PLAYSMET COPY

Metals Handbook Ninth Edition

Volume 2 Properties and Selection: Nonferrous Alloys and Pure Metals

Prepared under the direction of the ASM Handbook Committee

William H. Cubberly, Director of Reference Publications
Hugh Baker, Managing Editor
David Benjamin, Senior Editor
Paul M. Unterweiser, Manager, Publications Development
Craig W. Kirkpatrick, Chief Copy Editor
Vicki Knoll, Production Coordinator
Kathy Nieman, Editorial Assistant

AMERICAN SOCIETY FOR METALS
METALS PARK, OHIO 44073

First printing, November, 1979

Metals Handbook is a collective effort involving thousands of technical specialists. It brings together in one book a wealth of information from world-wide sources to help scientists, engineers and technicians solve current and long-range problems.

Great care is taken in the compilation and production of this volume, but it should be made clear that no warranties, express or implied, are given in connection with the accuracy or completeness of this publication, and no responsibility can be taken for any claims that may arise.

Nothing contained in the Metals Handbook shall be construed as a grant of any right of manufacture, sale, use or reproduction, in connection with any method, process, apparatus, product, composition or system, whether or not covered by letters patent, copyright or trademark, and nothing contained in the Metals Handbook shall be construed as a defense against any alleged infringement of letters patent, copyright or trademark, or as a defense against any liability for such infringement.

Comments, criticisms and suggestions are invited, and should be forwarded to the American Society for Metals.

Library of Congress Cataloging in Publication Data

American Society for Metals
Properties and selection: nonferrous alloys and pure metals

(Metals handbook; 9th ed., v. 2)
Includes bibliographical references and index.

1. Metals. 2. Alloys. I. Title. II. Series: American Society for
Metals, Metals handbook; 9th ed. v. 2.
TA459.A5 9th ed., vol. 2 669'.008s [669] 79-21644 ISBN 0-87170-008-5

Printed in the United States of America

Foreword

Reviewing the history and progress of Metals Handbook is always a rewarding experience because this unique and durable reference, with each succeeding edition, has consistently demonstrated a continuing upward trend in growth, in subject coverage, and—not least—in reader acceptance. It has been said, "nothing succeeds like success"—to which we might fittingly append—unless it is a long, uninterrupted series of successes!

Almost two decades have passed since the publication of Volume 1 (*Properties and Selection of Metals*) of the 8th edition of Metals Handbook in 1961. At that time, the forward noted that "although the subject matter dealt with in the area defined by this volume's title supersedes comparable material that occupied only two fifths of the 7th edition, greatly expanded coverage has resulted in a book almost as large as the previous single-volume edition". This expanded coverage was especially evident in the coverage accorded carbon and low-alloy steels and cast irons, where the number of volume pages devoted to these subjects increased from 108 in the 7th edition to a robust 344 in the 8th edition, an increase of 300%.

In 1961, the coverage accorded nonferrous alloys and pure metals, the principal subjects of the current volume, also reflected an increase when the 7th and 8th editions were compared, although the increase was a modest 15% when measured in total pages. The truly spectacular increase in these subject areas was yet to come. And now, with the publication of Volume 2 of the 9th edition, it can be viewed as *fait accomplis*—a volume of over 800 pages devoted exclusively to the properties and selection of nonferrous alloys and pure metals.

Over the intervening years, the special qualities of Metals Handbook have been retained—the standard for technical reliability and comprehensiveness for which the Handbook is internationally known, and the impeccable standards of its many contributors and benefactors.

To all of them, the tribute expressed in 1961 need not be revised nor amplified. The Handbook Committee and its subcommittees, the editorial staff of the Handbook, and contributors from industrial companies, research organizations, government establishments, and educational institutions have made a significant contribution to the metalworking industry—and have thereby earned its gratitude.

Elihu F. Bradley
President

Allan Ray Putnam
Managing Director

The Ninth Edition of Metals Handbook
is dedicated to the memory of
TAYLOR LYMAN, A. B. (Eng.), S. M., Ph. D.
(1917-1973)
Editor, Metals Handbook 1945-1973

Preface

This volume is the second of a three-volume set that, when completed in 1980, will supersede Volume 1 of the Eighth Edition of Metals Handbook. The need to present the information on composition, properties, selection and application of metals and alloys in three volumes instead of one reflects an awareness that knowledge about metals has expanded greatly over the past two decades. The extent of this expansion became acutely evident upon publication of Volume 1 of the Ninth Edition—a work of 793 pages that replaced 346 pages in Volume 1 of the Eighth Edition.

The comprehensive coverage of the properties and uses of nonferrous alloys and the properties of pure metals presented in this volume again emphasizes the tremendous growth in our knowledge of these materials. It has taken over 800 pages to cover essentially the same portion of metals technology that required only 339 pages in 1961.

This volume is designed to help readers quickly locate technical information needed to select the right nonferrous alloy for an application. To accomplish this, the information on properties and applications is presented in two distinct formats—descriptive articles and data compilations. The descriptive articles present information on the various alloy families—their designations, properties and characteristics. Also discussed are the product forms and sizes available in these alloy families, along with market information such as typical applications. This broad coverage allows the reader to select the individual metals and alloys that may be suitable for a specific application; these can then be further investigated by consulting the data compilations.

In the data compilations, or listings of property values for individual metals and alloys, information is provided on the mechanical and physical properties most often required for the solution of engineering problems. In many instances, these data are supplemented by information on properties less often of interest, but which could be of vital importance in certain applications. Coverage in the data compilations is not confined to the most popular alloys. Rather, properties are given for a wide range of alloys—the popular alloys plus many more that enjoy lesser degrees of use. Besides specific property values, the data compilations include listings of specifications, trade names, composition limits, foreign designations and in many instances, a summary of fabrication characteristics. These data compilations are not intended to provide extensive design information. Design criteria change and current issues of the applicable specifications should be consulted to obtain the values required for design.

By far the greater portion of this volume is devoted to alloys within the six most commercially important nonferrous metals: aluminum, copper, lead, magnesium, tin, and zinc. The seventh section contains information on the precious metals: silver, gold and the platinum-group metals. The eighth and last section consists of data compilations for pure metals.

Numerical values in the first seven sections are presented in dual engineering units. Those of Système Internationale d'Unités (SI) are the primary units of measure throughout the Ninth Edition of Metals Handbook. Equivalent units used in U.S. commercial practice are given where appropriate. (The Metals Handbook policy on units of measure is set forth more fully immediately following this preface.) Data on pure metals are given only in SI units.

This Handbook is the collective effort of the members of the many important technical committees, the individual contributors whose by-lines appear throughout the volume, and countless numbers of their colleagues, associates and contemporaries whose expertise was donated anonymously. We hereby acknowledge and commend each of them for their unselfish efforts. They deserve thanks from the entire metallurgical profession. Such efforts continually reinforce the original concept of Metals Handbook: a handbook written by members for the benefit of members.

The Editors

Policy on Units of Measure

By a resolution of its Board of Trustees, the American Society for Metals has adopted the practice of publishing data in both metric and customary U.S. units of measure. In preparing this Handbook, the editors have attempted to present data primarily in metric units based on Système Internationale d'Unités (SI), with secondary mention of the corresponding values in customary U.S. units. The decision to use SI as the primary system of units was based on the aforementioned resolution of the Board of Trustees, the widespread use of metric units throughout the world, and the expectation that the use of metric units in the United States will increase substantially during the anticipated lifetime of this Handbook.

For the most part, numerical engineering data in the text and in tables are presented in SI-based units with the customary U.S. equivalents in parentheses (text) or adjoining columns (tables). For example, pressure, stress and strength are shown in both SI units, which are pascals (Pa) with a suitable prefix (see the description of SI at the back of the volume), and in customary U.S. units, which are pounds per square inch (psi). To save space, large values of psi have been changed to kips per square inch (ksi), where one kip equals 1000 pounds. Some strictly scientific data are presented in SI units only.

To clarify some illustrations that depict machine parts described in the text, only one set of dimensions is presented on artwork. References in the accompanying text to dimensions in the illustrations are presented in both SI-based and customary U.S. units.

On graphs and charts, grids correspond to SI-based units, which appear along the left and bottom edges; where appropriate, corresponding customary U.S. units appear along the top and right edges. Some previously published histograms, particularly those illustrating the statistical distribution of values of mechanical properties, could not be redrawn because of the absence of the original data points; these have been reproduced in their original forms, with SI equivalents on the top and right edges.

Data pertaining to a specification published by a specification-writing group may be given in only the units used in that specification or in dual units, depending on the nature of the data. For example, the typical yield strength of aluminum sheet made to a specification written in customary U.S. units would be presented in dual units, but the thickness specified in that specification might be presented only in inches.

Data obtained according to specified test methods for which the specification implies a particular system of units are presented in the units of that system. Wherever feasible, equivalent units are also presented.

Conversions and rounding have been done in accordance with ASTM Standard E-380-76, with careful attention to the number of significant digits in the original data. For example, an annealing temperature of 1575 °F contains three significant digits (and possibly only two), because few commercial heat treatment systems can control the temperature of an entire load of parts within a spread of 10 °F. In this instance, the equivalent temperature would be given as 860 °C, or perhaps 850 °C depending on the degree of accuracy meant to be conveyed in the conversion, the exact conversion to 857.22 °C would not be appropriate. For an invariant physical phenomenon that occurs at a precise temperature (such as the melting of pure silver), it would be appropriate to report the temperature as 961.93 °C or 1763.5 °F. In many instances (especially in tables and data compilations), temperature values in °C and °F are alternatives rather than conversions.

The policy on units of measure in this Handbook contains several exceptions to strict conformance to ASTM E380; in each instance, the exception has been made to improve the clarity of the Handbook. Three examples of such exceptions are the use of "L", rather than "l" as the abbreviation for litre, reporting temperature in C rather than K and reporting stress intensity in $MPa\sqrt{m}$ rather than $MNm^{-3/2}$.

SI practice requires that only one virgule (diagonal) appears in units formed by combination of several basic units. Therefore, all of the units preceding the virgule are in the numerator and all units following the virgule are in the denominator of the expression (and no parentheses are required to prevent ambiguity).

Handbook Committee, Officers and Trustees

Members of the ASM Handbook Committee (1975-1979)

Author and Review Committees

ASM Committee on Aluminum and Aluminum Alloys

James L. McCall
Chairman, Manager
Materials Resources and Process
    Metallurgy Section
Battelle Columbus Laboratories

John F. Breedis
Supervisor Alloy Research Group
Olin Metals

Roy W. Brodie
Research Specialist-
    Materials and Processes
Lockheed-California Co.

Ernest D. Coberly, Jr.
Plant Metallurgist
Ross Aluminum Foundries

R. C. Cornell
Technical Director
American Die Casting Institute Inc.

Paul E. Fortin
Research Metallurgist
Alcan Research Centre

Harold Y. Hunsicker
Technical Advisor
Alcoa Laboratories

George C. Hsu
Manager, Industry Standards
Reynolds Metals Co.

Paul V. Mara
Vice President, Technical
The Aluminum Association

Peter A. Tomblin
Senior Materials Engineer
The De Havilland Aircraft Co.

Hugh Baker
Secretary, Managing Editor
Metals Handbook

ASM Review Committee on Magnesium and Magnesium Alloys

A. H. Braun
Chief, Research and Development
Wellman Dynamics Corp.

Fred H. Eckert
Extrusion Metallurgist
Consolidated Aluminum Corp.

P. A. Fisher
Consultant
Magnesium Elektron Ltd.

T. E. Leontis
Manager
Magnesium Research Center
Battelle Columbus Laboratories

Lloyd F. Lockwood
Magnesium, Business Section
The Dow Chemical Co.

H. J. Proffitt
Technical Director
Haley Industries Ltd.

H. G. Warrington
formerly with Chromasco Ltd.

ASM Review Committee on Copper and Copper Alloys

Ralph E. Ricksecker
Chairman, Consultant

A. W. Blackwood
Superintendent, Metals Applications
Central Research Laboratories
ASARCO Inc.

John F. Breedis
Supervisor, Alloy Research Group
Olin Metals

R. J. Cox
Vice President, Engineering
AMPCO Metals Inc.

Bruce A. Heyer
Manager, Non-Ferrous Research
Abex Corp.

John M. Kuzmech
Quality Assurance Metallurgist-Extrusion
Anaconda Industries
Brass Division

W. Stuart Lyman
Manager, Technical and Market Services
Copper Development Association Inc.

Robert S. Mroczkowski
Manager of Metallurgical and Polymeric Engineering
AMP Incorporated

Verne Pulsifer
Development Engineer
Essex Group of United Technologies

Fred L. Riddell
(retired), formerly Vice President
and Chief Metallurgist
H. Kramer and Co.

Donald G. Schmidt
Metallurgical Consultant
R. Lavin and Sons, Inc.

Richard Dale Smith
Senior Staff Metallurgist
Lexington Development Center
Kennecott Copper Corp.

Alfred Snowman
Research Manager-Industrial Products
Potters Industries Inc.

Andrew R. Somosi
Manager, Metallurgical Services
Brush Wellman, Inc.

Pierre W. Taubenblat
Associate Technical Director
AMAX Base Metals R&D Inc.

Keith G. Wikle
Senior Technical Specialist
Kawecki Berylco Industries

Ralph E. Willett
Senior Research Metallurgist
Anaconda Industries
Brass Divisions

David Benjamin
Secretary, Senior Editor
Metals Handbook

Contents

Aluminum 1

Introduction to Aluminum and
    Aluminum Alloys 3
Temper Designation System for
    Aluminum and Aluminum
    Alloys 24
Heat Treatment of Aluminum
    Alloys 28
Aluminum Mill Products.......... 44
Properties of Wrought Aluminums
    and Aluminum Alloys 63
Aluminum Foundry Products 140
Properties of Cast Aluminum
    Alloys 152
Stamping of Aluminum Alloy
    Sheet 180
Machinability of Aluminum
    Alloys 187
Joining Aluminum 191
Corrosion Resistance of Aluminum
    and Aluminum Alloys 204

Copper 237

Introduction to Copper and Copper
    Alloys 239
Temper Designations for Coppers
    and Copper Alloys 248
Heat Treating of Copper and Copper
    Alloys 252
Copper Tubular Products......... 261
Copper Wire and Cable 265
Properties of Wrought Coppers and
    Copper Alloys 275

Selection and Application of Copper
    Alloy Castings 383
Properties of Cast Copper Alloys.. 395
Joining of Copper and Its Alloys.. 440
Corrosion Characteristics of Copper
    and Its Alloys 458
Wrought Copper Alloys for Corrosion
    Service 466
Stress Relaxation in Copper and
    Copper Alloys 484

Lead 491

Lead and Lead Alloys 493
Properties of Leads and Lead
    Alloys 500
Corrosion Resistance of Lead..... 511

Magnesium 523

Selection and Application of
    Magnesium and Magnesium
    Alloys 525
Properties of Magnesium Alloys .. 553
Corrosion Resistance of Magnesium
    and Magnesium Alloys......... 596

Tin 611

Tin and Its Alloys 613
Properties of Tin and Tin Alloys.. 617

Zinc 627

Selection and Application of Zinc
    and Zinc Alloys 629

Properties of Zinc and Zinc
    Alloys 638
The Use of Zinc in Corrosion
    Service 646

Precious Metals 657

Precious Metals and Their Uses .. 659
Corrosion Resistance of Precious
    Metals 668
Properties of Silver and Silver
    Alloys 671
Properties of Gold and Gold
    Alloys 679
Gold in Dentistry 684
Properties of Platinum and
    Platinum Alloys 688
Properties of Palladium and
    Palladium Alloys 699

Pure Metals 707

Preparation and Characterization of
    Pure Metals 709
Properties of Pure Metals 714
Properties of the Actinide Metals . 832

**Periodic Table of the
Elements** 834

**Système Internationale d'Unités
(SI)** 835

Abbreviations and Symbols 836

Index 838

Aluminum

Introduction to Aluminum and Aluminum Alloys

By the ASM Committee on
Aluminum and Aluminum Alloys*

THE PROPERTIES of aluminum that make this metal and its alloys the most economical and attractive for a wide variety of uses are appearance, light weight, fabricability, physical properties, mechanical properties, and corrosion resistance, or a combination of these.

The sharp rise in the use of aluminum is shown by the world production record of 1966 to 1977 (Fig. 1) and the U.S. distribution of usage, in terms of general type of products for 1950 to 1977 (Fig. 2). The major applications for aluminum can be categorized, in decreasing order of current market size, into building and construction, containers and packaging, transportation, electrical, consumer durables, machinery and equipment, and other. The applications in each of these market categories are discussed at the end of this article.

Properties

Aluminum is probably best known for two properties—light weight and

*See page **X** for committee list.

corrosion resistance. It weighs only about 2.7 Mg/m^3 (about 0.1 lb/in^3), approximately one third as much as the same volume of steel, copper or brass. It has excellent resistance against corrosive elements in the atmosphere, water (including salt water), oils and many chemicals. (See the separate article on the corrosion resistance of aluminum and aluminum alloys in this volume.)

Aluminum also is highly reflective to radiant energy—visible light, radiant heat and electromagnetic waves. It has excellent electrical and thermal conductivity. It is nonferromagnetic, a property of importance in the electrical and electronics industries. It has nonsparking characteristics, which is important near inflammable or explosive materials. It is nontoxic, making it safe for use with foods and beverages. It has an attractive appearance in its natural finish, which can be soft and lustrous, or bright and shiny, or, if desired, it can be virtually any color or texture. Aluminum also is strong; some aluminum alloys exceed structural steel in strength. (The strengths of aluminum products are presented in the separate articles on aluminum mill and foundry

products, and in the data compilations for the individual alloys.) Finally, because of vast ore reserves, aluminum is the most abundant of all the structural metals, constituting over 8% of the earth's crust.

Electrical Conductivity. Aluminum is often selected for its high electrical conductivity. Its electrical conductivity is nearly twice that of copper on a weight basis. The success of aluminum in electrical applications depends largely on conductor design, which should incorporate high conductivity with adequate mechanical strength. An example of properly combined engineering features is to be found in long-line high-voltage transmission cable, which provides a steel core for strength with surrounding aluminum cable to transmit electricity. Another example is the "All Aluminum Alloy Conductor" (AAAC), which is used because of its improved corrosion resistance.

Electrical bus bar for feed-in and plug-in duct applications is normally made from rectangular extruded 6101 alloy aluminum bar, with ampere ratings ranging from 225 to 4000.

Fig. 1 Annual world production of primary aluminum

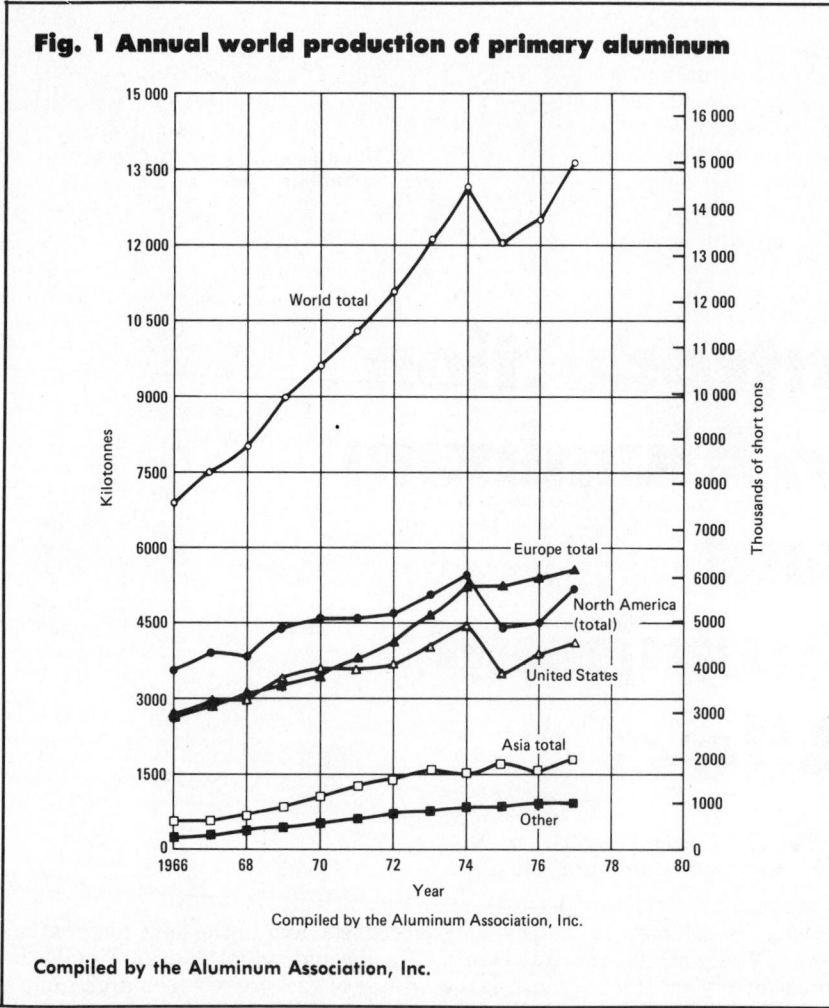

Compiled by the Aluminum Association, Inc.

Compiled by the Aluminum Association, Inc.

Fig. 2 Annual U.S. shipments of aluminum by product type

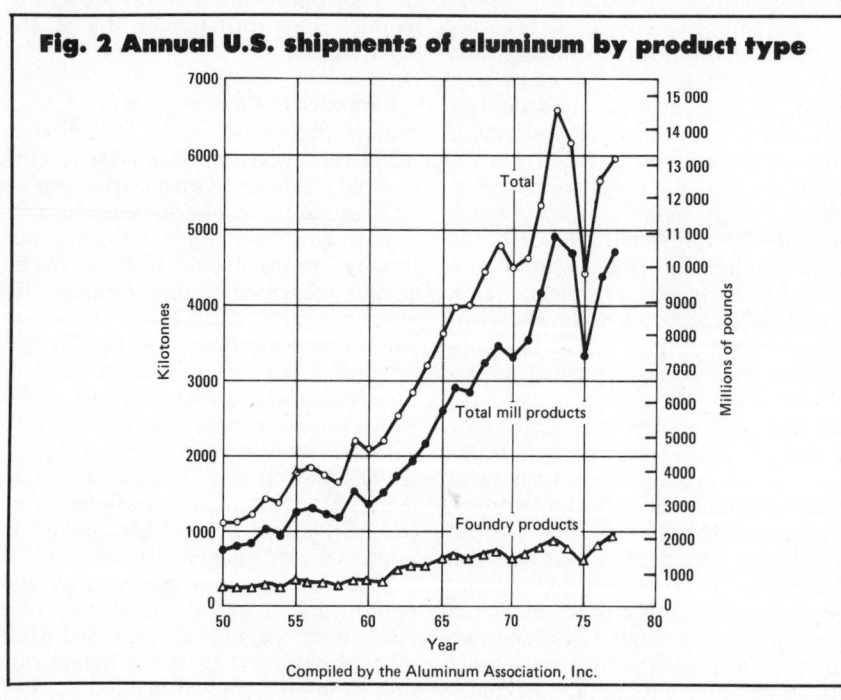

Compiled by the Aluminum Association, Inc.

Fig. 3 Auto emblem changed from an aluminum stamping to an aluminum extrusion

See text for discussion.

The thermal conductivity of aluminum (about 50 to 60% that of copper) is a decisive physical property in the selection of aluminum for refrigerator evaporators, electric range deep-well cooker pails and trivets, waffle-iron grids, sandwich grills, other food processing ware, and electric-iron sole plates. For rapid heat removal in the cylinder heads of high-compression engines, aluminum is often chosen over cast iron (conductivity about 10% that of copper).

The reflectance of polished aluminum, over a broad range of wave lengths, leads to its selection for such applications as the heater box in automatic dryers, adjustable reflecting shelves in electric range ovens, and reflectors beneath the resistance heating units in electric ranges. Its fabricability, low density, corrosion resistance, and reflectance make it a useful metal for low-cost lamp reflectors. Because aluminum reflects a high percentage of the sun's heat, it is used for roofing, reflective insulation, and the reflectors of heating panels.

Product Forms

Mill products made of aluminum and aluminum alloys are available in any standard form—flat-rolled products (sheet, plate and foil), rod, bar and wire, tubular products, and extruded standard and special shapes—and in a wide range of sizes. Flat-rolled products range from 0.005 mm or 0.00017 in. thick foil to 200 mm or 8 in. thick plate. Current facilities permit the production of a limited amount of extra large sheet, for example, up to 5 m (200 in.) wide by 25 m (1000 in.) long. Extra large plates, for example, up to 22 mm (⅞ in.) thick by 2¼ m (89 in.) wide by 32 m (105 ft) long, are supplied for the construction of wide-body aircraft. Extruded products range from 0.25 mm or 0.01 in. diam wire to shapes having a

circumscribing circle as large as 580 mm or 23 in. For a discussion of the alloys used in aluminum will products and their properties, see the separate article on this subject in this volume.

Engineered products are those products designed for one specific application, in contrast to "off-the-shelf" products such as standard size sheet, plate, foil, rod, bar, wire, tube, pipe and standard structural shapes. Therefore, engineered products include special extruded shapes, forgings, impacts, castings, powder metallurgy parts, ma-

chined parts and stampings, as well as special sizes of standard products.

Special Extruded Shapes. Designs that are symmetrical around one axis usually are adaptable to production from an extruded shape. With current technology, it is also possible to extrude many unbalanced shapes. The design of shapes is discussed in the article on aluminum mill products in this volume.

Most aluminum alloys can be obtained as precision extrusions with good as-extruded surfaces. Major di-

Fig. 4 Cost of an extruded fuel-tank attachment fitting as a function of quantity

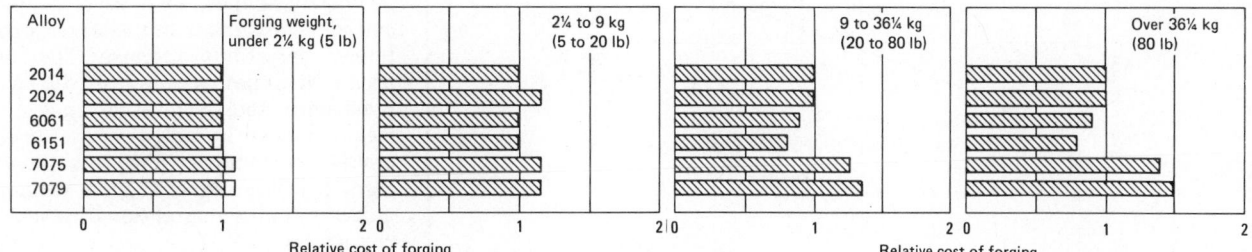

14.5   32.5

Dimensions in inches

7075-T6

Special extrusion

Relative cost per piece / Number of pieces

A part completely machined from bar stock is rated 100.

Table 1 Typical inch tolerances for aluminum die forgings

Weight kg	Weight lb(a)	Tolerance, in. Thickness(b)	Tolerance, in. Mismatch	Machining allowance per surface, in.	Width and length tolerances, in.(c) Dimension	Width and length tolerances, in.(c) Tolerance
0 to 0.5	0 to 1	+0.030 −0.015	0.015	0.090	0 to 8	+0.030 −0.015
0.5 to 1.8	1 to 4	+0.045 −0.030	0.020	0.090	8 to 12	+0.045 −0.020
1.8 to 7.7	4 to 17	+0.060 −0.030	0.025	0.090	12 to 16	+0.060 −0.025
7.7 to 10.9	17 to 24	+0.075 −0.030	0.030	0.120	16 to 20	+0.075 −0.030
10.9 to 22.7	24 to 50	+0.090 −0.030	0.035	0.120	20 to 24	+0.090 −0.040
22.7 to 45.4	50 to 100	+0.125 −0.045	0.055	0.150	24 to 28	+0.105 −0.050
45.4 to 113	100 to 250	+0.185 −0.060	0.090	0.150	28 to 32	+0.120 −0.060
Over 113	Over 250	+0.250 −0.060	0.125	0.190	32 to 36 (d)	+0.135 −0.070

(a) Maximum flash extension for forgings weighing up to 0.5 kg (1 lb) is 0.03 in.; 0.5 to 5.0 kg (1 to 12 lb), 0.06 in.; 5.4 to 10.9 kg (12 to 24 lb), 0.09 in.; over 10.9 kg (24 lb), 0.12 in. (b) Thickness tolerances apply to all dimensions to or across the parting plane and that lie approximately parallel to the forging stroke. (c) Width and length tolerances apply to all dimensions that are approximately parallel to the forging plane. Tolerance provides for the variation in shrinkage as well as a nominal amount for die allowance. (d) 36 to 40 in., to 0.150 and −0.080 in. For all dimensions above 40 in., add +0.004 and −0.002 per in. to tolerance for 40-in. dimension.

Fig. 5 Comparative die forging costs of different aluminum alloys as a function of forging weight

Alloy	Forging weight, under 2¼ kg (5 lb)	2¼ to 9 kg (5 to 20 lb)	9 to 36¼ kg (20 to 80 lb)	Over 36¼ kg (80 lb)
2014				
2024				
6061				
6151				
7075				
7079				

Relative cost of forging   Relative cost of forging

Although included in this comparison, alloy 2024 is not a standard forging alloy.

Fig. 6 Forgeability and forging temperatures of seven aluminum alloys

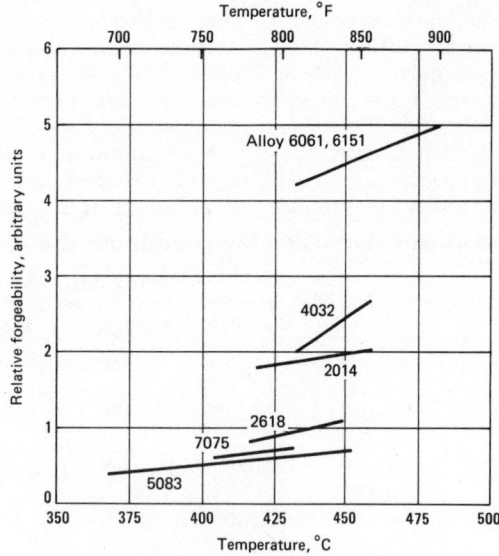

Forgeability increases as the arbitrary unit increases.

Fig. 7 Limiting web proportions for aluminum die forgings

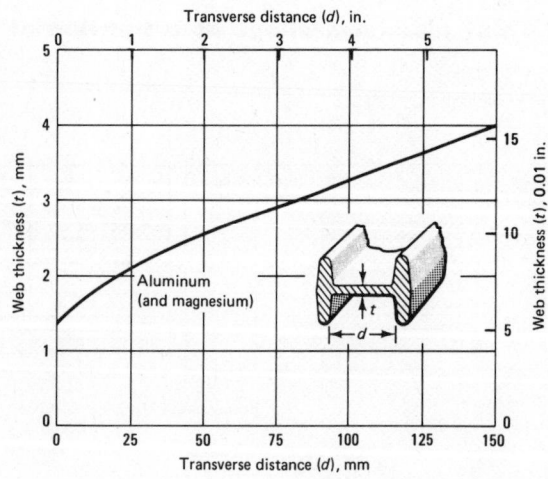

mensions usually do not need to be machined; tolerance of the as-extruded product often permits a manufacturer to complete the part with simple cutoff, drilling, broaching, or other minor machining operations. For any part for which the shape can be produced as an extrusion, the cost of the extrusion die is usually written off when a few parts have been produced.

Figure 3 shows an application in which several of the advantages of extrusions directly influenced the choice of that form. The auto emblem in this figure was originally an aluminum stamping of alloy 5357 in the O temper. Studs were resistance welded to the back of the emblem to allow for attachment. Rejection rate was high because of poor appearance and off-tolerance placement of the stud. Studs also failed in service. An extrusion, of the shape shown in the upper sketch in Fig. 3, was obtained in alloy 1100-O. The extrusion was then cold forged to emboss the name on the smooth face. Following this operation, horizontal blanking removed most of the material of the two upstanding legs, leaving only enough to form the fastening studs.

The surface finish obtained was superior to that of the original stamping for color anodizing. The integral studs were stronger, permitted easier handling, and resulted in fewer rejections from breakage in assembly and service.

Cost of machining may be the only selection consideration. This is illustrated by the cost figures for a fuel-tank attachment fitting (Fig. 4). The design permitted the use of an extrusion, which required very little machining compared with the same part fabricated from a solid bar. After about 100 pieces, the cost per piece was lowered substantially.

Forgings. Hand forgings are simple geometric shapes, such as rectangles, cylinders, and disks, or moderately contoured variations of these. They are produced on flat or contoured dies with blacksmith tools. These forgings fill a frequent need in industry when a few pieces are required for prototype designs and the expense and time required to make impression dies cannot be justified.

Forgings up to 450 kg (1000 lb) are made regularly, those between 450 and 900 kg (1000 and 2000 lb) are less common, and pieces over 900 kg (2000 lb) are specific items. Properties are lower in larger forgings.

Most aluminum forgings are pro-

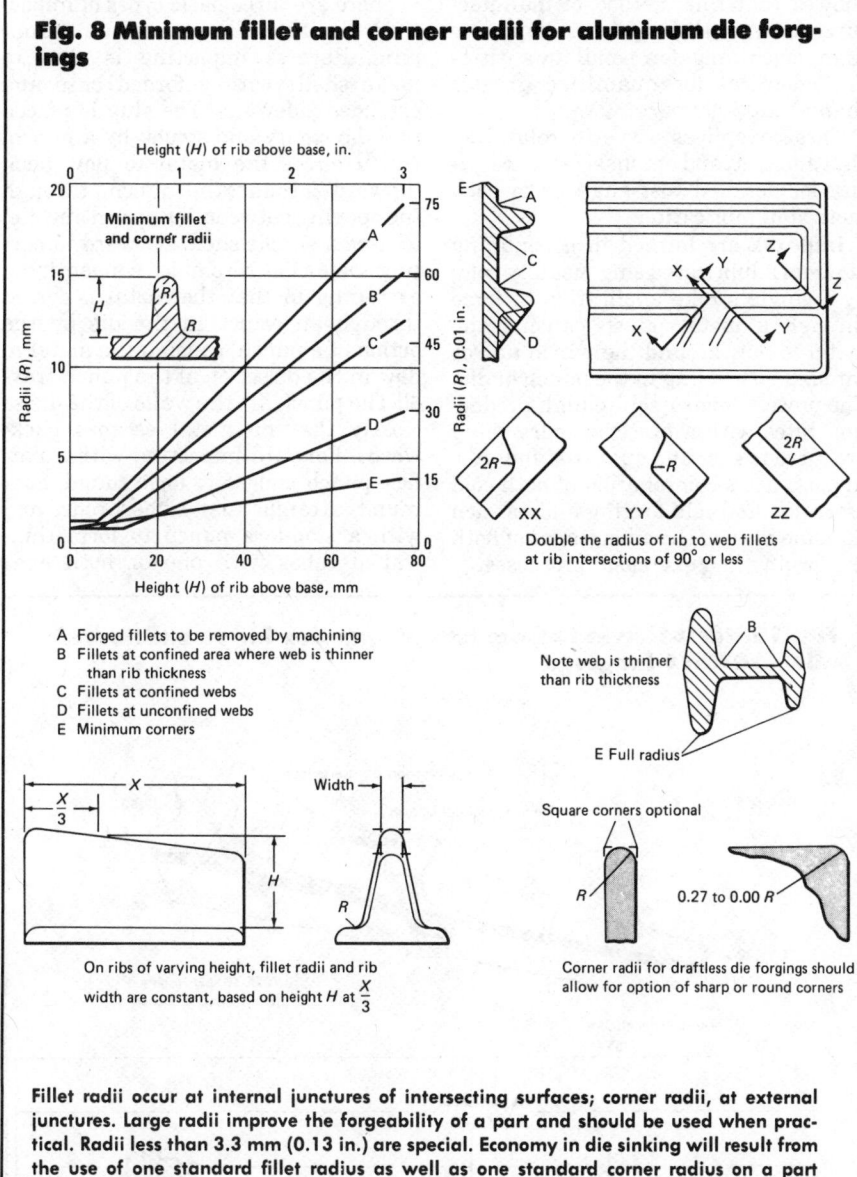

Fig. 8 Minimum fillet and corner radii for aluminum die forgings

A Forged fillets to be removed by machining
B Fillets at confined area where web is thinner than rib thickness
C Fillets at confined webs
D Fillets at unconfined webs
E Minimum corners

Double the radius of rib to web fillets at rib intersections of 90° or less

Note web is thinner than rib thickness

E Full radius

Square corners optional

0.27 to 0.00 R

Corner radii for draftless die forgings should allow for option of sharp or round corners

On ribs of varying height, fillet radii and rib width are constant, based on height H at $\frac{X}{3}$

Fillet radii occur at internal junctures of intersecting surfaces; corner radii, at external junctures. Large radii improve the forgeability of a part and should be used when practical. Radii less than 3.3 mm (0.13 in.) are special. Economy in die sinking will result from the use of one standard fillet radius as well as one standard corner radius on a part whenever possible.

Fig. 9 Straightness tolerances in inches for aluminum die forgings

Surface length, L, in.	Tolerance, in.
0 to 6	0.010
6 to 12	0.020
12 to 18	0.030
18 to 24	0.040
24 to 30	0.050

For each additional 6 in. in length, add 0.010 in. to tolerance.

Straightness is generally checked against a straightedge or surface plate, and the tolerance applies as a maximum deviation from the reference surface.

duced in closed dies. The cost of a raw aluminum die forging varies with alloy (see Fig. 5), reflecting the relative forgeability of the alloy. (The relative forgeabilities of seven aluminum alloys are presented in Fig. 6.) The absolute cost on a per piece basis varies with the weight of the piece, the complexity of the shape, the specified tolerances (standard, close tolerance, or low-draft), and the quantity. A ratio in price up to 10 to 1 for the same weight forging is common, depending on the above factors. Forging dies are more expensive than extrusion dies by a ratio varying from 4-to-1 to 3-to-1, for small and large sections. The more important general tolerances for aluminum die forgings are given in Table 1 and Fig. 7, 8 and 9.

The variation in mechanical properties with section size and flow condition is illustrated for two closed-die forgings in Fig. 10. Variation in properties is obtained because of the thickness variations, variations in reduction during forging, and variation in angle between the axis of the test specimen and the direction of grain flow. Both die forgings show markedly lower properties in the short transverse direction than in the longitudinal direction. Similar dif-

ferences prevail for longitudinal and transverse properties in premium strength spar forgings of 7175-T736 (Fig. 11). Such lower properties must be considered in design.

For alloys 2014, 7049, 7050, 7075 and 7175, properties are usually guaranteed in all three directions. For forging alloys and sizes where guaranteed values are not established, the user or producer must often test sample parts destructively to determine whether properties actually meet design requirements.

Fatigue strength of bars cut from forgings is affected in the same manner as tensile strength, yield strength, and elongation. Figure 12 shows the results of fatigue tests of specimens cut from two sample forgings in directions both parallel and transverse to the forging flow lines. (Note that fatigue strength is a function of both alloy and orientation.)

The fatigue strength of aluminum forgings also is affected by the extent of hot and cold working, and response to heat treatment. The curves shown in Fig. 12 are specific for the forging shown. Although alteration in design could change quantitative relationships, no change in qualitative relationships would be anticipated.

Tool costs play a major role in determining a choice between machining from bar stock and purchasing as die forgings. The cost of machining a few parts from a bar or slab is usually less than making a die and forging the part. As the number of parts to be produced increases, forging usually becomes the less expensive method. In borderline cases, a detailed study of machining and die costs is necessary to determine the crossover point. In determining this point, it is necessary to calculate only the original cost of the die, since the supplier of the forging is responsible for die replacements caused by breakage or wear. This replacement cost is included in the price of the forgings. The die cost varies with the size and intricacy of the part.

In Fig. 13, a die forging (Part A) is compared with a built-up design. Although 75% of the metal was machined away from the rough forging, the machined forging was more economical than the assembly for quantities greater than 125.

In some large, complicated forgings, the break-even point may be at the first or second forging. It also may be desirable to rough forge the part in relatively inexpensive roughing dies and complete the part by machining if only a few pieces are desired. Using this technique, the desirable flow of metal induced by forging and the consequent improvement in properties can be obtained at a lower price than would have to be paid for a part forged to final dimensions.

The curves in Fig. 13 compare costs of parts of different size and shape, produced by competitive methods. For the simpler Part B, the crossover occurs at about 100 pieces. For the complicated forging, Part C, the crossover is at 40 pieces. The items considered in determining the cost of these two parts include: fabrication-shop learning curves, unit-run labor, amortized setup, labor, tooling costs (including dies and fixtures), raw materials and overhead charges.

Die forging is not always the cheapest method of producing a large quantity of parts. A relatively simple fitting, Part D, was analyzed for production costs as a die forging, an impact extrusion, and as machined from plate. The machined fitting was more economical for all quantities, because cost of the finishing operations required on the die forging closely approached the cost of machining the part completely. The thin walls and deep crevices of this fitting should have made it ideally suited to impact extrusion. Analysis

showed that this method of manufacture was only slightly more expensive than machining for small quantities and identical for quantities greater than about 3000 parts.

These examples serve to relate design and cost, and emphasize the necessity for detailed cost analysis of each method of fabrication.

Impacts are formed in a confining die from a lubricated slug, usually cold, by a single stroke application of force through a metal punch causing the metal to flow around the punch and/or through an opening in the punch or die. The process lends itself to high production rates with a precision part being produced to exacting quality standards. Impacts are a combination of both cold extrusion and cold forging and as such combine most of the advantages of both the forging and extrusion processes.

There are three basic types of impacting, all of which are used on aluminum. Reverse impacting is used to make shells with a forged base and extruded sidewalls. The slug is placed in a die cavity and struck by a punch, which forces the metal to flow back (upward) around the punch, through the opening between the punch and die, to form a simple shell. Forward impacting somewhat resembles conventional extruding in that the metal is forced through an orifice in the die by the action of a punch, causing the metal to flow in the direction of the punch travel. The punch fits the walls of the die so closely that no metal escapes backwards. Forward impacting with a flat-face punch is used to form round, nonround, straight and ribbed rods, and with a stop-face punch to form thin-walled tubes with one or both ends

Fig. 10 Variation in mechanical properties for two different alloy 7075-T6 forgings

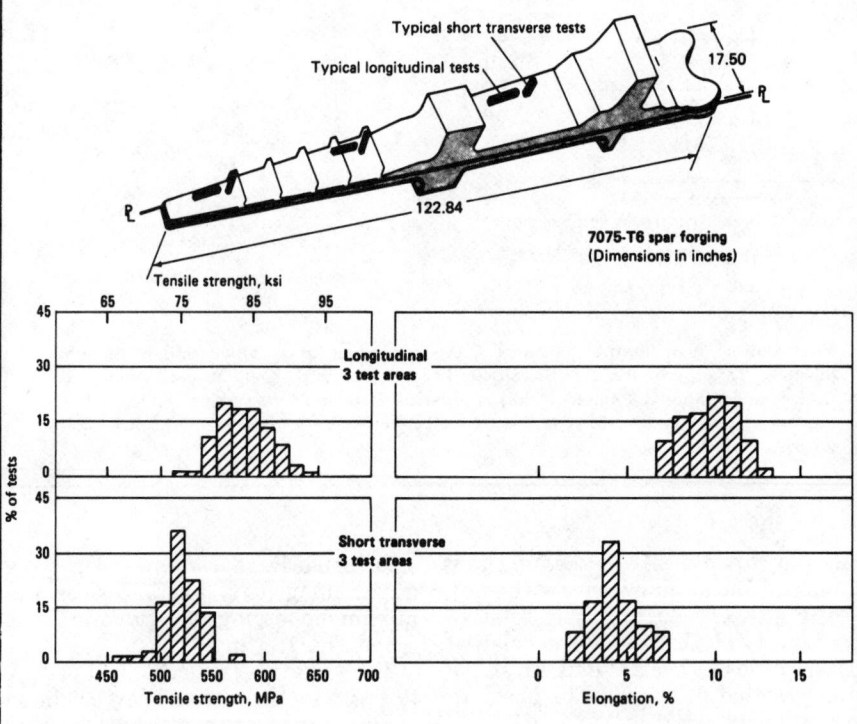

Both forgings were forged on a 31.75-kt (35,000-ton) press, solution treated at 470 °C (880 °F) for 3 h, quenched in water at 60 °C (140 °F) and aged at 120 °C (250 °F) for 24 h. **Spar forging.** Cast ingots were received in 17 lots from three sources. Two lots were in 19-in. rounds and 15 lots in 10 to 12-in. squares. Twenty-eight forgings were tested at the locations indicated. **Cylinder forging.** All cylinders were forged from 19-in. rounds. Six of the forgings were tested at locations shown.

Table 2 Practical minimum wall thicknesses for reverse aluminum impacts

Nominal outside diameter			Wall thickness for indicated alloy							
		1100		6061		2014		7075	
mm	in.	mm	in.	mm	in.	mm	in.	mm	in.
25	1	0.25	0.010	0.40	0.015	0.90	0.035	1.00	0.040
50	2	0.50	0.020	0.75	0.030	1.80	0.070	2.05	0.080
75	3	0.75	0.030	1.15	0.045	2.65	0.105	3.05	0.120
100	4	1.00	0.040	1.50	0.060	3.55	0.140	4.05	0.160
125	5	1.25	0.050	1.90	0.075	4.45	0.175	5.10	0.200
150	6	1.50	0.060	2.30	0.090	5.35	0.210	6.10	0.240
180	7	1.90	0.075	2.80	0.110	6.20	0.245	7.10	0.280
205	8	2.55	0.100	3.30	0.130	7.10	0.280	8.15	0.320
230	9	2.80	0.110	3.70	0.145	8.00	0.315	9.15	0.360
255	10	3.15	0.125	4.20	0.165	8.90	0.350	10.15	0.400

open, and with parallel or tapered side walls. If the punch is smaller than the die and the die contains an orifice, reverse and forward impacting can be combined to produce a combination impact.

A major consideration in designing aluminum impacts is the selection of the appropriate alloy. Alloys 1100, 2014, 3003, 6061, 6351 and 7075 are most often utilized in aluminum impacts. These alloys offer a range of mechanical properties that fit most applications. Generally, the stronger the alloy impacted, the shorter the life of the tools and the higher the production costs. Although each part must be considered individually, the stronger alloys generally require heavier minimum wall thicknesses (see Table 2).

Alloy 1100, which has excellent corrosion resistance in rural, industrial and marine atmospheres, is commonly impacted to form containers for liquid and semiliquid materials such as food preserves and products sprayed by aerosols. Alloy 3003 is used for many of the same applications as alloy 1100, but is selected when higher strength than that offered by 1100 is required. Alloy 6061, which is heat treatable and has excellent corrosion resistance, is widely used in the manufacture of parts for automotive, aircraft and marine applications, especially where welding is involved or high strength is required. Alloy 6351 is a medium-to-high strength heat treatable alloy with good corrosion resistance. Alloy 2014 is a heat treatable alloy used for general applications where high tensile and yield strength, combined with good ductility and good fatigue resistance are essential. It is widely used in structural applications and in aircraft, automobile and ordnance parts. Alloy 7075 has the highest strength and hardness of these alloys. This heat treatable alloy is used for many of the same applications as alloy 2014, but is selected where highest stresses are expected to be encountered or for maximum weight saving.

Impacts usually have properties in the longitudinal direction equal to those specified for other forms of similar composition. Figure 14 shows properties of a bomb ejector foot. Here, the properties of both the impact made at room temperature and the hot impact are equivalent and exceed the die forging properties for the same alloy. Typical mechanical properties in the longitudinal direction of aluminum impacts are given in Table 3.

Castings are used for products with intricate contours and hollowed or cored areas. Choice of castings over other product forms is usually based on lower final part cost rather than on mechanical properties. Thicker sections and more intricate designs are used to compensate for lower mechani-

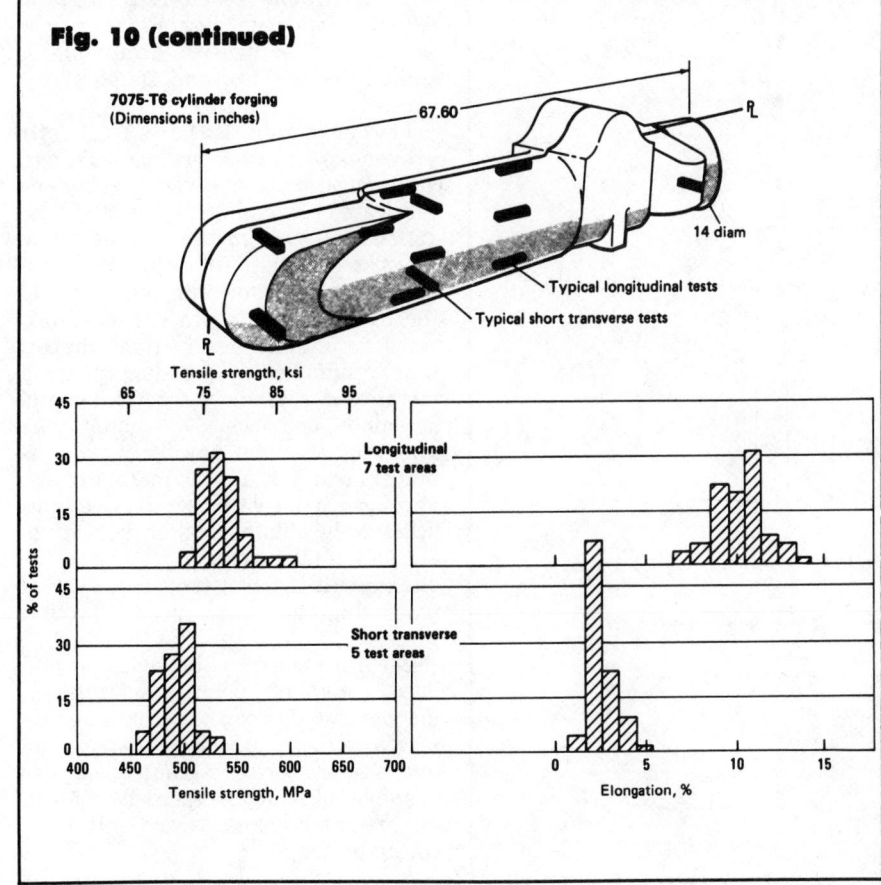

Fig. 10 (continued)

7075-T6 cylinder forging (Dimensions in inches)
67.60
14 diam
Typical longitudinal tests
Typical short transverse tests
Tensile strength, ksi
Longitudinal 7 test areas
Short transverse 5 test areas
% of tests
Tensile strength, MPa
Elongation, %

Table 3 Typical longitudinal mechanical properties of aluminum impacts

Alloy and temper	Tensile strength MPa	ksi	Yield strength MPa	ksi	Elongation in 50 mm or 2 in., %	Hardness, HB
1100-F	165	24	150	22	5	44
H185(a)	165	24	150	22	...	44
2014-T4	425	62	290	42	20	105
T6	485	70	415	60	11	135
3003-F	200	29	185	27	4	55
H185(a)	200	29	185	27	...	55
6061-F	195	28	170	25	5	52
H112	195	28	170	25	5	52
T4	240	35	145	21	22	65
T6	310	45	275	40	12	95
T84	295	43	270	39	6	85
6351-T6	340	49	295	43	13	95
7075-T6	570	83	505	73	10	145
T73	505	73	425	62	12	135

(a) Temper H185 applies to alloy 1100 and alloy 3003 parts where conformance to mechanical property limits throughout the side walls are confirmed by test data.

Fig. 11 Variation in mechanical properties of alloy 7175-T736 premium strength spar forging for F-5 aircraft

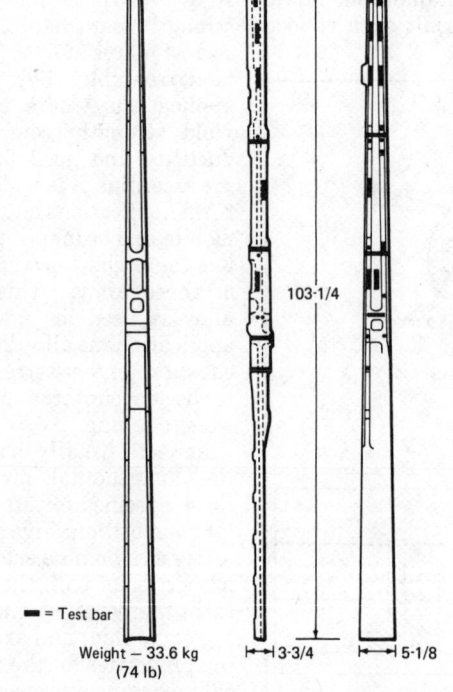

103-1/4

■ = Test bar

Weight — 33.6 kg (74 lb)   3-3/4   5-1/8

Test bar direction	Tensile strength MPa	ksi	Yield strength MPa	ksi	Elongation, %
Longitudinal					
average	549	79.6	502	72.8	13.8
minimum	524	76.0	455	66.0	7.0
Transverse					
average	519	75.3	456	66.1	9.9
minimum	490	71.0	427	62.0	4.0

200 production test specimens.

cal properties. Reinforcing ribs and bosses, which would be costly to machine in a part made from a wrought alloy, can be cast in place by appropriate construction of the pattern or mold. Aluminum die castings, permanent mold castings, sand castings and other types of castings are discussed in detail in a separate article in this volume.

Powder Metallurgy Parts. Powder metallurgy (P/M) consists of compressing metal powder in a shaped die to produce green compacts, and then sintering (diffusion bonding) the compacts at elevated temperature in a furnace with a protective atmosphere. During sintering, the compacts become consolidated and strengthened.

The density of sintered compacts may be increased by re-pressing. When re-pressing is performed primarily to improve the dimensional accuracy of a compact, it usually is termed "sizing"; when performed to improve configuration, it is termed "coining". Re-pressing may be followed by resintering, which relieves stress due to cold work in repressing and may further consolidate the compact. By pressing and sintering only, parts of over 80% theoretical density can be produced. By re-pressing, with or without resintering, parts of 90% theoretical density or more can be produced. The density attainable is limited by the size and shape of the compact.

Powder metallurgy parts frequently are competitive with forgings, castings, stampings, machined components and fabricated assemblies. Within the limitations inherent in the P/M processes (discussed in detail in the Metals Handbook volume on forming, Volume 4 of the 8th Edition), parts can be fabricated to final or nearly final shapes, thus eliminating or reducing scrap metal and secondary machining and assembly operations. Although the unit cost of metal powder usually is higher than that of solid metal in bars, forgings or castings, the savings achieved by eliminating fabricating operations and minimizing scrap losses often result in lower total cost for P/M parts than for parts made by other processes.

Certain metal products can be produced only by powder metallurgy; among these products are materials whose porosity (number distribution and size of pores) is controlled; two examples of controlled-porosity materials are filter elements and self-lubricating bearings.

Successful production by powder

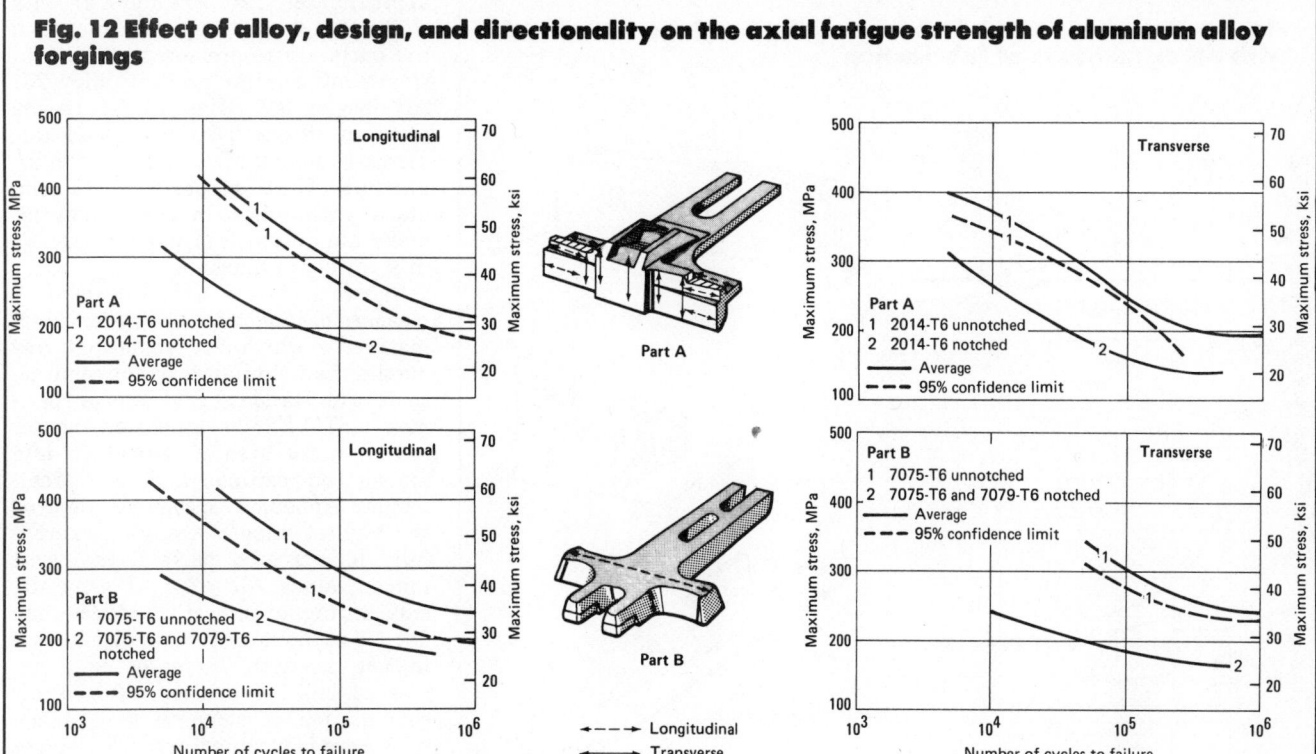

Fig. 12 Effect of alloy, design, and directionality on the axial fatigue strength of aluminum alloy forgings

Data apply to Parts A and B, as shown. Sheet-type fatigue specimens, 3.2 mm (0.125 in.) thick and 6.4 mm (0.250 in.) wide, were cut both parallel and transverse to the forging flow lines. Locations from which specimens were taken are shown on the drawings. (Mean stress was 91.7 MPa [13.3 ksi]; notch was 1.2 mm [0.047 in.] in diameter hole in center of specimen. K_t = 2.5.)

Table 4 Sintered properties of aluminum alloy 601AB(a)

Green density %	g/cm³	Temper	Tensile strength MPa	ksi	Yield strength MPa	ksi	Elongation, % in 25 mm or 1 in.	Apparent hardness, %
85	2.29	T1	110	16.0	50	7.0	6.0	55-60 HRH
		T4	140	20.5	95	14.0	5.0	80-85 HRH
		T6	185	26.5	...	...	1.0	70-75 HRE
90	2.42	T1	120	17.5	55	8.0	7.0	60-65 HRH
		T4	...	...	...	...	...	80-85 HRH
		T6	225	32.5	215	31.0	2.0	75-80 HRE
95	2.55	T1	125	18.0	60	8.5	8.0	65-70 HRH
		T4	150	22.0	105	15.0	5.0	85-90 HRH
		T6	250	36.5	240	35.0	2.0	80-85 HRE

(a) Sintered 30 min at 620 °C (150 °F) in N_2 with average dewpoint of 7 °C (45 °F).

metallurgy depends on proper selection and control of the following process variables (a) powder characteristics, (b) powder preparation, (c) type of compacting press, (d) design of compacting tools and dies, (e) type of sintering furnace, (f) composition of sintering atmosphere, and (g) choice of production cycle.

Two of the more common aluminum alloys used for P/M parts are alloys 601

AB and 201 AB. Their nominal compositions are:

	601 AB	201 AB
Cu	0.25	4.4
Si	0.60	0.8
Mg	1.0	0.5
Lubricant	1.5	1.5
Al	rem	rem

Aluminum P/M parts develop ultimate tensile strengths ranging from 75 to 350 MPa (11 to 50 ksi), depending on composition, density and thermal treatment. Tensile properties of sintered 601 AB and 201 AB after various thermal treatments are presented in Fig. 15 and 16, and in Tables 4 and 5. Because alloys 601 AB and 201 AB achieve properties by solution and precipitation of soluble alloying elements, the strengths of both compositions in the as-sintered (T1) temper are affected by the cooling rate from the sintering temperature. Parts cooled very slowly (about 30 °C or 50 °F per h) will develop the low strengths typical of the soft, annealed temper. Conversely, parts cooled rapidly, as by water quenching, will precipitation harden at room temperature to approach T4 temper properties. The actual cooling rate of parts in most sintering furnaces will lie somewhere between these extremes with the "as-sintered" properties reflecting the existing conditions.

In the solution and precipitation heat treated (T6) temper, high density 601 AB parts develop tensile strengths of

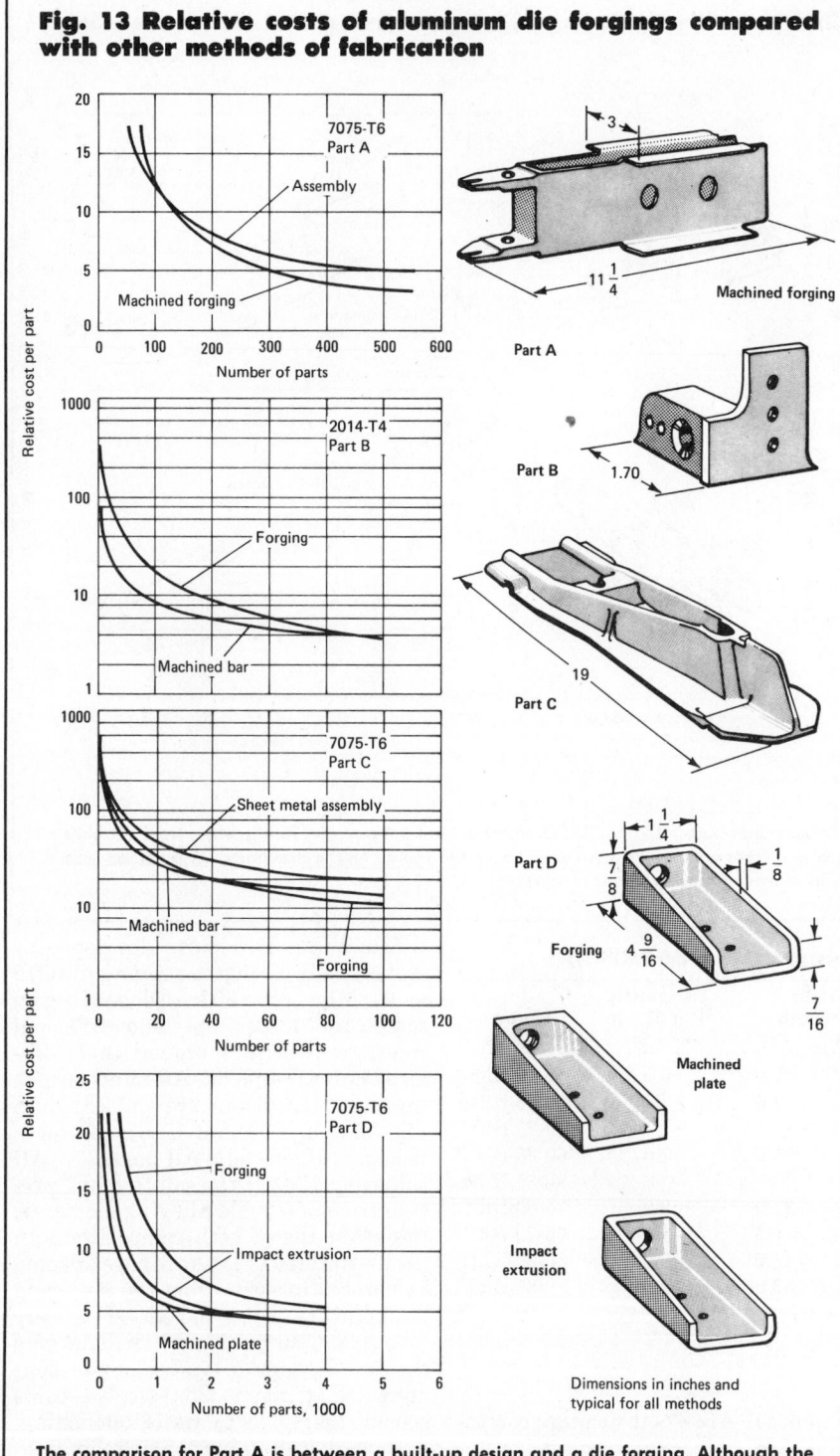

Fig. 13 Relative costs of aluminum die forgings compared with other methods of fabrication

Dimensions in inches and typical for all methods

The comparison for Part A is between a built-up design and a die forging. Although the rough forging was machined on all surfaces, a saving in fabrication cost was evident after about 125 forged fittings had been made. Part B is a simpler part, and the costs of forging compared with machining from a bar were the same at about the 100th piece. For a more complicated forging, Part C, the crossover point where the machined bar became more expensive than the forging occurred at 40 pieces. Part D, a relatively simple fitting, was made as a die forging, an impact extrusion, and a hollowed-out fitting. Forging was the most expensive approach. The cost of extrusion and of machining plate were about the same for 3000 fittings.

approximately 240 MPa (35 ksi), with slightly higher strengths achieved if the parts are re-pressed prior to heat treatment. Similar parts in alloy 201 AB develop 330 MPa (48 ksi) tensile strength in the T6 temper and this increases to 360 MPa (52 ksi) with re-pressing. These property levels combined with other inherent qualities make aluminum P/M parts attractive in a variety of markets.

Impact tests are used to provide a measure of toughness of powder metal materials, which are somewhat less ductile than similar wrought compositions. The standard P/M impact specimen (ASTM E23) is unnotched in order to achieve a greater spread of data among compositions and heat treatments. Annealed specimens develop the highest impact strength, whereas fully heat-treated parts have lowest impact values. Alloy 201 AB generally exhibits higher impact resistance than 601 AB at the same percent density and impact strength increases with increased 201 AB density. A desirable combination of strength and impact resistance is attained in the T4 temper for both alloys. In the T4 temper, 95% density 201 AB develops strength and impact properties exceeding those for as-sintered 99Fe-1C alloy, a P/M material frequently employed in applications requiring tensile strengths under 340 MPa (50 ksi).

Fatigue is an important design consideration for P/M parts subjected to dynamic stresses. The fatigue strength for aluminum alloys is defined as the maximum stress at which the material will withstand 500 million cycles of reversed bending without failure. Fatigue results were determined for 601 AB and 201 AB P/M aluminum alloys in both the as-sintered and heat-treated conditions using R. R. Moore-type specimens machined from sintered 95% density 16 by 16 by 89 mm (5/8 by 5/8 by 3½ in.) bars. The results are presented in Fig. 17 and 18 along with mean fatigue curves for wrought alloys 6061-T6 and 2014-T6. The fatigue strengths in the T6 temper are higher than in the as-sintered condition for both P/M alloys. However, beyond 1 million cycles the curves converge and the fatigue limits for both tempers are nearly the same. The average fatigue strengths for alloys 601 AB and 201 AB are 40 and 50 MPa (6 and 7 ksi), respectively. These values are about half those of wrought alloys 2014 and 6061, which also develop higher tensile

Fig. 14 Comparison of mechanical properties for bomb ejector feet made by forging and by impacting from 7075

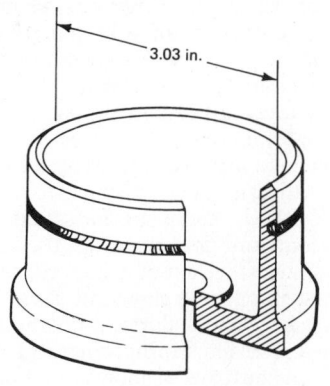

3.03 in.

Method	Tensile strength MPa ksi	Yield strength MPa ksi	Elongation, %
Die forging (a)	517 75.0	448 65.0	10.0
Impacting at room temperature	563 81.7	503 73.0	10.3
Impacting at 230 °C (450 °F)	561 81.3	492 71.3	10.0

(a) Specified properties for 7075-T6 in the required section size

Impacts were stress relieved at 340 °C (645 °F) after impacting, then solution treated and aged to T6 temper. Each value in the tabulation above is the average of five test specimens taken from the cylinder wall, parallel to the axis of impacting.

Fig. 15 Effect of density and thermal condition on tensile strength of alloy 601AB

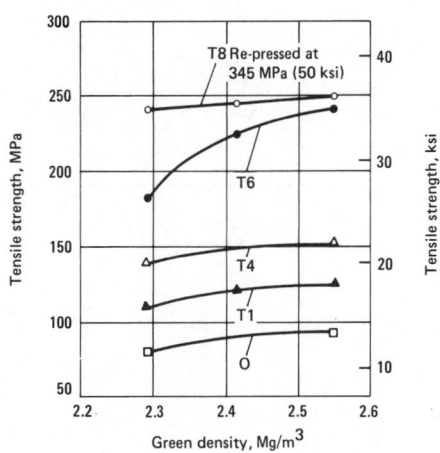

Sintered 30 min at 620 °C (1150 °F) in N_2 with average dew point of 7 °C (45 °F).

Table 5 Sintered properties of aluminum alloy 201AB(a)

Green density %	g/cm³	Temper	Tensile strength MPa	ksi	Yield strength MPa	ksi	Elongation, % in 25 mm or 1 in.	Apparent hardness, HRE
85	2.36	T1	170	24.5	145	21.0	2.0	60-65
		T4	210	30.5	180	26.0	3.0	70-75
		T6	250	36.0	...	...	...	75-80
90	2.50	T1	200	29.2	170	24.6	3.0	70-75
		T4	245	35.6	205	29.8	3.5	75-80
		T6	325	46.8	...	...	...	85-90
95	2.64	T1	210	30.3	180	26.2	3.0	70-75
		T4	260	38.0	215	31.0	5.0	80-85
		T6	330	48.1	325	47.5	2.0	85-90

(a) Sintered 30 min at 590 °C (1100 °F) in N_2 with average dewpoint of 7 °C (45 °F).

strengths than P/M materials because of the lower density of P/M parts.

In many materials, a relationship called fatigue ratio can be established between fatigue strength and tensile strength. This is not consistently true for aluminum, but for heat-treated aluminum alloys this ratio is about 0.3. The average fatigue ratio for aluminum P/M alloys is about 0.25 in the as-sintered condition and 0.18 in the T6 temper. The lower fatigue ratio for P/M parts is believed to result from porosity. However, this level of fatigue strength is suitable for many design applications.

Machined Parts. For some designs, machining may be the most economical production technique, because the set-up time may account for a large part of the cost of machining. In Fig. 19, for instance, the machined plate was the cheapest method of producing the alu-

minum part until production reached 600 pieces. At that point, the quantity would be sufficient to compensate for the cost of the extrusion die. The built-up design, with skin on both surfaces, was heavier and costlier.

Stampings are often chosen over other forms where intricate shapes must be held to close tolerances. The aluminum washing machine tub shown in Fig. 20 was originally designed as a sand casting, to be made in one piece from alloy 308.0 Need for light-weight and resistance to corrosion by washing detergents and tap water dictated the selection of aluminum. Although the engineering performance of the sand cast tub was satisfactory, the manufacturing costs were high and customer appeal was low.

The tub was redesigned to be made from two stamped parts welded together. The crown section was formed from alloy 5050 because it has good corrosion resistance and can be polished to a high luster. The bowl was deep drawn from alloy 3003. This alloy has excellent drawing characteristics and was completely compatible with the finishing requirements. Clear conversion coatings produced a bright lustrous finish for customer appeal. The two stamped portions were welded in an automatic machine in which the fixture and head moved in a manner to permit uninterrupted welding around the hub shape. The gas tungsten-arc process was used for welding.

The redesigned part was also more efficient to produce. The formed plate

Fig. 16 Effect of density and thermal condition on tensile strength of alloy 201AB

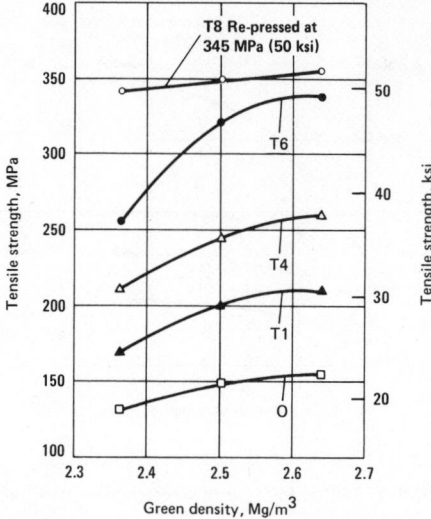

Sintered 30 min at 590 °C (1100 °F) in N_2 with average dew point of 7 °C (45 °F).

Fig. 17 Fatigue curves for P/M 601AB and wrought 6061

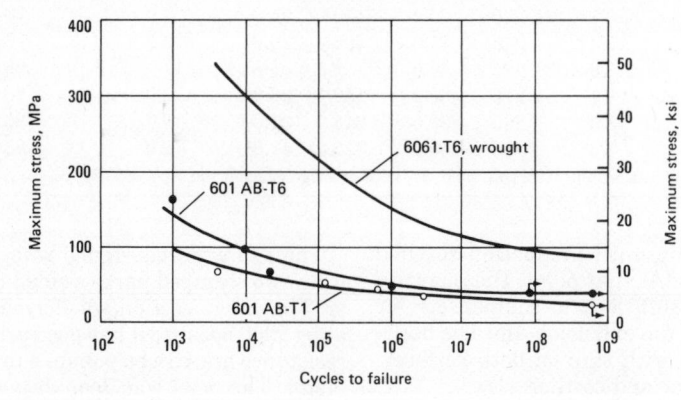

tub required 28 operations as compared with 57 for the cast tub. It permitted the detailed parts to be made on tooling that controlled finished dimensions accurately, reduced the cost of finishing operations, and simplified the manufacturing operations by processing the bowl without a top. Inventory control and scheduling problems were greatly reduced.

This comparison is valid only because enough parts were made to justify the more expensive tooling for drawing the tub parts from plate. The tooling was paid for with the savings from the manufacture of a few thousand tubs.

For further information on the use of aluminum in stampings, see the separate article on this subject that appears in this volume.

Fabrication Characteristics

Machinability of all aluminum alloys is excellent compared with steel or titanium. Among the various wrought and cast aluminum alloys and among the tempers in which they are produced, however, there is considerable variation in machinability. (See the separate article on the machinability of aluminum alloys in this volume.)

Chemical milling (removal of metal by dissolving in an alkaline or acid solution) is routine for specialized operations on aluminum. For flat parts on which large areas having complex or wavy peripheral outlines are to be reduced only slightly in thickness, chemical milling is usually the most economical method. A typical curve used by one company for choosing the most economical method of removing metal from such parts is shown in Fig. 21.

For sheet metal parts that cannot be formed after machining, chemical milling is the only practical way metal can be removed to obtain a waffle-type grid with uniform skin thickness. Even then, allowance must often be made for some spring back resulting from metal removal and the consequent redistribution of residual forming stresses.

Chemical milling can normally produce a stiffened skin to ±0.13 mm (±0.005 in.) thickness tolerance, with the cladding left intact on the unmilled surface. Mechanical milling of skins from sheet thicker than 3 mm (1/8 in.) requires a cleanup "skim" cut for flatness on the holddown surface, because of holddown limitations.

Formability. Aluminum and its alloys are among the most readily formable of the commonly fabricated metals. There are, of course, differences between aluminum alloys and other metals in the amount of permissible deformation, in some aspects of tool design, and in details of procedure. These differences stem primarily from the lower tensile and yield strengths of aluminum alloys, and from their comparatively low rate of work hardening. The compositions and tempers of aluminum alloys also affect their formability.

Ratings of comparable formability of the commercially available alloys in the various tempers depend on the forming process, and are presented in the article on forming of aluminum alloys in Volume 4 of the 8th edition of this handbook. Such ratings provide generally reliable comparisons of the working characteristics of metals, but at best are only an approximate guide to forming limits in any specific application. Trial runs and evaluative techniques developed for specialized ap-

Fig. 18 Fatigue curves for P/M 201AB and wrought 2014

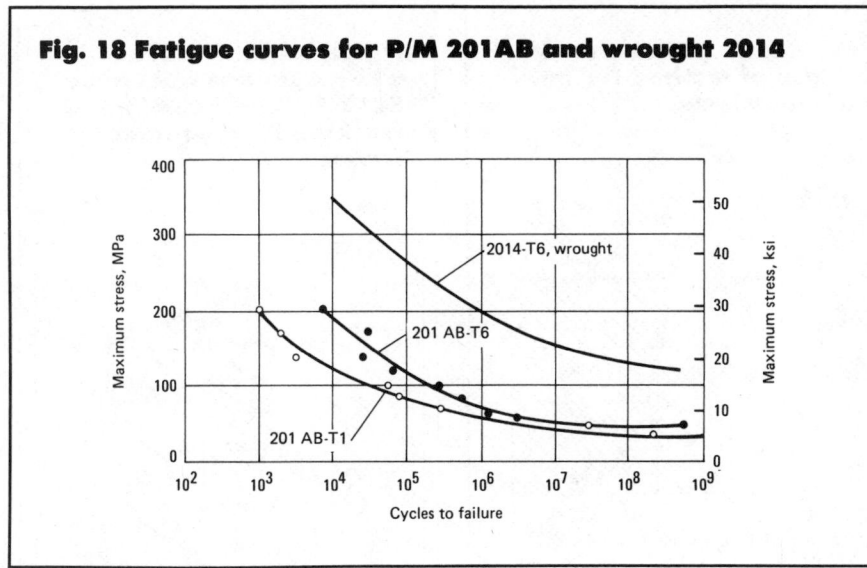

Fig. 19 Cost comparison of producing an alloy 2024-T4 part by three methods

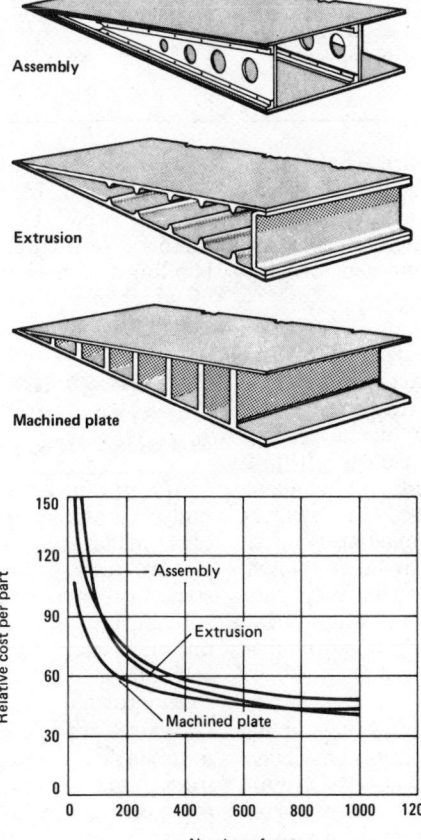

plications are needed in borderline or critical situations.

Choice of temper may depend on the severity of the forming operations. The annealed temper may be required for severed forming operations such as deep drawing, or for roll forming or bending to exceptionally small radii. Usually, the strongest temper that can be formed consistently is selected. For less severe forming operations, the intermediate tempers, or even full-hard work metal, can be used.

The non-heat-treatable alloys 1100 and 3003 are frequently used in forming applications, because of their excellent workability and low cost. If somewhat higher strength is required, alloys containing magnesium are commonly used (for example, in order of increasing strength, alloys 3004, 5052, 5154 and 5086).

If superior finishing characteristics are needed in addition to higher strength, alloy 5457, which contains a small amount of manganese in addition to magnesium can be used. Holding impurities at a low level in alloys used for decorative and finishing purposes also helps in developing bright, uniform finish.

Heat treatable alloys are used in applications for which a high strength-to-weight ratio is required. These include alloys 6061, 2014, 2024, 7075 and 7178, in approximate order of increasing strength.

The annealed temper (O) is the most workable condition for forming, but it entails the greatest expense in subsequent heat treating and straightening. Alloys that have been freshly solution heat treated and quenched (W temper) are nearly as formable as when annealed, and can be given increased strength after forming by natural or artificial aging, without reheating and consequent exposure of the finished part to warping. Alloys can be stored in the W temper for a reasonable period at a low temperature. (Almost no aging occurs in most alloys at −30 °C, or −20 °F.)

Material that has been solution heat treated at the mill, but not artificially aged (T3, T4 or W temper), is generally suitable only for mild forming operations such as bending, mild drawing, or moderate stretch forming.

Solution heat treated and artificially aged (T6 temper) alloys are seldom used for forming, other than bending to standard radii and forming of very shallow shapes. Although alloys in the

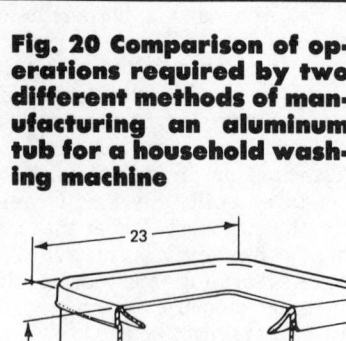

Fig. 20 Comparison of operations required by two different methods of manufacturing an aluminum tub for a household washing machine

Dimensions in inches

Type of operation	Number of operations
Sand cast (Alloy 308.0)	
Foundry	21
Machining	6
Polishing and finishing	26
Testing and inspection	4
Total	57
Formed sheet (5050 and 3003)	
Forming and finishing bowl	10
Forming and finishing crown	11
Welding	2
Machining weld and inspection	5
Total	28

See text for discussion.

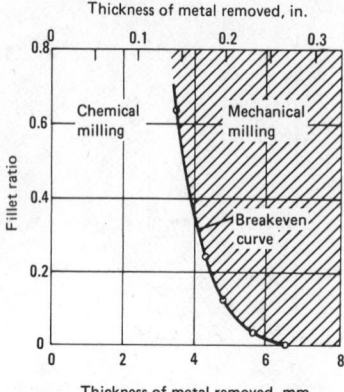

Fig. 21 A parameter for choice of method for milling aluminum

Thickness of metal removed, in.

Thickness of metal removed, mm

The parameter applies to the removal of metal from large areas having complex or wavy peripheral outlines. The curve shows that the chemical process should always be used for metal removal of 3 mm (0.125 in.) or less. The choice of method for thicknesses of 3 to 6 mm (0.125 to 0.250 in.) depends on fillet ratio or the amount of weight penalty. Metal thicknesses above 6 mm (0.250 in.) should be mechanically milled.

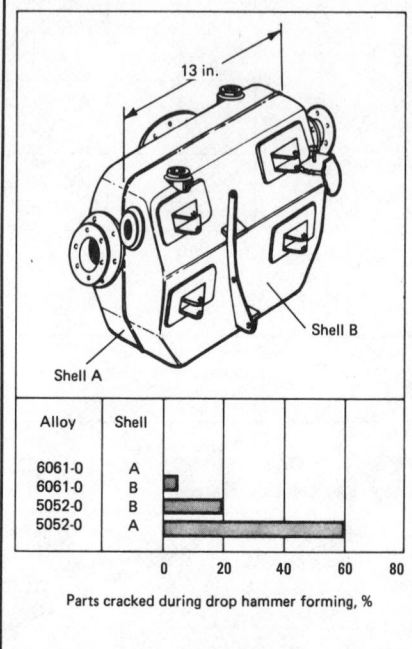

Fig. 22 Aircraft oil tank in which a change from alloy 5052 to 6061 eliminated cracking in drop hammer forming

Shell A

Shell B

Alloy	Shell
6061-0	A
6061-0	B
5052-0	B
5052-0	A

Parts cracked during drop hammer forming, %

T6 temper are much stronger, they have lost so much ductility in hardening that they are apt to fracture in even moderately severe forming.

Often a change in alloy will completely eliminate a forming problem. In the example of Fig. 22, producibility calculations indicated that the deformation needed to form the oil-tank shells did not exceed that expected for alloy 5052, and that this alloy, having excellent weldability, should be used, because the tank contained many welded bosses and fittings. However, in the shop, there was an excessive rejection rate in the drop hammer forming of the two halves of the shell. A change to alloy 6061 almost completely eliminated cracking in drop hammer forming and eliminated the need to revise tooling, even though the elongation of the two alloys is exactly the same as

measured by the standard tension test. The change to 6061 also permitted the final part to be solution heat treated and aged, so that a higher operating pressure could be imposed on the tank.

Economy is often realized in the fabrication of difficult parts by forming heat treatable alloys (especially the 2xxx and 7xxx series) in the freshly quenched (W) temper (usually within 2 h after quenching if material is held at room temperature, or longer if refrigerated to below 0 °F immediately after quenching). Forming in the W temper often eliminates the need for a separate straightening operation after heat treatment, to eliminate quenching distortion.

The use of aluminum alloy sheet for automobile stampings has been steadily increasing, because of aluminum's light weight, its excellent corrosion resistance, and its good formability. The formability of aluminum in such stampings is discussed in a separate article in this volume.

Joining. Most aluminum alloys are easily welded. In addition, aluminum can be joined by a great variety of methods—brazing, soldering, adhesive bonding and mechanical fastening—all of which are discussed in a separate article in this volume.

Buildings and Construction Applications

Aluminum is being used increasingly in static structures such as buildings, bridges, towers and large field storage tanks (see Fig. 23). Because structural steel shapes and plate are substantially lower in initial cost, aluminum is used only where light weight or the cost of maintenance is a major consideration.

Design and fabrication of aluminum static structures differ very little from practices used with steel. The most common alloys are 6061-T6 and 2014-T6. Tensile strengths are in the range of structural carbon steel and low-alloy steel, respectively; the modulus of elasticity is one third that of

Fig. 23 Annual U.S. shipments of aluminum by major market

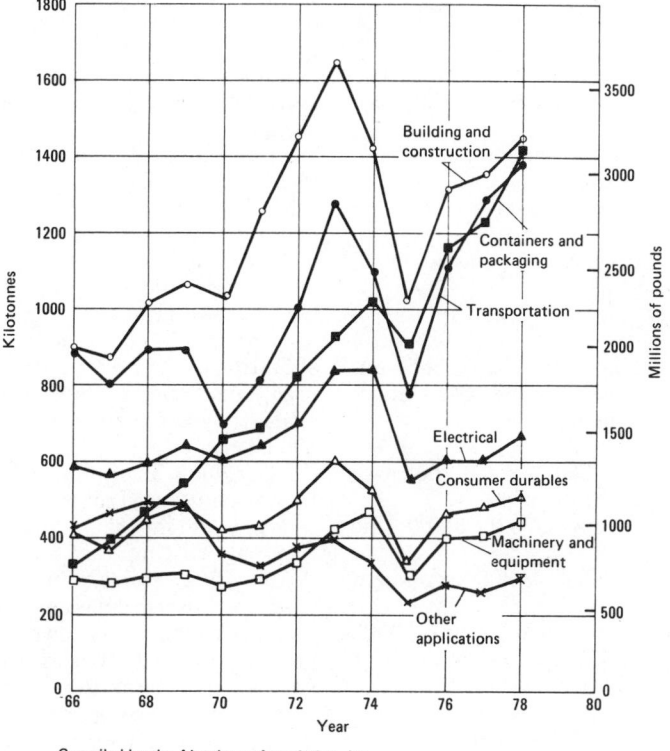

Building and construction

Containers and packaging

Transportation

Electrical

Consumer durables

Machinery and equipment

Other applications

Compiled by the Aluminum Association, Inc.

Compiled by the Aluminum Association, Inc.

steel, requiring special attention to compression members; it also provides a cushion against shock loads and minor foundation misalignment. Conditions of repeated loading require special consideration. High temperatures reduce strength and raise ductility; both are maintained or improved at subzero temperatures. When properly designed, aluminum typically saves over 50% of the weight of small structures compared to low-carbon steel; a savings of 75% or more may be possible in long-span or movable bridges. Savings also result from low maintenance costs and in slow deterioration from atmospheric corrosion.

Forming, shearing, sawing, punching and drilling are readily accomplished on the same equipment used for fabricating structural steel. Since the structural aluminum alloys owe their strength to properly controlled heat treatment, hot forming must be done

with caution, and sometimes reheat treatment is necessary. Burning or flame cutting should not be used, but arc cutting may be used and is as fast or faster. Special attention must be given to the strength requirements of welded areas, because of the annealing effects.

Buildings. The most important aluminum applications to farm buildings have been corrugated, or otherwise stiffened, sheet products. Alloy 3003 or similar compositions are employed, usually in the harder rolled tempers. Roofing, siding, ventilators and other components are made of similar materials.

The use of aluminum in industrial buildings has likewise been generally limited to portions exposed to weather. Roofing and siding are the most common applications; the alloys most used are 3003, 3004 and 5052, either bare or clad. Door and window frames are usu-

ally formed of 6063-T5 extruded shapes. A few industrial buildings, such as greenhouses, have 6061-T6 framework to resist corrosive conditions within.

Aluminum roofing, flashing, gutters and downspouts of 3003 or 3004, preferably clad, are used in homes, hospitals, schools, commercial establishments and office buildings. Exterior walls may be either 6063-T6 extrusions or formed sheet, bare, anodically treated, painted, or enameled. Curtain walls are customarily made from 6063-T5 extrusions and 3003, 3004 and 5005-H16 sheet. In addition, many interior applications, such as wiring, conduit, piping, ductwork, hardware, and railings, utilize aluminum in many forms and finishes.

Bridges and Highway Applications. A great deal of aluminum is used in bridge and highway accessories, such as bridge railings, highway guard rails, lighting standards, traffic control towers, traffic signs and chain-link fences. Aluminum also is economically feasible for bridge structures themselves, especially if extremely long-span or of the movable type (bascule and vertical-lift). Portable military bridges have been built in quantity as well as superhighway overpass bridges.

Scaffolding, ladders, electrical substation structures, and others of the same general type utilize aluminum, chiefly in the form of structural and special extruded shapes. Alloy 6061-T6 is the most widely used, because of strength and resistance to corrosion without paint.

Cranes, conveyors, and heavy-duty structures utilize smaller amounts of aluminum than the category just discussed. Welded construction, with 5083 and 5456 alloys, has been extensively used in the fabrication of heavy-duty cranes. For information on allowable stresses and design rules for riveted heavy-duty structures of this kind, the reader may refer to Trans ASCE, **117,** 1253 (1952). (Reference: Aluminum Construction Manual, Aluminum Association, Inc., 1976).

Containers and Packaging

Materials Handling Equipment. The food and drug industries use aluminum handling equipment more than other industries, because the metal is nontoxic, nonadsorptive and splinterproof, does not harbor bacteria, forms colorless salts, and can be steam cleaned. Low volumetric specific heat results in economies when containers or conveyors must be moved in and out of heated or refrigerated areas. The nonsparking property is valuable in flour mills and other plants subject to fire and explosion hazards.

Corrosion resistance is important in shipping fragile merchandise, valuable chemicals, and cosmetics. Large sealed aluminum containers, designed for either rail or truck shipments, are used for chemicals not suited to bulk shipment in box cars or hopper cars. Other examples of handling equipment include: airplane baggage carts, wheelbarrows, hand shovels, pallets, dock boards, bakery cabinets, tipping slings, electric hoist housings, lift-truck parts, beverage cases, bread handling racks, and liquid transfer pumps.

Packaging is one of the fastest growing markets for aluminum. Products include household wrap, flexible packaging and food containers, bottle caps and collapsible tubes. More than 80% of aluminum foil is used for packaging, for pouches and wraps for foodstuffs and drugs, as well as for household wrap. In the manufacture of cans, aluminum competes with tinplate, and a great many foodstuffs—soft drinks, beer, coffee, snack food and meat—are packaged in aluminum cans. Draft beer usually is shipped in alclad aluminum barrels. Currently, aluminum accounts for more than half of the production of collapsible tubes—for toothpaste, ointments, food and paints.

Transportation

Automobiles. Most pistons for automobile engines are permanent mold cast in aluminum alloy 332.0. Both aluminum die and permanent mold castings are utilized for other engine components in various automobile engines. These include cylinder heads, crankcases, and most structural parts. As discussed in the separate article on aluminum stampings, aluminum sheet metal parts are finding increased usage in automobiles.

A partial listing of other aluminum applications for various passenger car components follows:

Air-conditioning evaporators—No. 12 brazing sheet (4343 cladding on 3003 core) with 3003 or 3102 press fitted or solderd tube

Body panels—2036, 5182 sheet

Brake drums—355.0, 356.0, permanent mold cast

Bumpers—7016, 7029 extrusions

Bumper back-up bars—6061, 7021, 7146 sheet and extrusions

Carburetor bodies—360.0, 380.0, die cast

Compressor bodies, connecting rods and pistons—360.0, 380.0, die cast

Cranking motor end plates—360.0, 380.0, die cast

Distributor cap inserts—1100 or 1350 wire

Distributor housing base—360.0, 380.0, die cast

Door sill molding (scuff plates)—3003

Exterior trim parts—5252, 5657 anodized sheet or 6063, 6463 anodized extrusions

Fuel lines—3003 tube

Fuel pumps—360.0, 380.0, die cast

Horn coils—1350 strip

Horn trumpets—360.0, 380.0, die cast

Intake manifolds—355.0 permanent mold cast

Interior trim parts—1100, 3003 sheet or 5657 anodized sheet

Master cylinder body—360.0 die cast

Master cylinder pistons—360.0, 380.0, die cast, or 6061, impact extruded, or 2017, 2011, machined

Oil filter bases—360.0, 380.0, die cast

Oil filter bodies—1100 extrusions

Oil pumps—360.0, 380.0, die cast

Power brake housings—360.0, 380.0, die cast

Rocker arm covers—355.0 permanent mold cast

Spark plug well tubes—1100 extrusions

Stators—360.0, 380.0 die cast, or 319.0, 355.0, permanent or plaster mold cast

Timing chain covers—360.0, 380.0, die cast

Transmission housings—360.0, 380.0, die cast, or 355.0, 356.0 permanent mold cast

Transmission valve bodies—380.0, 413.0 die cast

Valve rocker arms—360.0, 380.0, die cast

Variable pitch stator blades—6063-T6 extrusions

Water pump bodies and covers—360.0, die cast

Trucks. Because of legal weight limitations and a desire to increase pay-load, truck manufacturers use aluminum for structural members and other truck components. Alloy 332.0 is widely used for truck engine pistons. Camshaft timing gears have been largely replaced by chain drives, but where camshaft timing gears are fabricated of aluminum, alloys 332.0 or alloy 355.0 permanent molded gear blanks are employed.

Sheet alloys 5052, in the H32 or H34 condition, or 6061-T4 are used in truck cabs. Dead weight is also reduced by making cab stringers of 6061-T6 or 6063-T6, and frame rails and cross members from 6061-T4 or 6061-T6 extrusions. Aluminum bumpers are also used on trucks. If formed from sheet, 5052-H34 or 6061-T4 alloys are usually used, but if the design requires extrusions, they are made from 6061-T4 or 6063-T4.

Alloy 5052, in the O or H34 condition, is being used for many truck fuel tanks for weight reduction as well as to eliminate rusting.

Alloys 2014 and 2024 have long been used for forged truck wheels. Good thermal conductivity lowers tire temperature, which increases tire life, and light weight increases fuel economy.

Mobile homes and travel trailers usually are constructed with alloy 3005-H25 or similar alloy sheet coverings, used bare or with a mill-applied baked-enamel finish. These coverings are either nailed to a wood frame or riveted to an extruded aluminum alloy frame.

Buses. Bus manufacturers are also concerned with minimizing dead weight, and manufacturers employ aluminum sheet and extrusions for body components. Body sheet metal is generally 5052-H34. Extrusions are usually made from 6061-T4 or T6, although 6063-T6 is sometimes used.

Truck Trailers are designed for maximum payload and operating economy, in consideration of legal weight requirements. Freight cargo trailers average about 1270 kg (2800 lb) of aluminum, and some may contain as much as 1590 kg (3500 lb). Aluminum is used for such items as the frames, floors, roofs, cross sills under the floor, and shelving. For high strength and rigidity, body side panels are usually made from Alclad 2024-T3, 6061-T6, or 5155.

Roof sheeting is usually 3003, in the H14 or H16 condition, while 6061, 6062 or 6063 extrusions, in the T6 condition, are utilized for stringers, top rails, bottom rails, and side posts. Trailer flooring is generally 6063-T6 alloy, while extruded or formed 6061-T6 is usually used for frames. Forged aluminum wheels are used on some trailers.

Tanker bodies are commonly made from 5052 or 5454, in the O or H34 temper, for both riveted and welded assemblies. Alloy 5454, 5083 and 5086 in the H32 or H34 temper and alloy 7005 in the T53 temper are most commonly used for welded dump bodies, but if they are constructed by riveting, 6061-T6 sheet and plate are used.

In investigating uses for aluminum die castings, vehicle manufacturers proceed on the premise that successful use depends on proper designing for aluminum, rather than mere substitution for other metals. Die casting has several well-recognized qualities, such as high production rate and lower assembly and unit costs. Lower final cost, resulting from several aluminum processing methods, tends to offset the higher cost of the aluminum metal as compared to ferrous materials. The amount of aluminum used and specific applications vary considerably from one manufacturer to another. Functional gains by the vehicle user are generally the primary motivation to substitute aluminum for ferrous materials. However, in many instances, cost reductions also result.

Bearings. Three aluminum-base alloys are used in some medium and heavy-duty gasoline and diesel engines for connecting-rod and main bearings. 852.0 (7.0Sn-1.0Cu-1.65Ni-0.6Si-1.0-Mg) and 850.0 (6.25Sn-1.0Cu-1.0Ni) are normally permanent mold castings. Another composition, 828.0 (similar to 850.0 but containing a nominal 1.5% Si), is a wrought alloy and is used in composite bearings with a steel backing and usually with a plated overlay. For a detailed discussion of aluminum alloy bearings, the reader may refer to the article on sleeve bearing materials, which will appear in Volume 3 of this handbook (9th edition).

Railroad Cars. Aluminum is used for railroad hopper cars, box cars, refrigerator cars and tank cars. Tank cars are of welded 1060, 1100, 3003, 5052, 5083, 5086, 5154, 5456 and 6061, in various tempers. These are built to specifications of either the Association of American Railroads or the Interstate Commerce Commission, depending on the commodity to be handled. In addition to freight cars, aluminum is used extensively in some passenger cars, mainly those for mass transit systems.

Marine Craft. Aluminum is commonly used for a large variety of marine applications, including main strength members. Some of these are hulls, deckhouses, stack enclosures, hatch covers, windows, air ports, accommodation ladders, gangways, bulkheads, deck plate, ventilation equipment, life-saving equipment, furniture, hardware and architectural trim. In addition, extensive use is now being made of ships fitted with large tanks of welded aluminum alloy plate for long-distance transportation of liquified natural gas.

Most structural plate used in marine applications is 5086, 5083 or 5456. Extrusions are from the same alloys or 6061. Alloy 5050 in sheet form is widely used for nonstructural applications.

The corrosion-resistant aluminum alloys in current use permit designs that save about 50% of the weight of similar designs in steel. Substantial savings of weight in deckhouses and topside equipment permit lighter supporting structures. The accumulative savings in weight improve the stability of the vessel and allow the beam to be decreased. For comparable speeds, the lighter, narrower craft will require a smaller power plant and will burn less fuel. Consequently, 1 kg of weight saved by the use of lighter structures or equipment frequently leads to an overall savings of 3 kg. Aluminum also saves on marine maintenance.

The low values of modulus of elasticity for aluminum alloys offer advantages in structures erected on a steel hull. Flexure of the steel hull results in low stresses in an aluminum superstructure, as compared with the stresses induced in a similar steel superstructure. Consequently, long continuous aluminum deckhouses can be built without expansion joints, thus eliminating a serious maintenance problem.

Die-cast alloy 413.0 predominates in outboard motor structural parts and housings subject to continuous immersion or to frequent wetting. Die-cast alloy 383.0 is used for motor hoods, shrouds and miscellaneous parts subject to less severe corrosive exposures. Wrought alloy 3003 is used for integral gas tanks, which are drawn from sheet stock.

Hull construction in most small craft, such as outboard motorboats, rowboats and canoes, including those designed for operation in fresh water, is shifting to 5052 and 6061-T4 or -T6 extrusions or sheet. A small amount of 3003 is used as casing for built-in safety buoys. Alloys 5005 and 5050 are finding use in nonstructural applications.

Aerospace Applications. Aluminum is used in many segments of the aircraft, missile, and spacecraft industry—in airframes, engines, propellers, accessories and tankage for liquid fuel and oxidizers. Aluminum is widely used because of its high strength-to-density ratio and weight efficiency in compressive designs. Minimum property values of these alloys for use when designing for aerospace applications are those listed in Military Handbook 5A, *Metallic Materials and Elements for Aerospace Vehicle Structures*, U.S. Department of Defense.

Corrosion resistance is important for many aerospace applications; it may be the primary design requirement, as in containers for fuming nitric acid, or a secondary attribute, as when resistance to corrosion in salt water is needed for carrier-based and water-based aircraft. Corrosion theory and corrosion behavior of aluminum alloys are described in the article on corrosion resistance of aluminum and aluminum alloys in this volume.

Storage and ground handling equipment for red and white fuming nitric acids, common oxidizers for rocket motors, is usually made from 1100 alloy; in flight vehicles, from 6061-T6.

Increased resistance to corrosion in salt water and other atmospheres is secured through the use of alclad alloys or anodic coatings. The exterior of aircraft exposed to salt water environment is usually fabricated from clad alloys. Anodized bare stock successfully resists corrosion when only occasional exposure to salt water is encountered. In these applications, the corrosion resistance may be further improved by organic finishes.

Water storage tanks are often made of alclad 6061 to improve resistance to pitting corrosion and to match the color of surrounding structural elements of alclad 2024 and 7075 in commercial transports.

Following is a partial list of aluminum usage in aircraft engines:

Baffles and deflectors—2024-T3
Castings in general—355.0-T6 or -T71, 295.0-T4 or -T6, 356.0-T6
Crankcases—6151-T6
Cylinder-barrell muff—6151-T6

Cylinder heads—2218-T6 or -T72, if forged; 242.0-T77, if cast

Oil lines and push-rod cover tubes—2024-T3

Oil pumps—6061-T6

Pistons—4032-T6, 2018-T61, 2618-T6

Supercharger impellers—2014-T6, 2025-T6

Following is a partial list of other usage of aluminum in aerospace applications:

Airplane skins and cowls—2024-T4, 7075-T6, 7475-T761

Control brackets, pulleys—295.0-T6, 356.0-T6, 355.0-T6, 712.0-F

Helicopter rotor hubs—2014-T6

Helicopter rotor skins—2024-T4, 6061-T651

Hydraulic line tubing—2024-T3, 5052-O, 6061-T4 or -T6

Instrument cases—518.0-F and 380.0-F die castings

Machined integrally stiffened skin—Alclad 7075-T7651, 7178-T51, 7475-T7651

Name plates—1100-O, 3003-O

Rivets—2117-T4, 2017-T4, 2014-T4, 2024-T4

Space vehicles—2014, 2024, 2124, 2219, 7075, 7178

Spun, drawn or cast pressure receptacles—3003, 5052, 5083, 5086, 5154, 5454, 5456

Water tanks—6061-T6 or Alclad 6061-T6

Wing spars and attachment forgings—2014-T6, 7049-T73, 7050-T736, 7075-T73

Electrical Applications

Conductor Alloys. Aluminum 1350 (formerly EC aluminum) accounts for all but a minor proportion of the aluminum now being used as conductor metal in the United States. Aluminum 1350 contains 99.50% Al min, with closely controlled impurities. It usually is treated with trace additions of boron to remove titanium, vanadium and zirconium, which are particularly harmful to electrical conductivity. Specifications are given in ASTM B230, B231, B232, B233, B236, B262, B314, B324, B400, B401 and B544.

The extensive use of aluminum 1350 rests on a combination of low cost, high electrical conductivity, adequate mechanical strength, low specific gravity, and excellent resistance to corrosion. Minimum conductivity of 61.8% of the International Annealed Copper Standard (IACS) and from 55 to 124 MPa (8 to 18 ksi) minimum tensile strength, depending on size, are readily maintained in commercial hard drawn aluminum 1350 wire. When compared with IACS on a basis of mass instead of volume, minimum conductivity of hard drawn aluminum 1350 is 204.6%.

Magnesium-silicide alloys are finding a growing use as alloy 6101 bus bar, as alloy 6063 tube and as alloy 6201 wire for service at slightly elevated temperatures. Alloys 8076 and X8176 are used for insulated wire for buildings.

Aluminum for cable sheathing varies widely from high-purity commercial aluminum to alloys of the 3003 type. Two methods are used: (a) extruding the sheath in final position and dimensions around the cable as it is fed through an axial orifice in the extrusion die, and (b) threading the cable through an oversized prefabricated tube and then squeezing the tube to final dimensions around the cable by tube reducers and draw dies.

Conductor accessories also vary widely from aluminum 1350 to the heat treatable, high-strength wrought alloys, and casting alloys such as 514.0 and 356.0.

Aluminum 1350 Wire Conductors. Common forms of aluminum 1350 conductors are single-wire and multiple-wire (stranded, bunched or rope layed). Both are used in overhead or other tensioned applications, as well as in nontensioned insulated applications. The tensioned conductors employ aluminum 1350 wire in the H19 temper; nontensioned, in the H26 temper.

Size for size, the direct-current resistance of an aluminum 1350 conductor is from about 1.6 to 2.0 times IACS. For equivalent direct-current resistance, an aluminum wire two American wire gage sizes larger than a copper wire must be used. Nevertheless, as a result of the lower specific gravity of aluminum, an aluminum 1350 conductor weighs only about half as much as an equivalent copper conductor.

Aluminum conductors, steel reinforced, consist of one or more layers of concentric-lay stranded aluminum 1350 wire around a high-strength galvanized or aluminized steel wire core, which itself may be a single wire, or a group of concentric-lay strands. Design and materials are specified by ASTM B232.

Constructions of aluminum conductors, steel reinforced (ACSR), are somewhat larger in diameter and far stronger than equivalent conductors employing only aluminum 1350. Electrical resistance is figured only on the aluminum cross section, whereas tensile strength is figured on the composite; the steel core provides 55 to 60% of the total strength.

ACSR constructions are used where great mechanical strength is advantageous. Their strength-to-weight ratio is usually about two times that of copper of equivalent direct-current resistance. ACSR cables, therefore, permit longer spans and fewer or shorter poles or towers.

Bus Bar Conductors. Commercial bus design in the United States utilizes four types of bus conductors: rectangular bar, solid round bar, tubular and structural shapes.

All types are supplied in aluminum 1350 and 6101, with tubular and structural shapes available in 6063 also.

Motors and Generators. Aluminum has long been used for cast rotor windings and for several structural parts. The end rings and usually the cooling fans are pressure cast integrally with the bars through the slots of the laminated core, from commercially pure aluminum ingot. The rotor thus formed stays permanently tight and quiet, in addition to its cost advantages.

Aluminum structural parts, such as stator frames and end shields, are often economically die cast. Their corrosion resistance may be necessary in specific environments—for example, in motors for spinning rayon (aluminum salts do not stain the fibers). They are also found in aircraft generators where light weight is paramount.

Aluminum 1350 has more recently been introduced for field coils on some of the direct-current machines, for stator windings in motors, and in transformer windings. Wire of the magnesium-silicide alloys is used in some of the extremely large turbogenerator field coils, where high operation temperatures and centrifugal forces cause creep failures in aluminum 1350 and other conductor wire.

Transformers. Aluminum windings have been extensively used in dry-type power transformers, and have been adapted to the secondary coil windings in the magnetic-suspension type of constant-current transformer to decrease weight and permit the coil to "float on magnetic suspension". In an application closely associated with transformers, aluminum is being used for concrete reactors (devices to protect transformers from overloads), where, as the name implies, conductors are cast in concrete.

Electronics applications, where aluminum is used primarily because of its electrical characteristics, include hollow shapes (both cast and wrought) in radar and sonar wave guides, copper-clad strips in printed circuits and other applications where weight is important, extruded shapes and punched sheet for radar antennas, extruded and roll formed tubing for television antennas, strips in lengths up to 300 ft for coiled line traps, drawn or impact-extruded cans for condensers and shields, and vaporized high-purity coatings inside cathode-ray tubes.

Examples where electrical properties, other than magnetic, are not dominant are chassis for electronic equipment, spun pressure receptacles for airborne equipment, etched name plates, and hardware, such as bolts, screws, and nuts. In addition, finned shapes are used to support electronic components to facilitate heat removal. A closely associated use for 6061-T63 is as a cell base for the deposition of selenium in the manufacture of selenium rectifiers.

Lighting. The use of 3004 aluminum for incandescent and fluorescent lamp bases, because of high melting temperature, and other sheet alloys for sockets are two established uses.

Capacitors. Aluminum, in the form of foil, dominates all other metals in the construction of capacitor electrodes. Dry electrolytic and nonelectrolytic capacitors are the only two basic types of condensers in extensive commercial use today. Dry electrolytic capacitors usually employ two parallel coiled or wrapped aluminum foil ribbons as electrodes. Paper wrapped into the coil mechanically separates the two ribbons. A glycol-borate paste, which is absorbed in the paper separators, functions as the operating electrolyte. In constructions designed for intermittent use in alternating circuits, both electrodes are anodized in a hot boric acid electrolyte, the thin anodic films constituting the dielectric element.

Only the anode foil is anodized in dry electrolytic assemblies intended for direct-current applications. Anodized electrodes are invariably made in aluminum of 99.8% or greater purity, whereas the nonanodized electrodes usually utilize foil ribbons of 99.4% minimum purity. Prior to anodizing, the foil is usually, but not always, etched to increase the effective surface area. Containers for dry electrolytic capacitors may be either drawn or impact-extruded cans of 1100 alloy.

Ordinary clean foil ribbons of 99.4% minimum purity usually serve as the electrodes in commercial nonelectrolytic capacitors. Oil-impregnated paper separates the electrodes and adjacent coils of the wrap. Nonelectrolytic foil assemblies are packed in either aluminum alloy or steel cans. Electrodes for variable air capacitors are usually made of 1100 or 3003 sheet in the intermediate tempers.

Consumer Durables

Household Appliances. Light weight, excellent appearance, adaptability to all forms of fabrication, and low fabricating costs are the reasons for the broad usage of aluminum in household electrical appliances. Light weight is an important sales characteristic in vacuum cleaners, electric irons, portable dishwashers, and portable food mixers which are continually moved about by the housewife.

Low fabricating costs depend on several properties, including adaptability to die casting and low finishing costs because of a natural pleasing appearance and good corrosion resistance, which eliminate the need for expensive finishing.

Aluminum is used for refrigerator and freezer evaporators because of its brazeability, in addition to other favorable characteristics. Tubing of 1100 and 3003 alloy is brazed to 3003 embossed or plain sheet, using an aluminum-silicon alloy for the braze metal. The tubing is placed on the embossed sheet over strips of brazing alloy with a suitable flux. The assembly is then furnace brazed, and the residual flux is removed by successive washes in boiling water, nitric acid and cold water. The result is an evaporator with high thermal conductivity and efficiency, good corrosion resistance and low manufacturing cost.

With the exception of a few permanent mold parts, virtually all of the aluminum castings in electrical appliances are die cast. The most popular alloy is 380.0, because of its good castability, adequate mechanical properties and low cost.

Most of the die castings are internal functional parts and are used without any finish. Organic finishes are usually applied to die-cast parts exposed to view, such as housings for food mixers.

Wrought forms, fabricated principally from sheet, tube and wire, are used in approximately the same quantities as die castings. The wrought alloys 1100, 3003, 5357, 5457, 5557, 5050, 5052 and a few 6063 or 6463 extrusions are selected because of corrosion resistance, anodizing characteristics and excellent formability. Alloys 5357, 5457 and 5557 are outstanding in their anodizing characteristics and are used with various colored anodized surfaces for trim on some electric ranges.

The natural colors that some of these alloys assume after anodizing are extremely important for food-handling equipment. Applications include refrigerator vegetable pans, ice-cube trays and wire shelves. The pans and trays are usually drawn from 5050-H38, anodized after fabrication. In the production of wire shelves, full-hard wire is cold headed over extruded strips, which form the borders.

An exception to the natural-color anodized surface for food-handling equipment occurs in waffle iron grids, which are given a silicone resin finish or are Teflon coated to facilitate removal of the waffle.

Furniture. Light weight, low maintenance and attractive appearance are the principal advantages of aluminum in furniture.

For office chairs, the most commonly used alloy is 6061-T6 in the form of drawn tube (round, square or rectangular), sheet or bar. Extruded tube and special shapes of 6063-T5 are also used. Frequently, the parts are formed in the annealed or partially heat treated tempers and are subsequently heat treated and aged. Designs are generally based on the service requirements, although styling may dictate overdesign or inefficient sections. Fabrication is conventional; joining is usually by welding or brazing. Various finishing procedures are used, such as mechanical, anodic oxide coatings or paint finishes.

Tubular sections, usually round and frequently formed and welded from flat strip, are the most popular form of aluminum in lawn furniture. Resistance to corrosion is an added advantage. Designs are established by structural engineering practices, with some deviations for appearance requirements. Alloys principally used are 3003, 6063 and 6061, the temper being selected by strength requirements and the degree of forming required. Conventional tube bending machines and mechanical joints are used. Finishing is usually by grinding and buffing, frequently followed by clear lacquers.

Cooking utensils may be cast, drawn, spun or drawn and spun from aluminum. Handles are often joined to the utensil by riveting or spot welding. In some utensils, an aluminum exterior is sometimes combined with a stainless steel interior; in others, the interior is coated with porcelain or Teflon.

Machinery and Equipment

Chemical Process Equipment. In general, the aluminum alloys most suitable for use in the chemical process industries are the lowest in cost. (For extensive corrosion data on aluminum in chemical process solutions, see the article on the corrosion resistance of aluminum and aluminum alloys, in this volume.)

In the petroleum industry, aluminum tops are used on steel storage tanks, exteriors are painted with aluminum paint, and aluminum pipelines are carriers of petroleum products. Aluminum is used extensively in the rubber industry because aluminum resists all types of corrosion that occur in rubber processing, and is nonadhesive with all forms of rubber. The nonsparking characteristic of aluminum is of particular importance in explosive atmospheres, and aluminum alloys find wide use in the manufacture of explosives. Strong oxidants are processed, stored and shipped in aluminum equipment. Sulfur, sulfuric acid, sulfides and sulfates are particularly well suited to processing in aluminum. In the nuclear energy industry, aluminum-jacketed fuel elements protect uranium from water corrosion, prevent the entry of fission products into the cooling water, transfer heat efficiently from uranium to water, and contribute to minimizing parasitic capture of neutrons. Aluminum tanks are used to hold heavy water.

Textile Equipment. Aluminum is used extensively in textile machinery and equipment in the form of extrusions, tube, sheet, castings and forgings. It is resistant to many corrosive agents encountered in textile mills and in the manufacture of yarns. High strength-weight ratio reduces inertia of high-speed machine parts. Permanent dimensional accuracy, with light weight, improves the dynamic balance of machine members running at high speeds and reduces fibration. Painting

is usually unnecessary. Alloys include:

3003-H14 for card rolls
3003-H18 for reed frames
6061-T6 for card rolls, roving frames, ring rails, spindle rails, roller beams, brush holders, bobbins, drying poles, finishing equipment, heddle frames, sewing machine parts, beam barrels, lay beams, and hand rails
2014-T4 for card plates and flats, spinning drive cylinders, heddle frames, hosiery knitting machines, guide bar hangers, and beam barrels
6063-T5 for card rolls, roving frames, spindle rails, roller beams, spinning mules, lay beams, hand rails, cloth rolls, yarn trays, plant construction, batten beams, and lap rolls
5052-H14 for separator blades
6053-T4 for spinning buckets
2011-T3 for tricot beams, spindles, spindle adaptors, and spindle sheaths
2024-T4 for bobbins and hosiery preboarding machines

Paper and Printing Industries. One of the principal applications is in returnable shipping cores. A tube of alloy 6063, fabricated in the same manner as irrigation tube, is used by some. The cores are reinforced with steel end-sleeves, which also constitute wear-resistant drive elements. Processing or rewinding cores are fabricated of the same alloy. Fourdrinier or table rolls for papermaking machines are made of 6061.

Curved aluminum sheet printing plates permit higher rotary press speeds and minimize misregister by decreasing centrifugal force. Aluminum lithographing sheet offers good reproduction and contributes greatly to ease of handling.

Coal Mine Machinery. The use of aluminum equipment in coal mines has increased in recent years; applications include cars, tubs and skips, roof props, nonsparking tools, portable jacklegs, and shaking conveyors. Aluminum is resistant to the corrosive conditions generally associated with coal mines; the metal is self-cleaning—wet coal will not stick to it—and it offers good resistance to abrasion, vibrations, splitting, and tearing.

Portable irrigation pipe, 2 to 8 in. OD, is extensively used for portable sprinkler irrigation systems. The alu-

minum pipe is extruded or drawn in 6063, or roll formed and welded from 3004, 5050, and other alloys.

Portable tools use large quantities of aluminum for motor housings. Precision cast housings are used for power drills, power saws, gasoline-driven chain saws, sanders, buffing machines, screw drivers, grinders, power shears, hammers, various impact tools and also for stationary bench tools.

Jigs, Fixtures and Patterns. Thick cast or rolled aluminum plates and bar, precisely machined to high finish and flatness, are used for tools and dies. The plate is suitable for hydropress form blocks, hydrostretch form dies, jigs, fixtures and other tooling. It is used in the aircraft industry for drill jigs, as formers, stiffeners and stringers for large assembly jigs, router bases and layout tables. Used in master tooling, cast aluminum eliminates warpage problems resulting from uneven expansion of the tool due to changes in ambient temperature. Large aluminum bars have been used to replace zinc alloys as a fixture base on spar mills with weight savings of two thirds.

Instruments. Aluminum alloys are used in the manufacture of clocks and instruments, where their light weight is an advantage. For cast parts, high-silicon alloys are preferred because of their castability, but anodizing must be done with care. For optical, telescopic, space guidance and other high precision devices, alloys 6061 or 2024 in the stress-relieved T651 temper are frequently preferred because of their good combinations of strength and dimensional stability. In manufacturing and assembling parts for such equipment, additional thermal stress-relief treatments are sometimes applied at stages of machining or after welding or mechanical assembly. Small quantities of aluminum strip are used in such items as pointers, where weight of the moving part is important.

Other Applications

Reflectors. Reflectivity of light is as high as 95% on specially prepared surfaces of high-purity aluminum. Aluminum is generally superior to other metals in its ability to reflect infrared or heat rays of the sun. It resists tarnish from sulfides, oxides and atmospheric contaminants, and has three to ten times the useful life of silver for mirrors in searchlights, telescopes and similar reflectors. Heat reflectivity may be as much as 98% for a highly polished surface; this is reduced only slightly as the metal weathers and loses its initial brilliance. When maximum reflectivity is desired, chemical or electrochemical brightening treatments are used. A short-time anodic treatment is usually given the parts, sometimes followed by a coat of clear lacquer. Reflectors which require less brightness may simply be buffed and lacquered. When a diffuse finish is desired, it may be obtained by etching in a mild caustic solution. This finish also is protected either by clear lacquer, an anodic coating or both.

Powders and Pastes. The addition of aluminum flakes to paint pigments utilizes the intrinsic advantages of this metal—high reflectance and durability, and low emissivity and moisture penetration. Other applications of powders and pastes involve printing inks, pyrotechnics, floating soap, aerated concrete, thermit welding, and additions to fuels to increase the amount of energy released.

Temper Designation System for Aluminum and Aluminum Alloys

THE TEMPER DESIGNATION SYSTEM used in the United States for aluminum and aluminum alloys is a part of the system adopted as an American National Standard (ANSI H35.1). It is used for all product forms, wrought and cast, except ingot. The system is based on the sequences of mechanical or thermal treatments, or both, used to produce the various tempers. The temper designation follows the alloy designation and is separated from it by a hyphen. Basic temper designations consist of individual capital letters. Major subdivisions of basic tempers, where required, are indicated by one or more digits following the letter. These digits designate specific sequences of treatments that produce specific combinations of characteristics in the product. Variations in treatment conditions within major subdivisions are identified by additional digits. The conditions during heat treatment (such as time, temperature and quenching rate) used to produce a given temper in one alloy may differ from those employed to produce the same temper in another alloy.

Designations for the common tempers, and descriptions of the sequences of operations used to produce these tempers, are given in the following paragraphs.

Basic Temper Designations

F   As fabricated. Applies to products shaped by cold working, hot working or casting processes in which no special control over thermal conditions or strain hardening is employed. For wrought products, there are no mechanical-property limits.

O   Annealed. Applies to wrought products that are annealed to obtain lowest strength temper, and to cast products that are annealed to improve ductility and dimensional stability. The O may be followed by a digit other than zero.

H   Strain hardened (wrought products only). Applies to products that have been strengthened by strain hardening, with or without supplementary heat treatment to produce some reduction in strength. The H is always followed by two or more digits, as discussed in the following section.

W   Solution heat treated. An unstable temper applicable only to alloys that naturally age (spontaneously age at room temperature) after solution heat treatment. This designation is specific only when the period of natural aging is indicated—for example, W ½ hr. (See also the discussion of the Tx51, Tx52 and Tx54 tempers, in the section on heat treatable alloys.)

T   Heat treated to produce stable tempers other than F, O, or H. Applies to products that are thermally treated, with or without supplementary strain hardening, to produce stable tempers. The T is always followed by one or more digits, as discussed in a later section.

System for Strain-hardened Products

Temper designations for wrought products that are strengthened by strain hardening consist of an H followed by two or more digits. The first digit following the H indicates the specific sequence of basic operations, as follows:

H1 Strain hardened only. Applies to products that are strain hardened to obtain the desired strength without supplementary thermal treatment. The digit following the H1 indicates the degree of strain hardening.

H2 Strain hardened and partially annealed. Applies to products that are strain hardened more than the desired final amount and then reduced in strength to the desired level by partial annealing. The digit following the H2 indicates the degree of strain hardening remaining after the product has been partially annealed.

H3 Strain hardened and stabilized. Applies to products that are strain hardened and whose mechanical properties are stabilized by a low-temperature thermal treatment that slightly decreases tensile strength and improves ductility. This designation is applicable only to those alloys that, unless stabilized, gradually age soften at room temperature. The digit following the H3 indicates the degree of strain hardening after stabilization.

For alloys that age soften at room temperature, each H2x temper has the same minimum ultimate tensile strength as the H3x temper with the same second digit. For other alloys, each H2x temper has the same minimum ultimate tensile strength as the H1x with the same second digit, and slightly higher elongation.

The digit following the designations H1, H2 and H3, which indicates the degree of strain hardening, is a numeral from 1 through 8. Numeral 8 indicates tempers with ultimate tensile strength equivalent to that achieved by about 75% cold reduction (temperature during reduction not to exceed 50 °C, or 120 °F) following full annealing. Tempers between 0 (annealed) and 8 are

designated by numerals 1 through 7. Material having an ultimate tensile strength approximately midway between that of the 0 temper and that of the 8 temper is designated by the numeral 4; approximately midway between the 0 and 4 tempers by the numeral 2; and approximately midway between the 4 and 8 tempers by the numeral 6. Numeral 9 designates tempers whose minimum ultimate tensile strength exceeds that of the 8 temper by 10 MPa or more (or 2.0 ksi or more when English unit strengths are used). For two-digit H tempers whose second digits are odd, the standard limits for strength are the arithmetic mean, rounded to the nearest multiple of 5 MPa or 0.5 ksi (in conformance with ASTM Recommended Practice E29), of the standard limits for the adjacent two-digit H tempers whose second digits are even.

For alloys that cannot be sufficiently cold reduced to establish an ultimate tensile strength applicable to the 8 temper (75% cold reduction after full annealing), the 6-temper tensile strength may be established by cold reduction of approximately 55% following full annealing, or the 4-temper tensile strength may be established by cold reduction of approximately 35% after full annealing.

When it is desirable to identify a variation of a two-digit H temper, a third digit (from 1 to 9) may be assigned. Zero has been assigned to indicate variations negotiated between the manufacturer and purchaser which are not used widely enough to justify registration. The third digit is used when the degree of control of temper or the mechanical properties are different from but close to those for the two-digit H temper designation to which it is added, or when some other characteristic is significantly affected. The minimum ultimate tensile strength of a

three-digit H temper is at least as close to that of the corresponding two-digit H temper as it is to either of the adjacent two-digit H tempers. Products in H tempers whose mechanical properties are below those of Hx1 tempers are assigned variations of Hx1. Some three-digit H temper designations have already been assigned; these are described below.

The following designations have been assigned for wrought products in all alloys:

Hx11 Applies to products that incur sufficient strain hardening after final annealing that they fail to qualify as O temper, but not so much or consistent amount of strain hardening that they qualify as Hx1 temper.

H112 Applies to products that may acquire some strain hardening during working at elevated temperature and for which there are mechanical-property limits.

The following designations have been assigned for wrought products in alloys with nominal magnesium contents greater than 4%.

H311 Applies to products that are strain hardened less than the amount required for a controlled H31 temper.

H321 Applies to products that are strain hardened less than the amount required for a controlled H32 temper.

H323,
H343 These designations apply to products that are specially fabricated to have acceptable resistance to stress-corrosion cracking.

The following three-digit H-temper designations have been assigned for patterned sheet or embossed sheet.

Patterned or embossed sheet	Fabricated from
H114	O temper
H124, H224, H324	H11, H21, H31 temper, respectively
H134, H234, H334	H12, H22, H32 temper, respectively
H144, H244, H344	H13, H23, H33 temper, respectively
H154, H254, H354	H14, H24, H34 temper, respectively
H164, H264, H364	H15, H25, H35 temper, respectively
H174, H274, H374	H16, H26, H36 temper, respectively
H184, H284, H384	H17, H27, H37 temper, respectively
H194, H294, H394	H18, H28, H38 temper, respectively
H195, H295, H395	H19, H29, H39 temper, respectively

System for Heat Treatable Alloys

The temper designation system for wrought and cast products that are strengthened by heat treatment employs the W and T designations described in the section on basic temper designations. The W designation denotes an unstable temper, whereas the T designation denotes a stable temper other than F, O or H. The T is followed by a number from 1 to 10; each number indicates a specific sequence of basic treatments, as follows:

T1 **Cooled from an elevated temperature shaping process and naturally aged to a substantially stable condition.** Applies to products that are not cold worked after an elevated temperature shaping process such as casting or extrusion, and for which mechanical properties have been stabilized by room-temperature aging. If the products are flattened or straightened after cooling from the shaping process, the effects of the cold work imparted by flattening or straightening are not accounted for in specified property limits.

T2 **Cooled from an elevated temperature shaping process, cold worked, and naturally aged to a substantially stable condition.** Applies to products that are cold worked specifically to improve strength after cooling from a hot working process such as rolling or extrusion, and for which mechanical properties have been stabilized by room-temperature aging. The effects of cold work, including any cold work imparted by flattening or straightening, are accounted for in specified property limits.

T3 **Solution heat treated, cold worked, and naturally aged to a substantially stable condition.** Applies to products that are cold worked specifically to improve strength after solution heat treatment, and for which mechanical properties have been stabilized by room-temperature aging. The effects of cold work, including any cold work imparted by flattening or straightening, are accounted for in specified property limits.

T4 **Solution heat treated and naturally aged to a substantially stable condition.** Applies to products that are not cold worked after solution heat treatment, and for which mechanical properties have been stabilized by room-temperature aging. If the products are flattened or straightened, the effects of the cold work imparted by flattening or straightening are not accounted for in specified property limits.

T5 **Cooled from an elevated temperature shaping process and artificially aged.** Applies to products that are not cold worked after an elevated temperature shaping process such as casting or extrusion, and for which mechanical properties or dimensional stability, or both, have been substantially improved by precipitation heat treatment. If the products are flattened or straightened after cooling from the shaping process, the effects of the cold work imparted by flattening or straightening are not accounted for in specified property limits.

T6 **Solution heat treated and artificially aged.** Applies to products that are not cold worked after solution heat treatment, and for which mechanical properties or dimensional stability, or both, have been substantially improved by precipitation heat treatment. If the products are flattened or straightened, the effects of the cold work imparted by flattening or straightening are not accounted for in specified property limits.

T7 **Solution heat treated and stabilized.** Applies to products that have been precipitation heat treated to the extent that they are overaged. Stabilization heat treatment carries the mechanical properties beyond the point of maximum strength to provide some special characteristic, such as enhanced resistance to stress corrosion cracking or exfoliation corrosion.

T8 **Solution heat treated, cold worked, and artificially aged.** Applies to products that are cold worked specifically to improve strength after solution heat treatment, and for which mechanical properties or dimen-

sional stability, or both, have been substantially improved by precipitation heat treatment. The effects of cold work, including any cold work imparted by flattening or straightening, are accounted for in specified property limits.

T9 **Solution heat treated, artificially aged, and cold worked.** Applies to products that are cold worked specifically to improve strength after they have been precipitation heat treated.

T10 **Cooled from an elevated temperature shaping process, cold worked, and artificially aged.** Applies to products that are cold worked specifically to improve strength after cooling from a hot working process such as rolling or extrusion, and for which mechanical properties or dimensional stability, or both, have been substantially improved by precipitation heat treatment. The effects of cold work, including any cold work imparted by flattening or straightening, are accounted for in specified property limits.

When it is desirable to identify a variation of one of the ten major T tempers described above, additional digits, the first of which cannot be zero, may be added to the designation.

The following specific sets of additional digits have been assigned to stress-relieved wrought products:

Tx51 **Stress relieved by stretching.** Applies to the following products when stretched to the indicated amounts after solution heat treatment or after cooling from an elevated-temperature shaping process:

Product form	Permanent set, %
Plate	1½ to 3
Rod, bar, shapes, extruded tube	1 to 3
Drawn tube	½ to 3

Applies directly to plate and to rolled or cold finished rod and bar. These products receive no further straightening after stretching. Applies to extruded rod, bar, shapes and tubing, and to drawn tubing, when designated as follows:

Tx510 Products that receive no further straightening after stretching

Tx511 Products that may receive minor straightening after stretching to comply with standard tolerances

Tx52 Stress relieved by compressing. Applies to products that are stress relieved by compressing after solution heat treatment, or after cooling from a hot working process to produce a permanent set of 1 to 5%

Tx54 Stress relieved by combining stretching and compressing. Applies to die forgings that are stress relieved by restriking cold in the finish die

(These same digits—and 51, 52 and 54—may be added to the designation W to indicate unstable solution heat treated and stress-relieved tempers.)

The following temper designations have been assigned to wrought products heat treated from the O or the F temper to demonstrate response to heat treatment:

T42 **Solution heat treated from the O or the F temper to demonstrate response to heat treatment, and naturally aged to a substantially stable condition**

T62 **Solution heat treated from the O or the F temper to demonstrate response to heat treatment, and artificially aged**

Temper designations T42 and T62 also may be applied to wrought products heat treated from any temper by the user when such heat treatment results in the mechanical properties applicable to these tempers.

System for Annealed Products

A digit following the "O", when used, indicates a product in annealed condition having special characteristics. For example, for heat treatable alloys, O1 indicates a product that has been heat treated at approximately the same time and temperature required for solution heat treatment and then air cooled to room temperature; this designation applies to products that are to be machined prior to solution heat treatment by the user.

Heat Treatment of Aluminum Alloys

By the ASM Committee on
Aluminum and Aluminum Alloys*

THE TERM "heat treatment", in its broadest sense, refers to any of the heating and cooling operations applied to change the mechanical properties, the metallurgical structure or the residual stress state of a metal product. When the term is applied to aluminum alloys, however, its use frequently is restricted to the specific operations employed to increase strength and hardness of the precipitation-hardenable wrought and cast alloys. These usually are referred to as the "heat treatable" alloys, to distinguish them from those alloys in which no significant strengthening can be achieved by heating and cooling. The latter, generally referred to as "non-heat-treatable" alloys, when in wrought form depend primarily on cold work to increase strength. Heating to decrease strength and increase ductility (annealing) is used with alloys of both types; metallurgical reactions may vary with type of alloy and with degree of softening desired. Except for the low-temperature stabilization treatment sometimes given 5xxx series alloys (which is a mill treatment and not discussed in this article), complete or partial annealing treatments are the

*See page **X** for committee list.

only ones used for non-heat-treatable alloys.

The annealing and heat treating schedules used for aluminum and aluminum alloys are given in *Aluminum Standards and Data* (for wrought alloys) and in *Standards for Aluminum Sand and Permanent Mold Castings,* both published by The Aluminum Association, Inc. ASTM B597 summarizes these schedules and presents recommended operating practices along with test requirements necessary to ensure proper physical and mechanical properties of the heat treated product.

Annealing

For both heat treatable and non-heat-treatable aluminum alloys, annealing to remove the effects of cold work is accomplished by heating within the temperature range from about 300°C (570 °F) (for batch treatment) to about 450 °C (840 °F) (for continuous treatment). The rate of softening is primarily temperature-dependent: the time required to soften a given material by a given amount can vary from hours at low temperatures to seconds at high temperatures. Products that can be heated and cooled very rapidly, such as wire, are annealed by continuous

processes that require a total heating and cooling time of only a few seconds. Continuous annealing of strip is accomplished in a total time at temperature of a few minutes for each portion of the strip. For these extremely rapid operations, maximum temperature occasionally may exceed 450 °C (840 °F).

Figure 1 shows changes in yield strength as functions of annealing temperature and time for sheet of two non-heat-treatable wrought alloys (1100 and 5052) initially in the highly cold worked condition (H18 temper). From these curves it is apparent that by selection of appropriate combinations of time and temperature, mechanical properties intermediate between those of cold worked material and those of fully annealed material can be obtained. It is evident also that yield strength depends much more strongly on temperature than on time of heating.

Annealing treatments employed for aluminum alloys are of several types that differ in objective. Annealing times and temperatures depend on alloy type as well as on initial microstructure and temper.

Full Annealing. The softest, most ductile and most workable condition of both non-heat-treatable and heat treat-

Table 1 Typical full annealing treatments for some common wrought aluminum alloys(a)

Alloy	Metal temperature °C	°F	Approximate time at temperature, h	Alloy	Metal temperature °C	°F	Approximate time at temperature, h
1060	345	650	(b)	5457	345	650	(b)
1100	345	650	(b)	5652	345	650	(b)
1350	345	650	(b)	6005	415	775(c)	2-3
2014	415	775(c)	2-3	6009	415	775(c)	2-3
2017	415	775(c)	2-3	6010	415	775(c)	2-3
2024	415	775(c)	2-3	6053	415	775(c)	2-3
2036	385	725(c)	2-3	6061	415	775(c)	2-3
2117	415	775(c)	2-3	6063	415	775(c)	2-3
2124	415	775(c)	2-3	6066	415	775(c)	2-3
2219	415	775(c)	2-3	7001	415	775(d)	2-3
3003	415	775	(b)	7005	345	650(e)	2-3
3004	345	650	(b)	7049	415	775(d)	2-3
3105	345	650	(b)	7050	415	775(d)	2-3
5005	345	650	(b)	7075	415	775(d)	2-3
5050	345	650	(b)	7079	415	775(d)	2-3
5052	345	650	(b)	7178	415	775(d)	2-3
5056	345	650	(b)	7475	415	775(d)	2-3
5083	345	650	(b)	**Brazing Sheet**			
5086	345	650	(b)	Nos. 11 and 12	345	650	(b)
5154	345	650	(b)				
5182	345	650	(b)	Nos. 21 and 22	345	650	(b)
5254	345	650	(b)				
5454	345	650	(b)	Nos. 23 and 24	345	650	(b)
5456	345	650	(b)				

(a) These treatments anneal the material to the "O" temper. The treatments listed in this table are typical for various sizes and methods of manufacture and may not exactly describe the optimum treatment for a specific item. (b) Time in the furnace need not be longer than necessary to bring all parts of load to annealing temperature. Rate of cooling is unimportant. (c) These treatments are intended to remove effects of solution heat treatment and include cooling at rate of about 30 °C (50 °F) per hour from the annealing temperature to 260 °C (500 °F). The rate of subsequent cooling is unimportant. Treatment at 345 °C (650 °F), followed by uncontrolled cooling, may be used to remove the effects of cold work, or to partially remove the effects of heat treatment. (d) This treatment is intended to remove the effects of solution heat treatment and includes cooling at an uncontrolled rate to 205 °C (400 °F) or less, followed by reheating to 230 °C (450 °F) for 4 h. Treatment at 345 °C (650 °F), followed by uncontrolled cooling, may be used to remove the effects of cold work, or to partially remove the effects of heat treatment. (e) Cooling rate to 205 °C (400 °F) or lower is less than or equal to 30 °C (50 °F) per hour.

able wrought aluminum alloys is produced by "full annealing" to the temper designated "O". Material in this temper normally is recrystallized. For heat treatable alloys, the phases formed by combination of solute elements with aluminum are sufficiently thoroughly precipitated by full annealing to stabilize them and prevent natural age hardening (see the section on precipitation hardening, which appears later in this article). Typical full annealing treatments for some wrought aluminum alloys in common use are listed in Table 1.

Partial Annealing. Annealing of cold worked non-heat-treatable wrought alloys to produce mechanical properties intermediate between those of material in the highly cold worked condition (H18 temper) and those of material in the fully annealed condition (O temper) is referred to as "partial annealing", "recovery annealing" and "temper annealing." Material in this intermediate condition is identified by an H2-type temper designation.

Temperatures used for partial annealing are below those that produce extensive recrystallization, and depending on the temperature, incomplete softening results from stress relief, substructural changes in dislocation density and rearrangement of dislocations into cellular patterns (polygonization), or partial recrystallization. Bendability and formability of an alloy annealed to an H2-type temper generally are significantly higher than those of the same alloy in which an equal strength level is developed by a final cold working operation (H1-type temper). Heat treating to H2-type tempers requires close control of temperature to achieve uniform and consistent mechanical properties.

Stress-relief Annealing. For heat treatable wrought alloys, annealing merely to remove the effects of strain hardening is referred to as "stress-relief annealing". Annealing treatments designed for stress relief employ temperatures up to about 345 °C (650 °F) and may result in recovery only, in partial recrystallization or even in full recrystallization. Although stress-relief annealing may fully recrystallize some alloys, it does not stabilize certain heat treatable alloys as does full annealing. Full stabilization (full annealing) of these alloys generally requires a higher maximum temperature, controlled cooling to a lower temperature and additional holding time at the lower temperature (see Table 1).

Annealing of castings for 2 to 4 h at temperatures from 315 to 345 °C (600 to 650 °F) provides the most complete relief of residual stresses and precipitation of phases formed by solute retained in solid solution in the as-cast condition. Such annealing treatments provide maximum dimensional stability for service at elevated temperatures. The annealed temper for castings is designated "O". This temper was designated "T2" prior to 1975.

Principles of Precipitation Hardening

Heat treatments that increase strength and hardness of either wrought or cast aluminum alloy products utilize the mechanism of precipitation hardening. One essential attribute of a precipitation-hardening alloy system is a temperature-dependent equilibrium solid solubility characterized by decreasing solubility with decreasing temperature, so that second-phase particles are formed by solid-state precipitation when the alloy is cooled from above the solvus. Although this condition is met by most binary aluminum systems, many exhibit very little precipitation hardening and ordinarily are not considered heat treatable. Alloys of the Al-Si and Al-Mn systems, for example, exhibit relatively insignificant changes in mechanical properties as a result of heat treatments that produce considerable precipitation. In contrast, alloys of the Al-Cu system exhibit significant changes.

In heat treatable aluminum alloys, such as the Al-Cu alloys, the first stages of precipitation can take place at room temperature; this is called "natural aging". Precipitation can be accelerated by heating the alloy to slightly

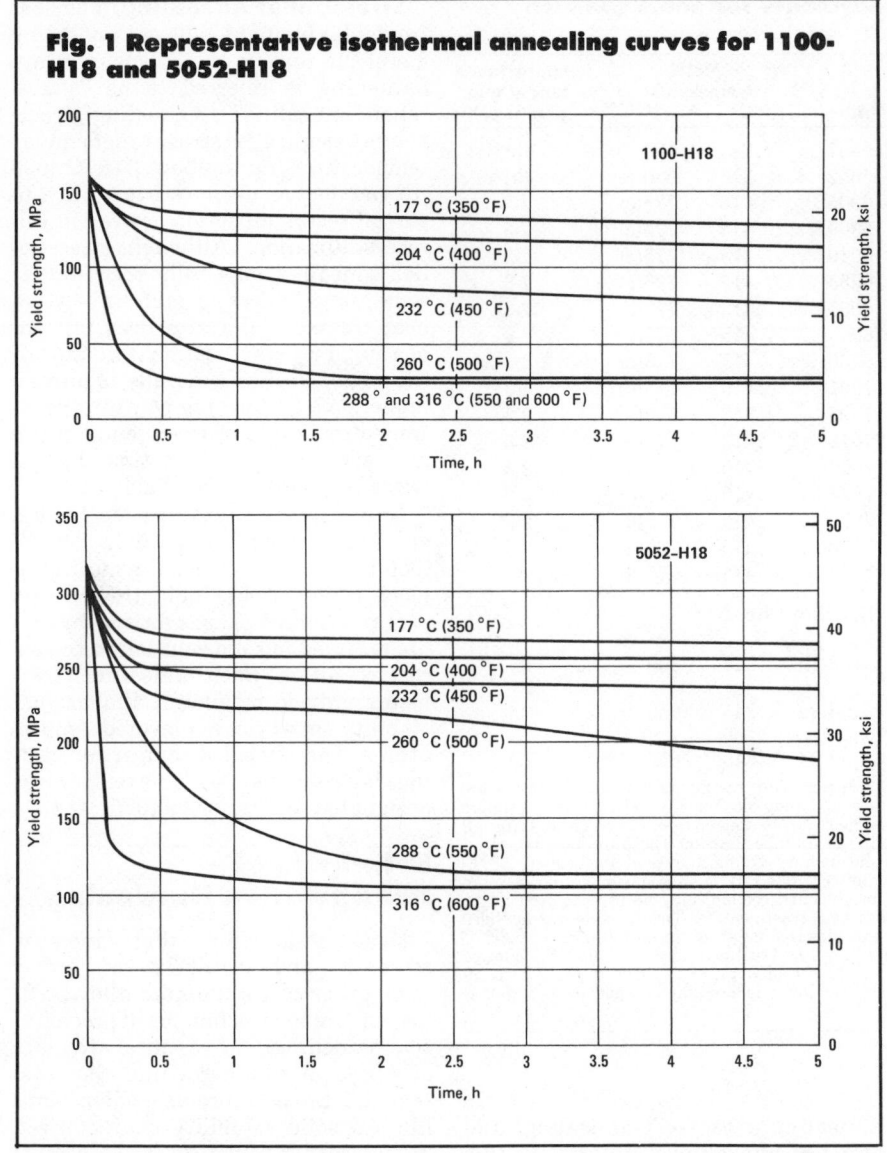

Fig. 1 Representative isothermal annealing curves for 1100-H18 and 5052-H18

elevated temperatures; this is called "precipitation heat treatment" or "artificial aging", and produces maximum strength and hardness. If the alloy is heated at too high a temperature or for too long a time, the increase in strength and hardness will be followed by a decrease of these properties; this is called "overaging".

The microstructural changes that occur during precipitation hardening to high-strength levels involve precipitate particles that are too small to be resolved by light microscopy. In certain instances, the presence of precipitates can be detected by etching metallographic specimens. In aluminum alloys in the solution heat treated and quenched condition, coloration contrast between grains of different orienta-tions is relatively high, particularly in 2xxx series wrought alloys and 2xx.0 series casting alloys. This contrast is noticeably decreased by precipitation heat treatment. Resolution of the precipitated particles, however, generally is possible only by electron microscopy.

Heat Treatable Alloys

Commercial alloys whose strength and hardness can be increased significantly by heat treatment include wrought alloys of the 2xxx, 6xxx and 7xxx series and casting alloys of the 2xx.0, 3xx.0 and 7xx.0 series. Some of these alloys contain only copper, or copper and silicon, as the primary strengthening alloy addition(s). Most of the heat treatable alloys, however, contain combinations of magnesium with one or more of the elements copper, silicon and zinc. Magnesium combines with copper, silicon and zinc to form complex compounds whose precipitation usually is faster and more pronounced than that of the simpler $CuAl_2$ and ZnAl compounds.

In the heat treatable wrought alloys, with some exceptions, such solute elements are present in amounts that are within the limits of mutual solid solubility at temperatures below the eutectic temperature (lowest melting temperature). In contrast, all of the casting alloys of the 3xx.0 series, and some 2xx.0 alloys, contain amounts of soluble elements that far exceed the solid-solubility limits. In these alloys, the phases formed by combination of the excess soluble elements with aluminum will never be dissolved, although the shapes of the undissolved particles may be changed by partial solution.

Aged Alloys. The first step in heat treatment of aluminum alloys is the production of a supersaturated solid solution and its retention at room temperature. This is usually accomplished by a high-temperature solution treatment followed by a quench. For some medium-strength wrought alloys (6063 and 7005 for example), and for many 3xx.0 and 7xx.0 series casting alloys, a sufficient degree of supersaturation is achieved by rapid cooling immediately following extrusion, hot rolling or casting. Natural aging, or preferably precipitation heat treating, can then be used to enhance strength and hardness.

Among the wrought alloys, only those that are relatively dilute (in which case the rate of cooling is not particularly critical) are well suited for this practice. Such wrought alloys are seldom left in a naturally aged (T1) temper but rather are precipitation heat treated to a T5-type temper. The mechanical properties of such alloys in T5-type tempers are lower than what would be obtained by solution heat treating and quenching prior to precipitation heat treating, however, the T5 procedure offers an economic advantage.

Many castings in T5-type tempers have properties that are well suited to their engineering applications and have better dimensional stability than castings of the same alloys in T6-type tempers. The wide usage of castings in T5-type tempers is thus attributable both to excellent service performance and to economics. High hardness, strength and dimensional stability at elevated temperatures account for the almost universal use of aluminum castings in T5-type tempers for pistons and other engine parts.

Castings of 7xx.0 series alloys may be used without heat treatment. The strength of such castings increases with time at room temperature at a diminishing rate but continuing over long periods of time. For example, the yield strength of sand cast 710.0 (formerly A712.0) alloy increases from about 85 MPa (12.5 ksi) shortly after removal from the mold to about 170 MPa (25 ksi) after 30 days of natural aging and to about 225 MPa (32.5 ksi) after two years of natural aging. Specification properties for these alloys in the T1 temper are based on testing after aging for 21 days. For most of these alloys, slow natural aging may be replaced by precipitation heat treating to a T5-type temper, which is more stable.

Solution Heat Treated and Naturally Aged Alloys. The more highly alloyed members of the 6xxx series of wrought alloys, the copper-containing members of the 7xxx series, and all of the 2xxx alloys are always solution heat treated and quenched. For some of these alloys, particularly the 2xxx alloys—the precipitation hardening that results from natural aging alone produces useful (T3- and T4-type) tempers that are characterized by high fracture toughness and resistance to fatigue. Natural aging is fairly rapid for the alloys of the 2xxx type, and in most of them 80 to 90% of the hardening is completed in 4 to 5 days. Tensile-property specifications for products in T3- and T4-type tempers are based on natural aging for a nominal time of 4 days. The changes that occur on further natural aging in alloys for which T3- or T4-type tempers are standard are relatively minor, and products of these combinations of alloy and temper are regarded as essentially stable after about 1 week.

Most wrought products undergo straightening or flattening operations after the quench, but the amount of deformation of these operations normally has a negligible effect on properties. After natural aging, these products are in a T4-type temper. If the amount of cold work after quench exceeds 1 to 2%, its effect on properties is significant and a different set of properties results; these are the T3-type tempers. (T3-type tempers are mill produced and cannot be duplicated by the user.) Further increases in strength can be obtained by cold rolling, additional stretching, combinations of these operations or, for products such as hand forgings, compressive deformation. Additional digits may be added to the T3 designation (T36 or T37) to indicate different degrees of strain hardening. These additional digits should not be confused with the specific combinations of the supplemental digits that are used to denote the tempers produced when mechanical deformation is used primarily to relieve residual stresses induced during quenching, and not to effect significant changes in mechanical properties (for example, the digits 51 following the basic T*x* designation, which denote products stress relieved by stretching).

In contrast to the relatively stable condition reached in a few days by the 2xxx alloys that are used in T3- or T4-type tempers, the 6xxx alloys, and to an even greater degree the 7xxx alloys, are considerably less stable at room temperature and continue to exhibit significant changes in mechanical properties for many years. The differences in rate and duration of changes in tensile yield strength among representative alloys of these three types are illustrated in Fig. 2. Because of the relative instability of the 7xxx alloys, the naturally aged temper (after solution heat treatment and quenching) is designated by the suffix letter W, rather than by a T3 or T4 type of designation. For specific description of the condition of the alloy, the time of natural aging should be included (for example, W ½ h).

Solution Heat Treated and Precipitation Heat Treated Alloys. Precipitation heat treatment (artificial aging) following solution heat treatment and quenching produces T6- and T7-type tempers. Alloys in T6-type tempers generally have the highest strength possible without sacrifice of the other properties and characteristics found by experience to be satisfactory and useful for engineering applications. Alloys in T7-type tempers are stabilized by "overaging", which means that some degree of strength has been sacrificed to improve one or more other characteristics. Strength may be sacrificed to improve dimensional stability, particularly in products intended for service at elevated temperatures, to lower residual stresses in order to reduce warpage or distortion in machining, or to develop better resistance to stress-corrosion cracking and exfoliation corrosion. For the most part, T7-type stabilization treatments are used for castings and forgings. T7-type tempers frequently are specified for cast or forged engine parts. To produce T7-type tempers primarily for dimensional stability, quenching media are used that provide moderate cooling rates, and the precipitation heat treating temperatures used generally are higher than those used to produce T6-type tempers in the same alloys.

Two other important groups of T7-type tempers—T73-type and T76-type tempers—have been developed for alloys of the 7xxx series that contain copper. These tempers were developed specifically to improve resistance of these wrought alloys to exfoliation corrosion and stress-corrosion cracking,

Fig. 2 Natural aging curves for three solution heat treated wrought aluminum alloys

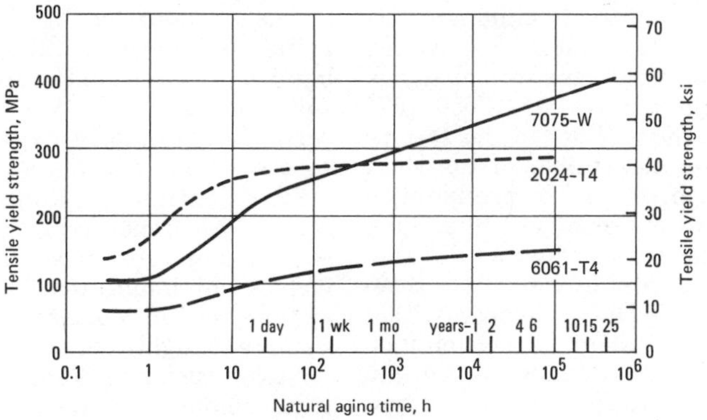

but as a result of overaging the alloys in these tempers also have increased fracture toughness and, under some conditions, reduced rates of fatigue-crack propagation. T73- and T76-type tempers have virtually eliminated exfoliation corrosion and stress-corrosion cracking of large and complex machined parts of alloys 7049, 7050, 7075 and 7475. These alloys in the T6-type tempers are susceptible in the short transverse direction. The precipitation heat treatments employed to produce T73-type tempers as well as the T76-type tempers are different from those used to produce T6-type tempers in that a two-stage isothermal precipitation heat treatment is employed. Certain properties in the T6, T76 and T73 type temper are:

| Property | Type of temper | | |
	T6	T76	T73
Strength	Highest	Intermediate	Lowest
Resistance to stress-corrosion cracking	Lowest	Intermediate	Highest
Resistance to exfoliation corrosion	Low	High	High

During the preliminary stage, a finer higher-density precipitate dispersion is nucleated, producing high strength. Then the second-stage heating is used to develop resistance to exfoliation and stress-corrosion cracking.

Cast products of heat treatable aluminum alloys have maximum combinations of strength, ductility and toughness when produced in T6-type tempers. In addition to type of heat treatment selected, properties of castings are affected by solidification conditions, as shown in Fig. 3. This illustration compares tensile properties developed during precipitation heat treating of solution heat treated and quenched alloy 356.0 sand and permanent mold castings at various temperatures. Because of the finer cast structure and greater supersaturation of the more rapidly solidified permanent mold castings, their tensile properties are superior to those of the similarly heat treated sand castings.

Solution Heat Treated, Cold Worked and Precipitation Heat Treated Alloys. The T8-type tempers, applicable only to alloys of the 2xxx series, are produced by solution heat treating, quenching and controlled plastic straining (cold working) to the T3-type temper, followed by precipitation heat treating. The plastic strain introduced by the producer prior to precipitation heat treatment nucleates a precipitate dispersion that is finer and denser than that found in material in T6-type tempers and that considerably increases strength. Alloys 2024, 2124 and 2219 in T8-type tempers are particularly suitable for applications in supersonic and military aircraft.

Specification Requirements. Quality assurance criteria that heat treated materials must meet always include minimum tensile properties and, for certain tempers, adequate resistance to highly detrimental forms of corrosion such as intergranular corrosion, exfoliation attack and stress-corrosion cracking. All processing steps through heat treatment must be controlled carefully to ensure high and reliable performance in all of these respects. For control of the corrosion characteristics of materials in certain tempers—notably, T73- and T76-type tempers—the materials must meet maximum yield strength and minimum electrical conductivity criteria as well as minimum tensile-property requirements. Although these criteria are based on indirect measurements of corrosion properties, their validity for ensuring the intended corrosion resistance has been established by extensive correlation testing.

In products of the newer high-strength alloys 2048, 2124, 2419, 7049, 7050, 7175 and 7475, which were developed to provide high fracture toughness, guaranteed minimum values of the applicable indices, K_{Ic} or K_c, are being established by accumulation of statistical data from production lots. Some existing specifications already contain minimum values of these fracture toughness indices for certain products, alloys and tempers.

Effects of Heat Treating Process Variables

Solution Heat Treating Temperature. The objective of solution heat treatment is to allow the maximum practical amounts of the soluble hardening elements of the alloy to dissolve into solid solution, thus obtaining a solid solution that is nearly homogeneous. Because solubility and diffusion rate both increase with temperature, it usually is desirable to use the highest treatment temperature that will not cause remelting. Nominal commercial solution heat treating temperatures are determined on the basis of alloy composition limits and allowances for unintended temperature variations. The ranges normally listed allow variations of ±5 °C (±10 °F) from the nominal temperature. These are the limits for some of the highly alloyed, controlled toughness, high strength alloys, whereas broader ranges may be allowable for alloys having greater intervals of temperature between their solvus and eutectic melting temperatures.

When the temperatures attained by the parts or pieces being heat treated are appreciably below the normal range, solution is incomplete, and strength somewhat lower than normal is expected. If eutectic melting discernible in the microstructure occurs as a result of overheating, strength, ductility and fracture toughness may be lowered. Specifications generally categorize as unacceptable those materials that exhibit microstructural evidence of overheating.

Solution Heat Treating Time. The time at the nominal solution heat treating temperature ("soak time") required to effect a satisfactory degree of solution of the undissolved or precipitated soluble phase constituents and to achieve good homogeneity of the solid solution is a function of microstructure before heat treatment. This time requirement can vary from a few minutes for thin sheet to as much as 20 h for large sand or plaster mold castings. The time required to attain the treatment temperature also increases with section thickness and furnace loading so that total cycle time increases with these factors.

Soak time for alclad sheet and for parts made from alclad sheet must be held to a minimum, because excessive diffusion of alloying elements from the core into the cladding reduces corrosion resistance. For the same reason, reheat treatment of alclad sheet less than 0.50 mm (0.020 in.) thick generally is prohibited, and the number of reheat treatments permitted for thicker alclad sheet is limited.

Quenching Rate. To avoid types of precipitation that are detrimental to mechanical properties or to corrosion resistance, the solid solution formed during solution heat treatment must be rapidly cooled (quenched) to produce a supersaturated solution at room temperature—the optimum condition for precipitation hardening. Most frequently, quenching is done by rapid immersion in cold water or, in continuous heat treating of sheet, plate or

Fig. 3 Comparison of the precipitation hardening characteristics of 356-T4 sand and permanent mold castings

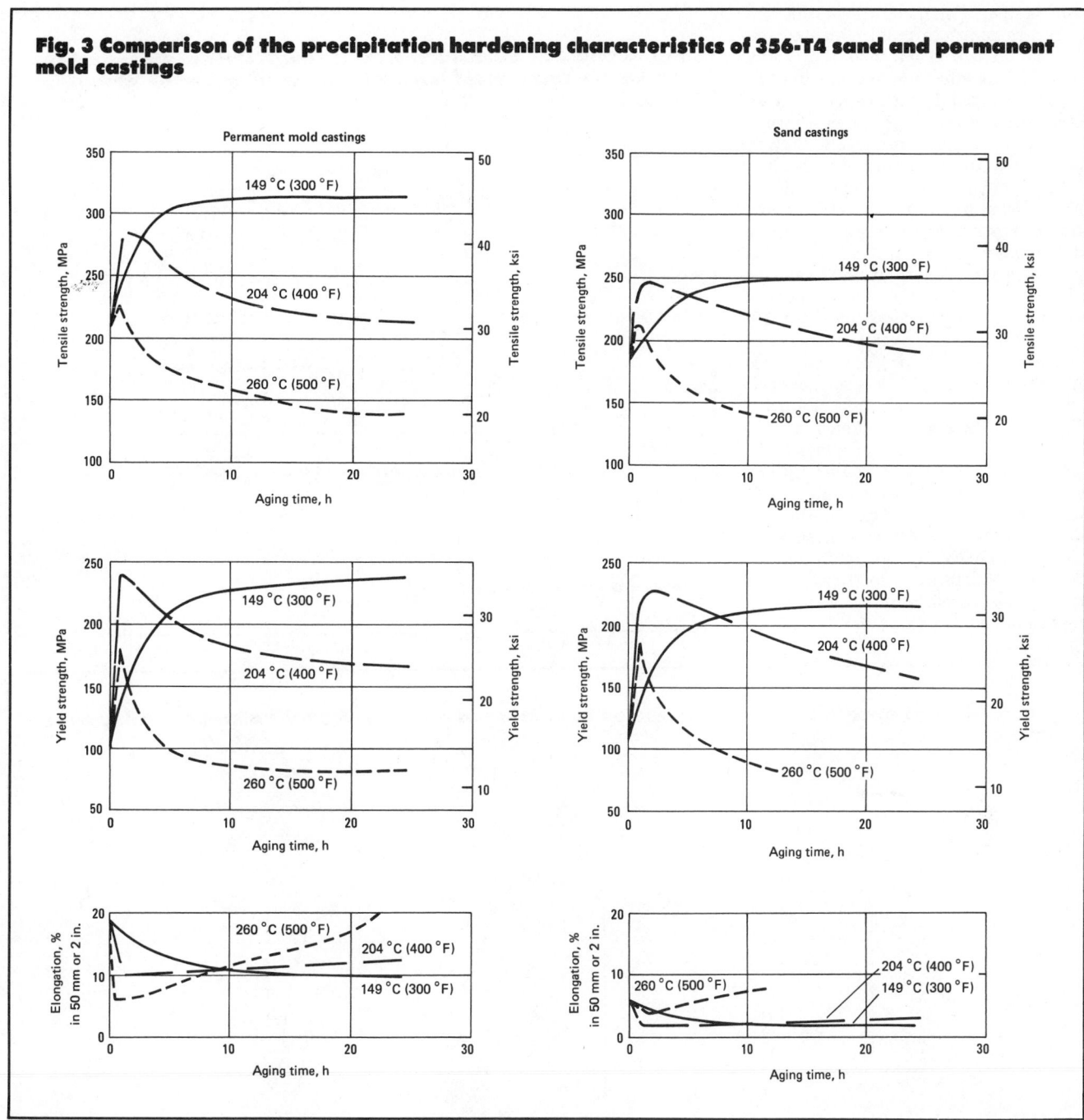

extrusions in primary fabrication plants, by progressive flooding or high-velocity spraying with cold water. However, parts of complex shape, often with both thin and thick sections (such as die forgings, most castings, or impact extrusions) and components formed from sheet, are commonly quenched in a medium that provides somewhat slower cooling. This medium may be hot water (65 to 80 °C, or 150 to 180 °F), boiling water, an aqueous solu-

tion or some other fluid medium. For maximum dimensional stability, some forged parts and castings are air quenched (fan cooled or still-air cooled); in such instances, precipitation-hardening response is limited, but strength and hardness satisfactory for the application are obtained. Extrusions produced without separate solution heat treatment usually are air quenched from the die. Alloys that are relatively dilute, such as 6063, are particularly

well suited for air quenching, and their mechanical properties are not greatly affected by the low quenching rate.

Low quenching rates are employed for forgings, castings and complex shapes to minimize warpage or other distortion and magnitude of residual stresses developed as a consequence of temperature nonuniformity from surfaces to interior. The effects of these low quenching rates on mechanical properties vary with alloy composition and temper. For most alloys, the temperature range through which cooling is most critical is 400 to 290 °C (750 to 550 °F); the effects of average quenching rate through this range on yield strength are shown for four alloys in Fig. 4. For alloys relatively high in sensitivity to quenching rate, such as 7075, quenching rates of about 330 °C/s (600 °F/s) or higher through the critical temperature range are required in order to obtain maximum strength after precipitation heat treating. The other alloys in Fig. 4 do not lose strength at quenching rates as low as about 100 °C/s (180 °F/s).

Corrosion resistance of some of the high-strength alloys is also markedly affected by quenching rate. Figure 5 illustrates the effects of quenching rate on corrosion resistance of alloys 2024-T4 and 7075-T6, as indicated by percentage loss in tensile strength due to corrosion and by changes in the predominant type of corrosive attack (pitting or intergranular). Maximum corrosion resistance requires rapid quenching—preferably at a rate not less than 280 °C/s (500 °F/s). The effects

Fig. 4 Effect of quenching rate through critical temperature range on final yield strength of four wrought aluminum alloys

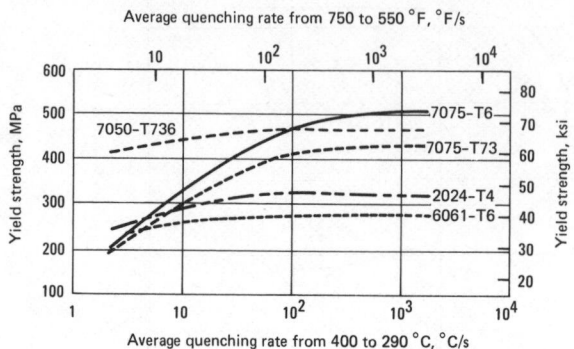

Tempers shown are those after aging.

Fig. 5 Effect of quenching rate on corrosion resistance of nonclad wrought aluminum alloy specimens

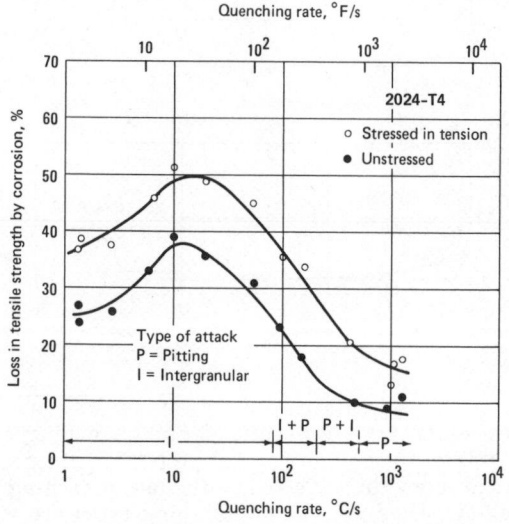
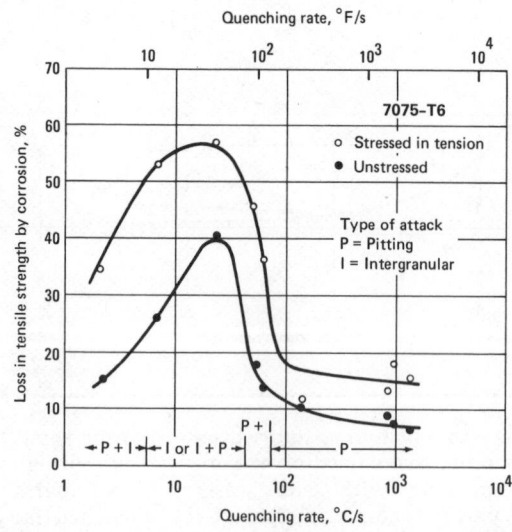

Corrosion resistance is indicated by percentage loss in tensile strength due to corrosion and by changes in predominant type of corrosive attack. Specimens of alloy 2024-T4 (left), 1.63 mm (0.064 in.) thick, were exposed to 48 h of alternate immersion in standard sodium chloride - hydrogen peroxide solution. Specimens of alloy 7075-T6 (right), also 1.63 mm thick, were exposed to three months of alternate immersion in a 3.5% aqueous solution of sodium chloride.

of corrosion on strength are accentuated when a sustained tensile stress is imposed during exposure.

As shown in Fig. 4, decreases in quenching rate reduce yield strength. However, the rate at which yield strength is reduced varies with alloy and temper. The strength of alloy 7075-T6 decreases significantly before the quenching rate becomes low enough to cause significant reduction in corrosion resistance. The opposite relationship applies for alloy 2024-T4: significant loss in corrosion resistance occurs at a higher quenching rate than significant loss in yield strength. Thus satisfactory strength does not necessarily ensure satisfactory corrosion resistance. Military specification MIL-H-6088E requires testing for evidence of excessive intergranular corrosion; acceptable material must exhibit a degree of corrosion no greater than that experienced when proper heat treating practices are followed.

In addition to sufficiently high quenching rates, prevention of losses in yield strength and/or corrosion resistance during quenching requires that atmospheric cooling during transfer from the solution heat treating furnace to the quenching medium be minimized.

Treatments After Quenching and Before Aging.

Most heat treatable aluminum alloys are nearly as ductile immediately after quenching as they are in the annealed condition. For this reason, forming of sheet metal parts and straightening of extrusions, forgings and castings frequently are done after quenching. Because natural aging increases strength and resistance to forming and straightening, these operations should be done as soon after quenching as possible.

The amounts of plastic strain involved in forming and in straightening range from small strains caused by minor corrections of warpage to strains approaching the limits of formability developed during production of complex parts from flat blanks. Although the most severe forming operations may have to be designed so that natural aging is avoided, aging prevents formation of Lüders lines, and thus some natural aging often is desirable. Formation of Lüders lines is most likely shortly after quenching and diminishes significantly after a few hours of natural aging. Complete freedom from Lüders lines, however, may require natural aging for 1 or 2 days

prior to forming. Therefore, timing of the forming operation may have to be selected to obtain the most appropriate combination of ability to be severely formed and freedom from Lüders lines for the specific parts involved. Lüders lines also are reduced by employing low strain rates or by forming at temperatures from 150 to 175 °C (300 to 350 °F).

Residual stresses in sheet metal parts formed in the solution heat treated and quenched condition are higher than those in parts formed after annealing but before solution heat treatment. Consequently, forming in the quenched condition should be selected judiciously for parts that must withstand fatigue or stress corrosion. (Figure 6 shows that alclad alloy 2024-T4 sheet has better fatigue characteristics if bent and flattened (unbent) in the annealed condition than if bent in the annealed condition but not flattened until after quenching following solution heat treatment.) However, forming in the annealed temper and then solution heat treating and quenching may not always be the better alternative; straightening usually is necessary to correct warpage caused by quenching, and the nonuniformity of residual strains introduced by this straightening may be as detrimental as forming in the quenched condition.

Re-solution heat treatment of parts formed after quenching is not recommended because it tends to cause excessive grain growth in critically strained regions.

Refrigeration.

Aging characteristics vary from alloy to alloy with respect to both time to initial change in mechanical properties and rate of change, but aging effects always are lessened by reductions in aging temperature (see Fig. 7). With some alloys, aging can be suppressed or delayed for several days by holding at a temperature of −18 °C (0 °F) or lower. It is usual practice to complete forming and straightening before aging changes mechanical properties appreciably. When scheduling makes this impractical, aging may be avoided in some alloys by refrigerating prior to forming. It is conventional practice to refrigerate alloy 2024-T4 rivets to maintain good driving characteristics. Full-size wing plates for current-generation jet aircraft have been solution heat treated and quenched at the primary fabricating mill, packed in dry ice in specially designed insulated shipping containers and transported by rail about 2000 miles to the aircraft manufacturer's plant for forming.

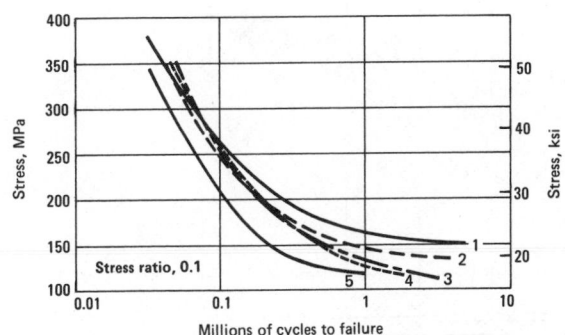

Fig. 6 Effects of flattening on fatigue characteristics of alclad alloy 2024-T4 sheet

Sheet 1.02 mm (0.040 in.) thick was annealed, solution heat treated and quenched, and then fatigue tested. The sheet represented by curve 1 was not bent. All other sheet was bent 90° in the annealed condition. Flattening (unbending) was done in either the annealed condition (curve 2) or the solution heat treated and quenched condition (curves 3, 4 and 5). Details of bending and flattening were as follows: (1) Not bent. (2) Bend radius, 3.18 mm (1/8 in.); flattened in annealed condition. (3) Bend radius, 3.18 mm (1/8 in.); flattened in quenched condition after 3 days of storage at −18 to −12 °C (0 to 10 °F). (4) Bend radius, 3.18 mm (1/8 in.); flattened in quenched condition after 14 days of storage at −18 to −12 °C (0 to 10 °F). (5) Bend radius, 1.59 mm (1/16 in.); flattened in quenched condition after 3 days of storage at −18 to −12 °C (0 to 10 °F).

Fig. 7 Aging characteristics of aluminum sheet alloys at room temperature at 0 °C (32 °F) and at −18 °C (0 °F)

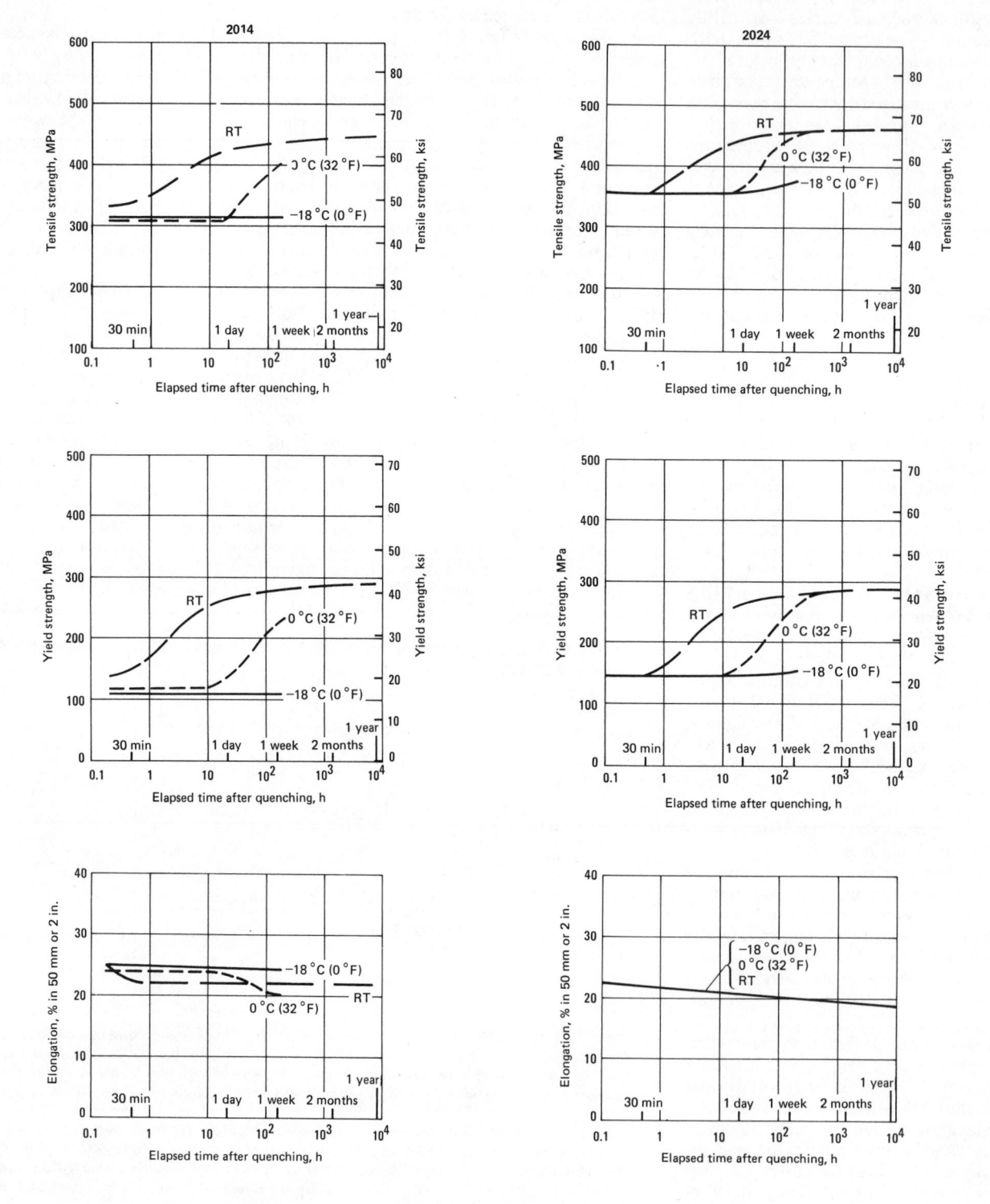

Fig. 7 (continued)

6061

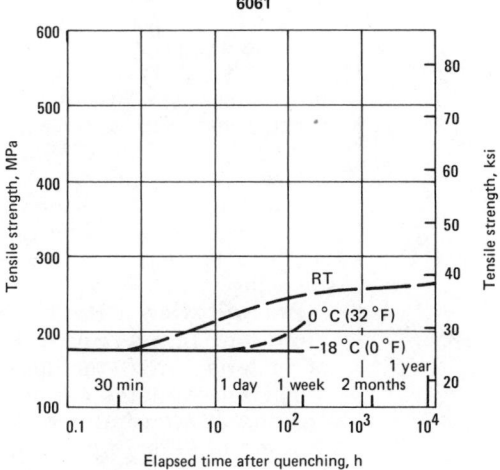

7050

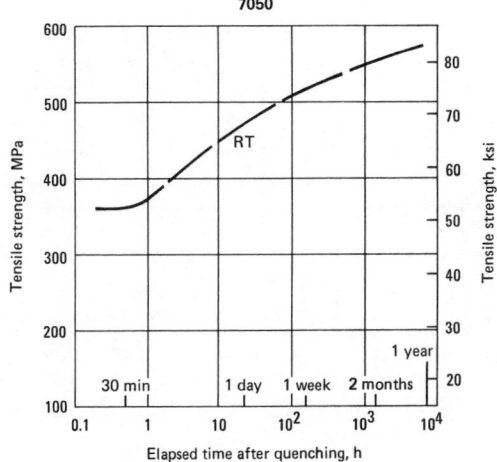

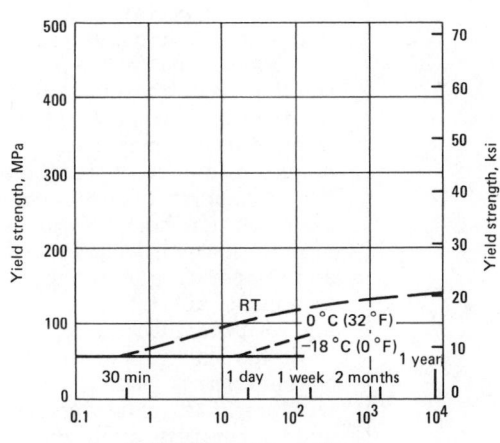

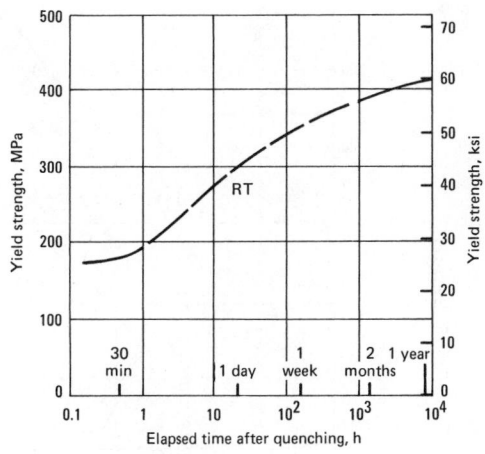

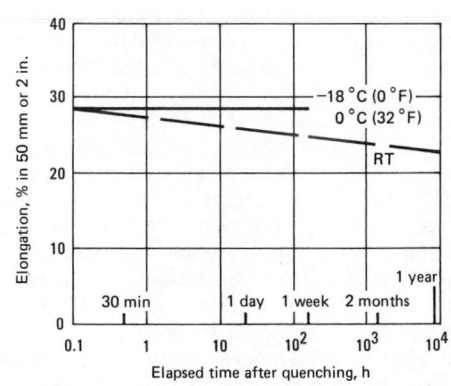

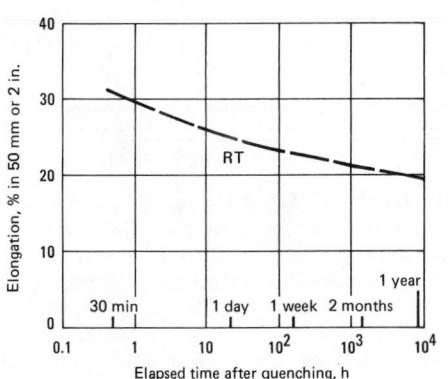

Fig. 7 (continued)

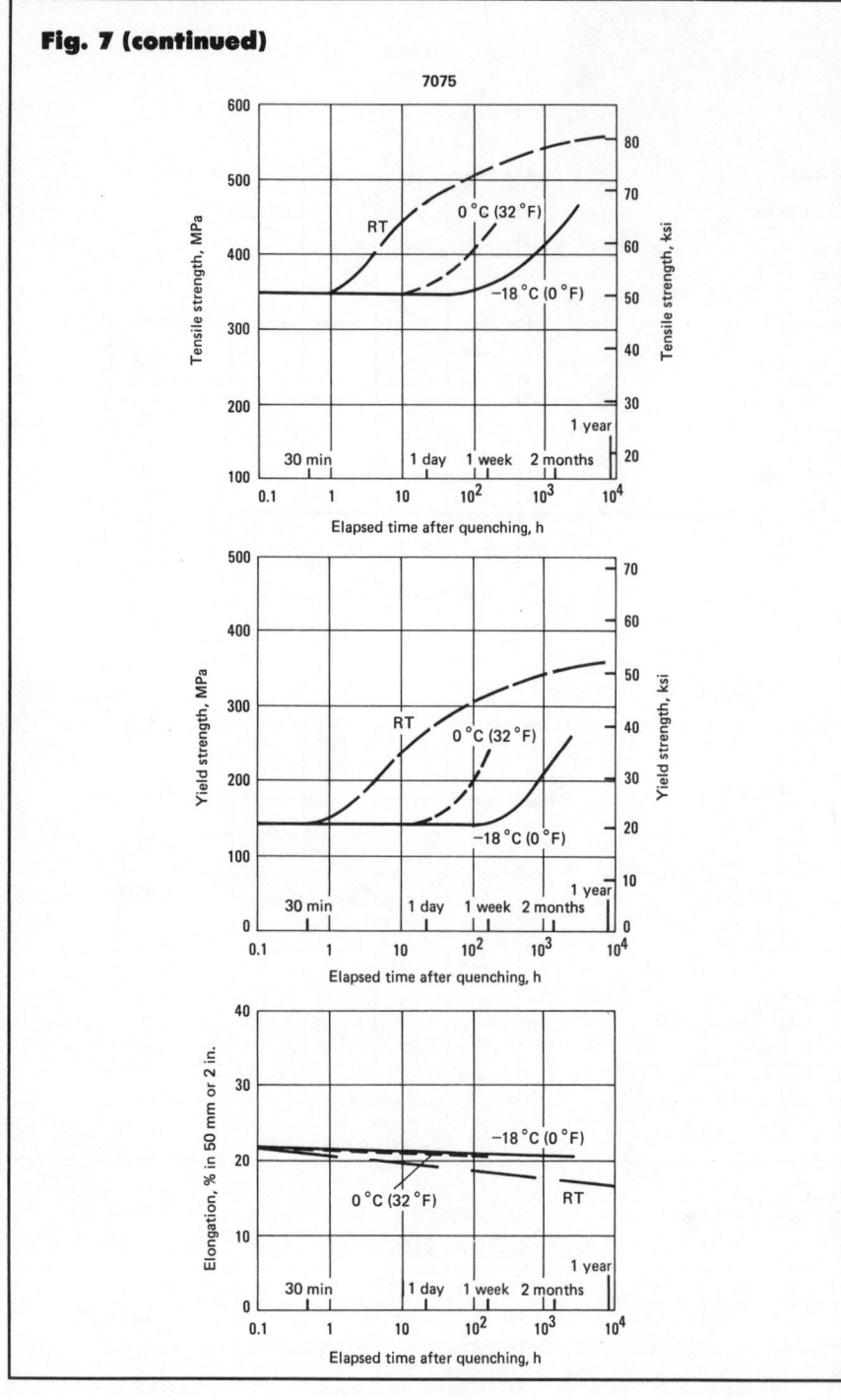

Mill Cold Work Processes. In some alloys—particularly certain alloys of the 2xxx series, strain introduced by cold working after solution heat treatment and quenching accelerates and greatly increases the changes in mechanical properties that occur during subsequent precipitation heat treatment, as illustrated for alloy 2024 in Fig. 8. This effect is the basis for the higher-strength T8-type tempers, used for alloys 2011, 2024, 2124, 2219 and 2419, which are produced by applying controlled amounts of cold rolling or stretching, or combinations of these operations. Because re-solution heat treatment of mill products supplied in T8-type tempers can result in grain growth and substantial lowering of strength, such reheat treatment is not recommended.

Alloys of the 7xxx series do not respond favorably to the sequence of operations used to produce T8-type tempers, and no such tempers are standard for these alloys. The strains associated with stretching or compressive stress relief of 7xxx alloys have relatively little effect on the mechanical properties of material precipitation heat treated to T6-type tempers. On the other hand, these operations have measurable detrimental effects on final strength when T73-, T736- or T76-type tempers are produced. Accordingly, specifications allow somewhat lower mechanical properties for the versions of these tempers that include stress relieving by cold work.

Precipitation Heat Treating Time and Temperature. Production of material in T5-through T10-type tempers necessitates precipitation heat treating at temperatures above room temperature. The type, size, distribution and amount of the precipitated particles in an aged alloy depend on precipitation heat treating time and temperature and on the initial state of the microstructure. Microstructure may vary in wrought products from unrecrystallized (hot worked prior to solution heat treatment) to recrystallized (cold worked prior to solution heat treatment) and may exhibit only modest strain due to quenching or somewhat higher strain due to cold working, after solution heat treatment. Therefore, these conditions, as well as time and temperature of precipitation heat treatment, affect final microstructure and resulting mechanical properties.

Mechanical properties and other characteristics change continuously with precipitation heat treatment time and temperature, as shown in Fig. 9 by typical curves for three wrought alloys. As may be seen in these graphs, maximum strength obtainable on aging is reduced as precipitation heat treatment temperature is raised. However, temperatures selected should be high enough to permit treatment times that are short and thus not too costly. Therefore, a temperature is selected that provides a compromise between high strength and short treatment time.

Good control of temperature and good temperature uniformity throughout the furnace and load are required for any precipitation heat treating operation. Recommended temperatures and times generally include adequate allowances for variations in composition within specified ranges, and variations in temperature in the furnace and load, while still ensuring a high probability that required minimum mechan-

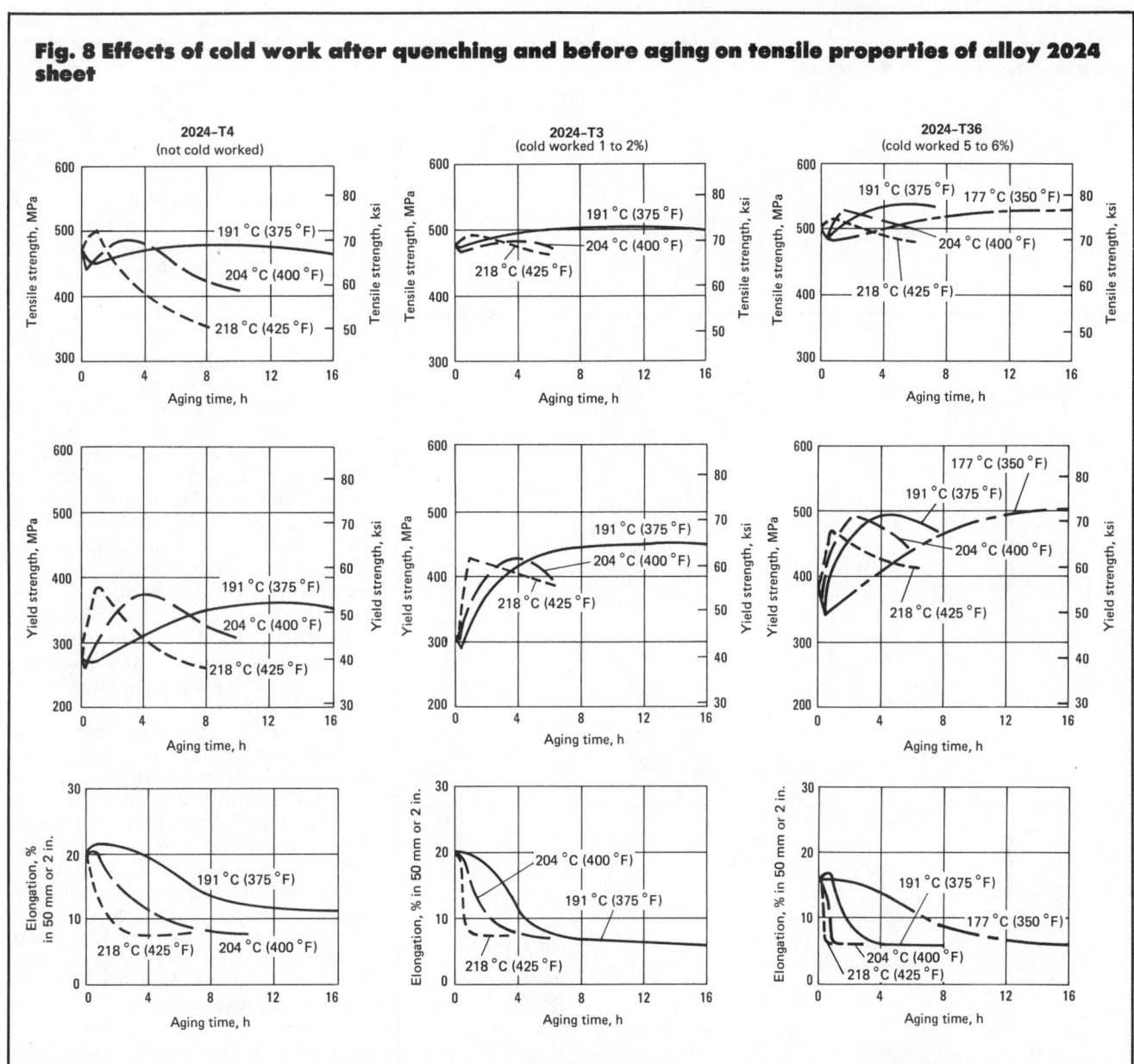

Fig. 8 Effects of cold work after quenching and before aging on tensile properties of alloy 2024 sheet

ical properties will be obtained. Treatments that incorporate such allowances include those recommended for development of T5- and T6-type tempers and of those tempers of the T7 type that are employed for dimensional and property stabilization. In heat treating of alloys 7049, 7050, 7075, 7175 and 7475 to T73-, T736- and T76-type tempers, on the other hand, relatively rapid overaging occurs at the temperatures employed in the second stage of the heat treating cycle (see Fig. 10), and obtaining the mechanical properties and corrosion resistance required of these tempers demands tight control of both time and temperature.

Dimensional Changes During Heat Treatment

In addition to the completely reversible changes in dimensions that are simple functions of temperature change and are caused by thermal expansion and contraction, dimensional changes of more permanent character are encountered during heat treatment. These changes are of several types, some of mechanical origin and others caused by changes in metallurgical structure. Changes of mechanical origin include those arising from stresses developed by gravitational or other applied forces, from thermally

induced stresses or from relaxation of residual stresses. Dimensional changes also accompany recrystallization and solution or precipitation of alloying elements.

Solution Heat Treatment. Distortion as a result of creep during solution heat treatment should be avoided by proper loading of parts in baskets, racks or fixtures or by provision of adequate support for long pieces heat treated in the horizontal position. Cyclically reversed lengthwise motion is employed in horizontal roller hearth furnaces to avoid sagging during heat treatment of plate, rod, bar and extrusions. Sheet is provided with air-pres-

Fig. 9a Typical artificial precipitation hardening aging curves for alloy 2024 sheet

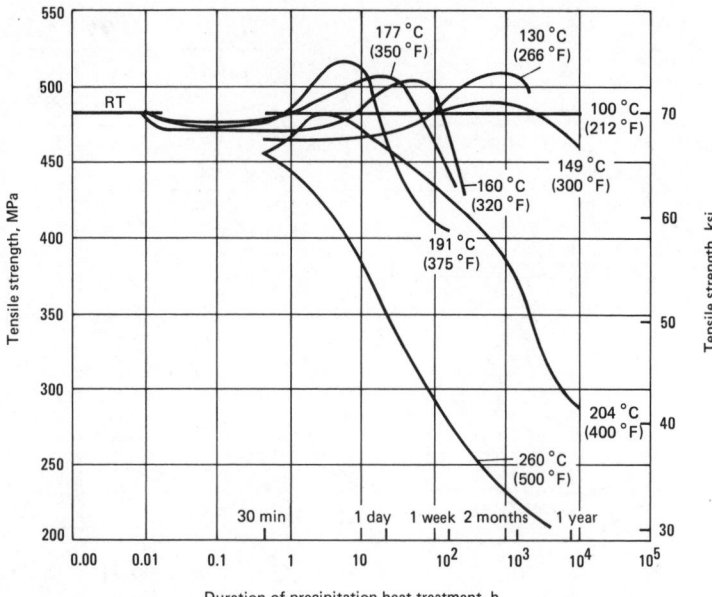

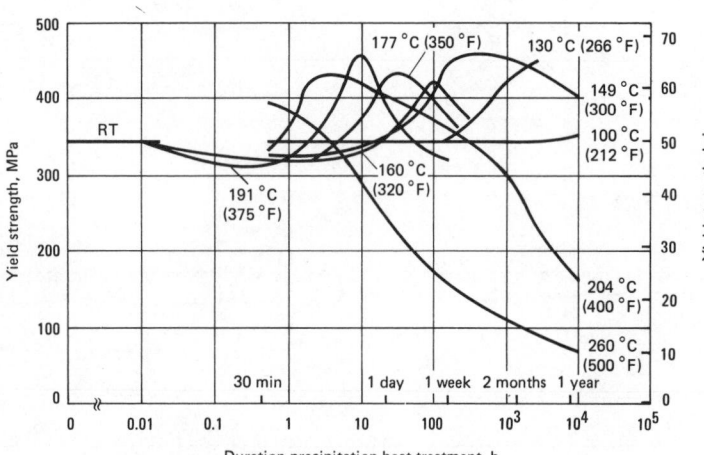

sure support in continuous heat treating furnaces to avoid scratching, gouging and distortion. For solution heat treatment fixtures or racks made of materials (such as steel) with coefficients of thermal expansion lower than that of the aluminum being treated, allowance should be made for differential expansion, if the free expansion of the aluminum is restricted. Straightening immediately after solution heat treating may be preferable to fixturing.

Solution of phases formed by major alloying elements causes volumetric expansion or contraction, depending on the alloy system, and this may have to be taken into account in heat treatment of long pieces. For example, solution heat treatment and quenching of alloy 2219 causes lengthwise contraction of about 2 mm/m (0.002 in./in.). Solution heat treatment and quenching of alloys of the 7xxx series is accompanied by lengthwise expansion—about 0.6 mm/m (0.0006 in./in.) for alloy 7075 rod or plate.

Quenching. The most troublesome changes in dimensions and shape are those that occur during quenching or that result from stresses induced by quenching. Because of its nonuniform cooling, quenching may produce warpage or distortion, particularly in thin material and in thin sections of parts that contain variations in thickness. For thick-section products or parts, changes in external shape may be small because of rigidity, but the interior-to-surface temperature gradients that form with rapid cooling create residual stresses; these stresses normally are compressive at the surfaces and tensile in the interior.

As previously discussed, warpage or distortion of thin-section material can be reduced by using a quenching medium that provides slower cooling; however, cooling must be sufficient to produce the required properties. Slower quenching can also reduce the magnitude of residual stresses in thicker parts or pieces, as shown in Fig. 11 for cylindrical specimens of alloy 6151 quenched in cold or boiling water. Stress range (maximum tensile stress plus maximum compressive stress) for a cylinder 89 mm (3.5 in.) in radius is about 205 MPa (30 ksi) when the cylinder is quenched in cold water but less than 70 MPa (10 ksi) when it is quenched in boiling water. The effects of average cooling rate through the temperature range from 400 to 290 °C (750 to 550 °F) on longitudinal stress

Fig. 9b Typical artificial precipitation hardening aging curves for alloy 2014 sheet

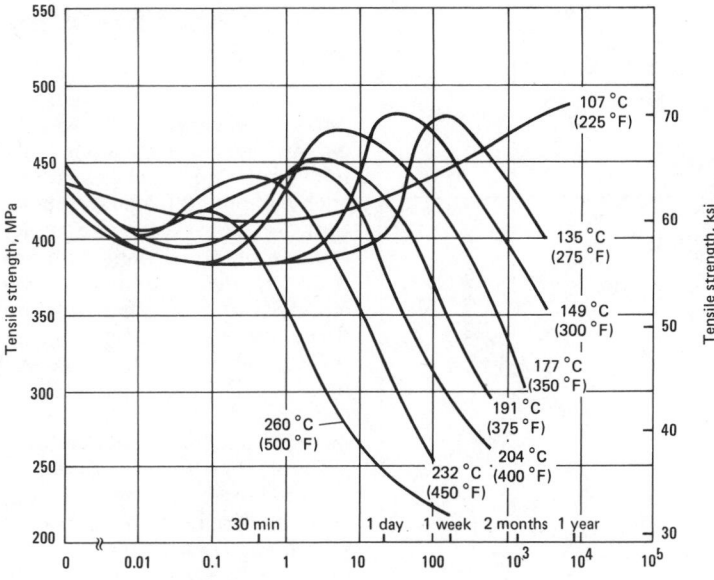

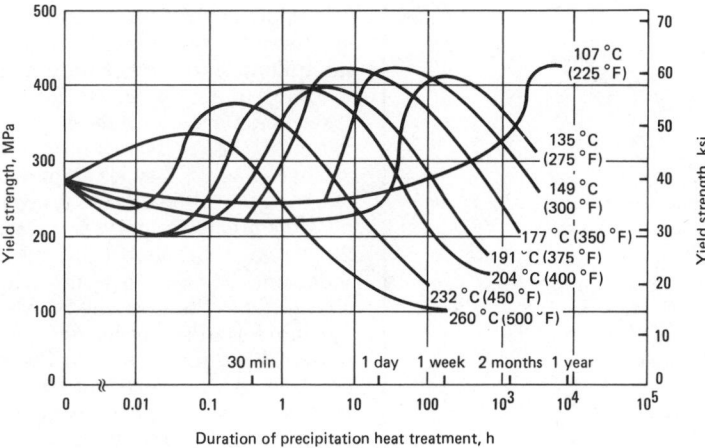

ranges developed in alloy 2014 cylinders 75 mm (3 in.) in diameter are shown in Fig. 12.

High stresses induced by rapid quenching generally are reduced only modestly by the precipitation heat treatments used to produce T6- or T8-type tempers. Consequently, for the alloys that require rapid cooling to develop the properties of these tempers, those incorporating mechanical stress relief (Tx51, Tx52) usually are specified when substantial metal must be removed to produce final shapes. Other T8-type tempers, such as T86 and T87, also have low residual stress as a result of the stretching required to produce them.

Heat Treatments for Precipitation and Stabilization. The most significant dimensional changes associated with precipitation heat treatments and stabilizing heat treatments arise from concurrent dilution of the solid solution (which changes lattice parameter) and formation of precipitate. Changes in density and specific volume resulting from these changes in metallurgical structure are the reverse of those caused by solution of the alloy phases. However, because the strongest tempers are those in which the precipitate is present in nonequilibrium transition forms, the amount of change during precipitation heat treatment does not totally compensate for the previous (and opposite) change that occurred during solution heat treatment. Most of the heat treatable alloys expand (grow) during precipitation heat treatment. Exceptions are alloys of the 7xxx wrought series and the 7xx.0 casting series, which exhibit contraction.

In alloys of the 2xxx series, the amount of growth decreases with increasing magnesium content. Thus, growth of about 1.5 mm/m (0.0015 in./in.) can be expected during precipitation heat treatment of alloy 2219-T87, about 0.5 mm/m (0.005 in./in.) for treatment of alloy 2014-T6 and less than 0.1 mm/m (0.0001 in./in.) for treatment of alloy 2024-T851. Alloys 7050 and 7075, on the other hand, contract about 0.3 mm/m (0.0003 in./in.) on precipitation heat treating from the W temper to the T6 temper and about 0.7 mm/m (0.0007 in./in.) on treating from the W temper to the T73 temper.

Stabilizing T7-type treatments cause greater amounts of growth than the T5-, T6- or T8-type treatments for the same alloys. This increased growth is associated either with formation of

Fig. 9c Typical artificial precipitation hardening aging curves for alloy 6061 sheet

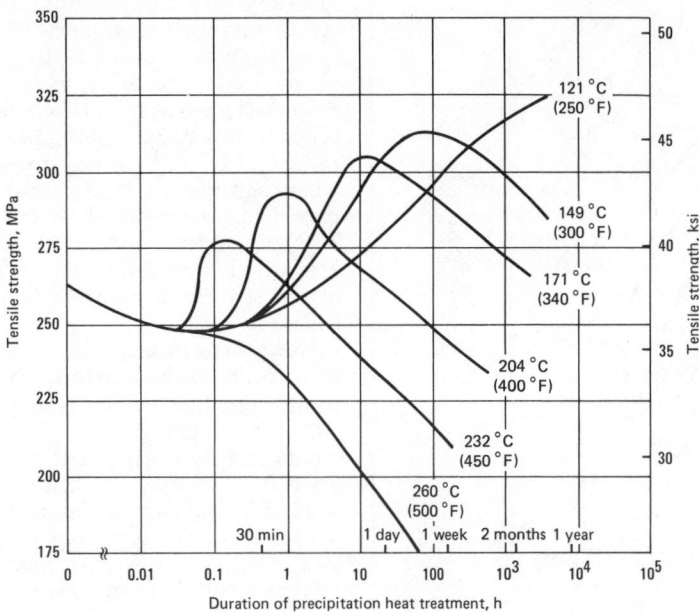

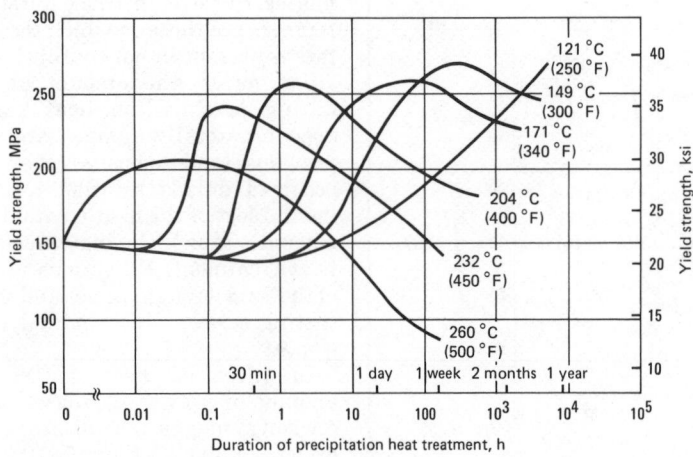

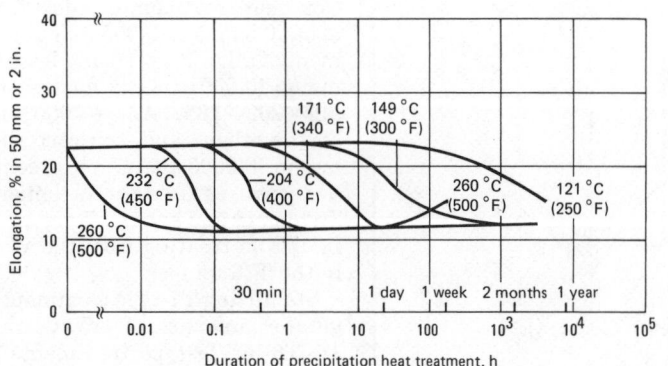

increased amounts of transition precipitates or with transformation of transition precipitates to equilibrium phases.

Dimensional Stability in Service

Dimensional stability of heat treated parts in service depends on alloy, temper and service conditions. Of the latter, excluding mechanical conditions such as applied loads, the most important is temperature range. Potential sources of dimensional change other than applied loading are residual stresses and further precipitation. Stress relief minimizes changes due to residual stresses, and most mill products usually are supplied in tempers that include stress relief. Potential dimensional change as a result of further precipitation in parts that operate at elevated temperatures is minimized for wrought products by use of T7-type stabilizing treatments and for castings by use of T5-type treatments. However, components of high-precision equipment, such as instruments for aerospace guidance systems and optical and telescopic devices, may require special supplementary treatments during manufacture to further reduce stresses or subsequent precipitation. (These treatments are discussed below, under "Stability of Precision Equipment".)

The T3- and T4-type tempers are the least stable dimensionally because of possible precipitation in service. Alloys 2024 and 2124 have the smallest dimensional change in aging; the total change from the quenched to the average state is of the order of 0.06 mm/m (0.00006 in./in.), less than the change due to a temperature variation of 3 °C (5 °F). These alloys therefore can be used in the T3- and T4-type tempers, except for precision equipment. For all other alloys, the T6- or T8-type tempers should be used, because in these tempers all the alloys have good dimensional stability.

Stability of Precision Equipment. Proper maintenance of high-precision devices, such as gyros, accelerometers and optical systems, requires use of materials in which dimensional changes from metallurgical instability are limited to 10 μm/m (10 μin./in.). Several laboratory investigations and considerable practical experience have shown that wrought alloys 2024 and 6061 and casting alloy 356.0 are well suited and generally preferred

Fig. 10 Effects of time and temperature in second stage of precipitation heat treatment of alloy 7075-T73 on resulting yield strength

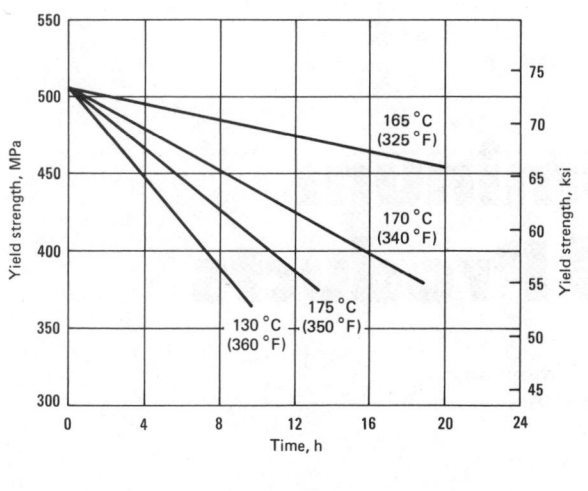

Fig. 12 Effect of quenching rate on longitudinal stress ranges in alloy 2014-T4 cylinders quenched in various media

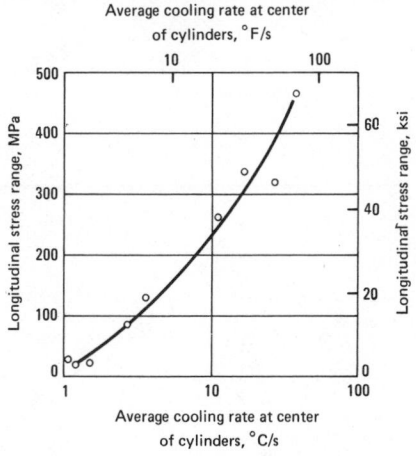

Cylinders were 75 mm (3 in.) in diameter by 230 mm (9 in.) long. Cooling rate was measured from 400 to 290 °C (750 to 555 °F). Stress range is maximum tensile stress plus maximum compressive stress.

Fig. 11 Effect of quenching from 540 °C (1000 °F) on residual stresses in solid cylinders of alloy 6151

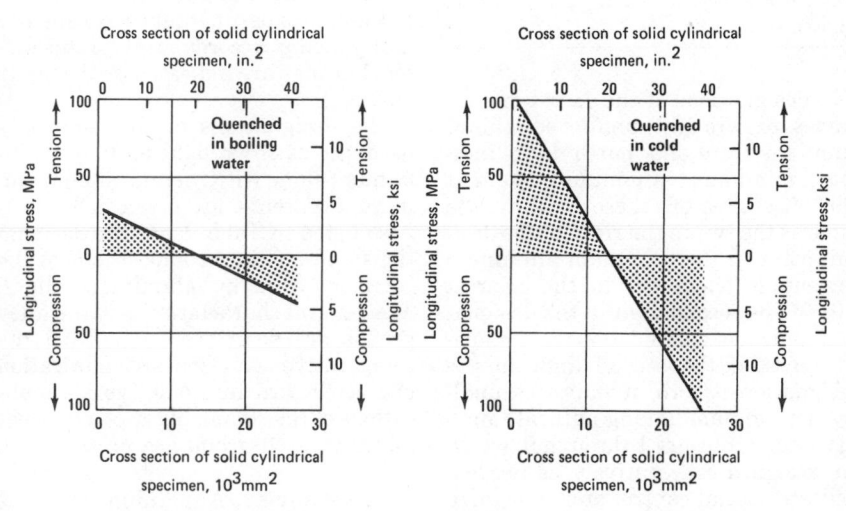

for such applications. Dimensional changes were no greater than 10 μm/m when alloys 2024-T851 and -T62, 6061-T651 and -T62, and 356.0-T51, -T6 and -T7 were tested for more than a year at room temperature and for several months at 70 °C (160 °F) and when the same alloys were tested with repeated thermal cycling between +20 and −70 °C (+68 and −94 °F).

Because stresses applied or induced by acceleration in such devices generally are not high, strength levels lower than those of the highest-strength tempers frequently are satisfactory. To increase precision of machining to intended dimensions, as well as to promote maximum stability, it is common practice to apply additional thermal treatments for stress relief and precipitation of 1 to 2 h at temperatures of 175 to 205 °C (350 to 400 °F) after rough machining. These additional treatments sometimes are repeated at successive stages of processing, and even after final machining. In addition, it has been claimed that one or two cyclic treatments consisting of cooling to −100 °C (−150 °F), holding for 2 h, heating to 232 to 240 °C (450 to 465 °F) and again holding for 2 h can improve dimensional stability of 356-T6 castings.

Aluminum Mill Products

By the ASM Committee on
Aluminum and Aluminum Alloys*

ALUMINUM mill products are those aluminum products that have been subjected to plastic deformation by hot and cold working mill processes such as rolling, extruding and drawing, either singly or in combination, that transform cast aluminum ingot into the desired product form. The microstructural changes associated with the working and with any accompanying thermal treatments are used to control certain properties and characteristics of the worked, or "wrought", product or alloy. Production of aluminum mill products has increased over the past quarter century at a fairly steady rate (see Fig. 1). Total production in the United States in 1978 (including forgings) was 5.1 Mt (5 600 000 tons), of which 65% was produced as rolled products (plate, sheet, strip and foil), 23.4% as extruded shapes and tube, 10.4% as rod, bar and wire, and 1.2% as forgings and impacts.

Wrought Alloy Series

Designation System. A standard system of four-digit numbers is used for designating wrought aluminum and aluminum alloys. The first digit indicates the alloy series, as follows:

*See page **X** for committee list.

Aluminum, 99.00% min Al or greater	1xxx
Aluminum alloys grouped by major alloying elements	
Copper	2xxx
Manganese	3xxx
Silicon	4xxx
Magnesium	5xxx
Magnesium and silicon	6xxx
Zinc	7xxx
Other element	8xxx
Unused series	9xxx

In designations of the 1xxx type (for grades of wrought unalloyed aluminum), the third and fourth digits indicate minimum aluminum content (99.00% or greater); these digits are the same as the two to the right of the decimal point in the minimum aluminum percentage expressed to the nearest 0.01%. The second digit in the designation indicates modifications in impurity limits. If the second digit in the designation is zero, it indicates unalloyed aluminum having natural impurity limits; integers 1 through 9, which are assigned consecutively as needed, indicate special control of one or more individual impurities or alloying elements.

In the 2xxx through 8xxx designations (for wrought aluminum alloys), the third and fourth digits in the designation have no numerical significance but only identify the different aluminum alloys in the group. The second digit in the designation indicates alloy modifications. If the second digit in the designation is zero, it indicates the original alloy; integers 1 through 9, which are assigned consecutively, indicate modifications of the original alloy (minor changes in alloy and/or impurity limits). Experimental alloys are also designated in accordance with this system but they are indicated by the prefix "X".

The designations of the commonly used grades of wrought unalloyed aluminum, along with their minimum aluminum contents, are given in Table 1. Also listed in Table 1 are the designations and nominal compositions of the common wrought aluminum alloys. The general characteristics of the alloy groups are described below, and the comparative corrosion and fabrication characteristics and some typical applications of the commonly used grades or alloys in each group are presented in Table 2.

1xxx Series. Aluminum of 99.00% or higher purity has many applications, especially in the electrical and chemical fields. These grades of aluminum are characterized by excellent corrosion resistance, high thermal and

Table 1 Product forms and nominal compositions of common wrought aluminum alloys

AA number	Product(a)	Al	Si	Cu	Mn	Mg	Cr	Zn	Others
1050	DT	99.50 min	...	...	...	...	...	...	...
1060	S, P, ET, DT	99.60 min	...	...	...	...	...	...	...
1100	S, P, F, E, ES, ET, C, DT, FG	99.00 min	...	0.12	...	...	...	...	...
1145	S, P, F	99.45 min	...	...	...	...	...	...	...
1199	F	99.99 min	...	...	...	...	...	...	...
1350	S, P, E, ES, ET, C	99.50 min	...	...	...	...	...	...	...
2011	E, ES, ET, C, DT	93.7	...	5.5	...	...	...	...	0.4Bi; 0.4Pb
2014	S, P, E, ES, ET, C, DT, FG	93.5	0.8	4.4	0.8	0.5	...	...	...
2024	S, P, E, ES, ET, C, DT	93.5	...	4.4	0.6	1.5	...	...	...
2036	S	96.7	...	2.6	0.25	0.45	...	...	...
2048	S, P	94.8	...	3.3	0.4	1.5	...	...	...
2124	P	93.5	...	4.4	0.6	1.5	...	...	...
2218	FG	92.5	...	4.0	...	1.5	...	...	2.0Ni
2219	S, P, E, ES, ET, C, FG	93.0	...	6.3	0.3	...	...	...	0.06Ti; 0.10V; 0.18Zr
2319	C	93.0	...	6.3	0.3	...	...	...	0.18Zn, 0.15Ti; 0.10V
2618	FG	93.7	0.18	2.3	...	1.6	...	...	1.1Fe; 1.0Ni; 0.07Ti
3003	S, P, F, E, ES, ET, C, DT, FG	98.6	...	0.12	1.2	...	...	...	...
3004	S, P, ET, DT	97.8	...	...	1.2	1.0	...	...	...
3105	S	99.0	...	...	0.55	0.50	...	...	...
4032	FG	85.0	12.2	0.9	...	1.0	...	...	0.9Ni
4043	C	94.8	5.2	...	...	...	...	...	...
5005	S, P, C	99.2	...	...	...	0.8	...	...	...
5050	S, P, C, DT	98.6	...	...	...	1.4	...	...	...
5052	S, P, F, C, DT	97.2	...	...	...	2.5	0.25	...	...
5056	F, C	95.0	...	...	0.12	5.0	0.12	...	...
5083	S, P, E, ES, ET, FG	94.7	...	...	0.7	4.4	0.15	...	...
5086	S, P, E, ES, ET, DT	95.4	...	...	0.4	4.0	0.15	...	...
5154	S, P, E, ES, ET, C, DT	96.2	...	...	...	3.5	0.25	...	...
5182	S	95.2	...	...	0.35	4.5	...	...	...
5252	S	97.5	...	...	...	2.5	...	...	...
5254	S, P	96.2	...	...	...	3.5	0.25	...	...
5356	C	94.6	...	...	0.12	5.0	0.12	...	0.13Ti
5454	S, P, E, ES, ET	96.3	...	...	0.8	2.7	0.12	...	...
5456	S, P, E, ES, ET, DT, FG	93.9	...	...	0.8	5.1	0.12	...	...
5457	S	98.7	...	...	0.3	1.0	...	...	...
5652	S, P	97.2	...	...	...	2.5	0.25	...	...
5657	S	99.2	...	...	...	0.8	...	...	...
6005	E, ES, ET	98.7	0.8	...	...	0.5	...	...	...
6009	S	97.7	0.8	0.35	0.5	0.6	...	...	...
6010	S	97.3	1.0	0.35	0.5	0.8	...	...	...
6061	S, P, E, ES, ET, C, DT, FG	97.9	0.6	0.28	...	1.0	0.2	...	...
6063	E, ES, ET, DT	98.9	0.4	...	...	0.7	...	...	...
6066	E, ES, ET, DT, FG	95.7	1.4	1.0	0.8	1.1	...	...	...
6070	E, ES, ET	96.8	1.4	0.28	0.7	0.8	...	...	...
6101	E, ES, ET	98.9	0.5	...	...	0.6	...	...	...
6151	FG	98.2	0.9	...	...	0.6	0.25	...	...
6201	C	98.5	0.7	...	...	0.8	...	...	...
6205	E, ES, ET	98.4	0.8	...	0.1	0.5	0.1	...	0.1Zr
6262	E, ES, ET, C, DT	96.8	0.6	0.28	...	1.0	0.09	...	0.6Bi; 0.6Pb
6351	E, ES	97.8	1.0	...	0.6	0.6	...	...	...
6463	E, ES	98.9	0.4	...	...	0.7	...	...	...
7005	E, ES	93.3	...	...	0.45	1.4	0.13	4.5	0.04Ti; 0.14Zr
7049	P, E, ES, FG	88.2	...	1.5	...	2.5	0.15	7.6	...
7050	P, E, ES, FG	89.0	...	2.3	...	2.3	...	6.2	0.12Zr
7072	S, F	99.0	...	...	...	...	...	1.0	...
7075	S, P, E, ES, ET, C, DT, FG	90.0	...	1.6	...	2.5	0.23	5.6	...
7175	S, P, FG	90.0	...	1.6	...	2.5	0.23	5.6	...
7178	S, P, E, ES, C	88.1	...	2.0	...	2.7	0.26	6.8	...
7475	S, P, FG	90.3	1.5	...	...	2.3	0.22	5.7	...

(a) S = sheet; P = plate; F = foil; E = extruded rod, bar and wire; ES = extruded shapes; ET = extruded tubes; C = cold finished rod, bar and wire; DT = drawn tube; FG = forgings.

Table 2 Comparative corrosion and fabrication characteristics and typical applications of wrought aluminum alloys

Alloy temper	Resistance to corrosion		Workability (cold)(e)	Machinability(e)	Weldability(f)			Brazeability(f)	Solderability(g)	Some typical applications of alloys
	General(a)	Stress-corrosion cracking(b)			Gas	Arc	Resistance spot and seam			
1050 0	A	A	A	E	A	A	B	A	A	Chemical equipment, railroad tank cars
H12	A	A	A	E	A	A	A	A	A	
H14	A	A	A	D	A	A	A	A	A	
H16	A	A	B	D	A	A	A	A	A	
H18	A	A	B	D	A	A	A	A	A	
1060 0	A	A	A	E	A	A	B	A	A	Chemical equipment, railroad tank cars
H12	A	A	A	E	A	A	A	A	A	
H14	A	A	A	D	A	A	A	A	A	
H16	A	A	B	D	A	A	A	A	A	
H18	A	A	B	D	A	A	A	A	A	
1100 0	A	A	A	E	A	A	B	A	A	Sheet-metal work, spun hollowware, fin stock
H12	A	A	A	E	A	A	A	A	A	
H14	A	A	A	D	A	A	A	A	A	
H16	A	A	B	D	A	A	A	A	A	
H18	A	A	C	D	A	A	A	A	A	
1145 0	A	A	A	E	A	A	B	A	A	Foil, fin stock
H12	A	A	A	E	A	A	A	A	A	
H14	A	A	A	D	A	A	A	A	A	
H16	A	A	B	D	A	A	A	A	A	
H18	A	A	B	D	A	A	A	A	A	
1199 0	A	A	A	E	A	A	B	A	A	Electrolytic capacitor foil, chemical equipment, railroad tank cars
H12	A	A	A	E	A	A	A	A	A	
H14	A	A	A	D	A	A	A	A	NA	
H16	A	A	B	D	A	A	A	A	A	
H18	A	A	B	D	A	A	A	A	A	
1350 0	A	A	A	E	A	A	B	A	A	Electrical conductors
H12, H111	A	A	A	E	A	A	A	A	A	
H14, H24	A	A	A	D	A	A	A	A	A	
H16, H26	A	A	B	D	A	A	A	A	A	
H18	A	A	B	D	A	A	A	A	A	
2011 T3	D(c)	D	C	A	D	D	D	D	C	Screw-machine products
T4, T451	D(c)	D	B	A	D	D	D	D	C	
T8	D	B	D	A	D	D	D	D	C	
2014 0	...	...	...	D	D	D	B	D	C	Truck frames, aircraft structures
T3, T4, T451	D(c)	C	C	B	D	B	B	D	C	
T6, T651, T6510, T6511	D	C	D	B	D	B	B	D	C	
2024 0	...	...	...	D	D	D	D	D	C	Truck wheels, screw-machine products, aircraft structures
T4, T3, T351, T3510, T3511	D(c)	C	C	B	C	B	B	D	C	
T361	D(c)	C	D	B	D	C	B	D	C	
T6	D	B	C	B	D	C	B	D	C	
T861, T81, T851, T8510, T8511	D	B	D	B	D	C	B	D	C	
T72	...	...	...	B	...	...	...	...	...	
2036 T4	C	...	B	C	...	B	B	D	...	Auto-body panel sheet
2124 T851	D	B	D	B	D	C	B	D	C	Military supersonic aircraft
2218 T61	D	C	...	...	...	...	C	...	C	Jet engine impellers and rings
T72	D	C	...	B	D	C	B	D	C	
2219 0	...	...	...	...	D	A	B	D		Structural uses at high temperatures (to 315 °C or 600 °F) high-strength weldments
T31, T351, T3510, T3511	D(c)	C	C	B	A	A	A	D	NA	
T37	D(c)	C	D	B	A	A	A	D		
T81, T851, T8510, T8511	D	B	D	B	A	A	A	D		
T87	D	B	D	B	A	A	A	D		
2618 T61	D	C	...	B	D	C	B	D	NA	Aircraft engines

(continued)

Table 2 (continued)

	General(a)	Stress-corrosion cracking(b)	Workability (cold)(e)	Machinability(e)	Gas	Arc	Resistance spot and seam	Brazeability(f)	Solderability(g)	Some typical applications of alloys
3003 0	A	A	A	E	A	A	B	A	A	Cooking utensils, chemical equipment, pressure vessels, sheet-metal work, builder's hardware, storage tanks
H12	A	A	A	E	A	A	A	A	A	
H14	A	A	B	D	A	A	A	A	A	
H16	A	A	C	D	A	A	A	A	A	
H18	A	A	C	D	A	A	A	A	A	
H25	A	A	B	D	A	A	A	A	A	
3004 0	A	A	A	D	B	A	B	B	B	Sheet-metal work, storage tanks
H32	A	A	B	D	B	A	A	B	B	
H34	A	A	B	C	B	A	A	B	B	
H36	A	A	C	C	B	A	A	B	B	
H38	A	A	C	C	B	A	A	B	B	
3105 0	A	A	A	E	B	A	B	B	B	Residential siding, mobile homes, rain-carrying goods, sheet-metal work
H12	A	A	B	E	B	A	A	B	B	
H14	A	A	B	D	B	A	A	B	B	
H16	A	A	C	D	B	A	A	B	B	
H18	A	A	C	D	B	A	A	B	B	
H25	A	A	B	D	B	A	A	B	B	
4032 T6	C	B	...	B	D	B	C	D	NA	Pistons
4043	B	A	NA	C	NA	NA	NA	NA	NA	Welding electrode
5005 0	A	A	A	E	A	A	B	B	B	Appliances, utensils, architectural, electrical conductors
H12	A	A	A	E	A	A	A	B	B	
H14	A	A	B	D	A	A	A	B	B	
H16	A	A	C	D	A	A	A	B	B	
H18	A	A	C	D	A	A	A	B	B	
H32	A	A	B	E	A	A	A	B	B	
H34	A	A	C	D	A	A	A	B	B	
H36	A	A	C	D	A	A	A	B	B	
H38	A	A		D	A	A	A	B	B	
5050 0	A	A	A	E	A	A	B	B	C	Builders' hardware, refrigerator trim, coiled tubes
H32	A	A	A	D	A	A	A	B	C	
H34	A	A	B	D	A	A	A	B	C	
H36	A	A	C	C	A	A	A	B	C	
H38	A	A	C	C	A	A	A	B	C	
5052 0	A	A	A	D	A	A	B	C	D	Sheet-metal work, hydraulic tube, appliances
H32	A	A	B	D	A	A	A	C	D	
H34	A	A	B	C	A	A	A	C	D	
H36	A	A	C	C	A	A	A	C	D	
H38	A	A	C	C	A	A	A	C	D	
5056 0	A(d)	B(d)	A	D	C	A	B	D	D	Cable sheathing, rivets for magnesium, screen wire, zippers
H111	A(d)	B(d)	A	D	C	A	A	D	D	
H12, H32	A(d)	B(d)	B	D	C	A	A	D	D	
H14, H34	A(d)	B(d)	B	C	C	A	A	D	D	
H18, H38	A(d)	C(d)	C	C	C	A	A	D	D	
H192	B(d)	D(d)	D	B	C	A	A	D	D	
H392	B(d)	D(d)	D	B	C	A	A	D	D	
5083 0	A(d)	B(d)	B	D	C	A	B	D	D	Unfired, welded pressure vessels, marine, auto aircraft cryogenics, TV towers, drilling rigs, transportation equipment, missile components
H321, H116	A(d)	B(d)	C	D	C	A	A	D	D	
H323	A(d)	B(d)	C	D	C	A	A	D	D	
H343	A(d)	B(d)	C	C	C	A	A	D	D	
H111	A(d)	B(d)	C	D	C	A	A	D	D	

(continued)

Table 2 (continued)

| | Resistance to corrosion | | Workability (cold)(e) | Machinability(e) | Weldability(f) | | | Brazeability(f) | Solderability(g) | Some typical applications of alloys |
	General(a)	Stress-corrosion cracking(b)			Gas	Arc	Resistance spot and seam			
5086 0	A(d)	A(d)	A	D	C	A	B	D	D	
H32, H116	A(d)	A(d)	B	D	C	A	A	D	D	
H34	A(d)	B(d)	B	C	C	A	A	D	D	
H36	A(d)	B(d)	C	C	C	A	A	D	D	
H38	A(d)	B(d)	C	C	C	A	A	D	D	
H111	A(d)	A(d)	B	D	C	A	A	D	D	
5154 0	A(d)	A(d)	A	D	C	A	B	D	D	Welded structures, storage tanks, pressure vessels, salt-water service
H32	A(d)	A(d)	B	D	C	A	A	D	D	
H34	A(d)	A(d)	B	C	C	A	A	D	D	
H36	A(d)	A(d)	C	C	C	A	A	D	D	
H38	A(d)	A(d)	C	C	C	A	A	D	D	
5182 0	A	A(d)	A	D	C	A	B	D	D	Automobile body sheet, can ends
H19	A	A(d)	D	B	C	A	A	D	D	
5252 H24	A	A	B	D	A	A	A	C	D	Automotive and appliance trim
H25	A	A	B	C	A	A	A	C	D	
H28	A	A	C	C	A	A	A	C	D	
5254 0	A(d)	A(d)	A	D	C	A	B	D	D	Hydrogen peroxide and chemical storage vessels
H32	A(d)	A(d)	B	D	C	A	A	D	D	
H34	A(d)	A(d)	B	C	C	A	A	D	D	
H36	A(d)	A(d)	C	C	C	A	A	D	D	
H38	A(d)	A(d)	C	C	C	A	A	D	D	
5356	A	A	NA	B	NA	NA	NA	NA	NA	Welding electrode
5454 0	A	A	A	D	C	A	B	D		Welded structures, pressure vessels, marine service
H32	A	A	B	D	C	A	A	D	NA	
H34	A	A	B	C	C	A	A	D		
H111	A	A	B	D	C	A	A	D		
5456 0	A(d)	B(d)	B	D	C	A	B	D		High-strength welded structures, storage tanks, pressure vessels, marine applications
H111	A(d)	B(d)	C	D	C	A	A	D		
H321, H115	A(d)	B(d)	C	D	C	A	A	D	NA	
H323	A(d)	B(d)	C	D	C	A	A	D		
H343	A(d)	B(d)	C	C	C	A	A	D		
5457 0	A	A	A	E	A	A	B	B	B	
5652 0	A	A	A	D	A	A	B	C	D	Hydrogen peroxide and chemical storage vessels
H32	A	A	B	D	A	A	A	C	D	
H34	A	A	B	C	A	A	A	C	D	
H36	A	A	C	C	A	A	A	C	D	
H38	A	A	C	C	A	A	A	C	D	
5657 H241	A	A	A	D	A	A	A	B		Anodized auto and appliance trim
H25	A	A	B	D	A	A	A	B		
H26	A	A	B	D	A	A	A	B	NA	
H28	A	A	C	D	A	A	A	B		
6005 T5	B	A	C	C	A	A	A	A	NA	Heavy-duty structures requiring good corrosion resistance applications, truck and marine, railroad cars, furniture, pipelines
6009 T4	A	A	A	C	A	A	A	A	B	Automobile body sheet
6010 T4	A	A	B	C	A	A	A	A	B	Automobile body sheet
6061 0	B	A	A	D	A	A	B	A	B	Heavy-duty structures requiring good corrosion resistance, truck and marine, railroad cars, furniture, pipelines
T4, T451, T4510, T4511	B	B	B	C	A	A	A	A	B	
T6, T651, T652, T6510, T6511	B	A	C	C	A	A	A	A	B	

(continued)

Table 2 (continued)

| | Resistance to corrosion | | Workability (cold)(e) | Machinability(e) | Weldability(f) | | | Brazeability(f) | Solderability(g) | Some typical applications of alloys |
	General(a)	Stress-corrosion cracking(b)			Gas	Arc	Resistance spot and seam			
6063 T1	A	A	B	D	A	A	A	A	B	Pipe railing, furniture, architectural extrusions
T4	A	A	B	D	A	A	A	A	B	
T5, T52	A	A	B	C	A	A	A	A	B	
T6	A	A	C	C	A	A	A	A	B	
T83, T831, T832	A	A	C	C	A	A	A	A	B	
6066 0	C	A	B	D	D	B	B	D		Forgings and extrusions for welded structures
T4, T4510, T4511	C	B	C	C	D	B	B	D	NA	
T6, T6510, T6511	C	B	C	B	D	B	B	D		
6070 T4, T4511	B	B	B	C	A	A	A	B	NA	Heavy-duty welded structures, pipelines
T6	B	B	C	C	A	A	A	B		
6101 T6, T63	A	A	C	C	A	A	A	A	NA	High-strength bus conductors
T61, T64	A	A	B	D	A	A	A	A		
6151 T6, T652	...	...	...	...	...	...	...	...	B	Moderate-strength, intricate forgings for machine and auto parts
6201 T81	A	A	...	C	A	A	A	A	NA	High-strength electric conductor wire
6262 T6, T651, T6510, T6511	B	A	C	B	A	A	A	A	NA	Screw-machine products
T9	B	A	D	B	A	A	A	A		
6351 T5, T6	B	A	C	C	A	A	A	A	B	Heavy-duty structures requiring good corrosion resistance, truck and tractor extrusions
6463 T1	A	A	B	D	A	A	A	A		Extruded architectural and trim sections
T5	A	A	B	C	A	A	A	A	NA	
T6	A	A	C	C	A	A	A	A		
7005 T53, T63	B	B	C	A	B	B	B	B	B	Heavy-duty structures requiring good corrosion resistance, trucks, trailers, dump bodies
7049 T73, T7351, T7352	C	B	D	B	D	C	B	D	D	Aircraft and other structures
T76, T7651	C	B	D	B	D	C	B	D	D	
7050 T736, T73651, T73652	C	B	D	B	D	C	B	D	D	Aircraft and other structures
T76, T761	C	B	D	B	D	C	B	D	D	
7072	A	A	A	D	A	A	A	A	A	Fin stock, cladding alloy
7075 0	...	...	...	D	D	C	B	D	D	Aircraft and other structures
T6, T651, T652, T6510, T6511	C(c)	C	D	B	D	C	B	D	D	
T73, T7351	C	B	D	B	D	C	B	D	D	
7175 T736, T73652	C	B	D	B	D	C	B	D	D	Aircraft and other structures, forgings
7178 0	...	...	...	D	C	B	D	D		Aircraft and other structures
T6, T651, T6510, T6511	C(c)	C	D	B	D	C	B	D	D	
7475 T6, T651	C	C	D	B	D	C	B	D	D	Aircraft and other structures
T73, T7351, T7352	C	B	D	B	D	C	B	D	D	
T76, T7651	C	B	D	B	D	C	B	D	D	

(a) Ratings A through E are relative ratings in decreasing order of merit, based on exposures to sodium chloride solution by intermittent spraying or immersion. Alloys with A and B ratings can be used in industrial and seacoast atmospheres without protection. Alloys with C, D and E ratings generally should be protected at least on faying surfaces. (b) Stress-corrosion cracking ratings are based on service experience and on laboratory tests of specimens exposed to the 3.5% sodium chloride alternate immersion test. A = No known instance of failure in service or in laboratory tests. B = No known instance of failure in service; limited failures in laboratory tests of short transverse specimens. C = Service failures with sustained tension stress acting in short transverse direction relative to grain structure; limited failures in laboratory tests of long transverse specimens. D = Limited service failures with sustained longitudinal or long transverse stress. (c) In relatively thick sections the rating would be E. (d) This rating may be different for material held at elevated temperature for long periods. (e) Ratings A through D for workability (cold), and A through E for machinability, are relative ratings in decreasing order of merit. (f) Ratings A through D for weldability and brazeability are relative ratings defined as follows: A = Generally weldable by all commercial procedures and methods. B = Weldable with special techniques or for specific applications which justify preliminary trials or testing to develop welding procedure and weld performance. C = Limited weldability because of crack sensitivity or loss in resistance to corrosion and mechanical properties. D = No commonly used welding methods have been developed. (g) Ratings A through D and NA for solderability are relative ratings defined as follows: A = Excellent. B = Good. C = Fair. D = Poor. NA = Not applicable.

Fig. 1 Annual U.S. shipments of aluminum mill products and forgings

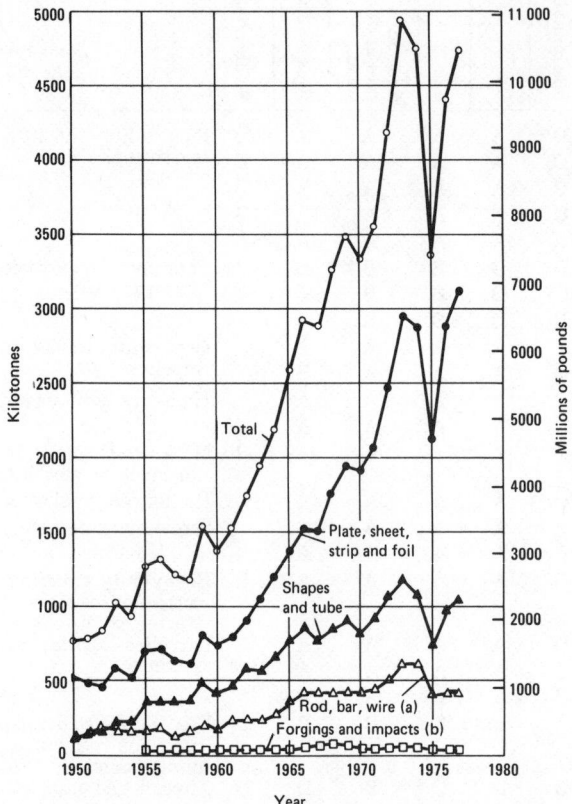

Compiled by the Aluminum Association, Inc. (a) A small amount of rolled structural shapes is included with rod and bar in 1946-1961, and with extruded shapes in subsequent years. (b) Forgings and impacts were not reported until 1955; instead, they were included in appropriate forging stock product categories, mainly rolled rod and bar.

electrical conductivities, low mechanical properties and excellent workability. Moderate increases in strength may be obtained by strain hardening. Iron and silicon are the major impurities. Typical uses include chemical equipment, reflectors, heat exchangers, electrical conductors and capacitors, packaging foil, architectural applications and decorative trim.

2xxx Series. Copper is the principal alloying element in 2xxx series alloys. These alloys require solution heat treatment to obtain optimum properties; in the solution heat treated condition, mechanical properties are similar to, and sometimes exceed, those of low-carbon steel. In some instances, precipitation heat treatment is employed to further increase mechanical properties. This treatment materially increases yield strength, with attendant loss in elongation; its effect on tensile strength is not as great.

The alloys in the 2xxx series do not have as good corrosion resistance as most other aluminum alloys, and under certain conditions they may be subject to intergranular corrosion. Therefore, these alloys in the form of sheet usually are clad with a high-purity aluminum or with a magnesium-silicon alloy of the 6xxx series, which provides galvanic protection of the core material and thus greatly increases resistance to corrosion.

Alloys in the 2xxx series are particularly well suited for parts and structures requiring high strength-to-weight ratios and are commonly used to make truck and aircraft wheels, truck suspension parts and aircraft fuselage and wing skins and structural parts and those parts requiring good strength at temperatures up to 150 °C (300 °F). These alloys have limited weldability, but some alloys in this series have superior machinability.

3xxx Series. Manganese is the major alloying element of 3xxx series alloys. These alloys generally are non-heat-treatable, but have about 20% more strength than 1xxx series alloys. Because only a limited percentage of manganese (up to about 1.5%) can be effectively added to aluminum, manganese is used as a major element in only a few alloys. However, three of them—3003, 3004 and 3105—are widely used as general-purpose alloys for moderate-strength applications requiring good workability. These applications include beverage cans, cooking utensils, heat exchangers, storage tanks, awnings, furniture, highway signs, roofing, siding and other architectural applications.

4xxx Series. The major alloying element in 4xxx series alloys is silicon, which can be added in sufficient quantities (up to 12%) to cause substantial lowering of the melting range without producing brittleness. For this reason, aluminum-silicon alloys are used in welding wire and as brazing alloys for joining aluminum, where a lower melting range than that of the base metal is required. Most alloys in this series are non-heat-treatable, but when used in welding heat treatable alloys, they will pick up some of the alloying constituents of the latter and so respond to heat treatment to a limited extent. The alloys containing appreciable amounts of silicon become dark gray to charcoal when anodic oxide finishes are applied, and hence are in demand for architectural applications. Alloy 4032 has a low coefficient of thermal expansion and high wear resistance, and thus is well suited to production of forged engine pistons.

5xxx Series. The major alloying element in 5xxx series alloys is magnesium, which is one of the most effective and widely used alloying elements for aluminum. When it is used as a major alloying element or with manganese, the result is a moderate-to-high-strength non-heat-treatable alloy. Magnesium is considerably more effective than manganese as a hardener, about 0.8% magnesium being equal to 1.25% manganese, and it can be added in considerably higher quantities. Alloys in this series possess good welding characteristics and good resistance to corrosion in marine atmospheres. However, certain limitations should be placed on the amount of cold work and the safe operating temperatures permissible for the higher-magnesium alloys (over about 3½% for operating temperatures above about 65 °C, or 150 °F) to avoid susceptibility to stress-corrosion cracking.

Uses include architectural, ornamental and decorative trim; cans and can ends; household appliances; streetlight standards; boats and ships, cryogenic tanks; and crane parts.

6xxx Series. Alloys in the 6xxx series contain silicon and magnesium approximately in the proportions required for formation of magnesium silicide (Mg_2Si), thus making them heat treatable. Although not as strong as most 2xxx and 7xxx alloys, 6xxx series alloys have good formability, weldability, machinability and corrosion resistance, with medium strength. Alloys in this heat treatable group may be formed in the T4 temper (solution heat treated but not precipitation heat treated) and strengthened after forming to full T6 properties by precipitation heat treatment. Uses include architectural applications, transportation equipment, bridge railings and welded structures.

7xxx Series. Zinc, in amounts of 1 to 8%, is the major alloying element in 7xxx series alloys, and when coupled with a smaller percentage of magnesium results in heat treatable alloys of moderate to very high strength. Usually other elements, such as copper and chromium, are also added in small quantities. 7xxx series alloys are used in airframe structures, mobile equipment and other highly stressed parts.

Types of Mill Products

Commercial wrought aluminum products are divided basically into five major categories based on production methods as well as geometric configurations. These are (a) flat rolled products (sheet, plate and foil); (b) rod, bar and wire; (c) tubular products; (d) shapes and (e) forgings. In the aluminum industry, rod, bar, wire, tubular products and shapes are termed "mill" products, as they are in the steel industry, even though they often are produced by extrusion rather than by rolling. Aluminum forgings usually are

not considered "mill" products, and are not discussed in this article. Production data for forgings, however, often are reported with data for the other wrought products (see Fig. 1).

In addition to production method and product configuration, wrought aluminum products also may be classified into heat treatable and non-heat-treatable alloys. Initial strength of non-heat-treatable (1xxx, 3xxx, 4043 and 5xxx) alloys depends on the hardening effects of elements such as manganese, silicon, iron and magnesium, singly or in various combinations. Because these alloys are work-hardenable, further strengthening is made possible by various degrees of cold working, denoted by the "H" series of tempers, as discussed earlier in this volume in the article on temper designations of aluminum and aluminum alloys. Alloys containing appreciable amounts of magnesium when supplied in strain-hardened tempers usually are given a final elevated-temperature treatment, called "stabilizing", to ensure stability of properties. Initial strength of heat treatable (2xxx, 4032, 6xxx and 7xxx) alloys is enhanced by addition of alloying elements such as copper, magnesium, zinc and silicon. Because these elements, singly or in various combinations, show increasing solid solubility in aluminum with increasing temperature, it is possible to subject them to thermal treatments that will impart pronounced strengthening. Heat treatment of these alloys is discussed in the preceding article.

Flat rolled products include sheet, plate and foil. They are manufactured by either hot or hot-and-cold rolling, are rectangular in cross section and form, and have uniform thickness. This category comprises about two-thirds of the total aluminum mill products purchased by U.S. fabricators.

"Plate" refers to a product whose thickness is greater than 6.3 mm (0.250 in. and greater). Plate up to 200 mm (8 in.) thick is available in some alloys. It usually has either sheared or sawed edges. Plate can be cut into circles, rectangles or odd-shape blanks. Plate of certain alloys—notably the high-strength 2xxx and 7xxx series alloys—also are available in alclad form, which comprises an aluminum alloy core having on one or both sides a metallurgically bonded aluminum or aluminum alloy coating that is anodic to the core, thus electrolytically protecting the core against corrosion. Most often, the coating consists of a high-purity aluminum,

a low magnesium-silicon alloy, or an alloy containing 1% zinc. Usually, coating thickness (one side) is from 2½ to 5% of the total thickness. The most commonly used plate alloys are 2024, 2124, 2219, 7075, 7475 and 7178 for aircraft structures; 5083, 5086 and 5456 for marine, cryogenics and pressure vessels; and 1100, 3003, 5052 and 6061 for general applications.

When the flat rolled product is over 0.15 through 0.63 mm (0.006 to 0.249 in.) in thickness, it is classified as "sheet". Sheet edges can be sheared, slit or sawed. Sheet is supplied in flat form, in coils or in pieces cut to length from coils. Current facilities permit production of a limited amount of extra-large sheet, for example, up to 5 m (200 in.) wide by 25 m (1000 in.) long. The term "strip", as applied to narrow sheet, is not used in the U.S. aluminum industry. Aluminum sheet usually is available in several surface finishes such as mill finish, one-side bright finish, or two-side bright finish. It may also be supplied embossed, perforated, corrugated, painted or otherwise surface treated; in some instances, it is edge-conditioned. As with aluminum plate, sheet made of the heat treatable alloys in which copper or zinc are the major alloying constituents, notably the high-strength 2xxx and 7xxx series alloys, also is available in alclad form for increased corrosion resistance. In addition, special composites may be obtained such as alclad non-heat-treatable alloys for extra corrosion protection, for brazing purposes, or for special surface finishes.

With a few exceptions, most alloys in the 1xxx, 2xxx, 3xxx, 5xxx and 7xxx series are available in sheet form. Along with alloy 6061, they cover a wide range of applications from builders' hardware to transportation equipment and from appliances to aircraft structures.

Foil is a product up through 0.15 mm (less than 0.006 in.) thick. Most foil is supplied in coils, although it is also available in rectangular form (sheets). One of the largest end uses of foil is household wrap. There is a wider variety of surface finishes for foil than for sheet. Foil often is treated chemically or mechanically to meet the needs of specific applications. Common foil alloys are limited to the higher-purity 1xxx series and 3003, 5052, 5056 and 8079 (Al-1.0Fe-0.15Si).

Bar, rod and wire are all solid products that are extremely long in

Table 3 Standard manufacturing limits (in inches) for aluminum extrusions

Diameter of circumscribing circle, in.	1060, 1100, 3003	6063	6061	2014, 5086, 5454	2024, 2219, 5083, 7001, 7075, 7079, 7178
		Minimum wall thickness, in.			
Solid and Semihollow Shapes, Rod, and Bar					
0.5 to 2 .... 0.040		0.040	0.040	0.040	0.040
2 to 3 .... 0.045		0.045	0.045	0.050	0.050
3 to 4 .... 0.050		0.050	0.050	0.050	0.062
4 to 5 .... 0.062		0.062	0.062	0.062	0.078
5 to 6 .... 0.062		0.062	0.062	0.078	0.094
6 to 7 .... 0.078		0.078	0.078	0.094	0.109
7 to 8 .... 0.094		0.094	0.094	0.109	0.125
8 to 10 .... 0.109		0.109	0.109	0.125	0.156
10 to 11 .... 0.125		0.125	0.125	0.125	0.156
11 to 12 .... 0.156		0.156	0.156	0.156	0.156
12 to 17 .... 0.188		0.188	0.188	0.188	0.188
17 to 20 .... 0.188		0.188	0.188	0.188	0.250
20 to 24 .... 0.188		0.188	0.188	0.250	0.500
Class 1 Hollow Shapes(a)					
1.25 to 3 .... 0.062		0.050	0.062	...	...
3 to 4 .... 0.094		0.050	0.062	...	...
4 to 5 .... 0.109		0.062	0.062	0.156	0.250
5 to 6 .... 0.125		0.062	0.078	0.188	0.281
6 to 7 .... 0.156		0.078	0.094	0.219	0.312
7 to 8 .... 0.188		0.094	0.125	0.250	0.375
8 to 9 .... 0.219		0.125	0.156	0.281	0.438
9 to 10 .... 0.250		0.156	0.188	0.312	0.500
10 to 12.75 ...... 0.312		0.188	0.219	0.375	0.500
12.75 to 14 .... 0.375		0.219	0.250	0.438	0.500
14 to 16 .... 0.438		0.250	0.375	0.438	0.500
16 to 20.25 ...... 0.500		0.375	0.438	0.500	0.625
Class 2 and 3 Hollow Shapes(b)					
0.5 to 1 .... 0.062		0.050	0.062	...	...
1 to 2 .... 0.062		0.055	0.062	...	...
2 to 3 .... 0.078		0.062	0.078	...	...
3 to 4 .... 0.094		0.078	0.094	...	...
4 to 5 .... 0.109		0.094	0.109	...	...
5 to 6 .... 0.125		0.109	0.125	...	...
6 to 7 .... 0.156		0.125	0.156	...	...
7 to 8 .... 0.188		0.156	0.188	...	...
8 to 10 .... 0.250		0.188	0.250	...	...

(a) Minimum inside diameter is one-half the circumscribing diameter, but never under 1 in. for alloys in first three columns or under 2 in. for alloys in last two columns. (b) Minimum hole size for all alloys is 0.110 sq in. in area or 0.375 in. in diam.

relation to their cross section. They differ from each other only in cross-sectional shape and in thickness or diameter. When the cross section is round or nearly round and over 10 mm (³⁄₈ in. or greater) in diameter, it is called "rod". It is called "bar" when the cross section is square, rectangular or in the shape of a regular polygon and when at least one perpendicular distance between parallel faces (thickness) is over 10 mm (³⁄₈ in. or greater). "Wire" refers to a product, regardless of its cross-sectional shape, whose diameter or greatest perpendicular distance

between parallel faces is 10 mm or less (less than ³⁄₈ in.).

Rod and bar can be produced by either hot rolling or hot extruding and brought to final dimensions with or without additional cold working. Wire usually is produced and sized by drawing through one or more dies, although roll flattening is also used. Alclad rod or wire for additional corrosion resistance is available only in certain alloys. Many aluminum alloys are available in bar, rod and wire; among these alloys 2011 and 6262 are specially designed for screw-machine products, 2117 and

6053 for rivets and fittings. Alloy 2024-T4 is a standard material for bolts and screws. Alloys 1350, 6101 and 6201 are extensively used as electrical conductors. Alloy 5056 is used for zippers and alclad 5056 for insect screen wire.

Tubular products include tube and pipe. They are hollow wrought products that are long in relation to their cross section and have uniform wall thickness except as affected by corner radii. Tube is round, elliptical, square, rectangular or regular polygonal in cross section. When round tubular products are in standardized combinations of outside diameter and wall thickness, commonly designated by "Nominal Pipe Sizes" and "ANSI Schedule Numbers", they are classified as "pipe".

Tube and pipe may be produced by using a hollow extrusion ingot, by piercing a solid extrusion ingot or by extruding through a porthole die or a bridge die. They also may be made by forming and welding sheet. Tube may be brought to final dimensions by drawing through dies. Tube (both extruded and drawn) for general applications is available in such alloys as 1100, 2014, 2024, 3003, 5050, 5086, 6061, 6063 and 7075. For heat-exchanger tube, alloys 1060, 3003, alclad 3003, 5052, 5454 and 6061 are most widely used. Clad tube is available only in certain alloys and is clad only on one side (either inside or outside). Pipe is available only in alloys 3003, 6061 and 6063.

Shapes. A "shape" is a product that is long in relation to its cross-sectional dimensions and has a cross-sectional shape other than that of sheet, plate, rod, bar, wire or tube. Most shapes are produced by extruding or by extruding plus cold finishing; shapes are now rarely produced by rolling because of economic disadvantages. Shapes may be solid, hollow (with one or more voids) or semihollow. The 6xxx series (Al-Mg-Si) alloys, because of their easy extrudability, are the most popular alloys for producing shapes. Some 2xxx and 7xxx series alloys are often used in applications requiring higher strength.

Standard structural shapes such as I beams, channels and angles produced in alloy 6061 are made in different and fewer configurations than similar shapes made of steel; the patterns especially designed for aluminum offer better section properties and greater structural stability than the steel design by using the metal more efficiently. The dimensions, weights and properties of the alloy 6061 standard structural shapes, along with other information needed by structural engineers and designers, are contained in the Aluminum Construction Manual, published by the Aluminum Association, Inc.

Most aluminum alloys can be obtained as precision extrusions with good as-extruded surfaces; major dimensions usually do not need to be machined because tolerances of the as-extruded product often permit manufacturers to complete the part with simple cutoff, drilling, or other minor operations.

In many instances, long aircraft structural elements involve large attachment fittings at one end. Such elements often are more economical to machine from stepped aluminum extrusions, with two or more cross sections in one piece, rather than from an extrusion having a uniform cross section large enough for the attachment fitting.

Design of Shapes

Aluminum shapes can be produced in a virtually unlimited variety of cross-sectional designs that place the metal where needed to meet functional and appearance requirements. Full utilization of this capability of the extrusion process depends principally on the ingenuity of designers in creating new and useful configurations. The cross-sectional design of an extruded shape, however, can have an important influence on its producibility, production rate, cost of tooling, surface finish and ultimate production cost. The optimum design of an extruded shape must take into account alloy thickness or thicknesses involved, and the size, type and complexity of the shape. Therefore, the extruder should be consulted during design to ensure adequate dimensional control, satisfactory finish and lowest cost while retaining the desired functional and appearance characteristics.

Classification of Shapes. The complexity of a shape producible as an extrusion is a function of metal-flow characteristics of the process and the means available to control flow. Con-

Table 4 Typical physical properties of wrought aluminum alloys

Alloy	Temper	Electrical conductivity(a) Volume	Weight	Electrical resistivity(b) nΩm	ohms(d)	Thermal conductivity(c) w/miK	Btu/ ft·h·°F
1050	O	61	190	28	17	231	133
1060	O	62	204	28	17	234	135
	H18	61	201	28	17	234	135
1100	O	59	194	29	18	222	128
	H18	57	187	30	18	218	126
1145	O	61	202	28	17	230	133
	H18	60	198	29	18	227	131
1199	O	65	215	27	16	243	140
1350	O	62	204	28	17	234	135
	H1x(e)	61	201	28	17	230	133
2011	T3, T4	39	123	44	27	152	88
	T8	45	142	38	23	173	100
2014	O	50	159	34	21	192	111
	T3, T4	34	108	51	31	134	77
	T6	40	127	43	26	155	89
2024	O	50	160	34	21	190	110
	T3, T4	30	96	57	35	120	69
	T6	37	119	46	28	145	84
	T8	39	125	44	27	152	88
2036	O	52	169	33	20	198	114
	T4	41	135	42	25	159	92
2048	T851	42	137	40	24	159	92
2124	O	50	161	35	21	191	110
	T851	39	126	44	27	152	88
2218	T61	38	121	45	27	148	86
	T72	40	128	43	26	155	90
2219	O	44	138	39	24	170	98
	T31, T37, T351	28	88	62	37	116	67
	T62, T81, T87, T851	30	95	57	35	130	75
2319	O	44	139	39	24	170	98
2618	T61	37	120	47	28	146	84
3003	O	47	154	37	22	180	104
	H12	42	138	41	25	162	94
	H14	41	134	42	25	159	92
	H18	40	130	43	26	155	90
3004	O (all)	42	137	41	25	162	94
3105	O (all)	45	148	38	23	173	100
4032	O	40	132	43	26	155	90
	T6	36	120	48	29	141	82
4043	O	42	140	41	25	163	94
5005	O, H38	52	172	33	20	205	118
5050	O, H38	50	165	34	21	191	110
5052	O, H38	35	116	49	30	137	79
5056	O	29	98	59	36	120	69
	H38	27	91	64	38	112	65
5083	All	29	98	60	36	120	69
5086	All	31	104	56	33	127	73
5154	All	32	108	54	32	127	73
5182	All	31	105	56	33	123	71
5252	All	35	117	49	30	138	80
5254	All	32	107	54	32	127	73
5356	O	29	98	59	36	116	67
5454	All	34	113	51	31	134	77
5456	All	29	98	60	36	116	67
5457	All	46	153	38	23	177	102
5652	All	35	116	49	30	137	79
5657	All	54	179	32	19	...	...

(continued)

Table 4 (continued)

Alloy	Temper	Electrical conductivity(a) Volume	Weight	Electrical resistivity(b) nΩm	ohms(d)	Thermal conductivity(c) w/miK	Btu/ ft·h·°F
6005	T5	49	162	35	21	167	97
6009	O	54	184	32	19	205	118
	T4	44	150	39	24	172	99
	T6	47	160	37	22	180	104
6010	O	53	175	33	20	202	117
	T4	39	129	44	27	151	87
	T6	44	146	39	24	180	104
6061	O	47	155	37	22	180	104
	T4	40	132	43	26	154	89
	T6	43	142	40	24	167	97
6063	O	58	191	30	18	218	126
	T1	50	165	35	21	193	112
	T5	55	181	32	19	209	121
	T6	53	175	33	20	201	116
6066	O	40	132	43	26	147	85
	T6	37	122	47	28	147	85
6070	T6	44	145	39	24	172	99
6101	T6	57	188	30	18	218	138
	T8	54	178	32	19	218	138
6151	O	54	178	32	19	205	118
	T4	42	138	41	25	163	94
	T6	45	148	38	23	175	101
6201	T81	54	179	32	19	205	118
6205	T1	45	149	37	22	172	99
	T5	49	162	35	21	188	109
6262	T9	44	145	39	24	172	99
6351	T6	46	152	38	23	176	102
6463	T1	50	165	34	21	192	111
	T5	55	181	31	19	209	121
	T6	53	175	33	20	201	116
7005	O	43	138	40	24	166	96
	T53	38	122	45	27	148	86
	T6	35	113	49	30	137	79
	T63	38	122	45	27	148	86
7049	T73	38	120	43	27	154	89
7050	O	47	148	37	22	180	104
	T73	40	127	43	26	157	91
	T76	40	125	44	26	154	89
7072	O	60	197	29	17	227	131
7075	T6	33	105	52	31	130	75
	T73	40	128	43	26	155	90
	T76	38	123	45	27	150	87
7175	O	46	147	38	23	177	102
	T66	36	115	48	29	142	82
	T73	40	128	43	26	155	90
7475	O	46	147	38	23	177	102
	T6	36	115	48	29	142	82
	T7351	40	128	43	26	155	90
	T76	42	134	41	25	163	94

(a) % IACS at 20 °C (68 °F). (b) At 20 °C (68 °F). (c) At 25 °C (77 °F). (d) Per circular mil/ft. (e) All H1x-type tempers.

trol of metal flow places a few limitations on the design features of the cross section of an extruded shape that affect production rate, dimensional and surface quality, and costs. Extrusions are classified by shape complexity from an extrusion-production viewpoint into solid, hollow and semihollow shapes. Each hollow shape—a shape with any part of its cross section completely enclosing a void—is further classified by increasing complexity as follows:

- Class 1—A hollow shape with a round void 25 mm (1 in.) or more in diameter and with its weight equally distributed on opposite sides of two or more equally spaced axes
- Class 2—Any hollow shape other than Class 1, not exceeding a 125-mm-diam (5-in.-diam) circle and having a single void of not less than 9.5 mm (0.375 in.) diam or 70 mm^2 (0.110 in.^{2}) area
- Class 3—Any hollow shape other than Class 1 or 2

A semihollow shape is a shape with any part of its cross section partly enclosing a void having the following ratios for the area of the void to the square of the width of the gap leading to the void:

Gap width		
mm	in.	Ratio
0.9 to 1.5	0.035 to 0.061	Over 2
1.6 to 3.1	0.062 to 0.124	Over 3
3.2 to 6.3	0.125 to 0.249	Over 4
6.4 to 12.6	0.250 to 0.499	Over 5
12.7 and greater	0.500 and greater	Over 6

Alloy Extrudability. Aluminum alloys differ in inherent extrudability. Alloy selection is important, because it establishes the minimum thickness for a shape and has a basic effect on extrusion cost. In general, the higher the alloy content and the strength of an alloy, the more difficult it is to extrude and the lower its extrusion rate.

The relative extrudabilities, as measured by extrusion rate, for several of the more important commercial extrusion alloys are given below.

Alloy	Extrudability, % of rate for 6063
1350	160
1060	135
1100	135
3003	120
6063	100
6061	60
2011	35
5086	25
2014	20
5083	20
2024	15
7075	9
7178	8

Actual extrusion rate depends on pressure, temperature and other require-

Fig. 2 Four examples of interconnecting extrusions that fit together or fit other products, and four examples of joining methods

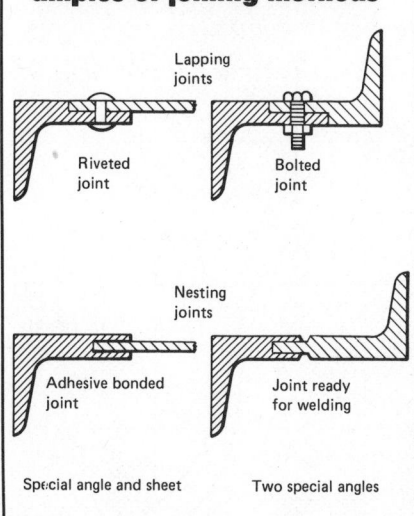

Lapping joints
Riveted joint
Bolted joint
Nesting joints
Adhesive bonded joint
Joint ready for welding
Special angle and sheet
Two special angles

Fig. 3 Two examples of extrusions with nonpermanent interconnections

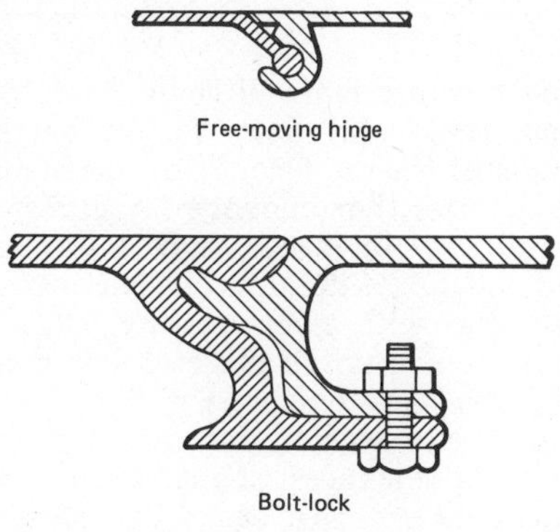

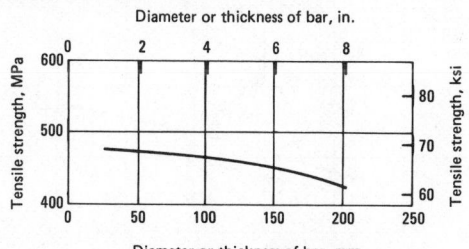

Free-moving hinge
Bolt-lock

Fig. 4 Six examples of interconnecting extrusions that lock together or lock to other products

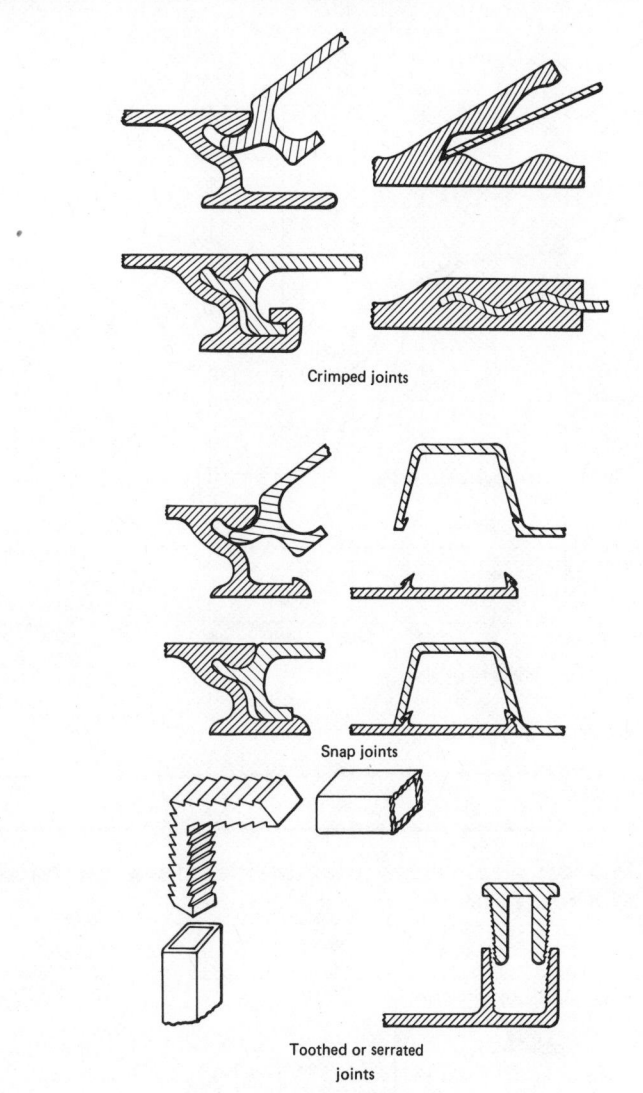

Crimped joints
Snap joints
Toothed or serrated joints

ments for the particular shape, as well as ingot quality.

Shape and Size Factors. The important shape factor of an extrusion is the ratio of its perimeter to its weight per unit length. For a single classification, increasing shape factor is a measure of increasing complexity. Designing for minimum shape factor promotes ease of extrusion.

The size of an extruded shape affects ease of extrusion and dimensional tolerances. As the circumscribing circle size (smallest diameter that completely encloses the shape) increases, extru-

Fig. 5 Variation of the tensile strength of extruded alloy 2014-T6 bars with bar size

Diameter or thickness of bar, in.
0
2
4
6
8
Tensile strength, MPa
600
500
400
0
50
100
150
200
250
Diameter or thickness of bar, mm
Tensile strength, ksi
80
70
60

Specimens were cut from the centers of bars of different thicknesses or diameters, after quenching in cold water.

Fig. 6 Effect of time and temperature on the tensile strength of two heat treated wrought aluminum alloys

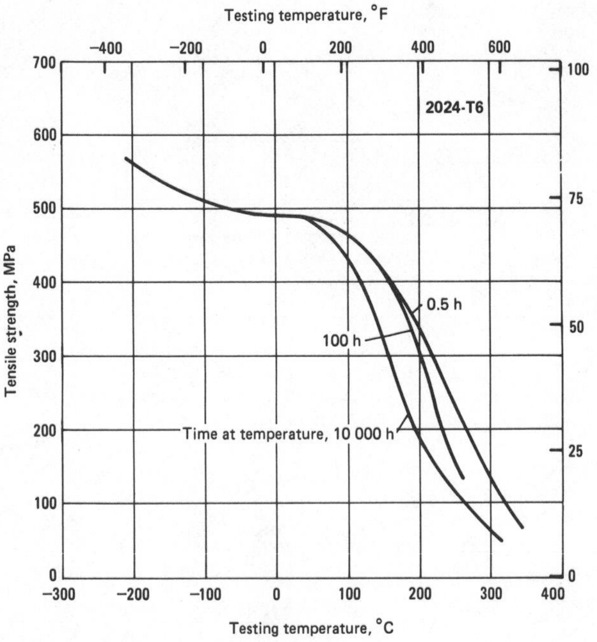

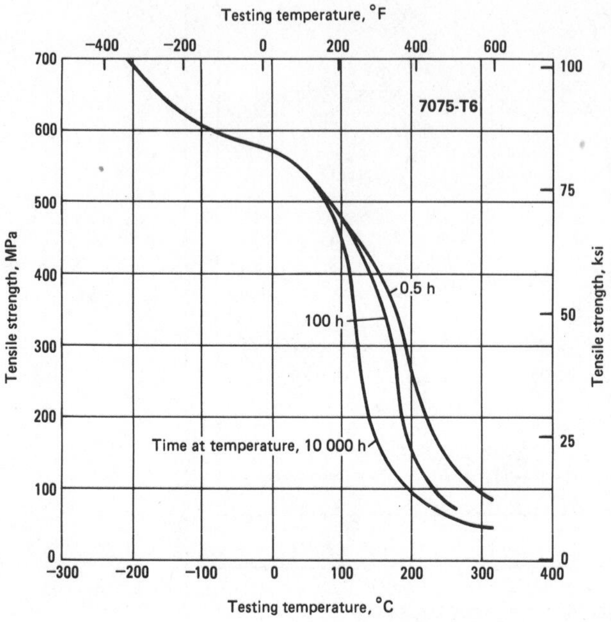

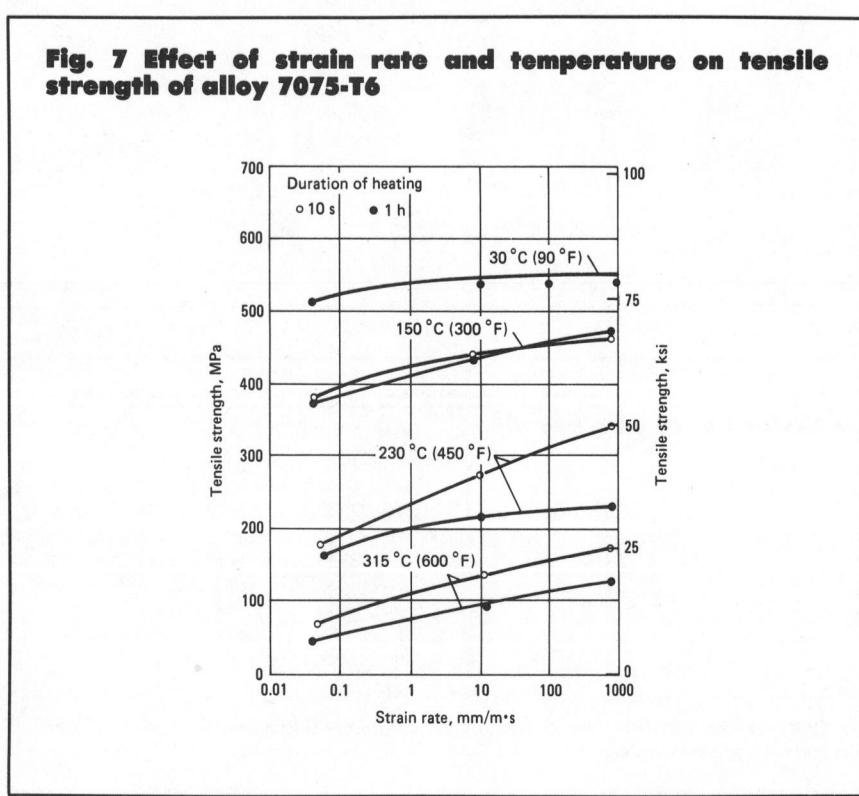

Strain rate held constant at 10 mm/m·s.

Fig. 7 Effect of strain rate and temperature on tensile strength of alloy 7075-T6

sion becomes more difficult. In extrusion, the metal flows fastest at the center of the die face. With increasing circle size, the tendency for different metal flow increases, and it is more difficult to design and construct extrusion dies with compensating features that provide uniform metal-flow rates to all parts of the shape.

Ease of extrusion improves with increasing thickness; shapes of uniform thickness are most easily extruded. A shape whose cross section has elements of widely differing thicknesses increases the difficulty of extrusion. The thinner a flange on a shape, the less the length of flange that can be satisfactorily extruded. Thinner elements at the ends of long flanges are difficult to fill properly and make it hard to obtain desired dimensional control and finish. Although it is desirable to produce the thinnest shape feasible for an application, reducing thickness can cause an increase in cost of extrusion that more than offsets the savings in metal cost. Extruded shapes 1 mm (0.040 in.) thick and even less can be produced, depending on alloy, shape, size and design. Manufacturing limits on mini-

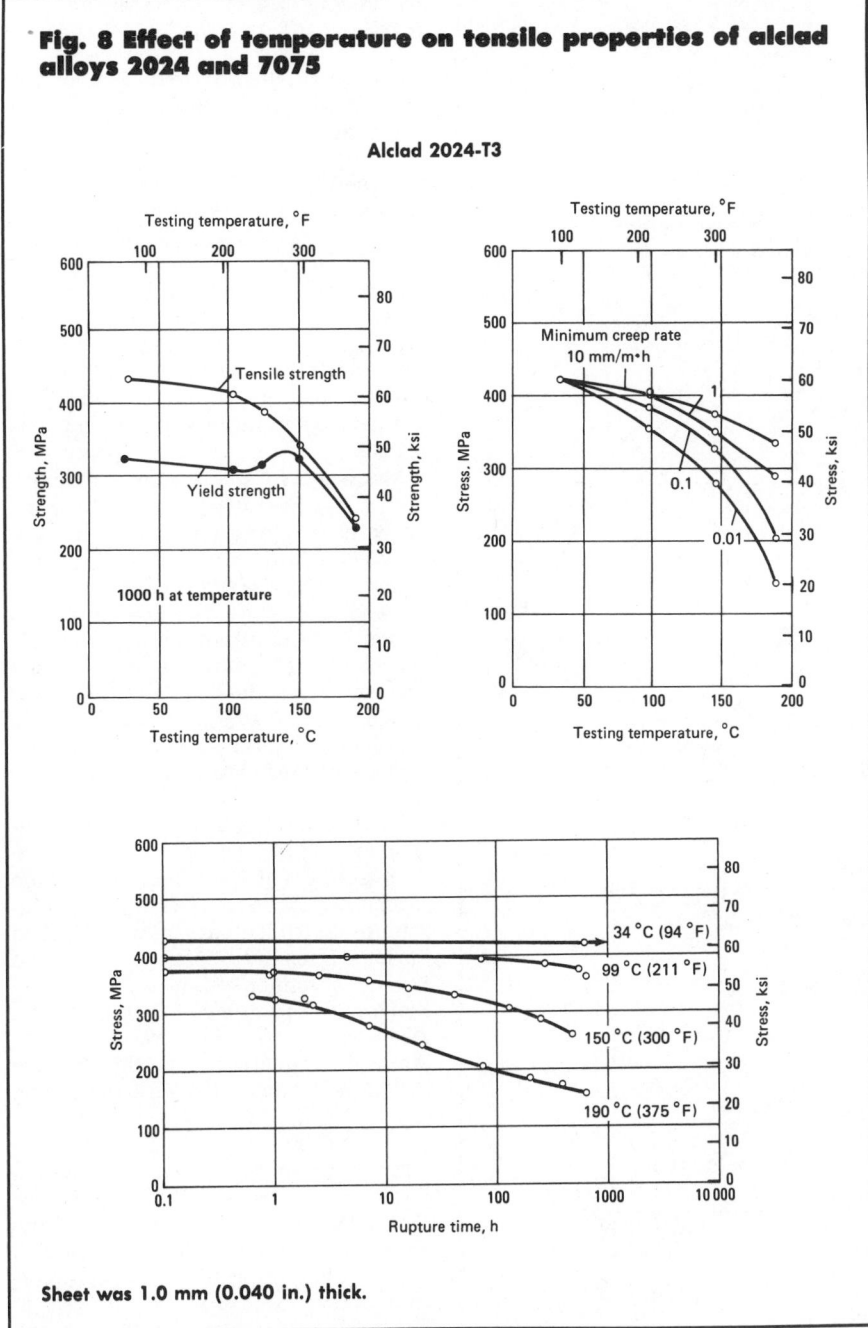

Fig. 8 Effect of temperature on tensile properties of alclad alloys 2024 and 7075

Alclad 2024-T3

1000 h at temperature

Sheet was 1.0 mm (0.040 in.) thick.

mum practical thickness of extruded shapes are given in Table 3.

Size and thickness relationships among the various elements of a shape can add to its complexity. Rod, bar and regular shapes of uniform thickness are easily produced. For example, a bar 3.2 mm (0.125 in.) thick, a rod 25 mm (1 in.) in diameter, and an angle 19 by 25 mm (0.75 by 1 in.) in cross section and 1.6 mm (0.0625 in.) thick are readily extruded, whereas extrusion of a 76-

mm (3-in.) bar-type shape with a 3.2-mm (0.125-in.) flange is more difficult.

Semihollow and channel shapes require a tongue in the extrusion die, which must have adequate strength to resist the extrusion force. Channel shapes become increasingly difficult to produce as the depth-to-width ratio increases. Wide, thin shapes are difficult to produce and make it hard to control dimension. Channel-type shapes and

wide, thin shapes may be fabricated if they are not excessively thin. Thin flanges or projections from a thicker element of the shape add to the complexity of an extruded design. On thinner elements at the extremities of high flanges, it is difficult to get adequate fill to obtain desired dimensions. The greater the difference in thickness of individual elements comprising a shape, the more difficult the shape is to produce. The effect of such thickness differences can be greatly diminished by blending one thickness into the other by tapered or radiused transitions. Sharp corners should be avoided wherever possible, because they reduce maximum extrusion speed and are locations of stress concentrations in the die opening that can cause premature die failure. Fillet radii of at least 0.8 mm (0.031 in.) are desirable, but corners with radii of only 0.4 mm (0.015 in.) are feasible.

In general, the more unbalanced and unsymmetrical an extruded-shape cross section, the more difficult that shape is to produce. Despite this, production of grossly unbalanced and unsymmetrical shapes is the basis of the great growth that has occurred in the use of aluminum extrusions, and such designs account for the bulk of extruded shapes produced today.

Interconnecting Shapes. It is becoming increasingly common to include an interconnecting feature in the design of an extruded shape to facilitate its assembly to a similar shape or to another product. This feature can be a simple step to provide a smooth lapping joint, or a tongue and groove for a nesting joint (see Fig. 2). Such connections can be secured by any of the common joining methods. Of special interest when the joint is to be arc welded is the fact that lapping and nesting types of interconnections can be designed to provide edge preparation and/or integral backing for the weld (see the sketch at bottom left in Fig. 2).

Interlocking joints can be designed to incorporate a free-moving hinge (see top sketch in Fig. 3) when one part is slid lengthwise into the mating portion of the next extrusion. Panel-type extrusions with hinge joints have found application in conveyor belts and roll-up doors.

A more common type of interlocking feature used in interconnecting extrusions is the nesting type that requires rotation of one part relative to the mating part for assembly (see bottom

Fig. 8 (continued)

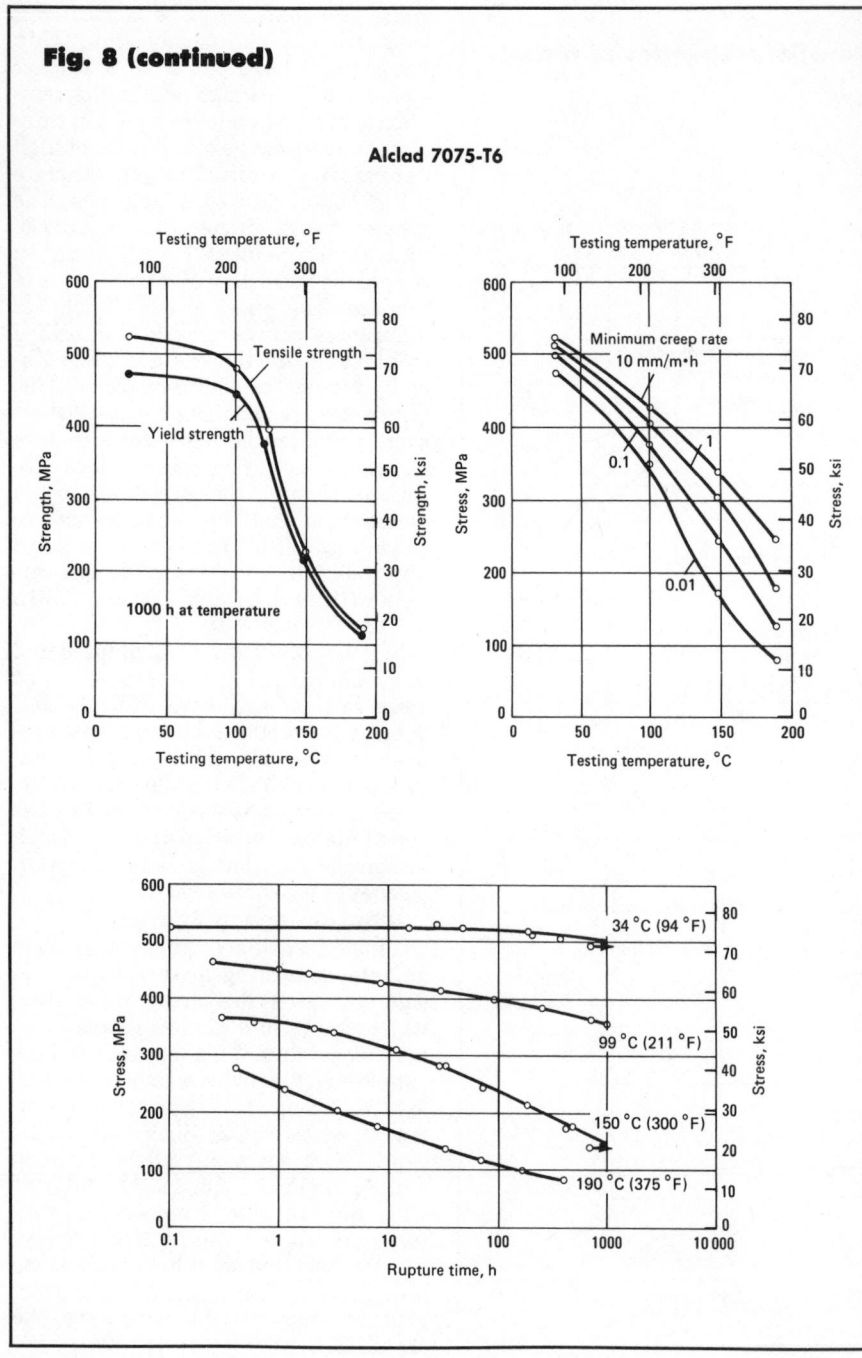

Alclad 7075-T6

Properties of Wrought Alloys

Physical Properties. Two of the most important physical properties in many applications are electrical conductivity and thermal conductivity, and the values of these properties can vary significantly from one temper to another. To aid in selection of alloy and temper for such applications, typical values of electrical resistivity are presented for most of the common wrought aluminum alloys in Table 4. These data, and the values of many other physical properties, also are listed in the data compilations that follow this article.

Mechanical properties of aluminum sheet are substantially the same in both longitudinal and transverse directions. Minimum mechanical properties usually are specified in the transverse direction for the heat treatable alloys and longitudinal direction for the non-heat-treatable alloys. Tensile properties specified for aluminum extrusions are usually parallel to the extrusion direction and represent the best properties obtainable. Transverse tests will show yield and ultimate strengths and markedly lower elongation.

In a study of the effect of mass on mechanical properties of wrought aluminum, extruded bars of alloy 2014 in various sizes were heat treated, and tensile specimens were subsequently machined from the center of each bar. As shown in Fig. 5, there was a substantial reduction in the tensile strength of alloy 2014 with increasing size.

Typical mechanical properties of wrought aluminum alloys at room temperature, which usually are reported for longitudinal specimens cut from shapes of common sizes, are presented in the data compilations that follow this article and are summarized in Table 5.

Properties at Elevated Temperature. Aluminum alloys are most widely used for structural parts within an operating temperature range of −240 to +200 °C (−400 to +400 °F). Certain alloys may be used at temperatures up to about 315 °C (600 °F) in special applications, whereas powder metallurgy alloys have been employed for short intervals up to about 480 °C (900 °F). The properties at elevated temperature, as tabulated in the data compilations in this handbook, represent typical values

sketch in Fig. 3). Such joints can be held together by gravity or by mechanical devices. If a nonpermanent joint is desired, a bolt or other fastener can be used, as illustrated in the bottom sketch in Fig. 3.

When a permanent joint is desired, a snapping or crimping feature can be added to interlocking extrusions (see Fig. 4). Crimping also can be used to make a permanent joint between an interlocking extrusion and sheet (Fig.

4). Extrusions also can be provided with longitudinal teeth or serrations, which will permanently grip smooth surfaces as well as surfaces provided with mating teeth or serrations; this is illustrated in the sketch at the bottom of Fig. 4.

Applications for interconnecting extrusions include doors, wall, ceiling and floor panels, pallets, aircraft landing mats, highway signs, window frames and large cylinders.

Table 5 Typical mechanical properties of wrought aluminum alloys

Alloy	Temper	Tensile strength MPa	ksi	Yield strength MPa	ksi	Elongation(a), % (b)	(c)	Hardness(d)	Shear strength MPa	ksi	Fatigue strength(e) MPa	ksi
1050	O	76	11	28	4	...	...	...	62	9	...	...
	H14	110	16	105	15	...	...	...	69	10	...	...
	H16	130	19	125	18	...	...	...	76	11	...	...
	H18	160	23	145	21	...	...	...	83	12	...	...
1060	O	69	10	28	4	43	...	19	48	7	21	3
	H12	83	12	76	11	16	...	23	55	8	28	4
	H14	97	14	90	13	12	...	26	62	9	34	5
	H16	110	16	105	15	8	...	30	69	10	45	6.5
	H18	130	19	125	18	6	...	35	76	11	45	6.5
1100	O	90	13	34	5	35	45	23	62	9	34	5
	H12	110	16	105	15	12	25	28	69	10	41	6
	H14	125	18	115	17	9	20	32	76	11	48	7
	H16	145	21	140	20	6	17	38	83	12	62	9
	H18	165	24	150	22	5	15	44	90	13	62	9
1350	O	83	12	28	4	23(f)	...	...	55	8	...	...
	H12	97	14	83	12	...	...	...	62	9	...	...
	H14	110	16	97	14	...	...	...	69	10	...	...
	H16	125	18	110	16	...	...	...	76	11	...	...
	H19	185	27	165	24	1.5(f)	...	...	105	15	48	7
2011	T3	380	55	295	43	...	15	95	220	32	125	18
	T8	405	59	310	45	...	15	100	240	35	125	18
2014	O	185	27	97	14	...	18	45	125	18	90	13
	T4	425	62	290(h)	42(h)	...	20	105	260	38	140	20
	T6(g)	485	70	415	60	...	13	135	290	42	125	18
Alclad 2014	O	170	25	69	10	21	...	...	125	18	...	...
	T3	435	63	275	40	20	...	...	255	37	...	...
	T4	420	61	255	37	22	...	...	255	37	...	...
	T6	470	68	415	60	10	...	...	285	41	...	...
2024	O	185	27	76	11	20	22	47	125	18	90	13
	T3	485	70	345	50	18	...	120	285	11	140	20
	T4, T351	470	68	325	47	20	19	120	285	41	140	20
	T361	495	72	395	57	13	...	130	290	42	125	18
Alclad 2024	O	180	26	76	11	20	...	...	125	18	...	...
	T	450	65	310	45	18	...	...	275	40	...	...
	T4, T351	440	64	290	42	19	...	...	275	40	...	...
	T361	460	67	365	53	11	...	...	285	41	...	...
	T81, T851	450	65	415	60	6	...	...	275	40	...	...
	T861	485	70	455	66	6	...	...	290	42	...	...
2036	T4	340	49	195	28	24	...	...	...	...	125(j)	18(j)
2048	...	455	66	415	60	8.3	...	...	...	...	220(k)	32(k)
2124	T851	490	71	440	64	9.4	...	...	...	...	...	...
2218	T61	405	59	305	44	...	13	115	...	...	...	...
	T71	345	50	275	40	...	11	105	...	...	...	...
	T72	330	48	255	37	...	11	95	205	30	...	...
2219	O	170	25	76	11	18	...	...	...	...	...	...
	T42	360	52	185	27	20	...	...	...	...	...	...
	T31, T351	360	52	250	36	17	...	...	...	...	...	...
	T37	395	57	315	46	11	...	...	...	...	...	...
	T62	415	60	290	42	10	...	...	...	...	105	15
	T81, T851	455	66	350	51	10	...	...	...	...	105	15
	T87	475	69	395	57	10	...	...	...	...	105	15
2618	All	440	64	370	54	10(c)	...	...	260	38	125	18
3003 and Alclad 3003(m)	O	110	16	42	6	30	40	28	76	11	48	7
	H12	130	19	125	18	10	20	35	83	12	55	8
	H14	150	22	145	21	8	16	40	97	14	62	9
	H16	180	26	170	25	5	14	47	105	15	69	10
	H18	200	29	185	27	4	10	55	110	16	69	10

(continued)

Table 5 (continued)

Alloy	Temper	Tensile strength MPa	ksi	Yield strength MPa	ksi	Elongation(a), % (b)	(c)	Hardness(d)	Shear strength MPa	ksi	Fatigue strength(e) MPa	ksi
3004 and Alclad 3004(m)	O..............	180	26	69	10	20	25	45	110	16	97	14
	H32.............	215	31	170	25	10	17	52	115	17	105	15
	H34.............	240	35	200	29	9	12	63	125	18	105	15
	H36.............	260	38	230	33	5	9	70	140	20	110	16
	H38.............	285	41	250	36	5	6	77	145	21	110	16
3105	O..............	115	17	55	8	24	...	...	83	12	...	...
	H12.............	150	22	130	19	7	...	...	97	14	...	...
	H14.............	170	25	150	22	5	...	...	105	15	...	...
	H16.............	195	28	170	25	4	...	...	110	16	...	...
	H18.............	215	31	195	28	3	...	...	115	17	...	...
	H25.............	180	26	160	23	8	...	...	105	15	...	...
4032	T6.............	380	55	315	46	...	9	120	260	38	110	16
4043	O..............	145	21	69	10	22	...	...	...	...	...	...
	H18.............	285	41	270	39	0.5	...	...	...	...	...	...
5005	O..............	125	18	41	6	25	...	28	76	11	...	...
	H12.............	140	20	130	19	10	...	...	97	14	...	...
	H14.............	160	23	150	22	6	...	...	97	14	...	...
	H16.............	180	26	170	25	5	...	...	105	15	...	...
	H18.............	200	29	195	28	4	...	...	110	16	...	...
	H32.............	140	20	115	17	11	...	36	97	14	...	...
	H34.............	160	23	140	20	8	...	41	97	14	...	...
	H36.............	180	26	165	24	6	...	46	105	15	...	...
	H38.............	200	29	185	27	5	...	51	110	16	...	...
5050	O..............	145	21	55	8	24	...	36	105	15	83	12
	H32.............	170	25	145	21	9	...	46	115	17	90	13
	H34.............	195	28	165	24	8	...	53	125	18	90	13
	H36.............	205	30	180	26	7	...	58	130	19	97	14
	H38.............	220	32	200	29	6	...	63	140	20	97	14
5052	O..............	195	28	90	13	25	27	47	125	18	110	16
	H32.............	230	33	195	28	12	16	60	140	20	115	17
	H34.............	260	38	215	31	10	12	68	145	21	125	18
	H36.............	275	40	240	35	8	9	73	160	23	130	19
	H38.............	290	42	255	37	7	7	77	165	24	140	20
5056	O..............	290	42	150	22	...	35	65	180	26	140	20
	H18.............	435	63	405	59	...	10	105	235	34	150	22
	H38.............	415	60	345	50	...	15	100	220	32	150	22
5083	O..............	290	42	145	21	...	22	...	170	25	...	...
	H112.............	305	44	195	28	...	16	...	...	...	...	...
	H113.............	315	46	230	33	...	16	...	...	...	...	...
	H321.............	315	46	230	33	...	16	...	...	...	160	23
	H323, H32........	325	47	250	36	...	10	...	...	...	...	...
	H343, H34........	345	50	285	41	...	9	...	...	...	...	...
5086	O..............	260	38	115	17	22	...	...	160	23	...	...
	H32, H116, H117..	290	42	205	30	12	...	...	...	...	...	...
	H34.............	325	47	255	37	10	...	...	185	27	...	...
	H112.............	270	39	130	19	14	...	...	...	...	...	...
5154	O..............	240	35	115	17	27	...	58	150	22	115	17
	H32.............	270	39	205	30	15	...	67	150	22	125	18
	H34.............	290	42	230	33	13	...	73	165	24	130	19
	H36.............	310	45	250	36	12	...	78	180	26	140	20
	H38.............	330	48	270	39	10	...	80	195	28	145	21
	H112.............	240	35	115	17	25	...	63	...	...	115	17
5182	O..............	275	40	140	19	25	...	58	150	22	140	20
	H32.............	315	46	235	34	12	...	...	...	...	...	...
	H34.............	340	49	285	41	10	...	...	...	...	...	...
	H19(n)	420	61	395	57	4	...	...	...	...	...	...
5252	H25.............	235	34	170	25	11	...	68	145	21	...	...
	H28, H38........	285	41	240	35	5	...	75	160	23	...	...
5254	O..............	240	35	115	17	27	...	58	150	22	115	17

(continued)

Table 5 (continued)

Alloy	Temper	Tensile strength MPa	ksi	Yield strength MPa	ksi	Elongation(a), % (b)	(c)	Hardness(d)	Shear strength MPa	ksi	Fatigue strength(e) MPa	ksi
5254	H32	270	39	205	30	15	...	67	150	22	125	18
	H34	290	42	230	33	13	...	73	165	24	130	19
	H36	310	45	250	36	12	...	78	180	26	140	20
	H38	330	48	270	39	10	...	80	195	28	145	21
	H112	240	35	115	17	25	...	63	...	...	115	17
5454	O	250	36	115	17	22	...	62	160	23	...	...
	H32	275	40	205	30	10	...	73	165	24	...	...
	H34	305	44	240	35	10	...	81	180	26	...	...
	H36	340	49	275	40	8	...	...	...	...	...	...
	H38	370	54	310	45	8	...	...	...	...	...	...
	H111	260	38	180	26	14	...	70	160	23	...	...
	H112	250	36	125	18	18	...	62	160	23	...	...
	H311	260	38	180	26	18	...	70	160	23	...	...
5456	O	310	45	160	23	...	24	...	...	...	...	...
	H111	325	47	230	33	...	18	...	...	...	...	...
	H112	310	45	165	24	...	22	...	...	...	...	...
	H321, H116	350	51	255	27	...	16	90	205	30	...	...
5457	O	130	19	48	7	22	...	32	83	12	...	...
	H25	180	26	160	23	12	...	48	110	16	...	...
	H28, H38	205	30	185	27	6	...	55	125	18	...	...
5652	O	195	28	90	13	25	30	47	125	18	110	16
	H32	230	33	195	28	12	18	60	140	20	115	17
	H34	260	38	215	31	10	14	68	145	21	125	18
	H36	275	40	240	35	8	10	73	160	23	130	19
	H38	290	42	255	34	7	8	77	165	24	140	20
5657	H25	160	23	140	20	12	...	40	97	14	...	...
	H28, H38	195	28	165	24	7	...	50	105	15	...	...
6005	T1	170	25	105	15	16	...	...	...	...	97	14
	T5	260	38	240	35	8	10	95	205	30	97	14
6009	T4	235	34	130	19	24	...	70(p)	150	22	115	17
	T6	345	50	325	47	12	...	...	...	...	...	...
6010	T4	255	37	170	25	24	...	76(p)	...	...	115	17
6061	O	125	18	55	8	25	30	30	83	12	62	9
	T4, T451	240	35	145	21	22	25	65	165	24	97	14
	T6, T651	310	45	275	40	12	17	95	205	30	97	14
Alclad 6061	O	115	17	48	7	25	...	...	76	11	...	...
	T4, T451	230	33	130	19	22	...	...	150	22	...	...
	T6, T651	290	42	255	37	12	...	...	185	27	...	...
6063	O	90	13	48	7	...	...	25	69	10	55	8
	T1	150	22	90	13	20	...	42	97	14	62	9
	T4	170	25	90	13	22	...	...	...	...	...	...
	T5	185	27	145	21	12	...	60	115	17	69	10
	T6	240	35	215	31	12	...	73	150	22	69	10
	T83	255	37	240	35	9	...	82	150	22	...	...
	T831	205	30	185	27	10	...	70	125	18	...	...
	T832	290	42	270	39	12	...	95	185	27	...	...
6066	O	150	22	83	12	...	18	43	97	14	...	...
	T4, T451	360	52	205	30	...	18	90	200	29	...	...
	T6, T651	395	57	360	52	...	12	120	235	34	110	16
6070	O	145	21	69	10	20	...	35	97	14	62	9
	T4	315	46	170	25	20	...	90	205	30	90	13
	T6	380	55	350	51	10	...	120	235	34	97	14
6101	H111	97	14	76	11	...	...	...	...	...	...	...
6151	T6	220	32	195	28	15(q)	...	71	140	20	...	...
6201	T6	330	48	300	43	17	...	90	...	...	...	...
	T81	330	48	310	45	6(f)	...	...	...	...	...	...
6205	T1	260	38	140	20	19	...	65	...	...	...	...
	T5	310	45	290	42	11	...	95	205	30	105	15
6262	T9	400	58	380	55	...	10	120	240	35	90	13

(continued)

Table 5 (continued)

Alloy	Temper	Tensile strength MPa	ksi	Yield strength MPa	ksi	Elongation(a), % (b)	(c)	Hardness(d)	Shear strength MPa	ksi	Fatigue strength(e) MPa	ksi
6351	T4	250	36	150	22	20	...	...	...	...	...	...
	T6	310	45	285	41	14	...	95	200	29	90	13
6463	T1	150	22	90	13	20	...	42	97	14	69	10
	T5	185	27	145	21	12	...	60	115	17	69	10
	T6	240	35	215	31	12	...	74	150	22	69	10
7005	O	193	28	83	12	20	...	...	117	17	...	...
	T53	393	57	345	50	15	...	...	221	32	140	20
	T6, T63, T6351	372	54	315	46	12	...	...	214	31	125	18
7049	T73	...	...	...	...	...	...	135	...	...	295(k)	43(k)
7050	T736	515	75	455	66	11	15	...	...	...	240(k)	35(k)
7072	O	...	...	...	...	...	...	20	55	8	...	...
	H12	...	...	...	...	...	...	28	62	9	...	...
	H14	...	...	...	...	...	...	32	69	10	...	...
7075	O	230	38	105	15	17	16	60	150	22	...	...
	T6, T651	570	83	505	73	11	11	150	330	48	160	23
	T73	505	73	435	63	13	...	...	...	...	...	...
Alclad 7075	O	220	32	95	14	17	...	...	150	22	...	...
	T6, T651	525	76	460	67	11	...	...	315	46	...	...
7175	T66	595	86	525	76	11	...	150	325	47	160	23
	T736	525	76	455	66	14	...	145	290	42	160	23
7475	T61	525	76	460	67	12	...	...	...	...	...	...
	T651	...	...	...	...	...	...	...	295	43	...	...
	T7351	...	...	...	...	...	...	...	270	9	220	32
	T7651	...	...	...	...	...	...	...	270	39	...	...

(a) In 50 mm or 2 in. (b) Specimen 1.6 mm (1/16 in.) thick. (c) Specimen 12.5 mm (0.50 in.) in diameter. (d) 500-kg load; 10-mm ball. (e) In 5×10^8 cycles; R. R. Moore — type test. (f) In 254 mm or 10 in. (g) Extruded products more than 19 mm (¾ in.) thick are 15 to 20% higher in strength. (h) Die forgings are about 20% lower in yield strength. (j) In 10^7 cycles using flexural-type testing of sheet specimens. (k) In 10^7 cycles; axially loaded specimens tested at $R = 0.1$. (m) No shear-strength or fatigue-strength values for Alclad. (n) Properties for this temper are those of container end stock 0.25 to 0.38 mm (0.010 to 0.015 in.) thick. (p) HR15T. (q) Specimen 6.35 mm (0.25 in.) thick.

for 10 000 h at test temperature.

The results of tension tests at elevated temperature for 2024-T6 and 7075-T6 are given in Fig. 6 and 7; effect of temperature on alclad alloys is shown in Fig. 8. Time at temperature varied from 10 s to 10 000 h. For heat treatable alloys, time at temperature has a marked effect. Such variations are less pronounced for non-heat-treatable alloys where no precipitation or overaging effects occur.

At high testing rates up to 690 MPa/s (100 ksi/s), marked increases in properties are observed for some alloys. The increases in yield strength at 260 °C (500 °F) when 2024-T4 and 7075-T6 are tested at 6.9 MPa/s (1 ksi/s) and 690 MPa/s (100 ksi/s) are 12.9 and 25.3%, respectively.

Low-Temperature Properties. Aluminum alloys are used for structural parts for operation at temperatures as low as −269 °C (−452 °F). As temperature is reduced, the strength of aluminum alloys increases — a characteristic that is sometimes used in design. Ductility also generally increases as temperature is reduced. Alloy 5083-O, which is the most widely used aluminum alloy for cryogenic applications, exhibits the following increases in tensile properties when cooled from room temperature to the boiling point of nitrogen (−195.8 °C, or −320.4 °F): about 40% in ultimate tensile strength, about 10% in yield strength, and 60% in elongation. Retention of toughness also is of major importance for equipment operating at low temperature. Aluminum alloys have no ductile-to-brittle transition; consequently, neither ASTM nor ASME specifications require low-temperature Charpy or Izod tests of aluminum alloys. Other tests, including notchtensile and tear tests, including notch-tensile and tear toughness of aluminum alloys at low temperatures. The low-temperature characteristics of welds in the weldable aluminum alloys parallel those described above for unwelded material.

Properties of Wrought Aluminums and Aluminum Alloys

By the ASM Committee on
Aluminum and Aluminum Alloys*

1050
99.5 min Al

Specifications

ASTM. B491
UNS number. A91050
Foreign. Canada: CSA 9950.
France: NF A5. United Kingdom: BS
1B. West Germany: DIN A199.5

Chemical Composition

Composition limits. 99.50 min Al;
0.25 max Si; 0.40 max Fe; 0.05 max
Cu; 0.05 max Mn; 0.05 max Mg; 0.05
max V; 0.03 max others (each)

Applications

Typical uses. Extruded coiled tube

*See page X for committee list.

for equipment and containers for
food, chemical and brewing industries; collapsible tubes; pyrotechnic
powder

Mechanical Properties

Tensile properties. See Table 1.

Mass Characteristics

Density. 2.705 Mg/m³ (0.0977 lb/in.³)

Table 1 Typical mechanical properties of 1050 aluminum

Temper	Tensile strength MPa	ksi	Yield strength MPa	ksi	Elonga-tion, %	Shear strength MPa	ksi
O	76	11	28	4	39	62	9
H14	110	16	103	15	10	69	10
H16	131	19	124	18	8	76	11
H18	159	23	145	21	7	83	12

at 20 °C (68 °F)

Thermal Properties

Liquidus temperature. 657 °C
(1215 °F)
Solidus temperature. 646 °C
(1195 °F)
Coefficient of thermal expansion.
Linear:

Temperature range		Average coefficient	
°C	°F	μm/m·K	μin./in.·°F
−50-+20	−58-+68	21.8	12.1
20-100	68-212	23.6	13.1
20-200	68-392	24.5	13.6
20-300	68-572	25.5	14.2

Volumetric: 68.1×10^{-6} m³/m³·K (3.78×10^{-5} in.³/in.³·°F)
Specific heat. 900 J/kg·K (0.215 Btu/lb·°F) at 20 °C (68 °F)
Thermal conductivity. O temper, 231 W/m·K (133 Btu/ft·h·°F) at 20 °C (68 °F)

Electrical Properties

Electrical conductivity. Volumetric. O temper, 61.3% IACS at 20 °C (68 °F)
Electrical resistivity. O temper: 28.1 nΩ·m at 20 °C (68 °F); Temperature coefficient, 0.1 nΩ·m per K at 20 °C (68 °F)

1060
99.60 min Al

Specifications

AMS. Sheet and plate: 4000
ASME. See Table 2.
ASTM. See Table 2.
SAE. J454
UNS number. A91060

Chemical Composition

Composition limits. 99.60 min Al;

Table 2 ASME and ASTM specifications for 1060 aluminum

Mill form and condition	Specification number	
	ASME	ASTM
Sheet and plate	SB209	B209
Wire, rod and bar (rolled or cold finished)	...	B211
Wire, rod, bar, shapes and tube (extruded)	SB221	B221
Pipe (gas and oil transmission)	...	B345
Tube (condenser)	SB234	B234
Tube (condenser with integral fins)	...	B404
Tube (drawn)	...	B483
Tube (drawn, seamless)	SB210	B210
Tube (extruded, seamless)	SB241	B241

Table 3 Typical mechanical properties of 1060 aluminum

Temper	Tensile strength		Yield strength		Elongation(a), %	Hardness, HB(b)	Shear strength		Fatigue limit(c)	
	MPa	ksi	MPa	ksi			MPa	ksi	MPa	ksi
O	69	10	28	4	43	19	48	7	21	3
H12 ...	83	12	76	11	16	23	55	8	28	4
H14 ...	97	14	90	13	12	26	62	9	34	5
H16 ...	110	16	103	15	8	30	69	10	45	6.5
H18 ...	131	19	124	18	6	35	76	11	45	6.5

(a) 1.6-mm (1/16 in.) thick specimens. (b) 500-kg load; 10-mm diam ball. (c) At 5×10^8 cycles; R. R. Moore type tests.

Table 4 Tensile-property limits for 1060 aluminum

Temper	Tensile strength				Yield strength (min)		Elongation (min), %(a)
	Minimum		Maximum				
	MPa	ksi	MPa	ksi	MPa	ksi	
Sheet and Plate							
O	55	8.0	95	14.0	15	2.5	15 to 25
H12	75	11.0	110	16.0	60	9.0	6 to 12
H14	85	12.0	115	17.0	70	10.0	1 to 10
H18	110	16.0	...	...	85	12.0	1 to 4
H112:							
0.250-0.499 in. thick ...	75	11.0	...	...	...	...	10
0.500-1.000 in. thick ...	70	10.0	...	...	...	...	20
1.001-3.000 in. thick ...	60	9.0	...	...	...	...	25
Drawn Tube (0.010-0.500 in. Wall Thickness)							
O	60	8.5	...	...	15	2.5	...
H12	70	10.0	...	...	30	4.0	...
H14	85	12.0	...	...	70	10.0	...
H18	110	16.0	...	...	90	13.0	...
H112	60	8.5	...	...	15	2.5	...
Extruded Tube							
O	60	8.5	95	14.0	15	2.5	...
H112	60	8.5	95(b)	14.0(b)	15	2.5	30(b)
Heat-exchanger Tube (0.010-0.200 in. Wall Thickness)							
H14	85	12.0	...	...	70	10.0	...

(a) In 2 in. or 4d, where d is diameter of reduced section of tensile test specimen. Where a range of values appears in this column, specified minimum elongation varies with thickness of the mill product. (b) Applicable only to tube 1.000 to 4.500 in. diam by 0.050 to 0.169 in. wall thickness.

0.25 max Si; 0.35 max Fe; 0.05 max Cu; 0.03 max Mn; 0.03 max Mg; 0.05 max Zn; 0.05 max V; 0.03 max Ti; 0.03 max others (each)

Applications

Typical uses. Applications requiring very good resistance to corrosion and good formability, but tolerate low strength. Chemical process equipment is typical

Mechanical Properties

Tensile properties. See Tables 3 and 4.
Hardness. See Table 3.
Poisson's ratio. 0.33 at 20 °C (68 °F)
Elastic modulus. Tension, 69 GPa (10×10^6 psi)
Fatigue strength. See Table 3.

Mass Characteristics

Density. 2.705×10^3 Mg/m³ (0.0977 lb/in.³) at 20 °C (68 °F)

Thermal Properties

Liquidus temperature. 657 °C (1215 °F)
Solidus temperature. 646 °C (1195 °F)
Coefficient of thermal expansion. Linear:

Temperature range		Average coefficient	
°C	°F	μm/m·K	μin./in.·°F
−50-+20	−58-+68	21.8	12.1
20-100	68-212	23.6	13.1
20-200	68-392	24.5	13.6
20-300	68-572	25.5	14.1

Volumetric: 68.1×10^{-6} m³/m³·K (3.8×10^{-5} in.³/in.³·°F)
Specific heat. 900 J/kg·K (0.215 Btu/lb·°F) at 20 °C (68 °F)
Thermal conductivity. 234 W/m·K (135 Btu/ft·h·°F) at 25 °C (77 °F)

Electrical Properties

Electrical conductivity. Volumetric at 20 °C (68 °F): O temper, 62% IACS; H18 temper, 61% IACS
Electrical resistivity. At 20 °C (68 °F): O temper, 27.8 nΩ·m; H18 temper, 28.3 nΩ·m. Temperature coefficient, O and H18 tempers, 0.1 nΩ·m per K at 20 °C (68 °F)
Electrolytic solution potential. -0.84 V vs 0.1N calomel electrode in aqueous solution containing 53 g NaCl plus 3 g H_2O_2 per litre

Fabrication Characteristics

Annealing temperature. 345 °C (650 °F)

1100
99.00 min Al-0.12Cu

Commercial Names

Common name. Aluminum

Specifications

AMS. See Table 5.
ASME. See Table 5.
ASTM. See Table 5.
SAE. J454
UNS number. A91100
Government. See Table 5.
Foreign. Canada: CSA 990C. France: NF A45. ISO: A199.0Cu

Chemical Composition

Composition limits. 99.00 min Al; 1.0 max Si + Fe; 0.05 to 0.20 Cu; 0.05 max Mn; 0.10 max Zn; 0.05 max others (each); 0.15 max others (total); 0.0008 max Be (welding electrode and filler wire only)

Applications

Typical uses. Applications requiring good formability and high resistance to corrosion where high strength is not necessary. Food and chemical handling and storage equipment, sheet metal work, drawn or spun hollow ware, welded assemblies, heat exchangers, litho plate, nameplates, light reflectors

Table 5 Standard specifications for 1100 aluminum

Mill form and condition	AMS	ASME	ASTM	Government
Sheet and plate	4001, 4003	SB209	B209	QQ-A-250/1
Wire, rod and bar (rolled or cold finished)	4102	···	B211	QQ-A-225/1
Wire, rod, bar, shapes and tube (extruded)	···	SB221	B221	···
Tube (extruded, seamless)	···	SB241	B241	···
Tube (extruded, coiled)	···	···	B491	···
Tube (drawn)	···	···	B483	···
Tube (drawn, seamless)	4062	···	B210	WW-T-700/1
Tube (welded)	···	···	B313, B547	···
Rivet wire and rod	···	···	B316	QQ-A-430
Spray gun wire	4180	···	···	MIL-W-6712
Forgings and forging stock	···	···	B247	···
Welding rod and electrodes (bare)	···	···	···	QQ-R-566, MIL-E-16053
Impacts	···	···	···	MIL-A-12545
Foil	···	···	···	QQ-A-1876

Table 6 Typical room-temperature mechanical properties of 1100 aluminum

Temper	Tensile strength MPa	ksi	Yield strength MPa	ksi	Elongation, % 1/16 in. thick specimens	1/2 in. thick specimens	Hardness, HB(a)	Shear strength MPa	ksi	Fatigue limit(b) MPa	ksi
O	90	13	34	5	35	45	23	62	9	34	5
H12	110	16	103	15	12	25	28	69	10	41	6
H14	124	18	117	17	9	20	32	76	11	48	7
H16	145	21	138	20	6	17	38	83	12	62	9
H18	165	24	152	22	5	15	44	90	13	62	9

(a) 500-kg load; 10-mm ball. (b) At 5×10^8 cycles; R. R. Moore type tests.

Fig. 1 Reflectivity of 1100 aluminum as a function of aluminum oxide coating thickness

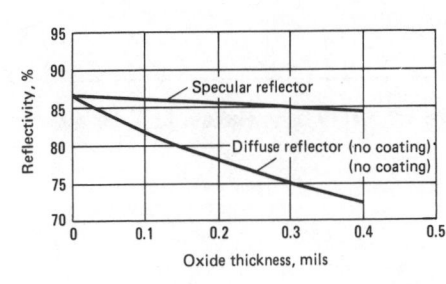

Mechanical Properties

Tensile properties. See Tables 6, 7 and 8.

Hardness. See Table 6.

Poisson's ratio. 0.33 at 20 °C (68 °F)

Elastic modulus. Tension, 69 GPa (10×10^6 psi); shear, 26 GPa (3.75×10^6 psi)
Fatigue strength. See Table 6.

Mass Characteristics

Density. 2.71 Mg/m³ (0.098 lb/in.³) at 20 °C (68 °F)

Table 7 Tensile-property limits for 1100 aluminum

Temper	Tensile strength Minimum MPa	ksi	Maximum MPa	ksi	Yield strength (min) MPa	ksi	Elongation (min), %(a)
Sheet and Plate							
O	75	11.0	105	15.5	25	3.5	15 to 28
H12	95	14.0	130	19.0	75	11.0	3 to 12
H14	110	16.0	145	21.0	95	14.0	1 to 10
H16	130	19.0	165	24.0	115	17.0	1 to 4
H18	15	22.0	...	...	...	...	1 to 4
H112:							
0.250-0.499 in. thick	90	13.0	...	...	50	7.0	9
0.500-2.000 in. thick	85	12.0	...	...	35	5.0	14
2.001-3.000 in. thick	80	11.5	...	...	30	4.0	20
Wire, Rod and Bar (Rolled or Cold Finished)							
O	75	11.0	105	15.5	20	3.0	25
H112	75	11.0	...	...	20	3.0	...
H12(b)	95	14.0	...	...	...	...	...
H14(b)	110	16.0	...	...	...	...	...
H16(b)	130	19.0	...	...	...	...	...
H18(b)	150	22.0	...	...	...	...	...
Wire, Rod, Bar and Shapes (Extruded)							
O	75	11.0	105	15.5	20	3.0	25
H112	75	11.0	...	...	20	3.0	...
Wire and Rod (Rivet and Cold Heading Grade)							
O(c)	...	...	105	15.5	...	...	...
H14(c)	110	16.0	145	21.0	...	...	...
Drawn Tube (0.014 to 0.500 in. Wall Thickness)							
O	...	...	105	15.5	...	...	...
H12	95	14.0	...	...	...	...	...
H14	110	16.0	...	...	...	...	...
H16	130	19.0	...	...	...	...	...
H18	150	22.0	...	...	...	...	...
Extruded Tube							
O	75	11.0	105	15.5	20	3.0	25
H112	75	11.0	...	...	20	3.0	25

(a) In 2 in. or 4d, where d is diameter of reduced section of tensile test specimen. Where a range of values appears in this column, the specified minimum elongation varies with thickness of the mill product. (b) Nominal thickness up thru 0.374 in. (c) Nominal diameter up thru 1.000 in.

Thermal Properties

Liquidus temperature. 657 °C (1215 °F)

Solidus temperature. 643 °C (1190 °F)

Coefficient of thermal expansion. Linear:

Temperature range °C	°F	Average coefficient μm/m·K	μin./in.·°F
−50-+20	−58-+68	21.8	12.1
20-100	68-212	23.6	13.1
20-200	68-392	24.5	13.6
20-300	68-572	25.5	14.1

Volumetric: 68.1×10^{-6} m³/m³·K (3.8×10^{-5} in.³/in.³·°F)

Specific heat. 904 J/kg·K (0.216 Btu/lb·°F) at 20 °C (68 °F)

Thermal conductivity. O temper, 222 W/m·K (128 Btu/ft·h·°F); H18 temper, 218 W/m·K (126 Btu/ft·h·°F)

Electrical Properties

Electrical conductivity. Volumetric at 20 °C (68 °F): O temper, 59% IACS; H18 temper, 57% IACS

Electrical resistivity. At 20 °C (68 °F): O temper, 29.2 nΩ·m; H18 temper, 30.2 nΩ·m. Temperature coefficient at 20 °C (68 °F): O and H18 tempers, 0.1 nΩ·m per K

Electrolytic solution potential. All tempers, −0.83 V vs 0.1N calomel electrode in aqueous solution containing 53 g NaCl plus 3 g H_2O_2 per litre at 25 °C (77 °F)

Optical Properties

Reflectance. Brightly polished or diffusely etched reflector: 86% for light from tungsten filament; 84% for light having a wavelength of 250 nm. See also Fig. 1.

Emittance. See Fig. 2.

Fabrication Characteristics

Annealing temperature. 343 °C (650 °F)

1145
99.45 min Al

Specifications

AMS. 4011
ASTM. B373
Government. QQ-A-1876

Fig. 2 Emissivity of 1100 aluminum foil as a function of coating thickness

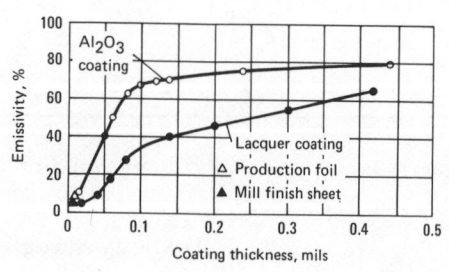

Table 8 Typical tensile properties of 1100 aluminum at various temperatures

Temperature °C	°F	Tensile strength MPa	ksi	Yield strength MPa	ksi	Elongation, %
O Temper						
−196	−320	170	25	41	6	50
−80	−112	105	15	38	5.5	43
−28	−18	97	14	34	5	40
+24	+75	90	13	34	5	40
100	212	69	10	32	4.6	45
149	300	55	8	29	4.2	55
204	400	41	6	24	3.5	65
260	500	28	4	18	2.6	75
316	600	20	2.9	14	2.0	80
371	700	14	2.1	11	1.6	85
H14 Temper						
−196	−320	205	30	140	20	45
−80	−112	140	20	125	18	24
−28	−18	130	19	115	17	20
+24	+75	125	18	115	17	20
100	212	110	16	105	15	20
149	300	97	14	83	12	23
204	400	69	10	52	7.5	26
260	500	28	4	18	2.6	75
316	600	20	2.9	14	2.0	80
371	700	14	2.1	11	1.6	85
H18 Temper						
−196	−320	235	34	180	26	30
−80	−112	180	26	160	23	16
−28	−18	170	25	160	23	15
+24	+75	165	24	150	22	15
100	212	145	21	130	19	15
149	300	125	18	97	14	20
204	400	41	6	24	3.5	65
260	500	28	4	18	2.6	75
316	600	20	2.9	14	2.0	80
371	700	14	2.1	11	1.6	85

Table 9 Tensile properties of 1145 aluminum foil

Temper	Tensile strength MPa	ksi	Yield strength MPa	ksi	Elongation, %
Typical Properties					
O	75	11	34	5	40
H18	145	21	117	17	5
Tensile Strength Limits(a)					
O	95 max	14 max	...	...	...
H19	140 min	20 min	...	...	...

(a) Unmounted foil 0.02 to 0.15 mm (0.0007 to 0.0059 in.) thick

Chemical Composition

Composition limits. 99.45 min Al, 0.55 max Si + Fe, 0.05 max Cu, 0.05 max Mn, 0.05 max Mg, 0.05 max Zn, 0.05 max V, 0.03 max Ti, 0.03 max others (each)

Applications

Typical uses. Foil for packaging, insulating and heat exchangers

Mechanical Properties

Tensile properties. See Table 9.

Mass Characteristics

Density. 2.705 Mg/m^3 (0.0977 lb/in.^{3}) at 20 °C (68 °F)

Thermal Properties

Liquidus temperature. 657 °C (1215 °F)
Solidus temperature. 646 °C (1195 °F)
Coefficient of thermal expansion. Linear:

Temperature range °C	°F	Average coefficient μm/m·K	μin./in.·°F
−50-+20	−58-+68	21.8	12.1
20-100	68-212	23.6	13.1
20-200	68-392	24.5	13.6
20-300	68-572	25.5	14.1

Volumetric: 68.1×10^{-6} m^3/m^3·K (3.78×10^{-5} in.^{3}/in.^{3}·°F)
Specific heat. 904 J/kg·K (0.216 Btu/lb·°F) at 20 °C (68 °F)
Thermal conductivity. At 20 °C (68 °F): O temper, 230 W/m·K (133 Btu/ft·h·°F); H18 temper, 227 W/m·K (131 Btu/ft·h·°F)

Electrical Properties

Electrical conductivity. Volumetric at 20 °C (68 °F): O temper, 61% IACS; H18 temper, 60% IACS
Electrical resistivity. At 20 °C (68 °F): O temper, 28.3 nΩ·m; H18 temper, 28.7 nΩ·m. Temperature coefficient at 20 °C: O and H18 tempers, 0.1 nΩ·m per K

Optical Properties

Reflectance. 95 to 97% for λ = 0.3 to 10 μm
Emittance. 3 to 5% for λ = 9.3 μm at 20 °C (68 °F)

Fabrication Characteristics

Annealing temperature. 343 °C (650 °F)

1199
99.99 min Al

Commercial Names

Trade name. Super-purity aluminum, Raffinal
Common name. Super-purity aluminum, refined aluminum

Chemical Composition

Composition limits. 99.99 min Al;

0.006 max Si; 0.006 max Fe; 0.006 max Cu; 0.002 max Mn; 0.006 max Mg; 0.006 max Zn; 0.002 max Ti; 0.005 max V; 0.005 max Ga; 0.002 max others (each)
Consequence of exceeding impurity limits. See Fig. 3.

Applications

Typical uses. Electrolytic capacitor foil, vapor deposited coatings for optically reflecting surfaces

Mechanical Properties

Tensile properties. See Table 10 and Fig. 3.
Hardness. O temper, 15 HB; H18 temper, 27 HB. (500-kg load; 10-mm diam ball) See also Fig. 3.
Elastic modulus. Tension, 62 GPa $(9.0 \times 10^6$ psi); shear, 25.0 GPa $(3.62 \times 10^6$ psi)

Mass Characteristics

Density. 2.705 Mg/m^3 (0.0975 $lb/in.^3$) at 20 °C (68 °F)

Thermal Properties

Melting point. 660 °C (1220 °F)
Coefficient of thermal expansion. Linear:

Temperature range		Average coefficient	
°C	°F	µm/m·K	µin./in.·°F
−50-+20	−58-+68	21.8	12.1
20-100	68-212	23.6	13.1
20-200	68-392	24.5	13.6
20-300	68-572	25.5	14.2

Specific heat. 900 J/kg·K at 25 °C (77 °F)
Heat of fusion. 390 kJ/kg·K
Thermal conductivity. O temper, 243 W/m·K (140 Btu/ft·h·°F) at 20 °C (68 °F)

Electrical Properties

Electrical conductivity. Volumetric, O temper: 64.5% IACS at 20 °C (68 °F)
Electrical resistivity. O temper: 2.67 nΩ·m at 20 °C (68 °F); temperature coefficient, O temper: 0.1 nΩ·m per K at 20 °C (68 °F)

Optical Properties

Reflectivity. 85 to 90% to visible light for an electrolytically brightened surface

Fig. 3 Effect of purity on strength and hardness of unalloyed aluminum

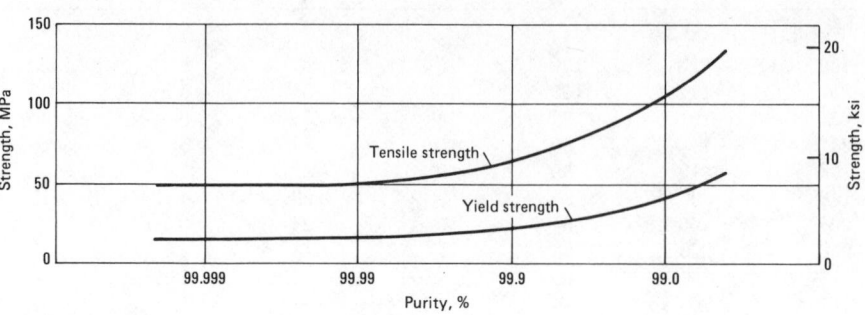

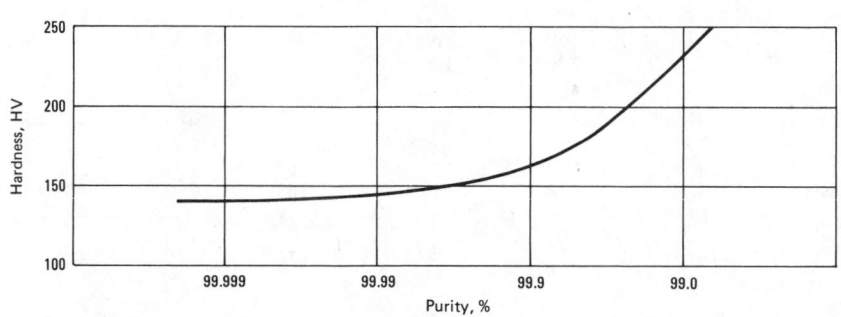

Table 10 Typical tensile properties of 1199 aluminum

Reduction by cold rolling, %	Tensile strength		Yield strength		Elongation, %
	MPa	ksi	MPa	ksi	
0 (annealed)	45	6.5	10	1.5	50
10	59	8.6	57	8.2	40
20	77	11.1	75	10.8	15
40	96	13.9	91	13.2	11
60	110	15.9	105	15.1	6
75	120	17.5	115	16.4	5

1350
99.50 min Al

Commercial Names

Common name. Electrical conductor grade (EC)

Specifications

ASTM. ACSR: B232, B401. Bus conductors: B236. Communication wire: B314. Rolled redraw rod: B233. Round wire: B230, B609. Wire, rectangular and square: B324. Round solid conductors: B544. Stranded conductors: B231, B400.

Foreign. France: NF A5/L. Spain: UNE AL99.5E. United Kingdom: BS 1E. West Germany: DIN E-A199.5

Chemical Composition

Composition limits. 99.50 min Al; 0.10 max Si; 0.40 max Fe; 0.05 max Cu; 0.01 max Mn; 0.01 max Cr; 0.05 max Zn; 0.03 max Ga; 0.02 max V + Ti; 0.05 max B; 0.03 max others (each); 0.10 max others (total)
Consequence of exceeding impurity limits. Impurity elements in excess of limits degrade electrical conductivity.

Applications

Typical uses. Wire, stranded conductors, bus conductors, transformer strip

Mechanical Properties

Tensile properties. Typical, see Table 11; property limits, see Tables 12 and 13.
Shear strength. See Table 11.
Poisson's ratio. 0.33 at 20 °C (68 °F)
Elastic modulus. Tension, 69 GPa (10×10^6 psi)
Fatigue strength. H19 temper, 48 MPa (7 ksi) at 5×10^8 cycles in an R.R. Moore type test

Mass Characteristics

Density. 2.705 Mg/m^3 (0.0977 $lb/in.^3$) at 20 °C (68 °F)

Thermal Properties

Liquidus temperature. 657 °C (1215 °F)
Solidus temperature. 646 °C (1195 °F)
Coefficient of thermal expansion. Linear:

Temperature range		Average coefficient	
°C	°F	μm/ m·K	μin./ in.·°F
−50-+20	−58-+68	21.8	12.1
20-100	68-212	23.6	13.1
20-200	68-392	24.5	13.6
20-300	68-572	25.5	14.2

Volumetric: 68.1×10^{-6} $m^3/m^3 \cdot K$ (3.8×10^{-5} $in.^3/in.^3 \cdot °F$)
Specific heat. 900 J/kg·K (0.215 Btu/lb·°F) at 20 °C (68 °F)
Thermal conductivity. O temper, 234 W/m·K (135 Btu/ft·h·°F); H19 temper, 230 W/m·K (133 Btu/ft·h·°F)

Electrical Properties

Electrical conductivity. Volumetric, at 20 °C (68 °F). O temper, 61.8% IACS min; H1x tempers, 61.0% IACS min
Electrical resistivity. O temper, 27.9 nΩ·m max at 20 °C (68 °F); H1x tempers, 28.2 nΩ·m max at 20 °C. Temperature coefficient, all tempers: 0.1 nΩ·m per K at 20 °C
Electrolytic solution potential. −0.84 V vs 0.1N calomel electrode in aqueous solution of 53 g NaCl plus 3 g H_2O_2 per litre at 25 °C (77 °F)

Fabrication Characteristics

Annealing temperature. 343 °C (650 °F)

Table 11 Typical mechanical properties of 1350 aluminum

Temper	Tensile strength MPa	ksi	Yield strength MPa	ksi	Elongation(a), %	Shear strength MPa	ksi
O	83	12	28	4	23	55	8
H12	97	14	83	12	...	62	9
H14	110	16	97	14	...	69	10
H16	124	18	110	16	...	76	11
H19	186	27	165	24	1.5	103	15

(a) In 250 mm or 10 in.; value applicable to wire only.

Table 12 Tensile-property limits for 1350 aluminum

Temper	Tensile strength Minimum MPa	ksi	Maximum MPa	ksi	Yield strength (min) MPa	ksi	Elongation (min), %(a)
Sheet and Plate							
O	55	8.0	95	14.0	...	...	15 to 28
H12	85	12.0	115	17.0	...	...	3 to 12
H14	95	14.0	130	19.0	...	...	1 to 10
H16	110	16.0	145	21.0	...	...	1 to 4
H18	125	18.0	...	...	...	...	1 to 4
H112:							
0.250-0.499 in.	75	11.0	...	...	...	...	10
0.500-1.000 in.	70	10.0	...	...	...	...	16
1.001-1.500 in.	60	9.0	...	...	...	...	22
Wire(b) and Redraw Rod(c)							
O	60	8.5	95	14.0	...	...	...
H12 and H22	85	12.0	115	17.0	...	...	...
H14 and H24	105	15.0	140	20.0	...	...	...
H16 and H26	115	17.0	150	22.0	...	...	...
Extrusions(d)							
H111	60	8.5	...	...	25	3.5	...
Rolled Bar(e)							
H12	85	12.0	...	...	55	8.0	...
Sawed-plate Bar							
H112							
0.125-0.499 in.	75	11.0	...	...	40	6.0	...
0.500-1.000 in.	70	10.0	...	...	30	4.0	...
1.001-1.500 in.	60	9.0	...	...	25	3.5	...

(a) In 2 in. or 4d, where d is diameter of reduced section of test specimen. Where a range of values appears in this column, specified minimum elongation varies with thickness of the mill product. (b) Up thru 9.50 mm (0.374 in.) diam. (c) 9.52 mm (0.375 in.) diam. (d) Bar, rod, tubular products and structural shapes. (e) 3 to 25 mm (0.125 to 1.0 in.) thick.

2011
5.5Cu-0.4Pb-0.4Bi

Specifications

ASTM. Drawn, seamless tube: B210. Rolled or cold finished wire, rod and bar: B211
SAE. J454
UNS number. A92011
Government. Rolled or cold finished wire, rod and bar: QQ-A-225/3
Foreign. Canada: CSA CB60. France: NF A-U4Pb. United Kingdom: BS FC1. Germany: DINAL Cu-BiPb

Chemical Composition

Composition limits. 0.40 max Si; 0.7 max Fe; 5.0 to 6.0 Cu; 0.30 max Zn; 0.20 to 0.6 Pb; 0.05 max others (each); 0.15 others (total); rem Al

Applications

Typical uses. Wire, rod and bar for screw machine products. Applications where good machinability and good strength are required

Mechanical Properties

Tensile properties. See Tables 14 and 15.

Table 13 Tensile-property limits for 1350 aluminum wire, H19 temper

Wire diameter, in.	Min tensile strength Individual(a) MPa	ksi	Average(b) MPa	ksi	Min elongation(c), % Individual(a)	Average(b)
0.0105-0.0500 160		23.0	170	25.0	...	...
0.0501-0.0600 185		27.0	200	29.0	1.2	1.4
0.0601-0.0700 185		27.0	195	28.5	1.3	1.5
0.0701-0.0800 185		26.5	195	28.0	1.4	1.6
0.0801-0.0900 180		26.0	190	27.5	1.5	1.6
0.0901-0.1000 180		25.5	185	27.0	1.5	1.6
0.1001-0.1100 170		24.5	180	26.0	1.5	1.6
0.1101-0.1200 165		24.0	175	25.5	1.6	1.7
0.1201-0.1400 160		23.5	170	25.0	1.7	1.8
0.1401-0.1500 160		23.5	170	24.5	1.8	1.9
0.1501-0.1800 160		23.0	165	24.0	1.9	2.0
0.1801-0.2100 160		23.0	165	24.0	2.0	2.1
0.2101-0.2600 155		22.5	160	23.5	2.2	2.3

(a) Minimum value for any test in a given lot. (b) Minimum value for average of all tests for a given lot. (c) In 10 in.

Table 14 Room-temperature mechanical properties of alloy 2011

Temper	Tensile strength MPa	ksi	Yield strength (a) MPa	ksi	Elon- gation, %	Hard- ness (b), HB	Shear strength MPa	ksi
Typical Properties								
T3(c) 379		55	296	43	15(d)	95	221	32
T8(c) 407		59	310	45	12(d)	100	241	35
Property Limits, Minimum Values for Rolled or Cold Finished Wire, Rod and Bar								
T3								
0.125 to 1.500 in. thick 310		45	260	38	10(a,e)	...	...	...
1.501 to 2.000 in. thick 295		43	235	34	12(a,e)	...	...	...
2.001 to 3.250 in. thick 290		42	205	30	14(a,e)	...	...	...
T4, T451								
0.375 to 8.000 in. thick 275		40	125	18	16	...	...	...
T8								
0.125 to 3.250 in. thick 370		54	275	40	10	...	...	...

(a) Yield strength and elongation limits not applicable to wire less than 0.125 in. in thickness or diameter. (b) 500-kg load; 10-mm diam ball. (c) Strengths and elongations generally unchanged or improved at low temperatures. (d) 12.7 mm (½ in.) diam specimen. (e) In 2 in. or 4d, where d is diameter of reduced section of tensile test specimen.

Compressive yield strength. Approximately equal to tensile yield strength
Hardness. See Table 14.
Poisson's ratio. 0.33 at 20 °C (68 °F)
Elastic modulus. Tension, 70.0 GPa (10.2 × 10⁶ psi); shear, 26.0 GPa (3.80 × 10⁶ psi)
Fatigue strength. At 5 × 10⁸ cycles,

R. R. Moore type test: T3 and T8 tempers, 124 MPa (18 ksi)

Mass Characteristics

Density. 2.82 Mg/m³ (0.102 lb/in.³) at 20 °C (68 °F)

Thermal Properties

Liquidus temperature. 638 °C (1180 °F)

Solidus temperature. 541 °C (1005 °F)
Incipient melting temperature. 535 °C (995 °F)
Coefficient of thermal expansion. Linear:

Temperature range °C	°F	Average coefficient μm/m·K	μin./in.·°F
−50-+20	−58-+68	21.4	11.9
20-100	68-212	23.1	12.8
20-200	68-392	24.0	13.3
20-300	68-572	25.0	13.9

Volumetric: 67 × 10⁻⁶ m³/m³·K (3.72 × 10⁻⁵ in.³/in.³·°F) at 20 °C (68 °F)

Specific heat. 864 J/kg·K (0.206 Btu/lb·°F) at 20 °C (68 °F)
Thermal conductivity. At 20 °C (68 °F): T3 and T4 tempers, 152 W/m·K (87.8 Btu/ft·h·°F); T8 temper, 173 W/m·K (99.9 Btu/f·h·°F)

Electrical Properties

Electrical conductivity. Volumetric, at 20 °C (68 °F): T3 and T4 tempers, 39% IACS; T8 temper, 45% IACS
Electrical resistivity. At 20 °C (68 °F): T3 and T4 tempers, 44.0 nΩ·m; T8 temper, 38.0 nΩ·m. Temperature coefficient. T3, T4, and T8 tempers, 0.1 nΩ·m per K at 20 °C (68 °F)
Electrolytic solution potential. At 25 °C (77 °F): −0.69 V (T3 and T4 tempers), −0.83 V (T8 temper) vs 0.1N calomel electrode in an aqueous solution containing 53 g NaCl plus 3 g H₂O₂ per litre

Fabrication Characteristics

Annealing temperature. 413 °C (775 °F)
Solution temperature. 524 °C (975 °F)
Aging temperature. T8 temper, 160 °C (320 °F); 14 h at temperature

2014, Alclad 2014
4.4Cu-0.8Si-0.8Mn-0.5Mg

Specifications

AMS. See Table 16.
ASME. Rolled or cold finished wire, rod, and bar: SB211. Forgings: SB247
ASTM. See Table 16.

SAE. J454
UNS number. A92014
Government. See Table 16.
Foreign. Canada: CSA CS41N.
France: NF A-U4SG. Germany: DIN
AlCuSiMn. ISO: AlCu4SiMg. United
Kingdom: BS H15

Chemical Composition

Composition limits. 3.9 to 5.0 Cu;
0.50 to 1.2 Si; 0.7 max Fe; 0.40 to 1.2
Mn; 0.20 to 0.8 Mg; 0.25 max Zn; 0.10
max Cr; 0.15 max Ti; 0.05 max others
(each); 0.15 max others (total); rem
Al. Alclad 2014: 6006 cladding—0.20
to 0.6 Si; 0.35 max Fe; 0.15 to 0.30
Cu; 0.05 to 0.20 Mn; 0.45 to 0.9 Mg;
0.10 max Cr; 0.10 max Zn; 0.10 max
Ti; 0.05 max others (each); 0.15 max
others (total); rem Al

Applications

Typical uses. Heavy-duty forgings,
plate and extrusions for aircraft fittings, wheels and major structure
components, space booster tankage
and structure, truck frame and suspension components. Applications requiring high strength and hardness
including service at elevated temperatures

Mechanical Properties

Tensile properties. See Tables 17 to
19.
Compressive yield strength. Approximately the same as tensile yield
strength.
Hardness. O temper: 87 to 98 HRH;
45 HB. T4 temper: 65 to 73 HRB; 105
HB. T6 temper: 80 to 86 HRB; 135
HB. HB values obtained using 500-
kg load and 10-mm diam ball.
Poisson's ratio. 0.33 at 20 °C (68
°F)
Elastic modulus. Tension: 2014,
72.4 GPa (10.5×10^6 psi); alclad
2014, 71.7 GPa (10.4×10^6 psi).
Shear: 2014 and alclad 2014, 28 GPa
(4.0×10^6 psi). Compression: 2014,
73.8 GPa (10.7×10^6 psi); alclad
2014, 73.1 GPa (10.6×10^6 psi)
Fatigue strength. O temper, 90
MPa (13 ksi); T4 temper, 140 MPa
(20 ksi); T6 temper, 125 MPa (18 ksi);
all at 5×10^8 cycles in a R. R. Moore
type test

Mass Characteristics

Density. 2.80 Mg/m³ (0.101 lb/in.³)
at 20 °C (68 °F)

Thermal Properties

Liquidus temperature. 638 °C
(1180 °F)
Solidus temperature. 507 °C (945
°F)
Coefficient of thermal expansion.
Linear:

Table 16 Standard specifications for alloy 2014

Mill form	Specification number		
	AMS	ASTM	Government
Sheet and plate	4014	B209	...
	4028	...	...
	4029	...	...
Rolled or cold finished wire, rod, and bar ...	4121	B211	QQ-A-225/4
Extruded wire, rod, bar, shapes and tube ..	4153	B221	QQ-A-200/2
Extruded seamless tube	...	B241	...
Drawn, seamless tube	...	B210	...
Forgings ...	4133	B247	QQ-A-367
	4134		MIL-A-22771
	4135		
Forging stock	4134	...	QQ-A-367
	4133	...	...
	4135	...	...
Impacts	...	...	MIL-A-12545
Sheet and plate (alclad) ...	...	B209	QQ-A-250/3

Table 15 Typical tensile properties of alloy 2011-T3

Temperature		Tensile strength (a)		Yield strength (a) (b)		Elon-gation, %
°C	°F	MPa	ksi	MPa	ksi	
24	75.........	379	55	296	43	15
100	212.........	324	47	234	34	16
149	300.........	193	28	131	19	25
204	400.........	110	16	76	11	35
260	500.........	45	6.5	26	3.8	45
316	600.........	21	3.1	12	1.8	90
371	700.........	16	2.3	10	1.4	125

(a) Lowest strength for exposures up to 10 000 h at temperature, no load; test loading applied at 5000 psi/min to yield strength and then at strain rate of 5%/min to fracture. (b) 0.2% offset.

Table 17 Typical tensile properties of alloy 2014

Temper	Tensile strength		Yield strength		Elongation, %	Hardness, HB	Shear strength		Fatigue strength	
	MPa	ksi	MPa	ksi			MPa	ksi	MPa	ksi
O	186	27	97	14	18(a)	45	125	18	90	13
T4	427	62	290(b)	42(a)	20(a)	105	260	38	140	20
T6(c)	483	70	414	60	13(a)	135	240	42	125	18
Alclad										
O	172	25	69	10	21(e)	...	125	18	...	...
T3(d)	434	63	276	40	20(e)	...	255	37	...	...
T4(d)	421	61	255	37	22(e)	...	255	37	...	...
T6(d)	469	68	414	60	10(e)	...	285	41	...	...

(a) Round bar, ½ in. diam. (b) Die forgings have about 20% lower yield strength. (c) Extruded products more than 19 mm (¾ in.) thick have 15 to 20% higher strengths. (d) Sheet less than 1 mm (0.04 in.) thick has slightly lower strength. (e) Sheet, 1/16 in. thick.

Temperature range		Average coefficient	
°C	°F	$\mu m/$ m·K	$\mu in./$ in.·°F
−50−+20	−58−+68	20.8	11.5
20−100	68−212	22.5	12.5
20−200	68−392	23.4	13.0
20−300	68−572	24.4	13.6

Volumetric: 65.1×10^{-6} m³/m³·K (3.62×10^{-5} in.³/in.³·°F) Btu/lb·°F) at 20 °C (68 °F)

Thermal conductivity. At 20 °C (68 °F): O temper, 192 W/m·K (111 Btu/ft·h·°F); T3, T4, T451 tempers, 134 W/m·K (77.4 Btu/ft·h·°F); T6, T651, T652 tempers, 155 W/m·K (89.5 Btu/ft·h·°F)

Electrical Properties

Electrical conductivity. At 20 °C (68 °F): O temper, 50% IACS, T3, T4, T451 tempers, 34% IACS; T6, T651, T652 tempers, 40% IACS

Electrical resistivity. At 20 °C (68 °F): O temper, 34 nΩ·m: T3, T4, T451 tempers, 51 nΩ·m: T6, T651, T652 tempers, 43 nΩ·m. Temperature coefficient: O, T3, T4, T451, T6, T651, T652 tempers, 0.1 nΩ·m per K at 20 °C (68 °F)

Electrolytic solution potential. At 25 °C (77 °F): −0.68 V (T3, T4, T451 tempers) or −0.78 V (T6, T651, T652 tempers) vs $0.1N$ calomel electrode in an aqueous solution containing 53 g NaCl plus 3 g H_2O_2 per litre

Fabrication Characteristics

Annealing temperature. 413 °C (775 °F)

Solution temperature. 502 °C (935 °F)

Aging temperature. T6 temper. Sheet, plate, wire, rod, bar, shapes and tube: 160 °C (320 °F) for 18 h at temperature. Forgings: 171 °C (340 °F) for 10 h at temperature

2024, Alclad 2024
4.4Cu-1.5Mg-0.6Mn

Specifications

AMS. See Table 20.
ASME. Rolled or drawn wire, rod and bar: SB211. Extrusions: SB221
ASTM. See Table 20.
SAE. J454

Table 18 Typical tensile properties of alloy 2014-T6 or -T651 at various temperatures(a)

Temperature		Tensile strength		Yield strength(b)		Elongation, %
°C	°F	MPa	ksi	MPa	ksi	
−196	−320	579	84	496	72	14
−80	−112	510	74	448	65	13
−28	−18	496	72	427	62	13
24	75	483	70	414	60	13
100	212	439	63	393	57	15
149	300	276	40	241	35	20
204	400	110	16	90	13	38
260	500	66	9.5	52	7.5	52
316	600	45	6.5	34	5	65
371	700	30	4.3	24	3.5	72

(a) Lowest strength for exposures up to 10 000 h at temperature under no load; test loading applied at 5 000 psi/min to yield strength and then at strain rate of 5%/min to fracture. (b) 0.2% offset.

Table 19 Tensile-property limits for alloy 2014

Temper	Tensile strength Minimum MPa	ksi	Maximum MPa	ksi	Yield strength Minimum MPa	ksi	Elongation(a), %
Flat Products, Bare							
Sheet and plate, O							
0.020–0.499 in. thick	···	···	220	32	110(max)	16(max)	16
0.500–1.000 in. thick	···	···	220	32	···	···	10
Flat sheet, T3							
0.020–0.039 in. thick	405	59	···	···	240	35	14
0.040–0.249 in. thick	405	59	···	···	250	36	14
Coiled sheet, T4							
0.020–0.249 in. thick	405	59			240	35	14
Plate, T451(b)							
0.250–2.000 in. thick	400	58	···	···	250	36	14–12
2.001–3.000 in. thick	395	57	···	···	250	36	8
Sheet and plate, T42							
0.020–1.000 in. thick	400	58	···	···	235	34	14
Sheet, T6, T62							
0.020–0.039 in. thick	440	64	···	···	395	57	6
0.040–0.249 in. thick	455	66	···	···	400	58	7
Plate, T62, T651							
0.250–2.000 in. thick	460	67	···	···	405	59	7–4
2.001–2.500 in. thick	450	65	···	···	400	58	2
2.501–3.000 in. thick	435	63	···	···	395	57	2
3.001–4.000 in. thick	405	59	···	···	380	55	1
Flat Products, Alclad							
Sheet and plate, O							
0.020–0.499 in. thick	···	···	205	30	95(max)	14(max)	16
0.500–1.000 in. thick	···	···	220	32	···	···	10
Flat sheet, T3							
0.020–0.024 in. thick	370	54	···	···	230	33	14
0.025–0.039 in. thick	380	55	···	···	235	34	14
0.040–0.249 in. thick	395	57	···	···	240	35	15
Coiled sheet, T4							
0.020–0.024 in. thick	370	54	···	···	215	31	14
0.025–0.039 in. thick	380	55	···	···	220	32	14
0.040–0.249 in. thick	395	57	···	···	235	34	15
Plate, T451(b)							
0.500–2.000 in. thick	400	58	···	···	250	36	12–14
0.250–0.499 in. thick	395	57	···	···	250	36	15
0.500–2.000 in. thick	400	58	···	···	250	36	12–14
2.001–3.000 in. thick	395	57	···	···	200	36	8
Sheet and plate, T4							
0.020–0.024 in. thick	370	54	···	···	215	31	14
0.025–0.039 in. thick	380	55	···	···	220	32	14
0.040–0.499 in. thick	395	57	···	···	235	34	15
0.500–1.000 in. thick	400	58	···	···	235	34	14
Sheet, T6							
0.020–0.024 in. thick	425	62	···	···	370	54	7
0.025–0.039 in. thick	435	63	···	···	380	55	7
0.040–0.249 in. thick	440	64	···	···	395	57	8

(continued)

Table 19 (continued)

Temper	Tensile strength Minimum MPa	ksi	Maximum MPa	ksi	Yield strength Minimum MPa	ksi	Elonga- tion(a), %
Plate, T62, T651							
0.250–0.499 in. thick 440	64	...	...		395	57	8
0.500–2.000 in. thick 460	67	...	...		405	59	6
2.001–2.500 in. thick 450	65	...	...		400	58	2
2.501–3.000 in. thick 435	63	...	...		395	57	2
3.001–4.000 in. thick 405	59	...	...		380	55	1
Rolled or Cold Finished Wire, Rod and Bar							
T4, T42, T451(b) 380	55	...	...		220	32	16
T6, T62, T651 450	65	...	...		380	55	8
Extruded Wire, Rod, Bar and Shapes							
O......................... ...	...	205	30		125(max)	12(max)	12
T4, T4510, T4511 345	50	...	...		240	35	12
T42.......................345	50	...	...		200	29	12
T6, T6510, T6511							
Up thru 0.499 in. thick ...415	60	...	...		365	53	7
0.500–0.749 in. thick 440	64	...	...		400	58	7
0.750 in. thick and over ...470	68	...	...		415	60	7
T62......................415	60	...	...		365	53	7(c)
Extruded Tube							
O......................... ...	...	205	30		125(max)	18(max)	12
T4, T4510, and T4511 345	50	...	...		240	35	12
T42.......................345	50	...	...		200	29	12
T6, T6510, T6511							
Up thru 0.499 in. thick ...415	60	...	...		365	53	7
0.500–0.749 in. thick 440	64	...	...		400	58	7
0.750 in. thick and over ...470	68	...	...		415(d)	60(d)	7(d)
T62......................415	60	...	...		365	53	7(c)
Drawn Tube							
O, 0.18–0.500 in. thick	...	220	32		110(max)	16(max)	...
T4, T42							
0.018–0.500 in. thick 370	54	...	...		205	30	10–16
Die forgings: Axis Parallel to Direction of Grain Flow							
T4, Up thru 4 in. thick 380	55	...	...		205	30	11(e)(f)
T6							
Up thru 2 in. thick 450	65	...	...		385	56	6(e)(g)
Over 2 thru 3 in. thick 450	65	...	...		380	55	6(e)(g)
Over 3 thru 4 in. thick 435	63	...	...		380	55	6(e)(g)
Die Forgings: Axis Not Parallel to Direction of Grain Flow							
T6, Up thru 2 in. thick 440	64	...	...		380	55	3(e)(h)
Over 2 thru 4 in. thick 435	63	...	...		370	54	2(e)
Hand Forgings							
T6							
Up thru 2.000 in. thick							
longitudinal,							
long transverse 450	65	...	...		385	56	3–8
2.001–3.000 in. thick							
Longitudinal 440	64	...	...		385	56	8
Long transverse 440	64	...	...		380	55	3
Short transverse 425	62	...	...		380	55	2
3.001–4.000 in. thick							
Longitudinal,							
long transverse.........435	63	...	...		380	55	3–8
Short transverse 420	61	...	...		370	54	2
4.001–5.000 in. thick							
Longitudinal,							
long transverse.........425	62	...	...		370	54	2–7
Short transverse 415	60	...	...		365	53	1
5.001–6.000 in. thick							
Longitudinal,							
long transverse.........420	61	...	...		365	53	2–7
Short transverse 405	59	...	...		365	53	1

(continued)

Table 20 Standard specifications for alloy 2024

Mill form and condition	Specification number AMS	ASTM	Government
Sheet and			
plate 4033	B209	QQ-A-250/4
4035	...	...
4037	...	...
4097	...	...
4098	...	...
4099	...	...
4103	...	...
4104	...	...
4105	...	...
4106	...	...
4192	...	...
4193	...	...
Wire, rod and			
bar (rolled			
or cold			
finished)....4112	B211	QQ-A-225/6
4119	...	...
4120	...	...
Wire, rod, bar,			
shapes and			
tube			
(extruded) ..4152	B221	QQ-A-200/3
4164	...	...
4165	...	...
Tube			
(extruded,			
seamless)	B241	...
Tube (drawn,			
seamless) ...4087	B210	WW-T-700/3
4088	...	MIL-T-50777
Tube			
(hydraulic) .4086	...	...
Rivet wire			
and rod	B316	QQ-A-430
Foil 4007	...	MIL-A-81596
Alclad 2024			
Sheet and			
plate 4034	B209	QQ-A-250/5
4040	...	...
4041	...	...
4042	...	...
4060	...	...
4061	...	...
4072	...	...
4073	...	...
4074	...	...
4075	...	...
4194	...	...
4195	...	...

UNS number. A92024
Government. See Table 20.
Foreign. Austria: Önorm AlCuMg2. Canada: CSA CG42. France: NF A-U4G1. Italy: UNI P-AlCu4.5MgMn; Alclad 2024, P-AlCu4.5MgMn placc. Spain: UNE L-314. Germany: DIN AlCuMg2

Chemical Composition

Composition limits. 0.05 max Si;

Table 19 (continued)

Temper	Tensile strength Minimum MPa	ksi	Maximum MPa	ksi	Yield strength Minimum MPa	ksi	Elonga-tion(a), %
6.001 – 7.000 in. thick							
Longitudinal,							
long transverse	415	60	...	...	360	52	2 – 7
Short transverse	400	58	...	...	360	52	1
7.001 – 8.000 in. thick							
Longitudinal,							
long transverse	405	59	...	...	350	51	2 – 7
Short transverse	395	57	...	...	350	51	1
T652							
Up thru 2.000 in. thick							
Longitudinal,							
long transverse	450	65	...	...	385	56	3 – 8
2.001 – 3.000 in. thick							
Longitudinal	440	64	...	...	385	56	8
Long transverse	440	64	...	...	380	55	3
Short transverse	425	62	...	...	360	52	2
3.001 – 4.000 in. thick							
Longitudinal,							
long transverse	435	63	...	...	380	55	3 – 8
Short transverse	420	61	...	...	350	51	2
4.001 – 5.000 in. thick							
Longitudinal,							
long transverse	425	62	...	...	370	54	2 – 7
Short transverse	415	60	...	...	345	50	1
5.001 – 6.000 in. thick							
Longitudinal,							
long transverse	420	61	...	...	365	53	2 – 7
Short transverse	405	59	...	...	345	50	1
6.001 – 7.000 in. thick							
Longitudinal,							
long transverse	415	60	...	...	360	52	2 – 6
Short transverse	400	58	...	...	340	49	1
7.001 – 8.000 in. thick							
Longitudinal,							
long transverse	405	59	...	...	350	51	2 – 6
Short transverse	395	57	...	...	331	48	1
Rolled Rings, T6, T652							
Up thru 2.500 in. thick							
Tangential	450	65	...	...	380	55	7
Axial	425	62	...	...	380	55	3
Radial	415	60	...	...	380	52	2
2.501 – 3.000 in. thick							
Tangential	450	65	...	...	380	55	6
Axial	425	62	...	...	360	52	2

(a) In 2 in. or 4d, where d is diameter of reduced section of tensile test specimen. Where a range of values appears in this column, specified minimum elongation varies with thickness of the mill product. (b) Upon artificial aging, T451 temper material develops properties applicable to T651 temper. (c) 6% elongation for products over 0.750 in. in diameter or thickness and over 25 thru 32 in.² in cross-sectional area. (d) Value slightly lower for material over 25 thru 32 in.² in cross-sectional area. (e) Test bar machined from sample forging. (f) 16% for test bar taken from separately forged coupon. (g) 8% for test bar taken from separately forged coupon. (h) 2% for forgings over 1 thru 2 in. thick.

0.50 max Fe; 3.8 to 4.9 Cu; 0.30 to 0.9 Mn; 1.2 to 1.8 Mg; 0.10 max Cr; 0.25 max Zn; 0.15 max Ti; 0.05 max others (each); 0.15 max others (total); rem Al. Alclad 2024: 1230 cladding— 99.30 min Al; 0.7 max Si + Fe; 0.10 max Cu; 0.05 max Mn; 0.05 max Mg; 0.10 max Zn; 0.05 max V; 0.03 max Ti; 0.03 max others (each)

Applications

Typical uses. Aircraft structures, rivets, hardware, truck wheels, screw machine products and other miscellaneous structural applications

Mechanical Properties

Tensile properties. See Tables 21, 22 and 23.
Shear strength. See Table 22.
Hardness. See Table 22.
Poisson's ratio. 0.33 at 20 °C (68 °F)
Elastic modulus. Tension, 72.4 GPa (10.5 × 10⁶ psi); shear, 28.0 GPa (4.0 × 10⁶ psi); compression, 73.8 GPa (10.7 × 10⁶ psi)
Fatigue strength. See Table 22.

Mass Characteristics

Density. 2.77 Mg/m³ (0.100 lb/in.³) at 20 °C (68 °F)

Thermal Properties

Liquidus temperature. 638 °C (1180 °F)
Solidus temperature. 502 °C (935 °F)
Incipient melting temperature. 502 °C (935 °F)
Coefficient of thermal expansion. Linear:

Temperature range °C	°F	Average coefficient μm/ m · K	μin./ in. · °F
−50 – +20	−58 – +68	21.1	11.7
20 – 100	68 – 212	22.9	12.7
20 – 200	68 – 392	23.8	13.2
20 – 300	68 – 572	24.7	13.7

Volumetric: 66.0 × 10⁻⁶ m³/m³·K (3.67 in.³/in.³·°F) at 20 °C (68 °F)
Specific heat. 875 J/kg·K (0.209 Btu/lb·°F) at 20 °C (68 °F)
Thermal conductivity:

Temper	Conductivity W/ m · K	Btu/ ft · h · °F
O	190	110
T3, T36, T351, T361,T4	120	69
T6, T81, T851, T861	151	88

Electrical Properties

Electrical conductivity. Volumetric, at 20 °C (68 °F):

Temper	Conductivity, % IACS
O	50
T3, T36, T351, T361, T4 ...	30
T6, T81, T851, T861	38

Electrical resistivity:

Temper	Resistivity, nΩ · m
O	34
T3, T36, T351, T361, T4	57
T6, T81, T851, T861	45

Temperature coefficient, 0.1 nΩ·m per K at 20 °C (68 °F)

Table 21 Typical tensile properties of alloy 2024

Temper	Temperature °C	°F	Tensile strength MPa	ksi	Yield strength(a) MPa	ksi	Elongation, %
T3 (sheet)	−196	−320	586	85	427	62	18
	−80	−112	503	73	359	52	17
	−28	−18	496	72	352	51	17
	24	75	483	70	345	50	17
	100	212	455	66	331	48	16
	149	300	379	55	310	45	11
	204	400	186	27	138	20	23
	260	500	76	11	62	9	55
	316	600	52	7.5	41	6	75
	371	700	34	5	28	4	100
T4, T351 (plate)	−196	−320	579	84	421	61	19
	−80	−112	490	71	338	49	19
	−28	−18	476	69	324	47	19
	24	75	469	68	324	47	19
	100	212	434	63	310	45	19
	149	300	310	45	248	36	17
	204	400	179	26	131	19	27
	260	500	76	11	62	9	55
	316	600	52	7.5	41	6	75
	371	700	34	5	28	4	100
T6, T651	−196	−320	579	84	469	68	11
	−80	−112	496	72	407	59	10
	−28	−18	483	70	400	58	10
	24	75	476	69	393	57	10
	100	212	448	65	372	54	10
	149	300	310	45	248	36	17
	204	400	179	26	131	19	27
	260	500	76	11	62	9	55
	316	600	52	7.5	41	6	75
	371	700	34	5	28	4	100
T81, T851	−196	−320	586	85	538	78	8
	−80	−112	510	74	476	69	7
	−28	−18	503	73	469	68	7
	24	75	483	70	448	65	7
	100	212	455	66	427	62	8
	149	300	379	55	338	49	11
	204	400	186	27	138	20	23
	260	500	76	11	62	9	55
	316	600	52	7.5	41	6	75
	371	700	34	5	28	4	100
T861	−196	−320	634	92	586	85	5
	−80	−112	558	81	531	77	5
	−28	−18	538	78	510	74	5
	24	75	517	75	490	71	5
	100	212	483	70	462	67	6
	149	300	372	54	331	48	11
	204	400	145	21	117	17	28
	260	500	76	11	62	9	55
	316	600	52	7.5	41	6	75
	371	700	34	5	28	4	100

(a) 0.2% offset.

Electrolytic solution potential. At 25 °C (77 °F) and vs 0.1N calomel electrode in an aqueous solution containing 53 g NaCl plus 3 g H_2O_2 per litre.

Temper	Volts
T3, T4, T361	−0.68
T6, T81, T861	−0.80
Alclad 2024	−0.83

Fabrication Characteristics

Annealing temperature. 413 °C (775 °F)

Solution temperature. 493 °C (920 °F)

Aging temperature. T6 and T8 tempers: 191 °C (375 °F) for 8 to 16 h at temperature

2036
2.6Cu-0.45Mg-0.25Mn

Specifications

UNS number. A92036

Chemical Composition

Composition limits. 0.50 max Si; 0.50 max Fe; 2.2 max Cu; 0.10 to 0.40 Mn; 0.30 to 0.6 Mg; 0.10 max Cr; 0.25 max Zn; 0.15 max Ti; 0.05 max others (each); 0.15 max others (total); rem Al

Applications

Typical uses. Sheet for auto body panels

Mechanical Properties

Tensile properties. Typical, for 0.64 to 3.18 mm (0.025 to 0.125 in.) flat sheet, T4 temper: tensile strength, 340 MPa (49 ksi); yield strength, 195 MPa (28 ksi); elongation, 24% in 50 mm or 2 in. Minimum, for 0.64 to 3.18 mm flat sheet, T4 temper: tensile strength, 290 MPa (42 ksi); yield strength, 160 MPa (23 ksi); elongation, 20% in 50 mm or 2 in.

Hardness. Typical, T4 temper: 80 HR15T

Strain-hardening exponent. 0.23

Elastic modulus. Tension, 70.3 GPa (10.2 × 10^6 ksi); compression, 71.7 GPa (10.4 × 10^6 ksi)

Fatigue strength. Typical, T4 temper: 124 MPa (18 ksi) at 10^7 cycles for flat sheet tested in reversed flexure

Mass Characteristics

Density. 2.75 Mg/m^3 (0.099 lb/in.^{3}) at 20 °C (68 °F)

Thermal Properties

Liquidus temperature. 649 °C (1200 °F)
Solidus temperature. 554 °C (1030 °F)
Incipient melting temperature. 510 °C (950 °F)
Coefficient of thermal expansion. Linear:

Temperature range		Average coefficient	
°C	°F	μm/ m·K	μin/ in.·°F
−50–+20	−58–+68	21.6	12.0
20–100	68–212	23.4	13.0
20–200	68–392	24.3	13.5
20–300	68–572	25.2	14.0

Volumetric: 67.5×10^{-6} m^3/m^3·K (3.75×10^{-5} in.^{3}/in.^{3}·°F) at 20 °C (68 °F)
Specific heat. 882 J/kg·K (0.211 Btu/lb·°F) at 20 °C (68 °F)
Thermal conductivity. At 20 °C (68 °F): O temper, 198 W/m·K (114 Btu/ft·h·°F); T4 temper, 159 W/m·K (91.8 Btu/ft·h·°F)

Electrical Properties

Electrical conductivity. Volumetric, at 20 °C (68 °F): O temper, 52% IACS; T4 temper, 41%
Electrical resistivity. At 20 °C (68 °F): O temper, 33.2 nΩ·m; T4 temper, 42.1 nΩ·m. Temperature coefficient. At 20 °C (68 °F): O and T4 tempers, 0.1 nΩ·m per K
Electrolytic solution potential. At 25 °C (77 °F): −0.75 V vs 0.1N calomel electrode in an aqueous solution containing 53 g NaCl plus 3 g H$_2$O$_2$ per litre

Fabrication Characteristics

Weldability. Arc welding with inert gas limited due to crack sensitivity, loss of mechanical properties, and/or loss in resistance to corrosion. When used for automotive parts, can be resistance welded with very good results.
Annealing temperature. 385°C (725 °F); hold 2 to 3 h at temperature for sheet
Solution temperature. 499 °C (930 °F)

Table 22 Typical mechanical properties of alloy 2024

Temper	Tensile strength MPa	ksi	Yield strength MPa	ksi	Elongation(a), %	Hardness(b), HB	Shear strength MPa	ksi	Fatigue strength(c) MPa	ksi
O	185	27	75	11	20	47	125	18	90	13
T3	485	70	345	50	18	120	285	41	140	20
T4, T351	470	68	325	47	20	120	285	41	140	20
T361	495	72	395	57	13	130	290	42	125	18
Alclad 2024										
O	180	26	75	11	20	···	125	18	···	···
T3	450	65	310	45	18	···	275	40	···	···
T4, T351	440	64	290	42	19	···	275	40	···	···
T361	460	67	365	53	11	···	285	41	···	···
T81, T851	450	65	415	60	6	···	275	40	···	···
T861	485	70	455	66	6	···	290	42	···	···

(a) 1.6-mm ($^1/_{16}$-in.) thick specimen. (b) 500-kg load; 10-mm ball. (c) At 5×10^8 cycles of completely reversed stress; R. R. Moore type test.

Table 23 Tensile property limits for alloy 2024

Temper	Tensile strength (min) MPa	ksi	Yield strength (min) MPa	ksi	Elongation (min)(a), %
Sheet and Plate					
O	220 (max)	32 (max)	95 (max)	14 (max)	12
T42					
0.010-0.499 in. thick	425	62	260	38	12-15
0.500-1.000 in. thick	420	61	260	38	8
1.001-2.000 in. thick	415	60	260	38	6-7
2.001-3.000 in. thick	400	58	260	38	4
T62					
0.010-0.499 in. thick	440	64	345	50	5
0.500-3.000 in. thick	435	63	345	50	5
T361					
0.020-0.062 in. thick	460	67	345	50	8
0.063-0.249 in. thick	470	68	350	51	9
0.250-0.500 in. thick	455	66	340	49	9-10
T861					
0.020-0.062 in. thick	485	70	425	62	3
0.063-0.249 in. thick	490	71	455	66	4
0.250-0.499 in. thick	485	70	440	64	4
Alclad O					
0.008-0.062 in. thick	205 (max)	30 (max)	95 (max)	14 (max)	10-12
0.063-1.750 in. thick(b)	220 (max)	32 (max)	95 (max)	14 (max)	12
Alclad T42					
0.008-0.009 in. thick	380	55	235	34	10
0.010-0.062 in. thick	395	57	235	34	12-15
0.063-0.499 in. thick	415	60	250	36	12-15
0.500-1.000 in. thick(b)	420	61	260	38	8
1.001-2.000 in. thick(b)	415	60	260	38	6-7
2.001-3.000 in. thick(b)	400	58	260	38	4
Alclad T62					
0.010-0.062 in. thick	415	60	325	47	5
0.063-0.499 in. thick	425	62	340	49	5
Alclad T361					
0.020-0.062 in. thick	420	61	325	47	8
0.063-0.499 in. thick	440	64	330	48	9
0.500 in. thick(b)	445	66	340	49	10
Alclad T861					
0.020-0.062 in. thick	440	64	400	58	3
0.063-0.249 in. thick	475	69	440	64	4
0.250-0.499 in. thick	470	68	425	62	4
0.500 in. thick(b)	485	70	440	64	4

(continued)

Table 23 (continued)

Temper	Tensile Strength (min) MPa	ksi	Yield strength (min) MPa	ksi	Elongation (min)(a), %
Flat Sheet					
T3					
0.008-0.128 in. thick	435	63	290	42	10-15
0.129-0.249 in. thick	440	64	290	42	15
T81	460	67	400	58	5
Alclad T3					
0.008-0.009 in. thick	400	58	270	39	10
0.010-0.062 in. thick	405	59	270	39	12-15
0.063-0.128 in. thick	420	61	275	40	15
0.129-0.249 in. thick	425	62	275	40	15
T81					
0.010-0.062 in. thick	425	62	370	54	5
0.063-0.249 in. thick	450	65	385	56	5
Sheet					
T72	415	60	315	46	5
Alclad T72					
0.010-0.062 in. thick	385	56	295	43	5
0.063-0.249 in. thick	400	58	310	45	5
Coiled Sheet					
T4	425	62	275	40	12-15
Alclad T4					
0.010-0.060 in. thick	400	58	250	36	12-15
0.063-0.128 in. thick	420	61	260	38	15
Plate					
T351					
0.250-0.499 in. thick	440	64	290	42	12
0.500-1.000 in. thick	435	63	290	42	8
1.001-2.000 in. thick	425	62	290	42	6-7
2.001-3.000 in. thick	415	60	290	42	4
3.001-4.000 in. thick	395	57	285	41	4
T851					
0.250-0.499 in. thick	460	67	400	58	5
0.500-1.000 in. thick	455	66	400	58	5
1.001-1.499 in. thick	455	66	395	57	5
Alclad T351					
0.250-0.499 in. thick	425	62	275	40	12
0.500-1.000 in. thick(b)	435	63	290	42	8
1.001-2.000 in. thick(b)	425	62	290	42	6-7
2.001-3.000 in. thick(b)	415	60	290	42	4
3.001-4.000 in. thick(b)	395	57	285	41	4
Alclad T851					
0.250-0.499 in. thick	450	65	385	56	5
0.500-1.000 in. thick(b)	455	66	400	58	5
Wire, Rod, and Bar (Rolled or Cold Finished)					
O	240 (max)	35 (max)	...	...	16
T36	475	69	360	52	10
T4					
Up thru 0.499 in. thick or in diam	425	62	310(c)	45(c)	10
0.500-4.500 in. thick or in diam	425	62	290(c)	42(c)	10
4.501-6.500 in. thick or in diam	425	62	275(c)	40(c)	10
6.501-8.00 in. in diam	400	58	260	38	10

(continued)

2048
3.3Cu-1.5Mg-0.40Mn

Specifications

UNS number. A92048

Chemical Composition

Composition limits. 0.15 max Si; 0.20 max Fe; 2.8 to 3.8 Cu; 0.20 to 0.6 Mn; 1.2 to 1.8 Mg; 0.25 max Zn; 0.10 max Ti; 0.05 max others (each); 0.15 max others (total); rem Al

Applications

Typical uses. Sheet and plate in structural components for aerospace application and military equipment

Mechanical Properties

Tensile properties. See Table 24 and Fig. 4.
Shear strength. Longitudinal, 271 MPa (39.3 ksi); transverse, 270 MPa (39.2 ksi)
Compressive properties. See Table 24 and Fig. 5.
Elastic modulus. See Fig. 4 and 5.
Impact strength. Charpy V-notch: longitudinal, 10.3 J (7.6 ft·lb); transverse, 6.1 J (4.5 ft·lb)
Fatigue strength. See Table 24 and Fig. 6 to 9.
Plane-strain fracture toughness. LT crack orientation, 35.2 MPa$\sqrt{m}$ (32.0 ksi $\sqrt{in}$.); TL crack orientation, 31.9 MPa$\sqrt{m}$ (29.1 ksi$\sqrt{in}$.)
Creep-rupture characteristics. See Table 24 and Fig. 10.

Mass Characteristics

Density. 2.75 Mg/m^3 (0.099 lb/in.^{3}) at 20 °C (68 °F)

Thermal Properties

Coefficient of thermal expansion. Linear, 23.5 μm/m·K (13.0 μin./in.·°F) at 21 to 104 °C (70 to 220 °F)
Specific heat. 926 J/kg·K (0.221 Btu/lb·°F) at 100 °C (212 °F)
Thermal conductivity. T851 temper, 159 W/m·K (92 Btu/ft·h·°F)

Electrical Properties

Electrical conductivity. Volumetric, T851 temper: 42% IACS at 20 °C (68 °F)
Electrical resistivity. T851 temper, 40.3 nΩ·m at 20 °C (68 °F)

Table 23 (continued)

Temper	Tensile Strength (min) MPa	ksi	Yield strength (min) MPa	ksi	Elongation (min)(a), %
T42	425	62	275	40	10
T351	425	62	310	45	10
T6	425	62	345	50	5
T62	415	60	315	46	5
T851	455	66	400	58	5
Wire, Rod, Bar and Shapes (Extruded)					
O	240 (max)	35 (max)	130 (max)	19 (max)	12
T3, T3510, T3511:					
Up thru 0.249 in. thick or in diam	395	57	290	42	12
0.250-0.749 in. thick or in diam	415	60	305	44	12
0.750-1.499 in. thick or in diam	450	65	315	46	10
1.5000 and over in. thick or in diam:					
Up thru 25 in.² area	485	70	360	52	10
Over 25 thru 32 in.² area	470	68	330	48	8
T42	395	57	260	38	8-12
T81, T851, T8510, T8511					
0.050-0.249 in. thick or in diam	440	64	385	56	4
0.250-1.500 and over in. thick or in diam: area up thru 32 in.²	455	66	400	58	5
Extruded Tube					
O	240 (max)	35 (max)	130 (max)	19 (max)	12
T3, T3510, T3511					
Up thru 0.249 in. thick ..	395	57	290	42	10
0.250-0.749 in. thick	415	60	305	44	10
0.750-1.499 in. thick	450	65	315	46	10
1.500 and over in. thick:					
Area up thru 25 in.² ...	485	70	330	48	10
Area over 25 thru 32 in.²	470	68	315	46	8
T42	395	57	260	38	12-8
T81, T8510, T8511					
0.050-0.249 in. thick	440	64	385	56	4
0.250-1.500 and over; area up thru 32 in.²	455	66	400	58	5
Drawn Tube					
O	220 (max)	32 (max)	105 (max)	15 (max)	...
T3	440	64	290	42	10-16(e)
T42	440	64	275	40	10-16(e)
Rivet and Cold-heading Wire and Rod					
O	240 (max)	35 (max)	...	...	...
H13	220	32	...	...	...
	290 (max)	42 (max)	...	...	...
T4	425	62	275	40	10

(a) In 2 in. or 4d, where d is diameter of reduced section of tension-test specimen. Where a range of values appears in this column, the specified minimum elongation varies with thickness of the mill product. (b) For plate 0.500 in. or over in thickness, listed properties apply to core material only. Tensile and yield strengths of composite plate are slightly lower than the listed value, depending on thickness of the cladding. (c) Minimum yield strength of coiled wire and rod, 276 MPa (40 ksi). (d) Applicable to rod only. (e) Full section specimen; min elongation is 10 to 12% for cut-out specimen.

2124
4.4Cu-1.5Mg-0.6Mn

Specifications

AMS. 4101
ASTM. B209
UNS number. A92124
Government. QQ-A-250/29

Chemical Composition

Composition limits. 0.20 max Si; 0.30 max Fe; 3.8 to 4.9 Cu; 0.30 to 0.9 Mn; 1.2 to 1.8 Mg; 0.10 max Cr; 0.25 max Zn; 0.15 max Ti; 0.05 max others (each); 0.15 max others (total); rem Al
Consequence of exceeding impurity limits. Degrades fracture toughness

Applications

Typical uses. Plate in thicknesses of 1.500 through 6.000 in. for aircraft structures

Mechanical Properties

Tensile properties. See Tables 25 and 26.
Poisson's ratio. 0.33 at 20 °C (68 °F)
Elastic modulus. See Table 26.
Plane-strain fracture toughness. T851 temper, plate: LT, 31.9 MPa $\sqrt{m}$ (29.0 ksi $\sqrt{in.}$); TL, 27.5 MPa $\sqrt{m}$ (25.0 ksi $\sqrt{in.}$); SL, 24.2 MPa $\sqrt{m}$ (22.0 ksi $\sqrt{in.}$)
Creep-rupture characteristics. See Table 27.

Mass Characteristics

Density. 2.77 Mg/m³ (0.100 lb/in.³) at 20 °C (68 °F)

Thermal Properties

Liquidus temperature. 638 °C (1180 °F)
Solidus temperature. 502 °C (935 °F)
Incipient melting temperature. 502 °C (935 °F)
Coefficient of thermal expansion. Linear:

Temperature range °C	°F	Average coefficient $\mu m/m \cdot K$	$\mu in./in. \cdot °F$
−50 − +20	−58 − +68	21.1	11.7
20 − 100	68 − 212	22.9	12.7
20 − 200	68 − 392	23.8	13.2
20 − 300	68 − 572	24.7	13.7

Volumetric: 66.0 × 10⁻⁶ m³/m³·K (3.6 × 10⁻⁵ in.³/in.³·°F) at 20 °C (68 °F)

Table 24 Typical mechanical properties of alloy 2048 plate, 75 mm (3 in.) thick

	Room temperature		120 °C (250 °F)		175 °C (350 °F)		260 °C (500 °F)	
Tensile Strength, MPa (ksi)								
Longitudinal	457	(66)	414	(60)	354	(51)	234	(34)
Transverse	465	(67)	414	(60)	347	(53)	230	(33)
Short transverse	463	(67)	...		...	...	...	...
Yield Strength, MPa (ksi)								
Longitudinal	416	(60)	392	(57)	338	(49)	219	(32)
Transverse	420	(61)	388	(56)	336	(49)	218	(32)
Short transverse	406	(59)	...		...	...	...	...
Elongation, %								
Longitudinal	8		13		14		10	
Transverse	7		13				8	
Short transverse	6		...		...		...	
Reduction in Area, %								
Longitudinal	16		32		37		23	
Transverse	12		28		34		15	
Short transverse	9		...		...		...	
Compressive Yield Strength, MPa (ksi)								
Longitudinal	420	(61)	391	(57)	349	(51)	243	(35)
Transverse	418	(61)	386	(56)	352	(51)	227	(33)
Elastic Moduli, GPa (10⁶ psi)								
In tension:								
Longitudinal	70.3	(10)	68.3	(10)	64	(9)	57	(8)
Transverse	72.4	(11)	67.6	(10)	64	(9)	53	(8)
Short transverse	76.5	(11)	...	...	...	...	...	...
In compression:								
Longitudinal	77.9	(11)	70.3	(10)	66	(10)	65	(9)
Transverse	76.5	(11)	71.0	(10)	67	(10)	66	(10)
Axial Fatigue (Longitudinal), MPa (ksi)								
Unnotched, R = 0.1:								
10³ cycles	469	(68)	469	(68)	469	(68)	...	...
10⁵ cycles	262	(38)	255	(37)	241	(35)	...	...
10⁷ cycles	221	(32)	193	(28)	172	(25)	...	...
Notched, K_t = 3.0, R = 0.1:								
10³ cycles	372	(54)	372	(54)	344	(50)	...	...
10⁵ cycles	152	(22)	145	(21)	131	(19)	...	...
10⁷ cycles	110	(16)	97	(14)	82	(12)	...	...
Creep Strength (Longitudinal)(a), MPa (ksi)								
100 h	...		303	(44)	241	(35)	59	(9)
1000 h	...		283	(41)	131	(19)	31	(5)
Rupture Strength (Longitudinal), MPa (ksi)								
100 h	...		345	(50)	269	(39)	90	(13)
1000 h	...		324	(47)	221	(32)	59	(9)

(a) Stress to produce 0.2% plastic extension in the indicated time.

Specific heat. 882 J/kg·K (0.210 Btu/lb·°F) at 20 °C (68 °F)
Thermal conductivity. At 20 °C (68 °F): O temper, 191 W/m·K (110 Btu/ft·h·°F); T851, 152 W/m·K (87.8 Btu/ft·h·°F)

Electrical Properties

Electrical conductivity. Volumetric, at 20 °C (68 °F): O temper, 50% IACS; T851, 39% IACS

Electrical resistivity. At 20 °C (68 °F): O temper, 34.5 nΩ·m; T851, 44.2 nΩ·m. Temperature coefficient, O and T851 tempers: 0.1 nΩ·m per K at 20 °C (68 °F)

Electrolytic solution potential. T851 temper, −0.80 V vs 0.1N calomel electrode in an aqueous solution containing 53 g NaCl plus 3 g H_2O_2 per litre at 25 °C (77 °F)

Fabrication Characteristics

Annealing temperature. 413 °C (775 °F)
Solution temperature. 493 °C (920 °F)
Aging temperature. 191 °C (375 °F)

2218
4.0Cu-2.0Ni-1.5Mg

Specifications

AMS. Forgings and forging stock: 4142
SAE. J454
UNS. A92218
Government. Forgings and forging stock: QQ-A-367
Foreign. France: NF A-U4N. Spain: UNE L-315. Switzerland: VSM Al-Cu-Ni

Chemical Composition

Composition limits. 0.9 max Si; 1.0 max Fe; 3.5 to 4.5 Cu; 0.20 max Mn; 1.2 to 1.8 Mg; 0.10 max Cr; 1.7 to 2.3 Ni; 0.25 max Zn; 0.05 max others (each); 0.15 max others (total); rem Al

Applications

Typical uses. Forgings; aircraft and diesel engine pistons; aircraft engine cylinder heads; jet engine impellers and compressor rings

Mechanical Properties

Tensile properties. See Tables 28 and 29.
Shear strength. T72 temper, 205 MPa (30 ksi)
Compressive yield strength. Approximately same as tensile yield strength
Hardness. See Table 28.
Poisson's ratio. 0.33 at 20 °C (68 °F)
Elastic modulus. Tension, 74.4 GPa (10.8 × 10⁶ psi); shear, 27.5 GPa (4.0 × 10⁶ psi)
Fatigue strength. See Table 30.
Creep-rupture characteristics. See Table 31.

Mass Characteristics

Density. 2.80 Mg/m³ (0.101 lb/in.³) at 20 °C (68 °F)

Thermal Properties

Liquidus temperature. 635 °C (1175 °F)

Fig. 4 Typical tensile properties of alloy 2048-T851 plate

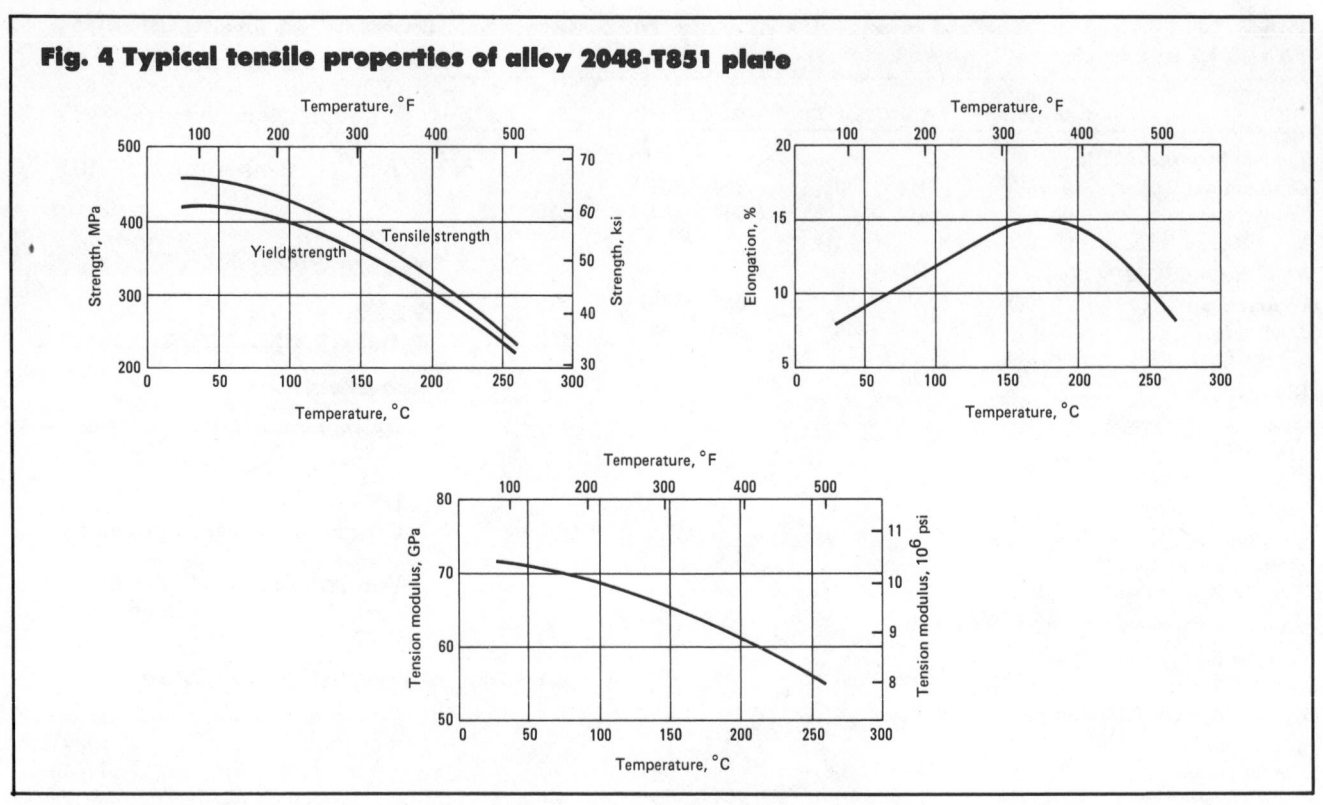

Fig. 5 Typical compressive properties of alloy 2048-T851 plate

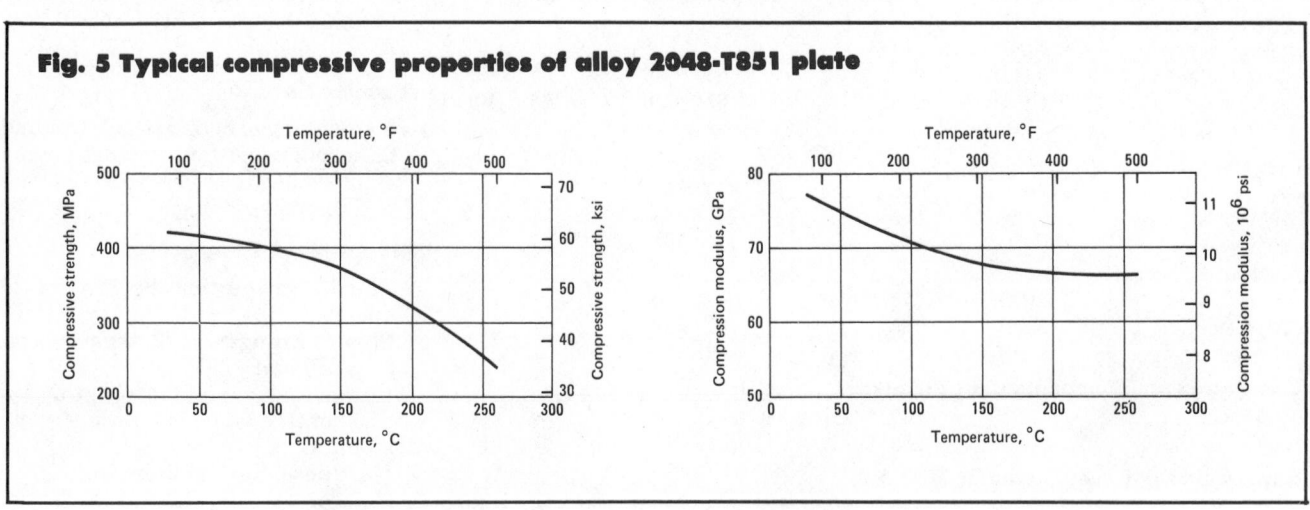

Fig. 6 Axial fatigue curves for unnotched specimens of alloy 2048-T851 plate

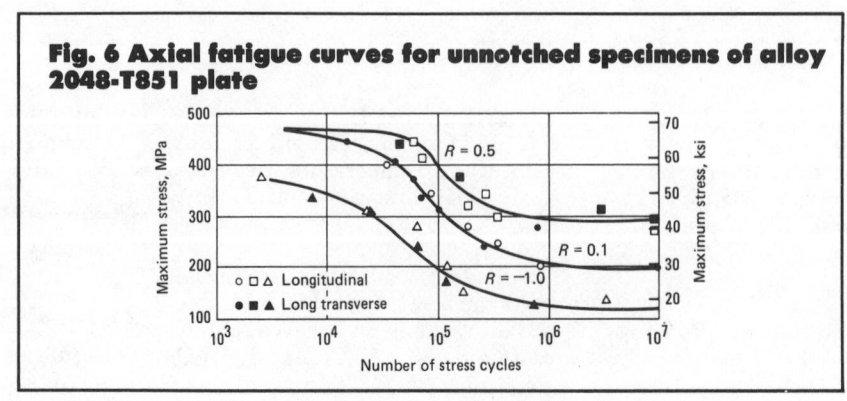

Fig. 7 Modified Goodman diagram for axial fatigue of unnotched specimens of alloy 2048-T851 plate

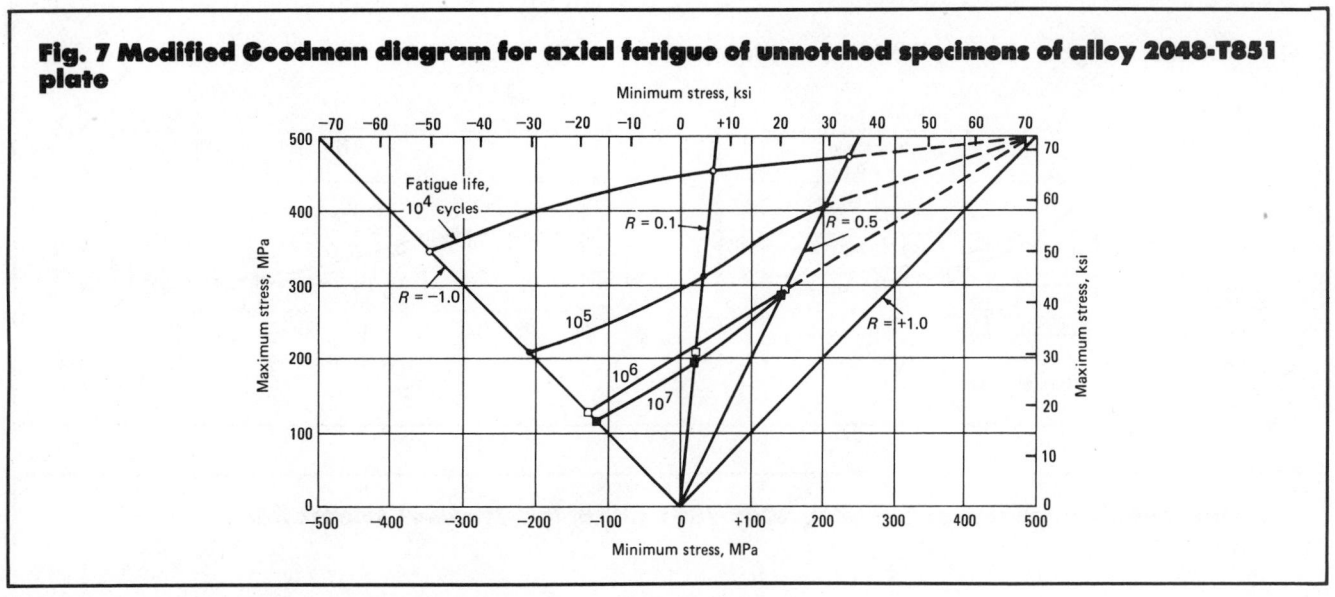

Fig. 8 Fatigue-crack propagation in alloy 2048-T851 plate

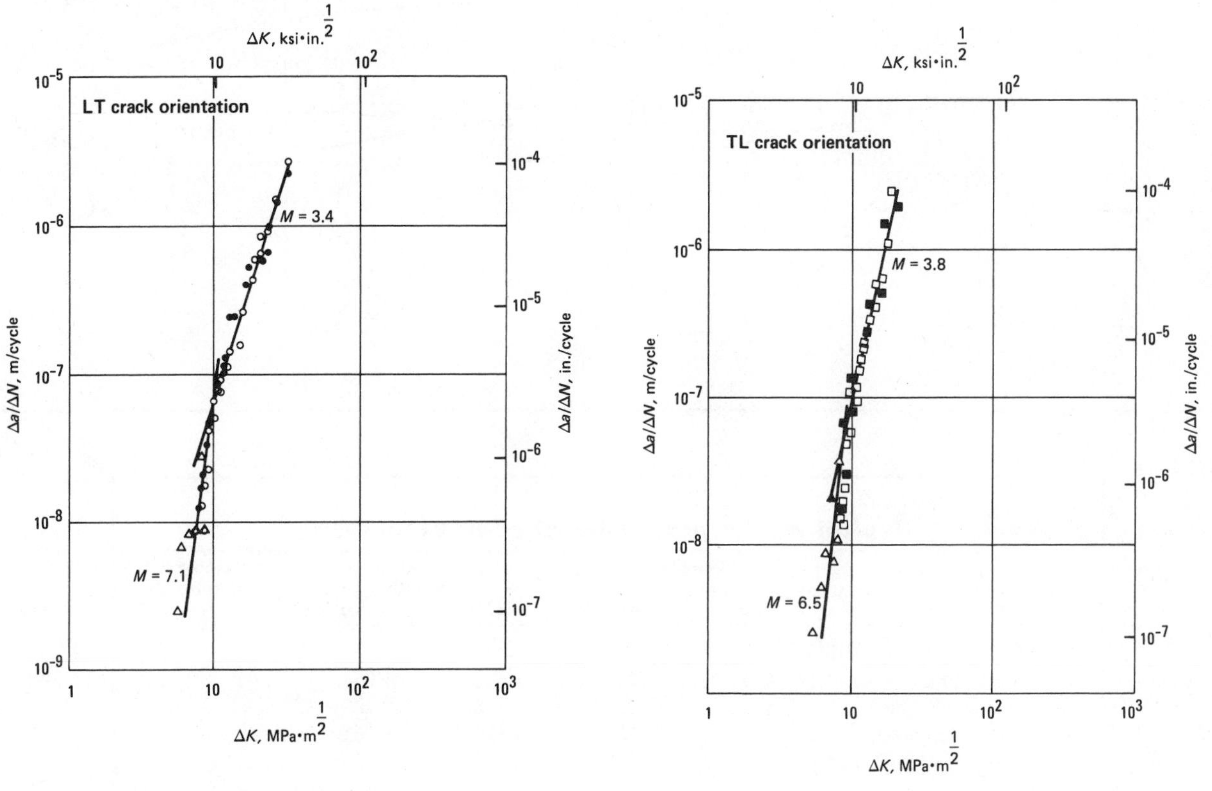

Fig. 9 Axial fatigue of alloy 2048-T851 plate

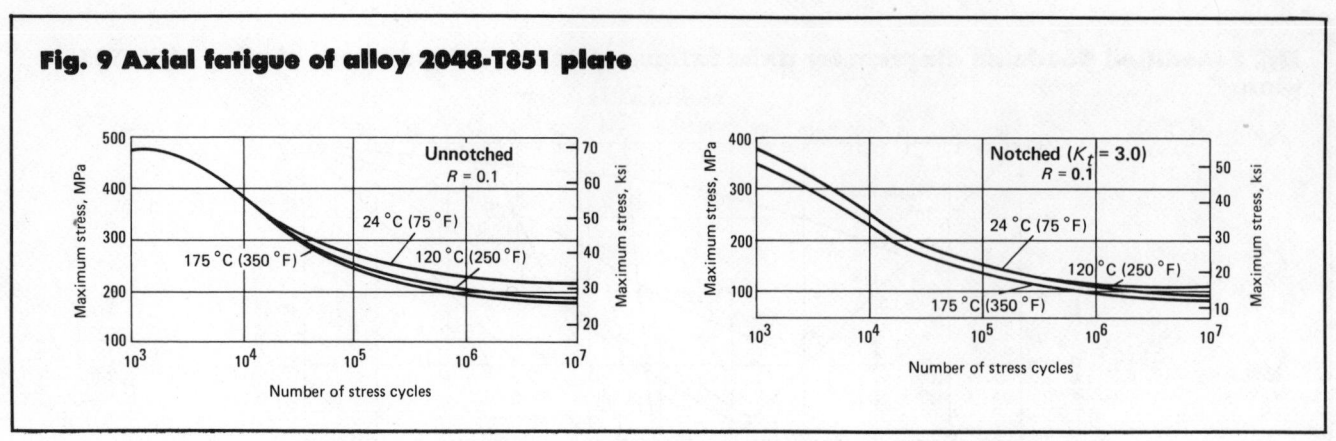

Fig. 10 Creep-rupture curves for alloy 2048-T851 plate, longitudinal orientation

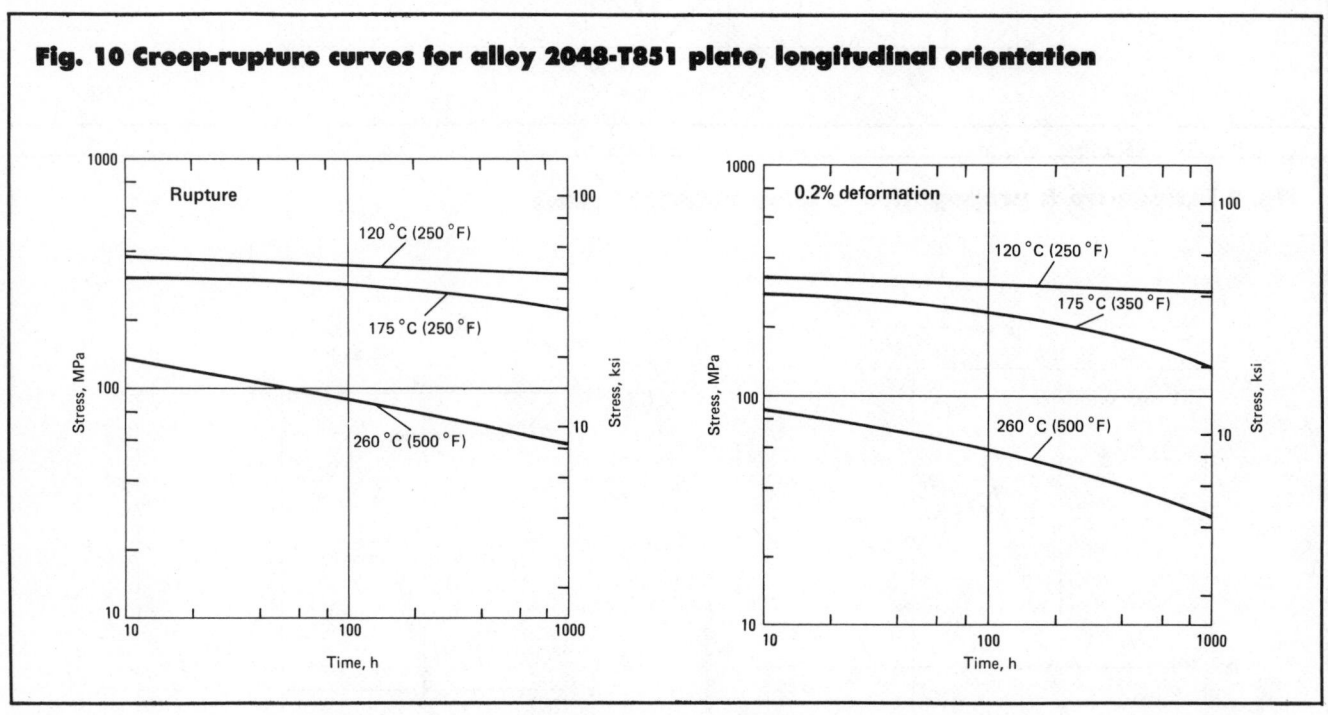

Table 25 Typical tensile properties of alloy 2124-T851

Specimen orientation	Tensile strength MPa	ksi	Yield strength MPa	ksi	Elongation, %
1.500 to 2.000 in. Thick					
Longitudinal	490	71	440	64	9
Long transverse	490	71	435	63	9
Short transverse	470	68	420	61	5
2.000 to 3.000 in. Thick					
Longitudinal	480	70	440	64	9
Long transverse	465	68	435	63	8
Short transverse	465	67	420	61	4

Table 26 Mechanical properties of alloy 2124-T851 plate, 70 mm (2.75 in.) thick

Temperature °C	°F	Time at temperature, h	At indicated temperature Tensile strength MPa	ksi	Yield strength MPa	ksi	Elonga-tion, %	Modulus of elasticity GPa	psi	At room temperature after heating Tensile strength MPa	ksi	Yield strength MPa	ksi	Elonga-tion, %
−269	−452	···	705	102	620	90	10	···	···	···	···	···	···	···
−195	−320	···	505	86	545	79	9	81	11.8	···	···	···	···	···
−80	−112	···	525	76	490	71	8	76	11.0	···	···	···	···	···
−28	−18	···	505	73	470	68	8	74	10.7	···	···	···	···	···
24	75	···	485	66	450	65	8	72	10.5	485	70	450	65	8
100	212	0.1 to 10 000	455	66	420	61	9	71	10.3	485	70	450	65	8
		100 000	450	65	415	60	9	71	10.3	···	···	···	···	···
150	300	0.1 to 10	415	60	395	57	10	68	9.9	185	70	450	65	8
		100	405	59	395	57	10	68	9.9	185	70	440	64	8
		1000	400	58	380	55	11	68	9.9	475	69	435	63	8
		10 000	370	54	330	48	13	68	9.9	460	67	405	59	8
		100 000	345	50	295	43	15	68	9.9	···	···	···	···	···
175	350	0.1	397	57	370	54	12	66	9.6	485	70	450	65	8
		0.5	385	56	365	53	12	66	9.6	485	70	450	65	8
		10	380	55	360	52	12	66	9.6	485	70	435	63	8
		100	360	52	340	49	12	66	9.6	470	68	420	61	8
		1000	330	48	305	44	14	66	9.6	455	66	400	58	8
		10 000	295	43	250	36	16	66	9.6	405	59	305	44	10
		100 000	220	32	180	26	23	66	9.6	···	···	···	···	···
205	400	0.1	365	53	340	49	13	63	9.2	···	···	···	···	···
		0.5	360	52	330	48	13	63	9.2	475	69	435	63	8
		10	330	48	310	45	14	63	9.2	460	67	405	59	8
		100	305	44	270	39	15	63	9.2	435	63	370	54	8
		1000	260	38	220	32	19	63	9.2	395	57	305	44	9
		10 000	185	27	140	20	28	63	9.2	290	42	165	24	12
		100 000	125	18	90	13	40	63	9.2	···	···	···	···	···
230	450	0.1	325	47	295	43	15	61	8.9	···	···	···	···	···
		0.5	310	45	285	41	15	61	8.9	470	68	425	62	8
		10	275	40	250	36	17	61	8.9	425	62	360	52	8
		100	235	34	200	29	20	61	8.9	370	54	275	40	10
		1000	170	25	125	18	30	61	8.9	290	42	170	25	12
		10 000	110	16	76	11	45	61	8.9	215	31	90	13	18
		100 000	83	12	59	8.5	55	61	8.9	···	···	···	···	···
260	500	0.1	270	39	240	35	17	59	8.5	···	···	···	···	···
		0.5	255	37	230	33	17	59	8.5	455	66	400	58	9
		10	205	30	185	27	20	59	8.5	385	56	295	43	10
		100	150	22	125	18	29	59	8.5	290	42	170	25	12
		1000	105	15	76	11	45	59	8.5	235	34	110	16	17
		10 000	76	11	55	8	60	59	8.5	195	28	83	12	22
		100 000	62	9	45	6.5	65	59	8.5	···	···	···	···	···
315	600	0.1	160	23	145	21	23	53	7.7	···	···	···	···	···
		0.5	140	20	115	17	26	53	7.7	340	49	230	33	10
		10	83	12	69	10	40	53	7.7	270	39	130	19	13
		100	69	10	55	8	50	53	7.7	240	35	105	15	17
		1000	62	9	45	6.5	65	53	7.7	215	31	83	12	22
		10 000	52	7.5	41	6	75	53	7.7	185	27	76	11	22
		100 000	45	6.5	38	5.5	80	53	7.7	···	···	···	···	···
370	700	0.1	76	11	69	10	35	45	6.5	···	···	···	···	···
		0.5	59	8.5	45	6.5	50	45	6.5	275	40	130	19	13
		10	48	7	34	5	75	45	6.5	255	37	105	15	18
		100	41	6	31	4.5	85	45	6.5	235	34	90	13	22
		1000	38	5.5	28	4.1	90	45	6.5	205	30	83	12	22
		10 000	34	5	28	4.1	95	45	6.5	185	27	76	11	22
		100 000	34	4.9	28	4.1	100	45	6.5	···	···	···	···	···
425	800	0.1	34	5	28	4.1	65	···	···	···	···	···	···	···
		0.5	30	4.4	24	3.5	85	···	···	···	···	···	···	···
480	900	···	16	2.3	12	1.8	65	···	···	···	···	···	···	···
535	1000	···	2	0.3	2	0.3	2	···	···	···	···	···	···	···

Table 27 Creep-rupture properties of alloy 2124-T851 plate, 70 mm thick

Temperature °C	°F	Time under stress, h	Rupture stress MPa	ksi	1.0% MPa	ksi	0.5% MPa	ksi	Stress for creep of 0.2% MPa	ksi	0.1% MPa	ksi
24	75	0.1	485	70	470	68	455	66	...	...	...	...
		1	475	69	460	67	450	65	...	...	...	...
		10	475	69	455	66	...	...	...	...	...	...
		100	470	68	...	...	...	...	...	...	...	...
		1000	470	68	...	...	...	...	...	...	...	...
100	212	0.1	455	66	435	63	425	62	420	61	415	60
		1	435	63	420	61	415	60	405	59	395	57
		10	420	61	405	59	395	57	380	55	370	54
		100	400	58	385	56	380	55	360	52	345	50
		1000	380	55	370	54	360	52	340	49	325	47
150	300	0.1	400	58	380	55	370	54	360	52	345	50
		1	370	54	360	52	345	50	330	48	310	45
		10	345	50	340	49	325	47	310	45	285	41
		100	315	46	310	45	305	44	290	42	250	36
		1000	290	42	285	41	270	39	235	34	205	30
		10 000	235	34	...	...	...	...	...	...	...	...
		100 000	170	25	...	...	...	...	...	...	...	...
175	350	0.1	365	53	345	50	340	49	325	47	305	44
		1	340	49	325	47	310	45	290	42	260	38
		10	305	44	290	42	275	40	255	37	230	33
		100	270	39	230	37	240	35	205	30	170	25
		1000	205	30	195	28	170	25	140	20	105	15
		10 000	145	21	...	...	...	...	...	...	...	...
		100 000	90	13	...	...	...	...	...	...	...	...
205	400	0.1	325	47	310	45	295	43	285	41	260	38
		1	290	42	275	40	270	39	250	36	220	32
		10	255	37	240	35	235	34	205	30	170	25
		100	200	29	185	27	180	26	150	22	115	17
		1000	130	19	125	18	115	17	90	13	52	7.5
		10 000	83	12	69	10	59	8.5	...	...	...	...
		100 000	52	7.5	...	...	...	...	...	...	...	...
230	450	0.1	275	40	260	38	250	36	235	34	215	31
		1	240	35	235	34	220	32	205	30	170	25
		10	195	28	185	27	180	26	150	22	115	17
		100	130	19	125	18	115	17	...	...	...	...
		1000	76	11	...	...	...	...	...	...	...	...
		10 000	48	7	...	...	...	...	...	...	...	...
		100 000	34	4.9	...	...	...	...	...	...	...	...
260	500	0.1	215	31	205	30	200	29	180	26	170	25
		1	185	27	180	26	170	25	150	22	130	19
		10	140	20	130	19	125	18	97	14	76	11
		100	83	12	76	11	69	10	...	...	...	...
		1000	48	7	...	...	...	...	...	...	...	...
		10 000	32	4.7	...	...	...	...	...	...	...	...
		100 000	23	3.4	...	...	...	...	...	...	...	...
315	600	0.1	110	16	110	16	105	15	97	14	90	13
		1	97	14	90	13	83	12	69	10	59	8.5
		10	59	8.5	55	8	52	7.5	45	6.5	38	5.5
		100	34	5	34	5	30	4.4	25	3.6	21	3
		1000	21	3	20	2.9	18	2.6	...	...	...	...

Solidus temperature. 532 °C (990 °F)

Incipient melting temperature. 504 °C (940 °F)

Coefficient of thermal expansion. Linear:

Temperature range		Average coefficient	
°C	°F	μm/ m·K	μin. in.·°F
−50−+20	−58−+68	20.7	11.5
20−100	68−212	22.4	12.4
20−200	68−392	23.3	12.9
20−300	68−572	24.2	13.4

Volumetric: 6.5×10^{-5} m³/m³·K (3.6×10^{-5} in./in.·°F) at 20 °C (68 °F)

Specific heat. 871 J/kg·K (0.208 Btu/lb·°F) at 20 °C (68 °F)

Thermal conductivity. At 20 °C (68 °F): T61 temper, 148 W/m·K (85.5 Btu/ft·h·°F); T72 temper, 155 W/m·K (89.6 Btu/ft·h·°F)

Electrical Properties

Electrical conductivity. Volumetric: T61 temper, 38% IACS; T72 temper, 40% IACS

Electrical resistivity. T61 temper, 45.0 nΩ·m; T72 temper, 43.0 nΩ·m. Temperature coefficient, T61 and T72 tempers: 0.1 nΩ·m per K at 20 °C (68 °F)

Fabrication Characteristics

Solution temperature. 510 °C (950 °F)

Aging temperature. T61 temper, 171 °C (340 °F) for 10 h at temperature; T72 temper, 238 °C (460 °F) for 6 h at temperature

Table 28 Typical mechanical properties of alloy 2218

Temper	Tensile strength MPa	ksi	Yield strength MPa	ksi	Elongation, %	Hardness(a), HB
T61	407	59	303	44	13	115
T71	345	50	276	40	11	105
T72	331	48	255	37	11	95

(a) 500-kg load; 10-mm diam ball.

Table 29 Tensile properties of alloy 2218-T61

Temperature °C	°F	Tensile strength(a) MPa	ksi	Yield strength(a) MPa	ksi	Elongation, %
−195	−320	495	72.0	360	52.0	15
−80	−112	420	61.0	310	45.0	14
−30	−18	405	59.0	305	44.0	13
+25	+75	405	59.0	305	44.0	13
100	212	385	56.0	290	42.0	15
150	300	285	41.0	240	35.0	17
205	400	150	22.0	110	16.0	30
260	500	70	10.0	40	6.0	70
315	600	40	5.5	20	3.0	85
370	700	30	4.0	15	2.5	100

(a) Lowest strength determined for representative lot during 10 000 h exposure at temperature under no load.

Table 30 Fatigue strength of alloy 2218-T61

Temperature °C	°F	No. of cycles	Fatigue strength MPa	ksi
23	75	10^5	270	39.0
		10^6	215	31.0
		10^7	170	25.0
		10^8	135	20.0
		5×10^8	125	18.0
150	300	10^5	...	...
		10^6	170	25.0
		10^7	130	19.0
		10^8	105	15.0
		5×10^8	100	14.0
205	400	10^5	...	...
		10^6	150	22.0
		10^7	105	15.0
		10^8	69	10.0
		5×10^8	59	8.5
260	500	10^5	145	21.0
		10^6	105	15.0
		10^7	72	10.0
		10^8	48	7.0
		5×10^8	41	6.0
315	600	10^5	90	13.0
		10^6	69	10.0
		10^7	48	7.0
		10^8	34	5.0
		5×10^8	31	4.5

(a) R. R. Moore type test.

Table 31 Creep-rupture properties of alloy 2218-T61

Temperature °C	°F	Time under stress, h	Rupture stress MPa	ksi	Stress for creep of: 1% MPa	ksi	0.5% MPa	ksi	0.2% MPa	ksi	0.1% MPa	ksi
100	212	Up to 1000	385	56.0	...	...	...	...	...	...	...	...
150	300	0.1	360	52.0	350	51.0	330	48.0	315	46.0	290	42.0
		1	350	51.0	345	50.0	325	47.0	310	45.0	285	41.0
		10	350	51.0	340	49.0	315	46.0	305	44.0	275	40.0
		100	330	48.0	325	47.0	310	45.0	295	43.0	230	33.0
		1000	290	42.0	290	42.0	290	42.0	270	39.0	140	20.0
205	400	0.1	325	47.0	315	46.0	290	42.0	275	40.0	255	37.0
		1	310	45.0	305	44.0	275	40.0	260	38.0	235	34.0
		10	255	37.0	250	36.0	240	35.0	220	32.0	160	23.0
		100	185	27.0	180	26.0	170	25.0	140	20.0	105	15.0
		1000	115	17.0	110	16.0	105	15.0	105	15.0	...	...
315	600	0.1	55	8.0	52	7.5	48	7.0	45	6.5	41	6
		1	48	7.0	48	7.0	45	6.5	41	6.0	38	5.5
		10	45	6.5	41	6.0	47	6.9	34	5.0	21	3.0
		100	27	3.9	23	3.4	21	3.0	14	2.1	...	...

2219, Alclad 2219
6.3Cu-0.3Mn-0.18Zr-0.10V-0.06Ti

Specifications

AMS. Sheet and plate: 4031. Extruded wire, rod, bar, shapes and tube: 4162, 4163. Forgings: 4143, 4144. Alclad 2219, sheet and plate: 4094, 4095, 4096.
ASTM. Sheet and plate: B209. Rolled or cold finished wire, rod, and bar: B211. Extruded wire, rod, bar, shapes and tube: B221. Extruded, seamless tube: B241. Forgings: B247. Alclad 2219, sheet and plate: B209
SAE. J454
UNS. A92219
Government. Sheet and plate: QQ-A-250/30. Forgings: QQ-A-367, MIL-A-22771. Armor plate: MIL-A-46118. Rivet wire and rod: QQ-A-430
Foreign. France: NF A-U6MT. United Kingdom: DTD 5004

Chemical Composition

Composition limits. 0.20 max Si; 0.30 max Fe; 5.8 to 6.8 Cu; 0.20 to 0.40 Mn; 0.02 max Mg; 0.10 max Zn; 0.05 to 0.15 V; 0.02 to 0.10 Ti; 0.10 to 0.25 Zr; 0.05 max others (each); 0.15 max others (total); rem Al. Alclad 2219: 7072 cladding—0.10 max Cu; 0.10 max Mn; 0.70 max Si + Fe; 0.80 − 1.3 Zn; 0.10 max Mg; 0.05 max others (each); 0.15 max others (total)

Applications

Typical uses. Welded space booster oxidizer and fuel tanks, supersonic aircraft skin and structure components. Readily weldable and useful for applications over temperature range of −269 to 300 °C (−452 to 600 °F). Has high fracture toughness, and the T8 temper is highly resistant to stress-corrosion cracking

Mechanical Properties

Tensile properties. See Tables 32 to 34.
Poisson's ratio. 0.33 at 20 °C (68 °F)

Elastic modulus. Tension, 73.8 GPa (10.7 × 10⁶ psi) compression, 75.2 GPa (10.9 × 10⁶ psi)

Fatigue strength. 103 MPa (15 ksi) at 5 × 10⁸ cycles, R. R. Moore type test

Table 32 Typical tensile properties of alloy 2219

Temper	Tensile strength MPa	ksi	Yield strength MPa	ksi	Elongation, %
O	172	25	76	11	18
T42	359	52	186	27	20
T31, T351	359	52	248	36	17
T37	393	57	317	46	11
T62	414	60	290	42	10
T81, T851	455	66	352	51	10
T87	476	69	393	57	10

Table 33 Tensile-property limits for alloy 2219

Temper	Min tensile strength MPa	ksi	Min yield strength MPa	ksi	Elongation(a), %
Sheet and Plate					
O	220 (max)	32 (max)	110 (max)	16 (max)	12
Alclad O	220 (max)	32 (max)	110 (max)	16 (max)	12
T31(b)					
0.020−0.039 in. thick	315	46	200	29	8
0.040−0.249 in. thick	315	46	195	28	10
Alclad T31(b)					
0.040−0.099 in. thick	290	42	170	25	10
0.100−0.249 in. thick	305	44	180	26	10
T351(c)					
0.250−2.000 in. thick	315	46	195	28	10
2.100−3.000 in. thick	305	44	195	28	10
3.100−4.000 in. thick	290	42	185	27	9
4.100−5.000 in. thick	275	40	180	26	9
5.001−6.000 in. thick	270	39	170	25	8
Alclad T351(c)	305	44	180	26	10
T37					
0.020−0.039 in. thick	340	49	260	38	6
0.040−2.500 in. thick	340	49	255	37	6
2.501−3.000 in. thick	325	47	250	36	6
3.001−4.000 in. thick	310	45	240	35	5
4.001−5.000 in. thick	295	43	235	34	4
Alclad T37					
0.040−0.099 in. thick	310	45	235	34	6
0.100−0.499 in. thick	325	47	240	35	6
T62	370	54	250	36	6-8
Alclad T62					
0.020−0.039 in. thick	305	44	200	29	6
0.040−0.099 in. thick	340	49	220	32	7
0.100−0.499 in. thick	350	51	235	34	7−8
0.500−2.000 in. thick(c)	370	54	250	36	7−8
T81(b)	425	62	315	46	6−7
Alclad T81(b)					
0.020−0.039 in. thick	340	49	255	37	6
0.040−0.099 in. thick	380	55	285	41	7
0.100−0.249 in. thick	400	58	295	43	7
T851(d)					
0.250−2.000 in. thick	425	62	315	46	7−8
2.001−3.000 in. thick	425	62	310	45	6
3.001−4.000 in. thick	415	60	305	44	5
4.001−5.000 in. thick	405	59	295	43	5
5.001−6.000 in. thick	395	57	290	42	4
Alclad T851(d)	400	58	290	42	8
T87					
0.020−0.249 in. thick	440	64	360	52	5−6
0.250−3.000 in. thick	440	64	350	51	6−7
3.001−4.000 in. thick	425	62	345	50	4
4.001−5.000 in. thick	420	61	340	49	3
Alclad T87					
0.040−0.099 in. thick	395	57	315	46	6
0.100−0.499 in. thick	415	60	330	48	6−7

(continued)

Table 33 (continued)

Temper	Min tensile strength MPa	ksi	Min yield strength MPa	ksi	Elongation(a), %
Wire, Rod, and Bar (Rolled or Cold Finished)					
T851					
0.500–2.000 in. thick or in diam 400		58	275	40	4
2.001–4.000 in. thick or in diam 395		57	270	39	4
Wire, Rod, Bar, and Shapes (Extruded)					
O ....221(max)		32(max)	125 (max)	18(max)	12
T31, T3510, T3511					
Up thru 0.499 in. thick or in diam 290		42	180	26	14
0.500–2.999 in. thick or in diam 310		45	185	27	14
T62 370		54	250	36	6
T81, T8510, T8511 400		58	290	42	6
Extruded Tube					
O ....220 (max)		32(max)	125 (max)	18(max)	12
T31, T3510, T3511					
Up thru 0.499 in. thick or in diam 290		42	180	26	14
0.500–2.999 in. thick or in diam 310		45	185	27	14
T62 370		54	250	36	6
T81, T8510, T8511 400		58	290	42	6
Die Forgings					
T6					
Specimen axis parallel to grain flow 400		58	260	38	8(e)(f)
Specimen axis not parallel to grain flow 385		56	250	36	4(e)
Hand Forgings(g)					
T6					
Longitudinal axis 400		58	275	40	6
Long transverse axis 380		55	255	37	4
Short transverse axis 365		53	240	35	2
Mechanical Property Limits					
T852					
Longitudinal axis 425		62	345	50	6
Long transverse axis 425		62	340	49	4
Short transverse axis 415		60	315	46	3
Rolled Rings(h)					
T6					
Tangential axis 385		56	275	40	6
Axial axis 380		55	255	37	4
Radial axis 365		53	240	35	2

(a) In 2 in. or 4d, where d is diameter of reduced section of tensile test specimen. Where a range of values appears in this column, specified minimum elongation varies with thickness of the mill product. (b) Sheet only. (c) For plate 12.7 mm (0.500 in.) or greater in thickness, property limits apply to core material only. Tensile and yield strengths of composite plate slightly lower depending on thickness of cladding. (d) Plate only. (e) Specimen taken from forging. (f) 10% for specimen taken from separately forged coupon. (g) Maximum cross-sectional area 256 in.² These properties not applicable to upset biscuit forgings or rolled rings. (h) Only applicable to rings having ratio of outside diameter to wall thickness equal to or greater than 10.

Creep-rupture characteristics. See Tables 35 and 36.

Mass Characteristics

Density. 2.84 Mg/m³ (0.103 lb/in.³) at 20 °C (68 °F)

Thermal Properties

Liquidus temperature. 643 °C (1190 °F)
Incipient melting temperature. 543 °C (1010 °F)
Coefficient of thermal expansion. Linear:

Temperature range °C	°F	Average coefficient μm/ m·K	μin./ in.·°F
−50– +20	−58– +68	20.8	11.5
20–100	68–212	22.5	12.5
20–200	68–392	23.4	13.0
20–300	68–572	24.4	13.6

Volumetric: 6.5×10^{-5} m³/m³·K (3.62×10^{-5} in.³/in.³·°F)
Specific heat. 864 J/kg·K (0.206 Btu/lb·°F) at 20 °C (68 °F)
Thermal conductivity. O temper, 170 W/m·K (98.2 Btu/ft·h·°F); T31, T37 tempers, 116 W/m·K (67.0 Btu/ft·h·°F); T62, T81, T87 tempers, 130 W/m·K (75.1 Btu/ft·h·°F)

Electrical Properties

Electrical conductivity. Volumetric, at 20 °C (68 °F): O temper, 44% IACS; T31, T37, T351 tempers, 28% IACS; T62, T81, T87, T851 tempers, 30% IACS
Electrical resistivity. At 20 °C (68 °F); O temper, 39 nΩ·m: T31, T37, T351 tempers, 62 nΩ·m: T62, T81, T87, T851 tempers, 57 nΩ·m. Temperature coefficient, all tempers: 0.1 nΩ·m per K at 20 °C (68 °F)
Electrolytic solution potential. T31, T37, T351 tempers, −0.64 V and T62, T81, T87, T851 tempers, −0.80 V vs 0.1N calomel electrode in an aqueous solution containing 53 g NaCl plus 3 g H₂O₂ per litre at 25 °C (77 °F)

Fabrication Characteristics

Annealing temperature. 413 °C (775 °F)
Solution temperature. 535 °C (995 °F)
Aging temperature. 165 to 190 °C (325 to 375 °F) from 18 to 36 h at temperature; appropriate combination of aging time and temperature is different for different tempers.

2319
5.3Cu-0.30Mn-0.18Zr-0.15Ti-0.10V

Specifications

UNS. A92319
Government. QQ-R-566, MIL-E-16053

Table 34 Typical tensile properties of alloy 2219 at various temperatures

Temper	Temperature °C	Temperature °F	Tensile strength(a) MPa	Tensile strength(a) ksi	Yield strength(a)(b) MPa	Yield strength(a)(b) ksi	Elongation, %
762	−196	−320	503	73	338	49	16
	−80	−112	434	63	303	44	13
	−28	−18	414	60	290	42	12
	24	75	400	58	276	40	12
	100	212	372	54	255	37	14
	149	300	310	45	227	33	17
	204	400	234	34	172	25	20
	260	500	186	27	133	20	21
	316	600	69	10	55	8	40
	371	700	30	4.4	26	3.7	75
T81, T851	−196	−320	572	83	421	61	15
	−80	−112	490	71	372	54	13
	−28	−18	476	69	359	52	12
	24	75	455	66	345	50	12
	100	212	414	60	324	47	15
	149	300	338	49	276	40	17
	204	400	248	36	200	29	20
	260	500	200	29	159	23	21
	316	600	48	7	41	6	55
	371	700	30	4.4	26	3.7	75

(a) Lowest strength for exposures up to 10 000 h at temperature under no load; test load applied at 5000 psi/min to yield strength and then at strain rate of 5% per min to fracture. (b) 0.2% offset.

Chemical Composition

Composition limits. 5.8 to 6.8 Cu; 0.20 to 0.40 Mn; 0.10 to 0.25 Zr; 0.10 to 0.20 Ti; 0.05 to 0.15 V; 0.20 max Si; 0.30 max Fe; 0.02 max Mg; 0.10 max Zn; 0.0008 max Be; 0.05 max others (each); 0.15 max others (total)

Applications

Typical uses. Electrodes and filler wire for welding 2219

Mass Characteristics

Density. 2.83 Mg/m³ (0.103 lb/in.³) at 20 °C (68 °F)

Thermal Properties

Liquidus temperature. 643 °C (1190 °F)
Incipient melting temperature. 543 °C (1010 °F)
Coefficient of thermal expansion. Linear:

Temperature range °C	Temperature range °F	Average coefficient μm/m · K	Average coefficient μin./in. · °F
−50– +20	−58–+68	20.8	11.5
20–100	68–212	22.5	12.5
20–200	68–392	23.4	13.0
20–300	68–572	24.4	13.6

Volumetric: 6.5 × 10⁻⁵ m³/m³·k (3.62 × 10⁻⁵ in.³/in.³·°F) at 20 °C (68 °F)

Specific heat. 864 J/kg·K (0.206 Btu/lb·°F) at 20 °C (68 °F)
Thermal conductivity. O temper: 170 W/m·k (98.2 Btu/ft·h·°F)

Electrical Properties

Electrical conductivity. Volumetric: O temper, 44% IACS at 20 °C (68 °F)
Electrical resistivity. O temper, 39 nΩ·m at 20 °C (68 °F)
Temperature coefficient. 2.94 × 10⁻³/k

Fabrication Characteristics

Annealing temperature. 413 °C (775 °F)

2618
2.3Cu-1.6Mg-1.1Fe-1.0Ni-0.18Si-0.07Ti

Specifications

AMS. Forgings and forging stock: 4132
ASTM. Forgings: B247
SAE. J454

Government. Forgings: QQ-A-367; MIL-A-22771
Foreign. France: NF A-U2GN. United Kingdom: BS H12

Chemical Composition

Composition limits. 0.10 to 0.25 Si; 0.9 to 1.3 Fe; 1.9 to 2.7 Cu; 1.3 to 1.8 Mg; 0.9 to 1.2 Ni; 0.10 max Zn; 0.04 to 0.10 Ti; 0.05 max others (each); 0.15 others (total); rem Al

Applications

Typical uses. Die and hand forgings. Pistons and rotating aircraft engine parts for operation at elevated temperatures. Tire molds

Mechanical Properties

Tensile properties. See Tables 37 and 38 and Fig. 11.
Shear strength. T61 temper, 260 MPa (38 ksi)
Compressive yield strength. Approximately the same as the tensile yield strength. See also Fig. 12.
Hardness. Die forgings, T61 temper: 115 HB min
Poisson's ratio. 0.33 at 20 °C (68 °F)
Elastic modulus. Tension, 74.4 GPa (10.8 × 10⁶ psi); shear, 28.0 GPa (4.0 × 10⁶ psi)
Fatigue strength. T61 temper, 125 MPa (18 ksi) at 5 × 10⁸ cycles; R. R. Moore type test
Creep-rupture characteristics. See Table 39.

Mass Characteristics

Density. 2.76 Mg/m³ (0.100 lb/in.³) at 20 °C (68 °F)

Thermal Properties

Liquidus temperature. 638 °C (1180 °F)
Solidus temperature. 549 °C (1020 °F)
Incipient melting temperature. 502 °C (935 °F)
Coefficient of thermal expansion. Linear:

Temperature range °C	Temperature range °F	Average coefficient μm/m · K	Average coefficient μin./in. · °F
−50– +20	−58–+68	20.6	11.4
20–100	68–212	22.3	12.4
20–200	68–392	23.2	12.9
20–300	68–572	24.1	13.4

Volumetric: 6.45 × 10⁻⁵ m³/m³·K (3.6 × 10⁻⁵ in.³/in.³·°F) at 20 °C (68 °F)

Table 35 Creep-rupture properties of alloy 2219-T851 plate

Temperature °C	°F	Time under stress, h	Rupture stress MPa	ksi	1.0% MPa	ksi	Stress for creep of: 0.5% MPa	ksi	0.2% MPa	ksi	0.1% MPa	ksi
24	75	0.1	455	66	435	63	415	60	365	53	350	51
		1	450	65	420	61	385	56	360	52	345	50
		10	435	63	400	58	365	53	345	50	330	48
		100	425	62	380	55	360	52	340	49	325	47
		1000	420	61	365	53	350	51	330	48	315	46
100	212	0.1	395	57	360	52	340	49	315	46	305	44
		1	370	54	340	49	325	47	305	44	285	41
		10	350	51	325	47	310	45	290	42	275	40
		100	330	48	310	45	295	43	275	40	270	39
		1000	315	46	295	43	285	41	270	39	260	38
150	300	0.1	340	49	305	44	295	43	275	40	260	38
		1	315	46	290	42	275	40	255	37	235	34
		10	290	42	270	39	255	37	235	34	205	30
		100	260	38	250	36	235	34	200	29	170	25
		1000	235	34	220	32	200	29	165	24	150	22
		10 000	205	30	...	...	...	...	...	...	...	...
		100 000	170	25	...	...	...	...	...	...	...	...
175	350	0.1	305	44	275	40	260	38	250	36	230	33
		1	275	40	255	37	240	35	220	32	200	29
		10	250	36	230	33	215	31	195	28	165	24
		100	220	32	200	29	185	27	160	23	130	19
		1000	185	27	170	25	160	23	140	20	105	15
		10 000	160	23	...	...	...	...	...	...	...	...
		100 000	130	19	...	...	...	...	...	...	...	...
205	400	0.1	270	39	240	35	235	34	215	31	195	28
		1	235	34	220	32	205	30	180	26	160	23
		10	205	30	195	28	180	26	150	22	130	19
		100	180	26	165	24	150	22	130	19	110	16
		1000	150	22	140	20	125	18	115	17	90	13
		10 000	125	18	125	18	...	...	...	...	...	...
		100 000	97	14	...	...	...	...	...	...	...	...
230	450	0.1	230	33	205	30	200	29	180	26	165	24
		1	200	29	185	27	170	25	150	22	140	20
		10	170	25	160	23	150	22	130	19	110	16
		100	150	22	140	20	130	19	110	16	90	13
		1000	125	18	115	17	110	16	90	13	69	10
		10 000	97	14	97	14	97	14	...	...	...	...
		100 000	66	9.5	...	...	...	...	...	...	...	...
260	500	0.1	180	26	170	25	165	24	160	23	145	21
		1	165	24	160	23	150	22	140	20	115	17
		10	150	22	140	20	130	19	110	16	90	13
		100	130	19	125	18	110	16	90	13	69	10
		1000	105	15	97	14	83	12	69	10	59	8.5
		10 000	69	10	69	10	...	...	...	...	...	...
		100 000	45	6.5	...	...	...	...	...	...	...	...
315	600	0.1	130	19	125	18	125	18	115	17	110	16
		1	115	17	115	17	110	16	105	15	90	13
		10	105	15	97	14	90	13	76	11	62	9
		100	69	10	69	10	62	9	52	7.5	38	5.5
		1000	41	6	41	6	38	5.5	28	4.1	23	3.4
		10 000	22	3.2	...	...	...	...	...	...	...	...
370	700	0.1	69	10	69	10	69	10	66	9.5	66	9.5
		1	62	9	62	9	59	8.5	45	6.5	32	4.7
		10	32	4.7	30	4.3	27	3.9	23	3.3	18	2.6
		100	22	3.2	20	2.9	18	2.6	13	1.9	...	...
		1000	14	2.1	13	1.9	11	1.6	...	...	...	...

Specific heat. 875 J/kg·K (0.209 Btu/lb·°F) at 20 °C (68 °F)

Thermal conductivity. T61 temper, 146 W/m·K (84 Btu/ft·h·°F) at 20 °C (68 °F)

Electrical Properties

Electrical conductivity. Volumetric, T61 temper, 37% IACS at 20 °C (68 °F)

Electrical resistivity. T61 temper, 47.0 nΩ·m at 20 °C (68 °F); temperature coefficient, T61 temper: 0.1 nΩ·m per K at 20 °C (68 °F)

Electrolytic solution potential. At 25 °C (77 °F): T61 temper, −0.80 V vs 0.1N calomel electrode in an aqueous solution containing 53 g NaCl plus 3 g H₂O₂ per litre

Fabrication Characteristics

Solution temperature. 529 °C (985 °F)

Aging temperature. T61, 199 °C (390 °F) for 20 h at temperature

Fig. 11 Influence of prolonged holding at elevated temperature on tensile properties of alloy 2618-T61 hand-forged billets

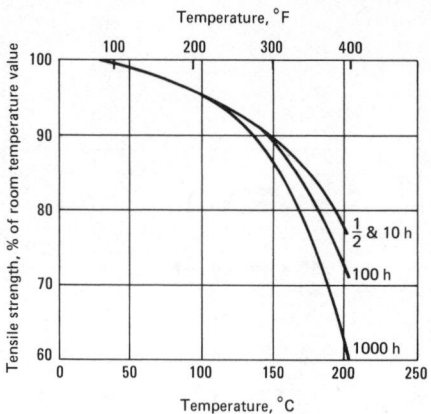

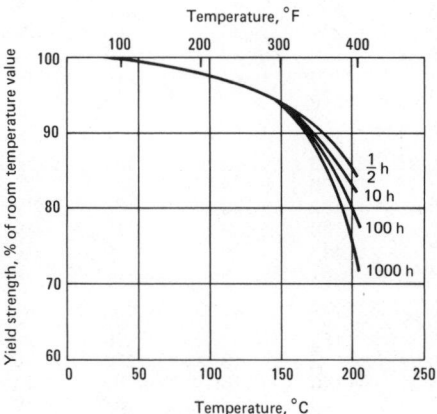

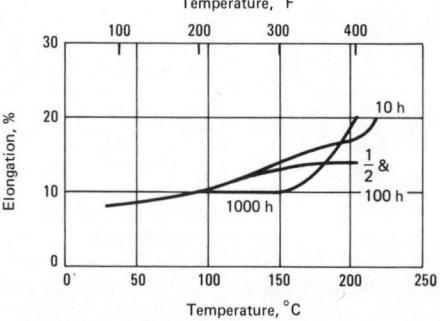

Properties determined at temperature after holding for the indicated time under no load. Tensile and yield strengths plotted as percentage of corresponding room-temperature value. Elongation plotted as value determined at temperature.

3003, Alclad 3003
1.2Mn-0.12Cu

Specifications

AMS. See Table 40.
ASME. See Table 40.
ASTM. See Table 40.
SAE. J454
UNS number. 3003: A93003.
Government. See Table 40.
Foreign. Canada: CSA MC10. France: NF A-M1. United Kingdom: BS N3. West Germany: DIN AlMn. ISO: AlMn1Cu

Chemical Composition

Composition limits. 3003: 0.6 max Si; 0.7 max Fe; 0.05 to 0.20 Cu; 1.0 to 1.5 Mn; 0.10 max Zn; 0.05 max others (each); 0.15 max others (total); rem Al. Alclad 3003: 7072 cladding - 0.10 max Cu; 0.10 max Mg; 0.10 max Mn; 0.7 max Fe plus Si; 0.8 to 1.3 Zn; 0.05 max others (each); 0.15 max others (total); rem Al

Applications

Typical uses. 3003: applications where good formability, very good resistance to corrosion or good weldability, or all three, are required, and where more strength is desired than is provided by unalloyed aluminum. Cooking utensils, food and chemical handling and storage equipment, tanks, trim in transportation equipment, lithographic sheet pressure vessels and piping. Alclad 3003: farm roofing and siding

Mechanical Properties

Tensile properties. See Tables 41 and 42. Directional characteristics: tensile strength and elongation of sheet in any of the H tempers are slightly lower in transverse direction.
Compressive yield strength. Approximately the same as tensile yield strength
Shear yield strength. Approximately 55% of the tensile strength
Hardness. See Table 41.
Poisson's ratio. 0.33 at 20 °C (68 °F)
Elastic modulus. Tension, 70 GPa (10.2 × 10⁶ psi); shear, 25 GPa (3.6 × 10⁶ psi)
Fatigue strength. See Table 41.

Table 36 Creep-rupture properties of alloy 2219-T87 plate

Temperature °C	°F	Time under stress, h	Rupture stress MPa	ksi	Stress for creep of: 1.0% MPa	ksi	0.5% MPa	ksi	0.2% MPa	ksi	0.1% MPa	ksi
24	75	0.1	460	67	450	65	420	61	385	56	370	54
		1	455	66	425	62	400	58	380	55	365	53
		10	450	65	405	59	385	56	370	54	360	52
		100	435	63	395	57	380	55	365	53	350	51
		1000	420	61	380	55	370	54	360	52	345	50
100	212	0.1	400	58	365	53	350	51	340	49	325	47
		1	380	55	345	50	340	49	325	47	310	45
		10	350	51	330	48	325	47	305	44	290	42
		100	330	48	315	46	310	45	290	42	260	38
		1000	315	46	305	44	295	43	260	38	240	35
150	300	0.1	345	50	315	46	310	45	290	42	270	39
		1	315	46	295	43	290	42	260	38	240	35
		10	290	42	275	40	260	38	235	34	205	30
		100	260	38	250	36	235	34	200	29	165	24
		1000	235	34	230	33	205	30	165	24	140	20
205	400	0.1	255	37	240	35	230	33	200	29	165	24
		1	230	33	215	31	200	29	165	24	140	20
		10	205	30	185	27	170	25	140	20	110	16
		100	180	26	165	24	145	21	115	17	97	14
		1000	150	22	145	21	130	19	97	14	83	12
230	450	0.1	206	30	195	28	180	26	165	24	140	20
		1	185	27	170	25	165	24	140	20	115	17
		10	170	25	160	23	145	21	115	17	97	14
		100	150	22	140	20	125	18	105	15	83	12
		1000	130	19	125	18	115	17	83	12	69	10
260	500	0.1	170	25	160	23	150	22	140	20	125	18
		1	160	23	145	21	140	20	125	18	105	15
		10	145	21	130	19	125	18	105	15	83	12
		100	125	18	115	17	110	16	90	13	69	10
		1000	105	15	105	15	97	14	69	10	59	8.5
315	600	0.1	115	17	110	16	105	15	97	14	83	12
		1	105	15	105	15	97	14	83	12	66	9.5
		10	90	13	83	12	76	11	62	9	52	7.5
		100	62	9	55	8	52	7.5	45	6.5	34	5
		1000	34	5	31	4.5	28	4	26	3.8	23	3.3
370	700	0.1	59	8.5	55	8	52	7.5	48	7	34	5
		1	48	7	45	6.5	41	6	32	4.7	18	2.6
		10	34	5	30	4.4	26	3.8	17	2.4	12	1.7
		100	23	3.4	20	2.9	17	2.5	11	1.6	8	1.2
		1000	17	2.4	13	1.9	11	1.6	8	1.2	6	0.9

Mass Characteristics

Density. 2.73 Mg/m³ (0.099 lb/in.³) at 20 °C (68 °F)

Thermal Properties

Liquidus temperature. 654 °C (1210 °F)

Solidus temperature. 643 °C (1190 °F)

Coefficient of thermal expansion. Linear:

Temperature range °C	°F	Average coefficient μm/m·K	μin/in.·°F
−50−+20	−58−+68	21.5	11.9
20−100	68−212	23.2	12.9
20−200	68−392	24.1	13.4
20−300	68−572	25.1	13.9

Volumetric: 67×10^{-6} m³/m³·K (3.72×10^{-5} in.³/in.³·°F) at 20 °C (68 °F)

Specific heat. 893 J/kg·K (0.213 Btu/lb·°F) at 20 °C (68 °F)

Thermal conductivity. At 20 °C (68 °F):

Temper	Conductivity W/m·K	Btu/ft·h·°F
O	193	112
H12	163	94.1
H14	159	91.9
H18	155	89.6

Electrical Properties

Electrical conductivity. Volumetric, at 20 °C (68 °F):

Temper	Conductivity, % IACS
O	50
H12	42
H14	41
H18	40

Electrical resistivity. At 20 °C (68 °F):

Temper	Resistivity, nΩ·m
O	34.0
H12	41.0
H14	42.0
H18	43.1

Temperature coefficient, all tempers: 0.1 nΩ·m per K at 20 °C (68 °F)

Electrolytic solution potential. 3003 and core of alclad 3003, −0.83

Table 37 Tensile properties of alloy 2618-T61

Product and orientation	Tensile strength MPa	ksi	Yield strength MPa	ksi	Elongation(a), %
Typical					
All products	441	64	372	54	10(b)
Property limits					
Die forgings, thickness up thru 4 in.(c)					
Axis parallel to grain flow	400	58	310	45	4(d)(e)
Axis not parallel to grain flow	380	55	290	42	4(d)
Hand forgings					
Thickness up thru 2.000 in.(c)(f)					
Longitudinal	400	58	325	47	7
Long transverse	380	55	290	42	5
Short transverse	360	52	290	42	4
2.001 to 3.000 in.					
Longitudinal	395	57	315	46	7
Long transverse	380	55	290	42	5
Short transverse	360	52	290	42	4
3.001 to 4.000 in.					
Longitudinal	385	56	310	45	7
Long transverse	365	53	275	40	5
Short transverse	350	51	270	39	4
Rolled rings, thickness up thru 2.500 in.(g)					
Tangential	380	55	285	41	6
Axial	380	55	285	41	5

(a) In 2 in. or 4 d, where d is diameter of reduced section of tensile test specimen. (b) 12.5 mm (½ in.) diameter specimen. (c) Properties also apply to forgings machined prior to heat treatment, provided machined thickness is not less than ½ original (as-forged) thickness. (d) Specimen taken from forging. (e) Elongation 6% min for specimen taken from separately forged coupon. (f) Maximum cross-sectional area 144 in.². Not applicable to upset biscuit forgings or to rolled rings. (g) Applicable only to rings having ratio of outside diameter to wall thickness equal to or greater than 10.

Fig. 12 Influence of temperature on compressive yield strength of alloy 2618-T61 hand-forged billets

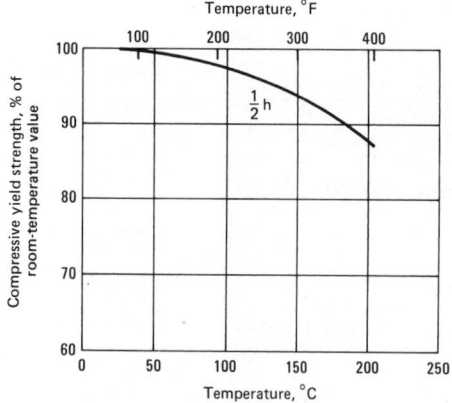

Compressive yield strength determined at temperature after holding ½ h under no load. Value plotted as percentage of corresponding room-temperature value.

V; 7072 cladding, -0.96 V vs $0.1N$ calomel electrode in an aqueous solution containing 53 g NaCl plus 3 g H_2O_2 per litre

Magnetic Properties

Magnetic susceptibility. Mass: 0.8 $\times 10^{-6}$ (cgs/g) at 25 °C (77 °C)

Fabrication Characteristics

Annealing temperature. 413 °C (775 °F). Commercial practice: 400 to 600 °C (750 to 1100 °F); higher temperatures used only for flash annealing.

3004, Alclad 3004
1.2Mn-1.0Mg

Specifications

ASTM. 3004: sheet and plate, B209; extruded tube, B221; welded tube, B313, B547. Alclad 3004: sheet and plate, B209; welded tube, B313; culvert pipe, B547
SAE. J454
UNS number. A93004
Government. Culvert pipe: WW-P-402
Foreign. Australia: A3004. France: NF A-M1G. West Germany: DIN AlMn1Mg1

Chemical Composition

Composition limits. 3004: 0.25 max Cu; 0.30 max Si; 0.70 max Fe; 1.0 to 1.5 Mn; 0.8 to 1.3 Mg; 0.25 max Zn; 0.05 max others (each); 0.15 max others (total); rem Al. Alclad 3004: 7072 cladding - 0.10 max Cu; 0.10 max Mg; 0.10 max Mn; 0.7 max Fe plus Si; 0.8 to 1.3 Zn; 0.05 max others (each); 0.15 max others (total); rem Al

Applications

Typical uses. 3004: drawn and ironed rigid containers (cans), chemical handling and storage equipment, sheet metal work, builders' hardware, incandescent and fluorescent lamp bases and similar applications requiring good formability and higher strength than provided by 3003. Alclad 3004: siding, culvert pipe, industrial roofing.

Mechanical Properties

Tensile properties. See Tables 43 and 44.

Compressive yield strength. Approximately the same as tensile yield strength
Shear yield strength. Approximately 55% of tensile strength
Hardness. See Table 43.
Poisson's ratio. 0.35 at 20 °C (68 °F)
Elastic modulus. Tension, 70 GPa (10.2×10^6 psi); shear, 25 GPa (3.6×10^6 psi)
Fatigue strength. See Table 43.

Mass Characteristics

Density. 2.72 Mg/m³ (0.098 lb/in.³) at 20 °C (68 °F)

Thermal Properties

Liquidus temperature. 654 °C (1210 °F)
Solidus temperature. 629 °C (1165 °F)
Coefficient of thermal expansion. Linear:

Temperature range		Average coefficient	
°C	°F	μm/ m·K	μin./ in.·°F
−50 – +20	−50 – +68 ..	21.5	11.9
20 – 100	68 – 212 ...	23.2	12.9
20 – 200	68 – 392 ...	24.1	13.4
20 – 300	68 – 572 ...	25.1	13.9

Volumetric: 67×10^{-6} m³/m³·K (3.72×10^{-5} in.³/in.³·°F) at 20 °C (68 °F)
Specific heat. 893 J/kg·K (0.213 Btu/lb·°F) at 20 °C (68 °F)
Thermal conductivity. O temper: 162 W/m·K (93.6 Btu/ft·h·°F) at 20 °C (68 °F)

Electrical Properties

Electrical conductivity. Volumetric, O temper: 42% IACS at 20 °C (68 °F)
Electrical resistivity. O temper: 41.0 nΩ·m at 20 °C (68 °F); temperature coefficient, 0.1 nΩ·m per K at 20 °C (68 °F)
Electrolytic solution potential. −0.84 V; 3004 and core of alclad 3004, 7072 cladding, −0.96 V (cladding) vs $0.1N$ calomel electrode in an aqueous solution containing 53 g NaCl plus 3 g H_2O_2 per litre

Table 38 Typical tensile properties of alloy 2618-T61 at various temperatures

Temperature		Tensile strength		Yield strength(a)		Elongation, %
°C	°F	MPa	ksi	MPa	ksi	
−196	−320	538	78.0	421	61.0	12
−80	−112	462	67.0	379	55.0	11
−28	−18	441	64.0	372	54.0	10
24	75	441	64.0	372	54.0	10
100	212	427	62.0	372	54.0	10
149	300	345	50.0	303	44.0	14
204	400	221	32.0	179	26.0	24
260	500	90	13.0	62	9.0	50
316	600	52	7.5	31	4.5	80
371	700	34	5.0	24	3.5	120

(a) 0.2% offset.

Table 39 Creep-rupture properties of alloy 2618

Temperature		Time under stress, h	Rupture stress		Stress for creep of:							
					1.0%		0.5%		0.2%		0.1%	
°C	°F		MPa	ksi	MPa	ksi	MPa	ksi	MPa	ksi	MPa	ksi
150	300	0.1	380	55	345	50	345	50	330	48	315	46
		1	360	52	340	49	330	48	315	46	290	42
		10	340	49	325	47	315	46	295	43	270	39
		100	305	44	305	44	290	42	270	39	240	35
		1000	255	37	255	37	250	36	240	35	205	30
177	350	0.1	340	49	325	47	315	46	295	43	285	41
		1	310	45	305	44	295	43	275	40	255	37
		10	285	41	275	40	260	38	250	36	220	32
		100	250	36	240	35	235	34	220	32	185	27
		1000	205	30	200	29	195	28	185	27	150	22
205	400	0.1	290	42	285	41	270	39	255	37	240	35
		1	260	38	255	37	250	36	235	34	205	30
		10	230	33	220	32	215	31	200	29	170	25
		100	195	28	185	27	180	26	165	24	140	20
		1000	160	23	150	22	145	21	130	19	90	13
260	500	0.1	185	27	170	25	165	24	160	23	145	21
		1	165	24	150	22	145	21	140	20	115	17
		10	140	20	130	19	125	18	110	16	83	12
		100	105	15	97	14	90	13	69	10	52	7.5
		1000	62	9	62	9	55	8	48	7	...	...
315	600	0.1	97	14	83	12	69	10	55	8	48	7
		1	69	10	62	9	55	8	45	6.5	41	6
		10	52	7.5	45	6.5	41	6	38	5.5	26	3.8
		100	32	4.6	28	4.1	26	3.7	19	2.8	15	2.2
		1000	20	2.9	17	2.5	14	2.1	...		...	...

Table 40 Standard specifications for alloy 3003

Mill form and condition	AMS	ASME	ASTM	Government
3003				
Sheet and plate	4006	SB209	B209	QQ-A-250/2
	4008	...	...	...
Wire, rod and bar (rolled or cold finished)	...	...	B211	QQ-A-225/2
Wire, rod, bar, shapes and tube (extruded)	...	SB221	B221	QQ-A-200/1
Tube				
Extruded, seamless	...	SB241	B241	...
Extruded, coiled	...	...	B491	...
Drawn	...	...	B483	...
Drawn, seamless	4065	SB210	B210	WW-T-700/2
	4067	...	...	...
Condenser	...	SB234	B234	...
Condenser with integral fins	...	...	B404	...
Welded	...	...	B313	...
	...	...	B547	...
Pipe: seamless	...	...	B241	MIL-P-25995
Gas and oil transmission	...	...	B345	...
Rivet wire and rod	...	...	B316	QQ-A-430
Forgings	...	SB247	B247	...
Foil	4010	...	...	MIL-A-81596
Alclad 3003				
Sheet and plate				B209
Tube				
Drawn, seamless				B210
Extruded				B221
Extruded, seamless				B241
Condenser				B234
Condenser with integral fins				B404
Welded				B547
Pipe (gas and oil transmission)				B345

Magnetic Properties

Magnetic susceptibility. Mass: 0.8 × 10^{-6} (cgs/g) at 25 °C (68 °F)

Fabrication Characteristics

Annealing temperature. 413 °C (775 °F)

3105
0.55Mn-0.50Mg

Specifications

ASTM. B209
SAE. J454

Chemical Composition

Composition limits. 0.6 max Si; 0.7 max Fe; 0.30 max Cu; 0.30 to 0.80 Mn; 0.20 to 0.80 Mg; 0.20 max Cr; 0.40 max Zn; 0.10 max Ti; 0.05 max others (each); 0.15 max others (total); rem Al

Table 41 Mechanical properties of alloy 3003(a)

Temper	Tensile strength MPa	ksi	MPa	ksi	Yield strength MPa	ksi	Elongation, %	Hardness HB(b)	HR	Shear strength MPa	ksi	Fatigue strength(c) MPa	ksi
Typical Properties													
O	110	16	...	...	42	6	30–40	28	45–65	76	11	48	7
H12	130	19	...	...	125	18	10–20	35	55–75	83	12	55	8
H14	150	22	...	...	145	21	8–16	40	70–90	97	14	62	9
H16	175	25	...	...	175	25	5–14	47	75–92	105	15	69	10
H18	200	29	...	...	185	27	4–10	55	84–95	110	16	69	10
Property Limits	**Minimum**		**Maximum**		**Minimum**								
O (0.006–3.000 in. thick)	97	14	130	19	34	5	14–25	...	...	...	...	...	...
H12 (0.017–2.000 in. thick)	115	17	160	23	83	12	3–10	...	...	...	...	...	...
H14 (0.009–1.000 in. thick)	140	20	180	36	115	17	1–10	...	...	...	...	...	...
H16 (0.006–0.162 in. thick)	165	24	205	30	145	21	1–4	...	...	...	...	...	...
H18 (0.006–0.128 in. thick)	185	27	...	...	165	24	1–4	...	...	...	...	...	...
H112													
(0.250–0.499 in. thick)	115	17	...	...	69	10	8	...	...	...	...	...	...
(0.500–2.000 in. thick)	105	15	...	...	41	6	12	...	...	...	...	...	...
(2.000–3.000 in. thick)	100	14.5	...	...	41	6	18	...	...	...	...	...	...
Property Limits, Alclad 3003													
O													
(0.006–0.499 in. thick)	90	13	125	18	31	4.5	14–25	...	...	...	...	...	...
(0.500–3.000 in. thick)	97	14	130	19	34	5.0	23	...	...	...	...	...	...
H12													
(0.017–0.499 in. thick)	110	16	150	22	77	11	4–9	...	...	...	...	...	...
(0.500–2.000 in. thick)	115	17	160	23	83	12	10	...	...	...	...	...	...
H14													
(0.009–0.499 in. thick)	130	19	170	25	110	16	1–8	...	...	...	...	...	...
(0.500–2.000 in. thick)	140	20	180	26	115	17	10	...	...	...	...	...	...
H16 (0.006–0.162 in. thick)	160	23	200	29	140	20	1–4	...	...	...	...	...	...
H18 (0.006–0.128 in. thick)	180	26	...	...	...	...	1–4	...	...	...	...	...	...
H112													
(0.250–0.499 in. thick)	110	16	...	...	62	9	8	...	...	...	...	...	...
(0.500–2.000 in. thick)	105	15	...	...	41	6	12	...	...	...	...	...	...
(2.000–3.000 in. thick)	100	14.5	...	...	41	6	18	...	...	...	...	...	...

(a) Mechanical properties of 3003 clad with 7072 are practically the same as for bare material, except that hardness and fatigue resistance tend to be slightly lower for the clad product. (b) 500-kg load, 10-mm ball, 30-s duration of loading. (c) At 5 x 10^8 cycles. R. R. Moore-type test.

Table 42 Typical mechanical properties of alloy 3003 at various temperatures

Temperature		Tensile strength(a)		Yield strength(a)		Elongation, %
°C	°F	MPa	ksi	MPa	ksi	
O temper						
−200	−328	250	33	60	8.6	46
−100	−148	150	22	52	7.5	43
−30	−22	115	17	45	6.5	41
+25	+77	110	16	41	6	40
100	212	80	13	38	5.5	43
200	392	60	8.6	30	4.3	60
300	572	29	4.2	17	2.5	70
400	752	10	2.6	12	1.7	75
H14 Temper						
−200	−328	250	36	170	25	30
−100	−148	175	25	155	22.5	19
−30	−22	150	22	145	21	16
+25	+77	150	22	145	21	16
100	212	145	21	130	19	16
200	392	96	14	62	9	20
300	572	29	4.2	17	2.5	70
400	752	18	2.6	12	1.7	75
H18 Temper						
−200	−328	290	42	230	33	23
−100	−148	230	33	210	30	12
−30	−22	210	30	190	38	10
+25	+77	200	29	185	27	10
100	212	180	26	145	21	10
200	392	96	14	62	9	18
300	572	29	4.2	17	2.5	70
400	752	18	2.6	12	1.7	75

(a) Lowest strengths for exposures up to 10 000 h at temperature, no load; test load applied at 5000 psi/min to yield strength and then at strain rate of 5%/min to fracture.

Applications

Typical uses. Residential siding, mobile home sheet, gutters and downspouts, sheet metal work, bottle caps and closures

Mechanical Properties

Tensile properties. See Table 45.
Poisson's ratio. 0.33
Elastic modulus. Tension, 69 GPa (10 × 10⁶ psi); shear, 25 GPa (3.6 × 10⁶ psi)

Mass Characteristics

Density. 2.71 Mg/m³ (0.098 lb/in.³) at 20 °C (68 °F)

Thermal Properties

Liquidus temperature. 657 °C (1215 °F)
Solidus temperature. 638 °C (1180 °F)
Coefficient of thermal expansion. Linear:

Temperature range		Average coefficient	
°C	°F	μm/ m · K	μin./ in. · °F
−50 – +20	−58 – +68	21.8	12.1
20 – 100	68 – 212	23.6	13.1
20 – 200	68 – 392	24.5	13.6
20 – 300	68 – 572	25.5	14.2

Volumetric: 68 × 10⁻⁶ m³/m³·K (3.77 × 10⁻⁵ in.³/in.³·°F) at 20 °C (68 °F)
Specific heat. 897 J/kg·K (0.214 Btu/lb·°F) at 20 °C (68 °F)
Thermal conductivity. 173 W/m·K (99.9 Btu/ft·h·°F) at 20 °C (68 °F)

Electrical Properties

Electrical conductivity. Volumetric, O temper: 45% IACS at 20 °C (68 °F)
Electrical resistivity. O temper: 38.3 nΩ·m at 20 °C (68 °F); temperature coefficient, 0.1 nΩ·m per K at 20 °C (68 °F)

Electrolytic solution potential. −0.84 V vs 0.1*N* calomel electrode in an aqueous solution containing 53 g NaCl plus 3 g H₂O₂ per litre

Magnetic Properties

Magnetic susceptibility. Mass: 0.7 × 10⁻⁶ (cgs/g) at 20 °C (68 °F)

Fabrication Characteristics

Annealing temperature. 343 °C (650 °F)

4032
12.2Si-1.0Mg-0.9Cu-0.9Ni

Specifications

AMS. Forgings and forging stock: 4145
ASTM. Forgings: B247
SAE. J454
UNS number. A94032
Government. Forgings: QQ-A-367
Foreign. Canada: CSA SG121. France: NF A-S12UN. Italy: UNI P-AlSi12MgCuNi

Chemical Composition

Composition limits. 11.0 to 13.5 Si; 1.0 max Fe; 0.50 to 1.30 Cu; 0.8 to 1.3 Mg; 0.10 max Cr; 0.50 to 1.3 Ni; 0.25 max Zn; 0.05 max others (each); 0.15 max others (total); rem Al

Applications

Typical uses. Pistons and other high-temperature service parts

Mechanical Properties

Tensile properties. T6 temper: tensile strength, 380 MPa (55 ksi); yield strength, 315 MPa (46 ksi); elongation, 9% in 50 mm or 2 in. For typical properties at various temperatures, see Table 46.
Hardness. T6 temper: 120 HB at 500-kg load, 10-mm ball
Poisson's ratio. 0.33
Elastic modulus. Tension, 79 GPa (11.4 × 10⁶ psi). Shear, 26 GPa (3.8 × 10⁶ psi)
Fatigue strength. T6 temper: 110 MPa (16 ksi) at 5 × 10⁸ cycles, R. R. Moore type test. At various temperatures, see Table 47.
Creep-rupture characteristics. See Table 48.

Table 43 Mechanical properties of alloy 3004(a)

Temper	Tensile strength MPa	ksi	MPa	ksi	Yield strength MPa	ksi	Elon-gation, %	Hardness, HB(b)	Shear strength MPa	ksi	Fatigue strength(c) MPa	ksi
Typical Properties												
O	180	26	...	...	69	10	20–25	45	110	16	97	14
H32	215	31	...	...	170	25	10–17	52	115	17	105	15
H34	240	35	...	...	200	29	9–12	63	125	18	105	15
H36	260	38	...	...	230	33	5–9	70	140	20	110	16
H38	285	41	...	...	250	36	4–6	77	145	21	110	16
Property Limits	Minimum		Maximum		Minimum							
O (0.006–3.000 in. thick)	150	22	200	29	59	8.5	10–18	...	...	...	...	...
H32 (0.017–2.000 in. thick)	195	28	240	35	145	21	1–6	...	...	...	...	...
H34 (0.009–1.000 in. thick)	220	32	260	38	170	25	1–5	...	...	...	...	...
H36 (0.006–0.162 in. thick)	240	35	285	41	195	28	1–4	...	...	...	...	...
H38 (0.006–0.128 in. thick)	260	38	...	...	215	31	1–4	...	...	...	...	...
H112 (0.250–3.000 in. thick)	160	23	...	...	62	9	7	...	...	...	...	...
Property Limits, Alclad 3004												
O												
(0.006–0.499 in. thick)	145	21	195	28	55	8	10–18	...	...	...	...	...
(0.500–3.000 in. thick)	150	22	200	29	59	8.5	16	...	...	...	...	...
H32												
(0.017–0.499 in. thick)	185	27	235	34	140	20	1–6	...	...	...	...	...
(0.500–2.000 in. thick)	195	28	240	35	145	21	6	...	...	...	...	...
H34												
(0.009–0.499 in. thick)	215	31	255	37	165	24	1–5	...	...	...	...	...
(0.500–1.000 in. thick)	220	32	260	38	170	25	5	...	...	...	...	...
H36 (0.006–0.162 in. thick)	235	34	275	40	185	27	1–4	...	...	...	...	...
H38 (0.006–0.128 in. thick)	255	37	...	...	...	...	1–4	...	...	...	...	...
H112												
(0.250–0.499 in. thick)	150	22	...	...	59	8.5	7	...	...	...	...	...
(0.500–3.000 in. thick)	160	23	...	...	62	9	7	...	...	...	...	...

(a) Mechanical properties of 3004 clad with 7072 are practically the same as for bare material, except that hardness and fatigue resistance tend to be slightly lower for the clad product. (b) 500-kg load, 10-mm ball, 30-s duration of loading. (c) At 5 x 10^8 cycles, R. R. Moore type test.

Mass Characteristics

Density. 2.68 Mg/m^3 (0.097 lb/in.^{3}) at 20 °C (68 °F)

Thermal Properties

Liquidus temperature. 571 °C (1060 °F)

Eutectic temperature. 532 °C (990 °F)

Incipient melting temperature. 532 °C (990 °F)

Coefficient of thermal expansion. Linear:

Temperature range °C	°F	Average coefficient μm/ m·K	μin./ in.·°F
−50–+20	−58–+68	18.0	10.0
20–100	68–212	19.5	10.8
20–200	68–392	20.2	11.2
20–300	68–572	21.0	11.7

Volumetric: 56 × 10^{-6} m^3/m^3·K (3.11 × 10^{-5} in.^{3}/in.^{3}·°F) at 20 °C (68 °F)

Specific heat. 864 J/kg·K (0.206 Btu/lb·°F) at 20 °C (68 °F)

Thermal conductivity. At 20 °C (68 °F): O temper, 155 W/m·K (89.6 Btu/ft·h·°F); T6 temper, 141 W/m·K (81.5 Btu/ft·h·°F)

Electrical Properties

Electrical conductivity. Volumetric, at 20 °C (68 °F): O temper, 40% IACS; T6 temper, 36% IACS

Electrical resistivity. At 20 °C (68 °F): O temper, 43.1 nΩ·m; T6 temper, 47.9 nΩ·m. Temperature coefficient, 0.1 nΩ·m per K at 20 °C (68 °F)

Fabrication Characteristics

Annealing temperature. 413 °C (775 °F); 2 to 3 h at temperature then furnace cooled to 260 °C (500 °F) at 25 °C (50 °F) per h max

Solution temperature. 504 to 516 °C (940 to 960 °F). Hold 4 min at temperature then quench in cold water; for heavy or complicated forgings, quench in water at 65 to 100 °C (150 to 212 °F)

Aging temperature. 168 to 174 °C (335 to 345 °F); 8 to 12 h at temperature

Hot working temperature. 316 to 480 °C (600 to 900 °F)

Table 44 Typical mechanical properties of alloy 3004 at various temperatures

Temperature		Tensile strength(a)		Yield strength(a)		Elongation, %
°C	°F	MPa	ksi	MPa	ksi	
O Temper						
−200	−328	290	42.5	90	13.2	38
−100	−148	200	29	80	11.5	31
−30	−22	175	26	69	10	26
+25	+77	175	26	69	10	25
100	212	175	26	69	10	25
200	392	96	14	65	9.5	55
300	572	50	7.2	34	4.9	80
400	752	30	4.4	9	2.8	90
H34 Temper						
−200	−328	360	52	235	34	26
−100	−148	270	39	212	31	17
−30	−22	245	36	200	29	13
+25	+77	240	35	200	29	12
100	212	240	35	200	29	12
200	392	145	21	105	15	35
300	572	50	7.2	34	4.9	80
400	752	30	4.4	19	2.8	90
H38 Temper						
−200	−328	400	58	295	43	20
−100	−148	310	45	267	39	10
−30	−22	290	42	245	36	7
+25	+77	280	41	245	36	6
100	212	275	40	245	36	7
200	392	150	22	105	15	30
300	572	50	7.2	34	4.9	80
400	752	30	4.4	19	2.8	90

(a) Lowest strength for exposures up to 10 000 h at temperature, no load; test loading applied at 5000 psi/min to yield strength and then at strain rate of 5%/min to fracture.

4043
5.2Si

Specifications

AMS. Bare welding rod and electrodes: 4190
SAE. J454
Government. Bare welding rod and electrodes: QQ-R-566, MIL-E-16053; spray gun wire: MIL-W-6712
Foreign. Australia: B4043. Canada: CSA S5. France: NF A-S5. United Kingdom: BS N21. Germany: DIN AlSi5, Werkstoff-Nr. 3.2245

Chemical Composition

Composition limits. 4.5 to 6.0 Si; 0.8 max Fe; 0.30 max Cu; 0.05 max Mn; 0.05 max Mg; 0.10 max Zn; 0.20 max Ti; 0.05 max others (each); 0.15 max others (total); 0.0008 max Be for welding electrode only; rem Al

Applications

Typical uses. General purpose weld filler alloy (rod or wire) for welding all wrought and foundry alloys except those rich in magnesium

Mechanical Properties

Tensile properties. See Table 49.

Mass Characteristics

Density. 2.68 Mg/m^3 (0.097 lb/in.^{3})

Table 45 Mechanical properties of alloy 3105 sheet

Temper	Tensile strength				Yield strength		Elongation, %	Shear strength	
	MPa	ksi	MPa	ksi	MPa	ksi		MPa	ksi
Typical Properties									
O	115	17	...	...	55	8	24	83	12
H12	150	22	...	...	130	19	7	97	14
H14	170	25	...	...	150	22	5	105	15
H16	195	28	...	...	170	25	4	110	16
H18	215	31	...	...	195	28	3	115	17
H25	180	26	...	...	160	23	8	105	15
Property Limits									
O (0.013 – 0.080 in. thick)	97	14	145	21	34	5	16 – 20	...	...
H12 (0.017 – 0.080 in. thick)	130	19	180	26	105	15	1 – 3	...	...
H14 (0.013 – 0.080 in. thick)	150	22	200	29	125	18	1 – 2	...	...
H16 (0.013 – 0.080 in. thick)	170	25	220	32	145	21	1 – 2	...	...
H18 (0.013 – 0.080 in. thick)	195	28	...	...	165	24	1 – 2	...	...
H25 (0.013 – 0.080 in. thick)	160	23	...	...	130	19	2 – 6	...	...

Table 46 Typical mechanical properties of alloy 4032-T6 at various temperatures

Temperature		Tensile strength		Yield strength		Elongation, %
°C	°F	MPa	ksi	MPa	ksi	
−200	−328 460		67	337	49	11
−100	−148 415		60	325	47	10
−30	−22 385		56	315	46	9
25	77 380		55	315	46	9
100	212 345		50	300	44	9
200	392 90		13	62	9	30
300	572 38		5.5	24	3.5	70
400	752 21		3.1	12	1.8	90

Table 48 Creep-rupture properties of alloy 4032

Temperature		Time under stress, h	Rupture stress		Stress for creep of:					
					1.0%		0.5%		0.2%	
°C	°F		MPa	ksi	MPa	ksi	MPa	ksi	MPa	ksi
100	212	0.1	331	48	283	41	269	39	...	...
		1	317	46	283	41	262	38	...	...
		10	303	44	283	41	262	38	...	...
		100	296	43	276	40	262	38	...	...
		1000	296	43	276	40	255	37	...	...
149	300	0.1	290	42	276	40	248	36	...	...
		1	276	40	269	39	241	35	...	...
		10	269	39	255	37	234	34	...	...
		100	248	36	241	35	221	32	...	...
		1000	207	30	200	29	186	27	...	...
204	400	0.1	234	34	228	33	221	32	138	20
		1	214	31	207	30	200	29	131	19
		10	186	27	179	26	165	24	103	15
		100	138	20	131	19	124	18	59	8.5
		1000	83	12	76	11	69	10	...	...

Table 49 Typical tensile properties of alloy 4043 welding wire

Wire diameter		Temper	Tensile strength		Yield strength(a)		Elongation, %
mm	in.		MPa	ksi	MPa	ksi	
5.0	0.20	H16 205		30	180	26	1.7
3.2	0.12	H14 170		25	165	24	1.3
1.6	0.06	H18 285		41	270	39	0.5
1.2	0.05	H16 200		29	185	27	0.4
5.0	0.20	O 130		19	50	7	25
3.2	0.12	O 115		17	55	8	31
1.6	0.06	O 145		21	65	10	22
1.2	0.05	O 110		16	55	8	29

(a) 0.2% offset.

Table 47 Fatigue strength of alloy 4032-T6 at various temperatures

Temperature		No. of cycles	Stress(a)	
°C	°F		MPa	ksi
24	75	10^4	359	52
		10^5	262	38
		10^6	207	30
		10^7	165	24
		10^8	124	18
		5×10^8	114	16.5
149	300	10^5	207	30
		10^6	165	24
		10^7	124	18
		10^8	90	13
		5×10^8	79	11.5
204	400	10^5	186	27
		10^6	138	20
		10^7	90	13
		10^8	55	8
		5×10^8	48	7
260	500	10^5	131	19
		10^6	83	12
		10^7	55	8
		10^8	34	5
		5×10^8	34	5

(a) Based on rotating beam tests at room temperature and cantilever beam tests at elevated temperatures.

Thermal Properties

Liquidus temperature. 630 °C (1170 °F)

Solidus temperature. 575 °C (1065 °F)

Coefficient of thermal expansion. Linear, 22.0 μm/m·K (12.2 μin./in.·°F) at 20 to 100 °C (68 to 212 °F)

Electrical Properties

Electrical conductivity. Volumetric, O temper: 42% IACS at 20 °C (68 °F)

Electrical resistivity. O temper: 41 nΩ·m at 20 °C (68 °F)

Fabrication Characteristics

Annealing temperature. 350 °C (660 °F)

5005
0.8Mg

Specifications

ASTM. Sheet and plate: B209. Wire, H19 temper: B396. Stranded conductor: B397. Rivet wire and rod: B316. Rolled rod: B531. Drawn tube: B210, B483

SAE. J454

UNS number. A95005

Government. Rivet wire and rod: QQ-A-430

Foreign. France: NF A-G0.6. United Kingdom: BS N41. Germany: DIN AlMg1. ISO: AlMg1

Chemical Composition

Composition limits. 0.30 max Si; 0.7 max Fe; 0.20 max Cu; 0.20 max Mn; 0.50 to 1.1 Mg; 0.10 max Cr; 0.25 max Zn; 0.05 max others (each); 0.15 max others (total); rem Al

Applications

Typical uses. Electrical conductor wire, cooking utensils, appliances, and architectural applications. Medium strength and good resistance to corrosion are two characteristics of

Table 50 Typical mechanical properties of alloy 5005

Temper	Tensile strength(a) MPa	ksi	Yield strength(a) MPa	ksi	Elongation(a)(b), %	Hardness(c), HB	Shear strength MPa	ksi
O	124	18	41	6	25	28	76	11
H12	138	20	131	19	10	...	97	14
H14	159	23	152	22	6	...	97	14
H16	179	26	172	25	5	...	103	15
H18	200	29	193	28	4	...	110	16
H32	138	20	117	17	11	36	97	14
H34	159	23	138	20	8	41	97	14
H36	179	26	165	24	6	46	103	15
H38	200	29	186	27	5	51	110	16

(a) Strengths and elongations unchanged or improved at low temperatures. (b) 1.6 mm (1/16 in.) thick specimen. (c) 500-kg load; 10-mm diam ball.

Table 51 Mechanical property limits for alloy 5005

Temper	Tensile strength Minimum MPa	ksi	Maximum MPa	ksi	Yield strength(min) MPa	ksi	Elongation(min), %(a)
Sheet and Plate							
O	105	15	145	21	35	5	12 to 22
H12	125	18	165	24	95	14	2 to 9
H14	145	21	185	27	115	17	1 to 8
H16	165	24	205	30	135	18	1 to 3
H18	185	27	...	...	...	...	1 to 3
H32	120	17	160	23	85	12	3 to 10
H34	140	20	180	26	105	15	2 to 8
H36	160	23	200	29	125	18	1 to 4
H38	180	26	...	...	...	...	1 to 4
H112							
0.250-0.492 in. thick	115	17	...	...	...	...	8
0.492-1.60 in. thick	105	15	...	...	...	...	10
1.60-3.20 in. thick	100	15	...	...	...	...	16

(a) In 2 in. or 5d, where d is diameter or reduced section of tensile test specimen. Where a range of values appears in this column, the specified minimum elongation varies with thickness of the mill product.

Table 52 Typical mechanical properties of alloy 5050

Temper	Tensile strength(a) MPa	ksi	Yield strength(a) MPa	ksi	Elongation(a)(b), %	Hardness(c), HB	Shear strength MPa	ksi	Fatigue strength(d) MPa	ksi
O	145	21	55	8	24	36	105	15	83	12
H32	170	25	145	21	9	46	115	17	90	13
H34	190	28	165	24	8	53	123	18	90	13
H36	205	30	180	26	7	58	130	19	97	14
H38	220	32	200	29	6	63	138	20	97	14

(a) Strengths and elongation generally unchanged or improved at low temperatures. (b) 1.6 mm (1/16 in.) thick sheet specimen. (c) 500-kg load; 10-mm diam ball. (d) At 5 × 10⁸ cycles; R.R. Moore type test.

5005 similar to those of 3003. When anodized, film on 5005 is clearer and lighter than on 3003 and gives better color match with 6063 architectural extrusions.

Mechanical Properties

Tensile properties. See Tables 50 and 51. Tensile strength and elongation are slightly lower in transverse direction than in longitudinal direction.

Shear yield strength. Approximately 55% of tensile yield strength.

Compressive yield strength. Approximately the same as tensile yield strength.

Hardness. See Table 50.

Poisson's ratio. 0.33

Elastic modulus. Tension, 68.2 GPa (9.90 × 10⁶ psi); shear, 25.9 GPa (3.75 × 10⁶ psi); compression, 69.5 GPa (10.1 × 10⁶ psi)

Mass Characteristics

Density. 2.70 Mg/m³ (0.097 lb/in.³) at 20 °C (68 °F)

Thermal Properties

Liquidus temperature. 652 °C (1205 °F)

Solidus temperature. 632 °C (1170 °F)

Coefficient of thermal expansion. Linear:

Temperature range °C	°F	Average coefficient μm/ m · K	μin./ in. · °F
−50 − +20	−58 − +68 ...	21.9	12.2
20 − 100	68 − 212 ...	23.7	13.2
20 − 200	68 − 392 ...	24.6	13.7
20 − 300	68 − 572 ...	25.6	14.2

Volumetric: 68 × 10⁻⁶ m³/m³·K (3.77 × 10⁻⁵ in.³/in.³·°F) at 20 °C (68 °F)

Specific heat. 900 J/kg·K (0.215 Btu/lb·°F) at 20 °C (68 °F)

Thermal conductivity. 205 W/m·K (118 Btu/ft·h·°F) at 20 °C (68 °F)

Electrical Properties

Electrical conductivity. Volumetric, O and H38 tempers: 52% IACS at 20 °C (68 °F)

Electrical resistivity. O and H38 tempers: 33.2 nΩ·m at 20 °C (68 °F); temperature coefficient, 0.1 nΩ·m per K at 20 °C (68 °F)

Electrolytic solution potential. −0.83 V vs 0.1N calomel electrode in an aqueous solution containing 53 g NaCl plus 3 g H₂O₂ per litre

Fabrication Characteristics

Annealing temperature. 343 °C (650 °F); holding at temperature not required

Hot working temperature. 260 to 510 °C (500 to 950 °F)

5050
1.4Mg

Specifications

ASTM. Sheet and plate: B209. Drawn, seamless tube: B210. Drawn

Table 53 Typical tensile properties of alloy 5050

Temperature		Tensile strength(a)		Yield strength(a)(b)	
°C	°F	MPa	ksi	MPa	ksi
−196	−320	255	37	70	10
−80	−112	150	22	60	8.5
−28	−18	145	21	55	8
24	75	145	21	55	8
100	212	145	21	55	8
149	300	130	19	55	8
204	400	95	14	50	7.5
260	500	60	9	41	6
316	600	41	6	29	4.2
371	700	27	3.9	18	2.6
−196	−320	305	44	205	30
−80	−112	205	30	170	25
−28	−18	195	28	165	24
24	75	195	28	165	24
100	212	195	28	165	24
149	300	170	25	150	22
204	400	95	14	50	7.5
260	500	60	9	41	6
316	600	41	6	29	4.2
371	700	27	3.9	18	2.6
−196	−320	315	46	250	36
−80	−112	235	34	205	30
−28	−18	220	32	200	29
24	75	220	32	200	29
100	212	215	31	200	29
219	300	185	27	170	25
204	400	95	14	50	7.5
260	500	60	9	41	6
316	600	41	6	29	4.2
371	700	27	3.9	18	2.6

(a) Lowest strengths for exposures up to 10 000 h at temperature; no load; test loading applied at 5000 psi/min to yield strength and then at strain rate of 5%/min to fracture. (b) 0.2% offset.

Table 54 Tensile-property limits for alloy 5050

Temper	Tensile strength(min)		Yield strength(min)		Elongation(min), %(a)
	MPa	ksi	MPa	ksi	
O	125	18	41	6	16 to 20
H32	150	22	110	16	4 to 6
H34	170	25	138	20	3 to 5
H36	185	27	151	22	2 to 4
H38	200	29	...	...	2 to 4

(a) Where a range of values appears in this column, specified minimum elongation varies with thickness of the mill product.

tube: B483. Welded tube: B313, B547
SAE. J454
UNS number. A95050
Foreign. France: NF A-Gl. Italy: P-AlMg 1.5. Switzerland: A11.5Mg. United Kingdom: BS 3L44. ISO: AlMg1.5

Chemical Composition

Composition limits. 0.40 max Si; 0.7 max Fe; 0.20 max Cu; 0.10 max Mn; 1.1 to 1.8 Mg; 0.10 max Cr; 0.25 max Zn; 0.05 max others (each); 0.15 max others (total); rem Al

Applications

Typical uses. Sheet used as trim in refrigerator applications; tube for automotive gas and oil lines; welded irrigation pipe; also available as plate, tube, rod, bar and wire

Mechanical Properties

Tensile properties. See Tables 52 to 54. Tensile strength and yield strength are approximately the same in both the transverse and longitudinal directions; however, elongation is slightly lower in the transverse direction than in the longitudinal direction.
Shear yield strength. Approximately 55% of the tensile yield strength
Compressive yield strength. Approximately the same as tensile yield strength
Hardness. See Table 52.
Poisson's ratio. 0.33
Elastic modulus. Tension, 68.9 GPa (10.0 × 10⁶ psi); shear, 25.9 GPa (3.75 × 10⁶ psi)

Mass Characteristics

Density. 2.69 Mg/m³ (0.097 lb/in.³) at 20 °C (68 °F)

Thermal Properties

Liquidus temperature. 652 °C (1205 °F)
Solidus temperature. 627 °C (1160 °F)
Coefficient of thermal expansion. Linear:

Temperature range		Average coefficient	
°C	°F	μm/ m · K	μin./ in. · °F
−50 − +20	−58 − +68 ...	21.8	12.1
20 − 100	68 − 212 ...	23.8	13.2
20 − 200	68 − 392 ...	24.7	13.7
20 − 300	68 − 572 ...	25.6	14.2

Specific heat. 900 J/kg·K (0.215 Btu/lb·°F) at 20 °C (68 °F)
Thermal conductivity. 191 W/m·K (110 Btu/ft·h·°F) at 20 °C (68 °F)

Electrical Properties

Electrical conductivity. Volumetric, O and H38 tempers: 50% IACS at 20 °C (68 °F)
Electrical resistivity. O and H38 tempers: 34 nΩ·m at 20 °C (68 °F); temperature coefficient, 0.1 nΩ·m per K at 20 °C (68 °F)
Electrolytic solution potential. −0.83 V vs 0.1N calomel electrode in an aqueous solution containing 53 g NaCl plus 3 g H₂O₂ per litre

Fabrication Characteristics

Annealing temperature. 343 °C (650 °F); holding at temperature not required
Hot working temperature. 260 to 510 °C (500 to 950 °F)

5052
2.5Mg-0.25Cr

Specifications

AMS. See Table 55.
ASTM. See Table 55.
SAE. J454
UNS number. A95052
Government. Sheet and plate: QQ-A-250/8. Foil: MIL-A-81596. Rolled or cold finished wire, rod, and bar: QQ-A-225/7. Drawn, seamless tube: WW-T-700/4. Rivet wire and rod: QQ-A430. Rivets: MIL-R-24243
Foreign. Canada: CSA GR20. France: NF A-G2.5C. Italy: UNI P-AlMg2.5. Germany: DIN AlMg2.5. ISO: AlMg2.5

Chemical Composition

Composition limits. 0.25 max Si; 0.40 max Fe; 0.10 max Cu; 0.10 max Mn; 2.2 to 2.8 Mg; 0.15 to 0.35 Cr; 0.10 max Zn; 0.05 max others (each); 0.15 max others (total); rem Al

Applications

Typical uses. Aircraft fuel and oil lines, fuel tanks, miscellaneous marine and transport applications, sheet metal work, appliances, street light standards, rivets and wire. Applications where good workability, very good resistance to corrosion, high fatigue strength, weldability, and moderate static strength are desired

Mechanical Properties

Tensile properties. See Tables 56 and 57.
Shear yield strength. Approximately 55% of tensile yield strength
Compressive yield strength. Approximately the same as tensile yield strength
Hardness. See Table 56.
Poisson's ratio. 0.33
Elastic modulus. Tension, 69.3 GPa (10.1 × 10⁶ psi); shear, 25.9 GPa (3.75 × 10⁶ psi); compression, 70.7 GPa (10.3 × 10⁶ psi)

Mass Characteristics

Density. 2.68 Mg/m³ (0.097 lb/in.³) at 20 °C (68 °F)

Thermal Properties

Liquidus temperature. 649 °C (1200 °F)
Solidus temperature. 607 °C (1125 °F)
Coefficient of thermal expansion. Linear:

Table 56 Typical mechanical properties of alloy 5052

Temper	Tensile strength(a) MPa	ksi	Yield strength(a) MPa	ksi	Elongation, %(a) 1.6 mm (¹/₁₆ in.) thick	12.5 mm (½ in.) diam	Hardness, HB(b)	Shear strength MPa	ksi	Fatigue strength(c) MPa	ksi
O	195	28	90	13	25	27	47	125	18	110	16
H32	230	33	195	28	12	16	60	140	20	115	17
H34	260	38	215	31	10	12	68	145	21	125	18
H36	275	40	240	35	8	9	73	160	23	130	19
H38	290	42	255	37	7	7	77	165	24	140	20

(a) Strengths and elongations unchanged or improved at low temperatures. (b) 500-kg load; 10-mm diam ball. (c) At 5 × 10⁸ cycles; R. R. Moore type test.

Table 57 Typical tensile properties of alloy 5052 at various temperatures

Temper	Temperature °C	°F	Tensile strength MPa	ksi	Yield strength(a) MPa	ksi	Elongation, %
O	−196	−320	303	44	110	16	46
	−80	−112	200	29	90	13	35
	−28	−18	193	28	90	13	32
	24	75	193	28	90	13	30
	100	212	193	28	90	13	36
	149	300	159	23	90	13	50
	204	400	117	17	76	11	60
	260	500	83	12	52	7.5	80
	316	600	52	7.5	38	5.5	110
	371	700	34	5	21	3	130
H34	−196	−320	379	55	248	36	28
	−80	−112	276	40	221	32	21
	−28	−18	262	38	214	31	18
	24	75	262	38	214	31	16
	100	212	262	38	214	31	18
	149	300	207	30	186	27	27
	204	400	165	24	103	15	45
	260	500	83	12	52	7.5	80
	316	600	52	7.5	38	5.5	110
	371	700	34	5	21	3	130
H38	−196	−320	414	60	303	44	25
	−80	−112	303	44	262	38	18
	−28	−18	290	42	255	37	15
	24	75	290	42	255	37	14
	100	212	276	40	248	36	16
	149	300	234	34	193	28	24

(a) 0.2% offset.

Table 55 Standard specifications for alloy 5052

Mill form	Specification No. AMS	ASTM
Sheet and plate	4015	B209
Sheet, plate, bar and shapes (extruded)	4016, 4017	B221
Wire, rod, and bar (rolled or cold finished)	4114	B211
Tube		
Drawn	4069	B483
Drawn, seamless	4070	B210
Hydraulic	4071	...
Extruded	...	B221
Extruded, seamless	...	B241
Condenser	...	B234
Condenser with integral fins	...	B404
Welded	...	B313, B547
Rivet wire and rod	...	B316
Foil	4004	...

Table 58 Typical mechanical properties of alloy 5056

Temper	Tensile strength(a) MPa	ksi	Yield strength(a) MPa	ksi	Elongation(a)(b), %	Hardness(c), HB	Shear strength MPa	ksi	Fatigue strength(d) MPa	ksi
O	290	42	152	22	35	65	179	26	138	20
H18	434	63	407	59	10	105	234	34	152	22
H38	414	60	345	50	15	100	221	32	152	22

(a) Strengths and elongations are unchanged or improved at low temperatures. (b) 12.5-mm (½-in.) diam; round specimen. (c) 500-kg load; 10-mm diam ball. (d) At 5 x 10^8 cycles, R. R. Moore type test.

Temperature range °C	°F	Average coefficient μm/ m·K	μin./ in.·°F
−50−+20	−58−+68	22.1	12.3
20−100	68−212	23.8	13.2
20−200	68−392	24.8	13.8
20−300	68−572	25.7	14.3

Volumetric: 69 × 10^{-6} m³/m³·K (3.83 × 10^{-5} in.³/in.³·°F) at 20 °C (68 °F)
Specific heat. 900 J/kg·K (0.215 Btu/lb·°F) at 20 °C (68 °F)
Thermal conductivity. 137 W/m·K (79.2 Btu/ft·h·°F) at 20 °C (68 °F)

Electrical Properties

Electrical conductivity. Volumetric, O and H38 tempers: 35% IACS at 20 °C (68 °F)
Electrical resistivity. O and H38 tempers: 49.3 nΩ·m at 20 °C (68 °F); temperature coefficient, 0.1 nΩ·m per K at 20 °C (68 °F)
Electrolytic solution potential. −0.85 V vs 0.1N calomel electrode in an aqueous solution containing 53 g NaCl plus 3 g H$_2$O$_2$ per litre

Fabrication Characteristics

Annealing temperature. 343 °C (650 °F); holding at temperature not required
Hot working temperature. 260 to 510 °C (500 to 950 °F)

5056, Alclad 5056
5.0Mg-0.1Mn-0.1Cr

Specifications

AMS. Rolled or cold finished wire, rod and bar: 4182. Foil: 4005
ASTM. Rivet wire and rod: B316. Rolled or cold finished wire, rod and bar: B211. Alclad, rolled or cold finished wire, rod and bar: B211
SAE. J454

Table 59 Typical tensile properties of alloy 5056

Temper	Temperature °C	°F	Tensile strength(a) MPa	ksi	Yield strength(a) MPa	ksi	Elongation, %
O	24	75	290	42	150	22	35
	149	300	214	31	117	17	55
	204	400	152	22	90	13	65
	260	500	110	16	69	10	80
	316	600	76	11	48	7	100
	371	700	41	6	28	4	130
H38	24	75	414	60	345	50	15
	149	300	262	38	214	31	30
	204	400	179	26	124	18	50
	260	500	110	16	69	10	80
	316	600	76	11	48	7	100
	371	700	41	6	28	4	130

(a) Lowest strengths for exposures up to 10 000 h at temperature, no load; test loading applied at 5000 psi/min to yield strength and then at strain rate of 5%/min to fracture.

UNS number. A95056
Government. Rivet wire and rod: QQ-A430. Foil: MIL-A-81596
Foreign. Austria: AlMg5. Canada: CSA-GM50R. United Kingdom: BS N6 2L.58. Germany: DIN AlMg5. ISO: AlMg5

Chemical Composition

Composition limits. 5056: 0.30 max Si; 0.40 max Fe; 0.10 max Cu; 0.05 to 0.20 Mn; 4.5 to 5.6 Mg; 0.20 max Cr; 0.10 max Zn; 0.05 max others (each); 0.15 max others (total); rem Al. Alclad 5056: 6253 cladding—Si, 45 to 65% of Mg content; 0.50 max Fe; 0.10 max Cu; 1.0 to 1.5 Mg; 0.15 to 0.35 Cr; 1.6 to 2.4 Zn; 0.05 max others (each); 0.15 max others (total); rem Al

Applications

Typical uses. Rivets for use with magnesium alloy and cable sheathing; zipper stock, nails; also Alclad wire is extensively used in fabrication of insect screens and other applications where wire products with good resistance to corrosion are required

Mechanical Properties

Tensile properties. See Tables 58, 59 and 60. Elongation, O temper:

Table 60 Mechanical-property limits for alloy 5056—rolled or cold finished wire, rod, and bar

Temper	Tensile strength (min) MPa	ksi
Bare Product		
O	315 (max)	46 (max)
H111	305	44
H12	315	46
H32	305	44
H14	360	52
H34	345	50
H18	400	58
H38	380	55
H192	415	60
H392	400	58
Alclad 5056		
H192	360	52
H392	345	50
H393	370(a)	54

(a) Yield strength (min), 325 MPa (47 ksi).

20% in 2 in. or 4d, where d is diameter of reduced section of tension test specimen
Shear yield strength. Approximately 55% of tensile yield strength
Compressive yield strength. Ap-

proximately the same as the tensile yield strength

Hardness. See Table 58.

Poisson's ratio. 0.33

Elastic modulus. Tension, 71.7 GPa $(10.4 \times 10^6$ psi); shear, 25.9 GPa $(3.75 \times 10^6$ psi); compression, 73.1 GPa $(10.6 \times 10^6$ psi)

Mass Characteristics

Density. 2.64 Mg/m³ (0.095 lb/in.³) at 20 °C (68 °F)

Thermal Properties

Liquidus temperature. 638 °C (1180 °F)

Solidus temperature. 568 °C (1055 °F)

Coefficient of thermal expansion. Linear, O temper:

Temperature range		Average coefficient	
°C	°F	µm/ m·K	µin./ in.·°F
−50 − +20	−58 − +65	...22.5	12.5
20 − 100	68 − 212	...24.1	13.7
20 − 200	68 − 392	...25.2	14.0
20 − 300	68 − 572	...26.1	14.5

Volumetric: 70×10^{-6} m³/m³·K $(3.89 \times 10^{-5}$ in.³/in.³·°F) at 20 °C (68 °F)

Specific heat. 904 J/kg·K (0.216 Btu/lb·°F) at 20 °C (68 °F)

Thermal conductivity. At 20 °C (68 °F): O temper, 120 W/m·K (69.3 Btu/ft·h·°F); H38 temper, 112 W/m·K (64.7 Btu/ft·h·°F)

Electrical Properties

Electrical conductivity. Volumetric, at 20 °C (68 °F): O temper, 29% IACS; H38 temper, 27% IACS

Electrical resistivity. At 20 °C (68 °F): O temper, 59 nΩ·m; H38 temper, 64 nΩ·m. Temperature coefficient, O and H38 tempers: 0.1 nΩ·m per K at 20 °C (68 °F)

Electrolytic solution potential. −0.87 V vs 0.1N calomel electrode in an aqueous solution containing 53 g NaCl plus 3 g H_2O_2 per litre

Fabrication Characteristics

Annealing temperature. 413 °C (775 °F); holding at temperature not required

Hot working temperature. 316 to 482 °C (600 to 900 °F)

Table 61 Typical tensile properties of alloy 5083

Temper	Tensile strength(a) MPa	ksi	Yield strength MPa	ksi	Elongation(a)(b), %
O	290	42	145	21	22
H112	303	44	193	28	16
H116	317	46	228	33	16
H321	317	46	228	33	16
H323, H32	324	47	248	36	10
H343, H34	345	50	283	41	9

(a) Strengths and elongations are unchanged or improved at low temperatures. (b) 1.6-mm (¹⁄₁₆-in.) thick specimen.

Table 62 Mechanical-property limits for alloy 5083

Temper	Tensile strength Minimum MPa	ksi	Maximum MPa	ksi	Yield strength Minimum MPa	ksi	Maximum MPa	ksi	Elongation (min), %(a)
O									
0.051-1.5000 in. thick	275	40	350	51	125	18	200	29	16
1.501-3.000 in. thick	270	39	345	50	115	17	200	29	16
3.001-5.000 in. thick	260	38	...	...	110	16	...	...	14-16
5.001-7.000 in. thick	255	37	...	...	105	15	...	...	14
7.001-8.000 in. thick	250	36	...	...	95	14	...	...	12
H112									
0.250-1.500 in. thick	275	40	...	...	125	18	...	...	12
1.501-3.000 in. thick	270	39	...	...	115	17	...	...	12
H116									
0.063-1.500 in. thick	305	44	...	...	215	31	...	...	12
1.501-3.000 in. thick	285	41	...	...	200	29	...	...	12
H321									
0.188-1.500 in. thick	305	44	385	56	215	31	295	43	12
1.501-3.000 in. thick	285	41	385	56	200	29	295	43	12
H323	310	45	370	54	235	34	305	44	8-10
H343	345	50	405	59	270	39	340	49	6-8

(a) In 2 in. or 4d, where d is diameter of reduced section of tensile test specimen. Where a range of values appears in this column, the specified minimum elongation varies with thickness of the mill product.

Table 63 Typical tensile properties of alloy 5083-O at various temperatures

Temperature °C	°F	Tensile strength(a) MPa	ksi	Yield strength(a)(b) MPa	ksi	Elongation, %
−195	−315	405	59	165	24	36
−80	−112	295	43	145	21	30
−30	−22	290	42	145	21	27
25	80	290	42	145	21	25
100	212	275	40	145	21	36
150	302	215	31	130	19	50
205	400	150	22	115	17	60
260	500	115	17	75	11	80
315	600	75	11	50	7.5	110
370	698	41	6	29	4.2	130

(a) Lowest strength for exposures up to 10 000 h at temperature, no load; test loading applied at 5000 psi/min to yield strength and then at strain rate of 10%/min to fracture. (b) 0.2% offset.

5083
4.4Mg-0.7Mn-0.15Cr

Specifications

AMS. Sheet and plate: 4056, 4057, 4058, 4059
ASTM. Sheet and plate: B209. Extruded wire, rod, bar, shapes and tube: B221. Extruded seamless tube: B241. Drawn seamless tube: B210. Welded Tube: B547. Forgings: B247. Gas and oil transmission pipe: B345.
SAE. J454
UNS number. A95083
Government. Sheet and plate: QQ-A-250/6. Extruded wire, rod, bar, shapes, and tube: QQ-A-200/4. Forgings: QQ-A-367. Armor plate: MIL-A-46027. Extruded armor: MIL-A-46083. Forged armor: MIL-A-45225
Foreign. Canada: CSA GM41. United Kingdom: BS N8. Germany: DIN AlMg4.5Mn; Werkstoff-Nr, 3.3547. ISO: AlMg4.5Mn

Chemical Composition

Composition limits. 0.40 max Si; 0.40 max Fe; 0.10 max Cu; 0.40 to 1.0 Mn; 4.0 to 4.9 Mg; 0.05 to 0.25 Cr; 0.25 max Zn; 0.15 max Ti; 0.05 max others (each); 0.15 max others (total); rem Al

Applications

Typical uses. Marine, auto and aircraft applications, unfired welded pressure vessels, cryogenics, TV towers, drilling rigs, transportation equipment, missile components, armor plate. Applications requiring a weldable moderate-strength alloy having good corrosion resistance

Mechanical Properties

Tensile properties. See Tables 61, 62 and 63.
Shear properties. O temper: shear strength, 172 MPa (25 ksi); shear yield strength, approximately 55% of tensile yield strength
Compressive yield strength. Approximately the same as tensile yield strength
Elastic modulus. Tension, 70.3 GPa (10.2×10^6 psi); shear, 26.4 GPa (3.83×10^6 psi); compression, 71.7 GPa (10.4×10^6 psi)
Fatigue strength. H321 and H116 tempers: 160 MPa (23 ksi) at 5×10^8 cycles; R. R. Moore type test

Table 64 Tensile properties of alloy 5086

Temper	Tensile strength MPa	ksi	MPa	ksi	Yield strength MPa	ksi	Elongation(a), %
Typical Properties							
O	260	38	...	...	115	17	22
H32, H116	290	42	...	...	205	30	12
H34	325	47	...	...	255	37	10
H112	270	39	...	...	130	19	14
Property Limits							
	Minimum		Maximum		Minimum		Minimum
O(0.020 – 2.000 in. thick)	240	35	305	44	95	14	15 to 18
H32(0.020 – 2.000 in. thick)	275	40	325	47	195	28	6 to 12
H34(0.009 – 1.000 in. thick)	305	44	350	51	235	34	4 to 10
H36(0.006 – 0.162 in. thick)	325	47	370	54	260	38	3 to 6
H38(0.006 – 0.020 in. thick)	345	50	...	...	285	41	3
H112							
(0.188 – 0.499 in. thick)	250	36	...	...	125	18	8
(0.500 – 1.000 in. thick)	240	35	...	...	110	16	10
(1.001 – 2.000 in. thick)	240	35	...	...	95	14	14
(2.001 – 3.000 in. thick)	235	34	...	...	95	14	14
H116(0.063 – 2.000 in. thick)	275	40	...	...	195	28	8 to 10

(a) In 2 in. or 4d, where d is diameter of reduced section of tensile test specimen. Where a range of values appears in this column, specified minimum elongation varies with thickness of the mill product.

Mass Characteristics

Density. 2.66 Mg/m³ (0.096 lb/in.³) at 20 °C (68 °F)

Thermal Properties

Liquidus temperature. 638 °C (1180 °F)
Solidus temperature. 574 °C (1065 °F)
Coefficient of thermal expansion. Linear:

Temperature range °C	°F	Average coefficient μm/ m·K	μin./ in.·°F
−50 – +20	−58 – +68	22.3	12.4
20 – 100	68 – 212	24.2	13.4
20 – 200	68 – 392	25.0	13.9
20 – 300	68 – 572	26.0	14.4

Volumetric: 70×10^{-6} m³/m³·K (3.89×10^{-5} in.³/in.³·°F) at 20 °C (68 °F)
Specific heat. 900 J/kg·K (0.215 Btu/lb·°F) at 20 °C (68 °F)
Thermal conductivity. 120 W/m·K (69.3 Btu/ft·h·°F) at 20 °C (68 °F)

Electrical Properties

Electrical conductivity. Volumetric, average of all tempers: 29% IACS at 20 °C (68 °F)
Electrical resistivity. 59.5 nΩ·m at 20 °C (68 °F); temperature coefficient, 0.1 nΩ·m per K at 20 °C (68 °F)
Electrolytic solution potential. −0.91 V vs 0.1N calomel electrode in an aqueous solution containing 53 g NaCl plus 3 g H_2O_2 per litre

Fabrication Characteristics

Annealing temperature. 413 °C (775 °F); holding at temperature not required
Hot working temperature. 316 to 482 °C (600 to 900 °F)

5086, Alclad 5086
4.0Mg-0.4Mn-0.15Cr

Specifications

ASTM. Sheet and plate: B209. Extruded wire, rod, bar, shapes and tube: B221. Extruded seamless tube: B241. Drawn, seamless tube: B210. Welded tube: B313, B547. Gas and oil transmission pipe: B345. Alclad 5086, sheet and plate: B209
SAE. J454
UNS number. A95086
Government. Sheet and plate: QQ-A-250/7, QQ-A-250/19. Extruded wire, rod, bar, shapes and tube: QQ-A-200/5. Drawn, seamless tube: WW-T-700/5.
Foreign. France: NF A-G4MC. Germany: DIN AlMg4. ISO: AlMg4

Chemical Composition

Composition limits. 5086: 0.40 max Si; 0.50 max Fe; 0.20 to 0.7 Mn; 3.5 to 4.5 Mg; 0.25 max Zn; 0.15 max Ti; 0.05 max others (each); 0.15 max others (total); rem Al

Applications

Typical uses. Marine, automotive and aircraft parts, cryogenics, TV towers, drilling rigs, transportation equipment, missile components, armor plate. Applications requiring weldable moderate-strength alloy having comparatively good corrosion resistance.

Mechanical Properties

Tensile properties. See Tables 64 and 65. Tensile strength and elongation are approximately equal in the longitudinal and transverse directions.

Shear properties. Shear strength: O temper, 159 MPa (23 ksi); H34 temper, 186 MPa (27 ksi). Shear yield strength: approximately 55% of tensile yield strength

Compressive yield strength. Approximately the same as tensile yield strength

Poisson's ratio. 0.33

Elastic modulus. Tension, 71.0 GPa $(10.3 \times 10^6$ psi); shear, 26.4 GPa $(3.83 \times 10^6$ psi); compression, 72.4 GPa $(10.5 \times 10^6$ psi)

Mass Characteristics

Density. 2.66 Mg/m^3 (0.096 lb/in.^{3}) at 20 °C (68 °F)

Thermal Properties

Liquidus temperature. 640 °C (1184 °F)

Solidus temperature. 585 °C (1085 °F)

Coefficient of thermal expansion. Linear:

Temperature range		Average coefficient	
°C	°F	µm/ m·K	µin./ in.·°F
−50 – +20	−50 – +68	22.0	12.2
20 – 100	68 – 212	23.8	13.2
20 – 200	68 – 392	24.7	13.7
20 – 300	68 – 572	25.8	14.3

Electrical Properties

Electrical conductivity. Volumetric, average of all tempers: 31% IACS at 20 °C (68 °F)

Electrical resistivity. Average of all tempers: 56.0 nΩ·m at 20 °C (68 °F); temperature coefficient, 0.1 nΩ·m per K at 20 °C (68 °F)

Electrolytic solution potential. −0.88 V vs 0.1N calomel electrode in an aqueous solution containing 53 g NaCl plus 3 g H$_2$O$_2$ per litre

Volumetric: 69×10^{-6} m^3/m^3·K $(3.83 \times 10^{-5}$ in.^{3}/in.^{3}·°F) at 20 °C (68 °F)

Specific heat. 900 J/kg·K (0.215 Btu/lb·°F) at 20 °C (68 °F)

Thermal conductivity. 127 W/m·K (73.4 Btu/ft·h·°F) at 20 °C (68 °F)

Table 65 Typical tensile properties of alloy 5086-O at various temperatures

Temperature		Tensile strength(a)		Yield strength(a)(b)		Elongation,
°C	°F	MPa	ksi	MPa	ksi	%
−196	−320	379	55	131	19	46
−80	−112	269	39	117	17	35
−28	−18	262	38	117	17	32
24	75	262	38	117	17	30
100	212	262	38	117	17	36
149	300	200	29	110	16	50
204	400	152	22	103	15	60
260	500	117	17	76	11	80
316	600	76	11	52	7.5	110
371	700	41	6	29	4.2	130

(a) Lowest strengths for exposures up to 10 000 h at temperature, no load; test loading applied at 5000 psi/min to yield strength and then at strain rate of 5%/min to fracture. (b) 0.2% offset.

Table 66 Mechanical properties of alloy 5154

Temper	Tensile strength MPa	ksi	MPa	ksi	Yield strength MPa	ksi	Elongation(a), %	Hardness(b), HB	Shear strength MPa	ksi	Fatigue strength(c) MPa	ksi
Typical Properties												
O	240	35	...	...	117	17	27	58	152	22	117	17
H32	270	39	...	...	207	30	15	67	152	22	124	18
H34	290	42	...	...	228	33	13	73	165	24	131	19
H36	310	45	...	...	248	36	12	78	179	26	138	20
H38	330	48	...	...	269	39	10	80	193	28	145	21
H112	240	35	...	...	117	17	25	63	...	...	117	17
Property limits												
	Minimum		Maximum		Minimum							
O(0.020 – 3.000 in. thick)	205	30	285	41	75	11	12 to 18	...	...	...	...	...
H32(0.020 – 2.000 in. thick)	250	36	295	43	180	26	5 to 12	...	...	...	...	...
H34(0.009 – 1.000 in. thick)	270	39	315	46	200	29	4 to 10	...	...	...	...	...
H36(0.006 – 0.162 in. thick)	290	42	340	49	220	32	3 to 5	...	...	...	...	...
H38(0.006 – 0.128 in. thick)	310	45	...	...	240	35	3 to 5	...	...	...	...	...
H112												
(0.250 – 0.499 in. thick)	220	32	...	...	125	18	8	...	...	...	...	...
(0.0500 – 3.000 in. thick)	205	30	...	...	75	11	11 to 15	...	...	...	...	...

(a) In 2 in. or 4d, where d is diameter of tensile test specimen. Where a range of values appears in this column, specified minimum elongation varies with thickness of the mill product. (b) 500-kg. load; 10-mm ball. (c) At 5×10^8 cycles of completely reversed stress; R. R. Moore type test.

Fabrication Characteristics

Annealing temperature. 343 °C (650 °F); holding at temperature not required

Hot working temperature. 316 to 482 °C (600 to 900 °F)

5154
3.5Mg-0.25Cr

Specifications

AMS. Sheet and plate: 4018, 4019
ASTM. Sheet and plate: B209. Rolled or cold finished wire, rod and bar: B211. Extruded wire, rod, bar, shapes and tube: B221. Drawn, seamless tube: B210. Welded tube: B313, B547
SAE. J454
UNS number. A95154
Foreign. Canada: CSA GR40. France: NF A-G3C. United Kingdom: BS N5. ISO: AlMg3.5

Chemical Composition

Composition limits. 0.25 max Si; 0.40 max Fe; 0.10 max Cu; 0.10 max Mn; 3.1 to 3.9 Mg; 0.15 to 0.35 Cr; 0.20 max Zn; 0.20 max Ti; 0.05 max others (each); 0.15 max others (total); rem Al

Applications

Typical uses. Welded structures, storage tanks, pressure vessels, marine structures, transportation trailer tanks

Mechanical Properties

Tensile properties. See Tables 66 and 67. Tensile strength and elongation are approximately equal in the longitudinal and transverse directions
Shear properties. Shear strength: see Table 66. Shear yield strength: approximately 55% of tensile yield strength
Compressive yield strength. Approximately the same as tensile yield strength
Hardness. See Table 66.
Poisson's ratio. 0.33
Elastic modulus. Tension, 69.3 GPa (10.1 × 10⁶ psi); shear, 25.9 GPa (3.75 × 10⁶ psi); compression, 70.7 GPa (10.3 × 10⁶ psi)
Fatigue strength. See Table 66.

Table 67 Typical tensile properties of alloy 5154-O at various temperatures

Temperature		Tensile strength		Yield strength(a)		Elongation(b), %
°C	°F	MPa	ksi	MPa	ksi	
−196	−320	360	52	130	19	46
−80	−112	250	36	115	17	35
−28	−18	240	35	115	17	32
24	75	240	35	115	17	30
100	212	240	35	115	17	36
149	300	200	29	110	16	50
204	400	150	22	105	15	60
260	500	115	17	75	11	80
316	600	75	11	50	7.5	110
371	700	41	6	29	4.2	130

(a) 0.2% offset. (b) In 2 in. or 4d, where d is diameter of reduced section of tensile test specimen.

Table 68 Typical tensile properties of alloy 5182

Temper	Tensile strength(a)		Yield strength(a)		Elongation(a)(b), %
	MPa	ksi	MPa	ksi	
O	276	40	138	19	25
H32	317	46	234	34	12
H34	338	49	283	41	10
H19(c)	421	61	393	57	4

(a) Strengths and elongations are unchanged or increased at low temperatures. (b) 1.6-mm (1/16-in.) thick specimen. (c) Properties of this temper are for container end stock 0.25 to 0.38 mm (0.010 to 0.015 in.) thick.

Mass Characteristics

Density. 2.66 Mg/m³ (0.096 lb/in.³) at 20 °C (68 °F)

Thermal Properties

Liquidus temperature. 643 °C (1190 °F)
Solidus temperature. 593 °C (1100 °F)
Coefficient of thermal expansion. Linear:

Temperature range		Average coefficient	
°C	°F	μm/ m·K	μin./ in.·°F
−50 – +20	−58 – +68	22.1	12.3
20 – 100	68 – 212	23.9	13.3
20 – 200	68 – 392	24.9	13.8
20 – 300	68 – 572	25.9	14.4

Volumetric: 69 × 10⁻⁶ m³/m³·K (3.83 × 10⁻⁵ in.³/in.³·°F) at 20 °C (68 °F)
Specific heat. 900 J/kg·K (0.215 Btu/lb·°F) at 20 °C (68 °F)
Thermal conductivity. 127 W/m·K (73.3 Btu/ft·h·°F) at 20 °C (68 °F)

Electrical Properties

Electrical conductivity. Volumetric, average of all tempers: 32% IACS at 20 °C (68 °F)
Electrical resistivity. Average of all tempers: 53.9 nΩ·m at 20 °C (68 °F); temperature coefficient, 0.1 nΩ·m per K at 20 °C (68 °F)
Electrolytic solution potential. −0.86 V vs 0.1N calomel electrode in an aqueous solution containing 53 g NaCl plus 3 g H₂O₂ per litre

Fabrication Characteristics

Annealing temperature. 343 °C (650 °F); holding at temperature not required
Hot working temperature. 260 to 510 °C (500 to 950 °F)

5182
4.5Mg-0.35Mg

Specifications

UNS. J95182

Chemical Composition

Composition limits. 0.20 max Si; 0.35 max Fe; 0.15 max Cu; 0.20 to

0.50 Mn; 4.0 to 5.0 Mg; 0.10 max Cr; 0.25 max Zn; 0.10 max Ti; 0.05 max others (each); 0.15 max others (total); rem Al

Applications

Typical uses. Sheet used for container ends, automotive body panels and reinforcement members, brackets and parts

Mechanical Properties

Tensile properties. See Table 68.
Shear properties. Shear strength: O temper, 152 MPa (22 ksi). Shear yield strength: approximately 55% of tensile yield strength
Compressive yield strength. Approximately the same as tensile yield strength.
Hardness. O temper, 58 HB with 500-kg load, 10-mm diam ball
Poisson's ratio. 0.33
Elastic modulus. Tension, 69.6 GPa (10.1×10^6 psi); compression, 70.9 GPa (10.3×10^6 psi)
Fatigue strength. O temper, 138 MPa (20 ksi) at 5×10^8 cycles in a R. R. Moore type rotating-beam test

Mass Characteristics

Density. 2.65 Mg/m³ (0.096 lb/in.³) at 20 °C (68 °F)

Thermal Properties

Liquidus temperature. 638 °C (1180 °F)
Solidus temperature. 577 °C (1070 °F)
Coefficient of thermal expansion. Linear:

Temperature		Average coefficient	
°C	°F	µm/ m·K	µin./ in.·°F
−50 – +20	−58 – +68...	22.2	12.3
20 – 100	68 – 212 ...	24.1	13.4
20 – 200	68 – 392 ...	25.0	13.9
20 – 300	68 – 572 ...	26.0	14.4

Volumetric: 70×10^{-6} m³/m³·K (3.89×10^{-5} in.³/in.³·°F) at 20 °C (68 °F)
Specific heat. 904 J/kg·K (0.216 Btu/lb·°F) at 20 °C (68 °F)
Thermal conductivity. 123 W/m·K (71.1 Btu/ft·h·°F) at 20 °C (68 °F)

Electrical Properties

Electrical conductivity. Volumetric, 31% IACS at 20 °C (68 °F)
Electrical resistivity. 55.6 nΩ·m at 20 °C (68 °F); temperature coefficient, 0.1 nΩ·m per K at 20 °C (68 °F)

Fabrication Characteristics

Annealing temperature. 343 °C (650 °F)
Hot working temperature. 260 to 510 °C (500 to 950 °F)

5252
2.5Mg

Specifications

ASTM. Sheet: B209
SAE. J454
UNS number. A95252

Chemical Composition

Composition limits. 0.08 max Si; 0.10 max Fe; 0.10 max Cu; 0.10 max Mn; 2.2 to 2.8 Mg; 0.05 max Zn; 0.05 max V; 0.03 max others (each); 0.10 max others (total); rem Al

Applications

Typical uses. Automotive and appliance trim where greater strength is required than in other trim alloys. Can be bright dipped or anodized to give a bright, clear finish

Mechanical Properties

Tensile properties. See Table 69.
Shear strength. H25 temper: 145 MPa (21 ksi); H28, H38 tempers: 160 MPa (23 ksi)
Compressive yield strength. Approximately the same as tensile yield strength
Hardness. H25 temper: 68 HB. H28, H38 tempers: 75 HB. Brinell hardness determined using 500-kg load, 10-mm ball, 30-s duration of loading
Elastic modulus. Tension, 68.3 GPa (9.90×10^6 psi); compression, 69.7 GPa (10.1×10^6 psi)

Mass Characteristics

Density. 2.67 Mg/m³ (0.097 lb/in.³) at 20 °C (68 °F)

Thermal Properties

Liquidus temperature. 649 °C (1200 °F)
Solidus temperature. 607 °C (1125 °F)
Coefficient of thermal expansion. Linear:

Temperature range		Average coefficient	
°C	°F	µm/ m·K	µin./ in.·°F
−50 – +20	−58 – +68	22.0	12.2
20 – 100	68 – 212	23.8	13.2
20 – 200	68 – 392	24.7	13.7
20 – 300	68 – 572	25.8	14.3

Volumetric: 69×10^{-6} m³/m³·K (3.83×10^{-5} in.³/in.³·°F) at 20 °C (68 °F)
Specific heat. 900 J/kg·K (0.215 Btu/lb·°F at 20 °C (68 °F)
Thermal conductivity. 138 W/m·K (80 Btu/ft·h·°F) at 20 °C (68 °F)

Electrical Properties

Electrical conductivity. Volumetric, average of all tempers: 35% IACS at 20 °C (68 °F)
Electrical resistivity. Average of all tempers: 49 nΩ·m at 20 °C (68 °F); temperature coefficient, 0.1 nΩ·m per K at 20 °C (68 °F)

Fabrication Characteristics

Annealing temperature. 343 °C (650 °F); holding at temperature not required
Hot working temperature. 260 to 510 °C (500 to 950 °F)

Table 69 Tensile properties of alloy 5252

Temper		Tensile strength			Yield strength		Elongation, %
	MPa	ksi	MPa	ksi	MPa	ksi	
Typical Properties							
H25............	235	34	···	···	170	25	11(a)
H28, H38.......	283	41	···	···	240	35	5(a)
Property Limits (0.030-0.090 in. thick sheet)							
	Minimum		Maximum				Minimum
H24............	205	30	260	38	···	···	10
H25............	215	31	270	39	···	···	9
H28............	260	38	···	···	···	···	3

(a) 1.6-mm (¹⁄₁₆-in.) thick specimen.

Table 70 Mechanical properties of alloy 5254

Temper	Tensile strength MPa	ksi	MPa	ksi	Yield strength MPa	ksi	Elon- gation, %	Hard- ness(a), HB	Shear strength MPa	ksi	Fatigue strength(b) MPa	ksi
Typical Properties(c)												
O	240	35	...	...	115	17	27	58	150	22	115	17
H32	270	39	...	...	205	30	15	67	150	22	125	18
H34	290	42	...	...	230	33	13	73	165	24	130	19
H36	310	45	...	...	250	36	12	78	180	26	140	20
H38	330	48	...	...	270	39	10	80	195	28	145	21
H112	240	35	...	...	115	17	25	63	...	...	115	17
Property Limits												
	Minimum		Maximum		Minimum		Minimum(d)					
O	205	30	285	41	75	11	12 to 18	...	...	...	...	...
H32	250	36	295	43	180	26	5 to 12	...	...	...	...	...
H34	270	39	315	46	200	29	4 to 10	...	...	...	...	...
H36	290	42	340	49	220	32	3 to 5	...	...	...	...	...
H38	310	45	...	...	240	35	3 to 5	...	...	...	...	...
H112												
(0.250–0.499 in. thick)	220	32	...	...	125	18	8	...	...	...	...	...
(0.500–3.000 in. thick)	205	30	...	...	75	11	11 to 15	...	...	...	...	...

(a) 500 kg load; 10 mm ball. (b) At 5×10^8 cycles; R. R. Moore type test. (c) Strengths and elongations are unchanged or increased at low temperatures. (d) In 2 in. or 4d, where d is diameter of reduced section of test specimen. Where a range of values appears in this column, specified minimum elongation varies with thickness of the mill product.

5254
3.5Mg-0.25Cr

Specifications

ASTM. Sheet and plate: B209. Extruded, seamless tube: B241.
SAE. J454
UNS number. A95254
Foreign. Canada: CSA GR40

Chemical Composition

Composition limits. 0.45 max Si + Fe; 0.05 max Cu; 0.01 max Mn; 3.1 to 3.9 Mg; 0.15 to 0.35 Cr; 0.20 max Zn; 0.05 max Ti; 0.05 max others (each); 0.15 max others (total); rem Al

Applications

Typical uses. Storage vessels for hydrogen peroxide and other chemicals

Mechanical Properties

Tensile properties. See Tables 70 and 71.
Shear yield strength. Approximately 55% of tensile yield strength
Compressive yield strength. Approximately the same as tensile yield strength
Hardness. See Table 70.
Elastic modulus. 69.6 GPa (10.2 $\times$ 10^6 psi); compression, 70.9 GPa (10.3 $\times$ 10^6 psi)
Fatigue strength. See Table 70.

Mass Characteristics

Density. 2.66 Mg/m^3 (0.096 lb/in.^{3}) at 20 °C (68 °F)

Table 71 Typical tensile properties of alloy 5254-O at various temperatures

Temperature °C	°F	Tensile strength(a) MPa	ksi	Yield strength(a) MPa	ksi	Elon- gation, %
−196	−320	360	52	130	19	46
−80	−112	250	36	115	17	35
−28	−18	240	35	115	17	32
24	75	240	35	115	17	30
100	212	240	35	115	17	36
149	300	200	29	110	16	50
204	400	150	22	105	15	60
260	500	115	17	75	11	80
316	600	75	11	50	7.5	110
371	700	41	6	29	4.2	130

(a) Lowest strengths for exposures up to 10 000 h at temperature, no load; test loading applied at 5 000 psi/min to yield strength and then at strain rate of 5%/min to fracture.

Thermal Properties

Liquidus temperature. 643 °C (1190 °F)
Solidus temperature. 593 °C (1100 °F)
Coefficient of thermal expansion. Linear:

Temperature range °C	°F	Average coefficient μm/ m·K	μin./ in.·°F
−50 – +20	−58 – +68	22.1	12.3
20–100	68–212	24.0	13.3
20–200	68–392	24.9	13.8
20–300	68–57	225.9	14.4

Volumetric: 69×10^{-6} m^3/m^3·K (3.83 $\times 10^{-5}$ in.^{3}/in.^{3}·°F) at 20 °C (68 °F)
Specific heat. 900 J/kg·K (0.215 Btu/lb·°F) at 20 °C (68 °F)
Thermal conductivity. 127 W/m·K (73.4 Btu/ft·h·°F) at 20 °C (68 °F)

Electrical Properties

Electrical conductivity. Volumetric, 32% IACS at 20 °C (68 °F)
Electrical resistivity. 54 nΩ·m at 20 °C (68 °F); temperature coefficient, 0.1 nΩ·m per K at 20 °C (68 °F)

Electrolytic solution potential. −0.86 V vs 0.1N calomel electrode in an aqueous solution containing 53 g NaCl plus 3 g H_2O_2 per litre

Fabrication Characteristics

Annealing temperature. 343 °C (650 °F); holding at temperature not required

Hot working temperature. 260 to 510 °C (500 to 950 °F)

5356
5.0Mg-0.12Mn-0.12Cr

Specifications

UNS number. A95356
Government. QQ-R-566, MIL-E-16053
Foreign. Canada: CSA GM50P. France: NF A-G5

Chemical Composition

Composition limits. 0.25 max Si; 0.40 max Fe; 0.10 max Cu; 0.05 to 0.20 Mn; 4.5 to 5.5 Mg; 0.05 to 0.20 Cr; 0.10 max Zn; 0.06 to 0.20 Ti; 0.05 max others (each); 0.15 max others (total); 0.0008 max Be; rem Al

Applications

Typical uses. Welding electrodes and filler wire for base metals with high magnesium content (> 3% Mg)

Mass Characteristics

Density. 2.64 Mg/m³ (0.0954 lb/in.³) at 20 °C (68 °F)

Thermal Properties

Liquidus temperature. 638 °C (1180 °F)
Solidus temperature. 574 °C (1065 °F)
Coefficient of thermal expansion. Linear:

Temperature range		Average coefficient	
°C	°F	μm/ m·K	μin./ in.·°F
−50 – +20	−58 – +68	22.3	12.3
20 – 100	68 – 212	24.2	13.4
20 – 200	68 – 392	25.1	13.9
20 – 300	68 – 572	26.1	14.5

Volumetric: 70×10^{-6} m³/m³·K (3.89×10^{-5} in.³/in.³·°F) at 20 °C (68 °F)

Specific heat. 904 J/kg·K (0.216 Btu/lb·°F) at 20 °C (68 °F)
Thermal conductivity. 116 W/m·K (67 Btu/ft·h·°F) at 20 °C (68 °F)

Electrical Properties

Electrical conductivity. Volumetric, O temper: 29% IACS at 20 °C (68 °F)
Electrical resistivity. O temper: 59.4 nΩ·m at 20 °C (68 °F). Temperature coefficient, 0.1 nΩ·m per K at 20 °C (68 °F)
Electrolytic solution potential. −0.87 V vs 0.1N calomel electrode in an aqueous solution containing 53 g NaCl plus 3 g H_2O_2 per litre

Fabrication Characteristics

Annealing temperature. 343 °C (650 °F); holding at temperature not required
Hot working temperature. 260 to 510 °C (500 to 950 °F)

5454
2.7Mg-0.8Mn-0.12Cr

Specifications

ASTM. Sheet and plate: B209. Extruded wire, rod, bar, shapes and tube: B221. Extruded seamless tube: B241. Condenser tube: B234. Condenser tube with integral fins: B404. Welded tube: B547
SAE. J454

UNS number. A95454
Government. Sheet and plate: QQ-A-250/10. Extruded wire, rod, bar, shapes and tube: QQ-A-200/6
Foreign. Canada: CSA GM31N. France: NF A-G2.5MC. United Kingdom: BS N51. Germany: DIN A1Mg2.7Mn. ISO: AlMg3Mn

Chemical Composition

Composition limits. 0.25 max Si; 0.40 max Fe; 0.10 max Cu; 0.50 to 1.0 Mn; 2.4 to 3.0 Mg; 0.05 to 0.20 Cr; 0.25 max Zn; 0.20 max Ti; 0.05 max others (each); 0.15 max others (total); rem Al

Applications

Typical uses. Welded structures, pressure vessels, tube for marine service

Mechanical Properties

Tensile properties. See Tables 72 and 73.
Shear yield strength. Approximately 55% of tensile yield strength
Compressive yield strength. Approximately the same as tensile yield strength
Hardness. See Table 72.
Elastic modulus. Tension, 69.6 GPa (10.1 × 10⁶ psi); compression, 71.0 GPa (10.3 × 10⁶ psi)

Mass Characteristics

Density. 2.68 Mg/m³ (0.097 lb/in.³) at 20 °C (68 °F)

Table 72 Mechanical properties of alloy 5454

Temper	Tensile strength MPa	ksi	Yield strength MPa	ksi	Elongation, %	Hardness(a), HB	Shear strength MPa	ksi
Typical properties								
O	250	36	⋯ ⋯		117	17	22	62	159	23
H32	275	40	⋯ ⋯		207	30	10	73	165	24
H34	305	44	⋯ ⋯		241	35	10	81	179	26
H36	340	49	⋯ ⋯		276	40	8	⋯	⋯ ⋯
H38	370	54	⋯ ⋯		310	45	8	⋯	⋯ ⋯
H111	260	38	⋯ ⋯		179	26	14	70	159	23
H112	250	36	⋯ ⋯		124	18	18	62	159	23
H311	260	38	⋯ ⋯		179	26	18	70	159	23
Property Limits								
	Minimum		Maximum		Minimum			
O	215	31	285	41	85	12	12 to 18(b)	⋯	⋯ ⋯
H32	250	36	305	44	180	26	5 to 12(b)	⋯	⋯ ⋯
H34	270	39	325	47	200	29	4 to 10(b)	⋯	⋯ ⋯
H112								
0.250-0.499 in. thick	220	32	⋯ ⋯		125	18	8	⋯	⋯ ⋯
0.500-3.00 in. thick	215	31	⋯ ⋯		85	12	11 to 15(b)	⋯	⋯ ⋯

(a) 500-kg load; 10-mm ball. (b) Range of values indicates that specified minimum elongation varies with thickness of mill product.

Thermal Properties

Liquidus temperature. 646 °C (1195 °F)
Solidus temperature. 602 °C (1115 °F)
Coefficient of thermal expansion. Linear:

Temperature range		Average coefficient	
°C	°F	μm/ m · K	μin/ in. · °F
−50 − + 20	−58 − +68 . . .	21.9	12.2
20 − 100	68 − 212 . . .	23.7	13.2
20 − 200	68 − 392 . . .	24.6	13.7
20 − 300	68 − 572 . . .	25.6	14.2

Volumetric: 68×10^{-6} m³/m³·K (3.77×10^{-5} in.³/in.³·°F) at 20 °C (68 °F)
Specific heat. 900 J/kg·K (0.215 Btu/lb·°F) at 20 °C (68 °F)
Thermal conductivity. 134 W/m·K (77.4 Btu/ft·h·°F) at 20 °C (68 °F)

Electrical Properties

Electrical conductivity. Volumetric, average of all tempers: 34% IACS at 20 °C (68 °F)
Electrical resistivity. Average of all tempers: 51 nΩ·m at 20 °C (68 °F). Temperature coefficient, 0.1 nΩ·m per K at 20 °C (68 °F)
Electrolytic solution potential. − 0.86 V vs 0.1N calomel electrode in an aqueous solution containing 53 g NaCl plus 3 g H_2O_2 per litre

Fabrication Characteristics

Annealing temperature. 343 °C (650 °F); holding at temperature not required
Hot working temperature. 260 to 510 °C (500 to 950 °F)

5456
5.1Mg-0.8Mn-0.12Cr

Specifications

ASTM. Sheet and plate: B209. Extruded wire, rod, bar, shapes and tube: B221. Extruded, seamless tube: B241. Drawn, seamless tube: B210
SAE. J454
UNS number. A95456
Government. Sheet and plate: QQ-A-250/9, QQ-A-250/20. Extruded wire, rod, bar, shapes and tube: QQ-A-200/7. Armor plate: MIL-A-46027.

Extruded armor: MIL-A-46083. Forged armor: MIL-A-45225

Chemical Composition

Composition limits. 0.25 max Si; 0.40 max Fe; 0.10 max Cu; 0.50 to 1.0 Mn; 4.7 to 5.5 Mg; 0.05 to 0.20 Cr; 0.25 max Zn; 0.20 max Ti; 0.05 max others (each); 0.15 max others (total); rem Al

Applications

Typical uses. Armor plate, high strength welded structures, storage tanks, pressure vessels, marine service

Mechanical Properties

Tensile properties. See Table 74.

Shear strength. H321, H116 tempers: 207 MPa (30 ksi)
Hardness. H321, H116 tempers: 90 HB
Elastic modulus. Tension, 70.3 GPa (10.2×10^6 psi); compression, 71.7 GPa (10.4×10^6 psi)

Mass Characteristics

Density. 2.66 Mg/m³ (0.096 lb/in.³) at 20 °C (68 °F)

Thermal Properties

Liquidus temperature. 638 °C (1180 °F)
Solidus temperature. 571 °C (1050 °F)
Coefficient of thermal expansion. Linear:

Table 73 Typical tensile properties of alloy 5454 at various temperatures

Temperature °C	°F	Tensile strength(a) MPa	ksi	Yield strength(a) MPa	ksi	Elongation, %
O Temper						
−196	−320	370	54	130	19	39
−80	−112	255	37	115	17	30
−28	−18	250	36	115	17	27
24	75	250	36	115	17	25
100	212	250	36	115	17	31
149	300	200	29	110	16	50
204	400	150	22	105	15	60
260	500	115	17	75	11	80
316	600	75	11	50	7.5	110
371	700	41	6	29	4.2	130
H32 Temper						
−196	−320	405	59	250	36	32
−80	−112	290	42	215	31	23
−28	−18	285	41	205	30	20
24	75	275	40	205	30	18
100	212	270	39	200	29	20
149	300	220	32	180	26	37
204	400	170	25	130	19	45
260	500	115	17	75	11	80
316	600	75	11	50	7.5	110
371	700	41	6	29	4.2	130
H34 Temper						
−196	−320	435	63	285	41	30
−80	−112	315	46	250	36	21
−28	−18	305	44	240	35	18
24	75	305	44	240	35	16
100	212	295	43	235	34	18
149	300	235	34	195	28	32
204	400	180	26	130	19	45
260	500	115	17	75	11	80
316	600	75	11	50	7.5	110
371	700	41	6	29	4.2	130

(a) Lowest strengths for exposures up to 10 000 h at temperature, no load, test loading applied at 5 000 psi/min to yield strength and then at strain rate of 5%/min to fracture.

Table 74 Tensile properties of alloy 5456

Temper	MPa	Tensile strength ksi	MPa	ksi	MPa	Yield strength ksi	MPa	ksi	Elongation, %
Typical properties									
O	310	45	· · ·	· · ·	159	23			24(a)
H111	324	47	· · ·	· · ·	228	33			18(a)
H112	310	45	· · ·	· · ·	165	24			22(a)
H321(b), H116(c)	352	51	· · ·	· · ·	255	37			16(a)

| **Property limits** | | | | | | | | Minimum(d) | |
	Minimum		Maximum		Minimum		Maximum	In 50 mm	In 5d(5.65√A)
O									
1.20 thru 6.30 mm thick	290	42	365	53	130	19	205	30	16	· · ·
6.30 thru 80.00 mm thick	285	41	360	53	125	18	205	30	16	14
80.00 thru 120.00 mm thick	275	40	· · ·	· · ·	120	17	· · ·	· · ·	· · ·	12
120.00 thru 160.00 mm thick	270	39	· · ·	· · ·	115	17	· · ·	· · ·	· · ·	12
160.00 thru 200.00 mm thick	265	38	· · ·	· · ·	105	15	· · ·	· · ·	· · ·	10
H112									
6.30 thru 40.00 mm thick	290	42	· · ·	· · ·	130	19	· · ·	· · ·	12	10
40.00 thru 80.00 mm thick	285	41	· · ·	· · ·	125	18	· · ·	· · ·	· · ·	10
H116(c)(e)									
1.60 thru 30.00 mm thick	315	46	· · ·	· · ·	230	33	· · ·	· · ·	10	10
30.00 thru 40.00 mm thick	305	44	· · ·	· · ·	215	31	· · ·	· · ·	· · ·	10
40.00 thru 80.00 mm thick	285	41	· · ·	· · ·	200	29	· · ·	· · ·	· · ·	10
80.00 thru 110.00 mm thick	275	40	· · ·	· · ·	170	25	· · ·	· · ·	· · ·	10
H321									
4.00 thru 12.50 mm thick	315	46	405	59	230	33	315	46	12	· · ·
12.50 thru 40.00 mm thick	305	44	385	56	215	31	305	44	· · ·	10
40.00 thru 80.00 mm thick	285	41	385	56	200	29	295	43	· · ·	10
H323									
1.20 thru 6.30 mm thick	330	48	400	58	250	36	315	46	6 to 8	· · ·
H343									
1.20 thru 6.30 mm thick	365	53	435	63	285	41	350	51	6 to 8	· · ·

(a) 12.5 mm (½ in.) diam specimen. (b) Material in this temper not recommended for applications requiring exposure to seawater. (c) H116 designation also applies to the condition previously designated H117. (d) Elongations in 50 mm apply to thicknesses thru 12.5 mm (½ in.); elongations in 5d (5.65 A), where d is diameter and A is cross-sectional area of tensile test specimen, apply to material over 12.5 mm thick. (e) Material in this temper required to pass an exfoliation corrosion test administered by the purchaser.

Temperature range °C	°F	Average coefficient μm/ m·K	μin./ in.·°F
−50– +20	−58– +68	22.1	12.3
20–100	68–212	23.9	13.3
20–200	68–392	24.8	13.8
20–300	68–572	25.9	14.4

Volumetric: 69×10^{-6} m³/m³·K (3.83×10^{-5} in.³/in.³·°F) at 20 °C (68 °F)
Specific heat. 900 J/kg·K (0.215 Btu/lb·°F) at 20 °C (68 °F)
Thermal conductivity. 116 W/m·K (67 Btu/ft·h·°F) at 20 °C (68 °F)

Electrical Properties

Electrical conductivity. Volumetric, average of all tempers: 29% IACS at 20 °C (68 °F)
Electrical resistivity. Average of all tempers: 59.5 nΩ·m at 20 °C (68 °F); temperature coefficient, 0.1 nΩ·m per K at 20 °C (68 °F)
Electrolytic solution potential. −0.87 V vs 0.1N calomel electrode in an aqueous solution containing 53 g NaCl plus 3 g H_2O_2 per litre

Fabrication Characteristics

Annealing temperature. 343 °C (650 °F); holding at temperature not required
Hot working temperature. 260 to 510 °C (500 to 950 °F)

5457
1.0Mg-0.30Mn

Specifications

ASTM. Sheet: B209
UNS number. A95457

Chemical Composition

Composition limits. 0.08 max Si; 0.10 max Fe; 0.20 max Cu; 0.15 to 0.45 Mn; 0.08 to 1.2 Mg; 0.05 max Zn; 0.05 max V; 0.03 max others (each); 0.10 max others (total); rem Al

Applications

Typical uses. Brightened and anodized automotive and appliance trim
Precautions in use. Fine grain size required for most applications of this alloy

Mechanical Properties

Tensile properties. O temper, thickness 0.030 to 0.090 in. Tensile strength: min, 110 MPa (16 ksi); max, 152 MPa (22 ksi). Elongation, 20% in 50 mm or 2 in. See also Table 75.
Shear strength. See Table 75.

Compressive yield strength. Approximately the same as tensile yield strength
Hardness. See Table 75.
Poisson's ratio. 0.33 at 20 °C (68 °F)
Elastic modulus. Tension, 68.2 GPa (10.0×10^6 psi); shear, 25.9 GPa (3.75×10^6 psi); compression, 69.6 GPa (10.1×10^6 psi)

Mass Characteristics

Density. 2.69 Mg/m³ (0.0972 lb/in.³) at 20 °C (68 °F)

Thermal Properties

Liquidus temperature. 654 °C (1210 °F)
Solidus temperature. 629 °C (1165 °F)
Coefficient of thermal expansion. Linear:

Temperature range		Average coefficient	
°C	°F	µm/ m·K	µin./ in.·°F
−50 − +20	−58 − +68	21.9	12.2
20 − 100	68 − 212	23.7	13.2
20 − 200	68 − 392	24.6	13.7
20 − 300	68 − 572	25.6	14.2

Volumetric: 68×10^{-6} m³/m³·K (3.77×10^{-5} in.³/in.³·°F) at 20 °C (68 °F)
Specific heat. 900 J/kg·K (0.215 Btu/lb·°F) at 20 °C (68 °F)
Thermal conductivity. 177 W/m·K (102 Btu/ft·h·°F) at 20 °C (68 °F)

Electrical Properties

Electrical conductivity. Volumetric, average of all tempers: 46% IACS at 20 °C (68 °F)
Electrical resistivity. 37.5 nΩ·m at 20 °C (68 °F); temperature coefficient, 0.1 nΩ·m per K at 20 °C (68 °F)

Electrolytic solution potential. −0.84 V vs 0.1N calomel electrode in an aqueous solution containing 53 g NaCl plus 3 g H_2O_2 per litre

Fabrication Characteristics

Formability. Readily formed in both annealed and H25 tempers
Annealing temperature. 343 °C (650 °F); holding temperature not required
Hot working temperature. 260 to 510 °C (500 to 950 °F)

5652
2.5Mg-0.25Cr

Specifications

ASTM. Sheet and plate: B209. Extruded, seamless tube: B241
SAE. J454
UNS number. A95652

Chemical Composition

Composition limits. 0.40 max Si + Fe; 0.04 max Cu; 0.01 max Mn; 2.2 to 2.8 Mg; 0.15 to 0.35 Cr; 0.10 max Zn; 0.05 max others (each); 0.15 max others (total); rem Al

Applications

Typical uses. Storage vessels for hydrogen peroxide and other chemicals

Mechanical Properties

Tensile properties. See Table 76.
Shear strength. See Table 76.
Compressive yield strength. Approximately the same as tensile yield strength
Hardness. See Table 76.
Poisson's ratio. 0.33
Elastic modulus. Tension, 68.2 GPa (9.89×10^6 psi); shear, 25.9 GPa (3.75×10^6 psi); compression, 69.6 GPa (10.1×10^6 psi)
Fatigue strength. See Table 76.

Mass Characteristics

Density. 2.68 Mg/m³ (0.097 lb/in.³) at 20 °C (68 °F)

Thermal Properties

Liquidus temperature. 649 °C (1200 °F)

Table 75 Typical mechanical properties of alloy 5457

Temper	Tensile strength(a) MPa	ksi	Yield strength(a) MPa	ksi	Elongation(a)(b), %	Hardness(c), HB	Shear strength MPa	ksi
O	130	19	50	7	22	32	85	12
H25	180	26	160	23	12	48	110	16
H38, H28	205	30	185	27	6	55	125	18

(a) Strengths and elongations are unchanged or improved at lower temperatures. (b) 1.6-mm (1/16-in.) thick specimen. (c) 500-kg load; 10-mm ball.

Table 76 Mechanical properties of alloy 5652

Temper	Tensile strength MPa	ksi	MPa	ksi	Yield strength MPa	ksi	Elongation(a), %	Hardness(b), HB	Shear strength MPa	ksi	Fatigue strength(c) MPa	ksi
Typical Properties												
O	195	28	···	···	90	13	25	47	124	18	110	16
H32	230	33	···	···	195	28	12	60	138	20	117	17
H34	260	38	···	···	215	31	10	68	145	21	124	18
H36	275	40	···	···	240	35	8	73	158	23	131	19
H38	290	42	···	···	255	37	7	77	165	24	138	20
Property Limits												
	Minimum		Maximum		Minimum		Minimum					
O	170	25	215	31	65	9.5	14 to 18	···	···	···	···	···
H32	215	31	260	38	160	23	4 to 12	···	···	···	···	···
H34	235	34	285	41	180	26	3 to 10	···	···	···	···	···
H36	255	37	305	44	200	29	2 to 4	···	···	···	···	···
H38	270	39	···	···	220	32	2 to 4	···	···	···	···	···
H112												
(0.250 to 0.499 in. thick)	195	28	···	···	110	16	7	···	···	···	···	···
(0.500 to 3.000 in. thick)	170	25	···	···	65	9.5	12 to 16	···	···	···	···	···

(a) In 2 in. or 4d, where d is diameter of reduced section of tension-test specimen. Where a range of values appears in this column, the specified minimum elongation varies with thickness of the mill product. (b) 500-kg load; 10-mm ball. (c) At 5×10^8 cycles; R.R. Moore type test.

Solidus temperature. 607 °C (1125 °F)

Coefficient of thermal expansion. Linear:

Temperature range		Average coefficient	
°C	°F	μm/ m·K	μin./ in.·°F
−50 − +20	−58 − +68	22.0	12.2
20 − 100	68 − 212	23.8	13.2
20 − 200	68 − 392	24.7	13.7
20 − 300	68 − 572	25.8	14.3

Volumetric: 69×10^{-6} m³/m³·K (3.83×10^{-5} in.³/in.³·°F) at 20 °C (68 °F)
Specific heat. 900 J/kg·K (0.215 Btu/lb·°F) at 20 °C (68 °F)
Thermal conductivity. 137 W/m·K (79.1 Btu/ft·h·°F) at 20 °C (68 °F)

Electrical Properties

Electrical conductivity. Volumetric, average of all tempers: 35% IACS at 20 °C (68 °F)
Electrical resistivity. 49.0 nΩ·m at 20 °C (68 °F); temperature coefficient, 0.1 nΩ·m per K at 20 °C (68 °F)
Electrolytic solution potential. −0.85 V vs 0.1N calomel electrode in an aqueous solution containing 53 g NaCl plus 3 g H_2O_2 per litre

Fabrication Characteristics

Annealing temperature. 343 °C (650 °F); holding at temperature not required
Hot working temperature. 260 to 510 °C (500 to 950 °F)

5657
0.8Mg

Specifications

ASTM. B209
UNS number. A95657
Foreign. Italy: P-AlMg0.9

Chemical Composition

Composition limits. 0.08 max Si; 0.10 max Fe; 0.10 max Cu; 0.03 max Mn; 0.6 to 1.0 Mg; 0.05 max Zn; 0.03 max Ga; 0.05 max V; 0.02 max others (each); 0.05 max others (total); rem Al

Applications

Typical uses. Brightened and anod-

ized automotive and appliance trim
Precautions in use. Fine grain size essential for almost all applications of this alloy

Mechanical Properties

Tensile properties. See Table 77.
Shear strength. H25 temper: 95 MPa (14 ksi); H28, H38 tempers: 105 MPa (15 ksi)
Compressive yield strength. Approximately the same as tensile yield strength
Hardness. H25 temper: 40 HB. H28 and H38 tempers: 50 HB. All hardness values obtained with 500-kg load, 10-mm diam ball and 30-s duration of loading
Poisson's ratio. 0.33
Elastic modulus. Tension, 68.2 GPa (9.89×10^6 psi); shear, 25.9 GPa (3.75×10^6 psi); compression, 69.6 GPa (10.1×10^6 psi)

Mass Characteristics

Density. 2.69 Mg/m³ (0.097 lb/in.³) at 20 °C (68 °F)

Thermal Properties

Liquidus temperature. 657 °C (1215 °F)
Solidus temperature. 638 °C (1180 °F)
Coefficient of thermal expansion. Linear:

Temperature range		Average coefficient	
°C	°F	μm/ m·K	μin./ in.·°F
−50 − +20	−58 − +68	21.9	12.2
20 − 100	68 − 212	23.7	13.2
20 − 200	68 − 392	24.6	13.7
20 − 300	68 − 572	25.6	14.2

Table 77 Tensile properties of alloy 5657

Temper	Tensile strength				Yield strength		Elongation(a), %
	MPa	ksi	MPa	ksi	MPa	ksi	
Typical Properties(b)							
H25	160	23	...	...	140	20	12
H28, H38	195	28	...	...	165	24	7
Property Limits							
	Minimum		Maximum				Minimum
H241(c)	125	18	180	26	...	...	13
H25	140	20	195	28	...	...	8
H26	150	22	205	30	...	...	7
H28	170	25	...	...	...	...	5

(a) In 2 in. or 4d, where d is diameter of reduced section of tension-test specimen. (b) Strengths and elongations are unchanged or increased at low temperatures. (c) Material in this temper subject to some recrystallization and attendant loss of brightness.

Volumetric: 68×10^{-6} m³/m³·K (3.77×10^{-5} in.³/in.³·°F) at 20 °C (68 °F)
Specific heat. 900 J/kg·K (0.215 Btu/lb·°F)

Electrical Properties

Electrical conductivity. Volumetric, 54% IACS at 20 °C (68 °F)
Electrical resistivity. 32 nΩ·m at 20 °C (68 °F); temperature coefficient, 0.1 nΩ·m per K at 20 °C (68 °F)

Fabrication Characteristics

Annealing temperature. 343 °C (650 °F); holding at temperature not required
Hot working temperature. 260 to 510 °C (500 to 950 °F)

6005
0.8Si-0.5Mg

Specifications

ASTM. Extruded wire, rod, bar, shapes and tube: B221
SAE. J454
UNS. A96005

Chemical Composition

Composition limits. 0.6 to 0.9 Si; 0.35 max Fe; 0.10 max Cu; 0.10 max Mn; 0.40 to 0.6 Mg; 0.10 max Cr; 0.10 max Zn; 0.10 max Ti; 0.05 max others (each); 0.15 max others (total); rem Al

Applications

Typical uses. Extruded shapes and tubing for commercial applications requiring strength greater than that of 6063; ladders and TV antennas are among the more common products **Precautions in use.** Not recommended for applications requiring resistance to impact loading

Mechanical Properties

Tensile properties. Tensile strength (minimum): T1 temper, 172 MPa (25 ksi); T5 temper, 262 MPa (38 ksi). Yield strength (minimum): T1 temper, 103 MPa (15 ksi); T5 temper: 241 MPa (35 ksi). Elongation (minimum): T1 temper, 16%; T5 temper, 8 to 10%, specific value varies with thickness of mill product.
Shear strength. T5 temper: 205 MPa (30 ksi)
Hardness. T5 temper: 95 HB
Elastic modulus. Tension, 69 GPa (10×10^6 psi)
Fatigue strength. (minimum). 97 MPa (14 ksi) at 5×10^8 cycles; R. R. Moore type test

Mass Characteristics

Density. 2.70 Mg/m^3 (0.098 lb/in.^{3}) at 20 °C (68 °F)

Thermal Properties

Liquidus temperature. 654 °C (1210 °F)
Solidus temperature. 607 °C (1125 °F)
Coefficient of thermal expansion. Linear, 23.4 μm/m·K (13.0 μin./in.·°F) at 20 to 100 °C (68 to 212 °F)
Thermal conductivity. T5 temper: 167 W/m·K (97 Btu/ft·h·°F) at 25 °C (77 °F)

Electrical Properties

Electrical conductivity. Volumetric, T5 temper: 49% IACS at 20 °C (68 °F)
Electrical resistivity. T5 temper: 40 nΩ·m at 20 °C (68 °F)

Fabrication Characteristics

Annealing temperature. 415 °C (778 °F); hold at temperature for 2 to 3 h
Solution temperature. 547 °C (1015 °F)
Aging temperature. 175 °C (346 °F), hold at temperature for 8 h

Table 78 Typical tensile properties of alloy 6009 automobile body sheet

Orientation	Tensile strength MPa	ksi	Yield strength MPa	ksi	Elongation, %
T4 Temper					
Longitudinal	234	34	131	19	24
Transverse and 45°	228	33	124	18	25
T6 Temper					
Longitudinal	345	50	324	47	12
Transverse and 45°	338	49	296	43	13

6009
0.80Si-0.60Mg-0.50Mn-0.35Cu

Specifications
UNS. A96009

Chemical Composition
Composition limits. 0.6 to 1.0 Si; 0.50 max Fe; 0.15 to 0.6 Cu; 0.20 to 0.8 Mn; 0.40 to 0.8 Mg; 0.10 max Cr; 0.25 max Zn; 0.10 max Ti; 0.05 max others (each); 0.15 max others (total); rem Al

Applications
Typical uses. Automobile body sheet

Mechanical Properties
Tensile properties. See Table 78
Yield stretch. Following simulated forming and a paint bake cycle consisting of 1 h at 175 °C (350 °F). T4 temper: no stretch, 228 MPa (33 ksi); 5% stretch, 262 MPa (38 ksi); 10% stretch, 290 MPa (42 ksi)
Shear strength. Auto body sheet, T4 temper: 152 MPa (22 ksi)
Hardness. T4 temper, auto body sheet: 70 HR15T
Poisson's ratio. 0.33
Elastic modulus. Tension, 69 GPa (10×10^6 psi); shear, 25.4 GPa (3.75×10^6 psi)
Fatigue strength. T4 temper: 117 MPa (17 ksi) at 10×10^6 cycles; sheet flexural specimens

Mass Characteristics
Density. 2.71 Mg/m^3 (0.098 lb/in.^{3}) at 20 °C (68 °F)

Thermal Properties
Liquidus temperature. 650 °C (1202 °F)
Solidus temperature. 560 °C (1040 °F)
Coefficient of thermal expansion. Linear:

Temperature range °C	°F	Average coefficient μm/ m·K	μin./ in.·°F
−50 − +20	−58 − +68	21.6	12.0
20 − 100	68 − 212	23.4	13.0
20 − 200	68 − 392	24.3	13.5
20 − 300	68 − 572	25.2	14.0

Volumetric: 67×10^{-6} m^3/m^3·K (3.72×10^{-5} in.^{3}/in.^{3}·°F) at 20 °C (68 °F)
Specific heat. 897 J/kg·K (0.214 Btu/lb·°F) at 20 °C (68 °F)
Thermal conductivity. At 20 °C (68 °F): O temper, 205 W/m·K (118 Btu/ft·h·°F); T4 temper, 172 W/m·K (99 Btu/ft·h·°F); T6 temper, 180 W/m·K (104 Btu/ft·h·°F)

Electrical Properties
Electrical conductivity. Volumetric, at 20 °C (68 °F): O temper, 54% IACS; T4 temper, 44% IACS; T6 temper: 47% IACS
Electrical resistivity. At 20 °C (68 °F): O temper, 31.9 nΩ·m; T4 temper, 39.2 nΩ·m; T6 temper, 36.7 nΩ·m. Temperature coefficient, 0.1 nΩ·m per K at 20 °C (68 °F)

Fabrication Characteristics
Formability. Auto body sheet, T4 temper. ½ *t* radius required for 90° bending or for flanging material 0.80 to 1.30 mm (0.032 to 0.050 in.) thick. Standard hems, which are made by bending 180° over 1 *t* interface thickness, also can be made in auto body sheet 0.80 to 1.30 mm thick. Olsen cup height, typically 0.38 in. when tested using a 1-in. diam top die, 2200 psi hold-down pressure and polyethylene film as a lubricant. Strain-hardening exponent (*n*) typically 0.23; plastic strain ratio (*r*) typically 0.70
Annealing temperature. 413 °C (775 °F)

Solution temperature. 554 °C (1030 °F)

Aging temperature. 177 °C (350 °F)

6010
1.0Si-0.8Mg-0.5Mn-0.35Cu

Specifications

UNS. A96010

Chemical Composition

Composition limits. 0.8 to 1.2 Si; 0.50 max Fe; 0.15 to 0.6 Cu; 0.20 to 0.8 Mn; 0.60 to 1.0 Mg; 0.10 max Cr; 0.25 max Zn; 0.10 max Ti; 0.05 max others (each); 0.15 max others (total); rem Al

Applications

Typical uses. Automobile body sheet

Mechanical Properties

Tensile properties. Typical. T4 temper: tensile strength, 290 MPa (42 ksi); yield strength, 172 MPa (25 ksi); elongation, 24% in 50 mm or 2 in. See also Table 79.

Yield stretch. Following simulated forming and a paint bake cycle consisting of 1 h at 175 °C (350 °F). T4 temper: no stretch, 255 MPa (37 ksi); 5% stretch, 296 MPa (43 ksi); 10% stretch, 324 MPa (47 ksi)

Hardness. T4 temper: 76 HR15T

Poisson's ratio. 0.33

Elastic modulus. Tension, 69 GPa (10×10^6 psi); shear, 25.4 GPa (3.75×10^6 psi)

Fatigue strength. T4 temper: 117 MPa (17 ksi) at 10×10^6 cycles; sheet flexural specimens

Mass Characteristics

Density. 2.70 Mg/m³ (0.098 lb/in.³) at 20 °C (68 °F)

Thermal Properties

Liquidus temperature. 649 °C (1200 °F)

Solidus temperature. 585 °C (1085 °F)

Incipient melting temperature. 577 °C (1070 °F)

Coefficient of thermal expansion. Linear:

Table 79 Typical tensile properties of alloy 6010 automobile body sheet

Orientation	Tensile strength		Yield strength		Elongation, %
	MPa	ksi	MPa	ksi	
T4 Temper					
Longitudinal	296	43	186	27	23
Transverse and 45°	290	42	172	25	24
T6 Temper					
Longitudinal	386	56	372	54	11
Transverse and 45°	379	55	352	51	12

Temperature range		Average coefficient	
°C	°F	μm/ m·K	μin./ in.·°F
−50 – +20	−58 – +68	21.5	11.9
20 – 100	68 – 21	223.2	12.9
20 – 200	68 – 392	24.1	13.4
20 – 300	68 – 572	25.1	13.9

Volumetric: 67×10^{-6} m³/m³·K (3.72×10^{-5} in.³/in.³·°F) at 20 °C (68 °F)

Specific heat. 897 J/kg·K (0.214 Btu/lb·°F) at 20 °C (68 °F)

Thermal conductivity. At 20 °C (68 °F): O temper, 202 W/m·K (117 Btu/ft·h·°F); T4 temper, 151 W/m·K (87.3 Btu/ft·h·°F); T6 temper, 180 W/m·K (104 Btu/ft·h·°F)

Electrical Properties

Electrical conductivity. Volumetric, at 20 °C (68 °F): O temper, 53% IACS; T4 temper, 39% IACS; T6 temper, 44% IACS

Electrical resistivity. At 20 °C (68 °F): O temper, 32.5 nΩ·m; T4 temper, 44.2 nΩ·m; T6 temper, 39.2 nΩ·m. Temperature coefficient, 0.1 nΩ·m per K at 20 °C (68 °F)

Fabrication Characteristics

Formability. Auto body sheet, T4 temper. 1 t radius required for 90° bending, 1 t for flanging material 0.80 to 1.30 mm (0.032 to 0.050 in.) thick. Only roped hems, which are made by bending 180° over 2 t interface thickness, can be made in auto body sheet 0.80 to 1.30 mm thick. Olsen cup height, typically 0.36 in. when tested using a 1-in. diam top die, 2200 psi hold-down pressure and polyethylene film as a lubricant. Strain-hardening exponent (n) typically 0.22; plastic strain ratio (r) typically 0.70

Annealing temperature. 413 °C (775 °F)

Solution temperature. 566 °C (1050 °F)

Aging temperature. 177 °C (350 °F)

6061, Alclad 6061
1.0Mg-0.6Si-0.30Cu-0.20Cr

Specifications

AMS. See Table 80.
ASTM. See Table 80.
UNS. A96061
Government. See Table 80.
Foreign. Canada: CSA GS11N. France: NF A-G5UC. United Kingdom: BS H20. ISO: AlMg1SiCu

Chemical Composition

Composition limits. 6061: 0.40 to 0.8 Si; 0.7 max Fe; 0.15 to 0.40 Cu; 0.15 max Mn; 0.8 to 1.2 Mg; 0.04 to 0.35 Cr; 0.25 max Zn; 0.15 max Ti; 0.05 max others (each); 0.15 max others (total); rem Al. Alclad 6061: 7072 cladding—0.7 max Si + Fe; 0.10 max Cu; 0.10 max Mn; 0.10 max Mg; 0.8 to 1.3 Zn; 0.05 max others (each); 0.15 max others (total); rem Al

Applications

Typical uses. Trucks, towers, canoes, railroad cars, furniture, pipelines and other structural applications where strength, weldability and corrosion resistance are needed

Mechanical Properties

Tensile properties. See Tables 81 and 82.
Shear strength. See Table 81.
Hardness. O temper: 30 HB; T4, T451 tempers: 65 HB; T6, T651 tempers: 95 HB. Data obtained using 500-kg load, 10 mm-diam ball and 30-s duration of loading
Elastic modulus. Tension, 68.3 GPa (10.0×10^6 psi); compression, 69.7 GPa (10.1×10^6 psi)

Fatigue strength. O temper: 62 MPa (9 ksi). T4, T451, T6 and T651 tempers: 97 MPa (14 ksi). Data correspond to 5×10^8 cycles of completely reversed stress in R. R. Moore type tests.

Mass Characteristics

Density. 2.70 Mg/m³ (0.098 lb/in.³) at 20 °C (68 °F)

Thermal Properties

Liquidus temperature. 652 °C (1206 °F)

Solidus temperature. 582 °C (1080 °F)

Coefficient of thermal expansion. Linear, 23.6 μm/m·K (13.1 μin./in.·°F) at 20 to 100 °C (68 to 212 °F)

Specific heat. 896 J/kg·K (0.214 Btu/lb·°F) at 20 °C (68 °F)

Thermal conductivity. At 25 °C (77 °F): O temper, 180 W/m·K (104 Btu/ft·h·°F); T4 temper, 154 W/m·K (89.0 Btu/ft·h·°F); T6 temper, 167 W/m·K (96.5 Btu/ft·h·°F)

Electrical Properties

Electrical conductivity. Volumetric at 20 °C (68 °F): O temper, 47% IACS; T4 temper, 40% IACS; T6 temper: 43% IACS

Electrical resistivity. At 20 °C (68 °F): O temper, 37 nΩ·m; T4 temper, 43 nΩ·m; T6 temper, 40 nΩ·m

Fabrication Characteristics

Solution temperature. 529 °C (985 °F)

Aging temperature. Rolled or drawn products: 160 °C (320 °F); hold at temperature for 18 h. Extrusions or forgings: 177 °C (350 °F); hold at temperature for 8 h

Table 80 Standard specifications for alloy 6061

Mill form and condition	AMS	Specification No. ASTM	Government
Sheet and plate	4025	B209	QQ-A-250/11
	4026	...	...
	4027	...	...
	4043	...	...
	4053	...	
Tread plate	...	B632	MIL-F-17132
Wire, rod, and bar (rolled or cold finished)	4115	B211	QQ-A-225/8
	4116	...	...
	4117	...	...
	4128	...	...
	4129	...	...
Rod, bar, shapes and tube (extruded)	4150	B221	QQ-A-200/8
	4160	...	...
	4161	...	...
	4172	...	...
	4173	...	...
Structural shapes	4113	B308	QQ-A-200/8
Tube (extruded, seamless)	...	B241	...
Tube (drawn)	...	B483	...
Tube (seamless)	4079	B210	WW-T-700/6
	4080	...	...
	4082	...	...
Tube (hydraulic)	4081	...	MIL-T-7081
	4083	...	...
Tube (condenser)	...	B234	...
Tube (condenser with integral fins)	...	B404	...
Tube (welded)	...	B313	...
	...	B549	...
Tube (wave guide)	...	...	MIL-W-85
	...	...	MIL-W-23068
	...	...	MIL-W-23351
Pipe	...	B241	MIL-P-25995
Pipe (gas and oil transmission)	...	B345	...
Forgings	4127	B247	QQ-A-367, MIL-A-22771
	4146		
Forging stock	4127	...	QQ-A-367
	4146	...	...
Rivet wire	...	B316	QQ-A-430
Impacts	...	...	MIL-A-12545
Structural pipe and tube (extruded)	...	B429	MIL-P-25995
Alclad			
Sheet and plate	4020	B209	...
	4021	...	...
	4022	...	...
	4023	...	...

Table 81 Typical mechanical properties of alloy 6061

Temper	Tensile strength MPa	ksi	Yield strength MPa	ksi	Elongation, % 1.6 mm (1/16 in.) thick specimen	13 mm (1/2 in.) diam specimen	Shear strength MPa	ksi
6061								
O	124	18	55	8	25	30	83	12
T4, T451	241	35	145	21	22	25	165	24
T6, T651	310	45	276	40	12	17	207	30
Alclad 6061								
O	117	17	48	7	25	...	76	11
T4, T451	228	33	131	19	22	...	152	22
T6, T651	290	42	255	37	12	...	186	27

Table 82 Typical tensile properties of alloy 6061-T6 or T651 at various temperatures

Temperature		Tensile strength(a)		Yield strength(a)(b)		Elon-gation, %
°C	°F	MPa	ksi	MPa	ksi	
−196	−320	414	60	324	47	22
−80	−112	338	49	290	42	18
−28	−18	324	47	283	41	17
24	75	310	45	276	40	17
100	212	290	42	262	38	18
149	300	234	34	214	31	20
204	400	131	19	103	15	28
260	500	51	7.5	34	5	60
316	600	32	4.6	19	2.7	85
371	700	24	3	12	1.8	95

(a) Lowest strength for exposures up to 10 000 h at temperature, no load; test loading applied at 5 000 psi/min to yield strength and then at strain rate of 5% per min to fracture. (b) 0.2% offset.

Table 84 Typical mechanical properties of alloy 6063

Temper	Tensile strength		Yield strength		Elon-gation, %	Hard-ness(a), HB	Shear strength		Fatigue strength	
	MPa	ksi	MPa	ksi			MPa	ksi	MPa	ksi
O	90	13	48	7	...	25	69	10	55	8
T1(c)	152	22	90	13	20	42	97	14	62	9
T4	172	25	90	13	22	...	...	...	...	...
T5	186	27	145	21	12	60	117	17	69	10
T6	241	35	214	31	12	73	152	22	69	10
T83	255	37	241	35	9	82	152	22	...	...
T831	207	30	186	27	10	70	124	18	...	...
T832	290	42	269	39	12	95	186	27	...	...

(a) 500-kg load; 10-mm diam ball. (b) At 5 x 10⁸ cycles; R. R. Moore type test. (c) Formerly T42 temper.

Table 83 ASTM specifications for alloy 6063

Mill form and condition	ASTM No.
Wire, rod, bar, shapes and tube (extruded)	B221
Tube (extruded, seamless); pipe	B241
Tube (extruded, coiled)	B491
Tube (drawn)	B483
Tube (drawn, seamless)	B210
Pipe (gas and oil transmission)	B345
Structural pipe and tube (extruded)	B429

Coefficient of thermal expansion. Linear:

Temperature range		Average coefficient	
°C	°F	µm/ m·K	µin./ in.·°F
−50 − +20	−58 − +68	21.8	12.1
20 − 100	68 − 212	23.4	13.0
20 − 200	68 − 392	24.5	13.6
20 − 300	68 − 572	25.6	14.2

Specific heat. 900 J/kg·K (0.215 Btu/lb·°F) at 20 °C (68 °F)
Thermal conductivity. At 25 °C (77 °F):

Temper	Conductivity	
	W/m·K	Btu/ft·h·°F
O	218	126
T1 (formerly T42)	193	112
T5	209	121
T6	201	116

6063
0.7Mg-0.4Si

Specifications

AMS. Extruded wire, rod, bar, shapes and tube: 4156
ASME. Extruded wire, rod, bar, shapes and tube: SB221. Pipe: SB241
ASTM. See Table 83.
SAE. J454
UNS. A96063
Government. QQ-A-200/9, MIL-P-25995
Foreign. Austria: Onorm AlMgSi0,5. Canada: CSA GS10. France: NF A-GS. Italy: UNI P-AlSi0.4Mg. United Kingdom: BS H19; DTD 372B. Germany: DIN AlMgSi0.5; Werkstoff-Nr. 3.3206. ISO: AlMgSi

Chemical Composition

Composition limits. 0.20 to 0.6 Si; 0.35 max Fe; 0.10 max Cu; 0.10 max Mn; 0.45 to 0.9 Mg; 0.10 max Cr; 0.10 max Zn; 0.10 max Ti; 0.05 max others (each); 0.15 max others (total); rem Al

Applications

Typical uses. Pipe, railings, furniture, architectural extrusions, truck and trailer flooring, doors, windows, irrigation pipes

Mechanical Properties

Tensile properties. See Tables 84 and 85.
Hardness. See Table 84.
Poisson's ratio. 0.33
Elastic modulus. Tension, 68.3 GPa (9.91 × 10⁶ psi); shear, 25.8 GPa (3.75 × 10⁶ psi); compression, 69.7 GPa (10.1 × 10⁶ psi)

Mass Characteristics

Density. 2.69 Mg/m³ (0.097 lb/in.³)

Thermal Properties

Liquidus temperature. 655 °C (1211 °F)
Solidus temperature. 615 °C (1139 °F)

Electrical Properties

Electrical conductivity. At 20 °C (68 °F):

Temper	Conductivity, % IACS	
	Equal volume	Equal weight
O	58	191
T1 (formerly T42)	50	165
T5	55	181
T6, T83	53	175

Electrical resistivity. At 20 °C (68 °F):

Temper	Resistivity, nΩ·m
O	30
T1 (formerly T42)	35
T5	32
T6, T83	33

Chemical Properties

General corrosion resistance. Highly resistant to all types of corrosion

Fabrication Characteristics

Machinability. Fair, depending on temper

Weldability. For all commercial processes, excellent weldability and brazability

Annealing temperature. 415 °C (775 °F); hold at temperature 2 to 3 h; cool at 28 °C (50 °F) per h from 415 °C (775 °F) to 260 °C (500 °F)

Solution temperature. 520 °C (970 °F)

Aging temperature. T5 temper: 205 °C (400 °F), hold at temperature for 1 h; or 182 °C (360 °F), hold at temperature for 1 h. All other artificially aged tempers: 175 °C (350 °F); hold at temperature for 8 h

6066
1.4Si-1.1Mg-1.0Cu-0.8Mn

Specifications

ASTM. Extruded wire, rod, bar, shapes and tube: B221
SAE. J454
UNS number. A96066
Government. Extruded wire, rod, bar, shapes and tube: QQ-A-200/10. Forgings: QQ-A-367
Foreign. United Kingdom: BS H11

Chemical Composition

Composition limits. 0.9 to 1.8 Si; 0.50 max Fe; 0.7 to 1.2 Cu; 0.6 to 1.1 Mn; 0.8 to 1.4 Mg; 0.40 max Cr; 0.25 max Zn; 0.20 max Ti; 0.50 max others (each); 0.15 max others (total); rem Al

Applications

Typical uses. Forgings and extrusions for welded structures

Mechanical Properties

Tensile properties. See Table 86.
Shear strength. Typical. O temper: 97 MPa (14 ksi); T4 and T451 tempers: 200 MPa (29 ksi); T6 and T651 tempers: 234 MPa (34 ksi)
Hardness. O temper: 43 HB; T4 and T451 tempers: 90 HB; T6 and T651 tempers: 120 HB

Table 85 Typical tensile properties of alloy 6063 at various temperatures

Temperature °C	°F	Tensile strength(a) MPa	ksi	Yield strength(b) MPa	ksi	Elongation, %
−196	−320	234	34	110	16	44
−80	−112	179	26	103	15	36
−28	−18	165	24	97	14	34
24	75	152	22	90	13	33
100	212	152	22	97	14	18
149	300	145	21	103	15	20
204	400	62	9	45	6.5	40
260	500	31	4.5	24	3.5	75
316	600	23	3.2	17	2.5	80
371	700	16	2.3	14	2	105
T5 Temper						
−196	−320	255	37	165	24	28
−80	−112	200	29	152	22	24
−28	−18	193	28	152	22	23
24	75	186	27	145	21	22
100	212	165	24	138	20	18
149	300	138	20	124	18	20
204	400	62	9	45	6.5	40
260	500	31	4.5	24	3.5	75
316	600	23	3.2	17	2.5	80
371	700	16	2.3	14	2	105
T6 Temper						
−196	−320	324	47	248	36	24
−80	−121	262	38	228	33	20
−28	−18	248	36	221	32	19
24	75	241	35	214	31	18
100	212	214	31	193	28	15
149	300	145	21	133	20	20
204	400	62	9	45	6.5	40
260	500	31	4.5	24	3.5	75
316	600	23	3.3	17	2.5	80
371	700	16	2.3	14	2	105

(a) Lowest strength for exposures up to 10 000 h at temperature, no load; test loading applied at 5 000 psi/min to yield strength and then at strain rate of 5%/min to fracture. (b) 0.2% offset.

Table 86 Tensile properties of alloy 6066

Temper	Tensile strength MPa	ksi	Yield strength(a) MPa	ksi	Elongation(b), %
Typical Properties					
O	150	22	83	12	18
T4, T451	360	52	207	30	18
T6, T651	395	57	359	52	12
Property Limits, Extrusions	Minimum		Minimum		Minimum
O	200 max	29 max	125 max	18 max	16
T4, T4510, T4511	275	40	170	25	14
T42	275	40	165	24	14
T6, T6510, T6511	345	50	310	45	8
T62	345	50	290	42	8
Property Limits, Die Forgings					
T6	345	50	310	45	...

(a) 0.2% offset. (b) In 2 in. or 4d, where d is diameter of reduced section of tensile test specimen.

Elastic modulus. Tension, 69 GPa (10 × 10⁶ psi)

Fatigue strength. T6 and T651 tempers, 110 MPa (16 ksi). Data correspond to 5 × 10⁸ cycles in R. R. Moore type tests.

Mass Characteristics

Density. 2.71 Mg/m³ (0.098 lb/in.³) at 20 °C (68 °F)

Thermal Properties

Liquidus temperature. 648 °C (1195 °F)

Solidus temperature. 563 °C (1045 °F)

Coefficient of thermal expansion. Linear, 23.2 μm/m·K (12.9 μin./in.·°F) at 20 to 100 °C (68 to 212 °F)

Specific heat. 887 J/kg·K (0.212 Btu/lb·°F) at 20 °C (68 °F)

Thermal conductivity. T6 temper, 147 W/m·K (85 Btu/ft·h·°F) at 20 °C (68 °F)

Electrical Properties

Electrical conductivity. Volumetric, at 20 °C (68 °F): O temper, 40% IACS; T6 temper, 37% IACS

Electrical resistivity. At 20 °C (68 °F): O temper, 43 nΩ·m; T6 temper, 47 nΩ·m

Fabrication Characteristics

Annealing temperature. 415 °C (778 °F); hold at temperature 2 to 3 h

Solution temperature. 530 °C (990 °F); followed by quenching

Aging temperature. 175 °C (350 °F); hold at temperature 8 h

6070
1.4Si-0.8Mg-0.7Mn-0.3Cu

Specifications

ASTM. Gas and oil transmission pipe: B345

SAE. J454

Government. Extruded rod, bar, shapes and tube: MIL-A-46104. Impacts: MIL-A-12545

Chemical Composition

Composition limits. 1.0 to 1.7 Si; 0.50 max Fe; 0.15 to 0.40 Cu; 0.40 to 1.0 Mn; 0.50 to 1.2 Mg; 0.10 max Cr; 0.25 max Zn; 0.15 max Ti; 0.05 max others (each); 0.15 max others (total); rem Al

Applications

Typical uses. Heavy duty welded structures, pipelines, extruded structural components for automobiles

Mechanical Properties

Tensile properties. Typical. Tensile strength: O temper, 145 MPa (21 ksi); T4 temper, 317 MPa (46 ksi); T6 temper, 379 MPa (55 ksi). Yield strength: O temper, 69 MPa (10 ksi); T4 temper, 172 MPa (25 ksi); T6 temper, 352 MPa (51 ksi). Elongation: O and T4 tempers, 20%; T6 temper, 10%

Shear strength. Typical. O temper: 97 MPa (14 ksi); T4 temper: 206 MPa (30 ksi); T6 temper: 234 MPa (34 ksi)

Hardness. O temper: 35 HB; T4 temper: 90 HB; T6 temper: 120 HB. Data obtained using 500-kg load, 10-mm diam ball and 30-s duration of loading.

Elastic modulus. Tension, 68 GPa (9.9 × 10⁶ psi)

Fatigue strength. O temper: 62 MPa (9 ksi); T4 temper: 90 MPa (13 ksi); T6 temper: 97 MPa (14 ksi). Data correspond to 5 × 10⁸ cycles of completely reversed stress in R. R. Moore type test

Mass Characteristics

Density. 2.71 Mg/m³ (0.098 lb/in.³)

Thermal Properties

Liquidus temperature. 649 °C (1200 °F)

Solidus temperature. 566 °C (1050 °F)

Specific heat. 891 J/kg·K (0.213 Btu/lb·°F) at 20 °C (68 °F)

Thermal conductivity. T6 temper: 172 W/m·K (99.1 Btu/ft·h·°F) at 20 °C (68 °F)

Electrical Properties

Electrical conductivity. Volumetric, T6 temper: 44% IACS at 20 °C (68 °F)

Electrical resistivity. 39 nΩ·m at 20 °C (68 °F)

Fabrication Characteristics

Solution temperature. 546 °C (1015 °F); followed by quenching

Annealing temperature. T4 temper: 545 °C (1015 °F)

Aging temperature. 160 °C (320 °F); hold at temperature for 18 h

---

6101
0.6Mg-0.5Si

Specifications

ASTM. Bus conductor: B317

SAE. J454

UNS number. A96101

Foreign. Austria: Önorm E-AlMgSi. France: NF A-GS/L. Italy: UNI P-AlSi0.5Mg. Switzerland: VSM Al-Mg-Si. United Kingdom: BS 91E. Germany: E-AlMgSi0.5; Werkstoff-Nr. 3.3207

Chemical Composition

Composition limits. 0.30 to 0.7 Si; 0.50 max Fe; 0.10 max Cu; 0.03 max Mn; 0.35 to 0.8 Mg; 0.03 max Cr; 0.10 max Zn; 0.06 max B; 0.03 max others (each); 0.10 max others (total); rem Al

Applications

Typical uses. High strength bus bars, electrical conductors, heat sinks

Mechanical Properties

Tensile properties. Typical. Tensile strength, 221 MPa (32 ksi); yield strength, 193 MPa (28 ksi); elongation, 15%. See also Tables 87 and 88.

Shear strength. 138 MPa (20 ksi)

Hardness. 71 HB with 500-kg load, 10-mm diam ball

Elastic modulus. Tension, 68.9 GPa (10.0 × 10⁶ psi); compression, 70.3 GPa (10.2 × 10⁶ psi)

Mass Characteristics

Density. 2.69 Mg/m³ (0.097 lb/in.³) at 20 °C (68 °F)

Thermal Properties

Liquidus temperature. 654 °C (1210 °F)

Solidus temperature. 621 °C (1150 °F)

Coefficient of thermal expansion. Linear:

Temperature range		Average coefficient	
°C	°F	μm/m·K	μin./in.·°F
−50 – +20	−58 – +68	21.7	12.0
20 – 100	68 – 212	23.5	13.0
−20 – 200	68 – 392	24.4	13.5
20 – 300	68 – 572	25.4	14.1

Table 87 Typical tensile properties of alloy 6101-T6 at various temperatures

Temperature		Tensile strength(a)		Yield strength(a)(b)		Elonga-tion(c),
°C	°F	MPa	ksi	MPa	ksi	%
−196	−320	296	43	228	33	24
−80	−112	248	36	207	30	20
−28	−18	234	34	200	29	19
24	75	221	32	193	28	19
100	212	193	28	172	25	20
149	300	145	21	131	19	20
204	400	69	10	48	7	40
260	500	33	4.8	23	3.3	80
316	600	24	3	16	2.3	100
371	700	17	2.5	12	1.8	105

(a) Lowest strength for exposures up to 10 000 h at temperature, no load; test loading applied at 5 000 psi/min to yield strength and then at strain rate of 5%/min to fracture. (b) 0.2% offset. (c) In 50 mm or 2 in.

Table 88 Property limits for alloy 6061 extrusions

Temper	Tensile strength(a)		Yield strength(a)		Electrical conductivity(a), % IACS
	MPa	ksi	MPa	ksi	
H111	83	12	55	8	59
T6	200	29	172	25	55
T61					
0.125-0.749 in. thick	138	20	103	15	57
0.750-1.499 in. thick	124	18	76	11	57
1.500-2.000 in. thick	103	15	55	8	57
T63	186	27	152	22	56
T64	103	15	55	8	59.5
T65	172-221	25-32	138-186	20-27	56.5

(a) Single entries are minimum values.

Specific heat. 895 J/kg·K (0.214 Btu/lb·°F) at 20 °C (68 °F)
Thermal conductivity. 218 W/m·K (138 Btu/ft·h·°F) at 25 °C (77 °F)

Electrical Properties

Electrical conductivity and resistivity at 20 °C (68 °F):

Temper	Electrical conductivity, % IACS	Electrical resistivity, nΩ·m
T6	57	30.2
T61	59	29.2
T63	58	29.7
T64	60	28.7
T65	58	29.7

Fabrication Characteristics

Solution temperature. 510 °C (950 °F); hold for 1 h at temperature
Aging temperature. 177 °C (349 °F); hold for 6 to 8 h at temperature
Hot working temperature. 260 to 510 °C (500 to 950 °F)

6151
0.9Si-0.6Mg-0.25Cr

Specifications

AMS. Forgings: 4125
SAE. J454
UNS number. A96151
Government. Forgings and forging

stock: QQ-A-367; MIL-A-22771
Foreign. Canada: CSA SG11P

Chemical Composition

Composition limits. 0.6 to 1.2 Si; 1.0 max Fe; 0.35 max Cu; 0.20 max Mn; 0.45 to 0.8 Mg; 0.15 to 0.35 Cr; 0.25 max Zn; 0.15 max Ti; 0.05 max others (each); 0.15 max others (total); rem Al

Applications

Typical uses. Die forgings and rolled rings for crank cases, fuses and machine parts. Applications requiring good forgeability, good strength and resistance to corrosion

Mechanical Properties

Tensile properties. See Tables 89 and 90.
Hardness. T6 temper: 90 HB with 500-kg load, 10-mm diam ball

Mass Characteristics

Density. 2.70 Mg/m³ (0.098 lb/in.³) at 20 °C (68 °F)

Thermal Properties

Liquidus temperature. 649 °C (1200 °F)
Solidus temperature. 588 °C (1090 °F)
Coefficient of thermal expansion. Linear:

Temperature range		Average coefficient	
°C	°F	μm/ m·K	μin./ in.·°F
−50 − +20	−58 − +68	21.8	12.1
20 − 100	68 − 212	23.0	12.8
20 − 200	68 − 392	24.1	13.4
20 − 300	68 − 572	25.0	13.9

Table 89 Tensile-property limits for alloy 6151

Temper	Tensile strength		Yield strength		Elon-gation(a),
	MPa	ksi	MPa	ksi	%
Die Forgings					
T6					
Axis parallel to grain flow	303	44	255	37	14 (coupon)
					10 (forging)
Axis not parallel to grain flow	303	44	255	37	6 (forging)
Rolled Rings					
T6 and T652					
Tangential	303	44	255	37	5
Axial	303	44	241	35	4
Radial	290	42	241	35	2

(a) In 2 in. or 4d, where d is diameter of reduced section of tensile test specimen.

Table 90 Typical tensile properties of alloy 6151

Temperature		Tensile strength(a)		Yield strength(a)(b)		Elongation, %
°C	°F	MPa	ksi	MPa	ksi	
−196	−321	395	57	345	50	20
−80	−112	345	50	315	46	17
−28	−18	340	49	310	45	17
24	76	330	48	298	43	17
100	212	295	43	275	40	17
149	300	195	28	185	27	20
204	400	95	14	85	12	30
260	500	45	6.5	34	5	50
316	600	34	5	27	3.9	43
371	700	28	4	22	3.2	35

(a) Lowest strength for exposures up to 10 000 h at temperature, no load; test loading applied at 5000 psi/min to yield strength and then at strain rate of 5%/min to fracture. (b) 0.2% offset.

Specific heat. 895 J/kg·K (0.214 Btu/lb·°F) at 20 °C (68 °F)

Thermal conductivity. At 20 °C (68 °F): O temper, 205 W/m·K (118 Btu/ft·h·°F); T4 temper, 163 W/m·K (94 Btu/ft·h·°F); T6 temper, 175 W/m·K (101 Btu/ft·h·°F)

Electrical Properties

Electrical conductivity. Volumetric, at 20 °C (68 °F): O temper, 54% IACS; T4 temper, 42% IACS; T6 temper, 45% IACS

Electrical resistivity. At 20 °C (68 °F): O temper, 32 nΩ·m; T4 temper, 41 nΩ·m; T6 temper, 38 nΩ·m

Electrolytic solution potential. −0.83 V vs 0.1N calomel electrode in an aqueous solution containing 53 g NaCl plus 3 g H_2O_2 per litre

Fabrication Characteristics

Annealing temperature. 413 °C (775 °F); hold at temperature 2 to 3 h; furnace cool to 260 °C (500 °F) at 27 °C (50 °F) per h max

Solution temperature. 510 to 525 °C (950 to 975 °F); hold at temperature 4 min, quench in cold water; heavy or complicated forgings, quench in water at 65 to 100 °C (150 to 212 °F)

Aging temperature. 165 to 175 °C (300 to 345 °F); hold at temperature 8 to 12 h

Hot working temperature. 260 to 480 °C (500 to 900 °F)

6201
0.7Si-0.8Mg

Specifications

ASTM. Wire, B398. Stranded conductor, T81 temper: B399

SAE. J454

UNS. A96201

Chemical Composition

Composition limits. 0.50 to 0.95 Si; 0.50 max Fe; 0.10 max Cu; 0.03 max Mn; 0.6 to 0.9 Mg; 0.03 max Cr; 0.10 max Zn; 0.06 max B; 0.03 max others (each); 0.10 max others (total); rem Al

Applications

Typical uses. Rod and wire for high strength electrical conductors

Mechanical Properties

Tensile properties. Typical. T81 temper: tensile strength, 331 MPa (48 ksi); yield strength, 310 MPa (45 ksi); elongation, 6% in 250 mm or 10 in.

Property limits for T81 temper wire. Specified diameter, 1/16 to 1/8 in.: min tensile strength (individual), 315 MPa (46 ksi); min tensile strength (average), 330 MPa (48 ksi). Specified diameter 1/8 to 3/16 in.: min tensile strength (individual), 305 MPa (44 ksi); min tensile strength (average), 315 MPa (46 ksi). Min elongation, 3% in 250 mm or 10 in. for all diameters.

Mass Characteristics

Density. 2.69 Mg/m³ (0.097 lb/in.³) at 20 °C (68 °F)

Thermal Properties

Liquidus temperature. 654 °C (1210 °F)

Solidus temperature. 607 °C (1125 °F)

Coefficient of thermal expansion.

Temperature range		Average coefficient	
°C	°F	µm/m·K	µin./in.°F
−50-+20	−58-+68	21.6	12.0
20-100	68-212	23.4	13.0
20-200	68-392	24.3	13.5
20-300	68-572	25.2	14.0

Specific heat. 895 J/kg·K (0.214 Btu/lb·°F) at 20 °C (68 °F)

Thermal conductivity. T8 temper: 205 W/m·K (118 Btu/ft·h·°F) at 25 °C (77 °F)

Electrical Properties

Electrical conductivity. Volumetric, T81 temper: 54% IACS at 20 °C (68 °F)

Electrical resistivity. T81 temper: 32 nΩ·m at 20 °C (68 °F)

Fabrication Characteristics

Solution temperature. 510 °C (950 °F)

Aging temperature. 150 °C (302 °F); hold at temperature approximately 4 h

6205
0.8Si-0.5Mg-0.10Mn-0.10Cr-0.10Zr

Specifications

UNS. A96205

Chemical Composition

Composition limits. 0.6 to 0.9 Si; 0.7 max Fe; 0.20 max Cu; 0.05 to 0.15 Mn; 0.40 to 0.6 Mg; 0.05 to 0.15 Cr; 0.25 max Zn; 0.05 to 0.15 Zr; 0.15 max Ti; 0.05 max others (each); 0.15 max others (total); rem Al

Applications

Typical uses. Plate, tread plate and extrusions for applications requiring high impact strength

Mechanical Properties

Tensile properties. Typical. T1 temper: tensile strength, 262 MPa (38 ksi); yield strength, 138 MPa (20 ksi); elongation, 19%. T5 temper: tensile strength, 310 MPa (45 ksi); yield strength, 290 MPa (42 ksi); elongation, 11%

Shear strength. T5 temper: 207 MPa (30 ksi)

Hardness. T1 temper: 65 HB; T5 temper: 95 HB

Fatigue strength. T5 temper: 103 MPa (15 ksi) at 5×10^8 cycles in R. R. Moore type test

Mass Characteristics

Density. 2.70 Mg/m³ (0.098 lb/in.³)

Thermal Properties

Liquidus temperature. 645 °C (1210 °F)

Solidus temperature. 613 °C (1135 °F)

Coefficient of thermal expansion. Linear, 23.0 µm/m·K (12.8 µin./in.·°F)

Thermal conductivity. At 25 °C (77 °F): T1 temper, 172 W/m·K (99.1 Btu/ft·h·°F); T5 temper, 188 W/m·K (109 Btu/ft·h·°F)

Electrical Properties

Electrical conductivity. Volumetric, At 20 °C (68 °F): T1 temper, 45% IACS; T5 temper, 49% IACS

Electrical resistivity. At 20 °C (68 °F): T1 temper, 37 nΩ·m per K; T5 temper, 35 nΩ·m

Fabrication Characteristics

Solution temperature. 527 °C (980 °F)

Aging temperature. 177 °C (350 °F); hold at temperature approximately 6 h

---

6262
1.0Mg-0.6Si-0.3Cu-0.09Cr-0.6Pb-0.6Bi

Specifications

ASTM. Rolled or cold finished wire, rod and bar: B211. Extruded wire, rod, bar, shapes and tube: B221. Drawn, seamless tube: B210. Drawn tube: B483

SAE. J454

UNS. A96262

Government. Rolled or cold finished wire, rod and bar: QQ-A-225/10

Chemical Composition

Composition limits. 0.40 to 0.8 Si; 0.7 max Fe; 0.15 to 0.40 Cu; 0.15 max Mn; 0.8 to 1.2 Mg; 0.04 to 0.14 Cr; 0.25 max Zn; 0.15 max Ti; 0.40 to 0.7 Bi; 0.40 to 0.7 Pb; 0.05 max others (each); 0.15 max others (total); rem Al

Table 91 Typical tensile properties of alloy 6262 at various temperatures

Temperature		Tensile strength(a)		Yield strength(a)(b)		Elongation, %
°C	°F	MPa	ksi	MPa	ksi	
T651 Temper						
−196	−320	414	60	324	47	22
−80	−112	338	49	290	42	18
−28	−18	324	47	283	41	17
24	75	310	45	276	40	17
100	212	290	42	262	38	18
149	300	234	34	214	31	20
T9 Temper						
−196	−320	510	74	462	67	14
−80	−112	427	62	400	58	10
−28	−18	414	60	386	56	10
24	75	400	58	379	55	10
100	212	365	53	359	52	10
149	300	262	38	255	37	14
204	400	103	15	90	13	34
260	500	59	8.5	41	6	48
316	600	32	4.6	19	2.7	85
371	700	24	3	12	1.8	95

(a) Lowest strength for exposures up to 10 000 h at temperature, no load; test loading applied at 5 000 psi/min to yield strength and then at strain rate of 5% per min to fracture. (b) 0.2% offset.

Applications

Typical uses. High-stress screw machine products requiring corrosion resistance superior to 2011 and 2017

Mechanical Properties

Tensile properties. Typical, T9 temper: tensile strength, 400 MPa (58 ksi); 0.2% yield strength, 379 MPa (55 ksi); see also Table 91.

Shear strength. Typical, T9 temper: 241 MPa (35 ksi)

Hardness. Typical, T9 temper: 120 HB with 500-kg load, 10-mm diam ball

Fatigue strength. Typical, T9 temper: 90 MPa (13 ksi) at 5×10^8 cycles; R. R. Moore type test

Mass Characteristics

Density. 2.71 Mg/m³ (0.098 lb/in.³) at 20 °C (68 °F)

Thermal Properties

Liquidus temperature. 652 °C (1204 °F)

Solidus temperature. 582 °C (1078 °F)

Coefficient of thermal expansion. Linear, 23.4 µm/m·K (13.0 µin./in.·°F) at 20 to 100 °C (68 to 212 °F)

Thermal conductivity. T9 temper: 172 W/m·K (99.1 Btu/ft.·h·°F) at 20 °C (68 °F)

Electrical Properties

Electrical conductivity. Volumetric, T9 temper: 44% IACS at 20 °C (68 °F)

Electrical resistivity. T9 temper: 39 nΩ·m at 20 °C (68 °F)

Fabrication Characteristics

Annealing temperature. 415 °C (780 °F); hold at temperature 2 to 3 h

Solution temperature. 540 °C (1000 °F); hold at temperature 8 to 12 h

Aging temperature. 170 °C (340 °F); hold at temperature 8 to 12 h

---

6351
1.0Si-0.6Mg-0.6Mn

Specifications

ASTM. Gas and oil transmission pipe: B345. Extruded wire, rod, bar, shapes and tube: B221

UNS. A96351

Chemical Composition

Composition limits. 0.7 to 1.3 Si; 0.50 max Fe; 0.10 max Cu; 0.40 to 0.8 Mn; 0.40 to 0.8 Mg; 0.20 max Zn; 0.20 max Ti; 0.05 max others (each); 0.15 others (total); rem Al

Applications

Typical uses. Extruded structures used in road vehicles and railroad stock; tubing and pipe for carrying water, oil or gasoline

Mechanical Properties

Tensile properties. Typical. T4 temper: tensile strength, 248 MPa (36 ksi); 0.2% yield strength, 152 MPa (22 ksi); elongation, 20%. T6 temper: tensile strength, 310 MPa (45 ksi); 0.2% yield strength, 283 MPa (41 ksi); elongation, 14%. Property limits for extrusions, T54 temper: tensile strength (min), 207 MPa (30 ksi); 0.2% yield strength (min), 138 MPa (20 ksi); elongation (min), 10%

Shear strength. T6 temper, 200 MPa (29 ksi)

Hardness. T6 temper, 95 HB with 500-kg load, 10-mm diam ball

Fatigue strength. Typical, T6 temper: 90 MPa (13 ksi) at 5×10^8 cycles in R. R. Moore type test

Mass Characteristics

Density. 2.71 Mg/m^3 (0.098 lb/in.^{3})

Thermal Properties

Liquidus temperature. 650 °C (1202 °F)

Solidus temperature. 555 °C (1030 °F)

Coefficient of thermal expansion. Linear, 23.4 µm/m·K (13.0 µin./in.·°F) at 20 to 80 °C (68 to 176 °F)

Thermal conductivity. 213 W/m·K (102 Btu/ft·h·°F) at 25 °C (77 °F)

Electrical Properties

Electrical conductivity. Volumetric, T6 temper: 46% IACS at 20 °C (68 °F)

Electrical resistivity. 38 nΩ·m at 20 °C (68 °F)

Fabrication Characteristics

Annealing temperature. 350 °C (662 °F); hold at temperature for about 4 h

Solution temperature. 505 °C (940 °F)

Aging temperature. 170 °C (338 °F); hold at temperature 6 h

6463
0.40Si-0.7Mg

Specifications

ASTM. Extruded wire, rod, bar, shapes and tube: B221
SAE. J454
UNS number. A96463
Foreign. United Kingdom: BS E6

Chemical Composition

Composition limits. 0.20 to 0.6 Si; 0.15 max Fe; 0.20 max Cu; 0.05 max Mn; 0.45 to 0.9 Mg; 0.05 max Zn; 0.05 max others (each); 0.15 max others (total); rem Al

Applications

Typical uses. Architectural, appliance, and bright anodized automotive extrusions

Mechanical Properties

Tensile properties. Typical. Tensile strength: T1 temper, 152 MPa (22 ksi); T5 temper, 186 MPa (27 ksi); T6 temper, 241 MPa (35 ksi). 0.2% yield strength: T1 temper, 90 MPa (13 ksi); T5 temper, 145 MPa (21 ksi); T6 temper: 214 MPa (31 ksi). Elongation: T1 temper, 20%; T5 and T6 tempers: 12%

Shear strength. T1 temper, 97 MPa (14 ksi); T5 temper, 117 MPa (17 ksi); T6 temper, 152 MPa (22 ksi)

Hardness. T1 temper, 42 HB; T5 temper, 60 HB; T6 temper, 74 HB. Values obtained with 500-kg load and 10-mm diam ball.

Fatigue strength. All tempers: 69 MPa (10 ksi) at 5×10^8 cycles; R. R. Moore type test

Mass Characteristics

Density. 2.69 Mg/m^3 (0.097 lb/in.^{3})

Thermal Properties

Liquidus temperature. 654 °C (1210 °F)

Solidus temperature. 621 °C (1150 °F)

Coefficient of thermal expansion. Linear, 23.4 µm/m·K (13.0 µin./in.·°F) at 20 to 100 °C (68 to 212 °F)

Thermal conductivity. At 25 °C (77 °F): T1 temper, 192 W/m·K (111 Btu/ft·h·°F); T5 temper, 209 W/m·K (121 Btu/ft·h·°F); T6 temper, 201 W/m·K (116 Btu/ft·h·°F)

Electrical Properties

Electrical conductivity. Volumetric, at 20 °C (68 °F): T1 temper, 50%

IACS; T5 temper, 55% IACS; T6 temper, 53% IACS

Electrical resistivity. At 20 °C (68 °F): T1 temper, 34 nΩ·m; T5 temper, 31 nΩ·m; T6 temper, 33 nΩ·m

Fabrication Characteristics

Annealing temperature. 415 °C (780 °F)

Solution temperature. 520 °C (968 °F)

Aging temperature. To produce T6 temper: 175 °C (350 °F), hold at temperature 8 h; can also use 182 °C (360 °F), hold at temperature 6 h. To produce T5 temper: 205 °C (400 °F), hold at temperature 1 h; can also use 182 °C (360 °F), hold at temperature 3 h

7005
4.6Zn-1.4Mg-0.5Mn-0.1Cr-0.1Zr-0.03Ti

Specifications

ASTM. Extruded wire, rod, bar, shapes and tube: B221
UNS number. A97005

Chemical Composition

Composition limits. 0.10 max Cu; 1.0 to 1.8 Mg; 0.20 to 0.70 Mn; 0.35 max Si; 0.40 max Fe; 0.06 to 0.20 Cr; 0.01 to 0.06 Ti; 4.0 to 5.0 Zn; 0.08 to 0.20 Zr; 0.05 max others (each); 0.15 max others (total); rem Al

Applications

Typical uses. Extruded structural members such as frame rails, cross members, corner posts, side posts and stiffeners for trucks, trailers, cargo containers and rapid transit cars. Welded or brazed assemblies requiring moderately high strength and high fracture toughness, such as large heat exchangers, especially where solution heat treatment after joining is impractical. Sports equipment such as tennis racquets and softball bats

Precautions in use. To avoid stress corrosion cracking, stresses in the transverse direction should be avoided at exposed machined or sawed surfaces. Parts should be cold formed in O temper, then heat treated; alternatively, parts may be cold formed in W temper, followed by

Table 92 Minimum mechanical properties of alloy 7005

Temper	Tensile strength MPa	ksi	Yield strength MPa	ksi	Elon-gation(a), %	Compres-sive yield strength MPa	ksi	Shear strength MPa	ksi	Shear yield strength MPa	ksi	Bearing strength MPa	ksi	Bearing yield strength MPa	ksi
Extrusions															
T53															
L direction...........	345	50	303	44	10	296	43	193	28	172	25	655(b)	95(b)	503(b)	73(b)
												496(c)	72(c)	407(c)	59(c)
LT direction.........	331	48	290	42	...	303	44	...	...	...	...	...	...	...	...
Sheet and Plate															
T6(d), T63(e), T6351(e) ..	324	47	262	38	...	269	39	186	27	152	22	634(b)	92(b)	448(b)	65(b)
												483(c)	70(c)	365(c)	53(c)

(a) In 2 in. or 4d, where d is diameter of reduced section of tensile test specimen. (b) e/d = 2.0, where e is edge distance and d is pin diameter. (c) e/d = 1.5. (d) Up to 0.250 in. thick. (e) 0.250 to 3.00 in. thick.

artificial aging. In parts intended for service in aggressive electrolytes such as seawater, selective attack along the heat affected zone in a weldment or torch-brazed assembly can be avoided by postweld aging. When the service environment is conducive to galvanic corrosion, 7005 should be coupled or joined only to aluminum alloy components having similar electrolytic solution potentials; alternatively, joint surfaces should be protected or insulated.

Mechanical Properties

Tensile properties. Typical. Tensile strength: O temper, 193 MPa (28 ksi); T53 temper, 393 MPa (57 ksi); T6, T63, T6351 tempers, 372 MPa (54 ksi). Yield strength: O temper, 83 MPa (12 ksi); T53 temper, 345 MPa (50 ksi); T6, T63, T6351 tempers, 317 MPa (46 ksi). Elongation in 2 in. or 4 d, where d is diameter of tensile test specimen: O temper, 20%; T53 temper, 15%; T6, T63, T6351 tempers, 12%. See also Tables 92 and 93.

Shear strength. Typical. O temper: 117 MPa (17 ksi); T53 temper: 221 MPa (32 ksi); T6, T63, T6351 tempers: 214 MPa (31 ksi); see also Table 92.

Compressive strength. See Table 92.

Elastic modulus. Tension, 71 GPa (10.3 × 10⁶ psi); shear, 26.9 GPa (3.9 × 10⁶ psi); compression, 72.4 GPa (10.5 × 10⁶ psi)

Fatigue strength. Rotating beam at 10⁸ cycles. T6351 plate: smooth specimens, 115 to 130 MPa (17 to 19 ksi); 60° notched specimens, 20 to 50 MPa (3 to 7 ksi). T53 extrusions: smooth specimens, 130 to 150 MPa (19 to 22 ksi); 60° notched specimens, 24 to 40 MPa (3.5 to 6 ksi). Axial (R = 0) at

Table 93 Typical tensile properties at various temperatures for alloy 7005-T53 extrusions

Temperature °C	°F	Tensile strength(a) MPa	ksi	Yield strength(a) MPa	ksi	Elon-gation, %
−269	−452	641	93	483	70	16
−196	−320	538	78	421	61	16
−80	−112	441	64	379	55	13
−28	−18	421	61	359	52	14
24	75	392	57	345	50	15
100	212	303	44	283	41	20
149	300	165	24	145	21	35
204	400	97	14	83	12	60
260	500	76	11	66	9.5	80

(a) Lowest strength for exposures up to 10 000 h at temperature, no load; test loading applied at 5000 psi/min to yield strength and then at strain rate of 5%/min to fracture.

10⁸ cycles, smooth specimens. T6351 plate: 195 MPa (28 ksi). T53 extrusions: 231 MPa (33.5 ksi)

Plane-strain fracture toughness. Typical, T6351 temper. LT orientation: 51.3 MPa$\sqrt{m}$ (46.7 ksi$\sqrt{in.}$); data from 3 in. thick notch bend specimens. TL orientation: 44 MPa$\sqrt{m}$ (40 ksi$\sqrt{in.}$); data from 3 in. thick notch bend specimens. SL orientation: 30.3 MPa$\sqrt{m}$ (27.6 ksi$\sqrt{in.}$); data from 1 to 1¼ in. thick compact tensile specimens.

Mass Characteristics

Density. 2.78 Mg/m³ (0.100 lb/in.³) at 20 °C (68 °F)

Thermal Properties

Liquidus temperature. 643 °C (1190 °F)

Solidus temperature. 604 °C (1120 °F)

Coefficient of thermal expansion. Linear:

Temperature range °C	°F	Average coefficient µm/ m·K	µin./ in.·°F
−50 – +20	−58 – +68 ..	21.4	11.9
20 – 100	68 – 212 ...	23.1	12.8
20 – 200	68 – 392 ...	24.0	13.3
20 – 300	68 – 572 ...	25.0	13.9

Volumetric: 67.0 × 10⁻⁶ m³/m³·K (3.72 × 10⁻⁵ in.³/in.³·°F) at 20 °C (68 °F)

Specific heat. 875 J/kg·K (0.209 Btu/lb·°F) at 20 °C (68 °F)

Thermal conductivity. At 20 °C (68 °F): O temper, 166 W/m·K (96 Btu/ft·h·°F); T53, T5351, T63, T6351 tempers, 148 W/m·K (86 Btu/ft·h·°F); T6 temper, 137 W/m·K (79 Btu/ft·h·°F)

Electrical Properties

Electrical conductivity. Volumetric, at 20 °C (68 °F): O temper, 43% IACS; T53, T5351, T63, T6351 tempers, 38% IACS; T6 temper, 35% IACS

Electrical resistivity. At 20 °C (68 °F): O temper, 40.1 nΩ·m; T53, T5351, T63, T6351 tempers, 45.4 nΩ·m; T6 temper, 49.3 nΩ·m. Temperature coefficient, all tempers: 0.1 nΩ·m per K at 20 °C (68 °F)

Fabrication Characteristics

Annealing temperature. 343 °C (650 °F)

Solution temperature. 399 °C (750 °F)

Heat treatment. T53: Press quench from hot working temperature, naturally age 72 h at room temperature, then two-stage artificially age 8 h at 100 to 110 °C (212 to 230 °F) plus 16 h at 145 to 155 °C (290 to 310 °F)

7049
7.6Zn-2.5Mg-1.5Cu-0.15Cr

Specifications

AMS. Extrusions: 4157, 4159. Forgings: 4111
UNS number. A97049

Government. Forgings: QQ-A-367, MIL-H-6088

Chemical Composition

Composition limits. 1.2 to 1.9 Cu; 2.0 to 2.9 Mg; 0.20 max Mn; 0.25 max Si; 0.35 max Fe; 0.10 to 0.22 Cr; 7.2 to 8.2 Zn; 0.10 max Ti; 0.05 max others (each); 0.15 max others (total); rem Al

Applications

Typical uses. Forged aircraft and missile fittings, landing gear cylinders, and extruded sections. Used where static strengths approximately the same as forged 7079-T6 and high resistance to stress corrosion cracking are required. Fatigue characteristics about equal to those of 7075-T6 products, toughness somewhat higher.

Precautions in use. Poor general corrosion resistance

Mechanical Properties

Tensile property limits. See Table 94.
Shear strength. See Table 94.
Compressive strength. See Table 94.

Bearing strength. See Table 94.
Hardness. 135 HB min; 500-kg load, 10-mm diam ball
Poisson's ratio. 0.33
Elastic modulus. Forgings, typical: tension, 70 GPa (10.2×10^6 psi). Extrusions, typical: tension, 72.5 GPa (10.5 ksi); shear, 27.6 GPa (4.0 ksi); compression, 76 GPa (11 ksi)
Fatigue strength. Axial fatigue at stress ratio R of 0.1 for material in the T73 temper. Smooth specimens from 125 mm (5 in.) thick forgings: 275 to 315 MPa (40 to 46 ksi) at 10^7 cycles for temperatures from room temperature to 175 °C (350 °F). Notched specimens from 75 mm (3 in.) thick forgings: 390 MPa (56 ksi) for K_t of 1.0; 115 MPa (17 ksi) for K_t of 3.0; both at 10^7 cycles
Plane-strain fracture toughness. K_Q values from compact tension tests of 7049 -T73 die forgings: LS orientation, 32 to 36 MPa $\sqrt{m}$ (29 to 33 ksi$\sqrt{in.}$); LT orientation, 31 to 40 MPa $\sqrt{m}$ (28 to 37 ksi $\sqrt{in.}$); SL orientation, 21 to 27 MPa $\sqrt{m}$ (19 to 25 ksi $\sqrt{in.}$)

Mass Characteristics

Density. 2.82 Mg/m³ (0.102 lb/in.³) at 20 °C (68 °F)

Table 94 Mechanical properties of alloy 7049

Size and direction	Tensile strength(a) MPa	ksi	Yield strength (a)(b) MPa	ksi	Elongation(a)(c), %	Compressive yield strength MPa	ksi	Shear strength MPa	ksi	Bearing strength(d) MPa	ksi	Bearing yield strength(a) MPa	ksi
Die Forgings (AMS 4111), T73 Temper													
Parallel to grain flow													
Up to 2 in., incl	496	72	427	62	7	441	64	283	41	917	133	662	96
Over 2 to 4 in., incl	490	71	421	61	7	434	63	276	40	903	131	655	95
Over 4 to 5 in., incl	483	70	414	60	7	427	62	269	39	890	129	641	93
Across grain flow													
Up to 1 in., incl	490	71	421	61	3	434	63	283	41	917	133	662	96
Over 1 to 4 in., incl	483	70	414	60	3 to 2	427	62	276	40	903	131	655	95
Over 4 to 5 in. incl	469	68	400	58	2	414	60	269	39	890	129	641	93
Extrusions (AMS 4157), T73511 Temper													
Up to 2.999 in., incl													
Longitudinal	510	74	441	64	7	448	65	276	40	758	110	⋯	⋯
Long transverse	483	70	414	60	5	420	61	276	40	993	144	⋯	⋯
Over 2.999 to 5.000 in., incl													
Longitudinal	496	72	427	62	7	435	63	269	39	738	107	⋯	⋯
Long transverse	469	68	400	58	5	407	59	269	39	965	140	⋯	⋯
Extrusions (AMS 4159), T75511 Temper													
Up to 2.999 in., incl													
Longitudinal	538	78	483	70	7	490	71	290	42	⋯	⋯	586	85
Long transverse	524	76	469	68	5	475	69	290	42	⋯	⋯	724	105
Over 2.999 to 5.000 in., incl													
Longitudinal	524	76	469	68	7	475	69	283	41	⋯	⋯	572	83
Long traverse	510	74	455	66	5	462	67	283	41	⋯	⋯	696	101

(a) Single values are minimum values. (b) 0.2% offset. (c) In 2 in. or 4d, where d is diameter of reduced section of tensile test specimen. Where a range appears in this column, the specified minimum elongation varies with thickness of mill product. (d) e/d = 2.0, where e is edge distance and d is pin diameter.

Thermal Properties

Liquidus temperature. 627 °C (1160 °F)

Solidus temperature. 477 °C (890 °F)

Coefficient of thermal expansion. Linear, 23.4 μm/m·K (13.0 μin./in.·°F) at 20 to 100 °C (68 to 212 °F)

Specific heat. 960 J/kg·K (0.23 Btu/lb·°F) at 100 °C (212 °F)

Thermal conductivity. 154 W/m·K (89 Btu/ft·h·°F) at 25 °C (77 °F)

Electrical Properties

Electrical conductivity. Volumetric, 38% IACS min at 20 °C (68 °F)

Electrical resistivity. 43 nΩ·m

7050
6.2Zn-2.3Mg-2.3Cu-0.12Zr

Specifications

AMS. 4050, 4107, 4108
UNS. number. A97050

Chemical Composition

Composition limits. 2.0 to 2.6 Cu; 1.9 to 2.6 Mg; 0.10 max Mn; 0.12 max Si; 0.15 max Fe; 0.04 max Cr; 0.08 to 0.15 Zr; 5.7 to 6.7 Zn; 0.06 max Ti; 0.05 max others (each); 0.15 max others (total)

Consequence of exceeding impurity limits. Excess Fe and Si degrade fracture toughness. Increased sensitivity to quenching rate due to excess Mn and Cr results in low strength in thick sections.

Applications

Typical uses. Plate, extrusions, hand and die forgings in aircraft structural parts. Other applications requiring very high strength coupled with high resistance to exfoliation corrosion and stress-corrosion crack-

Table 95 Minimum mechanical properties of alloy 7050-T736 die forgings

| | \multicolumn{8}{c}{Thickness, in.} |
Property	Up to 2.000 MPa	ksi	2.001-4.000 MPa	ksi	4.001-5.000 MPa	ksi	5.001-6.000 MPa	ksi
Tensile strength								
L direction	496	72	490	71	483	70	483	70
T direction	469	68	462	67	455	66	455	66
Yield strength								
L direction	427	62	421	61	414	60	467	59
T direction	386	56	379	55	372	54	372	54
Compressive yield strength								
L direction	434	63	434	63	434	63	427	62
T direction	400	58	393(a)	57(a)	379	55	372	54
Shear strength	290	42	283	41	283	41	283	41
Bearing strength								
e/d = 1.5	683	99	676	98	669	97	669	97
e/d = 2.0	903	131	889	129	876	127	876	127
Bearing yield strength								
e/d = 1.5	565	82	558	81	545	79	538	78
e/d = 2.0	662	96	655	95	641	93	634	92
Elongation(b), %								
L direction	7		7		7		7	
T direction	5		4		3		3	

(a) For material 3.001 to 4.000 in. thick, 386 MPa (56 ksi). (b) In 50 mm or 2 in.

Table 96 Minimum mechanical properties of alloy 7050-T73652 hand forgings

| | \multicolumn{14}{c}{Thickness, in.} |
Property	Up to 2.000 MPa	ksi	2.001-3.000 MPa	ksi	3.001-4.000 MPa	ksi	4.001-5.000 MPa	ksi	5.001-6.000 MPa	ksi	6.001-7.000 MPa	ksi	7.001-8.000 MPa	ksi
Tensile strength														
L direction	496	72	496	72	490	71	483	70	476	69	469	68	462	67
LT direction	490	71	483	70	483	70	476	69	469	68	462	67	455	66
ST direction	...	...	462	67	462	67	455	66	455	66	448	65	441	64
Yield strength														
L direction	434	63	427	62	421	61	414	60	407	59	400	58	393	57
LT direction	421	...	414	60	407	69	400	58	386	56	372	54	359	52
ST direction	...	...	379	55	379	55	372	54	365	53	352	51	345	50
Compressive yield strength														
L direction	441	64	434	63	427	62	421	61	414	60	407	59	400	58
LT directon	448	65	441	64	434	63	427	62	414	60	400	58	386	56
ST direction	...	...	421	61	421	61	414	60	407	59	393	57	379	55
Shear strength	290	42	283	41	283	41	283	41	276	40	269	39	269	39
Bearing strength														
e/d = 1.5	689	100	683	99	683	99	669	97	662	96	655	95	641	93
e/d = 2.0	903	131	896	130	896	130	883	128	869	126	855	124	841	122
Bearing yield strength														
e/d = 1.5	593	86	586	85	572	83	565	82	545	79	524	76	503	73
e/d = 2.0	696	101	689	100	676	98	662	96	641	93	621	90	593	86
Elongation, %														
L direction	9		9		9		9		9		9		9	
LT direction	5		5		5		4		4		4		4	
ST direction	...		4		4		3		3		3		3	

Table 97 Typical mechanical properties of alloy 7050

Temperature °C	°F	Time at temp, h	At indicated temperature Tensile strength MPa	ksi	Yield strength MPa	ksi	Elongation(a), %	At room temperature after heating Tensile strength MPa	ksi	Yield strength MPa	ksi	Elongation(a), %
T73651 Plate												
24	75	...	510	74	455	66	11	510	74	455	66	11
100	212	0.1 thru 10	441	64	427	62	13	510	74	455	66	11
		100	448	65	434	63	13	510	74	462	67	12
		1000	441	64	427	62	14	510	74	455	66	12
		10 000	441	64	421	61	15	510	74	441	64	12
149	300	0.1	393	57	386	56	16	510	74	455	66	11
		0.5	393	57	386	56	17	510	74	448	65	12
		10	393	57	386	56	18	503	74	441	64	12
		100	359	52	332	51	19	483	70	407	59	13
		1000	290	42	276	40	21	407	59	317	46	13
		10 000	221	32	193	28	29	331	48	228	33	14
177	350	0.1	359	52	345	50	19	510	74	448	65	12
		0.5	352	51	345	50	20	496	72	441	64	12
		10	324	47	310	45	22	469	68	400	58	13
		100	248	36	234	34	25	386	56	296	43	13
		1000	193	28	172	25	31	317	46	214	31	14
		10 000	159	23	124	18	40	248	36	152	22	15
204	400	0.1	303	44	290	42	22	490	71	434	63	12
		0.5	290	42	276	40	23	469	68	421	61	12
		10	221	32	207	30	27	386	56	283	41	13
		100	165	24	152	22	32	317	46	200	29	14
		1000	131	19	110	16	45	262	38	138	20	16
		10 000	117	17	90	13	54	234	34	117	17	19
T73652 Forgings												
−196	−320	...	662	96	572	83	13	...	...	...	...	
−80	−112	...	586	85	503	73	14	...	...	...	...	
−28	−18	...	552	80	476	69	15	...	...	...	...	
24	75	...	524	76	455	66	15	524	76	455	66	15
100	212	0.1 thru 10	462	67	427	62	16	524	76	455	66	15
		100	469	68	434	63	16	524	76	462	67	15
		1000	462	67	427	62	17	524	76	524	76	16
		10 000	462	67	421	61	17	517	75	517	75	16
149	300	0.1	414	60	386	56	17	517	75	455	66	15
		0.5	414	60	386	56	17	510	74	448	65	15
		10	407	59	386	56	18	503	73	441	64	16
		100	365	53	352	51	20	483	70	407	59	16
		1000	290	42	276	40	23	407	59	317	46	17
		10 000	221	32	193	28	29	331	48	228	33	17
177	350	0.1	379	55	345	50	19	510	74	448	65	15
		0.5	365	53	345	50	20	496	72	441	64	15
		10	324	47	310	45	22	469	68	400	58	16
		100	248	36	234	34	25	386	56	296	43	17
		1000	193	28	172	25	31	317	46	214	31	17
		10 000	159	23	124	18	40	248	36	152	22	18
204	400	0.1	324	47	290	42	22	503	73	434	63	15
		0.5	296	43	276	40	23	483	70	421	61	15
		10	221	32	207	30	27	386	56	283	41	16
		100	165	24	152	22	32	317	46	200	29	17
		1000	131	19	110	16	45	262	38	138	20	19
		10 000	117	17	90	13	54	234	34	117	17	22

(a) In 50 mm or 2 in.

ing, high fracture toughness and fatigue resistance.

Mechanical Properties

Tensile properties. See Tables 95 to 97.

Shear properties. See Tables 95 and 96.
Compressive properties. See Tables 95 and 96.
Bearing properties. See Tables 95 and 96.

Poisson's ratio. 0.33
Elastic modulus. Tension, 70.3 GPa (10.2×10^6 psi); shear, 26.9 GPa (3.9×10^6 psi); compression, 73.8 GPa (10.7×10^6 psi)

Fatigue strength. See Table 98.
Plane-strain fracture toughness. See Table 99.
Creep-rupture characteristics. See Table 100.

Mass Characteristics

Density. 2.83 Mg/m³ (0.102 lb/in.³) at 20 °C (68 °F)

Thermal Properties

Liquidus temperature. 635 °C (1175 °F)
Solidus temperature. 524 °C (957 °F)
Incipient melting temperature. 488 °C (910 °F) for homogenized (solution treated) wrought material
Eutectic temperature. 465 °C (870 °F) for unhomogenized wrought or as-cast material
Coefficient of thermal expansion. Linear:

Temperature range		Average coefficient	
°C	°F	µm/ m·K	µin./ in.·°F
−50 − +20	−58 − +68	21.7	12.1
20 − 100	68 − 212	23.5	13.1
20 − 200	68 − 392	24.4	13.6
20 − 300	68 − 572	25.4	14.1

Volumetric: 68.0×10^{-6} m³/m³·K (3.78×10^{-5} in.³/in.³·°F) at 20 °C (68 °F)
Specific heat. 860 J/kg·K (0.206 Btu/lb·°F) at 20 °C (68 °F)
Thermal conductivity. At 20 °C (68 °F): O temper, 180 W/m·K (104 Btu/ft·h·°F); T76, T7651 tempers, 154 W/m·K (89 Btu/ft·h·°F); T736, T73651 tempers, 157 W/m·K (91 Btu/ft·h·°F)

Electrical Properties

Electrical conductivity. Volumetric, at 20 °C (68 °F): O temper, 47% IACS; T76, T7651 tempers, 39.5% IACS; T736, T73651 tempers, 40.5% IACS
Electrical resistivity. At 20 °C (68 °F): O temper, 36.7 nΩ·m; T76, T7651 tempers, 43.6 nΩ·m; T736, T73651 tempers, 42.6 nΩ·m. Temperature coefficient, all tempers: 0.1 nΩ·m per K at 20 °C (68 °F)

Fabrication Characteristics

Annealing temperature. 413 °C (775 °F)
Solution temperature. 477 °C (890 °F)
Aging temperature. 121 to 177 °C (250 to 350 °F)

Table 98 Typical axial fatigue strength at 10⁷ cycles for alloy 7050

Product and temper	Stress ratio, R	Fatigue strength (max stress)			
		Smooth specimens		Notched specimens(a)	
		MPa	ksi	MPa	ksi
Plate, 25 to 150 mm (1 to 6 in.) Thick					
T6 type tempers	0.0	190-290	28-42	...	...
T73xxx tempers	0.0	170-300	24-44	50-90	7.5-13
Extrusions, 29.5 mm (1.16 in.) Thick					
T76511 temper	+0.5	320-340	46-50	110-125	16-18
	0.0	180-210	26-30	70-80	10-12
	−1.0	130-150	19-22	35-50	5-7
Die Forgings, 25 to 150 mm (1 to 6 in.) Thick					
T736 temper	0.0	210-275	30-40	75-115	11-17
Hand Forgings, 144 × 559 × 2130 mm (4½ × 22 × 84 in.)					
T73652 temper					
longitudinal	+0.5	325	47	145	21
	0.0	225	33	90	13
	−1.0	145	21	50	7
long transverse	+0.5	275	40	115	17
	0.0	170	25	90	13
	−1.0	125	18	50	7
short transverse	+0.5	260	38	115	17
	0.0	170	25	60	9
	−1.0	115	17	50	7

(a) Notch fatigue factor, K_t, of 3.0.

Table 99 Plane-strain fracture toughness of alloy 7050

Temper and orientation	Minimum		Average	
	MPa √m	ksi √in.	MPa √m	ksi √in.
Plate				
T73651				
LT	26.4	24	35.2	32
TL	24.2	22	29.7	27
SL	22.0	20	28.6	26
Extrusions				
T7651X				
LT	...	...	30.8	28
TL	...	...	26.4	24
SL	...	...	20.9	19
T7351X				
LT	...	...	45.1	41
TL	...	...	31.9	29
SL	...	...	26.4	24
Die Forgings				
T736				
LT	27.5	25	36.3	33
TL, SL	20.9	19	25.3	23
Hand Forgings				
T73652				
LT	29.7	27	36.3	33
TL	18.7	17	23.1	21
SL	17.6	16	22.0	20

7072
1.0Zn

Specifications

ASTM. B209

SAE. J454
UNS number. A97072

Chemical Composition

Composition limits. 0.10 max Cu;

Table 100 Creep and rupture properties of alloy 7050-T3651 plate

Tempera-ture °C	°F	Time under stress h	Rupture stress MPa	ksi	1.0% MPa	ksi	0.5% MPa	ksi	0.2% MPa	ksi	0.1% MPa	ksi
							Stress for creep of:					
24	75	0.1	510	74	496	72	476	69	455	66	448	65
		1	503	73	483	70	462	67	448	65	441	64
		10	490	71	469	68	455	66	441	64	441	64
		100	476	69	455	66	448	65	441	64	434	63
		1000	469	68	448	65	441	64	...	...	...	...
100	212	0.1	441	64	434	63	427	62	421	61	414	60
		1	427	62	414	60	407	59	400	58	386	56
		10	407	59	393	57	386	56	372	54	359	52
		100	379	55	372	54	365	53	345	50	331	48
		1000	359	52	352	51	345	50	317	46	...	...
149	300	0.1	372	54	365	53	359	52	345	50	324	47
		1	345	50	338	49	324	47	303	44	290	42
		10	310	45	303	44	290	42	269	39	228	33
		100	262	38	255	37	241	35	193	28	152	22
		1000	179	26	179	26	165	24	145	21	124	18

Table 101 Mechanical-property limits for alloy 7072 fin stock

Temper	Tensile strength Minimum MPa	ksi	Maximum MPa	ksi	Yield strength (min) MPa	ksi	Elon-gation (min), %(a)
O	55	8.0	90	13.0	21	3	15 to 20
H14	97	14.0	131	19.0	83	12	1 to 3
H18	131	19.0	...	...	...	...	1 to 2
H19	145	21.0	...	...	...	...	1
H25	107	15.5	148	21.5	83	12	2 to 3
H111, H211	62	9.0	97	14.0	41	6.0	12

(a) In 2 in. or 50 mm. where a range of values appears in this column, specified minimum elongation varies with thickness of the mill product.

0.10 max Mg; 0.10 max Mn; 0.7 max Si + Fe; 0.8 to 1.3 Zn; 0.05 max others (each); 0.15 max others (each); rem Al

Applications

Typical uses. Fin stock. Cladding alloy for alclad sheet, plate, and tube products with the following core alloys: 2219, 3003, 3004, 5050, 5052, 5154, 6061, 7075, 7475, 7178

Mechanical Properties

Tensile properties. See Table 101.
Shear strength. O temper, 55 MPa (8 ksi); H12 temper, 62 MPa (9 ksi); H14 temper, 69 MPa (10 ksi)
Hardness. O temper, 20 HB; H12 temper, 28 HB; H14 temper, 32 HB; all values obtained with 500-kg load, 10-mm diam ball and 30-s duration of loading
Poisson's ratio. 0.33
Elastic modulus. Tension, 68 GPa (9.9 × 10^6 psi); compression, 70 GPa (10.1 × 10^6 psi)

Mass Characteristics

Density. 2.72 Mg/m^3 (0.098 lb/in.^{3}) at 20 °C (68 °F)

Thermal Properties

Liquidus temperature. 657 °C (1215 °F)
Solidus temperature. 641 °C (1185 °F)
Coefficient of thermal expansion. Linear:

Temperature range °C	°F	Average coefficient μm/m·K	μin./in.·°F
−50 − +20	−58 − +68	21.8	12.1
20 − 100	68 − 212	23.6	13.1
20 − 200	68 − 392	24.5	13.6
20 − 300	68 − 572	25.5	14.2

Volumetric: 68 × 10^{-3} m^3/m^3·K (3.78 × 10^{-5} in.^{3}/in.^{3}·°F) at 20 °C (68 °F)
Specific heat. 893 J/kg·K (0.213 Btu/lb·°F) at 20 °C (68 °F)
Thermal conductivity. O temper: 227 W/m·K (131 Btu/ft·h·°F) at 20 °C (68 °F)

Electrical Properties

Electrical conductivity. Volumetric, O temper: 60% IACS at 20 °C (68 °F)
Electrical resistivity. 28.7 nΩ·m at 20 °C (68 °F); temperature coefficient, 0.1 nΩ·m per K at 20 °C (68 °F)
Electrolytic solution potential. −0.96 V vs 0.1N calomel electrode in an aqueous solution containing 53 g NaCl plus 3 g H$_2$O$_2$ per litre at 25 °C (77 °F)

Chemical Properties

General corrosion behavior. High resistance to general corrosion. Provides galvanic protection when used as cladding on several different alloys.

Fabrication Characteristics

Annealing temperature. 343 °C (650 °F)

7075, Alclad 7075
5.6Zn-2.5Mg-1.6Cu-0.23Cr

Specifications

AMS. See Table 102.
ASTM. See Table 102.
SAE. J454
UNS number. A97075
Government. See Table 102.
Foreign. Austria: Önorm AlZnMgCul.5. Canada: CSA ZG62, ZG62Alclad. France: NF A-Z5GU. Spain: UNE L-371. Switzerland: VSM Al-Zn-Mg-Cu; Alclad, Al-Zn-Mg-Cu-pl. United Kingdom: BS L.95, L.96. Germany: DIN AlZnMgCul.5; Werkstoff-Nr. 3.4365. ISO: AlZn6MgCu

Chemical Composition

Composition limits. 7075: 1.20 to 2.0 Cu; 2.1 to 2.9 Mg; 0.30 max Mn; 0.40 max Si; 0.50 max Fe; 0.18 to 0.28 Cr; 5.1 to 6.1 Zn; 0.20 max Ti; 0.05 max others (each); 0.15 max others (total); rem Al. Alclad 7075: 7072 cladding—0.10 max Cu; 0.10 max

Mg; 0.10 max Mn; 0.7 max Si + Fe; 0.8 to 1.3 Zn; 0.05 max others (each); 0.15 max others (total); rem Al

Applications

Typical uses. Aircraft structural parts and other highly stressed structural applications where very high strength and good resistance to corrosion are required

Precautions in use. Caution should be exercised in T6 temper applications where sustained tensile stresses are encountered, either residual or applied, particularly in the transverse grain direction. In such instances, the T73 temper should be considered, at some sacrifice in tensile strength

Mechanical Properties

Tensile properties. See Tables 103 and 104.

Shear strength. Bare and alclad products, O temper: 152 MPa (22 ksi). Bare products—T6, T651 tempers: 331 MPa (48 ksi); Alclad T6, T651: 317 MPa (46 ksi)

Hardness. O temper, 60 HB; T6, T651 temper, 150 HB; data obtained using 500-kg load, 10-mm diam ball and 30-s duration of loading

Poisson's ratio. 0.33

Elastic modulus. Tension, 71.0 GPa (10.3×10^6 psi); shear, 26.9 GPa (3.9×10^6 psi); compression, 72.4 GPa (10.5×10^6 psi)

Fatigue strength. T6, T651, T73 tempers: 159 MPa (23 ksi) at 5×10^8 cycles in R. R. Moore type test of smooth (unnotched) specimens

Plane-strain fracture toughness. See Table 105.

Directional properties. Transverse mechanical properties of many products, particularly tensile strength and ductility in the short transverse direction, are less than those in the longitudinal direction.

Mass Characteristics

Density. 2.80 Mg/m³ (0.101 lb/in.³) at 20 °C (68 °F)

Thermal Properties

Liquidus temperature. 635 °C (1175 °F)

Solidus temperature. 477 °C (890 °F); eutectic temperature for nonhomogeneous as-cast or wrought material that has not been solution heat treated

Incipient melting temperature. 532 °C (990 °F) for homogenized (so-

Table 102 Standard specifications for alloy 7075

Mill form and condition	AMS	ASTM	Government
Bare products			
Sheet and plate	4038	B209	QQ-A-250/2
	4044	...	...
	4045	...	...
	4078	...	...
Wire, rod, and bar (rolled or cold finished)	4122	B211	QQ-A-225/9
	4123	...	...
	4124	...	...
Rod, bar, shapes, and tube (extruded)	4154	B221	QQ-A-200/11
	4167	...	...
	4168	...	...
	4169	...	...
Tube (extruded, seamless)	...	B241	...
Tube (drawn, seamless)	...	B210	...
Forgings and forging stock	4139	B247	QQ-A-367
	...	...	MIL-A-22771
Impacts	4170	...	MIL-A-12545
Rivets	...	B316	QQ-A-430
Alclad Products			
Sheet and plate	4039	B209	QQ-A-250/13
	4048	...	...
	4049	...	...
Tapered sheet and plate	4047	...	...
Alclad One Side Products			
Sheet and plate	4046	B209	QQ-A-250/18

Table 103 Typical tensile properties for alloy 7075 at various temperatures

Temperature °C	°F	Tensile strength(a) MPa	ksi	Yield strength(a)(b) MPa	ksi	Elongation(c), %
T6, T651 Tempers						
−196	−320	703	102	634	92	9
−80	−112	621	90	545	79	11
−28	−18	593	86	517	75	11
24	75	572	83	503	73	11
100	212	483	70	448	65	14
149	300	214	31	186	27	30
204	400	110	16	87	13	55
260	500	76	11	62	9	65
316	600	55	8	45	6.5	70
271	700	41	6	38	4.6	70
T73, T7351 Tempers						
−196	−320	634	92	496	72	14
−80	−112	545	79	462	67	14
−28	−18	524	76	448	65	13
24	75	503	73	434	63	13
100	212	434	63	400	58	15
149	300	214	31	186	27	30
204	400	110	16	90	13	55
260	500	76	11	62	9	65
316	600	55	8	45	6.5	70
371	700	41	6	32	4.6	70

(a) Lowest strength for exposures up to 10 000 h at temperature, no load; test loading applied at 5000 psi/min to yield strength and then at strain rate of 5%/min to fracture. (b) 0.2% offset. (c) In 50 mm or 2 in.

lution heat treated) wrought material

Coefficient of thermal expansion.
Linear:

Temperature range		Average coefficient	
°C	°F	μm/ m·K	μin./ in.·°F
−50 − +20	−58 − +68 ..	21.6	12.0
20 − 100	68 − 212 ...	23.4	13.0
20 − 200	68 − 392 ...	24.3	13.5
20 − 300	68 − 572 ...	25.2	14.0

Volumetric, 68×10^{-6} m³/m³·K $(3.78 \times 10^{-5}$ in.³/in.³·°F) at 20 °C (68 °F)

Specific heat. 960 J/kg·K (0.23 Btu/ lb·°F) at 100 °C (212 °F)

Thermal conductivity. At 20 °C (68 °F). T6, T62, T651, T652 tempers: 130 W/m·K (75 Btu/ft·h·°F). T76, T7651 tempers: 150 W/m·K (87 Btu/ ft·h·°F). T73, T7351, T7352 tempers: 155 W/m·K (90 Btu/ft·h·°F)

Electrical Properties

Electrical conductivity. Volumetric, at 20 °C (68 °F). T6, T62, T651, T652 tempers: 33% IACS. T76, T7651 tempers: 38.5% IACS. T73, T7351, T7352 tempers: 40% IACS

Electrical resistivity. At 20 °C (68 °F). T6, T62, T651, T652 tempers: 52.2 nΩ·m. T76, T7651 tempers: 44.8 nΩ·m. T73, T7351, T7352 tempers: 43.1 nΩ·m. Temperature coefficient, all tempers: 0.1 nΩ·m per K at 20 °C (68 °F)

Fabrication Characteristics

Annealing temperature. 413 °C (775 °F)

Solution temperature. 466 to 482 °C (870 to 900 °F) depending on product

Aging temperature. T6 temper: 121 °C (250 °F); T7 temper: two-stage treatment—107 °C (225 °F) followed by 163 to 177 °C (325 to 350 °F), depending on product

7175
5.6Zn-2.5Mg-1.6Cu-0.23Cr

Commercial Names

Trade name. AA7175

Specifications

AMS. 4109, 4148, 4149, 4179

Table 104 Tensile properties of alloy 7075

Temper	Tensile strength MPa	ksi	Yield strength MPa	ksi	Elongation(a), %
Typical Properties					
O.....................228	38	103	15	17	
T6, T651 572	83	503	73	11	
T73.....................503	73	434	63	...	
Alclad O 221	32	97	14	17	
T6, T651 524	76	462	67	11	
Property Limits					
	Minimum		Minimum		Minimum
Sheet and Plate					
O.....................276 (max)	40 (max)	145 (max)	21 (max)	10	
Sheet					
T6, T62					
0.008-0.011 in. thick ...510	74	434	63	5	
0.012-0.039 in. thick ...524	76	462	67	7	
0.040-0.125 in. thick ...538	78	469	68	8	
0.126-0.249 in. thick ...538	78	476	69	8	
T73.....................462	67	386	56	8	
T76.....................503	73	427	62	8	
Plate					
T62, T651					
0.250-0.499 in. thick ...538	78	462	67	9	
0.500-1.000 in. thick ...538	78	469	68	7	
1.001-2.000 in. thick ...531	77	462	67	6	
2.001-2.500 in. thick ...524	76	441	64	5	
2.501-3.000 in. thick ...496	72	421	61	5	
3.001-3.500 in. thick ...490	71	400	58	5	
3.501-4.000 in. thick ...462	67	372	54	3	
T7351					
0.250-2.000 in. thick ...476	69	393	57	6-7	
2.001-2.500 in. thick ...455	66	359	52	6	
2.501-3.000 in. thick ...441	64	338	49	6	
T7651					
0.250-0.499 in. thick ...496	72	421	61	8	
0.500-1.000 in. thick ...490	71	414	60	6	
Alclad Sheet and Plate					
O					
0.008-0.062 in. thick ...248 (max)	36 (max)	138 (max)	20 (max)	9-10	
0.063-0.187 in. thick ...262 (max)	38 (max)	138 (max)	20 (max)	10	
0.188-0.499 in. thick ...269 (max)	39 (max)	145 (max)	21 (max)	10	
0.500-1.000 in. thick ...276 (max)	40 (max)	...	...	10	
Alclad Sheet					
T6, T62					
0.008-0.011 in. thick ...469	68	400	58	5	
0.012-0.039 in. thick ...483	70	414	60	7	
0.040-0.062 in. thick ...496	72	427	62	8	
0.063-0.187 in. thick ...503	73	434	63	8	
0.188-0.249 in. thick ...517	75	441	64	8	
T73					
0.040-0.062 in. thick ...434	63	352	51	8	
0.063-0.187 in. thick ...441	64	359	52	8	
0.188-0.249 in. thick ...455	66	372	54	8	
T76					
0.125-0.187 in. thick ...469	68	393	57	8	
0.188-0.249 in. thick ...483	70	407	59	8	

(continued)

Table 104 (continued)

Temper	Tensile strength MPa	ksi	Yield strength MPa	ksi	Elongation(a), %
Alclad Plate					
T62, T651					
0.250-0.499 in. thick ... 517	75		448	65	9
0.500-1.000 in. thick ... 538(b)	78(b)		469(b)	68(b)	7
1.001-2.000 in. thick ... 531(b)	77(b)		462(b)	67(b)	6
2.001-2.500 in. thick ... 524(b)	76(b)		441(b)	64(b)	5
2.501-3.000 in. thick ... 496(b)	72(b)		421(b)	61(b)	5
3.001-3.500 in. thick ... 490(b)	71(b)		400(b)	58(b)	5
3.501-4.000 in. thick ... 462(b)	67(b)		372(b)	54(b)	3
T7351					
0.250-0.499 in. thick ... 455	66		372	54	8
0.500-1.000 in. thick ... 476	69		393	57	7
T7651					
0.250-0.499 in. thick ... 476	69		400	58	8
0.500-1.000 in. thick ... 490(b)	71(b)		414(b)	60(b)	6

(a) In 2 in. or 4d, where d is diameter of reduced section of tensile test specimen. Where a range appears in this column, the specified minimum elongation varies with thickness of the mill product. (b) For plate 0.500 in. or over in thickness, listed properties apply to core material only. Tensile and yield strengths of composite plate are slightly lower than listed value, depending on thickness of cladding.

Table 105 Typical plane-strain fracture toughness of alloy 7075

Product and temper	Minimum MPa $\sqrt{m}$	ksi $\sqrt{in.}$	Average MPa $\sqrt{m}$	ksi $\sqrt{in.}$	Maximum MPa $\sqrt{m}$	ksi $\sqrt{in.}$
LT Orientation						
T651	27.5	25	28.6	26	29.7	27
T7351	...	...	33.0	30	...	...
Extruded shapes						
T6510,1	28.6	26	30.8	28	35.2	32
T7310,1	34.1	31	36.3	33	37.4	34
Forgings						
T652	26.4	24	28.6	26	30.8	28
T7352	29.7	27	34.1	31	38.5	35
TL Orientation						
Plate						
T651	22.0	20	24.2	22	25.3	23
T7351	27.5	25	31.9	29	36.3	33
Extruded shapes						
T6510,1	20.9	19	24.2	22	28.6	26
T7310,1	24.2	22	26.4	24	30.8	28
Forgings						
T652	...	...	25.3	23	...	...
T7352	25.3	23	27.5	25	28.6	26
SL Orientation						
Plate						
T651	16.5	15	17.6	16	19.8	18
T7351	20.9	19	22.0	20	23.1	21
Extruded shapes						
T6510,1	19.8	18	20.9	19	24.2	22
T7310,1	...	...	22.0	20	...	...
Forgings						
T651	...	...	18.7	17	...	...
T7351	20.9	19	23.1	21	27.5	25

UNS number. A97175

Chemical Composition

Composition limits. 1.2 to 2.0 Cu; 2.1 to 2.9 Mg; 0.10 max Mn; 0.15 max Si; 0.20 max Fe; 0.18 to 0.28 Cr; 5.1 to 6.1 Zn; 0.10 max Ti; 0.05 max others (each); 0.15 max others (total)
Consequence of exceeding impurity limits. Degraded fracture toughness

Applications

Typical uses. Die and hand forgings for structural parts requiring very high strength, such as aircraft components. T736 tempers supply high strength, resistance to exfoliation corrosion and stress-corrosion cracking, high fracture toughness and good fatigue resistance.

Mechanical Properties

Tensile properties. Typical. Tensile strength: T66 temper, 593 MPa (86 ksi); T736 temper, 524 MPa (76 ksi). Yield strength: T66 temper, 524 MPa (76 ksi); T736 temper, 455 MPa (66 ksi). Elongation: 11% in 50 mm or 2 in. See also Table 106.
Shear strength. Typical. T66 temper: 324 MPa (47 ksi); T736 temper: 290 MPa (42 ksi)
Hardness. Typical. T66 temper, 150 HB; T736 temper, 145 HB; data obtained with 500-kg load, 10-mm diam ball and 30-s duration of loading
Poisson's ratio. 0.33
Elastic modulus. Tension, 72 GPa (10.4×10^6 psi)
Fatigue strength. Typical. T66 and T736 tempers: 159 MPa (23 ksi)
Plane-strain fracture toughness. See Table 107.

Mass Characteristics

Density. 2.80 Mg/m^3 (0.101 lb/in.^{3})

Thermal Properties

Liquidus temperature. 635 °C (1175 °F)
Incipient melting temperature. 532 °C (990 °F) for homogenized (solution heat treated) wrought material
Eutectic temperature. 447 °C (890 °F) for nonhomogeneous as cast or wrought material that has not been solution heat treated
Coefficient of thermal expansion. Linear:

Table 106 Typical mechanical properties of alloy 7175-T736 die forgings up to 3 in. thick

Temperature °C	°F	Time at temperature, h	At indicated temperature Tensile strength MPa	ksi	Yield strength MPa	ksi	Elongation(a), %	At room temperature after heating Tensile strength MPa	ksi	Yield strength MPa	ksi	Elongation(a), %
−253	−423	⋯	876	127	745	108	12	⋯	⋯	⋯	⋯	⋯
−196	−320	⋯	731	106	676	98	13	⋯	⋯	⋯	⋯	⋯
−80	−112	⋯	621	90	572	83	14	⋯	⋯	⋯	⋯	⋯
−28	−18	⋯	600	87	552	80	16	⋯	⋯	⋯	⋯	⋯
24	75	⋯	552	80	503	73	14	552	80	503	73	14
100	212	0.1	490	71	476	69	14	552	80	503	73	14
		0.5	490	71	462	67	15	552	80	503	73	14
		10	496	72	476	69	16	552	80	510	74	14
		100	503	73	483	70	16	558	81	510	74	14
		1000	503	73	483	70	17	565	82	517	75	14
		10 000	496	72	476	69	17	558	81	503	73	14
149	300	0.1	427	62	414	60	20	552	80	503	73	15
		0.5	427	62	414	60	18	552	80	503	73	15
		10	427	62	414	60	20	552	80	496	72	15
		100	393	57	372	54	25	524	76	462	67	16
		1000	310	45	296	43	30	441	64	359	52	17
		10 000	241	35	214	31	30	352	51	248	36	18
176	350	0.1	365	53	345	50	20	538	78	490	71	14
		0.5	379	55	345	50	25	538	78	483	70	14
		10	338	49	324	47	25	496	72	427	62	16
		100	262	38	241	35	25	421	61	331	48	16
		1000	200	29	179	26	35	331	48	228	33	18
		10 000	165	24	131	19	55	262	38	152	22	20
204	400	0.1	324	47	303	44	20	524	76	469	68	16
		0.5	310	45	283	41	30	503	73	427	62	14
		10	228	33	214	31	35	393	57	296	43	16
		100	165	24	221	32	35	317	46	207	30	18
		1000	124	18	103	15	45	255	37	138	20	20
		10 000	124	18	90	13	65	234	34	110	16	25
232	450	0.1	262	38	241	35	20	510	74	441	64	16
		0.5	228	33	214	31	25	448	65	359	52	16
		10	159	23	145	21	35	338	49	228	33	17
		100	117	17	103	15	40	269	39	145	21	19
		1000	97	14	83	12	45	234	34	103	15	25
		10 000	90	13	76	11	50	221	32	97	14	25

(a) In 50 mm or 2 in.

Temperature range °C	°F	Average coefficient μm/m·K	μin./in.·°F
−50 – +20	−58 – +68 ..	21.6	12.0
20 – 100	68 – 212 ...	23.4	13.0
20 – 200	68 – 392 ...	24.3	13.5
20 – 300	68 – 572 ...	25.2	14.0

Volumetric: 68×10^{-6} $m^3/m^3 \cdot K$ (3.78×10^{-5} $in.^3/in.^3 \cdot °F$) at 20 °C (68 °F)

Specific heat. 864 J/kg·K (0.206 Btu/lb·°F) at 20 °C (68 °F)

Thermal conductivity. At 20 °C (68 °F): O temper, 177 W/m·K (102 Btu/ft·h·°F); T66 temper, 142 W/m·K (82 Btu/ft·h·°F) T736, T73652 tempers, 155 W/m·K (90 Btu/ft·h·°F)

Electrical Properties

Electrical conductivity. Volumetric, at 20 °C (68 °F): O temper, 46% IACS; T66 temper, 36% IACS; T736, T73652 tempers, 40% IACS

Electrical resistivity. At 20 °C (68 °F): O temper, 37.5 nΩ·m; T66 temper, 47.9 nΩ·m; T736, T73652 tempers, 43.1 nΩ·m. Temperature coefficient, all tempers: 0.1 nΩ·m per K at 20 °C (68 °F)

Table 107 Plane-strain fracture toughness of alloy 7175-T736 forgings

Temper and orientation	Plane-strain fracture toughness MPa √in.	ksi √in.	MPa √in.	ksi √in.
Die forgings	Minimum		Average	
T736				
LT 29.7		27	33.0	30
TL, SL 23.1		21	28.6	26
Hand Forgings				
T736				
LT 33.0		30	37.4	34
TL 27.5		25	29.7	27
SL 23.1		21	26.4	24

Fabrication Characteristics

Annealing temperature. 413 °C (775 °F)

Solution temperature. 516 °C (960 °F); must be preceded by soak at 477 to 485 °C (890 to 905 °F). Quench from lower temperature.

Aging temperature. 121 to 177 °C (250 to 350 °F)

7178, Alclad 7178
6.8Zn-2.7Mg-2.0Cu-0.3Cr

Specifications

AMS. Extruded wire, rod, bar, shapes and tube; 4158. Alclad 7178, sheet and plate: 4051, 4052

ASTM. See Table 108.

SAE. J454

Table 108 Standard specifications for alloy 7178

Mill form and condition	Specification number ASTM	Government
Sheet and plate	B209	QQ-A-250/14
	...	QQ-A-250/21
Wire, rod, bar, shapes and tube (extruded)	B221	QQ-A-200/13
	...	QQ-A-200/14
Rivet wire.........	B316	...
Tube (extruded, seamless)........	B241	...
Alclad sheet and plate.............	B209	QQ-A-250/15
	...	QQ-A-250/22
	...	QQ-A-250/28

UNS number. A97178

Government. See Table 108.

Chemical Composition

Composition limits. 7178: 1.6 to 2.4 Cu, 2.4 to 3.1 Mg, 0.30 max Mn, 0.40 max Si, 0.50 max Fe, 0.18 to 0.35 Cr, 6.3 to 7.3 Zn, 0.20 max Ti, 0.05 max others (each), 0.15 max others (total), rem Al. Alclad 7178: 7011 cladding— 0.05 max Cu, 1.0 to 1.6 Mg, 0.10 to 0.30 Mn, 0.15 max Si, 0.20 max Fe, 0.08 to 0.20 Cr, 4.0 to 5.5 Zn, 0.05 max Ti, 0.05 max others (each), 0.15 max others (total), rem Al; 7072 cladding—0.10 max Cu, 0.10 max Mg, 0.10 max Mn, 0.70 max Si + Fe, 0.8 to 1.3 Zn, 0.05 max others (each); 0.15 max others (total), rem Al

Table 109 Typical tensile properties of alloy 7178

Temperature °C	°F	Tensile strength(a) MPa	ksi	Yield strength(a)(b) MPa	ksi	Elongation(c), %
T6, T651 Tempers						
−196	−320	730	106	650	94	5
−80	−112	650	94	580	84	8
−28	−18	625	91	560	81	9
24	75	605	88	540	78	11
100	212	505	73	470	68	14
149	300	215	31	185	27	40
204	400	105	15	83	12	70
260	500	76	11	62	9	76
316	600	59	8.5	48	7	80
371	700	45	6.5	38	5.5	80
T76, T7651 Tempers						
−196	−320	730	106	615	89	10
−80	−112	625	91	540	78	10
−28	−18	605	88	525	76	10
24	75	570	83	505	73	11
100	212	475	69	440	64	17
149	300	215	31	185	27	40
204	400	105	15	83	12	70
260	500	76	11	62	9	76
316	600	59	8.5	48	7	80
371	700	45	6.5	38	5.5	80

(a) Lowest strength for exposures up to 10 000 h at temperature, no load; test loading applied at 5000 psi/min to yield strength and then at strain rate of 5%/min to fracture. (b) 0.2% offset. (c) In 50 mm or 2 in.

Table 110 Creep-rupture properties of alloy 7178-T6

Temperature °C	°F	Time under stress, h	Rupture stress MPa	ksi	Stress for creep of: 1.0% MPa	ksi	0.5% MPa	ksi	0.2% MPa	ksi	0.1% MPa	ksi
150	300......	0.1	440	64	420	61	415	60	395	57	365	53
		1	415	60	395	57	380	55	360	52	315	46
		10	370	54	345	50	340	49	310	45	250	36
		100	285	41	270	39	255	37	235	34	185	27
		1000	180	26	180	26	170	25	150	22	130	19
205	400......	0.1	275	40	260	38	255	37	235	34	205	30
		1	215	31	205	30	200	29	180	26	145	21
		10	150	22	145	21	145	21	130	19	97	14
		100	105	15	97	14	97	14	83	12	76	11
		1000	69	10	69	10	69	10	59	8.5	55	8
260	500......	0.1	110	16	110	16	110	16	105	15	97	14
		1	97	14	97	14	90	13	83	12	66	9.5
		10	69	10	69	10	66	9.5	55	8	41	6
		100	55	8	52	7.5	45	6.5	34	5	...	...
		1000	41	6	34	5	29	4.2	...	...	...	...
315	600......	0.1	62	9	52	7.5	48	7	45	6.5	38	5.5
		1	52	7.5	45	6.5	41	6	34	5	26	3.7
		10	41	6	38	5.5	34	4.9	26	3.8	...	...
		100	34	5	30	4.3	26	3.8	...	...	...	...
		1000	28	4	23	3.4	...	...	...	...	...	...

Applications

Typical uses. Aircraft and aerospace applications where high compressive yield is design criteria

Precautions in use. T6 temper highly susceptible to exfoliation corrosion. T76 temper has mechanical properties comparable to 7075-T6 and provides improved resistance to exfoliation corrosion

Mechanical Properties

Tensile properties. See Table 109.

Shear strength. T6, T6510, T6511 tempers: 305 MPa (44 ksi). T76, T76510, T76511 tempers: 295 MPa (43 ksi)

Compressive strength. T6, T6510, T6511 tempers: 530 MPa (77 ksi) at 0.1% permanent set. T76, T76510, T76511 tempers: 460 MPa (67 ksi) at 0.1% permanent set

Bearing properties. T6, T6510, T6511 tempers: bearing strength, 1035 to 1060 MPa (150 to 160 ksi); bearing yield strength, 680 to 730 MPa (99 to 106 ksi). T76, T76510, T76511 tempers: bearing strength, 965 MPa (140 ksi); bearing yield strength, 740 MPa (107 ksi). All data for e/d ratio of 2.0, where e is edge distance and d is pin diameter

Poisson's ratio. 0.33

Elastic modulus. Tension, 71.7 GPa (10.4 × 10^6 psi); shear, 27.5 GPa (4.0 × 10^6 psi); compression, 73.7 GPa (10.7 × 10^6 psi)

Fatigue strength. T76 type tempers: 200 to 290 MPa (29 to 42 ksi) at 10^7 cycles in axial fatigue tests ($R =$ 0.0) of smooth specimens; 130 to 195 MPa (19 to 28 ksi) at 10^8 cycles in rotating beam tests ($R = -1.0$) of polished specimens; 28 to 55 MPa (4 to 8 ksi) at 10^8 cycles in rotating beam tests ($R = -1.0$) of 60° V-notched specimens ($K_t = 3.0$)

Creep-rupture characteristics. See Table 110.

Mass Characteristics

Density. 2.83 Mg/m³ (0.102 lb/in.³) at 20 °C (68 °F)

Thermal Properties

Liquidus temperature. 629 °C (1165 °F)

Eutectic temperature. 477 °C (890 °F)

Coefficient of thermal expansion. Linear:

Temperature range		Average coefficient	
°C	°F	μm/m·K	μin./in.·°F
−50 – +20	−58 – +68	21.7	12.1
20 – 100	68 – 212	23.5	13.1
20 – 200	68 – 392	24.4	13.6
20 – 300	68 – 572	25.4	14.1

Volumetric: 68 × 10^{-6} m³/m³·K (3.78 × 10^{-5} in.³/in.³·°F) at 20 °C (68 °F)

Specific heat. 856 J/kg·K (0.20 Btu/lb·°F) at 20 °C (68 °F)

Thermal conductivity. At 20 °C (68 °F): O temper, 180 W/m·K (104 Btu/ft·h·°F); T6, T651 tempers, 127 W/m·K (73 Btu/ft·h·°F); T76, T7651 tempers, 152 W/m·K (88 Btu/ft·h·°F)

Electrical Properties

Electrical conductivity. Volumetric, at 20 °C (68 °F): O temper, 46% IACS; T6, T651 tempers, 32% IACS; T76, T7651 tempers, 39% IACS

Electrical resistivity. At 20 °C (68 °F): O temper, 37.5 nΩ·m; T6, T651 tempers, 53.9 nΩ·m; T76, T7651 tempers, 44.2 nΩ·m. Temperature coefficient, all tempers: 0.1 nΩ·m per K 20 °C (68 °F)

Electrolytic solution potential. T6 temper: −0.81 V vs 0.1 N calomel electrode in an aqueous solution containing 53 g NaCl plus 3 g H_2O_2 per litre

Fabrication Characteristics

Annealing temperature. 413 °C (775 °F)

Solution temperature. 468 °C (875 °F)

Aging temperature. T6 and T7 tempers, 121 °C (250 °F) for 24 h

7475
5.7Zn-2.3Mg-1.5Cu-0.22Cr

Specifications

AMS. 4084, 4085, 4089, 4090
UNS number. A94475

Chemical Composition

Composition limits. 1.2 to 1.9 Cu; 1.9 to 2.6 Mg; 0.06 max Mn; 0.18 to 0.25 Cr; 0.12 max Fe; 0.10 max Si; 5.2 to 6.2 Zn; 0.06 max Ti; 0.05 max others (each); 0.15 max others (total); rem Al

Consequence of exceeding impurity limits. Degrades fracture toughness

Applications

Typical uses. Bare and alclad sheet and plate for aircraft fuselage and wing skins, spars and bulkheads. Other structural applications requiring a combination of high strength and high fracture toughness

Mechanical Properties

Tensile properties. See Table 111.

Shear strength. Plate: T651 temper, 296 MPa (43 ksi); T7351, T7651 tempers, 269 MPa (39 ksi)

Compressive strength. At 0.1% permanent set. Plate: T651 temper, 476 MPa (69 ksi); T7351 temper, 379 MPa (55 ksi); T7651 temper, 414 MPa (60 ksi)

Bearing properties. Plate, all data for e/d ratio of 2.0, where e is edge distance and d is pin diameter. T761 temper: bearing strength, 990 MPa (144 ksi); bearing yield strength, 730 MPa (106 ksi). T7351 temper: bearing strength, 875 MPa (127 ksi); bearing yield strength, 640 MPa (93 ksi). T7651 temper: bearing strength, 925 MPa (134 ksi); bearing yield strength, 655 MPa (95 ksi)

Poisson's ratio. 0.33

Elastic modulus. Tension, 70 GPa (10.2 × 10^6 psi); shear, 27 GPa (3.9 × 10^6 psi); compression, 73 GPa (10.6 × 10^6 psi)

Fatigue strength. At 10^7 cycles in axial fatigue tests of smooth specimens from T7351 plate. Longitudinal or transverse orientation: 205 to 235 MPa (30 to 34 ksi) for $R = 0.0$. Transverse orientation: 315 MPa (46 ksi) for $R = +0.5$; 165 MPa (24 ksi) for $R = -1.0$

Plane-strain fracture toughness. See Table 112.

Creep-rupture characteristics. See Table 113.

Mass Characteristics

Density. 2.80 Mg/m³ (0.101 lb/in.³) at 20 °C (68 °F)

Thermal Properties

Liquidus temperature. 635 °C (1175 °F)

Incipient melting temperature. 538 °C (1000 °F) for homogenized (solution heat treated) wrought material

Eutectic temperature. 477 °C (890 °F) for as-cast or inhomogeneous wrought material that has not been solution heat treated

Table 111 Typical tensile properties of alloy 7475

Temperature °C	°F	Time at temperature, h	At indicated temperature Tensile strength MPa	ksi	Yield strength MPa	ksi	Elongation(a), %	At room temperature after heating Tensile strength MPa	ksi	Yield strength MPa	ksi	Elongation(a), %
T61 Sheet, 0.040 thru 0.249 in. Thick												
−196	−320	...	683	99	600	87	10	...	...	...	...	...
−80	−112	...	607	88	545	79	12	...	...	...	...	...
−28	−18	...	579	84	517	75	12	...	...	...	...	...
24	75	...	552	80	496	72	12	552	80	496	72	12
100	212	0.1-0.5	496	72	462	67	14	552	80	496	72	12
		10	496	72	462	67	14	558	81	496	72	12
		100	503	73	469	68	13	558	81	503	73	12
		1000	503	73	476	69	13	565	82	510	74	12
		10 000	483	70	448	65	14	552	80	490	71	13
149	300	0.1-0.5	434	63	414	60	18	552	80	496	72	12
		10	434	63	414	60	17	545	79	490	71	12
		100	379	55	372	54	19	510	74	434	63	12
		1000	262	38	255	37	23	400	58	310	45	13
		10 000	207	30	179	26	28	310	45	207	30	14
177	350	0.1	386	56	365	53	19	545	79	490	71	12
		0.5	379	55	365	53	19	538	78	483	70	12
		10	324	47	310	45	21	490	71	414	60	12
		100	228	33	221	32	23	386	56	290	42	12
		1000	172	25	159	23	30	303	44	193	28	14
		10 000	131	19	110	16	40	234	34	124	18	15
204	400	0.1	331	48	317	46	17	531	77	469	68	12
		0.5	296	43	283	41	19	496	72	427	62	12
		10	200	29	193	28	26	372	54	276	40	12
		100	145	21	138	20	35	296	43	186	27	13
		1000	110	16	97	14	45	234	34	117	17	15
		10 000	97	14	76	11	55	207	30	97	14	18
232	450	0.1	234	34	221	32	19	490	71	414	60	12
		0.5	200	29	186	27	21	421	61	331	48	12
		10	138	20	131	19	30	303	44	193	28	13
		100	97	14	90	13	45	241	35	124	18	14
		1000	83	12	76	11	60	214	31	97	14	18
		10 000	83	12	62	9	65	193	28	76	11	22
260	500	0.1	159	23	152	22	20	407	59	310	45	12
		0.5	131	19	124	18	25	338	49	221	32	12
		10	90	13	83	12	45	255	37	131	19	15
		100	76	11	69	10	60	228	33	97	14	19
		1000	69	10	59	8.5	70	207	30	83	12	21
		10 000	66	9.5	48	7	70	186	27	69	10	22
316	600	0.1	76	11	69	10	35	317	46	193	28	13
		0.5	69	10	62	9	45	269	39	131	19	15
		10	48	7	41	6	65	241	35	90	13	19
		100	45	6.5	38	5.5	75	221	32	83	12	20
		1000	45	6.5	38	5.5	80	207	30	76	11	21
		10 000	45	6.5	38	5.5	80	186	27	69	10	...
371	700	0.1	41	6	34	5	70	276	40	117	17	17
		0.5	38	5.5	32	4.7	70	...	...	...	...	...
		10-10 000	34	5	27	3.8	85	...	...	...	...	...
427	800	0.1	24	3.5	20	2.8	85	...	...	...	...	...
		0.5	23	3.3	19	2.7	85	...	...	...	...	...
482	900	...	18	2.6	15	2.2	50	...	...	...	...	...
538	1000	...	11	1.6	9	1.3	3	...	...	...	...	...

(continued)

Table 111 (continued)

Temperature °C	°F	Time at temperature, h	At indicated temperature					At room temperature after heating				
			Tensile strength		Yield strength		Elongation(a), %	Tensile strength		Yield strength		Elongation(a), %
			MPa	ksi	MPa	ksi	%	MPa	ksi	MPa	ksi	%
T761 Sheet, 0.040 thru 0.249 in. Thick												
−196	−320	⋯	655	95	565	82	11	⋯	⋯	⋯	⋯	⋯
−80	−112	⋯	579	84	503	73	12	⋯	⋯	⋯	⋯	⋯
−28	−18	⋯	552	80	483	70	12	⋯	⋯	⋯	⋯	⋯
24	75	⋯	524	76	462	67	12	524	76	462	67	12
100	212	0.1-10	455	66	434	63	14	524	76	462	67	12
		100-1000	455	66	434	63	13	531	77	469	68	12
		10 000	441	64	421	61	14	524	76	462	67	13
149	300	0.1-0.5	400	58	386	56	18	524	76	462	67	12
		10	393	57	379	55	17	524	76	455	66	12
		100	359	52	345	50	19	490	71	421	61	12
		1000	362	38	255	37	23	400	58	303	44	13
		10 000	207	30	179	26	28	310	45	207	30	14
177	350	0.1	352	51	338	49	19	517	75	455	66	12
		0.5	352	51	331	48	19	517	75	455	66	12
		10	303	44	290	42	21	469	68	393	57	12
		100	228	33	221	32	23	379	55	283	41	12
		1000	172	25	159	23	30	303	44	193	28	14
		10 000	131	19	110	16	40	234	34	124	18	15
204	400	0.1	290	42	269	39	17	503	73	434	63	12
		0.5	276	40	262	38	19	483	70	414	60	12
		10	200	29	193	28	26	372	54	276	40	12
		100	145	21	138	20	35	296	43	186	27	13
		1000	110	16	97	14	45	234	34	117	17	15
		10 000	97	14	76	11	55	207	30	97	14	18
232	450	0.1	221	32	207	30	19	462	67	386	56	12
		0.5	193	28	179	26	21	414	60	324	47	12
		10	138	20	131	19	30	303	44	193	28	13
		100	97	14	90	13	45	241	35	124	18	14
		1000	83	12	76	11	60	214	31	97	14	18
		10 000	83	12	62	9	65	193	28	76	11	22
260	500	0.1	159	23	152	22	20	386	56	283	41	12
		0.5	131	19	124	18	25	338	49	221	32	12
		10	90	13	83	12	45	255	37	131	19	15
		100	76	11	69	10	60	228	33	97	14	19
		1000	69	10	59	8.5	70	207	30	83	12	21
		10 000	66	9.5	48	7	70	186	27	69	10	22
316	600	0.1	76	11	69	10	35	310	45	186	27	13
		0.5	69	10	62	9	45	269	39	131	19	15
		10	48	7	41	6	65	241	35	90	13	19
		100	45	6.5	38	5.5	75	221	32	83	12	20
		1000	45	6.5	38	5.5	80	207	30	76	11	21
		10 000	45	6.5	38	5.5	80	186	27	69	10	⋯
371	700	0.1	41	6	34	5	70	276	40	117	17	17
		0.5	38	5.5	32	4.7	70	⋯	⋯	⋯	⋯	⋯
		10	34	5	27	3.9	80	⋯	⋯	⋯	⋯	⋯
		100-10 000	34	5	27	3.8	85	⋯	⋯	⋯	⋯	⋯

(a) In 50 mm or 2 in.

Coefficient of thermal expansion. Linear:

Temperature range		Average coefficient	
°C	°F	μm/ m·K	μin./ in.·°F
−50 − +20	−58 − +68	21.6	12.0
20−100	68−212	23.4	13.0
20−200	68−392	24.3	13.5
20−300	68−572	25.2	14.2

Volumetric: 68×10^{-6} m³/m³·K (3.78 $\times 10^{-5}$ in.³/in.³·°F) at 20 °C (68 °F)
Specific heat. 865 J/kg·K (0.215 Btu/lb·°F) at 20 °C (68 °F)
Thermal conductivity. At 20 °C (68 °F):

Temper	Conductivity	
	W/ m·K	Btu/ ft·h·°F
O	177	102
T61, T651	142	82
T761, T7651	155	90
T7351	163	94

Table 112 Typical fracture-toughness values for alloy 7475

Temper	LT MPa√m	LT ksi√in.	TL MPa√m	TL ksi√in.	SL MPa√m	SL ksi√in.
High-strength Plate (K_{Ic})(a)						
T651	42.9	39	37.4	34	29.7	24
T7651	47.3	43	38.5	35	30.8	28
T7351	52.7	48	41.8	38	35.2	32
High-strength Sheet (K_c)(b)						
T761						
0.047 in. thick, room temp	143	130	...	...	...	...
−54 °C (−65 °F)	90	82	...	...	...	...
0.055 in. thick, room temp	136	123	...	...	...	...
−54 °C (−65 °F)	87	79	...	...	...	...
0.063 in. thick, room temp	122	112	...	...	...	...
−54 °C (−65 °F)	102	93	...	...	...	...
0.063 in. thick, room temp	150	137	...	...	...	...
−54 °C (−65 °F)	111	101	...	...	...	...
0.063 in. thick, room temp	147	134	...	...	...	...
−54 °C (−65 °F)	109	99	...	...	...	...
0.071 in. thick, room temp	149	136	...	...	...	...
−54 °C (−65 °F)	125	114	...	...	...	...

(a) Determined using standard compact tension specimen. (b) Determined using 16 by 44 in. center cracked panel with anti-buckling guides.

Table 113 Creep-rupture properties of alloy 7475

Temperature °C	Temperature °F	Time under stress, h	Rupture stress MPa	Rupture stress ksi	1% MPa	1% ksi	0.5% MPa	0.5% ksi	0.2% MPa	0.2% ksi	0.1% MPa	0.1% ksi
T61 Sheet, 0.040 thru 0.249 in. Thick												
24	75	0.1	552	80	538	78	524	76	517	75	510	74
		1	545	79	531	77	517	75	510	74	503	73
		10	545	79	517	75	510	74	503	73	496	72
		100	538	78	510	74	503	73	496	72	...	...
		1000	524	76	503	73	496	72	...	...	...	...
100	212	0.1	490	71	476	69	469	68	455	66	448	65
		1	476	69	455	66	448	65	434	63	421	61
		10	455	66	434	63	427	62	414	60	393	57
		100	427	62	414	60	400	58	386	56	365	53
		1000	386	56	379	55	365	53	352	51	...	...
149	300	0.1	414	60	400	58	393	57	379	55	365	53
		1	386	56	372	54	365	53	345	50	310	45
		10	352	51	338	49	317	46	283	41	241	35
		100	262	38	248	39	241	35	214	31	193	28
		1000	186	27	179	26	179	26	165	24	159	23
T761 Sheet, 0.040 thru 0.249 in. Thick												
24	75	0.1	524	76	503	73	483	70	476	69	469	68
		1	517	75	490	71	476	69	469	68	462	67
		10	510	74	483	70	469	68	462	67	462	67
		100	496	72	476	69	469	68	462	67	455	66
		1000	490	71	462	67	462	67	455	66	448	65
100	212	0.1	441	64	421	61	414	60	414	60	400	58
		1	421	61	407	59	400	58	393	57	379	55
		10	400	58	386	56	386	56	372	54	359	52
		100	379	55	372	54	365	53	352	51	324	47
		1000	359	52	352	51	345	50	324	47	...	...
149	300	0.1	372	54	365	53	365	53	352	51	324	47
		1	345	50	338	49	331	48	310	45	276	40
		10	310	45	303	44	290	42	255	37	234	34
		100	248	36	234	34	228	33	207	30	193	28
		1000	186	27	179	26	179	26	165	24	159	23

Electrical Properties

Electrical conductivity. Volumetric, at 20 °C (68 °F):

Temper	Conductivity, % IACS
O	46
T61, T651	36
T761, T7651	40
T7351	42

Electrical resistivity. At 20 °C (68 °F):

Temper	Resistivity, nΩ·m
O	37.5
T61, T651	47.9
T761, T7651	43.1
T7351	41.1

Temperature coefficient, all tempers: 0.1 nΩ·m per K at 20 °C (68 °F)

Fabrication Characteristics

Annealing temperature. 413 °C (775 °F)

Solution temperature. 516 °C (960 °F); must be preceded by soak at 466 to 477 °C (870 to 890 °F)

Aging temperature. 121 to 177 °C (250 to 350 °F)

Aluminum Foundry Products

By the ASM Committee on
Aluminum and Aluminum Alloys*

PRODUCTION of aluminum foundry products (all types of castings exclusive of ingot) has increased over the past quarter century at a fairly steady rate (see Fig. 1). Total production in the United States in 1977 was 0.9 Mt (1 000 000 tons), of which 65% was accounted for by die castings, 22% by permanent mold castings, 11% by sand castings and 2% by castings of other types. Also increasing has been the proportion of castings produced by companies for use in products they manufacture rather than for use by other industrial consumers. In 1977, the proportion of castings used directly by producers in their own products was 64% by weight for die castings, 49% for permanent mold castings and 19% for sand castings.

Alloy Systems

Aluminum casting alloys must contain, in addition to strengthening elements, sufficient amounts of eutectic-forming elements (usually silicon) in order to have adequate fluidity to feed the shrinkage that occurs in all but the simplest castings. Required amounts of eutectic formers depend in part on casting process. Alloys for sand casting generally are lower in eutectics than those for casting in metal molds, because sand molds can tolerate a degree

*See page **X** for committee list.

of hot shortness that would lead to extensive cracking in nonyielding metal molds. The range of cooling rates characteristic of the casting process being used controls to some extent the distribution of alloying and impurity elements. For example, the extremely high cooling rates inherent in die casting result in fine dispersion of strengthening and eutectic-forming constituents, and reasonably good castings can

be obtained in spite of impurity contents that would render sand or plaster-mold castings unacceptable. However, with these minor exceptions, most aluminum foundry alloys can be cast by all processes, and choice of casting technique usually is controlled by factors other than alloy composition.

A large number of aluminum alloys has been developed for casting, but most of them are varieties of six basic

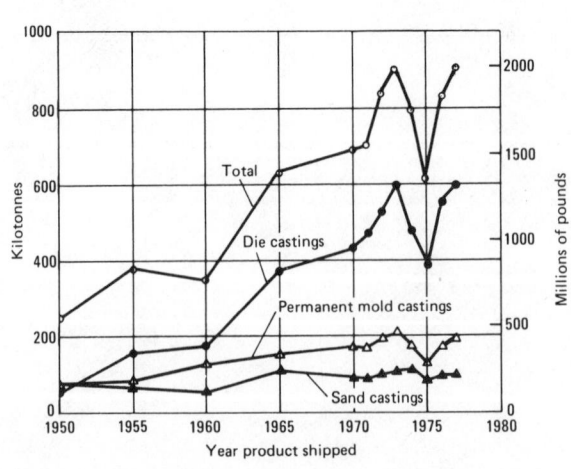

Fig. 1 Annual U.S. shipments of aluminum foundry products(a)

Compiled by the Aluminum Association, Inc.

types: aluminum-copper, aluminum-copper-silicon, aluminum-silicon, aluminum-magnesium, aluminum-zinc-magnesium and aluminum-tin alloys.

Aluminum-copper alloys that contain 4 to 5% Cu, with the usual impurities iron and silicon and sometimes with small amounts of magnesium, are heat treatable and can reach quite high strengths and ductilities, especially if prepared from ingot containing less than 0.15% iron. Manganese in small amounts also may be added, mainly to combine with the iron and silicon and reduce their embrittling effect. However, these alloys have poor castability and require very careful gating and risering if sound castings are to be obtained. Such alloys are used mainly in sand casting; when they are cast in metal molds, silicon must be added to increase fluidity and curtail hot shortness, and this addition of silicon substantially reduces ductility.

Al-Cu alloys with somewhat higher copper contents (7 to 8%), formerly the most commonly used aluminum casting alloys, have steadily been replaced by Al-Cu-Si alloys and today are used to a very limited extent. The best attribute of these higher-copper Al-Cu alloys is their insensitivity to impurities, but they have very low strength and only fair castability. Also in limited use are Al-Cu alloys that contain 9 to 11% Cu, whose high-temperature strength and wear resistance make them suitable for automotive pistons and cylinder blocks. These alloys usually contain manganese as an impurity because wrought metal scrap is used in preparing them. The manganese has little effect.

Very good high-temperature strength is an attribute of alloys containing copper, nickel and magnesium, sometimes with iron in place of part of the nickel.

Aluminum-Copper-Silicon Alloys. The most widely used aluminum casting alloys are those that contain silicon together with copper. The amounts of both additions vary widely, so that the copper predominates in some alloys and the silicon in others. In these alloys, the copper contributes to strength, and the silicon improves castability and reduces hot shortness; thus, the higher silicon alloys normally are used for more complex castings and for permanent mold and die casting processes, which cannot tolerate hot-short alloys.

Al-Cu-Si alloys with more than 3 to 4% Cu are heat treatable, but usually heat treatment is used only with those alloys that also contain magnesium, which enhances their response to heat treatment; without magnesium, response is too slow for heat treatment to be economical. High-silicon alloys (> 10% Si) have low thermal expansion, which makes them suitable for high-temperature operations. When silicon content exceeds 12 to 13% (silicon contents as high as 22% are typical), primary silicon crystals are present and, if properly distributed, impart excellent wear resistance. Automotive engine blocks and pistons are major uses of these alloys.

Aluminum-silicon alloys that do not contain copper additions are used when good castability and good corrosion resistance are needed. If high strength is also needed, magnesium additions make these alloys heat treatable. Alloys with silicon contents as low as 2% have been used for casting, but silicon content usually is between 5 and 13%. Strength and ductility of these alloys, especially the ones with higher silicon, can be substantially improved by "modification". Modification of the hypoeutectic alloys is particularly advantageous in sand castings, and can be effectively achieved through the addition of a controlled amount of sodium or strontium, which refines the silicon eutectic. Calcium and antimony additions are also used. Pseudomodification of sand castings, in which the size of the eutectic but not the structure is affected, may be achieved by solidification at high rates, such as occurs when chills are used. With permanent mold castings, modification of the eutectic also is advantageous, but the effect on properties is not as dramatic as with sand castings.

In hypereutectic alloys, the structure in sand and permanent mold castings is refined through modification of proeutectic silicon; here phosphorus additions are most effective. Phosphorus modification is required in only those die castings that have thick walls.

Aluminum-Magnesium Alloys. High corrosion resistance, especially to seawater and marine atmospheres, is the primary advantage of castings made of Al-Mg alloys. Best corrosion resistance requires low impurity content (both solid and gaseous), and thus alloys must be prepared from high-quality metals and handled with great care in the foundry. The relatively poor castability of Al-Mg alloys and the

tendency of the magnesium to oxidize increase handling difficulties and, therefore, cost.

Aluminum-zinc-magnesium alloys have the ability to naturally age, achieving full strength by 20 to 30 days at room temperature after casting. This strengthening process can be accelerated by furnace aging. The high-temperature solution heat treatment and drastic quenching required by other alloys (Al-Cu and Al-Si-Mg alloys, for example) is not necessary for optimum properties in most Al-Zn-Mg alloy castings. However, microsegregation of Mg-Zn phases can occur in these alloys, which reverses the accepted rule that faster solidification results in higher properties. When it is found in an Al-Zn-Mg alloy casting that the strength of the thin or highly chilled sections are lower than the thick or slowly cooled sections, the weaker sections can be strengthened to the required level by solution heat treatment and quenching, followed by natural or artificial (furnace) aging. Castability of Al-Zn-Mg alloys is poor, but they have good general corrosion resistance despite some susceptibility to stress corrosion.

Aluminum-tin alloys that contain about 6% Sn (and small amounts of copper and nickel for strengthening) are used for cast bearings because of the excellent lubricity imparted by tin. Bearing performance of Al-Sn alloys is strongly affected by casting method. Fine interdendritic distribution of tin, which is necessary for best bearing properties, requires small interdendritic spacing, and small spacing is obtained only with casting methods in which cooling is rapid.

Alloy Designation System

A standard system of four-digit numbers is used for designating aluminum and aluminum alloy castings and foundry ingot. The first digit indicates the major alloy group, as follows:

Aluminum (minimum Al content at least 99.00%	1xx.x
Al-Cu alloys	2xx.x
Al-Si alloys, with Cu and/or Mg	3xx.x
Al-Si alloys	4xx.x
Al-Mg alloys	5xx.x
Unused series	6xx.x
Al-Zn alloys	7xx.x
Al-Sn alloys	8xx.x

Table 1 Designations and nominal compositions of common aluminum alloys used for casting

AA number	Former AA designation	Former ASTM number	Product(a)	Cu	Mg	Mn	Si	Others
201.0	···	···	S	4.6	0.35	0.35	···	0.7 Ag, 0.25 Ti
206.0	···	···	S or P	4.6	0.25	0.35	0.10(b)	0.22 Ti, 0.15 Fe(b)
A206.0....	···	···	S or P	4.6	0.25	0.35	0.05(b)	0.22 Ti, 0.10 Fe(b)
208.0	108	CS43A	S	4.0	···	···	3.0	···
242.0	142	CN42A	S or P	4.0	1.5	···	···	2.0 Ni
295.0	195	C4A	S	4.5	···	···	0.8	···
296.0	B295.0, B195	···	P	4.5	···	···	2.5	···
308.0	A108	SC64A	S or P	4.5	···	···	5.5	···
319.0	319, Allcast	SC64D	S or P	3.5	···	···	6.0	···
336.0	A332.0, A132	SN122A	P	1.0	1.0	···	12.0	2.5 Ni
354.0	354	SC92A	P	1.8	0.50	···	9.0	
355.0	355	SC51A	S or P	1.2	0.50	0.50(b)	5.0	0.6 Fe(b), 0.35 Zn(b)
C355.0....	C355	SC51B	S or P	1.2	0.50	0.10(b)	5.0	0.20 Fe(b), 0.10 Zn(b)
356.0	356	SG70A	S or P	0.25(b)	0.32	0.35(b)	7.0	0.6 Fe(b), 0.35 Zn(b)
A356.0....	A356	SG70B	S or P	0.20(b)	0.35	0.10(b)	7.0	0.20 Fe(b), 0.10 Zn(b)
357.0	357	···	S or P	···	0.50	···	7.0	
A357.0....	A357	···	S or P	···	0.6	···	7.0	0.15 Ti, 0.005 Be
359.0	359	SG91A	S or P	···	0.6	···	9.0	···
360.0	360	SG100B	D	···	0.50	···	9.5	2.0 Fe(b)
A360.0....	A360	SG100A	D	···	0.50	···	9.5	1.3 Fe(b)
380.0	380	SC84B	D	3.5	···	···	8.5	2.0 Fe(b)
A380.0....	A380	SC84A	D	3.5	···	···	8.5	1.3 Fe(b)
383.0	···	SC102A	D	2.5	···	···	10.5	···
384.0	384	SC114A	D	3.8	···	···	11.2	3.0 Zn(b)
A384.0....	384	SC114A	D	3.8	···	···	11.2	1.0 Zn(b)
390.0	390	···	D	4.5	0.6	···	17.0	1.3 Zn(b)
A390.0....	A390	···	S or P	4.5	0.6	···	17.0	0.5 Zn(b)
413.0	13	S12B	D	···	···	···	12.0	2.0 Fe(b)
A413.0....	A13	S12A	D	···	···	···	12.0	1.3 Fe(b)
4430......	43	S5B	S	0.6(b)	···	···	5.2	···
A443.0....	43	···	S	0.30(b)	···	···	5.2	···
B443.0....	43	S5A	S or P	0.15(b)	···	···	5.2	···
C443.0....	A43	S5C	D	0.6(b)	···	···	5.2	2.0 Fe(b)
514.0	214	G4A	S	···	4.0	···	···	···
518.0	218	G8A	D	···	8.0	···	···	···
520.0	220	G10A	S	···	10.0	···	···	···
535.0	Almag 35	GM70B	S	···	6.8	0.18	···	0.18 Ti
A535.0....	A218	···	S	···	7.0	0.18	···	···
B535.0....	B218	···	S	···	7.0	···	···	0.18 Ti
712.0	D712.0, D612, 40E	ZG61A	S or P	···	0.6	···	···	5.8 Zn, 0.5 Cr, 0.20 Ti
713.0	613, Tenzaloy	ZC81A,B	S or P	0.7	0.35	···	···	7.5 Zn, 0.7 Cu
771.0	Precedent 71A	ZG71B	S	···	0.9	···	···	7.0 Zn, 0.13 Cr, 0.15 Ti
850.0	750	···	S or P	1.0	···	···	···	6.2 Sn, 1.0 Ni

(a) S = sand casting, P = permanent mold casting, D = die casting. (b) Maximum.

As discussed in the previous section, alloys in the Al-Cu ($2xx.x$) group may also contain other elements, such as magnesium and nickel. In addition, this group includes Al-Cu-Si alloys in which copper predominates over silicon. The remainder of the Al-Cu-Si alloys fall into the $3xx.x$ group. Those Al-Si alloys that contain magnesium also belong to the $3xx.x$ group; those that do not contain magnesium constitute the $4xx.x$ group. Aluminum-magnesium alloys comprise the $5xx.x$ group, Al-Zn-Mg alloys the $7xx.x$ group and Al-Sn alloys the $8xx.x$ group.

In designations of the $1xx.x$ type, the second and third digits indicate minimum aluminum content (99.00% or greater); these digits are the same as the two to the right of the decimal point in the minimum aluminum percentage expressed to the nearest 0.01%. The fourth digit in $1xx.x$ designations, which is to the right of the decimal point, indicates product form: 0 denotes castings (such as electric motor rotors), and 1 denotes ingot.

In $2xx.x$ through $8xx.x$ designations (for aluminum alloys), the second and third digits have no numerical significance but only identify the various alloys in the group. The digit to the right of the decimal point indicates product form: 0 denotes castings, 1 denotes standard ingot, and 2 denotes ingot having composition ranges narrower than but within those of standard ingot. Alloy modifications are identified by a capital letter preceding the numerical designation.

Alloying-element and impurity limits for ingot are the same as those for castings of the same alloy except that, when the ingot is remelted for making castings, iron and zinc contents tend to increase and magnesium content tends to decrease. Thus limits of these elements in ingot and in castings differ as follows:

Maximum content, %	
Element content in castings, %	Element content in ingot, %
Iron in Sand and Permanent Mold Castings (max)	
Through 0.15	0.03 less(a)
Over 0.15 through 0.25	0.05 less(a)
Over 0.25 through 0.6	0.10 less(a)
Over 0.6 through 1.0	0.2 less(a)
Over 1.0	0.3 less(a)
Iron in Die Castings (max)	
Through 1.3	0.3 less(a)
Over 1.3	1.1 max
Magnesium in All Types of Castings (min)	
To 0.50	0.05 more(b)
0.50 and greater	0.1 more(b)
Zinc in Die Castings (max)	
Over 0.25 through 0.6	0.10 less(a)
Over 0.6	0.1 less(a)

(a) Less than is allowed in castings. (b) More than is specified for castings; applicable only to the extent that the resulting range of Mg content in the ingot is at least 0.15%.

The designations of the commonly used cast aluminum alloys, along with their nominal compositions, are given in Table 1. The characteristics of the alloys commonly used for sand and permanent mold castings are listed in Table 2, whereas those of common die casting alloys are presented in Table 3.

Selection of Casting Alloy

The major factors that influence alloy selection for casting applications include casting process to be used, casting design, required properties, and economic (and availability) considerations.

Each casting process requires specific metal characteristics. For example,

die and permanent mold casting generally require alloys with good fluidity and resistance to hot tearing, whereas these properties are less critical in sand, plaster and investment casting, where molds and cores offer less resistance to shrinkage. Discussions of required alloy characteristics, and lists of alloys commonly used, are presented for the various casting processes in the section that follows.

The application for which a casting is to be made affects alloy selection by establishing requirements for strength and ductility, as well as special service requirements such as pressure characteristics, corrosion resistance and surface treatments.

Economic considerations also may be important in alloy selection. Total cost of making a casting is affected by required heat treatment and by weldability and machinability, in addition to ingot and melting costs.

Full development of the potential of any casting alloy depends in large part on foundry technique. Foundry personnel should be consulted on alloy selection; use of alloys with which such personnel are familiar often results in better and more economical castings.

Selection of the proper alloy requires careful consideration of all the factors discussed above, which are presented in the brief outline that follows.

Alloy characteristics necessary for casting process selected:
- (a) fluidity
- (b) resistance to hot tearing
- (c) solidification range

Casting design considerations:
- (a) solidification range
- (b) resistance to hot tearing
- (c) fluidity
- (d) die soldering (die casting)

Mechanical-property requirements:
- (a) strength and ductility
- (b) heat treatability
- (c) hardness

Service requirements:
- (a) pressure tightness characteristics
- (b) corrosion resistance
- (c) surface treatments
- (d) dimensional stability
- (e) thermal stability

Economics:
- (a) machinability
- (b) weldability
- (c) ingot and melting costs
- (d) heat treatment

Casting Processes

Aluminum is one of the few metals that can be cast by all of the processes used in casting metals. These processes, in decreasing order of amount of aluminum cast, are: die casting, permanent mold casting, sand casting (green sand and dry sand), plaster casting and investment casting. Aluminum also is continous cast. Each of these processes, and the castings produced by them, are discussed below.

Die Casting. Alloys of aluminum are used in die casting more extensively than alloys of any other base metal. In the United States alone, about 2.5 billion dollars worth of aluminum alloy die castings is produced each year. The die casting process consumes almost twice as many tonnes of aluminum alloys as all other casting processes combined (see Fig. 1).

Die casting is especially suited to production of large quantities of relatively small parts. Aluminum die castings weighing up to about 5 kg (10 lb) are common, but castings weighing as much as 50 kg (100 lb) are produced when the high tooling and casting-machine costs are justified. With die casting, it is possible to maintain close tolerances and produce good surface finishes; aluminum alloys can be die cast to basic linear tolerances of ±4 mm/m (±4 mils/in.) and commonly have finishes as fine as 1.3 μm (50 μin.). Die castings are best designed with uniform wall thickness; minimum practical wall thickness for aluminum alloy die castings is dependant on casting size. Small parts are cast as thin as 1.0 mm (0.040 in.). Cores, which are made of metal, are restricted to simple shapes that permit straight-line removal.

Die castings are made by injection of molten metal into metal molds under substantial pressure. Rapid injection (due to the high pressure) and rapid solidification under high pressure (due to the use of bare metal molds) combine to produce a dense, fine-grain surface structure, which results in excellent wear and fatigue properties. Air entrapment and shrinkage, however, may result in porosity, and machine cuts should be limited to 1.0 mm (0.040 in.) to avoid exposing it. Mold coatings are not practical in die casting, which is done at pressures of 2 MPa (300 psi) or higher, because the violence of the rapid injection of molten metal would remove the coating (production of thin-

Table 2 Characteristics of common aluminum alloys used in sand and permanent mold casting(a)(b)

Alloy	Type of mold(c)	Fluidity	Resistance to hot cracking	Pressure tightness	Heat treatment	Strength at elevated temperatures	General corrosion resistance	Machining	Polishing	Anodizing Appearance	Weldability
208.0	S	2	2	2	Optional	3	4	3	3	3	2
213.0	P	2	3	3	No	3	4	2	2	3	3
222.0	S or P	3	3	3	Yes	1	4	1	2	3	3
242.0	S or P	3	4	4	Yes	1	4	2	2	3	4
295.0	S	3	4	4	Yes	3	4	2	2	2	3
296.0	P	3	4	3	Yes	2	4	3	2	3	3
308.0	P	2	2	2	No	3	3	3	3	4	2
319.0	S or P	2	2	2	Optional	3	3	3	4	4	2
328.0	S	1	1	2	Optional	2	3	3	3	4	1
332.0	P	1	2	2	Yes	1	3	4	4	4	2
333.0	P	1	2	2	Yes	2	3	3	3	4	3
336.0	P	1	2	2	Yes	1	3	4	4	4	3
354.0	P	1	1	1	Yes	2	3	4	4	4	3
355.0	S or P	1	1	1	Yes	2	3	3	3	4	1
C355.0	S or P	1	1	1	Yes	2	3	3	3	4	1
356.0	S or P	1	1	1	Yes	3	2	3	4	4	1
A356.0	S or P	1	1	1	Yes	3	2	3	4	4	1
357.0	S or P	1	1	1	Yes	3	2	3	4	4	1
A357.0	S or P	1	1	1	Yes	2	2	3	4	4	1
359.0	S or P	1	2	2	Yes	2	2	4	4	4	1
B443.0	S or P	1	1	1	No	4	2	5	4	4	1
512.0	S	3	3	4	No	3	1	2	2	2	3
513.0	P	4	4	4	No	3	1	1	1	1	3
514.0	S	4	4	5	No	3	1	1	1	1	3
520.0	S	4	4	5	Yes	5	1	1	1	1	4
535.0	S	5	4	5	Optional	3	1	1	1	1	4
705.0	S or P	4	4	4	No	4	2	1	2	2	4
707.0	S or P	4	4	4	No	4	2	1	2	2	4
710.0	S	4	5	4	No	4	4	1	2	2	4
711.0	S	3	5	4	No	4	3	1	2	2	4
713.0	S or P	3	4	4	No	4	3	1	1	1	4
771.0	S	3	4	4	Yes	4	3	1	1	1	4
850.0	S or P	4	5	5	Yes	5	4	1	3	···	5
851.0	S or P	4	5	5	Yes	5	4	1	3	···	5
852.0	S or P	4	5	5	Yes	5	4	1	3	···	5

(a) From Standards for Aluminum Sand and Permanent Mold Castings, The Aluminum Association, 1977. (b) Characteristics are comparatively rated from 1 to 5; 1 is the highest or best possible rating. (c) S = sand; P = permanent.

section die castings may involve cavity fill times as brief as 20 ms).

Die castings are not easily welded or heat treated because of entrapped gases. Special techinques and care in production are required for pressure tight parts. The selection of an alloy with a narrow freezing range also is helpful. The use of vacuum for cavity venting is practiced in some die casting foundries for production of parts for some special applications. In the "pure free" process, the die cavity is purged with oxygen before injection. The entrapped oxygen reacts with the molten aluminum to form oxide particles rather than gas pores.

Approximately 85% of aluminum al-loy die castings is produced in alumi-num-silicon-copper alloys (alloy 380.0 and its several modifications). This family of alloys provides a good combi-nation of cost, strength and corrosion resistance, together with the high flu-idity and freedom from hot shortness that are required for ease of casting. Where better corrosion resistance is required, alloys lower in copper, such as 360.0 and 413.0, must be used.

Alloy 518.0 is occasionally specified when highest corrosion resistance is required. This alloy, however, has low fluidity and some tendency to hot short-ness. It is difficult to cast, which is reflected in higher cost per casting.

Aluminum alloy die castings usually are not heat treated, but occasionally are given dimensional and metallurgi-cal stabilization treatments.

The physical and mechanical proper-ties of the most commonly used alumi-num die casting alloys are given in the data compilations that follow this arti-cle. It should be noted that these prop-erties were obtained from test bars separately die cast under laboratory conditions, and they should be used only for assessing the suitability of an alloy for a particular application, and not for design purposes. Design-stress values are significantly below typical properties as discussed in the section on Mechanical Properties later in this ar-ticle. Actual design strength depends

Table 3 Characteristics of aluminum die casting alloys(a)

Alloy	Approximate melting temperature, °C	Hot cracking	Resistance to: Die soldering	Corrosion	Die filling capacity	Machining	Polishing	Electroplating	Anodized Appearance	surface Protection	Elevated temperature strength	Pressure tightness
360.0	557-596	1	2	2	3	3	3	2	3	3	1	2
A360.0	557-596	1	2	2	3	3	3	2	3	3	1	2
380.0	538-593	2	1	4	2	3	3	1	3	4	3	2
A380.0	583-593	2	1	4	2	3	3	1	3	4	3	2
383.0	516-582	1	2	3	1	2	3	1	3	4	2	2
384.0	516-582	2	2	5	1	3	3	2	4	5	2	2
413.0	574-582	1	1	2	1	4	5	3	5	3	3	1
A413.0	574-582	1	1	2	1	4	5	3	5	3	3	1
C443.0	574-632	3	4	2	4	5	4	2	2	2	5	3
518.0	535-621	5	5	1	5	1	1	5	1	1	4	5

(a) From ASTM B85. Relative rating of die casting alloys from 1 to 5; 1 is the highest or best possible rating. A rating of 5 in one or more categories does not rule an alloy out of commercial use if other attributes are favorable; however, ratings of 5 may present manufacturing difficulties.

on several factors, including (a) section size, (b) expected degree of porosity, (c) presence of sharp corners and (d) probability of cyclic loading in service.

Other characteristics of aluminum die casting alloys are presented in Table 3. Final selection of an aluminum alloy for a specific application can best be established by consultation with die casting suppliers.

The high cooling rates of the die casting process ensure extremely fine grain structure. During casting of sections no thicker than 1.6 mm (1/16 in.), heat is dissipated to the dies at 30 to 40 °C (55 to 70 °F) per second. The following tabulation lists some common aluminum die casting alloys and typical products cast from them.

Alloy 380.0	Lawnmower housings, gear cases, cylinder heads for air-cooled engines
Alloy A380.0	Streetlamp housings, typewriter frames, dental equipment
Alloy 360.0	Frying skillets, cover plates, instrument cases, parts requiring corrosion resistance
Alloy 413.0	Outboard-motor parts such as pistons, connecting rods and housings
Alloy 518.0	Escalator parts, conveyor components, aircraft and marine hardware and fittings

Permanent mold casting, like die casting, is suited to high-volume production. Permanent mold castings typically are larger than die castings. Maximum weight of permanent mold castings usually is about 10 kg (25 lb),

but much larger castings sometimes are made when costs of tooling and casting equipment are justified by the quality required for the casting.

Surface finish of permanent mold castings depends on whether or not a mold wash is used; generally, finishes range from 3.8 to 10 μm (150 to 400 μin.). Basic linear tolerances of about ±10 mm/m (±0.10 in./in.), and minimum wall thicknesses of about 3.6 mm (0.140 in.), are typical. Tooling costs are high, but less than those for die casting. Because sand cores can be used, internal cavities can be fairly complex. (When sand cores are used, the process usually is referred to as semipermanent mold casting.)

Permanent mold castings are gravity-fed and pouring rate is relatively low, but the metal mold produces rapid solidification. Permanent mold castings exhibit excellent mechanical properties. Castings are generally sound, provided that the alloys used exhibit good fluidity and resistance to hot tearing.

Mechanical properties of permanent mold castings can be further improved by heat treatment. If maximum properties are required, the heat treatment consists of a solution treatment at high temperature followed by a quench (usually in hot water) and then natural or artificial aging. For small castings in which the cooling rate in the mold is very rapid or for less critical parts, the solution treatment and quench may be eliminated and the fast cooling in the mold relied on to retain in solution the compounds that will produce age hardening.

In low-pressure casting (also called "low-pressure die casting" or "pressure

permanent mold casting"), molten metal is injected into the metal molds at pressures of 170 kPa (25 psi) or less. Filling of the mold and control of solidification are aided by application of refractory mold coating to selected areas of the die cavity, which slows down cooling in those areas. Thinner walls can be cast by low-pressure casting than by regular permanent mold casting. Low-pressure casting also has the economic advantage in that it can be highly automated.

Some common aluminum permanent mold casting alloys, and typical products cast from them, are presented below.

Alloy 366.0	Automotive pistons
Alloys 355.0, C355.0, A357.0	Timing gears, impellers, compressors, and aircraft and missile components requiring high strength
Alloys 356.0, A356.0	Machine tool parts, aircraft wheels, pump parts, marine hardware, valve bodies
Alloy B443.0	Carburetor bodies, waffle irons
Alloy 513.0	Ornamental hardware and architectural fittings

Other aluminum alloys commonly used for permanent mold castings include 296.0 and 319.0.

Sand Casting. In sand casting, the mold is formed around a pattern by ramming sand, mixed with the proper bonding agent, onto the pattern. Then the pattern is removed, leaving a cavity in the shape of the casting to be made. If the casting is to have internal cavities or undercuts, sand cores are used to make them. Molten metal is poured into the mold, and after it has solidified the mold is broken to remove the casting. In making molds and cores, various agents can be used for bonding the sand. The agent most often used is a mixture of clay and water. (Sand bonded with clay and water is called "green sand".) Sand bonded with oils or resins, which is very strong after baking, is used mostly for cores. Water glass (sodium silicate) hardened with CO_2 is used extensively as a bonding agent for both molds and cores.

The main advantages of sand casting are versatility (a wide variety of alloys, shapes and sizes can be sand cast) and low cost of minimum equipment when a small number of castings is to be made. Among its disadvantages are low dimensional accuracy and poor surface finish (basic linear tolerances of ± 30 mm/m (± 0.030 in./in.) and surface finishes of 7 to 13 μm, or 250 to 500 μin., are typical for aluminum sand castings), as well as low strength as a result of slow cooling. A minimum wall thickness of 4 mm (0.15 in.) normally is required for aluminum sand castings. Use of dry sands bonded with resins or water glass results in better surface finishes and dimensional accuracy, but with a corresponding decrease in cooling rate.

Casting quality is determined to a large extent by foundry technique. Proper metal-handling and gating practice is necessary for obtaining sound castings. Complex castings with varying wall thickness will be sound only if proper techniques are used. A minimum wall thickness of 3.8 mm (0.15 in.) normally is required for aluminum sand castings.

Typical products made from some common aluminum sand casting alloys include:

Alloy C355.0	Air-compressor fittings, crankcases, gear housings
Alloy A356.0	Automobile transmission cases, oil pans and rear-axle housings
Alloy 357.0	Pump bodies, cylinder blocks for water-cooled engines
Alloy 443.0	Pipe fittings, cooking utensils, ornamental fittings, marine fittings
Alloy 520.0	Aircraft fittings, truck and bus frame components, levers, brackets
Alloy 713.0	General-purpose casting alloy for applications that require strength without heat treatment or that involve brazing

Other aluminum alloys commonly used for sand castings include 319.0, 355.0, 356.0, 514.0 and 535.0.

Shell Mold Casting. In shell mold casting, the molten metal is poured into a shell of resin-bonded sand only 10 to 20 mm (0.4 to 0.8 in.) thick—much thinner than the massive molds commonly used in sand foundries. Shell mold castings surpass ordinary sand castings in surface finish and dimensional accuracy and cool at slightly higher rates; however, equipment and production costs are higher, and size and complexity of castings that can be produced are limited.

Plaster Casting. In this method, either a permeable (aerated) or impermeable plaster is used for the mold. The plaster in slurry form is poured around a pattern, the pattern is removed and the plaster mold is baked before the casting is poured. The high insulating value of the plaster allows castings with thin walls to be poured. Minimum wall thickness of aluminum plaster castings typically is 1.5 mm (0.060 in.). Plaster molds have high reproducibility, permitting castings to be made with fine details and close tolerances; basic linear tolerances of $\pm$ 5 mm/m are typical for aluminum castings. Surface finish of plaster castings also is very good; aluminum castings attain finishes of 1.3 to 3.2 μm (50 to 125 μin.). For castings of certain complex shapes, such as some precision impellers and electronic parts, mold patterns made of rubber are used because their flexibility makes them easier to withdraw from the molds than rigid patterns.

Mechanical properties and casting quality depend on alloy composition and foundry technique. Slow cooling due to the highly insulating nature of plaster molds tends to magnify solidification-related problems, and thus solidification must be controlled carefully to obtain good mechanical properties.

Plaster casting is sometimes used to make prototype parts before proceeding to make tooling for production die casting of the part.

Cost of basic equipment for plaster casting is low; however, because plaster molding is slower than sand molding, cost of operation is high. Aluminum alloys commonly used for plaster casting are 295.0, 355.0, C355.0, 356.0 and A356.0.

Investment casting of aluminum most commonly employs plaster molds and expendable patterns of wax or other fusible materials. A plaster slurry is "invested" around patterns for several castings, and the patterns are melted out as the plaster is baked.

Investment casting produces precision parts; aluminum castings can have walls as thin as 0.40 to 0.75 mm (0.015 to 0.030 in.), basic linear tolerances as narrow as ± 5 mm/m (± 5 mils/in.) and surface finishes of 1.5 to 2.3 μm (60 to 90 μin.). Some internal porosity usually is present, and it is recommended that machining be limited to avoid exposing it. However, investment molding is often used to produce large quantities of intricately shaped parts requiring no further machining so internal porosity seldom is a problem. Because of porosity and slow solidification, mechanical properties are low.

Investment castings usually are small, and thus gating techniques are limited. Christmas-tree gating systems often are employed to produce many parts per mold. Investment casting is especially suited to production of jewelry and parts for precision instruments.

Aluminum alloys commonly used for investment castings are 208.0, 295.0, 308.0, 355.0, 356.0, 443.0, 514.0, 535.0 and 712.0.

Centrifugal Casting. Centrifuging is another method of forcing metal into a mold. Steel, baked sand, plaster, cast iron, or graphite molds and cores are used for centrifugal casting of aluminum. Metal dies or molds provide rapid chilling, resulting in a level of soundness and mechanical properties comparable or superior to that of gravity-poured permanent mold castings. Baked sand and plaster molds are commonly used for centrifuge casting because multiple mold cavities can be arranged readily around a central pouring sprue. Graphite has two major advantages as a mold material: its high heat conductivity provides rapid chilling of the cast metal, and its low specific gravity, compared to ferrous mold materials, reduces the power required to attain the desired speeds.

Centrifugal casting has the advantage over other casting processes in that, if molds are properly designed, inclusions such as gases or oxides tend to be forced into the gates, and thus castings have properties that closely match those of wrought products. Limitations on shape and size are severe, and cost of castings is very high.

Wheels, wheel hubs, and papermaking or printing rolls are examples of aluminum parts produced by centrifugal casting. Aluminum alloys suitable for permanent mold, sand, or plaster casting can be cast centrifugally.

Continuous Casting. Long shapes of simple cross section (such as round, square and hexagonal rods) can be produced by continuous casting, which is done in a short, bottomless, water-cooled metal mold. The casting is continuously withdrawn from the bottom of the mold; because the mold is water cooled, cooling rate is very high. As a result of continuous feeding, castings generally are free of porosity. In most instances, however, the same product can be made by extrusion at approximately the same cost and with better properties, and thus use of continuous casting is limited. The largest application of continuous casting is production of ingot for rolling, extrusion or forging.

Composite-Mold Casting. Many of the molding methods described above can be combined to obtain greater flexibility in casting. Thus, dry sand cores often are used in green sand molds, and metal chills can be used in sand molds to accelerate local cooling. Semipermanent molds, which comprise metal molds and sand cores, take advantage of the better properties obtainable with metal molds and the greater flexibility in shape of internal cavities that results from use of cores that can be extracted piecemeal.

Selection of Casting Process

There are many factors that affect selection of a casting process for producing a specific aluminum alloy part. Some of the important factors in sand, permanent mold and die casting are discussed in Table 4. The most important factors for all casting processes are (a) feasibility and cost factors and (b) quality factors.

Feasibility and Cost. Many aluminum alloy castings can be produced by any of the available methods. Occasionally, minor design modifications may be necessary—for example, removal of undercuts from castings to be made in metal molds. For a considerable number of castings, however, dimensions or design features automatically determine the best casting method. Because metal molds weigh from 10 to 100 times as much as the castings they are used in producing, most very large cast products are made as sand castings rather than as die or permanent mold castings. Small castings usually are made with metal molds to ensure dimension-

Table 4 Factors affecting selection of casting process for aluminum alloys

Factor	Sand casting	Casting process Permanent mold casting	Die casting
Cost of equipment	Lowest cost if only a few items required	Less than die casting	Highest
Casting rate	Lowest rate	11 kg/h (25 lb/h) common; higher rates possible	4.5 kg/h (10 lb/h) common; 45 kg/h (100 lb/h) possible
Size of casting	Largest of any casting method	Limited by size of machine	Limited by size of machine
External and internal shape	Best suited for complex shapes where coring required	Simple sand cores can be used, but more difficult to insert than in sand castings	Cores must be able to be pulled because they are metal; undercuts can be formed only by collapsing cores or loose pieces
Minimum wall thickness	3.0-5.0 mm (0.125-0.200 in.) required; 4.0 mm (0.150 in.) normal	3.0-5.0 mm (0.125-0.200 in.) required; 3.5 mm (0.140 in.) normal	1.0-2.5 mm (0.100-0.040 in.); depends on casting size
Type of cores	Complex baked sand cores can be used	Reuseable cores can be made of steel, or nonreuseable baked cores can be used	Steel cores; must be simple and straight so they can be pulled
Tolerance obtainable	Poorest; best linear tolerance is 300 mm/m (300 mils/in.)	Best linear tolerance is 10 mm/m (10 mils/in.)	Best linear tolerance is 4 mm/m (4 mils/in.)
Surface finish	6.5-12.5 μm (250-500 μin.)	4.0-10 μm (150-400 μin.)	1.5 μm (50 μin.); best finish of the three casting processes
Gas porosity	Lowest porosity possible with good technique	Best pressure tightness; low porosity possible with good technique	Porosity may be present
Cooling rate	0.1-0.5 °C/s (0.2-0.9 °F/s)	0.3-1.0 °C/s (0.5-1.8 °F/s)	50-500 °C/s (90-900 °F/s)
Grain size	Coarse	Fine	Very fine on surface
Strength	Lowest	Excellent	Highest, usually used in the "as cast" condition
Fatigue properties	Good	Good	Excellent
Wear resistance	Good	Good	Excellent
Over-all quality	Depends on foundry technique	Highest quality	Tolerance and repeatability very good
Remarks	Very versatile as to size, shape, internal configurations	...	Excellent for fast production rates

al accuracy. Some parts can be produced much more easily if cast in two or more separate sections and bolted or welded together. Complex parts with many undercuts can be made easily by sand, plaster or investment casting, but may be practically impossible to cast in metal molds even if sand cores are used.

When two or more casting methods are feasible for a given part, the method used very often is dictated by costs. As a general rule, the cheaper the tooling (patterns, molds and auxiliary equipment) the greater the cost of producing each piece. Therefore, number of pieces is a major factor in the choice of a casting method. If only a few pieces are to be made, the method involving the least expensive tooling should be used, even if the cost of casting each piece is very high. For very large production runs, on the other hand, where cost of tooling is shared by a large number of castings, use of elaborate tooling usually decreases cost per piece and thus is justified. In mass production of small parts, for example, costs often are minimized by use of elaborate tooling that allows several castings to be poured simultaneously. Die castings are typical of this category.

Casting Quality. When applied to castings, the term "quality" refers to both degree of soundness (freedom from porosity, cracking and surface imperfections) and levels of mechanical properties (strength and ductility). The microstructural features that most strongly affect mechanical properties are (a) grain size and shape, (b) dendrite-arm spacing and (c) size and distribution of second-phase particles and inclusions.

Castings with fine, equiaxed grains have the best combinations of strength and ductility. To some extent, size and shape of grains can be controlled by addition of grain refiners, but use of low pouring temperatures and high cooling rates are the preferred methods.

All commercial aluminum alloys contain multiple phases, as well as oxide and gas inclusions. During freezing, inclusions and second-phase particles segregate to the spaces between dendrite arms. The farther apart the dendrite arms are, the coarser the distribution of microconstituents and the more pronounced their adverse effects on properties. Thus, small interdendritic spacing is necessary for high

casting quality. Although several factors affect spacing to some extent, the only efficient way of ensuring fine spacing is use of rapid cooling.

The finer the dispersion of inclusions and second-phase particles, the better the properties of the casting. Fine dispersion requires that particles be small; large masses of oxides or intermetallic compounds produce excessive brittleness. Controlling size and shape of microconstituents can be done to some extent by controlling composition, but is accomplished more efficiently by minimizing the period of time during which microconstituents can grow. Like minimizing grain size and interdendritic spacing, minimizing time for growth of microconstituents calls for rapid cooling. Thus, it is evident that high cooling rate is of paramount importance in obtaining good casting quality. The tabulation below presents characteristic ranges of cooling rate for the various casting processes.

Casting processes	Cooling rate, °C/s	Dendrite-arm spacing, mm
Plaster, dry sand. .	0.05-0.2	0.1-1
Green sand, shell .	0.1-0.5	0.05-0.5
Permanent mold . .	0.3-1	0.03-0.07
Die.	50-500	0.005-0.015
Continuous.	0.5-2	0.03-0.07

However, it should be kept in mind that in die casting, although cooling rates are very high, air tends to be trapped in the casting, which gives rise to appreciable amounts of porosity at the center. Extensive research has been conducted to find ways of reducing such porosity; however, it is impossible to eliminate completely, and die castings often are lower in strength than low-pressure or gravity-fed permanent mold castings, which are more sound in spite of slower cooling.

Mechanical Properties

Typical and minimum mechanical-property values commonly reported for castings of particular aluminum alloys are determined using separately cast test bars, that are 1/2-in. diam (for sand and permanent mold castings) or 1/4-in. diam (for die castings). As such, these values represent properties of sound castings, 13 or 6 mm (1/2 or 1/4 in.) in section thickness, made using normal casting practice; they do not represent properties in all sections and locations

of full-size production castings. Typical and minimum properties of test bars, however, are useful in determining relative strengths of the various alloy/temper combinations. Minimum properties—those values listed in applicable specifications—apply, except where otherwise noted, only to separately cast test bars. These values, unlike minimum values based on bars cut from production castings, are not usable as design limits for production castings. However, they can be useful in quality assurance. Actual mechanical properties, whether of separately cast test bars or of full-size castings, are dependent on two main factors: (a) alloy composition and heat treatment and (b) solidification pattern and casting soundness.

Test Specimens. Accurate determination of mechanical properties of aluminum alloy castings (or of castings of any other metal) requires proper selection of test specimens. For most wrought products, a small piece of the material often is considered typical of the rest, and mechanical properties determined from that small piece also are considered typical. Properties of castings, however, vary substantially from one area of a given casting to another, and may vary from casting to casting in a given heat.

If castings are small, one from each batch can be sacrificed and cut into test bars. If castings are too large to be economically sacrificed, test bars can be molded as an integral part of each casting, or can be cast in a separate mold.

Usually, test bars are cast in a separate mold. When this is done, care must be taken to ensure that the metal poured into the test-bar mold is representative of the metal in the castings that the test bars are supposed to represent. In addition, differences in pouring temperature and cooling rate, which can make the properties of separately cast test bars different from those of production castings, must be avoided.

For highly stressed castings, integrally cast test bars are preferable to separately cast bars. When integrally cast bars are selected, however, gating and risering must be designed carefully to ensure that test bars and castings have equivalent microstructure and integrity. Also, if there are substantial differences between test-bar diameter and wall thickness in critical areas of the casting, use of integrally cast test panels equal in thickness to those critical areas, instead of standard test bars, should be considered.

"Premium-quality" castings have guarantees for the tensile properties of test bars machined from the casting, either from designated areas in the casting or from any area. Premium-quality castings are discussed in a later section of this article.

The typical properties of aluminum sand casting alloys, as determined on separately cast test bars, are listed in Table 5.

ASTM E8 defines the test bars suitable for evaluation of aluminum castings. The use of test bars cut from die castings is not recommended; simulated service (proof) testing is considered more appropriate.

Chemical Composition and Heat Treatment. Mechanical properties of castings depend not only on choice of alloy but also depend somewhat on other considerations linked with the alloy. Variations in chemical composition, even within specified limits, can have measurable effects. Metallurgical considerations, such as coring, phase segregation and modification, also can alter properties.

Modification is commonly used for those aluminum alloys with 5% or more silicon. Figure 2 shows alloy 356 before and after modification.

In hypoeutectic Al-Si alloys, the coarse silicon eutectic has been refined and dispersed by modification. The modified structure increases both ductility and mechanical strength. Modification is accomplished by addition of small amounts (0.02%) of sodium or strontium. Making those additions often introduces gas into the melt. Their use, therefore, must be weighed against applicable radiographic specifications. In the hypereutectic aluminum-silicon alloys (silicon greater than 11.7%), refinement of primary silicon in sand and permanent mold castings is accomplished by adding 0.05% phosphorus. (Phosphorus modification is required in only those die castings that have thick walls.) In these alloys, the phosphorus addition provides moderate improvements in strength and machinability.

Where heat treatment is required, choice of temper affects properties. Heat treating variables, such as solution time and temperature, temperature of quenching medium and quench delay, also can alter properties.

Casting variables also contribute to mechanical-property variations. The

Table 5 Typical tensile properties for separately cast test bars of common aluminum casting alloys

Alloy	Product(a)	Temper	Tensile strength MPa	ksi	Yield strength(b) MPa	ksi	Elongation(c), %
201.0	S	T4	365	53	215	31	20
	S	T6	485	70	435	63	7
	S	T7	460	67	415	60	4.5
206.0, A206.0	S	T7	435	63	345	50	11.7
208.0	S	F	145	21	97	14	2.5
242.0	S	T21	185	27	125	18	1.0
	S	T571	220	32	205	30	0.5
	S	T77	205	30	160	23	2.0
	P	T571	275	40	235	34	1.0
	P	T61	325	47	290	42	0.5
295.0	S	T4	220	32	110	16	8.5
	S	T6	250	36	165	24	5.0
	S	T62	285	41	220	32	2.0
296.0	P	T4	255	37	130	19	9.0
	P	T6	275	40	180	26	5.0
	P	T7	270	39	140	20	4.5
308.0	P	F	195	28	110	16	2.0
319.0	S	F	185	27	125	18	2.0
	S	T6	250	36	165	24	2.0
	P	F	235	34	130	19	2.5
	P	T6	280	40	185	27	3.0
336.0	P	T551	250	36	195	28	0.5
	P	T65	325	47	295	43	0.5
354.0	P	T61	380	55	285	41	6.0
355.0	S	T51	195	28	160	23	1.5
	S	T6	240	35	175	25	3.0
	S	T61	270	39	240	35	1.0
	S	T7	265	38	250	36	0.5
	S	T71	175	35	200	29	1.5
	P	T51	210	30	165	24	2.0
	P	T6	290	42	190	27	4.0
	P	T62	310	45	280	40	1.5
	P	T7	280	40	210	30	2.0
	P	T71	250	36	215	31	3.0
356.0	S	T51	175	25	140	20	2.0
	S	T6	230	33	165	24	3.5
	S	T7	235	34	210	30	2.0
	S	T71	195	28	145	21	3.5
	P	T6	265	38	185	27	5.0
	P	T7	220	32	165	24	6.0
357.0, A357.0	S	T62	360	52	290	42	8.0
359.0	P	T61	330	48	255	37	6.0
		T62	345	50	290	42	5.5
360.0	D	F	325	47	170	25	3.0
A360.0	D	F	320	46	165	24	5.0
380.0	D	F	330	48	165	24	3.0
383.0	D	F	310	45	150	22	3.5
384.0, A384.0	D	F	330	48	165	24	2.5
390.0	D	F	280	41	240	35	1.0
	D	T5	300	43	260	38	1.0
A390.0	S	F, T5	180	26	180	26	<1.0
	S	T6	280	40	280	40	<1.0
	S	T7	250	36	250	36	<1.0
	P	F, T5	200	29	200	29	1.0
	P	T6	310	45	310	45	<1.0
	P	T7	260	38	260	38	<1.0
413.0	D	F	300	43	140	21	2.5
A413.0	D	F	290	42	130	19	3.5

(continued)

Table 5 (continued)

Alloy	Product(a)	Temper	Tensile Strength MPa	ksi	Yield strength(b) MPa	ksi	Elonga-tion(c), %
443.0	S	F	130	19	55	8	8.0
B443.0	P	F	159	23	62	9	10.0
C443.0	D	F	228	33	110	16	9.0
514.0	S	F	170	25	85	12	9.0
518.0	D	F	310	45	190	28	5.0-8.0
520.0	S	T4	330	48	180	26	16
535.0	S	F	275	40	140	20	13
712.0	S	F	240	35	170	25	5.0
713.0	S	T5	210	30	150	22	3.0
	P	T5	220	32	150	22	4.0
771.0	S	T6	345	50	275	40	9.0
850.0	P	T5	160	23	75	11	10.0

(a) S = sand casting, P = permanent mold casting, D = die casting. (b) 0.2% offset. (c) With 12.7-mm (½-in.) diam specimen.

Fig. 2 Effect of sodium modification on microstructure of sand cast alloy 356-F

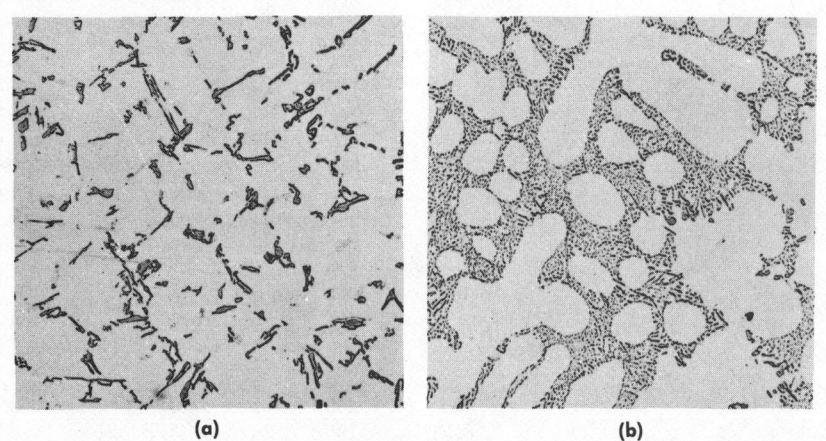

(a)   (b)

Both specimens were etched in 0.5% hydrofluoric acid and are shown at 100X. (a) As sand cast. Structure consists of a network of silicon particles (gray, sharp), which formed in the interdendritic Al-Si eutectic. (b) Modified by addition of 0.025% sodium to the melt. Constituents are the same as in (a), but the particles of silicon in the eutectic are smaller and less angular.

Fig. 3 Effect of gating system on formation of shrinkage cavities

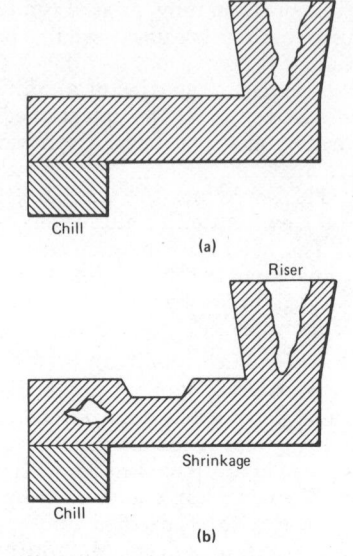

Solidification starts at the chill and progresses toward the riser. In (a), molten metal can easily feed from the riser into the entire length of the casting. In (b), the narrow portion of the casting can freeze shut before solidification of the left portion of the casting is completed, and thus the riser can no longer feed that portion and a shrinkage cavity develops.

differences between typical test-bar properties and mechanical properties of full-size castings result from differences in soundness and solidification characteristics. Casting soundness depends on the amount of porosity or other imperfections present in the casting. The presence of dross, shrinkage porosity and gas porosity will all decrease properties. Dross inclusions and gas porosity are minimized by proper melting and pouring techniques.

Properties of production castings vary depending on two aspects of the solidification characteristics in each section of the casting—solidification rate, and location and form of shrinkage. Shrinkage results when supply of molten metal is not adequate throughout solidification. Shrinkage may be apparent as sponge shrinkage, centerline shrinkage, shrinkage porosity or a large shrinkage cavity. Shrinkage, however, can be controlled by proper use of directional solidification.

Directional solidification in a casting section is accomplished by starting solidification at a selected point and allowing it to progress toward a riser. If solidification also starts at a second point (such as a thinner region), then shrinkage between these two points results, as illustrated in Fig. 3. In designing a gating system for a casting, each section is examined in an attempt to establish directional solidification by proper use of chills, risers and insulating materials. Casting design also is very important in ensuring that the necessary thermal gradients are established.

When the solidification characteristics of the entire casting are examined, it may be seen that solidification will be faster in some areas of the casting than in others; for example, the rate will be faster at a chilled area than at a riser area. Mechanical properties in castings are directly related to solidification rate; the more rapid the solidification rate, the higher the mechanical

Table 6 Mechanical properties and dendrite-arm spacing for an alloy C355.0—T6 test casting(a) (b)

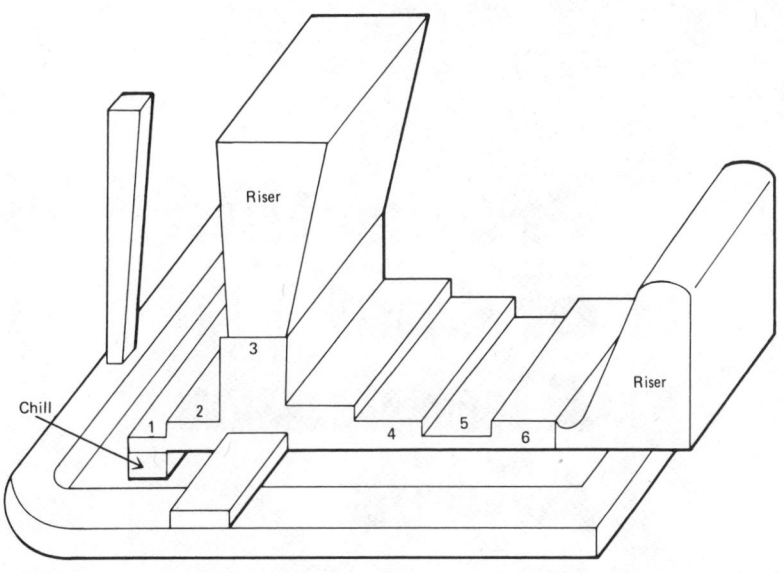

Location	Section thickness mm	in.	Description	Tensile strength MPa	ksi	Yield strength MPa	ksi	Elongation(c), %	Hardness, HB	Dendrite-arm spacing mm	in.
1	6.3	0.25	Chilled area	333	48.3	217	31.5	9.4	86	0.0279	0.0011
2	12.7	0.50	Sound cast section.....	272	39.5	217	31.5	3.8	86	0.0533	0.0021
3	51.0	2.0	Gate/riser area........	231	33.5	195	28.3	2.4	83	0.0762	0.0030
4	12.7	0.50	Typical area	236	34.2	205	29.7	2.4	83	0.0737	0.0029
5	6.3	0.25	Typical area	222	32.3	197	28.5	1.9	83	0.0889	0.0035
6	12.7	0.50	Gate/riser area........	224	32.5	199	28.9	1.8	83	0.0889	0.0035
...	...	...	Separately cast bar	279	40.5	215	31.2	2.9	89	0.0584	0.0023

(a) Casting illustrated is the test block proposed by The Aluminum Association. (b) Property values shown are not maximums or minimums. Properties may vary depending on casting design and other factors. (c) In 25 mm or 1 in.

properties. The solidification rate of a casting section can be determined using a metallographic technique that measures dendrite-arm spacing. High solidification rates produce relatively small dendrite-arm spacing. The data in Table 6 show that fine dendrite-arm spacing is accompanied by high mechanical properties.

As explained in the section on casting quality earlier in this article, casting processes all are characterized by solidification mode. In die casting, metal is rapidly injected and thus is subject to high chill rates. This results in rapid solidification and a fine metallographic structure at the surface. The fast chill rate and rapid injection, however, cause centerline shrinkage. It has generally been impractical to achieve directional solidification in die castings to overcome this centerline shrinkage. In permanent mold casting, chill rate

also is high but slower pouring allows longer feed times and more sound castings; design is very important in ensuring the proper solidification pattern. Because of high solidification rates, the properties of die and permanent mold castings are relatively insensitive to casting variables.

Due to relatively long solidification times, properties of sand and plaster castings are very sensitive to variations in casting technique; chills, risers and gates significantly affect properties. Plaster molds, with their insulating properties, keep solidification rate low and consequently produce castings with relatively low properties.

Premium-quality Castings. The production of "premium-quality" castings is an example of understanding and using the solidification process to

good advantage. In production of premium-quality castings, composite molds combining several mold materials are used to take advantage of the special properties of each casting process. High mechanical properties in designated areas are obtained by use of special chills. Plaster sections and risers may be used to extend the feeding range of the casting.

Premium-quality castings are made to the tight radiographic and mechanical specifications required for aerospace and other critical applications. Basic linear tolerances of ±15 mm/m (±15 mils/in.) are possible with aluminum alloy castings, depending on mold material, equipment and available fixtures, and minimum wall thickness of 3.8 mm (0.150 in.) is typical. Aluminum alloys commonly poured as premium-quality castings include C355.0, A356.0 and A357.

Properties of Cast Aluminum Alloys

By the ASM Committee on
Aluminum and Aluminum Alloys*

201.0
4.6Cu-0.7Ag-0.35Mn-0.35Mg-0.25Ti

Commercial Names
Trade name. KO-1

Specifications
Former ASTM. CQ 51A
SAE. 382
UNS number. A02010

Chemical Composition
Composition limits. 4.0 to 5.2 Cu;
0.15 to 0.55 Mg; 0.20 to 0.50 Mn; 0.10
max Si; 0.15 max Fe; 0.15 to 0.35 Ti;
0.40 to 1.0 Ag; 0.05 max others
(each); 0.10 max others (total); rem
Al
Consequence of exceeding impurity limits. High iron or silicon decreases tensile properties.

Applications
Typical uses. Sand castings, permanent mold and investment castings.
Structural casting members, aerospace housings, electrical transmission line fittings, insulator caps,
truck and trailer castings, other applications requiring highest tensile
and yield strengths with moderate
elongation. Gasoline engine cylinder
heads and pistons, turbine and supercharger impellers, rocker arms,
connecting rods, missile fins, other
applications where strength at elevated temperatures is important.
Structural gear housings, aircraft
landing gear castings, ordinance
castings, pump housings, other applications where high strength and
high energy absorption capacity are
required
Precautions in use. 201.0 castings
should be specified in the T7 temper
wherever resistance to stress corrosion cracking is an important consideration (exceeds requirement of 60
days alternate immersion exposure—10 min/h to 3.5% NaCl and
stressed to 75% of yield strength). T6
temper unsuitable for such applications.

Mechanical Properties
Tensile properties. See Table 1.
Hardness. T4 temper, 95 HB; T6
temper, 135 HB; T7 temper, 130 HB
(500-kg load, 10-mm ball)
Poisson's ratio. 0.33

Elastic modulus. Tension, 71 GPa
(10.3×10^6 psi); shear, 23 GPa (3.4×10^6 psi)
Impact strength. Charpy V-notch:
T4 temper, 21.7 J (16 ft·lb); T6 temper, 10.8 J (8 ft·lb)
Creep-rupture characteristics.
See Table 2.

Mass Characteristics
Density. 2.77 Mg/m³ (0.101 lb/in.³)
at 20 °C (68 °F)

Thermal Properties
Liquidus temperature. 650 °C
(1200 °F)
Solidus temperature. 535 °C (995 °F)
Coefficient of thermal expansion.
Linear, T6 and T7 conditions:

Temperature range		Average coefficient	
°C	°F	µm/ m·K	µin./ in.·°F
20–100	68–212	.... 19.3	10.7
20–200	68–392	.... 22.7	12.6
20–300	68–572	.... 24.7	13.7

Specific heat. 921 J/kg·K (0.220
Btu/lb·°F) at 100 °C (212 °F)

*See page X for committee list.

Table 1 Tensile properties of separately sand cast test bars of alloy 201.0

Condition	Exposure °C	°F	Exposure time, h	Tensile strength MPa	ksi	Yield strength(a) MPa	ksi	Elongation(b), %
Typical Properties								
T4......... ···	···	···	···	365	53	215	31	20
T6......... ···	···	···	···	485	70	435	63	7
T7......... ···	···	···	···	460	67	415	60	4.5
Short-timed Elevated Temperature Properties								
T7......... 150		300	0.5-100	380	55	360	52	6 to 8.5
			1000	360	52	345	50	8
			10 000	315	46	275	40	7
	205	400	0.5	325	47	310	45	9
			100	285	41	270	39	10
			1000	250	36	230	33	9
			10 000	185	27	150	22	14
	260	500	0.5	195	28	185	27	14
			100	150	22	140	20	17
			1000	125	18	110	16	18
	315	600	0.5	140	20	130	19	12
			100	85	12	75	11	30
			1000	70	10	60	9	39
			10 000	60	9	55	8	43

(a) 0.2% offset. (b) In 50 mm or 2 in. Where a range appears in this column, specified elongation varies with exposure time.

Table 2 Creep-rupture properties of separately sand cast test bars of alloy 201.0

Temperature °C	°F	Time under stress(a), h	Rupture stress MPa	ksi	Minimum creep rate at rupture stress, % per h	1.0% MPa	ksi	Stress for creep of: 0.5% MPa	ksi	0.25% MPa	ksi
150	300.......	10	Above yield		···	···	···	···	···	···	···
		100	Above yield		···	···	···	···	···	···	···
		1000	270	39	0.00013	260	38	250	36	250	36
		10 000	195	28	0.000023	195	28	185	27	180	76
175(b)	350(b)	10	Above yield		···	···	···	···	···	···	···
		100	250	36	0.00095	250	36	235	34	230	33
		1000	180	26	0.000175	180	26	170	25	170	24
		10 000	130	19	0.000035	130	19	125	18	125	18
205	400.......	10	250	36	0.0145	240	35	230	33	220	32
		100	180	26	0.0024	180	26	170	24	170	24
		1000	130	19	0.00046	130	19	125	18	110	16
		10 000	95	14	0.000088	95	14	90	13	85	12
230	450.......	10	185	27	0.028	185	27	170	25	170	24
		100	140	20	0.0048	140	20	130	19	115	17
		1000	95	14	0.00083	95	14	90	13	85	12
		10 000	70	10	0.000175	70	10	65	9.5	60	9
260	500.......	10	140	20	0.047	140	20	125	18	110	16
		100	95	14	0.0080	95	14	95	14	85	12
		1000	70	10	0.00130	70	10	70	9.8	60	8.6
		10 000	50	7.5	0.00028	50	7.5	50	7.2	45	6.3
290	550.......	10	90	13	0.062	83	12	83	12	69	10
		100	66	9.5	0.0118	63	9.2	58	8.4	54	7.8
		1000	48	7.0	0.0022	46	6.7	42	6.1	39	5.7
		10 000	34	5.0	0.00037	33	4.8	30	4.4	29	4.2
315	600.......	10	59	8.6	0.073	55	8.0	51	7.4	47	6.8
		100	43	6.2	0.0126	39	5.7	37	5.4	34	4.9
		1000	31	4.5	0.0023	29	4.2	27	3.9	24	3.5
		10 000	23	3.3	0.00043	21	3.0	20	2.9	18	2.6

(a) 10 000 h data are extrapolated. (b) For this temperature, properties are interpolations.

Latent heat of fusion. 389 kJ/kg (167 Btu/lb)

Thermal conductivity. 121 W/m·K (70 Btu/ft·h·°F) at 25 °C (77 °F)

Electrical Properties

Electrical conductivity. Volumetric, T6 condition: 27 to 32% IACS at 20 °C (68 °F)

Electrical resistivity. T6 condition: 54 to 64 nΩ·m at 20 °C (68 °F)

Electrolytic solution potential. vs 0.1N calomel electrode in an aqueous solution containing 53 g NaCl plus 3 g H_2O_2 per litre: as cast, -0.70 V; T4 condition, -0.59 V; T6 condition, -0.68 V; T7 condition, -0.73 V

Fabrication Characteristics

Solution temperature. See Table 3.

Aging temperature. See Table 3.

206.0, A206.0
4.5Cu-0.30Mn-0.25Mg-0.22Ti

Chemical Composition

Composition limits. 4.2 to 5.0 Cu; 0.15 to 0.35 Mg; 0.20 to 0.50 Mn; 0.10 max Si; 0.15 max Fe; 0.10 max Zn; 0.15 to 0.30 Ti; 0.05 max Ni; 0.05 max Sn; 0.05 max Sn; 0.05 max others (each); 0.15 max others (total); rem Al. A206.0: 4.2 to 5.0 Cu; 0.15 to 0.35 Mg; 0.20 to 0.50 Mn; 0.05 max Si; 0.10 max Fe; 0.10 max Zn; 0.15 to 0.30 Ti; 0.05 max Ni; 0.05 max Sn; 0.05 max others (each); 0.15 max Ni; 0.05 max Sn; 0.05 max others (each); 0.15 max others (total); rem Al

Applications

Typical uses. Structural castings in heat treated temper for automotive, aerospace, and other applications where high tensile and yield strength and moderate elongation are needed. Gear housings, truck spring hanger castings and other applications where high fracture toughness is required. Cylinder heads for gasoline and diesel motors, turbine and supercharger impellers, other applications where high strength at elevated temperatures and special aging treatment are required

Precautions in use. Subject to corrosion problems due to copper content of alloy. T4 and T7 heat treatments qualify and meet federal test requirements for stress corrosion cracking. T6 temper should not be used where stress corrosion cracking could be a problem.

Mechanical Properties

Tensile properties. Separately cast test bars. Tensile strength and yield strength, see Table 4. Elongation in 50 mm or 2 in. (typical for A206.0-T7): 11.7% at room temperature, 14.0% at 120 °C (250 °F), 17.7% at 175 °C (350 °F). Reduction in area (typical for A206.0-T7): 26.0% at room temperature, 40.4% at 120 °C (250 °F), 53.7% at 175 °C (350 °F)

Shear strength. See Table 4.

Compressive yield strength. See Table 4.

Bearing properties. See Table 4.

Hardness. T4 temper, 118 HV; T7 temper, 137 HV

Poisson's ratio. 0.33 at 20 °C (68 °F)

Elastic modulus. Tension: 70 GPa (10.2×10^6 psi) at room temperature, 69 GPa (10.0×10^6 psi) at 120 °C (250 °F), 65 GPa (9.4×10^6 psi) at 175 °C (350 °F)

Impact strength. Charpy V-notch, 9.5 J (7.0 ft·lb) at 20 °C (68 °F)

Table 3 Heat treatment practice for alloy 201.0(a)

Resulting temper	Temperature °C	°F	Time at temperature, h	Cooling
T4	490-500	910-930	2	None
	525-530(b)	980-990(b)	14-20(c)	Water(d)
T6(e)	Room temperature		12-24	None
	150-155	305-315	20	Not critical
T7(e)	Room temperature		12-24	None
	185-190	365-375	5	Not critical

(a) All treatments are performed in the two steps indicated for each resulting temper. (b) Careful composition and temperature control must be maintained during solution heat treatment in order to attain both adequate solution and to prevent incipient melting. (c) Soaking time periods required for average sand castings after load has reached specified temperature. Time changes may be required. Permanent mold and thin wall castings, in general, take less time. (d) At 65 to 100 °C (150 to 212 °F). (e) Start with T4 temper material.

Table 4 Typical mechanical properties for separately cast test bars of alloy A206.0-T7

Property	Room temperature MPa	ksi	120 °C MPa	250 °F ksi	175 °C MPa	350 °F ksi
Tensile strength	436	63	384	56	333	48
Tensile yield strength	347	50	316	46	302	44
Shear strength	257	37	232	34	208	30
Compressive yield strength	372	54	347	50	318	46
Bearing strength		100	632	92	545	79
$e/d = 1.5$(a)	692	131	784	114	635	92
$e/d = 2.0$(a)	960					
Bearing yield strength						
$e/d = 1.5$(a)	544	79	507	73	477	69
$e/d = 2.0$(a)	658	95	628	91	566	82
Axial fatigue strength						
Unnotched, $R = 0.1$						
10^3 cycles	435	63	...	...	350	51
10^5 cycles	290	42	...	...	250	36
10^7 cycles	205	30	...	...	160	23
Notched, $K_t = 3.0$, $R = 0.1$						
10^3 cycles	370	54	...	...	370	54
10^5 cycles	115	17	...	...	115	17
10^7 cycles	90	10	...	...	70	10

(a) Where e is edge distance and d is pin diameter.

Fatigue strength. See Table 4.
Plane-strain fracture toughness.
43.1 MPa$\sqrt{m}$ (39.2 ksi$\sqrt{in.}$) for
A206.0-T7 (not a true K_{Ic} value per
ASTM E399)

Mass Characteristics

Density. 2.80 Mg/m^3 (0.101 lb/in.^{3})
at 20 °C (68 °F)

Thermal Properties

Liquidus temperature. 650 °C
(1202 °F)
Solidus temperature. 570 °C (1058
°F)
Incipient melting temperature.
542 °C (1008 °F)
Coefficient of thermal expansion.
Linear: 19.3 μm/m·K (10.7 μin./
in.·°F) at 20 to 100 °C (68 to 212 °F)
Specific heat. 921 J/kg·K (0.20 Btu/
lb·°F) at 100 °C (212 °F)
Thermal conductivity. 121 W/m·K
(70.1 Btu/ft·h·°F)

Electrical Properties

Electrical resistivity. At 20 °C (68
°F): T6 temper, 54 to 64 nΩ·m; T7
temper, 50 to 54 nΩ·m

Chemical Properties

General corrosion behavior. Comparable to other wrought or cast aluminum alloys containing equivalent
amounts of copper

Fabrication Characteristics

Weldability. Fair repair welding
characteristics
Solution temperature. 530 °C (985
°F)
Aging temperature. T7 temper: 200
°C (390 °F), hold at temperature for 8
h. T6 temper: 155 °C (310 °F), hold at
temperature for 8 h. T4 temper: room
temperature

208.0
4Cu-3Si

Commercial Names

Former designation. 108

Specifications

Former ASTM. CS43A
UNS number. A 02080
Government. QQ-A-601, class 8

Chemical Composition

Composition limits. 3.5 to 4.5 Cu;
0.10 max Mg; 0.5 max Mn; 2.5 to 3.5
Si; 1.2 max Fe; 1.0 max Zn; 0.35 max

Ni; 0.25 max Ti; 0.50 max others
(total); rem Al
Consequence of exceeding impurity limits. High iron decreases mechanical properties, especially ductility. Zinc or tin decreases mechanical
properties. Magnesium reduces ductility.

Applications

Typical uses. Manifolds, valve
bodies, and similar castings requiring pressure tightness. Other applications where good casting characteristics, good weldability, pressure
tightness, and moderate strength are
required

Mechanical Properties

Tensile properties. Typical for separately cast test bars. F temper: tensile strength, 145 MPa (21 ksi); yield
strength, 95 MPa (14 ksi); elongation, in 50 mm or 2 in., 2.5%
Shear strength. 115 MPa (17 ksi)
Compressive yield strength. 105
MPa (15 ksi)
Hardness. 55 HB (500-kg load, 10-
mm ball)
Poisson's ratio. 0.33
Elastic modulus Tension, 71 GPa
(10.3 × 10^6 psi); shear, 26.5 GPa
(3.85 × 10^6 psi)
Fatigue strength. 76 MPa (11 ksi)
at 5 × 10^8 cycles (R. R. Moore type
test)

Mass Characteristics

Density. 2.79 Mg/m^3 (0.101 lb/in.^{3})
at 20 °C (68 °F)

Thermal Properties

Liquidus temperature. 625 °C
(1160 °F)
Solidus temperature. 520 °C (970
°F)
Coefficient of thermal expansion.
Linear:

| Temperature range | | Average coefficient | |
°C	°F	μm/ m·K	μin./ in.·°F
20–100	68–212	22.0	12.2
20–200	68–392	23.0	12.7
20–300	68–572	24.0	13.3

Specific heat. 963 J/kg·K (0.230
Btu/lb·°F) at 100 °C (212 °F)
Latent heat of fusion. 389 kJ/kg
(167 Btu/lb)
Thermal conductivity. At 25 °C (77

°F): as cast, 121 W/m·K (70.1 Btu/
ft·h·°F); annealed, 146 [/m·K (84.4
Btu/ft·h·°F)

Electrical Properties

Electrical conductivity. Volumet-
ric, at 20 °C (68 °F): as cast, 31%
IACS; annealed, 38% IACS
Electrical resistivity. At 20 °C (68
°F): as cast, 55.6 nΩ·m; annealed,
45.4 nΩ·m
Electrolytic solution potential.
−0.77 V vs 0.1N calomel electrode in
an aqueous solution containing 53 g
NaCl plus 3 g H_2O_2 per litre

Fabrication Characteristics

Melting temperature. 675 to 815 °C
(1250 to 1500 °F)
Casting temperature. 675 to 790 °C
(1250 to 1450 °F)
Joining. Rivet compositions: 2117-
T4, 2017-T4. Soft solder with Alcoa
No. 802; no flux. Rub-tin with Alcoa
No. 33 flux: flame either reducing
oxyacetylene or reducing ocyhydro-
gen. Metal-arc weld with 4043 alloy:
Alcoa No. 27 flux. Carbon-arc weld
with 4043 alloy: Alcoa No. 24 flux
(automatic), Alcoa No. 27 flux (manual). Tungsten-arc argon-atmosphere weld with 4043 alloy: no flux.
Resistance welding: spot, seam, and
flash methods
Aging temperature. T55 temper:
149 to 160 °C (300 to 320 °F) for 16
h

242.0
4Cu-2Ni-2.5Mg

Commercial Names

Former designation. 142

Specifications

AMS. 4220, 4222
Former ASTM. CN42A
SAE. 39
UNS number. A02420
Government. QQ-A-601, class 6
(snad); QQ-A-596, class 3 (perma-
nent mold)
Foreign. Canada: CSA CN42.
France: NF A-U4NT. ISO:
AlCu4Ni2Mg2

Chemical Composition

Composition limits. 3.5 to 4.5 Cu;
1.2 to 1.8 Mg; 0.35 max Mn; 0.7 max
Si; 1.0 max Fe; 0.25 max Cr; 0.35

Table 5 Typical mechanical properties of separately cast test bars of alloy 242.0

Temper	Tensile strength(a) MPa	ksi	Tensile yield strength(a) MPa	ksi	Elon-gation(a), %	Hard-ness(b), HB	Shear strength MPa	ksi	Fatigue strength(c) MPa	ksi	Compressive yield strength(a) MPa	ksi
Sand Cast												
T21	185	27	125	18	1.0	70	145	21	55	8.0	125	18
T571	220	32	205	30	0.5	85	180	26	75	11.0	235	34
T77	205	30	160	23	2.0	75	165	24	70	10.5	165	24
Permanent Mold Cast												
T571	275	40	235	34	1.0	105	205	30	70	10.5	235	34
T61	325	47	290	42	0.5	110	240	35	65	9.5	305	44

(a) Strengths and elongations remain unchanged or improve at low temperatures. (b) 500-kg load; 10-mm ball. (c) At 5 x 10^8 cycles; R. R. Moore type test.

max Zn; 0.25 max Ti; 1.7 to 2.3 Ni; 0.05 max others (each); 0.15 max others (total); rem Al

Consequence of exceeding impurity limits. High iron may cause shrinkage difficulties. High silicon decreases mechanical properties. Chromium decreases thermal conductivity.

Applications

Typical uses. Motorcycle, diesel and aircraft pistons, air-cooled cylinder heads, aircraft generator housings, other applications where excellent high-temperature strength is required

Mechanical Properties

Tensile properties. See Tables 5 and 6.
Shear strength. See Table 5.
Compressive yield strength. See Table 5.
Hardness. See Table 5.
Poisson's ratio. 0.33
Elastic modulus. Tension, 71 PGa (10.3 × 10^6 psi); shear, 26.5 GPa (3.85 × 10^6 psi)
Fatigue strength. See Table 5.

Mass Characteristics

Density. 2.823 Mg/m^3 (0.102 lb/in.^{3}) at 20 °C (68 °F)

Thermal Properties

Liquidus temperature. 635 °C (1175 °F)
Solidus temperature. 530 °C (990 °F)
Coefficient of thermal expansion. Linear:

Table 6 Typical tensile properties of separately cast test bars of alloy 242.0-T571 at elevated temperature

Temperature °C	°F	Tensile strength MPa	ksi	Yield strength MPa	ksi	Elon-gation(a), %
Sand Cast						
24	75	220	32	205	30	0.5
150	300	205	30	195	28	0.5
205	400	180	26	145	21	1.0
260	500	90	13	55	8	8.0
315	600	55	8	30	4	20.0
Permanent Mold Cast						
24	75	275	40	235	34	1.0
150	300	255	37	230	33	1.0
205	400	195	28	150	22	2.0
260	500	90	13	55	8	15.0
315	600	55	8	30	4	35.0

(a) In 50 mm or 2 in.

Table 7 Heat treatment for separately cast test bars of alloy 242.0

Purpose (and resulting temper)	Temperature °C	°F	Time, h
Sand Castings			
Annealing, T21	340 to 345	645 to 655	2 to 4
Aging, T571(a)	170 to 175	335 to 345	40 to 48
Solution heat treatment	520 to 525	965 to 975	6(b)(c)
Aging, T77(d)(e)	340 to 345	645 to 655	1 to 3
Permanent Mold Castings			
Aging, T571(b)	170 to 175	335 to 345	40 to 48
Solution heat treatment	515 to 520	955 to 965	4(b)(f)
Aging, T61(e)	200 to 205	395 to 405	3 to 5

(a) No solution heat treatment. (b) Soaking-time periods required for average castings after load has reached specified temperature. Time can be decreased or may have to be increased, depending on experience with particular castings. (c) Still-air cooling. (d) Start with solution heat treated material. (e) U.S. Patent 1,822,877. (f) Cool in water at 65 to 100 °C (150 to 212 °F).

Temperature range °C	°F	Average coefficient μm/ m·K	μin./ in.·°F
20–100	68–212	22.5	12.5
20–200	68–392	23.5	13.1
20–300	68–572	24.5	13.6

Specific heat. 963 J/kg·K (0.230 Btu/lb·°F) at 100 °C (212 °F)
Latent heat of fusion. 389 kJ/kg (167 Btu/lb)
Thermal conductivity. At 25 °C (77 °F):

Temper and product	Conductivity W/m·K	Btu/ft·h·°F
T21, sand	167	96.7
T571, sand	134	77.4
T77, sand	146	84.6
T61, permanent mold	130	74.9

Electrical Properties

Electrical conductivity. Volumetric:

Temper and product	Conductivity, % IACS
T21, sand	44
T571, sand	34
T77, sand	38
T61, permanent mold	33

Electrical resistivity. At 20 °C (68 °F):

Temper and product	Resistivity, nΩ·m
T21, sand	39.2
T571, sand	50.7
T77, sand	45.4
T61, permanent mold	52.2

Fabrication Characteristics

Melting temperature. 675 to 815 °C (1250 to 1500 °F)

Casting temperature. Sand and permanent mold castings: 675 to 790 °C (1250 to 1450 °F)

Annealing temperature. See Table 7.

Solution temperature. See Table 7.

Aging temperature. See Table 7.

Joining. Rivet compositions: 2117-T4, 2017-T4. Soft solder with Alcoa No. 802; no flux. Rub-tin with Alcoa No. 802. Metal-arc weld with 4043 alloy: Alcoa No. 27 flux. Carbon-arc weld with 4043 alloy: Alcoa No. 24 flux (automatic), Alcoa No. 27 flux (manual). Tungsten-arc argon-atmosphere weld with 4043 alloy: no flux. Resistance welding: spot, seam, and flash welds

295.0
4.5Cu-1.1Si

Commercial Names

Former disignation. 195

Table 8 Typical tensile properties for separately cast test bars of alloy 295.0 at elevated temperatures

Temperature °C	°F	Tensile strength MPa	ksi	Yield strength MPa	ksi	Elongation(a), %
T4 Temper						
24	75	220	32	110	16	8.5
150	300	195	28	105	15	5.0
205	400	105	15	60	9	15.0
260	500	60	9	40	6	25.0
315	600	30	4	20	3	75.0
T6 Temper						
24	75	250	36	165	24	5.0
150	300	195	28	140	20	5.0
205	400	105	15	60	9	15.0
260	500	60	9	40	6	25.0
315	600	30	4	20	3	75.0

(a) In 50 mm or 2 in.

Specifications

AMS. 4230, 4231
Former ASTM. C4A
SAE. 38
UNS number. A02950
Government. QQ-A-601, class 4
Foreign. Canada: CSA-C4. ISO: AlCu4Si

Chemical Composition

Composition limits. 4.0 to 5.0 Cu; 0.03 max Mg; 0.35 max Mn; 0.7 to 1.5 Si; 1.0 max Fe; 0.35 max Zn; 0.25 max Ti; 0.05 max others (each); 0.15 max others (total); rem Al

Consequence of exceeding impurity limits. High iron or silicon decreases tensile properties, especially ductility. Manganese or magnesium decreases ductility. Tin reduces strength, hardness, and resistance to corrosion.

Applications

Typical uses. Flywheel housings, rear-axle housings, bus wheels, aircraft wheels, fittings, crankcases and other applications where a combination of high tensile properties and good machinability is required, but pressure tightness is not needed.

Mechanical Properties

Tensile properties. Typical for separately cast test bars. Tensile strength: T4 temper, 220 MPa (32 ksi); T6 temper, 250 MPa (36 ksi); T62 temper, 285 MPa (41 ksi). Yield strength: T4 temper, 110 MPa (16 ksi); T6 temper, 165 MPa (24 ksi); T62 temper, 220 MPa (32 ksi). Elongation, in 50 mm or 2 in.: T4 temper, 8.5%; T6 temper, 5.0%; T62 temper,

2.0%. Strengths and elongations remain unchanged or improve at low temperatures. See also Table 8.

Shear strength. T4 temper: 180 MPa (26 ksi). T6 temper: 205 MPa (30 ksi). T62 temper: 230 MPa (33 ksi)

Compressive yield strength. T4 temper: 115 MPa (17 ksi); T6 temper: 170 MPa (25 ksi); T62 temper: 235 MPa (34 ksi)

Hardness. T4 temper, 60 HB; T6 temper, 75 HB; T62 temper, 90 HB (500-kg load, 10-mm ball)

Poisson's ratio. 0.33

Elastic modulus. Tension, 70 GPa $(10.0 \times 10^6$ psi); shear, 25.9 GPa $(3.75 \times 10^6$ psi)

Fatigue strength. At 5×10^8 cycles: T4 temper, 50 MPa (7.0 ksi); T6 temper, 50 MPa (7.5 ksi); T62 temper, 55 MPa (8.0 ksi) (R. R. Moore type test)

Mass Characteristics

Density. 2.823 Mg/m³ (0.102 lb/in.³) at 20 °C (68 °F)

Thermal Properties

Liquidus temperature. 645 °C (1190 °F)

Solidus temperature. 520 °C (970 °F)

Coefficient of thermal expansion. Linear:

Temperature range °C	°F	Average coefficient µm/m·K	µin./in.·F
20–100	68–212	23.0	12.8
20–200	68–392	24.0	13.3
20–300	68–572	25.0	13.9

Specific heat. 963 J/kg·K (0.230 Btu/lb·°F) at 100 °C (212 °F)

Latent heat of fusion. 389 kJ/kg (167 Btu/lb)

Thermal conductivity. T4 and T62 tempers: 138 W/m·K (79.8 Btu/ ft·h·°F) at 25 °C (77 °F)

Electrical Properties

Electrical conductivity. Volumetric: T4 temper, 35% IACS; T62 temper: 37% IACS

Electrical resistivity. T4 and T62 tempers: 49.3 nΩ·m at 20 °C (68 °F)

Electrolytic solution potential. vs 0.1N calomel electrode in an equeous solution containing 53 g NaCl plus 3 g H_2O_2 per litre: T4 temper, −0.70 V; T6 temper, −0.71 V; T62 temper, −0.73 V

Fabrication Characteristics

Melting temperature. 675 to 815 °C (1250 to 1500 °F)

Casting temperature. 675 to 790 °C (1250 to 1450 °F)

Solution temperature. 515 to 520 °C (955 to 965 °F); hold at temperature for 12 h; cool in water at 65 to 100 °C (150 to 212 °F)

Aging temperature. 150 to 155 °C (305 to 315 °F). To obtain T6 temper from solution heat treated material, hold at temperature for 3 to 5 h; for T62 temper, hold at temperature for 12 to 16 h

Joining. Rivet compositions: 2117-T4, 2017-T4. Soft solder with Alcoa No. 802; no flux. Rub-tin with Alcoa No. 802. Atomic-hydrogen weld with 4043 alloy; Alcoa No. 22 flux. Oxyacetylene weld with 4043 alloy; Alcoa No. 22 flux; flame neutral. Metal-arc weld with 4043 alloy; Alcoa No. 27 flux. Carbon-arc weld with 4043 alloy; Alcoa No. 24 flux (automatic), Alcoa No. 27 flux (manual). Tungsten-arc argon-atmosphere weld with 4043 alloy; no flux. Resistance welding: spot, seam, and flash methods

296.0
4.5Cu-2.5Si

Commercial Names

Former designations. B295.0, B195

Specifications

AMS. 4282, 4283

SAE. 380
UNS number. A-22950
Government. QQ-A-396, class 4

Chemical Composition

Composition limits. 4.0 to 5.0 Cu; 0.05 max Mg; 0.35 max Mn; 2.0 to 3.0 Si; 1.2 max Fe; 0.50 max Zn; 0.25 max Ti; 0.35 max Ni; 0.35 max others (total); rem Al

Consequence of exceeding impurity limits. High iron decreases tensile properties, especially ductility. Zinc or tin decreases tensile properties. Manganese or magnesium reduces ductility.

Applications

Typical uses. Aricraft fittings, aircraft gun control parts, aircraft wheels, railroad car seat frames, compressor connecting rods, full pump bodies, other applications requiring a combination of high tensile properties and good machinability.

Mechanical Properties

Tensile properties. Typical for separately cast test bars. Tensile strength: T4 temper, 255 MPa (37 ksi); T6 temper, 275 MPa (40 ksi); T7 temper, 270 MPa (39 ksi). Yield strength: T4 temper, 130 MPa (19 ksi); T6 temper, 180 MPa (26 ksi); T7 temper, 140 MPa (20 ksi). Elongation, in 50 mm or 2 in.: T4 temper, 9%; T6 temper, 5%; T7 temper, 4.5%. Strengths and elongations remain unchanged or improve at low temperatures. See also Table 9.

Shear strength. T4 and T7 tempers; 205 MPa (30 ksi). T6 temper: 220 MPa (32 ksi)

Compressive yield strength. T4 and T7 tempers: 140 MPa (20 ksi). T6 temper: 180 MPa (26 ksi)

Hardness. T4 temper: 75 HB. T6 temper: 90 HB. T7 temper: 80 HB (500-kg load, 10-mm ball)

Poisson's ratio. 0.33

Elastic modulus. Tension, 69 GPa (10.0×10^6 psi); shear, 26.2 GPa (3.80×10^6 psi)

Fatigue strength. At 5×10^8 cycles: T4 temper, 65 MPa (9.5 ksi); T6 temper, 70 MPa (10.0 ksi); T7 temper, 60 MPa (9.0 ksi) (R. R. Moore type test)

Mass Characteristics

Density. 2.796 Mg/m³ (0.101 lb/in.³) at 20 °C (68 °F)

Thermal Properties

Liquidus temperature. 635 °C (1170 °F)

Solidus temperature. 530 °C (990 °F)

Coefficient of thermal expansion. Linear:

| Temperature range | | Average coefficient | |
°C	°F	μm/ m·K	μin./ in.·°F
20–100	68–212	22	12.2
20–200	68–392	23	12.7
20–300	68–572	24	13.3

Specific heat. 963 J/kg·K (0.230 Btu/lb·°F) at 100 °C (212 °F)

Latent heat of fusion. 389 kJ/kg (167 Btu/lb)

Thermal conductivity. T4 and T6 tempers: 130 W/m·K (74.9 Btu/ ft·h·°F)

Table 9 Typical tensile properties of separately permanent mold cast test bars of alloy 296.00 at elevated temperature

| Temperature | | Tensile strength | | Yield strength | | Elongation(a), % |
°C	°F	MPa	ksi	MPa	ksi	
T4 Temper						
24	75	255	37	130	19	9
150	300	200	29	160	23	5
205	400	115	17	75	11	15
260	500	50	7	30	4	25
315	600	25	3.5	15	2.5	75
T6 Temper						
24	75	275	40	180	26	5
150	300	200	29	160	23	5
205	400	115	17	75	11	15
260	500	50	7	30	4	25
315	600	25	3.5	15	2.5	75

(a) In 50 mm or 2 in.

Electrical Properties

Electrical conductivity. Volumetric, T4 and T6 tempers: 33% IACS
Electrical resistivity. T4 and T6 tempers: 52.2 nΩ·m at 20 °C (68 °F)
Electrolytic solution potential. T4 temper: −0.71 V vs 0.1N calomel electrode in an aqueous solution containing 53 g NaCl plus 3 g H_2O_2 per litre

Fabrication Characteristics

Melting temperature. 675 to 815 °C (1250 to 1500 °F)
Casting temperature. 675 to 815 °C (1250 to 1500 °F)
Solution temperature. 505 to 515 °C (945 to 955 °F); hold at temperature for 8 h; cool in water at 65 to 100 °C (150 to 212 °F)
Aging temperature. To obtain T6 temper from solution heat treated material, 150 to 155 °C (305 to 315 °F) and hold at temperature 5 to 7 h; for T7 temper (U.S. Patent 1,822,877) 255 to 265 °C (495 to 505 °F) and hold at temperature 4 to 6 h
Joining. Same as alloy 295.0

308.0
5.5Si-4.5Cu

Commercial Names
Former designation. A108

Specifications
Former ASTM. SC64A
SAE. 330
UNS number. A03080
Government. QQ-A-596, class 6

Chemical Composition
Composition limits. 4.0 to 5.0 Cu; 0.10 max Mg; 0.50 max Mn; 5.0 to 6.0 Si; 1.0 max Fe; 1.0 max Zn; 0.25 max Ti; 0.50 max others (total); rem Al
Consequence of exceeding impurity limits. High iron, zinc, or tin decreases mechanical properties. Magnesium decreases ductility.

Applications
Typical uses. Ornamental grills, reflectors, general-purpose castings, and other applications where good casting characteristics, good weldability, pressure tightness, and moderate strength are required

Mechanical Properties
Tensile properties. Typical for separately cast test bars. F temper: tensile strength, 195 MPa (28 ksi); yield strength, 110 MPa (16 ksi); elongation in 50 mm or 2 in., 2.0%
Shear strength. 150 MPa (22 ksi)
Hardness. 70 HB (500-kg load, 10-mm ball)
Poisson's ratio. 0.33
Elastic modulus. Tension, 71 GPa (10.3 × 10⁶ psi); shear, 26.5 GPa (3.85 × 10⁶ psi)

Mass Characteristics
Density. 2.790 Mg/m³ (0.101 lb/in.³) at 20 °C (68 °F)

Thermal Properties
Liquidus temperature. 615 °C (1135 °F)
Solidus temperature. 520 °C (970 °F)
Coefficient of thermal expansion. Linear:

Temperature range		Average coefficient	
°C	°F	μm/m·K	μin./in.·°F
20–100	68–212	21.5	11.9
20–200	68–392	22.5	12.5
20–300	68–572	23.0	12.8

Specific heat. 963 J/kg·K (0.230 Btu/lb·°F) at 100 °C (212 °F)
Latent heat of fusion. 389 kJ/kg (167 Btu/lb)
Thermal conductivity. 142 W/m·K (82.2 Btu/ft·h·°F) at 25 °C (77 °F)

Electrical Properties
Electrical conductivity. Volumetric, 37% IACS at 20 °C (68 °F)
Electrical resistivity. 46.6 nΩ·m at 20 °C (68 °F)
Electrolytic solution potential. −0.75 V vs 0.1N calomel electrode in an aqueous solution containing 53 g NaCl plus 3 g H_2O_2 per litre

Fabrication Characteristics
Melting temperature. 675 to 816 °C (1250 to 1500 °F)
Casting temperature. 675 to 795 °C (1250 to 1450 °F)
Joining. Rivet compositions: 2117-T4, 2017-T4. Soft solder with Alcoa No. 802. Braze with Alcoa No. 717; Alcoa No. 33 flux; flame either reducing oxyacetylene or reducing oxyhydrogen. Metal-arc weld with 4043 alloy; Alcoa No. 27 flux. Carbon-arc weld with 4043 alloy; Alcoa No. 24 flux (automatic), Alcoa No. 27 flux (manual). Tungsten-arc argon-atmosphere weld with 4043 alloy, no flux. Resistance welding: spot, seam, and flash

319.0
6Si-3.5Cu

Commercial Names
Former designations. 319, Allcast

Specifications
Former ASTM. SC64D
SAE. 326
UNS number. A03190
Foreign. ISO: AlSi6Cu4

Chemical Composition
Composition limits. 3.0 to 4.0 Cu; 0.10 max Mg; 0.50 max Mn; 5.5 to 6.5 Si; 1.0 max Fe; 1.0 max Zn; 0.25 max Ti; 0.35 max Ni; 0.50 max others (total); rem Al
Consequence of exceeding impurity limits. Mechanical properties are relatively insensitive to impurities.

Applications
Typical uses. Automotive cylinder heads, internal combustion engine crankcases, typewriter frames, piano plates, and other applications where good casting characteristics and weldability, pressure tightness, and moderate strength are required

Mechanical Properties
Tensile properties. See Table 10.
Shear strength. See Table 10.
Compressive yield strength. See Table 10.
Hardness. See Table 10.
Poisson's ratio. 0.33
Elastic modulus. Tension, 74 GPa (10.7 × 10⁶ psi); shear, 28 GPa (4.0 × 10⁶ psi)
Fatigue strength. See Table 10.

Mass Characteristics
Density. 2.796 Mg/m³ (0.100 lb/in.³) at 20 °C (68 °F)

Thermal Properties
Liquidus temperature. 605 °C (1120 °F)
Solidus temperature. 515 °C (960 °F)
Coefficient of thermal expansion. Linear:

Temperature range		Average coefficient	
°C	°F	µm/ m·K	µin./ in.·°F
20–100	68–212 .... 21.5		11.9
20–200	68–392 .... 23.0		12.8
20–300	68–572 .... 23.5		13.1

Specific heat. 963 J/kg·K (0.230 Btu/lb·°F) at 100 °C (212 °F)

Latent heat of fusion. 389 kJ/kg (167 Btu/lb)

Thermal conductivity. 109 W/m·K (62.9 Btu/ft·h·°F) at 25 °C (77 °F)

Electrical Properties

Electrical conductivity. Volumetric, 27% IACS at 20 °C (68 °F)

Electrical resistivity. Sand: 63.9 nΩ·m at 20 °C (68 °F)

Electrolytic solution potential. −0.81 V (sand) and −0.76 V (permanent mold) vs 0.1N calomel electrode in an aqueous solution containing 53 g NaCl plus 3 g H_2O_2 per litre

Fabrication Characteristics

Melting temperature. 675 to 815 °C (1250 to 1500 °F)

Casting temperature. Sand: 675 to 790 °C (1250 to 1450 °F)

Solution temperature. 500 to 505 °C (935 to 945 °F); hold at temperature 12 h (sand), 8 h (permanent mold); cool in water at 65 to 100 °C (150 to 212 °F)

Aging temperature. To obtain T6 temper from solution treated material, 150 to 155 °C (305 to 315 °F) and hold at temperature 2 to 5 h

Joining. Same as for alloy 208.0

336.0
12Si-2.5Ni-1Mg-1Cu

Commercial Names

Former designations. A332.0, A132

Specifications

Former ASTM. SN122A
SAE. 321
UNS number. A13320
Government. QQ-A-596, class 9
Foreign. Canada: CSA SN122. France: NF A-S12N2G

Chemical Composition

Composition limits. 0.5 to 1.5 Cu; 1.3 max Mg; 0.35 max Mn; 11.0 to 13.0 Si; 1.2 max Fe; 0.35 max Zn; 0.25 max Ti; 0.05 max others (each); 0.15 max others (total); rem Al

Consequence of exceeding impurity limits. High iron or chromium promotes shrinkage difficulties

Applications

Typical uses. Automotive and diesel pistons, pulleys, sheaves, and other applications where good high-temperature strength, low coefficient of thermal expansion, and good resistance to wear are required

Mechanical Properties

Tensile properties. Typical for separately cast test bars. Tensile strength: T551 temper, 248 MPa (36 ksi); T65 temper, 324 MPa (47 ksi). Yield strength: T551 temper, 193 MPa (28 ksi); T65 temper, 296 MPa (43 ksi). Elongation, in 50 mm or 2 in.: T551 and T65 tempers, 0.5%. Strengths and elongations remain unchanged or improve at low temperatures. See also Table 11.

Shear strength. T551 temper, 193 MPa (28 ksi); T65 temper, 248 MPa (36 ksi)

Compressive yield strength. T551 temper, 193 MPa (28 ksi); T65 temper, 296 MPa (43 ksi)

Hardness. T551 temper, 105 HB; T65 temper, 125 HB (500-kg load, 10-mm ball)

Poisson's ratio. 0.33

Elastic modulus. Tension, 73 GPa (10.6 × 10⁶ psi; shear, 29.9 GPa (4.35 × 10⁶ psi)

Mass Characteristics

Density. 2.713 Mg/m³ (0.098 lb/in.³) at 20 °C (68 °F)

Thermal Properties

Liquidus temperature. 565 °C (1040 °F)

Solidus temperature. 540 °C (1000 °F)

Melting temperature. 677 to 816 °C (1250 to 1500 °F)

Coefficient of thermal expansion. Linear:

Temperature range		Average coefficient	
°C	°F	µm/ m·K	µin./ in.·°F
20–100	68–212 19		10.6
20–200	68–392 20		11.1
20–300	68–572 21		11.7

Specific heat. 963 J/kg·K (0.230 Btu/lb·°F) at 100 °C (212 °F)

Latent heat of fusion. 389 kJ/kg

Table 10 Typical mechanical properties for separately cast test bars of alloy 319.0

Temper	Tensile strength (a) MPa	ksi	Tensile yield strength (a) MPa	ksi	Elongation(a)(b), %	Hardness(c), HB	Shear strength MPa	ksi	Fatigue strength (d) MPa	ksi	Compressive yield strength (a) MPa	ksi
Sand Cast												
As cast..	185	27	125	18	2.0	70	150	22	70	10	130	19
T6......	250	36	165	24	2.0	80	200	29	75	11	170	25
Permanent Mold												
As cast..	235	34	130	19	2.5	85	165	24	70	10	130	19
T6......	280	40	185	27	3.0	95	185	27	...	...	...	...

(a) Strengths and elongations are unchanged or improved at low temperatures. (b) In 50 mm or 2 in. (c) 500-kg load; 10-mm ball. (d) At 5 x 10⁸ cycles; R. R. Moore type test.

Table 11 Typical tensile properties for separately cast test bars of alloy 336.0 at elevated temperature

Temperature °C	°F	Tensile strength MPa	ksi	Yield strength MPa	ksi	Elongation(a), %
25	75 250		36	195	28	0.5
150	300 215		31	150	22	1.0
205	400 180		26	105	15	2.0
260	500 125		18	70	10	5.0
315	600 70		10	30	4	10.0

(a) In 50 mm or 2 in.

(167 Btu/lb)

Thermal conductivity. T551 temper: 117 W/m·K (67.7 Btu/ft·h·°F) at 25 °C (77 °F)

Electrical Properties

Electrical conductivity. Volumetric, T551 temper: 29% IACS at 20 °C (68 °F)

Electrical resistivity. T551 temper: 59.5 nΩ·m at 20 °C (68 °F)

Fabrication Characteristics

Melting temperature. 675 to 816 °C (1250 to 1500 °F)

Casting temperature. 675 to 788 °C (1250 to 1450 °F)

Solution temperature. 515 to 520 °C (955 to 965 °F); hold 8 h at temperature; cool in water at 65 to 100 °C (150 to 212 °F)

Aging temperature. 170 to 175 °C (335 to 345 °F); hold at temperature 14 to 18 h to obtain T5 temper from as-cast material; 12 to 26 h to obtain T6 temper from solution heat treated material

Joining. Rivet compositions: 6053-T4, 6053-T6, 6053-T61. Soft solder with Alcoa No. 802; no flux. Rub-tin with Alcoa No. 802. Metal-arc wled with 4043 alloy; Alcoa No. 27 flux. Carbon-arc weld with 4043 alloy; Alcoa No. 24 flux (automatic); Alcoa No. 27 flux (manual). Tungsten-arc argon-atmosphere weld with 4043 alloys no flux. Resistance welding: spot, seam, and flash methods

354.0
9Si-1.8Cu-0.5Mg

Specifications

Former ASTM. SC92A
UNS number. AC3540
Government. MIL-A-21180

Chemical Composition

Composition limits. 1.6 to 2.0 Cu; 0.4 to 0.6 Mg; 0.10 max Mn; 8.6 to 9.5 Si; 0.2 max Fe; 0.1 max Zn; 0.2 max Ti; 0.05 max others (each); 0.15 max others (total); rem Al

Applications

Typical uses. Permanent mold castings used in applications requiring high strengths and heat treatability

Table 12 Minimum mechanical properties for castings of alloy 354.0-T61

Class(a)	Tensile strength(b) MPa	ksi	Tensile yield strength(b)(c) MPa	ksi	Elongation(b)(d), %	Compressive yield strength(e) MPa	ksi
1	324	47	248	36	3	248	36
2	345	50	290	42	2	290	42
10	324	47	248	36	3	248	36
11	296	43	227	33	2	227	33

(a) Classes 1 and 2 (levels of properties) obtainable only at designated areas of casting; classes 10 and 11 may be obtained at any location in casting. (b) Specified in MIL-A-21180. (c) 0.2% offset. (d) In 50 mm, 2 in. or 4*d*, where *d* is diameter of reduced section of tensile-test specimen. (e) Design values; not specified.

Mechanical Properties

Tensile properties. See Tables 12 and 13.

Compressive yield strength. See Table 12.

Elastic modulus. Tension, 73.1 GPa (10.6 × 10⁶ psi); shear, 27.6 GPa (4.0 × 10⁶ psi); compression, 74.5 GPa (10.8 × 10⁶ psi)

Fatigue strength. See Table 14.

Creep-rupture characteristics. See Table 15.

Mass Characteristics

Density. 2.71 Mg/m³ (0.098 lb/in.³)

Thermal Properties

Coefficient of thermal expansion. Linear: 20.9 μm/m·K (11.6 μin./in.·°F) at 20 to 100 °C (68 to 212 °F)

Specific heat. 963 J/kg·K (0.230 Btu/lb·°F) at 100 °C (212 °F)

Thermal conductivity. 128 W/m·K (74 Btu/ft·h·°F)

Fabrication Characteristics

Solution temperature. 525 °C (980 °F); hold at temperature 10 to 12 h; quench in hot water 60 to 80 °C (140 to 176 °F)

Aging temperature. To obtain T61 temper from solution heat treated material, room temperature for 8 to 16 h; 155 °C (310 °F); hold at temperature for 10 to 12 h

355.0, C355.0
5Si-1.3Cu-0.5Mg

Specifications

AMS. 4210, 4212, 4214, 4280, 4281
Former ASTM. 355.0: SC51A. C355.0: SC51B
SAE. 322

UNS number. A03550

Government. 355.0: sand castings, QQ-A-601, class 10; permanent mold castings, QQ-A-596, class 6. C355.0: MIL-A-21180

Foreign. Canada: CSA SC51

Chemical Composition

Composition limits. 355.0: 1.0 to 1.5 Cu; 0.40 to 0.60 Mg; 0.50 max Mn; 4.5 to 5.5 Si; 0.6 max Fe; 0.25 max Cr; 0.35 max Zn; 0.25 max Ti; 0.05 max others (each); 0.15 max others (total); rem Al (If Fe exceeds 0.45, Mn content may not be less than ½ Fe content.). C355.0: 1.0 to 1.5 Cu; 0.40 to 0.60 Mg; 0.10 max Mn; 4.5 to 5.5 Si; 0.20 max Fe; 0.10 max Zn; 0.20 max Ti; 0.05 max others (each); 0.15 max others (total); rem Al

Consequence of exceeding impurity limits. High iron decreases ductility. Nickel decreases resistance to corrosion. Tin reduces mechanical properties.

Applications

Typical uses. Aircraft supercharger covers, fuel-pump bodies, air-compressor pistons, liquid-cooled cylinder heads, liquid-cooled aircraft engine crankcases, water jackets, and blower housings. Other applications where good castability, weldability, and pressure tightness are required. The presence of copper in 355.0 increases strength but reduces corrosion resistance and ductility.

Mechanical Properties

Tensile properties. See Tables 16, 17 and 18.

Compressive yield strength. See Table 16.

Poisson's ratio. 0.33

Elastic modulus. 355.0: tension, 70.3 GPa (10.2 × 10⁶ psi) at 25 °C (75 °F), 67.6 GPa (9.8 × 10⁶ psi) at 149 °C

Table 13 Typical mechanical properties for separately cast test bars of alloy 354.0-T61 at various temperatures

Temperature °C	°F	Time at temperature, h	At indicated temperature						At room temperature after heating				
			Tensile strength MPa	ksi	Yield strength MPa	ksi	Elongation(a), %		Tensile strength MPa	ksi	Yield strength MPa	ksi	Elongation(a), %
−196	−320.........	...	470	68	340	49	5		...	...	...	...	...
−80, −28	−112, −18....	...	400	58	290	42	5		...	...	...	...	...
24	75...........	...	380	55	285	41	6		380	55	285	41	6
100	212..........	0.5	345	50	285	41	6		380	55	285	41	6
		10	350	51	285	41	6		385	56	290	42	6
		100	360	52	290	42	6		400	58	295	43	6
		1000	370	54	310	45	6		420	61	310	45	6
		10 000	415	60	340	49	6		435	63	350	51	5
150	300..........	0.5	325	47	275	40	6		380	55	290	42	6
		10	345	50	295	43	6		395	57	305	44	5
		100	350	51	315	46	6		425	62	345	50	5
		1000	340	49	305	44	6		405	59	360	52	4
		10 000	290	42	240	35	6		340	49	275	40	6
175	350..........	0.5	310	45	270	39	6		380	55	295	43	6
		10	325	47	290	42	6		405	59	340	49	4
		100	295	43	260	38	8		405	59	350	51	5
		1000	230	33	195	28	13		325	47	255	37	8
		10 000	130	19	95	14	24		205	30	115	17	16
205	400..........	0.5	290	42	270	39	6		405	59	340	49	5
		10	270	39	250	36	9		400	58	340	49	5
		100	205	30	180	26	17		330	48	255	37	7
		1000	130	19	105	15	30		220	32	125	18	14
		10 000	105	15	75	11	45		185	27	90	13	20
230	450..........	0.5	255	37	240	35	9		400	58	345	50	5
		10	195	28	170	25	15		315	46	250	36	8
		100	125	18	95	14	25		240	35	140	20	11
		1000	95	14	75	11	40		195	28	95	14	17
		10 000	80	12	60	8.5	55		170	25	75	11	22
260	500..........	0.5	195	28	170	25	16		360	52	290	42	6
		10	115	17	105	15	22		250	36	150	22	11
		100	80	12	65	9.5	35		205	30	105	15	15
		1000	65	9.5	50	7.5	50		185	27	80	12	19
		10 000	60	8.5	40	6	65		165	24	70	10	11
315	600..........	0.5	90	13	80	12	29		260	38	145	21	13
		10	60	8.5	50	7	60		205	30	90	13	17
		100	40	6	35	5	85		185	27	75	11	19
		1000	...	...	...	...	...		170	25	65	9.5	21
		10 000	...	...	...	...	...		160	23	60	8.5	23

(a) In 50 mm, 2 in. or 4*d*, where *d* is diameter of reduced section of tensile test specimen.

Table 14 Fatigue strengths for separately cast test bars of alloy 354.0-T61(a)

Temperature °C	°F	Cycles											
		10^4 MPa	ksi	10^5 MPa	ksi	10^6 MPa	ksi	10^7 MPa	ksi	10^8 MPa	ksi	5×10^8 MPa	ksi
24	75	345	50	275	40	215	31	175	25.5	145	21	135	19.5
150	300	...	...	255	37	200	29	150	21.5	115	17	110	16
205	400	...	...	215	31	150	22	105	15	70	10	60	9
260	500	195	28	140	20.5	96	14	60	9	40	6	40	6
315	600	...	...	75	11	55	8	40	6	30	4	30	4

(a) R. R. Moore type test.

Table 15 Creep-rupture properties for separately cast test bars of alloy 354.0-T61

| Temperature | | Time under stress, h | Rupture stress | | Stress for creep of: | | | | | | | |
| | | | | | 1% | | 0.5% | | 0.2% | | 0.1% | |
°C	°F		MPa	ksi	MPa	ksi	MPa	ksi	MPa	ksi	MPa	ksi
177	350	0.1	305	44	295	43	285	41	290	39	255	37
		1.0	295	43	290	42	285	41	290	39	255	37
		10	285	41	285	41	275	40	255	37	240	35
		100	240	35	235	34	230	33	205	30	115	17
		1000	170	25	165	24	165	24	138	20	76	11
205	400	0.1	285	41	275	40	270	39	255	37	240	35
		1.0	255	37	250	36	250	36	235	34	215	31
		10	220	32	215	31	205	30	180	26	125	18
		100	160	23	160	23	150	22	125	18	69	10
		1000	90	13	90	13	83	12	83	12	48	7

(300 °F), 64.1 GPa (9.3×10^6 psi) at 204 °C (400 °F), 56.5 GPa (8.2×10^6 psi) at 260 °C (500 °F); shear, 26.2 GPa (3.8×10^6 psi). C355.0: tension, 69.6 GPa (10.1×10^6 psi); shear, 26.5 GPa (3.85×10^6 psi); compression, 71 GPa (10.3×10^6 psi)

Fatigue strength. See Table 19.
Creep-rupture characteristics. See Table 20.

Mass Characteristics

Density. 2.71 Mg/m³ (0.098 lb/in.³) at 20 °C (68 °F)

Thermal Properties

Liquidus temperature. 620 °C (1150 °F)
Solidus temperature. 545 °C (1071 °F)
Coefficient of thermal expansion. Linear:

| Temperature range | | Average coefficient | |
°C	°F	μm/m·K	μin./in.·°F
20–100	68–212	22.4	12.4
20–200	68–392	23	12.8
20–300	68–572	24	13.3

Specific heat. 963 J/kg·K (0.230 Btu/lb·°F) at 100 °C (212 °F)
Thermal conductivity. At 25 °C (77 °F):

Temper and form	Conductivity W/m·K	Btu/ ft·h·°F
T51 (sand)	167	96
T6, T61 (sand)	152	88
T7 (sand)	163	94
T6 (permanent mold)	151	87

Table 16 Minimum mechanical properties for alloy C355.0-T61 castings

| Class(a) | Tensile strength(b) | | Tensile yield strength(b)(c) | | Elongation(d), % | Compressive yield strength(e) | |
	MPa	ksi	MPa	ksi		MPa	ksi
1	285	41	215	31	3	215	31
2	305	44	230	33	3	230	33
3	345	50	275	40	2	275	40
10	285	41	215	31	3	215	31
11	255	37	205	30	1	205	30
12	240	35	195	28	1	195	28

(a) Classes 1, 2 and 3 (levels of properties) obtainable only at designated areas of casting; classes 10, 11 and 12 may be obtained from any location in casting. (b) Specified in MIL-A-21180. High properties are obtained by advanced foundry techniques and by careful control of trace elements at lower levels than specified for alloy 355.0 castings. (c) 0.2% offset. (d) In 4d, where d is diameter of reduced section of tensile-test specimen. (e) Design values; not specified.

Table 17 Typical mechanical properties for separately cast test bars of alloy 355.0

| Temper | Tensile strength | | Tensile yield strength | | Elongation, % | Hardness(a) HB | Shear strength | | Fatigue strength(b) | | Compressive yield strength | |
	MPa	ksi	MPa	ksi			MPa	ksi	MPa	ksi	MPa	ksi
Sand Cast												
T51	195	28	160	23	1.5	65	150	22	55	8.0	165	24
T6	240	35	170	25	3.0	80	195	28	62	9.0	180	26
T61	270	39	240	35	1.0	90	215	31	66	9.5	255	37
T7	260	38	250	36	0.5	85	195	28	69	10.0	260	38
T71	172	35	200	29	1.5	75	180	26	69	10.0	205	30
Permanent Mold Cast												
T51	205	30	165	24	2.0	75	165	24	...	...	165	24
T6	290	42	185	27	4.0	90	235	34	69	10	185	27
T62	310	45	275	40	1.5	105	250	36	69	10	275	40
T7	275	40	205	30	2.0	85	205	30	69	10	205	30
T71	250	36	215	31	3.0	85	185	27	69	10	215	31

(a) 500-kg load; 10-mm ball. (b) At 5×10^8 cycles; R. R. Moore type test.

Electrical Properties

Electrical conductivity. Volumetric:

Temper and form	Conductivity, % IACS
T51 (sand)	43
T6 (sand)	36
T61 (sand)	39
T7 (sand)	42
T6 (permanent mold)	39

Electrical resistivity. At 20 °C (68 °F):

Temper and form	Resistivity, $n\Omega \cdot m$
T51 (sand)	40.1
T6 (sand)	47.9
T61 (sand)	44.2
T7 (sand)	41.0
T6 (permanent mold)	44.2

Electrolytic solution potential. T4 temper, -0.78 V and T6 temper, -0.79 V vs $0.1N$ calomel electrode in an aqueous solution containing 53 g NaCl plus 3 g H_2O_2 per litre

Fabrication Characteristics

Melting temperature. 675 to 815 °C (1250 to 1500 °F)
Casting temperature. 675 to 790 °C (1250 to 1450 °F)
Solution temperature. See Table 21.
Aging temperature. See Table 21.
Joining. Same as alloy 514.0

356.0, A356.0
7Si-0.3Mg

Specifications
AMS. 356.0: 4217, 4260, 4261, 4284, 4285, 4286. A356.0: 4218
Former ASTM. 356.0, SG70A; A356.0, SG70B
SAE. 356.0: J452, 323
UNS number. 356.0: A03560. A356.0: A13560
Government. 356.0: QQ-A-601, QQ-A-596. A356.0: MIL-C-21180 (class 12)
Foreign. ISO: AlSi7Mg

Chemical Composition
Composition limits. 356.0: 0.25 max Cu; 0.20 to 0.45 Mg; 0.35 max Mn; 6.5 to 7.5 Si; 0.6 max Fe; 0.35 max Zn; 0.25 max Ti; 0.05 max others (each); 0.15 max others (total); rem Al. A356.0: 0.20 max Cu; 0.25 to 0.45 Mg; 0.10 max Mn; 6.5 to 7.5 Si; 0.20 max Fe; 0.10 max Zn; 0.20 max Ti; 0.05 max others (each); 0.15 max others (total); rem Al
Consequence of exceeding impurity limits. High copper or nickel decreases ductility and resistance to corrosion. High iron decreases strength and ductility.

Applications
Typical uses. 356.0: aircraft pump parts, automotive transmission cases, aircraft fittings and control parts, water-cooled cylinder blocks. Other applications where excellent castability and good weldability, pressure tightness, and good resistance to corrosion are required. A356.0: aircraft structures and engine controls, nuclear energy installations, and other applications where high strength permanent mold or investment castings are required

Mechanical Properties
Tensile properties. See Tables 22, 23 and 24.
Compressive yield strength. See Table 22.
Poisson's ratio. 0.33
Elastic modulus. Tension, 72.4 GPa

Table 18 Typical tensile properties of separately cast test bars of alloy 355.0 at elevated temperatures

Temperature °C	°F	Tensile strength(a) MPa	ksi	Yield strength(a) MPa	ksi	Elongation(a)(b), %
T6 Temper, Sand Cast						
25	75	240	35	170	25	3
150	300	230	33	170	25	1.5
205	400	115	17	90	13	8
260	500	65	9.5	35	5	16
315	600	40	6	25	3	36
T6 Temper, Permanent Mold Cast						
25	75	290	42	180	27	4
150	300	220	32	170	25	10
205	400	130	19	90	13	20
260	500	65	9.5	35	5	40
315	600	40	6	25	3	50
T51 Temper, Sand Cast						
25	75	195	28	160	23	1.5
150	300	165	24	130	19	3
205	400	95	14	70	10	8
260	500	65	9.5	35	5	16
315	600	40	6	25	3	36

(a) Strengths and elongations remain unchanged or improve at low temperatures. (b) In 50 mm or 2 in.

Table 19 Fatigue properties for separately cast test bars of alloy C355.0-T61

Temperature °C	°F	No. of cycles	Fatigue strength(a) MPa	ksi
24	75	10^5	195	28.0
		10^6	130	19.0
		10^7	110	16.0
		10^8	100	14.5
		5×10^8	95	14.0
260	500	10^5	125	18.0
		10^6	80	11.5
		10^7	50	7.5
		10^8	40	5.5
		5×10^8	35	5.0

(a) Based on rotating-beam tests at room temperature and cantilever beam (rotating load) tests at elevated temperature.

Table 20 Creep-rupture properties for separately cast test bars of alloy C355.0-T61

Temperature		Time under stress, h	Rupture stress		Stress for creep of:							
					1%		0.5%		0.2%		0.1%	
°C	°F		MPa	ksi	MPa	ksi	Mpa	ksi	MPa	ksi	MPa	ksi
150	300	0.1	285	41	275	40	270	39	240	35	230	33
		1	285	41	270	39	260	38	235	34	220	32
		10	275	40	260	38	250	36	230	33	205	30
		100	260	38	250	36	235	34	215	31	170	25
		1000	220	32	215	31	206	30	185	27	140	20
205	400	0.1	250	36	250	36	240	35	230	33	170	25
		1	230	33	220	32	205	30	170	25	140	20
		10	180	26	120	25	160	23	130	19	110	16
		100	130	19	130	19	125	18	97	14	...	...
		1000	97	14	90	13	83	12	...	...	...	...
260	500	0.1	165	24	145	21	130	19	105	15	83	12
		1	125	18	110	16	97	14	83	12	59	8.5
		10	90	13	83	12	76	11	59	8.5	41	6
		100	62	9	62	9	55	8	41	6	...	...
		1000	45	6.5	45	6.5	41	6	...	...	...	...

(10.5 × 10⁶ psi); shear, 27.2 GPa (3.95 × 10⁶ psi)
Creep-rupture characteristics. See Table 25.

Mass Characteristics
Density. 2.685 Mg/m³ (0.097 lb/in.³) at 20 °C (68 °F)

Thermal Properties
Liquidus temperature. 615 °C (1135 °F)
Solidus temperature. 555 °C (1035 °F)
Coefficient of thermal expansion. Linear:

Temperature range		Average coefficient	
°C	°F	µm/ m·K	µin./ in.·°F
20–100	68–212	21.5	11.9
20–200	68–392	22.5	12.5
20–300	68–572	23.5	13.1

Specific heat. 963 J/kg·K (0.230 Btu/lb·°F) at 100 °C (212 °F)
Latent heat of fusion. 389 kJ/kg
Thermal conductivity. At 25 °C (77 °F):

Temper and form	Conductivity	
	W/m·K	Btu/ ft·h·°F
T51 (sand)	167	96
T6 (sand)	151	87
T7 (sand)	155	90
T6 (permanent mold)	159	92

Table 21 Heat treatments for separately cast test bars of alloy 355.0

Purpose (and resulting temper)	Temperature		Time at temperature, h
	°C	°F	
Sand Castings			
Solution	520 to 530	970 to 990	12(a)(b)
Aging			
T51(c)	225 to 230	435 to 445	7 to 9
T6(d)	150 to 155	305 to 315	3 to 5
T61(d)	150 to 160	300 to 320	8 to 10
T7(d)(e)	225 to 230	435 to 445	7 to 9
T71(d)(e)	245 to 250	470 to 480	4 to 6
Permanent Mold Castings			
Solution	520 to 530	970 to 980	8(a)(b)
Aging(f)			
T62(d)	170 to 175	335 to 345	14 to 18

(a) Soaking time periods required for average castings after load has reached specified temperature. Time can be decreased or may have to be increased, depending on experience with particular castings. (b) Cool in water at 65 to 100 °C (150 to 212 °F). (c) No solution heat treatment. (d) Start with solution heat treated material. (e) U. S. Patent 1,822,877. (f) Except for temper listed under this head, temperature values for all tempers are the same as for sand castings.

Table 22 Minimum mechanical properties for alloy A356.0-T61 castings

Class(a)	Tensile strength(b)		Tensile yield strength(b)(c)		Elon- gation(d), %	Compressive yield strength(e)	
	MPa	ksi	MPa	ksi		MPa	ksi
1	260	38	195	28	5	195	28
2	275	40	205	30	3	205	30
3	310	45	235	34	3	235	34
10	260	38	195	28	5	195	28
11	230	33	185	27	3	185	27
12	220	32	150	22	2	150	22

(a) Classes 1, 2, and 3 (levels of properties) obtainable only at designated areas of casting; classes 10, 11, and 12 may be specified at any location in casting. (b) Specified in MIL-A-21180. (c) 0.2% offset. (d) In 4d, where d is diameter of reduced section of tensile test specimen. (e) Design values; not specified.

Table 23 Typical mechanical properties for separately cast test bars of alloy 356.0

Temper	Tensile strength MPa	ksi	Yield strength MPa	ksi	Elon-gation(a), %	Hard-ness(b), HB	Shear strength MPa	ksi	Fatigue strength(c) MPa	ksi	Compressive yield strength MPa	ksi
Sand Cast												
T51	172	25	140	20	2.0	60	140	20	55	8.0	145	21
T6	228	33	165	24	3.5	70	180	26	60	8.5	170	25
T7	234	34	205	30	2.0	75	165	24	60	9.0	215	31
T71	193	28	145	21	3.5	60	140	20	60	8.5	150	22
Permanent Mold												
T6	262	38	185	27	5.0	80	205	30	90	13	185	27
T7	221	32	165	24	6.0	70	170	25	75	11	165	24

(a) In 50 mm or 2 in. (b) 500-kg load; 10-mm ball. (c) At 5×10^8 cycles; R. R. Moore type test.

Electrical Properties

Electrical conductivity. Volumetric:

Temper and form	IACS, %
T51 (sand)	43
T6 (sand)	39
T7 (sand)	40
T6 (permanent mold)	41

Electrical resistivity. At 20 °C (68 °F):

Temper and form	Resistivity, nΩ·m
T51 (sand)	40.1
T6 (sand)	44.2
T7 (sand)	43.1
T6 (permanent mold)	42.1

Electrolytic solution potential. T6 temper (sand): -0.82 V vs $0.1N$ calomel electrode in an aqueous solution containing 53 g NaCl plus 3 g H_2O_2 per litre

Nuclear Properties

Effect of neutron irradiation. See Table 26.

Fabrication Characteristics

Melting temperature. 675 to 815 °C (1250 to 1500 °F)
Casting temperature. 675 to 790 °C (1250 to 1450 °F)
Solution temperature. See Table 27.
Aging temperature. See Table 27.
Joining. Same as alloy 514.0

Table 24 Typical tensile properties of separately cast test bars of alloy 356.0-T6

Temperature °C	°F	Tensile strength(a) MPa	ksi	Yield strength(a) Mpa	ksi	Elon-gation(a)(b), %
24	75	230	33	165	24	3.5
150	300	160	23	140	20	6.0
205	400	85	12	60	8.5	18
260	500	50	7.5	35	5.0	35
315	600	30	4.0	25	3.0	60

(a) Strengths and elongations remain unchanged or improve at low temperatures. (b) In 50 mm or 2 in.

Table 25 Creep-rupture properties for separately cast test bars of alloy A356.0-T61

Tempera-ture °C	°F	Time under stress, h	Rupture stress MPa	ksi	1% MPa	ksi	Stress for creep of: 0.5% MPa	ksi	0.2% MPa	ksi	0.1% MPa	ksi
150	300 . . .	0.1	235	34	215	31	205	30	195	28	185	27
		1	235	34	215	31	200	29	185	27	180	26
		10	230	33	205	30	195	28	180	26	170	25
		100	200	29	195	28	185	27	170	25	165	24
		1000	165	24	165	24	160	23	. . .	. . .	. . .	. . .

Table 26 Effect of neutron radiation on tensile properties of alloy A356.0-T61(a)

Fast neutron flux, n/cm²	Tensile strength MPa	ksi	Yield strength MPa	ksi	Elon-gation, %
Control sample	230	33	180	26	4
2.0×10^{19}	255	37	200	29	6
1.2×10^{20}	290	42	230	33	6
5.6×10^{20}	315	46	290	42	6
9.8×10^{20}	375	54	360	52	3

(a) Separately cast test bars; irradiation temperature, 50 °C (120 °F).

357.0, A357.0
7Si-0.5Mg

Specifications

UNS number. 357.0: A03570.
A357.0: A13570
Government. A357.0: MIL-A-
21180

Chemical Composition

Composition limits. 357.0: 0.05
max Cu; 0.45 to 0.6 Mg; 0.03 max
Mn; 6.5 to 7.5 Si; 0.15 max Fe; 0.05
max Zn; 0.20 max Ti; 0.05 max others
(each); 0.15 max others (total); rem
Al. A357.0: 0.20 max Cu; 0.40 to 0.7
Mg; 0.10 max Mn; 6.5 to 7.5 Si; 0.20
max Fe; 0.10 max Zn; 0.10 to 0.20 Ti;
0.04 to 0.07 Be; 0.05 max others
(each); 0.15 max others (total); rem
Al

Applications

Typical uses. Critical aerospace applications and other uses requiring
heat treatable permanent mold casting that combines ready weldability
with high strength and good toughness

Mechanical Properties

Tensile properties. See Tables 28
and 29.
Compressive yield strength. See
Table 28.
Hardness. A357.0, T61 temper: 100
HB
Elastic modulus. A357.0: tension,
71.7 GPa (10.4 × 10⁶ psi); shear, 26.8
GPa (3.9 × 10⁶ psi); compression,
72.4 GPa (10.5 × 10⁶ psi)
Fatigue strength. A357.0-T62 (ro-
tating-beam tests):

| No. of | Stress | |
cycles	MPa	ksi
10⁵	255	37
10⁶	195	28
10⁷	145	21
10⁸	115	17
5 x 10⁸	110	16

Mass Characteristics

Density. 2.68 Mg/m³ (0.097 lb/in.³)

Thermal Properties

Liquidus temperature. A357.0: 615
°C (1135 °F)
Solidus temperature. 555 °C (1035
°F)
Coefficient of thermal expansion.

Linear, 21.6 μm/m·K (12.0 μin./
in.·°F) at 17 to 100 °C (63 to 212 °F)
Specific heat. 963 J/kg·K (0.230
Btu/lb·°F) at 100 °C (212 °F)
Thermal conductivity. 152 W/m·K
(88 Btu/ft·h·°F) at 25 °C (77 °F)

Fabrication Characteristics

Solution temperature. 540 °C (1005
°F); hold at temperature for 8 h; hot
water quench
Aging temperature. T6 temper: 170
°C (340 °F); hold at temperature 3 to
5 h
Joining. Because of the beryllium
content, care should be taken not to
inhale fumes during welding

Table 27 Heat treatments for separately cast test bars of alloys 356.0 and A356.0

Purpose (and resulting temper)	Temperature °C	°F	Time at temperature, h
Sand Castings			
Solution	535 to 540	995 to 1005	12(a)(b)
Aging			
T51(c)	225 to 230	435 to 445	7 to 9
T6(d)	150 to 155	305 to 315	2 to 5
T7(d)(e)	225 to 230	435 to 445	7 to 9
T71(d)	245 to 250	470 to 480	2 to 4
Permanent Mold Castings			
Solution	535 to 540	995 to 1005	8(a)(b)
Aging(f)			
T6(d)	150 to 155	305 to 315	3 to 5

(a) Soaking-time periods required for average casting after load has reached specified temperature. Time can be decreased or may have to be increased, depending on experience with particular castings. (b) Cool in water at 65 to 100 °C (150 to 212 °F). (c) No solution heat treatment. (d) Start with solution heat treated material. (e) U. S. Patent 1,822,877. (f) Except for temper listed under this head, temperature values for all tempers are the same as for sand castings.

Table 28 Minimum mechanical properties for alloy A357.0 castings

Class(a)	Tensile strength(b) MPa	ksi	Tensile yield strength(b)(c) MPa	ksi	Elon- gation(b)(d), %	Compressive yield strength(e) MPa	ksi
T61, Permanent Mold Castings							
1	317	46	248	36	3	...	...
10	283	41	214	31	3	...	...
T62 Castings							
1	310	45	241	35	3	241	35
2	345	50	276	40	5	276	40
10	262	38	193	28	5	193	28
11	283	41	214	31	3	214	31

(a) Classes 1 and 2 (levels of properties obtainable only at designated areas of casting; classes 10 and 11 may be obtained from any location in castings. (b) Specified in MIL-A-21180. (c) 0.2% offset. (d) In 4d, where d is diameter of reduced section of tensile-test specimen. (e) Design values; not specified.

359.0
9Si-0.6Mg

Specifications
Former ASTM. SG91A
UNS number. A03590
Government. MIL-A-21180

Chemical Composition
Composition limits. 0.20 max Cu;
0.50 to 0.7 Mg; 0.10 max Mn; 8.5 to
9.5 Si; 0.20 max Fe; 0.10 max Zn;
0.20 max Ti; 0.05 max others (each);
0.15 max others (total); rem Al

Applications
Typical uses. A moderately high
strength permanent mold casting alloy having superior casting characteristics

Table 29 Typical mechanical properties of separately cast test bars of alloy A357.0-T62 at various temperatures

Temperature °C	°F	Time at temperature, h	Tensile strength MPa	ksi	Yield strength MPa	ksi	Elongation(a), %
-196	-320	...	425	62	330	48	6
-80	-112	...	380	55	310	45	6
-28	-18	...	370	54	305	44	6
24	75	...	360	52	290	42	8
100	212	0.5 thru 100	315	46	270	39	10
		1000	315	46	275	49	8
		10 000	330	48	310	45	6
150	300	0.5	270	39	240	35	10
		10	285	41	255	37	9
		100	290	42	275	40	7
		1000	260	38	250	36	7
		10 000	160	23	145	21	20
175	350	0.5	255	37	235	34	7
		10	275	40	260	38	6
		100	240	35	230	33	7
		1000	150	22	140	20	19
		10 000	90	13	75	11	35
205	400	0.5	250	36	240	35	6
		10	205	30	195	28	7
		100	160	23	145	21	23
		1000	85	12	70	10	40
		10 000	70	10	50	7.5	50
230	450	0.5	215	31	105	30	9
		10	130	19	125	18	13
		100	95	14	90	13	45
260	500	0.5	160	23	150	22	16
		10	85	12	75	11	23
		100	55	8	50	7	55
315	600	0.5	70	10	65	9.5	35

(a) In 4d, where d is diameter of reduced section of tensile test specimen.

Table 30 Minimum mechanical properties for alloy 359.0-T61

Class(a)	Tensile strength(b) MPa	ksi	Tensile yield strength(b)(c) MPa	ksi	Elongation(b)(d), %	Compressive yield strength(e) MPa	ksi
1	310	45	241	35	4	241	35
2	324	47	262	38	3	262	38
10	310	45	234	34	4	234	34
11	276	40	207	30	3	207	30

(a) Classes 1 and 2 (levels of properties) obtainable only from designated areas of casting; classes 10 and 11 may be obtained from any location in casting. (b) Specified in MIL-A-21180. (c) 0.2% offset. (d) In 4d, where d is diameter of reduced section of tensile-test specimen. (e) Design values; not specified.

Mechanical Properties

Tensile properties. See Tables 30 and 31.

Compressive yield strength. See Table 30.

Elastic modulus. Tension, 72.4 GPa (10.5 × 10⁶ psi); shear, 27.6 GPa (4.0 × 10⁶ psi); compression, 73.8 GPa (10.7 × 10⁶ psi)

Fatigue strength. Rotating-beam tests, T61 temper:

No. of cycles	Stress MPa	ksi
10⁵	255	37
10⁶	195	28
10⁷	145	21
10⁸	115	17
5 x 10⁸	110	16

Mass Characteristics

Density. 2.685 Mg/m³ (0.097 lb/in.³)

Thermal Properties

Liquidus temperature. 615 °C (1135 °F)

Solidus temperature. 555 °C (1035 °F)

Coefficient of thermal expansion. Linear, 20.9 μm/m·K (11.6 μin./in.·°F) at 20 to 100 °C (68 to 212 °F)

Specific heat. 963 J/kg·K (0.230 Btu/lb·°F)

Thermal conductivity. 138 W/m·K (80 Btu/ft·h·°F)

Fabrication Characteristics

Solution temperature. 540 °C (1000 °F); hold at temperature 10 to 14 h; hot water quench 60 to 80 °C (140 to 175 °F)

Aging temperature. Room temperature for 8 to 16 h after solution treatment then 155 °C (310 °F) for 10 to 12 h (T61 temper) or 170 °C (340 °F) for 6 to 10 h (T62 temper)

360.0, A360.0
9.5Si-0.5Mg

Specifications

AMS. 360.0: 4290F

Former ASTM. 360.0: SG100B. A360.0: SG100A

SAE. A360.0: J452, 309

UNS number. 360.0: A03600. A360.0: A13600

Government. 360.0: QQ-A-591

Chemical Composition

Composition limits. 360.0: 0.6 max Cu; 0.40 to 0.6 Mg; 0.35 max Mn; 9.0 to 10.0 Si; 2.0 max Fe; 0.50 max Ni; 0.50 max Zn; 0.15 max Sn; 0.25 max others (total); rem Al. A360.0: 0.6 max Cu; 0.40 to 0.6 Mg; 0.35 max Mn; 9.0 to 10.0 Si; 1.3 max Fe; 0.50 max Ni; 0.50 max Zn; 0.15 max Sn; 0.25 max others (total); rem Al

Consequence of exceeding impurity limits. Increasing copper limits lowers resistance to corrosion; increasing iron lowers ductility. Decreasing silicon reduces castability.

Applications

Typical uses. Die castings requiring improved corrosion resistance compared to 3800. Other applications where excellent castability, pressure tightness, resistance to hot cracking, strength at elevated temperatures and ability to be electroplated are required. Poor weldability and brazeability. General purpose casting alloy for such items as cover plates and instrument cases

Mechanical Properties

Tensile properties. Typical for separately cast test bars, as cast. 360.0: tensile strength, 305 MPa (44 ksi); yield strength, 170 MPa (25 ksi); elongation, 2.5% in 50 mm or 2 in. A360.0: tensile strength, 320 MPa (46 ksi); yield strength, 170 MPa (25 ksi); elongation, 3.5% in 50 mm or 2 in. See also Table 32.
Shear strength. 360.0: 190 MPa (28 ksi). A360.0: 180 MPa (26 ksi)
Poisson's ratio. 0.33
Elastic modulus. Tension, 71.0 GPa (10.3 × 10^6 psi); shear, 26.5 GPa (3.85 × 10^6 psi)
Fatigue strength. At 5 × 10^8 cycles, 360.0: 140 MPa (20 ksi). A360.0: 120 MPa (18 ksi) (R. R. Moore type test)

Mass Characteristics

Density. 2.630 Mg/m^3 (0.095 lb/in.^{3}) at 20 °C (68 °F)

Thermal Properties

Liquidus temperature. 595 °C (1105 °F)
Solidus temperature. 555 °C (1035 °F)
Coefficient of thermal expansion. Linear:

Temperature range		Average coefficient	
°C	°F	μm/ m·K	μin./ in.·°F
20–100	68–212	21	11.6
20–200	68–392	22	12.2
20–300	68–572	23	12.8

Specific heat. 963 J/kg·K (0.230 Btu/lb·°F) at 100 °C (212 °F)
Latent heat of fusion. 389 kJ/kg (167 Btu/lb)
Thermal conductivity. 113 W/m·K (65.3 Btu/ft·h·°F) at 25 °C (77 °F)

Electrical Properties

Electrical conductivity. Volumetric: 360.0, 28% IACS; A360.0, 30% IACS
Electrical resistivity. 61.6 nΩ·m at 20 °C (68 °F)

Fabrication Characteristics

Melting temperature. 650 to 760 °C (1200 to 1400 °F)
Die casting temperature. 635 to 705 °C (1175 to 1300 °F)
Joining. Same as alloys 413.0 and A413.0

Table 31 Typical tensile properties of separately cast test bars of alloy 359.0-T6 at various temperatures

Temperature		Time at temperature, h	Tensile strength		Yield strength		Elongation(a), %
°C	°F		MPa	ksi	MPa	ksi	
−196	−320	⋯	435	63	325	47	4
−80	−112	⋯	380	55	325	47	5
−28	−18	⋯	360	52	310	45	6
150	300	100	290	42	260	38	10
		1000	250	36	235	34	11
		10 000	125	18	95	14	30
260	500	0.5	125	18	115	17	25
		10	65	9.5	60	8.5	40
		100	60	8.5	50	7	50
		1000	50	7.5	40	6	55
		10 000	50	7	35	5	60
315	600	0.5	50	7.5	45	6.5	50
		10	40	6	40	5.5	60
		100	40	5.5	30	4.4	65
370	700	0.5	30	4.4	30	4	55

(a) In 4d, where d is diameter of reduced section of tensile test specimen.

Table 32 Typical tensile properties for separately cast test bars of alloys 360.0-F and A360.0-F at elevated temperature

Temperature		Tensile strength		Yield strength(a)		Elongation(b), %
°C	°F	MPa	ksi	MPa	ksi	
360.0 Aluminum						
24	75	325	47	170	25	3
100	212	305	44	170	25	2
150	300	240	35	165	24	4
205	400	150	22	95	14	8
250	500	85	12	50	7.5	20
315	600	50	7	30	4.5	35
370	700	30	4.5	25	3	40
A360.0 Aluminum						
24	75	315	46	165	24	5
100	212	295	43	165	24	3
150	300	235	34	160	23	5
205	400	145	21	90	13	14
250	500	75	11	45	6.5	30
315	600	45	6.5	30	4	45
370	700	30	4	15	2.5	45

(a) 0.2% offset. (b) In 50 mm or 2 in.

Table 33 Typical tensile properties of separately cast test bars of alloy 380.0-F at elevated temperature

Temperature		Tensile strength		Yield strength		Elongation, %
°C	°F	MPa	ksi	MPa	ksi	
24	75	330	48	165	24	3
100	212	310	45	165	24	4
150	300	235	34	150	22	5
205	400	165	24	110	16	8
260	500	90	13	55	8	20
315	600	50	7	30	4	30
370	700	30	4	15	2.5	35

380.0, A380.0
8.5Si-3.5Cu

Specifications

AMS. A380.0: 4291
Former ASTM. 380.0 SC84B. A380.0: SC84A
SAE. 380.0: 308. A380.0: 306
UNS number. 380.0: A03800. A380.0: A13800
Government. A380.0: QQ-A-591
Foreign. 380.0: Canada, CSA SC84

Chemical Composition

Composition limits. 380.0: 3.0 to 4.0 Cu; 0.10 max Mg; 0.50 max Mn; 7.5 to 9.5 Si; 2.0 max Fe; 0.50 max Ni; 3.0 max Zn; 0.35 max Sn; 0.50 max others (total); rem Al. A380.0: 3.0 to 4.0 Cu; 0.10 max Mg; 0.50 max Mn; 7.5 to 9.5 Si; 1.3 max Fe; 0.50 max Ni; 3.0 max Zn; 0.35 max Sn; 0.50 max others (total); rem Al
Consequence of exceeding impurity limits. Increasing iron will lower ductility. Relatively large quantities of impurities may be present before serious effects are detected.

Applications

Typical uses. Vacuum cleaners, floor polishers, parts for automotive and electrical industries such as motor frames and housings. Most widely used aluminum die casting alloy. Poor weldability and brazeability; fair strength at elevated temperatures.

Mechanical Properties

Tensile properties. Typical for separately cast test bars, as cast. 380.0: tensile strength, 315 MPa (46 ksi); yield strength, 160 MPa (23 ksi); elongation, 3.5% in 50 mm or 2 in. A380.0: tensile strength, 325 MPa (47 ksi); yield strength, 160 MPa (23 ksi); elongation, 3.5% in 50 mm or 2 in. See also Table 33.
Shear strength. 380.0: 195 MPa (28 ksi). A380.0: 185 MPa (27 ksi)
Poisson's ratio. 0.33
Elastic modulus. 71.0 GPa (10.3 × 10^6 psi); shear, 26.5 GPa (3.85 × 10^6 psi)
Fatigue strength. At 5 × 10^8 cycles, 380.0 and A380.0: 138 MPa (20 ksi) (R. R. Moore type test)

Mass Characteristics

Density. 2.740 Mg/m^3 (0.098 lb/in.^{3}) at 20 °C (68 °F)

Thermal Properties

Liquidus temperature. 595 °C (1100 °F)
Solidus temperature. 540 °C (1000 °F)
Coefficient of thermal expansion. Linear, at 20 to 200 °C (68 to 392 °F). 380.0: 22.0 μm/m·K (12.2 μin./in.·°F). A380.0: 21.8 μm/m·K (12.1 μin./in.·°F)
Specific heat. 963 J/kg·K (0.230 Btu/lb·°F) at 100 °C (212 °F)
Latent heat of fusion. 389 kJ/kg. (167 Btu/lb)
Thermal conductivity. 96.2 W/m·K (55.6 Btu/ft·h·°F) at 25 °C (77 °F)

Electrical Properties

Electrical conductivity. Volumetric, 27% IACS at 20 °C (68 °F)
Electrical resistivity. 75 nΩ·m at 20 °C (68 °F)

Fabrication Characteristics

Melting temperature. 650 to 760 °C (1200 to 1400 °F)
Die casting temperature. 635 to 704 °C (1175 to 1300 °F)
Annealing temperature. For increased ductility, 260 to 370 °C (500 to 700 °F); hold at temperature 4 to 6 h; furnace cool or cool in still air
Stress relief temperature. 175 to 260 °C (350 to 500 °F); hold at temperature 4 to 6 h; cool in still air
Joining. Same as alloy 413.0 and A413.0.

383.0
10.5Si-2.5Cu

Specifications

Former ASTM. SC102A
SAE. 383
UNS number. A03830

Chemical Composition

Composition limits. 2.0 to 3.0 Cu; 0.10 max Mg; 0.50 max Mn; 9.5 to 11.5 Si; 1.3 max Fe; 0.30 max Ni; 3.0 max Zn; 0.15 max Sn; 0.50 max others (total); rem Al

Applications

Typical uses. Applications requiring good die filling capacity, fair pressure tightness, electroplating and machining characteristics and strength at elevated temperature,

poor weldability and brazeability; anodizing quality is poor

Mechanical Properties

Tensile properties. Typical for separately cast test bars, as cast: tensile strength, 310 MPa (45 ksi); yield strength, 150 MPa (22 ksi); elongation, 3.5% in 50 mm or 2 in.
Hardness. 75 HB (500-kg load, 10-mm ball)
Poisson's ratio. 0.33
Fatigue strength. 145 MPa (21 ksi) at 5 × 10^8 cycles
Impact strength. Charpy V-notch: 4 J (3 ft·lb)

Mass Characteristics

Density. 2.740 Mg/m^3 (0.099 lb/in.^{3})

Thermal Properties

Liquidus temperature. 580 °C (1080 °F)
Solidus temperature. 515 °C (960 °F)
Coefficient of thermal expansion. Linear, 21.1 μm/m·K (11.7 μin./in.·°F) at 20 to 100 °C (68 to 212 °F)
Thermal conductivity. 96.2 W/m·K (55.6 Btu/ft·h·°F)

Electrical Properties

Electrical conductivity. Volumetric, 23% IACS at 20 °C (68 °F)

Fabrication Characteristics

Die casting temperature. 615 to 700 °C (1140 to 1292 °F)
Stress relief temperature. 175 to 260 °C (350 to 500 °F); hold at temperature 4 to 6 h; cool in still air
Annealing temperature. For increased ductility, 260 to 370 °C (500 to 700 °F); hold at temperature 4 to 6 h; furnace cool or cool in still air

384.0, A384.0
11.2Si-3.8Cu

Specifications

Former ASTM. SC114A
SAE. 303
UNS number. 384.0: A03840. A384.0: A13840
Government. 384.0: QQ-A-591

Chemical Composition

Composition limits. 384.0: 3.0 to 4.5 Cu; 0.10 max Mg; 0.5 max Mn; 10.5 to 12.0 Si; 1.3 max Fe; 0.50 max Ni; 3.0 max Zn; 0.35 max Sn; 0.50

max others (total); rem Al. A384.0: 3.0 to 4.5 Cu; 0.10 max Mg; 0.50 max Mn; 10.5 to 12.0 Si; 1.3 max Fe; 0.50 max Ni; 1.0 max Zn; 0.35 max Sn; 0.50 max others (total); rem Al

Consequence of exceeding impurity limits. Generally insensitive to minor variations in composition, but resistance to corrosion is reduced and lowers as copper increases

Applications

Typical uses. Die casting applications where fair pressure tightness and fair strength at elevated temperatures are required. Better die filling than 380.0. Poor weldability and brazeability

Mechanical Properties

Tensile properties. Typical for separately cast test bars, as cast, 384.0 and A384.0: tensile strength, 330 MPa (48 ksi); yield strength, 165 MPa (24 ksi); elongation, 2.5% in 50 mm or 2 in.

Shear strength. 384.0: 200 MPa (29 ksi)

Hardness. 384.0 and A384.0: 85 HB (500-kg load, 10-mm ball)

Fatigue strength. 384.0: 140 MPa (20 ksi)

Mass Characteristics

Density. 384.0: 2.823 Mg/m^3 (0.102 lb/in.^{3}) A384.0: 2.768 Mg/m^3 (0.100 lb/in.^{3})

Thermal Properties

Liquidus temperature. 580 °C (1080 °F)

Solidus temperature. 515 °C (960 °F)

Coefficient of thermal expansion. Linear. 384.0: 20.8 μm/m·K (11.6 μin./in.·°F). A384.0: 20.7 μm/m·K (11.5 μin./in.·°F)

Thermal conductivity. 384.0: 92 W/m·K (53 Btu/ft·h·°F). A384.0: 96 W/m·K (56 Btu/ft·h·°F)

Electrical Properties

Electrical conductivity. Volumetric, at 20 °C (68 °F). 384.0: 22% IACS. A384.0: 23% IACS

Fabrication Characteristics

Die casting temperature. 615 to 700 °C (1140 to 1280 °F)

Stress relief temperature. 175 to 260 °C (350 to 500 °F); hold at temperature 4 to 6 h; cool in still air

Annealing temperature. For increased ductility, 260 to 370 °C (500 to 700 °F); hold at temperature 4 to 6 h; furnace cool or cool in still air

Table 34 Typical room-temperature mechanical properties for separately cast test bars of alloys 390.0 and A390.0

Temper	Tensile strength(a) MPa	ksi	Yield strength(b) MPa	ksi	Hardness(a)(c), HB	Fatigue strength(d) MPa	ksi
A390.0, Sand Castings							
F, T5	180	26	180	26	100	...	...
T6	275	40	275	40	140	105	15
T7	250	36	250	36	115	...	...
A390.0, Permanent Mold Castings							
F, T5	200	29	200	29	110	...	...
T6	310	45	310	45	145	115	17
T7	260	38	260	38	120	100	14.5
390.0, Conventional Die Castings							
F	280	40.5	240	35	120	140	20
T5	295	43	260	38	125	...	...
390.0, Acurad Castings							
F	205	30	195	28	110	90	13
T5	205	30	200	29	110	95	14
T6	365	53	365	53	150	115	17
T7	275	40	275	40	125	110	16

(a) Tensile properties and hardness are determined from standard cast-to-size tensile specimens 12.7-mm (½-in.) diameter for sand, permanent mold and Acurad castings and 6.4-mm (¼-in.) diameter for die castings and tested without machining the surface. (b) 0.2% offset. For sand and permanent mold castings, yield strength normally equals tensile strength because 0.2% offset is not reached prior to fracture. (c) 500-kg load; 10-mm ball. (d) At 5 x 10^8 cycles; R. R. Moore type test.

Table 35 Typical elevated-temperature tensile yield strength for separately cast test bars of alloy 390.0

Temper	Temperature °C	°F	Yield strength(a) MPa	ksi
Acurad Castings				
F	38	100	195	28
	95	200	195	28
	150	300	180	26
	205	400	155	22
	260	500	100	14
T5	38	100	210	30
	95	200	225	32
	150	300	195	28
	205	400	160	23
	260	500	85	12
T6	38	100	365	52
	95	200	335	48
	150	300	305	44
	205	400	235	34
	260	500	70	10
T7	38	100	280	40
	95	200	270	39
	150	300	245	35
	205	400	195	28
	260	500	70	10
Die Castings				
F	38	100	260	37
	95	200	285	41
	150	300	265	38
	205	400	210	30
	260	500	125	18

(a) Based on cast-to-size test specimens tested after 1000 h holding at test temperature.

390.0, A390.0
17.0Si-4.5Cu-0.6Mg

Specifications

UNS number. 390.0 die castings, A03900. A390.0: sand and permanent mold castings, A13900

Chemical Composition

Composition limits. 390.0: 4.0 to 5.0 Cu; 0.45 to 0.65 Mg; 0.10 max Mn; 16.0 to 18.0 Si; 1.3 max Fe; 0.10 max Zn; 0.20 max Ti; 0.10 max others (each); 0.20 max others (total); rem Al. A390.0: 4.0 to 5.0 Cu; 0.45 to 0.65 Mg; 0.10 max Mn; 16.0 to 18.0 Si; 0.5 max Fe; 0.10 max Zn; 0.20 max Ti; 0.10 max others (each); 0.20 max others (total); rem Al

Applications

Typical uses. Automotive cylinder block, four cycle air-cooled engines, air compressors, Freon compressors, pumps requiring abrasive resistance, pulleys and brake shoes. Other applications where high wear resistance, low coefficient of thermal expansion, good elevated temperature strength and good fluidity are required

Mechanical Properties

Tensile properties. See Tables 34 and 35. Typical elongation. 390.0: die and Acurad castings (F and T5 tempers), 1.0% in 50 mm or 2 in.; Acurad castings (T6 and T7 tempers), <1.0% in 50 mm or 2 in. A390.0: sand castings (all tempers) and permanent mold castings (T6 and T7 tempers), <1.0% in 50 mm or 2 in.; permanent mold castings (F and T5 tempers), 1.0% in 50 mm or 2 in.

Hardness. See Table 34.

Elastic modulus. Tension, 81.2 GPa (11.8 × 10⁶ psi); compression, 82.8 GPa (12.0 × 10⁶ psi)

Fatigue strength. See Table 34.

Mass Characteristics

Density. 2.730 Mg/m³ (0.099 lb/in.³) at 20 °C (68 °F)

Thermal Properties

Liquidus temperature. 650 °C (1200 °F)

Solidus temperature. 505 °C (945 °F)

Coefficient of thermal expansion. Linear, 18.0 μm/m·K (10.0 μin./in.·°F) at 20 to 100 °C (68 to 212 °F)

Thermal conductivity. 134 W/m·K (77.4 Btu/ft·h·°F) at 25 °C (77 °F)

Electrical Properties

Electrical conductivity. Volumetric, at 20 °C (68 °F). F temper: 27% IACS. T5 temper: 25% IACS

Fabrication Characteristics

Solution temperature. 495 °C (925 °F)

Aging temperature. T5 and T7 tempers: 230 °C (450 °F). T6 temper: 175 °C (350 °F); hold at temperature for 8 h

413.0, A413.0
12Si

Commercial Names

Former designation. 413.0: 13. A413.0: A13

Specifications

Former ASTM. 413.0: S12B. B85 S12A

SAE. A413.0: J453, 305

UNS number. 413.0: A04130. A413.0: A14130

Government. A413.0: QQ-A-591 (class 2)

Foreign. Canada: A413.0, CSA S12P. France: NF A-S13. ISO: AlSi12

Table 36 Typical tensile properties for separately cast test bars of alloy 413.0-F at elevated temperature

Temperature		Tensile strength		Yield strength(a)		Elongation(b), %
°C	°F	MPa	ksi	MPa	ksi	
24	75	295	43	145	21	2.5
100	212	255	37	140	20	5
150	300	220	32	130	19	8
205	400	165	24	105	15	15
260	500	90	13	60	9	30
315	600	50	7	30	4.5	35
370	700	30	4.5	15	2.5	40

(a) 0.2% offset. (b) In 50 mm or 2 in.

Table 37 Typical tensile properties for separately cast test bars of alloys 443.0, 443.0-F, B443.0-F and C443.0-F

Temperature		Tensile strength		Yield strength(a)		Elongation(b), %
°C	°F	MPa	ksi	MPa	ksi	
443.0-F Sand Castings						
24	75	130	19	55	8	8
B443.0-F Permanent Mold Castings						
24	75	160	23	60	9	10
C443.0-F Die Castings						
24	75	230	33	110	16	9
100	212	195	28	110	16	9
150	300	150	22	105	15	10
205	400	110	16	85	12	25
260	500	60	9	40	6	30
315	600	35	5	25	3.5	35
370	700	25	3.5	15	2.5	35

(a) 0.2% offset. (b) In 50 mm or 2 in.

Chemical Composition

Composition limits. 413.0: 1.0 max Cu; 0.10 max Mg; 0.35 max Mn; 11.0 to 13.0 Si; 2.0 max Fe; 0.50 max Ni; 0.50 max Zn; 0.15 max Sn; 0.25 max others (total); rem Al. A413.0: 1.0 maxCu; 0.10 max Mg; 0.35 max Mn; 11.0 to 13.0 Si; 1.3 max Fe; 0.50 max Ni; 0.50 max Zn; 0.15 max Sn; 0.25 max others (total); rem Al

Consequence of exceeding impurity limits. Content of impurities may be quite high before serious effects are detected. Increasing copper lowers corrosion resistance; increasing iron and magnesium lowers ductility; increasing silicon content may lead to machining problems

Applications

Typical uses. Miscellaneous thin-walled and intricately designed castings. Other applications where excellent castability, resistance to corrosion and pressure tightness are required

Mechanical Properties

Tensile properties. Typical for separately cast test bars, as cast. 413.0: tensile strength, 300 MPa (43 ksi); yield strength, 140 MPa (21 ksi); elongation, 2.5% in 50 mm or 2 in. A413.0: tensile strength, 290 MPa (42 ksi); yield strength, 130 MPa (19 ksi); elongation, 3.5% in 50 mm or 2 in. See also Table 36.

Shear strength. 170 MPa (25 ksi)

Fatigue strength. At 5 × 10⁸ cycles, 130 MPa (19 ksi) (R. R. Moore type test)

Mass Characteristics

Density. 2.657 Mg/m³ (0.096 lb/in.³) at 20 °C (68 °F)

Thermal Properties

Coefficient of thermal expansion. Linear:

Temperature range		Average coefficient	
°C	°F	µm/ m·K	µin./ in.·°F
20 – 100	68 – 212 .... 20.4		11.3
20 – 200	68 – 392 .... 21.4		11.8
20 – 300	68 – 572 .... 22.4		12.4

Specific heat. 963 J/kg·K (0.230 Btu/lb·°F)
Latent heat of fusion. 389 kJ/kg (167 Btu/lb)
Thermal conductivity. 121 W/m·K (70.1 Btu/ft·_6°F) at 25 °C (77 °F)

Electrical Properties

Electrical conductivity. Volumetric, 31% IACS at 20 °C (68 °F)
Electrical resistivity. 55.6 nΩ·m at 20 °C (68 °F)

Fabrication Characteristics

Melting temperature. 650 to 760 °C (1200 to 1400 °F)
Die casting temperature. 635 to 704 °C (1175 to 1300 °F)
Joining. Rivet compositions: 6053-T4, 6053-T6, 6053-T61. Soft solder: After copper plating, then use methods applicable to copper-base alloys. Resistance welding: flash method

---

443.0, A443.0, B443.0, C443.0
5.2Si

Commercial Names

Former designation. 43

Specifications

Former ASTM. 443.0: S5B. B443.0: S5A. C443.0: S5C
SAE. C443.0: 304
UNS number. 443.0: A04430. A443.0: A14430. B443.0: A24430 C443.0: A34430
Government. B443.0: QQ-A-601 (class 2). C443.0: QQ-A-591
Foreign. Canada: CSA S5

Chemical Composition

Composition limits. 443.0: 0.6 max Cu; 0.05 max Mg; 0.50 max Mn; 4.5 to 6.0 Si; 0.8 max Fe; 0.25 max Cr; 0.50 max Zn; 0.25 max Ti; 0.35 max others (total); rem Al. A443.0: 0.30 max Cu; 0.05 max Mg; 0.50 max Mn; 4.5 to 6.0 Si; 0.8 max Fe; 0.25 max Cr; 0.50 max Zn; 0.25 max Ti; 0.35 max others (total); rem Al. B443.0: 0.15 max Cu; 0.05 max Mg; 0.35 max Mn; 4.5 to 6.0 Si; 0.8 max Fe; 0.35 max Zn; 0.25 max Ti; 0.25 max others (total); rem Al. C443.0: 0.6 max Cu; 0.10 max Mg; 0.35 max Mn; 4.5 to 6.0 Si; 2.0 max Fe; 0.50 max Ni; 0.50 max Zn; 0.15 max Sn; 0.25 max others (total); rem Al

Consequence of exceeding impurity limits. For die cast alloy, relatively large quantities of impurities may be present before serious effects are detected. Increasing copper tends to lower resistance to corrosion; increasing iron and magnesium tends to lower ductility. For sand and permanent mold cast alloys, high copper, iron, or nickel decrease ductility and resistance to corrosion. Increasing magnesium reduces ductility.

Applications

Typical uses. Cooking utensils, food-handling equipment, marine fittings, miscellaneous thin-section castings. Die castings: applications where good pressure tightness, above average ductility, and excellent resistance to corrosion are required. Sand and permanent mold castings: applications where very good castability and resistance to corrosion with moderate strength are required.

Mechanical Properties

Tensile properties. See Table 37.
Shear strength. F temper: 443.0 (sand castings): 95 MPa (14 ksi). B443.0 (permanent mold castings): 110 MPa (16 ksi). C443.0 (die castings): 145 MPa (21 ksi)
Hardness. F temper: 443.0 (sand castings): 40 HB. B443.0 (permanent mold castings): 45 HB. C443.0 (die castings): 65 HB (500-kg load, 10-mm ball)
Poisson's ratio. 0.33
Elastic modulus. Tension, 71.0 GPa (10.3 × 10^6 psi); shear 26.5 GPa (3.85 × 10^6 psi)
Fatigue strength. F temper, at 5 × 10^8 cycles. 443.0 (sand castings) and B443.0 (permanent mold castings): 55 MPa (8 ksi). C443.0 (die castings): 115 MPa (17 ksi) (R. R. Moore type test)

Mass Characteristics

Density. 2.690 Mg/m^3 (0.097 lb/in.^{3}) at 20 °C (68 °F)

Thermal Properties

Liquidus temperature. 630 °C (1170 °F)

Solidus temperature. 575 °C (1065 °F)
Coefficient of thermal expansion. Linear:

Temperature range		Average coefficient	
°C	°F	µm/ m·K	µin./ in.·°F
20 – 100	68 – 212 22		12.2
20 – 200	68 – 392 23		12.8
20 – 300	68 – 572 24		13.3

Specific heat. 963 J/kg·K (0.230 Btu/lb·°F) at 100 °C (212 °F)
Latent heat of fusion. 389 kJ/kg (167 Btu/lb)
Thermal conductivity. As cast: 142 W/m·K (82.2 Btu/ft·h·°F). Annealed: 163 W/m·K (94.3 Btu/ft·h·°F)

Electrical Properties

Electrical conductivity. Volumetric at 20 °C (68 °F). As cast (sand, permanent mold, and die castings): 37% IACS
Electrical resistivity. At 20 °C (68 °F). As cast (sand, permanent mold, and die castings): 46.6 nΩ·m. Annealed (sand and permanent mold): 41.0 nΩ·m
Electrolytic solution potential. –0.83 V (sand cast) and –0.82 V (permanent mold cast) vs 0.1N calomel electrode in an aqueous solution containing 53 g NaCl plus 3 g H$_2$O$_2$ per litre

Fabrication Characteristics

Melting temperature. Die castings: 650 to 760 °C (1200 to 1400 °F). Sand and permanent mold castings: 675 to 816 °C (1250 to 1500 °F)
Casting temperature. Die castings: 635 to 705 °C (1175 to 1300 °F). Sand and permanent mold castings: 675 to 790 °C (1250 to 1450 °F)
Joining. Rivet compositions: 6053-T4, 6053-T6, 6053-T61. Soft solder with copper plate and use methods applicable to copper-base alloys for die castings. Use Alcoa No. 802, no flux or rub-tin with Alcoa No. 802 for sand and permanent mold castings. Sand and permanent mold casting alloys (unless otherwise noted): braze with Alcoa No. 717; Alcoa No. 33 flux; flame either reducing oxyacetalene or reducing oxyhydrogen. Atomic-hydrogen weld with 4043 alloy; Alcoa No. 22 flux. Oxyacetylene weld with 4043 alloy; Alcoa No. 22 flux; neutral flame. Metal-arc weld with

4043 alloy; Alcoa No. 27 flux. Carbon-arc weld with 4043 alloy; Alcoa No. 24 flux (automatic), Alcoa No. 27 flux (manual). Tungsten-arc argon-atmosphere weld with 4043 alloy; no flux. Resistance weld: flash method for die cast alloys; spot, seam, and flash methods for sand and permanent mold cast alloys

Table 38 Typical tensile properties for separately cast test bars of alloy 514.0-F

Temperature		Tensile strength		Yield strength		Elongation, %
°C	°F	MPa	ksi	MPa	ksi	
24	75	170	25	85	12	9
150	300	150	22	85	12	7
205	400	125	18	85	12	9
260	500	90	13	55	8	12
315	600	60	9	30	4	17

514.0
4Mg

Commercial Names

Former designation. 214

Specifications

Former ASTM. G4A
SAE. 320
UNS number. A05140
Government. QQ-A-601 (class 5)
Foreign. Canada: CSA G4. United Kingdom: DTD 165. ISO: AlMg3

Chemical Composition

Composition limits. 0.15 max Cu; 3.5 to 4.5 Mg; 0.35 max Mn; 0.35 max Si; 0.50 max Fe; 0.15 max Zn; 0.25 max Ti; 0.05 max others (each); 0.15 max others (total); rem Al
Consequence of exceeding impurity limits. High copper or nickel greatly decreases resistance to corrosion and decreases ductility. High iron, silicon, or manganese decreases strength and ductility. Tin reduces resistance to corrosion.

Applications

Typical uses. Dairy and food-handling applications, cooking utensils, fittings for chemical and sewage use. Other applications where excellent resistance to corrosion and tarnish are required

Mechanical Properties

Tensile properties. Typical, F temper. Tensile strength, 145 MPa (21 ksi); yield strength, 95 MPa (14 ksi); elongation, 3.0%. See also Table 38.
Shear strength. 140 MPa (20 ksi)
Compressive yield strength. 85 MPa (12 ksi)
Hardness. 50 HB (500-kg load, 10-mm ball)
Poisson's ratio. 0.33
Elastic modulus. Tension, 71.0 GPa (10.3 × 10^6 psi); shear, 26.5 GPa (3.85 × 10^6 psi)
Fatigue strength. 50 MPa (7 ksi) at 5 × 10^8 cycles (R. R. Moore type test)

Mass Characteristics

Density. 2.650 Mg/m^3 (0.096 lb/in.^{3}) at 20 °C (68 °F)

Thermal Properties

Liquidus temperature. 630 °C (1170 °F)
Solidus temperature. 585 °C (1090 °F)
Coefficient of thermal expansion. Linear:

Temperature range		Average coefficient	
°C	°F	µm/ m·K	µin./ in.·°F
20–100	68–212	24	13.3
20–200	68–392	25	13.9
20–300	68–572	26	14.4

Specific heat. 963 J/kg·K (0.230 Btu/lb·°F) at 100 °C (212 °F)
Latent heat of fusion. 389 kJ/kg (167 Btu/lb)
Thermal conductivity. 146 W/m·K (84.6 Btu/ft·h·°F) at 25 °C (77 °F)

Electrical Properties

Electrical conductivity. Volumetric, 38% IACS at 20 °C (68 °F)
Electrical resistivity. 49.3 nΩ·m at 20 °C (68 °F)
Electrolytic solution potential. -0.87 V vs 0.1N calomel electrode in an aqueous solution containing 53 g NaCl plus 3 g H_2O_2 per litre

Fabrication Characteristics

Melting temperature. 675 to 815 °C (1250 to 1500 °F)
Casting temperature. 675 to 790 °C (1250 to 1450 °F)
Joining. Rivet compositions: 6053-T4, 6053-T6, 6053-T61. Soft solder with Alcoa No. 802; no flux. Rub-tin with Alcoa No. 802. Braze with Alcoa No. 717; Alcoa No. 33 flux; flame either reducing oxyacetylene or reducing oxyhydrogen. Atomic-hydrogen weld with 4043 alloy; Alcoa No. 22 flux. Oxyacetylene weld with 4043 alloy; Alcoa No. 22 flux; flame neutral. Metal-arc weld with 4043 alloy; Alcoa No. 27 flux. Carbon-arc weld with 4043 alloy; Alcoa No. 24 flux (automatic), Alcoa 27 flux (manual). Tungsten-arc argon-atmosphere weld with 4043; no flux. Resistance welding: spot, seam, and flash welds.

518.0
8Mg

Commercial Names

Former designation. 218

Specifications

Former ASTM. G8A
UNS number. A05180
Government. QQ-A-591

Chemical Composition

Composition limits. 0.25 max Cu; 7.5 to 8.5 Mg; 0.35 max Mn; 0.35 max Si; 1.8 max Fe; 0.15 max Ni; 0.15 max Zn; 0.15 max Sn; 0.25 max others (total); rem Al

Applications

Typical uses. Alloy has excellent corrosion resistance and machinability; high ductility; poor castability (is hot short). Takes a high polish; difficult to attain a uniform appearance after anodizing. Nonheat treatable. Poor weldability and brazeability. Used for die cast marine fittings, ornamental hardware, ornamental automotive parts and other applications requiring the highest corrosion resistance.

Mechanical Properties

Tensile properties. Typical, F temper. Tensile strength, 310 MPa (45 ksi); yield strength, 190 MPa (28 ksi); elongation, 5 to 8% in 50 mm or 2 in.

Shear strength. 205 MPa (30 ksi)

Hardness. 80 HB (500-kg, 10-mm load)

Impact strength. Charpy V-notch: 9 J (7 ft·lb)

Fatigue strength. 159 MPa (23 ksi) at 5×10^8 cycles (R. R. Moore type test)

Mass Characteristics

Density. 2.574 Mg/m³ (0.093 lb/in.³)

Thermal Properties

Liquidus temperature. 620 °C (1150 °F)

Solidus temperature. 535 °C (995 °F)

Coefficient of thermal expansion. Linear, 24.1 μm/m·K (13.4 μin./in.·°F) at 20 to 100 °C (68 to 212 °F)

Thermal conductivity. 96.2 W/m·K (55.6 Btu/ft·h·°F)

Electrical Properties

Electrical conductivity. Volumetric, 25% IACS at 20 °C (68 °F)

520.0
10Mg

Commercial Names

Former designation. 220

Specifications

AMS. 4240
Former ASTM. G10A
SAE. 324
UNS number. A05200
Government. QQ-A-601 (class 16)
Foreign. Canada: CSA G10. France: NF A-G10. ISO: AlMg10

Chemical Composition

Composition limits. 0.25 max Cu; 9.5 to 10.6 Mg; 0.15 max Mn; 0.25 max Si; 0.30 max Fe; 0.15 max Zn; 0.25 max Ti; 0.05 max others (each); 0.15 max others (total); rem Al

Consequence of exceeding impurity limits. High copper or nickel greatly decreases resistance to corrosion. High iron, silicon, or manganese contents adversely affect mechanical properties

Table 39 Typical tensile properties for separately cast test bars of alloy 520.0-F at elevated temperature

Temperature		Tensile strength		Yield strength		Elongation(a), %
°C	°F	MPa	ksi	MPa	ksi	
24	75	315	46	170	25	14
150	300	240	35	130	19	16
205	400	150	22	80	11.5	40
260	500	105	15	50	7.5	55
315	600	70	10.5	25	3.5	70

(a) In 50 mm or 2 in.

Applications

Typical uses. Aircraft fittings, railroad passenger-car frames, miscellaneous castings, requiring strength and shock resistance. Other applications where excellent machinability and resistance to corrosion with highest strength and elongation of any aluminum sand casting alloy are desired

Mechanical Properties

Tensile properties. Typical. T4 temper: tensile strength, 330 MPa (48 ksi); yield strength, 180 MPa (26 ksi); elongation in 50 mm or 2 in., 16%. See also Table 39.

Shear strength. 235 MPa (34 ksi)

Compressive yield strength. 2 MPa (27 ksi)

Hardness. 75 HB (500-kg load, 10-mm ball)

Poisson's ratio. 0.33

Elastic modulus. Tension, 66 GPa (9.5×10^6 psi); shear, 24.5 GPa (3.55×10^6 psi)

Fatigue strength. 55 MPa (8 ksi) at 5×10^8 cycles (R. R. Moore type test)

Mass Characteristics

Density. 2.570 Mg/m³ (0.093 lb/in.³) at 20 °C (68 °F)

Thermal Properties

Liquidus temperature. 605 °C (1120 °F)

Solidus temperature. 450 °C (840 °F)

Coefficient of thermal expansion. Linear:

Temperature range		Average coefficient	
°C	°F	μm/m·K	μin./in.·°F
20–100	68–212	25	13.9
20–200	68–392	26	14.4
20–300	68–572	27	15.0

Specific heat. 963 J/kg·K (0.230 Btu/lb·°F) at 100 °C (212 °F)

Latent heat of fusion. 389 kJ/kg (167 Btu/lb)

Thermal conductivity. T4 temper: 87.9 W/m·K (50.8 Btu/ft·h·°F) at 25 °C (77 °F)

Electrical Properties

Electrical conductivity. Volumetric, T4 temper: 21% IACS at 20 °C (68 °F)

Electrical resistivity. T4 temper: 82.1 nΩ·m at 20 °C (68 °F)

Electrolytic solution potential. T4 temper: −0.89 V vs 0.1N calomel electrode in an aqueous solution containing 53 g Na©l plus 3 g H₂O₂ per litre

Fabrication Characteristics

Melting temperature. 675 to 816 °C (1250 to 1500 °F)

Casting temperature. 675 to 788 °C (1250 to 1450 °F)

Joining. Rivet compositions: 6053-T4, 6053-T6, 6053-T61. Soft solder with Alcoa No. 802; no flux. Rub-tin with Alcoa No. 802. Resistance welding: spot, seam, and flash methods

535.0, A535.0, B535.0
7Mg

Commercial Names

Former designations. 535.0: Almag35. A535.0: A218. B535.0: B218

Specifications

Former AMS. 4238A, 4239
Former ASTM. 535.0: GM70B
UNS number. 535.0: A05350. A535.0: A15350. B535.0: A25350.
Government. 535.0: QQ-A-601, QQ-A-371

Chemical Composition

Composition limits. 535.0. 0.05 max Cu; 6.2 to 7.5 Mg; 0.10 to 0.25 Mn; 0.15 max Si; 0.15 max Fe; 0.10 to 0.25 Ti; 0.003 to 0.007 Be; 0.002 max B; rem Al. A535.0: 0.10 max Cu; 6.5 to 7.5 Mg; 0.10 to 0.25 Mn; 0.20 max Si; 0.20 max Fe; 0.25 max Ti; 0.05 max others (each); 0.15 max others (total); rem Al. B535.0: 0.10 max Cu; 6.5 to 7.5 Mg; 0.05 max Mn; 0.15 max Si; 0.15 max Fe; 0.10 to 0.25 Ti; 0.05 max others (each); 0.15 max others (total); rem Al

Applications

Typical uses. Maximum properties are available immediately after casting without the aid of heat treatment or natural aging. Used in parts in computing devices, aircraft and missile guidance systems, and electric equipment where dimensional stability is essential. Highly useful in marine and other corrosive-prone applications

Mechanical Properties

Tensile properties. F and T5 tempers: 535.0: Tensile strength: typical, 275 MPa (40 ksi); minimum, 240 MPa (35 ksi). Yield strength: typical, 140 MPa (20 ksi); minimum, 125 MPa (18 ksi). Elongation in 50 mm or 2 in.: typical, 13%; minimum, 8.0%. See also Table 40.
Shear strength. 190 MPa (27.5 ksi)
Compressive yield strength. Typical: 162 MPa (23.5 ksi)
Hardness. Typical: 60 HB. Minimum: 70 HB
Elastic modulus. Tension, 71.0 GPa (10.3×10^6 psi)
Impact strength. Charpy: 90 notch specimen, 14.2 J (10.5 ft·lb); keyhole specimen, 6.7 J (4.95 ft·lb); unnotched specimen, 77.0 J (56.8 ft·lb)

Mass Characteristics

Density. 2.620 Mg/m³ (0.095 lb/in.³)

Thermal Properties

Liquidus temperature. 630 °C (1165 °F)
Solidus temperature. 550 °C (1020 °F)
Coefficient of thermal expansion. Linear:

Table 40 Typical tensile properties for separately cast test bars of alloy 535.0-F at elevated temperature

Temperature		Tensile strength		Elongation(a),
°C	°F	MPa	ksi	%
150	300	260	37.8	11
175	350	235	34.4	14
205	400	220	32.3	14
260	500	180	26.7	13
315	600	140	20.8	13
370	700	105	15.5	12

(a) In 50 mm or 2 in.

Table 41 Typical tensile properties for separately cast test bars of alloy 712.0-F at elevated temperature

Temperature		Tensile strength		Yield strength		Elongation(a),
°C	°F	MPa	ksi	MPa	ksi	%
79	175	235	33.8	210	30.7	3
120	250	205	29.5	175	25.2	2
175	350	135	19.7	115	17.0	6

(a) In 50 mm or 2 in.

Temperature range		Average coefficient	
°C	°F	µm/ m·K	µin./ in.·°F
−60– 20	−76–68	21.6	12.0
20–100	68–212	23.6	13.1
20–200	68–392	25.6	14.2
20–300	68–572	26.6	14.8

Electrical Properties

Electrical conductivity. Volumetric, 23.2% IACS at 20 °C (68 °F)
Electrical resistivity. 44.7 nΩ·m at 20 °C (68 °F)

Chemical Properties

General corrosion behavior. 535.0 has the highest resistance to corrosion of any of the common aluminum casting alloys.

Fabrication Characteristics

Machinability. Superior, can be milled at speeds four times faster than other aluminum casting alloys. High microfinishes can be achieved at high speeds. 535.0 takes a very high mirror polish. Normally this alloy used as sand and permanent mold castings, but it can also be used for die casting. Where high dimensional tolerance is required, the following procedure should be used: rough machine parts; heat at 200 °C (400 °F) for 14 h; cycle between −73 to 100 °C (+100 to +212 °F) five times (30 h/cycle); finish machine; heat 10 h at 200 °C (400 °F); cycle between −73 to +100 °C (−100 to +212 °F) 25 times (30 h/cycle). 535.0 may be stress relieved at approximately 370 °C (700 °F) for 5 h; air cool. Creep resistance at 370 °C (700 °F) is very low, permitting plastic flow under the load of locked up stresses and resulting in stress-free castings. On air cooling from 370 °C (700 °F), 535.0 will have full hard and physical properties and will be stable. After being stress relieved, most castings from 535.0, A535.0, B535.0 can be rough and finish machined without breaking into the machining sequence.
Weldability. Can be welded by any inert gas shielded arc systems using filler material of 5356 or 535.0 aluminum. Welding fluxes should be avoided if possible. Because of the beryllium content in alloy 535.0, care should be taken not to inhale fumes during welding.
Anodizing. Use sulfuric acid process to produce a pure satin white finish capable of being dyed to brilliant pastel colors.

712.0
5.8Zn-0.6Mg-0.5Cr-0.2Ti

Commercial Names

Former designations. D712.0, D612, 40E

Specifications

Former ASTM. ZG61A
SAE. 310
UNS number. A47120
Government. QQ-A-601 (class 17)

Chemical Composition

Composition limits. 0.25 max Cu; 0.50 to 0.65 Mg; 0.10 max Mn; 0.30 max Si; 0.50 max Fe; 0.40 to 0.6 Cr; 5.0 to 6.5 Zn; 0.15 to 0.25 Ti; 0.05 max others (each); 0.20 max others (total); rem Al

Applications

Typical uses. Applications where a good combination of mechanical properties is required without heat treatment: shock and corrosion resistance, machinability, dimensional stability, no distortion in heat treating

Mechanical Properties

Tensile properties. F or T5 temper. Typical tensile strength, 240 MPa (35 ksi); yield strength, 170 MPa (25 ksi); elongation, 5%. Low temperature strength after 24 h at −94 °F: tensile strength, 265 MPa (38.4 ksi); elongation in 50 mm or 2 in., 5%. See also Table 41.
Shear strength. 180 MPa (26 ksi)
Compressive proportional limit. 96 MPa (14 ksi)
Hardness. 70 HB (500-kg load, 10-mm ball)
Poisson's ratio. 0.33
Elastic modulus. Tension, 71.0 GPa (10.3×10^6 psi); shear, 26.5 GPa (3.85×10^6 psi)
Impact strength. Charpy V-notch: 2.7 to 4.0 J (2 to 3 ft·lb)
Fatigue strength. 62 MPa (9 ksi) at 5×10^8 cycles (R. R. Moore type test)

Mass Characteristics

Density. 2.810 Mg/m^3 (0.101 lb/in.^{3}) at 20 °C (68 °F)

Thermal Properties

Liquidus temperature. 615 °C (1140 °F)
Solidus temperature. 570 °C (1060 °F)
Coefficient of thermal expansion. Linear, 24.7 μm/m·K (13.7 μin./in.·°F) at 20 to 93 °C (68 to 199 °F)
Specific heat. 963 J/kg·K (0.230 Btu/lb·°F) at 100 °C (212 °F)
Latent heat of fusion. 389 kJ/kg (167 Btu/lb)

Thermal conductivity. 138 W/m·K (79.8 Btu/ft·h·°F) at 25 °C (77 °F)

Electrical Properties

Electrical conductivity. Volumetric, 35% IACS at 20 °C (68 °F)
Electrical resistivity. 49.3 nΩ·m at 20 °C (68 °F)

Fabrication Characteristics

Melting temperature. F temper: 610 to 648 °C (1132 to 1200 °F)
Aging temperature. T5 temper: room temperature for 21 days or at 157 °C (315 °F) for 6 to 8 h

713.0
7.5Zn-0.7Cu-0.35Mg

Commercial Names

Former designation. 613, Tenzaloy

Specifications

Former ASTM. Sand castings, B26 ZC81A. Permanent mold castings, B108 ZC81B
Former SAE. 315
UNS number. A07130
Government. Sand castings, QQ-A-601 (class 22). Permanent mold castings: QQ-A-596 (class 12)

Chemical Composition

Composition limits. 0.40 to 1.0 Cu; 0.20 to 0.50 Mg; 0.6 max Mn; 0.25 max Si; 1.1 max Fe; 0.35 max Cr; 0.15 max Ni; 7.0 to 8.0 Zn; 0.25 max Ti; 0.10 max others (each); 0.25 max others (total); rem Al

Applications

Typical uses. Cast aluminum furniture and other very large casting applications that require high strength without heat treatment. 713.0 ages at room temperature to produce mechanical properties equivalent to those of common heat treated aluminum cast alloys. These properties develop in 10 to 14 days at room temperature or in 12 h at 120 °C (250 °F).

Mechanical Properties

Tensile properties. Typical for T5 temper, aged at room temperature for 21 days or artificially aged at 120 ± 5.5 °C (250 ± 10 °F) for 16 h. Sand casting: tensile strength, 205 MPa

(30 ksi); yield strength: 150 MPa (22 ksi); elongation, 4.0% in 50 mm or 2 in. Permanent mold casting: tensile strength, 220 —pa (32 ksi); yield strength: 150 MPa (22 ksi); elongation, 3.0% in 50 mm or 2 in.
Shear strength. 180 MPa (26 ksi)
Compressive yield strength. 170 MPa (25 ksi)
Impact strength. Charpy V-notch: sand castings, 3.4 J (2.5 ft·lb); permanent mold castings, 4 J (3 ft·lb). Unnotched: sand castings, 16.3 J (12 ft lb); permanent mold castings, 27.1 J (20 ft·lb)
Fatigue strength. 60 MPa (9 ksi) at 5×10^8 cycles (R. R. Moore type test)

Mass Characteristics

Density. 2.810 Mg/m^3 (0.102 lb/in.^{3})

Thermal Properties

Coefficient of thermal expansion. Linear, 24.1 μm/m·K (13.4 μin./in.·°F) at 20 to 200 °C (68 to 392 °F)
Thermal conductivity. 140 W/m·K (80 Btu/ft·h·°F) at 25 °C (77 °F)

Electrical Properties

Electrical conductivity. Volumetric, 35% IACS at 20 °C (68 °F)

Chemical Properties

General corrosion behavior. Good resistance to corrosion, equivalent to aluminum-silicon alloys. A typical corrosion test showed no loss in mechanical properties after immersion for 90 days in aerated 3% salt water solution. Not subject to acceleration of corrosion by stress or to stress-corrosion cracking as determined by the standard test of exposure for 14 days to the corrosive medium while under a continuous load of 75% of yield strength.

Fabrication Characteristics

Melting temperature. Approximate, 595 to 640 °C (1100 to 1185 °F)
Machinability. Good machinability and polishing characteristics. Very good dimensional stability. Fully aged material shows a decrease in length of less than 0.1 min/in. of length. If 713.0 is given a stress relief treatment of 6 h at 450 °C (850 °F) and air cooled, it ages naturally. The resulting product is a stress-free, full strength casting. This is not possible with any heat-treatable aluminum alloy.

Weldability. For high strength welds, shielded-arc methods can be used with filler alloys 5154 and 5356.

Brazeability. Readily brazed at 537 to 593 °C (1000 to 1100 °F) using any of the common brazing methods.

771.0
7Zn-0.9Mg-0.13Cr

Commercial Names

Former designation. Precedent 71A

Specifications

Former ASTM. 771.0: ZG71B
UNS number. A07710
Government. 771.0: QQ-A-601E

Chemical Composition

Composition limits. 0.10 max Cu; 0.8 to 1.0 Mg; 0.10 max Mn; 0.15 max Si; 0.15 max Fe; 0.06 to 0.20 Cr; 6.5 to 7.5 Zn; 0.10 to 0.20 Ti; 0.05 max others (each); 0.15 others (total); rem Al

Applications

Typical uses. Applications where free machine and dimension stability are important. Polishes to a high lustre; anodizes with good clean appearance. Good corrosion resistance

Mechanical Properties

Tensile properties. See Table 42.
Compressive properties. Compressive strength, T71 temper, 925 MPa (134 ksi); compressive yield strength, 370 MPa (54 ksi)
Elastic modulus. Tension, 71.0 GPa (10.3 × 10^6 psi)

Mass Characteristics

Density. 2.823 Mg/m^3 (0.102 lb/in.^{3})

Thermal Properties

Liquidus temperature. 645 °C (1190 °F)
Solidus temperature. 605 °C (1120 °F)
Coefficient of thermal expansion. Linear, 24.7 µm/m·K (13.7 µin./in.·°F) at 20 to 100 °C (68 to 212 °F)
Thermal conductivity. 138 W/m·K (79.8 Btu/ft·h·°F)

Electrical Properties

Electrical conductivity. Volumetric, 27% IACS at 20 °C (68 °F)

Table 42 Minimum mechanical properties for separately cast test bars of alloy 771.0

Temper	Tensile strength (min) MPa	ksi	Yield strength (min)(a) MPa	ksi	Elongation(b), %	Hardness(c), HB
T5	290	42	260	38	1.5	100
T51	220	32	185	27	3.0	85
T52	250	36	205	30	1.5	85
T6	290	42	240	35	5.0	90
T71	330	48	310	45	2.0	120

(a) 0.2% offset. (b) In 50 mm or 2 in. (c) 500-kg load, 10-mm ball.

Table 43 Heat treatments for alloy 771.0

Temper	Treatment
T2	Hold at 415 ± 14 °C (775 ± 25 °F) for 5 h; cool outside furnace in still air to room temperature; harden by reheating to 180 ± 3 °C (360 ± 5 °F) for 4 h; cool in air
T5	Hold at 180 ± 3 °C (355 ± 5 °F) for 3 to 5 h; cool outside furnace in still air to room temperature
T6	Hold at 580 to 595 °C (1080 to 1100 °F) for 6 h; cool outside furnace to room temperature in still air; age by holding for 3 h at 130 °C (265 °F) followed by cooling in still air
T51	Age by holding at 205 °C (405 °F) for 6 h; cool in still air
T52	Hold at 415 °C (775 ± 25 °F) for 5 h; cool from 415 to 345 °C (775 to 650 °F) in 2 h or more; cool from 345 to 230 °C (650 to 450 °F) in not more than ½ h (20 min desirable); cool from 230 to 120 °C (450 to 250 °F) in approximately 2 h; cool from 120 °C (250 °F) to room temperature in still air outside of furnace; harden by reheating to 165 °C (330 °F) for 6 to 16 h and cooling outside of furnace in still air
T71	Hold at 580 to 595 °C (1080 to 1100 °F) for 6 h; cool outside furnace to room temperature in still air; age by holding at 140 °C (285 °F) for 15 h followed by cooling in still air. Similar properties can be obtained by aging at 155 °C (310 °F) for 3 h

Fabrication Characteristics

Machinability. 771.0-T5 has good stability and machinability. It can be milled five times faster and hole worked at twice the speed of alloys such as 356.0 and 319.0. It can be finished machined in one clamping operation to flatness tolerance of 0.001 in. This reduces total cost of machining over most casting alloys, which require two clamping operations to obtain this type of flatness tolerance.

Welding. Can be welded by either gas tungsten-arc or gas metal-arc welding using 5356 rod or wire. Special procedure should be followed in welding to ensure good results.

If parts are to be welded, the operation should be made part of the heat treating cycle. If welding is to be done on T6 or T71 parts, the castings are heated to 580 °C (1080 °F), removed from the heat treating furnace and welded while hot. The parts are then returned to the furnace and the T6 and T71 heat treatments continued. If the parts are to be used in the T52 or T2 temper, they are heat-ed to 415 °C (775 °F), taken from the furnace, welded hot, then returned to the furnace and the heat treatment continued. Items to be used in the T51 temper are heated to 205 °C (405 °F), taken from the furnace, welded hot, then returned to the furnace and T51 treatment continued. Repair weld parts should be heated and welded as described above.

The T5 temper should not be welded but can be welded if the procedure for T51 is used.

Heat treatments. See Table 43.

850.0
6.2Sn-1Cu-1Ni

Commercial Names

Former designation. 750

Specifications

AMS. Permanent mold casting: 4275
UNS number. A08500
Government. QQ-A-596 (class 15)

Chemical Composition

Composition limits. 0.7 to 1.3 Cu; 0.10 max Mg; 0.10 max Mn; 0.7 max Si; 0.7 max Fe; 5.5 to 7.0 Sn; 0.7 to 1.3 Ni; 0.20 max Ti; 0.30 max others (total); rem Al

Consequence of exceeding impurity limits. High iron, manganese, or magnesium decreases ductility and increases hardness. High silicon modifies bearing characteristics.

Applications

Typical uses. Applications where excellent bearing qualities are required

Mechanical Properties

Tensile properties. Typical for T5 temper: tensile strength, 160 MPa (23 ksi); yield strength, 75 MPa (11 ksi); elongation in 50 mm or 2 in., 10%
Shear strength. 103 MPa (15 ksi)

Compressive yield strength. 75 MPa (11 ksi)

Hardness. T5 temper: 45 HB (500-kg load, 10-mm ball)
Poisson's ratio. 0.33

Elastic modulus. Tension, 71.0 GPa (10.3×10^6 psi); shear, 26.5 GPa (3.85×10^6 psi)

Fatigue strength. 60 MPa (9 ksi) at 5×10^8 cycles (R. R. Moore type test)

Mass Characteristics

Density. 2.880 Mg/m^3 (0.104 lb/in.^{3}) at 20 °C (68 °F)

Thermal Properties

Liquidus temperature. 650 °C (1200 °F)

Solidus temperature. 225 °C (435 °F)

Coefficient of thermal expansion. Linear:

Temperature range		Average coefficient	
°C	°F	μm/ m·K	μin./ in.·°F
20–100	68–212 .... 23.1		12.8
20–200	68–392 .... 24.3		13.5

Specific heat. 963 J/kg·K (0.230 Btu/lb·°F) at 100 °C (212 °F)
Latent heat of fusion. 389 kJ/kg (167 Btu/lb)
Thermal conductivity. 180 W/m·K (104 Btu/ft·h·°F)

Electrical Properties

Electrical conductivity. Volumetric, 47% IACS at 20 °C (68 °F)
Electrical resistivity. 36.7 nΩ·m at 20 °C (68 °F)

Fabrication Characteristics

Melting temperature. 650 to 730 °C (1200 to 1350 °F)
Casting temperature. 650 to 705 °C (1200 to 1300 °F)
Aging temperature. 230 °C (450 °F); hold at temperature for 8 h

Stamping of Aluminum Alloy Sheet

By R. J. Traficante
and
W. L. Weeks
Chrysler Corporation

ALUMINUM SHEET has only recently been used in applications that require high-volume forming techniques such as mechanical stamping with hard tooling. High strength-to-weight ratio and excellent corrosion resistance are the primary engineering advantages of aluminum over low-carbon steel in such applications. These advantages have made aluminum sheet attractive to the automobile industry, where obligations to meet governmental mileage standards now have largely overcome the higher initial cost of aluminum.

In evaluating the ease with which a particular stamping can be formed from aluminum sheet, three basic forming parameters—the shape of the part, the specific alloy and the tooling (or process)—should be considered. Complex forming can be analyzed by evaluating severely formed regions as individual symmetrical shapes and then relating these shapes to laboratory cup testing to generate an over-all severity rating for the part. Because most automobile panels are formed in one operation (except for subsequent trimming and flanging), attention in this article is focused primarily on single-stage forming.

Types of Forming

The principal types of forming encountered in high-volume production stamping are:

1 *Pure bending*—such as simple bending in a press brake
2 *Stretch bending*—used for parts featuring ribs and troughs
3 *Flanging*—such as wiping an overhanging edge down over a bend line, either straight or curved; the flange edge is in either stretch or compression in a curved bend line
4 *Biaxial stretching*—in which metal is strained under tension in two directions in the plane of the sheet, as occurs over a punch nose
5 *Deep drawing*—in which blank flanges outside a die cavity are drawn into the cavity, which places flanges in compression.

Typically, high-volume stamping involves combinations of these types of forming. For instance, in forming a cup shape, the material over the punch nose will be stretched, while the rest of the material will be subjected to drawing in. A bracket may combine bending with flange stretching or compression. The relative percentages of each type of forming will be determined by the shape of the part and the process chosen to arrive at that shape. The shapes of most complex parts can be broken down into components having more simple shapes producible by the types of forming listed above. Analyses of the simplified components can be made and then added to obtain a final analysis of the part. A more detailed discussion of this approach to assessment of formability is presented later in this article, with examples of its application to production parts.

Material Properties and Tests

To conduct a complete analysis of a part, the required material properties, as determined by several standard tests, must be considered. These properties include those determined by the tensile test and by other tests designed to simulate various production forming processes. These simulative tests include the cup test, the bend test and the hole-expansion test.

The tensile test is used to determine the commonly reported properties—(ultimate) tensile strength, (tensile) yield strength and (total) elongation—as well as two properties especially important in forming (Ref 1). These latter two are the strain-hardening exponent, n, and the plastic strain ratio, r.

The strain-hardening exponent of a material is determined from the true stress–true strain curve for that material according to the formula:

$$\sigma = K\epsilon^n \qquad \text{(Eq 1)}$$

where σ is true stress, ϵ is true strain and K is the constant of proportionality.

The plastic strain ratio describes the resistance of the material to thinning during forming operations and equals the ratio of the true strain in the width direction (ϵ_w) to the true strain in the thickness direction (ϵ_t) of plastically strained sheet.

$$r = \frac{\epsilon_w}{\epsilon_t} \qquad \text{(Eq 2)}$$

A standard method of determining r using a tension specimen is given in ASTM E517.

The tensile properties (and other mechanical properties) of many aluminum sheet alloys in medium and hard tempers exhibit directional sensitivity. The test direction should be reported along with test results. Directional sensitivity is important in analysis of forming operations that involve bending, flange stretching or plane straining, which are encountered in forming of ribs and troughs. Orientation of the rolling direction of the sheet relative to the direction of critical strain in the part often can mean the difference between producing a good part and producing scrap. For circular-cup analysis of stretch/draw forming (forming involving combined stretching and drawing), it has been found that planar averages of mechanical properties work best. The planar average for any property is equal to the sum of the values of that property in the longitudinal and transverse directions plus twice the value in the diagonal direction (45° to the longitudinal), all divided by four:

$$\text{planar average} = \frac{\text{long.} + \text{trans} + 2\,\text{diag}}{4} \qquad \text{(Eq 3)}$$

A planar average value often is indicated by a bar over the symbol for the property (for example, $\bar{r}$). However, the planar average of the plastic strain ratio also is indicated by r_m.

Typical tensile properties of several aluminum sheet alloys are presented in Table 1.

The Olsen cup test is a biaxial-stretch-forming test that has been in use for 75 years (Ref 2). A specimen is stretched over a 7/8-in.-diam ball that is lubricated by a small disk of oiled polyethylene. The flange of the specimen is tightly clamped. Maximum cup height is measured when necking occurs. The cup-test value reported is the ratio of cup height to cup diameter and is given the symbol O_d. Typical values of O_d are given in Table 1.

The Swift Cup Test. In the Swift cup test, a deep drawn cup is used to determine the limiting draw ratio, LDR, of blank size to cup diameter. It is obtained with a 2.0-in.-diam flat-bottom punch and a draw die appropriate for the thickness of the specimen. A circular blank is cut to a diameter smaller than the expected draw limit. Lubrication is provided by two oiled polyethylene disks, one on each side of the blank. The blank is drawn to maximum punch load, which occurs before the cup is fully formed. Successively larger blanks are drawn until one fractures before being drawn completely through the die. The diameter of the largest blank that can be drawn without fracturing, divided by cup diameter, determines the limiting draw ratio (LDR). Typical values of LDR are given in Table 1.

A predictive version of this test uses a single blank size, which is drawn to maximum forming load, flange clamped and then drawn to fracture (Ref 3). The limiting blank diameter (LBD) is calculated as follows:

Table 1 Typical mechanical properties of several aluminum sheet alloys (a)

Alloy and temper	Thickness mm	in.	Ultimate MPa	ksi	Yield MPa	ksi	$\bar{E}$, %	$\bar{n}$	$\bar{r}$	Olsen, O_d	Swift, LDR	Longi-tudinal	Trans-verse
2036-T4	0.99	0.039	349	50.6	203	29.5	23	0.226	0.75	0.306	2.10	1t	2t
3003-O	0.89	0.035	107	15.5	52	7.6	33	0.235	0.66	0.300	2.04	0	0
5052-O	0.86	0.034	211	30.6	97	14.0	22	0.282	0.62	0.316	2.08	0	0
5056-O	0.86	0.034	283	41.0	147	21.3	27	0.279	0.79	0.325	2.16	0	0
5086-O	1.63	0.064	284	41.2	136	19.7	24	0.359	0.68	0.365	2.06	2t	2t
5086-H32	1.63	0.064	303	43.9	203	29.5	16	0.291	0.71	0.340	2.01	2t	2t
5182-O	1.02	0.040	280	40.6	132	19.2	26	0.340	0.78	0.340	2.09	0	½t
5252-H25	0.79	0.031	247	35.8	192	27.8	14	0.140	1.04	0.216	1.98	⋯	⋯
6009-T4	1.32	0.052	261	37.8	150	21.7	24	0.264	0.68	0.330	2.04	0	½t
6010-T4	0.81	0.032	296	42.9	170	24.6	27	0.260	0.61	0.320	2.04	½t	½t
6061-T4	1.02	0.040	301	43.6	193	28.0	23	0.194	0.71	0.290	2.08	2t	2t
6151-T4	1.02	0.040	279	40.5	180	26.1	21	0.195	0.62	0.305	2.07	2t	2t
7021-O	2.36	0.093	195	28.3	132	19.1	20	0.202	0.990	0.345	2.15	2t	2t
7021-O	4.75	0.187	215	31.2	140	20.3	20	0.176	0.372	⋯	⋯	2t	2t
7029-O	2.56	0.101	273	39.6	123	17.9	22	0.298	0.611	0.293	2.15	2t	2t
7029-O	5.08	0.200	228	33.0	110	15.9	25	0.264	0.544	0.377	⋯	2t	2t
7029-W	2.54	0.100	222	32.2	79	11.4	26	0.428	⋯	0.387	2.30	2t	2t
7146-O	4.72	0.186	174	25.2	133	19.3	21	0.122	0.680	0.370	⋯	2t	2t
7146-O	2.54	0.100	169	24.5	137	19.8	19	0.123	0.642	0.347	2.26	2t	2t

(a) See text for explanation of column headings. (b) Values shown are planar averages of tensile strength ($\bar{T}$), yield strength ($\bar{Y}$), elongation ($\bar{E}$), strain-hardening exponent ($\bar{n}$) and plastic strain ratio ($\bar{r}$).

$$LBD = \frac{\text{fracture load}}{\text{maximum drawing load}} \times$$

$$\text{(blank diam } - \text{ die diam)} +$$

$$\text{die diam} \quad \text{(Eq 4)}$$

The limiting draw ratio can then be calculated from the limiting blank diameter as follows:

$$LDR = \frac{\text{limiting blank diam}}{\text{punch diam}} \quad \text{(Eq 5)}$$

This predictive test is valid for annealed tempers, but loses accuracy with increasingly harder tempers.

The Bend Test. In the bend test, strips of material are bent around mandrels with different tip radii. Often, mandrels are in the form of pins or rods. The mandrel is forced against one side of the specimen strip and the other side is simply supported at the end points. The value reported is the minimum radius of mandrel, in multiples of the material thickness (t), around which the material can be bent 180° without cracking. The direction of the bend relative to the rolling (or extrusion) direction should be indicated along with the test results. Typical minimum radii are given in Table 1.

The Hole-Expansion Test. Limits of flange-edge strain are influenced by plastic strain ratio (r value) and tensile elongation. The laboratory test used to determine flange stretch limits is the hole-expansion test, in which a cleanly blanked hole is expanded by a conical punch until cracking occurs around the hole. The value reported (e_f) is related to the ratio of final hole radius (R_2) to initial hole radius (R_1) according to the expression:

$$e_f = \frac{R_2}{R_1} - 1 \quad \text{(Eq 6)}$$

Correlation Between Test Results. It has been found that results of simulative forming tests correlate quite well with results of tension tests. Specifically, results of cup ductility tests, such as the Olsen and Swift tests, show good correlation with values of tensile elongation, strain-hardening exponent and plastic strain ratio. Olsen cup values correlate well with tensile elongation, and Swift cup values with plastic strain ratio. The closest correlations, however, are between cup values and combined tensile-test values, as shown in Equations 7 and 8 (Ref 4).

The equation for the Swift cup parameter is:

$$LDR = 1.93 + 0.00216 \, \bar{E} +$$

$$0.226 \, \bar{r} \quad \text{(Eq 7)}$$

The correlation coefficient of this equation is 0.835 and is valid for LDR values up to 2.5, which was the limit of the study.

The equation for the Olsen cup parameter is:

$$O_d = 0.217 + 0.0074 \, \bar{E} +$$

$$0.00392 \, \bar{r} \quad \text{(Eq 8)}$$

The correlation coefficient of this equation is 0.925.

Although the correlations of these equations are good, and they can be used to some extent to predict cup-test values, they are not presented as a substitute for direct determination of cup-test values, which should be performed if at all possible.

Correlation with Production Forming. The material properties that apply to flange stretching are tensile elongation and plastic strain ratio. Close correlation between these tensile properties and the flange stretch limit, e_f, has not yet been achieved due to the strong effects of burring and work hardening of the sheared edge during blanking. Flanging tests were performed on blanked specimens of several 2xxx, 3xxx, 5xxx and 6xxx sheet alloys used for automotive stamping, which contained burrs with a height/stock ratio of 0.10 max, at stock thicknesses below 1.3 mm (0.050 in.). All heat treatable alloys were in the T4 temper (solution heat treated and naturally aged); the 5xxx and 3xxx alloys were in the O (annealed) temper. Unit edge strains of up to 0.35 were achieved, with a working average of 0.20. 7xxx series alloys in thick stock (thicker than 2.54 mm, or 0.100 in.) showed an average of 0.35.

Limiting edge strain, like tensile elongation, is lower for harder tempers. For example, a fully hard 5xxx alloy has a strain limit of 0.03. Programs are currently underway to find better correlations between tensile-test results and flange stretch limits.

Ease of forming ribs and troughs in parts can best be predicted from results of tests that produce conditions of plane strain. Plane strain is the condition of straining in which the minor strain is zero. This condition is found at or close to the low midpoint on forming-limit diagrams, which are described below.

Forming-limit diagrams, also known as forming-limit curves, are direct and useful representations of the formability of aluminum sheet. These diagrams illustrate the biaxial combinations of strain that can occur without failure.

SAE Recommended Practice J863c describes one method of constructing forming-limit diagrams. An array of circles, which often are 2.5 mm (0.1 in.) in diameter, is imprinted by photoprinting, photoetching or electroetching on the surface of the steel sheet

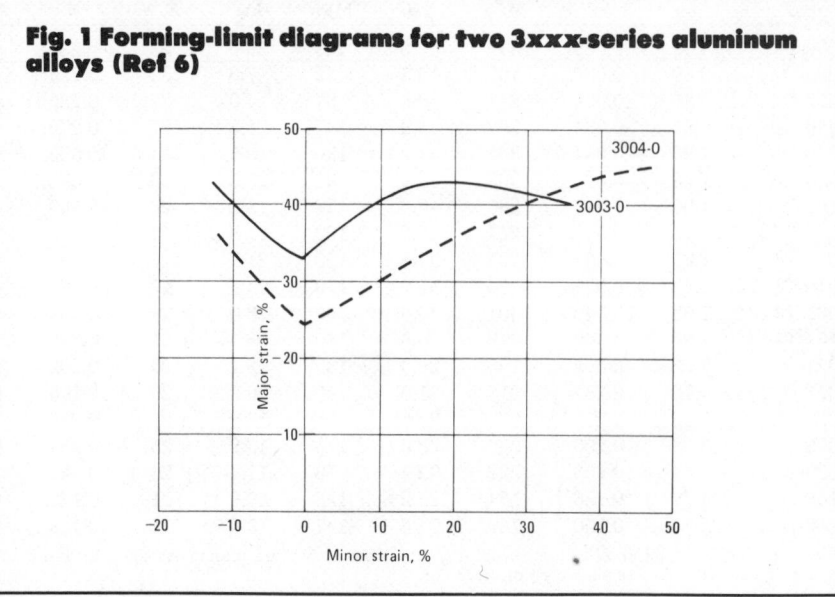

Fig. 1 Forming-limit diagrams for two 3xxx-series aluminum alloys (Ref 6)

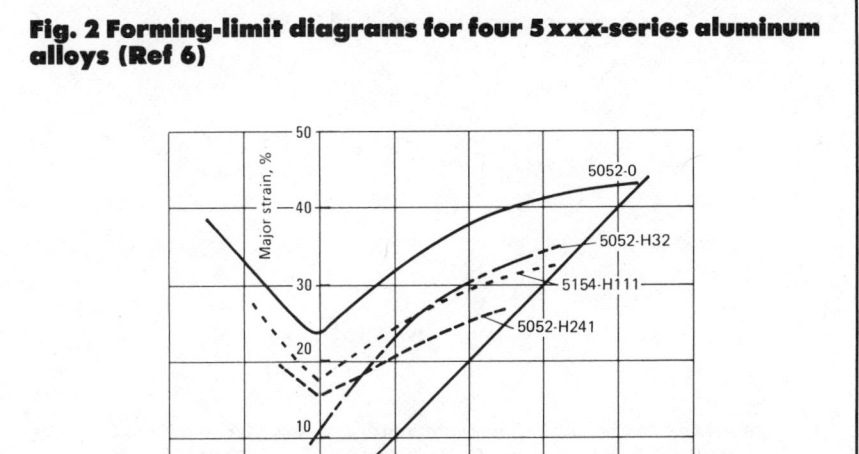

Fig. 2 Forming-limit diagrams for four 5xxx-series aluminum alloys (Ref 6)

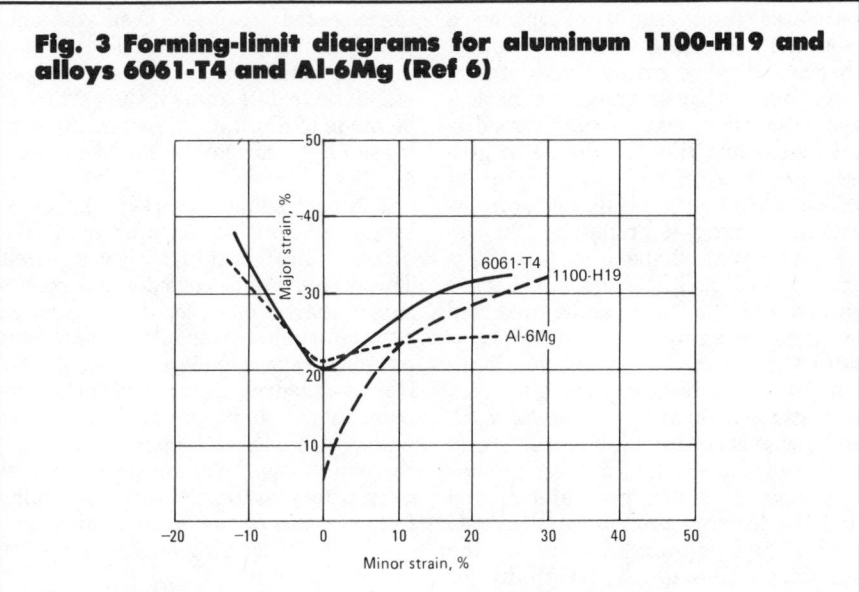

Fig. 3 Forming-limit diagrams for aluminum 1100-H19 and alloys 6061-T4 and Al-6Mg (Ref 6)

seen from the curves in these diagrams that a universal forming-limit curve for aluminum alloys does not exist.

Forming Characteristics of Aluminum

Aluminum stampings often are considered replacements for stampings of low-carbon steel. Therefore, the forming characteristics of the two metals often are compared. Choosing between an aluminum alloy and a low-carbon steel for a particular application requires detailed analysis, but several general comments can be made:

1   Formability of medium-strength aluminum alloys in deep draw and biaxial-stretch cup-type forming operations is about two-thirds that of low-carbon steel.

2   Minimum bend radii are approximately three times those for steel. This lower bendability is related to aluminum's characteristically low reduction in area; it cannot be severely strained in local areas that have sharp formations.

3   Aluminum's high notch sensitivity requires that blanking tools—particularly those with sharp edges—be designed for close tolerances. The blanking and shearing characteristics of 7xxx series aluminum alloys require greater clearances than those for steel. Tools must be sharp and precise to minimize formation of burrs and thus reduce edge-splitting tendencies in subsequent bending or flange-stretching operations. Lancing of blanks, to improve interior metal flow, should be avoided.

4   In mechanical presses, the highest speed in the stroke cycle occurs during workpiece contact. The low strain-rate sensitivity of aluminum creates high stresses in the metal during initial metal movement, especially during deep drawing. It is compensated for by lower blank-hold-down pressure, increased draw-ring and punch-nose radii, and by use of lubricants formulated for aluminum. Aluminum is sensitive to lubrication as a result of its abrasive oxide layer and the fact that sheet products are received dry.

5   Because the elastic modulus of aluminum is lower than that of steel, formed aluminum panels exhibit more elastic recovery, or springback, than formed steel panels.

before it is formed. The individual circles become ellipses wherever deformation occurs, except in areas where pure biaxial stretching occurs. The major and minor axes of the ellipses are compared with the circles of the original grid to determine the major and minor strains at each location. The areas immediately adjacent to failures are of particular concern in evaluating the forming capabilities of the metal. Failure can be defined by several criteria, but the onset of a visible neck is the most widely used criterion. The loci of strain combinations that produce failures define the forming-limit curve. The area below this curve encompasses all the combinations of strain that the metal can withstand.

Forming-limit diagrams for two annealed Al-Mn alloys (3003-O and 3004-O) are presented in Fig. 1; the diagrams for four Al-Mg alloys with different levels of work hardening (5052-O, 5154-H111, 5052-H241 and 5052-H32) are shown in Fig. 2; and the diagrams for fully hardened aluminum (1100-H19), a lightly rolled Al-6Mg alloy and a heat treated Cu-Mg-Si alloy (6001-T4) are presented in Fig. 3. It may be

Fig. 4 Stress-strain forming path

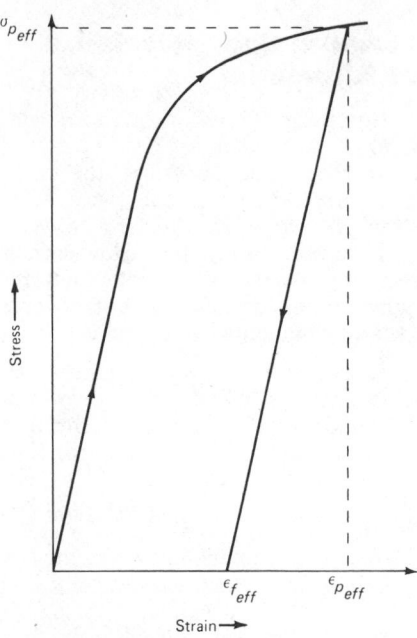

See text for discussion.

Fig. 5 Fuel-tank upper stamping (Ref 8, 9)

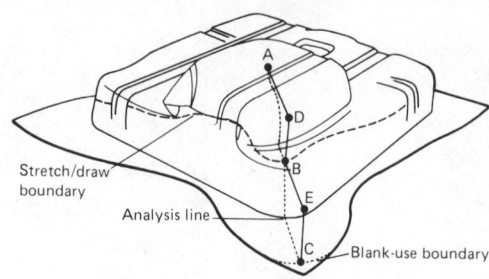

I_t is point A, intersection of analysis line with stretch/draw boundary is point B, O_t is point C, L_t is distance AB, L_d is distance BC, C_{hs} is distance BD, C_{hd} is distance CE, L_{os} is distance AD, and L_{od} is distance BE. See Table 2 for identification of symbols.

This must be compensated for by increasing overcrown in the draw die and/or incorporating locking beads in the binder, to ensure that all material has been "set" by plastic deformation. When the metal is plastically deformed, there will always be a component of elastic recovery. The amount of recovery depends on total plastic strain, as shown in Fig. 4. Of the total strain induced by tooling during forming, $\epsilon_{p_{eff}}$, there is elastic recovery back to $\epsilon_{f_{eff}}$ after forming. Elastic recovery is discussed in further detail in Ref 7.

Predicting Formability of Production Parts

The technique of analyzing forming severity of sheet-metal stampings by use of data from simulative laboratory tests is termed "shape analysis" (Ref 7). Such analysis is applied to those critically formed areas of the stamping that establish its over-all formability. Generally, critical locations are those that involve combinations of stretch and draw. In shape analysis, these stretch/draw formations are broken down into combinations of spherical and cylindrical sections whose formabilities can be estimated from results of optimized laboratory forming tests of similar shapes. Forming ratios that describe the shapes of these simpler spherical and cylindrical sections can be readily calculated and then combined to give an over-all severity rating.

This technique recognizes the interaction of material properties, die design, and part shape in producing a final stamping. Given any two of these parameters, the third can be predicted using shape analysis. For example, it can be assumed that a specific aluminum alloy has been chosen, after considerations of such factors as strength and stress-corrosion resistance. It also can be assumed that die design has been optimized for the material and that the forming process has been selected. The task then is to evaluate the adequacy of the alloy's formability for producing the part shape. This approach is illustrated in the section that follows, in which a stretch/draw application is discussed.

There are, however, limitations to shape analysis. When the shape deviates greatly from simple spherical or cylindrical sections, analysis is much more difficult. Multiple-draw forming is significantly more complex to analyze than single-stage forming. Shape analysis cannot predict local wrinkling and buckling phenomena; it can, however, be used to evaluate the effects on forming severity by tooling adjustments made to control wrinkling.

Stretch/Draw Forming. The upper half of a fuel tank (Fig. 5) was made of terne-coated steel and then was subjected to shape analysis to optimize forming conditions. A second analysis was done to determine if the part could be made of aluminum. Development of these analyses may be found in Ref 8 and 9.

The part shape was reviewed to establish the critical forming area. The corner indicated in Fig. 5 was selected. This corner can be considered to constitute a quarter of a deep drawn circular cup. (A shelf, or shoulder, part way down the corner made determination of the stretch/draw boundary more complex, but the shape was considered to be within a reasonable approximation of the cup models.) An analysis line was drawn diagonally across the stamping from crest to corner so that (ideally) a symmetry of forming strain existed on either side of it.

The inner terminal of the analysis line, I_t (point A in Fig. 5), was chosen as the effective cup center. The analysis defines it as the point on the analysis line where the forming strain approaches zero. It will be found inside the horizontal corner radius. As the punch shoulder radius increases (more domed) it will move inward toward the geometric center of the dome. Due to the large radii of this part's domed shoulder and corner, a point on the analysis line close to the geometric center was chosen. Large flat areas on the noses of punches show areas of zero forming strain, moving the inner terminal toward the center of the horizontal corner radius.

Table 2 Identification of symbols used in shape analysis of fuel-tank stamping

Symbol	Identification
I_t	Analysis-line inner terminal
O_t	Analysis-line outer terminal
L_s	Final analysis-line length in stretch
L_d	Final analysis-line length in draw
L_{os}	Original analysis-line length in stretch
L_{od}	Original analysis-line length in draw
C_{hs}	Stretch cup height
C_{hd}	Draw cup height
L_{ds}	Final analysis-line length
L_o	Original analysis-line length
STR	Amount of forming contributed by stretching
$DRAW$	Amount of forming contributed by drawing
C_{ws}	Stretch cup width at stretch/draw boundary
C_{wd}	Draw cup width at die ring
E_{bw}	Effective blank width
R_s	Stretch cup ratio
R_d	Draw cup ratio
O_d	Olsen cup-test value
LDR	Swift limiting draw ratio
SEV	Forming severity

The outer terminal, O_t (point C in Fig. 5) was located at the effective outer periphery of the blank, rather than at the tip of the corner. In rectangular blanks, a region at the tip of each corner undergoes no measurable forming strain and the stresses are low and balanced. Therefore, this region is considered to have an insignificant effect on the forming operation. This region can be defined by a blank-use boundary, constructed by projecting the radius at the part base to the tangent points at the flange edges, as shown in Fig. 5.

The stretch/draw boundary, which is the boundary between the metal initially suspended over the die cavity (and thus stretched over the punch nose) and the metal in the flange that moved into the die cavity over the die ring, was readily seen as an impact mark caused by movement against the die surface. This boundary also is the transition between biaxial stretch and draw.

A series of measurements was then made on the steel stamping (see Tables 2 and 3, and Fig. 5). The distance, L_s, along the analysis line from the inner terminal to the stretch/draw boundary

and the distance, L_d, from the stretch/draw boundary to the outer terminal, are measured. These distances represent final formed lengths. The original values of these distances, L_{os} and L_{od}, are determined by horizontal projection of the appropriate sections of the analysis line (see Fig. 5). By vertical projection of these sections to vertical lines, the stretch cup height, C_{hs}, and the draw cup height, C_{hd}, can be determined, the stretch cup being measured from inner terminal to stretch/draw boundary and the draw cup from stretch/draw boundary to die ring.

The relative contributions of stretch (STR) and draw ($DRAW$) to lengthening of the analysis line were calculated by comparing the lengths of lines on the part with their equivalent original lengths from the projections. The equations for these calculations are:

$$STR = \frac{L_s - L_{os}}{L_{ds} - L_o} \qquad (Eq\ 9)$$

and

$$DRAW = \frac{L_d - L_{od}}{L_{ds} - L_o} \qquad (Eq\ 10)$$

where

$$L_{ds} = L_d + L_s \qquad (Eq\ 11)$$

and

$$L_o = L_{od} + L_{os} \qquad (Eq\ 12)$$

The cup-forming ratios in stretch, R_s, and in draw, R_d are then calculated:

$$R_s = \frac{C_{hs}}{C_{ws}} \qquad (Eq\ 13)$$

$$R_d = \frac{E_{bw}}{C_{wd}} \qquad (Eq\ 14)$$

where C_{ws} is stretch cup width at the stretch/draw boundary, C_{wd} is draw cup width at the die ring, and E_{bw} is effective blank width. Because the analysis line shows the profile of only half a cup, the measured value of C_{ws} for the entire cup is approximately twice L_{os}. Similarly, the measured value of C_{wd} is approximately twice the value of L_{od}, and the measured value of E_{bw} is approximately twice the value of L_o.

Finally, the forming severity, SEV, is calculated by a two-step formula. First, the amounts of forming contributed by stretching and drawing are combined in proportion to their line-lengthening ratios. Then, the combined value is compared with the results of the laboratory cup tests that give the limits of stretch and draw for the mate-

rial under consideration. The laboratory cup-test values, which were discussed earlier, are the Olsen cup-test value, O_d, which measures stretch, and the Swift limiting draw ratio, LDR. Forming severity is calculated as follows:

$$SEV = [(R_s \times STR) + (R_d \times DRAW) - DRAW] \div [DRAW (LDR - O_d) + O_d - DRAW] \qquad (Eq\ 15)$$

For the terne-coated steel, for which O_d = 0.375 and LDR = 2.28, forming severity was calculated to be 0.65. If the stamping were to be made of aluminum alloy 5182-O, for which O_d = 0.340 and LDR = 2.09 (see Table 1), forming severity would be 0.76. (The measured and calculated shape-analysis values for the stamping are listed in Table 3.) For this value of forming severity to be valid, process variables such as blankholder pressure and lubrication would have to be adjusted so that the stretch/draw boundary would be in the same location in the aluminum as in the steel. If the boundary is not in the same location, a new set of calculations must be performed on the basis of the new boundary, and these calculations would indicate the change

Table 3 Shape-analysis values for steel and aluminum fuel-tank upper stampings(a)

Symbol	Value
L_s	10.75 in.
L_d	10.35 in.
L_{os}	9.60 in.
L_{od}	8.43 in.
C_{hs}	3.34 in.
L_{ds}	21.10 in.
L_o	18.03 in.
STR	0.375
$DRAW$	0.625
C_{ws}	17.48 in.
C_{wd}	19.80 in.
E_{bw}	37.02 in.
R_s	0.191
R_d	1.87
Steel	
O_d	0.375
LDR	2.28
SEV	0.65
Aluminum	
O_d	0.340
LDR	2.09
SEV	0.76

(a) Terne-coated steel and 5182-O aluminum.

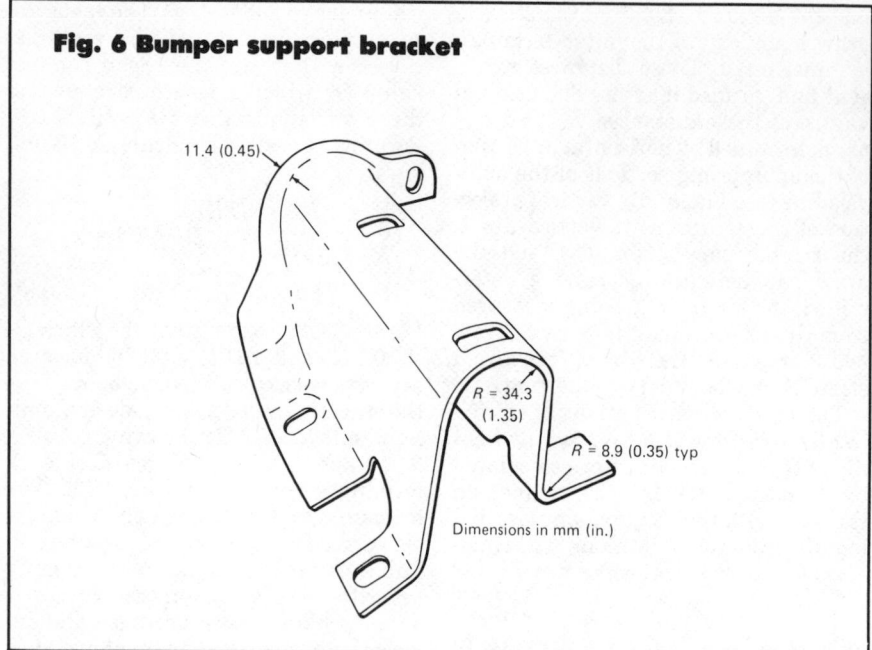

Fig. 6 Bumper support bracket

11.4 (0.45)

R = 34.3 (1.35)

R = 8.9 (0.35) typ

Dimensions in mm (in.)

in severity due to the changes in tooling variables.

Shape analysis indicated clearly that the aluminum alloy was suitable for forming the fuel-tank stamping, and stampings were satisfactorily formed.

Bending and Flange Stretching. An application of bending and flange stretching is the bumper-support bracket shown in Fig. 6, which was formed from 4.33-mm (0.170-in.) thick alloy 7021-O sheet. Shape requirements for the design of this part were established by the bend-radius limits and flange-stretch limits. The forming process consisted of bending a flat blank into a U shape. At one end, a flange was stretched to provide stiffness, and at the other, attaching tabs were bent. The radius of the U bend was 34.3 mm (1.35 in.). Flange stretch limits for the alloy were determined to be $e_f = 0.35$, a value common to thick sheets of the aluminum alloys evaluated to date,

$$e_f = \frac{R_2}{R_1} - 1 \qquad \text{(Eq 16)}$$

where R_2 is the flange edge radius and R_1 is the flange bend-line radius. With R_1 the radius of the U bend, and e_f of 0.35, then the highest flange cannot exceed a radius, R_2, of 46.2 mm (1.82 in.), or a height of 11.9 mm (0.47 in.). A recommendation was made to the de-

signers to limit flange height to less than 11.4 mm (0.45 in.). A typical minimum bend radius for 7021 alloy is $2t$, as shown in Table 1. For a stock thickness of 4.32 mm (0.170 in.), a typical minimum bend radius of 8.64 mm (0.340 in.) was calculated and a design minimum inside bend radius of 8.9 mm (0.35 in.) was recommended for the part. Subsequent production experience verified both recommendations.

Production Forming

It is possible to form aluminum into complex stampings such as the fuel-tank half in Fig. 5 and the bumper bracket in Fig. 6 if the steps necessary to accommodate the material's specific properties (discussed earlier) are taken. Radii must be generous, and flange heights must not be excessive. The significance of good blanking procedures and proper lubrication with uniform coverage cannot be overemphasized. The lubricant applied, the surface roughness of both the sheet and the die, and any dirt present in the tooling will all influence drawing friction. A local increase in friction can significantly alter the strain pattern in the stamping.

Variations in mechanical properties occur due to standard tolerances in material characteristics such as compo-

sition, temper, thickness, and rolling and coiling parameters. The effects of such factors must be considered when stampings are analyzed using cup-test and tensile-test values determined for "typical" materials. The closer the strains in these critical areas approach the forming limits of the typical material the greater the expected scrap rate.

When any two of the basic parameters have been selected, such as alloy and forming process, the third variable (in this instance, shape) can be predicted. It often is possible to modify the design slightly to alleviate these local difficulties. By considering severity of both local and over-all forming early in the design program, aluminum can be used more efficiently, saving both material and costly development time.

REFERENCES

1. A. S. Kasper, Forming Sheet Metal Parts, Paper M. F. 69-516, *ASTME Creative Manufacturing Seminar,* Chicago, Jan 1969
2. "Comparison of Olsen Cup Values on Aluminum Alloys", Aluminum Association Publication T13, Feb 1975
3. D. V. Wilson, B. J. Sunter and D. F. Martin, A Single-Blank Test For Drawability, *Sheet Metal Industries,* Vol 43 (No. 470), 1966, p 465-476
4. A. S. Kasper, How We Will Predict Sheet Metal Formability, *Metal Progress,* Vol 96 (No. 5), Oct 1969, p 159-164
5. P. N. Richards, Deep Drawing—A Sketch of Modern Development, *Proceedings of 1975 Metals Congress,* Australian Institute of Metals, p 27-62
6. S. S. Hecker, A Simple Forming Limit Curve Technique & Results on Aluminum Alloys, *IDDRG Congress,* Oct 1972
7. D. G. Adams, A. S. Kasper and G. M. Kurajian, "Springback Analysis of Biaxially Stretched Panels", SAE Publication 730529
8. A. S. Kasper and P. J. Vanderveen, "A New Method of Predicting the Formability of Materials", SAE Publication 720019
9. A. S. Kasper, D. G. Adams and J. A. DiCello, Sheet Metal Forming Limits with Manufacturing Applications, *Proceedings of Sagamore Army Materials Research Conference,* Aug 1974

Machinability of Aluminum Alloys

By B. Chamberlain
Aluminum Company of Canada, Ltd.

MACHINABILITY of aluminum alloys varies widely with composition and temper. However, all aluminum alloys, wrought or cast, have far better machinability compared with steel or titanium, and they can be machined under relatively simple cutting conditions. It is necessary, however, to select cutting conditions that will make machining of the alloy as economical as possible.

In evaluations of the machinability of aluminum alloys, three factors usually are considered—tool life (tool wear), chipping characteristics, and surface finish. Unfortunately, changing cutting conditions to improve one factor may be detrimental to one or both of the other factors.

Many of the chipping and surface-finish difficulties encountered in machining aluminum alloys can be eliminated by increasing cutting speed. For some applications, speeds of 35 to 50 m/s (7000 to 10 000 ft/min) are used to obtain smooth surfaces. In special cases, speed has been increased to 75 m/s (15 000 ft/min) to eliminate tearing. Tool wear, however, increases with cutting speed and can become intolerable at very high speeds. Usually, cutting conditions must be selected that result in reasonable compromises among the three machinability factors.

Tool Life

Condition of tool cutting surfaces is very important in controlling tool wear. In general, the softer aluminum alloys—and, to a lesser extent, some of the harder alloys—tend to weld to the tool face and cause a built-up edge. This built-up edge forms a new cutting surface, resulting in erratic chip formation and poor surface finish and causing overheating and premature failure of the tool. For best results, the tool is ground and lapped in the direction of chip travel so that the chips slide easily over the tool face. Wide flutes are also essential to give the chips unimpeded passage.

The balance between tool finishing cost and machining cost must be established for each operation on the basis of total operation economy. In one instance, the life of a high speed steel broaching tool was increased an average of 270% (7400 compared with 2000 pieces) when the cutting edge was finished by a wet, superfine sandblasting. In another instance, life of a carbide-tipped grooving tool lapped with a 350-grit wheel instead of with a 200-grit wheel increased 140 to 180%.

Lubrication is an important factor in tool life. The lubricant serves three functions: it lubricates the tool to provide ease of chip travel, cools both the tool and workpiece, and flushes the broken chips away from the workpiece. It is essential to use a good lubricant and to ensure that it is directed to the critical areas.

It is essential that feed be sufficiently heavy to ensure that the tool cuts, rather than rubs against, the workpiece. The workpiece, tools and tool holders must be rigidly mounted, and the cutting tool must contact the centerline of the workpiece wherever possible.

Choice of alloy and temper play an important part in tool life. Alloying elements in aluminum alloys can be either in solution, or out of solution as discrete particles. The constituents out of solution usually are harder than the aluminum matrix, and these harder constituents can cause tool wear. Constituents that are dissolved in the aluminum by solution heat treatment, and held there by rapid quenching, have little abrasive effect on tools. However, if they are allowed to grow as a result of slow solidification, poor quenching from solution heat treatment, or overaging, the tool is abraded by the constituent. For example, a little less tool wear occurs in machining alloy 2011 when it is in the T3 temper than when it is in the T8 temper. This is because the T8 temper necessitates an artificial aging treatment, which precipitates $CuAl_2$. Alloy 6061 in the O temper may contain relatively coarse Mg_2Si constituent; solution heat treating and quenching will take the Mg_2Si into solution and reduce tool wear. In 6061 in the solution heat treated and aged condition (T6 temper), the Mg_2Si is precipitated as fine constituent particles and tool wear is less than for the O temper.

In aluminum alloy castings, shape and size of constituents out of solution likewise are important, and increased tool life is obtained by proper control of solidification rate and heat treatment. The faster the solidification rate, the finer the constituents and the greater the tool life. (Solidification rates of the various types of casting methods are discussed in the article on aluminum foundry products in this volume.)

Silicon is one of the major alloying ingredients in many aluminum casting alloys (Al-Si, Al-Si-Cu, Al-Si-Mg and Al-Si-Mg-Cu types). Although the hard particles of free silicon in these alloys have no detrimental effects on chipping characteristics, they cause rapid tool wear. Both silicon content and silicon particle size directly influence the amount of tool wear. Increasing silicon particle size without changing silicon content, and increasing silicon content without changing particle size, both increase tool wear. Silicon particle size in hypoeutectic alloys (containing less than 11.6% Si) can be decreased by addition of a modifier such as sodium, strontium or calcium. In the hypereutectic silicon alloys, phosphorus is the most effective primary silicon nucleant, and its addition to the melt results in more numerous, but smaller, silicon particles. In both instances, refinement of silicon particles is very effective in reducing tool wear.

Iron, which normally is present in all aluminum casting alloys, forms hard particles, and increasing iron content increases tool wear. Deliberate iron additions made to overcome die soldering in the die casting process can be made less harmful to tool life by addition of up to 0.20% Mn; the presence of manganese both refines the iron constituent and inhibits die soldering.

Table 1 Machinability ratings of aluminum alloys

Alloy(a)	Temper(a)	Rating(b)
Casting Alloys		
208.0	F	C
242.0	T21, T571, T61, T77	B
295.0	T4, T6, T7, T62	B
308.0	F	C
319.0	F	C
	T5, T6, T7	C
354.0	T61, T62	C
355.0	T51, T6, T61, T62, T7, T71	C
C355.0	T61	B
356.0	T51, T6, T7, T71	C
A356.0, A357.0	T61	C
357.0	T6	C
359.0	T61, T62	C
360.0	F	C
A360.0	F	C
380.0, A380.0	F, T5	B
390.0, A390.0	T5	E
413.0, 443.0	F	E
514.0	F	A
518.0, B535.0	F	B
520.0	T4	A
712.0	F	A
713.0	F	A
850.0	T5	A
Wrought Alloys		
1100	O, H112, H12	E
	H14 to H18	D
1350	O, H111, H112, H12	E
	H14 to H19	D
2011	T3, T4, T6, T8	A
2014(c), 2124	O	C
	T3, T4, T6	B
2017	O	C
	T4	B
2024(c)	T3, T4, T6, T8	B
2036	T4	C
2219	T3, T6, T8	B
2618	T61	B
3003(c)	O, H112, H12	E
	H14 to H18	D
3004(c)	O, H112, H32	D
	H34 to H38	C
5005	O, H112, H12, H32	E
	H14 to H18	D
	H34 to H38	D
5050	O, H112, H32	D
	H34 to H38	C
5052	O, H112, H32	D
	H34 to H38	C
5056	O	D
	H18, H38	C
5083	O, H111, H116, H321, H323	D
	H131, H343	C
5086, 5154	O, H111, H116, H32	D
	H34 to H38	C
5182	O	D
	H19	C
5454, 5456	O, H112, H311	D
	H343	C
(continued)		

Table 1 (continued)

Alloy(a)	Temper(a)	Rating(b)
5457	O	D
	H25, H28, H38	C
5657	O	E
	H25, H28, H38	D
6061(c)	O	D
	T4, T6	C
6009, 6010	T4	D
6063	O, T2, T4	D
	T5, T6, T8	C
6262	T4, T9	B
6463	O, T1	D
	T4, T5, T6	C
7049, 7050	T73, T76	B
	T736, T76	B
7075(c)	T6, T73, T76	B
7175	T736	B
7178(c)	T6	B
7475	T6, T73, T76	B

(a) Alloys and tempers are those commonly used. Alloy modifications designated by other second digits and temper variations designated by added numerals will have the same ratings. (b) A, B, C, D and E are relative ratings in increasing order of chip length (see Fig. 1) and decreasing order of quality of finish and are defined as: A—Free cutting, very small broken chips and excellent finish; B—Curled or easily broken chips and good-to-excellent finish; C—Continuous chips and good finish; D—Continuous chips and satisfactory finish; E—Optimum tool design and machine settings required to obtain satisfactory control of chip and finish. (c) Includes clad alloys and tempers.

As in wrought aluminum alloys, the presence of either magnesium or copper in aluminum casting alloys increases tool wear through formation of relatively hard constituents. Constituents of either type (Mg or Cu) can be taken into solution by heat treatment, and in this manner tool wear can be decreased.

Chipping Characteristics

The chips that are easiest to handle are tight, easily broken individual coils that come cleanly from the workpiece and tool (see the photographs marked "A" in Fig. 1). Proper tool angles, tool finishes, cutting speeds, and feeds all are essential ingredients in machining operations that produce such chips, but these cutting conditions only assist the inherent nature of the alloy. Basic chipping characteristics are determined by alloying ingredients and to some degree by alloy temper. As in tool life, alloying elements in or out of solution play an important part.

In the O temper, the non-heat-treatable alloys (such as 1100, 3003 and 5052) offer little resistance to cutting, but form continuous chips and may require special lubricants or larger top and side rakes on tools to obtain good finish. Such alloys are easier to machine in the full-hard (cold worked) temper than in the annealed condition. Most heat treatable alloys also produce

continuous chips but usually are easily machined to good finishes.

Copper content of either wrought or cast alloys is a good indication of ease of chipping, and the 2xxx series alloys have the best relative machinability. One common 2xxx alloy—2011, which has a nominal copper content of 5.5% and both lead and bismuth additions—has been designed specifically for excellent machinability. Alloy 2011 gives the best performance of all aluminum alloys on automatic screw machine lathes, but all 2xxx alloys have superior chipping characteristics and are considered suitable for continuous machining on such equipment.

The 7xxx series alloys, which have high magnesium and zinc contents, are easily machined; however, because these alloys contain no copper, machining them produces continuous chips. The 5xxx series alloys, which have relatively high magnesium contents, are the next most easily machined. The 6xxx series alloys are relatively difficult to machine because the magnesium they contain is essentially tied up with silicon to form Mg$_2$Si particles. Alloy 6262, which was designed specifically for improved machinability and has a nominal composition of 0.6 Si, 0.28 Cu, 1.0 Mg, 0.09 Cr, 0.6 Bi and 0.6 Pb, offers some ease of machinability, but a magnesium-bismuth phase may form in this alloy and hamper its free-cutting characteristics. The alloys in

the 4xxx series contain high levels of silicon and are relatively difficult to machine; however, these are not the common structural wrought alloys and are used only for forgings and for welding and brazing filler metals.

The alloys in the 1xxx and 3xxx series are similar in their machining characteristics. These alloys are not heat treatable and have relatively soft, aluminum matrixes. Under normal machining conditions, the material appears to be torn rather than cut. Thus, for machining these alloys, it is essential to have sharp tools and highly polished faces to prevent buildup on the tool. These alloys are best machined at very low speeds, or at very high speeds (on the order of 10 m/s or 2 000 ft/min) using light cuts. Under either low or high speeds, excellent surface finishes can be obtained; with high-speed cutting, fine chips can be produced.

As in wrought aluminum alloys, copper is the favored addition to aluminum casting alloys to improve their chipping characteristics, and the 2xx.x alloys have very good machinability. The best machinability of the common casting alloys is found in alloy 850.x, which contains tin as well as copper and nickel.

The 7xx.x series casting alloys, which have high magnesium and zinc contents, have very good chipping characteristics. Alloys of the high-magnesium 5xx.x series, with the exception of alloys 511.0, 512.0 and 515.0 have almost as good machinability as the 7xx.x series alloys, as does alloy 222.0 (which has a high copper content). The remainder of the 2xx.x series alloys, and alloys 511.0, 512.0 and 515.0, have slightly reduced machinabilities, the amount of the reduction being directly proportional to their silicon contents.

Silicon does not directly adversely affect the chipping characteristics of aluminum casting alloys (silicon is a major constituent in alloys of the 3xx.x and 4xx.x series, and some alloys in the 2xx.x, 5xx.x and 8xx.x series). Silicon, however, is detrimental to tool life, and as a tool becomes dull, its ability to produce good chips and good surface finishes is decreased. Alloys containing more than 10% Si are the most difficult to machine, and none containing more than 5% Si will develop a bright lustrous surface finish. As a result of tool wear, chips in such alloys are likely to

Fig. 1 Typical chips for machinability ratings A to E (Table 1) for aluminum alloys

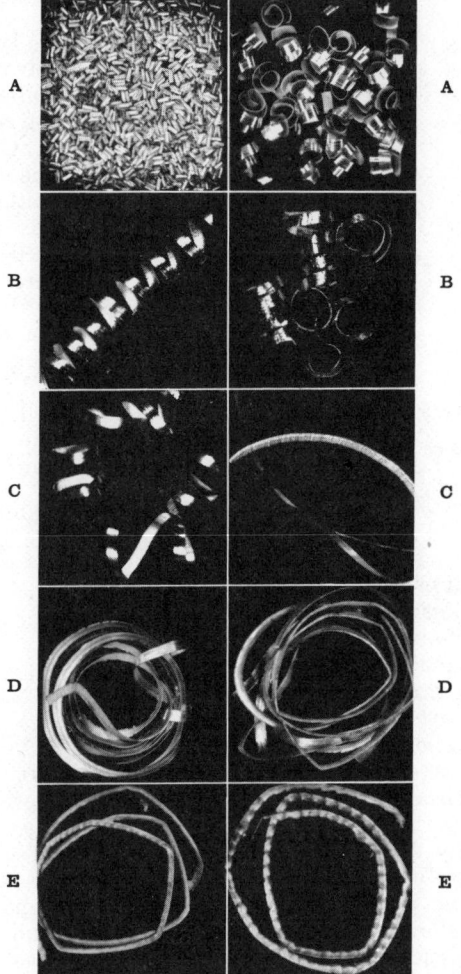

| Machinability rating | Alloy | Speed | | Feed | | | |
| | | m/s | ft/min | Left photo | | Right photo | |
				mm/rev	in./rev	mm/rev	in./rev
A	2011-T3	2.0	400	0.66	0.0026	0.15	0.0060
B	2024-T4	0.5	100	0.15	0.0060	0.26	0.0104
C	6061-T6	2.0	400	0.15	0.0060	0.26	0.0104
D	3004-H32	2.0	400	0.15	0.0060	0.26	0.0104
E	1100-H12	2.0	400	0.15	0.0060	0.26	0.0104

All chips were made with 20°-rake tool and 2.5-mm (0.100-in.) depth of cut.

tear, rather than shear, from the work.

The machinabilities of many of the common aluminum alloys, both cast and wrought, have been classified into five groups: A, B, C, D and E, in increasing order of chip length and in decreasing order of finish quality, as defined in the footnotes in Table 1. Typical chips for each rating are illustrated in Fig. 1.

Surface Finish and Integrity of Machined Parts

The softness and low modulus of elasticity of aluminum alloys, relative to the steels and brasses, introduces two characteristics that have a considerable influence on their machinability. Unless the cutting tools are sharp and the workpiece well supported, aluminum tends to move away from the cutting tool. A machined surface showing bright burnished areas and irregular gouges usually indicates chatter caused by inadequate clamping; it also, however, could be caused by irregular pressure of the tool against the workpiece. With chattering, the tool alternately rubs against the workpiece, giving the burnished appearance and powder-like chips, and then when the slack or elasticity is exhausted, the tool bites the metal, gouging it and producing thick chips. With aluminum, adequate clamping and support of both the tool and workpiece is essential.

The thermal conductivity of aluminum is relatively high, and with high cutting speeds, the heat generated in cutting can be carried away in the chips. Under very slow cutting speeds, little heat is generated unless the cut is heavy. At intermediate speeds (typical of those used for free machining steels) and with inadequate cooling, the heat is maintained in the workpiece; the resulting expansion of the workpiece will result in off-tolerances parts. These conditions (intermediate speeds and inadequate cooling) are also conducive to the formation of a built-up edge on the tool and torn machined surfaces.

For the best surface finishes and wherever possible, there should be a roughing cut and a finishing cut. The oxide coating, which is always present on as-received stock, causes rapid tool wear; and when the tool becomes dull, holding tight tolerances and a good surface is extremely difficult. A light finishing cut with sharp lapped tools, however, will enable the tolerance to be met and a much finer finished surface to be produced. With wrought stock, the material should be ordered so that either no material is removed from the surface or there is sufficient material to allow the second cut. With castings, the same conditions should apply, especially when drilling blind or cast-in holes is required.

Joining Aluminum

By the ASM Committee on
Aluminum and Aluminum Alloys*

ALUMINUM can be joined by a wide variety of methods, including fusion and resistance welding, brazing and soldering, adhesive bonding, and mechanical methods such as riveting and bolting. These various methods are presented in Fig. 1. When proper techniques are used, joints have more than adequate strength for the many demanding applications of aluminum. This article discusses the various types of joints that can be made in aluminum and how joining method affects choice of alloy, temper and product form. For detailed information on the specific techniques used in welding and brazing aluminum, see Vol 6 of the 8th Edition of this Handbook.

Factors That Affect Welding

Aluminum oxide immediately forms on aluminum surfaces exposed to air. This layer of aluminum oxide increases in thickness with increasing time and temperature, and it is quite thick on heat treated aluminum. Before aluminum can be welded by fusion methods, thick oxide layers must be removed mechanically by machining, filing, wire brushing, scraping or chemical cleaning. During welding, the oxide must be prevented from re-forming by shielding the joint area with a nonoxidizing gas such as argon, helium or hydrogen, or chemically by use of

fluxes. In some methods (ultrasonic, friction, pressure and explosive welding), the oxide film is broken up and dispersed mechanically during the operation.

Thermal conductivity is a measure of the rate at which a material will conduct heat from one region of itself to another and is the physical property that most affects weldability. As can be

seen in Table 1, thermal conductivity of aluminum alloys is about one-half that of copper and four times that of low-carbon steel. This means that heat must be supplied four times as fast to aluminum alloys as to steel to raise the temperature locally by the same amount. However, the high thermal conductivity of aluminum alloys helps to solidify the molten weld pool of

*See page X for committee list.

Table 1 Nominal values of physical properties important in welding

Property	Aluminum alloy 1100-O	Aluminum alloy 5454-H32	Aluminum alloy 6061-T6	Low-carbon steel	Copper C11000
Thermal conductivity					
W/m·K..............	223	134	167	52	388
Btu/ft·h. °F.........	129	77.4	96.5	30	224
Specific heat (volumetric)					
MJ/m³·K............	2.45	2.41	2.60	3.99	3.38
Btu/in.³·°F..........	0.0212	0.0209	0.0225	0.034	0.0295
Heat of fusion (volumetric)					
GJ/m³..............	1.05	...	...	2.14	1.82
Btu/in.³..............	16.4	...	...	33.1	10.6
Coefficient of linear thermal expansion(a)					
µm/m·K.............	23.6	23.6	23.6	12.6	17.0
µin./in.·°F..........	13.1	13.1	13.1	7.0	9.4
Melting temperature					
°C..................	645-655	600-645	580-650	1450-1520	1065-1083
°F..................	1190-1215	1115-1195	1080-1205	2640-2770	1949-1981
Electrical conductivity, % IACS..............	59	34	43	10	100
Density					
Mg/m³..............	2.71	2.68	2.70	7.85	8.89
lb/in.³	0.098	0.097	0.098	0.283	0.321

(a) At 20 to 100 °C (68 to 392 °F).

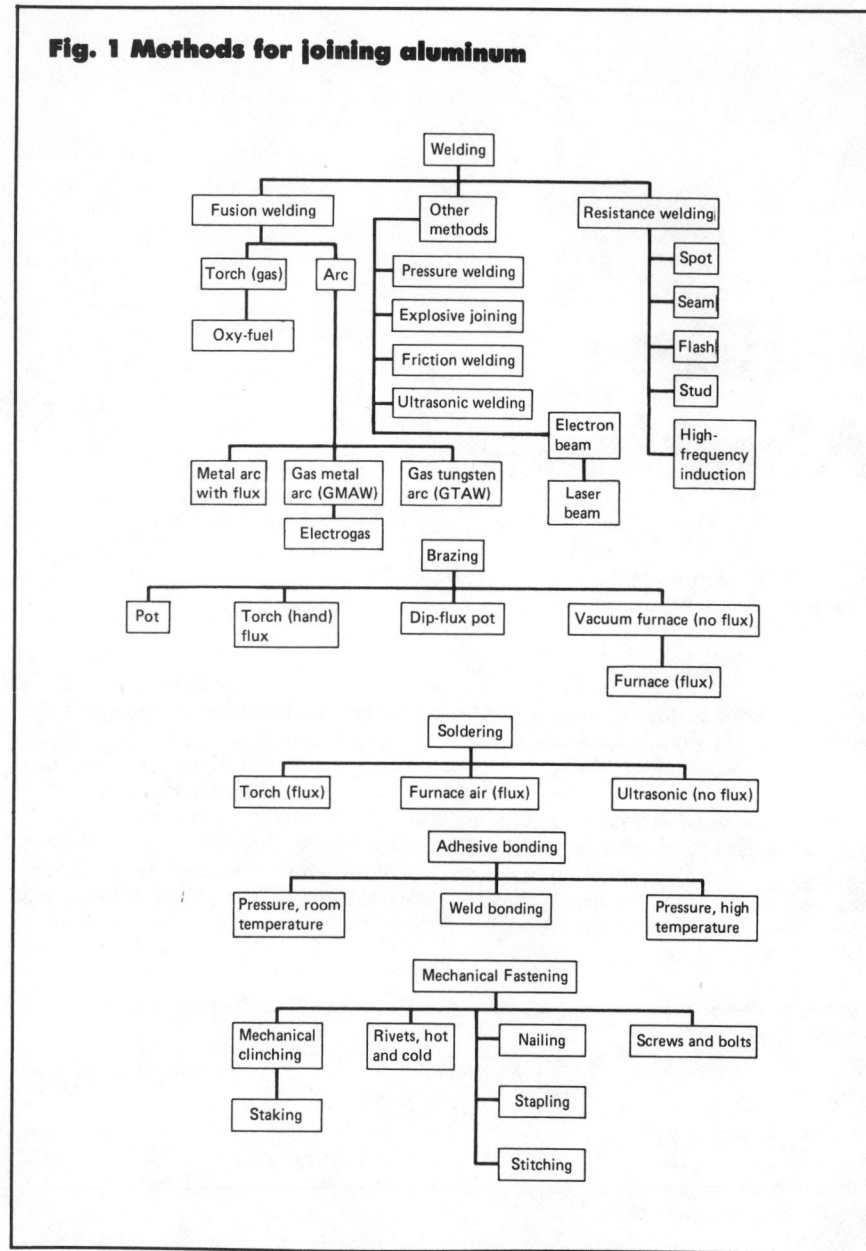

Fig. 1 Methods for joining aluminum

Melting range is the temperature range over which a solid material turns to a liquid. The melting ranges for aluminum alloys are considerably lower than those for copper or steel (Table 1). This fact, combined with the lower volumetric specific heats of aluminum alloys, means that the amount of heat required to reach the melting range is much lower for aluminum alloys than for copper or steel.

As the temperature of copper or steel nears the melting point there is a change in color, which indicates that melting is near. Aluminum alloys, because of their lower melting ranges, do not exhibit changes in color prior to melting, and care must be taken during welding not to melt too great an area.

Electrical conductivity has little influence on fusion welding but is a very important property for materials that are to be resistance welded. In resistance welding, resistance of the metal to the flow of welding current produces heat, which causes the portion of the metal through which the current flows to approach or reach its melting point. Aluminum has higher conductivity than steel (see Table 1), which means that much higher currents are required to produce the same heating effect as for steel. Consequently, resistance welding machines for aluminum must have much higher output capabilities than those normally used for steel, for welding comparable sections.

Weldability of Aluminum Alloys

All aluminum alloys can be resistance welded and most aluminum alloys can be joined by either gas metal-arc or gas tungsten-arc welding. The weldabilities of aluminum alloys are essentially the same for both gas-shielded arc processes. The most common alloys are grouped by gas-shielded arc weldability rating in Table 2.

Wrought alloys most easily welded by gas-shielded arc processes are those of the non-heat-treatable 1xxx, 3xxx and 5xxx series; the alloys in the heat treatable 6xxx series are also easily welded. Alloys of the high-strength, heat treatable 2xxx series also can be arc welded, but special techniques may be required and somewhat lower ductility may be obtained. Of the high-strength heat treatable 7xxx series, alloys 7049, 7050, 7075, 7079 and 7178 are weldable, but are more prone to cracking, and therefore welding usual-

aluminum and, consequently, makes out-of-position welding easier by reducing sag.

Specific heat is the amount of heat required to raise the temperature of a specified amount of a material one degree on a specified scale. As may be seen in Table 1, it takes only 65p as much heat to raise the temperature of aluminum a certain amount as it does to raise the temperature of an equal volume of steel the same amount.

Coefficient of linear thermal expansion is a measure of the change in length of a material with a change in

its temperature. As shown in Table 1, the coefficient of linear thermal expansion for aluminum is twice that for steel. This means that extra care must be taken in welding aluminum to ensure that the joint space remains uniform. This may necessitate preliminary joining of the parts of the assembly by tack welding prior to the main welding operation.

The combination of high coefficient of thermal expansion and high thermal conductivity would cause considerable distortion of aluminum during welding were it not for the high welding speed possible.

Table 2 Weldability of aluminum alloys by the gas metal-arc and gas tungsten-arc processes

Readily Weldable

Wrought alloys
  Unalloyed aluminum, 1060, 1100, 1350
  2219
  3003, 3004, 3105
  5005, 5050, 5052, 5056, 5083, 5086,
   5154, 5252, 5254, 5454, 5456, 5457,
   5652, 5657
  6061, 6063, 6070, 6101, 6201, 6262, 6463
  7005
Casting alloys
  328.0, 355.0, C355.0, 356.0, A356.0,
   357.0, A357.0, 359.0
   443.0, A443.0, B443.0

Weldable in Most Applications(a)

Wrought alloys: 2014, 4032, 6066
Casting alloys
  208.0, 308.0, 319.0, 332.0
   413.0, 712.0

Limited Weldability(b)

Wrought alloys: 2024, 2218, 2618
Casting alloys
  213.0, 222.0, 295.0, 296.0
   333.0, 336.0, 354.0
   512.0, 513.0, 514.0
   Die casting alloys

Welding Not Recommended

Wrought alloys: 2011, 7075, 7178
Casting alloys
  242.0, 520.0, 535.0
   705.0, 707.0, 710.0, 711.0, 713.0, 771.0

(a) May require special techniques for some applications. (b) Require special techniques.

Fig. 2 The basic types of joints used in welding aluminum

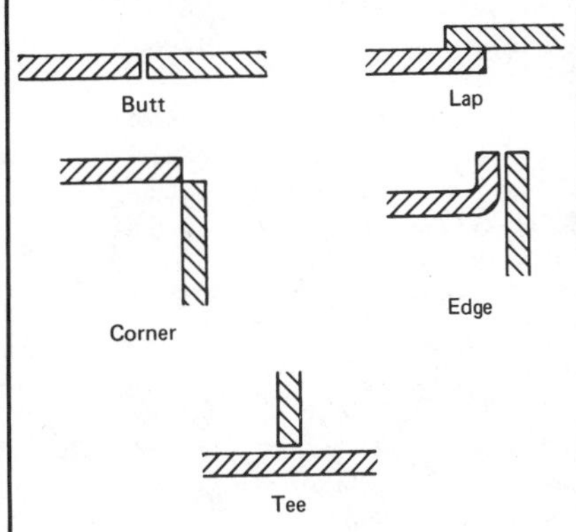

Butt

Lap

Corner

Edge

Tee

ly is not recommended for these alloys; however, alloys 7005 and 7039 were developed specifically for welding and have good weldability. Alloys 7005 and 7039 are of special interest for large structures in which the welds must be of high strength, because welds will age naturally to 70 to 90% of the strength of the heat treated base metal (depending on the chemical composition of the weld deposit) within 30 to 90 days after welding. The gas-shielded arc weldability and resistance weldability ratings of wrought alloys are given in Table 2 in the article entitled "Aluminum Mill Products" in this volume.

The heat of welding removes part or all of the effects of strain hardening adjacent to the fusion zone; in consequence, the yield strength of the heat-affected zone of a weld in a non-heat-treatable alloy may not exceed that of the annealed alloy. The size of the low-strength zone will depend primarily on the speed of welding and the welding method employed. On the whole, weldments exhibit good joint efficiency and ductility.

When a heat treated alloy (T4 or T6 condition) is arc welded, its strength in the as-welded condition is slightly less than that of the unwelded alloy in the T4 condition. This decrease in strength is attributed to the comparative weakness of the heat-affected zone. The zone normally consists of an area of resolution material adjacent to the weld, an area where partial annealing has occurred, and an overaged area. Because of the high strength of the heat-affected zone, weldments of alloys in the T6 condition have low as-welded joint efficiency and often reduced weld ductility. Solution heat treatment and aging after welding may restore much of the strength, but little improvement in weld ductility may occur. After post-weld aging alone, strength may be no higher, and ductility can be lower, than those of some non-heat-treatable alloys, so that there is no advantage in using the heat treatable alloy. To obtain optimum properties in weldments, it may be preferable to postweld solution heat treat and age.

Weld strengths are the same as base-metal strengths when heat treatable or non-heat-treatable alloys are welded in the annealed condition.

Casting Alloys. Most aluminum casting alloys can be gas-shielded arc welded. Aluminum sand and permanent mold castings are welded to repair foundry defects, to repair items broken in service, or to join cast fittings to wrought members. Gas-shielded arc

weldability ratings for sand and permanent mold castings are given in Table 2 in the article entitled "Aluminum Foundry Products" in this volume.

Formerly, die cast fittings were seldom used where gas-shielded arc welded construction was required, because they often contained porosity, but recent advances in casting technique, such as vacuum die casting, have resulted in improved quality; die castings are now satisfactorily welded for some applications, such as irrigation tubing.

Filler Metals

The aluminum alloys commonly used as filler metals for welding aluminum are 1100, 4043, 5183, 5356, 5554, 5556 and 5654. The filler metal used in the gas shielded arc welding processes is made with special care to avoid imperfections, oil and dirt. All arc welding filler rod and electrode are packaged to keep out dirt and prevent oxidation. Standard sizes of filler rod for gas tungsten-arc welding are: 1.6 mm (1/16 in.), 2.4 mm (3/32 in.), 3.2 mm (1/8 in.), 4.0 mm (5/32 in.), 4.8 mm (3/16 in.) and 6.4 mm (1/4 in.). Standard electrode sizes for gas metal-arc welding are: 0.8 mm (0.30 in.), 0.9 mm (0.035 in.), 1.2 mm (3/64 in.), 1.6 mm (1/16 in.), 2.4 mm (3/32 in.) and 3.2 mm (1/8 in.). It is important that welding rod and wire be stored in a clean, dry, temperature-controlled atmosphere. Many problems in welding can be traced to poor choice of filler alloy and/or to poor filler-metal surface conditions.

The composition of the base metal governs choice of filler-metal alloy. An incorrect choice of filler metal can cause weld cracking, low strength, poor corrosion resistance and poor color match after chemical treatments. Table 3 gives the filler metals recommended for joining various combinations of common wrought and cast aluminum alloys by gas-shielded arc welding.

Preweld Operations

Preparation of aluminum and its alloys for joining by the various fusion welding and brazing processes includes several common but important operations:

1 Selection of type of joint
2 Edge preparation
3 Cleaning of the joint area
4 Spacing

Table 3 Filler metals suitable for gas-shielded arc welding of various combinations of wrought and cast aluminum alloys

Base metals to be welded (in column below, and in column heads at right)	319.0, 355.0, C355.0	356.0, 413.0, A443.0	514.0	7005, 712.0, A712.0, C712.0	6061, 6063, 6101, 6151	5456	5454	5154, 5254a
1060, 1350	4145(b)(c)	4043(e)(d)	4043(c)(e)	4043(c)	4043(c)	5356(b)	4043(c)(e)	4043(c)(e)
1100, 3003, alclad 3003	4145(b)(c)	4043(c)(d)	4043(c)(e)	4043(c)	4043(c)	5356(b)	4043(c)(e)	4043(c)(e)
2014, 2024	4145(b)	4145	...	...	4145	...	...	...
2219	4145(b)(c)(f)		4043(c)	4043(c)	4043(c)(d)	4043	4043(c)	4043(c)
3004, alclad 3004	4043(c)	4043(c)	5654(g)	5356(e)	4043(g)	5356(e)	5654(g)	5654(g)
5005, 5050	4043(c)	4043(c)	5654(g)	5356(e)	4043(g)	5356(e)	5654(g)	5654(g)
5052, 5652a	4043(c)	4043(c)(g)	5654(g)	5356(e)	5356(b)(g)	5356(g)	5654(g)	5654(a)(g)
5083	...	5356(b)(c)(e)	5356(e)	5183(e)	5356(e)	5183(c)	5356(e)	5356(e)
5086	...	5356(b)(c)(e)	5356(e)	5356(e)	5356(e)	5356(e)	5356(g)	5356(g)
5154, 5254a	...	4043(c)(g)	5654(g)	5356(g)	5356(b)(g)	5356(g)	5654(g)	5654(a)(g)
5454	4043(c)	4043(c)(g)	5654(g)	5356(g)	5356(b)(g)	5356(g)	5554(b)(e)	
5456	...	5356(b)(c)(e)	5356(e)	5556(e)	5356(e)	5556(e)		
6061, 6063, 6101, 6151	4145(b)(c)	4043(c)(g)	5356(b)(g)	5356(b)(c)(g)	4043(c)(g)			
7005, 712.0, A712.0, C712.0	4043(c)	4043(c)(g)	5356(g)	5183(e)				
514.0	...	4043(c)(g)	5654(g)(h)					
356.0, 413.0, A443.0	4145(b)(c)	4043(c)(h)						
319.0, 355.0, C355.0	4145(b)(c)(h)							

Base metals to be welded (in column below, and in column heads at right)	5086	5083	5052, 5652a	5005, 5050	3004, alclad 3004	2219	2014, 2024	1100, 3003, alclad 3003	1060, 1350
1060, 1350	5356(b)	5356(b)	4043(c)	1100(b)	4043	4145	4145	1100(b)	1100(b)
1100, 3003, alclad 3003	5356(b)	5356(b)	4043(c)(e)	4043(e)	4043(e)	4145	4145	1100(b)	
2014, 2024	...	...	...	...	...	4145(h)	4145(f)		
2219	4043	4043	4043(c)	4043	4043	2319(b)(c)(d)			
3004, alclad 3004	5356(e)	5356(e)	4043(c)(e)	4043(e)	4043(e)				
5005, 5050	5356(e)	5356(e)	4043(c)(e)	4043(e)(h)					
5052, 5652a	5356(e)	5356(e)	5654(a)(b)(g)						
5083	5356(e)	5183(e)							
5086	5356(e)								
5154, 5254a									
5454									
5456									
6061, 6063, 6101, 6151									
7005, 712.0, A712.0, C712.0									
514.0									
356.0, 413.0, A443.0									
319.0, 355.0, C355.0									

Note: All filler metals shown here are covered by AWS specification A5.10-69, prefixed by the letters "ER". Throughout this table, the prefix has been omitted, to conserve space. Filler metals 5356, 5556 and 5654 are not suitable for sustained service at temperatures higher than 65 °C (150 °F). Other service conditions, such as immersion in fresh or salt water or exposure to specific chemicals, may also limit the choice of filler metal. Where no filler metal is listed, the base-metal combination is not recommended for welding. (a) Base metals 5254 and 5652 are used for hydrogen peroxide service. Filler metal 5654 is used for welding both alloys for service at temperatures of 65 °C (150 °F) and below. (b) 4043 may be used for some jobs. (c) 4047 may be used for some jobs. (d) 4145 may be used for some jobs. (e) 5183, 5356 or 5556 may be used. (f) 2319 may be used for some jobs. (g) 5183, 5356, 5554, 5556 and 5654 may be used. In some instances they provide better color match after anodizing treatment, highest weld ductility, and higher weld strength. Filler metal 5554 is suitable for service at elevated temperatures. (h) Filler metal of the same composition as the base metal is sometimes used.

5 Jigging
6 Preheating (if required).

Attention to the above operations will ensure good welds.

Material to be welded by resistance processes usually requires only degreasing, oxide film treatment and jigging. The above points are covered in more detail under resistance welding.

Types of Joints. The five basic types of weld joints for aluminum alloys are similar to those used in welding steel (see Fig. 2). The joints may need an edge preparation, such as beveling, depending on how they are used. In practice, many variations and combinations of these types of joints are used.

Edge preparation for welding is very important if good welds are to be made. The type of edge preparation required is governed largely by base alloy composition, material thickness, type and location of joint, welding process used and required weld quality.

Aluminum sheet less than 6.4 mm (1/4 in.) thick may be sawed or cut on ordinary shears with good set and sharp blades. No further dressing of sheared edges is usually necessary. However, filing sometimes is needed to remove any fold. Material over 6.4 mm (1/4 in.) thick also may be sheared, but further edge preparation usually is required for fit-up and welding.

Aluminum may be readily cut by hacksaw, band saw, jigsaw or circular saw. Usually, no further dressing is required if the cut is smooth. Sawing lubricants must be removed.

Band saws are one of the most versatile tools for cutting aluminum. A tilting table is often used so the band saw can be used effectively for a beveled edge preparation. Aluminum is best sawn at a blade speed of at least 30 m/s (6000 ft/min).

Pneumatic tools are very effective for removing defective metal and, if the

Table 4 Typical surface preparations for fusion welding of aluminum

Methods and conditions	Remarks
Hand wiping with clean rags	Applicable to all alloys for all types of metallurgical joining
Degreasing with commercial solvents by wiping, spraying or dipping, or with vapor degreasing	Removes oil, grease, dirt and loose particles; applicable to all alloys for all types of metallurgical joining
Mechanical removal of oxide by wire brushing, filing, milling, rubbing with steel wool, sanding, routing or rotary planing	Sufficient treatment where edges are not heavily oxidized; applicable to all alloys for fusion welding, but not usually employed for resistance welding
Chemical surface treatment by immersion in 50% solution of concentrated nitric acid at room temperature for 15 min, followed by cold-water rinsing, hot-water rinsing and then drying	For removing thin oxide film for all fusion welding processes
Chemical removal of oxide by immersion in 5% sodium hydroxide (caustic soda) solution at 70 °C (160 °F) for 10 to 60 s followed in turn by cold-water rinsing, rinsing in concentrated nitric acid at room temperature, hot-water rinsing and then drying	For removing thick oxide film from all alloys for all welding and brazing processes
Surface treatment by proprietary products	Consult supplier for use and recommended procedures

Fig. 3 Fatigue characteristics of arc welded butt joints in three aluminum alloys

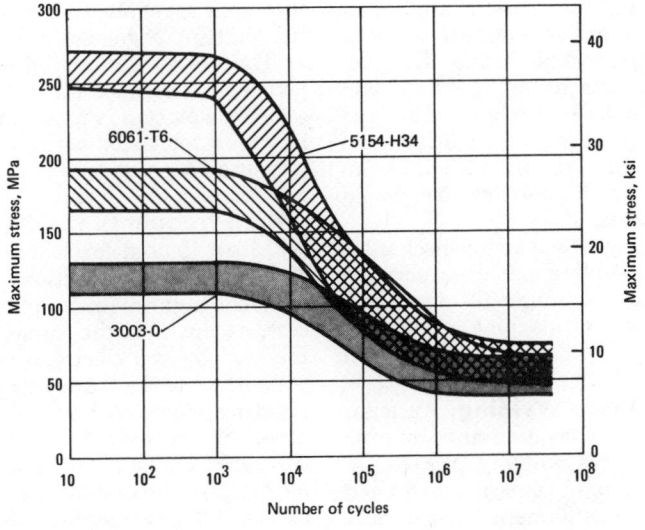

9.5-mm (³⁄₈-in.) plate; weld reinforcement removed; axial loading; $R = 0$.

object is too large or otherwise impossible to place on a band-saw table, for edge preparation.

For large thick plates, a shop rotary planer or (in the field) a portable hand planer can be used. Routers and edge millers are useful in shaping and preparing edges for welding.

Surface and Joint Cleaning.
Cleanness of joint surfaces is a prerequisite for sound welded, brazed and soldered joints in aluminum. All grease, oil, dirt, finger prints, water and loose particles of metal must be removed. (The fluxes used in welding and brazing will remove only aluminum oxides.)

Table 4 lists the typical methods of preparing aluminum for fusion welding. The method of cleaning depends on material condition, welding process and required weld quality.

Joint Spacing. In order to ensure good penetration and to minimize distortion, joint spacing must be carefully controlled. Correct edge preparation and joint spacing are imperative if cracking of butt joints in heat treatable alloys is to be avoided. With proper edge preparation and joint spacing, the weld metal will have sufficient plasticity or ductility to absorb stresses and avoid cracking. The heat of welding causes expansion of the base metal, which may change the joint spacing as welding progresses. Care should be taken to maintain joint spacing during welding by means of jigs, clamps or tack welds.

Jigging. Jigs are tools used to hold parts in position for welding, brazing or soldering. Properly designed jigs provide easy handling of the work and accurate alignment of the edges to be joined, thus providing greater production, economy and precision.

Thin material must be supported during joining, because strength of aluminum alloys drops rapidly above 150 °C (300 °F). Jigs help support the aluminum and assist in avoiding sag.

When a temporary backing strip is part of the jigging for GMAW (MIG) welding, care must be taken to see that the backing is grooved to allow an underbead of metal to protrude below the plane of the bottom surface of the components being welded; this will ensure adequate penetration. The groove should be quite shallow (0.25 to 0.75 mm, or 0.010 to 0.030 in.), and wide enough so that it does not confine the edge of the penetration bead.

Preheating depends on welding process, base metal, material thickness, nature of the structure, and ambient temperature. It is best carried out in a furnace, but when this is not practical, a preheating torch can be used. Care must be taken to ensure uniform heating. Careful preheating reduces distortion.

Finishing of Weldments

Weldments made by tungsten-arc and metal-arc inert-gas processes usually do not require finishing except when a flush appearance is required. If a flux process has been used, the flux must be removed for appearance and for prevention of corrosion. Flux removal should be done immediately after welding or brazing, by immersion in hot water and vigorous scrubbing with a fiber brush. Successive rinsing operations are recommended to completely remove all traces of the flux. For large assemblies it may only be possible to clean by means of a jet of steam and hot water. Complete removal of flux may require dipping in a chemical bath followed by rinsing in cold water and then in hot water.

Mechanical finishing may be required for certain weldments, such as those in food-handling and chemical equipment. In such applications, mechanical finishing may consist of chipping, filing, planing and grinding. Some aluminum weldments may be finished by chemical and electrochemical means to provide additional protection from corrosion or to form a base for decorative finishing. Aluminum weldments that are to be anodized for decorative work should be made with the filler metals recommended in Table 3.

After being cleaned by mechanical or chemical means, aluminum weldments may be readily painted. The recommended paint system on aluminum consists of a coat of wash and primer conforming to U. S. Military specification MIL-P-15328 or a zinc chromate primer conforming to U. S. Military specification MIL-P-6889. Almost any good-quality finishing paint is suitable, provided that it is compatible with the primer. Paints provide excellent finishes on aluminum.

Welding Processes

The two main types of processes used for welding aluminum are fusion welding and resistance welding. The various welding processes are listed in Fig. 1.

Gas tungsten-arc welding, known as GTAW or TIG welding, uses no flux and usually employs alternating current. In the alternating-current GTAW process, an arc is struck between a nonconsumable tungsten electrode and the workpiece in a shield of inert gas, usually argon. The arc rids the area of oxide film and the inert gas shields the molten weld pool of molten aluminum from the atmosphere to prevent the oxide film from re-forming. The GTAW process is suitable for manual welding in all positions and for machine welding with or without addition of filler metal. Gas tungsten-arc welding is slower than gas metal-arc welding but is more adaptable to weld-

ing aluminum pipe and aluminum parts less than 3 mm (1/8 in.) thick.

Gas metal-arc welding, known as GMAW or MIG welding, was developed from the GTAW process and combines most of its advantages with much higher welding speeds. In GMAW, a reverse-polarity, direct-current arc is established between a consumable aluminum wire electrode (which becomes the filler metal) and the workpiece in a shield of inert gas (argon or helium). The GMAW process is faster than GTAW, produces less distortion, and is primarily used to weld 3-mm (1/8-in.) and thicker aluminum. Both processes usually employ water-cooled torches.

Both GMAW and GTAW can be done as hand or machine operations. GMAW is the most widely used process for welding aluminum and is used for welding truck bodies, rail cars, storage tanks and other large aluminum structures.

Stud welding is a means of affixing studlike attachments to the surfaces of aluminum products. These studs may be used for attaching handles and legs to aluminum cooking utensils or for holding architectural curtain walling in place.

When using a "drawn-arc" stud welding process, a gas-shielded arc is struck between the workpiece and the stud, which first acts as an electrode and causes the joint area to melt. The stud is then plunged into the molten weld pool. Aluminum stud from 1.6 to 13 mm (1/16 to 1/2 in.) in diameter can be applied by this process.

When using a "capacitor-discharge" type of stud welding machine, inert-gas shielding is not commonly employed. With this type of machine, aluminum studs from 1.6 to 6.4 mm (1/16 to 1/4 in.) in diameter can be applied.

Oxy-fuel Gas Welding. An older process used for welding aluminum is the oxy-fuel gas welding process. In this process, an oxygen and hydrogen or acetylene gas flame is used to heat the metal locally until a molten weld pool is developed; the filler alloy is added to the pool. An active flux is required for removal of the oxide film from the aluminum before welding. This flux must be removed after welding to prevent severe corrosion.

Welding with an oxy-fuel gas torch is relatively slow and can cause overheating and distortion of the workpiece. This process is seldom used today except by those who do not frequently weld aluminum or who do not have

GTAW or GMAW equipment available. This process, however, is useful for welding up to 25-mm (1-in.) thick aluminum and for repair work in the field.

Resistance welding is a process in which the heat for welding is generated in the parts to be joined by the resistance they offer to the controlled flow of an electric current. In addition, pressure is applied to forge the heated parts together. Several types of resistance welding are used in making lap joints in aluminum: spot welding, seam welding, flash welding and resistance stud welding.

Spot welding is used extensively in aircraft and related industries to replace riveting in fabrication of assemblies from aluminum up to about 5 mm (3/16 in.) thick. Little loss of strength occurs during spot welding.

High-frequency induction seam welding is used to join the longitudinal seam in tube roll formed from coiled strip. The welding equipment is included in the roll form system, just beyond the last rolls. The formed tube passes through an induction coil, which heats the abutting edges to welding temperature. The formed tube then passes through a set of pressure rolls, in which the abutting edges are pressed together. The molten metal is forced out of the joint and is scraped off with a cutting tool. This process is used to make such items as television-aerial tubing and irrigation pipe at speeds that can exceed 2.5 m/s (500 ft/min).

High-frequency resistance seam welding is used for tubular sections, Tee sections and H sections. It is similar to the high-frequency induction process, except that the induction coil is replaced by two electrical contacts on the surface of the formed strip near the abutting edges. As the edges of the sheet are pressed together, a high-frequency current flows and a weld is formed. This process also can operate in excess of 2.5 m/s (500 ft/min).

Pressure welding is a process of joining metals with homogeneous welds produced by pressure at room temperature or, in certain cases, elevated temperature. No electricity, flux, chemical or gas is used. The joint area must be clean of oil, grease and dirt. Prior to welding, the oxide film is broken up and removed by wire brushing. Welding pressure should be applied soon after the oxide film has been removed. This process can be used to make lap joints in aluminum of foil

Table 5 Minimum expected properties at room temperature for butt welded aluminum alloys(a)(b)

Alloy and temper	Product forms	Thickness range mm	Thickness range in.	Tensile Ultimate(b) MPa	Tensile Ultimate(b) ksi	Tensile Yield(c) MPa	Tensile Yield(c) ksi	Compressive yield strength(c) MPa	Compressive yield strength(c) ksi	Shear Ultimate MPa	Shear Ultimate ksi	Shear Yield MPa	Shear Yield ksi	Bearing Ultimate MPa	Bearing Ultimate ksi	Bearing Yield MPa	Bearing Yield ksi
1100-H12, H14	All	All	All	76	11	31	4.5	31	4.5	55	8	17	2.5	160	23	55	8
3003-H12, H14, H16, H18	All	All	All	97	14	48	7	48	7	69	10	28	4	205	30	83	12
Alclad 3003-H12, H14, H16, H18	All	All	All	90	13	41	6	41	6	69	10	24	3.5	205	30	76	11
3004-H32, H34, H36, H38	All	All	All	150	22	76	11	76	11	97	14	45	6.5	315	46	140	20
Alclad 3004-H32, H34, H36, H38, H14, H16	All	All	All	145	21	76	11	76	11	90	13	45	6.5	305	44	130	19
3003-H25	Sheet	All	All	115	17	62	9	62	9	83	12	34	5	250	36	105	15
5005-H12, H14, H32, H34	All	All	All	97	14	48	7	48	7	62	9	28	4	195	28	69	10
5050-H32, H34	All	All	All	125	18	55	8	55	8	83	12	30	4.5	250	36	83	12
5052-H32, H34	All	All	All	170	25	90	13	90	13	110	16	52	7.5	345	50	130	19
5083-H111	Extrusions	All	All	270	39	145	21	140	20	160	23	83	12	540	78	220	32
H321	Sheet and plate	4.7-38.1	0.188-1.5	275	40	165	24	165	24	165	24	97	14	550	80	250	36
H321	Plate	38.1-76.2	1.501-3.0	270	39	160	23	160	23	165	24	90	13	540	78	235	34
H323, H343	Sheet	All	All	275	40	165	24	165	24	165	24	97	14	550	80	250	36
5086-H111	Extrusions	All	All	240	35	125	18	115	17	145	21	69	10	485	70	195	28
H112	Plate	6.4-12.7	0.25-0.499	240	35	115	17	115	17	145	21	66	9.5	485	70	195	28
H112	Plate	12.7-25.4	0.50-1.0	240	35	110	16	110	16	145	21	62	9	485	70	195	28
H112	Plate	25.4-50.8	1.001-2.0	240	35	97	14	97	14	145	21	55	8	485	70	195	28
H32, H34	Sheet and plate	All	All	240	35	130	19	130	19	145	21	76	11	485	70	195	28
5154-H38	Sheet	All	All	205	30	105	15	105	15	130	19	59	8.5	415	60	160	23
5454-H111	Extrusions	All	All	215	31	110	16	105	15	130	19	66	9.5	425	62	165	24
H112	Extrusions	All	All	215	31	83	12	83	12	130	19	49	7	425	62	165	24
H32, H34	Sheet and plate	All	All	215	31	110	16	110	16	130	19	66	9.5	425	62	165	24
5456-H111	Extrusions	All	All	285	41	165	24	150	22	165	24	97	14	565	82	260	38
H112	Extrusions	All	All	285	41	130	19	130	19	165	24	76	11	565	82	260	38
H321	Sheet and plate	4.7-38.1	0.188-1.5	290	42	180	26	165	24	170	25	105	15	580	84	260	38
H321	Plate	38.1-76.2	1.501-3.0	285	41	165	24	160	23	170	25	97	14	565	82	260	38
H323, H343	Sheet	All	All	290	42	180	26	180	26	170	25	105	15	580	84	260	38
6061-T6, T651, T6510, T6511(d)	All	Over 9.5	Over 0.375	165	24	140	20	140	20	105	15	83	12	345	50	205	30
T6, T651, T6510, T6511(e)	All	Over 9.5	Over 0.375	165	24	105	15	105	15	105	15	62	9	345	50	205	30
6063-T5, T6	All	All	All	115	17	76	11	76	11	76	11	45	6.5	235	34	150	22
6351-T5(d)	Extrusions	Over 9.5	Over 0.375	165	24	140	20	140	20	105	15	83	12	345	50	205	30
T5(e)	Extrusions	Over 9.5	Over 0.375	165	24	105	15	105	15	105	15	62	9	345	50	205	30

(a) Gas tungsten-arc or gas metal-arc welding with no postweld heat treatment. (b) Filler wires used are those recommended in Table 3. Ultimate tensile values are ASME weld-qualification-test values. (c) 0.2% offset in 250-mm (10-in.) gage length across a butt weld. (d) Values are for welding with 5183, 5356 or 5556 filler wire, regardless of thickness. Values also apply to thicknesses less than 9.5 mm (0.375 in.) when welding is done with 4043, 5554 or 5654 filler wire. (e) Values are for welding with 4043, 5554 or 5654 filler wire.

thickness to 6 mm (1/4 in.) thick, or butt joints in aluminum wire and rod from 0.6 to 10 mm (0.025 to 3/8 in.) in diameter, and for joining aluminum to other metals.

Explosive welding is accomplished by the energy released from detonation of high explosives. This is a lap welding process that allows aluminum to be joined to itself or to other metals without loss of strength, and is used to produce transition joints between aluminum and dissimilar metals. Explosive welding has been useful in joining aluminum pipes and bus bars in the field.

Friction (inertia) welding is a butt welding process in which a spinning part is pushed against a stationary part and the frictional heat and force generated produce a solid-state bond at the interface. Aluminum alloys can be joined by friction welding to similar or dissimilar aluminum alloys and to other metals. When alloy 6061-T6 is fusion welded to 304 stainless steel, a brittle intermetallic compound is formed, but when 6061-T6 is friction welded to 304 it forms a satisfactory joint for cryogenic and heat-transfer applications.

Ultrasonic welding resembles resistance welding in the general configuration of equipment used. Overlapping parts to be joined are clamped between tips to weld them at a spot, or between rolls to weld a seam. No heat energy, as such, is supplied. The weld is made without fusion in the ordinary sense of the term. An intense, localized, mechanical wiping does occur where the parts are pressed together. This action scrubs away oxides or other impurities, rearranges the metallic structural configuration at both surfaces, and merges them to form a sound metallurgical bond.

Ultrasonic welding is well suited for handling material having thicknesses in the lower part of resistance welding's range—for aluminum, the range is from foil thicknesses (less than 0.005 mm, or 0.0002 in.) up to about 2.4 mm (3/32 in.). Furthermore, with ultrasonic welding very thin aluminum foil can be joined to very thick aluminum, and aluminum can be joined to many dissimilar metals.

Although it is not formed by general fusion of metal, the bond achieved by ultrasonic welding is an intimate metallurgical union. Strength is proportional to weld size. When tested to destruction, ultrasonic welds in hard alloys usually fail in shear. Those in softer alloys usually pull out a button. Weld strength data for electric-resistance spot welds may be used for design. Testing done so far indicates this would be conservative. Some test results are published in ASME paper No. 60-WA-332 in the table entitled "Comparison of Ultrasonic Strength with Resistance Welds on Aluminum Alloy Applications for Ultrasonic Welding of Aluminum."

Indentation caused by clamping force seldom exceeds 5% of metal thickness. It is greater with the softer aluminum alloys than with the harder materials.

Adjacent or overlapping welds can be made, and proximity to an already completed weld spot does not downgrade the quality or strength of the next one. When lap seams are made to join foil, the joints can be trimmed simply by tearing off the excess lap adjacent to the weld seam.

Resistance of aluminum alloys to corrosion is not adversely affected by ultrasonic welding, but the usual precautions should be taken whenever dissimilar metals are joined.

Electron beam welding applies heat by means of a concentrated beam of high-velocity electrons. Magnetic fields focus and accelerate electrons from an incandescent filament. The beam and the work are usually under high vacuum—in the order of one micron. While the size of the work is limited by the available vacuum chamber, this environment is ideal for preventing weld contamination. Equipment for welding in low vacuum and in air also has been developed.

Well-fitted joint preparation is essential, and high concentration of energy in a focused beam makes very narrow, single-pass welds possible. Welding has little or no effect on the heat-treated or cold-worked structure of the base metal, and weld contamination is nil.

Because equipment cost is high, this method is useful only where sound, high-strength joints are more important than low welding cost. Apparatus is avalable to handle parts 9 m (30 ft) or more in diameter, and aluminum alloy thicknesses up to 100 mm (4 in.) have been welded.

Laser Beam Welding. The word "laser" means light amplification by stimulated emission of radiation. Laser welding has developed into a high-volume production process for joining and cutting metals. The high-intensity light beam from a fused laser is highly collimated and has good wavelength purity. Because of the resultant high power density, material in a local area of a workpiece can be melted and/or vaporized with little or no effect on the adjacent material. One of the disadvan-

Table 6 Minimum expected tensile strengths at various temperatures for butt welded aluminum alloys

Alloy and temper	Filler-metal alloy		Tensile strength in MPa (ksi) at:				
		−185 °C (−300 °F)	−129 °C (−200 °F)	73 °C (−100 °F)	38 °C (100 °F)	150 °C (300 °F)(a)	260 °C (500 °F)(a)
2219-T37(b)	2319	334 (48.5)	275 (40)	250 (36)	240 (35)	215 (31)	130 (19)
2219(c)	2319	445 (64.5)	410 (59.5)	380 (55)	345 (50)	260 (38)	150 (22)
3003	1100	190 (27.5)	148 (21.5)	121 (17.5)	97 (14)	66 (9.5)	34 (5)
5052	5356	260 (38)	215 (31)	183 (26.5)	170 (25)	145 (21)	72 (10.5)
5083	5183	375 (54.5)	315 (46)	280 (40.5)	275 (40)	...	...
5086	5356	330 (48)	280 (40.5)	245 (35.5)	240 (35)	...	...
5454	5554	305 (44)	255 (37)	220 (32)	215 (31)	180 (26)	105 (15)
5456	5556	385 (56)	328 (47.5)	293 (42.5)	290 (42)	...	...
6061-T6(b)	4043	238 (34.5)	205 (30)	183 (26.5)	165 (24)	150 (20)	41 (6)
6061(c)	4043	380 (55)	341 (49.5)	315 (46)	290 (42)	217 (31.5)	48 (7)

(a) Alloys not listed at 150 °C (300 °F) and 260 °C (500 °F) are not recommended for use at sustained operating temperatures of over 65 °C (150 °F). (b) As welded. (c) Heat treated and aged after welding.

Table 7 Minimum expected shear strengths of fillet welds in aluminum alloys

Filler-metal alloy	Shear strength MPa	ksi
1100	52	7.5
2319(a)(b)	110	16
2319(c)	150	22
4043(a)	79	11.5
5052	52	7.5
5154	83	12
5554	115	17
5556	140	20
5654	83	12

(a) Naturally aged (two to three months). (b) Artificially aged after welding. (c) Heat-treated and aged after welding.

Fig. 4 Range of typical shear strengths for single spot welds in three high-strength aluminum alloys

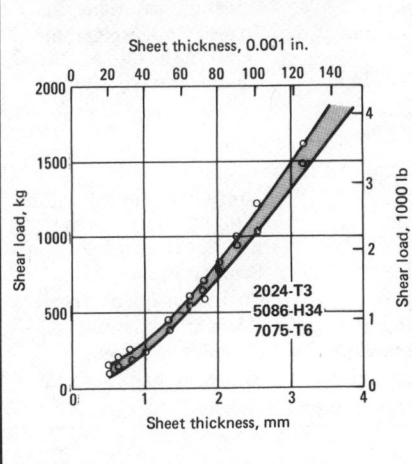

2024-T3
5086-H34
7075-T6

Fig. 5 Allowable shear strengths for single spot welds in aluminum alloys of different tensile strengths

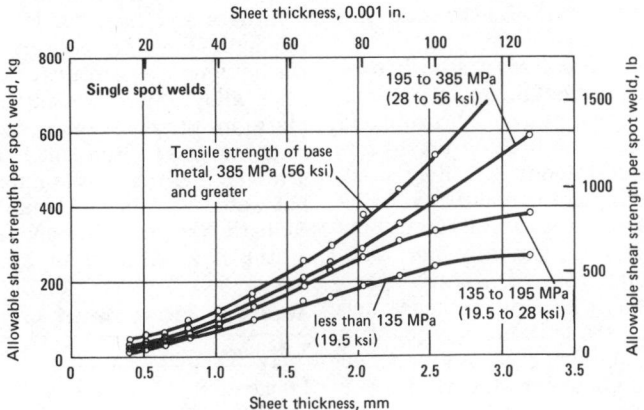

In welding two sheets of unequal thickness, allowable shear strength is based on the thickness of the thinner sheet.

tages of this process is the initial equipment cost, which can go as high as $1 000 000.

Strength of Welds

The strength of sound, normal, as-deposited aluminum alloy weld metal is approximately 80 to 280 MPa (12 to 40 ksi). Joints in non-heat-treatable alloys can be made so that they will fail in the base metal, rather than in the weld metal, by proper selection of the filler alloy. For hard-rolled tempers, the base metal in the heat-affected zone on either side of the joint is softened by the heat of welding, precluding the possibility of obtaining high joint efficiency. With heat treatable alloys of

the 6xxx series, 100% efficiency can be obtained when the welded structure can be solution and precipitation heat treated after welding. Nearly 100% efficiency can be obtained when welding in the T4 temper by using one of the high-speed welding techniques, such as inert-gas-shielded metal-arc welding or electron beam welding, in which the amount of heat flowing into the base metal of the joint is limited, and by precipitation heat treating after welding. In the 2xxx and 7xxx series, such practice produces less improvement in joint efficiency. Alloy 2014 and especially alloy 2219 are widely welded for spacecraft boosters, where machined or chemically milled lands are provided to achieve adequate strength across the welded joints. Summaries of minimum expected butt-welded properties of commonly welded aluminum alloys are given in Tables 5 and 6. The minimum expected shear strength of fillet welded aluminum is given in Table 7.

The fatigue life of welded joints at high loads varies with the alloy. As the load is decreased, differences disappear until, at about one to ten million cycles of axial loading ($R = 0$), the fatigue strength of an arc-welded joint is approximately the same regardless of alloy and is 50 to 70% that of the unwelded alloy. Typical data are given in Fig. 3 for alloys 3003-O, 5154-H34 and 6061-T6.

Design criteria for resistance spot welded aluminum usually are based

on the minimum reproducible shear strength (of single-spot specimens) obtainable under the established manufacturing conditions. Figure 4 presents such test data for three high-strength alloys. Figure 5 shows design-allowable shear strength in relation to the tensile strength and thickness of the alloy being welded.

Design allowables developed in single-spot shear specimens are applicable to multispot patterns, provided the distance between adjacent welds is not less than two nugget diameters. Closer distances will result in shunting current losses, which will reduce current density in the spot weld and reduce the size of the weld and its shear strength. Such losses, which increase with the thickness of the material, must be taken into account in design.

Brazing Processes

Brazing is a type of joining process in which coalescence is obtained by heating the base metal to a temperature below its solidus temperature and adding a filler alloy, using flux. The flux promotes wetting of the joint surface by the filler alloy, which is selected to have a liquidus temperature slightly below the solidus of the base metal. Although the base metal is not melted, there is definite diffusion between the base and filler metals.

Brazing can be used successfully on very thin material and on assemblies that have relatively inaccessible joints. Because very active fluxes frequently are used, flux must be removed after brazing to prevent subsequent corrosion. A noncorrosive flux, which does not require removal after brazing, has recently been developed.

The brazing processes presently used for joining aluminum are torch brazing, vacuum or air furnace brazing and dip brazing. Torch brazing (oxyhydrogen or oxyacetylene) is widely used, because equipment is readily available and costs are kept to a minimum. Furnace and dip brazing require more elaborate equipment and fixtures than those for torch brazing. Furnace and dip brazing are suitable for quantity production and are used in production of aluminum heat exchangers. Table 8 lists the procedures used in torch, furnace and dip brazing.

Aluminum brazing filler metals, which are available in the forms of wire, rod, powder, sheet, and cladding on sheet, are listed in Tables 9 and 10, along with brazing temperature ranges. The brazeabilities of wrought aluminum alloys are given in Table 2 in the article on "Aluminum Mill Products" in this volume.

Vacuum Furnace Brazing. Several vacuum furnace brazing processes have been developed for brazing aluminum heat exchangers and automobile radiators. The reason for brazing in a vacuum furnace is to reduce or eliminate the need for the corrosive flux used in air furnace brazing. There are many problems associated with vacuum furnace brazing, but they are gradually being eliminated. There are

several high-production furnaces in operation today, mainly producing parts in the heat-transfer field.

Dip brazing is a process whereby a cleaned and jigged unit is preheated in a furnace and then quickly immersed in a bath of molten salts maintained at the liquidus temperature of the brazing filler alloy. After brazing has taken place in the salt bath, the unit is removed, the molten salt is allowed to drain off and the unit is immersed in a hot-water wash bath. Usually a chemical bath is required to remove all traces of the flux. Aluminum heat-transfer

units for automobile air-conditioning systems are brazed by this method, using halide salts (usually chloride-based salts).

Soldering

Soldering is a metallurgical bonding process in which the filler metal is a dissimilar metal or alloy having a melting temperature range below 425 °C (800 °F) and appreciably below that of the alloy being joined. Solders have not only appreciably lower melting temperatures than brazing alloys but also

Table 8 Flow sheet of procedures for brazing of aluminum

Torch brazing	Furnace brazing	Dip brazing
Preparation of materials (cutting, bending, drawing and other forming)	Preparation of materials (cutting, bending, drawing and other forming)	Preparation of materials (cutting, bending, drawing and other forming)
Preparation of torch equipment	Preparation of equipment (furnace at proper temperature: cleaning solutions ready)	Preparation of equipment (furnace and flux bath at proper temperature: cleaning solutions ready)
Degreasing or cleaning of all material	Degreasing or cleaning of all material	Degreasing or cleaning of all material
Chemical etch cleaning or wire brushing	Chemical etch cleaning	Chemical etch cleaning
Flux preparation and application	Flux preparation and application(a)	. . .
Assembly (jigging if necessary)	Assembly (jigging if necessary)	Assembly (preferably self-jigging or minimum jigging)
. . .	. . .	Preheat in furnace
Brazing with torch	Brazing in furnace	Brazing in flux bath
Cooling or quenching	Cooling or quenching	Cooling or quenching
Hot-water washing	Hot-water washing(a)	Hot-water washing
Chemical flux removal	Chemical flux removal(a)	Chemical flux removal
Finishing and inspection	Finishing and inspection	Finishing and inspection

(a) Required for air furnace brazing, not for vacuum furnace brazing.

Table 9 Brazing filler metals for joining aluminum alloys

Designations AWS A5.8	AA	Product forms	Nominal composition, % Si	Cu	Mg	Zn	Approximate melting range °C	°F	Nominal brazing range °C	°F	Applicable brazing processes	Remarks
BAlSi-2	4343	Sheet, cladding	7.5	. . .	. . .	. . .	577-613	1070-1135	599-621	1110-1150	Furnace, dip	
BAlSi-3	4145	Wire, rod, sheet	10	4	. . .	. . .	521-585	970-1085	571-604	1060-1120	Torch, furnace, dip	Desirable where control of fluidity is necessary
BAlSi-4	4047	Wire, rod, sheet, powder	12	. . .	. . .	. . .	577-582	1070-1080	582-604	1080-1120	Torch, furnace, dip	Fluid in entire brazing range
BAlSi-5	4045	Sheet, cladding	10	. . .	. . .	. . .	577-591	1070-1095	588-604	1090-1120	Furnace, dip	. . .
BAlSi-6	. . .	Sheet	7.5	. . .	2.5	. . .	559-607	1038-1125	599-621	1110-1150	Furnace, dip	. . .
BAlSi-7	. . .	Sheet	10.0	. . .	1.5	. . .	559-591	1038-1105	588-604	1090-1120	Furnace, dip	. . .
BAlSi-8	. . .	Wire, sheet	12.0	. . .	1.5	. . .	559-579	1038-1075	582-604	1080-1120	Torch, furnace, dip	. . .

Table 10 Types of clad aluminum brazing sheet

Brazing sheet(a)	Optimum brazing range °C	°F	Core alloy	Cladding alloy	Sides clad	Cladding on each side,* % of sheet thickness
11	600 to 620	1110 to 1150	3003	.4343	1	10% for 1.6 mm (0.063 in.) and less; 5% for 1.6 mm (0.064 in.) and over
12	600 to 620	1110 to 1150	3003	.4343	2	10% for 1.6 mm (0.063 in.) and less; 5% for 1.6 mm (0.064 in.) and over
21	600 to 615	1110 to 1140	6951	.4343	1	10% for 2.3 mm (0.090 in.) and less; 5% for 2.3 mm (0.091 in.) and over
22	600 to 615	1110 to 1140	6951	.4343	2	10% for 2.3 mm (0.090 in.) and less; 5% for 2.3 mm (0.091 in.) and over
23	590 to 605	1090 to 1120	6951	.4045	1	10% for 2.3 mm (0.090 in.) and less; 5% for 2.3 mm (0.091 in.) and over
24	590 to 605	1090 to 1120	6951	.4045	2	10% for 2.3 mm (0.090 in.) and less; 5% for 2.3 mm (0.091 in.) and over

(a) Designations registered with the Aluminum Association.

Table 11 Properties and typical applications of solders used in joining aluminum

Nominal composition	Solidus °C	°F	Liquidus °C	°F	Width of melting range °C	°F	Typical applications
Tin-Zinc							
91Sn-9Zn	199	390	199	390	0	0	Aluminum to itself
80Sn-20Zn	199	390	269	518	70	128	Aluminum to itself
70Sn-30Zn	199	390	311	592	112	202	Aluminum to itself
60Sn-40Zn	199	390	340	645	141	255	Aluminum to itself
30Sn-70Zn	199	390	375	708	176	318	Aluminum to itself
Cadmium-Silver and Cadmium-Zinc							
95Cd-5Ag	338	640	393	740	55	100	. . .
82.5Cd-17.5Zn	265	509	265	509	0	0	Aluminum to itself and other metals
40Cd-60Zn	265	09	35	635	70	126	Aluminum to itself and other metals
10Cd-90Zn	265	09	99	750	134	241	Aluminum to itself and other metals
Zinc-Aluminum							
95Zn-5Al	382	720	382	720	0	0	Aluminum to itself and to copper
Indium							
50Sn-50In	117	243	125	257	8	14	Cryogenic assemblies; good for glass
7.5Pb-37.5Sn-25In	138	230	138	230	0	0	Cryogenic assemblies
50Pb-50In	180	356	209	408	29	52	Cryogenic assemblies; good resistance to alkalis

much wider melting ranges. Various soldering alloys for joining aluminum are described in Table 11.

Solders can be obtained in various forms such as rod, wire, powders, sheet and reaction fluxes. There are many processes for soldering aluminum. Large numbers of aluminum heat ex-changers of various types are soldered by hand torch soldering, machine torch soldering, air furnace soldering, dip-solder pot soldering and ultrasonic hand or dip soldering.

If a flux has been used in soldering, it must be removed to prevent subsequent corrosion. The flux is best removed immediately after soldering, while the assembly is still hot, by immersion in a hot water tank. If tubular assemblies are being produced, hot water should be recirculated through the assembly to completely remove the flux.

Because solders are dissimilar to the metal being joined, galvanic cells can exist in the presence of an electrolyte, resulting in corrosive attack. Therefore, soldered joints in applications where moisture is expected should be protected from the moisture.

The solderabilities of various wrought aluminum alloys are given in Table 2 in the article entitled "Aluminum Mill Products" in this volume.

Adhesive Bonding

There are many adhesives used for bonding aluminum alloys. Among the more common types are those made from various natural and synthetic elastomers such as natural rubber; butadiene-styrene, butyl, acrylonitride, polysulfide and neoprene synthetic rubbers; modified epoxy resin adhesives; and vinyl plastisol adhesives, the latter two being most extensively investigated. These types of adhesives cover a range of tensile-shear bond strengths from about 3 to 35 MPa (400 to 5000 psi), and have curing schedules that generally fit the paint baking schedules used in finishing automobile body assemblies. A list of adhesives used in the automotive industry are given in Aluminum Association publication T14, "Adhesive Bonding of Aluminum Automotive Body Sheet Alloys".

Adhesive-bonded joints have several advantages:

1 The load is more evenly distributed, and stress concentrations are minimized. As a result, bonded joints have longer fatigue lives than joints made by spot welding or mechanical fastening.

2 Adhesive-bonded joints help in damping vibrations.

3 Dissimilar materials, with different coefficients of thermal expansion, can be joined with adhesives

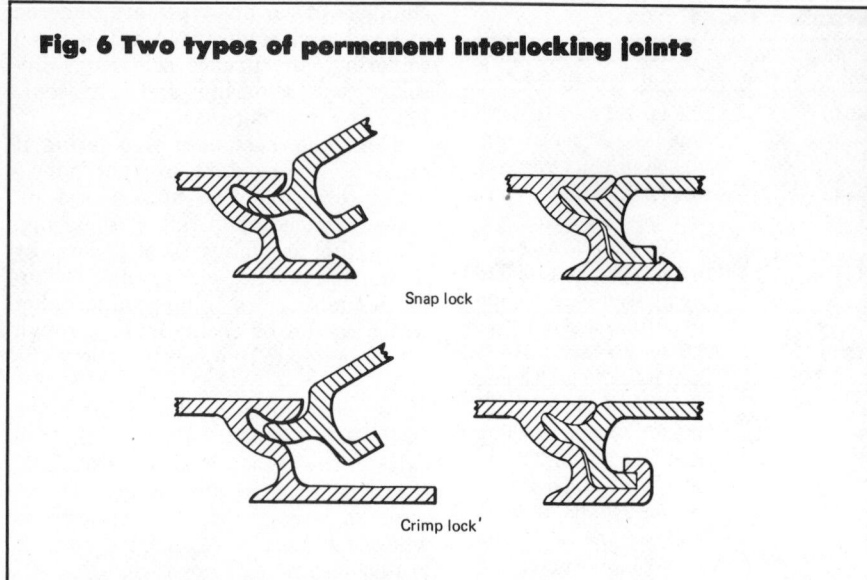

Fig. 6 Two types of permanent interlocking joints

Snap lock

Crimp lock'

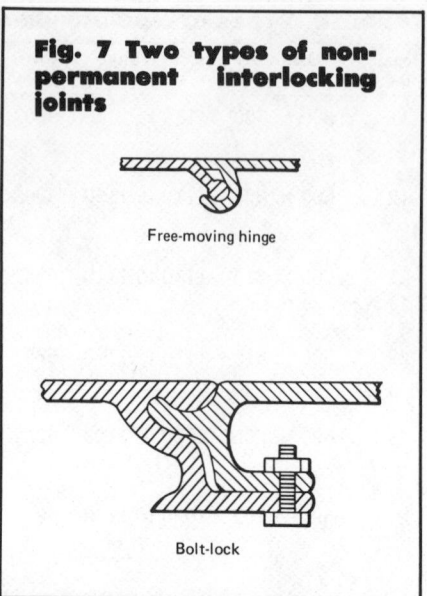

Fig. 7 Two types of non-permanent interlocking joints

Free-moving hinge

Bolt-lock

having selected flexibility characteristics.

4 In contrast to fusion joining processes, metals of greatly different thicknesses can be assembled easily at low, room or elevated temperatures.

5 Adhesive bonding is the only process for continuous joining of metals to nonmetals. However, adhesive joints generally are not satisfactory where peel loads are applied between the two joined members. Adhesive bonding of alclad alloys should be avoided because the cladding, which is sacrificial to the core, may be attacked in corrosive environments, resulting in debonds.

Weldbonding

Weldbonding produces a joint by using a combination of resistance spot welding and adhesive bonding. Usually, the joint is made by resistance spot welding through an uncured adhesive, although a capillary-application procedure is possible. Further information is available in the Aluminum Association publication T17, "Weldbonding—An Alternative Joining Method for Aluminum Auto Body Alloys". The advantages of weldbonded joints are:

1 Resistance to peel forces is better than that obtained by spot welding or adhesive bonding alone.
2 Fatigue life and durability are superior to those of spot welded or mechanical joints.
3 The adhesive acts as a seal at the

joint to provide better corrosion protection.

The main disadvantage of weldbonding is the problem of producing good spot welds through an adhesive and keeping the electrodes free of adhesive.

Mechanical Fastening

Aluminum assemblies can be joined by the same mechanical fastening processes used for other metals. These include riveting, bolting, nailing, stitching and stapling, and clinching. Adhesives can be used with mechanical fastening to improve over-all strength of the joint. These combination joining processes produce joints with characteristics similar to those of well-made adhesive-bonded joints.

Riveting. Although welding has superseded riveting for many fastening applications, riveting is still used to make joints in such structures as aircraft, small boats and van trailers. Rivets produce high-strength joints without the distortion and loss of strength that result from high welding temperatures, and less skill is required to make a riveted joint. Corrosion can be prevented in the lap by sealing the faying surfaces with special joint compounds.

Aluminum alloys should be joined with aluminum rivets. Small aluminum rivets are driven cold whereas large aluminum rivets are driven hot. Special rivets are available for blind riveting, where only one side of the joint is accessible. Some of the special

rivets available are pop rivets, cherry rivets, Chobert rivets and explosive rivets.

In high-strength applications such as aircraft, heat treated rivets in alloys 2017-T4 and 2024-T4 are often used. These rivets are refrigerated immediately after solution heat treating and quenching, to prevent natural aging, and are driven immediately after they are taken from refrigeration, while they are still soft. Rivets of alloy 2219 are also used in high-strength applications. These rivets do not require refrigeration to prevent natural aging.

Bolting. Bolted joints have high joint strength and efficiency, are free from heat effects, are easy to assemble and inspect, and can be disassembled into reusable components. Aluminum parts may be joined with either steel or aluminum threaded fasteners—steel for high-performance structural connections and aluminum for moderate-strength connections requiring good appearance and freedom from corrosion. For joining aluminum to steel, galvanized or cadmium-plated steel or stainless steel threaded fasteners are recommended.

Nailing. Aluminum nails are widely used to attach aluminum siding and roofing. These nails are made from alloys 6061 or 5056 and are slightly larger than steelnails having the same size number.

Stitching and stapling are rapid methods of connecting thin sheets of metal to each other or to other materials. Stitching wire and staples for use on aluminum sheet are made from high-strength stainless or galvanized steel wire.

Clinching offers an attractive alternative to spot welding of sheet. Machines in current use with aluminum typically make an indentation 6 to 13 mm (0.25 to 0.50 in.) in diameter through the overlapping sheets. This indentation is flattened slightly on the opposite side to make a good mechanically locked joint. These joints exhibit tensile and peel strengths similar to those of spot welded joints.

Interlocking joints can be designed into aluminum extrusions for rapid and inexpensive connection to sheet or to other extrusions. The connections provided by such joints have good appearance and good strength. In addition, interlocking joints can be designed as permanent joints, such as those shown in Fig. 6, or can be hinged for free movement or semipermanent rigid connection (see Fig. 7). Interlocking joints constitute only one of several types of interconnection that can be designed into extrusions. For a more comprehensive discussion of interconnecting extrusions, see the article entitled "Aluminum Mill Products".

Minimizing Corrosion

Most aluminum alloys have good to excellent resistance to corrosion. Because there are differences in corrosion resistance, the properties of individual alloys should be considered for each application. Corrosion usually will not undercut protective and decorative coatings on aluminum, even when the coating is scratched. If bright-finished, corrosion-resistant products are required, anodizing can be specified.

Galvanic corrosion may be encountered where aluminum is coupled to steel or to any other dissimilar metal in the presence of an electrolyte. Corrosion of this type usually can be reduced or eliminated by taking the following precautions:

1 Avoid designs that include crevices and spaces in which dirt and debris can collect and form a poultice.
2 Keep electrolyte out of the joint. This can be done by placing sealants in hemmed joints and in bolted or riveted connections.
3 Replace bare steel with galvanized steel, aluminized steel or stainless steel when joining aluminum to steel.
4 Proper control of quenching and precipitation heat treating practices is especially important for those heat treatable alloys and special tempers susceptible to stress-corrosion cracking.

In attaching corrugated aluminum sheet to a steel-panel building, one or more of the following precautions must be taken to prevent galvanic corrosion:

1 The steel should be coated with zinc chromate or other suitable primer and at least one coat of aluminum metal-and-masonry paint or other suitable protective coating, excluding those containing lead pigments.
2 The steel should be painted with a coating of heavy-bodied bituminous paint.
3 A good-quality sealant should be placed between the aluminum and the steel.
4 A nonabsorptive tape or gasket should be placed between the aluminum and the steel.
5 When aluminum sheet is attached to wood, special precautions should be followed to prevent corrosion.
6 Aluminum or stainless steel bolts, screws or nails with aluminum-neoprene composite washers should be used to both avoid galvanic corrosion and to keep out moisture.

Corrosion Resistance of Aluminum and Aluminum Alloys

By E. H. Hollingsworth
and
H. Y. Hunsicker
Aluminum Company of America

As indicated by its position in the electromotive force series, aluminum is a thermodynamically reactive metal; among structural metals, only beryllium and magnesium and zinc are more reactive. Aluminum owes its excellent resistance to corrosion and its usage as one of the primary metals of commerce to the aluminum oxide film that is bonded strongly to its surface and that, if damaged, re-forms immediately in most environments. On a surface freshly abraded and then exposed to air, the oxide film is only 5 to 10 nm (0.2 to 0.4 μin.) thick but is highly effective in protecting the aluminum from corrosion.

The conditions for thermodynamic stability of the oxide film are expressed by the Pourbaix diagram of potential vs pH shown in Fig. 1. As shown by this diagram, aluminum is passive (is protected by its oxide film) in the pH range of about 4 to 8.5. The limits of this range, however, vary somewhat with temperature, the form of oxide film present, and the rate at which this film dissolves in the electrolyte. The relative inertness in the passive range is further illustrated in Fig. 2, which gives results of weight-loss measurements for alloy 3004-H14 specimens exposed in water and in solutions at various pH values (see Fig. 2).

Beyond the limits of its passive range, aluminum corrodes, by simple chemical reactions, in both acidic and alkaline environments, yielding Al^{+++} ions in the former and AlO_2^- (aluminate) ions in the latter. There are a few instances where corrosion does not occur—either where the oxide film is not soluble or where it is maintained by the oxidizing nature of the solution.

Pitting Corrosion

Corrosion of aluminum in the passive range is localized, usually manifested by random formation of pits. The pitting-potential principle establishes the conditions under which metals in the passive state are subject to corrosion by pitting. Simply stated, pitting potential is that potential in a particular solution above which pits will initiate and below which they will not.

Pitting potential of a metal is determined by the following procedure. A variable external potential is impressed between the metal and cathode, which are immersed in an electrolyte free of cathodic reactant. The potential is increased until the anode (the metal) is polarized, and the resulting current is measured. In one experiment, an aluminum 1100 specimen was immersed in neutral deaerated sodium chloride solution, and the relationship between anode potential and current was plotted (solid line in Fig. 3). When anode polarization occurred, the potential leveled off at the value of the pitting potential, E_p. The potential-current relationships for various cathodic reactions are indicated by the dashed curves in Fig. 3. Only when cathodic potential is sufficient to polarize the metal to its pitting potential will significant current flow and pitting occur.

Fig. 1 Pourbaix diagram for aluminum (Ref 1)

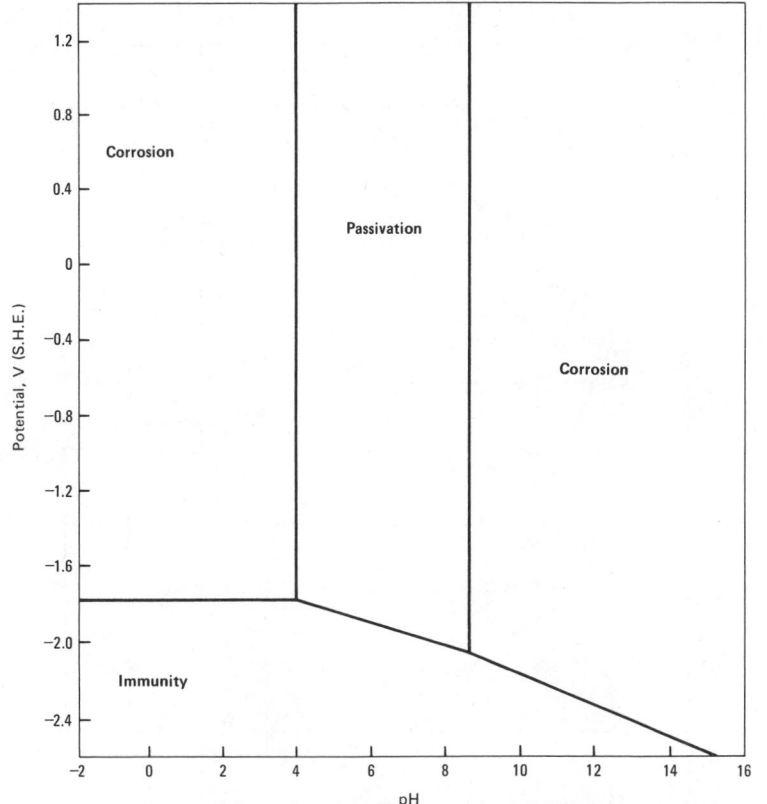

Diagram is for aluminum with an oxide film of hydragillite (Al$_2$O$_3$, 3H$_2$O) at 25 °C (77 °F). Potential values are for the standard hydrogen electrode scale.

Fig. 2 Weight loss of alloy 3004-H14 exposed 1 week in distilled water and in solutions of various pH values

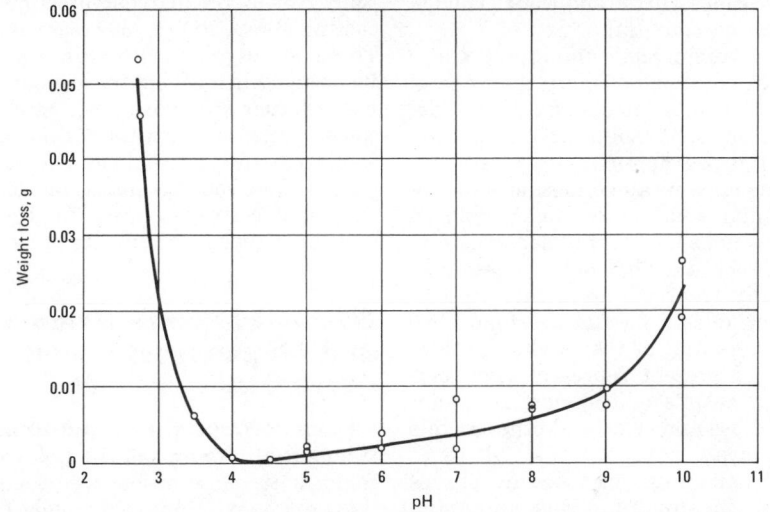

Specimens were 1.6 × 13 × 75 mm (0.06 × 0.5 × 3.0 in.). The pH values of solutions were adjusted with HCl and NaOH. Test temperature was 60 °C (140 °F).

Fig. 3 Anodic polarization curve for aluminum 1100

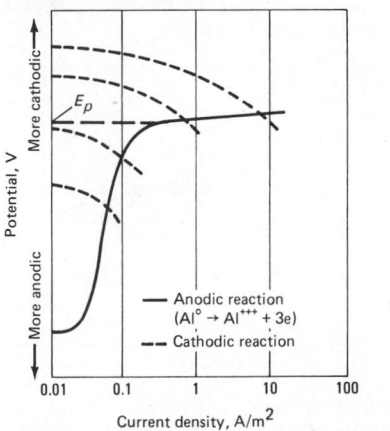

Specimens were immersed in neutral deaerated NaCl solution free of cathodic reactant. Pitting develops only at potentials more cathodic than the pitting potential, E$_p$. The intersection of the anodic curve for aluminum (solid line) with a curve for the applicable cathodic reaction (one of the representative dashed lines) determines the potential to which the aluminum is polarized, either by cathodic reaction on the aluminum itself, or on another metal connected to it electrically. The potential to which the aluminum is polarized by a specific cathodic reaction determines corrosion current density and corrosion rate.

For aluminum, pitting corrosion is most commonly produced by halide ions, of which chloride ions are most frequently encountered in service. The effect of chloride-ion activity on the pitting potential of aluminum 1199 (99.99+% Al) is shown in Fig. 4. Pitting of aluminum in aerated halide solutions occurs because, in the presence of the cathodic reactant O$_2$, the metal is readily polarized to its pitting potential—that is, it develops an electrode potential in these solutions that equals or exceeds the pitting potential. In deaerated solutions of equal chloride concentration, in the absence of the cathodic reactant O$_2$, aluminum may not corrode by pitting because it is not polarized to its pitting potential. Generally, aluminum does not develop pitting in aerated solutions of most nonhalide salts because its pitting potentials in these solutions are considerably more cathodic than those in halide solutions, and it is not polarized to these potentials.

Pitting potentials for several aluminum alloys in several electrolytes are reported in Ref 2 and 3. Examples of

Fig. 4 Effect of chloride-ion activity on pitting potential of aluminum 1199 (99.99 + % Al) in NaCl solutions

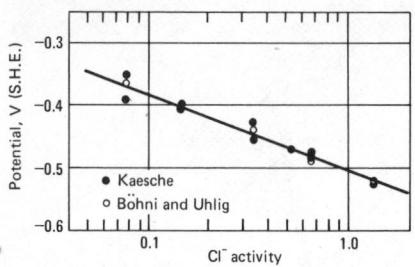

Fig. 5 Effects of principal alloying elements on electrolytic solution potential of aluminum

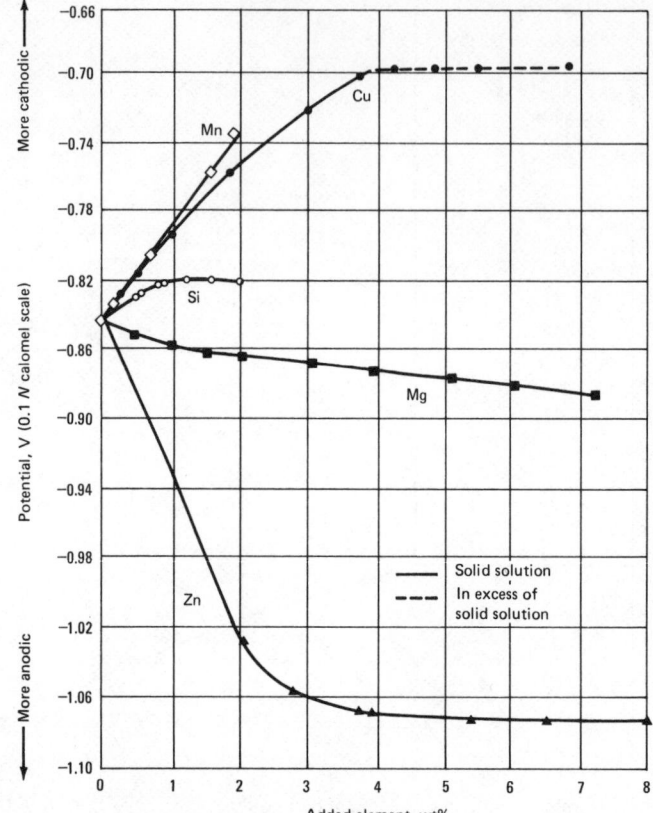

Potentials are for high-purity binary alloys solution heat treated and quenched. Measured in a solution of 53 g/L NaCl plus 3 g/L H₂O₂ maintained at 25 °C (77 °F).

application of pitting-potential analysis to particular corrosion problems are presented in Ref 4 and 5.

Electrolytic Solution Potentials

Because of the electrochemical nature of most corrosion processes, relationships among solution potentials of different aluminum alloys, and between potentials of aluminum alloys and those of other metals, are of considerable importance. Furthermore, the solution-potential relationships among the microstructural constituents of a particular alloy significantly affect its corrosion behavior. Compositions of solid solutions and additional phases, as well as amounts and spatial distributions of the additional phases, may affect both type and extent of corrosion.

The solution potential of an aluminum alloy is determined primarily by the composition of the aluminum-rich solid solution, which constitutes the predominant volume fraction and area fraction of the alloy microstructure. Solution potential is not affected significantly by second-phase particles of microscopic size, but because these particles frequently have solution potentials differing from that of the solid-solution matrix in which they occur, localized galvanic cells may be formed between them and the matrix.

The effects of principal alloying elements on solution potential of high-purity aluminum are shown in Fig. 5. For each element, the significant changes that occur do so within the range in which the element is completely in solid solution. Further addition of the same element, which forms a second

phase, causes little additional change in solution potential.

Most commercial aluminum alloys contain additions of more than one of these elements; effects of multiple elements in solid solution on solution potential are approximately additive. The amounts retained in solid solution, particularly for more highly alloyed compositions, are highly dependent on fabrication and thermal processing so that heat treatment and other processing variables influence the final electrode potential of the product. Tables 1a to 1d present representative solution potentials of commercial aluminum alloys and of several other metals and alloys.

The amounts of second phases present in aluminum and aluminum alloy products vary from nearly zero in those of aluminum 1199 and some others that also are nearly pure solid solu-

tions to over 20% in hypereutectic Al-Si casting alloys, such as 392.0 and 393.0. These phases generally are intermetallic compounds of binary, ternary or higher-order compositions, although some elements in excess of their solid solubility are present as elemental phases. Electrode potentials of some of the simpler second-phase constituents have been measured and are presented in Table 2.

Effects of Composition and Microstructure on Corrosion

1xxx Wrought Aluminums. Wrought aluminums of the 1xxx series conform to composition specifications that set maximum individual, combined and total contents for several elements present as natural impurities in the smelter-grade or refined alumi-

Table 1a Solution potentials of non-heat-treatable commercial wrought aluminum alloys(a)

Alloy	Potential(b), V
1060	−0.84
1100	−0.83
3003	−0.83
3004	−0.84
5050	−0.84
5052	−0.85
5154	−0.86
5454	−0.86
5056	−0.87
5456	−0.87
5182	−0.87
5083	−0.87
5086	−0.85
7072	−0.96

(a) Values are the same for all tempers of each alloy. (b) Potential vs standard calomel electrode.

Table 1b Solution potentials of heat treatable commercial wrought aluminum alloys

Alloy	Temper	Potential (a), V
2014	T4	−0.69(b)
	T6	−0.78
2219	T3	−0.64(b)
	T4	−0.64(b)
	T6	−0.80
	T8	−0.82
2024	T3	−0.69(b)
	T4	−0.69(b)
	T6	−0.81
	T8	−0.82
2036	T4	−0.72
6009	T4	−0.80
6010	T4	−0.79
6151	T6	−0.83
6351	T5	−0.83
6061	T4	−0.80
	T6	−0.83
6063	T5	−0.83
	T6	−0.83
7005	T6	−0.94
X7016	T6	−0.86
X7021	T6	−0.99
X7029	T6	−0.85
X7146	T6	−1.02
7049	T73	−0.84(c)
	T76	−0.84(c)
7050	T73	−0.84(c)
	T76	−0.84(c)
7075	T6	−0.83(c)
	T73	−0.84(c)
	T76	−0.84(c)
7475	T6	−0.83(c)
	T73	−0.84(c)
	T76	−0.84(c)
7178	T6	−0.83(c)

(a) Potential vs standard calomel electrode. (b) Varies ±0.01 V with quenching rate. (c) Varies ±0.02 V with quenching rate.

Table 1c Solution potentials of cast aluminum alloys

Alloy	Temper	Type of mold(a)	Potential (b), V
208.0	F	S	−0.77
238.0	F	P	−0.74
295.0	T4	S or P	−0.70
	T6	S or P	−0.71
	T62	S or P	−0.73
296.0	T4	S or P	−0.71
308.0	F	P	−0.75
319.0	F	S	−0.81
	F	P	−0.76
355.0	T4	S or P	−0.78
	T6	S or P	−0.79
356.0	T6	S or P	−0.82
443.0	F	S	−0.83
	F	P	−0.82
514.0	F	S	−0.87
520.0	T4	S or P	−0.89
710.0	F	S	−0.99

(a) S, sand; P, permanent. (b) Potential vs standard calomel electrode.

Table 1d Solution potentials of some nonaluminum base metals

Metal	Potential(a), V
Magnesium	−1.73
Zinc	−1.10
Cadmium	−0.82
Mild carbon steel	−0.58
Lead	−0.55
Tin	−0.49
Copper	−0.20
Bismuth	−0.18
Stainless steel(b)	−0.09
Silver	−0.08
Nickel	−0.07
Chromium	−0.40 to +0.18

(a) Potential vs standard calomel electrode. (b) Series 300, type 430.

Table 2 Electrode potentials of some second-phase constituents in aluminum alloys

Phase	Potential(a), V
Si	−0.26
Al_3Ni	−0.52
Al_3Fe	−0.56
Al_2Cu	−0.73
Al_6Mn	−0.85
Al_8Mg_5	−1.24

(a) Potential vs standard calomel electrode.

the ratio of iron to silicon and on thermal history. The microstructural particles of these phases are cathodic to the aluminum solid solution, and exposed surfaces of these particles are covered by an oxide film thinner than that covering exposed areas of the solid solution. Corrosion may be initiated earlier and progress more rapidly in the aluminum solid solution immediately surrounding the particles. The number and/or size of such corrosion sites is proportional to the area fraction of the second-phase particles.

It should be noted that not all impurity elements are detrimental to corrosion resistance of lxxx series aluminums and that detrimental elements may reduce resistance of some types of alloys but have no ill effects in others. Therefore, specification limitations established for impurity elements often are based on maintaining consistent and predictable levels of corrosion resistance in a variety of applications rather than on their effects in any specific application.

2xxx wrought alloys and 2$xx.x$ casting alloys, in which copper is the major alloying element, are less resistant to corrosion than alloys of other series, which contain much lower amounts of copper. Alloys of this type were the first heat treatable, high-strength aluminum-base materials, and have been used for more than 50 years in structural applications—most prominently in aircraft and aerospace applications. Much of the thin sheet made of these alloys is produced as an alclad composite, but thicker sheet and other products in many applications require no protective cladding.

Electrochemical effects on corrosion can be stronger in these alloys than in alloys of many other types because of two factors: greater change in electrode potential with variations in amount of copper in solid solution (see Fig. 5) and, under some conditions, the presence of

num used to produce these products. Aluminums 1100 and 1135 differ somewhat from the others in this series in having minimum as well as maximum specified copper contents. Corrosion resistance of all lxxx compositions is very high, but under many conditions decreases slightly with increasing alloy content. Iron, silicon and copper are the elements present in the largest percentages. The copper and part of the silicon are in solid solution. The second-phase particles present contain either iron or iron and silicon—Al_6Fe, Al_3Fe and $Al_{12}Fe_3Si$. The specific phase present, or the relative amounts when more than one are present, depend on nonuniformities in solid-solution concentration. However, the fact that general resistance to corrosion decreases with increasing copper content is not primarily attributable to these solid-solution or second-phase-solution potential relationships, but to galvanic cells created by formation of minute copper particles or films deposited on the alloy surface as a result of corrosion. As corrosion progresses, copper ions, which initially go into solution, replate onto the alloy to form metallic copper cathodes. Reduction of copper ions and increased efficiency of O_2 and H^+ reduction reactions in the presence of copper increase corrosion rate.

These alloys invariably are solution heat treated and are used in either the naturally aged or the precipitation heat treated temper. Development of these tempers using good heat treating practice can minimize electrochemical effects on corrosion resistance.

3xxx Wrought Alloys. Wrought alloys of the 3xxx series (Al-Mn and Al-Mn-Mg) also have very high resistance to corrosion. The manganese is present in the aluminum solid solution, in submicroscopic particles of precipitate, and in larger particles of Al_6 (Mn, Fe) or Al_{12} (Mn, Fe)$_3$ Si phases, both of which have solution potentials almost the same as that of the solid-solution matrix. Such alloys are widely used for cooking and food-processing equipment, chemical equipment and a variety of architectural products requiring high resistance to corrosion.

4xxx Wrought Alloys and 3xx.x and 4xx.x Casting Alloys. Elemental silicon is present as second-phase constituent particles in wrought alloys of the 4xxx series, in brazing and welding alloys, and in casting alloys of the 3xx.x and 4xx.x series. Silicon is cathodic to the aluminum solid-solution matrix by several hundred millivolts and accounts for a considerable area fraction of most of the silicon-containing alloys. However, the effects of silicon on the corrosion resistance of these alloys are minimal because of low corrosion current density resulting from the fact that the silicon particles are highly polarized.

Corrosion resistance of 3xx.x casting alloys is strongly affected by copper content, which can be as high as 5% in some compositions, as well as by impurity levels. Modifications of certain basic alloys have more restrictive limits on impurities, which benefit corrosion resistance and mechanical properties.

5xxx Wrought Alloys and 5xx.x Casting Alloys. Wrought alloys of the 5xxx series (Al-Mg-Mn, Al-Mg-Cr and Al-Mg-Mn-Cr) and casting alloys of the 5xx.x series (Al-Mg) have high resistance to corrosion, and this accounts in part for their use in a wide variety of building products and chemical-processing and food-handling equipment, as well as applications involving exposure to seawater.

Alloys in which the magnesium is present in amounts that remain in solid solution, or is partially precipitated as Al_8Mg_5 particles dispersed uniformly throughout the matrix, are as generally resistant to corrosion as commercially pure aluminum, and more resistant to salt water and some alkaline solutions such as those of sodium carbonate and amines. The wrought alloys containing about 3% or more magnesium under conditions that lead to an almost continuous intergranular Al_8Mg_5 precipitate, with very little precipitate within the grains, may be susceptible to exfoliation or stress-corrosion cracking. Tempers have been developed for these higher-magnesium wrought alloys to produce microstructures having extensive Al_8Mg_5 precipitate within the grains, thus eliminating such susceptibility.

In the 5xxx alloys that contain chromium, this element is present as a submicroscopic precipitate, $Al_{12}Mg_2Cr$. Manganese in these alloys is in the form of Al_6 (Mn, Fe), both submicroscopic and larger particles. Such precipitates and particles do not adversely affect corrosion resistance of these alloys.

6xxx Wrought Alloys. Moderately high strength and very good resistance to corrosion combine to make the heat treatable wrought alloys of the 6xxx series (Al-Mg-Si) highly suitable in a wide variety of structural, building, marine, machinery, process-equipment and other applications. The Mg_2Si phase, which is the basis for precipitation hardening, is unique in that it is an ionic compound, and is not only anodic to aluminum but also reactive in acidic solutions. However, either in solid solution or as submicroscopic precipitate, Mg_2Si has a negligible effect on electrode potential. Because these alloys normally are used in the heat treated condition, no detrimental effects derive from the major alloying elements or from the supplementary chromium, manganese or zirconium, which are added for control of grain structure. Copper additions,

which augment strength in many of these alloys are limited to small amounts to minimize effects on corrosion resistance. In general, the level of resistance decreases somewhat with increasing copper content.

7xxx wrought alloys and 7xx.x casting alloys contain major additions of zinc along with magnesium or magnesium plus copper in combinations that develop various levels of strength. Those containing copper have the highest strengths and have been used as constructional materials—principally in aircraft applications—for more than 30 years. The copper-free alloys of the series have many desirable characteristics: moderate to high strength, excellent toughness, and good workability, formability and weldability. Use of these copper-free alloys has increased in recent years and now includes automotive applications (such as bumpers), structural members and armor plate for military vehicles, and components of other transportation equipment.

The 7xxx wrought and 7xx.x casting alloys, because of their zinc contents, are anodic to 1xxx wrought aluminums and to other aluminum alloys. They are among the aluminum alloys most susceptible to stress-corrosion cracking. However, stress-corrosion cracking can be avoided by proper alloy and temper selection and observance of appropriate design, assembly and application precautions. Stress-corrosion cracking of aluminum alloys is discussed in greater detail in a subsequent section.

Resistance to general corrosion of the copper-free wrought 7xxx alloys is good, approaching those of wrought 3xxx, 5xxx and 6xxx alloys. The copper-containing alloys of the 7xxx series, such as 7049, 7050, 7075 and 7178, have lower resistance to general corrosion than those of the same series that do not contain copper. All 7xxx alloys are more resistant to general corrosion than 2xxx alloys, but less resistant than wrought alloys of other groups.

Although the copper in both wrought and cast alloys of the Al-Zn-Mg-Cu type has a detrimental effect on resistance to general corrosion, it is beneficial from the standpoint of resistance to stress-corrosion cracking. Copper allows these alloys to be precipitated at higher temperatures without excessive loss in strength, and thus makes possible the development of T73-type tempers, which couple high strength with excellent resistance to stress-corrosion cracking.

Table 3 Relative ratings of resistance to general corrosion and to stress-corrosion cracking of wrought aluminum alloys

Alloy	Temper	Resistance to corrosion General(a)	Stress-corrosion cracking(b)	Alloy	Temper	Resistance to corrosion General(a)	Stress-corrosion cracking(b)
1060	All	A	A	5083	All	A(d)	B(d)
1100	All	A	A	5086	O, H32, H116	A(d)	A(d)
1350	All	A	A		H34, H36, H38, H111	A(d)	A(d)
2011	T3, T4, T451	D(c)	D	5154	All	A(d)	A(d)
	T8	D	B	5252	All	A	A
2014	O	...	...	5254	All	A(d)	A(d)
	T3, T4, T451	D(c)	C	5454	All	A	A
	T6, T651, T6510, T6511	D	C	5456	All	A(d)	B(d)
2017	T4, T451	D(c)	C	5457	O	A	A
2018	T61	...	...	5652	All	A	A
2024	O	...	...	5657	All	A	A
	T4, T3, T351, T3510, T3511, T361	D(c)	C	6053	O	...	...
	T6, T861, T81, T851, T8510, T8511	D	B		T6, T61	A	A
	T72	...	...	6061	O	B	A
2025	T6	D	C		T4, T451, T4510, T4511	B	B
2036	T4	C	...		T6, T651, T652, T6510, T6511	B	A
2117	T4	C	A	6063	All	A	A
2218	T61, T72	D	C	6066	O	C	A
2219	O	...	...		T4, T4510, T4511, T6, T6510, T6511	C	B
	T31, T351, T3510, T3511, T37	D(c)	C	6070	T4, T4511, T6	B	B
	T81, T851, T8510, T8511, T87	D	B	6101	T6, T63, T61, T64	A	A
2618	T61	D	C	6151	T6, T652	...	...
3003	All	A	A	6201	T81	A	A
3004	All	A	A	6262	T6, T651, T6510, T6511, T9	B	A
3105	All	A	A	6463	All	A	A
4032	T6	C	B	7001	O	C(c)	C
5005	All	A	A	7075	T6, T651, T652, T6510, T6511	C(c)	C
5050	All	A	A		T73, T7351	C	B
5052	All	A	A	7178	T6, T651, T6510, T6511	C(c)	C
5056	O, H11, H12, H32, H14, H34	A(d)	B(d)				
	H18, H38	A(d)	C(d)				
	H192, H392	B(d)	D(d)				

(a) Ratings A through E are relative ratings in decreasing order of merit, based on exposures to NaCl solution by intermittent spraying or immersion. Alloys with A and B ratings can be used in industrial and seacoast atmospheres without protection. Alloys with C, D and E ratings generally should be protected, at least on faying surfaces. (b) Stress-corrosion cracking ratings are based on service experience and on laboratory tests of specimens exposed to alternate immersion in 3.5% NaCl solution. A—no known instance of failure in service or in laboratory tests. B—no known instance of failure in service; limited failures in laboratory tests of short transverse specimens. C—service failures when sustained tension stress acts in short transverse direction relative to grain structure; limited failures in laboratory tests of long transverse specimens. D—limited service failures when sustained stress acts in longitudinal or long transverse direction relative to grain structure. (c) In relatively thick sections the rating would be E. (d) This rating may be different for material held at elevated temperatures for long periods.

Effects of Additional Alloying Elements. In addition to the major elements that define the various alloy systems discussed above, commercial aluminum alloys may contain other elements that provide special characteristics. Lead and bismuth are added to alloys 2011 and 6262 to improve chip breakage and other machining characteristics. Nickel is added to wrought alloys 2018, 2218 and 2618, which were developed for elevated-temperature service, and to certain 3xx.x cast alloys used for pistons, cylinder blocks and other engine parts subjected to high temperatures. Cast aluminum bearing alloys of the 850.0 group contain tin. In all instances, these alloying additions introduce constituent phases that are cathodic to the matrix and decrease resistance to corrosion in aqueous saline media. In many instances, however, these alloys are used in environments in which they are not subject to corrosion.

Corrosion Ratings of Alloys and Tempers

Simplified ratings of resistance to general corrosion and to stress-corrosion cracking for wrought and cast aluminum alloys are presented in Tables 3 and 4. These ratings may be of assistance in making rough evaluations and comparisons of alloy/temper combinations for corrosion service. (More detailed ratings of resistance to stress-corrosion cracking for high-strength wrought aluminum alloys are given in Table 6, which appears later in this article, in the section on stress-corrosion cracking.)

Galvanic Corrosion and Protection

The solution-potential values in Tables 1a to 1d, as measured against a standard calomel electrode, form a galvanic series for aluminum alloys and other metals. The galvanic relationships indicated by these values have wide applicability because of the similarity of the electrochemical behavior of these metals in the NaCl solution to that in marine and other saline environments. This galvanic series, however, is not necessarily valid in nonsaline solutions. For example, aluminum

Table 4a Relative ratings of resistance to general corrosion and to stress-corrosion cracking of aluminum casting alloys

Alloy	Temper	Resistance to corrosion	
		General(a)	Stress-corrosion cracking(b)
Sand Castings			
208.0	F	B	B
224.0	T7	C	B
240.0	F	D	C
242.0	All	D	C
A242.0	T75	D	C
249.0	T7	C	B
295.0	All	C	C
319.0	F, T5	C	B
	T6	C	C
355.0	All	C	A
C355.0	T6	C	A
356.0	T6, T7, T71, T51	B	A
A356.0	T6	B	A
443.0	F	B	A
512.0	F	A	A
513.0	F	A	A
514.0	F	A	A
520.0	T4	A	C
535.0	F	A	A
B535.0	F	A	A
705.0	T5	B	B
707.0	T5	B	C
710.0	T5	B	B
712.0	T5	B	C
713.0	T5	B	B
771.0	T6	C	C
850.0	T5	C	B
851.0	T5	C	B
852.0	T5	C	B

(a) Relative ratings of general corrosion resistance A through E are in decreasing order of merit, based on exposures to NaCl solution by intermittent spray or immersion. (b) Relative ratings of resistance to stress-corrosion cracking are based on service experience and on laboratory tests of specimens exposed to alternate immersion in 3.5% NaCl solution. A—no known instance of failure in service when properly manufactured. B—failure not anticipated in service from residual stresses or from design and assembly stresses below about 45% of the minimum guaranteed yield strength given in applicable specifications. C—failures have occurred in service with either this specific alloy/temper combination or with alloy/temper combinations of this type; designers should be aware of the potential stress-corrosion cracking problem that exists when these alloys and tempers are used under adverse conditions.

is anodic to zinc in an aqueous $1M$ Na_2CrO_4 solution and cathodic to iron in an aqueous $1M$ Na_2SO_4 solution.

Under most environmental conditions frequently encountered in service, aluminum and its alloys are the anodes in galvanic cells with most other metals, protecting them by corroding sacrificially. Only magnesium and zinc are more anodic. Sacrificial corrosion of aluminum or cadmium when these two metals are coupled in a galvanic cell is slight because of the small difference in electrode potential between them.

Contact of aluminum with more cathodic metals should be avoided in any environment in which aluminum by itself is subject to pitting corrosion. Where such contact is necessary, protective measures should be employed to minimize sacrificial corrosion of the aluminum. In such an environment, aluminum is already polarized to its pitting potential, and the additional potential imposed by contact with the more cathodic metal greatly increases the corrosion current. In many environments, aluminum can be used in contact with chromium or stainless steel with only slight acceleration of corrosion; chromium and stainless steel become highly polarized in mild environments, where they are passive, so that the corrosion current is small despite the large electrode potentials between these metals and aluminum.

To minimize corrosion of aluminum wherever contact with more cathodic metals cannot be avoided, the ratio of the exposed surface area of the aluminum to that of the more cathodic metal should be as high as possible in order to minimize the current density at the aluminum and hence the rate of corrosion. The area ratio may be increased by painting the cathodic metal or both metals, but painting only the aluminum is not effective and may even accentuate corrosion.

Corrosion of aluminum in contact with more cathodic metals is much less severe in solutions of most nonhalide salts, in which aluminum alone normally is not polarized to its pitting potential, than in solutions of halide salts, in which it is. (As shown in Fig. 3, increases in potential, as long as the value does not reach the pitting potential, have small effects on current density.)

Regardless of the solution, galvanic current between aluminum and another metal can be reduced to a low value by removal of the cathodic reactant. Thus, corrosion rate of aluminum coupled to copper in seawater is greatly reduced wherever the seawater is deaerated. In closed multimetallic systems, the corrosion rate of aluminum, although initially high, decreases to a low value whenever the cathodic reactant is depleted. Galvanic current also is low in solutions having high electrical resistivity, such as high-purity water.

Some semiconductors, such as graphite and magnetite, are cathodic to aluminum, and when in contact with them aluminum corrodes sacrificially.

Alclad Products. In alclad aluminum products, the difference in solution potential between the core alloy and the cladding alloy is used to provide cathodic protection to the core (Ref 6). These products, primarily sheet and tube, consist of a core coated on one or both surfaces with a metallurgically bonded layer of an alloy that is anodic to the core alloy. The thickness of the cladding layer usually is less than 10% of the over-all thickness of the product. Cladding alloys generally are of the non-heat-treatable type, although, for higher strength, heat treatable alloys sometimes are used. Composition relationships of core and cladding alloys generally are designed so that the cladding is 80 to 100 mV more anodic than the core. Several core alloy/cladding alloy combinations for common alclad products are listed in Table 5. Because of the cathodic protection provided by the cladding, corrosion progresses only to the core/cladding interface, and then spreads laterally. This is highly effective in eliminating perforation of thin products.

Table 4b Relative ratings of resistance to general corrosion and to stress-corrosion cracking of aluminum casting alloys

Alloy	Temper	Resistance to corrosion General(a)	Stress-corrosion cracking(b)
Permanent Mold Castings			
242.0	T571, T61	D	C
308.0	F	C	B
319.0	F	C	B
	T6	C	C
332.0	T5	C	B
336.0	T551, T65	C	B
354.0	T61, T62	C	A
355.0	All	C	A
C355.0	T61	C	A
356.0	All	B	A
A356.0	T61	B	A
F356.0	All	B	A
A357.0	T61	B	A
358.0	T6	B	A
359.0	All	B	A
B443.0	F	B	A
A444.0	T4	B	A
513.0	F	A	A
705.0	T5	B	B
707.0	T5	B	C
711.0	T5	B	A
713.0	T5	B	B
850.0	T5	C	B
851.0	T5	C	B
852.0	T5	C	B
Die Castings			
360.0	F	C	A
A360.0	F	C	A
364.0	F	C	A
380.0	F	E	A
A380.0	F	E	A
383.0	F	E	A
384.0	F	E	A
390.0	F	E	A
392.0	F	E	A
413.0	F	C	A
A413.0	F	C	A
C443.0	F	B	A
518.0	F	A	A
Rotor Metal(c)			
100.1		A	A
150.1		A	A
170.1		A	A

(a) Relative ratings of general corrosion resistance A through E are in decreasing order of merit, based on exposures to NaCl solution by intermittent spray or immersion. (b) Relative ratings of resistance to stress-corrosion cracking are based on service experience and on laboratory tests of specimens exposed to alternate immersion in 3.5% NaCl solution. A—no known instance of failure in service when properly manufactured. B—failure not anticipated in service from residual stresses or from design and assembly stresses below about 45% of the minimum guaranteed yield strength given in applicable specifications. C—failures have occurred in service with either this specific alloy/temper combination or with alloy/temper combinations of this type; designers should be aware of the potential stress-corrosion cracking problem that exists when these alloys and tempers are used under adverse conditions. (c) For electric motor rotors.

Table 5 Combinations of aluminum alloys used in some alclad products

Core alloy	Cladding alloy
2014	6003 or 6053
2024	1230
2219	7072
3003	7072
3004	7072 or 7013
6061	7072
7075	7072, 7008 or 7011
7178	7072

these alloys should not be high enough to make the solution sufficiently alkaline to cause significant corrosion.

Deposition Corrosion

In designing aluminum and aluminum alloys for satisfactory corrosion resistance, it is important to keep in mind that ions of several metals have reduction potentials more cathodic than the electrode potential of aluminum and therefore can be reduced to metallic form by aluminum. For each chemical equivalent of so-called "heavy metal" ions reduced, a chemical equivalent of aluminum is oxidized. Reduction of only a small amount of these ions can lead to severe corrosion of aluminum, because the metal reduced from them plates onto the aluminum and sets up galvanic cells. The more important heavy metals are copper, lead, mercury, nickel and tin. The effects of these metals on aluminum are of greatest concern in acidic solutions; in alkaline solutions, they have much lower solubilities and thus much less severe effects.

Copper is the heavy metal most commonly encountered in applications of aluminum. A copper-ion concentration of 0.02 ppm in neutral or acidic solutions generally is accepted as the threshold concentration for reduction of copper by aluminum.

Contamination of solutions in contact with aluminum by ions of copper or other heavy metals should always be avoided or minimized. The amount of corrosion resulting from such contamination depends to a considerable degree on other components of the solution. For example, copper ions are reduced much more rapidly in aerated halide solutions than in aerated non-halide solutions, because in halide solutions aluminum is polarized to its pitting potential and thus corrosion current is high.

Cathodic Protection. In some applications, aluminum alloy parts, assemblies, structures and pipelines are cathodically protected externally by anodes either made of more anodic metals or made anodic to the aluminum by means of impressed potentials. In either instance, because the usual cathodic reaction on aluminum alloys produces hydroxyl ions, the current on

As discussed previously, the relatively low corrosion resistance of aluminum-copper alloys results from reduction of copper ions present in the corrosion product of the alloy.

Aluminum reduces ions of ferric iron, but these ions rarely are encountered in service because they react with oxygen and water to form insoluble oxides and hydroxides, except in acidic solutions outside the passive range of aluminum. At room temperature, the most anodic aluminum alloys (those with electrode potentials approaching -1 V as measured against a standard calomel electrode) reduce ions of ferrous iron, but presence of these ions also is unlikely because they exist only in deaerated or other solutions free of oxidizing agents. With increasing temperature, more cathodic aluminum alloys become capable of reducing such ions (Ref 7).

Mercury amalgamates with aluminum with difficulty, because the natural oxide film on aluminum prevents metal-to-metal contact. However, when the film is broken by mechanical or chemical action, mercury is the most damaging to aluminum of all the heavy metals. This effect can be catastrophic when stress is present. For example, attack by mercury and zinc amalgam combined with residual stresses from welding caused cracking of the weldment shown in Fig. 6. The damaging effect of mercury is severe with or without stress, because amalgamation, once initiated, progresses for long periods of time; continued amalgamation occurs because the aluminum in the amalgam oxidizes immediately in the presence of moisture, and thus the mercury is continuously regenerated. It is immaterial, therefore, whether metallic mercury is reduced from its ions in solution or is introduced directly. It is difficult to determine the safe level of mercury that can be tolerated in a solution or atmosphere because of the difficulty with which amalgamation initiates, but in solutions any concentration greater than a few ppb should be viewed with suspicion, and in atmospheres where attack is initiated less readily, any concentration exceeding that allowed by EPA regulations is suspect. No amount of metallic mercury should ever be allowed to contact aluminum.

Intergranular Corrosion

Under certain metallurgical conditions and in certain environments, any

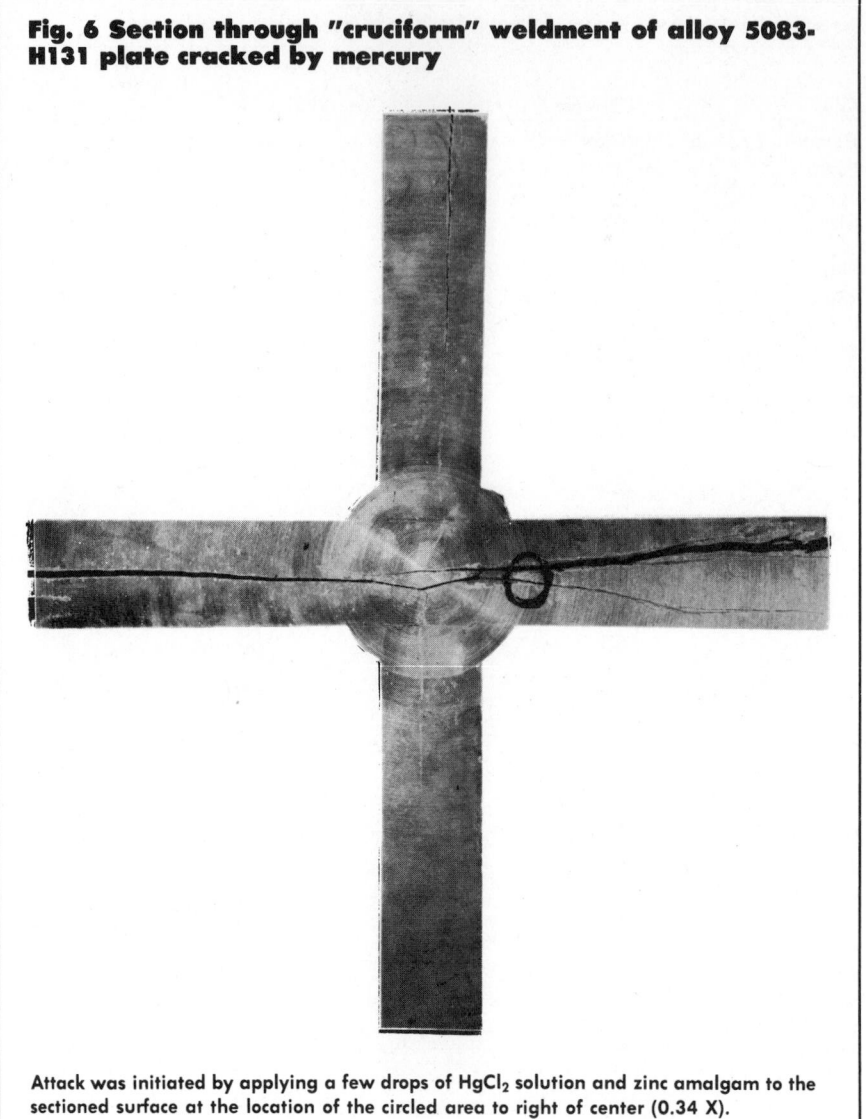

Fig. 6 Section through "cruciform" weldment of alloy 5083-H131 plate cracked by mercury

Attack was initiated by applying a few drops of HgCl₂ solution and zinc amalgam to the sectioned surface at the location of the circled area to right of center (0.34 X).

aluminum alloy may exhibit intergranular corrosion. Even super-purity aluminum, heat treated at high temperature and rapidly quenched, corrodes intergranularly in hydrochloric acid. Because intergranular corrosion is involved in stress-corrosion cracking, it frequently is presumed more deleterious than pitting or general corrosion. However, in alloys that are not susceptible to stress-corrosion cracking, intergranular corrosion usually is no more severe than pitting corrosion, tends to decrease with time; and for equal depths of corrosion, its effect on strength is no greater than that of pitting corrosion (although fatigue cracks may be more likely to initiate at

areas of intergranular corrosion than at random pits).

Stress-Corrosion Cracking

Only aluminum alloys that contain appreciable amounts of soluble alloying elements (primarily copper, magnesium, silicon and zinc) are susceptible to stress-corrosion cracking. For most commercial alloys, tempers have been developed that provide a high degree of immunity to stress-corrosion cracking in most environments.

The electrochemical theory of stress corrosion, developed about 1940, describes certain conditions required for stress-corrosion cracking of aluminum alloys (Ref 8 to 10). Subsequent research has demonstrated inadequacies in this theory, and the complex interactions among factors that lead to stress-corrosion cracking of aluminum alloys are not yet fully understood. However, there is general agreement that the electrochemical factor is significant, and the electrochemical theory continues to be an important basis for developing aluminum alloys and tempers resistant to stress-corrosion cracking.

Stress-corrosion cracking in aluminum alloys characteristically is intergranular. According to the electrochemical theory, this requires a condition along grain boundaries that makes them anodic to the rest of the microstructure so that corrosion propagates selectively along them. Such a condition is produced by localized decomposition of solid solution, with a high degree of continuity of decomposition products, along the grain boundaries. The most anodic regions may be either the boundaries themselves (most commonly, the precipitate formed in them) or regions adjoining the boundaries that have been depleted of solute.

In 2xxx alloys, it is the solute-depleted regions that are the most anodic, whereas in 5xxx alloys, it is the Al_8Mg_5 precipitate along the boundaries. The most anodic grain-boundary regions in other alloys have not been identified with certainty. Strong evidence for the presence of anodic regions, and of the electrochemical nature of their corrosion in aqueous solutions, is provided by the fact that stress-corrosion cracking can be greatly retarded, if not eliminated, by cathodic protection (Ref 10).

Figure 7 shows four different microstructures in an alloy containing 5% magnesium. These microstructures represent degrees of susceptibility to stress-corrosion cracking ranging from high susceptibility to high resistance, depending on heat treatment. It can be seen that treatments providing high resistance to cracking are those that produce microstructures either free of precipitate along grain boundaries (Fig. 7a) or with precipitate distributed as uniformly as possible within grains (Fig. 7d). In the latter instance, corrosion along boundaries is minimized because the presence of precipitate or depleted regions throughout the microstructure increases the ratio of the total area of anodic regions to that of cathod-

Table 6 Relative stress-corrosion (cracking) ratings for wrought products of high-strength aluminum alloys(a)

Alloy and temper(b)	Test direction(c)	Rolled plate	Rod and bar(d)	Extruded shapes	Forgings
2011-T3, -T4	L	(e)	B	(e)	(e)
	LT	(e)	D	(e)	(e)
	ST	(e)	D	(e)	(e)
2011-T8	L	(e)	A	(e)	(e)
	LT	(e)	A	(e)	(e)
	ST	(e)	A	(e)	(e)
2014-T6	L	A	A	A	B
	LT	B(f)	D	B(f)	B(f)
	ST	D	D	D	D
2024-T3, -T4	L	A	A	A	(e)
	LT	B(f)	D	B(f)	(e)
	ST	D	D	D	(e)
2024-T6	L	(e)	A	(e)	A
	LT	(e)	B	(e)	A(f)
	ST	(e)	B	(e)	D
2024-T8	L	A	A	A	A
	LT	A	A	A	A
	ST	B	A	B	C
2048-T851	L	A	(e)	(e)	(e)
	LT	A	(e)	(e)	(e)
	ST	B	(e)	(e)	(e)
2124-T851	L	A	(e)	(e)	(e)
	LT	A	(e)	(e)	(e)
	ST	B	(e)	(e)	(e)
2219-T3, -T37	L	A	(e)	A	(e)
	LT	B	(e)	B	(e)
	ST	D	(e)	D	(e)
2219-T6, -T8	L	A	A	A	A
	LT	A	A	A	A
	ST	A	A	A	A
6061-T6	L	A	A	A	A
	LT	A	A	A	A
	ST	A	A	A	A
7005-T53, -T63	L	(e)	(e)	A	A
	LT	(e)	(e)	A(f)	A(f)
	ST	(e)	(e)	D	D
7039-T63, -T64	L	A	(e)	A	(e)
	LT	A(f)	(e)	A(f)	(e)
	ST	D	(e)	D	(e)
7049-T73	L	A	(e)	A	A
	LT	A	(e)	A	A
	ST	A	(e)	B	A
7049-T76	L	(e)	(e)	A	(e)
	LT	(e)	(e)	A	(e)
	ST	(e)	(e)	C	(e)
7149-T73	L	(e)	(e)	A	A
	LT	(e)	(e)	A	A
	ST	(e)	(e)	B	A
7050-T736	L	A	(e)	A	A
	LT	A	(e)	A	A
	ST	B	(e)	B	B
7050-T76	L	A	A	A	(e)
	LT	A	B	A	(e)
	ST	C	B	C	(e)
7075-T6	L	A	A	A	A
	LT	B(f)	D	B(f)	B(f)
	ST	D	D	D	D
7075-T73	L	A	A	A	A
	LT	A	A	A	A
	ST	A	A	A	A
7075-T736	L	(e)	(e)	(e)	A
	LT	(e)	(e)	(e)	A
	ST	(e)	(e)	(e)	B
7075-T76	L	A	(e)	A	(e)
	LT	A	(e)	A	(e)
	ST	C	(e)	C	(e)

(continued)

Table 6 (continued)

Alloy and temper(b)	Test direction(c)	Rolled plate	Rod and bar(d)	Extruded shapes	Forgings
7175-T736	L	(e)	(e)	(e)	A
	LT	(e)	(e)	(e)	A
	ST	(e)	(e)	(e)	B
7475-T6	L	A	(e)	(e)	(e)
	LT	B(f)	(e)	(e)	(e)
	ST	D	(e)	(e)	(e)
7475-T73	L	A	(e)	(e)	(e)
	LT	A	(e)	(e)	(e)
	ST	A	(e)	(e)	(e)
7475-T76	L	A	(e)	(e)	(e)
	LT	A	(e)	(e)	(e)
	ST	C	(e)	(e)	(e)
7178-T6	L	A	(e)	A	(e)
	LT	B(f)	(e)	B(f)	(e)
	ST	D	(e)	D	(e)
7178-T76	L	A	(e)	A	(e)
	LT	A	(e)	A	(e)
	ST	C	(e)	C	(e)
7079-T6	L	A	(e)	A	A
	LT	B(f)	(e)	B(f)	B(f)
	ST	D	(e)	D	D

(a) Resistance ratings are as follows: A—very high; B—high; C—intermediate; D—low. See text for more detailed explanation of these ratings. (b) Ratings apply to standard mill products in the types of tempers indicated and also in Tx5x and Tx5xx (stress-relieved) tempers, and may be invalidated in some instances by use of nonstandard thermal treatments, or mechanical deformation at room temperature, by the user. (c) Test direction refers to orientation of direction in which stress is applied relative to the directional grain structure typical of wrought alloys, which for extrusions and forgings may not be predictable on the basis of the cross-sectional shape of the product: L—longitudinal; LT—long transverse; ST—short transverse. (d) Sections with width-to-thickness ratios equal to or less than two, for which there is no distinction between LT and ST properties. (e) Rating not established because product not offered commercially. (f) Rating is one class lower for thicker sections: extrusions, 25 mm (1 in.) and thicker; plate and forgings, 40 mm (1.5 in.) and thicker.

ic ones, thereby reducing the corrosion current on each anodic region. For alloys requiring microstructural control to avoid susceptibility, resistance is obtained by using treatments that produce precipitate throughout the microstructure, because precipitate always forms first along boundaries, and its formation there usually cannot be prevented.

According to electrochemical theory, susceptibility to intergranular corrosion is a prerequisite for susceptibility to stress-corrosion cracking, and treatment of aluminum alloys to improve resistance to stress-corrosion cracking also improves their resistance to intergranular corrosion. For most alloys, however, optimum levels of resistance to these two types of failure require different treatments, and resistance to intergranular corrosion is not a reliable indication of resistance to stress-corrosion cracking.

In many instances, susceptibility to stress-corrosion cracking of an aluminum alloy cannot be predicted reliably by examining its microstructure. Many observations have been made of the progressive changes in dislocation network, precipitation pattern and other microstructural features that occur as an alloy is treated to improve its resistance to stress-corrosion cracking, but these changes have not been correlated quantitatively with susceptibility.

Effect of Stress. Whether or not stress-corrosion cracking develops in a susceptible aluminum alloy product depends on both magnitude and duration of tensile strength acting at the surface. The effects of these factors have been established most commonly by means of accelerated laboratory tests; results of one set of such tests are reflected in the shaded bands in Fig. 8. Despite introduction of fracture mechanics techniques capable of determining crack-growth rates, such tests continue to be the basic tools used in evaluating resistance of aluminum alloys to stress-corrosion cracking. They suggest a minimum (threshold) stress that is required for cracking to develop. This value has questionable significance in a fundamental sense and, because it implies the existence of a "safe" stress, questionable validity in an empirical sense. Despite these limitations, the threshold value does provide a valid measure of the relative susceptibilities of aluminum alloys to stress-corrosion cracking under the specific conditions of a particular test or environment. Also, for some alloy/temper combinations, results of accelerated laboratory tests reliably predict stress-corrosion performance in service environments; for example, results of an 84-day alternate immersion test of alloy 7075 and alloy 7178 products correlated well with performance of these products in a seacoast environment.

Effects of Grain Structure and Stress Direction. Many wrought aluminum alloy products have highly directional grain structures (see Fig. 9). Such products are highly anisotropic with respect to resistance to stress-corrosion cracking (see Fig. 8). Resistance, which is measured by magnitude of tensile stress required to cause cracking, is highest when the stress is applied in the longitudinal direction, lowest in the short-transverse direction and intermediate in other directions. These directional differences are most marked in the more susceptible tempers, but are usually much lower in tempers produced by extended precipitation treatments, such as T6-type and T8-type tempers for 2xxx alloys and T73-type, T736-type and T76-type tempers for 7xxx alloys.

Thus, direction and magnitude of stresses anticipated under conditions of assembly and service may govern alloy and temper selection. For products of thin section, applied in ways that induce little or no tensile stress in the short-transverse direction, resistance of 2xxx alloys in T3-type or T4-type tempers or of 7xxx alloys in T6-type tempers may be satisfactory. Resistance in the short-transverse direction usually controls application of products that are of thick section or are machined or applied in ways that result in sustained tensile stresses in the short-transverse direction. More resistant tempers are preferred for these applications.

Effects of Environment. Much research indicates that water or water vapor is the key environmental factor required to produce stress-corrosion cracking in aluminum alloys. Halide ions have the greatest effects in accelerating attack. Chloride is the most important halide ion because it is a natural constituent of marine environ-

Fig. 7 Microstructures of alloy 5356-H12 after treatment to produce various degrees of susceptibility to stress-corrosion cracking

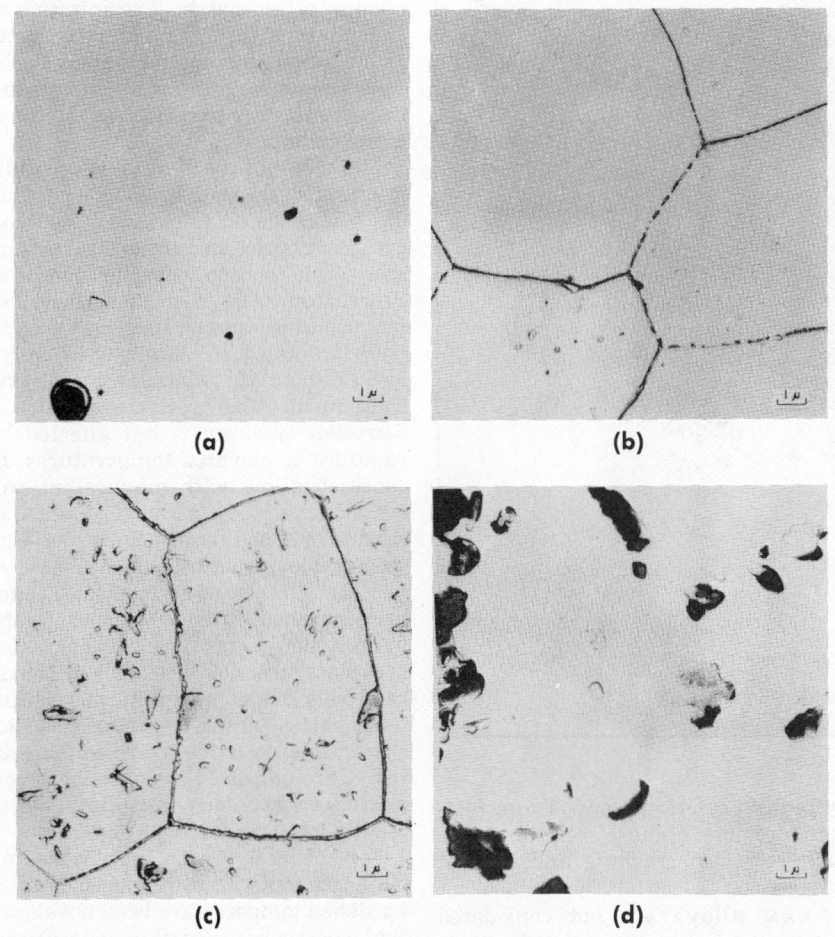

(a)

(b)

(c)

(d)

(a) Highly resistant—cold rolled 20%. (b) Highly susceptible—cold rolled 20% and heated 1 yr at 100 °C (212 °F). (c) Slightly susceptible—cold rolled 20% and heated 1 yr at 149 °C (300 °F). (d) Highly resistant—cold rolled 20% and heated 1 yr at 204 °C (400 °F).

Fig. 8 Stress-corrosion cracking behavior of alloy 7075-T651 plate

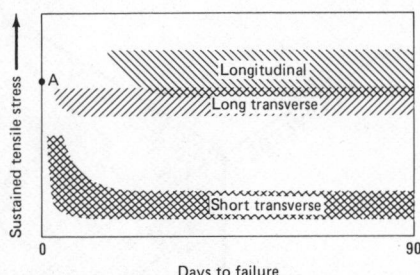

Shaded bands indicate combinations of stress and time known to produce stress-corrosion cracking in specimens intermittently immersed in 3.5% NaCl solution. Point A is minimum yield strength in the long transverse direction for plate 76 mm (3 in.) thick.

minimum specified yield strength) produced by any combination of sources including heat treatment, straightening, forming, fit-up and sustained service loading.

C—**Intermediate.** Stress-corrosion cracking not anticipated if total sustained tensile stress is maintained below 25% of minimum specified yield strength. This rating is designated for the short transverse direction in products used primarily for high resistance to exfoliation corrosion in relatively thin structures, where appreciable stresses in the short transverse direction are unlikely.

D—**Low.** Failure from stress-corrosion cracking is anticipated in any application involving sustained tensile stress in the designated test direction. This rating currently is designated only for the short transverse direction in certain products.

The stress levels mentioned in the above descriptions are not to be interpreted as "threshold" stresses and are not recommended for design. Documents such as MIL-HANDBOOK-5, MIL-STD-1568, NASC SD-24 and MSFC-SPEC-522A should be consulted for design recommendations.

The relative ratings of resistance to stress-corrosion cracking for high-strength wrought aluminum alloys are presented in Table 6. These ratings, assigned primarily by alloy and temper, also make distinctions among test directions and product types.

ments and is present in other environments as a contaminant. Because it accelerates stress-corrosion cracking, chloride is the principal component of environments used in laboratory tests to determine susceptibility of aluminum alloys to this type of attack. In general, susceptibility is greater in neutral solutions than in alkaline solutions, and greater still in acidic solutions.

Stress-Corrosion Ratings. A system of ratings of resistance to stress-corrosion cracking for high-strength aluminum alloy products has been developed by a joint task group of ASTM and the Aluminum Association to aid in alloy and temper selection. Definitions of these ratings, which range from "A" (highest resistance) to "D" (lowest resistance), are as follows:

A—**Very high.** No record of service problems; stress-corrosion cracking not anticipated in general applications

B—**High.** No record of service problems; stress-corrosion cracking not anticipated at stresses of the magnitude caused by solution heat treatment. Precautions must be taken to avoid high sustained tensile stresses (exceeding 50% of the

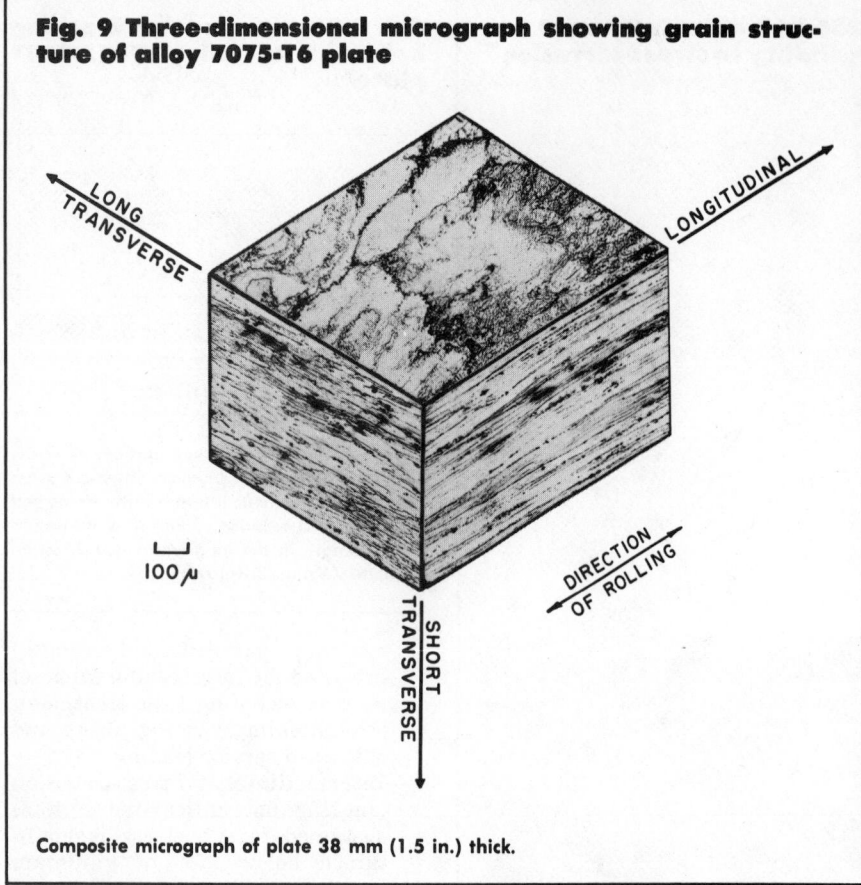

Fig. 9 Three-dimensional micrograph showing grain structure of alloy 7075-T6 plate

LONG TRANSVERSE

LONGITUDINAL

100 μ

DIRECTION OF ROLLING

SHORT TRANSVERSE

Composite micrograph of plate 38 mm (1.5 in.) thick.

2xxx Alloys. Thick-section products of 2xxx alloys in the naturally aged T3- and T4-type tempers have low ratings of resistance to stress-corrosion cracking in the short-transverse direction. Ratings of such products in other directions are higher, as are ratings of thin-section products in all directions. These differences are related to effects of quenching rate (largely determined by section thickness) on amount of precipitation occurring during quenching. If 2xxx alloys in T3- and T4-type tempers are heated for short periods in the temperature range used for artificial aging, selective precipitation along grain or subgrain boundaries may further impair their resistance.

Longer heating, as specified for T6- and T8-type tempers, produces more general precipitation and significant improvements in resistance to stress-corrosion cracking. Precipitates are formed within grains at a greater number of nucleation sites during treatment to T8-type tempers, which requires stretching, or cold working by other means, after quenching following solution heat treatment and before pre-

cipitation heat treatment. These tempers provide highest resistance to stress-corrosion cracking and highest strength.

5xxx alloys are not considered heat treatable and do not develop their strength through heat treatment. However, these alloys are processed to H3-type tempers, which require a final thermal stabilizing treatment to eliminate age-softening, or to H2-type tempers, which require final partial annealing. The H116 temper also is used for high-magnesium 5xxx alloys and involves special temperature control during fabrication to achieve a defined microstructural pattern of precipitate. The alloys of the 5xxx series span a wide range of magnesium contents, and the tempers that are standard for each alloy are established primarily by its magnesium content and the desirability of microstructures highly resistant to stress-corrosion cracking and other forms of corrosion.

Although 5xxx alloys are not heat treatable, they develop good strength through solution hardening by the magnesium retained in solid solution,

dispersion hardening by precipitates, and strain-hardening effects. Because the solid solutions in the higher-magnesium alloys are more highly supersaturated, the excess magnesium tends to precipitate out as Al_8Mg_5, which is anodic to the matrix. Precipitation of this phase with high selectivity along grain-boundary sites, accompanied by little or no precipitation within grains, may result in susceptibility to stress-corrosion cracking.

The probability that a susceptible microstructure will develop in a 5xxx alloy depends on magnesium content, grain structure, amount of strain hardening and subsequent time/temperature history. Alloys with relatively low magnesium contents, such as 5052 and 5454 (2.5 and 2.75% Mg, respectively), are only mildly supersaturated, and consequently their resistance to stress-corrosion cracking is not affected by exposure to elevated temperatures. In contrast, alloys with magnesium contents higher than about 3%, when in strain-hardened tempers, may develop susceptible structures as a result of heating, or even after very long times at room temperature. For example, the microstructure of alloy 5083-O (4.5% Mg) plate stretched 1% (see Fig. 10a) is relatively free of precipitate (no continuous paths), and the material is not susceptible to stress-corrosion cracking. Prolonged heating, however, produces continuous precipitate, which results in susceptibility (Fig. 10b).

For the alloys with higher magnesium contents, mill-produced strain-hardened tempers have been developed that decrease susceptibility to stress-corrosion cracking. These tempers (H116, H321, H323 and H343) are designed to develop uniform precipitate structures that are unaltered by subsequent exposure to slightly elevated temperatures, or to room temperature for long times, and thus maintain resistance to cracking.

6xxx Alloys. The service record of 6xxx alloys shows no reported instances of stress-corrosion cracking. In laboratory tests, however, at high stresses and in aggressive solutions, cracking has been demonstrated in 6xxx alloys of particularly high alloy content, containing silicon in excess of the Mg_2Si ratio and/or high percentages of copper.

7xxx Alloys Containing Copper. The alloy of 7xxx series that has been used most extensively and for the longest period of time is 7075, an Al-Zn-Mg-

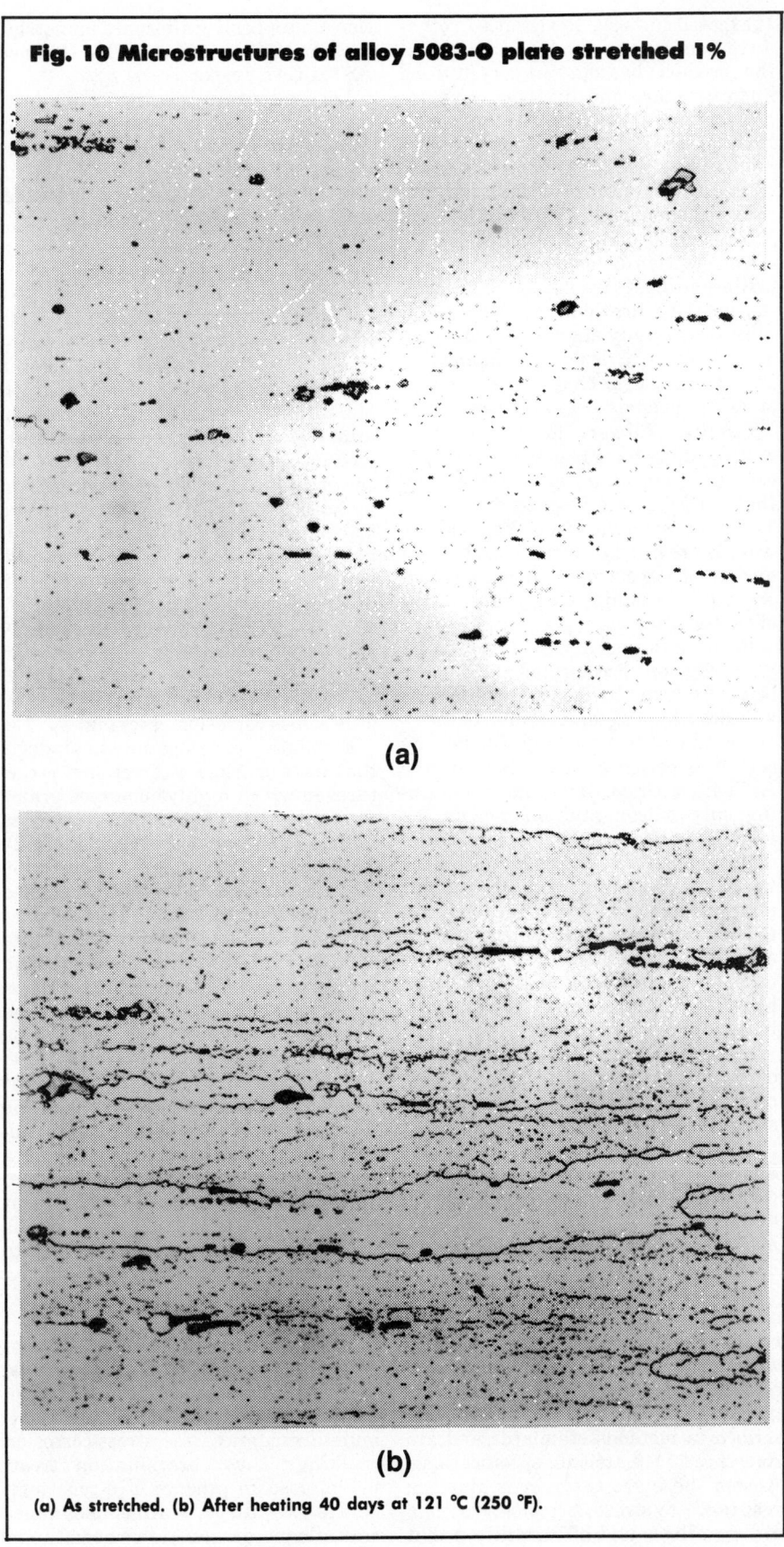

Fig. 10 Microstructures of alloy 5083-O plate stretched 1%

(a)

(b)

(a) As stretched. (b) After heating 40 days at 121 °C (250 °F).

Cu-Cr alloy. Introduced in 1943, this aircraft-construction alloy initially was used for products with thin sections, principally sheet and extrusions. In these products, quenching rate normally is very high and tensile stresses are not encountered in the short-transverse direction, and thus stress-corrosion cracking is not a problem for material in the highest-strength (T6-type) tempers. When 7075 was used in products of greater size and thickness, however, it became apparent that such products heat treated to T6-type tempers often were unsatisfactory. Parts extensively machined from large forgings, extrusions or plate frequently were subjected to continuous stresses, arising from interference misfit during assembly or from service loading, that were tensile at exposed surfaces and aligned in unfavorable orientations. Under such conditions, stress-corrosion cracking was encountered in service with significant frequency.

This problem resulted in introduction (in about 1960) of the T73-type tempers for thick-section 7075 products. The precipitation treatment used to develop these tempers requires two-stage artificial aging, the second stage of which is done at a higher temperature than that used to produce T6-type tempers. During the preliminary stage, a fine, high-density precipitation dispersion is nucleated, producing high strength. The second stage is then used to develop resistance to stress-corrosion cracking (and exfoliation). Extensive accelerated and environmental testing has demonstrated that 7075-T73 resists stress-corrosion cracking even when stresses are oriented in the least favorable direction, at stress levels of at least 300 MPa (44 ksi), while under similar conditions except for stress level, 50 MPa (7 ksi) is about the maximum stress at which 7075-T6 resists cracking. The excellent test results for 7075-T73 have been confirmed by extensive service experience in a variety of applications.

The additional precipitation required to produce 7075 in T73-type tempers, which have high resistance to stress-corrosion cracking, reduces strength to levels below those of 7075 in T6-type tempers. Alloy 7175, a variant of 7075, was developed for forgings. In the T736 temper, 7175 has strength nearly comparable to that of 7075-T6, and better resistance to stress-corrosion cracking. Other newer alloys—such as 7049 and 7475, which are used in the T73 temper, and 7050, which is used in the T736 temper—couple high strength with very high resistance and improved fracture toughness.

The T76-type tempers, which also require two-stage artificial aging and which are intermediate to the T6- and T73-type tempers in both strength and resistance to stress-corrosion cracking, are developed in copper-containing 7xxx alloys for certain products. Comparative ratings of resistance for various products of all these alloys, as well as for products of 7178, are given in Table 6.

The microstructural differences among the T6-, T73- and T76-type tempers of these alloys are differences in size and type of precipitate, which changes from predominantly GP zones in T6-type tempers to η', the metastable transition form of $\eta(MgZn_2)$, in T73- and T76-type tempers. None of these differences can be detected by light microscopy. In fact, even the resolutions possible in transmission electron microscopy are insufficient for determining whether the precipitation reaction has been adequate to ensure the expected level of resistance to stress-corrosion cracking. For quality assurance, copper-containing 7xxx alloys in T73- and T76-type tempers are required to have specified minimum values of electrical conductivity and, in some instances, tensile yield strengths that fall within specified ranges. The validity of these properties as measures of resistance to stress-corrosion cracking is based on many correlation studies involving these measurements, laboratory and field stress-corrosion tests, and service experience.

Copper-free 7xxx Alloys. Wrought alloys of the 7xxx series that do not contain copper are of considerable interest because of their good resistance to general corrosion, moderate to high strength, and good fracture toughness and formability. Alloys 7004 and 7005 have been used in extruded form, and to a smaller extent in sheet form, for structural applications. More recently introduced compositions, including 7016, X7021, X7029 and X7146, have been used in automobile bumpers formed from extrusions or sheet.

As a group, copper-free 7xxx alloys are less resistant to stress-corrosion cracking than other types of aluminum alloys when tensile stresses are developed in the short transverse direction

at exposed surfaces. Resistance in other directions may be good, particularly if the product has an unrecrystallized microstructure and has been properly heat treated. Products with recrystallized grain structures generally are more susceptible to cracking as a result of stresses induced by forming or mechanical damage after heat treatment. When cold forming is required, subsequent solution heat treatment or precipitation heat treatment is recommended. Applications of these alloys must be carefully engineered, and consultation among designers, application engineers and product producers or suppliers is advised in all cases.

Casting Alloys. Most aluminum casting alloys have resistance to stress-corrosion cracking sufficiently high that cracking rarely occurs in service. The microstructures of these alloys usually are nearly isotropic, and resistance to stress-corrosion cracking, consequently, is unaffected by orientation of tensile stresses.

Relative ratings of cast alloys, based primarily on accelerated laboratory tests, are listed in Table 4. It has been indicated by accelerated and natura - environment testing, and verified by service experience, that alloys of the Al-Si 4xx.x series, 3xx.x alloys containing only silicon and magnesium as alloying additions, and 5xx.x alloys with magnesium contents of 8% or lower have virtually no susceptibility to stress-corrosion cracking. Alloys of the 3xx.x group that contain copper are rated as less resistant, although the numbers of castings of these alloys that have failed by stress-corrosion cracking have not been significant.

Significant stress-corrosion cracking of aluminum alloy castings in service has occurred only in the highest-strength Al-Zn-Mg 7xx.x alloys and in the Al-Mg alloy 520.0 in the T4 temper. For such alloys, factors that require careful consideration include casting design, assembly and service stresses, and anticipated environmental exposure.

Specifications and Tests. Several aluminum alloy product specifications require defined levels of performance with respect to resistance to stress-corrosion cracking. Standard tests used to measure such performance are described in methods standards and are referenced in materials specifications. Among these are tests for evaluating resistance to stress-corrosion cracking of 2xxx alloys, and of 7xxx alloys that

contain copper, by alternate immersion in 3.5% NaCl solution (ASTM G44 and ASTM G47, respectively).

Lot acceptance criteria for products of 7xxx copper-containing alloys in T76-, T73- and T736-type tempers are based on combined requirements for tensile yield strength and electrical conductivity.

Exfoliation Corrosion

In certain tempers, wrought products of aluminum alloys are subject to corrosion by exfoliation, which sometimes is described as lamellar, layer or stratified corrosion. In this type of corrosion, attack proceeds along selective subsurface paths parallel to the surface. As shown in Fig. 11a, layers of uncorroded metal between the selective paths are split apart and raised above the original surface; this delamination is promoted by the voluminous corrosion product formed along the paths of attack. Because it can be detected readily at an early stage and is restricted in depth, exfoliation does not cause sudden, unexpected structural failure as does stress-corrosion cracking.

Exfoliation develops only in products that have markedly directional structures in which highly elongated grains form platelets that are thin relative to their length and width (see Fig. 12). Susceptibility to this type of corrosion may result from the presence of aligned intergranular or subgrain-boundary precipitates, or from aligned strata that differ slightly from each other in composition. Alloys most subject to exfoliation are those of the 2xxx, 5xxx and 7xxx series. In these alloys, exfoliation is caused primarily by unfavorable precipitate distribution, and processing to eliminate this form of attack is of the type that promotes either more uniform precipitation within grains or a more advanced stage of precipitation. Thus, increases in precipitation heat treating time or temperature are as effective in reducing susceptibility to exfoliation, as they are in reducing susceptibility to stress-corrosion cracking.

During long-duration or high-temperature precipitation treatments, maximum resistance to exfoliation usually is achieved sooner than maximum resistance to stress-corrosion cracking. Thus, precipitation treatments used to produce T76-type tempers in 7xxx alloys, which employ times and temperatures intermediate to

Fig. 11 Effect of temper on exfoliation resistance of an alloy 7075 extrusion exposed in a seacoast environment

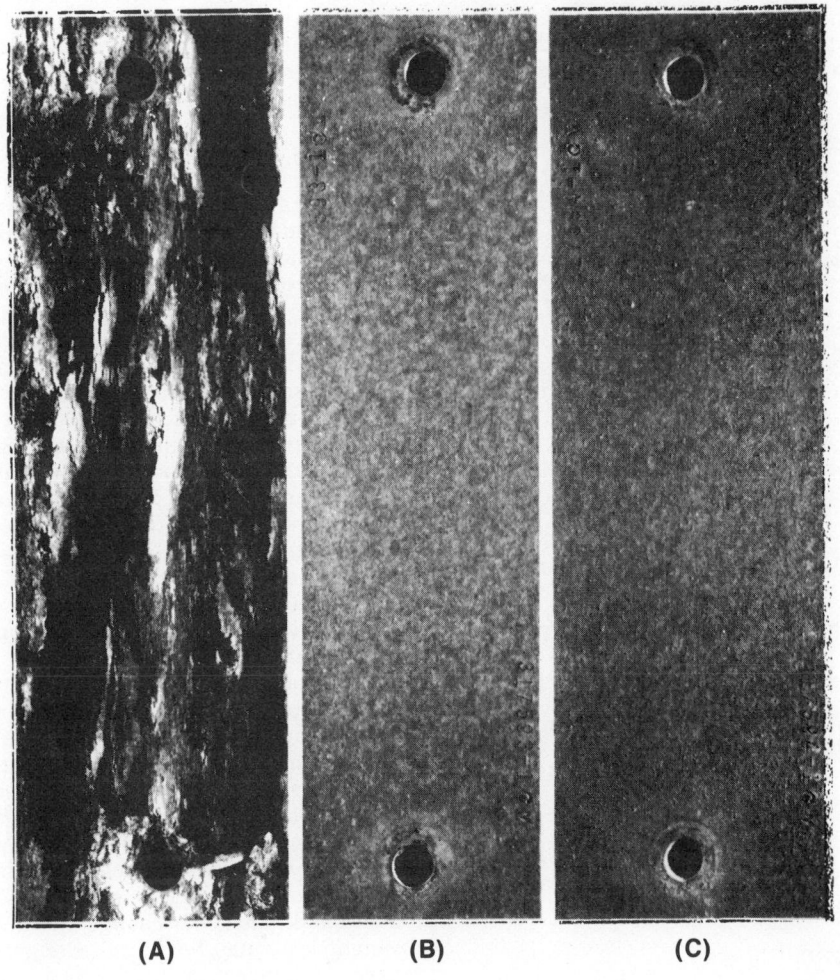

(A)   (B)   (C)

Specimens were exposed for 4 years. The specimen in the T6510 temper (a) developed exfoliation after only 5 months; those in the T76510 and T73510 tempers (b and c) were unaffected after 4 years.

be demonstrated by testing (see ASTM B209, Annex Al) that the material has acceptable resistance to exfoliation.

Corrosion Fatigue

Fatigue strengths of aluminum alloys are lower in corrosive environments such as seawater and other salt waters than they are in air, especially when evaluated by low-stress, long-period tests. As shown in Fig. 13, such corrosive environments produce smaller reductions in fatigue strength in alloys of the more corrosion-resistant types, such as 5xxx and 6xxx alloys, than in less resistant alloys, such as those of the 2xxx and 7xxx series. Like stress-corrosion cracking of aluminum alloys, corrosion fatigue requires the presence of water. In contrast to stress-corrosion cracking, corrosion fatigue is not appreciably affected by test direction, because fracture resulting from this type of attack is predominantly transgranular.

Erosion-Corrosion

In noncorrosive environments, such as high-purity water, the stronger aluminum alloys have the greatest resistance to erosion-corrosion because resistance in such an environment is controlled almost entirely by the mechanical component of the environment. In a corrosive environment, such as seawater, the corrosion component becomes the controlling factor, and thus resistance may be greater for the more corrosion-resistant alloys even though they are lower in strength.

Atmospheric Corrosion

Most aluminum alloys have excellent resistance to atmospheric corrosion (often called "weathering"), and in many outdoor applications such alloys do not require shelter, protective coatings or maintenance. Aluminum alloy products that have no external protection and therefore depend critically on this property include electrical conductors, outdoor lighting poles, ladders and bridge railings. Such products often retain a bright metallic appearance for many years, but their surfaces may become dull, gray or even black as a result of accumulation of pollutants. Corrosion of most aluminum alloys by weathering is restricted to mild surface roughening by shallow pitting, with no general thinning. However, such at-

those of T6- and T73-type treatments, provide excellent resistance to exfoliation (see Fig. 11b) but only intermediate resistance to stress-corrosion cracking. The T73-type tempers provide highest resistance to both types of corrosion (see Fig. 11c), but at a sacrifice in strength compared to T76-type tempers.

Among the standard tests for evaluating resistance to exfoliation of 5xxx and 7xxx alloys are those that require total immersion in aggressive acidified solutions of mixed salts (Ref 12). Such

tests are described in ASTM B209, Annex Al (5xxx alloys) and in ASTM G34 (7xxx alloys).

Acceptability of Al-Mg alloys 5083, 5086 and 5456 in the H116 temper is based on comparison of the microstructure disclosed by etching in a defined manner with a reference microstructure that is predominantly free from continuous grain-boundary network of Al_8Mg_5 precipitate particles. Material containing such precipitate in amounts exceeding that shown by the reference standard is unacceptable unless it can

Fig. 12 Exfoliation corrosion in alloy 7178-T651 plate exposed in a seacoast environment

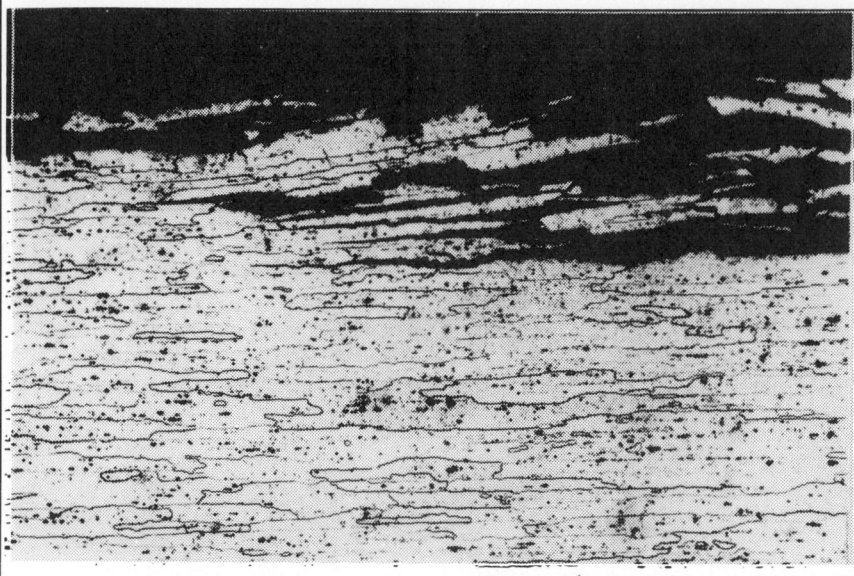

Cross section of plate, showing how exfoliation develops by corrosion along boundaries of thin, elongated grains.

Fig. 13 Ratio of axial-stress fatigue strength of aluminum alloy sheet in 3% NaCl solution to that in air

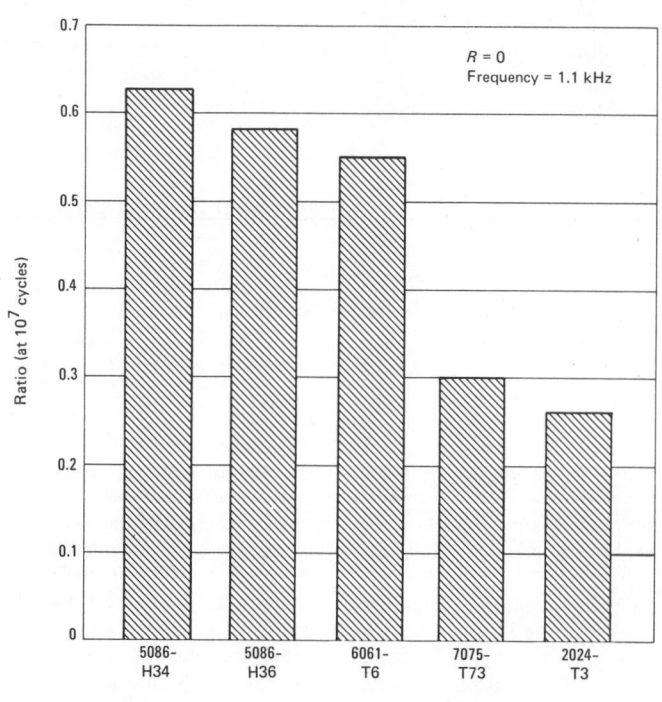

tack is more severe for alloys with higher copper contents, and such alloys seldom are used in outdoor applications without protection.

Corrosivity of the atmosphere to metals varies greatly from one geographic location to another, depending on meteorological factors such as wind direction; precipitation and temperature changes; amount and type of urban and industrial pollutants (if any); and proximity to natural bodies of water. Laboratory exposure tests (such as salt-spray, total-immersion and alternate-immersion tests) provide useful comparative information, but have limited value for predicting actual service performance and sometimes exaggerate differences among alloys that are negligible under atmospheric conditions. Consequently, extensive long-term evaluations of the effects of exposure in different industrial, chemical, seacoast, tropical and rural environments have been made. Large programs have been arranged and conducted cooperatively by industry under ASTM sponsorship, in addition to those carried out by producers, users and organizations on a world-wide basis, so that a large volume of information is available.

Data collected in these programs include measurements of maximum and mean depth of attack, weight loss and changes in tensile properties. Because of the localized nature of the prevalent pitting corrosion, which leaves some (in many instances, most) of the original surface intact even after many years of weathering, weight loss, or calculated average dimensional change based on weight loss, may have limited significance. Changes in tensile strength, which reflect the effects of size, number, distribution and acuity of pits, generally are most significant from a structural standpoint, while depth-of-attack determinations provide realistic measures of penetration rate.

Effect of Exposure Time. A very important characteristic of weathering of aluminum, and of corrosion of aluminum under many other environmental conditions, is that corrosion rate decreases with time (Ref 13 to 21). This deceleration of corrosion, which is illustrated in Fig. 14 to 16, occurs regardless of alloy composition, type of environment or the parameter by which the corrosion is measured. However, loss in tensile strength, which is influenced somewhat by pit acuity and distribution but is basically a result of loss of effective cross section, decelerates more

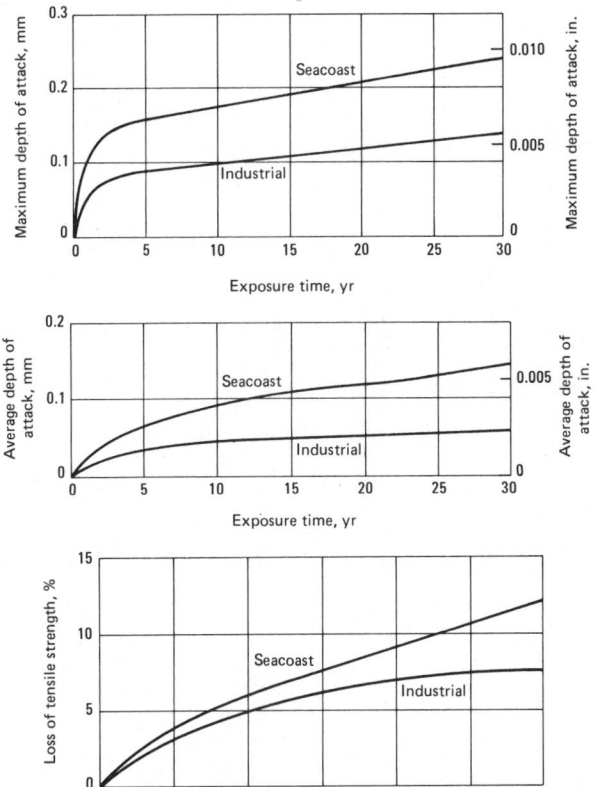

Fig. 14 Effects of weathering depth of corrosion and loss of tensile strength for alloys 1100, 3003 and 3004

Shown is the average performance of the three alloys, all in H14 temper. Seacoast exposure was at a severe location (Pt. Judith, RI); industrial exposure was at New Kensington, PA. Tensile strengths were computed using original cross-sectional area, and loss in strength is expressed as percentage of original tensile strength.

gradually than depth of attack (see Fig. 14).

The decrease in rate of penetration of corrosion is dramatic. In general, rate of attack at discrete locations, which initially is about 0.1 mm/yr (4 mils/yr), decreases to much lower and nearly constant rates within a period of about 6 months to 2 years. For the deepest pits, the maximum rate after about 2 years does not exceed about 0.003 mm/yr (0.11 mil/yr) for severe seacoast locations, and may be as low as 0.0008 mm/yr (0.03 mil/yr) in rural or arid climates. The dramatic deceleration in penetration is illustrated by the specimen cross sections shown in Fig. 15, and by the depth-of-attack curves shown in Fig. 16, both of which are from the same 30-year test program.

Also shown in Fig. 16 are results (shown as vertical bars) from other test programs in which various articles made of aluminum alloys were exposed continuously for various periods and in different locations, many of which are less severe than the relatively aggressive industrial environment of New Kensington.

Data for Wrought Alloys. Two major test programs investigating weathering of aluminum alloy sheet have been conducted under the supervision of ASTM Committee B03. The first program, started in 1931, was somewhat limited in variety of alloys tested but included desert, rural, seacoast and industrial exposures. Data

obtained after 20 years of exposure are listed in Table 7. Corrosion rates were calculated from cumulative weight loss after 20 years, and average and maximum depths of attack were measured microscopically. In aggressive (seacoast and industrial) environments, the bare (non-alclad) heat treated alloys—2017-T3 and, to a lesser extent, 6051-T4—exhibited more severe corrosion, and greater resulting loss in tensile strength, than the non-heat-treatable alloys. Alclad 2017-T3, although as severely corroded as the non-heat-treatable materials, did not show measurable loss in strength; in fact, some specimens of this alloy were 2 to 3% higher in strength after 20 years because of long-term natural aging.

Data from a comprehensive program initiated in 1958 were compiled from examinations and tests performed after 7 years of exposure. Thirty-four combinations of alloy and temper in the form of 1.27-mm (0.050-in.) thick sheet were exposed at four sites—two seacoast, one industrial and one rural; Table 8 presents average values of measurements reported at two of the more aggressive sites. In another ASTM program, 10 years of weathering produced the changes in tensile strength reported in Table 9.

The information from these and many other weathering programs (Ref 17 to 21) demonstrate that differences in resistance to weathering among non-heat-treatable alloys are not great, that alclad products retain their strength quite well because corrosion penetration is confined to the cladding layer, and that corrosion and resulting strength loss tend to be greater for bare (nonalclad) heat treatable 2xxx series alloys.

Data for Casting Alloys. The testing program that was the source of the strength-change data for wrought alloys given in Table 9 also provided weathering data for casting alloys exposed for the same period of time and at the same sites. Specimens were separately sand cast and permanent mold cast tensile bars, each with a reduced section 12.7 mm (0.5 in.) in diameter. Strength-change data for these alloys are summarized in Table 10. Alloys with relatively high copper contents, such as 295.0-T6, 208.0-F, 319.0-T6

Fig. 15 Sectioned specimens cut from alloy 3003-H14 panels 1.62 mm (0.064 in.) thick after exposure in two atmospheric environments

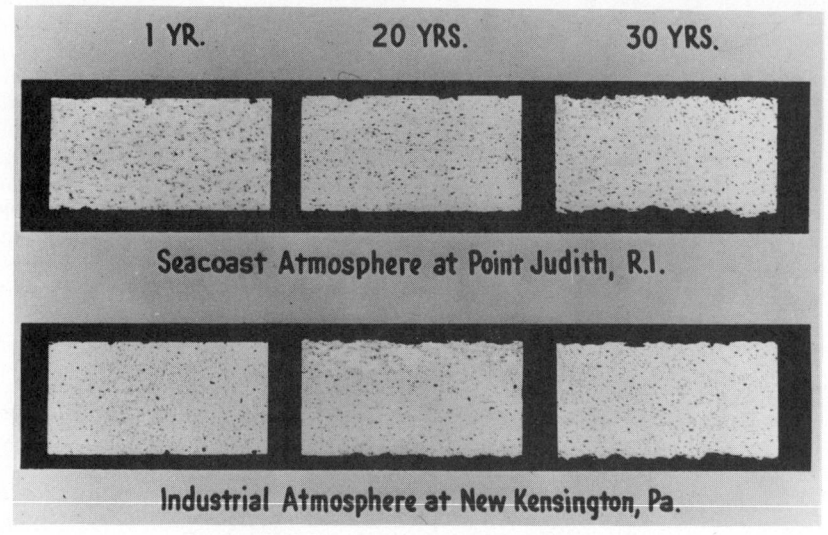

Fig. 16 Correlation of weathering data for specimens of alloys 1100, 3003 and 3004, all in H14 temper, exposed in industrial atmosphere (curves) with service experience with aluminum alloys in various locations (bars)

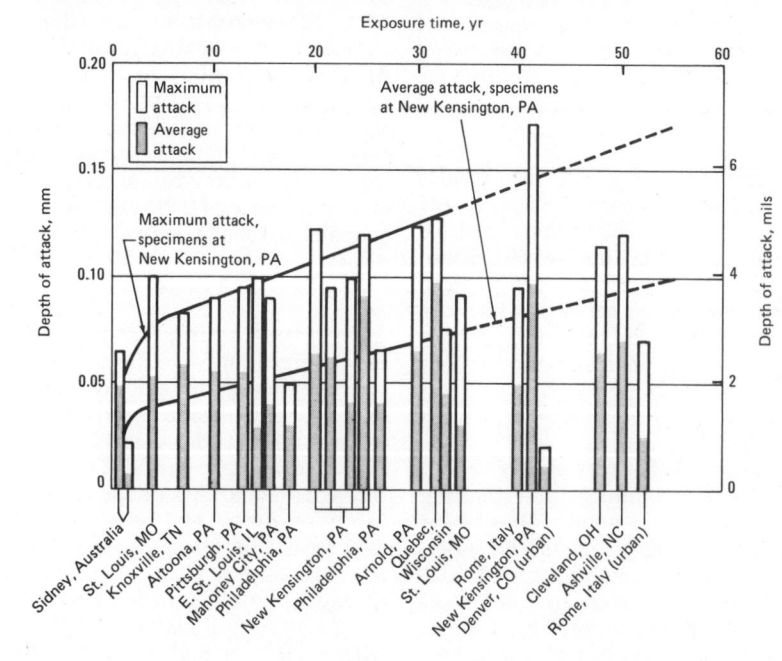

and 319.0-T61, showed the greatest losses. Alloys of the zinc-containing 7xx.x series generally exhibited larger strength losses than alloys having low zinc or copper contents. In all instances, as for wrought materials, severity of corrosion varied widely depending on environmental conditions.

Comparison with Other Metals. Other metals were exposed to the same weathering environments over the same time periods employed in evaluating corrosion of aluminum alloys. Comparative corrosion rates (average loss in thickness per side calculated from weight losses measured after exposures of 10 and 20 years) are shown in Table 11 for aluminum, copper, lead and zinc panels. Losses in tensile strengths at several weathering sites for unprotected low-carbon steel (0.09C, 0.07Cu) and for aluminum alloys are presented in Fig. 17.

Corrosion in Waters

High-Purity Water. Suitability of the more corrosion-resistant aluminum alloys for use with high-purity water at room temperature is well established by both laboratory testing and service experience. The slight reaction with the water that occurs initially ceases almost completely within a few days following development of a protective oxide film of equilibrium thickness. After this conditioning period, the amount of metal dissolved by the water becomes negligible.

Corrosion resistance of aluminum alloys in high-purity water is not significantly decreased by dissolved carbon dioxide or oxygen in the water or, in most cases, by the various chemicals added to high-purity water in the steam power industry to provide the required compatibility with steel. These additives include ammonia and neutralizing amines for pH adjustment to control carbon dioxide, hydrazine and sodium sulfate to control oxygen, and filming amines (long-chain polar compounds) to produce nonwettable surfaces.

Somewhat surprisingly, effects of alloying elements on corrosion resistance of aluminum alloys in high-purity water at elevated temperatures are opposite to their effects at room temperature; elements (including impurities) that decrease resistance at room temperature improve it at elevated temperatures.

Table 7 Weathering data for aluminum alloy sheet 0.89 mm (0.035 in.) thick after 20-year exposure (ASTM program started in 1931) (Ref 13)

Alloy and temper	Corrosion rate nm/yr	μin./yr	Average depth of attack μm	mils	Maximum depth of attack μm	mils	Loss in tensile strength, %
Phoenix, AZ (Desert)							
1100-H14	76	3.0	8	0.3	18	0.7	0
2017-T3	76	3.0	23	0.9	51	2.0	0
2017-T3, alclad	13	0.5	10	0.4	23	0.9	0
3003-H14	13	0.5	5	0.2	10	0.4	0
6051-T4	13	0.5	28	1.1	74	2.9	0
State College, PA (Rural)							
1100-H14	76	3.0	36	1.4	89	3.5	3
2017-T3	102	4.0	25	1.0	81	3.2	2
2017-T3, alclad	76	3.0	10	0.4	25	1.0	0
3003-H14	89	3.5	23	0.9	56	2.2	3
6051-T4	76	3.0	23	0.9	96	3.8	0
Sandy Hook, NJ (Seacoast)							
1100-H14	279	11.0	96	3.8	231	9.1	3
2017-T3	⋯	⋯	43	1.7	132	5.2	10
2017-T3, alclad	⋯	⋯	23	0.9	33	1.3	⋯
3003-H14	356	14.0	36	1.4	84	3.3	⋯
6051-T4	343	13.5	58	2.3	137	5.4	9
La Jolla, CA (Seacoast)							
1100-H14	584	23.0	102	4.0	356	14.0	8
2017-T3	2260	89.0	147	5.8	515	20.3	20
2017-T3, alclad	584	23.0	33	1.3	74	2.9	0
3003-H14	610	24.0	107	4.2	259	10.2	7
6051-T4	775	30.5	84	3.3	307	12.1	20
New York, NY (Industrial)							
1100-H14	749	29.5	89	3.5	213	8.4	7
2017-T3	1260	49.6	51	2.0	180	7.1	7
2017-T3, alclad	762	30.0	28	1.1	36	1.4	0
3003-H14	965	38.0	51	2.0	163	6.4	8
6051-T4	914	36.0	74	2.9	170	6.7	12

At 200 °C (390 °F), high-purity aluminum of sheet thickness disintegrates completely within a few days by reaction with high-purity water to form aluminum oxide. In contrast, Al-Ni-Fe alloys have the best elevated-temperature resistance to high-purity water of all aluminum metals; for example, alloy X8001 (1.0Ni, 0.5Fe) has good resistance at temperatures as high as 315 °C (600 °F).

Natural Waters. Aluminum alloys of the 1xxx, 3xxx, 5xxx and 6xxx series are resistant to corrosion by many natural waters. As described in an earlier section, the more important factors controlling corrosivity of natural waters to aluminum include water temperature, pH and conductivity; availability of cathodic reactant; presence or absence of heavy metals; and the free-corrosion and pitting potentials of the specific alloys. Various correlations among the corrosivities of natural waters to aluminum alloys have been developed (see Ref 23), but none predicts the corrosivities of all natural waters with complete reliability.

Seawater. Service experience with 1xxx, 3xxx, 5xxx and 6xxx wrought aluminum alloys in marine applications, including structures, pipelines, boats and ships, demonstrates their good resistance and long life under conditions of partial, intermittent or total immersion. Casting alloys of the 356.0 and 514.0 types also show high resistance to seawater corrosion, and these alloys are used widely for fittings, housings and other marine parts.

Fig. 17 Tensile strength losses for low-carbon steel (0.09% C, 0.07% Cu) and representative non-heat-treatable aluminum alloys at several atmospheric exposure sites (Ref 22)

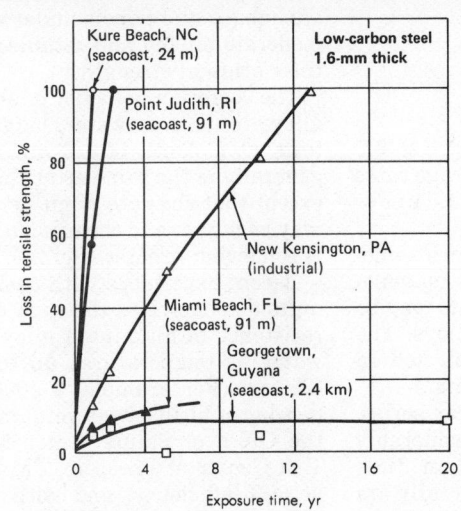
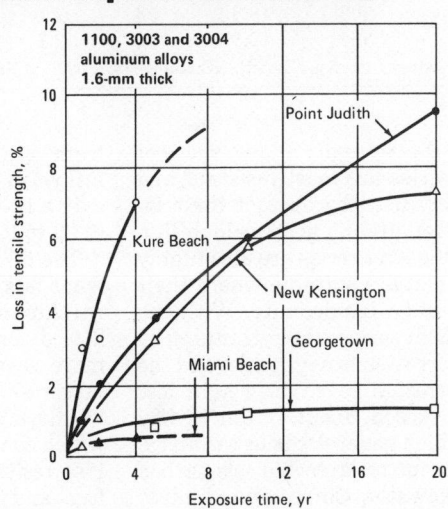

Note that the strength losses of the aluminum alloys are less than one tenth of the low-carbon steel.

Table 8 Weathering data for aluminum alloy sheet 1.27 mm (0.05 in.) thick after 7-year exposure (ASTM program started in 1958) (Ref 14)(a)

Alloy and temper	Corrosion rate(b)		Maximum depth of attack in 7 years		Average depth of attack in 7 years		Loss in tensile strength in 7 years, %
	nm/yr	μin./yr	μm	mils	μm	mils	
Non-Heat-Treatable Alloys							
1100-H14	345	13.6	70	2.6	29	1.1	0
1135-H14	321	12.6	83	3.3	37	1.5	0.4
1188-H14	250	9.8	121	4.8	46	1.8	0
1199-H18	205	8.1	96	3.8	57	2.2	3.9
3003-H14	295	11.6	86	3.4	52	2.0	1.1
3004-H34	414	16.3	119	4.7	44	1.7	1.1
4043-H14	335	13.2	105	4.1	34	1.3	2.8
5005-H34	373	14.7	76	3.0	27	1.1	0.9
5050-H34	349	13.7	107	4.2	58	2.3	0.5
5052-H34	362	14.3	62	2.4	43	1.7	0.8
5154-H34	326	12.8	91	3.6	65	2.6	0.9
5454-O	348	13.7	95	3.7	41	1.6	1.5
5454-H34	342	13.5	105	4.1	30	1.2	0.5
5456-O	381	15.0	104	4.1	37	1.5	0.4
5357-H34	292	11.5	138	5.4	102	4.0	0.4
5083-O	469	18.5	102	4.0	52	2.0	1.8
5083-H34	375	14.8	88	3.5	56	2.2	2.2
5086-H34	436	17.2	105	4.1	76	3.0	1.9
Heat Treatable Alloys							
2014-T6	644	25.4	77	3.0	50	2.0	1.7
2024-T3	1022	40.2	76	3.0	67	2.6	2.0
2024-T81	725	28.5	97	3.8	76	3.0	6.0
2024-T86	806	31.7	77	3.0	58	2.3	6.2
6061-T4	378	14.9	57	2.2	38	1.5	0.4
6061-T6	422	16.6	98	3.9	42	1.7	0.7
7075-T6	688	27.1	119	4.7	71	2.8	1.7
7079-T6	635	25.0	65	2.6	37	1.5	0.5
Alclad Alloys—Heat Treatable and Non-Heat Treatable							
2014-T6	358	14.1	43	1.7	28	1.1	0
2024-T3	264	10.4	46	1.8	27	1.1	0
3003-H14	345	13.6	128	5.0	117	4.6	0
5155-H34	345	13.6	53	2.1	35	1.4	0
6061-T6	356	14.0	98	3.9	25	1.0	0.7
7075-T6	502	19.8	53	2.1	41	1.6	0.1
7079-T6	324	12.8	72	2.8	36	1.4	0

(a) Average values from Kure Beach, NC and Newark, NJ. (b) Based on weight change.

Among the wrought alloys, those of the 5xxx series are most resistant and most widely used because of their favorable strength and good weldability. Alloys of the 3xxx series are also highly resistant and are suitable where their strength range is adequate. With the 3xxx and 5xxx series alloys, thinning by uniform corrosion is negligible, and the rate of corrosion based on weight loss does not exceed about 5 μm/yr (0.2 mil/yr), which generally is less than 5% of the rate for unprotected low-carbon steel in seawater. Corrosion is mainly of the pitting type, characterized by deceleration of penetration with time from rates of 3 to 6 μm/yr (0.1 to 0.2 mil/yr) in the first year to average rates over a 10-year period of 0.8 to 1.5 μm/yr (0.03 to 0.06 mil/yr).

The Al-Mg-Si 6xxx alloys are somewhat less resistant; although no general thinning occurs, weight loss may be 2 to 3 times that for 5xxx alloys. The more severe corrosion is reflected in larger and more numerous pits.

Alloys of the 2xxx and 7xxx series, which contain copper, are considerably less resistant to seawater than 3xxx, 5xxx and 6xxx alloys and generally are not used unprotected. Protective measures, such as use of alclad products and coating by metal spraying or by painting, provide satisfactory service in certain situations.

The literature on corrosion testing of aluminum alloys in seawater is extensive. Summaries of information are provided in Ref 19 and 24. Results of 10-year immersion testing of a variety of alloys in the form of rolled plate exposed in three locations are presented in Table 12. The relationships among the types of alloys that have been discussed and a comparison with unprotected low-carbon steel are apparent. Similar data for extruded products of several 6xxx alloys and one 5xxx alloy are given in Table 13. Direct comparison of the data in Tables 12 and 13 is provided in Table 14, where corrosion is expressed in terms of average weight loss, and in Fig. 18, which illustrates the deceleration of corrosion rate with time that is characteristic of aluminum alloys. Data on corrosion rates, maximum and average depths of pitting and changes in tensile strength compiled during 10-year tidal and full-immersion exposure of seven 5xxx alloys and super-purity aluminum 1199 are summarized in Table 15. Full immersion generally resulted in more extensive corrosion than tidal exposure, although the reverse relationship also has been observed. Tensile-strength losses were 5% or less, and yield-strength losses were less than 5% in the panels completely immersed and generally lower in those exposed to tidal immersion.

The data in Table 16 illustrate the corrosion resistance of aluminum alloy plates, with and without riveted or welded joints, in flowing seawater. All assemblies and panels underwent only moderate pitting and retained most of their original strength.

The corrosion behavior of aluminum alloys in deep seawater, judging from tests at a depth of 1.6 km (1 mile), generally is the same as at the surface except that the rate of pit penetration may be higher and the effect of crevices is somewhat greater (Ref 27).

Recent experience with desalination units demonstrates the high degree of resistance of aluminum alloys to seawater at temperatures up to 120 °C (250 °F). For example, a 3000-gallon-per-day multiflash aluminum unit at the Office of Saline Water Materials Test Center at Freeport, TX, operated at 99% efficiency, and with minimal corrosion, for 38 months under process conditions selected to match those of a

Table 9a Loss in tensile strength for wrought aluminum alloys during various atmospheric exposures (ASTM program) (Ref 16)(a)(b)

Alloy and temper	State College, PA					New York, NY					Kure Beach, NC				
	6 mo	1 yr	3 yr	5 yr	10 yr	6 mo	1 yr	3 yr	5 yr	10 yr	6 mo	1 yr	3 yr	5 yr	10 yr
1.62-mm (0.064-in.) Sheet															
2024-T3	+8	+1	+2	0	+1	+2	−8	−7	−11(c)	−11(c)	+6	−3	−4	−6	−4
3003-H14	+6	0	−2	0	+1	+4	−4	−5	−8	−6	+5	0	−2	−4	0
3004-H34	+6	−1	0	0	+1	+7	−2	−5	−5	−7	+6	+2	−2	−2	−1
5050-H34	+6	0	−1	0	−1	+4	−2	−1	−8	−4	+5	−1	−1	−1	−2
5052-H34	+9	0	−1	−1	0	⋯	−1	−6(c)	−5(c)	−7(c)	+6	0	−2	−3(c)	−1
6061-T6	+5	−2	−2	−3	0	⋯	−3	−7	−8	−11	+4	−1	−1	−1	−4
7075-T6	+5	−1	−3	0	−1	+3	−1	−5	−6(c)	−8(c)	+4	−2	−2	−4	−4
1.62-mm (0.064-in.) Alclad Sheet															
2014-T6	+5	−1	−1	−2	+2	+4	+1	−2	−4	−4	−2(c)	−1	−1	−4	−2
2024-T3	+7	−1	+1	+1	0	+8	−2	−1	−3	−3	+6	+1	0	0	−1
7075-T6	+6	0	+6	−2	−2	+5	+1	−2	−5	−5	+6	+2	+2	−1	0
6.35-mm (0.25-in.) Plate															
2014-T4	−3	0	0	0	0	−5	0	−2	−1	−4	−4	+1	0	0	−12
2014-T6	0	−1	0	0	+1	0	−2	−1	−1	−1	−2	−2	−1	−1	−1
6061-T6	−4	0	−2	−1	−5	+7	−1	−2	+4	+3	−4	−1	0	−1	−8
6.35-mm (0.25-in.) Alclad Plate															
2014-T6	0	−1	0	+1	−1	0	0	+1	−1	−2	−1	0	0	0	0
2024-T3	0	0	0	−1	+1	0	−2	−2	−2	−2	+2	0	+1	+2	+1
7075-T6	0	0	0	0	0	0	+1	−1	0	+1	0	+1	0	0	−11(c)
6.35-mm (0.25-in.) Extruded Bar															
2014-T4	+2	+3	+1	−1	−4	+1	+1	0	+1	−2	−0	0	−1	−1	−13
2014-T6	−1	0	0	−1	0	−1	+1	−2	−1	−2	−1	+2	−1	−2	−1
6061-T6	0	0	0	−1	+7	−2	−1	0	−3	−3	−1	−1	−2	−1	+6
6063-T5	+1	−1	−1	−1	+1	+1	−1	−2	+9	+11	−1	+8	+3	+6	+2
7075-T6	−1	−1	−3	−2	−3	−1	−2	−2	−1	−4	−2	−1	0	+1	−2

(a) Exposed as 102-by-203-mm (4-by-8-in.) panels. (b) Calculated from average tensile strength of several specimens (usually four). (c) Average tensile-strength values were below required minimum.

commercial installation. Such experience has shown, however, that galvanic attack of aluminum alloys by contact with dissimilar metals is more severe at elevated than at room temperature.

Corrosion in Soils

Soils differ widely in mineral content, texture and permeability, moisture, pH and aeration, presence of organic matter and microorganisms, and electrical resistivity. Because of these variations, corrosion performance of unprotected buried aluminum alloys, like that of other metals, varies considerably. In many environments where carbon steel requires protective coating, unprotected aluminum alloys have performed well; however, because corrosivity of soils is highly variable, protection is recommended for such applications.

Corrosion of the copper-bearing 2xxx and 7xxx alloys in moist, low-resistivity soils, measured by weight loss and pitting depth, is several times more rapid than corrosion of the more resistant 1xxx, 3xxx, 5xxx and 6xxx series alloys, and application of the copper-bearing alloys is limited accordingly. Use of alclad products or cathodic protection effectively reduces or limits penetration.

Aluminum alloys 3003, 6063, and 6061 are satisfactory for several types of surface and underground pipelines for irrigation, petroleum and mining applications. In several instances, buried pipelines have been examined after periods of 3 to 11 years (Ref 19 and 26). Unprotected sections exhibited corrosion attack ranging from almost none to deep pitting, depending on the type of soil. Cathodically protected sections of some of the same pipes in aggressive soils showed either no attack or only mild etching.

Resistance of Anodized Aluminum

Anodizing is an electrolytic oxidation process employed to produce on a metal surface an integral coating of oxide that is much thicker than the natural film. The anodic coatings used for decoration and/or protection of aluminum have a thin, nonporous barrier portion adjacent to the metal interface and a porous outer portion that is sealed by hydrothermal treatment in water or a metal salt solution to increase its protective value. The entire coating tightly adheres to the aluminum, resists abrasion and, when of adequate thickness, provides greatly improved protection against weathering and other corrosive conditions.

For outdoor applications of aluminum parts, a coating thickness of 5 to 7.6 μm (0.2 to 0.3 mil) is normally specified for bright automotive trim and 17 to 30 μm (0.7 to 1.2 mils) for architectural product finishes. Dichromate sealing affords added protection in severe saline environments. Because coatings can be attacked and stained by alkaline building materials (such as mortar, cement and plaster), a clear, nonyellowing lacquer often is applied

Table 9b Loss in tensile strength for wrought aluminum alloys during various stmospheric exposures (ASTM program) (Ref 16)(a)(b)

Alloy and temper	Change in strength, %, during exposure of indicated length at:									
	Point Reyes, CA					Freeport, TX				
	6 mo	1 yr	3 yr	5 yr	10 yr	6 mo	1 yr	3 yr	5 yr	10 yr
1.62-mm (0.064-in.) Sheet										
2024-T3	···	−13(c)	−19(c)	−19(c)	−23(c)	+3	−2	−9(c)	−8	−13(c)
3003-H14	···	+1	−3	−1	−4	+3	0	−5	+1	−4
3004-H34	···	−3	−1	−1	+1	+5	−1	−4	0	−2
5050-H34	···	+2	−1	0	−2	+5	0	−4	0	−3
5052-H34	···	−1	−2	0	−1	+4	−1	−7(c)	0	−1
6061-T6	···	−3	−4	−5	−5	+1	−3	−4	−1	−3
7075-T6	···	−3	−4	−4	−11(c)	+1	−1	−5	−3	−8(c)
1.62-mm (0.064-in.) Alclad Sheet										
2014-T6	···	−3	−1	−4	−4	+3	−1	−3	−3	−2
2024-T3	···	−1	−1	−1	−3	+6	−1	−2	0	−3
7075-T6	···	+3	−2	−3	−6	+5	+4	−1	−1	−2
6.35-mm (0.25-in.) Plate										
2014-T4	···	−1	−3	−6	−5	+1	···	−2	−1	−22(c)
2014-T6	···	−13(c)	−4	−8(c)	−8(c)	0	···	−2	0	−2
6061-T6	···	+1	0	+2	0	−4	0	−2	0	−2
6.35-mm (0.25-in.) Alclad Plate										
2014-T6	···	0	−1	0	−1	−1	+2	−1	0	−2
2024-T3	···	+2	0	−1	+1	+1	0	−1	0	0
7075-T6	···	+1	−1	0	−1	0	+2	−1	+1	0
6.35-mm (0.25-in.) Extruded Bar										
2014-T4	···	+3	−6	−3	−8	+1	+3	−2	+2	−5
2014-T6	···	···	−4	−3	−7	+1	+1	−1	−2	−3
6061-T6	···	−1	−1	−1	···	0	0	−2	−1	−2
6063-T5	···	+3	+3	+3	+7	+11	+2	0	+8	−1
7075-T6	···	−3	−3	−4	0	0	0	−1	−1	−4

(a) Exposed as 102-by-203-mm (4-by-8-in.) panels. (b) Calculated from average tensile strength of several specimens (usually four). (c) Average tensile-strength values were below required minimum.

to anodized aluminum architectural parts to protect the finish during construction. An added advantage of lacquer coatings is that they minimize soil accumulation during service.

In general, chemical resistance of anodic coatings is greatest in approximately neutral solutions, but such coatings usually are serviceable and protective if the pH is between 4 and 8.5. More acidic and more alkaline solutions attack anodic coatings.

Under atmospheric weathering, the number of pits developed in the basis metal decreases exponentially with increasing coating thickness, as shown in Fig. 19. The pits may form at minute discontinuities or voids in the coating, some of which result from large second-phase particles in the microstructure. The pit density was determined by dissolving the anodic coating in a stripping solution that does not attack the metal substrate. After the 8½-year ex-posure, the pits were of pin-point size and had penetrated less than 50 μm (2.0 mils). Specimens with coatings at least 22 μm (0.9 mil) thick were practically free of pitting.

Weathering of anodic coatings involves relatively uniform erosion of the coating by wind-borne solid particles, rainfall and some chemical reaction with pollutants. The available information indicates that such erosion occurs at a reasonably constant rate, which averaged 0.33 μm/yr (0.013 mil/yr) for several alloys exposed to an industrial atmosphere for 18 years (see Fig. 20).

Three-year seacoast exposure of specimens of several alloys with sulfuric acid coatings 23 μm (0.9 mil) thick caused no visible pitting except in several alloys of the 7xxx series and in a 2xxx alloy (see Table 17). Alloys that exhibited pitting were not protected any more effectively by coatings 51 μm (2.0 mils) thick. This confirms a gener-al observation that optimum protection against atmospheric corrosion is achieved in the coating-thickness range of 18 to 30 μm (0.7 to 1.2 mils), and that use of thicker coatings adds little more protection.

Anodized aluminum exterior automotive parts, such as bright trim and bumpers, exhibit good resistance to deicing salts and other ingredients of road splash despite the limited thickness applied to maintain brightness and image clarity. Development of a hazy coating appearance is considered more of a problem than pitting during service in these applications. The hazy appearance results from scattering of light from a coating surface that has been microroughened as a result of inadequate sealing or use of excessively harsh alkaline cleaners.

Anodic coatings, unless used as part of a protective system that includes other measures such as shot peening or painting, are not reliable for protection against stress-corrosion cracking of susceptible alloys. Data obtained with short-transverse-direction specimens from plate of alloy 7075-T651 and other susceptible alloys show that the anodic coating may retard, have no effect, or even accelerate stress-corrosion cracking, depending on level of stress and, to some extent, on whether or not the stress was present before anodizing. High stresses applied after anodizing crack the coating. The effects of several applied protective measures on lifetimes of specimens in industrial and seacoast environments under relatively high elastic strain are shown in Fig. 21, where the relatively small protective value of anodic coatings is apparent.

Effects of Nonmetallic Building Materials

Many nonmetallic building materials that contact aluminum during and after construction, either intentionally or accidentally, have been evaluated to determine their corrosive effects (Ref 30). Many of these materials that contain calcium or magnesium hydroxides are alkaline and, when wet, may cause over-all surface attack of bare aluminum. This early reaction produces protective films of limited solubility that resist further corrosion. Such materials cause only superficial or mild surface attack, most of which occurs during initial stages of exposure.

Drainage from freshly applied con-

Table 10a Loss in tensile strength for cast aluminum alloys during various atmospheric exposures (ASTM program) (Ref 16)(a)(b)

Alloy and temper	State College, PA					New York, NY					Kure Beach, NC				
	6 mo	1 yr	3 yr	5 yr	10 yr	6 mo	1 yr	3 yr	5 yr	10 yr	6 mo	1 yr	3 yr	5 yr	10 yr
Sand Castings															
208.0-F	−1	−2	−2	−1	−2	−1	−4	−4	−3	0	−2	−5	−7	−6	−4
295.0-T6	+1	−3	−2	−4	−2	−2	−6	−6	−5	−5	−1	−5	−9	−10	−9
319.0-T6	0	−1	−3	0	−3	+1	−2	−6	−8	−5	−1	−5	−7	−6	−4
355.0-T6	0	+1	−2	+1	−3	+1	0	−3	−1	−3	+2	+2	0	−1	−3
356.0-T6	+1	−1	0	−1	−1	+1	−1	−2	−2	−3	+1	−1	0	−2	−2
443.0-F	+3	0	−2	−2	−2	0	+3	−2	−4	−3	−2	0	0	−1	−2
520.0-T4	+1	−5	−4	−6	⋯	+2	−1	−1	−2	⋯	−2	−2	−5	−6	⋯
705.0-T5	+1	−2	−6	−4	−1	0	0	−4	−3	−10	+1	−2	−3	−3	−4
707.0-T5	+1	−2	−1	−3	0	+2	+1	−5	−9	−15	+2	−3	−9	−13	−18
710.0-T5	+2	−2	−2	−1	−5	+1	−3	−3	−2	−1	+2	−1	−1	−2	−1
712.0-T5	0	−8	−3	−2	−7	0	−2	−4	−5	−2	−4	−3	−8	−2	−8
713.0-T5	+1	+3	−2	+1	−1	−3	−4	−1	−1	−5	−5	−3	−8	−1	−3
Permanent Mold Castings															
319.0-T61	+1	−2	−1	−2	−2	+1	−3	0	−4	−4	−5	−3	−4	−7	−5
355.0-T6	+3	0	+7	+2	−4	+1	−2	+8	−2	−7	+2	−7	+5	−1	−5
443.0-F	+3	0	−1	−1	−2	+1	−3	−1	+1	0	−1	0	−6	+2	0
705.0-T5	−1	−2	−3	−5	−3	−2	0	−2	−3	−7	−3	−3	−5	−9	−5
707.0-T5	+2	−2	−3	−3	−4	0	−2	−1	−4	−7	+1	−2	−4	−7	−12
711.0-T5	−8	−11	−7	−6	−8	+2	−4	−5	−2	−6	−2	−6	−6	−6	−11
713.0-T5	−2	−2	0	−1	−2	−1	−11	−2	−7	−2	−11	−12	−6	−4	−1

(a) Exposed as separately cast tensile specimens. (b) Calculated from average tensile strength of several specimens (usually six). (c) Average tensile-strength values were below required minimum.

Table 10b Loss in tensile strength for cast aluminum alloys during various atmospheric exposures (ASTM program) (Ref 16)(a)(b)

Alloy and temper	Point Reyes, CA					Freeport, TX				
	6 mo	1 yr	3 yr	5 yr	10 yr	6 mo	1 yr	3 yr	5 yr	10 yr
Sand Castings										
208.0-F	⋯	−11	−13	−11	−10	−4	−5	−5	−9	−6
295.0-T6	⋯	−13	−15	−17	−16	−2	−9	−10	−10	−12
319.0-T6	⋯	−9	−14	−11	−10	−2	−1	−7	−6	−4
355.0-T6	⋯	−4	−8	−7	−10	+1	−1	−4	−3	−7
356.0-T6	⋯	0	−1	−2	−5	+2	−3	0	−3	−4
443.0-F	⋯	−7	−10	−10	−10	0	−1	−2	−4	−6
520.0-T4	+1	−3	−6	−7	⋯	+1	−4	−7	−11	⋯
705.0-T5	⋯	+3	−8	−6	−4	+6	+3	−5	−4	−8
707.0-T5	⋯	−5	−8	−7	−9	−1	−5	−15	−16	−32(c)
710.0-T5	⋯	−1	−3	−4	−3	+4	−1	−1	0	−2
712.0-T5	⋯	−7	−7	−8	−14	+1	−7	−6	−9	−9
713.0-T5	⋯	−3	−6	0	−3	−4	−6	−7	−6	−9
Permanent Mold Castings										
319.0-T61	⋯	−7	−15(c)	−14(c)	−16(c)	0	−7	−4	−5	−5
355.0-T6	⋯	−6	−2	−8	−13	+4	−4	+5	−2	−7
443.0-F	⋯	−7	−11	−8	−10	0	−1	−3	−2	−2
705.0-T5	⋯	−5	−6	−3	−4	−3	−5	−5	−8	−14
707.0-T5	⋯	−3	−2	−2	−9	+1	−3	−6	−10	−24(c)
711.0-T5	⋯	−5	−9	−6	−9	+1	−4	−3	−1	−8
713.0-T5	⋯	−9	−6	−4	−9	−6	−9	−2	0	−6

(a) Exposed as separately cast tensile specimens. (b) Calculated from average tensile strength of several specimens (usually six). (c) Average tensile-strength values were below required minimum.

crete, plaster, mortar or stucco is highly alkaline and causes slight attack and discoloration. This is most likely to occur during or shortly after construction, and leaching by subsequent rains, as well as conversion to carbonates, reduces the alkalinity and further attack. Staining can be effectively prevented by organic coatings.

Some insulating materials that are porous and absorbent may cause corrosion when wet. If more cathodic metals, such as steel or copper alloys, are electrically coupled with the aluminum through these materials, galvanic attack may occur. Protective paint films on the cathodic metal, moisture barriers or chemical inhibition are required for optimum performance under these conditions.

Concrete, plaster, mortar and cements also cause superficial etching of aluminum, most of which occurs during the curing period. The surface attack involves dissolution of the natural oxide film and some of the metal, but a new film is formed that prevents further corrosion. Coupling with more cathodic metals has little effect on aluminum embedded in these materials except in those that contain certain curing or antifreeze additives.

When partly embedded in concrete,

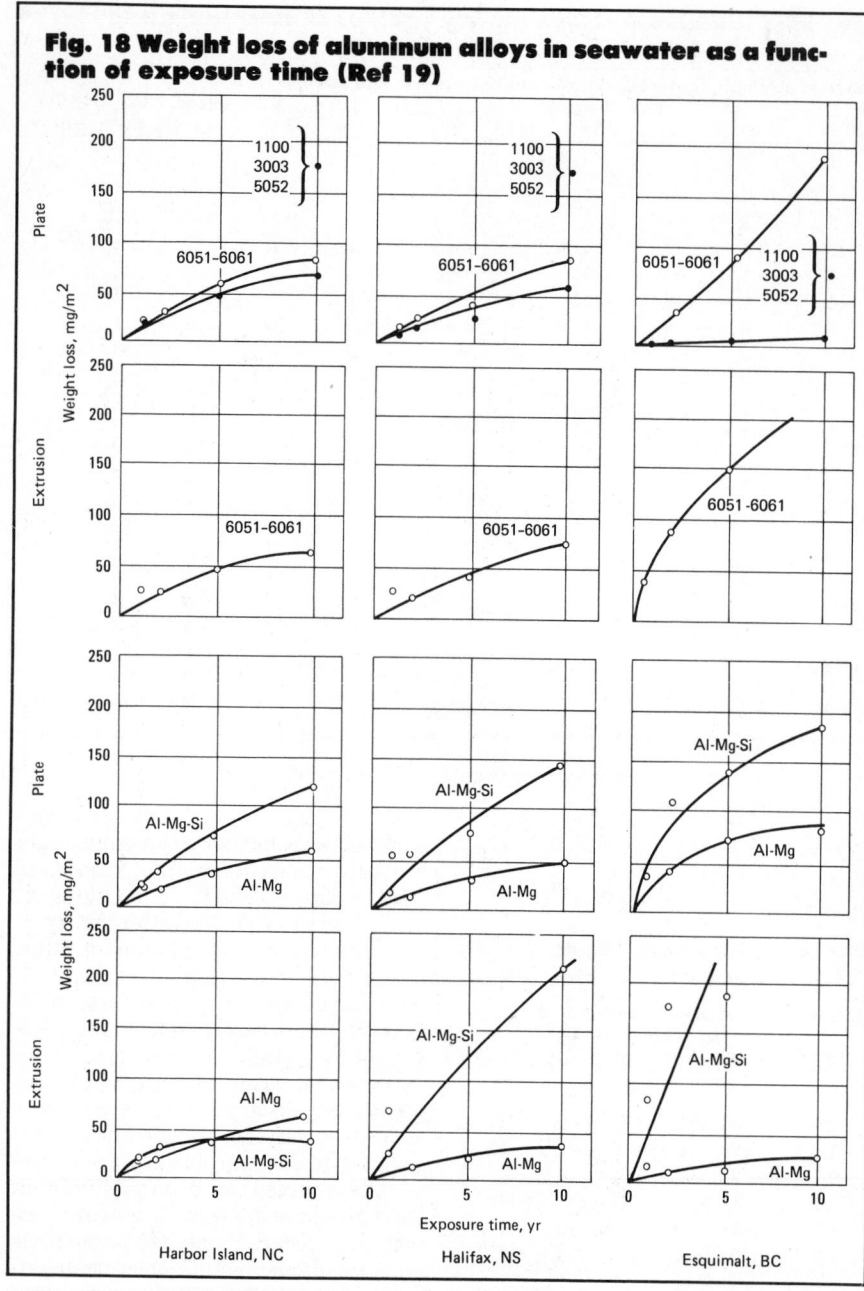

Fig. 18 Weight loss of aluminum alloys in seawater as a function of exposure time (Ref 19)

some metals undergo accelerated corrosion where the metal intersects the exposed surface of the concrete. This effect usually is not important for aluminum, but special consideration must be given to protection of faying surfaces or crevices between the aluminum and the concrete, which may entrap environmental contaminants. For example, highway railings and streetlight standards and stanchions usually are coated with a sealing compound where they are fastened to concrete, to prevent entry of salt-laden road splash into crevices.

Contact with Foods, Pharmaceuticals and Chemicals

The widespread usage of aluminum in processing, handling and packaging of foods, beverages, and pharmaceutical and chemical products is based on economic factors and the excellent compatibility of aluminum with many of these products. In addition to high corrosion resistance in contact with such products, many of these applications depend on the nontoxicity of aluminum and its salts, as well as its freedom from catalytic effects that cause product discoloration.

Application of aluminum for packaging foods and pharmaceutical products has grown phenomenally since 1970 so that this usage now accounts for about 20% of the aluminum marketed in the United States. The largest amount is used in beverage cans, and a smaller amount is used for foods. These cans generally have both internal and external organic coatings, not for corrosion protection but for decoration and for prevention of effects on product taste.

Large quantities of aluminum foil, either uncoated or with plastic coatings, are used in flexible packages.

Table 11 Atmospheric corrosion rates for aluminum and other nonferrous metals at several exposure sites (Ref 18)

Location	Type of atmosphere	Aluminum(b) 10 yr	Aluminum(b) 20 yr	Copper(c) 10 yr	Copper(c) 20 yr	Lead(d) 10 yr	Lead(d) 20 yr	Zinc(e) 10 yr	Zinc(e) 20 yr
Phoenix, AZ	Desert	0.000	0.076	0.13	0.13	0.23	0.10	0.25	0.18
State College, PA	Rural	0.025	0.076	0.58	0.43	0.48	0.30	1.07	1.09
Key West, FL	Seacoast	0.10	...	0.51	0.56	0.56	...	0.53	0.66
Sandy Hook, NJ	Seacoast	0.20	0.28	0.66	...	...	...	1.40	...
La Jolla, CA	Seacoast	0.71	0.63	1.32	1.27	0.41	0.53	1.73	1.73
New York, NY	Industrial	0.78	0.74	1.19	1.37	0.43	0.38	4.8	5.6
Altoona, PA	Industrial	0.63	...	1.17	1.40	0.69	...	4.8	6.9

Depth of metal removed per side(a), in μm/yr, during exposure of indicated length for specimens of:

(a) Calculated from weight loss, assuming uniform attack, for panels 0.89 mm (0.035 in.) thick. (b) Aluminum 1100-H14. (c) Tough pitch copper (99.9% Cu). (d) Commerical lead (99.92% Pb). (e) Prime western zinc (98.9% Zn).

Fig. 19 Density of corrosion pits in anodized aluminum 1100 as a function of coating thickness (Ref 28)

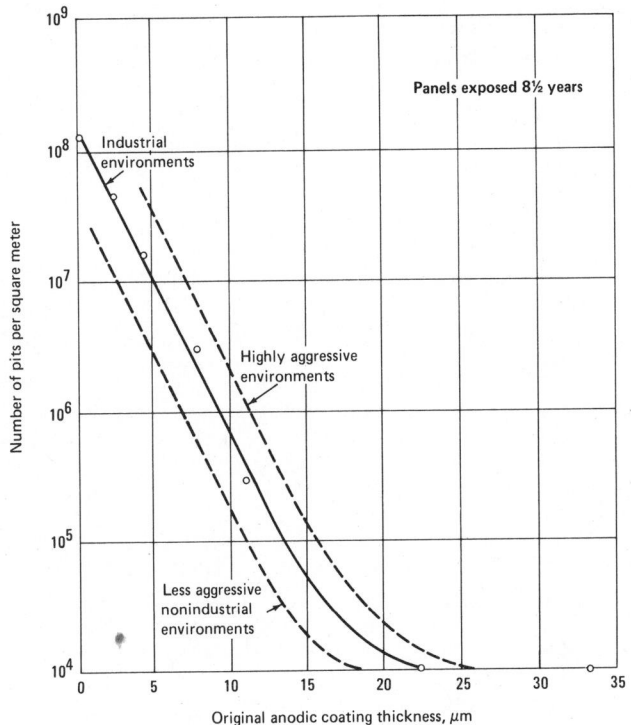

Fig. 21 Relative effectiveness of various protective systems in preventing stress-corrosion cracking of susceptible aluminum alloys

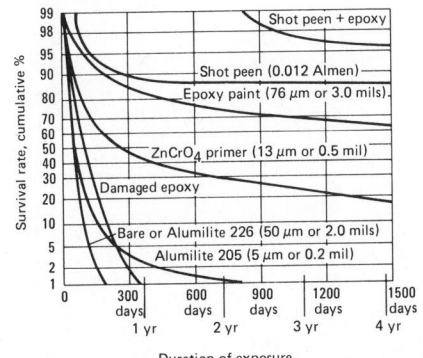

Combined data for highly elastically strained specimens of alloys 2014-T651 and 7079-T651, exposed at Pt. Judith, RI; Pt. Comfort, TX; and New Kensington, PA. (Reproduced from Ref 29 with permission of the National Academy of Sciences, Washington, DC.)

Coated foil also is used with fiber board in construction of rigid containers. Foil in such rigid containers, because of its extreme thinness, must be coated; only the slightest degree of corrosion can be tolerated, and perforation must not occur even during long periods of storage.

Packaging foils are produced from unalloyed aluminum corresponding to composition limits for aluminum 1230, beverage can bodies generally from alloy 3004, food can bodies from alloy 5352 or alloy 5050, and can ends from alloy 5182. These alloys have very high corrosion resistance and are not subject to corrosion problems in such applications.

Aluminum alloy household cooking utensils, usually made of alloy 3003, have been used for many years. These utensils, and commercial food-processing equipment, do not require protective coatings, but ceramic coatings often are applied to the exteriors of cooking utensils for aesthetic reasons and plastic coatings to the food-contacting surfaces for nonsticking characteristics. Alloys employed in commercial food processing include alloy 3003, 5xxx alloys, and casting alloy 514.0. Unsatisfactory performance sometimes is caused by use of improper cleaners, but

Fig. 20 Weathering data for anodically coated aluminum in an industrial atmosphere

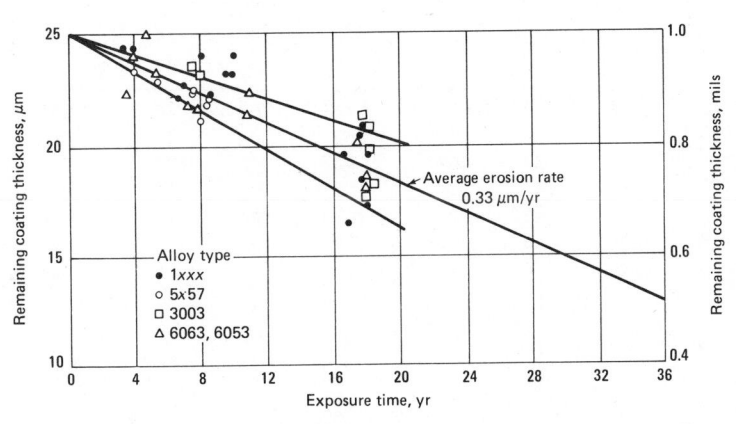

Table 12 Average weight loss and maximum depth of pitting for aluminum alloy plate specimens after immersion in seawater (Ref 19)(a)

Test series	Alloy and temper	Harbor Island, NC(b)				Halifax, NS				Esquimalt, BC(b)			
		1 yr	2 yr	5 yr	10 yr	1 yr	2 yr	5 yr	10 yr	1 yr	2 yr	5 yr	10 yr
Weight Loss, grams													
1	1100-H14	4.4	5.4	10.3	11.1	1.9	3.5	5.3	12.7	0.0	2.4	1.3	2.3
	3003-H14	4.1	6.4	9.3	11.2	0.0	3.3	4.6	7.5	0.0	0.0	3.0	2.2
	5052-H34	4.5	6.5	9.0	14.9	2.8	3.3	...	14.2	1.7	0.0	0.0	0.6
	6051-T4	3.7	4.9	9.9	12.3	0.0	0.7	3.5	8.0	1.9	7.8	19.0	14.6
	6051-T6	4.4	5.7	10.3	13.1	2.1	5.5	6.1	19.5	22.5	13.8	19.9	27.3
	6061-T4	4.8	6.6	12.4	18.6	4.4	6.0	8.0	15.6	0.9	2.3	28.2	62.0
	6061-T6	5.5	7.7	14.0	21.5	4.3	7.3	12.7	22.8	6.7	7.1	11.1	44.3
	7072	...	...	...	10.2	...	...	...	15.9	...	...	...	3.1
	7075-T6	...	...	...	149.0	...	...	...	242.6	...	...	...	246.5
2	5083	2.5	3.7	...	7.3	2.8	0.0	6.1	8.5	1.3	1.9	2.7	3.3
	5083	4.7	3.4	5.7	8.1	2.6	3.2	5.2	7.5	15.3	16.3	36.3	31.1
	5056	3.7	4.7	6.0	9.2	2.5	3.3	5.7	10.4	10.7	16.5	19.5	28.9
	5056	4.5	5.2	...	16.7	4.0	4.1	5.5	11.1	7.0	6.0	11.0	11.4
	6051-T4	3.9	4.2	12.1	9.1	3.6	3.1	5.5	9.2	9.1	18.8	15.3	51.0
	6051-T6	4.1	4.5	7.7	10.6	5.3	4.1	8.4	18.6	17.2	23.3	30.6	33.5
	6053-T6	4.1	4.5	6.6	9.7	3.0	3.3	5.6	14.8	15.7	25.1	19.3	25.8
	6061-T6	7.6	13.4	29.4	51.6	9.8	11.2	33.2	48.5	12.3	26.8	48.7	48.0
	6061-T6	5.5	6.5	15.4	34.2	10.0	9.4	19.1	54.1	7.3	7.0	21.3	18.6
	Al-7Mg	4.1	4.1	6.5	9.4	2.4	2.4	4.6	8.0	1.6	2.9	2.1	3.3
	Low-carbon steel(c)	219.0	294.0	471.3	979.8	208.0	292.6	761.1	1450.0	277.0	455.4	1012.4	2240.8
3	5154	2.8	5.2	6.0	...	2.4	2.6	3.8	...	1.4	2.1	2.6	...
	5083	3.5	4.6	6.0	...	2.0	2.8	3.6	...	0.2	2.2	2.8	...
	6053-T6	3.8	6.6	25.9	...	19.3	29.2	4.7	...	45.6	80.4	86.0	...
	7075-T6	60.4	49.3	74.8	...	44.8	66.1	116.0	...	50.9	71.3	153.5	...
	3003, alclad ...	4.3	12.0	...	...	1.6	2.3	...	...	...	1.9	...	...
	6061, alclad ...	4.3	3.9	5.7	...	8.4	3.3	6.5	...	20.8	15.8	34.3	...
	7075, alclad ...	4.4	5.2	6.1	...	2.8	3.6	6.8	...	8.5	14.5	16.6	...
Maximum Depth of Pitting, mils													
1	1100-H14	0	0	40	0	17	32	0	29	30	26	15	0
	3003-H14	0	0	13	21	13	15	21	22	5	20	0	10
	5052-H34	0	0	0	0	5	20	6	12	16	6	0	5
	6051-T4	0	0	5	0	0	10	0	62	10	65	51	37
	6051-T6	2	0	5	0	19	56	15	64	70	60	181	238
	6061-T4	0	13	2	14	12	18	21	33	15	50	20	28
	6061-T6	36	24	60	95	36	43	43	54	30	25	80	116
	7072	...	...	...	56	...	...	...	150	...	...	...	26
	7075-T6	...	...	...	66	...	...	...	(d)	...	...	...	(d)
2	5083	12	9	6	0	3	0	12	7	13	5	0	6
	5083	16	13	6	10	4	23	16	22	29	38	47	55
	5056	7	10	7	5	3	0	12	15	20	39	34	35
	5056	10	10	5	28	0	0	10	24	20	1	0	11
	6051-T4	16	4	7	15	3	0	9	35	25	47	109	170
	6051-T6	11	17	9	15	5	8	25	60	55	34	184	200
	6053-T6	30	15	14	58	28	34	66	95	93	126	165	105
	6061-T4	67	100	144	130	50	67	90	122	60	100	125	125
	6061-T6	15	27	36	40	38	47	58	67	35	48	60	55
	Al-7 Mg	12	7	8	8	3	0	12	14	8	12	0	7
3	5154	12	9	5	...	0	12	15	...	0	0	3	...
	5083	22	1	7	...	0	11	7	...	0	0	5	...
	6053-T6	28	150	186	...	93	91	34	...	81	118	118	...
	7075-T6	25	25	25	...	18	17	(e)	...	15	11	(e)	...
	3003, alclad ...	0	12	...	...	0	13	...	...	...	13	...	...
	6061, alclad ...	10	10	9	...	9	11	9	...	11	13	9	...
	7075, alclad ...	10	15	13	...	13	14	12	...	12	14	15	...

(a) Specimens were 6.35 x 305 x 305 mm (0.250 x 12 x 12 in.) and weighed approximately 1.6 kg (3.5 lb). (b) Harbor Island is near Wilmington, NC; Esquimalt is near Victoria, BC. (c) Original weight about 4.8 kg (10.6 lb). (d) Plate was perforated. (e) Could not determine because no original surface left.

Table 13 Average weight loss and maximum depth of pitting for aluminum alloy extruded specimens after immersion in seawater (Ref 19)(a)

Test series	Alloy and temper	Harbor Island, NC(b) 1 yr	2 yr	5 yr	10 yr	Halifax, NS 1 yr	2 yr	5 yr	10 yr	Esquimalt, BC(b) 1 yr	2 yr	5 yr	10 yr
Weight Loss, grams													
1	6051-T4	2.9	0.0	8.0	8.2	7.8	1.5	4.5	6.3	2.8	0.0	...	...
	6051-T6	6.2	0.0	...	14.6	9.0	10.5	12.8	23.4	15.4	40.7	29.7	83.0
	6061-T4	4.7	10.9	...	7.4	0.0	0.4	4.3	7.4	0.2	10.0	29.8	38.3
	6061-T6	3.0	6.9	8.1	16.4	3.8	5.0	7.1	15.5	15.5	14.1	25.2	59.2
2	5056	3.2	2.7	6.3	9.9	6.3	2.8	4.2	6.2	5.0	1.9	3.6	4.8
	6051-T4	3.0	3.4	6.5	8.0	13.1	9.2	7.8	12.0	16.0	16.7	18.0	35.4
	6051-T6	4.9	11.1	5.7	9.4	19.9	...	23.0	78.9	23.0	30.2	41.3	122.8
	6053-T6	3.3	4.9	6.9	10.6	2.6	3.0	4.5	8.9	4.5	43.5	35.3	99.9
3	5056	2.0	5.2	5.1	...	1.3	1.3	2.1	...	3.5	9.4	2.4	...
	6063-T5	2.6	3.3	6.5	...	2.4	3.5	4.9	...	6.6	13.4	13.1	...
	6053-T6	2.8	3.0	5.3	...	25.7	12.0	30.6	...	43.5	29.9	77.1	...
Maximum Depth of Pitting, mils													
1	6051-T4	0	0	27	20	27	27	14	32	35	23	65	72
	6051-T6	70	40	46	67	52	68	125	(c)	70	(c)	160	(c)
	6061-T4	23	23	27	12	25	20	12	32	33	45	56	70
	6061-T6	13	13	10	15	15	9	14	27	20	30	46	45
2	5056	13	7	35	32	60	0	16	41	30	17	99	50
	6051-T4	57	5	20	15	34	65	30	74	66	65	90	115
	6051-T6	58	>100	34	45	100	...	84	(c)	64	85	107	>200(d)
	6053-T6	13	25	93	46	0	0	7	34	80	110	175	210
3	5056	28	68	17	...	0	3	15	...	37	72	63	...
	6063-T5	42	35	45	...	27	25	30	...	70	66	136	...
	6053-T6	28	1	20	...	185	90	(c)	...	178	(c)	(c)	...

(a) Specimens were 6.35 mm (0.250 in.) thick, 0.170 m² (1.83 ft²) in area, and weighed approximately 1.2 kg (2.6 lb). (b) Harbor Island is near Wilmington, NC; Esquimalt is near Victoria, BC. (c) Plate was perforated. (d) In thick web of angle.

Table 14 Average weight loss (mg/m²) for aluminum alloys in seawater (Ref 19) (from Tables 12 and 13)

Test series	Harbor Island, NC 1 yr	2 yr	5 yr	10 yr	Halifax, NS 1 yr	2 yr	5 yr	10 yr	Esquimalt, BC 1 yr	2 yr	5 yr	10 yr
Series 1: Plate(a)												
1100, 3003, 5052	22	32	49	64	9	18	26	60	3	4	7	9
6051, 6061	24	32	60	85	12	25	39	85	41	40	101	191
Series 1: Extrusions(b)												
6051, 6061	25	26	47	68	30	26	42	78	50	95	166	354
Series 2: Plate(a)												
Al-Mg	20	22	32	52	15	13	28	47	37	39	74	81
Al-Mg-Si	26	34	75	119	55	54	75	149	64	111	140	183
Series 2: Extrusions(b)												
Al-Mg	19	19	38	62	25	16	21	37	18	11	15	26
Al-Mg-Si	22	38	38	41	70	24	70	196	85	177	185	506

(a) Plate surface area, 0.193 m² (2.08 ft²). (b) Extrusion surface area, 0.170 m² (1.83 ft²).

is not caused by incompatibility with the food product. Some alkaline cleaners cause excessive corrosion and should not be used unless they are inhibited effectively.

Aluminum alloys are used in processing, handling and packaging a wide variety of chemical products. Aluminum alloys are compatible with dry salts of most inorganic chemicals. Factors controlling compatibility of aluminum alloys with aqueous solutions are described in earlier sections. Within their passive pH range (about 4 to 9), aluminum alloys are resistant to corrosion by solutions of most inorganic chemicals, but they are subject to pitting in aerated solutions—particularly halide solutions, in which they are polarized to their pitting potentials by free corrosion.

Figure 22 illustrates the corrosion behavior of aluminum in several acids and bases. Aluminum alloys are not suitable for handling mineral acids, with the exception of nitric acid in concentrations above 82 weight percent and sulfuric acid from 98 to 100 weight percent. Aluminum alloys resist most alcohols; however, some alcohols may cause corrosion when extremely dry and at elevated temperatures (Fig. 22d). The same characteristics are associated with phenol (Fig. 22f). Aldehydes have little or no action on aluminum (Fig. 22e). Under most conditions, particularly at room temperature, aluminum alloys are resistant to halogenated organic compounds, but under some conditions, they may react rapidly or violently with these chemicals. If water is present, these chemicals may hydrolyze to yield mineral acids that destroy aluminum's protective oxide film. Such corrosion by mineral acids may in turn promote further reaction with the chemicals themselves, because the aluminum halides formed by this corrosion are catalysts for some

Table 15 Summary of data from 10-year seawater exposures at Wrightsville Beach, NC (Ref 25)

Alloy and temper	% Mg	Thickness mm	Thickness in.	Corrosion rate based on weight change μm/yr	Corrosion rate based on weight change mil/yr	Maximum depth of attack in 10 years mm	Maximum depth of attack in 10 years mil	Average depth of attack in 10 years mm	Average depth of attack in 10 years mil	Change in tensile strength in 10 years, %
Half-Tide Exposure										
1199	...	1.27	0.050	0.91	0.036	0.99	0.039	0.07	0.003	0
5154-H38	3.5	1.27	0.050	0.94	0.037	0.50	0.020	0.13	0.005	−2.1
5454-H34	2.7	6.35	0.250	1.04	0.041	0.39	0.015	0.07	0.003	−0.7
5457-H34	1.0	1.02	0.040	0.91	0.036	0.56	0.022	0.03	0.001	−4.2
5456-O	5.1	6.17	0.243	0.36	0.014	1.74	0.069	0.32	0.013	−0.4
5456-H321	5.1	6.17	0.243	1.29	0.051	1.83	0.072	0.34	0.013	−4.5
5083-O	4.5	6.35	0.250	0.91	0.036	0.97	0.038	0.31	0.012	0
5086-O	4.0	2.03	0.080	0.89	0.035	0.69	0.027	0.06	0.002	−2.7
Full-Immersion Exposure										
1199	...	1.27	0.050	1.55	0.061	...	...	...	...	0
5154-H38	3.5	1.27	0.050	1.40	0.055	...	...	...	...	−5.1
5454-H34	2.7	6.35	0.250	1.50	0.059	0.51	0.020	0.10	0.004	−0.5
5457-H34	1.0	1.02	0.040	1.42	0.056	...	...	...	...	−5.2
5456-O	5.1	6.17	0.243	2.95	0.116	3.33	0.131	1.01	0.040	−3.0
5456-H321	5.1	6.17	0.243	1.62	0.064	1.12	0.044	0.31	0.012	−1.1
5083-O	4.5	6.35	0.250	1.50	0.059	0.61	0.024	0.03	0.001	0
5086-O	4.0	2.03	0.080	1.45	0.057	...	...	...	...	−3.7

Table 16 Corrosion resistance of aluminum alloy plate, with and without joints, partially immersed in flowing seawater at Kure Beach, NC (Ref 26)

Alloy and temper	Type of joint	Exposure period, yr	Maximum depth of attack, mils Plate Outside surface	Maximum depth of attack, mils Faying surface	Rivet or weld	Change in tensile strength due to corrosion(a), %
Continuously Immersed						
6053-T6	Riveted(b)	6	1.4	3.0	8.4	0
6061-T6	Riveted(c)	1	1.4	2.8	2.8	0
6053-T6	Welded(d)	2	5.0	...	4.2	...
6061-T6	Welded(d)	1	5.0	...	9.8	...
6061-T4	None	3	2.1	...	...	+2
6061-T6	None	3	1.4	...	...	+1
2024-T4 alclad(e)	None	5	4.2	...	...	0
3004-H14 alclad(f)	None	5	1.4	...	...	−5
520.0-T4(g)	None	3	4.2	...	...	−4
Not Immersed (Atmospheric Exposure)						
6053-T6	Riveted(d)	6	5.6	5.6	11.7	−1
6061-T6	Riveted(e)	1	5.6	2.1	8.5	0
6053-T6	Welded(f)	2	3.3	...	9.8	...
6061-T6	Welded(f)	1	7.0	...	9.8	...
6061-T4	None	3	2.1	...	...	+1
6061-T6	None	3	4.2	...	...	+1
2024-T4 alclad(e)	None	5	8.4	...	...	0
3003-H14 alclad(f)	None	5	7.0	...	...	−5
520.0-T4(g)	None	3	1.4	...	...	+4

(a) Results of testing ASTM tensile specimens 6.4 mm (0.25 in.) thick cut from indicated location in test plate. (Generally, two specimens were cut from each test plate and the results were averaged.) (b) 6053-T6 rivets. (c) 6061-T43 rivets. (d) 4043 filler metal. (e) Average thickness of cladding on each surface, 297 μm (11.7 mils). (f) Average thickness of cladding on each surface, 307 μm (12.1 mils). (g) Sand cast.

Table 17 Results of 3-year seacoast exposure testing of anodized aluminum alloys(a)

Alloy and temper	Results
Sheet	
1100	No visible pitting
2024-T3, alclad	Edge pitting only
5456-H343	No visible pitting
5086-H34	No visible pitting
6061-T6	No visible pitting
7039-T6	No visible pitting
7075-T6	Edge pitting only
7075-F, alclad	Edge pitting only
7079-T6	Edge pitting only
Extrusions	
6351-T6	No visible pitting
6061-T6	No visible pitting
6063-T5	No visible pitting
6070-T6	No visible pitting
7039-T6	Scattered small pits

(a) H_2SO_4 anodic coatings 23 μm (0.9 mil) thick, sealed in boiling water on test panels 100 by 150 mm (4 by 6 in.) cut from sheet and extrusions.

such reactions. To ensure safety, service conditions should be ascertained before aluminum alloys are used with these chemicals, and the most stringent precautions should be exercised before they are used in finely divided form.

Reactivity of aluminum alloys with halogenated organic chemicals is inversely related to the chemical stability of these reagents. Thus, they are most resistant to chemicals containing fluorine and decreasingly resistant to those containing chlorine, bromine and iodine. Aluminum alloys are resistant to highly polymerized halogenated chemicals, reflecting the high degree of stability of these chemicals.

Resistance of aluminum and its alloys to a large number of foods and chemicals, representing practically all classifications, has been established in laboratory tests, and in many instances by service experience as well. Data are readily available from several handbooks, from proprietary literature and from publications of trade associations. Especially useful is "Aluminum with Food and Chemicals", which is published, with frequent revisions, by the Aluminum Association (Ref 31).

Much of the data from laboratory tests are for chemicals of high purity. Caution should be exercised in using these data to predict performance of aluminum alloys with commercial grades of chemicals. Corrosion of aluminum alloys by inorganic chemicals frequently is caused by impurities such as copper, lead, mercury and nickel, and corrosion by organic chemicals often results from the presence of other organic chemicals. The combined effect of impurities may be greater than the sum of their individual effects.

REFERENCES

1. M. Pourbaix *et al*, *Atlas of Electrochemical Equilibria in Aqueous Solutions*, Pergamon Press, Oxford, NY, 1966

2. H. Bohni and H. H. Uhlig, Environmental Factors Affecting the Critical Pitting Potential of Aluminum, *Journal of the Electrochemical Society*, Vol 116, 1969, p 906

3. J. R. Galvele *et al*, Critical Potentials for Localized Corrosion of Aluminum Alloys, *Localized Corrosion*, National Association of Corrosion Engineers, 1974, p 580

4. R. L. Horst and G. C. English, Corrosion Evaluation of Aluminum Easy-Open Ends on Tinplate Cans, *Corrosion*, Vol 16, 1977, p 23

5. R. A. Bonewitz and E. D. Verink, Jr., Correlation Between Long Term Testing of Aluminum Alloys for Desalination and Electrochemical Methods of Evaluation, *Materials Performance*, Vol 14, 1975, p 16

6. R. H. Brown, Aluminum Alloy Laminates; Alclad and Clad Aluminum Alloy Products, *Engineering Laminates*, edited by A. G. H. Dietz, Wiley, 1949

7. E. H. Cook and F. L. McGeary, Electrodeposition of Iron From Aqueous Solutions Onto an Aluminum Alloy, *Corrosion*, Vol 20, 1964, p 11

8. E. H. Dix, Jr., Acceleration of the Rate of Corrosion by High Constant Stresses, *Transactions of the American Institute of Mining, Metallurgical and Petroleum Engineers*, Vol 137, 1940, p 11

9. R. B. Mears, R. H. Brown and E. H. Dix, Jr., A Generalized Theory of Stress-Corrosion Cracking of Alloys, *Symposium on Stress-Corrosion Cracking of Alloys*, published jointly by the American Society for Testing and Materials and the American Institute of Mining, Metallurgical and Petroleum Engineers, 1944, p 329

10. D. O. Sprowls and R. H. Brown, Stress-Corrosion Mechanisms for Aluminum Alloys, *Fundamental Aspects of Stress-Corrosion Cracking*, edited by R. W. Staehle, A. J. Forty and D. Van Rooyen, National Association of Corrosion Engineers, Houston, 1969, p 466–512

11. M. O. Speidel, Hydrogen Embrittlement of Aluminum Alloys, *Hydrogen in Metals*, edited by L. M. Bernstein and A. W. Thompson,

Fig. 22 Corrosion of aluminum 1100-H14 in various chemical solutions

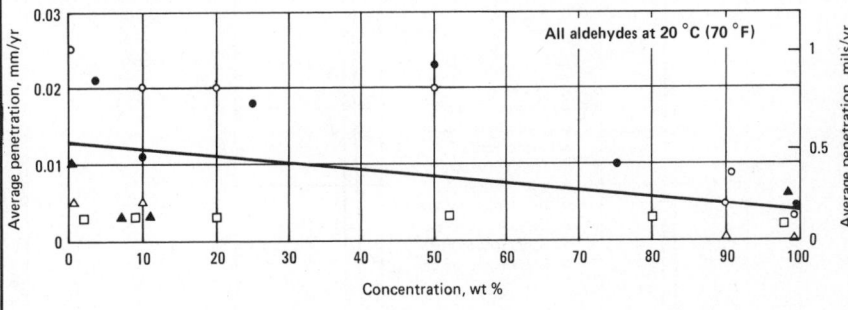

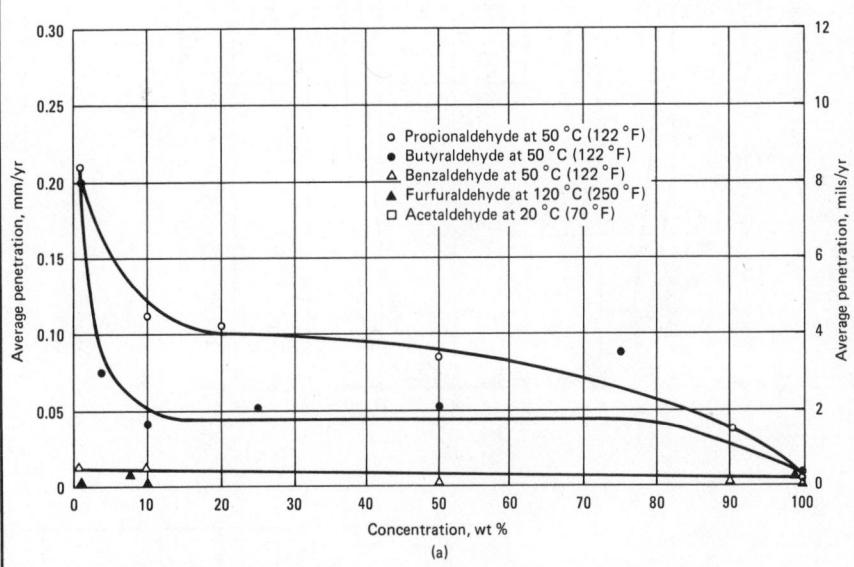

Note: average penetration calculated from weight-loss data in short-time tests.

(a) **Effect of concentration of aqueous solutions of several aldehydes. The rates of attack are low enough so that aluminum should be satisfactory for handling all of these solutions.** (b) **Effect of pH. The concentration of all the solutions ranged from 0.00001 to 0.1 N, except the acetic acid (0.00001 to 17.4 N), the ammonium hydroxide (0.00001 to 15 N) and the sodium disilicate solutions (0.00001 to 1.0 N). (c) Effect of concentration of nitric acid solutions at room temperature. (d) Effect of concentration and temperature of acetic acid. (e) Effect of concentration of boiling aqueous solutions of three alcohols. (f) Effect of temperature of phenol. Rapid reaction above 120 °C (250 °F) can be stopped by small additions of steam or water.**

Fig. 22 (continued)

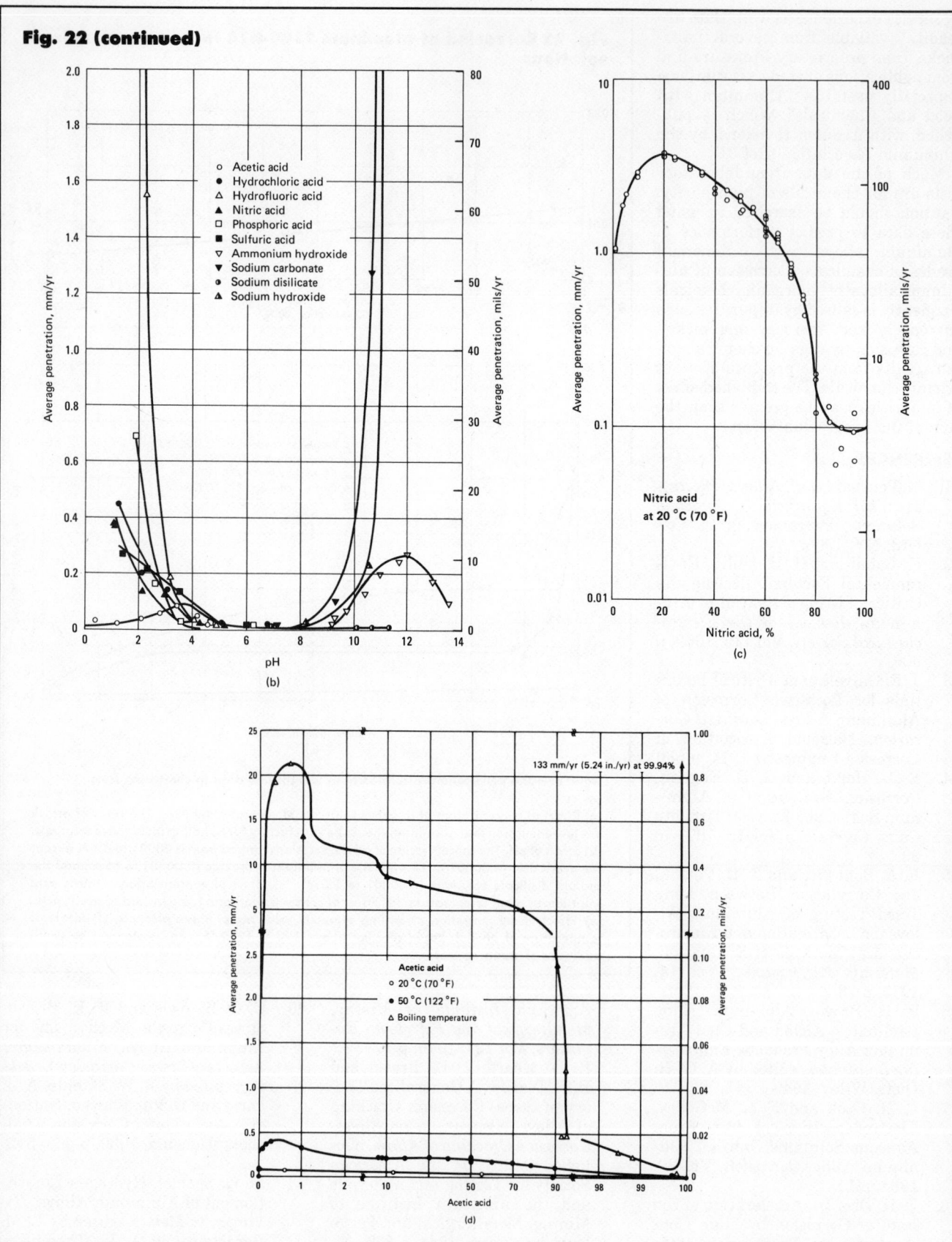

(b)

(c)

(d)

Fig. 22 (continued)

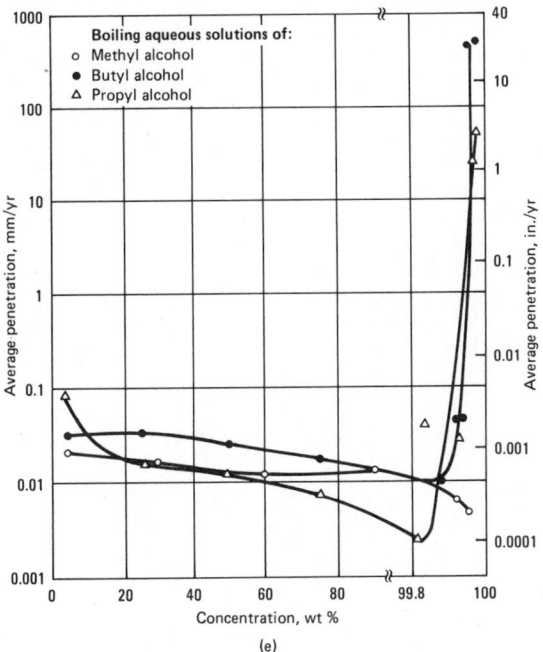

(e)

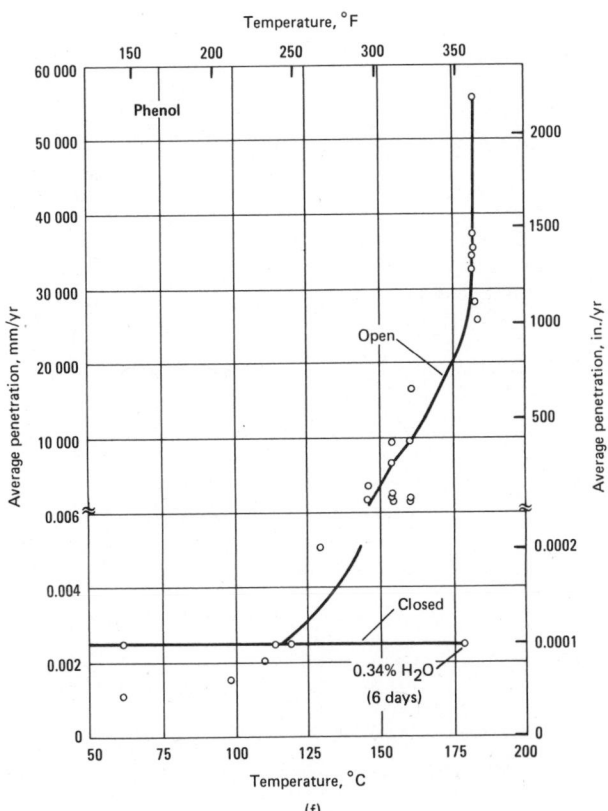

(f)

American Society for Metals, 1974, p 249

12. B. W. Lifka and D. O. Sprowls, Relationship of Accelerated Test Methods for Exfoliation Resistance of 7XXX Series Aluminum Alloys with Exposure to a Seacoast Atmosphere, *Corrosion in Natural Environments*, STP 558, American Society for Testing and Materials, 1974, p 306

13. C. J. Walton and W. King, Resistance of Aluminum-Base Alloys to 20-year Atmospheric Exposure, *Symposium on Metallic Materials for Service at Temperatures Above 1600 °F*, STP 174, American Society for Testing and Materials, 1956, p 21

14. F. L. McGeary, E. T. Englehart and P. J. Ging, Weathering of Aluminum, *Materials Protection*, Vol 6, June 1967, p 33

15. F. L. McGeary, T. J. Summerson and W. H. Ailor, Jr., Atmospheric Exposure of Nonferrous Metals and Alloys—Aluminum: Seven-year Data, *Symposium on Metal Corrosion in the Atmosphere*, STP 435, American Society for Testing and Materials, 1968, p 141

16. S. M. Brandt and L. H. Adam, Atmospheric Exposure of Light Metals, *Symposium on Metal Corrosion in the Atmosphere*, STP 435, American Society for Testing and Materials, 1968, p 95

17. W. H. Ailor, Jr., Performance of Aluminum Alloys at Other Test Sites, *Symposium on Metal Corrosion in the Atmosphere*, STP 435, American Society for Testing and Materials, 1968, p 285

18. E. Mattsson and S. Lindgren, Hard Rolled Aluminum Alloys, *Symposium on Metal Corrosion in the Atmosphere*, STP 435, American Society for Testing and Materials, 1968, p 240

19. H. P. Godard *et al*, *The Corrosion of Light Metals*, Wiley, 1967

20. V. E. Carter, Atmospheric Corrosion of Aluminum and Its Alloys: Results of Six-year Exposure Tests, *Symposium on Metal Corrosion in the Atmosphere*, STP 435, American Society for Testing and Materials, 1968, p 257

21. W. K. Boyd and F. W. Fink, "Corrosion of Metals in the Atmosphere", Report MCIC-74-33, Battelle Memorial Institute, Columbus, OH, 1974

22. E. H. Dix, Jr., R. H. Brown and W. W. Binger, The Resistance of Aluminum Alloys to Corrosion, *Metals Handbook,* 8th Ed., Vol 1, American Society for Metals, 1961, p 916–935

23. B. R. Pathak and H. P. Godard, Equations for Predicting the Corrosivity of Natural Fresh Waters to Aluminum, *Nature,* Vol 218 (No. 5144), June 1968, p 893

24. W. K. Boyd and F. W. Fink, "Corrosion of Metals in Marine Environments", Report MCIC-74-245 R, Battelle Memorial Institute, Columbus, OH, May 1975

25. W. H. Ailor, Jr., Ten-year Seawater Tests on Aluminum, *Corrosion in Natural Environments,* STP 558, American Society for Testing and Materials, 1974, p 117

26. W. W. Binger, E. H. Hollingsworth and D. O. Sprowls, Resistance to Corrosion and Stress Corrosion, *Aluminum,* Vol 1, American Society for Metals, 1967, p 209

27. F. M. Reinhart, "Corrosion of Metals and Alloys in the Deep Ocean", U.S. Naval Engineering Laboratory Report R834, Port Hueneme, CA, February 1976

28. W. C. Cochran and D. O. Sprowls, Anodic Coatings for Aluminum, *Conference on Corrosion Control by Coatings,* Lehigh University, Bethlehem, PA, November 1978

29. D. O. Sprowls *et al,* "Investigation of the Stress-Corrosion Cracking of High-Strength Aluminum Alloys", Final Report, Contract NAS-8-5340, for the period May 1963 to October 1966, Accession No. NASA CR88110, available from National Technical Information Center, Springfield, VA

30. C. J. Walton, F. L. McGeary and E. T. Englehart, The Compatibility of Aluminum with Alkaline Building Products, *Corrosion,* Vol 13, 1957, p 807t–816t

31. *Aluminum with Food and Chemicals,* 3rd Ed., The Aluminum Association, Washington, 1975

Copper

Introduction to Copper and Copper Alloys

COPPER and its alloys constitute one of the major groups of commercial metals. They are widely used because of their excellent electrical and thermal conductivities, outstanding resistance to corrosion, and ease of fabrication, together with good strength and fatigue resistance. They are generally nonmagnetic. They can be readily soldered and brazed, and many coppers and copper alloys can be welded by various gas, arc and resistance methods. For decorative parts, standard alloys having specific colors are readily available. Copper alloys can be polished and buffed to almost any desired texture and luster. They can be plated, coated with organic substances or chemically colored to further extend the variety of available finishes.

Pure copper is used extensively for cables and wires, electrical contacts, and a wide variety of other parts that are required to pass electrical current. Coppers and certain brasses, bronzes and cupronickels are used extensively for automobile radiators, heat exchangers, home heating systems, panels for absorbing solar energy and various other applications requiring rapid conduction of heat across or along a metal section. Because of their outstanding ability to resist corrosion, coppers, brasses, some bronzes, and cupronickels are used for pipes, valves and fittings in systems carrying potable water, process water or other aqueous fluids.

In all classes of copper alloys, certain alloy compositions for wrought products have counterparts among the cast alloys, which enables the designer to make an initial alloy selection before deciding on the manufacturing process. Most wrought alloys are available in various cold worked conditions, which have room temperature strengths and fatigue resistances that depend on the amount of cold work more than on alloy content. Typical applications of cold worked conditions (cold worked tempers) include springs, fasteners, hardware, small gears, and cams. Certain types of parts—most notably plumbing fittings and valves—are produced by hot forging simply because no other fabrication process can produce the required shapes and properties as economically.

Copper alloys containing 1 to 6% Pb are free machining grades, and are used widely for machined parts—especially those produced in screw machines.

Certain compositions in all classes of wrought copper alloys have cast counterparts. Properties and applications of cast copper alloys are described in greater detail in the article in this volume entitled "Copper Alloy Castings".

During the 1930's, it was not uncommon for an integrated brass mill to list 300 to 400 wrought copper alloys. In present-day commercial production, remain chiefly those alloys whose unique properties or combinations of properties have proved them superior for one or more major applications. Alloys of marginal utility have been weeded out, and new alloys of commercial significance are likely to appear only occasionally.

Properties of Importance

Good resistance to corrosion, good electrical conductivity, good thermal conductivity, color and ease of fabrication are the chief reasons why copper or one of its alloys is chosen over other metals. Strength, resistance to fatigue and ability to take a good finish also are criteria for selection, but are of lesser importance.

Corrosion Resistance. Copper is a member of the noble metal group in the periodic table. However, copper is not inert, like gold and other precious metals, because common reagents and environments can attack copper.

Copper resists this attack quite well under most corrosive conditions. A detailed description of the corrosion resistance of copper and its alloys can be found in the articles "Corrosion Resistance of Wrought Coppers and Copper Alloys" and "Corrosion Resistance of Copper Alloy Castings" in this volume. Although they resist corrosion, copper and its alloys sometimes have limited usefulness in certain environments because of hydrogen embrittlement or stress-corrosion cracking.

Hydrogen embrittlement is most often observed in tough pitch coppers and those alloys containing cuprous oxide. Consequently, deoxidized coppers are generally preferred over tough pitch coppers for applications where exposure to a reducing atmosphere is likely. Most of the copper alloys are deoxidized, and thus are not subject to hydrogen embrittlement.

Stress-corrosion cracking (season cracking) most commonly occurs in brass that is exposed to ammonia or amines. Brasses containing more than 15% zinc are the most susceptible. Copper and most copper alloys that either do not contain zinc or are low in zinc content usually are not susceptible to stress-corrosion cracking. Because stress-corrosion cracking requires both tensile stress and a specific chemical species to be present at the same time, removal of either the stress or the chemical species can prevent cracking. Annealing or stress relieving after forming frequently prevents stress-corrosion cracking because residual stresses from cold working are a major source of the stress present in the part in service. Stress relieving is effective only if the parts are not bent or strained in service, for bending or straining reintroduce tensile stresses.

Electrical and Thermal Conductivity. Copper and its alloys are relatively good conductors of electricity and heat. In fact, copper is used for these purposes more often than any other metal. Alloying invariably decreases electrical conductivity and, to a lesser extent, thermal conductivity. For this reason, coppers and high copper alloys are preferred over copper alloys containing more than a few percent total alloy content when high electrical or thermal conductivity is required for the application. The amount of reduction due to alloying does not depend on conductivity or any other bulk property of the alloying element, but only on the effect that the particular foreign atoms have on the copper lattice.

Color. Copper and certain copper alloys are used for decorative purposes alone, or when a particular color and finish is combined with a desirable mechanical or physical property of the alloy. Table 1 lists the standard copper alloys that can be obtained with consistent and reproducible colors.

Ease of Fabrication. Although there may be restrictions on the extent to which any given fabrication process can be used for a given copper or copper alloy, copper and its alloys are generally capable of being shaped to the required form and dimensions by any of the common fabricating processes. Copper metals are routinely rolled, stamped, drawn and headed cold; they are rolled, extruded, forged and formed at elevated temperature; there are casting alloys for all of the generic families of coppers and copper alloys.

Copper metals can be polished, textured, plated or coated to provide a wide variety of functional or decorative surfaces.

Electrical Coppers

Commercially pure copper is represented by UNS numbers C10100 to C13000. The various coppers within this group have different degrees of purity, and therefore different metal characteristics. Fire refined tough pitch copper C12500 is made by deoxidizing anode copper until the oxygen content has been lowered to the optimum value of 0.02 to 0.04%. Both the traditional method of "poling" (or "pitching") a bath of molten anode copper and the more modern method of deoxidizing with hydrocarbons produce metal with essentially the same high ductility and excellent electrical conductivity. Fire-refined tough pitch copper contains a small amount of residual sulfur, normally 10 to 30 ppm, and a somewhat larger amount of cuprous oxide, normally 500 to 3000 ppm. Traditionally, microscopic examination to detect the presence of a very small amount of cuprous oxide was the means of ensuring electrical conductivity above 95% IACS. The metallographic test has been largely replaced by actual conductivity testing for quality control purposes, especially because the metallographic test is invalid for high conductivity coppers other than the tough pitch coppers or cadmium copper C16200.

Electrolytic tough pitch copper C11000 is made from cathode copper—that is, copper that has been refined electrolytically. C11000 is the most common of all the electrical coppers. It has high electrical conductivity, in excess of 100% IACS. It has the same oxygen content as C12500, but differs in sulfur content and in over-all purity. C11000 has less than 50 ppm total metallic impurities (including sulfur).

Oxygen-free coppers C10100 and C10200 are made by induction melting prime-quality cathode copper under nonoxidizing conditions produced by a granulated graphite bath covering and a protective reducing atmosphere that is low in hydrogen. Oxygen-free coppers are particularly suitable for applications requiring high conductivity coupled with exceptional ductility, low gas permeability, freedom from hydrogen embrittlement or low out-gassing tendency.

If resistance to softening at slightly elevated temperature is required, C11100 is often specified. This copper contains a small amount of cadmium, which raises the temperature at which recovery and recrystallization occurs. Oxygen-free copper, electrolytic tough pitch copper and fire refined tough pitch copper are available as silver-bearing coppers having specific minimum silver contents. The silver, which may be present as an impurity in anode copper or may be intentionally added to

Table 1 Standard color controlled wrought copper alloys

UNS number	Common name	Color description
C11000	Electrolytic tough pitch copper	Soft pink
C21000	Gilding, 95%	Red brown
C22000	Commercial bronze, 90%	Bronze gold
C23000	Red brass, 85%	Tan gold
C26000	Cartridge brass, 70%	Green gold
C28000	Muntz metal, 60%	Light brown gold
C61200	Aluminum bronze	Brown gold
C65500	High-silicon bronze, A	Lavender-brown
C70600	Copper-nickel, 10%	Soft lavender
C74500	Nickel silver, 65-10	Gray white
C75200	Nickel silver, 65-18	Silver

molten cathode copper, also imparts resistance to softening to cold worked metal. Silver-bearing coppers and cadmium-bearing coppers are used for applications such as automotive radiators and electrical conductors that must operate at temperatures above about 200 °C (400 °F).

If good machinability is required, C14500 (tellurium-bearing copper) or C14700 (sulfur-bearing copper) can be selected. As might be expected, machinability is gained at a modest sacrifice in electrical conductivity.

Mechanical Working of Copper

High purity copper is a very soft metal. It is softest in its undeformed, single-crystal form, requiring a shear stress of only 3.9 MPa (570 psi) on {111} crystal planes for slip. Annealed tough pitch copper is almost as soft as high purity copper, but many of the copper alloys are much harder and stiffer, even in annealed tempers.

Copper is easily deformed cold. Once flow has been started it takes little energy to continue, and thus extremely large changes in shape or reductions in section are possible in a single pass. The only limitation appears to be the ability to design and build the necessary tools. Very heavy reductions are possible, especially with continuous flow. Rolling reductions of more than 90% in one pass are used for rolling strip.

Copper and many of its alloys also respond well to sequential cold working. Tandem rolling and gang-die drawing are common. Some copper alloys work harden rapidly, so there is a limit to the number of operations that can be performed before annealing to resoften the metal.

Copper can be cold reduced almost limitlessly without annealing but heavy deformation (more than about 80 to 90%) may induce preferred crystal orientation, or texturing. Textured metal has different properties in different directions, which is undesirable for some applications.

Cold working increases both tensile strength and yield strength, but the effect is more pronounced on the latter. For most coppers and copper alloys, the tensile strength of the hardest cold worked temper is approximately twice the tensile strength of the annealed temper. For the same alloys, the yield strength of the hardest cold worked temper may be as much as five to six times that of the annealed temper.

Hardness as a measure of temper is inaccurate—the relation between hardness and strength is different for different alloys. Usually, hardness and strength for a given alloy can be correlated only over a rather narrow range of conditions. Also, the range of correlation is often different for different methods of hardness determination.

Hot Working. Not all shaping is confined to cold deformation. Hot working is commonly used for alloys that remain ductile above the recrystallization temperature. Hot working permits more extensive changes in shape than cold working, so that a single operation can replace a sequence of forming and annealing operations. To avoid preferred orientation and textures, as well as to achieve processing economy, copper and many of its alloys are hot worked to nearly finished size. Hot working reduces as-cast grain size from about 1 to 10 mm to about 0.1 mm or less, and yields a soft, texture-free structure suitable for cold finishing.

Some hot working operations may produce strengths that exceed that of the annealed temper. However, property control by hot working is very difficult, and is rarely attempted.

Annealing

Work hardened metal can be returned to a soft state by heating, or annealing. During annealing, deformed and highly stressed crystals are transformed into unstressed crystals by recovery, recrystallization and grain growth. In severely deformed metal, recrystallization occurs at lower temperatures than in lightly deformed metal. Also, the grains are smaller and more uniform in size when severely deformed metal is recrystallized.

Grain size can be controlled by proper selection of cold working and annealing practices. Large amounts of prior cold work, fast heating to annealing temperature and short annealing times favor fine grain sizes. Larger grain sizes are normally produced by a combination of limited deformation and long annealing times. In normal commercial practice, annealed grain sizes are controlled about a median value in the range 0.01 to 0.10 mm. Total variations in grain size of two orders of magnitude are not uncommon, even though the control range is usually much narrower.

Variations in annealed grain size produce variations in hardness and other mechanical properties that are smaller than the variations that occur in cold worked material, but by no means negligible. Coarse-grained metal is somewhat softer than fine-grained metal and therefore somewhat more easily cold worked. Fine grain sizes often are required to enhance end-product characteristics such as load-carrying capacity, fatigue resistance, resistance to stress-corrosion cracking, and surface quality for polishing or buffing of either annealed or cold formed parts.

Anneal-resistant Coppers. Addition of small amounts of elements such as silver and cadmium to deoxidized copper imparts resistance to softening at times and temperatures encountered in soldering operations such as those used to join components of automobile and truck radiators.

The thermal and electrical conductivities of copper are relatively unaffected by small amounts of either silver or cadmium. Room temperature mechanical properties also are unchanged. C11100, C14300 and C16200 (cadmium-bearing coppers) work harden at higher rates than either C11400 (10 oz/ton silver-bearing copper) or C11000 (electrolytic tough pitch copper), as shown in Fig. 1.

Cold rolled silver-bearing copper is used extensively for automobile-radiator fins. Usually such strip is only moderately cold rolled, because heavy cold rolling makes silver-bearing copper more likely to soften during soldering or baking operations. Some manufacturers prefer cadmium copper C14300, because it can be severely cold rolled without making it susceptible to softening during soldering. Figure 2 illustrates the softening characteristics of C14300 and C11400 as measured for several temperatures and two tempers. As illustrated in Fig. 2(b), C14300 cold rolled to a tensile strength of 440 MPa (64 ksi) retains 91% of its strength after a typical core bake of 3 min at 345 °C (650 °F). Silver-bearing copper C11400 given the same cold reduction retains only 60% of its tensile strength after the same baking schedule.

Copper Alloys

The most common way to catalog copper and its alloys is to divide them

Fig. 1 Tensile strength versus reduction in area during rolling for cadmium-bearing copper and tough pitch copper

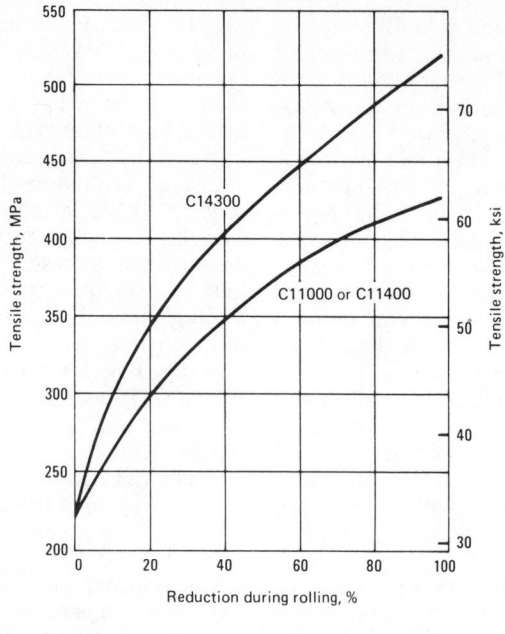

Table 2 Classification of copper and copper alloys

Family	Principal alloying element	Solid solubility, at. %(a)	UNS numbers(b)
Coppers, high copper alloys .... (c)		...	C10000
Brasses .... Zn		37	C20000, C30000, C40000, C66400 to C69800
Phosphor bronzes .... Sn		9	C50000
Aluminum bronzes .... Al		19	C60600 to C64200
Silicon bronzes .... Si		8	C64700 to C66100
Copper nickels, nickel silvers .... Ni		100	C70000

(a) At 20 °C (68 °F). (b) Wrought alloys. (c) Various elements having less than 8 at. % solid solubility at 20 °C (68 °F).

into six families: coppers, dilute copper alloys, brasses, bronzes, copper nickels and nickel silvers. The first family, the coppers, is essentially commercially pure copper, which ordinarily is soft and ductile and contains less than about 0.7% total impurities. The dilute copper alloys contain small amounts of various alloying elements that modify one or more of the basic properties of copper. Each of the remaining families contains one of five major alloying elements as its primary alloying ingredient (see Table 2). All five of the major alloying elements have room-temperature solid solubility in copper of at least 8 at. %.

Solid Solution Alloys. The most compatible alloying elements with copper are those that form solid-solution fields. These include all elements forming useful alloy families (see Table 2) plus manganese. Hardening in these systems is great enough to make useful objects without encountering brittleness associated with second phases or compounds.

Cartridge brass is typical of this group, consisting of 30% Zn in copper and exhibiting no beta phase except an occasional small amount due to segregation, which normally disappears after the first anneal. Provided that there are no tramp elements, such as Fe, cold

working and grain growth relationships are easily reproduced in practice.

Age-hardenable Alloys. Age hardening produces very high strengths, but is limited to those few copper alloys in which the solubility of the alloying element decreases sharply with decreasing temperature. The beryllium coppers can be considered typical of the age-hardenable copper alloys. Other age-hardenable alloys include C15000 (zirconium copper); C18200, C18400 and C18500 (chromium coppers); C19000 and C19100 (copper nickel phosphorus alloys); and C64700 (copper nickel silicon alloy).

By combining cold working with heat treatment, higher strengths can be obtained than can be achieved by either cold working or age hardening alone. Beryllium copper illustrates well the effects of heat treatment and cold working: in the soft, solution treated condition, the tensile strength is about 500 MPa (70 ksi); solution treated and aged, about 1000 MPa (150 ksi); solution treated, cold worked and aged, about 1400 MPa (200 ksi).

Some age-hardening alloys have different desirable characteristics, such as high strength combined with better electrical conductivity than the beryllium coppers.

Other Hardenable Alloys. Certain aluminum bronzes, most notably those containing more than about 9% Al, can be hardened by quenching from above a critical temperature. The hardening process is a martensitic-type process, similar to the martensitic hardening that occurs when iron-carbon alloys are quenched. Mechanical properties of aluminum bronzes can be varied somewhat by temper annealing after quenching or by using an interrupted quench instead of a standard quench.

C71900 and similar copper alloys also can be hardened, but the process is spinodal decomposition rather than martensite formation.

Insoluble Alloying Elements. Lead, tellurium and selenium are added to copper and its alloys to improve machinability. They, along with bismuth, make hot rolling and hot forming nearly impossible and severely limit the useful range of cold working.

An exception here are the high zinc brasses, which become fully beta phase at high temperature. The beta phase can dissolve lead—thus avoiding a liquid grain-boundary phase at hot forg-

Fig. 2 Softening characteristics of cadmium-bearing copper and silver-bearing tough pitch copper

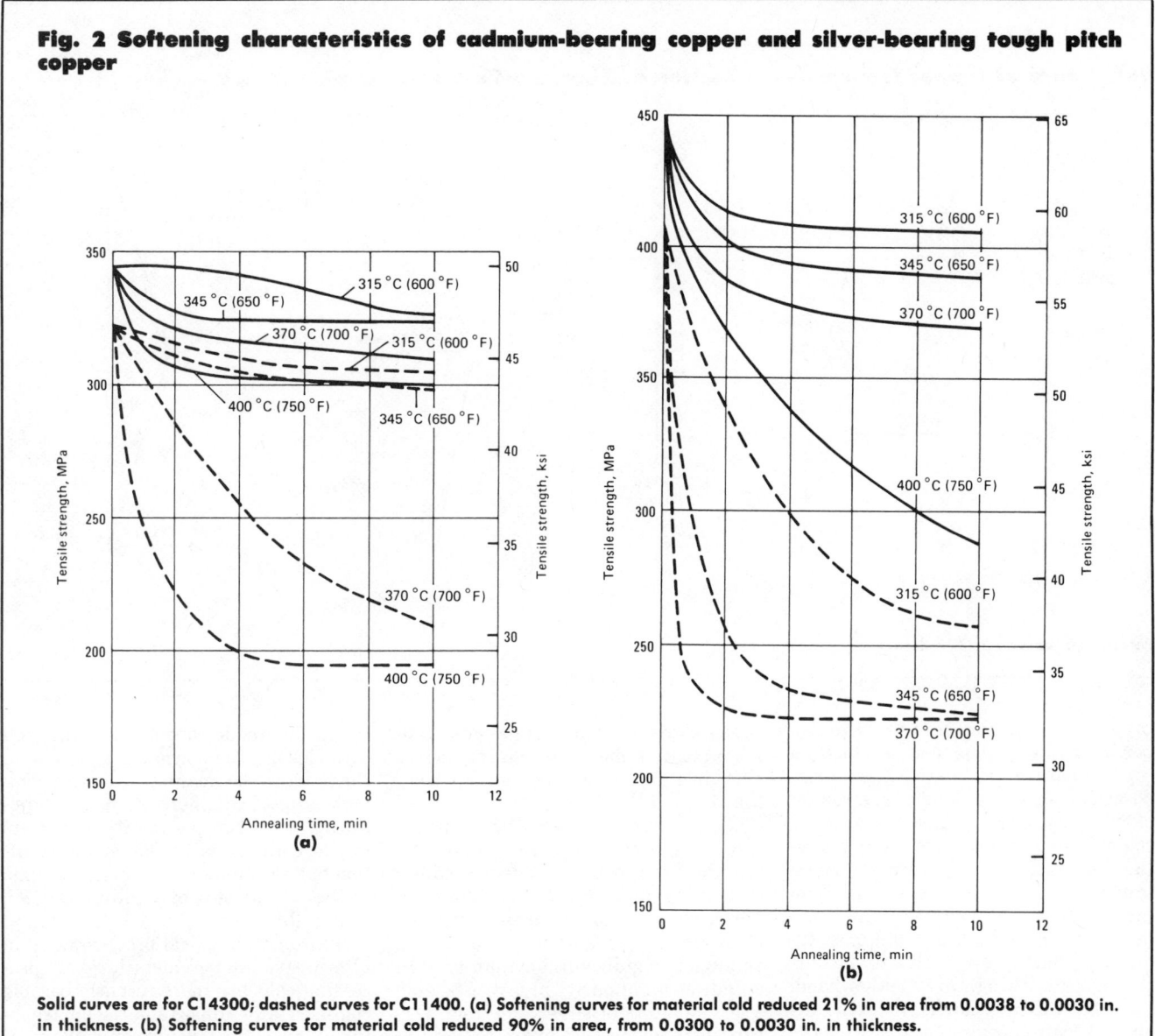

Solid curves are for C14300; dashed curves for C11400. (a) Softening curves for material cold reduced 21% in area from 0.0038 to 0.0030 in. in thickness. (b) Softening curves for material cold reduced 90% in area, from 0.0300 to 0.0030 in. in thickness.

ing or extrusion temperatures. Most free-cutting brass rod is made by beta extrusion. C37700, one of the leaded high zinc brasses, is so readily hot forged that it is the standard alloy against which the forgeability of all copper alloys is judged.

Deoxidizers

Li, Na, Be, Mg, B, Al, C, Si and P can be used to deoxidize copper. Ca, Mn and Zn can sometimes be considered deoxidizers, although they normally fulfill different roles.

The first requirement of a deoxidizer is that it have an affinity for oxygen in molten copper. Probably the second most important requirement is that it be relatively inexpensive compared to copper and any other additions. Thus, although zinc normally functions as a solid-solution strengthener, it is sometimes added in small amounts to function as a deoxidizer, because it has high affinity for oxygen and is relatively low in cost. In tin bronze, phosphorus has traditionally been the deoxidizer, hence the name "phosphor bronzes" for these alloys. Silicon instead of phosphorus is the deoxidizer for chromium coppers because phosphorus severely reduces electrical conductivity. Most

deoxidizers contribute to hardness and other qualities, which often makes classification as a deoxidizer indistinct.

Production of Copper Metals

The copper industry in the United States, broadly speaking, is composed of two segments: producers (mining, smelting and refining companies) and fabricators (wire mills, brass mills, foundries and powder plants). The end products of copper producers, the most important of which is refined cathode

Fig. 3 Flow of copper from mine production and scrap collection through end use

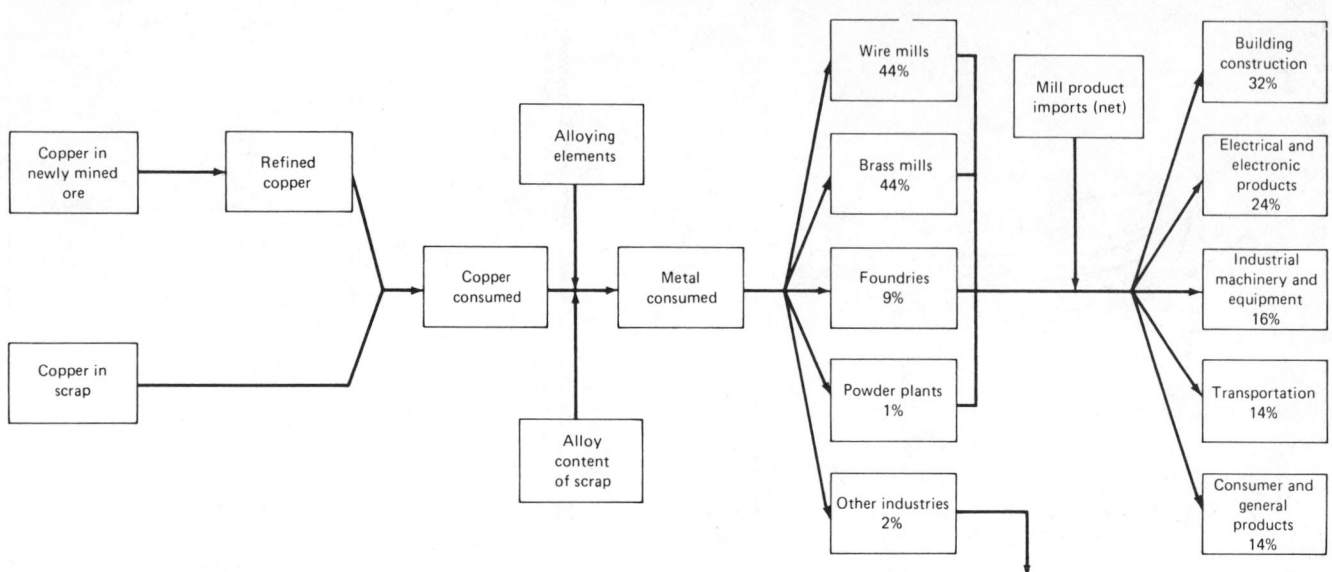

Percentages based on 1977 data.

copper, are sold almost entirely to copper fabricators. The end products of copper fabricators may be generally described as mill products, and consist of wire and cable, sheet, strip, plate, rod, bar, mechanical wire, tubing, forgings, extrusions and castings. These products are sold to a wide variety of industrial users. Certain mill products—chiefly wire, cable and most tubular products—are used without further metalworking. On the other hand, most flat-rolled products, rod, bar, mechanical wire, forgings and castings go through multiple metalworking, machining, finishing and/or assembly operations before emerging as finished products.

Copper Producers. Figure 3 is a simplified flow diagram of the copper industry. The box at upper left represents mining companies which remove vast quantities of low-grade material, mostly from open-pit mines, in order to extract copper from the earth's crust. Approximately 2.5 t (tonnes) of overburden must be removed along with each tonne of copper ore. (The ratio of overburden to ore is sometimes as high as 3 to 1.) The ore itself averages only about 0.5% copper. The importance of efficient materials handling by copper producers is dramatized by the fact

that, in the United States, more tonnage must be moved in mining copper than in mining all other metals combined (see Table 3).

Copper ore normally is crushed, ground and concentrated (usually by flotation) to produce a beneficiated ore containing 20 to 25% copper. Most copper ores are sulfide ores, composed chiefly of copper sulfide but also containing a significant amount of iron sulfide as an impurity plus recoverable trace amounts of silver and gold. Ore concentrates most often are reduced to the metallic state by a pyrometallurgical process. Traditionally, the concentrated ore is smelted in a reverberatory furnace to produce a copper sulfide–iron sulfide matte, and the matte is oxidized in a converter to convert the FeS to iron oxides, which separate out in a slag, and to burn off the sulfur from the CuS, leaving blister copper, which contains at least 98.5% copper. Fire refining of blister copper removes most of the oxygen and other impurities, leaving a product at least 99.5% pure, which is cast into anodes. Finally, most anode copper is electrolytically refined, usually to a purity of at least 99.95%. Gold and silver slimes are a byproduct of electrorefining, which helps defray the cost of the electrical energy used to

refine the anode copper. The resulting cathodes are the normal end product of the producer companies, and are a common item of commerce. The consumption of refined copper (mostly cathodes) in the United States was about 2 million t (4.4 billion lb) in 1977, about 22% of the world total of 9 million t (19.8 billion lb).

Hydrometallurgical processing is an alternative to pyrometallurgical processing that has more recently become commercially important. Heap leaching followed by cementation is the most common hydrometallurgical process, although others such as vat leaching or agitation leaching, both of which require richer ores, have also become important. In heap leaching, sulfuric acid is percolated through waste dumps formed from the rejected material (tailings) of the flotation process to leach out (dissolve) most of the remaining copper. The copper in the pregnant leach liquor is recovered by a process called cementation; in this process, the liquor is passed over scrap steel, upon which the copper precipitates from solution. Impure copper precipitates then are sent to a converter or smelting furnace for pyrometallurgical refining.

The liquor from heap leaching can be

Table 3 Total material moved(a) in mining of metals in the United States(b)

Metal mined	Ore	Overburden	Total
Copper	282	687	969
Iron	250	246	496
Other metals	114	331	445

(a) Millions of short tons. (b) U.S. Bureau of Mines data for 1976.

concentrated to produce an electrolyte suitable for use in electrowinning, whereas vat leaching and agitation leaching produce such electrolytes directly. In electrowinning, copper is extracted electrolytically from the electrolyte much as anode copper is electrorefined. The chief difference is that the copper is present in an enriched electrolyte instead of as impure copper anodes. Electrowon copper (about 99.9% pure) is not quite as pure as electrorefined copper (ordinarily about 99.97% (pure).

The box at lower left in Fig. 3 represents the portion of copper supply provided by scrap. This portion is substantial: nearly half of the copper consumed in the United States each year is derived from recycled scrap. (Runaround scrap, which is scrap recycled within a particular plant, is not included in these statistics.) About 25 to 30% of the scrap is fed into the smelting or refining stream and thus quickly loses its identity. The remainder is consumed directly by brass mills, by ingot makers (whose main function is to melt scrap into alloy ingot for use by foundries) and by foundries themselves.

The box labeled "Copper consumed" in Fig. 3 represents the total tonnage of refined copper plus the copper content of scrap consumed directly by fabricator companies. To this sum are added the amounts of various alloying elements used in producing copper alloys, and the alloy content of the directly consumed scrap, to obtain "Metal consumed".

Copper Fabricators. The four classes of copper fabricators together consume about 98% of the total output of the copper producers. Other industries, such as the steel, aluminum and chemical industries, consume the remaining 2%. As shown in Fig. 3, wire mills and brass mills are roughly equal in output: on the average, each produces 40 to 50% of total mill products. Foundries account for roughly 10% of the fabricated mill products, and powder plants about 1%.

Wire mills make copper wire and cable. The starting material is refined copper cathodes. Traditionally, cathodes have been melted and then cast into wirebars, followed by hot rolling to wire rod. More recently, continuous casting has been chosen for new plants or for major renovation of older plants. Facilities for producing wire rod may be integrated into a refinery, may be a separate operation or may be integrated into a wiredrawing plant.

Wire rod is cold drawn to final dimensions through a series of dies. The cold drawn wire may or may not be annealed, depending on requirements. Wire may be used as a single conductor, but more often is fabricated into a stranded conductor; most copper wire is insulated. Various types of electrical cable are produced from individual conductors, each of which may be stranded and/or insulated separately before being incorporated into the finished cable.

Brass mills melt and alloy feedstock to make sheet, strip, plate, tubing, rod, bar, mechanical wire, forgings and extrusions. Of the feedstock used by brass mills, about half is scrap and about half is virgin metal. Fabricating processes such as hot rolling, cold rolling, extrusion and drawing are employed to convert the melted and cast feedstock into wrought mill products. Some brass mills are secondary mills that do not melt and cast the feedstock they fabricate into mill products, but merely reroll strip to thinner gages or redraw tubing or mechanical wire to smaller dimensions.

About one-third of the output of U. S. brass mills is unalloyed copper and high copper alloys—chiefly in such forms as plumbing and air-conditioning tube, busbar and roofing sheet. Copper alloys comprise the remaining two-thirds, and are distributed approximately as indicated in Table 4. The several varieties of leaded brass rod (which exhibits outstanding machinability and good corrosion resistance) and unleaded brass strip (which has high strength, corrosion resistance, excellent formability and good electrical properties) constitute about three-fourths of the total tonnage of wrought copper alloys shipped from U. S. brass mills. Other alloy types of commercial significance include tin bronzes, which are noted for their excellent cold forming behavior; tin brasses, known for outstanding corrosion resistance; copper nickels, which are particularly resistant to seawater; nickel silvers, which combine a silvery appearance with good formability and corrosion resistance; beryllium coppers, which can provide outstanding strength when hardened; and aluminum bronzes,

Table 4 Major wrought copper and copper-base alloy systems

Copper or copper alloy group	UNS numbers	Approximate U.S. shipments in 1977, millions of pounds	Remarks
Wire Mill Shipments			
Coppers	C10000 to C15900	2691	C11000 is predominant material.
Brass Mill Shipments			
Coppers	C10000 to C15900, C16000 to C16900, C18000 to C18900	952	Includes modified coppers, cadmium copper, and chromium coppers.
Common brasses	C20000 to C29900	648	Of this, 89% is sheet, strip and plate.
Leaded brasses	C30000 to C39900	797	Of this, 95% is rod.
Tin bronzes	C50000 to C53900	41	Unleaded only; also known as phosphor bronzes.
Aluminum bronzes, silicon bronzes and manganese bronzes	C60000 to C68400	32	
Copper nickels	C70000 to C72900	71	
Nickel silvers	C73000 to C79900	17	
Others	C17000 to C17900, C19000 to C19900, C40000 to C49900, C54000 to C54900, C68500 to C69900	112	Includes beryllium coppers, copper-iron alloys, tin brasses, leaded tin bronzes, aluminum brasses, and silicon brasses.

which have high strength along with good resistance to both chemical attack and mechanical abrasion.

Foundries use prealloyed ingot, scrap and virgin metal as raw materials. Their chief products are shape castings for many different industrial and consumer goods, the most important of which are plumbing products, industrial valves and bearings.

Powder plants produce powder metallurgy parts, chiefly sintered bronze bushings.

Applications of Copper and Its Alloys

The five major market categories at far right in Fig. 3 constitute the chief customer industries of the copper fabricators. Of the chief customer industries, the largest is building construction, which purchases large quantities of electrical wire, tubing and parts, for builders' hardware and for electrical, plumbing, heating and air-conditioning systems. Next are electrical and electronic products, including those for telecommunications, electronics, wiring devices, electric motors and power utilities. The industrial machinery and equipment category includes industrial valves and fittings; industrial, chemical and marine heat exchangers; and various other types of heavy equipment, off-road vehicles and machine tools. Transportation applications include road vehicles, railroad equipment and aircraft parts; automobile radiators and wiring harnesses are the most important products in this category. Finally, consumer and general products include electrical appliances, fasteners, ordnance, coinage and jewelry.

About 90% of the total tonnage of wrought copper alloys sold by U. S. fabricator plants is represented by the 16 application categories listed in Table 5. In the three categories that account for the greatest tonnages—telecommunications, automotive, and plumbing and heating—a continuing effort has been made to conserve materials and to manufacture products more efficiently. Most often this effort involves redesign of components, and is accomplished through reductions in material gage. In some instances, such as small motors for appliances and other devices, there is a trend toward using more copper for each unit to increase the energy efficiency of the end product.

Table 5 also shows the mill products

Table 5 Major end-use applications for copper and copper alloys in the United States

Application	% of total	Mill products	Principal reason(s) for using copper
Telecommunications	13.9	Copper wire	Electrical properties
Automotive: automobiles, trucks and buses	13.8	Brass and copper strip, copper wire	Corrosion resistance, heat transfer, electrical properties
Plumbing and heating	12.2	Copper tube, brass rod, castings	Corrosion resistance, mechinability
Building wiring	10.4	Copper wire	Electrical properties
Heavy industrial equipment	7.5	All	Corrosion resistance, wear resistance, electrical properties, heat transfer, machinability
Air conditioning and commercial refrigeration	6.4	Copper tube	Heat transfer, formability
Industrial valves and fittings	5.1	Brass rod, castings	Corrosion resistance, machinability
Power utilities	4.1	Copper wire and bar	Electrical properties
Appliances	3.8	Copper wire and tube	Electrical properties, heat transfer
Lighting and wiring devices	3.1	Alloy strip, copper wire	Electrical properties
Electronics	2.3	Alloy strip, copper wire	Electrical properties
Fasteners	2.2	Brass wire	Machinability, corrosion resistance
Military and commercial ordnance	1.7	Brass strip and tube	Ease of fabrication
Coinage	1.3	Alloy and copper strip	Ease of fabrication, corrosion resistance, electrical properties, aesthetics
Builders' hardware	1.2	Brass rod and strip	Corrosion resistance, formability, aesthetics
Heat exchangers	1.0	Alloy tube and plate	Heat transfer, corrosion resistance

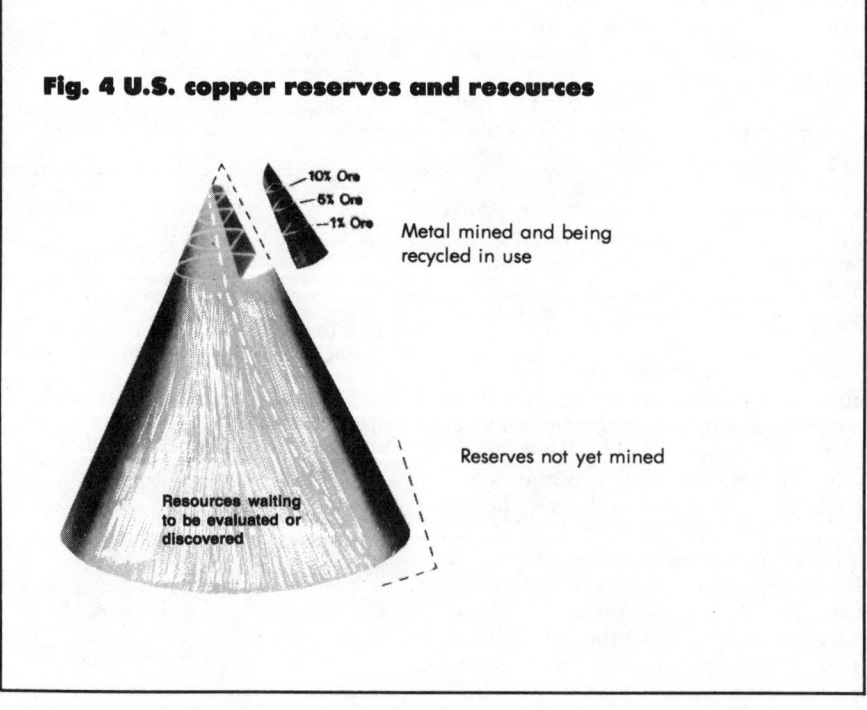

Fig. 4 U.S. copper reserves and resources

10% Ore
5% Ore
1% Ore

Metal mined and being recycled in use

Reserves not yet mined

Resources waiting to be evaluated or discovered

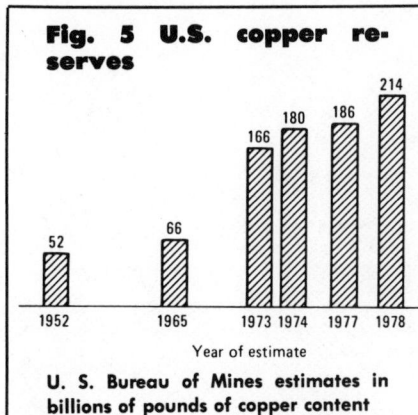

Fig. 5 U.S. copper reserves

Year of estimate

U. S. Bureau of Mines estimates in billions of pounds of copper content

used in each application category, plus the property or properties that dictate the use of copper or its alloys. High electrical conductivity is a major reason for choosing copper in nine of the sixteen categories, corrosion resistance a major reason in eight, ease of fabrication (including good machinability) in six (although ease of fabrication is a significant factor in all categories), and good heat-transfer properties in five.

Copper Supply and Resources

All of the copper in the earth's crust is represented in Fig. 4 as a hypothetical "cone of resources". In this representation, the richest ores are at the tip of the cone. Up to now, man has discovered, extracted and used only a slice off the tip of the cone. Behind that slice are the known copper reserves that have been discovered but not yet mined. Below the known reserves are the rest of the earth's resources, only a fraction of which are known to be economically recoverable using current methods of extractive metallurgy. In 1978, the U. S. Bureau of Mines estimated that there were 214 billion lb of known copper reserves in the United States, plus another 640 billion lb of resources. By comparison, only about 3 billion lb of newly mined copper were consumed in 1978.

Not only are there substantial copper reserves in the United States (enough to ensure self-sufficiency for about the next half-century even if no new re-

serves are added), but there also are continual improvements in discovery and extractive metallurgy. As shown in Fig. 5, U. S. copper reserves increased from 52 billion lb to 214 billion lb from 1952 to 1978—a rate of increase in reserves which far outstripped the total consumption of newly mined copper over the same period of time.

The U.S. copper industry relies heavily on scrap copper as a raw material. For instance, in 1978 the copper industry used 2.9 billion lb of scrap, or about 49% of the total amount of copper used in that year. Such reliance is typical of a mature economy, and recycling of products such as building wire and plumbing, even after a 40- to 50-yr service life, is not uncommon. In fact, by percentage, copper is recycled more than any other engineering metal.

Substantial reserves and extensive use of scrap both contribute to the self-sufficiency of the United States, and therefore to the reliability of supply of copper to manufacturing industries. Over the period 1970 to 1977, the United States was more than 90% self-sufficient in copper.

Temper Designations for Coppers and Copper Alloys

By Richard Dale Smith
Kennecott Copper Corp.

DESIGNATION of tempers for copper and its alloys must allow for highly diversified and complex variation in product forms and processing methods. Out of this diversity, one group of temper designations can be specified on the basis of cold reduction by rolling or drawing. Such a scheme is presented in Table 1, where the nominal temper designations are related directly to the amount of reduction stated in Browne & Sharpe (B&S) gage numbers. In the B&S gage system, each gage is related to both adjacent gages by a fixed ratio. As the gage numbers increase, the thickness (or diameter) decreases by 10.95% for each step. For example, a strip of flat rolled material 0.032 in. thick or drawn wire 0.032 in. in diameter is designated B&S gage 20; cold reduction one B&S number (to 21 gage) reduces the thickness or diameter 10.95% (to 0.0285 in.), and a further reduction to 22 gage reduces the diameter or thickness another 10.95% (to 0.0254 in.). Because the reductions are based on thickness for flat material and on diameter for wire, the reduction in cross-sectional area is less for strip

than for wire for a given gage interval.

Temper designations for coppers and copper alloys require a broad-based classification scheme because (a) some alloys have much higher work-hardening rates than the coppers and brasses for which the scheme summarized in Table 1 was developed and (b) a temper-designation scheme based on cold

reduction cannot be applied to product forms such as rod, tube, extrusions and castings or to heat treatable copper alloys. For example, some manufacturers of high-strength alloys with high work-hardening rates have established nominal temper designations (1/4 hard, 1/2 hard, etc.) based on strength level rather than on dimensional reduction. In an extreme case for rod, "hard"

Table 1 Scheme of temper designations for wrought copper and brass based on cold reduction

Nominal temper designation	Increase in B&S gage numbers	Rolled sheet		Drawn wire		
		Reduction in thickness and area, %	True strain(a)	Reduction in diameter, %	Reduction in area, %	True strain(a)
1/4 hard	1	10.9	0.116	10.9	20.7	0.232
1/2 hard	2	20.7	0.232	20.7	37.1	0.463
3/4 hard	3	29.4	0.347	29.4	50.1	0.694
Hard	4	37.1	0.463	37.1	60.5	0.926
Extra hard	6	50.1	0.696	50.1	75.1	1.39
Spring	8	60.5	0.928	60.5	84.4	1.86
Extra spring	10	68.6	1.16	68.6	90.2	2.32
Special spring	12	75.1	1.39	75.1	93.8	2.78
Super spring	14	80.3	1.62	80.3	96.1	3.25

(a) True strain equals $\ln A_0/A$, where A_0 is initial cross-sectional area and A is final area.

temper can mean anything from "annealed to a fine grain size" to a cold reduction of 10, 15, 20, 35 or even 40%. "Half hard" and "quarter hard" tempers are also inconsistent with the scheme of temper designations based on dimensional reduction.

Similar problems exist for annealed tempers. With simple single-phase alloys, annealed tempers are generally specified on the basis of grain size of recrystallized material. This system cannot be applied to grain-size-stabilized alloys because there may be considerable variation in properties for material annealed at various temperatures up to several hundred degrees Fahrenheit above the recrystallization temperature, even though the grain size remains quite stable. For these alloys, the terms "light annealed", "soft annealed" and "annealed to temper" are generally used instead of grain-size designations.

In order to clarify this complex situation, the American Society for Testing and Materials has published Standard Recommended Practice ASTM B601, "Temper Designation for Copper and Copper Alloys—Wrought and Cast", which is intended to be unambiguous and applicable to all product forms currently in widespread use. An alphanumeric code has been assigned to each of the standard descriptive temper designations (see Table 2). Wherever possible in this Handbook, data for different tempers have been listed using these codes for identification of material condition.

Table 2 ASTM B601 temper designation codes for copper and copper alloys

Temper designation	Temper name or material condition
Cold Worked Tempers	
H00	1/8 hard
H01	1/4 hard
H02	1/2 hard
H03	3/4 hard
H04	Hard
H06	Extra hard
H08	Spring
H10	Extra spring
H12	Special spring
H13	Ultra spring
H14	Super spring
H50	Extruded and drawn
H52	Pierced and drawn
H55	Light drawn; light cold rolled
H58	Drawn general purpose
H60	Cold heading; forming
H63	Rivet
H64	Screw
H66	Bolt
H70	Bending
H80	Hard drawn
H85	Medium-hard-drawn electrical wire
H86	Hard-drawn electrical wire
Cold Worked and Stress-relieved Tempers	
HR01	H01 and stress relieved
HR02	H02 and stress relieved
HR04	H04 and stress relieved
HR06	H06 and stress relieved
HR08	H08 and stress relieved
HR10	H10 and stress relieved
HR50	Drawn and stress relieved
Cold Worked and Order-strengthened Tempers	
HT04	H04 and order heat treated
HT06	H06 and order heat treated
HT08	H08 and order heat treated
As-manufactured Tempers	
M01	As sand cast
M02	As centrifugal cast
M03	As plaster cast
M04	As pressure die cast
M05	As permanent mold cast
M06	As investment cast

Table 2 ASTM B601 temper designation codes for copper and copper alloys (continued)

Temper designation	Temper name or material condition
M07	As continuous cast
M10	As hot forged and air cooled
M11	As hot forged and quenched
M20	As hot rolled
M30	As hot extruded
M40	As hot pierced
M45	As hot pierced and rerolled
Annealed Tempers(a)	
O10	Cast and annealed(b)
O20	Hot forged and annealed
O25	Hot rolled and annealed
O30	Hot extruded and annealed
O40	Hot pierced and annealed
O50	Light annealed
O60	Soft annealed
O61	Annealed
O65	Drawing annealed
O68	Deep drawing annealed
O70	Dead soft annealed
O80	Annealed to temper—1/8 hard
O81	Annealed to temper—1/4 hard
O82	Annealed to temper—1/2 hard
Annealed Tempers(c)	
OS005	Average grain size 0.005 mm
OS010	Average grain size 0.010 mm
OS015	Average grain size 0.015 mm
OS025	Average grain size 0.025 mm
OS035	Average grain size 0.035 mm
OS050	Average grain size 0.050 mm
OS070	Average grain size 0.070 mm
OS100	Average grain size 0.100 mm
OS120	Average grain size 0.120 mm
OS150	Average grain size 0.150 mm
OS200	Average grain size 0.200 mm
Solution-treated Temper	
TB00	Solution heat treated
Solution-treated and Cold Worked Tempers	
TD00	TB00 cold worked to 1/8 hard
TD01	TB00 cold worked to 1/4 hard
TD02	TB00 cold worked to 1/2 hard
TD03	TB00 cold worked to 3/4 hard
TD04	TB00 cold worked to full hard
Precipitation-hardened Temper	
TF00	TB00 and precipitation hardened
Cold Worked and Precipitation-hardened Tempers	
TH01	TD01 and precipitation hardened
TH02	TD02 and precipitation hardened
TH03	TD03 and precipitation hardened
TH04	TD04 and precipitation hardened
Precipitation-hardened and Cold Worked Tempers	
TL00	TF00 cold worked to 1/8 hard
TL01	TF00 cold worked to 1/4 hard
TL02	TF00 cold worked to 1/2 hard
TL04	TF00 cold worked to full hard
TL08	TF00 cold worked to spring
TL10	TF00 cold worked to extra spring

Table 2 ASTM B601 temper designation codes for copper and copper alloys (continued)

Temper designation	Temper name or material condition
TR01	TL01 and stress relieved
TR02	TL02 and stress relieved
TR04	TL04 and stress relieved
Mill-hardened Tempers	
TM00	AM
TM01	1/4 HM
TM02	1/2 HM
TM04	HM
TM06	XHM
TM08	XHMS
Quench-hardened Tempers	
TQ00	Quench hardened
TQ50	Quench hardened and temper annealed
TQ75	Interrupted quench hardened
Tempers of Welded Tubing(d)	
WH00	Welded and drawn to 1/8 hard
WH01	Welded and drawn to 1/4 hard
WM00	As welded from H00 strip
WM01	As welded from H01 strip
WM02	As welded from H02 strip
WM03	As welded from H03 strip
WM04	As welded from H04 strip
WM06	As welded from H06 strip
WM08	As welded from H08 strip
WM10	As welded from H10 strip
WM15	WM50 and stress relieved
WM20	WM00 and stress relieved
WM21	WM01 and stress relieved
WM22	WM02 and stress relieved
WM50	As welded from O60 strip
WO50	Welded and light annealed
WR00	WM00; drawn and stress relieved
WR01	WM01; drawn and stress relieved

(a) To produce specified mechanical properties. (b) Homogenization anneal. (c) To produce prescribed average grain size. (d) Tempers of fully finished tubing that has been drawn or annealed to produce specified mechanical properties or that has been annealed to produce a prescribed average grain size are commonly identified by the appropriate H, O or OS temper designation.

Heat Treating of Copper and Copper Alloys

By the ASM Committee on Copper and Copper Alloys*

COPPER AND COPPER ALLOYS may be heat treated for any of these purposes:

1 Homogenizing—to dissolve and absorb segregation and coring found in some cast and hot worked materials, chiefly those containing tin and nickel. Homogenizing is done at the mill before shipment to a fabricator.

2 Softening, or annealing—to allow additional cold work or provide ease of forming in a cold worked product. Recrystallization normally occurs during annealing.

3 Stress relieving—to reduce or eliminate residual stress, thereby reducing the likelihood that the part will fail by season cracking or corrosion fatigue in service. Parts are stress relieved at temperatures below the normal annealing range that do not cause recrystallization and consequent softening of the metal.

4 Solution treating and precipitation hardening—to strengthen special types of copper alloys above the levels ordinarily obtained by cold working. Examples of precipitation

*See page XI for committee list.

hardening copper alloys include the beryllium coppers, some of which also contain nickel, cobalt or chromium; the copper-chromium alloys; the copper-zirconium alloys; the copper-nickel-silicon alloys; and the copper-nickel-phosphorus alloys.

5 Transformation hardening—to strengthen certain alloys by inducing a phase change to a harder and stronger phase. Two-phase aluminum bronzes and some manganese bronzes are given quench-and-temper treatments to increase strength without unduly sacrificing ductility.

6 Hardening by spinodal decomposition—to strengthen alloys such as C71900 by inducing formation of a spinodal structure.

7 Order hardening—to strengthen certain alloys by inducing a short-range ordering reaction.

8 Microduplex hardening—to strengthen certain alloys by controlled thermomechanical processing that results in optimum distribution of a very fine two-phase structure.

Homogenizing

In homogenizing, high temperatures and relatively long times are employed to eliminate or decrease coring (segregation) in cast metal that is to be hot or cold worked. The various wrought brasses are seldom homogenized; however, because repeated process annealing between reductions is necessary when heavy stock is cold worked to finished size, this repeated alternate cold working and annealing generally ensures that finished brass mill products will be free of all evidence of segregation and elimates the need for homogenization.

Diffusion and homogenization are slower and more difficult in tin bronzes, silicon bronzes and copper nickels than in most other copper alloys. Therefore, these alloys usually are subjected to prolonged homogenizing treatments before hot or cold working operations.

The high-tin phosphor bronzes (above 8% Sn) are noted for extreme segregation. Although these alloys sometimes are hot worked, usual practice is to roll them cold, making it necessary to first diffuse the brittle segregated tin phase, thereby increasing strength and ductility and decreasing hardness before rolling. These objectives are accomplished by homogenizing at about 760 °C (1400 °F).

Annealing

Softening (annealing) of cold worked metal is accomplished by heating to a temperature that causes recrystallization and, if maximum softening is desired, by heating well above the recrystallization temperature to cause grain growth. Temperatures commonly used for annealing cold worked coppers and copper alloys are given in Table 1.

Annealing is controlled primarily by controlling metal temperature, time at temperature and amount of prior cold work. Rates of heating and cooling, except in multi-phase alloys and hardenable alloys, are relatively unimportant. Method of heating, furnace design, furnace atmosphere, and shape of workpiece are important, because they affect uniformity of results, finish, and cost of annealing.

For copper and brass mill alloys, grain size is the standard means of evaluating a recrystallizing anneal. Because many interreacting variables influence the annealing process, it is difficult to predict a specific combination of time and temperature that will always produce a given grain size in a given metal. Figure 1 presents typical tensile strength, elongation and grain size for C27000 wire initially in the H80 temper after annealing at different temperatures. The curves in Fig. 1 apply only to material with a specific prior history; quite different curves might result if C27000 (H80 temper) with a different prior history were annealed under the same conditions. Similar data for other coppers and copper alloys may be found in the section of this volume devoted to data compilations for wrought coppers and copper alloys. Figure 2 gives time-temperature combinations that produce different recrystallized grain sizes for one production lot of C27000 strip, H80 temper.

Increasing the amount of cold work prior to annealing lowers the recrystallization temperature. The smaller the degree of prior deformation, the larger the grain size after annealing. For fixed temperature and duration of annealing, the larger the original grain size before working, the larger the grain size after recrystallization.

In commercial mill practice, copper alloys usually are annealed at progres-

Table 1 Temperatures commonly used for annealing cold worked copper metals

UNS numbers	Alloy type	Annealing temperature °C	°F
Wrought Coppers			
C11000	Electrolytic tough pitch	250-650	500-1200
C12000	Deoxidized, low residual P	325-650	600-1200
C12200	Deoxidized, high residual P	375-650	700-1200
C10200, C14500, C18700	Oxygen-free; free-machining	425-650	800-1200
C11300 to C11600, C12700 to C13000	Silver-bearing coppers	400-475	750-900
Wrought Copper Alloys			
C21000, C22000	Gilding; commercial bronze	425-800	800-1450
C22600, C26000, C60600	Jewelry bronze; cartridge brass; aluminum bronze, 5% Al	425-750	800-1400
C23000	Red brass	425-725	800-1350
C24000, C27000, C35300	Low brass; yellow brass high-leaded brass	425-700	800-1300
C31400 to C35600, C37000	Leaded brasses; free-cutting Muntz metal	425-650	800-1200
C28000, C36500 to C38500, C44300 to C48500, C66700, C67400, C67500, C68700	Muntz metal; leaded high-zinc brasses; admiralty metals; naval brasses; manganese bronzes; arsenical aluminum brass	425-600	800-1100
C51000 to C54400, C65100	Phosphor bronzes (except low Sn alloys); low silicon bronze B	475-675	900-1250
C50500	Phosphor bronze, 1.25% E	475-650	900-1200
C71500	Copper nickel, 30%	650-815	1200-1500
C70600, C74500 to C78200	Copper nickel, 10%; nickel silvers	600-815	1100-1500
C65500	High silicon bronze A	475-700	900-1300
C61300, C61400	Alpha aluminum bronze	815	1500
C63800	Aluminum-silicon bronze	Above 650	Above 1200
C63000, C64200	Aluminum bronze	575-650	1050-1200
C17000 to C17600	Beryllium coppers	775-1050(a)	1425-1900(a)

(a) Solution treating temperature.

sively lower temperatures, with intermediate cold reductions of at least 35% and as high as 50 to 60% wherever practicable. The higher initial temperatures accelerate homogenization, and the resulting large grains permit more economical reduction during the early stages of mill reduction.

In successive process anneals, grain size should be decreased gradually to approximate the final grain size required. This point is usually reached one or two anneals before the final anneal. With such a sequence and with sufficiently severe intermediate reductions, it is then possible to obtain uniform final grain size within a lot and from lot to lot.

Grain-size Stabilized Alloys. Several copper alloys have been developed in which the grain size is stabilized by the presence of a finely distributed second phase. Examples include copper-iron alloys such as C19200, C19400 and C19500, and aluminum-

containing brasses and bronzes such as C61500, C63800, C68800 and C69000. These alloys will maintain an extremely fine grain size at temperatures well beyond their recrystallization temperature, up to the temperature where the second phase finally dissolves or coarsens, which allows grain growth to proceed. Specification of grain size as a means of approximating properties is not practical for the grain-size stabilized alloys. Generally, two annealed tempers are available: light anneal, which is performed at a temperature slightly above the recrystallization temperature, and soft anneal, which is performed several hundred degrees higher, at a temperature just below the point at which rapid grain growth begins.

Correlation with Mechanical Properties. The grain size and mechanical properties required for cold forming of parts will vary considerably with the alloy and the amount and kind of further cold work to be done. A

Fig. 1 Annealing curves for copper alloy C27000

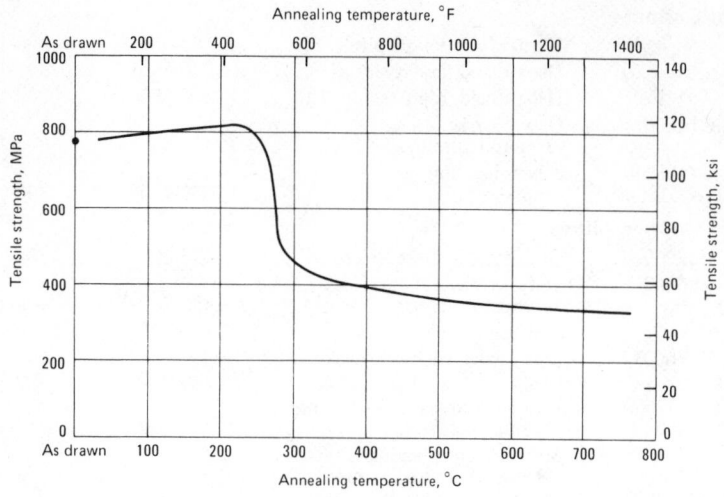

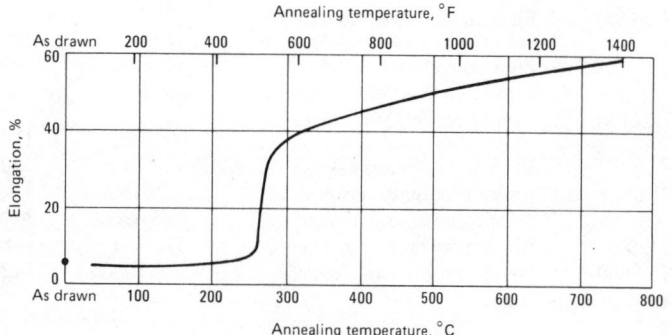

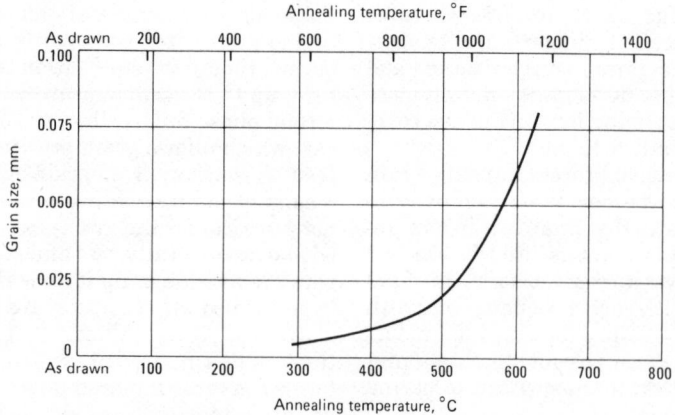

Data are for yellow brass wire drawn 63% to H80 temper and annealed 1 h at temperature.

generalized relationship between grain size and cold formability for coppers and copper alloys is shown below, along with a general rating of ability to take a fine polished finish.

Nominal grain size, mm	Typical use
0.015	Simple forming operations
0.025	Shallow drawing, excellent polishing capability
0.035	Best combination of drawing capability and polishing capability
0.050	Deep drawing operations, fair polishing capability
0.070	Heavy drawing or thick gages, difficult to polish

This table is particularly applicable to low-copper brasses such as C26000 and C27000. The higher-copper alloys are more ductile and do not harden so rapidly with cold work. The 0.035-mm nominal grain size is generally acceptable for deep drawing these high-copper alloys.

The objectives of annealing in preparation for cold forming is to obtain an optimum combination of ductility and strength and to keep the surface texture refined sufficiently to minimize the amount of buffing required to achieve an acceptable finish. Annealing must be governed by definite specifications, and must be coordinated with cold working operations to meet these objectives.

General Precautions. For best results in annealing copper and copper alloys, the following precautions should be observed:

Sampling and Testing. Test specimens must represent extremes of temperature in the furnace load. For simple single-phase alloys, the best and most accurate measure of the extent of annealing is average grain size. Grain size usually is the basis for acceptance or rejection of the material, but determination requires special equipment not always available. For convenience, Rockwell hardness can be used to approximate grain size. (ASTM specifications correlate Rockwell hardness with grain size for many copper alloys.)

Effect of Pretreatment. Because the extent of cold working and the grain structure resulting from the last prior anneal greatly affect results obtained in any given anneal,

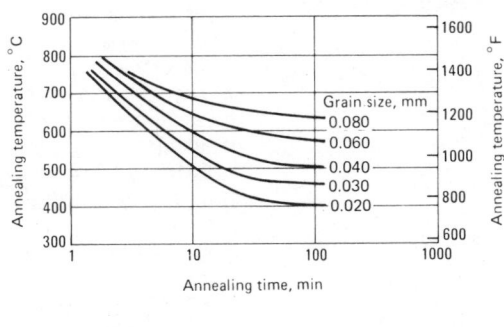

Fig. 2 Time-temperature combinations that produce various recrystallized grain sizes in 1.3-mm (0.05-in.) C27000 strip in H80 temper

annealing schedules must take such pretreatment into account. Once a schedule has been established, both the anneal and the pretreatment must be adhered to for consistent results.

Effect of Time. In most furnaces, there is an appreciable difference between the temperature of the metal and that of the furnace; consequently, time in the furnace has an important effect on final temperature of the metal. For a specific annealed grain size and furnace temperature, time must vary with the size and type of furnace load. Thermocouples placed strategically within the furnace load are a common means of monitoring metal temperature.

Oxidation should be minimized to reduce loss of metal, to lower pickling costs and to improve finish. In some instances, specially prepared atmospheres such as exothermic gas or dissociated ammonia are used for producing brightly annealed material. Control of furnace atmosphere usually also results in better furnace economy.

Hydrogen Embrittlement. When annealing copper that contains oxygen, the hydrogen in the atmosphere must be kept to a minimum to avoid embrittlement. For temperatures lower than about 480 °C (about 900 °F), hydrogen preferably should not exceed 1%, and as the temperature is increased above 480 °C, the hydrogen content of the furnace atmosphere should approach zero.

Impurities in the alloy sometimes interfere with the ability to obtain a specific grain size when annealing under standard conditions. If annealing behavior is erratic from lot to lot and a specific grain size is necessary, it may be advisable to specifically limit certain impurities.

Annealing for Specific Tensile Properties. Because there is an extremely rapid rate of change of tensile properties with temperature in the vicinity of the recrystallization temperature, it is difficult to reliably obtain specific values of tensile strength or hardness intermediate between heavily cold worked tempers and fully recrystallized tempers. Control of both heating rates and temperature is not accurate or uniform enough to permit consistent reproduction of any desired strength or corresponding mechanical property by batch annealing. As a result, batch annealing to temper should be specified only when accurate reproduction of mechanical properties is not critical. When using batch annealing, the best practice is to anneal completely and then to obtain desired intermediate properties by controlled cold working. Closer control of the thermal cycle is possible in continuous strand annealing, and this annealing method can be used to achieve more consistent (and frequently better) combinations of properties than is usual for batch annealing.

Loading. It is usually inadvisable to anneal a variety of different sizes or kinds of material in the same charge, because different rates of heating produce different thermal cycles, which result in corresponding property variations.

Fire cracking is a type of cracking that occurs when some alloys that contain residual stresses are heated too rapidly. Leaded alloys are particularly susceptible to fire cracking. The remedy is to heat slowly until the stresses are relieved. Mechanical stress relief, such as by roller leveling or straightening, can aid considerably in preventing fire cracking.

Thermal Shock. Hot short materials such as leaded brasses should be heated slowly and uniformly to avoid cracking due to thermal stress.

Sulfur Stains. Excessive sulfur in furnace fuel or in lubricant residue on formed parts will cause red stains on yellow brass and reddish-brown stains on copper-rich alloys. Regardless of the workpiece or furnace, it is advisable to eliminate as much lubricant as possible before the metal is heated.

Stress Relieving

Some copper alloys fail by stress-corrosion cracking (also referred to as "season cracking") as a result of residual surface stresses induced by forming at temperatures below the recrystallization temperature. Stress-corrosion cracking can take place at ordinary temperatures and in common environments. Residual stresses contribute to this type of failure, which is frequently seen in brasses containing 15% zinc or more. Even higher-copper alloys such as aluminum bronzes and silicon bronzes may crack under critical combinations of stress and specific corrodent, and all copper alloys are susceptible to more rapid corrosion attack when in the stressed condition.

Stressed phosphor bronzes and copper nickels have comparatively slight tendencies toward stress-corrosion cracking; these alloys are more susceptible to fire cracking, which is cracking caused when stressed metal is heated too rapidly to the annealing temperature. Slow heating provides a measure of stress relief and minimizes nonuniform temperature distributions, which lead to thermal stress.

Mill practice for relieving stress incidental to plastic deformation involves either mechanical relieving, such as bending or straightening, or thermal stress relieving below the recrystallization temperature—or a combination of both. Thermal stress relieving frequently is applied to finished parts to prevent failure from exposure to corrosive conditions—even such mild ones

Table 2 Typical stress-relieving temperatures for 12 wrought copper alloys

UNS number	Common name	Stress-relieving temperature(a) °C	°F
C21000	Gilding, 95%.........................190	190	375
C22000	Commercial bronze, 90% 205	205	400
C23000	Red brass, 85%.......................230	230	450
C24000	Low brass, 80% 245	245	475
C26000	Cartridge brass, 70% 260	260	500
C27000	Yellow brass, 65%....................260	260	500
C28000	Muntz metal, 60%.....................205	205	400
C36000	Free-cutting brass....................245	245	475
C44300	Inhibited admiralty metal..............290	290	550
C51000	Phosphor bronze 190	190	375
C71500	Copper nickel, 30%...................245	245	475
C75200	Nickel-silver, 65-18..................245	245	475

(a) Time at temperature, 1 h.

as ammonia in the atmosphere or contact with soapy water. Typical stress-relieving temperatures for 12 copper alloys are given in Table 2.

Using a high stress-relieving temperature for a short time is generally considered best for keeping processing time and cost to a practical minimum, even though there is usually some sacrifice in mechanical properties. Using a lower temperature for a longer time will provide complete stress relief with no decrease in mechanical properties. Actually, the hardness and strength of severely cold worked alloys will increase slightly when low stress-relieving temperatures are used.

An additional benefit of a thermal stress-relieving is dimensional stability of cold formed parts. Also, it is often advisable to stress relieve welded or cold formed structures. For these structures, stress-relieving temperature is 85 to 110 °C (150 to 200 °F) above that used for mill products of the same alloy.

Precipitation Hardening

High strength in most coppers and copper alloys is achieved by cold working. But for certain copper alloys containing small amounts of beryllium, chromium or zirconium, or nickel in combination with silicon or phosphorus, unusually high strength and hardness can be obtained by precipitation hardening.

All precipitation-hardening copper alloys have similar metallurgical characteristics: they can be solution treated to a soft condition by quenching from a high temperature, and then subse-

quently precipitation hardened by aging at a moderate temperature for a time usually not exceeding 3 h. The chief advantages of these alloys are:

- Customer fabrication is easily performed in the soft, solution-annealed condition.
- The precipitation-hardening heat treatment performed by the fabricator is relatively simple. It is carried out at moderate temperatures, usually in air; controlled cooling is not needed, and time of treatment is not of critical importance.
- Different combinations of properties—including strength, hardness, ductility, conductivity, impact resistance and anelasticity—can be obtained by varying hardening times and temperatures. The particular requirements of the application determine the type of hardening treatment.

Age-hardenable alloys are furnished in the solution-treated condition, in the solution treated and cold worked condition or in the age-hardened condition. When the mill performs the age-hardening operation, further age hardening of the alloy after fabrication of parts is not required. However, it may be desirable to stress relieve the parts to remove residual stresses induced during fabrication. This treatment is particularly desirable for highly formed cantilever-type springs and intricate machined shapes that require maximum resistance to relaxation at moderately elevated temperatures.

Beryllium Coppers

Wrought beryllium coppers can de-

velop wide ranges of mechanical properties, depending on solution treating and aging conditions, on the amount of cold work imparted to the alloy and on whether the alloy is cold worked after solution treating and before aging or is cold worked after aging. Cast beryllium coppers generally are not cold worked, so variations in properties can be developed only by varying solution treating and aging conditions.

Solution Treating. Wrought beryllium copper mill products are supplied solution treated or solution treated and cold worked. Material in these conditions can be fabricated directly into parts. Solution treating by the fabricator generally is not done unless it is needed to fulfill a special requirement, such as softening a semifinished product for additional forming, or to salvage parts that have been heated incorrectly for precipitation hardening. Solution treating must be carefully controlled to produce the desired grain size, dimensional tolerances and mechanical properties, and to prevent surface oxidation.

Temperature limits must be adhered to if optimum properties are to be obtained during subsequent aging. Exceeding the upper limit causes grain coarsening in wrought material and may cause overheating in wrought and cast materials. Coarse grain size impairs formability; overheating results in brittleness that prevents full response to precipitation hardening. Solution treating below the specified minimum temperature results in insufficient solution of the beryllium-rich phase, which in turn results in lower hardness of the material after aging.

Time at the solution treating temperature depends on the amount of beryllium-rich phase that must be dissolved. Solution of this phase must be complete to produce maximum strength after precipitation hardening.

In cast products, the as-cast structure usually contains a large amount of microsegregation within the dendritic pattern. Therefore, castings must be heated for a time sufficient to homogenize the structure (a minimum of 3 h at temperature is recommended).

Solution treating of wrought material also removes the effects of cold working and permits additional forming. Some grain growth will occur during softening for additional forming, because the solution treating temperature is above the recrystallization temperature. Therefore, to minimize grain growth, excessive time at temperature must be avoided.

Water quenching is the most common method for retaining the solid-solution condition in both wrought and cast products; however, depending on shape, some castings may crack as the result of such rapid cooling. These castings may be quenched in oil or forced air; slower cooling rates may result in lower properties after precipitation hardening because slow rates lead to coarse precipitation.

When beryllium coppers are solution treated in air or an oxidizing atmosphere, two types of oxidation are encountered. A continuous and tenacious oxide surface layer forms on alloys with high beryllium contents. Low-beryllium alloys form a loosely adhering scale and are subject to internal oxidation.

The oxide layer on high-beryllium alloys does not significantly affect the mechanical properties of the precipitation-hardened material, but it is abrasive and causes severe wear of tools and dies. In addition to the abrasive effect of the oxide, oxidation of low-beryllium alloys decreases apparent mechanical properties because the internal oxidation reduces the effective section thickness of the material.

Oxide on both types of alloys may be removed by chemical or abrasive cleaning methods.

Aging. Cold working of solution treated (annealed) beryllium copper influences the strength attained when the material is subsequently aged; the highest response to aging occurs in material that has been cold rolled to at least TD04 (hard) temper. Generally, there is no advantage to work hardening beyond TD04 temper, because formability becomes undesirably poor and control of the aging treatment for maximum strength requires exceptional precision. Yet, for some applications, wire is drawn to higher levels of cold work prior to aging.

Table 3 lists the properties usually developed in strip of three common beryllium coppers—C17000, C17200 and C17500—using standard aging treatments. Figure 3 illustrates the relations between time and temperature for developing maximum strength in C17200 in each of the standard aged tempers. For mill products other than strip, standard aging treatments vary slightly from those given in Table 3, and so do the corresponding optimum properties. Special combinations of

properties can be obtained by varying either the aging time or the aging temperature. Recommended aging cycles for four solution treated cast beryllium coppers are given in Table 4.

Beryllium copper may be aged at any temperature within the normal aging range, but a change in temperature affects the time required to reach maximum strength. Although high temperatures promote rapid aging, they can result in lower maximum strength than can be obtained with the optimum treatment. Temperature should be controlled within ±5 °C (±10 °F) about the selected aging temperature, which normally is between 315 and 370 °C (600 and 700 °F).

The effect of grain size on properties is less significant for beryllium coppers than it is for solid-solution alloys such as brasses. The relatively high temperatures required for solution treating of beryllium coppers usually override the effects of cold work and time at temperature. Low solution treating temperatures result in fine grain size, but if the temperature is too low to completely dissolve the beryllium-rich phase, response to aging is adversely affected and the benefits obtained from fine grain size are nullified. For this reason, grain sizes below about 0.015 mm are not practical for most beryllium copper products. With normal commercial practice, grain size of solution treated material usually ranges from about 0.015 to 0.060 mm.

Copper-Nickel-Phosphorus Alloys

Alloys containing about 1% nickel and about 0.25% phosphorus, typified by C19000, are used for a wide variety of small parts requiring high strength, such as springs, clips, electrical connectors and fasteners. C19000 is solution treated at 700 to 800 °C (1300 to 1450 °F). A reducing or neutral atmosphere should be used (especially when heating thin sections) to prevent internal oxidation. Water quenching is the preferred cooling method, although individually handled small parts may be air cooled.

If the metal must be softened between cold working steps prior to aging, it may be satisfactorily annealed at temperatures as low as 620 °C (1150 °F). Rapid cooling from the annealing temperature is not necessary.

For aging, the material is held at 425 to 475 °C (800 to 900 °F) for 1 to 3 h.

Chromium Coppers

Chromium coppers containing about 1% Cr, such as C18200, C18400 and C18500, are solution treated at 950 to 1010 °C (1750 to 1850 °F) and rapidly quenched. Solution treating usually is done in molten salt, but may be done in a controlled-atmosphere furnace to prevent surface scaling and internal oxidation.

Solution treated chromium copper is

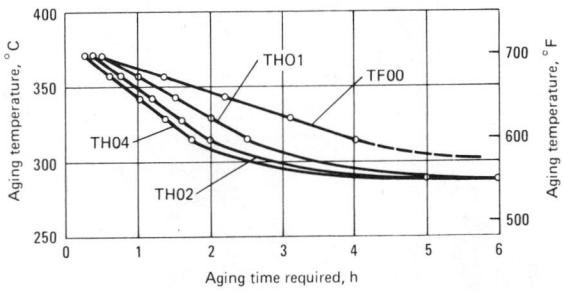

Fig. 3 Typical time-temperature relations in aging C17200

Aging time required to develop maximum strength in annealed, 1/4-hard, 1/2-hard and hard beryllium copper strip (1.9Be-0.2Co) aged at various temperatures in a recirculating air furnace to produce TF00, TH01, TH02 and TH04 tempers, respectively.

Table 3 Properties corresponding to usual standard aging treatments for strip of three beryllium coppers

Initial condition	Time, h	Standard aging treatment Temperature °C	°F	Final temper	Tensile strength MPa	ksi	Yield strength(a) MPa	ksi	Elongation, %	Hardness	Electrical conductivity, % IACS
C17000 (97.9Cu-1.7Be-0.4Co + Ni)											
Annealed(b) ..	···	None		TB00	415-540	60-78	170-240	25-35	35-60	45-78 HRB	17-19
¼ hard.......	···	None		TD01	515-605	75-88	310-515	45-75	10-35	68-90 HRB	16-18
½ hard.......	···	None		TD02	585-690	85-100	450-620	65-90	5-25	88-96 HRB	15-17
Hard	···	None		TD04	690-825	100-120	550-760	80-110	2-8	96-102 HRB	15-17
Annealed	3	315	600	TF00.....	1035-1240	150-180	825-1105	120-160	4-10	33-39 HRC	22-25
Annealed	3	345	650	TF00(c) ..	1105-1275	160-185	860-1140	125-165	4-10	34-40 HRC	22-25
¼ hard.......	2	315	600	TH01	1105-1275	160-185	860-1140	125-165	3-6	34-40 HRC	22-25
¼ hard.......	3	330	625	TH01(c) ..	1170-1345	170-195	895-1170	130-170	3-6	36-41 HRC	22-25
½ hard.......	2	315	600	TH02	1170-1345	170-195	895-1170	130-170	2-5	36-41 HRC	22-25
½ hard.......	2	330	625	TH02(c) ..	1240-1380	180-200	965-1240	140-180	2-5	38-42 HRC	22-25
Hard	2	315	600	TH04	1240-1380	180-200	965-1240	140-180	2-5	38-42 HRC	22-25
Hard	2	330	625	TH04(c) ..	1275-1415	185-205	1070-1345	155-195	2-5	39-43 HRC	22-25
C17200 (97.7Cu-1.9Be-0.4Co + Ni)											
Annealed	···	None		TB00	415-540	60-78	195-250	28-36	35-60	45-78 HRB	17-19
¼ hard.......	···	None		TD01	515-605	75-88	415-550	60-80	10-35	68-90 HRB	16-18
½ hard.......	···	None		TD02	585-690	85-100	515-620	75-90	5-25	88-96 HRB	15-17
Hard	···	None		TD04	690-825	100-120	620-770	90-112	2-8	96-102 HRB	15-17
Annealed	3	315	600	TF00(c) ..	1140-1310	165-190	965-1205	140-175	4-10	35-40 HRC	22-25
Annealed	½	370	700	TF00.....	1105-1310	160-190	895-1205	130-175	3-10	34-40 HRC	22-25
¼ hard.......	2	315	600	TH01(c) ..	1205-1380	175-200	1035-1275	150-185	3-6	37-42 HRC	22-25
¼ hard.......	¼	370	700	TH01	1170-1380	170-200	965-1275	140-185	2-6	36-42 HRC	22-25
½ hard.......	2	315	600	TH02(c) ..	1275-1450	185-210	1105-1345	160-195	2-5	39-44 HRC	22-25
½ hard.......	¼	370	700	TH02	1240-1450	180-210	1035-1345	150-195	2-5	38-44 HRC	22-25
Hard	2	315	600	TH04(c) ..	1310-1480	190-215	1140-1415	165-205	1-4	40-45 HRC	22-25
Hard	¼	370	700	TH04	1275-1480	185-215	1105-1415	160-205	1-4	39-45 HRC	22-25
C17500 (97Cu-0.5Be-2.5Co)											
Annealed	···	None		TB00	240-380	35-55	140-205	20-30	20-35	20-43 HRB	25-30
Hard	···	None		TD04	515-585	75-85	380-550	55-80	5-10	78-88 HRB	22-27
Annealed(b) ..	3	480	900	TF00.....	690-760	100-110	550-690	80-100	8-12	92-100 HRB	48-52
Annealed(b) ..	3	455	850	TF00(c) ..	725-825	105-120	550-725	80-105	8-12	93-100 HRB	45-52
Hard	2	480	900	TH04	760-860	110-125	690-825	100-120	5-8	95-103 HRB	48-52
Hard	2	455	850	TH04(c) ..	792-950	115-138	725-860	105-125	5-8	97-104 HRB	45-52

(a) At 0.2% offset. (b) All annealing of these alloys is solution treating. (c) Heat treatment that provides optimum strength.

Table 4 Typical aging cycles and properties of four beryllium copper casting alloys

UNS number	Nominal composition	Aging treatment(a)	Tensile strength MPa	ksi	Elongation, %	Electrical conductivity, % IACS
C81800 ..	97Cu-0.4Be-1.5Co-1Ag	3 h at 480 °C (900 °F)	705	102	10	45
C82000 ..	96.9Cu-0.6Be-2.5Co	3 h at 480 °C (900 °F)	705	102	9	45
C82400 ..	98Cu-1.7Be-0.3Co	3 h at 345 °C (650 °F)	1100	160	2.5	18
C82500 ..	97.5Cu-2Be-0.5Co	3 h at 345 °C (650 °F)	1140	165	2	18

(a) Typical aging conditions for solution treated castings.

soft and ductile; therefore, it can be cold worked in a manner similar to that used for unalloyed copper.

Solution treated chromium copper is aged at 400 to 500 °C (750 to 930 °F) for several hours to produce the desired mechanical and physical properties. A typical aging cycle is 455 °C (850 °F) for 4 h or more.

Typical effects of heat treatment and cold work on properties of chromium copper are shown in Table 5. The hard drawn samples were obtained by reducing the cross-sectional area of the solution annealed samples by approximately 40%.

Zirconium Copper

Zirconium copper C15000 (99.8Cu-0.2Zr) is solution treated at 900 to 925 °C (1650 to 1700 °F), then quenched in water. Time at the solution treating temperature should be minimized to limit grain growth and possible internal oxidation by reaction of zirconium with the furnace atmosphere. Because solution and diffusion of the zirconium occur rapidly at the solution treating temperature, holding at temperature is not required.

Aging is done at 500 to 550 °C (930 to 1020 °F) for 1 to 4 h. If the material has been cold worked following solution treating, aging temperature may be reduced to 375 to 475 °C (700 to 900 °F).

Maximum mechanical properties and resistance to softening are developed when all of the zirconium present in the alloy is dissolved during solution treating. If material containing 0.15% zirconium or more is heated above 975 °C (1790 °F), the Cu_3Zr phase will begin to melt. A slight amount of melting will not affect mechanical properties; but if excessive melting occurs, ductility of the alloy will decrease.

Normally, as the solution temperature is increased above 900 °C (1600 °F), the aging temperature should also be increased to maintain high electrical conductivity. The following aging treatments will produce the best combination of mechanical properties and electrical conductivity:

Condition	Aging treatment
Solution treated at 900 °C (1600 °F)	3 h at 500 °C (930 °F)
Solution treated at 900 °C and cold worked	3 h at 400 °C (750 °F)
Solution treated at 975 °C (1790 °F)	3 h at 550 °C (1020 °F)
Solution treated at 975 °C and cold worked	3 h at 450 °C (840 °F)

High strength in zirconium copper depends primarily on cold work. Although aging results in some added strength, its chief effect is to increase electrical conductivity.

Alpha Aluminum Bronzes

The structure and consequent heat treatability of aluminum bronze varies greatly with composition.

Single-phase (alpha) aluminum bronzes that contain only copper and aluminum (up to about 10% Al) can be strengthened only by cold working. They can be softened by annealing at 425 to 760 °C (800 to 1400 °F).

Although single-phase binary alloys such as C60600 and C61000 cannot be age hardened, additions of certain elements, such as cobalt and nickel, produce alloys that are age hardenable. The strength of some age-hardenable alpha aluminum bronzes can be increased by prior cold work.

Transformation Hardening

The transformation hardening mech-

Table 5 Typical properties of chromium coppers

	Tensile strength		Yield strength(a)		Elonga-tion(b),		Electrical conductivity,
Temper	MPa	ksi	MPa	ksi	%	Hardness	% IACS
TB00	240	35	105	15	42	50 HRF	40
TF00	410	60	345	50	20	60 to 65 HRB	80 to 85
TD04	435	63	310	45	15	65 HRF	40
TH04	480	70	425	62	18	80 to 85 HRB	80

(a) 0.5% extension under load. (b) In 50 mm or 2 in.

anism most often observed in copper alloys is associated with two-phase aluminum bronzes. These alloys are hardened by cooling rapidly from a high temperature to produce a martensitic type of structure, and then are tempered at a lower temperature to stabilize the structure and partly restore ductility and toughness.

Two-Phase Aluminum Bronzes. Binary copper-aluminum alloys have two stable phases at room temperature when the aluminum content is 9.5 to 16%. When other elements (most notably about 1 to 5% iron) are added, the corresponding aluminum content for two-phase alloys is 8 to 14%. Any of the two-phase alloys can be strengthened by quenching and tempering. At temperatures of 815 to 1010 °C (1500 to 1850 °F), the two room-temperature phases transform to beta in the same manner that alpha plus Fe_3C in steel transforms to austenite. Rapid quenching produces a hard, brittle structure due to formation of metastable, ordered, close-packed-hexagonal beta, which is referred to as martensitic beta. Both oil and water quenching are used commercially.

Tempering for 2 h at 595 to 650 °C (1100 to 1200 °F) causes reprecipitation of fine acicular alpha in a tempered beta-martensite structure, reducing hardness while increasing ductility and toughness. For large sections, rapid cooling (by fan or water spray) from the tempering temperature is advisable to avoid transformation of residual tempered beta to the embrittling eutectoid structure of lamellar or nodular alpha–gamma-2.

Nickel aluminum bronzes, although more complex, respond to quench-and-temper treatments in a similar manner. An additional phase, ordered body-centered-cubic kappa, generates structures resembling coarse pearlite in the alpha crystals, in addition to stabilizing the quenched beta. Nickel-bearing alloys such as C95500 and C63000 quench to a higher hardness

and are more susceptible to quench cracking in heavy and/or complex sections, making oil quenching desirable.

Cast two-phase aluminum bronzes often are normalized by heating to 815 °C (1500 °F), furnace cooling to about 550 °C (1000 °F) and then cooling in air to room temperature. This treatment produces uniform hardness and improves machinability.

Process Control. Temperature variations of ±10 °C (±20 °F) during heat treating of aluminum bronzes do not materially affect final properties. Excessively high annealing temperatures increase grain size and thus decrease strength. For single-phase alloys, the critical annealing temperature is about 650 °C (1200 °F).

In general, atmospheres are not used for heat treating aluminum bronzes, because these alloys form protective aluminum oxide surface films.

Spinodal Decomposition

Spinodal decomposition is a hardening mechanism that occurs in some copper alloys, such as C71900 (67.2Cu-30Ni-2.8Cr). Spinodal structures can be formed in alloy systems that exhibit a miscibility gap (either stable or unstable), and in which atoms of the component metals possess sufficient mobility at the heat treating temperature.

Spinodal structures are formed when an appropriate alloy is homogenized at a temperature above the miscibility gap, for a sufficient time that only statistical variations in composition exist. If the alloy is then cooled rapidly to a temperature within the miscibility gap and held at that temperature, spinodal decomposition proceeds at a rate controlled by the diffusion rates of the two metals. Spinodal decomposition also may occur on continuous cooling from the homogenization temperature to room temperature, provided the cooling rate is low enough and the diffusion rates of the metals are fast enough.

Order Hardening

Certain alloys, generally those that are nearly saturated with an alloying element dissolved in the alpha phase, will undergo an ordering reaction when highly cold worked material is annealed at a relatively low temperature. Alloys C61500, C63800, C68800 and C69000 are examples of copper alloys that exhibit this behavior. Strengthening is attributed to short range ordering of the solute atoms within the copper matrix, which greatly impedes the motion of dislocations through the crystals.

The low temperature order annealing treatment also acts as a stress relieving treatment, which raises the yield strength by reducing stress concentrations in the lattice at the focuses of dislocation pileups. As a result, order annealed alloys exhibit improved stress relaxation characteristics.

Order annealing is done for relatively short times at relatively low temperatures, generally in the range 150 to 400 °C (300 to 750 °F). Because of the low temperature, no special protective atmosphere is required. Order hardening is frequently done after the final fabrication step to take full advantage of the stress relieving aspect of the treatment, especially where resistance to stress relaxation is desired.

Microduplex Hardening

Microduplexing is a hardening mechanism applicable to certain copper alloys that can be processed to achieve a two phase structure in which both phases are stable or metastable at room temperature and slightly elevated temperatures. In microduplexing, the thermomechanical history of the alloy is closely controlled so that there is an optimum intermingling and dispersion of the two phases. The resultant structure is very fine (grain size usually less than 0.010 mm), and there is a marked improvement in strength with practically no change in elongation compared to the same alloy that has not undergone microduplex hardening.

Copper Tubular Products

By the ASM Committee on Copper and Copper Alloys*

TUBE AND PIPE made of copper or copper alloys are used extensively for carrying potable water in buildings and homes. These products also are used throughout the oil, chemical and process industries to carry diverse fluids, ranging from various natural and process waters to seawater to an extremely broad range of strong and dilute organic and inorganic chemicals. In the automotive and aerospace industries, copper tube is used for hydraulic lines, heat exchangers (such as automotive radiators), air conditioning systems and various formed or machined fittings. In marine service, copper tube and pipe are used to carry potable water, seawater and other fluids, but their chief application is in tube bundles for condensers, economizers and auxiliary heat exchangers. Copper tube and pipe are used in food and beverage industries to carry process fluids for beet and cane sugar refining, for brewing of beer and for many other food processing operations. In the building trades, copper tube is used widely for heating and air conditioning systems in homes, commercial buildings, and industrial plants and offices. Table 1 summarizes the copper alloys that are standard tube alloys, and gives ASTM

*See page **XI** for committee list.

specifications and typical uses for each of the alloys.

Frequently, resistance to corrosion is a critical factor in selecting a tube alloy for a specific application. Information that can help determine the alloy(s) most suitable for a given type of service can be found in the articles "Corrosion of Copper and Copper Alloys" and "Wrought Copper Alloys for Corrosion Service" elsewhere in this volume.

Joints in copper tube and pipe are made in various ways. Permanent joints can be made by brazing or welding. Semipermanent joints are made most often by soldering, usually in conjunction with standard socket-type solder fittings, but threaded joints also can be considered semipermanent joints for pipe. Detachable joints are almost always some form of mechanical joint—flared joints, flange-and-gasket joints, and joints made using any of a wide variety of specially designed compression fittings are all common.

Properties of Tube. As with most wrought products, the mechanical properties of copper tube depend on prior processing. With copper, it is not so much the methods used to produce tube, but rather the resulting metallurgical condition that has the greatest bearing on properties. Table 2 summarizes tensile properties for the standard

tube alloys in their most widely used conditions. Information on other properties of tube alloys can be found in the data compilations for the individual alloys, see the section of this volume entitled "Properties of Wrought Coppers and Copper Alloys".

Production of Tube Shells

Copper tubular products are produced from shells made by extruding or piercing copper billets.

Extrusion of copper and copper alloy tube shells is done by heating a billet of material above the recrystallization temperature, and then forcing material through an orifice in a die and over a mandrel held in position within the die orifice. The clearance between mandrel and die determines the wall thickness of the extruded tube shell.

In extrusion, the die is located at one end of the container section of an extrusion press; the metal to be extruded is driven through the die by a ram, which enters the container from the end opposite the die. Tube shells are produced either by starting with a hollow billet or by a two-step operation in which a solid billet is first pierced and then extruded.

Extrusion pressure varies with alloy

Table 1 Copper tube alloys and typical applications

UNS number	Alloy type	ASTM specifications	Typical uses
C10200	Oxygen-free copper	B68, B75, B88, B111, B188, B280, B359, B372, B395, B447	Bus tube, conductors, wave guides
C12200	Phosphorus deoxidized copper	B68, B75, B88, B111, B280, B306, B359, B360, B395, B447, B543	Water tubes; condenser, evaporator and heat exchanger tubes; air conditioning and refrigeration, gas, heater and oil burner lines; plumbing pipe and steam tubes; brewery and distillery tubes; gasoline, hydraulic and oil lines; rotating bands
C19200	Copper	B111, B359, B395, B469	Automotive hydraulic brake lines; flexible hose
C23000	Red brass, 85%	B111, B135, B359, B395, B543	Condenser and heat exchanger tubes, flexible hose; plumbing pipe; pump lines
C26000	Cartridge brass, 70%	B135	Plumbing brass goods
C33000	Low-leaded brass (tube)	B135	Pump and power cylinders and liners; plumbing brass goods
C36000	Free-cutting brass		Screw machine parts; plumbing goods
C43500	Tin brass		Bourdon tubes; musical instruments
C44300	Inhibited admiralty metal	B111, B359, B395	Condenser, evaporator and heat exchanger tubes; distiller tubes
C44400			
C44500			
C46400	Naval brass		Marine hardware, nuts
C46500			
C46600			
C46700			
C60800	Aluminum bronze, 5%	B111, B359, B395	Condenser, evaporator and heat exchanger tubes; distiller tubes
C65100	Silicon bronze B	B315	Heater exchanger tubes; electrical conduits
C65500	Silicon bronze A	B315	Chemical equipment, heat exchanger tubes; piston rings
C68700	Arsenical aluminum brass	B111, B359, B395	Condenser, evaporator and heat exchanger tubes; distiller tubes
C70600	Copper nickel, 10%	B111, B359, B395, B466, B467, B543, B552	Condenser, evaporator and heat exchanger tubes; salt water piping; distiller tubes
C71500	Copper nickel, 30%	B111, B359, B395, B446, B467, B543, B552	Condenser, evaporator and heat exchanger tubes; distiller tubes; salt water piping

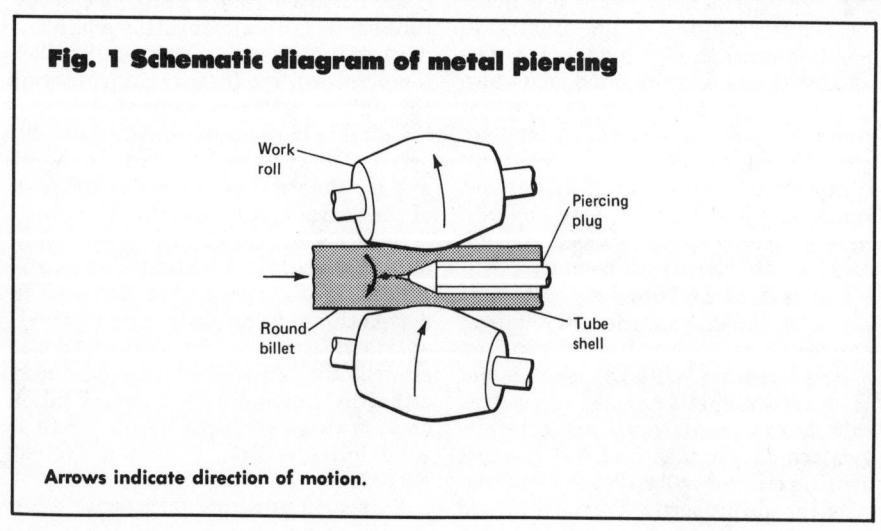

Fig. 1 Schematic diagram of metal piercing

Work roll

Piercing plug

Round billet

Tube shell

Arrows indicate direction of motion.

composition. C36000 (61.5Cu-3Pb-35.5Zn) requires a relatively low pressure, whereas C26000 (70Cu-30Zn) and C44300 (71.5Cu-1Sn-27.5Zn-0.06As) require the highest pressure of all the brasses. Most of the coppers require an extrusion pressure intermediate between those for C26000 and C36000. C71500 (70Cu-30Ni) requires a very high extrusion pressure.

Extrusion pressure also depends on billet temperature, extrusion ratio (the ratio of the cross-sectional area of the billet to that of the extruded section), speed of extrusion and degree of lubrication. The flow of metal during extrusion depends on many factors, including copper content of the metal, amount of lubricant, and die design.

Table 2 Typical mechanical properties for copper alloy tube(a)

Temper	Tensile strength MPa	ksi	Yield strength(b) MPa	ksi	Elongation(c), %
C10200					
OS050	220	32	69	10	45
OS025	235	34	76	11	45
H55	275	40	220	32	25
H80	380	55	345	50	8
C12200					
OS050	220	32	69	10	45
OS025	235	34	76	11	45
H55	275	40	220	32	25
H80	380	55	345	50	8
C19200					
H55(d)	290	42	205(e)	30(e)	35
C23000					
OS050	275	40	83	12	55
OS015	305	44	125	18	45
H55	345	50	275	40	30
H80	485	70	400	58	8
C26000					
OS050	325	47	105	15	65
OS025	360	52	140	20	55
H80	540	78	440	64	8
C33000					
OS050	325	47	105	15	60
OS025	360	52	140	20	50
H80	515	75	415	60	7
C43500					
OS035	315	46	110	16	46
H80	515	75	415	60	10
C44300, C44400, C44500					
OS025	365	53	150	22	65
C46400, C46500, C46600, C46700(f)					
H80	605	88	455	66	18
C60800					
OS025	415	60	185	27	55
C65100					
OS015	310	45	140	20	55
H80	450	65	275	40	20
C65500					
OS050	395	57	. . .	. . .	70
H80	640	93	. . .	. . .	22
C68700					
OS025	415	60	185	27	55
C70600					
OS025	305	44	110	16	42
H55	415	60	395	57	10
C71500					
OS025	415	60	170	25	45

(a) Tube size: 25 mm (1 in.) OD by 1.65 mm (0.065 in.) wall. (b) 0.5% extension under load. (c) In 50 mm or 2 in. (d) Tube size: 4.8 mm (0.1875 in.) OD by 0.76 mm (0.030 in.) wall. (e) 0.2% offset. (f) Tube size: 9.5 mm (0.375 in.) OD by 2.5 mm (0.097 in.) wall.

Rotary piercing on a Mannesmann mill is another method commonly used to produce seamless pipe and tube from copper and certain copper alloys. Piercing is the most severe forming operation customarily applied to metals. The process takes advantage of tensile stresses that develop at the center of a billet when it is subjected to compressive forces around its periphery. In rotary piercing, one end of a heated cylindrical billet is fed between rotating work rolls that lie in a horizontal plane and are inclined at an angle to the axis of the billet, see Fig. 1. Guide rolls beneath the billet prevent it from dropping from between the work rolls. Because the work rolls are set at an angle to each other as well as to the billet, the billet is simultaneously rotated and driven forward toward the piercing plug, which is held in position between the work rolls.

The opening between work rolls is set smaller than the billet, and the resultant pressure acting around the periphery of the billet opens up tensile cracks, and then a rough hole, at the center of the billet just in front of the piercing plug. The piercing plug assists in further opening the axial hole in the center of the billet, smooths the wall of the hole and controls the wall thickness of the formed tube.

Coppers and plain alpha brasses can be pierced, provided the lead content is held to less than 0.01%. Alpha-beta brasses can tolerate higher levels of lead without adversely affecting their ability to be pierced.

When piercing brass, close temperature control must be maintained because the range in which brass can be pierced is narrow. Each alloy has a characteristic temperature range within which it is sufficiently plastic for piercing to take place. Below this range, the central hole does not open up properly under the applied peripheral forces. Overheating may lead to cracked surfaces. Suggested piercing temperatures for various alloys are given below:

UNS number	Piercing temperature °C	°F
C11000	815-870	1500-1600
C12200	815-870	1500-1600
C22000	815-870	1500-1600
C23000	815-870	1500-1600
C26000	760-790	1400-1450
C28000	705-760	1300-1400
C46400	730-790	1350-1450

Table 3 ASTM and ASME specifications for copper tube and pipe

Tubular product	ASTM	ASME
Seamless pipe and tube, copper-nickel alloy	B466	SB466
Seamless pipe and tube, copper-silicon alloy	B315	SB315
Seamless pipe and tube, for electrical conductors	B188	...
Seamless pipe, standard sizes	B42	...
Seamless pipe, threadless	B302	...
Seamless tube	B75	SB75
Seamless tube, brass	B135	SB135
Seamless tube, bright annealed	B68	...
Seamless tube, capillary, hard-drawn	B360	...
Seamless tube, condenser and heat exchanger	B111, B395	SB111, SB395
Seamless tube, condenser and heat exchanger, with integral fins	B359	SB359
Seamless tube, for air conditioning and refrigeration service	B280	...
Seamless tube, drainage	B306	...
Seamless tube, general requirements	B251	...
Seamless tube, rectangular waveguide	B372	...
Seamless tube, water	B88	...
Welded pipe and tube, copper-nickel alloy	B467	SB467
Welded tube, C10800 and 12000	B543	SB543
Welded tube, all other coppers	B447	...

Production of Finished Tubes

Cold drawing of extruded or pierced tube shells to smaller sizes is done on draw blocks for coppers and on draw benches for brasses and other alloys. With either type of machine, the metal is cold worked by pulling the tube through a die that reduces the diameter. Concurrently, wall thickness is reduced by drawing over a plug or mandrel that may be either fixed or floating. Cold drawing increases the strength of the material and simultaneously reduces ductility. Tube size is reduced—outside diameter, inside diameter, wall thickness and cross-sectional area all are smaller after drawing. Because the metal work hardens, tubes may be annealed at intermediate stages when drawing to small sizes. However, coppers are so ductile that they frequently can be drawn to finished size without intermediate annealing.

Tube reducing is an alternative process for cold sizing of tube. In tube reducing, semicircular grooved dies are rolled or rocked back and forth along the tube while a tapered mandrel inside the tube controls the inside diameter and wall thickness. The process yields tube having very accurate dimensions and better concentricity than can be achieved by tube drawing.

The grooves in the tube reducing dies are tapered, one end of the grooved section being somewhat larger than the outside diameter of the tube to be sized. As the dies are rocked back and forth, the tube is pinched against the tapered mandrel, which reduces wall thickness and increases tube length. The tube is fed longitudinally, and rotated on its axis to distribute the cold work uniformly around the entire circumference. Feeding and rotating are synchronized with die motion and take place after the dies have completed their forward stroke.

Tube reducing may be used for all alloys that can be drawn on draw benches. Slight changes in die design and operating conditions may be required to accommodate different alloys.

Small diameter tube may be produced by block or bench drawing following tube reducing.

Product Specifications

Copper tube and pipe are available in a wide variety of nominal diameters and wall thicknesses, from small diameter capillary tube to 12-in. nominal diameter pipe. To a certain extent, dimensions and tolerances for copper tube and pipe depend on the type of service for which it is intended. The standard dimensions and tolerances for several different kinds of copper tube and pipe are given in the ASTM specifications listed in Table 3, along with other requirements for the tubular products. Seamless copper tube for automotive applications (1/8 to 3/4 in. nominal diameter) is covered by SAE J528. Requirements for copper tube and pipe to be used in condensers, heat exchangers, economizers and similar unfired pressure vessels are given in the ASME specifications listed in Table 3. (ASME materials specifications are almost always identical to ASTM specifications having the same numerical designation; for example, ASME SB111 is identical to ASTM B111.) Certain tube alloys are covered in AMS specifications, which apply to materials for aerospace applications. These are given below:

AMS specification	Product	Copper alloy
4555	Seamless brass tube, light annealed	C26000, C33000
4558	Seamless brass tube drawn	C33200
4625	Phosphor bronze, hard temper	C51000
4640	Aluminum bronze	C63000
4665	Seamless silicon bronze tube, annealed	C65500

Copper Wire and Cable

By the ASM Committee on Copper and Copper Alloys*

WIRE made from copper and its alloys has been used since about 2000 to 3000 BC. According to archaeological evidence, the ancient Assyrians, Babylonians and Egyptians were skilled in producing of copper wire for ornamental purposes. Drawing wire through a die is a much more modern development. The earliest evidence of drawn wire comes from sixth century AD Venetian and French artifacts. Theophilus, a German monk, produced the first written records in a treatise on metalworking circa 1110 to 1140 AD. His description of wiredrawing reads, in part, "Two pieces of iron three or four fingers wide, smaller at the top and bottom, rather thin, pierced with three or four holes through which wire may be drawn. . . ." By 1270, a set of rules had been passed to govern wiredrawing in Paris and at least nine wiredrawers were at work in that city.

Development of wiredrawing processes during the Middle Ages concentrated to a large extent on drawing iron and steel wires to make pins and instrument strings. But with the invention of the electric telegraph in 1847 came the requirement for long continuous lengths of electric conductor wire made of copper. In 1850, copper wire was used to make a submarine cable connecting England and France.

*See page XI for committee list.

At the beginning of the twentieth century, wire was still being drawn through single dies—a process commonly known as "bull-block" drawing. Dies were made by punching a series of holes in a steel plate. These holes were then trimmed to final size with a master punch. Rows of single capstans, power driven from a common drive shaft, were used for drawing single lengths of wire. As each reduction was completed, the steel-plate die was replaced with one containing smaller holes until the final diameter was achieved.

Multiple wire-drawing machines were introduced about 1900. As a result, chilled cast iron plates and dies that could be reamed to size replaced the punch-sized steel-plate dies. Lubricants were introduced because the considerable heat generated by friction between the wire and draw-capstan and by successive reductions through progressively smaller dies. In turn, use of lubricants permitted wire to be drawn at faster speeds.

The prime development during the 1920's was the introduction of drawing dies made of tungsten carbide. High hardness and lack of porosity made tungsten carbide dies ideal for high-speed wiredrawing and provided longer die life than was usually possible with dies made of chilled cast iron. Tungsten carbide dies are standard today. For very fine wire sizes, below about 1.3 mm (0.05 in.), diamond dies are used because they are harder and last longer than tungsten carbide dies. Some wire mills use diamond dies for high speed drawing of larger wires, up to 8 mm (0.32 in.) in diameter, to reduce the frequency of shutdowns for die replacement.

Classification of Copper for Conductors

Copper metals used for electrical conductors fall into three general categories: high-conductivity coppers, high-copper alloys and electrical bronzes.

High-conductivity coppers are covered by ASTM specifications B4, B5, B170, B442 and B623. ASTM B4 covers both high-resistance Lake copper and low-resistance Lake copper. Lake copper is fire refined from local Lake Superior ore deposits. ASTM B5 covers copper electrolytically refined from blister copper, converter copper, black copper or Lake copper. ASTM B170 covers oxygen-free electrical copper.

Oxygen-free copper is produced by special manufacturing techniques and is used to avoid embrittlement where conductors are subjected to hydrogen or other reducing gases at elevated temperatures.

Some specialty coppers are produced by adding minimal amounts of hardening agents (such as chromium, tellurium, beryllium, cadmium or zirconium). These are used in applications where high anneal resistance is required.

A series of bronzes has been developed for use as conductors; these alloys are covered by ASTM B105. These bronzes are intended to provide better corrosion resistance and higher tensile strengths than standard conductor coppers. There are nine conductor bronzes, designated 8.5 to 85 in accordance with their electrical conductivities, as given below:

ASTM B105 alloy designation	Alternative alloy types
8.5	Cu-Si-Fe
	Cu-Si-Mn
	Cu-Si-Zn
	Cu-Si-Sn-Fe
	Cu-Si-Sn-Zn
13	Cu-Al-Sn
	Cu-Al-Si-Sn
	Cu-Si-Sn
15	Cu-Al-Si
	Cu-Al-Sn
	Cu-Al-Si-Sn
	Cu-Si-Sn
20	Cu-Sn
30	Cu-Sn
	Cu-Zn-Sn
40(a)	Cu-Sn
	Cu-Sn-Cd
55(a)	Cu-Sn-Cd
65(a)	Cu-Sn
	Cu-Sn-Cd
80(a)	Cu-Cd
85	Cu-Cd

(a) Normally used for trolley-wire applications in either a round or grooved cross-sectional configuration, as set forth in ASTM B9.

The compositions of these alloys must be within the total limits prescribed below, and no alloy may contain more than the allowed maximum of any constituent other than copper.

Element	Composition limit, % max
Fe	0.75
Mn	0.75
Cd	1.50
Si	3.00
Al	3.50
Sn	5.00
Zn	10.50
Cu	89.00 min
Sum of above elements	99.50 min

Classification of Wire and Cable

Round Wire. Standard nominal diameters and cross-sectional areas of solid round copper wires used as electrical conductors are prescribed in ASTM B258. Wire sizes have almost always been designated in the American Wire Gauge (AWG) system. This system is based on fixed diameters for two wire sizes (4/0 and 36 AWG, respectively) with a geometric progression of wire diameters for the thirty-eight intermediate gages and for gages smaller than 36 AWG (see Table 1). This is an inverse series in which a higher number denotes a smaller wire diameter. Each increase of one AWG number is approximately equivalent to a 20.7% reduction in cross-sectional area.

ASTM B1 specifies hard drawn round wire that has been reduced at least four AWG numbers (60% reduction in area) and ASTM B3 specifies soft (or annealed) copper wire. Table 1 gives the sizes and properties of wires specified in ASTM B1, B3 and B258.

Square and Rectangular Wire. ASTM B48 specifies soft (annealed) square and rectangular copper wire.

Stranded wire is normally used in electrical applications where some degree of flexing is encountered either in service or during installation. In order of increasing flexibility, the common forms of stranded wire are: concentric lay, unilay, rope lay and bunched.

Concentric-lay stranded wire and cable are composed of a central wire surrounded by one or more layers of helically laid wires, with the direction of lay reversed in successive layers, and with the length of lay increased for each successive layer. The outer layer usually has a left-hand lay.

ASTM B8 establishes five classes of concentric-lay stranded wire and cable, from AA (the coarsest) to D (the finest). Details of concentric-lay constructions are given in Table 2.

Unilay stranded wire is composed of a central core surrounded by more than one layer of helically laid wires, all layers having a common lay length and direction. This type of wire sometimes is referred to as "smooth bunch". The layers usually have a left-hand lay.

Rope lay stranded wire and cable are composed of a stranded member (or members) as a central core, around which are laid one or more helical layers of similar stranded members. The members may be concentric or bunch-stranded. ASTM B173 and B172 establish five classes of rope-lay stranded conductors: classes G and H, which have concentric members; and classes I, K and M, which have bunched members. Construction details are shown in Tables 3 and 4. These cables are normally used to make large, flexible conductors for portable service, such as mining cable or apparatus cable.

Bunch stranded wire is composed of any number of wires twisted together in the same direction without regard to geometric arrangement of the individual strands. ASTM B174 provides for five classes (I, J, K, L and M); these conductors are commonly used in flexible cords, hookup wires and special flexible welding conductors. Typical construction details are given in Table 5.

Tin-coated Wire. Solid and stranded wires are available with tin coatings. These are manufactured to the latest revisions of ASTM B33, which covers soft or annealed tinned copper wires, and B246, which covers hard drawn or medium hard drawn tinned copper wires. Characteristics of tinned round solid wire are given in Table 6.

Fabrication of Wire Rod

Wire rod is the intermediate product in the manufacture of wire. Although "wire rod" is the term used in the U.S. for the intermediate product, the term "drawing stock" is used in international standards and customs documents.

Rolling. The traditional process for converting prime copper into wire rod involves hot rolling of cast wirebar. Almost all drawing stock is rolled to 8 mm (0.32 in.) diameter. Larger sizes, up to 22 mm (0.87 in.) or more in diameter are available on special order.

Some special oxygen-free copper wirebar is produced by vertical casting, but most wirebar is produced by horizontal casting of tough-pitch copper into open molds. The oxygen content is controlled at 0.03 to 0.06% to give a level surface. Cast wirebars weigh about 110 to 135 kg (250 to 300 lb) each. Their ends are tapered to facilitate entry into the first pass of the hot rolling mill.

Prior to rolling, bars are heated to

Table 1 Characteristics of solid round copper wire: ASTM B1, B3, B258

Conductor size, AWG	Conductor diameter, mils	Conductor area, circular mils	Net weight, lb/1000 ft	Soft (annealed) wire		Hard drawn wire		
				Minimum elongation(a), %	Nominal resistance, Ω/1000 ft	Nominal breaking strength, lb	Nominal tensile strength, ksi	Nominal resistance, Ω/1000 ft
4/0......460.0		211 600	640.5	35	0.0491	8143	49.0	0.05044
3/0......409.6		167 800	507.8	35	0.06180	6720	51.0	0.06361
2/0......364.8		133 100	402.8	35	0.07791	5519	52.8	0.08019
1.0......324.9		105 600	319.5	35	0.09821	4518	54.5	0.1011
1 ......289.3		83 690	253.3	30	0.1239	3888	56.1	0.1289
2 ......257.6		66 360	200.9	30	0.1563	3002	57.6	0.1625
3 ......229.4		52 620	159.3	30	0.1971	2439	59.0	0.2050
4 ......204.3		41 740	126.3	30	0.2485	1970	60.1	0.2584
5 ......181.9		33 090	100.2	30	0.3134	1590	61.2	0.3259
6 ......162.0		26 240	79.44	30	0.3952	1280	62.1	0.4110
7 ......144.3		20 820	63.03	30	0.4981	1030	63.1	0.5180
8 ......128.5		16 510	49.98	30	0.6281	826.1	63.7	0.6532
9 ......114.4		13 090	39.62	30	0.7923	660.9	64.3	0.8239
10 ......101.9		10 380	31.43	25	0.9991	529.3	64.9	1.039
11 ......90.7		8 230	24.9	25	1.26	423	65.4	1.31
12 ......80.8		6 530	19.8	25	1.59	337	65.7	1.65
13 ......72.0		5 180	15.7	25	2.00	268	65.9	2.08
14 ......64.1		4 110	12.4	25	2.52	214	66.2	2.62
15 ......57.1		3 260	9.87	25	3.18	170	66.4	3.31
16 ......50.8		2 580	7.81	25	4.02	135	66.6	4.18
17 ......45.3		2 050	6.21	25	5.06	108	66.8	5.26
18 ......40.3		1 620	4.92	25	6.40	85.5	67.0	6.66
19 ......35.9		1 290	3.90	25	8.04	68.0	67.2	8.36
20 ......32.0		1 020	3.10	25	10.2	54.2	67.4	10.6
21 ......28.5		812	2.46	25	12.8	43.2	67.7	13.3
22 ......25.3		640	1.94	25	16.2	34.1	67.9	16.9
23 ......22.6		511	1.55	25	20.3	27.3	68.1	21.1
24 ......20.1		404	1.22	20	25.7	21.7	68.3	26.7
25 ......17.9		320	0.970	20	32.4	17.3	68.6	33.7
26 ......15.9		253	0.765	20	41.0	13.7	68.8	42.6
27 ......14.2		202	0.610	20	51.4	10.9	69.0	53.4
28 ......12.6		159	0.481	20	65.2	8.64	69.3	67.8
29 ......11.3		128	0.387	20	81.0	6.96	69.4	84.3
30 ......10.0		100	0.303	15	104.0	5.47	69.7	108.0
31 ......8.9		79.2	0.240	15	131.0	4.35	69.9	136.0
32 ......8.0		64.0	0.194	15	162.0	3.53	70.2	169.0
33 ......7.1		50.4	0.153	15	206.0	2.79	70.4	214.0
34 ......6.3		39.7	0.120	15	261.0	2.20	70.6	272.0
35 ......5.6		31.4	0.0949	15	330.0	1.75	70.9	343.0
36 ......5.0		25.0	0.0757	15	415.0	1.40	71.1	431.0
37 ......4.5		20.2	0.0613	15	513.0	1.13	71.3	534.0
38 ......4.0		16.0	0.0484	15	648.0	0.898	71.5	674.0
39 ......3.5		12.2	0.0371	15	850.0	0.691	71.8	884.0
40 ......3.1		9.61	0.0291	15	1079.0	0.543	72.0	1122.0
41 ......2.8		7.84	0.0237	15	1323.0	0.443	72.0	1376.0
42 ......2.5		6.25	0.0189	15	1659.0	0.353	72.0	1726.0
43 ......2.2		4.48	0.0147	15	2143.0	0.274	72.0	2228.0
44 ......2.0		4.00	0.0121	15	2593.0	0.226	72.0	2696.0

(a) In 10 in.

about 925 °C (1700 °F) in a neutral atmosphere and then rolled on a continuous mill through a series of reductions to yield round rod about 6 to 22 mm (¼ to ⅞ in.) in diameter. The hot rolled rod is coiled, water quenched, and then pickled to remove the black cupric oxide that forms during rolling. This method can produce rod at rates up to 7.5 kg/s (30 tons/h).

Disadvantages of this process include (a) high capital investment to achieve low operating cost; (b) relatively small coils that must be welded together for efficient production, where the welded junctions present potential sources of weakness in subsequent wiredrawing operations; and (c) unsuitability of rod

Table 2 Characteristics of concentric-lay stranded copper conductors: ASTM B8

Conductor size, circular mils or AWG	Nominal weight lb/1000 ft	Nominal resistance(a), Ω/1000 ft	Class AA No. of wires	Class AA Diameter of individual wires, mils	Class A No. of wires	Class A Diameter of individual wires, mils	Class B No. of wires	Class B Diameter of individual wires, mils	Class C No. of wires	Class C Diameter of individual wires, mils	Class D No. of wires	Class D Diameter of individual wires, mils
5 000 000	15 890	0.002 178	…	…	169	172.0	217	151.8	271	135.8	271	135.8
4 500 000	14 300	0.002 420	…	…	169	163.2	217	144.0	271	128.9	271	128.9
4 000 000	12 590	0.002 696	…	…	169	153.8	217	135.8	271	121.5	271	121.5
3 500 000	11 020	0.003 082	…	…	127	166.0	169	143.9	271	127.0	271	113.6
3 000 000	9 353	0.003 561	…	…	127	153.7	169	133.2	271	117.6	271	105.2
2 500 000	7 794	0.004 278	…	…	91	165.7	127	140.3	217	121.6	217	107.3
2 000 000	6 175	0.005 289	…	…	91	148.2	127	125.5	217	108.8	217	96.0
1 900 000	5 886	0.005 568	…	…	91	144.5	127	122.3	217	106.0	217	93.6
1 800 000	5 558	0.005 877	…	…	91	140.6	127	119.1	217	103.2	217	91.1
1 750 000	5 404	0.006 045	…	…	91	138.7	127	117.4	217	101.8	217	89.8
1 700 000	5 249	0.006 223	…	…	91	136.7	127	115.7	217	100.3	217	88.5
1 600 000	4 940	0.006 612	…	…	91	132.6	127	112.2	217	97.3	217	85.9
1 500 000	4 631	0.007 052	…	…	61	156.6	91	128.4	169	108.7	169	94.2
1 400 000	4 323	0.007 556	…	…	61	151.5	91	124.0	169	105.0	169	91.0
1 300 000	4 014	0.008 137	…	…	61	146.0	91	119.5	169	101.2	169	87.7
1 250 000	3 859	0.008 463	…	…	61	143.1	91	117.2	169	99.2	169	86.0
1 200 000	3 705	0.008 815	…	…	61	140.3	91	114.8	169	97.2	169	84.3
1 100 000	3 396	0.009 617	…	…	61	134.3	91	109.9	169	93.1	169	80.7
1 000 000	3 088	0.010 88	37	164.4	61	128.0	61	128.0	127	104.8	127	88.7
900 000	2 779	0.011 75	37	156.0	61	121.5	61	121.5	127	99.4	127	84.2
800 000	2 470	0.013 22	37	147.0	61	114.5	61	114.5	127	93.8	127	79.4
750 000	2 316	0.014 10	37	142.4	61	110.9	61	110.9	127	90.8	127	76.8
700 000	2 161	0.015 11	37	137.5	61	107.1	61	107.1	127	87.7	127	74.2
650 000	2 007	0.016 27	37	132.5	61	103.2	61	103.2	127	84.5	127	71.5
600 000	1 853	0.017 63	37	127.3	37	127.3	61	99.2	127	81.2	127	68.7
550 000	1 698	0.019 23	37	121.9	37	121.9	61	95.0	127	77.7	127	65.8
500 000	1 544	0.021 16	19	162.2	37	116.2	37	116.2	61	90.5	91	74.1
450 000	1 389	0.023 51	19	153.9	37	110.3	37	110.3	61	85.9	91	70.3
400 000	1 235	0.026 45	19	145.1	19	145.1	37	104.0	61	81.0	91	66.3
350 000	1 081	0.030 22	12	170.8	19	135.7	37	97.3	61	75.7	91	62.0
300 000	926.3	0.035 26	12	158.1	19	125.7	37	90.0	61	70.1	91	57.4
250 000	771.9	0.042 31	12	144.3	19	114.6	37	82.2	61	64.0	91	52.4
4/0	653.3	0.049 99	7	173.9	7	173.9	19	105.5	37	75.6	61	58.9
3/0	518.1	0.063 04	7	154.8	7	154.8	19	94.0	37	67.3	61	52.4
2/0	410.9	0.079 48	7	137.9	7	137.9	19	83.7	37	60.0	61	46.7
1/0	326.0	0.100 2	7	122.8	7	122.8	19	74.5	37	53.4	…	…
1	258.4	0.126 4	3	167.0	7	109.3	19	66.4	37	47.6	…	…
2	204.9	0.159 4	3	148.7	7	97.4	7	97.4	19	59.1	…	…
3	162.5	0.201 0	3	132.5	7	86.7	7	86.7	19	52.6	…	…
4	128.9	0.253 4	3	118.0	…	…	7	77.2	19	46.9	…	…
5	102.2	0.319 7	…	…	…	…	7	68.8	19	41.7	…	…
6	81.05	0.403 1	…	…	…	…	7	61.2	19	37.2	…	…
7	64.28	0.508 1	…	…	…	…	7	54.5	19	33.1	…	…
8	50.98	0.640 7	…	…	…	…	7	48.6	19	29.5	…	…
9	40.42	0.808 1	…	…	…	…	7	43.2	19	26.2	…	…

(a) Uncoated wire.

Table 3 Characteristics of rope-lay stranded copper conductors having uncoated or tinned concentric members: ASTM B173

Conductor sizes, circular mils or AWG	Class G				Class H			
	Diameter of individual wires, mils	No. of ropes	No. of wires each rope	Net weight, lb/1000 ft	Diameter of individual wires, mils	No. of ropes	No. of wires each rope	Net weight, lb/1000 ft
5 000 000 . . . . 65.7		61	19	16 052	53.8	91	19	15 057
4 500 000 . . . . 62.3		61	19	14 434	51.0	91	19	14 429
4 000 000 . . . . 58.7		61	19	12 814	48.1	91	19	12 835
3 500 000 . . . . 55.0		61	19	11 249	45.0	91	19	11 234
3 000 000 . . . . 50.9		61	19	9 635	41.7	91	19	9 647
2 500 000 . . . . 59.6		37	19	8 012	46.4	61	19	8 006
2 000 000 . . . . 53.3		37	19	6 408	41.5	61	19	6 405
1 900 000 . . . . 52.0		37	19	6 099	40.5	61	19	6 100
1 800 000 . . . . 50.6		37	19	5 775	39.4	61	19	5 773
1 750 000 . . . . 49.9		37	19	5 617	38.9	61	19	5 627
1 700 000 . . . . 49.2		37	19	5 460	38.3	61	19	5 455
1 600 000 . . . . 47.7		37	19	5 132	37.2	61	19	5 146
1 500 000 . . . . 59.3		61	7	4 772	46.2	37	19	4 815
1 400 000 . . . . 57.3		61	7	4 456	44.6	37	19	4 487
1 300 000 . . . . 55.2		61	7	4 135	43.0	37	19	4 171
1 250 000 . . . . 54.1		61	7	3 972	42.2	37	19	4 017
1 200 000 . . . . 53.0		61	7	3 814	41.3	37	19	3 847
1 100 000 . . . . 50.8		61	7	3 502	39.6	37	19	3 537
1 000 000 . . . . 48.4		61	7	3 179	37.7	37	19	3 206
900 000 . . . . 45.9		61	7	2 859	35.8	37	19	2 891
800 000 . . . . 43.3		61	7	2 544	33.7	37	19	2 562
750 000 . . . . 41.9		61	7	2 383	32.7	37	19	2 412
700 000 . . . . 40.5		61	7	2 226	31.6	37	19	2 252
650 000 . . . . 39.0		61	7	2 064	30.4	37	19	2 085
600 000 . . . . 37.5		61	7	1 908	29.2	37	19	1 923
550 000 . . . . 35.9		61	7	1 749	28.0	37	19	1 768
500 000 . . . . 43.9		37	7	1 579	34.2	61	7	1 587
450 000 . . . . 41.7		37	7	1 425	32.5	61	7	1 433
400 000 . . . . 39.3		37	7	1 265	30.6	61	7	1 271
350 000 . . . . 36.8		37	7	1 109	28.6	61	7	1 110
300 000 . . . . 34.0		37	7	947.1	26.5	61	7	953.0
250 000 . . . . 31.1		37	7	792.4	24.2	61	7	794.8
4/0 . . . . 39.9		19	7	666.6	28.6	37	7	670.1
3/0 . . . . 35.5		19	7	527.7	25.5	37	7	532.7
2/0 . . . . 31.6		19	7	418.1	22.7	37	7	422.2
1/0 . . . . 28.2		19	7	333.0	20.7	37	7	334.3
1 . . . . 25.1		19	7	263.8	18.0	37	7	265.4
2 . . . . 36.8		7	7	206.9	22.3	19	7	208.2
3 . . . . 37.8		7	7	164.4	19.9	19	7	165.8
4 . . . . 29.2		7	7	130.3	17.7	19	7	131.2
5 . . . . 26.0		7	7	103.3	15.8	19	7	104.5
6 . . . . 23.1		7	7	81.52	14.0	19	7	82.06
7 . . . . 20.6		7	7	64.83	12.5	19	7	65.42
8 . . . . 18.4		7	7	51.72	11.1	19	7	51.59
9 . . . . 15.3		7	7	40.59	9.9	19	7	41.04
10 . . . . 14.6		7	7	32.57	. . .	. . .	. . .	. . .
12 . . . . 11.5		7	7	20.20	. . .	. . .	. . .	. . .
14 . . . . 9.2		7	7	12.93	. . .	. . .	. . .	. . .

rolled from cast wirebars for certain specialized wire applications.

Because of the disadvantages inherent in producing rolled rod from conventionally case wirebars processes have been developed for continuously converting liquid metal directly into wire rod, thus avoiding the intermediate wirebar stage.

Continuous Casting. In 1963, the first plant for continuous casting and rolling of copper wirebar went into operation. This process, which was developed by the Southwire Co. in conjunction with Western Electric Co., was essentially an adaptation of a process developed by Ilario Properzi prior to 1950—a process that had been used for many years by aluminum and zinc producers for conversion of prime metal or scrap into wire rod.

Continuously cast wire rod has come to dominate the copper wire rod market and now accounts for more than 1.8 billion kg (2 million tons) annually, or about 50% of the total amount of wire rod produced.

Other casting systems that have been developed include the Properzi, Hazelett and General Electric (G.E.) dip-form systems. Smaller-capacity machines have been developed by Outokumpu, Davy, Wertli, Lamitref and others.

Advantages of continuous casting and rolling include (a) large coil weights, up to 10 Mg (11 tons); (b) ability to reprocess scrap at considerable savings; (c) improved rod quality and surface condition; (d) homogeneous metallurgical conditions and close process control; and (e) low capital investment and low operating costs for moderate production rates.

The standard feed for continuous casting processes is cathode copper, which is charged directly into a melting furnace. An ASARCO shaft furnace is used for the Southwire, Properzi and Hazelett systems, but an electric furnace is preferred for the smaller systems, such as the G. E. dip-form process.

Southwire Continuous Rod System. In the Southwire system, the casting machine produces a cast copper bar 2500 to 3000 mm² (4 to 5 in.²) in cross-sectional area by pouring molten copper between a grooved wheel made of steel or copper and a steel band. The cast bar is cooled by water sprays as the wheel rotates, and is withdrawn by pinch rolls as it exits from the wheel. Next it goes to a bar-conditioning unit, and then to the rolling mill. After passing through a series of reductions in the mill, the rod enters an in-line pickling system, which quenches and cleans the rod prior to final coiling.

Properzi System. The Properzi system is basically the same as the Southwire process except for slight differences in design of the casting wheel and configuration of the rod mill.

G.E. Dip-form Process. The General Electric dip-form process was introduced in 1964. A seed rod approximately 9 mm (0.35 in.) in diameter is passed at a controlled rate through a bath of molten C10100. Copper freezes onto the seed rod, thickening its diameter to about 16.5 mm (0.65 in.). The rod emerging from the copper bath is hot rolled on a 2-stand mill, cooled and

Table 4 Characteristics of rope-lay stranded copper conductors having uncoated or tinned bunched members: ASTM B172

Conductor size, circular mils or AWG	Class of strand	Construction and wire size, AWG	Total No. of wires	Approx diameter, in.	Net weight, lb/1000 ft
1 000 000	I	19x7x19/24	2 527	1.290	3306
	K	37x7x39/30	10 101	1.329	3272
	M	61x7x59/34	25 193	1.353	3239
900 000	I	19x7x17/24	2 261	1.217	2959
	K	37x7x35/30	9 065	1.255	2936
	M	61x7x53/34	22 631	1.279	2909
800 000	I	19x7x15/24	1 995	1.140	2611
	K	19x7x60/30	7 980	1.174	2585
	M	61x7x47/34	20 069	1.200	2580
750 000	I	19x7x14/24	1 862	1.099	2437
	K	19x7x57/30	7 581	1.143	2455
	M	61x7x44/34	18 788	1.160	2415
700 000	I	19x7x13/24	1 729	1.057	2262
	K	19x7x52/30	6 916	1.089	2240
	M	61x7x41/34	17 507	1.117	2251
650 000	I	19x7x12/24	1 596	1.014	2088
	K	19x7x49/30	6 517	1.056	2111
	M	61x7x38/34	16 226	1.074	2086
600 000	I	7x7x30/24	1 470	0.971	1906
	K	19x7x45/30	5 985	1.010	1938
	M	61x7x35/34	14 945	1.028	1921
550 000	I	7x7x28/24	1 372	0.936	1779
	K	19x7x41/30	5 453	0.961	1766
	M	61x7x32/34	13 664	0.981	1757
500 000	I	7x7x25/24	1 225	0.882	1588
	K	19x7x38/30	5 054	0.924	1637
	M	37x7x49/34	12 691	0.900	1631
450 000	I	7x7x23/24	1 127	0.845	1461
	K	19x7x34/30	4 522	0.871	1465
	M	37x7x44/34	11 396	0.892	1465
400 000	I	7x7x20/24	980	0.785	1270
	K	19x7x30/30	3 990	0.816	1292
	M	37x7x39/34	10 101	0.837	1298
350 000	I	7x7x18/24	882	0.743	1143
	K	19x7x26/30	3 458	0.757	1120
	M	37x7x34/34	8 806	0.779	1132
300 000	I	7x7x15/24	735	0.675	953
	K	7x7x61/30	2 989	0.701	959
	M	19x7x57/34	7 581	0.720	975
250 000	I	7x7x13/24	637	0.626	826
	K	7x7x61/30	2 499	0.638	802
	M	19x7x48/34	6 384	0.658	821
4/0	I	19x28/24	532	0.569	683
	K	7x7x43/30	2 107	0.584	676
	M	19x7x40/34	5 320	0.598	684
3/0	I	19x22/24	418	0.502	537
	K	7x7x34/30	1 666	0.516	535
	M	19x7x32/34	4 256	0.532	547
2/0	I	19x18/24	342	0.452	439
	K	7x7x27/30	1 323	0.457	424
	M	19x7x25/34	3 325	0.467	427
1/0	I	19x14/24	266	0.396	342
	K	19x56/30	1 064	0.408	338

(continued)

Table 4 (continued)

Conductor size, circular mils or AWG	Class of strand	Construction and wire size, AWG	Total No. of wires	Approx diameter, in.	Net weight, lb/1000 ft
1/0.............M	M	7x7x54/34	2 646	0.414	337
1.............I	I	7x30/24	210	0.350	267
	K	19x44/30	836	0.359	266
	M	7x7x43/34	2 107	0.368	268
2.............I	I	7x23/24	161	0.304	205
	K	19x35/30	665	0.319	211
	M	7x7x34/34	1 666	0.325	212
3.............I	I	7x19/24	133	0.275	169
	K	19x28/30	532	0.283	169
	M	7x7x27/34	1 323	0.288	168
4.............I	I	7x15/24	105	0.243	134
	K	7x60/30	420	0.250	132
	M	19x56/34	1 064	0.257	134
5.............I	I	7x12/24	84	0.216	107
	K	7x48/30	336	0.223	106
	M	19x44/34	836	0.226	105
6.............I	I	7x9/24	63	0.186	80
	K	7x38/30	266	0.197	84
	M	19x35/34	665	0.201	84
7.............K	K	7x30/30	210	0.174	66
	M	19x28/34	532	0.178	67
8.............K	K	7x/30	168	0.155	53
	M	7x60/34	420	0.158	53
9.............K	K	7x19/30	133	0.137	42
	M	7x48/34	336	0.140	42
10.............M	M	7x37/34	259	0.122	33
12.............M	M	7x24/34	168	0.097	21

coiled. The entire operation is performed in a controlled atmosphere, from the time C10100 cathodes enter the melting furnace until the rod emerges. This rod is generally used for production of fine and ultrafine wires. Production rates up to 2.5 kg/s (10 tons/h) can be obtained.

Hazelett Process. This process was originally developed in 1957 for zinc slab, and was further refined in conjunction with Metallurgic Hoboken Overpelt (Belgium) to make it applicable to copper wire rod. Molten copper is passed between two water-cooled, counter rotating steel belts having specially cooled side dams. The resulting bar is sent through a conventional Krupp rolling mill similar to that used in the Southwire system. Production rates of about 6.3 to 7.5 kg/s (25 to 30 tons/h) have been achieved.

Outokumpu Process. The Outokumpu process is similar to horizontal continuous casting. Metal is pulled through a graphite die where it solidifies. One end of the die extends into the melt and the other is surrounded by a water cooled jacket. The Outokumpu process differs from horizontal continuous casting in that the direction of withdrawal is vertically upwards. A 12-strand plant can produce up to 12.7 Gg (14 000 tons) of oxygen-free rod annually.

Wiredrawing and Wire Stranding

Preparation of Rod. In order to provide a wire of good surface quality, it is necessary to have clean wire rod with a smooth, oxide-free surface. Conventional hot rolled rod must be cleaned in a separate operation, but with the advent of continuous casting, which provides better surface quality, a separate cleaning operation is not required. Instead, the rod passes through a cleaning station as it exits from the rolling mill.

The standard method for cleaning copper wire rod is pickling in hot 20% sulfuric acid followed by rinsing in water. When fine wire is being produced, it is necessary to provide rod of even better surface quality. This can be achieved in a number of ways. One is by open-flame annealing of cold drawn rod—that is, heating to 700 °C (1300 °F) in an oxidizing atmosphere. This eliminates shallow discontinuities. A more common practice, especially for fine magnet-wire applications, is die shaving, in which rod is drawn through a circular cutting die made of steel or carbide to remove approximately 0.13 mm (0.005 in.) from the entire surface of the rod. A further refinement of this cleaning operation for rod made from conventionally cast wirebar involves scalping the top surface of cast wirebar and subsequently die shaving the hot rolled rod.

Wiredrawing. Single-die machines called "bull blocks" are used for drawing special heavy sections such as trolley wire. Drawing speeds range from about 1 to 2.5 m/s (200 to 500 ft/min). Tallow is generally used as the lubricant and the wire is drawn through hardened steel or tungsten carbide dies. In some instances, multiple-draft tandem bull blocks (in sets of 3 or 5 passes) are used instead of single-draft machines.

Tandem drawing machines having 10 to 12 dies for each machine are used for breakdown of hot rolled or continuous cast copper rod. The rod is reduced in diameter from 8.3 mm (0.325 in.) to about 2 mm (0.08 in.) by drawing it through dies at speeds up to 25 m/s (5000 ft/min). The drawing machine operates continuously; the operator merely welds the end of each rod coil to the start of the next coil.

Intermediate and fine wires are drawn on smaller machines that have 12 to 20 or more dies each. The wire is reduced in steps of 20 to 25% in cross-sectional area. Intermediate machines can produce wire as small as 0.5 mm (0.020 in.) in diameter, and fine wire machines can produce wire in diameters from 0.5 mm less than 0.25 mm (0.010 in.). Drawing speeds are typically 25 to 30 m/s (5000 to 6000 ft/min) and may be even higher.

All drawing is performed with a copious supply of lubricant to cool the wire and prevent rapid die wear. Traditional lubricants are soap and fat emulsions, which are fed to all machines from a central reservoir. Breakdown of rod

Table 5 Characteristics of bunch stranded copper conductors having uncoated or tinned members: ASTM B174

Conductor size, AWG	Class of strand	Number and size of wire, AWG	Approx. diameter, in.	Approx. weight, lb/1000 ft
7	I	52/24	0.168	64.9
8	I	41/24	0.148	51.1
9	I	33/24	0.132	41.2
10	I	26/24	0.117	32.4
	J	65/28	0.118	31.9
	K	104/30	0.120	32.1
12	J	41/28	0.093	20.1
	K	65/30	0.094	20.1
	L	104/32	0.096	20.6
14	J	26/28	0.073	12.7
	K	41/30	0.074	12.7
	L	65/32	0.075	12.8
	M	104/34	0.076	12.7
16	J	16/28	0.057	7.84
	K	26/30	0.058	8.03
	L	41/32	0.059	8.10
	M	65/34	0.059	7.97
18	J	10/28	0.044	4.90
	K	16/30	0.045	4.94
	L	26/32	0.046	5.14
	M	41/34	0.046	5.02
20	J	7/28	0.038	3.43
	K	10/30	0.035	3.09
	L	16/32	0.036	3.16
	M	26/34	0.037	3.19

usually requires a lubricant concentration of about 7%; drawing of intermediate and fine wires, concentrations of 2 to 3%. Today, synthetic lubricants are becoming more widely accepted.

Drawn wire is collected on reels or stem packs, depending on the next operation. Fine wire is collected on reels carrying as little as 4.5 kg (10 lb); large-diameter wire, on stem packs carrying up to 450 kg (1000 lb). To ensure continuous operation, many drawing machines are equipped with dual take-up systems. When one reel is filled, the machine automatically flips the wire onto an adjacent empty reel and simultaneously cuts the wire. This permits the operator to unload the full reel and replace it with an empty one without stopping the wiredrawing operation.

Until the early 1970's, hydrostatic extrusion was essentially a batch process and was not considered a competitor to conventional wiredrawing. In 1970, Western Electric patented a "viscous drag machine" that uses a pressurized, flowing viscous fluid to feed wire rod into an extrusion chamber and through a die for continuous extrusion of wire. By forcing the viscous fluid to flow along the surface of the wire rod, shear stresses between fluid and rod are used to move the rod through the die. In 1973, a refinement of this process was announced; in the refined system, shear forces transmitted through a viscous medium were used to feed the rod toward the extrusion die. This process has not yet proved economical enough to be a significant commercial process.

Several newer continuous extrusion processes for production of wire are currently under development. These include the U. K. Atomic Energy Authority helical extrusion process and the Conform process. Data indicate so far that copper wire can be produced successfully by these techniques.

Production of Flat or Rectangular Wire. Depending on size and quantity, flat or rectangular wire is drawn on bull block machines or Turk's Head machines, or is rolled on tandem rolling mills with horizontal and vertical rolls. Larger quantities are produced by rolling, smaller quantities by drawing.

Annealing. Wiredrawing, like any other cold working operation, increases tensile strength and reduces ductility of copper. Although it is possible to cold work copper up to 99% reduction in area, copper wire usually is annealed after 90% reduction.

In some plants, electrical-resistance heating methods are used to fully anneal copper wire as it exits from the drawing machines. Wire coming directly from drawing passes over suitably spaced contact pulleys that carry the electrical current necessary to heat the wire above its recrystallization temperature in less than a second.

In plants where batch annealing is practiced, drawn wire is treated either in a continuous tunnel furnace, where reels travel through a neutral or slightly reducing atmosphere and are annealed during transit, or in batch bell furnaces under a similar protective atmosphere. Annealing temperatures range from 400 to 600 °C (750 to 1100 °F) depending chiefly on wire diameter and reel weight.

Wire Coating. Four basic coatings are used on copper conductors for electrical applications:

1 Lead, or lead alloy
   (80Pb-20Sn) ...... (ASTM B189)
2 Nickel .... (ASTM B355)
3 Silver .... (ASTM B298)
4 Tin .... (ASTM B33)

Coatings are applied to (a) retain solderability for hookup-wire applications, (b) provide a barrier between the copper and insulation materials such as rubber, that would react with the copper and adhere to it (thus making it difficult to strip insulation from the wire to make an electrical connection) or (c) prevent oxidation of the copper during high-temperature service.

Tin-lead alloy coatings and pure tin coatings are the most common; nickel and silver are used for specialty and high-temperature applications.

Copper wire can be coated by hot dipping in a molten metal bath, electroplating or cladding. With the advent of continuous processes, electroplating has become the dominant process, especially because it can be done "on line" following the wiredrawing operation.

Stranded wire is produced by twisting or braiding several wires together to provide a flexible cable. (For a description of various strand constructions, see the section of this article entitled "Classification of Wire and Cable".) Different degrees of flexibility for a given current-carrying capacity can be achieved by varying the number, size and arrangement of individual wires. Solid wire, concentric strand, rope strand and bunched strand provide increasing degrees of flexibility; within the last three categories, a larger number of finer wires provides greater flexibility.

Table 6 Characteristics of tinned solid round copper wire: ASTM B33, B246, B258

Conductor size, AWG	Net weight, lb/1000 ft	Soft (annealed) wire Nominal resistance, Ω/1000 ft	Minimum elonga-tion(a), %	Hard drawn wire Nominal resistance, Ω/1000 ft	Minimum breaking strength, lb
2	200.9	0.1609	25	...	...
3	159.3	0.2028	25	...	...
4	126.3	0.2557	25	0.2680	1773
5	100.2	0.3226	25	0.3380	1432
6	79.44	0.4067	25	0.4263	1152
7	63.03	0.5127	25	0.5372	927.3
8	49.98	0.6465	25	0.6776	743.1
9	39.62	0.8154	25	0.8545	595.1
10	31.43	1.039	20	1.087	476.1
11	24.9	1.31	20	1.37	381.0
12	19.8	1.65	20	1.73	303.0
13	15.7	2.08	20	2.18	241.0
14	12.4	2.62	20	2.74	192.0
15	9.87	3.31	20	3.46	153.0
16	7.81	4.18	20	4.37	121.0
17	6.21	5.26	20		
18	4.92	6.66	20		
19	3.90	8.36	20		
20	3.10	10.6	20		
21	2.46	13.3	20		
22	1.94	16.9	20		
23	1.55	21.1	20		
24	1.22	26.7	15		
25	0.970	34.4	15		
26	0.765	43.5	15		
27	0.610	54.5	15		
28	0.481	69.3	15		
29	0.387	86.1	15		
30	0.303	110.0	10		
31	0.204	141.0	10		
32	0.194	174.0	10		
33	0.153	221.0	10		
34	0.120	281.0	10		

(a) In 10 in.

Stranded copper wire and cable are made on machines known as "bunchers" or "stranders". Conventional bunchers are used for stranding small-diameter wires (34 AWG up to 10 AWG). Individual wires are payed off reels located alongside the equipment and are fed over flyer arms that rotate about the take-up reel to twist the wires. The rotational speed of the arm relative to the take-up speed controls the length of lay in the bunch. For small, portable, flexible cables, individual wires are usually 30 to 34 AWG, and there may be as many as 150 wires in each cable.

A tubular buncher has up to 18 wire-payoff reels mounted inside the unit. Wire is taken off each reel while it remains in a horizontal plane, is threaded along a tubular barrel and is twisted together with other wires by a rotating action of the barrel. At the take-up end, the strand passes through a closing die to form the final bunch configuration. The finished strand is wound onto a reel that also remains within the machine.

Supply reels in conventional stranders for large-diameter wire are fixed onto a rotating frame within the equipment and revolve about the axis of the finished conductor. There are two basic types of machines. In one, known as a rigid-frame strander, individual supply reels are mounted in such a way that each wire receives a full twist for every revolution of the strander. In the other, known as a planetary strander, the wire receives no twist as the frame rotates.

These types of stranders are comprised of multiple bays, with the first bay carrying six reels and subsequent bays carrying increasing multiples of six. The core wire in the center of the strand is payed off externally. It passes through the machine center and individual wires are laid over it. In this manner, strands with up to 127 wires are produced in one or two passes through the machine depending on its capacity for stranding individual wires.

Normally, hard-drawn copper wire is stranded on a planetary machine so that the strand will not be as springy and will tend to stay bunched rather than spring open when it is cut off. The finished product is wound onto a power-driven external reel that maintains a prescribed amount of tension on the stranded wire.

Insulation and Jacketing

Of the three broad categories of insulation—polymeric, enamel and paper-and-oil—polymeric insulation is the most widely used.

Polymeric Insulation. The most common polymers are polyvinyl chloride (PVC), polyethylene, ethylene propylene rubber (EPR), silicone rubber, polytetrafluoroethylene (PTFE) and fluorinated ethylene propylene (FEP). Polymide coatings are used where fire resistance is of prime importance, such as in wiring harnesses for manned space vehicles. Until a few years ago, natural rubber was used, but this has now been supplanted by synthetics such as butyl rubber and EPR. Synthetic rubbers are used wherever good flexibility must be maintained, such as in welding or mining cable.

Many varieties of PVC are made, including several that are flame resistant. PVC has good dielectric strength and flexibility, and is one of the least expensive conventional insulating and jacketing materials. It is used mainly for communication wire, control cable, building wire and low-voltage power cables. PVC insulation is normally selected for applications requiring continuous operation at temperatures up to 75 °C (165 °F).

Polyethylene, because of its low and stable dielectric constant, is specified when better electrical properties are required. It resists abrasion and solvents. It is used chiefly for hookup wire, communication wire and high-voltage cable. Cross-linked polyethylene (XLPE), which is made by adding organic peroxides to polyethylene and then vulcanizing the mixture, yields better heat resistance, better mechanical properties, better aging characteristics, and freedom from environmental stress cracking. Special compounding can provide flame resistance in cross-linked polyethylene. Typical uses include building wire, control cables and power cables. The usual maximum sustained operating temperature is 90 °C (200 °F).

PTFE and FEP are used to insulate jet aircraft wire, electronic equipment wire and specialty control cables, where heat resistance, solvent resistance and high reliability are important. These electrical cables can operate at temperatures up to 250 °C (480 °F).

All of the polymeric compounds are applied over copper conductors by hot extrusion. The extruders are machines that convert pellets or powders of thermoplastic polymers into continuous covers. The insulating compound is loaded into a hopper that feeds it into a long, heated chamber. A continuously revolving screw moves the pellets into the hot zone, where the polymer softens and becomes fluid. At the end of the chamber, molten compound is forced out through a small die over the moving conductor, which also passes through the die opening. As the insulated conductor leaves the extruder it is water cooled and taken up on reels. Cables jacketed with EPR and XLPE go through a vulcanizing chamber prior to cooling to complete the cross-linking process.

Enamel Film Insulation. Film-coated wire, usually fine magnet wire, is composed of a metallic conductor coated with a thin, flexible enamel film. These insulated conductors are used for electromagnetic coils in electrical devices, and must be capable of withstanding high breakdown voltages. Temperature ratings range from 105 to 220 °C (220 to 425 °F), depending on enamel composition. The most commonly used enamels are based on polyvinyl acetals, polyesters and epoxy resins.

Equipment for enamel coating of wire often is custom built, but standard lines are available. Basically, systems are designed to insulate large numbers of wires simultaneously. Wires are passed through an enamel applicator that deposits a controlled thickness of liquid enamel onto the wire. Then the wire travels through a series of ovens to cure the coating, and finished wire is collected on spools. In order to build up a heavy coating of enamel, it may be necessary to pass wires through the system several times. In recent years, some manufacturers have experimented with powder-coating methods. These avoid evolution of solvents, which is characteristic of curing conventional enamels, and thus make it easier for the manufacturer to meet OSHA and EPA standards. Electrostatic sprayers, fluidized beds and other experimental devices are used to apply the coatings.

Paper-and-oil Insulation. Cellulose is one of the oldest materials for electrical insulation and is still used for certain applications. Oil-impregnated cellulose paper is used to insulate high-voltage cables for critical power-distribution applications. The paper, which may be applied in tape form, is wound helically around the conductors using special machines in which six to twelve paper-filled pads are held in a cage that rotates around the cable. Paper layers are wrapped alternately in opposite directions, free of twist. Paper-wrapped cables then are placed inside special impregnating tanks to fill the pores in the paper with oil and to ensure that all air has been expelled from the wrapped cable.

The other major use of paper insulation is for flat magnet wire. In this application, magnet-wire strip (with a width-to-thickness ratio greater than 50 to 1) is helically wrapped with one or more layers of overlapping tapes. These may be bonded to the conductor with adhesives or varnishes. The insulation provides highly reliable mechanical separation under conditions of electrical overload.

Properties of Wrought Coppers and Copper Alloys

By the ASM Committee on Copper
and Copper Alloys*

C10100, C10200

Commercial Names

Previous trade names. C10100: Oxygen-free electronic copper. C10200: Oxygen free copper
Common name. Oxygen-free copper
Designations. C10100: OFE. C10200: OF

Specifications

ASTM. C10100. Flat products: B48, B133, B152, B187, B272, B432, F68. Pipe: B42, B188, F68. Rod: B12, B49, B133, B187, F68. Shapes: B133, B187, F68. Tubing: B372, B68, B75, B188, B280, F68. Wire: B1, B2, B3, F68. **C10200.** Flat products: B48, B133, B152, B187, B272, B370, B432. Pipe: B42, B188, Rod: B12, B49, B124, B133, B187. Tubing: B68, B75, B88, B111, B188, B280, B359, B372, B395, B447. Wire: B1, B2, B3, B33, B47, B116, B189, B246, B286, B298, B355. Shapes: B124, B133, B187

*See page **XI** for committee list.

Government. C10100. Rod: QQ-C-502. **C10200.** Flat products: QQ-C-576. Rod and shapes: QQ-C-502. Tubing: WW-T-775. Wire: QQ-C-502, QQ-W-343, MIL-W-3318

Chemical Composition

Composition limits. C10100: 99.99 min Cu (There are specific limits in ppm for 17 named elements; refer to ASTM B170 or CDA Standards Handbook.) C10200: 99.95 min Cu + Ag
Consequence of exceeding impurity limits. C10100 and C10200 are high-conductivity electrolytic coppers produced without use of metal or metalloid deoxidizers. Excessive amounts of impurities reduce conductivity. Excessive oxygen causes the metal to fail the ASTM B170 bend test after being heated 30 min at 850 °C in pure hydrogen.

Applications

Typical uses. Bus bars, waveguides, lead-in wire, anodes, vacuum seals, transistor components, glass-to-metal seals, coaxial cables, klystrons, microwave tubes
Precautions in use. Avoid heating in oxidizing atmospheres.

Mechanical Properties

Tensile properties. See Table 1, Fig. 1 and Fig. 2.
Shear strength. See Table 1.
Hardness. See Table 1.
Elastic modulus. Tension, 115 GPa (17×10^6 psi); shear, 44 GPa (6.4×10^6 psi)
Impact resistance. See Fig. 2.
Fatigue strength. See Table 1.
Creep-rupture characteristics. See Tables 2 and 3.

Mass Characteristics

Density. 8.94 Mg/m³ (0.323 lb/in.³) at 20 °C (68 °F)

Thermal Properties

Melting point. 1083 °C (1981 °F)
Coefficient of thermal expansion. Linear, 17.0 μm/m·K (9.4 μin./in.·°F) at 20 to 100 °C (68 to 212 °F); 17.3 μm/m·K (9.6 μin./in. ·°F) at 20 to 200 °C (68 to 392 °F); 17.7 μm/m·K (9.8

Table 1 Typical mechanical properties of C10100 and C10200

Temper	Tensile strength MPa	ksi	Yield strength(a) MPa	ksi	Elongation in 50 mm or 2 in., %	Hardness HRF	HRB	HR30T	Shear strength MPa	ksi	Fatigue strength(b) MPa	ksi
Flat Products, 1-mm (0.04-in.) Thick												
M20	235	34	69	10	45	45	...	...	160	23	...	...
OS025	235	34	76	11	45	45	...	...	160	23	76	11
OS050	220	32	69	10	45	40	...	...	150	22	...	...
H00	250	36	195	28	30	60	10	25	170	25	...	...
H01	260	38	205	30	25	70	25	36	170	25	...	...
H02	240	42	250	36	14	84	40	50	180	26	90	13
H04	345	50	310	45	6	90	50	57	195	28	90	13
H08	380	55	345	50	4	94	60	63	200	29	95	14
H10	395	57	360	53	4	95	62	64	200	29	...	...
Flat Products, 6-mm (0.25-in.) Thick												
M20	220	32	69	10	50	40	...	...	150	22	...	...
OS050	220	32	69	10	50	40	...	...	150	22	...	...
H00	250	36	195	28	40	60	10	...	170	25	...	...
H01	260	38	205	30	35	70	25	...	170	25	...	...
H04	345	50	310	45	12	90	50	...	195	28	...	...
Flat Products, 25-mm (1-in.) Thick												
H04	310	45	275	40	20	85	45	...	180	26	...	...
Rod, 6-mm (0.25-in.) Diam												
H80 (40%)	380	55	345	50	10	94	80	...	200	29	...	...
Rod, 25-mm (1-in.) Diam												
M20	220	32	69	10	55(c)	40	...	...	150	22	...	...
OS050	220	32	69	10	55(c)	40	...	...	150	22	...	...
H80 (35%)	330	48	305	44	16(d)	87	47	...	185	27	115	17
Rod, 50-mm (2-in.) Diam												
H80 (16%)	310	45	275	40	20	85	45	...	180	26	...	...
Wire, 2-mm (0.08-in.) Diam												
OS050	240	35	...	...	35(e)	45	...	...	165	24	...	...
H04	380	55	...	...	1.5(f)	...	...	...	200	29	...	...
H08	455	66	...	...	1.5(f)	...	...	...	230	33	...	...
Tubing, 25-mm (1-in.) OD by 1.65 mm (0.065 in.) Wall Thickness												
OS025	235	34	76	11	45	45	...	...	160	23	...	...
OS050	220	32	69	10	45	40	...	...	150	22	...	...
H55 (15%)	275	40	220	32	25	77	35	45	180	26	...	...
H80 (40%)	380	55	345	50	8	95	60	63	200	29	...	...
Shapes, 13-mm (0.50-in.) Diam												
M20	220	32	69	10	50	45	...	...	150	22	...	...
M30	220	32	69	10	50	45	...	...	150	22	...	...
OS050	220	32	69	10	50	45	...	...	150	22	...	...
H80 (15%)	275	40	220	32	30	...	35	...	180	26	...	...

(a) At 0.5% extension under load. (b) At 10^8 cycles. (c) 70% reduction in area. (d) 55% reduction in area. (e) Elongation in 254 mm (10 in.). (f) Elongation in 150 cm (60 in.).

μin./in. ·°F) at 20 to 300 °C (68 to 572 °F)

Specific heat. 385 J/kg·K (0.092 Btu/lb·°F) at 20 °C (68 °F)

Thermal conductivity. 391 W/m·k (226 Btu/ft·h·°F) at 20 °C (68 °F)

Electrical Properties

Electrical conductivity. Annealed: volumetric, 101% IACS min at 20 °C (68 °F)

Electrical resistivity. 17.1 nΩ·m at 20 °C (68 °F)

Chemical Properties

General corrosion behavior. Copper is cathodic to hydrogen in the electromotive series and therefore is the cathode in galvanic couples with other base metals such as iron, aluminum, magnesium, lead, tin and zinc. C10100 and C10200 have excellent resistance to atmospheric corrosion and to corrosion by most waters, including brackish water and seawater. They have good resistance to nonoxidizing acids but poor resistance to oxidizing acids, moist ammonia, moist halogens, sulfides, and solutions containing ammonium ions.

Fabrication Characteristics

Machinability. 20% of C36000, free-cutting brass

Forgeability. 65% of C37700, forging brass

Formability. Readily formed by a wide variety of hot and cold methods.

Table 2 Creep properties of C10100 and C10200

Condition and grain size	Test temperature °C	°F	Stress(a) for creep rate of: 10^{-6}%/h MPa	ksi	10^{-5}%/h MPa	ksi	10^{-4}%/h MPa	ksi	10^{-3}%/h MPa	ksi	10^{-2}%/h MPa	ksi	10^{-1}%/h MPa	ksi
OS025(b)	43	110	...	...	...	...	...	...	170	25	185	27	200	29
	120	250	...	...	...	...	...	...	125	18	150	22	165	24
	150	300	11	1.6	25	3.6	55	8.0	110	16	130	19	150	22
	205	400	3	0.5	10	1.5	33	4.8	...	...	...	...	...	...
	260	500	0.7	0.1	3	0.4	12	1.7	...	...	...	...	...	...
	370	700	...	...	...	...	...	...	...	...	21	3.1	(40)	(5.8)
	480	900	...	...	...	...	...	...	...	...	9.9	1.45	(23)	(3.3)
Cold drawn 40%(c) ...	43	110	...	...	...	...	...	...	310	45	330	48	...	...
	120	250	...	...	...	...	...	...	240	35	270	39	(295)	(43)
	150	300	...	...	...	...	...	...	200	29	235	34	250	36
	370	700	...	...	...	...	...	...	11	1.6	26	3.8	(39)	(5.6)
	480	900	...	...	...	...	...	...	...	...	8.3	1.2	(17)	(2.4)
	650	1200	...	...	...	...	...	...	...	...	3	0.5	6	0.9
Cold drawn 84%(d) ...	150	300	...	...	55	8.0	89.6	13.0	...	...	...	...	...	...
	205	400	(4.5)	(0.65)	12	1.7	35	5.0	...	...	...	...	...	...

(a) Parentheses indicate extrapolated values. (b) Tensile strength, 220 MPa (31.9 ksi) at 21 °C (70 °F). (c) Tensile strength, 352 MPa (51.1 ksi) at 21 °C. (d) Tensile strength, 376 MPa (54.5 ksi) at 21 °C.

Can be easily stamped, bent, coined, sheared, spun, upset, swaged, forged, roll threaded and knurled.

Weldability. Can be readily soldered, brazed, gas tungsten-arc welded, gas metal-arc welded or upset welded. Its capacity for being oxyfuel gas welded is fair. Shielded metal-arc welding and most resistance welding methods are not recommended.

Annealing temperature. 375 to 650 °C (700 to 1200 °F)

Hot working temperature. 750 to 875 °C (1400 to 1600 °F)

Table 3 Stress-rupture properties of C10100 and C10200

Temper or condition	Test temperature °C	°F	Stress(a) for rupture in: 10 h MPa	ksi	100 h MPa	ksi	1000 h MPa	ksi
OS025(b)	150	300	...	...	161	23.4	147	21.3
	200	380	...	...	130	18.9	106	15.3
Cold drawn 40%(c)	120	250	...	...	272	39.4	(245)	(35.6)
	150	300	...	...	241	35.0	(215)	(31.2)
H80(d)	450	840	33	4.8	17	2.4	...	...
	650	1200	9.7	1.4	5.2	0.75	...	...

(a) Parentheses indicate extrapolated values. (b) Tensile strength, 238 MPa (34.5 ksi) at 21 °C (70 °F). (c) Tensile strength, 352 MPa (51.1 ksi) at 21 °C. (d) Tensile strength, 426 MPa (61.8 ksi) at 21 °C.

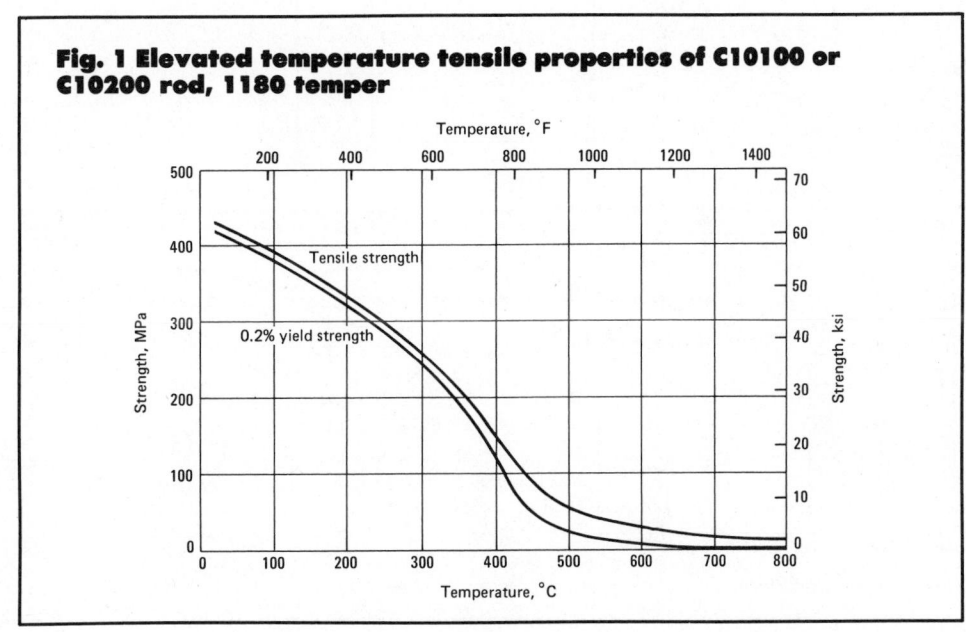

Fig. 1 Elevated temperature tensile properties of C10100 or C10200 rod, 1180 temper

Fig. 1 (continued)

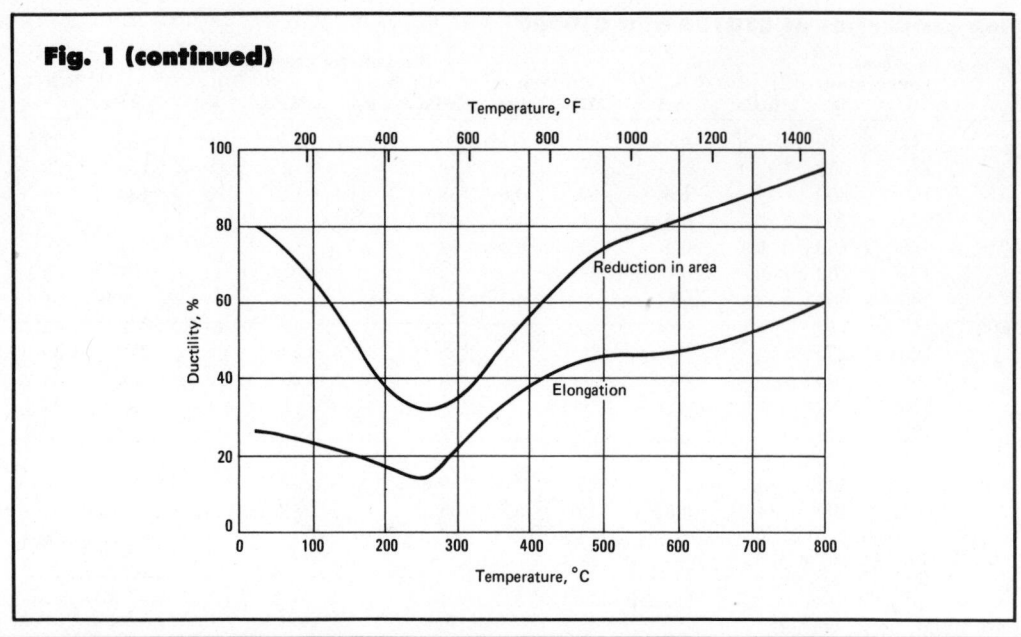

Fig. 2 Low temperature mechanical properties of C10100 or C10200 bar

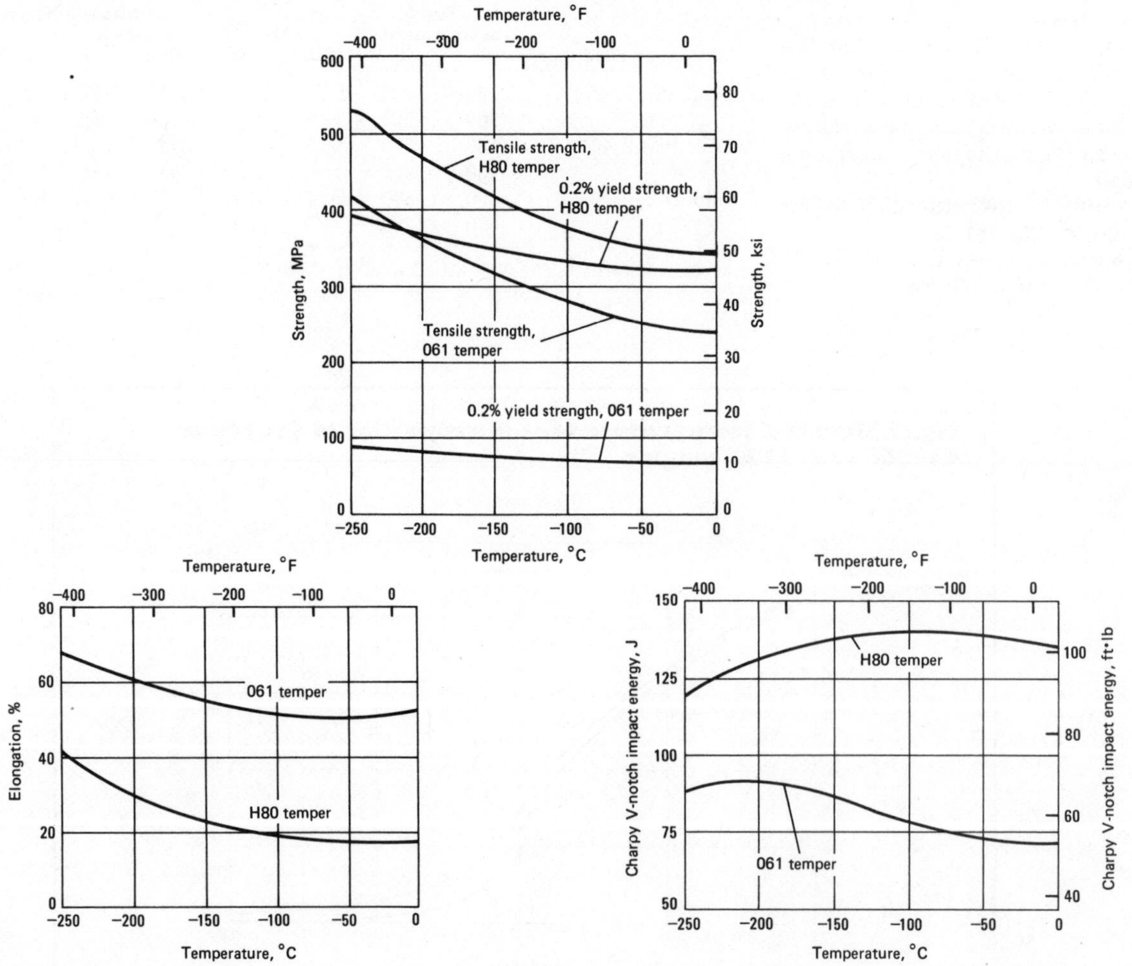

C10300

Commercial Names

Common name. Oxygen-free extra low phosphorus copper
Designation. OFXLP

Specifications

ASTM. Flat products: B133, B152, B187, B272, B432. Pipe: B42, B188, B302, Rod: B12, B133, B187. Shapes: B133, B187. Tubing: B68, B75, B88, B111, B188, B251, B280, B306, B359, B372, B395, B447

Chemical Composition

Composition limits. 99.95 min Cu + Ag + P; 0.001 to 0.005 P; 0.05 max others (total)

Applications

Typical uses. Bus bars, electrical conductors and terminals, commutators, tubular bus bars, clad products, waveguide tubing, thermostatic control tubing

Mechanical Properties

Tensile properties. See Table 4.
Shear strength. See Table 4.
Hardness. See Table 4.
Elastic modulus. Tension, 115 GPa (17×10^6 psi); shear, 44 GPa (6.4×10^6 psi).
Fatigue strength. 1-mm (0.04-in.) thick strip: OS025 temper, 76 MPa (11 ksi); H02 or H04 temper, 90 MPa (13 ksi).

Mass Characteristics

Density. 8.94 Mg/m^3 (0.323 $lb/in.^3$) at 20 °C (68 °F)

Thermal Properties

Liquidus temperature. 1083 °C (1981 °F)
Solidus temperature. 1083 °C (1981 °F)

Coefficient of thermal expansion. Linear, 17.0 μm/m·K (9.4 μin./in.· °F) at 20 to 100 °C (68 to 212 °F); 17.3 μm/m·K (9.6 μin./in.· °F) at 20 to 200 °C (68 to 392 °F); 17.7 μm/m·K (9.8 μin./in.·°F) at 20 to 300 °C (68 to 572 °F)
Specific heat. 385 J/kg·K (0.092 Btu/lb.· °F) at 20 °C (68 °F)
Thermal conductivity. 386 W/m·K (223 Btu/ft·h· °F) at 20 °C (68 °F)

Electrical Properties

Electrical conductivity. O61 temper: volumetric, 99% IACS at 20 °C (68 °F)
Electrical resistivity. 17.4 nΩ·m at 20 °C (68 °F)

Fabrication Characteristics

Machinability. 20% of C36000, free-cutting brass
Annealing temperature. 375 to 650 °C (700 to 1200 °F)
Hot working temperature. 750 to 875 °C (1400 to 1600 °F)

Table 4 Typical mechanical properties of C10300

Temper	Tensile strength MPa	ksi	Yield strength(a) MPa	ksi	Elongation(b), %	HRF	Hardness HRB	HR30T	Shear strength MPa	ksi
Flat Products, 1-mm (0.04-in.) Thick										
OS050	220	32	69	10	45	40	...	...	150	22
OS025	235	34	76	11	45	45	...	...	160	23
H00	250	36	195	28	30	60	10	25	170	25
H01	260	38	205	30	25	70	25	36	170	25
H02	290	42	250	36	14	84	40	50	180	26
H04	345	50	310	45	6	90	50	57	195	28
Flat Products, 6-mm (0.25-in.) Thick										
OS050	220	32	69	10	50	40	...	...	150	22
H00	250	36	195	28	40	60	10	...	170	25
H04	345	50	310	45	12	90	50	...	195	28
M20	220	32	69	10	50	40	...	...	150	22
Flat Products, 25-mm (1-in.) Thick										
H04	310	45	275	40	20	85	45	...	180	26
Rod, 6-mm (0.25 in.) Diam										
H80 (40%)	380	55	345	50	20	94	60	...	200	29
H80 (35%)	330	48	305	44	16	87	47	...	185	27
H80 (16%)	310	45	275	40	20	85	45	...	180	26
Tubing, 25-mm (1-in.) OD by 1.65-mm (0.65-in.) Wall										
OS050	220	32	69	10	45	40	...	...	150	22
OS025	235	34	76	11	45	45	...	...	160	23
H80 (15%)	275	40	220	32	25	77	35	45	180	26
H80 (40%)	380	55	345	50	8	95	60	63	200	29
Pipe, ¾ SPS										
H80 (30%)	345	50	310	45	10	90	50	...	195	28
Shapes, 13-mm (0.50-in.) Section Size										
OS050	220	32	69	10	50	40	...	...	150	22
H80 (15%)	275	40	220	32	30	...	35	...	180	26

(a) At 0.5% extension under load. (b) In 50 mm or 2 in. (c) at 10^8 cycles.

Table 5 Summary of ASTM and government specifications for C10400, C10500 and C10700

Mill product	C10400	C10500	C10700
ASTM Numbers			
Flat products	B48, B133, B152, B187, B272	B152, B187, B272	B152, B187, B272
Pipe	B42, B188	B188	B188
Rod	B12, B49, B133, B187	B12, B49, B133, B187	B12, B49, B133, B187
Shapes	B133, B187	B187	B187
Tube	B188	B188	B188
Wire	B1, B2, B3	B1, B2, B3	B1, B2, B3
Government Numbers			
Flat products	QQ-C-502, QQ-C-576	QQ-C-502, QQ-C-576	QQ-C-576
Rod	QQ-C-502	QQ-C-502	QQ-C-502
Shapes	QQ-C-502, QQ-B-825	QQ-C-502	QQ-B-825, MIL-B-19231
Tube	QQ-B-825	QQ-B-825	QQ-B-825
Wire	QQ-W-343, MIL-W-3318	QQ-W-343, MIL-W-3318	QQ-W-343, MIL-W-3318

Table 6 Typical mechanical properties of C10400, C10500 and C10700

Temper	Tensile strength MPa	ksi	Yield strength(a) MPa	ksi	Elongation(b), %	Hardness HRF	HRB	HR30T	Shear strength MPa	ksi
Flat Products, 1-mm (0.04-in.) Thick										
OS025	235	34	76	11	45	45	...	...	160	23
H00	250	36	195	28	30	60	10	25	170	25
H01	260	38	205	30	25	70	25	36	170	25
H02	290	42	250	36	14	84	40	50	180	26
H04	345	50	310	45	6	90	50	57	195	28
H08	380	55	345	50	4	94	60	63	200	29
H10	395	57	365	53	4	95	62	64	200	29
M20	235	34	69	10	45	45	...	...	160	23
Flat Products, 6-mm (0.25-in.) Thick										
OS050	220	32	69	10	50	40	...	...	150	22
H00	250	36	195	28	40	60	10	...	170	25
H01	260	38	205	30	35	70	25	...	170	25
H04	345	50	310	45	12	90	50	...	195	28
M20	220	32	69	10	50	40	...	...	150	22
Flat Products, 25-mm (1-in.) Thick										
H04	310	45	275	40	20	85	45	...	180	26
Rod, 6-mm (0.25-in.) Diam										
H80 (40%)	380	55	345	50	10	94	60	...	200	29
Rod, 25-mm (1-in.) Diam										
OS050	220	32	69	10	55	40	...	...	150	22
H80 (35%)	330	48	305	44	16	87	47	...	185	27
M20	220	32	69	10	55	40	...	...	150	22
Rod, 50-mm (2-in.) Diam										
H80 (16%)	310	45	275	40	20	85	45	...	180	26
Wire, 2-mm (0.08-in.) Diam										
OS050	240	35	...	...	35(c)	...	...	...	165	24
H04	380	55	...	...	1.5(d)	...	...	...	200	29
H08	455	66	...	...	1.5(d)	...	...	...	230	33
Shapes, 13-mm (0.50-in.) Section Size										
OS050 mm	220	32	69	10	50	40	...	...	150	22
H80 (15%)	275	40	220	32	30	...	35	...	180	26
M20	220	32	69	10	50	40	...	...	150	22
M30	220	32	69	10	50	40	...	...	150	22
Tubing, 25-mm (1.0-in.) Diam by 1.65-mm (0.065-in.) Wall										
OS050	220	32	69	10	45	40	...	...	150	22
OS025	235	34	76	11	45	45	...	...	160	23
H80 (15%)	275	40	220	32	25	77	35	45	180	26
H80 (50%)	380	55	345	50	8	95	60	62	200	29

(a) At 0.5% extension under load. (b) In 50 mm or 2 in. (c) Elongation in 25 mm (10 in.) (d) Elongation in 150 cm (60 in.).

C10400, C10500, C10700

Commercial Names

Trade name. AMSIL copper
Common name. Oxygen-free silver copper

Specifications

ASTM. See Table 5.

Chemical Composition

Composition limits. C10400: 99.95 min Cu + Ag; 0.027 min Ag. C10500: 99.95 min Cu + Ag; 0.034 min Ag. C10700: 99.95 min Cu + Ag; 0.085 min Ag.

Applications

Typical uses. Bus bars, conductivity wire, contacts, radio parts, windings, switches, commutator segments, automotive gaskets and radiators, chemical-plant equipment, printing rolls, printed-circuit foil. Many uses are based on the good creep strength at elevated temperatures and the high softening temperature of these alloys.

Mechanical Properties

Tensile properties. See Table 6 and Fig. 3.
Shear strength. See Table 6.
Hardness. See Table 6.
Elastic modulus. Tension, 115 GPa (17×10^6 psi); shear, 44 GPa (6.4×10^6 psi)

Mass Characteristics

Density. 8.94 Mg/m³ (0.323 lb/in.³) at 20 °C (68 °F)

Thermal Properties

Liquidus temperature. 1083 °C (1981 °F)
Solidus temperature. 1083 °C (1981 °F)
Coefficient of thermal expansion. Linear, 17.0 μm/m·K (9.4 μin./·°F) at 20 to 100 °C (68 to 212 °F); 17.3 μm/m·K (9.6 μin./in.· °F) at 20 to 200 °C (68 to 392 °F); 17.7 μm/m·K (9.8 μin./in.· °F) at 20 to 300 °C (68 to 572 °F)
Specific heat. 385 J/kg·K (0.092 Btu/1b· °F) at 20 °C (68 °F)
Thermal conductivity. 388 W/m·K (224 Btu/ft·h· °F) at 20 °C (68 °F)

Electrical Properties

Electrical conductivity. O61 temper: volumetric, 100% IACS at 20 °C (68 °F)
Electrical resistivity. 17.2 nΩ·m at 20 °C (68 °F)

Fabrication Characteristics

Machinability. 20% of C36000, free-cutting brass
Annealing temperature. 475 to 750 °C (900 to 1400 °F). See also Fig. 3.
Hot working temperature. 750 to 875 °C (1400 to 1600 °F)

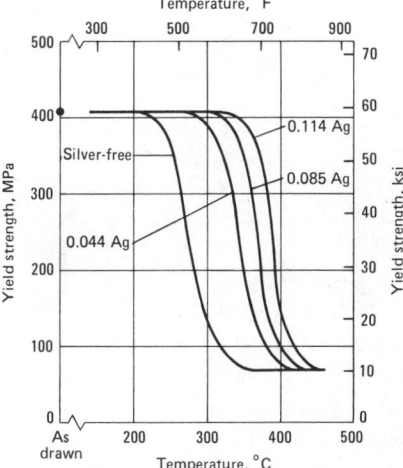

Fig. 3 Softening characteristics of oxygen-free copper containing various amounts of silver

Data are for copper wire cold worked 90% to 2 mm (0.08 in.) diam, and then annealed ½ h at various temperatures.

C10800

Commercial Names

Trade name. AMAX-LP copper
Common name. Oxygen-free low phosphorus copper

Specifications

ASTM. Flat products: B113, B152, B187, B432. Pipe: B42, B302. Rod: B12, B133. Shapes: B133. Tubing: B68, B75, B88, B111, B188, B251, B280, B306, B357, B360, B372, B395, B447, B543

Chemical Composition.

Composition limits. 99.95 min Cu + Ag + P; 0.005 to 0.012 P

Applications

Typical uses. Refrigerator and air-conditioner tubing and terminals, commutators, clad products, gas and burner lines and units, oil-burner tubes, condenser and heat-exchanger tubes, pulp and paper lines, steam and water lines, tank gage lines, plumbing pipe and tubing, thermostatic control tubing, plate for welded continuous casting molds, tanks, kettles, rotating bands and similar uses.

Mechanical Properties

Tensile properties. See Table 7.
Shear strength. See Table 7.
Hardness. See Table 7.
Elastic modulus. Tension, 115 GPa (17×10^6 psi); shear, 44 GPa (6.4×10^6 psi)
Fatigue strength. See Table 7.

Mass Characteristics

Density. 8.94 Mg/m³ (0.323 1b/in.³) at 20 °C (68 °F)

Thermal Properties

Liquidus temperature. 1083 °C (1981 °F)
Solidus temperature. 1083 °C (1981 °F)
Coefficient of thermal expansion. Linear, 17.0 μm/m·K (9.4 μin./in.· °F) at 20 to 100 °C (68 to 212 °F); 17.3 μm/m·K (9.6 μin./in.· °F) at 20 to 200 °C (68 to 392 °F); 17.7 μm/m·K (9.8 μin./in.· °F) at 20 to 300 °C (68 to 572 °F)
Specific heat. 385 J/kg·K (0.092 Btu/1b· °F) at 20 °C (68 °F)
Thermal conductivity. 350 W/m·K (202 Btu/ft·h· °F) at 20 °C (68 °F)

Electrical Properties

Electrical conductivity. O61 temper: volumetric, 92% IACS at 20 °C (68 °F)
Electrical resistivity. 18.7 nΩ·m at 20 °C (68 °F)

Table 7 Typical mechanical properties of C10800

Temper	Tensile strength MPa	ksi	Yield strength(a) MPa	ksi	Elongation(b), %	Hardness HRF	HRB	HR30T	Shear strength MPa	ksi	Fatigue strength(c) MPa	ksi
Flat Products, 1-mm (0.04-in.) Thick												
OS025	235	34	76	11	45	45	...	...	160	23	76	11
H00	250	36	195	28	30	60	10	25	170	25	...	...
H01	260	38	205	30	25	70	25	36	170	25	...	...
H02	290	42	250	36	14	84	40	50	180	26	90	13
H04	345	50	310	45	6	90	50	57	195	28	90	13
H08	380	55	345	50	4	94	60	63	200	29	97	14
Flat Products, 6-mm (0.25-in.) Thick												
OS050	220	32	69	10	50	40	...	...	150	22	...	...
H00	250	36	195	28	40	60	10	...	170	25	...	...
H04	345	50	310	45	12	90	50	...	195	28	...	...
M20	220	32	69	10	50	40	...	...	150	22	...	...
Flat Products, 25-mm (1-in.) Thick												
H04	310	45	275	40	20	85	45	...	180	26	...	...
Rod, 6-mm (0.25-in.) Diam												
H80 (40%)	380	55	345	50	20	94	60	...	200	29	...	...
Rod, 25-mm (1-in.) Diam												
H80 (35%)	330	48	305	44	16	87	47	...	185	27	115	17
Rod, 50-mm (2-in.) Diam												
H80 (16%)	310	45	275	40	20	85	45	...	180	26	...	...
Tubing, 25-mm (1-in.) OD by 1.65-mm (0.065-in.) Wall												
OS050	220	32	69	10	45	40	...	...	150	22	...	...
OS025	235	34	76	11	45	45	...	...	160	23	...	...
H55 (15%)	275	40	220	32	25	77	35	45	180	26	...	...
H80 (40%)	380	55	345	50	8	95	60	63	200	29	...	...
Pipe, ¾ SPS												
H80 (30%)	345	50	310	45	10	90	50	...	195	28	...	...

(a) At 0.5% extension under load. (b) In 50 mm or 2 in. (c) At 10^8 cycles.

Fabrication Characteristics

Machinability. 20% of C36000, free-cutting brass

Annealing temperature. 375 to 650 °C (700 to 1200 °F)

Hot working temperature. 750 to 875 °C (1400 to 1600 °F)

C11000
99.95Cu-0.040

Commercial Names

Common name. Electrolytic tough pitch copper

Designation. ETP

Specifications

AMS. Sheet and strip: 4500. Wire: 4701

ASME. Plate for locomotive fireboxes: SB11. Rod for locomotive staybolts: SB12

ASTM. See Table 8.

SAE. J463

Government. Federal specifications: See Table 8. Military specifications: Rod, MIL-C-12166; wire, MIL-W-3318, MIL-W-6712

Chemical Composition

Composition limits. 99.90 min Cu (silver counted as copper)

Consequence of exceeding impurity limits. *Silver* has little effect on mechanical and electrical properties, but does raise the recrystallization temperature and tends to produce a fine-grained copper. *Iron,* as present in commercial copper, has no effect on mechanical properties, but even traces of iron may cause C11000 to be slightly ferromagnetic. *Sulfur* causes spewing and unsoundness, and is kept below 0.003% in ordinary refinery practice. *Selenium* and *tellurium* are usually considered undesirable impurities, but may be added to improve machinability. *Bismuth* creates brittleness in amounts greater than 0.001%. *Lead* should not be present in amounts greater than 0.005% if copper is to be hot rolled. *Cadmium* is rarely present: its effect is to toughen copper without much loss in conductivity. *Arsenic* decreases the conductivity of copper noticeably, although it is often added intentionally to copper not used in electrical service because it increases the toughness and heat resistance of the metal. *Antimony* is sometimes added to copper when high recrystallization temperature is desired.

Applications

Typical uses. Produced in all forms except pipe, and used for building fronts, downspouts, flashing, gutters, roofing, screening, spouting, gaskets, radiators, bus bars, electrical wire, stranded conductors, contacts, radio parts, switches, terminals, ball floats, butts, cotter pins, nails, rivets, soldering copper, tacks, chemical process equipment, kettles, pans, printing rolls, rotating bands, road-bed expansion plates, vats

Precautions in use. C11000 is subject to embrittlement when heated at 370 °C (700 °F) or above in a reducing atmosphere, as in annealing, brazing or welding. If hydrogen or carbon monoxide is present in the reducing atmosphere, embrittlement can be rapid.

Mechanical Properties

Tensile properties. See Table 9 and Fig. 4 to 8.
Shear strength. See Table 9.
Hardness. See Table 9 and Fig. 9.
Poisson's ratio. 0.33
Elastic modulus. O60 temper: tension, 115 GPa (17×10^6 psi); shear, 44 GPa (6.4×10^6 psi). Cold worked (H) tempers: tension, 115 to 130 GPa (17×10^6 to 19×10^6 psi); shear, 44 to 49 GPa (6.4×10^6 to 7.1×10^6 psi)
Impact strength. See Table 10.
Fatigue strength. See Table 9; values shown there are typical of all tough pitch, oxygen-free, phosphorus deoxidized and arsenical coppers. Copper does not exhibit an "endurance limit" under fatigue loading and, on the average, will fracture in fatigue at the stated number of cycles when subjected to an alternating stress equal to the corresponding fatigue strength, see Fig. 10.
Creep-rupture characteristics. See Table 11.
Specific damping capacity. The damping capacity of coppers and brasses depends on the amplitude and, in some instances, on the frequency of vibration; it is also affected by the condition of the metal. Up to a point, damping capacity increases with increasing cold work; for instance, the damping capacity of 70-30 brass has been reported to increase for reductions up to 60%. When subjected to the same conditions, coppers have about three times the damping capacity of C21000 or C22000. A specific damping capacity of 5×10^{-5} has been recorded for single-crystal annealed copper. Log decrement: O60 temper, 3.2; cold rolled (H) tempers, 5.0
Coefficient of friction. Values given below apply to any of the unalloyed coppers in contact with the indicated materials without lubrication of any kind between the contacting surfaces.

Table 8 ASTM and federal specifications for C11000

Product and condition	Specification number	
	ASTM	Federal
Flat Products		
General requirements for Cu and Cu alloy plate, sheet, strip and rolled bar	B248	...
Sheet, strip, plate and rolled bar	B152	QQ-C-576
Sheet, lead coated	B101	...
Sheet and strip for building construction	B370	...
Strip and flat wire	B272	QQ-C-502
Foil, strip and sheet for printed circuits	B451	...
Rod, Bar and Shapes		
General requirements for Cu and Cu alloy rod, bar and shapes	B249	...
Rod, bar and shapes	B133	QQ-C-502, QQ-C-576
Rod, hot rolled	B49	...
Rod, bar and shapes for forging	B124	QQ-C-502
Bus bars, rods and shapes	B187	QQ-B-825
Wire		
General requirements for Cu and Cu alloy wire	B250	...
Hard drawn	B1	QQ-W-343
tinned	B246	...
Medium hard drawn	B2	QQ-W-343
tinned	B246	...
Soft	B3	QQ-W-343
Lead-alloy coated	B189	...
Nickel-coated	B355	...
Rectangular and square	B48, B272	...
Tinned	B33	...
Silver coated	B298	...
Trolley	B47, B116	...
Conductors		
Bunch stranded	B174	...
Concentric-lay stranded	B8, B226, B496	...
Conductors for electronic equipment	B286, B470	...
Rope-lay stranded	B172, B173	...
Composite conductors (Cu plus Cu-clad steel)	B229	...
Tubular Products		
Bus pipe and tube	B188	QQ-B-825
Pipe	...	WW-P-377
Welded copper tube	B447	...
Miscellaneous		
Standard classification of coppers	B224	...
Electrolytic Cu wirebars, cakes, slabs, billets, ingots and ingot bars	B5	...
Anodes	...	QQ-A-673
Die forgings	B283	...

Opposing material	Coefficient of friction	
	Static	Sliding
Carbon steel	0.53	0.36
Cast Iron	1.05	0.29
Glass	0.68	0.53

Mass Characteristics

Density. Solid: 8.89 Mg/m³ (0.321 lb/in.³) at 20 °C (68 °F); 8.32 Mg/m³ (0.301 lb/in.³) at 1083 °C (1981 °F); see also Fig. 11. Liquid: 7.93 Mg/m³ (0.286 lb/in.³) at 1083 °C

Thermal Properties

Liquidus temperature. 1083 °C (1981 °F)
Solidus temperature. Eutectic point, 1065 °C (1950 °F)
Coefficient of thermal expansion. Linear, 17 μm/m·K (9.4 μin./in.·°F) at 20 to 100 °C (68 to 212 °F); 17.3 μm/m·K (9.6 μin./in.·°F) at 20 to 200 °C (68 to 392 °F); 17.7 μm/m·K

Table 9 Typical mechanical properties of C11000

Temper	Tensile strength MPa	ksi	Yield strength(a) MPa	ksi	Elongation(b), %	HRF	Hardness HRB	HR30T	Shear strength MPa	ksi	Fatigue strength(c) MPa	ksi
Flat Products, 1 mm (0.04 in.) Thick												
OS050	220	32	69	10	45	40	...	...	150	22	...	...
OS025	235	34	76	11	45	45	...	...	160	23	76	11
H00	250	36	195	28	30	60	10	25	170	25	...	...
H01	260	38	205	30	25	70	25	36	170	25	...	...
H02	290	42	250	36	14	84	40	50	180	26	90	13
H04	345	50	310	45	6	90	50	57	195	28	90	13
H08	380	55	345	50	4	94	60	63	200	29	97	14
H10	395	57	365	53	4	95	62	64	200	29	...	...
M20	235	34	69	10	45	45	...	...	160	23	...	...
Flat Products, 6 mm (0.25 in.) Thick												
OS050	220	32	69	10	50	40	...	...	150	22	...	...
H00	250	36	195	28	40	60	10	...	170	25	...	...
H01	260	38	205	30	35	70	25	...	170	25	...	...
H04	345	50	310	45	12	90	50	...	195	28	...	...
M20	220	32	69	10	50	40	...	...	150	22	...	...
Flat Products, 25 mm (1.0 in.) Thick												
H04	310	45	275	40	20	85	45	...	180	26	...	...
Rod, 6 mm (0.25 in.) Diam												
H80(40%) ...	380	55	345	50	10	94	60	...	200	29	...	...
Rod, 25 mm (1.0 in.) Diam												
OS050	220	32	69	10	55	40	...	...	150	22	...	...
H80(35%) ...	330	48	305	44	16	87	47	...	185	27	115(d)	17(d)
M20	220	32	69	10	55	40	...	...	150	22	...	...
Rod, 50 mm (2.0 in.) Diam												
H80(16%) ...	310	45	275	40	20	85	45	...	180	26	...	...
Wire, 2 mm (0.08 in.) Diam												
OS050	240	35	...	...	35(e)	...	...	...	165	24	...	...
H04	380	55	...	...	1.5(f)	...	...	...	200	29	...	...
H08	455	66	...	...	1.5(f)	...	...	...	230	33	...	...
Tube, 25 mm (1.0 in.) Diam by 1.65 mm (0.065 in.) Wall												
OS050	220	32	69	10	45	40	...	...	150	22	...	...
OS025	235	34	76	11	45	45	...	...	160	23	...	...
H55(15%) ...	275	40	220	32	25	77	35	45	180	26	...	...
H80(40%) ...	380	55	345	50	8	95	60	63	200	29	...	...
Shapes, 13 mm (0.50 in.) Section Size												
OS050	220	32	69	10	50	40	...	...	150	22	...	...
H80(15%) ...	275	40	220	32	30	...	35	...	180	26	...	...
M20	220	32	69	10	50	40	...	...	150	22	...	...
M30	220	32	69	10	50	40	...	...	150	22	...	...

(a) At 0.5% extension under load. (b) In 50 mm or 2 in. (c) At 10^8 cycles in a reversed bending test. (d) At 3×10^8 cycles in a rotating beam test. (e) Elongation in 250 mm (10 in.). (f) Elongation in 150 cm (60 in.).

(9.8 μin./in.·°F) at 20 to 300 °C (68 to 572 °F). See also Fig. 12.
Specific heat. 380 J/kg·K (0.092 Btu/lb·°F) at 20 °C (68 °F)
Enthalpy. See Fig. 12.
Latent heat of fusion. 205 kJ/kg
Thermal conductivity. 388 W/m·K (224 Btu/ft·h·°F) at 20 °C (68 °F). For high conductivity coppers, a value of 387 W/m·K (223 Btu/ft·h·°F) is an adjusted value corresponding to an electrical conductivity of 101% IACS.

Temperature K	°C	Thermal conductivity, W/m · K
4.2	−268.8	300
20	−253	1300
77	−196	550
194	−79	400
273	0	390
373	+100	380
573	300	370
973	700	300

Electrical Properties

Electrical conductivity. Volumetric: O60 temper, 100 to 101.5% IACS; H14 temper, 97% IACS. See also Fig. 13.
Electrical resistivity. O60 temper: 17.00 to 17.24 nΩ·m; temperature coefficient, 0.00393/K at −100 to +200 °C (−148 to +392 °F) for 100% IACS material, 0.00397/K at −100 to +200 °C for 101% IACS material. H14 temper: 1.78 nΩ·m; temperature

coefficient, 0.00381/K at 0 to 100 °C (32 to 212 °F) for 97% IACS material. See also Fig. 13.

Thermoelectric potential. See Fig. 14.

Electrochemical equivalent. Cu^{++}, 0.329 mg/C; Cu^+, 0.659 mg/C

Electrolytic solution potential. Cu^{++}, −0.344 V vs standard hydrogen electrode; Cu^+, −0.470 V vs standard hydrogen electrode; temperature coefficient, −0.01 mV/K at 20 to 50 °C (68 to 122 °F)

Hydrogen overvoltage. Approximately 0.23 V in dilute sulfuric acid; specific value varies with current density

Hall effect. Hall coefficient, −52 pV·m/A·T

Optical Properties

Color. Reddish metallic
Spectral reflectivity. 32.7% for λ of 420 nm; 43.7% for λ of 500 nm; 71.8% for λ of 600 nm; 83.4% for λ of 700 nm. See also Fig. 15 and 16.

Chemical Properties

General corrosion behavior. Although many factors influence the corrosion resistance of copper under specific conditions of service, copper is generally less subject to corrosion than other engineering metals. Often, copper is used where resistance to corrosion is of prime importance. Sometimes, it is better to use a copper alloy rather than an unalloyed copper. In general, copper resists nonoxidizing mineral and organic acids, caustic solutions, saline solutions and various natural waters or process waters. It is suitable for underground service because it resists soil corrosion. Copper is not suitable for service in oxidizing acids such as nitric acid. Also, copper may corrode in aerated nonoxidizing acids such as sulfuric or acetic acids, even though it is practically immune to these acids in the complete absence of air. Tough pitch copper is considered to be immune from stress-corrosion cracking in ammonia and the other agents that induce season cracking of brasses. However, tough pitch copper is susceptible to embrittlement in reducing atmospheres, especially those containing hydrogen.

Resistance to specific corroding agents. Depending on concentration and specific conditions of exposure, copper generally resists the following

Fig. 4 Variation of tensile properties and grain size with similar coppers

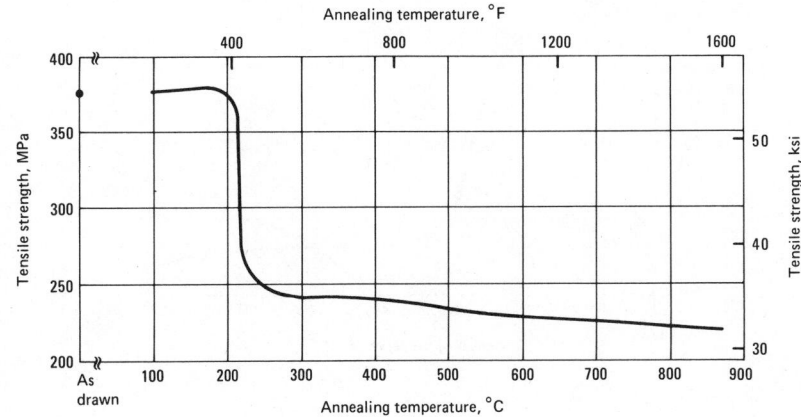

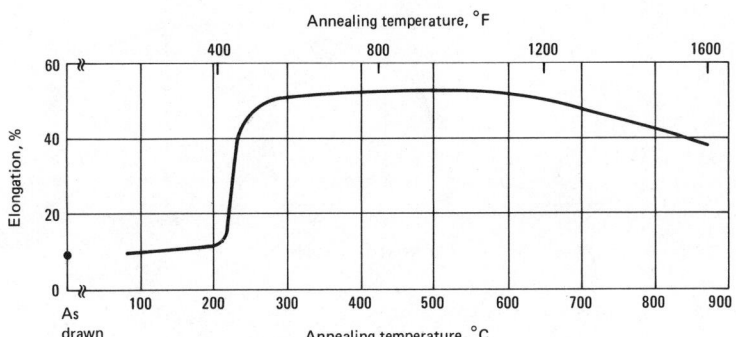

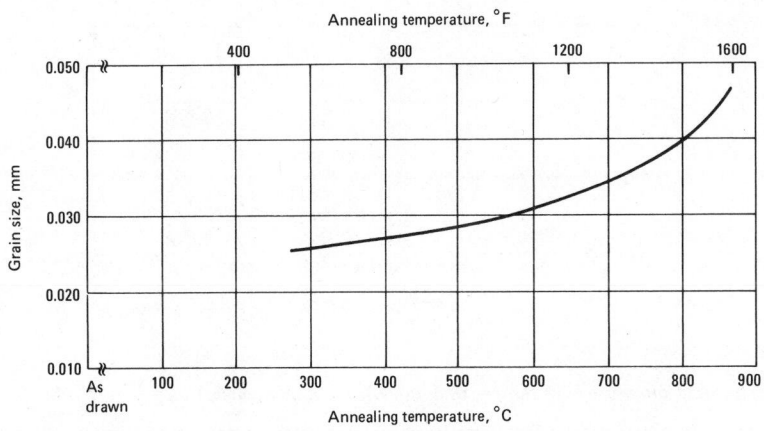

Material cold drawn to H10 temper and annealed ½ h at temperature.

Fig. 4 (continued)

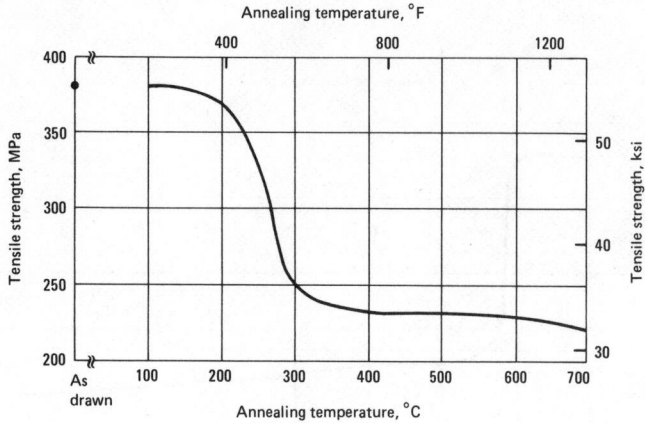

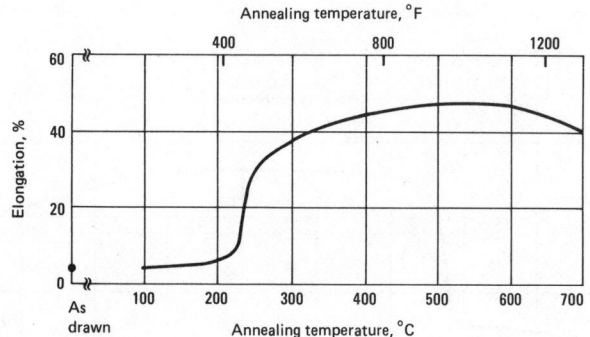

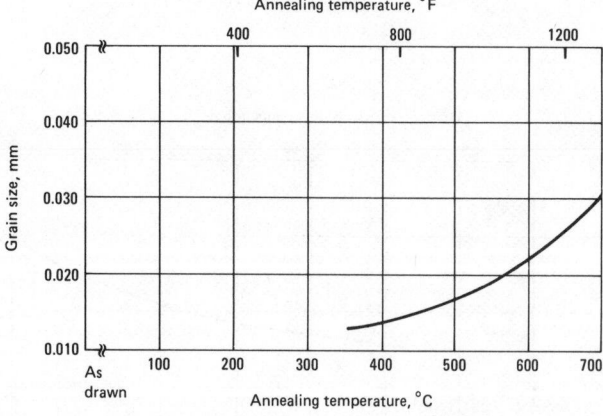

Material cold drawn to H04 temper and annealed 1 h at temperature

Table 10 Typical impact strength of C11000

Product and condition	Impact strength	
	J	ft·lb
Charpy V-notch		
Hot rolled, annealed........	96	71
Charpy keyhole-notch		
As-cast....................	11	8
As hot rolled	43	32
Rod		
annealed	52	38
commercial temper	35	26
Izod		
Rod		
annealed and drawn 30%..	54	40
drawn 30%	45	33
Plate		
as hot rolled	52	38
annealed	53(a)	39(a)
	39(b)	29(b)
cold rolled 50%...........	26(a)	19(a)
	12(b)	9(b)

(a) Parallel to rolling direction. (b) Transverse to rolling direction.

hydroxide; concentrated or dilute caustic solutions. *Salt Solutions:* aluminum chloride, aluminum sulfate, calcium chloride, copper sulfate, sodium carbonate, sodium nitrate, sodium sulfate and zinc sulfate. *Waters:* all types of potable water, many industrial and mine waters, seawater and brackish water. *Copper* is not recommended for use with ammonia, nitric acid, acid chromate solutions, ferric chloride, mercury salts, perchlorates or persulfates.

Fabrication Characteristics

Machinability. 20% of C36000, free-cutting brass

Forgeability. 65% of C37700, forging brass

Formability. Excellent for cold working and hot forming

Weldability. Soldering: excellent. Brazing and resistance butt welding: good. Gas-shielded arc welding: fair. Oxyfuel gas, shielded metal-arc, resistance spot and resistance seam welding: not recommended.

Annealing temperature. 475 to 750 °C (900 to 1400 °F). See also Fig. 4 and 17.

Hot working temperature. 750 to 875 °C (1400 to 1600 °F)

Typical softening temperature. 360 °C (675 °F)

agents. *Acids:* mineral acids such as hydrochloric and sulfuric acids; organic acids such as acetic acid (including vinegar and acetates), carbolic acid, citric acid, formic acid,

oxalic acid and tartaric acid; fatty acids; and acidic solutions containing sulfur, such as the sulfurous acid and sulfite solutions used in pulp mills. *Alkalies:* fused sodium or potassium

Table 11 Creep properties of copper(a)

Testing temperature °C	°F	Stress MPa	ksi	Duration of test, h	Total extension(b), %	Intercept, %	Min creep rate, % per 1000 h
Strip (c), OS030 Temper							
130	265	55	8	2500	2.6	2.0	0.15
		100	14.5	2600	10.0	7.6	1.2
		140	20	170	29.8 (d)	···	39
175	345	55	8	2000	3.3	2.3	0.65
		100	14.5	350	15(d)	8.0	6.3
Strip (c), H01 Temper							
130	265	55	8	8250	0.20	0.15	0.01
		100	14.5	8600	0.67	0.26	0.042
		140	20	1750	2.4(d)	0.32	0.45
175	345	55	8	6850	1.14	0.14	0.088
		100	14.5	1100	2.0	0.22	0.66
Strip (c), H02 Temper							
130	265	55	8	7200	0.24	0.13	0.01
		100	14.5	8600	1.02	0.25	0.054
		140	20	4680	3.4(d)	0.36	0.27
175	345	55	8	1050	3.3(d)	···	0.6
Strip (c), H06 Temper							
130	265	55	8	8250	1.58	0.08	0.035
		100	14.5	8700	7.31	0.16	0.055
		140	20	4030	11(d)	0.24	0.17
Rod (e), OS025 Temper							
260	500	2.5	0.36	6000	0.08	0.016	0.011
		4.1	0.60	6000	0.19	0.010	0.030
		7.2	1.05	6500	0.64	0.113	0.080
		13.8	2.0	6500	2.88	0.87	0.306
Rod (e), H08 Temper							
205	400	7.2	1.05	6500	0.06	0.045	0.011
		14.5	2.1	6500	0.20	0.112	0.012
		28	4.05	6500	1.08	0.41	0.097
		50	7.25	6500	5.42	2.47	0.44

(a) Values shown are typical for the tough pitch grades of copper. Oxygen-free, phosphorus deoxidized and arsenical coppers have marginally greater resistance to creep deformation. (b) Total extension is initial extension (not given in table) plus intercept (column 7) plus the product of min creep rate (column 8) and duration (column 5). (c) 2.5 mm (0.10 in.) thick. (d) Rupture test. (e) 3.2 mm (0.13 in.) diam.

Fig. 5 Variation of tensile properties with amount of cold reduction by rolling for C11000 and similar coppers

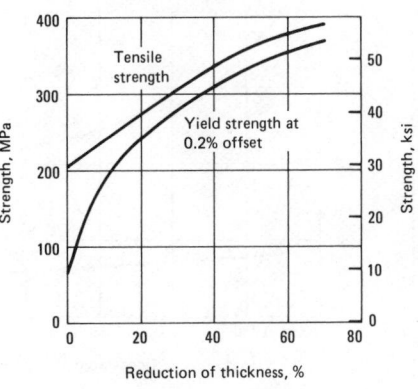

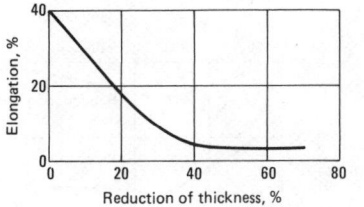

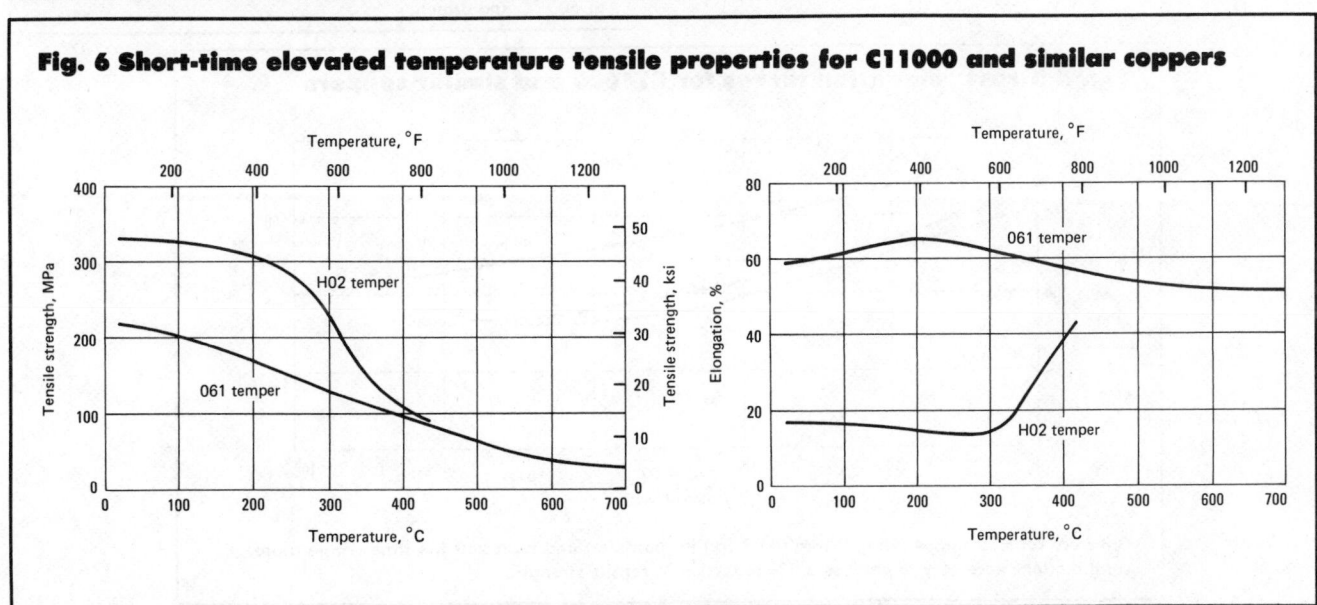

Fig. 6 Short-time elevated temperature tensile properties for C11000 and similar coppers

Fig. 7 Low temperature tensile properties of C11000 and similar coppers

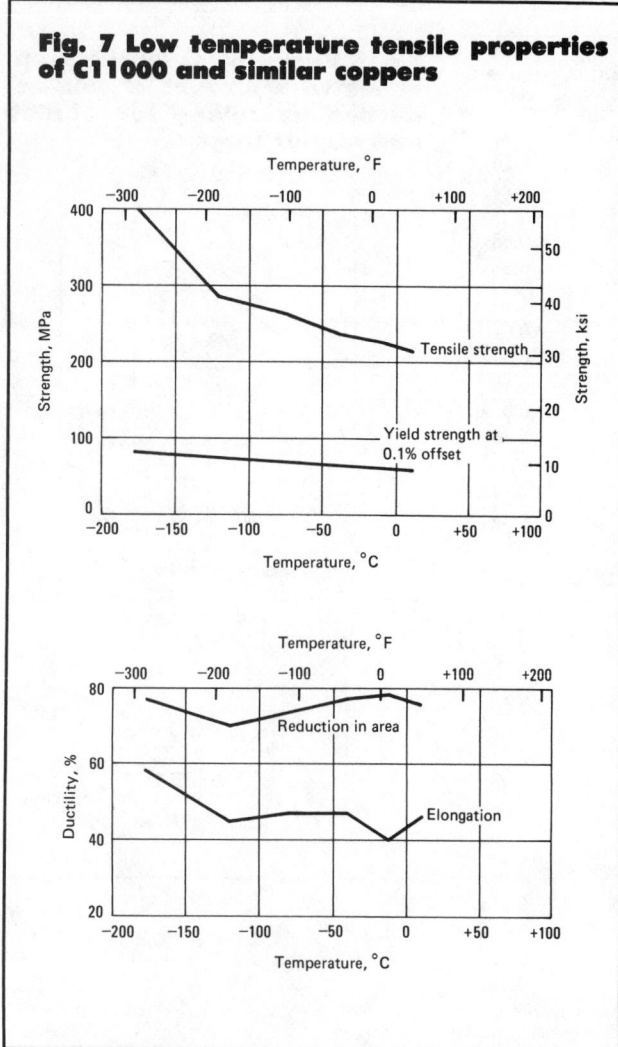

Fig. 9 Variation of hardness with amount of cold reduction by rolling for C11000 and similar coppers.

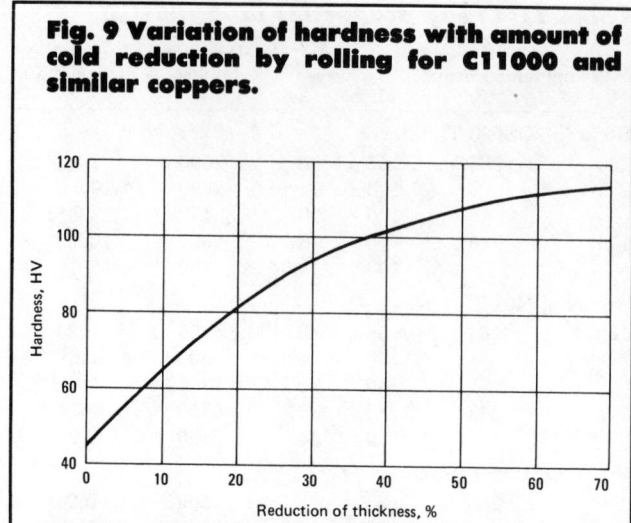

Fig. 11 Variation of density with amount of cold reduction by rolling for C11000 and similar coppers

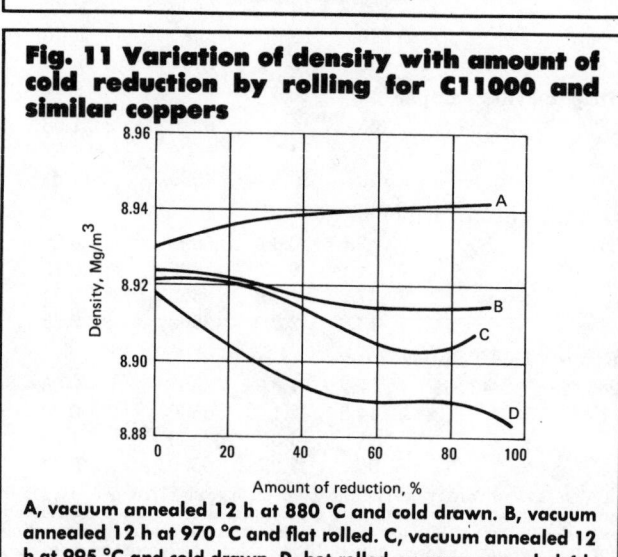

A, vacuum annealed 12 h at 880 °C and cold drawn. B, vacuum annealed 12 h at 970 °C and flat rolled. C, vacuum annealed 12 h at 995 °C and cold drawn. D, hot rolled, vacuum annealed 4 h at 600 °C and drawn.

Fig. 8 Stress relaxation curves for C11000 and similar coppers

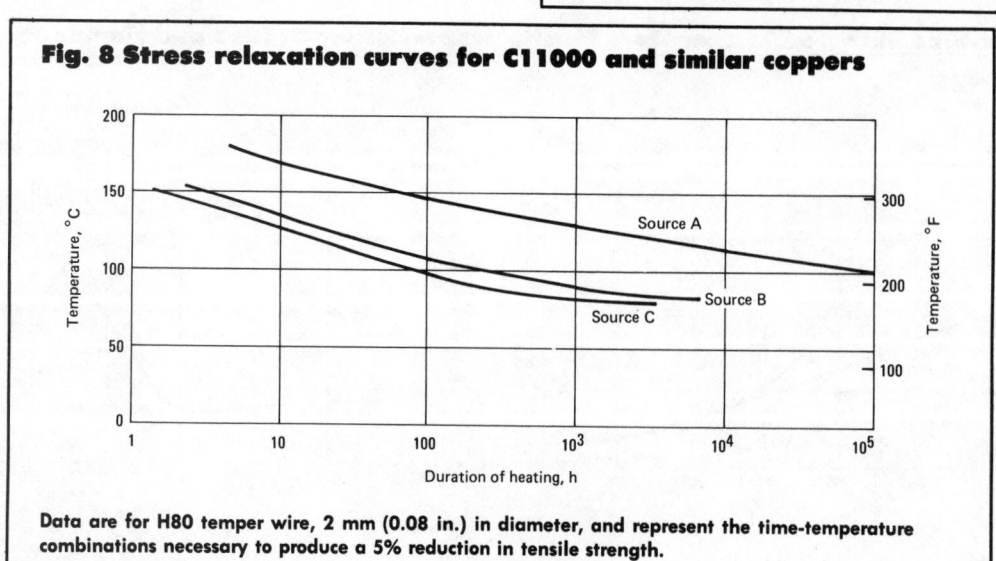

Data are for H80 temper wire, 2 mm (0.08 in.) in diameter, and represent the time-temperature combinations necessary to produce a 5% reduction in tensile strength.

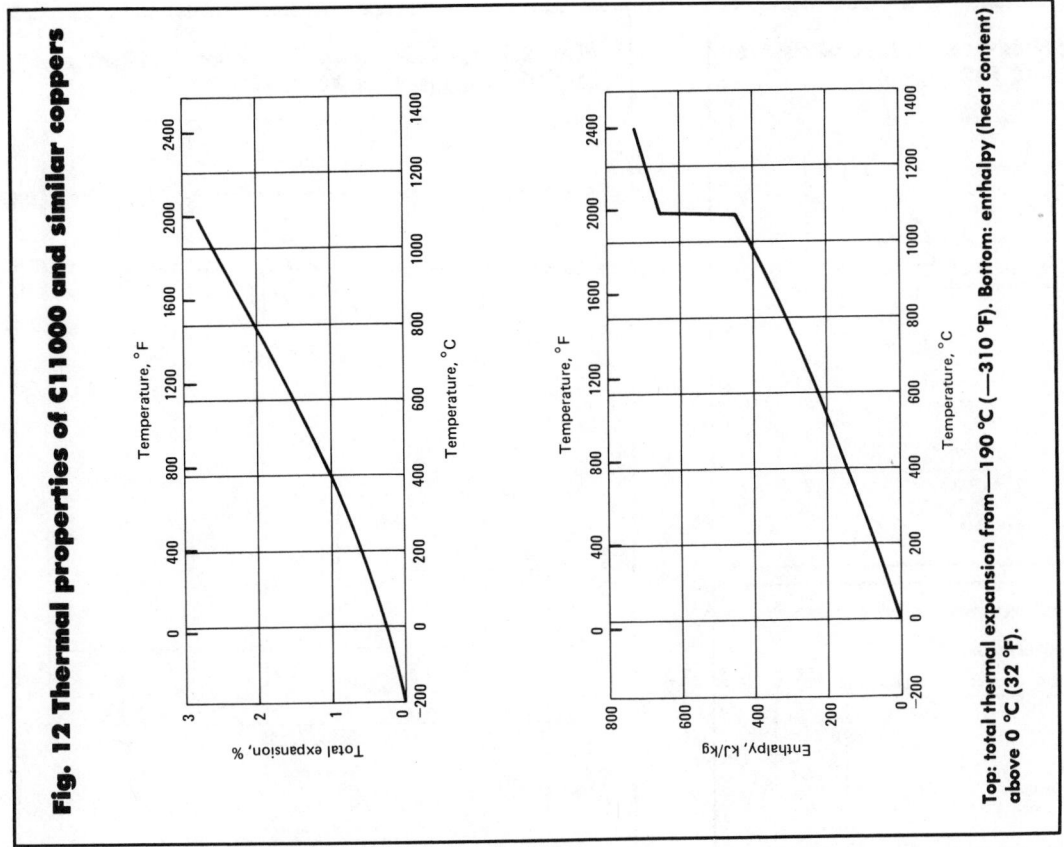

Fig. 12 Thermal properties of C11000 and similar coppers

Top: total thermal expansion from —190 °C (—310 °F). Bottom: enthalpy (heat content) above 0 °C (32 °F).

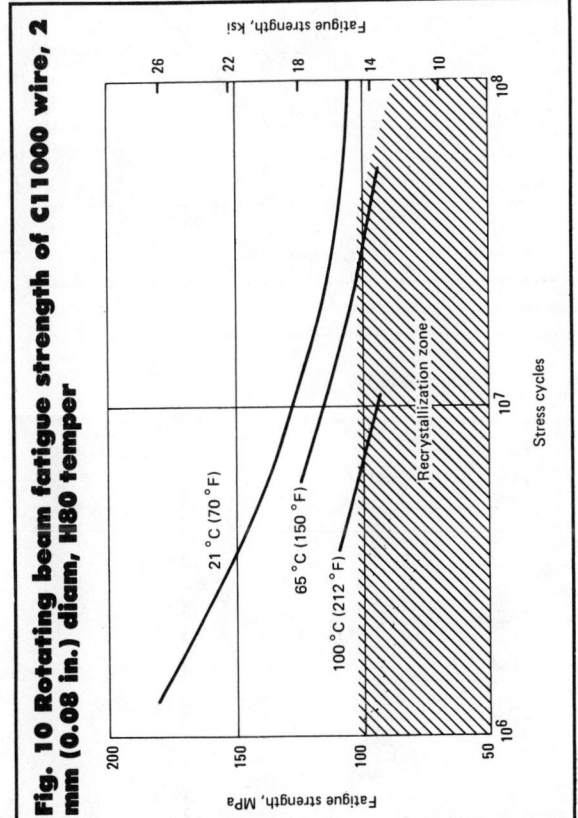

Fig. 10 Rotating beam fatigue strength of C11000 wire, 2 mm (0.08 in.) diam, H80 temper

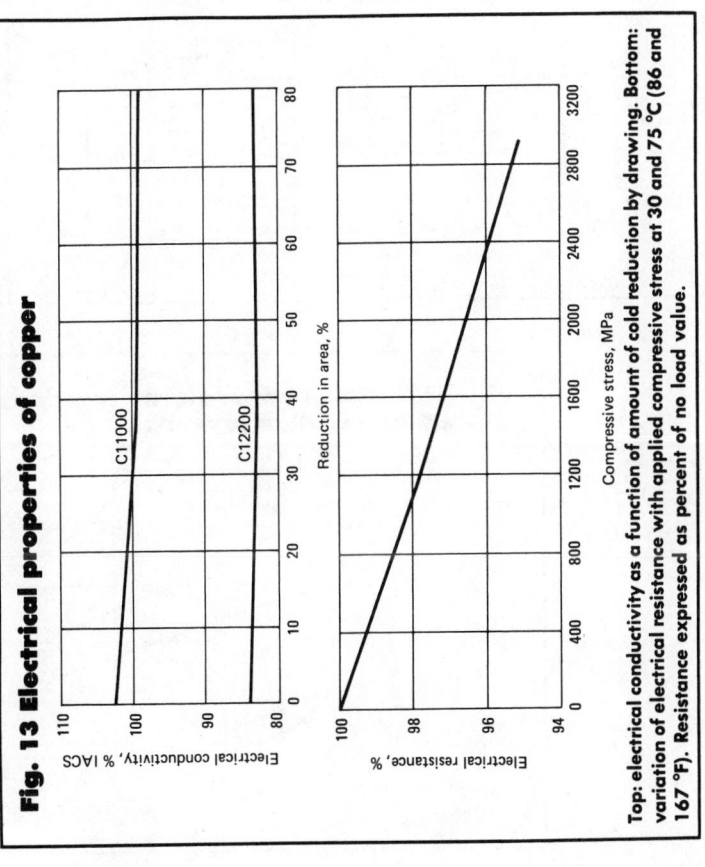

Fig. 13 Electrical properties of copper

Top: electrical conductivity as a function of amount of cold reduction by drawing. Bottom: variation of electrical resistance with applied compressive stress at 30 and 75 °C (86 and 167 °F). Resistance expressed as percent of no load value.

Fig. 14 Thermoelectric properties of copper for cold junctions 0 °C (32 °F)

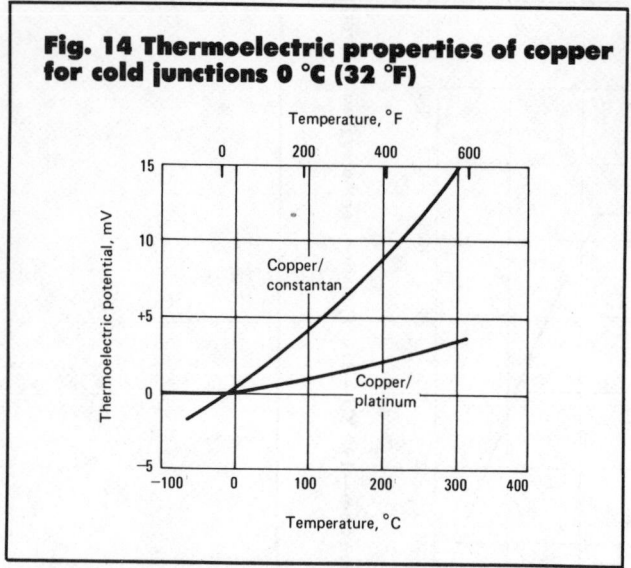

Fig. 15 Optical properties of C11000 and similar coppers at 21 °C (70 °F)

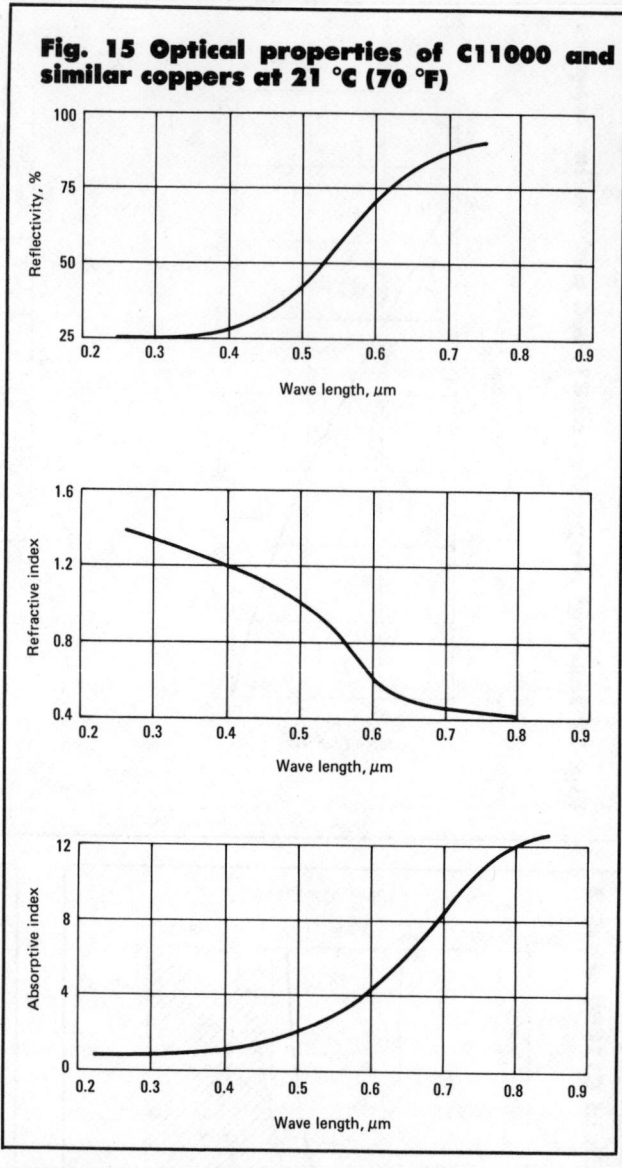

Fig. 16 Emissivity of commercial coppers

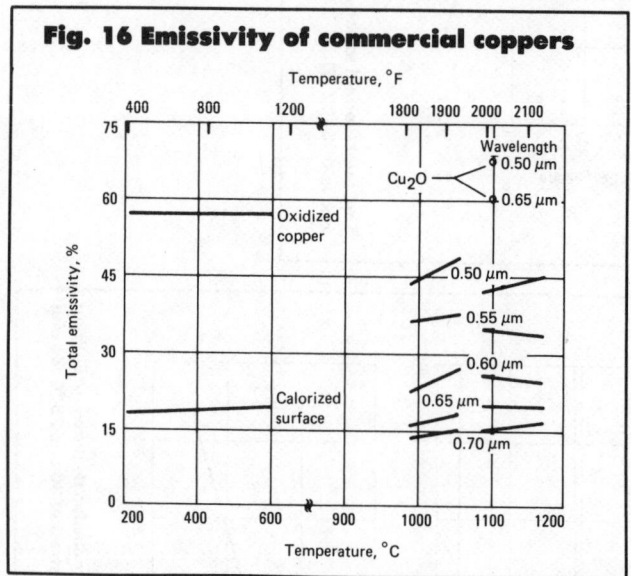

Fig. 17 Time-temperature relationships for annealing C11000 and similar coppers

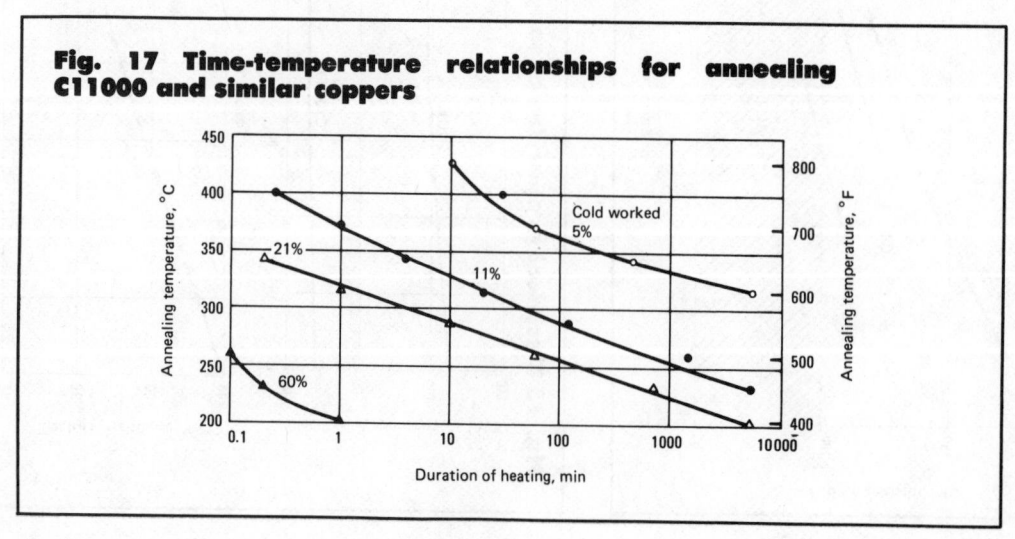

C11100
99.95Cu-0.04O-0.01Cd

Commercial Names

Previous trade name. Electrolytic tough pitch copper, anneal resistant
Common name. Anneal-resistant electrolytic copper

Specifications

ASTM. See Table 12.
SAE. Bar, plate, sheet, strip: J461, J463
Government. See Table 12.

Chemical Composition

Composition limits. 99.90 min Cu. Limits on O and Cd or other elements present to make this copper anneal resistant are established by conductivity tests and/or stress relaxation tests rather than by chemical analysis.

Applications

Typical uses. Produced mainly as wire for electrical power transmission where resistance to softening under overloads is desired

Mechanical Properties

Tensile properties. (Typical) Tensile strength, 455 MPa (66 ksi); elongation, 1.5% in 150 cm or 60 in.
Elastic modulus. Tension, 115 GPa (17×10^6 psi); shear, 44 GPa (6.4×10^6 psi)

Mass Characteristics

Density. 8.89 to 8.94 Mg/m^3 (0.321 to 0.323 lb/in.^{3}) at 20 °C (68 °F)

Thermal Properties

Liquidus temperature. 1085 °C (1980 °F)
Solidus temperature. 1065 °C (1950 °F)
Coefficient of thermal expansion. Linear, 17.0 μm/m·K (9.4 μin./in.·°F) at 20 to 100 °C (68 to 212 °F); 17.3 μm/m·K (9.6 μin./in.·°F) at 20 to 200 °C (68 to 392 °F); 17.7 μm/m·K (9.8 μin./in.·°F) at 20 to 300 °C (68 to 572 °F)
Specific heat. 385 J/kg·K (0.092 Btu/lb·°F) at 20 °C (68 °F)
Thermal conductivity. 388 W/m·K (224 Btu/ft·h·°F)

Electrical Properties

Electrical conductivity. Volumetric, 100% IACS at 20 °C (68 °F)
Electrical resistivity. 17.2 nΩ·m at 20 °C (68 °F)

Fabrication Characteristics

Machinability. 20% of C36000, free-cutting brass
Forgeability. 65% of C37700, forging brass
Formability. Excellent for cold working and hot forming. Common processes include drawing, stranding and stamping.
Weldability. Soldering: excellent. Brazing and resistance butt welding: good. Gas-shielded arc welding: fair. Oxyacetylene coated metal-arc and resistance spot and seam welding: not recommended
Annealing temperature. 475 to 750 °C (900 to 1400 °F)
Hot working temperature. 750 to 875 °C (1400 to 1600 °F)
Typical softening temperature. 355 °C (675 °F)

Table 12 Specifications for C11100

Product	Federal	ASTM
Bar	QQ-C-502, QQ-C-576	...
Bar, bus	QQ-B-865	...
Pipe, bus	QQ-B-825	...
Plate	QQ-C-576	...
Rod	QQ-B-502	B49, B133
Rod, bus	QQ-B-825	...
Shapes	QQ-C-502	...
Shapes, bus	QQ-B-825	...
Sheet	QQ-C-576	...
Strip	QQ-C-502, QQ-C-576	...
Tubing, bus	QQ-B-825	...
Wire, coated	...	B246
with tin	...	B334
with lead alloy	...	B189
with nickel	...	B355
with silver	...	B298
Wire, flat	QQ-C-502	...
Wire, hard drawn	QQ-W-343	B1
Wire, medium hard drawn	...	B2
Wire, stranded	...	B8, B172, B173, B174, B226, B228, B229, B286
Wire, rod	...	B47
Wire, trolley	...	B116

C11300, C11400, C11500, C11600
99.96Cu + Ag-0.40

Commercial Names

Previous trade name. Tough pitch copper with silver
Common name. Silver-bearing tough pitch copper
Designation. STP

Specifications

AMS. Soft wire (all alloys) and trolley wire (C11300 only): 4701
ASME. Strip (C11300 only): SB152
ASTM. See Table 13.
SAE. Bar, sheet, strip (C11300, C11400 and C11600) and plate (C11300 and C11400): J463
Federal. See Table 13.
Military. Soft wire (all alloys) and trolley wire (C11300 only): MIL-W-3318. Commutator bar (11600 only): MIL-B-19231

Chemical Composition

Composition limits. 99.0 to 99.9 Cu; 0.027 to 0.85 Ag(a); 0.04 O

(a) C11300, 0.027 Ag; C11400, 0.034 Ag; C11500, 0.054 Ag; C11600, 0.085 Ag. These coppers may be Low-Resistance Lake Copper or electrolytic copper to which Ag has been intentionally added.

Applications

Typical uses. All forms except pipe and tubing: gaskets, radiators, bus bars, conductivity wire, contacts, radio parts, windings, switches, terminals, commutator segments, chemical process equipment, printing rolls, clad metals, printed circuit foil

Mechanical Properties

Tensile properties. See Table 14.
Shear strength. See Table 14.
Hardness. See Table 14.
Elastic modulus. Tension, 115 GPa (17 × 10^6 psi); shear, 44 GPa (6.4 × 10^6 psi)

Mass Characteristics

Density. 8.89 to 8.94 Mg/m^3 (0.321 to 0.323 $lb/in.^3$) at 20 °C (68 °F)

Thermal Properties

Liquidus temperature. 1080 °C (1980 °F)
Coefficient of thermal expansion. Linear, 17.7 μm/m·K (9.8 μin./in.·°F) at 20 to 300 °C (68 to 572 °F)
Specific heat. 385 J/kg·K (0.092 Btu/lb·°F) at 20 °C (68 °F)
Thermal conductivity. 388 W/m·K (224 Btu/ft·h·°F) at 20 °C (68 °F)

Electrical Properties

Electrical conductivity. Volumetric, 100% IACS at 20 °C (68 °F)
Electrical resistivity. 17.2 nΩ·m at 20 °C (68 °F)

Fabrication Characteristics

Machinability. 20% of C36000, free-cutting brass
Forgeability. 65% of C37700, forging brass
Formability. Excellent for cold working and hot forming.
Weldability. Soldering: excellent. Brazing and resistance butt welding: good. Gas-shielded arc welding: fair. Oxyacetylene coated metal-arc and resistance spot and seam welding: not recommended.

Annealing temperature. 475 to 750 °C (900 to 1400 °F)
Hot working temperature. 750 to 875 °C (1400 to 1600 °F)

Table 13 Specifications for C11300, C11400, C11500 and C11600

Product	ASTM	Federal
Bar	B152(a)	QQ-C-576(a), QQ-C-502(b)
Bar, bus	B187(a)	QQ-B-825(a)
Pipe, bus	B188(a)	QQ-B-825(c)
Plate	B152(a)	QQ-B-825(a)
Rod	B49(e)	QQ-C-502(f)
Rod, bus	B187(a)	QQ-B-825(f)
Shapes	. . .	QQ-C-502(f)
Shapes, bus	B187(a)	QQ-B-825(f)
Sheet	B152(a)	QQ-C-576(f)
Sheet, clad	B506(a)	QQ-C-502(c)
Strip	B152(a), B272(e)	QQ-C-502(b), QQ-C-576(f)
Strip, clad	B506(a)	. . .
Tube, bus	B188(a)	QQ-B-825(f)
Wire, coated with		
Tin	B246, B334(e)	. . .
Lead alloy	B189(e)	. . .
Nickel	B355(e)	. . .
Silver	B298(e)	. . .
Wire, flat	B272(e)	QQ-C-502(a)
Wire, hard drawn	B1(e)	QQ-W-343(e)
Wire, medium hard drawn	B2(e)	QQ-W-343(e)
Wire, rod	B49(e)	. . .
Wire, soft	B3(a), B48(e)	QQ-W-343(e)
Wire, stranded	B8, B172, B173, B174, B226, B228, B229, B286(e)	. . .
Wire, trolley	B47, B116(e)	QQ-W-343(g)

(a) C11300, C11400 and C11600. (b) C11600 only. (c) C11400 only. (d) C11300 and C11400. (e) C11300, C11400, C11500 and C11600. (f) C11400 and C11600. (g) C11300 only.

C12500, C12700, C12800, C12900, C13000

Commercial Names

Previous trade name. Fire refined tough pitch copper (12500); fire refined tough pitch copper with silver (12700, C12800, C12900, C13000)
Common name. Fire refined copper
Designation. C12500; FRTP. Others: FRTSP

Specifications

ASTM. Flat products: B11, B124, B133, B152, B272. Rod: B12, B124, B133. Shapes: B124, B133, B216. Lake Cu wirebar, cake, slab, billet and ingot: B4

Government. MIL-W-3318

Chemical Composition

Composition limits. ASTM B216: 99.88 min Cu + Ag(a); 0.012 max As; 0.003 max Sb; 0.025 max Se + Te; 0.05 max Ni; 0.003 max Bi; 0.004 max Pb

(a) Minimum Ag content may be specified by agreement.

Consequence of exceeding impurity limits. Bi and Pb can cause hot workability problems if composition limits are exceeded. Se and Te greatly affect recrystallization and grain growth.

Applications

Typical uses. Architectural: Building fronts, downspouts, flashing, gutters, roofing, screening, spouting. Automotive: gaskets, radiators. Electrical: bus bars, contacts, radio parts, commutator segments, switches, terminals. Miscellaneous: anodes, chemical process equipment, kettles, pans, printing rolls, rotating bands, road bed expansion plates, vats. This copper is suitable for use where the high conductivity and low annealing temperature of electrolytic tough pitch copper are not required.
Precautions in use. This copper is subject to embrittlement when heated in a reducing atmosphere, as in annealing, brazing, or welding at temperatures of 370 °C (700 °F) or above. If hydrogen or carbon monoxide is present, embrittlement can be rapid.

Table 14 Typical mechanical properties of C11300, C11400, C11500 and C11600

Size	Temper	Tensile strength MPa	ksi	Yield strength(a) MPa	ksi	Elongation in 50 mm or 2 in., %	HRF	Hardness HRB	HR30T	Shear strength MPa	ksi
Flat Products											
1-mm (0.04-in.) thick	OS025	235	34	72	11	45	45	...	...	160	23
	H00	250	36	195	28	30	60	10	25	170	25
	H01	260	38	205	30	25	70	25	36	170	25
	H02	290	42	250	36	14	84	40	50	180	26
	H04	345	50	310	45	6	90	50	57	195	28
	H08	380	55	345	50	4	94	60	63	200	29
	H10	395	57	365	53	4	95	62	64	200	29
	M20	235	34	69	10	45	45	...	...	160	23
6-mm (0.25-in.) thick	OS050	220	32	69	10	50	40	...	...	150	22
	H00	250	36	195	28	40	60	10	...	170	25
	H01	260	38	205	30	35	70	25	...	170	25
	H04	345	50	310	45	12	90	50	...	195	28
	M20	220	32	69	10	50	40	...	...	150	22
25-mm (1.0-in.) thick	H04	310	45	275	40	20	85	45	...	180	26
Rod											
6-mm (0.25-in.) diam	H80(40%)	380	55	345	50	10	94	60	...	200	29
25-mm (1.0-in.) diam	OS050	220	32	69	10	55	40	...	...	150	22
	H80(35%)	330	48	305	44	16	87	47	...	185	27
	M20	220	32	69	10	55	40	...	...	150	22
51-mm (2.0-in.) diam	H80(16%)	310	45	275	40	20	85	45	...	180	26
Wire											
2-mm (0.08-in. diam)	OS050	240	35	...	...	35(b)	...	...	...	165	24
	H04	380	55	...	...	1.5(c)	...	...	...	200	29
	H08	455	66	...	...	1.5(c)	...	...	...	230	33
Shapes											
13-mm (0.50-in.) diam	OS050	220	32	69	10	50	40	...	...	150	22
	H80(15%)	275	40	220	32	30	...	35	...	180	26
	M20	220	32	69	10	50	40	...	...	150	22
	M30	220	32	69	10	50	40	...	...	150	22

(a) At 0.5% extension under load. (b) Elongation in 250 mm (10 in.). (c) Elongation in 150 cm (60 in.).

Mechanical Properties

Tensile properties. See Table 15.
Shear strength. See Table 15.
Hardness. See Table 15.
Fatigue strength. Strip, OS025 temper, 76 MPa (11 ksi)

Mass Characteristics

Density. 8.89 Mg/m^3 (0.321 lb/in.^{3}) at 20 °C (68 °F)

Thermal Properties

Liquidus temperature. 1085 °C (1980 °F)
Coefficient of thermal expansion. Linear, 16.8 μm/m·K (9.3 μin./in.·°F) at 20 to 100 °C (68 to 212 °F); 17.4 μm/m·K (9.7 μin./in.·°F) at 20 to 200 °C (68 to 392 °F); 17.7 μm/m·K (9.8 μin./in.·°F) at 20 to 300 °C (68 to 572 °F)
Specific heat. 385 J/kg·K (0.092 Btu/lb·°F) at 20 °C (68 °F)
Thermal conductivity. 377 W/m·K (218 Btu/ft·h·°F) at 20 °C (68 °F)

Electrical Properties

Electrical conductivity. Volumetric, 98% IACS at 20 °C (68 °F), annealed
Electrical resistivity. 17.6 nΩ·m at 20 °C (68 °F)

Fabrication Characteristics

Machinability. 20% of C36000, free-cutting brass
Formability. Excellent for hot or cold forming, but should not be heated for forming or annealed in a reducing atmosphere.
Joining. Riveting: use copper rivets. Pressure welding: use Koldweld proprietary method. Soft solder with all grades of solder, commercial solder fluxes or rosin. Silver braze with all types of flame using copper phosphorus, silver, or copper zinc (see ASTM B260). Satisfactory fluxes are commercially available. Use gas-shielded arc welding processes with recommended filler metals, depending on application. Other welding methods are not generally recommended.
Annealing temperature. 400 to 650 °C (750 to 1200 °F)
Hot working temperature. 750 to 950 °C (1400 to 1750 °F)

Table 15 Typical mechanical properties of C12500, C12700, C12800, C12900 and C13000

Temper	Tensile strength MPa	ksi	At 0.5% extension under load MPa	ksi	At 0.2% offset MPA	ksi	Elongation in 50 mm or 2 in., %	HRF	HRB	H30T	Shear strength MPa	ksi
Flat Products: 1 mm (0.04 in.) Thick												
OS025	235	34	76	11	...	...	45	45	...	...	160	23
H00	250	36	195	28	235	34	30	60	10	25	170	25
H02	290	42	250	36	270	39	14	84	40	50	180	26
H04	345	50	310	45	327	47.5	6	90	50	57	195	28
H08	380	55	345	50	360	52	4	94	60	63	200	29
H10	395	57	365	53	370	54	4	95	62	64	200	29
M20	235	34	69	10	...	...	45	45	...	...	160	23
Rod: 25-mm (1-in.) Diam												
OS050	220	32	69	10	...	...	55	40	...	...	150	22
H80(35%)	330	48	305	44	...	...	16	87	47	...	185	25
M20	220	32	69	10	...	...	55	40	...	...	150	22
Wire: 2-mm (0.08-in.) Diam												
OS050	240	35	...	...	...	...	35(b)	...	...	...	165	24
H04	380	55	...	...	...	...	1.5(c)	...	...	...	200	29
H08	455	66	...	...	...	...	1.5(c)	...	...	...	230	33
Shapes: 13-mm (0.50-in.) Diam												
OS050	220	32	69	10	...	...	50	40	...	...	150	22
H80(15%)	275	40	220	32	...	...	30	...	35	...	180	26
M20	220	32	69	10	...	...	50	40	...	...	150	22
M30	220	32	69	10	...	...	50	40	...	...	150	22

(a) At 10^8 cycles. (b) Elongation in 250 mm (10 in.). (c) Elongation in 150 cm (60 in.).

C14300, C14310
99.9Cu-0.1Cd;
99.8Cu-0.2Cd

Commercial Names

Previous trade names. C14300, cadmium copper, deoxidized

Chemical Composition

Composition limits. 99.8 to 99.9 Cu, 0.05 to 0.3 Cd(a)

(a) C14300, 0.05 to 0.15 Cd; C14310, 0.1 to 0.3 Cd

Applications

Typical uses. Rolled strip for anneal-resistant electrical applications: applications requiring thermal softening and embrittlement resistance, such as lead frames, contacts, terminals, solder-coated and solder-fabricated parts. Furnace brazed assemblies and welded components such as tube or cable wrap.

Mechanical Properties

Tensile properties. See Table 16.

Shear strength. See Table 16.
Hardness. See Table 16.
Elastic modulus. Tension, 115 GPa (17×10^6 psi); shear, 44 GPa (6.4×10^6 psi)

Mass Characteristics

Density. 8.94 Mg/m³ (0.323 lb/in.³) at 20 °C (68 °F)

Thermal Properties

Liquidus temperature. 1080 °C (1976 °F)
Solidus temperature. 1052 °C (1926 °F)
Coefficient of thermal expansion. Linear, 17.0 μm/m·K (9.4 μin./in.·°F) at 20 to 100 °C (68 to 212 °F); 17.3 μm/m·K (9.6 μin./in.·°F) at 20 to 200 °C (68 to 392 °F); 17.7 μm/m·K (9.8 μin./in.·°F) at 20 to 300 °C (68 to 572 °F)
Specific heat. 385 J/kg·K (0.092 Btu/lb·°F) at 20 °C (68 °F)
Thermal conductivity. C14300, 377 W/m·K (218 Btu/ft·h·°F) at 20 °C (68 °F); C14310, 343 W/m·K (198 Btu/ft·h·°F)

Electrical Properties

Electrical conductivity. Volumetric, C14300, 96% IACS at 20 °C (68 °F); C14310, 85% IACS at 20 °C (68 °F)
Electrical resistivity. C14300, 18 nΩ·m at 20 °C (68 °F); C14310, 20.3 nΩ·m at 20 °C (68 °F)

Fabrication Characteristics

Machinability. 20% of C36000, free-cutting brass
Forgeability. 65% of C37700, forging brass
Formability. Excellent capacity for cold working and hot forming.
Weldability. Soldering, brazing and gas-shielded arc welding: excellent. Oxyacetylene welding and resistance butt welding: good. Coated metal-arc and resistance seam and spot welding: not recommended.
Annealing temperature. 535 to 750 °C (1000 to 1400 °F)
Hot working temperature. 750 to 875 °C (1400 to 1600 °F)

Table 16 Typical mechanical properties of C14300 and C14310

Temper	Tensile strength MPa	ksi	Yield strength (a) MPa	ksi	Elongation in 50 mm or 2 in., %
Strip: 0.3- to 1-mm (0.01- to 0.04-in.) Diam					
OS025	220	32	75	11	42
H04	310	45	275	40	14
H08	350	51	330	48	7
H10	400	58	385	56	3
	400 min	58 min	385 min	56 min	1 min

(a) At 0.2% offset.

C14500
99.5Cu-0.5Te

Commercial Names

Previous trade name. Phosphorus-deoxidized, tellurium-bearing copper
Common name. Free-machining copper
Designation. DPTE

Specifications

ASTM. Flat products and rod: B124, B301. Shapes: B124, B283

Chemical Composition

Composition limits. 99.90 min Cu + Ag + Te, 0.004 to 0.012 P, 0.40 to 0.60 Te

Applications

Typical uses. Forgings and screw machine products requiring high conductivity, extensive machining, corrosion resistance, copper color, or a combination of these qualities; typical parts include electrical connectors, motor parts, switch parts, plumbing fixtures, soldering tips, welding-torch tips, transistor bases and parts that are assembled by furnace brazing.
Precautions in use. Carbide-tipped tools should be used for machining C14500.

Mechanical Properties

Tensile properties. See Table 17.
Shear strength. See Table 17.
Hardness. See Table 17.
Elastic modulus. Tension, 115 GPa (17×10^6 psi); shear, 44 GPa (6.4×10^6 psi)

Mass Characteristics

Density. 8.94 Mg/m^3 (0.323 lb/in.^{3}) at 20 °C (68 °F)

Thermal Properties

Liquidus temperature. 1075 °C (1967 °F)
Solidus temperature. 1051 °C (1924 °F)
Coefficient of thermal expansion. Linear, 17.1 μm/m·K (9.5 μin./in.·°F) at 20 to 100 °C (68 to 212 °F); 17.4 μm/m·K (9.7 μin./in.·°F) at 20 to 200 °C (68 to 392 °F); 17.8 μm/m·K (9.9 μin./in.·°F) at 20 to 300 °C (68 to 572 °F)
Specific heat. 415 J/kg·K (0.092 Btu/lb·°F) at 20 °C (68 °F)
Thermal conductivity. 355 W/m·K (205 Btu/ft·h·°F) at 20 °C (68 °F)

Electrical Properties

Electrical conductivity. Volumetric, 93% IACS at 20 °C (68 °F)
Electrical resistivity. 18.6 nΩ·m at 20 °C (68 °F)

Fabrication Characteristics

Machinability. 85% of C36000, free-cutting brass
Formability. Good capacity for being cold worked, usually by drawing, rolling or swaging. Excellent capacity for being hot formed, most often by extrusion, forging or rolling.
Weldability. Soldering or brazing: excellent. Arc welding, oxyfuel gas welding and most resistance welding processes are not recommended.
Annealing temperature. 425 to 650 °C (800 to 1200 °F)
Hot working temperature. 750 to 875 °C (1400 to 1600 °F)

C14700
99.6Cu-0.4S

Commercial Names

Previous trade name. Sulfur-bearing copper
Common name. Free-machining copper, sulfur copper

Specifications

ASTM. Flat products and rod: B301

Chemical Composition

Composition limits. 0.20 to 0.50 S, 0.10 max others (total); rem Cu + Ag

Applications

Typical uses. Screw machine products and parts requiring high con-

Table 17 Typical mechanical properties of C14500, C14700 and C18700 rod

Temper	Tensile strength MPa	ksi	Yield strength(a) MPa	ksi	Elongation(b), %	Hardness, HRB	Shear strength MPa	ksi
6 mm (0.25 in.) diam								
H02	295	43	275	40	18	43	180	26
H04	365	53	340	49	10	54	200	29
13 mm (0.50 in.) diam								
OS015	230	33	76	11	46	43 HRF	150	22
H02	295	43	275	40	20	43	180	26
H04	330	48	305	44	15	48	185	27
25 mm (1 in.) diam								
OS050	220	32	69	10	50	40 HRF	150	22
H02	290	42	275	40	25	42	170	25
H04	330	48	305	44	20	48	185	27
50 mm (2 in.) diam								
H02	290	42	270	39	35	42	170	25

(a) At 0.5% extension under load. (b) In 50 mm or 2 in.

ductivity, extensive machining, corrosion resistance, copper color or a combination of these; electrical connectors, motors and switch components, plumbing fittings, furnace-brazed articles, screws, soldering coppers, rivets and welding torch tips

Mechanical Properties

Tensile properties. See Table 17, C14500.
Hardness. See Table 17, C14500.
Elastic modulus. Tension, 115 GPa (17×10^6 psi); shear, 44 GPa (6.4×10^6 psi)

Mass Characteristics

Density. 8.94 Mg/m³ (0.323 lb/in.³) at 20 °C (68 °F)

Thermal Properties

Liquidus temperature. 1076 °C (1969 °F)
Solidus temperature. 1067 °C (1953 °F)
Coefficient of thermal expansion. Linear, 17.0 μm/m·K (9.4 μin./in.·°F) at 20 to 100 °C (68 to 212 °F); 17.3 μm/m·K (9.6 μin./in.·°F) at 20 to 200 °C (68 to 392 °F); 17.7 μm/m·K (9.8 μin./in.·°F) at 20 to 300 °C (68 to 572 °F)
Specific heat. 385 J/kg·K (0.092 Btu/lb·°F) at 20 °C (68 °F)
Thermal conductivity. 374 W/m·K (216 Btu/ft·h·°F) at 20 °C (68 °F)

Electrical Properties

Electrical conductivity. Volumetric: O61 temper, 95% IACS at 20 °C (68 °F)
Electrical resistivity. 18.1 nΩ·m at 20 °C (68 °F)

Fabrication Characteristics

Machinability. 85% of C36000, free-cutting brass
Annealing temperature. 425 to 650 °C (800 to 1200 °F)
Hot working temperature. 750 to 875 °C (1400 to 1600 °F)

C15000
99.85Cu-0.15Zr

Commercial Names

Trade name. Amzirc Brand copper; N-4 alloy
Common name. Zirconium copper

Chemical Composition

Composition limits. 99.95 min Cu+Ag+Zr; 0.13 to 0.20 Zr

Applications

Typical uses. Stud bases for power transmitters and rectifiers, switches and circuit breakers for high-temperature service, commutators, resistance welding tips and wheels, solderless wrapped connectors. Zirconium copper is heat treatable and retains much of its room temperature strength up to 450 °C (840 °F).
Precautions in use. During hot working, forging should be discontinued if the temperature falls below 800 °C (1470 °F). The part must be reheated to at least 900 °C (1650 °F) before forging can be resumed.

Mechanical Properties

Tensile properties. See Tables 18 and 19, plus Fig. 18.
Hardness. Rod, up to 16 mm (0.62 in.) diam, TB04 or TH04 temper: 72 HRB. Wire: 6 mm (0.25 in.) diam, OS025 temper, 40 HRB; 13 mm (0.50 in.) diam, temper, 90 HRB.
Elastic modulus. Tension, 129 GPa (18.7×10^6 psi)
Fatigue strength. TH04 temper: 180 MPa (26 ksi) at 10^8 cycles
Impact strength. See Table 19.
Creep-rupture characteristics. See Table 20 and Fig. 19.

Mass Characteristics

Density. 8.89 Mg/m³ (0.321 lb/in.³) at 20 °C (68 °F)

Thermal Properties

Liquidus temperature. 1080 °C (1976 °F)
Solidus temperature. 980 °C (1796 °F)
Coefficient of thermal expansion. Linear, 16.9 μm/m·K (9.4 μin./in.·°F) at 20 to 100 °C (68 to 212 °F); 17.6

Table 18 Typical mechanical properties of C15000

Section size mm	in.	Cold work, %, after: Solution treating(c)	Aging(d)	Tensile strength MPa	ksi	Yield strength(a) MPa	ksi	Elongation(b), %
Rod								
5	0.20	...	76	430	62	385	56	8
6	0.25	10(e)	...	285	41	250	36	34
9.5	0.37	80	44	470	68	440	64	11
13	0.50	56	47	460	67	435	63	15
16	0.62	61	31	440	64	430	62	15
19	0.75	50	34	435	63	420	61	15
22	0.87	48	52	430	62	415	60	15
25	1.0	48	47	430	62	415	60	15
32	1.25	32	17	430	60	400	58	18
Wire								
1	(0.04)	...	98(f)	525	76	495	72	1.5
2.3	(0.09)	...	62(f)	495	72	470	68	3
		0	...	200	29	40	6	54
		...	0	205	30	90	13	49
6	(0.25)	0(e)(g)	...	255	37	75	11	50
13	(0.50)	30(e)	...	365	53	340	49	23

(a) At 0.5% extension under load. (b) In 50 mm or 2 in. (c) At 900 to 925 °C (1650 to 1695 °F). (d) For 1 h or more at 400 to 425 °C (750 to 795 °F). (e) Mill annealed. (f) Solution treated, cold worked the stated amount, then aged. (g) OS025 temper.

Table 19 Typical low temperature mechanical properties of C15000(a)

Test temperature °C	°F	Tensile strength MPa	ksi	Notch tensile strength(b) MPa	ksi	Yield strength(c) MPa	ksi	Elongation(d), %	Reduction in area, %	Impact strength(e) J	ft·lb
22	72	445	64.5	673	97.6	411	59.6	16	62	121	89
−78	−108	463	67.2	711	103.1	423	61.3	20	66	142	105
−197	−323	534	77.4	775	112.4	453	65.7	26	71	155	114
−253	−423	587	85.2	820	119.0	458	66.4	37	72	155	114
−269	−452	591	85.7	838	121.6	446	64.7	36	69	...	...

(a) Data are for TH04 temper — solution treated at 950 °C (1740 °F), cold worked 85 to 90%, and aged 1 h at 450 °C (840 °F). (b) For K_t of 5.0. (c) At 0.2% offset. (d) In 2 diameters. (e) Charpy V-notch, standard 10 mm square specimen.

μm/m·K (9.8 μin./in.·°F) at 20 to 300 °C (68 to 572 °F); 20.2 μm/m·K (11.2 μin./in.·°F) at 20 to 650 °C (68 to 1200 °F)

Specific heat. 385 J/kg·K (0.092 Btu/lb·°F) at 20 °C (68 °F)

Thermal conductivity. Solution treated, cold worked 84% and aged material, 367 W/m·K (212 Btu/ft·h·°F) at 20 °C (68 °F)

Electrical Properties

Electrical conductivity. Volumetric, 93% IACS at 20 °C (68 °F)

Electrical resistivity. Solution treated, cold worked 84% and aged material, 18.6 nΩ·m at 20 °C (68 °F)

Fabrication Characteristics

Machinability. 20% of C36000, free-cutting brass

Formability. Excellent capacity for being cold worked or hot formed. Most often fabricated by swaging, bending, heading or forging.

Weldability. Soldering: excellent. Brazing or resistance butt welding: good. Other welding processes not recommended.

Heat treating. Solution treat 5 to 30 min at temperature, then age 1 to 4 h. Aging time and temperature depends on section size and amount of previous cold work.

Annealing temperature. 600 to 700 °C (1112 to 1292 °F)

Solution temperature. 900 to 925 °C (1650 to 1700 °F)

Aging temperature. Aged only, 500 to 550 °C (930 to 1020 °F); cold worked and aged, 375 to 475 °C (705 to 885 °F)

Hot working temperature. 900 to 950 °C (1650 to 1740 °F)

Table 20 Typical creep strength of C15000(a)

| Test temperature | | Stress for 1% creep in | | | | | |
| | | 1000 h | | 10 000 h | | 100 000 h | |
°C	°F	MPa	ksi	MPa	ksi	MPa	ksi
TH01 Temper (17% Cold Work)							
300	570	277	40.2	241	35.0	208	30.2
350	660	217	31.5	166	24.0	185	26.8
400	750	150	21.7	123	17.9	102	14.8
450	840	98	14.2	70	10.2	51	7.4
500	930	88	12.7	39	5.6	16	2.3
600	1110	28	4.1	15	2.2	7.5	1.1
TH02 Temper (43% Cold Work)							
250	480	343	49.7	330	47.8	317	46.0
300	570	325	47.2	297	43.1	272	39.5
350	660	247	35.8	212	30.7	181	26.2
400	750	176	25.6	142	20.6	114	16.5
450	840	100	14.5	74	10.7	51	7.4
500	930	74	10.7	53	7.7	39	5.6
600	1110	18	2.6	12	1.8	8.3	1.2
TH04 Temper (82% Cold Work)							
250	480	321	46.5	312	45.2	303	44.0
300	570	305	44.2	271	39.3	240	34.8
350	660	257	37.3	238	34.5	219	31.8
400	750	201	29.2	161	23.4	139	20.2
450	840	77	11.1	53	7.7	44	6.4
500	930	63	9.2	41	6.0	28	4.0
600	1100	5.2	0.75	2.8	0.41	1.5	0.22
650	1200	3.0	0.44	1.7	0.25	1.0	0.14

(a) Solution treated, cold worked the indicated amount, then aged 1 h at 425 °C (795 °F).

Fig. 18 Short time elevated temperature tensile properties of C15000

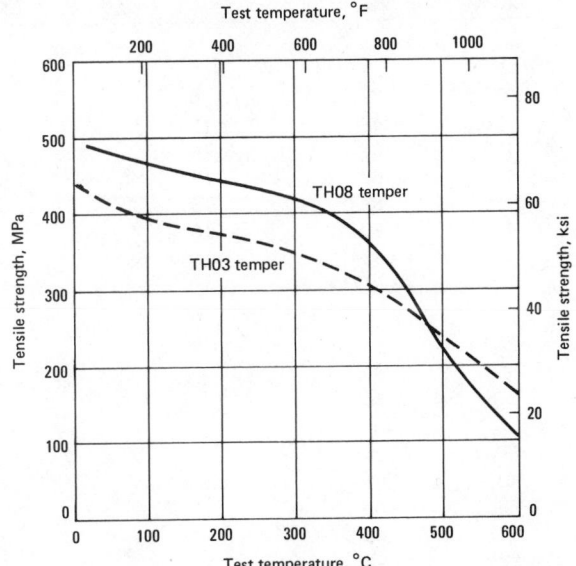
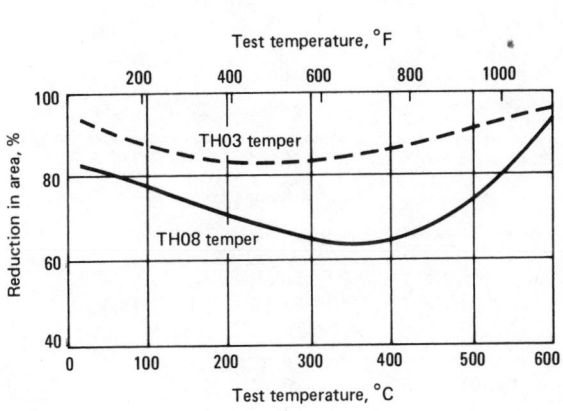

Material was solution treated 15 min at 900 °C (1650 °F), quenched, cold worked and aged TH03 temper material was cold worked 54%, then aged 1 h at 400 °C (750 °F); TH08 temper material was cold worked 84%, then aged 1 h at 375 °C (705 °F).

Fig. 19 Stress rupture properties of C15000, TH08 temper

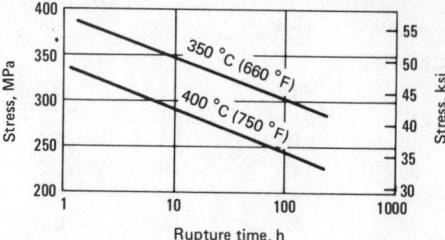

Material was solution treated 1 h at 950 °C (1740 °F), quenched, cold worked 85% and aged 1 h at 425 °C (795 °F).

C15710
99.8Cu-0.2Al$_2$O$_3$

Chemical Composition

Composition limits. 99.69 to 99.85 Cu; 0.15 to 0.25 Al$_2$O$_3$; 0.01 max Fe; 0.01 max Pb; 0.04 max O

Applications

Typical uses. Rolled strip, rolled flat wire, rod and wire for electrical connectors, light-duty current-carrying springs, inorganic insulated wire, thermocouple wire, lead wire resistance welding electrodes for aluminum, heat sinks

Mechanical Properties

Tensile properties. See Table 21. Hardness. See Table 21. Elastic modulus. Tension, 105 GPa (15 × 10^6 psi)

Mass Characteristics

Density. 8.82 Mg/m^3 (0.319 lb/in.^{3}) at 20 °C (68 °F)

Thermal Properties

Liquidus temperature. 1080 °C (1980 °F)
Coefficient of thermal expansion. Linear, 19.5 μm/m·K (10.8 μin./in.·°F) at 20 to 300 °C (68 to 572 °F)
Specific heat. 380 J/kg·K (0.09 Btu/lb·°F) at 20 °C (68 °F)
Thermal conductivity. 360 W/m·K (208 Btu/ft·h·°F)

Electrical Properties

Electrical conductivity. Volumetric, 90% IACS at 20 °C (68 °F)
Electrical resistivity. 19.2 nΩ·m at 20 °C (68 °F); temperature coefficient, 5.22 nΩ·m per K at 20 °C (68 °F)

Fabrication Characteristics

Formability. Excellent for cold working; poor for hot forming
Weldability. Soldering: excellent. Brazing: good. Resistance butt welding: fair. Resistance spot and seam welding: poor. Oxyacetylene, gas-shielded arc and coated metal-arc welding: not recommended.
Annealing temperature. 650 to 875 °C (1200 to 1600 °F)

Table 21 Typical mechanical properties of C15710

Size	Temper	Tensile strength(a) MPa	ksi	Yield strength(a)(b) MPa	ksi	Elongation in 50 mm or 2 in., %	Hardness, HRB
Rod							
	Cold worked						
24-mm (0.94-in.) diam	0%	325	47	270	39	20	60
22-mm (0.88-in.) diam	13%	345	50	330	48	18	65
19-mm (0.75-in.) diam	39%	415	60	400	58	16	70
16-mm (0.63-in.) diam	56%	450	65	425	62	12	70
10-mm (0.38-in.) diam	82%	510	74	470	68	10	72
6-mm (0.25-in.) diam	93%	530	77	485	70	10	74
	O61	325	47	275	40	20	60
Wire							
	Cold worked						
2-mm (0.09-in.) diam	98.5%	565	82	540	78	...	...
1-mm (0.05-in.) diam	99.5%	650	94	620	90	...	...
	O61	325	47	275	40	...	...
	Cold worked						
0.8-mm (0.03-in.) diam	99.8%	685	99	650	94	...	...
	65%	455	66	420	61	...	...
0.5-mm (0.02-in.) diam	99.9%	725	105	690	100	...	...
	85%	475	69	450	65	...	...
	O61	345	50	290	42	...	...

(a) Properties will vary, depending on extrusion ratio and temperature. (b) At 0.2% offset.

Table 22 Typical mechanical properties of C15720

Size	Temper	Tensile strength(a)		Yield strength(a)(b)		Elongation in 50 mm or 2 in., %	Hardness, HRB
		MPa	ksi	MPa	ksi		
Flat Products							
	Cold worked						
0.76-mm (0.03-in.) thick	91% 570		83	545	79	7	...
0.51-mm (0.02-in.) thick	95% 585		85	565	82	6	...
0.25-mm (0.01-in.) thick	97% 605		88	580	84	5	...
0.152-mm (0.006-in.) thick	98% 615		89	585	85	3.5	...
	O61 485		70	380	55	13	...
Rod							
	Cold worked						
24-mm (0.94-in.) diam	0% 470		68	365	53	19	74
21-mm (0.81-in.) diam	26% 495		72	470	68	16	77
18-mm (0.72-in.) diam	42% 510		74	485	70	14	78
16-mm (0.63-in.) diam	56% 530		77	495	72	13	79
13-mm (0.50-in.) diam	72% 540		78	505	73	11	79
10-mm (0.38-in.) diam	82% 550		80	510	74	10	80
76-mm (3.0-in.) diam	M30 525		76	510	74	13	78
102-mm (4.0-in.) diam	M30 460		67	395	57	20	68

(a) Properties will vary, depending on extrusion ratio and temperature. (b) At 0.2% offset.

C15720
99.6Cu-0.4Al$_2$O$_3$

Chemical Composition

Composition limits. 99.49 to 99.6 Cu; 0.35 to 0.45 Al$_2$O$_3$; 0.01 max Pb; 0.01 max Fe; 0.04 max O

Applications

Typical uses. Rolled and drawn strip, rolled flat wire, drawn bar, rod, wire and shapes for relay and switch springs, lead frames, contact supports, heat sinks, circuit breaker parts, rotor bars, resistance welding electrodes and wheels, and connectors. Parts requiring a combination of high strength and conductivity, particularly after exposure to high manufacturing or operating temperatures.

Mechanical Properties

Tensile properties. See Table 22.
Hardness. See Table 22.
Elastic modulus. Tension, 112 GPa (16.4 × 10^6 psi)

Mass Characteristics

Density. 8.81 Mg/m^3 (0.319 lb/in.^{3}) at 20 °C (68 °F)

Thermal Properties

Liquidus temperature. 1080 °C (1980 °F)
Coefficient of thermal expansion. Linear, 19.6 μm/m·K (10.9 μin./in.·°F) at 20 to 300 °C (68 to 572 °F)

Specific heat. 380 J/kg·K (0.09 Btu/lb·°F) at 20 °C (68 °F)
Thermal conductivity. 353 W/m·K (204 Btu/ft·h·°F)

Electrical Properties

Electrical conductivity. Volumetric, 89% IACS at 20 °C (68 °F)
Electrical resistivity. 19.4 nΩ·m at 20 °C (68 °F)

Fabrication Characteristics

Formability. Excellent for cold working; poor for hot forming
Weldability. Soldering: excellent. Brazing: good. Resistance butt welding: fair. Resistance spot and seam welding: poor. Oxyacetylene, gas-shielded arc and coated metal-arc welding: not recommended.
Annealing temperature. 650 to 925 °C (1200 to 1700 °F)

C15735
99.3Cu-0.7Al$_2$O$_3$

Chemical Composition

Composition limits. 99.19 to 99.35 Cu; 0.65 to 0.75 Al$_2$O$_3$; 0.01 max Fe; 0.01 max Pb; 0.04 max O

Applications

Typical uses. Rod for resistance welding electrodes, circuit breakers, feed-through conductors, heat sinks, motor parts, parts requiring retention of high strength and conductivity after high-temperature exposure

Mechanical Properties

Tensile properties. See Table 23.
Hardness. See Table 23.
Elastic modulus. Tension, 142 GPa (17.8 × 10^6 psi)

Mass Characteristics

Density. 8.80 Mg/m^3 (0.318 lb/in.^{3}) at 20 °C (68 °F)

Thermal Properties

Liquidus temperature. 1080 °C (1980 °F)
Coefficient of thermal expansion. Linear, 20 μm/m·K (11.1 μin./in.·°F) at 20 to 300 °C (68 to 572 °F)
Specific heat. 422 J/kg·K (0.10 Btu/lb·°F) at 20 °C (68 °F)
Thermal conductivity. 339 W/m·K (196 Btu/ft·h·°F)

Electrical Properties

Electrical conductivity. Volumetric, 85% IACS at 20 °C (68 °F)
Electrical resistivity. 20.3 nΩ·m at 20 °C (68 °F)

Fabrication Characteristics

Formability. Excellent for cold working; poor for hot forming
Weldability. Soldering: excellent. Brazing: good. Resistance butt welding: fair. Resistance spot and seam welding: poor. Oxyacetylene, gas-shielded arc and coated metal-arc welding: not recommended.
Annealing temperature. 650 to 925 °C (1200 to 1700 °F)

Table 23 Typical mechanical properties of C15735

Size	Temper	Tensile strength(a) MPa	ksi	Yield strength(a)(b) MPa	ksi	Elongation in 50 mm or 2 in., %	Hardness, HRB
Rod	Cold worked						
24-mm (0.94-in.) diam	0%	485	70	420	61	16	77
19-mm (0.75-in.) diam	39%	550	80	540	78	13	80
16-mm (0.63-in.) diam	56%	585	85	565	82	10	83
64-mm (2.5-in.) diam	M30	490	71	415	60	16	76
76-mm (3.0-in.) diam	M30	565	82	540	78	11	78
102-mm (4.0-in.) diam	M30	515	75	485	70	13	75

(a) Properties will vary, depending on extrusion ratio and temperature. (b) At 0.2% offset.

C16200
99Cu-1Cd

Commercial Names

Previous trade name. Cadmium copper

Specifications

ASTM. Wire: B9, B105
SAE. J463

Chemical Composition

Composition limits. 98.78 to 99.3 Cu; 0.7 to 1.2 Cd; 0.02 max Fe

Applications

Typical uses. Rolled strip, rod and wire for trolley wire, heating-pad and electric-blanket elements, spring contacts, railbands, high-strength transmission lines, connectors, cable wrap, switch gear components, wave guide cavities

Mechanical Properties

Tensile properties. See Table 24.
Shear strength. Rod, 13 mm (0.50 in.) diam: OS050 temper, 185 MPa (27 ksi); H04 temper, 385 MPa (56 ksi)
Hardness. See Table 24.
Elastic modulus. Tension, 11 GPa (17×10^6 psi); shear, 44 GPa (6.4×10^6 psi)
Fatigue strength. At 10^8 cycles for rod, 13 mm (0.50 in.) diam: OS050 temper, 100 MPa (14.5 ksi); H04 temper 205 MPa (30 ksi)

Mass Characteristics

Density. 8.89 Mg/m^3 (0.321 $lb/in.^3$) at 20 °C (68 °F)

Thermal Properties

Liquidus temperature. 1076 °C (1969 °F)

Solidus temperature. 1030 °C (1886 °F)
Coefficient of thermal expansion. Linear, 17.0 μm/m·K (9.4 μin./in.·°F) at 20 to 100 °C (68 to 212 °F); 17.3 μm/m·K (9.6 μin./in.·°F) at 20 to 200 °C (68 to 392 °F); 17.7 μm/m·K (9.8 μin./in.·°F) at 20 to 300 °C (68 to 572 °F)
Specific heat. 380 J/kg·K (0.09 Btu/lb·°F) at 20 °C (68 °F)
Thermal conductivity. 360 W/m·K (208 Btu/ft·h·°F)

Electrical Properties

Electrical conductivity. Volumetric, 90% IACS at 20 °C (68 °F)

Electrical resistivity. 19.2 nΩ·m at 20 °C (68 °F)

Fabrication Characteristics

Machinability. 20% of C36000, free-cutting brass
Formability. Excellent for cold working; good for hot forming
Weldability. Soldering and brazing: excellent. Oxyfuel gas, gas shielded arc and resistance butt welding: good. Shielded metal arc, resistance spot and resistance seam welding: not recommended.
Annealing temperature. 425 to 750 °C (800 to 1400 °F)
Hot working temperature. 750 to 875 °C (1400 to 1600 °F)

Table 24 Typical mechanical properties of C16200

Temper	Tensile strength MPa	ksi	Yield strength(a) MPa	ksi	Elongation(b), %	Hardness
Flat products, 1 mm (0.04 in.) Thick						
0.025 mm	240	35	76	11	52	54 HRF
Hard	415	60	310	45	5	64 HRB
Spring	440	64	...	...	3	73 HRB
Extra spring	495	72	405	59	1	75 HRB
Rod, 13 mm (0.50 in.) Diam						
0.050 mm	240	35	48	7	56	46 HRF
0.025 mm	250	36	83	12	57	46 HRF
Half hard (25%)	400	58	310	45	12	65 HRB
Hard	505	73	474	68.7	9	73 HRB
Wire, 0.25 mm (0.01 in.) Diam						
Drawn (> 99%)	690	100	...	...	1.0(c)	...
Wire, 2 mm (0.08 in.) Diam						
0.025 mm	260	38	83	12	50	...
Hard	485	70	380	55	6	...
Spring	550	80	455	66	2	...
Drawn (> 96%)	605	88	...	...	1.5(c)	...

(a) At 0.5% extension under load. (b) In 50 mm or 2 in. (c) In 1.5 m (60 in.).

C17000
98Cu-1.7Be-0.3Co

Commercial Names

Trade name. Berylco 165
Common name. Beryllium copper; 165 alloy

Specifications

ASTM. Flat products: B194. Rod, bar: B196. Forgings and extrusions: B570
SAE. J463
Government. QQ-C-533

RWMA. Class IV

Chemical Composition

Composition limits. 1.60 to 1.79 Be; 0.20 min Ni + Co; 0.6 max Ni + Fe + Co; rem Cu

Applications

Typical uses. Bellows, Bourdon tubing, diaphragms, fuse clips, fasteners, lock washers, springs, switch and relay parts, electrical and electronic components, retaining rings, roll pins, valves, pumps, spline shafts, rolling mill parts, welding equipment, nonsparking safety tools
Precautions in use. *Health hazard.* Because this alloy contains beryllium, ventilation must be provided for dry sectioning and grinding, machining, melting, welding and any other process that produces metal dust or fumes.

Mechanical Properties

Tensile properties. See Table 25 and 26.
Hardness. See Tables 26 and 27.
Poisson's ratio. 0.30
Elastic modulus. Tension, 115 GPa

Table 25 Typical mechanical properties and electrical conductivity of C17000 strip

Temper	Tensile strength MPa	ksi	Proportional limit(a) MPa	ksi	Yield strength(b) MPa	ksi	Elongation(c), %	Electrical conductivity, % IACS	Fatigue strength(d) MPa	ksi
TB00	410-540	60-78	100-140	15-20	190-370	28-53	35-60	17-19	190-230	28-33
TD01	520-610	75-88	280-410	40-60	310-520	45-75	10-35	16-18	200-230	29-34
TD02	590-690	85-100	380-480	55-70	450-620	65-90	5-25	15-17	220-260	32-38
TD04	690-825	100-120	480-590	70-85	550-760	80-110	2-8	15-17	240-270	35-39
TF00(e)	1030-1240	150-180	550-760	80-110	895-1140	130-165	4-10	22-25	240-270	35-39
TH01(f)	1100-1280	160-185	620-795	90-115	930-1170	135-170	3-6	22-25	250-280	36-41
TH02(f)	1170-1340	170-195	660-860	95-125	1000-1210	145-175	2-5	22-25	250-290	36-42
TH04(f)	1240-1380	180-200	690-930	100-135	1070-1240	155-180	1-4	22-25	260-310	38-45
TM00(g)	690-760	100-110	480-590	70-85	520-620	75-90	18-22	20-33	230-260	33-37
TM01(g)	760-825	110-120	520-660	75-95	620-760	90-110	15-19	20-33	230-260	34-38
TM02(g)	825-930	120-135	550-690	80-100	690-860	100-125	12-16	20-33	240-270	35-39
TM04(g)	930-1030	135-150	590-725	85-105	760-930	110-135	9-13	20-33	250-280	36-40
TM06(g)	1030-1100	150-160	590-760	85-110	860-965	125-140	9-12	20-33	260-290	37-40
TM08(g)	1100-1210	160-175	620-795	90-115	965-1140	140-165	3-7	20-33	230-310	33-45

(a) At 0.002% offset. (b) At 0.2% offset. (c) In 50 mm or 2 in. (d) Rotating beam at 10^8 cycles. (e) Aged 3 h at 315 °C (600 °F). (f) Aged 2 h at 315 °C (600 °F). (g) Proprietary mill heat treatment intended to produce the stated tensile properties.

Table 26 Typical mechanical properties and electrical conductivity of C17000 rod, bar, plate, tubing, billets and forgings

Size	Temper	Tensile strength MPa	ksi	Yield strength(a) MPa	ksi	Elongation(b), %	Hardness HRB	HRC	Electrical conductivity, % IACS
Rod, Bar, Plate, Tubing									
All sizes	TB00	415-585	60-85	140-205	20-30	35-60	45-85	···	17-19
	TF00(c)	1035-1240	150-180	860-1070	125-155	4-10	···	32-39	22-25
<10 mm (<⅜ in.)	TD04	655-895	95-130	515-725	75-105	10-20	92-103	···	15-17
	TH04(d)	1205-1380	175-200	930-1140	135-165	2-5	···	36-41	22-25
10-25 mm (⅜-1 in.)	TD04	620-825	90-120	515-725	75-105	10-20	91-102	···	15-17
	TH04(d)	1170-1345	170-195	930-1140	135-165	2-5	···	35-40	22-25
>25 mm (>1 in.)	TD04	585-795	85-115	515-725	75-105	10-20	88-101	···	15-17
	TH04(d)	1140-1310	165-190	930-1140	135-165	2-5	···	34-39	22-25
Billet									
···	As cast	515-585	75-85	275-345	40-50	15-30	80-85	···	16-22
	Cast and aged(c)	655-690	95-100	485-515	70-75	10-25	···	18-25	18-23
	TB00	415-515	60-75	170-205	25-30	25-45	65-75	···	13-18
	TF00(c)	965-1170	140-170	725-930	105-135	1-4	···	30-38	18-25
Forgings									
···	TB00	415-585	60-85	140-205	20-30	35-60	45-85	···	17-19
	TF00(c)	1035-1240	150-180	860-1070	125-155	4-10	···	32-39	22-25

(a) At 0.2% offset. (b) In 50 mm or 2 in. (c) Aged 3 h at 350 °C (625 °F). (d) Aged 2 to 3 h at 330 °C (625 °F).

(17×10^6 psi); shear, 50 GPa (7.3×10^6 psi)

Fatigue strength. See Table 25.

Mass Characteristics

Density. 8.26 Mg/m^3 (0.298 lb/in.^{3}) at 20 °C (68 °F)

Volume change on phase transformation. During age hardening: 0.2% max decrease in length; 0.6% max increase in density

Thermal Properties

Liquidus temperature. 980 °C (1800 °F)

Solidus temperature. 865 °C (1590 °F)

Coefficient of thermal expansion. Linear, 16.7 µm/m·K (9.3 µin./in.·°F) at 20 to 100 °C (68 to 212 °F); 17.0 µm/m·K (9.4 µin./in.·°F) at 20 to 200 °C (68 to 392 °F); 17.8 µm/m·K (9.9 µin./in.·°F) at 20 to 300 °C (68 to 572 °F)

Specific heat. 420 J/kg·K (0.10 Btu/lb·°F) at 20 °C (68 °F)

Thermal conductivity. 118 W/m·K (69 Btu/ft·h·°F) at 20 °C (68 °F); 145 W/m·K (84 Btu/ft·h·°F) at 200 °C (392 °F)

Electrical Properties

Electrical conductivity. Volumetric, 15 to 33% IACS at 20 °C (68 °F), depending on heat treatment. See also Tables 25 and 26.

Electrical resistivity. Typical, 76.8 nΩ·m at 20 °C (68 °F), but varies with heat treatment

Chemical Properties

General corrosion behavior. Similar to that of other high-copper alloys, and basically the same as that of pure copper

Resistance to specific corroding agents. Essentially the same as that of C17200. See also Table 28.

Fabrication Characteristics

Machinability. 20% of C36000, free-cutting brass

Formability. This alloy can be formed, drawn, blanked, pierced and machined in the unhardened condition.

Weldability. Soldering, brazing, gas-shielded arc welding, shielded metal-arc welding and resistance spot welding: good. Resistance seam and resistance butt welding: fair. Oxyfuel gas welding: not recommended.

Recrystallization temperature. 730 °C (1350 °F)

Annealing temperature. Strip, thin rod, wire: 775 to 800 °C (1425 to 1475 °F) for 10 min, water quench.

Larger sections: 1 h for each 25 mm (1 in.) of thickness

Solution temperature. 760 to 790 °C (1400 to 1450 °F). All annealing of this material is a solution treatment.

Aging temperature. 260 to 425 °C (500 to 800 °F). Maximum strength is obtained by aging 1 to 3 h at 315 to 345 °C (600 to 650 °F), depending on amount of cold work preceding the aging treatment.

Hot working temperature. 650 to 825 °C (1200 to 1500 °F)

Hot-shortness temperature. 845 °C (1550 °F)

Table 27 Typical hardnesses of C17000 strip

Temper(a)	HV	Standard Rockwell	Superficial Rockwell
TB00	90-160	45-78 HRB	45-67 HR30T
TD01	150-190	68-90 HRB	62-75 HR30T
TD02	185-225	88-96 HRB	74-79 HR30T
TD04	200-260	96-102 HRB	79-83 HR30T
TF00	320 min	33-38 HRC	55-58 HR30N
TH01	343 min	35-39 HRC	55-59 HR30N
TH02	360 min	37-40 HRC	56-80 HR30N
TH04	370 min	39-41 HRC	58-61 HR30N
TM00	200-235	18-23 HRC	37-42 HR30N
TM01	230-265	21-26 HRC	42-46 HR30N
TM02	260-295	25-30 HRC	46-50 HR30N
TM04	290-325	30-35 HRC	50-54 HR30N
TM06	320-350	31-37 HRC	52-56 HR30N
TM08	434-375	32-38 HRC	55-58 HR30N

(a) For explanation of tempers, see footnotes in **Table 1**.

Table 28 Approximate corrosion resistance of C17000

Good resistance	Fair resistance	Poor resistance
Acetate solvents	Acetic acid, cold, aerated	Acetic acid, hot
Acetic acid, cold, unaerated	Acetic anhydride	Ammonia, moist
Alcohols	Acetylene	Ammonium hydroxide
Ammonia, dry	Ammonium chloride	Ammonium nitrate
Atmosphere, rural, industrial, marine	Ammonium sulfate	Bromine, aerated or hot
Benzine	Aniline	Chlorine, moist or warm
Borax	Bromine, dry	Chromic acid
Boric acid	Carbonic acid	Ferric chloride
Brine	Copper nitrate	Ferric sulfate
Butane	Ferrous chloride	Fluorine, moist or warm
Carbon dioxide	Ferrous sulfate	Hydrochloric acid, over 0.1%
Carbon tetrachloride	Fluorine, dry	Hydrocyanic acid
Chlorine, dry	Hydrochloric acid, up to 0.1%	Hydrofluoric acid, concentrate
Freon	Hydrofluoric acid, dilute	Hydrogen sulfide, moist
Gasoline	Hydrofluosilicic acid	Lactic acid, hot or aerated
Hydrogen	Hydrogen peroxide	Mercuric chloride
Nitrogen	Nitric acid, up to 0.1%	Mercury
Oxalic acid	Phenol	Mercury salts
Potassium chloride	Phosphoric acid, unaerated	Nitric acid, over 0.1%
Potassium sulfate	Potassium hydroxide	Phosphoric acid, aerated
Propane	Sodium hydroxide	Picric acid
Rosin	Sodium hypochlorite	Potassium cyanide
Sodium bicarbonate	Sodium peroxide	Silver chloride
Sodium chloride	Sodium sulfide	Sodium cyanide
Sodium sulfate	Sulfur	Stannic chloride
Sulfur dioxide	Sulfur chloride	Sulfuric acid, aerated
Sulfur trioxide	Sulfuric acid, unaerated	Sulfurous acid
Water, fresh or salt	Zinc chloride	Tartaric acid, hot or aerated

Note: Good: less than 0.25 mm/year (0.01 in./year) penetration. Fair: 0.025 to 2.5 mm/year (0.001-0.10 in./year) penetration. Poor: more than 0.25 mm/year (0.01 in./year) penetration.

C17200, C17300

Commercial Names

Previous trade names. C17200: 25 alloy, alloy 25. C17300: alloy M25
Common name. Beryllium copper

Specifications

AMS. Flat products: 4530, 4532. Rod, bar and forgings: 4650. Wire: 4725
ASTM. Flat products: B194 (C17200 only), B196. Rod and bar: B196. Wire: B197 (C17200 only). Forgings and extrusions: B570 (C17200 only)
SAE. J463 (C17200)
Government. Strip: QQ-C-533 (C17200 only). Rod and bar: MIL-C-21657, QQ-C-530. Wire: QQ-C-530
RWMA. Class IV

Chemical Composition

Composition limits. 1.80 to 2.00 Be; 0.20 min Ni + Co; 0.6 max Ni + Co + Fe; 0.10 max Pb (C17200) or 0.20 to 0.6 Pb (C17300); 0.5 max others (total); rem Cu
Consequence of exceeding impurity limits. Excessive P and Si decrease electrical conductivity. Excessive Sn and Pb cause hot shortness.

Applications

Typical uses. C17200 and C17300 are used in parts that are subject to severe forming conditions but require high strength, anelasticity, and fatigue and creep resistance (a wide variety of springs, flexible metal hose, Bourdon tubing, bellows, clips, washers, retaining rings); in parts that require high strength or wear resistance along with good electrical conductivity and/or magnetic characteristics (navigational instruments, nonsparking safety tools, firing pins, bushings, valves, pumps, shafts, rolling mill parts); and in parts requiring high strength and good corrosion resistance and electrical conductivity (electromechanical springs, diaphragms, contact bridges, bolts, screws).
Precautions in use. *Health hazard;* this alloy contains beryllium. Adequate ventilation should be provided for dry sectioning, melting, grinding, machining, welding and any other fabrication or testing process that produces dust or fumes.

Table 29 Tensile property ranges for C17200 and C17300 strip of various tempers(a)

Temper	Tensile strength MPa	ksi	Proportional limit(b) MPa	ksi	Yield strength(c) MPa	ksi	Elongation(d), %
TB00......	415-540	60-78	105-140	15-20	195-380	28-55	35-60
TD01	515-605	75-88	275-415	40-60	415-605	60-88	10-36
TD02	585-690	85-100	380-485	55-70	515-655	75-95	5-25
TD04	690-825	100-120	485-585	70-85	620-770	90-112	2-8
TF00(e) ...	1140-1310	165-190	690-860	100-125	965-1205	140-175	4-10
TH01(f) ...	1205-1380	175-200	760-930	110-135	1035-1275	150-185	3-6
TH02(f) ...	1275-1450	185-210	825-1000	120-145	1105-1345	160-195	2-5
TH04(f) ...	1310-1480	190-215	860-1070	125-155	1140-1415	165-205	1-4
TM00(g) ..	690-760	100-110	450-585	65-85	515-620	75-90	18-23
TM01(g) ..	760-825	110-120	515-655	75-95	620-760	90-110	15-20
TM02(g) ..	825-930	120-135	585-725	85-105	690-860	100-125	12-18
TM04(g) ..	930-1035	135-150	655-795	95-115	795-930	115-135	9-15
(g)	1035-1105	150-160	725-825	105-120	860-965	125-140	9-14
TM06(g) ..	1105-1205	160-175	760-860	110-125	1000-1170	145-170	4-10
TM08(g) ..	1205-1310	175-190	795-895	115-130	1070-1240	155-180	3-9

(a) For hardness, conductivity and fatigue strength, see **Table 3.** (b) At 0.002% offset. (c) At 0.2% offset. (d) In 50 mm or 2 in. (e) Solution treated and aged 3 h at 315 °C (600 °F). (f) Cold rolled and aged 2 h at 315 °C (600 °F). (g) Proprietary mill treatment to produce the indicated tensile properties.

Mechanical Properties

Tensile properties. See Tables 29 and 30.
Hardness. See Tables 29 and 30.
Poisson's ratio. 0.30
Elastic modulus. Tension, 125 to 130 GPa (18 to 19 × 10^6 psi); shear, 50 GPa (7.3 × 10^6 psi)
Fatigue strength. Rotating beam: 380 to 480 MPa (55 to 70 ksi) at 10^7 cycles for both TF00 temper rod having a tensile strength of 1140 to 1310 MPa (165 to 190 ksi) and TH04 temper rod having a tensile strength of 1280 to 1480 MPa (185 to 215 ksi). Reversed torsion: 170 to 275 MPa (25 to 40 ksi). See also Table 31.

Structure

Crystal structure. α-Cu solid solution is face-centered cubic, disordered. At 20 °C (68 °F), the lattice parameter of the parent phase with about 1.8% Be, homogenized at 815 °C (1500 °F) and quenched in water, is 0.3570 nm. The lattice parameter decreases strongly with increasing beryllium content. Age hardening begins with the formation of coherent Guinier-Preston (G-P) zones on {100} planes. The intermediate precipitate gamma-prime may be nucleated either from the G-P zones or discontinuously at the grain boundaries. In either case, it has a B2 superlattice structure, a lattice parameter of 0.270 nm, and the orientation $(\bar{1}13)_\alpha \parallel (130)_{\gamma'}$, $[110]_\alpha \parallel [001]_{\gamma'}$.

The equilibrium precipitate gamma, which requires longer aging times than are normally used commercially, is body-centered cubic of the CsCl type with a B2 superlattice structure, a lattice parameter of 0.270 nm, and the orientation $(\bar{1}11)_\alpha \parallel (110)_\gamma$, $[110]_\alpha \parallel [001]_\gamma$.
Microstructure. Small, mainly spheroidal, uniformly dispersed (Cu,Co)Be beryllides (bluish gray) in a matrix of equiaxed α-Cu. (Typical grain size is 0.012 to 0.030 nm in wrought product.) There is a strong tendency to form mechanical and annealing twins. In the age-hardened condition, the matrix shows pronounced striations (the so-called "tweed structure") caused by G-P zone formation on {110} planes. At long aging times (>8 h) or high aging temperatures (>315 °C, or >600 °F), there is a strong tendency to form continuous bands of cellular precipitate at the grain boundaries and along twin boundaries. Conventional metallographic techniques may be used. Dry sectioning and grinding should be done in a ventilated area. One of the common etchants for immersion etching is ammonium persulfate—3 parts concentrated NH_4OH, 1 part 3% H_2O_2, 2 parts 10% $(NH_4)_2S_2O_3$ and 7 to 10 parts H_2O. This etchant reveals general details of the microstructure. The matrix is stained blue to deep lavender, depending on the state of heat treatment, etchant concentration and

Table 30 Property ranges for various mill products of C17200 and C17300

Temper	Thickness or diameter	Tensile strength MPa	ksi	Yield strength(a) MPa	ksi	Elonga-tion(b), %	Hardness		Electrical conductivity, % IACS
Rod, Bar, Plate and Tubing									
TB00	All sizes	415-585	60-85	140-205	20-30	35-60	45-85	HRB	17-19
TD04	Up to 9.5 mm (⅜ in.)	655-900	95-130	515-725	75-105	10-20	92-103	HRB	15-17
	9.5-25 mm (⅜-1 in.)	620-825	90-120	515-725	75-105	10-20	91-102	HRB	15-17
	Over 25 mm (1 in.)	585-790	85-115	515-725	75-105	10-20	88-102	HRB	15-17
TF00(c)	All sizes	1140-1310	165-190	1000-1210	145-175	3-10	36-40	HRC	22-25
TH04(d) ...	Up to 9.5 mm (⅜ in.)	1280-1480	185-215	1140-1380	165-200	2-5	39-45	HRC	22-25
	9.5-25 mm (⅜-1 in.)	1240-1450	180-210	1140-1380	165-200	2-5	38-44	HRC	22-25
	Over 25 mm (1 in.)	1210-1410	175-205	1030-1340	150-200	2-5	37-43	HRC	22-25
Wire									
TB00	All sizes	400-540	58-78	140-240	20-35	35-55	· · ·		17-19
TD04	Up to 2 mm (0.08 in.)	895-1070	130-155	760-930	110-135	2-8	· · ·		15-17
	2-9.5 mm (0.08-0.38 in.)	655-900	95-130	515-725	75-105	10-35	· · ·		15-17
	Over 9.5 mm (0.38 in.)	620-825	90-120	515-725	75-105	10-35	· · ·		15-17
TF00(c)	All sizes	1140-1310	165-190	1000-1210	145-175	3-8	· · ·		22-25
TH04(e)	Up to 2 mm (0.08 in.)	1310-1590	190-230	1240-1410	180-205	1-3	· · ·		22-25
TH04(f)	2-9.5 mm (0.08-0.38 in.)	1280-1480	185-215	1210-1380	175-200	2-5	· · ·		22-25
	Over 9.5 mm (0.38 in.)	1240-1450	180-210	1140-1380	165-200	2-5	· · ·		22-25
Billets									
As cast	· · ·	515-585	75-85	275-345	40-50	15-30	80-85	HRB	16-22
Cast and aged(c) ...	· · ·	725-760	105-110	515-550	75-80	10-20	20-25	HRC	18-23
TB00	· · ·	415-515	60-75	170-205	25-30	25-45	65-75	HRB	13-18
TF00(c)	· · ·	1070-1210	155-175	860-1030	125-150	1-3	36-42	HRC	18-25
Forgings									
TB00	· · ·	415-585	60-85	140-205	20-30	35-60	45-85	HRB	17-19
TF00(c) · · · ·	· · ·	1140-1310	165-190	1000-1210	145-175	3-10	36-41	HRC	22-25

(a) At 0.2% offset. (b) In 50 mm or 2 in. (c) Aged 3 h at 330 °C (625 °F). (d) Aged 2 to 3 h at 330 °C (625 °F). (e) Aged 1 h at 330 °C (625 °F). (f) Aged 1½ to 3 h at 330 °C (625 °F)

etching time. The etchant should be freshly made. A common etchant for swabbing is potassium dichromate—$K_2Cr_2O_7$, 1.5 g NaCl, 8 ml H_2SO_4 and 100 ml H_2O. This etchant emphasizes grain boundaries, particularly when heavily decorated with discontinuous precipitate. A very effective procedure for studying grain boundaries and discontinuous precipitation is to first etch with ammonium persulfate, then remove the stain with a single wipe of the dichromate etchant.

Mass Characteristics

Density. 8.25 Mg/m^3 (0.298 $lb/in.^3$) at 20 °C (68 °F)

Volume change on phase transformation. During age hardening, there is a maximum decrease in length of 0.2% and a maximum increase in density of 0.6%

Table 31 Hardness, conductivity and fatigue strength for C17200 and C17300 strip of various tempers(a)

Temper	HV	Hardness HRC		HR30N	Electrical conductivity, % IACS	Fatigue strength(b) MPa	ksi
TB00	90-160	45-78	HRB	45-67 HR30T	17-19	205-240	30-35
TD01	150-190	68-90	HRB	62-75 HR30T	16-18	215-250	31-36
TD02	185-225	88-96	HRB	74-79 HR30T	15-17	220-260	32-38
TD04	200-260	96-102	HRB	79-83 HR30T	15-17	240-270	35-39
TF00	343 min	31-41		56-61	22-25	240-260	35-38
TH01	370 min	38-42		58-63	22-25	240-270	35-39
TH02	380 min	39-44		59-65	22-25	270-295	39-43
TH04	385 min	40-45		60-65	22-25	285-315	41-46
TM00	200-235	18-23		37-42	20-28	230-255	33-37
TM01	230-265	21-26		42-47	20-28	235-260	34-38
TM02	260-295	25-30		45-51	20-28	240-295	35-43
TM04	290-325	30-35		50-55	20-28	260-310	38-45
(c)	320-350	31-37		52-56	20-28	260-310	38-45
TM06	343-375	32-38		55-58	20-28	260-310	38-45
TM08	370-400	33-42		56-63	20-28	275-330	40-48

(a) For tensile properties, see Table 29. (b) In reversed bending at 10^8 cycles. (c) Proprietary mill heat treatment to produce tensile strength of 1030 to 1100 MPa.

Table 32 Approximate corrosion resistance of C17200 and C17300

Good resistance	Fair resistance	Poor resistance
Acetate solvents	Acetic acid, cold, aerated	Acetic acid, hot
Acetic acid, cold, unaerated	Acetic anhydride	Ammonia, moist
Alcohols	Acetylene	Ammonium hydroxide
Ammonia, dry	Ammonium chloride	Ammonium nitrate
Atmosphere, rural, industrial, marine	Ammonium sulfate	Bromine, aerated or hot
Benzene	Aniline	Chlorine, moist or warm
Borax	Bromine, dry	Chromic acid
Boric acid	Carbonic acid	Ferric chloride
Brine	Copper nitrate	Ferric sulfate
Butane	Ferrous chloride	Fluorine, moist or warm
Carbon dioxide	Ferrous sulfate	Hydrochloric acid, over 0.1%
Carbon tetrachloride	Fluorine, dry	Hydrocyanic acid
Chlorine, dry	Hydrochloric acid, up to 0.1%	Hydrofluoric acid, concentrated
Freon	Hydrofluoric acid, dilute	Hydrogen sulfide, moist
Gasoline	Hydrofluosilicic acid	Lactic acid, hot or aerated
Hydrogen	Hydrogen peroxide	Mercuric chloride
Nitrogen	Nitric acid, up to 0.1%	Mercury
Oxalic acid	Phenol	Mercury salts
Potassium chloride	Phosphoric acid, unaerated	Nitric acid, over 0.1%
Potassium sulfate	Potassium hydroxide	Phosphoric acid, aerated
Propane	Sodium hydroxide	Picric acid
Rosin	Sodium hypochlorite	Potassium cyanide
Sodium bicarbonate	Sodium peroxide	Silver chloride
Sodium chloride	Sodium sulfide	Sodium cyanide
Sodium sulfate	Sulfur	Stannic chloride
Sulfur dioxide	Sulfur chloride	Sulfuric acid, aerated
Sulfur trioxide	Sulfuric acid, unaerated	Sulfurous acid
Water, fresh or salt	Zinc chloride	

Note: Good: less than 0.25 mm/year (0.01 in./year) attack. Fair: 0.025 to 2.54 mm/year (0.001 to 0.10 in./year) attack. Poor: more than 0.25 mm/year (0.01 in./year) attack.

Thermal Properties

Liquidus temperature. 980 °C (1800 °F)
Solidus temperature. 865 °C (1590 °F)

Coefficient of thermal expansion. Linear, 16.7 μm/m·K (9.3 μin./in.·°F) at 20 to 100 °C (68 to 212 °F); 17.0 μm/m·K (9.4 μin./in.·°F) at 20 to 200 °C (68 to 392 °F); 17.8 μm/m·K (9.9 μin./in.·°F) at 20 to 300 °C (68 to 572 °F)

Specific heat. 420 J/kg·K (0.10 Btu/lb·°F) at 20 to 100 °C (68 to 212 °F); 425 J/kg·K (0.11 Btu/lb·°F) at 100 to 300 °C (212 to 572 °F)
Thermal conductivity. 105 to 130 W/m·K (60 to 75 Btu/ft·h·°F) at 20 °C (68 °F); 130 to 133 W/m·K (75 to 77 Btu/ft·h·°F) at 200 °C (392 °F)

Electrical Properties

Electrical conductivity. Volumetric, 15 to 30% IACS at 20 °C (68 °F), depending on heat treatment. See also Tables 30 and 31.
Electrical resistivity. 57 to 115 nΩ·m at 20 °C (68 °F), depending on heat treatment

Chemical Properties

General corrosion behavior. Similar to that of other high-copper alloys; basically the same as that of pure copper
Resistance to specific corroding agents. See Table 32.

Fabrication Characteristics

Machinability. C17200: 20% of C36000, free-cutting brass. C17300: 50% of C36000. Both alloys can be readily machined by all conventional methods. Specific machining parameters depend on shapes, machining method and temper or condition of the metal. The leaded version of this alloy, C17300, is especially intended for machined parts. Other properties are unchanged by the addition of lead to enhance machinability.
Recrystallization temperature. Approximately 730 °C (1350 °F)

Annealing temperature. Strip, thin rod and wire: 760 to 790 °C (1400 to 1450 °F)/10 min/water quench. Larger sections: 1 h per inch or fraction of an inch of cross section

Solution temperature. 760 to 790 °C (1400 to 1450 °F). All annealing of this material is a solution treatment.

Aging temperature. 260 to 425 °C (500 to 800 °F). Maximum strength is obtained by aging material 1 to 3 h at 315 to 345 °C (600 to 650 °F), depending on the amount of cold work.

Hot working temperature. 650 to 800 °C (1200 to 1475 °F). C17300 cannot be hot rolled or forged, but can be hot extruded.
Hot-shortness temperature. 845 °C (1580 °F)

C17500
97Cu-0.5Be-2.5Co

Commercial Names

Trade names. 10 alloy, Alloy 10, Berylco 10

Common name. Low-beryllium copper

Specifications

ASTM. Flat products: B534. Rod, bar: B441.
SAE. J463
Government. Rod, bar: MIL-C-46087. Strip: MIL-C-81021.
RWMA. Class III

Chemical Composition

Composition limits. 0.40 to 0.7 Be; 2.4 to 2.7 Co; 0.10 max Fe; 0.5 max others (total); rem Cu

Applications

Typical uses. Strip, wire: fuse clips, fasteners, springs, switch parts, electrical connectors and conductors. Rod, plate: resistance spot welding tips, seam welding discs, die casting plunger tips, tooling for plastic molding.

Precautions in use. *Health hazard.* Because this alloy contains beryllium, adequate safety precautions are mandatory for all melting, welding, grinding and machining operations.

Mechanical Properties

Tensile properties. See Table 33 and Fig. 20.

Hardness. See Table 33.

Elastic modulus. Tension, 125 to 130 GPa (18 to 19 × 10^6 psi)

Fatigue strength. Rod, TF00 temper (rotating beam tests): 275 to 310 MPa (40 to 45 ksi) at 10^7 cycles: see Table 33.

Structure

Crystal structure. The α-Cu solid solution is face-centered cubic. The beryllide, (Cu,Co)Be, is ordered body-centered cubic of the CsCl (B2) type.

Microstructure. α-Cu with beryllium in solid solution, and (Cu,Co)Be beryllide inclusions. The appearance of the matrix of the beryllides depends on the extent of deformation and the state of heat treatment. In the cast condition, the matrix is essentially like pure copper; the beryllides, which are blue-gray, are large and sharply angular in the grain boundaries and small with Widmanstätten orientation within the grains. When such cast shapes are annealed, the cored appearance is reduced slightly and the matrix becomes slightly "cleaner" as small amounts of the beryllides are dissolved. As the cast product is reduced by either hot or cold working, the beryllides are broken up and uniformly distributed. For such products as strip or rod, the microstructure is fine-grained, equiaxed α-Cu with small, mainly spherical, uniformly dispersed beryllides. For all product types, there is little difference in microstructure between the annealed and the aged conditions. Metallography is by ordinary metallographic techniques except that grinding must be performed in a vented area, and all other appropriate OSHA requirements should be strictly observed.

Mass Characteristics

Density. 8.75 Mg/m^3 (0.316 lb/in.^{3}) at 20 °C (68 °F)

Volume change on phase transformation. Slight contraction during age hardening; exact amount depends on starting condition of material, and on time and temperature of aging.

Thermal Properties

Liquidus temperature. 1070 °C (1955 °F)

Solidus temperature. 1030 °C (1885 °F)

Coefficient of thermal expansion. Linear, 17.6 μm/m·K (9.8 μin./in.·°F) at 20 to 200 °C (68 to 392 °F)

Specific heat. 420 J/kg·K (0.10 Btu/lb·°F) at 20 °C (68 °F)

Electrical Properties

Electrical conductivity. See Table 33.

Electrical resistivity. 29 to 86 nΩ·m at 20 °C (68 °F), depending on heat treatment

Chemical Properties

General corrosion behavior. Comparable to that of other high-copper

Table 33 Typical mechanical properties and electrical conductivity of C17500

Temper	Tensile strength MPa	ksi	Proportional limit(a) MPa	ksi	Yield strength(b) MPa	ksi	Elongation(c), %	Hardness, HRB	Electrical conductivity, % IACS	Fatigue strength(d) MPa	ksi
Strip											
TB00	240-380	35-55	69-140	10-20	140-205	20-30	20-35	28-50	20-30	...	...
H04	485-585	70-85	240-450	35-65	380-550	55-80	3-10	70-80	20-30	205	30
TF00	690-825	100-120	380-515	55-75	550-690	80-100	10-20	92-100	45-60	205	30
TH04	760-895	110-130	485-655	70-95	690-825	100-120	8-15	98-102	50-60	240	35
HTR(e)	825-1035	120-150	550-760	80-110	760-965	110-140	1-4	98-103	45-60	240-260	35-38
HTC(f)	515-585	75-85	205-415	30-60	345-515	50-75	8-15	79-88	60 min	205-240	30-35
Rod, Bar, Plate, Tubing											
TB00	240-380	35-55	...	...	140-205	20-30	20-35	20-50	20-30	...	...
H04	450-550	65-80	...	...	380-515	55-75	10-15	60-80	20-30	...	...
TF00	690-825	100-120	...	...	550-690	80-100	10-25	92-100	45-60	...	...
TH04	760-895	110-130	...	...	690-825	100-120	10-20	95-102	50-60	...	...
Forged Products											
TB00	240-380	35-55	...	...	140-205	20-30	20-35	20-50	20-30	...	...
TF00	690-825	100-120	...	...	550-690	80-100	10-25	92-100	45-60	...	...

(a) At 0.002% offset. (b) At 0.2% offset. (c) In 50 mm or 2 in. (d) Reversed bending; at 10^8 cycles. (e) Proprietary mill hardening for maximum strength. (f) Proprietary mill hardening for maximum electrical conductivity.

alloys. May tarnish in humid or sulfur-bearing atmospheres.

Fabrication Characteristics

Machinability. Readily machinable by all common methods. Recommended machining conditions depend greatly on shape of part, on state of heat treatment of material and on type of machining operation. Because this alloy contains beryllium, OSHA requirements must be strictly observed. Normally, these requirements include flooding and/or special ventilation to prevent personnel from inhaling or ingesting metal dust.

Annealing temperature. All annealing of this alloy is a solution treatment.

Solution treatment. Strip, rod, bar, tubing, wire: 10 min at 900 to 955 °C (1650 to 1750 °F), water quench. Large sections: 1 h per inch or fraction of an inch at 900 to 925 °C (1650 to 1700 °F), water quench.

Aging temperature. For maximum strength: 3 to 6 h at 425 °C (800 °F), depending on degree of cold work. Commercial practice: 2 to 3 h at 468 to 495 °C (875 to 925 °F) to provide a combination of high strength and electrical conductivity. Cooling rate after aging is not critical. See also Fig. 20.

Hot working temperature. 700 to 925 °C (1300 to 1700 °F)

Hot-shortness temperature. 980 °C (1800 °F)

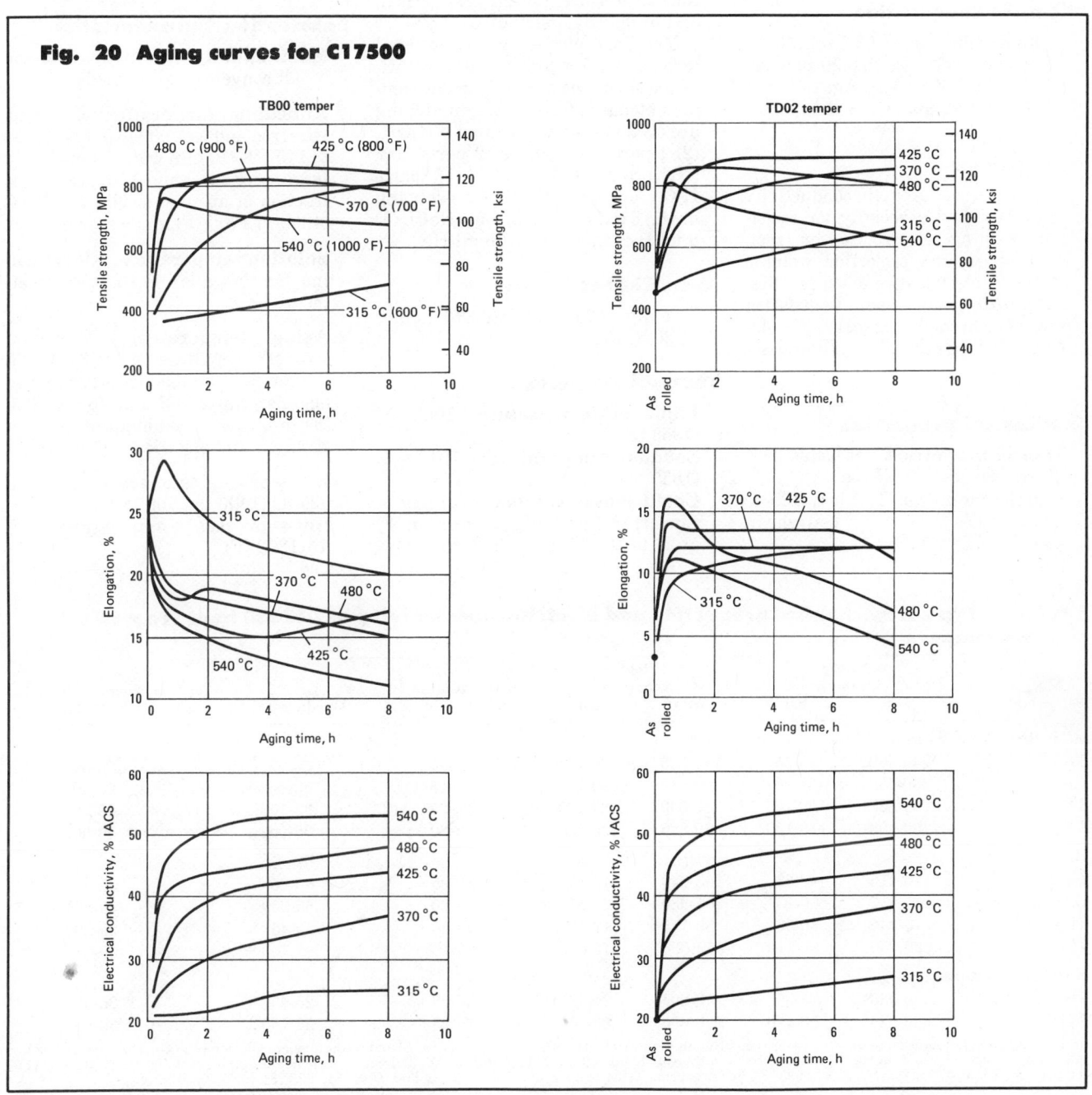

Fig. 20 Aging curves for C17500

C17600

Commercial Names

Previous trade name. 50 alloy, alloy 50

Common name. Beryllium copper

Specifications

SAE. J463 (CA176)
RWMA. Class III

Chemical Composition

Composition limits. 99.5 min Cu + Be + additives; 0.25 to 0.50 Be; 1.40 to 1.70 Co; 0.90 to 1.10 Ag; 1.40 min Co + Ni; 1.90 max Co + Ni + Fe

Applications

Typical uses. A high conductivity alloy designed especially for resistance-welding electrodes for spot, seam, flash and projection welding methods; electrical connectors, clips

Precautions in use. Ventilation should be used during melting, welding, grinding and all machining operations.

Mechanical Properties

Tensile properties. See Table 34.
Hardness. See Table 34.
Elastic modulus. Tension, 125 to 130 GPa (18 to 19 × 10^6 psi); shear, 44 GPa (6.8 × 10^6 psi)

Structure

Crystal structure. α-Copper solid solution is face-centered cubic; the beryllide, (Cu,Co)Be, is ordered body-centered cubic of the CsCl (B2) type.

Microstructure. Matrix of α-copper; large and sharply angular blue-gray beryllide inclusions in grain boundaries of cast product, smaller Widmanstatten beryllides within the grain. In wrought product with large amounts of deformation, the beryllides are small, mainly spherical, and uniformly distributed.

Metallography is by conventional techniques. For dry grinding, ventilation should be provided. Some common etchants for immersion etching are (1) 3 parts concentrated NH_4OH, (2) 1 part 3% H_2O_2, (3) 2 parts 10% $(NH_4)_2S_2O_3$, and (4) 7 to 10 parts H_2O. Common etchants for swabbing are (1) 3 g $K_2Cr_2O_7$, (2) 1.5 g NaCl, (3) 8 ml H_2SO_4, and (4) 100 ml H_2O.

Mass Characteristics

Density. 8.75 Mg/m^3 (0.316 $lb/in.^3$) at 20 °C (68 °F)

Thermal Properties

Liquidus temperature. 1068 °C (1955 °F)
Solidus temperature. 1013 °C (1855 °F)
Coefficient of thermal expansion. Linear, 16.7 μm/m·K (9.3 μin./in.·°F) at 20 to 200 °C (68 to 392 °F)

Specific heat. 420 J/kg·K (0.10 Btu/lb·°F) at 20 °C (68 °F)
Thermal conductivity. 215 to 245 W/m·K (125 to 140 Btu/ft·h·°F) at 20 °C (68 °F)

Electrical Properties

Electrical conductivity. See Table 34.

Electrical resistivity. 28.7 to 86.2 nΩ·m at 20 °C (68 °F), depending strongly on heat treatment

Fabrication Characteristics

Machinability. Readily machinable by all conventional methods

Annealing temperature. For strip, wire, rod and bar, 900 to 950 °C (1650 to 1750 °F)/10 min/water quench. For larger sections, anneal 1 h per inch or fraction of an inch at 900 to 925 °C (1650 to 1700 °F) and water quench.

Solution temperature. All annealing for this alloy is solution treatment.

Aging temperature. Maximum strength is obtained by 3 to 6 h at 425 °C (800 °F). Commercial practice is to age material 2 to 3 h at 480 °C (900 °F) to obtain a combination of high strength and electrical conductivity.

Hot working temperature. 750 to 925 °C (1400 to 1700 °F)
Hot-shortness temperature. 975 °C (1800 °F)

Table 34 Typical mechanical properties and electrical conductivity of C17600 heat treated to various tempers

Temper(a)	Tensile strength MPa	ksi	Yield strength(b) MPa	ksi	Elongation in 50 mm or 2 in., %	Hardness, HRB	Electrical conductivity, % IACS
Rod, Bar, Wire, Tubing, Plate							
TB00	240-380	35-55	140-205	20-30	20-35	20-50	20-30
H04	450-550	65-80	380-515	55-75	10-15	60-80	20-30
TF00	690-825	100-120	550-690	80-100	10-25	92-100	45-60
TH04	760-900	110-130	690-825	100-120	10-20	95-102	50-60
Billet							
As cast	310-415	45-60	105-240	15-35	15-25	60-65	32-37
Cast and aged	415-515	60-75	205-380	30-55	10-20	65-90	40-50
TB00	275-345	40-50	69-115	10-17	20-40	10-45	22-28
TF00	655-760	95-110	515-550	75-80	3-15	92-100	50-60
Forged Products							
TB00	240-380	35-55	140-205	20-30	20-35	25-45	20-30
TF00	690-825	100-120	550-690	80-100	10-25	92-100	50-60

(a) For TB00 temper: solution treat strip, bar, rod and tubing 10 min at 900 to 955 °C (1650 to 1750 °F) and water quench; solution treat thicker products such as billet 1 h for each 25 mm (1 in.) of thickness or fraction thereof at 900 to 925 °C (1650 to 1700 °F) and water quench. For aging cast billets or producing TF00 temper, age 3 h at 470 to 500 °C (875 to 925 °F). For producing TH04 temper, age 2 h at 470 to 500 °C (875 to 925 °F). (b) At 0.2% offset.

C18200, C18400, C18500
99Cu-1Cr

Commercial Names

Previous trade name. CA182, CA184, CA185; Chrome Copper 999 (C18200)

Common name. Chromium copper

Specifications

ASTM. Wire: F9
SAE. J463 (C18400 only)
Government. Bar, forgings, rod, strip: MIL-C-19311 (C18400, C18500)

Chemical Composition

Composition limits. C18200: 0.6 to 1.2 Cr; 0.10 max Fe; 0.10 max Si; 0.05 max Pb; 0.5 max others (total); rem Cu + Ag. C18400: 0.40 to 1.2 Cr; 0.7 max Zn; 0.15 max Fe; 0.10 max Si; 0.05 max P; 0.05 max Li; 0.005 max As; 0.005 max Ca; 0.2 max others (total); rem Cu + Ag. C18500: 0.40 to 1.0 Cr; 0.08 to 0.12 Ag; 0.04 max P; 0.04 max Li; 0.015 max Pb; rem Cu + Ag.

Applications

Typical uses. Applications requiring excellent cold workability and good hot workability, coupled with medium to high conductivity. Uses include resistance welding electrodes, seam welding wheels, switch gears, electrode holder jaws, cable connectors, current-carrying arms and shafts, circuit-breaker parts, molds, spot welding tips, flash welding electrodes, electrical and thermal conductors requiring more strength than that provided by unalloyed coppers, and switch contacts.

Mechanical Properties

Tensile properties. See Table 35.
Hardness. See Table 35.
Elastic modulus. Tension, 130 GPa (19×10^6 psi); shear, 50 GPa (7.2×10^6 psi)

Mass Characteristics

Density. 8.89 Mg/m^3 (0.321 lb/in.^{3}) at 20 °C (68 °F)

Thermal Properties

Liquidus temperature. 1075 °C (1965 °F)
Solidus temperature. 1070 °C (1960 °F)
Coefficient of thermal expansion. Linear, 17.6 μm/m·K (9.8 μin./in.·°F) at 20 to 100 °C (68 to 212 °F)
Specific heat. 385 J/kg·K (0.092 Btu/lb·°F) at 20 °C (68 °F)

Table 35 Typical mechanical properties of C18200, C18400 and C18500

Temper	Tensile strength MPa	ksi	Yield strength(a) MPa	ksi	Elongation(b), %	Hardness, HRB
Flat Products: 1 mm (0.04 in.) Thick						
TB00 235	34	130	19	40	16
TF00(c) 350	51	250	36	22	59
TD04 365	53	350	51	6	66
TH04(d) 460	67	405	59	14	79
Plate: 50 mm (2.0 in.) Thick						
TF00 400	58	290	42	25	70
Plate: 75 mm (3.0 in.) Thick						
TF00 385	56	275	40	30	68
Rod: 4-mm (0.156-in.) Diam						
TD08 510	74	505	73	5	. . .
TH08 595	86	530	77	14	. . .
Rod: 13-mm (0.50-in.) Diam						
TB00 310	45	97	14	40	. . .
TF00(c) 485	70	380	55	21	70
TD04 395	57	385	56	11	65
TH04(d) 530	77	450	65	16	82
TH03, cold worked 6% 530	77	460	67	19	83
Rod: 25-mm (1.0-in.) Diam						
TF00 495	72	450	65	18	80
Rod: 50-mm (2.0-in.) Diam						
TF00 485	70	450	65	18	75
Rod: 75-mm (3.0-in.) Diam						
TF00 450	65	380	55	18	70
Rod: 100-mm (4.0-in.) Diam						
TF00 380	55	295	43	25	68
Tube: 9.5-mm (3/8-in.) OD, 2.4-mm (0.094-in.) Wall						
O60 275	40	105	15	50	59 HRF
Tube: 31.8-mm (1¼-in.) OD, 5.4-mm (0.212-in.) Wall						
TD04 405	59	395	57	21	67
TH04, cold worked 28% 475	69	435	63	26	84

(a) At 0.5% extension under load. (b) In 50 mm or 2 in. (c) Aged 3 h at 500 °C (930 °F). (d) Aged 3 h at 450 °C (840 °F).

Thermal conductivity. TB00 temper: 171 W/m·K (99 Btu/ft·h·°F) at 20 °C (68 °F). TH04 temper: 324 W/m·K (187 Btu/ft·h·°F) at 20 °C (68 °F)

Electrical Properties

Electrical conductivity. Volumetric. TB00 temper: 40% IACS at 20 °C (68 °F). TH04 temper: 80% IACS at 20 °C (68 °F)
Electrical resistivity. TH04 temper: 21.6 nΩ·m at 20 °C (68 °F)

Fabrication Characteristics

Machinability. 20% of C36000, free-cutting brass
Formability. Suited for hot working by extrusion, rolling and forging (subsequent solution treatment required) and for cold working (in soft, solution annealed or suitable drawn temper) by drawing, rolling, impacting, heading, bending or swaging
Weldability. Welding and brazing temperatures lower the properties developed by heat treatment; such processes are normally applied to material in the soft condition, followed by necessary heat treatment. Soldering: good. Oxyfuel gas, shielded metal-arc, resistance spot and resistance seam welding: not recommended
Solution treatment. 980 to 1000 °C (1800 to 1850 °F) for 10 to 30 min, water quench
Aging temperature. 425 to 500 °C (800 to 930 °F) for 2 to 4 h
Hot working temperature. 800 to 925 °C (1500 to 1700 °F)

C18700
99Cu-1Pb

Commercial Names

Previous trade name. Leaded copper

Common name. Free-machining copper

Specifications

ASTM. Flat products, rod: B301
SAE. Rod: J463

Chemical Composition

Composition limits. 0.8 to 1.5 Pb; 0.10 max others (total); rem Cu. Oxygen-free grades or grades containing deoxidizers such as P, B or Li may be specified.

Applications

Typical uses. Electrical connectors, motor parts, switch parts, and screw machine parts requiring high conductivity

Precautions in use. Unless specifically deoxidized, this copper is subject to embrittlement when heated in a reducing atmosphere (as in annealing or brazing), at temperatures of 350 °C (660 °F) or higher. If hydrogen or carbon monoxide is present, embrittlement can be rapid.

Mechanical Properties

Tensile properties. See Tables 17 and 36.

Shear strength. See Tables 17 and 36.

Hardness. See Tables 17 and 36.

Elastic modulus. Tension, 115 GPa (17×10^6 psi); shear, 44 GPa (6.4×10^6 psi)

Mass Characteristics

Density. 8.94 Mg/m^3 (0.323 $lb/in.^3$) at 20 °C (68 °F)

Thermal Properties

Liquidus temperature. 1080 °C (1975 °F)

Solidus temperature. 950 °C (1750 °F)

Coefficient of thermal expansion. Linear, 17.6 μm/m·K (9.8 μin./in.·°F) at 20 to 300 °C (68 to 572 °F)

Specific heat. 385 J/kg·K (0.092 Btu/lb·°F) at 20 °C (68 °F)

Thermal conductivity. 377 W/m·K (218 Btu/ft·h·°F) at 20 °C (68 °F)

Electrical Properties

Electrical conductivity. Volumetric, 96% IACS at 20 °C (68 °F)

Electrical resistivity. 17.9 nΩ·m at 20 °C (68 °F)

Fabrication Characteristics

Machinability. 85% of C36000, free-cutting brass

Formability. Good for cold working; poor for hot forming

Weldability. Soldering: excellent. Brazing: good. Most arc, gas and resistance welding processes not recommended

Annealing temperature. 425 to 650 °C (800 to 1200 °F)

Hot working temperature. 750 to 875 °C (1400 to 1600 °F)

Table 36 Typical mechanical properties of C18700 rod, H04 temper

Size	Tensile strength MPa	ksi	Yield strength MPa	ksi	Elongation in 50 mm or 2 in., %	Hardness, HRB	Shear strength MPa	ksi
6-mm (0.25-in.) diam	415	60	380	55	10	55	200	32
13-mm (0.50-in.) diam	380	55	345	50	11	50	205	30
19-mm (0.75-in.) diam	365	53	330	48	12	50	200	29
25-mm (1.0-in.) diam	350	51	315	46	14	50	195	28

Table 37 Typical mechanical properties of C19200

Size	Temper	Tensile strength MPa	ksi	Yield strength At 0.5% extension under load MPa	ksi	At 0.2% offset MPa	ksi	Elongation in 50 mm or 2 in., %	Hardness, HRB
Strip									
1-mm (0.04-in.) diam	O60	310	45	· · ·	· · ·	140 min	20 min	25 min	38
	O82	395	57	· · ·	· · ·	305	44	20	55
	H02	395	57	· · ·	· · ·	305	44	9	55
	H04	450	65	· · ·	· · ·	415	60	7	72
	H06	485	70	· · ·	· · ·	460	67	3	75
	H08	510	74	· · ·	· · ·	490	71	2 min	76
	H10	530	77	· · ·	· · ·	510	74	2 min	77
Tubing									
48-mm (1.88-in.) OD, 3-mm (1.88-in.) wall	O50	290	42	160	23	150	22	30	· · ·
	O60	255	37	83	12	76	11	40	· · ·
	H80 (40%)	385	56	360	52	360	52	7	· · ·
5-mm (0.19-in.) OD, 0.8-mm (0.03-in.) wall	H55	290	42	215	31	205	30	35	· · ·

C19200
98.97Cu-1.0Fe-0.03P

Specifications

ASTM. Tubing: B111, B359, B395, B469

Chemical Composition

Composition limits. 98.7 to 99.19 Cu; 0.8 to 1.2 Fe; 0.01 to 0.04 P

Applications

Typical uses. Rolled strip and tubing for air conditioning and heat exchanger tubing, applications requiring resistance to softening and stress corrosion, automotive hydraulic brake lines, cable wrap, circuit breaker components, contact springs, electrical connectors and terminals, eyelets, flexible hose, fuse clips, gaskets, gift hollow ware, lead frames

Mechanical Properties

Tensile properties. See Table 37.
Hardness. See Table 37.
Elastic modulus. Tension, 115 GPa (17×10^6 psi); shear, 44 GPa (6.4×10^6 psi)

Mass Characteristics

Density. 8.87 Mg/m^3 (0.320 $lb/in.^3$) at 20 °C (68 °F)

Thermal Properties

Liquidus temperature. 1084 °C (1983 °F)

Solidus temperature. 1077 °C (1973 °F)
Coefficient of thermal expansion. Linear, 16.2 μm/m·K (9.0 μin./in.·°F) at 20 to 100 °C (68 to 212 °F)
Specific heat. 380 J/kg·K (0.09 Btu/lb·°F) at 20 °C (68 °F)
Thermal conductivity. Strip, 251 W/m·K (145 Btu/ft·h·°F) at 20 °C (68 °F); tubing, 216 W/m·K (125 Btu/ft·h·°F) at 20 °C (68 °F)

Electrical Properties

Electrical conductivity. Strip, 60% IACS at 20 °C (68 °F); tubing, 50% IACS at 20 °C (68 °F)
Electrical resistivity. Strip, 28.8 nΩ·m at 20 °C (68 °F); tubing, 34.5 nΩ·m at 20 °C (68 °F)

Fabrication Characteristics

Machinability. 20% of C36000, free-cutting brass
Forgeability. 65% of C37700, forging brass
Weldability. Soldering, brazing and gas-shielded arc welding: excellent. Oxyacetylene welding: good. Coated metal-arc and resistance seam, spot and butt welding: not recommended.
Annealing temperature. 700 to 825 °C (1300 to 1500 °F)
Hot working temperature. 825 to 950 °C (1500 to 1750 °F)

C19400
Cu-2.35Fe-0.03P-0.12Zn

Commercial Names

Previous trade name. High-strength modified copper, HSM copper

Specifications

ASME. Welded tubing: SB543
ASTM. Flat products: B465. Welded tubing: B543, B586

Chemical Composition

Composition limits. 2.1 to 2.6 Fe; 0.05 to 0.20 Zn; 0.015 to 0.15 P; 0.03 max Pb; 0.03 max Sn; 0.15 max others (total); rem Cu

Applications

Typical uses. Applications requiring excellent hot and cold workability as well as high strength and conductivity. Specific uses include circuit breaker components, contact springs, electrical clamps, springs and terminals, flexible hose, fuse clips, gaskets, gift hollow ware, plug contacts, rivets, welded condenser tubes, semi-conductor lead frames and cable shielding.

Mechanical Properties

Tensile properties. See Tables 38, 39, 40 and 41.
Hardness. See Table 38.
Elastic modulus. Tension, 121 GPa

Table 38 Typical mechanical properties of C19400

Temper	Tensile strength MPa	ksi	Yield strength(a) MPa	ksi	Elongation(b), %	Hardness HRB	HR30T	Fatigue strength(c) MPa	ksi
Flat Products: 0.64-mm (0.025-in.) Thick									
O60	310	45	150	22 max	29 min	38	...	110	16
O50	345	50	160	23	28	45	...	...	...
O82	400	58	255	37	15	...	...	...	...
Flat Products: 1-mm (0.04-in.) Thick									
H02	400	58	315(d)	46(d)	18	68	66	...	...
H04	450	65	380	55	7	73	69	145	21
H06	485	70	465	67.5	3	74	71	...	...
H08	505	73	486	70.5	3	75	72	148	21.5
H10	530	77	506	73.5	2 max	77	74	141	20.5
H14	550 min	80 min	530 min	77 min	2 max	...	>73	...	...
Tubing: 25-mm (1-in.) OD by 0.9-mm (0.035-in.) Wall									
O60	310	45	165	24	28	38	...	...	...
O50	345	50	205	30	16	45	...	...	...
WM02	400	58	365	53	9	61	60	...	...
WM04	450	65	435	63	4	73	66	...	...
WM06	485	70	465	67.5	3	74	68	...	...
WM08	505	73	486	70.5	2	75	69	...	...
WM10	525	76	505	73	1	76	69	...	...
H55 (15%)	400	58	380	55	9	61	60	...	...
H80 (35%)	470	68	455	66	2	73	66	...	...

(a) At 0.2% offset. (b) In 50 mm or 2 in. (c) At 10^8 cycles as rotating beam test. (d) At 0.5% extension under load.

Table 40 Typical elevated-temperature properties of annealed C19400 strip

Test temperature	Tensile strength, min		Yield strength, min(a)		Creep strength, min(b)		Stress-rupture stress, min(c)	
	MPa	ksi	MPa	ksi	MPa	ksi	MPa	ksi
Ambient 341	341	49.5	150	22.0	. . .	. . .	. . .	. . .
65 °C (150 °F) 324	324	47.0	144	20.9	. . .	. . .	. . .	. . .
95 °C (200 °F) 313	313	45.4	144	20.9	. . .	. . .	. . .	. . .
120 °C (250 °F) 300	300	43.5	144	20.9	190	27.6	. . .	. . .
150 °C (300 °F) 289	289	41.9	139	20.2	171	24.8	171	24.9
175 °C (350 °F) 276	276	40.1	135	19.6	143	20.8	148	21.4
205 °C (400 °F) 266	266	38.6	131	19.0	124	18.0	125	18.1
230 °C (450 °F) 253	253	36.8	131	19.0	110	16.0	105	15.2
260 °C (500 °F) 235	235	34.1	127	18.4	96	13.9	82	11.9
290 °C (550 °F) 219	219	31.8	123	17.8	84	12.2	65	9.4
315 °C (600 °F) 203	203	29.5	116	16.8	74	10.8	47	6.8

(a) At 0.2% offset. (b) Stress causing secondary creep of 0.01% per 1000 h in a 10 000-h test. (c) Stress causing rupture in 100 000 h (extrapolated from 10 000 h).

(17.5 × 10⁶ psi); shear, 45.5 GPa (6.6 × 10⁶ psi)

Impact strength. (Charpy) Plate, O61 temper: longitudinal, 144 J (106 ft·lb) at −196 °C (−320 °F); transverse, 99 J (73 ft·lb at −196 °C)

Fatigue strength. See Table 38.

Creep and stress-rupture properties. See Table 40.

Mass Characteristics

Density. 8.78 Mg/m³ (0.317 lb/in.³) at 20 °C (68 °F)

Thermal Properties

Liquidus temperature. 1090 °C (1990 °F)

Solidus temperature. 1080 °C (1980 °F)

Coefficient of thermal expansion. Linear, 16.3 μm/m·K (9.0 μin./in.·°F) at 20 to 300 °C (68 to 572 °F)

Specific heat. 385 J/kg·K (0.092 Btu/lb·°F) at 20 °C (68 °F)

Thermal conductivity. 260 W/m·K (150 Btu/ft·h·°F) at 20 °C (68 °F)

Electrical Properties

Electrical conductivity. Volumetric, at 20 °C (68 °F). O60 temper: 40% IACS nominal. H14 temper: 50% IACS min. All other tempers: 65% IACS nominal, 60% IACS min. In O50, O82 and H02 tempers, 75% IACS min conductivity may be available depending on mill processing restrictions.

Electrical resistivity. At 20 °C (68 °F). O60 temper: 43.1 nΩ·m nominal. H14 temper: 34.5 nΩ·m max. All other tempers: 26.6 nΩ·m nominal, but may be only 23.0 nΩ·m max under certain circumstances.

Magnetic Properties

Magnetic permeability. 1.1

Table 41 Annealing response of C19400 strip(a)

Annealing Temperature	Tensile strength		Yield strength(b)		Elongation(c) %	Electrical conductivity, % IACS
	MPa	ksi	MPa	ksi		
H04 Temper						
100 °C (212 °F) 460	460	67	450	65	3	66
205 °C (400 °F) 450	450	65	435	63	5	67
315 °C (600 °F) 440	440	64	415	60	9	68
370 °C (700 °F) 415	415	60	385	56	12	68
425 °C (800 °F) 415	415	60	360	52	14	72
480 °C (900 °F) 400	400	59	345	50	16	71
540 °C (1000 °F) 385	385	56	310	45	17	64(d)
595 °C (1100 °F) 350	350	51	220	32	23	52
650 °C (1200 °F) 315	315	46	140	20	33	51
705 °C (1300 °F) 310	310	45	115	17	34	49
760 °C (1400 °F) 305	305	44	110	16	36	48
815 °C (1500 °F) 305	305	44	110	16	36	48
H10 Temper						
100 °C (212 °F) 510	510	74	490	71	3	65
205 °C (400 °F) 495	495	72	460	67	5	66
315 °C (600 °F) 485	485	70	415	60	8	67
370 °C (700 °F) 330	330	48	170	25	25	71
425 °C (800 °F) 325	325	47	145	21	27	74
480 °C (900 °F) 315	315	46	140	20	28	69
540 °C (1000 °F) 315	315	46	140	20	31	64(d)
595 °C (1100 °F) 310	310	45	130	19	33	58
650 °C (1200 °F) 305	305	44	130	19	34	52
705 °C (1300 °F) 295	295	43	115	17	34	49
760 °C (1400 °F) 290	290	42	110	16	35	48
815 °C (1500 °F) 285	285	41	105	15	35	48

(a) Typical ambient-temperature properties after 1 h at temperature. (b) At 0.2% offset. (c) In 50 mm or 2 in. (d) Conductivity may be restored to about 70% IACS by holding at 500 °C (925 °F) for 1 h.

Chemical Properties

General corrosion behavior. Very corrosion resistant and essentially immune to stress corrosion cracking

Fabrication Characteristics

Machinability. 20% of C36000, free-cutting brass

Formability. Suited to forming by blanking, coining, coppersmithing, drawing, bending, heading and upsetting, hot forging and pressing, piercing and punching, roll threading and knurling, shearing, spinning, squeezing and stamping

Weldability. Joining by soldering, brazing and TIG welding: excellent

Annealing temperature. See Table 41.

Table 39 Typical low-temperature (cryogenic) properties of C19400

Temper	Tensile strength MPa	ksi	Yield strength(a) MPa	ksi	Elongation(b), %
Room-Temperature Properties					
O61	325	47	170	25	28
H02	405	59	360	52	15
H04	455	66	405	59	10
Cryogenic Properties: −196 °C (−320 °F)					
O61	475	69	195	28	38
H02	570	83	425	62	30
H04	615	89	485	70	23

(a) At 0.2% offset. (b) In 50 mm or 2 in.

Table 42 Typical mechanical properties of C19500

Temper	Tensile strength MPa	ksi	Yield strength(a) MPa	ksi	Elongation in 50 mm or 2 in., %	Hardness, HRB
O61	360 min	52 min	170 min	25 min	25 min	· · ·
O50	520-590	75-85	395-530	57-77	11-17	81-89
H02	565-620	82-90	505-605	73-88	3-13	85-88
H08	605-670	88-97	585-650	85-94	2-5	87-90
H10	670 min	97 min	650 min	94 min	2 max	90 min

(a) At 0.2% offset.

C19500
97Cu-1.5Fe-0.1P-0.8Co-0.6Sn

Commercial Names

Trade name. Strescon

Chemical Composition

Composition limits. 1.3 to 1.7 Fe; 0.6 to 1.0 Co; 0.08 to 0.12 P; 0.40 to 0.7 Sn; 0.20 max Zn; 0.02 max Al; 0.02 max Pb; 0.05 max others (each); 0.10 max others (total); rem Cu

Applications

Typical uses. Electrical springs, sockets, terminals, connectors, clips and other current carrying parts requiring strength and exceptional softening resistance. Applications requiring excellent hot and cold workability, as well as high strength and high conductivity.

Mechanical Properties

Tensile properties. See Table 42.
Hardness. See Table 42.
Elastic modulus. Tension, 119 GPa (17.3×10^6 psi)

Mass Characteristics

Density. 8.92 Mg/m^3 (0.322 lb/in.^{3}) at 20 °C (68 °F)

Thermal Properties

Liquidus temperature. 1090 °C (1995 °F)
Solidus temperature. 1085 °C (1985 °F)
Coefficient of thermal expansion. Linear, 16.9 μm/m·K (9.4 μin./in.·°F) at 20 to 300 °C (68 to 572 °F)
Thermal conductivity. 199 W/m·K (115 Btu/ft·h·°F) at 20 °C (68 °F)

Electrical Properties

Electrical conductivity. Volumetric, 50% IACS at 20 °C (68 °F), annealed
Electrical resistivity. 34.4 nΩ·m at 20 °C (68 °F)

Fabrication Characteristics

Machinability. 20% of C36000, free-cutting brass
Formability. Suited to forming by bending, coining, drawing and stamping

Variation of properties with Zn content for wrought Cu-Zn alloys

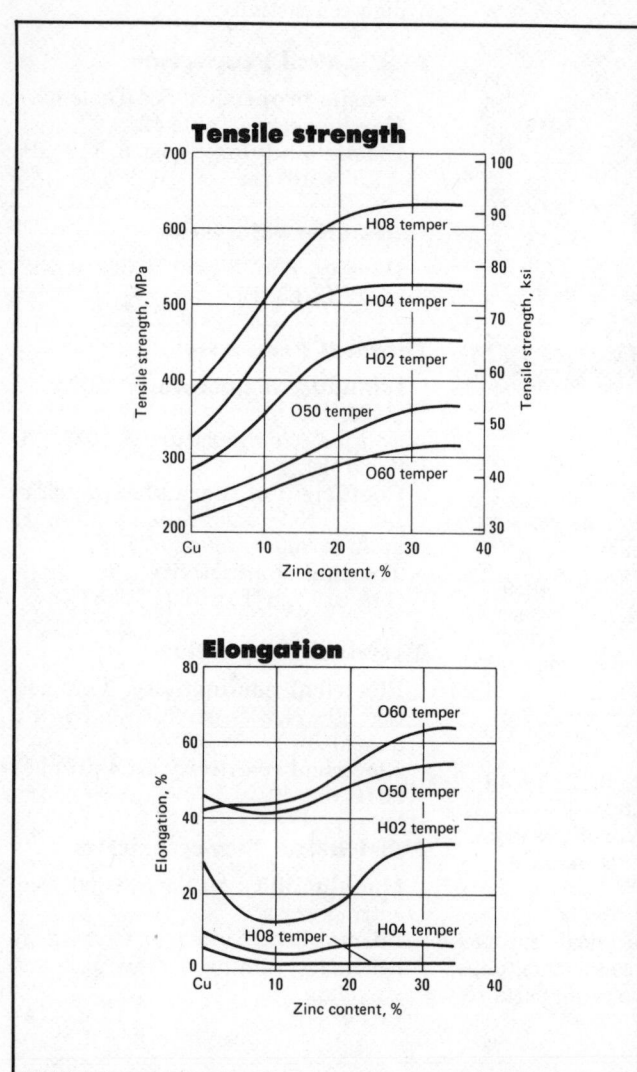

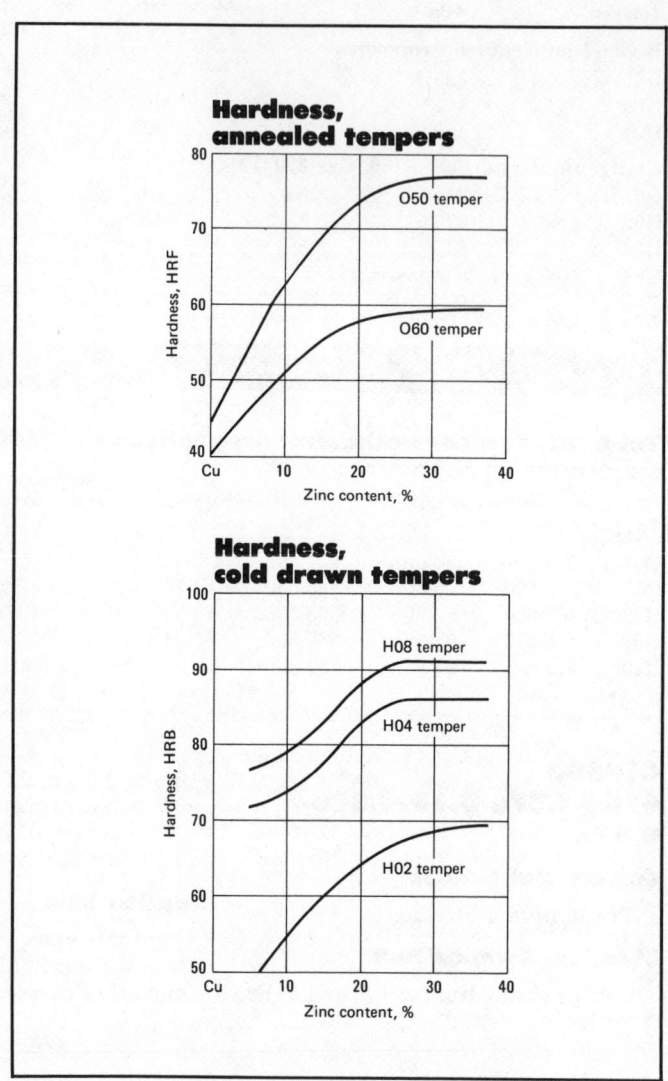

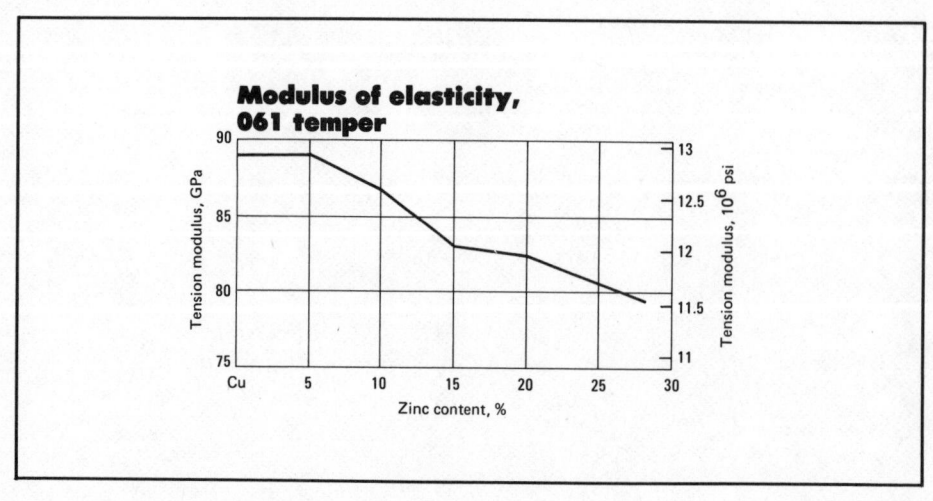

Variation of properties with Zn content for wrought Cu-Zn alloys

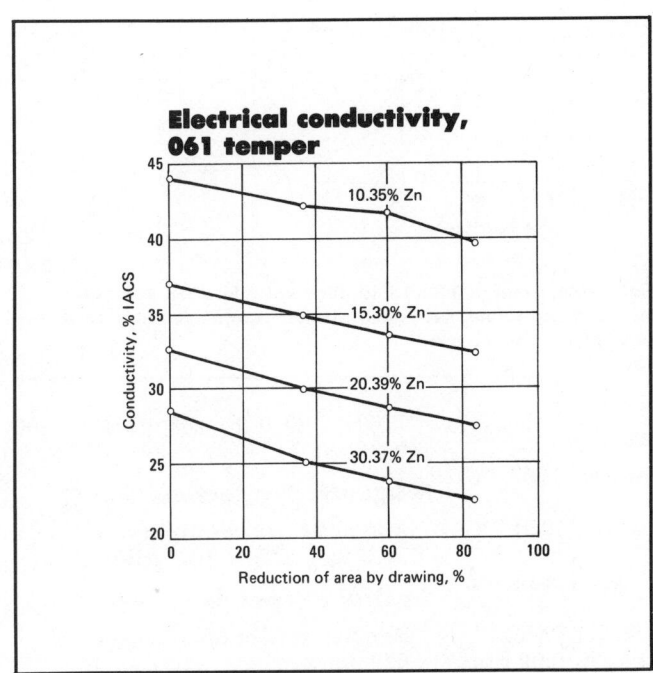

Electrical conductivity, 061 temper

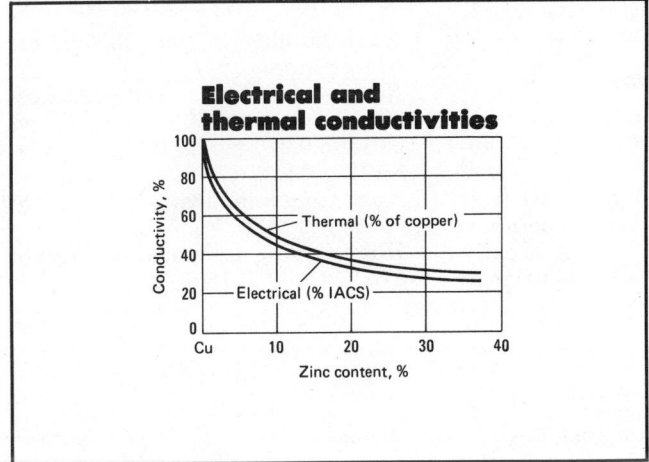

Electrical and thermal conductivities

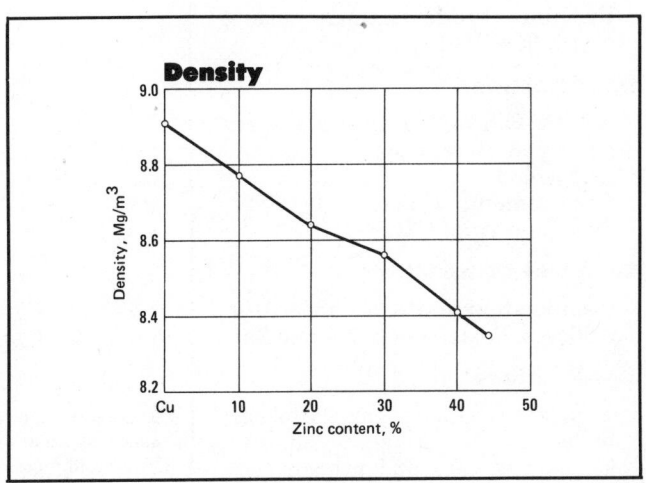

Density

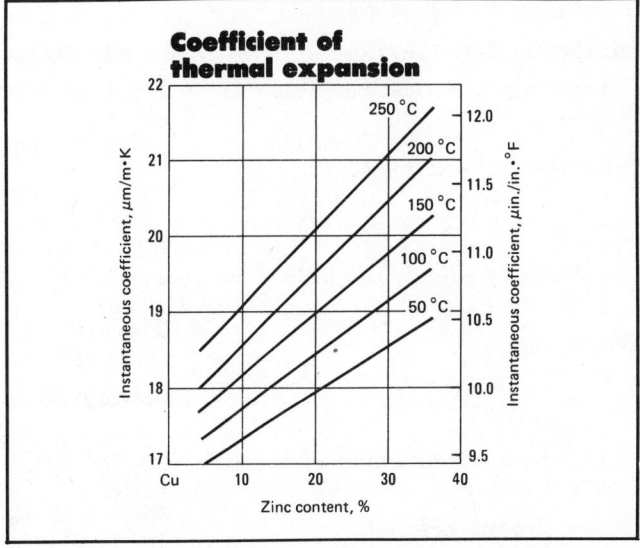

Coefficient of thermal expansion

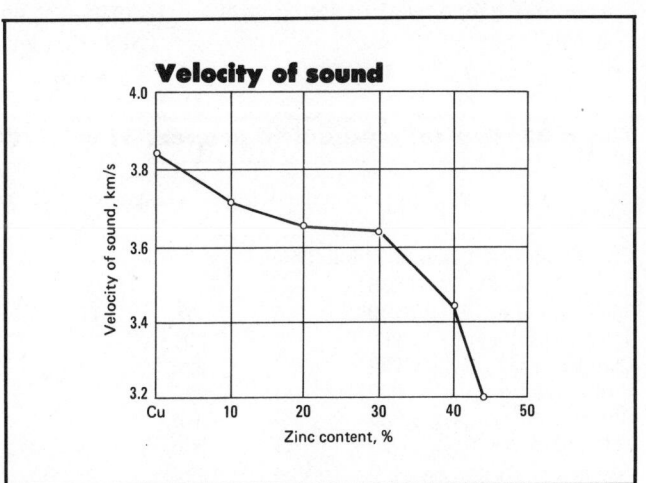

Velocity of sound

C21000
95Cu-5Zn

Commercial Names

Previous trade name. Gilding metal, 95%; CA210

Specifications

ASTM. Rolled bar, plate, sheet and strip: B36. Wire: B134
SAE. J463
Government. Wire: QQ-W-321. Sheet and strip: MIL-C-21768

Chemical Composition

Composition limits. 94.0 to 96.0 Cu; 0.05 max Pb; 0.05 max Fe; rem Zn

Applications

Typical uses. Coins, medals, tokens, bullet jackets, firing-pin supports, shells, fuse caps and primers, emblems, jewelry plaques, base for gold plate, base for vitreous enamel

Mechanical Properties

Tensile properties. See Table 43 and Fig. 21.
Shear strength. See Table 43.
Hardness. See Table 43.
Elastic modulus. Tension, 115 GPa (17×10^6 psi); shear, 44 GPa (6.4×10^6 psi)
Velocity of sound. 3.78 km/s at 20 °C (68 °F)

Structure

Crystal structure. Face-centered cubic alpha; lattice parameter, 0.3627 nm
Minimum interatomic distance. 0.2564 nm

Mass Characteristics

Density. 8.86 Mg/m³ (0.320 1b/in.³) at 20 °C (68 °F)

Thermal Properties

Liquidus temperature. 1065 °C (1950 °F)
Solidus temperature. 1050 °C (1920 °F)
Coefficient of thermal expansion. Linear, 18.1 μm/m·K (10.0 μin./in.· °F) at 20 to 300 °C (68 to 572 °F)
Specific heat. 380 J/kg·K (0.09 Btu/1b. °F) at 20 °C (68 °F)
Thermal conductivity. 234 W/m·K (135 Btu/ft·h· °F) at 20 °C (68 °F)

Electrical Properties

Electrical conductivity. Volumetric, 56% IACS at 20 °C (68 °F) annealed
Electrical resistivity. 31 nΩ·m at 20 °C (68 °F), annealed; temperature coefficient, 0.0231 nΩ·m per K at 20 °C (68 °F). Liquid phase: 244 nΩ·m at 1100 °C (2010 °F), 266 nΩ·m at 1300 °C (2370 °F).

Magnetic Properties

Magnetic susceptibility. -1.0×10^{-6} to -12.5×10^{-6} (SI units)

Optical Properties

Special reflectance. 90% for λ = 578 nm

Fabrication Characteristics

Machinability. 20% of C36000, free-cutting brass
Recrystallization temperature. 370 °C (700 °F) for 50% reduction and 0.015 to 0.070 mm initial grain size. See Fig. 21.
Annealing temperature. 425 to 800 °C (800 to 1450 °F)
Hot working temperature. 750 to 875 °C (1400 to 1600 °F)

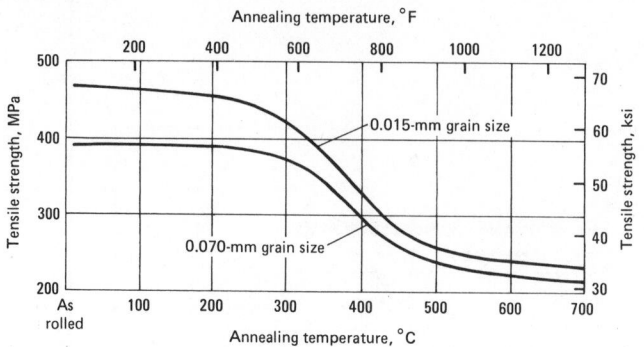

Fig. 21 Variation of tensile strength with annealing temperature for C21000

Data are for ready-to-finish strip 1-mm (0.04-in.) thick that was cold rolled 50%, then annealed 1 h at the indicated temperature. Recrystallization temperature is 370 °C (700 °F) for initial grain sizes of 0.015 to 0.070 mm.

Table 43 Typical mechanical properties of C21000

Temper	Tensile strength MPa	ksi	Yield strength(a) MPa	ksi	Elongation in 50 mm or 2 in., %	HRF	Hardness HRB	HR30T	Shear strength MPa	ksi
Flat Products: 1-mm (0.04-in.) Thick										
0.050-mm anneal	235	34	69	10	45	46	...	...	...	...
0.035-mm anneal	240	35	76	11	45	52	...	4	195	28
0.015-mm anneal	260	38	97	14	42	60	...	15	205	30
Quarter hard	290	42	220	32	25	...	38	44	220	32
Half hard	330	48	275	40	12	...	52	54	235	34
Hard	385	56	345	50	5	...	64	60	255	37
Extra hard	420	61	380	55	4	...	70	64	270	39
Spring	440	64	400	58	4	...	73	66	275	40
(a) At 0.5% extension under load.										

C22000
90Cu-10Zn

Commercial Names

Previous trade name. Commercial bronze, 90%; CA220

Specifications

ASTM. Rolled bar, plate and sheet: B36. Strip: B36 and B130. Cups, bullet jacket: B131. Tube, rectangular wave guide: B372. Seamless tube: B135. Wire: B134

SAE. Rolled bar, plate, sheet, strip and seamless tube: J 463 (CA220)

Government. Wire: QQ-W-321; MIL-W-6712. Bands, projectile rotating: MIL-B-18907. Blanks, rotating band for projectiles: MIL-B-20292. Cups, bullet jacket: MIL-C-3383. Sheet and strip: MIL-C-21768. Tube, rectangular wave guide: MIL-W-85. Seamless tube for microwave use: MIL-T-52069

Chemical Composition

Composition limits. 89.0 to 91.0 Cu; 0.05 max Pb; 0.05 max Fe; rem Zn

Consequence of exceeding impurity limits. See general statement for brasses under C26000.

Applications

Typical uses. Architectural: etching bronze, grillwork, screen cloth, weather stripping. Hardware: escutcheons, kick plates, line clamps, marine hardware, rivets, screws, screw shells. Munitions: primer caps, rotating bands. Miscellaneous: compacts, lipstick cases, costume jewelry, ornamental trim, screen wire, base for vitreous enamel, wave guides

Mechanical Properties

Tensile properties. See Table 44 and Fig. 22.

Shear strength. See Table 44.

Hardness. See Table 44.

Elastic modulus. Tension, 115 GPa (17×10^6 psi); shear, 44 GPa (6.4×10^6 psi)

Fatigue strength. Spring temper, flat product 1.016 mm (0.40 in.) thick: 145 MPa (21 ksi) at 15×10^6 cycles; hard wire 2.032 mm (0.080 in.) in diameter: 160 MPa (23 ksi) at 10^8 cycles

Velocity of sound. 3720 m/s (12 200 ft/s) at 20 °C (68 °F)

Structure

Crystal structure. Face-centered cubic alpha; lattice parameter, 0.364 nm

Minimum interatomic distance. 0.257 nm

Mass Characteristics

Density. 8.80 Mg/m³ (0.318 1b/in.³) at 20 °C (68 °F)

Thermal Properties

Liquidus temperature. 1045 °C (1910 °F)

Solidus temperature. 1020 °C (1870 °F)

Boiling point. About 1400 °C (2550 °F) at 101 kPa (1 atm)

Coefficient of thermal expansion. Linear, 18.4 μm/m·K (10.2 μin./in.·°F) at 20 to 300 °C (68 to 572 °F), cold rolled

Specific heat. 376 J/kg·K (0.09 Btu/1b·°F) at 20 °C (68 °F)

Thermal conductivity. 189 W/m·K (109 Btu/ft·h·°F) at 20 °C (68 °F)

Electrical Properties

Electrical conductivity. Volumetric, 44% IACS at 20 °C (68 °F), annealed

Electrical resistivity. 39.1 nΩ·m at 20 °C (68 °F) Liquid phase, 272 nΩ·m at 1100 °C (2012 °F). Temperature coefficient, 0.00186 nΩ·m per K at 20 °C (68 °F)

Magnetic Properties

Magnetic susceptibility. -0.086×10^{-6} to -1.00×10^{-6} (cgs units)

Fabrication Characteristics

Machinability. 20% of C36000, free-cutting brass

Recrystallization temperature. 370 °C (700 °F) for 37% reduction and 0.050 mm (0.002 in.) initial grain size. See Fig. 22.

Annealing temperature. 425 to 800 °C (800 to 1450 °F)

Hot working temperature. 750 to 875 °C (1400 to 1600 °F)

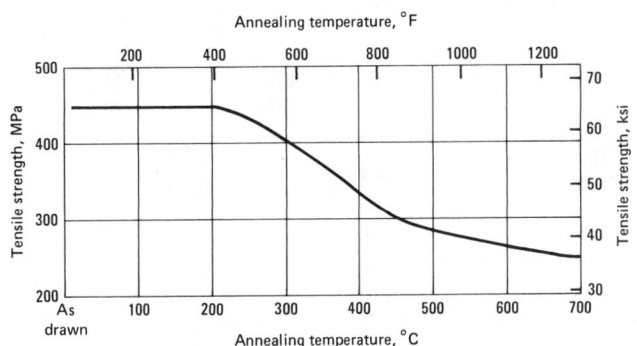

Fig. 22 Variation of tensile strength and grain size with annealing temperature for C22000

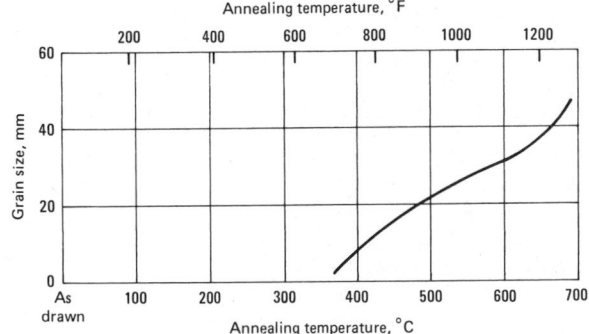

Data are for rod less than 25 mm (1 in.) in diameter that was cold drawn to a 37% reduction in area, then annealed 1 h at the indicated temperature. Grain size before annealing was 0.050 mm.

Table 44 Typical mechanical properties of C22000

Temper	Tensile strength MPa	ksi	Yield strength(a) MPa	ksi	Elongation in 50 mm or 2 in., %	HRF	Hardness HRB	HR30T	Shear strength MPa	ksi
Flat Products: 1-mm (0.040-in.) Thick										
OS050	255	37	69	10	45	53	...	6	195	28
OS035	260	38	83	12	45	57	...	12	205	30
OS025	270	39	97	14	44	60	...	16	215	31
OS015	280	41	105	15	42	65	...	26	220	32
H01	310	45	240	35	25	...	42	44	230	33
H02	360	52	310	45	11	...	58	56	240	35
H04	420	61	370	54	5	...	70	63	260	38
H06	460	67	400	58	4	...	75	67	275	40
H08	495	72	425	62	3	...	78	69	290	42
M20	270	39	97	14	44	60	...	...	215	31
Flat Products: 6-mm (0.250-in.) Thick										
OS035	260	38	83	12	50	57	...	...	205	30
H02	360	52	310	45	15	...	58	...	240	35
M20	255	37	69	10	45	53	...	...	195	28
Wire: 2-mm (0.080-in.) Diam										
OS035	275	40	...	...	50	...	...	...	205	30
OS015	290	42	...	...	48	...	...	...	220	32
H00	305	44	...	...	27	...	...	...	230	33
H01	345	50	...	...	13	...	...	...	235	34
H02	415	60	...	...	6	...	...	...	255	37
H04	510	74	...	...	4	...	...	...	290	42
H06	570	83	...	...	3	...	...	...	...	...
H08	620	90	...	...	3	...	...	...	...	...
Tubing: 25-mm (1-in.) OD, 1.65-mm (0.065-in.) Wall										
OS025	260	38	83	12	50	57	...	12	...	...
H80(b)	415	60	365	53	6	...	69	62	...	...
Rod: 12.7-mm (0.500 in.) Diam										
OS035	275	40	...	...	50	55	...	...	220	32
H00	310	45	...	...	25	...	42	...	230	33

(a) At 0.5% extension under load. (b) Drawn 35%.

C22600
87.5Cu-12.5Zn

Commercial Names

Previous trade name. Jewelry bronze, 87½%; CA226
Common name. Jewelry bronze

Chemical Composition

Composition limits. 86.0 to 89.0 Cu; 0.05 max Pb; 0.005 max Fe; rem Zn

Applications

Typical uses. Architectural: angles, channels. Hardware: chain, eyelets, fasteners, slide fasteners. Novelties: compacts, costume jewelry, emblems, etched articles, lipstick containers, plaques, base for gold plate.

Mechanical Properties

Tensile properties. See Table 45 and Fig. 23.

Shear strength. See Table 45.
Elastic modulus. Tension, 115 GPa (17×10^6 psi); shear, 44 GPa (6.4×10^6 ksi)

Mass Characteristics

Density. 8.78 Mg/m³ (0.317 lb/in.³) at 20 °C (68 °F)

Thermal Properties

Liquidus temperature. 1035 °C (1895 °F)
Solidus temperature. 1005 °C (1840 °F)
Coefficient of thermal expansion. Linear, 18.6 μm/m·K (10.3 μin./in.·°F) at 20 to 300 °C (68 to 572 °F)
Specific heat. 380 J/kg·K (0.09 Btu/lb·°F) at 20 °C (68 °F)
Thermal conductivity. 173 W/m·K (100 Btu/ft·h·°F) at 20 °C (68 °F)

Electrical Properties

Electrical conductivity. Volumetric, 40% IACS at 20 °C (68 °F), annealed
Electrical resistivity. 43 nΩ·m at 20 °C (68 °F), annealed

Fabrication Characteristics

Machinability. 30% of C36000, free-cutting brass

Recrystallization temperature. About 330 °C (625 °F) for 1-mm (0.04-in.) strip rolled six B and S numbers hard from a 0.035-mm (0.001-in.) grain size. See also Fig. 23.

Annealing temperature. 425 to 750 °C (800 to 1400 °F)
Hot working temperature. 750 to 900 °C (1400 to 1650 °F)

Table 45 Typical mechanical properties of C22600

Temper	Tensile strength MPa	ksi	Yield strength(a) MPa	ksi	Elongation in 50 mm or 2 in., %	Hardness HRF	HRB	Shear strength MPa	ksi
Flat Products: 1-mm (0.04-in.) Thick									
OS050 270	270	39	76	11	46	55	...	200	29
OS035 275	275	40	90	13	45	59	...	205	30
OS025 290	290	42	105	15	44	64	...	215	31
OS015 305	305	44	110	16	42	68	...	220	32
H01 325	325	47	255	37	25	...	47	235	34
H02 370	370	54	325	47	12	...	61	250	36
H04 455	455	66	385	56	5	...	73	275	40
H06 495	495	72	415	60	4	...	78	290	42
H08 545	545	79	425	62	4	...	82	305	44
Wire: 2-mm (0.08-in.) Diam									
OS050 275	275	40	...	13	44	...	...	200	29
OS035 285	285	41	...	15	42	...	...	205	30
OS025 295	295	43	...	17	40	...	...	215	31
OS015 310	310	45	...	18	38	...	...	220	32
H00 325	325	47	...	35	26	...	...	235	34
H01 385	385	56	...	52	12	...	...	250	36
H02 470	470	68	...	60	7	...	70	275	40
H04 570	570	83	...	64	5	...	...	...	...
H06 615	615	89	...	65	4	...	...	...	...
H08 670	670	97	...	66	3	...	...	...	...

(a) At 0.5% extension under load.

Fig. 23 Annealing characteristics of C22600

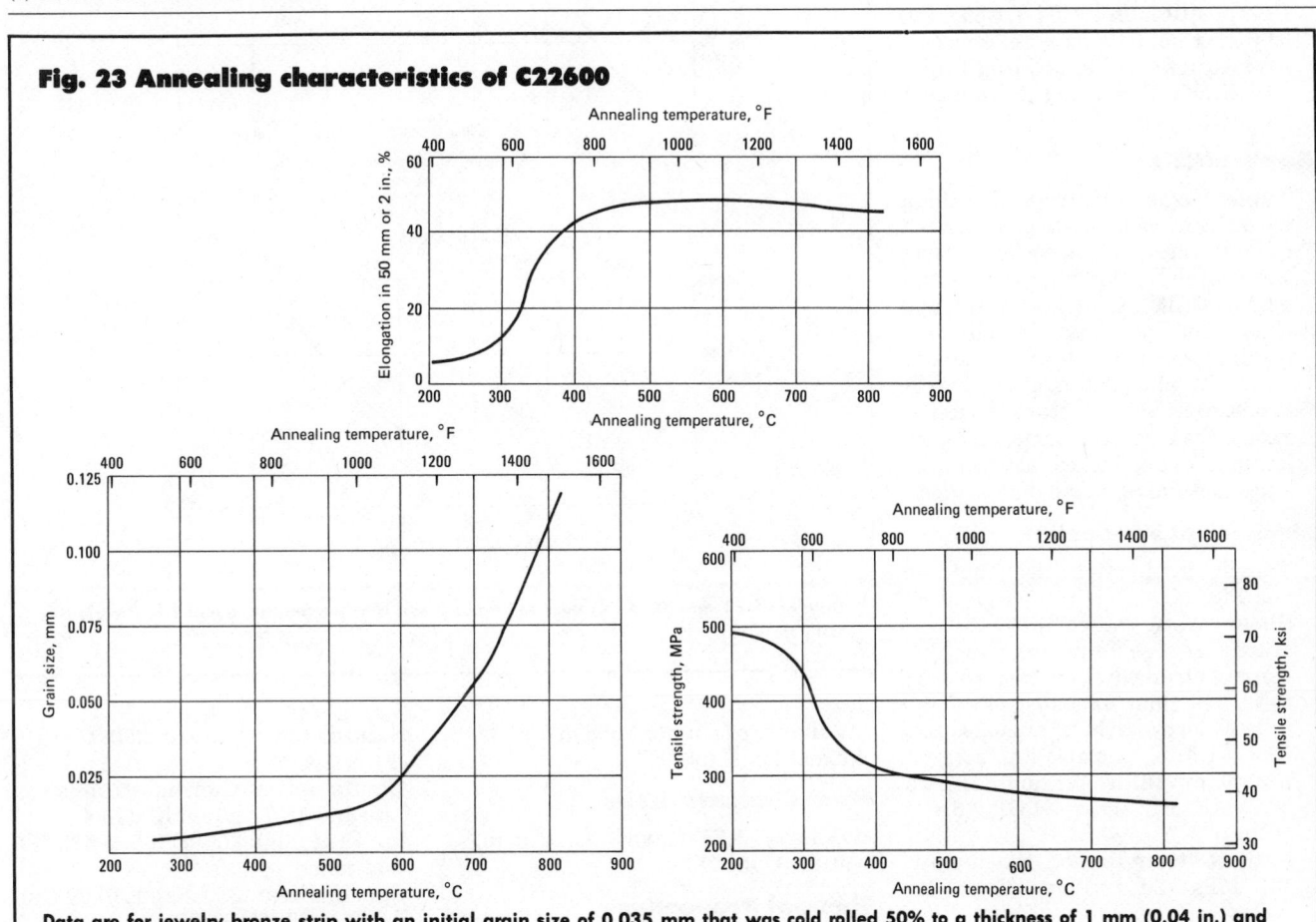

Data are for jewelry bronze strip with an initial grain size of 0.035 mm that was cold rolled 50% to a thickness of 1 mm (0.04 in.) and annealed 1 h at various temperatures.

C23000
85Cu-15Zn

Commercial Names

Previous trade name. Red brass, 85%; CA230

Common name. Red brass

Specifications

ASME. Pipe: SB43. Condenser tubing: SB111. Finned tubing: SB359. U-band tubing: SB395

ASTM. Plate, sheet, strip, hot rolled bar: B36. Pipe: B43. Condenser tubing: B111. Finned tubing: B359. Seamless tubing: B135. U-bend tubing: B395. Wire: B134

SAE. Sheet, strip, seamless tubing: J463 (CA230)

Government. Bar, forgings, rod, shapes, strip: QQ-B-626. Plate, sheet, strip, hot rolled bar: QQ-B-613. Pipe: WW-P-351. Seamless tubing: WW-T-791; MIL-T-20168. Wire: QQ-W-321

Chemical Composition

Composition limits. 84.0 to 86.0 Cu; 0.06 max Pb; 0.05 max Fe; rem Zn

Consequence of exceeding impurity limits. See general statement for cartridge brass, C27000.

Applications

Typical uses. Architectural: etching parts, trim, weather strip. Electrical: conduit, screw shells, sockets. Hardware: eyelets, fasteners, fire extinguishers. Industrial: condenser and heat exchanger tubes, flexible hose, pickling crates, pump lines, radiator cores. Plumbing: plumbing pipe, J-bends, service lines, traps. Miscellaneous: badges, compacts, costume jewelry, dials, etched articles, lipstick containers, name plates, tags.

Mechanical Properties

Tensile properties. See Table 46 and Fig. 24.

Shear strength. See Table 46.

Hardness. See Table 46.

Impact strength. Izod: cast, 45 J (33 ft·lb); cast and annealed, 43 J (32 ft·lb). Charpy keyhole: annealed rod, 69 J (51 ft·lb). See also Fig. 25.

Elastic modulus. Tension, 115 GPa (17×10^6 psi); shear, 44 GPa (6.4×10^6 psi)

Fatigue strength. Rod, H00 temper, 140 MPa (20 ksi) at 300×10^6 cycles

Creep-rupture characteristics. See Fig. 26.

Fig. 24 Annealing characteristics of C23000

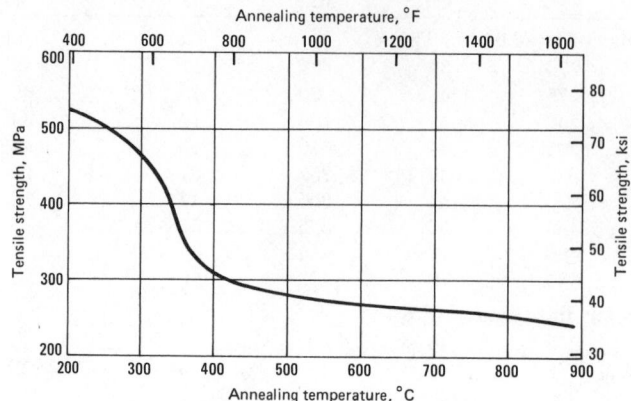

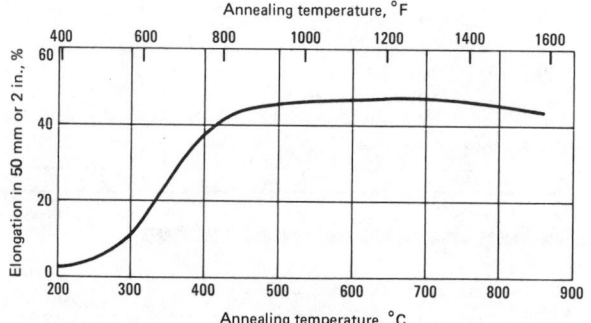

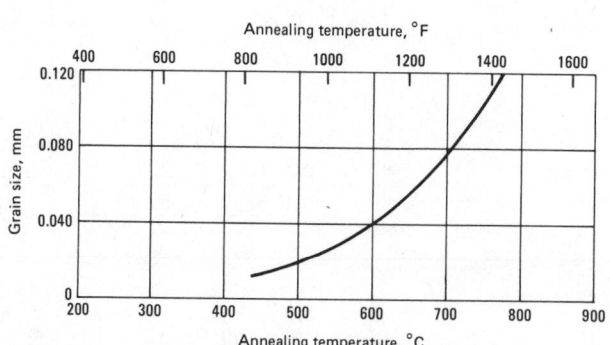

Data are for 1 mm (0.04 in.) thick red brass sheet, H06 temper, annealed 1 h at various temperatures.

Velocity of sound. 3658 m/s (12 000 ft/s at 20 °C (68 °F)

Mass Characteristics

Density. 8.75 Mg/m³ (0.316 lb/in.³) at 20 °C (68 °F)

Thermal Properties

Liquidus temperature. 1025 °C (1880 °F)

Solidus temperature. 990 °C (1810 °F)

Coefficient of thermal expansion. Linear, 18.7 μm/m·K (10.4 μin./in.·°F) at 20 to 300 °C (68 to 572 °F), cold rolled

Specific heat. 380 J/kg·K (0.09 Btu/lb·°F) at 20 °C (68 °F)

Thermal conductivity. 159 W/m·K (92 Btu/ft·h·°F) at 20 °C (68 °F)

Electrical Properties

Electrical conductivity. Volumetric, 37% IACS at 20 °C (68 °F), annealed

Electrical resistivity. 47 nΩ·m at 20 °C (68 °F), annealed. Liquid: 299 nΩ·m at 1100 °C (2012 °F); 304 nΩ·m at 1200 °C (2192 °F). Temperature coefficient, 0.0016/°C at 20 °C (68 °F).

Magnetic Properties

Magnetic susceptibility. Approximately -1.00×10^{-6} cgs

Fabrication Characteristics

Machinability. 30% of C26000, free-cutting brass

Recrystallization temperature. About 350 °C (660 °F) for 1-mm (0.04-in.) sheet rolled six B and S numbers hard, 50% reduction and 0.035-mm (0.001-in.) initial grain size

Annealing temperature. 425 to 725 °C (800 to 1350 °F) See also Fig. 24.

Hot working temperature. 800 to 900 °C (1450 to 1650 °F)

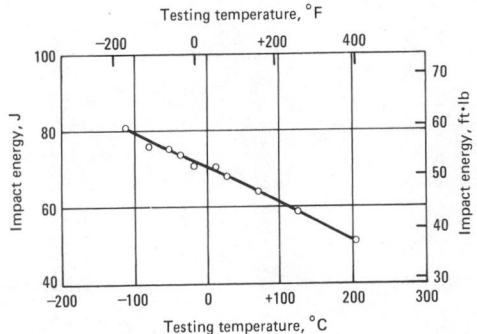

Fig. 25 Impact strength of C23000

Charpy keyhole specimens were machined from O61 temper material, then tested at the indicated temperatures. Impact strengths represent energy absorbed without fracture.

Table 46 Typical mechanical properties of C23000

Temper	Tensile strength MPa	ksi	Yield strength(a) MPa	ksi	Elongation in 50 mm or 2 in., %	Hardness HRB	HRF	HR30T	Shear strength MPa	ksi
Flat Products: 1 mm (0.04 in.) Thick										
OS070 270	39	69	10	48	...	56	10	215	31	
OS050 275	40	83	12	47	...	59	14	215	31	
OS035 285	41	97	14	46	...	63	22	215	31	
OS025 295	43	110	16	44	...	66	28	220	32	
OS015 310	45	125	18	42	...	71	38	230	33	
H01 345	50	270	39	25	55	...	54	240	35	
H02 395	57	340	49	12	65	...	60	255	37	
H04 485	70	395	57	5	77	...	68	290	42	
H06 540	78	420	61	4	83	...	72	305	44	
H08 580	84	435	63	3	86	...	74	315	46	
Wire: 2-mm (0.08-in.) Diam										
OS035 285	41	...	...	48	...	...	...	215	31	
OS025 295	43	...	...	...	...	...	...	220	32	
OS015 310	45	...	...	...	...	...	...	230	33	
H00 345	50	...	...	25	...	...	...	240	35	
H01 405	59	...	...	11	...	...	...	260	38	
H02 495	72	...	...	8	...	...	...	295	43	
H04 605	88	...	...	6	...	...	...	330	48	
H08 725	105	...	...	...	...	...	...	370	54	
Tubing: 25-mm (1.0-in.) OD, 1.65-mm (0.065-in.) Wall										
OS050 275	40	83	12	55	...	60	15	...	...	
OS015 305	44	125	18	45	...	71	38	...	...	
H55(15%) 345	50	275	40	30	55	...	54	...	...	
H80(35%) 485	70	365	53	8	77	...	68	...	...	
Pipe: 19-mm (0.75-in.) SPS										
OS015 305	44	125	18	45	...	71	...	...	...	

(a) At 0.5% extension under load.

Fig. 26 Minimum creep rates for C23000 wire

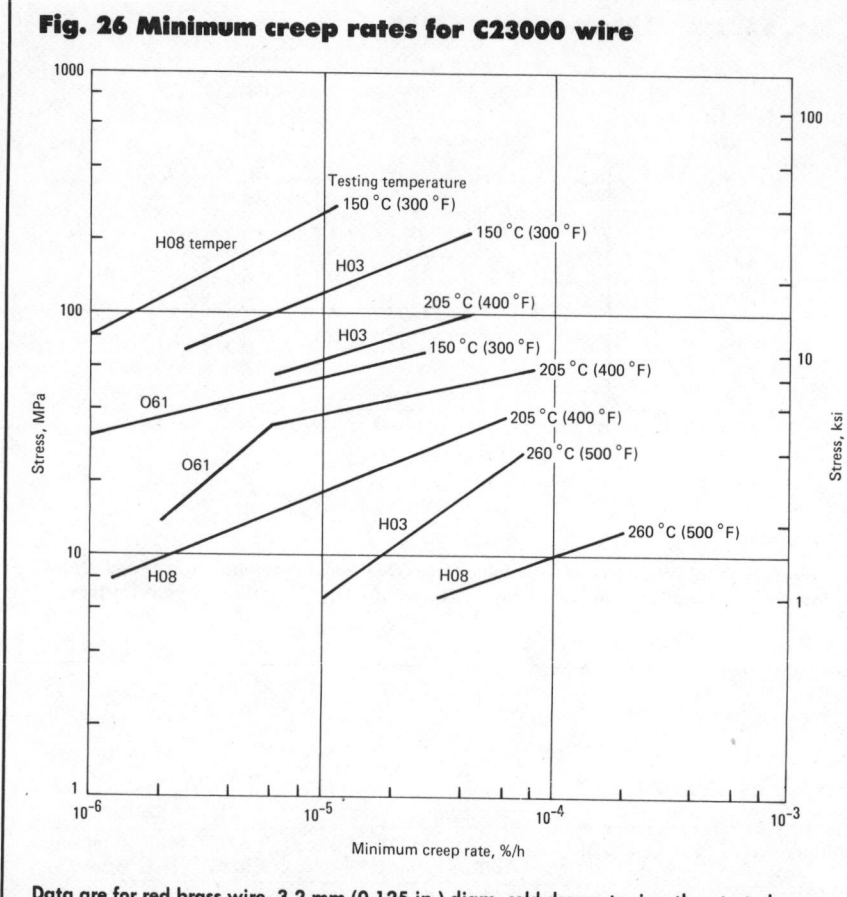

Data are for red brass wire, 3.2 mm (0.125 in.) diam, cold drawn to size, then tested as drawn or annealed prior to testing.

C24000
80Cu-20Zn

Commercial Names

Trade name. Low brass, 80%; CA240

Common name. Low brass

Specifications

ASTM. Flat products: B36. Wire: B134

SAE. Sheet, strip: J463 (CA240)

Government. Finished-edge bar and strip, forgings, rod, shapes: QQ-B-626. Rolled bar, plate, sheet, strip: QQ-B-613. Wire: QQ-W-321. Brazing alloy wire: QQ-B-650

Chemical Composition

Composition limits. 78.5 to 81.5 Cu; 0.05 max Pb; 0.05 max Fe; rem Zn

Applications

Typical uses. Ornamental metal work, medallions, spandrels, electrical battery caps, bellows and musical instruments, clock dials, flexible hose, pump lines, tokens

Mechanical Properties

Tensile properties. See Table 47 and Fig. 27.

Shear strength. See Table 47.

Hardness. See Table 47.

Elastic modulus. Tension, 110 GPa (16×10^6 psi); shear, 40 GPa (6×10^6 psi)

Fatigue strength. 1 mm (0.04 in.) thick strip, H08 temper: 165 MPa (24 ksi) at 20×10^6 cycles

Fig. 27 Tensile strength and grain size vs annealing temperature for C24000, annealed from HC2 temper

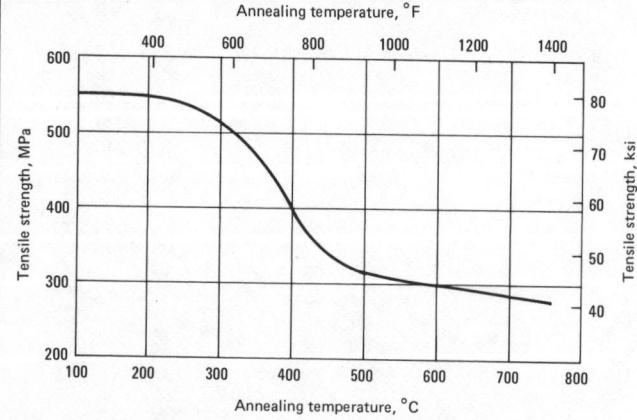

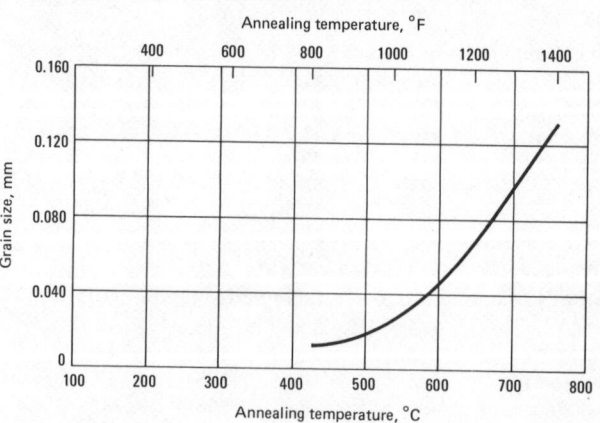

Data are for low brass and with an initial grain size of 0.060 mm, cold drawn 37% to a diameter less than 25 mm (1 in.) and annealed 1 h at the indicated temperature.

Table 47 Typical mechanical properties of C24000

Temper	Tensile strength MPa	ksi	Yield strength(a) MPa	ksi	Elongation(b), %	Hardness HRB	HRF	HR30T	Shear strength MPa	ksi
Flat products: 1-mm (0.04-in.) Thick										
OS070	290	42	83	12	52	...	57	8	...	...
OS050	305	44	97	14	50	...	61	16	220	32
OS035	315	46	105	15	48	...	66	28	...	...
OS025	330	48	115	17	47	...	69	32	...	...
OS015	345	50	140	20	46	...	75	42	230	33
H01	365	53	275	40	30	55	...	54	250	36
H02	420	61	345	50	18	70	...	64	270	39
H04	510	74	405	59	7	82	...	71	295	43
H08	625	91	450	65	3	91	...	77	330	48
Wire: 2-mm (0.08-in.) Diam										
OS050	305	44	...	...	55	...	...	...	220	32
OS035	315	46	...	...	50	...	...	...	...	...
OS015	345	50	...	...	47	...	...	...	230	33
H00	385	56	...	...	27	...	...	...	255	37
H01	470	68	...	...	12	...	...	...	290	42
H02	565	82	...	...	8	...	...	...	325	47
H04	740	107	...	...	5	...	...	...	365	53
H06	800	116	...	...	4	...	...	...	...	...
H08	860	125	...	...	3	...	...	...	415	60

(a) At 0.5% extension under load. (b) In 50 mm or 2 in.

Structure

Crystal structure. Face-centered cubic alpha; lattice parameter, 0.366 nm
Minimum interatomic distance. 0.259 nm

Mass Characteristics

Density. 8.67 Mg/m^3 (0.313 lb/in.^{3}) at 20 °C (68 °F)
Solidification shrinkage. 5 to 6%

Thermal Properties

Liquidus temperature. 1000 °C (1830 °F)
Solidus temperature. 965 °C (1770 °F)
Coefficient of thermal expansion. Linear, 19.1 μm/m·K (10.6 μin./in.·°F) at 20 to 300 °C (68 to 572 °F)
Specific heat. 380 J/kg·K (0.09 Btu/lb·°F) at 20 °C (68 °F)
Thermal conductivity. 140 W/m·K (81 Btu/ft·h·°F) at 20 °C (68 °F)

Electrical Properties

Electrical conductivity. Volumetric, 061 temper: 32% IACS at 20 °C (68 °F)
Electrical resistivity. 061 temper: 54 nΩ·m at 20 °C (68 °F). Liquid: 330 nΩ·m at 1000 °C (1830 °F); 338 nΩ·m at 1200 °C (2190 °F). Temperature coefficient, 0.00154/°C at 20 °C (68 °F).

Magnetic Properties

Magnetic susceptibility. Approximately -1.00×10^{-6} cgs

Fabrication Characteristics

Machinability. 30% of C36000, free-cutting brass
Recrystallization temperature. About 400 °C (750 °F) for 37% reduction and 0.060 mm initial grain size.
Annealing temperature. 425 to 700 °C (800 to 1300 °F). See also Fig. 27.
Hot working temperature. 825 to 900 °C (1500 to 1650 °F)

C26000
70Cu-30Zn

Commercial Names

Previous trade name. Cartridge brass, 70%; CA260
Common name. Cartridge brass, 70-30 brass, spinning brass, spring brass, extra quality brass

Specifications

AMS. Flat products: 4505, 4507. Tube: 4555
ASTM. Flat products: B19, B36, B569. Cups for cartridge cases: B129. Tube: B135, B587. Wire: B134
SAE. J463
Government. Flat products: QQ-B-613, QQ-B-626, MIL-C-50. Rod, bar, shapes, forgings: QQ-B-626. Tube: MIL-T-6945, MIL-T-20219. Wire: QQ-W-321, QQ-B-650. Shim stock, laminated: MIL-S-22499. Cups for cartridge cases: MIL-C-10375

Chemical Composition

Composition limits. 68.5 to 71.5 Cu; 0.07 max Pb; 0.05 max Fe; 0.15 max others (total); rem Zn
Consequence of exceeding impurity limits. *Lead* should be kept under 0.01% for hot rolling, although additions of lead up to 4% improve machinability in material processed by extrusion and cold working. Lead lowers room temperature ductility in brass, and also leads to hot shortness above 315 °C (600 °F). *Aluminum* as high as 2% has no adverse effect on hot or cold working. However, annealing and grain size are affected. *Arsenic* does not affect hot or cold working, but tends to refine the grain size, and thus to lower the ductility. *Cadmium* is controversial; some claim as much as 0.10% has little effect, others would keep it under 0.05%. *Chromium* affects temperature of anneal and grain size. This condition is aggravated when iron is present. *Iron* chiefly affects annealing and magnetic properties. *Nickel*

Table 48 Typical mechanical properties of C26000

Temper	Tensile strength MPa	ksi	Yield strength(a) MPa	ksi	Elongation (b), %	Hardness HRF or HRB	HR30T	Shear strength MPa	ksi	Fatigue strength(a) MPa	ksi
Flat Products, 1 mm (0.04 in.) Thick											
O5100	300	44	75	11	68	F 54	11	215	31	90	13
O5070	315	46	95	14	65	F 58	15	220	32	90	13
O5050	325	47	105	15	62	F 64	26	230	33	...	...
O5035	340	49	115	17	57	F 68	31	235	34	...	...
O5025	350	51	130	19	55	F 72	36	235	34	95	14
O5015	365	53	150	22	54	F 78	43	240	35	105	15
H01	370	54	275	40	43	B 55	54	250	36	...	...
H02	425	62	360	52	23	B 70	65	275	40	125	18
H04	525	76	435	63	8	B 82	73	305	44	145	21
H06	595	86	450	65	5	B 83	76	315	46	...	...
H08	650	94	...	...	3	B 91	77	330	48	160	23
H10	680	99	...	...	3	B 93	78	...	...	...	...
Wire, 2 mm (0.08 in.) Diam											
O5050	330	48	110	16	64	...	...	230	33	...	...
O5035	345	50	125	18	60	...	...	235	34	...	...
O5025	360	52	145	21	58	...	...	240	35	...	...
O5015	370	54	160	23	58	...	...	250	36	...	...
H00	400	58	315	46	35	...	...	260	38	...	...
H01	485	70	395	57	20	...	...	290	42	...	...
H06	855	124	...	...	4	...	...	...	...	...	...
H08	895	130	...	...	3	...	...	415	60	150	22
Tube, 25 mm (1 in.) OD by 1.6 mm (0.065 in.) Wall											
O5050	325	47	105	15	65	F 64	26	...	...	...	...
O5025	360	52	140	20	55	F 75	40	...	...	...	...
H80	540	78	440	64	8	B 82	73	...	...	...	...
Rod, 25 mm (1.0 in.) Diam											
O5050	330	48	110	16	65	F 65	...	235	34	...	...
H00	380	55	275	40	48	B 60	...	260	38	...	...
H02	480	70	360	52	30	B 80	...	290	42	22(d)	150(d)

(a) At 0.5% extension under load. (b) In 50 mm or 2 in. (c) Reverse bending, at 10^8 cycles. (d) Reverse bending, at 5×10^7 cycles.

Table 49 Typical tensile properties of cold rolled and annealed C26000 sheet

Direction in sheet	Tensile strength MPa	ksi	Elongation, %
Parallel to RD	330	48	59
45° to RD	305	44	66
90° to RD	325	47	61

Note: Values are approximate for material given a ready-to-finish anneal at 400 °C (750 °F), then cold rolled 70% and annealed 1 h at 575 °C (1070 °F).

restrains grain growth. *Phosphorus* has no adverse effect up to 0.04%; it does, however, restrain grain growth, increase tensile strength, and lower ductility to some extent.

Applications

Typical uses. *Architectural:* grillwork. *Automotive:* radiator cores and tanks. *Electrical:* bead chain, flashlight shells, reflectors, lamp fixtures, socket shells, screw shells. *Hard-ware:* eyelets, fasteners, pins, hinges, kick plates, locks, rivets, springs, stampings, tubes, etched articles. *Munitions:* ammunition components, particularly cartridge cases. *Plumbing:* accessories, fittings. *Industrial:* pump and power cylinders, cylinder liners

Precautions in use. Highly susceptible to season cracking in ammoniacal environments

Mechanical Properties

Tensile properties. See Tables 48 and 49 and Fig. 28 to 30.
Hardness. See Table 48 and Fig. 30.
Elastic modulus. Tension, 110 GPa (16×10^6 psi); shear, 40 GPa (6×10^6 psi)
Fatigue strength. See Table 48.
Impact strength. Charpy V-notch: O61 temper, 60 J (44 ft·lb); M20 temper, 19 J (14 ft·lb). Izod: O61 temper, 89 J (66 ft·lb) for notched round specimen

Creep-rupture properties. See Fig. 31.
Velocity of sound. 3660 m/s (12 000 ft/s) at 20 °C (68 °F)

Structure

Crystal structure. Face-centered cubic; lattice parameter, 0.3684 nm; **Minimum interatomic distance.** 0.2605 nm
Microstructure. Single-phase alpha, usually with extensive pattern of annealing twins
Damping capacity. See Fig. 32.

Mass Characteristics

Density. 8.53 Mg/m³ (0.308 lb/in.³) at 20 °C (68 °F)

Thermal Properties

Liquidus temperature. 955 °C (1750 °F)
Solidus temperature. 915 °C (1680 °F)

Fig. 28 Tensile strength and grain size as a function of annealing temperature for C26000 rod

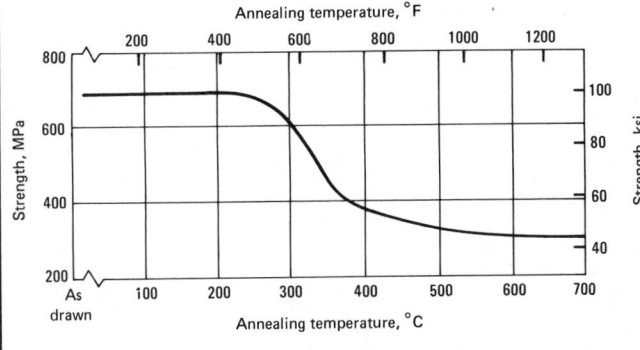
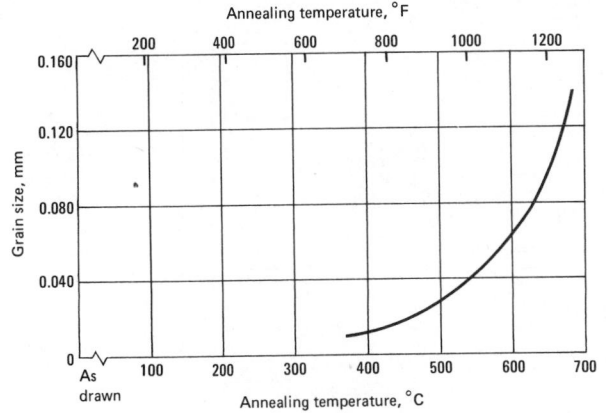

Data are for cartridge brass rod less than 25 mm (1 in.) diam that was cold drawn 50% from starting material having a grain size of 0.045 mm, and then annealed 1 h at the indicated temperature.

Fig. 29 Low-temperature tensile properties of C26000 rod, 061 temper

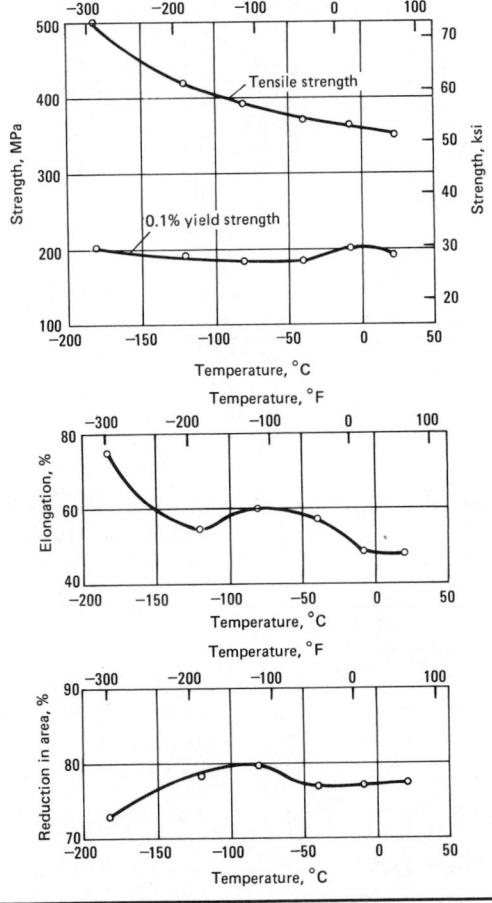

Coefficient of thermal expansion. Linear, cold rolled stock: 19.9 μm/m·K (11.1 μin./in.·°F) at 20 to 300 °C (68 to 572 °F). Equation for 20 to 300 °C:

$$L_t = L_0 [1 + (17.75t + 0.00653t^2) \times 10^{-6}]$$

where t is temperature difference from 20 °C

Specific heat. 375 J/kg·K (0.09 Btu/lb·°F) at 20 °C (68 °F)

Thermal conductivity. 120 W/m·K (70 Btu/ft·h·°F) at 20 °C (68 °F)

Electrical Properties

Electrical conductivity. Volumetric, O61 temper, 28% IACS at 20 °C (68 °F)

Electrical resistivity. O61 temper, 62 nΩ·m at 20 °C (68 °F); temperature coefficient, 0.092 nΩ·m per K at 20 °C

Hall coefficient. 25 pV·m/A·T

Magnetic Properties

Iron in excess of 0.03% can precipitate from C26000 during suitable low-temperature anneals. Precipitation is slow and occurs chiefly in a nonmagnetic form, which is converted to a ferromagnetic structure on subsequent cold working.

Magnetic susceptibility. -8×10^{-8} to -16×10^{-8} mks units; susceptibility in alpha brasses decreases with increasing zinc content

Fig. 30 Typical distribution of tensile properties and hardness for C26000 strip, HO1 temper

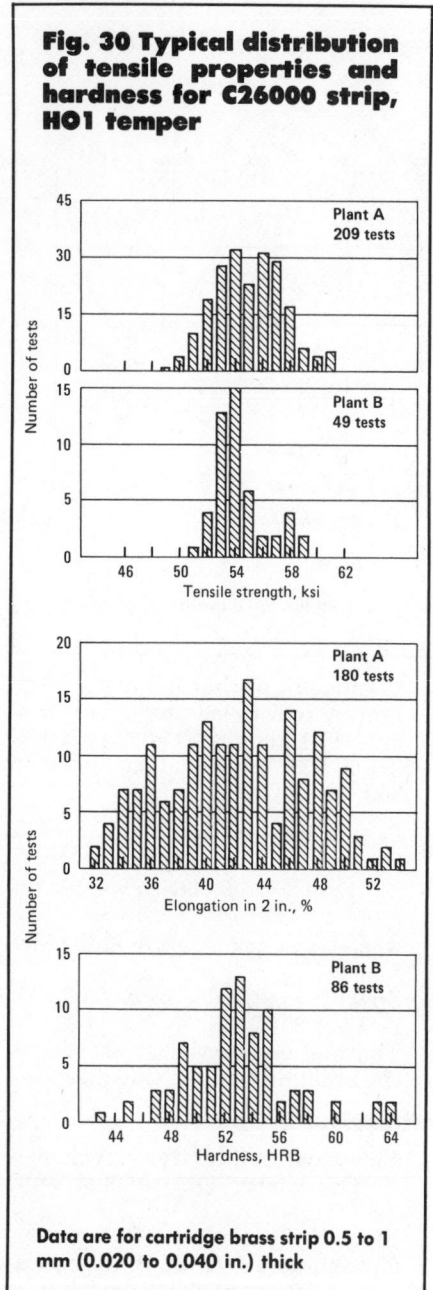

Data are for cartridge brass strip 0.5 to 1 mm (0.020 to 0.040 in.) thick

Fig. 31 Minimum creep rates for C26000

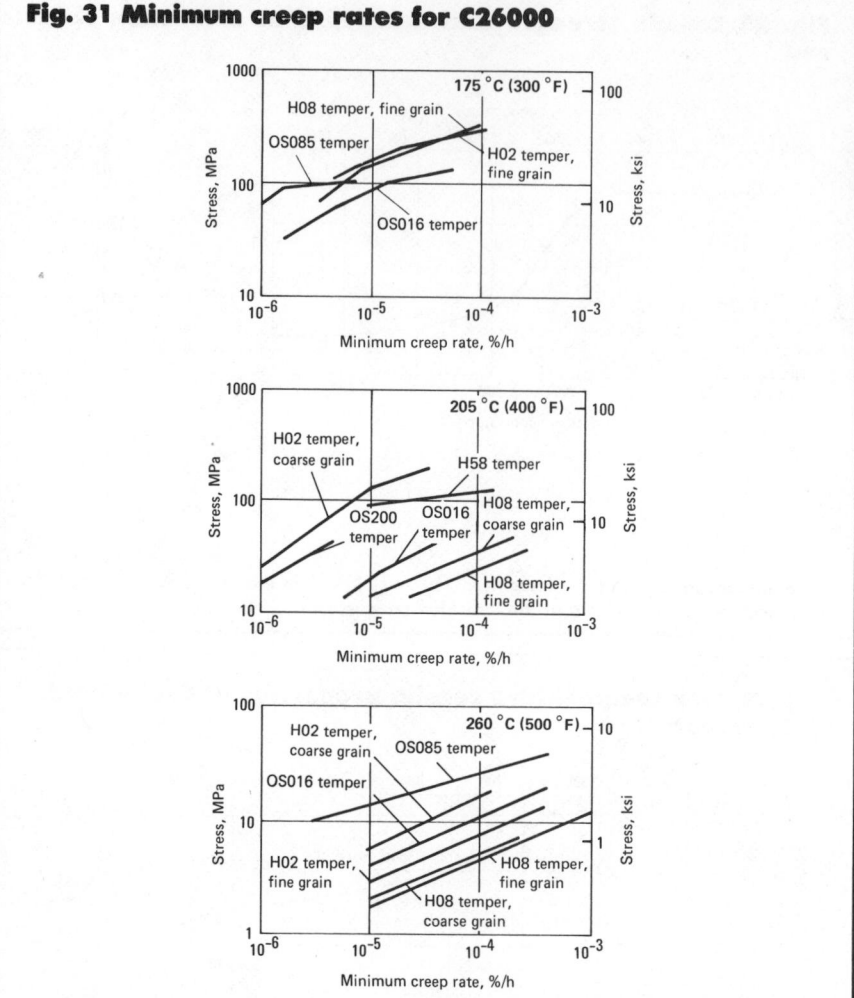

Chemical Properties

General corrosion behavior. Resists corrosion in a wide variety of waters and chemical solutions, but may undergo dezincification in stagnant or slowly moving salt solutions, brackish water, or mildly acidic solutions. Is susceptible to stress-corrosion cracking (season cracking), especially in ammoniacal environments

Fabrication Characteristics

Machinability. 30% of C36000, free-cutting brass

Formability. Excellent for cold working and forming; fair for hot forming. Directionality in brass is more readily developed with high zinc content, such as in C26000 and higher zinc brasses. Earing usually occurs as 45° to the direction of rolling and is aggravated by heavy final reductions, low ready-to-finish annealing temperatures and high finish annealing temperatures

Weldability. Soldering and brazing: excellent. Oxyfuel gas, resistance spot and resistance butt welding: good. Gas metal arc welding: fair. Other welding processes not recommended

Recrystallization temperature. About 300 °C (575 °F) for 0.045 mm initial grain size and a cold reduction of 50%

Annealing temperature. 425 to 750 °C (800 to 1400 °F)

Hot working temperature. 725 to 850 °C (1350 to 1550 °F)

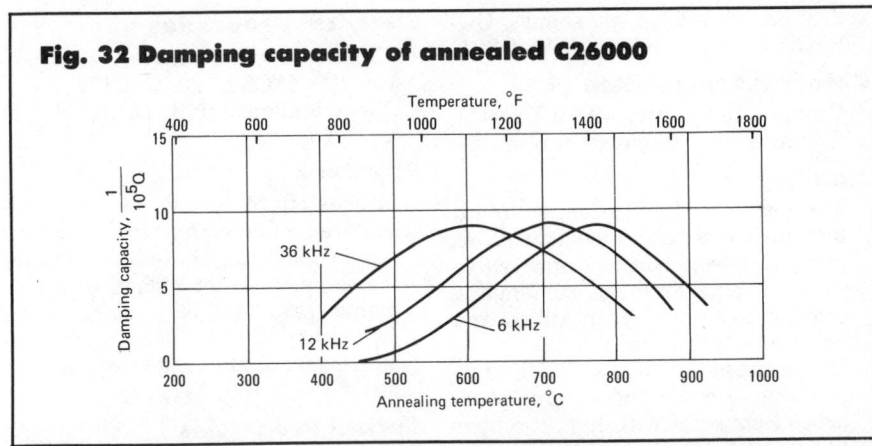

Fig. 32 Damping capacity of annealed C26000

B-626. Wire: QQ-W-321, MIL-W-6712

Chemical Composition

Composition limits. C26800: 64.0 to 68.5 Cu; 0.15 max Pb; 0.05 max Fe; rem Zn. C27000: 63.0 to 68.5 Cu; 0.10 max Pb; 0.07 max Fe; rem Zn

Applications

Typical uses. Architectural grillwork, radiator cores and tanks, reflectors, flashlight shells, lamp fixtures, screw shells, socket shells, bead chain, chain, eyelets, fasteners, grommets, kick plates, push plates, stencils, plumbing accessories, sink strainers, wire, pins, rivets, screws, springs

C26800, C27000
65Cu-35Zn

Commercial Names

Previous trade name. C26800: Yellow brass, 66%. C27000: Yellow brass, 65%
Common name. Yellow brass

Specifications

AMS. Wire: 4710, 4712
ASTM. Flat products: B36 (C26800). Tube: B135 (C27000), B587 (26800 and C27000). Wire: B134
SAE: J463
Government. Flat products: QQ-B-613. Bar, rod, forgings, shapes: QQ-

Mechanical Properties

Tensile properties. See Table 50.
Shear strength. See Table 50.
Hardness. See Table 50.
Elastic modulus. Tension, 105 GPa (15 × 10⁶ psi); shear, 35 GPa (5 × 10⁶ psi)
Fatigue strength. Rotating beam tests. At 10⁸ cycles, for strip 1 mm (0.04 in.) thick: OS070 temper, 83

Table 50 Typical mechanical properties of C26800 and C27000

Temper	Tensile strength MPa	ksi	Yield strength(a) MPa	ksi	Elongation(b), %	Hardness HRF	HRB	HR30T	Shear strength MPa	ksi
Flat Products: 1 mm (0.04-in.) Thick										
OS070	315	46	97	14	65	58	...	15	220	32
OS050	325	47	105	15	62	64	...	26	230	33
OS035	340	49	115	17	57	68	...	31	235	34
OS025	350	51	130	19	55	72	...	36	240	35
OS015	365	53	150	22	54	78	...	43	250	36
H01	370	54	275	40	43	...	55	54	250	36
H02	420	61	345	50	23	...	70	65	275	40
H04	510	74	415	60	8	...	80	70	295	43
H06	585	85	425	62	5	...	87	74	310	45
H08	625	91	425	62	3	...	90	76	325	47
H10	675	98	435	63	3	...	91	77	...	...
Rod: 25 mm (1.0-in.) Diam										
OS050	330	48	110	16	65(c)	65	...	...	235	34
H00 (6%)	380	55	275	40	48(d)	...	...	55	...	36
Wire: 2 mm (0.08-in.) Diam										
OS050	330	48	110	16	64	...	...	...	230	33
OS035	345	50	125	18	60	...	...	...	235	34
OS025	360	52	145	21	58	...	...	...	240	35
OS015	370	54	160	23	55	...	...	...	250	36
H00	400	58	315	46	35	...	...	...	260	38
H01	485	70	395	57	20	...	...	...	290	42
H02	605	88	420	61	15	...	...	...	...	...
H04	760	110	...	...	8	...	...	...	380	55
H06	825	120	...	...	4	...	...	...	...	...
H08	885	128	...	...	3	...	...	...	415	60

(a) At 0.5% extension under load. (b) In 50 mm or 2 in. (c) 75% reduction in area. (d) 70% reduction in area.

MPa (12 ksi); H04 temper, 97 MPa (14 ksi); H08 temper, 140 MPa (20 ksi)

Mass Characteristics

Density. 8.47 Mg/m³ (0.306 lb/in.³) at 20 °C (68 °F)

Thermal Properties

Liquidus temperature. 930 °C (1710 °F)

Solidus temperature. 905 °C (1660 °F)

Coefficient of thermal expansion. Linear, 20.3 µm/m·K (11.3 µin./ in.·°F) at 20 to 300 °C (68 to 572 °F)

Specific heat. 380 J/kg·K (0.09 Btu/ lb·°F) at 20 °C (68 °F)

Thermal conductivity. 116 W/m·K (67 Btu/ft·h·°F) at 20 °C (68 °F)

Electrical Properties

Electrical conductivity. Volumetric, 27% IACS at 20 °C (68 °F), annealed

Electrical resistivity. 64 nΩ·m at 20 °C (68 °F), annealed

Structure

Crystal structure. Face-centered cubic alpha

Microstructure. Single-phase alpha

Fabrication Characteristics

Machinability. 30% of C36000, free-cutting brass

Recrystallization temperature. About 290 °C (550 °F) for strip cold rolled 50% to 1 mm (0.04 in.) thickness and having an initial grain size of 0.035 mm.

Maximum cold reduction between anneals. 90%

Annealing temperature. 425 to 700 °C (800 to 1300 °F)

Hot working temperature. 700 to 820 °C (1300 to 1500 °F)

C28000
60Cu-40Zn

Commercial Names

Previous trade name. Muntz metal, 60%; CA280

Common name. Muntz metal

Specifications

ASME. Condenser tubing: SB111

ASTM. Tubing: B111, B135

Government. Flat products: QQ-B-613. Bar, rod, forgings, shapes: QQ-B-626. Seamless tubing: WW-T-791

Chemical Composition

Composition limits. 59.0 to 63.0 Cu; 0.30 max Pb; 0.07 max Fe; rem Zn

Applications

Typical uses. Decoration, as architectural panel sheets; structural, as heavy plates; bolting and valve stems; tubing for heat exchangers; brazing rod (for copper alloys and cast iron); hot forgings

Precautions in use. C28000 has poor cold drawing and forming properties (compared with higher-copper alloys), but excellent hot working properties. It is the strongest of the copper-zinc alloys but is less ductile than higher-copper alloys. It is subject to dezincification and stress-corrosion cracking under certain conditions.

Mechanical Properties

Tensile properties. See Table 51 and Fig. 33 and 34.

Shear strength. See Table 51.

Hardness. See Table 51.

Elastic modulus. Tension, 105 GPa (15 × 10⁶ psi); shear, 39 GPa (5.6 × 10⁶ psi)

Mass Characteristics

Density. 8.39 Mg/m³ (0.303 lb/in.³) at 20 °C (68 °F)

Thermal Properties

Liquidus temperature. 905 °C (1660 °F)

Solidus temperature. 900 °C (1650 °F)

Coefficient of thermal expansion. Linear, 20.8 µm/m·K (11.6 µin./ in.·°F) at 20 to 300 °C (68 to 572 °F)

Specific heat. 375 J/kg·K (0.09 Btu/ lb·°F) at 20 °C (68 °F)

Thermal conductivity. 123 W/m·K (71 Btu/ft·h·°F) at 20 °C (68 °F)

Electrical Properties

Electrical conductivity. Volumetric, 28% IACS at 20 °C (68 °F)

Electrical resistivity. 61.6 nΩ·m at 20 °C (68 °F)

Structure

Microstructure. Two phase—face-centered cubic alpha plus body-centered cubic beta. Beta phase appears lemon yellow when etched with ammonia peroxide; it is dark when etched with ferric chloride. In grain size determination, ignore the beta phase.

Optical Properties

Color. Reddish compared to C26000 (70-30 cartridge brass). C28000 is used as a good match to the color of C23000 (85-15 red brass).

Chemical Properties

General corrosion behavior. Generally good; similar to copper except as noted below.

Resistance to specific corroding agents. Better resistance to sulfur-bearing compounds than higher copper alloys.

Fabrication Characteristics

Machinability. 40% of C36000, free-cutting brass

Forgeability. 90% of C37700, forging brass

Formability. Fair capacity for cold working; excellent capacity for hot forming

Weldability. Soldering or brazing: excellent. Oxyfuel gas welding, resistance spot welding or resistance butt welding: good. Gas shielded arc welding: fair

Annealing temperature. 425 to 600 °C (800 to 1100 °F). See also Fig. 34.

Hot working temperature. 625 to 800 °C (1150 to 1450 °F)

Table 51 Typical mechanical properties of C28000

Temper	Tensile strength MPa	ksi	Yield strength(a) MPa	ksi	Elongation(b), %	Hardness, HRF	Shear strength MPa	ksi
Flat Products, 1-mm (0.04-in.) Thick								
M20	370	54	145	21	45	85	275	40
O61	370	54	145	21	45	80	275	40
H00	415	60	240	35	30	55 HRB	290	42
H02	485	70	345	50	10	75 HRB	305	44
Rod, 25-mm (1-in.) Diam								
M30	360	52	140	20	52	78	270	39
O61	370	54	145	21	50	80	275	40
H01	495	72	345	50	25	78	310	45

(a) 0.5% extension under load. (b) In 50 mm or 2 in.

Fig. 33 Typical mechanical properties of extruded and drawn C28000

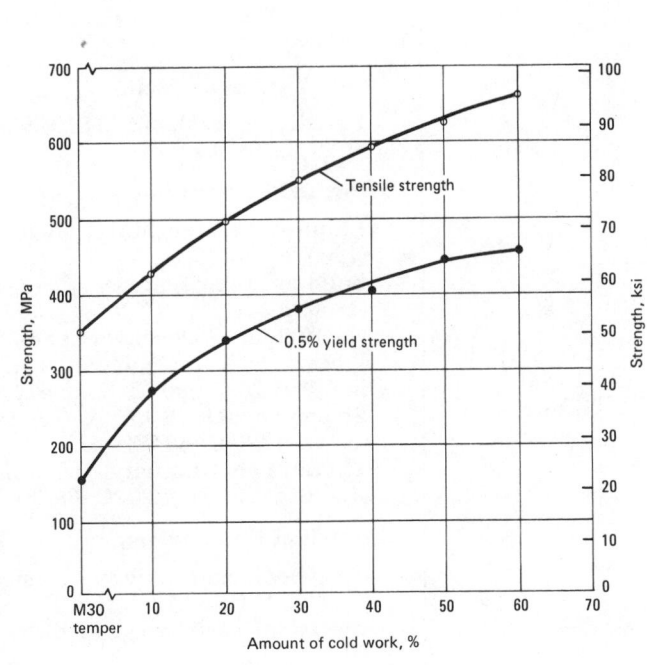

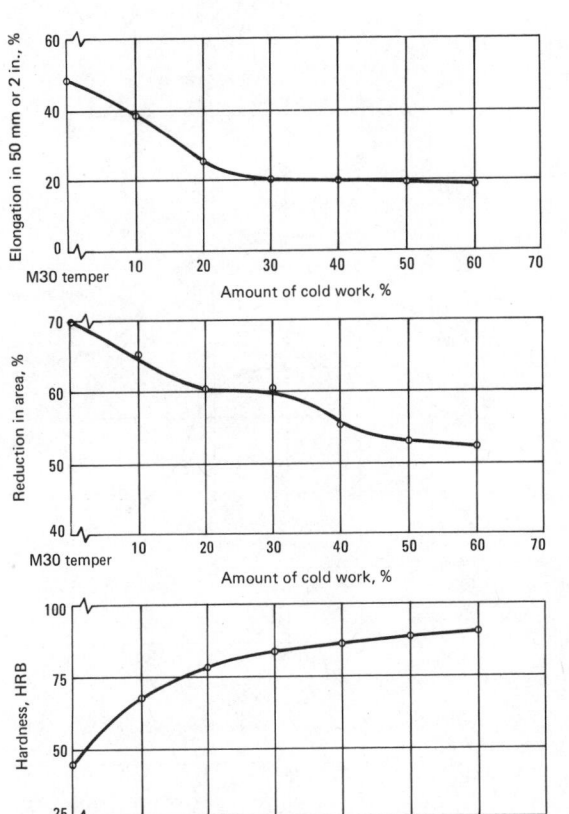

Data are for Muntz metal, rod, less than 25 mm (1 in.) diam, that was extruded, then cold drawn to various percentages reduction in area.

Fig. 34 Annealing curves for C28000

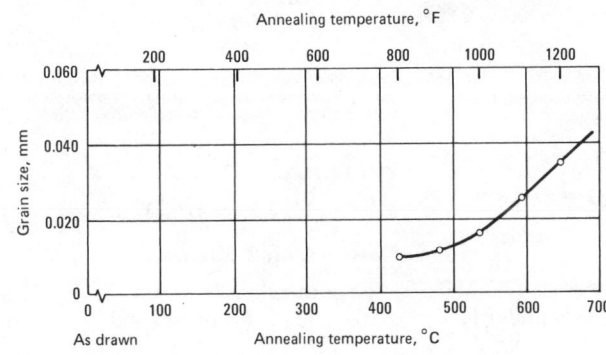

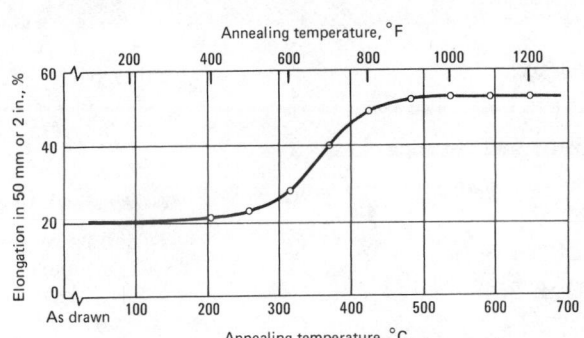

Data are for Muntz metal, rod, less than 25 mm (1 in.) diam, that was extruded, cold drawn 30%, and annealed 1 h at various temperatures.

Fig. 34 (continued)

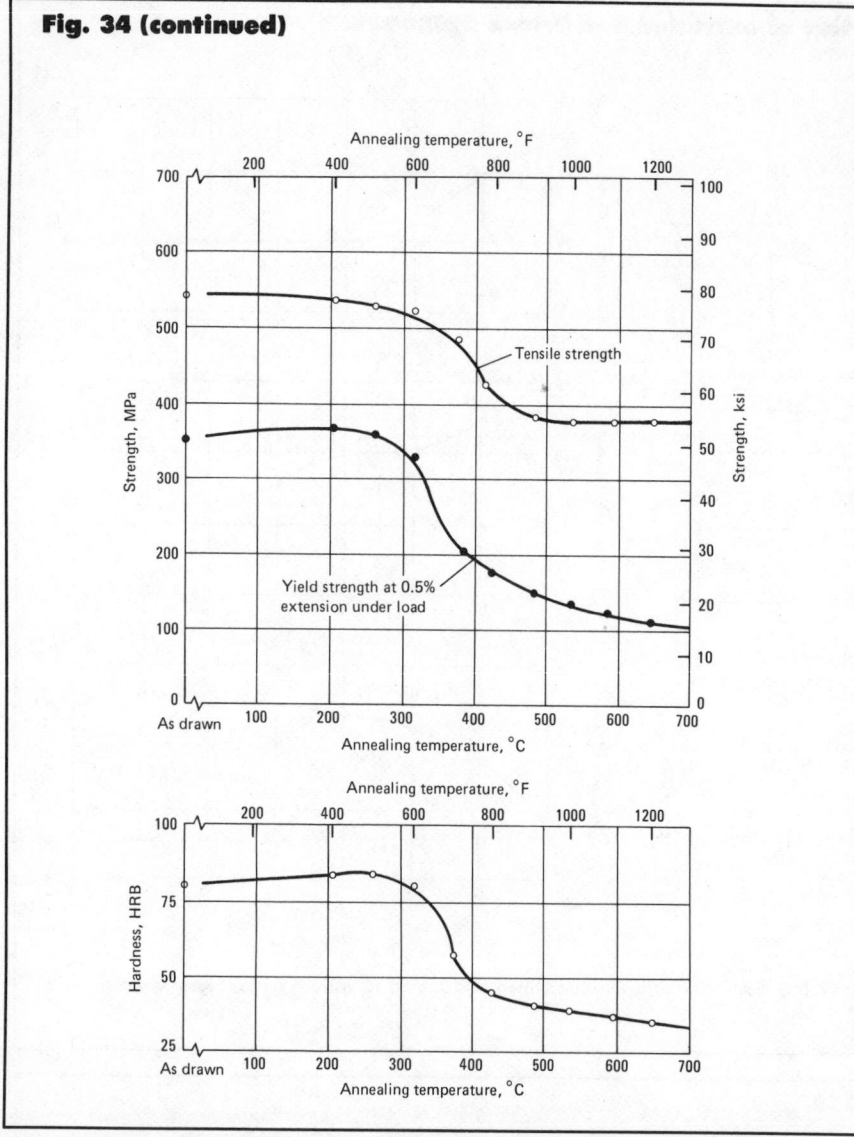

temper, 165 MPa (24 ksi); H02 temper, 205 MPa (30 ksi)

Hardness. O61 temper, 55 HRF; H02 temper, 58 HRB; H04 temper, 61 to 65 HRB

Elastic modulus. Tension, 115 GPa (17×10^6 psi); shear, 45 GPa (6.4×10^6 psi)

Mass Characteristics

Density. 8.83 Mg/m^3 (0.319 $lb/in.^3$) at 20 °C (68 °F)

Thermal Properties

Liquidus temperature. 1040 °C (1900 °F)

Solidus temperature. 1010 °C (1850 °F)

Coefficient of thermal expansion. Linear, 18.4 μm/m·K (10.2 μin./in.·°F) at 20 to 300 °C (68 to 572 °F)

Specific heat. 375 J/kg·K (0.09 Btu/lb·°F) at 20 °C (68 °F)

Thermal conductivity. 180 W/m·K (104 Btu/ft·h·°F) at 20 °C (68 °F)

Electrical Properties

Electrical conductivity. Volumetric, 42% IACS at 20 °C (68 °F)

Electrical resistivity. 41 nΩ·m at 20 °C (68 °F)

Optical Properties

Color. Rich bronze

Fabrication Characteristics

Machinability. 80% of C36000, free-cutting brass

Formability. Cold working, good; hot forming, poor

Weldability. Soldering: excellent. Brazing: good. Resistance butt welding: fair. All other welding processes not recommended

Annealing temperature. 425 to 650 °C (800 to 1200 °F)

C31400
89Cu-9.1Zn-1.9Pb

Commercial Names

Previous trade name. Leaded commercial bronze; CA314

Specifications

ASTM. B140

Chemical Composition

Composition limits. 87.5 to 90.5 Cu; 1.3 to 2.5 Pb; 0.10 max Fe; 0.7 max Ni; 0.5 max total others, rem Zn

Applications

Typical uses. Screws, screw-machine parts, pickling racks and fixtures, electrical plug-type connectors, builders' hardware

Mechanical Properties

Tensile properties. Rod, typical. O61 temper: tensile strength, 255 MPa (37 ksi); yield strength, 83 MPa (12 ksi) at 0.5% extension under load; elongation, 45% in 50 mm or 2 in.; reduction in area, 70%. H02 temper: tensile strength, 360 MPa (52 ksi); yield strength, 310 MPa (45 ksi); elongation, 18%; reduction in area, 60%

Shear strength. Rod, typical: O61

C31600
89Cu-8.1Zn-1.9Pb-1Ni

Commercial Names

Previous trade name. Leaded commercial bronze-nickel bearing; CA316

Specifications

ASTM. B140

Chemical Composition

Composition limits. 87.5 to 90.5 Cu;

1.3 to 2.5 Pb; 0.7 to 1.2 Ni; 0.1 max Fe; 0.04 to 0.10 P; 0.5 max total others; rem Zn

Applications

Typical uses. Electrical connectors, fasteners, hardware, nuts, screws, screw machine parts. Most commonly used as rod or drawn bar

Mechanical Properties

Tensile properties. See Table 52.
Hardness. See Table 52.
Elastic modulus. Tension, 115 GPa (17×10^6 psi)

Mass Characteristics

Density. 8.86 Mg/m³ (0.320 lb/in.³) at 20 °C (68 °F)

Thermal Properties

Liquidus temperature. 1040 °C (1900 °F)
Solidus temperature. 1010 °C (1850 °F)
Coefficient of thermal expansion. Linear, 18.4 μm/m·K (10.2 μin./in.·°F) at 20 to 300 °C (68 to 572 °F)

Specific heat. 380 J/kg·K (0.09 Btu/lb·°F) at 20 °C (68 °F)
Thermal conductivity. 140 W/m·K (81 Btu/ft·h·°F) at 20 °C (68 °F)

Electrical Properties

Electrical conductivity. Volumetric, 32% IACS at 20 °C (68 °F)
Electrical resistivity. 54 nΩ·m at 20 °C (68 °F)

Optical Properties

Color. Rich bronze

Fabrication Characteristics

Machinability. 80% of C36000, free-cutting brass

Formability. Cold working: good. Hot forming: poor

Weldability. Soldering: excellent. Brazing: good. Resistance butt welding: fair. All other welding processes not recommended.
Annealing temperature. 425 to 650 °C (800 to 1200 °F)

Table 52 Typical mechanical properties of C31600

Temper	Tensile strength MPa	ksi	Yield strength(a) MPa	ksi	Elongation(b), %	Hardness, HRB	Shear strength MPa	ksi
Drawn Bar, 6-mm (0.25-in.) Diam								
H04	435	63	385	56	12	70	...	...
Rod, 13-mm (0.50-in.) Diam								
H04	460	67	405	59	13	72	275	40
Rod, 25-mm (1-in.) Diam								
OS050	255	37	83	12	45	55 HRF	165	24
H04	450	65	395	57	15	70	270	39

(a) 0.5% extension under load. (b) In 50 mm or 2 in.

C33000
66Cu-33.5Zn-0.5Pb

Commercial Names

Previous trade name. Low leaded brass (tube)
Common name. High brass; yellow brass

Specifications

AMS. 4555
ASTM. B135
SAE. J463
Government. WW-T-791, MIL-T-46072

Chemical Composition

Composition limits. 65 to 68 Cu; 0.2 to 0.8 Pb; 0.07 max Fe; 0.5 max total others; rem Zn. For tubing more than 125 mm (5 in.) OD, Pb content may be less than 0.2%.

Applications

Typical uses. General purpose use: Where some degree of machinability is required together with moderate cold working properties; for instance, primers for munitions. Plumbing: J-bends, pump lines, trap lines

Mechanical Properties

Tensile properties. See Table 53.

Hardness. See Table 53.
Elastic modulus. Tension, 105 GPa (15×10^6 psi); shear, 39 GPa (5.6×10^6 psi)

Mass Characteristics

Density. 8.50 Mg/m³ (0.31 lb/in.³) at 20 °C (68 °F)

Thermal Properties

Liquidus temperature. 940 °C (1720 °F)
Solidus temperature. 905 °C (1660 °F)
Coefficient of thermal expansion. Linear, 20.2 μm/m·K (11.2 μin./in.·°F) at 20 to 300 °C (68 to 572 °F)
Specific heat. 380 J/kg·K (0.09 Btu/lb·°F) at 20 °C (68 °F)
Thermal conductivity. 115 W/m·K (67 Btu/ft·h·°F) at 20 °C (68 °F)

Electrical Properties

Electrical conductivity. Volumetric. O61 temper, 26% IACS at 20 °C (68 °F)
Electrical resistivity. 66 nΩ·m at 20 °C (68 °F)

Fabrication Characteristics

Machinability. 60% of C36000, free-cutting brass.
Formability. Cold working, excellent; hot forming, poor
Weldability. Soldering: excellent. Brazing: good. Oxyfuel gas, gas-shielded arc, resistance spot, and resistance butt welding: fair. All other welding processes not recommended
Recrystallization temperature. 290 °C (550 °F)
Annealing temperature. 425 to 650 °C (800 to 1200 °F)

Table 53 Typical mechanical properties of C33000 tubing(a)

Temper	Tensile strength MPa	ksi	Yield strength(b) MPa	ksi	Elongation(c), %	HRF	Hardness HRB	HR30T
OS050	325	47	105	15	60	64	...	26
OS025	360	52	135	20	50	75	...	36
H58	450	65	345	50	32	100	70	66
H80	515	75	415	60	7	...	85	76

(a) 25 mm (1.0 in.) OD by 1.65 mm (0.065 in.) wall. (b) 0.5% extension under load. (c) In 50 mm or 2 in.

C33200
66Cu-32.4Zn-1.6Pb

Commercial Names

Previous trade name. High-leaded brass (tube)

Common name. Free-cutting tube brass

Specifications

AMS. 4558
ASTM. B135
Government. MIL-T-46072

Chemical Composition

Composition limits. 65.0 to 68.0 Cu; 1.3 to 2.0 Pb; 0.07 max Fe; 0.5 max others (total); rem Zn

Applications

Typical uses. General purpose screw-machine products

Mechanical Properties

Tensile properties. See Table 54 and Fig. 35.
Hardness. See Table 54 and Fig. 35.
Elastic modulus. Tension, 105 GPa (15×10^6 psi); shear, 39 GPa (6.5×10^6 psi)

Mass Characteristics

Density. 8.53 Mg/m^3 (0.31 lb/in.^{3}) at 20 °C (68 °F)

Thermal Properties

Liquidus temperature. 930 °C (1710 °F)
Solidus temperature. 900 °C (1650 °F)
Coefficient of thermal expansion. Linear, 20.3 μm/m·K (11.3 μin./in.·°F) at 20 to 300 °C (68 to 572 °F)
Specific heat. 380 J/kg·K (0.09 Btu/lb·°F) at 20 °C (68 °F)
Thermal conductivity. 115 W/m·K (67 Btu/ft·h·°F) at 20 °C (68 °F)

Electrical Properties

Electrical conductivity. Volumetric. O61 temper, 26% IACS at 20 °C (68 °F)
Electrical resistivity. 66 nΩ·m at 20 °C (68 °F)

Fabrication Characteristics

Machinability. 80% of C36000, free-cutting brass
Formability. Cold working, fair; hot forming, poor
Weldability. Soldering: excellent. Brazing: good. Resistance butt welding: fair. All other welding processes not recommended.
Recrystallization temperature. 288 °C (550 °F)
Annealing temperature. 425 to 650 °C (800 to 1200 °F)

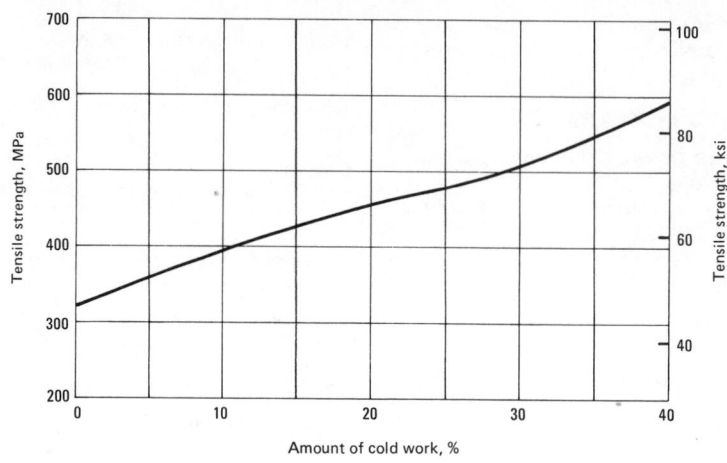

Fig. 35 Typical mechanical properties of cold drawn C33200 copper alloy tubing

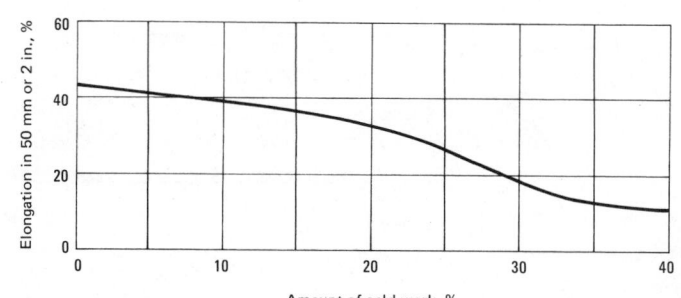

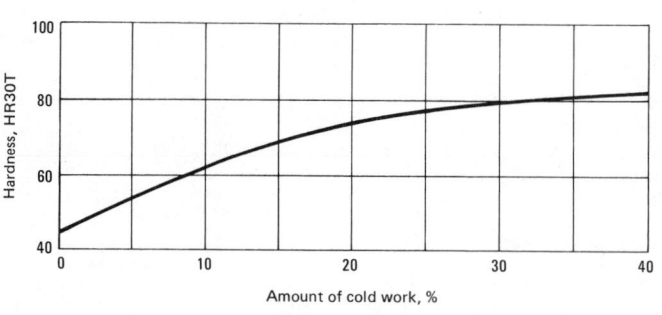

Table 54 Typical mechanical properties of C33200 tubing(a)

Temper	Tensile strength		Yield strength(b)		Elongation(c), %	Hardness		
	MPa	ksi	MPa	ksi		HRF	HRB	HR30T
OS050	325	47	105	15	60	64	...	26
OS025	360	52	135	20	50	75	...	36
H58	450	65	345	50	32	100	70	66
H80	515	75	415	60	7	...	85	76

(a) 25 mm (1 in.) OD by 1.65 mm (0.065 in.) wall. (b) 0.5% extension under load. (c) In 50 mm or 2 in.

C33500
65Cu-34.5Zn-0.5 Pb

Commercial Names

Previous trade name. Low-leaded brass

Specifications

ASTM. Flat products: B121. Rod: B453

Government. Flat products: QQ-B-613. Bar, forgings, rod, shapes, strip: QQ-B-626

Chemical Composition

Composition limits. 62.5 to 66.5 Cu; 0.3 to 0.8 Pb; 0.1 max Fe; 0.5 max others (total); rem Zn

Applications

Typical uses. Hardware such as butts and hinges; watch backs

Mechanical Properties

Tensile properties. See Table 55.
Shear strength. See Table 55.
Hardness. See Table 55.
Elastic modulus. Tension, 105 GPa (15×10^6 psi); shear, 39 GPa (5.6×10^6 psi)

Mass Characteristics

Density. 8.47 Mg/m^3 (0.306 lb/in.^{3}) at 20 °C (68 °F)

Table 55 Typical mechanical properties of C33500

Temper	Tensile strength MPa	ksi	Yield strength(a) MPa	ksi	Elon-gation(b), %	Hardness HRF	HRB	HR30T	Shear strength MPa	ksi
Flat Products, 1-mm (0.04-in.) Thick										
OS070	315	46	97	14	65	58	⋯	15	220	32
OS050 ...	325	47	105	15	62	64	⋯	26	⋯	⋯
OS035 ...	340	49	115	17	57	68	⋯	31	235	34
OS025 ...	350	51	130	19	55	72	⋯	36	⋯	⋯
H01	370	54	275	40	43	⋯	55	54	250	36
H02	420	61	345	50	23	⋯	70	65	275	40
H04	510	74	415	60	8	⋯	80	69	295	43
H06	580	84	⋯	⋯	⋯	⋯	86	74	⋯	⋯

(a) 0.5% extension under load. (b) In 50 mm or 2 in.

Thermal Properties

Liquidus temperature. 925 °C (1700 °F)
Solidus temperature. 900 °C (1650 °F)

Coefficient of thermal expansion. Linear, 20.3 μm/m·K (11.3 μin./in.·°F) at 20 to 300 °C (68 to 572 °F)
Specific heat. 380 J/kg·K (0.09 Btu/lb·°F) at 20 °C (68 °F)
Thermal conductivity. 115 W/m·K (67 Btu/ft·h·°F) at 20 °C (68 °F)

Electrical Properties

Electrical conductivity. Volumetric, 26% IACS at 20 °C (68 °F)

Electrical resistivity. 66 nΩ·m at 20 °C (68 °F)

Fabrication Characteristics

Machinability. 60% of C36000, free-cutting brass
Formability. Cold working, good; hot forming, poor. Commonly fabricated by blanking, drawing, machining, piercing, punching and stamping
Weldability. Soldering: excellent. Brazing: good. Oxyfuel gas, gas shielded arc, resistance spot and resistance butt welding: fair. Shielded metal arc and resistance seam welding not recommended.
Annealing temperature. 425 to 700 °C (800 to 1300 °F)

C34000
65Cu-34Zn-1Pb

Commercial Names

Previous trade name. Medium-leaded brass, 64.5%

Specifications

ASTM. Flat products: B121. Rod: B453

Government. Flat products: QQ-B-613. Bar, forgings, rod, shapes, strip: QQ-B-626

Chemical Composition

Composition limits. 62.5 to 66.5 Cu; 0.8 to 1.4 Pb; 0.10 max Fe; 0.5 max others (total); rem Zn

Applications

Typical uses. Flat products: butts, dials, engravings, gears, instrument plates, nuts or drawn shells, all involving piercing, threading or ma-

Table 56 Typical mechanical properties of C34000

Temper	Tensile strength MPa	ksi	Yield strength(a) MPa	ksi	Elon-gation,(b) %	Hardness HRB	HR30T	Shear strength MPa	ksi
Flat Products, 1-mm (0.04-in.) Thick									
OS035	340	49	115	17	54	68 HRF	31	225	33
OS025	350	51	130	19	53	72 HRF	36	235	34
H01	370	54	275	40	41	55	54	250	36
H02	420	61	345	50	21	70	63	275	40
H04	510	74	415	60	7	80	70	295	43
H06	585	85	425	62	5	87	73	310	45
Rod, 25-mm (1.0-in.) Diam									
OS025	345	50	135	20	60	70 HRF	⋯	235	34
H03	380	55	290	42	40	60	⋯	250	36
H02	435	63	330	48	30	68	⋯	275	40
Wire, 2-mm (0.08-in.) Diam									
OS025	345	50	⋯	⋯	50	⋯	⋯	235	34
H00	400	58	⋯	⋯	30	⋯	⋯	260	38
H01	485	70	⋯	⋯	13	⋯	⋯	290	42
H02	605	88	⋯	⋯	7	⋯	⋯	315	46

(a) 0.5% extension under load. (b) In 50 mm or 2 in.

chining. Rod, bar and wire: couplings, free-machining screws and rivets, gears, nuts, tire valve stems, screw-machine products involving severe knurling and roll threading or moderate cold heading, flaring, spinning or swaging.

Mechanical Properties

Tensile properties. See Table 56.
Shear strength. See Table 56.
Hardness. See Table 56.
Elastic modulus. Tension, 105 GPa (15×10^6 psi); shear, 39 GPa (5.6×10^6 psi)

Mass Characteristics

Density. 8.47 Mg/m^3 (0.306 lb/in.^{3}) at 20 °C (68 °F)

Thermal Properties

Liquidus temperature. 925 °C (1700 °F)
Solidus temperature. 885 °C (1630 °F)
Coefficient of thermal expansion. Linear, 20.3 μm/m·K (11.3 μin./in.·°F) at 20 to 300 °C (68 to 572 °F)
Specific heat. 380 J/kg·K (0.09 Btu/lb·°F) at 20 °C (68 °F)
Thermal conductivity. 115 W/m·K (67 Btu/ft·h·°F) at 20 °C (68 °F)

Electrical Properties

Electrical conductivity. Volumetric. O61 temper, 26% IACS at 20 °C (68 °F)
Electrical resistivity. 66 nΩ·m at 20 °C (68 °F)

Fabrication Characteristics

Machinability. 60% of C36000, free-cutting brass
Formability. Cold working, good; hot forming, poor
Weldability. Soldering: excellent. Brazing: good. Resistance butt welding: fair. All other welding processes not recommended.
Recrystallization temperature. 288 °C (550 °F)
Annealing temperature. 425 to 650 °C (800 to 1200 °F)

C34200
64.5Cu-33.5Zn-2Pb
C35300
62Cu-36.2Zn-1.8Pb
Commercial Names

Previous trade name. High leaded brass

Common name. Clock brass, engraver's brass, heavy-leaded brass

Specifications

ASTM. Flat products: B121. Rod: B453
SAE. J463
UNS number. C34200, C35300
Government. Flat products: QQ-B-613. Bar, forgings, rod, shapes, strip: QQ-B-626

Chemical Composition

Composition limits. C34200: 62.5 to 66.5 Cu; 1.5 to 2.5 Pb; 0.1 max Fe; 0.05 max others (total); rem Zn. C35300: 59.0 to 64.5 Cu; 1.3 to 2.3 Pb; 0.1 max Fe; 0.5 max others (total); rem Zn

Applications

Typical uses. Flat products: gears, wheels, nuts, plates for clocks, keys, bearing cages, engraver's plates. Rod: gears, pinions, valve stems, automatic screw-machine parts that necessitate more severe cold working than can be tolerated with free-cutting brass—for example, knurling, moderate staking

Mechanical Properties

Tensile properties. See Table 57.
Hardness. See Table 57.
Elastic modulus. Tension, 105 GPa (15×10^6 psi); shear, 39 GPa (5.6×10^6 psi)

Structure

Crystal structure. Face-centered cubic alpha
Microstructure. Two phase, alpha and lead

Mass Characteristics

Density. 8.5 Mg/m^3 (0.307 lb/in.^{3}) at 20 °C (68 °F)

Thermal Properties

Liquidus temperature. 910 °C (1670 °F)
Solidus temperature. 885 °C (1630 °F)
Coefficient of thermal expansion. Linear, 20.3 μm/m·K (11.3 μin./in.·°F) at 20 to 300 °C (68 to 572 °F)
Specific heat. 380 J/kg·K (0.09 Btu/lb·°F) at 20 °C (68 °F)
Thermal conductivity. 115 W/m·K (67 Btu/ft·h·°F) at 20 °C (68 °F)

Electrical Properties

Electrical conductivity. Volumetric. O61 temper, 26% IACS at 20 °C (68 °F)
Electrical resistivity. 66 nΩ·m at 20 °C (68 °F)

Fabrication Characteristics

Machinability. 90% of C36000, free-cutting brass
Formability. Cold working, fair; hot forming, poor
Weldability. Soldering: excellent. Brazing: good. Resistance butt welding: fair. All other welding processes not recommended.
Recrystallization temperature. 320 °C (600 °F)
Annealing temperature. 425 to 600 °C (800 to 1100 °F). See also Fig. 36.
Hot working temperature. 785 to 815 °C (1445 to 1500 °F)

Table 57 Typical mechanical properties of C34200

Temper	Tensile strength MPa	ksi	Yield strength(a) MPa	ksi	Elongation(b), %	Hardness HRB	HR30T	Shear strength MPa	ksi
Flat Products, 1-mm (0.04-in.) Thick									
OS015	370	54	165	24	45	78 HRF	41	255	37
OS025	360	52	140	20	48	76 HRF	37	250	36
OS035	340	49	115	17	52	68 HRF	32	235	34
OS050	325	47	105	15	55	66 HRF	28	225	33
H01	370	54	275	40	38	55	54	250	36
H02	420	61	345	50	20	70	63	275	40
H04	510	74	415	60	7	80	71	295	43
H06	585	85	425	62	5	87	75	310	45
Rod, 25-mm (1.0-in.) Diam									
O50	325	47	125	18	50(c)	66 HRF	...	...	...
H55	400	58	270	39	28(d)	65	...	...	...
H02	450	65	310	45	23(e)	72	...	...	...

(a) 0.5% extension under load. (b) In 50 mm or 2 in. (c) Reduction in area 65%. (d) Reduction in area 50%. (e) Reduction in area 35%.

Fig. 36 Annealing behavior of C34200

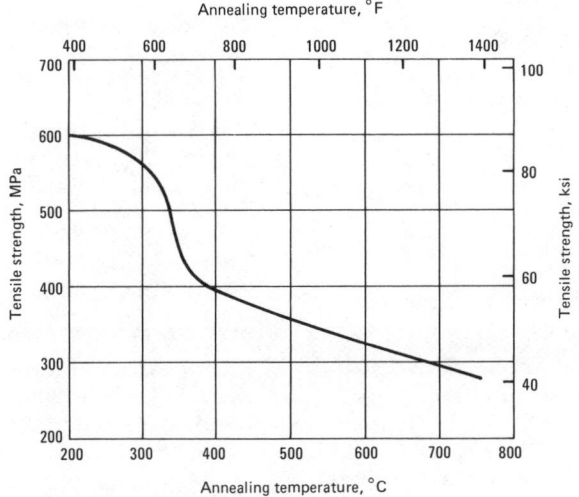

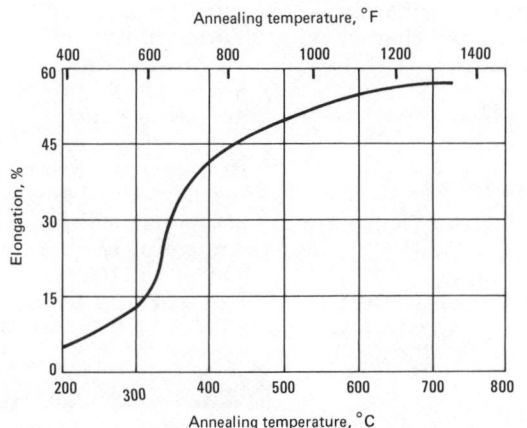

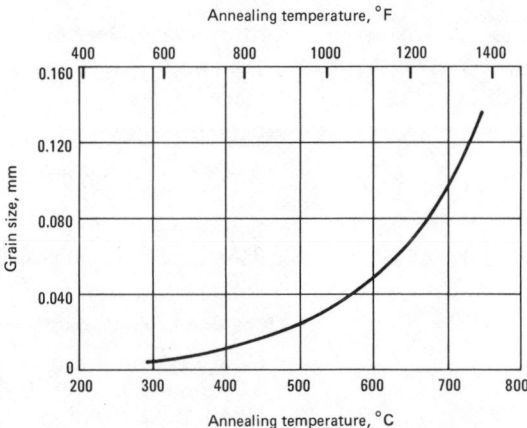

Curves are for 1 mm (0.04 in.) thick strip cold rolled from OS035 temper starting stock.

C34900
62Cu-37.5Zn-0.3Pb

Chemical Composition

Composition limits. 61.0 to 64.0 Cu; 0.1 to 0.5 Pb; 0.1 max Fe; 0.5 max others (total); rem Zn

Applications

Typical uses. Building hardware, drilled and tapped rivets, plumbing goods, saw nuts and parts requiring moderate cold working combined with some machining

Mechanical Properties

Tensile properties. See Table 58.
Shear strength. See Table 58.
Hardness. See Table 58.
Elastic modulus. Tension, 105 GPa (15×10^6 psi); shear, 39 GPa (5.6×10^6 psi)

Mass Characteristics

Density. 8.44 Mg/m³ (0.305 lb/in.³) at 20 °C (68 °F)

Thermal Properties

Liquidus temperature. 910 °C (1670 °F)
Solidus temperature. 895 °C (1640 °F)
Coefficient of thermal expansion. Linear, 20.3 μm/m·K (11.3 μin./in.·°F) at 20 to 300 °C (68 to 572 °F)
Specific heat. 380 J/kg·K (0.09 Btu/lb·°F) at 20 °C (68 °F)
Thermal conductivity. 115 W/m·K (67 Btu/ft·h·°F) at 20 °C (68 °F)

Electrical Properties

Electrical conductivity. Volumetric, 26% IACS at 20 °C (68 °F)
Electrical resistivity. 66 nΩ·m at 20 °C (68 °F)

Fabrication Characteristics

Machinability. 50% of C36000, free-cutting brass
Formability. Cold working, good; hot forming, fair
Weldability. Soldering: excellent. Brazing: good. Oxyfuel gas, gas-shielded arc and resistance spot and butt welding: fair. Shielded metal-arc and resistance seam welding: not recommended.
Annealing temperature. 425 to 650 °C (800 to 1200 °F)
Hot working temperature. 675 to 800 °C (1250 to 1450 °F)

Table 58 Typical mechanical properties of C34900

Temper	Tensile strength MPa	ksi	Yield strength(a) MPa	ksi	Elongation(b), %	Hardness, HRF	Shear strength MPa	ksi
Rod: 6-mm (0.25-in.) Diam								
OS035	365	53	165	24	50	75	235	34
Rod: 25-mm (1.0-in.) Diam								
H01	385	56	290	42	42	70 HRB	250	36
Wire: 6-mm (0.25-in.) Diam								
OS015	380	55	150	22	48	70	240	35
H01	470	68	380	55	18	72 HRB	285	410
Wire: 19-mm (0.75-in.) Diam								
OS050	330	48	110	16	72	67	220	32

(a) At 0.5% extension under load. (b) In 50 mm or 2 in.

C35000
62.5Cu-36.4Zn-1.1Pb

Commercial Names

Previous trade name. Medium-leaded brass, 62%

Specifications

ASTM. Flat products: B121. Rod: B453
SAE. J463
Government. Flat products: QQ-B-613. Bar, forgings, rod, shapes, strip: QQ-B-626

Chemical Composition

Composition limits. 59.0 to 64.0 Cu; 0.8 to 1.4 Pb; 0.1 max Fe; 0.5 max others (total); rem Zn

Applications

Typical uses. Bearing cages, book dies, clock plates, engraving plates, gears, hinges, hose couplings, keys, lock parts, lock tumblers, meter parts, sink strainers, strike plates, templates, nuts, type characters, washers, wear plates

Mechanical Properties

Tensile properties. See Table 59.
Shear strength. See Table 59.
Hardness. See Table 59.
Elastic modulus. Tension, 105 GPa (15×10^6 psi); shear, 39 GPa (5.6×10^6 psi)

Mass Characteristics

Density. 8.47 Mg/m^3 (0.306 lb/in.^{3}) at 20 °C (68 °F)

Thermal Properties

Liquidus temperature. 915 °C (1680 °F)
Solidus temperature. 895 °C (1640 °F)
Coefficient of thermal expansion. Linear, 20.3 μm/m·K (11.3 μin./in.·°F) at 20 to 300 °C (68 to 572 °F)
Specific heat. 380 J/kg·K (0.09 Btu/lb·°F) at 20 °C (68 °F)
Thermal conductivity. 115 W/m·K (67 Btu/ft·h·°F) at 20 °C (68 °F)

Electrical Properties

Electrical conductivity. Volumetric, 26% IACS at 20 °C (68 °F)
Electrical resistivity. 66 nΩ·m at 20 °C (68 °F)

Fabrication Characteristics

Machinability. 70% of C36000, free-cutting brass
Forgeability. 50% of C37700, forging brass
Formability. Fair for cold working and hot forming
Weldability. Soldering: excellent. Brazing: good. Resistance butt welding: fair. All other welding processes not recommended.
Annealing temperature. 425 to 600 °C (800 to 1100 °F)
Hot working temperature. 760 to 800 °C (1400 to 1500 °F)

Table 59 Typical mechanical properties of C35000

Temper	Tensile strength MPa	ksi	Yield strength 0.5% extension under load MPa	ksi	0.2% offset MPa	ksi	Elongation(a), %	Hardness HRB	HR30T	Shear strength MPa	ksi
Flat Products: 1-mm (0.04-in.) Thick											
OS050 ...	310	45	90	13	90	13	57	61 HRF	...	...	...
OS035 ...	325	47	110	16	110	16	54	67 HRF	...	...	...
OS025 ...	330	48	135	20	135	20	50	70 HRF	...	...	...
OS015 ...	350	51	170	25	170	25	46	74 HRF	52	...	...
H01	370	54	220	32	235	34	43	66	60	...	...
H02	415	60	310	45	310	45	29	75	68	...	...
H03	460	67	365	53	380	55	17	80	71	...	...
H04	505	73	415	60	415	60	10	86	75	...	...
H06	580	84	450	65	475	69	5				
Rod: 12-mm (0.5-in.) Diam											
OS050 ...	330	48	110	16	...	...	56	65 HRF	25	235	34
OS015 ...	380	55	170	25	...	...	46	85 HRF	50	250	36
H01	400	58	305	44	...	...	42	60	57	260	38
H02	485	70	360	52	...	...	22	80	70	290	42

(a) In 50 mm or 2 in.

C35600
62Cu-35.5Zn-2.5Pb

Commercial Names

Previous trade name. Extra-high leaded brass

Specifications

ASTM. Flat products: B121. Rod: B453
Government. Flat products: QQ-B-613. Bar, rod, shapes, strip: QQ-B-626

Chemical Composition

Composition limits. 59.0 to 64.5 Cu; 2.0 to 3.0 Pb; 0.1 max Fe; 0.5 max others (total); rem Zn

Applications

Typical uses. Hardware: clock plates and nuts, clock and watch backs, clock gears and wheels. Industrial: channel plate

Mechanical Properties

Tensile properties. See Table 60.
Hardness. See Table 60.
Elastic modulus. Tension, 97 GPa (14×10^6 psi); shear, 37 GPa (5.3×10^6 psi)

Mass Characteristics

Density. 8.5 Mg/m^3 (0.307 $lb/in.^3$) at 20 °C (68 °F)

Thermal Properties

Liquidus temperature. 905 °C (1660 °F)
Solidus temperature. 885 °C (1630 °F)
Coefficient of thermal expansion. Linear, 20.5 μm/m·K (11.4 μin./in.·°F) at 20 to 300 °C (68 to 572 °F)

Specific heat. 380 J/kg·K (0.09 Btu/lb·°F) at 20 °C (68 °F)
Thermal conductivity. 115 W/m·K (67 Btu/ft·h·°F) at 20 °C (68 °F)

Electrical Properties

Electrical conductivity. Volumetric. O61 temper, 26% IACS at 20 °C (68 °F)
Electrical resistivity. 66 nΩ·m at 20 °C (68 °F)

Table 60 Typical mechanical properties of C35600 sheet and strip(a)

Temper	Tensile strength MPa	ksi	Yield strength(b) MPa	ksi	Elongation(c), %	Hardness HRB	HR30T
OS035	340	49	115	17	50	68 HRF	31
H01	370	54	275	40	35	55	54
H02	420	61	345	50	20	70	65
H04	510	74	415	60	7	80	69

(a) 1 mm (0.04 in.) thick. (b) At 0.5% extension under load. (c) In 50 mm or 2 in.

Fabrication Characteristics

Machinability. 100% of C36000, free-cutting brass
Formability. Cold working, poor; hot forming, fair
Weldability. Soldering: excellent. Brazing: good. Resistance butt welding: fair. All other welding processes not recommended.
Annealing temperature. 425 to 600 °C (800 to 1100 °F)
Hot working temperature. 700 to 800 °C (1300 to 1450 °F)

C36000
61.5Cu-35.5Zn-3Pb

Commercial Names

Previous trade name. Free-cutting brass
Common name. Free turning brass, free-cutting yellow brass, high-leaded brass

Specifications

AMS. 4610
ASTM. B16
SAE. J463
Government. Flat products: QQ-B-613. Bar, forgings, rod, shapes, strip: QQ-B-626

Chemical Composition

Composition limits. 60.0 to 63.0 Cu; 2.5 to 3.7 Pb; 0.35 max Fe; 0.5 max others (total); rem Zn

Applications

Typical uses. Hardware: gears, pinions. Industrial: automatic high-speed screw-machine parts

Mechanical Properties

Tensile properties. See Table 61 and Fig. 37.
Hardness. See Table 61.
Elastic modulus. Tension, 97 GPa

(14×10^6 psi); shear, 37 GPa (5.3×10^6 psi)
Fatigue strength. Rotating beam tests on 0.350 in. diam specimens taken from 50 mm (2 in.) diam rod. H02 temper (cold drawn 15%): 140 MPa (20 ksi) at 10^8 cycles; 97 MPa (14 ksi) at 3×10^8 cycles

Structure

Microstructure. Generally three-phase—alpha, beta and lead

Mass Characteristics

Density. 8.5 Mg/m^3 (0.307 $lb/in.^3$) at 20 °C (68 °F)

Table 61 Typical mechanical properties of C36000

Temper	Tensile strength MPa	ksi	Yield strength(a) MPa	ksi	Elongation(b), %	Reduction in area, %	Hardness, HRB	Shear strength MPa	ksi
Rod, 6 mm (0.25 in.) Diam									
H02(c)	470	68	360	52	18	48	80	260	38
Rod, 25 mm (1 in.) Diam									
O61	340	49	125	18	53	58	68 HRF	205	30
H02(d)	400	58	310	45	25	50	78	235	34
Rod, 50 mm (2 in.) Diam									
H02(e)	380	55	305	44	32	52	75	220	32
Shapes									
M30	340	49	125	18	50	· · ·	68 HRF	205	30
H01(f)	385	56	310	45	20	· · ·	62	230	33

(a) 0.5% extension under load. (b) In 50 mm or 2 in. (c) Cold drawn 25%. (d) Cold drawn 20%. (e) Cold drawn 18%. (f) Cold drawn 15%.

Fig. 37 Annealing curves for C36000

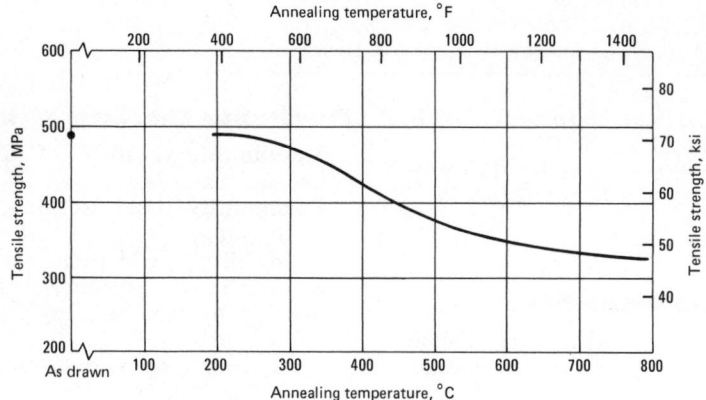

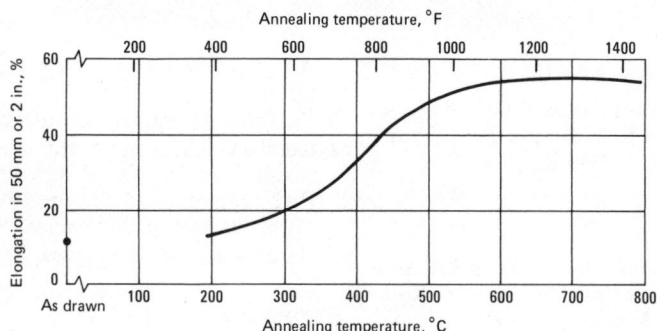

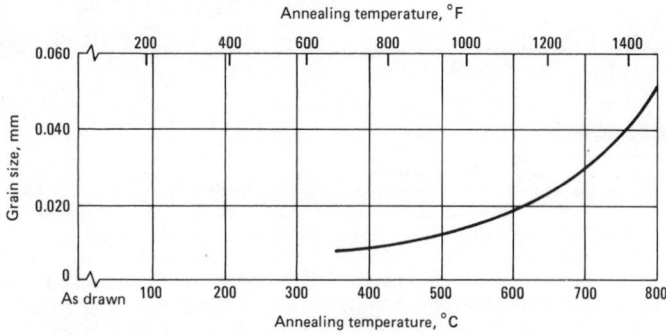

Curves are for free-cutting brass rod, cold drawn 30% to 19 mm (0.75 in.) diam from M30 temper (as extruded) starting stock, then annealed 1 h at temperature.

Thermal Properties

Liquidus temperature. 900 °C (1660 °F)
Solidus temperature. 885 °C (1630 °F)
Coefficient of thermal expansion. Linear, 20.5 µm/m·K (11.4 µin./in.·°F) at 20 to 300 °C (68 to 572 °F)
Specific heat. 380 J/kg·K (0.09 Btu/lb·°F) at 20 °C (68 °F)
Thermal conductivity. 115 W/m·K (67 Btu/ft·h·°F) at 20 °C (68 °F)

Electrical Properties

Electrical conductivity. Volumetric. O61 temper, 26% IACS at 20 °C (68 °F)
Electrical resistivity. 66 nΩ·m at 20 °C (68 °F)

Fabrication Characteristics

Machinability. 100% of C36000, free-cutting brass
Formability. Cold working, poor; hot forming, fair
Weldability. Soldering: excellent. Brazing: good. Resistance butt welding: fair. All other welding processes not recommended.
Recrystallization temperature. 330 °C (625 °F)
Annealing temperature. 425 to 600 °C (800 to 1100 °F). See also Fig. 37.
Hot working temperature. 700 to 800 °C (1300 to 1450 °F)

C36500, C36600, C36700, C36800
60Cu-39.4Zn-0.6Pb

Commercial Names

Previous trade name. C36500, uninhibited leaded Muntz metal; C36600, arsenical leaded Muntz metal; C36700, antimonial leaded Muntz metal; C36800, phosphorized leaded Muntz metal
Common name. Leaded Muntz metal; inhibited leaded Muntz metal

Specifications

ASME. Plate, condenser tube: SB171
ASTM. Plate, condenser tube: B171. Plate, clad: B432

Chemical Composition

Composition limits. 58.0 to 61.0 Cu;

0.4 to 0.9 Pb; 0.15 max Fe; 0.25 max Sn; As, Sb or P(a); 0.1 max others (total); rem Zn

(a) C36500, none specified; C36600, 0.02 to 0.1 As; C36700, 0.02 to 0.1 Sb; C36800, 0.02 to 0.1 P

Applications

Typical uses. Main tube sheets for condensers and heat exchangers; support sheets; baffles

Mechanical Properties

Tensile properties. 25 mm (1 in.) plate, M20 temper: tensile strength, 370 MPa (54 ksi); yield strength, 140 MPa (20 ksi); elongation, 45% in 50 mm or 2 in.
Shear strength. M20 temper: 275 MPa (40 ksi)
Hardness. M20 temper: 80 HRF
Elastic modulus. Tension, 105 GPa (15 × 10^6 psi); shear, 39 GPa (5.6 × 10^6 psi)

Structure

Crystal structure. Face-centered cubic
Microstructure. Alpha and beta with undissolved lead. Beta phase appears lemon yellow with ammonia peroxide etch; it may be darkened with ferric chloride etch. Lead appears as insoluble gray particles randomly distributed throughout the structure.

Mass Characteristics

Density. 8.41 Mg/m^3 (0.304 lb/in.^{3}) at 20 °C (68 °F)

Thermal Properties

Liquidus temperature. 900 °C (1650 °F)
Solidus temperature. 885 °C (1630 °F)
Coefficient of thermal expansion. Linear, 20.8 μm/m·K (11.6 μin./in.·°F) at 20 to 300 °C (68 to 572 °F)
Specific heat. 380 J/kg·K (0.09 Btu/lb·°F) at 20 °C (68 °F)
Thermal conductivity. 123 W/m·K (71 Btu/ft·h·°F) at 20 °C (68 °F)

Electrical Properties

Electrical conductivity. Volumetric, O61 temper: 28% IACS at 20 °C (68 °F)
Electrical resistivity. 62 nΩ·m at 20 °C (68 °F)

Chemical Properties

General corrosion behavior. Good resistance to corrosion in both fresh and salt water. C36500 is the uninhibited alloy and is subject to dezincification; the inhibited alloys each contain 0.02 to 0.10% of an inhibitor element (As, Sb or P), which imparts high resistance to dezincification.

Fabrication Characteristics

Machinability. 60% of C36000, free-cutting brass
Formability. Cold working, fair; hot working, excellent
Weldability. Soldering: excellent. Brazing: good. Oxyfuel gas, gas shielded arc and resistance butt welding: fair. All other welding processes not recommended.
Annealing temperature. 425 to 600 °C (800 to 1100 °F)
Hot working temperature. 625 to 800 °C (1150 to 1450 °F)

C37000
60Cu-39Zn-1Pb

Commercial Names

Previous trade name. Free-cutting Muntz metal

Specifications

ASTM. Tube: B135
Government. Flat products: QQ-B-613. Bar, forgings, rod, strip: QQ-B-626. Tube: MIL-T-46072

Chemical Composition

Composition limits. 59.0 to 62.0 Cu; 0.9 to 1.4 Pb; 0.15 max Fe; 0.5 max others (total); rem Zn

Applications

Typical uses. Automatic screw machine parts

Mechanical Properties

Tensile properties. See Table 62.
Hardness. See Table 62.
Elastic modulus. Tension, 105 GPa (15 × 10^6 psi); shear, 39 GPa (5.6 × 10^6 psi)

Mass Characteristics

Density. 8.41 Mg/m^3 (0.304 lb/in.^{3}) at 20 °C (68 °F)

Thermal Properties

Liquidus temperature. 900 °C (1650 °F)
Solidus temperature. 885 °C (1630 °F)
Coefficient of thermal expansion. Linear, 20.8 μm/m·K (11.6 μin./in.·°F) at 20 to 300 °C (68 to 572 °F)
Specific heat. 375 J/kg·K (0.09 Btu/lb·°F) at 20 °C (68 °F)
Thermal conductivity. 120 W/m·K (69 Btu/ft·h·°F) at 20 °C (68 °F)

Electrical Properties

Electrical conductivity. Volumetric, O61 temper: 27% IACS at 20 °C (68 °F)
Electrical resistivity. 63.9 nΩ·m at 20 °C (68 °F)

Fabrication Characteristics

Machinability. 70% of C36000, free-cutting brass
Formability. Cold working, fair; hot forming, excellent
Weldability. Soldering: excellent. Brazing: good. Resistance butt welding: fair. All other welding processes not recommended.
Annealing temperature. 425 to 600 °C (800 to 1100 °F)
Hot working temperature. 625 to 800 °C (1150 to 1450 °F)

Table 62 Typical mechanical properties of C37000

Temper	Tensile strength MPa	ksi	Yield strength(a) MPa	ksi	Elongation(b), %	Hardness HRB	HR30T
Tube: 38-mm (1.5-in.) OD by 3-mm (0.125-in.) Wall							
O50	370	54	140	20	40	80 HRF	43
H80(c)	550	80	415	60	6	85	74
Tube: 50-mm (2-in.) OD by 6-mm (0.25-in.) Wall							
H80(d)	485	70	310	45	10	75	67

(a) At 0.5% extension under load. (b) In 50 mm or 2 in. (c) Cold drawn 35%. (d) Cold drawn 25%.

C37700
60Cu-38Zn-2Pb

Commercial Names

Previous trade name. Forging brass

Specifications

AMS. Die forgings, forging rod: 4614
ASME. Die forgings: SB283
ASTM. Bar, forging, rod, shapes: B124. Die forgings: B283
SAE. Die forgings: J463
Government. QQ-B-626. Die forgings: MIL-C-13351

Chemical Composition

Composition limits. 58.0 to 62.0 Cu; 1.5 to 2.5 Pb; 0.3 max Fe; 0.5 max others (total); rem Zn

Applications

Typical uses. Forgings and pressings of all kinds

Mechanical Properties

Tensile properties. M30 temper: tensile strength, 360 MPa (52 ksi); yield strength, 140 MPa (20 ksi); elongation, 45% in 50 mm or 2 in. See also Fig. 38 and 39.
Hardness. M30 temper: 78 HRF. See also Fig. 38 and 39.
Elastic modulus. Tension, 105 GPa (15×10^6 psi); shear, 39 GPa (5.6×10^6 psi)

Structure

Crystal structure. Face-centered cubic
Microstructure. Two phase, alpha and beta, with undissolved lead. Beta phase appears lemon yellow with ammonia-peroxide etch. Ferric chloride darkens beta phase. Lead appears as gray particles.

Mass Characteristics

Density. 8.44 Mg/m³ (0.305 lb/in.³) at 20 °C (68 °F)

Thermal Properties

Liquidus temperature. 895 °C (1640 °F)
Solidus temperature. 880 °C (1620 °F)
Coefficient of thermal expansion. Linear, 20.7 μm/m·K (11.5 μin./in.·°F) at 20 to 300 °C (68 to 572 °F)
Specific heat. 380 J/kg·K (0.09 Btu/lb·°F) at 20 °C (68 °F)
Thermal conductivity. 120 W/m·K (69 Btu/ft·h·°F) at 20 °C (68 °F)

Fig. 38 Typical mechanical properties of extruded and drawn C37700

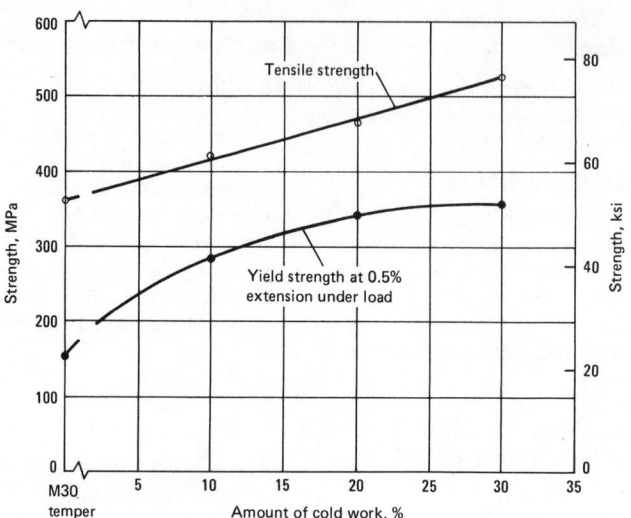
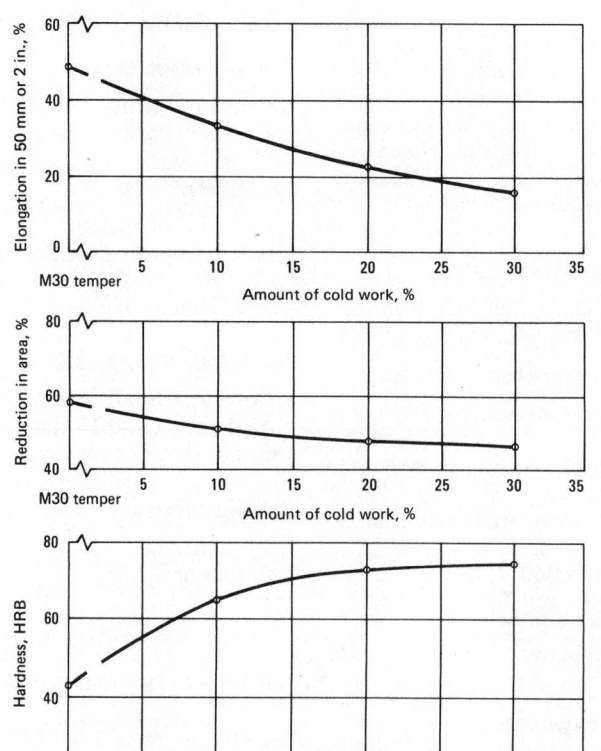

Data are for forging brass rod, less than 25 mm (1 in.) diam, that was extruded, then cold drawn to various percentages reduction in area.

Fig. 39 Annealing curves for C37700

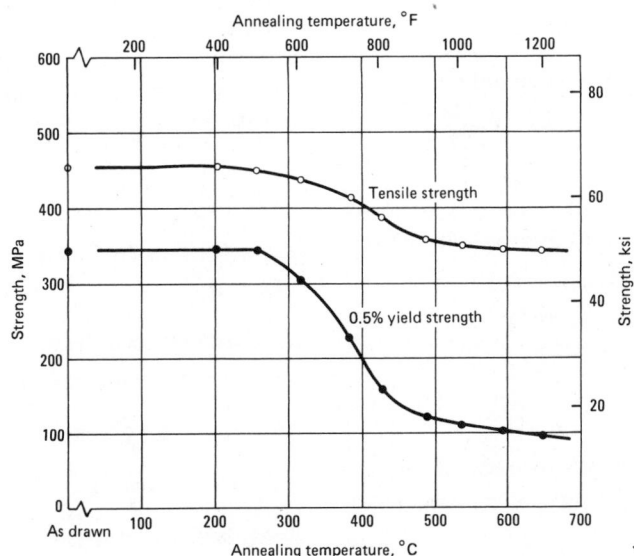

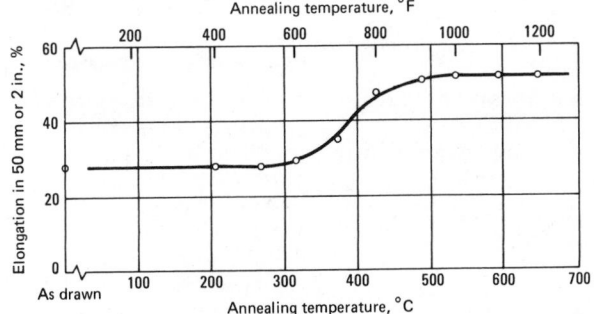

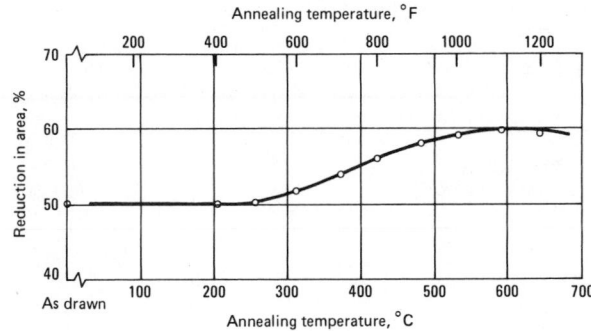

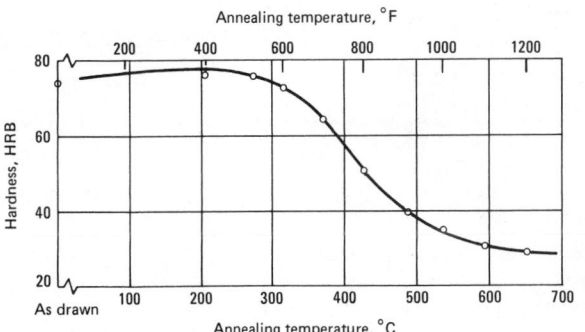

Electrical Properties

Electrical conductivity. Volumetric, 27% IACS at 20 °C (68 °F)
Electrical resistivity. 64 nΩ·m at 20 °C (68 °F)

Magnetic Properties

Magnetic susceptibility. Nonmagnetic

Optical Properties

Color. Golden hue compared to yellow of C26000, cartridge brass

Fabrication Characteristics

Machinability. 80% of C36000, free-cutting brass
Forgeability. 100%—This is the standard material against which the forging qualities of all other copper alloys are judged.
Formability. Cold working, poor; hot forming, excellent.
Weldability. Soldering: excellent. Brazing: good. Resistance butt welding: fair. All other welding processes not recommended.
Annealing temperature. 425 to 600 °C (800 to 1100 °F). See also Fig. 39.
Hot working temperature. 650 to 825 °C (1200 to 1500 °F)

Data are typical for forging brass rod, less than 25 mm (1 in.) diam, that was extruded, cold drawn 18%, and annealed 1 h at various temperatures.

C38500
57Cu-40Zn-3Pb

Commercial Names

Previous trade name. Architectural bronze

Specifications

ASTM. Shapes: B455

Chemical Composition

Composition limits. 55.0 to 60.0 Cu; 2.0 to 3.8 Pb; 0.35 max Fe; 0.5 max others (total); rem Zn

Applications

Typical uses. Architectural extrusions, store fronts, thresholds and trim; hardware butts, hinges and lock bodies; industrial forgings

Mechanical Properties

Tensile properties. M30 temper: tensile strength, 415 MPa (60 ksi); yield strength, 140 MPa (20 ksi); elongation, 30% in 50 mm or 2 in.
Shear strength. M30 temper: 240 MPa (35 ksi)
Hardness. M30 temper: 65 HRB
Elastic modulus. Tension, 97 GPa (14×10^6 psi); shear, 37 GPa (5.3×10^6 psi)

Mass Characteristics

Density. 8.47 Mg/m^3 (0.306 lb/in.^{3}) at 20 °C (68 °F)

Thermal Properties

Liquidus temperature. 890 °C (1630 °F)
Solidus temperature. 875 °C (1610 °F)
Coefficient of thermal expansion. Linear, 20.9 μm/m·K (11.6 μin./in.·°F) at 20 to 300 °C (68 to 572 °F)
Specific heat. 380 J/kg·K (0.09 Btu/lb·°F) at 20 °C (68 °F)
Thermal conductivity. 123 W/m·K (71 Btu/ft·h·°F) at 20 °C (68 °F)

Electrical Properties

Electrical conductivity. Volumetric, O61 temper: 28% IACS at 20 °C (68 °F)
Electrical resistivity. 62 nΩ·m at 20 °C (68 °F)

Fabrication Characteristics

Machinability. 90% of C36000, free-cutting brass
Formability. Cold working, poor; hot forming, excellent
Weldability. Soldering: excellent. Brazing: good. Resistance butt welding: fair. All other welding processes not recommended.
Annealing temperature. 425 to 600 °C (800 to 1100 °F)
Hot working temperature. 625 to 725 °C (1150 to 1350 °F)

C40500
95Cu-4Zn-1Sn

Commercial Names

Trade name. High conductivity bronze
Common name. Penny bronze

Specifications

ASTM. B591

Chemical Composition

Composition limits. 94 to 96 Cu; 0.7 to 1.3 Sn; 0.05 max Pb; 0.05 max Fe; rem Zn

Applications

Typical uses. Meter clips, terminals, fuse clips, contact springs, relay springs, washers from rolled strip, rolled bar, sheet

Mechanical Properties

Tensile properties. See Table 63.
Shear strength. See Table 63.
Hardness. See Table 63.
Elastic modulus. Tension: hard, 110 GPa (16×10^6 psi); annealed, 125 GPa (18×10^6 psi). Shear: see Table 63.

Mass Characteristics

Density. 8.83 Mg/m^3 (0.319 lb/in.^{3}) at 20 °C (68 °F)

Thermal Properties

Liquidus temperature. 1060 °C (1940 °F)
Solidus temperature. 1025 °C (1875 °F)
Thermal conductivity. 165 W/m·K (95 Btu/ft·h·°F) at 20 °C (68 °F)

Electrical Properties

Electrical conductivity. Volumetric, 41% IACS at 20 °C (68 °F)
Electrical resistivity. 42 nΩ·m at 20 °C (68 °F)

Fabrication Characteristics

Machinability. 20% of C36000, free-cutting brass
Formability. Excellent for cold working; good for hot forming
Weldability. Soldering and brazing: excellent. Gas-shielded arc and resistance spot and butt welding: good. Oxyacetylene and resistance seam welding: fair. Coated metal-arc welding: not recommended.
Annealing temperature. 510 to 670 °C (950 to 1240 °F)
Hot working temperature. 830 to 890 °C (1525 to 1635 °F)

Table 63 Typical mechanical properties of C40500

Temper	Tensile strength MPa	ksi	At 0.5% extension under load MPa	ksi	At 0.2% offset MPa	ksi	Elongation in 50 mm or 2 in., %	HRF	Hardness HRB	HR30T	Shear strength MPa	ksi
Flat Products: 1 mm (0.04 in.) Thick												
OS035	270	39	83	12	69	10	49	55	···	10	215	31
OS025	280	41	83	12	76	11	48	58	···	13	230	33
OS015	290	42	90	13	97	14	47	64	···	24	230	33
H01	325	47	250	36	250	36	30	···	46	47	240	35
H02	360	52	295	43	345	50	15	···	60	56	255	37
H03	400	58	340	49	385	56	12	···	67	62	260	38
H04	440	64	380	55	425	62	10	···	72	65	270	39
H06	475	69	415	60	460	67	7	···	76	69	280	41
H08	510	74	435	63	495	72	4	···	79	71	295	43
H10	540	78	485	70	525	76	3	···	82	72	310	45

Table 64 Typical mechanical properties of C40800

Temper	Tensile strength MPa	ksi	Yield strength At 0.5% extension under load MPa	ksi	At 0.2% offset MPa	ksi	Elongation in 50 mm or 2 in., %	HRF	Hardness HRB	HR30T	Shear strength MPa	ksi
Flat Products: 1 mm (0.04 in.) Thick												
OS035	290	42	90	13	...	...	43	60	...	22	230	33
OS025	305	44	97	14	...	...	43	65	...	26	235	34
OS015	310	45	105	15	...	...	42	69	...	31	235	34
H01	345	50	270	39	310	45	24	...	50	54	250	36
H02	370	54	315	46	380	55	12	...	65	62	260	38
H03	425	62	360	52	415	60	6	...	72	67	280	41
H04	460	67	395	57	485	70	5	...	76	70	295	43
H06	505	73	420	61	540	78	4	...	82	73	310	45
H08	545	79	455	66	565	82	3	...	85	77	330	48
H10	545 min	79 min	515	75	580	84	3	...	84 min	75 min	340	49

C40800
95Cu-2Sn-3Zn

Specifications

ASTM. Flat products: B591

Chemical Composition

Composition limits. 94 to 96 Cu; 1.8 to 2.2 Sn; 0.05 max Pb; 0.05 max Fe; rem Zn

Applications

Typical uses. Rolled strip for electrical connectors

Mechanical Properties

Tensile properties. See Table 64.
Shear strength. See Table 64.
Hardness. See Table 64.
Elastic modulus. Tension: hard, 110 GPa (16 × 10⁶ psi); Shear, 41 GPa (6 × 10⁶ psi).

Mass Characteristics

Density. 8.86 Mg/m³ (0.320 lb/in.³) at 20 °C (68 °F)

Thermal Properties

Liquidus temperature. 1054 °C (1930 °F)
Solidus temperature. 1038 °C (1900 °F)
Coefficient of thermal expansion. Linear, 18.2 μm/m·K (10.1 μin./in.·°F) at 20 to 300 °C (68 to 572 °F)
Specific heat. 380 J/kg·K (0.09 Btu/lb·°F) at 20 °C (68 °F)
Thermal conductivity. 160 W/m·K (92 Btu/ft·h·°F)

Electrical Properties

Electrical conductivity. Volumetric, 37% IACS at 20 °C (68 °F)
Electrical resistivity. 46.6 nΩ·m at 20 °C (68 °F)

Fabrication Characteristics

Machinability. 20% of C36000, free-cutting brass
Formability. Excellent for cold working; fair for hot forming
Weldability. Soldering and brazing: excellent. Oxyacetylene, gas shielded arc and resistance butt welding: good. Resistance spot and seam welding: not recommended.
Annealing temperature. 450 to 675 °C (850 to 1250 °F)
Hot working temperature. 830 to 890 °C (1525 to 1635 °F)

C41100
91Cu-8.5Zn-0.5Sn

Commercial Names

Previous trade name. Lubaloy

Specifications

ASTM. Flat products: B508, B591. Wire: B105

Chemical Composition

Composition limits. 89 to 93 Cu; 0.3 to 0.7 Sn; 0.1 max Pb; 0.05 max Fe; rem Zn

Applications

Typical uses. Rolled strip, rolled bar, rod and sheet for bushings, bearing sleeves, thrust washers, terminals, connectors, flexible metal hose and electrical conductors

Mechanical Properties

Tensile properties. See Table 65.
Hardness. See Table 65.

Elastic modulus. Tension: hard, 115 GPa (16.7 × 10⁶ psi); annealed, 125 GPa (18 × 10⁶ psi). Shear, 46 GPa (6.7 × 10⁶ psi).

Mass Characteristics

Density. 8.80 Mg/m³ (0.318 lb/in.³) at 20 °C (68 °F)

Thermal Properties

Liquidus temperature. 1040 °C (1905 °F)
Solidus temperature. 1020 °C (1870 °F)
Coefficient of thermal expansion. 18 μm/m·K (10.2 μin./in.·°F) at 20 to 100 °C (68 to 212 °F)
Thermal conductivity. 130 W/m·K (75 Btu/ft·h·°F) at 20 °C (68 °F)
Specific heat. 380 J/kg·K (0.09 Btu/lb·°F) at 20 °C (68 °F)

Electrical Properties

Electrical conductivity. Volumetric, 32% IACS at 20 °C (68 °F)
Electrical resistivity. 53.2 nΩ·m at 20 °C (68 °F)

Fabrication Characteristics

Machinability. 20% of C36000, free-cutting brass
Formability. Excellent for cold working; good for hot forming
Weldability. Soldering: excellent. Gas-shielded arc welding and resistance butt welding: good. Brazing, oxyacetylene and resistance spot welding: fair. Coated metal-arc and resistance seam welding: not recommended.
Annealing temperature. 500 to 700 °C (930 to 1290 °F)
Hot working temperature. 830 to 890 °C (1525 to 1635 °F)

Table 65 Typical mechanical properties of C41100

Size	Temper	Tensile strength MPa	ksi	Yield strength At 0.5% extension under load MPa	ksi	At 0.2% offset MPa	ksi	Elongation in 50 mm or 2 in., %	HRF	Hardness HRB	HR30T	Shear strength MPa	ksi
Flat Products													
1 mm (0.04 in.) thick	OS050	260	38	76	11	62	9	44	58	...	...	220	32
	OS035	270	39	76	11	83	12	43	60	...	...	230	33
	OS025	280	41	83	12	97	14	41	68	...	...	...	...
	OS015	290	42	83	12	105	15	40	71	...	...	235	34
	H01	330	48	260	38	280	41	23	...	52	58	...	...
	H02	380	55	325	47	365	53	14	...	62	60	250	36
	H03	415	60	360	52	400	58	6	...	70	66	...	...
	H04	455	66	380	55	440	64	5	...	76	69	275	40
	H06	495	72	415	60	485	70	4	...	78	71	...	...
	H08	540	78	485	70	515	75	3	...	81	72	...	...
	H10	550	80	495	72	525	76	2	...	83	73	...	...
Wire	H80												
6-mm (0.25-in.) diam	70%	560	81	...	...	...	...	2(a)	...	...	...	...	...
3-mm (0.10-in.) diam	95%	705	102	...	...	...	...	1(b)	...	...	...	...	...
1-mm (0.05-in.) diam	98.7% ...	730	106	...	...	...	...	0.9(b)	...	...	...	...	...

(a) Elongation in 254 mm (10 in.). (b) Elongation in 150 cm (60 in.).

Table 66 Typical mechanical properties of C41500

Temper	Tensile strength MPa	ksi	Yield strength At 0.5% extension under load MPa	ksi	At 0.2% offset MPa	ksi	Elongation in 50 mm or 2 in., %	HRF	Hardness HRB	HR30T	Fatigue strength MPa	ksi
Flat Products: 1 mm (0.04 in.) Thick												
OS035	315	46	115	17	125	18	44	64	...	24	240	35
OS025	...	...	...	...	...	...	...	68	...	29	...	...
OS015	345	50	180	26	185	27	42	74	...	36	250	36
H01	345	50	280	41	...	...	28	...	62	58	...	...
H02	385	56	365	53	370	54	16	...	74	65	280	41
H03	435	63	...	...	...	...	...	...	78	68	290	...
H04	485	70	450	65	455	66	5	...	83	72	305	42
H06	525	76	490	71	515	75	4	...	86	73	305	44
H08	560	81	505	73	570	83	3	...	90	75	345	50
H10	560 min	81 min	515	75	605	88	2	...	89 min	74 min	360	52

C41500
91Cu-7.2Zn-1.8Sn

Specifications

ASTM. B591

Chemical Composition

Composition limits. 89 to 93 Cu; 1.5 to 2.2 Sn; 0.1 max Pb; 0.05 max Fe; rem Zn

Applications

Typical uses. Rolled strip for spring applications for electrical switches

Mechanical Properties

Tensile properties. See Table 66.
Hardness. See Table 66.
Elastic modulus. Tension: hard, 110 GPa (16×10^6 psi); annealed, 125 GPa (18×10^6 psi). Shear, 46 GPa (6.7×10^6 psi).

Mass Characteristics

Density. 8.80 Mg/m^3 (0.318 lb/in.^{3}) at 20 °C (68 °F)

Thermal Properties

Liquidus temperature. 1032 °C (1890 °F)

Solidus temperature. 1010 °C (1850 °F)
Coefficient of thermal expansion. Linear, 18.6 μm/m·K (10.3 μin./ in.·°F) at 20 to 300 °C (68 to 572 °F)
Specific heat. 380 J/kg·K (0.09 Btu/ lb·°F) at 20 °C (68 °F)
Thermal conductivity. 123 W/m·K (71 Btu/ft·h·°F) at 20 °C (68 °F)

Electrical Properties

Electrical conductivity. Volumetric, 28% IACS at 20 °C (68 °F)
Electrical resistivity. 62 nΩ·m at 20 °C (68 °F)

Fabrication Characteristics

Machinability. 30% of C36000, free-cutting brass

Formability. Excellent for cold working; fair for hot forming

Weldability. Soldering and brazing: excellent. Oxyacetylene, gas-shielded arc and resistance butt welding: good. Resistance spot welding: fair. Coated metal-arc and resistance seam welding: not recommended.

Annealing temperature. 400 to 705 °C (750 to 1300 °F)

Hot working temperature. 730 to 845 °C (1350 to 1550 °F)

C41900
90.5Cu-4.35Zn-5.15Sn

Commercial Names

Previous trade name. CA419
Common name. Tin brass

Chemical Composition

Composition limits. 89 to 92 Cu; 4.8 to 5.5 Sn; 0.10 max Pb; 0.05 max Fe; rem Zn

Applications

Typical uses. Electrical connectors

Mechanical Properties

Tensile properties. See Table 67.
Hardness. See Table 67.
Elastic modulus. Tension, 120 GPa (18 × 10⁶ psi)

Mass Characteristics

Density. 8.80 Mg/m³ (0.318 lb/in.³) at 20 °C (68 °F)

Thermal Properties

Liquidus temperature. 1025 °C (1880 °F)
Solidus temperature. 1000 °C (1830 °F)
Coefficient of thermal expansion. Linear, 18.7 μm/m·K (10.4 μin./in.·°F) at 20 to 300 °C (68 to 572 °F)
Thermal conductivity. 100 W/m·K (58 Btu/ft·h·°F) at 20 °C (68 °F)

Electrical Properties

Electrical conductivity. Volumetric, 22% IACS at 20 °C (68 °F)
Electrical resistivity. 76 nΩ·m at 20 °C (68 °F)

Fabrication Characteristics

Annealing temperature. 480 to 680 °C (900 to 1250 °F)

Table 67 Typical mechanical properties of C41900 strip

	Tensile strength		Yield strength(a)		Elongation(b),	Hardness
Temper	MPa	ksi	MPa	ksi	%	HRB
O61	340	49	130	19	42	67 HRF
H01	400	58	315	46	25	64
H02	470	68	395	57	14	73
H03	515	75	450	65	5	78
H04	565	82	510	74	4	87
H06	640	93	530	77	3	92
H08	705	102	550	80	2	95

(a) At 0.2% offset. (b) In 50 mm or 2 in.

Table 68 Typical mechanical properties of C42200

	Tensile strength		Yield strength At 0.5% extension under load		At 0.2% offset		Elongation in 50 mm or 2 in.,		Hardness	
Temper	MPa	ksi	MPa	ksi	MPa	ksi	%	HRF	HRB	HR30T
Flat Products: 1 mm (0.04 in.) Thick										
OS035 ..	295	43	105	15	97	14	46	65	···	27
OS025 ..	305	44	110	16	105	15	45	70	···	31
OS015 ..	315	46	115	17	130	19	44	75	···	40
H01	360	52	275	40	270	39	30	···	56	54
H02	415	60	350	51	395	57	12	···	70	64
H03	455	66	380	55	440	64	6	···	77	68
H04	505	73	450	65	485	70	4	···	81	70
H06	550	80	470	68	525	76	3	···	84	72
H08	600	87	505	73	560	81	2	···	87	73
H10	605 min	88 min	515	75	580	84	2	···	86 min	74 min

C42200
87.5Cu-11.4Zn-1.1Sn

Commercial Names

Previous trade name. Lubronze

Specifications

ASTM. B591

Chemical Composition

Composition limits. 86.0 to 89.0 Cu; 0.8 to 1.4 Sn; 0.35 max P; 0.05 max Pb; 0.05 max Fe; rem Zn

Applications

Typical uses. Rolled strip, rolled bar and sheet for sash chains, terminals, fuse clips, spring washers, contact springs and electrical connectors

Mechanical Properties

Tensile properties. See Table 68.
Hardness. See Table 68.
Elastic modulus. Tension: hard, 110 GPa (16 × 10⁶ psi); annealed, 125 GPa (18 × 10⁶ psi)

Mass Characteristics

Density. 8.80 Mg/m³ (0.318 lb/in.³) at 20 °C (68 °F)

Thermal Properties

Liquidus temperature. 1040 °C (1905 °F)
Solidus temperature. 1020 °C (1870 °F)
Thermal conductivity. 130 W/m·K (75 Btu/ft·h·°F)

Electrical Properties

Electrical conductivity. Volumetric, 31% IACS at 20 °C (68 °F)
Electrical resistivity. 55 nΩ·m at 20 °C (68 °F)

Fabrication Characteristics

Machinability. 30% of C36000, free-cutting brass

Formability. Excellent for cold working, good for hot forming

Weldability. Soldering and gas-shielded arc welding: excellent. Resistance spot and butt welding: good. Resistance seam welding and brazing: fair. Oxyacetylene welding: not recommended.

Annealing temperature. 500 to 674 °C (930 to 1250 °F)

Hot working temperature. 830 to 890 °C (1525 to 1635 °F)

Table 69 Typical mechanical properties of C42500

Temper	Tensile strength MPa	ksi	Yield strength At 0.5% extension under load MPa	ksi	At 0.2% offset MPa	ksi	Elongation in 50 mm or 2 in., %	HRF	Hardness HRB	HR30T
Flat Products: 1 mm (0.04 in.) Thick										
OS035	310	45	125	18	105	15	49	70	...	32
OS025	315	46	125	18	125	18	48	72	...	36
OS015	325	47	135	20	130	19	47	79	...	45
H01	370	54	310	45	315	46	35	...	60	56
H02	435	63	345	50	405	59	20	...	75	68
H03	470	68	395	57	450	65	15	...	80	70
H04	525	76	435	63	505	73	9	...	86	73
H06	565	82	485	70	545	79	7	...	90	74
H08	615	89	515	75	585	85	4	...	92	76
H10	635 min	92 min	525	76	615	89	2	...	92 min	76 min

C42500
88.5Cu-9.5Zn-2Sn

Specifications

ASTM. B591

Chemical Composition

Composition limits. 87 to 90 Cu; 1.5 to 3.0 Sn; 0.35 max P; 0.05 max Pb; 0.05 max Fe; rem Zn

Applications

Typical uses. Rolled strip, rolled bar and sheet for electrical switch springs, terminals, connectors, fuse clips, pen clips and weather stripping

Mechanical Properties

Tensile properties. See Table 69.
Hardness. See Table 69.
Elastic modulus. Tension: hard, 110 GPa (16×10^6 psi); annealed, 125 GPa (18×10^6 psi)

Mass Characteristics

Density. 8.78 Mg/m^3 (0.317 lb/in.^{3}) at 20 °C (68 °F)

Thermal Properties

Liquidus temperature. 1030 °C (1890 °F)
Solidus temperature. 1010 °C (1850 °F)
Coefficient of thermal expansion. Linear, 18.4 μm/m·K (10.2 μin./in.·°F) at 20 to 100 °C (68 to 212 °F)
Specific heat. 380 J/kg·K (0.09 Btu/lb·°F) at 20 °C (68 °F)
Thermal conductivity. 120 W/m·K (69 Btu/ft·h·°F) at 20 °C (68 °F)

Electrical Properties

Electrical conductivity. Volumetric, 28% IACS at 20 °C (68 °F)

Electrical resistivity. 62 nΩ·m at 20 °C (68 °F)

Fabrication Characteristics

Machinability. 30% of C36000, free-cutting brass
Formability. Excellent for cold working; fair for hot forming
Weldability. Soldering and brazing: excellent. Oxyacetylene, gas-shielded arc and resistance butt welding: good. Coated metal-arc and resistance spot and seam welding: not recommended.
Annealing temperature. 425 to 700 °C (800 to 1300 °F)
Hot working temperature. 790 to 840 °C (1455 to 1545 °F)

C43000
87Cu-10.8Zn-2.2Sn

Specifications

ASTM. Flat products: B591

Chemical Composition

Composition limits. 84 to 87 Cu; 1.7 to 2.7 Sn; 0.10 max Pb; 0.05 max Fe; rem Zn

Applications

Typical uses. Rolled strip and sheet for electrical switches, springs, fuse and pen clips and weather stripping

Mechanical Properties

Tensile properties. See Table 70.
Hardness. See Table 70.

Elastic modulus. Tension, 110 GPa (16×10^6 psi); shear, 119 GPa (17.3×10^6 psi)

Mass Characteristics

Density. 8.75 Mg/m^3 (0.316 lb/in.^{3}) at 20 °C (68 °F)

Thermal Properties

Liquidus temperature. 1025 °C (1877 °F)
Solidus temperature. 1000 °C (1832 °F)
Coefficient of thermal expansion. Linear, 18.4 μm/m·K (10.2 μin./in.·°F) at 20 to 100 °C (68 to 212 °F)
Specific heat. 380 J/kg·K (0.09 Btu/lb·°F) at 20 °C (68 °F)
Thermal conductivity. 119 W/m·K (69 Btu/ft·h·°F)

Electrical Properties

Electrical conductivity. Volumetric, 27% IACS at 20 °C (68 °F)
Electrical resistivity. 63.8 nΩ·m at 20 °C (68 °F)

Fabrication Characteristics

Machinability. 30% of C36000, free-cutting brass
Formability. Excellent for cold working; good for hot forming
Weldability. Soldering and brazing: excellent. Oxyacetylene, gas shielded arc and resistance butt welding: good. Resistance spot welding: fair. Coated metal arc and resistance seam welding: not recommended.
Annealing temperature. 425 to 700 °C (800 to 1300 °F)
Hot working temperature. 790 to 840 °C (1455 to 1545 °F)

Table 70 Typical mechanical properties of C43000

Temper	Tensile strength MPa	ksi	Yield strength(a) MPa	ksi	Elongation in 50 mm or 2 in., %	HRF	Hardness HRB	HR30T
Flat Products: 1 mm (0.04 in.) Thick								
OS035 . . . 315	46	125	18	55	69	30	. . .
OS025 . . . · · ·	. . .	. . .	. . .	. . .	72	34	. . .
OS015 . . . · · ·	. . .	. . .	. . .	. . .	77	39	. . .
H01 365	53	275	40	44	. . .	57	57
H02 425	62	380	55	25	. . .	73	65
H03 495	72	450	65	13	. . .	79	69
H04 540	78	460	67	10	. . .	84	73
H06 605	88	485	70	5	. . .	81	75
H08 650	94	495	72	4	. . .	91	77
H10 . . . . . 620 min	90 min	505	73	3	. . .	90 min	75 min

(a) At 0.5% extension under load.

Table 71 Typical mechanical properties of C43400

Temper	Tensile strength MPa	ksi	Yield strength(a) MPa	ksi	Elongation in 50 mm or 2 in., %	HRF	Hardness HRB	HR30T	Shear strength MPa	ksi
Flat Products: 1 mm (0.04 in.) Thick										
OS035 . . 310	45	105	15	49	64	. . .	22	250	36
OS025 . . 315	46	110	16	48	65	. . .	26	255	37
OS015 . . 330	48	115	17	47	70	. . .	30	255	37
H01 360	52	280	41	28	. . .	54	55	275	40
H02 405	59	350	51	18	. . .	66	63	290	42
H03 470	68	405	59	10	. . .	73	68	310	45
H04 510	74	460	67	7	. . .	80	71	340	49
H06 580	84	490	71	5	. . .	83	74	360	52
H08 620	90	510	74	4	. . .	86	76	370	54
H10 . . . . 605 min	88 min	515	75	3	. . .	84 min	74 min	385	56

(a) At 0.5% extension under load.

C43400
85Cu-14.3Zn-0.7Sn

Specifications

ASTM. Flat products: B591

Chemical Composition

Composition limits. 84 to 87 Cu; 0.4 to 1.0 Sn; 0.05 max Pb; 0.05 max Fe; rem Zn

Applications

Typical uses. Rolled strip for electrical uses: switch parts, blades, relay springs, contacts

Mechanical Properties

Tensile properties. See Table 71.
Shear strength. See Table 71.
Hardness. See Table 71.

Elastic modulus. Tension: hard, 110 GPa (16×10^6 psi); annealed, 40 GPa (6×10^6 psi)

Mass Characteristics

Density. 8.75 Mg/m^3 (0.316 lb/in.^{3}) at 20 °C (68 °F)

Thermal Properties

Liquidus temperature. 1020 °C (1870 °F)
Solidus temperature. 990 °C (1810 °F)
Coefficient of thermal expansion. Linear, 18.9 μm/m·K (10.5 μin./in.·°F) at 20 to 300 °C (68 to 572 °F)
Specific heat. 380 J/kg·K (0.09 Btu/lb·h·°F) at 20 °C (68 °F)
Thermal conductivity. 137 W/m·K (79 Btu/ft·h·°F) at 20 °C (68 °F)

Electrical Properties

Electrical conductivity. Volumetric, 31% IACS at 20 °C (68 °F)
Electrical resistivity. 55 nΩ·m at 20 °C (68 °F)

Fabrication Characteristics

Machinability. 30% of C36000, free-cutting brass
Formability. Excellent for cold working; fair for hot forming
Weldability. Soldering and brazing: excellent. Oxyacetylene, gas-shielded arc and resistance spot and butt welding: good. Resistance seam welding: not recommended.
Annealing temperature. 425 to 675 °C (800 to 1250 °F)
Hot working temperature. 815 to 870 °C (1500 to 1600 °F)

C43500
81Cu-18.1Zn-0.9Sn

Chemical Composition

Composition limits. 79 to 83 Cu; 0.6 to 1.2 Sn; 0.1 max Pb; 0.05 max Fe; 0.15 max others (total); rem Zn

Applications

Typical uses. Rolled strip and tubing for Bourdon tubing and musical instruments

Mechanical Properties

Tensile properties. See Table 72.
Shear strength. See Table 72.
Elastic modulus. Tension, 110 GPa (16×10^6 psi); shear, 40 GPa (6×10^6 psi)

Mass Characteristics

Density. 8.66 Mg/m^3 (0.313 lb/in.^{3}) at 20 °C (68 °F)

Thermal Properties

Liquidus temperature. 1005 °C (1840 °F)
Solidus temperature. 965 °C (1770 °F)
Coefficient of thermal expansion. Linear, 19.4 μm/m·K (10.8 μin./in.·°F) at 20 °C (68 °F)
Specific heat. 380 J/kg·K (0.09 Btu/lb·°F) at 20 °C (68 °F)

Electrical Properties

Electrical conductivity. Volumetric, 28% IACS at 20 °C (68 °F)

Table 72 Typical mechanical properties of C43500

Temper	Tensile strength MPa	ksi	Yield strength(a) MPa	ksi	Elongation in 50 mm or 2 in., %	HRF	Hardness HRB	HR30T	Shear strength MPa	ksi
Flat Products: 1 mm (0.04 in.) Thick										
OS025	340	49	125	18	46	70	...	31	250	36
H02	450	65	370	54	16	...	72	...	286	41.5
H04	550	80	470	68	7	...	85	...	310	45
Tubing: 25-mm (1.0-in.) OD, 1.65-mm (0.065-in.) Wall										
OS035	315	46	110	16	46	69	...	40	...	...
H80(35%)	515	75	415	60	10	...	...	...	...	...

(a) At 0.5% extension under load.

Electrical resistivity. 61.6 nΩ·m at 20 °C (68 °F)

Fabrication Characteristics

Machinability. 30% of C36000, free-cutting brass

Formability. Excellent for cold working; good for hot forming

Weldability. Soldering and brazing: excellent. Oxyacetylene and resistance spot and butt welding: good. Gas-shielded metal-arc welding: fair. Coated metal-arc and resistance seam welding: not recommended.

C44300, C44400, C44500
71Cu-28Zn-1Sn

Commercial Names

Previous trade names. C44300, arsenical admiralty metal; C44400, antimonial admiralty metal; C44500, phosphorized admiralty metal

Common names. Inhibited admiralty metal; admiralty brass

Specifications

ASME. Condenser plate: SB171. Tubing: SB111, SB359, SB395, SB543

ASTM. Condenser plate: B171. Tubing: B111, B359, B395, B543

Chemical Composition

Composition limits. 70.0 to 73.0 Cu; 0.07 max Pb; 0.06 max Fe; 0.9 to 1.2 Sn(a); As, Sb or P(b); rem Zn

(a) For flat rolled products, 0.8 to 1.2 Sn. (b) C44300, 0.02 to 0.10 As; C44400, 0.02 to 0.10 Sb; C44500, 0.02 to 0.10 P.

Applications

Typical uses. Condenser, distiller and heat-exchanger tubes, ferrules, strainers, condenser-tube plates

Precautions in use. These three alloys are susceptible to stress corrosion cracking. Whenever possible, they should be used in the annealed condition. Where fabrication results in residual stresses, a suitable stress-relieving heat treatment should be applied.

Mechanical Properties

Tensile properties. See Table 73 and Fig. 40.

Hardness. See Table 73.

Elastic modulus. Tension, 110 GPa (16 × 10^6 psi); shear, 40 GPa (6 × 10^6 psi)

Impact strength. See Table 74.

Fatigue strength. 115 to 125 MPa (17 to 18 ksi) at 10^7 cycles

Creep-rupture characteristics. See Table 75.

Mass Characteristics

Density. 8.53 Mg/m³ (0.308 1b/in.³) at 20 °C (68 °F)

Table 73 Typical mechanical properties of C44300, C44400 and C44500

Temper	Tensile strength MPa	ksi	Yield strength(a) MPa	ksi	Elongation(b), %	HRF	Hardness HR15T	HR30T
Tubing: 25-mm (1-in.) OD, 1.65-mm (0.065-in.) Wall								
OS025	365	53	152	22	65	75	...	37
H01	434	63	...	...	45	...	86.5	...
H02	503	73	...	...	29	...	90	78
H03	565	82	...	...	15	...	90	81
H04	669	97	...	...	4	...	93	84
Plate: 25-mm (1-in.) Diam								
M20	330	48	124	18	65	70	...	...
Strip: 1-mm (0.04-in.) Diam								
O60 (0.080 mm) ...	310	45	90	13	69	59	9	20
O50 (0.015 mm) ...	330	48	97	14	62	60	9	20
H04	607	88	496	72	4	109	90	76

(a) At 0.5% extension under load. Apparent elastic limit (tubing), 125 MPa (18 ksi). (b) In 50 mm or 2 in.

Table 74 Typical Charpy impact strength of C44300, C44400 or C44500(a)

Test temperature °C	°F	Impact strength J	ft·lb
68	20	82.4	60.8
38	3	82.2	60.6
0	−18	79.7	58.8
−25	−30	82.4	60.8
−60	−50	79.9	58.9
−110	−80	83.4	61.5
−175	−115	80.3	59.2

(a) Annealed specimens, cut from 19-mm-diam (0.75-in.-diam) rod into keyhole-notch bars. Values are averages of data from three tests (specimens did not fracture). Tensile strength at 20 °C (68 °F), 320 MPa (46.5 ksi); yield strength, 92 MPa (13.3 ksi); elongation, 83.5%; hardness, 64 HRF.

Thermal Properties

Liquidus temperature. 935 °C (1720 °F)

Solidus temperature. 900 °C (1650 °F)

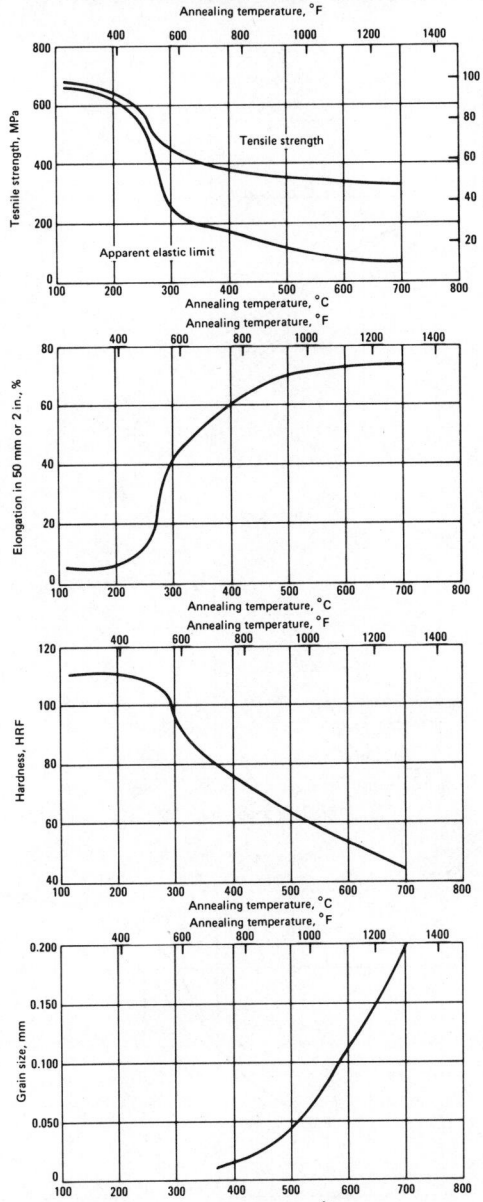

Fig. 40 Variation of properties and grain size with annealing temperature for C44300, C44400 or C44500

Inhibited admiralty metal tubing (71Cu-28Zn-1Sn), cold drawn 50% and annealed 1 h at temperature.

Coefficient of thermal expansion. Linear, 20.2 μm/m·K (11.2 μin./in.· °F) at 20 to 300 °C (68 to 572 °F)
Specific heat. 380 J/kg·K (0.09 Btu/lb·°F) at 20 °C (68 °F)
Thermal conductivity. 110 W/m·K (64 Btu/ft·h·°F) at 20 °C (68 °F)

Electrical Properties

Electrical conductivity. Volumetric, 25% IACS at 20 °C (68 °F)
Electrical resistivity. 69 nΩ·m at 20 °C (68 °F)

Chemical Properties

General corrosion behavior. Good resistance to salt and fresh waters at low velocities. Water velocities above 6 fps give rise to impingement attack. A different inhibitor is added to each alloy to protect against dezincification.

Fabrication Characteristics

Machinability. 30% of C36000, free-cutting brass
Formability. Excellent for cold working; fair for hot forming
Weldability. Soft soldering: excellent. Silver alloy brazing, oxyfuel-gas welding, resistance spot welding and flash welding: good. Gas-shielded arc welding: fair. Shielded metal-arc welding and resistance seam welding: not recommended.
Recrystallization temperature. 300 °C (575 °F) for 1-mm (0.04-in.) strip cold rolled hard (50% reduction) from a grain size of 0.015 mm. See also Fig. 40.
Annealing temperature. 425 to 600 °C (800 to 1100 °F)
Hot working temperature. 650 to 800 °C (1200 to 1450 °F)

C46400, C46500, C46600, C46700
60Cu-39.2Zn-0.8Sn

Commercial Names

Previous trade names. C46400, uninhibited naval brass; C46500, arsenical naval brass; C46600, antimonial naval brass; C46700, phosphorized naval brass
Common names. Naval brass; inhibited naval brass

Specifications

AMS. Bar and rod (C46400 only): 4611, 4612

Table 75 Typical creep data for C44300, C44400 or C44500(a)

Temperature		Stress required to produce designated creep in 1000 h							
		Nil(b)		0.01%		0.10%		1.00%	
°C	°F	MPa	ksi	MPa	ksi	MPa	ksi	MPa	ksi
205	400	 69	10	90	13	117	17	130	19
315	600	..... (c)	(c)	6.9	1.0	13.4	1.95	26	3.8
425	800	..... (c)	(c)	0.37	0.054	1.1	0.16	3.4	0.5

(a) Rod, hot rolled to 22.2 mm (0.875 in.) then cold drawn to 19.0 mm (0.750 in.). (b) No measurable flow. (c) Nearly zero.

ASME. Condenser plate: SB171
ASTM. Bar, rod and shapes (C46400 only): B21, B124, Forgings (C46400 only): B283. Condenser plate: B171.
SAE. Bar, rod and shapes (C46400 only): J461, J463
Government. QQ-B-626. Bar, rod, shapes, forgings and wire (C46400 only): QQ-B-637. Bar and flat products (C46400 only): QQ-B-639. Tubing (C46400 only): MIL-T-6945.

Chemical Composition

Composition limits. 59.0 to 62.0 Cu; 0.50 to 1.0 Sn; 0.20 max Pb; 0.10 max Fe; As, Sb or P(a); rem Zn

(a) C46400, none specified; C46500, 0.02 to 0.10 As; C46600, 0.02 to 0.10 Sb; C46700, 0.2 to 0.10 P.

Applications

Typical uses. Condenser plates, welding rod, marine hardware, propeller shafts, valve stems, airplane turnbuckle barrels, balls, nuts, bolts, rivets, fittings

Mechanical Properties

Tensile properties. See Table 76 and Fig. 41.
Shear strength. See Table 76.
Hardness. See Table 76.
Elastic modulus. Tension, 100 GPa (15×10^6 psi); shear, 39 GPa (5.6×10^6 psi)
Impact strength. 43 J (32 ft·lb) at 21 °C (70 °F) for Charpy keyhole specimens 10 mm (0.4 in.) square machined from annealed plate 13 mm (0.5 in.) thick; plate hardness, 96 HRF
Fatigue strength. 100 MPa (15 ksi) at 3×10^8 cycles

Structure

Crystal structure. Face-centered-cubic alpha and body-centered-cubic beta
Microstructure. Generally two phases—alpha and beta

Mass Characteristics

Density. 8.41 Mg/m³ (0.304 1b/in.³) at 20 °C (68 °F)

Thermal Properties

Liquidus temperature. 900 °C (1650 °F)
Solidus temperature. 885 °C (1630 °F)
Coefficient of thermal expansion. Linear, 21.2 μm/m·K (11.8 μin./in.· °F) at 20 to 300 °C (68 to 572 °F)

Specific heat. 380 J/kg·K (0.09 Btu/1b· °F) at 20 °C (68 °F)
Thermal conductivity. 116 W/m·K (67 Btu/ft·h· °F) at 20 °C (68 °F)

Electrical Properties

Electrical conductivity. Volumetric, 26% IACS at 20 °C (68 °F), annealed

Electrical resistivity. 66.3 nΩ·m at 20 °C (68 °F), annealed

Chemical Properties

General corrosion behavior. Good resistance to corrosion in both fresh and salt water, different inhibitor elements are added to C46500,

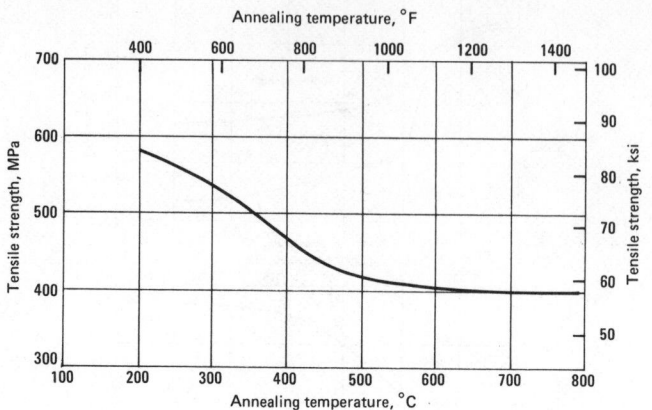

Fig. 41 Variation of strength, ductility and grain size with annealing temperature for C46400, C46500, C46600 or C46700

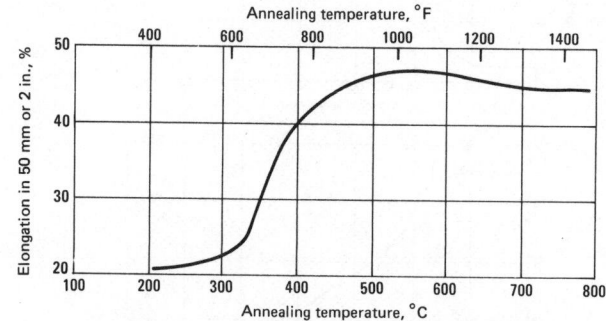

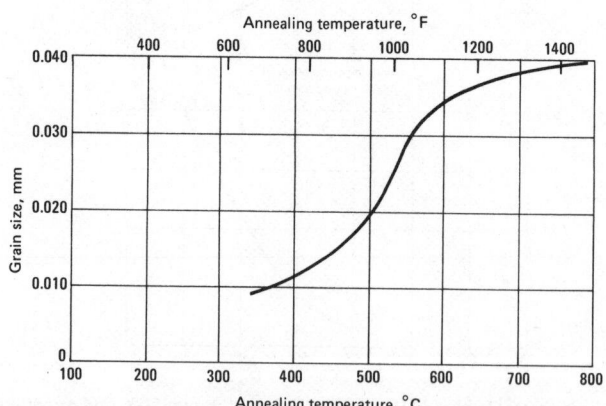

Naval brass rod (60 Cu-39.2 Zn-0.8Sn), 19-mm (0.75-in.) diam, cold drawn 30% and annealed 1 h at temperature. Grain size before cold drawing, 0.025 mm.

Table 76 Typical mechanical properties of C46400, C46500, C46600 or C46700

Size	Temper	Tensile strength MPa	ksi	Yield strength(a) MPa	ksi	Elongation in 50 mm or 2 in., %	Reduction in area, %	Hardness, HRB	Shear strength MPa	ksi
Flat Products										
1 mm (0.04 in.) thick	O50	427	62	207	30	40	· · ·	60	283	41
	H01	483	70	400	58	17	· · ·	75	296	43
6 mm (0.25 in.) thick	O60	400	58	172	25	49	· · ·	56	276	40
	O50	414	60	193	28	45	· · ·	58	283	41
25 mm (1.0 in.) thick	M20	379	55	172	25	50	· · ·	55	276	40
Rod and Bar										
6 mm (0.25-in.) diam	O60	400	58	186	27	45	60	56	276	40
	O50	434	63	207	30	40	55	60	290	42
	H01 (10%) ...	482	70	331	48	25	50	80	296	43
	H02 (20%) ...	552	80	393	57	20	45	85	310	45
25-mm (1.0-in.) diam	O60	393	57	172	25	47	60	55	276	40
	O50	434	63	207	30	40	55	60	290	42
	H01 (8%)	476	69	317	46	27	50	78	296	43
	H02 (20%) ...	517	75	365	53	20	45	82	303	44
51-mm (2.0-in.) diam	O60	386	56	172	25	47	60	55	276	40
	O50	427	62	193	28	43	55	60	290	42
	H01 (8%)	462	67	276	40	35	50	75	296	43
Tubing										
9.5-mm (0.375-in.) OD, 2.5-mm (0.097-in.) wall	H80 (35%) ...	607	88	455	66	18	40	95	· · ·	· · ·
	O61	427	62	207	30	45	· · ·	25	· · ·	· · ·
Extruded Shapes										
· · ·	M30	400	58	170	25	40	· · ·	· · ·	275	40

(a) At 0.5% extension under load.

C46600, C46700 to protect against dezincification.

Fabrication Characteristics

Machinability. 30% of C36000, free-cutting brass
Forgeability. 90% of C37700, forging brass

Formability. Excellent for hot forming; fair for cold working

Weldability. Soft soldering and silver alloy brazing: excellent. Oxyfuel-gas welding, resistance spot welding and flash welding: good. Gas-shielded arc welding and resistance seam welding: fair. Shielded arc welding: not recommended.

Recrystallization temperatures. About 350 °C (660 °F) for 19-mm-diam (0.75-in.- diam) rod cold drawn 30%. See also Fig. 41.
Annealing temperature. 425 to 600 °C (800 to 1100 °F)
Maximum cold reduction between anneals. 30%
Hot working temperature. 650 to 825 °C (1200 to 1500 °F)

C48200
60.5Cu-38Zn-0.8Sn-0.7 Pb

Commercial Names

Previous trade names. Naval brass, medium leaded; CA482
Common name. Leaded naval brass

Specifications

ASTM. Rod, bar and shapes: B21 (CA482), B124 (C48200)
Government. QQ-B-626. Bar, rod, shapes, forgings and wire: QQ-B-637. Bar and plate: QQ-B-639

Chemical Composition

Composition limits. 59.0 to 62.0 Cu; 0.40 to 1.0 Pb; 0.10 max Fe; 0.50 to 1.0 Sn; rem Zn

Applications

Typical uses. Marine hardware, screw-machine products, valve stems

Mechanical Properties

Tensile properties. See Table 77.
Shear strength. See Table 77.
Hardness. See Table 77.
Elastic modulus. Tension, 100 GPa (15 × 10⁶ psi); shear, 39 GPa (5.6 × 10⁶ psi)

Structure

Microstructure. Generally three phases—alpha, beta and lead

Mass Characteristics

Density. 8.44 Mg/m³ (0.305 1b/in.³) at 20 °C (68 °F)

Thermal Properties

Liquidus temperature. 900 °C (1650 °F)
Solidus temperature. 885 °C (1625 °F)

Coefficient of thermal expansion. Linear, 21.2 μm/m·K (11.8 μin./in.· °F) at 20 to 300 °C (68 to 572 °F)

Specific heat. 380 J/kg·K (0.09 Btu/1b· °F) at 20 °C (68 °F)
Thermal conductivity. 116 W/m·K (67 Btu/ft·h· °F)

Table 77 Typical mechanical properties of C48200

Temper	Tensile strength MPa	ksi	Yield strength(a) MPa	ksi	Elongation in 50 mm or 2 in., %	Hardness, HRB	Shear strength MPa	ksi
Rod: 25-mm (1.0-in.) Diam								
O60	395	57	170	25	40	55	260	38
O50	435	63	205	30	35	60	270	39
H01 (8%)	475	69	315	46	20	78	275	40
H02 (20%)	515	75	365	53	15	82	285	41
Rod: 51-mm (2.0-in.) Diam								
O60	385	56	170	25	40	55	260	38
O50	425	62	195	28	37	60	270	39
H01 (8%)	460	67	275	40	30	75	275	40
H02 (15%)	485	70	360	52	17	78	285	41
Rod: 76-mm (3.0-in.) Diam								
H01 (4%)	435	63	230	33	43	78	275	40
Bar: 10-mm (0.38-in.) Diam								
M30	435	63	230	33	34	60	270	39
Bar: 38-mm (1.5-in.) Diam								
H01	455	66	275	40	32	75	275	40

(a) At 0.5% extension under load.

Electrical Properties

Electrical conductivity. Volumetric, 26% IACS at 20 °C
Electrical resistivity. 66.3 nΩ·m at 20 °C (68 °F)

Chemical Properties

General corrosion behavior. Good resistance to seawater and marine atmospheres

Fabrication Characteristics

Machinability. 50% of C36000, free-cutting brass
Forgeability. 90% of C37700, forging brass
Formability. Good for hot working; poor for cold working
Weldability. Soft soldering: excellent. Silver alloy brazing: good. Flash welding: fair. Oxyfuel-gas welding, arc welding and most resistance welding processes: not recommended.
Recrystallization temperature. About 360 °C (680 °F) for 19-mm-diam (0.75-in.-diam) rod cold drawn 30%
Annealing temperature. 425 to 600 °C (800 to 1100 °F)
Hot working temperature. 650 to 760 °C (1200 to 1400 °F)

C48500
60Cu-37.5Zn-1.8Pb-0.7Sn

Commercial Names

Previous trade names. High-leaded naval brass; CA485
Common name. Leaded naval brass

Specifications

ASTM. Rod, bar, and shapes: B21 (CA485), B124 (C48500). Forgings: B283 (CA485)
Government. QQ-B-626. Bar, rod, shapes, forgings and wire: QQ-B-637. Bar and flat products: QQ-B-639

Chemical Composition

Composition limits. 59.0 to 62.0 Cu; 1.3 to 2.2 Pb; 0.10 max Fe; 0.50 to 1.0 Sn; rem Zn

Applications

Typical uses. Marine hardware, screw-machine products, valve stems

Mechanical Properties

Tensile properties. See Table 78.

Shear strength. See Table 78.
Hardness. See Table 78.
Elastic modulus. Tension, 100 GPa (15 × 10⁶ psi); shear, 39 GPa (5.6 × 10⁶ psi)

Structure

Microstructure. Generally three phases—alpha, beta and lead

Mass Characteristics

Density. 8.44 Mg/m³ (0.305 1b/in.³) at 20 °C (68 °F)

Thermal Properties

Liquidus temperature. 900 °C (1650 °F)
Solidus temperature. 885 °C (1625 °F)
Coefficient of thermal expansion. Linear, 21,2 μm/m·K (11.8 μin./in.·°F) at 20 to 300 °C (68 to 572 °F)
Specific heat. 380 J/kg·K (0.09 Btu/1b·°F) at 20 °C (68 °F)
Thermal conductivity. 116 W/m·K (67 Btu/ft·h·°F)

Electrical Properties

Electrical conductivity. Volumetric, 26% IACS at 20 °C (68 °F)
Electrical resistivity. 66.3 nΩ·m at 20 °C (68 °F)

Chemical Properties

General corrosion behavior. Good resistance to seawater and marine atmospheres

Fabrication Characteristics

Machinability. 70% of C36000, free-cutting brass
Forgeability. 90% of C37700, forging brass
Formability. Good for hot working; poor for cold working
Weldability. Soft soldering: excellent. Silver alloy brazing: good. Flash welding: fair. Oxyfuel-gas welding, arc welding and most resistance welding processes: not recommended.

Table 78 Typical mechanical properties of C48500(a)

Temper	Tensile strength MPa	ksi	Yield strength(b) MPa	ksi	Elongation in 50 mm or 2 in., %	Hardness, HRB	Shear strength MPa	ksi
O60	393	57	172	25	40	55	248	36
H01 (8%)	476	69	317	46	20	78	269	39
H02 (20%)	517	75	365	53	15	82	276	40

(a) Rod, 25-mm diam (1.0-in. diam). (b) At 0.5% extension under load.

Fig. 42 Variation of strength, ductility and grain size for C48500

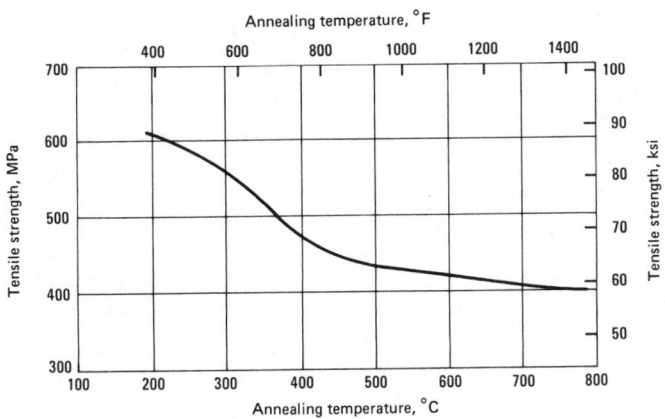

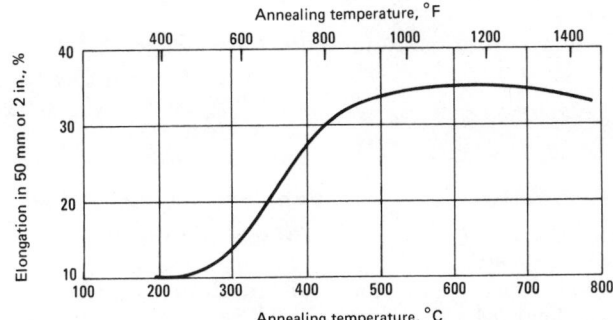

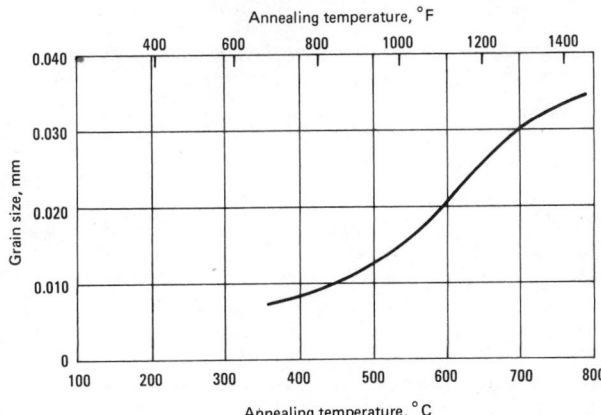

High leaded naval brass (60Cu-37.5Zn-1.8Pb-0.7Sn) rod, 19 mm diam (0.75 in. diam) cold drawn 30% and annealed 1 h at temperature. Grain size before cold drawing, 0.025 mm.

Recrystallization temperature. About 360 °C (680 °F) for 19-mm-diam (0.75-in-diam) rod cold worked 30%. See also Fig. 42.
Annealing temperature. 425 to 600 °C (800 to 1100 °F)

Maximum cold reduction between anneals. 20%

Hot working temperature. 650 to 760 °C (1200 to 1400 °F)

C50500
98.7Cu-1.3Sn

Commercial Names

Previous names. Phosphor bronze, 1.25% E; CA505
Common name. Phosphor bronze (1.25% Sn)

Specifications

ASTM. Strip: B105. Wire: B105.

Chemical Composition

Composition limits. 1.0 to 1.7 Sn, 0.05 max Pb, 0.10 max Fe, 0.30 max Zn, 0.35 max P, rem Cu; 99.5 min Cu + Sn + P

Applications

Typical uses. Electrical contacts, flexible hose, pole line hardware.

Mechanical Properties

Tensile properties. See Table 79.
Hardness. See Table 79.
Elastic modulus. Tension, 117 GPa (17×10^6 psi); shear, 44 GPa (6.4×10^6 psi)
Fatigue strength. See Table 79.

Mass Characteristics

Density. 8.89 Mg/m^3 (0.321 1b/in.^{3}) at 20 °C (68 °F)

Thermal Properties

Liquidus temperature. 1075 °C (1970 °F)
Solidus temperature. 1035 °C (1900 °F)
Coefficient of thermal expansion. Linear, 17.8 μm/m·K (9.9 μin./in.·°F) at 20 to 300 °C (68 to 572 °F)
Specific heat. 380 J/kg·K (0.09 Btu/1b·°F) at 20 °C (68 °F)
Thermal conductivity. 208 W/m·K (120 Btu/ft·h·°F) at 20 °C (68 °F)

Electrical Properties

Electrical conductivity. Volumetric, 48% IACS at 20 °C (68 °F)
Electrical resistivity. 35.9 nΩ·m at 20 °C (68 °F)

Fabrication Characteristics

Machinability. 20% of C36000, free-cutting brass
Formability. Cold: excellent. Hot: good. Commonly fabricated by blanking, forming, bending, heading, upsetting, shearing, squeezing and swaging.
Weldability. Flash welding, soldering and brazing: excellent. Gas

Table 79 Typical mechanical properties of C50500(a)

Temper	Grain size, mm	Tensile strength(a) MPa	ksi	Yield strength(b) MPa	ksi	Elongation in 50 mm or 2 in., %	Hardness, HRB	Fatigue strength(c) MPa	ksi
OS035	0.035	276	40	76	11	47.0	...	114	16.5
OS075	0.015	290	42	90	13	47.0	...	121	17.5
H02	0.035	365	53	352	51	12.0	59.0	162	23.5
	0.015	372	54	359	52	13.0	60.0	172	25
H04	0.035	421	61	414	60	5.0	67.0	179	26
	0.015	441	64	434	63	5.0	69.0	190	27.5
H06	0.035	462	67	455	66	3.0	73.0	172	25
	0.015	483	70	476	69	3.0	75.0	193	28
H08	0.035	483	70	476	69	3.0	76.0	197	28.5
	0.015	510	74	503	73	3.0	78.0	203	29.5
H10	0.035	510	74	503	73	3.0	79.0	197	28.5
	0.015	524	76	517	75	3.0	80.0	210	30.5

(a) Flat products 1 mm (0.040 in.) thick. Data in this table were interpolated from ASTM STP 183. (b) At 0.2% offset. (c) At 10⁸ cycles of fully reversed stress.

metal-arc welding: good. Oxyfuel-gas welding and shielded metal-arc welding: fair. Other processes not recommended.

Annealing temperature. 475 to 650 °C (900 to 1200 °F)
Hot working temperature. 800 to 875 °C (1450 to 1600 °F)

Fatigue strength. At 10^8 cycles. Flat products: H04 temper, 170 MPa (25 ksi); H08 temper, 150 MPa (22 ksi). Wire: H04 temper, 185 MPa (27 ksi); H06 temper, 205 MPa (30 ksi)

Mass Characteristics

Density. 8.86 Mg/m^3 (0.320 1b/in.^{3}) at 20 °C (68 °F)

Thermal Properties

Liquidus temperature. 1060 °C (1945 °F)
Solidus temperature. 975 °C (1785 °F)
Coefficient of thermal expansion. Linear, 17.8 μm/m·K (9.9 μin./in. °F) at 20 to 300 °C (68 to 572 °F)
Specific heat. 380 J/kg·K (0.09 Btu/1b· °F)
Thermal conductivity. 84 W/m·K (48.4 Btu/ft·h·°F) at 20 °C (68 °F)

Electrical Properties

Electrical conductivity. Volumetric, 20% IACS at 20 °C (68 °F)
Electrical resistivity. 87 nΩ·m at 20 °C (68 °F)

C51000
94.8Cu-5Sn-0.2P

Commercial Names

Previous trade name. Phosphor bronze, 5% A

Specifications

AMS. Flat products: 4510. Bar, rod, tubing: 4625. Wire: 4720
ASTM. Flat products: B100, B103. Bar: B103, B139. Rod, shapes: B139. Wire: B159
SAE. J463
Government. Flat products, bar, shapes: QQ-B-750. Rod: QQ-B-750, MIL-B-13501. Bearings: MIL-B-13501. Wire: QQ-B-750, QQ-W-321, MIL-W-6712

Chemical Composition

Composition limits. 93.6 to 95.6 Cu, 4.2 to 5.8 Sn, 0.03 to 0.35 P, 0.05 max Pb, 0.1 max Fe, 0.3 max Zn

Applications

Typical uses. Architectural: bridge bearing plates. Hardware: beater bars, bellows, Bourdon tubing, clutch disks, cotter pins, diaphragms, fuse clips, fasteners, lockwashers, sleeve bushings, springs, switch parts, truss wire, wire brushes. Industrial: chemical hardware, perforated sheets, textile machinery, welding rods.

Mechanical Properties

Tensile properties. See Table 80.
Hardness. See Table 80.
Elastic modulus. Tension, 110 GPa (16 × 10⁶ psi); shear, 41 GPa (6 × 10⁶ psi)

Fabrication Characteristics

Machinability. 20% of C36000, free-cutting brass
Formability. Excellent capacity for cold working by blanking, drawing, forming, bending, roll threading,

Table 80 Typical mechanical properties of C51000

Temper	Tensile strength MPa	ksi	Yield strength(a) MPa	ksi	Elongation(b), %	Hardness, HRB
Flat Products, 1-mm (0.04-in.) Thick						
OS050	325	47	130	19	64	26
OS035	340	49	140	20	58	28
OS025	345	50	145	21	52	30
OS015	365	53	150	22	50	34
H02	470	68	380	55	28	78
H04	560	81	515	75	10	87
H06	635	92	550	80	6	93
H08	690	100	...	...	4	95
H10	740	107	...	...	3	97
Rod, 13-mm (0.05-in.) Diam						
H02	515	75	450	65	25	80
Rod, 25-mm (1-in.) Diam						
H02	480	70	400	58	25	78
Wire, 2-mm (0.08-in.) Diam						
OS035	345	50	140	20	58	...
H01	470	68	415	60	24	...
H02	585	85	550	80	8	...
H04	760	110	...	...	5	...
H06	895	130	...	...	3	...
H08	965	140	...	...	2	...

(a) 0.5% extension under load. (b) In 50 mm or 2 in.

knurling, shearing and stamping. Poor capacity for hot forming.
Weldability. Soldering, brazing and resistance butt welding: excellent. Gas metal-arc and resistance spot welding: good. Oxyfuel gas, shielded metal-arc and resistance seam welding: fair
Annealing temperature. 475 to 675 °C (900 to 1250 °F)

C51100
95.6Cu-4.2Sn-0.2P

Specifications

ASTM. Flat products: B100, B103

Chemical Composition

Composition limits. 94.5 to 96.3 Cu, 3.5 to 4.9 Sn, 0.003 to 0.35 P, 0.05 max Pb, 0.1 max Fe, 0.3 max Zn

Applications

Typical uses. Architectural: bridge bearing plates. Hardware: beater bars, bellows, clutch disks, connectors, diaphragms, fuse clips, fasteners, lockwashers, sleeve bushings, springs, switch parts, terminals. Industrial: chemical hardware, perforated sheets, textile machinery

Mechanical Properties

Tensile properties. See Table 81.
Hardness. See Table 81.
Elastic modulus. Tension, 110 GPa (16 × 10⁶ psi); shear, 41 GPa (6 × 10⁶ psi)

Mass Characteristics

Density. 8.86 Mg/m³ (0.32 lb/in.³) at 20 °C (68 °F)

Thermal Properties

Liquidus temperature. 1060 °C (1945 °F)
Solidus temperature. 975 °C (1785 °F)
Coefficient of thermal expansion. Linear, 17.8 μm/m·K (9.9 μin./in.·°F) at 20 to 300 °C (68 to 572 °F)
Specific heat. 380 J/kg·K (0.09 Btu/lb·°F) at 20 °C (68 °F)
Thermal conductivity. 84 W/m·K (48.4 Btu/ft·h·°F) at 20 °C (68 °F)

Electrical Properties

Electrical conductivity. Volumetric, 20% IACS at 20 °C (68 °F)
Electrical resistivity. 87 nΩ·m at 20 °C (68 °F)

Table 81 Typical mechanical properties of C51100 strip, 1-mm (0.04-in.) thick

Temper	Tensile strength MPa	ksi	Yield strength(a) MPa	ksi	Elongation(b), %	HRF	Hardness HRB	HR30T
OS050	315	46	110	16	48	70	...	...
OS035	330	48	130	19	47	73	...	...
OS025	345	50	145	21	46	75	...	...
OS015	350	51	160	23	46	76	...	...
H01	380	55	295	43	36	...	48	45
H02	425	62	385	56	19	...	70	65
H03	510	74	495	72	11	...	84	72
H04	550	80	530	77	7	...	86	74
H06	635	92	615	89	4	...	91	78
H08	675	98	655	95	3	...	93	79
H10	710	103	675	98	2	...	95	80

(a) 0.2% offset. (b) In 50 mm or 2 in.

Fabrication Characteristics

Machinability. 20% of C36000, free-cutting brass
Formability. Excellent capacity for cold working by blanking, drawing, forming, bending, roll threading, knurling, shearing and stamping. Poor capacity for hot forming
Weldability. Soldering, brazing and resistance butt welding: excellent. Gas metal-arc and resistance spot welding: good. Oxyfuel gas, shielded metal-arc and resistance seam welding: fair
Annealing temperature. 475 to 675 °C (900 to 1250 °F)

C52100
92Cu-8Sn

Commercial Names

Previous trade name. Phosphor bronze, 8% C

Specifications

ASTM. Flat products: B103. Bar: B103, B139. Rod, shapes: B139. Wire: B159
SAE. J463
Government. MIL-E-23765

Chemical Composition

Composition limits. 90.5 to 92.8 Cu,

Table 82 Typical mechanical properties of C52100

Temper	Tensile strength MPa	ksi	Yield strength(a) MPa	ksi	Elongation(b), %	HRF	Hardness HRB	HR30T
Flat products, 1-mm (0.04-in.) Thick								
OS050	380	55	...	...	70	75	...	...
OS035	400	58	...	...	65	80	...	...
OS025	415	60	165	24	63	82	50	...
OS015	425	62	...	...	60	85	...	...
H02	525	76	380	55	32	...	84	73
H04	640	93	495	72	10	...	93	78
H06	730	106	550	80	4	...	96	80
H08	770	112	...	...	3	...	98	81
H10	825	120	...	...	2	...	100	82
Rod, 13-mm (0.5-in.) Diam								
H02	550	80	450	65	33	...	85	...
Wire, 2-mm (0.08-in.) Diam								
OS035	415	60	165	24	65	...	...	...
H01	560	81	...	...	...	...	...	...
H02	725	105	...	...	...	...	...	...
H04	895	130	...	...	...	...	...	...
H06	965	140	...	...	...	...	...	...

(a) 0.5% extension under load. (b) In 50 mm or 2 in.

7.0 to 9.0 Sn, 0.03 to 0.35 P, 0.05 max Pb, 0.1 max Fe, 0.2 max Zn

Applications

Typical uses. For more severe service conditions than C51000. Architectural: bridge bearing plates. Hardware: beater bars, bellows, Bourdon tubing, clutch disks, cotter pins, diaphragms, fuse clips, fasteners, lockwashers, sleeve bushings, springs, switch parts, truss wire, wire brushes. Industrial: chemical hardware, perforated sheets, textile machinery, welding rods

Mechanical Properties

Tensile properties. See Table 82.
Hardness. See Table 82.
Elastic modulus. Tension, 110 GPa (16×10^6 psi); shear, 41 GPa (6×10^6 psi)
Fatigue strength. Strip, 1 mm (0.04 in.) thick, H04 temper: 150 MPa (22 ksi) at 10^8 cycles

Mass Characteristics

Density. 8.8 Mg/m³ (0. 318 lb/in.³) at 20 °C (68 °F)

Thermal Properties

Liquidus temperature. 1020 °C (1880 °F)
Solidus temperature. 880 °C (1620 °F)
Coefficient of thermal expansion. Linear, 18.2 μm/m·K (10.1 μin./in.·°F) at 20 to 300 °C (68 to 572 °F)
Specific heat. 380 J/kg·K (0.09 Btu/lb·°F) at 20 °C (68 °F)
Thermal conductivity. 62 W/m·K (36 Btu/ft·h·°F) at 20 °C (68 °F)

Electrical Properties

Electrical conductivity. Volumetric, 13% IACS at 20 °C (68 °F)
Electrical resistivity. 133 nΩ·m at 20 °C (68 °F)

Fabrication Characteristics

Machinability. 20% of C36000, free-cutting brass
Formability. Good capacity for cold working by blanking, drawing, forming, bending, shearing and stamping: Poor capacity for hot forming
Weldability. Soldering, brazing and resistance butt welding: excellent. Gas metal-arc and resistance spot welding: good. Oxyfuel gas, shielded metal-arc and resistance seam welding: fair
Annealing temperature. 475 to 675 °C (900 to 1250 °F)

C52400
90Cu-10Sn

Commercial Names

Previous trade name. Phosphor bronze, 10% D

Specifications

ASTM. Flat products: B103. Bar: B103, B139. Rod, shapes: B139. Wire: B159
Government. Flat products, wire: QQ-B-750

Chemical Composition

Composition limits. 88.3 to 90.07 Cu, 9.0 to 11.0 Sn, 0.03 to 0.35 P, 0.05 max Pb, 0.1 max Fe, 0.2 max Zn

Applications

Typical uses. Heavy bars and plates for severe compression, good wear and corrosion resistance; bridge and expansion plates and fittings; and articles requiring extra spring qualities, greatest resiliency, particularly in fatigue

Mechanical Properties

Tensile properties. Tensile strength and elongation, See Table 83. Yield strength, typical, OS035 temper: 195 MPa (28 ksi) at 0.5% extension under load
Hardness. See Table 83.
Elastic modulus. Tension, 110 GPa (16×10^6 psi); shear, 41 GPa (6×10^6 psi)

Mass Characteristics

Density. 8.78 Mg/m³ (0.317 lb/in.³) at 20 °C (68 °F)

Thermal Properties

Liquidus temperature. 1000 °C (1830 °F)
Solidus temperature. 845 °C (1550 °F)
Coefficient of thermal expansion. Linear, 18.4 μm/m·K (10.2 μin./in.·°F) at 20 to 300 °C (68 to 572 °F)
Specific heat. 380 J/kg·K (0.09 Btu/lb·°F) at 20 °C (68 °F)
Thermal conductivity. 50 W/m·K (29 Btu/ft·h·°F) at 20 °C (68 °F)

Electrical Properties

Electrical conductivity. Volumetric, 11% IACS at 20 °C (68 °F)
Electrical resistivity. 157 nΩ·m at 20 °C (68 °F)

Fabrication Characteristics

Machinability. 20% of C36000, free-cutting brass
Formability. Good capacity for cold working by blanking, forming, bending and shearing. Poor capacity for hot forming
Weldability. Soldering, brazing and resistance butt welding: excellent. Gas metal-arc and resistance spot welding: good. Oxyfuel gas, shielded metal-arc and resistance seam welding: fair.
Annealing temperature. 475 to 675 °C (900 to 1250 °F)

Table 83 Typical mechanical properties of C52400

Temper	Tensile strength MPa	ksi	Elongation(a), %	Hardness, HRB
Flat Products, 1-mm (0.04-in.) Thick				
OS035 ..	455	66	68	55
H02	570	83	32	92
H04	690	100	13	97
H06	795	115	7	100
H08	840	122	4	101
H10	885	128	3	103
Wire, 2-mm (0.08-in.) Diam				
OS035 ..	455	66	70	...
H01	640	93	...	...
H02	815	118	...	...
H04	1013	147	...	...

(a) In 50 mm or 2 in.

C54400
88Cu-4Pb-4Sn-4Zn

Commercial Names

Previous trade name. Phosphor bronze B-2
Common names. Free-cutting phosphor bronze; 444 bronze; bearing bronze

Specifications

AMS. Strip: 4520
ASTM. B103, B139
SAE. J463. Bearing alloy: J460 (791)
Government. Bar and rod: QQ-B-750

Chemical Composition

Composition limits. 3.5 to 4.5 Pb, 3.5 to 4.5 Sn, 1.5 to 4.5 Zn, 0.10 max Fe, 0.01 to 0.50 P, rem Cu; 99.5 min Cu + Pb + Sn + Zn + P

Applications

Typical uses. Bearings (sleeve and thrust), bushings, gears, pinions, screw-machine products, shafts, thrust washers, valve parts

Mechanical Properties

Tensile properties. See Table 84.
Hardness. See Table 84.
Elastic modulus. Tension, 103 GPa $(15 \times 10^6$ psi); shear, 39 GPa $(5.6 \times 10^6$ psi)

Mass Characteristics

Density. 8.89 Mg/m^3 (0.321 lb/in.^{3}) at 20 °C (68 °F)

Thermal Properties

Liquidus temperature. 1000 °C (1830 °F)
Solidus temperature. 930 °C (1700 °F)
Coefficient of thermal expansion. Linear, 17.3 μm/m·K (9.6 μin./in.·°F) at 20 to 300 °C (68 to 572 °F)
Specific heat. 380 J/kg·K (0.09 Btu/lb·°F) at 20 °C (68 °F)
Thermal conductivity. 87 W/m·K (50 Btu/ft·h·°F) at 20 °C (68 °F)

Electrical Properties

Electrical conductivity. Volumetric, 19% IACS at 20 °C (68 °F)

Electrical resistivity. 90.8 nΩ·m at 20 °C (68 °F)

Fabrication Characteristics

Machinability. 80% of C36000, free-cutting brass
Formability. Good cold working characteristics; commonly fabricated by machining, shearing, blanking, drawing, forming, bending. Hot working and hot forming not recommended.
Weldability. Soldering: excellent. Brazing: good. Flash welding: fair. Other welding processes not recommended.

Table 84 Typical mechanical properties of C54400

Temper	Tensile strength MPa	ksi	Yield strength(a) MPa	ksi	Elongation(b), %	Hardness HRB
Sheet and strip, 1-mm (0.04-in.) Thick						
OS050 315	46	...	...	48	70 HRF
OS035 330	48	...	...	47	73 HRF
OS025 345	50	...	...	46	75 HRF
OS015 350	51	...	...	46	76 HRF
H02 425	62	370	54	19	70
H04 550	80	510	74	7	86
H06 635	92	...	...	4	91
H08 675	98	550	80	3	93
H10 710	103	...	...	2	95
Flat Products, 8-mm (0.38-in.) Thick						
H04 415	60	310	45	20	70
Flat Products, 19-mm (0.75-in.) Thick						
H04 380	55	240	35	25	...
Rod, 13-mm (0.50-in.) Diam						
H04 515	75	435	63	15	83
Rod, 25-mm (1.0-in.) Diam						
H04 470	68	385	57	20	80

(a) At 0.5% extension under load. (b) In 50 mm or 2 in.

C60600
95Cu-5Al

Commercial Names

Previous trade name. Aluminum bronze A; CA606
Common name. Aluminum bronze, 5%

Specifications

ASTM. Flat products: B169
Government. Bar, rod, forgings, shapes: QQ-C-645. Sheet and plate: QQ-C-450. Strip: QQ-C-450, QQ-C-465

Chemical Composition

Composition limits. 92.0 to 96.0 Cu; 4.0 to 7.0 Al; 0.50 max Fe; 0.50 max others (total)
Consequence of exceeding impurity limits. Excessive amounts of Pb, Zn and P will cause hot shortness and difficulties in welding.

Applications

Typical uses. Produced as sheet, strip and rolled bar; used to make fasteners, deep drawn "gold" decoration and parts requiring corrosion resistance
Precautions in use. Not suitable for use in oxidizing acids

Mechanical Properties

Tensile properties. Typical data for 13 mm (0.5 in.) thick plate. Tensile strength: O61 temper, 310 MPa (45 ksi); H04 temper, 415 MPa (60 ksi). Yield strength: O60 temper, 115 MPa (17 ksi); H04 temper, 165 MPa (24 ksi). Elongation: O60 temper, 40% in 50 mm or 2 in.; H04 temper, 25% in 50 mm or 2 in.
Hardness. O60 temper, 42 HRB; H04 temper, 55 HRB
Poisson's ratio. 0.326
Elastic modulus. Tension, 121 GPa $(17.5 \times 10^6$ psi); shear, 46 GPa $(6.6 \times 10^6$ psi)
Fatigue strength. Rotating beam, 169 MPa (24.5 ksi) at 10^8 cycles

Structure

Microstructure. Alpha structure, face-centered cubic

Mass Characteristics

Density. 8.17 Mg/m^3 (0.295 lb/in.^{3}) at 20 °C (68 °F)
Volume change on freezing. Approximately 1.6% contraction

Thermal Properties

Liquidus temperature. 1065 °C (1945 °F)
Solidus temperature. 1050 °C (1920 °F)

Coefficient of thermal expansion. Linear, 18.1 μm/m·K (10.1 μin./in.·°F) at 20 to 300 °C (68 to 572 °F)
Specific heat. 375 J/kg·K (0.09 Btu/lb·°F) at 20 °C (68 °F)
Thermal conductivity. 79.5 W/m·K (45.9 Btu/ft·h·°F) at 20 °C (68 °F)

Electrical Properties

Electrical conductivity. Volumetric, 17% IACS at 20 °C (68 °F)
Electrical resistivity. 100 nΩ·m at 20 °C (68 °F)

Magnetic Properties

Magnetic permeability. 1.01

Chemical Properties

General corrosion resistance. See C61400.
Resistance to specific agents. Has been used in sulfuric acid pickling applications where oxygen content is low. Has been used for anhydrous NH4OH, but the presence of moisture leads to season cracking. Not suitable for use with nitric acid. Oxidizing salts such as chromates, and metal salts such as ferric chloride, are generally corrosive to C60600.

Fabrication Characteristics

Machinability. 20% of C36000, free-cutting brass. Tends to form tough, stringy chips. Good lubrication and cooling essential for good finish. Carbide or tool steel cutters may be used.
Recrystallization temperature. 350 °C (660 °F) at 44% reduction and 0.075 mm (0.003 in.) initial grain size
Annealing temperature. 550 to 650 °C (1020 to 1200 °F)
Hot working temperature. 815 to 870 °C (1500 to 1600 °F)

C60800
95Cu-5Al

Commercial Names

Previous trade name. 5% aluminum bronze
Common name. Aluminum bronze, 5%

Specifications

ASME. Tubing: SB111, SB359, SB395
ASTM. Tubing: B111, B359, B395

Chemical Composition

Composition limits. 92.5 to 94.8 Cu; 5.0 to 6.5 Al; 0.02 to 0.35 As; 0.10 max Pb; 0.10 max Fe
Consequence of exceeding impurity limits. Excessive amounts of Pb, Zn and P will cause difficulties in welding and hot working

Applications

Typical uses. Produced as seamless tubing and ferrule stock—for heat exchanger tubes, condenser tubes and other applications requiring corrosion-resistant seamless tubing
Precautions in use. Not suitable for use in oxidizing acids

Mechanical Properties

Tensile properties. Typical data for 25-mm (1.0-in.) OD by 1.65-mm (0.065-in.) wall tubing, OS025 temper: tensile strength, 415 MPa (60 ksi); yield strength, 185 MPa (27 ksi); elongation, 55% in 50 mm or 2 in.
Hardness. OS025 temper: 77 HRF
Poisson's ratio. 0.325
Elastic modulus. Tension, 121 GPa (17.5 × 10^6 psi); shear, 46 GPa (6.6 × 10^6 psi)

Structure

Microstructure. Alpha structure, face-centered cubic

Mass Characteristics

Density. 8.17 Mg/m³ (0.295 lb/in.³) at 20 °C (68 °F)
Volume change on freezing. Approximately 1.6% contraction

Thermal Properties

Liquidus temperature. 1065 °C (1945 °F)
Solidus temperature. 1050 °C (1920 °F)
Coefficient of thermal expansion. Linear, 18.1 μm/m·K (10.1 μin./in.·°F) at 20 to 300 °C (68 to 572 °F)
Specific heat. 380 J/kg·K (0.09 Btu/lb·°F) at 20 °C (68 °F)
Thermal conductivity. 79.5 W/m·K (45.9 Btu/ft·h·°F) at 20 °C (68 °F)

Electrical Properties

Electrical conductivity. Volumetric, 17% IACS at 20 °C (68 °F)
Electrical resistivity. 100 nΩ·m at 20 °C (68 °F)

Magnetic Properties

Magnetic permeability. 1.01

Chemical Properties

General corrosion resistance. See C61400.
Resistance to specific agents. Has been used in sulfuric acid pickling applications where oxygen content is low. Has been used for anhydrous NH4OH, but the presence of moisture leads to season cracking. Not suitable for use with nitric acid. Oxidizing salts such as chromates, and metal salts such as ferric chloride, are generally corrosive to C60800.

Fabrication Characteristics

Machinability. 20% of C36000, free-cutting brass. Tends to form tough, stringy chips. Good lubrication and cooling essential for good finishes. Carbide or tool steel cutters may be used.
Formability. Good for cold working; fair for hot forming
Weldability. Arc and resistance welding: good. Brazing: fair. Soldering and oxyfuel gas welding: not recommended.
Recrystallization temperature. 350 °C (660 °F) at 44% reduction and 0.075 mm (0.003 in.) initial grain size
Annealing temperature. 550 to 650 °C (1020 to 1200 °F)
Hot working temperature. 800 to 875 °C (1470 to 1610 °F)

C61000
92Cu-8Al

Commercial Names

Common name. 8% aluminum bronze

Specifications

ASME. SB169
ASTM. B169
Government. QQ-C-450; MIL-E-23765

Chemical Composition

Composition limits. 6.0 to 8.5 Al, 0.50 max Fe, 0.02 max Pb, 0.20 max Zn, 0.10 max Si, 0.50 max others (total), rem Cu

Applications

Typical uses. Produced as rod or wire and used to make bolts, shafts, tie rods and pump parts. Also used as welded overlay on steel to give a surface with improved wear resistance.

Mechanical Properties

Tensile properties. Typical for rod, 25 mm (1 in.) diam. O60 temper: tensile strength, 480 MPa (70 ksi); yield strength (0.5% extension under load), 205 MPa (30 ksi); elongation, 65% in 50 mm or 2 in. H04 temper: tensile strength, 550 MPa (80 ksi); yield strength, 380 MPa (55 ksi); elongation, 25%
Hardness. O60 temper: 60 HRB. H04 temper: 85 HRB
Elastic modulus. Tension, 117 GPa (17×10^6 psi); shear, 44 GPa (6.4×10^6 psi)

Mass Characteristics

Density. 7.78 Mg/m³ (0.281 lb/in.³) at 20 °C (68 °F)

Thermal Properties

Liquidus temperature. 1040 °C (1905 °F)
Coefficient of thermal expansion. Linear, 17.9 μm/m·K (9.9 μin./in.·°F) at 20 to 300 °C (68 to 572 °F)
Specific heat. 375 J/kg·K (0.09 Btu/lb·°F) at 20 °C (68 °F)
Thermal conductivity. 69 W/m·K (40 Btu/ft·h·°F) at 20 °C (68 °F)

Electrical Properties

Electrical conductivity. Volumetric, 15% IACS at 20 °C (68 °F)
Electrical resistivity. 115 nΩ·m at 20 °C (68 °F)

Fabrication Characteristics

Machinability. 20% of C36000, free-cutting brass
Forgeability. 70% of C37700, forging brass
Formability. Good capacity for being hot formed or cold worked. Common fabrication processes include blanking, drawing, forming, bending, cold heading and roll threading.
Weldability. Arc welding, resistance spot welding and resistance butt welding: good. Soldering and resistance seam welding: fair. Brazing and oxyfuel gas welding: not recommended.
Annealing temperature. 600 to 675 °C (1100 to 1250 °F)
Hot working temperature. 760 to 875 °C (1400 to 1600 °F)

C61300
90Cu-7Al-2.7Fe-0.3Sn

Commercial Names

Common name. Aluminum bronze, 7%

Specifications

Government. Flat products: QQ-C-450

Chemical Composition

Composition limits. 88.5-91.5 Cu; 6.0-7.5 Al; 0.20-0.50 Sn; 2.0-3.0 Fe; 0.10 max Mn; 0.15 max Ni(+Co); 0.01 max Pb; 0.05 max Zn; 0.05 max others
Consequence of exceeding impurity limits. Excessive amounts of Pb, Zn, P or Si will cause hot-shortness, which can lead to problems during hot working or welding.

Applications

Typical uses. Produced as rod, bar, sheet, plate, seamless tubing and pipe, and welded pipe, fasteners, tube sheets, heat exchanger tubes, acid-resistant piping, columns, water boxes and corrosion-resistant vessels
Precautions in use. Not suitable for use in oxidizing acids

Mechanical Properties

Tensile properties. Typical data for 13-mm (0.50-in.) thick plate Tensile strength: O60 temper, 540 MPa (78 ksi); H04 temper, 585 MPa (85 ksi). Yield strength: O60 temper, 240 MPa (35 ksi); H04 temper, 400 MPa (58 ksi). Elongation: O60 temper, 42% in 50 mm or 2 in.; H04 temper, 35% in 50 mm or 2 in. Reduction in area: O60 temper, 32%; H04 temper, 25%. See also Table 86.
Compressive properties. Typical data for 13-mm (0.50-in.) thick plate. Compressive strength, ultimate: O60 temper, 825 MPa (120 ksi); H04 temper, 860 MPa (125 ksi)
Hardness. O60 temper, 82 HRB; H04 temper, 91 HRB. See also Table 86.
Poisson's ratio. 0.312
Elastic modulus. Tension, 115 GPa (17×10^6 psi); shear, 44 GPa (6.4×10^6 psi). See also Table 86.
Impact strength. Charpy keyhole, 81 to 88 J (60 to 65 ft·lb) at −30 to +150 °C (−20 to +300 °F); Izod, 54 to 66 J (40 to 49 ft·lb) at −30 to +150

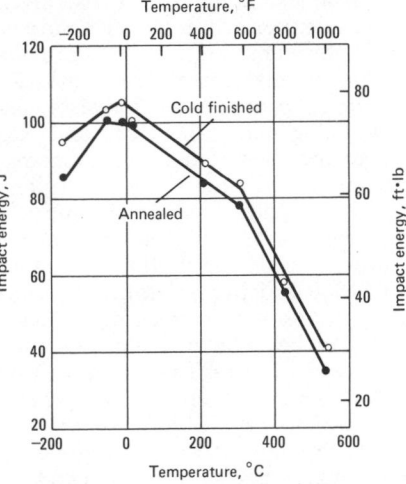

Fig. 43 Variation of Charpy V-notch impact strength with temperature for C61300 and C61400

°C (−20 to +300 °F). See also Fig. 43.
Fatigue strength. Reverse bending, 180 MPa (26 ksi) at 10^8 cycles

Structure

Microstructure. Alpha structure, single phase, with iron-rich precipitates

Mass Characteristics

Density. 7.89 Mg/m³ (0.285 lb/in.³) at 20 °C (68 °F)
Volume change on freezing. Approximately 1.8% contraction

Thermal Properties

Liquidus temperature. 1045 °C (1915 °F)
Solidus temperature. 1040 °C (1905 °F)
Coefficient of thermal expansion. Linear, 16.2 μm/m·K (9.0 μin./in.·°F) at 20 to 300 °C (68 to 572 °F)
Specific heat. 375 J/kg·K (0.09 Btu/lb·°F) at 20 °C (68 °F)
Thermal conductivity. 56.5 W/m·K (32.7 Btu/ft·h·°F) at 20 °C (68 °F); temperature coefficient, 0.12 W/m·K per K at 20 °C (68 °F)

Electrical Properties

Electrical conductivity. Volumetric, 12% IACS at 20 °C (68 °F)
Electrical resistivity. 144 nΩ·m at 20 °C (68 °F)

Magnetic Properties

Magnetic permeability. 1.16

Chemical Properties

General corrosion resistance. See C61400.

Resistance to specific agents. C61300 is very resistant to neutral and nonoxidizing salts. It has given extended service in potash solutions of potassium chloride, sodium chloride, magnesium chloride and calcium chloride. The alloy resists nonoxidizing mineral acids, and has been used successfully for tanks containing hydrofluoric acid in glassetching applications. In organic acid service, it has been used to make acetic acid distillation columns. C61300 is highly resistant to dealloying and to season cracking in steam and in hot oxidizing aqueous solutions and vapors. The presence of tin in 7% aluminum bronze (C61300 contains 0.3% Sn, C61400 does not contain Sn) evidently renders the alloy immune to stress-corrosion cracking in these environments. Like many copper alloys, C61300 is susceptible to season cracking in moist ammonia and mercurous nitrate solutions. However, it is highly resistant to season cracking in anhydrous ammonia, especially when the moisture content is below 500 ppm and the temperature is below 85 °C (180 °F). Because of its high resistance to corrosion in salt water, C61300 has been specified for a wide variety of components for marine and desalting-plant service. Typical uses of C61300 include tube sheets for condensers in both nuclear and fossil-fueled power stations, cooling tower transfer piping, seawater piping for secondary cooling systems in nuclear power plants, and piping for geothermal heat transfer systems.

Fabrication Characteristics

Machinability. Fair to poor, with chips tending to be stringy and gummy. Good lubrication and cooling are essential. Tool steel cutters: roughing speed, 90 m/min (300 ft/min) with a feed of 0.3 mm/rev (0.011 in./rev); finishing speed, 350 m/min (1150 ft/min) with a feed of 0.3 mm/rev (0.011 in./rev)

Forgeability. 50% of C37700, forging brass

Formability. Good for cold working and hot forming

Weldability. Arc and resistance welding: good. Brazing: fair. Soldering and oxyfuel gas welding: not recommended

Recrystallization temperature. 785 to 870 °C (1450 to 1600 °F)

Annealing temperature. 600 to 875 °C (1125 to 1600 °F)

Hot working temperature. 800 to 925 °C (1450 to 1700 °F)

Hot-shortness temperature. 1010 °C (1850 °F)

C61400
91Cu-7Al-2Fe

Commercial Names

Previous trade name. Aluminum bronze D

Common name. Aluminum bronze, 7%

Specifications

ASME. Flat products: SB169, SB171. Bar, rod, shapes: SB150

ASTM. Flat products: B169, B171. Bar, rod, shapes: B150

SAE. J463

Government. Flat products: QQ-C-450, QQ-C-465. Bar, rod, shapes, forgings: QQ-C-465. Flat wire: QQ-C-465

Chemical Composition

Composition limits. 88.0 to 92.5 Cu; 6.0 to 8.0 Al; 1.5 to 3.5 Fe; 1.0 max Mn; 0.20 max Zn; 0.01 max Pb; 0.015 max P; 0.5 max others (total)

Consequence of exceeding impurity limits. Excessive amounts of Pb, Zn, Si or P will cause hot shortness and cracking during hot working and welding

Applications

Typical uses. Produced as seamless tubing, welded and seamless pipe, sheet, plate, rod and bar for condenser and heat exchanger tubes, fasteners, tube sheets and corrosion-resistant vessels

Precautions in use. Not suitable for use in oxidizing acids. Is susceptible to stress-corrosion cracking in moist ammonia or in steam environments, especially when stress levels are high.

Mechanical Properties

Tensile properties. See Tables 85 and 86.

Shear strength. See Table 85.

Compressive properties. Compressive strength, ultimate: O60 temper, 825 MPa (120 ksi); H04 temper, 860 MPa (125 ksi)

Hardness. O60 temper, 80 to 84 HRB; H04 temper, 84 to 91 HRB. See also Table 86.

Poisson's ratio. 0.312

Elastic modulus. Tension, 115 GPa (17×10^6 psi); shear, 44 GPa (6.4×10^6 psi). See also Table 86.

Impact strength. Charpy keyhole, 81 to 88 J (60 to 65 ft·lb); Izod, 54 to 61 J (40 to 45 ft·lb). See Fig. 43.

Fatigue strength. Reverse bending, 180 MPa (26 ksi) at 10^8 cycles

Structure

Microstructure. Alpha solid solution with precipitates of iron-rich phase

Mass Characteristics

Density. 7.89 Mg/m³ (0.285 lb/in.³) at 20 °C (68 °F)

Volume change on freezing. Approximately 1.8% expansion

Thermal Properties

Liquidus temperature. 1045 °C (1915 °F)

Solidus temperature. 1040 °C (1905 °F)

Coefficient of thermal expansion. Linear, 16.2 μm/m·K (9.0 μin./in.·°F) at 20 to 300 °C (68 to 572 °F)

Specific heat. 375 J/kg·K (0.09 Btu/lb·°F) at 20 °C (68 °F)

Thermal conductivity. 56.5 W/m·K (32.6 Btu/ft·h·°F) at 20 °C (68 °F); temperature coefficient, 0.12 W/m·K per K at 20 °C (68 °F)

Electrical Properties

Electrical conductivity. Volumetric, 14% IACS at 20 °C (68 °F)

Electrical resistivity. 123 nΩ·m at 20 °C (68 °F)

Magnetic Properties

Magnetic permeability. 1.16

Chemical Properties

General corrosion behavior. The aluminum bronzes resist nonoxidizing mineral acids such as sulfuric, hydrochloric and phosphoric. Resistance tends to decrease with increasing concentration of dissolved oxygen or oxidizing agents, particularly as temperatures increase above 55 °C (130 °F). Aluminum bronzes are generally suited for service in alkalies, neutral salts, nonoxidizing acid salts and many organic acids and compounds. Oxidizing acids, oxidizing salts and heavy-metal salts are corrosive. Aluminum bronzes resist waters, whether potable water, brackish water or seawater. Softened water tends to be more corrosive than hard water. Aluminum bronzes resist dealloying, but to different degrees depending on alloy composition. In general, corrosion resistance is influenced most by solution concentration, aeration, temperature, velocity and the type and amount of any impurities in the solution. Like many other copper alloys, the aluminum bronzes are susceptible to stress-corrosion cracking in moist ammonia and mercury compounds. When stress levels are high, they may also be susceptible to stress-corrosion cracking in purified steam or in steam containing acidic or salt vapors.

Resistance to specific agents. C61400 has been used successfully to contain mineral acids, alkalies such as sodium or potassium hydroxide, neutral salts such as sodium chloride, and organic acids such as acetic, lactic or oxalic. C61400 resists anhydrous ammonia, but precautions must be taken to exclude moisture and thus avoid season cracking. Similarly, this alloy resists anhydrous chlorinated hydrocarbons such as carbon tetrachloride, but the presence of moisture makes those chemicals corrosive.

Fabrication Characteristics

Machinability. 20% of C36000, free-cutting brass. Tendency to form continuous, stringy chips. Good lubrication and cooling essential. Tool steel or carbide cutters may be used. Typical conditions, using tool steel cutters: roughing speed, 90 m/min (300 ft/min) with a feed of 0.3 mm/rev (0.011 in./rev); finishing speed, 350 m/min (1150 ft/min) with a feed of 0.3 mm/rev (0.011 in./rev)

Formability. Fair for cold working; good for hot forming

Weldability. Gas-shielded arc, coated metal-arc and resistance welding: good. Brazing: fair. Soldering, oxyacetylene and carbon arc welding: not recommended

Recrystallization temperature. 785 to 870 °C (1450 to 1600 °F)

Annealing temperature. 600 to 900 °C (1125 to 1650 °F)

Hot working temperature. 800 to 925 °C (1450 to 1700 °F)

Hot-shortness temperature. 1010 °C (1850 °F)

Table 85 Typical mechanical properties of C61400

Size	Tensile strength MPa	ksi	Yield strength(a) MPa	ksi	Elongation(b), %	Shear strength MPa	ksi
Flat Products: O60 Temper							
3-mm (0.12-in.) thick	565	82	310	45	40	310	45
8-mm (0.31-in.) thick	550	80	275	40	40	290	42
13-mm (0.50-in.) thick ...	535	78	240	35	42	275	40
25-mm (1.00-in.) thick ...	525	76	230	33	45	275	40
Flat Products: H04 Temper							
3-mm (0.12-in.) thick	615	89	415	60	32	...	...
8-mm (0.31-in.) thick	585	85	400	58	35	...	...
13-mm (0.50-in.) thick....	550	80	370	54	38	...	...
25-mm (1.00-in.) thick ...	535	78	310	45	40	...	...
Rod: H04 Temper							
13-mm (0.50-in.) diam ...	585	85	310	45	35	330	48
25-mm (1.00-in.) diam ...	565	82	275	40	35	310	45
51-mm (2.00-in.) diam ...	550	80	240	35	35	275	40

(a) At 0.5% extension under load. (b) In 50 mm or 2 in.

Table 86 Typical mechanical properties of C61300 and C61400 rod at various temperatures

Temperature °C	°F	Tensile strength MPa	ksi	Yield strength(a) MPa	ksi	Elongation(b), %	Reduction in area, %	Modulus of elasticity GPa	10^6 psi	Hardness(c), HB
Cold Finished										
−182	−295	718	104.1	397	57.6	50	49	156	22.7	186
−60	−75	611	88.6	335	48.7	45	55	149	21.6	170
−29	−20	606	87.9	339	49.2	44	58	172	25.0	162
+20	+70	590	85.5	318	46.1	42	59	126	18.3	157
+204	+400	532	77.2	298	43.3	35	32	128	18.5	144
+316	+600	432	62.6	271	39.3	22	24	88	12.8	137
+427	+800	170	24.6	105	15.2	52	41	48	6.9	83
+538	+1000	88	12.8	71	10.3	27	26	45	6.6	49
Annealed										
−182	−295	707	102.6	347	50.3	52	51	139	20.2	185
−60	−75	610	88.4	305	44.3	47	57	176.5	25.6	162
−29	+20	600	87.1	303	43.9	44	56	172	24.9	161
+20	+70	583	84.6	288	41.8	45	56	136.5	19.8	155
+204	+400	522	75.7	276	40.1	34	32	130	18.8	142
+316	+600	427	62.0	256	37.1	30	27	81	11.7	134
+427	+800	174	25.3	123	17.8	60	55	67	9.7	84
+538	+1000	92	13.3	68	9.9	36	33	45	6.5	50

(a) At 0.5% extension under load. (b) In 50 mm or 2 in. (c) 3000-kg load.

C61500
90Cu-8Al-2Ni

Commercial Names

Previous trade name. Lusterloy

Chemical Composition

Composition limits. 89.0 to 90.5 Cu; 7.7 to 8.3 Al; 1.8 to 2.2 Ni; 0.015 max Pb

Applications

Typical uses. Hardware, decorative metal trim, interior furnishings, giftware, springs, fasteners, architectural panels and structural sections, deep drawn articles, tarnish-resistant articles

Mechanical Properties

Tensile properties. See Table 87.
Hardness. See Table 87.
Elastic modulus. Tension, 112 GPa (16.6 × 10⁶ psi)
Fatigue strength. See Table 87.

Mass Characteristics

Density. 7.65 Mg/m³ (0.278 lb/in.³) at 20 °C (68 °F)

Thermal Properties

Liquidus temperature. 1040 °C (1904 °F)
Solidus temperature. 1030 °C (1890 °F)
Coefficient of thermal expansion. Linear, 16.8 μm/m·K (9.3 μin./in.·°F) at 20 to 300 °C (68 to 572 °F)
Specific heat. 380 J/kg·K (0.09 Btu/lb·°F) at 20 °C (68 °F)
Thermal conductivity. 58 W/m·K (33.6 Btu/ft·h·°F) at 20 °C (68 °F)

Electrical Properties

Electrical conductivity. Volumetric, 12.6% IACS at 20 °C (68 °F)
Electrical resistivity. 136 nΩ·m at 20 °C (68 °F)

Optical Properties

Color. Gold

Chemical Properties

General corrosion behavior. Excellent; similar to other aluminum bronzes

Fabrication Characteristics

Machinability. 30% of C36000, free-cutting brass
Forgeability. 50% of C37700, forging brass
Formability. Suitable for forming by bending, drawing, deep drawing, forging, extrusion, blanking and stamping; only slight directionality in bending. Good for cold working and hot forming.
Weldability. Gas-shielded arc welding, shielded metal-arc welding and resistance welding: excellent. Soldering and brazing: easily done using mildly aggressive fluxes. Oxyfuel gas welding: not recommended.
Annealing temperature. 620 to 675 °C (1150 to 1250 °F)
Aging temperature. Order strengthening, 300 °C (575 °F) for 1 h
Hot working temperature. 815 to 870 °C (1500 to 1600 °F)

Table 87 Typical mechanical properties of C61500 sheet and strip(a)

Temper	Tensile strength MPa	ksi	Yield strength(b) MPa	ksi	Elongation(c), %	Hardness HR30T	Fatigue strength(d) MPa	ksi
O60	485	70	150	22	55	42	...	...
O50	585	85	345	50	36	70	260	38
H02	725	105	515	75	15	81	...	...
H04	860	125	620	90	5	83	...	...
H06	930	135	690	100	4	84	270	39
H08	965	140	725	105	3	84.5	...	...
HR06(e)	1000	145	965	140	1	86.5	275	40

(a) 1 mm (0.04 in.) thick. (b) At 0.2% offset. (c) In 50 mm or 2 in. (d) At 10⁸ cycles. (e) Cold worked 50%, then stress relieved for 1 h at 300 °C (570 °F).

C62300
87Cu-10Al-3Fe

Commercial Names

Common name. Aluminum bronze, 9%

Specifications

ASME. Bar, rod, shapes: SB150
ASTM. Bar, rod, shapes: B150. Forgings: B283
SAE. J463
Government. Forgings: MIL-B-16166

Chemical Composition

Composition limits. 82.2 to 89.5 Cu; 8.5 to 11.0 Al; 2.0 to 4.0 Fe; 1.0 max Ni (+ Co); 0.6 max Sn; 0.50 max Mn; 0.25 max Si; 0.5 max others (total)
Consequence of exceeding impurity limits. An excessive amount of Pb will cause hot shortness, and excessive Si will cause the alloy to lose ductility. Excessive Al will reduce ductility and corrosion resistance.

Applications

Typical uses. Produced as rod and bar for bearings, bushings, bolts, nuts, gears, valve guides, pump rods, cams and applications requiring corrosion resistance
Precautions in use. Not suitable for use in oxidizing acids

Fig. 44 Variation of Charpy V-notch impact strength with temperature for C62300

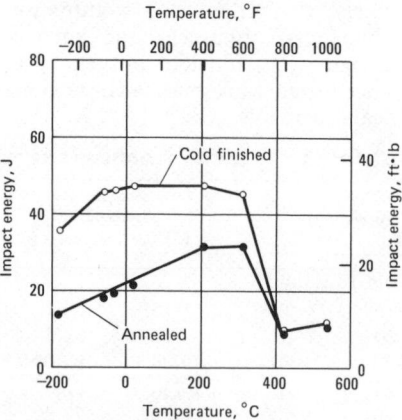

Mechanical Properties

Tensile properties. Typical. Tensile strength, 605 MPa (88 ksi); yield strength, 305 MPa (44 ksi); elongation, 15% in 50 mm or 2 in.; reduction in area, 15%. See also Table 88.
Compressive properties. See Table 89.
Hardness. 89 HRB. See also Table 88.
Poisson's ratio. 0.328
Elastic modulus. Tension, 115 GPa

Table 88 Typical mechanical properties of C62300 rod at various temperatures

Temperature °C	°F	Tensile strength MPa	ksi	Yield strength(a) MPa	ksi	Elongation(b), %	Reduction in area, %	Modulus of elasticity(c) GPa	10⁶ psi	Hardness(d), HB
Cold Finished										
−182	−295	778	112.8	390	56.5	37	41	127	18.4	193
−60	−75	682	98.9	340	49.3	34	41	108	15.7	170
−29	−20	663	96.2	326	47.3	34	44	114	16.5	168
+20	+70	652	94.5	320	46.4	34	44	110	16.1	165
+204	+400	550	79.8	296	43.0	22	22	114	16.5	152
+316	+600	465	67.5	296	43.0	10	13	85	12.4	148
+427	+800	196	28.5	138	20.0	32	33	54	7.9	98
+538	+1000	103	15.0	92	13.3	18	29	41	5.9	54
Annealed										
−182	−295	762	110.5	377	54.7	35	38	125	18.1	195
−60	−75	664	96.3	330	47.9	33	39	121	17.5	171
−29	−20	647	93.8	323	46.9	31	38	124	18.0	168
+20	+70	620	90.0	294	42.6	32	39	120	17.4	161
+204	+400	534	77.5	302	43.8	20	21	138	20.0	151
+316	+600	448	65.0	288	41.8	10	13	79	11.5	146
+427	+800	210	30.5	153	22.2	46	39	73	10.6	97
+538	+1000	94	13.6	84	12.2	27	32	51	7.4	46

(a) At 0.5% extension under load. (b) In 50 mm or 2 in. (c) In tension. (d) 3000-kg load.

(17 × 10⁶ psi); shear, 44 GPa (6.4 × 10⁶ psi). See also Table 88.
Impact strength. Charpy V-notch, 25 to 40 J (18 to 30 ft·lb); Izod, 43 to 47 J (32 to 35 ft·lb). See also Fig. 44.
Fatigue strength. Reverse bending, 200 MPa (29 ksi) at 10⁸ cycles

Structure

Microstructure. Duplex structure of face-centered cubic alpha plus metastable body-centered cubic beta with iron-rich precipitates

Mass Characteristics

Density. 7.65 Mg/m³ (0.276 lb/in.³) at 20 °C (68 °F)
Volume change on freezing. Approximately 2% expansion

Thermal Properties

Liquidus temperature. 1045 °C (1915 °F)
Solidus temperature. 1040 °C (1905 °F)
Phase-transformation temperature. Eutectoid transformation, 563 to 570 °C (1045 to 1055 °F)
Coefficient of thermal expansion. Linear, 16.2 µm/m·K (9.0 µin./in.·°F) at 20 to 300 °C (68 to 572 °F)
Specific heat. 375 J/kg·K (0.09 Btu/lb·°F) at 20 °C (68 °F)
Thermal conductivity. 54.4 W/m·K (31.4 Btu/ft·h·°F) at 20 °C (68 °F); temperature coefficient, 0.12 W/m·K per K at 20 °C (68 °F)

Table 89 Typical compressive properties for C62300 rod, H50 temper

Rod diameter	Compressive strength at permanent set of: 0.1% MPa	ksi	1% MPa	ksi	10% MPa	ksi	20% MPa	ksi
Up to 25 mm (1 in.)	369	52	485	70	825	120	965	140
Over 25 to 50 mm (1 to 2 in.)	345	50	450	65	675	98	930	135
Over 50 to 75 mm (2 to 3 in.)	315	46	415	60	620	90	895	130

Electrical Properties

Electrical conductivity. Volumetric, 12% IACS at 20°C (68 °F)
Electrical resistivity. 144 nΩ·m at 20 °C (68 °F)

Magnetic Properties

Magnetic permeability. 1.17

Chemical Properties

General corrosion behavior. See C61400.
Resistance to specific agents. C62300 resists nonoxidizing mineral acids, but hydrochloric acid is more corrosive than other nonoxidizing mineral acids. C62300 resists dealloying, but to a lesser extent than C61300 or C61400. Like other aluminum bronzes, C62300 is not suitable for use in an oxidizing acid such as nitric acid.

Fabrication Characteristics

Machinability. Fair, with good surface finish possible. Carbide or tool steel cutters may be used. Typical conditions using tool steel cutters: roughing speed, 107 m/min (350 ft/min) with a feed of 0.3 mm/rev (0.011 in./rev); finishing speed, 350 m/min (1150 ft/min) with a feed of 0.15 mm/rev (0.006 in./rev)
Forgeability. 75% of C37700, forging brass
Formability. Good for cold working and hot forming
Weldability. Gas-shielded arc, shielded metal-arc and all types of resistance welding: good. Brazing: fair. Soldering and oxyfuel gas welding: not recommended
Annealing temperature. 600 to 650 °C (1110 to 1200 °F)
Hot working temperature. 700 to 800 °C (1290 to 1605 °F)
Hot-shortness temperature. 1010 °C (1850 °F)

C62400
86Cu-11Al-3Fe

Commercial Names

Common name. Aluminum bronze, 11%

Specifications

SAE. J463

Chemical Composition

Composition limits. 82.8 to 88.0 Cu; 10.0 to 11.5 Al; 2.0 to 4.5 Fe; 0.30 max Mn; 0.25 max Si; 0.20 max Sn; 0.5 max others (total)

Consequence of exceeding impurity limits. Excessive amounts of Si and Al decrease ductility.

Applications

Typical uses. Produced as rod and bar for gears, wear plates, cams, bushings, nuts, drift pins and tie rods

Precautions in use. May lose ductility upon prolonged heating in range of from 370 to 565 °C (700 to 1050 °F). Not suitable for use in oxidizing acids.

Mechanical Properties

Tensile properties. Typical data for 25-mm (1-in.) diam round rod. Tensile strength, 655 MPa (95 ksi); yield strength, 310 MPa (45 ksi); elongation, 12% in 50 mm or 2 in.; reduction in area, 11%

Compressive properties. See Table 90.

Hardness. 92 HRB

Poisson's ratio. 0.318

Elastic modulus. Tension, 115 GPa (17×10^6 psi); shear, 44 GPa (6.4×10^6 psi)

Impact strength. Charpy keyhole, 15 J (11 ft·lb) at -23 to $+27$ °C (-10 to $+80$ °F); Izod, 23 J (17 ft·lb) at -23 to $+27$ °C (-10 to $+80$ °F)

Fatigue strength. Reverse bending, 235 MPa (34 ksi) at 10^8 cycles

Structure

Microstructure. Duplex-structure alpha plus metastable beta phases and iron-rich precipitates

Mass Characteristics

Density. 7.45 Mg/m³ (0.269 lb/in.³) at 20 °C (68 °F)

Volume change on freezing. Approximately 2% contraction

Table 90 Typical compressive properties for C62400 rod, H50 temper

| Rod diameter | Compressive strength at permanent set of: | | | | | | | |
| | 0.1% | | 1% | | 10% | | Ultimate | |
	MPa	ksi	MPa	ksi	MPa	ksi	MPa	ksi
Up to 25 mm (1 in.)	290	42	470	68	885	128	1140	165
Over 25 to 50 mm (1 to 2 in.) ...	220	32	400	58	825	120	1090	158
Over 50 to 75 mm (2 to 3 in.) ...	175	25	330	48	795	115	1090	158

Thermal Properties

Liquidus temperature. 1040 °C (1900 °F)

Solidus temperature. 1025 °C (1880 °F)

Phase transformation temperature. Eutectoid, 560 to 570 °C (1045 to 1055 °F)

Coefficient of thermal expansion. Linear, 16.5 μm/m·K (9.2 μin./in.·°F) at 20 to 300 °C (68 to 572 °F)

Specific heat. 375 J/kg·K (0.09 Btu/lb·°F) at 20 °C (68 °F)

Thermal conductivity. 58.6 W/m·K (33.9 Btu/ft·h·°F) at 20 °C (68 °F); temperature coefficient, 0.12 W/m·K per K at 20 °C (68 °F)

Electrical Properties

Electrical conductivity. Volumetric, 12% IACS at 20 °C (68 °F)

Electrical resistivity. 144 nΩ·m at 20 °C (68 °F)

Magnetic Properties

Magnetic permeability. 1.34

Chemical Properties

General corrosion behavior. See C61400.

Resistance to specific agents. C62400 resists nonoxidizing mineral acids, but hydrochloric acid is more corrosive than other nonoxidizing mineral acids. C62400 is susceptible to dealloying, but proper heat treatment increases resistance to this type of corrosion. (For further information on the influence of heat treatment on resistance to dealloying, refer to the section on aluminum bronzes in the article "Corrosion of Copper and Its Alloys" in this volume.) Like other aluminum bronzes, C62400 is not suitable for use in an oxidizing acid such as nitric acid.

Fabrication Characteristics

Machinability. 50% of C36000, free-cutting brass; chips break readily. Carbide or tool steel cutters may be used. Typical conditions using tool steel cutters: roughing speed, 90 m/min (300 ft/min) with a feed of 0.3 mm/rev (0.011 in./rev); finishing speed, 290 m/min (950 ft/min) with a feed of 0.1 mm/rev (0.004 in./rev). Using carbide cutters (2.3- to 6.4-mm cut, or 0.09- to 0.25-in. cut): roughing speed, 53 m/min (175 ft/min) with a feed of 0.3 mm/rev (0.011 in./rev); finishing speed, 36 to 46 m/min (125 to 150 ft/min) with a feed of 0.3 mm/rev (0.011 in./rev)

Weldability. Similar to C62300.

Annealing temperature. 600 to 700 °C (1100 to 1300 °F)

Hot working temperature. 760 to 925 °C (1400 to 1700 °F)

C62500
82.7Cu-4.3Fe-13Al

Commercial Names

Trade name. Ampco 21, Wearite 4-13

Chemical Composition

Composition limits. 12.5 to 13.5 Al; 3.5 to 5.0 Fe; 2.0 max Mn; 0.5 max others (total); rem Cu

Consequence of exceeding impurity limits. Possibility of hot shortness, reduced wear resistance, increased spalling tendency and lower strength when elements such as Pb, Zn, P and Si are present in more than trace quantities

Applications

Typical uses. Guide bushings, wear strips, cams, sheet metal forming dies, forming rolls

Precautions in use. Low ductility and impact resistance make it advisable to provide adequate structural support for components made of C62500 that will be subjected to shock loads or high stress. Corrosion resistance is inferior to aluminum bronzes containing less aluminum.

Mechanical Properties

Tensile properties. Typical M30 and O61 tempers: tensile strength, 690 MPa (100 ksi); yield strength, 380 MPa (55 ksi); elongation, 1% in 50 mm or 2 in.; reduction in area, 1%

Compressive properties. Compressive strength, 450 MPa (65 ksi) at a permanent set of 0.1%; 880 MPa (128 ksi) at a permanent set of 1%.

Hardness. 27 HRC

Poisson's ratio. 0.312

Elastic modulus. Tension, 110 GPa (16×10^6 psi); shear, 42.3 GPa (6.13×10^6 psi)

Impact strength. Izod or Charpy keyhole, 3 J (2 ft·lb) at −18 to 100 °C (0 to 212 °F)

Fatigue strength. Rod, M30 temper: 460 MPa (67 ksi) at 10^8 cycles

Structure

Microstructure. Primarily metastable bcc beta phase with small crystals of ordered cph gamma phase

Magnetic Properties

Magnetic permeability. 1.2

Mass Characteristics

Density. 7.21 Mg/m³ (0.260 lb/in.³) at 20 °C (68 °F)

Thermal Properties

Liquidus temperature. 1052 °C (1925 °F)

Solidus temperature. 1047 °C (1917 °F)

Coefficient of thermal expansion. Linear, 16.2 μm/m·K (9.0 μin./in.·°F) at 20 to 300 °C (68 to 572 °F)

Specific heat. 380 J/kg·K (0.09 Btu/lb·°F) at 20 °C (68 °F)

Thermal conductivity. 38.9 W/m·K (22.5 Btu/ft·h·°F) at 20 °C (68 °F); temperature coefficient, 0.093 W/m·K per K at −100 to +150 °C (−150 to +300 °F)

Electrical Properties

Electrical conductivity. Volumetric, 8% IACS at 20 °C (68 °F)

Electrical resistivity. 172 nΩ·m at 20 °C (68 °F)

Chemical properties

General corrosion behavior. Adequate corrosion resistance to ambient moisture and industrial atmospheres. C62500 is rarely used for its corrosion characteristics in strongly corrosive environments. General corrosion characteristics are inferior to those of C62400 and C62300.

Fabrication Characteristics

Machinability. 20% of C36000, free-cutting brass

Formability. Not recommended for cold working; excellent for hot forming

Weldability. Gas-shielded arc and shielded metal-arc welding: good. Brazing and resistance welding: fair. Oxyfuel gas welding and soldering: not recommended.

Annealing temperature. 600 to 650 °C (1100 to 1200 °F)

Hot working temperature. 745 to 850 °C (1375 to 1550 °F)

C63000
82Cu-10Al-5Ni-3Fe

Commercial Names

Previous trade name. Aluminum bronze E

Common name. Nickel aluminum bronze

Specifications

AMS. Bar, shapes: 4640

ASME. Bar, rod, shapes: SB150. Condenser tube plate: SB171

ASTM. Bar, rod, shapes: B124, B150. Condenser tube plate: B171. Forgings: B283

SAE: J463

Government. Flat products, rod, shapes: QQ-C-465. Forgings: QQ-C-465, MIL-B-16166

Chemical Composition

Composition limits. 78.0 to 85.0 Cu; 9.0 to 11.0 Al; 2.0 to 4.0 Fe; 4.0 to 5.5 Ni (+Co); 1.5 max Mn; 0.30 max Zn; 0.25 max Si; 0.20 max Sn; 0.5 max others (total)

Consequence of exceeding impurity limits. Excessive amounts of Zn, Sn and Pb will cause cracking during hot working and joining. Excessive Si will result in machining difficulties.

Applications

Typical uses. Produced as rod, bar and forgings for nuts, bolts, shafting, pump parts, valve seats, faucet balls, gears, cams, structural members and tube sheets for condensers in power stations and desalting units.

Precautions in use. Not suitable for use in oxidizing acids

Mechanical Properties

Tensile properties. Typical data for 25-mm (1-in.) diam round rod, HR50 temper: tensile strength, 760 MPa (110 ksi); yield strength, 470 MPa (68 ksi); elongation, 10% in 50 mm or 2 in.; reduction in area, 10%. See also Table 91.

Compressive properties. HR50

Table 91 Typical mechanical properties of C63000 rod at various temperatures

Temperature °C	°F	Tensile strength MPa	ksi	Yield strength(a) MPa	ksi	Elongation(b), %	Reduction in area, %	Modulus of elasticity GPa	10⁶ psi	Hardness(c), HB
Cold Finished										
−182	−295	845	122.5	469	68.1	8	10	128	18.5	238
−60	−75	774	112.3	443	64.3	26	28	131	19.0	216
−29	−20	784	113.7	463	67.1	24	29	132	19.1	209
+20	+70	776	112.5	407	59.1	20	21	117	16.9	200
+204	+400	694	100.7	403	58.5	13	15	136	19.7	188
+316	+600	582	84.4	373	54.1	8	9	85	12.3	181
+427	+800	245	35.5	166	24.1	51	56	57	8.2	98
+538	+1000	107	15.5	88	12.8	41	59	44	6.4	47
Annealed										
−182	−295	867	125.8	431	62.5	12	12	130	18.8	235
−60	−75	784	113.7	379	54.9	24	26	123	17.9	212
−29	−20	781	113.3	384	55.7	23	23	139	20.1	209
+20	+70	766	111.1	370	53.6	21	21	125	18.1	200
+204	+400	706	102.4	348	50.4	16	15	107	15.5	189
+316	+600	605	87.7	337	48.9	10	11	110	15.9	176
+427	+800	232	33.7	158	22.9	41	46	64	9.3	101
+538	+1000	95	13.7	78	11.3	39	46	48	7.0	50

(a) At 0.5% extension under load. (b) In 50 mm or 2 in. (c) 3000-kg load.

Fig. 45 Variation of Charpy V-notch impact strength with temperature for C63000

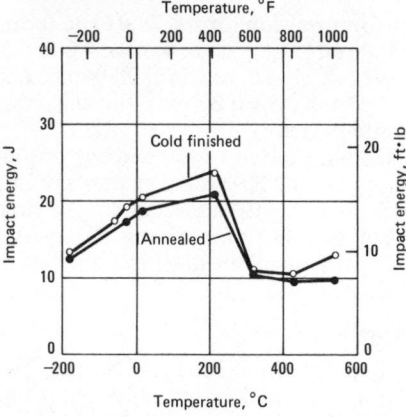

temper. Compressive strength, ultimate: 1035 MPa (150 ksi)

Hardness. HR50 temper: 94 HRB. See also Table 91.

Poisson's ratio. 0.328

Elastic modulus. Tension, 115 GPa (17×10^6 psi); shear, 44 GPa (6.4×10^6 psi). See also Table 91.

Impact strength. Charpy V-notch, 16 to 21 J (12 to 15 ft·lb) at 20 °C (68 °F). See also Fig. 45.

Fatigue strength. Reverse bending, 255 MPa (37 ksi) at 10^8 cycles

Structure

Microstructure. Alpha, kappa and metastable beta phases in various structures, depending on heat treatment and/or thermal history and composition. Normally, alpha plus alpha-kappa lamellar structure with areas of beta.

Mass Characteristics

Density. 7.58 Mg/m³ (0.274 lb/in.³) at 20 °C (68 °F)

Volume change on freezing. Approximately 2% contraction

Thermal Properties

Liquidus temperature. 1055 °C (1930 °F)

Solidus temperature. 1035 °C (1895 °F)

Coefficient of thermal expansion. Linear, 16.2 μm/m·K (9.0 μin./in.·°F) at 20 to 300 °C (68 to 572 °F).

Specific heat. 375 J/kg·K (0.09 Btu/lb·°F) at 20 °C (68 °F)

Thermal conductivity. 37.7 W/m·K (21.8 Btu/ft·h·°F) at 20 °C (68 °F);

temperature coefficient, 0.09 W/m·K per K at 20 °C (68 °F)

Electrical Properties

Electrical conductivity. Volumetric, 9% IACS at 20 °C (68 °F)

Electrical resistivity. 192 nΩ·m at 20 °C (68 °F)

Magnetic Properties

Magnetic permeability. 1.05

Chemical Properties

General corrosion behavior. See C61400.

Fabrication Characteristics

Machinability. 30% of C36000, free-cutting brass, with breaking to slightly stringy chips. Carbide or tool steel cutters may be used, and good lubrication and cooling are essential. Typical conditions using tool steel cutters: roughing speed, 20 m/min (250 ft/min) with a feed of 0.3 mm/rev (0.011 in./rev); finishing speed, 290 m/min (950 ft/min) with a feed of 0.1 mm/rev (0.004 in./rev). Using carbide cutters: roughing speed, 53 m/min (175 ft/min) with a feed of 0.3 mm/rev (0.011 in./rev); finishing speed, 38 m/min (125 ft/min) with a feed of 0.3 mm/rev (0.011 in./rev).

Forgeability. 75% of C37700, forging brass

Formability. Poor for cold working; good for hot forming

Weldability. Gas-shielded arc, coated metal-arc and spot, seam and butt resistance welding: good. Brazing: fair. Soldering and oxyacetylene welding: not recommended.

Annealing temperature. 600 to 700 °C (1100 to 1300 °F)

Hot working temperature. 800 to 925 °C (1450 to 1700 °F)

C63200
82Cu-9Al-5Ni-4Fe

Commercial Names

Common name. Nickel aluminum bronze

Chemical Composition

Composition limits. 75.9 to 84.5 Cu; 8.5 to 9.5 Al; 3.0 to 5.0 Fe(a); 4.0 to 5.5 Ni (+Co); 3.5 max Mn; 0.10 max Si; 0.02 max Pb; 0.5 max others (total)

(a) Fe content shall not exceed Ni content.

Consequence of exceeding impurity limits. Excessive Pb and Si will cause hot shortness in weld joints. Excessive Mn will reduce corrosion resistance.

Applications

Typical uses. Produced as rod, bar and forgings for nuts, bolting, shafts, pump parts, propellers and miscellaneous uses for corrosion-resistant, spark-resistant parts for industrial, marine and submarine applications

Precautions in use. Not suitable for use in oxidizing acids

Mechanical Properties

Tensile properties. (Depending on amount of cold work or heat treatment) Tensile strength, 640 to 725 MPa (93 to 105 ksi); yield strength, 330 to 380 MPa (48 to 55 ksi); elongation, 18% in 50 mm or 2 in.; reduction in area, 18%

Compressive properties. Compressive strength, ultimate: 760 MPa (110 ksi)

Hardness. 92 to 97 HRB

Poisson's ratio. 0.320

Elastic modulus. Tension, 115 GPa (17×10^6 psi); shear, 44 GPa (6.4×10^6 psi)

Impact strength. Charpy V-notch, 23 to 27 J (17 to 20 ft·lb) at −29 to +20 °C (−20 to +78 °F); Charpy keyhole, 13 to 16 J (10 to 12 ft·lb) at −29 to +20 °C (−20 to +78 °F)

Structure

Microstructure. Alpha, kappa and metastable beta phases in various structures, depending on heat treatment and/or thermal history and composition. Normally, alpha plus alpha-kappa lamellar structure with or without areas of beta phase.

Mass Characteristics

Density. 7.64 Mg/m³ (0.276 lb/in.³) at 20 °C (68 °F)

Volume change on freezing. Approximately 2% contraction

Thermal Properties

Liquidus temperature. 1060 °C (1940 °F)

Solidus temperature. 1040 °C (1905 °F)

Coefficient of thermal expansion. Linear, 16.2 μm/m·K (9.0 μin./in.·°F) at 20 to 300 °C (68 to 572 °F)

Specific heat. 439 J/kg·K (0.105 Btu/lb·°F) at 20 °C (68 °F)

Thermal conductivity. 36 W/m·K (21 Btu/ft·h·°F) at 20 °C (68 °F)

Electrical Properties

Electrical conductivity. Volumetric, 7% IACS at 20 °C (68 °F)
Electrical resistivity. 246 nΩ·m at 20 °C (68 °F)

Magnetic Properties

Magnetic permeability. 1.04

Fabrication Characteristics

Machinability. Fair. Tendency to form stringy chips and to gall makes good lubrication and cooling essential. Tool steel or carbide cutters may be used. Good finish and fine thread tapping possible. Typical conditions using tool steel cutters: roughing speed, 76 m/min (250 ft/min) with a feed of 0.3 mm/rev (0.011 in./rev); finishing speed, 290 m/min (950 ft/min) with a feed of 0.1 mm/rev (0.004 in./rev). Using carbide cutters: roughing speed, 53 m/min (175 ft/min) with a feed of 0.3 mm/rev (0.011 in./rev); finishing speed, 38 m/min (125 ft/min) with a feed of 0.3 mm/rev (0.011 in./rev).
Annealing temperature. 705 to 880 °C (1300 to 1615 °F)
Hot working temperature. 705 to 925 °C (1300 to 1700 °F)

C63600
95.5Cu-3.5Al-1.0Si

Chemical Composition

Composition limits. 93.5 to 96.3 Cu; 3.0 to 4.0 Al; 0.7 to 1.3 Si; 0.50 max Zn, 0.20 max Sn; 0.15 max Ni; 0.15 max Fe; 0.05 max Pb

Applications

Typical uses. Rod and wire for components for pole line hardware; cold-headed nuts for wire and cable connectors; bolts; screw machine products

Mechanical Properties

Tensile properties. See Table 92.
Hardness. See Table 92.
Elastic modulus. Tension, 110 GPa (16 × 10^6 psi)

Mass Characteristics

Density. 8.33 Mg/m^3 (0.301 lb/in.^3) at 20 °C (68 °F)

Thermal Properties

Liquidus temperature. 1035 °C (1890 °F)

Table 92 Typical mechanical properties of C63600

Size	Temper	Tensile strength MPa	ksi	Elongation(a), %	Hardness, HRB
Rod					
16-mm (0.63-in.) diam O61		415	60	64	...
14-mm (0.56-in.) diam H01		510	74	31	...
Wire					
10-mm (0.40-in.) diam O61		415	60	67	...
11-mm (0.42-in.) diam .... H00(7%)		470	68	52	71
12-mm (0.49-in.) diam .... H01(21%)		580	84	29	84

(a) In 50 mm or 2 in.

Coefficient of thermal expansion. Linear, 17.2 μm/m·K (9.4 μin./in.·°F) at 20 to 300 °C (68 to 572 °F)
Thermal conductivity. 57 W/m·K (33 Btu/ft·h·°F) at 20 °C (68 °F)

Electrical Properties

Electrical conductivity. Volumetric, 12% IACS at 20 °C (68 °F)
Electrical resistivity. 143 nΩ·m at 20 °C (68 °F)

Fabrication Characteristics

Machinability. 40% of C36000, free-cutting brass
Formability. Excellent for cold working; fair for hot forming
Weldability. Gas-shielded arc, shielded metal-arc and resistance welding: fair. Soldering, brazing and oxyfuel gas welding: not recommended.
Hot working temperature. 760 to 875 °C (1400 to 1600 °F)

C63800
95Cu-2.8Al-1.8Si-0.40Co

Commercial Names

Trade name. Coronze

Chemical Composition

Composition limits. 2.5 to 3.1 Al; 1.5 to 2.1 Si; 0.25 to 0.55 Co; 0.80 max Zn; 0.10 max Ni; 0.05 max Pb; 0.10 max Fe; 0.10 max Mn; rem Cu

Applications

Typical uses. Springs, switch parts, contacts, relay springs, glass sealing and porcelain enameling

Mechanical Properties

Tensile properties. See Table 93 and Fig. 46 and 47.
Hardness. See Table 93.
Poisson's ratio. 0.312
Elastic modulus. Tension, 11.7 GPa (16.7 × 10^6 psi)

Mass Characteristics

Density. 8.28 Mg/m^3 (0.299 lb/in.^3) at 20 °C (68 °F)

Thermal Properties

Liquidus temperature. 1030 °C (1885 °F)
Solidus temperature. 1000 °C (1830 °F)
Coefficient of thermal expansion. Linear, 17.1 μm/m·K (9.5 μin./in.·°F) at 20 to 300 °C (68 to 572 °F)
Specific heat. 375 J/kg·K (0.09 Btu/lb·°F) at 20 °C (68 °F)
Thermal conductivity. 42 W/m·K (24 Btu/ft·h·°F) at 20 °C (68 °F)

Electrical Properties

Electrical conductivity. Volumetric, 10% IACS at 20 °C (68 °F), annealed
Electrical resistivity. 174 nΩ·m at 20 °C (68 °F), annealed

Chemical Properties

General corrosion behavior. C63800 is more resistant to stress corrosion than the nickel silvers, approaching the performance of the highly resistant phosphor bronzes. This alloy is far superior to most other copper alloys in resistance to crevice corrosion. At elevated temperature, the oxidation resistance of C63800 is excellent. For instance,

Fig. 46 Typical short-time tensile properties of C63800, H02 temper

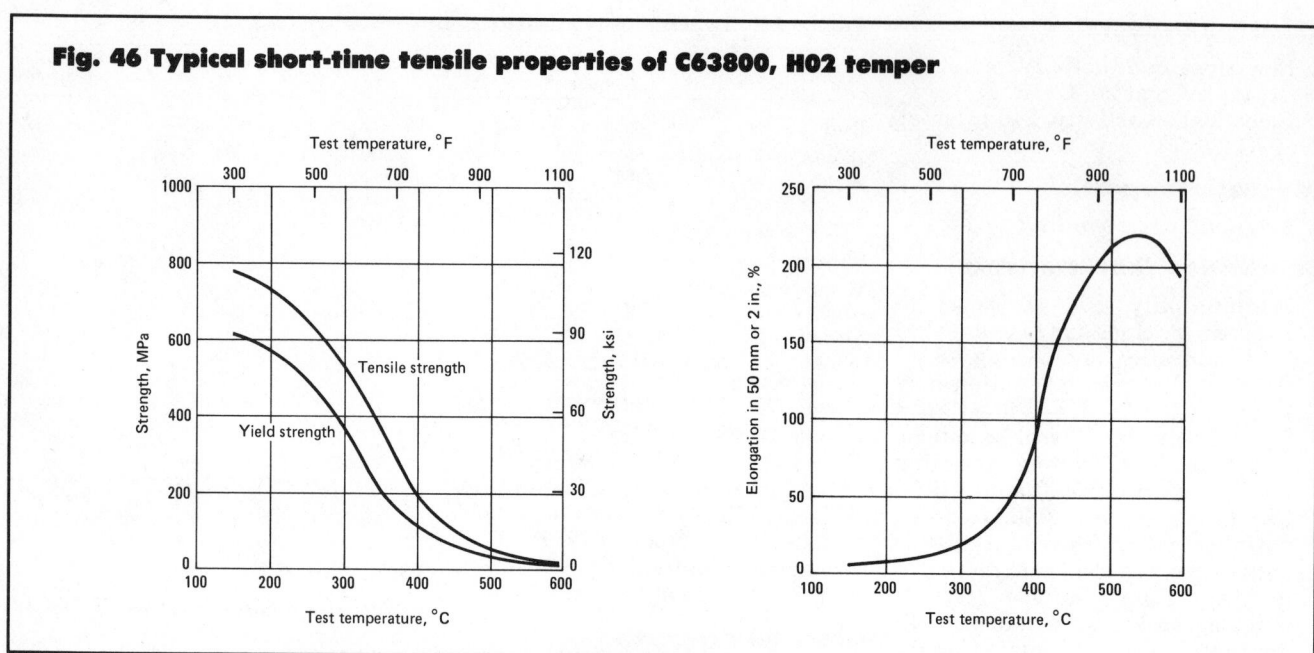

Fig. 47 Anneal resistance of C63800 strip, H08 temper

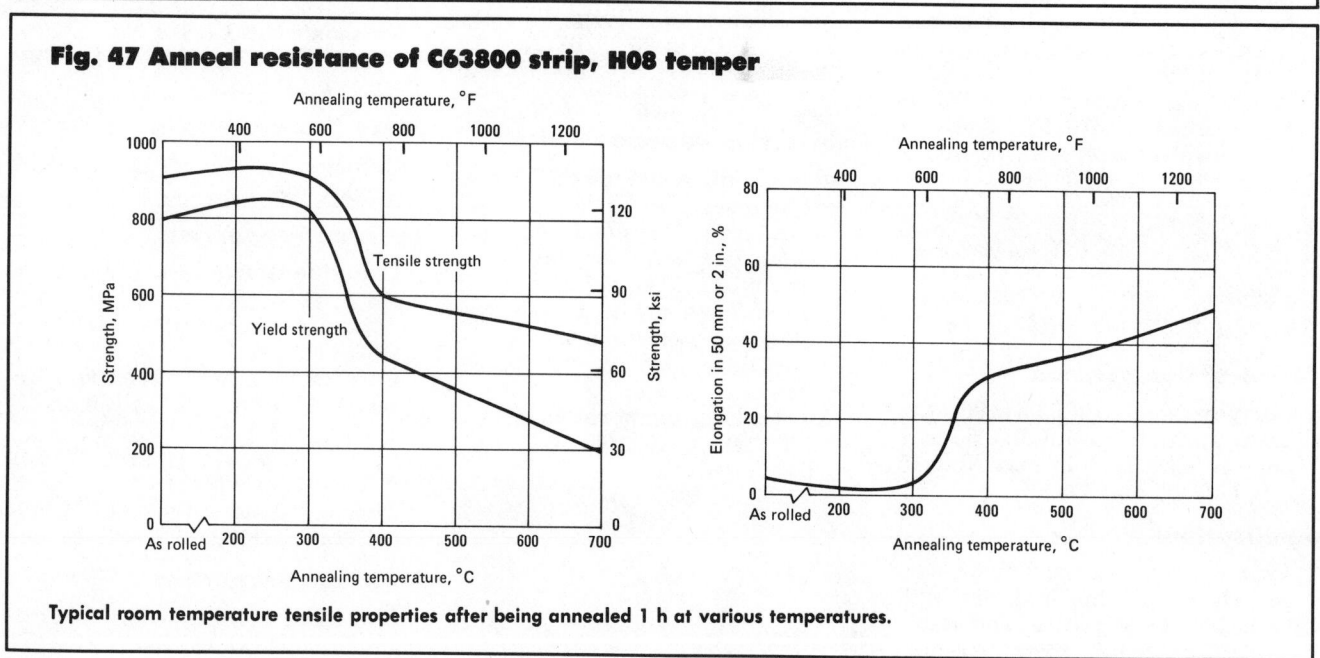

Typical room temperature tensile properties after being annealed 1 h at various temperatures.

Table 93 Typical mechanical properties of C63800 sheet and strip

Temper	Tensile strength MPa	ksi	Yield strength(a) MPa	ksi	Elongation(b), %	Hardness HRB	HR30T
O61	565	82	385	56	33	. . .	74
H01	660	96	565	82	17	94	78
H02	730	106	640	93	10	97	80
H03	765	111	680	99	8	98	81
H04	825	120	750	109	5	99	82
H06	855	124	780	113	4	100	82
H08	895	130	800	116	3	100	83
H10	895 min	130 min	820 min	119 min	2 max	100 min	83 min

(a) At 0.2% offset. (b) In 50 mm or 2 in.

after heating in air for 2 h, the film thickness on C83800 was 7 nm (0.26 μin.) at 450 °C, 12 nm (0.47 μin.) at 600 °C, 24 nm (0.94 μin.) at 700 °C and 24 nm at 800 °C. On the basis of weight gain after heating 2 to 24 h in air at temperatures of 600 to 800 °C, C63800 was consistently superior to Nickel 270, Nichrome (80Ni-20Cr), type 301 stainless steel, Incoloy 800 (ASTM B408), and C60600. The superiority of C63800 was especially evident at 800 °C.

Fabrication Characteristics

Formability. Suitable for blanking, drawing, bending, shearing and stamping. Excellent for cold working and hot forming

Weldability. Soft soldering utilizing standard fluxes normally employed. Brazing, gas-shielded arc welding and all forms of resistance welding also commonly used

Annealing temperature. 400 to 600 °C (750 to 1100 °F). See also Fig. 47.

C65100
98.5Cu-1.5Si

Commercial Names

Previous trade name. Low-silicon bronze B

Common name. Low-silicon bronze

Specifications

ASME. Bar, rod, shapes: SB98. Tubular products: SB315

ASTM. Flat products: B97. Bar, rod, shapes: B98. Tubular products: B315. Wire: B99

Government. QQ-C-591

Chemical Composition

Composition limits. 0.8 to 2.0 Si; 0.05 max Pb; 0.8 max Fe; 1.5 max Zn; 0.7 max Mn; rem Cu

Applications

Typical uses. Aircraft: hydraulic pressure lines. Hardware: anchor screws, bolts, cable clamps, cap screws, machine screws, marine hardware, nuts, pole line hardware, rivets, U-bolts. Industrial: electrical conduits, heat exchanger tubes, welding rod.

Mechanical Properties

Tensile properties. See Table 94.
Shear strength. See Table 94.
Hardness. See Table 94.
Elastic modulus. Tension, 115 GPa (17×10^6 psi); shear, 44 GPa (6.4×10^6 psi)
Fatigue strength. Reverse bending, H04 temper, 170 MPa (25 ksi) at 10^8 cycles; H06 temper, 195 MPa (28 ksi) at 10^8 cycles

Mass Characteristics

Density. 8.75 Mg/m^3 (0.316 lb/in.^{3}) at 20 °C (68 °F)

Table 94 Typical mechanical properties of C65100

Temper	Tensile strength		Yield strength(a)		Elonga- tion(b), %	Hardness	Shear strength	
	MPa	ksi	MPa	ksi			MPa	ksi
Rod: 25-mm (1-in.) Thick								
OS035	275	40	105	15	50	55 HRF	...	...
H04(36%)	485	70	380	55	15	80 HRB	310	45
H06(50%)	620	90	460	67	12	90 HRB	345	50
Wire: 2-mm (0.08-in.) Diam								
H00	380	55	275	40	40	...	250	36
H01	450	65	345	50	25	...	275	40
H02	550	80	435	63	15	...	310	45
H04	690	100	485	70	11	...	345	50
H06	725	105	490	71	10	...	365	53
Wire: 11-mm (0.44-in.) Diam								
H00(21%)	435	63	...	...	30	...	...	...
H02(37%)	550	80	...	...	20	...	...	...
H04(60%)	655	95	...	...	12	...	...	...
Tubing: 25-mm (1.0-in.) OD by 1.65-mm (0.065-in.) Wall								
OS015	310	45	140	20	55	68 HRF	...	...
H80(35%)	450	65	275	40	20	75 HRB	...	...

(a) At 0.5% extension under load. (b) In 50 mm or 2 in.

Thermal Properties

Liquidus temperature. 1060 °C (1940 °F)
Solidus temperature. 1030 °C (1890 °F)
Coefficient of thermal expansion. Linear, 18 μm/m·K (9.9 μin./in.·°F) at 20 to 300 °C (68 to 572 °F)
Specific heat. 380 J/kg·K (0.09 Btu/lb·°F) at 20 °C (68 °F)
Thermal conductivity. 57 W/m·K (33 Btu/ft·h·°F) at 20 °C (68 °F)

Electrical Properties

Electrical conductivity. Volumetric, 12% IACS at 20 °C (68 °F)
Electrical resistivity. 144 nΩ·m at 20 °C (68 °F)

Fabrication Characteristics

Machinability. 30% of C36000, free-cutting brass
Formability. Excellent for cold working and hot forming
Weldability. Soldering, brazing, gas-shielded arc, resistance spot and resistance butt welding: excellent. Oxyfuel gas and resistance seam welding: good. Shielded metal-arc welding: fair.
Annealing temperature. 475 to 675 °C (900 to 1250 °F)
Hot working temperature. 700 to 875 °C (1300 to 1600 °F)

C65500
97Cu-3Si

Commercial Names

Previous trade name. High-silicon bronze A
Common name. High-silicon bronze

Specifications

AMS. Bar, rod: 4615. Tubing: 4665
ASME. Flat products: SB96. Bar, rod, shapes: SB98. Tubular products: SB315
ASTM. Flat products: B96, B97, B100. Bar, rod, shapes: B98, B124. Forgings: B283. Tubular products: B315. Wire: B99
SAE. J463
Government. QQ-C-591. Tubing: MIL-T-8231

Chemical Composition

Composition limits. 2.8 to 3.8 Si; 0.5 max Pb; 0.8 max Fe; 1.5 max Zn; 1.5 max Mn; 0.6 max Ni; rem Cu

Applications

Typical uses. Aircraft: hydraulic pressure lines. Hardware: bolts, burrs, butts, clamps, cotter pins, hinges, marine hardware, nails, nuts, pole line hardware, screws. Industrial: bearing plates, bushings,

Table 95 Typical mechanical properties of C65500

Temper	Tensile strength MPa	ksi	Yield strength(a) MPa	ksi	Elongation(b), %	Hardness, HRB	Shear strength MPa	ksi
Flat Products: 1-mm (0.04-in.) Thick								
OS070	385	56	145	21	63	40	290	42
OS035	415	60	170	25	60	62	295	43
OS015	435	63	205	30	55	66	310	45
H01	470	68	240	35	30	75	325	47
H02	540	78	310	45	17	87	345	50
H04	650	94	400	58	8	93	390	57
H06	715	104	415	60	6	96	415	60
H08	760	110	427	62	4	97	435	63
Rod: 25-mm (1.0-in.) Diam								
OS050	400	58	150	22	60	60	295	43
H02(20%)	540	78	310	45	35	85	360	52
H04(36%)	635	92	380	55	22	90	400	58
H06(50%)	745	108	415	60	13	95	425	62
Wire: 2-mm (0.08-in.) Diam								
OS035	415	60	170	25	60	...	295	43
H00	485	70	275	40	35	...	330	48
H01	550	80	330	48	20	...	360	52
H02	675	98	395	57	8	...	400	58
H04	860	125	450	65	5	...	450	65
H08(80%)	1000	145	485	70	3	...	485	70
Tubing: 25-mm (1.0-in.) OD, 1.65-mm (0.065-in.) Wall								
OS050	395	57	...	...	70	45	...	...
H80(35%)	640	93	...	...	22	92	...	...

(a) At 0.5% extension under load. (b) In 50 mm or 2 in.

cable, channels, chemical equipment, heat exchanger tubes, kettles, piston rings, tanks, rivets, screen cloth and wire, screen plates, shafting. Marine: propeller shafts.

Mechanical Properties

Tensile properties. See Table 95.
Shear strength. See Table 95.
Hardness. See Table 95.
Elastic modulus. Tension, 105 GPa (15 × 10^6 psi); shear, 39 GPa (5.6 × 10^6 psi)
Fatigue strength. Reverse bending, H04 temper, 200 MPa (29 ksi) at 10^8 cycles; H08 temper, 205 MPa (30 ksi) at 10^8 cycles

Mass Characteristics

Density. 8.53 Mg/m^3 (0.308 lb/in.^{3}) at 20 °C (68 °F)

Thermal Properties

Liquidus temperature. 1025 °C (1880 °F)
Solidus temperature. 970 °C (1780 °F)
Coefficient of thermal expansion.
Linear, 18 μm/m·K (10 μin./in.·°F) at 20 to 300 °C (68 to 572 °F)
Specific heat. 380 J/kg·K (0.09 Btu/lb·°F) at 20 °C (68 °F)
Thermal conductivity. 36 W/m·K (21 Btu/ft·h·°F) at 20 °C (68 °F)

Electrical Properties

Electrical conductivity. Volumetric, 7% IACS at 20 °C (68 °F)
Electrical resistivity. 246 nΩ·m at 20 °C (68 °F)

Fabrication Characteristics

Machinability. 30% of C36000, free-cutting brass
Forgeability. 40% of C37700, forging brass
Formability. Excellent for cold working and hot forming
Weldability. Brazing, gas-shielded arc and all forms of resistance welding: excellent. Soldering and oxyfuel gas welding: good. Shielded metal-arc welding: fair.
Annealing temperature. 475 to 700 °C (900 to 1300 °F)
Hot working temperature. 700 to 875 °C (1300 to 1600 °F)

C66400
86.5Cu-1.5Fe-0.5Co-11.5Zn

Commercial Names

Previous trade name. Cobron

Chemical Composition

Composition limits. 1.3 to 1.7 Fe; 0.30 to 0.70 Co; 11.0 to 12.0 Zn; 0.05 max Sn; 0.05 max Ni; 0.05 max Al; 0.05 max Mn; 0.05 max Si; 0.05 max Ag; 0.02 max P; 0.015 max Pb; rem Cu. Note: the Fe + Co content shall be 1.8 to 2.0 (total).

Applications

Typical uses. Spring washers, switch blades, fuse clips, contact springs, socket contacts, connectors, terminals and similar parts for electronic and electromechanical assemblies

Mechanical Properties

Tensile properties. See Table 96.
Elastic modulus. Tension, 11.5 GPa (16.3 × 10^6 psi)
Fatigue strength. Reverse bending, O60 temper, 165 MPa (24 ksi) at 10^8 cycles; H04 temper, 185 MPa (27 ksi) at 10^8 cycles

Mass Characteristics

Density. 8.74 Mg/m^3 (0.317 lb/in.^{3}) at 20 °C (68 °F)

Thermal Properties

Liquidus temperature. 1055 °C (1930 °F)
Solidus temperature. 1035 °C (1895 °F)
Thermal conductivity. 116 W/m·K (67 Btu/ft·h·°F) at 20 °C (68 °F)

Electrical Properties

Electrical conductivity. Volumetric, O61 temper: 30% IACS at 20 °C (68 °F)
Electrical resistivity. O61 temper: 57.5 nΩ·m at 20 °C (68 °F)

C68800
73.5Cu-22.7Zn-3.4Al-0.4Co

Commercial Names

Trade name. Alcoloy

Specifications

ASTM. Flat products: B592

Table 96 Typical mechanical properties of C66400

Temper	Tensile strength MPa	ksi	Yield strength(a) MPa	ksi	Elongation(b), %
O60	435	63	310	45	25
H01	495	72	455	66	13
H02	545	79	525	76	7
H03	570	83	560	81	6
H04	605	88	585	85	5
H06	650	94	615	89	4
H08	670	97	635	92	3
H10	690	100	640	93	3

(a) At 0.2% offset. (b) In 50 mm or 2 in.

Table 97 Typical mechanical properties of C68800 strip(a)

Temper	Tensile strength MPa	ksi	Yield strength(b) MPa	ksi	Elongation(c), %	Hardness HRB	HR30T
O60(d)	565	82	365	53	35	78	69
O50	615	89	475	69	30	...	...
H01	650	94	525	76	20	90.5	78
H02	725	105	635	92	9	95	81
H04	780	113	705	102	5	97	82.5
H06	825	120	750	109	3	98	83
H08	885	128	785	114	2	99	83.5
H10	895 min	130 min	805 min	117 min	2 max	99 max	84 min

(a) 1 mm (0.04 in.) thick. (b) At 0.2% offset. (c) In 50 mm or 2 in. (d) Normally, annealed C68800 is very fine grained—0.010 mm or less.

Table 98 Typical mechanical properties of C68800 after low temperature thermal treatment

	As rolled				Stabilization treated(a)			
Temper	Tensile strength MPa	ksi	Yield strength(b) MPa	ksi	Tensile strength MPa	ksi	Yield strength(b) MPa	ksi
H02(c)	695	101	640	93	725	105	690	100
H04	750	109	670	97	780	113	740	107
H06	840	122	760	110	910	132	895	130
H08	890	129	785	114	960	139	925	134
H10	895 min	130 min	805 min	117 min	965	140	945	137

(a) Heated 1 h at 205 to 230°C (400 to 250°F). (b) At 0.2% offset. (c) Stabilization treatment is not effective on H00 or H01 temper material.

Chemical Composition

Composition limits. 72.3 to 74.7 Cu; 3.0 to 3.8 Al; 0.25 to 0.55 Co; 0.05 max Pb; 0.05 max Fe; 0.010 max others (total); rem Zn(a)

(a) 25.1 to 27.1 Al + Zn

Applications

Typical uses. Springs, switches, contacts, relays, terminals, plug recepticals, connectors

Mechanical Properties

Tensile properties. See Tables 97 and 98.
Hardness. See Table 97.
Elastic modulus. Tension, 11.8 GPa (16.8 × 10^6 psi)

Mass Characteristics

Density. 8.20 Mg/m³ (0.296 lb/in.³) at 20 °C (68 °F)

Thermal Properties

Liquidus temperature. 965 °C (1765 °F)
Solidus temperature. 950 °C (1740 °F)

Coefficient of thermal expansion. Linear, 18.2 μm/m·K (10.1 μin./in.·°F) at 20 to 300 °C (68 to 572 °F)
Specific heat. 375 J/kg·K (0.09 Btu/lb·°F) at 20 °C (68 °F)
Thermal conductivity. 69 W/m·K (40 Btu/ft·h·°F) at 20 °C (68 °F)

Electrical Properties

Electrical conductivity. Volumetric: O61 temper, 18% IACS at 20 °C (68 °F) H08 temper, 16.6% IACS at 20 °C (68 °F)
Electrical resistivity. O61 temper, 96 nΩ·m at 20 °C (68 °F); H08 temper, 104 nΩ·m

Magnetic Properties

Magnetic permeability. 1.003

Chemical Properties

General corrosion behavior. C68800 is more resistant than C26000 to both corrosion and stress-corrosion cracking.

Fabrication Characteristics

Formability. Suitable for blanking, drawing, bending, shearing and stamping. Bending characteristics are nearly nondirectional for all annealed and rolled tempers. Excellent for cold working and hot forming.
Weldability. Can be joined by soft soldering when mildly activated commercial fluxes are used and exhibits substantially better tarnish resistance than most other copper alloys. Can also be joined by brazing and resistance welding.
Annealing temperature. 400 to 600 °C (750 to 1100 °F)
Order strengthening. When heated to 220 °C (425 °F), temper rolled C68800 undergoes an ordering reaction that increases strength (see Table 98) and decreases ductility. Because this decrease in ductility would adversely affect formability, parts should be order strengthened after forming. Specific times and temperatures for thermal treatment may vary, depending on cold worked temper. Susceptibility to stress-corrosion cracking increases dramatically with an increase in the degree of ordering.
Stabilization treatment. Stabilization treatment is performed when enhanced stress relaxation is desired. This treatment causes little change in the 0.2% offset yield strength. Temperature and time of the treatment vary, depending on cold worked temper; the ranges are 280 to 320 °C (535 to 610 °F) and 10 min to 2 h, respectively. To gain the maximum benefit from a stabilization treatment, parts should be stabilized after forming. There is no increase in stress corrosion susceptibility as a result of this treatment.

C69000
73.3Cu-22.7Zn-3.4Al-0.6Ni

Chemical Composition

Composition limits. 72 to 74.5 Cu; 3.3 to 3.5 Al; 0.50 to 0.70 Ni; 0.05 max Fe; 0.025 max Pb; rem Zn

Applications

Typical uses. Electrical component parts, contacts, connectors, switches, relays, springs, high-strength shells

Mechanical Properties

Tensile properties. See Table 99.
Hardness. See Table 99.
Elastic modulus. Tension, 115 GPa $(16.7 \times 10^6$ psi)

Mass Characteristics

Density. 8.19 Mg/m^3 (0.296 lb/in.^{3}) at 20 °C (68 °F)

Thermal Properties

Liquidus temperature. 960 °C (1760 °F)
Solidus temperature. 950 °C (1745 °F)
Coefficient of thermal expansion. Linear, 18 μm/m·K (10 μin./in.·°F) at 20 to 300 °C (68 to 572 °F)
Specific heat. 380 J/kg·K (0.09 Btu/lb·°F) at 20 °C (68 °F)
Thermal conductivity. 40 W/m·K (23 Btu/ft·h·°F) at 20 °C (68 °F)

Electrical Properties

Electrical conductivity. Volumetric, O61 temper: 18% IACS at 20 °C (68 °F)
Electrical resistivity. O61 temper: 96 nΩ·m

Chemical Properties

General corrosion behavior. Significantly better corrosion performance than C26000, both in uniform corrosion rate and stress-corrosion resistance

Fabrication Characteristics

Formability. Behavior in blanking, drawing, forming, bending, stamping and other cold forming operations is similar to C26000, but with lower directionality in bending of cold worked tempers. Excellent for cold working and hot forming.
Weldability. Resistance welding: good. Soldering and brazing: fair pro-vided that active flux is used. Oxy-fuel gas and arc welding: not recommended.
Annealing temperature. 400 to 590 °C (750 to 1100 °F); stress relief anneal, 225 °C (435 °F) for 1 h
Hot working temperature. 790 to 840 °C (1450 to 1550 °F)

Table 99 Typical mechanical properties of C69000

Temper	Tensile strength MPa	ksi	Yield strength(a) MPa	ksi	Elongation(b), %	Hardness HRB	HR30T
OS025	565	82	360	52	35	...	69
H01	650	94	525	76	19.5	90.5	...
H02	715	105	635	92	9	95	...
H04	780	113	700	102	4.5	97	...
H06	825	120	750	109	2.5	98	...
H08	875	126	785	114	1.5	96	...
H10	895 min	130 min	805	117	2 max	99 min	...
EHT(c)	930	135	880	127	1	...	84.5

(a) At 0.2% offset. (b) In 50 mm or 2 in. (c) Cold rolled 50% and stress relief annealed 1 h at 220°C (425°F).

C69400
81.5Cu-14.5Zn-4Si

Commercial Names

Previous trade name. Silicon red brass, CA694

Specifications

ASTM. Rod: B371 (CA694)

Chemical Composition

Composition limits. 80.0 to 83.0 Cu, 0.30 max Pb, 0.20 max Fe, 3.5 to 4.5 Si, rem Zn

Applications

Typical uses. Valve stems requiring a combination of corrosion resistance and high strength; forged or screw-machined parts

Mechanical Properties

Tensile properties. See Table 100.
Hardness. See Table 100.
Elastic modulus. Tension, 110 GPa $(16 \times 10^6$ psi)

Mass Characteristics

Density. 8.19 Mg/m^3 (0.296 lb/in.^{3}) at 20 °C (68 °F)

Thermal Properties

Liquidus temperature. 920 °C (1685 °F)
Solidus temperature. 820 °C (1510 °F)
Coefficient of thermal expansion. Linear, 20.2 μm/m·K (11.2 μin./in.·°F) at 20 to 300 °C (68 to 572 °F)
Specific heat. 380 J/kg·K (0.09 Btu/lb·°F) at 20 °C (68 °F)
Thermal conductivity. 26 W/m·K (15 Btu/ft·h·°F)

Electrical Properties

Electrical conductivity. Volumetric, 6.2% IACS at 20 °C (68 °F)
Electrical resistivity. 280 nΩ·m at 20 °C (68 °F)

Fabrication Characteristics

Machinability. 30% of C36000, free-cutting brass
Forgeability. 80% of C37700, forging brass
Formability. Excellent capacity for being hot formed; poor for being cold formed
Weldability. Soft soldering and silver alloy brazing: excellent. Oxyfuel-gas welding and resistance welding: good. Arc welding: not recommended.
Annealing temperature. 425 to 650 °C (800 to 1200 °F)
Hot working temperature. 650 to 875 °C (1200 to 1600 °F)

Table 100 Typical mechanical properties of copper alloy C69400 rod

Temper	Section size mm	in.	Tensile strength MPa	ksi	Yield strength (0.5% ext. under load) MPa	ksi	Elongation(a), %	Hardness, HRB
O60	13	0.5	621	90	310	45	20	85
	25	1.0	586	85	296	43	25	85
	51	2.0	532	80	276	40	25	85
H00	19	0.75	689	100	393	57	21	95

(a) In 50 mm or 2 in.

Table 101 Typical mechanical properties of C70400

Temper	Tensile strength MPa	ksi	Yield strength At 0.5% extension under load MPa	ksi	At 0.2% offset MPa	ski	Elongation in 50 mm or 2 in., %	HRF	Hardness HRB	HR30T
Strip										
O61 260	38	83		12	. . .	. . .	41	. . .	8	. . .
H01 350	51	. . .		. . .	275	40	21	. . .	54	57
H02 395	57	. . .		. . .	380	55	11	. . .	67	65
H04 440	64	. . .		. . .	435	63	5	. . .	72	68
H06 485	70	. . .		. . .	475	69	3	. . .	75	69
H08 530	77	. . .		. . .	525	76	2 min	. . .	76 min	70 min
Tubing: 25-mm (1.0-in.) OD, 1.65-mm (0.065-in.) Wall										
OS015 285	41	97		14	. . .	. . .	46	58	. . .	. . .
H55 330	48	250		36	. . .	. . .	18	67	. . .	. . .

C70400
92.4Cu-5.5Ni-1.5Fe-0.6Mn

Specifications

ASTM. Pipe: B466. Tubing: B111, B359, B395, B466, B543.

Chemical Composition

Composition limits. 91.2 min Cu; 4.8 to 6.2 Ni; 1.3 to 1.7 Fe; 0.3 to 0.8 Mn; 1.0 max Zn; 0.05 max Pb

Applications

Typical uses. Rolled strip, sheet and tubing for industrial uses: condensers, condenser plates, evaporator and heat exchanger tubes, ferrules, salt water piping, lithium bromide absorption system tubing, shipboard condenser intake systems

Mechanical Properties

Tensile properties. See Table 101.
Hardness. See Table 101.
Elastic modulus. Tension, 115 GPa (17 × 10⁶ psi); shear, 44 GPa (6.4 × 10⁶ psi)

Mass Characteristics

Density. 8.94 Mg/m³ (0.323 lb/in.³) at 20 °C (68 °F)

Thermal Properties

Liquidus temperature. 1125 °C (2050 °F)
Coefficient of thermal expansion. Linear, 17.5 μm/m·K (9.7 μin./in.·°F) at 20 to 300 °C (68 to 572 °F)
Specific heat. 380 J/kg·K (0.09 Btu/lb·°F) at 20 °C (68 °F)
Thermal conductivity. 64 W/m·K (37 Btu/ft·h·°F) at 20 °C (68 °F)

Electrical Properties

Electrical conductivity. Volumetric, 14% IACS at 20 °C (68 °F)

Electrical resistivity. 117 nΩ·m at 20 °C (68 °F)

Fabrication Characteristics

Machinability. 20% of C36000, freecutting brass
Formability. Excellent for cold working; good for hot forming
Weldability. Soldering, brazing and gas-shielded arc welding: excellent. Coated metal-arc and resistance spot, seam and butt welding: good. Oxyacetylene welding: fair.

Annealing temperature. 575 to 825 °C (1050 to 1500 °F)
Hot working temperature. 825 to 950 °C (1500 to 1750 °F)

C70600
90Cu-10Ni

Commercial Names

Previous trade names. Copper nickel, 10%; CA706
Common name. 90-10 cupronickel

Specifications

ASME. Flat products: SB171, SB402. Pipe: SB466, SB 467. Tubing: SB111, SB359, SB395, SB466, SB467, SB543
ASTM. Flat products: B122, B171, B402, B432. Pipe: B466, B467. Rod: B151. Tubing: B111, B359, B395, B466, B467, B543, B552
SAE. Plate and tubing: J463
Government. Bar, flat products, forgings, rod: MIL-C-15726 E(2). Tubing: MIL-T-16420 J(3), MIL-T-1368 C(2), MIL-T-23520 A(4). Condenser tubing: MIL-T-15005 F

Chemical Composition

Composition limits. 0.05 max Pb, 1 to 1.8 Fe, 1.0 max Zn, 9 to 11 Ni, 1.0 max Mn, 0.5 max others (total), rem Cu

Applications

Typical uses. Condensers, condenser plates, distiller tubes, evaporator and heat-exchanger tubes, ferrules, salt water piping, boat hulls

Mechanical Properties

Tensile properties. See Table 102 and Fig. 48.
Elastic modulus. Tension, 140 GPa (20 × 10⁶ psi); shear, 52 GPa (7.5 × 10⁶ psi)
Fatigue strength. Tubing, H55 temper: 138 MPa (20 ksi) at 10⁸ cycles

Mass Characteristics

Density. 8.94 Mg/m³ (0.323 lb/in.³) at 20 °C (68 °F)

Thermal Properties

Liquidus temperature. 1150 °C (2100 °F)
Solidus temperature. 1100 °C (2010 °F)
Coefficient of thermal expansion. 17.1 μm/m·K (9.5 μin./in.·°F) at 20 to 300 °C (68 to 572 °F)
Specific heat. 380 J/kg·K (0.09 Btu/lb·°F) at 20 °C (68 °F).
Thermal conductivity. 40 W/m·K (23 Btu/ft·h·°F) at 20 °C (68 °F)

Electrical Properties

Electrical conductivity. Volumetric, 9.1% IACS
Electrical resistivity. 191 nΩ·m at 20 °C (68 °F)

Fig. 48 Mechanical properties of cold drawn C70600 tubing

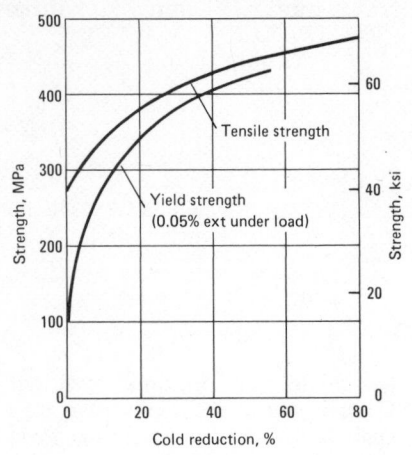

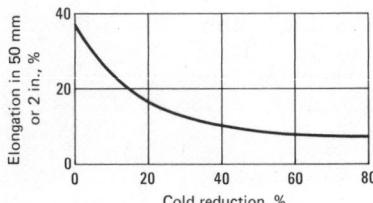

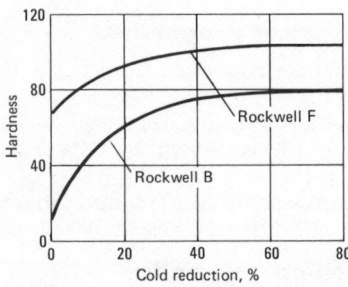

Variation of mechanical properties with amount of cold reduction for C70600 tubing. 60 mm (2³⁄₈ in.) diam by 4.8 mm (³⁄₁₆) wall.

Table 102 Typical mechanical properties of C70600 and C71000

Temper	Tensile strength MPa	ksi	Yield strength 0.5% extension under load MPa	ksi	0.2% offset MPa	ksi	Elongation, %(a)	Hardness HRF	HRB
Flat products, 1 mm (0.04 in.) Thick									
OS050	350	51	90	13	90	13	35	72	25
OS035	358	52	98	14	98	14	35	73	27
OS025	365	53	110	16	110	16	35	75	30
H01	415	60	330	48	338	49	20	92	58
H02	468	68	430	62	433	63	8	100	75
H04	518	75	490	71	500	72	5	...	80
H06	540	78	518	75	525	76	4	...	82
H08	565	82	540	78	545	79	3	...	84
H10	585	85	540	78	545	79	3	...	86
Tubing, 25 mm (1 in.) OD by 1.65 mm (0.065 in.) Wall									
OS025	338	49	125	18	...	...	40	72	25
H55	468	68	430	62	...	...	14	...	76
Wire, 2 mm (0.080 in.) Diam									
H10	655	95	585	85	...	...	5	...	...

(a) In 50 mm or 2 in.

welding: excellent. Shielded metal arc and resistance spot and seam welding: good. Oxyfuel gas welding: fair.

Annealing temperature. 600 to 825 °C (1100 to 1500 °F)
Hot working temperature. 850 to 950 °C (1550 to 1750 °F)

Optical Properties

Color. Pink silver

Chemical Properties

Resistance to specific agents. Excellent resistance to seawater

Fabrication Characteristics

Machinability. 20% of C36000, free-cutting brass
Formability. Good capacity for being both cold worked and hot formed
Weldability. Soldering, brazing, gas shielded arc and resistance butt

C71000
80Cu-20Ni

Commercial Names

Previous trade names. Copper nickel, 20%; CA710
Common name. 80-20 cupronickel

Specifications

ASME. Pipe: SB466, SB467. Tubing: SB111, SB359, SB395, SB466, SB467
ASTM. Bar and flat products: B122. Pipe: B466, B467. Tubing: B111, B359, B395, B466, B467. Wire: B206
SAE. Bar, flat products and tubing: J463

Chemical Composition

Composition limits. 0.05 max Pb, 1.00 max Fe, 1.00 max Zn, 19 to 23 Ni, 1.00 max Mn, 0.5 max others (total), rem Cu

Applications

Typical uses. Communication relays, condensers, condenser plates, electrical springs, evaporator and heat-exchanger tubes, ferrules, resistors

Mechanical Properties

Tensile properties. See Table 102, C70600
Fatigue strength. Tubing, H55 temper: 138 MPa (20 ksi) at 10^8 cycles
Elastic modulus. Tension, 140 GPa (20×10^6 psi); shear, 52 GPa (7.5×10^6 psi)

Mass Characteristics

Density. 8.94 Mg/m³ (0.323 lb/in.³) at 20 °C (68 °F)

Thermal Properties

Liquidus temperature. 1200 °C (2190 °F)
Solidus temperature. 1150 °C (2100 °F)
Coefficient of thermal expansion. 16.4 μm/m·K (9.1 μin./in.·°F) at 20 to 300 °C (68 to 572 °F)
Specific heat. 380 J/kg·K (0.09 Btu/lb·°F) at 20 °C (68 °F)
Thermal conductivity. 36 W/m·K (21 Btu/ft·h·°F) at 20 °C (68 °F)

Electrical Properties

Electrical conductivity. Volumetric, O61 temper: 6.5% IACS at 20 °C (68 °F)

Electrical resistivity. O61 temper: 266 nΩ·m at 20 °C (68 °F)

Optical Properties

Color. Pale silver

Fabrication Characteristics

Machinability. 20% of C36000, free-cutting brass

Formability. Good capacity for being cold worked by blanking, forming and bending, and for being hot formed.

Weldability. Soldering, brazing, gas shielded arc welding and resistance welding (all forms): excellent. Shielded metal arc welding: good. Oxyfuel gas welding: fair

Annealing temperature. 650 to 825 °C (1200 to 1500 °F)

Hot working temperature. 875 to 1050 °C (1600 to 1900 °F)

C71500
70Cu-30Ni

Commercial Names

Previous trade names. Copper nickel, 30%; CA715

Common name. 70-30 cupronickel

Specifications

ASME. Flat products: SB171, SB402. Pipe: SB466, SB467. Tubing: SB111, SB359, SB395, SB466, SB467, SB543

ASTM. Flat products: B122, B151, B171, B402. Pipe: B466, B467. Rod: B151. Tube: B111, B359, B395, B466, B467, B543, B552

SAE. Bar, flat products, tubing. J463

Government. Bar, flat products, forgings, rod, wire: MIL-C-15726. Tubing: MIL-T-15005, MIL-T-16420, MIL-T-22214

Chemical Composition

Composition limits. 0.05 max Pb, 0.4 to 0.7 Fe, 1.0 max Zn, 29 to 33 Ni, 1.0 max Mn, 0.5 max others (total), rem Cu

Applications

Typical uses. Condensers, condenser plates, distiller tubes, evaporator and heat-exchanger tubes, ferrules, salt water pipe

Precautions in use. Stress relieving or full annealing should precede exposure to solders of all kinds.

Mechanical Properties

Tensile properties. See Table 103 and Fig. 49.

Elastic modulus. Tension, 150 GPa (22×10^6 psi) at 20 °C (68 °F); shear, 57 GPa (8.3×10^6 psi) at 20 °C.

Hardness. See Table 103.

Impact strength. Charpy keyhole data: 107 J (79 ft·lb) at 21 °C (70 °F); 88 to 93 J (65 to 69 ft·lb) at −73 °C (−100 °F) after holding 30 to 140 days at −87 °C (−125 °F). Data are for 10-mm square specimens machined from 25-mm (1-in.) thick plate having a room temperature hardness of 88 HRF. See Table 104 for additional impact data.

Fatigue strength. Rod, H80 temper (drawn 50% to 1 in. diam): 220 MPa (32 ksi) at 10^8 cycles. Rod, O61 temper (drawn 50% to 1 in. diam, then annealed at 760 °C, or 1400 °F): 150 MPa (22 ksi) at 10^8 cycles.

Creep-rupture characteristics. Creep strength for 3.2-mm (0.125-in.) diam wire, OS020 temper: for a creep rate of 0.001% in 1000 h, 165 MPa (24 ksi) at 150 °C (300 °F) or 110 MPa (16 ksi) at 260 °C (500 °F); for a creep rate of 0.01% in 1000 h, 240 MPa (35 ksi) at 150 °C or 205 MPa (30 ksi) at 260 °C. Creep strength for 19-mm (³⁄₄-in.) diam rod, O61 temper (drawn to size and annealed at 550 °C, or 1020 °F): 63 MPa (9.1 ksi) for a creep rate of 0.01% in 1000 h at 400 °C; 130 MPa (18.8 ksi) for a creep rate of 0.1% in 1000 h at 400 °C.

Mass Characteristics

Density. 8.94 Mg/m³ (0.323 lb/in.³) at 20 °C (68 °F)

Thermal Properties

Liquidus temperature. 1240 °C (2260 °F)

Solidus temperature. 1170 °C (2140 °F)

Coefficient of thermal expansion. 16.2 μm/m·K (9 μin./in.·°F) at 20 to 300 °C (68 to 572 °F)

Specific heat. 380 J/kg·K (0.09 Btu/lb·°F) at 20 °C (68 °F)

Thermal conductivity. 29 W/m·K (17 Btu/ft·h·°F) at 20 °C (68 °F)

Electrical Properties

Electrical conductivity. Volumetric, O61 temper: 4.6% IACS at 20 °C (68 °F)

Electrical resistivity. O61 temper: 375 nΩ·m at 20 °C (68 °F); temperature coefficient, 4.8×10^{-5}/K (2.6×10^{-5}/°F) at 20 to 200 °C (68 to 392 °F)

Table 103 Typical mechanical properties of C71500

Size and temper	Tensile strength MPa	ksi	Yield strength(a) MPa	ksi	Elongation(b), %	Hardness, HRB
Flat Products						
25 mm (1 in.) plate, M20 temper	380	55	140	20	45	36
1 mm (0.04 in.) strip, O61 temper(c)	380	55	125	18	36	40
1 mm (0.04 in.) strip, H80 temper	580	84	545	79	3	86
Rod						
Up to 25 mm (1 in.) diam, O61 temper(d)	380	55	140	20	45	37
Up to 25 mm (1 in.) diam, H80 temper(e)	585	85	540	78	15	81
25 mm (1 in.) diam, H02 temper(f)	515	75	485	70	15	80
Tubing						
19 mm (0.75 in.) OD by 1.25 mm (0.049 in.) wall, O61 temper(c)	340	49			50	⋯
19 mm OD by 1.25 mm wall, H80 temper	580	84			4	⋯
25 mm (1 in.) OD by 1.65 mm (0.065 in.) wall, OS025 temper	415	60	170	25	45	45
114 mm (4.5 in.) OD by 2.75 mm (0.109 in.) wall, OS035 temper	370	54			45	36

(a) 0.5% extension under load. (b) In 50 mm or 2 in. (c) Annealed at 705 °C (1300 °F). (d) Annealed at 760 °C (1400 °F). (e) Cold drawn 50%. (f) Cold drawn 20%.

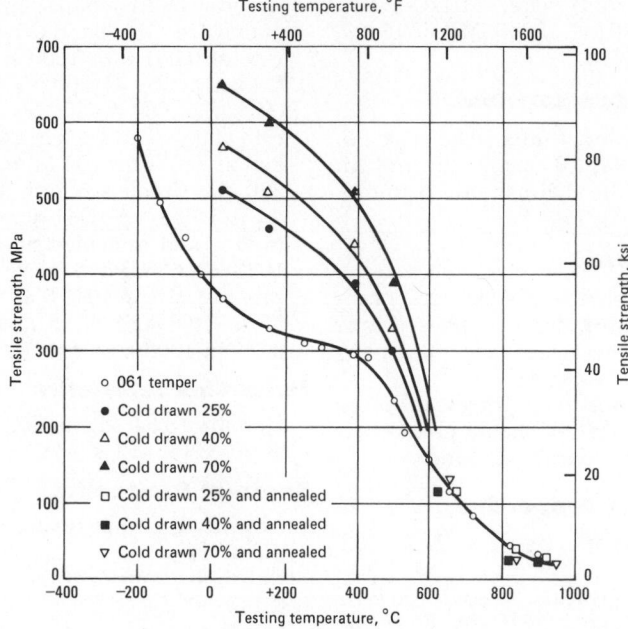

Fig. 49 Typical tensile and yield strengths of C71500 rod

Legend (upper chart):
- ○ 061 temper
- ● Cold drawn 25%
- △ Cold drawn 40%
- ▲ Cold drawn 70%
- □ Cold drawn 25% and annealed
- ■ Cold drawn 40% and annealed
- ▽ Cold drawn 70% and annealed

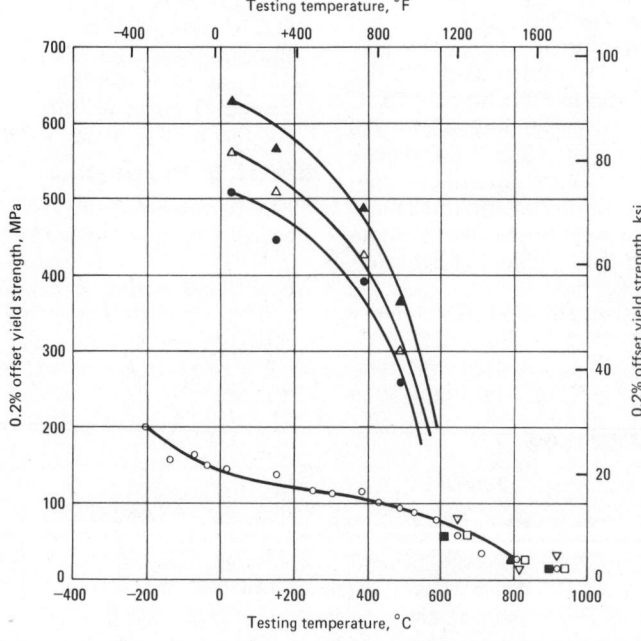

Table 104 Typical Charpy impact strengths for C71500

Testing temperature		Charpy impact strength(a)	
°C	°F	J	ft · lb
−115	−175	81	60
−18	0	81	60
3	38	87	64
20	68	89	66
65	150	72	53
120	250	72	53
205	400	68	50

(a) For 10-mm square keyhole specimens machined from annealed rod.

C71900
67.2Cu-30Ni-2.8Cr

Commercial Names

Previous trade names. Copper nickel, chromium-bearing; CA719
Common name. Cupronickel with Cr

Chemical Composition

Composition limits. 28 to 32 Ni, 2.4 to 3.2 Cr, 0.5 max Fe, 0.2 to 1.0 Mn, 0.01 to 0.20 Ti, 0.02 to 0.25 Zr, 0.04 max C, 0.25 max Si, 0.5 max others (total), rem Cu

Applications

Typical uses. Heat-exchanger tubes, tube sheets, water boxes, ferrules, salt water pipe

Mechanical Properties

Spinodally Decomposed Condition
Tensile properties. Tensile strength, 550 MPa (78 ksi); yield strength, 330 MPa (47 ksi) at 0.2% offset; elongation, 25%. See also Table 105.
Elastic modulus. Tension, 150 GPa (22×10^6 psi); shear, 59 GPa (8.5×10^6 psi)
Fatigue strength. Smooth bar, rotating beam: 275 MPa (40 ksi) at 10^8 cycles for both spinodally decomposed condition and half hard temper (spinodally decomposed plus 44% cold work).

Mass Characteristics

Density. 8.85 Mg/m³ (0.319 lb/in.³) at 20 °C (68 °F)

Thermal Properties

Liquidus temperature. 1220 °C (2225 °F)

Optical Properties

Color. White

Fabrication Characteristics

Machinability. 20% of C36000, free-cutting brass
Formability. Good capacity for being both cold worked and hot formed by bending and forming and welding processes
Weldability. Soldering, brazing, all forms of arc welding and all forms of resistance welding: excellent. Oxy-fuel gas welding: good.
Annealing temperature. 650 to 825 °C (1200 to 1500 °F)
Hot working temperature. 925 to 1050 °C (1700 to 1900 °F)

Solidus temperature. 1170 °C (2140 °F)
Coefficient of thermal expansion. 16.8 μm/m·K (9.3 μin./in.·°F) at 20 to 200 °C (68 to 392 °F); 17.1 μm/m·K (9.5 μin./in.·°F) at 20 to 300 °C (68 to 572 °F)
Thermal conductivity. 29 W/m·K (16.5 Btu/ft·h·°F) at 20 °C (68 °F)

Electrical Properties

Spinodally Decomposed Condition
Electrical conductivity. Volumetric, 4.4% IACS at 20 °C (68 °F)
Electrical resistivity. 395 nΩ·m at 20 °C (68 °F)

Magnetic Properties

Magnetic permeability. 1.0003 at magnetic field strength of 16 kA/m

Chemical Properties

Resistance to specific corroding agents. Seawater: C71900 resists both general and localized attack. The corrosion rate is very low (generally less than 0.1 mm/yr, or less than 4 mils/yr) in seawater flowing at velocities above about 1.8 m/s (6 ft/s). This level of corrosion resistance is comparable to or slightly less than the resistance of C71500 exposed to the same conditions. At low velocities or in stagnant seawater, C71900 exhibits slightly higher general weight loss than C71500, and corrosion is a broad, uniform type of attack. C71900 is not quite as resistant to crevice corrosion as C71500 at all velocities: for instance, in a 3-month test in seawater flowing at intermediate velocity C71900 incurred 0.2 to 0.33 mm (8 to 13 mils) penetration compared to nil penetration for C71500. Welding does not have an adverse effect on corrosion resistance; corrosion in seawater is about the same for the weld zone, heat-affected zone and unaffected base metal. C71900 appears to be immune to stress-corrosion cracking in seawater, even when the seawater is contaminated with 5 ppm H_2S. C71900 is cathodic to carbon steel and Ni-Resist type cast iron, slightly cathodic to C71500, and anodic to austenitic stainless steels.

Fabrication Characteristics

Formability. Can be cold worked in a manner similar to C71500, although C71900 has higher tensile and yield strengths at any given reduction. Hot working is readily accomplished from a starting temperature of 1040 to 1065 °C (1900 to 1950 °F), but working should not be continued below 840 °C (1550 °F) because of reduced ductility. About 25% more extrusion pressure is required for C71900 than for C71500. Because of microsegregation, cast billets should be homogenized 3 to 4 h at 1040 to 1065 °C before being extruded.
Weldability. Soldering, brazing and all forms of arc welding: excellent. Resistance welding not normally used for this alloy. Oxyfuel gas welding not recommended. Material thick enough to require multipass welds develops a minimum yield strength of 345 MPa (50 ksi) as welded. Single-pass welds develop a minimum yield strength of 275 MPa (40 ksi) as welded. The yield strength can be raised to 345 MPa by a post-weld heat treatment consisting of 1 h at 480 °C (900 °F).
Heat treatment. Full properties of the spinodally decomposed condition can be achieved by slow cooling (furnace cooling or still air cooling) through the temperature range 760 to 425 °C (1400 to 800 °F) from a soaking temperature of 900 to 1000 °C (1650 to 1850 °F).
Hot working temperature. 900 to 1065 °C (1650 to 1950 °F)

C72200
83Cu-16.5-Ni-0.5Cr

Commercial Names

Previous trade names. Copper nickel, chromium-bearing; CA722
Common name. Cupronickel with Cr

Chemical Composition

Composition limits. 15 to 18 Ni, 0.3 to 0.7 Cr, 0.5 to 1.0 Fe, 0.4 to 0.9 Mn, 0.03 max Si, 0.03 max Ti, 0.03 max C, 0.5 max others (total), rem Cu

Applications

Typical uses. Condenser and heat-exchanger tubing, salt water pipe

Mechanical Properties

Tensile properties. O61 temper: tensile strength, 315 MPa (46 ksi); yield strength, 125 MPa (18 ksi) at 0.2% offset; elongation, 46%. H04 temper: tensile strength, 475 MPa (70 ksi); yield strength, 445 MPa (66 ksi) at 0.2% offset; elongation, 6%
Elastic modulus. Tension, 135 GPa (20 × 10⁶ psi); shear, 55 GPa (8.2 × 10⁶ psi)

Mass Characteristics

Density. 8.94 Mg/m³ (0.323 lb/in.³) at 20 °C (68 °F)

Thermal Properties

Liquidus temperature. 1176 °C (2148 °F)
Solidus temperature. 1122 °C (2052 °F)
Coefficient of thermal expansion. 15.8 μm/m·K (8.8 μin./in.·°F) at 20 to 300 °C (68 to 572 °F)
Specific heat. 396 J/kg·K (0.094 Btu/lb·°F) at 20 °C (68 °F)
Thermal conductivity. 34.5 W/m·K (19.9 Btu/ft·h·°F) at 20 °C (68 °F)

Electrical Properties

Electrical conductivity. Volumetric, 6.53% IACS
Electrical resistivity. 264 nΩ·m at 20 °C (68 °F)

Fabrication Characteristics

Formability. Good capacity for being cold worked or hot formed
Weldability. Gas shielded arc welding: excellent. Soldering, brazing, shielded metal arc welding and resistance welding (all forms): good. Oxyfuel gas welding: fair

Table 105 Typical mechanical properties of C71900 strip

Condition	Tensile strength MPa	ksi	Yield strength(a) MPa	ksi	Elongation(b), %	Hardness, HRB
Heat treated(c)	600	87	365	53	32	87
Half hard temper(d)	730	106	685	99	14	100
Hard temper(e)	780	113	740	107	8	100
Spring temper(f)	835	121	800	116	6	101

(a) 0.2% offset. (b) In 50 mm or 2 in. (c) Spinodally decomposed by air cooling from 900 °C (1650 °F). (d) Spinodally decomposed, then cold rolled 20%. (e) Spinodally decomposed, then cold rolled 37%. (f) Spinodally decomposed, then cold rolled 60%.

Annealing temperature. 730 to 800 °C (1350 to 1500 °F)

Hot working temperature. 900 to 1040 °C (1650 to 1900 °F)

C72500
88.2Cu-9.5Ni-2.3Sn

Commercial Names

Previous trade names. Copper nickel, tin-bearing; CA725
Common name. Cupronickel with Sn

Chemical Composition

Composition limits. 0.05 max Pb, 0.6 max Fe, 0.5 max Zn, 0.2 max Mn, 8.5 to 10.5 Ni, 1.8 to 2.8 Sn, 0.2 max others (total), rem Cu

Applications

Typical uses. Relay and switch springs, connectors, lead frames, control and sensing bellows, brazing alloy

Mechanical Properties

Tensile properties. See Table 106.
Elastic modulus. Tension, 137 GPa (20 × 10⁶ psi); shear, 52 GPa (7.5 × 10⁶ psi)

Mass Characteristics

Density. 8.89 Mg/m³ (0.321 lb/in.³) at 20 °C (68 °F)

Thermal Properties

Liquidus temperature. 1130 °C (2065 °F)
Solidus temperature. 1060 °C (1940 °F)
Coefficient of thermal expansion. 16.5 μm/m·K (9.2 μin./in.·°F) at 20 to 300 °C (68 to 572 °F)
Thermal conductivity. 55 W/m·K (31 Btu/ft·h·°F) at 20 °C (68 °F)

Electrical Properties

Electrical conductivity. Volumetric, 11% IACS at 20 °C (68 °F)
Electrical resistivity. O61 temper, 157 nΩ·m at 20 °C (68 °F)

Optical Properties

Color. Silver

Chemical Properties

Resistance to specific agents. Excellent resistance to seawater

Table 106 Tensile properties of C72500

Temper	Tensile strength MPa	ksi	Yield strength 0.5% ext. under load MPa	ksi	0.2% offset MPa	ksi	Elongation, %	Hardness, HRB
Flat Products, 1-mm (0.04-in.) Thick								
Annealed(a)	380	55	150	22	150	22	35	42
Quarter Hard	450	65	365	53	400	58	18	71
Half Hard	490	71	450	65	475	69	6	78
Hard	570	83	515	75	555	81	3	85
Extra Hard	600	87	555	81	590	86	2	88
Spring	625	91	570	83	620	90	1	90
Super Spring	780	112	570	83	740	108	1	99
Wire, 2-mm (0.08-in.) Diam								
Annealed(a)	415	60	170	25	...	...	...	...

(a) Grain size, 0.015 mm.

Fabrication Characteristics

Machinability. 20% of C36000, free-cutting brass
Formability. Excellent capacity for being both cold worked and hot formed by blanking, coining, drawing, forming, bending, heading, upsetting, roll threading, knurling, shearing, spinning, squeezing, stamping and swaging
Weldability. Soldering, brazing and resistance spot and resistance butt welding: excellent. Gas shielded arc, shielded metal arc and resistance seam welding: good. Oxyfuel gas welding: fair
Annealing temperature. 650 to 800 °C (1200 to 1475 °F)
Hot working temperature. 850 to 950 °C (1550 to 1750 °F)

C74500
65Cu-25Zn-10Ni

Commercial Names

Common name. Nickel silver, 65-10

Specifications

ASTM. Flat products: B122. Bar: B122, B151. Rod: B151. Wire: B206
Government. Flat products: QQ-C-585. Bar: QQ-C-585, QQ-C-586. Rod, shapes, flat wire: QQ-C-586. Wire: QQ-W-321

Chemical Composition

Composition limits. 63.5 to 68.5 Cu; 9.0 to 11.0 Ni; 0.10 max Pb; 0.25 max Fe; 0.5 max Mn; 0.5 max others; rem Zn

Applications

Typical uses. Hardware: rivets, screws, slide fasteners. Optical goods: optical parts. Miscellaneous: etching stock, holloware, nameplates, platers' bars

Mechanical Properties

Tensile properties. See Table 107.
Hardness. See Table 107.
Elastic modulus. Tension, 120 GPa (17.5 × 10⁶ psi); shear, 46 GPa (6.6 × 10⁶ psi)

Mass Characteristics

Density. 8.69 Mg/m³ (0.314 lb/in.³) at 20 °C (68 °F)

Thermal Properties

Liquidus temperature. 1020 °C (1870 °F)
Coefficient of thermal expansion. Linear, 16.4 μm/m·K (9.1 μin./in.·°F) at 20 to 300 °C (68 to 572 °F)
Specific heat. 380 J/kg·K (0.09 Btu/lb·°F) at 20 °C (68 °F)
Thermal conductivity. 45 W/m·K (26 Btu/ft·h·°F)

Electrical Properties

Electrical conductivity. Volumetric, 9.0% IACS at 20 °C (68 °F)
Electrical resistivity. 192 nΩ·m at 20 °C (68 °F)

Fabrication Characteristics

Machinability. 20% of C36000, free-cutting brass
Formability. Excellent for cold working; poor for hot forming
Weldability. Soldering and brazing:

Table 107 Typical mechanical properties of C74500

Temper	Tensile strength MPa	ksi	Yield strength(a) MPa	ksi	Elongation in 50 mm or 2 in., %	HRF	Hardness HRB	HR30T	Shear strength MPa	ksi
Flat Products: 1 mm (0.04 in.) Thick										
OS070 340	340	49	125	18	49	67	22	30	...	...
OS050 350	350	51	130	19	46	71	28	34	...	...
OS035 365	365	53	140	20	43	76	35	38	285	41
OS025 385	385	56	160	23	40	80	42	44	...	...
OS015 415	415	60	195	28	36	85	52	51	...	...
H00 415	415	60	240	35	34	...	60	55	295	43
H01 450	450	65	310	45	25	...	70	63	310	45
H02 505	505	73	415	60	12	...	80	70	345	50
H04 590	590	86	515	75	4	...	89	76	380	55
H06 655	655	95	525	76	3	...	92	78	405	59
Wire: 2-mm (0.08-in.) Diam										
OS070 345	345	50	...	...	50	...	...	...	...	...
OS050 360	360	52	...	...	48	...	...	...	...	...
OS035 385	385	56	...	...	45	...	...	...	...	...
OS025 400	400	58	...	...	40	...	...	...	...	...
OS015 435	435	63	...	...	35	...	...	...	...	...
H00(10%) 450	450	65	...	...	25	...	...	...	...	...
H01(20%) 495	495	72	...	...	10	...	...	...	...	...
H02(37%) 585	585	85	...	...	7	...	...	...	...	...
H04(60%) 725	725	105	...	...	5	...	...	...	...	...
H06(75%) 825	825	120	...	...	3	...	...	...	...	...
H08(84%) 895	895	130	...	...	1	...	...	...	...	...

(a) At 0.5% extension under load.

excellent. Oxyfuel gas, resistance spot and resistance butt welding: good. Gas metal-arc and resistance seam welding: fair. Shielded metal-arc welding: not recommended
Annealing temperature. 600 to 750 °C (1100 to 1400 °F)

C75200
65Cu-18Ni-17Zn

Commercial Names

Common name. Nickel silver, 65-18

Specifications

ASTM. Flat products: B122. Bar: B122, B151. Rod: B151. Wire: B206
SAE. J463
Government. Flat products: QQ-C-585. Bar: QQ-C-585, QQ-C-586. Rod, shapes, flat wire: QQ-C-586. Wire: QQ-W-321

Chemical Composition

Composition limits. 63.0 to 66.5 Cu; 16.5 to 19.5 Ni; 0.1 max Pb; 0.25 max Fe; 0.5 max Mn; 0.5 max others (total); rem Zn

Table 108 Typical mechanical properties of C75200

Temper	Tensile strength MPa	ksi	Yield strength(a) MPa	ksi	Elongation in 50 mm or 2 in., %	HRF	Hardness HRB	HR30T
Flat Products: 1 mm (0.04 in.) Thick								
OS035 400	400	58	170	25	40	85	40	...
OS015 415	415	60	205	30	32	90	55	...
H01 450	450	65	345	50	20	...	73	65
H02 510	510	74	427	62	8	...	83	72
H04 585	585	85	510	74	3	...	87	75
Rod: 13-mm (0.5-in.) Diam								
OS035 385	385	56	170	25	42	...	...	...
H02(20%) ... 485	485	70	415	60	20	...	78	...
Wire: 2-mm (0.08-in.) Diam								
OS035 400	400	58	170	25	45	...	...	...
OS015 415	415	60	205	30	35	...	...	...
H01 505	505	73	450	65	16	...	...	...
H02 590	590	86	550	80	7	...	...	...
H04 710	710	103	620	90	3	...	...	...

(a) At 0.5% extension under load.

Applications

Typical uses. Hardware: rivets, screws, table flatware, truss wire, zippers. Optical goods: bows, camera parts, core bars, temples. Miscellaneous: base for silver plate, costume jewelry, etching stock, hollowware, nameplates, radio dials

Mechanical Properties

Tensile properties. See Table 108.
Hardness. See Table 108.
Elastic modulus. Tension, 125 GPa (18×10^6 psi); shear, 47 GPa (6.8×10^6 psi)

Mass Characteristics

Density. 8.73 Mg/m^3 (0.316 lb/in.^{3}) at 20 °C (68 °F)

Thermal Properties

Liquidus temperature. 1110 °C (2030 °F)

Solidus temperature. 1070 °C (1960 °F)

Coefficient of thermal expansion. Linear, 16.2 µm/m·K (9.0 µin./in.·°F) at 20 to 300 °C (68 to 572 °F)

Specific heat. 380 J/kg·K (0.09 Btu/lb·°F) at 20 °C (68 °F)

Thermal conductivity. 33 W/m·K (19 Btu/ft·h·°F) at 20 °C (68 °F)

Electrical Properties

Electrical conductivity. Volumetric, 6% IACS at 20 °C (68 °F)

Electrical resistivity. 287 nΩ·m at 20 °C (68 °F)

Fabrication Characteristics

Machinability. 20% of C36000, free-cutting brass

Formability. Excellent for cold working; poor for hot forming

Weldability. Soldering and brazing: excellent. Oxyfuel gas, resistance spot and resistance butt welding: good. Gas metal-arc and resistance seam welding: fair. Shielded metal-arc welding: not recommended.

C75400
65Cu-20Zn-15Ni

Commercial Names

Common name. Nickel silver, 65-15

Chemical Composition

Composition limits. 63.5 to 66.5 Cu; 14.0 to 16.0 Ni; 0.1 max Pb; 0.25 max Fe; 0.5 max Mn; 0.5 max others (total); rem Zn

Applications

Typical uses. Camera parts, optical equipment, etching stock, jewelry

Mechanical Properties

Tensile properties. See Table 109.
Hardness. See Table 109.
Elastic modulus. Tension, 125 GPa (18 × 10⁶ psi); shear, 47 GPa (6.8 × 10⁶ psi)

Mass Characteristics

Density. 8.70 Mg/m³ (0.314 lb/in.³) at 20 °C (68 °F)

Thermal Properties

Liquidus temperature. 1075 °C (1970 °F)

Solidus temperature. 1040 °C (1900 °F)

Coefficient of thermal expansion. Linear, 16.2 µm/m·K (9.0 µin./in.·°F) at 20 to 300 °C (68 to 572 °F)

Specific heat. 380 J/kg·K (0.09 Btu/lb·°F) at 20 °C (68 °F)

Thermal conductivity. 36 W/m·K (21 Btu/ft·h·°F) at 20 °C (68 °F)

Electrical Properties

Electrical conductivity. Volumetric, 7% IACS at 20 °C (68 °F)

Electrical resistivity. 246 nΩ·m at 20 °C (68 °F)

Fabrication Characteristics

Machinability. 20% of C36000, free-cutting brass

Formability. Excellent for cold working by blanking, drawing, forming, bending, heading, upsetting, roll threading, knurling, shearing, spinning, squeezing or swaging; poor for hot forming

Weldability. Soldering and brazing: excellent. Oxyfuel gas, resistance spot and resistance butt welding:

good. Gas metal-arc and resistance seam welding: fair. Shielded metal-arc welding: not recommended.

Annealing temperature. 600 to 825 °C (1100 to 1500 °F)

C75700
65Cu-23Zn-12Ni

Commercial Names

Common name. Nickel silver, 65-12

Specifications

ASTM. Bar, rod: B151. Wire: B206.
Government. Wire: QQ-W-321

Chemical Composition

Composition limits. 63.5 to 66.5 Cu; 11.0 to 13.0 Ni; 0.05 max Pb; 0.25 max Fe; 0.5 max Mn; 0.5 max others (total); rem Zn

Applications

Typical uses. Slide fasteners, camera parts, optical parts, etching stock, nameplates

Mechanical Properties

Tensile properties. See Table 110.
Hardness. See Table 110.
Elastic modulus. Tension, 125 GPa (18 × 10⁶ psi); shear, 47 GPa (6.8 × 10⁶ psi)

Mass Characteristics

Density. 8.69 Mg/m³ (0.314 lb/in.³) at 20 °C (68 °F)

Thermal Properties

Liquidus temperature. 1050 °C (1900 °F)

Coefficient of expansion. Linear,

Table 109 Typical mechanical properties of C75400 sheet or strip, 1 mm (0.04 in.) thick

Temper	Tensile strength MPa	ksi	Yield strength(a) MPa	ksi	Elongation in 50 mm or 2 in., %	Hardness HRF	HRB	HR30T	Shear strength MPa	ksi
OS070	365	53	125	18	43	69	...	27	...	...
OS050	380	55	130	19	42	73	...	33	...	...
OS035	395	57	145	21	40	79	...	41	285	41
OS025	405	59	165	24	37	82	...	46	...	...
OS015	420	61	195	28	34	89	...	53	...	...
H00	425	62	240	35	30	...	60	55	295	43
H01	450	65	340	49	21	...	70	63	305	44
H02	510	74	425	62	10	...	80	70	325	47
H04	585	85	515	75	3	...	87	75	360	52
H06	635	92	545	79	2	...	90	77	385	54

(a) At 0.5% extension under load.

Table 110 Typical mechanical properties of C75700 sheet or strip, 1 mm (0.04 in.) thick

Temper	Tensile strength MPa	ksi	Yield strength(a) MPa	ksi	Elongation in 50 mm or 2 in., %	Hardness HRF	HRB	HR30T	Shear strength MPa	ksi
OS070	360	52	125	18	48	69	22	27	...	...
OS050	370	54	130	19	45	73	30	33	...	...
OS035	385	56	145	21	42	78	37	38	285	41
OS025	405	59	165	24	38	82	45	44	...	...
OS015 ....·...	420	61	195	28	35	88	55	51	...	...
H00	415	60	240	35	32	...	60	55	295	43
H01	450	65	310	45	23	...	70	63	305	44
H02	505	73	415	60	11	...	80	70	325	47
H04	585	85	515	75	4	...	89	75	360	52
H06	640	93	545	79	2	...	92	77	385	56

(a) At 0.5% extension under load.

16.2 μm/m·K (9.0 μin./in.·°F) at 20 to 300 °C (68 to 572 °F)
Specific heat. 380 J/kg·K (0.09 Btu/lb·°F) at 20 °C (68 °F)
Thermal conductivity. 40 W/m·K (23 Btu/ft·h·°F) at 20 °C (68 °F)

Electrical Properties

Electrical conductivity. Volumetric, 8% IACS at 20 °C (68 °F)
Electrical resistivity. 216 nΩ·m at 20 °C (68 °F)

Fabrication Characteristics

Machinability. 20% of C36000, free-cutting brass
Formability. Excellent for cold working by blanking, drawing, etching, forming, bending, heading, upsetting, roll threading, knurling, shearing, spinning, squeezing or swaging; poor for hot forming
Weldability. Soldering and brazing: excellent. Oxyfuel gas, resistance spot and resistance butt welding: good. Gas metal-arc and resistance seam welding: fair. Shielded metal-arc welding: not recommended.
Annealing temperature. 600 to 825 °C (1100 to 1500 °F)

C77000
55Cu-27Zn-18Ni

Commercial Names

Common name. Nickel silver, 55-18

Specifications

ASTM. Flat products: B122. Bar: B122, B151. Rod: B151. Wire: B206
SAE. J463
Government. Flat products: QQ-C-585. Bar: QQ-C-585, QQ-C-586. Rod,

Table 111 Typical mechanical properties of C77000

Temper	Tensile strength MPa	ksi	Yield strength(a) MPa	ksi	Elongation in 50 mm or 2 in., %	Hardness HRF	HRB	HR30T
Flat Products: 1 mm (0.04 in.) Thick								
OS035	415	60	185	27	40	90	55	...
H04	690	100	585	85	3	...	91	77
H06	745	108	620	90	2.5	...	96	80
H08	795	115	...	...	2.5	...	99	81
Wire: 2-mm (0.08-in.) Diam								
OS035	415	60	...	...	40	...	...	...
H08(68%) ...	1000	145	...	...	2	...	...	...

(a) At 0.5% extension under load.

shapes, flat wire: QQ-C-586. Wire: QQ-W-321

Chemical Composition

Composition limits. 53.5 to 56.5 Cu; 16.5 to 19.5 Ni; 0.1 max Pb; 0.25 max Fe; 0.5 max Mn; 0.5 max others (total); rem Zn

Applications

Typical uses. Optical goods, springs, resistance wire

Mechanical Properties

Tensile properties. See Table 111.
Hardness. See Table 111.
Elastic modulus. Tension, 125 GPa (18 × 10⁶ psi); shear, 47 GPa (6.8 × 10⁶ psi)

Mass Characteristics

Density. 8.70 Mg/m³ (0.314 lb/in.³) at 20 °C (68 °F)

Thermal Properties

Liquidus temperature. 1055 °C (1930 °F)

Coefficient of thermal expansion. Linear, 16.7 μm/m·K (9.3 μin./in.·°F) at 20 to 300 °C (68 to 572 °F)
Specific heat. 380 J/kg·K (0.09 Btu/lb·°F) at 20 °C (68 °F)
Thermal conductivity. 29 W/m·K (17 Btu/ft·h·°F) at 20 °C (68 °F)

Electrical Properties

Electrical conductivity. Volumetric, 5.5% IACS at 20 °C (68 °F)
Electrical resistivity. 314 nΩ·m at 20 °C (68 °F)

Fabrication Characteristics

Machinability. 30% of C36000, free-cutting brass
Formability. Good for cold working by blanking, forming, bending and shearing; poor for hot forming
Weldability. Soldering and brazing: excellent. Oxyfuel gas, resistance spot and resistance butt welding: good. Gas metal-arc and resistance seam welding: fair. Shielded metal-arc welding: not recommended.
Annealing temperature. 600 to 825 °C (1100 to 1500 °F)

C78200
65Cu-25Zn-8Ni-2Pb

Chemical Composition

Composition limits. 63.0 to 67.0 Cu; 1.5 to 2.5 Pb; 7.0 to 9.0 Ni; 0.35 max Fe; 0.50 max Mn; 0.10 max others (total); rem Zn

Applications

Typical uses. Key blanks, watch plates, watch parts

Mechanical Properties

Tensile properties. See Table 112.
Shear strength. See Table 112.
Hardness. See Table 112.
Elastic modulus. Tension, 117 GPa (17×10^6 psi); shear, 44 GPa (6.4×10^6 psi)

Mass Characteristics

Density. 8.69 Mg/m^3 (0.314 $lb/in.^3$) at 20 °C (68 °F)

Thermal Properties

Liquidus temperature. 1000 °C (1830 °F)
Solidus temperature. 970 °C (1780 °F)

Table 112 Typical mechanical properties of C78200 sheet(a)

Temper	Tensile strength MPa	ksi	Yield strength(b) MPa	ksi	Elongation(c), %	Hardness, HRB	Shear strength MPa	ksi
OS035	365	53	160	23	40	78 HRF	275	40
OS015	405	59	185	27	32	85 HRF	295	43
H01	425	62	290	42	24	65	305	44
H02	475	69	400	58	12	78	325	47
H03	540	78	435	63	5	84	350	51
H04	585	85	505	73	4	87	370	54
H06	625	91	525	76	3	90	400	58

(a) 1 mm (0.04 in.) thick. (b) 0.5% extension under load. (c) In 50 mm or 2 in.

Coefficient of thermal expansion. Linear, 18.5 μm/m·K (10.3 μin./in.·°F) at 20 to 100 °C (68 to 212 °F)
Specific heat. 380 J/kg·K (0.09 Btu/lb·°F) at 20 °C (68 °F)
Thermal conductivity. 48 W/m·K (28 Btu/ft·h·°F)

Electrical Properties

Electrical conductivity. Volumetric, 10.9% IACS at 20 °C (68 °F)
Electrical resistivity. 172 nΩ·m at 20 °C (68 °F)

Fabrication Characteristics

Machinability. 60% of C36000, free cutting brass
Formability. Cold working, good; hot forming, poor. Commonly fabricated by blanking, milling and drilling.

Weldability. Soldering: excellent. Brazing: good. Oxyfuel gas, arc and resistance welding generally not recommended.
Annealing temperature. 500 to 620 °C (930 to 1150 °F)

Selection and Application of Copper Alloy Castings

COPPER ALLOY CASTINGS are used in applications that require superior corrosion resistance, high thermal or electrical conductivity, good bearing-surface qualities or other special properties. Casting makes it possible to produce parts whose shape cannot be easily obtained by fabricating methods such as forming or machining. Often, it is more economical to produce a part as a casting than to fabricate it by other means.

Compositions of copper casting alloys (see Table 1) may differ from those of their wrought counterparts for various reasons. Generally, casting permits greater latitude in the use of alloying elements, because the effects of composition on hot or cold working properties are not important. However, imbalances among certain elements, and trace amounts of certain impurities in some alloys, will diminish castability and may result in castings of questionable quality. Nevertheless, certain cast alloys may be more suitable than wrought alloys for specific applications.

For example, nineteen of the alloys listed in Table 1 have lead contents of 5% or more. Alloys containing such high percentages of lead are not suited to hot working, but are ideal for low-1 to medium-speed bearings, where the lead prevents galling and excessive wear under boundary-lubrication conditions.

The tolerance for impurities is normally greater in castings than in their wrought counterparts—again because of the adverse effects certain impurities have on hot or cold workability. On the other hand, impurities that inhibit response to heat treatment must be avoided in both castings and wrought products.

The choice of an alloy for any casting usually depends on four factors: metal cost, castability, properties and final cost.

Metal cost is a minor consideration if only a few castings are to be made. However, when the product is a mass-produced or highly competitive item, or when metal cost is a major portion of the final cost of the castings, this factor becomes of prime importance.

Castability

Castability should not be confused with "fluidity", which is the ability of a molten alloy to fill a mold cavity completely in every detail. Castability, on the other hand, is the ease with which an alloy responds to ordinary foundry practice, without requiring special techniques for gating, risering, melting, sand conditioning or any of the other factors involved in making good castings. High fluidity often ensures good castability, but is not solely responsible for that quality in a casting alloy.

Foundry alloys generally are classed as "high shrinkage" or "low shrinkage". To the former class belong the manganese bronzes, aluminum bronzes, silicon bronzes, silicon brasses, and some nickel silvers. They are more fluid than the low-shrinkage red brasses, more easily poured, and give high-grade castings in sand, permanent mold, plaster, die and centrifugal casting processes. With high-shrinkage alloys, careful design is necessary to (a) promote directional solidification, (b) avoid abrupt changes in cross section, (c) avoid notches (by using generous fillets), and (d) properly place gates and risers, all of which help avoid internal shrinks and draws. Turbulent pouring must be avoided to prevent formation of dross. Liberal use of risers or exothermic compounds ensures adequate molten metal to feed all sections of the casting. Table 2 presents foundry characteristics of nineteen standard alloys, including a comparative ranking for both fluidity and overall castability for sand casting; number 1 represents the highest castability or fluidity ranking.

All copper alloys can be successfully cast in sand. Sand casting is the most economical casting method and allows the greatest flexibility in casting size and shape.

Permanent mold casting is best suited for tin, silicon, aluminum and manganese bronzes, and for yellow brasses. Die casting is well suited for yellow brasses, but increasing amounts of permanent mold alloys are also being die cast. Size is a definite limitation for both methods, although large slabs weighing as much as 4500 kg (10 000 lb) have been cast in permanent molds. Brass die castings generally weigh less than 0.2 kg (0.5 lb) and seldom exceed 0.9 kg (2 lb). The limitation of size is due to reduced die life with larger castings.

Virtually all copper alloys can be cast successfully by the centrifugal casting process. Castings of virtually any size from less than 100 g to more than 22 000 kg (less than 0.25 to more than 50 000 lb) have been made.

Because of their low lead contents, aluminum bronzes, yellow brasses, manganese bronzes, low-nickel bronzes, and silicon brasses and bronzes are best adapted to plaster mold casting. For most of these alloys, lead should be held to a minimum because it reacts with the calcium sulfate in the plaster, resulting in discoloration of the surface of the casting and increased cleaning and machining costs. Size is a limitation on plaster mold casting, although aluminum bronze castings that weigh as little as 100 g (0.22 lb) have been made by the

Table 1 Nominal compositions of principal copper casting alloys

UNS number	Common name	Previous ASTM designation	Cu	Sn	Pb	Zn	Fe	Al	Others
ASTM B22									
C86300	Manganese bronze	B22-E	63	...	...	25	3	6	3 Mn
C90500	Tin bronze	B22-D	88	10	...	2	...	...	...
C91100	Tin bronze	B22-B	84	16	...	...	...	...	...
C91300	Tin bronze	B22-A	81	19	...	...	...	...	...
C93700	High-lead tin bronze	B22-C	80	10	10	...	...	...	...
ASTM B61									
C92200	Valve bronze	...	88	6	1.5	4	...	...	1 max Ni
ASTM B62									
C83600	Leaded red brass	...	85	5	5	5	...	...	...
ASTM B66									
C93800	High-lead tin bronze	...	78	7	15	...	...	...	...
C94300	High-lead tin bronze	...	70	5	25	...	...	...	...
C94400	Leaded phosphor bronze	...	81	8	11	...	...	...	0.35 P
C94500	High-lead tin bronze	...	73	7	19	1	...	...	...
ASTM B67									
C94100	High-lead tin bronze	...	70	5.5	18.5	3 max	...	...	...
ASTM B148									
C95200	Aluminum bronze	B148-9A	88	...	...	...	3	9	...
C95300	Aluminum bronze	B148-9B	89	...	...	...	1	10	...
C95400	Aluminum bronze	B148-9C	85	...	...	...	4	11	...
C95500	Nickel-aluminum bronze	B148-9D	81	...	...	...	4	11	4 Ni
C95800	Nickel-aluminum bronze	...	81.3	...	...	...	4	9	4.5 Ni, 1.2 Mn
ASTM B176 (Die Casting Alloys)									
C85800	Yellow brass	Z30A	58	1	1	40	...	...	...
C87800	Silicon brass	ZS144A	82	...	...	14	...	...	4 Si
C87900	Silicon brass	ZS331A	65	...	...	33	...	...	1 Si
ASTM B584									
C83600	Leaded red brass	B145-4A	85	5	5	5	...	...	...
C83800	Leaded red brass	B145-4B	83	4	6	7	...	...	...
C84400	Leaded semi-red brass	B145-5A	81	3	7	9	...	...	...
C84800	Leaded semi-red brass	B145-5B	76	3	6	15	...	...	...
C85200	Leaded yellow brass	B146-6A	72	1	3	24	...	...	...
C85400	Leaded yellow brass	B146-6B	67	1	3	29	...	...	...
C85700	Leaded naval brass	B146-6C	63	1	1	34.7	0.03	...	...
C86200	High-strength manganese bronze	B147-8B	64	...	...	26	3	4	3 Mn
C86300	High-strength manganese bronze	B147-8C	63	...	...	25	3	6	3 Mn
C86400	Leaded manganese bronze	B147-7A or B132-A	59	...	1	38	1	0.5	0.5 Mn
C86500	Manganese bronze	B147-8A	58	...	...	39	1	1	1 Mn
C86700	Leaded manganese bronze	B132-B	58	1	1	34	2	2	2 Mn
C87200	Silicon bronze	B198-12A	Several nominal compositions available						
C87400	Silicon brass	B198-13A	82	...	0.5	14	...	...	3.5 Si
C87500	Silicon brass	B198-13B	82	...	...	14	...	...	4 Si
C87600	Silicon brass	B198-13C	89	...	...	6	...	...	5 Si
C90300	Modified G bronze	B143-1B	88	8	...	4	...	...	...
C90500	G bronze	B143-1A	88	10	...	2	...	...	...
C92200	Steam bronze (a)	B143-2A	88	6	1.5	4.5	...	...	...
C92300	Leaded tin bronze	B143-2B	87	8	1	4	...	...	...
C93200	High-lead tin bronze	B144-3B	83	7	7	3	...	...	...

(continued)

Table 1 (continued)

UNS number	Common name	Previous ASTM designation	Cu	Sn	Pb	Zn	Fe	Al	Others
C93500	High-lead tin bronze	B144-3C	85	5	9	1	...	...	...
C93700	High-lead tin bronze	B144-3A	80	10	10	...	...	...	...
C93800	High-lead tin bronze	B144-3D	78	7	15	...	...	...	...
C94300	High-lead tin bronze	B144-3E	70	5	25	...	...	...	...
C94700	Nickel-tin bronze	B292-A	88	5	...	2	...	...	5 Ni
C94800	Leaded nickel-tin bronze	B292-B	87	5	1	2	...	...	5 Ni
C94900	Leaded nickel-tin bronze	...	80	5	5	5	...	...	5 Ni
C97300	Leaded nickel silver	B149-10A	56	2	10	20	...	...	12 Ni
C97600	Leaded nickel silver	B149-11A	64	4	4	8	...	...	20 Ni
C97800	Leaded nickel silver	B149-11B	66	5	2	2	...	...	25 Ni

(a) Also known as valve bronze or Navy M bronze.

Table 2 Foundry properties for principal copper alloys for sand casting

UNS number	Common name	Shrinkage allowance, %	Approx liquidus temperature °C	°F	Castability rating (a)	Fluidity rating (a)
C83600	Leaded red brass	5.7	1010	1850	2	6
C84400	Leaded semi-red brass	2.0	980	1795	2	6
C84800	Leaded semi-red brass	1.4	955	1750	2	6
C85400	Leaded yellow brass ...	1.5 to 1.8	940	1725	4	4
C85800	Yellow brass..............	2.0	925	1700	4	4
C86300	Manganese bronze........	2.3	920	1690	6	2
C86500	Manganese bronze........	1.9	880	1615	6	2
C87200	Silicon bronze.........	1.8 to 2.0			8	3
C87500	Silicon brass........	1.9	915	1680	7	1
C90300	Tin bronze............	1.5 to 1.8	980	1795	3	6
C92200	Leaded tin bronze	1.5	990	1810	3	6
C93700	High-lead tin bronze......	2.0	930	1705	1	6
C94300	High-lead tin bronze......	1.5	925	1700	1	6
C95300	Aluminum bronze	1.6	1045	1910	8	5
C95800	Aluminum bronze	1.6	1060	1940	8	5
C97600	Nickel silver.............	2.0	1145	2090	5	7
C97800	Nickel silver.............	1.6	1180	2160	5	7

(a) Relative rating for casting in sand molds. The alloys are ranked from 1 to 8 in both over-all castability and fluidity; 1 is the highest or best possible rating.

lost-wax process and castings that weigh more than 150 kg (330 lb) have been made by conventional plaster molding.

Control of Solidification. Production of consistently sound castings requires an understanding of the solidification characteristics of the alloys as well as knowledge of relative magnitudes of shrinkage. The actual amount of contraction during solidification does not differ greatly from alloy to alloy. Its distribution, however, is a function of the freezing range and the temperature gradient in critical sections. Manganese and aluminum bronzes are similar to steel in that their freezing ranges are

quite narrow—about 40 and 14 °C (70 and 25 °F), respectively. Large castings can be made by the same conventional methods used for steel, as long as proper attention is given to placement of gates and risers—both those for controlling directional solidification and those for feeding the primary central shrinkage cavity.

Tin bronzes have wider freezing ranges (about 165 °C or 300 °F for C83600). Alloys with such wide freezing ranges form a mushy layer during solidification, resulting in interdendritic shrinkage or microshrinkage. Because feeding cannot take place properly under these conditions, open

grain or porosity results in the affected sections. The only practical means of overcoming interdendritic shrinkage is to maintain close temperature control of the metal during pouring and to provide for rapid solidification. These requirements limit thickness of section and pouring temperatures, and require a gating system that will ensure directional solidification. Sections up to 25 mm (1 in.) in thickness are routinely cast. Sections up to 50 mm (2 in.) thick may be cast, but only with difficulty and under carefully controlled conditions. A bronze with a narrow solidification range and good directional solidification characteristics is recommended for castings having section thickness greater than about 25 mm.

It is difficult to achieve directional solidification in complex castings. The most effective and most easily used device is the chill, which can be employed to initiate or accelerate solidification, and thus promote soundness. For irregular sections, chills must be shaped to fit the contour of the section of the mold in which they are placed. Insulating pads and riser sleeves sometimes are effective in slowing down the solidification rate in certain areas to maintain directional solidification.

Mechanical Properties

Most copper-base casting alloys containing tin, lead or zinc have only moderate tensile and yield strengths, low to medium hardness, and high elongation. When higher tensile or yield strength is required, the aluminum bronzes, manganese bronzes, silicon brasses, silicon bronzes, and some nickel silvers are used instead. Most of the higher strength alloys have better-than-average resistance to corrosion and wear. Mechanical and physical properties of copper-base casting alloys are presented in Table 3. (Throughout this discussion, as well as in Table 3, mechanical properties quoted are for test bars. Properties of the castings themselves are almost always lower and depend on section size and process variables.)

Tensile strengths for cast test bars of aluminum bronzes and manganese bronzes range from 450 to 900 MPa (65 to 130 ksi), depending on composition; aluminum bronzes attain maximum tensile strength only after heat treatment.

Table 3 Typical properties of copper casting alloys

UNS number	Tensile strength MPa	ksi	Yield strength(a) MPa	ksi	Compressive yield strength(b) MPa	ksi	Elon-gation, %	Hardness HB(c)	Electrical conductivity, % IACS
ASTM B22									
C86300	820	119	570(d)	82(d)	490	71	18	177	9.05
C90500	275-345	40-50	140-160	20-23	...	...	24-43	75-85	10.5-11.5
C93700	270	39	125	18	125	18	30	67	10.0
ASTM B61									
C92200	280	41	110	16	105	15	45	64	14.5
ASTM B62									
C83600	240	35	105	15	100	14	32	62	15.0
ASTM B66									
C94300	160-205	23-30	75-105	11-15	80-95	12-14	7-16	42-55	...
ASTM B147									
C86200	625-670	91-97	315-345	46-50	345	50	19-25	170-195(e)	7-8
C86300	820	119	570(d)	82(d)	490	71	18	177	9.0
C86400	415-540	60-78	170-275	25-40	140-180	20-26	15-30	80-95	20-24
C86500	490	71	180	26	165	24	40	98	20.5
ASTM B148									
C95200	480-600	70-87	170-205	25-30	185-215	27-31	22-38	110-140(e)	12-14
C95300(f)	480-585	70-85	205-240	30-35	110-140	16-20	20-35	110-160(e)	12-15
C95300(g)	550-655	80-95	275-380	40-55	240-275	35-45	12-16	160-225(e)	13.8
C95400(f)	515-655	75-95	205-285	30-41	...	...	12-20	150-185(e)	13-15
C95400(g)	620-690	90-100	310-360	45-52	...	...	6-15	190-235(e)	...
C95500(f)	620-725	90-105	275-345	40-50	...	...	7-20	175-210(e)	8-9.5
C95500(g)	760-855	110-124	415-550	60-80	...	...	5-12	215-260(e)	...
ASTM B176									
C85800(h)	380	55	205(d)	30(d)	...	...	15	...	22
C87800(h)	620	90	205(d)	30(d)	...	...	25	...	...
C87900(h)	400	58	205(d)	30(d)	...	...	15	...	...
ASTM B584									
C83600	243	35	105	15	100	14	32	62	15
C83800	205-260	30-38	85-115	12-17	76-83	11-12	15-27	50-60	...
C84400	200-270	29-39	90-115	13-17	...	...	18-30	50-60	18
C84800	260	38	105	15	85	12	37	59	16.5
C85200	240-275	35-40	85-95	12-14	55-70	8-10	25-40	40-55	15-22
C85400	205-260	30-38	75-105	11-15	62	9	20-35	40-60	18-25
C85700	275-310	40-45	95-140	14-20	...	...	15-25	50-75	20-26
C86200	625-670	91-97	315-345	46-50	345	50	19-25	170-195(e)	7-8
C86300	820	119	573	82(d)	490	71	18	177	9.0
C86400	415-540	60-78	170-275	25-40	...	20-26	15-30	80-95	20-24
C86500	490	71	180	26	140-180	24	40	98	20.5
C86700	550	80	220	32	165	...	15	...	...
C87200	380-450	55-65	150-205	22-30	105-150	15-22	25-55	85-120	4.5-6.4
C87400	345-485	50-70	145-225	21-33	...	...	20-50	70-130	...
C87500	470	68	207	30	185	27	20-50	115	6.0
C87600	414 min	60 min	207 min	30 min	...	...	17	...	...
C90300	275-345	40-50	125-150	18-22	...	...	16 min	...	...
C90500	275-345	40-50	140-160	20-23	...	...	25-50	60-75	12-13
C92200	280	41	110	16	105	15	24-43	75-85	10.5-11.5
C92300	225-295	33-43	110-165	16-24	62-76	9-11	45	64	14.5
C93200	205-260	30-38	115-145	17-21	...	...	18-30	60-75	10-12
C93500	195-240	28-35	83-105	12-15	90	13	12-20	55-65	...
C93700	270	39	125	18	125	18	20-35	55-65	15
C93800	170-225	25-33	95-140	14-20	90-110	13-16	30	67	10.0
C94300	160-205	23-30	76-105	11-15	83-97	12-14	10-18	50-60	...
C94700(f)	310	45	140	20	...	...	7-16	42-55	...
C94700(j)	515	75	345	50	...	...	25	...	...
C94800(f)	275	40	140	20	...	...	5	...	...
C94900	262 min	38 min	97 min	15 min	...	...	20	...	...
C97300	205-275	30-40	105-140	15-20	...	...	15 min	...	...
C97600	325	47	180	26	168	24	10-25	50-60	5.7
C97800	345-450	50-65	180-275	26-40	...	...	22	85	4.8
							15-25	120-150	4-5

(a) At 0.5% extension under load. (b) At permanent set of 0.1%. (c) 500 kg load; 10 mm diam ball. (d) At 0.2% offset. (e) 3000 kg load. (f) MO1 temper. (g) TQ00 temper. (h) MO4 temper. (j) TF00 temper.

Although manganese and aluminum bronzes often are used for the same applications, the manganese bronzes are handled more easily in the foundry. As-cast tensile strengths as high as 800 MPa (115 ksi) and elongations of 15 to 20% can be obtained readily in sand castings, and slightly higher values in centrifugal castings. Stresses may be relieved at 175 to 200 °C (350 to 400 °F). Lead may be added to the lower-strength manganese bronzes to increase machinability, but at the expense of decreased tensile strength and elongation. Lead content should not exceed 0.1% in high-strength manganese bronzes. Although manganese bronzes range in hardness from 125 to 250 HB, they are readily machined with proper tools.

Tin is added to the low-strength manganese bronzes to enhance resistance to dezincification, but should be limited to 0.1% in high-strength manganese bronzes unless great sacrifices in strength and ductility can be accepted.

Manganese bronzes are specified for marine propellers and fittings, pinions, ball-bearing races, worm wheels, gearshift forks and architectural work. Manganese bronzes also are used for rolling-mill screwdown nuts and slippers, bridge trunnions, gears and bearings, all of which require high strength and hardness.

Various cast aluminum bronzes contain 9 to 14% Al and lesser amounts of iron, manganese or nickel. They have a very narrow solidification range and, because of the greater need for adequate gating and risering compared to most other copper casting alloys, are more difficult to cast. A wide range of properties is obtainable, especially after heat treatment, but close control of composition is necessary. Like the manganese bronzes, aluminum bronzes can develop tensile strengths well over 700 MPa (100 ksi).

Most aluminum bronzes contain from 0.75 to 4% Fe to refine grain structure and increase strength. Alloys containing from 8 to 9.5% Al cannot be heat treated unless other elements (such as nickel or manganese) in amounts over 2% are added as well. They have higher tensile strength and greater ductility and toughness than any of the ordinary tin bronzes. Appli-

cations include valve nuts, cam bearings, impellers, hangers in pickling baths, agitators, crane gears, and connecting rods.

The heat treatable aluminum bronzes contain from 9.5 to 11.5% Al, in addition to iron, with or without nickel or manganese. These alloys resemble heat treated steels in structure and in response to quenching and tempering; castings are quenched in water or oil from temperatures between 760 and 925 °C (1400 and 1700 °F) and tempered at 425 to 650 °C (800 to 1200 °F), depending on exact composition and required properties.

From the range of properties shown in Table 3, it can be seen that all the maximum properties cannot be obtained in any one aluminum bronze. In general, alloys with higher tensile strength, yield strength and hardness have lower values of elongation. Typical applications of the higher-hardness alloys are rolling-mill screwdown nuts and slippers, worm gears, bushings, slides, impellers, nonsparking tools, valves and dies.

Aluminum bronzes resist corrosion in many substances, including pickling solutions. When corrosion occurs, it often proceeds by dealuminification, a form of dealloying in which aluminum is lost preferentially. Duplex alpha-plus-beta aluminum bronzes are more susceptible to dealloying than the all-alpha aluminum bronzes.

Aluminum bronzes have a high fatigue limit, considerably greater than that of manganese bronze or any other cast copper alloy. Unlike those of Cu-Zn and Cu-Sn-Pb-Zn alloys, the mechanical properties of aluminum and manganese bronzes do not decrease much with increases in casting cross section. This is because of a narrow freezing range, which results in a denser structure when castings are properly designed and properly fed.

Whereas manganese bronzes become hot short above 230 °C (450 °F), aluminum bronzes can be used at temperatures as high as 400 °C (750 °F) for short periods of time without appreciable loss in strength. For example, room-temperature tensile strengths of 540 MPa (78 ksi) decline to 530 MPa (76.7 ksi) at 260 °C (500 °F), 460 MPa (67 ksi) at 400 °C (750 °F) and 400 MPa (58 ksi) at 540 °C (1000 °F). Corresponding elongation values change from 28% to 32, 35 and 25%, respectively.

Unlike manganese bronzes, many

aluminum bronzes increase in yield strength and hardness, but decrease in tensile strength and elongation, on slow cooling in the mold. While some manganese bronzes precipitate a relatively soft phase during slow cooling, aluminum bronzes precipitate a hard constituent rather rapidly within the narrow temperature range 565 to 480 °C (1050 to 900 °F). Hence, large castings, or smaller castings that are cooled slowly, will have properties different from small castings cooled relatively rapidly. The same phenomenon occurs on heat treating the hardenable aluminum bronzes. Cooling slowly through the critical temperature range after quenching, or tempering at temperatures within this range, will decrease elongation. Addition of 2 to 5% Ni greatly diminishes this effect.

Nickel brasses, silicon brasses and silicon bronzes, although generally higher in strength than red metal alloys, are used more for their corrosion resistance and are discussed under "Selection of Alloys for Corrosion Service", below.

Distributions of hardness and tensile-strength data for separately cast test bars of three different alloys are shown in Fig. 1.

Properties of Test Bars. Mechanical properties of separately cast test bars often differ widely from those of production castings poured at the same time, particularly when the thickness of the casting differs markedly from that of the test bar.

Variation in casting section size particularly affects mechanical properties of tin bronzes. With increasing section size up to about 50 mm (2 in.), mechanical properties—both strength and elongation—of the castings themselves are progressively lower than the corresponding properties of separately cast test bars. Elongation is particularly affected; for some tin bronzes, elongation of a 50-mm section may be as little as 1/10 that of a 10-mm (0.4-in.) section or of a separately cast test bar.

The metallurgical behavior of many copper alloy systems is complex. Cooling rate (a function of casting section size) directly influences grain size, segregation and intergranular shrinkage; these factors, in turn, affect the mechanical properties of the cast metal. Therefore, molding and casting techniques are based on metallurgical characteristics as well as casting shape.

Fig. 1 Distribution of hardness for three copper casting alloys of different tensile strengths

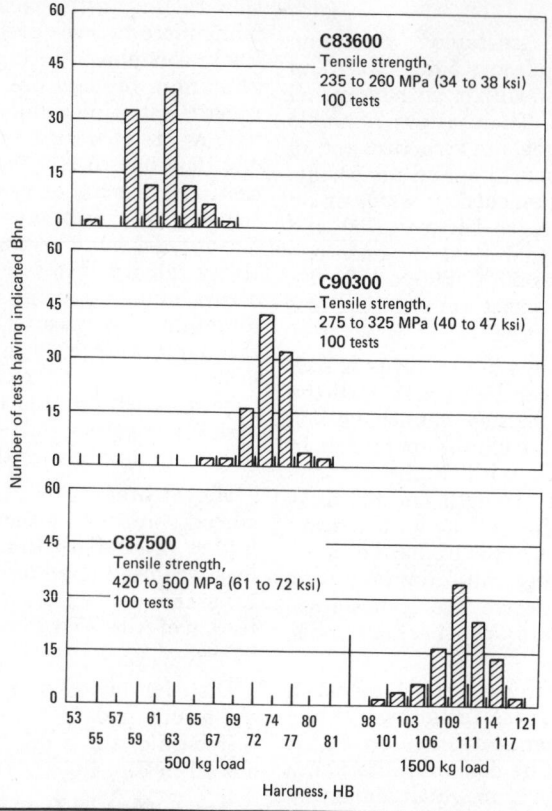

Fig. 2 Variations from design dimensions for a typical red brass casting

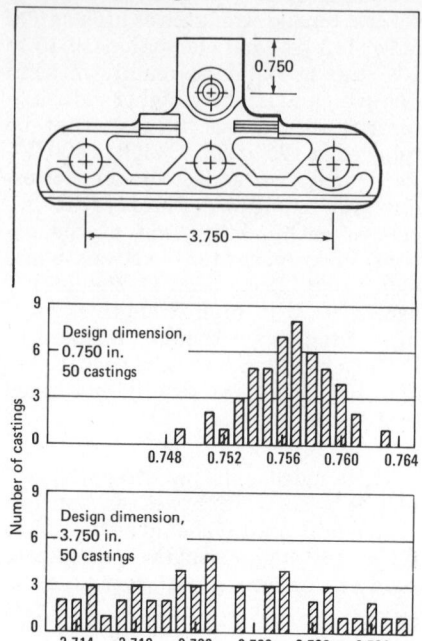

Parts were cast in green sand molds made using the same pattern.

Dimensional Tolerances

Typical dimensional tolerances are different for castings produced by different molding methods. A molding process involving two or more mold parts requires greater tolerances for dimensions that cross the parting line than for dimensions wholly within one mold part. For castings made in green sand molds, tolerance across the parting line depends on the accuracy of pins and bushings that align the cope with the drag.

Figure 2 shows variations in two important dimensions for 50 production castings of red brass. The 3.750-in. dimension presented the greatest difficulty: none of the 50 production castings had an actual dimension as large as the nominal design value. Figure 3 shows dimensional variations in two similar cored valve castings. For each design, both the cores and the corresponding cavities in the castings were measured for about 100 castings. For both designs, the castings had actual dimensions less than those of the cores.

This indicates that cores may need to have a slightly larger nominal size than is desired in the finished casting in order to ensure proper as-cast hole sizes.

Machinability

Machinability ratings of copper casting alloys are similar to those of their wrought counterparts. The cast alloys can be separated into three groups. The first group includes only those containing a single, copper-rich phase plus lead. Whether present for some other purpose or merely to improve machinability, lead facilitates chip breakage, thus allowing higher machining speeds with decreased tool wear and improved surface finishes.

Alloys of the second group contain two or more phases. Generally, the secondary phases are harder or more brittle than the matrix. Silicon bronzes, several aluminum bronzes and the high tin bronzes belong to this group. Hard and brittle secondary phases act as internal chip breakers, resulting in short chips and easier ma-

chining. Based on metallographic appearance, manganese bronzes would seem likely to produce a short chip, but they produce a long spiral chip, smooth on both sides, which does not break. Some aluminum bronzes, on the other hand, produce a long spiral chip that is rough on the underside and that breaks, thus acting like a short chip. Some of the alloys in the second group are classified "moderately machinable" because tools wear more rapidly when these alloys are machined, even though chip formation is entirely adequate.

The third group, the most difficult to machine, is composed mainly of the high-strength manganese bronzes and aluminum bronzes that are high in iron or nickel content. Relative machinability of alloys belonging to the three groups is shown in Table 4.

General-Purpose Alloys

General-purpose copper casting alloys are often classified as either red or yellow alloys. Nominal compositions and general properties of these alloys are shown in Tables 1 and 3.

Fig. 3 Variations from design dimensions for two typical cast red brass valve bodies

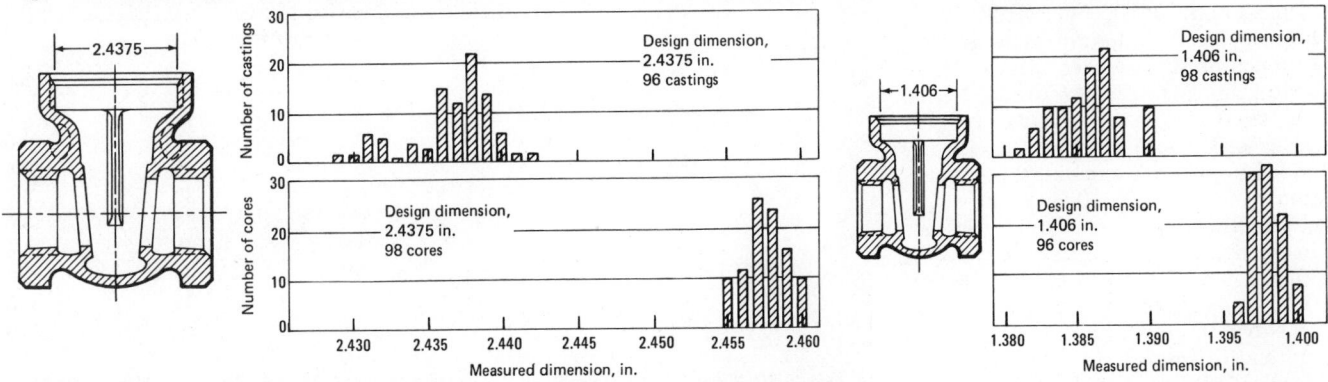

Valve bodies, similar in design but of different sizes, were made using dry sand cores to shape the internal cavities. Upper histograms indicate dimensional variations for the castings; lower histograms, variations for the corresponding cores.

Table 4 Machinability ratings of several copper casting alloys

UNS number	Common name	Machinability rating (a), %
Group 1—Free-cutting Alloys		
C83600	Leaded red brass	90
C83800	Leaded red brass	90
C84400	Leaded semi-red brass	90
C84800	Leaded semi-red brass	90
C94300	High-leaded tin bronze	90
C85200	Leaded yellow brass	80
C85400	Leaded yellow brass	80
C93700	High-leaded tin bronze	80
C93800	High-leaded tin bronze	80
C93200	High-leaded tin bronze	70
C93500	High-leaded tin bronze	70
C97300	Leaded nickel brass	70
Group 2—Moderately Machinable Alloys		
C86400	Leaded high-strength manganese bronze	60
C92200	Leaded tin bronze	60
C92300	Leaded tin bronze	60
C90300	Tin bronze	50
C90500	Tin bronze	50
C95600	Silicon-aluminum bronze	50
C95300	Aluminum bronze	35
C86500	High-strength manganese bronze	30
Group 3—Hard-to-Machine Alloys		
C86300	High-strength manganese bronze	20
C95200	9% aluminum bronze	20
C95400	11% aluminum bronze	20
C95500	Nickel-aluminum bronze	20

(a) Machinability rating expressed as a percentage of the machinability of C36000, free-cutting brass. The rating is based on relative speed for equivalent tool life. For instance, a material having a rating of 50 should be machined at about half the speed that would be used to make a similar cut in C36000.

The leaded red and leaded semi-red brasses respond readily to ordinary foundry practice and are rated very high in castability. C83600 is the best known of this group and usually is referred to by one of its common names—85-5-5-5 or ounce metal. C83600 and its modification, C83800 (83-4-6-7), constitute the largest tonnage of copper-base foundry alloys.

They are used where moderate corrosion resistance, good machinability, moderate strength and ductility, and good castability are required. C83800 has lower mechanical properties, but better machinability and lower initial metal cost than C83600.

Both C83600 and C83800 are used for plumbing goods, stopcocks, faucets, flanges, feed pumps, meter casings and parts, general household and machinery hardware and fixtures, paper machinery, hydraulic and steam valves, valve disks and seats, impellers, injectors, memorial markers, plaques, statuary and similar products.

C84400 and C84800 are higher in lead and zinc, and lower in copper and tin, than C83600 and C83800. They are lower in price, tensile strength and hardness. Their widest application is in the plumbing goods industry.

The leaded yellow brasses C85200 and C85400 are still lower in price and mechanical properties. Their main applications are die castings for plumbing goods and accessories, low-pressure valves, air and gas fittings, general hardware and ornamental castings. In general, they are best suited for small parts; for larger parts, thick sections should be avoided, and phosphorus may be added to increase fluidity if pressure tightness is required. Aluminum (0.15 to 0.25%) is sometimes added to yellow brasses to increase fluidity and to give a smoother surface, but aluminum should not be added if the casting must be pressure-tight or if phosphorus is to be added as well.

All the red and yellow general-purpose alloys, when properly made and cleaned, can be plated with nickel or chromium.

Alloys that do not contain lead, such as the tin bronzes C90500 (Navy G bronze) and C90300 (modified Navy G bronze), are considerably more difficult to machine than leaded alloys. Alloys containing 10 to 12% Sn, 1 to 2% Ni and 0.1 to 0.3% P are known as gear bronzes. Up to 1.5% Pb is frequently added to increase machinability. Addition of lead to C90300 increases machinability, but a concurrent decrease in tin is needed to maintain elongation. The leaded tin bronzes include C92200 (known as steam bronze, valve bronze or Navy M bronze) and C92300 (commercial G bronze).

All of the tin bronzes are suitable wherever corrosion resistance, leak tightness or greater strength is required at higher operating temperatures than can be tolerated with leaded red or semi-red brasses. The limiting temperature for long-time operation of C92200 is 290 °C (550 °F); for C90300, C90500 and C92300, it is 260 °C (500 °F) because of embrittlement caused by precipitation of a high-tin phase. This reaction does not occur in tin bronzes with tin contents less than about 8%.

For elevated-temperature service in handling fluids and gases, the ASME Boiler and Pressure Vessel Code (in Table UNF-23) defines allowable working stresses for C92200 (leaded tin bronze ASTM B61) and C83600 (leaded red brass ASTM B62) at different temperatures, as given in Table 5.

Nickel frequently is added to tin bronzes to increase density and leak tightness. Alloys containing more than 3% Ni are heat treatable, but must contain less than 0.01% Pb for optimum properties; one example is C94700 (88 Cu - 5 Sn - 2 Zn - 5 Ni).

Selection of Alloys for Corrosion Service

In Table 6, relative corrosion resistance in a wide variety of liquids and gases is given for 14 different classes of copper casting alloys. Certain generalizations may be drawn from examination of the table. The yellow brasses do not have as high corrosion resistance as the other copper alloys in many liquids. The high strength of these alloys may make them desirable selections even though some corrosion may be encountered. Corrosion of these alloys often takes place by dezincification; if this is a problem, alloys with copper contents of 80% or more must be selected.

Table 5 Allowable working stresses for C92200 and C83600 castings (a)

Temperature °C	°F	ASTM B61 MPa	ksi	ASTM B62 MPa	ksi
38	100	47	6.8	41	6.0
65	150	47	6.8	41	6.0
93	200	47	6.8	40	5.8
120	250	47	6.8	38	5.5
150	300	45	6.5	34	5.0
175	350	41	6.0	31	4.5
205	400	38	5.5	24	3.5
230	450	34	5.0	24	3.5
260	500	28	4.0	24	3.5
290	550	23	3.3	24	3.5

(a) Tensile strength of 235 MPa (34 ksi) min. is specified for C92200 in ASTM B61; 205 MPa (30 ksi) min. for C83600 in ASTM B62. (b) From ASME Boiler and Pressure Vessel Code, Table UNF-23.

Table 6 Corrosion ratings of cast copper metals in various media

Corrosive medium	Copper	Tin bronze	Leaded tin bronze	High-leaded tin bronze	Leaded red brass	Leaded semi-red brass	Leaded yellow brass	Leaded high-strength yellow brass	High-strength yellow brass	Aluminum bronze	Leaded nickel brass	Leaded nickel bronze	Silicon bronze	Silicon brass
Acetate solvents	B	A	A	A	A	A	B	A	A	A	A	A	A	B
Acetic acid														
20%	A	C	B	C	B	C	C	C	C	A	C	A	A	B
50%	A	C	B	C	B	C	C	C	C	A	C	B	A	B
Glacial	A	A	A	C	A	C	C	C	C	A	B	B	A	A
Acetone	A	A	A	A	A	A	A	A	A	A	A	A	A	A
Acetylene (a)	C	C	C	C	C	C	C	C	C	C	C	C	C	C
Alcohols (b)	A	A	A	A	A	A	A	A	A	A	A	A	A	A
Aluminum chloride	C	C	C	C	C	C	C	C	C	B	C	C	C	C
Aluminum sulfate	B	B	B	B	B	C	C	C	C	A	C	C	A	A
Ammonia, moist gas	C	C	C	C	C	C	C	C	C	C	C	C	C	C
Ammonia, moisture-free	A	A	A	A	A	A	A	A	A	A	A	A	A	A
Ammonium chloride	C	C	C	C	C	C	C	C	C	C	C	C	C	C
Ammonium hydroxide	C	C	C	C	C	C	C	C	C	C	C	C	C	C
Ammonium nitrate	C	C	C	C	C	C	C	C	C	C	C	C	C	C
Ammonium sulfate	B	B	B	B	B	C	C	C	A	C	C	A	A	
Aniline and aniline dyes	C	C	C	C	C	C	C	C	C	B	C	C	C	C
Asphalt	A	A	A	A	A	A	A	A	A	A	A	A	A	A
Barium chloride	A	A	A	A	C	C	C	C	C	A	A	A	A	C
Barium sulfide	C	C	C	C	C	C	C	C	C	B	C	C	C	C
Beer (b)	A	A	B	B	B	C	C	C	A	A	C	A	A	B
Beet sugar syrup	A	A	B	B	B	A	A	A	B	A	A	A	B	B
Benzine	A	A	A	A	A	A	A	A	A	A	A	A	A	A
Benzol	A	A	A	A	A	A	A	A	A	A	A	A	A	A
Boric acid	A	A	A	A	A	A	A	B	A	A	A	A	A	A
Butane	A	A	A	A	A	A	A	A	A	A	A	A	A	A
Calcium bisulfite	A	A	B	B	B	C	C	C	C	A	B	A	A	B
Calcium chloride (acid)	B	B	B	B	B	C	C	C	C	A	C	C	A	C
Calcium chloride (alkaline)	C	C	C	C	C	C	C	C	C	A	C	A	C	B
Calcium hydroxide	C	C	C	C	C	C	C	C	C	B	C	C	C	C

(continued)

Table 6 (continued)

Corrosive medium	Copper	Tin bronze	Leaded tin bronze	High-leaded tin bronze	Leaded red brass	Leaded semi-red brass	Leaded yellow brass	Leaded high-strength yellow brass	High-strength yellow brass	Aluminum bronze	Leaded nickel brass	Leaded nickel bronze	Silicon bronze	Silicon brass
Calcium hypochlorite	C	C	B	B	B	C	C	C	C	B	C	C	C	C
Cane sugar syrups	A	A	B	A	B	A	A	A	A	A	A	A	A	B
Carbonated beverages (b)	A	C	C	C	C	C	C	C	C	A	C	C	A	C
Carbon dioxide, dry	A	A	A	A	A	A	A	A	A	A	A	A	A	A
Carbon dioxide, moist (b)	B	B	B	C	B	C	C	C	C	A	C	A	A	B
Carbon tetrachloride, dry	A	A	A	A	A	A	A	A	A	A	A	A	A	A
Carbon tetrachloride, moist	B	B	B	B	B	B	B	B	B	B	B	A	A	A
Chlorine, dry	A	A	A	A	A	A	A	A	A	A	A	A	A	A
Chlorine, moist	C	C	B	B	B	C	C	C	C	C	C	C	C	C
Chromic acid	C	C	C	C	C	C	C	C	C	C	C	C	C	C
Citric acid	A	A	A	A	A	A	A	A	A	A	A	A	A	A
Copper sulfate	B	A	A	A	A	C	C	C	C	B	B	B	A	A
Cottonseed oil (b)	A	A	A	A	A	A	A	A	A	A	A	A	A	A
Creosote	B	B	B	B	B	C	C	C	C	A	B	B	B	B
Ethers	A	A	A	A	A	A	A	A	A	A	A	A	A	A
Ethylene glycol	A	A	A	A	A	A	A	A	A	A	A	A	A	A
Ferric chloride, sulfate	C	C	C	C	C	C	C	C	C	C	C	C	C	C
Ferrous chloride, sulfate	C	C	C	C	C	C	C	C	C	C	C	C	C	C
Formaldehyde	A	A	A	A	A	A	A	A	A	A	A	A	A	A
Formic acid	A	A	A	A	A	B	B	B	B	A	B	B	B	C
Freon	A	A	A	A	A	A	A	A	A	A	A	A	A	B
Fuel oil	A	A	A	A	A	A	A	A	A	A	A	A	A	A
Furfural	A	A	A	A	A	A	A	A	A	A	A	A	A	A
Gasoline	A	A	A	A	A	A	A	A	A	A	A	A	A	A
Gelatin (b)	A	A	A	A	A	A	A	A	A	A	A	A	A	A
Glucose	A	A	A	A	A	A	A	A	A	A	A	A	A	A
Glue	A	A	A	A	A	A	A	A	A	A	A	A	A	A
Glycerin	A	A	A	A	A	A	A	A	A	A	A	A	A	A
Hydrochloric or muriatic acid	C	C	C	C	C	C	C	C	C	B	C	C	C	C
Hydrofluoric acid	B	B	B	B	B	B	B	B	B	A	B	B	B	B
Hydrofluosilicic acid	B	B	B	B	B	C	C	C	C	B	C	C	B	C
Hydrogen	A	A	A	A	A	A	A	A	A	A	A	A	A	A
Hydrogen peroxide	C	C	C	C	C	C	C	C	C	C	C	C	C	C
Hydrogen sulfide, dry	C	C	C	C	C	C	C	C	C	B	C	C	B	C
Hydrogen sulfide, moist	C	C	C	C	C	C	C	C	C	B	C	C	C	C
Lacquers	A	A	A	A	A	A	A	A	A	A	A	A	A	A
Lacquer thinners	A	A	A	A	A	A	A	A	A	A	A	A	A	A
Lactic acid	A	A	A	A	A	C	C	C	C	A	C	C	A	C
Linseed oil	A	A	A	A	A	A	A	A	A	A	A	A	A	A
Liquors														
Black liquor	B	B	B	B	B	C	C	C	C	B	C	C	B	B
Green liquor	C	C	C	C	C	C	C	C	C	B	C	C	C	B
White liquor	C	C	C	C	C	C	C	C	C	A	C	C	C	B
Magnesium chloride	A	A	A	A	A	C	C	C	C	A	C	C	A	B
Magnesium hydroxide	B	B	B	B	B	B	B	B	B	A	B	B	B	B
Magnesium sulfate	A	A	A	A	B	C	C	C	C	A	C	B	A	B
Mercury, mercury salts	C	C	C	C	C	C	C	C	C	C	C	C	C	C
Milk (b)	A	A	A	A	A	A	A	A	A	A	A	A	A	A

(continued)

Often, experience must be relied on for proper selection of alloys. Although laboratory tests can be very useful for selection, many tests are valueless because they fail to duplicate or approximate the conditions to be encountered in service. When used in a "recommended" service application (see Table 6), copper metals generally give the greatest service life per dollar. However, the table can only serve as a guide, and it should be used judiciously.

Additional data are given in two other articles in this volume, "Corrosion of Copper and Copper Alloys" and "Wrought Copper Alloys for Corrosion Service".

Atmospheric Corrosion. Copper alloy castings have been used for centuries for their superior resistance to atmospheric corrosion. Resistance is afforded by the formation of a coating or patina of basic copper sulfate, which ultimately reacts further to form some basic copper carbonate. The sulfate is virtually insoluble in water and so affords good protection.

Liquid Corrosion. Copper alloy castings are widely used for their superior corrosion resistance in many liquid media. Their resistance to corrosion in liquids, like their resistance to atmospheric corrosion, is increased by formation of a stable, adherent reaction product. If the coating is removed by chemical or mechanical means, corrosion resistance is reduced—often severely reduced. Thus rapid corrosion takes place in aerated mineral acids, or under conditions of severe agitation, impingement or high-velocity flow.

Copper metals are attacked by strong organic and inorganic acids and, to some extent, by weak organic acids. Although a copper metal may not visibly corrode, even a minute quantity of copper ions in the solution may not be acceptable in certain applications. This is particularly true for food products, in which adverse color or taste may develop.

Ammonium hydroxide attacks all copper alloys severely, and these alloys are not recommended where ammonium ions may be formed.

Copper metals are generally satisfactory for applications involving neutral organic compounds, including petroleum products, solvents, and animal and vegetable products. However, in the presence of moisture, certain of these materials may form acids, which in turn may attack a copper metal.

In aqueous solutions, attack is accelerated by dissolved oxygen and carbon dioxide. Thus, although copper alloys are widely used for plumbing goods, they are attacked by many natural waters, especially the very soft waters with high oxygen and carbon dioxide contents. Here, carbonic acid is formed, which prevents development of a resistant layer or dissolves any layer formed previously. Dezincification of high-zinc alloys frequently results if they are indiscriminately used in fresh water service.

Bearing and Wear Properties

Copper alloys have long been used for bearings because of their combination of moderate to high strength, corrosion resistance, and either wear resistance or self-lubrication properties. The choice of an alloy depends on required corrosion resistance and fatigue strength, rigidity of backing material, lubrication, thickness of bearing material, load, speed of rotation, atmospheric conditions and other factors. Copper alloys may be cast into plain bearings, cast on steel backs, cast on rolled strip, made into sintered powder-metallurgy shapes, or pressed and sintered onto a backing material.

Three groups of alloys are used for bearing and wear-resistant applications: phosphor bronzes (Cu-Sn); copper-tin-lead (low-zinc) alloys; and manganese, aluminum and silicon bronzes (see Table 1).

Phosphor bronzes (Cu-Sn-P or Cu-Sn-Pb-P alloys) have residual phosphorus ranging from a few hundredths of 1% (for deoxidation and slight hardening) to a maximum of 1%, which imparts great hardness. Nickel often is added to refine grain size and disperse the lead. Copper-tin bearings have high resistance to wear, high hardness and moderately high strength. C90700 is so widely used for gears that it is called gear bronze.

Phosphor bronzes of higher tin content, such as C91100 and C91300, are used in bridge turntables, where loads are high and rotational movement is slow. The maximum load permitted for C91100 (16% Sn) is 17 MPa (2500 psi), and for C91300 (19% Sn), 24 MPa (3500 psi). These bronzes are high in phosphorus (1% max) to impart high hardness, and low in zinc (0.25% max) to prevent seizing. They are very brittle,

Table 6 (continued)

Corrosive medium	Copper	Tin bronze	Leaded tin bronze	High-leaded tin bronze	Leaded red brass	Leaded semi-red brass	Leaded yellow brass	Leaded high-strength yellow brass	High-strength yellow brass	Aluminum bronze	Leaded nickel brass	Leaded nickel bronze	Silicon bronze	Silicon brass
Molasses (b)	A	A	A	A	A	A	A	A	A	A	A	A	A	A
Natural gas	A	A	A	A	A	A	A	A	A	A	A	A	A	A
Nickel chloride	A	A	A	A	A	C	C	C	C	B	C	C	A	C
Nickel sulfate	A	A	A	A	A	C	C	C	C	A	C	C	A	C
Nitric acid	C	C	C	C	C	C	C	C	C	C	C	C	C	C
Oleic acid	A	A	B	B	B	C	C	C	C	A	C	A	A	B
Oxalic acid	A	A	B	B	B	C	C	C	C	A	C	A	A	B
Phosphoric acid	A	A	A	A	A	C	C	C	C	A	C	A	A	A
Picric acid	C	C	C	C	C	C	C	C	C	C	C	C	C	C
Potassium chloride	A	A	A	A	A	C	C	C	C	A	C	A	A	C
Potassium cyanide	C	C	C	C	C	C	C	C	C	C	C	C	C	C
Potassium hydroxide	C	C	C	C	C	C	C	C	C	A	C	C	C	C
Potassium sulfate	A	A	A	A	A	C	C	C	C	A	C	C	A	C
Propane gas	A	A	A	A	A	A	A	A	A	A	A	A	A	A
Sea water	A	A	A	A	A	C	C	C	C	A	C	C	B	B
Soap solutions	A	A	A	A	B	C	C	C	C	A	C	C	A	C
Sodium bicarbonate	A	A	A	A	A	A	A	A	A	A	A	A	A	B
Sodium bisulfate	C	C	C	C	C	C	C	C	C	A	C	C	C	C
Sodium carbonate	C	A	A	A	A	C	C	C	C	A	C	C	C	A
Sodium chloride	A	A	A	A	B	C	C	C	C	A	C	A	C
Sodium cyanide	C	C	C	C	C	C	C	C	C	B	C	C	C	C
Sodium hydroxide	C	C	C	C	C	C	C	C	C	A	C	C	C	C
Sodium hypochlorite	C	C	C	C	C	C	C	C	C	C	C	C	C	C
Sodium nitrate	B	B	B	B	B	B	B	B	B	A	B	A	B	A
Sodium peroxide	B	B	B	B	B	B	B	B	B	B	B	B	B	A
Sodium phosphate	A	A	A	A	A	A	A	A	A	A	A	A	A	A
Sodium sulfate, silicate	A	A	B	B	B	C	C	C	C	A	C	C	A	B
Sodium sulfide, thiosulfate	C	C	C	C	C	C	C	C	C	B	C	C	C	C
Stearic acid	A	A	A	A	A	A	A	A	A	A	A	A	A	A
Sulfur, solid	C	C	C	C	C	C	C	C	C	A	C	C	C	C
Sulfur chloride	C	C	C	C	C	C	C	C	C	C	C	C	C	C
Sulfur dioxide, dry	A	A	A	A	A	A	A	A	A	A	A	A	A	A
Sulfur dioxide, moist	A	A	A	B	B	C	C	C	C	A	C	C	A	B
Sulfur trioxide, dry	A	A	A	A	A	A	A	A	A	A	A	A	A	A
Sulfuric acid														
78% or less	B	B	B	B	B	C	C	C	C	A	C	C	B	B
78% to 90%	C	C	C	C	C	C	C	C	C	B	C	C	C	C
90% to 95%	C	C	C	C	C	C	C	C	C	B	C	C	C	C
Fuming	C	C	C	C	C	C	C	C	C	A	C	C	C	C
Tannic acid	A	A	A	A	A	A	A	A	A	A	A	A	A	A
Tartaric acid	B	A	A	A	A	A	A	A	A	A	A	A	A	A
Toluene	B	B	A	A	A	B	B	B	B	B	B	B	B	A
Trichlorethylene, dry	A	A	A	A	A	A	A	A	A	A	A	A	A	A
Trichlorethylene, moist	A	A	A	A	A	A	A	A	A	A	A	A	A	A
Turpentine	A	A	A	A	A	A	A	A	A	A	A	A	A	A
Varnish	A	A	A	A	A	A	A	A	A	A	A	A	A	A
Vinegar	A	A	B	B	B	C	C	C	C	B	C	C	A	B
Water, acid mine	C	C	C	C	C	C	C	C	C	C	C	C	C	C

(continued)

Table 6 (continued)

Corrosive medium	Copper	Tin bronze	Leaded tin bronze	High-leaded tin bronze	Leaded red brass	Leaded semi-red brass	Leaded yellow brass	Leaded high-strength yellow brass	High-strength yellow brass	Aluminum bronze	Leaded nickel brass	Leaded nickel bronze	Silicon bronze	Silicon brass
Water, condensate............	A	A	A	A	A	A	A	A	A	A	A	A	A	A
Water, potable	A	A	A	A	A	B	B	B	A	A	A	A	A	A
Whiskey (b)	A	A	C	C	C	C	C	C	C	A	C	C	A	C
Zinc chloride................	C	C	C	C	C	C	C	C	C	B	C	C	B	C
Zinc sulfate.................	A	A	A	A	A	C	C	C	C	B	C	A	A	C

Note: ratings: A, recommended; B, acceptable; C, not recommended.
(a) Acetylene forms an explosive compound with copper when moist or when certain impurities are present and the gas is under pressure. Alloys containing less than 65% Cu are satisfactory under this use. When gas is not under pressure other copper alloys are satisfactory. (b) Copper and copper alloys resist corrosion by most food products. Traces of copper may be dissolved and affect taste or color. In such cases, copper metals are often tin coated.

Table 7 Composition and typical properties of heat treated copper casting alloys of high strength and conductivity

UNS number	Nominal composition	Tensile strength MPa	ksi	Yield strength MPa	ksi	Elongation, %	Hardness	Electrical conductivity, %IACS
C81400	99Cu-0.8Cr-0.06Be	365	53	250	36	11	69 HRB	70
C81500	99Cu-1Cr	350	51	275	40	17	105 HB	85
C81800	97Cu-1.5Co-1Ag-0.4Be	705	102	515	75	8	96 HRB	48
C82000	97Cu-2.5Co-0.5Be	660	96	515	75	6	96 HRB	48
C82200	98Cu-1.5Ni-0.5Be	655	95	515	75	7	96 HRB	48
C82500	97Cu-2Be-0.5Co-0.3Si	1105	160	1035	150	1	43 HRC	20
C82800	96.6Cu-2.6Be-0.5Co-0.3Si	1140	165	1070	155	1	46 HRC	18

and because of this brittleness are sometimes replaced by manganese bronzes or aluminum bronzes.

High-leaded tin bronzes are used where a softer metal is required at slow to moderate speeds and at loads not exceeding 5.5 MPa (800 psi). Alloys of this type include C93200, C93500 and C93700. The last of these alloys, also known as 80-10-10, is an excellent general bearing alloy, especially well-suited for applications where lubrication may be deficient. C93700 is widely used in machine tools, electrical and railroad equipment, steel-mill machinery and automotive applications. C93200 and C93500 are less costly than C93700 and are used chiefly for replacement bearings in machinery. C93800 (15% Pb) and C94300 (24% Pb) are used where high loads are encountered under conditions of poor or nonexistent lubrication, or under corrosive conditions, such as in mining equip-

ment (pumps and car bearings), or in dusty atmospheres, as in stone-crushing and cement plants. These alloys replace the tin bronzes or low-leaded tin bronzes where operating conditions are unsuitable for alloys containing little or no lead.

High-strength manganese bronzes have high tensile strength, hardness and resistance to shock. Large gears, bridge turntables (slow motion and high compression), roller tracks for antiaircraft guns, and recoil parts of cannons are typical applications.

Aluminum bronzes with 8 to 9% Al are used widely for bushings and bearings in light-duty or high-speed machinery. Aluminum bronzes containing 11% Al, either as cast or heat treated, are suitable for heavy-duty service (such as valve guides, rolling-mill bearings, screwdown nuts and slippers) and precision machinery. As aluminum content increases above 11%, hardness

increases and elongation decreases to low values. Such bronzes are well suited for guides and aligning plates, where wear is excessive. Above 13% Al, aluminum bronzes exceed 300 HB in hardness but are brittle. Such alloys are suitable for dies and other parts not subjected to impact loads.

Aluminum bronze generally has a considerably higher fatigue limit and freedom from galling than manganese bronze. On the other hand, manganese bronze has great toughness for equivalent tensile strength and does not need to be heat treated.

Electrical and Thermal Conductivity

Electrical and thermal conductivity of any casting will invariably be lower than for wrought metal of the same composition. Copper castings are used in the electrical industry for their current-carrying capacity and as water-cooled parts of melting and refining furnaces for their high thermal conductivity. However, for a copper casting to be sound and have electrical or thermal conductivity of at least 85%, care must be taken in melting and casting. The ordinary deoxidizers (silicon, tin, zinc, aluminum and phosphorus) cannot be used, because small residual amounts lower electrical and thermal conductivity drastically. Calcium boride or metallic lithium will help to produce sound castings with high conductivity.

Cast copper is soft and low in strength. Increased strength and hardness and good conductivity can be obtained with heat treated alloys containing silicon, cobalt, chromium, nickel and beryllium in various combinations.

These alloys, however, are expensive and less readily available than the standardized alloys. Table 7 presents some of the properties of these alloys after heat treatment.

Cost Considerations

During design of a copper alloy casting, foundry personnel or the design engineer must choose a method of producing internal cavities. There is no general rule for choosing between cored and coreless designs. A cost analysis will determine which is the more economical method of producing the casting, although frequently the choice can be decided by experience.

In one instance, costs were compared

for producing a small (½ in.) valve disk both as a cored casting and as a machined casting (internal cavities made without cores). The machined casting could be produced for about 78% of the cost of making the identical casting using dry sand cores—a savings of 22% in favor of the machined casting. In a similar instance, producing a larger (1½ in.) valve disk as a cored casting requiring only a minimal amount of machining saved more than 8% in overall cost compared to producing the same valve disk without cores. Thus, for two closely related parts, there may be a decisive difference in manufacturing economy when all cost factors are taken into account.

Properties of Cast Copper Alloys

By the ASM Committee on
Copper and Copper Alloys*

C81100

Commercial Names

Previous trade name. CA811

Chemical Composition

Composition limits. 99.70 min Cu + Ag; 0.30 max others (total); (0.01 max P + Si to achieve a conductivity of 92% IACS)

Applications

Typical uses. Electrical and thermal conductors, applications requiring resistance to corrosion and oxidation

Mechanical Properties

Tensile properties. Typical data for sand cast test bars: tensile strength, 170 MPa (25 ksi); yield strength, 62 MPa (9 ksi) at 0.5% extension under load; elongation, 40% in 50 mm or 2 in.
Hardness. 44 HB
Elastic modulus. Tension, 115 GPa (17×10^6 psi)
Fatigue strength. 62 MPa (9 ksi) at 10^8 cycles

Mass Characteristics

Density. 8.94 Mg/m^3 (0.323 lb/in.^{3}) at 20 °C (68 °F)
Volume change on freezing. 4.92% contraction

*See page XI for committee list.

Patternmaker's shrinkage. 21 mm/m (¼ in./ft)

Thermal Properties

Liquidus temperature. 1083 °C (1981 °F)
Solidus temperature. 1065 °C (1948 °F)
Coefficient of thermal expansion. Linear, 16.9 μm/m·K (9.4 μin/in.·°F) at 20 to 300 °C (68 to 572 °F)
Specific heat. 380 J/kg·K (0.09 Btu/lb·°F) at 20 °C (68 °F)
Thermal conductivity. 346 W/m·K (200 Btu/ft·h·°F) at 20 °C (68 °F)

Electrical Properties

Electrical conductivity. Volumetric, 92% IACS at 20 °C (68 °F)

Magnetic Properties

Magnetic permeability. 1.0

Fabrication Characteristics

Machinability. 10% of C36000, free-cutting brass

C81400
99Cu-0.8Cr-0.06Be

Commercial Names

Previous trade name. Beryllium copper 70C, CA814
Common name. Be-modified chrome copper

Specifications

RWMA. Class II

Chemical Composition

Composition limits. 98.5 min Cu; 0.6 to 1.0 Cr; 0.02 to 0.10 Be

Applications

Typical uses. Electrical parts that meet RWMA Class II standards. The beryllium content of this alloy ensures that the chromium content will be kept under control during melting and casting, thus allowing the production of chrome copper castings of consistently high quality.
Precautions in use. Health hazard: see C82500.

Mechanical Properties

Tensile properties. Typical as cast: tensile strength, 205 MPa (30 ksi); yield strength, 83 MPa (12 ksi) at 0.2% offset; elongation, 35% in 50 mm or 2 in. TF00 temper: tensile strength, 365 MPa (53 ksi); yield strength, 250 MPa (36 ksi) at 0.2% offset; elongation, 11% in 50 mm or 2 in.
Hardness. As cast: 62 HRB. TF00 temper: 69 HRB
Elastic modulus. Tension, 110 GPa (16×10^6 psi); shear, 41 GPa (5.9×10^6 psi)

Mass Characteristics

Density. 8.81 Mg/m^3 (0.317 lb/in.^{3}) at 20 °C (68 °F)

Patternmaker's shrinkage. 1.96%

Thermal Properties

Liquidus temperature. 1100 °C (2000 °F)

Solidus temperature. 1075 °C (1950 °F)

Coefficient of thermal expansion. Linear, 18 μm/m·K (10 μin./in.·°F) at 20 to 300 °C (68 to 572 °F)

Specific heat. 389 J/kg·K (0.093 Btu/lb·°F) at 20 °C (68 °F)

Thermal conductivity. 259 W/m·K (150 Btu/ft·h·°F) at 20 °C (68 °F)

Electrical Properties

Electrical conductivity. Volumetric, 70% IACS at 20 °C (68 °F)

Electrical resistivity. 246 μΩ·m at 20 °C (68 °F)

Fabrication Characteristics

Machinability. As cast or TB00 temper: 30% of C36000, free-cutting brass; TF00 temper: 40% of C36000

Melting temperature. 1065 to 1095 °C (1950 to 2000 °F)

Casting temperature. Light castings, 1200 to 1260 °C (2200 to 2300 °F); heavy castings, 1175 to 1230 °C (2150 to 2250 °F)

Solution temperature. 1000 to 1010 °C (1830 to 1850 °F)

Aging temperature. 480 °C (900 °F)

C81500
99Cu-1Cr

Commercial Names

Previous trade names. Chromium copper; CA815

Common name. Chrome copper

Chemical Composition

Composition limits. 98.0 to 99.6 Cu; 0.40 to 1.50 Cr; 0.015 max Pb; 0.04 max P; 0.15 max others (total)

Consequence of exceeding impurity limits. Elements that contribute to hot shortness must be avoided. Because of the high solution temperatures necessary to develop the desired mechanical properties, elements that enter into solid solution must be held to close limits.

Applications

Typical uses. Electrical and/or thermal conductors used as structural members in applications requiring greater strength and hardness than that of cast coppers C80100 to C81100

Mechanical Properties

Tensile properties. Typical data for sand cast test bars, heat treated: tensile strength, 350 MPa (51 ksi); yield strength, 275 MPa (40 ksi) at 0.5% extension under load; elongation, 17% in 50 mm or 2 in.

Hardness. Heat treated, 105 HB

Poisson's ratio. 0.32

Elastic modulus. Tension, 115 GPa (17 × 10^6 psi)

Impact strength. Izod, 41 J (30 ft·lb); Charpy V-notch, 27 J (20 ft·lb)

Fatigue strength. 105 MPa (15 ksi) at 10^8 cycles

Mass Characteristics

Density. 8.82 Mg/m^3 (0.319 lb/in.^{3}) at 20 °C (68 °F)

Patternmaker's shrinkage. 21 mm/m (¼ in./ft)

Thermal Properties

Liquidus temperature. 1085 °C (1985 °F)

Solidus temperature. 1075 °C (1967 °F)

Coefficient of thermal expansion. Linear, 17.1 μm/m·K (9.5 μin./in.·°F) at 20 to 300 °C (68 to 572 °F)

Specific heat. 376 J/kg·K (0.09 Btu/lb·°F) at 20 °C (68 °F)

Thermal conductivity. 315 W/m·K (182 Btu/ft·h·°F) at 20 °C (68 °F)

Electrical Properties

Electrical conductivity. Volumetric: solution heat treated, 40 to 50% IACS at 20 °C (68 °F); precipitation hardened, 80 to 90% IACS at 20 °C (68 °F)

Electrical resistivity. Solution heat treated, 38.3 nΩ·m at 20 °C (68 °F); precipitation hardened, 21 nΩ·m at 20 °C (68 °F). Temperature coefficient: solution heat treated, 0.08 nΩ·m per K at 20 °C (68 °F); precipitation hardened, 0.06 nΩ·m per K at 20 °C (68 °F).

Magnetic Properties

Magnetic permeability. 1.0

Fabrication Characteristics

Machinability. 20% of C36000, free-cutting brass

Weldability. Chromium copper can be silver soldered, soft soldered or brazed; it can be carbon arc welded with copper-chromium filler rod and fused-borax flux.

Solution temperature. 1000 to 1010 °C (1830 to 1850 °F)

Aging temperature. 480 °C (900 °F)

C81800
97Cu-1.5Co-1Ag-0.4Be

Commercial Names

Previous trade name. Beryllium copper alloy 50C, CA818

Specifications

RWMA. Class III

Chemical Composition

Composition limits. 0.30 to 0.55 Be; 1.4 to 1.7 Co; 0.8 to 0.12 Ag; 0.15 max Si; 0.20 max Ni; 0.10 max Fe; 0.10 max Al; 0.10 max Sn; 0.002 max Pb; 0.10 max Zn; 0.10 max Cr; rem Cu

Consequence of exceeding impurity limits. See C82500.

Applications

Typical uses. The silver content of C81800 provides an improved surface conductivity over other RWMA Class III alloys. Typical uses are resistance welding electrode tips and holders and arms.

Precautions in use. *Health hazard: see C82500.*

Mechanical Properties

Tensile properties. See Table 1.

Hardness. See Table 1 and Fig. 1.

Poisson's ratio. 0.33

Elastic modulus. Tension, 110 GPa (16 × 10^6 psi); shear, 41 GPa (6 × 10^6 psi)

Mass Characteristics

Density. 8.62 Mg/m^3 (0.311 lb/in.^{3}) at 20 °C (68 °F)

Patternmaker's shrinkage. 1.56%

Thermal Properties

Liquidus temperature. 1070 °C (1955 °F)

Solidus temperature. 1010 °C (1855 °F)

Coefficient of thermal expansion. Linear, 18 μm/m·K (10 μin./in.·°F) at 20 to 300 °C (68 to 572 °F)

Specific heat. 420 J/kg·K (0.10 Btu/lb·°F) at 20 °C (68 °F)

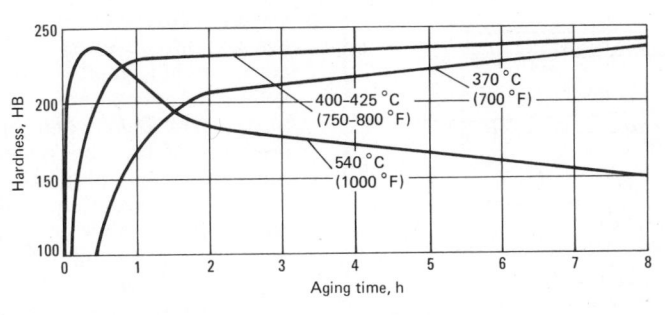

Fig. 1 Aging curves for cast-and-solution-treated C81800

Table 1 Typical mechanical properties of C81800

Temper	Tensile strength MPa	ksi	Yield strength(a) MPa	ksi	Elongation(b), %	Hardness, HRB
As cast	345	50	140	20	20	50
Cast and aged(c)	450	65	275	40	15	70
TB00(c)	310	45	83	12	25	40
TF00(d)(c)	705	102	515	75	8	96

(a) At 0.2% offset. (b) In 50 mm or 2 in. (c) Aged 3 h at 480 °C (900 °F). (d) Solution treated at 900 to 950 °C (1650 to 1750 °F).

Thermal conductivity. 218 W/m·K (126 Btu/ft·h·°F) at 20 °C (68 °F)

Electrical Properties

Electrical conductivity. Volumetric, 48% IACS at 20 °C (68 °F)
Electrical resistivity. 359 μΩ·m at 20 °C (68 °F)

Magnetic Properties

Magnetic susceptibility. See C82000.

Nuclear Properties

Effect of neutron irradiation. See C82500.

Chemical Properties

See C82000.

Fabrication Characteristics

Machinability. As cast or TB00 temper: 30% of C36000, free-cutting brass. TF00 temper: 40% of C36000.
Melting temperature. 1010 to 1070 °C (1855 to 1955 °F)
Casting temperature. Light castings, 1175 to 1230 °C (2150 to 2250 °F); heavy castings, 1120 to 1175 °C (2050 to 2150 °F)
Solution temperature. 900 to 925 °C (1650 to 1700 °F)
Aging temperature. 480 °C (900 °F) See also Fig. 1.

C82000
97Cu-2.5Co-0.5Be

Commercial Names

Previous trade name. Beryllium copper alloy 10C, CA820
Common name. Beryllium copper casting alloy 10C

Specifications

Government. QQ-C-390 (CA820), MIL-C-19464 (Class I)

Chemical Composition

Composition limits. 0.45 to 0.8 Be; 2.4 to 2.7 Co; 0.15 max Si; 0.20 max Ni; 0.10 max Fe; 0.10 max Al; 0.10 max Sn; 0.02 max Pb; 0.10 max Zn; 0.10 max Cr; rem Cu
Consequence of exceeding impurity limits. See C82500.

Applications

Typical uses. C82000 castings are used when a combination of high conductivity and high strength is required. Applications include resistance welding tips, holders and arms, circuit breaker parts, switch gear parts, plunger tips for die casting, con-casting molds, for continuous casting installations, soldering-iron tips, brake drums, and whenever RWMA Class III properties are required.
Precautions in use. *Health hazard:* see C82500.

Mechanical Properties

Tensile properties. See Table 2.
Hardness. See Table 2 and Fig. 2 and 3.
Poisson's ratio. 0.33
Elastic modulus. Tension, 115 GPa (17×10^6 psi); shear, 44 GPa (6.4×10^6 psi)
Fatigue strength. Rotating beam, 125 MPa (18 ksi) at 5×10^7 cycles

Mass Characteristics

Density. 8.62 Mg/m^3 (0.311 lb/in.^{3}) at 20 °C (68 °F)
Patternmaker's shrinkage. 1.56%

Thermal Properties

Liquidus temperature. 1090 °C (1990 °F)
Solidus temperature. 970 °C (1780 °F)
Coefficient of thermal expansion. Linear, 17.8 μm/m·K (9.9 μin./in.·°F) at 20 to 300 °C (68 to 572 °F)
Specific heat. 420 J/kg·K (0.10 Btu/lb·°F) at 20 °C (68 °F)

Table 2 Typical mechanical properties of C82000

Temper	Tensile strength MPa	ksi	Yield strength(a) MPa	ksi	Elongation(b), %	Hardness, HRB
As cast	345	50	140	20	20	52
Cast and aged(c)	450	65	255	37	12	70
TB00(d)	325	47	105	15	25	40
TF00(d)(e)	660	96	515	75	6	96

(a) At 0.2% offset. (b) In 50 mm or 2 in. (c) Aged 2 h at 480 °C (900 °F). (d) Solution treated at 900 to 950 °C (1650 to 1750 °F). (e) Aged 3 h at 480 °C.

Fig. 2 Aging curves for cast-and-solution-treated C82000

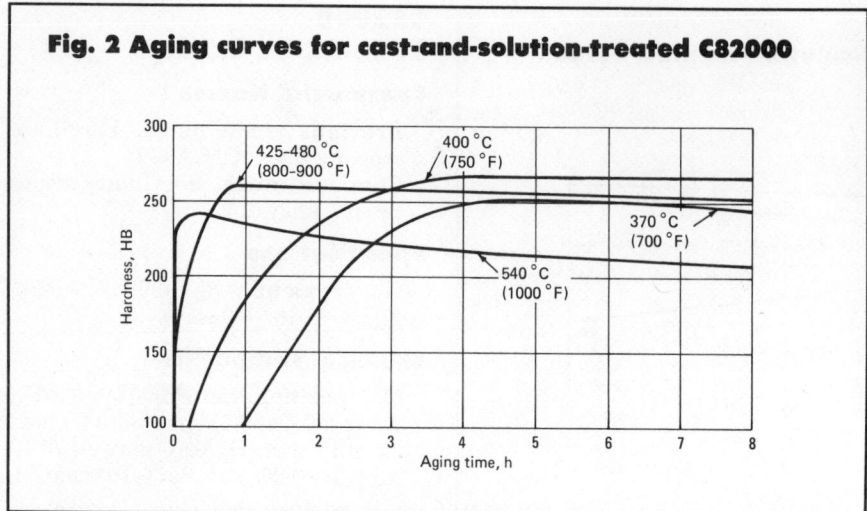

Fig. 3 Hot hardness of C82000, TF00 temper

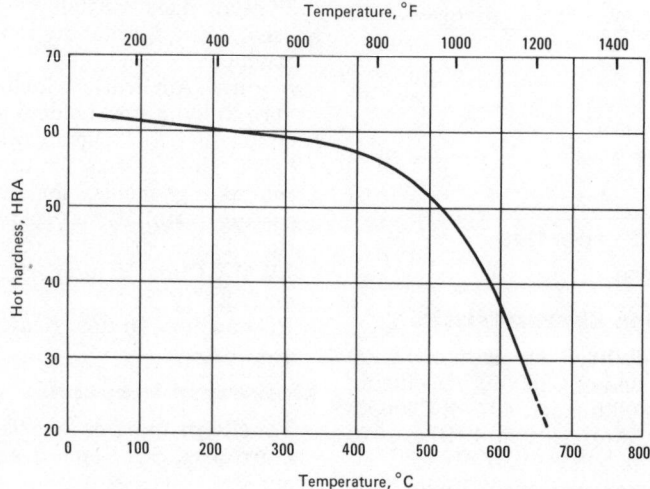

Cast specimens were solution treated, then aged at 480 °C (900 °F). Useful design range, up to about 400 °C (750 °F).

Thermal conductivity. 218 W/m·K (126 Btu/ft·h·°F) at 20 °C (68 °F)

Electrical Properties

Electrical conductivity. Volumetric, 48% IACS at 20 °C (68 °F)
Electrical resistivity. 359 μΩ·m at 20 °C (68 °F)

Magnetic Properties

Magnetic susceptibility. Commercial beryllium copper casting alloys containing 0.02 to 0.8% Be exhibit magnetic susceptibility of +0.001 cgs units or less. Magnetic suscepti-bility varies principally with iron content—on the high side of the commercial range of iron content (0.10%), magnetic susceptibility is approximately 0.001 cgs units; on the low side of the range (0.05%), the value will be much lower (0.0001 cgs units or less). A high-temperature solution treatment of 950 °C (1750 °F) and a low aging temperature of 425 to 450 °C (800 to 850 °F) will tend to keep iron in solution, and keep the alloy nonmagnetic—that is, with a magnetic susceptibility less than 0.0001 cgs units.

Nuclear Properties

Effect of neutron irradiation. See C82500.

Chemical Properties

General corrosion behavior. At elevated temperatures, beryllium is an active oxide former. Beryllium in beryllium copper alloys will preferentially form BeO when low partial pressures of oxygen are present. This causes intergranular oxidation during solution treating in air, often resulting in surface deterioration up to 0.05 mm (0.002 in.) deep. For resistance to water, chemical solutions, organic chemicals and chemical gases, and for resistance to stress-corrosion cracking, see C82500.

Fabrication Characteristics

Machinability. As cast or TB00 temper: 30% of C36000, free-cutting brass. TF00 temper: 40% of C36000.
Melting temperature. 970 to 1090 °C (1780 to 1990 °F)
Casting temperature. Light castings, 1175 to 1230 °C (2150 to 2250 °F); heavy castings, 1120 to 1175 °C (2050 to 2150 °F)
Solution temperature. 900 to 925 °C (1650 to 1700 °F)
Aging temperature. 480 °C (900 °F) See also Fig. 2.

C82200
98Cu-1.5Ni-0.5Be

Commercial Names

Previous trade name. Beryllium copper alloy 30C, CA822
Common name. Beryllium copper casting alloy 30C, 35C or 53B

Specifications

RWMA. Class III

Chemical Composition

Composition limits. 0.35 to 0.8 Be; 1.0 to 2.0 Ni; 0.15 max Si; 0.20 max Co; 0.10 max Fe; 0.10 max Al; 0.10 max Sn; 0.02 max Pb; 0.10 max Zn; 0.10 max Cr; rem Cu
Consequence of exceeding impurity limits. See C82500.

Applications

Typical uses. Seam welder electrodes, projection welder dies, spot welding tips, beam welder shapes,

Fig. 4 Hot hardness of C82200, TF00 temper

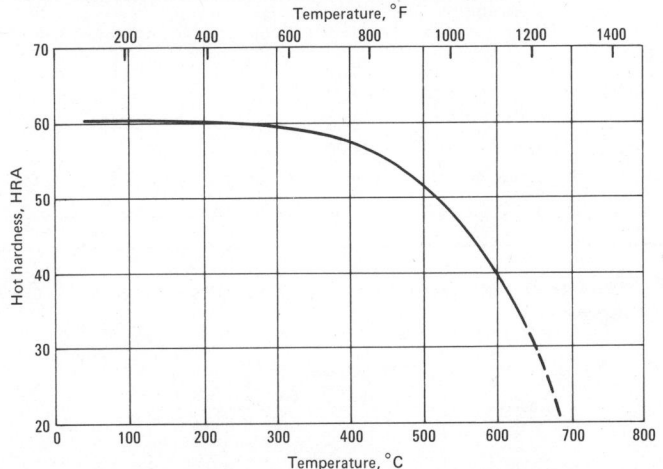

Aged at 480 °C (900 °F). Useful design range, up to 370 °C (700 °F).

Table 3 Typical mechanical properties of C82200

Temper	Tensile strength MPa	ksi	Yield strength(a) MPa	ksi	Elongation(b), %	Hardness, HRB
As cast	345	50	170	25	20	55
Cast and aged(c)	450	65	275	40	15	75
TB00(d)	310	45	85	12	30	30
TF00(d)(c)	655	95	515	75	7	96

(a) At 0.2% offset. (b) In 50 mm or 2 in. (c) Aged 3 h at 480 °C (900 °F). (d) Solution treated at 900 to 955 °C (1650 to 1750 °F).

water-cooled holders, arms bushings for resistance welding, clutch rings, brake drums
Precautions in use. *Health hazard:* see C82500.

Mechanical Properties

Tensile properties. See Table 3.
Hardness. See Table 3 and Fig. 4.
Poisson's ratio. 0.33
Elastic modulus. Tension, 114 GPa (16.5×10^6 psi); shear, 43 GPa (6.2×10^6 psi)

Mass Characteristics

Density. 8.75 Mg/m³ (0.316 lb/in.³) at 20 °C (68 °F)
Patternmaker's shrinkage. 1.56%

Thermal Properties

Liquidus temperature. 1115 °C (2040 °F)
Solidus temperature. 1040 °C (1900 °F)
Coefficient of thermal expansion. Linear, 16.2 μm/m·K (9 μin./in.·°F) at 20 to 200 °C (68 to 392 °F)
Specific heat. 420 J/kg·K (0.10 Btu/lb·°F) at 20 °C (68 °F)
Thermal conductivity. 183 W/m·K (106 Btu/ft·h·°F) at 20 °C (68 °F)

Electrical Properties

Electrical conductivity. Volumetric, 48% IACS at 20 °C (68 °F)
Electrical resistivity. 359 μΩ·m at 20 °C (68 °F)

Magnetic Properties

Magnetic susceptibility. See C82000.

Nuclear Properties

Effect of neutron irradiation. See C82500.

Chemical Properties

See C82000.

Fabrication Characteristics

Machinability. As cast or TB00 temper: 30% of C36000, free-cutting brass. TF00 temper: 40% of C36000.
Melting temperature. 1035 to 1115 °C (1900 to 2040 °F)
Casting temperature. Light castings, 1200 to 1260 °C (2200 to 2300 °F); heavy castings, 1150 to 1200 °C (2000 to 2200 °F)
Solution temperature. 900 to 955 °C (1650 to 1750 °F)
Aging temperature. 445 to 455 °C (835 to 850 °F)

C82400
98Cu-1.7Be-0.3Co

Commercial Names

Previous trade name. Beryllium copper alloy 165C; CA824
Common name. Beryllium copper casting alloy 165C

Specifications

Government. QQ-C-390 (CA824)

Chemical Composition

Composition limits. 1.65 to 1.75 Be; 0.20 to 0.40 Co; 0.10 max Ni; 0.20 max Fe; 0.15 max Al; 0.10 max Sn; 0.02 max Pb; 0.10 max Zn; 0.10 max Cr; rem Cu
Consequence of exceeding impurity limits. See C82500.

Applications

Typical uses. C82400 was developed for use in marine service as a corrosion-resistant, pressure-tight casting material. Its lower beryllium content compared to C82500 makes this alloy the least expensive of the commercial high-strength beryllium copper alloys. When its hardness is relatively low, C82400 exhibits greater than normal toughness. Typical uses include various parts for the submarine telephone cable repeater system and hydrophone, molds for forming plastics, safety tools, plunger tips for die casting, cams, bushings, bearings, valves, pump parts and gears.
Precautions in use. *Health hazard:* see C82500.

Mechanical Properties

Tensile properties. See Table 4.
Hardness. See Table 4.
Poisson's ratio. 0.30
Elastic modulus. Tension, 128 GPa (18.5×10^6 psi); shear, 50 GPa (7.3×10^6 psi)
Fatigue strength. Rotating beam, 160 MPa (23 ksi) at 5×10^7 cycles

Structure

Crystal structure. See C82500.

Mass Characteristics

Density. 8.31 Mg/m³ (0.301 lb/in.³) at 20 °C (68 °F)
Patternmaker's shrinkage. 1.56%
Dilation during aging. Linear, 0.2%
Change in density during aging. 0.6% increase

Thermal Properties

Liquidus temperature. 1000 °C (1825 °F)

Solidus temperature. 900 °C (1650 °F)

Incipient melting temperature. 865 °C (1585 °F)

Coefficient of thermal expansion. Linear, 17.0 μm/m·K (9.4 μin./in.·°F) at 20 to 200 °C (68 to 392 °F)

Specific heat. 419 J/kg·K (0.10 Btu/lb·°F) at 20 °C (68 °F)

Thermal conductivity. 109 W/m·K (61 Btu/ft·h·°F) at 20 °C (68 °F)

Electrical Properties

Electrical conductivity. Volumetric, 25% IACS at 20 °C (68 °F)

Electrical resistivity. 690 μΩ·m at 20 °C (68 °F)

Magnetic Properties

Magnetic susceptibility. See C82500.

Nuclear Properties

Effect of neutron irradiation. See C82500.

Chemical Properties

See C82500.

Fabrication Characteristics

Machinability. As cast or TB00 temper: 30% of C36000, free-cutting brass. Cast and aged or TF00 temper: 10 to 20%

Melting temperature. 900 to 1000 °C (1650 to 1825 °F)

Casting temperature. Light castings, 1080 to 1135 °C (1975 to 2075 °F); heavy castings, 1025 to 1080 °C (1875 to 1975 °F)

Solution temperature. 790 to 815 °C (1450 to 1500 °F)

Aging temperature. 345 °C (650 °F)

C82500
97.2Cu-2Be-0.5Co-0.25Si

Commercial Names

Previous trade name. Beryllium copper 20C, CA825

Common name. Standard beryllium copper casting alloy

Specifications

AMS. Investment castings: 4890
Government. Sand castings: QQ-C-390, MIL-C-19464 (class 2). Centrifugal castings: QQ-C-390. Precision castings: MIL-C-11866 (composition 17), MIL-C-17324. Investment castings: MIL-C-22087

Other. ICI-Cu-2-10780

Chemical Composition

Composition limits. 95.5 min Cu; 1.90 to 2.15 Be; 0.35 to 0.7 Co; 0.20 to

Table 4 Typical mechanical properties of C82400

Temper	Tensile strength MPa	ksi	Yield strength(a) MPa	ksi	Elongation(b), %	Hardness
As cast	485	70	275	40	15	78 HRB
Cast and aged(c)	690	100	550	80	3	21 HRC
TB00(d)	415	60	140	20	40	59 HRB
TF00(d)(c)	1070	155	1000	145	1	38 HRC

(a) At 0.2% offset. (b) In 50 mm or 2 in. (c) Aged 3 h at 345 °C (650 °F) (d) Solution treated at 800 to 815 °C (1475 to 1500 °F).

Fig. 5 Elevated temeprature tensile properties of C82500, TF00 temper

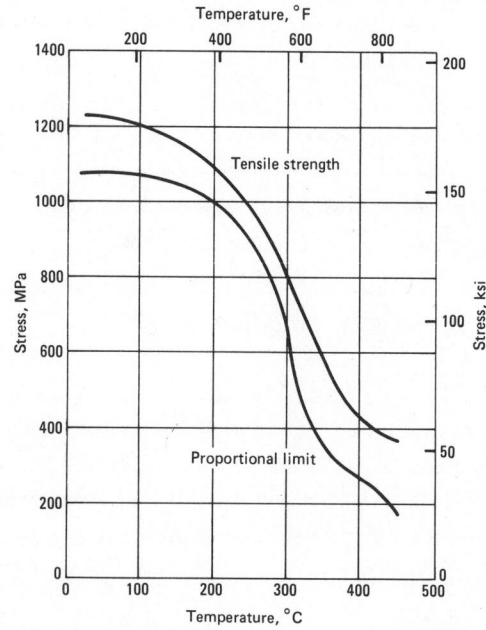

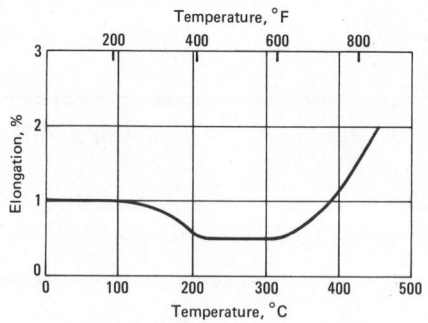

Sand cast test bars were solution treated, then aged at 345 °C (650 °F). Useful design range limited to about 220 °C (425 °F).

Table 5 Typical mechanical properties of C82500

Temper	Tensile strength		Yield strength(a)		Elongation(b), %	Hardness
	MPa	ksi	MPa	ksi		
As cast	515	75	275	40	15	81 HRB
Cast and aged(c)	825	120	725	105	2	30 HRC
TB00(d)	415	60	170	25	35	63 HRB
TF00(d)(c)	1105	160	1035	150	1	43 HRC

(a) At 0.2% offset. (b) In 50 mm or 2 in. (c) Aged 3 h at 345 °C (650 °F). (d) Solution treated at 790 to 800 °C (1450 to 1475 °F).

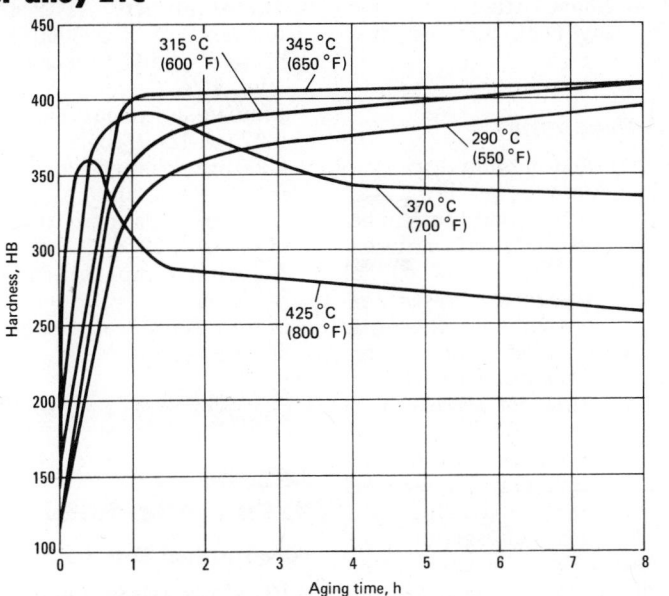

Fig. 6 Aging curves for solution treated C82500 or beryllium copper alloy 21C

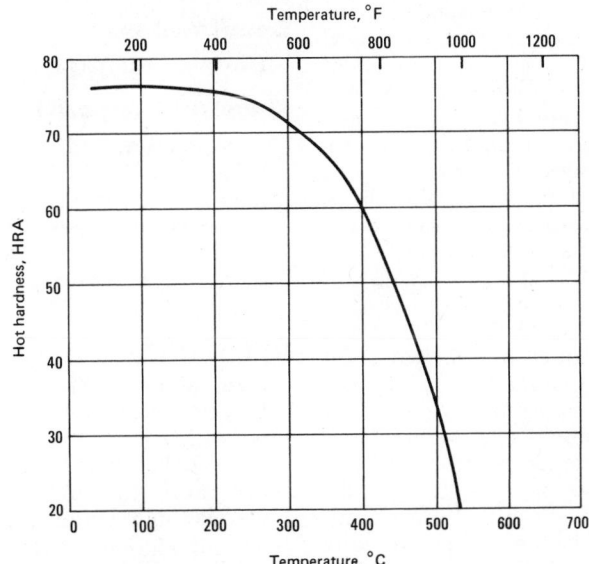

Fig. 7 Hot hardness of C82500, TF00 temper

Specimens were solution treated, then aged at 345 °C (650 °F).

0.35 Si; 0.20 max Ni; 0.25 max Fe; 0.15 max Al; 0.10 max Sn; 0.02 max Pb; 0.10 max Zn; 0.10 max Cr. Available with or without 0.02 to 0.10% Ti added as a grain refiner.

Consequence of exceeding impurity limits. Generally, electrical conductivity is lowered. High Fe raises magnetic susceptibility. High Sn, Zn or Pb causes hot shortness. High Cr diminishes response to precipitation hardening.

Applications

Typical uses. Molds for forming plastics, die casting plunger tips, safety tools, cams, bushings, bearings, gears, sleeves, valves, wear parts, structural parts, resistance welding electrodes and inserts, holders and structural members. Exhibits its low casting temperature, good castability, excellent ability to reproduce fine detail in the pattern, high strength, high electrical and thermal conductivity and excellent resistance to corrosion and wear. Can be sand, shell, ceramic, investment, permanent, pressure and die cast. Especially suited for investment casting and often replaces ferrous castings having similar mechanical properties. Investment castings are used for communication, textile, aerospace, business machine, firearm, instrument and ordnance parts.

Precautions in use. *Health hazard.* Melting, casting, abrasive-wheel operations, abrasive blasting, welding, arc cutting, flame cutting, grinding, polishing and buffing under improper conditions may raise the concentration of beryllium in the air to levels above the limits prescribed by OSHA, thus creating a potential for personnel to contract berylliosis, a chronic lung disease. Exhaust ventilation, the principal means of achieving compliances with these limits, is a specific OSHA requirement for processes involving beryllium alloys. Careful attention to the exhaust-ventilation requirements of these and any other effluent-producing operations is essential. Actual exposure of workers should be continually monitored using prescribed air-sampling and calculation methods to determine compliance or noncompliance with OSHA limits.

Mechanical Properties

Tensile properties. See Table 5 and Fig. 5.

Compressive properties. Compressive yield strength, 1030 to 1200 MPa (150 to 175 ksi) at a permanent set of 0.1%.

Hardness. See Table 5 and Fig. 6 and 7.

Poisson's ratio. 0.30

Elastic modulus. Tension, 128 GPa (18.5×10^6 psi); shear, 50 GPa (7.3×10^6 psi)

Fatigue strength. Rotating beam, 165 MPa (24 ksi) at 5×10^7 cycles

Tensile properties and hardness vs temperature. See Fig. 6.

Structure

Crystal structure. Alpha phase, face-centered cubic. Lattice parameter, a: solution treated (2.1% Be in solid solution), 0.357 nm; precipitation hardened, 0.361 nm

Mass Characteristics

Density. 8.26 Mg/m³ (0.298 lb/in.³) at 20 °C (68 °F)

Patternmaker's shrinkage. 1.56%

Dilation during aging. Linear, 0.2%

Change in density during aging. 0.6% increase

Thermal Properties

Liquidus temperature. 975 °C (1800 °F)

Solidus temperature. 850 °C (1575 °F)

Incipient melting temperature. 835 °C (1535 °F)

Coefficient of thermal expansion. Linear, 17 μm/m·K (9.4 μin./in.·°F) at 20 to 200 °C (68 to 392 °F)

Specific heat. 420 J/kg·K (0.10 Btu/lb·°F) at 20 °C (68 °F)

Thermal conductivity. 105 W/m·K (61 Btu/ft·h·°F) at 20 °C (68 °F)

Electrical Properties

Electrical conductivity. Volumetric, 20% IACS at 20 °C (68 °F)

Electrical resistivity. 862 μΩ·m at 20 °C (68 °F)

Magnetic Properties

Magnetic susceptibility. Commercial beryllium copper casting alloys with 1.6 to 2.7% Be content exhibit magnetic susceptibility of approximately +0.002 cgs units. Magnetic susceptibility varies principally with iron content. On the high side of the range for iron content (0.25%), magnetic susceptibility is greater than 0.002 cgs units. On the low side of the commercial range (0.05%), the value is much less than 0.001 cgs units. A high-temperature solution treatment of 815 °C (1500 °F) and a low aging temperature of 315 to 345 °C (600 to 650 °F) will tend to keep iron in solution and keep magnetic susceptibility below 0.002 cgs units.

Nuclear Properties

Effect of neutron irradiation. Neutron irradiation causes precipitation hardening because of thermal spikes and induced vacancies. This will affect as-cast or solution treated tempers but have little effect on material already peak aged or overaged.

Chemical Properties

General corrosion behavior. At elevated temperatures, beryllium is an active oxide former. Beryllium in beryllium copper alloys will preferentially form BeO when low partial pressures of oxygen are present, especially when the environment is reducing with respect to copper. This leads to preferential formation of BeO films during hot processing of alloys containing 1.6% Be or more. BeO films may be abrasive to fabricating tools, and may be removed mechanically or by pickling. The general corrosion resistance of beryllium copper alloys is similar to that of deoxidized copper, except as indicated above.

Resistance to specific agents. Beryllium copper alloys possess excellent resistance to atmospheric corrosion in marine, industrial and rural environments. They have excellent resistance to organic chemicals such as alcohols, aldehydes, esters and ketones. They are slightly more resistant to seawater than tough pitch or deoxidized copper. Resistance is good with respect to: fresh water; most organic acids, hot or cold dilute sulfuric acid, cold concentrated sulfuric acid and cold dilute hydrochloric acid; hot or cold dilute alkalis and cold concentrated alkalis; salts, including most sulfates and chlorides. Resistance is only fair towards sulfides, especially at elevated temperatures. Resistance is poor towards: mercury and mercury compounds; nitric acid; ferric chloride, ferric sulfate and other heavy-metal salts with oxidizing cations and strong acid anions; acid chromates; and halogens (fluorine, chlorine, bromine and iodine), particularly at elevated temperatures.

Stress-corrosion cracking. Beryllium copper alloys resist stress-corrosion cracking in marine and most chemical environments, even when stressed up to 90% of their 0.2% offset yield strengths. They are susceptible to stress-corrosion cracking in ammonia and halogen gas environments, especially at elevated temperature.

Fabrication Characteristics

Machinability. As cast or TB00 temper: 30% of C36000, free-cutting brass. Cast and aged or TF00 temper: 10 to 20%

Melting temperature. 850 to 980 °C (1575 to 1800 °F)

Casting temperature. Light castings, 1065 to 1175 °C (1950 to 2150 °F); heavy castings, 1010 to 1065 °C (1850 to 1950 °F)

Solution temperature. 790 to 800 °C (1450 to 1475 °F)

Aging temperature. 345 °C (650 °F). See also Fig. 6.

C82600
97Cu-2.4Be-0.5Co

Commercial Names

Previous trade name. Beryllium copper 245C

Common name. Beryllium copper casting alloy 245C

Specifications

Government. QQ-C-390

Chemical Composition

Composition limits. 2.25 to 2.45 Be; 0.35 to 0.7 Co; 0.20 to 0.35 Si; 0.20 max Ni; 0.25 max Fe; 0.15 max Al; 0.10 max Sn; 0.02 max Pb, 0.10 max Zn; 0.10 max Cr; rem Cu

Consequence of exceeding impurity limits. See C82500.

Applications

Typical uses. C82600 is a beryllium copper casting alloy intermediate in beryllium content between C82500 and C82800. It exhibits better fluidity, castability and hardness than C82500 and better toughness and lower cost than C82800. C82600 is used primarily to produce molds for plastic parts. In pressure castings, the lower pouring temperature results in longer tool life than for similar castings of C82500.

Precautions in use. *Health hazard:* see C82500.

Mechanical Properties

Tensile properties. See Table 6.
Hardness. See Table 6.
Poisson's ratio. 0.30
Elastic modulus. Tension, 130 GPa (19×10^6 psi); shear, 50 GPa (7.3×10^6 psi)

Structure

Crystal structure. Alpha phase, face-centered cubic

Mass Characteristics

Density. 8.16 Mg/m³ (0.295 lb/in.³) at 20 °C (68 °F)
Patternmaker's shrinkage. 1.56%
Dilation during aging. Linear, 0.2%
Change in density during aging. 0.6% increase

Thermal Properties

Liquidus temperature. 950 °C (1750 °F)
Solidus temperature. 850 °C (1575 °F)
Incipient melting temperature. 825 °C (1535 °F)
Coefficient of thermal expansion. Linear, 17 μm/m·K (9.4 μin./in.·°F) at 20 to 200 °C (68 to 392 °F)
Specific heat. 420 J/kg·K (0.10 Btu/lb·°F) at 20 °C (68 °F)
Thermal conductivity. 100 W/m·K (58 Btu/ft·h·°F) at 20 °C (68 °F)

Electrical Properties

Electrical conductivity. Volumetric, 19% IACS at 20 °C (68 °F)
Electrical resistivity. 907 μΩ·m at 20 °C (68 °F)

Magnetic Properties

Magnetic susceptibility. See C82500.

Nuclear Properties

Effect of neutron irradiation. See C82500.

Chemical Properties

See C82500.

Fabrication Characteristics

Machinability. As cast or TB00 temper: 30% of C36000, free-cutting brass. Cast and aged or TF00 temper: 10 to 20% of C36000.

Table 6 Typical mechanical properties of C82600

Temper	Tensile strength		Yield strength(a)		Elongation(b), %	Hardness
	MPa	ksi	MPa	ksi		
As cast	550	80	345	50	10	86 HRB
Cast and aged(b)	825	120	725	105	2	31 HRC
TB00(c)	485	70	205	30	12	75 HRB
TF00(d)(c)	1140	165	1070	155	1	45 HRC

(a) At 0.2% offset. (b) In 50 mm or 2 in. (c) Aged 3 h at 345 °C (650 °F). (d) Solution treated at 790 to 800 °C (1450 to 1475 °F).

Melting temperature. 860 to 955 °C (1575 to 1750 °F)
Casting temperature. Light castings, 1040 to 1150 °C (1900 to 2100 °F); heavy castings, 980 to 1040 °C (1800 to 1900 °F)
Solution temperature. 790 to 800 °C (1450 to 1475 °F)
Aging temperature. 345 °C (650 °F)

C82800
96.6Cu-2.6Be-0.5Co-0.3Si

Commercial Names

Previous trade name. Beryllium copper alloy 275C, CA828
Common name. Beryllium copper casting alloy 275C

Specifications

Government. QQ-C-390, MIL-T-16243, MIL-C-19464 (Class IV)
Other. ICI-Cu-2-10785

Chemical Composition

Composition limits. 94.8 min Cu; 2.50 to 2.75 Be; 0.37 to 0.7 Co; 0.20 to 0.35 Si; 0.20 max Ni; 0.25 max Fe; 0.15 max Al; 0.10 max Sn; 0.02 max Pb; 0.10 max Zn; 0.10 max Cr
Consequence of exceeding impurity limits. See C82500.

Applications

Typical uses. C82800 is a special-purpose, high-fluidity casting alloy developed for molds for forming plastics and other applications where the casting process should replicate finest detail with maximum fidelity and the resultant part must exhibit maximum hardness and wear resistance for a cast beryllium copper alloy. The relative slow pouring temperature results in increased tool life during pressure casting and permanent molding. Typical uses are molds for forming plastics, cams, bushings, bearings, valves, pump parts, sleeves and precision cast parts for the communications, textile, aerospace, business machine, firearm, instrument, ordnance and other industries.
Precautions in use. See C82500.

Mechanical Properties

Tensile properties. See Table 7 and Fig. 8.
Hardness. See Table 7 and Fig. 9 and 10.
Poisson's ratio. 0.30
Elastic modulus. Tension, 133 GPa (19.3×10^6 psi); shear, 51 GPa (7.4×10^6 psi)

Structure

Crystal structure. See C82500.

Mass Characteristics

Density. 8.09 Mg/m³ (0.292 lb/in.³) at 20 °C (68 °F)
Patternmaker's shrinkage. 1.56%
Linear dilation during aging. 0.2%

Table 7 Typical mechanical properties of C82800 sand cast test bars

Temper	Tensile strength		Yield strength(a)		Elongation(b), %	Hardness
	MPa	ksi	MPa	ksi		
As cast	550	80	345	50	10	88 HRB
Cast and aged(c)	860	125	760	110	2	31 HRC
TB00(d)	550	80	240	35	10	85 HRB
TF00(d)(c)	1140	165	1070	155	1	46 HRC

(a) At 0.2% offset. (b) In 50 mm or 2 in. (c) Aged 3 h at 345 °C (650 °F). (d) Solution treated at 790 to 800 °C (1450 to 1475 °F).

Fig. 8 Typical tensile properties of C82800, TF00 temper

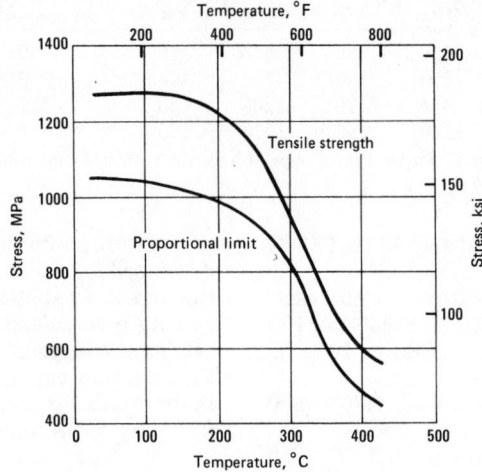

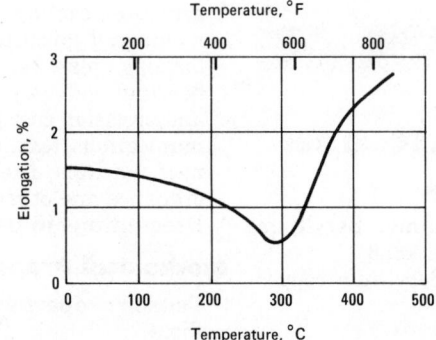

Sand cast test bars were solution treated, then aged at 345 °C (650 °F).

Fig. 9 Aging curves for solution treated C82800

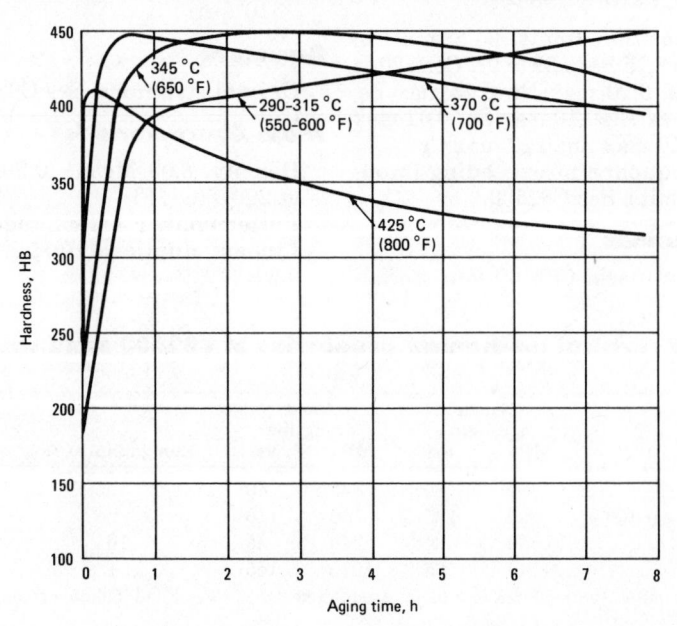

Change in density during aging. 0.6% increase

Thermal Properties

Liquidus temperature. 930 °C (1710 °F)
Solidus temperature. 835 °C (1535 °F)
Incipient melting temperature. 855 °C (1575 °F)
Coefficient of thermal expansion. Linear, 17 μm/m·K (9.4 μin./in.·°F) at 20 to 200 °C (68 to 392 °F)
Specific heat. 420 J/kg·K (0.10 Btu/lb·°F) at 20 °C (68 °F)
Thermal conductivity. 95 W/m·K (55 Btu/ft·h·°F) at 20 °C (68 °F)

Electrical Properties

Electrical conductivity. Volumetric, 18% IACS at 20 °C (68 °F)
Electrical resistivity. 958 μΩ·m at 20 °C (68 °F)

Magnetic Properties

Magnetic susceptibility. See C82500.

Nuclear Properties

Effect of neutron irradiation. See C82500.

Chemical Properties

See C82500.

Fabrication Characteristics

Machinability. As cast or TB00 temper: 30% of C36000, free-cutting brass. Cast and aged or TF00 temper: 10 to 20% of C36000.
Melting temperature. 860 to 930 °C (1575 to 1710 °F)
Casting temperature. Light castings, 1040 to 1150 °C (1900 to 2100 °F); heavy castings, 965 to 1040 °C (1770 to 1900 °F)
Solution temperature. 790 to 800 °C (1450 to 1475 °F)
Aging temperature. 345 °C (650 °F). See also Fig. 9.

C83600
85Cu-5Sn-5Pb-5Zn

Commercial Names

Previous trade names. Leaded red brass; CA836
Common names. Ounce metal; 85-5-5-5; composition metal

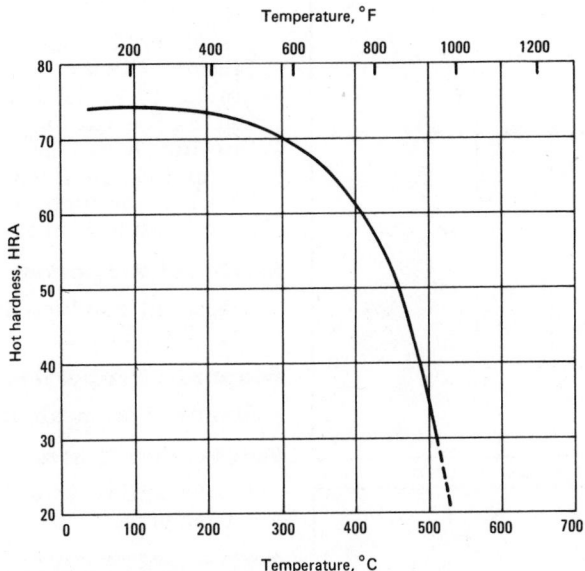

Fig. 10 Hot hardness of C82800, TF00 temper

Specimens were solution treated, then aged at 345 °C (650 °F).

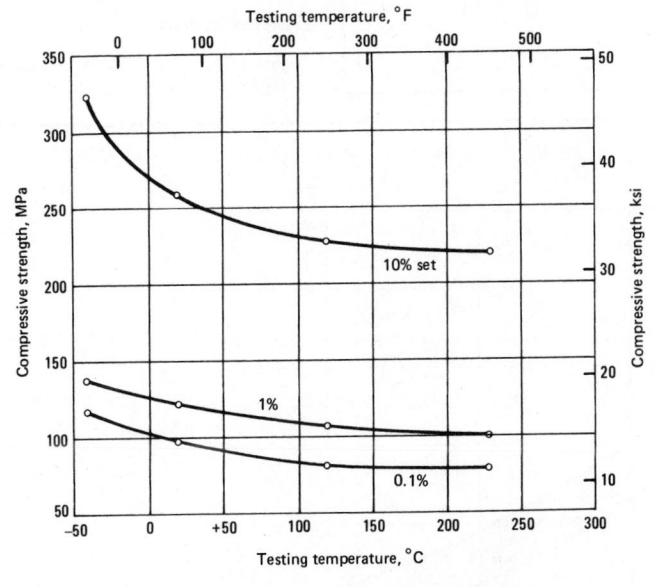

Fig. 11 Typical compressive strength for C83600

Chemical Composition

Composition limits. 84.0 to 86.0 Cu, 4.0 to 6.0 Sn, 4.0 to 6.0 Pb, 4.0 to 6.0 Zn, 0.30 max Fe, 0.25 max Sb, 1.0 max Ni, 0.05 max P (1.5 max for continuous castings), 0.08 max S, 0.005 max Al, 0.005 max Si. In determining min Cu, Cu may be calculated as Cu + Ni.

Consequence of exceeding impurity limits. Aluminum and/or silicon in excess of 0.005% will adversely affect mechanical properties and pressure tightness.

Applications

Typical uses. Good general-purpose casting alloy. For castings requiring moderate strength, soundness and good machinability, such as low-pressure valves, pipe fittings, gasoline- and oil-line fittings, fire-equipment fittings, small gears, small pump parts, general plumbing hardware.

Mechanical Properties

Tensile properties. Typical data for separately cast test bars: tensile strength, 255 MPa (37 ksi); yield strength, 117 MPa (17 ksi) at 0.5% extension under load; elongation, 30% in 50 mm or 2 in.

Compressive properties. Compressive strength at room temperature: 97 MPa (14 ksi) at permanent set of 0.1%; 120 MPa (17.4 ksi) at permanent set of 1%; 258 MPa (37.5 ksi) at permanent set of 10%. See also Fig. 11.

Hardness. 60 HB, typical

Elastic modulus. Tension, 83 GPa (12×10^6 psi) at 20 °C (68 °F). See also Fig. 12.

Impact strength. Izod, 14 J (10 ft·lb); Charpy V-notch, 15 J (11 ft·lb)

Fatigue strength. 76 MPa (11 ksi) at 10^8 cycles. See also Fig. 13.

Creep strength. For 0.1% creep in 10 000 h: 86 MPa (12.5 ksi) at 180 °C (350 °F); 77 MPa (11.1 ksi) at 230 °C (450 °F); 48 MPa (7 ksi) at 290 °C (550 °F)

Mass Characteristics

Density. 8.83 Mg/m³ (0.318 lb/in.³) at 20 °C (68 °F)

Volume change on freezing. 10.6% contraction. Patternmaker's shrinkage, 13 to 16 mm/m (5/32 to 3/16 in./ft)

Specifications

AMS. 4855
ASTM. B30, B62, B271, B505, B584

SAE. J462 (CA836)
Ingot identification number. 115
Government. QQ-C-390 (CA836), MIL-C-15345 (Alloy 1)

Fig. 12 Typical modulus of elasticity in tension for C83600

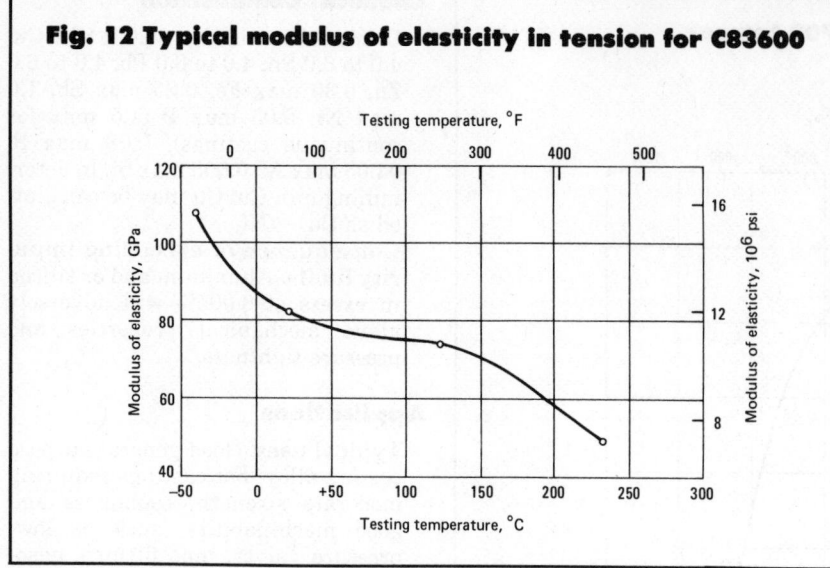

Fig. 14 Mean thermal expansion of C83600

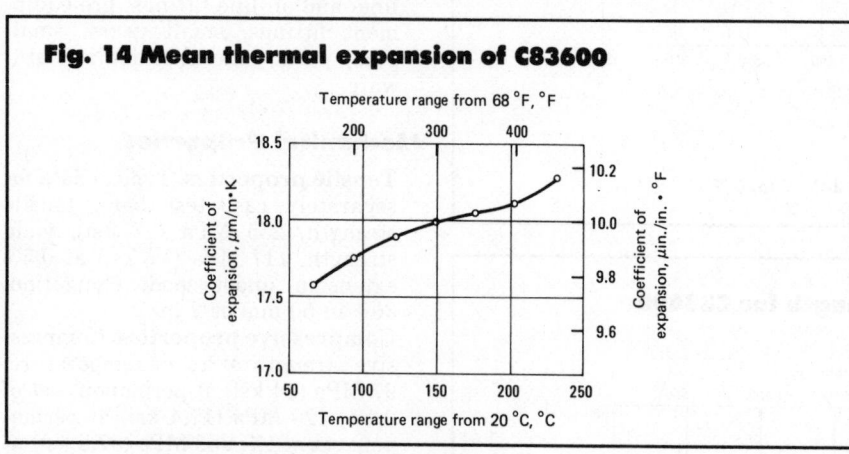

Thermal Properties

Liquidus temperature. 1010 °C (1850 °F)
Solidus temperature. 855 °C (1570 °F)
Coefficient of thermal expansion. Linear, 18.0 μm/m·K (10.0 μin./in.·°F) at 20 to 205 °C (68 to 400 °F). See also Fig. 14.
Specific heat. 380 J/kg·K (0.09 Btu/lb·°F) at 20 °C (68 °F)
Thermal conductivity. 72.0 W/m·K (41.6 Btu/ft·h·°F) at 20 °C (68 °F)

Electrical Properties

Electrical conductivity. Volumetric, 15% IACS

Magnetic Properties

Magnetic permeability. 1.0

Fabrication Characteristics

Machinability. 84% of C36000, free-cutting brass

C83800
83Cu-4Sn-6Pb-7Zn

Commercial Names

Previous trade name. CA838
Common names. Hydraulic bronze; 83-4-6-7

Specifications

ASTM. B30 (CA838), B271 (CA838), B505 (CA838), B584 (CA838)

Fig. 13 Fatigue strength of C83600

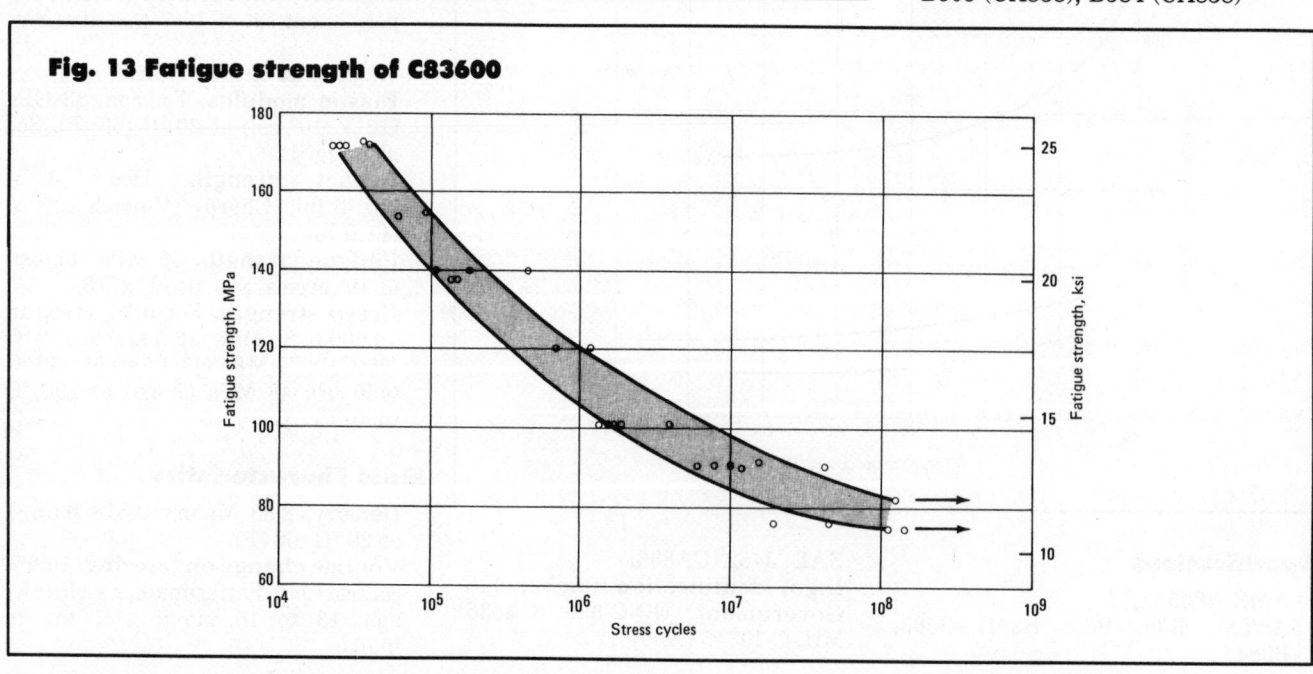

SAE. J462
Ingot identification number. 120
Government. QQ-C-390

Chemical Composition

Composition limits. 82.0 to 83.8 Cu, 3.3 to 4.2 Sn, 5.0 to 7.0 Pb, 5.0 to 8.0 Zn, 0.30 max Fe, 0.25 max Sb, 1.0 max Ni, 0.03 max P (1.5 max for continuous castings), 0.08 max S, 0.005 max Al, 0.005 max Si. In determining min Cu, Cu may be calculated as Cu + Ni

Consequence of exceeding impurity limits. Aluminum and/or silicon in excess of 0.005% will adversely affect mechanical properties and pressure tightness.

Applications

Typical uses. General-purpose free-machining alloy. For air, gas and water fittings, plumbing supplies and fittings, pumps and pump fittings, hardware, carburetors, injectors, railroad catenary and overhead fittings.

Mechanical Properties

Tensile properties. Typical data for separately cast test bars: tensile strength, 240 MPa (35 ksi); yield strength, 110 MPa (16 ksi) at 0.5% extension under load; elongation, 25% in 50 mm or 2 in.
Compressive properties. Compressive strength: 79 MPa (11.5 ksi) at permanent set of 0.1%; 200 MPa (29 ksi) at permanent set of 10%
Hardness. 60 HB
Elastic modulus. Tension, 92 GPa $(13.3 \times 10^6$ psi)
Impact strength. Izod, 11 J (8 ft·lb)

Mass Characteristics

Density. 8.6 Mg/m³ (0.318 lb/in.³) at 20 °C (68 °F)
Patternmaker's shrinkage. 15.6 mm/m (³/16 in./ft)

Thermal Properties

Liquidus temperature. 1005 °C (1840 °F)
Solidus temperature. 845 °C (1550 °F)
Coefficient of thermal expansion. Linear, 18 μm/m·K (10 μin./in.·°F) at 20 to 232 °C (68 to 450 °F)
Specific heat. 380 J/kg·K (0.09 Btu/lb·°F) at 20 °C (68 °F)
Thermal conductivity. 72.5 W/m·K (41.9 Btu/ft·h·°F) at 20 °C (68 °F)

Electrical Properties

Electrical conductivity. Volumetric, 15% IACS

Magnetic Properties

Magnetic permeability. 1.0

Fabrication Characteristics

Machinability. 90% of C36000, free-cutting brass

C84400
81Cu-3Sn-7Pb-9Zn

Commercial Names

Previous trade names. Leaded semi-red brass; CA844
Common names. Valve metal; 81-3-7-9

Specifications

ASTM. B30 (CA844), B271 (CA844), B505 (CA844), B584 (CA844)
Ingot identification number. 123
Government. QQ-C-390

Chemical Composition

Composition limits. 78.0 to 82.0 Cu, 2.3 to 3.5 Sn, 6.0 to 8.0 Pb, 7.0 to 10.0 Zn, 0.40 max Fe, 0.25 max Sb, 1.0 max Ni, 0.02 max P (1.5 max for continuous castings), 0.08 max S, 0.005 max Al, 0.005 max Si. In determining min Cu, Cu may be calculated as Cu + Ni.
Consequence of exceeding impurity limits. Aluminum and/or silicon in excess of 0.005% will adversely affect mechanical properties and pressure tightness.

Applications

Typical uses. Low-pressure valves and fittings, general hardware fittings, plumbing supplies and fixtures, ornamental fixtures

Mechanical Properties

Tensile properties. Typical data for separately cast test bars: tensile strength, 235 MPa (34 ksi); yield strength, 105 MPa (15 ksi) at 0.5% extension under load; elongation, 26% in 50 mm or 2 in.
Hardness. 55 HB
Elastic modulus. Tension, 90 GPa $(13.0 \times 10^6$ psi)
Impact strength. Izod, 11 J (8 ft·lb)

Mass Characteristics

Density. 8.70 Mg/m³ (0.314 lb/in.³) at 20 °C (68 °F)
Patternmaker's shrinkage. 15.6 mm/m (³/16 in./ft)

Thermal Properties

Liquidus temperature. 1005 °C (1940 °F)
Solidus temperature. 840 °C (1540 °F)
Coefficient of thermal expansion. Linear, 18 μm/m·K (10 μin./in.·°F) at 20 to 260 °C (68 to 500 °F)
Specific heat. 380 J/kg·K (0.09 Btu/lb·°F) at 20 °C (68 °F)
Thermal conductivity. 72.5 W/m·K (41.9 Btu/ft·h·°F) at 20 °C (68 °F)

Electrical Properties

Electrical conductivity. Volumetric, 16.4% IACS

Magnetic Properties

Magnetic permeability. 1.0

Fabrication Characteristics

Machinability. 90% of C36000, free-cutting brass

C84800
76Cu-2½Sn-6½Pb-15Zn

Commercial Names

Common name. Leaded semi-red brass, plumbing goods brass, 76-2½-6½-15

Specifications

ASTM. B30, B271, B505, B584
Government. QQ-C-390, CA848
Other. Ingot Code No. 130

Chemical Composition

Composition limits. 75.0 to 77.0 Cu(a); 2.0 to 3.0 Sn; 5.5 to 7.0 Pb; 13.0 to 17.0 Zn; 0.40 max Fe; 0.25 max Sb; 1.0 max Ni; 0.02 max P(b); 0.08 max S; 0.005 max Al; 0.005 max Si

(a) In determining Cu, minimum may be calculated as Cu + Ni. (b) 1.5 max P for continuous castings.

Consequence of exceeding impurity limits. Aluminum and/or silicon in excess of 0.005% will adversely affect mechanical properties and pressure tightness.

Applications

Typical uses. Plumbing fixtures,

cocks, faucets, stops, wastes, air and gas line fittings, general hardware fittings, low-pressure valves and fittings

Mechanical Properties

Tensile properties. Typical data for separately cast test bars: tensile strength, 255 MPa (37 ksi); yield strength, 97 MPa (14 ksi) at 0.5% extension under load; elongation, 35% in 50 mm or 2 in.

Compressive properties. Typical compressive strength: 88.3 MPa (12.8 ksi) at a permanent set of 0.1%; 109 MPa (15.8 ksi) at a permanent set of 1%; 236 MPa (34.3 ksi) at a permanent set of 10%.

Hardness. 55 HB.

Elastic modulus. Tension, 105 GPa $(15 \times 10^6$ psi).

Impact strength. Charpy V-notch, 16 J (12 ft·lb).

Fatigue strength. 76 MPa (11 ksi) at 10^8 cycles.

Creep rupture characteristics. Limiting creep stress for 10^{-5}%/h: 82.0 MPa (11.9 ksi) at 177 °C (350 °F); 55 MPa (8 ksi) at 204 °C (400 °F); 20 MPa (3 ksi) at 288 °C (550 °F)

Mass Characteristics

Density. 8.58 Mg/m^3 (0.310 lb/in.^{3}) at 20 °C (68 °F)

Patternmaker's shrinkage. 16 mm/m (3/16 in./ft)

Thermal Properties

Liquidus temperature. 954 °C (1750 °F)

Solidus temperature. 832 °C (1530 °F)

Coefficient of thermal expansion. Linear, 18.7 µm/m·K (10.4 µin./in.·°F) at 20 to 260 °C (68 to 500 °F).

Specific heat. 376 J/kg·K (0.09 Btu/lb·°F) at 20 °C (68 °F)

Thermal conductivity. 72.0 W/m·K (41.6 Btu/ft·h·°F) at 20 °C (68 °F).

Electrical Properties

Electrical conductivity. Volumetric, 16.4% IACS at 20 °C (68 °F)

Magnetic Properties

Magnetic permeability. 1.0

Fabrication Characteristics

Machinability. 90% of C36000, free-cutting brass

C85200
72Cu-1Sn-3Pb-24Zn

Commercial Names

Previous trade names. Leaded yellow brass; CA852

Common names. High copper yellow brass; 72-1-3-24

Specifications

ASTM. B30 (CA852), B271 (CA852), B584 (CA852)

SAE. J462

Ingot identification number. 400

Government. QQ-C-390 (CA852), MIL-C-15345 (Alloy 28)

Chemical Composition

Composition limits. 70.0 to 74.0 Cu, 0.7 to 2.0 Sn, 1.5 to 3.8 Pb, 20.0 to 27.0 Zn, 0.6 max Fe, 0.20 max Sb, 1.0 max Ni, 0.02 max P, 0.05 max S, 0.005 max Al, 0.05 max Si

Applications

Typical uses. Plumbing fittings and fixtures, ferrules, low-pressure valves, hardware fittings, ornamental brass, chandeliers, andirons

Mechanical Properties

Tensile properties. Typical data for separately cast test bars: tensile strength, 260 MPa (38 ksi); yield strength, 90 MPa (13 ksi) at 0.5% extension under load; elongation, 35% in 50 mm or 2 in.

Hardness. 45 HB

Elastic modulus. Tension, 76 GPa $(11 \times 10^6$ psi)

Mass Characteristics

Density. 8.50 Mg/m^3 (0.307 lb/in.^{3}) at 20 °C (68 °F)

Volume change on freezing. 12.4% contraction. Patternmaker's shrinkage, 15.6 mm/m (3/16 in./ft)

Thermal Properties

Liquidus temperature. 940 °C (1725 °F)

Solidus temperature. 925 °C (1700 °F)

Coefficient of thermal expansion. Linear, 21 µm/m·K (11.5 µin./in.·°F) at 20 to 100 °C (68 to 212 °F)

Specific heat. 380 J/kg·K (0.09 Btu/lb·°F) at 20 °C (68 °F)

Thermal conductivity. 83.9 W/m·K (48.5 Btu/ft·h·°F) at 20 °C (68 °F)

Electrical Properties

Electrical conductivity. Volumetric, 18.6% IACS

Magnetic Properties

Magnetic permeability. 1.0

Fabrication Characteristics

Machinability. 80% of C36000, free-cutting brass

C85400
67Cu-1Sn-3Pb-29Zn

Commercial Names

Previous trade names. Leaded yellow brass; CA854

Common names. No. 1 yellow brass; 67-1-3-29

Specifications

ASTM. B30 (CA854), B271 (CA854), B584 (CA854)

SAE. J462 (CA854)

Ingot identification number. 403

Government. QQ-C-390 (CA854), MIL-C-15345 (Alloy 23)

Chemical Composition

Composition limits. 65.0 to 70.0 Cu, 0.50 to 1.5 Sn, 1.5 to 3.5 Pb, 24.0 to 32.0 Zn, 0.7 max Fe, 1.0 max Ni, 0.35 max Al (a), 0.05 max Si

(a) Addition of 0.20 to 0.30% Al improves castability.

Applications

Typical uses. General-purpose casting alloy. For lightweight castings not subject to high internal pressure, such as furniture hardware, ornamental castings, radiator fittings, ship trimmings, gas cocks, light fixtures, battery clamps

Mechanical Properties

Tensile properties. Typical data for separately cast test bars: tensile strength, 235 MPa (34 ksi); yield strength, 83 MPa (12 ksi) at 0.5% extension under load; elongation, 35% in 50 mm or 2 in.

Hardness. 50 HB

Mass Characteristics

Density. 8.45 Mg/m^3 (0.305 lb/in.^{3}) at 20 °C (68 °F)

Patternmaker's shrinkage. 15.6 mm/m (3/16 in./ft)

Thermal Properties

Liquidus temperature. 940 °C (1725 °F)

Solidus temperature. 925 °C (1700 °F)

Coefficient of thermal expansion. Linear, 20.2 μm/m·K (11.2 μin./in.·°F) at 20 to 100 °C (68 to 212 °F)

Specific heat. 380 J/kg·K (0.09 Btu/lb·°F) at 20 °C (68 °F)

Thermal conductivity. 88 W/m·K (51 Btu/ft·h·°F) at 20 °C (68 °F)

Electrical Properties

Electrical conductivity. Volumetric, 19.6% IACS

Magnetic Properties

Magnetic permeability. 1.0

Fabrication Characteristics

Machinability. 80% of C36000, free-cutting brass

C85700, C85800
63Cu-1Sn-1Pb-35Zn

Commercial Names

Previous trade names. CA857, CA858

Common names. Leaded yellow brass; 63-1-1-35

Specifications

ASTM. B30 (CA857, CA858), B176 (CA858), B271 (CA857), B584 (CA857)

SAE. J462

Ingot identification number. 406

Government. QQ-C-390 (CA857), MIL-C-15345 (Alloy 3)

Chemical Composition

Composition limits. See Table 8. Addition of 0.20 to 0.30% Al improves castability.

Applications

Typical uses. Bushings, hardware fittings, ornamental castings, lock hardware

Mechanical Properties

Tensile properties. Typical data for separately cast test bars. Sand castings or centrifugal castings (C85700): tensile strength, 345 MPa (50 ksi); yield strength, 125 MPa (18 ksi) at 0.5% extension under load; elongation, 40% in 50 mm or 2 in. Die castings (C85800): tensile strength, 380 MPa (55 ksi); yield strength, 201 MPa (30 ksi) at 0.5% extension under

Table 8 Composition limits of C85700 and C85800

Sand Castings or Centrifugal Castings (C85700)

Cu	58.0-64.0	Fe	0.7 max
Sn	0.50-1.50	Ni	1.0 max
Pb	0.80-1.50	Al	0.55 max
Zn	32.0-40.0	Si	0.05 max

Die Castings (C85800)

Cu	58.0 min	Fe	0.50 max
Sn	1.5 max	Al	(b)
Pb	1.5 max	Mn	0.25 max
Zn	(a)	Others	0.50 max (c)

(a) ASTM B176, 31-41; SAE J462, 31.0-34.0. (b) ASTM B176, 0.25 max; SAE J462, 0.50 max. (c) SAE J462 allows 0.05 max Sb, 0.50 max Ni, 0.05 max As, 0.05 max S, 0.01 max P and 0.25 max Si before determination of total unnamed elements.

load; elongation, 15% in 50 mm or 2 in.

Hardness. Sand castings or centrifugal castings (C85700), 75 HB; die castings (C85800), 102 HB

Elastic modulus. Tension: sand castings or centrifugal castings (C85700), 97 GPa (14 × 10⁶ psi); die castings (C85800), 105 GPa (15 × 10⁶ psi)

Mass Characteristics

Density. 8.41 Mg/m³ (0.304 lb/in.³) at 20 °C (68 °F)

Patternmaker's shrinkage. 15.6 mm/m (³/₁₆ in./ft)

Thermal Properties

Liquidus temperature. 920 °C (1688 °F)

Solidus temperature. 903 °C (1657 °F)

Coefficient of thermal expansion. Linear, 22 μm/m·K (12 μin./in.·°F) at 20 to 260 °C (68 to 500 °F)

Specific heat. 376 J/kg·K (0.09 Btu/lb·°F) at 20 °C (68 °F)

Thermal conductivity. 83.9 W/m·K (48.5 Btu/ft·h·°F) at 20 °C (68 °F)

Electrical Properties

Electrical conductivity. Volumetric, 22% IACS

Magnetic Properties

Magnetic permeability. 1.0

Fabrication Characteristics

Machinability. 80% of C36000, free-cutting brass

C86100, C86200
64Cu-24Zn-3Fe-5Al-4Mn

Commercial Names

Common names. Manganese

bronze (90 000 psi); High strength yellow brass; CA861; CA862

Specifications

ASTM. C86100: none. C86200: Ingot, B30; centrifugal castings, B271; sand castings, B584; continuous castings, B505

SAE. J462. (former alloy number: 430A)

Government. QQ-C-390, QQ-C-523. C86100: centrifugal castings, MIL-C-15345 (alloy 5); investment castings, MIL-C-22087 (composition 7); sand castings, MIL-C-22229 (composition 10). C86200: investment castings, MIL-C-22087 (composition 9); precision castings, MIL-C-11866 (composition 20); sand castings, MIL-C-22229 (composition 9)

Ingot identification number. 423

Chemical Composition

Composition limits. C86100: 66.0 to 68 Cu; 4.5 to 5.5 Al; 2.0 to 4.0 Fe; 2.5 to 5.0 Mn; 1.0 max Ni; 0.2 max Sn; 0.2 max Pb; rem Zn. C86200: 60.0 to 68.0 Cu; 3.0 to 7.5 Al; 2.0 to 4.0 Fe; 2.5 to 5.0 Mn; 1.0 max Ni; 0.2 max Sn; 0.2 max Pb; rem Zn

Applications

Typical uses. Marine castings, gears, gun mounts, bushings and bearings

Mechanical Properties

Tensile properties. Nominal. Tensile strength, 655 MPa (95 ksi); yield strength, 330 MPa (48 ksi); elongation, 20% in 50 mm or 2 in.

Compressive properties. Compressive strength, 345 MPa (50 ksi) at a permanent set of 0.1%

Hardness. 180 HB

Elastic modulus. Tension, 105 GPa (15 × 10⁶ psi)

Impact strength. Izod, 16 J (12 ft·lb)

Mass Characteristics

Density. 7.9 Mg/m³ (0.285 lb/in.³) at 20 °C (68 °F)
Volume change on freezing. 2%

Thermal Properties

Liquidus temperature. 941 °C (1725 °F)
Solidus temperature. 899 °C (1650 °F)
Coefficient of thermal expansion. Linear, 22 µm/m·K (12 µin./in.·°F) at 20 to 260 °C (68 to 500 °F)
Specific heat. 376 J/kg·K (0.09 Btu/lb·°F) at 20 °C (68 °F)
Thermal conductivity. 35 W/m·K (20 Btu/ft·h·°F) at 20 °C (68 °F)

Electrical Properties

Electrical conductivity. Volumetric, 7.5% IACS at 20 °C (68 °F)

Magnetic Properties

Magnetic permeability. 1.24 at field strength of 16 kA/m

Fabrication Characteristics

Machinability. 30% of C36000, free-cutting brass
Annealing temperature. 260 °C (500 °F)

C86300
64Cu-26Zn-3Fe-3Al-4Mn

Commercial Names

Common names. Manganese bronze (110 000 psi); High strength yellow brass; CA863

Specifications

AMS. 4862
ASTM. Sand castings: B22, B584. Centrifugal castings: B271. Continuous castings: B505. Ingot: B30
SAE. J462
Government. QQ-C-390, QQ-C-523. Centrifugal castings: MIL-C-15345, alloy 6. Investment castings: MIL-C-22087, composition 9. Precision castings: MIL-C-11866, composition 21. Sand castings: MIL-C-22229, composition 8
Ingot identification number. 424

Chemical Composition

Composition limits. 60.0 to 68.0 Cu; 2.5 to 5.0 Mn; 3.0 to 7.5 Al; 2.0 to 4.0 Fe; 0.2 max Pb; 0.2 max Sn; rem Zn
Consequence of exceeding impurity limits. Excessive Sn causes brittleness; excessive Pb or Ni decreases elongation

Applications

Typical uses. Extra-heavy duty, high strength alloy for gears, cams, bearings, screw-down nuts, bridge parts, hydraulic cylinder parts
Precautions in use. Not to be used in marine atmospheres, ammonia or high corrosive atmospheres.

Mechanical Properties

Tensile properties. Nominal. Tensile strength, 820 MPa (119 ksi); yield strength, 460 MPa (67 ksi); elongation, 18% in 50 mm or 2 in.

Compressive properties. Compressive strength: 415 MPa (60 ksi) at permanent set of 0.1%; 670 MPa (97 ksi) at permanent set of 1%
Hardness. 225 HB

Elastic modulus. Tension, 105 GPa (15.5 × 10⁶ psi)
Fatigue strength. Rotating beam: 170 MPa (25 ksi) at 100 million cycles
Impact strength. Izod: 20 J (15 ft·lb). Charpy V-notch: 16 J (12 ft·lb)

Creep-rupture characteristics. Stress for 0.17% creep in 10 000 h: 390 MPa (56.5 ksi) at 120 °C (250 °F); 225 MPa (32.5 ksi) at 150 °C (300 °F); 130 MPa (19 ksi) at 175 °C (350 °F); 3 MPa (0.5 ksi) at 230 °C (450 °F). See also Fig. 15 and 16.

Mass Characteristics

Density. 7.7 Mg/m³ (0.278 lb/in.³) at 20 °C (68 °F)
Volume change on freezing. 2%

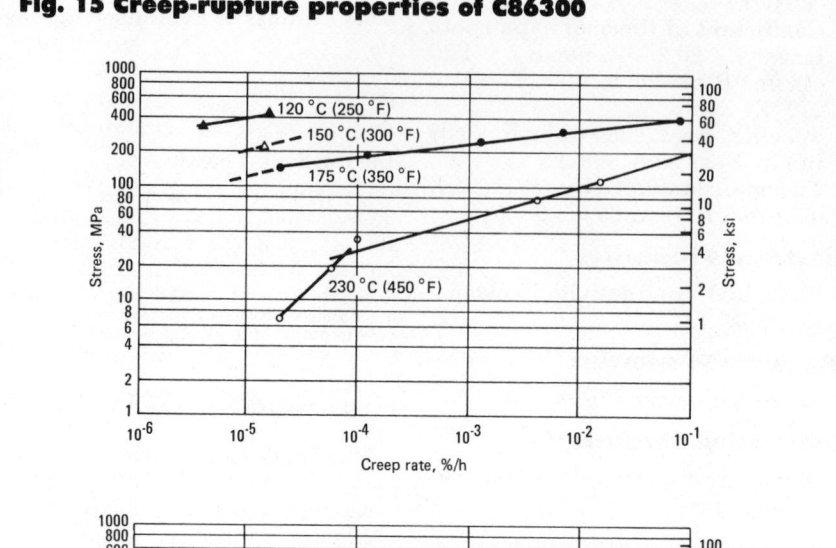

Fig. 15 Creep-rupture properties of C86300

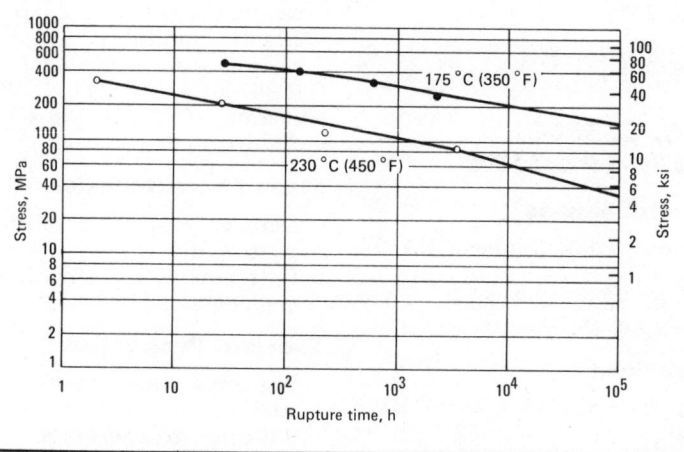

Fig. 16 Isochrenous stress-strain curves for C86300

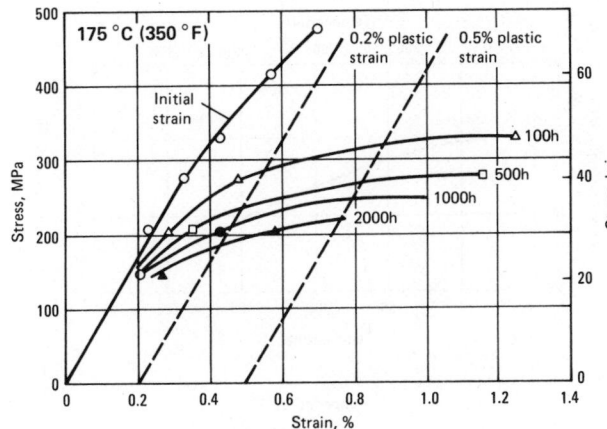

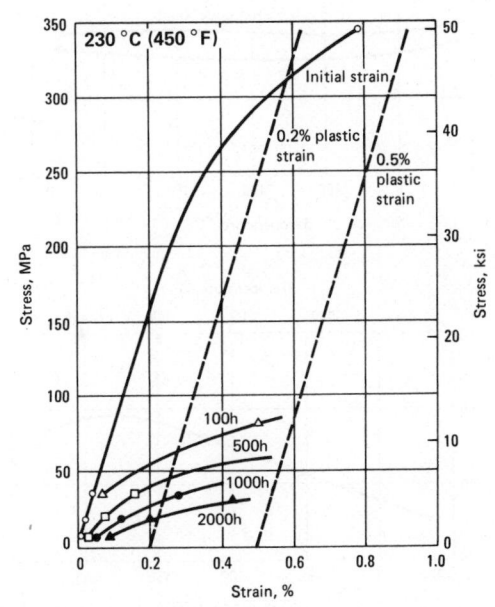

C86400
59Cu-0.75Sn-0.75Pb-37Zn-1.25Fe-0.75Al-0.5Mn

Commercial Names

Previous trade name. Leaded high strength yellow brass; stem manganese bronze
Common name. Manganese bronze (60 000 psi)

Specifications

ASTM. Sand castings: B584. Centrifugal castings: B271. Ingot: B30
Government. QQ-C-390, QQ-C-523
Ingot identification number. 420

Chemical Composition

Composition limits. 56.0 to 62.0 Cu; 1.5 max Sn; 0.5 to 1.5 Pb; 2.0 max Fe; 1.5 max Al; 1.5 max Mn; 1.0 max Ni; rem Zn

Applications

Typical uses. Free-machining manganese bronze for valve stems, marine castings and fittings, pump bodies

Mechanical Properties

Tensile properties. Typical tensile strength, 450 MPa (65 ksi); yield strength, 170 MPa (25 ksi); elongation, 20% in 50 mm or 2 in.
Compressive properties. Compressive strength: 150 MPa (22 ksi) at 0.1% permanent set; 600 MPa (87 ksi) at 10% permanent set
Hardness. 105 HB
Elastic modulus. Tension, 96 GPa (14×10^6 psi)
Impact strength. Izod: 40 J (30 ft·lb). Charpy V-notch: 34 J (25 ft·lb)

Mass Characteristics

Density. 8.32 Mg/m³ (0.301 lb/in.³) at 20 °C (68 °F)
Volume change on freezing. 2%

Thermal Properties

Liquidus temperature. 880 °C (1615 °F)
Solidus temperature. 860 °C (1585 °F)
Coefficient of thermal expansion. Linear, 20 μm/m·K (11.4 μin./in.·°F) at 21 to 204 °C (70 to 400 °F)
Specific heat. 376 J/kg·K (0.09 Btu/lb·°F) at 20 °C (68 °F)
Thermal conductivity. 88 W/m·K (51 Btu/ft·h·°F) at 20 °C (68 °F)

Thermal Properties

Liquidus temperature. 923 °C (1693 °F)
Solidus temperature. 885 °C (1625 °F)
Coefficient of thermal expansion. Linear, 22 μm/m·K (12 μin./in.·°F) at 20 to 260 °C (68 to 500 °F)
Specific heat. 376 J/kg·K (0.09 Btu/lb·°F) at 20 °C (68 °F)
Thermal conductivity. 36 W/m·K (21 Btu/ft·h·°F) at 20 °C (68 °F)

Electrical Properties

Electrical conductivity. Volumetric, 9% IACS at 20 °C (68 °F)

Magnetic Properties

Magnetic permeability. 1.09 at field strength of 16 kA/m

Fabrication Characteristics

Machinability. 8% of C36000, free-cutting brass
Annealing temperature. 260 °C (500 °F)

Electrical Properties

Electrical conductivity. Volumetric, 22% IACS at 20 °C (68 °F)

Fabrication Characteristics

Machinability. 60% of C36000, free-cutting brass

Casting temperature range. Light castings: 1038 to 1121 °C (1900 to 2050 °F). Heavy castings: 954 to 1038 °C (1750 to 1900 °F)

Annealing temperature. 260 °C (500 °F)

C86500
58Cu-39Zn-1.3Fe-1A1-0.5Mn

Commercial Names

Previous trade name. High strength yellow brass

Common name. Manganese bronze (65 000 psi)

Specifications

AMS. 4860A

ASTM. Sand castings: B584. Centrifugal castings: B271. Ingot: B30

SAE. J462

Government. QQ-C-390. Sand castings: MIL-C-22229, composition 7. Centrifugal castings: MIL-C-15345, alloy 4. Investment castings: MIL-C-22087, composition 5

Ingot identification number. 421

Chemical Composition

Composition limits. 55.0 to 60.0 Cu; 0.4 to 2.0 Fe; 0.5 to 1.5 Al; 1.5 max Mn; 0.4 max Pb; 1.0 max Sn; 1.0 max Ni; rem Zn

Applications

Typical uses. Propeller hubs, blades and other parts in contact with salt and fresh water, gears, liners

Mechanical Properties

Tensile properties. Typical. Tensile strength, 490 MPa (71 ksi); yield strength, 195 MPa (28 ksi); elongation, 30% in 50 mm or 2 in. See also Fig. 17.

Compressive properties. Compres-

Fig. 17 Typical tensile properties of C86500

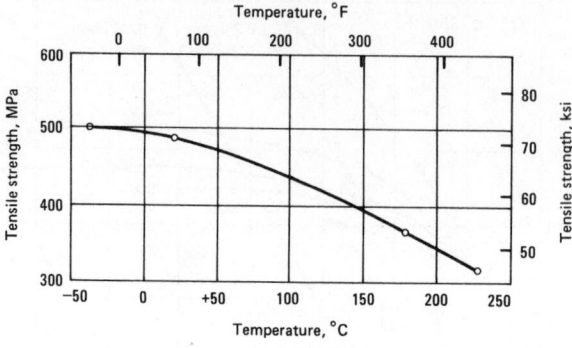

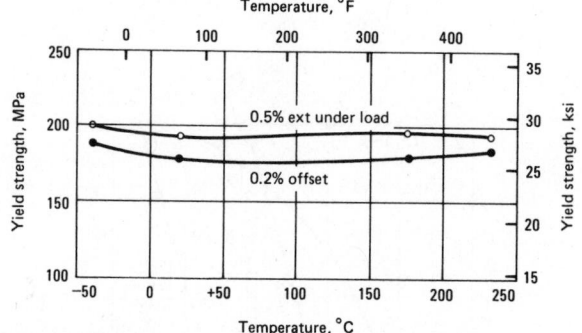

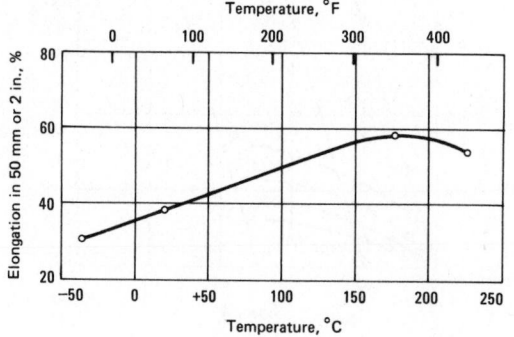

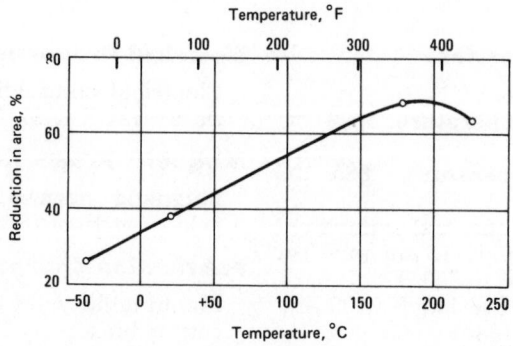

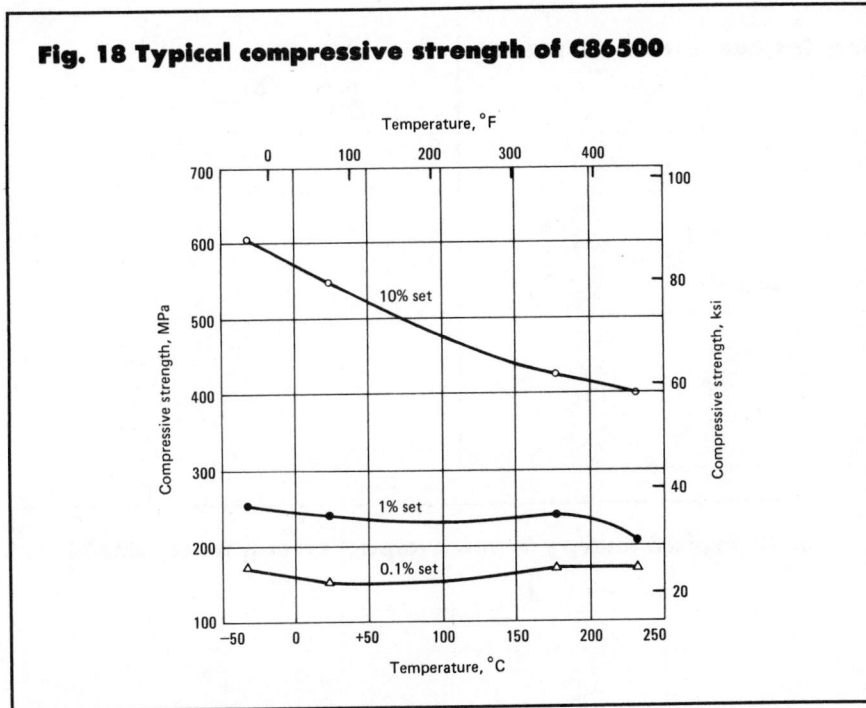

Fig. 18 Typical compressive strength of C86500

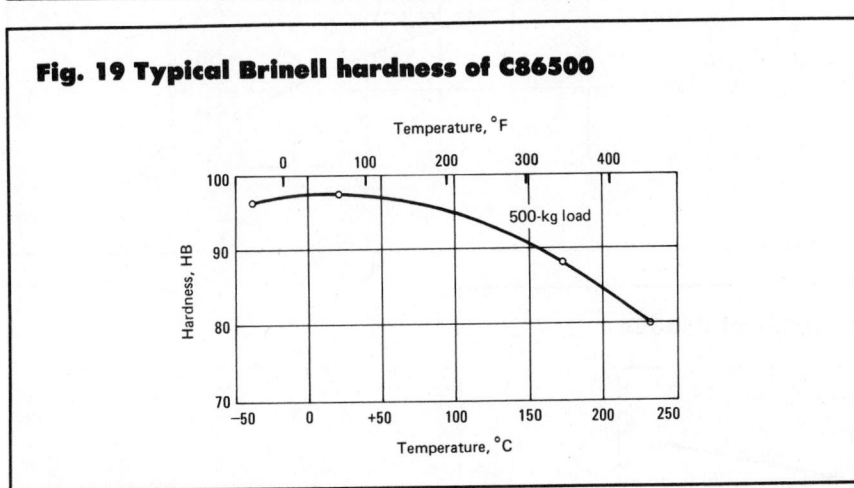

Fig. 19 Typical Brinell hardness of C86500

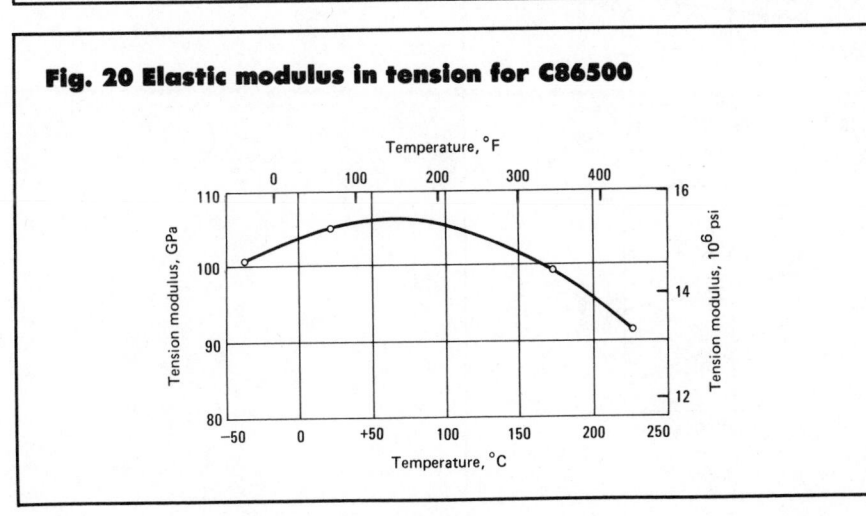

Fig. 20 Elastic modulus in tension for C86500

sive strength: 165 MPa (24 ksi) at permanent set of 0.1%; 240 MPa (35 ksi) at permanent set of 1%; 545 MPa (79 ksi) at permanent set of 10%. See also Fig. 18.

Hardness. 130 HB. See also Fig. 19.

Elastic modulus. Tension, 105 GPa (15×10^6 psi). See also Fig. 20.

Fatigue strength. Reverse bending: 145 MPa (21 ksi) at 10^8 cycles. See also Fig. 21.

Impact strength. Charpy, 43 J (31 ft·lb). See also Fig. 22.

Creep-rupture characteristics. Stress for 0.1% creep in 10 000 h: 190 MPa (28 ksi) at 120 °C (250 °F); 43 MPa (6.2 ksi) at 175 °C (350 °F); 12 MPa (1.7 ksi) at 230 °C (450 °F). See also Fig. 23 and 24.

Mass Characteristics

Density. 8.3 Mg/m³ (0.299 lb/in.³) at 20°C (68 °F)

Patternmaker's shrinkage. 1.65 to 2.15% for pouring temperature of 905 °C (1665 °F)

Thermal Properties

Liquidus temperature. 880 °C (1616 °F)

Solidus temperature. 862 °C (1583 °F)

Coefficient of thermal expansion. Linear, 21.6 μm/m·K (11.3 μin./in.·°F) at 21 to 93 °C (70 to 200 °F). See also Fig. 25.

Specific heat. 373 J/kg·K (0.089 Btu/lb·°F) at 20 °C (68 °F)

Thermal conductivity. 87 W/m·K (50.2 Btu/ft·h·°F) at 20 °C (68 °F). See also Fig. 25.

Electrical Properties

Electrical conductivity. Volumetric, 20.5% IACS at 20 °C (68 °F). See also Fig. 26.

Electrical resistivity. See Fig. 26.

Magnetic Properties

Magnetic permeability. 1.09 at field strength of 16 kA/m

Fabrication Characteristics

Machinability. 26% of C36000, free-cutting brass

Annealing temperature. 260 °C (500 °F)

Fig. 21 Typical reverse bending fatigue curve at room temperature for C86500

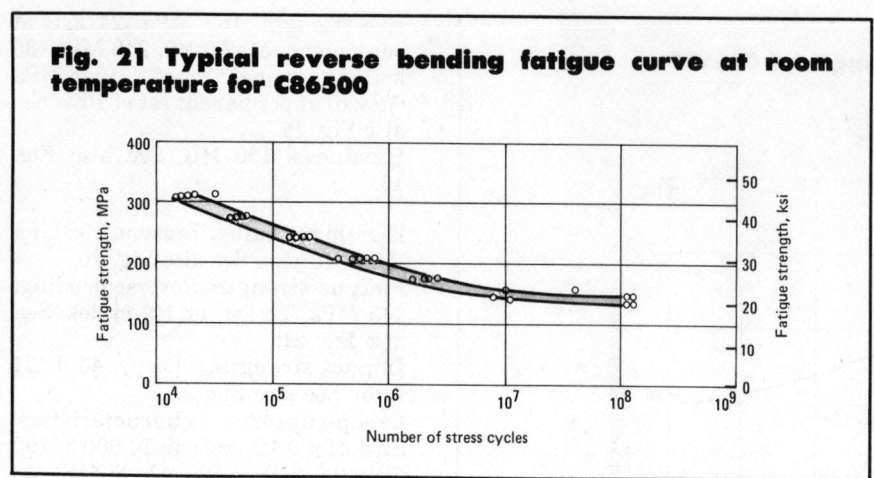

Fig. 22 Typical Charpy V-notch impact strength for C86500

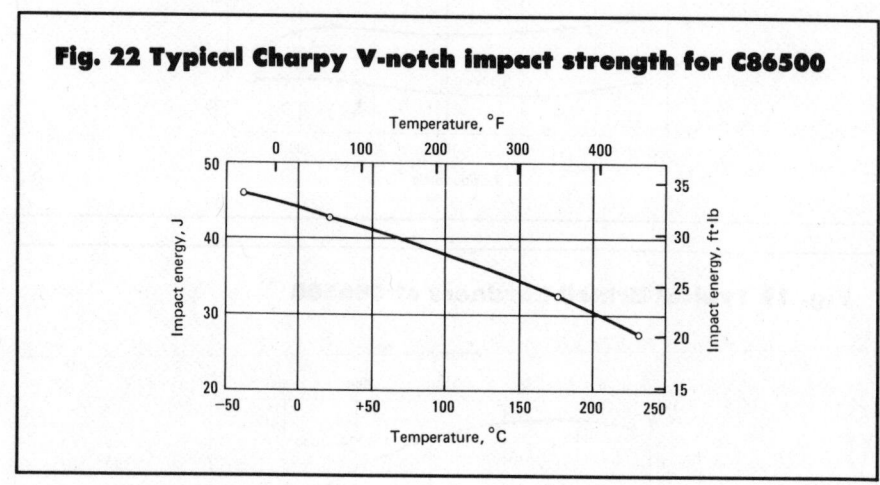

Fig. 23 Typical creep-rupture properties of C86500

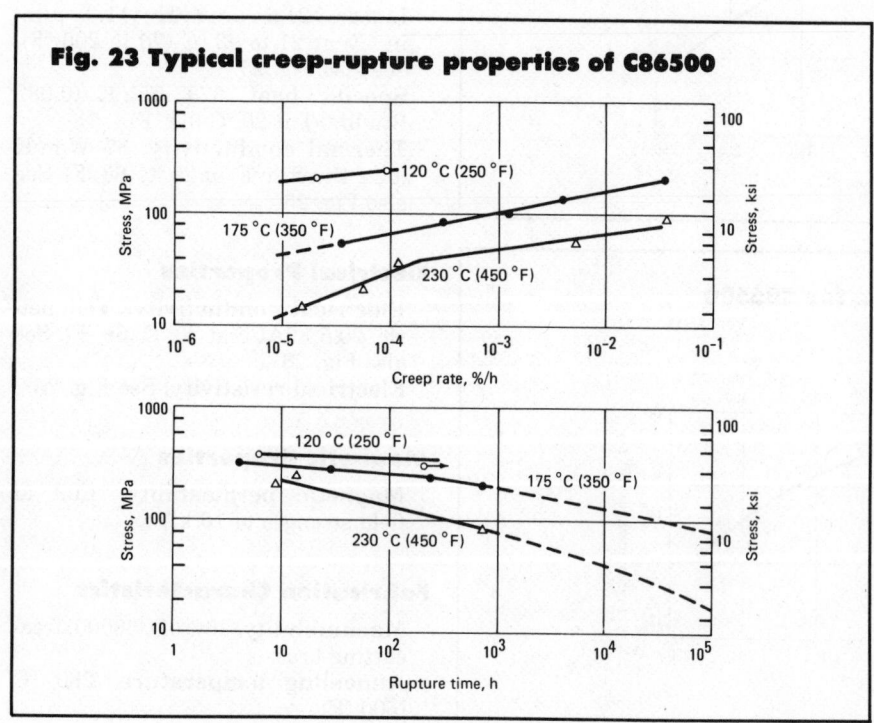

Fig. 24 Isochronous stress-strain curves for C86500

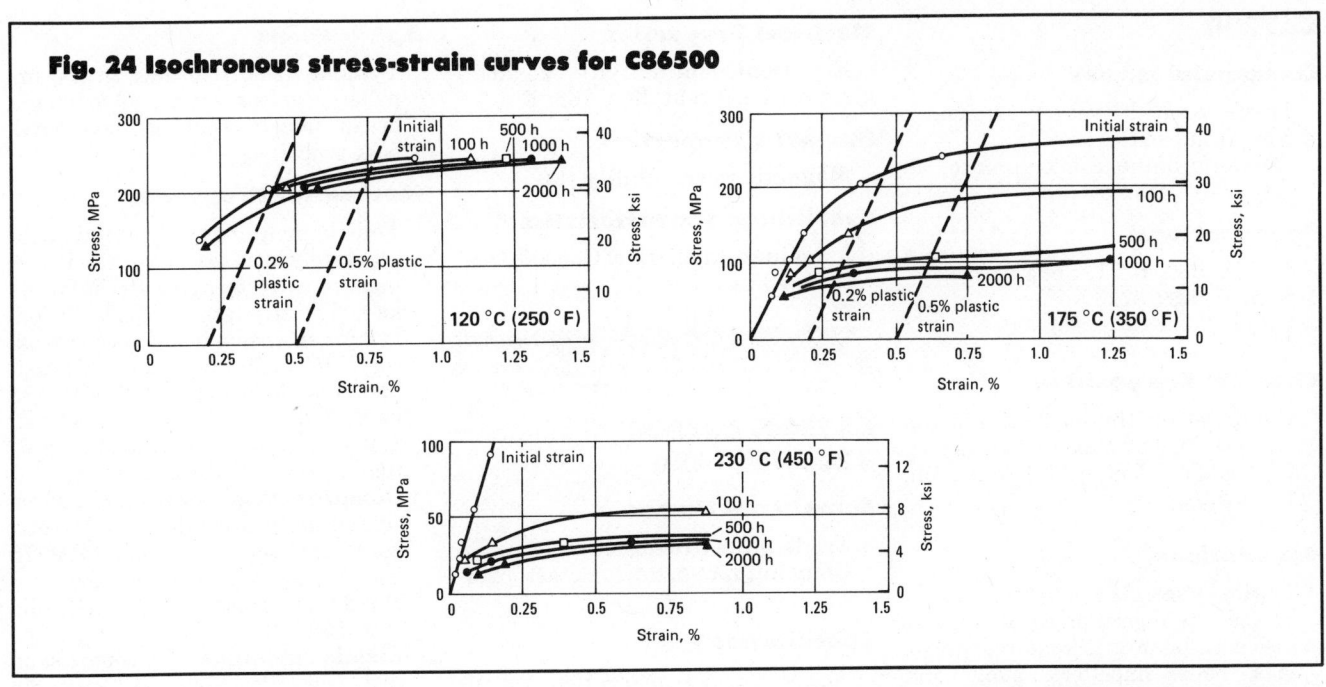

Fig. 25 Typical thermal properties of C86500

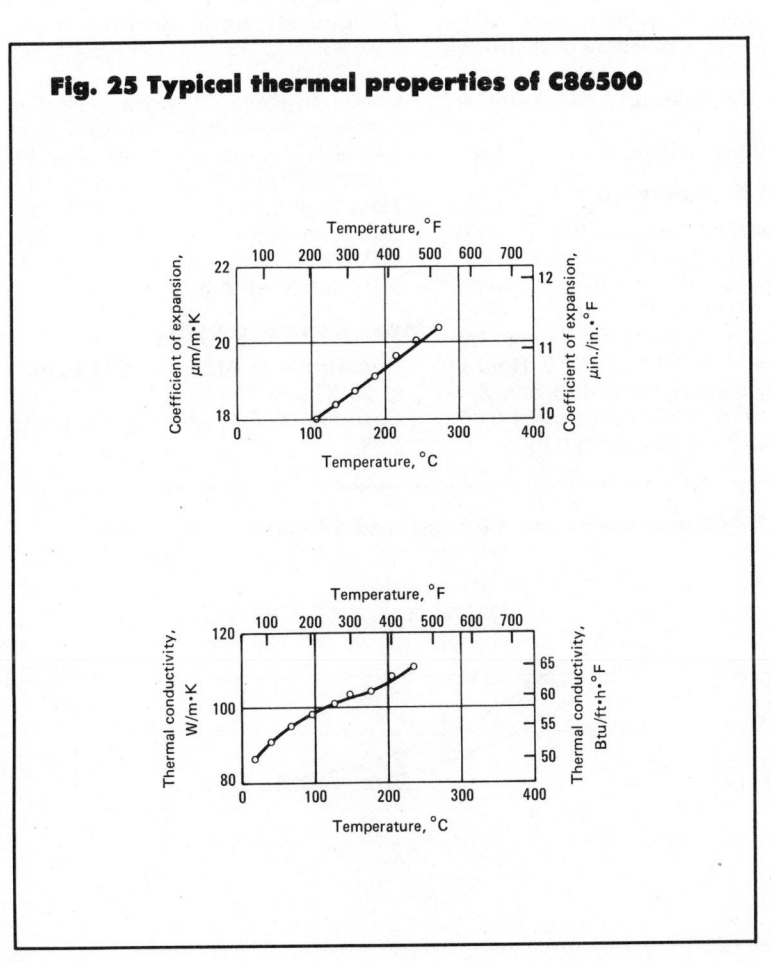

Fig. 26 Variation of electrical properties with temperature for C86500

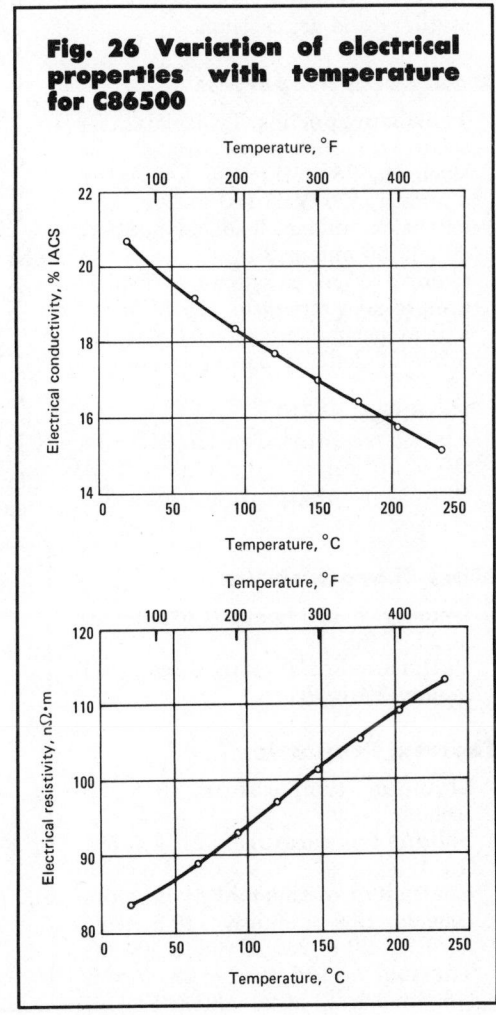

C87200

Commercial Names

Trade name. Everdur, Herculoy, Navy Tombasil
Common name. Silicon bronze, 95-1-4, 92-4-4, 89-6-5

Specifications

ASTM. B30, B271, B584 (CA872)
SAE. J462 (C87200)
Other. Ingot Code No. 500

Chemical Composition

Composition limits. 89.0 min Cu; 1.0 max Sn; 0.50 max Pb; 5.0 max Zn; 2.5 max Fe; 1.5 max Al; 1.5 max Mn; 1.0 to 5.0 Si

Applications

Typical uses. As a substitute for tin bronze where good physical and corrosion resistance is required. Bearings, bells, impellers, pump and valve components, marine fittings, statuary and art castings.

Mechanical Properties

Tensile properties. Typical data for separately cast test bars: tensile strength, 380 MPa (55 ksi); yield strength, 170 MPa (25 ksi) at 0.5% extension under load; elongation, 30% in 50 mm or 2 in.
Compressive properties. Typical compressive strength: 125 MPa (18 ksi) at permanent set of 0.1%; 415 MPa (60 ksi) at permanent set of 10%
Hardness. 85 HB
Elastic modulus. Tension, 105 GPa (15×10^6 psi)
Impact strength. Izod, 45 J (33 ft·lb)

Mass Characteristics

Density. 8.36 Mg/m³ (0.302 lb/in.³) at 20 °C (68 °F)
Patternmaker's shrinkage. 21 mm/m (1/4 in./ft)

Thermal Properties

Liquidus temperature. 916 °C (1680 °F)
Solidus temperature. 821 °C (1510 °F)
Coefficient of thermal expansion. Linear, 19.6 μm/m·K (10.9 μin./in.·°F) at 20 to 260 °C (68 to 500 °F)
Thermal conductivity. 28 W/m·K (16 Btu/ft·h·°F) at 20 °C (68 °F)

Electrical Properties

Electrical conductivity. Volumetric, 6.7% IACS at 20 °C (68 °F)

Magnetic Properties

Magnetic permeability. 1.0

Fabrication Characteristics

Machinability. 50% of C36000, free-cutting brass

C87500, C87800
82Cu-4Si-14Zn

Commercial Names

Trade name. Tombasil
Common name. Silicon brass, 82-4-14

Specifications

ASTM. C87500: ingots, B30; centrifugal castings, B271; sand castings, B584. C87800: die castings, B176
SAE. J462
Government. C87500: sand castings, QQ-C-390; investment castings, MIL-C-22087, (composition 4). C87800: die castings, MIL-B-15894 (class 3)
Other. Ingot code 500T

Chemical Composition

Composition limits. C87500: 79.0 min Cu, 0.50 max Pb; 12.0 to 16.0 Zn; 0.50 max Al; 3.0 to 5.0 Si. C87800: 80.0 to 83.0 Cu; 0.25 max Sn; 0.15 max Pb; 0.15 max Fe; 0.15 max Mn; 0.15 max Al; 3.75 to 4.25 Si; 0.01 max Mg; 0.25 max others (total) rem Zn, but As, Sb and S not to exceed 0.05 each, and P not to exceed 0.01

Applications

Typical uses. Bearings, gears, impellers, rocker arms, valve stems, brush holders, bearing races, small boat propellers

Mechanical Properties

Tensile properties. Typical data for separately cast test bars. Sand castings: tensile strength, 460 MPa (67 ksi); yield strength, 205 MPa (30 ksi) at 0.5% extension under load; elongation, 21% in 50 mm or 2 in. Die castings: tensile strength, 585 MPa (85 ksi); yield strength, 310 MPa (45 ksi) at 0.5% extension under load; elongation, 25% in 50 mm or 2 in.
Compressive properties. Compressive strength, 183 MPa (26.5 ksi) at a permanent set of 0.1%; 515 MPa (75 ksi) at a permanent set of 10%
Hardness. Sand cast, 134 HB; die cast, 163 HB
Elastic modulus. Tension: sand cast, 106 GPa (15.4×10^6 psi); die cast, 138 GPa (20.0×10^6 psi)
Impact strength. Charpy V-notch, 43 J (32 ft·lb)
Fatigue strength. Rotating beam: 150 MPa (22 ksi) at 10^8 cycles. See also Fig. 27.
Creep-rupture characteristics. Limiting creep stress for 10^{-5} %/h: 195 MPa (28 ksi) at 180 °C (350 °F); 75 MPa (11 ksi) at 230 °C (450 °F); 9.5 MPa (1.4 ksi) at 290 °C (550 °F). Stress for rupture in 100 000 h: 125 MPa (18 ksi) at 230 °C (450 °F); 20 MPa (3 ksi) at 290 °C (550 °F)

Mass Characteristics

Density. 8.28 Mg/m³ (0.299 lb/in.³) at 20 °C (68 °F)
Patternmaker's shrinkage. 1.5 to 1.9%

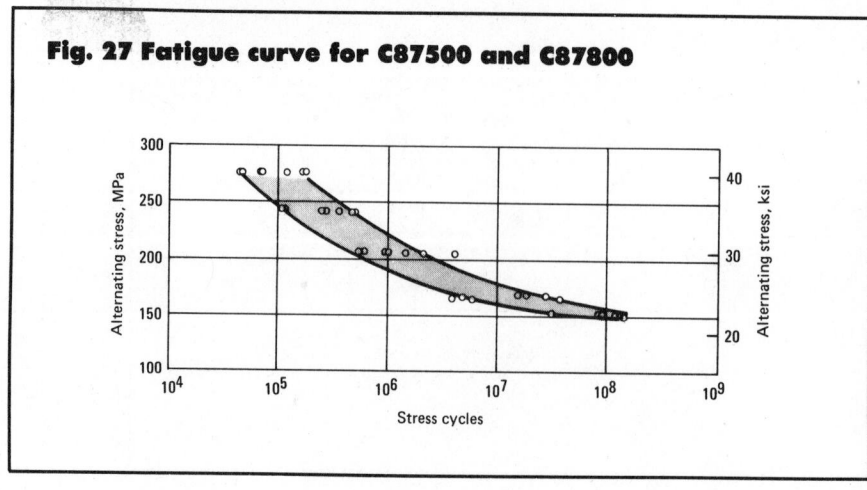

Fig. 27 Fatigue curve for C87500 and C87800

Fig. 28 Selected thermal properties of C87500 and C87800

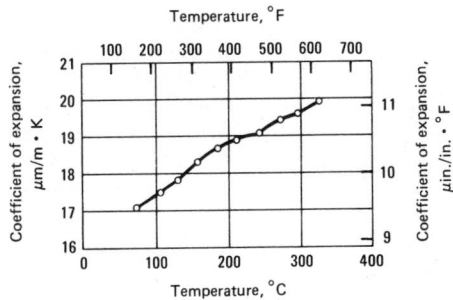

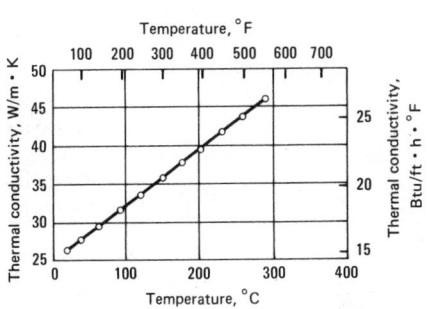

Fig. 29 Electrical conductivity and resistivity of C87500 and C87800

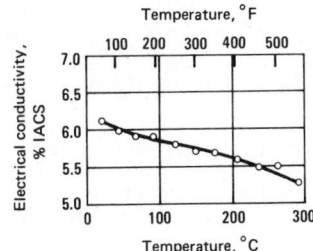

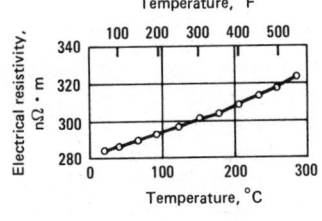

Thermal Properties

Liquidus temperature. 917 °C (1683 °F)

Solidus temperature. 821 °C (1510 °F)

Coefficient of thermal expansion. Linear, 19.6 μm/m·K (10.9 μin./in.·°F) at 20 to 260 °C (68 to 500 °F). See also Fig. 28.

Specific heat. 375 J/kg·K (0.09 Btu/lb·°F) at 20 °C (68 °F)

Thermal conductivity. 28 W/m·K (16 Btu/ft·h·°F) at 20 °C (68 °F). See also Fig. 28.

Electrical Properties

Electrical conductivity. Volumetric, 6.7% IACS at 20 °C (68 °F). See also Fig. 29.

Electrical resistivity. 284 nΩ·m at 20 °C (68 °F). See also Fig. 29.

Magnetic Properties

Magnetic permeability. 1.0

Fabrication Characteristics

Machinability. C87500: 50% of C36000, free-cutting brass. C87800: 40% of C36000

Casting temperature. 980 to 955 °C (1800 to 1750 °F)

Stress relieving temperature. 260 °C (500 °F)

C90300
88Cu-8Sn-4Zn

Commercial Names

Common name. Tin bronze; 88-8-0-4; "G"-bronze

Specifications

ASTM. Sand castings: B584. Centrifugal castings: B271. Continuous castings: B505. Ingot: B30

SAE. J462

Government. QQ-C-390, QQ-C-525. Sand castings: MIL-C-22229, composition 1. Centrifugal castings: MIL-C-15345, alloy 8. Investment castings: MIL-C-22087, composition 3. Precision castings: MIL-C-11866, composition 26

Ingot identification number. 225

Chemical Composition

Composition limits. 86.0 to 89.0 Cu; 7.5 to 9.0 Sn; 3.0 to 5.0 Zn; 1.0 max Ni; 0.30 max Pb; 0.15 max Fe; 0.05 max P(a); 0.2 max Sb; 0.05 max S; 0.005 max Si; 0.005 max Al

(a) For continuous castings, 1.5 max P.

Applications

Typical uses. Bearings, bushings, pump impellers, piston rings, valve components, seal rings, steam fittings, gears

Mechanical Properties

Tensile properties. Typical tensile strength, 310 MPa (45 ksi); yield strength, 145 MPa (21 ksi); elongation, 30% in 50 mm or 2 in.

Compressive properties. Compressive strength, 90 MPa (13 ksi)

Hardness. 70 HB

Elastic modulus. Tension, 97 GPa (14 × 10⁶ psi)

Impact strength. Charpy V-notch, 19 J (14 ft·lb)

Mass Characteristics

Density. 8.80 Mg/m³ (0.318 lb/in.³) at 20 °C (68 °F)

Volume change on freezing. 1.6%

Thermal Properties

Liquidus temperature. 1000 °C (1832 °F)

Solidus temperature. 854 °C (1570 °F)

Coefficient of thermal expansion. Linear, 18 μm/m·K (10 μin./in.·°F) at 20 to 177 °C (68 to 340 °F)

Specific heat. 376 J/kg·K (0.09 Btu/lb·°F) at 20 °C (68 °F)

Thermal conductivity. 74 W/m·K (43 Btu/ft·h·°F)

Electrical Properties

Electrical conductivity. Volumetric, 12% IACS at 20 °C (68 °F)

Magnetic Properties

Magnetic permeability. 1.0

Fabrication Characteristics

Machinability. 30% of C36000, free-cutting brass

C90500
88Cu-10Sn-2Zn

Commercial Names

Common name. Tin bronze; Gun metal; 88-10-0-2

Specifications

AMS. 4845
ASTM. Sand castings: B22, B584. Centrifugal castings: B271. Continuous castings: B505. Ingot: B30
SAE. J462
Government. QQ-C-390
Ingot identification number. 210

Chemical Composition

Composition limits. 86.0 to 89.0 Cu; 9.0 to 11.0 Sn; 1.0 to 3.0 Zn; 1.0 max Ni; 0.3 max Pb; 0.15 max Fe; 0.05 max P(a); 0.2 max Sb; 0.05 max S; 0.005 max Si; 0.005 max Al

(a) For continuous castings, 1.5 max P.

Applications

Typical uses. Bearings, bushings, pump impellers, piston rings, pump bodies, valve components, steam fittings, gears

Mechanical Properties

Tensile properties. Typical tensile strength, 310 MPa (45 ksi); yield strength, 150 MPa (22 ksi); elongation, 25% in 50 mm or 2 in.; reduction in area, 40%
Compressive properties. Compressive strength, 275 MPa (40 ksi)
Elastic modulus. Tension, 105 GPa (15×10^6 psi)
Fatigue strength. Rotating beam: 90 MPa (13 ksi) at 10^8 cycles
Impact strength. Izod: 14 J (10 ft·lb)

Mass Characteristics

Density. 8.72 Mg/m³ (0.315 lb/in.³) at 20 °C (68 °F)
Volume change on freezing. 1.6%

Thermal Properties

Liquidus temperature. 999 °C (1830 °F)
Solidus temperature. 854 °C (1570 °F)
Coefficient of thermal expansion. Linear, 20 μm/m·K (11 μin./in.·°F) at 20 to 300 °C (68 to 572 °F)
Specific heat. 376 J/kg·K (0.09 Btu/lb·°F) at 20 °C (68 °F)
Thermal conductivity. 74 W/m·K (43 Btu/ft·h·°F)

Electrical Properties

Electrical conductivity. Volumetric, 11% IACS at 20 °C (68 °F)

Magnetic Properties

Magnetic permeability. 1.0

Fabrication Characteristics

Machinability. 30% of C36000, free-cutting brass

C90700
89Cu-11Sn

Commercial Names

Common name. Tin bronze, 65; Phosphor gear bronze

Specifications

ASTM. Continuous castings: B505. Ingot: B30
Ingot identification number. 205

Chemical Composition

Composition limits. 88.0 to 90.0 Cu; 10.0 to 12.0 Sn; 0.15 max Fe; 0.1 to 0.3 P; 0.005 max Al; 0.50 max Pb(a); 0.50 max Zn(a)

(a) Pb + Zn + Ni, 1.0 max

Consequence of exceeding impurity limits. Ductility decreases rapidly with tin contents over 12%, with 13% a practical limit for gear applications

Applications

Typical uses. Worm wheels and gears; bearings expected to carry heavy loads at relatively low speeds

Mechanical Properties

Tensile properties. Typical. Sand

castings: tensile strength, 305 MPa (44 ksi); yield strength, 150 MPa (22 ksi); elongation, 20% in 50 mm or 2 in. Permanent mold castings: tensile strength, 380 MPa (55 ksi); yield strength, 205 MPa (30 ksi); elongation, 16% in 50 mm or 2 in.
Hardness. Sand castings, 80 HB; permanent mold castings, 102 HB
Elastic modulus. Tension, 105 GPa (15×10^6 psi)
Fatigue strength. Rotating beam: 170 MPa (25 ksi) at 10^8 cycles

Mass Characteristics

Density. 8.77 Mg/m³ (0.317 lb/in.³) at 20 °C (68 °F)
Volume change on freezing. 1.6%

Thermal Properties

Liquidus temperature. 1000 °C (1830 °F)
Solidus temperature. 832 °C (1528 °F)
Coefficient of thermal expansion. Linear, 18 μm/m·K (10 μin./in.·°F) at 20 to 200 °C (68 to 392 °F)
Specific heat. 376 J/kg·K (0.09 Btu/lb·°F) at 20 °C (68 °F)
Thermal conductivity. 71 W/m·K (41 Btu/ft·h·°F)

Electrical Properties

Electrical conductivity. Volumetric, 9.6% IACS at 20 °C (68 °F)
Electrical resistivity. 15 nΩ·m at 20 °C (68 °F)

Magnetic Properties

Magnetic permeability. 1.0

Fabrication Characteristics

Machinability. 20% of C36000, free-cutting brass

C91700
86½Cu-12Sn-1½Ni

Commercial Names

Common name. Nickel gear bronze, 86½-12-0-0-1½

Specifications

ASTM. Ingot: B30. Sand castings: B427
Other. Ingot Code No. 205

Chemical Composition

Composition limits. 85.0 to 87.5 Cu; 11.3 to 12.5 Sn; 0.25 max Pb; 1.3 to 2.0 Ni; 0.30 max P

Applications

Typical uses. Worm wheels and gears, bearings with heavy loads and relatively low speeds

Mechanical Properties

Tensile properties. Typical data for sand cast test bars: tensile strength, 305 MPa (44 ksi); yield strength, 150 MPa (22 ksi) at 0.5% extension under load; elongation, 16% in 50 mm or 2 in. Typical data for centrifugal or permanent mold test bars: tensile strength, 415 MPa (60 ksi); yield strength, 220 MPa (32 ksi) at 0.5% extension under load; elongation, 16% in 50 mm or 2 in.

Hardness. Sand cast, 85 HB; centrifugal or permanent mold cast, 106 HB

Elastic modulus. Tension, 105 GPa (15 × 10⁶ psi)

Mass Characteristics

Density. 8.75 Mg/m³ (0.316 lb/in.³) at 20 °C (68 °F)

Patternmaker's shrinkage. 16 mm/m (³/₁₆ in./ft)

Thermal Properties

Liquidus temperature. 1015 °C (1860 °F)

Solidus temperature. 850 °C (1565 °F)

Coefficient of thermal expansion. Linear, 16.2 μm/m·K (9.0 μin./in.·°F) at 20 to 200 °C (68 to 392 °F)

Specific heat. 376 J/kg·K (0.09 Btu/lb·°F) at 20 °C (68 °F)

Thermal conductivity. 71 W/m·K (41 Btu/ft·h·°F) at 20 °C (68 °F)

Electrical Properties

Electrical conductivity. Volumetric, 10% IACS at 20 °C (68 °F)

Magnetic Properties

Magnetic permeability. 1.0

Fabrication Characteristics

Machinability. 20% of C36000, free-cutting brass

C92200
88Cu-6Sn-1½Pb-4½Zn

Commercial Names

Common name. Navy "M" bronze, steam bronze, 88-6-1½-4½

Fig. 30 Tensile properties of C92200

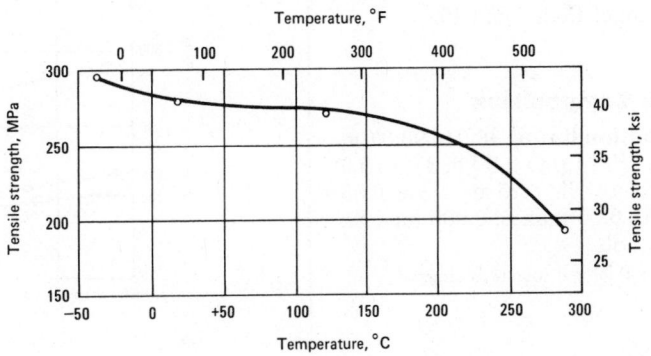

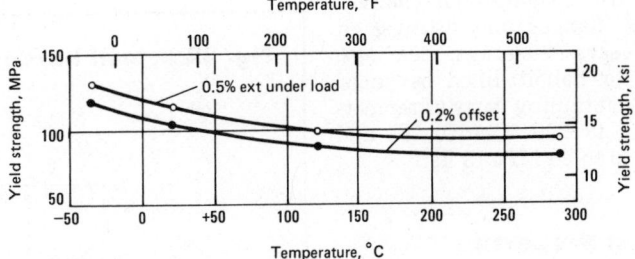

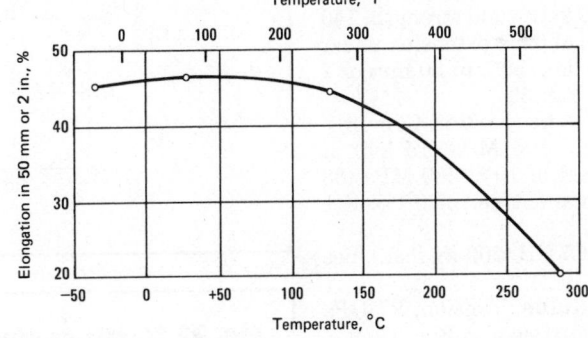

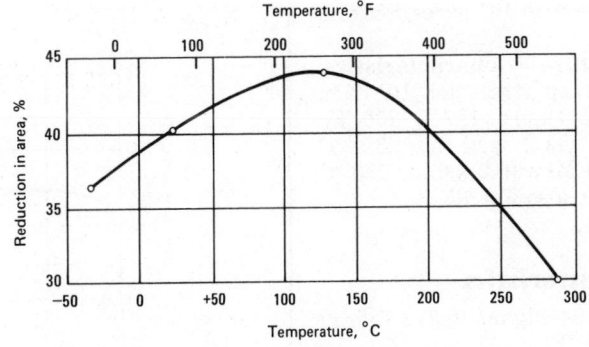

Specifications

ASTM. B584, B61, B271, B505, B30
SAE. J462 (C92200)
Government. CA922, QQ-B-225 (Alloy No. 1), MIL-B-16541, MIL-B-15345
Other. Ingot Code No. 245

Chemical Composition

Composition limits. 86.0 to 90.0 Cu; 5.5 to 6.5 Sn; 1.0 to 2.0 Pb; 3.0 to 5.0 Zn; 1.0 max Ni; 0.25 max Fe; 0.05 max P(a); 0.05 max S; 0.005 max Si; 0.25 max Sb

(a) 1.5 max P for continuous castings.

Applications

Typical uses. Component castings of valves, flanges and fittings, oil pumps, gears, bushings, bearings, backing for babbitt-lined bearings, pressure-containing parts at temperatures up to 290 °C (550 °F) and stresses up to 20 MPa (3 ksi)

Mechanical Properties

Tensile properties. Typical data for sand cast test bars: tensile strength, 275 MPa (40 ksi); yield strength, 140 MPa (20 ksi) at 0.5% extension under load; elongation, 30% in 50 mm or 2 in. See also Fig. 30.

Compressive properties. Compressive strength, 105 MPa (15 ksi) at permanent set of 10%; 260 MPa (38 ksi) at permanent set of 0.1%. See also Fig. 31.

Hardness. 65 HB (500-kg load). See also Fig. 32.

Elastic modulus. Tension, 97 GPa (14 × 10⁶ psi). See also Fig. 33.

Fatigue strength. Rotating beam, 76 MPa (11 ksi) at 10^8 cycles. See also Fig. 34.

Creep-rupture characteristics. Limiting creep stress for 10^{-5}%/h: 110 MPa (16.0 ksi) at 177 °C (350 °F); 77.2 MPa (11.2 ksi) at 232 °C (450 °F); 43 MPa (6.2 ksi) at 288 °C (550 °F). See also Fig. 35.

Mass Characteristics

Density. 8.64 Mg/m³ (0.312 lb/in.³) at 20 °C (68 °F)

Patternmaker's shrinkage. 16 mm/m (³⁄₁₆ in./ft)

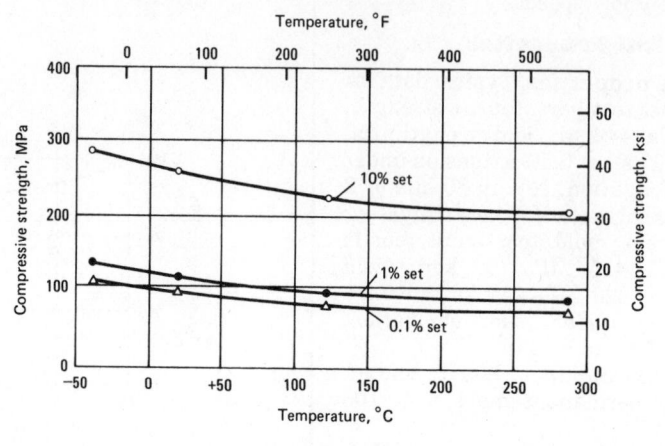

Fig. 31 Compressive strength of C92200

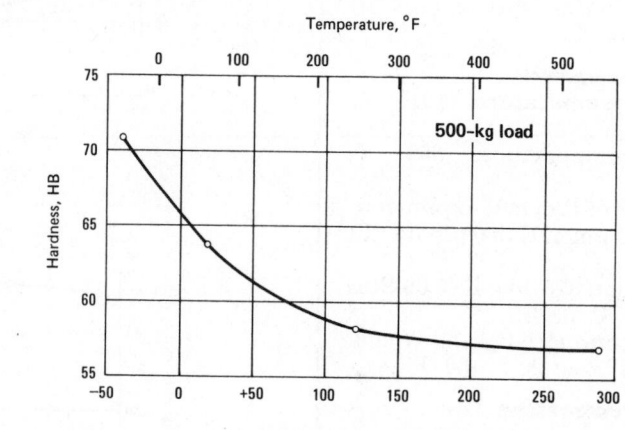

Fig. 32 Brinell hardness of C92200

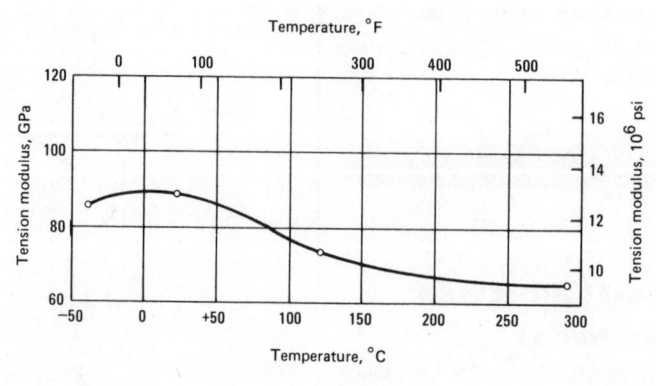

Fig. 33 Elastic modulus in tension for C92200

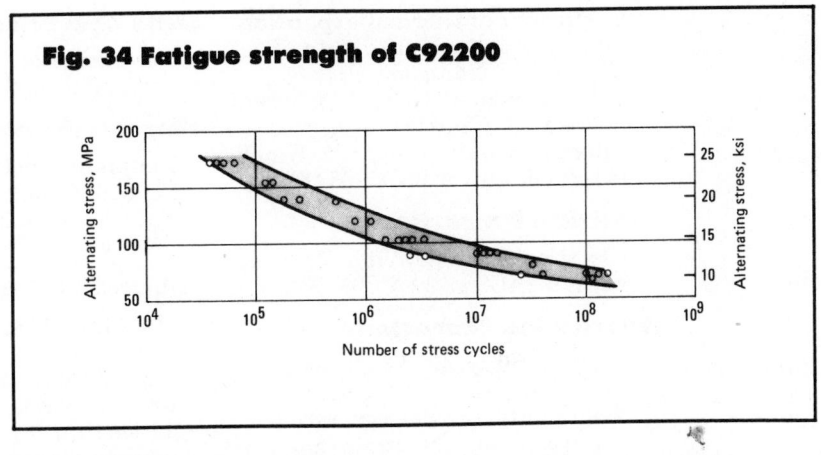

Fig. 34 Fatigue strength of C92200

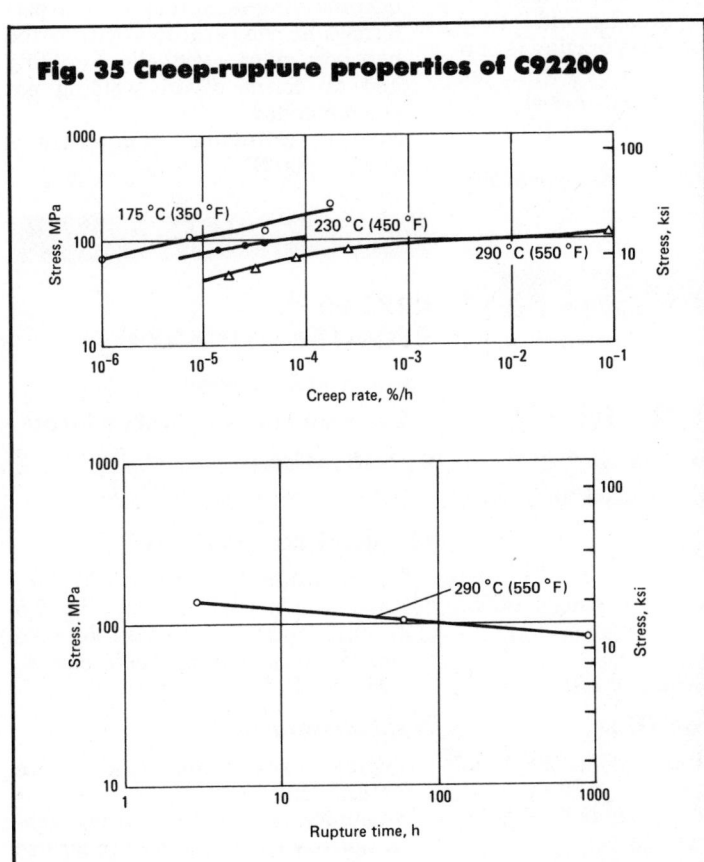

Fig. 35 Creep-rupture properties of C92200

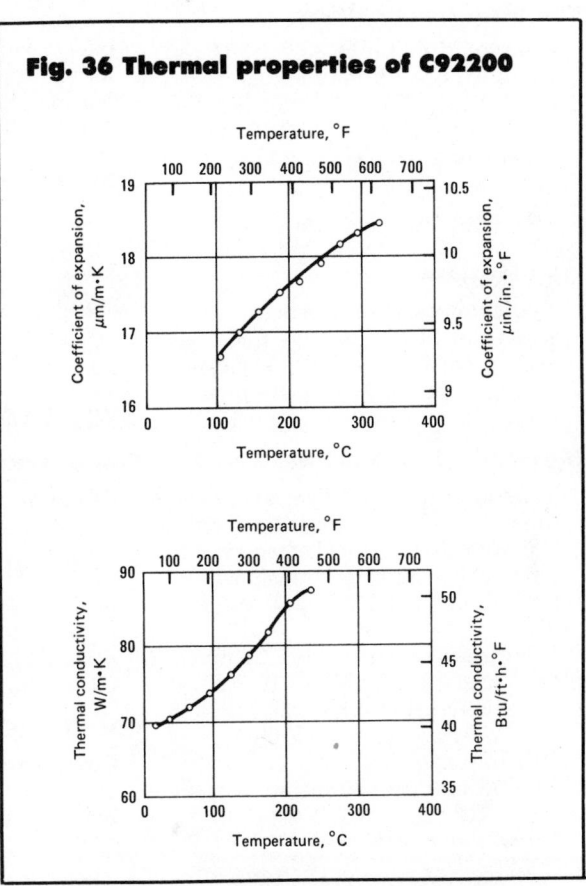

Fig. 36 Thermal properties of C92200

Thermal Properties

Liquidus temperature. 990 °C (1810 °F)
Solidus temperature. 825 °C (1520 °F)

Incipient melting temperature. Pb, 315 °C (600 °F)
Coefficient of thermal expansion. See Fig. 36, top.
Specific heat. 376 J/kg·K (0.09 Btu/lb·°F) at 20 °C (68 °F)

Thermal conductivity. 70 W/m·K (40 Btu/ft·h·°F) at 20 °C (68 °F). See also Fig. 36, bottom.

Electrical Properties

Electrical conductivity. Volumetric, 14.3% IACS at 20 °C (68 °F)
Electrical resistivity. 120 nΩ·m at 20 °C (68 °F)

Magnetic Properties

Magnetic permeability. 1.0

Fabrication Characteristics

Machinability. 42% of C36000, free-cutting brass
Weldability. Soldering: excellent. Brazing: excellent, but strain must be avoided during brazing and subsequent cooling because brazing is performed at temperatures within the hot short range. Oxyfuel gas welding and all forms of arc welding not recommended
Stress relieving temperature. 260 °C (500 °F)

C92300
87Cu-8Sn-1Pb-4Zn

Commercial Names

Common names. Leaded tin bronze, leaded Navy "G" bronze, 87-8-1-4

Specifications

ASTM. Sand castings: B584. Centrifugal castings: B271. Continuous castings: B505. Ingot: B30
SAE. J462
Government. QQ-C-390. Centrifugal castings: MIL-C-15345 (Alloy 10)
Other. Ingot Code No. 230

Chemical Composition

Composition limits. 85.0 to 89.0 Cu; 7.0 to 9.0 Sn; 1.0 max Pb; 2.5 to 5.0 Zn; 1.0 max Ni; 0.25 max Fe; 0.05 max P(a); 0.25 max Sb; 0.05 max S; 0.005 max Si; 0.005 max Al

(a) 1.5 max P for continuous castings

Applications

Typical uses. Strong general-utility structural bronze for use under severe conditions; valves, expansion joints, special high-pressure pipe fittings, steam pressure castings

Mechanical Properties

Tensile properties. Typical data for sand cast test bars: tensile strength, 275 MPa (40 ksi); yield strength, 140 MPa (20 ksi) at 0.5% extension under load; elongation, 25% in 50 mm or 2 in.
Compressive properties. Compressive strength, 69 MPa (10 ksi) at permanent set of 0.1%; 250 MPa (35 ksi) at permanent set of 10%
Hardness. 70 HB
Elastic modulus. Tension, 97 GPa (14×10^6 psi)
Impact strength. Izod, 18.3 J (13.5 ft·lb)

Mass Characteristics

Density. 8.80 Mg/m^3 (0.317 lb/in.^{3}) at 20 °C (68 °F)
Patternmaker's shrinkage. 16 mm/m (³/₁₆ in./ft)

Thermal Properties

Liquidus temperature. 1000 °C (1830 °F)
Solidus temperature. 855 °C (1570 °F)
Incipient melting temperature. Pb, 315 °C (600 °F)

Coefficient of thermal expansion. Linear, 18 μm/m·K (10 μin./in.·°F) at 20 to 177 °C (68 to 350 °F)
Specific heat. 376 J/kg·K (0.09 Btu/lb·°F) at 20 °C (68 °F)
Thermal conductivity. 75 W/m·K (43 Btu/ft·h·°F) at 20 °C (68 °F)

Electrical Properties

Electrical conductivity. Volumetric, 12% IACS at 20 °C (68 °F)

Fabrication Characteristics

Machinability. 42% of C36000, free-cutting brass
Weldability. Soldering: excellent. Brazing: good, but strain must be avoided during brazing and subsequent cooling because brazing is performed at temperatures within the hot short range. Oxyfuel gas welding and all forms of arc welding not recommended
Stress relieving temperature. 260 °C (500 °F)

C92500
87Cu-11Sn-1Pb-1Ni

Commercial Names

Common name. Leaded tin bronze, 640; 87-11-1-0-1

Specifications

ASTM. Continuous castings: B505. Ingot: B30
SAE. J462
Other. Ingot Code No. 250

Chemical Composition

Composition limits. 85.0 to 88.0 Cu; 10.0 to 12.0 Sn; 1.0 to 1.5 Pb; 0.5 max Zn; 0.8 to 1.5 Ni; 0.15 max Fe; 0.20 to 0.30 P; 0.005 max Al

Applications

Typical uses. Gears, automotive synchronizer rings

Mechanical Properties

Tensile properties. Typical data for sand cast test bars: tensile strength, 305 MPa (44 ksi); yield strength, 140 MPa (20 ksi) at 0.5% extension under load; elongation, 20% in 50 mm or 2 in.
Hardness. 80 HB
Elastic modulus. Tension, 110 GPa (16×10^6 psi)

Mass Characteristics

Patternmaker's shrinkage. 16 mm/m (³/₁₆ in./ft)

Thermal Properties

Incipient melting temperature. Pb, 316 °C (600 °F)
Specific heat. 376 J/kg·K (0.09 Btu/lb·°F) at 20 °C (68 °F)

Fabrication Characteristics

Machinability. 30% of C36000, free-cutting brass
Weldability. Soldering: excellent. Brazing: good, but strain must be avoided during brazing and subsequent cooling because brazing is performed at temperatures within the hot short range. Oxyfuel gas welding and all forms of arc welding not recommended
Stress relieving temperature. 260 °C (500 °F)

C92600
87Cu-10Sn-1Pb-2Zn

Commercial Names

Common name. Leaded tin bronze

Specifications

Other. Ingot Code No. 215

Chemical Composition

Composition limits. 86.0 to 88.5 Cu; 9.3 to 10.5 Sn; 0.8 to 1.2 Pb; 1.3 to 2.5 Zn; 0.75 max Ni; 0.15 max Fe; 0.25 max Sb; 0.05 max S; 0.005 max Si; 0.03 max P; 0.005 max Al

Applications

Typical uses. Commercial bronze for high-duty bearings where wear resistance is essential; strong general-utility structural bronze for use under severe conditions; bolts, nuts, gears; heavy-pressure bearings and bushings to use against hardened steel; valves, expansion joints, special high-pressure pipe fittings; pump pistons; elevator components; steam pressure castings

Mechanical Properties

Tensile properties. Typical data for sand cast test bars: tensile strength, 305 MPa (44 ksi); yield strength, 140 MPa (20 ksi) at 0.5% extension under load; elongation, 30% in 50 mm or 2 in.

Compressive properties. Compressive strength, 85 MPa (12 ksi) at permanent set of 0.1%; 275 MPa (40 ksi) at permanent set of 10%
Hardness. 78 HRF, 72 HB
Elastic modulus. Tension, 105 GPa (15×10^6 psi)
Impact strength. Izod, 9 J (7 ft·lb)

Mass Characteristics

Density. 8.70 Mg/m^3 (0.315 lb/in.^{3}) at 20 °C (68 °F)
Patternmaker's shrinkage. 16 mm/m (³/₁₆ in./ft)

Thermal Properties

Liquidus temperature. 890 °C (1800 °F)
Solidus temperature. 850 °C (1550 °F)
Incipient melting temperature. Pb, 315 °C (600 °F)
Specific heat. 376 J/kg·K (0.09 Btu/lb·°F) at 20 °C (68 °F)

Electrical Properties

Electrical conductivity. Volumetric, 9% IACS at 20 °C (68 °F)

Fabrication Characteristics

Machinability. 40% of C36000, free-cutting brass
Weldability. Soldering: excellent. Brazing: good, but strain must be avoided during brazing and subsequent cooling because brazing is done at temperatures within the hot short range. Oxyfuel gas welding and all forms of arc welding not recommended
Stress relieving temperature. 260 °C (500 °F)

C92700
88Cu-10Sn-2Pb

Commercial Names

Common name. Leaded tin bronze, 88-10-2-0

Specifications

ASTM. Continuous castings: B505. Ingot: B30
SAE. J462
Other. Ingot Code No. 206

Chemical Composition

Composition limits. 86.0 to 89.0 Cu; 9.0 to 11.0 Sn; 1.0 to 2.5 Pb; 0.7 max Zn; 1.0 max Ni; 0.15 max Fe; 0.25 max P; 0.005 max Al

Applications

Typical uses. Bearings, bushings, pump impellers, piston rings, valve components, steam fittings, gears

Mechanical Properties

Tensile properties. Typical data for sand cast test bars: tensile strength, 290 MPa (42 ksi); yield strength, 145 MPa (21 ksi) at 0.5% extension under load; elongation, 20% in 50 mm or 2 in.
Hardness. 77 HB
Elastic modulus. Tension, 110 GPa (16×10^6 psi)

Mass Characteristics

Density. 8.80 Mg/m^3 (0.317 lb/in.^{3}) at 20 °C (68 °F)
Patternmaker's shrinkage. 16 mm/m (³/₁₆ in./ft)

Thermal Properties

Liquidus temperature. 980 °C (1800 °F)
Solidus temperature. 850 °C (1550 °F)
Incipient melting temperature. Pb, 315 °C (600 °F)
Coefficient of thermal expansion. Linear, 18 µm/m·K (10 µin./in.·°F) at 20 to 177 °C (68 to 350 °F)
Specific heat. 376 J/kg·K (0.09 Btu/lb·°F) at 20 °C (68 °F)

Electrical Properties

Electrical conductivity. Volumetric, 11% IACS at 20 °C (68 °F)

Fabrication Characteristics

Machinability. 45% of C36000, free-cutting brass
Weldability. Soldering: excellent. Brazing: good, but parts must not be strained during brazing or subsequent cooling because brazing is done at temperatures within the hot short range. Oxyfuel gas welding and all forms of arc welding not recommended
Stress relieving temperature. 260 °C (500 °F)

C92900
84Cu-10Sn-2¹/₂Pb-3¹/₂Ni

Commercial Names

Common name. Leaded nickel-tin bronze, 84-10-2¹/₂-0-3¹/₂

Specifications

ASTM. Sand and centrifugal castings: B427. Continuous castings: B505. Ingot. B30
SAE. J462

Chemical Composition

Composition limits. 81.0 to 85.5 Cu; 9.0 to 11.0 Sn; 2.0 to 3.2 Pb; 2.8 to 4.0 Ni; 0.50 max P; 0.50 max others (total)

Applications

Typical uses. Gears, wear plates and guides, cams

Mechanical Properties

Tensile properties. Typical data for sand cast test bars: tensile strength, 325 MPa (47 ksi); yield strength, 180 MPa (26 ksi) at 0.5% extension under load; elongation, 20% in 50 mm or 2 in.
Hardness. 80 HB
Elastic modulus. Tension, 97 GPa (14×10^6 psi)
Impact strength. Izod, 16 J (12 ft·lb)

Mass Characteristics

Density. 8.79 Mg/m^3 (0.320 lb/in.^{3}) at 20 °C (68 °F)
Patternmaker's shrinkage. 16 mm/m (³/₁₆ in./ft)

Thermal Properties

Liquidus temperature. 1030 °C (1887 °F)
Solidus temperature. 860 °C (1575 °F)
Incipient melting temperature. Pb, 315 °C (600 °F)
Coefficient of thermal expansion. Linear, 17 µm/m·K (9.5 µin./in.·°F) at 20 to 200 °C (68 to 392 °F)
Specific heat. 376 J/kg·K (0.09 Btu/lb·°F) at 20 °C (68 °F)
Thermal conductivity. 58.2 W/m·K (33.6 Btu/ft·h·°F) at 20 °C (68 °F)

Electrical Properties

Electrical conductivity. Volumetric, 9.2% IACS at 20 °C (68 °F)

Fabrication Characteristics

Machinability. 40% of C36000, free-cutting brass
Weldability. Soldering: excellent. Brazing: good, but parts must not be strained during brazing or subsequent cooling because brazing is done at temperatures within the hot short range. Oxyfuel gas welding and all forms of arc welding not recommended
Stress relieving temperature. 260 °C (500 °F)

C93200
83Cu-7Sn-7Pb-3Zn

Commercial Names

Common name. High leaded tin bronze; bearing bronze 660; 83-7-7-3

Specifications

ASTM. Sand castings: B584. Centrifugal castings: B271. Continuous castings: B505. Ingot: B30
SAE. J462
Government. QQ-C-390; QQ-C-525; QQ-L-225 (alloy 12); MIL-C-15345 (alloy 17); MIL-C-11553 (alloy 12); MIL-B-16261 (alloy VI)
Other. Ingot Code No. 315

Chemical Composition

Composition limits. 81.0 to 85.0 Cu(a); 6.3 to 7.5 Sn; 6.0 to 8.0 Pb; 2.0 to 4.0 Zn; 0.50 max Ni; 0.20 max Fe; 0.15 max P(b); 0.35 max Sb; 0.08 max S; 0.003 max Si

(a) In determining Cu, minimum may be calculated as Cu + Ni. (b) 1.5 max P for continuous castings; 0.50 max P for permanent mold castings.

Applications

Typical uses. General-utility bearings and bushings, automobile fittings

Mechanical Properties

Tensile properties. Typical data for sand cast test bars: tensile strength, 240 MPa (35 ksi); yield strength, 125 MPa (18 ksi) at 0.5% extension under load; elongation, 20% in 50 mm or 2 in.
Compressive properties. Compressive strength, 315 MPa (46 ksi) at permanent set of 10%
Hardness. 65 HB
Elastic modulus. Tension, 100 GPa (14.5×10^6 psi)
Impact strength. Izod, 8 J (6 ft·lb)
Fatigue strength. Reverse bending, 110 MPa (16 ksi) at 10^8 cycles

Mass Characteristics

Density. 8.93 Mg/m^3 (0.322 lb/in.^{3}) at 20 °C (68 °F)
Patternmaker's shrinkage. 18 mm/m ($^7/_{32}$ in./ft)

Thermal Properties

Liquidus temperature. 975 °C (1790 °F)
Solidus temperature. 855 °C (1570 °F)
Incipient melting temperature. Pb, 315 °C (600 °F)
Coefficient of thermal expansion. Linear, 18 µm/m·K (10 µin./in.·°F) at 0 to 100 °C (32 to 212 °F)
Specific heat. 376 J/kg·K (0.09 Btu/lb·°F) at 20 °C (68 °F)
Thermal conductivity. 59 W/m·K (34 Btu/ft·h·°F) at 20 °C (68 °F)

Electrical Properties

Electrical conductivity. Volumetric, 12% IACS at 20 °C (68 °F)

Fabrication Characteristics

Machinability. 70% of C36000, free-cutting brass
Weldability. Soldering: excellent. Brazing: good, but parts must not be strained during brazing or subsequent cooling because brazing is done at temperatures within the hot short range. Oxyfuel gas welding and all forms of arc welding not recommended
Stress relieving temperature. 260 °C (500 °F)

C93500
85Cu-5Sn-9Pb-1Zn

Commercial Names

Common name. High leaded tin bronze, 85-5-9-1

Specifications

ASTM. Sand castings: B584. Centrifugal castings: B271. Continuous castings: B505. Ingot: B30
SAE. J462
Government. QQ-C-390; QQ-L-225 (alloy 14); MIL-B-11553B (alloy 14)
Other. Ingot Code No. 326

Chemical Composition

Composition limits. 83.0 to 86.0 Cu(a); 4.5 to 6.0 Sn; 8.0 to 10.0 Pb; 2.0 max Zn; 0.50 max Ni; 0.20 max Fe; 0.02 max P(b); 0.30 max Sb; 0.08 max S; 0.003 max Si

(a) In determining Cu, minimum may be calculated as Cu + Ni. (b) 1.5 max P for continuous castings.

Applications

Typical uses. Small bearings and bushings, bronze backings for babbitt-lined automotive bearings

Mechanical Properties

Tensile properties. Typical data for sand cast test bars: tensile strength, 220 MPa (32 ksi); yield strength, 110 MPa (16 ksi) at 0.5% extension under load; elongation, 20% in 50 mm or 2 in.
Compressive properties. Compressive strength, 90 MPa (13 ksi) at permanent set of 0.1%
Hardness. 60 HB
Elastic modulus. Tension, 100 GPa (14.5×10^6 psi)
Impact strength. Charpy V-notch or Izod, 11 J (8 ft·lb)

Mass Characteristics

Density. 8.87 Mg/m^3 (0.320 lb/in.^{3}) at 20 °C (68 °F)
Patternmaker's shrinkage. 16 mm/m ($^3/_{16}$ in./ft)

Thermal Properties

Liquidus temperature. 1000 °C (1830 °F)
Solidus temperature. 855 °C (1570 °F)
Incipient melting temperature. Pb, 316 °C (600 °F)
Coefficient of thermal expansion. Linear, 18 µm/m·K (10 µin./in.·°F) at 20 to 200 °C (68 to 392 °F)
Specific heat. 376 J/kg·K (0.09 Btu/lb·°F) at 20 °C (68 °F)
Thermal conductivity. 71 W/m·K (41 Btu/ft·h·°F) at 20 °C (68 °F)

Electrical Properties

Electrical conductivity. Volumetric, 15% IACS at 20 °C (68 °F)

Magnetic Properties

Magnetic permeability. 1.0

Fabrication Characteristics

Machinability. 70% of C36000, free-cutting brass
Weldability. Soldering: good. Brazing: good, but parts must not be strained during brazing or subsequent cooling because brazing is done at temperatures within the hot short range. Oxyfuel gas welding and all forms of arc welding not recommended
Stress relieving temperature. 260 °C (500 °F)

C93700
80Cu-10Sn-10Pb

Commercial Names

CDA and UNS number. C93700
Common names. High leaded tin bronze; bushing and bearing bronze; 80-10-10

Specifications

AMS. Sand and centrifugal castings: 4842

ASTM. Sand castings: B22, B584. Centrifugal castings: B271. Continuous castings: B505. Ingot: B30

SAE. J462

Government. QQ-C-390; MIL-B-13506 (alloy A2)

Other. Ingot Code No. 305

Chemical Composition

Composition limits. 78.0 to 82.0 Cu; 9.0 to 11.0 Sn; 8.0 to 11.0 Pb; 0.70 max Zn; 0.70 max Ni; 0.15 max Fe; 0.05 max P; 0.50 max Sb; 0.08 max S; 0.003 max Si

Applications

Typical uses. Bearings for high speed and heavy pressure, pumps, impellers, applications requiring corrosion resistance, pressure-tight castings

Mechanical Properties

Tensile properties. Typical data for sand cast test bars: tensile strength, 240 MPa (35 ksi); yield strength, 125 MPa (18 ksi) at 0.5% extension under load; elongation, 20% in 50 mm or 2 in. See also Fig. 37.

Compressive properties. Compressive strength, 90 MPa (13 ksi) at permanent set of 0.1%; 325 MPa (47 ksi) at permanent set of 10%. See also Fig. 38.

Hardness. 60 HB

Elastic modulus. See Fig. 39.

Impact strength. Izod, 7 J (5 ft·lb); Charpy V-notch, 15 J (11 ft·lb)

Fatigue strength. Reverse bending: 90 MPa (13 ksi) at 10^8 cycles. See also Fig. 40.

Creep-rupture characteristics. Limiting creep stress for 10^{-5}%/h: 71.7 MPa (10.4 ksi) at 177 °C (350 °F); 51 MPa (7.4 ksi) 232 °C (450 °F); 12 MPa (1.8 ksi) at 288 °C (550 °F). See also Fig. 41.

Mass Characteristics

Density. 8.95 Mg/m³ (0.322 lb/in.³) at 20 °C (68 °F)

Volume change on freezing. 7.3%

Patternmaker's shrinkage. 11 mm/m (⅛ in./ft)

Thermal Properties

Liquidus temperature. 928 °C (1705 °F)

Solidus temperature. 762 °C (1403 °F)

Fig. 37 Typical tensile properties of C93700 at various temperatures

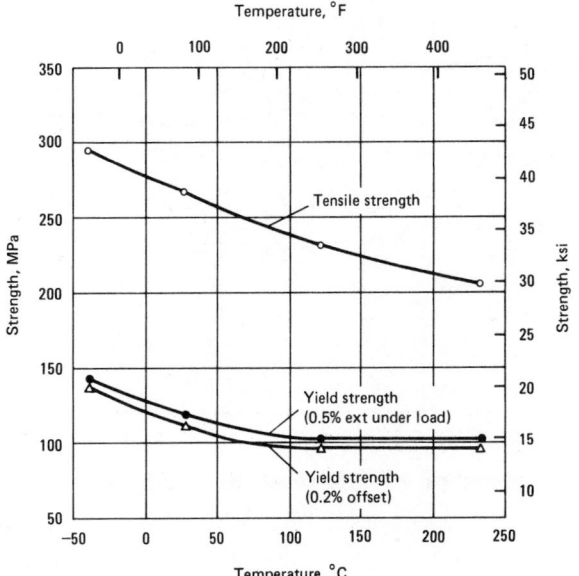

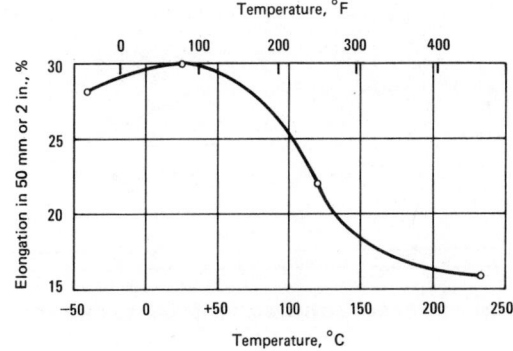

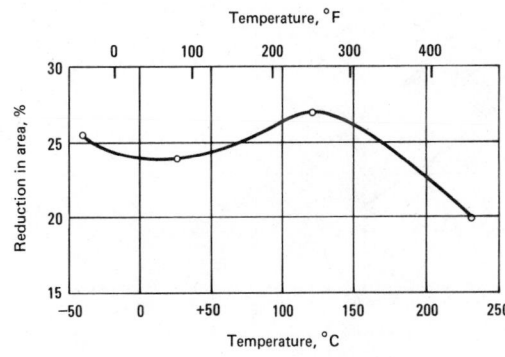

Fig. 38 Variation of compressive strength with temperature for C93700

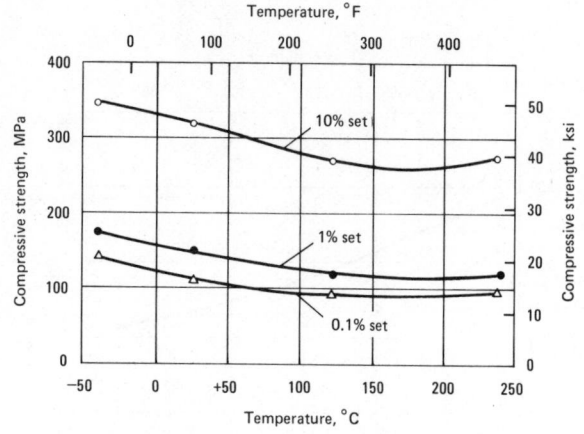

Fig. 39 Variation of elastic modulus with temperature for C93700

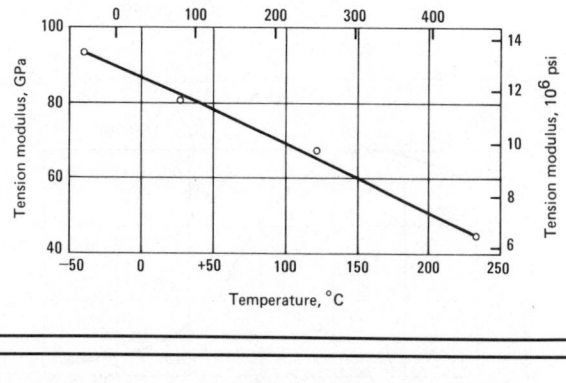

Fig. 40 Typical reverse bending fatigue curve for C93700

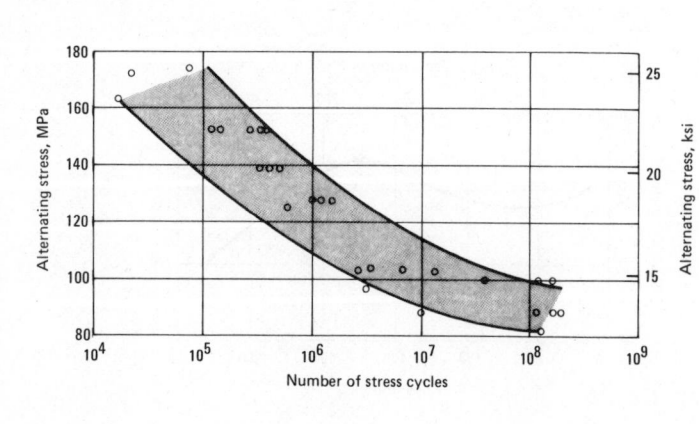

Incipient melting temperature. Pb, 315 °C (600 °F)

Coefficient of thermal expansion. Linear, 18.5 μm/m·K (10.3 μin./ in.·°F) at 20 to 200 °C (68 to 392 °F). See also Fig. 42.

Specific heat. 376 J/kg·K (0.09 Btu/ lb·°F) at 20 °C (68 °F)

Thermal conductivity. 46.9 W/m·K (27.1 Btu/ft·h·°F) at 20 °C (68 °F). See also Fig. 42.

Electrical Properties

Electrical conductivity. See Fig. 43.

Electrical resistivity. 170 nΩ·m at 20 °C (68 °F)

Magnetic Properties

Magnetic permeability. 1.0

Fabrication Characteristics

Machinability. 80% of C36000, free-cutting brass

Weldability. Soldering: good. Brazing: good, but parts must not be strained during brazing or subsequent cooling because brazing is done at temperatures in the hot short range. Oxyfuel gas welding and all forms of arc welding not recommended

Stress relieving temperature. 260 °C (500 °F)

Fig. 43 Variation of electrical conductivity with temperature for C93700

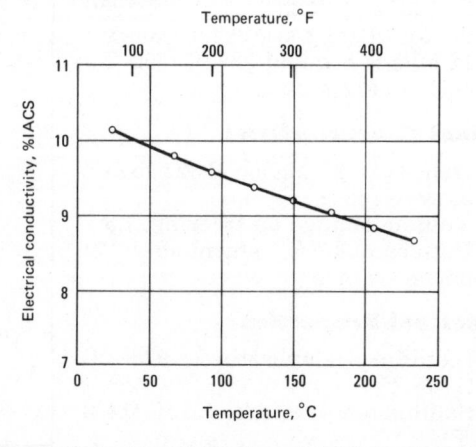

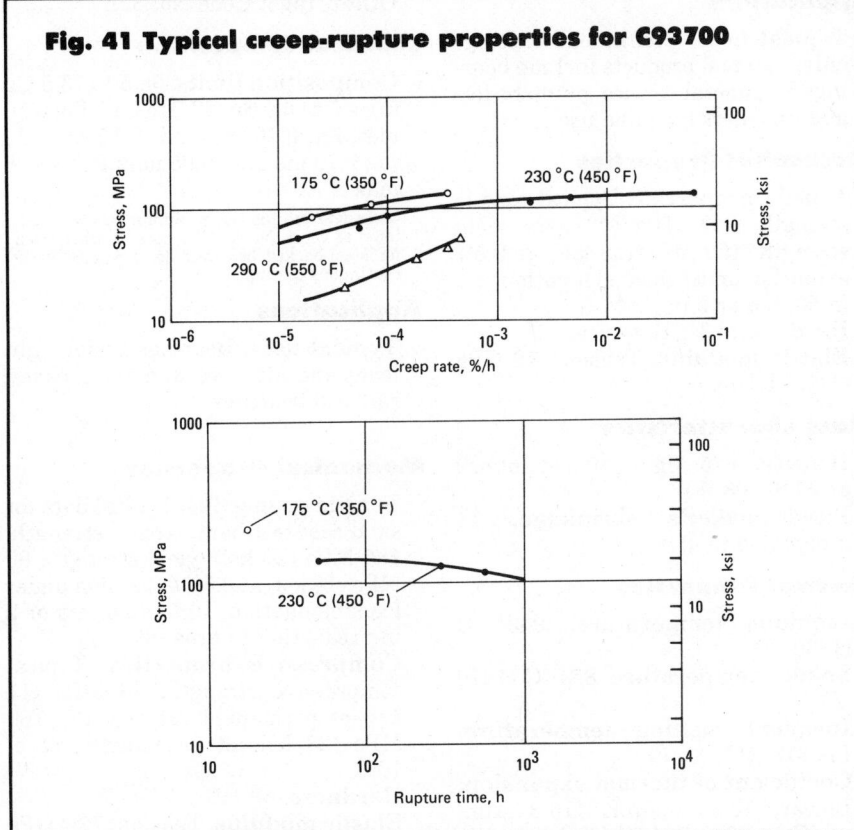

Fig. 41 Typical creep-rupture properties for C93700

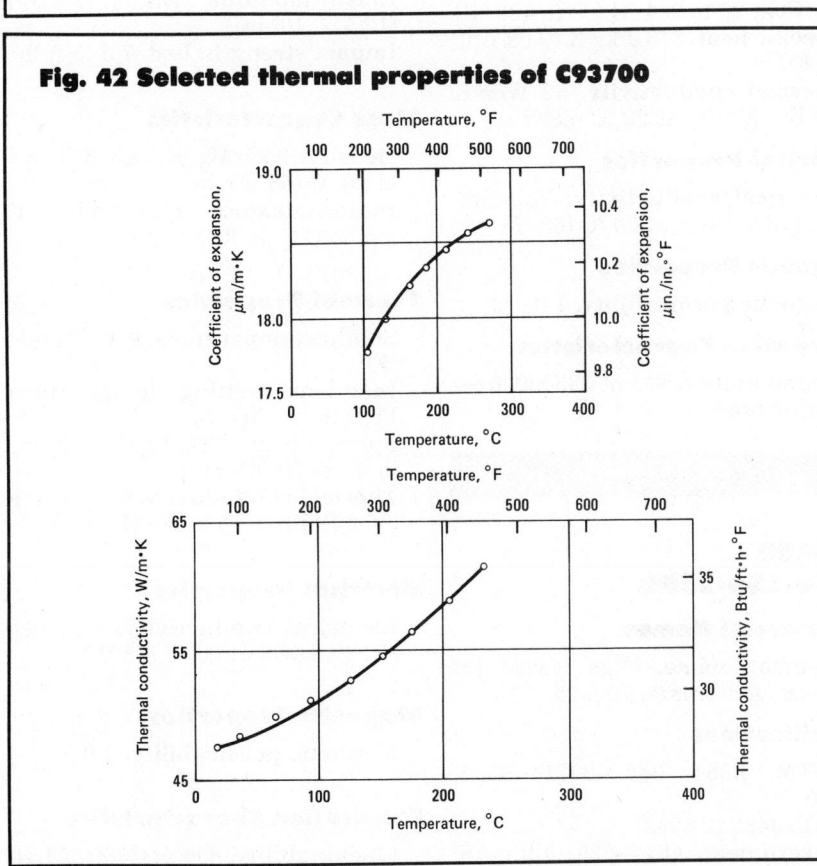

Fig. 42 Selected thermal properties of C93700

C93800
78Cu-7Sn-15Pb

Commercial Names

Common names. High leaded tin bronze, anti-acid metal, 78-7-15

Specifications

ASTM. Sand castings: B66, B584. Centrifugal castings: B271. Continuous castings: B505. Ingot: B30
SAE. J462
Government. QQ-C-390; QQ-C-525 (alloy 7); QQ-L-225 (alloys 19 and 7); MIL-B-16261 (alloy IV)
Other. Ingot Code No. 319

Chemical Composition

Composition limits. 75.0 to 79.0 Cu(a); 6.3 to 7.5 Sn; 13.0 to 16.0 Pb; 0.70 max Zn; 0.70 max Ni; 0.15 max Fe; 0.05 max P; 0.70 max Sb; 0.08 max S; 0.003 max Si; 0.005 max Al

(a) In determining Cu, minimum may be calculated as Cu + Ni.

Consequence of exceeding impurity limits. Aluminum or silicon causes lead sweating during solidification and may cause a substantial portion of castings to be unsound.

Applications

Typical uses. Locomotive engine castings and general-service bearings for moderate pressure; general-purpose wearing metal for rod bushings, shoes and wedges; freight car bearings; backs for lined journal bearings for locomotive tenders and passenger cars; pump impellers and bodies for use in acid mine water

Mechanical Properties

Tensile properties. Typical data for sand cast test bars: tensile strength, 205 MPa (30 ksi); yield strength, 110 MPa (16 ksi) at 0.5% extension under load; elongation, 18% in 50 mm or 2 in. Typical data for chilled centrifugally cast test bars: tensile strength, 230 MPa (33 ksi); yield strength, 140 MPa (20 ksi) at 0.5% extension under load; elongation, 12% in 50 mm or 2 in.
Shear strength. 105 MPa (15 ksi)
Compressive properties. Compressive strength. Sand cast: 83 MPa (12 ksi) at permanent set of 0.1%; 260 MPa (38 ksi) at permanent set of 10%. Centrifugally cast: 130 MPa (19 ksi) at permanent set of 0.1%
Hardness. Sand cast: 55 HB
Elastic modulus. Sand cast test

bars: tension, 72.4 GPa (10.5×10^6 psi)

Impact strength. Sand cast: Charpy V-notch or Izod, 7 J (5 ft·lb)

Fatigue strength. Reverse bending, sand cast test bars: 69 MPa (10 ksi) at 10^8 cycles

Mass Characteristics

Density. 9.25 Mg/m³ (0.334 lb/in.³) at 20 °C (68 °F)

Patternmaker's shrinkage. 11 mm/m (⅛ in./ft)

Thermal Properties

Liquidus temperature. 945 °C (1730 °F)

Solidus temperature. 855 °C (1570 °F)

Incipient melting temperature. Pb, 315 °C (600 °F)

Coefficient of thermal expansion. Linear, 18.5 μm/m·K (10.2 μin./in.·°F) at 20 to 205 °C (68 to 400 °F)

Specific heat. 376 J/kg·K (0.09 Btu/lb·°F) at 20 °C (68 °F)

Thermal conductivity. 52 W/m·K (30 Btu/ft·h·°F) at 20 °C (68 °F)

Electrical Properties

Electrical conductivity. Volumetric, 11.5% IACS at 20 °C (68 °F)

Magnetic Properties

Magnetic permeability. 1.0

Fabrication Characteristics

Machinability. 80% of C36000, free-cutting brass

Weldability. Soldering: good. Brazing: poor. Oxyfuel gas welding and all forms of arc welding not recommended

Stress relieving temperature. 260 °C (500 °F)

C93900
79Cu-6Sn-15Pb

Commercial Names

Common name. High leaded tin bronze, 79-6-15

Specifications

ASTM. B505, B30

Chemical Composition

Composition limits. 76.5 to 79.5 Cu; 5.0 to 7.0 Sn; 14.0 to 18.0 Pb; 1.5 max Zn; 0.80 max Ni; 0.40 max Fe; 0.05 max P (a)

(a) 1.5 max P for continuous castings

Applications

Typical uses. Continuous castings only; common products include bearings for general service, pump bodies and impellers for mine use.

Mechanical Properties

Tensile properties. Typical tensile strength, 220 MPa (32 ksi); yield strength, 150 MPa (22 ksi) at 0.5% extension under load; elongation, 7% in 50 mm or 2 in.

Hardness. 63 HB, typical

Elastic modulus. Tension, 76 GPa (11×10^6 psi)

Mass Characteristics

Density. 9.25 Mg/m³ (0.334 lb/in.³) at 20 °C (68 °F)

Patternmaker's shrinkage. 11 mm/m (1/8 in./ft)

Thermal Properties

Liquidus temperature. 943 °C (1730 °F)

Solidus temperature. 854 °C (1570 °F)

Incipient melting temperature. Pb, 316 °C (600 °F)

Coefficient of thermal expansion. Linear, 18.3 μm/m·K (10.3 μin./in.·°F) at 20 to 204 °C (68 to 400 °F)

Specific heat. 376 J/kg·K (0.09 Btu/lb·°F)

Thermal conductivity. 52 W/m·K (30 Btu/ft·h·°F) at 20 °C (68 °F)

Electrical Properties

Electrical conductivity. Volumetric, 11.5% IACS at 20 °C (68 °F)

Magnetic Properties

Magnetic permeability. 1.0

Fabrication Characteristics

Machinability. 80% of C36000, free-cutting brass

C94300
70Cu-5Sn-25Pb

Commercial Names

Common name. High leaded tin bronze, soft bronze, 70-5-25

Specifications

ASTM. B584, B66, B271, B505, B30

SAE. J462 (CA943)

Government. QQ-L-225, alloy 18; MIL-B-16261, alloy V

Other. Ingot Code No. 322

Chemical Composition

Composition limits. 68.5 to 73.5 Cu (a); 4.5 to 6.0 Sn; 22.0 to 25.0 Pb; 0.50 max Zn; 0.70 max Ni; 0.15 max Fe (b); 0.70 max Sb; 0.05 max P (c); 0.08 max S

(a) In determining Cu, minimum may be calculated as Cu + Ni. (b) 0.35 max Fe when used for steel-backed bearings. (c) 1.5 max P for continuous castings.

Applications

Typical uses. Bearings under light loads and high speed, driving boxes, railroad bearings

Mechanical Properties

Tensile properties. Typical data for sand cast test bars: tensile strength, 185 MPa (27 ksi); yield strength, 90 MPa (13 ksi) at 0.5% extension under load; elongation, 10% in 50 mm or 2 in.; reduction in area, 8%

Compressive properties. Typical compressive strength: 76 MPa (11 ksi) at permanent set of 0.1%; 160 MPa (23 ksi) at permanent set of 10%

Hardness. 48 HB

Elastic modulus. Tension, 72.4 GPa (10.5×10^6 psi)

Impact strength. Izod, 7 J (5 ft·lb)

Mass Characteristics

Density. 9.29 Mg/m³ (0.336 lb/in.³) at 20 °C (68 °F)

Patternmaker's shrinkage. 11 mm/m (1/8 in./ft)

Thermal Properties

Solidus temperature. 900 °C (1650 °F)

Incipient melting temperature. Pb, 316 °C (600 °F)

Specific heat. 376 J/kg·K (0.09 Btu/lb·°F) at 20 °C (68 °F)

Thermal conductivity. 62.7 W/m·K (36.2 Btu/ft·h·°F) at 20 °C (68 °F)

Electrical Properties

Electrical conductivity. Volumetric, 9% IACS at 20 °C (68 °F)

Magnetic Properties

Magnetic permeability. 1.0

Fabrication Characteristics

Machinability. 80% of C36000, free-cutting brass

C94500
73Cu-7Sn-20Pb

Commercial Names

Common name. Medium bronze

Specifications

ASTM. Sand castings: B66. Ingot: B30

Government. QQ-L-225, alloy 15; MIL-B-16261, alloy I

Chemical Composition

Composition limits. 6.0 to 8.0 Sn; 16 to 22 Pb; 1.2 max Zn; 1.0 max Ni; 0.8 max Sb; 0.005 max Al; 0.15 max Fe; 0.5 max P(a); 0.08 max S; 0.005 max Si; rem Cu

(a)1.5 max P for continuous castings.

Applications

Typical uses. Locomotive wearing parts, high-load low-speed bearings

Mechanical Properties

Tensile properties. Typical. Tensile strength, 170 MPa (25 ksi); yield strength, 83 MPa (12 ksi); elongation, 12% in 50 mm or 2 in.

Compressive properties. Compressive strength, 250 MPa (36 ksi)

Hardness. 50 HB

Elastic modulus. Tension, 72 GPa (10.5×10^6 psi); shear, 90 GPa (13×10^6 psi)

Fatigue strength. Rotating beam: 69 MPa (10 ksi) at 10^8 cycles

Impact strength. Izod, 5.4 J (4.0 ft·lb)

Mass Characteristics

Density. 9.4 Mg/m³ (0.34 lb/in.³) at 20 °C (68°F)

Volume change on freezing. 1.1%

Thermal Properties

Liquidus temperature. 940 °C (1725 °F)

Solidus temperature. 800 °C (1475 °F)

Incipient melting temperature. 316 °C (600 °F)

Coefficient of thermal expansion. Linear, 18.5 µm/m·K (10.3 µin./in.·°F) at 20 to 200 °C (68 to 392 °F)

Specific heat. 376 J/kg·K (0.09 Btu/lb·°F) at 20 °C (68 °F)

Thermal conductivity. 52 W/m·K (30 Btu/ft·h·°F) at 20 °C (68 °F)

Electrical Properties

Electrical conductivity. Volumetric, 10% IACS at 20 °C (68 °F)

Magnetic Properties

Magnetic permeability. 1.0

Fabrication Characteristics

Machinability. 80% of C36000, free-cutting brass

C95200
88Cu-3Fe-9Al

Commercial Names

Previous trade name. Ampco A1
Common name. Aluminum bronze 9A; 88-3-9

Specifications

ASME. Sand castings: SB148. Centrifugal castings: SB271.

ASTM. Sand castings: B148. Centrifugal castings: B271. Continuous castings: B505. Ingot: B30

SAE. J462

Government. Centrifugal, sand and continuous castings: QQ-C-390. Sand castings: MIL-C-22229

Other. Ingot Code No. 415

Chemical Composition

Composition limits. 86 min Cu; 8.5 to 9.5 Al; 2.5 to 4.0 Fe; 1.0 max others (total)

Consequence of exceeding impurity limits. Possible hot shortness and/or hot cracking, embrittlement and reduced soundness of castings

Applications

Typical uses. Acid-resisting pumps, bearings, bushings, gears, valve seats, guides, plungers, pump rods, pickling hooks, nonsparking hardware

Precautions in use. Not suitable for use in oxidizing acids

Mechanical Properties

Tensile properties. Typical data for sand cast test bars: tensile strength, 550 MPa (80 ksi); yield strength, 185 MPa (27 ksi); elongation, 35% in 50 mm or 2 in. See also Fig. 44.

Hardness. 64 HRB; 125 HB (3000-kg load)

Poisson's ratio. 0.31

Elastic modulus. Tension, 105 GPa (15×10^6 psi); shear, 39 GPa (5.7×10^6 psi)

Impact strength. Charpy keyhole, 27 J (20 ft·lb) at -18 to $+38$ °C (0 to 100 °F); Izod, 40 J (30 ft·lb) at -18 to $+38$ °C (0 to 100 °F)

Fatigue strength. Rotating beam: 150 MPa (22 ksi) at 10^8 cycles

Creep-rupture characteristics. Limiting creep stress for 10^{-5}%/h: 145 MPa (21 ksi) at 230 °C (450 °F); 54 MPa (7.9 ksi) at 315 °C (600 °F). See also Fig. 45.

Structure

Microstructure. As cast, the microstructure is primarily fcc alpha, with precipitates of iron-rich alpha in the form of rosettes and spheres. Depending on the cooling rate, small amounts of metastable cph beta or alpha-gamma eutectoid decomposition products may be present. Annealing followed by rapid cooling reduces the amount of residual beta to about 5% of the apparent volume.

Metallographic etchant. Acid ferric chloride (10% HCl, 5% $FeCl_3$)

Mass Characteristics

Density. 7.64 Mg/m³ (0.276 lb/in.³) at 20 °C (68 °F)

Volume change on freezing. Approximately 1.7% contraction

Patternmaker's shrinkage. 2%

Thermal Properties

Liquidus temperature. 1045 °C (1915 °F)

Solidus temperature. 1040 °C (1905 °F)

Coefficient of thermal expansion. Linear, 16.2 µm/m·K (9.0 µin./in.·°F) at 20 to 300 °C (68 to 572 °F)

Specific heat. 377 J/kg·K (0.091 Btu/lb·°F) at 20 °C (68 °F)

Thermal conductivity. 50 W/m·K (29.1 Btu/ft·h·°F) at 20 °C (68 °F)

Electrical Properties

Electrical conductivity. Volumetric, 12% IACS at 20 °C (68 °F)

Electrical resistivity. 144 nΩ·m at 20 °C (68 °F)

Magnetic Properties

Magnetic permeability. 1.20 at 16 000 A/m (200 oersteds)

Chemical Properties

General corrosion behavior. C95200 has generally fair resistance to attack in nonoxidizing mineral acids such as sulfuric, hydrochloric and phosphoric, and in alkalies such as sodium and potassium hydroxide. Cast components are used successfully in systems for seawater, brackish water and potable water. The alloy resists many organic acids, including acetic and lactic, plus all

Fig. 44 Typical short-time tensile properties of C95200, as cast

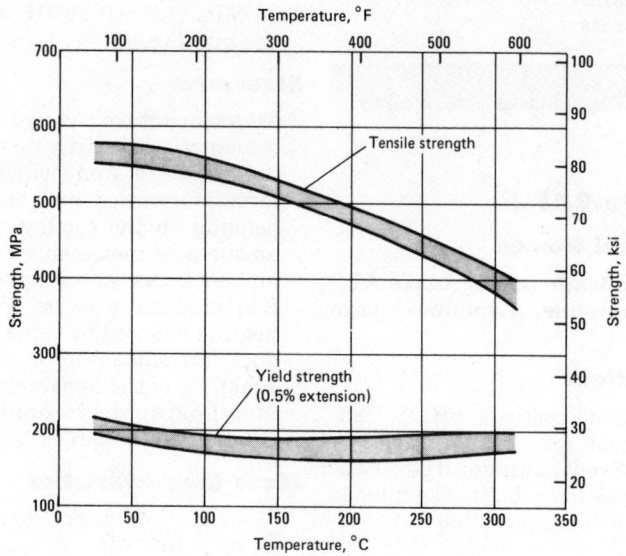

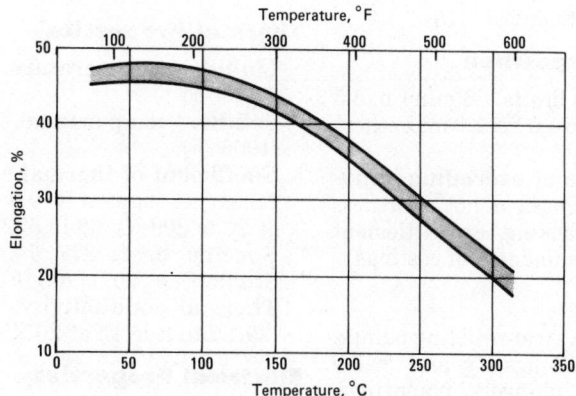

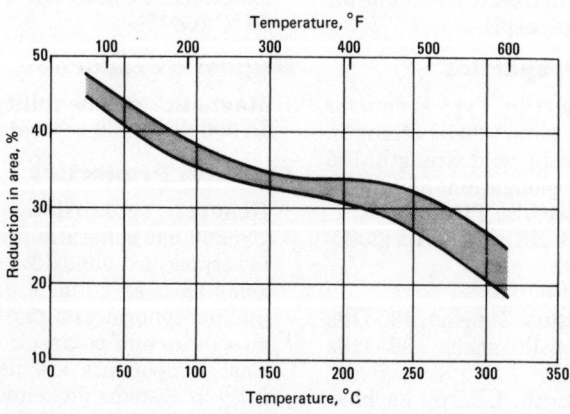

esters and ethers. Moist ammonia atmospheres can cause stress-corrosion cracking.

Fabrication Characteristics

Machinability. 20% of C36000, free-cutting brass. Carbide or tool steel cutters may be used. Good surface finish and precision attainable with all conventional methods. Typical conditions using tool steel cutters: roughing speed, 105 m/min (350 ft/min) with a feed of 0.3 mm/rev (0.011 in./rev); finishing speed, 350 m/min (1150 ft/min) with a feed of 0.15 mm/rev (0.006 in./rev)

Annealing temperature. 650 to 745 °C (1200 to 1375 °F)

C95300
89Cu-1Fe-10Al

Commercial Names

Trade name. Ampco B2
Common names. Aluminum bronze 9B; 89-1-10

Specifications

ASTM. Sand castings: B148. Centrifugal castings: B271. Continuous castings: B505. Ingots: B30
SAE. J462
Government. Centrifugal and sand castings: QQ-C-390. Precision castings: MIL-C-11866, composition 22
Ingot identification number. 415

Chemical Composition

Composition limits. 86 min Cu; 9.0 to 11.0 Al; 0.8 to 1.5 Fe; 1.0 max others (total)
Consequence of exceeding impurity limits. Possible hot shortness, loss of casting soundness, embrittlement, reduced response to heat treatment

Applications

Typical uses. Pickling baskets, nuts, gears, steel mill slippers, marine equipment, welding jaws, nonsparking hardware
Precautions in use. Not suitable for exposure to oxidizing acids. Prolonged heating in the 320 to 565 °C (610 to 1050 °F) range can result in a loss of ductility and notch toughness.

Mechanical Properties

Tensile properties. Minimum val-

Fig. 45 Typical creep properties of C95200, as cast

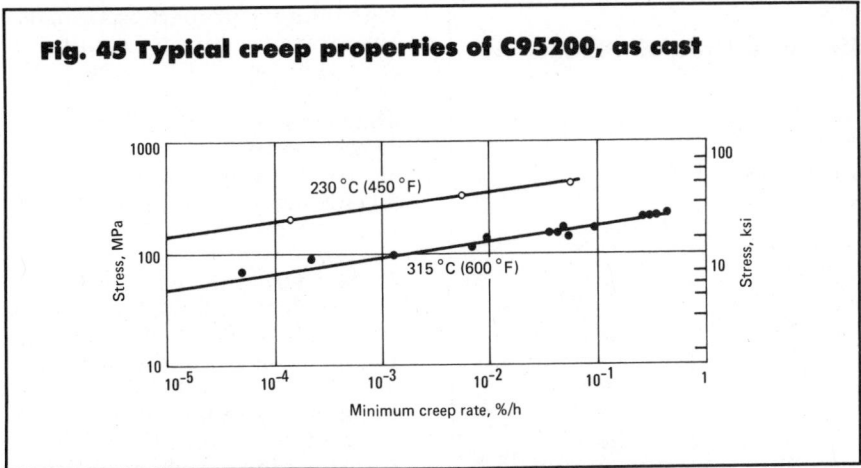

ues. As cast: tensile strength, 450 MPa (65 ksi); yield strength, 170 MPa (25 ksi); elongation, 20% in 50 mm or 2 in.; reduction in area, 25%. TQ50 temper: tensile strength, 550 MPa (80 ksi); yield strength, 275 MPa (40 ksi); elongation, 12% in 50 mm or 2 in.; reduction in area, 14%

Compressive properties. Compressive ultimate strength: as cast, 760 MPa (110 ksi); TQ50 temper, 825 MPa (120 ksi). Elastic limit: as cast, 125 MPa (18 ksi); TQ50 temper, 205 MPa (30 ksi)

Hardness. As cast, 67 HRB; TQ50 temper, 81 HRB

Poisson's ratio. 0.314

Elastic modulus. Tension, 110 GPa (16×10^6 psi); shear, 42 GPa (6.1×10^6 psi)

Impact strength. Cast and annealed: Charpy keyhole, 31 J (23 ft·lb); Izod, 38 J (28 ft·lb) at −20 to +100 °C (−5 to +212 °F). TQ50 temper: Charpy keyhole, 37 J (27 ft·lb) at −20 to +100 °C (−5 to +212 °F)

Structure

Crystal structure. Alpha phase, face-centered cubic; beta phase, close-packed hexagonal

Microstructure. As cast and properly cooled or annealed, the structure is approximately 70% alpha and 30% metastable beta. Quenched and tempered (TQ50 temper), the structure is largely tempered metastable beta martensite, but also contains both primary alpha and reprecipitated acicular alpha

Mass Characteristics

Density. 7.53 Mg/m³ (0.272 lb/in.³) at 20 °C (68 °F)

Patternmaker's shrinkage. 1.6%

Thermal Properties

Liquidus temperature. 1045 °C (1915 °F)

Solidus temperature. 1040 °C (1905 °F)

Coefficient of thermal expansion. Linear, 16.2 μm/m·K (9.0 μin./in.·°F) at 20 to 300 °C (68 to 572 °F)

Specific heat. 375 J/kg·K (0.09 Btu/lb·°F) at 20 °C (68 °F)

Thermal conductivity. 63 W/m·K (36 Btu/ft·h·°F) at 20 °C (68 °F); temperature coefficient, 0.12 W/m·K per K at 20 °C (68 °F)

Electrical Properties

Electrical conductivity. Volumetric, 13% IACS at 20 °C (68 °F)

Electrical resistivity. 133 nΩ·m at 20 °C (68 °F)

Magnetic Properties

Magnetic permeability. 1.07 at field strength of 8 kA/m

Chemical Properties

General corrosion behavior. Corrosion characteristics of C95300 are slightly inferior to those of C95200, primarily because C95300 has more and larger beta areas. Heat treatment enhances corrosion resistance, particularly in mediums that promote dealloying. The alloy shows characteristic resistance to nonoxidizing mineral acids, neutral salt solutions, seawater, brackish water and some organic acids

Fabrication Characteristics

Machinability. 55% of C36000, free-cutting brass. Tool steel or carbide cutters may be used. Good surface and precision finish may be obtained in the as cast, cast and annealed and TQ50 tempers. Typical conditions us-

ing tool steel cutters: roughing speed, 90 m/min (300 ft/min) at a feed of 0.2 mm/rev (0.009 in./rev); finishing speed, 290 m/min (950 ft/min) at a feed of 0.1 mm/rev (0.004 in./rev)

Annealing temperature. 595 to 650 °C (1100 to 1200 °F)

C95400
85Cu-4Fe-11Al

Commercial Names

Trade name. Ampco C3

Common names. Aluminum bronze 9C; G5; 85-4-11

Specifications

ASME. Sand castings: SB148

ASTM. Sand castings: B148. Centrifugal castings: B271. Continuous castings: B505. Ingots: B30

Government. QQ-C-390. Sand castings: MIL-C-22229, composition 6. Investment castings: MIL-C-15345, alloy 13. Centrifugal castings: MIL-C-22087, composition 8

Ingot identification number. 415

Chemical Composition

Composition limits. 83 min Cu; 10.0 to 11.5 Al; 3.0 to 5.0 Fe; 0.50 max Mn; 2.5 max Ni(+Co); 0.5 max others (total)

Consequence of exceeding impurity limits. Possible hot shortness, reduced casting soundness, embrittlement and loss of heat treating response

Applications

Typical uses. Pump impellers, bearings, gears, worms, bushings, valve seats and guides, rolling mill slippers, slides, nonsparking hardware

Precautions in use. Not suitable for use in oxidizing acids. Prolonged heating in the 320 to 565 °C (610 to 1050 °F) range can result in loss of ductility and notch toughness

Mechanical Properties

Tensile properties. Minimum values. As cast: tensile strength, 515 MPa (75 ksi); yield strength, 205 MPa (30 ksi); elongation, 12% in 50 mm or 2 in.; reduction in area, 12%. TQ50 temper: tensile strength, 620 MPa (90 ksi); yield strength, 310 MPa (45 ksi); elongation, 6% in 50 mm or 2 in., reduction in area, 6%. See also Fig. 46.

Fig. 46 Typical short-time tensile properties of C95400, as cast

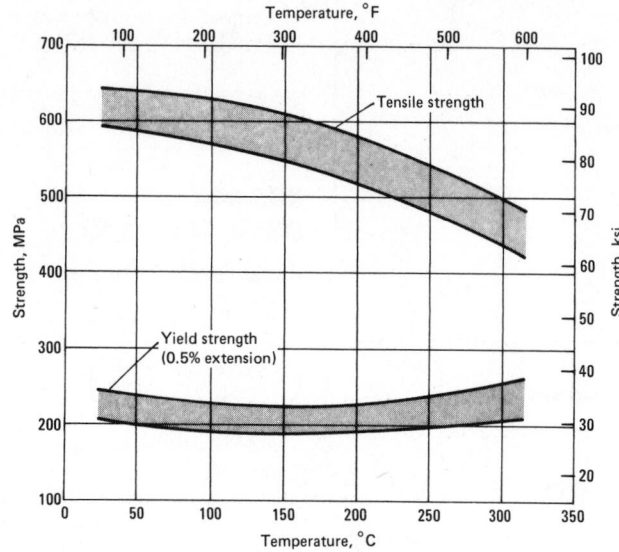

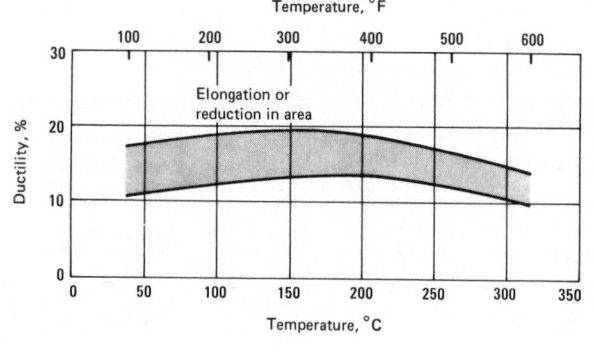

Fig. 47 Typical creep properties for C95400, as cast

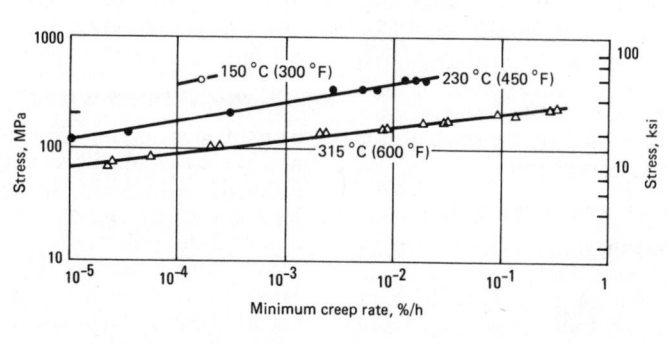

Compressive properties. Compressive strength, ultimate: as cast, 940 MPa (136 ksi); TQ50 temper, 1070 MPa (155 ksi)

Hardness. As cast, 83 HRB; TQ50 temper: 94 HRB

Poisson's ratio. 0.316

Elastic modulus. Tension, 110 GPa (16×10^6 psi); shear, 42 GPa (6.1×10^6 psi)

Impact strength. As cast: Charpy keyhole, 15 J (11 ft·lb); Izod, 22 J (16 ft·lb) at 20 °C (68 °F). TQ50 temper: Charpy keyhole, 9 J (7 ft·lb); Izod, 15 J (11 ft·lb) at 20 °C (68 °F)

Fatigue strength. Reverse bending: 355 MPa (36 ksi) at 10^8 cycles

Creep-rupture characteristics. Limiting creep stress at a strain rate of 10^{-5}%/h: 115 MPa (17 ksi) at 230 °C (450 °F); 51 MPa (7.4 ksi) at 315 °C (600 °F); 30 MPa (4.4 ksi) at 370 °C (700 °F); 20 MPa (2.9 ksi) at 425 °C (800 °F). See also 47.

Structure

Crystal structure. Alpha, face-centered cubic; beta, close-packed hexagonal

Microstructure. As-cast and annealed material normally consists of approximately 50% alpha and 50% metastable beta. Under some conditions, eutectoid decomposition may produce an alpha–gamma–2 structure instead of the beta phase. Quenched and tempered structures consist of fine acicular alpha crystals in a tempered beta matrix.

Mass Characteristics

Density. 7.45 Mg/m³ (0.269 lb/in.³) at 20 °C (68 °F)

Patternmaker's shrinkage. 1.6%

Thermal Properties

Liquidus temperature. 1040 °C (1900 °F)

Solidus temperature. 1025 °C (1880 °F)

Coefficient of thermal expansion. Linear, 16.2 μm/m·K (9.0 μin./in.·°F) at 20 to 300 °C (68 to 572 °F)

Specific heat. 420 J/kg·K (0.10 Btu/lb·°F) at 20 °C (68 °F)

Thermal conductivity. 59 W/m·K (34 Btu/ft·h·°F) at 20 °C (68 °F); temperature coefficient, 0.117 W/m·K per K at 20 °C (68 °F)

Electrical Properties

Electrical conductivity. Volumetric, 13% IACS at 20 °C (68 °F)

Electrical resistivity. 133 nΩ·m at 20 °C (68 °F)

Magnetic Properties

Magnetic permeability. As cast, 1.27 at field strength of 16 kA/m; TQ50 temper, 1.20 at field strength of 16 kA/m

Chemical Properties

General corrosion behavior. C95400 has fair resistance to attack by nonoxidizing solutions of mineral acids such as sulfuric and phosphoric, as well as to neutral salts such as sodium chloride. The alloy also resists acetic, lactic and oxalic acids; organic solvents such as esters and ethers; and seawater, brackish water and potable waters. In some environments, C95400 can undergo dealloying caused by corrosive attack on the beta phase. Heat treatment improves resistance to dealloying. Moist ammonia environments may cause stress corrosion cracking under high levels of applied stress.

Fabrication Characteristics

Machinability. 60% of C36000, freecutting brass. C95400, in either ascast or TQ50 temper, is easily machined by all standard operations using high-strength tool steel or carbide cutters. Typical conditions using tool steel cutters: roughing speed, 90 m/min (300 ft/min) at a feed of 0.3 mm/rev (0.011 in./rev); finishing speed, 290 m/min (950 ft/min) at a feed of 0.1 mm/rev (0.004 in./rev)
Annealing temperature. 620 °C (1150 °F)

C95500
81Cu-4Fe-4Ni-11Al

Commercial Names

Previous trade name. Ampco D4
Common names. Aluminum bronze 9D; 415; 81-4-4-11

Specifications

AMS. 4880
ASTM. Sand castings: B148. Centrifugal castings: B271. Continuous castings, B505. Ingots: B30
SAE. J462
Government. QQ-C-390. Centrifugal castings: MIL-C-15345, alloy 14. Sand castings: MIL-C-22229, composition 6. Investment castings: MIL-C-22087, composition 8
Ingot identification number. 415

Chemical Composition

Composition limits. 78 min Cu; 10.0 to 11.5 Al; 3.0 to 5.0 Fe; 3.5 max Mn; 3.0 to 5.5 Ni(+Co); 0.5 max others (total)
Consequence of exceeding impurity limits. Possible hot shortness in welding, embrittlement, increased quench cracking susceptibility, possible loss of heat treating response. Excessive Si can cause machining difficulties.

Applications

Typical uses. Valve guides and seats in aircraft engines, corrosion-resistant parts, bushings, gears, worms, pickling hooks and baskets, agitators
Precautions in use. Not suitable for use in strong oxidizing acids

Mechanical Properties

Tensile properties. (Typical) As cast: tensile strength, 620 MPa (90 ksi); yield strength, 275 MPa (40 ksi); elongation, 6% in 50 mm or 2 in.; reduction in area, 7%. TQ50 temper: tensile strength, 760 MPa (110 ksi); yield strength, 415 MPa (60 ksi); elongation, 5% in 50 mm or 2 in.; reduction in area, 5%
Compressive properties. As cast: compressive strength, 895 MPa (130 ksi); compressive yield strength, 825 MPa (120 ksi) at a permanent set of 10%; elastic limit, 310 MPa (45 ksi). TQ50 temper: compressive strength, 1140 MPa (165 ksi); compressive yield strength, 1030 MPa (150 ksi) at a permanent set of 10%; elastic limit, 415 MPa (60 ksi)
Hardness. As cast, 87 HRB; TQ50 temper, 96 HRB
Poisson's ratio. 0.32
Elastic modulus. Tension: as cast, 110 GPa (16×10^6 psi); TQ50 temper, 115 GPa (17×10^6 psi). Shear: as cast, 42 GPa (6.1×10^6 psi); TQ50 temper, 44 GPa (6.4×10^6 psi)
Impact strength. Charpy keyhole, 14 J (10 ft·lb); Izod, 18 J (13 ft·lb) at 20 °C (68 °F)
Fatigue strength. Rotating beam: as cast, 215 MPa (31 ksi) at 10^8 cycles; TQ50 temper, 260 MPa (38 ksi) at 10^8 cycles
Creep-rupture characteristics. Limiting creep stress at a strain rate of 10^{-5}%/h: 72 MPa (10.5 ksi) at 315 °C (600 °F); 38 MPa (5.5 ksi) at 370 °C (700 °F); 17 MPa (2.5 ksi) at 425 °C (800 °F)

Structure

Crystal structure. Alpha, face-centered cubic; beta, close-packed hexagonal; kappa, ordered face-centered cubic
Microstructure. As-cast or annealed structures consist of alpha crystals plus kappa precipitates, forming a pearlitic appearance. Small areas of metastable beta may exist. Heat treated structures consist of tempered beta martensite with very fine reprecipitated alpha needles. Some undissolved equiaxed alpha crystals may be evident, depending on the actual composition and quenching temperature.

Mass Characteristics

Density. 7.53 Mg/m³ (0.272 lb/in.³) at 20 °C (68 °F)
Patternmaker's shrinkage. 1.6%

Thermal Properties

Liquidus temperature. 1055 °C (1930 °F)
Solidus temperature. 1040 °C (1900 °F)
Coefficient of thermal expansion. Linear, 16.2 µm/m·K (9.0 µin./in.·°F) at 20 to 300 °C (68 to 572 °F)
Specific heat. 418 J/kg·K (0.10 Btu/lb·°F) at 20 °C (68 °F)
Thermal conductivity. 42 W/m·K (24 Btu/ft·h·°F) at 20 °C (68 °F)

Electrical Properties

Electrical conductivity. Volumetric, 8.5% IACS at 20 °C (68 °F)
Electrical resistivity. 203 nΩ·m at 20 °C (68 °F)

Magnetic Properties

Magnetic permeability. As cast, 1.30 at field strength of 16 kA/m; TQ50 temper, 1.20 at field strength of 16 kA/m

Chemical Properties

General corrosion behavior. Good cavitation resistance in salt water and fresh tap water. Avoid nitric acid and strong aeration when using other acids.

Fabrication Characteristics

Machinability. 50% of C36000, freecutting brass. Heat treating reduces machinability in drilling and tapping operations. Tool steel or carbide cutters may be used. Typical conditions using tool steel cutters follow. Roughing speed: as cast, 76 m/min (250 ft/min) at a feed of 0.3 mm/rev

(0.011 in./rev); TQ50 temper, 90 m/min (300 ft/min) at a feed of 0.2 mm/rev (0.009 in./rev). Finishing speed: as cast and TQ50 temper, 290 m/min (950 ft/min) at a feed of 0.1 mm/rev (0.004 in./rev)

Annealing temperature. 620 to 705 °C (1150 to 1300 °F)

C95700
75Cu-3Fe-8Al-2Ni-12Mn

Commercial Names

Previous trade name. Superstone 40, Novoston, Ampcoloy 495
Common name. Manganese aluminum bronze; 75-3-8-2-12

Specifications

ASTM. Sand castings: B148. Ingot: B30
Government. Sand castings: MIL-B-24480

Chemical Composition

Composition limits. 71.0 min Cu; 11.0 to 14.0 Mn; 7.0 to 8.5 Al; 2.0 to 4.0 Fe; 1.5 to 3.0 Ni; 0.10 max Si; 0.03 max Pb; 0.5 max others (total)
Consequence of exceeding impurity limits. Possible hot shortness and reduced cast strength

Applications

Typical uses. Propellers, impellers, stator clamp segments, safety tools, welding rods, valves, pump casings, marine fittings
Precautions in use. Slow cooling or prolonged heating in the 350 to 560 °C (660 to 1050 °F) range may cause embrittlement. Not suitable for use in oxidizing acids.

Mechanical Properties

Tensile properties. Typical data for sand cast test bars: tensile strength, 620 MPa (90 ksi); yield strength, 275 MPa (40 ksi); elongation, 20% in 50 mm or 2 in.; reduction in area, 24%
Compressive properties. Compressive strength, as cast: 1035 MPa (150 ksi) at a permanent set of 0.1%
Hardness. As cast, or cast and annealed: 85 to 90 HRB
Poisson's ratio. 0.326
Elastic modulus. Tension, 125 GPa (18 × 10^6 psi); shear, 44 GPa (6.4 × 10^6 psi)
Impact strength. Izod, 27 J (20 ft·lb) at 20 °C (68 °F)

Fatigue strength. Reverse bending, 231 MPa (33.5 ksi) at 10^8 cycles
Creep-rupture characteristics. Limiting creep stress for 10^{-5}%/h: 66 MPa (9.6 ksi) at 205 °C (400 °F); 31 MPa (4.5 ksi) at 290 °C (550 °F). Rupture stress for 10^5 h life: 470 MPa (68 ksi) at 205 °C (400 °F); 232 MPa (33.6 ksi) at 260 °C (500 °F); 39 MPa (5.7 ksi) at 370 °C (700 °F)

Structure

Microstructure. As cast and annealed tempers: fcc alpha crystals with cph beta phase in various amounts, typically 25% by volume

Mass Characteristics

Density. 7.53 Mg/m³ (0.272 lb/in.³) at 20 °C (68 °F)
Patternmaker's shrinkage. 1.6%

Thermal Properties

Liquidus temperature. 990 °C (1815 °F)
Solidus temperature. 950 °C (1740 °F)
Coefficient of thermal expansion. Linear, 17.6 μm/m·K (9.8 μin./in.·°F) at 20 to 300 °C (68 to 572 °F)
Specific heat. 440 J/kg·K (0.105 Btu/lb·°F) at 20 °C (68 °F)
Thermal conductivity. 12.1 W/m·K (7.0 Btu/ft·h·°F) at 20 °C (68 °F)

Electrical Properties

Electrical conductivity. Volumetric, 3.1% IACS at 20 °C (68 °F)
Electrical resistivity. 556 nΩ·m at 20 °C (68 °F)

Magnetic Properties

Magnetic permeability. Magnetic condition (as cast, slow cooled): 2.2 to 15.0. Demagnetized (annealed, fast cooled): 1.03

Chemical Properties

General corrosion behavior. Generally comparable to that of the aluminum bronzes and nickel-aluminum bronzes. See C95200.

Fabrication Characteristics

Machinability. 50% of C36000, free-cutting brass. Tool steel or carbide cutters may be used. Good surface finishes and tolerance are possible in all conventional machining operations. Typical conditions using tool steel cutters: roughing speed, 75 m/min (250 ft/min) with a feed of 0.3 mm/rev (0.011 in./rev); finishing speed, 290 m/min (950 ft/min) with a feed of 0.1 mm/rev (0.004 in./rev)
Annealing temperature. 620 °C (1150 °F)

C95800
82Cu-4Fe-9Al-4Ni-1Mn

Commercial Names

Common names. Alpha nickel aluminum bronze; propeller bronze

Specifications

ASTM. Sand castings: B148. Centrifugal castings: B271. Continuous castings: B505. Ingots: B30
SAE. J462
Government. Sand and centrifugal castings: QQ-C-390; MIL-B-24480. Centrifugal castings only: MIL-C-15345, alloy 28
Ingot identification number. 415

Chemical Composition

Composition limits. 79.0 min Cu; 0.03 max Pb; 3.5 to 4.5 Fe; 4.0 to 5.0 Ni(+ Co); 0.8 to 1.5 Mn; 8.5 to 9.5 Al; 0.10 max Si
Consequence of exceeding impurity limits. Hard spots, embrittlement, possible hot shortness, possible weld cracking

Applications

Typical uses. Propeller blades and hubs for fresh- and salt-water service, fittings, gears, worm wheels, valve guides and seals, structural applications
Precautions in use. Not suitable for use in oxidizing acids or strong alkalies

Mechanical Properties

Tensile properties. (Typical) Cast and annealed: tensile strength, 585 MPa (85 ksi); yield strength, 240 MPa (35 ksi); elongation, 15% in 50 mm or 2 in.; reduction in area, 16%
Compressive properties. Compressive strength, cast and annealed: 240 MPa (35 ksi) at a permanent set of 0.1%; 330 MPa (48 ksi) at a permanent set of 1%; 690 MPa (100 ksi) at a permanent set of 10%
Hardness. Cast and annealed, 84 to 89 HRB
Poisson's ratio. 0.32
Elastic modulus. Tension, 110 GPa (16 × 10^6 psi); shear, 42 GPa (6.1 × 10^6 psi)
Impact strength. Charpy keyhole, 13 J (10 ft·lb) at −23 to +66 °C (−10 to +150 °F); Charpy V-notch, 22 J (16 ft·lb) at −23 to +66 °C (−10 to +150 °F)
Fatigue strength. Rotating beam: 230 MPa (33 ksi) at 10^8 cycles

Structure

Crystal structure. Alpha, face-centered cubic; beta, close-packed hexagonal; kappa, ordered face-centered cubic

Microstructure. As-cast or annealed structures are generally continuous equiaxed alpha crystals with small areas of metastable beta phase. Kappa phase precipitates are found in the alpha phase, in grain boundaries and in beta areas. Quench-and-temper treatments result in refinement and redistribution of the kappa phase throughout a matrix of tempered beta martensite and alpha-kappa eutectoid decomposition product. Some undissolved primary alpha crystals may also be present.

Mass Characteristics

Density. 7.64 Mg/m^3 (0.276 lb/in.^{3}) at 20 °C (68 °F)
Patternmaker's shrinkage. 1.6%

Thermal Properties

Liquidus temperature. 1060 °C (1940 °F)
Solidus temperature. 1045 °C (1910 °F)
Coefficient of thermal expansion. Linear, 16.2 μm/m·K (9.0 μin./in.·°F) at 20 to 300 °C (68 to 572 °F)
Specific heat. 440 J/kg·K (0.105 Btu/lb·°F) at 20 °C (68 °F)
Thermal conductivity. 36 W/m·K (21 Btu/ft·h·°F) at 20 °C (68 °F)

Electrical Properties

Electrical conductivity. Volumetric, 7.1% IACS at 20 °C (68 °F)
Electrical resistivity. 243 nΩ·m at 20 °C (68 °F)

Magnetic Properties

Magnetic permeability. 1.05 at field strength of 16 kA/m

Chemical Properties

General corrosion behavior. Corrosion properties of C95800 are similar to those of other nickel aluminum bronzes, except that C95800 has better resistance to cavitation and seawater fouling attack. Resists dealloying in most mediums.

Fabrication Characteristics

Machinability. 50% of C36000, free-cutting brass. Excellent surface finish and tolerances possible in all standard machining operations. Carbide or tool steel cutters may be used.

Typical conditions using tool steel cutters: roughing speed, 76 m/min (250 ft/min) at a feed of 0.3 mm/rev (0.011 in./rev); finishing speed, 290 m/min (950 ft/min) at a feed of 0.1 mm/rev (0.004 in./rev)
Annealing temperature. 650 to 705 °C (1200 to 1300 °F)

C96400
70Cu-30Ni

Commercial Names

Previous trade name. 70–30 copper nickel

Specifications

ASTM. Centrifugal castings: B369. Continuous castings: B505. Sand castings, ingot: B30
Government. Centrifugal castings: MIL-C-15345 (alloy 24). Sand castings: QQ-C-390, MIL-C-20159 (type 1)

Chemical Composition

Composition limits. 65.0 to 69.0 Cu; 28.0 to 32.0 Ni; 0.50 to 1.5 Nb; 0.25 to 1.5 Fe; 1.5 max Mn; 0.50 max Si; 0.15 max C; 0.03 max Pb (0.01 max Pb for welding applications)

Applications

Typical uses. Centrifugal, continuous and sand castings for valves, pump bodies, flanges, and elbows for applications requiring resistance to seawater corrosion

Mechanical Properties

Tensile properties. Typical data for sand cast test bars: tensile strength, 470 MPa (68 ksi); yield strength, 255 MPa (37 ksi) at 0.5% extension under load; elongation, 28% in 50 mm or 2 in.
Hardness. Typical, 140 HB using 3000-kg load
Elastic modulus. Tension, 145 GPa (21 × 10^6 psi)
Impact strength. Charpy V-notch, 106 J (78 ft·lb)
Fatigue strength. Reverse bending: 125 MPa (18 ksi) at 10^8 cycles

Mass Characteristics

Density. 8.94 Mg/m^3 (0.323 lb/in.^{3}) at 20 °C (68 °F)
Patternmaker's shrinkage. 19 mm/m (7/32 in./ft)

Thermal Properties

Liquidus temperature. 1238 °C (2260 °F)

Solidus temperature. 1171 °C (2140 °F)
Coefficient of thermal expansion. Linear, 16 μm/m·K (9.0 μin./in.·°F) at 20 to 300 °C (68 to 572 °F)
Specific heat. 375 J/kg·K (0.09 Btu/lb·°F) at 20 °C (68 °F)
Thermal conductivity. 29 W/m·K (17 Btu/ft·h·°F) at 20 °C (68 °F)

Electrical Properties

Electrical conductivity. Volumetric, as cast tempers: 5% IACS at 20 °C (68 °F)

Fabrication Characteristics

Machinability. 20% of C36000, free-cutting brass
Weldability. Soldering, brazing: excellent. Gas-shielded arc and shielded metal-arc welding: good, using RCuNi or ECuNi filler metal oxyfuel gas and carbon arc welding: not recommended

C96600
69.5Cu-30Ni-0.5Be

Commercial Names

Previous trade name. Beryllium cupro-nickel alloy 71C; CA966
Common name. Beryllium cupro-nickel

Specifications

Government. Sand castings: MIL-C-81519

Chemical Composition

Composition limits. 0.40 to 0.7 Be; 29.0 to 33.0 Ni; 0.8 to 1.1 Fe; 1.0 max Mn; 0.15 max Si; 0.01 max Pb; rem Cu
Consequence of exceeding impurity limits. An excessive amount of Si will increase as-cast hardness and lower ductility. High Pb will cause hot shortness.

Applications

Typical uses. C96600 is a high-strength version of the well-known cupro-nickel alloy C96400, possessing twice the strength. Like C96400, C96600 exhibits excellent corrosion resistance to seawater. Typical uses are high-strength constructional parts for marine service; pressure housings for long, unattended submergence; pump bodies; valve bodies; seawater line fittings;

marine tow-line hardware; gimbal assemblies; and release mechanisms **Precautions in use.** See C82500.

Mechanical Properties

Tensile properties. Typical data for separately cast test bars. TB00 temper: tensile strength, 515 MPa (75 ksi); yield strength, 260 MPa (38 ksi); elongation in 50 mm or 2 in., 12%. TF00 temper: tensile strength, 825 MPa (120 ksi); yield strength, 515 MPa (75 ksi); elongation, 12%
Hardness. TB00 temper: 74 HRB. TF00 temper: 24 HRC
Poisson's ratio. 0.33
Elastic modulus. Tension, 150 GPa (22×10^6 psi); shear, 57 GPa (8.3×10^6 psi)

Mass Characteristics

Density. 8.80 Mg/m^3 (0.320 lb/in.^{3}) at 20 °C (68 °F)
Patternmaker's shrinkage. 1.8%

Thermal Properties

Liquidus temperature. 1180 °C (2160 °F)
Solidus temperature. 1095 °C (2010 °F)
Coefficient of thermal expansion. Linear, 16 μm/m·K (9 μin./in.·°F) at 20 to 300 °C (68 to 572 °F)
Specific heat. 377 J/kg·K (0.091 Btu/lb·°F) at 20 °C (68 °F)
Thermal conductivity. 30 W/m·K (17.3 Btu/ft·h·°F) at 20 °C (68 °F)

Electrical Properties

Electrical conductivity. Volumetric, 4.3% IACS at 20 °C (68 °F)
Electrical resistivity. 4 nΩ·m at 20 °C (68 °F)

Chemical Properties

General corrosion behavior. Essentially identical to that of C96400

Fabrication Characteristics

Machinability. TF00 temper, 40% of C36000, free-cutting brass
Melting temperature. 1090 to 1180 °C (2010 to 2160 °F)
Casting temperature. 1260 to 1370 °C (2300 to 2500 °F)
Solution temperature. 995 °C (1825 °F)
Aging temperature. 510 °C (950 °F). Typical aging time, 3 h

C97300
56Cu-2Sn-10Pb-20Zn-12Ni

Commercial Names
Previous trade name. 12% nickel silver
Common name. Leaded nickel brass; 56-2-10-20-12

Specifications
ASTM. Centrifugal castings: B271. Sand castings: B584. Ingot: B30

Chemical Composition
Composition limits. 53.0 to 58.0 Cu; 1.5 to 3.0 Sn; 8.0 to 11.0 Pb; 17.0 to 25.0 Zn; 11.0 to 14.0 Ni; 1.5 max Fe; 0.50 max Mn; 0.35 max Sb; 0.15 max Si; 0.08 max S; 0.05 max P; 0.005 max Al

Applications
Typical uses. Investment, centrifugal, permanent mold and sand castings for hardware fittings, valves and valve trim, statuary, and ornamental castings

Mechanical Properties
Tensile properties. Typical data for sand cast test bars: tensile strength, 240 MPa (35 ksi); yield strength, 115 MPa (17 ksi) at 0.5% extension under load; elongation, 20% in 50 mm or 2 in.
Hardness. Typical, 55 HB using 500-kg load
Elastic modulus. Tension, 110 GPa (16×10^6 psi)

Mass Characteristics
Density. 8.95 Mg/m^3 (0.321 lb/in.^{3}) at 20 °C (68 °F)
Patternmaker's shrinkage. 16 mm/m (³⁄₁₆ in./ft)

Thermal Properties
Liquidus temperature. 1040 °C (1904 °F)
Solidus temperature. 1010 °C (1850 °F)
Coefficient of thermal expansion. Linear, 16.2 μm/m·K (9.0 μin./in.·°F) at 20 to 260 °C (68 to 500 °F)
Specific heat. 375 J/kg·K (0.09 Btu/lb·°F) at 20 °C (68 °F)
Thermal conductivity. 29.6 W/m·K (16.5 Btu/ft·h·°F) at 20 °C (68 °F)

Electrical Properties
Electrical conductivity. Volumetric, as cast tempers: 5.7% IACS at 20 °C (68 °F)

Fabrication Characteristics
Machinability. 70% of C36000, free-cutting brass

Weldability. Soldering, brazing: excellent. Welding: not recommended
Stress-relieving temperature. 260 °C (500 °F), 1 h for each 25 mm (1 in.) of section thickness
Casting temperature. Light castings, 1200 to 1315 °C (2200 to 2400 °F); heavy castings, 1090 to 1200 °C (2000 to 2200 °F). Melt rapidly at no more than 55 to 85 °C (100 to 150 °F) above maximum casting temperature

C97600
64Cu-4Sn-4Pb-8Zn-20Ni

Commercial Names
Previous trade name. 20% nickel silver
Common name. Dairy metal, leaded nickel bronze, 64-4-4-8-20

Specifications
ASME. Sand castings: SB584
ASTM. Centrifugal castings: B271. Sand castings: B584. Ingot: B30
Government. Sand castings: MIL-C-17112
Other. Ingot Code No. 412

Chemical Composition
Composition limits. 63.0 to 67.0 Cu; 3.5 to 4.5 Sn; 3.0 to 5.0 Pb; 3.0 to 9.0 Zn; 19.0 to 21.5 Ni; 1.5 max Fe; 1.0 max Mn; 0.25 max Sb; 0.15 max Si; 0.08 max S; 0.05 max P; 0.005 max Al

Applications
Typical uses. Centrifugal, investment and sand castings for marine castings, sanitary fittings, ornamental hardware, valves and pumps

Mechanical Properties
Tensile properties. Typical data for sand cast test bars: tensile strength, 310 MPa (45 ksi); yield strength, 165 MPa (24 ksi) at 0.5% extension under load; elongation, 20% in 50 mm or 2 in.
Compressive properties. Compressive strength, 205 MPa (30 ksi) at a permanent set of 1%; 395 MPa (57 ksi) at a permanent set of 10%
Hardness. Typical, 80 HB using 500-kg load
Elastic modulus. Tension, 130 GPa (19×10^6 psi)
Impact strength. Charpy V-notch, 15 J (11 ft·lb)
Fatigue strength. Reverse bending: 107 MPa (15.5 ksi) at 10^8 cycles
Creep-rupture characteristics.

Limiting stress for creep of 10^{-5} %/h: 224 MPa (32.5 ksi) at 230 °C (450 °F); 153 MPa (22.2 ksi) at 290 °C (550 °F)

Mass Characteristics
Density. 8.90 Mg/m³ (0.321 lb/in.³) at 20 °C (68 °F)
Patternmaker's shrinkage. 11 mm/m (⅛ in./ft)

Thermal Properties
Liquidus temperature. 1142 °C (2089 °F)
Solidus temperature. 1108 °C (2027 °F)
Coefficient of thermal expansion. Linear, 17 μm/m·K (9.3 μin./in.·°F) at 20 to 300 °C (68 to 572 °F)
Specific heat. 375 J/kg·K (0.90 Btu/lb·°F) at 20 °C (68 °F)
Thermal conductivity. 22 W/m·K (13 Btu/ft·h·°F) at 20 °C (68 °F)

Electrical Properties
Electrical conductivity. Volumetric, as cast tempers: 5% IACS at 20 °C (68 °F)

Fabrication Characteristics
Machinability. 70% of C36000, free-cutting brass
Weldability. Soldering, brazing: excellent. Welding: not recommended
Stress-relieving temperature. 260 °C (500 °F), 1 h for each 25 mm (1 in.) of section thickness
Casting temperature. Light castings, 1260 to 1430 °C (2300 to 2600 °F); heavy castings, 1230 to 1320 °C (2250 to 2400 °F). Melt rapidly at no more than 55 to 85 °C (100 to 150 °F) above casting temperature range.

C97800
66.5Cu-5Sn-1.5Pb-2Zn-25Ni

Commercial Names
Previous trade name. 25% nickel silver
Common name. Leaded nickel bronze; 66-5-2-2-25

Specifications
ASTM. Centrifugal castings: B271. Sand castings: B584. Ingot: B30

Chemical Composition
Composition limits. 64.0 to 67.0 Cu; 4.0 to 5.5 Sn; 1.0 to 2.5 Pb; 1.0 to 4.0 Zn; 24.0 to 27.0 Ni; 1.5 max Fe; 1.0 max Mn; 0.20 max Sb; 0.15 max Si; 0.08 max S; 0.05 max P; 0.005 max Al

Applications
Typical uses. Investment, permanent mold and sand castings for ornamental castings, sanitary fittings, valve bodies, valve seats, and musical instrument components

Mechanical Properties
Tensile properties. Typical data for sand cast test bars: tensile strength, 380 MPa (55 ksi); yield strength, 205 MPa (30 ksi) at 0.5% extension under load; elongation, 15% in 50 mm or 2 in.
Hardness. Typical, 130 HB using 3000-kg load
Elastic modulus. Tension, 130 GPa (19 × 10⁶ psi)

Mass Characteristics
Density. 8.86 Mg/m³ (0.320 lb/in.³) at 20 °C (68 °F)
Patternmaker's shrinkage. 16 mm/m (3/16 in./ft)

Thermal Properties
Liquidus temperature. 1180 °C (2156 °F)
Solidus temperature. 1140 °C (2084 °F)
Coefficient of thermal expansion. Linear, 17.5 μm/m·K (9.7 μin./in.·°F) at 20 to 260 °C (68 to 500 °F)
Specific heat. 375 J/kg·K (0.09 Btu/lb·°F) at 20 °C (68 °F)
Thermal conductivity. 25.4 W/m·K (14.7 Btu/ft·h·°F) at 20 °C (68 °F)

Electrical Properties
Electrical conductivity. Volumetric, as-cast tempers: 4.5% IACS at 20 °C (68 °F)

Fabrication Characteristics
Machinability. 60% of C36000, free-cutting brass
Weldability. Soldering, brazing: excellent. Welding: not recommended
Stress-relieving temperature. 260 °C (500 °F)

C99400
90.4Cu-2.2Ni-2.0Fe-1.2Al-1.2Si-3.0Zn

Commercial Names
Common name. Nondezincification alloy, NDZ

Chemical Composition
Composition limits. 0.25 max Pb; 1.0 to 3.5 Ni; 1.0 to 3.0 Fe; 0.50 to 2.0 Al; 0.50 to 2.0 Si; 0.50 to 5.0 Zn; 0.50 max Mn; rem Cu

Applications
Typical uses. Centrifugal, continuous, investment and sand castings for valve stems, propeller wheels, electrical parts, gears for mining equipment, outboard motor parts, marine hardware and other environmental uses where resistance to dezincification and dealuminification is required.

Mechanical Properties
Tensile properties. Typical. MO1 temper: tensile strength, 455 MPa (66 ksi); yield strength, 235 MPa (34 ksi) at 0.5% extension under load; elongation, 25% in 50 mm or 2 in. TF00 temper: tensile strength, 545 MPa (79 ksi); yield strength, 370 MPa (54 ksi) at 0.5% extension under load
Shear strength. MO1 temper: 330 MPa (48 ksi)
Hardness. MO1 temper: 125 HB; TF00 temper, 170 HB. Determined using 3000-kg load
Elastic modulus. Tension, 133 GPa (19.3 × 10⁶ psi)

Mass Characteristics
Density. 8.30 Mg/m³ (0.30 lb/in.³) at 20 °C (68 °F)
Patternmaker's shrinkage. 16 mm/m (3/16 in./ft)

Electrical Properties
Electrical conductivity. Volumetric, TF00 temper: 16.8% IACS at 20 °C (68 °F)

Fabrication Characteristics
Machinability. 50% of C36000, free-cutting brass
Weldability. Shielded metal-arc welding: poor
Solution temperature. 885 °C (1625 °F), 1 h for each 25 mm (1 in.) of section thickness
Aging temperature. 480 °C (900 °F), 1 h at temperature
Stress-relieving temperature. 315 °C (600 °F), 1 h for each 25 mm (1 in.) of section thickness

C99700
56.5Cu-5Ni-1Al-1.5Pb-12Mn-24Zn

Commercial Names
Common name. White manganese brass

Trade name. White Tombasil

Chemical Composition

Composition limits. 54.0 min Cu; 19.0 to 25.0 Zn; 11.0 to 15.0 Mn; 4.0 to 6.0 Ni; 2.0 max Pb; 1.0 max Sn; 1.0 max Fe; 0.50 to 3.0 Al

Applications

Typical uses. Building hardware (interior and exterior), architectural and ornamental fittings, marine hardware, floor drain covers, food handling equipment, swimming pool hardware, valves

Mechanical Properties

Tensile properties. Typical data for separately cast test bars. Sand cast: tensile strength, 380 MPa (55 ksi); yield strength, 170 MPa (25 ksi) at 0.5% extension under load; elongation, 25% in 50 mm or 2 in. Die cast: tensile strength, 450 MPa (65 ksi); yield strength, 185 MPa (27 ksi) at 0.5% extension under load; elongation, 15% in 50 mm or 2 in.
Hardness. Sand cast: 110 HB (3000-kg load). Die cast: 125 HB
Elastic modulus. Tension, 114 GPa (16.5 × 10^6 psi)

Mass Characteristics

Density. 8.19 Mg/m^3 (0.296 lb/in.^{3}) at 20 °C (68 °F)
Patternmaker's shrinkage. 21 mm/m (0.25 in./ft)

Thermal Properties

Liquidus temperature. 900 °C (1655 °F)
Solidus temperature. 880 °C (1615 °F)

Electrical Properties

Electrical conductivity. Volumetric, 3% IACS at 20 °C (68 °F)

Fabrication Characteristics

Machinability. 80% of C36000, free-cutting brass

Beryllium copper 21C
97Cu-2Be-1Co

Commercial Names

Common name. Grain-refined beryllium copper casting alloy 21C

Chemical Composition

Composition limits. 2.00 to 2.25 Be;

Table 9 Typical mechanical properties of beryllium-copper alloy 21C

Temper	Tensile strength MPa	ksi	Yield strength MPa	ksi	Elongation in 50 mm or 2 in., %	Hardness HRB	HRC
As cast	515	75	275	40	25	75	...
Cast and aged(a)	825	120	725	105	5	...	30
Solution treated(b)	415	60	170	25	40	63	...
Solution treated(b) and aged(a)	1105	160	1035	150	1	...	42

(a) Aged 3 h at 345 °C (650 °F). (b) At 790 to 800 °C (1450 to 1475 °F).

1.0 to 1.2 Co; 0.20 to 0.40 Si; 0.20 max Ni; 0.25 max Fe; 0.15 max Al; 0.10 max Sn; 0.02 max Pb; 0.10 max Zn; 0.10 max Cr
Consequence of exceeding impurity limits. See C82500.

Applications

Typical uses. The 1% Co content is a strong grain refiner, and as a result, this alloy is used instead of beryllium copper alloys C82500 and C82400 when thin sections must be cast at high temperatures or when thick and thin sections are present within the same casting in order to achieve a uniform fine-grained structure. The higher cobalt content imparts better wear resistance but less desirable polishability and machinability. Typical uses are comparable to those of beryllium-copper alloys C82400 and C82500.
Precautions in use. See C82500.

Mechanical Properties

Tensile properties. See Table 9.
Hardness. See Table 9.
Poisson's ratio. 0.30
Elastic modulus. Tension, 128 GPa (18.5 × 10^6 psi); shear, 50 GPa (7.3 × 10^6 psi)

Mass Characteristics

Density. 8.26 Mg/m^3 (0.298 lb/in.^{3}) at 20 °C (68 °F)
Dilation during aging. Linear, 0.2%
Change in density during aging. 0.6% increase
Patternmaker's shrinkage. 1.56%

Thermal Properties

Liquidus temperature. 980 °C (1800 °F)
Solidus temperature. 860 °C (1575 °F)
Incipient melting temperature. 835 °C (1535 °F)
Coefficient of thermal expansion.

Linear, 8.6 μm/m·K (5.5 μin./in.·°F) at 20 to 200 °C (68 to 392 °F)
Specific heat. 419 J/kg·K (0.10 Btu/lb·°F) at 20 °C (68 °F)
Thermal conductivity. 105 W/m·K (61 Btu/ft·h·°F) at 20 °C (68 °F)

Electrical Properties

Electrical conductivity. Volumetric, 20% IACS at 20 °C (68 °F)
Electrical resistivity. 862 μΩ·m at 20 °C (68 °F)

Magnetic Properties

Magnetic susceptibility. See C82500.

Nuclear Properties

Effect of irradiation. See C82500.

Chemical Properties

Same as C82500

Fabrication Characteristics

Machinability. As cast or solution treated, 30% of C36000, free-cutting brass. Cast and aged, or solution treated and aged, 10 to 20% of C36000.
Solution temperature. 790 to 800 °C (1450 to 1475 °F)
Aging temperature. 340 °C (650 °F).
Melting temperature. 850 to 980 °C (1575 to 1800 °F)
Casting temperature. Light castings, 1065 to 1175 °C (1950 to 2150 °F); heavy castings, 1000 to 1065 °C (1850 to 1950 °F)

Beryllium copper nickel 72C
68.8Cu-30Ni-1.2Be

Commercial Names

Common name. Modified beryllium cupro-nickel alloy 72C

Chemical Composition

Composition limits. 1.1 to 1.2 Be; 29.0 to 33.0 Ni; 0.7 to 1.0 Fe; 0.10 to 0.20 Zr; 0.10 to 0.20 Ti; 0.7 max Mn; 0.15 max Si; 0.1 max Pb; rem Cu

Consequence of exceeding impurity limits. High silicon will raise as-cast hardness and lower ductility. High lead will cause hot shortness. High carbon will result in undesirable carbides.

Applications

Typical uses. Alloy 72C is a modified version of beryllium cupro-nickel alloy 71C, its increased beryllium content providing improved castability. Its field of application is the plastic tooling industry. Alloy 72C generally is ceramic mold cast into tooling used for molding flame-retardant plastics containing bromine, bromine-boron, chlorinated paraffins and phosphates and other halogens. Additionally, alloy 72C tooling is resistant to corrosion by the foaming agents used in structural plastics that generate ammonia at elevated temperatures, as well as to decompositional products of PVC that contain HCl. The good castability of 72C allows it to be cast into tooling of fine detail.

Precautions in use. See C82500.

Table 10 Typical mechanical properties of cast beryllium cupro-nickel alloy 72C

Temper	Tensile strength MPa	ksi	Yield strength MPa	ksi	Elongation in 50 mm or 2 in., %	Hardness HRB	HRC
As cast and aged(a)	555	81	310	45	15	90	...
Solution treated(b) and aged(a)	860	125	550	80	70	...	26

(a) Aged 3 h at 510 °C (950 °F) (b) Water quenched from 995 °C (1825 °F).

Mechanical Properties

Tensile properties. See Table 10.
Hardness. See Table 10.
Poisson's ratio. 0.33
Elastic modulus. Tension, 150 GPa (22×10^6 psi); shear, 57 GPa (8.3×10^6 psi)

Mass Characteristics

Density. 8.60 Mg/m³ (0.311 lb/in.³) at 20 °C (68 °F)
Patternmaker's shrinkage. 1.8%

Thermal Properties

Liquidus temperature. 1150 °C (2110 °F)
Solidus temperature. 1070 °C (1950 °F)
Coefficient of thermal expansion. Linear, 16 μm/m·K (9 μin./in.·°F) at 20 to 300 °C (68 to 572 °F)
Specific heat. 337 J/kg·K (0.08 Btu/lb·°F) at 20 °C (68 °F)
Thermal conductivity. 30 W/m·K (17 Btu/ft·h·°F) at 20 °C (68 °F)

Electrical Properties

Electrical conductivity. Volumetric, 43% IACS at 20 °C (68 °F)
Electrical resistivity. 4 mΩ·m at 20 °C (68 °F)

Chemical Properties

General corrosion behavior. Essentially the same as C96400

Fabrication Characteristics

Machinability. Solution treated and aged, 40% of C36000, free-cutting brass
Solution temperature. 996 °C (1825 °F)
Aging temperature. 510 °C (950 °F)
Melting temperature. 1066 to 1154 °C (1950 to 2110 °F)
Casting temperature. 1150 to 1300 °C (2200 to 2400 °F)

Joining of Copper and Its Alloys

By the Cleveland Chapter of ASM*

ASSEMBLIES in which some or all of the components are made of copper or copper alloys can be produced using any of the various types of mechanical or bonding processes commonly used to join metals. Mechanical joining processes include any of the methods that depend on an applied force to maintain joint integrity. Crimping, staking, riveting and bolting are common examples of mechanical processes. Bonding processes include any of the methods in which joint integrity depends solely on atomic or molecular attraction. Soldering, welding and silver alloy brazing are the bonding processes most often used for copper metals, although processes such as adhesive bonding and diffusion bonding can be used when appropriate from both engineering and economic standpoints.

Selection of the best joining process is governed chiefly by service requirements, joint configuration, thickness of components and alloy composition. The ability of various joining processes to satisfy these four selection criteria is summarized in Tables 1 to 4. These tables provide a broad basis for process selection, and the remainder of this article serves mainly as a guide for interpreting the tables.

Mechanical Joining

Copper and copper alloys can be readily joined to other copper metals, to dissimilar metals or to nonmetals by mechanical joining methods—that is, with mechanical fasteners, by crimping or staking, or by use of interlocking joints. When properly made, the resulting assemblies are both long-lasting and reliable.

Fasteners are well suited for joining noncritical structural components as well as parts to be used at low to moderately high temperatures and pressures. Critical structural members made of copper alloys, and components intended for service at temperatures above room temperature or under pressure or extremely high vacuum, should not be joined with mechanical fasteners. Joint designs in mechanically fastened copper and copper alloy components are generally restricted to flange, lap, scarf, and strap butt configurations, but all such joints may be subjected to either shear or tensile loads. Minimum thickness for copper alloy parts to be mechanically fastened is about 0.25 mm (0.010 in.).

In all mechanically fastened joints, the size, strength and number of fasteners, rather than the alloy or alloys selected for the components being joined, usually dictate whether the joints will sustain the loads applied in service. The main requirement of the component material is that it be able to withstand the forces applied through the fasteners without deforming during assembly and without cracking in service due to stress concentrations at the fastener locations.

Mechanical fasteners may be fabricated from any wrought copper or copper alloy. Mechanical properties specified for externally threaded fasteners made of the copper alloys most widely used for fasteners are shown in Table 5. The primary criteria for selection of a copper alloy fastener material are fastener strength requirements, fastener design (which places demands on material formability), and to a limited extent, typically in architectural applications, color match with the components to be joined. The strength and color matching characteristics of the most common copper-base fastener alloys are shown in Table 6. Where copper alloy components are mechanically fastened to other metals, resistance to galvanic corrosion must be given special consideration in selection of a fastener material. For example, copper metals can be fastened to stainless steel with copper alloy or stainless steel fasteners. Similarly, carbon steel and

*Editorial Committee—Warren J. Haws, Metallurgist, Glidden Metals Division, SCM Corp., *Chairman*; Peter I. Basalyk, Product Manager—New Products, Glidden Metals Division, SCM Corp.; John C. Harkness, Staff Metallurgist, Brush-Wellman Corp.; R. M. Hawson, Manager, Technical Service, Chase Brass and Copper Co.; Albert J. Mastrangelo, Glidden Metals Division, SCM Corp.; Willis J. Resiner, Consultant; John Snyder, Glidden Metals Division, SCM Corp.

Table 1 Joining-process selection guide based on service requirements

| Service requirement | Fusion welding | | | | | | | Resistance welding | | | | Solid-state welding | | | | Brazing | Soldering | Adhesive bonding | Mechanical fastening |
| | Arc welding | | | | Other welding processes | | | | | | | | | | | | | | |
	Gas metal-arc	Gas tungsten-arc	Plasma-arc	Shielded metal-arc	Electron beam	Laser beam	Oxyfuel gas	Resistance spot	Resistance seam	Projection	Flash	Diffusion	Explosive	Ultrasonic	Friction	Brazing	Soldering	Adhesive bonding	Mechanical fastening
Primary structural																			
Elevated temperature	B	A	E	B	A	E	C	A	A	D	A	A	C	C	D	B	D	C	B
Ambient temperature	A	A	E	B	A	E	C	A	A	D	A	A	C	C	D	A	C	A	A
Cryogenic	B	A	E	C	A	E	D	A	A	D	A	A	C	C	D	B	D	D	B
Vacuum	B	A	E	D	A	E	D	D	B	D	C	A	C	C	D	B	D	C	C
Atmospheric pressure	A	A	E	B	A	E	C	B	A	D	B	A	C	C	D	A	D	C	C
High pressure	B	A	E	D	A	E	D	D	C	D	C	A	C	C	D	C	D	D	C
Secondary structural	A	A	E	A	A	E	B	A	A	C	A	A	C	C	C	A	D	A	A
Noncritical	A	A	C	A	A	E	B	A	A	A	A	A	C	B	B	A	B	A	A
Dissimilar metal joining	C	C	C	B	C	E	C	D	D	D	C	A	B	A	B	B	B	A	A

Note: A, most satisfactory; B, satisfactory; C, restricted use; D, prohibited use; E, experimental.

Table 2 Joining-process selection guide based on joint configuration

| Joint configuration | Fusion welding | | | | | | | Resistance welding | | | | Solid-state welding | | | | Brazing | Soldering | Adhesive bonding | Mechanical fastening |
| | Arc welding | | | | Other welding processes | | | | | | | | | | | | | | |
	Gas metal-arc	Gas tungsten-arc	Plasma-arc	Shielded metal-arc	Electron beam	Laser beam	Oxyfuel gas	Resistance spot	Resistance seam	Projection	Flash	Diffusion	Explosive	Ultrasonic	Friction	Brazing	Soldering	Adhesive bonding	Mechanical fastening
Butt joint	A	A	A	A	A	A,E	B	D	D	D	A	C	D	D	B	D	D	D	D
Tee-joint	A	A	B	A	C	C,E	B	D	D	D	C	B	D	D	D	B	B	C	D
Edge joint	A	A	B	B	A	A,E	B	D	D	D	D	C	D	D	D	C	D	D	D
Corner joint	A	A	B	B	A	A,E	B	D	D	D	D	C	D	D	D	B	B	D	D
Flange joint	A	A	B	B	A	A,E	B	D	D	C	D	C	D	D	B	C	C	C	A
Scarf joint	D	D	D	D	C	D	D	D	D	D	D	C	B,E	D	D	B	B	B	B
Strap butt joint (splice joint)	C	C	C	C	C	B,E	D	B	B	C	D	C	C	B	D	A	B	A	A
Lap joint																			
Shear load	B	B	B	B	A	A,E	D	A	A	C	D	A	B	B	D	A	B	A	A
Tensile load	B	B	B	B	A	A,E	D	B	B	C	D	A	B	B	D	D	D	C	A

Note: A, most satisfactory; B, satisfactory; C, restricted use; D, prohibited use; E, experimental.

copper metals can be fastened together with copper alloy or plated carbon steel fasteners. But if copper and aluminum are to be fastened together, nonmagnetic stainless steel fasteners are recommended.

Crimping and Staking. As an alternative to using fasteners, copper and copper alloy components can be mechanically joined to themselves or to other materials by crimping or staking, in which one or both components are physically deformed to lock the parts together. The component(s) must have sufficient ductility to survive the assembly process without cracking or breaking, and the final assembly must be strong enough to withstand service loads. In cold rolled copper alloy strip, for example, formability (usually defined as the minimum radius around which the strip can be bent 90° or 180° without cracking) worsens and becomes more anisotropic with greater amounts of cold work in the strip. Softer, more ductile material can be selected for improved formability, but section size may have to be increased to provide sufficient strength. Where high-strength alloys must be used, crimped or staked joints may have to be rede-

Table 3 Joining-process selection guide based on thickness of parts being joined

Metal thickness mm	in.	Gas metal-arc	Gas tungsten-arc	Plasma-arc	Shielded metal-arc	Electron beam	Laser beam	Oxyfuel gas	Resistance spot	Resistance seam	Projection	Flash	Diffusion	Explosive	Ultrasonic	Friction	Brazing	Soldering	Adhesive bonding	Mechanical fastening
0.025–0.25	0.001–0.010	D	B	A	D	A	B,E	D	C	C	D	D	A	D	A	D	B	C	B	C
0.25–0.50	0.010–0.020	D	A	B	D	A	C,E	D	A	A	B	D	A	D	A	D	B	B	A	B
0.50–1.25	0.020–0.050	C	A	B	D	A	D	B	A	A	A	D	A	D	C	B	B	A	A	A
1.25–2.50	0.050–0.100	A	A	A	B	A	D	B	A	A	A	A	A	D	D	B	B	C	A	A
2.50–3.75	0.100–0.150	A	A	A	A	A	D	B	A	A	C	A	A	C	D	B	B	C	A	A
3.75–6.25	0.150–0.250	A	A	A	A	D	C	C	C	C	C	A	A	C	D	B	C	D	B	A
6.25–12.50	0.250–0.500	A	A	C	A	A	D	D	D	D	D	A	A	C	D	B	C	D	C	A
12.50–25.0	0.500–1.00	A	A	D	A	A	D	D	D	D	D	A	A	C	D	B	D	D	D	A
25.0–62.5	1.00–2.50	B	B	D	A	A	D	D	D	D	D	C	A	C	D	C	D	D	D	A
Over 62.5	Over 2.50	C	B	D	A	A	D	D	D	D	D	C	A	C	D	C	D	D	D	A
Thick to thin		A	A	C	B	A	B,E	D	C	C	C	C	A	A	A	···	B	A	A	A

Note: A, most satisfactory; B, satisfactory; C, restricted use; D, prohibited use; E, experimental.

Table 4 Joining-process selection guide based on alloy composition

Copper alloy family	Gas metal-arc	Gas tungsten-arc	Submerged arc	Shielded metal-arc	Stud	Electron beam	Electroslag	Laser beam	Oxyfuel gas	Spot	Seam	Projection	Flash butt	Brazing	Soldering	Adhesive bonding	Mechanical fastening
Coppers																	
Oxygen free	B	B	D	D	D	B,E	D	E	C	D	D	D	B	A	A	C	A
Deoxidized	A	A	D	D	D	B,E	D	E	B	D	D	D	B	A	A	C	A
Tough pitch	C	C	D	D	D	B,E	D	E	D	D	D	D	B	B	A	C	A
High Copper Alloys																	
Cadmium	B	B	D	D	D	···	D	···	B	D	D	D	B	A	A	C	A
Beryllium	B	B	D	B	D	···	D	···	D	B	C	B	C	B	B	C	A
Chromium	B	B	D	D	D	···	D	···	D	D	D	D	C	B	B	C	A
Leaded	D	D	D	D	D	···	D	···	D	D	D	D	C	B	A	C	A
Oxide dispersion strengthened	D	D	D	D	D	···	D	···	D	C	C	C	C	B	A	C	A
Brasses																	
Red	B	B	D	D	D	···	D	···	B	C	D	C	B	A	A	C	A
Yellow	C	C	D	D	D	···	D	···	B	B	D	B	B	A	A	C	A
Leaded	D	D	D	D	D	···	D	···	D	D	D	D	C	B	A	C	A
Tin	B	B	D	C	D	···	D	···	B	B	C	B	B	A	A	C	A
Bronzes																	
Phosphor	B	B	D	C	D	···	D	···	C	B	C	B	A	A	A	C	A
Leaded phosphor	D	D	D	D	D	···	D	···	D	D	D	D	C	B	A	C	A
Aluminum	A	A	D	B	D	···	D	···	D	B	B	B	B	C	C	C	A
Silicon	A	A	D	C	D	···	D	···	B	A	A	A	A	A	A	C	A
Copper Nickels																	
10% Ni	A	A	D	B	D	···	D	···	C	B	B	B	A	A	A	C	A
30% Ni	C	C	D	C	D	···	D	···	B	A	A	A	A	A	A	C	A
Nickel silvers	C	C	D	D	D	···	D	···	B	B	C	B	B	A	A	C	A

Note: A, most satisfactory; B, satisfactory; C, restricted use; D, prohibited use; E, experimental.

Table 5 Mechanical-property requirements for metric bolts, screws and studs

| Copper alloy | Nominal diam | Full-size bolts, screws and studs | | | Machined test specimens of bolts, screws and studs | | | | |
		Yield strength, MPa min	Tensile strength, MPa min	max	Yield strength, MPa min	Tensile strength, MPa min	Elongation, % min	Hardness, HRB min	max
C11000	All	69	207	345	69	207	15	65 HRF	90 HRF
C27000	All	345	415	620	345	400	35	55 HRF	30 HRF
C46200	All	172	345	550	172	345	20	65	90
C46400	All	103	345	550	103	345	25	55	75
C51000	All	240	415	620	207	380	15	60	95
C61400	All	240	520	760	240	520	30	70	95
C63000	All	345	690	900	345	690	5	85	100
C64200	All	240	520	760	240	520	10	75	95
C65100	Thru M20	380	485	690	365	485	8	75	95
	Over M20 to M36	275	380	620	260	370	8	70	95
C65500	All	138	345	550	103	345	20	60	80
C66100	All	240	495	690	240	485	15	75	95
C67500	All	172	330	590	172	380	20	60	90
C71000	All	103	310	520	103	310	40	50	85
C71500	All	138	330	590	138	380	45	50	95

signed to accommodate lower material ductility.

Interlocking Joint Designs. A specialized area of mechanical joining relevant to copper and copper alloys (as well as to other metals) is the use of mechanically interlocking joint designs. This technique is typically employed in the assembly of large or irregular architectural shapes from smaller or simpler extruded forms having designed-in dovetails, clinch locks, or splines such as those sketched in Fig. 1. Such designs are useful where the final assembled shape has a cross section too complicated, or of too great an area, to extrude in one piece; or where dissimilar metal shapes, or a copper alloy and a non-metal are to be joined.

Interaction of Alloy Selection and Design. In all mechanical joints, regardless of the joining method used, joint design must take into account special service conditions or material characteristics. Joints subject to vibration must be engineered to accommodate the displacements expected in service in order to avoid loosening and subsequent chafing or fretting. Flexible washers, spring-loaded fasteners and Belleville washers provide means of accommodating movement in mechanical joints. Where joint integrity depends on elastic behavior of fasteners or wedged interlocking parts, compensation must be made for stress relaxation in the stressed member, especially above room temperature.

Corrosive environments pose a different set of problems. Alloys prone to

Fig. 1 Three examples of mechanically interlocking shapes commonly produced as copper alloy extrusions

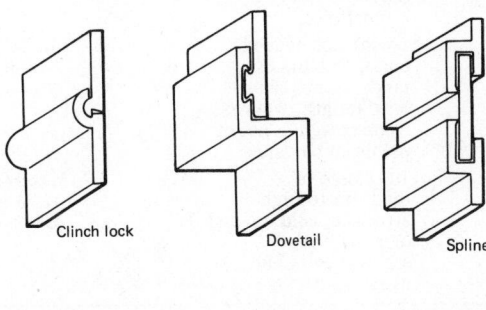

Clinch lock   Dovetail   Spline

stress-corrosion cracking in the service environment must almost always be avoided for both the fasteners and the components to be joined. All mechanical joints contain inherent crevices. Hence any tendency of materials toward galvanic corrosion in oxygen-starvation cells must be considered. In all mechanical joints, appropriate means of preventing corrosion should be incorporated in the joint design. A more detailed discussion of this subject may be found in the article "Wrought Copper Alloys for Corrosion Service" in this volume.

Soldering

Copper and copper alloys are among the most readily soldered engineering materials. Soldering is used to join copper components in such diverse applications as plumbing, aerospace hardware, automotive radiators and printed circuits. In all cases, users can expect long, reliable performance.

Soldering is a particularly useful process for joining electrical conductors. Table 7 illustrates the many advantages of soldering over brazing, welding, mechanical joining and chemical bonding for electrical applications. Soldering is quite cost-effective for joints exposed to service temperatures below 120 °C (250 °F.)

Solderability of copper alloys ranges from excellent to poor. In order of decreasing solderability, copper metals may be roughly classified as follows: coppers, copper-tin alloys, copper-zinc

Table 6 Criteria for selecting copper alloy fasteners for joining copper metals

Fastener material	Fastener types	Color matching
C26000	Full range of medium-strength, cold headed, roll threaded (coarse thread) screws, bolts and nuts	Matches C26000 sheet, tube and rod
C28000	Screws, nuts and bolts of relatively low strength where color match is of prime importance	Excellent match with C38500 products
C36000	Full range of medium-to-low-strength machine screws, bolts and nuts	Good match with C26000
C46400 thru C46700	Full range of medium-strength screws, bolts and nuts	Fair to good color match, although slightly yellower than C38500
C48500	Machine screws, bolts and nuts of medium-to-low strength	Yellow color is only a fair to good match with C28000 and C38500
C65100	Full range of medium-to-high-strength nails, screws, bolts and nuts where color match is not of prime importance	Slightly redder than C38500
C65500	Special hot headed bolts, or bolts of large diameter or long length, where color match is not of prime importance	Slightly redder than C38500
C74500	Full range of medium-to-high-strength, cold headed (coarse thread) screws, bolts and nuts	Matches C74500 sheet, tube and rod, plus C79600 extrusions

alloys, copper-nickel alloys, copper-chromium alloys, copper-beryllium alloys, copper-silicon alloys and copper-aluminum alloys. Soldering of most copper-base metals presents no serious problems, but copper alloys containing beryllium, silicon or aluminum require special fluxes.

The high thermal conductivity of copper and many of its alloys makes a high rate of heat input necessary when localized heating is used to complete the joint.

Solders. Limitations on use of any particular solder generally are imposed by production methods and final performance requirements. Factors to be considered include maximum allowable soldering temperature, cost of solder, joint strength and other physical characteristics.

The most widely used solders are alloys of tin and lead. Tin, the active component, readily reacts with and diffuses into copper, and an intermetallic phase, Cu_6Sn_5, is created during soldering. This phase is formed at the interface while the solder is still liquid; however, aging of the soldered joint promotes growth of Cu_6Sn_5 and formation of Cu_3Sn, which results in decreased joint strength. Elevated temperature accelerates aging.

The 50Sn-50Pb composition is the most common solder used for copper. The eutectic composition (63Sn-37Pb) provides the lowest melting temperature at 183 °C (361 °F), well below that of either tin or lead. Other elements contained in the solder are either impurities that can be detrimental to the process, or elements added to improve

bond strength or alter some other characteristic in a beneficial manner. A representative listing of alloys for soldering copper metals is presented in Table 8, along with typical applications.

ASTM class A tin-lead solders contain less than 0.12% antimony as an impurity. Class B tin-lead solders contain 0.20 to 0.50% antimony as an intentional alloying element to improve the reliability of soldered joints at low temperatures. Small amounts of antimony inhibit the phase transformation from beta tin to alpha tin below about 0 °C (32 °F), which is accompanied by a 26% increase in volume and may result in powdery disintegration of the solder.

Solders are commercially available in various sizes, shapes and forms. They can be grouped into several major classifications, as follows:

- *Pig*: Available in 20-, 40-, 50- and 100-lb sizes
- *Cake* or *ingot*: Rectangular or circular in shape; 3-, 5- and 10-lb sizes
- *Bar*: Available in weights from ½ to 2 lb
- *Paste*: A mixture of powdered metal, flux, solvent, and a gelling or suspending agent that prevents the metal from settling
- *Segment* or *drop*: Wire or triangular bar cut into pieces or lengths
- *Foil, sheet* and *ribbon*: Supplied in various thicknesses and widths
- *Solid wire*: Supplied in diameters of 0.010 to 0.250 in., on spools weighing 1, 5, 20, 25 and 50 lb or in bulk packs
- *Flux-cored wire*: Tubular wire cored with organic, inorganic or rosin flux, supplied in same diameters and spool weights as solid wire
- *Preforms*: A wide range of custom-designed preform shapes is available. Each shape is a derivative of one or more of the following: wire, punched parts, spheres and flux-coated metal forms.

Among these commercial forms, solder preforms and specially formulated solder pastes have undergone substantial growth in commercial usage, even though they are more expensive than most other forms. Solder paste can be especially profitable because it (a) allows closer control over deposit thickness and area covered, (b) eliminates separate fluxing operations and (c) can

Table 7 Comparison of soldering with other bonding methods for electrical applications

Factor	Soldering	Metallurgical bonding Brazing	Welding	Mechanical bonding Crimping	Screwing	Wrapping	Adhesive bonding Conductive cement
Temperature limit of joint, °C (°F)	73-460 (100-800)	460-900 (800-1600)	Conductor melting temperature	No limit except wire	No limit except wire	No limit except wire	100-180 (160-300)
Heating effect on assembly...............	Small	Large	Small (quick)	None	None	None	Cures at ambient to 120 °C (250 °F)
Ease of rework and rebonding.............	Simple	Simple	Not practical	Not practical	Simple	Simple	Not practical
Process economy Equipment cost........	Low	Medium	High	Low	Low	Low	Low
Ease of automation.....	Easiest	More difficult	More difficult	More difficult	More difficult	More difficult	More difficult
Extra hardware........	No	No	No	Yes	Yes	No	No
Joint stable under vibration...............	Yes	Yes	Yes	Yes	No	Yes	Yes
Joint stable in oxidizing environment	Yes	Yes	Yes	No	No	Yes	Yes

Table 8 Representative list of solders used for joining copper metals

ASTM classification	Nominal composition	Solidus °C	°F	Liquidus °C	°F	Melting range °C	°F	Density Mg/m³	lb/in.³	Typical applications
Tin-Lead(a)										
5A	95Pb-5Sn	308	586	312	594	4	8	11.3	0.408	Coating and joining
10B	90Pb-10Sn	268(b)	514(b)	301	573	33	59	10.8	0.389	Coating and joining
15B	85Pb-15Sn	225(b)	437(b)	290	553	65	116	10.5	0.379	Coating and joining
20A	80Pb-20Sn	183	361	280	535	97	174	10.2	0.368	Coating and joining
25A	75Pb-25Sn	183	361	267	511	84	150	9.99	0.361	Machine and torch soldering
30A	70Pb-30Sn	183	361	255	491	72	130	9.69	0.350	Machine and torch soldering
35A	65Pb-35Sn	183	361	247	477	64	116	9.69	0.350	General purpose; wiping
40A	60Pb-40Sn	183	361	235	455	52	94	9.27	0.335	Wiping; auto radiators
45A	55Pb-45Sn	183	361	228	441	45	80	8.97	0.324	Auto radiators; roofing seams
50A	50Pb-50Sn	183	361	217	421	34	60	8.83	0.319	General purpose; most widely used on copper
60A	40Pb-60Sn	183	361	190	374	7	13	8.64	0.312	"Fine solder"; general purpose, especially where low soldering temperature is essential
63A	37Pb-63Sn	183	361	183	361	0	0	8.40	0.303	"Eutectic solder"; lowest-melting lead-tin solder
70A	30Pb-70Sn	183	361	192	378	9	17	8.32	0.301	
Tin-Lead-Antimony and Tin-Antimony										
20C	79Pb-20Sn-1.0Sb	184	363	270	517	86	154	10.2	0.367	Machine soldering and coating(c)
25C	73.7Pb-25Sn-1.3Sb	185	364	262	504	77	140	9.94	0.359	Torch and machine soldering(c)
30C	68.4Pb-30Sn-1.6Sb	185	364	250	482	65	118	9.63	0.348	Torch and machine soldering(c)
35C	63.2Pb-35Sn-1.8Sb	186	365	243	470	57	105	9.44	0.341	Wiping(c)
40C	58Pb-40Sn-2.0Sb	186	365	231	448	45	83	9.22	0.333	General purpose(c)
95TA........	95Sn-5Sb	232	450	240	464	8	14	7.80	0.260	Copper joints in electrical, plumbing and heating systems
Lead-Silver, Lead-Tin-Silver and Tin-Silver										
2.5S.........	97.5Pb-2.5Ag	304	579	579	579	275	0	11.3	0.409	Torch soldering of copper and brass
5.5S.........	94.5Pb-5.5Ag	304	579	343	689	39	110	...	...	Torch soldering of copper and brass
1.5S........	97.5Pb-1.0Sn-1.5Ag	313	588	313	588	0	0	11.3	0.409	Torch soldering of copper and brass
... .	97Pb-2.5Sn-0.5Ag	303	577	310	590	7	13	...	...	Torch soldering of copper and brass
... .	94.5Pb-5Sn-0.5Ag	294	561	301	574	7	13	...	...	Torch soldering of copper and brass
... .	36Pb-62Sn-2Ag	180	354	190	372	10	18	...	...	...
96TS	96Sn-4Ag	221	430	221	430	0	0	10.4	0.375	Delicate instruments; electrical conductors for use at high temperature

(a) Most class A solders also available as antimonial class B solders (0.20 to 0.50% Sb). (b) This alloy has virtually no strength at temperatures above 183 °C (361 °F). (c) Not recommended for soldering to galvanized iron.

Table 9 Characteristics of fluxes commonly used for soldering copper metals

Type of flux	Active chemicals	Vehicles	Type of joint	Temperature stability	Tarnish removal	Corrosiveness	Postcleaning methods
Inorganic							
Acids	Hydrochloric, hydrofluoric, orthophosphoric	Water, petrolatum paste	Structural	Good	Very good	High	Hot-water rinsing and neutralizing; organic solvents; degreasing
Salts..............	Zinc chloride, ammonium chloride, tin chloride	Water, petrolatum paste, polyethylene glycol	Structural	Excellent	Very good	High	Hot-water rinsing and neutralizing; 2% HCl, hot-water rinsing and neutralizing; organic solvents; degreasing(a)
Gases	Hydrogen-forming gas, dry HCl	None	Electrical	Excellent	Very good, at high temperatures	None, normally	None required
Organic Nonrosin							
Acids	Lactic, oleic, stearic, glutamic, phthalic	Water, organic solvents, petrolatum paste, polyethylene glycol	Structural, electrical	Fairly good	Fairly good	Moderate	Hot-water rinsing and neutralizing; organic solvents; degreasing(a)
Halogens..........	Aniline hydrochloride, glutamic hydrochloride, bromide derivatives of palmitic acid, hydrazine hydrochloride, hydrazine hydrobromide	Water, organic solvents, petrolatum paste, polyethylene glycol	Structural, electrical	Fairly good	Fairly good	Moderate	Hot-water rinsing and neutralizing; organic solvents; degreasing(a)
Amines and amides..	Urea, ethylene diamine	Water, organic solvents, petrolatum paste, polyethylene glycol	Structural, electrical	Poor	Fair	None, normally	Hot-water rinsing and neutralizing; organic solvents; degreasing(a)
Organic Rosin							
Activated..........	Water-white rosin plus activators	Isopropyl alcohol, organic solvents, polyethylene glycol	Electrical	Poor	Fair	None, normally	Water-base detergents; isopropyl alcohol; organic solvents; degreasing(a)
Water white	Water-white rosin alone	Isopropyl alcohol, organic solvents, polyethylene glycol	Electrical	Poor	Poor	None	Does not normally require postcleaning; when cleaning is required, same as above

(a) For optimum cleansing, follow by rinsing with demineralized and distilled water.

be satisfactorily dispensed onto the joint by use of relatively high mechanical or air pressure.

Fluxes react with, dislodge and dissolve metallic oxides on copper surfaces. There are three basic types of soldering flux: (a) inorganic (acid), (b) organic and (c) rosin. Table 9 summarizes the characteristics and uses of the chief types of flux. Acid and organic nonrosin fluxes generally are not appropriate for soldering electrical connections. Both produce corrosive hydroscopic residues that could damage an electrical circuit unless thoroughly washed away. Acid and organic fluxes are used extensively in applications where the assembly can be washed after soldering, such as automotive radiators.

Rosin in its natural, solid state is noncorrosive but is a poor flux. It is activated by dissolution in organic solvents. The residue after soldering has a very high resistivity and is not hygroscopic. Activated rosin fluxes are sometimes considered corrosive, but actually are not if properly exposed to heat.

Noncorrosive fluxes are excellent for soldering coppers, and may be used with some success in soldering copper alloys containing tin and zinc, depending on how clean the surfaces are initially. The flux should be applied to clean surfaces and only enough should be used to lightly coat the areas to be joined.

Intermediate fluxes are used in soldering coppers and copper-tin, copper-zinc, copper-beryllium and copper-chromium alloys. Some of the more active fluxes may be adequate for copper nickels and silicon bronzes, but a generalization in this respect could be misleading.

Corrosive fluxes can be used with all copper-base metals, but they are necessary only for those that develop refractory oxides, such as silicon bronzes and aluminum bronzes. Soldering of aluminum bronzes is especially difficult and requires special fluxes or copper plating. Chloride fluxes are useful for soldering silicon bronzes and copper nickels.

Oxide films may re-form quickly on copper and copper alloys after they have been cleaned. Therefore, flux should be applied as soon as possible after cleaning.

The fluxes best suited for use with 50Sn-50Pb and 95Sn-5Sb solders on copper plumbing systems are mildly corrosive liquid or petrolatum pastes containing chlorides of zinc and ammonium. Many liquid fluxes for plumbing applications are self-cleaning (that is, they will wash away when the system is put into service), but there still is a risk that they will corrode the joint if they are not washed away completely. Highly corrosive flux can effectively remove heavy oxides and dirty films, but when a highly corrosive flux is used as an alternative to proper cleaning, there is always a chance that corrosive action will continue after soldering. It is always best to clean the surface first and then apply the minimum amount

of the least-active flux that is appropriate for the alloy being soldered.

Surface preparation plays a vital role in producing high-quality soldered joints. Dirt, oil, grease and oxides may be removed by chemical and/or mechanical means. Solvent or alkali degreasing generally is recommended for removing dirt, oil and grease, and acid immersion for removing most oxides. Possible mechanical methods for removing both dirt and oxides include sand blasting, grit blasting, wire brushing, filing, scraping, abrasive cleaning (sanding) and machining. The method used to obtain clean, uncontaminated surfaces should be chosen carefully, keeping in mind both the characteristics of the material(s) and the expected service conditions.

Plating with other metals sometimes is used to preserve clean surfaces for soldering, especially when components must retain solderability even after long-term storage.

Solvent or alkaline degreasing procedures are suitable for removing organic residues from copper-base metals; mechanical methods such as wire brushing and sanding, may be used to remove oxides. Chemical removal of oxides requires proper choice of a pickling solution (usually a sulfuric acid solution) followed by thorough rinsing. Mechanical cleaning, rather than pickling, is preferred for arsenical and antimonial brasses to avoid development of surface contamination (slimes), which may interfere with soldering and produce brittle joints. After being heat treated, beryllium coppers are coated with oxide, which requires two-stage pickling—first with hot 20% sulfuric acid and then with cold 30% nitric acid. Following this treatment, beryllium

copper can be soldered using a plain or activated rosin flux. Mechanical cleaning is a recommended alternative cleaning procedure for beryllium coppers.

Soldering Methods. Soldering by any method requires sufficient heat to melt the solder and to warm the surface, which facilitates good wetting and flowing of the molten solder. Table 10 lists some of the advantages and disadvantages of the most common methods.

The soldering equipment most commonly used includes soldering irons, solder pots (including baths, waves, jets and cascades), torches, ovens, induction heaters, oil baths, electrical resistance heaters and electromagnetic radiation (infrared) heaters.

With few exceptions, rapid heating and cooling are desirable. Flux tends to degrade when hot, and can lose its effectiveness before soldering is completed unless the joint is completed quickly. Base-metal surfaces may oxidize and become difficult to solder if they are heated for a prolonged period of time. Prolonged contact with molten solder can cause formation of intermetallic compounds, or erosion or dissolution of the base metal, any of which may be unacceptable. Degradation of desirable characteristics (such as electrical properties of electronic devices) may occur if too much heat is allowed to accumulate in the joint. A pin wave soldered into a circuit board is an example of a rapidly soldered component. A wave temperature of approximately 270 °C (515 °F) and an immersion time of about 1.25 s are typical for this application.

In dip soldering of copper and brass, contamination of the solder bath with copper and zinc is always a problem.

Table 10 Comparison of various soldering methods

Method	Advantages	Disadvantages
Soldering iron	Easily applied, inexpensive	Slow, only small parts
Dip or wave	High volume rates with many connections simultaneously	Solder contamination, close control of process
Flame	For large masses, portability	Open fire, overheating, little control of solder flow
Induction heating	For large masses, localized heat control, high volume and high-quality joints	Requires extreme cleanness, good part clearances
Electrical resistance	Localized heating, useful when soldering irons are not suitable	Small volume, small assemblies
Oven heating	For large, complicated assemblies and mass production	Expensive to set up
Ultrasonic	Removes surface oxides, thus eliminates fluxes	Only for small areas, cannot be used for lap or crimp joints

The degree to which contamination is controlled has a direct bearing on quality of the soldered joint. Low bath temperatures keep contamination low, but the bath should have sufficient capacity to heat the parts to temperature rapidly. A bath temperature no more than 85 °C (150 °F) above the liquidus of the solder is recommended for printed circuit boards, and as much as 200 °C (350 °F) above the liquidus for heat exchangers such as automotive radiators. Contamination is commonly held to less than 0.3% Cu for wave soldering of electronic components.

Solderability testing is often subjective. No standard, recognized universal method has yet emerged. Two tests that are used widely are a spread test and a dip test. In the spread test, a flat piece of material is fluxed and heated to a prescribed temperature; then a prescribed amount of solder is placed on it, melted and allowed to spread; after the test piece has cooled, the area covered by the solder is measured. In the dip test, the sample material is immersed in flux and then in a solder bath, from which it is slowly withdrawn. The coverage and appearance are visually rated from "good" to "bad" in five steps (Classes I to V) as described in Table 11. The spread test eliminates operator judgment and correlates with direct-heat soldering (such as use of a soldering iron) whereas the dip test correlates with dip or wave soldering.

Although clean copper is the easiest of all metals to solder, its solderability, like those of copper alloys, degrades with time due to gradual surface oxidation and tarnishing. The solderability after a given period of storage, known as "shelf-life solderability", is a common criterion for rating both bare and coated metals for soldering applications. All copper alloys tend to tarnish when exposed to the atmosphere, so none exhibits outstanding shelf-life solderability if stored in a bare condition; soldering must be done in the presence of a flux that removes tarnish. Plating with a good tarnish-resistant metal such as gold, tin or solder enhances shelf-life solderability.

The shelf-life solderability of copper-zinc alloys varies inversely with zinc content, degrading more rapidly for alloys of high zinc content. Zinc exhibits high diffusion mobility and high oxidation potential, so it can diffuse through plating and oxidize on the exposed surface when the plating is thin and storage time long. Although copper-zinc alloys present the greatest soldering problems after long periods of storage, copper also oxidizes as do alloying elements like tin and aluminum. Table 12 describes relative solderabilities among the major families of copper metals.

Soldering of Coated Copper Alloys. Except for thermal conductivity, only characteristics of the coating —usually tin, lead, tin-lead alloys, nickel, chromium or silver—affect the solderability of coated copper metals. None of the coatings other than chromium offer any serious problem. Chromium plating should be removed from faying surfaces before soldering chromium-plated copper articles.

Mechanical properties of soldered joints are different from the properties of the bulk solder and depend on both solder composition and process variables. Good design practice generally requires soldered copper joints to be stressed primarily in shear. Also, the area of the joint must be large enough to ensure that service stresses will be below the levels that can result in creep or fatigue failure. Shear strength of soldered copper joints varies with solder composition. An example is given in Fig. 2 for binary tin-lead solders; maximum strength is obtained with solders of approximately eutectic composition (63Sn-37Pb). Aging at room or slightly elevated temperature reduces shear strength. For instance, joints soldered with tin-lead solders may lose as much as 30% in measured shear strength if they are aged for several weeks before they are tested. Shear strength decreases with temperature, as shown in Fig. 3 for 95Sn-5Sb and 50Sn-50Pb, which are the solders most often used in plumbing systems.

Burst pressure and working-pressure ratings for soldered joints in copper tubes of different sizes are given in Table 13. (Copper tubes and socket joints are designed so that, when the assembly is subjected to increasing internal pressure, the tube splits before a properly soldered joint fails.) The calculated burst pressures in Table 13 vary with tube size, but are the same

Table 11 Dip-test solderability rating system

Class I
An "ideal" coating. The solder layer is smooth and of uniform thickness with no surface irregularities or gaps.

Class II
An essentially continuous solder coating without solder dewetting or solder pullback. Less than 1% of the surface may be taken up by "pinholes" (small areas where the solder did not wet the surface). The solder coating may be slightly rougher than a class I coating; it need not be of equal thickness at all points on the specimen.

Class III
Solder pullback and/or dewetting is evident on up to 50% of the specimen surface. Up to 10% of the surface may be bare.

Class IV
Dewetting and/or pullback occurs over 50% or more of the surface. More than 10% of the surface is bare.

Class V
No adhesion to the surface, except for small amounts usually in the form of small droplets which can be readily removed mechanically.

Table 12 Relative solderability of various types of copper metals

Type of copper metal	Solderability and remarks
Coppers(a)	Excellent. Need only rosin or other noncorrosive flux
Copper-tin alloys	Good. Easily soldered with activated rosin and intermediate fluxes
Copper-zinc alloys	Good. Easily soldered with activated rosin and intermediate fluxes
Copper-nickel alloys	Good. Easily soldered with intermediate and corrosive type fluxes
Chromium copper and beryllium copper	Good. Require intermediate and corrosive type fluxes
Copper-silicon alloys	Fair. Silicon produces refractory oxides that require use of corrosive fluxes
Copper-aluminum alloys	Difficult. May be soldered with help of very corrosive fluxes
High-strength manganese bronze	Not recommended. Should be plated to ensure consistent solderability

(a) Includes tough-pitch, oxygen-free, phosphorized, arsenical, silver-bearing, leaded, tellurium, and selenium coppers.

Fig. 2 Variation of shear strength with tin content for copper joints soldered with tin-lead alloys

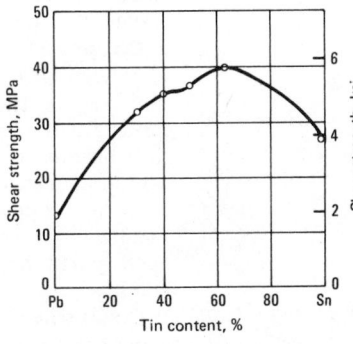

Fig. 3 Shear strengths at elevated temperature for copper joints soldered with 95Sn-5Sb and 50Sn-50Pb

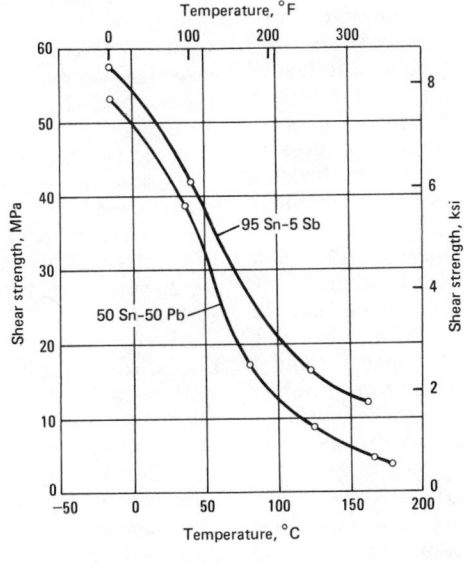

joints soldered with tin-lead filler metals.

It is important to keep temperature, time at temperature and volume of solder to a minimum, especially with solders containing large proportions of tin, because the intermetallic compounds formed by reaction of the molten solder with the base metal reduce the strength of the joint. High temperatures, prolonged contact with molten solder, and presence of large volumes of molten solder in the joint all lead to thicker reaction layers and lower strength. With high-tin solders, time at temperature is the most critical factor.

Brazing

Strong, high-quality permanent joints can be produced in coppers and copper alloys by brazing. Brazed joints exhibit strength equal to or better than the strengths of the base metals, and retain joint strength and integrity at temperatures from 200 to 260 °C (400 to 500 °F) and higher. Brazing is relatively easy to automate, is suitable for joining dissimilar metals and produces assemblies having good appearance.

The diverse applications of brazed copper alloy assemblies include electrical conductors, heat exchangers, jewelry, marine fittings and chemical process equipment. Most copper alloys can be brazed to one another, but compromise in selection of temperature, process technique, filler metal, and/or flux may be required. It is not uncommon for three or more alloys to be joined in a single assembly.

Coppers. High-conductivity oxygen-free coppers and deoxidized coppers can be joined readily by furnace or torch brazing. Tough pitch (oxygen-bearing) grades are subject to oxide migration and hydrogen embrittlement at elevated temperatures, and should be furnace brazed in inert atmospheres or vacuum. Sometimes, tough pitch coppers can be torch brazed using a neutral or slightly oxidizing flame.

The BCuP filler metals are self fluxing on copper, but flux may be used to help inhibit oxidation when prolonged heating is required. Joints made with phosphorus-bearing filler metals are subject to corrosive attack when exposed to sulfurous atmospheres.

The RBCuZn filler metals should not be overheated, because volatilization of zinc will cause voids in joints. Use of an oxidizing flame reduces zinc fuming

regardless of whether the joint is soldered with 95Sn-5Sb or 50Sn-50Pb filler metal, both of which have shear strengths of about 31 MPa (4.5 ksi). Pressure ratings are different, however, because they are based on creep resistance rather than shear strength. Solder alloy 95Sn-5Sb has considerably greater creep resistance than 50Sn-

50Pb; tin-base solders containing 3.5 to 5% silver have creep resistances similar to that of 95Sn-5Sb.

Besides joint area, joint spacing is the chief design factor that influences the strength of a soldered joint. Most investigators agree that a spacing of about 0.10 to 0.15 mm (0.004 to 0.006 in.) provides optimum strength for

Table 13 Typical burst pressures and pressure ratings for soldered copper tube

Nominal tube size, in.	Burst pressure(a) MPa	ksi	Pressure rating(b) kPa	psi
50Sn-50Pb Solder				
1/2	99	14.4	1380	200
3/4	106	15.4	1380	200
1	110	15.9	1380	200
2	80	11.6	1200	175
95Sn-5Sb Solder				
1/2	99	14.4	3450	500
3/4	106	15.4	3450	500
1	110	15.9	3450	500
2	80	11.6	2760	400

(a) Calculated for standard circular lap (socket) joint, assuming 31-MPa (4.5-ksi) shear strength for solder alloy. (b) At 38 °C (100 °F).

during torch brazing. Corrosion resistance of these filler metals is rated inferior to that of copper.

Materials containing no more than 1% silver, selenium, sulfur, lead or tellurium are readily brazed with BCuP filler metals. Use of flux improves wetting and flow.

High-strength beryllium coppers must be solution treated and aged to obtain optimum mechanical properties. By brazing a beryllium copper assembly at the solution treating temperature, possible adverse effects of reheating a brazed part to a high temperature for solution treating are avoided. Typically, parts are furnace brazed and solution treated simultaneously by heating to 790 °C (1450 °F), after which the temperature is lowered to 760 °C (1400 °F) to solidify the filler metal, and the assembly is rapidly quenched in water and then aged at 315 to 345 °C (600 to 650 °F). BAg-8, the copper-silver eutectic filler metal, and AWS type 3A flux generally are used.

High-conductivity beryllium coppers, which are solution treated at 927 °C (1700 °F), cannot be brazed simultaneously. Instead, they are brazed with BAg-1 or BAg-1a filler metal in the age-hardened condition, losing hardness (about 10 to 15 points HRB) in the process due to overaging.

Chromium copper and zirconium copper can be brazed with silver-bearing filler metals after solution annealing and cold working but before age hardening.

Cadmium-bearing copper is brazed in the same manner as deoxidized copper, but with added safety precautions due to the toxicity of cadmium.

Copper-Zinc Alloys. BAg and BCuP filler metals are commonly used to braze all brasses. Higher-melting (low-zinc) alloys can be brazed with RBCuZn filler metals. AWS type 3 flux is used with BAg and BCuP, type 5 flux with RBCuZn. Flux is recommended, even in protective atmospheres, to promote good wetting.

Zinc volatilization above 400 °C (750 °F) can be reduced by fluxing parts during furnace brazing and by using an oxidizing flame for torch brazing.

Aluminum brasses and silicon brasses require treatments similar to those for corresponding aluminum bronzes and silicon bronzes. The lead in leaded brasses may alloy with filler metal, and can embrittle it if the concentration of lead in the filler metal becomes greater than 2 to 3%. Slow uniform heating plus liberal use of flux is recommended in brazing leaded brasses.

Copper-Tin Alloys. Phosphor bronzes are brazed with BAg and BCuP filler metals; low-tin alloys can be brazed with RBCuZn. Prior to brazing, powder metallurgy compacts require pretreatment with a colloidal graphite suspension to seal pores, followed by baking at low temperatures, cleaning and degreasing.

Copper-Aluminum Alloys. AWS type 4 flux and silver-bearing filler metals are recommended. Formation of refractory oxides at aluminum concentrations above 8% can be avoided by electroplating copper 0.013 mm (0.5 mil) thick on the surface to be brazed. In furnace brazing, flux should be used in addition to a protective atmosphere.

Copper-Silicon Alloys. Mechanical cleaning, plus use of either AWS type 3 flux coating or copper electroplating, is recommended to prevent formation of refractory silicon oxides. Sil-

ver-bearing filler metals generally are used for brazing these alloys.

Copper-Nickel Alloys. BAg filler metals and AWS type 3 flux usually give best results. BCuP filler metals may be used, but formation of brittle nickel phosphide may present problems. Assemblies should be stress relieved before being brazed.

Copper-Nickel-Zinc Alloys. Nickel silvers are brazed with procedures similar to those used for brasses. RBCuZn filler metals may not be suitable because they require high brazing temperatures. Nickel silver parts must be stress relieved before brazing, and localized overheating may occur in such parts because of poor thermal conductivity.

Dissimilar metals can be brazed to copper metals quite readily. Only aluminum and magnesium alloys are not brazed to copper; soldering is preferred for these dissimilar metal combinations.

Filler Metals. Major factors for choosing a filler metal are liquidus temperature, which must be sufficiently below the solidus temperature of the base metal, and limitations imposed by production methods. Other filler-metal properties that may influence the choice for a specific application are electrical conductivity, corrosion resistance, cost, flow characteristics and strength.

Table 14 summarizes the specific filler metals, atmospheres and fluxes commonly used for brazing copper metals of the various classes. Table 15 lists the filler metals most often recommended for dissimilar metal combinations with copper.

All copper and copper alloys may be brazed with BAg filler metals. They are well suited for brazing zinc-bearing alloys because they have relatively low liquidus temperatures. Cadmium-bearing filler metals should be avoided for applications in the food, dairy and pharmaceutical industries. The major disadvantage of BAg filler metals is high cost.

BAu filler metals are suitable only for coppers and copper nickels because they have a high liquidus temperature. Because of high cost, BAu filler metals generally are restricted to applications requiring superior corrosion resistance.

Brazing Fluxes and Atmospheres. The common function of fluxes and atmospheres is to promote flow of filler metal and wetting of the

Table 14 Guide to selecting proper filler metals, atmospheres and fluxes for brazing copper metal assemblies

Alloy type	Brazing filler metals	AWS brazing atmospheres(a)	AWS brazing flux	Remarks
Coppers	BCuP-2(b), BCuP-3(b), BCuP-5(b), RBCuZn, BAg-1a, BAg-6, BAg-18, BAg-1, BAg-2, BAg-5	1 or 2 or 5	3 or 5	Oxygen-bearing coppers should not be brazed in hydrogen-containing atmospheres
High copper alloys	BAg-8, BAg-1	...	3A	...
Red brasses	BAg-1a, BAg-1, BAg-2, BCuP-5, BCuP-3, BAg-5, BAg-6, RBCuZn	1 or 2 or 5	3; or for RBCuZn, 5	...
Yellow brasses	BCuP-4, BAg-1a, BAg-1, BAg-5, BAg-6, BCuP-5, BCuP-3	3 or 4 or 5	3	Keep brazing cycle short
Leaded brasses	BAg-1a, BAg-1, BAg-2, BAg-7, BAg-18, BCuP-5	3 or 4 or 5	3	Keep brazing cycle short and stress relieve before brazing
Tin brasses	BAg-1a, BAg-1, BAg-2, BAg-5, BAg-6, BCuP-5, BCuP-3 (RBCuZn for low tin)	3 or 4 or 5	3	...
Phosphor bronzes	BAg-1a, BAg-1, BAg-2, BCuP-5, BCuP-3, BAg-5, BAg-6	1 or 2 or 5	3	Stress relieve before brazing
Silicon bronzes	BAg-1a, BAg-1, BAg-2	4 or 5	3	Stress relieve before brazing. Abrasive cleaning may be helpful
Aluminum bronzes	BAg-3, BAg-1a, BAg-1, BAg-2	4 or 5	4	...
Copper nickel	BAg-1a, BAg-1, BAg-2, BAg-18, BAg-5, BCuP-5, BCuP-3	1 or 2 or 5	3	Stress relieve before brazing
Nickel silver	BAg-1a, BAg-1, BAg-2, BAg-5, BAg-6, BCuP-5, BCuP-3	3 or 4 or 5	1	Stress relieve before brazing and heat uniformly

(a) Inert gas, including hydrogen, or vacuum atmospheres are usually acceptable (AWS type 6, 9 or 10). (b) Protective atmospheres are not required for brazing copper with these filler metals.

Table 15 Guide to selecting proper filler metals for brazing copper metals to dissimilar metals

Dissimilar metal	Preferred filler metal
Carbon steels, low-alloy steels	BAg, BAu, RBCuZn
Cast iron	BAg, BAu, RBCuZn
Stainless steels	BAg, BAu
Nickel and nickel alloys	BAg, BAu, RBCuZn
Titanium and titanium alloys	BAg
Reactive metals	BAg
Refractory metals	BAg
Tool steels	BAg, BAu, RBCuZn, BNi

Note: Aluminum alloys and magnesium alloys are not brazed to copper metals.

base metal. This is accomplished primarily by removal of oxides, coatings and other foreign matter from the base-metal surface. Because there are so many variables in brazing, no universal flux exists.

AWS types 3A, 3B, 4 and 5 generally are recommended for brazing copper metals. Only type 4 fluxes and a few type 3A fluxes are suitable for aluminum bronzes. In cases involving high-volume production, the optimum flux should be determined through experiments. Criteria for maximum effectiveness of a specific flux within any given AWS type number are:

1 For dip brazing, water (including water of hydration) must be avoided.
2 For resistance brazing, the flux mixture must allow electric current to pass. This generally requires a dilute, wet flux.
3 The effective temperature range for the flux must span the brazing temperature range for the specific brazing filler metal involved. This is important on both the high and low sides of the brazing temperature range.
4 If the time during brazing is long (many minutes), a less active and longer-lived flux is desirable; a quick heating cycle (such as in induction brazing) needs a more active flux but does not require long flux life.
5 If a nonoxidizing atmosphere is used, the life of the flux can be prolonged.
6 Flux should leave a residue that is easy to remove if the part must be cleaned after brazing.
7 Corrosive action on the base metal or filler metal should be minimized.

Fluxes are available commercially as liquids, powders and pastes, with pastes being the most common.

Flux residues should be removed after brazing if exposure to moisture is anticipated in service. Residues, which are amorphous if considerable oxide has been removed from joint surfaces by the flux, will be easier to remove if the part is immersed in cold water before the joint has cooled from the brazing temperature. This technique is most useful in torch brazing, but removal solely by mechanical or chemical means may be more appropriate if other brazing processes have been employed.

Controlled atmospheres are commonly used in furnace, induction and resistance brazing to produce high-quality joints. The atmosphere prevents formation of oxides during brazing and often chemically reduces surface oxides, allowing filler metal to flow onto clean base metal. An advantage of atmosphere brazing is that it removes oxide and/or inhibits further oxide formation over the whole part or assembly, thus reducing or eliminating postbraze cleaning. Atmospheres of AWS types 2 through 6 (which contain hydrogen or dissociated ammonia) cannot be used with tough pitch coppers because they induce hydrogen embrittlement. Inert atmospheres such as argon, nitrogen and helium are suitable for all copper metals. If base or filler metals are essentially free of elements having high vapor pressure (such as Zn, P or Cd), vacuum is also suitable.

Surface Preparation. Joint surfaces must be clean and free from dirt, grease, oxides, scale and other foreign substances if high-quality brazed joints are to be obtained. Chemical degreasing (solvent or alkali) as well as mechanical methods (sanding, wire brushing, grinding, scraping or filing) may be employed. To remove oxide films completely, chemical pickling solutions are recommended.

For alloys that react rapidly with air, plating with copper after pickling may increase shelf life prior to brazing.

Joint clearance greatly affects the mechanical strength of a brazed joint. Filler-metal distribution depends on capillary action, and maintaining proper clearance reduces development of voids, which degrade joint strength. It is recommended that a part or assembly to be joined by brazing be designed specifically for brazing. Among factors to be considered in part design are (a)

compositions of filler and base metals and (b) service requirements.

Brazing Processes. Among factors that influence choice of a brazing process are size, shape and quantity of parts or assemblies, cost, and available manpower and equipment.

- *Furnace brazing* is preferred for batch or continuous processing of large numbers of parts, and generally low unit cost. Equipment consists of a furnace, which may be electric, gas fired or oil fired, plus temperature and protective-atmosphere controls. Continuous processing is done in conveyor furnaces. Filler metal usually is preplaced wire, preforms, slugs, cemented powder or paste. For high-volume production, brazing paste is often the most economical form of brazing alloy. Aside from cost effectiveness, specific advantages of furnace brazing include the ability to braze several joints simultaneously on the same assembly, and protection against surface oxidation by use of an enclosed container and a protective atmosphere. With regard to specific benefits for copper-base alloys, the slower and more uniform heating and cooling rates normally employed in furnace brazing are desirable for copper alloys susceptible to fire cracking. Limitations of furnace brazing include initial cost of equipment, floor-space requirements, and frequency of required maintenance of conveyor belts and muffles. As a general rule, maintenance costs increase with furnace temperature.

- *Torch brazing* is used most often for low to moderate production rates, and for parts of medium size and/or complex shape. It is of critical importance that joint members achieve temperature uniformity so that filler metal will melt freely and fill the entire joint. An active flux that melts not too far below the filler-metal liquidus can be used to signal when the temperature approaches the brazing range. Advantages of torch brazing include low equipment cost; flexibility in terms of assembly size; and selective heating, which facilitates bringing unequal masses to temperature uniformity. Copper-base alloys most often torch brazed are oxygen-free copper, deoxidized copper and copper-zinc alloys. In brazing of tough-pitch coppers, the torch should be adjusted to give a neutral or slightly oxidizing flame; a reducing flame promotes hydrogen

embrittlement. Alloys containing refractory-forming elements (aluminum, beryllium, chromium and silicon) should be protected by appropriate fluxes and should be brazed with a reducing flame. Brasses containing zinc are subject to volatilization when overheated or held at brazing temperature for extended periods of time. Phosphor bronzes and other alloys subject to fire cracking must be heated slowly. Filler metals commonly used in torch brazing include the relatively low melting BAg, BCuP and RBCuZn-A filler metals.

- *Induction brazing* usually is selected where localized application of heat constitutes a distinct advantage. Considerations that make induction brazing attractive are: (a) the localized heating can sometimes prevent overheating and distortion, (b) the localized heating may be more economical than heating the entire assembly and (c) the rapid heating characteristic of induction heating may be more desirable than the slower heating of other methods. The general advantage of induction brazing copper metals is uniform localized heating. Limitations include relatively high initial cost of equipment and lower efficiency of heating copper metals (compared to heating steels). Control of heating rate is poor, hence alloys susceptible to fire cracking and volatilization may not be suitable for induction brazing. Filler metals usually are preplaced rings, wire, shims, washers, paste or powder. Moist or liquid fluxes are not recommended; moisture is driven off as steam during rapid heating cycles, and may cause spattering of flux and/or molten filler metal. Induction brazing with active or protective atmospheres can be performed satisfactorily, provided a closed vessel or tube is used to contain the parts during brazing.

- *Resistance brazing* is used primarily to join copper conductors and other parts to make lap-joint electrical connections. In this type of brazing, it is difficult to obtain uniform current distribution, hence if the area to be brazed is large or discontinuous heating will not be uniform. The filler metal most frequently used is BCuP-5, which is normally used without flux. Preplaced filler metal in the form of wire, shims, washers, rings, paste or powder is used.

- *Dip brazing* is done by dipping loosely assembled joints into molten filler metal or a molten chemical (flux) bath. With the latter method, filler metal is preplaced in the joint. Molten metal bath dip brazing usually is limited to small wire connections or metal strips. Molten filler metal is contained in an externally heated crucible, and is covered with flux. Molten chemical bath dip brazing requires a heated vessel to maintain temperature. In production brazing with molten chemical fluxes, parts usually are preheated close to the melting temperature of the flux before dipping. Preheating prevents flux from freezing on cold surfaces and avoids spattering, which can occur when moist parts are dipped in molten flux.

- *Infrared brazing* is considered a form of furnace brazing in which heat is supplied by radiation from high-intensity quartz lamps. Quarts-lamp heaters of up to 5 kW capacity are available, and reflectors are used to concentrate the heat. The position of the assembly with respect to the heaters can be adjusted to obtain optimum, uniform heating of the joint. Because the process is similar to furnace brazing, selection of filler metal, flux, atmosphere and other process variables is based on the same considerations as for furnace brazing. The major advantages of infrared brazing are energy-related. In atmosphere brazing, a large amount of energy goes into heating the atmosphere and the furnace chamber. In infrared brazing, only the parts are heated; thus, time and energy are not lost in bringing the furnace to temperature, the furnace can be turned on and off as needed, switchover from one assembly to another is faster, and downtime is minimized because furnace components are more accessible. Obviously, these energy-related advantages are of more benefit in intermittent and one-shift operations than in continuous, three-shift operations.

Postbraze treatment usually involves removal of flux or oxide. These operations are best achieved by: (a) mechanical abrasion such as wire brushing or blast cleaning; (b) hot-water rinsing for soluble fluxes; (c) acid pickling with a suitable commercial solution (for copper alloys this must not contain nitric acid); or (d) strong, hot, caustic solutions, which generally react more slowly than acid solutions. Thorough water rinsing is necessary after all chemical treatment.

Welding

Copper and its alloys are readily welded, the choice of welding method being determined by the alloy, the application and the configuration of the joint.

Oxyfuel-gas welding and shielded metal-arc welding (SMAW) once were the preferred processes for welding copper. Gas-shielded arc processes—gas metal-arc welding (GMAW) and gas tungsten-arc welding (GTAW)—now have largely replaced oxyfuel-gas welding and SMAW because the gas-shielded processes produce superior welds. The primary criteria for choosing between GMAW and GTAW are thickness of the metal to be welded and amount of welding to be performed. GMAW is generally preferred for thicknesses greater than about 6 mm (¼ in.), and GTAW for thicknesses less than about 2 mm (0.08 in.). For thicknesses between 2 and 6 mm, both processes will work satisfactorily.

Table 16 gives recommended filler metals for GMAW and GTAW processes for welding copper and copper alloys to other copper metals and to dissimilar metals. This table is only a guide, because many factors are involved in choosing the right filler metal.

Resistance welding is also used extensively because it is readily adapted to assembly-line production. It can be used for joining parts less than 3.8 mm (0.15 in.) thick, although production of good welds in highly conductive alloys (>30% IACS) requires strict control of operating procedures. Table 17 lists conductivity values and relative ratings of performance in resistance welding for several copper alloy classes.

Coppers. Typical applications of welded copper include bus bars, electrical conductors, wave guides, air conditioners, plumbing lines and chemical process vessels.

The easiest coppers to arc weld are oxygen-free or deoxidized coppers. Copper oxides cause the most trouble in welding because they migrate rapidly to grain boundaries at temperatures above 1065 °C (1950 °F) and reduce ductility and strength. Oxygen-bearing (tough pitch) coppers are very difficult to fusion weld; they should be used only for weldments that will not be subjected to structural loads. Several processes—GTAW, GMAW, SMAW, oxyfuel-gas welding or even electron-beam welding—are all satisfactory for coppers. Selection of a process is usually governed by joint configuration, number of joints to be welded, section thickness and welding position. The presence of hydrogen and copper oxides should be avoided during welding; above 370 °C (700 °F), they react to form steam, which results in a porous, brittle weld. Welds in phosphorus-deoxidized copper have the best ductility and strength because the phosphorus ensures freedom from copper oxides.

The GMAW process is preferred for welding parts thicker than about 6 mm (about ¼ in.); GTAW is more effective on light gages; deoxidized filler metals are used. Shielded metal-arc welding is limited to minor repair welds on light gage stock, fillet welds in locations that are difficult to reach, and welds involving dissimilar metals. Gas welding should be done using a neutral or slightly oxidizing flame, a deoxidized filler metal, and flux. Resistance welding is extremely difficult because coppers have very high conductivities. Spot welding of stock less than 1.5 mm (0.060 in.) thick can be done with tungsten- or molybdenum-tipped electrodes to prevent sticking.

High-copper alloys such as cadmium coppers, zirconium coppers, chromium coppers and beryllium coppers are typically used to make springs, electrical connectors, valves, pump parts and welding equipment.

Except for beryllium coppers, procedures for arc or gas welding of deoxidized coppers are a good starting point for developing welding procedures for high-copper alloys. The GTAW process is recommended for beryllium coppers of any thickness, with filler rod being used on thick sections. On sections no thicker than 1.0 mm (0.040 in.), excellent welds have been obtained without addition of filler metal. Alloys having high beryllium contents are more readily welded than alloys with low beryllium contents, because the latter are susceptible to weld cracking and to

Table 16 Filler-metal selection guide for arc welding copper metals

Weld to	Coppers	Low-zinc brasses	High-zinc brasses; tin brasses; special brasses	Phosphor bronzes	Aluminum bronzes	Silicon bronzes	Copper nickels
Low-zinc brasses	Cu-CuSn-C	CuSi, CuSn-A, CuSn-C	...	...	...	...	...
High-zinc brasses, tin and special brasses	Cu, CuSi, CuSn-C	CuSn-C	CuSn-C, CuAl-A2, RCuSi-A	...	...	...	...
Phosphor bronzes	Cu, CuSn-C	CuSn-C	CuSn-C	CuSn-A, CuSn-C, RCuSi-A	...	...	...
Aluminum bronzes	CuAl-A2	CuAl-A2	CuAl-A2	CuSn-C, CuAl-A2	CuAl-A1, CuAl-A2, CuAl-B	...	...
Silicon bronzes......	Cu, CuSn-C	CuAl-A2	CuSi, CuAl-A2	CuSi	CuAl-A2	CuSi	...
Copper nickels......	Cu, CuAl-A2, CuNi	CuAl-A2	CuAl-A2	CuSn-C	CuAl-A2	CuAl-A2	CuNi
Nickel and Ni-Cu alloys.............	CuNi, RNi-3	(a)	(a)	(a)	(a)	(a)	CuNi, NiCu-7
Ni-Cr, Ni-Fe and Ni-Cr-Fe alloys	RNi-3	(a)	(a)	(a)	(a)	(a)	RNi-3
Low-carbon steel	Cu, CuAl-A2, CuNi, Ni-3	CuSn-C	CuAl-2	CuSn-C	CuAl-A2	CuAl-A2	CuAl-A2, Ni-3
Medium-carbon steel	Cu, CuAl-A2, Ni-3	CuAl-A2	CuAl-A2	CuSn-C	CuAl-A2	CuAl-A2	CuAl-A2, Ni-3
High-carbon steel ...	Cu, CuAl-A2, Ni-3	CuAl-A2	CuAl-A2	CuSn-C	CuAl-A2	CuAl-A2	CuAl-A2, Ni-3
Low-alloy steel......	Cu, CuAl-A2, Ni-3	CuAl-A2	CuAl-A2	CuSn-C	CuAl-A2	CuAl-A2	CuAl-A2, Ni-3
Stainless steel	Cu, CuAl-A2, Ni-3	CuAl-A2	CuAl-A2	CuSn-C	CuAl-A2	CuAl-A2	CuAl-A2, Ni-3
Gray and malleable irons	Cu, CuAl-A2	CuSn-C, CuAl-A2	CuAl-A2	CuSn-C	CuAl-A2	CuSi, CuAl-A2	CuAl-A2, CuNi
Ductile iron	Cu, CuAl-A2	CuAl-A2	CuAl-A2	CuSn-C	CuAl-A2	CuSi, CuAl-A2	CuAl-A2, CuNi

(a) This metal combination is seldom welded.

cracking during postweld heat treatment. When any welding is done on beryllium coppers, rigid safety precautions must be followed; fumes must not be inhaled because of their extremely toxic nature. High-copper alloys should be welded under protective atmosphere to prevent preferential oxidation of the alloying elements.

Chromium coppers and beryllium coppers are precipitation-hardening alloys and require heat treatment after welding. Cadmium coppers and zirconium coppers are strengthened principally by cold working, and as-welded joints in these alloys will not be as strong as the parent metal. Arc-welding processes are not recommended for oxide-dispersion-strengthened copper alloys, because melting in the weld zone would destroy the oxide dispersion.

Spot welding of high-copper alloys is difficult because of their high conductivities. High-beryllium alloys are spot welded best in the solution-annealed condition, when electrical conductivity is lower.

Brasses are typically used for architectural applications, plumbing fittings, automotive radiators, evaporators, marine hardware, and tubes for condensers and heat exchangers.

Compared to coppers, brasses require less preheating for fusion welding and

Table 17 Comparative behavior of copper metals in resistance welding

Alloy type	Common name	Conductivity, % IACS	Rating
Cu-Si	Silicon bronzes	7 to 12	Excellent
Cu-Ni	Copper nickels	4 to 10	Excellent
Cu-Ni-Zn	Nickel silvers	5 to 9	Good
Cu-Al	Aluminum bronzes	7 to 18	Good
Cu-Sn	Phosphor bronzes	10 to 22	Fair
Cu-Zn-Mn	Manganese brass	15 to 16	Fair
Cu-Zn-Si	Silicon red brass	15 to 16	Fair
Cu-Zn (high Zn)	Yellow brasses	22 to 28	Fair
Cu-Zn-Mn-Fe	Manganese bronzes	22 to 28	Fair
Cu-Zn (low Zn)	Red brasses	32 to 43	Poor
Cu	High-copper alloys	50 to 65	Poor

less power for resistance welding. The presence of oxygen and hydrogen during welding does not pose problems, because hydrogen has a low solubility in brass and zinc is an effective deoxidizer. Restraint during postweld cooling should be minimized to avoid cracking.

For low-zinc brasses (those containing less than 20% Zn), GTAW is the process most commonly used, particularly for light gages. The GMAW process can be used for heavy gages. Light gages can be welded without filler metal, but a zinc-free filler metal such as RCuSn-C or RCuSi is recommended for thicknesses greater than about 1.5 mm (1/16 in.) to improve color match and flow, respectively. Shielded metal-arc welding can be done in the flat position using the same types of filler metals as for GMAW. Oxyfuel-gas welding is done on piping using RCuSi-A filler, and the resulting welds have excellent corrosion resistance. The process should be controlled to avoid zinc fuming, which constitutes a serious health hazard to personnel if the fumes are inhaled. Zinc fuming increases with zinc content.

Like low-zinc brasses, high-zinc brasses can be satisfactorily welded by any fusion welding method. Again, zinc-free filler metals are used with GTAW. ERCuSi-A filler metal is effective in controlling zinc fuming and produces porosity-free welds with good corrosion resistance. Very high zinc contents (above about 35% Zn) require other filler metals to match the strength of the base metal, but these are not as effective in controlling zinc fuming. High-zinc brasses can be welded by GMAW and SMAW, but fuming is greater than in GTAW. Oxyfuel-gas welding is done in a slightly oxidizing flame to control zinc vapor-

ization; filler metals RBCuZn-A or RCuZn-C are used with AWS type 5 flux.

Tin brasses can be welded by the same methods used for plain brasses of similar zinc content. Leaded brasses are not suitable for welding because the lead causes hot shortness and cracking.

The conductivities of all brasses are high enough to make resistance welding difficult. RWMA class I electrodes help achieve acceptable weld quality. Friction welding can be used to join brasses, but only when the joint is of an appropriate (chiefly circular) configuration.

Tin Bronzes. Because of their wide freezing ranges and large dendritic grain structures, tin bronzes (phosphor bronzes) have a tendency to crack during welding if proper procedures are not followed. Hot peening and preheating often are used to refine the grain structure for better weldability. Tin bronzes are welded by SMAW, GMAW and GTAW with CuSn-A or CuSn-C filler metal. Filler metals are always used because they contain sufficient phosphorus to deoxidize the molten metal and prevent porosity. The GTAW process is recommended for welding light gages, for minor repair of castings and for surfacing, whereas GMAW and SMAW are recommended for welding heavy gages and for fabricating structures and chemical process vessels. Oxyfuel-gas welding is not recommended because excessive contraction, cracking and porosity cannot be adequately controlled in these hot short materials. If oxyfuel-gas welding must be used, a neutral flame and ERCuSn-A filler metal are employed. Improved ductility and resistance to stress-corrosion cracking are obtained by annealing

and rapidly cooling after welding. Leaded phosphor bronzes are not welded for the same reason leaded brasses are not welded. Resistance welding is seldom used for tin bronzes except to butt weld rod.

Aluminum Bronzes. Of the alpha aluminum bronzes, those binary alloys containing less than 6% aluminum (C60600 and C60800, for example) are hot short and tend to crack in the heat-affected zone, and are therefore difficult to fabricate by welding. The higher-aluminum alpha alloys (such as C61000, C61300 and C61400) are less difficult to weld using conventional methods and equipment. Unfired pressure vessels and components are commonly fabricated from C61400 by welding, and such welded assemblies are capable of meeting requirements of the ASME Boiler and Pressure Vessel Code, sections III and VIII.

Both cast and wrought aluminum bronzes with duplex alpha-beta structures are readily welded by conventional methods. They also can be successfully welded to dissimilar metals such as steels, nickel alloys and high-copper alloys using copper-aluminum filler metals.

Various compositions of aluminum bronzes are deposited as welded overlays on steels. In this form, the aluminum bronze serves as a wear resistant or corrosion resistant surface layer.

GTAW and GMAW can produce high-quality welds, and are used for both joining and repair welding of aluminum bronzes. SMAW also can be used, but requires careful cleaning to remove slag residue. Low thermal and electrical conductivities make it relatively easy to join aluminum bronzes by resistance welding. Current densities should be about 25% higher than those used for steel. Careful surface preparation to remove oxides prior to welding is essential for both arc and resistance welding.

Silicon Bronzes. Of all the copper alloys, silicon bronzes are generally considered the easiest to weld—a result of low thermal conductivity, the deoxidizing effect of silicon and a relatively narrow hot shortness range. They are readily welded by GTAW (using the proper filler metal), although GMAW also can be used readily. Filler metals should have the same composition as the base metal. Silicon bronzes are hot short and require fast cooling through the hot shortness range. Oxyfuel-gas welding can be done using a neutral or slightly oxidizing flame and a flux of high acid content. A small weld puddle should be maintained to minimize contraction strains, and high restraint should be avoided in joint design. Silicon bronzes are easily joined by resistance butt and resistance seam welding. Removal of surface oxides is critical prior to resistance welding.

Copper nickels are used extensively for desalination equipment, marine piping and hardware, antifouling surfaces, and tubing for marine condensers and heat exchangers. Copper nickels have thermal and electrical conductivities close to those of carbon steel, which makes them relatively easy to weld. The GTAW, GMAW and SMAW processes are used regularly, with deoxidized 70Cu-30Ni filler metal for all copper nickels regardless of composition. Lead, sulfur or phosphorus contamination of the weld metal can cause intergranular hot cracking in restrained weldments. Quality of oxyfuel-gas welds is poor except in thin sheet. A reducing flame and liberal use of flux are necessary. Alloys containing more than 10% nickel have good to excellent resistance welding properties, provided surfaces are clean and free of contaminants.

Nickel Silvers. The welding metallurgy of nickel silvers is similar to that of brasses. Arc welding is seldom used because nickel silvers frequently are used in decorative rather than critical structural applications. If arc welding is necessary, GTAW is preferred. Filler metals containing silicon are used to reduce porosity. Lead, bismuth or sulfur contamination of weld metal leads to hot cracking. In gas welding, flux is used to reduce nickel oxides, and RBCuZn-D filler metal is employed. Resistance welding is very easy because the conductivities of nickel silvers are among the lowest of all copper alloys.

Diffusion Bonding

Diffusion bonding refers to any of several joining methods in which metals are bonded together as a result of interdiffusion of interfacial atoms across the joint interface. The process may be used for making joints between large, flat surfaces; butt joints between tubes or bars; line joints, such as the joints that bond a corrugated core to the skin sheets of a reinforced panel; or concentric lap (socket) joints between

tubes. Metal may be molten at the interface, but usually it is not; most often, diffusion-bonding temperatures are about one-third to one-half the melting temperature of the base metal.

Chief advantages of diffusion bonding are: ability to produce continuous, leaktight, completely metallurgical bonds without voids, brittle phases or discontinuities; ability to make bonds between metals that otherwise cannot be metallurgically bonded; joint strength equal to that of the lower-strength member; freedom from limitations regarding complexity of part shape or difference in metal thickness between members; exceptional electrical and thermal continuity across the joint; and the capacity to produce bonded composites that can be formed or welded after bonding without affecting the quality of the bond. Limitations are as follows: costs are high, both for development and for production of certain bonded assemblies; certain applications may require use of an intermediate layer at the interface to act as a filler metal or barrier film; some bonding methods may require precision machining to prepare the joint for bonding; and inspection to ensure reliable bonding may be difficult for some assemblies. In addition, diffusion bonding requires use of a very clean atmosphere, such as inert gas, hydrogen or vacuum, during the portion of the process when diffusion across the joint interface takes place at elevated temperature, but this usually is not considered a limitation.

The most advanced and most commonly employed diffusion bonding process is isostatic gas pressure bonding. In this process, pressure is applied at the interface by placing the assembly in a closed container and then introducing high-pressure gas into the container. Other diffusion bonding processes include roll bonding, coextrusion, press bonding and creep-controlled bonding. Diffusion brazing and eutectic bonding are processes that make use of an intermediate liquid phase at the interface to transport metal atoms across the interface and thereby speed diffusion. Although not strictly a diffusion bonding process, a diffusion heat treatment often is employed following explosive welding, in order to further consolidate the bond.

Diffusion bonds are readily produced between copper metals. Typical conditions are 1 min at a temperature of 315

°C (600 °F) under hydrogen gas pressurized to about 140 MPa (20 000 psi). Copper also can be bonded to dissimilar metals such as aluminum, titanium, Kovar, Cb-1Zr and type 316 stainless steel, but under vacuum rather than under pressurized hydrogen. Temperatures used for welding copper to these metals range from about 510 °C (950 °F) for aluminum to about 980 °C (1800 °F) for 316 stainless; times, from 10 min for Kovar to 4 h for Cb-1Zr. No intermediate layer of material at the bond line is needed for any of these combinations, but closely mating surfaces with finishes of about 4 to 6 μin. rms are essential for production of reliable bonds.

Adhesive Bonding

Application of adhesive bonding to copper metals most often involves bonding of thin copper sheet to substrates such as steel, plywood, cement-asbestos board, masonite, urethane foam and particle board to produce bonded laminates for architectural applications. The copper is almost always applied for purely decorative purposes, although such laminates may also serve as structural members.

There are no restrictions on thickness or on composition of the alloy, although for economic reasons a very thin gage usually is desirable. The copper sheet most often is only 0.05 to 0.8 mm (0.002 to 0.032 in.) thick.

Successful adhesive bonding depends mainly on selection of the proper adhesive, careful surface preparation of both surfaces, and joint design compatible with both the adhesive and the application. For example, a flexible adhesive is required for applications where the temperature varies over a significant range so that the adhesive will give rather than debond under the joint stresses caused by differential thermal expansion and contraction between the substrate and the copper overlay. When the laminate is to be used outdoors, the edges must be protected so that moisture does not penetrate along the bond line and cause the copper to separate from the substrate.

Many commercial adhesives have been used to bond copper; thermoplastic resins, thermosetting resins and epoxies are the most common. Most of the applicable adhesives have only limited temperature stability, which limits use of copper-faced laminates to temperatures below 100 °C (212 °F).

Special Joining Techniques

The processes described in this section are very limited in application because of their high cost. For certain applications, the advantages outweigh the disadvantages and make the particular process feasible and/or economical compared to other alternatives. In many instances, special techniques are limited to high-quality precision parts because extremely close tolerances and precise fit-ups are required for successful and reliable joining.

Electroplating has been used to join copper to aluminum for electrical applications. The process is performed at room temperature, which eliminates distortion, shrinkage stresses and recrystallization problems associated with conventional processes that use heat and pressure. Electroplating has also been used to make vacuum-tight ceramic-to-metal seals. In joining of dissimilar metals, formation of brittle intermetallics can be prevented by using this process.

Roll bonding can be used to join copper sheet or plate to other materials, thus making layered, composite, metallurgically bonded sheet or plate. Applications include cookware and coins. Roll bonding is most suitable where the joint surface is so large in area that welding or brazing would not be feasible.

Friction welding and explosive bonding have not been very successful in producing copper-to-copper joints. Either method can be used to join copper to other metals such as steel or brass. Joints are not very strong, and in general these processes are useful only in joining thick to thin materials.

Ultrasonic welding also is most effective in joining copper to other metals. Weld joints are weak, and the process is best suited for joining two thin materials, or for joining thick materials to thin materials.

Electron beam welding is a fusion process where a concentrated high-energy electron beam bombards the workpiece. Sound, porosity-free welds with excellent weld strengths are obtained. The process is limited to copper alloys that can be joined by other fusion processes. Alloys containing lead or zinc are to be specifically avoided. The process is usually carried out in a relatively hard vacuum, but equipment that allows the workpiece to be under only soft vacuum (or no vacuum at all) has been developed to make electron beam welding more flexible.

Laser welding is chiefly an experimental process at this time. The process is currently limited to materials no thicker than 0.8 mm (0.032 in.), but as higher-power lasers are developed this range could expand. Because it is a fusion process, laser welding cannot be used for copper alloys that contain lead or zinc.

Other Welding Processes. Plasma-arc welding is seldom used for copper metals. Stud welding, submerged arc welding and electroslag welding are prohibited for welding copper and its alloys.

Corrosion Characteristics of Copper and Its Alloys

By the ASM Committee on Copper and Copper Alloys*

COPPER and its alloys are used in many applications that require service for extended periods in environments that can be aggressive to other metals. Copper metals resist the atmosphere, fresh and salt waters, alkaline solutions (except those containing ammonia) and many organic chemicals. Their resistance to corrosion by oxidizing acids depends mainly on the severity of oxidizing conditions in the acid solution. Copper metals are suitable for use with many salt solutions.

Copper reacts with sulfur and sulfides to form copper sulfide. As a result, coppers and copper alloys normally are not selected for service in environments known to contain sulfur or sulfides.

This article describes the chief causes and forms of corrosion affecting copper metals. Detailed information on the resistance of copper and its alloys to specific environments can be found in the articles "Wrought Copper Alloys for Corrosion Service" and "Copper Alloy Castings" in this volume.

Coppers and copper alloys, like most other metals and alloys, are susceptible to several forms of corrosion, depending mainly on environmental conditions.

*See page **XI** for committee list.

Table 1 presents identifying characteristics of the forms of corrosion that commonly attack copper metals, and the most effective means of combating each.

General corrosion is well-distributed attack of an entire surface with little or no localized penetration, and it is the least damaging of all forms of attack. It is the only form of corrosion for which weight-loss data can be used to accurately estimate penetration rates.

General or uniform corrosion results from prolonged contact with environments in which the corrosion rate is very low, such as fresh, brackish and salt waters; many types of soil; neutral, alkaline and acid salt solutions; organic acids; and sugar juices. Many other substances that bring about uniform thinning at a faster rate include oxidizing acids, sulfur-bearing compounds, ammonia and cyanides.

Galvanic Corrosion. An electrochemical potential almost always exists between two dissimilar metals when they are immersed in a conductive solution. If two dissimilar metals are in electrical contact with each other and immersed in a conductive solution, a potential results which enhances corrosion of the more electronegative

member of the couple (the anode) and partly or completely protects the more electropositive member (the cathode). Copper metals are almost always cathodic to other common structural metals such as steel and aluminum. When steel or aluminum is put in contact with a copper metal, the corrosion rate of the steel or aluminum increases whereas that of the copper metal decreases. The common grades of stainless steel exhibit variable behavior: that is, copper metals may be either anodic or cathodic to the stainless steel, depending on conditions of exposure. Copper metals usually corrode preferentially when coupled with high-nickel alloys, titanium or graphite.

Corrosion potentials of copper metals generally range from -0.2 to -0.4 mV when measured against a saturated calomel reference electrode; the potential of pure copper is about -0.3 mV. Alloying additions of zinc or aluminum move the potential toward the anodic (more electronegative) end of the range; additions of tin or nickel move the potential toward the cathodic (less electronegative) end. Galvanic corrosion between two copper metals is seldom a significant problem, because the potential difference is so small.

Table 2 lists a galvanic series of

Table 1 Guide to corrosion of copper metals

Form of corrosion	Identifying characteristics	Techniques for preventing corrosion
General thinning	Uniform metal removal	Select proper alloy for environmental conditions based on weight-loss data.
Galvanic corrosion	Corrosion preferentially near a more cathodic metal	Avoid electrically coupling dissimilar metals. Maintain optimum ratio of anode area to cathode area. Maintain optimum concentration of oxidizing constituent in corroding medium.
Pitting Concentration cell	Water-line pitting Crevice corrosion Pitting under foreign objects or dirt	Design out crevices. Keep metal clean
Impingement	Erosion attack from turbulent flow plus dissolved gases, generally as lines of pits in direction of fluid flow	Design for streamline flow. Keep velocity low. Remove gases from liquid phase.
Fretting	Chafing or galling, often occurring during shipment	Lubricate contacting surfaces. Interleave sheets of paper between sheets of metal. Decrease load on bearing surfaces.
Dealloying	Preferential dissolution of zinc or nickel, resulting in layer of sponge copper	Select proper alloy for environmental conditions based on metallographic examination of corrosion specimens.
Stress-corrosion cracking	Cracking, usually intercrystalline but sometimes transcrystalline, that is often fairly rapid	Select proper alloy based on stress-corrosion tests. Reduce applied or residual stress. Remove mercury compounds or ammonia from environment.
Corrosion fatigue	Several transcrystalline cracks	Select proper alloy based on fatigue tests in service environment. Reduce mean or alternating stress.
Intercrystalline corrosion	Corrosion along grain boundaries without visible signs of cracking	Select proper alloy for environmental conditions based on metallographic examination of corrosion specimens.

Fig. 1 Galvanic attack between steel and red brass

Galvanic corrosion of a 38-mm (1½-in.) steel coupling joining two sections of red brass (C23000) pipe. Note that the brass pipes (cathodic) have not corroded, whereas the steel coupling was severely attacked. About ⅔ size.

metals and alloys valid for dilute aqueous solutions such as seawater and weak acids. The metals that are grouped together may be coupled to each other without significant galvanic damage. However, connecting metals from different groups leads to damage of the more anodic metal; the larger the difference in galvanic potential between groups, the greater the corrosion.

Accelerated damage due to galvanic effects is usually greatest near the junction, where electrochemical current density is the highest. Figure 1 shows a common example of galvanic corrosion of a steel coupling that joined two lengths of red brass pipe.

Another factor that affects galvanic corrosion is area ratio. An unfavorable area ratio exists when the cathodic area is large and the anodic area is small. The corrosion rate of the small anodic area may be several hundred times greater than if the anodic and cathodic areas were equal in size. Conversely, when a large anodic area is coupled to a small cathodic area, current density and damage due to galvanic corrosion are much less. For example, copper rivets (cathodic) used to fasten steel plates together lasted longer than 1.5 years in seawater, but steel rivets used to fasten copper plates were completely destroyed during the same period.

There are five major methods of eliminating or significantly reducing galvanic corrosion:

1 Selecting dissimilar metals that are as close as possible to each other in the galvanic series.
2 Avoiding coupling of small anodes to large cathodes.
3 Insulating dissimilar metals completely wherever practicable.
4 Applying coatings and keeping them in good repair, particularly on the cathodic member.
5 Using a sacrificial anode—that is, coupling the system to a third metal that is anodic to both structural metals.

Pitting. As with most commercial metals, corrosion of copper metals results in pitting under certain conditions. Sometimes pitting is general over the entire surface, giving the metal an irregular and roughened appearance. In other instances, pits are concentrated in specific areas and are of various sizes and shapes.

Localized pitting is the most damaging form of corrosive attack because it reduces load carrying capacity and increases stress concentration by creating depressions or holes in the metal. Pitting is the usual form of corrosive

Table 2 Galvanic series in seawater

Anodic End

Magnesium	Lead
Magnesium alloys	Tin
Zinc	Muntz metal (C28000)
Galvanized steel	Manganese bronze (C67500)
Aluminum 5052H	Naval brass (C46400)
Aluminum 3004	Nickel (active)
Aluminum 3003	Inconel (active)
Aluminum 1100	Cartridge brass (C26000)
Aluminum 6053	Admiralty metal (C44300)
Alclad aluminum alloys	Aluminum bronze (C61400)
Cadmium	Red brass (C23000)
Aluminum 2017	Copper (C11000)
Aluminum 2024	Silicon bronze (C65100)
Low-carbon steel	Copper nickel, 30% (C71500)
Wrought iron	Nickel (passive)
Cast iron	Inconel (passive)
Ni-Resist	Monel
Type 410 stainless steel (active)	Type 304 stainless steel (passive)
50Pb-50Sn solder	Type 316 stainless steel (passive)
Type 304 stainless steel (active)	Silver
Type 316 stainless steel (active)	Gold
	Platinum
	Cathodic End

Fig. 2 Impingement attack of a condenser tube

Flattened section of admiralty metal (C44300) tube taken from a ship condenser. High velocity and turbulence of the fluid stream caused impingement attack, which can be seen as a series of horseshoe-shape pits at top of flattened tube section. Shown at about 2X.

attack at surfaces on which there are incomplete protective films, nonprotective deposits of scale or extraneous deposits of dirt or other foreign substances.

Crevice corrosion, water-line attack, deposit attack, impingement attack, concentration-cell action and *fretting* are some of the names used for special types of pitting. Pitting results from formation of local electrolytic cells due to differences in metal-ion or oxygen concentration at adjacent areas of the metal surface. Crevice corrosion results from a depletion of oxygen in crevices so that the metal in a crevice becomes anodic to metal outside the crevice, which is exposed to an oxygen-bearing solution. Water-line attack is a term used to describe pitting due to a differential oxygen cell functioning between the well-aerated surface layer of a liquid and the oxygen-starved layer immediately beneath it. The pitting occurs just below the water line.

Local cell action may also result from the presence of foreign objects or debris such as dirt, pieces of shell, or vegetation, or it may result from rust, permeable scales or uneven accumulation of corrosion product on the metallic surface. Such materials screen the affected area, causing oxygen deficiency by making it difficult for fresh oxygen-bearing solution to diffuse to the site—

hence the name deposit attack. Sometimes this type of attack can be controlled by cleaning the surfaces. For instance, condensers and heat exchangers are cleaned periodically to prevent deposit attack.

Impingement attack (sometimes called erosion-corrosion) occurs where gases, vapors or liquids impinge on metal surfaces at high velocities, such as in condensers or heat exchangers. It is most often found with waters containing low compounds of sulfur and with polluted, contaminated or silty salt water or brackish water. The erosive action locally removes protective films, thereby contributing to the formation of concentration cells and to localized pitting of anodic sites.

Impingement attack is characterized by undercut grooves, waves, ruts, gullies and rounded holes; it usually exhibits a directional pattern. Pits are elongated in the direction of flow and are undercut on the downstream side. When the condition becomes severe, it may result in a pattern of horseshoe-shaped grooves or pits with their open ends pointing downstream. As attack progresses, the pits may join, forming fairly large patches of undercut pits (see Fig. 2). When this form of corrosion occurs in a condenser tube, it usually is confined to a region near the inlet end of the tube where fluid flow is rapid and

turbulent. If some of the tubes in a bundle become plugged, the velocity is increased in the remaining tubes; therefore, the unit should be kept as clean as possible.

Impingement attack can be reduced, and the life of the unit extended, by decreasing fluid velocity, streamlining the flow and removing entrained air. This usually is accomplished by redesigning water boxes, injector nozzles and piping to reduce or eliminate low-pressure pockets, obstructions to smooth flow, abrupt changes in flow direction and other features that cause local regions of high-velocity or turbulent flow. Condensers and heat exchangers are less susceptible to impingement attack if they are made of one of the aluminum brasses or copper nickels, which are more erosion resistant than the brasses or tin brasses. When contaminated waters are involved, filtering or screening of the liquids and frequent cleaning of surfaces can be very effective in minimizing impingement attack.

Another form of attack, called *fretting* or *fretting corrosion*, appears as pits or grooves in the metal surface surrounded or filled with corrosion product. Fretting is sometimes referred to as *chafing, road burn, friction oxidation, wear oxidation* or *galling*.

The basic requirements for fretting are as follows:

1 Repeated relative (sliding) motion between two surfaces must occur. The relative amplitude of the motion may be very small—motion of only a few tenths of a millimetre is typical.

2 The interface must be under load.

3 Both load and relative motion must be sufficient to produce deformation of the interface.

4 Oxygen and/or moisture must be present.

Fretting does not occur on lubricated surfaces in continuous motion such as axle bearings, but rather on dry interfaces subject to repeated small relative displacements. A classic type of fretting occurs during shipment of bundles of mill products having flat faces, such as hexagonal rod, octagonal rod and rectangular bar.

Two major mechanisms have been proposed to explain fretting: the wear-oxidation theory and the oxidation-wear theory. Both theories account for the fact that in fretting attack pitting occurs at the contacting interfaces, accompanied by production of oxide debris. Fretting is not confined to coppers and copper alloys, but has been recognized on almost every kind of surface—steel, aluminum, noble metals, mica and glass.

Fretting can be controlled, and sometimes eliminated, by (a) lubricating with low-viscosity, high-tenacity oils to reduce friction at the interface between the two metals and to exclude oxygen from the interface; (b) separating the faying surfaces by interleaving an insulating material; (c) increasing the load to reduce motion between faying surfaces (may be difficult in practice, because only a minute amount of relative motion is necessary to produce fretting); or (d) decreasing the load at bearing surfaces to increase the relative motion between parts, which reduces fretting.

Dealloying is a corrosion process whereby the more active metal is selectively removed from an alloy, leaving behind a weak deposit of the more noble metal.

Copper-zinc alloys containing more than 15% Zn are susceptible to a dealloying process called dezincification. In dezincification of brass, selective removal of zinc leaves a relatively porous and weak layer of copper and copper oxide. Corrosion of a similar nature continues beneath the primary corrosion layer, resulting in gradual replacement of sound brass by weak, porous copper. Unless arrested, dealloying eventually penetrates the metal, weakening it structurally and allowing liquids or gases to leak through the porous mass in the remaining structure. The term *plug-type dealloying* re-

fers to the dealloying that occurs in local areas; surrounding areas usually are unaffected or only slightly corroded. An early stage of plug-type dezincification is illustrated in Fig. 3. In *uniform-layer dealloying,* the active component of the alloy is leached out over a broad area of the surface. Dezincification is the usual form of corrosion for uninhibited brasses in prolonged contact with waters high in oxygen and carbon dioxide. It is frequently encountered with quiescent or slowly moving solutions. Slightly acidic water, low in salt content and at room temperature, is likely to produce uniform attack, whereas neutral or alkaline water, high in salt content and above room temperature, often produces plug-type attack.

Brasses with copper contents of 85% or more resist dezincification. Brasses with two-phase structures usually dezincify in two stages: the high-zinc beta phase first and then the lower-zinc alpha phase.

Tin tends to inhibit dealloying, especially in cast alloys. C46400 (naval brass) and C67500 (manganese bronze), which are alpha-beta brasses containing about 1% Sn and are widely used for naval equipment, have reasonably good resistance to dezincification. Addition of a small amount of phosphorus, arsenic or antimony to admiralty metal (an all-alpha 71Cu-28Zn-1Sn

brass) very effectively inhibits dezincification. Inhibitors are not entirely effective in preventing dezincification of the alpha-beta brasses, because they do not prevent dezincification of the beta phase.

Where dezincification is a problem, red brass, commercial bronze, inhibited admiralty metal and inhibited aluminum brass can be used successfully. In some instances, the economic penalty of avoiding dealloying by selecting a low-zinc alloy may be unacceptable. (Low-zinc alloy tube requires fittings that are available only as sand castings, whereas fittings for higher zinc tube can be die cast or forged much more economically.) Where selection of a low-zinc alloy is unacceptable, inhibited yellow brasses generally are preferred.

Attack similar to dezincification occurs in other alloys. Dealuminification occurs in some copper-aluminum alloys. Decobaltification of cobalt-base alloys has been reported, and denickelification of copper nickels may occur under special conditions. "Parting" of gold-silver alloys is a form of dealloying.

Intercrystalline corrosion is an infrequently encountered form of attack that occurs most often in applications involving high-pressure steam. This type of corrosion penetrates the metal along grain boundaries—often to

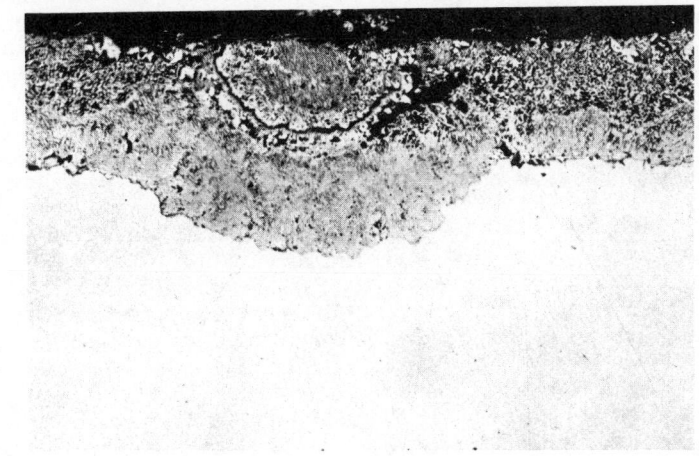

Fig. 3 Dezincification of C27000 innercooler tubes from an air compressor

Combined uniform-layer and plug-type dezincification that occurred over 17 years of service; about 75X, unetched.

Fig. 4 Intercrystalline corrosion of C44500

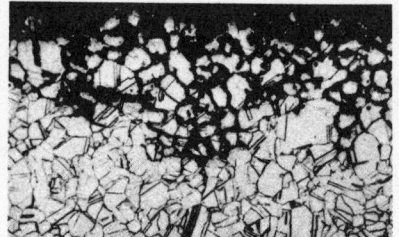

Longitudinal section of an arsenical admiralty metal tube; about 150X, etched.

Fig. 5 Typical appearance of stress-corrosion cracking in copper alloys

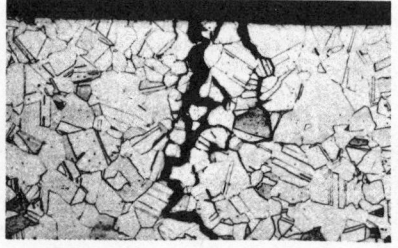

Intercrystalline cracking shown at about 60X in an etched specimen.

a depth of several grains (see Fig. 4)—which distinguishes it from surface roughening. Mechanical stress apparently is not a factor in intercrystalline corrosion. The alloys that appear to be the most susceptible to this form of attack are Muntz metal, admiralty metal, aluminum brasses and silicon bronzes.

Stress-Corrosion Cracking

Stress-corrosion cracking and season cracking are two names for the same phenomenon—the apparently spontaneous cracking of stressed metal. Stress-corrosion cracking is largely intercrystalline (see Fig. 5), but sometimes transcrystalline cracking also may be detected. Stress-corrosion cracking occurs only if a susceptible alloy is subjected to the combined effects of sustained stress and certain chemical substances.

Ammonia and ammonium compounds are the corrosive substances most often associated with stress-corrosion cracking of copper alloys. Sometimes these compounds are present in the atmosphere; in other instances, they are in cleaning compounds or in chemicals used for the treatment of boiler water. Both oxygen and moisture must be present for ammonia to be corrosive to copper alloys; other compounds, such as carbon dioxide, are thought to accelerate stress-corrosion cracking in ammoniacal atmospheres. Moisture films on metal surfaces will dissolve significant quantities of ammonia, even from atmospheres that are low in ammonia concentration.

A specific corrosive environment and sustained stress are the primary causes of stress-corrosion cracking; microstructure and alloy composition may affect the rate of crack propagation in susceptible alloys. Microstructure and composition can be controlled most effectively by selecting the right combination of alloy, forming process, thermal treatment and metal-finishing process. Although test results may indicate that a finished part is not susceptible to stress-corrosion cracking, such an indication does not ensure complete freedom from cracking, particularly where service stresses are high.

Both applied and residual stresses can lead to failure by stress-corrosion cracking. Susceptibility is largely a function of stress magnitude. Stresses near the yield strength are usually required, but parts have failed under lower stresses. In general, the higher the stress, the weaker the corroding medium must be to cause cracking. The reverse also is true: the stronger the corroding medium, the lower the required stress.

Sources of Stress. Applied stresses result from ordinary service loading or from fabricating techniques such as riveting, bolting, shrink fitting, brazing and welding.

Residual stresses are of two types: differential-strain stresses, resulting from nonuniform plastic strain during cold forming; and differential-thermal-contraction stresses, resulting from nonuniform heating and/or cooling.

Residual stresses induced by nonuniform straining are influenced chiefly by the method of fabrication. In some fabricating processes, it is possible to cold work a metal extensively and yet produce only a low level of residual stress. For example, residual stress in a drawn tube is influenced by die angle and amount of reduction. Wide-angle dies (about 32°) produce higher residual stresses than narrow-angle dies (about 8°). Light reductions yield high residual stresses because only the surface of the alloy is stressed; heavy reductions yield low residual stresses because the region of cold working extends deeper into the metal. Most drawing operations can be planned so that residual stresses are low and susceptibility to stress-corrosion cracking is negligible.

Residual stresses resulting from upsetting, stretching or spinning are harder to evaluate and harder to control by varying tooling and process conditions. For these operations, stress-corrosion cracking can be prevented more effectively by selecting a resistant alloy or by treating the metal after fabrication.

Alloy Composition. Brasses containing less than 15% Zn are highly resistant to stress-corrosion cracking. Phosphorus-deoxidized copper and tough pitch copper rarely exhibit stress-corrosion cracking, even under severe conditions. On the other hand, brasses containing 20 to 40% Zn are highly susceptible to stress-corrosion cracking. Susceptibility increases only slightly as zinc content is increased from 20 to 40%.

There is no indication that other elements commonly added to brasses increase the probability of stress-corrosion cracking. Phosphorus, arsenic, magnesium, tellurium, tin, beryllium and manganese are thought to decrease susceptibility under some conditions. Addition of 1.5% silicon is known to decrease the probability of cracking.

Altering the microstructure cannot make a susceptible alloy totally resistant to stress-corrosion cracking. However, the rapidity with which susceptible alloys crack appears to be affected by grain size and structure. All other factors being equal, rate of cracking increases as grain size increases. The effects of structure on stress-corrosion cracking are not sharply defined, chiefly because they are interrelated with effects of both composition and stress.

Control Measures. Stress-corrosion cracking can be controlled, and sometimes prevented, by (a) selecting copper alloys that have high resistance to cracking (notably those with less than 15% Zn); (b) reducing residual stress to a safe level by thermal stress relief, which usually can be applied without significantly decreasing strength; or (c) altering the environment, such as by changing the predominant chemical species present or introducing a corrosion inhibitor.

Residual and assembly stresses may be eliminated by recrystallization annealing after forming or assembly. Recrystallization annealing cannot be used when the integrity of the structure depends on the higher strength of strain-hardened metal, which always contains a certain amount of residual stress. Thermal stress relief (sometimes called relief annealing) can be specified when the higher strength of a cold worked temper must be retained. Thermal stress relief consists of heating the part for a relatively short time at low temperature. Specific times and temperatures depend on alloy composition, severity of deformation, prevailing stresses and size of load being heated. Usually, time is from 30 min to 1 h and temperature is from 150 to 425 °C (300 to 800 °F). Typical stress-relieving times and temperatures for some of the more common copper alloys follow.

UNS number	Common name	Temperature °C	Temperature °F	Time, h
C22000	Commercial bronze	205	400	1
C26000	Cartridge brass	260	500	1
C28000	Muntz metal	190	375	½
C44300, C44400, C44500	Admiralty metal	300	575	1
C51000, C52400	Phosphor bronze, 5 or 10%	190	375	1
C65500	Silicon bronze	370	700	1
C61300, C61400	Aluminum bronze	400	750	1
C71500	Copper nickel, 30%	425	800	1

The exact thermal treatment should be established by examining specific parts for residual stress, as described in the next section of this article. If such examination indicates that a thermal treatment is insufficient, temperature and/or time should be adjusted until satisfactory results are obtained. Parts in the center of a furnace load may not reach the desired temperature as soon as parts around the periphery. Because of this, it may be necessary to compensate for furnace loading when setting

process controls or to limit the number of parts that can be stress relieved together.

Mechanical methods such as stretching, flexing, bending, straightening between rollers, peening and shot blasting also may be used to reduce residual stresses to a safe level. These methods depend on plastic deformation to either decrease dangerous tensile stresses or convert them to less objectionable compressive stresses.

Corrosion Fatigue

The combined action of corrosion (usually pitting corrosion) and cyclic stress may result in corrosion-fatigue cracking. Like ordinary fatigue cracks, corrosion-fatigue cracks generally propagate at right angles to the maximum tensile stress in the affected region. However, cracks resulting from simultaneous alternating stress and corrosion propagate much more rapidly than cracks caused by alternating stress alone. Also, corrosion-fatigue failure usually involves several parallel cracks, whereas it is rare for more than one crack to be found in a part that has failed as a result of simple fatigue. The cracks shown in Fig. 6 are characteristic of service failures resulting from corrosion fatigue.

Ordinarily, corrosion fatigue can be readily identified by the presence of several cracks emanating from corrosion pits. Cracks not visible to the unaided eye or at low magnification can be made visible by deep etching or plastic deformation or detected by eddy-current inspection. Corrosion-fatigue cracking is often transcrystalline, but there is evidence that certain environments induce intercrystalline fatigue cracking in copper metals.

In addition to effectively resisting corrosion, copper and copper alloys also resist corrosion fatigue in many applications that involve repeated stress and corrosion. These applications include such parts as springs, switches, diaphragms, bellows, aircraft and automotive gasoline and oil lines, tubes for condensers and heat exchangers, and fourdrinier wire for the paper industry.

Copper alloys high in both fatigue limit and resistance to corrosion in the service environment are more likely to have good resistance to corrosion fatigue. Alloys frequently used in applications involving both cyclic stress and corrosion include beryllium coppers, phosphor bronzes, aluminum bronzes and copper nickels.

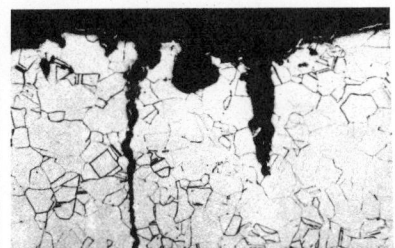

Fig. 6 Typical appearance of corrosion-fatigue cracking

Transcrystalline cracks originating at the base of corrosion pits on the roughened inner surface of a tube; about 150X, etched.

Effect of Alloy Composition

Brasses are basically copper-zinc alloys and are the most widely used group of copper alloys. Resistance of brasses to corrosion by aqueous solutions does not change markedly as long as the zinc content does not exceed about 15%; above 15% Zn, dezincification may occur. Quiescent or slowly moving saline solutions, brackish waters and mildly acidic solutions are environments that often lead to dezincification.

As shown in Fig. 7, resistance to pitting is almost total when zinc content exceeds 15%. But the brasses that resist pitting are severely degraded by dezincification, which causes them to lose a substantial portion of their strength.

Where exposure to sulfur compounds is involved, brasses containing the highest amounts of zinc have the best resistance (see Tables 3 and 4).

Susceptibility to stress-corrosion cracking is significantly affected by zinc content: alloys containing more zinc are more susceptible. Resistance increases substantially as zinc content decreases from 15% to zero; stress-corrosion cracking in commercial coppers is practically unknown.

Elements such as lead, tellurium, beryllium, chromium, phosphorus and manganese have little or no effect on corrosion resistance of coppers and binary copper-zinc alloys. These elements are added to enhance mechanical properties such as machinability, strength and hardness.

Tin-bearing Brasses. Tin additions significantly increase the corro-

Fig. 7 Effect of zinc content on corrosion of brasses

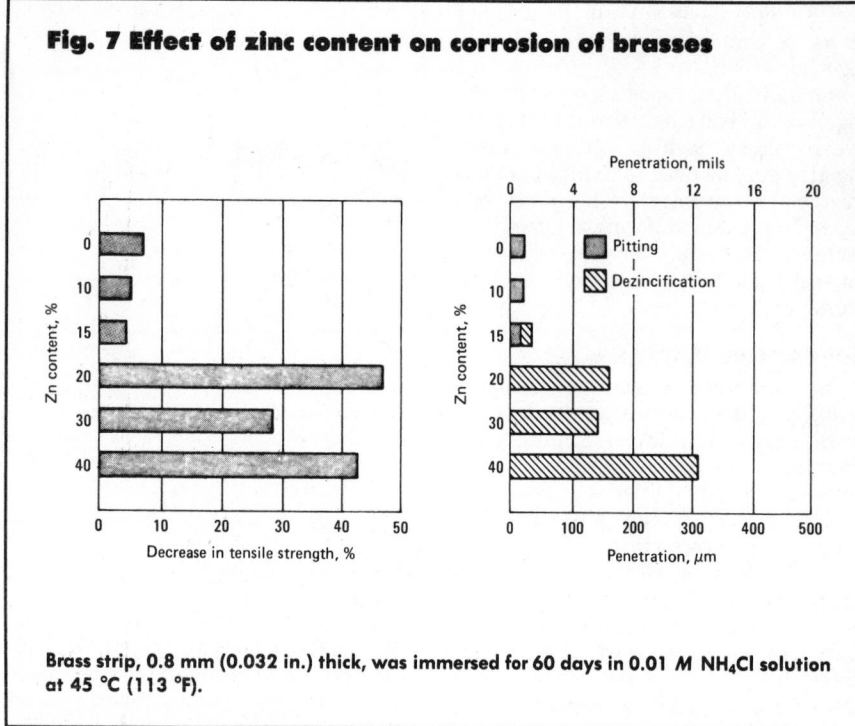

Brass strip, 0.8 mm (0.032 in.) thick, was immersed for 60 days in 0.01 M NH₄Cl solution at 45 °C (113 °F).

Table 4 Corrosion of copper alloys in hot paper-mill vapor containing SO₂ (a)

UNS number	Alloy type	Weight loss, $g/m^2 \cdot d$
⋯	Bronze (90Cu-10Sn)	22.03
C61800	Aluminum bronze	26.42
C51100	Phosphor bronze	28.58
C73200	Nickel silver, 75-20	35.56
C52100	Phosphor bronze, 8% C	39.43
C65800	Silicon bronze	50.22
C77000	Nickel silver, 55-18	63.78
C75200	Nickel silver, 65-18	67.42
⋯	Nickel bronze (88.5Cu-5Sn-5Ni-1.5Si)	70.49

(a) Temperature, 200 to 220 °C (390 to 430 °F); atmosphere, 17 to 18% SO₂ plus 1 to 2% O₂; test duration, mainly 30 days, but some longer.

Table 3 Corrosion of selected copper alloys in cracked oil containing 1.4% S

UNS number	Alloy type	Exposure time, days	Percent loss in tensile strength (a) at: 360 °C (680 °F)	315 °C (600 °F)	285 °C (545 °F)	255 °C (490 °F)
C23000	Red brass, 85%	27	100(b)	100(c)	100	100
C28000	Muntz metal	27	12(b)	7.5(d)	1	1.5
C46400	Naval brass	24	⋯	1.5	0	2
⋯	Uninhibited admiralty metal	27	13(b)	6(c)	3	2
C44400	Antimonial admiralty metal	27	16.5(b)	6(c)	4	2.5
⋯	Aluminum brass	24	⋯	7	16	10
C71500	Copper nickel, 30%	24	⋯	100	100	57
⋯	Silicon bronze, 3%	34	⋯	100	100	100

(a) Specimens 0.8 by 13 mm (0.032 by 0.50 in.) in cross section were exposed at different locations within a high-pressure fractionating column, each location having a characteristic average temperature. (b) 115-day exposure. (c) 26-day exposure. (d) Length of exposure unavailable.

sion resistance of some brasses, especially resistance to dezincification. Prime examples of this effect are two tin-bearing brasses: uninhibited admiralty metal (no active UNS number) and naval brass (C46400). Uninhibited admiralty metal was once widely used to make heat-exchanger tubes; it has largely been replaced by inhibited grades of admiralty metal (C44300, C44400 and C44500), which have even greater resistance to dealloying. Admiralty metal is a variation of cartridge brass (C26000) produced by adding about 1% Sn to the basic 70Cu–30Zn composition. Similarly, naval brass is the alloy resulting from addition of 0.75% Sn to the basic 60Cu–40Zn composition of Muntz metal (C28000).

Aluminum Brasses. An important constituent of the corrosion film on a brass that contains a few percent aluminum in addition to copper and zinc is aluminum oxide, which markedly increases resistance to impingement attack in turbulent, high-velocity saline water. For example, the arsenical aluminum brass C68700 (76Cu-22Zn-2Al) is frequently used for marine condensers and heat exchangers where impingement attack is likely to pose a serious problem. Aluminum brasses are susceptible to dezincification unless they are inhibited, which usually is done by adding 0.02 to 0.10% As.

Inhibited Alloys. Addition of phosphorus, arsenic or antimony (typically, 0.02 to 0.10%) to admiralty metal, naval brass or aluminum brass effectively produces high resistance to dezincification at alpha service temperatures. Inhibited alloys have been used extensively for components such as condenser tubes, which must accumulate several years of continuous service between shutdowns for repair or replacement.

Phosphor Bronzes. Addition of tin and phosphorus to copper produces good resistance to flowing seawater and to most nonoxidizing acids except hydrochloric. Alloys containing 8 to 10% Sn have high resistance to impingement attack. Phosphor bronzes are much less susceptible to stress-corrosion cracking than brasses and are similar to copper in resistance to sulfur attack.

Copper Nickels. C71500 (copper nickel, 30%) has the best general resistance to aqueous corrosion of all the commercially important copper alloys, but C70600 (copper nickel, 10%) is often selected because it offers good resistance at lower cost. Both of these alloys, although well suited to applications in the chemical industry, have been used most extensively for condenser tubes and heat-exchanger tubes in recirculating steam systems. They are superior to coppers and to other copper alloys in resisting acid solutions and are highly resistant to stress-corrosion cracking and impingement corrosion.

Nickel Silvers. The two most common nickel silvers are C75200 (nickel silver, 65-18) and C77000 (nickel silver, 55-18). They have good resistance to corrosion in both fresh and salt waters. Chiefly because their relatively high nickel contents inhibit dezincification, C75200 and C77000 usually are much more resistant to corrosion in saline solutions than are brasses of similar copper content.

Copper-silicon alloys generally have the same corrosion resistance as copper, but higher mechanical properties and superior weldability. These alloys appear to be much better than the common brasses in resistance to stress-corrosion cracking. Silicon bronzes are susceptible to embrittlement by high-pressure steam and should be tested for suitability in the service environment before being specified for components to be used at elevated temperature.

Aluminum bronzes containing 5 to 12% Al have excellent resistance to impingement corrosion and high-temperature oxidation. Aluminum bronzes are used for beater bars and for blades in wood pulp machines because of their ability to withstand mechanical abrasion and chemical attack by sulfite solutions.

In most practical commercial applications, the corrosion characteristics of aluminum bronzes are primarily related to aluminum content. Alloys with up to 8% Al normally have completely face-centered-cubic alpha structures and good resistance to corrosion attack. Alloys C60600, C60800, C61000, C61300, C61400 and C61500 are of this type. As aluminum content increases above 8%, alpha-beta duplex structures appear. The beta phase is a high-temperature phase retained at room temperature on fast cooling from 565 °C (1050 °F) or above. Slow cooling, or long exposure at temperatures from 320 to 565 °C (610 to 1050 °F), tends to decompose the beta phase into a brittle alpha + gamma-2 eutectoid having either a lamellar or a nodular structure. The beta phase is less resistant to corrosion than the alpha phase, and eutectoid structures are even more susceptible to attack.

Depending on specific environmental conditions, beta phase or eutectoid structure in aluminum bronze can be selectively attacked by a mechanism similar to dezincification of brasses. The use of proper quench-and-temper treatments on duplex alloys such as C62400 and C95400 produces a tempered beta structure with reprecipitated acicular alpha crystals, a combination often superior in corrosion resistance to the normal annealed structures.

Iron-rich particles are distributed as small round or rosette particles throughout the structures of aluminum bronzes containing more than about 0.5% Fe. These particles sometimes impart a rusty tinge to the surface, but have no known effect on corrosion rates.

Nickel-aluminum bronzes are more complex in structure, with the introduction of the kappa phase. Nickel appears to alter corrosion characteristics of the beta phase to give greater resistance to dealloying and cavitation-erosion in most liquids. There is some evidence that for C63200, and perhaps for C95800, quench-and-temper treatments yield even greater resistance to dealloying. C95700, a high-manganese cast aluminum bronze, is somewhat inferior in corrosion resistance to C95500 and C95800, which are low in manganese and slightly higher in aluminum.

Aluminum bronzes are generally suitable for service in nonoxidizing mineral acids, such as phosphoric, sulfuric and hydrochloric: organic acids, such as lactic, acetic, or oxalic; neutral salt solutions, such as sodium or potassium chloride; alkalies, such as sodium hydroxide, potassium hydroxide and anhydrous ammonium hydroxide; and various natural waters, including sea, brackish and potable waters. Environments to be avoided include nitric acid, some metallic salts such as ferric chloride and chromic acid, moist chlorinated hydrocarbons, and moist ammonia. Aeration can result in accelerated corrosion in many media that appear compatible.

Exposure under high tensile stress to moist ammonia can result in stress-corrosion cracking. In certain environments, corrosion can lower the fatigue limit to 25 to 50% of the normal atmospheric value.

Protective Coatings

Copper metals resist corrosion in many environments because on initial exposure they react with one or more constituents of the environment, thereby forming an inert surface layer of protective reaction products.

In certain applications, corrosion resistance of copper metals may be increased by applying metallic or organic protective coatings. Provided the coating material is able to resist corrosion adequately, service life may depend on impermeability, continuity, and adhesion to the basis metal. The electropotential relationship of the coating to the basis metal may be important, especially with metallic coatings and at uncoated edges.

Tin, lead and solder, used extensively as coatings, ordinarily are applied by hot dipping, although electroplating is used also.

Tin arrests corrosion caused by sulfur; it is most effective as a coating for copper wire and cable insulated by rubber that contains sulfur. Lead-coated copper is used chiefly for roofing applications where contact with flue gases or other products that contain dilute sulfuric acid is likely. Tin or lead coatings sometimes are applied to copper intended for ordinary atmospheric exposure, but this is done primarily for architectural effect; the atmospheric corrosion resistance of bare copper is excellent in rural, urban, marine and most industrial locations.

Electroplated chromium is used for decoration, for improvement of wear resistance or for reflectivity. Because it is somewhat porous, it is not effective for corrosion protection. Where corrosion protection is important, electroplated nickel is used most often as a protective coating under electroplated chromium.

Clear lacquer sometimes is applied to preserve a bright natural color for decorative reasons. For instance in one particular architectural application, a commercial lacquer specifically developed for use on copper effectively preserved the warm color of a copper roof.

Wrought Copper Alloys for Corrosion Service

By the ASM Committee on Copper and Copper Alloys*

COPPER and its alloys have been used extensively in applications requiring resistance to many different corrosive substances. The commercially important copper alloys vary widely in composition and hence vary significantly in resistance to corrosion. Many of the alloying elements added to copper give it improved corrosion resistance and enhanced mechanical properties compared to those of a related binary alloy or unalloyed copper. The wrought coppers and copper alloys whose compositions are given in Table 1 are noted for their corrosion resistance and are used extensively in chemical process equipment.

Copper and copper alloys are not merely suitable but often superior for many applications included in the following broad classifications:

1 Applications requiring resistance to atmospheric exposure, such as roofing and other architectural uses, hardware, building fronts, grille work, hand rails, butts, lock bodies, doorknobs and kick plates
2 Fresh-water supply lines and plumbing fittings, where superior resistance to corrosion by various

*See page **XI** for committee list.

types of water and soils are important
3 Marine applications—most often fresh-water and seawater supply lines, heat exchangers, condensers, shafting, valve stems and marine hardware—where resistance to seawater, hydrated salt deposits and biofouling from marine organisms are important
4 Heat exchangers and condensers in marine service, in steam power plants and in chemical process applications, and liquid-to-gas or gas-to-gas heat exchangers where either process stream may contain a corrosive contaminant
5 Industrial and chemical-plant process equipment involving exposure to a wide variety of organic and inorganic chemicals

Selection for a Specific Environment

Selection of a suitably resistant material requires consideration of the many factors that influence corrosion. Operating records are the most reliable guidelines as long as the data are accu-

rately interpreted. Some of the information in this article has been collected over a period of 20 years or more. Results of short-term laboratory and field testing also are described, but these data may not be as reliable for solving certain problems. Laboratory corrosion tests often do not duplicate operating factors such as stress, velocity, galvanic coupling, concentration cells, initial surface condition and contamination of the surrounding medium. If damage occurs by pitting, intergranular corrosion or dealloying (as in dezincification), or if a thick adherent scale forms, corrosion rates calculated from a change in weight may be misleading. For these forms of corrosion, estimates of reduction in mechanical strength often are more meaningful. Corrosion fatigue and stress–corrosion cracking also are potential sources of failure that cannot be predicted from routine measurements of weight loss or dimensional change.

Over the years, experience has been the best criterion for selecting the most suitable alloy for a given environment. The Copper Development Association has compiled much field experience in the form of the ratings shown in Table 2. The table should be used only as a

Table 1 Corrosion-resistant wrought coppers and copper alloys

CDA or UNS number	Common name	Nominal composition, %
C11000	Electrolytic tough pitch copper	99.90Cu + Ag(min)
C12200	Phosphorus deoxidized copper, high residual phosphorus	99.90Cu + Ag(min)-0.02P
C22000	Commercial bronze, 90%	90Cu-10Zn
C23000	Red brass, 85%	85Cu-15Zn
C26000	Cartridge brass, 70%	70Cu-30Zn
C28000	Muntz metal, 60%	60Cu-40Zn
C36000	Free-cutting brass	61.5Cu-35.5Zn-3Pb
C38500	Architectural bronze	57Cu-40Zn-3Pb
C44300	Arsenical admiralty metal	71Cu-28Zn-1Sn-0.06As
C44400	Antimonial admiralty metal	71Cu-28Zn-1Sn-0.06Sb
C44500	Phosphorized admiralty metal	71Cu-28Zn-1Sn-0.06P
C46400	Naval brass	60Cu-39.25Zn-0.75Sn
C46500	Arsenical naval brass	60Cu-39.2Zn-0.75Sn-0.06As
C46600	Antimonial naval brass	60Cu-39.2Zn-0.75Sn-0.06Sb
C46700	Phosphorized naval brass	60Cu-39.2Zn-0.75Sn-0.06P
C51000	Phosphor bronze, 5%A	95Cu-5Sn-0.2P
C52100	Phosphor bronze, 8%C	92Cu-8Sn-0.2P
C61300	Aluminum bronze	91Cu-7Al-2.5Fe-0.5Sn
C63700	Aluminum bronze	90.5Cu-7.5Al-1.7Si
C68700	Arsenical aluminum brass	77.5Cu-20.5Zn-2Al-0.06As
C65100	Low-silicon bronze B	98.5Cu-1.5Si
C65500	High-silicon bronze A	97Cu-3Si
C70600	Copper nickel, 10%	88.7Cu-1.3Fe-10Ni
C71500	Copper nickel, 30%	70Cu-0.6Fe-30Ni
C75200	Nickel silver 65-18	65Cu-17Zn-18Ni

guide: small changes in the environmental conditions sometimes change the performance of a given alloy from "suitable" to "not suitable".

Whenever there is a lack of operating experience, whenever reported test conditions do not closely match conditions for which alloy selection is being made, and whenever there is doubt as to the applicability of published data, it is always best to conduct an independent test program. Field tests are the most reliable. Laboratory tests can be equally valuable, but only if operating conditions are precisely defined and then accurately simulated in the laboratory. Long-term tests generally are preferred because the reaction that dominates the initial stages of corrosion may be quite different from the reaction that dominates later on. If short-term tests must be used as the basis for alloy selection, the test program should be supplemented with field tests so that the laboratory results can be re-evaluated in light of true operating experience.

Erroneous conclusions based on laboratory results also may be reached by inaccurately measuring corrosion damage, especially when corrosion is slight. It has become common practice to express test results in terms of penetration or average reduction in metal thickness, even when corrosion was actually measured by weight loss. Weight-loss or average-penetration data are valid only when corrosion is uniform. When corrosion occurs predominantly by pitting or some other localized form, or when corrosion is intergranular or involves formation of a thick, adherent scale, direct measurement of the extent of corrosion provides the most reliable information. (A common technique is to measure the maximum depth of penetration observed on a metallographic cross section through the region of interest.) Statistical averaging of repeated measurements on one or more specimens may or may not be warranted. Despite the deficiencies in laboratory testing, information gained in this way serves as a useful starting point for alloy selection. Operating experience may later indicate the need for a more discriminating selection.

Atmospheric Exposure

Comprehensive tests conducted over a 20-year period under the supervision of The American Society for Testing and Materials, plus many service records, have confirmed the suitability of copper and copper alloys for atmospheric exposure (see Table 3). Copper and copper alloys resist corrosion by industrial, marine and rural atmospheres except atmospheres containing ammonia or certain other agents where stress-corrosion cracking has been observed in high-zinc alloys (greater than 20% Zn). The copper metals most widely used in atmospheric exposure are C11000, C22000, C23000, C38500 and C75200. C11000 is a most satisfactory material for roofing, flashings, gutters and downspouts.

Colors of different copper alloys are often important in architectural applications, and color may be the main criterion for selection of a specific alloy. After surface preparation such as sanding or polishing, different copper alloys vary in color from silver to yellow to gold to reddish shades. Different alloys having the same initial color may show differences in color after weathering under similar conditions. Therefore, alloys having the same or nearly the same composition are usually used together for consistent appearance in a specific structure.

Soils

Copper, zinc, lead and iron are the metals most commonly used in underground construction. Data compiled by the National Bureau of Standards (Circular 579) compare the behavior of these materials in soils of the following four types: (a) well-aerated acid soils low in soluble salts, (b) poorly aerated soils, (c) alkaline soils high in soluble salts (Docas clay), and (d) soils high in sulfides.

Corrosion data as a function of time for copper, iron, lead and zinc exposed to these four types of soil are given in Fig. 1. Copper shows high resistance to corrosion by these soils, which are representative of most soils found in the United States. Where local soil conditions are unusually corrosive, it may be necessary to employ some means of protection, such as cathodic protection, neutralizing backfill (limestone, for example), protective coating or wrapping.

For many years, the National Bureau of Standards has conducted studies on corrosion of underground structures to determine the specific behavior of metals and alloys when exposed for long periods in a wide range of soils. Results indicate that tough pitch coppers, deoxidized coppers, silicon bronzes and low-zinc brasses behave essentially alike. Soils containing cinders with high concentrations of sulfides, chlorides, or hydrogen ions cor-

Table 2 Corrosion ratings of copper and copper alloys in various corrosive media

This table is intended to serve only as a general guide to the behavior of copper and copper alloys in corrosive environments. It is impossible to cover in a simple tabulation the performance of a material for all possible variations of temperature, concentration, velocity, impurity content, degree of aeration and stress. The ratings are based on general performance; they should be used with caution, and then only for the purpose of screening candidate alloys.

The letters E, G, F and P have the following significance:

E—Excellent. Resists corrosion under almost all conditions of service.

G—Good. Some corrosion will take place, but satisfactory service can be expected under all but the most severe conditions.

F—Fair. Corrosion rates are higher than for the "G" classification, but the metal can be used if needed for a property other than corrosion resistance and if either the amount of corrosion does not cause excessive maintenance expense or the effects of corrosion can be lessened, such as by use of coatings or inhibitors.

P—Poor. Corrosion rates are high, and service is generally unsatisfactory.

Corrosive medium	Coppers	Low-zinc brasses	High-zinc brasses	Special brasses	Phosphor bronzes	Aluminum bronzes	Silicon bronzes	Copper nickels	Nickel silvers
Acetate solvents	E	E	G	E	E	E	E	E	E
Acetic acid(a)	E	E	P	P	E	E	E	E	G
Acetone	E	E	E	E	E	E	E	E	E
Acetylene(b)	P	P(b)	P	P	P	P	P	P	P
Alcohols(a)	E	E	E	E	E	E	E	E	E
Aldehydes	E	E	F	F	E	E	E	E	E
Alkylamines	G	G	G	G	G	G	G	G	G
Alumina	E	E	E	E	E	E	E	E	E
Aluminum chloride	G	G	P	P	G	G	G	G	G
Aluminum hydroxide	E	E	E	E	E	E	E	E	E
Aluminum sulfate and alum	G	G	P	G	G	G	G	E	G
Ammonia, dry	E	E	E	E	E	E	E	E	E
Ammonia, moist(c)	P	P	P	P	P	P	P	F	P
Ammonium chloride(c)	P	P	P	P	P	P	P	F	P
Ammonium hydroxide(c)	P	P	P	P	P	P	P	F	P
Ammonium nitrate(c)	P	P	P	P	P	P	P	F	P
Ammonium sulfate(c)	F	F	P	P	F	F	F	G	F
Aniline and aniline dyes	F	F	F	F	F	F	F	F	F
Asphalt	E	E	E	E	E	E	E	E	E
Atmosphere: Industrial(c)	E	E	E	E	E	E	E	E	E
Atmosphere: Marine	E	E	E	E	E	E	E	E	E
Atmosphere: Rural	E	E	E	E	E	E	E	E	E
Barium carbonate	E	E	E	E	E	E	E	E	E
Barium chloride	G	G	F	F	G	G	G	G	G
Barium hydroxide	E	E	G	E	E	E	E	E	E
Barium sulfate	E	E	E	E	E	E	E	E	E
Beer(a)	E	E	G	E	E	E	E	E	E
Beet-sugar syrup(a)	E	E	G	E	E	E	E	E	E
Benzene, benzine, benzol	E	E	E	E	E	E	E	E	E
Benzoic acid	E	E	E	E	E	E	E	E	E
Black liquor, sulfate process	P	P	P	P	P	P	P	G	P
Bleaching powder (wet)	G	G	P	G	G	G	G	G	G
Borax	E	E	E	E	E	E	E	E	E
Bordeaux mixture	E	E	G	E	E	E	E	E	E
Boric acid	E	E	G	E	E	E	E	E	E
Brines	G	G	P	G	G	G	G	E	E
Bromine, dry	E	E	E	E	E	E	E	E	E
Bromine, moist	G	G	P	F	G	G	G	G	G
Butane(d)	E	E	E	E	E	E	E	E	E
Calcium bisulfate	G	G	P	G	G	G	G	G	G
Calcium chloride	G	G	F	G	G	G	G	G	G
Calcium hydroxide	E	E	G	E	E	E	E	E	E
Calcium hypochlorite	G	G	P	G	G	G	G	G	G
Cane-sugar syrup(a)	E	E	E	E	E	E	E	E	E
Carbolic acid (phenol)	F	G	P	G	G	G	G	G	G
Carbonated beverages(a)(e)	E	E	E	E	E	E	E	E	E
Carbon dioxide, dry	E	E	E	E	E	E	E	E	E
Carbon dioxide, moist(a)(e)	E	E	E	E	E	E	E	E	E
Carbon tetrachloride (dry)	E	E	E	E	E	E	E	E	E
Carbon tetrachloride (moist)	G	G	F	G	E	E	E	E	E
Castor oil	E	E	E	E	E	E	E	E	E
Chlorine, dry(f)	E	E	E	E	E	E	E	E	E
Chlorine, moist	F	F	P	F	F	F	F	G	F
Chloracetic acid	G	F	P	F	G	G	G	G	G
Chloroform, dry	E	E	E	E	E	E	E	E	E
Chromic acid	P	P	P	P	P	P	P	P	P
Citric acid(a)	E	E	F	E	E	E	E	E	E
Copper chloride	F	F	P	F	F	F	F	F	F
Copper nitrate	F	F	P	F	F	F	F	F	F
Copper sulfate	G	G	P	G	G	G	G	E	G
Corn oil(a)	E	E	G	E	E	E	E	E	E
Cottonseed oil(a)	E	E	G	E	E	E	E	E	E
Creosote	E	E	G	E	E	E	E	E	E
Dowtherm "A"	E	E	E	E	E	E	E	E	E
Ethanol amine	G	G	G	G	G	G	G	G	G
Ethers	E	E	E	E	E	E	E	E	E
Ethyl acetate (esters)	E	E	G	E	E	E	E	E	E
Ethylene glycol	E	E	G	E	E	E	E	E	E
Ferric chloride	P	P	P	P	P	P	P	P	P
Ferric sulfate	P	P	P	P	P	P	P	P	P
Ferrous chloride	G	G	P	G	G	G	G	G	G
Ferrous sulfate	G	G	P	G	G	G	G	G	G
Formaldehyde (aldehydes)	E	E	G	E	E	E	E	E	E
Formic acid	G	G	P	F	G	G	G	G	G
Freon, dry	E	E	E	E	E	E	E	E	E
Freon, moist									
Fuel oil, light	E	E	E	E	E	E	E	E	E
Fuel oil, heavy	E	E	G	E	E	E	E	E	E
Furfural	E	E	F	E	E	E	E	E	E
Gasoline	E	E	E	E	E	E	E	E	E
Gelatin(a)	E	E	E	E	E	E	E	E	E
Glucose(a)	E	E	E	E	E	E	E	E	E
Glue	E	E	G	E	E	E	E	E	E
Glycerin	E	E	G	E	E	E	E	E	E
Hydrobromic acid	F	F	P	F	F	F	F	F	F
Hydrocarbons	E	E	E	E	E	E	E	E	E
Hydrochloric acid (muriatic)	F	F	P	F	F	F	F	F	F
Hydrocyanic acid, dry	E	E	E	E	E	E	E	E	E
Hydrocyanic acid, moist	P	P	P	P	P	P	P	P	P
Hydrofluoric acid, anhydrous	G	G	P	G	G	G	G	G	G
Hydrofluoric acid, hydrated	F	F	P	F	F	F	F	F	F

(Table 2 continued on next page.)

Table 2 (continued)

Corrosive medium	Coppers	Low-zinc brasses	High-zinc brasses	Special brasses	Phosphor bronzes	Aluminum bronzes	Silicon bronzes	Copper nickels	Nickel silvers
Hydrofluosilicic acid	G	G	P	G	G	G	G	G	G
Hydrogen(d)	E	E	E	E	E	E	E	E	E
Hydrogen peroxide up to 10%	G	G	F	G	G	G	G	G	G
Hydrogen peroxide over 10%	P	P	P	P	P	P	P	P	P
Hydrogen sulfide, dry	E	E	E	E	E	E	E	E	E
Hydrogen sulfide, moist	P	P	F	F	P	P	P	F	F
Kerosine	E	E	E	E	E	E	E	E	E
Ketones	E	E	E	E	E	E	E	E	E
Lacquers	E	E	E	E	E	E	E	E	E
Lacquer thinners (solvents)	E	E	E	E	E	E	E	E	E
Lactic acid(a)	E	E	F	E	E	E	E	E	E
Lime	E	E	E	E	E	E	E	E	E
Lime sulfur	P	P	F	F	P	P	P	F	F
Linseed oil	G	G	G	G	G	G	G	G	G
Lithium compounds	G	G	P	F	G	G	G	E	E
Magnesium chloride	G	G	F	F	G	G	G	G	G
Magnesium hydroxide	E	E	G	E	E	E	E	E	E
Magnesium sulfate	E	E	G	E	E	E	E	E	E
Mercury or mercury salts	P	P	P	P	P	P	P	P	P
Milk(a)	E	E	G	E	E	E	E	E	E
Molasses	E	E	G	E	E	E	E	E	E
Natural gas(d)	E	E	E	E	E	E	E	E	E
Nickel chloride	F	F	P	F	F	F	F	F	F
Nickel sulfate	F	F	P	F	F	F	F	F	F
Nitric acid	P	P	P	P	P	P	P	P	P
Oleic acid	G	G	F	G	G	G	G	G	G
Oxalic acid(g)	E	E	P	P	E	E	E	E	E
Oxygen(h)	E	E	E	E	E	E	E	E	E
Palmitic acid	G	G	F	G	G	G	G	G	G
Paraffin	E	E	E	E	E	E	E	E	E
Phosphoric acid	G	G	P	F	G	G	G	G	G
Picric acid	P	P	P	P	P	P	P	P	P
Potassium carbonate	E	G	E	E	E	E	E	E	E
Potassium chloride	G	G	P	F	G	G	G	E	E
Potassium cyanide	P	P	P	P	P	P	P	P	P
Potassium dichromate (acid)	P	P	P	P	P	P	P	P	P
Potassium hydroxide	G	G	F	G	G	G	G	E	E
Potassium sulfate	E	E	G	E	E	E	E	E	E
Propane(d)	E	E	E	E	E	E	E	E	E
Rosin	E	E	E	E	E	E	E	E	E
Sea water	G	G	F	E	G	E	G	E	E
Sewage	E	E	F	E	E	E	E	E	E
Silver salts	P	P	P	P	P	P	P	P	P
Soap solution	E	E	E	E	E	E	E	E	E
Sodium bicarbonate	E	E	G	E	E	E	E	E	E
Sodium bisulfate	G	G	F	G	G	G	G	E	E
Sodium carbonate	E	E	G	E	E	E	E	E	E
Sodium chloride	G	G	P	F	G	G	G	E	E
Sodium chromate	E	E	E	E	E	E	E	E	E
Sodium cyanide	P	P	P	P	P	P	P	P	P
Sodium dichromate (acid)	P	P	P	P	P	P	P	P	P
Sodium hydroxide	G	G	F	G	G	G	G	E	E
Sodium hypochlorite	G	G	P	G	G	G	G	G	G
Sodium nitrate	G	G	P	F	G	G	G	E	E
Sodium peroxide	F	F	P	F	F	F	F	G	G
Sodium phosphate	E	E	G	E	E	E	E	E	E
Sodium silicate	E	E	G	E	E	E	E	E	E
Sodium sulfate	E	E	G	E	E	E	E	E	E
Sodium sulfide	P	P	F	F	P	P	P	F	F
Sodium thiosulfate	P	P	F	F	P	P	P	F	F
Steam	E	E	F	E	E	F	E	E	E
Stearic acid	E	E	F	E	E	E	E	E	E
Sugar solutions	E	E	G	E	E	E	E	E	E
Sulfur, solid	G	G	E	G	G	G	G	E	G
Sulfur, molten	P	P	P	P	P	P	P	P	P
Sulfur chloride (dry)	E	E	E	E	E	E	E	E	E
Sulfur chloride (moist)	P	P	P	P	P	P	P	P	P
Sulfur dioxide (dry)	E	E	E	E	E	E	E	E	E
Sulfur dioxide (moist)	G	G	P	G	G	G	G	F	F
Sulfur trioxide (dry)	E	E	E	E	E	E	E	E	E
Sulfuric acid 80-95%(j)	G	G	P	F	G	G	G	G	G
Sulfuric acid 40-80%(j)	F	F	P	F	F	F	F	F	F
Sulfuric acid 40%(j)	G	G	P	F	G	G	G	G	G
Sulfurous acid	G	G	P	G	G	G	G	F	F
Tannic acid	E	E	E	E	E	E	E	E	E
Tartaric acid(a)	E	E	G	E	E	E	E	E	E
Toluene	E	E	E	E	E	E	E	E	E
Trichloracetic acid	G	G	P	F	G	G	G	G	G
Trichlorethylene (dry)	E	E	E	E	E	E	E	E	E
Trichlorethylene (moist)	G	G	F	G	E	E	E	E	E
Turpentine	E	E	E	E	E	E	E	E	E
Varnish	E	E	E	E	E	E	E	E	E
Vinegar(a)	E	E	P	F	E	E	E	E	G
Water, acidic mine	F	F	P	F	G	F	F	P	F
Water, potable	E	E	G	E	E	E	E	E	E
Water, condensate(c)	E	E	E	E	E	E	E	E	E
Wetting agents(k)	E	E	E	E	E	E	E	E	E
Whiskey(a)	E	E	E	E	E	E	E	E	E
White water	G	G	G	E	E	E	E	E	E
Zinc chloride	G	G	P	G	G	G	G	G	G
Zinc sulfate	E	E	P	E	E	E	E	E	E

(a) Copper and copper alloys are resistant to corrosion by most food products. Traces of copper may be dissolved and affect taste or color of the products. In such cases, copper alloys often are tin coated. (b) Acetylene forms an explosive compound with copper when moisture or certain impurities are present and the gas is under pressure. Alloys containing less than 65% copper are satisfactory; when the gas is not under pressure, other copper alloys are satisfactory. (c) Precautions should be taken to avoid stress-corrosion cracking. (d) At elevated temperatures, hydrogen will react with tough pitch copper, causing failure by embrittlement. (e) Where air is present, corrosion rate may be increased. (f) Below 150 °C (300 °F), corrosion rate is very low; above this temperature, corrosion is appreciable and increases rapidly as the temperature rises. (g) Aeration and elevated temperature may increase corrosion rate substantially. (h) Excessive oxidation may begin above 120 °C (250 °F). If moisture is present, oxidation may begin at lower temperatures. (j) Use of high-zinc brasses should be avoided in acids because of the likelihood of rapid corrosion by dezincification. Copper, low-zinc brasses, phosphor bronzes, silicon bronzes, aluminum bronzes and copper nickels offer good resistance to corrosion by hot and cold dilute sulfuric acid and to corrosion by cold concentrated sulfuric acid. Intermediate concentrations of sulfuric acid sometimes are more corrosive to copper alloys than either concentrated or dilute acid. Concentrated sulfuric acid may be corrosive at elevated temperatures due to breakdown of acid and formation of metallic sulfides and sulfur dioxide, which cause localized pitting. Tests indicate that copper alloys may undergo pitting in 90 to 95% sulfuric acid at about 50 °C (122 °F), in 80% acid at about 70 °C (160 °F), and in 60% acid at about 100 °C (212 °F). (k) Wetting agents may increase corrosion rates of copper and copper alloys slightly to substantially when carbon dioxide or oxygen is present, by preventing formation of a film on the metal surface and by combining (in some instances) with the dissolved copper to produce a green, insoluble compound.

Table 3 Atmospheric corrosion of selected copper alloys

	Corrosion rates(a) at indicated locations												
	Altoona, PA		New York, NY		Key West, FL		La Jolla, CA		State College, PA		Phoenix, AZ	
Alloy	μm/yr	mils/yr	μm/yr	mils/yr	μm/yr	mils/yr	μm/yr	mils/yr	μm/yr	mils/yr	μm/yr	mils/yr
C11000	1.40	0.055	1.38	0.054	0.56	0.022	1.27	0.050	0.43	0.017	0.13	0.005
C12000	1.32	0.052	1.22	0.048	0.51	0.020	1.42	0.056	0.36	0.014	0.08	0.003
C23000	1.88	0.074	1.88	0.074	0.56	0.022	0.33	0.013	0.46	0.018	0.10	0.004
C26000	3.05	0.120	2.41	0.095	0.20	0.008	0.15	0.006	0.46	0.018	0.10	0.004
C52100	2.24	0.088	2.54	0.100	0.71	0.028	2.31	0.091	0.33	0.013	0.13	0.005
C61000	1.63	0.064	1.60	0.063	0.10	0.004	0.15	0.006	0.25	0.010	0.51	0.002
C65500	1.65	0.065	1.73	0.068	...	...	1.38	0.054	0.51	0.020	0.15	0.006
C44200	2.13	0.084	2.51	0.099	...	...	0.33	0.013	0.53	0.021	0.10	0.004
70Cu-29Ni-1Sn(b) ...	2.64	0.104	2.13	0.084	0.28	0.011	0.36	0.014	0.48	0.019	0.10	0.004

(a) Derived from 20-yr exposure tests. Types of atmospheres: Altoona, industrial; New York City, industrial marine; Key West, tropical rural marine; La Jolla, humid marine; State College, northern rural; Phoenix, dry rural. (b) This alloy is obsolete, but it indicates the corrosion resistance expected of C71500.

Fig. 1 Corrosion of copper, iron, lead and zinc in four different types of soil

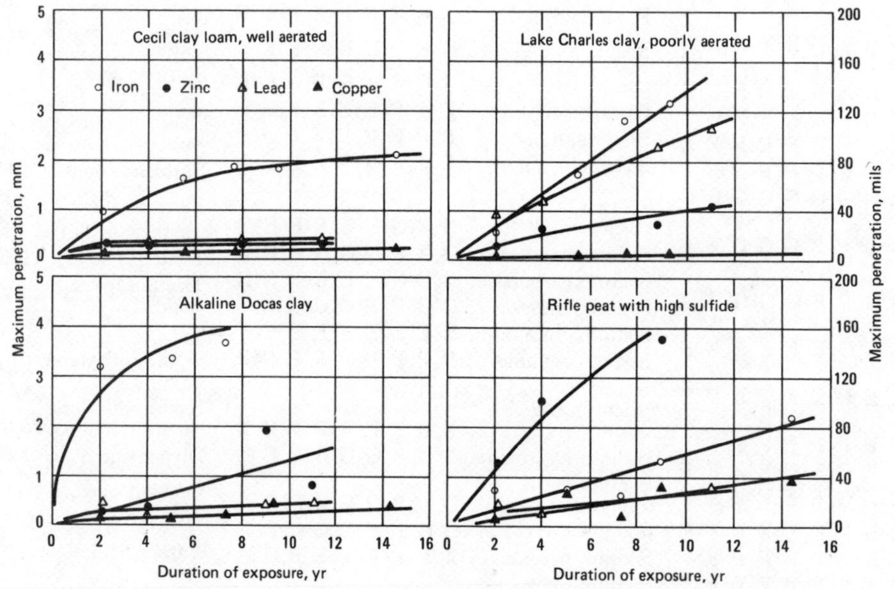

rode these materials. In this type of contaminated soil, corrosion rates of copper-zinc alloys containing more than about 22% zinc increase as zinc content increases. Corrosion generally results from dezincification. In soils that contain only sulfides, corrosion rates of the copper-zinc alloys decline with increasing zinc content, and no dezincification occurs. Although not included in these tests, inhibited admiralty metals would offer significant resistance to dezincification.

Exposure to Fresh Water

Copper is used extensively for handling fresh water. Copper tube in the K-gage range with flared fittings was designed for underground water service, and along with type L tube has now become standard for this application.

The greatest single application of copper tube is for hot- and cold-water distribution lines in homes and other buildings, although considerable quantities are also used in heating lines (including radiant heating lines for homes), drain tubes and fire safety systems.

Copper. Minerals in water combine with dissolved carbon dioxide and oxygen and react with copper to form a protective film. Therefore, the rate of corrosion is low—5 to 25 μm/yr (0.2 to 1.0 mil/yr) in most exposures. In distilled water or very soft water, protective films are less likely to form; therefore, the corrosion rate may vary from less than 2.5 μm/yr (0.1 mil/yr) to 125 μm/yr (5 mils/yr) or more, depending on oxygen and carbon dioxide contents.

Copper-Zinc Alloys. Corrosion resistance of the brasses is good in unpolluted fresh water—normally 2.5 to 25 μm/yr (0.1 to 1.0 mil/yr). In nonscaling water containing carbon dioxide and oxygen, corrosion rates are somewhat higher. Uninhibited brasses of high zinc content (35 to 40% Zn) are subject to dezincification when used with stagnant or slowly moving brackish or slightly acid waters. On the other hand, inhibited admiralty metal and brasses containing 15% zinc or less are highly resistant to dezincification and are used very successfully in these waters. Inhibited yellow brasses are widely used for dezincification resistant service in Europe and are gaining acceptance in North America. C68700 (arsenical aluminum brass, an inhibited 77Cu-21Zn-2Al alloy) has been used successfully for condenser and heat-exchanger tubes.

Copper nickels generally have corrosion rates less than 25 μm/yr (1 mil/yr) in unpolluted water and are sometimes used to resist impingement attack where severe velocity and entrained-air conditions cannot be overcome by changes in either operating conditions or equipment design.

Copper-silicon alloys (silicon bronzes) also have excellent corrosion resistance and for these alloys the amount of dissolved oxygen in the water does not significantly influence corrosion. If carbon dioxide is also present, there is an increase in corrosion rate (but not an excessive increase), particularly at temperatures above 60 °C (140 °F). Corrosion rates for the silicon bronzes are similar to those for copper.

Copper-Aluminum Alloys. The aluminum bronzes have been used in many waters, from potable water to brackish water to seawater. Softened waters are usually more corrosive than hard waters. C61300 and C63200 are used in cooling tower hardware where the make-up water is sewage effluent. Aluminum bronzes resist oxidation and impingement corrosion because of the aluminum in the surface film.

Exposure to Steam. Copper and copper alloys are resistant to attack by pure steam, but if much carbon dioxide, oxygen or ammonia is present, the condensate is corrosive. Even though wet steam at high velocities can cause severe impingement attack, copper alloys are used extensively in condensers and heat exchangers. Copper alloys also are used for feedwater heaters, although their use in such applications is somewhat limited because of their rapid decline in strength and creep resistance at moderately elevated temperatures. Copper nickels are the preferred copper alloys for the higher temperatures and pressures.

Use of copper in systems handling hot water and steam is limited by working pressures of tubes and joints. For example, copper tube ¼ to 1 in. in nominal diameter joined with 50Sn-50Pb solder may be used at 120 °C (250 °F) and 585 kPa (85 psi). The working pressure at this temperature in tube of the same size can be increased to 1380 kPa (200 psi) when the system is joined using 95Sn-5Sb solder. When the joining material is a silver-base brazing alloy having a melting point above 540 °C (1000 °F), working pressure at 120 °C for tube in this size range can be increased to 2070 kPa (300 psi). A few copper alloys have shown a tendency to fail by stress-corrosion cracking when highly stressed and exposed to steam. Alpha aluminum bronzes that do not contain tin are among the susceptible alloys.

Steam condensate that has been properly treated so that it is relatively free of noncondensable gases, as in a power-generating station, is relatively noncorrosive to copper and copper alloys. Rates of attack in most such exposures are less than 2.5 μm/yr (0.1 mil/yr). Copper and its alloys are not attacked by condensate that contains a significant amount of oil (such as condensate from a reciprocating steam engine).

The rate of attack is significantly increased by dissolved carbon dioxide or oxygen, or both. For example, condensate containing 4.6 ppm oxygen and 14 ppm carbon dioxide, having a pH of 5.5 at 68 °C (154 °F), caused an average penetration of 175 to 350 μm/yr (7 to 14 mils/yr) when in contact with C12200 (phosphorus-deoxidized copper), C14200 (arsenical copper), C23000 (red brass), C44300 to C44500 (admiralty metal) and C71000 (copper nickel, 20%).

Steel tested under the same conditions was penetrated at about twice the rate given for the copper alloys listed above, whereas tin-coated copper proved much more resistant and was attacked at a rate of less than 25 μm/yr (1 mil/yr).

Steps that may be taken to attain the best life in condensate systems are (a) ensure that tubes are installed with enough slope to allow proper drainage, (b) reduce the quantity of corrosive agents (usually carbon dioxide and oxygen) at the source by either mechanical or chemical treatment of feedwater, or (c) chemically treat the steam.

The quantity of ammonia present, particularly in a condensing system, significantly influences service life of copper alloys. Boilerhouse operators should be cautioned about using ammonia to make boiler water more alkaline. This practice can lead to contamination and subsequent corrosion or stress-corrosion cracking in condensers and heat exchangers.

Salt Water

An important use of copper alloys is in handling seawater in ships and tidewater power stations. Copper itself, although fairly useful, is usually less resistant than C44300 to C44500 (inhibited admiralty metal), C61300 (aluminum bronze), C68700 (aluminum brass), C70600 (copper nickel, 10%) or C71500 (copper nickel, 30%). The superior performance of these alloys is partly the result of their inherent insolubility in seawater, but more because of their ability to form films of corrosion products that resist erosion by turbulently flowing seawater carrying entrained air. The velocity limits of copper and some copper alloy tubes in clean seawater are tabulated below:

Alloy	Performance rating	Velocity limits	
		m/s	ft/s
Copper	Fair	0.9	3
C44300, C44400, C44500	Good	1.8	6
C60800, C61300	Excellent	2.7	9
C68700	Excellent	2.4	8
C65100, C65500	Fair	0.9	3
C70600	Excellent	3.0	10
C71500	Excellent	4.5	15

Copper nickels are superior to copper and other copper alloys in resisting erosion in high-velocity salt water. Addition of iron to the copper nickels (or to the fluid stream) significantly improves erosion resistance. For this reason, standard 10% and 30% copper nickels contain small amounts of iron. Iron can be added to the water by treating it with ferrous sulfate or by placing steel waster plates in the system. Corrosion of copper alloys in a well-controlled desalting plant is described in a paper by A. Cohen and P. F. George: "Copper Alloys in the Desalting Environment" (NACE, 1974).

Corrosion rates of copper and its alloys in relatively quiescent seawater are typically less than 50 μm/yr (2 mils/yr). Copper nickels, aluminum bronzes and aluminum brass often show corrosion rates of less than 25 μm/yr (1 mil/yr). The instantaneous corrosion rate of all copper alloys tends to decrease as duration of exposure increases. Localized corrosion (pitting), with penetration rates on the order of 125 μm/yr (5 mils/yr), can occur under deposits or debris, or adjacent to crevices. These higher corrosion rates will also be seen if the velocity limits given above are exceeded. Plug-type dealloying of high-zinc brasses and certain aluminum bronzes can occur at rates often exceeding 250 μm/yr (10 mils/yr).

Biofouling. The copper alloys as a group tend to resist fouling by marine organisms, although the degree of resistance varies among the various alloys. The alloys most resistant to biofouling are those containing more than 85% copper (except the aluminum bronzes) and are the ones usually selected when this property is of prime importance. In order to maintain fouling resistance, copper alloys must be allowed to corrode freely. Traditionally, the minimum corrosion rate thought necessary to provide fouling resistance was about 25 μm/yr (1 mil/yr). However, copper nickels resist fouling when the instantaneous corrosion rate is on the order of 2.5 μm/yr (0.1 mil/yr), which implies that the corrosion-product film itself may resist fouling. The corrosion rates of copper alloys are such that they can protect only themselves from fouling. They do not release enough copper ions into seawater to prevent adjacent noncopper surfaces from becoming fouled.

Heat Exchangers and Condensers

The choice of material for condenser and heat-exchanger tubes requires a survey of service conditions, examination of tube previously used and evaluation of its service life, and a review of the type, form and location of corrosion experienced in the unit or in similar units. Types of water and operating conditions vary widely, and any estimate of probable tube performance must be based on specific operating factors. Tubes of the various alloys discussed in this section have been found to give satisfactory and economical performance for the services described.

Inhibited admiralty metal (C44300, C44400 and C44500) has good corrosion resistance and is used extensively for tubes in a variety of services, especially steam condensers cooled with fresh, salt or brackish water. Admiralty metal tubes also are used for heat exchangers in oil refineries, where corrosion from sulfur compounds and

contaminated water may be very severe, and for feedwater heaters and heat-exchanger equipment and in other industrial processes. Admiralty metal tubes often are used in equipment operating at temperatures of 200 °C (400 °F) or higher. The 0.02 to 0.06% of phosphorus, antimony or arsenic in inhibited admiralty metal markedly increases dezincification resistance.

Inhibited aluminum brass (C68700) resists the action of high-velocity salt and brackish water and is commonly used for condenser tubes. The outstanding characteristic of C68700 is its high resistance to impingement attack. Tubes of this alloy are frequently recommended for use in marine and land power stations where cooling-water velocities are high and where inhibited admiralty metal tubes have failed because of impingement attack.

Aluminum Bronzes. Tube sheets made of C61300 and C63200 have been specified for coastal power station condensers. C61300 also is used for emergency raw seawater cooling system piping in coastal nuclear power plants.

Copper nickel, 10% (C70600) has shown excellent resistance to impingement attack; it appears to be inferior only to copper nickel, 30%. It is also highly resistant to stress-corrosion cracking. This alloy is suitable for marine condenser-tube installations in place of aluminum brass, especially where higher water velocities are encountered.

Copper nickel, 30% (C71500) has, in general, the best resistance of any of the copper alloys to impingement attack and to corrosion by most acids and waters. It is being used in increasing quantities under severely corrosive conditions where longer life is desired than that given by other copper alloys. It is used by the U.S. Navy for most shipboard condensers and heat exchangers.

Phosphorus-deoxidized coppers (C12000 to C12300) are used extensively in beet-sugar and cane sugar refineries for condensers and evaporators.

Deoxidized coppers are standard materials for use in the refrigeration industry, and for transferring heat from steam to water or air, because of their excellent resistance to corrosion by fresh water and their high thermal conductivities.

Bimetal tubes sometimes are used to meet severe corrosion problems not handled adequately by tubes of a single metal or alloy. Two tubes of different

alloys, one inside the other, form one integral tube. Copper may be the inner or outer layer, depending on the application.

Drain Tubes. Copper has been used successfully for waste and vent lines in drains. The first such installations were made in the mid 1930's and since then many municipalities have approved the use of copper drain lines. Development of "Sovent" fittings now enables construction of a single-stack drain system in high-rise buildings instead of the two-stack system formerly used.

Acids

Copper is widely used for industrial equipment handling acid solutions. A fairly definite separation exists between those acids that can be handled by copper and those that cannot. In general, copper alloys are successfully used with nonoxidizing acids such as acetic, sulfuric, hydrochloric and phosphoric, as long as the concentration of oxidizing agents such as dissolved oxygen (air) and ferric or bichromate ions is low. Broadly speaking, a thoroughly agitated or stirred solution, or one into which a stream of air has been bubbled, approaches air saturation and thus is not a suitable acid medium for copper. Acids that are oxidizing agents in themselves (such as nitric, sulfurous and not concentrated sulfuric acids, and acids carrying oxidizing agents such as ferric salts, bichromate ions or permanganate ions) cannot be handled in equipment made of copper or its alloys.

In dilute solutions (up to 1% acid), the corrosive action of a nonoxidizing acid on copper is relatively low; corrosion rates are usually below 6 g/m²·d (60 mdd) or 250 μm/yr (10 mils/yr). This is only true of oxidizing acids when the concentration does not exceed 0.01%. At such low acid concentrations, aeration has little effect in either oxidizing or nonoxidizing acids.

Nonoxidizing acids with near-zero aeration have virtually no corrosive effect. Rates in 1.2N sulfuric, hydrochloric and acetic acids are less than 0.1 g/m²·d in the absence of air. Figure 2 illustrates the general effect of various concentrations of oxygen on the corrosion rate of copper in these acids.

Except for hydrochloric acid, nonoxidizing acids that contain as much air as is absorbed in quiet contact with the atmosphere are weakly corrosive. Rates generally range from 0.5 to 6 g/

Acid	g/m²·d	Corrosion rate mm/yr	mils/yr
32% nitric	5700	240	9500
Concentrated hydrochloric	18	0.75	30
17% sulfuric	2	0.1	4

Fig. 2 Effect of oxygen on corrosion rates for copper in 1.2N solutions of nonoxidizing acids

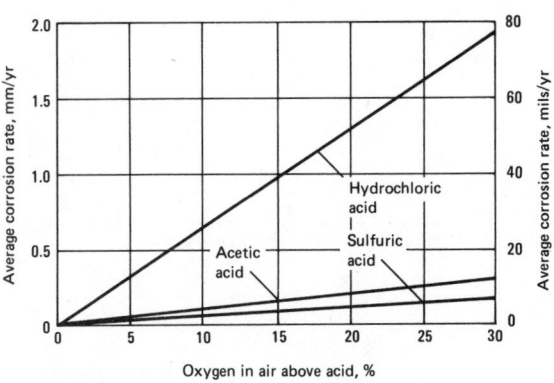

Specimens were immersed for 24 h at 24 °C (75 °F) in 1.2N acid solutions. Oxygen content of the solutions varied from test to test, depending on the concentration of oxygen in the atmosphere above the solutions.

Fig. 3 Corrosion of C65500 in sulfuric acid solutions

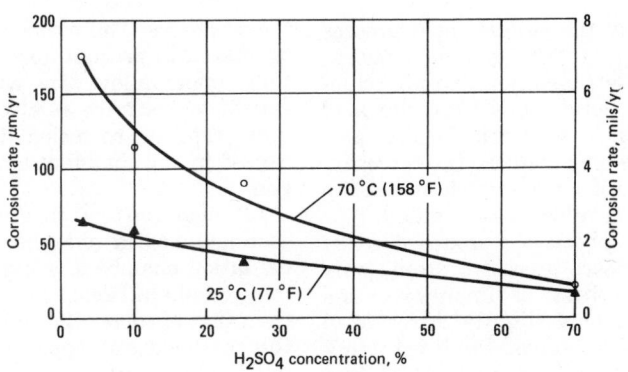

Specimens were immersed for 48 h at the indicated temperatures. The solution was not agitated or intentionally aerated.

Phosphoric, acetic, tartaric, formic, oxalic, malic and similar acids normally react in a manner comparable to sulfuric acid.

Many of the copper alloys can be brazed with brazing rod of the same composition, which provides a joint about as corrosion resistant in acids as the base metal.

Factors that may accelerate corrosion vary from one plant to another, and it is advisable to conduct preliminary service or field tests under actual operating conditions before purchasing large quantities of any alloy. Corrosion-accelerating factors then can be evaluated. Selection of the most suitable material for use in a chemical process depends not only on corrosion resistance but also on such factors as continuing availability of the alloy in the desired form and size (which should be ensured before any alloy is given serious consideration).

The following corrosion data were obtained in tests made under various conditions for handling different acids and acid solutions. Because of the variety of factors affecting all chemical reactions, the values shown cannot be taken as absolute, and should be considered only as trends.

Sulfuric Acid. The corrosion rate of C65500 (3% silicon bronze) in sulfuric acid indicates that this alloy can be used successfully with solutions of 3 to 70% sulfuric acid (by weight) at temperatures of 25 to 70 °C (77 to 158 °F). Laboratory test results are shown in Fig. 3.

Rate of attack by sulfuric acid varies with concentration as shown in Table 4. The presence of copper or iron salts in acid solutions accelerates the corrosion rate of copper, as shown in Table 5.

Phosphoric Acid. Copper and copper alloys are used in heat-exchanger tube, pipe and fittings for handling phosphoric acid, although corrosion rates of some of these alloys may be comparatively high.

Laboratory tests were performed on eight groups of copper alloys in aerated and unaerated acid, with specimens at

m²·d, or an equivalent of about 20 to 250 μm/yr (0.8 to 10 mils/yr).

Air-saturated solutions of nonoxidizing acids are likely to be strongly corrosive, with rates of 5 to 30 g/m²·d, which is equivalent to 0.2 to 1.25 mm/yr (8 to 50 mils/yr). This rate is higher for hydrochloric acid. The actual corrosion in any aerated acid depends on acid concentration, temperature and other factors difficult to classify. Except in very dilute solutions, oxidizing acids corrode copper rapidly—at rates usually above 50 g/m²·d or 2.1 mm/yr (85 mils/yr). The reaction is independent of aeration.

Corrosion rates of three common acids are compared below (temperature and aeration are not specified):

Table 4 Corrosion of copper alloys completely immersed in sulfuric acid of various strengths

Alloy	30%		40%		50%	
	μm/yr	mils/yr	μm/yr	mils/yr	μm/yr	mils/yr
Average penetration for sulfuric acid concentration of:						
Exposure Time 24 to 48 h, Boiling at a Pressure of 100 mm Hg						
C11000	670-700	26.4-27.6	487-700	19.2-27.6	660-792	26.0-31.2
C14200	640-670	25.2-26.4	487-548	19.2-21.6	610	24.0
C51000	640	25.2	395-457	15.6-18.0	915	36.0
C26000	...	...	...	...	...	...
Exposure Time 16 to 24 h, Solution Agitated						
C11000	60-245	2.4-9.6	18-60	0.7-2.4	60	2.4
C14200	92-335	3.6-13.2	Nil	Nil	50-60	2.0-2.4

Alloy	60%		70%		80%	
	μm/yr	mils/yr	μm/yr	mils/yr	μm/yr	mils/yr
Average penetration for sulfuric acid concentration of:						
Exposure Time 24 to 48 h, Boiling at a Pressure of 100 mm Hg						
C11000	2195-2255	86.4-88.8	853-1067	33.6-42.0	39 630-166 420	1560-6552
C14200	2285-2377	90.0-93.6	945	37.2	67 310-527 300	650-20,760
C51000	2957-3385	116.4-133.2	945-1067	37.2-42.0	60 660-62 080	2388-2444
C26000	...	...	580-793	22.8-31.2	72 850-206 050	2868-8112
Exposure Time 16 to 24 h, Solution Agitated						
C11000	60-92	2.4-3.6	1830-2745	72.0-108.0	39 370-40 890	1550-1610
C14200	15-60	0.6-2.4	2135	84.0	39 370-50 550	1550-1990

the waterline, in quiet immersion and totally submerged.

Acid concentrations ranged from 5 to 90% and temperatures ranged from 20 to 85 °C (68 to 185 °F) except for the Cu-Al-Si, which was tested only in 6.5% acid at 20 °C (68 °F) with specimens at the waterline and in quiet immersion. Corrosion rates for the eight alloy groups were as follows:

Alloy type	Corrosion rate	
	mm/yr	mils/yr
Copper	0.55-3.7	22-148
Copper-zinc (70% Cu min)	0.13-7.0	5-280
Copper-tin	0.025-1.3	1-52
Copper-nickel	0.025-0.63	1-25
Copper-silicon	0.13-0.93	5-37
Copper-aluminum-iron	0.13-0.25	5-10
Copper-aluminum-silicon	0.28-2.4	11-97

On the whole, copper and copper alloys give satisfactory service in handling pure phosphoric acid solutions in various concentrations. The acid concentration seems to have less effect on the corrosion rate than the amount of impurities. The impure phosphoric acid

produced by the sulfuric acid process may contain a markedly higher concentration of ferric, sulfate, sulfite, chloride and fluoride ions than the acid produced by the electric furnace process. These ions increase the corrosion rate up to 150 times, which makes the life of copper alloys quite limited.

Pure phosphoric acid produced by the electric furnace process contains only small quantities of impurities and therefore is only slightly corrosive to copper and its alloys. Inhibited admiralty metals C44300, C44400 and C44500 are suggested for solutions of pure phosphoric acid.

Accumulation of corrosion products on metal surfaces may also increase both the rate of corrosion and the possibility of pitting. Low-copper alloys such as C46400 (naval brass) appear to form thin, adherent films of corrosion products. Copper, copper-silicon alloys, and other high-copper alloys form more voluminous, porous films or scales beneath which roughened or pitted surfaces are likely to be found.

The phosphoric acid vapors that condense in electrostatic precipitators at about 120 °C (250 °F) are noticeably more corrosive than solutions of pure phosphoric acid at the same or lower

temperatures. The corrosion rates encountered in precipitators are so high that copper alloy wire will not give satisfactory service as electrodes. The high rate of corrosion is probably caused by an abundant supply of oxygen.

Although the corrosion rate of copper cooling tubes in phosphoric acid condensation chambers is high, about 10 mm/yr (400 mils/yr), the rate is lower than that of some other materials. For this reason, use of copper tubes is feasible for this purpose.

The preceding discussion on the effect of phosphoric acid on copper and its alloys emphasizes the value of keeping service records. Such records are valuable for anticipating repairs, making changes to minimize the effect of various factors, and selecting materials for replacement parts.

Hydrochloric acid is one of the most corrosive of the nonoxidizing acids when in contact with copper and its alloys, and is successfully handled only in dilute concentrations. The rates for C65800 in HCl of various concentrations are shown in Table 6. Corrosion rates for two nonstandard silicon bronzes were about the same as for C65800.

Table 5 Corrosion of copper in boiling 30% sulfuric acid containing Cu and Fe salts

Copper, ppm	Average penetration		Iron, ppm	Average penetration		Iron and copper, ppm	Average penetration rate	
	μm/yr	mils/yr		μm/yr	mils/yr		μm/yr	mils/yr
0	60	2.4	0	122	4.8	0	13	0.5
20	183	7.2	28	122	4.8	20Cu+28Fe	152	6.0
40	213	8.4	58	245	9.6	40Cu+56Fe	244	9.6
80	243	9.6	112	427	16.8	80Cu+112Fe	457	18.0
200	335	13.2	196	782	30.8	200Cu+196Fe	730	28.8
280	360	14.2	280	975	38.4	280Cu+280Fe	1005	39.6
360	427	16.8	364	1097	43.2	360Cu+364Fe	1250	49.2
440	457	18.0	447	1280	50.4	440Cu+447Fe	1525	60.0

Table 6 Corrosion of C65800 totally submerged in hydrochloric acid

HCl concentration, wt%	Corrosion rate(a)		
	g/m²·d	μm/yr	mils/yr
At 25 °C (77 °F)			
3	2.3	99	3.9
10	2.3	99	3.9
20	1.8	79	3.1
35	12.3	526	20.7
At 70 °C (158 °F)			
3	18.3	780	30.7
10	13.7	508	20.0
20	23.8	102	40.1
35	160.8	6860	270.1

(a) Size of specimens, 50 by 25 by 1.3 mm (2 by 1 by 0.050 in.); surface condition, pickled; velocity of solution, natural convection; aeration, none; duration of test, 48 h.

Table 7 Corrosion of wrought copper alloys in anhydrous hydrofluoric acid

Temperature		Corrosion rate(a)					
		C51000		C44400		C71500	
°C	°F	μm/yr	mils/yr	μm/yr	mils/yr	μm/yr	mils/yr
16–27	60–80	510	20	255	10	180	7
27–38	80–100	480	18.8	480	18.8	...	...
82–88	180–190	1525	60	510	20	255	10

(a) These values are representative of results on copper alloys having high copper content such as copper, aluminum bronze, silicon bronze, and inhibited admiralty metal. Corrosion rates for C23000 are between C44400 and C51000.

Table 8 Corrosion of copper in acetic acid – acetic anhydride mixtures

Copper alloy	Exposure time, h	Test conditions	Average penetration rate	
			μm/yr	mils/yr
C11000	1115	Acetic acid – acetic anhydride – acetone mixture, 110 to 140 °C (230 to 284 °F)	483	19.0
	2952	Same as above	66-70	2.6-2.8
C65500	1115	Same as above	213	8.4
	2952	Same as above	70-90	2.7-3.6
C11000	1115	1:1 acetic acid – acetic anhydride mixture, 130 to 145 °C (266 to 293 °F)	120-533	4.7-21.0
C65500	1115	Same as above	116-236	4.6-9.3
C11000	865	95% acetic acid – 5% acetic anhydride, liquid phase, 120 °C (248 °F)	97-116	3.8-4.4
C11000 coupled to type 316 stainless steel	865	Same as above	102-216	4.0-8.5
C11000	865	95% acetic acid – 5% acetic anhydride, vapor phase, 120 °C (248 °F)	102-104	4.0-4.1
C11000 coupled to type 316 stainless steel	865	Same as above	94-213	3.7-8.4
C11000	2448	50:50 acetic acid – acetic anhydride, 150 °C (302 °F)	84-90	3.3-3.6
C11000	2448	Essentially pure acetic acid	5	0.2

The corrosion rate of copper nickels in 2N HCl at 24 °C (75 °F) may range from 2.3 to 7.6 mm/yr (90 to 300 mils/yr) depending on the degree of aeration and other factors. Specimens of C71000 (copper nickel, 20%) corrode as follows in stagnant solutions at room temperature: in 1% HCl, 305 μm/yr (12 mils/yr); in 10% HCl, 790 μm/yr (31 mils/yr).

Hydrofluoric acid is less corrosive than hydrochloric acid and can be successfully handled by C71500 (copper nickel, 30%), which has good resistance to both the aqueous and anhydrous acids. C71500 is not sensitive to velocity effects as are some other copper alloys. Laboratory tests in conjunction with the hydrofluoric acid alkylation process in anhydrous acid provided the data given in Table 7.

Acetic Acid and Acetic Anhydride. Copper and copper alloys are used successfully in commercial processes involving exposure to acetic acid and related chemical compounds or in the manufacture of this acid.

One plant kept records concerning the corrosion rate of C11000 used in two different acetic acid still systems. One still operated at 115 to 140 °C (240 to 285 °F) and handled a solution containing 50% acetic acid and about 50% anhydride with some esters also present. After operating for 663 h, the kettle showed an average penetration rate of 210 μm/yr (8.4 mils/yr). The rate was lower, 60 μm/yr (2.4 mils/yr), for the bottom column and lower yet, only 30 μm/yr (1.2 mils/yr), for the middle and top columns. A second still, operating at 60 to 140 °C (140 to 285 °F) contained a 70% solution of acetic acid, the remainder being anhydride, esters and ketones. After 1464 h, the kettle showed a corrosion rate of 120 μm/yr (4.8 mils/yr). The rate was only 30 μm/ yr (1.2 mils/yr) for the middle and top columns.

In another field test, C11000 and C65500 coupons were placed in an acetic acid storage tank at ambient temperature. The stored solution contained 27% acetic acid, 1% butyl acetate, 70% water, and small amounts of acetates, aldehydes, and other acids. During the 3984-h exposure, the specimens were immersed in the liquid phase 80% of the time and were in the vapor phase 20% of the time. The C11000 specimens showed a corrosion rate of 38 to 53 μm/ yr (1.5 to 2.1 mils/yr) and the C65500 specimens, 30 to 45 μm/yr (1.2 to 1.8 mils/yr).

The results of other field tests for C11000 and C65500 exposed in acetic acid mixtures are given in Tables 8 to 10. Test conditions included a variety of temperatures, concentrations, exposure times and locations in equipment as well as the presence of other chemicals, as indicated in Tables 9 and 10.

In laboratory tests at room temperature, C61300 and C62300 exhibit typical corrosion rates of 65 to 80 μm/yr (2.5 to 3.2 mils/yr) in 10 to 40% acetic acid.

Hydrocyanic acid can be handled successfully by copper and copper alloys. Results of field tests for C11000 and C65500 are given in Tables 11 and 12.

Fatty Acids. Under severe service conditions, fatty acids attack copper alloys at somewhat higher rates than do other organic acids, such as acetic or citric.

Tests were conducted for 400 h in a copper-lined wooden splitting tank containing a mixture of about 60% fatty acids, 39% water and 1.17% sulfuric acid heated to 100 °C (212 °F) and agitated violently with an open steam jet. Specimens of C71000 (copper nickel, 20%) showed a corrosion rate of 64 μm/yr (2.6 mils/yr), and specimens of C71500 (copper nickel, 30%) a rate of 59 μm/yr (2.4 mils/yr), when submerged just below the liquid level in the tank. Similar specimens submerged 150 mm (6 in.) from the bottom of the tank showed corrosion rates of 178 and 185 μm/yr (7.0 and 7.3 mils/yr) for C71000 and C71500, respectively.

Oleic Acid. Copper and copper-zinc alloys are highly resistant to attack by pure oleic acid. However, oleic acid will attack these alloys when air and water are present. Temperature also influences the rate of attack. Copper and several copper alloys were tested in oleic acid at 24 °C (75 °F); C51000 and C61300 corroded at less than 50 μm/yr (2 mils/yr) compared with about 500 μm/yr (20 mils/yr) for C26000 and C65500.

Stearic acid, like all other fatty acids, attacks copper and copper alloys when moisture and air are present. Temperature and impurities also influence the rate of attack. Tests made at 24 to 100 °C (75 to 212 °F) in stearic

Table 9 Corrosion of C11000 in isopropyl ether – acetic acid mixtures

Concentration, %		Average penetration rate	
Isopropyl ether	Acetic acid	μm/yr	mils/yr
Exposed 72 h at 60-65 °C (140-150 °F)			
93	7	40-50	1.6-2.0
85	15	18-20	0.7-0.8
Exposed 328 h at 20 °C (68 °F)			
93	7	100	4.0
85	15	13	0.5

Table 10 Corrosion of copper metals in acetic acid

Alloy	Exposure time, h	Average penetration rate μm/yr	mils/yr
Acetic Anhydride(a)			
C11000	2448(b)	60	2.4
	2448(c)	915-1100	36.0-43.2
C65500	2448(b)	60	2.4
	2448(c)	488-732	19.2-28.8
90% Acetic Acid(d)			
C11000, annealed	672	60	2.4
	816	30	1.2
C11000, cold worked	672	90	3.6
	792	90	3.6
Copper joint(e)	1512	183	7.2
	4000	120	4.8
Copper joint(f)	1512	183	7.2
	4000	120	4.8
45% Acetic Acid(g)			
C11000	1038	30 max	1.2 max
C65500	1038	30 max	1.2 max
Copper joint(f)	1038	30 max	1.2 max
25% Acetic Acid(h)			
C11000	432	274	10.8
	792	152	6.0

(a) Test specimens were exposed in stills separating acetic acid from acetic anhydride. (b) Top of column. (c) Kettle. (d) Test specimens were exposed in cycle feed lines at 30 to 50 °C (86 to 122 °F). (e) Joint brazed with BCuP-5 filler metal. (f) BAg filler metal. (g) Test specimens were exposed in the acetic acid recovery column, where concentration of the acetic acid was 45% max. (h) Test specimens were exposed to crude by-product acetic acid (approximately 25% concentration) in pump suction line from storage tank.

Table 11 Corrosion of copper and copper alloys in production of hydrogen cyanide

Alloy	Exposure time, h	Average penetration rate(a)							
		Stripping still μm/yr	mils/yr	Top of HCN refining still μm/yr	mils/yr	Base of HCN stripping still μm/yr	mils/yr	Base of partial condenser μm/yr	mils/yr
C11000	573	173-218	6.8-8.6	54-60	2.1-2.4	1033-1186	40.7-46.7	1534-14 170	60.4-558
	671	155-609	6.1-24.0	18-25	0.7-1.0	Nil	Nil	478	18.8
C65500	573	229-244	9.0-9.6	18-25	0.7-1.0	777-1145	30.6-45.1	1138-5385	44.8-212
	671	137-503	5.4-19.8	...	...	275	10.8	343	13.5

(a) All data from separate specimens; differences at similar locations imply expected variability.

Table 12 Corrosion of C11000 and C65500 in hydrocyanic acid solutions

Alloy	Exposure time, h	Test conditions	Average penetration rate μm/yr	mils/yr
C11000	3144	Ethylene cyanohydrin residues, 70 °C (158 °F)	5-35	0.2-1.4
C11000	2232	Ethylene cyanohydrin residues, 30 to 90 °C (86 to 194 °F)	13	0.5
C65500	2232	Same as above	40	1.6
C11000	1621	Cyanohydrin stripping still products (kettle).	690	27
C65500	1621	Same as above	35	1.4

Table 14 Corrosion of copper and brass in ammonia

Alloy	Average penetration rate(a) Liquid μm/yr	mils/yr	Vapor μm/yr	mils/yr
Anhydrous Ammonia				
C11000	2.5	0.1	<2.5	<0.1
C26000	<2.5	<0.1	<2.5	<0.1
Anhydrous Ammonia plus 1% H_2O(b)				
C11000	<2.5	<0.1	<2.5	<0.1
C26000	2.5	0.1	<2.5	<0.1
Anhydrous Ammonia plus 2% H_2O(b)				
C11000	2.5	0.1	2.5	0.1
C26000	5.0	0.2	2.5	0.1

(a) Atmospheric temperature and 345 to 1035 kPa (50 to 150 psi) pressure for 1600-h exposure. Specimens were placed at the top and bottom of two-litre bombs that were charged with ammonia. Pressure varied throughout the test, depending on temperature. Water was added to two of the bombs prior to charging with ammonia. (b) Any air present probably was depleted rapidly during initial stages of test.

Table 13 Corrosion of copper alloys in contact with tartaric acid at 24 °C (75 °F)

Solution strength, %	Corrosion rate μm/yr	mils/yr
C26000 and C23000		
10	50 max	2 max
30	500-1250	20-50
50	500-1250	20-50
100	50 max	2 max
C71000		
5	25 max	1 max
C71300		
2	40	1.6

acid showed corrosion rates of C11000, C26000 and C65500 to be in the range of 500 to 1250 μm/yr (20 to 50 mils/yr).

Tartaric Acid. Copper and its alloys corrode rather slowly when exposed to various concentrations of tartaric acid, as indicated by the laboratory test data in Table 13.

Alkalis

Copper and its alloys resist alkaline solutions except those containing ammonium hydroxide, or compounds that hydrolyze to ammonium hydroxide or cyanides. Ammonium hydroxide reacts with copper to form soluble complex copper cations whereas the cyanides react to form soluble complex copper anions. The rate of attack for copper-zinc alloys exposed to alkalis other than those specified above is about 50 to 500 μm/yr (2 to 20 mils/yr) at room temperature under stagnant conditions, but is about 500 to 1750 μm/yr (20 to 70 mils/yr) in aerated boiling solutions.

C71500 corrodes at less than 5 μm/yr (0.2 mil/yr) in 1N to 2N NaOH solutions at room temperature, and the degree of aeration usually has no significant effect. This rate is two to three times as great as the rate in boiling solutions.

Copper-tin alloys (phosphor bronzes) corrode at less than 250 μm/yr (10 mils/yr) in 1N to 2N NaOH solutions at room temperature and are apparently unaffected by aeration.

Copper and two grades of silicon bronze were tested in a 50% solution of sodium hydroxide at 60° C (140 °F) for four weeks. The specimens were bright rolled and degreased sheet, about 25 by 50 by 1.3 mm (1 by 2 by 0.05 in.). The solution was exposed to air (no additional aeration), and velocity was limited to natural convection. C11000 showed a corrosion rate of 1.7 g/m²·d or 70 μm/yr (2.8 mils/yr); C65100, a rate of 1.5 g/m²·d or 63 μm/yr (2.5 mils/yr); and C65500, a rate of 1.1 g/m²·d or 47 μm/yr (1.85 mils/yr).

Ammonium Hydroxide. Strong NH_4OH solutions attack copper and copper alloys rapidly, compared with the rates of attack by metallic hydroxides, because of the formation of a soluble complex copper-ammonium compound. However, in some applications the corrosion of copper exposed to dilute solutions of ammonium hydroxide is low. For example, copper specimens submerged in 0.01N NH_4OH solution at room temperature for one week lost

weight at a rate of 1.5 g/m²·d, which is equivalent to about 60μm/yr (2.5 mils/yr).

Ammonium hydroxide solutions also attack copper-zinc alloys. Alloys containing more than 15% zinc are susceptible to stress-corrosion cracking when stressed and exposed to ammonium hydroxide. The stress may be due to applied service loads or to unrelieved residual stresses.

In quiescent 2N solutions at room temperature, copper-zinc alloys corrode at 1.8 to 6.6 mm/yr (70 to 260 mils/yr), copper-nickel alloys at 0.25 to 0.50 mm/yr (10 to 20 mils/yr), copper-tin alloys at 1.3 to 2.5 mm/yr (50 to 100 mils/yr) and copper-silicon alloys at 0.75 to 5 mm/yr (30 to 200 mils/yr).

Anhydrous Ammonia. Copper and its alloys are suitable for handling anhydrous ammonia provided that the ammonia remains anhydrous and is not contaminated with water and oxygen. In one test conducted for 1200 h, C11000 and C26000 each showed an average penetration of 5 μm/yr (0.2 mil/yr) in contact with anhydrous ammonia at atmospheric temperature and pressure. Tests showed the rates of corrosion to be low in the presence of small amounts of water, but probably when oxygen was excluded. Data on exposure for 1600 h are given in Table 14. For any new installation, tests simulating the expected conditions are recommended.

Salts

Copper metals are widely used in equipment for handling salt solutions of various kinds, particularly those that are nearly neutral. Among these are the nitrates, sulfates, and chlorides of sodium and potassium. Chlorides usually are more corrosive than the

Table 15 Corrosion of C11000 in 30% CaCl₂ refrigeration brine

Inhibitor	Corrosion rate µm/yr	mils/yr
None(a)	10	0.4
K₂Cr₂O₇(b)	6	0.23
Aluminum foil(c)	119	4.7

(a) Exposed for 325 days at −12 °C (+10 °F). (b) Exposed for 372 days, cold. (c) Exposed for 50 days with slight agitation in brine with a pH of 9.0.

Table 16 Corrosion of C11000 in a sodium chloride brine refrigeration system

Location in equipment	Corrosion rate(a) µm/yr	mils/yr
With Na₂Cr₂O₇ Inhibitor; pH, 6.0 to 6.5		
Brine tank near main outlet	5	0.2
Top of brine pump, high agitation	10	0.4
Inside cooler tube	15	0.6
Return line to storage tank	2.5	0.1
Brine tank near agitator	2.5	0.1
Without Inhibitor; pH, 10.5		
Open brine tank	160	6.3
Brine cooler outlet, rapid flow	360	14.2
Cooler inlet	157	6.2
Cooler outlet	250	9.8

(a) Field test; 98 days at −15 °C (+4 °F).

Table 17 Corrosion of copper alloys in alkaline salt solutions

Alloy family	Common name	Corrosion rate µm/yr	mils/yr
Sodium Silicate, Phosphate or Carbonate			
Cu-Zn	Brasses	50 to 125	2 to 5
Cu-Sn	Phosphor bronzes	<50	<2
Cu-Ni	Copper nickels	2.5 to 40	0.1 to 1.5
Sodium Cyanide			
Cu-Zn	Brasses	250 to 500	10 to 20
Cu-Sn	Phosphor bronzes	875	35
Cu-Ni	Copper nickels	500 to 2500	20 to 100

Table 18 Corrosion of copper metals in amine-system service

Alloy	Exposure time, h	Test conditions	Average penetration rate µm/yr	mils/yr
C11000	1622	Coupons exposed in ethylenediamine refining still	Nil to 180	Nil to 7
C11000	1580	Aqueous ethylenediamine	25	1
C71500	1580	Same as above	75	3
C11000	806	Liquid-vapor containing ammonia and mono-, di- and triethanolamines; 90 to 156 °C (194 to 313 °F)	760	30
C65500	806	Same as above	790	31.2
C11000	1437	Liquid-vapor containing ammonia and mono-, di- and triethanolamines; 180 to 195 °C (356 to 383 °F)	28	1.1
C65500	1437	Same as above	48	1.9
C11000	2622	Vapor phase of diethanolamine still containing mono-, di- and triethanolamines; 180 to 195 °C	28	1.1
C26000	887	Denuded monoethanolamine (20%)	Nil	Nil
C44200(a)	887	Same as above	Nil	Nil
C26000 coupled to carbon steel	168	20% monoethanolamine (4 mol CO₂ per mol MEA); 60 °C (140 °F)	4 550	179
C44200(a)	900	Lean solution of diethanolamine containing impurities	50	2
C26000	1440	Rich solution of monoethanolamine	330	13
C11000	1440	Same as above	Dissolved	Dissolved
C26000	1440	Lean solution of monoethanolamine	3 000	118
C11000	1440	Same as above	11 500	454

(a) Uninhibited admiralty metal, now an obsolete alloy.

other salts, especially in strongly agitated, aerated solutions.

The nonoxidizing acid salts, such as the alums and certain metal chlorides (magnesium and calcium chlorides) that hydrolyze in water to produce an acidic pH, behave essentially the same as dilute solutions of the corresponding acids. Corrosion rates generally range from 2.5 to 1500 µm/yr (0.1 to 60 mils/yr) at room temperature, depending on the degree of aeration and the acidity. Test data for corrosion of copper in 30% calcium chloride refrigeration brine with and without inhibitors are given in Table 15.

Neutral salt solutions can be handled successfully by copper alloys. Consequently, these alloys are used in heat-exchanger and condenser equipment exposed to seawater. Corrosion rates of copper in sodium chloride brine are given in Table 16. These rates are not necessarily the same as those in seawater.

Such alkaline salts as sodium silicate, sodium phosphate and sodium carbonate attack copper alloys at low but different rates at room temperature. On the other hand, alkali cyanide is aggressive and attacks copper alloys fairly rapidly because it forms a soluble complex copper anion. Specific rates are given in Table 17.

Oxidizing salts corrode copper and copper alloys rapidly; therefore, copper metals should not be used with oxidizing salt solutions except those that are very dilute. Aqueous sodium dichromate solutions can be safely handled by copper alloys, but the presence of a highly ionized acid such as chromic or sulfuric acid may increase the corrosion rate several hundred times, because the dichromate acts as an oxidizing agent in acidic solutions. In one test, a copper nickel corroded at 2.5 to 250 µm/yr (0.1 to 10 mils/yr), and a copper-tin alloy (phosphor bronze) at 5 µm/yr (0.2 mil/yr), when handling an aqueous sodium dichromate solution. The rate increased 200 to 300 times for both metals when chromic acid was added to the solution. In solutions containing ferric, mercuric or stannic ions, a copper nickel showed a corrosion rate of 75 µm per *day* (3 mils per *day*), while copper-zinc and copper-tin alloys showed a still greater rate of 625 μm per *day* (25 mils per *day*).

Salts of metals more noble than copper (such as the nitrates of mercury and silver) corrode copper alloys rapidly, concurrently plating out the noble metal on the copper surface. Rate of attack is influenced by temperature and acidity. A film of mercury on a high-zinc brass (more than 15% Zn) may cause intercrystalline cracking by liquid-metal embrittlement if the alloy is under tensile stress, either residual or applied.

Organic Compounds

Copper and many of its alloys resist corrosive attack by organic compounds such as amines, alkanolamines, esters, glycols, ethers, ketones, alcohols, aldehydes, naphtha and gasoline, and by most organic solvents.

Although corrosion rates of copper and copper alloys in pure alkanolamines and amines are low, they can be significantly increased if these compounds are contaminated with water, acids, alkalis or salts, or with combinations of these impurities, particularly at high temperatures.

Tables 18 to 24 report the results of corrosion testing of copper and a limited but representative variety of copper alloys in contact with various amines, esters, ethers, ketones, aldehydes, ethylene glycols, and alcohols under many conditions.

Gasoline, naphtha and other related hydrocarbons in pure form will not attack copper or any of the copper alloys. However, in manufacture of hydrocarbon materials, process streams are likely to be contaminated with one or more substances such as water, sulfides, acids and various organic compounds. These contaminants attack copper and its alloys. Corrosion rates for C44300 and C71500 exposed to gasoline are low (see Table 25), and these two alloys are successfully used in equipment for refining gasoline.

Corrosion rates for copper and for alloys exposed to contaminated naphtha in two different environments are given in Table 26.

Creosote. Copper and copper alloys are generally suitable for use with creosote, although creosote attacks some high-zinc brasses. C11000, C23000, C26000, C51000 and C65500 typically corrode at rates less than 500 μm/yr (20 mils/yr) when exposed to creosote at 24 °C (75 °F).

Table 19 Corrosion of copper and copper alloys in ester solutions

Alloy	Exposure time, h	Test conditions	Average penetration rate μm/yr	mils/yr
Acetates				
C11000	400	Alkenyl acetate plus H$_2$SO$_4$	6100	240
C65500	400	Same as above	3050	120
	257	Allylidene diacetate; 110 °C (230 °F)	183-213	7.2-8.4
C11000	240	Butyl acetate plus 1% H$_2$SO$_4$ ...	1625-4090	64-161
C65500	240	Same as above	2870	113
C11000	2328	2-chloroallylidene diacetate.....	5	0.2
C11000	250	Crude vinyl acetate; 110 to 150 °C (230 to 302 °F).........	25	1.0
C71500	250	Same as above	7.5-125	0.3-5
C11000	550	Ethyl acetate plus 1.0% H$_2$SO$_4$	483	19
C65500	550	Same as above	400	16
C11000	991	Ethyl acetate reaction mixture; liquid; 90 °C (194 °F)	550	21.6
C62300	991	Same as above	395	15.6
C65500	991	Same as above	518	20.4
C11000	991	Ethyl acetate reaction mixture; vapor; 90 °C (194 °F)..........	5	0.2
C62300	991	Same as above	15	0.6
C65500	991	Same as above	13	0.5
C11000	2976	Ethyl acetoacetate.............	10	0.4
C65500				
Cold worked	216	Isopropyl acetate	6700	264
Annealed	216	Isopropyl acetate	6100	240
	480	Isopropyl acetate process; liquid; 120 °C (248 °F)...............	300	12
C11000	519	Methylamyl acetate process; batch still coils; 115 °C (239 °F)	500-685	22-27
C65500	519	Same as above	280-300	11-12
C11000	519	Methylamyl acetate process; batch still down pipe; 115 °C (239 °F)	330	13
C65500	51	Same as above	300	12
C11000	1345	Methylamyl acetate process; batch still condenser 30 °C (86 °F)	840-940	33-37
C65500	1345	Same as above	1400-1575	55-62
C63600	3312	Methylamyl acetate process; batch still coils; 95 °C (203 °F)	483	19
C51000	3312	Same as above	430-483	17-19
C60800	3312	Same as above	330	13
C51000	3312	Methylamyl acetate process; batch still downpipe; 95 °C (203 °F)	380-483	15-19
C63600	3312	Same as above	400-460	16-18
C60800	3312	Same as above	280	11
C11000	217	Refined isopropenylacetate; 98 °C (208 °F).................	60	2.4
	2784	Vinyl acetate, inhibited	2.5	0.1
C11000	250	Vinyl acetate, process; 150 to 190 °C (302 to 374 °F).........	355-400	14-16
C71500	250	Same as above	685-1250	27-49
C11000	768	Vinyl acetate process; batch still kettle..................	685-1170	27-46
C65500	768	Same as above	150-483	6-19
C11000	864	Same as above	2290-3500	90-138
C65500	864	Same as above	660-2160	26-85
		(continued)		

Table 19 (continued)

Alloy	Exposure time, h	Test conditions	Average penetration rate μm/yr	mils/yr
Acrylates				
C11000	240	Acidified sodium acrylate containing 5% H_2SO_4; 49 °C (120 °F)	945	37.2
	254	Ethyl acrylate process; 130 to 150 °C (266 to 302 °F)	1220	48
C65500	254	Same as above	430	16.8
C11000	240	Isopropyl ether solution of acrylic acid (18%); 49 °C (120 °F)	18	0.7
	240	Sodium acrylate solution containing 1% NaOH; 49 °C (120 °F)	5	0.2
	240	Washings from isopropyl ether solution of acrylic acid; 49 °C (120 °F)	210	8.3
	240	Wet calcium acrylate	240	9.4
	504	2-ethylhexylacrylate process; 95 °C (203 °F)	230-275	9.0-10.8
C65500	504	Same as above	220-275	8.6-10.8
C11000	566	2-ethylhexyl acrylate process; condensate tank; 30 °C (86 °F)	66-74	2.6-2.9
C51000	566	Same as above	60-86	2.4-3.4
C65500	566	Same as above	114-122	4.5-4.8
C11000	566	2-ethylhexyl acrylate process; 120 °C (248 °F)	236-239	9.3-9.4
C51000	566	Same as above	264	10.4
C65500	566	Same as above	328-360	12.9-14.2
Benzoates				
C11000	1680	Butyl benzoate	Nil	Nil
	1296	Butyl benzoate process; circulating line; 40 °C (104 °F)	800-1025	31.4-40.4
C60800	1296	Same as above	1060	37.7-41.8
C65500	1296	Same as above	843-1090	33.2-42.8
C23000	1296	Same as above	790-1085	31.2-42.7
C22000	1296	Same as above	900-985	35.6-38.8
C11000	1296	Butyl benzoate process; 40 °C (104 °F)	280	11.1
C65500	1296	Same as above	350-400	13.7-15.7
C11000	1296	Butyl benzoate process; batch still kettle; 185 °C (365 °F)	7.5-38	0.3-1.5
C65500	1296	Same as above	7.5-25	0.3-1.0
C11000	1680	Methyl benzoate (refined)	2.5	0.1
C11000	1680	Methyl benzoate (Cu-free)	7.5	0.3

Linseed Oil. Copper and its alloys are fairly resistant to corrosion by linseed oil. All of the alloys show some attack, but none exhibits corrosion severe enough to make it unsuitable for this application. C11000, C51000 and C65500 showed corrosion rates less than 500 μm/yr (20 mils/yr) in linseed oil at 24 °C (75 °F). C26000 had a rate of 500 to 1250 μm/yr (20 to 50 mils/yr).

Benzol and Benzene. C11000, C23000, C26000, C51000 and C65500

tested in these two materials at 24 °C had corrosion rates less than 500 μm/yr (20 mils/yr).

Sugar. Copper is used successfully for vacuum-pan heating coils, evaporators, and juice extractors in manufacture of both cane and beet sugar. Inhibited admiralty metals, aluminum brass, aluminum bronzes and copper nickels are also used for tubes in juice heaters and evaporators. Bimetal tubes of copper and steel have been used by

manufacturers of beet sugar to counteract stress-corrosion cracking of copper tubes caused by ammonia from beets grown in fertilized soil. Results of tests run on copper and copper alloys in a beet-sugar refinery are presented in *Table 27*.

Beer. Copper is used extensively in brewing of beer. In one installation, the wall thickness of copper kettles thinned from an original thickness of 16 mm (5/8 in.) to 10 mm (3/8 in.) in a 30-year period. Brazing with BAg filler metals eliminates the possibility that the alkaline compounds used for cleaning copper equipment will destroy joints by attacking tin-lead solders. Steam coils require more frequent replacement than any other brewery equipment. They have a service life of 15 to 20 years. The service life of other copper items exposed to process streams in a brewery is 30 to 40 years.

Sulfur compounds free to react with copper (hydrogen sulfide, sodium sulfide or potassium sulfide) form copper sulfide. Reaction rates depend on alloy composition; the alloys of highest resistance are those of high zinc content.

Strip tensile specimens of seven copper alloys were exposed in a fractionating tower in which oil containing 1.4% S was being processed. The results of this accelerated test are presented in Table 3 in the article "Corrosion Characteristics of Copper and Its Alloys".

These data show the suitability of the higher zinc alloys for use with sulfur-bearing compounds. C28000 (60Cu-40Zn) showed good corrosion resistance; in contrast, C23000 (85Cu-15Zn) was completely destroyed.

Inhibited admiralty metals also are excellent alloys for use in heat exchangers and condensers handling sulfur-bearing petroleum products and using water as the coolant. C44300, C44400 and C44500, which are inhibited toward dezincification by addition of arsenic, antimony or phosphorus to the basic 70Cu-29Zn-1Sn composition, offer good resistance to corrosion from sulfur plus excellent resistance to the water side of the heat exchanger.

Gases

Carbon dioxide and carbon monoxide in dry forms are usually inert to copper and its alloys; however, when moisture is present some corrosion takes place. The rate of reaction depends on the amount of moisture.

Table 20 Corrosion of copper and copper alloys in various ethers

Alloy	Exposure time, h	Test conditions	Average penetration rate μm/yr	mils/yr
C11000	2784	γ-methylbenzyl ether, N₂ atmosphere	2.5 max	0.1 max
C11000	2784	γ-methylbenzyl ether, air atmosphere	2.5 max	0.1 max
C11000	288	Recovered butyl ether	Nil	Nil
C65500	288	Same as above	2.5	0.1
C11000	94	Dichloro ethyl ether residues, 80 °C (176 °F)	183-915	7.2-36
C65500	94	Same as above	61-245	2.4-9.6
C11000	71	Crude dichloro ethyl ether, 80 °C (176 °F)	2130-3050	84-120
C65500	71	Same as above	1220-3050	48-120
C11000	70	Dichloro ethyl ether, 80 °C (176 °F)	150	6
C65500	70	Same as above	120	4.8
C11000	70	Dichloro ethyl ether, 100 °C (212 °F)	610	24
C65500	70	Same as above	245	9.6
C11000	70	Dichloro ethyl ether, boiling	183	7.2
C65500	70	Same as above	213	8.4

Table 21 Corrosion of copper and copper alloys in various ketones

Alloy	Exposure time, h	Test conditions	Average penetration rate μm/yr	mils/yr
C11000	138	Phenylxylol ketone mixture	41-43	1.6-1.7
C65500	138	Same as above	76	3.0
C11000	163	Pentanedione mixture	46-91	1.8-3.6
C65500	163	Same as above	33-84	1.3-3.3
C12000	43	Diethyl ketone, 30 °C (86 °F)	Nil	Nil
C12000	42	Diethyl ketone, boiling	Nil to 7.6	Nil to 0.3
C12000	43	Methyl n-propyl ketone, 30 °C (86 °F)	Nil	Nil
C12000	42	Methyl n-propyl ketone, boiling	Nil	Nil
C11000	216	Methylamyl ketone, boiling	2.5	0.1
C11000	353	Methyl ethyl ketone, boiling	12.7	0.5
C11000	409	Phenylxylol ketone containing NaOH	457-518	18-20.4
C65500	409	Same as above	701-823	27.6-32.4
C11000	165	Acetone dispersion of cellulose acetate, 56 °C (133 °F)	10.2	0.4
C26000	165	Same as above	5.1	0.2

Table 22 Corrosion of copper metals in aldehydes

Alloy	Exposure time, h	Test conditions	Average penetration rate μm/yr	mils/yr
C11000	49	Boiling 2-ethylbutyraldehyde	33	1.3
C11000	112	Boiling butyraldehyde	33	1.3
C11000	1752	2-hydroxyadipaldehyde	20-23	0.8-0.9
C11000	168	Diethyl acetal mixture, 45 °C (113 °F)	60-120	2.4-4.8
C65500	168	Same as above	90-150	3.6-6.0
C26000	168	Same as above	90-150	3.6-6.0
C11000	70	2-ethyl – 3 – propylacrolein, 98 °C (208 °F)	33	1.3
C11000	168	Diacetoxybutyraldehyde, 160 °C (320 °F)	230-240	9.0-9.4
C65500	168	Same as above	75	3.0
C51000	168	Same as above	18-20	0.7-0.8
C11000	540	Propionaldehyde	1420-1550	56.0-61.0
C11000	216	Propionaldehyde, 190 °C (374 °F)	610-1220	24.0-48.0
C11000	443	Butylaldehyde	310	12.2
C51000	443	Same as above	360	14.2
C11000	2374	Same as above	165	6.5
C65500	2374	Same as above	20	0.8
C51000	2374	Same as above	10	0.4

Because some alloy steels are attacked by carbon monoxide, high pressure equipment used for handling this gas is often lined with copper or copper alloys.

Sulfur Dioxide. Gases containing SO₂ attack copper in a manner similar to oxygen. The dry gas does not corrode copper or copper alloys, but the moist gas reacts to produce a mixture of oxide and sulfide scale. The rates of corrosion of some copper alloys in hot paper-mill vapors containing sulfur dioxide are given in Table 28.

Hydrogen Sulfide. Moist hydrogen sulfide gas reacts with copper and copper-zinc alloys to form copper sulfide. Alloys containing more than 20% zinc have considerably better resistance than the alloys having less zinc, or copper itself. Hot wet vapors of hydrogen sulfide corrode C26000, C28000 or C44300 at the rate of only 50 to 75 μm/yr (2 to 3 mils/yr), whereas the rate for C11000 and C23000 for the same conditions is 1250 to 1625 μm/yr (50 to 65 mils/yr).

Halogen Gases. When dry, fluorine, chlorine and bromine, and their hydrogen compounds are not corrosive to copper and its alloys. When moisture is present, however, they are aggressive. Corrosion rates of copper metals in wet hydrogen compounds are comparable to those given for hydrofluoric and hydrochloric acids in Tables 6 and 7.

Hydrogen. Copper and its alloys are not susceptible to attack by hydrogen unless they contain copper oxide in their microstructures. Tough pitch coppers such as C11000 contain small quantities of cuprous oxide. Deoxidized coppers with low residual deoxidizer contents (C12000, for instance) may contain cuprous oxide, but less than the tough pitch coppers; these coppers are not immune from hydrogen embrittlement. Deoxidized coppers with high residual deoxidizer contents (such as C12200, between the oxide and the low amount which has a phosphorus content of 0.015 to 0.040%) are not considered susceptible to hydrogen embrittlement because the oxygen is tied up in complex oxides that do not react appreciably with hydrogen.

When oxygen-bearing copper is heated in hydrogen or hydrogen-bearing gases, the hydrogen diffuses into the metal and reacts with the oxide to form water, which is converted to high-pressure steam if the temperature is above 374 °C (705 °F). The steam produces fissures, which decrease the ductility of

Table 23 Corrosion of copper metals in ethylene glycol solutions

Alloy	Exposure time, h	Test conditions	Average penetration rate μm/yr	mils/yr
C11000	1344	Triethylene glycol solution, aerated; room temperature	Nil	Nil
C11000	2560	Triethylene glycol air-conditioning system; 175 °C (347 °F)	40	1.6
C26000	2560	Same as above	50	2.0
C11000	3320	Same as above	10	0.4
C26000	3320	Same as above	15	0.6
C11000	8328	Same as above	25	1.0
C26000	8328	Same as above	35	1.4
C11000	2880	Triethylene glycol air-conditioning system(a); 160 °C (320 °F)	7.5	0.3
C26000	2880	Same as above	7.5	0.3
C11000	5760	Same as above	2.5	0.1
C26000	5760	Same as above	2.5	0.1
C51000	2880	Ethylene glycol solution(b) plus 0.03% H_2SO_4; 99 °C (210 °F)	7.5-10	0.3-0.4
C60800	2880	Same as above	2.5-7.5	0.1-0.3
C63000	2880	Same as above	2.5-18	0.1-0.7
C65500	2880	Same as above	20-25	0.8-1.0
C11000	2400	Ethylene glycol solution(b) plus 0.03 to 0.05% H_2SO_4; second run; 99 °C (210 °F)	580	23
C44200(c)	2400	Same as above	530	21
C61800	2400	Same as above	380	15
C70600	2400	Same as above	480	19
C71500	2400	Same as above	460	18
C11000	305	Glycol maleate, 79 °C (175 °F)	20	0.8

(a) 87 to 95% glycol. (b) 15% glycol, 85% H_2O. (c) Uninhibited admiralty metal, now an obsolete alloy.

Table 24 Corrosion of copper metals in various alcohols

Alloy	Exposure time, h	Test conditions	Average penetration rate μm/yr	mils/yr
C11000	503	Crude C-5 alcohols; 126 to 140 °C (259 to 284 °F)	7.5	0.3
C11000	210	Crude decyl alcohol; 175 °C (347 °F)	3 to 5	0.1 to 0.2
C11000	288	Primary decyl alcohol; 175 °C	15 to 45	0.6 to 1.8
C65500	288	Same as above	20 to 60	0.8 to 2.4
C44400	8160	Isopropanol and water; 118 to 145 °C (244 to 293 °F)	10 to 38	0.4 to 1.5
C23000	8160	Same as above	10 to 56	0.4 to 2.2
C11000	8160	Same as above	8 to 75	0.3 to 3.0
C65500	8160	Same as above	10 to 63	0.4 to 2.5
C11000	264	Allyl alcohol; refluxed at 88 °C (190 °F)	25	1
C11000	94	Methanol; boiling	Nil	Nil
C11000	46	Denaturing grade ethanol; boiling	25	1
C23000	165	2-ethyl—2-butyl—1,3 propanediol; 45 °C (113 °F)	5	0.2

the metal. This condition is generally known as "hydrogen embrittlement." Any degree of embrittlement can lead to catastrophic failure, and should be avoided; there is no "safe" depth of attack.

Figure 4 shows the depth of damage or "embrittlement" of C11000 after it has been heated in hydrogen at about 600 °C (1100 °F) for varying times. The reaction is especially important when oxygen-containing copper is "bright annealed" in reducing atmospheres containing relatively small amounts of hydrogen (1 to 1.5%). Annealing of tough pitch coppers in such atmo- spheres at temperatures much above 475 °C (900 °F) may lead to severe embrittlement, especially when annealing times are long. In fact, tough pitch coppers should not be exposed to hydrogen at any temperature if they will subsequently be exposed to temperatures above 370 °C (700 °F).

When tough pitch coppers are welded or brazed, the possibility of hydrogen embrittlement must be anticipated and hydrogen atmospheres must not be used. Where copper must be heated in hydrogen atmospheres, an oxygen-free copper or deoxidized copper with high residual deoxidizer content should be selected. No hydrogen embrittlement problems have been encountered with these materials.

Dry Oxygen. Copper and copper alloy tubes are used to convey oxygen at room temperature, as in hospital oxygen service systems. When heated in air, copper develops a cuprous oxide film that exhibits a series of interference tints (temper colors) as it increases in thickness. The colors associated with different oxide-film thicknesses are shown in the following table.

Color	Film thickness, nm
Dark brown	37-38
Very dark purple	45-46
Violet	48
Dark blue	50-52
Yellow	94-98
Orange	112-120
Red	124-126

Black cupric oxide forms over the cuprous oxide layer as the film thickness increases above the interference-color range.

When copper is used at high temperatures in air or oxygen scaling results. At low temperatures up to 100 °C (212 °F), the oxide film increases in thickness logarithmically with time. There is an irregular increase of scaling rate with further increases in temperature, and a rapid increase with pressure up to 1.6 kPa (12 mm of Hg). Above 20 kPa (150 mm of Hg), the rate of increase is steady. Beyond the interference-color range, the rate of growth of oxide film is approximately defined by:

$$W^2 = kt \qquad \text{(Eq 1)}$$

Where W is weight gain (or increase in equivalent thickness) per unit area, t is

Table 25 Corrosion of C44300 and C71500 exposed to gasoline in a refinery

Service condition(a)	Temperature °C	°F	Average penetration rate µm/yr	mils/yr
C44300				
Straight-run (untreated)				
Tower liquid(b)	121	250	1270 min	50 min
Storage(c)	4-27	40-80	63	2.5
Distilled tops from straight-run gasoline(d)	35	95	1270	50
Cracked gasoline (top tray in tower)(e)	204	400	15	0.6
Sweet gasoline vapor(f)	177	350	7.5	0.3
C71500				
Straight-run (untreated)				
Tower liquid(b)	121	250	180	7
Storage(c)	4-27	40-80	180	7
Distilled tops from straight-run gasoline(d)	35	95	1140	45
Cracked gasoline (top tray in tower)(e)	204	400	200	8
Sweet gasoline vapor(f)	177	350	10	0.4
Aviation gasoline (top of column)	121	250	2.5	0.1

(a) Gasoline or related hydrocarbons will not attack copper or its alloys. Attack depends on the type and amount of impurities in the gasoline, such as water, sulfides, mercaptans, aliphatic acids, naphthenic acids, phenols, nitrogen bases, and dissolved gases. (b) 100 lb of H_2S present per 1000 bbl of gasoline. (c) 0.02 to 0.03 g H_2S per litre of gasoline. (d) pH controlled by NH_3. (e) H_2S and HCl present. (f) Vacuum operation.

Table 26 Corrosion of copper metals in contaminated naphtha

Alloy	Corrosion rate µm/yr	mils/yr
At 21 °C (70 °F)(a)		
C23000	230	9
C46400	50	2
C28000	75	3
C44200	200	8
C11000	1270	50
At 177 °C (350 °F)(b)		
C23000	2030	80
C46400	10	0.4
C28000	10	0.4
C44200	200	8

(a) The naphtha contained H_2S, H_2O and HCl. (b) The naphtha contained H_2S, mercaptans and naphthenic acids.

Table 27 Corrosion of copper metals in beet-sugar solution

Alloy	Decrease in tensile strength, %, for test rack number(a): 1	2	3	4
C11000	0	4.0	3.5	0
C44300	2.0	9.5	11.5	2.5
C44400	0	3.0	6.0	0
C44500	4.5	9.0	12.5	5.5
C71000	1.0	4.5	7.0	0
C71500	0	5.0	8.0	0

(a) Corrosion specimens (0.8-mm-thick strips) were exposed in contact with beet-sugar solution for 100 days in normal refinery operations. Test racks 1 and 4 were at the finishing pan containing Steffen's filtrate; rack 2 was in the first-effect thin-juice evaporator; rack 3 was at the third body of the triple-effect evaporator.

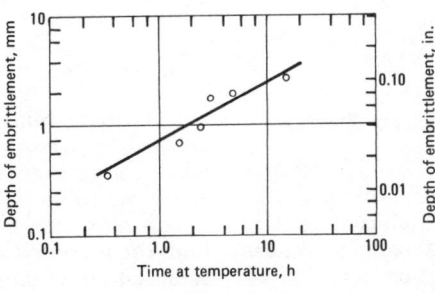

Fig. 4 Hydrogen embrittlement of tough pitch coppers heated in pure hydrogen at 600 °C (1100 °F)

time and k is a constant of proportionality. Values for k are given in Table 29. Different investigators report different oxidation rates; however, values reported by N. B. Pilling and R. E. Bedworth (*J Inst Metals*, Vol 29, 1923, p 529-582) seem to be as reliable as any.

Low concentrations of lead, oxygen, zinc, nickel and phosphorus in copper have little influence on oxidation rate. Silicon, magnesium, beryllium and aluminum form very thin insulating (nonconductive) oxide films on copper, which protect the metal surface and retard oxidation.

Table 29 Values of rate constant for oxide growth on unalloyed copper

Temperature °C	°F	Rate constant, k (a) Pure O_2	Air
400	750	4.4×10^{-8}	...
500	950	4.4×10^{-7}	...
600	1100	3.24×10^{-6}	...
700	1300	1.6×10^{-5}	8.03×10^{-6}
800	1475	8.69×10^{-5}	7.97×10^{-5}
900	1650	3.49×10^{-4}	3.36×10^{-4}
950	1750	7.30×10^{-4}	...
1000	1850	1.78×10^{-3}	1.35×10^{-3}

(a) For calculation of weight gain in g/m^2 from Eq 1 when time is measured in seconds.

Table 28 Corrosion of copper alloys in hot paper-mill vapor containing SO_2(a)

UNS number	Alloy type	Weight loss, $g/m^2 \cdot d$
...	Bronze (90Cu-10Sn)	22.0
C61800	Aluminum bronze	26.4
C51100	Phosphor bronze	28.6
C73200	Nickel silver, 75-20	35.6
C52100	Phosphor bronze, 8% C	39.4
C65800	Silicon bronze	50.2
C77000	Nickel silver, 55-18	63.8
C75200	Nickel silver, 65-18	67.4
...	Nickel bronze (88.5Cu-5Sn-5Ni-1.5Si)	70.5

(a) Temperature, 200 to 220 °C (390 to 430 °F); atmosphere, 17 to 18% SO_2 plus 1 to 2% O_2; test duration, mainly 30 days, but some longer.

Stress Relaxation in Copper and Copper Alloys

By A. Fox
Bell Laboratories

COPPER and copper alloys are used extensively in structural applications in which they are subject to moderately elevated temperatures. Examples include automotive radiators, solar heating panels, communications cable and electrical connectors. At relatively low operating temperatures, these alloys can undergo stress relaxation (decrease in stress resulting from transformation of elastic strain into plastic strain in a constrained solid), which can lead to service failures. Because of the wide variations in composition and processing among commercial copper alloys, resistance to stress relaxation varies considerably. Of course, selection of an alloy for a given application is based not only on stress-time-temperature response but also on such factors as cost, basic mechanical and physical properties, operating temperature, service environment and formability. For many applications, electrical conductivity is a primary consideration.

Stress Relaxation Data

Unalloyed copper C11000 (electrolytic tough pitch copper) is probably the most inexpensive high-conductivity copper and is used extensively because of its ease of fabrication. The stress-relaxation behavior of this material is rather poor, as demonstrated in Fig. 1, in which relaxed stress is plotted as a function of time and temperature for 0.25-mm (0.010-in.) C11000 wire initially stressed in tension to 89 MPa (13 ksi). Comparison of stress values at a given time for different temperatures illustrated the very sharp dependence of stress relaxation on temperature for this copper. At 93 °C (200 °F), for example, no tension remains after 10^5 h (11.4 years), whereas 40% of the initial stress remains after 40 years at room temperature. For C11000 and for many other copper metals, stress relaxation in a given time period is inversely proportional to absolute temperature (Ref 1).

The stress-relaxation behavior of C10200 (oxygen-free copper) is somewhat better than that of C11000, as shown in Table 1, which also presents stress-relaxation data for many other high-conductivity copper metals. (For compositions of these metals, see Table 2; basic mechanical properties are given in Table 3.) A more extensive comparison of the mechanical behavior of C10200 and C11000 has been presented by Opie, Taubenblat and Hsu (Ref 2).

Among the high-conductivity coppers, relaxation is greatest in very-high-purity copper (99.999 + %)—a material used mainly in research. Improvement in the stress-relaxation behavior of high-conductivity copper can be achieved by adding alloying elements that cause solid-solution strengthening, age hardening or dispersion hardening (Ref 3). For example, minute additions of silver sig-

Table 1 Tensile-stress-relaxation data for selected types of copper wire

Material	Temper	Length of test, h	Temperature °C	°F	Initial stress MPa	ksi	Percent of initial stress remaining after: 10 000 h	40 years
0.25 mm (0.01-in.) Diam Wire								
C10200, tinned	O61	10 000	27	80	41.0	5.95	72	55
		10 000	27	80	61.5	8.92	70	53
		10 000	27	80	82.0	11.9	69	50
		2850	121	250	82.0	11.9	15	0
		2850	149	300	82.0	11.9	6	0
C10200, tinned	H04	10 000	27	80	79.9	11.6	82	68
		8600	66	150	88.9	12.9	78	68
		9300	93	200	88.9	12.9	67	42
		2850	121	250	88.9	12.9	55	37
		2850	149	300	88.9	12.9	42	18
		10 000	27	80	160	23.2	80	68
		8600	66	150	160	23.2	69	57
		9300	93	200	160	23.2	59	43
		2850	121	250	160	23.2	40	14
		2850	149	300	160	23.2	22	0
C11000, tinned	O61	10 000	23	73	44.8	6.5	60	41
		9300	66	150	44.8	6.5	47	22
		9700	93	200	44.8	6.5	32	3
		2850	121	250	44.8	6.5	12	0
		2850	149	300	44.8	6.5	12	0
		10 000	23	73	88.9	12.9	60	38
		9300	66	150	88.9	12.9	30	6
		9700	93	200	88.9	12.9	20	0
		2850	121	250	88.9	12.9	8	0
		2850	149	300	88.9	12.9	8	0
C12000, tinned	O61	10 000	27	80	52.4	7.6	86	80
		10 000	27	80	77.9	11.3	85	79
		10 000	27	80	104	15.1	84	78
C13400, tinned(a)	H00	2833	93	200	88.9	12.9	50	27
		2833	93	200	101	14.7	49	28
		2833	93	200	152	22.1	45	25
		2833	93	200	203	29.5	42	19
C13700, tinned(b)	H00	9700	23	73	88.9	12.9	88	83
		9300	66	150	88.9	12.9	78	67
		9700	93	200	88.9	12.9	70	52
		2850	121	250	88.9	12.9	51	27
		2760	149	300	88.9	12.9	41	8
		9700	23	73	136	19.7	86	81
		9300	66	150	136	19.7	77	64
		9700	93	200	136	19.7	67	48
		2850	121	250	136	19.7	42	19
		2760	149	300	136	19.7	28	0
C15000, tinned	H04(c)	9700	23	73	88.9	12.9	93	92
		9300	66	150	88.9	12.9	93	89
		9700	93	200	88.9	12.9	92	82
		2850	121	250	88.9	12.9	82	78
		2850	149	300	88.9	12.9	80	76
		9700	23	73	203	29.5	93	92
		9300	66	150	203	29.5	93	87
		9700	93	200	203	29.5	92	82
		2850	121	250	203	29.5	80	76
		2850	149	300	203	29.5	78	74

(continued)

Table 1 (continued)

Material	Temper	Length of test, h	Temperature °C	Temperature °F	Initial Stress MPa	Initial Stress ksi	Percent of initial stress remaining after: 10 000 h	Percent of initial stress remaining after: 4 years
C15000, bare	H04(c)	9700	23	73	88.9	12.9	96	95
		9600	66	150	88.9	12.9	96	95
		9700	93	200	88.9	12.9	96	95
		9700	23	73	203	29.5	96	95
		9600	66	150	203	29.5	96	95
		9700	93	200	203	29.5	86	79
C15000, tinned	H00	2800	93	200	88.9	12.9	96	91
		2800	93	200	128	18.6	95	90
		2800	93	200	192	27.9	94	89
C15000, silver plated		2800	93	200	256	37.2	93	89
	(d)	9800	27	80	74.4	10.8	97.9	95
		9800	27	80	112	16.2	98.8	94
		9800	27	80	149	21.6	96.7	93
C16200, tinned	H04(e)	9700	23	73	88.9	12.9	97	94
		9700	66	150	88.9	12.9	93	92
		9700	93	200	88.9	12.9	92	87
		2800	121	250	88.9	12.9	79	71
		2800	149	300	88.9	12.9	62	40
		9700	23	73	226	32.8	95	92
		9700	66	150	226	32.8	91	88
		9700	93	200	226	32.8	88	84
		2800	121	250	226	32.8	77	64
		2800	149	300	226	32.8	60	34
C16200, tinned	H00	2800	93	200	88.9	12.9	91	85
		2800	93	200	114	16.6	91	84
		2800	93	200	172	24.9	91	84
		2800	93	200	229	33.2	91	84
0.5-mm (0.02-in.) Diam Wire								
C10200	O61	22 600	27	80	58.6	8.5	81	71
		22 600	27	80	75.8	11.0	81	71
		22 600	27	80	86.2	12.5	81	71
		22 600	27	80	103	15.0	79	70
		22 600	27	80	110	16.0	78	67
C10200	H00	4060	93	200	68.9	10.0	48	9
		4060	93	200	142	20.6	42	0
C11000	O61	35 000	27	80	34.5	5.0	60	43
		35 000	27	80	68.9	10.0	55	39
C1100	O61	24 500	27	80	34.5	5.0	60	38
		24 500	27	80	41.2	6.0	60	38
		24 500	27	80	51.7	7.5	59	38
		24 500	27	80	68.9	10.0	57	38
		24 500	27	80	82.7	12.0	56	38
		24 500	27	80	96.5	14.0	55	37
C11000	H00	4100	93	200	68.9	10.0	35	6
		4100	93	200	121	17.5	23	0
C11600	H00	4100	93	200	68.9	10.0	50	20
		4100	93	200	143	20.7	43	18
C13400	H00	4100	93	200	68.9	10.0	53	27
		4100	93	200	148	21.4	38	14
C15500, bare	H00	4060	93	200	68.9	10.0	78	62
		4060	93	200	164	23.8	74	60
C16200	H00	4100	93	200	68.9	10.0	88	82
		4100	93	200	158	22.9	80	69

(a) Boron-deoxidized copper containing 0.027% Ag. (b) Boron-deoxidized copper containing 0.085% Ag. (c) In-process strand annealed. (d) Proprietary mill processing. (e) Batch annealed.

Fig. 1 Tensile-stress-relaxation characteristics of C11000

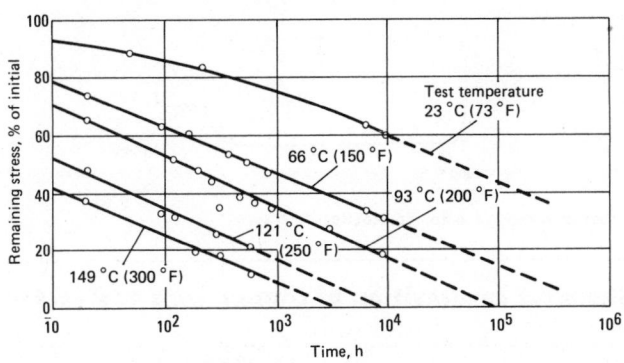

Data are for tinned 30 AWG (0.25-mm-diam) annealed ETP copper wire; initial elastic stress, 89 MPa (13 ksi).

Fig. 2 Stress relaxation in C17200 at two levels of initial stress

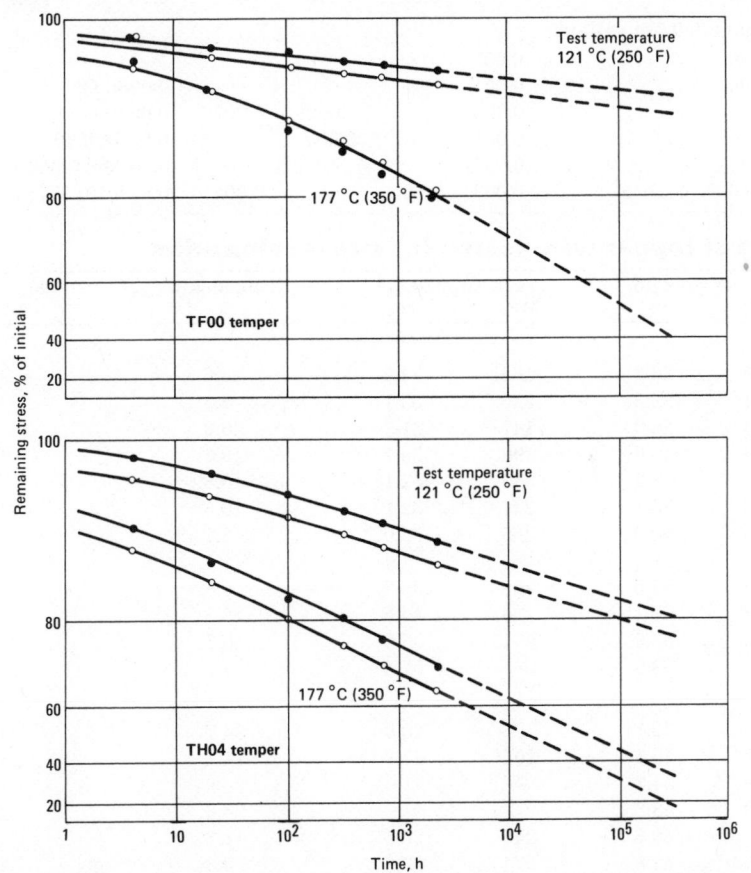

Data are for beryllium copper (1.9% Be) strip, 0.38 mm (0.015 in.) thick. Open symbols are for an initial test stress equal to 80% of the monotonic bending yield stress; solid symbols, for an initial stress 50% of the bending yield stress.

nificantly reduce stress relaxation in copper (Ref 4, 5).

Besides the strengthened high-conductivity coppers for which stress-relaxation data are given in Table 1, proprietary coppers strengthened with small amounts of cadmium, chromium, zirconium or a combination of one or more of these elements have been found to have good to superior stress relaxation resistance (Ref 1). Processing variations that strengthen copper metals, even including internal oxidation, are almost always beneficial.

As with high-conductivity copper metals, low-conductivity copper metals that are strengthened by precipitation hardening or spinodal decomposition exhibit improved resistance to stress relaxation (Ref 7). C17200 is a typical example of a low-conductivity copper alloy strengthened by precipitation hardening. Stress-relaxation data for this alloy are shown in Fig. 2 for two levels of cold work. C17200 has adequate resistance to stress relaxation at temperatures up to 120 °C (250 °F), provided the initial stress is below the elastic limit. Other resistant alloys include some of the phosphor bronzes, nickel silvers, copper nickels, beryllium coppers, and copper-nickel-tin alloys.

Stress Relaxation in Mechanical Components

A solderless wrapped connection such as the one shown in Fig. 3, in which electrical contact is made by wrapping a wire around a terminal, is a typical application for high-conductivity copper metals where stress relaxation is of concern. Typical operating temperatures can be as high as 85 °C (185 °F); generally, conductivities higher than 98% IACS at 20 °C (68 °F) are desirable. After the connection is made, it is maintained by elastic stresses in the two members. If the wire undergoes stress relaxation, electrical contact between the wire and the terminal may be lost.

The spring in the alarm fuse shown in Fig. 4 is a typical application for copper alloys with room-temperature conductivities of 55 to 85% IACS. This spring conducts relatively high electrical currents and also triggers an alarm circuit if the fuse blows. To perform the latter function reliably, the spring must retain spring force for extended periods of time. But if the spring material undergoes stress relaxation, the device may fail to trigger the alarm when the fuse blows. C16200 (Cu-1Cd) has been used successfully in spring loaded alarm fuses operating at temperatures below 95 °C (200 °F). (This alloy has an electrical conductivity of 80 to 85% IACS at 20 °C). For higher operating temperatures up to 165 °C (330 °F), C19000 (Cu-Ni-P alloy) springs have performed adequately, provided the ratio of nickel to phosphorus is at least 5 to 1 (Ref 6). (The nominal ratio for this alloy is about 3.5 to 1.) For applications where lower electrical conductivity can be tolerated, C17500 (Cu-2.5Co-0.6Be; conductivity, 45% IACS at 20 °C, is a satisfactory alternative for temperatures up to 165 °C. Both C19000 and C17500 must be age hardened after forming.

A typical application of the lower-conductivity, high-strength copper alloys is the pressure-type, split-beam connector shown in Fig. 5. The knife edges of the connector first cut through the insulation on the conductor and then must maintain electrical contact with it. Materials used for connectors of this type, depending on operating stress and temperature, include some of the phosphor bronzes, nickel silvers, copper nickels, beryllium coppers and some of the newer copper-nickel-tin alloys (Ref 8) strengthened by spinodal

Fig. 3 Typical solderless wrapped connection

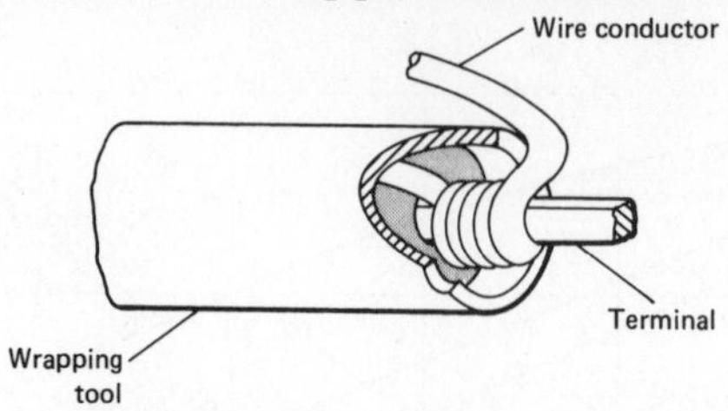

Wrapping tool is removed after connection is made.

Table 2 Chemical composition of copper wire tested for stress relaxation

Material	Ag	Pb	Composition, % Fe	Ni	Others
0.25-mm (0.01-in.) Diam					
C10200	0.002	...	...	...	...
C11000	0.002	0.001	0.002	0.001	...
C12000	0.002	...	...	...	...
C13400	0.031	...	0.003	...	0.02 B, 0.001 Si
C13700	0.090	0.001	0.004	...	0.01 B, 0.001 Si
C15000	0.003	0.001	0.002	...	0.001 Mg, 0.15 Zr
C16200	0.005	0.01	0.015	0.005	0.75 Cd, 0.01 Sn
0.05-mm (0.02-in.) Diam					
C10200	0.002	...	...	...	...
C11000	0.001	...	...	...	0.0355 O
C13000	0.083	...	0.0029	0.0079	0.0310 O
C13400	0.031	...	0.003	...	0.02 B, 0.001 Si
C15500	0.037	...	...	...	0.10 Mg, 0.063 P
C16200	0.003	0.005	...	0.005	0.97 Cd, 0.007 Sn, 0.07 Zn

Table 3 Typical mechanical properties of copper wire tested for stress relaxation

Material	Temper	Tensile strength MPa	ksi	Yield strength(a) MPa	ksi	Elongation(b), %	Conductivity, % IACS
0.25-mm (0.01-in.) Diam							
C10200, tinned	O61	250	36.3	160	23.2	24.6	99
C10200, tinned	H04	271	39.3	228	33.1	2.3	99
C11000, tinned	O61	254	36.9	147	21.3	30.0	99
C12000, tinned	O61	276	40.0	198	28.7	18.1	98
C13400, tinned	H00	365	53.0	359	52.1	0.82	98
C13700, tinned	H00	242	35.1	221	32.1	10.5	99
C15000, tinned	H04	414	60.0	365	53.0	3.7	90
C15000, bare	H04	...	...	...	...	...	...
C15000, tinned	H00	393	57.0	388	56.3	0.82	93
C15000, silver plated	...	347	50.3	298	43.3	13.5	...
C16200, tinned	H04	473	68.6	404	58.6	6.6	85
C16200, tinned	H00	403	58.5	395	57.3	0.91	85
0.05-mm (0.02-in.) Diam							
C10200, tinned	O61	249	36.1	138	20.0	34.7	101
C10200, bare	H00	278	40.3	261	37.8	11.1	101
C11000, tinned	O61	239	34.7	...	...	28.8	101
C11000, silver plated	O61	233	33.8	...	...	24.6	101
C11000, bare	H00	280	40.6	245	35.5	11.7	101
C11600	H00	295	42.8	274	39.8	9.5	98
C13400	H00	258	37.5	240	34.8	9.4	98
C15500, bare	H00	367	53.2	332	48.1	1.9	93.8
C16200	H00	312	45.3	276	40.1	10.9	85

(a) 0.2% offset. (b) In 254 mm (10 in.).

Fig. 4 Typical spring-type alarm fuse

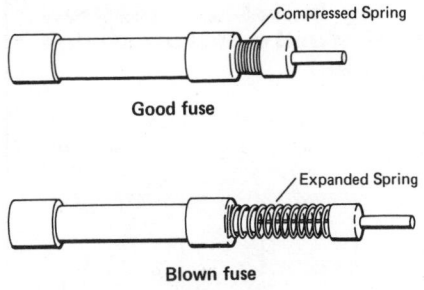

Good fuse

Blown fuse

Fig. 5 Typical quick clip connection

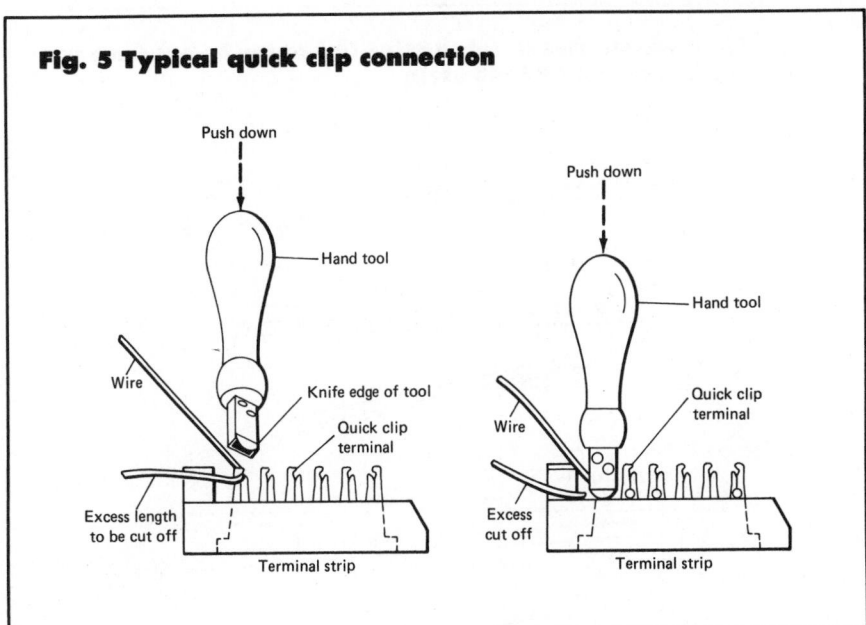

decomposition. For copper alloys in which high strength is achieved through cold rolling, resistance to stress relaxation can be significantly improved by stress-relief aging (Ref 9, 10). This is illustrated for C51000 in Fig. 6 and for C72500 in Fig. 7.

Stress relaxation can produce mechanical or thermal ratcheting, which sometimes occurs in multilayer circuit boards such as the one illustrated in Fig. 8. A multilayer board usually consists of an epoxy-glass composite substrate with several lands of electroplated copper and including a plated-through hole. When a leadwire is soldered into the plated-through hole, differential thermal expansion causes the hole to expand more than the substrate, and a tensile stress is applied to the copper barrel. While at temperature, the stressed electroplated copper relaxes according to behavior that varies with the plating system and bath used (Ref 11). If the board is repeatedly heated and cooled, the electroplated copper alternately expands and relaxes during each heating cycle, and permanent strain accumulates in the copper barrel until the assembly fails by buckling or by low-cycle fatigue.

REFERENCES

1. A. Fox, Stress-Relaxation Characteristics in Tension of High-Strength, High Conductivity Copper and High Copper Alloy Wires, *Journal of Testing and Evaluation,* Vol 2, No. 1 (Jan 1974), p 32–39

2. W. R. Opie, P. W. Taubenblat and Y. T. Hsu, A Fundamental Comparison of the Mechanical Behavior of Oxygen-Free and Tough Pitch Coppers, *Journal of the Institute of Metals,* Vol 98, 1970, p 245

Fig. 6 Anisotropic stress-relaxation behavior in bending for highly cold worked C51000 strip

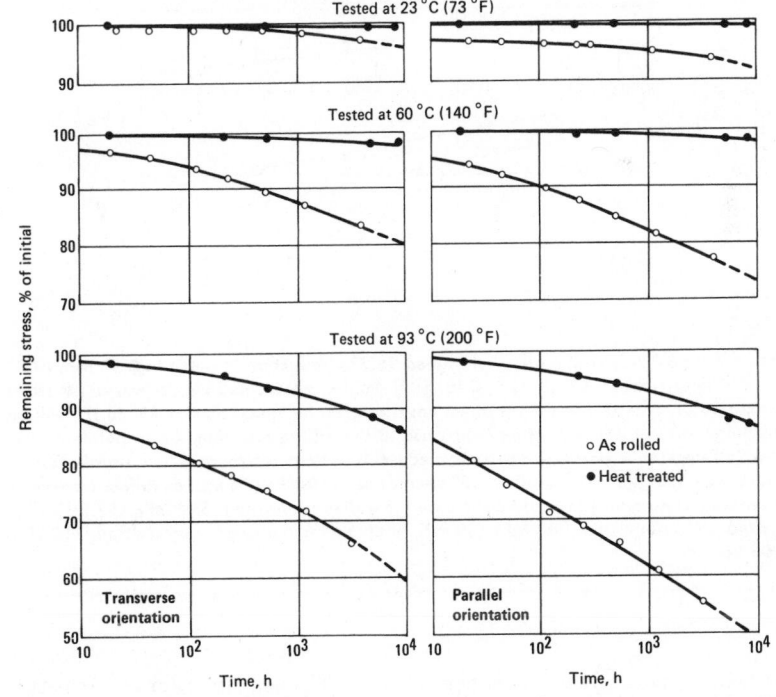

Data are for 5% Sn phosphor bronze cold rolled 93% (reduction in area) to 0.25 mm (0.01 in.) and heat treated 2 h at 260 °C (500 °F). Graphs at left are for stress relaxation transverse to the rolling direction; graphs at right, for stress relaxation parallel to the rolling direction. Initial stresses: as rolled, parallel orientation, 607 MPa (88 ksi); as rolled, transverse orientation, 634 MPa (92 ksi); heat treated, parallel orientation, 641 MPa (93 ksi); heat treated, transverse orientation, 738 MPa (107 ksi).

Fig. 7 Anisotropic stress-relaxation behavior in bending for highly cold worked C72500 strip

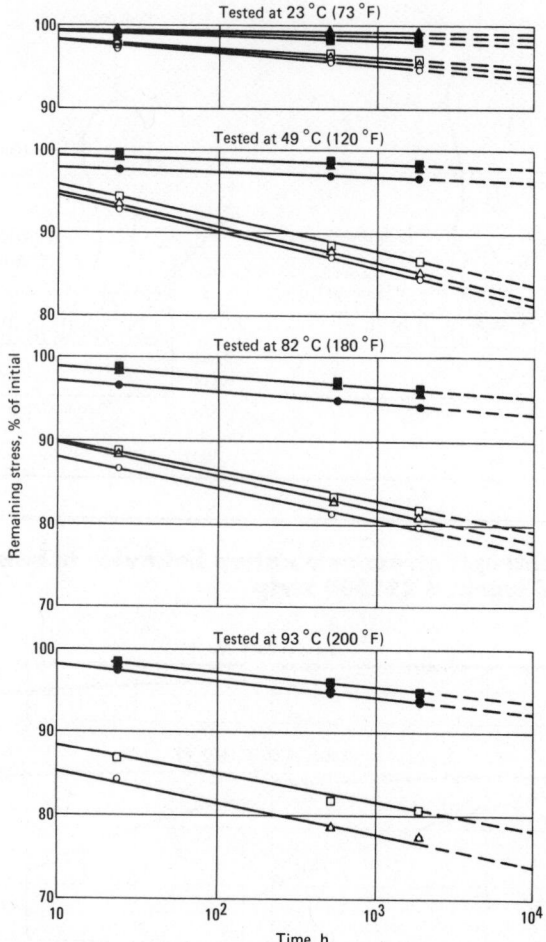

Data are for 89Cu-9Ni-2Sn alloy cold rolled 98.7% (reduction in area) to 0.25 mm (0.01 in.) and heat treated 2 h at 357 °C (675 °F). Points represented by circles are for stress relaxation parallel to the rolling direction; triangles, for relaxation at 45° to the rolling direction; squares, for relaxation transverse to the rolling direction. Open points are for as-rolled stock; solid points, for heat treated stock. Initial stresses: as rolled, parallel orientation, 524 MPa (76 ksi); as rolled, 45° orientation, 510 MPa (74 ksi); as rolled, transverse orientation, 586 MPa (85 ksi); heat treated, parallel orientation, 669 MPa (97 ksi); heat treated, 45° orientation, 552 MPa (80 ksi); heat treated transverse orientation, 710 MPa (103 ksi).

Fig. 8 Cutaway view through a typical multilayer circuit board showing barrel-land construction and a plated-through hole

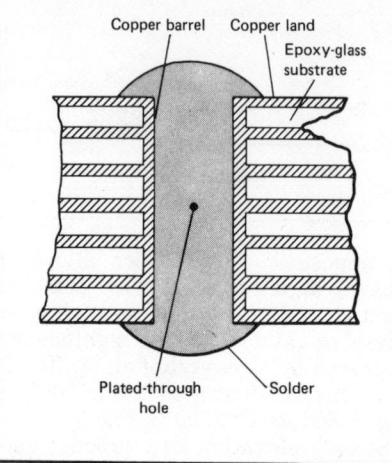

3. A. Fox and J. J. Swisher, Superior Hook-Up Wires for Miniaturized Solderless Wrapped Connections, *Journal of the Institute of Metals*, Vol 100, 1972, p 30

4. A. Fox, "The Creep and Stress Relaxation Behavior of Silver Bearing Copper Wire", Master's Thesis, New Jersey Institute of Technology, 1972

5. W. L. Finlay, *Silver-Bearing Copper: A Compendium of the Origin, Characteristics, Uses and Future of Copper Containing 12 to 25 Ounces per Ton of Silver*, Corinthian Editions, New York, 1968

6. A. Fox and R. C. Stoffers, High Conductivity, High Strength Wire Springs for Fuse Applications, *Electrical Contacts/1977 Proceedings of the Annual Holm Conference on Electrical Contacts*, p 233–240

7. A. Fox and E. O. Fuchs, Mechanical Properties in Bending at Elevated Temperature of High Strength Copper Alloy Flat Spring Materials, *Journal of Testing and Evaluation*, Vol 6, No. 3 (May 1978), p 211–220

8. J. T. Plewes, Spinodal Cu-Ni-Sn Alloys Are Strong and Superductile, *Metal Progress*, July 1974, p 46

9. A. Fox, The Effect of Extreme Cold Rolling on the Stress Relaxation Characteristics of CDA Copper Alloy 510 Strip, *Journal of Materials*, Vol 6, No. 2 (June 1971), p 422–435

10. A. Fox, Stress Relaxation and Fatigue of Two Electromechanical Spring Materials Strengthened by Thermomechanical Processing, *IEEE Transactions on Parts, Materials and Packaging*, Vol PMP-7, No. 1 (March 1971), p 34–47

11. A. Fox, Mechanical Properties at Elevated Temperature of Cu Bath Electroplated Copper for Multilayer Boards, *Journal of Testing and Evaluations*, Vol 4, No. 1 (Jan 1976), p 74–84

Lead

Lead and Lead Alloys

By Jerome F. Smith
Vice President and Secretary
Lead Industries Assoc.
and
Raymond R. Kubalak (deceased)
Manager
Market Development—Lead
St. Joe Minerals Corp.

LEAD was one of the first metals known to man. Probably the oldest lead artifact is a figure made about 3000 BC, which was found in the temple of Osiris by Abydos. All civilizations beginning with the ancient Egyptians, Assyrians and Babylonians have used lead for many ornamental and structural purposes. Many magnificent buildings erected in the 15th and 16th centuries still stand under their original lead roofs.

Pipe was one of the earliest applications of lead. The Romans produced 15 standard sizes of water pipe in regular 10-ft lengths. In the old Roman baths at Bath, England, the lead pipe installed by the Romans 1900 years ago is still in use. The Romans made their pipe by folding heavy sheets of cast lead and fusing the seams together.

Sources of Lead. Although there are at least 60 known lead-containing minerals, by far the most important as a source of primary lead is galena (PbS). Recycling of scrap lead (from batteries, lead sheet and cable sheathing) is also a major source, providing nearly half the lead used in the United States. Large amounts of antimonial lead and some soft lead are produced from scrap. Considerable tonnages of scrap solder and bearing metals are recovered and used again, generally after addition of sufficient new metal to rectify the composition.

Except in the Missouri lead belt and a few other base-metal mining areas, galena generally is associated with substantial amounts of zinc and copper minerals and is not mined independently. Whenever possible, galena is separated from other minerals through differential flotation. Sometimes, however, the mineral particles are so small and so tightly locked together that separation by flotation is not possible. In such instances, bulk concentrates containing lead, zinc and copper are produced, which subsequently are separated by the Imperial Smelting Process.

More lead ore is mined in the United States than in any other country; Canada is second, followed by Australia, Peru and Mexico. Total annual world production of lead from ores is about 2.4 million tonnes (2.6 million tons).

Smelting and Refining. When they leave the concentrating mill, lead concentrates contain more than 40% lead—generally about 70%. They are roasted to remove sulfur and to agglomerate the fine flotation product, which is not physically desirable in a blast furnace. The roasted sinter is charged into the top of a blast furnace along with suitable fluxes and coke; the resulting impure lead-base bullion, containing silver, gold and various impurities, is shipped to the refinery for further treatment before the metal is suitable for industrial use.

In initial refining, copper is removed by addition of sulfur or sulfur-bearing material to molten furnace bullion (usually in the smelter). The resulting copper-rich drosses are retreated in a reverberatory furnace to produce a high-copper matte, which is further treated elsewhere to recover the copper.

In the next step, antimony, tin and arsenic are removed—usually by the Harris process. In this process, additions of sodium hydroxide and sodium nitrate react with the antimony, tin and arsenic to form sodium salts, which then are removed and treated to recover the metals. Alternatively, the antimony, tin and arsenic can be oxidized in a reverberatory furnace (by air blowing) and later recovered from the resulting oxide dross or slag.

Precious metals are removed next, by either electrolytic or nonelectrolytic processing. In Park's process, a non-electrolytic method, zinc is added in small amounts and forms intermetallic compounds with the precious metals. These compounds are skimmed from the molten bullion and further processed to recover the gold and silver. Residual zinc generally is removed by the vacuum dezincing process of W. T. Isbell. In electrolytic practice, the lead bullion is cast into anodes, which are placed in electrolytic cells. Under the action of the cell current, purified lead is deposited at the cathode, and precious metals precipitate to the bottom of the cell as a slimy residue, which is reworked to recover them.

Bismuth, if present, can be removed either by electrolytic refining or by the Betterton process, in which a Ca-Mg alloy is added to the molten lead and selectively reacts with the bismuth to form an intermetallic, floating dross.

The purity of commercial lead refined by these processes varies with ore composition and processing conditions, but with final purification using caustic or nitre, a purity of 99.99+% is commonly attained. Pig lead generally is cast into bars weighing roughly 30 to 45 kg (60 to 100 lb), although blocks weighing about 900 kg (1 ton) have become more popular in recent years, particularly among large users.

Grades of Lead. Composition limits of the four grades of pig lead covered in ASTM B29 are given in Table 1. These grades are corroding lead and common lead (both containing 99.94% min lead), and chemical lead and copper-bearing lead (both containing 99.90% min lead). Lead of higher specified purity (99.99%) also is available in commercial quantities.

Most lead produced in the United States is corroding lead (99.94% min Pb). This grade, which exhibits the outstanding corrosion resistance typical of lead and its alloys, is named not for a characteristic of the metal but rather for a process by which it was formerly produced. Corroding lead is used in making pigments, lead oxides, and a wide variety of other lead chemicals. The required high purity is necessary to avoid problems caused by impurities during and after processing, such

as unwanted colors in white lead pigment.

Refined lead with a residual copper content of 0.04 to 0.08% and a residual silver content of 0.002 to 0.02% is particularly desirable in the chemical industries and thus is called "chemical lead". (Most chemical lead produced in the United States is refined from the lead ores of southeastern Missouri, which contain small amounts of copper and silver.) The copper content significantly improves corrosion resistance and mechanical strength, making chemical lead the next most frequently used grade after corroding lead. The silver content also improves corrosion resistance in some applications.

Copper-bearing lead provides corrosion protection comparable to that of chemical lead in most applications that require high corrosion resistance. This grade is made by adding copper to fully refined lead and differs primarily from chemical lead in its higher bismuth contents.

Lead of the common grade, which contains lower amounts of silver and copper than either chemical lead or copper-bearing lead, has more than adequate corrosion resistance for the many uses of lead in which high corrosion resistance is not required.

Specifications other than ASTM B29 for grades of pig lead include federal specification QQ-L-171, German standard DIN1719 and British specification BS334.

There are two grades of lead besides the pig lead grades: arsenical lead and calcium lead. Arsenical lead, used for cable sheathing, contains about 0.15% arsenic, 0.10% tin and 0.10% bismuth.

Calcium lead, used for lead-acid storage batteries and for casting applications, contains 0.03 to 0.11% calcium.

Lead-base alloys are used commercially in very large amounts; antimony and tin are the most common alloying elements. Antimony generally is used to give greater hardness and strength, as in storage-battery grids, sheet, pipe and castings. Antimony contents of Pb-Sb alloys range from 1 to 25%, usually 4 to 7%. (Within the past decade, calcium leads have become prominent in a number of the same applications for which antimonial lead is used—particularly, storage-battery grids.) Adding tin to lead or to lead alloys increases hardness and strength, but Pb-Sn alloys are more commonly used for their good melting and casting properties, as in type metals and solders. In solders and terne metal, tin imparts the ability to wet and bond with metals such as steel and copper (unalloyed lead has poor wetting characteristics). Lead combined with tin and bismuth forms the principal ingredient of many low-melting alloys.

Properties of Lead

The properties of lead that make it useful in a wide variety of applications are density, malleability, lubricity, flexibility and coefficient of thermal expansion, all of which are quite high; and elastic modulus, elastic limit, strength, hardness, melting point and electrical conductivity, all of which are quite low. Lead also has good resistance to corrosion under a wide variety of conditions, is easily alloyed with many other metals and is easily cast.

Table 1 Composition limits of pig leads (ASTM B29)

| Element | Composition(a), % | | | |
	Corroding lead(b)	Common lead(c)	Chemical lead(d)	Acid-copper lead(e)
Silver, max	0.0015	0.005	0.020	0.002
Silver, min	...	...	0.002	...
Copper, max	0.0015	0.0015	0.080	0.080
Copper, min	...	...	0.040	0.040
Silver + copper, max	0.0025	...	...	...
Arsenic + antimony + tin, max	0.002	0.002	0.002	0.002
Zinc, max	0.001	0.001	0.001	0.001
Iron, max	0.002	0.002	0.002	0.002
Bismuth, max	0.050	0.050(f)	0.005	0.025
Lead(g), min	99.94	99.94	99.90	99.90

(a) By agreement between the purchaser and the supplier, analyses may be required and limits established for elements (or compounds) not specified here. (b) "Corroding lead" is a designation used in the trade to describe lead that has been refined to a high degree of purity. (c) "Common lead" is fully refined desilverized lead. (d) "Chemical lead" designates the undesilverized lead produced from southeastern Missouri ores. (e) "Copper-bearing lead" is made by adding copper to fully refined lead. (f) By agreement between the purchaser and the supplier, bismuth levels of up to 0.150% may be allowed. (g) By difference.

Density. The high density of lead makes it very effective in shielding against x-rays and gamma radiation. In very large installations, it is often used for lining concrete structures to greatly reduce the mass of concrete that otherwise would be required.

The combination of high density, high "limpness" (low stiffness) and high damping capacity make lead an excellent material for deadening sound and for isolating equipment and structures from mechanical vibrations.

Because of its high density, lead generally is excluded from use in applications where light weight is important. However, even when light weight is desirable, the high density of lead sometimes can be used to advantage. For example, use of lead in aircraft counterweights often reduces total weight, because lead's high density allows more mass to be concentrated at the point of greatest effect. In addition, lead's low melting point and ease of casting make it possible to fit lead weights into irregular and out-of-the-way spaces.

Malleability, softness and lubricity are three related properties that account for the extensive use of lead in many applications. For example, high malleability is largely responsible for the value of lead as a calking material, enabling it to fill calked joints completely. The softness and self-lubricating properties of lead account in substantial part for its use in bearing alloys, gaskets, washers and lead-headed nails. As a coating on wire or sheet metal, lead acts as a drawing lubricant, and in the form of powder or wire it imparts lubricity to friction materials such as brake linings. The commercial application in which the malleability of lead is used to greatest advantage is manufacture of foil; lead foil often is rolled as thin as 0.01 mm (0.0005 in.), and sometimes even thinner.

On the other hand, the softness of lead requires that care be taken in designing for many applications. For example, excessive stream velocity in lead pipes may result in severe erosion if proper precautions are not taken in system design.

Strength. The fact that lead has low tensile strength and low creep strength must always be considered in designing lead components. However, even when good strength is an essential design criterion, the low strength of lead does not necessarily preclude its use. Lead products can be designed to be self-supporting, or inserts or supports of other materials can be provided. Alloying with certain other metals—notably antimony—is a common method of strengthening lead for many applications. In general, consideration should always be given to supporting lead structures by lead-covered steel straps. When lead is used as a lining in a structure made of a stronger material, the lining can be supported by bonding it to the structure.

Thermal Expansion. The relatively high coefficient of thermal expansion of lead is another important design parameter. In lead roofing and flashing, thermal expansion is always a factor, and is provided for by using small sheets and loose-locking each sheet to the next, thus minimizing both individual and cumulative expansion. In pipelines subject to wide variations in temperature, provision must be made for free expansion. The excellent flexibility of lead can be used to advantage in designing such systems.

Corrosion Resistance. Lead is highly resistant to corrosion by the atmosphere, by waters and by a wide range of chemicals in common use. Where resistance to corrosion must be combined with long service life, the limitations imposed by the mechanical properties of lead, which were discussed previously, must be carefully considered in final design. For a detailed description of the corrosion resistance of lead and lead alloys, see the following article, "Corrosion Resistance of Lead".

Products and Applications

The most significant applications of lead and lead alloys are lead-acid storage batteries (grid plates, posts and support straps), ammunition, cable sheathing and building construction (such as sheet, pipe, solder, and wool for calking). Other important uses include counterweights, battery clamps and other cast products, bearings, ballast, gaskets, type metal, terneplate and foil. Lead in various forms and combinations is finding increased application as a material for controlling sound and mechanical vibrations. Also, in many forms it is important as shielding against x-rays and (in the nuclear industry) gamma radiation. In addition, lead is used as an alloying element in steel and in copper alloys to improve machinability and other characteristics, and is used in fusible (low-melting) alloys for fire sprinkler systems.

Although most of the lead used is in metallic form, substantial amounts are used in the form of lead compounds. These include tetraethyl and tetramethyl lead (used as antiknock compounds in gasoline), litharge (PbO), and various corrosion-inhibiting lead pigments such as red lead (Pb_3O_4), lead chromates, lead silicochromates and lead silicates. Litharge is used in paste mixtures for grid plates of lead-acid storage batteries, in cements, glasses and ceramics, and as a starting material for preparation of many other lead compounds.

Red lead has long been one of the most important rust-inhibiting pigments used in primers and undercoats for protection of steel structures. Commercially important white corrosion-inhibiting pigments are basic lead carbonate, dibasic lead phosphite, dibasic lead phosphosilicate and basic lead silicate. The most important colored pigments are triberic lead chromosilicate, basic lead silicochromate and normal lead silicochromate. The color of the first pigment varies from red to orange, whereas the color of the second is generally yellow. The last is used in yellow paint for marking pavement.

Battery Grids. The largest use of lead is in manufacture of lead-acid storage batteries. These batteries consist of a series of grid plates made from either cast or wrought calcium lead or antimonial lead, pasted with a mixture of lead oxides and immersed in sulfuric acid. The lead from discarded batteries is reclaimed in secondary smelter operations.

The length of the active life of a battery depends on the resistance of the lead alloy grids to corrosion under repeated cycling (charge and discharge) in the sulfuric acid. Automotive batteries usually are made from antimony-lead alloys containing 4 to 6% antimony and various other elements such as tin and calcium. The exact composition used varies with manufacturer. "Maintenance-free" automotive batteries are made from 2% antimonial lead and from lead-calcium alloys containing 0.04 to 0.06% calcium and about 0.1% tin. Industrial batteries usually are made from alloys containing about 8% antimony and various other elements. For all of these alloys, long battery life requires close control of impurities. Large standby stationary batteries may be made with grids of relatively pure lead sheet. Lead-calcium alloys also have been used in this type of battery with considerable success. It is important that support straps and connector posts in any battery be made of the same alloy as the grid plates.

Type metals, a class of metals used in the printing industry, generally are alloys of lead, antimony and tin; small amounts of copper are added to increase hardness for some applications. Compositions of type metals in present commercial use are given in Table 2. The lead base provides low cost, low melting point and ease of casting—properties desired in all type metals. Additions of antimony not only harden the alloy and make it more resistant to compressive impact and wear, but also lower the casting temperature and minimize contraction during freezing. Tin adds fluidity and greater ease of casting, reduces brittleness, and imparts a finer structure—a characteristic helpful in reproducing fine detail.

Electrotype metal contains the lowest percentages of tin and antimony (see Table 2) because it is used as a backing metal only, and is not required to resist wear. Unlike electrotype metal, stereotype metal ordinarily is used directly for printing; hence it must be harder and more wear resistant than electrotype metal, necessitating higher contents of tin and antimony. Sometimes, for greater resistance to wear, stereotypes are lightly electroplated with chromium or nickel.

Linotype, or slug-casting metal, is used for high-speed composition of newspaper type. For this purpose, a low melting point and short temperature range during solidification are of greatest importance, and the ternary eutectic alloy containing 84% Pb, 4% Sn and 12% Sb, or an alloy of similar composition, is favored.

Like linotype metal, monotype metal is machine die cast. In monotype casting, because only one type character is cast at a time, a rapid cooling rate is possible, permitting the use of harder alloys of higher melting range than linotype metal, which is die cast an entire line at a time.

Foundry type metal is used exclusively to cast type for hand composition. The cast type is used over and over again instead of being melted before reuse, as other type metals are. If not used to print directly, this type is subjected to heavy pressure in forming molds for electrotypes, stereotypes and other duplicate plates. Such service requires the hardest, most wear-resistant alloy that is practical to use. Small

Table 2 Typical compositions and properties of type metals

Item	Composition, % Pb	Sn	Sb	Hardness, HB(a)	Liquidus °C	°F	Solidus °C	°F
Electrotype								
General....................	95	2.5	2.5	...	303	578	246	475
General....................	94	3	3	12.4	298	568	246	475
Curved plates	93	4	3	12.5	294	561	245	473
Stereotype								
Flat plate	80	6	14	23	256	493	239	462
General....................	80.5	6.5	13	22	252	485	239	462
Curved plates	77	8	15	25	263	505	239	462
Linotype								
Standard...................	86	3	11	19	247	477	239	462
Special	84	5	11	22	246	475	239	462
Ternary eutectic alloy	84	4	12	22	239	463	239	462
Monotype								
Ordinary..................	78	7	15	24	262	503	239	462
Display...................	75	8	17	27	271	520	239	462
Case type(b)...............	72	9	19	28.5	286	546	239	462
Case type	64	12	24	33	330	626	239	462
Rules.....................	75	10	15	26	270	518	239	462
Foundry Type								
Hard (1.5% Cu)............	60.5	13	25	...	...	...	...	...
Hard (1.5% Cu)............	58.5	20	20	...	...	...	...	...
Hard (2.0% Cu)............	61	12	25	...	...	...	...	...

(a) 10-mm ball, 250-kg load. (b) Lanston standard.

additions of copper as a hardener are feasible for foundry type.

Because type metals other than foundry type are remelted and recast repeatedly, there is always a possibility of contamination by unwanted metals, as well as by oxide and dross formed during melting and handling of molten metal. Also, because new type metal usually comes from secondary metal refineries, it is seldom as pure as the virgin constituent metals. Copper, zinc, nickel, aluminum, and arsenic are the principal metallic impurities that may impair castability of type metals. Iron also is present in very small amounts, principally because of the action of molten tin on the steel equipment used in melting and casting; but iron usually is not considered a harmful impurity, except that it increases the amount of dross.

Cable Sheathing. Lead sheathing extruded around electrical power and communication cables gives the most durable protection against moisture and corrosion damage, as well as mechanical protection of the insulation. Chemical lead, 1% antimonial lead and arsenical lead are most commonly employed for this purpose. The additional stiffness imparted to lead by antimony is advantageous for overhead cables.

The additional resistance to bending and creep imparted by arsenic is desirable in applications involving severe vibration. Lead alloyed with 0.03% calcium and with tellurium also has been used with satisfactory results.

Lead-sheathed cables used underground or under water usually are protected against mechanical damage to the sheathing. Sheathing on underground cable generally is protected from contact with the ground by wood, cement, clay or fiber. Where scoring of the sheathing or a severely corrosive environment is likely to be encountered, a polyethylene or neoprene jacket is applied over the lead sheathing. Underwater lead-sheathed cables are protected with asphalt-impregnated jute and galvanized steel wire.

Sheet. Lead sheet is a constructional material of major importance in chemical and related industries because lead resists attack by a wide range of chemicals. This is discussed in detail in the following article, "Corrosion Resistance of Lead." Lead sheet is also used in building construction—for roofing and flashing, shower pans, flooring, x-ray and gamma-ray protection, and vibration damping and soundproofing. Sheet for use in chemical industries and building construction is made from either lead or 6% antimonial lead. Calcium-lead alloys are also suitable for many of these applications.

Lead sheet is rolled in widths up to 3.6 m (11¾ ft) and in any thickness desired. Thickness is designated by weight per square foot; lead weighs approximately 1 lb/ft² for each ¹⁄₆₄ in. of thickness. (This approximation should be used with care for thicknesses exceeding ¼ in.; lead sheet ½ in. thick weighs 30 lb/ft².)

Roofing and flashing for general purposes are made of 3-lb antimonial lead sheet (³⁄₆₄ in. thick) containing 6 to 7% antimony. Flashing installed in contact with fresh cement, mortar or concrete should be coated with black asphalt.

Pans placed beneath the concrete flooring of shower and bath stalls are made of at least 4-lb lead sheet. They should be coated on both sides with asphalt or covered with tar paper. As flooring, lead sheet offers a corrosion-resistant surface as well as being non-sparking, which is required for some specialized applications. Because of its excellent absorption characteristics, lead sheet is widely used as radiation shielding for medical and industrial installations.

Lead sheet is used in many applications where its vibration-damping characteristics are advantageous. For example, vibration-damping pads of lead sheet, asbestos board and steel are placed under column footings of buildings to prevent transmission of underground vibrations, such as originate from subway and railroad trains. The lead serves as a moistureproof envelope in addition to absorbing vibration. Hangers for rigid pipes often are lined with lead, which acts as a vibration and movement absorber. The soundproofing abilities of lead sheet are discussed at the end of this article.

Pipe. Seamless pipe made from lead and lead alloys is readily fabricated by extrusion. Because of its corrosion resistance and flexibility, lead pipe finds many uses in the chemical industry and in plumbing and water distribution. Pipe for these applications is made from either lead or 6% antimonial lead. Calcium-lead alloys are also suitable in many instances. Sizes range from fine tubing to pipes 12 in. or more in diameter, with almost any wall thickness.

In the chemical industry, horizontal runs of exposed lead pipe usually are supported continuously in troughs or sheet metal shells. Use of unbonded

lead-lined steel pipe sometimes is a simple solution to the problem of pipe support. Lengths of pipe may also be fabricated with welded-on lead hanger bars through which hanger hooks may be attached. Vertical runs are supported at intervals of approximately 18 in. Lengths of pipe are joined by welding or by bolting through welded-on flanges. Expansion bends are provided for pipe that will operate at elevated or fluctuating temperatures.

Heating and cooling coils are important uses of lead pipe in the chemical industry. They usually are in the form of helixes or return-bend banks of coils. Lead spacer supports are welded between turns at about 18-in. intervals.

For lead pipe used in the chemical industry, appropriate wall thickness depends on operating pressures and allowances for corrosion and abrasion. Pipe 1½ in. in diameter with ½-in. walls is commonly used with steam pressures up to 310 kPa (45 psi). For nonpressure service, pipes up to 2 in. in diameter usually are no lighter than the class known as "B" or "M" weight, and minimum wall thickness of larger pipes generally is ¼ in. For pressurized lead pipe, safe working pressure is calculated using the formula:

$$P = \frac{2St}{D}$$

Where P is working pressure, in psi; S is maximum allowable fiber stress, in psi; t is wall thickness, in inches; and D is the inside diameter of the pipe, in inches. For chemical lead, maximum allowable fiber stress ranges from 200 psi (1400 kPa) at room temperature to 80 psi (600 kPa) at 150 °C (300 °F). Proper design of steam lines allows for condensate drainage, thus eliminating water-hammer damage.

When greatest strength is essential, lead-lined steel pipe may be used, or coils may be made of copper tubing completely covered with an adherent layer of lead.

Lead pipes and traps have had a long history of use in water and waste service because of lead's excellent corrosion resistance and ability to adjust to ground settlement without damage. Joints in service pipes have been successfully made by wiping, by welding, by cupping and soldering, and by means of compression-type couplings. Service pipe should be laid with goosenecks to allow for settlement, and a

cast iron sleeve should be provided where the pipe passes through foundation walls. Where electrolysis, free lime, or cinder fill is encountered, lead pipe should be suitably protected.

Solders are used in many industries, of which electronics, construction and automobile manufacturing are important examples. Solders are basically lead-tin alloys, sometimes containing other elements such as antimony and silver, which are added to improve strength and corrosion resistance. One of the most common solders contains 50% lead and 50% tin by weight—the "half-and-half" grade. Other solders contain from 0 to 98% lead and from 96 to 0% tin. Plumber's wiping solder contains 38 to 42% Sn, 0 to 2% Sb, remainder Pb. ASTM B32 lists 26 solders that contain at least 50% lead. Properties of four of the most common lead-base solders are listed in the lead alloy data compilations, which follow this article.

Bearings. Lead-base bearing alloys, which are called lead-base babbitt metals, vary widely in composition. An alloy commonly used for railroad-car journal bearings contains 86% Pb, 9% Sb and 5% Sn. Copper-lead bearing alloys contain up to 40% lead. Leaded bronzes contain 4 to 25% lead along with the copper and tin. Numerous alloys of lead and alkaline earth metals such as calcium and sodium are used as bearing materials. Four lead-base bearing alloys are covered by ASTM B23, and four are covered by SAE J459. Properties of five of the most common lead-base bearing alloys are listed in the data compilations that follow this article. Additional information may be found in the article on sleeve bearings that will appear in Vol 3 of this Handbook.

Ammunition. Large quantities of lead are used in ammunition for both military and sporting purposes. Alloys used for shot contain up to 1% arsenic; those used for bullet cores, up to 2% antimony.

Terne metal, an alloy of lead and tin (usually 8 to 12% tin), is used for coating steel sheet to produce terneplate. Because terneplate is sturdy and highly resistant to corrosion, it is widely used for automotive gasoline tanks, filter covers and similar parts. It also is used for roofing, where the attractive patina and durability of lead are desired but weight must be limited.

For further information on terneplate, see the article on precoated steel sheet in Vol 1 of this Handbook.

Lead foil, generally known as composition metal foil, is usually made by rolling a sandwich of lead between two sheets of tin, producing a tight union of the metals. Thicknesses of 0.01 mm (0.0005 in.) or less are common. Lead foil is used for moisture protection in the construction industry and for decoration of wine and champagne bottles.

Fusible Alloys. Many lead alloys melt at relatively low temperatures and are used for electric fuses, automatic sprinkler systems and boiler plugs. Compositions of fusible lead alloys will be presented in the article on low-melting alloys in Vol 3 of this Handbook.

Anodes made of lead or lead alloys are used in electrolytic refining and plating of metals, such as manganese and zinc. They are preferred over other metals in these applications, usually because of their high resistance to the sulfuric acid used in electrolytic solutions.

Lead anodes also have high resistance to corrosion by seawater, making them economical to use in systems for cathodic protection of ships and offshore rigs.

Anodes for these purposes usually are made of unalloyed lead but sometimes are of lead alloyed with silver, tin or antimony. These anodes are made not only in cast form but also as extruded bars or supported sheet.

Structures

In many applications, lead is combined with stiffer and stronger materials to make structures that have the best qualities of both materials. An example of this type of structure is the series of lead and lead-coated structures described below, which are used for corrosion-resistant equipment. In contrast, lead is combined with plastics having relatively low stiffness and strength to make structures with superior sound-control characteristics.

The plumbum series is a group of material combinations of which lead is a major constituent, used in applications requiring good strength and high resistance to corrosion. Lead or lead alloy coatings provide corrosion resistance, and strength is provided by steel, concrete, wood, brick or other suitable material.

Table 3 Sound-control materials containing lead

Material	Description	Uses
Sheet lead	Usual weight, 0.25 to 2 kg (½ to 4 lb)	Used alone or laminated to substrates of various types
Lead foam composites	Usual weight, 0.25- or 0.5-kg (½- or 1-lb) lead sheet sandwiched between layers of polyurethane foam	Laminated to enclosures
Leaded plastic sheets	Lead-loaded vinyl or neoprene sheet with or without fabric reinforcement	As a curtain or to line enclosures
Damping tile	Lead-loaded epoxy or urethane tiles	Damping heavy machinery
Casting compounds	Lead-loaded epoxy	Potting, filling complex voids
Troweling compounds	Lead-loaded epoxy or urethane	Damping enclosures, surfaces, resonating members and rattling panels

The six members of the plumbum series are described below, along with examples of typical uses, in roughly increasing order of cost and strength.

1 *Basic Plumbum (Ba.P.).* Lead or lead alloys in cast or extruded form with limited support. Uses: cast antimonial lead valves, pipe fittings, pumps, anodes and vessels
2 *Supported Plumbum (Su.P.).* Lead or lead alloys in sheet, pipe or other extruded form mechanically fastened to supporting structures of steel, wood, concrete, copper or other metals. Uses: concrete cells lined with lead sheet (loosely lined or cage supported) used in electrolytic refining of metals, flues, ducts, towers, floors, expanded lead-lined pipe, cable sheathing, roofing and anodes
3 *Adhesive Plumbum (Ad.P.).* Lead or lead alloys in sheet, pipe or other form joined with an adhesive to steel, concrete, wood or any other material for extensive support. Uses: acid storage tanks made of lead sheet joined with an adhesive to a steel outer shell
4 *Bonded Plumbum (Bd.P.).* A heavy lead or lead alloy layer metallurgically bonded to steel, copper or other metal. Uses: homogeneously bonded, lead-lined steel reaction vessels and lead-clad copper heating and cooling coils
5 *Brick Plumbum (Bk.P.).* Lead or lead alloy sheet sandwiched between an outer shell of concrete or steel and an inner layer of chemical-resistant ceramic brick or masonry (usually acid-brick); the

sheet is metallurgically or chemically bonded to the outer shell and the inner layer; sometimes a layer of cushioning material is placed between the sheet and the inner layer. Uses: sulfuric acid mist scrubbers, precipitators, concentrators, and storage tanks
6 *Plumbum Coatings (Co.P.).* Thin lead or lead alloy coatings metallurgically or mechanically bonded to equipment to protect it from corrosion. This is the lead-tin alloy coating used on steel for roofing, gutters and downspouts.

The five major characteristics of the plumbum series are relatively low material costs, low to moderate installation and maintenance costs, inherently high corrosion resistance, long service life and adaptability to a wide range of operating conditions. The relatively low material costs of plumbum series equipment are due to the fact that the series' three basic components—steel, lead and concrete—are comparatively inexpensive. A fourth component, wood, is used only in applications such as manufacture of explosives, where its nonsparking property makes it essential. The fifth major component, chemical-resistant masonry, is not low in cost. However, it is used only where high-temperature strength or abrasion resistance is required. Under those conditions, the cost of using the only other suitable materials or material combinations is usually comparable or higher.

All plumbum series equipment may be used in cold climates without failure due to embrittlement. Temperatures as high as 1000 °C (1830 °F) have been handled successfully by Bk.P. Both Bk.P. and Bd.P. have high resistance to damage by thermal shock due to large and rapid fluctuations in temperature. High heat conductivity is a normal feature of Bd.P., Ba.P. and most types of Co.P. A strong barrier to heat transfer is provided by Bk.P., Ad.P. and a few types of Su.P. Thus, Bd.P. is used to make heating and cooling coils, and Bk.P. is widely used as insulation in ducts handling very hot gases.

Ba.P. performs reliably under pressures up to 3 atm. All other plumbums can withstand substantially higher pressures, with Bd.P. especially resistant to damage. Both Bd.P. and Bk.P. can be used to handle vacuums as well as pressures that fluctuate both above and below atmospheric. Adequate abrasion resistance is often provided by using the harder alloys of lead. However, for extremely abrasive conditions, the hard masonry of Bk.P. is required. The pliability and malleability of soft lead make possible the manufacture of a wide variety of intricately shaped items such as corrugated helical heating and cooling coils. These qualities also allow manufacture of glassy smooth lead pipes that minimize friction energy losses.

Ba.P., Bk.P., Su.P. and Co.P. are electrically conductive. However, they can be made into insulating barriers by incorporating a layer of nonconducting material. By choosing the appropriate adhesive, Ad.P. can be made either insulating or conductive. Bd.P. is always electrically conductive. Equipment with nonsparking surfaces can be constructed using Su.P. or Ad.P. combinations that contain only wood and lead as major components.

Sound-Control Materials. Lead is an excellent barrier to sound transmission. Essentially, a good sound barrier should have high density and low stiffness, and should be impermeable. Lead and lead composites more than satisfy these requirements. In addition, the high internal damping capacity of lead and lead composites make them even more effective in controlling sound. Several examples of sound-control materials that contain lead are described in Table 3. An advantage of these products is that they are unaffected by coolants, cutting oils, drawing compounds and similar industrial fluids.

Properties of Leads and Lead Alloys

Corroding Lead 99.94 + % Pb

Compiled by R. R. Kubalak
(deceased)
St. Joe Lead Co.

Commercial Names

Common name. Corroding lead, common desilverized lead, soft unde-silverized lead

Specifications

ASTM. B29 (corroding lead)
Government. QQ-L-171

Chemical Composition

Composition limits. 99.94 min Pb, 0.0015 max Ag, 0.0015 max Cu, 0.0025 max Ag + Cu, 0.002 max As + Sb + Sn, 0.001 max Zn, 0.002 max Fe, 0.050 max Bi

Applications

Typical uses. Storage batteries, sheath for electric cable, paint, ammunition, antiknock fluid, sheet, calking, insecticides, type metal, solder, bearing metals, foil, ceramics, liquid medium for heat treating, collapsible tubes, coatings, seals, powder

Precautions in use. Lead presents a health hazard. Avoid inhalation of lead dust and fumes. Do not use lead to conduct very soft water for drinking, or in contact with foods.

Mechanical Properties

Tensile properties. Tensile strength: sand cast, 12 to 13 MPa (1.7 to 1.9 ksi); chill cast, 14 MPa (2.0 ksi). Yield strength: sand cast, 55 MPa (8.0 ksi). Elongation: sand cast, 30%; chill cast, 47%. Reduction in area: sand cast and chill cast, 100%
Shear strength. Sand cast and chill cast, 12.5 MPa (1.82 ksi)
Hardness. Sand cast, 3.2 to 4.5 HB; chill cast, 4.2 HB (using 10-mm ball, 100-kg load and 30-s duration)
Poisson's ratio. Chill cast, 0.40 to 0.45
Elastic modulus. Tension: sand cast, 14 GPa (2×10^6 psi)
Impact strength. Charpy V-notch: chill cast, 14.1 J (10.4 ft·lb)
Fatigue strength. Sand cast, 32 MPa (4.6 ksi) at 10^7 cycles
Damping capacity. See Fig. 1.

Mass Characteristics

Density. At 20 °C (68 °F): cast, 11.34 Mg/m^3 (0.4097 lb/in.^{3}); rolled, 11.35 Mg/m^3 (0.4100 lb/in.^{3})

X-ray absorption characteristics. See Fig. 2.

Thermal Properties

Melting point. 327.4 °C (621.0 °F)
Coefficient of thermal expansion. Linear, 29.3 μm/m·K (16.3 μin./in.·°F) from 17 to 100 °C (63 to 212 °F)
Specific heat. 129 J/kg·K (0.0309 Btu/lb·°F) at 0 °C (32 °F)
Latent heat of fusion. 23 kJ/kg (10 Btu/lb)
Thermal conductivity. 35 W/m·K (20 Btu/ft·h·°F)

Electrical Properties

Electrical conductivity. Volumetric, 8.3% IACS
Electrical resistivity. 206.43 nΩ·m at 20 °C (68 °F)

Chemical Properties

Resistance to specific corroding agents. Similar to chemical lead; see Table 2.

REFERENCES

1. Bolt, Beranek and Newman, "Improved Sound Barriers Employing Lead," Lead Industries Association, 1960
2. Lead in Modern Industry, Second International Congress of Radiology, Lead Industries Association, 1952, p 104

Fig. 1 Damping capacity of lead compared with that of other materials (Ref 1)

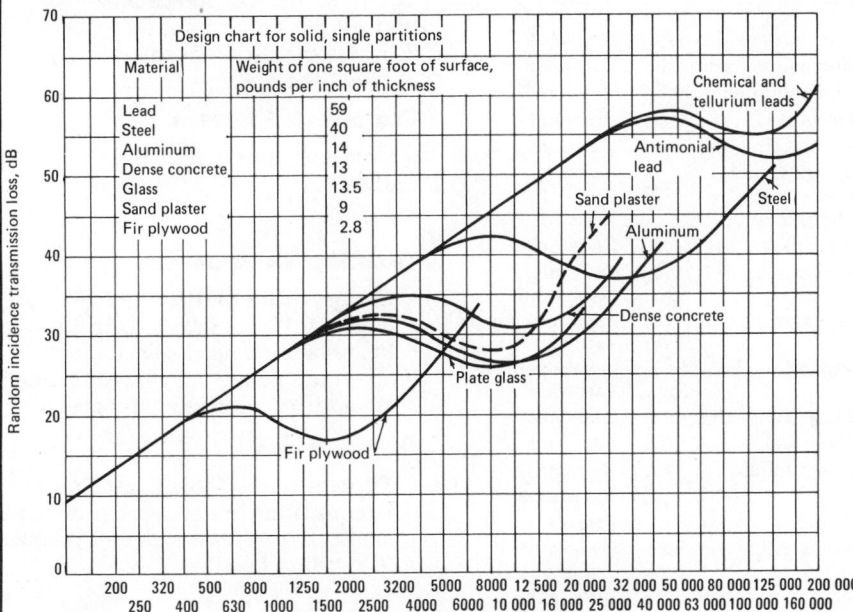

Fig. 2 X-ray absorption characteristics of lead (Ref 2)

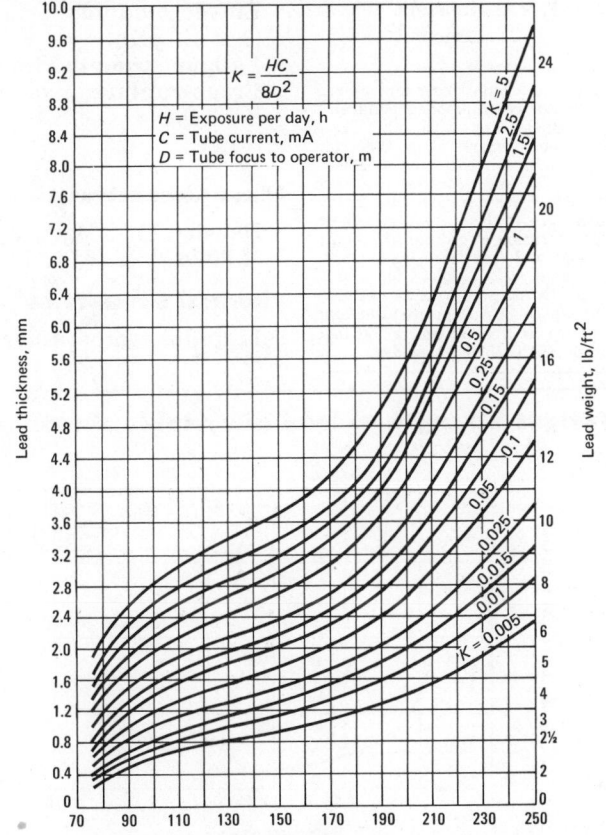

$$K = \frac{HC}{8D^2}$$

H = Exposure per day, h
C = Tube current, mA
D = Tube focus to operator, m

The thickness of lead barriers required to reduce x-ray dosage rate of the useful beam to 6.25 mR/h under indicated conditions.

Chemical Lead 99.90+% Pb

Compiled by F. W. Ling
and
R. W. Balliett
St. Joe Minerals Corp.

Specifications

ASTM. B29 (chemical lead)

Chemical Composition

Composition limits. 99.90 min Pb, 0.002 to 0.020 Ag, 0.040 to 0.080 Cu, 0.002 max As + Sb + Sn, 0.001 max Zn, 0.002 max Fe, 0.005 max Bi

Applications

Typical uses. Used principally as a corrosion resistant material in the chemical industry. Also used as x-ray shielding.

Precautions in use. Lead presents a health hazard; see this entry under "Corroding Lead."

Mechanical Properties

Tensile properties. Tensile strength, 16 to 19 MPa (2.3 to 2.7 ksi); yield strength, 6 to 8 MPa (0.9 to 1.2 ksi); proportional limit, 7 MPa (1 ksi); elongation, 30 to 60%. See also Table 1.

Hardness. 4 to 6 HB (using 10-mm ball, 100-kg load and 30-s duration), 80 to 85 HRR (30-s duration)

Elastic modulus. Tension, 13 to 14 GPa (1.9 to 2.1 × 10^6 psi)

Impact strength. Charpy V-notch: 11 to 16 J (8 to 12 ft·lb)

Creep characteristics. Creep rate at 20 °C (68 °F) and 2.07 MPa (300 psi): 3.0% per year

Damping capacity. See Fig. 1.

Mass Characteristics

Density:

°C	°F	Mg/m³	lb/in.³
20	68 ..... 11.340		0.4097
325.6(a)	618(a) ... 11.005		0.3976
325.6(b)	618(b) ... 10.686		0.3854
550	1022 ..... 10.418		0.3754
(a) Solid. (b) Liquid.			

X-ray absorption characteristics. See Fig. 2.

Thermal Properties

Melting point. 325.6 °C (618 °F)

Coefficient of thermal expansion. Linear, 29.3 µm/m·K (16.3 µin./in.·°F)

Latent heat of fusion. 22.9 to 26.2 kJ/kg (9.85 to 11.26 Btu/lb)

Thermal conductivity. At 20 °C (68 °F), 35 W/m·K (20.1 Btu/ft·h·°F); at 105 °C (221 °F), 34 W/m·K (19.6 Btu/ft·h·°F)

Electrical Properties

Electrical resistivity. 206 nΩ·m at 20 °C (68 °F)

Chemical Properties

Resistance to specific corroding agents. See Table 2.

Fabrication Characteristics

Casting temperature. 421 to 443 °C (790 to 830 °F)

Recrystallization temperature. Below 0 °C (32 °F) for 99.999% Pb

Table 1 Elevated temperature tensile properties of chemical lead

Temperature		Tensile strength(a)		Elongation(b), %
°C	°F	MPa	ksi	
38	100	15.5	2.25	45
66	150	13.1	1.90	45
93	200	10.3	1.50	52
121	250	9.3	1.35	52
149	300	7.2	1.05	52
177	350	4.8	0.70	69

(a) Mean value; range, ±1.4 MPa (±200 psi).
(b) Mean value; range, ±10%.

Table 2 Resistance of chemical lead to specific corroding agents

Corrosive agent	Resistance
Acetone	Resistant
Acetylene	Resistant
Acid, acetic(a)	Moderate general attack
Acid, chromic	Resistant
Acid, citric	Moderate general attack
Acid, hydrochloric(b)	Moderate general attack
Acid, hydrofluoric	Resistant
Acids, mixed(c)	Resistant
Acid, nitric(d)	Severe general attack
Acid, phosphoric(e)	Resistant
Acid, sulfuric(f)	Resistant
Acid sulfurous	Resistant
Acid, tartaric	Moderate general attack
Air	Resistant
Alcohol, ethyl	Resistant
Alcohol, methyl	Resistant
Aluminum sulfate	Resistant
Ammonia(g)	Resistant
Ammonium azide	Resistant

Table 2 (continued)

Corrosive agent	Resistance
Ammonium chloride(h)	Resistant
Ammonium hydroxide	Resistant
Ammonium phosphate	Resistant
Benzol	Resistant
Bromine(i)	Resistant
Carbon dioxide	Resistant
Carbon tetrachloride(j)	Resistant
Chlorine(k)	Resistant
Dyestuffs	Generally resistant
Formaldehyde	Moderate general attack
Magnesium chloride	Severe general attack
Magnesium sulfate	Resistant
Motor fuel	Resistant
Nickel sulfate	Resistant
Oxygen	Resistant
Phenols	Resistant
Photographic solutions	Generally resistant
Sodium carbonate	Resistant
Sodium chloride(m)	Resistant
Sodium hydroxide(n)	Resistant
Sodium sulfate(p)	Resistant
Sulfur dioxide(q)	Resistant
Water, chlorinated	Resistant
Water, sea	Resistant

(a) Used to handle acetic anhydride and glacial acetic acid. (b) Use not recommended generally. (c) At ordinary temperatures with 30% H_2O. (d) Used at normal temperatures above 80% concentration. (e) Up to 80% concentration at 200 °C (390 °F). (f) Up to 96% concentration at room temperature or 85% concentration at 220 °C (430 °F). (g) Unless Na or K is dissolved in it. (h) Up to 10% concentration at ordinary temperatures. (i) When cold and free from acid. (j) At ordinary temperatures. (k) Moist to 110 °C (230 °F) or dry. (m) Dilute solutions at ordinary temperatures. (n) Up to 26% concentration and 80 °C (175 °F). (p) Up to 10% concentration boiling. (q) Moist up to 200 °C (390 °F), or dry.

Arsenical Lead
Pb-0.15As-0.10Sn-0.10Bi

Compiled by A. T. Balcerzak
and
R. R. Kubalak (deceased)
St. Joe Lead Co.

Commercial Names

Common name. Arsenical lead; F-3 alloy

Chemical Composition

Composition limits. 0.12 to 0.20 As, 0.08 to 0.12 Sn, 0.05 to 0.15 Bi, rem Pb. (Arsenical lead sold under various trade names may contain small amounts of tellurium or copper.)

Applications

Typical uses. Cable sheathing

Precautions in use. Lead presents a health hazard; see this entry under "Corroding Lead".

Mechanical Properties

Tensile properties. See Table 3.

Hardness. See Table 3.

Elastic modulus. Tension, 21 GPa (3×10^6 psi)

Fatigue strength. See Fig. 3.

Creep-rupture characteristics. See Table 4.

Mass Characteristics

Density. 11.33 Mg/m³ (0.4092 lb/in.³) at 20 to 25 °C (68 to 77 °F)

Thermal Properties

Liquidus temperature. 327 °C (621 °F)

Fig. 3 Fatigue strength of lead alloy cable sheath in bending

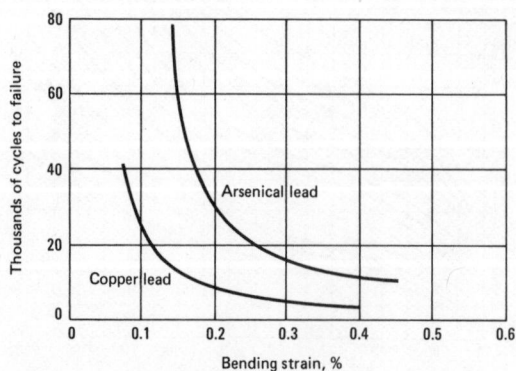

Bending was at 25 °C (77 °F) and one cycle per minute.

Table 3 Typical mechanical properties of arsenical lead extruded cable sheath and pipe

Extrusion temperature		Tensile strength				Elon-gation, %	Hardness (pipe only), HB
		Cable sheath only		Pipe only			
°C	°F	MPa	ksi	MPa	ksi		
(a)	(a) 17.2		2.5	16.2	2.35	40	4.9
795(b)	425(b) 19.9		2.9	20.6	3.0	35	6.6
842(b)	450(b) 20.6		3.0	22.0	3.2	35	7.0
1022(b)	550(b) ..... ···		···	31.7	4.6	25	10.0

(a) Extruded at any temperature and air cooled. (b) Water quenched after extruding.

Solidus temperature. Approximately 302 °C (575 °F)

Chemical Properties

Resistance to specific corroding agents. Similar to chemical lead; see Table 2.

Fabrication Characteristics

Casting temperature. 400 °C (750 °F)

Hot working temperature. Extrusion, 200 to 230 °C (400 to 450 °F)

Table 4 Typical creep-rate data for arsenical lead cable sheath

Testing temperature		Applied load		Creep rate % per year
°C	°F	MPa	psi	
43	110	1.41	205	.... 0.1
43	110	2.34	340	.... 0.5
43	110	2.93	425	.... 1.0
25	76	2.07	300	.... 0.13
43	110	2.07	300	.... 0.32
65	149	2.07	300	.... 1.8

Calcium Lead Pb − 0.07Ca

Compiled by A. T. Balcerzak
and
R. R. Kubalak (deceased)
St. Joe Lead Co.

Chemical Composition

Composition limits. 0.06 to 0.08 Ca; 0.001 max Ag; 0.001 max Bi; 0.0005 max each for Sn, Sb, Cu, As, Zn, Cd, Ni and Fe; rem Pb

Applications

Typical uses. Cast grids for maintenance-free and standby batteries
Precautions in use. Lead presents a health hazard; see this entry under "Corroding Lead".

Mechanical Properties

Tensile properties. Fully aged, air cooled castings: tensile strength, 36 to 39 MPa (5.2 to 5.6 ksi); elongation, 30 to 45%
Hardness. 70 to 80 HR

Mass Characteristics

Density. 11.34 Mg/m^3 (0.409 lb/in.^{3}) at 20 °C (68 °F)

Thermal Properties

Solidus temperature. 328 °C (623 °F)
Coefficient of thermal expansion. Linear, 30.2 μm/m·K (16.7 μin./in.·°F)

Electrical Properties

Electrical resistivity. 218 nΩ·m at 20 °C (68 °F)

Chemical Properties

General corrosion behavior. Excellent anodic corrosion resistance in sulfuric acid
Resistance to specific corroding agents. Similar to chemical lead; see Table 2.

Calcium Lead Pb-0.09Ca-0.3Sn

Compiled by A. T. Balcerzak
and
R. R. Kubalak (deceased)
St. Joe Lead Co.

Chemical Composition

Composition limits. 0.08 to 0.11 Ca; 0.2 to 0.4 Sn; 0.001 max Ag; 0.001

max Bi; 0.0005 max each for Sb, Cu, As, Zn, Cd, Ni and Fe; rem Pb

Applications

Typical uses. Cast parts requiring high strength and cast grids for maintenance-free and standby batteries
Precautions in use. Lead presents a health hazard; see this entry under "Corroding Lead."

Mechanical Properties

Tensile properties. Fully aged, air cooled casting: 41 to 45 (6.0 to 6.5 ksi); elongation, 20 to 35% in 50 mm or 2 in.
Hardness 90 to 95 HR

Mass Characteristics

Density. 11.34 Mg/m^3 (0.409 lb/in.^{3}) at 20 °C (68 °F)

Thermal Properties

Liquidus temperature. 328 °C (622 °F)
Solidus temperature. 338 °C (640 °F)

Electrical Properties

Electrical resistivity. 219 nΩ·m at 20 °C (68 °F)

Chemical Properties

Resistance to specific corroding agents. Similar to chemical lead; see Table 2.

Calcium Lead Pb-0.09Ca-0.5Sn

Compiled by A. T. Balcerzak
and
R. R. Kubalak (deceased)
St. Joe Lead Co.

Chemical Composition

Composition limits. 0.08 to 0.11 Ca; 0.4 to 0.6 Sn; 0.001 max Ag; 0.001 max Bi; 0.005 max each for Sb, Cu, As, Zn, Cd, Ni and Fe; rem Pb

Applications

Typical uses. Cast parts requiring improved strength and surface finish properties. Cast grids for maintenance-free and standby batteries
Precautions in use. Lead presents a health hazard; see this entry under "Corroding Lead."

Mechanical Properties

Tensile properties. Fully aged, air cooled castings: 44.8 to 51.7 MPa (6.5 to 7.5 ksi); elongation, 25 to 35% in 50 mm or 2 in.
Hardness. 85 to 90 HR

Mass Characteristics

Density. 11.34 Mg/m³ (0.409 lb/in.³) at 20 °C (68 °F)

Thermal Properties

Liquidus temperature. 327 °C (621 °F)
Solidus temperature. 336 °C (637 °F)

Electrical Properties

Electrical resistivity. 219 nΩ·m at 20 °C (68 °F)

Chemical Properties

General corrosion behavior. Excellent anodic corrosion resistance in sulfuric acid
Resistance to specific corroding agents. Similar to chemical lead; see Table 2.

Calcium Lead
Pb-0.065Ca-0.7Sn
Compiled by A. T. Balcerzak
and
R. R. Kubalak (deceased)
St. Joe Lead Co.

Chemical Composition

Composition limits. 0.06 to 0.07 Ca; 0.6 to 0.8 Sn; 0.001 max Ag; 0.001 max Bi; 0.0005 max Sb; 0.0005 max other elements; rem Pb

Applications

Typical uses. Strip, sheet and wire applications requiring strength and creep resistance as well as conventional properties of lead; maintenance-free batteries
Precautions in use. Lead presents a health hazard: see this entry under "Corroding Lead."

Mechanical Properties

Tensile properties. Longitudinal properties of rolled strip: tensile strength, 62 MPa (9.0 ksi); yield strength (0.2% offset), 55 MPa (8.0 ksi); elongation, 10%

Creep-rupture characteristics. Rolled strip: rupture life, 50 h when stressed at 28 MPa (4.0 ksi); 500 h when stressed at 21 MPa (3.0 ksi)

Mass Characteristics

Density. 11.34 Mg/m³ (0.409 lb/in.³) at 20 °C (68 °F)

Thermal Properties

Solidus temperature, 327 °C (621 °F)

Electrical Properties

Electrical resistivity. 219 nΩ·m at 20 °C (68 °F)

Chemical Properties

General corrosion behavior. Excellent anodic corrosion resistance in sulfuric acid
Resistance to specific corroding agents. Similar to chemical lead; see Table 2.

Calcium Lead
Pb-0.09Ca-1.0Sn

Compiled by A. T. Balcerzak
and
R. R. Kubalak (deceased)
St. Joe Lead Co.

Chemical Composition

Composition limits. 0.08 to 0.10 Ca; 0.9 to 1.1 Sn; 0.001 max Ag; 0.001 max Bi; 0.0005 max each for Sb, Cu, As, Zn, Cd, Ni and Fe; rem Pb

Applications

Typical uses. Cast parts requiring high strength and cast grids for maintenance-free and standby batteries.
Precautions in use. Lead presents a health hazard; see this entry under "Corroding Lead."

Mechanical Properties

Tensile properties. Fully aged, air cooled casting: 52 to 55 MPa (7.5 to 8.0 ksi); elongation, 20 to 35% in 50 mm or 2 in.
Hardness 90 to 95 HR

Mass Characteristics

Density. 11.34 Mg/m³ (0.409 lb/in.³) at 20 °C (68 °F)

Thermal Properties

Liquidus temperature. 325 °C (617 °F)
Solidus temperature. 332 °C (630 °F)

Electrical Properties

Electrical resistivity. 212 nΩ·m at 20 °C (68 °F)

Chemical Properties

Resistance to specific corroding agents. Similar to chemical lead; see Table 2

Calcium Lead
Pb-0.065Ca-1.3Sn
Compiled by A. T. Balcerzak
and
R. R. Kubalak (deceased)
St. Joe Lead Co.

Chemical Composition

Composition limits. 0.06 to 0.07 Ca; 1.2 to 1.4 Sn; 0.001 max Ag; 0.001 max Bi; 0.0005 max each for Cu, As, Zn, Cd, Ni and Fe; rem Pb

Applications

Typical uses. Strip, sheet, wire applications requiring strength and creep resistance as well as conventional properties of lead
Precautions in use. Lead presents a health hazard; see this entry under "Corroding Lead."

Mechanical Properties

Tensile properties. Longitudinal properties of rolled strip: tensile strength, 70 MPa (10 ksi); yield strength (0.2% offset), 66 MPa (9.5 ksi); elongation, 10%
Creep-rupture characteristics. Rolled strip: minimum creep rate at 20 °C (68 °F), 0.06% in 10 000 h when stressed at 6.9 MPa (1.0 ksi); rupture life, 1000 h when stressed at 28 MPa (4.0 ksi)

Mass Characteristics

Density. 11.34 Mg/m³ (0.409 lb/in.³) at 20 °C (68 °F)

Thermal Properties

Solidus temperature. 323 °C (613 °F)
Coefficient of thermal expansion. Linear, 26.6 μm/m·K (14.7 μin./in.·°F)

Electrical Properties

Electrical resistivity. 220 nΩ·m at 20 °C (68 °F)

Chemical Properties

General corrosion behavior. Excellent anodic corrosion resistance in sulfuric acid
Resistance to specific corroding agents. Similar to chemical lead; see Table 2.

Silver Lead Solder
97.5Pb-1.5Ag-1Sn

Specifications

ASTM. B32, alloy 1.5S

Chemical Composition

Composition limits. 97.5 Pb nominal; 0.75 to 1.25 Sn; 1.3 to 1.7 Ag; 0.40 max Sb; 0.25 max Bi; 0.08 max Cu; 0.02 max Fe; 0.005 max Al; 0.005 max Zn; 0.02 max As

Applications

Precautions in use. Lead presents a health hazard; see this entry under "Corroding Lead."

Mechanical Properties

Hardness. 13.0 HB

Thermal Properties

Melting point. 310 °C (590 °F)

Chemical Properties

Resistance to specific corroding agents. The tin in this alloy is required to make it resistant to humid air.

5-95 Solder
95Pb-5Sn

Specifications

ASTM. B32, alloy 5B

Chemical Composition

Composition limits. 95 Pb nominal; 4.5 to 5.5 Sn; 0.20 to 0.50 Sb; 0.25 max Bi; 0.08 max Cu; 0.02 max Fe; 0.005 max Al; 0.005 max Zn; 0.02 max As

Applications

Precautions in use. Lead presents a health hazard; see this entry under "Corroding Lead."

Mechanical Properties

Tensile properties. Tensile strength, 23 MPa (3.4 ksi); yield strength, 10 MPa (1.5 ksi); elongation, 50%; reduction in area, 80%
Hardness. 8 HB

Mass Characteristics

Density. 11.0 Mg/m³ (0.397 lb/in.³) at 20 °C (68 °F)
Volume change at freezing. 3.6%

Thermal Properties

Liquidus temperature. 312 °C (594 °F)
Solidus temperature. 270 °C (518 °F)
Coefficient of thermal expansion. Linear, 28.7 μm/m·K (15.9 μin./in.·°F) at 15 to 110 °C (60 to 230 °F)
Thermal conductivity. 36 W/m·K (21 Btu/ft·h·°F) at 54 °C (129 °F)

Electrical Properties

Electrical conductivity. Volumetric, 8.8% IACS
Electrical resistivity. 195 nΩ·m

Magnetic Properties

Magnetic susceptibility. -0.1×10^{-6} mks

Chemical Properties

Resistance to specific corroding agents. 5-95 solder is resistant to acetone, milk, mineral oils, motor fuel and petroleum products, and sea water. It is resistant to air, but does tarnish. It is subject to moderate general attack by ammonium sulfate, photographic solutions, potassium chloride, sodium chloride and sodium sulfate.

20-80 Solder
80Pb-20Sn

Specifications

ASTM. B32, alloy 20B

Chemical Composition

Composition limits. 80 Pb nominal; 20 Sn desired; 0.20 to 0.50 Sb; 0.25 max Bi; 0.08 max Cu; 0.02 max Fe; 0.005 max Zn; 0.005 max Al; 0.02 max As

Applications

Typical uses. Coating and joining metals; filling dents or seams in automobile bodies (body solder)
Precautions in use. Lead presents a health hazard; see this entry under "Corroding Lead."

Mechanical Properties

Tensile properties. Tensile strength, 40 MPa (5.8 ksi); yield strength, 25 MPa (3.6 ksi); elongation, 16%; reduction in area, 50%
Shear strength. 32.5 MPa (4.71 ksi)
Hardness. 11.3 HB
Impact strength. Charpy V-notch, 16.5 J (12.2 ft·lb)

Mass Characteristics

Density. 10.2 Mg/m³ (0.368 lb/in.³) at 20 °C (68 °F)

Thermal Properties

Liquidus temperature. 277 °C (531 °F)
Solidus temperature. 183 °C (361 °F)
Coefficient of thermal expansion. Linear, 26.5 μm/m·K (14.7 μin./in.·°F) at 15 to 110 °C (60 to 230 °F)
Thermal conductivity. 37 W/m·K (22 Btu/ft·h·°F) at 54 °C (129 °F)

Electrical Properties

Electrical conductivity. Volumetric, 9.8% IACS
Electrical resistivity. 175 nΩ·m

Magnetic Properties

Magnetic susceptibility. -1.0×10^{-6} mks

Chemical Properties

Resistance to specific corroding agents. 20-80 solder is resistant to milk, mineral oils, motor fuel and petroleum. It is fairly resistant to air, but does tarnish.

50-50 Solder
50Pb-50Sn

Specifications

ASTM. B32, alloys 50A and 50B

Chemical Composition

Composition limits. Alloy 50A: 50 Pb nominal; 50 Sn desired; 0.12 max Sb; 0.25 max Bi; 0.08 max Cu; 0.02 max Fe; 0.005 max Al; 0.005 max Zn; 0.025 max As. Alloy 50B: same as alloy 50A except Sb range is 0.20 to 0.50

Applications

Typical uses. The most popular of all lead-tin solders for general-purpose applications

Precautions in use. Lead presents a health hazard; see this entry under "Corroding Lead."

Mechanical Properties

Tensile properties. Tensile strength, 42 MPa (6.1 ksi); yield strength, 33 MPa (4.8 ksi); elongation, 60%; reduction in area, 70%

Shear strength. 40.7 MPa (5.9 ksi)

Hardness. 14.5 HB

Impact strength. Charpy V-notch, 21.6 J (15.9 ft·lb)

Dynamic liquid viscosity. 1.27 mP·s at 350 °C (662 °F)

Liquid surface tension. 472 mN/m

Mass Characteristics

Density. 8.89 Mg/m³ (0.321 lb/in.³) at 20 °C (68 °F)

Volume change on freezing. 2.3%

Thermal Properties

Liquidus temperature. 216 °C (421 °F)

Solidus temperature. 183 °C (361 °F)

Coefficient of thermal expansion. Linear, 23.4 μm/m·K (13 μin./in.·°F) at 15 to 110 °C (60 to 230 °F)

Specific heat. Solid: 210 J/kg·K (0.051 Btu/lb·°F) at 16 to 183 °C (60 to 361 °F); liquid: 190 J/kg·K (0.046 Btu/lb·°F) at 216 to 300 °C (421 to 572 °F)

Latent heat of fusion. 536 J/kg

Thermal conductivity. 46.5 W/m·K (26.8 Btu/ft·h·°F) at 54 °C (129 °F)

Enthalpy. 33.73 kJ/kg (14.50 Btu/lb) at 183 °C (361 °F); 103.51 kJ/kg (44.50 Btu/lb) at 300 °C (572 °F)

Electrical Properties

Electrical conductivity. Volumetric, 11% IACS

Electrical resistivity. 156 nΩ·m

Magnetic Properties

Magnetic susceptibility. -0.5×10^{-6} mks

Chemical Properties

Resistance to specific corroding agents. 50-50 solder is resistant to the following chemicals: carbonic and phosphoric acids; ammonium chloride, nitrate and sulfate; magnesium sulfate, milk, mineral oils, motor fuel, petroleum products, potassium chloride, sodium carbonate, chloride, nitrate and sulfate; and sea water. It is moderately resistant to air but does tarnish. Subject to severe attack by nitric acid.

1% Antimonial Lead
99Pb-1Sb

Applications

Typical uses. Cable sheathing

Precautions in use. Lead presents a health hazard; see this entry under "Corroding Lead."

Mechanical Properties

Tensile properties. Extruded and aged 1 month: tensile strength, 20 MPa (3.0 ksi); elongation, 50%

Hardness. 7 HB

Elastic modulus. Tension, 14 GPa (2×10^6 psi)

Fatigue strength. 7.93 MPa (1.15 ksi) at 2×10^7 million cycles

Creep-rupture characteristics. Minimum creep rate at 10^{-5}%/h for extruded cable sheath: 4.5 MPa (650 psi) at 0 °C (32 °F); 2.4 MPa (350 psi) at 30 °C (86 °F); 1.0 MPa (150 psi) at 66 °C (150 °F)

Mass Characteristics

Density. 11.2 Mg/m³ (0.406 lb/in.³) at 18 °C (64 °F)

Volume change on freezing. 3.72%

Thermal Properties

Liquidus temperature. 320 °C (608 °F)

Solidus temperature. 312 °C (595 °F)

Coefficient of thermal expansion. Linear, 28.8 μm/m·K (16 μin./in.·°F) at 20 to 100 °C (68 to 212 °F)

Specific heat. 131 J/kg·K (0.0312 Btu/lb·°F) at 20 to 100 °C (68 to 212 °F)

Thermal conductivity. 33 W/m·K (19 Btu/ft·h·°F) at 0 °C (32 °F)

Electrical Properties

Electrical conductivity. Volumetric, 7.88% IACS at 20 °C (68 °F) for material heat treated at 235 °C (455 °F), quenched, and then aged 150 days

Fabrication Characteristics

Casting temperature. Ingot, 400 to 500 °C (750 to 930 °F)

4% Antimonial Lead
96Pb-4Sb

Commercial Names

Common name. Hard lead (96-4)

Applications

Typical uses. Storage battery grids

Precautions in use. Lead presents a health hazard; see this entry under "Corroding Lead."

Mechanical Properties

Tensile properties. Sheet, cold rolled 95%: tensile strength, 27.6 MPa (4.0 ksi); elongation, 48.3%. Heat treated at 235 °C (455 °F), quenched, and then aged 150 days: tensile strength, 80.7 MPa (11.7 ksi); elongation, 6.3%

Hardness. Sheet, cold rolled 95%: 8.1 HB. Heat treated at 235 °C (455 °F), quenched, and then aged 150 days: 24 HB. (Both tests with 1/16-in. ball, 9.85-kg load and 30-s duration.)

Fatigue strength. Sheet, cold rolled 95%: 10.3 MPa (1.50 ksi) at 2×10^7 cycles

Creep-rupture characteristics. Cold rolled sheet minimum creep rate at 30 °C (86 °F), 10^{-5} %/h

Mass Characteristics

Density. 11.02 Mg/m³ (0.398 lb/in.³) at 18 °C (64 °F)

Volume change on freezing. 3.36%

Thermal Properties

Liquidus temperature. 299 °C (570 °F)
Solidus temperature. 252 °C (486 °F)
Coefficient of thermal expansion. Linear, 27.8 μm/m·K (15.4 μin./in.·°F) at 20 to 100 °C (68 to 212 °F)
Specific heat. 133 J/kg·K (0.0318 Btu/lb·°F) at 20 to 100 °C (68 to 212 °F)
Thermal conductivity. 31 W/m·K (18 Btu/ft·h·°F) at 20 to 100 °C (68 to 212 °F)

Electrical Properties

Electrical conductivity. Volumetric, 7.7% IACS at 20 °C (68 °F) for material heat treated at 235 °C (455 °F), quenched, and then aged 150 days

Fabrication Characteristics

Casting temperature. Ingot, 400 to 500 °C (750 to 930 °F)

6% Antimonial Lead 94Pb-6Sb

Commercial Names

Common name. Hard lead (94-6)

Applications

Typical uses. Rolled sheet, extruded pipe for industrial applications requiring higher strength than that of soft lead, but similar corrosion resistance
Precautions in use. Lead presents a health hazard; see this entry under "Corroding Lead."

Mechanical Properties

Tensile properties. See Table 5.
Hardness. See Table 5.
Fatigue strength. At 2×10^7 cycles: chill cast, 17.2 MPa (2.50 ksi); cold rolled 95%, 10.3 MPa (1.50 ksi); extruded, 8.3 MPa (1.20 ksi)
Creep-rupture characteristics. Minimum creep rate at 10^{-5}%/h for cold rolled sheet: 2.8 MPa (400 psi) at 30 °C (86 °F); 340 kPa (50 psi) at 100 °C (212 °F)

Mass Characteristics

Density. 10.88 Mg/m^3 (0.393 lb/in.^{3}) at 18 °C (64 °F)
Volume change on freezing. 3.11%

Thermal Properties

Liquidus temperature. 285 °C (545 °F)
Solidus temperature. 252 °C (486 °F)
Coefficient of thermal expansion. Linear, 27.2 μm/m·K (15.1 μin./in.·°F) at 20 to 100 °C (68 to 212 °F)
Specific heat. 135 J/kg·K (0.0322 Btu/lb·°F) at 20 to 100 °C (68 to 212 °F)
Thermal conductivity. 29 W/m·K (17 Btu/ft·h·°F) at 20 to 100 °C (68 to 212 °F)

Electrical Properties

Electrical conductivity. Volumetric, 7.6% IACS for material heat treated at 235 °C (455 °F), quenched, and then aged 150 days
Electrical resistivity. 253 nΩ·m at 20 °C (68 °F)

Chemical Properties

Resistance to specific corroding agents.

Corrosive agent	Resistance
Acetone	Good
Acid	
Acetic	General attack
Carbonic	Severe general attack
Citric	Severe general attack
Fatty	Satisfactory in absence of O_2
Formic	Severe general attack
Hydrochloric	General attack

Corrosive agent	Resistance
Hydrofluoric	General attack
Nitric	Severe general attack
Oxalic	Severe general attack
Phosphoric	Resistant
Sulfuric	Resistant
Tartaric	Moderate attack
Air	Resistant
Alcohol	
Ethyl	Resistant
Methyl	Resistant
Higher	Resistant
Aluminum sulfate	Resistant
Ammonia	Resistant
Ammonium sulfate	Resistant
Bromine	General attack
Calcium chloride	General attack
Carbon dioxide	Moderate attack in H_2O solution
Carbon tetrachloride	Resistant in anhydrous solutions
Copper sulfate	Resistant
Ferric chloride	Severe attack
Hydrogen sulfide	General attack
Iodine	General attack
Magnesium sulfate	Resistant
Potassium nitrate	General attack
Refrigerants	Resistant in anhydrous solutions
Sodium sulfate	Resistant
Water	
Distilled	Considerable attack
Sea	Resistant

Fabrication Characteristics

Casting temperature. Ingot, 400 to 500 °C (750 to 930 °F)

Table 5 Typical mechanical properties of 6% antimonial lead

Condition and temperature	Tensile strength MPa	ksi	Elongation(a), %	Hardness, HB(b)
Chill cast				
Room temperature	47.2	6.84	24	13.0
100 °C (212 °F)	24.1	3.50	...	6.8
200 °C (392 °F)	5.9	0.85	...	2.0
Cold rolled 95%				
Room temperature	28.3	4.10	47	...
100 °C (212 °F)	12.8	1.85	...	3.9
200 °C (392 °F)	4.1	0.60	...	1.6
Extruded				
Room temperature	22.8	3.30	65	10.7

(a) In 50 mm or 2 in. (b) Tested with 1/16-in. ball, 9.85-kg load and 30-s duration.

8% Antimonial Lead
92Pb-8Sb

Applications

Typical uses. Similar to the uses of 9% antimonial lead
Precautions in use. Lead presents a health hazard; see this entry under "Corroding Lead."

Mechanical Properties

Tensile properties. Rolled 95%: tensile strength, 32.1 MPa (4.65 ksi); elongation, 31.3%. Heat treated at 235 °C (455 °F), quenched, and then aged 1 day at room temperature: 85.15 MPa (12.35 ksi); elongation, 4.7%.
Hardness. Rolled 95%: 9.5 HB. Heat treated and aged: 26.3 HB. (Both tests with 1/16-in. ball, 9.85-kg load and 30-s duration.)
Fatigue strength. Rolled 95%: 12.1 MPa (1.75 ksi) at 2×10^7 cycles
Creep-rupture characteristics. Sheet, cold rolled at 30 °C (86 °F): minimum creep rate, 10^{-5}%/h for stress of 2.93 MPa (425 psi)

Mass Characteristics

Density. 10.74 Mg/m³ (0.388 lb/in.³) at 18 °C (64 °F)
Volume change on freezing. 2.88%

Thermal Properties

Liquidus temperature. 271 °C (520 °F)
Solidus temperature. 252 °C (486 °F)
Coefficient of thermal expansion. Linear, 26.7 μm/m·K (14.8 μin./in.·°F) at 20 to 100 °C (68 to 212 °F)
Specific heat. 136 J/kg·K (0.0326 Btu/lb·°F) at 20 to 100 °C (68 to 212 °F)
Thermal conductivity. 27 W/m·K (16 Btu/ft·h·°F) at 20 to 100 °C (68 to 212 °F)

Electrical Properties

Electrical conductivity. Volumetric, 7.5% IACS for material heat treated at 235 °C (455 °F), quenched, and then aged 150 days
Electrical resistivity. 265 nΩ·m at 20 °C (68 °F)

Fabrication Characteristics

Casting temperature. Ingot, 400 to 500 °C (750 to 930 °F)

9% Antimonial Lead
91Pb-9Sb

Applications

Typical uses. Heavy-duty motive-power battery grids
Precautions in use. Lead presents a health hazard; see this entry under "Corroding Lead." Repeated charging and discharging of the battery can cause prohibitive corrosion unless the grid metal is adequately pure.

Mechanical Properties

Tensile properties. Chill cast: tensile strength, 52 MPa (7.5 ksi); elongation, 17%
Hardness. 15.4 HB
Fatigue strength. 19 MPa (2.7 ksi) at 20×10^6 cycles

Mass Characteristics

Density. 10.65 Mg/m³ (0.385 lb/in.³) at 18 °C (64 °F)
Volume change on freezing. 2.76%

Thermal Properties

Liquidus temperature. 265 °C (509 °F)
Solidus temperature. 252 °C (486 °F)
Coefficient of thermal expansion. Linear, 26.4 μm/m·K (14.6 μin./in.·°F) at 20 to 100 °C (68 to 212 °F)
Specific heat. 137 J/kg·K (0.0328 Btu/lb·°F) at 100 °C (212 °F)
Thermal conductivity. 27 W/m·K (15 Btu/ft·h·°F) at 20 to 100 °C (68 to 212 °F)

Electrical Properties

Electrical conductivity. Volumetric, 7.4% IACS for material heat treated at 235 °C (455 °F), quenched, and then aged 150 days
Electrical resistivity. 271 nΩ·m at 20 °C (68 °F)

Lead-base Babbitt
(Alloy 7)
75Pb-15Sb-10Sn

Specifications

ASTM. B23, alloy 7
SAE. J460, No. 14

Chemical Composition

Composition limits. ASTM: 9.3 to 10.7 Sn; 14.0 to 16.0 Sb; 0.30 to 0.60 As; 0.50 max Cu; 0.10 max Fe; 0.10 max Bi; 0.005 max Zn; 0.005 max Al; 0.50 max Cd; rem Pb

Applications

Typical uses. Sleeve bearings operating at moderate loads and speeds, such as bearings for blowers, pumps, electric motors and machine tools
Precautions in use. Lead presents a health hazard; see this entry under "Corroding Lead."

Mechanical Properties

Tensile properties. See Table 6.
Hardness. See Table 6.
Elastic modulus. Chill cast: tension, 29 GPa (4.2×10^6 psi)
Fatigue strength. Chill cast: 28 MPa (4.0 psi) at 20×10^6 cycles

Mass Characteristics

Density. 9.7 Mg/m³ (0.35 lb/in.³) at 20 °C (68 °F)
Volume change on freezing. 2.3%

Thermal Properties

Liquidus temperature. 268 °C (514 °F)
Solidus temperature. 240 °C (464 °F)
Coefficient of thermal expansion. Linear, 19.6 μm/m·K (10.8 μin./in.·°F) at 20 to 100 °C (68 to 212 °F)
Specific heat. 160 J/kg·K (0.038 Btu/lb·°F) at 20 to 100 °C (68 to 212 °F)
Latent heat of fusion. 110 kJ/kg
Thermal conductivity. 24 W/m·K (14 Btu/ft·h·°F) at 20 °C (68 °F)

Table 6 Mechanical properties of alloy 7 lead-base babbitt, chill cast

Temperature		Tensile strength		Elongation, %	Hardness, HB
°C	°F	MPa	ksi		
25	77	72	10.5	4	22
100	212	38	5.5	25	10.5
150	302	21	3.0	52	8

Electrical Properties

Electrical conductivity. Volumetric, 6.0%
Electrical resistivity. 286 nΩ·m at 20 °C (68 °F)

Fabrication Characteristics

Casting temperature. 325 to 400 °C (617 to 750 °F)

Lead-base Babbitt (Alloy 8) 80Pb-15Sb-5Sn

Specifications

ASTM. B23, alloy 8
Government. QQ-T-390, grade 6

Chemical Composition

Composition limits. 4.5 to 5.5 Sn; 14.0 to 16.0 Sb; 0.30 to 0.60 As; 0.50 max Cu; 0.10 max Fe; 0.10 max Bi; 0.005 max Zn; 0.005 max Al; 0.05 max Cd; rem Pb

Applications

Typical uses. Sleeve bearings operating at light loads and moderate speeds, such as car journals and bearings for mining and transmission machinery; used with all kinds of shafting
Precautions in use. Lead presents a health hazard; see this entry under "Corroding Lead."

Mechanical Properties

Tensile properties. See Table 7.
Hardness. See Table 7.
Elastic modulus. Chill cast: tension, 29 GPa (4.2×10^6 psi)
Fatigue strength. Chill cast: 27 MPa (3.9 ksi) at 20×10^6 cycles

Mass Characteristics

Density. 9.96 Mg/m^3 (0.36 lb/in.^{3}) at 20 °C (68 °F)
Volume change on freezing. Approximately 2%

Thermal Properties

Liquidus temperature. 272 °C (522 °F)
Solidus temperature. 240 °C (464 °F)
Coefficient of thermal expansion. Linear, 24 μm/m·K (13.3 μin./in.·°F) at 20 to 100 °C (68 to 212 °F)
Specific heat. 150 J/kg·K (0.036 Btu/lb·F) at 20 to 100 °C (68 to 212 °F)

Latent heat of fusion. 110 kJ/kg

Electrical Properties

Electrical conductivity. Volumetric, 6.1% IACS

Table 7 Mechanical properties of alloy 8 lead-base babbitt, chill cast

Temperature		Tensile strength		Elongation,	Hardness,
°C	°F	MPa	ksi	%	HB
25	77	69	10.0	5	20
100	212	38	5.4	27	10
150	302	20	2.9	55	7.3

Lead-base Babbitt (Alloy 13) 85Pb-10Sb-5Sn

Specifications

ASTM. B23, alloy 13
SAE. J460, No. 13
Other. Similar to AAR (car and tender bearings)

Chemical Composition

Composition limits. ASTM: 5.5 to 6.5 Sn; 9.5 to 10.5 Sb; 0.50 max Cu; 0.10 max Fe; 0.25 max As; 0.10 max Bi; 0.005 max Zn; 0.005 max Al; 0.05 max Cd; rem Pb

Applications

Typical uses. Sleeve bearings operating at light loads and low speeds, such as car journal bearings
Precautions in use. Lead presents a health hazard; see this entry under "Corroding Lead."

Mechanical Properties

Tensile properties. See Table 8.
Hardness. See Table 8.

Elastic modulus. Chill cast: tension, 29 GPa (4.2×10^6 psi)
Fatigue strength. Chill cast: 26 MPa (3.7 ksi) at 20×10^6 cycles

Mass Characteristics

Density. 10.5 Mg/m^3 (0.38 lb/in.^{3}) at 20 °C (68 °F)
Volume change on freezing. 2+%

Thermal Properties

Liquidus temperature. 256 °C (493 °F)
Solidus temperature. 240 °C (464 °F)
Specific heat. 150 J/kg·K (0.036 Btu/lb·°F) at 20 to 237 °C (68 to 459 °F)
Latent heat of fusion. 900 J/kg

Electrical Properties

Electrical resistivity. 282 nΩ·m

Fabrication Characteristics

Casting temperature. 340 to 425 °C (645 to 800 °F)

Electrical Properties

Electrical conductivity. Volumetric, 6.0% IACS
Electrical resistivity. 287 nΩ·m at 20 °C (68 °F)

Fabrication Characteristics

Casting temperature. 325 to 400 °C (617 to 750 °F)

Table 8 Mechanical properties of alloy 13 lead-base babbitt, chill cast

Temperature		Tensile strength		Elongation,	Hardness,
°C	°F	MPa	ksi	%	HB
25	77	69	10	5	19
100	212	34	4.9	30	8.5

Lead-base Babbitt (Alloy 15) 83Pb-15Sb-1Sn-1As

Specifications

ASTM. B23, alloy 15
SAE. J460, No. 15
Government. QQ-T-390, grade 10

Chemical Composition

Composition limits. ASTM: 0.8 to 1.2 Sn; 14.5 to 17.5 Sb; 0.8 to 1.4 As; 0.6 max Cu; 0.10 max Fe; 0.10 max Bi; 0.005 max Zn; 0.005 max Al; 0.05 max Cd; rem Pb

Applications

Typical uses. Sleeve bearings operating at high loads and speeds, such as bearings for diesel engines, automotive engines, steamships, steel mills, and all kinds of machinery
Precautions in use. Lead presents a health hazard; see this entry under "Corroding Lead."

Mechanical Properties

Tensile properties. See Table 9.
Hardness. See Table 9.
Elastic modulus. Chill cast: tension, 29 GPa (4.2×10^6 psi)
Fatigue strength. Chill cast: 30 MPa (4.3 ksi) at 20×10^6 cycles

Mass Characteristics

Density. 10.1 Mg/m^3 (0.365 lb/in.^{3}) at 20 °C (68 °F)

Volume change on freezing. Approximately 2.5%

Thermal Properties

Liquidus temperature. 353 °C (667 °F)
Solidus temperature. 247 °C (595 °F)

Fabrication Characteristics

Casting temperature. 480 to 540 °C (900 to 1000 °F)

Table 9 Mechanical properties of alloy 15 lead-base babbitt, chill cast

Temperature °C	°F	Tensile strength MPa	ksi	Elongation, %	Hardness, HB
25	77	71	10.4	2	20
100	212	41	6.4	9	12.5
150	302	26	3.7	26	7.4

Corrosion Resistance of Lead*

By Jerome F. Smith
Vice President and Secretary
Lead Industries Association, Inc.

LEAD has such a successful record of service in exposure to the atmosphere and to water that its resistance to corrosion by these media is often taken for granted. Underground, thousands of kilometres of lead-sheathed cable and lead pipe give reliable, long-term performance all over the world. In the chemical industry, lead is a major constituent in the strong, corrosion-resistant equipment necessary for handling many chemicals.

The Nature of Lead Corrosion

Corrosion of lead in aqueous electrolytes is an electrochemical process. The metal either enters the solution at anodic sites as metallic cations or is converted anodically to solid compounds. Both corrosion reactions can be represented by the reaction:

$$Pb - 2e^- \rightarrow Pb^{++} \qquad (Eq\ 1)$$

This oxidation reaction, which takes place at anodic sites, is accompanied by a reduction of some constituent in the electrolyte at cathodic sites. In neutral salt solutions, the cathodic reaction is the reduction of dissolved oxygen:

$$\tfrac{1}{2}O_2 + H_2O + 2e^- \rightarrow 2OH^- \qquad (Eq\ 2)$$

In acid solutions free of oxygen, the corresponding cathodic reaction is:

$$2H^+ + 2e^- \rightarrow H_2 \qquad (Eq\ 3)$$

Rate of corrosion is a function of the current flowing between the anodes and cathodes of the corrosion cell. Many factors and conditions can initiate or influence this flow of current. In corrosion of a single metal such as lead, local anodes and cathodes may be set up as a result of inclusions, inhomogeneities, stress variations and differences in temperature. In bimetallic (galvanic) corrosion, the anodic and cathodic sites are on different metals, with the less noble metal (anode) corroding in preference to the more noble metal (cathode).

In most environments, lead is cathodic to steel, aluminum, zinc, cadmium and magnesium, and thus will accelerate corrosion of these metals. With titanium and passivated stainless steels, lead is the anode of the cell and suffers accelerated attack. In either instance, rate of corrosion is governed by the difference in potential between the two metals, the ratio of their areas and their polarization characteristics.

The corrosion rate of lead usually is under anodic control, because the most important determinant usually is the solubility and other physical character-

istics of the corrosion products formed at anodic sites. The majority of these products are relatively insoluble lead salts that are deposited on the lead surface as impervious films, which tend to stifle further attack. The formation of such insoluble protective films is responsible for the high resistance of lead to corrosion by sulfuric, chromic and phosphoric acids.

In general, anything causing injury to the protective film increases corrosion rate. Factors that help create or strengthen the film reduce corrosion rate. Therefore, the life of lead-protected equipment can be extended, for example, by washing it with film-forming aqueous solutions containing sulfates, carbonates or silicates. This procedure is suggested for protecting lead when it will be in contact with corrosives that do not form protective films.

Corrosion in Water

Distilled water free of oxygen and carbon dioxide does not attack lead. Distilled water containing carbon dioxide but not oxygen also has little effect on lead. Corrosion behavior of lead in distilled water containing dissolved carbon dioxide and dissolved oxygen depends on CO_2 concentration. This dependency, which causes many differ-

Table 1 Corrosion of chemical lead in industrial and domestic waters(a)

Type of water	Temperature °C	Temperature °F	Aeration	Agitation	Corrosion rate μm/yr	Corrosion rate mils/yr
Condensed steam, traces of acid	21-38	70-100	None	Slow	21.59	0.85
Mine water:						
ph 8.3, 110 ppm hardness	20	68	Yes	Slow	6.60	0.26
160 ppm hardness	19	67	Yes	Slow	7.11	0.28
110 ppm hardness	22	72	Yes	Slow	6.35	0.25
Cooling-tower water, oxygenated, from Lake Erie	16-29	60-85	Complete	None	134.6	5.3
Los Angeles aqueduct water, treated with chlorine and copper sulfate ..	Ambient		...	0.5 ft/s	9.65	0.38
Spray cooling water, chromate treated.........................	16	60	Yes	...	9.4	0.37

(a) Total immersion.

Table 2 Corrosion of lead in natural waters

Location and type of water	Type of test	Agitation	Corrosion rate μm/yr	Corrosion rate mils/yr	Ref
Bristol Channel; seawater	Immersion about 93% of the time	...	12.7	0.50	1
Southhampton Docks; seawater ..	Half tide level	...	2.79	0.11	2
Gatun Lake, CZ; tropical fresh water	Immersion	None	2.03	0.08	3
Fort Amador, CZ; tropical Pacific Ocean	Immersion	Flowing(a)	9.14	0.36	
Fort Amador, CZ; tropical Pacific Ocean	Mean tide level	Flowing(a)	5.08	0.20	
San Francisco Harbor; seawater..	Mean tide level	Flowing	10.67	0.42	
Port Hueneme Harbor, CA; seawater	Immersion	Flowing(b)	5.59	0.22	4
Kure Beach, NC; seawater	Immersion	...	15.24	0.60	

(a) At 150 mm/s (0.5 ft/s). (b) At 60 mm/s (0.2 ft/s).

ent reactions to take place in a narrow range of concentration, explains the contradictory nature of much of the corrosion data reported in the literature.

For instance, lead steam coils that handle pure water condensate are not severely corroded in systems where all condensate is returned to the boiler and negligible make-up water is used. However, if make-up water is used, dissolved oxygen can be introduced to the condensate and corrosion can be severe. Carbon dioxide also can be generated from the breakdown of carbonates and bicarbonates in boiler water, decreasing the severity of corrosion of lead. The oxygen level in the make-up water usually is controlled by adding oxygen scavengers, such as hydrazine or sodium sulfite.

In general, corrosion rate in natural and domestic waters depends on degree of water hardness. Water hardness is caused mainly by calcium and magnesium salts in the water. These salts, if present in at least moderate amounts

(greater than 125 ppm), form films on lead that adequately protect it against corrosive attack. Silicate salts present in the water increase both hardness and the protective value of the film. In contrast, nitrate and chloride ions either interfere with the formation of the protective film or penetrate it, and thus increase corrosion.

In soft, aerated natural and domestic waters, corrosion rate depends on both hardness and oxygen content of the water. When water hardness is less than 125 ppm, corrosion rate, like the rate in distilled water, depends on the relative proportions of dissolved carbon dioxide and dissolved oxygen. Potable waters, in which lead content is not permitted to exceed 0.10 ppm, often have hardness below 125 ppm and often contain considerable amounts of carbon dioxide and oxygen; thus lead frequently cannot be used for pipe or containers that handle potable waters. This problem of contamination limits the use of lead in many applications

where, from a service point of view, corrosion rate is negligible.

Corrosion rates of chemical lead in several industrial and domestic waters are presented in Table 1. It should be noted that corrosion rate is relatively low, even where water hardness is below 125 ppm. (A corrosion rate for a fresh water is also included among the data for seawater in Table 2.)

Corrosion of lead in seawater is relatively slight and may be retarded by incrustations of lead salts. Data on performance of lead in seawater at several locations are given in Table 2. A comparison of two of the entries in this table shows that at the same tropical location (Panama Canal Zone), corrosion rate of lead in fresh water is about one-fourth the rate in seawater.

Extensive service experience and laboratory testing have indicated that corrosion rate of lead generally is quite low in a wide variety of waters. The only major applications where lead cannot be used are those involving some pure waters containing oxygen and soft natural waters, especially if contamination is of concern. In contrast, as discussed above, addition of calcium and magnesium salts further enhances resistance of lead to corrosion by water.

Atmospheric Corrosion

Lead in most of its forms exhibits consistent durability in all types of atmospheric exposure, including industrial, rural and marine (see Table 3). These three atmospheric environments are distinct, because each involves different factors that promote corrosion. In rural areas, which are relatively free of pollutants, the only important environmental factors influencing corrosion rate are humidity, rainfall and air flow. However, near or on the sea, chlorides entrained in marine air often exert a strong effect on corrosivity. In industrial environments, sulfur oxide gases and the minerals in solid emissions considerably change patterns of corrosion behavior. However, the protective films that form on lead and its alloys are so effective that corrosion is insignificant in most natural atmospheres. The extent of this protection is demonstrated by the survival of lead roofing and auxiliary products after hundreds of years of atmospheric exposure. In fact, according to Burns (Ref 10), the metal is preserved permanently if these films are not damaged.

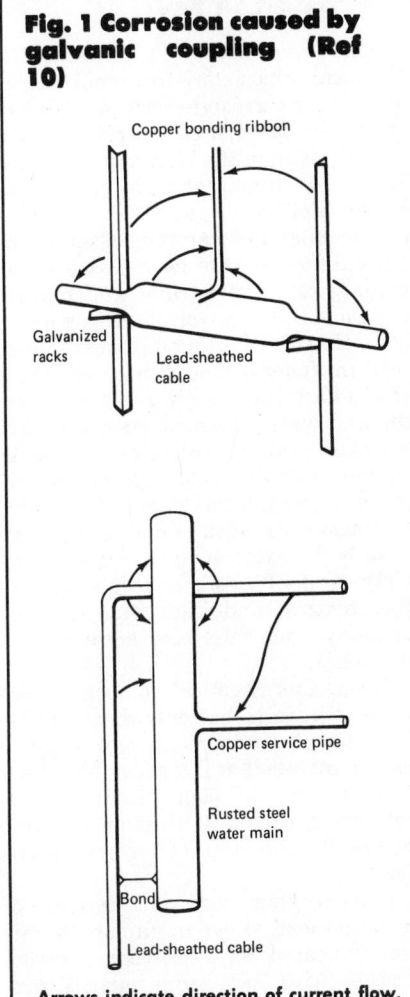

Antimonial lead exhibits approximately the same corrosion rate in atmospheric environments as chemical lead. However, its greater hardness, strength and resistance to creep often make it more desirable for use in roofs and reflecting pools. The ability of some antimonial leads to retain this greater mechanical strength in atmospheric environments has been demonstrated in exposure tests in which sheets containing 4% antimony and smaller amounts of arsenic and tin were placed in semirestricted positions for 3 years. They showed less tendency to buckle than chemical lead, indicating that their greater resistance to creep had been retained.

Painting of lead coatings, especially terne metal (a coating containing 8 to 12% Sb, rem Pb), further raises their resistance to corrosion in outdoor environments. Terne metal has such good paint retention that one coat will far outlast two separate coats on plain steel.

Corrosion in Underground Ducts

Lead is used extensively in the form of sheathing for power and communications cables because of its impermeability to water and its excellent resistance to corrosion in a wide variety of soil conditions. Cables are either buried directly in the ground or installed in ducts or conduits. In the United States, the preferred method is to lay cable in ducts or conduits made of materials such as cement, vitrified clay or wood.

Severe corrosion of lead in underground service (in ducts or directly in the soil) is the exception rather than the rule. However, because repair or replacement of underground components is difficult and expensive, proper corrosion protection is recommended in any underground service. Although the discussion that follows is based on preventive methods used for lead-sheathed cables, in many ways it is directly applicable to underground behavior of other lead products, such as water service pipe.

The environment within ducts often is quite complex (Ref 9). It can include combinations of highly humid manhole and soil atmospheres, free lime leached from concrete, and alkalies formed by electrolysis of salts in the water that seeps into ducts. Compton (Ref 10) describes some of the factors involved in corrosion of lead cable sheathing and how they relate to cable assembly and installation. He discusses their influence in initiating or accelerating corrosion and uses simple sketches for illustration. Two of these factors—galvanic coupling and differential aeration—are discussed below.

Galvanic Coupling. Figure 1, from Compton (Ref 10), illustrates two typical examples of contact between lead and other metals. In the presence of an electrolyte, such a dissimilar-metal couple forms a galvanic cell in which the more anodic metal is corroded. A difference in potential sufficient to cause corrosion may also arise when

Table 3 Corrosion of lead in various natural outdoor atmospheres

Location	Type of atmosphere	Duration of test, years	Type of lead	Corrosion rate µm/yr	mils/yr	Ref
Altoona, PA	Industrial	10	Chemical	0.737	0.029	5,6
			Pb-1Sb	0.584	0.023	5,6
New York City	Industrial	20	Chemical	0.381	0.015	5,6
			Pb-1Sb	0.330	0.013	5,6
Sandy Hook, NJ	Seacoast	20	Chemical	0.533	0.021	5,6
			Pb-1Sb	0.508	0.020	5,6
Key West, FL	Seacoast	10	Chemical	0.584	0.023	5,6
			Pb-1Sb	0.559	0.022	5,6
LaJolla, CA	Seacoast	20	Chemical	0.533	0.021	5,6
			Pb-1SB	0.584	0.023	5,6
State College, PA	Rural	20	Chemical	0.330	0.013	5,6
			Pb-1Sb	0.356	0.014	5,6
Phoenix, AZ	Semi-arid	20	Chemical	0.102	0.004	5,6
			Pb-1Sb	0.308	0.012	5,6
Kure Beach, NC (80 ft site)	East coast, marine	2	Chemical	1.321	0.052	7
			Pb-6Sb	1.041	0.041	7
Newark, NJ	Industrial	2	Chemical	1.473	0.058	7
			Pb-6Sb	1.067	0.042	7
Point Reyes, CA	West coast, marine	2	Chemical	0.914	0.036	7
			Pb-6Sb	0.660	0.026	7
State College, PA	Rural	2	Chemical	1.397	0.055	7
			Pb-6Sb	0.991	0.039	7
Birmingham, England	Urban	7	99.96%Pb	0.939	0.037	8
			Pb-1.6Sb	0.102	0.004	8
Wakefield, England	Industrial	1	99.995%Pb	1.879	0.074	8
Southport, England	Marine	1	99.995%Pb	1.778	0.070	8
Bourneville, England	Suburban	1	99.995%Pb	1.956	0.077	8
Cardington, England	Rural	1	99.995%Pb	1.422	0.056	8
Cristobal, CZ	Tropical, marine	8	Chemical	1.346	0.053	3
Miraflores, CZ	Tropical, marine	8	Chemical	0.762	0.030	3

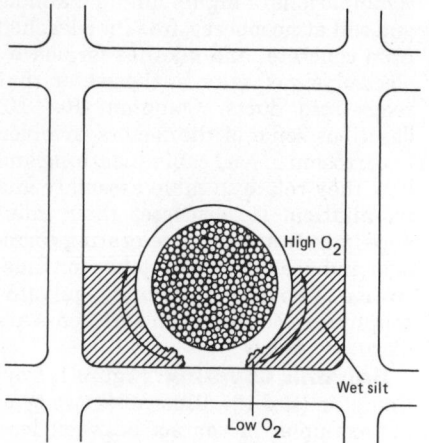

Fig. 2 Corrosion caused by differential aeration in a duct (Ref 10)

High O₂

Wet silt

Low O₂

Arrows indicate direction of current flow.

the surface of the lead is scratched to expose bright, active metal. In such instances, the exposed metal is the anode and is attacked.

Differential Aeration. Figure 2 is the illustration used by Compton (Ref 10) to show differential-aeration corrosion. In this type of corrosion cell, areas exposed to low oxygen concentration tend to become anodic to areas exposed to higher oxygen concentrations. As shown, the amount of air able to penetrate the silt and reach the crevice where the cable sheath and the duct meet is less than the amount available at the upper surface of the sheath; this results in corrosion.

An actual example of differential-aeration corrosion is described in a NACE Technical Committee Report (Ref 11). Lead-sheathed cable was pressed tightly against the inner surface of a tile duct, and water formed a meniscus extending from the sheathing surface to the tile. The area that was pressed against the tile did not corrode. However, an adjacent area, where the water was farthest from contact with air, corroded severely. The lead surface in contact with water closer to the air in the duct was the cathode.

Alkalinity. Another factor causing corrosion of cable sheathing has been described by Perry (Ref 12). Sheathing on cable installed in continuous concrete or asbestos cement ducts in concrete tunnels under waterways was found to be severely corroded. Analysis

of water samples from these locations revealed that the corrosion had resulted from the presence of up to 1000 ppm of hydroxides. These alkaline water samples (pH 10.9 to 12.2) contained mainly calcium hydroxide. Sodium hydroxide was also found in some tunnels.

The source of the calcium hydroxide was incompletely cured concrete. Electrolysis of solutions of deicing salts that had seeped into the tunnels was believed to be the source of sodium hydroxide. The buildup in NaOH concentration occurred because seepage water was not being removed (the ducts had been designed to function without removal of seepage water). Proper drainage and use of completely cured, impervious concrete were suggested as corrective measures.

Stray currents can cause severe corrosion of lead pipe or lead cable sheathing. Stray currents are those that follow paths outside intended circuits. They may also be minor earth currents. Stray currents cause corrosion at the point where they leave the metal. Sources of stray currents include electric railway systems, grounded electric dc power, electric welders, cathodic protection systems and electroplating plants. Stray alternating currents are much less damaging than stray direct currents.

It has also been found that corrosion of lead cable sheathing in manhole waters depends more on the magnitude and polarity of the potential between the ground and the lead sheathing than it does on the natural dissolved salts in the water. Corrosion is at a minimum when the sheathing is cathodic to the ground (Ref 13 and 14). Grounding prevents this type of corrosion.

Other factors that can initiate corrosion of lead sheathing include contact with acetic acid (in wooden ducts), microorganisms and corroded steel-tape armor. Bacterial corrosion usually occurs when aeration is poor and mud, water or organic matter is present. Bacteria capable of reducing sulfates to sulfides are the principal cause of attack. Microbial decomposition of the hydrocarbons present in cable coatings also may produce organic acids corrosive to lead. Corrosion of lead by corroded steel-tape armor can occur when the oxide coating formed on the steel is cathodic to lead.

Corrosion in Soil

Soils vary widely in physical and chemical characteristics and, consequently, in corrosive effect. A summary of the characteristics and properties of soils is given in the National Bureau of Standards circular on underground corrosion (Ref 15).

More than 200 varieties of soil in the United States have been classified according to texture, color and natural drainage. The physical properties of soils that most influence corrosion of lead in underground service are those that affect permeability of the soil to air and water, because good drainage tends to minimize corrosion. Soils with coarse textures, such as sands and gravels, permit free circulation of air. Corrosion in such soils is approximately the same as in the atmosphere. Clays and silty soils generally exhibit fine texture and high water-holding capacity, and thus poor aeration and drainage.

Numerous chemical compounds are present in soils, but only those soluble in water play important roles in corrosion of metals. For instance, the calcareous nature of some Indiana soils influences corrosion through alkaline attack or promotion of bacterial activity.

Considerable corrosion testing of lead and lead alloys in numerous soils has indicated that corrosion rate decreases with increasing particle size, and that distribution of anodic and cathodic areas depends on soil particle size, water-activation value of the soil, soil pH and duration of exposure. Test results also show that lead tends to become passive in soils regardless of water content; however, addition of sodium bicarbonate reactivates it.

The data in Table 4 show that in most soils the average corrosion rate of lead is low—from less than 2.5 to 10 μm (0.1 to 0.4 mil) per year. It should be noted, however, that depth of pitting often is a more important measure of underground corrosion behavior than corrosion rate.

The most comprehensive investigation of corrosion of metals buried in soils was conducted by the National Bureau of Standards from 1910 to 1955. This investigation included lead alloy pipe of three different compositions buried in 14 soils. Specimens were removed periodically; maximum exposure time was 11 years (Table 4).

Analysis of the data in Table 4 indicates that, in general, weight loss and

Table 4 Corrosion of lead alloys in various soils(a) (Ref 15)

	Chemical lead(b)				Tellurium lead(c)				Antimonial lead(d)			
	Corrosion rate		Max pit depth		Corrosion rate		Max pit depth		Corrosion rate		Max pit depth	
Type of soil	μm/yr	mils/yr	μm	mils	μm/yr	mils/yr	μm	mils	μm/yr	mils/yr	μm	mils
Cecil clay loam	<2.54	<0.1	457	18	<2.54	<0.1	406	16	<2.54	<0.1	229	9
Hagerstown loam	<2.54	<0.1	787	31	<2.54	<0.1	762	30	<2.54	<0.1	406	16
Lake Charles clay	7.62	0.3	2540	100	10.16	0.4	2718	107	10.16	0.4	2642	104
Muck	7.62	0.3	1321	52	7.62	0.3	1346	53	7.62	0.3	1295	51
Carlisle muck	5.08	0.2	508	20	5.08	0.2	533	21	2.54	0.1	305	12
Rifle peat	<2.54	<0.1	838	33	<2.54	<0.1	584	23	<2.54	<0.1	711	28
Sharkey clay	7.62	0.3	1778	70	7.62	0.3	1854	73	10.16	0.4	2261	89
Susquehanna clay	<2.54	<0.1	864	34	2.54	0.1	1016	40	2.54	0.1	356	14
Tidal marsh	<0.25	<0.01	305	12	<0.25	<0.01	203	8	<0.25	<0.01	152	6
Docas clay	<2.54	<0.01	635	25	<2.54	<0.1	432	17	<2.54	<0.1	483	19
Chino silt loam	<2.54	<0.1	381	15	<2.54	<0.1	508	20	<2.54	<0.1	178	7
Mohave fine gravelly clay . .	<2.54	<0.1	610	24	<2.54	<0.1	584	23	2.54	<0.1	406	16
Cinders	7.62	0.3	2159	85	7.62	0.3	1549	61	10.16	0.4	1168	46
Merced silt loam	<2.54	<0.1	610	24	<2.54	<0.1	406	16	<2.54	<0.1	229	9

(a) Maximum exposure time, 11 years. (b) 0.056 Cu, 0.002 Bi, 0.001 Sb. (c) 0.08 Cu, 0.01 Sb, 0.043 Te. (d) 0.036 Cu, 5.3 Sb, 0.016 Bi.

maximum pit depth decrease with increasing aeration of the soil. For example, poor aeration caused severely deep pitting of the lead buried in Sharkey clay, in Lake Charles clay and in cinders, whereas pitting of pipe buried in the well-aerated Cecil clay loam was shallow.

Resistance to Chemicals

The excellent resistance of lead and lead alloys to corrosion by a wide variety of chemicals is attributed to the polarization of local anodes caused by formation of a relatively insoluble surface film of lead corrosion products (Ref 16). The extent of protection depends on the compactness, adherence and solubility of these films.

Solubilities of various lead compounds in water at room temperature are given in Table 5. These data are general indicators of the behavior of lead in solutions that promote formation of these compounds. The solubility of a lead corrosion product, however, depends on the solution in which the lead is immersed. Therefore, the solubility of that corrosion product in water is not always an adequate indicator of its behavior in another solution. This fact is illustrated by the variation in solubility of lead sulfate in sulfuric acid as acid concentration and temperature change (Fig. 3). The lead sulfate film is less soluble in sulfuric acid solutions than it is in water. Solubility drops to a minimum value at acid concentrations of 30 to 60% and then increases at higher concentrations. At intermediate

Table 5 Solubility of lead compounds (Ref 17)

Lead compound	Formula	Temperature °C	°F	Solubility (a), Kg/m³
Acetate	$Pb(C_2H_3O_2)_2$	20	68	433
Bromide	$PbBr_2$	20	68	8.441
Carbonate	$PbCO_3$	20	68	0.0011
Basic carbonate	$2PbCO_3, Pb(OH)_2$	. . .	. . .	Insoluble
Chlorate	$Pb(ClO_3)_2, H_2O$	18	64	0.513
Chloride	$PbCl_2$	20	68	9.9
Chromate	$PbCrO_4$	25	77	0.000058
Fluoride	PbF_2	18	64	0.64
Hydroxide	$Pb(OH)_2$	18	64	0.155
Iodide	PbI_2	18	64	0.63
Nitrate	$Pb(NO_3)_2$	18	64	565
Oxalate	PbC_2O_4	18	64	0.0016
Oxide	PbO	18	64	0.017
Orthophosphate	$Pb_3(PO_4)_2$	18	64	0.00014
Sulfate	$PbSO_4$	25	77	0.0425
Sulfide	PbS	18	64	0.1244
Sulfite	$PbSO_3$	. . .	. . .	Insoluble

(a) In water at room termperature.

concentrations, the sulfate film is so insoluble that corrosion is negligible.

Another example of the importance of the solubility relationship of the lead film to its environment is shown in Fig. 4. Lead nitrate is quite soluble in dilute and intermediate-strength solutions of nitric acid at room temperature. Lead is not resistant to corrosion under such conditions. However, above a nitric acid concentration of 50%, lead nitrate is only slightly soluble, and lead is quite resistant to attack.

Increases in temperature generally increase corrosion rate (Fig. 3). This effect is primarily due to increases in film solubility.

Galvanic Corrosion. When lead is anodic to a metal to which it is coupled

and a firm film develops on the lead, galvanic corrosion of the lead will be negligible. For example, when lead is galvanically connected to a copper or a copper alloy in a sulfuric, chromic or phosphoric acid solution, the lead is protected by a firm film even though it is the anode in the galvanic cell. However, when the other metal is the anode, galvanic corrosion of the lead may occur and sometimes is severe. Aluminum or magnesium will be severely corroded if coupled with lead in the presence of an electrolyte. If lead is coupled with Monel in a 6% sulfuric acid solution, corrosion of the Monel will be accelerated.

A factor to be kept in mind is that environmental changes can reverse the

Fig. 3 Solubility of lead sulfate in sulfuric acid

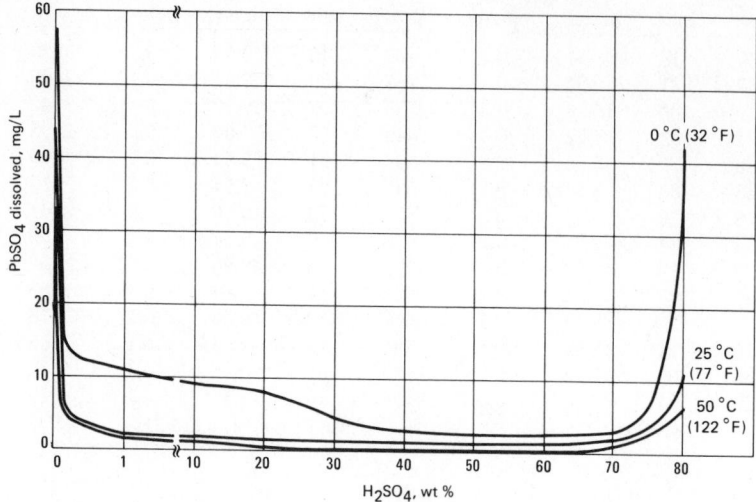

H_2SO_4, wt %	PbSO$_4$ dissolved, mg/L, at:		
	0 °C	25 °C	50 °C
0	33.0	44.5	57.7
0.005	8.0	10.0	24.0
0.01	7.0	8.0	21.0
0.10	4.6	5.2	13.0
1.0	1.8	2.2	11.3
10.0	1.2	1.6	9.6
20.0	0.5	. . .	8.0
30.0	0.4	1.2	4.6
60.0	0.4	1.2	2.8
70.0	1.2	1.8	3.0
75.0	2.8	3.0	6.6
80.0	6.5	11.5	42.0

Fig. 4 Solubility of lead nitrate in nitric acid

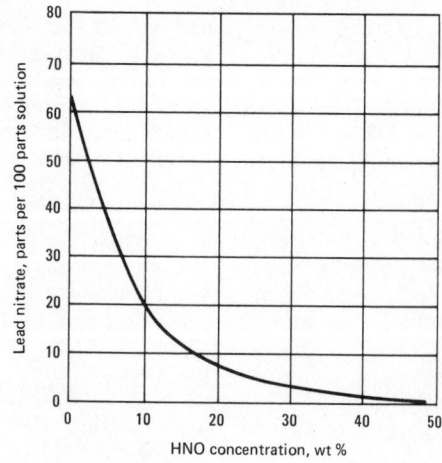

galvanic positions of two metals. For example, iron is anodic to lead in acids and cathodic to it in alkalies.

In general, galvanic corrosion of lead is significant only when the lead is coupled with a metal to which it is anodic, when an electrolyte is present and when a firm film cannot be maintained; corrosion of metals anodic to lead seldom occurs. When galvanic corrosion of either lead or the dissimilar metal does occur, it is unlikely to be severe, because lead occupies a central position in the galvanic series.

Physical variables also influence corrosion rate of lead in many situations. For instance, if the flow velocity of a solution is above a critical point, it can completely erode the protective film, leaving only a clean lead surface exposed for continued attack. This is demonstrated by the rapid increase in corrosion of lead in 20% sulfuric acid at velocities greater than 1.5 m/s (300 ft/min). Presence of foreign insoluble matter further aggravates this condition.

Fatigue stresses also may break the protective film on lead, repeatedly exposing lead to environmental attack. However, applied fatigue stresses no greater than the creep strength of lead do not significantly affect corrosion properties.

Quantitative Corrosion Data. It is important to remember when evaluating quantitative corrosion data that lead weighs more per unit of volume and is normally used in greater thicknesses than most other metals. The effects of these two factors should be considered in evaluating the data presented in this section.

Lead has high corrosion resistance to chromic, sulfurous, sulfuric and phosphoric acids and is widely used in their manufacture and handling. Lead satisfactorily resists all but the most dilute solutions of sulfuric acid. It performs well at acid concentrations up to 95% at ambient temperatures, up to 85% at 220 °C (428 °F) and up to 93% at 150 °C (302 °F) (Fig. 5). Below a concentration of 5%, the corrosion rate increases, but it is still relatively low. In the lower range of concentration, antimonial lead is recommended.

Lead exhibits the same excellent corrosion resistance to higher concentrations of chromic, sulfurous and phosphoric acids at elevated temperatures. Lead also is generally resistant to solutions of salts formed by each of these acids. However, the reaction to mix-

Table 6 Corrosion of chemical lead in commercial phosphoric acid at 21 °C (70 °F)

Solution	Corrosion rate	
	μm/yr	mils/yr
20% H₃PO₄	86.4	3.4
30% H₃PO₄	124.5	4.9
40% H₃PO₄	144.8	5.7
50% H₃PO₄	162.6	6.4
85% H₃PO₄	40.6	1.6
80% H₃PO₄(a)	325.1	12.8
(a) Pure grade.		

Table 7 Corrosion of lead in hydrochloric acid at 24 °C (75 °F)

HCl concen- tration, %	Chemical lead		6% Antimonial lead	
	μm/yr	mils/yr	μm/yr	mils/yr
1%	610	24	840	33
5%	410	16	510	20
10%	560	22	1 090	43
15%	790	31	3 810	150
20%	1880	74	4 060	160
25%	4830	190	5 080	200
35%(a)..	8890	350	13 720	540
(a) Commercially concentrated HCl.				

Fig. 5 Corrosion rate of lead in sulfuric acid (Ref 19)

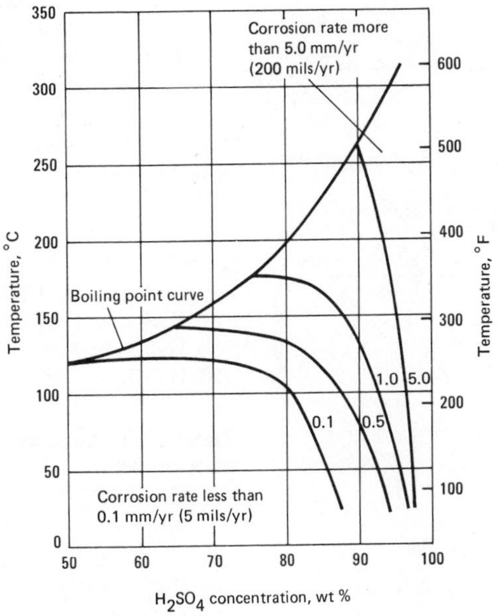

Table 8 Corrosion of lead in hydrochloric acid-ferric chloride mixtures at 24 °C (75 °F)

Solution	Chemical lead		6% Antimonial lead	
	μm/yr	mils/yr	μm/yr	mils/yr
5% HCl + 5% FeCl₃	711	28	940	37
10% HCl + 5% FeCl₃	1041	41	1930	76
15% HCl + 5% FeCl₃	2235	88	4064	160
20% HCl + 5% FeCl₃	3810	150	4826	190

Table 9 Corrosion of lead in nitric acid

Solution	Corrosion rate			
	24°C (75°F)		50°C (122°F)	
	μm/yr	mils/yr	μm/yr	mils/yr
1% HNO₃...	3556	140	15 240	600
5% HNO₃..	41 910	1650	46 990	1850
10% HNO₃.	86 360	3400	88 646	3490

tures of these salts is much more complex.

Lead finds especially wide application in manufacture of phosphoric acid from phosphate rock when sulfuric acid is used in the process. Corrosion rates are low for all acid concentrations up to 85% (see Table 6). Six percent antimonial lead has been reported to have a lower corrosion rate than chemical lead

in a plant test using a solution containing 32% phosphoric acid, 0.4% sulfuric acid and 1% chlorides at 88 °C (190 °F). In pure acid manufactured from elemental phosphor, lead corrodes at a higher rate due to the absence of sulfates.

Lead has fair corrosion resistance to dilute hydrochloric acid (up to 15%) at 24 °C (75 °F); corrosion rate increases at higher concentrations and at higher temperatures (Table 7). The presence of 5% ferric chloride also accelerates corrosion (Table 8).

The resistance of lead to corrosion by hydrofluoric acid is only fair. However, lead is used to handle hydrofluoric acid because it is the only low-priced metal that has adequate corrosion resistance. The corrosion rate in this acid (if it is free of air) is less than 510 μm/yr (20

mils/yr) for a wide range of temperatures and concentrations (see Fig. 6).

Nitric, acetic and formic acids in most concentrations corrode lead at rates high enough to preclude its use in these acids. However, although nitric acid rapidly attacks lead when dilute, it has little effect at strengths of 52 to 70%. The same is true of hydrofluoric acid, acetic acid and acid sodium sulfate.

Addition of sulfuric acid to acids corrosive to lead often lowers corrosion rate. For example, although nitric acid in concentrations less than 50% is quite corrosive to lead (see Table 9), in the presence of 54% sulfuric acid the corrosion rate in 1% and 5% nitric acid is quite low even at 118 °C or 245 °F (see Table 10). Other concentrations of sulfuric acid also lower corrosion rate in nitric acid (see Table 11). The composition range of mixed H₂SO₄ and HNO₃ solutions for which chemical lead has a corrosion rate of less than 500 μm (20 mils) per year is shown in Fig. 7. Chemical lead is preferred over 6% antimonial lead for handling these mixtures of acids.

Corrosion rates of chemical lead and 6% antimonial lead in hydrochloric acid and in fluosilicic acid are retarded by the presence of sulfuric acid (see Tables 12 and 13). Data on corrosion of

Fig. 6 Resistance of lead to corrosion in air-free hydrofluoric acid

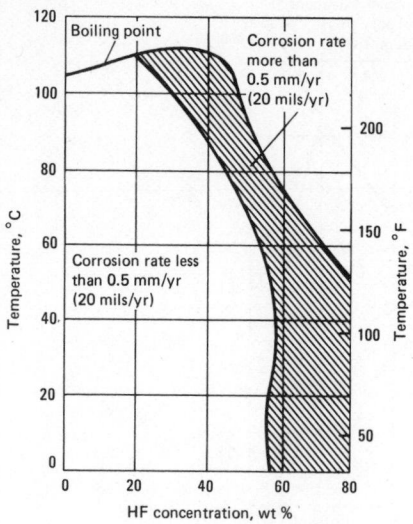

lead in chemical process fluids containing sulfuric acid or closely related compounds are presented in Table 14.

Qualitative corrosion data serve to provide guidelines for screening suitable metals for chemical equipment. Laboratory test environments may not always simulate actual plant conditions, and there may be significant variations among plants manufacturing the same product. Therefore, it often is more helpful to be less specific when categorizing corrosion rates of lead in various chemicals. Table 15 presents such less-specific information, and should be used only as a guide for determining whether further tests are warranted. Most of the data in Table 15 are for chemical lead. The corrosion rates of different grades of lead in the same chemical all normally fall within the same category. Therefore, no mention is made of variations in corrosion rate for other grades of lead.

Table 10 Effect of nitric acid in sulfuric acid on the corrosion of lead at 118 °C (245 °F)

Solution	Chemical lead		6% Antimonial lead	
	µm/yr	mils/yr	µm/yr	mils/yr
78% H_2SO_4 + 0% HNO_3	188	7.4	356	14
78% H_2SO_4 + 1% HNO_3	150	5.9	559	22
78% H_2SO_4 + 5% HNO_3	213	8.4	2896	114

Table 11 Corrosion of chemical lead with sulfuric-nitric mixed acids

Solution	Corrosion rate			
	24 °C (75 °F)		50 °C (122 °F)	
	µm/yr	mils/yr	µm/yr	mils/yr
78% H_2SO_4 + 0% HNO_3........	25.4	1	50.8	2
78% H_2SO_4 + 1% HNO_3........	76.2	3	304.8	12
78% H_2SO_4 + 3.5% HNO_3	91.4	3.6	457.2	18
78% H_2SO_4 + 7.5% HNO_3	101.6	4	889	35

Table 12 Corrosion of lead in hydrochloric acid–sulfuric acid mixtures

Solution	Chemical lead				6% Antimonial lead			
	24 °C (75 °F)		66 °C (150 °F)		24 °C (75 °F)		66 °C (150 °F)	
	µm/yr	mils/yr	µm/yr	mils/yr	µm/yr	mils/yr	µm/yr	mils/yr
1% HCl + 9% H_2SO_4	130	5	230	9	130	5	300	12
3% HCl + 7% H_2SO_4	360	14	810	32	530	21	1040	41
5% HCl + 5% H_2SO_4	360	14	1070	42	530	21	1650	65
7% HCl + 3% H_2SO_4	410	16	1140	45	560	22	1880	74
9% HCl + 3% H_2SO_4	460	18	1190	47	760	30	2130	84
5% HCl + 25% H_2SO_4 ...	250	10	560	22	560	22	860	34
10% HCl + 20% H_2SO_4 ...	430	17	1070	42	2030	80	1470	58
15% HCl + 15% H_2SO_4 ...	1040	41	1880	74	2290	90	4570	180
20% HCl + 10% H_2SO_4 ...	2180	86	3050	120	2790	110	4570	180
25% HCl + 5% H_2SO_4	3560	140	4060	160	3810	150	5330	210
5% HCl + 45% H_2SO_4 ...	1580	62	...	...	1350	53	...	...
10% HCl + 40% H_2SO_4 ...	1650	65	...	...	2130	84	...	...
15% HCl + 35% H_2SO_4 ...	1680	66	...	...	3050	120	...	...
20% HCl + 30% H_2SO_4 ...	2130	84	...	...	3300	130	...	...
25% HCl + 25% H_2SO_4 ...	3050	120	...	...	5330	210	...	...

Table 13 Effect of sulfuric acid on the corrosion of lead by fluosilicic acid at 45 °C (113 °F)

Solution	Chemical lead		6% Sb lead	
	µm/yr	mils/yr	µm/yr	mils/yr
5% H_2SiF_6	1346	53	1956	77
5% H_2SiF_6 + 5% H_2SO_4	229	9	356	14
10% H_2SiF_6	1626	64	2921	115
10% H_2SiF_6 + 1% H_2SO_4	2235	88	1930	76
1% H_2SiF_6 + 10% H_2SO_4	102	4	229	9

REFERENCES

1. J. N. Friend, The Relative Corrodibilities of Ferrous and Nonferrous Metals and Alloys—Part 1—The Results of Four Years Exposure in Bristol Channel, *Journal of the Institute of Metals*, Vol. 39, 1928, p 111–143
2. J. N. Friend, The Relative Corrodibilities of Ferrous and Nonferrous Metals and Alloys—Part III—Results of Three Years Exposure at Southampton Docks, *Journal of the Institute of Metals*, Vol 48, 1932, p 109–120
3. B. W. Forgeson *et al*, Corrosion of Metals in Tropical Environments, *Corrosion*, Vol 14, 1958, p 73t–81t
4. C. V. Brouilette, Corrosion Rates in Port Hueneme Harbor, *Corrosion*, Vol 14, 1958, p 352t–356t
5. G. O. Hiers and E. J. Minarcik, "The Use of Lead and Tin Outdoors", STP 175, American Society for Testing and Materials, 1955, p 135–140
6. Report of Subcommittee VI of ASTM Committee B-3 on Atmospheric Corrosion Tests of Nonferrous Metals and Alloys, Proceedings of the American Society for Testing and Materials, Vol 44, 1944, p 224
7. Report of Subcommittee VI of ASTM Committee B-3 on Atmo-

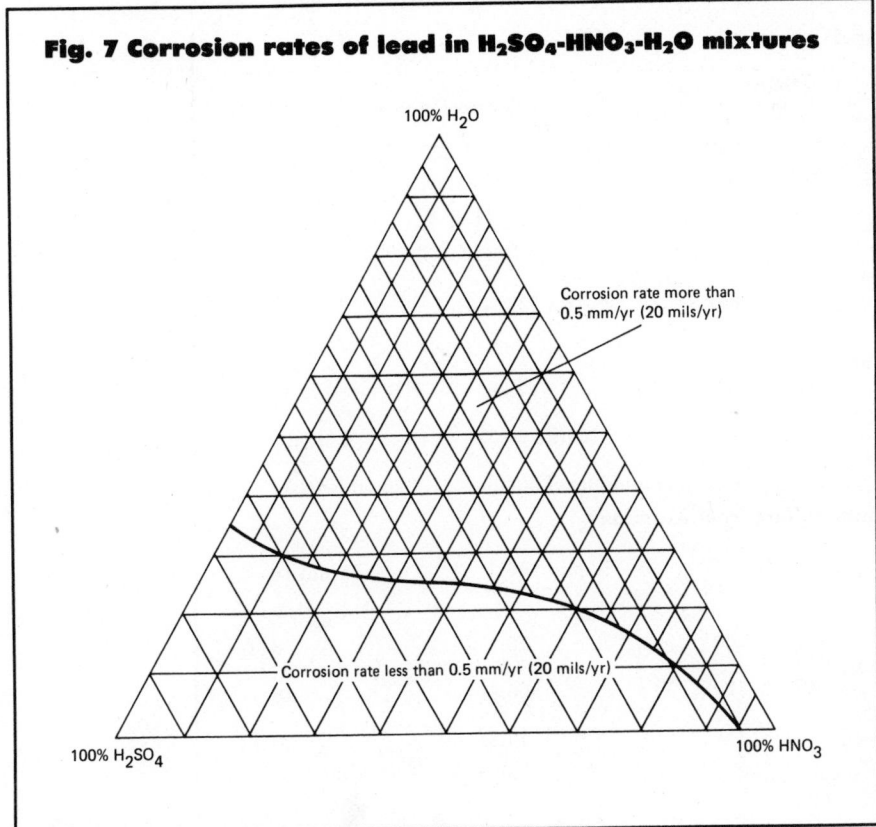

Fig. 7 Corrosion rates of lead in H₂SO₄-HNO₃-H₂O mixtures

100% H₂O

Corrosion rate more than
0.5 mm/yr (20 mils/yr)

Corrosion rate less than 0.5 mm/yr (20 mils/yr)

100% H₂SO₄

100% HNO₃

Table 14 Corrosion of lead in chemical process fluids (Ref 20)

Fluids	Temperature °C	°F	Corrosion rate µm/yr	mils/yr
Sulfation of Oils with 25% Sulfuric Acid (66° Be) — 140 °F (60 °C)				
Castor	. . .	. . .	76.2	3
Tallow.	. . .	. . .	304.8	12
Olive.	. . .	. . .	76.2	3
Cod Liver	. . .	. . .	152.4	6
Neatsfoot	. . .	. . .	279.4	11
Fish.	. . .	. . .	279.4	11
Vegetable.	. . .	. . .	584.2	23
Peanut	. . .	. . .	457.2	18
Sulfonation with 93% Sulfuric Acid (66° Be)				
Naphthalene	166	330	1143	45
Phenol.	120	248	76.2	3
Washing and Neutralization of Sulfated and Sulfonated Compounds				
Sulfated vegetable oil + water wash-neutralized with sodium hydroxide	60	140	228.6	9
Naphthalene sulfonic acid + water wash-neutralized with caustic soda pH 3	70	158	990.6	39
Washing tallow with 2% by wt 60° Be sulfuric acid	121	250	127	5
Storage of liquid alkyl detergent	. . .	. . .	7.62	0.3

(continued)

spheric Corrosion Tests of Nonferrous Metals and Alloys, Proceedings of the American Society for Testing and Materials, Vol 62, 1962, p 216

8. J. N. Friend, The Relative Corrodibilities of Ferrous and Nonferrous Alloys—Part II—The Results of Seven Years Exposure to Air at Birmingham, *Journal of the Institute of Metals,* Vol 42, 1929, p 149–155

9. R. M. Burns, Corrosion of Metals II—Lead and Lead Alloy Cable Sheathing, *Bell System Technical Journal,* Vol 15, 1936, p 603–625

10. K. G. Compton, Factors Involved in Corrosion of Lead Cable Sheath, *Corrosion,* Vol 17, 1961, p 409t–412t

11. NACE Task Group T-4B-1, Cell Corrosion on Lead Cable Sheaths, *Corrosion,* Vol 12, 1956, p 257t–259t

12. R. I. Perry, Preventing Corrosion of Lead-Sheathed Power Cables in Concrete Tunnels, *Corrosion,* Vol 12, 1956, p 207t–212t

13. Y. Yamaguchi *et al,* Studies on Corrosion of Communication Cable Lead Sheath by Manhole Water, *Corrosion Engineering* (Japan), Vol 5, 1956, p 302–306

14. NACE Technical Unit Committee T-4B, Corrosion of Lead Sheath in Manhole Water, *Corrosion,* Vol 14, 1958, p 85t–87t

15. M. Romanoff, "Underground Corrosion", National Bureau of Standards Circular 579, April 1957, p 227

16. E. L. Littauer and H. C. Wesson, Lead and Lead Alloys, chapter 4.3 in *Corrosion,* Vol 1, edited by L. L. Shrier, Wiley, New York, 1963, p 4:68–4:85

17. *Handbook of Chemistry and Physics* (45th Ed.), edited by R. C. Weast, CRC Press, Cleveland, 1964

18. H. D. Crockford and D. J. Brawley, Solubility of Lead Sulfate in Water and Aqueous Solutions of Sulfuric Acid, *Journal of the American Chemical Society,* Vol 56, 1934, p 2600

19. M. G. Fontana, *Industrial and Engineering Chemistry,* Vol 43, 1951, p 105A

20. Modified and taken from: G. A. Nelson, *Corrosion Data Survey* (1967 Ed.), National Association of Corrosion Engineers

Table 14 (continued)

Fluids	Temperature °C	°F	Corrosion rate μm/yr	mils/yr
Storage of 50% chlorosulfonic acid-50% sulfur trioxide	. . .	. . .	15.24	0.6
Mixing tank and crystallizer-saturated ammonium sulfate–5% sulfuric acid solution	47	116	25.4 to 127	1 to 5
Splitting				
Olive oil and 0.5% sulfuric acid (66° Be)	88	190	279.4	11
Storage of split fatty acids	. . .	. . .	Liquid 20.32	Liquid 0.8
Storage of split fatty acids	. . .	. . .	Liquid level 304.8	Liquid level 12
Extraction of Aluminum Sulfate from Alumina				
Bauxite + sulfuric acid–boiling	. . .	. . .	Liquid 406.4	Liquid 16
Bauxite + sulfuric acid–boiling	. . .	. . .	Vapor 127	Vapor 5
Alum evaporator	116	240	76.2	3
Tank for dissolving alum paper mill	49	120	406.4	16
Storage of 24% alum solution	. . .	. . .	15.24	0.6
Dorr Settling Tank				
19.5 sulfuric acid, 20% ferrous sulfate, 10% titanium oxide as TiSO4 .	70	158	254	10
Evaporator				
Nickel sulfate solution	100	212	152.4	6
Zinc sulfate solution	107	225	152.4	6
Ammonium Sulfate Production				
Solution-saturated ammonium sulfate + 5% sulfuric acid	47	116	Mixing tank 25.4	Mixing tank 1
Solution-saturated ammonium sulfate + 5% sulfuric acid	47	116	Crystallizer 127	Crystallizer 5
Acid Washing				
Lube oil-treatment with 25% sulfuric acid	104	220	635	25
Sludge oil + 15% sulfuric acid-steam treatment	. . .	. . .	508	20
Benzol (crude)-treatment with 3% sulfuric acid washed with water, neutralized with lime	60	140	152.4	6
Tar oil-treatment with 25% sulfuric acid, washed with water, neutralized with sodium hydroxide . .	77	170	609.6	24
Wet acid gases from regeneration of sulfuric acid	121	250	152.4	6
Polymerization				
Polymerization of butenes with 72% sulfuric acid . . .	80	175	12.7	0.5
Polymerization of butenes with 72% sulfuric acid . . .	80	175	356 pits	14 pits

(continued)

Table 14 (continued)

Fluids	Temperature °C	°F	Corrosion rate µm/yr	mils/yr
Viscose Rayon Spinning Bath				
Evaporator—6% sulfuric acid, 17% sodium sulfate, 30% other inorganic sulfates	40	104	127	5
Evaporator—concentrated bath of 20% sulfuric acid 30% sodium sulfate	55	130	101.6	4
Vapors from spin bath evaporator	49	120	127	5
Spinning bath drippings	46	115	203.2	8
Storage-reclaimed spinning bath liquor	...	...	50.8	2
Pickling Solution				
Brass and copper-sulfuric acid + 5% cupric sulfate	71	160	127	5

Table 15 Corrosion rate of lead in chemical environments(a)

Chemical	Temperature °C	°F	Concentration, %	Corrosion class(b)
Acetic acid	24	75	Glacial	B
Acetic anhydride	24	75	...	A
Acetone	24-100	75-212	10-90	A
Alcohol, ethyl	24-100	75-212	10-100	A
Alcohol, methyl	24-100	75-212	10-100	A
Aluminum chloride	24	75	0-10	B
Aluminum potassium sulfate	24-100	75-212	10-20	A
Aluminum potassium sulfate	24-100	75-212	20-100	B
Ammonia	24-100	75-212	10-30	B
Ammonium chloride	24	75	0-10	B
Ammoniun hydroxide	27	80	3.5-40	A
Ammonium nitrate	24-49	75-120	10-30	D
Ammonium sulfate	24	75	...	B
Amyl acetate	24	75	80-100	B
Aniline	20	68	...	A
Antimony chloride	24	75	...	C
Arsenic trichloride	100-149	212-300	...	B
Barium chloride	24-100	75-212	10	B
Benzaldehyde	24	75	10-100	D
Benzene	24	75	...	B
Benzoic acid	24	75	...	D
Benzyl alcohol	24-100	75-212	...	B
Benzyl chloride	24-100	75-212	...	B
Beryllium chloride	100	212	...	D
Boric acid	24-149	75-300	10-100	B
Bromine	24	75	...	B
Butyric acid	24	75	10-100	D
Cadmium sulfate	24-100	75-212	10-30	A
Calcium bicarbonate	24	75	...	C
Calcium chloride	24	75	20	A
Calcium fluoride	24-100	75-212	...	B
Calcium nitrate	24	75	10	D
Calcium sulfate	24-100	75-212	10	B
Carbon disulfide	24-100	75-212	...	A
Carbonic acid	24	75	...	D

(continued)

Table 15 (continued)

Chemical	Temperature °C	°F	Concentration, %	Corrosion class (b)
Cellulose acetate	24	75	...	A
Cellulose nitrate	24-100	75-212	...	B
Chloroacetic acid	24	75	...	B
Chloric acid	24	75	10	D
Chlorine	38	100	...	B
Chloroform	24-62	75-143	...	B
Chromic acid	24	75	...	B
Copper chloride	24	75	10-40	D
Creosote	24	75	90	D
Dichlorobenzene	24-100	75-212	10-100	B
Diethyl ether	24	75	...	B
Dioxane	24-100	75-212	...	B
Ethyl acetate	24-79	75-175	...	B
Ferric ammonium sulfate	24-100	75-212	10-20	A
Ferric chloride	24	75	20-30	D
Ferric sulfate	24-79	75-175	10-20	A
Ferrous chloride	24	75	10-30	C
Ferrous sulfate	24-100	75-212	10	B
Fluosilicic acid	45	113	10	D
Formaldehyde	24-52	75-125	20-100	B
Formic acid	24-100	75-212	10-100	D
Glycerol	24	75	...	B
Hydrazine	24	75	20-100	D
Hydriodic acid	24	75	10-50	D
Hydrobromic acid	24	75	10-70	D
Hydrochloric acid	24	75	0-10	C
Hydrogen peroxide	24	75	10-30	D
Isopropanol	24	75	...	A
Lead acetate	24	75	10-30	D
Lead choloride	24-100	75-212	...	B
Lithium hydroxide	24	75	...	D
Magnesium chloride	24	75	10-100	D
Magnesium sulfate	24-100	75-212	10-60	B
Mercury	24	75	100	D
Methyl ethyl ketone	24-100	75-212	10-100	B
Nitrobenzene	24-52	75-125	...	B
Oxalic acid	24	75	20-100	D
Phenol	24	75	90	B
Phosphoric acid	24-93	75-200	...	B
Potassium chloride	8	47	0.25-8.0	B
Potassium hydroxide	24-60	75-140	0-50	B
Pyridine	24-100	75-212	10	B
Sodium acetate	25	77	4	B
Sodium bicarbonate	24	75	10	B
Sodium chloride	25	77	0.5-24	A
Sodium hydroxide	26	79	0-30	B
Sodium nitrate	24	75	10	D
Sodium sulfate	24	75	2-20	A
Stannous chloride	24	75	10-50	D
Zinc sulfate	35	95	...	B
Zinc chloride	79	175	25	B

(a) For corrosion rate information on lead in other chemical environments, see the more extensive tables contained in Lead for Corrosion Resistant Applications—A Guide, Lead Industries Association, Inc. (b) The four categories of the Table are: A <2 mils/year: negligible corrosion—lead recommended for use. B <20 mpy: practically resistant—lead recommended for use. (When the only information available is that "lead is resistant" to a certain chemical, that chemical was arbitrarily placed in this category.) C is 20-50 mpy: lead may be used where this effect on life can be tolerated. D >50 mpy: corrosion rate too high to merit any consideration of lead.

Magnesium

Selection and Application of Magnesium and Magnesium Alloys

By the ASM Review Committee on
Magnesium and Magnesium Alloys*

MAGNESIUM and magnesium alloys are used in a wide variety of structural and nonstructural applications. Structural applications include industrial, materials-handling, commercial and aerospace equipment. In industrial machinery, such as textile and printing machines, magnesium alloys are used for parts that operate at high speeds and thus must be lightweight to minimize inertial forces. Materials-handling equipment includes dockboards, grain shovels and gravity conveyors. Commercial applications include luggage and ladders. Good strength and stiffness at both room and elevated temperatures combined with light weight make magnesium alloys valuable for aerospace applications.

Magnesium is also employed in various nonstructural applications. It is used as an alloying element in alloys of aluminum, zinc, lead and other nonferrous metals. It is used as an oxygen scavenger and desulfurizer in the manufacture of nickel and copper alloys, as a desulfurizer in the iron and steel

*See page X for committee list.

industry, and as a reducing agent in the production of beryllium, titanium, zirconium, hafnium and uranium. Another important nonstructural use of magnesium is in the Grignard reaction in organic chemistry. In finely divided form, magnesium finds some use in pyrotechnics, both as pure magnesium and alloyed with 30% or more aluminum. It is also used for cathodic protection of other metals from corrosion and in construction of dry-cell and reserve cell batteries. Gray iron foundries use magnesium and magnesium-containing alloys as ladle-addition agents introduced just before the casting is poured. The magnesium makes the graphite particles nodular and greatly improves the properties of the cast iron.

Because of its rapid but controllable response to etching as well as its light weight, magnesium is used increasingly in photoengraving.

Table 1 presents data on consumption of primary magnesium in the United States in 1965, 1970, 1972, 1974, 1976 and 1977. Primary magnesium is furnished to ASTM B92, grade

9980A, with a specified minimum magnesium content of 99.8%. Also available are special grades of primary magnesium in which manganese, aluminum and iron impurities are held to especially low levels. These special grades are employed in chemical and metallurgical applications, such as preparation of uranium metal and other reactive metals.

Aluminum and zinc are relatively soluble in solid magnesium, but their solubilities decrease at low temperatures. The solubility of aluminum is 12.7% by weight at 437 °C (819 °F) and 3.0% at 93 °C (200 °F); solubility of zinc is 6.2% at 340 °C (644 °F) and 2.8% at 204 °C (400 °F). Solubilities of manganese, zirconium and cerium are less than 1.0% by weight at 482 °C (900 °F). At the eutectic temperature, 4.5% thorium is soluble in magnesium. Manganese is effective in improving corrosion stability of magnesium alloys that contain aluminum and zinc.

Designations. A standard system of alloy and temper designations, adopted in 1948, is explained in Table 2. As an example of how the system works, con-

Table 1 U.S. consumption of primary magnesium(a)

Use	Metric tons consumed					
	1965	1970	1972	1974	1976	1977
Structural						
Castings						
Sand	2684	1574	635	1244	1121	1038
Die	5078(b)	8165	8459	10 707	4326	4555
Permanent mold	738	236	668	907	963	953
Wrought products						
Sheet and plate	4478	(c)	3462	(d)	(d)	(d)
Extrusions, shapes, tubing	5438	11 111	7029	6642	5863	(d)
Other (including forgings)	(c)	(c)	1253	5465	3447	11 484
Total structural	18 416	21 086	21 506	24 965	15 720	18 030
Nonstructural						
Alloying ingredient in						
Aluminum	23 824	33 145	39 418	56 374	49 382	50 987
Zinc	123	32	25	22	26	21
Copper		(d)	34	17	13	9
Other metals	2010	(d)	99	15	9	7
Anodes for cathodic protection	4170	5241	5935	9468	7099	3712
Constituent in chemicals	3452	7605	8827	8348	9218	9037
Powder for pyrotechnics	(d)	5121	(d)	(d)	(d)	(d)
Nodulizing agent for ductile iron	(d)	4281	6896	9617	6894	6634
Agent for scavenging and deoxidizing	154	(d)	297	259	(d)	(d)
Reducing agent for titanium, zirconium, hafnium, uranium and beryllium	7680	5714	5523	6865	5441	4759
Other	3320	2577	5492	2007	1155	964
Total nonstructural	44 733	63 716	72 546	92 992	79 237	76 130
Total all uses	63 149	84 802	94 052	117 957	94 957	94 160

(a) Source: U.S. Bureau of Mines. (b) Includes investment castings. (c) Included with extrusions. (d) Included with Other.

sider magnesium alloy AZ91C-T6, the nominal composition and typical properties of which are given in Table 3. The first part of the designation, AZ, signifies that aluminum and zinc are the two principal alloying elements. The second part of the designation, 91, means that aluminum and zinc are present in rounded-off percentages of 9 and 1, respectively. The third part, C, indicates that this is the third alloy standardized with 9% Al and 1% Zn as the principal alloying additions. The fourth part, T6, denotes that the alloy is solution treated and artificially aged.

Casting Alloys

There are several systems of magnesium alloys for sand and permanent mold castings: magnesium-aluminum-manganese with and without silicon or zinc (AM, AS and AZ), magnesium-zirconium (K), magnesium-zinc-zirconium with and without rare earths (ZK, ZE and EZ), magnesium-thorium-zirconium with and without zinc (HK, HZ and ZH), and magnesium-silver-zirconium with rare earths or thorium (QE and QH). Nominal compositions and typical properties of these alloys are given in Table 3.

AZ91C and AZ81A have almost completely replaced AZ63A where good ductility and moderately high yield strength are required at temperatures up to 120 °C (250 °F). In similar fashion, AZ92A has virtually replaced AM100A. In any of these Mg-Al-Zn alloys, an increase in aluminum content raises yield strength but reduces ductility for comparable heat treatment. The castability of these alloys is good, and final selection of the specific composition may be based on tests of the finished castings.

The difference between die-casting alloys AZ91A and AZ91B is maximum copper content, which does not affect mechanical properties. Because of its higher maximum copper content (0.30%), AZ91B has lower resistance to corrosion. The reason for the higher maximum copper is to allow the alloy to be made from secondary metal, which

will reduce the cost of the alloy. Die castings are used in the as-cast condition.

Die-cast alloy AM60A has better elongation and toughness, but lower tensile and yield strengths, than AZ91A or AZ91B. It is used in the production of die-cast automotive wheels and in some archery equipment. Die-cast alloy AS41A has creep strength much superior to that of AZ91A, AZ91B or AM60A up to 175 °C (350 °F), and good elongation, yield strength and tensile strength. One use of AS41A is in crankcases of air-cooled automotive engines.

KIA is primarily used where high damping capacity is required. It has low tensile and yield strength.

The ZK and ZH alloys develop the highest yield strengths of the casting alloys and can be cast into complicated shapes. However, these grades are more costly than the alloys of the AZ series.

The two ZK casting alloys in use are ZK51A and ZK61A. The latter, which has a slightly higher zinc content, has significantly greater strength than ZK51A (Table 3). Both alloys maintain high ductility after an artificial aging treatment (T5). The strength of ZK61A can be further increased (3 to 4%) by solution treatment plus artificial aging (T6), without impairing ductility. Both of these alloys have fatigue strengths equal to those of the Mg-Al-Zn alloys, but they are more susceptible to microporosity and hot cracking and are less weldable. Addition of either thorium or rare-earth metals overcomes these deficiencies. The strength properties of ZE63A are equivalent to those of ZK61A, those of ZH62A are equivalent to or better than those of ZK51A, but those of ZE41A are somewhat lower than those of ZK51A (Table 3).

ZE41A alloy was developed to meet the growing need for an alloy with medium strength, good weldability, and improved castability over AZ91C and AZ92A. It has good fatigue and creep properties and maximum freedom from microshrinkage. Unlike the AZ alloys, there is a very close relationship between separately-cast test bar properties and those obtained from the casting itself, even where relatively thick cast sections are involved. ZE41A is used up to 160 °C (320 °F) in such applications as aircraft engines, helicopter and airframe components, and wheels and gear boxes.

ZE63A is a high strength alloy with excellent tensile and yield strength,

Table 2 Standard four-part ASTM system of alloy and temper designations for magnesium alloys(a)(b)

First part	Second part	Third part	Fourth part
Indicates the two principal alloying elements	Indicates the amounts of the two principal	Distinguishes between different alloys with the same percentages of the two principal alloying elements	Indicates condition (temper)
Consists of two code letters representing the two main alloying elements arranged in order of decreasing percentage (or alphabetically if percentages are equal)	Consists of two numbers corresponding to rounded-off percentages of the two main alloying elements and arranged in same order as alloy designations in first part	Consists of a letter of the alphabet assigned in order as compositions become standard	Consists of a letter followed by a number (separated from the third part of the designation by a hyphen)
A-Aluminum E-Rare Earth H-Thorium K-Zirconium M-Manganese Q-Silver S-Silicon T-Tin Z-Zinc	Whole numbers	Letters of alphabet except I and O	F-As fabricated O-Annealed H10 and H11-Slightly strain hardened H23, H24 and H26-Strain hardened and partially annealed T4-Solution heat treated T-5-Artificially aged only T6-Solution heat treated and artificially aged T8-Solution heat treated, cold worked and artificially aged

(a) As an example of a typical four-part designation, AZ91C-T6 is explained in the text. (b) This system is now also used by SAE.

which is achieved by heat treating in a hydrogen atmosphere. Because hydriding proceeds from the surface, heat treating time and penetrability are limiting factors. This alloy has excellent casting characteristics.

The Mg-RE-Zr alloys are used at temperatures of from 175 to 260 °C (350 to 500 °F). Because their high-temperature strengths exceed those of the Mg-Al-Zn alloys, a savings in weight is possible.

The Mg-RE-Zn-Zr alloy EZ33A has good strength stability when exposed to elevated temperatures. (Strength stability is the ability to resist deterioration of strength from extended exposure to elevated temperature.) EZ33A castings usually are quite free from porosity, but are more susceptible to inclusions of dross than are the Mg-Al-Zn alloys. For these reasons, they are more difficult to cast in some designs than Mg-Al-Zn alloys. EZ33A castings

have excellent pressure tightness. ZE41A, discussed earlier, is similar to EZ33A, but with higher tensile and yield strength due to the higher zinc content. Some sacrifice is made in castability and weldability of ZE41A for the higher mechanical properties.

When the operating temperature of an engine housing was increased from 120 to 205 °C (250 to 400 °F), alloy EZ33A-T5 was successfully substituted for AZ92A-T6. The change was based on creep tests of separately cast bars of the two alloys; stress values at three temperatures, for 0.1% creep in 1000 h, were as follows:

Alloy	Temperature °C	°F	Stress MPa	ksi
AZ92A-T6	205	400	6.9	1.0
	260	500	2.1	0.3
EZ33A-T5	205	400	58	8.4
	260	500	26	3.7
	315	600	8.3	1.2

The Mg-Th-Zr alloys HK31A and HZ32A are intended primarily for use at temperatures of 200 °C (400 °F) and higher, for which properties superior to those of EZ33A are required. For full development of properties, HK31A requires the T6 treatment (solution heat treatment plus artificial aging), whereas HZ32A, which contains zinc, requires only the T5 treatment (artificial aging). HK31A and HZ32A castings have been used at temperatures as high as 345 to 370 °C (650 to 700 °F) in a few applications. The Mg-Zn-Th-Zr alloy ZH62A differs from other Mg-Th-Zr alloys in that it is intended primarily for use at room temperature.

Mg-Th-Zr alloys are more difficult to cast than EZ33A because they are more subject to formation of inclusions and defects as a result of gating turbulence. The tendency for inclusions to form in the Mg-Th-Zr alloys is particularly marked in thin-wall parts that require rapid pouring rates. These alloys have adequate castability for production of complex parts of moderate to heavy wall thickness.

At 260 °C (500 °F) and slightly higher, HZ32A is equal to or better than HK31A in short-time and long-time creep strength at all extensions. HK31A has higher tensile, yield and short-time creep strengths up to 370 °C (700 °F). However, HZ32A has greater strength stability at elevated temperatures, and much better foundry characteristics, than does HK31A.

QE22A is a high tensile strength and yield strength alloy with fairly good properties at temperatures up to 204 °C (400 °F). QH21A has similar properties to QE22A at room temperature but superior properties at temperatures at 204 °C (400 °F) up to 260 °C (500 °F). Both QE22A and QH21A have good castability and weldability. They do require solution and aging heat treatments to achieve the higher mechanical properties. Also, they are relatively expensive because of their silver contents.

Wrought Alloys

Wrought magnesium alloys are produced as bars, billets and shapes, wire, sheet, plate and forgings.

Extruded bars and shapes are made of several types of magnesium alloys (see Table 3). For normal strength requirements, one of the Mg-Al-Zn (AZ) alloys is usually selected. The strength of these alloys increases

Table 3 Nominal compositions and typical room-temperature mechanical properties of magnesium alloys

Alloy	Al	Mn(a)	Th	Zn	Zr	Others	Tensile strength MPa	ksi	Yield strength Tensile MPa	ksi	Compressive MPa	ksi	Bearing MPa	ksi	Elongation in 50 mm or 2 in., %	Shear strength MPa	ksi	Hardness, HRB(b)
Sand and Permanent Mold Castings																		
AM100A-T61	10.0	0.1	...	...	...	...	275	40	150	22	150	22	...	...	1	...	...	69
AZ63A-T6	6.0	0.15	...	3.0	...	...	275	40	130	19	130	19	360	52	5	145	21	73
AZ81A-T4	7.6	0.13	...	0.7	...	...	275	40	83	12	83	12	305	44	15	125	18	55
AZ91C-T6	8.7	0.13	...	0.7	...	...	275	40	195	21	145	21	360	52	6	145	21	66
AZ92A-T6	9.0	0.10	...	2.0	...	...	275	40	150	22	150	22	450	65	3	150	22	84
EZ33A-T5	...	...	...	2.7	0.6	3.3 RE	160	23	110	16	110	16	275	40	2	145	21	50
HK31A-T6	...	...	3.3	...	0.7	...	220	32	105	15	105	15	275	40	8	145	21	55
HZ32A-T5	...	...	3.3	2.1	0.7	...	185	27	90	13	90	13	255	37	4	140	20	57
K1A-F	...	...	...	...	0.7	...	180	26	55	8	...	...	125	18	1	55	8	...
QE22A-T6	...	...	...	...	0.7	2.5 Ag, 2.1 Di	260	38	195	28	195	28	...	...	3	...	...	80
QH21A-T6	...	...	60	...	0.7	2.5 Ag, 1.0 Di	275	40	205	30	...	...	...	...	4	...	...	...
ZE41A-T5	...	...	...	4.2	0.7	1.2 RE	205	30	140	20	140	20	350	51	3.5	160	23	62
ZE63A-T6	...	...	...	5.8	0.7	2.6 RE	300	44	190	28	195	28	...	...	10	...	...	60-85
ZH62A-T5	...	...	1.8	5.7	0.7	...	240	35	170	25	170	25	340	49	4	165	24	70
ZK51A-T5	...	...	...	4.6	0.7	...	205	30	165	24	165	24	325	47	3.5	160	23	65
ZK61A-T5	...	...	...	6.0	0.7	...	310	45	185	27	185	27	...	...	...	170	25	68
ZK61A-T6	...	...	...	6.0	0.7	...	310	45	195	28	195	28	...	...	10	180	26	70
Die Castings																		
AM60A-F	6.0	0.13	...	...	...	...	205	30	115	17	115	17	...	...	6	...	...	...
AS41A-F(d)	4.3	0.35	...	...	...	1.0 Si	220	32	150	22	150	22	...	...	4	...	...	...
AZ91A and B-F(e)	9.0	0.13	...	0.7	...	...	230	33	150	22	165	24	...	...	3	140	20	63
Extruded Bars and Shapes																		
AZ10A-F	1.2	0.2	...	0.4	...	...	240	35	145	21	69	10	...	...	10	...	...	...
AZ21X1-F	1.8	0.02	...	1.2	...	...	...	...	...	...	...	...	...	...	...	...	...	...
AZ31 B and C-F(d)	3.0	...	...	1.0	...	...	260	38	200	29	97	14	230	33	15	130	19	49
AZ61A-F	6.5	...	...	1.0	...	...	310	45	230	33	130	19	285	41	16	140	20	60
AZ80A-T5	8.5	...	...	0.5	...	...	380	55	275	40	240	35	...	...	7	165	24	82
HM31A-F	...	1.2	3.0	...	...	...	290	42	230	33	185	27	345	50	10	150	22	...
M1A-F	...	1.2	...	...	...	...	255	37	180	26	83	12	195	28	12	125	18	44
ZK21A-F	...	...	...	2.3	0.45(a)	...	260	38	195	28	135	20	...	...	4	...	...	...
ZK40A-T5	...	...	...	4.0	0.45(a)	...	276	40	255	37	140	20	...	...	4	...	...	...
ZK60A-T5	...	...	...	5.5	0.45(a)	...	365	53	305	44	250	36	405	59	11	180	26	88
Sheet and Plate																		
AZ31B-H24	3.0	...	...	1.0	...	...	290	42	220	32	180	26	325	47	15	160	23	73
HK31A-H24	...	...	3.0	...	0.6	...	255	33	200	29	160	23	285	41	9	140	20	68
HM21A-T8	...	0.6	2.0	...	...	...	235	34	170	25	130	19	270	39	11	125	18	...
PE(f)	3.3	...	...	0.7	...	...	...	...	...	...	...	...	...	...	...	...	...	...

(a) Minimum. (b) 500-kg load, 10-mm ball. (c) A and B are identical except that 0.30% max residual Cu is allowable in AZ91B. (d) For battery applications. (e) Properties of B and C are identical, but AZ31C has 0.15 min Mn, 0.1 max Cu and 0.03 max Ni. (f) Photoengraving grade.

as aluminum content increases. AZ31B is a widely-used, moderate-strength alloy with good formability. AZ31B is also used extensively for cathodic protection. AZ31C is a lower purity, commercial variation of AZ31B for lightweight structural applications. M1A and ZM21A are alloys that can be extruded at higher speeds than AZ31B but have limited use due to their lower strength. Alloy AZ10A, because of its low aluminum content, has lower strength than AZ31B, but it can be welded without subsequent stress relief. AZ61A and AZ80A can be artificially aged for additional strength (with a sacrifice in ductility). AZ80A is not available in hollow shapes.

AZ21X1 is an alloy designed specially for use in battery applications.

Alloy ZK60A is used where high strength and good toughness are required. This alloy is heat treatable and normally is used in the artificially aged (T5) condition. ZK21A and ZK40A are lower strength and more readily extrudable than ZK60A and have had limited use in hollow tubular strength requirements.

HM31A has moderate strength and is suitable for use in applications requiring good strength and creep resistance at temperatures in the range of 150 to 425 °C (300 to 800 °F).

Forgings are made of alloys AZ31B, AZ61A, AZ80A, M1A and ZK60A, the compositions and properties of which are listed under extruded bars and shapes in Table 3. Alloy HM21A, which is listed under sheet and plate alloys in Table 3, is also a good forging alloy. Alloys M1A and AZ31B may be used for hammer forgings, whereas the other alloys are almost always press forged. However, there has been a gradual decline in the use of the Mg-Mn alloy M1A. AZ80A has greater strength than AZ61A and requires the slowest rate of deformation of the Mg-Al-Zn alloys. ZK60A has essentially the same strength as AZ80A and greater ductility. To develop maximum properties, both AZ80A and ZK60A are heat treated to the T5 (artificially aged) condition. AZ80A may be given the T6 solution heat treatment, followed by artificial aging to provide maximum creep stability. HM21A is used in the T5 temper, and is useful at elevated temperatures up to 370 to 425 °C (700 to 800 °F) where good creep resistance is needed.

The forgeability of four magnesium alloys measured by transverse ductility, assuming a web thickness of approximately 3 mm (⅛ in.) and a minimum draft angle of 1°, is given in Table 4. Here, ZK60A shows slightly higher forgeability than the other three high-strength alloys.

Sheet and plate are rolled from Mg-Al-Zn (AZ and photoengraving grade alloy, or "PE") and Mg-Th (HK and HM) alloys (see Table 3).

AZ31B is the alloy most widely used for sheet and plate and is available in several grades and tempers. It can be used at temperatures up to 100 °C (200 °F). HK31A and HM21A are suitable up to 315 and 345 °C (600 and 650 °F), respectively. HM21A has superior strength and creep resistance. For example, an air impeller manufactured from thick plate of alloy HK31A failed as a result of excessive creep. A change to HM21A led to satisfactory performance and service life. Test coupons machined from the two materials gave the following stress values:

Alloy	Stress for 0.1% creep in 100 h MPa	ksi
At 205 °C (400 °F)		
HM21A	86.2	12.5
HK31A	41	6.0
At 260 °C (500 °F)		
HM21A	72.4	10.5
HK31A	28	4.0
At 315 °C (600 °F)		
HM21A	52	7.5
HK31A	14	2.0

Alloy PE is a special-quality sheet, with excellent flatness, corrosion resistance and etchability, used in photoengraving.

Good formability is an important requirement for most sheet materials. The approximate formability of magnesium alloy sheet is indicated by its ability to withstand 90° bending over a mandrel without cracking. The minimum-size mandrel (minimum radius) over which the sheet can be bent without cracking depends on alloy composition and temper, material thickness and temperature (see Table 5).

Table 4 Forgeability of four magnesium alloys(a)

Alloy	Transverse ductility, %	Forging characteristics
ZK60A	7.0	Excellent on hydraulic or mechanical presses for small forgings; large forgings confined to hydraulic press. Properties nearly equivalent to AZ80A.
AZ80A	5.0	Forgings have maximum strength. Forging limited to hydraulic presses.
HK31A	5.0	Readily forged if proper temperature is maintained. Recommended for elevated-temperature applications.
HM21A	5.0	Rolled ring and die forgings. Recommended for elevated-temperature applications.

(a) For minimum web thickness of approximately 3 mm (⅛ in.) and minimum draft of 1°.

Table 5 Recommended minimum radii for 90° bends in magnesium sheet(a)

Alloy and temper	Forming temperature(b) 20 °C (70 °F)	95 °C (200 °F)	150 °C (300 °F)	205 °C (400 °F)	260 °C (500 °F)	315 °C (600 °F)	370 °C (700 °F)	425 °C (800 °F)
AZ31B-O	5.5t	5.5t	4t	3t	2t	⋯	⋯	⋯
AZ31B-H24	8t	8t	6t	3t	2t	⋯	⋯	⋯
HK31A-O	6t	6t	6t	5t	4t	3t	2t	1t
HK31A-H24	13t	13t	13t	9t	8t	5t	3t	⋯
HM21A-T8	9t	9t	9t	9t	9t	8t	6t	4t

(a) Numerical values of bend radii are given as multiples of sheet thickness. See Table 14 for maximum time at temperature.

When correct temperatures and forming conditions are employed, all magnesium alloys can be deep drawn to about equal reduction.

Mechanical Properties

The mechanical and physical properties of magnesium alloys are given in the data compilations at the end of this section. These are typical values and, for castings, are obtained by testing separately cast specimens. Tensile strengths of investment mold and shell mold castings compare favorably with those of sand and permanent mold castings: yield strength, tensile strength and percentage elongation may vary with cooling rate and generally are lower than those of separately cast sand mold test bars.

The effect of specimen location on the tensile properties of specimens machined from two representative sand castings having sections of varying thickness is shown in Fig. 1. Some specifications permit a 25% reduction in tensile strength and a 75% reduction in elongation for specimens machined from castings, compared with requirements for separately cast bars.

Most magnesium alloys have ratios of tensile strength to density and tensile yield strength to density that are comparable to those of other common structural metals.

Compressive Strength. Compressive yield strength is defined as the stress required to produce a deviation or offset of 0.2% from the modulus line. For castings, compressive yield strength is approximately equal to tensile yield strength. For wrought alloys, however, yield strength in compression may be considerably less than yield strength in tension. The ratio of yield strength in compression to yield strength in tension varies from about 0.4 for alloy M1A to an average value of about 0.7 for the other wrought magnesium alloys. Typical compressive yield strength values for various magnesium alloys are given in Table 3.

Maximum design stresses for magnesium-alloy columns that are loaded axially and that have sufficient stability to prevent local failure may be determined, for columns in the long-column range, by using the Euler column formula. (A long column is one whose length and cross section are such that the stress at which it will buckle does not exceed the elastic limit of the column material.) Maximum design stresses for magnesium-alloy columns in the short-column range are dependent on strengths and forms of the alloys being tested. (A short column is

Fig. 1 Effect of specimen location on tensile properties of two AZ91A-T6 production sand castings

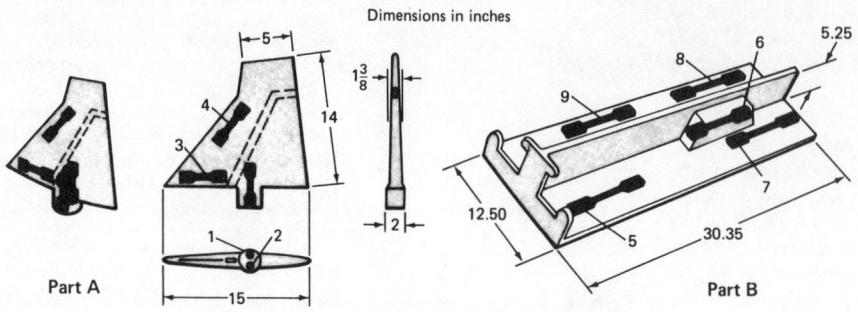

Dimensions in inches

Part A

Part B

(Part A) Average mechanical properties for parts made from 11 different heats. Specimens 1 and 2 were machined 13 mm (0.50 in.) in diameter; specimens 3 and 4, 6.4 by 13 mm (0.25 by 0.50 in.) flat. (Part B) Average mechanical properties for parts made from 16 heats. Castings weighed approximately 7 kg (15 lb), and all sections were 6.4 to 9.5 mm (¼ to ⅜ in.) thick, except specimen 6, which was taken from a section about 25 mm (1.0 in.) thick. Test specimens were 6.4 by 13 mm (¼ by ½ in.) flat, except No. 6, which was 13 mm (0.50 in.) in diameter. Customer required a minimum of 176 MPa (25.5 ksi) tensile strength, 110 MPa (16.0 ksi) yield strength and 1.0% elongation in 50 mm or 2 in.

| | Specimen location | | | | | | | | |
| | Part A | | | | Part B | | | | |
Property	1	2	3	4	5	6	7	8	9
Tensile strength									
MPa	202.4	199.6	227.5	265.4	242.7	214.8	248.6	250.6	244.4
ksi	29.35	28.95	33.00	38.50	35.20	31.15	36.05	36.35	35.45
Yield strength									
MPa	116.5	123.1	133.1	141.0	140.8	135.5	142.1	121.9	140.0
ksi	16.90	17.85	19.30	20.45	20.42	19.65	20.61	20.57	20.30
Elongation in 50 mm or 2 in., %	2.8	2.8	2.5	4.0	2.8	3.1	2.5	2.2	22.8

any column of such length and cross section that it fails under compressive loading by plastic yielding and/or crushing, rather than by buckling.) In practical application, the maximum design stress of a column is considered to be the minimum compressive yield stress of the material. Various formulas have been developed for deriving maximum design stresses for columns of intermediate length (those that fail by a combination of elastic buckling and plastic yielding and/or crushing). Column-strength curves for several magnesium extrusion alloys are shown in Fig. 2.

Bearing strength is particularly important in the design of bolted and riveted joints. Bearing yield strength is defined as the stress required to produce an offset from the initial straight portion of the curve equal to 2% of hole diameter. Bearing-strength values listed in Table 3 were determined using specimens with an edge distance (from the center of the hole) of 2½ times the pin diameter and a width of 8 times the pin diameter. Increasing edge distance to more than about twice the pin diameter has little effect on bearing-strength values. Sheet thicknesses in a wide range have been tested, and no effect of the ratio of pin diameter to sheet thickness has been observed, except when buckling occurred. A pin diameter not greater than four times sheet thickness prevents buckling.

Shear strength is important in design of joints in magnesium parts, such as threaded joints and spot welds. Values for castings and extrusions given in Table 3 were obtained by the conventional double-shear method, using solid rods. Values for sheet and AZ80A-T5 structural shapes were obtained by the punch method, using flat specimens.

Hardness and Wear Resistance. Magnesium alloys have sufficient hardness for all structural applications except those involving severe abrasion. Hardness values are given in the data compilations and in Table 3. Although rather wide variations in hardness are observed, resistance of the alloys to abrasion varies by only about 15 to 20%. When subjected to wear by rubbing, by frequent removal of studs, or by heavy bearing loads, magnesium may be protected by inserts of steel, bronze or nonmetallic materials, attached as sleeves, liners, plates or bushings. Such inserts may be attached mechanically by pressing, shrinking, riveting, bolting or bonding; in castings, inserts may be cast in place.

Fig. 2 Minimum column-strength curves for several magnesium extrusion alloys

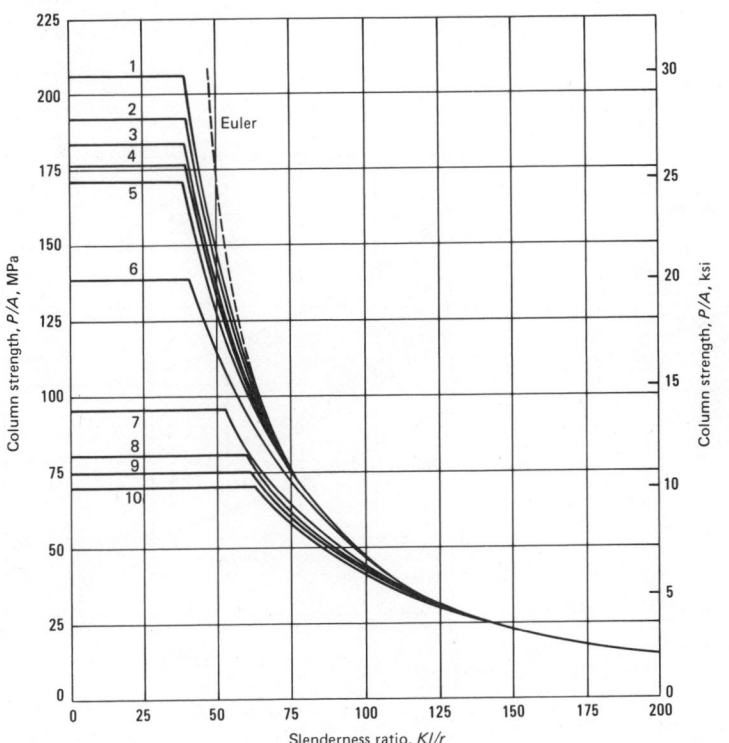

P = ultimate column load, A = cross-sectional area, K = constant that depends on end conditions, l = column length, and r = minimum radius of gyration of column cross section.

Alloy	Curve No.	Least dimension of area range	
ZK60A-T5	1	under 1209 mm^2	under 2.000 in.^{2}
	2	1290 to 1935 mm^2	2.000 to 2.999 in.^{2}
	5	1936 to 3225 mm^2	3.000 to 4.999 in.^{2}
AZ80A-T5	2	6.35 to 38.09 mm	0.250 to 1.499 in.
	3	38.10 to 63.49 mm	1.500 to 2.499 in.
	4	63.50 to 127.00 mm	2.500 to 5.000 in.
ZK21A-F	6	under 3226 mm^2	under 5.000 in.^{2}
AZ61A-F	7	6.35 to 127.00 mm	0.250 to 5.000 in.
AZ31B-F	8	up to 6.35 mm	up to 0.250 in.
	9	6.35 to 63.49 mm	0.250 to 2.499 in.
	10	63.50 to 127.00 mm	2.500 to 5.000 in.

Magnesium alloys perform satisfactorily as bearing materials where loads do not exceed 14 MPa (2 ksi), shafts are hardened (350 to 600 HB), lubrication is ample, speeds are low (5 m/s, or 1000 ft/min, max) and operating temperature does not exceed 105 °C (220 °F).

Fatigue strength of magnesium al-

loys, as determined using laboratory test samples, covers a relatively wide scatter band such as is characteristic of other metals. The S-N curves have a gradual change in slope and become essentially parallel to the horizontal axis at 10 to 100 million cycles.

The fatigue strengths are higher for

wrought products than for cast test bars. Fatigue strengths of several alloys are listed in the data compilations. Increasing surface smoothness improves resistance to fatigue failure. For example, removing the relatively rough as-cast surfaces of castings by machining improves the fatigue properties of the castings (see Fig. 3). Sharp notches, small radii, fretting and corrosion are more likely to reduce fatigue life than are variations in chemical composition or heat treatment.

A specimen of alloy ZK60A-T5 (static yield strength, 290 MPa or 42 ksi) with a machined 60° notch of 0.025-mm (0.001-in.) radius has a fatigue limit of 28 MPa (4 ksi) at 500 million cycles, compared with 110 MPa (16 ksi) for the unnotched specimen. This is a notch factor of 0.25. For a shorter life of 100 000 cycles, the notch factor is about 0.48. As the severity of the notch decreases, its effect on fatigue limit decreases rapidly. For instance, a semi-circular notch with radius of 1.2 mm (0.047 in.) reduces fatigue strength by only 20%, compared with 75% for the sharp V-notch cited above.

When fatigue is the controlling factor in design, every effort should be made to decrease the severity of stress raisers. Use of generous fillets in re-entrant corners and gradual changes of section greatly increase fatigue life. Situations in which the effects of one stress raiser overlap those of another should be eliminated. Further improvement in fatigue strength can be obtained by inducing stress patterns conducive to long life. Cold working the surfaces of critical regions by rolling or peening to achieve appreciable plastic deformation produces residual compressive surface stress and increases fatigue life.

Surface rolling of radii is especially beneficial to fatigue resistance, because radii generally are the locations of higher-than-normal stresses. In surface rolling, size and shape of the roller, as well as feed and pressure, are controlled to obtain definite plastic deformation of the surface layers for an appreciable depth (0.25 to 0.38 mm, or 0.010 to 0.015 in.). In all surface working processes, caution must be exercised to avoid surface cracking, which decreases fatigue life. For example, if shot peening is used, the shot must be smooth and round. Use of broken shot or grit may result in surface cracks.

Low-temperature Properties. With decreasing temperature, magne-

Fig. 3 Effect of type of surface on fatigue properties of cast Mg-Al-Zn alloys

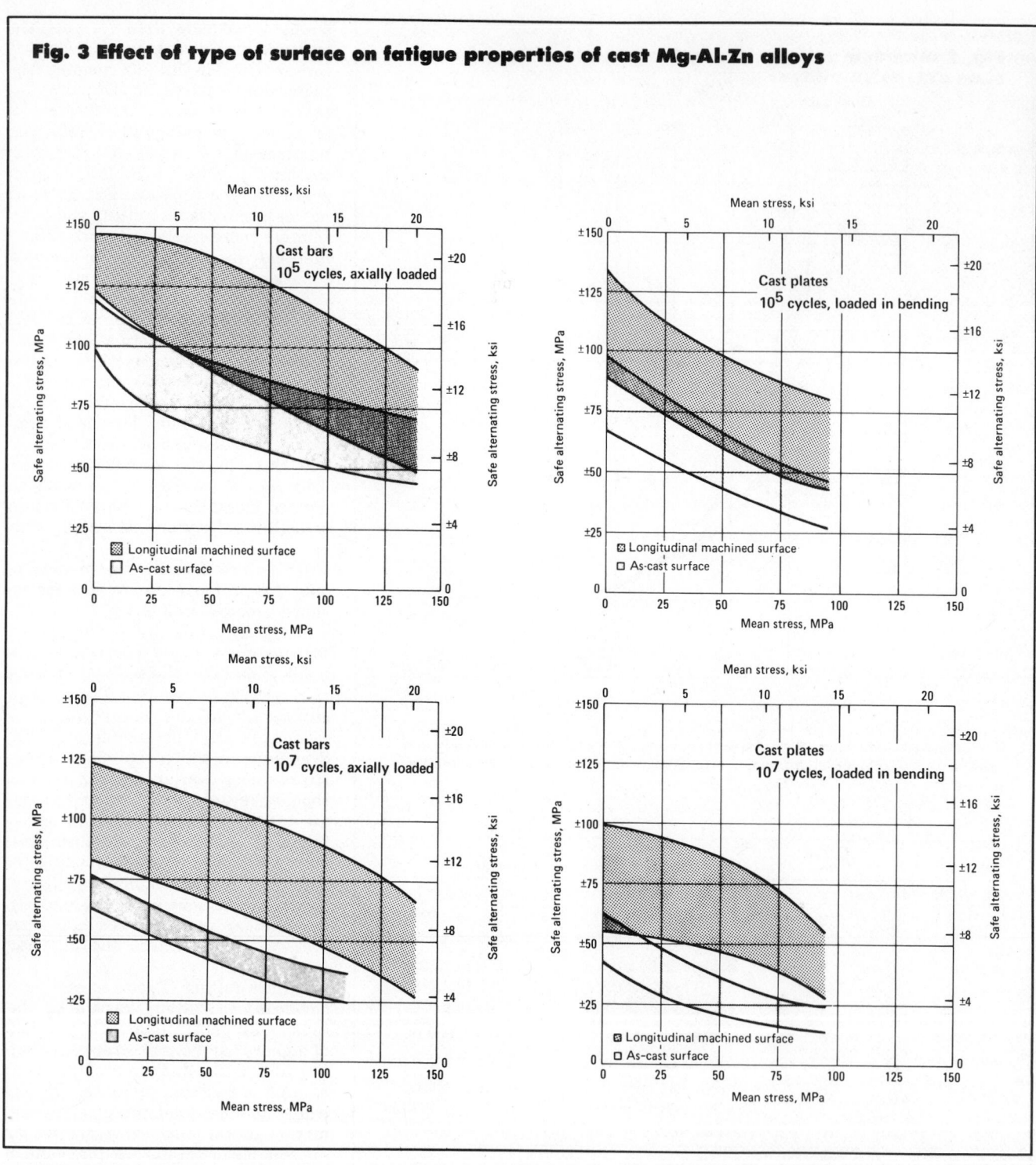

sium alloys increase in tensile strength, yield strength and hardness, but generally decrease in ductility (see Tables 6a and b for the results of tensile tests at both room and low temperatures).

Elevated temperatures have ad-

verse effects on ultimate and yield strengths, as exemplified in the bearing-strength data given for two magnesium alloys in Table 7. The effect of elevated temperatures on the mechanical properties of magnesium alloys is evaluated by considering: (a) the strength as determined by bringing the test specimen up to temperature and

testing immediately (short-time test); (b) the strength at temperature after prolonged heating at elevated temperature; (c) the effect on room-temperature properties of heating at elevated temperature for short and long times; and (d) the deformation produced by prolonged heating under load (creep test).

Table 6a Some low-temperature tensile properties of various magnesium alloys

Alloy	Thickness mm	in.	Tensile strength MPa	ksi	Yield strength MPa	ksi	Elongation, %
Transverse Tests of Plate Alloys at 24 °C (75 °F)							
HK31A-H24	6.35	0.250	240	35.2	180	25.9	21.0
HK31A-O	6.35	0.250	200	29.0	125	18.0	30.5
HM21A-T8	6.35	0.250	240	35.0	170	24.8	13.7
Longitudinal Tests of Sheet and Plate Alloys at 24 °C (75 °F)							
HK31A-H24	1.63	0.064	250	36.3	200	29.0	7.5
HK31A-H24	6.35	0.250	240	34.5	190	27.3	14.2
Welded(a)	6.35	0.250	200	28.8	150	21.7	2.4
HK31A-O	1.63	0.064	205	29.7	125	17.9	27.5
HK31A-O	6.35	0.250	200	28.9	120	17.7	29.7
Welded(a)	6.35	0.250	160	23.4	120	17.3	3.2
HM21A-T5	(b)	(b)	210	30.4	105	15.5	8.0
HM21A-T8	1.63	0.064	220	32.2	160	23.1	7.2
HM21A-T8	6.35	0.250	235	32.4	175	25.1	5.6
Welded(a)	6.35	0.250	195	28.6	130	18.6	2.7
Longitudinal Tests of Sheet and Plate Alloys at −54 °C (−65 °F)							
HK31A-H24	1.63	0.064	300	43.3	220	32.0	5.0
....	6.35	0.250	280	40.8	230	33.4	9.0
HK31A-O	1.63	0.064	275	39.9	150	21.4	20.7
....	6.35	0.250	265	38.3	150	21.5	18.0
HM21A-T5	(b)	(b)	270	39.5	110	15.8	9.3
HM21A-T8	1.63	0.064	275	39.6	175	25.6	6.2
....	6.35	0.250	265	38.4	205	29.7	4.7
Longitudinal Tests of Sheet and Plate Alloys at −72 °C (−98 °F)							
HK31A-H24	1.63	0.064	295	42.7	210	30.6	4.2
HK31A-H24	6.35	0.250	290	42.4	235	33.8	11.5
Welded(a)	6.35	0.250	195	28.1	165	23.6	0.5
HK31A-O	1.63	0.064	285	41.3	145	21.1	17.5
HK31A-O	6.35	0.250	275	40.2	150	21.9	20.2
Welded(a)	6.35	0.250	205	29.4	145	21.0	2.2
HM21A-T5	(b)	(b)	275	40.0	110	16.3	8.3
HM21A-T8	1.63	0.064	280	40.8	150	22.1	17.5
....	6.35	0.250	275	40.1	215	31.3	5.0
....	6.35	0.250	200	29.2	120	17.5	1.5
Longitudinal Tests of Sheet and Plate Alloys at −196 °C (−320 °F)							
HK31A-H24	1.63	0.064	370	54.0	225	33.0	6.2
HK31A-H24	6.35	0.250	365	52.9	240	34.7	8.0
Welded(a)	6.35	0.250	230	33.7	180	25.9	1.5
HK31A-O	1.63	0.064	330	47.9	170	24.3	12.7
HK31A-O	6.35	0.250	325	47.2	170	24.7	12.5
Welded(a)	6.35	0.250	205	29.7	150	21.6	2.2
HM21A-T5		(b)	320	46.6	125	18.1	8.0
HM21A-T8	1.63	0.064	330	47.6	170	24.9	4.0
HM21A-T8	6.35	0.250	325	47.3	210	30.6	4.2
Welded(a)	6.35	0.250	330	33.1	145	20.9	1.5

(a) Welding rod was EZ33A; weld bead intact. (b) Specimen machined from a forging.
NOTE: Values for wrought alloys are averages of two to four tests at room temperature (2-in. gage length). Values of duplicate tests at low temperatures are also averages (1-in. gage length). Values for cast alloys are averages of two to four tests on separately cast bars.

Data showing the effects of elevated temperature on the mechanical properties of several magnesium alloys are presented in Tables 8 and 9. Elevated-temperature properties, including isochronous stress-strain curves, are given in greater detail in the data compilations for magnesium alloys.

Designs of many parts for use at elevated temperatures under continuous loads are based on maximum allowable deformation. The limiting creep-stress values given in Table 9 are based on 0.2% total extension. The isochronous curves in the data compilations cover a wide range of deformations. The

alloys that contain thorium have the greatest resistance to creep at 205 and 315 °C (400 and 600 °F), and the Mg-Al-Zn alloys have the lowest resistance.

The decrease in modulus of elasticity with increase in temperature, a characteristic of the Mg-Al-Zn alloys, is considerably less for thorium-containing alloys.

Selection of Product Form

Selection of a particular product form for a structural application is based on mechanical-property requirements and on cost, availability and fabricability. Requirements for production and design may change under operating conditions or as need arises. A part originally machined from bar stock may subsequently be made by extrusion or forging. Assemblies built up by joining sheets and extrusions may be redesigned as castings with equivalent performance at lower cost.

Castings. Parts too intricate to fabricate economically by other methods can be produced as castings. Sand, permanent mold and die castings are more widely used than investment and shell mold castings. The choice of casting method is determined primarily by size, shape, quantity, cost and desired mechanical properties of the casting. Cost of magnesium alloy castings is governed largely by ingot price, alloy castability and required heat treatment. Ingot price increases with additions of rare-earth metals, zirconium and thorium. Small changes in composition may affect cost of heat treatment. Comparative costs for casting an aircraft engine part from three different magnesium alloys are given in Table 10.

Magnesium alloys are cast by the permanent mold process when the number of parts required justifies the very high cost of equipment. The mechanical properties of sand and permanent mold castings are comparable, but the permanent mold process normally provides closer control of dimensions and produces better cast surfaces.

Cost of castings also is influenced by such factors as required tolerances, mold and die costs, and machining costs. The quantity of a part to be produced is an important factor affecting cost and must be considered in seeking the most economical method of production. For example, in making the part illustrated in Fig. 4, tooling

Table 6b Some low-temperature tensile properties of various magnesium alloys

Alloy	Tensile strength MPa	ksi	Yield strength MPa	ksi	Elon- gation %	Charpy impact (a) J	ft · lb	(b) J	ft · lb
Cast Alloys at 24 °C (75 °F)									
AZ91C-T6 ...	290	41.8	130	19.2	6.3	7.96	5.87	1.36	1.00
AZ92A-T6 ...	290	41.8	160	23.4	4.0	7.62	5.62	0.68	0.50
EZ33A-T5 ...	190	27.5	115	16.9	7.6	7.46	5.50	0.84	0.62
HK31A-T6...	225	32.7	110	16.3	9.5	16.61	12.25	3.80	2.81
ZH62A-T5 ...	275	39.9	190	27.9	5.7	15.02	11.08	1.02	0.75
Cast Alloys at −78 °C (−109 °F)									
AZ91C-T6 ...	305	44.3	150	21.6	5.1	6.26	4.62	1.36	1.00
AZ92A-T6 ...	295	42.7	170	24.6	2.3	6.44	4.75	0.76	0.56
EZ33A-T5 ...	190	27.6	125	18.0	3.1	4.83	3.56	0.68	0.50
HK31A-T6...	300	43.3	120	17.5	8.6	16.43	12.12	3.21	2.37
ZH62A-T5 ...	330	47.6	200	29.2	2.7	18.99	14.00	1.02	0.75
Cast Alloys at −196 °C (−321 °F)									
AZ91C-T6 ...	310	44.9	180	26.0	1.7	4.06	3.00	1.02	0.75
AZ92A-T6 ...	320	46.5	195	28.5	0.8	4.57	3.37	0.68	0.50
EZ33A-T5 ...	200	29.0	140	20.3	2.2	5.00	3.69	0.68	0.50
HK31A-T6...	330	48.1	135	19.6	6.1	13.72	10.12	3.05	2.25
ZH62A-T5 ...	320	46.6	235	34.1	1.0	8.56	6.31	1.02	0.75

(a) Unnotched specimens. (b) Notched specimens.
NOTE: Values for wrought alloys are averages of two to four tests at room temperature (2-in. gage length). Values of duplicate tests at low temperatures are also averages (1-in. gage length). Values for cast alloys are averages of two to four tests on separately cast bars.

Table 7 Typical bearing strengths of two magnesium alloys at elevated temperature

Temperature °C	°F	Bearing strength Ultimate MPa	ksi	Yield MPa	ksi
HK31A-H24 Sheet					
20	70	420	61	285	41
205	400	285	41.4	210	30.3
260	500	225	32.4	190	27.8
315	600	170	25	145	21
HM21A-T8 Sheet					
20	70	415	60	275	40
205	400	270	39	185	27
260	500	250	36	180	26
315	600	200	29	150	22
425	800	160	23	115	17

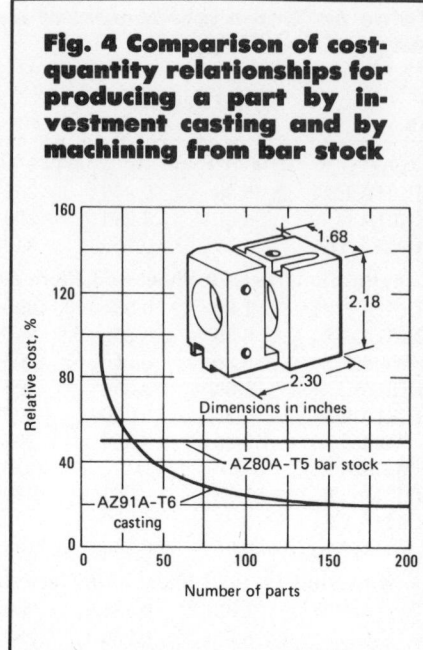

Fig. 4 Comparison of cost-quantity relationships for producing a part by investment casting and by machining from bar stock

costs would have made investment casting more expensive if only 30 pieces were produced; but because more than 30 pieces were produced, investment casting was cheaper than machining from bar stock. Although rejection rates vary from one casting method to another (see Fig. 5), they are high for all methods when tolerances are close.

Figure 6 illustrates a casting that was produced at lower cost by shell mold casting than by sand casting and machining when more than 700 castings were made. The sand mold casting required one extra machining opera-tion to obtain a dimension that could be held within the tolerance of the shell mold.

Die castings made of magnesium al-loys may be selected in preference to aluminum die castings of the same design because of the savings in weight. Service requirements and size may govern whether a magnesium al-loy is selected for use in a die casting, but quantity is the most important factor, because die castings are high-production items. The effects of both quantity and weight are shown in Fig. 7. Magnesium alloy die castings, like castings in general, are always priced and purchased on a per piece basis. Cost per pound varies, depending pri-marily on complexity of design, wall thickness, number of cavities in the mold, and quality level.

Extrusions. Magnesium alloys are extruded as round rods and a variety of bars, tubes and shapes. A wide variety of special shapes also can be extruded. Extrusion is selected as a means of producing certain shapes when (a) sev-eral small extrusions or a combination of extrusions and sheet can be joined to form an assembly, (b) shapes are de-sired that are uneconomical to machine from castings, and (c) pieces cut from extrusions can replace individually cast or forged parts.

The extrusion process offers many design possibilities not economically attainable by other production meth-ods. These include re-entrant angles and undercuts, thin-wall tubing of large diameter and variations in sec-tion thickness, almost without restric-tion. Probably the most important factor in determining whether a mag-nesium alloy shape will extrude well is good symmetry, preferably around both axes.

Very thin and wide sections with large circumscribing circles should be avoided. The optimum width-to-thick-ness (w/t) ratio for magnesium extru-sions normally is less than 20. Parts with higher ratios can be extruded, but require more generous tolerances. A thick section tapering to a thin wedge

Table 8 Effect of elevated temperature on tensile strength of magnesium alloys

| Alloy | 20 °C MPa | (70 °F) ksi | Tested at exposure temperature | | | | | | | | Tested at room temperature | | | |
| | | | Exposed 10 min at | | | | | Exposed 1000 h at | | | Exposed 1000 h at | | | |
			150 °C MPa	(300 °F) ksi	315 °C MPa	(600 °F) ksi	205 °C MPa	(400 °F) ksi	315 °C MPa	(600 °F) ksi	205 °C MPa	(400 °F) ksi	315 °C MPa	(600 °F) ksi
Castings														
AZ63A-T6 ..	275	40	165	24	55	8	110	16	...	...	255	37	...	...
AZ92A-T6 ..	275	40	195	28	55	8	115	17	...	...	270	39	...	...
EZ33A-T5 ...	160	23	145	21	83	12	130	19	76	11	170	25	180	26
HK31A-T6 ..	215	31	195	28	125	18	180	26	62	9	240	35	180	26
HZ32A-T5 ..	200	29	145	21	83	12	115	17	76	11	220	32	235	34
ZH62A-T5 ..	290	42	195	28	69	10	...	...	...	...	...	...	...	...
QH21A-T6 ..	275	40	235	34	97	14	...	...	...	...	...	...	...	...
Extrusions														
AZ80A-T5 ..	380	55	235	34	69	10	...	...	...	...	...	...	...	...
ZK60A-T5 ..	365	53	180	26	41	6	...	...	...	...	315	46	315	46
HM31A-F ...	275	40	195(a)	28(a)	115	17	...	...	...	...	...	...	...	...
Sheet														
AZ31B-H24 .	285	41	145	21	48	7	90	13	62(a)	9(a)	255	37	260	38
HK31A-T6 ..	255	37	180	26	115	17	...	...	55	8	255	37	215	31
HM21A-T8 ..	235	34	140	20	97	14	...	...	...	...	...	...	...	...

(a) Tested at 260 °C (500 °F).

Table 9 Effect of elevated temperature on values of creep stress and elastic modulus for magnesium alloys

| Alloy | Creep stress(a) at | | | | Elastic modulus at | | | |
	205 °C (400 °F) MPa	ksi	315 °C (600 °F) MPa	ksi	205 °C (400 °F) GPa	10^6 psi	315 °C (600 °F) GPa	10^6 psi
Castings								
AZ92A-T6	3.4	0.5	...	...	31	4.5	21	3.0
EZ33A-T5	38	5.5	6.9	1.0	40	5.8	38	5.5
HK31A-T6	64	9.3	14	2.0	40	5.8	39	5.6
HZ32A-T5	52	7.5	22	3.2	40	5.8	39	5.6
ZH62A-T5	17	2.5	...	...	40	5.8	38	5.5
Extrusions								
ZK60A-T5	7	1.0(b)	...	...	...	...	...	...
HM31A-F	83	12.0	41	6.0	40	5.8	38	5.5
Sheet								
AZ31B-H24 ...	7	1.0(b)	...	...	30	4.3	17	2.5
HK31A-T6	69	10.0	17	2.5	40	5.8	25	3.6
HM21A-T8	76	11.0	34	5.0	40	5.8	34	5.0

(a) Stress to produce 0.2% total extension in 1000 h for cast alloys and 100 h for wrought alloys.
(b) Tested at 150 °C (300 °F).

Table 10 Comparison of costs for making a typical aircraft engine casting from three different magnesium alloys

Item	AZ91C	ZE41A	QE22A
Cores	$ 41.18	$ 41.18	$ 41.18
Molding	79.44	79.44	79.44
Metal	76.33	125.83	310.77
Cleaning	56.66	56.66	56.66
Heat treatment	18.28	1.39	8.03
Visual inspection	11.00	11.00	11.00
Nondestructive testing	94.68	94.68	94.68
Fixturing	8.80	8.80	8.80
Total cost per casting ...	$386.37	$418.98	$610.56

must always be modified by rounding the edge, or the die may not fill properly. A thin leg attached to a thick body of an extrusion should be limited to a length not exceeding ten times the leg thickness. Semiclosed shapes requiring long, thin die tongues should be avoided. For best extrudability, the length of the tongue should not exceed three times its width, although it is possible to extrude lengths five times the width. Similarly, shapes requiring unbalanced die tongues do not constitute good extrusion design. Hollow shapes that contain unsymmetrical voids, or voids separated by sections of inadequate thickness, are undesirable.

Sharp outside corners result in excessive stress concentration and die breakage and should be avoided. Inside corners should be filleted to reduce stress concentration in the part and to ensure complete filling of the die during extrusion. Regardless of the shape being extruded, it is difficult to hold distances between thin sections to close tolerances.

Many shapes can be extruded economically. Extrusion dies are relatively inexpensive, and dimensions can be held closely enough so that often machining is unnecessary. Part A in Fig. 8 is one that lends itself to extrusion. However, because more metal had

to be machined from this part as an extrusion than as a die casting, die casting proved to be cheaper.

A slight design change in the bracket shown as part B in Fig. 8 permitted the part to be made more economically by extrusion than by casting.

Impact extrusions are tubular parts of symmetrical shape. The impact type of hot extrusion is particularly applicable when:

1 It is not practical to make the part by any other method, such as with parts requiring very thin walls, where thin walls having high strength are essential or where irregular profiles must be incorporated in the part.

Fig. 5 Dimensional variations, tolerances and rejection rates for two Mg-Al-Zn alloy castings

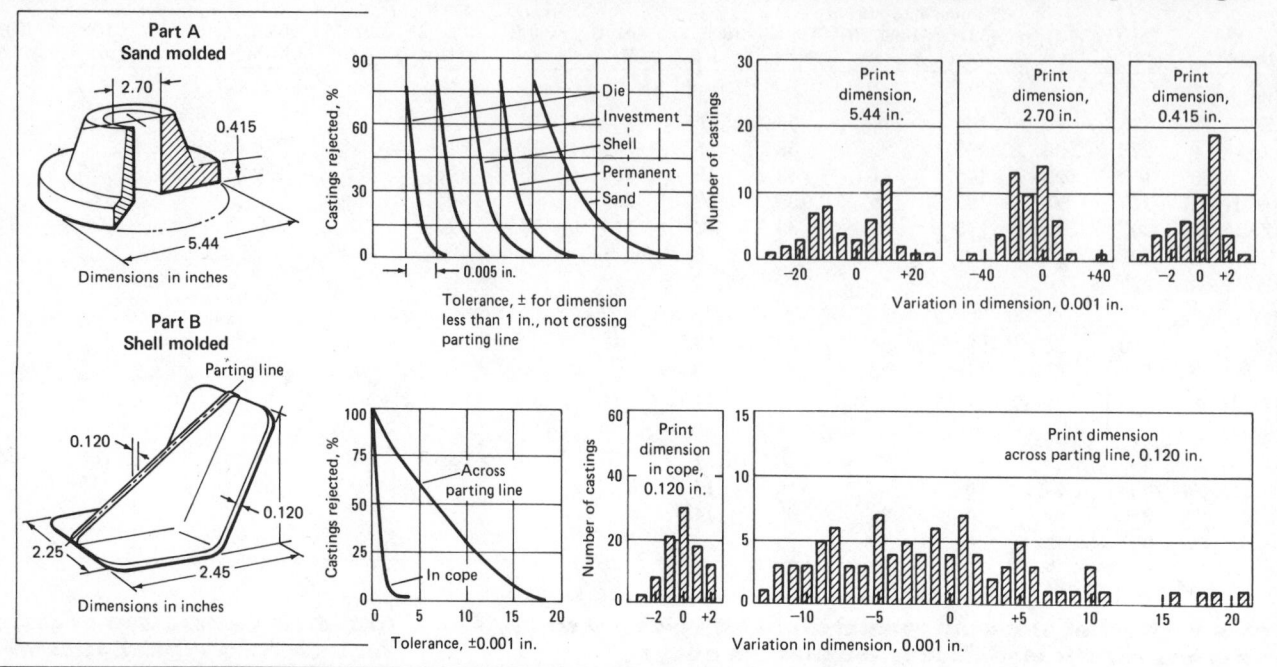

Fig. 6 Comparison of cost-quantity relationships for two methods of casting a Mg-Al-Zn alloy part

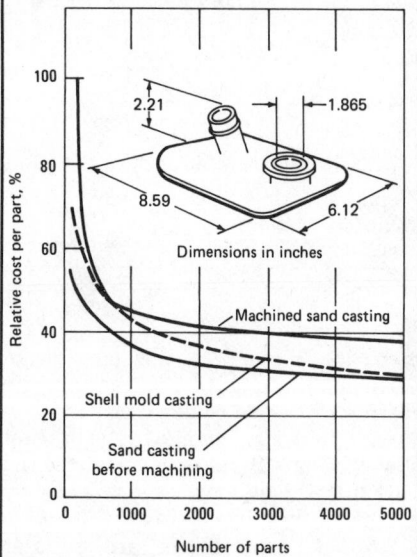

The sand casting required an extra machining operation to meet a dimensional limit that could be held in the shell mold casting without machining. Thus, the curve labeled "Machined sand casting" should be used in comparing total costs of the two casting methods.

Fig. 7 Typical effects of weight and quantity on cost of magnesium die castings

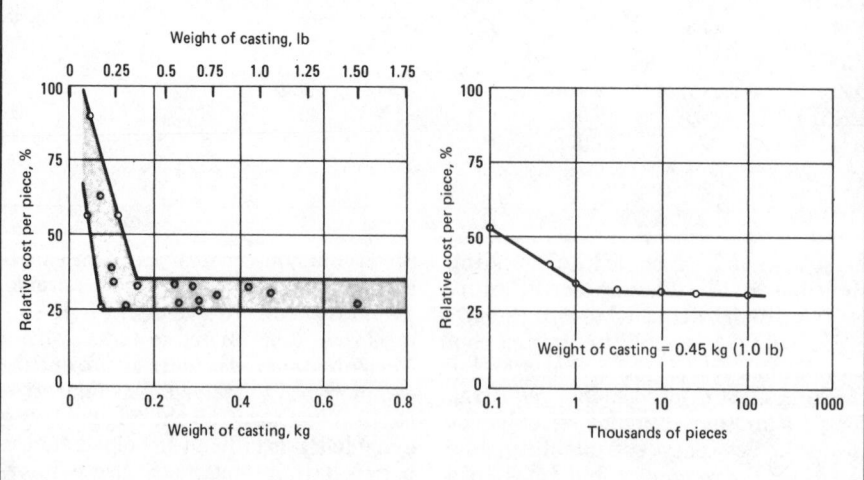

On a weight basis, the cost of die castings heavier than about 0.15 kg (⅓ lb) is shown to be fairly constant. For lighter castings, weighing less than about 0.10 or 0.15 kg (4 or 5 oz), cost per pound may vary by a factor of three or four, depending on design, wall thickness and quality level.

Fig. 8 Effect of manufacturing method on cost of two magnesium-alloy parts

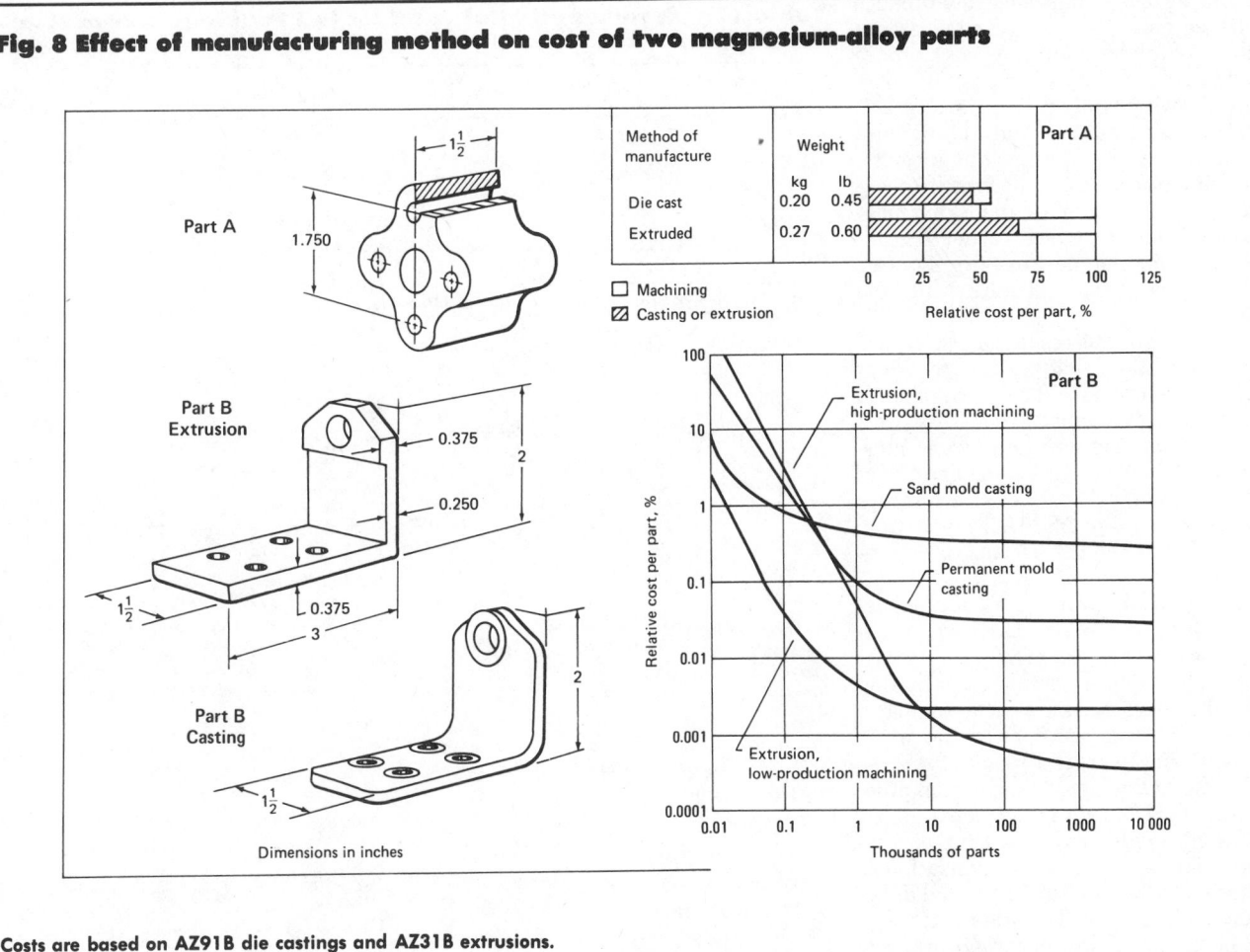

Method of manufacture	Weight		
	kg	lb	
Die cast	0.20	0.45	
Extruded	0.27	0.60	

Part A

☐ Machining
▨ Casting or extrusion

Relative cost per part, %

Dimensions in inches

Costs are based on AZ91B die castings and AZ31B extrusions.

2 High production rates are required, where scrap loss from machining would be excessive if the part were made by other means, where strength requirements cannot be met by die castings, where the number of manufacturing operations or the number of parts in an assembly can be reduced by the use of impact extrusions, where portions of the part require zero draft, and where closer tolerances are required.

In designing impact extrusions, the following factors should be taken into consideration:

1 A wide variety of symmetrical shapes is possible.
2 Variations in wall thickness are possible (thin sidewall, thick bottom; thick sidewall, thin bottom).
3 Ribs, flanges, bosses and indentations can be incorporated.
4 Length-to-diameter ratios may range from 1.5:1 to 15:1. Ratios from 6:1 to 8:1 are considered good working ratios.
5 Reduction in area varies with the alloy being extruded and is limited by the size of available equipment. Parts with reduction in area up to 95% have been made. In general, extrusions of alloys M1A and AZ31B c n have thinner walls than AZ61A, AZ80A and ZK60A extrusions.
6 Sharp corner radii are possible in some areas of impact extrusions. This is not true of other product forms.
7 Average properties of impact extrusions are slightly higher than typical properties of the hot extruded stock from which the parts are made.

Forgings. Magnesium forgings can be produced in the same variety of shapes and sizes as forgings of other metals. Maximum size is limited primarily by size of available equipment. Tolerances can be held to the same values as in normal forging of other metals and vary somewhat with forging size and design.

Forgings have the best combination of strength characteristics of all forms of magnesium. They are used where light weight coupled with rigidity and high strength are required. Magnesium forgings are sometimes used because of their pressure tightness, machinability and lack of warpage rather than because of their high strength-to-weight ratio.

Forging is used for parts to be produced in quantities sufficient to amortize die costs and for parts requiring high strength and ductility and greater uniformity and soundness than can be obtained with castings. For small quantities, hand forgings may be used, but die forgings have better mechanical properties and are less expensive in larger quantities.

The ease with which magnesium can be worked greatly reduces the number of forging operations needed to produce finished parts. Many of the steps commonly required in forging brass, bronze and steel (such as punching, planishing, drawing and ironing, sizing and coining, and edging and rolling) are unnecessary in forging magnesium. Bending, blocking and finishing are the principal steps used in forging magnesium. Recommended corner and fillet radii for magnesium forgings (see Fig. 9) are given in Tables 11(a) and 11(b).

Die design and resulting metal flow cause variations in tensile properties at different sections of large forgings. An example involving aircraft wheels forged from magnesium alloys is illustrated in Fig. 10. Test specimens were taken at several locations in forgings made from alloy AZ80A-T5 and from alloy ZK60A in the T5 and T6 conditions.

Mechanical properties of magnesium-alloy forgings or castings, as determined by testing of separately cast or forged bars, are useful for evaluating certain characteristics on a comparative basis and serve as a means of control. However, test results for these bars may vary significantly from properties of specimens taken from various locations in production castings or forgings. The amount of this variation is affected by section thickness and direction of metal flow. The wide variations in properties between forged brackets of the same design made from two different heat treated alloys are shown in Fig. 11, and a comparison of the properties of separately cast and forged test specimens with those of specimens cut from castings and forgings are presented in Fig. 12. Forgings made from Mg-Th alloys are considerably more expensive than those of Mg-Zn-Zr alloys.

Comparison of the costs of forging and casting requires careful analysis for each design and specification.

Inserts

Magnesium surfaces that are subjected to heavy bearing loads or severe wear require protection. Inserts that provide such protection to the surfaces of holes can be made of various materials and can be attached in numerous ways.

Cast-in Inserts. Inserts in magnesium may be fixed in place by casting the magnesium around them. Cast-in inserts for use in magnesium may be made of steel, brass, bronze or other metals. Nonferrous inserts may be plated with chromium to prevent alloying with magnesium, although such plating is seldom used. Tinning of ferrous inserts prevents galvanic corrosion of the magnesium.

Cast-in inserts become securely fixed when the cast metal shrinks around them. The insert is even more securely fixed if the outside of the insert is knurled or grooved. Care should be taken to ensure that sharp corners or insufficient metal around the insert does not set up concentrations of stress.

Shrinkage of the magnesium alloy around cast-in inserts may cause high residual stress in the metal surround-

Table 11a Corner and fillet radii (in in.) for large magnesium forgings(a)

D, in.	r_1, in. min	r_2, in. min
Under 3/16	3/32	1/4
3/16 to 13/32	3/16	13/32
13/32 to 1	5/16	5/8
1 to 1¾	½	1
1¾ to 2½	¾	1¼
2½ to 4	1 1/16	1 5/8
Over 4	1¼	2

D, in.	r_3, in.	r_4, in.	D, in.	r_3, in.	r_4, in.
3/8	1/4	1/16	1½	5/8	3/16
½	9/32	1/16	1¾	23/32	7/32
5/8	11/32	1/16	2	13/16	1/4
¾	3/8	5/64	2¼	7/8	9/32
7/8	7/16	3/32	2½	15/16	5/16
1	15/32	7/64	2¾	1 1/16	11/32
1 1/8	½	1/8	3	1 1/8	3/8
1¼	9/16	5/32			

D, in.	r_5, in. min
W, 5/8 in. max	
13/32 max	7/64
W, 5/8 to 13/16 in.	
13/32 max	7/64
13/32 to ½	5/32
W, 13/16 to 1 in.	
13/32 max	5/32
13/32 to ½	5/32
½ to 5/8	5/32
W, 1 to 1 9/32 in.	
13/32 max	5/32
13/32 to ½	5/32
½ to 5/8	5/32
5/8 to 13/16	1/4
W, 1 9/32 to 1 5/8 in.	
13/32 max	5/32
13/32 to ½	1/4
½ to 5/8	1/4
5/8 to 13/16	1/4
13/16 to 1	1/4
W, 1 5/8 to 2 in.	
13/32 max	1/4
13/32 to ½	1/4
½ to 5/8	1/4
5/8 to 13/16	13/32
13/16 to 1	13/32
1 to 1 5/16	13/32
W, 2 to 2½ in.	
13/32 max	1/4
13/32 to ½	1/4
½ to 5/8	1/4
5/8 to 13/16	13/32

D, in.	r_5, in. min
13/16 to 1	13/32
1 to 1 5/16	13/32
W, 2½ to 3 in.	
13/32 max	13/32
13/32 to ½	13/32
½ to 5/8	13/32
5/8 to 13/16	13/32
13/16 to 1	13/32
1 to 1 5/16	13/32
1 5/8 to 2	5/8
W, 3 to 4 in.	
13/32 max	13/32
13/32 to ½	13/32
½ to 5/8	13/32
5/8 to 13/16	13/32
13/16 to 1	13/32
1 to 1 5/16	5/8
1 5/8 to 2	5/8
W, 4 to 5 in.	
13/32 max	13/32
13/32 to ½	13/32
½ to 5/8	13/32
5/8 to 13/16	5/8
13/16 to 1	5/8
1 to 1 5/16	5/8
1 5/8 to 2	13/16

D, in.	r_6, in.
13/32 max	1/16
13/32 to ½	1/16
½ to 5/8	7/64
5/8 to 13/16	7/64
13/16 to 1	7/64
1 to 1 5/16	5/32
1 5/8 to 2	1/4

D, in.	r_8, in.
3/8 max	1/16
½	1/4
5/8	5/16
¾	11/32
7/8	3/8
1	13/32
1 1/8	7/16
1¼	½
1½	19/32
1¾	21/32
2	¾
2¼	27/32
2½	29/32
2¾	1
3	1 1/16

(a) See Fig. 9 for identification of symbols.
Note: The values given for r_3 and r_4, for corner and fillet radii for I-sections in magnesium forgings apply when the W/D ratio is approximately 2 to 1. For W/D ratios greater than 2 to 1 the radii should be increased and for lower W/D ratios the radii may be decreased. In corner and fillet radii for channel sections, r_7 equals r_5 plus t_1.

Fig. 9 Large magnesium forgings for which corner and fillet radii are given in Tables 11a and 11b

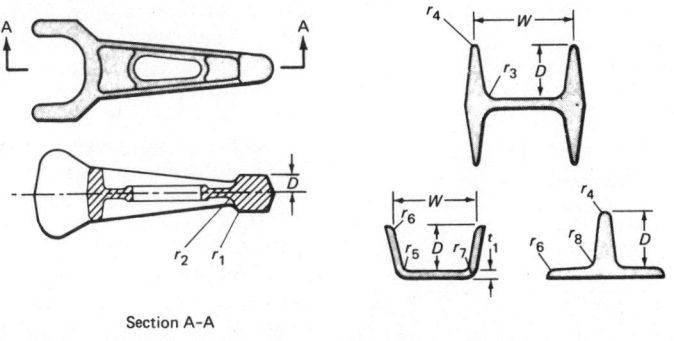

Section A-A

Fig. 11 Comparison of mechanical properties of forged brackets made of two different magnesium alloys

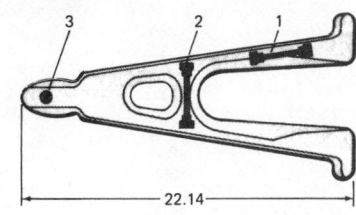

Dimensions in inches

Specimen location	Tensile strength MPa	ksi	Yield strength MPa	ksi	Elongation in 50 mm or 2 in., %
ZK60A-T					
1.....	305	44.1	238	34.5	8.5
2.....	302	43.8	203	29.5	23.5
3.....	260	37.7	131	19.0	12.5
AZ80A-T5					
1.....	368	53.4	279	40.5	4.0
2.....	333	48.3	243	35.2	2.7
3.....	283	41.1	163	23.7	3.2

Fig. 10 Effect of alloy, heat treatment and specimen location on mechanical properties of forged magnesium aircraft wheels

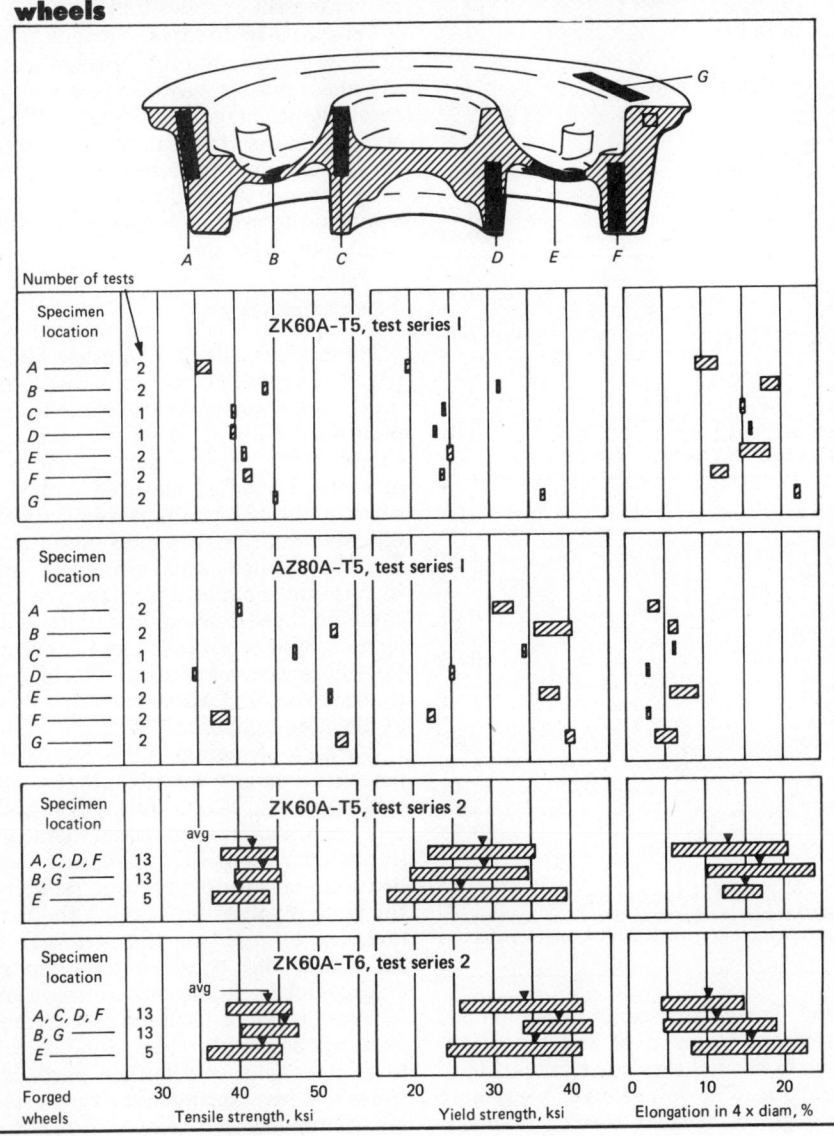

ing the insert. This possibility can be significantly reduced by preheating the insert, although inserts with wall thickness of 1.3 mm (0.050 in.) or less will preheat sufficiently from contact with the hot die and the molten magnesium. If these design and manufacturing conditions are not followed, the high residual stresses may lead to failure of the part in service due to stress-corrosion cracking. Safe design dimensions for cast-in steel inserts with wall thickness greater than 1.3 mm (0.050 in.) are shown in Fig. 13.

In highly stressed castings, it may be desirable to cast in a pilot to which the insert may be attached, thus avoiding high shrink stresses around the insert. Cast-in inserts may also be used for providing design details not otherwise feasible, such as lubrication lines of steel or copper, or appendages that could not be attached conveniently otherwise. Cast-in inserts complicate manufacture of castings and should be used only if other methods are not feasible.

Press-fit and Shrink-fit Inserts. Because of the low modulus of elasticity of magnesium, greater interference must be used for press-fit or shrink-fit inserts in magnesium than for inserts in other metals in order to obtain suffi-

Table 11b Corner and fillet radii (in millimetres) for large magnesium forgings(a)

D, mm	r_1, mm min	r_2, mm min
Under 4.76	2.38	6.35
4.76 to 10.32	4.76	10.32
10.32 to 25.4	7.94	15.87
25.4 to 44.45	12.70	25.4
44.45 to 63.50	19.04	31.75
63.50 to 101.6	26.99	41.28
over 101.6	31.75	50.80

D, mm	r_3, mm	r_4, mm
9.52	6.35	1.59
12.70	7.14	1.59
15.87	8.73	1.59
19.04	9.52	1.98
22.22	11.11	2.38
25.40	11.91	2.78
28.58	12.70	3.17
31.75	14.28	3.97
38.10	15.87	4.76
44.45	18.25	5.56
50.80	20.64	6.35
57.15	22.22	7.14
63.50	23.81	7.93
69.85	26.99	8.73
76.20	28.57	9.52

D, mm	r_5, mm min
W, 15.87 mm max	
10.32 max	2.78
W, 15.87 to 4.76	
10.32 max	2.78
10.32 to 12.70	3.97
W, 476 to 25.4 mm	
10.32 max	3.97
10.32 to 12.70	3.97
12.70 to 15.87	3.97
W, 25.4 to 32.54 mm	
10.32 max	3.97
10.32 to 12.70	3.97
12.70 to 15.87	3.97
15.87 to 20.64	6.35
W, 32.54 to 41.28 mm	
10.32 max	3.97
10.32 to 12.70	6.35
12.70 to 15.87	6.35
15.87 to 20.64	6.35
20.64 to 25.4	6.35
W, 41.28 to 50.80 mm	
10.32 max	6.35
10.32 to 12.70	6.35
12.70 to 15.87	6.35
15.87 to 20.64	10.32
20.64 to 25.4	10.32
25.4 to 33.34	10.32
W, 50.80 to 63.50 mm	
10.32 max	6.35

D, mm	r_5, mm min
10.32 to 12.70	6.35
12.70 to 15.87	6.35
15.87 to 20.64	10.32
20.64 to 25.4	10.32
25.4 to 33.34	10.32
W, 63.50 to 76.2 mm	
10.32 max	10.32
10.32 to 12.70	10.32
12.70 to 15.87	10.32
15.87 to 20.64	10.32
20.64 to 25.4	10.32
25.4 to 33.34	10.32
33.34 to 50.8	15.87
W, 76.2 to 101.6 mm	
10.32 max	10.32
10.32 to 12.70	10.32
12.70 to 15.87	10.32
15.87 to 20.64	10.32
20.64 to 25.4	10.32
25.4 to 33.34	15.87
33.34 to 50.8	15.87
W, 101.6 to 127 mm	
10.32 max	10.32
10.32 to 12.70	10.32
12.70 to 15.87	10.32
15.87 to 20.64	15.87
20.64 to 25.4	15.87
25.4 to 33.34	15.87
33.34 to 50.8	20.64

D, mm	r_6, mm
10.32 max	1.59
10.32 to 12.70	1.59
12.70 to 15.87	2.78
15.87 to 20.64	2.78
20.64 to 25.4	2.78
25.4 to 33.34	3.97
33.34 to 50.8	6.35

D, mm	r_8, mm
9.52 max	1.59
12.70	6.35
15.87	7.94
19.04	8.73
22.22	9.52
25.40	10.32
28.58	11.11
31.75	12.70
38.10	15.08
44.45	16.67
50.80	19.04
57.15	21.43
63.50	23.02
69.85	25.40
76.20	26.99

(a) See Fig. 9 for identification of symbols.
Note: The values given for r_3 and r_4, for corner and fillet radii for I-sections in magnesium forgings apply when the W/D ratio is approximately 2 to 1. For W/D ratios greater than 2 to 1 the radii should be increased and for lower W/D ratios the radii may be decreased. In corner and fillet radii for channel sections, r_7 equals r_5 plus t_1.

cient gripping force. An interference of 0.5 to 1.0 mm/m (0.0005 to 0.001 in./in.) is usually satisfactory, but may be increased appreciably where high torque loads are likely to be encountered. Table 12 presents recommended interferences for steel and bronze inserts in normal service at various temperatures. The values given serve only as a guide, because service tempera-ture, type of insert material, severity of service, thickness of insert, and sensitivity of the alloy to stress-corrosion cracking all influence the correct amount of interference. Differences in thermal expansion, yield strength and modulus of elasticity must also be considered for service at elevated temperature.

Inserts are more easily assembled by shrinking than by pressing. It is relatively easy to heat a magnesium part and thus expand a hole in the part sufficiently to receive an insert. Where necessary, the insert may be cooled to facilitate insertion. On the other hand, assembly by pressing requires careful machining of both insert and hole, and proper lubrication. When large interferences are required, it is best to specify shrink-fit inserts, because press-fits may score the magnesium.

Screwed-in Inserts. Various types of screwed-in (and other mechanically attached) inserts may also be used successfully in magnesium parts. When screwed-in inserts are used, locations of threaded holes must be chosen carefully in order to avoid stress concentrations and to provide sufficient engagement length for the thread.

Formability

Magnesium alloys, like other alloys with hexagonal crystal structures, are much more workable at elevated temperatures than at room temperature. Consequently, magnesium alloys usually are formed at elevated temperatures, and cold forming is used only for mild deformations around generous radii. The methods and equipment used in forming magnesium alloys are the same as those commonly employed in forming alloys of other metals, except for differences in tooling and technique that are required when forming is done at elevated temperatures.

Working of metals at elevated temperature has several advantages over cold working. Magnesium parts usually are drawn at elevated temperature in one operation without repeated annealing and redrawing, thus reducing the time involved for making the part and also eliminating the necessity of additional die equipment for extra stages. Hardened dies are unnecessary for most types of forming. Hot formed parts can be made to closer dimensional tolerances than cold formed parts because of less springback. Suggested maximum forming temperatures and times for various wrought magnesium alloys are given in Table 13. The column headed "Time" gives the maximum time the alloy can be held at temperature without adverse effects on properties.

Sheet and Plate. Rolled magnesium alloy products include flat sheet and plate, coiled sheet, circles, tooling plate and tread plate. These products are supplied in a variety of standard and nonstandard sizes (see Table 14).

The ability to use increased section thickness without weight penalty is of particular importance in designs that employ magnesium sheet. Thick-sheet construction provides the rigidity necessary in a structure, without the need for costly assembly of ribs and similar reinforcing members.

Rolled magnesium-alloy products can be worked by most conventional methods. For severe forming, sheet in the annealed (O temper) condition is preferred. However, sheet in the partially annealed (H24 temper) condition can be formed to a considerable extent. Because heat has significant effects on properties of hard rolled magnesium, properties of the metal after exposure to elevated temperature must be considered in forming. The design curves shown in Fig. 14 give minimum values suitable for design use. Although the curves are based primarily on tests of sheet 1.63 mm (0.064 in.) thick or less, check tests indicate reasonable applicability for gages up to 6.35 mm (0.250 in.).

Figure 14 shows how the properties of AZ31B-H24 vary with exposure time at several temperatures. The curves have been extrapolated above the typical property levels of AZ31B-H24 sheet. Thus, if the value selected from a

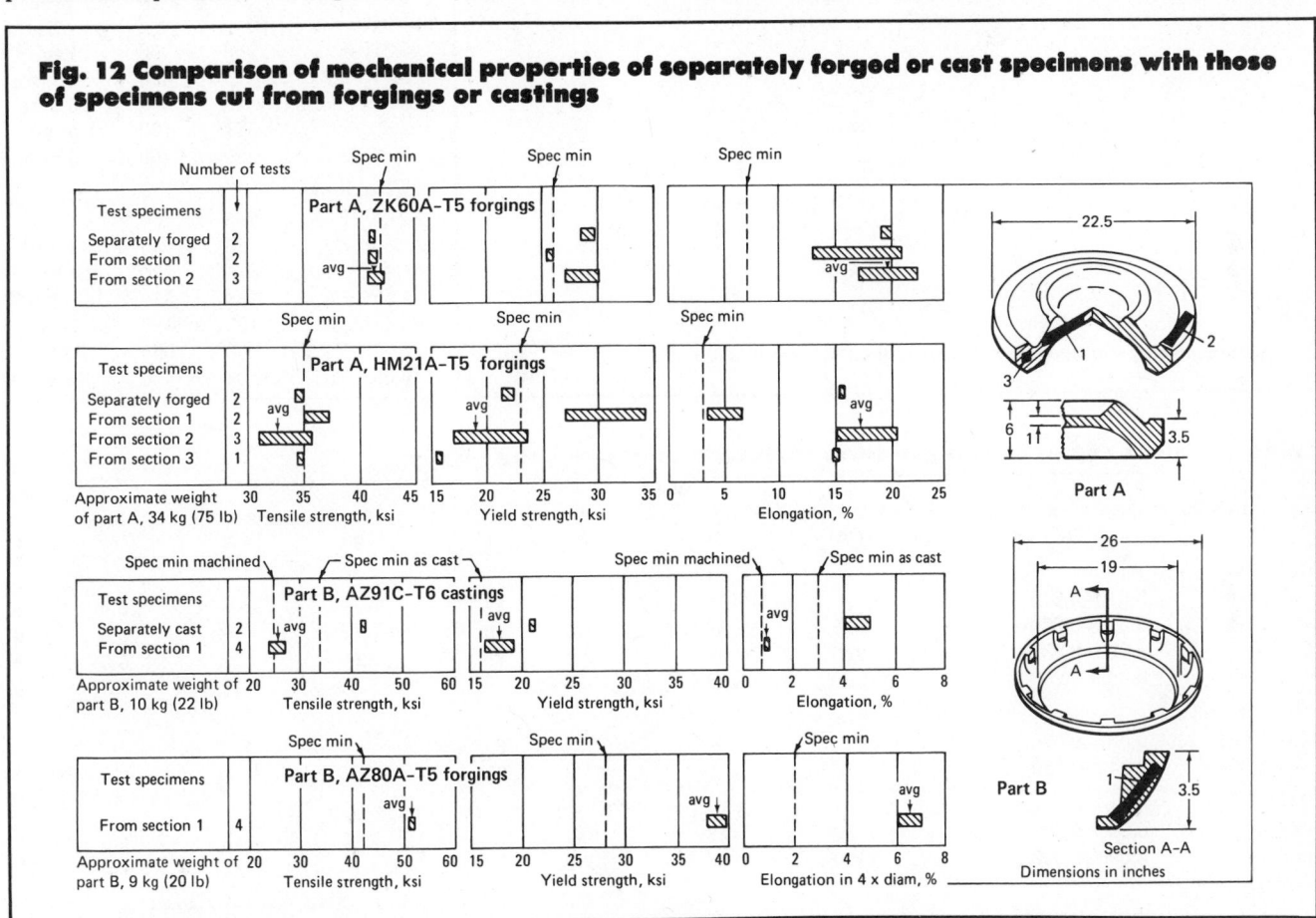

Fig. 12 Comparison of mechanical properties of separately forged or cast specimens with those of specimens cut from forgings or castings

Table 12 Approximate interferences suggested for press-fit and shrink-fit inserts in magnesium

| Outside diam. | | Wall thickness | | | Steel inserts Operating temperatures | | | | Bronze inserts Operating temperatures | | |
mm	in.	mm	in.	21 °C (70 °F)	38 °C (100 °F)	93 °C (200 °F)	149 °C (300 °F)	21 °C (70 °F)	38 °C (100 °F)	93 °C (200 °F)	149 °C (300 °F)
12.70	0.50	1.59	0.06	0.0004	0.0005	0.0009	0.0013	0.0004	0.0005	0.0007	.0009
25.40	1.0	2.38	0.09	0.0006	0.0010	0.0017	0.0025	0.0006	0.0008	0.0012	.0016
50.80	2.0	3.17	0.13	0.0010	0.0015	0.0031	0.0048	0.0010	0.0013	0.0019	.0028
76.20	3.0	3.97	0.16	0.0015	0.0024	0.0047	0.0072	0.0015	0.0019	0.0028	.0044
101.6	4.0	4.76	0.19	0.0021	0.0034	0.0062	0.0096	0.0021	0.0026	0.0039	.0060
127.0	5.0	5.56	0.22	0.0028	0.0044	0.0080	0.0121	0.0028	0.0034	0.0053	.0079
152.4	6.0	6.35	0.25	0.0036	0.0053	0.0098	0.0147	0.0036	0.0043	0.0070	.0098

Fig. 13 Safe design dimensions for cast-in steel inserts in magnesium alloy castings

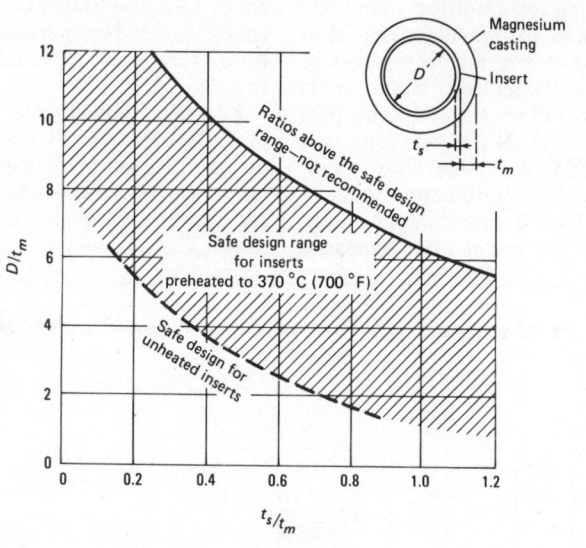

For inserts having wall thickness (t_s) greater than 1.3 mm (0.050 in.).

curve exceeds the actual property level of the material before exposure, the actual figure must be used.

Tests indicate that the effects of multiple exposures at elevated temperature are cumulative. Using Fig. 14 as an example, suppose that a part is held at 195 °C (380 °F) for 10 min; the resul-

Table 13 Maximum forming temperatures and times for wrought magnesium alloys

Alloy	Temperature °C	°F	Time
Sheet			
AZ31B-O	288	550	1 h
AZ31B-H24	163	325	1 h
HK31A-H24	343	650	15 min
	371	700	5 min
	399	750	3 min
Extrusions			
AZ61A-F	288	550	1 h
AZ31B-F	288	550	1 h
M1A-F	371	700	1 h
AZ80A-F	288	550	½ h
AZ80A-T5	193	380	1 h
ZK60A-F	288	550	½ h
ZK60A-T5	204	400	½ h

Table 14 Sizes of flat-rolled products available in magnesium alloys

Thickness range mm	in.	Width Standard mm	in.	Maximum mm	in.	Length Standard m	ft	Maximum(a) m	ft
Flat Sheet									
0.25–0.41	0.010–0.016	...	...	610	24	...	...	5.5	18
0.41–0.51	0.016–0.020	915	36	915	36	3.66	12	5.5	18
0.51–0.81	0.020–0.032	1220	48	1220	48	3.66	12	5.5	18
0.81–6.35	0.032–0.250	1220	48	1525	60	3.66	12	5.5	18
Coiled Sheet(b)									
0.81–1.02	0.032–0.040	...	...	1525	60	...	...	...	...
1.02–6.35	0.040–0.250	...	...	1830	60	...	...	...	...
Plate									
6.35–50.8	0.250–2.000	1220	48	1830	72	3.66	12	5.5	18
50.8–76.2	2.000–3.000	1220	48	1830	72	3.66	12	3.9	13
Tooling Plate									
6.35, 9.52	0.250, 0.375	1220	48	1830	72	2.44, 3.66	8, 12	5.5	18
12.7–50.8	0.500–2.000(c)	1220, 1525	48, 60	1830	72	2.44, 3.66	8, 12	5.5	18
63.5	2.500	1220	48	1830	72	2.44	8	4.9	16
76.2	3.000	1220	48	1830	72	2.44	8	3.9	13
88.9	3.500	1220	48	1830	72	2.44	8	3.4	11
101.6	4.000	1245	49	1830	72	...	7	3.0	10
127.0	5.000	1245	49	1830	72	...	6	2.4	8
152.4	6.000	1245	49	1830	72	...	5	1.6	6.5
Tread Plate (Raised-Pattern Floor Plate)									
3.18	0.125	1220, 1525	48, 60	1830	72	3.66	12	5.5	18
4.76–9.52	0.188–0.375(d)	1220, 1525	48, 60	1830	72	3.66	12	5.5	18
11.11–15.88	0.438–0.625(d)	1220, 1525	48, 60	1830	72	3.66	12	5.5	18
19.05	0.750	1220, 1525	48, 60	1830	72	3.66	12	5.5	18

(a) Maximum length for maximum width. Size of a single piece is limited to 998 kg (2200 lb). (b) Coiled sheet up to 610 mm (24 in.) wide can be produced in thicknesses down to 0.25 mm (0.10 in.). (c) In steps of 6.35 mm (0.250 in.). (d) In steps of 1.588 mm (0.0625 in.).

tant design compressive yield stress is 159 MPa (23.1 ksi). Then suppose the part is subsequently exposed for 300 min at 170 °C (340 °F). Because 200 min at 170 °C and 10 min at 195 °C both result in a minimum compressive yield stress of 159 MPa, they are equivalent exposures. The total equivalent exposure time at 170 °C then is 300 plus 200, or 500 min, which according to the data given in Fig. 14 results in a compressive yield stress of 155 MPa (22.5 ksi).

AZ31B-H24 sheet is commonly hot formed at temperatures below 160 °C (325 °F) to avoid annealing it to room-temperature property levels lower than the specified minimums. Annealing is a function of both time and temperature of exposure; thus, temperatures higher than 160 °C can be tolerated if exposure is carefully controlled. Table 15 shows the maximum permissible combination of time and temperature that will ensure that the specified minimum room-temperature properties of AZ31B-H24, HK31A-H24 and HM21A-T8 can be retained. This table is used for establishing limits of time and temperature for single exposures in normal forming operations. Whenever the sheet must endure multiple exposures, or whenever time of exposure at a given temperature must exceed the value indicated in Table 15, the data compilations at the end of the magnesium section of this Handbook should be consulted.

Deep Drawing. Magnesium alloys can be cold drawn to a maximum reduction of 15 to 25% in the annealed condition. The cold drawability limit of alloy AZ31B-O is about 20%. The drawability, or percentage reduction in blank diameter, is calculated by the formula:

$$\text{Percentage reduction} = \frac{D - d}{D} \times 100$$

where D is blank diameter before drawing and d is diameter of punch.

The time and temperature for annealing AZ31B-O are 15 min at 260 °C (500 °F). Cold drawn parts are stress relieved at 150 °C (300 °F) for 1 h after the final draw, to eliminate the danger of cracking from residual stresses.

The technique for hot drawing of magnesium alloy sheet has been developed largely so that drawing can be completed in a single operation. Heating magnesium alloys increases their drawability to such an extent that most parts can be made in a single draw. The amount of possible reduction increases as temperature increases up to about 230 °C (450 °F) for alloy AZ31B. It is usual practice to draw annealed AZ31B sheet to reductions as high as 68% in a single draw. When heated, magnesium can be drawn to higher reduction in a single draw than other metals. Maximum single-draw reduction as a function of temperature is illustrated in Fig. 15. These values were determined for 1.63-mm (0.064-in.) sheet using a cupping die 38 mm (1½ in.) in diameter. The radii on the draw ring and punch were both 5t, and drawing speeds were 38 mm (1½ in.) and 508 mm (20 in.) per minute. Note in Fig. 15 that the amount of reduction possible at a given temperature also is influenced by drawing speed.

The possibilities inherent in two-step draws are illustrated by the following parts: In the first operation, 610-mm (24-in.) blanks of 0.64-mm (0.025-in.) annealed sheet were drawn to a cup 200 mm (8 in.) in diameter by 400 mm (16 in.) deep; they were redrawn to a cup 140 mm (5½ in.) in diameter by 585 mm (23 in.) deep. Starting with a rectangular blank of 1.3-mm (0.051-in.) AZ31B-O, 455 by 485 mm (18 by 19 in.), a rectangular box 111 by 273 by 165 mm (4⅜ by 10¾ by 6½ in.) deep was drawn in the first operation. This box was then redrawn into a rectangular box 89 by 254 by 171 mm (3½ by 10 by 6¾ in.) deep, having 5.6-mm (7/32-in.) corner radii.

For most parts, however, depth of draw is not a primary consideration, and usually no trouble is experienced in drawing to the depth required. More trouble is encountered in keeping the metal free from puckers in parts with rounded corners or contours. Temperatures above those required for maximum drawability often are necessary to eliminate these puckers. On unusual or difficult jobs, it may be necessary to vary the procedure to obtain minimum scrap.

Choice of die materials is influenced chiefly by severity of the operation and number of parts to be produced. For most applications, unhardened low-carbon steel boiler plate or cast iron is satisfactory. For runs of 10 000 parts or more, for maximum surface smoothness, or for close tolerances where no significant die wear can be tolerated, hardened tool steels are recommended. W1 or O1 tool steels are satisfactory for extremely long runs (one million parts). For the most severe draws, however, the more abrasion-resistant tool

Table 15 Maximum time at temperature to maintain properties of magnesium alloy sheet(a)

Max time, min	Temperature	
	°C	°F
AZ31B-H24		
0.3	260	500
1	224	435
2	210	410
3	202	395
4	196	385
5	188	370
10	182	360
30	174	345
60	163	325
HK31A-H24		
15	343	650
5	371	700
3	385	725
3	399	750
HM21A-T8(b)		
60	399	750
10	427	800

(a) Annealed sheet will endure much higher temperatures than heat treated sheet for short periods without significant reduction of properties at room temperature. (b) Based on limited data obtained in laboratory tests.

steels, such as A2 or D2, will probably be more satisfactory and more economical. For room-temperature drawing, it is usually desirable that die steels be heat treated to obtain near maximum hardness in service. However, for elevated-temperature drawing, the maximum temperature to which the dies will be exposed in drawing must also be considered. In this situation, the dies must be tempered slightly above the maximum service temperature, even though some hardness may be sacrificed.

Parts with smooth bottoms are usually drawn with open dies or dies in which the punch does not bottom on the female, or bottom, plate. These parts are formed by the pull exerted on the blank by the pressure ring and the draw ring. Mating dies are used with magnesium only in forming parts having re-entrant portions that cannot be made by other means.

Stretch Forming. Both magnesium sheet and magnesium extrusions can be stretch formed. The temper of the alloy has no effect on techniques employed. Sheet usually is heated to 165 to 290 °C (325 to 550 °F) and slowly stretched to the desired contour. Annealed sheet can be stretched at room temperature to a limited extent. How-

Fig. 14 Effect of exposure time at elevated temperature on mechanical properties of AZ31B-H24 at room temperature

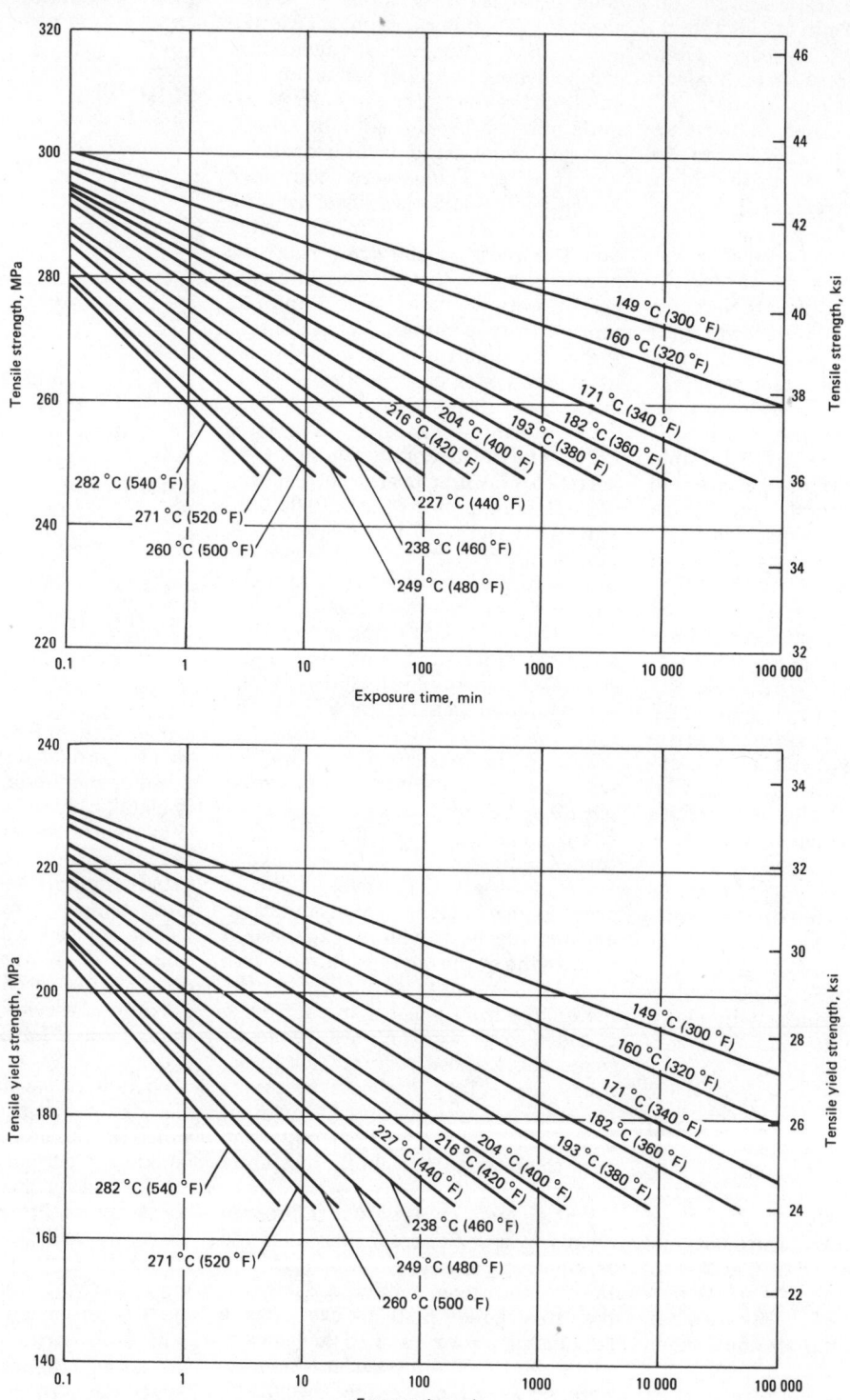

Data are based on sheet 1.63 mm (0.064 in.) thick. Check tests indicate reasonable applicability for thickness up to 6.35 mm (0.250 in.).

Fig. 14 (continued)

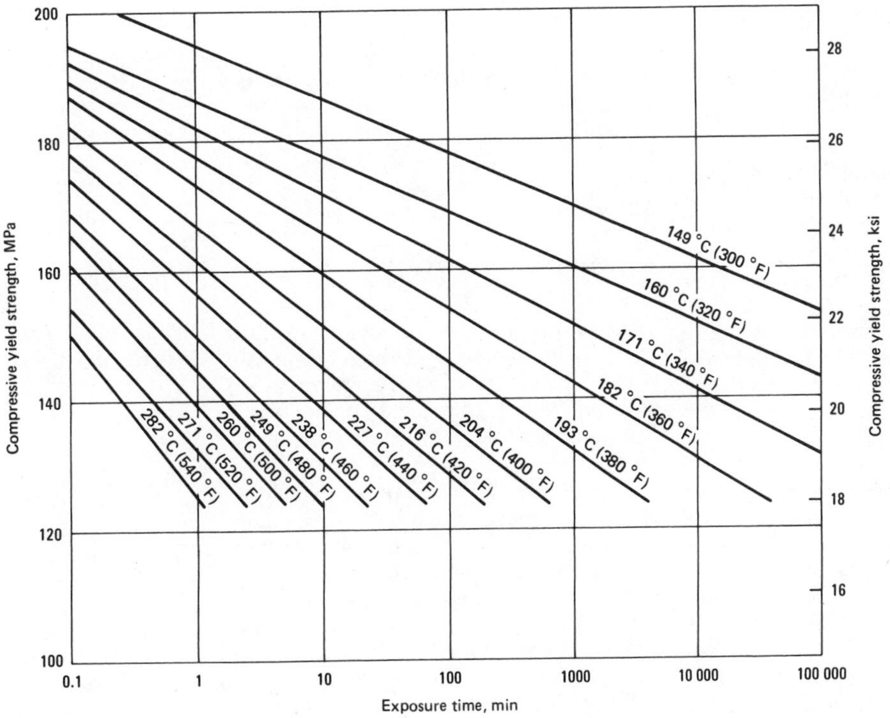

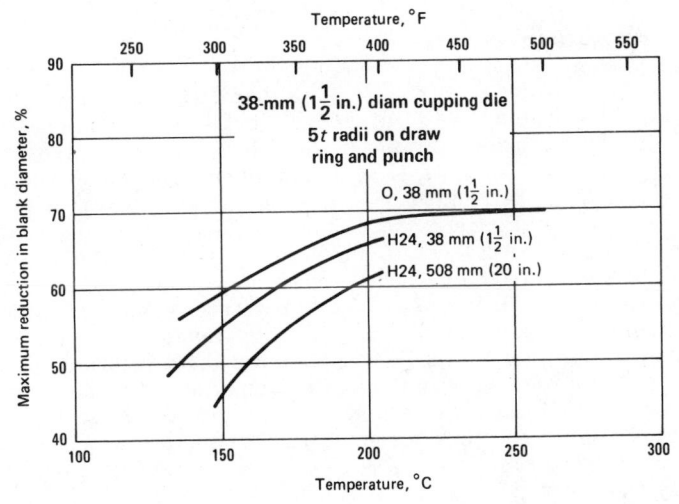

Fig. 15 Effect of temperature on the drawability of 1.63-mm (0.064-in.) AZ31B sheet

ever, the formabilities of alloys of any temper are so much greater at elevated temperatures than at room temperature that elevated-temperature forming is preferred for most operations. The percentage of differential stretching (P) during stretch forming is expressed as follows:

$$P = \frac{L - S}{S} \times 100$$

where L is the longest stretched length in the part and S is the shortest comparable length parallel to the longest stretched length. L and S are assumed to be of equal length before stretching.

Hot stretch forming results in minimum springback; the little springback that may occur is controlled by adding about 1% to the total stretch. Contour and springback control are good for positive contour radii curvature up to 6 m (20 ft).

For differential stretching of sheet over dies of low curvature, the maximum practical limit is about 15%. A 12% maximum is considered more desirable, however, because it permits enough overstretch for springback control.

Wrinkling often is a problem in stretch forming magnesium sheet, particularly in making asymmetrical parts of low curvature. The best method of controlling wrinkling is by building the proper restraints in the dies.

Dies may be made of a variety of materials, including magnesium alloys, aluminum alloys, iron, steel or zinc alloys. One aircraft company has reported the use of dies made of concrete cast over a wire mesh and heated by electrical resistance. Zinc alloy blocks should not be used above 230 °C (450 °F). Grippers should not have sharp serrated edges, which tear magnesium alloys; the use of emery paper between grips and magnesium sheet helps to reduce the possibility of tearing.

Heat control is important, and proper arrangement of heating units provides correct heat distribution. In addition to electrical resistance heating, infrared radiant heating units can be employed. Thermostatic control is essential and should be used on all heaters. The resistance heaters can be placed at various points where critical forming occurs. It is important that heating be sufficient to induce plastic flow, yet not great enough to cause excessive elongation and rupture of the part.

Die temperature varies with the material being formed. AZ31B-O sheet is usually formed at 290 °C (550 °F) without loss of mechanical properties. Hard rolled AZ31B-H24 can withstand a temperature of 165 °C (325 °F) for 1 h and higher temperatures for shorter periods. The maximum time at temperature to maintain original properties is given in Table 15. Extruded magnesium alloys may be heated to 315 °C (600 °F) in most instances. For most contours, only heating of the die is necessary, because the sheet will pick up heat quickly from the hot die. More complicated parts may require additional heating of the sheet by radiant heat or thermal blankets. Thick material may require the same treatment.

Lubricants are normally recommended, colloidal graphite being best for high-temperature forming. Lower temperature operation (up to 260 °C, or 500 °F) may permit the use of heat-resisting waxes and greases, or a dry-film stearate-type lubricant instead of colloidal graphite. Some shops have used heat-resisting synthetic rubber or fiberglass cloth between the magnesium sheet and the forming block.

Annealed magnesium sheet may be shrunk satisfactorily at room temperature. The amount of shrinkage possible can be greatly increased by heating. Standard shrinking machines are employed.

Extrusion Formability. Production bending of extrusions may be done on standard angle rolls, in mating dies, in stretch-forming machines or in other specialized bending equipment. If the forming is severe, extrusions are heated to approximately 260 to 345 °C (500 to 650 °F) and formed hot.

Some bend-radii data for flat extruded magnesium strip, as well as maximum forming temperatures and times, are given in Table 16. Bending of more complex extrusions is more difficult, and the bend radii required must be established for a given shape.

The minimum bend radii for round magnesium tubing varies with the ratio of outer diameter to wall thickness (Table 17).

Joining of Magnesium Alloys

Welding. Magnesium alloys are welded readily by gas metal-arc welding and by resistance spot welding. Rods of approximately the same composition as the base metal generally are satisfactory. With alloys HM21A and HM31A, EZ33A rods give higher joint efficiencies (see Table 18).

Butt and fillet joints are preferred in magnesium because they are the easiest to make by arc welding, and they provide more consistent results than other types of joints. Lap joints sometimes are used, but generally are less satisfactory than butt joints for load-carrying applications.

Arc-welded joints in annealed magnesium-alloy sheet and plate have room-temperature tensile strengths less than 10% lower than those of the base metal (joint efficiencies greater than 90%). Tensile strengths of arc welds in hard rolled material, however, are significantly lower than those of the base metal (joint efficiencies of only 60 to 85%) as a result of the annealing effect of welding. Consequently, room-temperature strengths of arc-welded joints in magnesium-alloy sheet and plate are about the same regardless of the temper of the base metal.

Joint efficiencies also are affected by service temperatures. For example, arc welds in HK31A-H24 sheet exhibit joint efficiencies of 75 to 80% at room temperature, but nearly 100% at 260 °C (500 °F). Joint efficiencies of arc welds in HM21A-T8 sheet range from about 80% at room temperature to 100% at

Table 16 Suggested limits for bending flat magnesium alloy extrusions

| Alloy | Typical bend radius at 21 °C (70 °F) | Limits for hot bending | | | Typical bend radius |
| | | At temperature | | | |
		°C	°F	Time, h	
Extruded Flat Strip: 2.29 by 22.2 mm (0.090 by 0.875 in.)					
AZ61A-F	1.9t	288	550	1	1.0t
AZ80A-F	2.4t	288	550	½	0.7t
AZ31B-F	2.4t	288	550	1	1.5t
M1A-F	4.8t	371	700	1	2.0t
AZ80A-T5	8.3t	193	380	1	1.7t
ZK60A-F	12t	288	550	½	2.0t
ZK60A-T5	12t	204	400	½	6.6t

Table 17 Form bending of magnesium tubing

| Alloy | Forming temperature | | Bend radius, D(a) |
	°C	°F	
AZ31B-F	21	70	4D
	93	200	3D
AZ61A-F	21	70	4D
	−7	20	3D
M1A-F	21	70	6D
	204	400	4D
ZK60A-F	21	70	5D

| | Min bend radius at 21 °C (70 °F)(b) | | |
	D/t = 17	D/t = 6	D/t = 3
AZ61A-F(c)	5D	2½D	2D
AZ61A-F(d)	2½D	2½D	2½D
AZ31B-F(c)	6D	4D	3D
AZ31B-F(d)	3D	2D	2D
M1A-F(c)	6D	3D	2½D
M1A-F(d)	6D	6D	2½D

(a) D = tube outside diameter. Bend radius taken to axis of tube. (b) Minimum bend radius for various D/t ratios at 21 °C (70 °F). D = tube outside diameter; t = wall thickness. (c) Tubing unfilled before bending. (d) Tubing filled with low-melting alloy (50 Bi, 26.7 Pb, 13.3 Sn, 10 Cd) before bending.

Table 18 Weldability of magnesium alloys

Alloy	Thickness mm	in.	Welding rod	Joint efficiency, %	Joint ductility(a)
AZ31B-O	1.63	0.064	AZ61A, AZ92A	97	12.0
AZ31B-H24	1.63	0.064	AZ61A, AZ92A	88	10.0
ZE10A-O	1.63	0.064	AZ61A, AZ92A	94	7.0
ZE10A-H24	1.63	0.064	AZ61A, AZ92A	87	3.0
M1A-F	3.17	0.125	M1A	55	2.0
AZ31B-F	3.17	0.125	AZ61A, AZ92A	92	12.0
AZ61A-F	3.17	0.125	AZ61A, AZ92A	89	8.0
AZ80A-F	3.17	0.125	AZ61A, AZ92A	86	4.0
AZ63A-F	12.70	0.5	AZ63A	83	2.5
AZ63A-T4	12.70	0.5	AZ63A	70	5.0
AZ63A-T6	12.70	0.5	AZ63A	75	2.0
AZ92A-F	12.70	0.5	AZ92A	100	2.5
AZ92A-T4	12.70	0.5	AZ92A	70	4.0
AZ92A-T6	12.70	0.5	AZ92A	75	2.0
AZ91C-F	12.70	0.5	AZ92A	100	2.5
AZ91C-T4	12.70	0.5	AZ92A	78	4.0
AZ91C-T6	12.70	0.5	AZ92A	75	2.0
AZ81A-F	12.70	0.5	AZ92A	100	2.5
AZ81A-T4	12.70	0.5	AZ92A	85	8.0
EK41A-T5	12.70	0.5	EK41A	100	1.0
EK41A-T6	12.70	0.5	EK41A	93	6.2
EZ33A-T5	12.70	0.5	EZ33A	100	1.1
HK31A-T6	12.70	0.5	HK31A	100	9.5
HK31A-H24	...	...	EZ33A	83	1.0
HZ32A-T5	12.70	0.5	HZ32A	93	3.8
HM21A-T8	1.63	0.064	EZ33A	88	1.5
	...	...	HM31A	74	1.5
HM31A-F	15.88	0.625	EZ33A	71	1.8
			HM31A	58	2.5

(a) Percentage elongation across the weld over a 50-mm or 2-in. gage length from tension tests.

Table 19 Times and temperatures for stress relieving arc welds in magnesium alloys

Alloy	Stress relief Temperature °C	°F	Time, min
Sheet			
AZ31B-H24 ...	149	300	60
AZ31B-O	260	500	15
Extrusions(a)			
AZ31B-F	260	500	15
AZ61A-F	260	500	15
AZ80A-F	260	500	15
AZ80A-T5	204	400	60
Castings			
AZ63A	260	500	60
AZ81A	260	500	60
AZ91C	260	500	60
AZ92A	260	500	60

(a) When extrusions are welded to sheet, distortion may be minimized by using a lower stress-relieving temperature and longer time. For example, 60 min at 150 °C (300 °F) instead of 15 min at 260 °C (500 °F).

200 °C (400 °F). HM31-T5 extrusions exhibit joint efficiencies of 75 to 85% at room temperature to about 370 °C (700 °F), and 100% at 425 °C (800 °F) and above. There are no appreciable differences in properties between welds made with alternating current and those made with direct current.

Stress Relieving. Arc welds in some magnesium alloys—specifically the Mg-Al-Zn series and alloys containing more than 1% Al—are subject to stress-corrosion cracking, and thermal treatment must be used to remove the residual stresses that cause this condition. The parts are placed in a jig or clamping plate and heated at the temperatures indicated in Table 19 for the specified times. After heating, the parts are cooled in still air. The use of jigs is sometimes necessary so that relief of stresses does not result in warpage of the assembly.

The other types of magnesium alloys, including those containing manganese, rare earths, thorium, zinc or zirconium, are not sensitive to stress corrosion and normally do not require stress relief after welding.

Spot welds in magnesium have good static strength, but fatigue strength is lower than for either riveted or adhesive-bonded joints. Spot welded assemblies are used mainly for low-stress applications and are not recommended where joints are subject to vibration. Typical shear strengths of spot welds in three alloys are given in Table 20. Shear strengths of welds AZ61A and HK31A are about the same as those in AZ31B.

Recommended spot spacings and edge distances for spot welds are given in Table 21. Where magnesium sheets of unequal thickness are to be spot welded, the thickness ratio should not exceed 2½ to 1.

Seam welds of the continuous or intermittent types have strength properties comparable to those of spot welds. Shear strengths of about 19.2 to 40.2 kg/linear mm (1075 to 2250 lb/linear in.) of welded seam can be obtained in AZ31B sheet from 1.0 to 3.2 mm (0.040 to 0.12 in.) thick.

Cost of weldments is less likely to vary significantly with quantity than cost of other methods of fabrication. Therefore, weldments are used most often where quantities are small or where fabrication of specific designs is impractical or impossible by other methods. For a dozen parts of the design shown in Fig. 16, sand castings cost twice as much as weldments; at about 35 pieces, the tooling cost for casting was absorbed, and casting was more economical for larger lots.

For the electronic mounting base shown in Fig. 17, the die casting was superior to the weldment in mechanical properties, although properties of both were above minimum requirements. Die castings were less expensive than weldments in quantities of 5000 (including cost of tooling); in quantities of 100, weldments were less expensive.

Adhesive bonding of magnesium has become an important fabrication technique. The fatigue characteristics of adhesive-bonded lap joints are better than those of other types of joints. The probability of stress-concentration failure in adhesive-bonded joints is minimal. Adhesive bonding permits the use of thinner materials than can be effectively riveted. The adhesive fills the spaces between the contacting surfaces and thus acts as an insulator between any dissimilar metals in the joint. It also permits manufacture of assemblies having surfaces smoother than those associated with riveting.

Fig. 16 Effect of quantity on cost of magnesium-alloy sand castings compared with the same parts made as weldments

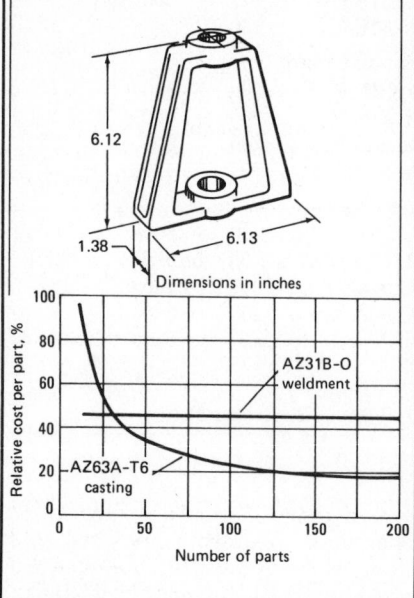

Fig. 17 Comparison of a magnesium-alloy electronic mounting base as manufactured by welding and by casting

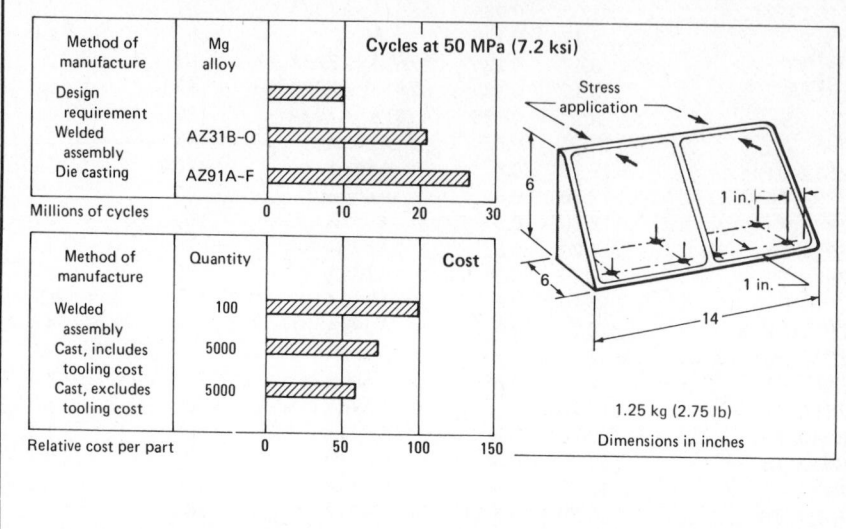

Table 20 Typical shear strengths of spot welds in magnesium alloys

Material thickness		Average spot diameter		Single spot shear strength			
				AZ31B-O		HK31A-H24	
mm	in.	mm	in.	kg	lb	kg	lb
Sheet							
0.508	0.020	3.56	0.14	100	220	...	...
0.635	0.025	4.06	0.16	120	270	...	...
0.813	0.032	4.57	0.18	150	330	135	300
1.016	0.040(a)	5.08	0.20	185	410	170	375
1.270	0.050	5.84	0.23	240	530	250	550
1.600	0.063(b)	6.86	0.27	340	750	325	720
2.032	0.080	7.87	0.31	405	890	...	...
2.540	0.100	8.64	0.34	535	1180	...	...
3.175	0.125(c)	9.65	0.38	695	1530	675	1490
				M1A-F			
				kg	lb		
Extrusions							
0.508	0.020	3.05	0.12	50	105		
0.635	0.025	3.56	0.14	70	150		
0.813	0.032	4.06	0.16	95	210		
1.016	0.040	4.57	0.18	130	285		
1.295	0.051	5.33	0.21	175	385		
1.626	0.064	6.10	0.24	225	500		
2.057	0.081	7.11	0.28	305	670		
2.591	0.102	7.87	0.31	400	885		
3.175	0.125	8.89	0.35	515	1135		

(a) Single-spot shear strength for HM21A-T8 alloy is 165 kg (360 lb). (b) Single-spot shear strength for HM21A-T8 alloy is 300 kg (660 lb). (c) Single-spot shear strength for HM21A-T8 alloy is 555 kg (1220 lb).

Adhesive bonding has been limited almost exclusively to lap joints. The following are a few of the general factors that should be considered when designing adhesively bonded joints:

1 Joint strengths vary with lap width, metal thickness, direction in which loads are applied and type of adhesive used.
2 The joint should be designed so that it provides a sufficiently large bonded area.
3 The adhesive layer should be uniform in thickness.
4 The adhesive layer should be as thin as possible, yet applied in sufficient quantity so that no joints are "starved".
5 Joints should be designed so that pressure and heat can be readily applied.
6 The curing temperatures of the common structural adhesives are below the temperatures at which the properties of hard rolled magnesium sheet are affected, and thus they do not significantly reduce the properties of magnesium alloys in the annealed (O) condition.

The effect of lap width on shear strength of joints bonded with phenolic rubber-base resin adhesive is shown at left in Fig. 18. The effect of temperature on the shear strength of adhesive-bonded joints in magnesium and aluminum is shown in the chart at right in Fig. 18.

The characteristics and properties of some adhesives used with magnesium are shown in Table 22. These adhesives cannot be utilized in assemblies operating above 80 °C (180 °F) because of low shear strength.

Riveting. Essentially the same procedures employed in riveting other materials are used in riveting magnesium alloys. Standard procedures are used for drilling and countersinking holes. Both dimpling and machine countersinking are used in flush riveting. With machine countersinking, it is desirable to have a cylindrical land with a minimum depth of 0.38 mm (0.015 in.) at the bottom of the hole. Thus, machine countersinking is limited to sheet thick enough to permit lands of this depth

with a given size of rivet. Dimpling of magnesium alloy sheet is a hot forming operation; to prevent reduction of properties during dimpling, the sheet must not be heated to excessively high temperatures or for long periods.

Only aluminum rivets should be used if galvanic incompatibility is to be minimized, and those up to 8 mm (5/16 in.) in diameter can be driven cold. The ease of driving rivets of alloy 5056 will vary with the temper. Quarter-hard temper (5056-H32) is satisfactory for all normal riveting.

Machinability

Magnesium and its alloys can be machined at extremely high speeds using greater depths of cut and higher rates of feed than can be used in machining other structural metals. There

are no significant differences in machinability among magnesium alloys. Therefore, a specific magnesium alloy rarely, if ever, is selected in place of another magnesium alloy solely on the basis of machinability.

Because of the free-cutting characteristic of magnesium, chips produced in machining it are well broken. Dimensional tolerances of about ±0.1 mm (a few thousandths of an inch) can be obtained using standard operations.

The power required to remove a given amount of metal is lower for magnesium than for any other commonly machined metal. The tabulation below compares power requirements for various metals, based on volume of metal removed per minute:

Metal	Relative power
Magnesium alloys	1.0
Aluminum alloys	1.8
Brass	2.3
Cast iron	3.5
Low-carbon steel	6.3
Nickel alloys	10.0

An outstanding machining characteristic of magnesium alloys is their ability to acquire an extremely fine finish. Often, it is unnecessary to grind and polish magnesium to obtain a smooth finished surface. Surface smoothness readings of about 0.1 μm (3 to 5 μin.) have been reported for machined magnesium and are attainable at both high and low speeds, with or without cutting fluids.

Table 21 Recommended spot spacing and edge distance for spot welds in magnesium alloy sheet

Sheet thickness		Spot spacing				Edge distance			
		mm		in.		mm		in.	
mm	in.	Min	Nom	Min	Nom	Min	Nom	Min	Nom
0.508	0.020	... 6.35	12.70	0.25	0.50	3.81	6.35	0.15	0.25
0.635	0.025	... 6.35	12.70	0.25	0.50	4.06	6.35	0.16	0.25
0.813	0.032	... 7.87	15.75	0.31	0.62	4.57	6.35	0.18	0.25
1.015	0.040	... 9.65	19.05	0.38	0.75	5.08	6.35	0.20	0.25
1.296	0.051	...10.41	19.05	0.41	0.75	5.84	7.87	0.23	0.31
1.626	0.064	...12.70	25.40	0.50	1.00	6.85	9.65	0.27	0.38
2.057	0.081	...15.75	31.75	0.62	1.25	7.87	10.41	0.31	0.41
2.591	0.102	...15.75	31.75	0.62	1.25	9.40	12.70	0.37	0.50
3.175	0.125	...19.05	38.10	0.75	1.50	11.18	15.75	0.44	0.62

Table 22 Characteristics of adhesives used for bending magnesium (a)

General type of composition	Temperature °C	°F	Curing conditions Time, min	Pressure MPa	ksi	Adhesive thickness mm	in.	Shear strength MPa	ksi
Phenol formaldehyde plus polyvinyl formal powder(b)	132	270	32	0.34–3.44	0.05–0.5	0.0–0.152	0.0–0.006	11–18	1.6–2.6
Phenolic rubber-base resin(b)	163	325	20	1.38	0.2	0.076–0.152	0.003–0.006	15–18	2.2–2.6
Phenolic synthetic rubber base plus thermosetting resin	177	350	10	0.048–0.310	0.007–0.045	0.127–0.508	0.005–0.020	7–17	1.0–2.5
			60	0.689	0.1	0.127–0.508	0.005–0.020	14–20	2.1–2.9
Ethoxyline resin-liquid, powder or stick used like solder	199	390	60	Contact	Contact	0.076–0.152	0.003–0.006	10–15	1.5–2.2
Ethoxyline resin (two liquids)	Room	Room	24 h	Contact	Contact	0.025–0.152	0.001–0.006	...	...
Epoxy type resin paste plus liquid activator	93	200	60	Contact	Contact	0.076–0.127	0.003–0.005	21 max	3.0 max
Rubber base	204	400	8	1.38	0.2	0.254–0.381	0.010–0.015	12–16	1.7–2.3
Vinyl phenolic	149	300	8 preheat	1.38	0.2	0.102–0.305	0.004–0.012	7–12	1.0–1.7
	135–204	275–400	70–4	...	...	...	...	...	...
Epoxy-type resin	93	200	45	Contact	Contact	0.254–0.762	0.010–0.030	8–12	1.2–1.8
	93	200	45	0.096	0.014	(tape)	(tape)	10–12	1.4–1.7
Phenolic	149	300	15	0.193	0.028	0.051–0.102	0.002–0.004	17 max	2.4 max

(a) These adhesives are for service at temperatures up to 82 °C (180 °F). (b) Known to meet USAF specifications.

Fig. 18 Effect of lap width and temperature on shear strength of adhesive bonded joints

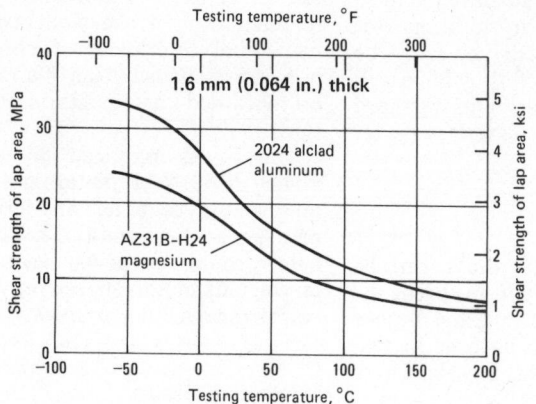

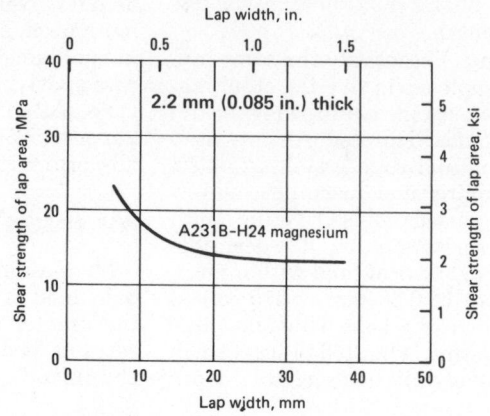

Phenolic rubber-base resin adhesive.

Fig. 19 Effect of plate-buckling index and temperature on structural efficiency of magnesium, aluminum and titanium alloys

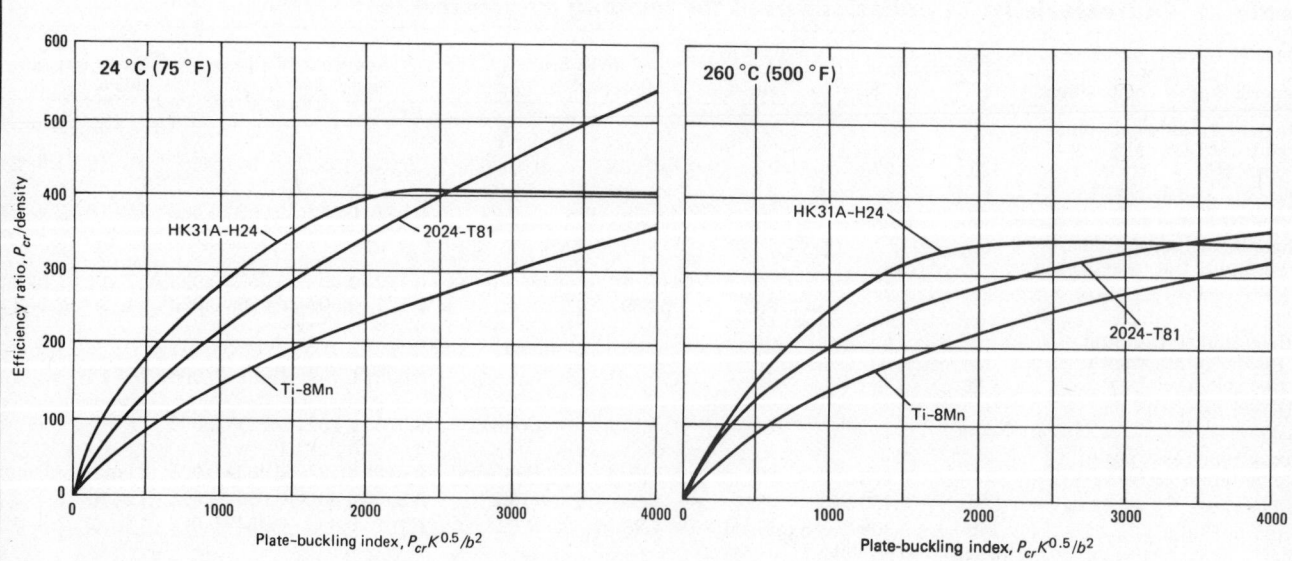

See text for discussion.

Cutting Fluids ("Coolants"). In machining of magnesium alloys, cutting fluids provide far smaller reductions in friction than they provide in machining of other metals and thus are of little use in improving surface finish and tool life. Most machining of magnesium alloys is done dry, but cutting fluids sometimes are used for cooling the work.

Although less heat is generated during machining of magnesium alloys than during machining of other metals, higher cutting speeds and magnesium's low heat capacity and relatively high thermal expansion characteristics may make it necessary to dissipate the small amount of heat that is generated. Generation of heat can be minimized by use of correct tooling and machining techniques, but sometimes cutting fluids are needed to reduce the possibilities of distortion of the work and ignition of fine chips. Because they are used primarily to dissipate heat, cutting fluids are referred to as "coolants" when used in machining of magnesium alloys.

Numerous mineral-oil cutting fluids of relatively low viscosity are satisfactory for use as coolants in machining magnesium. Suitable coolants represent a compromise between cooling power and flash point. Additives designed to increase wetting power are usually beneficial. Only mineral oils should be used as coolants; animal and vegetable oils are not recommended.

Water-soluble oils, oil-water emulsions or water solutions of any kind should not be used on magnesium. Water reduces the scrap value of magnesium turnings and introduces potential fire hazards during shipment and storage of machine shop scrap.

Safe Practice. The possibility of chips or turnings catching fire must be considered when magnesium is to be machined. Chips must be heated close to their melting point before ignition can occur. Roughing cuts and medium finishing cuts produce chips too large to be readily ignited during machining. Fine finishing cuts, however, produce fine chips that can be ignited by a spark. Stopping the feed and letting the tool dwell before disengagement, and letting the tool or tool holder rub on the work, produce extremely fine chips and should be avoided.

Factors that increase the probability of chip ignition are (a) extremely fine feeds, (b) dull or chipped tools, (c) improperly designed tools, (d) improper

machining techniques and (e) sparks caused by tools hitting iron or steel inserts. Feeds less than 0.02 mm (0.001 in.) per revolution and cutting speeds higher than 5 m/s (1000 ft/min) increase the risk of fire. Even under the most adverse conditions—with dull tools and fine feeds—chip fires are very unlikely at cutting speeds below 3.5 m/s (700 ft/min).

Any fire hazard connected with machining of magnesium is easy to control, and large quantities of magnesium are machined without difficulty. Following the rules given below will reduce the fire hazard:

1 Keep all cutting tools sharp and ground with adequate relief and clearance angles.
2 Use heavy feeds to produce thick chips.
3 Use mineral-oil coolants (15 to 19 L/min, or 4 to 5 gal/min) whenever possible; when not possible, avoid fine cuts.
4 Do not allow chips to accumulate on machines or on clothing of operators. Remove dust and chips at frequent intervals and store in clean, plainly labeled, covered metal cans.
5 Keep an adequate supply of a recommended magnesium fire extinguisher within reach of operators.

If dry chips are ignited, they will burn with a brilliant white light, but the fire will not flare up unless disturbed. Burning chips should be extinguished as follows:

1 Scatter a generous layer of clean, dry cast iron chips or metal extinguishing powder over the burning magnesium.
2 Cover actively burning fires on combustible surfaces like wood floors with a layer of the extinguishant, then shovel the entire mass into an iron container or onto a piece of iron plate.
3 Do not use water or any of the common liquid or foam-type extinguishers, which intensify magnesium-chip fires.

Distortion of magnesium parts during machining occurs rarely and usually can be attributed to excessive heating or improper chucking or clamping.

Heating of the work is increased by use of dull or improperly designed tools or very fine cuts. Because magnesium has a relatively high coefficient of thermal expansion, such excessive heating results in substantial increases in di-

mensions—particularly in thin sections, where heating causes relatively large increases in temperature. Use of sharp, properly designed tools, mineral-oil coolants and relatively coarse feeds and depths of cut reduces excessive heating. Wide variations in room temperature during machining also can cause sufficient dimensional change to affect machining tolerances.

Clamping should always be done on heavier sections of magnesium castings, and clamping pressures should not be high enough to cause distortion. Special care should be taken with light parts that could be distorted easily by the chuck or by use of heavy cuts.

Distortion of magnesium parts seldom is caused by stresses during casting, forging or extruding, but may result from stresses caused by straightening or welding. Such stresses can be relieved prior to machining by heating at 260 °C (500 °F) for 2 h and slowly cooling. However, such treatment causes some loss of strength in AZ31B-H24 sheet products. If distortion of parts is observed after rough machining, the cutting tool should be inspected to ensure that it is sharp and properly ground. If so, size of cut should be decreased. With complex parts or parts machined to extremely close tolerances, it may be advisable to stress relieve or, if time permits, to store parts for 2 or 3 days, between rough machining and finishing.

Design and Weight Reduction

By substitution of magnesium alloys for heavier metals such as steel and aluminum alloys, many structural parts can be substantially reduced in weight with little or no redesign. This is possible because manufacturing limitations make many parts heavier than necessary. For example, a casting, for successful filling of the mold, may require minimum wall thickness greater than that dictated by service requirements and the strength of the metal used. Similarly, forgings and extrusions sometimes must be made thicker than necessary, and the light weight of magnesium can be used to advantage with these product forms also. In many instances, a casting, forging or extrusion for which magnesium is substituted for a heavier metal can have adequate strength with no increase in wall thickness.

In other parts, substitution of magnesium may require greater wall thickness, and substantial redesign may be

Table 23 Relative bending strength, stiffness and weight of some structural metals (a)

Material	Thickness	Bending strength	Stiffness	Weight
For Equal Thickness				
1025 steel	100	100.0	100.0	100.0
6061-T6 aluminum sheet and extrusions............................	100	97.2	34.5	34.5
AZ31B magnesium extrusions	100	47.2	22.4	22.5
ZK60A-T5 magnesium extrusions	100	88.9	22.4	22.5
AZ31B-H24 magnesium sheet	100	73.4	22.4	22.5
For Equal Bending Strength				
1025 steel	100	100	100.0	100.0
6061-T6 aluminum sheet and extrusions............................	101	100	35.8	34.8
AZ31B magnesium extrusions	146	100	69.2	32.9
ZK60A-T5 magnesium extrusions	106	100	26.7	23.9
AZ31B-H24 magnesium sheet	117	100	35.6	26.3
For Equal Stiffness				
1025 steel	100	100	100	100.0
6061-T6 aluminum sheet and extrusions............................	143	199	100	49.3
AZ31B magnesium extrusions	165	129	100	37.2
ZK60A-T5 magnesium extrusions	165	242	100	37.2
AZ31B-H24 magnesium sheet	165	200	100	37.2
For Equal Weight				
1025 steel	100	100	100	100
6061-T6 aluminum sheet and extrusions............................	290	817	841	100
AZ31B magnesium extrusions	444	930	1962	100
ZK60A-T5 magnesium extrusions	444	1753	1962	100
AZ31B-H24 magnesium sheet	444	1451	1962	100

(a) Comparison made at room temperature for rectangular beams of constant width with the following minimum yield strengths: 1025 steel, 250 MPa (36 ksi); 6061-T6 aluminum, 240 MPa (35 ksi); magnesium alloys, average of minimum tensile yield and compressive yield strengths.

necessary in order to realize maximum saving of weight. Because strength and stiffness in bending of many structural sections increase approximately as the square and cube of the section depth, respectively, it is possible to obtain large increases in strength and stiffness with moderate increases in depth and cross-sectional area. When such increases in depth are permissible, it usually is economical to redesign the part for magnesium. The greater bulk of the redesigned part reduces local instability, and although the saving in weight is less than maximum, the reduction in instability allows design simplification and thus reduces manufacturing costs.

Magnesium alloys are compared with aluminum alloys and steel on the bases of thickness, strength, stiffness and weight, at room temperature, in Table 23. Bending strength is defined as the product of yield strength and section modulus.

Bending. Rectangular steel, aluminum and magnesium sections of equal thickness have rigidities in the ratio of their moduli of elasticity. The magnesium section weighs about 63% as much as the aluminum section and about 22% as much as the steel section.

The rigidity in bending of a rectangular section is proportional both to the cube of its depth and to its modulus of elasticity. If the section thicknesses of a magnesium section, an aluminum section and a steel section are adjusted until their rigidities are equal, the magnesium section will weigh about 71% of the aluminum and about 40% of the steel. If the section thickness of the magnesium is increased to about twice that of the steel, the magnesium will be more than 70% more rigid than the steel and less than 50% as heavy. Magnesium supporting its own weight shows no more deflection than other metals under the same conditions.

At high temperatures, the difference between short-time ultimate and yield strengths of certain magnesium alloys decreases significantly. Creep properties that depend on time must also be considered in evaluating materials for long-time operation at elevated temperature. Creep-strength values of several magnesium alloys are given in the data compilations starting on page 553.

Plate Buckling. Structures subjected to compressive loads may be limited in efficiency (load carried versus weight of structure) by buckling at relatively low stresses.

A valuable aid to designers in selection of optimum materials for plate structures that are critical in compression loading is a structural index. A structural index is nondimensional—that is, equivalent designs give the same value of structural index regardless of size of the actual part. The plate-buckling index is computed from the maximum edge load P_{cr} that will not cause crippling, the width b of the plate and a factor K (for a simply supported edge, $K = 4.0$) determined by the amount of restraint or clamping along the unloaded edges. The formula is

$$\text{Index} = P_{cr}K^{0.5}/b^2$$

Using this index, the efficiency of various structural materials can be compared directly for given conditions of loading and structural configuration. For example, the efficiencies of three materials at room temperature and at 260 °C (500 °F) for plate-buckling indexes up to 4000 are shown in Fig. 19. Comparisons are based on typical properties after short-time exposure to temperature.

A low value of plate-buckling index means either that the critical edge load is low or that the plate is wide, corresponding in either instance to a more lightly stressed structure. As the index value increases, it represents a transition to narrower plates and/or heavier edge loads and, at high values, corresponds to a condition of pure prismatic compression.

The ratio of working stress to density is an inverse measure of structural weight—the higher the ratio, the lighter the structure. Figure 19 shows the expected advantage in efficiency of the lower density magnesium alloy HK31A-H24 over the higher density aluminum and titanium alloys. This advantage fades as the index increases, and the stress condition moves from elastic buckling toward prismatic compression. Comparison of the two charts shows how the range over which the magnesium alloy is the most efficient of the three alloys (magnesium, aluminum and titanium) is higher at 260 °C (500 °F) than at room temperature.

Properties of Magnesium Alloys

By the ASM Review Committee on
Magnesium and Magnesium Alloys*

AZ10A

Specifications

UNS. M11100
Government. Extruded rods, bars
and shapes: QQ-M-31

Chemical Composition

Composition limits. 1.0 to 1.5 Al;
0.2 to 0.6 Zn; 0.2 min Mn; 0.1 max Si;
0.1 max Cu; 0.005 max Ni; 0.005 max
Fe; 0.04 max Ca; rem Mg

Applications

Typical uses. Low cost extrusion
alloy with moderate mechanical
properties and high elongation. Used
in as-extruded (F) temper.

Mechanical Properties

Tensile properties. See Table 1.
Compressive properties. See Table
1.
Poisson's ratio. 0.35
Elastic modulus. Tension, 45 GPa
(6.5 × 10^6 psi)

*See page X for committee list.

Table 1 Typical mechanical properties of AZ10A at room temperature

Size and shape	Tensile strength MPa	ksi	Yield strength MPa	ksi	Elonga-tion, %	Compressive yield strength MPa	ksi
Solid shapes with least dimension up to 6.4 mm (0.025 in.)	240	35	145	21	10	69	10
Solid shapes with least dimension to 6.4 to 38 mm (0.025 to 1.5 in.)	240	35	150	22	10	76	11
Hollow and semi-hollow shapes	230	33	145	21	8	69	10
Tube (152 mm or 6 in. (O.D. max) with 0.7 to 6.4 mm (0.028 to 0.25 in.) wall	230	33	145	21	8	69	10

Mass Characteristics

Density. 1.76 Mg/m^3 (0.064 lb/in.^{3})
at 20 °C (68 °F)

Thermal Properties

Liquidus temperature. 645 °C
(1190 °F)
Solidus temperature. 630 °C
(1170 °F)
Coefficient of thermal expansion.
Linear, 26.6 μm/m·K (14.8 μin./
in.·°F) at 21 to 204 °C (70 to 400 °F)
Thermal conductivity. 110 W/m·K
(64 Btu/ft·h·°F) at 20 °C (68 °F)

Electrical Properties

Electrical resistivity. 64 nΩ·m at
20 °C (68 °F)

Fabrication Characteristics

Weldability. Good. Does not require
stress relief after welding.

AZ21X1

Specifications

UNS. M11210

Chemical Composition

Composition limits. 1.6 to 2.5 Al; 0.8 to 1.6 Zn; 0.1 to 0.25 Ca; 0.15 max Mn; 0.05 max Si; 0.05 max Cu; 0.005 max Fe; 0.002 max Ni; 0.3 max others; rem Mg

Applications

Typical uses. Impact extruded battery anodes. Used in as-extruded (F) temper.

AZ31B, AZ31C

Specifications

AMS. Sheet, AZ31B: O temper, 4357; H24 temper, 4376
ASTM. Sheet: B90. Extruded rod, bar, shapes, tubing and wire: B107. Forgings, AZ31B: B91
SAE. AZ31B: J466. Former SAE alloy number: 510
UNS numbers. AZ31B: M11311. AZ31C: M11312
Government. AZ31B: forgings, sheet and plate, QQ-M-40; extruded bar, rod and shapes, QQ-M-31B; extruded tubing, WW-T-825B
Foreign. Elektron AZ31 (extruded bar and tubing). British: sheet, BS3370 MAG111; extruded bar and tubing, BS3373 MAG111. German: DIN9715 3.5312. French: AFNOR G-A371

Chemical Composition

Composition limits. AZ31B: 2.5 to 3.5 Al; 0.20 min Mn; 0.60 to 1.4 Zn; 0.04 max Ca; 0.10 max Si; 0.05 max Cu; 0.005 max Ni; 0.005 max Fe; 0.30 max others (total); rem Mg. AZ31C: 2.4 to 3.6 Al; 0.15 min Mn; 0.50 to 1.5 Zn; 0.10 max Cu; 0.03 max Ni; 0.10 max Si; rem Mg
Consequence of exceeding impurity limits. Excessive Cu, Ni or Fe degrades corrosion resistance.

Applications

Typical uses. AZ31B and AZ31C: forgings and extruded bar, rod, shapes, structural sections and tubing, with moderate mechanical properties and high elongation; AZ31C is the commercial grade, with the same properties as AZ31B but higher impurity limits. AZ31B only: sheet and plate with good formability and strength, high resistance to corrosion and good weldability. AZ31B and AZ31C are used in the as-fabricated (F), annealed (O) and hard rolled (H24) tempers.

Mechanical Properties

Tensile properties. See Tables 2 and 3.
Shear strength. See Table 2.
Compressive yield strength. See Table 2.
Bearing properties. See Table 2.
Hardness. See Table 2.
Poisson's ratio. 0.35
Elastic modulus. Tension, 45 GPa (6.5 × 10^6 psi); shear, 17 GPa (2.4 × 10^6 psi)
Impact strength. Forgings and extruded bar, rod and solid shapes: Charpy V-notch, 4 J (3.2 ft·lb)
Directional properties. See Table 4.

Mass Characteristics

Density. 1.77 Mg/m³ (0.064 lb/in.³) at 20 °C (68 °F)

Table 3 Typical tensile properties of AZ31B at various temperatures

Testing temperature		Tensile strength		Yield strength		Elongation(a), %
°C	°F	MPa	ksi	MPa	ksi	
Sheet, Hard Rolled						
−80	−112	331	48.0	234	34.0	...
−27	−18	310	45.0	234	34.0	...
21	70	290	42.0	221	32.0	15
100	212	207	30.0	145	21.0	30
149	300	152	22.0	90	13.0	45
204	400	103	15.0	59	8.5	55
260	500	76	11.0	31	4.5	75
316	600	41	6.0	21	3.0	125
371	700	28	4.0	14	2.0	140
Extrusions, As Fabricated						
−185	−300	434	63.0	338	49.0	6.0
−129	−200	359	52.0	303	44.0	7.5
−73	−100	314	45.5	262	38.0	9.5
−18	0	283	41.0	228	33.0	12.5
21	70	262	38.0	200	29.0	15.0
93	200	238	34.5	148	21.5	23.5
121	250	217	31.5	117	17.0	29.5
149	300	179	26.0	100	14.5	37.5

(a) In 50 mm or 2 in.

Table 2 Typical mechanical room-temperature mechanical properties of AZ31B

Product form	Tensile strength MPa	ksi	Tensile yield strength(a) MPa	ksi	Elongation(b), %	Hardness HBN(c)	HBE	Shear strength MPa	ksi	Compressive yield strength(a) MPa	ksi	Ultimate bearing strength(d) MPa	ksi	Bearing yield strength(d) MPa	ksi
Sheet, annealed	255	37	150	22	21	56	67	145	21	110	16	485	70	290	42
Sheet, hard rolled	290	42	220	32	15	73	83	160	23	180	26	495	72	325	47
Extruded bar, rod and solid shapes	255	37	200	29	12	49	57	130	19	97	14	385	56	230	33
Extruded hollow shapes and tubing	241	35	165	24	16	46	51	...	...	83	12	...	...	...	...
Forgings	260	38	170	25	15	50	59	130	19	...	...	...	...	...	...

(a) At 0.2% offset. (b) In 50 mm or 2 in. (c) 500-kg load, 10-mm ball. (d) ³⁄₁₆-in. pin diameter.

Thermal Properties

Liquidus temperature. 630 °C (1170 °F)

Solidus temperature. 605 °C (1120 °F)

Coefficient of thermal expansion. Linear, 26 μm/m·K (14 μin./in.·°F)

Specific heat vs temperature. $C_p = 0.2441 + 0.000105T - 2783T^{-2}$

Latent heat of fusion. 331 to 348 kJ/kg (142 to 149 Btu/lb)

Thermal conductivity. 96 W/m·K (56 Btu/ft·h·°F) at 100 to 300 °C (212 to 572 °F)

Electrical Properties

Electrical conductivity. 18.5% IACS

Electrical resistivity. 92 nΩ·m at 20 °C (68 °F)

Electrolytic solution potential. 1.59 V vs saturated calomel electrode

Fabrication Characteristics

Weldability. Gas-shielded arc welding with AZ61A or AZ92A rod (AZ61A preferred), excellent; stress relief required. Resistance welding, excellent.

Recrystallization temperature. Recrystallizes after 1 h at 205 °C (400 °F) following 15% cold work

Annealing temperature. 345 °C (650 °F)

Hot working temperature. 230 to 425 °C (450 to 800 °F)

Table 4 Typical directional properties of AZ31B

Condition	Tensile strength MPa	ksi	Yield strength MPa	ksi	Elongation(a), %
Parallel to Rolling Direction					
Annealed....	255	37	150	22	21
Hard rolled..	290	42	220	32	15
Perpendicular to Rolling Direction					
Annealed....	270	39	170	25	19
Hard rolled..	295	43	235	34	19

(a) In 50 mm or 2 in.

PE

Chemical Composition

Composition limits. 2.5 to 4.0 Al; 0.08 max Mn; 0.7 to 1.6 Zn; 0.05 max Si; 0.05 max Cu, 0.005 max Ni; 0.005 max Fe; 0.04 max Ca; 0.03 max other impurities (total); rem Mg

Consequence of exceeding impurity limits. Poor etch quality

Applications

Typical uses. Photoengraving

Mass Characteristics

Density. 1.76 Mg/m³ (0.064 lb/in.³) at 20 °C (68 °F)

Thermal Properties

Liquidus temperature. 632 °C (1170 °F)

Solidus temperature. 605 °C (1120 °F)

Incipient melting temperature. 532 °C (990 °F)

Coefficient of thermal expansion. Linear, 26 μm/m·K (14 μin./in.·°F)

Specific heat. 1047 J/kg·K (0.25 Btu/lb·°F) at 20 °C (68 °F)

Latent heat of fusion. 330, 757 to 347, 504 J/kg

Fabrication Characteristics

Weldability.

Annealing temperature. 345 °C (650 °F)

Hot working temperature. 230 to 425 °C (450 to 800 °F)

Shortness temperature. 345 °C (650 °F)

AZ61A

Specifications

AMS. Extrusions: 4350. Forgings: 4358

ASTM. Extrusions: B107. Forgings: B91

SAE. J466. Former SAE alloy numbers: 520 (extrusions) and 531 (forgings)

UNS number. M11610

Government. Extruded bar, rod and shapes: QQ-M-31B. Extruded tubing: WW-T-825A. Forgings: QQ-M-40B

Foreign. Elektron AZ61 (extruded bar, sections and tubing). British: extruded bar, sections and tubing, BS3373 MAG121; forgings, BS3372 MAG121. German: DIN9715 3.5612; casting, DIN1729 3.5612 French: AFNOR G-A6Z1

Chemical Composition

Composition limits. 5.8 to 7.2 Al; 0.15 min Mn; 0.40 to 1.5 Zn; 0.10 max Si; 0.05 max Cu; 0.005 max Ni; 0.005 max Fe; 0.30 max others (total); rem Mg

Consequence of exceeding impurity limits. Excessive Cu, Ni or Fe degrades corrosion resistance.

Applications

Typical uses. General-purpose extrusions with good properties and moderate cost, and forgings with good mechanical properties, used in the as-fabricated (F) temper. This alloy is used in sheet form for battery applications only.

Mechanical Properties

Tensile properties. See Tables 5 and 6.

Shear strength. See Table 5.

Compressive yield strength. See Table 5.

Bearing properties. See Table 5.

Hardness. See Table 5.

Poisson's ratio. 0.35

Elastic modulus. Tension, 45 GPa (6.5 × 10⁶ psi); shear, 17 GPa (2.4 × 10⁶ psi)

Impact strength. Charpy V-notch: forgings, 3 J/kg (2.2 ft·lb); extruded rod, bar and shapes, 4.1 J/kg (3.0 ft·lb)

Table 6 Typical properties of AZ61A-F extrusions at various temperatures

Temperature °C	°F	Tensile strength MPa	ksi	Yield strength MPa	ksi	Elongation(a), %
−184	−300	379	55.0	317	46.0	4
−129	−200	355	51.5	296	43.0	6.5
−73	−100	331	48.0	265	38.5	9.5
−18	0	317	46.0	238	34.5	13
21	70	310	45.0	228	33.0	16
93	200	286	41.5	179	26.0	23
149	300	217	31.5	134	19.5	32
204	400	145	21.0	97	14.0	48.5
316	600	52	7.5	34	5.0	70

(a) In 50 mm or 2 in.

Mass Characteristics

Density. 1.8 Mg/m³ (0.065 lb/in.³) at 20 °C (68 °F)

Thermal Properties

Liquidus temperature. 620 °C (1145 °F)
Solidus temperature. 525 °C (975 °F)
Incipient melting temperature. 418 °C (785 °F)
Coefficient of thermal expansion. Linear, 26 μm/m·K (14 μin./in.·°F) at 20 °C (68 °F)

Specific heat. 1.05 kJ/kg·K (0.25 Btu/lb·°F) at 25 °C (78 °F)
Latent heat of fusion. 373 kJ/kg (160 Btu/lb)
Thermal conductivity. 80 W/m·K (46 Btu/ft·h·°F)

Electrical Properties

Electrical conductivity. 11.6% IACS at 20 °C (68 °F)
Electrical resistivity. 125 nΩ·m at 20 °C (68 °F)
Electrolytic solution potential. 1.58 V vs saturated calomel electrode

Fabrication Characteristics

Weldability. Gas-shielded arc welding with AZ61A or AZ92A rod (AZ61A preferred), good; stress relief required. Resistance welding, excellent
Recrystallization temperature. Recrystallizes after 1 h at 288 °C (550 °F) following 20% cold work
Annealing temperature. 345 °C (650 °F)
Hot working temperature. 230 to 400 °C (450 to 750 °F)
Hot-shortness temperature. 415 °C (780 °F)

Table 5 Typical room-temperature mechanical properties of AZ61A-F

Form and condition	Tensile strength MPa	ksi	Tensile yield strength(a) MPa	ksi	Elonga-tion(b), %	Hardness HBN(c)	HBE	Shear strength MPa	ksi	Compressive yield strength(a) MPa	ksi	Ultimate bearing strength(d) MPa	ksi	Bearing yield strength(d) MPa	ksi
Forgings	295	43	180	26	12	55	66	145	21	125	18	...	...	...	...
Extruded bar, rod and shapes ...	305	44	205	30	16	60	72	140	20	130	19	470	68	285	41
Extruded tubing and hollow shapes	285	41	165	24	14	50	60	...	...	110	16	...	...	...	...
Sheet	305	44	220	32	8	...	...	...	...	150	22	...	...	...	...

(a) At 0.2% offset. (b) In 50 mm or 2 in. (c) 500-kg load, 10-mm ball. (d) 3/16-in. pin diameter.

AZ80A

Specifications

AMS. Forgings: 4360
ASTM. Extruded rod, bar and shapes: B107. Forgings: B91
SAE. J466. Former SAE alloy numbers: 523 (extrusions) and 532 (forgings)
UNS number. M11800
Government. Extruded bar, rod and shapes: QQ-M-31B. Extruded tubing: WW-T-825. Forgings: QQ-M-40B

Chemical Composition

Composition limits. 7.8 to 9.2 Al; 0.20 to 0.80 Zn; 0.12 min Mn; 0.10 max Si; 0.05 max Cu; 0.005 max Ni; 0.005 max Fe; 0.30 max others (total); rem Mg
Consequence of exceeding impurity limits. Excessive Si, Cu, Ni or Fe degrades corrosion resistance.

Applications

Typical uses. Extruded products and press forgings. This alloy can be heat treated.

Mechanical Properties

Tensile properties. See Tables 7 and 8.
Shear strength. See Table 7.
Compressive yield strength. See Table 7.
Bearing properties. See Table 7.
Hardness. See Table 7.
Poisson's ratio. 0.35
Elastic modulus. Tension, 45 GPa (6.5 × 10⁶ psi); shear, 17 GPa (2.4 × 10⁶ psi)

Mass Characteristics

Density. 1.8 Mg/m³ (0.065 lb/in.³) at 20 °C (68 °F)

Thermal Properties

Liquidus temperature. 610 °C (1130 °F)
Solidus temperature. 490 °C (915 °F)
Incipient melting temperature. 427 °C (800 °F)
Coefficient of thermal expansion. Linear, 26 μm/m·K (14 μin./in.·°F) at 20 °C (68 °F)
Specific heat. 1.05 kJ/kg·K (0.25 Btu/lb·°F) at 25 °C (78 °F)

Thermal conductivity. 78 W/m·K (44 Btu/ft·h·°F) at 100 to 300 °C (212 to 572 °F)

Electrical Properties

Electrical conductivity. Extruded condition, 10.6% IACS at 20 °C (68 °F)
Electrical resistivity. 145 nΩ·m at 20 °C (68 °F)
Electrolytic solution potential. 1.57 V vs saturated calomel electrode

Fabrication Characteristics

Weldability. Gas-shielded arc welding with AZ61A or AZ92A rod (AZ61A preferred), good; stress relief required. Resistance welding, excellent
Recrystallization temperature. Recrystallizes after 1 h at 345 °C (650 °F) following 10% cold work
Annealing temperature. 385 °C (725 °F)
Hot working temperature. 320 to 400 °C (600 to 700 °F)
Hot-shortness temperature. 420 °C (775 °F)

Table 7 Typical room-temperature mechanical properties of AZ80A

Form and condition	Tensile strength MPa	ksi	Tensile yield strength(a) MPa	ksi	Elongation (b), %	Hardness HBN(c)	HBE	Shear strength MPa	ksi	Compressive yield strength MPa	ksi	Ultimate bearing strength MPa	ksi	Bearing yield strength MPa	ksi
Forgings:															
as forged	330	48	230	33	11	69	80	150	22	170	25	...	...	...	...
aged (T5 temper)	345	50	250	36	6	72	82	160	23	195	28	...	...	...	...
Bar, rod and shapes:															
as extruded	340	49	250	36	11	67	77	150	22	...	...	550	80	350	51
aged (T5 temper)	380	55	275	40	7	80	88	165	24	240	35	...	...	...	...

(a) At 0.2% offset. (b) In 50 mm or 2 in. (c) 500-kg load, 10-mm ball.

Table 8 Typical mechanical properties of AZ80A-F at various temperatures

Testing temperature °C	°F	Tensile strength MPa	ksi	Yield strength MPa	ksi	Elongation(a), %
−73	−100	386	56.0	269	39.0	8.5
18	0	355	51.5	252	36.5	10.5
21	70	338	49.0	248	36.0	11.0
93	200	307	44.5	221	32.0	18.0
149	300	241	35.0	176	25.5	25.5
204	400	197	28.5	121	17.5	35.0
260	500	110	16.0	76	11.0	57.0

(a) In 50 mm or 2 in.

Table 10 Thermal conductivity of HK31A sheet and plate at various temperatures

Testing temperature °C	°F	Thermal conductivity W/m·K	Btu/ft·h·°F
H24 Temper			
18	65 114		66
38	100 114		66
93	200 119		69
149	300 123		71
204	400 128		74
260	500 132		76
O Temper			
18	65 107		62
38	100 107		62
93	200 110		64
149	300 114		66
204	400 119		69
260	500 123		71

HK31A

See also cast alloy HK31A.

Specifications

AMS. Annealed sheet and plate: 4384E
ASTM. Sheet and plate: B90
SAE. J465. Former SAE alloy number: 507
UNS number. M13310
Government. Sheet and plate: MIL-M-26075

Chemical Composition

Composition limits. 2.5 to 4.0 Th; 0.4 to 1.0 Zr; 0.3 max Zn; 0.1 max Cu; 0.01 max Ni; 0.3 max others (total); rem Mg

Applications

Typical uses. Sheet and plate with excellent weldability and formability, and with high strength up to 315 °C (600 °F)

Mechanical Properties

Tensile properties. Tensile strength: H24 temper, 260 MPa (38 ksi); O temper, 230 MPa (33 ksi). Yield strength: H24 temper, 205 MPa (30 ksi); O temper, 140 MPa (20 ksi). Elongation in 50 mm or 2 in.: O temper, 23%; H24 temper, 9%

Tensile properties vs temperature. See Table 9 and Fig. 1 and 2.

Compressive yield strength. O temper: 97 MPa (14 ksi) at 21 °C (70 °F). H24 temper: 160 MPa (23 ksi) at 21 and 149 °C (70 and 300 °F); 150 MPa (22 ksi) at 204 °C (400 °F). See also Fig. 1 and 2.

Bearing properties. H24 temper: ultimate bearing strength, 420 MPa (61 ksi); bearing yield strength, 285 MPa (41 ksi)

Hardness. H24 temper, 68 HRE; O temper, 55 HRE

Poisson's ratio. 0.35

Elastic modulus. Tension, 45 GPa (6.5×10^6 psi); shear, 17 GPa (2.4×10^6 psi)

Impact strength. Charpy V-notch, at 20 °C (68 °F): H24 temper, 4.1 J (3.0 ft·lb); O temper, 5.4 J (4.0 ft·lb)

Creep characteristics. See Fig. 3 and 4.

Mass Characteristics

Density. 1.79 Mg/m³ (0.065 lb/in.³) at 20 °C (68 °F)

Thermal Properties

Liquidus temperature. 650 °C (1200 °F)

Solidus temperature. 590 °C (1090 °F)

Incipient melting temperature. 627 to 632 °C (1160 to 1170 °F) in circulating air

Specific heat vs temperature. $C_p = 1374 + 0.0002306T + 3370T^{-2}$

Latent heat of fusion. 318 to 335 kJ/kg (137 to 144 Btu/lb)

Thermal conductivity. See Table 10.

Electrical Properties

Electrical resistivity. At 20 °C (68 °F): H24 temper, 61 nΩ·m; O temper, 60 nΩ·m

Fabrication Characteristics

Weldability. Gas-shielded arc welding with HK31A or EZ33A rod (EZ33A preferred), excellent; stress relief can be used for sheet and plate, but is not required. Resistance welding, excellent.

Fig. 1 Typical stress-strain curves for HK31A-H24 sheet 1.63 mm (0.064 in.) thick

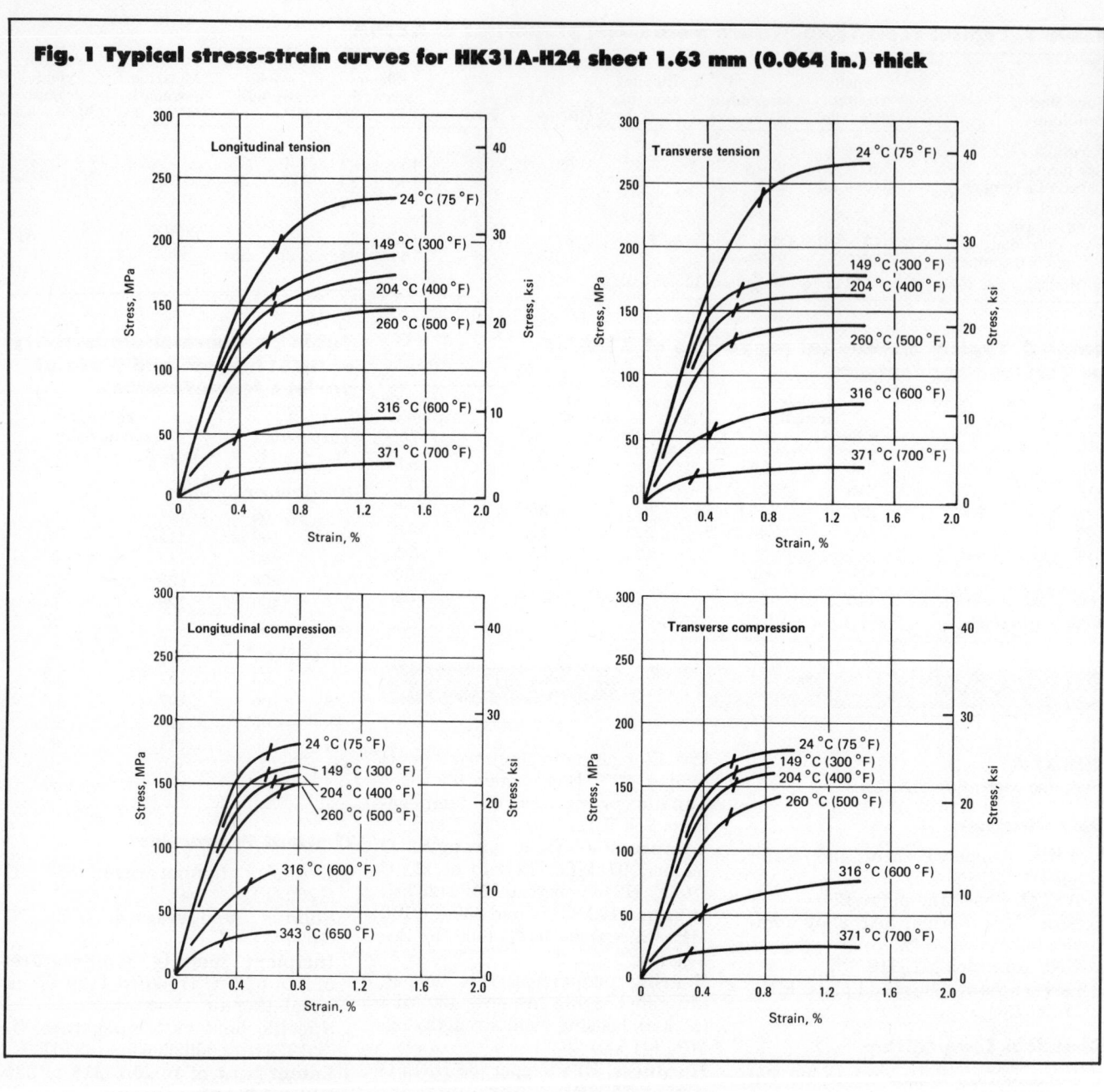

Table 9 Typical tensile properties of HK31A-H24 sheet at elevated temperatures

Testing temperature		Tensile strength		Yield strength		Elongation(a),
°C	°F	MPa	ksi	MPa	ksi	%
21	70	260	38	205	30	8
149	300	180	26	165	24	20
204	400	165	24	145	21	21
260	500	140	20	115	17	19
316	600	89	13	48	7	70
343	650	55	8	28	4	>100

(a) In 50 mm or 2 in.

Fig. 2 Typical stress-strain curves for HK31A-0 sheet 1.63 mm (0.064 in.) thick

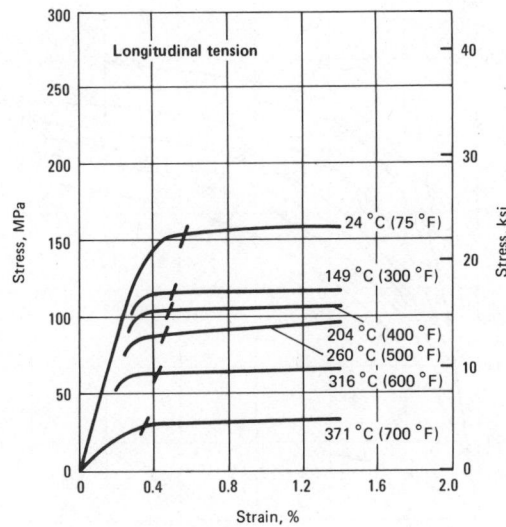

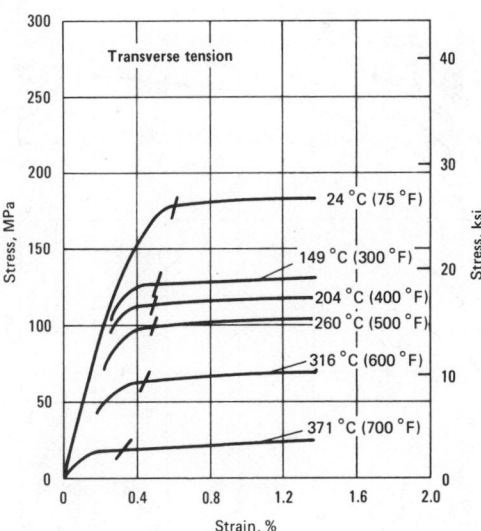

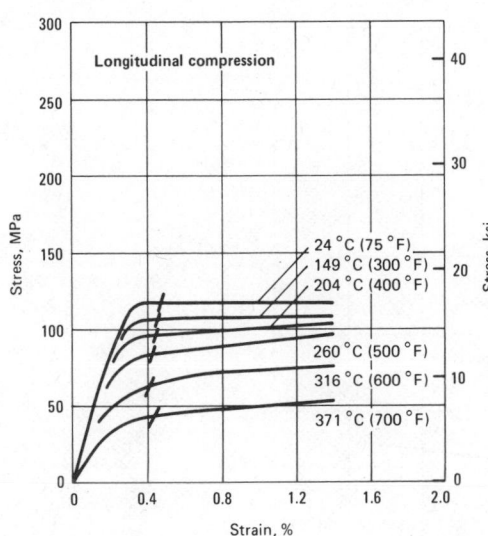

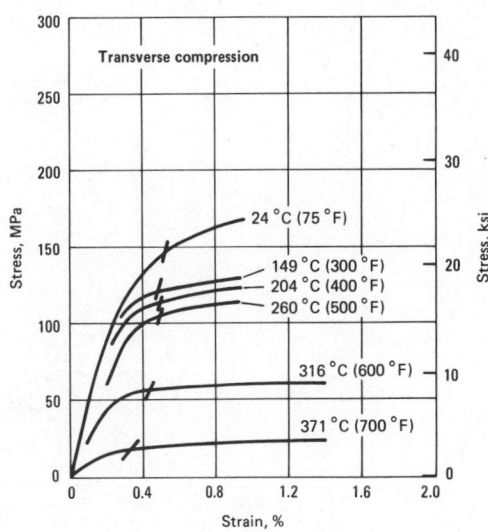

Fig. 3 Isochronous stress-strain curves for HK31A-H24 sheet 1.63 mm (0.064 in.) thick

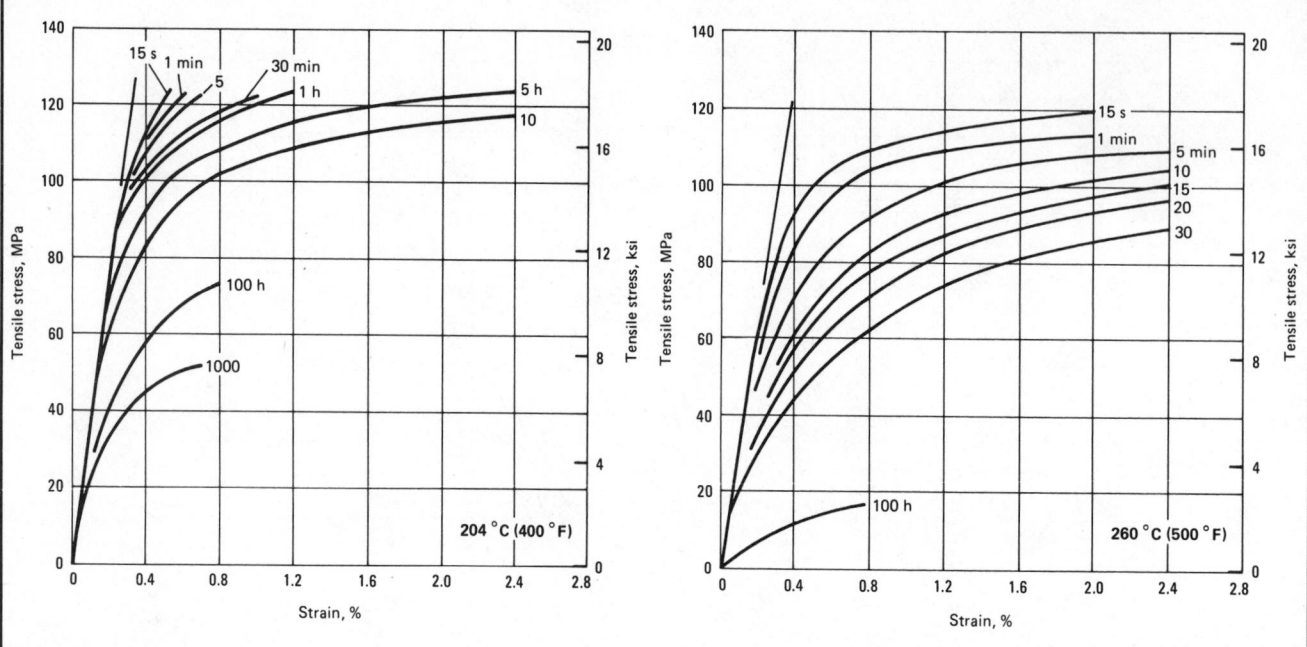

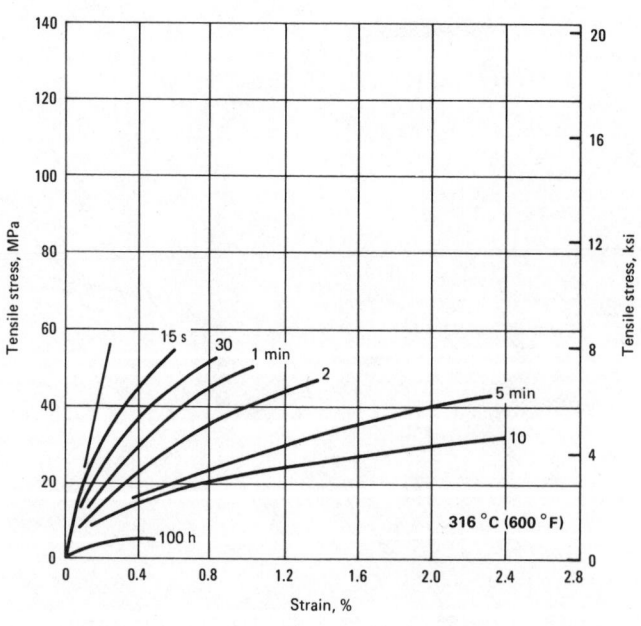

Specimens exposed at testing temperatures for 3 h before loading.

Fig. 4 Isochronous stress-strain curves for HK31A-0 sheet 1.63 mm (0.064 in.) thick

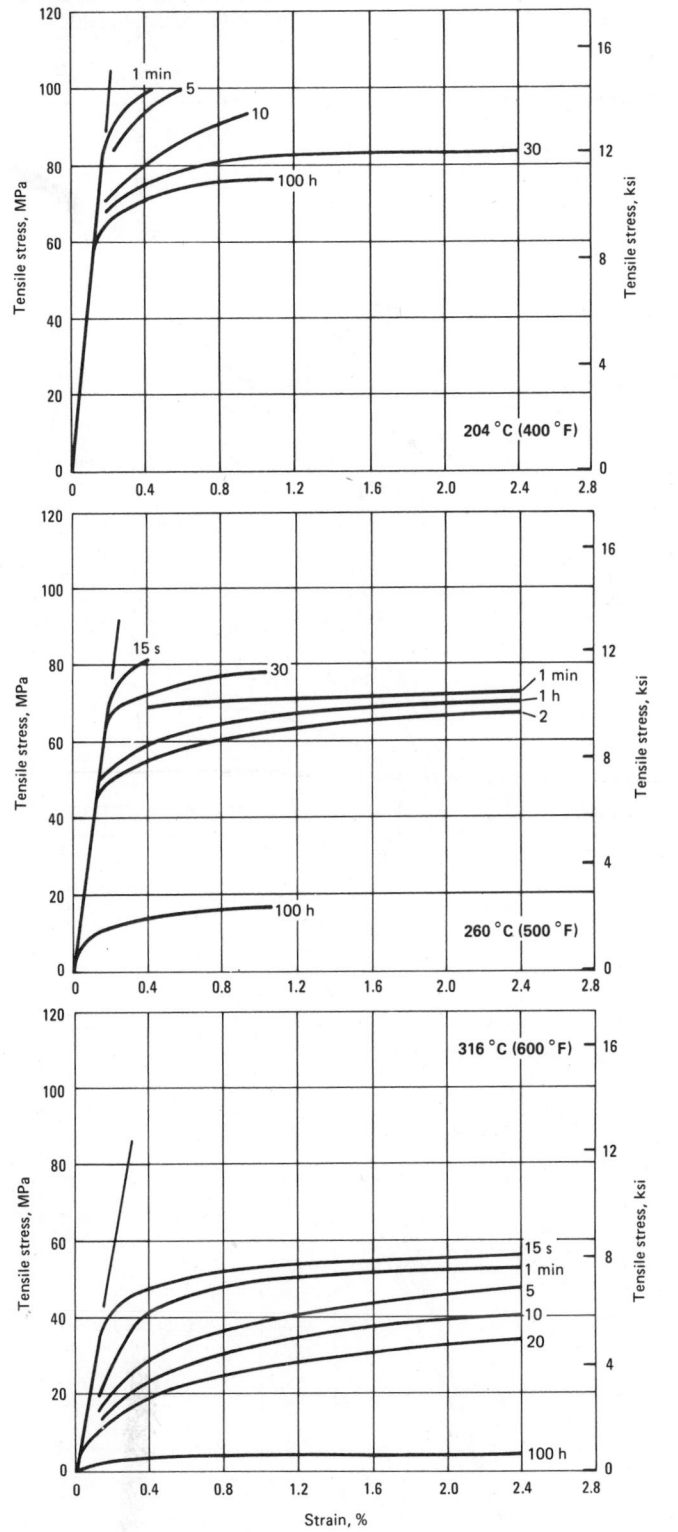

Specimens exposed at testing temperatures for 3 h before loading.

HM21A

Specifications

AMS. Sheet and plate: 4390. Forgings: 4363
ASTM. Sheet and plate: B90. Forgings: B91
UNS number. M13210
Government. Sheet and plate: MIL-M-8917. Forgings: QQ-M-40

Chemical Composition

Composition limits. 1.5 to 2.5 Th; 0.45 to 1.1 Mn; 0.30 max others (total); rem Mg

Applications

Typical uses. Sheet, plate and forgings in solution heat treated, cold worked and annealed condition (T8 temper), usable to 343 °C (650 °F) and above

Mechanical Properties

Tensile properties. T8 temper: tensile strength, 235 MPa (34 ksi); yield strength at 0.2% offset, 170 MPa (25 ksi)
Shear strength. 125 MPa (18 ksi)
Compressive yield strength. 130 MPa (19 ksi)
Tensile and compressive properties vs temperature. See Table 11 and Fig. 5.
Bearing properties. Ultimate bearing strength, 415 MPa (60 ksi); bearing yield strength, 270 MPa (39 ksi)
Poisson's ratio. 0.35
Elastic modulus. Tension, 45 GPa (6.5×10^6 psi); shear, 17 GPa (2.4×10^6 psi)
Creep characteristics. See Table 12 and Fig. 6.

Mass Characteristics

Density. 1.78 Mg/m³ (0.064 lb/in.³) at 20 °C (68 °F)

Thermal Properties

Liquidus temperature. 650 °C (1200 °F)
Solidus temperature. 605 °C (1120 °F)
Specific heat vs temperature. $C_p = 0.1412 + 0.0002294T + 3068T^{-2}$
Latent heat of fusion. 343 kJ/kg
Thermal conductivity. H24 temper, 134 W/m·K (77 Btu/ft·h·°F); O temper, 138 W/m·K (80 Btu/ft·h·°F)

Electrical Properties

Electrical resistivity. At 20 °C (68 °F): H24 temper, 52 nΩ·m; O temper, 50 nΩ·m

Fabrication Characteristics

Weldability. Gas-shielded arc welding with EZ33A rod, excellent; resistance welding, very good

Annealing temperature. 455 °C (850 °F)

Hot working temperature. 455 to 595 °C (850 to 1100 °F)

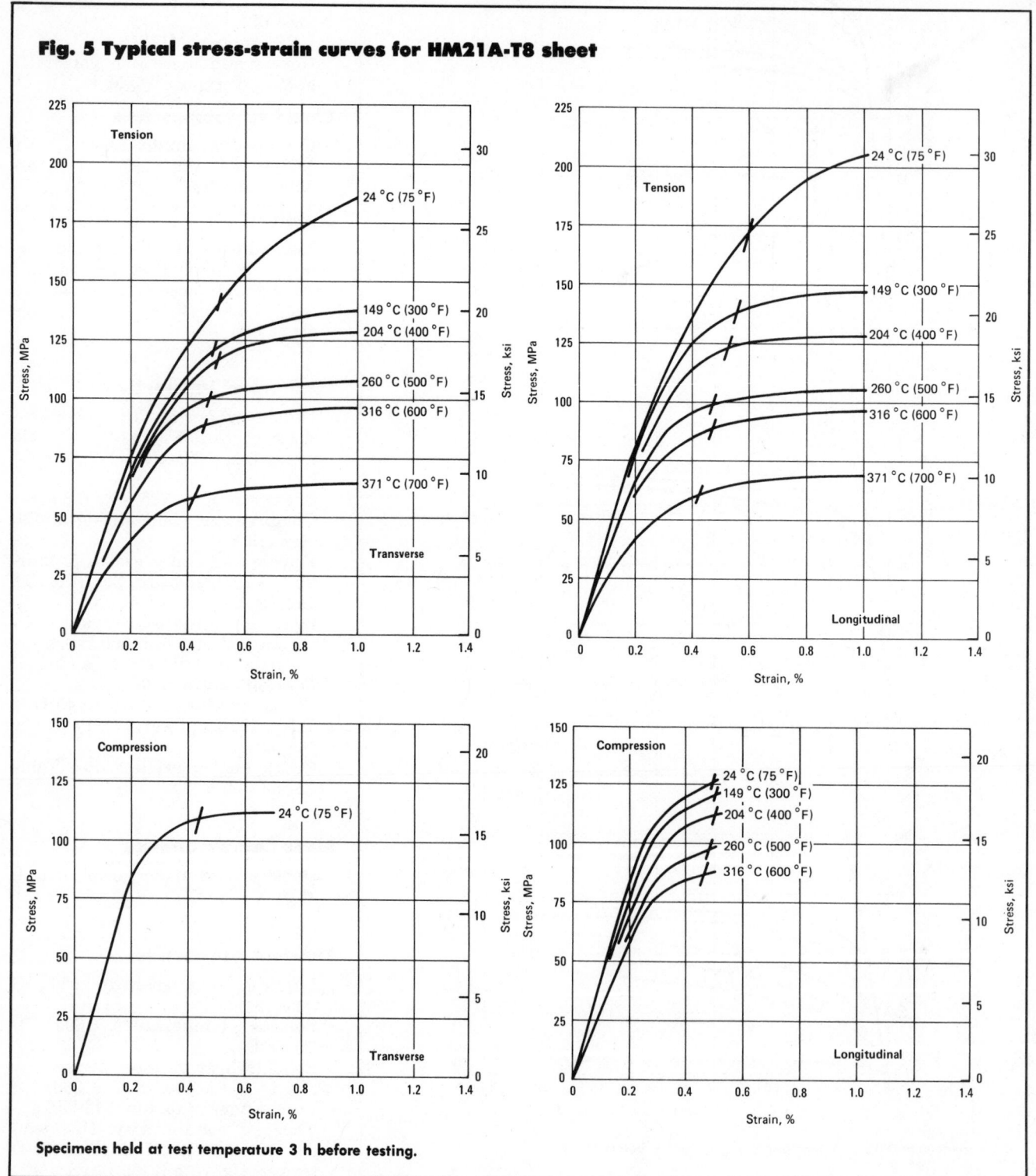

Fig. 5 Typical stress-strain curves for HM21A-T8 sheet

Specimens held at test temperature 3 h before testing.

Fig. 6 Isochronous stress-strain curves for HM21A-T8 sheet

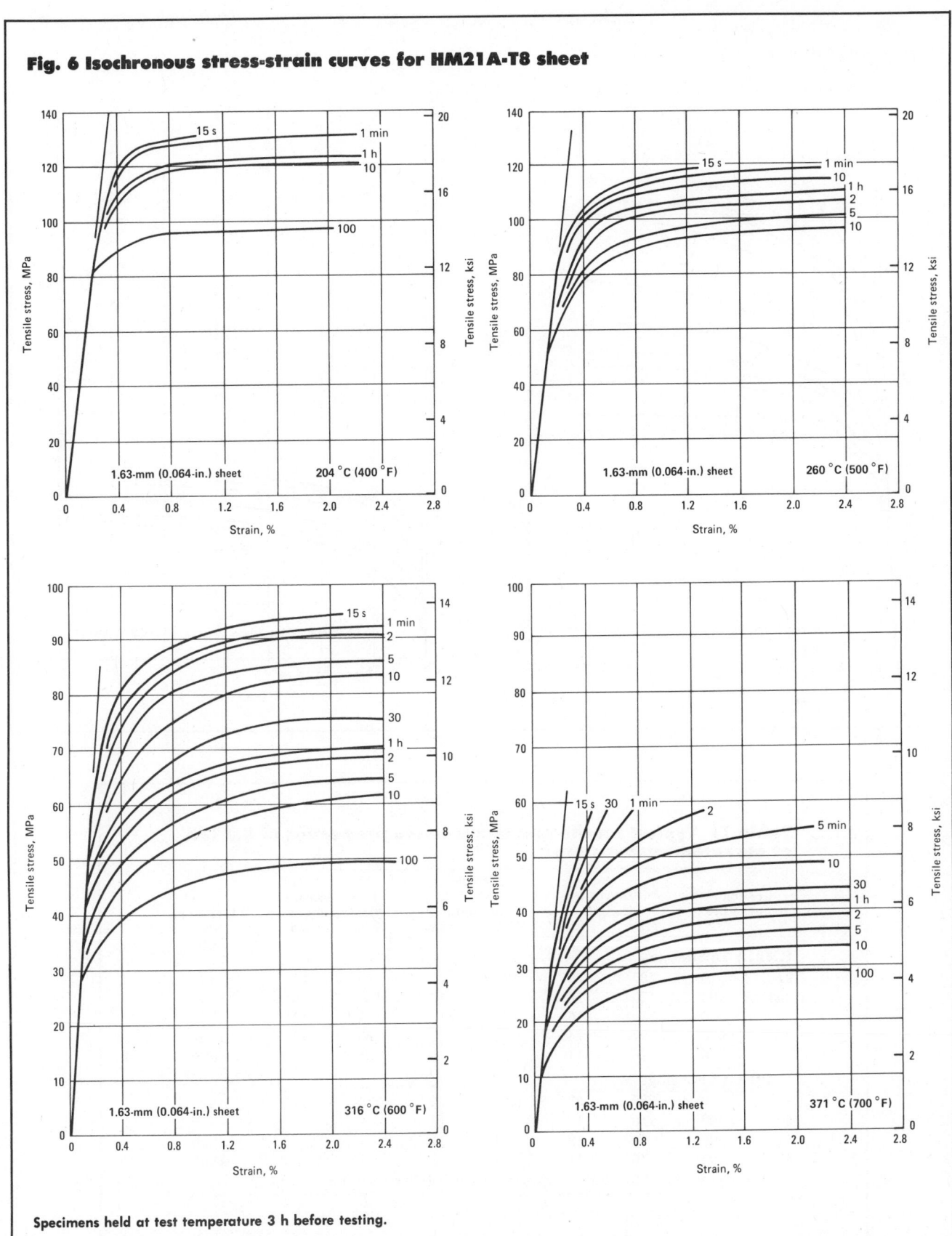

Specimens held at test temperature 3 h before testing.

Fig. 6 (continued)

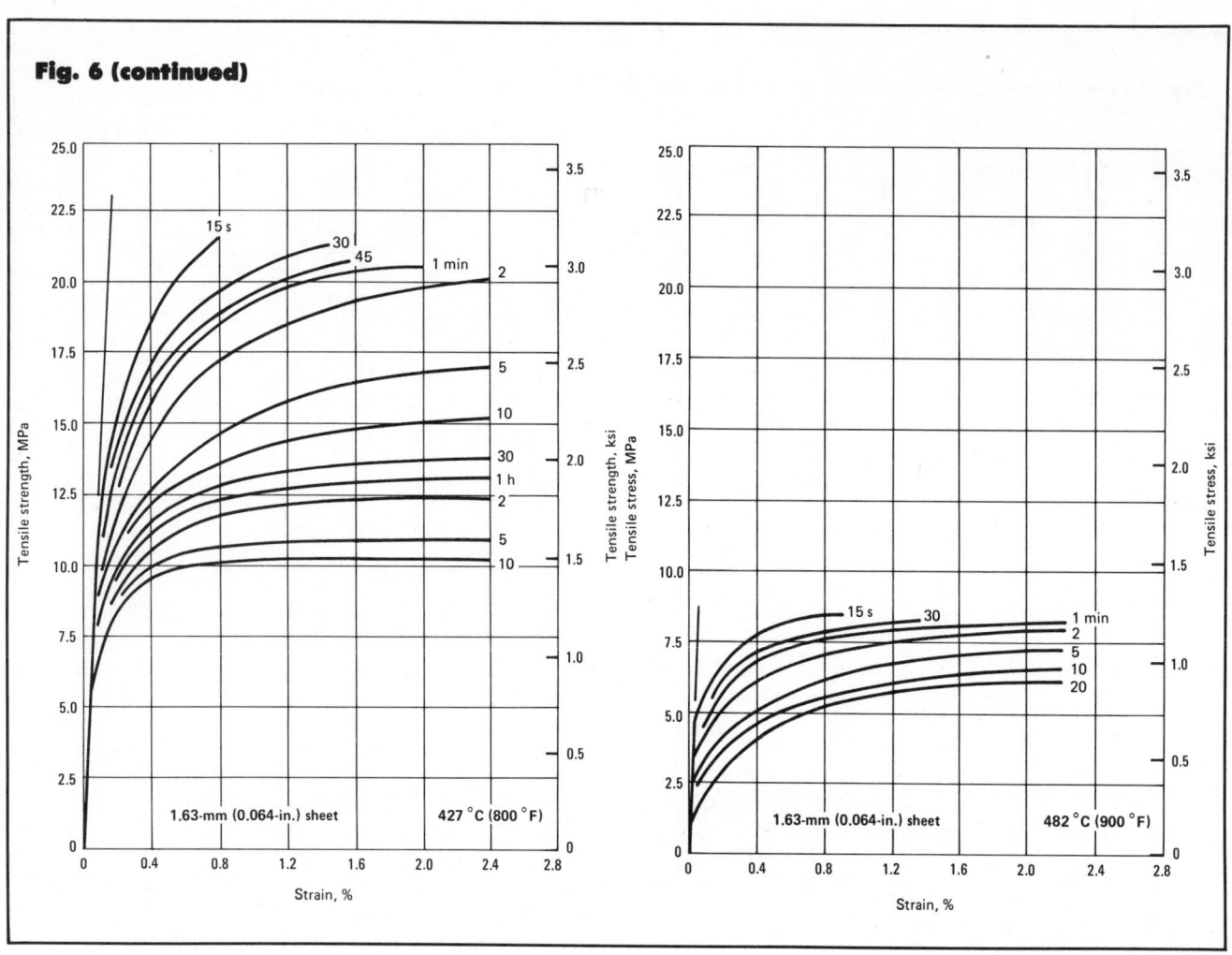

Table 11 Typical tensile and compressive properties of HM21A at elevated temperatures

Testing temperature °C	°F	Tensile strength MPa	ksi	Tensile yield strength MPa	ksi	Compressive yield strength MPa	ksi	Elongation, %
HM21A-T8 Sheet(a)								
21	70	235	34	170	25	130	19	8(b)
204	400	125	18	115	17	105	15	30(b)
260	500	110	16	105	15	105	15	25(b)
316	600	97	14	83	12	83	12	15(b)
370	700	76	11	55	8	55	8	50(b)
HM21A-T5 Forgings(c)								
21	70	230	33	140	20	115	17	15
204	400	110	16	90	13	...	...	49(d)
316	600	90	13	76	11	...	...	37(d)
370	700	76	11	55	8	...	...	43(d)

(a) In 50 mm or 2 in. (b) Under 100-h exposure. (c) Rapid heating. (d) In 1 in.

Table 12 Typical creep properties of HM21A-T8 sheet

Testing temperature		Stress to produce, in 100 h extension of					
		0.1% (creep)		0.2% (total)		0.5% (total)	
°C	°F	MPa	ksi	MPa	ksi	MPa	ksi
149	300	105	14.9	80	11.5	110	15.6
204	400	92	13.3	72	10.5	93	13.5
260	500	55	8.0	48	7.0	62	9.0
316	600	34	5.0	34	5.0	41	6.0
370	700	16	2.3	18	2.6	24	3.5

HM31A

Specifications

AMS. As extruded: 4388. Extruded and aged: 4389
SAE. J466
UNS number. M13312
Government. MIL-M-8916

Chemical Composition

Composition limits. 2.5 to 3.5 Th; 1.2 min Mn; 0.30 max others (total); rem Mg

Applications

Typical uses. Weldable alloy developed primarily for elevated-temperature structural service in the form of extruded bar, rod, shapes and tubing. Exposure to temperatures up to 316 °C (600 °F) for 1000 h causes virtually no change in short-time room- and elevated-temperature properties. Superior elastic modulus, particularly at elevated temperatures. Although certain extruded sections develop optimum properties in the as-extruded (F) temper, other sections require aging to the T5 temper.

Mechanical Properties

Tensile properties. See Table 13 and Fig. 7.
Shear strength. Punch, 150 MPa (22 ksi) at 21 °C (70 °F)
Compressive yield strength. See Table 13.
Bearing properties. At 21 °C (70 °F): ultimate bearing strength, 480 MPa (70 ksi); bearing yield strength, 345 MPa (50 ksi)
Poisson's ratio. 0.35
Elastic modulus. Tension: 45 GPa (6.5 × 10⁶ psi) at 21 °C (70 °F); 42 GPa (6.1 × 10⁶ psi) at 149 °C (300 °F); 40 GPa (5.9 × 10⁶ psi) at 204 °C (400 °F); 39 GPa (5.6 × 10⁶ psi) at 316 °C (600 °F). Shear: 17 GPa (2.4 × 10⁶ psi) at 21 °C (70 °F)
Creep characteristics. See Table 14.

Mass Characteristics

Density. 1.81 Mg/m³ (0.065 lb/in.³)

Thermal Properties

Liquidus temperature. 650 °C (1200 °F)
Solidus temperature. 605 °C (1120 °F)
Incipient melting temperature. 482 °C (900 °F)
Coefficient of thermal expansion. Linear: 26 μm/m·K (14.5 μin./in.·°F) at 20 to 93 °C (68 to 200 °F); 28 μm/ m·K (15.6 μin./in.·°F) at 20 to 316 °C (68 to 600 °F); 30 μm/m·K (16.8 μin./ in.·°F) at 20 to 538 °C (68 to 1000 °F)
Specific heat vs temperature. $Cp = 0.0982 + 0.0002894T + 5300T^{-2}$
Latent heat of fusion. 331 kJ/kg
Thermal conductivity. 104 W/m·K (60 Btu/ft·h·°F)

Electrical Properties

Electrical conductivity. F temper at 20 °C (68 °F): volumetric, 26% IACS; mass, 135% IACS
Electrical resistivity. F temper: 66 nΩ·m at 20 °C (68 °F); 79 nΩ·m at 93 °C (200 °F); 97 nΩ·m at 204 °C (400 °F); 115 nΩ·m at 316 °C (600 °F)
Temperature coefficient of electrical resistivity. 0.18 nΩ·m per K

Fabrication Characteristics

Weldability. Gas-shielded arc welding with EZ33A rod, excellent; no stress relief is necessary. Resistance welding, very good
Recrystallization temperature. Recrystallizes after 1 h at 398 °C (750 °F) following 50% cold work
Hot working temperature. 370 to 538 °C (700 to 1000 °F)

Table 13 Typical tensile and compressive properties of HM31A extrusions up to 4 in.² in area

Testing temperature		Tensile strength		Tensile yield strength		Compressive yield strength		Elongation(a),
°C	°F	MPa	ksi	MPa	ksi	MPa	ksi	%
21	70	283	41	230	33	165	24	10
149	300	195	28	180	26	170	25	30
204	400	165	24	160	23	160	23	32
260	500	145	21	140	20	140	20	25
316	600	115	17	110	16	110	16	22
370	700	90	13	83	12	...	...	35
426	800	55	8	48	7	...	...	60
482	900	14	2	7	1	...	...	100

(a) In 50 mm or 2 in.

Table 14 Typical creep properties of HM31A extrusions

Testing temperature		Stress to produce, in 100 h, extension of					
		0.1% (creep)		0.2% (total)		0.5% (total)	
°C	°F	MPa	ksi	MPa	ksi	MPa	ksi
204	400	110	16	83	12	115	17
260	500	76	11	69	10	83	12
316	600	41	6	41	6	48	7

Fig. 7 Typical stress-strain curves for HM31A extrusions

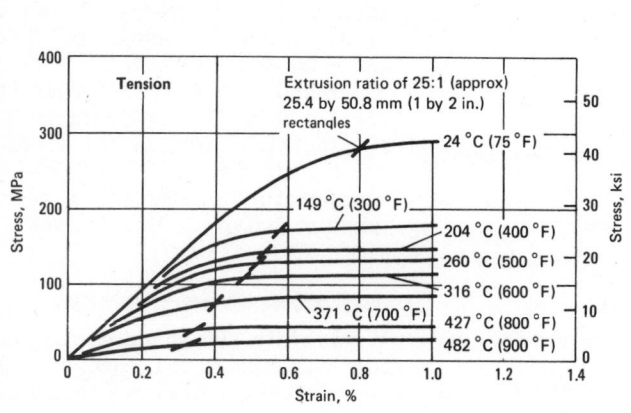

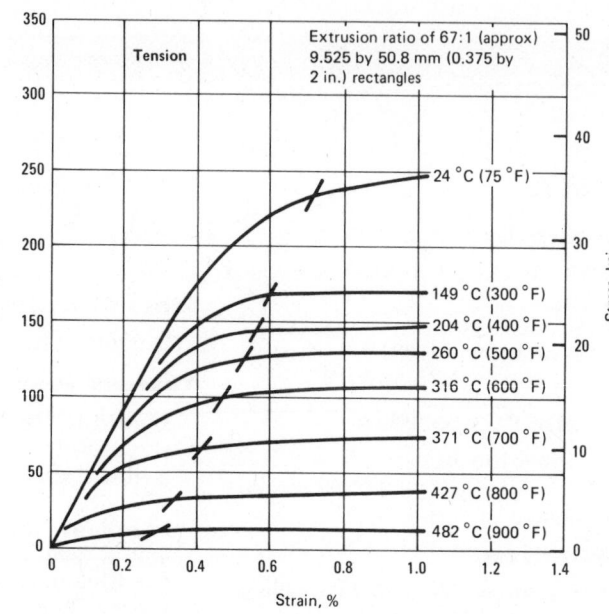

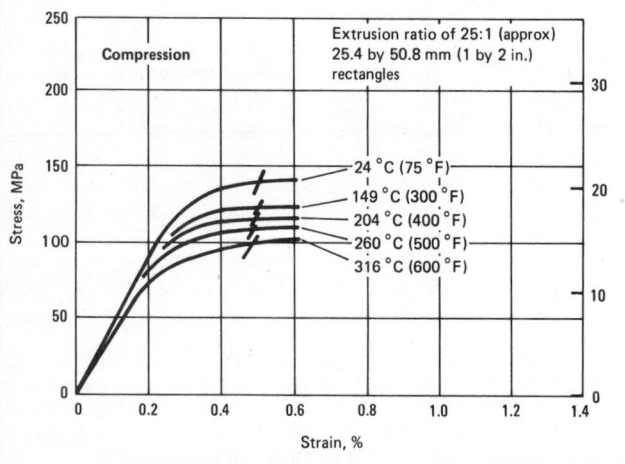

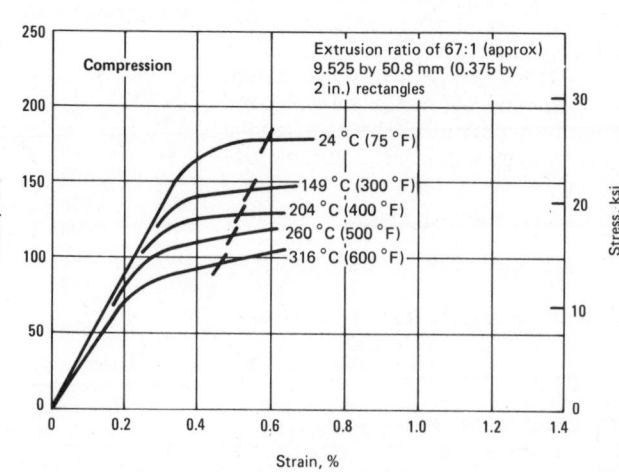

Tested in the longitudinal direction.

Table 15 Typical room-temperature mechanical properties of M1A

Product form	Tensile strength MPa	ksi	Tensile yield strength(a) MPa	ksi	Elonga- tion(b), %	Hardness HBN(c)	HBE	Shear strength MPa	ksi	Compressive yield strength(a) MPa	ksi	Ultimate bearing strength(d) MPa	ksi	Bearing yield strength(d) MPa	ksi
Sheet, annealed	230	33	125	18	17	48	55	115	17	76	11	350	51	200	29
Sheet, hard rolled	240	35	180	26	7	54	65	115	17	125	18	395	57	270	39
Extruded bar and shapes	255	37	180	26	12	44	45	125	18	83	12	350	51	195	28
Extruded tubing and hollow shapes ...	240	35	145	21	9	42	41	...	...	62	9	...	...	...	...
Forgings	250	36	160	23	7	47	54	110	16	...	...	...	...	...	...

(a) At 0.2% offset. (b) In 50 mm or 2 in. (c) 500-kg load, 10-mm ball. (d) 3/16-in. pin diameter.

M1A

Specifications

ASTM. Extruded rod, bar, shapes and tubing: B107
SAE. J466. Former SAE alloy numbers: 522 (extrusions) and 533 (forgings)
UNS number. M15100
Government. Extruded bar, rod and shapes: QQ-M-31. Extruded tubing: WW-T-825. Forgings: QQ-M-40. Sheet and plate: QQ-M-54
Foreign. Elektron AM503. British: BS3370 MAG101. German: DIN9715 3.5200

Chemical Composition

Composition limits. 1.2 min Mn; 0.30 max Ca; 0.05 max Cu; 0.01 max Ni; 0.10 max Si; 0.30 max others (total); rem Mg
Consequence of exceeding impurity limits. Excessive Si tends to precipitate Mn. Excessive Cu or Ni degrades corrosion resistance in salt water.

Applications

Typical uses. Wrought products with moderate mechanical properties, and excellent weldability, corrosion resistance and hot formability; not heat treatable

Mechanical Properties

Tensile properties. See Tables 15 and 16.
Shear strength. See Table 15.
Compressive properties. See Table 15.
Bearing properties. See Table 15.
Hardness. See Table 15.
Directional properties. See Table 17.
Poisson's ratio. 0.35
Elastic modulus. Tension, 45 GPa (6.5 × 10⁶ psi); shear, 17 GPa (2.4 × 10⁶ psi)

Table 16 Typical tensile properties of M1A at elevated temperatures

Testing temperature °C	°F	Tensile strength MPa	ksi	Yield strength MPa	ksi	Elonga- tion, %
Bar and Shapes, Extruded						
93	200	186	27.0	145	21.0	16
121	250	165	24.0	131	19.0	18
149	300	145	21.0	110	16.0	21
204	400	117	17.0	83	12.0	27
316	600	62	9.0	34	5.0	53
Sheet, Annealed						
93	200	169	24.5	110	16.0	31
121	250	148	21.5	100	14.5	41
149	300	133	19.3	86	12.5	44
Sheet, Hard Rolled						
93	200	203	29.5	183	26.5	11
121	250	190	27.5	169	24.5	13
149	300	172	25.0	145	21.0	15
Forgings						
93	200	165	24.0	121	17.5	25
121	250	145	21.0	107	15.5	26
149	300	131	19.0	93	13.5	31
204	400	114	16.5	69	10.0	34
260	500	83	12.0	45	6.5	67
316	600	41	6.0	28	4.0	140

Mass Characteristics

Density. 1.76 Mg/m³ (0.064 lb/in.³) at 20 °C (68 °F)

Thermal Properties

Liquidus temperature. 649 °C (1200 °F)
Solidus temperature. 648 °C (1198 °F)
Coefficient of thermal expansion. Linear, 26 μm/m·K (14 μin./in.·°F) at 20 to 100 °C (68 to 212 °F)
Specific heat. 1.05 kJ/kg·K (0.25 Btu/1b·°F)
Latent heat of fusion. 373 kJ/kg (160 Btu/lb)
Thermal conductivity. 138 W/m·K (79.8 Btu/ft·h·°F)

Electrical Properties

Electrical conductivity. 34.5% IACS at 20 °C (68 °F)

Electrical resistivity. 50 nΩ·m at 20 °C (68 °F)
Electrolytic solution potential. 1.64 V vs saturated calomel electrode

Fabrication Characteristics

Weldability. Gas-shielded arc welding with AZ61A, AZ92A or M1A rod (AZ61A preferred), excellent; stress relief not required, but may be used. Resistance welding, good. Oxyacetylene welding, if necessary, can be done with M1A rod, magnesium flux and neutral flame.
Recrystallization temperature. Recrystallizes after 1 h at 260 °C (500 °F) following 20% cold work
Annealing temperature. 370 °C (700 °F)
Hot working temperature. 295 to 540 °C (560 to 1000 °F)

Table 17 Typical directional properties of M1A sheet

Condition	Tensile strength MPa	ksi	Yield strength MPa	ksi	Elongation, %
Parallel to Rolling Direction					
Annealed	230	33	125	18	17
Hard rolled	250	36	180	26	7
Perpendicular to Rolling Direction					
Annealed	220	32	115	17	17
Hard rolled	255	37	185	27	13

ZK21A

Specifications

AMS. Extruded tubes, bars, rods and shapes: 4387
UNS. M16210
Government. Extrusions: MIL-M-46039

Chemical Composition

Composition limits. 2.0 to 2.6 Zn, 0.45 to 0.8 Zr, 0.3 max impurities (total), rem Mg

Applications

Typical uses. Moderate-strength extrusion alloy with good weldability. Stress relief is not required. Used in as-extruded (F) temper.

Mechanical Properties

Tensile properties. See Table 18.
Compressive properties. See Table 18.

Fabrication Characteristics

Weldability. Gas shielded arc welding with AZ61A or AZ92A rod is satisfactory. Resistance welding is satisfactory.

Table 18 Minimum mechanical properties at room temperature of ZK21A-F extrusions

Form	Tensile strength MPa	ksi	Yield strength MPa	ksi	Compressive yield strength MPa	ksi	Elongation, %
Rods, bars and shapes	260	38	195	28	135	20	4
Tubing	235	34	180	26	97	14	4

Table 19 Minimum mechanical properties of ZK40A-T5 at room temperature

Form	Tensile strength MPa	ksi	Yield strength MPa	ksi	Elongation, %	Compressive yield strength MPa	ksi
Extruded bars and shapes	276	40	255	37	4	140	20
Extruded tubes	276	40	248	36	4	140	20

ZK40A

Specifications

ASTM. Extrusions: B107
UNS. M16400
Foreign. Canadian, CSA HG.5 ZK40A

Chemical Composition

Composition limits. 3.5 to 4.5 Zn; 0.45 min Zr; 0.30 max others (total); rem Mg

Applications

Typical uses. High yield strength extrusion alloy, available in as extruded (F) and artificially aged (T5) tempers. Not as sensitive to stress concentration at thread roots as other high strength alloys. Can be heat treated. Can replace ZK60A, especially for diamond drill rod, and is more readily extruded.

Mechanical Properties

Tensile properties. See Table 19.
Poisson's ratio. 0.35
Elastic modulus. Tension, 45 GPa $(6.5 \times 10^6$ psi); shear, 17 GPa $(2.4 \times 10^6$ psi)

Mass Characteristics

Density. 1.82 Mg/m^3 (0.066 lb/in.^{3}) at 20 °C (68 °F)

ZK60A

Specifications

AMS. Extrusions: 4352. Forgings: 4362
ASTM. Extrusions: B107. Forgings: B91
SAE. J466. Former SAE alloy number: 524
UNS number. M16600
Government. Extruded rod, bar and shapes: QQ-M-31. Extruded tubing: WW-T-825. Forgings: QQ-M-40
Foreign. Elektron ZW6. British: BS3373 MAG161. German: DIN9715 3.5161. French: AFNOR G-Z5Zr

Chemical Composition

Composition limits. 4.8 to 6.2 Zn; 0.45 min Zr; 0.30 max others (total); rem Mg

Applications

Typical uses. Extruded products and press forgings with high strength and good ductility; can be artificially aged to T5 temper

Mechanical Properties

Tensile properties. See Table 20.
Shear strength. See Table 20.
Compressive yield strength. See Table 20.
Bearing properties. See Table 20.
Hardness. See Table 20.
Poisson's ratio. 0.35
Elastic modulus. Tension, 45 GPa (6.5 × 10⁶ psi); shear, 17 GPa (2.4 × 10⁶ psi)

Mass Characteristics

Density. 1.83 Mg/m³ (0.066 lb/in.³) at 20 °C (68 °F)

Thermal Properties

Liquidus temperature. 635 °C (1175 °F)

Solidus temperature. 520 °C (970 °F)
Incipient melting temperature. 515 °C (965 °F)
Coefficient of thermal expansion. Linear, 26 μm/m·K (14 μin./in.·°F) at 20 °C (68 °F)
Specific heat vs temperature. $C_p = 0.1233 + 0.0002566T + 3939T^{-2}$
Latent heat of fusion. 301 to 335 kJ/kg (129 to 144 Btu/lb)
Thermal conductivity. F temper 117 W/m·K (68 Btu/ft·h·°F) at 20 °C (68 °F); T5 temper, 121 W/m·K (70 Btu/ft·h·°F) at 20 °C (68 °F)

Electrical Properties

Electrical conductivity. At 20 °C

(68 °F): F temper, 29% IACS; T5 temper, 30% IACS

Fabrication Characteristics

Weldability. Gas-shielded arc welding with AZ92A welding rod is possible, but is not recommended because these alloys are prone to hot-shortness cracking; when welds free of cracks are obtained, they exhibit high weld efficiencies. Resistance welding, excellent
Aging temperature. 150 °C (300 °F) for 24 h in the air, followed by air cooling
Hot working temperature. 315 to 400 °C (600 to 750 °F)
Hot-shortness temperature. Cast, 315 °C (600 °F); wrought, 510 °C (950 °F)

Table 20 Typical mechanical properties of ZK60A at room temperature

Form and condition	Tensile strength MPa	ksi	Tensile yield strength(a) MPa	ksi	Elongation, %	Hardness HBN(b)	HBE	Shear strength MPa	ksi	Compressive yield strength MPa	ksi	Ultimate bearing strength MPa	ksi	Bearing yield strength MPa	ksi
Extruded Bars, Rod and Shapes															
ZK60A-F	340	49	260	38	11	75	84	185	27	230	33	550	80	380	55
ZK60A-T5 ...	350	51	285	41	11	82	88	180	26	250	36	585	85	405	59
Extruded Hollow Shapes and Tubing															
ZK60A-F	315	46	235	34	12	75	84	...	...	170	25	...	...	...	...
ZK60A-T5 ...	345	50	275	40	11	82	88	...	...	200	29	...	...	...	...
Forgings															
ZK60A-T5 ...	305	44	215	31	16	65	77	165	24	160	23	420	61	285	41

(a) 0.2% offset. (b) 500-kg load, 10-mm ball.

AM60A

Specifications

ASTM. Die casting: B94
UNS. M10600
Foreign. German: DIN1729 3.5662

Chemical Composition

Composition limits. 92.27 to 93.37 Mg; 5.5 to 6.5 Al; 0.13 min Mn; 0.50 max Si; 0.35 max Cu; 0.22 max Zn; 0.03 max Ni

Applications

Typical uses. Die casting alloy used

in as-cast (F) temper for production of automotive wheels and other parts requiring good elongation and toughness combined with reasonable yield and tensile properties

Mechanical Properties

Tensile properties. F temper: tensile strength, 220 MPa (32 ksi); yield strength, 130 MPa (19 ksi); elongation, 6% in 50 mm or 2 in.
Compressive properties. F temper: compressive yield strength, 130 MPa (19 ksi)
Poisson's ratio. 0.35
Elastic modulus. Tension, 45 GPa (6.5 × 10⁶ psi)

Mass Characteristics

Density. 1.79 Mg/m³ (0.065 lb/in.³) at 20 °C (68 °F)

Thermal Properties

Liquidus temperature. 615 °C (1140 °F)
Solidus temperature. 540 °C (1005 °F)
Coefficient of thermal expansion. Linear, 25.6 μm/m·K (14.2 μin./in.·°F) at 20 to 100 °C (68 to 212 °F)
Thermal conductivity. 61 W/m·K (36 Btu/ft·h·°F) at 20 °C (68 °F)

Fabrication Characteristics

Casting temperature. 650 to 695 °C (1200 to 1280 °F)
Weldability. Not weldable

AM100A

Specifications

AMS. Permanent mold castings: 4483. Investment castings: 4455
ASTM. Sand castings: B80. Ingot for sand, permanent mold and die castings: B93. Permanent mold castings: B199. Investment castings: B403
SAE. J465. Former SAE alloy number: 502
UNS number. M10100
Government. Permanent mold castings: QQ-M-55

Chemical Composition

Composition limits. 9.3 to 10.7 Al; 0.10 min Mn; 0.30 max Zn; 0.30 max Si; 0.10 max Cu; 0.01 max Ni; 0.30 max others (total); rem Mg
Consequence of exceeding impurity limits. Corrosion resistance decreases with increasing amounts of Cu, Ni and Fe. Increased amounts of Zn decreases pressure tightness. More than 0.5% Si decreases elongation.

Applications

Typical uses. Pressure-tight sand and permanent mold castings with good combination of tensile strength, yield strength and elongation

Mechanical Properties

Tensile properties. See Tables 21 and 22, and Fig. 8.
Shear strength. See Table 21.
Compressive yield strength. See Table 21.
Hardness. At room temperature: see Table 21. At -78 °C (-108 °F): F temper, 63 HB or 75 HRE; T4 temper, 60 HB or 73 HRE; T6 temper, 85 HB or 90 HRE
Bearing properties. Ultimate bearing strength: T4 temper, 475 MPa (69 ksi); T61 temper, 560 MPa (81 ksi). Bearing yield strength: T4 temper, 310 MPa (45 ksi); T61 temper, 470 MPa (68 ksi)
Poisson's ratio. 0.35
Impact strength. Charpy V-notch. At 20 °C (68 °F): F temper, 0.8 J (0.6 ft·lb); T4 temper, 2.7 J (2.0 ft·lb); T61 temper, 0.9 J (0.7 ft·lb). At -78 °C (-108 °F): F temper, 1.1 J (0.8 ft·lb); T4 temper, 3.4 J (2.5 ft·lb); T6 temper, 1.1 J (0.8 ft·lb)
Fatigue strength. At 5×10^8 cycles: F and T61 tempers, 69 MPa (10 ksi); T4 temper, 76 MPa (11 ksi) (R. R. Moore type test)

Table 21 Typical mechanical properties of AM100A sand castings at room temperature

Temper	Tensile strength MPa	ksi	Tensile or compressive yield strength(a) MPa	ksi	Elongation(b), %	Hardness HB	HRE	Shear strength MPa	ksi
F	150	22	83	12	2	53	61	125	18
T4	275	40	90	13	10	52	62	140	20
T61	275	40	150	22	1	69	80	145	21
T5	150	22	110	16	2	58	70	...	...
T7	260	38	125	18	1	67	78	...	...

(a) Values are the same for tensile and compressive yield strengths. (b) In 50 mm or 2 in.

Table 22 Typical tensile properties of AM100A sand castings at elevated and subzero temperatures

Testing temperature °C	°F	Tensile strength MPa	ksi	Tensile yield strength MPa	ksi	Elongation(a), %
F Temper						
-78	-108	150	22	125	18.0	1
T4 Temper						
-78	-108	260	38	125	18.0	7
93	200	235	34	...	...	1.5
149	300	160	23	...	...	9
260	500	83	12	...	...	22
T6 Temper(b)						
-78	-108	270	39	180	26.0	2
149	300	165	24	62	9.0	4
204	400	115	17	45	6.5	25
260	500	83	12	28	4.0	45
316	600	59	8.5	17	2.5	60
370	700	38	5.5	10	1.5	100

(a) In 50 mm or 2 in. (b) Elevated-temperature properties were determined after prolonged heating.

Elastic modulus. Tension, 45 GPa (6.5×10^6 psi); shear, 17 GPa (2.4×10^6 psi)

Mass Characteristics

Density. 1.81 Mg/m³ (0.066 lb/in.³) at 20 °C (68 °F)

Thermal Properties

Liquidus temperature. 595 °C (1100 °F)
Solidus temperature. 465 °C (865 °F)
Incipient melting temperature. 430 °C (810 °F)
Coefficient of thermal expansion. Linear, 25 µm/m·K (14 µin./in.·°F) at 18 to 100 °C (65 to 212 °F)
Specific heat. 1.05 kJ/kg·K (0.25 Btu/lb·°F) at 25 °C (77 °F)
Thermal conductivity. 73 W/m·K (42 Btu/ft·h·°F) at 100 to 300 °C (212 to 572 °F)

Latent heat of fusion. 373 kJ/kg (89 Btu/ft)

Electrical Properties

Electrical conductivity. F temper, 11.5% IACS; T4 temper, 9.9% IACS; T6 temper, 12.3% IACS
Electrical resistivity. F temper, 150 nΩ·m; T4 temper, 175 nΩ·m; T6 temper, 140 nΩ·m at 20 °C (68 °F)
Electrolytic solution potential. 1.57 V vs saturated calomel electrode
Hydrogen overvoltage. 0.27 V for extrusions; 0.06 V for castings

Fabrication Characteristics

Weldability. Gas-shielded arc welding wiht AM100A rod, very good
Casting temperatures. Sand castings, 735 to 845 °C (1350 to 1550 °F); permanent mold castings, 650 to 815 °C (1200 to 1500 °F); ingot, 650 to 705 °C (1200 to 1300 °F)

Fig. 8 Distribution of tensile properties for AM100A separately sand cast test bars

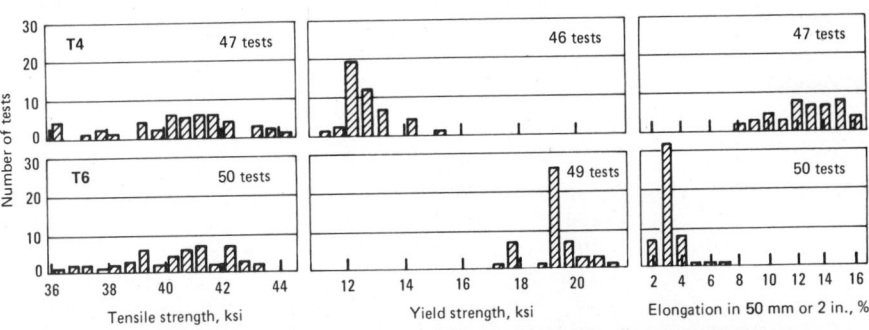

Applications

Typical uses. Sand castings with good strength, ductility and toughness

Mechanical Properties

Tensile properties. Tensile strength: F and T5 tempers, 200 MPa (29 ksi); T4, T6 and T7 tempers, 275 MPa (40 ksi). Yield strength: F and T4 tempers, 97 MPa (14 ksi); T5 temper, 105 MPa (15 ksi); T6 temper, 130 MPa (19 ksi); T7 temper, 115 MPa (17 ksi). Elongation in 50 mm or 2 in.: F and T7 tempers, 6%; T4 temper, 12%; T5 temper, 4%; T6 temper, 5%. See also Fig. 9.

Tensile properties vs temperature. See Table 23.

Shear strength. F and T4 tempers, 125 MPa (18 ksi); T5 temper, 130 MPa (19 ksi); T6 and T7 tempers, 140 MPa (20 ksi)

Compressive yield strength. F, T4 and T5 tempers, 97 MPa (14 ksi); T6 temper, 130 MPa (19 ksi); T7 temper, 115 MPa (17 ksi)

Bearing properties. Ultimate bearing strength: F, T4 and T6 tempers, 415 MPa (60 ksi); T5 temper, 455 MPa (66 ksi); T7 temper, 515 MPa (75 ksi). Bearing yield strength: F and T5 tempers, 275 MPa (40 ksi); T4 temper, 305 MPa (44 ksi); T6 temper, 360 MPa (52 ksi); T7 temper, 325 MPa (47 ksi)

Hardness. F temper, 50 HB, 59 HRE; T4 and T5 tempers, 55 HB, 66 HRE; T6 temper, 73 HB, 83 HRE; T7 temper, 64 HB, 76 HRE

Poisson's ratio. 0.35

Elastic modulus. Tension, 45 GPa (6.5×10^6 psi); shear, 17 GPa (2.4×10^6 psi)

Impact strength. Charpy V-notch: F temper, 1.4 J (1.0 ft·lb); T4 temper, 3.4 J (2.5 ft·lb); T5 temper, 3.5 J (2.6 ft·lb); T6 temper, 1.5 J (1.1 ft·lb)

Fatigue strength. At 5×10^8 cycles: F, T5 and T6 tempers, 76 MPa (11 ksi); T4 temper, 83 MPa (12 ksi); T7 temper, 115 MPa (17 ksi) (R.R. Moore type test)

Mass Characteristics

Density. 1.84 Mg/m^3 (0.066 lb/in.^{3}) at 20 °C (68 °F)

Thermal Properties

Liquidus temperature. 610 °C (1130 °F)

Solidus temperature. 455 °C (850 °F)

AS41A

Specifications

ASTM. Die castings: B94
UNS number. M10410
Foreign. German: DIN1729 3.5470

Chemical Composition

Composition limits. 3.5 to 5.0 Al; 0.50 to 1.5 Si; 0.20 to 0.50 Mn; 0.12 max Zn; 0.06 max Cu; 0.03 max Ni; 0.30 max others; rem Mg

Applications

Typical uses. Die castings used in the as-cast condition (F temper), with creep resistance superior to that of AZ91A, AZ91B or AM60A up to 175 °C (350 °F), and good tensile strength, tensile yield strength and elongation

Mechanical Properties

Tensile properties. F temper: tensile strength, 210 MPa (31 ksi); yield strength, 140 MPa (20 ksi); elongation, 6% in 50 mm or 2 in.

Compressive yield strength. F temper, 140 MPa (20 ksi)

Poisson's ratio. 0.35

Elastic modulus. Tension, 45 GPa (6.5×10^6 psi)

Mass Characteristics

Density. 1.77 Mg/m^3 (0.064 lb/in.^{3}) at 20 °C (68 °F)

Thermal Properties

Liquidus temperature. 620 °C (1150 °F)

Solidus temperature. 565 °C (1050 °F)

Coefficient of thermal expansion. Linear, 26.1 μm/m·K (14.5 μin./in.·°F) at 20 to 100 °C (68 to 212 °F)

Specific heat. 1.02 kJ/kg·K (0.242 Btu/lb·°F) at 20 °C (68 °F)

Thermal conductivity. 68 W/m·K (40 Btu/ft·h·°F) at 20 °C (68 °F)

Fabrication Characteristics

Casting temperature. 660 to 695 °C (1220 to 1280 °F)

Weldability. Not weldable

AZ63A

Specifications

AMS. Sand castings: F temper, 4420; T4 temper, 4422; T5 temper, 4424
ASTM. Ingot: B93. Sand castings: B80
SAE. J465. Former SAE alloy number: 50
UNS number. M11630
Government. Sand castings: QQ-M-56. Permanent mold castings: QQ-M-55

Foreign. Elektron AZG

Chemical Composition

Composition limits. 5.3 to 6.7 Al; 2.5 to 3.5 Zn; 0.15 min Mn; 0.30 max Si; 0.25 max Cu; 0.01 max Ni; 0.30 others (total); rem Mg

Consequence of exceeding impurity limits. Excessive Si causes brittleness. Excessive Cu degrades mechanical properties and corrosion resistance. Excessive Ni degrades corrosion resistance.

Coefficient of thermal expansion. Linear, 26.1 μm/m·K (14.5 μin./in.·°F) at 20 to 100 °C (68 to 212 °F)

Specific heat. 1.05 kJ/kg·K (0.25 Btu/lb·°F) at 25 °C (78 °F)

Latent heat of fusion. 373 kJ/kg (160 Btu/lb)

Thermal conductivity. 77 W/m·K (44.3 Btu/ft·h·°F) at 100 to 300 °C (212 to 572 °F)

Electrical Properties

Electrical conductivity. At 20 °C (68 °F): F temper, 15% IACS; T4 temper, 12.3% IACS; T5 temper, 13.8% IACS

Electrical resistivity. At 20 °C (68 °F): F temper, 115 nΩ·m; T4 temper, 140 nΩ·m; T5 temper, 125 nΩ·m

Electrolytic solution potential. 1.57 V vs saturated calomel electrode

Hydrogen overvoltage. As cast, 0.34 V

Fabrication Characteristics

Casting temperature. Sand castings, 705 to 845 °C (1300 to 1550 °F)

Weldability. Gas-shielded arc welding with AZ63A or AZ92A rod (AZ63A preferred), fair

Table 23 Typical tensile properties of AZ63A sand castings at elevated temperatures(a)

Testing temperature °C	°F	Tensile strength MPa	ksi	Yield strength MPa	ksi	Elongation(b), %
F Temper						
24	75	197	28.6	94	13.7	4.5
66	150	210	30.5	...	...	3.0
93	200	208	30.1	...	...	4.5
121	250	191	27.7	...	...	7.5
149	300	166	24.1	...	...	20.5
204	400	105	15.3	...	...	50.5
260	500	71	10.3	...	...	38.0
T4 Temper						
24	75	254	36.8	94	13.6	10.0
66	150	253	36.7	...	...	9.0
93	200	236	34.3	...	...	7.0
121	250	207	30.0	...	...	9.0
149	300	154	22.4	...	...	33.2
204	400	101	14.6	...	...	38.0
260	500	75	10.9	...	...	26.0
T6 Temper						
35	95	232	33.7	122	17.7	5.5
93	200	248	36.0	119	17.3	11.0
121	250	223	32.4	114	16.5	11.0
149	300	169	24.5	103	15.0	15.0
204	400	121	17.5	83	12.0	17.0
260	500	83	12.0	61	8.8	15.0
316	600	57	8.2	39	5.6	20.0

(a) Tested as soon as specimens reached testing temperature. (b) In 50 mm or 2 in.

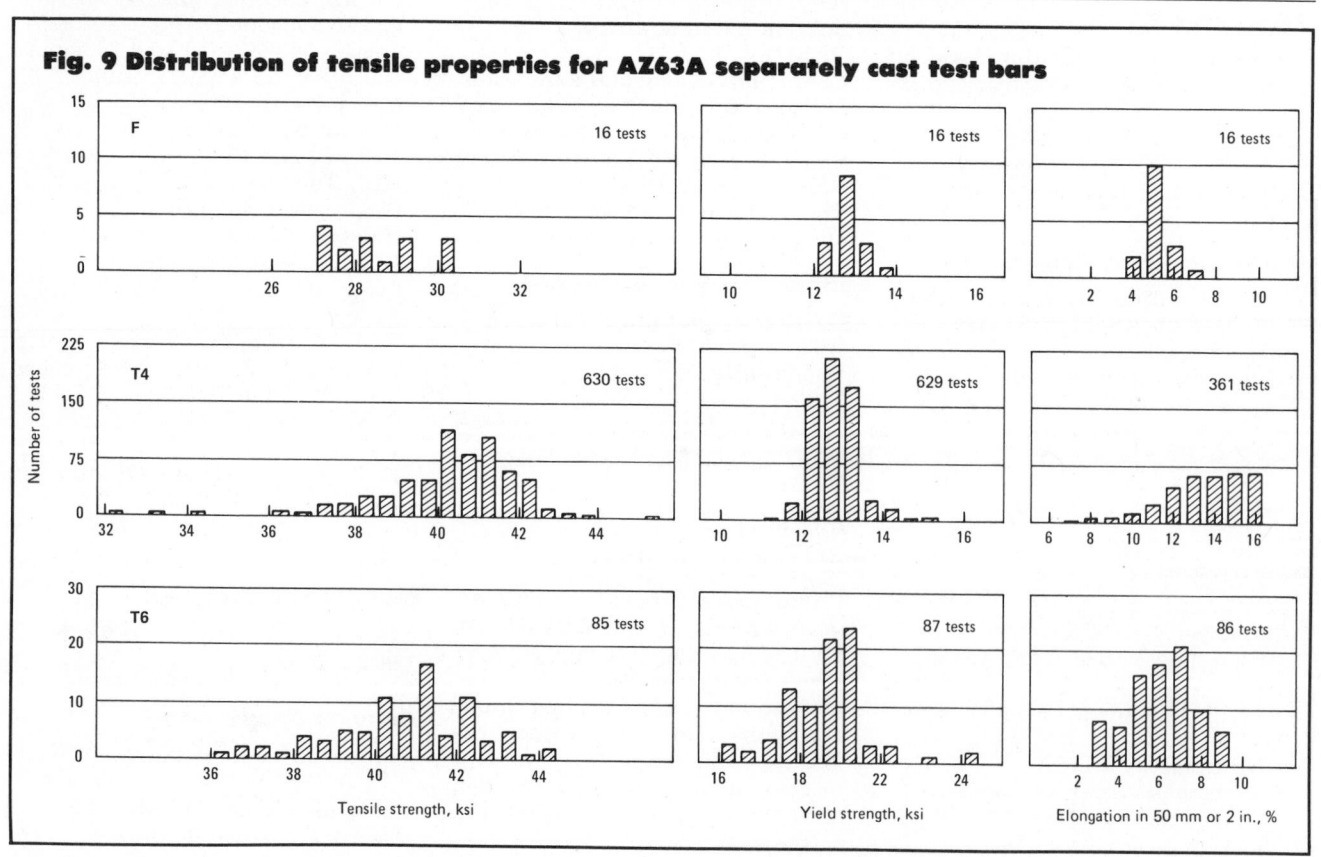

Fig. 9 Distribution of tensile properties for AZ63A separately cast test bars

AZ81A

Specifications

ASTM. Sand castings: B80. Ingot: B93. Permanent mold castings: B199. Investment castings: B403

SAE. J465. Former SAE alloy number: 505

UNS number. M11810

Government. Sand castings: QQ-M-56. Permanent mold castings: QQ-M-55

Foreign. Elektron A8. British: BS2970 MAG1. German: DIN1729 3.5812. French: AIR3380 G-A9

Chemical Composition

Composition limits. 7.0 to 8.1 Al; 0.4 to 1.0 Zn; 0.13 min Mn; 0.30 max Si; 0.10 max Cu; 0.01 max Ni; 0.30 max others (total); rem Mg

Consequence of exceeding impurity limits. Excessive Si causes brittleness. Excessive Cu degrades mechanical properties and corrosion resistance. Excessive Ni degrades corrosion resistance.

Applications

Typical uses. Sand and permanent mold castings used in the solution treated condition (T4 temper), with good strength and excellent ductility and toughness. This alloy is readily castable, with a low microshrinkage tendency.

Mechanical Properties

Tensile properties. T4 temper: tensile strength, 275 MPa (40 ksi); yield strength, 83 MPa (12 ksi); elongation, 15% in 50 mm or 2 in. See also Fig. 10.

Tensile properties vs temperature. See Table 24.

Shear strength. T4 temper, 145 MPa (21 ksi)

Bearing properties. Ultimate bearing strength, 400 MPa (58 ksi); bearing yield strength, 240 MPa (35 ksi)

Compressive yield strength. 83 MPa (12 ksi)

Hardness. 55 HB, 66 HRE

Poisson's ratio. 0.35

Elastic modulus. Tension, 45 GPa (6.5×10^6 psi); shear, 17 GPa (2.4×10^6 psi)

Impact strength. Charpy V-notch, 6.1 J (4.5 ft·lb)

Creep characteristics. See Table 25.

Mass Characteristics

Density. 1.80 Mg/m³ (0.065 lb/in.³) at 20 °C (68 °F)

Thermal Properties

Liquidus temperature. 610 °C (1130 °F)

Solidus temperature. 490 °C (915 °F)

Coefficient of thermal expansion. 25 μm/m·°C (14 μin./in.·°F)

Thermal conductivity, 51.1 W/m·K (29.5 Btu/ft·h·°F) at 20 °C (68 °F)

Electrical Properties

Electrical conductivity. 12% IACS at 20 °C (68 °F)

Electrical resistivity. 13 nΩ·m

Fabrication Characteristics

Casting temperature. 705 to 845 °C (1300 to 1550 °F)

Weldability. Gas-shielded arc welding with AZ92A rod, very good

Table 24 Typical tensile properties of AZ81A-T4 sand castings at elevated temperatures(a)

Testing temperature		Tensile strength		Yield strength		Elongation(b), %
°C	°F	MPa	ksi	MPa	ksi	
21	70	275	40.0	83	12.0	15.0
93	200	260	37.5	83	12.0	20.0
149	300	190	27.5	80	11.5	24.5
204	400	140	20.0	76	11.0	29.0
260	500	97	14.0	72	10.5	35.0

(a) Properties determined using separately cast test bars. (b) In 50 mm or 2 in.

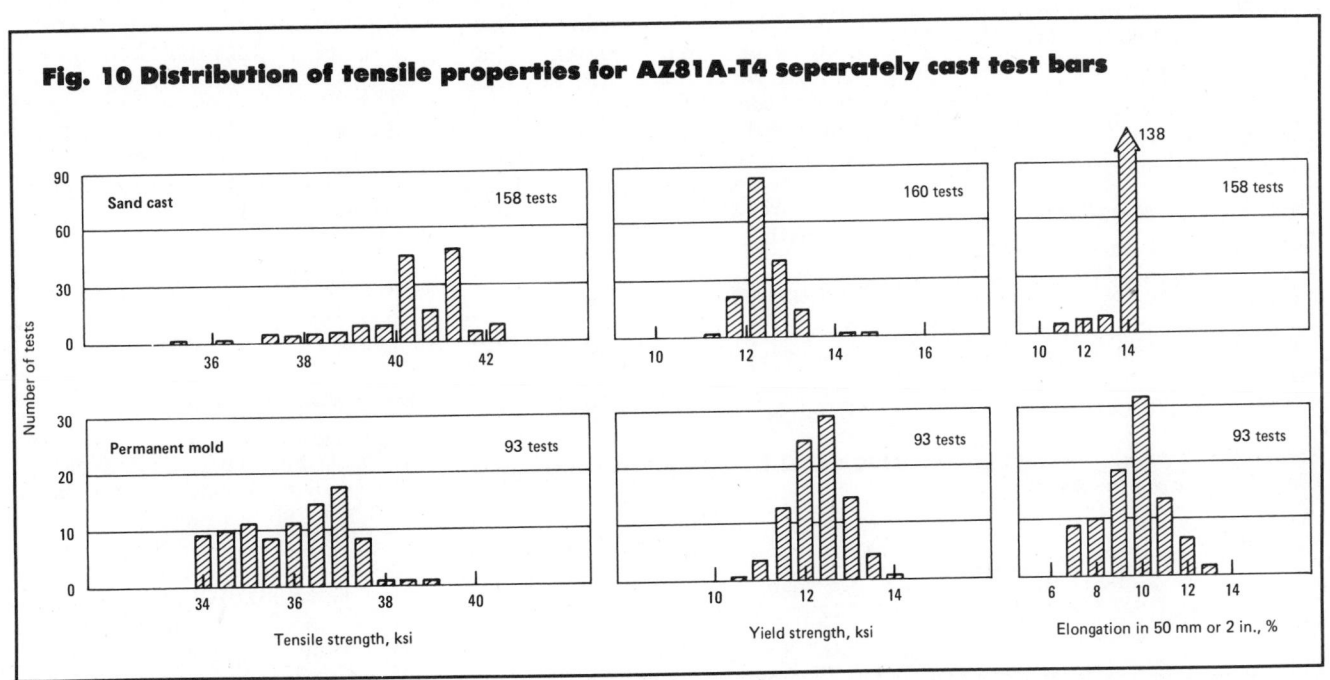

Fig. 10 Distribution of tensile properties for AZ81A-T4 separately cast test bars

Table 25 Typical creep properties of AZ81A-T4 sand castings(a)

Time under load, h	0.1%		0.2%		0.5%	
	MPa	ksi	MPa	ksi	MPa	ksi
At 93 °C (200 °F)						
1	39	5.6	58	8.4	86	12.5
10	37	5.4	55	8.0	83	12.0
100	36	5.2	51	7.4	81	11.8
At 149 °C (300 °F)						
1	37	5.4	53	7.7	...	...
10	28	4.0	45	6.5	62	9.0
100	15	2.2	24	3.5	46	6.6
At 204 °C (400 °F)						
1	23	3.4	41	6.0	...	...
10	12	1.7	21	3.1	...	...
100	7	1.0	12	1.7	21	3.0

(a) Properties determined using separately cast test bars. (b) Total extension equals initial extension plus creep extension.

AZ91A, AZ91B, AZ91C

Specifications

AMS. Die castings, AZ91A: 4490. Sand castings, AZ91C: 4437
ASTM. Die castings, AZ91A and AZ91B: B94. Sand castings, AZ91C: B80. Permanent mold castings, AZ91C: B199. Investment castings, AZ91C: B403. Ingot: B93
SAE. J465. Former SAE alloy numbers: AZ91A, 501; AZ91B, 501A; AZ91C, 504
UNS numbers. AZ91A: M11910. AZ91B: M11912. AZ91C: M11914
Government. Die castings, AZ91A: QQ-M-38. Permanent mold castings, AZ91C: QQ-M-55, MIL-M-46062. Sand castings, AAZ91C: QQ-M-56, MIL-M-46062
Foreign. Elektron AZ91. British: BS2970 MAG3. French: AIR3380 G-AZ91. German: DIN1729 3.5912

Chemical Composition

Composition limits. AZ91A: 8.3 to 9.7 Al; 0.13 min Mn; 0.35 to 1.0 Zn; 0.50 max Si; 0.10 max Cu; 0.03 max Ni; 0.30 max others; rem Mg. AZ91B: 8.3 to 9.7 Al; 0.13 min Mn; 0.35 to 1.0 Zn; 0.50 max Si; 0.35 max Cu; 0.03 max Ni; 0.30 max others; rem Mg

AZ91C: 8.1 to 9.3 Al; 0.13 min Mn; 0.40 to 1.0 Zn; 0.30 max Si; 0.10 max Cu; 0.01 max Ni; 0.03 max others (total); rem Mg
Consequence of exceeding impurity limits. Corrosion resistance decreases with increasing Cu or Ni content. More than 0.5% Si decreases elongation.

Applications

Typical uses. AZ91A and AZ91B, which have the same nominal composition except for copper content, are die casting alloys generally used in the as-cast condition (F temper). AZ91B is the most commonly used magnesium die casting alloy; AZ91A, which has a lower allowable copper content, is used when corrosion resistance is required. AZ91C is used in pressure-tight sand and permanent mold castings with high tensile strength and moderate yield strength.

Mechanical Properties

Tensile properties. See Tables 26 and 27, and Fig. 11, 12, and 13.
Shear strength. AZ91A and AZ91B, F temper: 140 MPa (20 ksi)

Compressive yield strength. See Table 26.
Bearing properties. See Table 26.
Hardness. See Table 26.
Poisson's ratio. 0.35
Elastic modulus. Tension, 45 GPa (6.5×10^6 psi); shear, 17 GPa (2.4×10^6 psi)
Impact strength. See Table 26.
Fatigue strength. At 5×10^8 cycles: AZ91A and AZ91B, F temper: 97 MPa (14 ksi) at 5×10^8 cycles (R. R. Moore type tests)

Mass Characteristics

Density. 1.81 Mg/m³ (0.066 lb/in.³) at 20 °C (68 °F)

Thermal Properties

Liquidus temperature. 595 °C (1105 °F)
Solidus temperature. 470 °C (875 °F)
Coefficient of thermal expansion. Linear, 26 μm/m·K (14 μin./in.·°F) at 20 to 100 °C (68 to 212 °F)
Specific heat. 1.05 kJ/kg·K (0.25 Btu/lb·°F) at 20 °C (68 °F)
Latent heat of fusion. 373 kJ/kg (89 Btu/lb)
Thermal conductivity. 72 W/m·K (41.8 Btu/ft·h·°F) at 100 to 300 °C (212 to 572 °F)
Incipient melting temperature. 421 °C (790 °F)

Electrical Properties

Electrical conductivity. AZ91A: F temper, 10.1% IACS. AZ91C: F temper, 11.5% IACS; T4 temper, 9.9% IACS; T6 temper, 11.2% IACS
Electrical resistivity. AZ91A: F temper, 170 nΩ·m. AZ91C: F temper, 150 nΩ·m; T4 temper, 175 nΩ·m; T6 temper, 151.5 nΩ·m
Electrolytic solution potential. 1.58 V vs saturated calomel electrode
Hydrogen overvoltage. As cast, 0.40 V

Fabrication Characteristics

Casting temperature. AZ91C: sand castings, 705 to 845 °C (1300 to 1550 °F); permanent mold castings, 650 to 815 °C (1200 to 1500 °F)
Weldability. AZ91C can be readily welded by the gas-shielded arc process using AZ91C or AZ92A rod; stress relief required. AZ91A and AZ91B not weldable
Hot-shortness temperature. 400 °C (750 °F)

Table 27 Typical tensile properties of AZ91C-T6 sand castings at elevated temperatures

Testing temperature		Tensile strength		Tensile yield strength		Elongation(a), %
°C	°F	MPa	ksi	MPa	ksi	
149	300	185	27	97	14	40
204	400	115	17	83	12	40

(a) In 50 mm or 2 in.

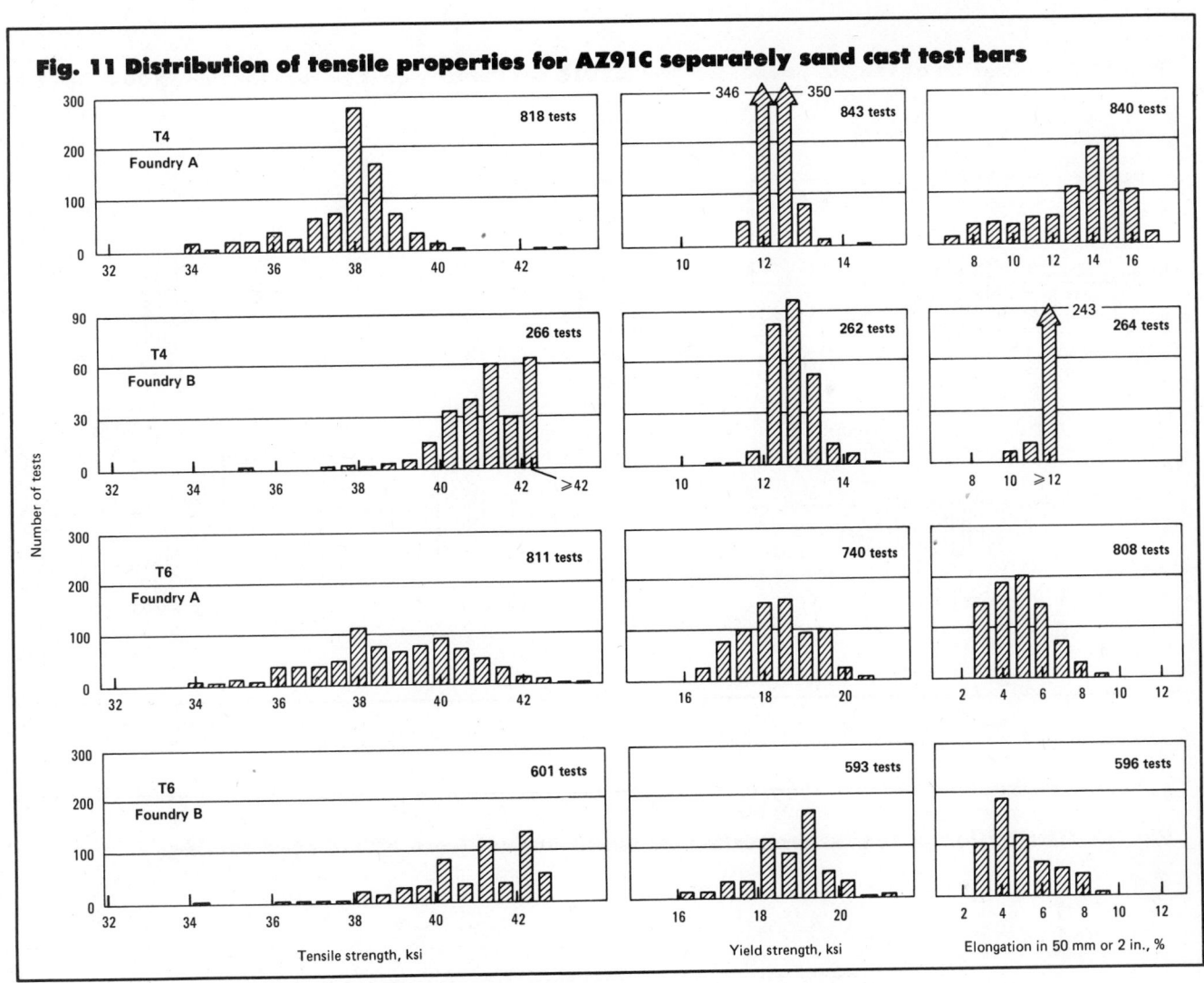

Fig. 11 Distribution of tensile properties for AZ91C separately sand cast test bars

Tensile strength, ksi

Yield strength, ksi

Elongation in 50 mm or 2 in., %

Table 26 Typical room-temperature mechanical properties of AZ91A, AZ91B and AZ91C castings

Property	AZ91A and AZ91B, F temper	F temper	AZ91C T4 temper	T6 temper
Tensile strength, MPa (ksi)	230(33)	165(24)	275(40)	275(40)
Tensile yield strength, MPa (ksi)	150(22)	97(14)	90(13)	145(21)
Elongation(a), %	3	2.5	15	6
Compressive yield strength(b), MPa (ksi)	165(24)	97(14)	90(13)	130(19)
Ultimate bearing strength, MPa (ksi)	...	415(60)	415(60)	515(75)
Bearing yield strength, MPa (ksi)	...	275(40)	305(44)	360(52)
Hardness: HB	63	60	55	70
HRE	75	66	62	77
Charpy V-notch impact strength, J (ft·lb)	2.7(2.0)	0.79(0.58)	4.1(3.0)	1.4(1.0)

(a) In 50 mm or 2 in. (b) At 0.2% offset.

Fig. 12 Distribution of tensile properties for AZ91C separately cast permanent-mold test bars

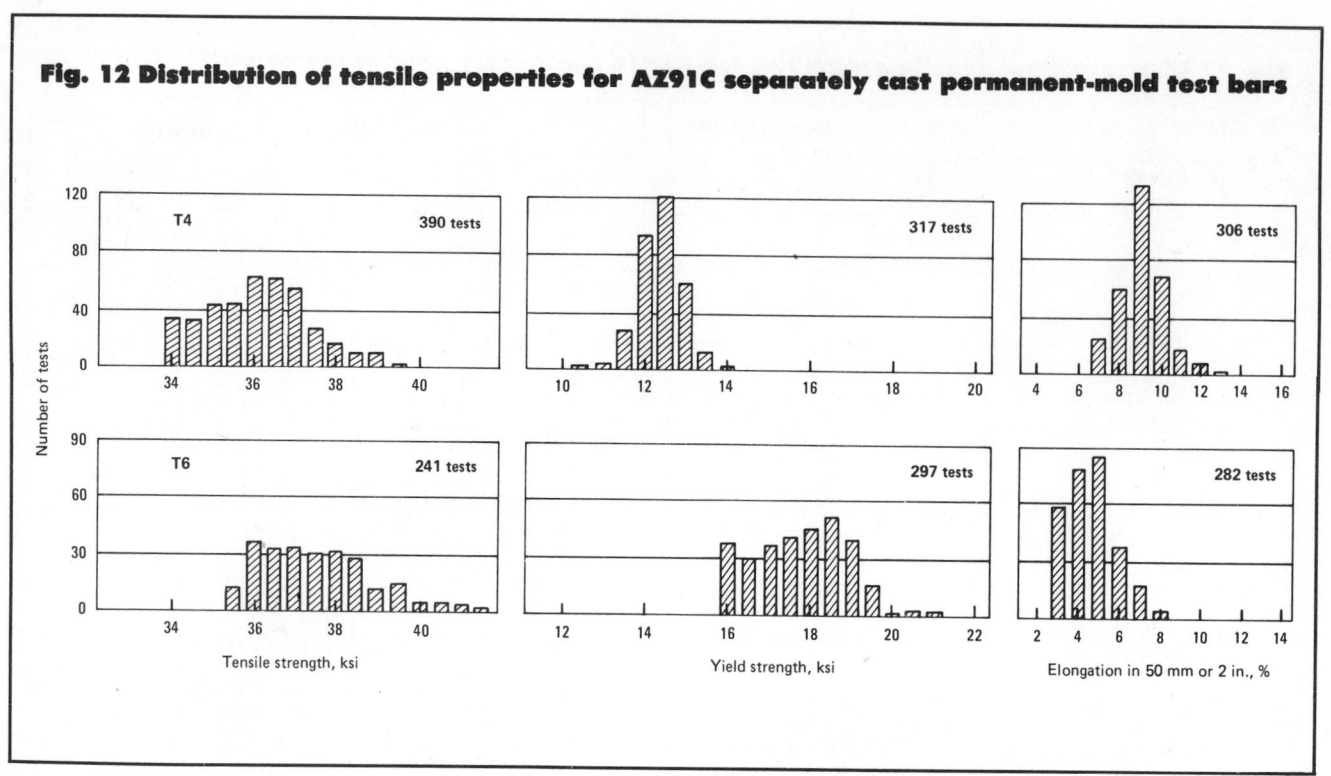

Fig. 13 Distribution of tensile properties for specimens cut from AZ91C sand castings

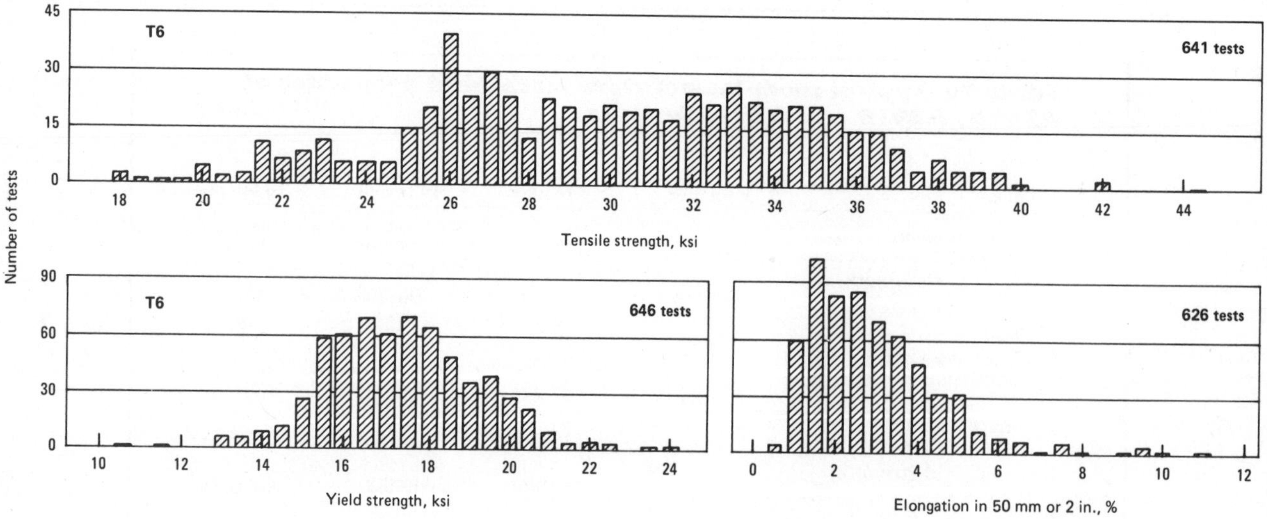

AZ92A

Specifications

AMS. Sand castings: 4434. Investment castings: 4453. Permanent mold castings: 4484.
ASTM. Ingot: B93. Sand castings: B80. Permanent mold castings: B199. Investment castings: B403
SAE. J465. Former SAE alloy number: 500
UNS number. M11920
Government. Sand castings: QQ-M-56 and MIL-M-46062. Permanent mold castings: QQ-M-55 and MIL-M-46062

Chemical Composition

Composition limits. 8.3 to 9.7 Al; 0.10 min Mn; 1.6 to 2.4 Zn; 0.30 max Si; 0.25 max Cu; 0.01 max Ni; 0.30 max others (total); rem Mg
Consequence of exceeding impurity limits. Excessive Cu or Ni degrades corrosion resistance. More than 0.5% Si decreases elongation.

Applications

Typical uses. Pressure-tight sand and permanent mold castings with high tensile strength and good yield strength

Mechanical Properties

Tensile properties. See Table 28 and Fig. 14.
Tensile properties vs temperature. See Table 29.
Shear strength. F temper, 125 MPa (18 ksi); T4 and T5 tempers, 140 MPa (20 ksi); T6 temper, 145 MPa (21 ksi); T7 temper, 150 MPa (22 ksi)
Compressive yield strength. F and

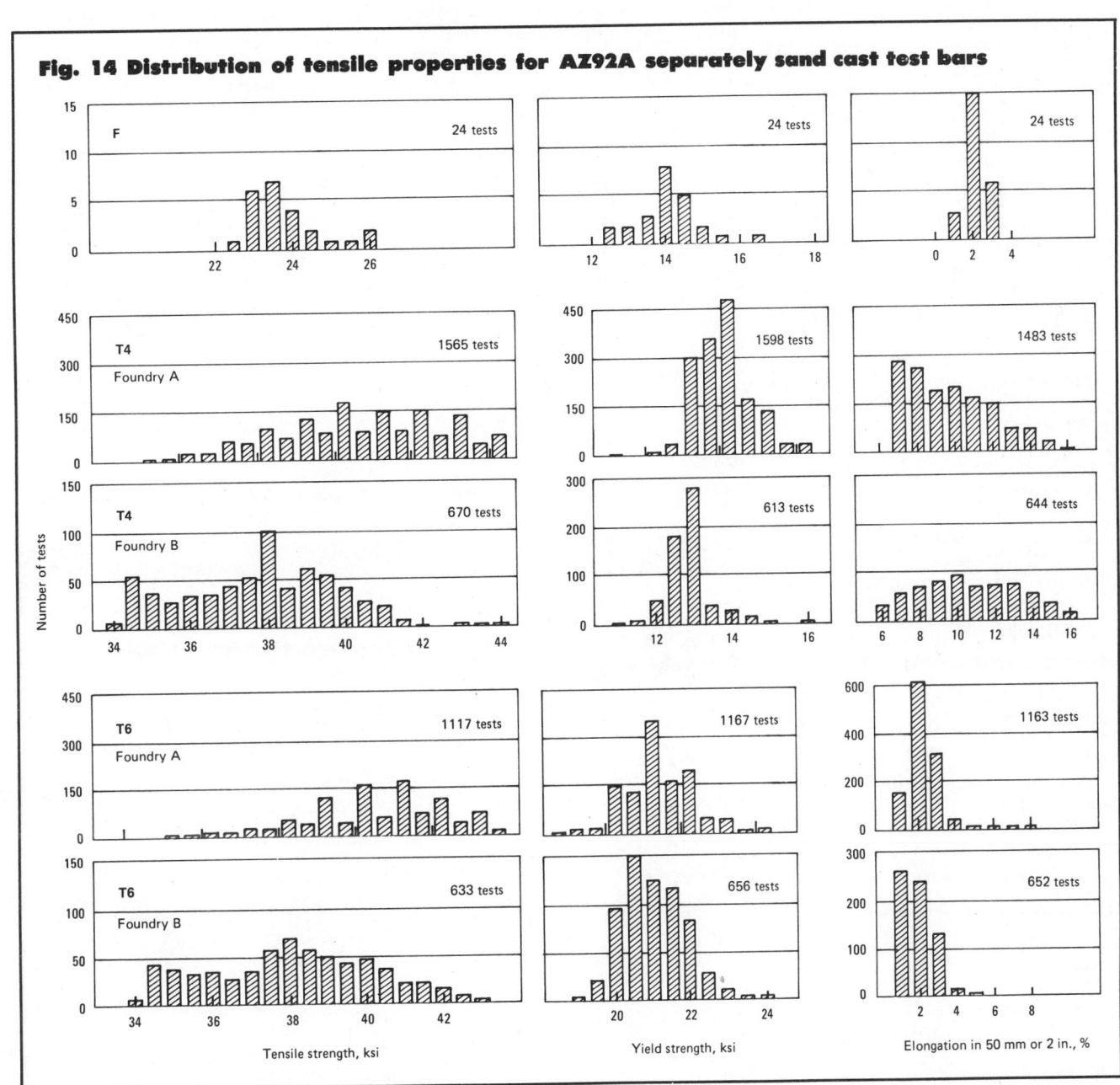

Fig. 14 Distribution of tensile properties for AZ92A separately sand cast test bars

Table 28 Typical tensile properties of AZ92A sand castings(a)

Temper	Tensile strength MPa	ksi	Yield strength MPa	ksi	Elongation(b), %
F	170	25	97	14	2
T4	275	40	97	14	10
T5	170	25	115	17	1
T6	275	40	150	22	3
T7	275	40	145	21	3

(a) Properties determined using separately cast test bars. (b) In 50 mm or 2 in.

T4 tempers, 97 MPa (14 ksi); T5 temper, 115 MPa (17 ksi); T6 temper, 150 MPa (22 ksi); T7 temper, 145 MPa (21 ksi)

Bearing properties. Ultimate bearing strength: F and T5 tempers, 345 MPa (50 ksi); T4 temper, 470 MPa (68 ksi); T6 temper, 550 MPa (80 ksi). Bearing yield strength: F, T4 and T5 tempers, 315 MPa (46 ksi); T6 temper, 450 MPa (65 ksi)

Hardness. F temper: 65 HB, 76 HRE. T4 temper: 63 HB, 75 HRE. T5 temper: 69 HB, 80 HRE. T6 temper: 81 HB, 88 HRE. T7 temper 78 HB, 86 HRE

Poisson's ratio. 0.35

Elastic modulus. Tension, 45 GPa $(6.5 \times 10^6$ psi); shear, 17 GPa $(2.4 \times 10^6$ psi)

Impact strength. Charpy V-notch: F temper, 0.7 J (0.5 ft·lb); T4 temper, 2.7 J (2.0 ft·lb); T6 temper, 1.1 J (0.8 ft·lb)

Fatigue strength. At 5×10^8 cycles: F and T6 tempers, 83 MPa (12 ksi); T4 and T7 tempers, 90 MPa (13 ksi); T5 temper, 76 MPa (11 ksi) (R.R. Moore type test)

Mass Characteristics

Density. 1.82 Mg/m³ (0.066 lb/in.³) at 20 °C (68 °F)

Thermal Properties

Liquidus temperature. 595 °C (1100 °F)

Solidus temperature. 445 °C (830 °F)

Coefficient of thermal expansion. Linear, 26 µm/m·K (14 µin./in.·°F) at 18 to 100 °C (65 to 212 °F)

Incipient melting temperature. 410 °C (770 °F)

Specific heat. 1.05 kJ/kg·K (0.25 Btu/lb·°F) at 25 °C (78 °F)

Latent heat of fusion. 373 kJ/kg (160 Btu/lb)

Thermal conductivity. 72 W/m·K (41.8 Btu/ft·h·°F) at 100 to 300 °C (212 to 572 °F)

Electrical Properties

Electrical conductivity. At 20 °C (68 °F): F temper, 12.3% IACS; T4 temper, 10.5% IACS; T6 temper, 12.3% IACS

Electrical resistivity. At 20 °C (68 °F): F temper, 140 nΩ·m; T4 temper, 165 nΩ·m; T6 temper, 140 nΩ·m

Electrolytic solution potential. 1.56 V vs saturated calomel electrode

Hydrogen overvoltage. As cast, 0.3 V

Fabrication Characteristics

Casting temperature. Sand castings, 705 to 845 °C (1300 to 1550 °F); permanent mold castings, 650 to 815 °C (1200 to 1500 °F)

Weldability. Gas-shielded arc welding with AZ92A rod, good; stress relief required

Table 29 Typical tensile properties of AZ92A sand castings at elevated temperatures(a)

Testing temperature °C	°F	Tensile strength MPa	ksi	Elongation(b), %	Time at temperature(c), days
F Temper					
93	200	170	25	2	80
149	300	150	22	3	160
204	400	110	16	36	160
260	500	83	12	34	40
T4 Temper					
93	200	275	40	8	160
149	300	180	26	40	160
204	400	115	17	41	160
260	500	76	11	52	40
T6 Temper					
93	200	260	38	7	160
149	300	170	25	40	160
204	400	115	17	43	160
260	500	76	12	47	40

(a) Tested after prolonged heating at testing temperature. (b) In 50 mm or 2 in. (c) Prior to testing.

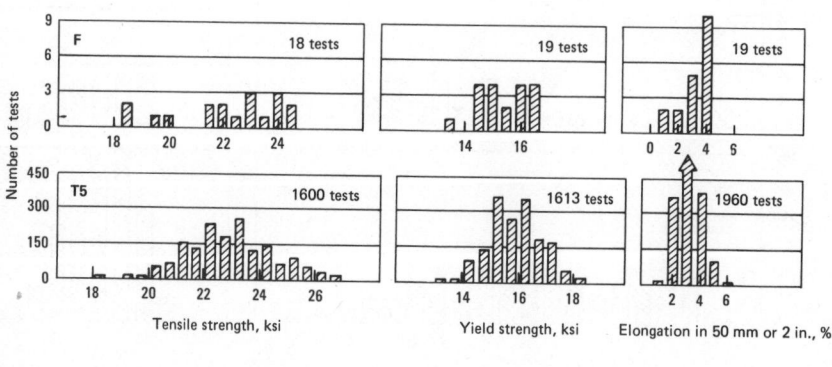

Fig. 15 Distribution of tensile properties for EZ33A separately sand cast test bars

EZ33A

Specifications

AMS. Sand castings: 4442
ASTM. Sand castings: B80. Permanent mold castings: B199. Investment castings: B403
SAE. J465. Former SAE alloy number: 506
UNS number. M12330
Government. Sand castings: QQ-M-56. Permanent mold castings: QQ-M-55. Welding rod: MIL-R-6944
Foreign. Elektron ZRE1. British: BS2970 MAG6. German: DIN1729 3.5103. French: AIR3380 ZRE1

Chemical Composition

Composition limits. 2.5 to 4.0 rare earths; 2.0 to 3.1 Zn; 0.50 to 1.0 Zr; 0.10 max Cu; 0.01 max Ni; 0.30 max others (total); rem Mg

Applications

Typical uses. Pressure-tight sand and permanent mold castings relatively free from microporosity, used in T5 condition for applications requiring good strength properties up to 260 °C (500 °F)

Mechanical Properties

Tensile properties. T5 temper: tensile strength, 160 MPa (23 ksi); yield strength, 110 MPa (16 ksi); elongation, 3% in 50 mm or 2 in. See also Fig. 15.
Tensile properties vs temperature. See Table 30 and Fig. 16.
Shear strength. T5 temper, 135 MPa (19.8 ksi)
Compressive yield strength. 110 MPa (16 ksi)
Bearing properties. T5 temper: ultimate bearing strength, 395 MPa (57 ksi); bearing yield strength, 275 MPa (39.9 ksi)
Hardness. 50 HB, 59 HRE
Creep characteristics. See Table 31 and Fig. 17.

Mass Characteristics

Density. 1.83 Mg/m³ (0.066 lb/in.³) at 20 °C (68 °F)

Thermal Properties

Liquidus temperature. 645 °C (1190 °F)
Solidus temperature. 545 °C (1010 °F)
Coefficient of thermal expansion.
Linear, 26.1 μm/m·K (14.5 μin./in.·°F) at 20 to 100 °C (68 to 212 °F)
Specific heat. 1.05 kJ/kg·K (0.25 Btu/lb·°F) at 20 °C (68 °F)
Latent heat of fusion. 373 kJ/kg (160 Btu/lb)
Thermal conductivity. 100 W/m·K (58 Btu/ft·h·°F)

Electrical Properties

Electrical conductivity. 25% IACS at 20 °C (68 °F)
Electrical resistivity. 70 nΩ·m at 20 °C (68 °F)

Fabrication Characteristics

Casting temperature. Sand castings, 705 to 815 °C (1300 to 1500 °F); permanent mold castings, 675 to 785 °C (1250 to 1450 °F)
Weldability. Gas-shielded arc welding with EZ33A rod, excellent; preheating not necessary, but may be used; postweld heat treatment required

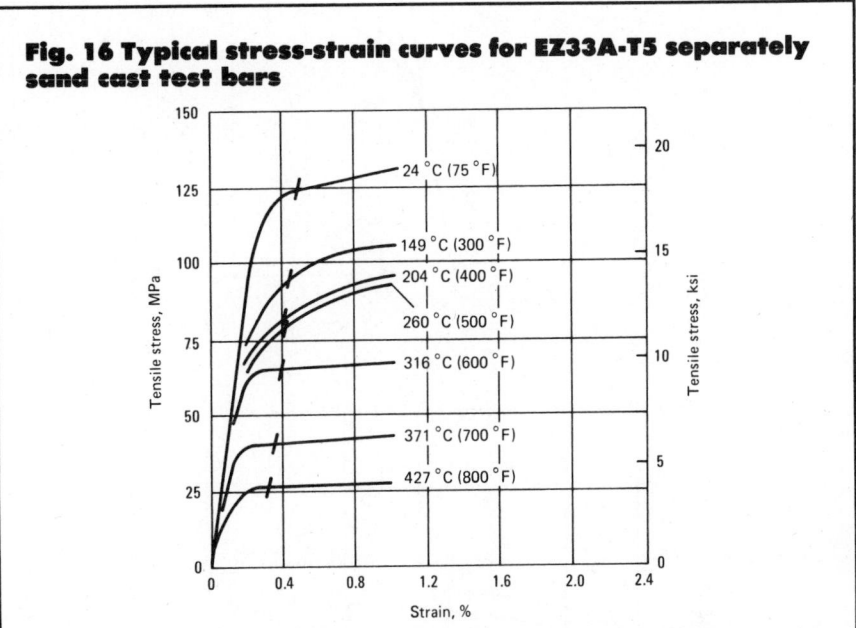

Fig. 16 Typical stress-strain curves for EZ33A-T5 separately sand cast test bars

Table 31 Typical creep properties of EZ33A-T5 sand castings(a)

| Time under load, h | \multicolumn{8}{c}{Tensile stress resulting in total extension(b) of:} | | | | | | | |
| | 0.1% | | 0.2% | | 0.5% | | 1.0% | |
	MPa	ksi	MPa	ksi	MPa	ksi	MPa	ksi
At 204 °C (400 °F)								
1	41	6	69	10	89	13	105	15
10	41	6	62	9	83	12	89	13
100	34	5	55	8	69	10	76	11
1000	28	4	41	6	48	7	55	8
At 260 °C (500 °F)								
1	34	5	55	8	69	10	83	12
10	28	4	34	5	48	7	55	8
100	14	2	21	3	28	4	34	5
1000	14	2	14	2	14	2	21	3
At 316 °C (600 °F)								
1	14	2	21	3	28	4	34	5
10	14	2	14	2	21	3	21	3
100	14	2	7	1	14	2	14	2
1000	7	1	7	1	7	1	7	1

(a) Properties determined using separately cast test bars. (b) Total extension equals initial extension plus creep extension.

Fig. 17 Isochronous stress-strain curves for EZ33A-T5 separately sand cast test bars.

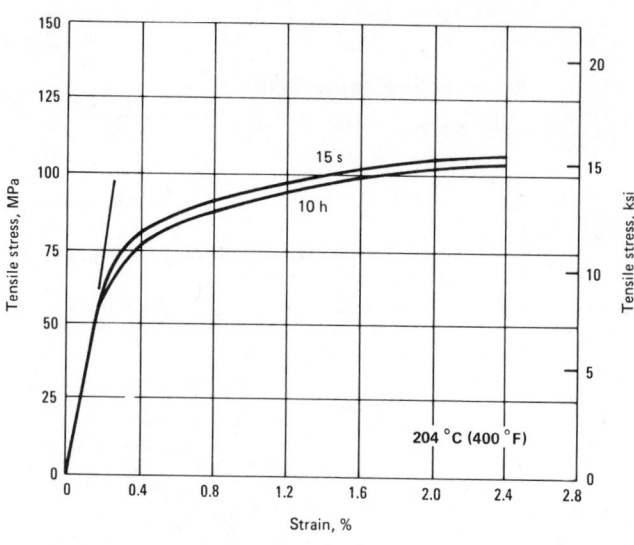

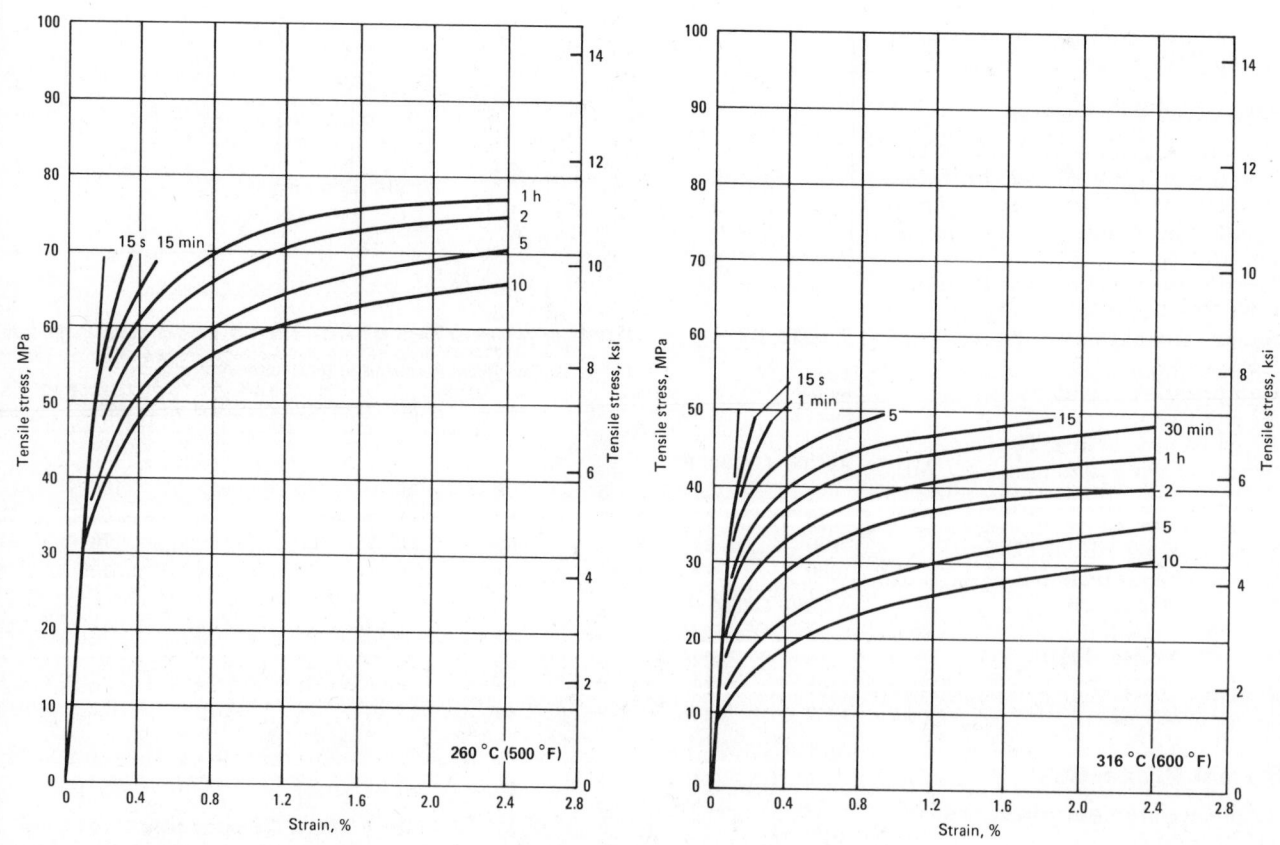

Specimens exposed at testing temperatures for 3 h before loading.

Fig. 17 (continued)

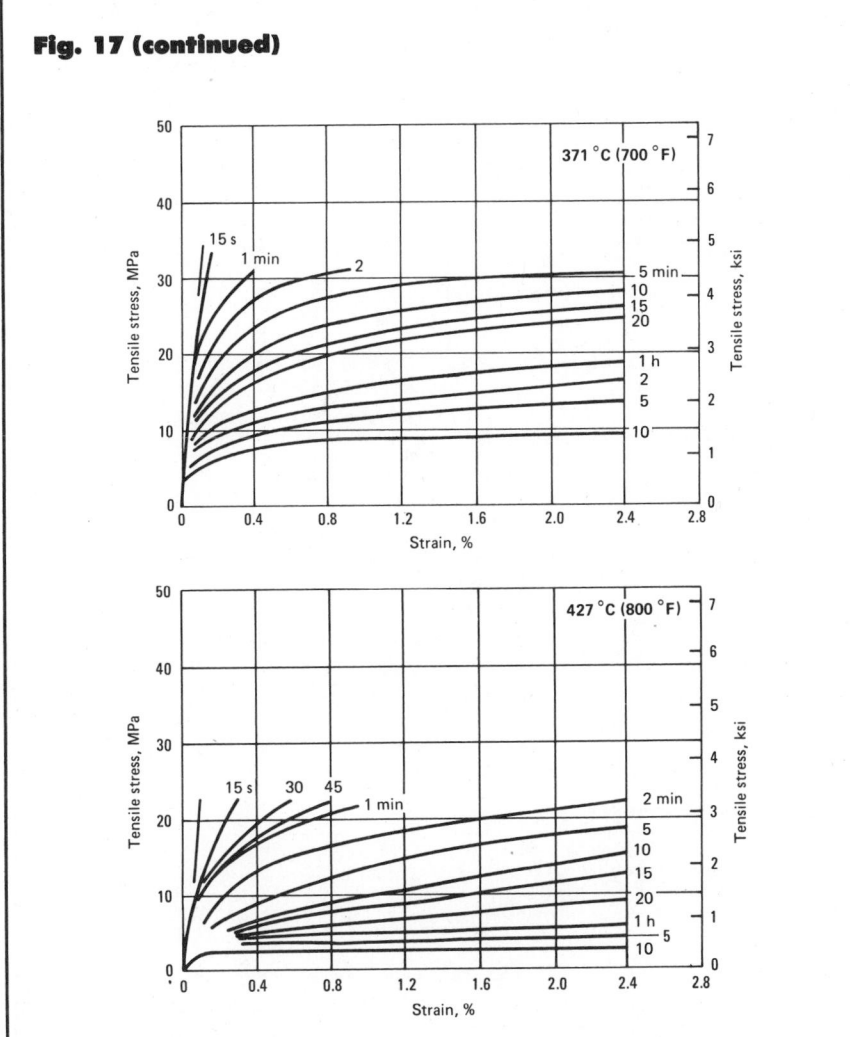

Table 30 Typical tensile properties of EZ33A-T5 sand castings at elevated temperatures(a)

Testing temperature		Tensile strength		Yield strength		Elongation(b), %
°C	°F	MPa	ksi	MPa	ksi	
24	75	160	23	110	16	3
149	300	150	22	97	14	10
204	400	145	21	76	11	20
260	500	125	18	69	10	31
316	600	83	12	55	8	50

(a) Properties determined using separately cast test bars. (b) In 50 mm or 2 in.

HK31A

See also wrought alloy HK31A.

Specifications

AMS. Sand castings: 4445
ASTM. Sand castings: B80. Permanent mold castings: B199. Investment castings: B403
SAE. J465. Former SAE alloy number: 507
UNS number. M13310
Government. Sand castings: QQ-M-
56 and MIL-M-46062. Permanent
mold castings. QQ-M-55 and MIL-M-
46062

Chemical Composition

Composition limits. 2.5 to 4.0 Th;
0.40 to 1.0 Zr; 0.30 max Zn; 0.10 max
Cu; 0.01 max Ni; 0.30 max others
(total); rem Mg

Applications

Typical uses. Sand castings for use
at temperatures up to 345 °C
(650 °F)

Mechanical Properties

Tensile properties. T6 temper: tensile strength, 220 MPa (32 ksi); yield
strength, 105 MPa (15 ksi); elongation, 8% in 50 mm or 2 in. See also
Fig. 18.
**Tensile properties vs tempera-
ture.** See Table 32 and Fig. 19.
Compressive yield strength. T6
temper, 105 MPa (15 ksi)
Bearing properties. T6 temper: ultimate bearing strength, 420 MPa
(61 ksi); bearing yield strength, 275
MPa (40 ksi)
Hardness. T6 temper, 66 HRE
Poisson's ratio. 0.35
Elastic modulus. Tension, 45 GPa
(6.5×10^6 psi); shear, 17 GPa (2.4×10^6 psi)
Creep characteristics. See Table
33 and Fig. 20.

Mass Characteristics

Density. 1.79 Mg/m³ (0.065 lb/in.³)
at 20 °C (68 °F)

Thermal Properties

Liquidus temperature. 650 °C
(1205 °F)
Solidus temperature. 590 °C
(1090 °F)
Incipient melting temperature.
627 to 632 °C (1160 to 1170 °F) in
circulating air
Specific heat vs temperature. C_p
$= 1374 + 0.0002306T + 3370T^{-2}$
Latent heat of fusion. 318 to 335
kJ/kg (137 to 144 Btu/lb)
Thermal Conductivity. At 20 °C (68
°F); T6 temper, 92 W/m·K (53 Btu/
ft·h·°F); H24 temper, 113 W/m·K (65
Btu/ft·h·°F); O temper, 105 W/m·K
(61 Btu/ft·h·°F)
**Thermal conductivity vs tempera-
ture.** See Table 34.

Electrical Properties

Electrical conductivity. T6 temper,
22% IACS at 20 °C (68 °F)
Electrical resistivity. At 20 °C (68

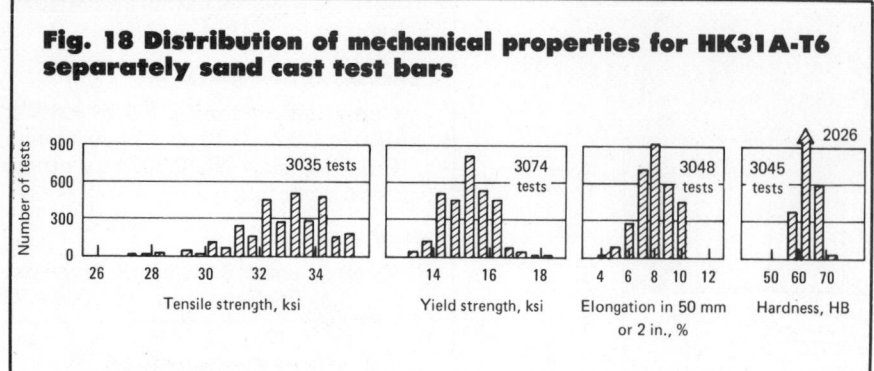

Fig. 18 Distribution of mechanical properties for HK31A-T6 separately sand cast test bars

°F): T6 temper, 77 nΩ·m; H24 temper, 61 nΩ·m; O temper, 60 nΩ·m

Fabrication Characteristics

Casting temperature. Sand and permanent mold castings, 675 to 800 °C (1250 to 1470 °F)

Weldability. Gas-shielded arc welding with EZ33A or HK31A rod (EZ33A preferred), very good; stress relief required for sand castings

Table 32 Typical tensile properties of HK31A-T6 sand castings at elevated temperatures(a)

Testing temperature		Tensile strength		Yield strength		Elonga- tion(b), %
°C	°F	MPa	ksi	MPa	ksi	
24	75	215	31	110	16	6
204	400	165	24	97	14	17
260	500	160	23	89	13	19
316	600	140	20	83	12	22
371	700	89	13	55	8	26

(a) Properties determined using separately cast test bars. (b) In 50 mm or 2 in.

Table 34 Thermal conductivity of HK31A-T6 sand castings at various temperatures

Temperature		Thermal conductivity	
°C	°F	W/m · K	Btu/ft · h · °F
20	68	92	53
38	100	92	53
93	200	100	58
149	300	105	61
204	400	109	63
260	500	113	65

Table 33 Creep properties of HK31A-T6 sand castings(a)

Time under load h	0.1%		0.2%		0.5%		1.0%	
	MPa	ksi	MPa	ksi	MPa	ksi	MPa	ksi
At 204 °C (400 °F)								
1	41	6.0	71	10.3	103	15.0	110	16.0
10	40	5.8	68	9.8	103	15.0	110	16.0
100	39	5.6	66	9.5	103	15.0	110	16.0
1000	37	5.4	63	9.1	97	14.0	109	15.8
t 260 °C (500 °F)								
1	36	5.25	69	10.0	97	14.0	107	15.5
10	30	4.4	59	8.6	88	12.7	100	14.4
100	24	3.5	43	6.3	67	9.7	84	12.2
1000	21	3.1	29	4.2	47	6.8	52	7.6
At 288 °C (550 °F)								
1	···	···	54	7.8	85	12.3	···	···
10	···	···	44	6.4	66	9.5	···	···
100	···	···	31	4.5	43	6.3	···	···
1000	···	···	17	2.5	22	3.2	···	···
At 316 °C (600 °F)								
1	29	4.15	43	6.2	72	10.4	85	12.3
10	22	3.25	33	4.75	50	7.2	60	8.7
100	15	2.15	20	2.9	24	3.5	28	4.1
1000	6	0.94	8	1.1	10	1.4	11	1.55
At 349 °C (660 °F)								
1	···	···	30	4.4	41	6.0	···	···
10	···	···	16	2.3	22	3.2	···	···
100	···	···	7	1.0	9	1.3	···	···
1000	···	···	4	0.63	5	0.72	···	···

Tensile stress resulting in total extension(b) of:

(a) Properties determined using separately cast test bars. (b) Total extension equals initial extension plus creep extension.

Fig. 20 Isochronous stress-strain curves for HK31A-T6 separately cast test bars

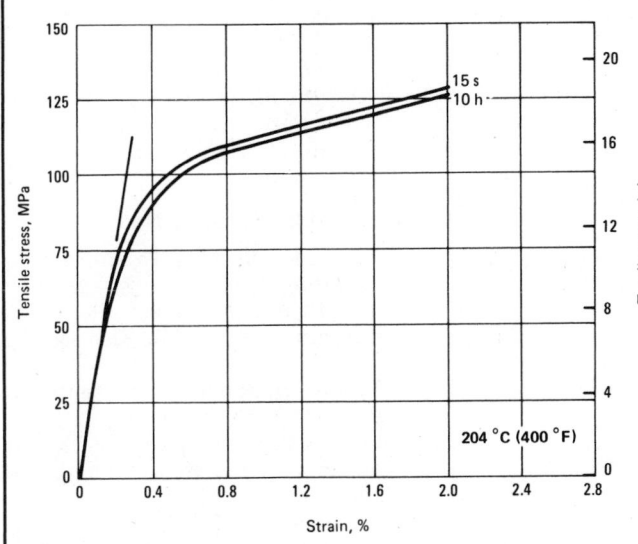

15 s
10 h

204 °C (400 °F)

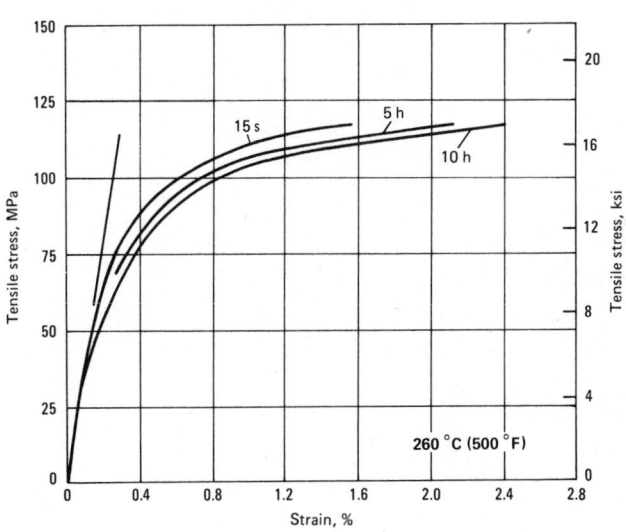

15 s
5 h
10 h

260 °C (500 °F)

Fig. 19 Typical stress-strain curves for HK31A separately cast test bars

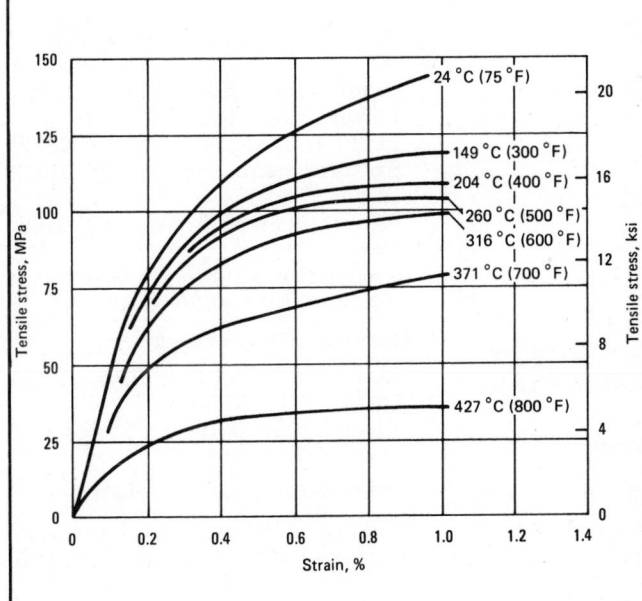

24 °C (75 °F)
149 °C (300 °F)
204 °C (400 °F)
260 °C (500 °F)
316 °C (600 °F)
371 °C (700 °F)
427 °C (800 °F)

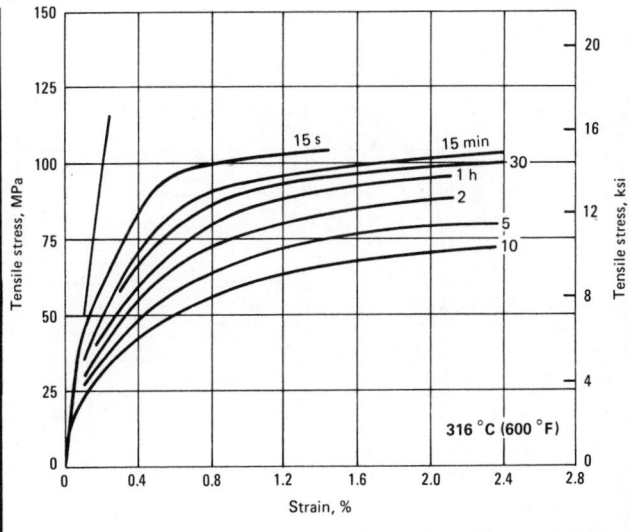

15 s
15 min
1 h
2
5
10
30

316 °C (600 °F)

Specimens exposed at testing temperature for 3 h before loading.

Fig. 20 (continued)

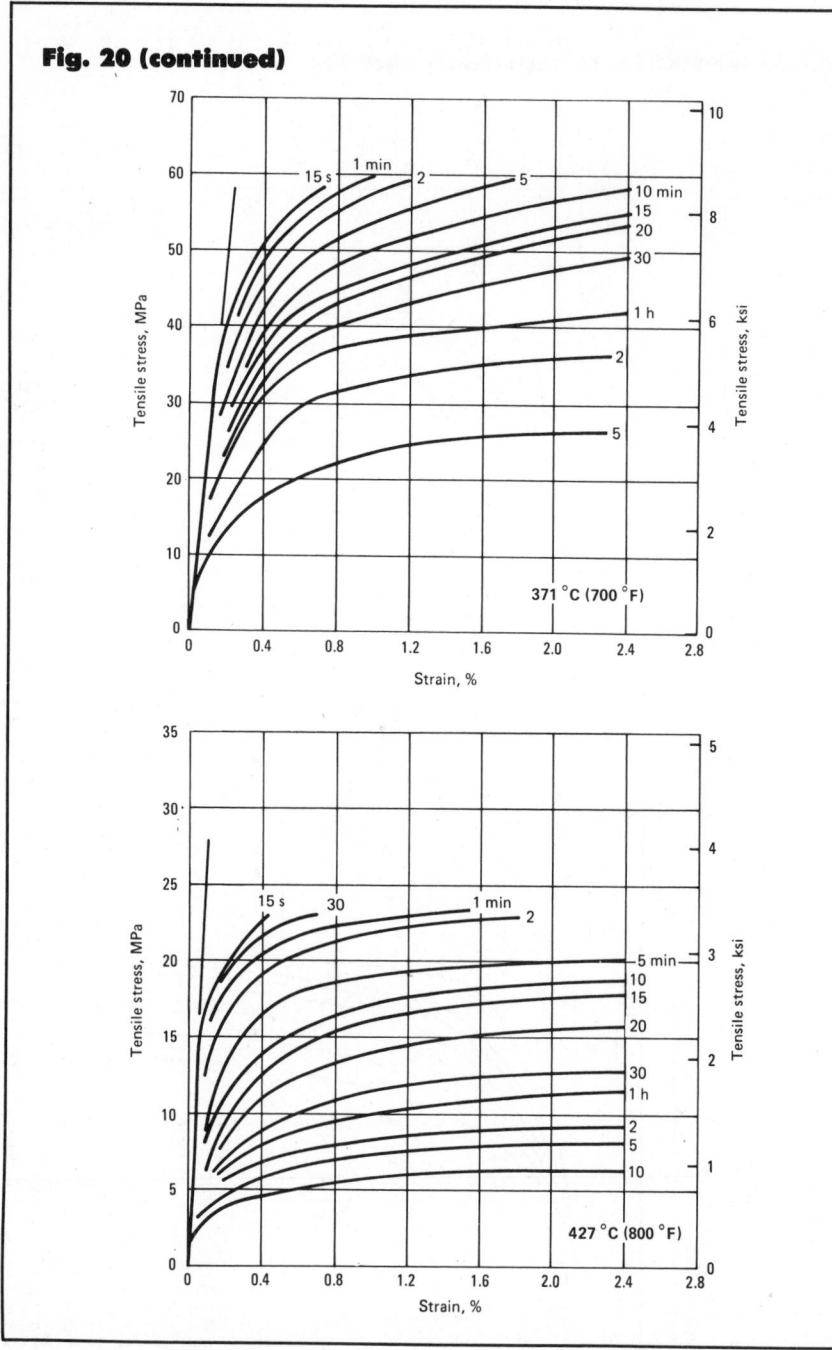

Applications

Typical uses. Sand castings used in the artificially aged condition (T5 temper), with moderate strength and an optimum combination of properties for medium- and long-time exposure at temperatures above 260 °C (500 °F). Castings are pressure tight, and under long-time exposure can withstand higher stresses and higher temperatures than any other commercially available magnesium alloy.

Mechanical Properties

Tensile properties. T5 temper: tensile strength, 185 MPa (27 ksi); yield strength, 90 MPa (13 ksi); elongation in 50 mm or 2 in., 4%. Also see Fig. 21.

Tensile properties vs temperature. See Table 35 and Fig. 22.

Compressive properties. T5 temper: compressive yield strength, 110 MPa (16 ksi)

Hardness. 55 HB

Poisson's ratio. 0.3

Elastic modulus. Tension, 45 GPa (6.5×10^6 psi); shear, 17 GPa (2.4×10^6 psi)

Creep characteristics. See Fig. 23.

Mass Characteristics

Density. 1.83 Mg/m³ (0.066 lb/in.³) at 20 °C (68 °F)

Thermal Properties

Liquidus temperature. 650 °C (1200 °F)

Solidus temperature. 550 °C (1025 °F)

Coefficient of thermal expansion. Linear, 26.7 μm/m·K (14.8 μin./in.·°F) at 20 to 200 °C (68 to 330 °F)

Specific heat. 0.96 kJ/kg·K (0.23 Btu/lb·°F)

Latent heat of fusion. 373 kJ/kg (160 Btu/lb)

Thermal conductivity. 110 W/m·K (64 Btu/ft·h·°F) at 20 °C (68 °F)

Electrical Properties

Electrical conductivity. 26.5% IACS at 20 °C (68 °F)

Electrical resistivity. T5 temper: 65 nΩ·m at 20 °C (68 °F)

Fabrication Characteristics

Casting temperature. Sand castings, 675 to 800 °C (1250 to 1470 °F)

Weldability. Gas-shielded arc welding with HZ32A or EZ33A welding rod, fair; heavy-section castings require stress relief after welding.

HZ32A

Specifications

AMS. Sand castings: 4447
ASTM. Sand castings: B80
UNS number. M13320
Government. Sand castings: QQ-M-56, MIL-M-46062
Foreign. Elektron ZT1. British: BS2970 MAG 8. German: DIN1729-3.5105

Chemical Composition

Composition limits. 1.7 to 2.5 Zn; 2.5 to 4.0 Th; 0.10 max rare earths; 0.50 to 1.0 Zr; 0.10 max Cu; 0.01 max Ni; 0.30 max others (total); rem Mg

Consequence of exceeding impurity limits. More than 0.1% rare earths causes loss in creep resistance

Table 35 Typical tensile properties of HZ32A-T5 sand castings at elevated temperatures

Testing temperature °C	°F	Tensile strength MPa	ksi	Yield strength MPa	ksi	Elongation(a), %
24	75	200	29	105	15	6
93	200	180	26	97	14	15
149	300	150	22	83	12	23
204	400	115	17	69	10	33
260	500	97	14	63	9	33
316	600	83	12	55	8	28
371	700	69	10	48	7	29

(a) In 50 mm or 2 in.

Fig. 21 Distribution of tensile properties for HZ32A-T5 separately sand cast test bars

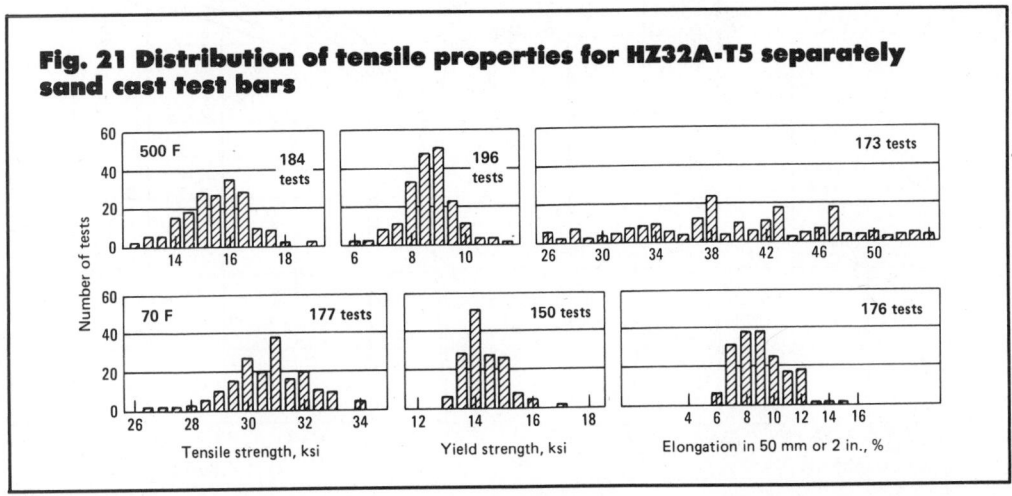

Fig. 22 Typical stress-strain curves for HZ32A-T5 separately sand cast test bars

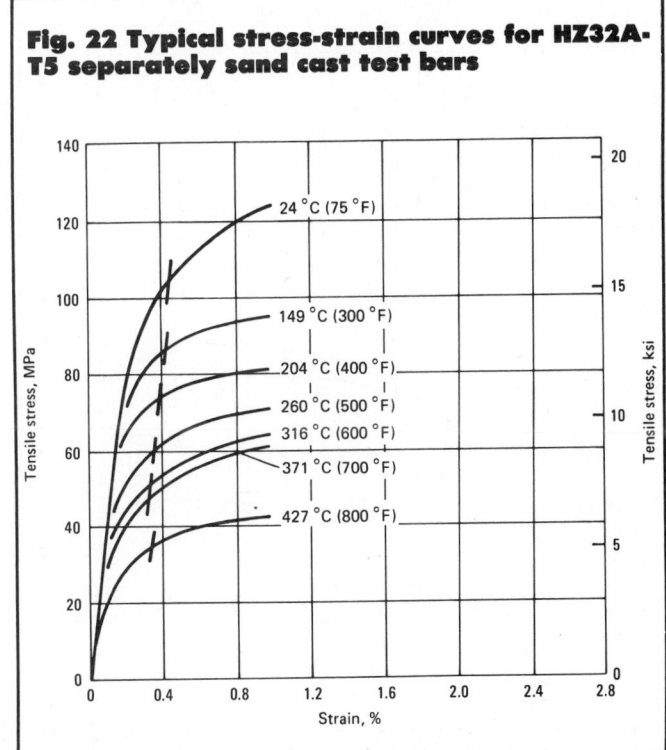

Fig. 23 Isochronous stress-strain curves for HZ32A-T5 separately sand cast test bars

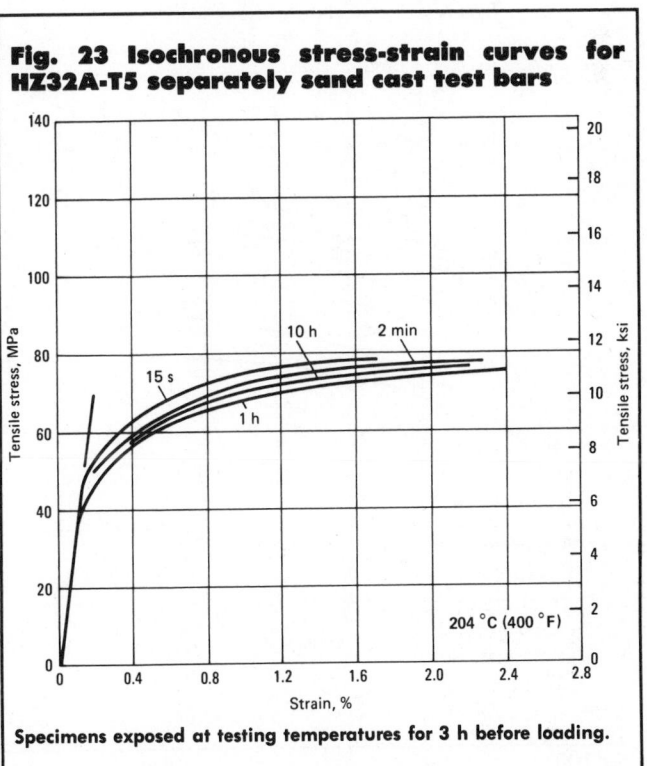

Specimens exposed at testing temperatures for 3 h before loading.

Fig. 23 (continued)

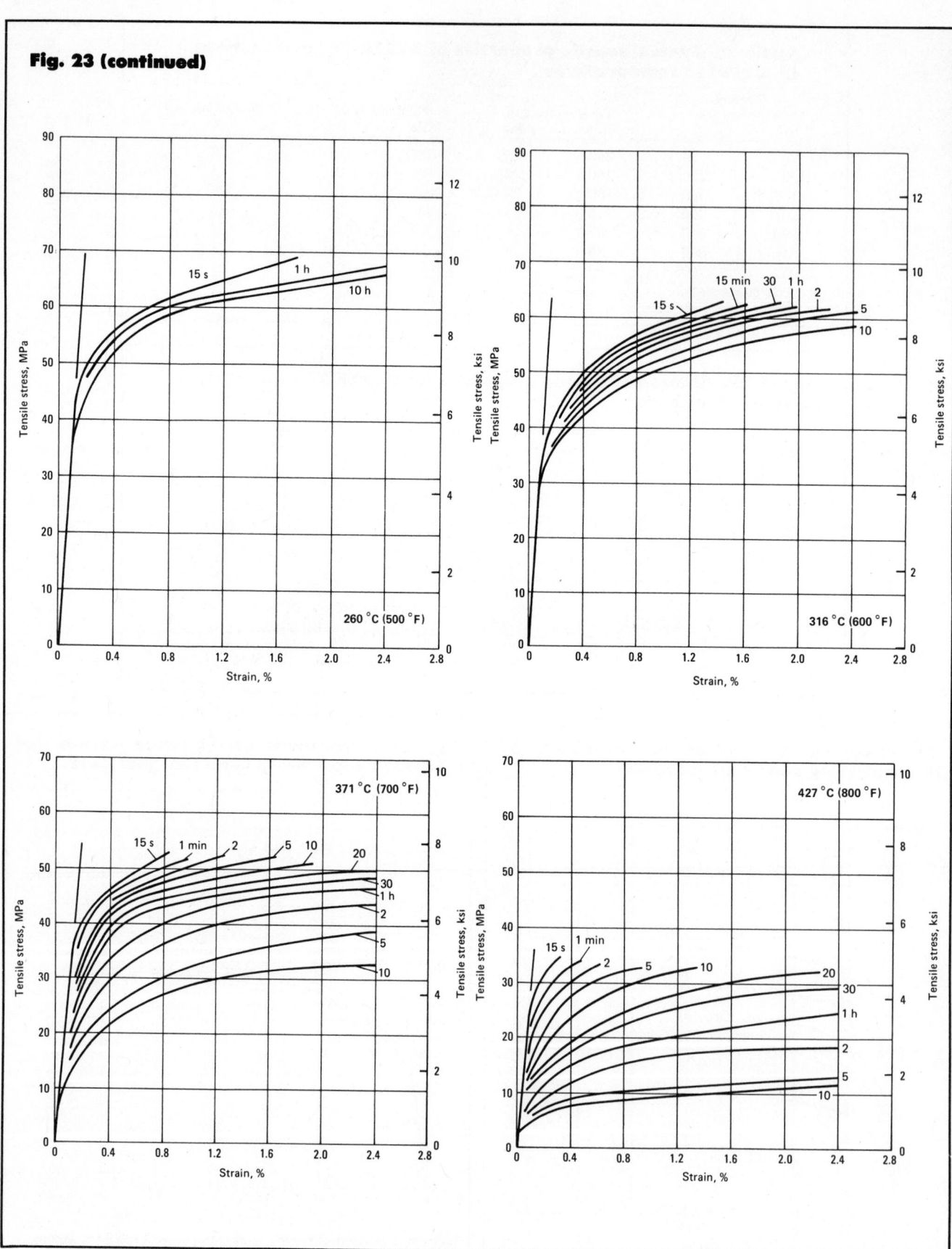

K1A

Specifications

ASTM. Sand castings: B80
UNS number. M18010

Chemical Composition

Composition limits. 0.40 to 1.0 Zr;
0.30 max others (total); rem Mg

Applications

Typical uses. K1A is used in the as-cast condition (F temper) for its high damping capacity. It has slightly better mechanical properties as die cast than as sand cast.

Mechanical Properties

Tensile properties. Sand castings, F temper: tensile strength, 180 MPa (26 ksi); yield strength, 55 MPa (8 ksi); elongation, 19%. Die castings, F temper: tensile strength, 165 MPa (24 ksi); yield strength, 83 MPa (12 ksi); elongation, 8%.
Tensile properties vs temperature. See Table 36.
Shear strength. Sand castings, F temper: 55 MPa (8 ksi).
Bearing properties. Sand castings, F temper: ultimate bearing strength, 317 MPa (46 ksi); bearing yield strength, 125 MPa (18 ksi)

Mass Characteristics

Density. 1.74 Mg/m³ (0.063 lb/in.³) at 20 °C (68 °F)

Thermal Properties

Liquidus temperature. 649 °C (1200 °F)
Solidus temperature. 648 °C (1200 °F)
Coefficient of thermal expansion. Linear, 27 μm/m·K (15 μin./in.·°F) at 21 to 204 °C (70 to 400 °F)
Thermal conductivity. 122 W/m·K (71 Btu/ft·h·°F) at 20 °C (68 °F)
Latent heat of fusion. 343 to 359 kJ/kg

Electrical Properties

Electrical resistivity. 57 nΩ·m

Fabrication Characteristics

Casting temperature. Sand castings, 705 to 800 °C (1300 to 1470 °F)
Weldability. Can be readily welded and soldered.

Table 36 Typical tensile properties of K1A-F sand castings at elevated temperatures

Testing temperature		Tensile strength		Tensile yield strength		Elongation, %
°C	°F	MPa	ksi	MPa	ksi	
93	200	115	17	48	7	30
204	400	55	8	34	5	71
316	600	28	4	14	2	78

QE22A

Specifications

AMS. Sand castings: 4418C
ASTM. Sand castings: B80. Permanent mold castings: B199. Investment castings: B403
UNS number. M18220
Government. Sand castings: QQ-M-56B. Sand and permanent mold castings: MIL-M-46062B. Permanent mold castings: QQ-M-55
Foreign. Elektron MSR-B. British: DTD 5055. French: MSR-B AECMA MG-C-51. German: DIN1729 3.5164

Chemical Composition

Composition limits. 2.0 to 3.0 Ag; 1.75 to 2.5 Nd-rich rare earths; 0.4 to 1.0 Zr; 0.1 max Cu; 0.01 max Ni; 0.3 max others (total); rem Mg
Consequence of exceeding impurity limits. Zr content below 0.5% may result in somewhat coarser as-cast grains, lower mechanical properties.

Applications

Typical uses. Sand and permanent mold castings used in the solution treated and artificially aged condition (T6 temper), with high yield strength up to 200 °C (390 °F). Castings have excellent short-time elevated-temperature mechanical properties and are pressure tight and weldable.

Mechanical Properties

Tensile properties. T6 temper: tensile strength, 260 MPa (38 ksi); yield strength, 195 MPa (28 ksi); elongation, 3% in 50 mm or 2 in.
Tensile properties vs temperature. See Table 37 and Fig. 24 and 25.
Compressive properties. T6 temper: compressive strength, 345 MPa (50 ksi); compressive yield strength, 195 MPa (28 ksi)
Hardness. 65 to 85 HB
Poisson's ratio. 0.35
Elastic modulus. Tension, 45 GPa (6.5 × 10⁶ psi) (see also Fig. 26); shear, 17 GPa (2.5 × 10⁶ psi); compression, 44 GPa (6.4 × 10⁶ psi)
Impact strength. Charpy, at 20 °C (68 °F): unnotched, 6.8 to 13.6 J (5 to 10 ft·lb); V-notched, 1.4 to 2.7 J (1 to 2 ft·lb)
Fatigue strength. See Fig. 27.
Creep-rupture characteristics. See Table 38 and Fig. 28.

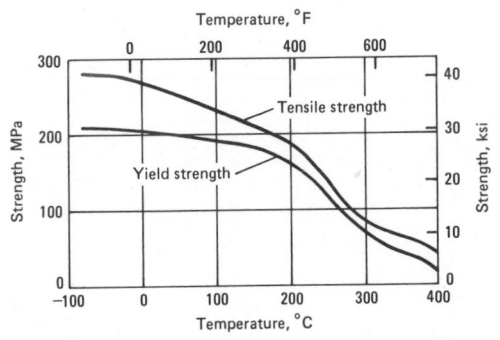

Fig. 24 Effect of temperature on strength of QE22A-T6 sand castings

Specific damping capacity. 0.4 at stress equal to 10% of tensile yield strength

Mass Characteristics

Density. 1.81 Mg/m³ (0.065 lb/in.³) at 20 °C (68 °F)

Thermal Properties

Liquidus temperature. 645 °C (1190 °F)

Solidus temperature. 545 °C (1020 °F)

Coefficient of thermal expansion. Linear, 26.7 μm/m·K (14.8 μin./in.·°F) from 20 to 200 °C (68 to 390 °F)

Specific heat. 1.00 kJ/kg·K (0.24 Btu/lb·°F) at 20 to 100 °C (68 to 212 °F)

Latent heat of fusion. 373 kJ/kg (160 Btu/lb)

Thermal conductivity. 113 W/m·K (65.3 Btu/ft·h·°F)

Electrical Properties

Electrical conductivity. 25.2% IACS at 20 °C (68 °F)

Electrical resistivity. 68.5 nΩ·m at 20 °C (68 °F)

Fabrication Characteristics

Casting temperature. Sand castings, 705 to 800 °C (1300 to 1470 °F); permanent mold castings, 705 to 785 °C (1300 to 1450 °F)

Weldability. Gas-shielded arc welding with welding rod of base-metal composition, good

Solution temperature. 520 to 530 °C (970 to 990 °F)

Aging temperature. 200 °C (390 °F)

Table 37 Typical tensile properties of QE22A sand castings at various temperatures

Testing temperature		Tensile strength		Yield strength	
°C	°F	MPa	ksi	MPa	ksi
20	68	... 263	38.1	208	30.2
100	212	... 235	34.1	193	28.0
200	392	... 193	28.0	166	24.0
300	57	... 83	12.0	69	10

Table 38 Long-time creep properties of QE22A sand castings

| Time under load, h | Tensile stress resulting in creep extension(a) of: | | | | | | | | | |
| | 0.05% | | 0.1% | | 0.2% | | 0.5% | | 1.0% | |
	MPa	ksi	MPa	ksi	MPa	ksi	MPa	ksi	MPa	ksi
At 150 °C (302 °F)										
10	150	21.6	...	...	...	...	...	...	...	...
100	120	17.4	140	20.5	165	23.8	...	...	...	...
1000	90	13.0	105	15.5	125	18.0	150	21.7		
At 200 °C (392 °F)										
10	83	12.0	105	15.0	...	...	...	...	...	...
100	55	8.0	73	10.6	87	12.6	105	15.0	110	16.0
1000	...	...	...	...	55	8.0	72	10.5	78	11.3
At 250 °C (482 °F)										
10	32	4.7	41	6.0	...	...	...	...	...	...
100	17	2.5	26	3.7	32	4.7	40	5.8	...	...
1000	...	...	10	1.4	16	2.3	22	3.2	26	3.8
(a) Does not include initial extension.										

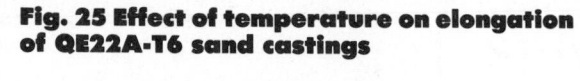

Fig. 25 Effect of temperature on elongation of QE22A-T6 sand castings

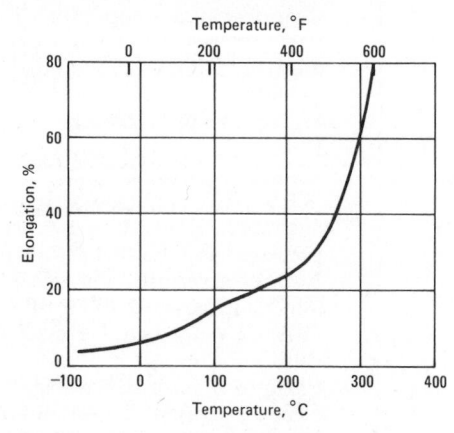

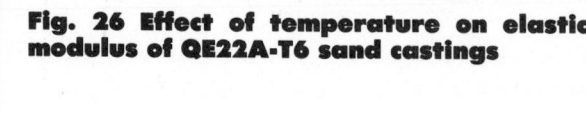

Fig. 26 Effect of temperature on elastic modulus of QE22A-T6 sand castings

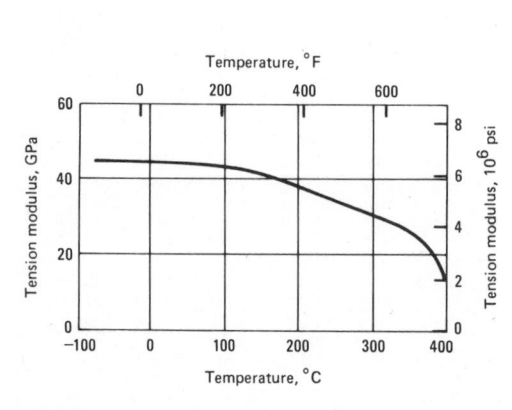

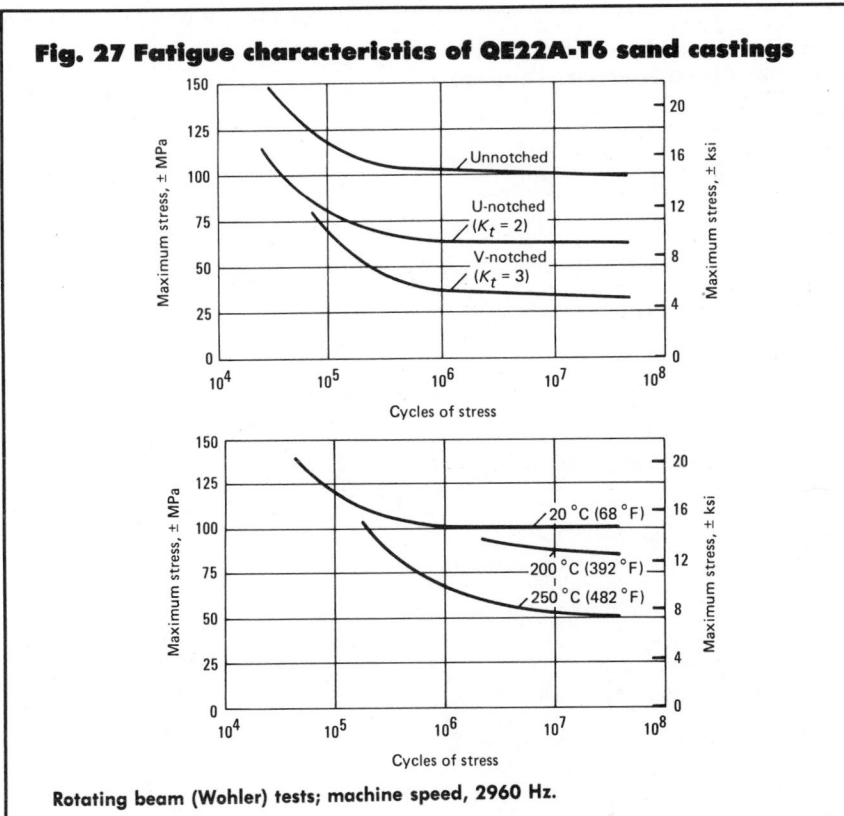

Fig. 27 Fatigue characteristics of QE22A-T6 sand castings

Rotating beam (Wohler) tests; machine speed, 2960 Hz.

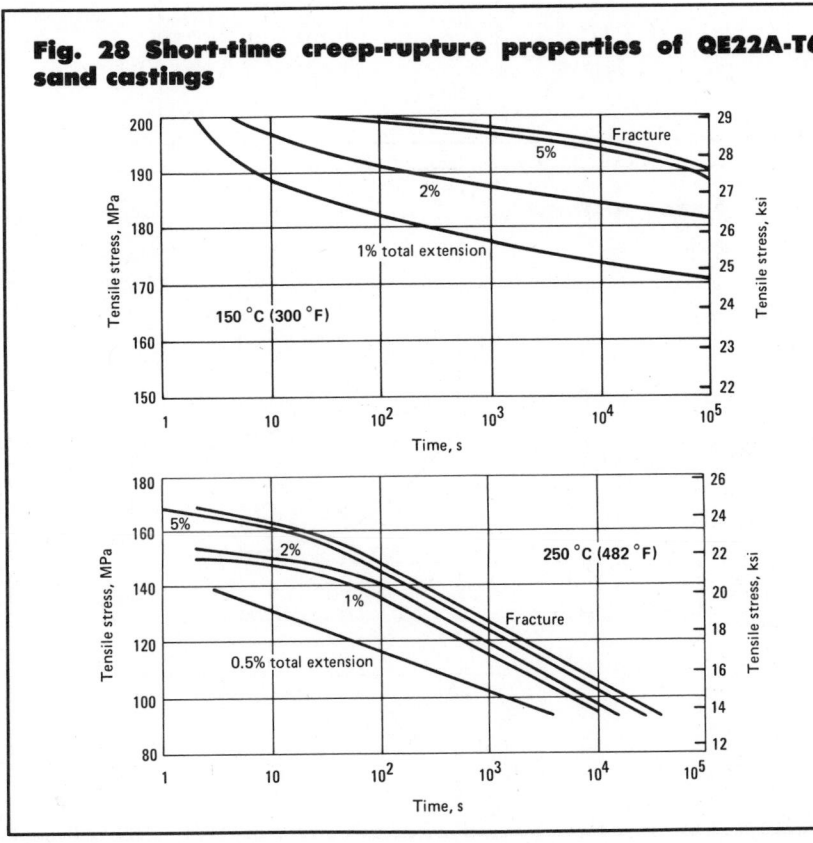

Fig. 28 Short-time creep-rupture properties of QE22A-T6 sand castings

QH21A

Specifications

Foreign. Elektron QH21A

Chemical Composition

Composition limits. 2.0 to 3.0 Ag, 0.6 to 1.6 Th; 0.6 to 1.5 rare earths (composed of at least 70% Nd); 0.40 to 1.0 Zr; 0.20 max Zn; 0.10 max Cu; 0.01 max Ni; 0.30 max others (total); rem Mg (Th plus rare earths, 1.6 to 2.2 optimum)

Applications

Typical uses. Castings used in solution heat treated and artificially aged condition (T6 temper). Ideally suited for aircraft and aerospace components, especially where pressure tightness is required. This alloy is of particular interest to designers and stress engineers for highly stressed components operating at temperatures up to 249 °C (480 °F).

Mechanical Properties

Tensile properties. See Table 39 and Fig. 29.
Fatigue strength. See Fig. 30.
Creep characteristics. See Fig. 31.

Mass Characteristics

Specific gravity. 1.82 Mg/m³ (0.066 lb/in.³) at 21 °C (70 °F)

Thermal Properties

Liquidus temperature. 640 °C (1185 °F)
Solidus temperature. 535 °C (1000 °F)
Coefficient of thermal expansion. Linear, 26.7 μm/m·K (14.8 μin/in.·°F) at 20 to 100 °C (68 to 212 °F)
Specific heat. 1.00 kJ/kg·K (0.23 Btu/lb·°F)
Thermal conductivity. 113 W/m·K (65 Btu/ft·h·°F) at 21 °C (70 °F)

Electrical Properties

Electrical resistivity. 68.5 nΩ·m

Fabrication Characteristics

Castability. Fine-grain alloy with good casting characteristics
Casting temperature. Sand castings, 675 to 815 °C (1250 to 1500 °F)
Weldability. Fully weldable by the gas tungsten-arc process, using rod of the base-metal composition

Table 39 Typical tensile properties of QH21A-T6 castings before and after exposure to elevated temperature(a)

Exposure condition	Tensile strength		Tensile yield strength		Elonga-tion(b), %
	MPa	ksi	MPa	ksi	
Unexposed	276	40.0	207	30.0	4
Exposed 500 h at 199°C (390°F)	284	41.2	205	29.7	8
Exposed 1000 h at 199°C (390°F)	282	40.9	200	29.0	8

(a) Room-temperature values determined using separately cast test bars. (b) In 50 mm or 2 in.

Fig. 29 Effect of temperature on strength of QH21A-T6 sand castings

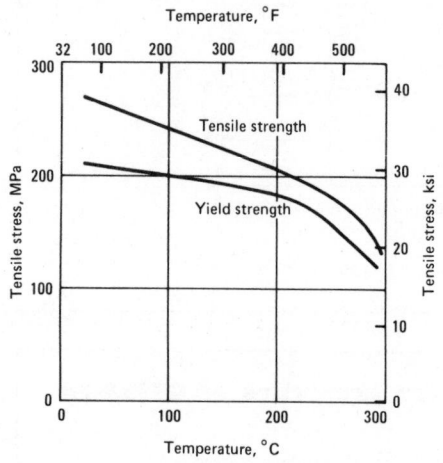

Fig. 30 Fatigue characteristics of QH21A-T6 sand castings

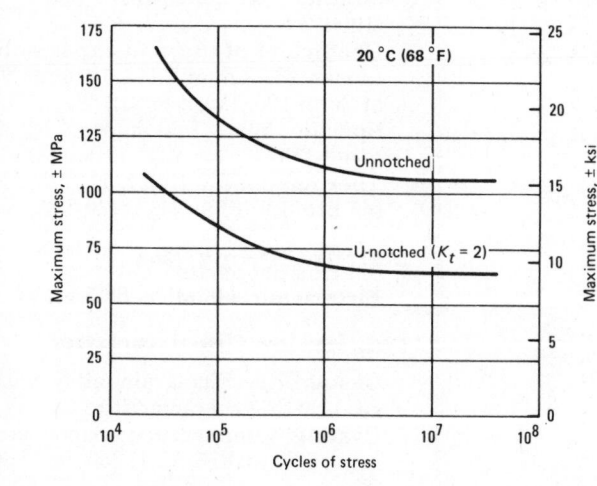
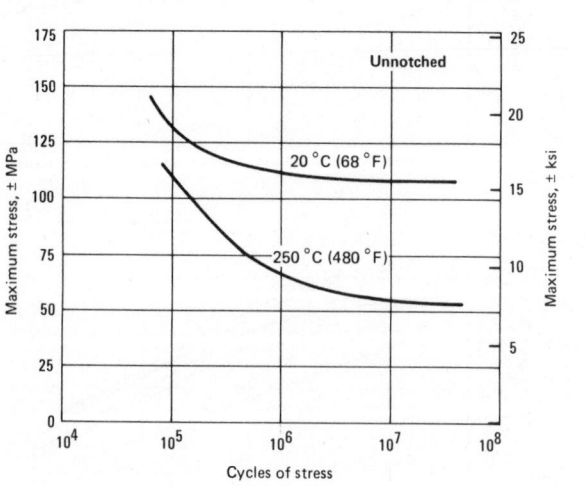

Rotating beam (Wohler) tests; machine speed 2960 Hz.

Fig. 31 Creep properties of QH21A-T6 sand castings

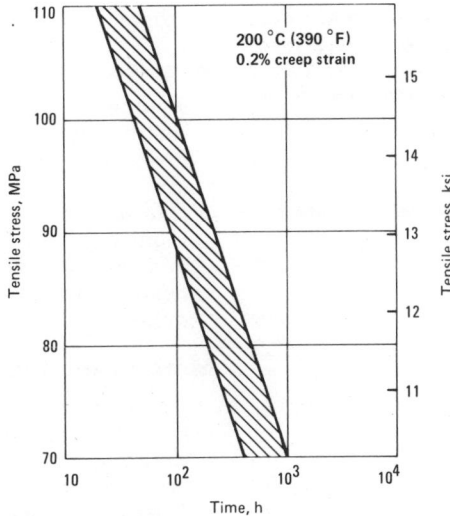

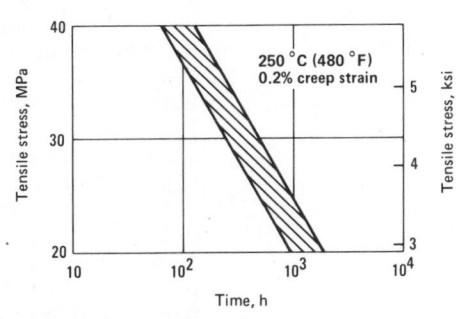

Fig. 32 Typical stress-strain curves for ZE41A-T5 separately cast test bars

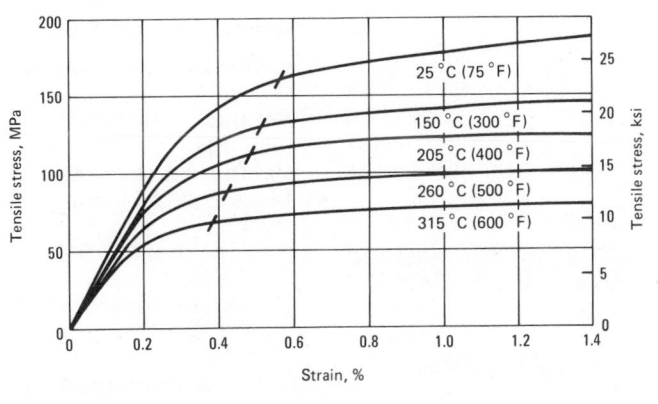

ZE41A

Specifications

ASTM. Sand castings: B80
UNS number. M16410
Foreign. Elektron RZ5. British: BS 2970 MAG5. German: DIN1729 3.5101. French: AIR3380 RZ5

Chemical Composition

Composition limits. 3.5 to 5.0 Zn; 0.75 to 1.75 rare earths (as mischmetal); 0.40 to 1.0 Zr; 0.15 max Mn; 0.10 max Cu; 0.01 max Ni; 0.30 max others (total); rem Mg
Consequence of exceeding impurity limits. 0.6 minimum soluble Zr. Lower content may increase grain size, consequently reducing mechanical properties; weldability also may decrease

Applications

Typical uses. Sand castings used in the artificially aged condition (T5 temper), with better castability than ZK51A and good strength up to 93 °C (200 °F). Useful in pressure-tight applications. Can be welded. Stress relieved at 343 °C (650 °F)

Mechanical Properties

Tensile properties. T5 temper: tensile strength, 205 MPa (30 ksi); yield strength, 140 MPa (20 ksi); elongation, 3.5%
Tensile properties vs temperature. See Table 40 and Fig. 32.
Compressive properties. T5 temper: compressive strength, 345 MPa (50 ksi); compressive yield strength, 140 MPa (20 ksi)
Impact strength. Charpy V-notch, 1.4 J (1 ft·lb)
Elastic modulus. Tension, 45 GPa (6.5 × 10⁶ psi); shear, 17 GPa (2.4 × 10⁶ psi)
Creep characteristics. See Table 41.
Poisson's ratio. 0.35
Bearing properties. Ultimate bearing strength, 485 MPa (70 ksi); bearing yield strength, 350 MPa (51 ksi)
Hardness. 62 HB, 72 HRE

Mass Characteristics

Density. 1.82 Mg/m³ (0.066 lb/in.³) at 20 °C (68 °F)

Thermal Properties

Liquidus temperature. 645 °C (1190 °F)

Solidus temperature. 525 °C (975 °F)

Thermal conductivity. 113 W/m·K (65.3 Btu/ft·h·°F)

Electrical Properties

Electrical resistivity. T5 temper: 60 nΩ·m at 20 °C (68 °F)

Fabrication Characteristics

Casting temperature. Sand castings, 675 to 815 °C (1250 to 1500 °F)

Weldability. Gas-shielded arc welding with welding rod of base metal composition, good; complete all welding before hydrogen treatment; stress relief required

Chemical Composition

Composition limits. 5.5 to 6.0 Zn; 2.1 to 3.0 rare earths; 0.40 to 1.0 Zr; 0.10 max Cu; 0.01 max Ni; 0.30 max others (total); rem Mg

Applications

Typical uses. Sand and investment castings used in solution heat treated and artificially aged condition (T6 temper). Especially useful in thin-section castings for applications requiring high mechanical strength and freedom from porosity. Special heat treatment in hydrogen is required to develop properties.

Mechanical Properties

Tensile properties. T6 temper: tensile strength, 300 MPa (44 ksi); yield strength, 190 MPa (28 ksi); elongation, 10% in 5.65 $\sqrt{A}$

Tensile properties vs temperature. See Table 42.

Compressive properties. T6 temper: compressive strength, 450 MPa (65 ksi); compressive yield strength, 195 MPa (28 ksi)

Hardness. 60 to 85 HB

Poisson's ratio. 0.35

Elastic modulus. Tension, 45 GPa (6.5×10^6 psi); shear, 17 GPa (2.5×10^6 psi); compression, 44 GPa (6.4×10^6 psi)

Impact strength. Unnotched, 0.33 to 0.55 J (0.24 to 0.41 ft·lb); notched, 0.063 to 0.084 J (0.046 to 0.062 ft·lb)

Plane-strain fracture toughness. 168 MPa$\sqrt{m}$ (24.5 ksi$\sqrt{in.}$)

Creep characteristics. See Table 43.

Mass Characteristics

Density. 1.87 Mg/m³ (0.067 lb/in.³) at 20 °C (68 °F)

Thermal Properties

Liquidus temperature. 635 °C (1175 °F)

Solidus temperature. 510 °C (950 °F)

Coefficient of thermal expansion. Linear, 26.5 μm/m·K (14.7 μin./in.·°F)

Specific heat. 0.96 kJ/kg·K (0.23 Btu/lb·°F)

Thermal conductivity. 109 W/m·K (63 Btu/ft·h·°F)

Electrical Properties

Electrical conductivity. 30.9% IACS at 20 °C (68 °F)

Table 40 Typical tensile properties of ZE41A-T5 sand castings at elevated temperatures(a)

Testing temperature		Tensile strength		Tensile yield strength		Elongation(b), %
°C	°F	MPa	ksi	MPa	ksi	
93	200	193	28.0	138	20.0	8
149	300	172	25.0	130	18.8	12
204	400	141	20.5	114	16.5	31
260	500	106	15.4	88	12.7	40
316	600	82	11.9	69	10.0	45

(a) Properties determined on separately cast test bars. (b) In 50 mm or 2 in.

Table 41 Creep properties of ZE41A-T5 sand castings(a)

Time under load, h	Tensile stress resulting in total extension(b) of:							
	0.1%		0.2%		0.5%		1.0%	
	MPa	ksi	MPa	ksi	MPa	ksi	MPa	ksi
At 93 °C (200 °F)								
1	47	6.8	85	12.3	135	20.0	...	...
10	46	6.6	83	12.0	130	19.0	...	...
100	42	6.1	76	11.0	125	18.1	...	...
1000	37	5.4	68	9.8	115	16.5	...	...
At 149 °C (300 °F)								
1	43	6.3	74	10.7	110	16.3	...	...
10	43	6.2	71	10.3	105	15.2	...	...
100	41	6.0	68	9.9	99	14.3	...	...
1000	34	5.0	63	9.1	86	12.5	...	...
At 204 °C (400 °F)								
1	38	5.5	67	9.7	105	15.1	...	...
10	33	4.8	56	8.1	91	13.2	...	...
100	23	3.4	41	6.0	74	10.7	...	...
1000	14	2.1	23	3.3	37	5.4	...	...
At 260 °C (500 °F)								
1	28	4.1	39	5.6	55	8.0	66	9.5
10	16	2.3	23	3.4	35	5.1	43	6.2
100	7	1.0	12	1.8	21	3.0	25	3.6
1000	6	0.84	7	1.0	10	1.4	12	1.7

(a) Properties determined using separately cast test bars. (b) Total extension equals initial extension plus creep extension.

ZE63A

Specifications

AMS. Sand castings: 4425
UNS number. M16630

Government. Sand castings: MIL-M-46062B
Foreign. Elektron ZE63A. British: DTD 5045

Electrical resistivity. 56 nΩ·m at 20 °C (68 °F)

Fabrication Characteristics

Casting temperature. Sand castings, 705 to 815 °C (1300 to 1500 °F)

Weldability. Gas-shielded arc welding with ZE63A welding rod, very good. Must be welded prior to heat treatment.

Table 42 Typical tensile properties of ZE63A sand castings at various temperatures

Testing temperature		Tensile strength		Tensile yield strength	
°C	°F	MPa	ksi	MPa	ksi
20	68	289	41.9	173	25.1
100	212	235	34.1	131	19.0
150	302	187	27.1	111	16.1
200	392	131	19.0	97	14.1

Table 43 Creep properties of ZE63A sand castings

Stress				Time, h, to reach total extension(a) of:							
MPa	ksi	0.15%	0.2%	0.25%	0.3%	0.5%	0.75%	1.0%	2.0%	3.0%	4.0%
At 100 °C (212 °F)											
46	6.7	25	650	1440	...	...	...	...	...	...	...
62	9.0	...	120	480	960	...	...	...	...	...	...
77	11.1	...	...	30	135	1250	...	...	...	...	...
92	13.3	...	...	...	15	280	1400	...	...	...	...
At 150 °C (302 °F)											
39	5.7	70	530	...	...	...	...	...	...	...	...
46	6.7	20	156	...	912	...	...	...	...	...	...
54	7.8	...	8	...	50	720	...	...	...	...	...
62	9.0	...	5	...	35	335	840	1200	...	...	...
70	10.1	...	...	...	12	135	350	550	920	...	...
77	11.1	...	...	...	5	35	90	145	290	350	390

(a) Total extension equals initial extension plus creep extension.

ZH62A

Specifications

AMS. Sand castings: 4448
ASTM. Sand castings: B80
SAE. J465. Former SAE alloy number: 508
UNS number. M16620
Government. Sand castings: QQ-M-56, MIL-M-46062
Foreign. Elektron TZ6. British: BS2970 MAG9. German: DIN1729-3.5102. French: AIR 3380 TZ6

Chemical Composition

Composition limits. 5.2 to 6.2 Zn; 1.4 to 2.2 Th; 0.50 to 1.0 Zr; 0.10 max Cu; 0.01 max Ni; 0.30 max others (total); rem Mg

Applications

Typical uses. Sand and permanent mold castings used in artificially aged condition (T5 temper); for room-temperature service. Highest in yield strength of all magnesium casting alloys except ZK61A-T6 and QE22A-T6

Mechanical Properties

Tensile properties. T5 temper: tensile strength, 240 MPa (35 ksi); yield strength, 150 MPa (22 ksi); elongation, 4% in 50 mm or 2 in. See also Fig. 33.

Compressive yield strength. T5 temper, 150 MPa (22 ksi)
Poisson's ratio. 0.3
Elastic modulus. Tension 45 GPa (6.5 × 10^6 psi); shear, 17 GPa (2.5 × 10^6 psi)
Hardness. 70 HB
Impact strength. Notched Izod, 3.4 J (2.5 ft·lb) at 20 °C (68 °F)

Mass Characteristics

Density. 1.86 Mg/m³ (0.067 lb/in.³) at 20 °C (68 °F)

Thermal Properties

Liquidus temperature. 630 °C (1170 °F)
Solidus temperature. 520 °C (970 °F)
Coefficient of thermal expansion. Linear, 27.1 μm/m·K (15 μin./in.·°F) at 20 to 200 °C (68 to 390 °F)
Specific heat. 0.963 kJ/kg·K (0.41 Btu/lb·°F)
Latent heat of fusion. 373 kJ/kg (160 Btu/lb)
Thermal conductivity. 110 W/m·K (62.9 Btu/ft·h·°F) at 20 °C (68 °F)

Electrical Properties

Electrical conductivity. 26.5% IACS at 20 °C (68 °F)
Electrical resistivity. 65 nΩ·m at 20 °C (68 °F)

Fabrication Characteristics

Casting temperature. Sand castings, 705 to 815 °C (1300 to 1500 °F)

Weldability. Gas-shielded arc welding with EZ33A or ZH62A welding rod, poor; castings should be heat treated after welding.

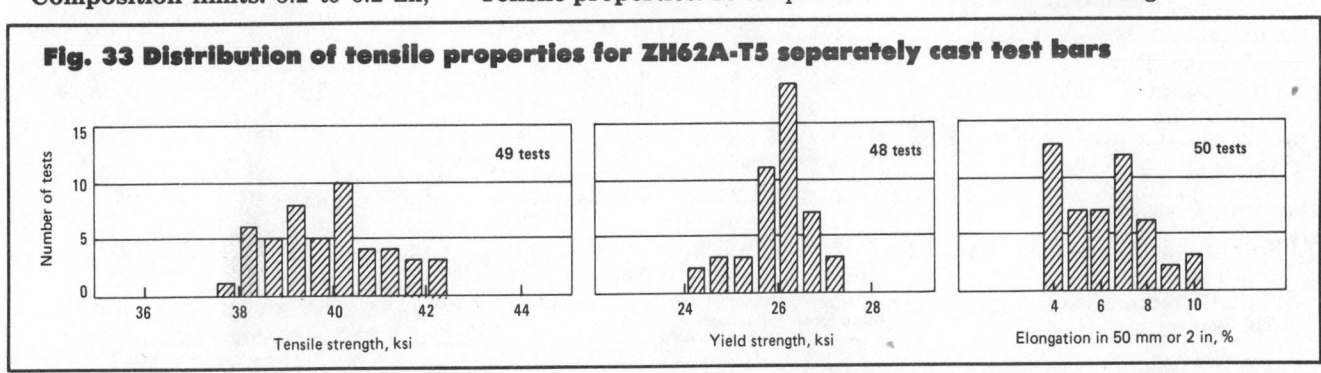

Fig. 33 Distribution of tensile properties for ZH62A-T5 separately cast test bars

ZK51A

Specifications

AMS. Sand castings: 4443
ASTM. Sand castings: B80
SAE. J465. Former SAE alloy number: 509
UNS number. M16510
Government. Sand castings: QQ-M-56a, MIL-M-46062
Foreign. Elektron Z5Z. British: BS2970 MAG4. French: AIR3380 Z5Z

Chemical Composition

Composition limits. 3.6 to 5.5 Zn; 0.50 to 1.0 Zr; 0.10 max Cu; 0.01 max Ni; 0.30 max others (total); rem Mg

Applications

Typical uses. Sand castings used in artificially aged condition (T5 temper), with high yield strength and good ductility. This alloy is suggested for highly stressed parts that are small or relatively simple in design. Solution treatment not required.

Mechanical Properties

Tensile properties. T5 temper: tensile strength, 205 MPa (30 ksi); yield strength, 140 MPa (20 ksi); elongation, 3.5% in 50 mm or 2 in.
Tensile properties vs temperature. See Table 44.
Shear strength. T5 temper, 150 MPa (22 ksi)
Compressive properties. T5 temper: compressive strength, 345 MPa (50 ksi); compressive yield strength, 140 MPa (20 ksi)
Bearing properties. T5 temper: ultimate bearing strength, 485 MPa (70 ksi); bearing yield strength, 350 MPa (51 ksi)
Hardness. 62 HB, 72 HRE
Creep characteristics. See Table 45.

Mass Characteristics

Density. 1.81 Mg/m³ (0.066 lb/in.³) at 20 °C (68 °F)
Solidification shrinkage. 4.2%
Volume change during cooling. 5% contraction from 600 to 20 °C (1110 to 68 °F)

Thermal Properties

Liquidus temperature. 640 °C (1185 °F)
Solidus temperature. 560 °C (1040 °F)
Nonequilibrium solidus temperature. 550 °C (1020 °F)
Coefficient of thermal expansion. Linear, 26 μm/m·K (14.5 μin./in.·°F) at 20 °C (68 °F)
Specific heat. 1.02 kJ/kg·K (0.244 Btu/lb·°F) at 20 °C (68 °F)
Latent heat of fusion. 318 kJ/kg (76 Btu/lb)
Thermal conductivity. 110 W/m·K (63 Btu/ft·h·°F) at 20 °C (68 °F)

Electrical Properties

Electrical conductivity. 28% IACS at 20 °C (68 °F)
Electrical resistivity. 62 nΩ·m at 20 °C (68 °F). Temperature coefficient, 0.16 nΩ·m per K at 20 °C (68 °F)

Fabrication Characteristics

Casting temperature. Sand castings, 705 to 815 °C (1300 to 1500 °F)
Weldability. Gas-shielded arc welding with EZ33A or ZK51A rod (EZ33A preferred), limited; preheating not necessary, but may be used; postweld heat treatment required.

Table 44 Typical tensile properties of ZK51A-T5 sand castings at elevated temperatures(a)

Testing temperature		Tensile strength		Yield strength		Elongation(b), %
°C	°F	MPa	ksi	MPa	ksi	
25	75	275	40	180	26	8
95	200	205	30	145	21	12
150	300	160	23	115	17	14
205	400	115	17	90	13	17
260	500	83	12	62	9	16
315	600	55	8	41	6	16

(a) Properties determined using separately cast test bars. (b) In 50 mm or 2 in.

Table 45 Creep properties of ZK51A-T5 sand castings(a)

Time under load, h	Tensile stress resulting in total extension(b) of:							
	0.1%		0.2%		0.5%		1.0%	
	MPa	ksi	MPa	ksi	MPa	ksi	MPa	ksi
At 95 °C (200 °F)								
1	47	6.8	85	12.3	138	20.0	...	...
10	46	6.6	83	12.0	131	19.0	...	...
100	42	6.1	76	11.0	125	18.1	...	...
1000	37	5.4	68	9.8	114	16.5	...	...
At 150 °C (300 °F)								
1	43	6.3	74	10.7	112	16.3	...	...
10	43	6.2	71	10.3	105	15.2	...	...
100	41	6.0	68	9.9	99	14.3	...	...
1000	34	5.0	63	9.1	86	12.5	...	...
At 205 °C (400 °F)								
1	38	5.5	67	9.7	104	15.1	...	...
10	33	4.8	56	8.1	91	13.2	...	...
100	23	3.4	41	6.0	74	10.7	...	...
1000	14	2.1	23	3.3	37	5.4	...	...
At 260 °C (500 °F)								
1	28	4.1	39	5.6	55	8.0	66	9.5
10	16	2.3	23	3.4	35	5.1	43	6.2
100	7	1.0	12	1.8	21	3.0	25	3.6
1000	6	0.84	7	1.0	10	1.4	12	1.7

(a) Properties determined using separately cast test bars. (b) Total extension equals initial extension plus creep extension.

ZK61A

Specifications

ASTM. Sand castings: B80
SAE. J465. Former SAE alloy number: 513
UNS number. M16610
Government. Sand castings: QQ-M-56B

Chemical Composition

Composition limits. 5.5 to 6.5 Zn; 0.6 to 1.0 Zr; 0.10 max Cu; 0.01 max Ni; 0.30 max others (total); rem Mg

Applications

Typical uses. Simple, highly stressed castings of uniform cross section. High in cost. Intricate castings subject to microporosity and cracking due to shrinkage. Not readily welded. Sometimes used in the artificially aged condition (T5 temper), but usually in the solution heat treated and artificially aged condition (T6 temper) to develop properties fully

Mechanical Properties

Tensile properties. T6 temper: tensile strength, 310 MPa (45 ksi); yield strength, 195 MPa (28 ksi); elongation in 50 mm or 2 in., 10%
Fatigue properties. At least equal to those of the Mg-Al-Zn alloys

Mass Characteristics

Density. 1.83 Mg/m^3 (0.066 lb/in^3) at 20 °C (68 °F)

Thermal Properties

Liquidus temperature. 635 °C (1175 °F)
Solidus temperature. 530 °C (985 °F)
Coefficient of thermal expansion. Linear, 27.0 to 27.1 μm/m·K (15.0 to 15.1 μin./in.·°F) at 20 to 200 °C (68 to 390 °F)

Fabrication Characteristics

Weldability. Not readily weldable. Addition of thorium or rare earths decreases porosity and improves weldability.
Casting temperature. Sand castings, 705 to 815 °C (1300 to 1500 °F)

Corrosion Resistance of Magnesium and Magnesium Alloys

By the ASM Committee on
Magnesium and Magnesium Alloys*

CORROSION RESISTANCE of a magnesium or magnesium-alloy part depends on environmental conditions and on the chemical composition, thermal and mechanical history and surface condition of the part. The factors affecting corrosion resistance that are discussed in this article are the presence of heavy-metal impurities (iron, copper, etc.) in the magnesium or magnesium alloy, the type of environment (rural atmosphere, marine atmosphere, elevated temperature, etc.), the surface condition of the part (bare, treated, painted, etc.), and whether the part is in contact with other parts.

In some environments, contact with parts made of dissimilar metals can lead to severe damage of the magnesium part due to galvanic corrosion, unless the joint is properly designed and protected. Therefore, a separate section on magnesium-to-dissimilar-metal assemblies has been included near the end of this article.

Unalloyed magnesium is not extensively used for structural purposes.

Consequently, it is the corrosion resistance of the alloys of magnesium that usually are of concern. Magnesium alloys, when properly made and applied, are corrosion-resistant and are used successfully in a wide variety of commercial, industrial and aerospace applications. The corrosion that is encountered usually is a result of improper design or application, or inadequate protective finish.

Metallurgical Factors

Chemical Composition. Figure 1 shows the effects of 17 alloying elements on corrosion of magnesium-base binary alloys in 3% NaCl solution. Four elements—iron, nickel, cobalt and copper—have extremely deleterious effects on corrosion resistance; these elements are considered impurities in magnesium, with definite tolerance limits for good corrosion resistance.**

Magnesium-manganese binary alloy M1A, which contains about 1.2% Mn, has comparitively good tolerance for both iron and nickel impurities; the iron tolerance limit is 0.017%, the same as for pure magnesium. Greater amounts of iron are less detrimental

when manganese is present. Manganese also increases the tolerance limit for nickel.

Magnesium-aluminum binary alloys have tolerance limits for iron that are significantly lower than that of unalloyed magnesium. With as little as a few hundredths percent aluminum, the tolerance limit for iron is decreased from 0.017% to a few thousandths percent. With 7% Al, it is about 0.0005%. With 10% Al, the limit is too low to be determined, possibly because of the formation of an iron-aluminum phase even more active than discrete iron particles alone. The addition of even a few hundredths percent manganese to the magnesium-aluminum binaries is sufficient to raise their tolerance limit for iron to 0.002%, probably because of the formation of the compound $(Fe, Mn) Al_3$.

Figure 2 shows how additions of 0.5 and 3% zinc affect the relationship of iron content to corrosion of Mg-6Al-0.2Mn alloys in 3% NaCl solutions. Addition of 0.5% Zn does not shift the position of the tolerance limit for iron (0.002%); it does somewhat reduce corrosion rates at higher percentages of iron. Addition of 3% Zn raises the toler-

---

*See page **X** for committee list.
**J. D. Hanawalt, C. E. Nelson and J. A. Peloubet, Transactions of AIME, 147 (1942), p 273–299.

Fig. 1 Effects of alloy content on corrosion rates of magnesium-base binary alloys tested in 3% NaCl solution

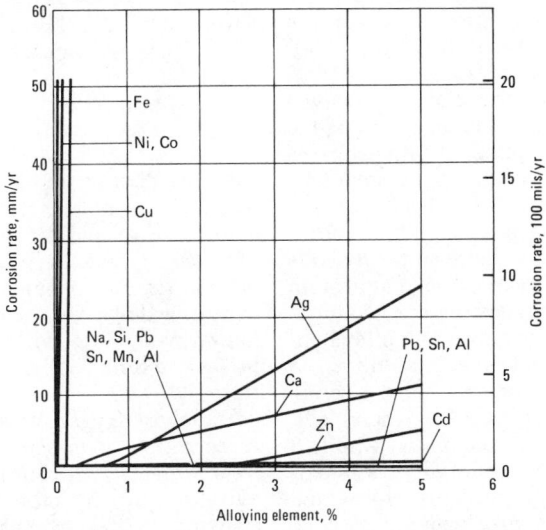

Fig. 2 Effect of zinc content on relationship between iron content and corrosion rate for Mg-6A1-0.2Mn alloys in 3% NaCl solution

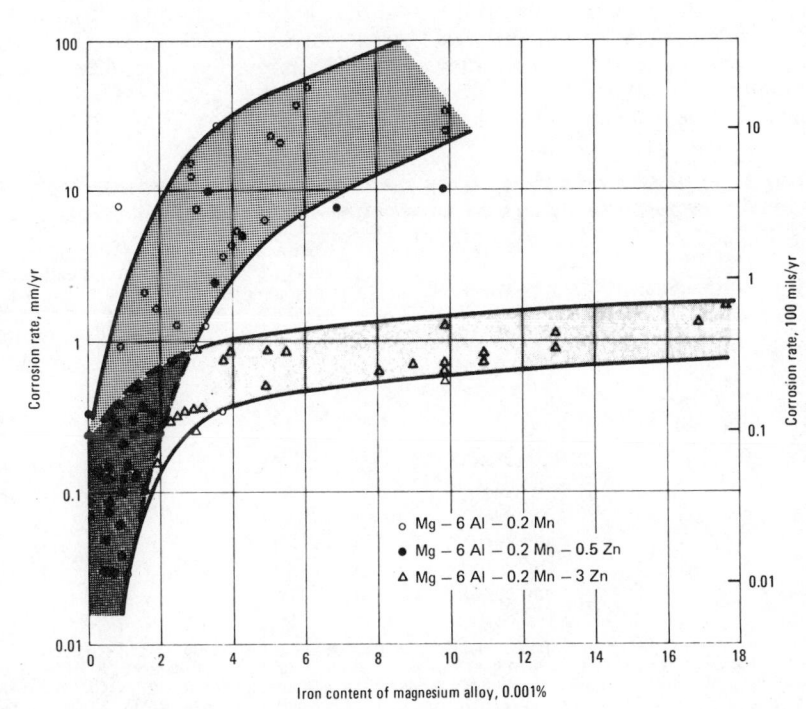

ance limit for iron to 0.003% and reduces the corrosion rate when iron content is higher than 0.003%. Zinc additions show somewhat similar effects in raising the tolerance limits for nickel and copper.

In outdoor urban exposures, corrosion of magnesium alloys containing aluminum decreases as aluminum content increases, up to about 9%. However, additions of cerium (mischmetal), thorium or zinc usually result in increased rates of atmospheric corrosion.

Heat Treatment and Cold Work. With controlled-purity Mg-Al-Mn alloys containing 0 to 1% Zn, there is little if any difference in corrosion rate between as-fabricated material and solution heat treated and aged material, provided the heat treated material is air cooled from the solution heat treating temperature. With similar cooling, controlled-purity alloys containing 2 to 3% Zn corrode at slightly higher rates in the solution heat treated or solution heat treated and aged condition than in the as-cast condition. In alloys containing iron above the tolerance limit, solution heat treatment will increase the corrosion rate by a factor of 2 to 5. This factor is lower if aging follows solution treatment. Data in Table 1 indicate the effect of heat treatment on rates of corrosion of alloys AZ63A and AZ92A in contact with a 3% NaCl solution.

The effect of heat treatment on corrosion rate in 3% NaCl is particularly noticeable adjacent to welds in alloys containing aluminum and zinc (severe attack may occur in the rapidly cooled heat-affected zone.)

Cold working of magnesium alloys, such as stretching or bending, has no appreciable effect on corrosion rate. Corrosion rate is increased by shot or grit blasting because of surface contamination. Acid pickling to a depth of 0.04 to 0.05 mm (0.0015 to 0.0021 in.) often is used to remove the corrosion-active contamination, but re-precipitation of the contaminant is possible with acid pickling. Therefore, fluoride anodizing is used instead when complete removal of the contaminant is essential.

Environmental Factors

Atmospheres. A clean unprotected magnesium-alloy surface exposed to indoor or outdoor atmospheres free from salt spray will develop a gray film (discussed later) that protects the metal from corrosion while causing only negligible losses in mechanical properties. Materials-handling equipment is an example of successful application of unfinished magnesium. Chlorides, sulfates and foreign materials that hold moisture on the surface will promote corrosion and pitting unless the metal is protected by properly applied coatings.

The surface film that ordinarily forms on magnesium alloys exposed to the atmosphere does not protect the metal from further attack. If the film is tight and adherent, however, it decreases the rate of further attack. Unprotected magnesium and magnesium-alloy parts are resistant to rural atmospheres and moderately resistant to industrial and mild marine atmospheres, provided they do not contain joints or recesses that entrap water and are not in contact with dissimilar metals. Magnesium alloys generally do lose strength in industrial and mild marine atmospheres, whether exposed in a stressed or an unstressed condition (see Tables 2 and 3). However, except for exposure close to seawater, the conductivity of any entrapped water present during atmospheric exposure is low. Consequently, the rate of corrosion is low and is less influenced by alloy composition and impurities.

Compositions of the corrosion products that form on magnesium vary from one geographical location to another and from indoor to outdoor exposure. In one urban location (Teddington, England), the compositions of the corrosion products that formed on magnesium exposed to an indoor atmosphere for 412 days and to an outdoor atmosphere for 217 days were as follows:

Indoor exposure
$MgCO_3$ 46.5%
$MgSO_4$ 9.8
$Mg(OH)_2$ 16.3
H_2O 27.4

Outdoor exposure
$MgCO_3 \cdot 3H_2O$ 61.5%
$MgSO_4 \cdot 7H_2O$ 26.7
$Mg(OH)_2$ 6.4
Carbonaceous matter 2.5
$Fe_2O_3 + Al_2O_3$ 2.9

X-ray diffraction analyses of corrosion products scraped from magnesium ingots after 18 months of exposure in a rural atmosphere have shown the presence of $MgCO_3 \cdot H_2O$, $MgCO_3 \cdot 5H_2O$ and $3MgCO_3 \cdot Mg(OH)_2 \cdot 3H_2O$. In an industrial atmosphere, hydrated and basic carbonates were found, together with $MgSO_3 \cdot 6H_2O$ and $MgSO_4 \cdot 7H_2O$.

These analyses, in addition to similar analyses after shorter periods of exposure, indicate that the primary reaction in corrosion of magnesium is formation of magnesium hydroxide, followed by a secondary reaction with carbonic acid (carbon dioxide dissolved in moisture) to convert the hydroxide to hydrated carbonate. In atmospheres contaminated with sulfur compounds, sulfites or sulfates may also be present in the corrosion product. The sulfates may be formed by the reaction of acidic sulfur-bearing gases with magnesium hydroxide or magnesium carbonate.

Corrosion of magnesium alloys increases with relative humidity. At 9.5% humidity, neither pure magnesium nor any of its alloys shows evidence of surface corrosion after 18 months. At 30% humidity, a small amount of an amorphous phase and slight corrosion are evident. At 80% humidity, an amorphous phase is present over about 30% of the surface, and the surface exhibits considerable corrosion. Crystalline magnesium hydroxide is formed only when relative humidity is at or above 93%.

In marine atmospheres heavily loaded with salt spray, magnesium alloys require protection for prolonged survival. Magnesium alloys are less resistant to atmospheric corrosion than aluminum alloys but considerably more resistant than low-carbon steel. Table 4 compares a magnesium alloy, an aluminum alloy and a low-carbon steel on the bases of loss in weight and loss in tensile strength in three different atmospheres. None of the metals was given surface protection. Table 5 gives data on corrosion resistance for bare specimens of one common magnesium alloy (AZ31B-H24) exposed to three different atmospheric conditions; these data are based on losses in mechanical properties.

Fresh Water. In stagnant distilled water at room temperature, magnesium alloys rapidly form a protective film that prevents further corrosive action. Small amounts of dissolved salts in water, particularly chlorides or heavy-metal salts, will break down the protective film locally, resulting in pitting.

Dissolved oxygen plays no major role in corrosion of magnesium, in either fresh water or salt solutions. However, agitation, or any other means of destroying or preventing formation of a protective film, leads to corrosion. When magnesium is immersed in a small volume of stagnant water, its corrosion rate is negligible. When the water is constantly replenished or circulated through an ion-exchange bed so that the solubility limit of magnesium hydroxide is never reached, the corrosion rate becomes significant. For example, AZ31B showed little attack after 35 days in stagnant distilled water at 52 °C (125 °F), but when the water was continuously replenished so that the pH remained at about 6.8, the rate of attack increased to 0.18 mm/yr (7 mils/yr).

Table 1 Typical corrosion rates and electrode potentials for several magnesium alloys in contact with 3% NaCl solution

Alloy	Condition(a)	Form	Corrosion rate in 3% NaCl, g/m²·d(b)	Potential vs saturated calomel(c) 7 days	Potential vs saturated calomel(c) 1 h
AZ31B	F	Sheet, extrusion	3	−1.55	−1.59
HK31A	H24	Sheet	2	−1.70	−1.74
HM21A	T8	Sheet	2	−1.59	...
AZ61A	H24	Sheet	2	...	...
M1A	F	Extrusion	4	−1.59	...
ZK60A	T5	Extrusion	8	−1.50	−1.60
AZ63A	F	Sand casting	15	...	−1.57
AZ63A	T4	Sand casting	40	...	...
AZ63A	T6	Sand casting	20	...	...
AZ92A	F	Sand casting	40	...	...
AZ92A	T4	Sand casting	100	...	...
AZ92A	T6	Sand casting	60	...	...

(a) F: as-cast, as-extruded or as-rolled. T4: solution heat treated. T5: artificially aged. T6: solution heat treated and artificially aged. T8: solution heat treated, strain hardened and artificially aged. H24: strain hardened and partially annealed. (b) Fourteen days alternate immersion in 3% NaCl. Surface of specimen ground with Al_2O_3. (c) Measured in stagnant 3% NaCl solution at 25 °C (77 °F). Surface of specimen ground with Al_2O_3.

Table 2 Effect of industrial atmosphere on tensile properties of sand cast magnesium alloys(a)

Alloy	Original properties Tensile strength MPa	ksi	Yield strength MPa	ksi	Elongation, %	% change in 5 years Tensile strength	Unstressed Yield strength	Elongation	Stress of 69 MPa (10 ksi)(b) Tensile strength	Yield strength	Elongation
AZ92A-T6........	301	43.6	173	25.1	1.6	−5	−4	−20	−11	−7	...
AM100A-T61.....	288	41.8	159	23.0	2.0	−3	−2	...	−7	+2	+10

(a) Average of several 13-mm (½-in.) test bars, exposed and tested in Cleveland, with surface as cast. (b) Static stress applied during exposure.

Table 3 Percentage changes in tensile strength and elongation of magnesium alloys exposed unstressed to the industrial atmosphere of New Kensington, PA

Alloy	Product	Tensile strength MPa	ksi	Original properties Yield strength MPa	ksi	Elongation, %	% change in 4 years Tensile strength	Elongation
AZ31B-F	Sheet(a)	356	51.6	285	41.3	11.8	−15	−16
AZ31B-O	Sheet(a)	248	36.0	141	20.4	22.2	−19	−15
AZ31B-H24	Sheet(a)	285	41.3	188	27.2	20.0	−13	−25
	Sheet(a)	304	44.1	239	34.7	12.7	−17	−45
AZ31B-H24(b)	Sheet(a)	312	45.2	250	36.2	9.5	−17	−37
	Sheet(a)	307	44.5	248	35.9	7.8	−15	−33
AZ31B-F(c)	Strip, extruded(d)	268	38.9	159	23.1	17.7	−14	−12
AZ31B-F(c)(e)	Extrusion(f)	262	38.0	204	29.6	15.7	−11	−12
AZ61A-F(c)	Tubing(g)	294	42.7	105	15.2	19.8	−16	−39
AZ61A-F(e)	Rod, extruded(h)	338	49.0	170	24.6	11.2	−11	+4
AZ61A-H24	Sheet(a)	383	55.5	303	43.9	4.7	−19	−34
AZ61A-H24(b)......	Sheet(a)	233	33.8	165	24.0	13.3	−12	−19
AZ61A-H24	Sheet(a)	391	56.7	307	44.5	6.3	−14	−37
AZ61A-H24(b)......	Sheet(a)	364	52.8	280	40.6	6.7	−13	−20
AZ80A(e)	Rod, extruded(i)	343	49.8	221	32.1	14.8	−4	−27
AZ80A-T5(e)	Rod, extruded(i)	392	56.8	279	40.5	8.6	−10	−61
AZ80A-F(e)	Rod, extruded(j)	339	49.2	220	31.9	13.2	−11	−24
MIA-F(e)...........	Rod, extruded(k)	282	40.9	195	28.3	4.4	−11	−9
MIA-H24	Sheet(a)	252	36.5	176	25.5	11.8	−26	−40
	Sheet(a)	252	36.6	181	26.3	13.0	−18	−91
MIA-H24	Sheet(a)	240	34.8	194	28.2	10.2	−19	−82

Note: All test bars except those indicated by footnote (b) were tested without surface treatment. (a) 1.6 mm (0.064 in.) thick. (b) Chrome pickle (Dow 1). (c) With grain. Unless indicated otherwise, all specimens are across the grain. (d) 2 by 4.5 mm (0.080 by 1¾ in.). (e) Cylindrical test bar, 11.1-mm (0.437-in.) diam. (f) 38 mm (1½ in.) square. (g) 64 mm (2½ in.) OD; 2.4-mm (0.095-in.) wall. (h) 76 mm (3 in.). (i) 19 mm (¾ in.). (j) 50 mm (2 in.). (k) 60 mm (2⅜ in.).

Table 4 Results of a 2.5-year atmospheric exposure of an aluminum alloy, a magnesium alloy and a low-carbon steel

Alloy	Corrosion rate µm/yr	mils/yr	% change in tensile strength
Marine Atmosphere(a)			
Aluminum, 2024-T3	2	0.06	2.5
Magnesium, AZ31B-H24	18	0.70	7.4
Low-carbon steel (0.27% Cu)	150	5.91	75.4
Industrial Atmosphere(b)			
Aluminum, 2024-T3	2	0.08	1.5
Magnesium, AZ31B-H24	27.7	1.09	11.2
Low-carbon steel (0.27% Cu)	25.4	1.00	11.9
Rural Atmosphere(c)			
Aluminum, 2024-T3	0.1	0.005	0.4
Magnesium, AZ31B-H24	13	0.53	5.9
Low-carbon steel (0.27% Cu)	15	0.59	7.5

(a) At Kure Beach, NC. (b) At Madison, IL. (c) Near Midland, MI.

Salt Solutions. Severe corrosion may occur in neutral solutions of salts of heavy metals such as copper, iron and nickel. Such corrosion occurs when the heavy metal or the heavy-metal basic salts, or both, plate out to form active cathodes on the anodic magnesium surface.

Table 6 shows the effects of representative oxidizing and nonoxidizing salt solutions on corrosion of a commercial magnesium alloy. Chloride solutions are corrosive because chlorides, even in small amounts, usually break down the protective film on magnesium, although sodium chloride is less corrosive than sodium iodide (see Table 6). Fluorides form insoluble magnesium fluoride and consequently are not

Table 5 Effect of exposure on tensile properties of bare magnesium alloy AZ31B-H24

Exposure time, years	Tensile strength			Yield strength			Elongation(a)	
	MPa	ksi	% change	MPa	ksi	% change	%	% change
Original Properties (Longitudinal Direction)								
0 283	283	41.1	···	226	32.8	···	6.2	···
Tidewater Exposure(b)								
0.375 229	229	33.2	−19.2	167	24.2	−26.2	4.1	−33.9
1 253	253	36.7	−10.7	···	···	···	2.0	−67.7
3 146	146	21.2	−48.4	155	22.5	−31.4	1.2	−80.6
Marine-Atmosphere Exposure(b)								
0.375 289	289	41.9	+1.9	224	32.5	−0.9	12.7	−104.8
3 265	265	38.4	−6.6	196	28.4	−13.4	7.9	−27.4
Urban-Atmosphere Exposure(c)								
1 277	277	40.2	−2.2	219	31.7	−3.4	8.4	−35.5
3 261	261	37.9	−7.8	197	28.5	−13.1	9.6	−54.8
Original Properties (Transverse to Direction of Rolling)								
0 293	293	42.5	···	229	33.2	···	10.0	···
Tidewater Exposure(b)								
0.375 276	276	40.0	−5.9	219	31.8	−4.2	4.0	−60.0
1 249	249	36.1	−15.1	···	···	···	2.1	−79.0
3 123	123	17.8	−58.1	56	8.1	−75.6	1.0	−90.0
Marine-Atmosphere Exposure(b)								
0.375 285	285	41.4	−2.6	226	32.8	−1.2	10.0	0
3 275	275	39.9	−6.1	204	29.6	−10.8	12.0	+20.0
Urban-Atmosphere Exposure(c)								
1 288	288	41.8	−1.6	225	32.7	−1.5	11.3	+13.0
3 278	278	40.3	−5.2	206	29.9	−9.9	9.6	−4.0

Note: Each value represents an average of at least four specimens.
(a) In 50 mm or 2 in. (b) Exposed at the Naval Air Station, Norfolk, VA. (c) Exposed at Washington, DC.

appreciably corrosive. Oxidizing salts—especially those containing chlorine or sulfur atoms—are more corrosive than nonoxidizing salts; but chromates, vanadates, phosphates and many others are film-forming, and thereby retard corrosion, except at elevated temperatures.

The amounts and distribution of impurities in magnesium alloys strongly affect results of corrosion tests in salt solutions. For example, chloride impurities from included flux particles can cause catastropic breakdown of a magnesium-alloy part because magnesium chloride will react with water to form magnesium hydroxide and release chlorine for further chloride formation, which starts a chain reaction.

Acids and Alkalis. Magnesium is attacked rapidly by all mineral acids except hydrofluoric and chromic acids. Hydrofluoric acid does not attack magnesium to an appreciable extent because it forms an insoluble, protective magnesium fluoride film on the magnesium; however, pitting develops at low acid concentrations. With increasing temperature, the rate of attack increases at the liquid line, but to a negli-gible extent elsewhere. Magnesium has been used commercially for float-control units, tanks and piping for closed units handling concentrated hydrofluoric acid.

Pure chromic acid attacks magnesium and its alloys at a very low rate. However, traces of chloride ion in the acid will markedly increase this rate. A boiling solution of 20% chromic acid in water is widely used to remove corrosion products from magnesium alloys without attacking the base metal.

Aqueous solutions of organic acids attack magnesium and magnesium alloys at various rates. Dilute acetic or glycolic acid, with sodium or magnesium nitrate added to reduce the rate of attack, is used as a pickling treatment for wrought magnesium products.

Magnesium and its alloys are not significantly attacked by fatty acids, with or without water, at room or elevated temperature. A superficial layer of the corresponding magnesium soap forms and prevents continued attack.

Magnesium is resistant to dilute alkalis and even to 50% caustic liquors at temperatures as high as 60 °C (140 °F), but the corrosion rate increases rapidly with temperatures above 60 °C. Therefore, hot alkaline cleaners suitable for cleaning steel are recommended for cleaning magnesium. Such cleaners should have a pH of 11 to 13.

Magnesium-alloy fabricated assemblies have been used extensively for concrete wall forms and buckets. Concrete does not adhere to magnesium as well as it does to wood or steel, minimizing the need to use parting oils for stripping.

Even where calcium chloride has been added to the concrete mixture for low-temperature pouring, magnesium forms have been used for over ten years with no failures due to corrosion. Steel fasteners used to hold magnesium forms together, however, should be tin plated to minimize galvanic corrosion.

Magnesium alloys also are used successfully in tools for handling plaster and cement.

Organic Compounds. Aliphatic and aromatic hydrocarbons, ketones, ethers, glycols and higher alcohols are not corrosive to magnesium and its alloys. Ethyl alcohol causes slight attack. Anhydrous methyl alcohol causes

Table 6 Corrosion of magnesium alloy AZ61A in salt solutions

Salt	Corrosion rate, g/m²·d
Nonoxidizing Acid Salts	
Aluminum sulfate	11.2
Zinc chloride	77
Sodium acid tartrate	15.5
Sodium dihydrogen phosphate	8.5
Oxidizing Acid Salts	
Ammonium persulfate	46.5
Sodium dichromate	0.4
Ferric sulfate	29.7
Nonoxidizing Neutral Salts	
Sodium chloride	1.4
Sodium bromide	0.6
Sodium iodide	2.9
Sodium fluoride	0.3
Sodium sulfate	0.8
Sodium nitrate	0.5
Oxidizing Neutral Salts	
Sodium chromate	0.2
Sodium chlorate	5.9
Sodium pyrophosphate	5.1
Nonoxidizing Alkaline Salts	
Sodium metasilicate	0.07
Sodium sulfite	0.2
Sodium metaborate	0.5
Sodium phosphate	0.3
Oxidizing Alkaline Salts	
Calcium hypochlorite	15.5
Sodium hypochlorite	0.5
Sodium iodate	4

Test conditions: Specimen size, 75 by 25 by 1.5 mm (3 by 1 by 0.06 in.); surface preparation, HNO₃ pickling; temperature, 35 °C (95 °F); duration, 7 days; volume of testing solution, 100 ml; concentration of solution, 3% by weight. Specimens were alternately immersed 30 s in solution and held 2 min in air.

severe attack. The rate of attack in the latter is reduced by the presence of water.

Pure halogenated organic compounds do not attack magnesium at ambient temperatures. At elevated temperatures, or if water is present, such compounds may cause severe corrosion—particularly compounds whose final products are acidic.

Dry fluorinated hydrocarbons, such as the freon refrigerants, usually do not attack magnesium alloys at room temperature, but when water is present they may stimulate significant attack. At elevated temperatures, fluorinated hydrocarbons may react violently with magnesium alloys.

In aqueous organic acids, such as fruit juices, attack of magnesium is slow but measurable. Milk causes attack, particularly when souring.

At room temperature, ethylene glycol solutions produce negligible corrosion of magnesium used alone or galvanically connected to steel; at elevated temperatures (such as 115 °C or 240 °F), the rate increases, and galvanic corrosion occurs unless inhibitors such as those used for aircraft engines are added. Alloy M1A has been used extensively for aircraft oil and gasoline tanks. However, tetraethyl lead and ethylene dibromide (added to raise the antiknock rating of gasoline) will cause severe pitting at the liquid line in gasoline-water mixtures, unless slushing compounds are used or special capsules containing slightly soluble chromates or other inhibitors are placed in the sump of the tank. Alloy AZ91C is used in applications involving contact with oil, such as accessory housings.

Gases. Dry chlorine, iodine, bromine and fluorine cause little or no corrosion of magnesium at room temperature. Even when it contains 0.02% water, dry bromine causes no more attack at its boiling temperature (58 °C or 136 °F) than at room temperature. The presence of a small amount of water causes pronounced attack by chlorine, some attack by iodine, and negligible attack by bromine or fluorine. Dry, gaseous sulfur dioxide or ammonia causes no attack at ordinary temperatures; if water vapor is present, some corrosion may occur.

Soils. Except when used as galvanic anodes, magnesium alloys have good corrosion resistance in clay or nonsaline sandy soils but poor resistance in saline sandy soils. Bituminous paints over primed surfaces retard attack and provide protection against corrosion equivalent to or better than comparable treatments on low-carbon steel.

Elevated Temperature. The effect of elevated temperature on corrosion of magnesium depends largely on the purity of the metal. The corrosion rates of controlled-purity magnesium and its alloys immersed in 3% sodium chloride solution at 100 °C (212 °F) are approximately twice the rates at room temperature. When cathodic impurities, such as iron and nickel, are present in amounts greater than their tolerance limits, corrosion and pitting increase as temperature increases.

Water vapor in air or in oxygen sharply increases the rates of oxidation of magnesium and its alloys at temperatures above 100 °C (212 °F), but boron trifluoride, sulfur dioxide and sulfur hexafluoride are effective in reducing them. The presence of boron trifluoride or sulfur hexafluoride in the ambient atmosphere is particularly effective in suppressing high-temperature oxidation up to and including the temperature at which the alloy normally ignites.

The corrosion rate of magnesium in ordinary tap water containing approximately 70 ppm chloride is low, but increases with temperature. Figure 3 shows comparative data for four alloys. Alloy M1A shows a greater tendency toward severe pitting with increasing temperature than magnesium alloys containing aluminum. Soluble fluorides are effective inhibitors of corrosion of magnesium in hot water. For example, addition of 0.10% sodium fluoride (by weight) reduced the corrosion rate of alloy AZ31B in boiling distilled water from 0.41 to 0.02 mm/yr (16 to 0.9 mil/yr).

The effects of temperature on corrosion rates of typical magnesium alloys in steam are shown in Fig. 4. Alloy

Fig. 3 Effects of temperature on corrosion rates of magnesium alloys M1A, AZ61A, AZ92A and A10 in tap water

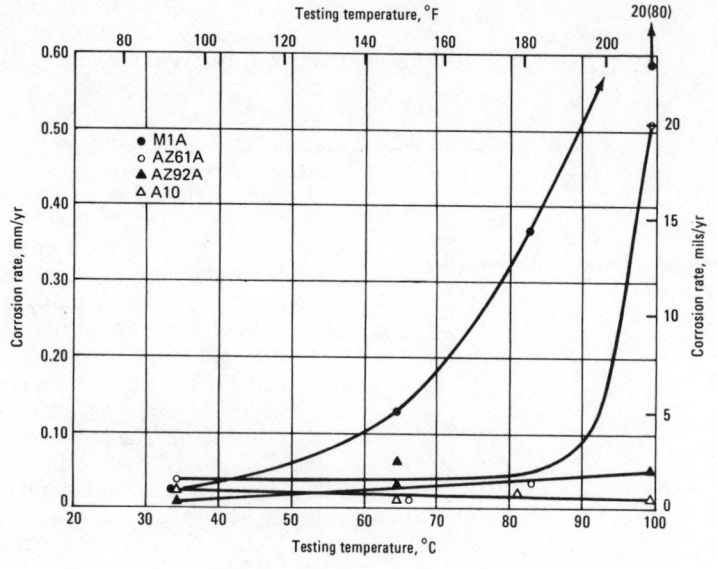

Fig. 4 Effects of temperature on corrosion rates of magnesium alloys M1A, AZ63A and A4 in steam

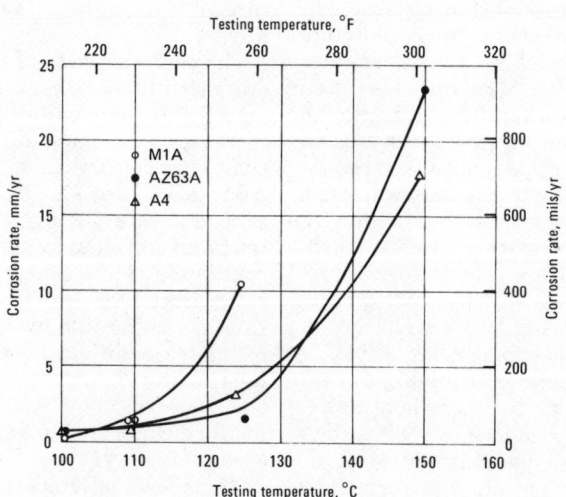

M1A exhibits a corrosion rate greater than those of alloys containing aluminum or aluminum and zinc. None of the magnesium alloys is resistant to steam at temperatures above 120 °C (250 °F).

The rate of oxidation of magnesium in oxygen increases with temperature. At elevated temperature, the oxidation rate is a linear function of time. Most common alloying elements, such as aluminum and zinc, increase the rate of oxidation at a given temperature. Cerium, lanthanum, calcium, and beryllium reduce the rate of oxidation below that of pure magnesium. Beryllium additions have the most striking effects, protecting some alloys at temperatures up to the melting point over extended periods of time.

Protective Surface Treatments

Chemical treatments of magnesium alloys are used extensively to provide bases for paint and to protect the metal. (See Volume 2 of the 8th Edition of Metals Handbook for details of these treatments.) The dip coatings (chrome pickle and dichromate treatments; see Table 7) are very thin coatings used primarily for protection during shipment and storage and as primers for subsequent painting. These coatings should not be heated above 260 °C (500 °F). Discretion should be exercised when using chromate coating without subsequent painting because some chromate coatings are pyrophoric—they spark when hit.

Anodic coatings are thicker and harder than dip coatings. Flash (thin) coats often are used as paint bases. Thicker (heavier) coats (without paint) give some protection against corrosion in moderately corrosive atmospheres, but sealing the coating by impregnation or painting is usually desirable. In marine atmospheres, or in any location conducive to galvanic corrosion, resin impregnation followed by painting with chromate-pigmented primer and then with a glossy-finish top coat is necessary for maximum serviceability. Several commercially available primers and finishing lacquers are satisfactory. Anodic coatings can be heated to 345 °C (650 °F) with no reduction in corrosion resistance. With suitable heat-resistant paints, exposure times of 1/2 h at 425 °C (800 °F) have been survived by anodically treated ramjet parts made of alloys HK31A and HM21A.

Table 7 Some protective chemical treatments for magnesium alloys

Common names	Military specifications	Type of treatment	Alloys on which treatment may be used	Type of solution	Time, min	Temperature °C	°F	Current density, A/m²	Remarks
Chrome pickle, Dow 1	MIL-M-3171C, type I	Dip	All alloys	Chromic acid, nitric acid, water	1 to 15	88-100	190-212	None	Protection during shipment and storage; good paint base; slight dimensional etching loss
Dichromate, Dow 7	MIL-M-3171C, type III (DOW 7)	Dip	All wrought alloys except those containing thorium	Sodium dichromate, calcium or magnesium fluoride, water	30	100	212	None	Good combination of paint-base and protective qualities; requires acid fluoride pickle pretreatment
Dow 17	MIL-M-45202B, type I, class C (thin) type II, class D (heavy)	Anodic	All alloys	Ammonium acid fluoride, sodium dichromate, phosphoric acid, water	1 to 30	71-82	160-180	55 to 540 at 60 to 95 v	Thin coatings (0.005 mm or 0.0002 in.) for flexibility and paint base; heavy coatings (0.03 mm or 0.001 in.) for maximum corrosion and abrasion resistance
HAE	MIL-M-45202B, type I, class A (thin) type II, class A (heavy)	Anodic	All alloys	Potassium hydroxide, aluminum hydroxide, trisodium phosphate, potassium fluoride, potassium manganate, water	60 to 90	24-29	75-85	160 at 110 v	Thin coatings (0.005 mm or 0.0002 in.) for flexibility and paint base; heavy coatings (0.03 mm or 0.001 in.) for maximum corrosion and abrasion resistance

Environmental Corrosion Testing

Accelerated corrosion tests such as alternate (intermittent) immersion in salt water and salt spray often are used to compare the corrosion resistance of magnesium alloys relative to each other as well as to other metals. In addition, such tests are used to determine the relative merits of protective chemical treatments and the galvanic compatibilities of dissimilar metals coupled to magnesium. However, the results of such tests must be used with caution because there usually is poor correlation with environmental exposure in rural, urban, industrial and marine atmospheres. Table 1 presents typical corrosion rates for several commercial magnesium alloys in a 3% NaCl solution; data on electrolytic-solution potential for some of these alloys are also given. In Table 4, a magnesium alloy, an aluminum alloy and a low-carbon steel are compared on the basis of actual performance in rural, industrial and marine atmospheres.

Corrosion of magnesium alloys in salt solutions is strongly affected by the amounts and distribution of impurities in the alloy. In the atmosphere, except close to seawater, the electrolytic conductivity is low. Therefore, the rate of atmospheric corrosion of magnesium alloys (except near seawater) is lower and less influenced by alloy composition and impurities than the rate of corrosion in salt solutions.

Corrosion rates of alloys AZ31B, HK31A and HM21A in 3% NaCl solutions are given in Table 1. These data illustrate the difficulty of comparing the corrosion resistance of various magnesium alloys on this basis. For example, salt-water corrosion rates of clean specimens indicate that HK31A and HM21A alloys have better corrosion resistance than AZ31B. However, if the same amount of corrosion-active surface contamination, such as rolled-in iron-bearing mill scale, is present in specimens of all three alloys, the HK31A and HM21A alloys will suffer more severe pitting in a salt solution than AZ31B, because they have higher electrode potentials. Therefore, the

electromotive force between the cathodic contamination and the metal will be greater with the former alloys than with AZ31B.

Results of Testing in Salt Spray and a Marine Environment. Specimens of magnesium alloy AZ31B-H24 sheet and of sand cast alloys AZ63A-F and AZ91C-F were tested in salt spray and in two types of marine environment. Detailed test results are given in Table 8. Compositions (by weight) of the three alloys tested were as follows:

	AZ31B-H24	AZ63A-F	AZ91C-F
Al	2.6	5.8	8.8
Zn	1.0	2.9	0.68
Be	...	<0.0003	<0.0003
Ca	0.115	...	...
Cu	0.0019	0.015	0.013
Fe	0.0007	0.005	0.006
Mn	0.51	0.25	0.22
Ni	0.0005	<0.001	<0.01
Pb	0.0039	...	...
Si	0.0017	<0.05	<0.05
Sn	0.001	...	...

Table 8 Weight loss of bare and coated magnesium alloys after exposure in corrosive environments

Exposure			Surface treatment			
Salt spray, h	Tidewater, days	Marine atmosphere, days	None, g/m²·d	MIL-M-3171C, type III (Dow 7), g/m²·d	MIL-M-45202B, type II, class D (heavy Dow 17), g/m²·d	MIL-M-45202B, type II, class A (heavy HAE), g/m²·d
Alloy AZ31B-H24						
1000	...	...	5.39	0.54	0.35	0.003
1000	...	...	2.34	0.40	0.70	0.57
...	3	...	17.4	...	...	...
...	98	...	1.36	0.10	0.36	0.12
...	111	...	1.63	...	...	...
...	181	...	1.48	0.35	0.44	0.13
...	182	...	1.08	...	...	...
...	273	...	1.60	0.14	0.14	0.06
...	363	...	1.19	0.18	0.17	0.12
...	...	364	0.09	0.06	0.039	0.03
...	...	1052	0.09	0.07	0.065	0.02
Alloy AZ63A-F						
1000	...	...	52.5	35.0	9.40	1.99
1000	...	...	52.1	15.4	4.88	2.92
...	98	...	6.89	1.85	1.03	0.18
...	181	...	11.5	3.23	2.76	0.24
...	273	...	9.27	1.84	1.37	...
...	363	...	7.97	1.41	0.87	0.09
...	...	364	0.08	0.02	0.00	0.00
...	...	1052	0.08	0.04	0.012	0.00
Alloy AZ91C-F						
1000	...	...	61.9	52.0	31.8	2.87
1000	...	...	78.3	51.1	24.6	21.5
...	98	...	22.3	7.17	5.78	0.59
...	181	...	38.1	30.1	23.2	3.37
...	211	...	...	23.0	7.67	0.85
...	302	...	...	12.8	3.52	0.91
...	...	364	0.10	0.004	0.013	0.00
...	...	1052	0.15	0.029	0.013	0.00

The AZ31B-H24 test specimens were panels 1.63 by 100 by 200 mm (0.064 by 4 by 8 in.); the cast alloy specimens were panels 9.52 by 100 by 200 mm (0.375 by 4 by 8 in.). The cast panels were sand blasted and all panels were acid pickled, removing 0.05 mm (0.002 in.) from the cast and 0.03 mm (0.001 in.) from the wrought panels. Groups of 30 panels of each alloy were surface treated to specifications MIL-M-3171C type III (Dow 7); MIL-M-45202B, type II, class D (heavy Dow 17); and MIL-M-45202B, type II, class A (heavy HAE).

The salt-spray panels were subjected to the mist formed by atomizing a 20% NaCl solution in a chamber maintained at 35 +1, −2 °C (95 +2, −3 °F) at the National Bureau of Standards. The panels were inclined 15° from the vertical.

The marine-atmosphere panels were exposed at the Naval Air Station at Norfolk, VA, directly over the water, 3 m (10 ft) above mean tide level, at an angle of 45° from the horizontal and facing east-southeast. The tidewater panels were placed on edge, with the 200-mm (8-in.) dimension parallel to the surface of the water. These panels were totally immersed in the water during high tide and totally exposed to the marine atmosphere during low tide.

Panels were removed from the salt-spray cabinet after 1000 h of exposure, from the tidewater after 98, 181, 273 and 363 days, and from the marine atmosphere after 364 and 1052 days.

Galvanic Corrosion

Magnesium in electrical contact with other metals can suffer severe galvanic corrosion in high-conductivity environments, such as salt solutions, unless one or more of the following measures are taken:

1 Selection of the dissimilar metal for galvanic compatibility with the magnesium, or electroplating the magnesium with a metal having galvanic compatibility with the dissimilar metal.

2 Protection of the magnesium and the dissimilar metal by suitable surface treatments.

3 Use of an insulating washer or gasket between the dissimilar metals to prevent the completion of an electrical circuit.

4 Inhibition of the galvanic cell by using chromates in the sealing compounds or primers.

High-purity (99.99%) aluminum is galvanically compatible with magnesium, but the small amounts of iron and copper normally present in commercial aluminum alloys effectively destroy such compatibility. In addition, the high pH (10.5) of water in contact with magnesium can lead to attack of the aluminum.

Figure 5 illustrates the effects of iron and magnesium in aluminum on the galvanic compatibility of aluminum with magnesium alloy AZ31B. The effects were determined using actual couples immersed in stagnant 3% NaCl solutions. High-purity aluminum causes almost no galvanic corrosion of magnesium, but when the iron content of the aluminum exceeds about 200 ppm (0.02%), cathodic activity becomes significant. Compatibility then diminishes rapidly with increasing iron content. Figure 5 also shows that addition of magnesium to aluminum markedly improves the tolerance limit for iron. This is why rivets made of aluminum alloy 5056, which contains about 5% Mg, have been used successfully in magnesium-alloy structures without causing significant galvanic corrosion.

Table 9 shows the effects of other elements in high-purity aluminum on its galvanic compatibility with magnesium alloy AZ31B. As much as 0.27% Si and 0.30% Mn are not detrimental to galvanic compatibility of aluminum in 3% NaCl or in simulated seawater, but 0.16% Zn is mildly harmful. The activating influence of nickel in aluminum is almost as great as those of iron and copper. The extreme galvanic corrosion of magnesium coupled with a dissimilar metal that occurs in sodium chloride solutions is no criterion for the activity of the same couple in seawater. Marked suppression of such activity can arise because of the magnesium-ion content of seawater. For this reason, the practical value, if any, of tests in sodium chloride solutions is associated with prediction of service life of couples in such saline environments as road splash from sodium chloride deicing salts, rather than in marine environments.

The ratio of cathode area to anode area has a significant effect on the corrosion rate of magnesium alloys in galvanic couples, as illustrated in Table 10.

Magnesium ions in a sodium chloride solution may suppress the galvanic corrosion of magnesium–aluminum alloy couples in such a solution. However, the addition of magnesium chloride to a sodium chloride solution actually increases the dissolution rate of uncoupled commercial-purity magnesium in the sodium chloride. Uncoupled magnesium is attacked less by natural seawater than by a sodium chloride solution of similar concentration, because of the sulfate content of the seawater, which probably stimulates flocculation of magnesium hydroxide sol on the metal surface to form a semiprotective film.

Figure 6 illustrates how suitable control of iron content can improve the compatibility of a commercial aluminum fastener alloy (6061) with magnesium alloy AZ31B. In seawater, 6061 is compatible with AZ31B, but in applications involving exposure to chlorides (such as road splash), AZ31B suffers severe galvanic corrosion. As shown in Fig. 6, aluminum alloy 6061 never becomes completely compatible with AZ31B even when the iron content of the 6061 is very low. An improvement is realized by holding the iron to a maximum of 0.1% (1000 ppm).

Ferrous alloys, nonferrous alloys containing nickel and/or copper, and

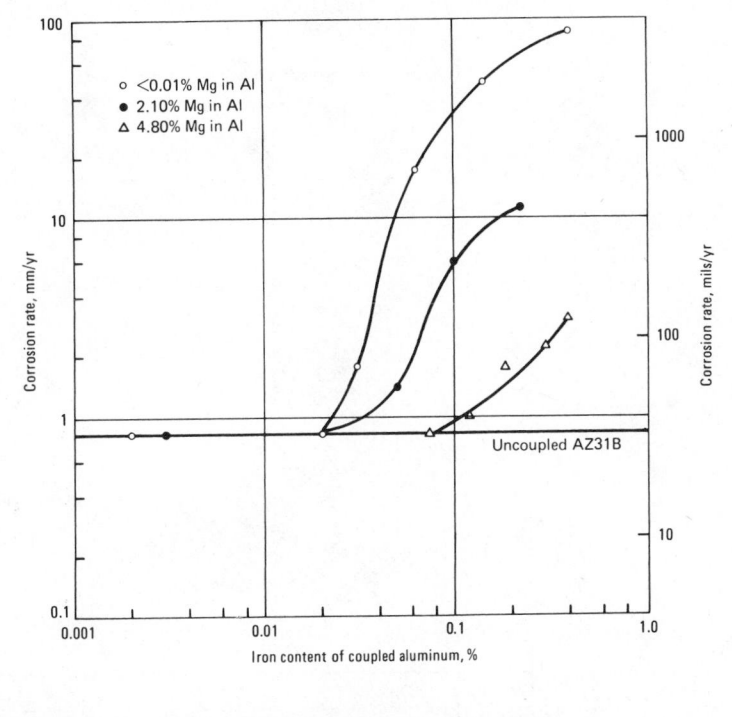

Fig. 5 Corrosion rates of magnesium alloy AZ31B coupled with aluminum containing varying amounts of iron and magnesium in 3% NaCl solution

- ○ <0.01% Mg in Al
- ● 2.10% Mg in Al
- △ 4.80% Mg in Al

Uncoupled AZ31B

Corrosion rate, mm/yr

Corrosion rate, mils/yr

Iron content of coupled aluminum, %

Corrosion rate of uncoupled AZ31B also shown for comparison.

Table 9 Effects of various constituents on galvanic compatibility of high-purity aluminum with magnesium alloy AZ31B

Aluminum	Weight loss in AZ31B, g/m²·d	
	In 3% NaCl	In 3% NaCl + 0.3% MgCl₂
99.9+% pure:		
Unalloyed	9.4	2.4
Alloyed with 0.16% Zn	40.0	5.9
Alloyed with 0.27% Si	7.8	3.3
99.99+% pure:		
Unalloyed	3.6	2.2
Alloyed with 0.015% Ni	12.7	...
Alloyed with 0.30% Mn	3.4	2.3

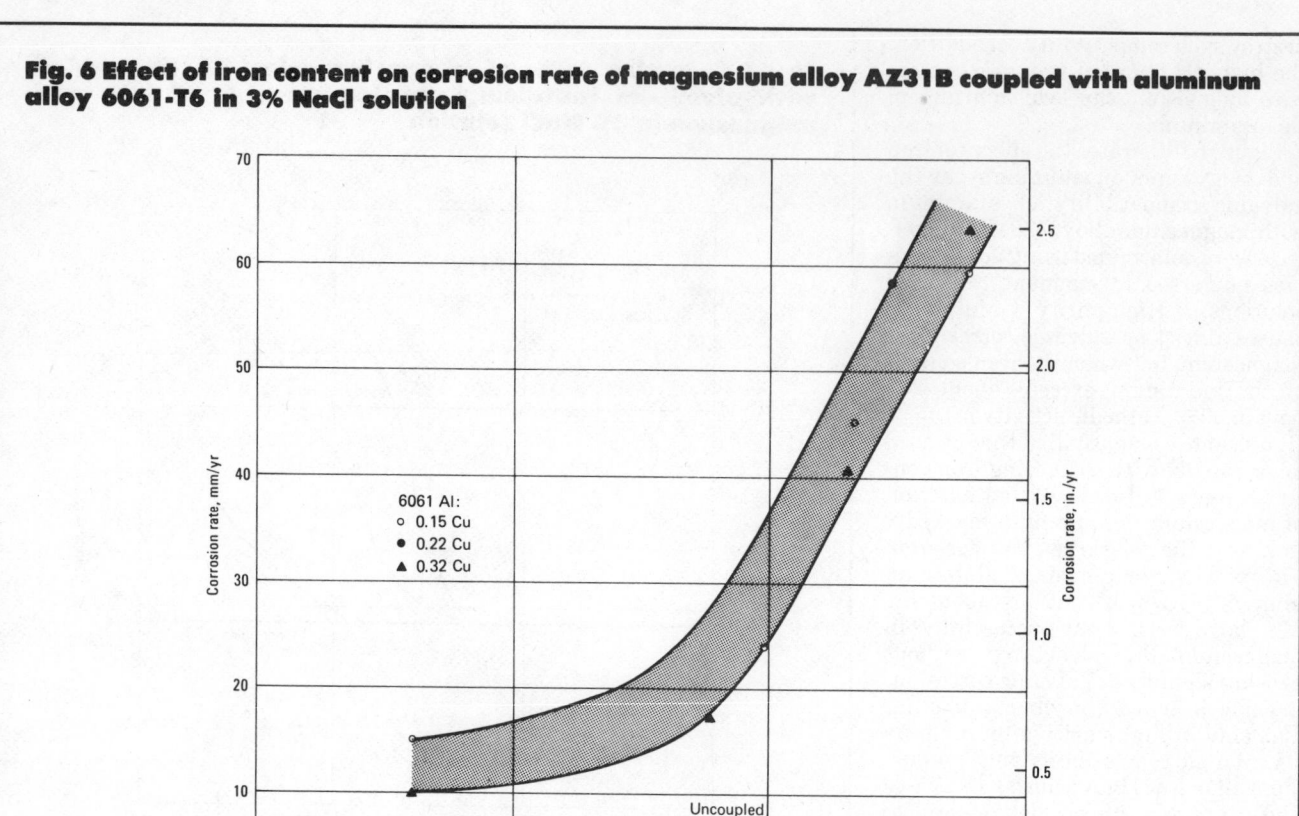

Fig. 6 Effect of iron content on corrosion rate of magnesium alloy AZ31B coupled with aluminum alloy 6061-T6 in 3% NaCl solution

Corrosion rate of uncoupled AZ31B also shown for comparison.

titanium-base alloys have poor galvanic compatibility with magnesium alloys. The compatibility of steel fasteners with magnesium alloys is increased by electroplating the fasteners with tin, cadmium, zinc or chromium. Tin is the best coating for steel fasteners in contact with magnesium alloys, but better still are fasteners made of high-purity aluminum alloys containing magnesium. It should be noted that when magnesium-aluminum couples are immersed in seawater, the alkaline magnesium corrosion products induced by contact with aluminum in turn corrode the aluminum.

Metals compatible with one magnesium alloy may be incompatible with another. For example, reducing the iron content of aluminum alloy 6061 provides a significant reduction in galvanic corrosion when the 6061 is coupled with Mg-Al-Zn alloys, but a much smaller reduction when it is coupled

Table 10 Corrosion rates of galvanic couples of magnesium alloy AZ31B-H24 with commercially pure titanium

Exposure, days	Uncoupled AZ31B-H24	Corrosion rate, $g/m^2 \cdot d$ Cathode-anode area ratio(a) 1:6	6:1
Tidewater Environment(b)			
3	17.4	26.5	88.7
Marine Atmosphere(b)			
358	0.106	0.171	0.372
715	0.095	0.156	0.235
1087	0.082	0.125	0.207
2563	0.077	0.115	0.204
Average	0.090	0.142	0.255
Urban Atmosphere(c)			
368	0.096	0.120	0.148
722	0.101	0.120	0.173
1087	0.096	0.120	0.161
2575	0.078	0.099	0.130
Average	0.093	0.112	0.153

Note: Each value represents an average of at least four specimens. (a) The cathode was commercially pure titanium, and the anode was AZ31B-H24. (b) At Naval Air Station, Norfolk, VA. (c) At Washington, DC.

with Mg-Th alloys. The improvement with Mg-Th alloys is small because these alloys, especially HK31A, develop appreciably higher potentials than do Mg-Al-Zn alloys and thus require a higher degree of galvanic compatibility for metals coupled with them.

In actual practice, magnesium-alloy parts are most widely used under conditions of atmospheric exposure where the corrosive environment is a poor electrical conductor, resulting in very small galvanic current. For example, die-cast magnesium-alloy generator parts with steel inserts and bolted connections showed little galvanic attack after 15 years of exposure in an industrial atmosphere, whereas parts made entirely of steel rusted under the same conditions.

Galvanic Corrosion Tests. At Kure Beach, NC, the effects of a wide variety of metals on promotion of galvanic corrosion of magnesium alloys was determined by fastening sheets of dissimilar metals directly to panels of magnesium alloys AZ31B and AZ61A by means of cadmium-plated steel bolts or aluminum alloy rivets. The magnesium alloy panels were acid pickled before assembly. All galvanic assemblies exposed were in the bare condition, except that rivets of aluminum alloy 2017 used for fastening some assemblies were painted. The results of these tests are summarized in Table 11.

Alloys in group 1 in Table 11 caused the least galvanic corrosion; each successive group of alloys caused a greater amount of such corrosion. The severity of attack was much greater at a site 24 m (80 ft) from the ocean (high-water mark) than at one 240 m (800 ft) from the ocean, but the relative effects of the various metals were not substantially changed.

Assemblies of magnesium-alloy parts with the metals in group 1 require little or no insulation, depending on the environment. The need for protective insulation increases for each of the succeeding groups.

Protection of Assemblies

Contact of magnesium with magnesium, with dissimilar metals and with wood provides potential sources of corrosion of various types unless the magnesium is adequately protected. The protection required varies with the severity of the corrosive environment and the material with which the magne-

sium is coupled. Although corrosive attack from any source can jeopardize the satisfactory performance of magnesium, attack resulting from metal-to-metal contact is probably the most detrimental. Applications conducive to severe attack include those involving continuous outdoor exposure or wetting of the magnesium with corrosive materials such as chloride-laden splash. For indoor use where condensation is not likely, no protection is necessary; there are many such applications. Ordinary paint coatings applied for decorative purposes provide adequate protection under normal indoor conditions.

Magnesium-to-Magnesium Assemblies. For all practical purposes, galvanic corrosion between magnesium alloys is negligible. However, good assembly practice dictates that the magnesium faying surfaces be given one or more coats of a chromate-pigmented primer. Alternatively, or in addition, a technique known as "wet assembly" can be used. A sealing compound is placed between the surfaces at the time of assembly. This ensures that there is no crevice at the joint into which water can gain access by capillary action. Figure 7 shows a properly assembled magnesium-to-magnesium riveted joint. The same procedure should be used for bolted magnesium joints.

Magnesium-to-wood assemblies present an unusual problem because of the absorbency of wood. If wood is wetted, it will hold the water in contact with the magnesium. As the water is absorbed, the natural acids of the wood tend to leach out, and continuous contact of the acids with the magnesium causes corrosion. To protect the magnesium from attack, the wood should be sealed with paint or varnish to prevent absorption of moisture and the faying surface of the magnesium should be treated as described above under Magnesium-to-Magnesium Assemblies.

Table 11 Relative effects of various metals on galvanic corrosion of magnesium alloy AZ31B and AZ61A sheet exposed at the 80- and 800-ft stations, Kure Beach, NC

Group 1 (least effect)	Group 2	Group 3	Group 4	Group 5 (greatest effect)
5052	6063	Alclad 2024	Zinc-plated steel	Low-carbon steel
5056	Alclad 7075	2017	Cadmium-plated steel	Stainless steel
6061	3003	2024		Monel
	7075	Zinc		Titanium
				Lead
				Copper
				Brass

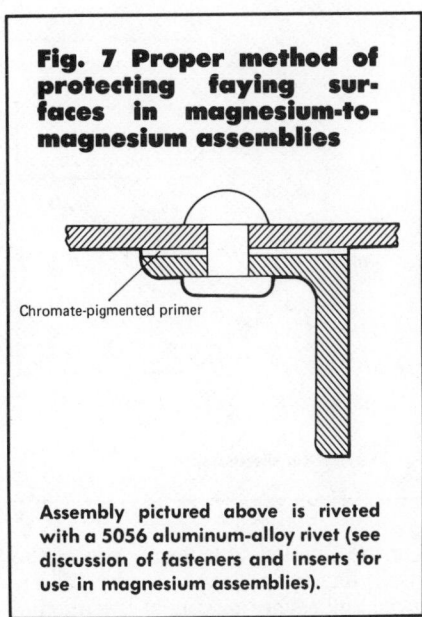

Fig. 7 Proper method of protecting faying surfaces in magnesium-to-magnesium assemblies

Chromate-pigmented primer

Assembly pictured above is riveted with a 5056 aluminum-alloy rivet (see discussion of fasteners and inserts for use in magnesium assemblies).

Magnesium-to-Dissimilar-Metal Assemblies. In order to minimize or prevent galvanic action between magnesium and dissimilar metals, it is necessary to employ one of the following procedures:

1 Protect the dissimilar metal as well as the magnesium.

2 Use dissimilar metals that are compatible with magnesium.

3 Separate one from the other so that the corroding medium cannot complete an electrical circuit.

Moisture-impervious films should be used to separate magnesium from dissimilar metals (often steel or aluminum). If either the dissimilar metal or the magnesium is coated with an unbroken film (see Fig. 8a), no galvanic corrosion occurs. However, if the film is broken, such as at a scratch, corrosion of the magnesium can begin. When the broken film is on the magnesium (as in Fig. 8b), a very small area of magne-

Fig. 8 Effects of moisture-impervious films on corrosion rates of magnesium in assemblies with dissimilar metals

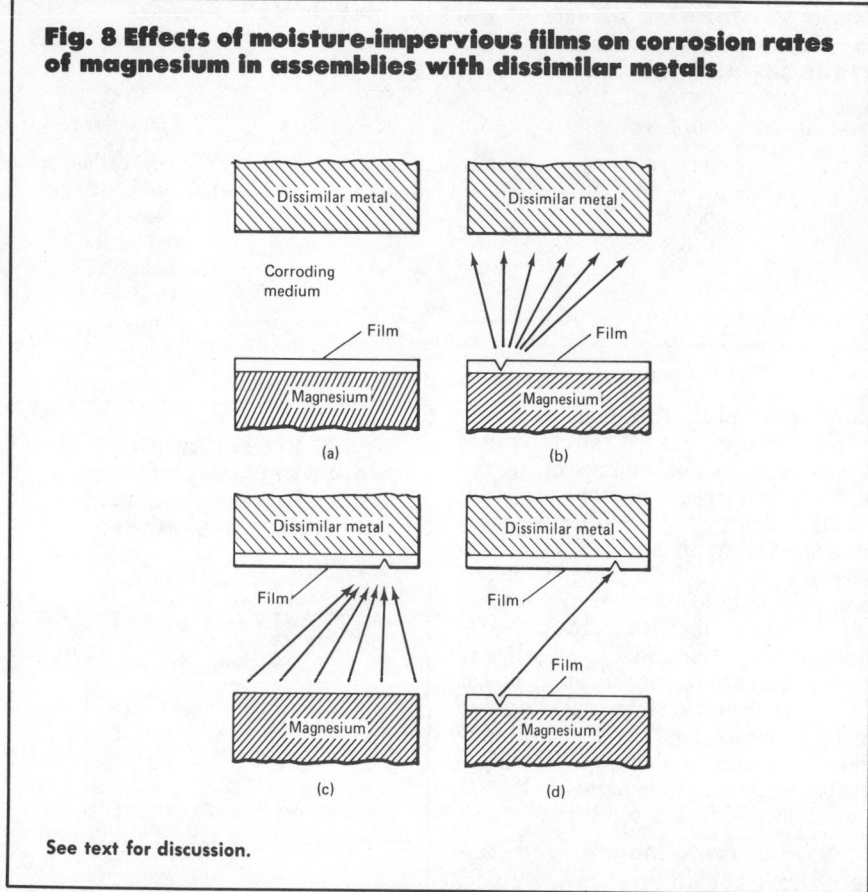

See text for discussion.

Fig. 9 A recommended bolt-and-washer combination for fastening a magnesium member

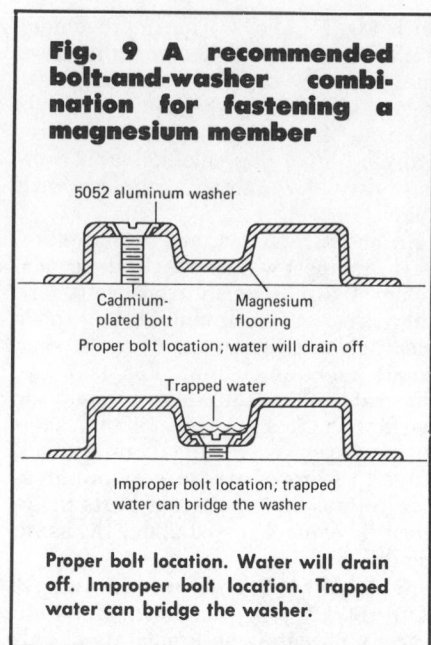

Proper bolt location. Water will drain off. Improper bolt location. Trapped water can bridge the washer.

sium is coupled to a large area of the dissimilar metal, as shown by the arrows, and severe pitting of the magnesium occurs. Therefore, it is preferable that the film be applied to the dissimilar metal, as shown in Fig. 8(c). In this instance, a large area of magnesium is coupled to a very small area of the dissimilar metal, and corrosion of the magnesium is minor. For best protection, both the magnesium and the dissimilar metal should be coated with moisture-impervious films (see Fig. 8d); the probability of breaks in both films lining up closely enough for significant galvanic corrosion to occur is extremely small. In any event, the film or films used should be alkaline-resistant so that they will not break down once corrosion begins and magnesium hydroxide is formed.

Dissimilar metals that are compatible with magnesium, such as aluminum alloys 5052, 6053, 6061 and 6063, should be used for washers, shims, fasteners (rivets and bolts) and structural members. Figure 9 illustrates a recommended method of minimizing attack, in which an aluminum alloy 5052 washer is used with a cadmium-plated steel bolt. To be effective, however, the assembly must be located so that liquid does not readily bridge the washer, as in the groove in the flooring.

Tin is the preferred coating material for steel and brass surfaces in contact with magnesium; it is more effective than cadmium and zinc coatings. However, tin, cadmium and zinc coatings only retard, and do not prevent, galvanic corrosion under the worst conditions. In some milder environments they are satisfactory.

Stainless steel, titanium, copper, Monel, and aluminum alloys such as 2024, Alclad 2024, 7075, Alclad 7075, and 3003 will corrode magnesium when coupled with it under corrosive conditions. Therefore, protection is required.

Proper design also calls for use of nonabsorbent tapes or sealing compounds in joints. Vinyl tapes as thin as 0.08 mm (0.003 in.) and other nonabsorbent tapes, such as rubber tape, can be used. (Cloth-supported tape is not recommended because the cloth acts as a wick.) Choice of tape or sealing compound is governed by the particular assembly and its protective requirements. Figures 10 and 11 illustrate methods of separating magnesium from dissimilar materials in riveted and bolted assemblies. In highly stressed joints, sealing materials must remain nonconductive in use and must extend beyond the joint far enough to prevent bridging of an electrolyte (see Fig. 10 and 11). Figure 12 shows two satisfactory and two unsatisfactory ways of draining and sealing a joint where collection of salt water could not be avoided by other design changes. The magnesium sheet at upper left will corrode where it contacts the aluminum extrusion. At upper right, there is a satisfactory tape separator in the joint. At left in lower sketch, the tape should extend farther beyond the joint so that the drain hole can be drilled through both it and the magnesium sheet. The sketch at lower right shows a drain hole drilled through the tape and the magnesium in a satisfactory manner. A similar effect can be obtained by wrapping the tape up the sides of the extrusion (as in sketch at upper right).

Service experience has demonstrated that thin vinyl tape, with virtually no overlap, suffices to separate magnesium from dissimilar metals and prevent galvanic corrosion in structures that do not trap water, such as those shown in Fig. 13. The lower sketch in Fig. 13 shows how the tape can be extended, insulating the other metal from the magnesium.

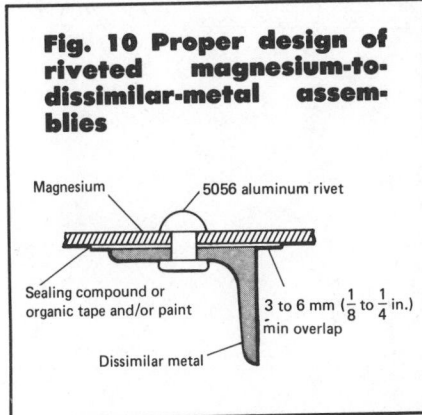

Fig. 10 Proper design of riveted magnesium-to-dissimilar-metal assemblies

Magnesium
5056 aluminum rivet
Sealing compound or organic tape and/or paint
3 to 6 mm ($\frac{1}{8}$ to $\frac{1}{4}$ in.) min overlap
Dissimilar metal

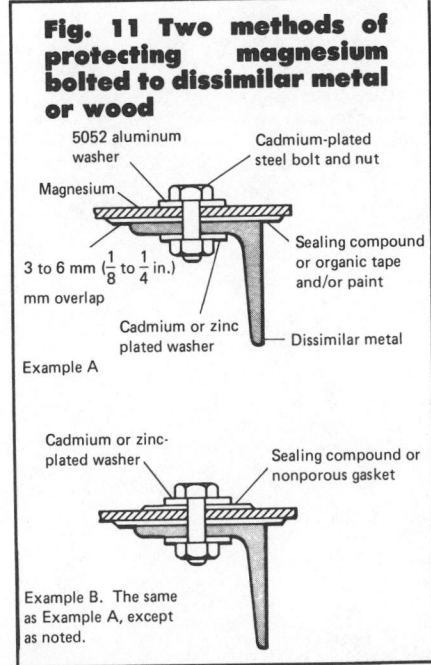

Fig. 11 Two methods of protecting magnesium bolted to dissimilar metal or wood

5052 aluminum washer
Cadmium-plated steel bolt and nut
Magnesium
3 to 6 mm ($\frac{1}{8}$ to $\frac{1}{4}$ in.) mm overlap
Sealing compound or organic tape and/or paint
Cadmium or zinc plated washer
Dissimilar metal
Example A

Cadmium or zinc-plated washer
Sealing compound or nonporous gasket
Example B. The same as Example A, except as noted.

Fig. 12 Four designs of a magnesium-to-dissimilar-metal assembly in which salt water was allowed to collect

Salt water
Aluminum extrusion
Magnesium sheet
Unsatisfactory

Vinyl tape
Drain hole
Satisfactory

Vinyl tape
Drain hole
Unsatisfactory

Vinyl tape
Drain hole
Satisfactory

Sealing tapes and drain holes must be properly used so that the conductive liquid (salt water) does not bridge between the magnesium and the dissimilar metal.

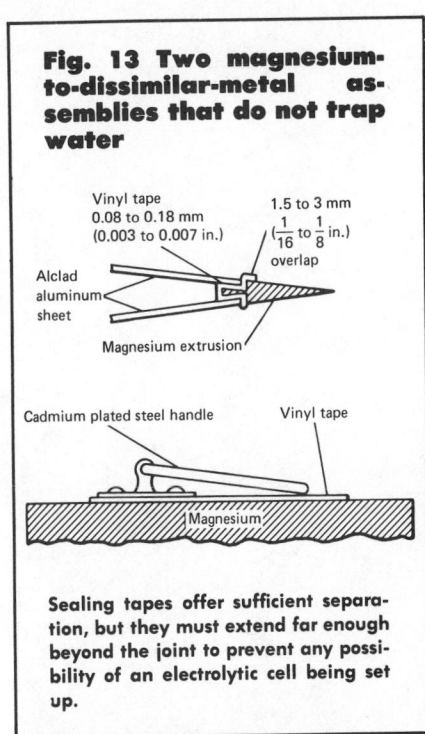

Fig. 13 Two magnesium-to-dissimilar-metal assemblies that do not trap water

Vinyl tape 0.08 to 0.18 mm (0.003 to 0.007 in.)
1.5 to 3 mm ($\frac{1}{16}$ to $\frac{1}{8}$ in.) overlap
Alclad aluminum sheet
Magnesium extrusion

Cadmium plated steel handle
Vinyl tape
Magnesium

Sealing tapes offer sufficient separation, but they must extend far enough beyond the joint to prevent any possibility of an electrolytic cell being set up.

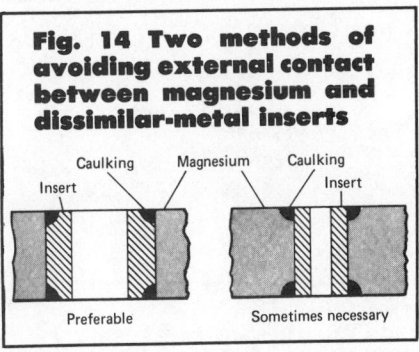

Fig. 14 Two methods of avoiding external contact between magnesium and dissimilar-metal inserts

Caulking
Magnesium
Caulking
Insert
Insert
Preferable
Sometimes necessary

Fasteners and Inserts. Most assemblies require fasteners such as rivets, bolts or screws. Magnesium alloys are not used for making such fasteners, and thus choice of fastener material is important.

Very pure aluminum is almost completely compatible with magnesium. Alloy additions of magnesium, manganese or silicon to aluminum do not affect this compatibility. Alloy additions or impurities such as copper, nickel and iron, and to a lesser degree zinc, can adversely affect the compatibility. However, the adverse effect of these elements is suppressed to a large degree by the presence of magnesium in the aluminum alloy. As a consequence, aluminum alloy 5056 rivets are recommended for use with magnesium. Rivets of aluminum alloys 6053 and 6061 are almost as compatible with magnesium as 5056 rivets and may be used as substitutes (6053 rivets are preferable to 6061 rivets). Aluminum rivets should be anodized or chemically treated before use. Washers of alloy 5052 under cadmium-plated steel bolts are also recommended. Bolts and hardware of compatible aluminum alloys are preferred to coated steel bolts and hardware wherever it is possible to use them.

Steel and copper rivets, as well as steel, nickel, aluminum (other than alloys 5056, 6061 and 6053) and brass bolts and screws, should not be used bare in magnesium assemblies, because they are not compatible with the magnesium. In general, tin, zinc or cadmium plating of these parts is recommended, followed by chemical treatment to provide the plating with better paint-adhesion characteristics. However, such plating only retards, and does not prevent, galvanic corrosion.

Special organic coatings, such as baked vinyl plastisols, epoxies and high-temperature-resistant fluorinated hydrocarbon resin coatings, can be used for insulation of fasteners and inserts.

Most steel inserts used in magnesium parts should be plated with tin, cadmium or zinc. If severe service conditions are expected, annular grooves 3.2 mm (1/8 in.) or more in width should be provided and filled with caulking compound, as shown in Fig. 14. Preferably, these grooves should be formed by machining the inserts, but usually the small size of the inserts makes it necessary to counterbore the hole in the magnesium. Inserts that are well covered with grease or oil during service usually do not require additional protection.

The choice of materials for bimetallic joints is given in the following tabulation, listed in order of preference:

Magnesium-to-aluminum assemblies
1 5056 (wire and rivets)
2 6061 (extrusions and sheet)
3 5052 (sheet)
4 6053 (extrusions and rivets)

Magnesium-to-steel assemblies
1 Tin-coated steel
2 Cadmium-coated steel
3 Zinc-coated steel

Tin

Tin and Its Alloys

By Joseph B. Long
Tin Research Institute, Inc.

TIN was one of the first metals known to man. Throughout ancient history, various cultures recognized the virtues of tin in coatings, alloys and compounds, and use of the metal increased with advancing technology. Today, tin is an important metal in industry even though the annual tonnage used is much smaller than those of many other metals. One reason for the small tonnage is that, in most applications, only very small amounts of tin are used at a time.

Production of tin ore generally is centered in areas far distant from centers of consumption. The leading tin-producing countries (excluding the USSR and China) are, in descending order, Malaysia, Bolivia, Indonesia, Thailand, Australia, Zaire and Nigeria. These countries supply over 85% of total world production.

Cassiterite, a naturally occurring oxide of tin, is by far the most economically important tin mineral. The bulk of the world's tin ore is obtained from low-grade placer deposits of cassiterite derived from primary ore bodies or veins associated with granites or rocks of granitic composition.

Primary ore deposits can contain very low percentages of tin (0.01%, for example), and thus large amounts of soil or rock must be worked to provide recoverable amounts of tin minerals. Unlike ores of other metals, cassiterite is very resistant to chemical and mechanical weathering, but extended erosion of primary lodes by air and water has resulted in deposition of the ore as eluvial and alluvial deposits.

Underground lode deposits of tin ores are worked by sinking shafts and driving adits, the rock being broken from the working face by drilling and blasting. Cassiterite is recovered from eluvial and alluvial deposits by dredging, gravel pumping and hydraulicking. In open-pit mining, a much less widely employed mining method, mechanical and manual methods are used to move tin-bearing materials. After ball-mill concentration of the ore, a final culling is provided at dressing stations.

The final concentrates, which contain 70 to 77% tin, are then sent to the smelter where they are mixed with anthracite and limestone. This charge is heated in a reverberatory furnace to about 1400 °C (2550 °F) to reduce the tin oxide to impure tin metal, which is again heated in huge cast iron melting pots to refine the metal. Steam or compressed air is introduced into the molten metal, and this treatment plus addition of controlled amounts of other elements which combine with the impurities, results in tin of high purity (99.75 to 99.85%). This high-purity tin often is treated again by liquating or electrolytic refining, which provides tin with a purity level approaching 99.99%.

After the tin is refined, it is cast into ingots weighing 12 to 25 kg (26 to 56 lb) or bars in weights of 1 kg (2 lb) and upwards. Tin normally is sold by brand name, and the choice of brand is determined largely by the amounts of impurities that can be tolerated in each end product. High-purity brands of tin may contain small amounts of lead, antimony, copper, arsenic, iron, bismuth, nickel, cobalt and silver. Total impurities in commercially pure tin rarely exceed 0.25%.

Tin in Coatings

Tinplate. The largest single application of tin is in manufacture of tinplate (steel sheet coated with tin), which accounts for about 40% of total world tin consumption. Since 1940, the traditional hot dip method of making tinplate has been largely replaced by electrodeposition of tin on continuous strips of rolled steel. Electrolytic tinplate can be produced with either equal or unequal amounts of tin on the two surfaces of the steel base metal. Nominal coating thicknesses for equally coated tinplate range from 0.38 to 1.54 μm (15 to 60 μin.) on each surface. The thicker coating on tinplate with unequal coatings (differential tinplate) rarely exceed 2.0 μm (80 μin.). Tinplate is produced in thicknesses from 0.15 to 0.60 mm (0.006 to 0.024 in.).

Over 90% of world production of tinplate is used for containers (tin cans). Traditional tinplate cans are made of three pieces of tin-coated steel: two ends and a body with a soldered side-seam. Innovations in can manufacture have produced two-piece cans made by drawing and ironing. Tinplate cans find their most important use in packaging of food products, beer and soft drinks, but also are used for holding paint, motor oil, disinfectants, detergents and polishes. Other applications of tinplate include fabrication of signs, filters, batteries, toys, gaskets, and containers for pharmaceuticals, cosmetics, fuels, tobacco and numerous other commodities.

Electroplating accounts for one of the major uses of tin and tin chemicals. Tin is used in anodes, and tin chemicals are used in formulating various electrolytes, for coating a variety of substrates. Tin electroplating can be performed in either acid or alkaline solutions. Sodium or potassium stannates form the bases of alkaline tin plating electrolytes that are very efficient and capable of producing high-quality deposits. Advantages of these alkaline stannate baths are that they are not corrosive to steel and that they do not require additional agents. Acid electrotinning solutions operate at higher current densities and higher plating rates and require additions of organic compounds.

A number of alloy coatings can be electroplated from mixed stannate-cyanide baths, including coatings of tin-zinc and tin-cadmium alloys and a wide range of tin-copper alloys (bronzes). The bronzes range in tin content from 7 to 98%. Red bronze deposits contain up to 20% tin; high-tin bronzes, called speculum, usually contain about 40% tin.

Tin-nickel and tin-lead electrodeposits are plated from acid electrolytes and are important coatings for printed circuits and electronic components. Tin-cobalt plate is used in applications requiring an attractive finish and good corrosion resistance.

Two ternary alloy electrodeposits are used by industry. These are the copper-tin-lead for bearing surfaces and the copper-tin-zinc alloy for coatings in certain electronic applications.

Hot Dip Coatings. Coating of steel with lead-tin alloys produces a material called terneplate. Terneplate is easily formed and easily soldered and is used as a roofing and weather sealing material and in construction of automotive gasoline tanks, signs, radiator header tanks, brackets, chassis and covers for electronic equipment and sheathing for cable and pipe.

Hot dip tin coatings are used on wire for component leads as well as food handling and processing equipment. In addition, hot dip tin coatings are used to provide the bonding layer for babbitting of bearing shells.

Unalloyed Tin

There are only a few applications where tin is used unalloyed with other metals. Unalloyed tin is well recognized as the most practical lining material for handling high-purity water in distillation plants because it is chemically inert to pure water and will not contaminate the water in any way.

In the manufacture of plate glass, the molten glass is fed from the furnace on to the surface of a molten tin bath, which is protected from oxidation by an atmosphere of nitrogen containing some hydrogen. The natural forces of surface tension and gravity within the bath ordinarily produce plate glass about 6 mm (¼ in.) thick, but thickness of the glass can be varied by the speed at which the molten glass is drawn from the float bath and the temperature of the tin. Glass ribbons are formed with surfaces flat and parallel. Surfaces of the glass are so smooth that surface polishing is not required.

Tin in Alloys

Solders account for the second largest use of tin (after tinplate). Tin is an important constituent in solders because it wets and adheres to many common base metals at temperatures considerably below their melting points. Tin is alloyed with lead to produce solders with melting points lower than those of either tin or lead. Small amounts of various metals, notably antimony and silver, are added to tin-lead solders to increase their strength. These solders can be used for joints subjected to high or even subzero service temperatures.

Both solder compositions and applications of joining by soldering are many and varied. Commercially pure tin is used for soldering sideseams of cans for special food products and aerosol sprays. The electronics and electrical industries employ solders containing 40 to 70% tin, which provide strong and reliable joints under a variety of environmental conditions. General-purpose solders (50Sn-50Pb and 40Sn-60Pb) are used for light engineering applications, plumbing and sheet metal work. Lower-tin solders (20 to 35% Sn, remainder Pb) are used in joining cable and in production of automobile radiators and heat exchangers. Low-tin solders are used in large amounts to fill crevices at seams and welds in automotive bodies, thereby providing smooth joints and contours. Solders containing about 2% tin (remainder lead) are used for can sideseams to provide hermetic seals. Tin-zinc solders are used to join aluminum, while tin-antimony and tin-silver solders are employed in applications requiring joints with high creep resistance.

Alloys for Organ Pipes. Tin-lead alloys are used in the manufacture of organ pipes. These materials commonly are named "spotted metal" because they develop large nucleated crystals or "spots" when solidified as strip on casting tables. The pipes that produce the diapason tones of organs generally are made of alloys with tin contents varying from 20 to 90% according to the tone required. Broad tones generally are produced by alloys rich in lead; as tin content increases, the tone becomes brighter. Cold rolled tin-copper-antimony alloys (95% Sn) also have been used successfully in the manufacture of pipes, and adoption of these alloys has improved the efficiency and speed of fabrication of finished pipes. This composition provides for a bright appearance which is more tarnish resistant than the tin-lead alloys.

Pewter is a tin-base white metal containing antimony and copper. Originally, pewter was defined as an alloy of tin and lead, but to avoid toxicity and dullness of finish, lead is excluded from modern pewter. These modern compositions contain 1 to 8% antimony and 0.25 to 3.0% copper. Pewter casting alloys usually are lower in copper than pewters used for spinning hollowares and thus have greater fluidity at casting temperatures.

Pewter is malleable and ductile and is easily spun or formed into intricate designs and shapes. Pewter parts do not require annealing during fabrication. Much of the costume jewelry produced today is made of pewter alloys centrifugally cast in rubber or silicone molds.

Bearing Materials. Tin has a low coefficient of friction, which is the first consideration in its use as a bearing material. Tin is structurally a weak metal, and when used in bearing applications it is alloyed with copper and antimony for increased hardness, tensile strength and fatigue resistance. Normally, the quantity of lead in these alloys, called tin-base babbitts, is limited to 0.35 to 0.5% to avoid formation of the tin-lead eutectic, which would significantly reduce strength properties at operating temperatures.

Lead-base bearing alloys, called lead-base babbitts, contain up to 10% tin and 12 to 18% antimony. In general, these alloys are inferior in strength to tin-base babbitts, and this must be equated with their lower cost. Segregation of the constituents of these alloys may provide some difficulties during centrifugal casting of linings. During casting, careful selection of rotational speed in relation to bearing size is necessary. Additions of cerium, arsenic, or nickel also assists in controlling segregation of these alloys.

In addition to the tin-base and lead-base babbitts, there is a series of intermediate lead-tin bearing alloys. These alloys have tin and lead contents between 20 and 65% plus various amounts of antimony and copper. Increasing the tin content of these alloys provides higher hardness and greater ease of casting. These alloys are less prone to segregation during melting than lead-base babbitts. Cast intermediate bearing alloys, however, exhibit lower strength values than either tin-base or lead-base babbitts.

Bearing alloys must maintain a balance between softness and strength. Aluminum-tin bearing alloys represent an excellent compromise between the requirements for high fatigue strength and the need for good surface properties such as softness, seizure resistance and embeddability. Aluminum-tin bearing alloys are usually employed in conjunction with hardened steel or ductile iron crankshafts and allow significantly higher loading than tin- or lead-base bearing alloys.

Low-tin aluminum-base alloys (5 to 7% Sn) containing small amounts of strengthening elements, such as copper and nickel, are often used for connecting-rod and thrust bearings in high-duty engines. Strict dimensional tolerances must be adhered to and oil contamination should be avoided. Alloys containing 20 to 40% tin, remainder aluminum, show excellent resistance to corrosion by products of oil breakdown and good embeddability, particularly in dusty environments. The higher-tin alloys have adequate strength and better surface properties, which make them useful for crosshead bearings in high-power marine diesel engines.

Type metals are cast alloys containing various proportions of lead, antimony and tin. They do not readily segregate on solidification from the melt, but they are subject to porosity in the central regions of type characters and slugs because air in molds escapes with difficulty. Good fill of the mold should be ensured by rapid injection, and temperature of the metal should be high enough to avoid premature solidification and entrapment of gases.

Battery-grid Alloys. Lead-calcium-tin alloys have been developed for storage-battery grids—largely as replacements for antimonial lead alloys. Use of ternary lead-base alloys containing up to 1.3% tin has substantially reduced gassing, and therefore batteries whose grids are made of these alloys do not require periodic water additions during their working life. Two chief methods of grid manufacture are casting and fabrication of wrought alloys including punching, roll forging and expanded metal processes.

Copper Alloys. Copper-tin bronzes were some of the first alloys used by man, and these alloys continue to be used for structural and decorative purposes. True bronzes contain tin in amounts up to 10% as well as very small amounts of phosphorus. Quaternary bronzes containing 5 Sn, 5 Zn, 5 Pb, remainder Cu are used for general-purpose castings for applications requiring reasonable strength and soundness, such as gears, pumps, and automotive fittings. Special copper-base alloys with 20 to 24% tin have been used historically for cast bells of excellent tonal quality. Spinodal Cu-Ni-Sn alloys containing 2 to 8.5% tin have excellent elastic properties and have replaced tin-free Cu-Ni alloys in some spring and electrical-contact applications. In addition to these uses in copper-base alloys, small quantities of tin (0.75 to 1.0%) are added to copper-zinc alloys (brasses) for increased corrosion resistance. Cast leaded brasses may contain up to 4% tin.

Dental alloys for making amalgams contain silver, tin, mercury, and some copper and zinc. The copper increases hardness and strength and the zinc acts as a scavenger during alloy manufacture, protecting major constituents from oxidation. Most dental alloys presently available contain 25 to 27% tin and consist mainly of the intermetallic compound Ag_3Sn. When porcelain veneers are added to gold alloys for high-grade dental restoration, 1% tin is added to the gold alloy to ensure bonding with the porcelain.

Cast Irons. The presence of about 0.1% tin in flake or ductile iron castings ensures a completely pearlitic structure, and this pearlite is retained even at elevated temperatures. Commercially pure tin is added to the cast iron in the form of shot, bars or cast pieces; in cupola melting, the tin is commonly added to the ladle or to the cupola spout during tapping. Tin is also added to special mixing chambers along with suitable inoculant materials in the production of ductile iron castings. Because the mixing chambers are an integral part of the mold, this technique allows one-step treatment of the molten metal as it enters the mold and overcomes "fading" (loss of effectiveness of inoculating additions before the metal is cast). In addition, the mixing chamber provides immediate dissolution of the tin in the iron and ensures uniform distribution in the casting.

Titanium Alloys. Tin strengthens titanium alloys by forming solid solutions. Titanium can exist in the low-temperature alpha phase or the higher-temperature beta phase, which remains stable up to the melting point. In titanium alloys, relative amounts of alpha and beta phases present at the service temperature have profound effects on properties. Aluminum additions raise the transformation temperature and stabilize the alpha phase, but may cause embrittlement in amounts greater than 7%. However, with tin additions, increased strength without embrittlement can be obtained in aluminum-stabilized alpha titanium alloys. Optimum strength and workability can be obtained with 5% aluminum and 2.5% tin; in addition, this alloy has the advantage of being weldable. Alpha-beta titanium alloys contain aluminum as an alpha stabilizer and combinations of beta stabilizers (such as chromium, iron, molybdenum, manganese or vanadium), as well as tin and zirconium as substitutional solid-solution strengthening elements. Such alloys have good strength and creep-resistance at elevated temperatures. Strength and forming properties of many of these alloys can be optimized by various heat treatments.

Zirconium alloys are similar to titanium alloys in that the elements they contain can be divided into two classes: alpha stabilizers, which raise the transformation temperature, and beta stabilizers, which lower it. Tin and aluminum are alpha stabilizers in zirconium alloys and enhance high-temperature strength. A commercial series of corrosion-resistant zirconium alloys containing 0.15 to 2.5% tin has been developed for nuclear service.

Powder Applications. Most of the supply of tin powders is used in making sintered bronze or sintered iron parts. However, tin powders are also employed in making paste solders and creams used in the plumbing and electronic manufacturing industries. A minor use of tin and tin alloy powders is in sprayed coatings for food-handling equipment, metallizing of nonconductors and bearing repairs. Tin particles are also used in food can lacquers to decrease dissolution of iron and any exposed lead-base solder by the food product.

Additions of 2% tin powder and 3% copper powder aid sintering of iron compacts. The tin provides a low-melting-point phase which in turn provides diffusion paths for the iron. Iron-tin-copper compacts sintered at 950 °C (1740 °F) have mechanical properties comparable to those of iron-copper powder metallurgy parts containing 7 to 10% copper sintered at 1150 °C (2100 °F). In addition, closer control of finished dimensions is afforded by the iron-tin-copper mixture, which results in improved quality and cost effectiveness.

Sintered compacts made from mixtures of iron and tin-lead solder powders are suitable for certain low-stress engineering applications. Warm pressing of these compacts provides cohesion of the iron-solder mixtures. Warm pressing of these compacts (at about 450 °C or 840 °F) provides cohesion of the iron-solder mixtures, but avoids recrystallizing the iron powder so that any work hardening obtained during compaction is retained. Different properties in pressed-and-sintered compacts can be obtained by varying the pressing conditions and the relative amounts of iron and solder powders.

Tin in Chemicals

The manufacture of inorganic and organic chemicals containing tin constitutes one of the major uses of metallic tin. The use of tin compounds has grown so rapidly over the past quarter century that the tin chemicals industry has been transformed from one based mainly on recovered secondary tin to one that consumes significant amounts of primary ingot tin.

Tin chemicals are used for such widely diversified purposes as: electrolyte solutions for depositing tin and its alloys; pigments and opacifiers for ceramics and glazes; catalysts and stabilizers for plastics; pesticides, fungicides and antifouling agents in agricultural products, paints and adhesives; and corrosion-inhibiting additives for lubricating oils.

Properties of Tin and Tin Alloys

Compiled by Joseph B. Long
Manager
Tin Research Institute, Inc.

Table 1 Designations, chemical compositions and applications of commercially pure tins

Grade designation			Composition, % (a)											General applications
ASTM B339	Designation	Class	Sn max	Sb max	As max	Bi max	Cd max	Cu max	Fe max	Pb max	Ni + Co max	S max	Zn max	
AAA	Electrolytic	Extra-high purity	99.98	0.008	0.0005	0.001	0.001	0.002	0.005	0.010	0.005	0.002	0.001	Analytical standards, research
AA	Electrolytic	High purity	99.95	0.02	0.01	0.01	0.001	0.02	0.01	0.02	0.01	0.01	0.005	Research, pharmaceuticals, fine chemicals
A(b)	A, Straits	High purity; commercial	99.80	0.04	0.05	0.015	0.001	0.04	0.015	0.05	0.01	0.01	0.005	Tinplate, foil, collapsible tubes, block tin products, pewter
B(c)	B	General purpose	99.80	···	0.05	···	···	···	···	···	···	···	···	Less exacting, general purpose
C	C	Intermediate grade	99.65	···	···	···	···	···	···	···	···	···	···	General purpose alloys
D	D	Lower intermediate grade	99.50	···	···	···	···	···	···	···	···	···	···	General purpose alloys
E	E	Common	99.00	(d)	···	···	···	(d)	···	(d)	···	···	···	Cast bronze, bearing metal, general purpose alloys, lead base alloys

(a) The maximum impurity limits listed below, which are from ASTM Standard Classification B339, are not specification limits, but simply guides to the maximum impurity contents commonly found in the various brands of tin that fall into these grades. (b) ASTM Grade A includes about 80 to 90% of the refined tin produced. (c) Grade B is intended for those uses where the specific impurity limitations of Grade A are not critical. (d) Limits of these impurities may be specified for some uses.

Commercially Pure Tins 99+%Sn

Commercial Names

Designations. See Table 1.
Common names. Electrolytic tin, Straits tin. Commercial tin (Grade A) is considered to be "pure tin" with a minimum purity of 99.8%. However, some brands have been consistently purer than 99.8% tin. Many of the properties listed here for pure tin were determined on tin of 99.95% purity.

Specifications

ASTM number. See Table 1.
Government. QQ-T-371, grade A (99.75%)
Foreign. British: BS3252, grade T (99.8%). German: DIN1704, grade A2 (99.75%)

Chemical Composition

Composition limits. See Table 1.

Applications

Typical uses. See Table 1.

Mechanical Properties

Tensile properties. See Table 2.
Hardness:

Temperature		Hardness, HB
°C	°F	
0	32	4.2
20	68	3.9
60	140	3.0
100	212	2.3
140	284	1.7
180	356	1.2
220	428	0.7

Poisson's ratio. 0.33
Elastic modulus. Tension, at room temperature: as cast (coarse grained), 41.6 GPa (6.03×10^6 psi); self annealed (fine grained), 44.3 GPa (6.43×10^6 psi)
Impact strength:

Temperature		Charpy V-notch energy	
°C	°F	J	ft·lb
−80	−112	3.7	2.75
−60	−76	11.5	8.5
−15	+5	28.5	21.0
0	32	44.1	32.5
150	302	22.7	16.75
190	374	20.3	15.0
215	419	2.7	2.0

Fatigue strength. Rotating cantilever tests: at 15 °C (59 °F), 2.9 MPa (430 psi) for 10^7 cycles and 2.6 MPa (380 psi) for 10^8 cycles; at 100 °C (212 °F), 2.4 MPa (340 psi) for 10^7 cycles
Creep-rupture characteristics. At room temperature:

Initial stress		Time,	Extension,
MPa	psi	days	%
1.083	157.0	..551	3.5
1.351	196.0	..551	7
2.256	327.1	..173(a)	101
2.772	402.1	.. 79(a)	132
3.227	468.1	.. 21(a)	119
4.214	611.2	.. 4.6	105
7.069	1025.2	.. 0.5(a)	78

(a) Specimen failed.

Specific damping capacity. Tests on bars vibrating at audio frequencies in the "free-free" mode:

Temperature		Logarithmic decrement	
°C	°F	Polycrystalline	Single crystals
25	77....	0.022	0.0010
50	122....	0.045	0.0013
75	167....	0.060	0.0015
100	212....	0.054	0.0018
125	257....	0.045	0.0024
150	302....	0.060	0.0032

Mass Characteristics

Density. β phase, 5.765 Mg/m³ (0.2083 lb/in.³) at 1 °C (33.8 °F); α phase, 7.2984 Mg/m³ (0.2636 lb/in.³) at 15 °C (59 °F)
Volume change on freezing. 2.8%
Volume change on phase transformation. 27% expansion from β phase to α phase (from tetragonal to cubic crystal structure) during cooling below 13.2 °C (55.8 °F)

Table 2 Tensile properties of commercially pure tin

Temperature		Yield strength		Elongation in 25 mm or 1 in., %	Reduction in area, %
°C	°F	MPa	ksi		
Strained at 0.2 mm/m · min (0.0002 in./in. · min)					
−200	−328	36.2	5.25	6	6
−160	−256	90.3	13.10	15	10
−120	−184	87.6	12.71	60	97
−80	−112	38.9	5.64	89	100
−40	−40	20.1	2.92	86	100
0	32	12.5	1.81	64	100
23	73	11.0	1.60	57	100
Strained at 0.4 mm/m · min (0.0004 in./in. · min)					
15	59	14.5	2.10	75	...
50	122	12.4	1.80	85	...
100	212	11.0	1.60	55	...
150	302	7.6	1.10	55	...
200	392	4.5	0.65	45	...

Note: It is uncertain if the inconsistencies among these data are due to differences in purity or the difference in straining rate.

Thermal Properties

Melting point. 231.9 °C (449.4 °F)
Boiling point. 2770 °C (4118 °F)
Phase transformation temperature. On cooling (β phase to α phase), 13.2 °C (55.8 °F)
Coefficient of thermal expansion. Linear:

Temperature		Coefficient	
°C	°F	μm/m·K	μin./in.·F
50	122	23.1	12.8
100	212	23.8	13.2
150	302	26.7	14.8

Specific heat. 22 J/kg·K (0.053 Btu/lb·°F) at 25 °C (77 °F)
Thermal conductivity. β phase; 60.7 W/m·K (35.1 Btu/ft·h·°F) at 100 °C (212 °F), 56.5 W/m·K (32.6 Btu/ft·h·°F) at 200 °C (392 °F)

Electrical Properties

Electrical conductivity. Volumetric, 15.6% IACS at 20 °C (68 °F)

Chemical Properties

General corrosion behavior. Tin reacts with both strong acids and strong alkalies, but is relatively resistant to nearly neutral solutions. Oxygen greatly accelerates corrosion in aqueous solutions. In general, with mineral acids the rate of attack increases with the temperature and concentration. Dilute solutions of weak alkalies have little effect on tin, but strong alkalies are corrosive even in cold dilute solutions. Salts with an acid reaction attack tin in the presence of oxidizers or air. Tin resists demineralized waters but is slightly attacked near the water line by hard tap waters.

Resistance to specific corroding agents.

Corrosive agent	Resistance	Remarks
Acid, acetic	Slight attack	(increased by air)
Acid, butyric	Resistant	. . .
Acid, citric	Moderate attack	At water line
Acids, fatty	Moderate attack	. . .
Acid, hydrochloric	Severe attack	In presence of air
Acid, hydrofluoric	Severe attack	In presence of air
Acid, lactic	Moderate attack	Increased by air
Acid, nitric	Severe attack	. . .
Acid, oxalic	Moderate attack	(a)
Acid, phosphoric	Resistant	. . .
Acid, salts	Severe attack	Air present
Acid, sulfuric	Severe attack	(b)
Acid, tartaric	Slight attack	. . .
Air	Resistant	. . .
Ammonia	Resistant	. . .
Bromine	Severe attack	. . .
Carbon tetrachloride	Resistant	. . .
Chlorine	Severe attack	. . .
Iodine	Severe attack	. . .
Milk	Resistant	. . .
Motor fuel	Resistant	. . .
Petroleum products	Resistant	. . .
Potassium hydroxide	Severe attack	Increased by air
Sodium carbonate	Slight attack	. . .
Sodium hydroxide	Severe attack	Increased by air
Water, distilled	Resistant	. . .
Water, sea	Slight attack	. . .

(a) Most corrosive of common organic acids. (b) Increased with concentration and in presence of air.

Fabrication Characteristics

Machinability. Excellent

Hard Tin 99.6Sn-0.4Cu

Applications

Typical uses. Collapsible tubes and foil

Mechanical Properties

Tensile properties. Typical tensile strength. Strip, 2.5-mm (0.1-in.) thick: annealed 3 h at 100 °C (212 °F),

23 MPa (3.3 ksi); annealed 3 h at 200 °C (392 °F), 21 MPa (3.1 ksi); hard rolled (80% reduction), 28 MPa (4.0 ksi)

Bursting strength. A tube 25 mm (1 in.) in diameter and 0.1 mm (0.004 in.) in wall thickness burst under an internal pressure of 320 kPa (46 psi)

Bend test. A flattened impact extruded collapsible tube 0.1 mm (0.004 in.) in wall thickness survived 21 bends over 90° jaws (1-kg load)

Thermal Properties

Liquidus temperature. 230 °C (446 °F)

Solidus temperature. 227 °C (441 °F)

Chemical Properties

General corrosion behavior. Resistant to attack by foodstuffs, medicinal products, cosmetics and artist's colors

Antimonial Tin Solder 95Sn-5Sb

Specifications

ASTM. B32, alloy grade 95TA

Chemical Composition

Composition limits. 95 Sn desired, 0.20 max Pb, 4.5 to 5.5 Sb, 0.15 max Bi, 0.08 max Cu, 0.04 max Fe, 0.005 max Al, 0.005 max Zn, 0.05 max As

Applications

Typical uses. Soldering of electrical equipment, joints in copper tubing and cooling coils for refrigerators. High-tin solders are used for joining parts of electrical apparatus because they have higher electrical conductivity than high-lead solders. High-tin solders are also used where lead may be a hazard—for instance, in contact with foodstuffs. Tin solders that contain 5% Sb (or 5% Ag) are suitable for use at higher temperatures than the tin-lead solders.

Mechanical Properties

Tensile properties. Cast: typical tensile strength, 40.7 MPa (5.9 ksi); elongation in 100 mm or 4 in., 38%. Soldered copper joint: typical tensile strength, 97.9 MPa (14.2 ksi)

Shear strength. Cast, 41.4 MPa (6.0 ksi). Soldered copper joint, 76.5 MPa (11.1 ksi)

Impact strength. Cast (Izod test), 27 J (20 ft·lb)

Mass Characteristics

Density. 7.25 Mg/m^3 (0.262 lb/in.^{3})

Thermal Properties

Liquidus temperature. 240 °C (464 °F)

Solidus temperature. 234 °C (452 °F)

Electrical Properties

Electrical conductivity. Volumetric, 11.9% IACS at 20 °C (68 °F)

Electrical resistivity. 145 nΩ·m at 25 °C (77 °F)

Chemical Properties

Resistance to specific corroding agents. Resistant to SO$_2$

Tin-Silver Solder 95Sn-5Ag

Applications

Typical uses. Soldering of components for electrical and high-temperature service. See Antimonial Tin Solder.

Mechanical Properties

Tensile properties. Sheet, 1.02-mm (0.040-in.) thick, aged 14 days at room temperature: typical tensile strength, 31.7 MPa (4.6 ksi); yield strength, 24.8 MPa (3.6 ksi); elongation in 50 mm or 2 in., 49%. Soldered copper joint: typical tensile strength, 96.5 MPa (14 ksi)

Shear strength. Soldered copper joint, 73.1 MPa (10.6 ksi)

Thermal Properties

Liquidus temperature. 245 °C (473 °F)

Solidus temperature. 221 °C (430 °F)

Electrical Properties

Electrical conductivity. Volumetric, 16.6% IACS at 20 °C (68 °F)

Electrical resistivity. 104 nΩ·m at 0 °C (32 °F)

Temperature coefficient of electrical resistivity. 0 to 100 °C (32 to 212 °F), 42.3 pΩ·m per K

Soft Solder 70Sn-30Pb

Commercial Names

Common name. 70-30 solder

Specifications

ASTM. B32, alloy grades 70A and 70B
Government. QQ-S-571, grade Sn70

Chemical Composition

Composition limits. Grade 70A: 70 Sn desired, 30 Pb nominal, 0.12 max Sb, 0.25 max Bi, 0.08 max Cu, 0.02 max Fe, 0.005 max Al, 0.005 max Zn, 0.03 max As. Grade 70B: same as grade 70A except limits for Sb are 0.20 to 0.50
Consequence of exceeding impurity limits. See Soft Solder (60Sn-40Pb).

Applications

Typical uses. Joining and coating of metals (see Antimonial Tin Solder). Except in very special situations, grades 70A and 70B are used for the same applications. Grade 70B is specified when the presence of antimony is required to ensure that the change from beta tin to alpha tin (called the "tin pest"), with the accompanying change in volume and drastic loss in solder strength, does not occur.

Mechanical Properties

Tensile properties. Cast: typical tensile strength, 46.9 MPa (6.8 ksi)
Hardness. 12 HB

Mass Characteristics

Density. 8.32 Mg/m^3 (0.300 lb/in.^{3})

Thermal Properties

Liquidus temperature. 192 °C (378 °F)
Solidus temperature. 183 °C (361 °F)
Coefficient of thermal expansion. Linear: 15 to 110 °C (59 to 230 °F), 21.6 μm/m·K (12.0 μin./in.·°F)

Electrical Properties

Electrical conductivity. Volumetric, 11.8% IACS
Electrical resistivity. 146 nΩ·m

Soft Solder 63Sn-37Pb

Commercial Names

Previous trade name. Eutectic solder
Common name. 63-37 solder
Government. QQ-S-571, grade Sn63

Specifications

ASTM. B32, alloy grades 63A and 63B

Chemical Composition

Composition limits. Grade 63A: 63 Sn desired, 37 Pb nominal, 0.12 max Sb, 0.25 max Bi, 0.08 max Cu, 0.02 max Fe, 0.005 max Al, 0.005 max Zn, 0.03 max As. Grade 63B: same as grade 63A except limits for Sb are 0.20 to 0.50
Consequence of exceeding impurity limits. See Soft Solder (60Sn-40Pb).

Applications

Typical uses. Soldering of electrical components (see Antimonial Tin Solder). Except in very special situations, grades 63A and 63B are used for the same applications. Grade 63B is specified when the presence of antimony is required to ensure that the change from alpha tin to beta tin (called the "tin pest"), with accompanying change in volume and drastic loss in solder strength, does not occur.

Mechanical Properties

Tensile properties. Cast: typical tensile strength, 51.7 MPa (7.5 ksi); elongation in 100 mm or 4 in., 32%. Soldered copper joint: typical tensile strength, 200 MPa (20 ksi)
Shear strength. Cast; 42.7 MPa (6.2 ksi); soldered copper joint, 55.2 MPa (8 ksi)
Hardness. Cast, 14 HB
Impact strength. Cast (Izod test), 20 J (15 ft·lb)
Creep characteristics. Minimum creep rate: at room temperature and 2.3 MPa (335 psi), 0.1 mm/m (100 μin./in.) per day; at 80 °C (176 °F) and 467 MPa (68 psi), 0.1 mm/m (100 μin./in.) per day
Dynamic viscosity. 1.33 mPa·s (0.0133 poise) at 280 °C (536 °F)
Liquid surface tension. 0.490 N/m at 280 °C (536 °F)

Mass Characteristics

Density. 8.42 Mg/m^3 (0.304 lb/in.^{3})

Thermal Properties

Melting point. 183 °C (361 °F)
Coefficient of thermal expansion. Linear: at 15 to 110 °C (59 to 230 °F), 24.7 μm/m·K (13.7 μin./in.·°F)
Thermal conductivity. 50 W/m·K (29 Btu/ft/h·°F) at 0 to 180 °C (32 to 356 °F)

Electrical Properties

Electrical conductivity. Volumetric, 11.9% IACS
Electrical resistivity. 145 nΩ·m

Soft Solder 60Sn-40Pb

Commercial Names

Common name. 60-40 solder

Specifications

ASTM. B32, alloy grades 60A and 60B
Government. QQ-S-571, grade Sn60
Foreign. British: BS219, grade K. German: DIN1707, LSn60Pb, 2.3660; and DIN1707, LSn60Pb (Sb), 2.3665

Chemical Composition

Composition limits. Grade 60A: 60 Sn desired, 40 Pb nominal, 0.12 max Sb, 0.25 max Bi, 0.08 max Cu, 0.02 max Fe, 0.005 max Al, 0.005 max Zn, 0.03 max As. Grade 60B: same as grade 60A except limits for Sb are 0.20 to 0.50
Consequence of exceeding impurity limits. Antimony is slightly detrimental to wetting properties; antimony is an intentional addition to grade 60B, as noted below under typical uses. Bismuth causes some discoloration of solder surface. Copper levels above about 0.25% and iron levels above about 0.1% cause grittiness of solder. Excessive zinc causes oxidation of solder to be more noticeable. Excessive aluminum causes appreciable oxidation of solder. Even at the maximum allowable level of 0.03%, arsenic may cause dewetting problems when soldering brass.

Applications

Typical uses. Solder for electronic and electrical work, especially mass-soldering of printed circuits (see Antimonial Tin Solder). Except in very special situations, grades 60A and 60B are used for the same applications. Grade 60B is specified when the presence of antimony is required to ensure that the change from alpha tin to beta tin (called the "tin pest"), with accompanying change in volume and drastic loss in solder strength, does not occur.

Mechanical Properties

Tensile properties. Bulk solder at room temperature (measurements depend greatly on conditions of casting and testing): tensile strength (mean), 52.5 MPa (7.61 ksi); elongation (range), 30 to 60%. Effect of temperature on properties of specimens cast at 300 °C (570 °F) steel molds (specimens not machined):

Temperature °C	°F	Tensile strength MPa	ksi	Elongation, %
Cast into 150 °C (300 °F) Molds				
19	66 ..	56.4	8.18	60(a)
50	122 ..	45.4	6.58	80(a)
75	167 ..	41.7	6.05	90(a)
100	212 ..	30.9	4.48	110(a)
125	257 ..	19.3	2.80	180(a)
150	302 ..	12.4	1.80	180(a)
Cast into 200 °C (390 °F) Molds				
0	+32 ..	59	8.6	50(b)
−40	−40 ..	76	11.0	50(b)
−80	−112 ..	97	14.1	55(b)
−120	−184 ..	119	17.3	30(b)
−160	−256 ..	112	16.2	10(b)
−200	−328 ..	109	15.8	5(b)

(a) In 22.5 mm (0.89 in.). (b) In 25.4 mm (1.00 in.).

Superplasticity was examined for eutectic and near-eutectic solder compositions. It was found that the strain-rate sensitivity m has a value of about 0.4 at a strain rate of 10^{-4} m/m·s, increasing to a relative maximum of about 0.5 at a strain rate of 10^{-3} m/m·s, then decreasing to a value near 0.2 at a strain rate of 10^{-1} m/m·s

Shear strength. Mean, 37.1 MPa (5.38 ksi) (depends greatly on conditions of casting and testing)

Hardness. 16 HV (depends on casting conditions)

Elastic modulus. Tension (bulk solder), 30.0 GPa (4.35×10^6 psi)

Creep-rupture characteristics. Limiting creep stress, 2.2 to 3.0 MPa (320 to 430 psi) for a strain rate of 10^{-4} m/m per day at room temperature. Rupture life: 1000 h under stress of 4.5 MPa (650 psi) at 26 °C (79 °F); 1000 h under stress of 1.4 MPa (200 psi) at 80 °C (176 °F)

Dynamic liquid viscosity. Estimated, 2.0 mPa·s (0.020 poise) at the liquidus

Liquid surface tension. Estimated: 468 mN/m at 330 °C (626 °F), 461 mN/m at 430 °C (806 °F)

Mass Characteristics

Density. 8.520 Mg/m³ (0.3078 lb/in.³) at 25 °C (77 °F). Temperature dependence, $\rho = 9.079 + (9.708 \times 10^{-4}) T$ Mg/m³, where $T = °C$

Volume change on freezing. 2.4% contraction

Atomic volume. Liquid, 18.26 cm³/mol at 350 °C (660 °F)

Thermal Properties

Liquidus temperature. 189 °C (372 °F)

Solidus temperature. 183 °C (361 °F)

Coefficient of thermal expansion. Linear: 15 to 110 °C (59 to 230 °F), 24 μm/m·K (13 μin./in.·°F)

Specific heat. Estimated: 150 J/kg·K (0.036 Btu/lb·°F)

Latent heat of fusion. Estimated: 37 kJ/kg (16 Btu/lb)

Thermal conductivity. Estimated: 50 W/m·K (28 Btu/ft·h·°F)

Electrical Properties

Electrical conductivity. Volumetric, 11.5% IACS

Electrical resistivity. 149.9 nΩ·m

Thermoelectric potential. Same as pure tin when measured against copper

Temperature of superconductivity. 7.05 K. Critical field 83.2 mT at 1.3 K

Fabrication Characteristics

Dissolution rate of copper. In liquid solder:

Temperature °C	°F	Dissolution rate μm/s	nin./s
250	482	0.47	18.5
275	527	0.70	27.6
300	572	0.88	34.6
350	662	1.10	43.3

REFERENCES

1. L. T. Greenfield and P. G. Forrester, "The Properties of Tin Alloys," Tin Research Institute, Publication No. 155, 1947
2. H. Thresh, A. Crawley and D. White, Trans Met Soc, AIME, 1968, p 242, 819 and 859
3. H. H. Manko, Product Engineering, March 6, 1961, p 39
4. Fidos and Schreiner, Z Metallkunde, 1970, 61, p 273
5. H. Charnock and S. A. Yeo, S Sci Instruments, 1959, Vol 36, Issue 11, p 478-9
6. W. H. Warren and W. G. Bader, Rev Sci Instruments, 1969, Vol 40, Issue 1, p 180
7. C. J. Thwaites and B. T. K. Barry, "Soldering," Oxford Univ Press, 1975
8. H. S. Kalish and F. J. Dunkerley, Trans AIME, 1949, Vol 180, p 637
9. S. W. Zehr and W. A. Backofen, Trans ASM, 1968, Vol 61, p 300
10. M. L. Ackroyd, C. A. MacKay and C. J. Thwaites, Metals Technology, 1975, Vol 2, part 2, p 73-85

Tin Babbitt Alloy 1
91Sn-4.5Sb-4.5 Cu

Specifications

AMS. 4800
ASTM. Bearings: B23, 1. Die castings: B102, alloy CY44A
SAE. 10
Government. QQ-M-161, grade 1. Navy 46M2, No. 1

Chemical Composition

Composition limits. Bearings (ASTM B23: 90 to 92 Sn, 4 to 5 Sb, 4 to 5 Cu, 0.35 max Pb, 0.08 max Fe, 0.10 max As. Die castings (ASTM B102): 90 to 92 Sn, 4 to 5 Sb, 4 to 5 Cu, 0.35 max Pb, 0.08 max Fe, 0.08 max As, 0.01 max Zn, 0.01 max Al

Applications

Typical uses. Sleeve bearings, die castings

Mechanical Properties

Tensile properties. Chill cast: typical tensile strength, 64 MPa (9.3 ksi), elongation in 50 mm or 2 in., 2%. Die cast: typical tensile strength, 62 MPa (9 ksi); elongation in 50 mm or 2 in., 2%

Compressive properties. See Table 3.

Hardness. Chill cast, 17 HB
Elastic modulus. Tension (chill cast), 50 GPa (7.3×10^6 psi)
Impact strength. See Table 3.

Fatigue strength. Chill cast, 26 MPa (3.8 ksi) at 2×10^{-7} cycles, R. R. Moore-type test

Mass Characteristics

Density. 7.34 Mg/m^3 (0.265 lb/in.^{3})

Thermal Properties

Liquidus temperature. 371 °C (700 °F)

Solidus temperature. 223 °C (433 °F)

Fabrication Characteristics

Casting temperature. Chill casting, 400 to 440 °C (750 to 825 °F)

Table 3 Mechanical properties of chill cast tin babbitt alloy 1

| Temperature | | Compressive yield strength | | | | Impact energy (Izod) | |
| | | 0.125% set | | 0.3% set | | | |
°C	°F	MPa	ksi	MPa	ksi	J	ft · lb
20(a)	68(a)	30.3	4.4	43.4	6.3	3.4	2.5
60	140	...	...	35.8	5.2	...	...
100(b)	212(b)	17.9	2.6	...	...	4.1	3.0
150	302	...	...	...	...	4.5	3.3
200	392	...	...	6.9	1.0	3.1	2.3

(a) Compressive strength at 25% set, 88.2 MPa (12.8 ksi); hardness, 17 HB. (b) Compressive strength at 25% set, 47.6 MPa (6.9 ksi), hardness, 8 HB.

Tin Babbitt Alloy 2
89Sn-7.5Sb-3.5Cu

Specifications

ASTM. B23, alloy 2
SAE. 12
Government. QQ-M-161, grade 2. Navy 46M2, grade 2

Chemical Composition

Composition limits. 88 to 90 Sn, 7 to 8 Sb, 3 to 4 Cu, 0.35 max Pb, 0.08 max Fe, 0.10 max As, 0.08 max Bi, 0.005 max Zn, 0.005 max Al, 0.05 max Cd

Applications

Typical uses. Sleeve bearings

Mechanical Properties

Tensile properties. Typical: tensile strength, cast from 315 °C (600 °F) into mold at 150 °C (300 °F): 87 MPa (12.6 ksi). Proportional limit, cast from 400 °C (750 °F) into mold at 100 °C (212 °F): 10.1 MPa (1.46 ksi). For tensile properties of chill castings, see Table 4.

Hardness. Chill cast, 24 HB. Cast from 315 °C (600 °F) into mold at 150 °C (300 °F), 24 HB. Cast from 400 °C (750 °F) into mold at 100 °C (212 °F), 22 HB

Elastic modulus. Tension, cast from 400 °C (750 °F) into mold at 100 °C (212 °F): 52 GPa (7.6×10^6 psi)

Fatigue strength. Chill cast, 33 MPa (4.8 ksi) at 2×10^7 cycles, R. R. Moore-type test

Mass Characteristics

Density. 7.39 Mg/m^3 (0.267 lb/in.^{3})

Thermal Properties

Liquidus temperature. 354 °C (669 °F)
Solidus temperature. 241 °C (466 °F)

Fabrication Characteristics

Casting temperature. Chill castings, 425 °C (795 °F)

Tin Babbitt Alloy 3
84Sn-8Sb-8Cu

Specifications

ASTM. B23, alloy 3
Government. QQ-M-161, grade 3. Navy 46M2, grade 3

Chemical Composition

Composition limits. 83 to 85 Pb, 7.5 to 8.5 Sb, 7.5 to 8.5 Cu, 0.35 max Pb, 0.08 max Fe, 0.10 max As, 0.08 max Bi, 0.005 max Zn, 0.005 max Al, 0.05 max Cd

Applications

Typical uses. Sleeve bearings

Mechanical Properties

Tensile properties. Die cast, typical: tensile strength, 69 MPa (10 ksi); elongation, 1%
Compressive properties. See Table 5.
Hardness. Chill cast, 27 HB at 20 °C (68 °F); 14 HB at 100 °C (212 °F). Die cast, 30 HB at 20 °C (68 °F)

Mass Characteristics

Density. 7.45 Mg/m^3 (0.269 lb/in.^{3})

Thermal Properties

Liquidus temperature. 422 °C (792 °F)
Solidus temperature. 240 °C (464 °F)

Fabrication Characteristics

Casting temperature. Chill castings, 490 °C (915 °F)

Table 4 Mechanical properties of chill cast tin babbitt alloy 2

| Temperature | | Tensile strength | | Elongation, %(a) | Reduction in area, % |
°C	°F	MPa	ksi		
20	68(b)	77	11.2	18	25
49	120	63	9.2	24	27
100	212(c)	45	6.5	23	28
149	300	28	4.0	32	38
175	345	20	2.9	38	44

(a) Gage length equals $4\sqrt{area}$. (b) Compressive yield strength, 0.125% set, 42 MPa (6.1 ksi); compressive strength, 25% set, 103 MPa (14.9 ksi). (c) Compressive yield strength, 0.125% set, 21 MPa (3.0 ksi); compressive strength, 25% set, 60 MPa (8.7 ksi).

Table 5 Mechanical properties of chill cast tin babbitt alloy 3

| Temperature | | Compressive yield strength | | | | Compressive strength 25% set | | Impact energy (Izod) | |
| | | 0.125% set | | 0.3% set | | | | | |
°C	°F	MPa	ksi	MPa	ksi	MPa	ksi	J	ft·lb
20	68	46	6.6	55	8	117	17	1.2	0.9
60	140	...	...	47	6.8	...	...	1.3	1
100	212	21	3.1	35	5.1	68	9.9	1.3	1
150	302	...	...	...	...	...	...	1.5	1.1
200	392	...	...	7	1	...	...	0.9	0.7

Bearing Alloy 75Sn-12Sb-10Pb-3Cu

Chemical Composition

Composition limits. 74 to 76 Sn, 11 to 13 Sb, 9.3 to 10.7 Pb, 2.5 to 3.5 Cu, 0.08 max Fe, 0.15 max As

Applications

Typical uses. Sleeve bearings

Mechanical Properties

Compressive properties. See Table 6.
Hardness. Chill cast, 27 HB. See also Table 6.

Mass Characteristics

Density. 7.53 Mg/m^3 (0.272 lb/in.^{3})

Casting Alloy 65Sn-18Pb-15Sb-2Cu

Specifications

ASTM. B102, alloy PY1815A

Chemical Composition

Composition limits. 64 to 66 Sn, 14 to 16 Sb, 17 to 19 Pb, 1.5 to 2.5 Cu, 0.08 max Fe, 0.15 max As, 0.01 max Zn, 0.01 max Al

Applications

Typical uses. Sleeve bearings, die castings

Mechanical Properties

Tensile properties. Die cast, typical: tensile strength, 54 MPa (7.8 ksi); elongation in 50 mm or 2 in., 1.5%
Compressive properties. See Table 7.
Hardness. Die cast: 23 HB. See also Table 7.

Thermal Characteristics

Liquidus temperature. 306 °C (583 °F)
Solidus temperature. 184 °C (363 °F)

Table 6 Mechanical properties of 75Sn-12Sb-10Pb-3Cu

| Temperature | | Compressive yield strength (0.125% set) | | Compressive strength (25% set) | | Hardness, HB |
°C	°F	MPa	ksi	MPa	ksi	
20	68	38.3	5.55	111.4	16.15	24
100	212	14.8	2.15	47.6	6.9	12

Mass Characteristics

Density. 7.75 Mg/m^3 (0.280 lb/in.^{3})

Thermal Properties

Liquidus temperature. 296 °C (565 °F)
Solidus temperature. 181 °C (358 °F)

Table 7 Mechanical properties of chill cast 65Sn-18Pb-15Sb-2Cu

| Temperature | | Compressive yield strength (0.125% set) | | Compressive strength (25% set) | | Hardness, HB |
°C	°F	MPa	ksi	MPa	ksi	
20	68	34	5	103	15	22.5
100	212	14	2.1	46	6.7	10

Tin Die-Casting Alloy 82Sn-13Sb-5Cu

Specifications

ASTM. B102 YC135A

Chemical Composition

Composition limits. 80 to 84 Sn, 12 to 14 Sb, 4 to 6 Cu, 0.35 max Pb, 0.08 max Fe, 0.08 max As, 0.01 max Zn, 0.01 max Al

Applications

Typical uses. Die castings

Mechanical Properties

Tensile properties. Die cast, typical: tensile strength, 69 MPa (10 ksi); elongation in 50 mm or 2 in., 1%
Hardness. 29 HB
Creep rupture strength. 17 MPa (2.5 ksi) for 1-yr life, 12.93 MPa (1.875 ksi) for 10-yr life

Tin Foil 92Sn-8Zn

Applications

Typical use. Foil for packaging food

Mechanical Properties

Tensile properties. Typical: tensile strength, 60 MPa (8.7 ksi); yield strength, 41 MPa (6.0 ksi); elongation, 40%
Bending properties. Force-deformation-time characteristics of various foils using static and dynamic bend tests reported in Ref 1.

Thermal Properties

Solidus temperature. 199 °C (390 °F)
Thermal conductivity. 59 W/m·K (34 Btu/ft·h·°F) at 0 to 200 °C (32 to 390 °F)

Electrical Properties

Electrical conductivity. Volumetric, 14.2% IACS
Electrical resistivity. 121.0 nΩ·m

Chemical Properties

Resistance to specific corroding agents. Immersion and bottle-capping tests with milk showed that this alloy is only slightly soluble and has no effect on the milk (Ref 2).

REFERENCES

1. B. Chalmers and P. W. Seddon, The Mechanical Properties of Metal Foils, *J. Inst. Metals*, 68, 283 (1942)
2. R. Kerr, The Behavior of Some Metal Foils in Contact with Milk, *J. Soc. Chem. Ind.*, 61, 128 (1942)

White Metal 92Sn-8Sb

Applications

Typical uses. Costume jewelry

Mechanical Properties

Tensile properties. See Table 8.
Shear properties. See footnote with Table 8.
Hardness. Effect on rolling: the alloy hardens at first on cold rolling; the maximum hardness is reached at a reduction of about 40 to 45%; further working causes progressive softening until, at about 80% reduction, the hardness approaches that of the cast alloy; annealing at 200 to 225 °C (392 to 437 °F) causes the severely worked alloy to harden slightly. See also footnotes with Table 8.
Impact strength. See footnote with Table 8.
Creep rupture characteristics. See Table 9.

Mass Characteristics

Density. 7.28 Mg/m³ (0.263 lb/in.³)

Thermal Properties

Solidus temperature. 246 °C (475 °F)

Electrical Properties

Electrical conductivity. Volumetric, 11.1% IACS
Electrical resistivity. 154.8 nΩ·m at 25 °C (77 °F)

Table 9 Creep rupture characteristics of white metal (a)

Stress		Time to fracture, days	Final extension, %
MPa	ksi		
9.7	1.4	19	66
8.3	1.2	54	54
7.6	1.1	71	37
6.9	1.0	155	42
6.2	0.9	198	49
5.5	0.8	360	98
4.1	0.6	339(a)	4.12(b)
2.8	0.4	337(a)	1.06(b)

(a) Tests conducted at room temperatures of 9 to 27 °C (48 to 81 °F) on rolled material 2.5 mm (0.1 in.) thick. (b) Specimen did not fracture.

Pewter 91Sn-7Sb-2Cu

Commercial Names

Common names. Modern pewter, Britannia metal

Specifications

ASTM. B560
Foreign. British: BS5140. German: DIN17810

Chemical Composition

Nominal composition. Although a wide range of compositions has been called pewter, the usual modern alloys contain 90 to 95% Sn, 1 to 3% Cu and the remainder Sb. Some "pewter-like" materials are sand cast or spun aluminum alloys, which are traditionally not considered to be pewter. Although some pewter contains lead as an alloying constituent, a considerable portion of lead is undesirable where the material may be in contact with food or beverages. In addition, lead may impart a dullness to the ware.
Composition limits. See Table 10.

Applications

Typical uses. Coffee and tea services, trays, steins, mugs, candy dishes, jewelry, bowls, plates, vases, candlesticks, compotes, decanters, and cordial cups

Mechanical Properties

Tensile properties. See Tables 11 and 12.
Hardness. See Tables 11 and 12.
Elastic modulus. Tension, 53 GPa (7.7 × 10⁶ psi)

Structure

Microstructure. Modern pewter consists of a cored solid solution of antimony in tin within which are distributed fine crystals of η (Cu₆Sn₅) phase

Mass Characteristics

Density. 7.28 Mg/m³ (0.263 lb/in.³)

Thermal Properties

Liquidus temperature. 295 °C (563 °F)
Solidus temperature. 244 °C (471 °F)

Chemical Properties

Resistance to specific corroding agents. Pewter tarnishes in soft water with the production of a visible

Table 8 Mechanical properties of white metal

Form and condition	Section size		Tensile strength		Elongation, %
	mm	in.	MPa	ksi	
Chill cast, tested 2 months after casting (a)	5 by 1.3	2 by ½	50	7.2	...
Chill cast, annealed at 225 °C (437 °F) (b)	5 by 1.3	2 by ½	45	6.5	50
Cast (c)	...	...	...	...	...
Annealed sheet	2.5	0.1	46	6.7	70(d)
Sheet, quenched from 220 °C (428 °F)	2.5	0.1	51	7.4	28(d)
Sheet, aged 150 °C (302 °F)	2.5	0.1	61	8.8	28(d)
Wire, extruded	0.35	0.14	59	8.5	63
Wire, extruded and annealed 24 h at 225 °C (437 °F)	0.35	0.14	54	7.8	10

(a) Brinell hardness, 20. (b) Brinell hardness, 17. (c) Izod impact value, 30 J (22 ft·lb); shear strength, 46 MPa (6.7 ksi) (d) in 50 mm or 2 in.

film of interference-tint thickness. In hard water, there was no tarnishing but localized attack at the water line and sometimes elsewhere if a chalky deposit was formed from the water. Pewter is attacked by dilute hydrochloric and citric acids in the presence of air.

Fabrication Characteristics

Casting temperature. 315 to 330 °C (600 to 625 °F)

Formability. Can be formed by rolling, hammering, spinning or drawing. Erichsen cup height: annealed sheet, 9.6 mm; sheet cold rolled to 32% reduction, 11.6 mm. It has been demonstrated that the "earing" of pewter sheet can be reduced by an intermediate cross-rolling operation or heat treatment, rolling then being continued down to final thickness. (Table 3).

Solderability. Good

REFERENCE

1. R. Duckett and P. A. Ainsworth, Sheet Metal Ind, 50, 7, 412 (1973)

Table 10 Chemical composition limits for modern pewter

Specification	Sn	Sb	Cu	Composition, % Pb max	As max	Fe max	Zn max	Cd max
ASTM B560								
type 1(a)	91-93	6-8	0.25-2.0	0.05	0.05	0.015	0.005	...
type 2(b)	90-93	5-7.5	1.5-3.0	0.05	0.05	0.015	0.005	...
type 3(c)	95-98	1.0-3.0	1.0-2.0	0.05	0.05	0.015	0.005	...
BS5140	rem	5-7	1.0-2.5	0.5	...	...	...	0.05
	rem	3-5	1.0-2.5	0.5	...	...	...	0.05
DIN17810	rem	1-3	1-2	0.5	...	...	...	...
	rem	3.1-7.0	1-2	0.5	...	...	...	...

(a) Casting alloy, nominal composition 92Sn-7.5Sb-0.5Cu. (b) Sheet alloy, nominal composition 91Sn-7Sb-2Cu. (c) Special-purpose alloy.

Table 11 Typical mechanical properties of pewter

Form and condition	Section thickness mm	in.	Tensile strength MPa	ksi	Elongation in 50 mm or 2 in., %	Hardness, HB
Chill cast(a)	19.05	0.750	...	...	...	23.8
Sheet, annealed 1 h at 205 °C (400 °F), air cooled(b)	6.12	0.241	59	8.6	40	9.5
Sheet, cold rolled, 32% reduction	6.12	0.241	52	7.6	50	8.0

Table 12 Effect of processing variables on mechanical properties of pewter sheet and on amount of earing during drawing (a) (Ref 1)

Processing	Delay between processing and testing	Tensile strength, MPa (ksi), at angle to rolling direction of: 0°	55°	90°	Elongation, % at angle to rolling direction of: 0°	55°	90°	Hardness, HV	Earing, %
Cross rolling from intermediate thickness	12 months	64(9.3)	62(9.0)	64(9.3)	56	49	53	15	10
	24 h	48(7.0)	48(7.0)	50(7.3)	92	136	122	13	4½
Unidirectional rolling, with heat treatment(b) at intermediate thickness	24 h	68(9.9)	69(10.0)	73(10.6)	47	36	17	20	2½

(a) Properties are mean values of three determinations each on sheets of Sn-6Sb-2Cu alloy, 1-mm (0.04 in.) thick, cold rolled from 25-mm (1.00-in.) thick cast slabs. About 150 to 200 °C (302 to 392 °F).

Zinc

Selection and Application of Zinc and Zinc Alloys

By Ernest W. Horvick
Zinc Institute, Inc.

THIS ARTICLE deals primarily with zinc and zinc alloys for structural applications. Uses based on corrosion resistance are discussed in a subsequent article in this volume.

Slab zinc is available in three grades, as specified by the American Society for Testing and Materials (Table 1). The kinds and amounts of impurities in the metal have important influences on physical and chemical properties.

Zinc coating, which constitutes the largest single use of zinc, is accomplished mainly by dipping the iron in molten zinc or by electroplating (electrogalvanizing). Annual consumption of slab zinc used in various galvanized products in the United States during recent years is summarized in Fig. 1. Hot dip galvanizing employs chiefly the less pure grades of zinc. The life of galvanized material is proportional to the thickness of the coating, and recent developments in both electrogalvanizing and hot dip galvanizing have been toward application of heavy coatings that will withstand deformation without peeling.

All grades of zinc are used in brass; the specific grade has some effect on alloy properties.

The use of zinc-base alloys for die casting is important; for this work, it is essential to use alloys based on Special High Grade zinc.

All grades of zinc are used for rolling. In general, the presence of natural impurities or alloying additions increases the hardness and stiffness of rolled zinc. Special combinations of lead, iron and cadmium are used to control the chemical characteristics of zinc for dry batteries and photoengraving.

Superplastic zinc is a fairly recent development that is eliciting considerable interest because of its properties.

Superplastic zinc can be formed like plastics. Forming processes, including vacuum forming, blow molding and compression forming, have been successfully applied in manufacture of a variety of parts from superplastic zinc alloys.

All commercial superplastic zinc alloys are based on the eutectoid composition of 78% Zn and 22% Al (by weight) with various minor additions, generally to increase strength and creep resistance.

The superplastic structure of these alloys consists of fine, equiaxed grains (about 1 to 2 μm in diameter) of the aluminum-rich α phase and of the zinc-rich β phase in nearly equal proportions. This structure usually is pro-

Table 1 Grades and compositions of slab zinc (ASTM B6)

Grade	Composition(a), %			Zn, min (by difference)
	Pb, max	Fe, max	Cd, max	
Special High Grade(b)	0.003	0.003	0.003	99.990
High Grade	0.03	0.02	0.02	99.90
Prime Western(c)	1.4	0.05	0.20	98.0

(a) When specified for use in manufacture of rolled zinc or brass, aluminum is held to 0.005% max. (b) Tin in Special High Grade Zinc is held to 0.001% max. (c) Aluminum in Prime Western Zinc is held to 0.05% max.

Fig. 1 Annual consumption of slab zinc used in galvanized products in the United States

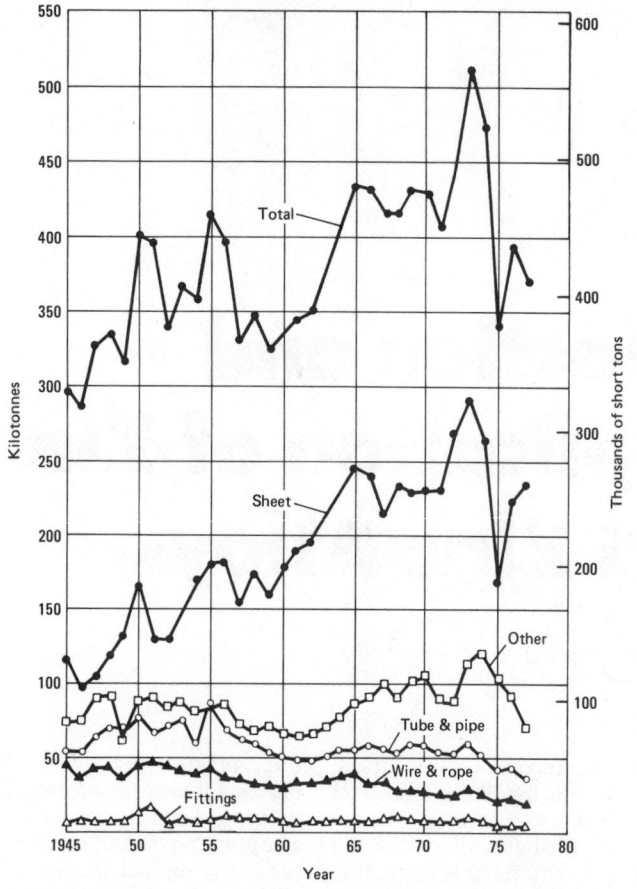

duced by heating the material in the range 275 to 375 °C (525 to 705 °F) to convert it completely to the α′ solid-solution phase before quenching in water. Quenching initially retains the α′ solid solution, but after a short incubation period this structure transforms to the very fine mixture of α and β phases. This α–β structure is superplastic when worked within the temperature range 250 to 270 °C (480 to 520 °F) with a strain-sensitivity index (m) of about 0.5.

Although material with this structure has excellent superplastic properties, its room-temperature properties are similar to those of lead. To achieve a usable engineering condition, the material requires further heat treatment—annealing at about 340 °C (645 °F) followed by slow cooling. The mechanical properties are, however, sensitive to this slow cooling rate so that two grades of usable materials may be identified, depending on the rate of cooling after annealing. These two grades are the "air cooled" and "furnace cooled" grades.

Other forms of zinc include zinc oxide and zinc dust. Zinc oxide is widely used in paint, rubber and copier paper. Zinc dust is used as a pigment in primers and finish paints for metal, as a reducing agent in chemical processes, and as a component in mechanical plating and lubricating compounds.

Die Castings

The major use of zinc as a structural material is in alloys for pressure die casting. Development of the modern zinc die casting alloys was directly related to use of Special High Grade zinc (99.99 + % Zn), addition of particular alloying constituents held within close limits, and control of impurities.

For die castings it is essential to use this extra-pure grade of zinc to ensure extremely low iron, lead, cadmium and tin contents. It is also necessary to limit these same impurities in the metals added to make the desired zinc alloy composition. Only by such control can zinc die castings be produced that are stable in dimensions and properties.

The impurities lead, cadmium and tin, if present in castings in amounts greater than the established maximums (0.005% Pb, 0.004% Cd, 0.003% Sn), cause subsurface network corrosion. These limits are close to critical values, and thus lead, cadmium and tin contents must be strictly controlled in order to get maximum utility from the die cast part. Iron is held to 0.10% max to prevent excessive skimming losses and machining problems.

Standardization of zinc-base die casting alloys by ASTM and SAE followed the commercial development of these alloys. The composition limits given in ASTM B240 (zinc-base alloys in ingot form for die castings) and those given in ASTM B86 and specified for Alloy 7 (zinc alloy die castings) are presented in Table 2.

Alloys. The zinc die casting alloys presently in use are known as ASTM AG40A, AC41A and Alloy 7 or, respectively, SAE 903, 925 and 903. These alloys fulfill the requirements of good die casting metal. AG40A and AC41A contain about 4% Al and 0.04% Mg. Alloy 7 contains about 4% Al and 0.01 to 0.02% Mg. Alloy AG40A is essentially copper-free; AC41A contains approximately 1% Cu. Alloy AG40A is more often specified, has slightly higher ductility, and retains its impact strength better at elevated temperature than alloy AC41A. Alloy AC41A is somewhat harder and stronger, and has slightly better castability. Alloy 7 is superior to the other two in casting properties, surface finish, and ductility. Otherwise, these three alloys are similar in properties, and for service at normal temperatures the differences among them are slight.

The International Lead Zinc Research Organization has recently developed an additional zinc die casting alloy known as ILZRO 16, which has superior creep resistance at both ambient and elevated temperatures. ILZRO 16 was designed for processing in cold-chamber die casting machines and has the composition limits given in Table 2.

The zinc alloys are low in cost of metal per casting, are easy to die cast, are cast at low temperatures, have strength greater than all other die casting metals except the copper alloys, lend themselves to casting within close dimensional limits, permit the thinnest sections yet produced, and are machined at minimum cost. Their resistance to surface corrosion is adequate in a wide range of applications. Prolonged contact with moisture results in formation of white corrosion products, but surface treatments can be applied that largely prevent formation of such products.

Limiting service conditions for standard zinc-base die casting alloys are as follows. At temperatures slightly above 95 °C (200 °F) their tensile strength is reduced 30% and their hardness 40%. At subzero temperatures some embrittlement occurs, but impact strength is still in the same range as that of aluminum and magnesium die casting alloys at normal service temperatures. At room temperature, impact strength of zinc die castings is much higher than that of aluminum or magnesium die castings or iron sand castings. Tensile and other properties of alloys AG40A and AC41A after various aging treatments are given in Fig. 2; average properties of these alloys are summarized in Table 3. Also listed in Table 3 are average properties for the two newer alloys ILZRO 16 and Alloy 7.

Stabilization. Changes brought about by aging produce a maximum shrinkage of 0.7 mm/m (0.0007 in./in.) for alloy AG40A and 0.9 mm/m (0.0009 in./in.) for alloy AC41A. About two thirds of the shrinkage occurs in 4 to 5 weeks; the remainder takes place over a period of years. Because these changes can be accelerated by annealing treatments at low temperatures, it is customary to use one of the following heat treatments when castings are required to maintain rigid dimensional tolerances: (a) 3 to 6 h at 100 °C (212 °F), (b) 5 to 10 h at 85 °C (185 °F) or (c) 10 to 20 h at 70 °C (158 °F). On aging, castings that have been given one of these treatments will exhibit additional shrinkage equal to about one third of that occurring in untreated castings.

Assembly. All die castings have at least a light flash at the die parting, and those requiring movable cores will have some flash around the cores. Flash is also formed around ejector pins at the points at which they make contact with the casting.

Although it often is cheaper to cut threads than to cast them, for many pieces cast threads are more economical. Male threads usually are made with a parting parallel to the axis; this leaves a flash at the parting. The flash can be removed with a shaving tool in some instances, but in others a chasing operation is necessary.

Zinc die castings are invariably cast within quite close dimensional limits, but some machining is commonly required in addition to removal of flash, even though it may consist only of such simple operations as punching, drilling, reaming or tapping of holes.

Zinc die castings can be soldered or welded, but ordinarily neither is used except for special applications or repair work.

Soldered joints have low strength. Soldering with the usual soft solder is possible only on nickel-plated surfaces. A more satisfactory procedure is to use an 82.5Cd-17.5Zn solder, without flux.

Gas welding is best done with a reducing flame and with welding rods made from the same alloy as the die casting. Because zinc die castings have a low melting point, the flame must be held almost parallel to the casting.

Pulsed-arc welding also may be used in assembly work.

Finishing. Many of the finishes applied to other types of metal products can be applied also to zinc die castings, although some differences in formulation, as well as occasional differences in the method of application, are desirable. The types of finishes applicable to zinc die castings are: (a) mechanical finishes—buffed, polished, brushed and tumbled; (b) electrodeposited finishes—copper, nickel, chromium, brass, silver, and black nickel; (c) chemical finishes—chromate, phosphate, molybdate and black nickel; (d) organic finishes—enamel, lacquer, paint and varnish; and (e) plastic finishes.

Zinc die castings can be mechanically finished rather easily; however, un-

Table 2a Designations of zinc die casting alloys

Alloy	UNS number	SAE number	Government specification	ASTM specification
AG40A	Z33520	903	QQ-Z-363	B240 B86
AC41A	Z35530	925	QQ-Z-363	B240 B86
Alloy 7	...	...	...	...
ILZRO 16..	...	...	...	...

Table 2b Compositions of zinc die casting alloys

Alloy	Form	Cu	Al	Mg	Pb, max	Cd, max	Sn, max	Fe, max	Others	Zn
AG40A	Ingot	0.10 max	3.9-4.3	0.025-0.05	0.004	0.003	0.002	0.075	(a)	rem
AC41A	Die castings	0.25 max(b)	3.5-4.3	0.020-0.05(c)	0.005	0.004	0.003	0.100	(a)	rem
Alloy 7	Ingot	0.75-1.25	3.9-4.3	0.03-0.06	0.004	0.003	0.002	0.075	(a)	rem
ILZRO 16 ...	Die castings	0.75-1.25	3.5-4.3	0.03-0.08(c)	0.005	0.004	0.003	0.100	(a)	rem
	Die castings	0.25 max	3.5-4.3	0.010-0.02	0.0020	0.0020	0.0010	0.050	...	rem
	Die castings	1.0 to 1.5	1.01-0.04	...	...	...	...	...	(d)	rem

(a) May contain nickel, chromium, silicon and manganese in amount of 0.02, 0.02, 0.035 and 0.5%, respectively. No harmful effects have ever been noted due to the presence of these elements in these concentrations; therefore, analyses are not required for these elements. (b) For the majority of commercial applications, a copper content in the range of 0.25 to 0.75% will not adversely affect the serviceability of die castings and should not serve as a basis for rejection. (c) Magnesium content may be as low as 0.015% provided that lead, cadmium and tin contents do not exceed 0.003, 0.003 and 0.001%, respectively. (d) 0.15 to 0.25 Ti, 0.10 to 0.20 Cr, 0.30 to 0.40 Ti + Cr.

Fig. 2 Mechanical properties and dimensional changes characteristic of zinc die casting alloys

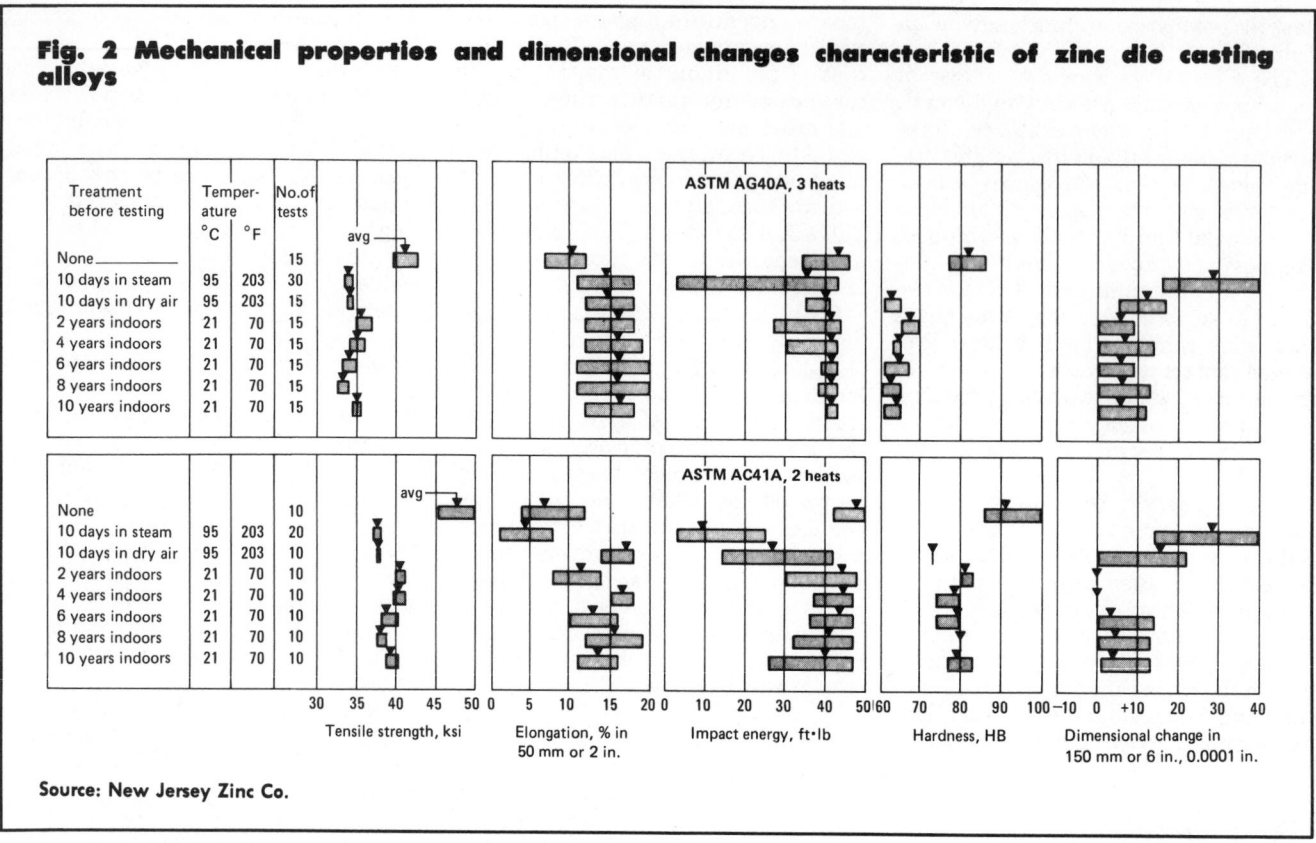

Treatment before testing	Temperature °C	°F	No. of tests
None			15
10 days in steam	95	203	30
10 days in dry air	95	203	15
2 years indoors	21	70	15
4 years indoors	21	70	15
6 years indoors	21	70	15
8 years indoors	21	70	15
10 years indoors	21	70	15
None			10
10 days in steam	95	203	20
10 days in dry air	95	203	10
2 years indoors	21	70	10
4 years indoors	21	70	10
6 years indoors	21	70	10
8 years indoors	21	70	10
10 years indoors	21	70	10

Tensile strength, ksi — Elongation, % in 50 mm or 2 in. — Impact energy, ft·lb — Hardness, HB — Dimensional change in 150 mm or 6 in., 0.0001 in.

ASTM AG40A, 3 heats — ASTM AC41A, 2 heats

Source: New Jersey Zinc Co.

less the surface is protected, tarnishing will occur more quickly and will be more noticeable than on untreated castings. For this reason, mechanical finishing of zinc die castings is seldom done except as a preparatory step for other types of finishes.

Electrodeposited coatings of virtually any metal capable of being electrodeposited can be applied to zinc die castings. Coatings of nickel, or of copper and nickel, finished with a thin layer of chromium are used very extensively. Plated coatings are most commonly applied for decorative purposes, but they also improve resistance to corrosion and to abrasion. All plated coatings are somewhat porous; when moisture penetrates through the pores, some corrosion of the basis metal usually results. If the plated coatings are too thin, such corrosion will destroy the appearance of the casting in a short time; therefore, the performance and useful life to be expected from such coatings depends on the thickness of the deposit and the method of application. The American Society for Testing and Materials has adopted specifications covering three classes of commercial coatings for zinc die casting

alloys—namely, coatings for mild, ordinary and severe exposures.

In addition to the electrodeposited coatings described above, some newer plating methods, which provide better results than the older methods have been developed. These involve the use of (a) conventional chromium over duplex nickel, (b) bright, crack-free chromium over conventional nickel or (c) bright, crack-free chromium over duplex nickel. Dual chromium also is applied over conventional nickel. (All of these nickel-chromium combinations are plated over a conventional copper coat.)

Chemical treatments, although sometimes used in place of other applied finishes, are more often applied as bases for enamels and lacquers. Certain chemical treatments are used alone for inhibiting corrosion.

Organic finishes of many types are available, each offering some particular advantage, such as color, hardness, gloss, adhesion or resistance to weathering. Most of these finishes are baked for greater durability and adhesion. When finishes are baked on zinc alloy die castings, the properties of the castings may be adversely affected if the

castings are heated for more than: ½ h at 220 °C (428 °F); 1 h at 190 °C (374 °F); 2 h at 165 °C (329 °F); or 3 h at 150 °C (302 °F). If a casting is copper-nickel-chromium plated and then given a baked paint finish, the maximum temperature that can be used without blistering the paint is 125 °C (257 °F).

Plastic finishes are applied to some high-production zinc die castings, often by repeated dipping in or spraying with nitrocellulose or ethyl cellulose solutions. These finishes are several times as thick as other organic coatings and differ from similar-base finishes not only in thickness but also in composition. In another instance of the use of plastic over die cast zinc, only a core of the part is die cast, and plastic (cellulose-acetate or acetate-butyrate) is injection molded around this core; the plastic gives the part its final shape, and the die cast core provides strength.

Applications. The automotive industry uses by far the largest number of zinc alloy die castings. Some of the most important mechanical parts made as zinc alloy die castings include carburetors, bodies for fuel pumps, windshield-wiper parts, speedometer

Table 3 Average properties of zinc die casting alloys

Properties	ASTM AG40A; SAE 903	ASTM AC41A; SAE 925	Alloy 7	ILZRO 16
Mechanical Properties				
Charpy impact strength, ¼-by-¼-in. bar:				
As cast, J(ft·lb)	58 (43)	65 (48)	. . .	. . .
After aging indoors 10 yr, J(ft·lb)	56 (41)	54 (40)	275 (40)	. . .
Tensile strength:				
As cast, MPa (ksi)	285(41)	330 (47.6)	285 (41)	230 to 235 (33 to 34)
After aging indoors 10 yr, MPa (ksi)	240 (35)	270 (39.3)	. . .	. . .
Elongation, % in 50 mm or 2 in.:				
As cast	10	7	14	5
After aging indoors 10 yr	16	13	. . .	. . .
Expansion, after aging indoors 10 yr at room temperature, μm/m	80	70	. . .	. . .
Other Properties and Constants of As-Cast Alloys				
Brinell hardness (HB)	82	91	76	75 to 77
Compressive strength, MPa (ksi)	415 (60)	600 (87)	. . .	. . .
Electrical conductivity, % IACS	27.5	26.5	. . .	. . .
Liquidus temperature, °C (°F)	387 (728)	386 (727)	. . .	417 (785)
Solidus temperature, °C (°F)	381 (717)	380 (716)	. . .	415 (780)
Modulus of rupture, MPa (ksi)	655 (95)	725 (105)	. . .	. . .
Shear strength, MPa (ksi)	215 (31)	260 (38)	. . .	. . .
Specific heat, J/kg (Btu/lb)	420 (0.10)	420 (0.10)	. . .	. . .
Thermal conductivity, W/m·K (Btu/ft·h·°F)(a)	113 (65.3)	109 (62.9)	. . .	. . .
Thermal expansion, μm/m·K (μin./in.·°F)	27.4 (15.2)	27.4 (15.2)	. . .	. . .
Transverse deflection, mm (in.)	6.9 (0.27)	4.1 (0.16)	. . .	. . .
Density, Mg/m³ (lb/in.³)	6.6 (0.238)	6.7 (0.242)	. . .	. . .
(a) At 18 °C (64 °F).				

frames, grilles, horns, heaters, and parts for hydraulic brakes. Parts that perform both structural and decorative functions include grilles for radiators and radios, lamp and instrument bezels, hubs for steering wheels, brackets, horn rings, exterior and interior hardware, instrument panels and body moldings.

The electrical industry probably uses a larger diversity of die castings than the automotive industry, because of the range of different products manufactured. Zinc alloy die castings are employed for numerous parts of washing machines, oil burners, stokers, motor housings, vacuum cleaners, electric clocks, and kitchen equipment and utensils.

Zinc die castings are used in business machines and other light machines of all types—for example, in typewriters, recording machines, picture projectors, vending machines, accounting machines, cash registers, cameras, slicing machines, garbage disposers, gasoline pumps, hoists, and drink mixers.

Many tools are made with die cast parts. Zinc die castings are not hard enough to serve as cutters, but steel inserts are often cast in place as parts of the tools, thus saving appreciable assembly costs. Die castings are not limited to portable tools but are also used on such large equipment as drill presses and lathes.

Building hardware, padlocks, toys and novelties consume a substantial percentage of the total production of zinc-base die castings.

Corrosion Resistance. The resistance of zinc die castings to corrosion, as well as that of rolled zinc and galvanized products, is discussed in the article on use of zinc and zinc alloys in corrosion service.

Engineering of Die Castings

Engineering of die castings involves two problems—design of dies and design of castings. Because the manufacturer of the castings is responsible for die design, and also should be consulted on casting design, the following discussion of this subject is intended merely to identify some of the necessary considerations, not to explain them.

Die Design. In die design, after the parting line and the general die construction have been planned, it is highly important to design proper gating and venting of the casting cavity. The gates and vents should be as thick as can be tolerated and must be placed at proper positions. In addition, large overflow wells should be connected to the cavity at the vent end of the die and at other points where entrapment of air is likely. Because loss of metal is small during remelting of overflow wells, the cost of such wells is low, considering the improvement in casting quality they provide.

Design of Die Castings. The rules for designing die castings have many exceptions, but the following generally assist designers in arriving at designs that are both rational and helpful in attaining low cost:

1 Keep sections at minimum thickness and use ribs for added strength where necessary. Make a gradual transition when section thickness varies. Use cores to keep sections uniform and use fillets to avoid sharp corners.
2 Incorporate studs or bosses for economical fastening. Bosses provide for stronger joints than do studs.
3 Avoid plain flat surfaces when such areas must be cast smooth for finishing purposes.
4 Allow ample drafts on cavity walls and on cores, and never specify draft tolerances that are closer than necessary.
5 Employ inserts whenever their advantages cannot be realized at equal cost by other means.
6 Take advantage of the ductility of zinc alloys by designing parts for bending or forming after casting.
7 Avoid undercuts except where their advantages, such as metal

savings, offset their disadvantages of added cost and slower casting cycle.

8 Linear tolerances should not be specified closer than are essential to meet service requirements.

9 If areas of a die casting are to be polished or buffed, the configuration of the casting must be such that these areas are readily accessible to polishing or buffing wheels or belts; deep recesses and sharp edges must be avoided.

10 Location and size of ejector pins should be determined by locations of shrinkage stresses resulting from cooling of metal in the die. Pins should be positioned so as not to have adverse effects on the appearance and finishing of the die casting.

11 Keep casting size and weight at minimum values consistent with other requirements.

12 Design castings so as to minimize machining and trimming costs.

In designing zinc alloy die castings to withstand constant loads, it must be remembered that these alloys do not have recognized elastic moduli or yield strengths. Hence, standard engineering formulas used in calculating cross sections of elastic materials cannot be used for zinc alloys. However, data developed from creep tests may be applied with satisfactory results.

The creep characteristics of zinc die casting alloys subjected to constant transverse loads can be calculated from graphs (a) and (b) in Fig. 3. The method of using graph (b) is as follows:

Determine the percentage of elongation that is permissible in the particular beam member in some definite period of time—for instance, five years.

In applications involving beams, the following approximate formulas have been found useful in determining the relations between deflection and elongation in the outer fiber of such members:

For cantilever beams,

$$e = \frac{150dt}{l^2} \qquad \text{(Eq 1)}$$

and for simple beams (concentrated load),

$$e = \frac{600dt}{l^2} \qquad \text{(Eq 2)}$$

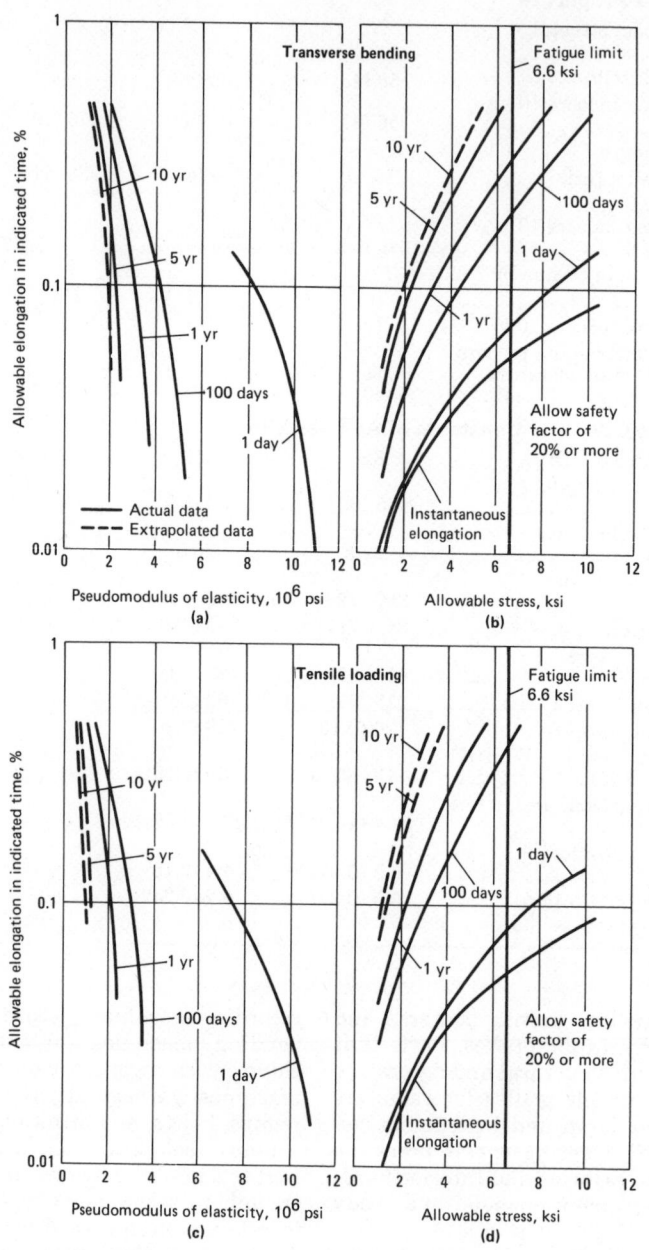

Fig. 3 Room-temperature creep data for AG40A and AC41A zinc die casting alloys

Source: E. H. Keltner and B. D. Grissinger, Trans. AIME, Vol 161, 1945, p 466.

where e is percentage elongation in the outer fiber of the beam; d is deflection of the beam, in inches; t is thickness of the beam, in inches; and l is length of beam, in inches. Find this percentage of elongation on the ordinate and move across the graph horizontally until the chosen time curve is intercepted. From this point, move vertically down to the abscissa, which will indicate the permissible stress. For many applications, a safety factor of 20% is ordinarily applied. Thus, if the permissible stress for the particular application is 2.5 ksi, the cross section would be designed to permit a stress no greater than 2.0 ksi.

Curves for determining the "pseudomodulus" in transverse bending are presented in graph (a) in Fig. 3. Pseudomoduli are not true elastic moduli but the instantaneous moduli calculated from the stress-strain curves in graph (b) at any particular time. A pseudomodulus may be employed in standard engineering formulas instead of the true modulus, provided it is understood that the value applies only for one particular load and one particular time at the temperature specified.

Graph (b) in Fig. 3 contains a curve labeled "instantaneous elongation." This is simply the stretching that takes place immediately upon application of load and is similar to "initial stretch" or "initial elongation," terms applied by other investigators. In this instance, the elongation is largely elastic and hence recoverable by unloading. Curves labeled "1 day," "100 days" and so on, include this "instantaneous elongation;" hence they show total deformation, part of which is elastic.

In graph (b), the vertical line indicating the fatigue limit is at 6.6 ksi. Thus, if the particular application involves cyclic loading, a working stress of 6.6 ksi should not be exceeded.

Zinc and zinc alloys are influenced in their creep properties by temperature. The magnitude of this effect may be estimated by the following empirical formula derived from temperature studies:

$$S_t = \frac{(S_{25} \times 28)}{T} \qquad \text{(Eq 3)}$$

where S_t is permissible stress at any temperature from 25 to 100 °C (77 to 212 °F); S_{25} is permissible stress at 25 °C (77 °F); and T is given temperature, in degrees Celsius. Thus, for an increase in temperature from 25 to 56 °C (77 to 133 °F), the permissible stress is reduced by about 50%. Zinc alloy die castings are not suggested for uses involving continuous stress at temperatures above 100 °C (212 °F).

Graphs (a) and (b) in Fig. 3 deal only with deformation in bending. Similar data for tensile loading are shown in graphs (c) and (d).

Gravity Casting Alloys

A recent alloy development has been the generation of a family of zinc foundry alloys suitable for sand, permanent mold, plaster mold, shell mold and investment casting. Many of these alloys also can be die cast in cold chamber machines where strength and/or hardness beyond the properties of AG41A are required. Currently, two alloys are finding increasing application, a 12%-Al alloy (ILZRO 12) and a 27%-Al alloy (Zn-27Al). These two zinc gravity casting alloys have been designed for structural applications and should not be confused with die casting and slush casting alloys, which often are used for decorative gravity cast parts.

The mechanical properties of these zinc alloys make them attractive substitutes for cast iron and copper alloys in many structural and pressure-tight applications. Because zinc is less costly than copper, these zinc alloys have a distinct cost advantage over copper-base alloys. The ease of machining of zinc and its inherent corrosion resistance give it advantages over cast iron. In situations where zinc plating or some other protective finish is required to prevent atmospheric corrosion of cast iron, direct substitution of a zinc foundry alloy may be possible.

Zinc gravity casting alloys have attractive foundry properties. Due to their low melting temperatures (below 540 °C or 1000 °F) and casting temperatures, energy requirements are low. They are readily cast in thin sections—less than 2.5 mm (0.10 in.) in sand molds. Melting and casting of these alloys are virtually pollution-free. No fluxing or degassing is required, and because of the low casting temperatures minimal pollution from the sand mold results. The 12%-Al alloy is preferred for heavy sections and is suitable for permanent mold casting in both metal and graphite molds. Its permanent mold casting characteristics are similar to those of aluminum permanent mold alloys. The 27%-Al alloy should be specified when higher mechanical properties are required in thin-section sand castings. Care should be taken to prevent hot spots in the mold, which contribute to underside shrinkage.

Applications. Zinc gravity casting alloys can be used for general industrial applications where strength, hardness, wear resistance or good pressure tightness is required. They often are used as direct substitutes for conventional sand casting alloys (such as cast iron or 85-5-5-5 bronze). The patterns and rigging used for the conventional alloys normally can be used, without modification, with zinc foundry alloys.

Zinc alloys often are employed to replace cast iron because of their similar properties and higher machinability ratings. The good bearing and wear characteristics of zinc alloys permit them to be used for bearing bushings and flanges. Other applications in which zinc alloys have been successfully substituted for cast iron or copper alloys include fuel-handling components, pulleys, electrical fittings and hardware components.

One of the original uses of zinc gravity casting alloys was prototype manufacturing. Because of their similarity in strength to zinc die castings, zinc gravity castings are suitable for short runs of parts to be trial marketed prior to production die casting. They also are suitable for production runs too short for die casting to be economical. In certain applications, permanent mold cast zinc components can compete favorably with structural plastics and metal stampings when the production quantity is low.

Components in a wide range of sizes have been successfully cast in either sand or permanent molds. Castings range from a few grams up to about 50 kg (100 lb) and from 1.5 to 50 mm (0.060 to 2 in.) in section thickness.

Wrought Zinc and Zinc Alloys

Wrought zinc and zinc alloys may be obtained as rolled strip, sheet and foil; extruded rod and shapes; and drawn rod and wire. These metals exhibit good resistance to corrosion in many types of service, and because the corrosion products that may form on them are white, other materials are not stained by them. Wrought zinc has chemical characteristics particularly adapted to certain uses, such as dry batteries and photoengraver's plate, and offers combinations of desirable physical and mechanical properties at relatively low cost. In common with many other metals and alloys, wrought zinc creeps under constant loads that are substantially less than its ultimate strength; that is, wrought zinc does not have clearly defined elastic moduli, and hence creep data from service tests must be used in designing for strength

Table 4 Classification of wrought zinc alloys

Pb	Fe	Cd	Composition, % Cu	Mg	Al	Other	Characteristics	Typical uses
0.05 to 0.10	0.012 max	0.005 max	0.001 max	...	...	...	.... High ductility with low hardness and stiffness. Very little work hardening possible	Drawn battery cans, eyelets, fuse links, and a variety of articles drawn, formed and spun. Address plates
0.05 to 0.10	0.012 max	0.06	0.005 max	...	...	...	.... High ductility with low hardness. Can be work hardened slightly	Drawn battery cans, eyelets and grommets. Extruded battery cans. Address plates, laundry tags
0.15 to 0.35	0.017 max	0.15 to 0.30	0.005 max	...	...	...	.... High hardness and stiffness. Uniform etching quality. Can be work hardened	Soldered battery cans, photoengraver's plate, lithographer sheet, boiler and ship plates, weather-strips
0.05 to 0.10	0.012 max	0.005 max	0.85 to 1.25	...	...	...	.... High hardness and stiffness. Good ductility. Good creep resistance. Work hardens easily	Weatherstrips and drawn and formed articles requiring stiffness
0.05 to 0.10	0.015 max	0.005 max	0.85 to 1.25	0.007 to 0.02	...	...	.... High stiffness and creep resistance. Can be severely work hardened	Flat or formed articles requiring high stiffness and strength
0.005 to 0.10	0.012 max	0.05 max	0.50 to 1.50	...	...	0.12 to 1.50 Ti(a)	.... Outstanding creep resistance. Can be severely work hardened. Lowest thermal expansivity with the grain. Very high resistance to grain growth during annealing	Corrugated roofing, leaders and gutters, and other uses requiring maximum creep resistance
0.15 to 0.35	0.014 to 0.025	0.15 to 0.30	0.005 max	0.005 to 0.025	...	...	.... High hardness. Can be baked without severe softening. Good etching characteristics	Photoengraver's sheet
...	...	...	...	0 to 0.025	0.25 to 0.60	...	.... High hardness. Can be baked without severe softening. Good etching characteristics	Photoengraver's sheet
0.007 max	0.10 max	0.007 max	0 to 3.5	0.02 to 0.10	3.5 to 4.5	0.005 max Sn	.... High strength and hardness	Shearing and forming dies. Extruded rod, tube and moldings
0.005 to 0.10	0.012 max	0.05 max	0.50 to 1.5	...	...	0.12 to 0.50 Ti	.... Good creep resistance	Corrugated roofing, leaders and gutters, and formed articles requiring maximum creep resistance

(a) U.S. Patent 2472402.

and rigidity under conditions of continuous stress.

Rolled Products. Zinc usually is cast into flat slabs 25 to 100 mm (1 to 4 in.) thick and in widths and lengths that are convenient for rolling. The slabs are preheated and roughed with a leveling pass of about 10%; subsequent reductions may be increased according to the capacity of the mill. Slabs for strip are reduced by longitudinal rolling to a thickness convenient for coiling—generally less than 2.5 mm (0.1 in.). In production of sheet, slabs are rolled laterally until the desired width is obtained. Then the sheets are turned 90° and rolled to a thickness two to four times the final gage.

Schedules for finish rolling of zinc strip vary widely, depending on the ultimate product desired. Strip is produced in widths as great as 2 m (84 in.) and in thicknesses as low as 0.1 mm (0.004 in.) in regular mills. Foil in thicknesses of 0.025 mm (0.001 in.) or less is produced in special mills. For a bright surface combined with high duc-tility, finish rolling is performed hot; this requires that the rolls be held at temperatures ranging from approximately 120 to 150 °C (250 to 300 °F).

Zinc sheet is produced by the pack-rolling method, in which as many as 30 rough-rolled sheets are stacked together and rolled simultaneously. Packs must be "split" frequently, interchanging inner and outer sheets to equalize temperatures and reductions.

Extrusion. Although it lacks a well-defined elastic modulus at low rates of deformation, zinc has a high apparent modulus under rapid deformation; consequently, pressures required for extrusion of zinc and zinc alloys are higher than those for most nonferrous metals. Ease of extrusion increases with temperature; hence, billets for extrusion usually are heated to the maximum temperature that will not cause hot shortness. This temperature will vary from 250 °C (480 °F) to more than 300 °C (570 °F), depending on composition. Speeds of extrusion of zinc are low compared with those for brasses be-cause, as speed of extrusion increases, the temperature in the product rapidly increases to the point of hot shortness, and cracking and crumbling may result.

Drawing of Rod and Wire. Rolled or extruded zinc or zinc alloy rod may be drawn to wire in operations quite similar to those used for other nonferrous metals.

Fabrication. All severe fabrication of wrought zinc should be done at temperatures above 20 °C (70 °F). Warm, soapy water (made with a neutral soap to avoid staining) is the cheapest lubricant available and works as well as any other.

Rolled zinc of the proper grade is readily drawn into a great variety of articles such as batter cups, eyelets, meter cases, novelties, flashlight reflectors and fruit-jar caps. Suitable grades of rolled zinc also are readily rolled, press formed, stamped or spun into items such as plates for addressing machines, buckles, ferrules, ornaments, nameplates, gaskets, weatherstripping and lamp parts.

The 4Al-0.04Mg alloy that contains as much as 3.5% Cu has been used widely in the aircraft industry, in the form of heavy rolled plate, for dies employed in blanking aluminum alloy sheet and thin steel sheet. Similar compositions are available as extruded rod, tube and moldings.

The softer grades of wrought zinc are self-annealing at temperatures developed during fabrication; hence no annealing is required between operations. The harder grades work harden somewhat and may require intermediate annealing. Temperatures slightly above 100 °C (212 °F) are ordinarily sufficient for annealing the usual commercial grades, and care must be taken not to raise the temperature much above the minimum required and not to leave the article exposed to the annealing temperature for more than a few minutes; otherwise, coarse brittle structures will develop. The 1%-Cu alloys should be annealed at temperatures between 175 and 205 °C (350 and 400 °F).

Classification. The classifications of wrought zinc presented in Table 4 are arbitrary. The compositions shown are typical but do not include every composition supplied by the various companies. To accommodate special applications, many producers supply special grades that are intermediate to or quite different from those listed.

Wrought zinc alloys containing titanium have far greater resistance to creep and grain growth and a lower coefficient of thermal expansion than any other zinc alloy in strip and sheet form.

Mechanical Properties. Among the specific uses listed for rolled zinc in Table 4, few ever require high strength under constant load. When wrought zinc is employed for structural purposes, creep data from service tests must be applied during design, and properties such as tensile strength and tensile elongation have little significance. The control tests—dynamic ductility, hardness, temper, and dynamic bend—permit comparison with other shipments of similar material and general identification of the type and the rolling treatment of samples of unknown history.

Soldering and Welding. The ordinary grades of wrought zinc can be soldered easily by conventional methods. The usual precautions should be observed regarding proper cleaning and fluxing; the metal must not be overheated to the point where it melts. Pulsed-arc welding may be used for joining; gas welding of zinc is used only for repair work.

Machining. Wrought zinc is easily machined using standard methods and tools. However, if it is necessary to machine zinc containing exceedingly coarse grains, the metal should be heated to a temperature between 70 and 100 °C (160 and 210 °F) in order to avoid cleavage of crystals.

Finishing. In general, the same methods are used for finishing wrought zinc and zinc alloys as are used for finishing zinc-base die castings. However, in bake-enameling of wrought zinc, greater caution should be exercised in avoiding temperatures high enough to impair mechanical properties. For this reason air-dried finishes are preferable.

Properties of Zinc and Zinc Alloys

AG40A Alloy
Zn-4A1-0.04Mg

Commercial Names

Trade name. Number 3 Die Casting Alloy
Previous trade name. Zamak-3 (die casting)
Foreign. Mazak-3, Gomak-3

Specifications

ASTM. B86, alloy AG40A (die castings); B240, alloy AG40A (ingot)
SAE. J468, alloy 903
UNS. Z33520
Government. QQ-Z-363
Foreign. BS1400, alloy A; CSA, AG40; DIN, Z400; J15, type I

Chemical Composition

Composition limits. ASTM B86: 3.5 to 4.3 Al, 0.020 to 0.05 Mg, 0.25 max Cu, 0.100 max Fe, 0.005 max Pb, 0.004 max Cd, 0.003 max Sn, rem Zn. (Special high grade zinc must be used as the basic material in making this alloy.)
Consequence of exceeding impurity limits. Alloy becomes subject to intergranular attack and fails prematurely by warping and cracking.

Applications

Typical uses. For die castings, such as automotive parts, household utensils, office equipment, building hardware, padlocks, toys and novelties

Mechanical Properties

Tensile properties. Die cast specimen, 6-mm or ¼-in. diam: tensile strength, 283 MPa (41 ksi); elongation, 10% in 50 mm or 2 in.
Shear strength. 214 MPa (31 ksi)
Compressive properties. Compressive strength, 414 MPa (60 ksi)
Hardness. 82 HB (500 kg load, 10-mm hardened steel ball, 30-s duration of loading)
Impact strength. Charpy (¼ by ¼ by 3 in., unnotched), 58 J (43 ft·lb)

Mass Characteristics

Density. 6.6 Mg/m³ (0.24 lb/in.³) at 21 °C (70 °F)
Volume change on freezing. 1.17% shrinkage (Use 0.3 to 0.6% die design.)

Thermal Properties

Liquidus temperature. 387 °C (728 °F)
Solidus temperature. 381 °C (717 °F)
Coefficient of thermal expansion. Linear, 27.4 μm/m·K (15.2 μin./in.·°F) at 20 to 100 °C (68 to 212 °F)
Specific heat. 418.7 J/kg·K (0.10 Btu/lb·°F) at 20 to 100 °C (68 to 212 °F)
Thermal conductivity. 113.0 W/m·K (65.3 Btu/ft·h·°F) at 70 to 140 °C (160 to 285 °F)

Electrical Properties

Electrical conductivity. Volumetric, 27% IACS at 20 °C (68 °F)

Electrical resistivity. 63.694 nΩ·m at 20 °C (68 °F); temperature coefficient, 0.03774 nΩ·K from 0 to 100 °C (32 to 212 °F)

Chemical Properties

General corrosion behavior. This alloy can be used safely wherever zinc, zinc-coated iron or zinc-coated steel has been used successfully in the past. Atmospheric corrosion rate is 78 μm/yr (3.1 μin./yr) in Palmerton, PA and 220 μm/yr (8.7 μin./yr) in New York City

Fabrication Characteristics

Joining. Solder with Pb-Sn solder over nickel-plated surface; acidulated zinc chloride flux. Oxyacetylene weld with Zamak-3, no flux, soft flame. Resistance welding: pulsation technique is usable.
Casting temperature range. 393 to 427 °C (740 to 800 °F)
Precautions in melting. Allow a 15-min holding period after the aluminum has become molten, during which time fine particles of aluminum and silicon (believed to be oxides) will rise to the surface of the melt and must be removed by skimming. Skimming will also remove all iron-aluminum compounds. To avoid depletion of alloying elements, molten metal must not be held at high temperature.
Precautions in casting. Strict adherence to chemical composition limits and proper die design to ensure sound castings having uniformly high strength

AC41A Alloy
Zn-4Al-1Cu-0.04Mg

Commercial Names

Trade name. Number 5 Die Casting Alloy
Previous trade name. Zamak-5 (die casting and sand casting)
Foreign. Mazak-5, Gomak-5

Specifications

ASTM. B86, alloy AC41A (die castings); B240, alloy AC41A (ingot for die castings)
SAE. J468, alloy 925
UNS. Z35530
Government. QQ-Z-363
Foreign. BS1400, alloy B; CSA, AC41; DIN, Z410; J15, type II

Chemical Composition

Composition limits. ASTM B86: 3.5 to 4.3 Al; 0.75 to 1.25 Cu, 0.03 to 0.08 Mg, 0.100 max Fe, 0.005 max Pb, 0.004 max Cd, 0.003 max Sn, rem Zn. (Special high grade zinc must be used as the basic material in making this alloy.)
Consequence of exceeding impurity limits. Alloy becomes subject to intergranular attack and fails prematurely by warping and cracking

Applications

Typical uses. For die castings, such as automotive parts, household utensils, office equipment, building hardware, padlocks, toys and novelties. Sand castings used for drop hammer dies.

Mechanical Properties

Tensile properties. Die cast specimen, 6-mm or 1/4-in. diam: (typical) tensile strength, 328 MPa (47.6 ksi); elongation, 7% in 50 mm or 2 in.
Shear strength. 262 MPa (38 ksi)
Compressive properties. Compressive strength 600 MPa (87 ksi)
Hardness. 91 HB
Impact strength. Charpy (unnotched), 65 J (48 ft·lb)
Fatigue strength. Reverse bending, 56.36 MPa (8.175 ksi) at 10^8 cycles

Mass Characteristics

Density. 6.6 Mg/m^3 (0.24 lb/in.^{3}) at 21 °C (70 °F)
Volume change on freezing. 1.17% shrinkage (Use 0.3 to 0.6% for die design.)

Thermal Properties

Liquidus temperature. 386 °C (727 °F)

Solidus temperature. 380 °C (717 °F)
Coefficient of thermal expansion. Linear, 27.4 μm/m·K (15.2 μin./in.·°F) at 20 to 100 °C (68 to 212 °F)
Specific heat. 419 J/kg·K (0.10 Btu/lb·°F) at 20 to 100 °C (68 to 212 °F)
Thermal conductivity. 109 W/m·K (63 Btu/ft·h·°F) at 70 to 140 °C (160 to 285 °F)

Electrical Properties

Electrical conductivity. Volumetric, 26% IACS at 20 °C (68 °F)
Electrical resistivity. 65.359 nΩ·m at 20 °C (68 °F). Temperature coefficient, 3.527 pΩ·m/K at 0 to 100 °C (32 to 212 °F)

Chemical Properties

General corrosion behavior. This alloy can be used safely wherever zinc, zinc-coated iron or zinc-coated steel has been used successfully in the past. Atmospheric corrosion rate is 63 μm/yr (2.5 μin./yr) in Palmerton, PA and 280 μm/yr (11 μin./yr) in New York City

Fabrication Characteristics

Joining. Solder with Pb-Sn solder over nickel-plated surface; acidulated zinc chloride flux. Oxyacetylene weld with Zamak-3, no flux, soft flame. Resistance welding: pulsation technique is usable.
Die casting temperature. 393 to 427 °C (740 to 800 °F)
Precautions in melting. Allow a 15 min holding period after the aluminum has become molten, during which time fine particles of aluminum and silicon (believed to be oxides) will rise to the surface of the melt and must be removed by skimming. Skimming will also remove all iron-aluminum compounds.
Precautions in casting. Strict adherence to chemical composition limits and proper die design to ensure sound castings having uniformly high strength

Zinc-base Slush-casting Alloy
Zn-4.75Al-0.25Cu

Chemical Composition

Composition limits. 4.50 to 5.00 Al, 0.2 to 0.3 Cu, 0.007 max Pb, 0.005 max Cd, 0.005 max Sn, 0.100 max Fe, rem Zn. (Special high grade zinc

must be used as the basic material in making this alloy.)
Consequence of exceeding impurity limits. Alloy becomes hot short, subject to intergranular attack and fails prematurely by warping and cracking

Applications

Typical uses. Slush and permanent mold castings, principally parts for lighting fixtures

Mechanical Properties

Tensile properties. Chill cast specimen, 12 mm or 1/2 in. diam: tensile strength, 193 MPa (28 ksi)
Impact strength. Charpy (1/4 by 1/4 by 3 in., unnotched), 4 J (3 ft·lb)

Thermal Properties

Liquidus temperature. Approx 390 °C (734 °F)
Solidus temperature. 380 °C (716 °F)

Fabrication Characteristics

Precautions in casting. Avoid contamination with lead, tin or cadmium

Zinc-base Slush-casting Alloy
Zn-5.5Al

Commercial Names

Trade name. Unbreakable Metal

Chemical Composition

Composition limits. 5.25 to 5.75 Al, 0.1 max Cu, 0.007 max Pb, 0.005 max Cd, 0.005 max Sn, 0.100 max Fe, rem Zn. (Special high grade zinc must be used as the basic material in making this alloy.)
Consequence of exceeding impurity limits. Alloy becomes hot short, subject to intergranular attack and fails prematurely by warping and cracking

Applications

Typical uses. For all slush and permanent mold castings, chiefly in the manufacture of lighting fixtures

Mechanical Properties

Tensile properties. Chill cast specimen, 12 mm or 1/2 in. diam: tensile strength, 172 MPa (25 ksi); elongation, 1% in 50 mm or 2 in.

Impact strength. Charpy (¼ by ¼ by 3 in., unnotched), 1 J (1 ft·lb)

Thermal Properties

Liquidus temperature. Approx 395 °C (745 °F)

Solidus temperature. 380 °C (715 °F)

Fabrication Characteristics

Precautions in casting. Avoid contamination with Pb, Sn, or Cd

Zinc Foundry Alloy ZA-12 Zn-11Al

Compiled by Michael L. Bess
Eastern Alloys, Inc.
and Ernest W. Horvick
Zinc Institute, Inc.

Commercial Names

Previous trade name. Korloy 2570, ZA-12

Foreign. Kayem 12, EZDA 12, Korloy 2570

Chemical Composition

Composition limits. 10.5 to 11.5 Al, 0.5 to 1.0 Cu, 0.01 to 0.02 Mg, 0.005 max Pb, 0.004 max Cd, 0.003 max Sn, 0.075 max Fe, 0.05 max others (total), rem Zn

Consequence of exceeding impurity limits. Alloy becomes subject to intergranular corrosion and premature warping and cracking result from high Pb, Cd, Sn, Bi. High Fe and Si cause excessive tool wear.

Applications

Typical uses. For sand and permanent mold gravity castings such as commercial hardware, fluid handling components, hose couplings, electrical conduit fittings, industrial hardware, hydraulic and pneumatic control equipment, transmission line components, machinery components, and bearing and bushing applications. Normally supplied as alloy ingot.

Precautions in use. Do not use under highly acidic or caustic conditions.

Mechanical Properties

Tensile properties. Typical. Tensile strength: sand, 310 MPa (42 ksi); permanent mold, 360 MPa (50 ksi). Yield strength: 210 MPa (30.5 ksi). Elongation: 1 to 3%

Shear strength. 220 MPa (32 ksi)

Compressive properties. Compressive strength, 585 MPa (85 ksi); compressive yield strength (0.1% offset), 145 MPa (21 ksi)

Hardness. 105 to 125 HB (500 kg load, 10 mm ball)

Elastic modulus. Tension, 83×10^6 GPa (12×10^6 psi)

Impact strength. Charpy (unnotched): sand, 8 J (6 ft·lb); permanent mold, 17.6 J (13 ft·lb)

Fatigue strength. Rotating beam, 55 MPa (8 ksi) at 10^8 cycles

Structure

Microstructure. Primary Al rich, α phase in a matrix of eutectic $\alpha + \beta$

Mass Characteristics

Density. 6.03 Mg/m³ (0.218 lb/in.³) at 21 °C (70 °F)

Volume change on freezing. 1.9% shrinkage

Thermal Properties

Liquidus temperature. 432 °C (810 °F)

Solidus temperature. 380 °C (715 °F)

Melting temperature. 380 to 430 °C (715 to 810 °F)

Coefficient of thermal expansion. Linear, 28 μm/m·K (15.5 μin./in.·°F) at 20 to 100 °C (68 to 212 °F)

Thermal conductivity. 87 to 92 W/m·K (50 to 53 Btu/ft·h·°F) at 20 °C (68 °F)

Electrical Properties

Electrical conductivity. Volumetric, 25% IACS at 20 °C (68 °F)

Chemical Properties

General corrosion behavior. Alloy can be used safely wherever zinc-coated steel or zinc-coated iron has been used successfully in the past.

Good resistance to atmospheric corrosion.

Fabrication Characteristics

Machinability. 80% of C36000, free-cutting brass

Casting temperature. Sand, 441 to 538 °C (825 to 1000 °F); permanent mold, 482 to 580 °C (900 to 1075 °F)

Joining. Mechanical fasteners, adhesives, TIG or MIG welding, Zn-Cd soft solder, Zn-Al-Cu soft solder

Precautions in casting. Strict adherence to chemical composition limits is necessary. The alloy should be melted and held in clay-graphite, ceramic or silicon carbide crucibles rather than iron crucibles.

ILZRO 16 Zn-1.2 Cu

Compiled by Ernest W. Horvick
Zinc Institute, Inc.

Chemical Composition

Composition limits. 1.0 to 1.5 Cu, 0.15 to 0.25 Ti, 0.10 to 0.20 Cr, 0.30 to 0.40 Ti + Cr, 0.01 to 0.04 Al, rem Zn

Applications

Typical uses. Lead-bearing components for elevated temperature service

Mechanical Properties

Tensile properties. Typical. Tensile strength, 225 to 235 MPa (33 to 34 ksi); yield strength, 135 to 145 MPa (20 to 21 ksi); elongation, 5 to 6% in 50 mm or 2 in.

Hardness. 75 to 77 HB

Elastic modulus. Tension, 97 GPa (14×10^6 psi)

Stress-rupture characteristics. Stress to produce 0.1% plastic strain in 500 h:

Temperature		Stress	
°C	°F	MPa	ksi
20	68	108	15.7
75	167	101	14.6
125	257	80	11.6
150	302	46	6.7

Thermal Properties

Liquidus temperature. 417 °C (785 °F)

Solidus temperature. 415 °C (780 °F)

Electrical Properties

Electrical resistivity. 84 nΩ·m at 20 °C (68 °F)

Fabrication Characteristics

Casting temperature. 460 to 470 °C (860 to 880 °F)

Zn-27Al

Compiled by Michael L. Bess
Eastern Alloys, Inc.

Commercial Names

Common name. Zinc foundry alloy ZA-27

Chemical Composition

Composition limits. 25 to 28 Al, 0.01 to 0.02 Mg, 2.0 to 2.5 Cu, bal Zn

Consequence of exceeding impurity limits. Intergranular corrosion results from high Pb, Cd, Sn, Bi. High Fe and Si cause excessive tool wear.

Applications

Typical uses. Thin-section castings that require higher strength than Zinc Foundry Alloy ZA-12. Normally supplied as alloy ingot.

Precautions in use. Do not use under highly acidic or caustic conditions.

Mechanical Properties

Tensile properties. Typical. Tensile strength: cast, 400 to 440 MPa (58 to 64 ksi); heat treated, 310 to 325 MPa (45 to 47 ksi). Yield strength: cast, 365 MPa (53 ksi); heat treated, 275 MPa (40 ksi). Elongation: cast, 3 to 6%; heat treated, 8 to 11%

Hardness. As cast, 110 to 120 HB; heat treated, 90 to 100 HB (500 kg load, 100 mm ball)

Elastic modulus. Tension, 145 GPa (21×10^6 psi)

Fatigue strength. Rotating beam at 10^8 cycles: as cast, 170 MPa (25 ksi); heat treated, 105 MPa (15 ksi)

Impact strength. Charpy (¼ in. square specimen) as cast, 15 J (11 ft·lb); heat treated, 26 J (19 ft·lb)

Structure

Microstructure. As cast, primary Al rich α dendrites surrounded by α + β eutectic; heat treated, primary Al rich α dendrites surrounded by lamellar α + β eutectic

Mass Characteristics

Density. 5.01 Mg/m³ (0.181 lb/in.³) at 20 °C (68 °F)

Volume change on freezing. 2.2% shrinkage

Thermal Properties

Liquidus temperature. 492 °C (919 °F)

Solidus temperature. 378 °C (714 °F)

Coefficient of thermal expansion. Linear, 25.9 μm/m·K (14.4 μin./in.·°F) at 20 to 100 °C (68 to 212 °F)

Electrical Properties

Electrical conductivity. Volumetric, 28% IACS at 20 °C (68 °F)

Chemical Properties

General corrosion behavior. Same as Zinc Foundry Alloy ZA-12

Fabrication Characteristics

Joining. Zn-Cd soft solder; Zn-Al-Cu soft solder, TIG welding

Casting range. 537 to 593 °C (1000 to 1100 °F)

Precautions in casting. Strict adherence to chemical composition limits is necessary. The alloy should be melted and held in clay-graphite, ceramic or silicon carbide crucibles rather than iron crucibles.

Commercial Rolled Zinc Zn-0.08Pb

Commercial Names

Previous trade name. Deep drawing zinc

Chemical Composition

Composition limits. Strip: 0.10 max Pb, 0.012 max Fe, 0.005 max Cd, 0.001 max Cu, 0.001 max Sn, 0.001 max Al

Consequence of exceeding impurity limits. Iron or cadmium increases hardness and reduces ductility; copper alters chemical characteristics; tin causes hot shortness; aluminum promotes intergranular corrosion.

Applications

Typical uses. Generally drawn, formed or spun articles requiring some rigidity, such as drawn battery cans and formed eyelets and grommets

Precautions in use. Deforms under light continuous load, particularly at 38 °C (100 °F) or higher

Mechanical Properties

Tensile strength. See Table 1.

Hardness. Hot rolled strip, 38 HB

Fatigue strength. Rotating beam at 10^8 cycles; hot rolled strip, 17 MPa (2.5 ksi)

Mass Characteristics

Density. 7.14 Mg/m³ (0.258 lb/in.³) at 21 °C (70 °F)

Thermal Properties

Melting point. 419 °C (786 °F)

Coefficient of thermal expansion. Linear, at 20 to 40 °C (70 to 105 °F): 32.5 μm/m·K (18.05 μin./in.·°F) longitudinal, 23 μm/m·K (12.7 μin./in.·°F) transverse

Specific heat. 395 J/kg·K (0.094 Btu/lb·°F) at 20 to 100 °C (68 to 212 °F)

Thermal conductivity. Longitudinal: 108 W/m·K (62.4 Btu/ft·h·°F) at 18 °C (64 °F)

Electrical Properties

Electrical conductivity. Volumetric: 28.44% IACS for hot rolled; 28.27% IACS for cold rolled

Electrical resistivity. At 20 °C (68 °F): 60.6 nΩ·m for hot rolled, 61.0 nΩ·m for cold rolled

Chemical Properties

General corrosion behavior. Excellent resistance to atmospheric corrosion. Penetration same as for Zn-0.06Pb-0.06Cd alloy.

Fabrication Characteristics

Suitable forming methods. Suited to forming by drawing, bending, roll forming, spinning, swaging, impact extrusion

Precautions in forming. Keep temperature above 21 °C (70 °F). Use pure soapy water or noncorrosive mineral oil as lubricant.

Standard finishes. Precautions in finishing. Air-dried coatings are preferred over baked finishes because many enamels must be baked at temperatures high enough to impair mechanical properties of the zinc.

Joining. Soft solder with Pb-Sn, acidulated zinc chloride flux

Hot working temperature. 120 to 275 °C (250 to 525 °F)

Hot-shortness temperature. 300 to 420 °C (570 to 785 °F)

Table 1 Typical tensile properties of commercial rolled zinc (Zn-0.08Pb) strip(a)

Orientation	Tensile strength MPa	ksi	Elongation, %
Hot Rolled			
Longitudinal	134	19.5	65
Transverse	159	23.0	50
Cold Rolled			
Longitudinal	145	21.0	50
Transverse	186	27.0	40
(a) Finish rolled to 0.6 mm (0.024 in.).			

Commercial Rolled Zinc Zn-0.06Pb-0.06Cd

Chemical Composition

Composition limits. Strip: 0.05 to 0.10 Pb, 0.05 to 0.08 Cd, 0.012 max Fe, 0.005 max Cu, 0.001 max Sn, 0.001 max Al

Consequence of exceeding impurity limits. Iron increases hardness and reduces ductility; copper alters chemical characteristics; tin causes hot shortness; aluminum promotes intergranular corrosion.

Applications

Typical uses. Generally drawn, formed or spun articles requiring some rigidity, such as drawn battery cans and formed eyelets and grommets

Precautions in use. Deforms under light continuous load, particularly at temperatures slightly above room temperature (38 °C or 100 °F) or higher

Mechanical Properties

Tensile strength. See Table 2.
Hardness. Hot rolled strip, 43 HB
Fatigue strength. Rotating beam at 10^8 cycles; hot rolled strip, 26 MPa (3.8 ksi)

Mass Characteristics

Density. 7.14 Mg/m^3 (0.258 lb/in.^{3}) at 21 °C (70 °F)

Thermal Properties

Melting point. 419 °C (786 °F)
Coefficient of thermal expansion. Linear, at 20 to 40 °C (70 to 105 °F): 32.5 μm/m·K (18.05 μin./in.·°F) longitudinal, 23 μm/m·K (12.7 μin./in.·°F) transverse
Specific heat. 395 J/kg·K (0.094 Btu/lb·°F) at 20 to 100 °C (68 to 212 °F)
Thermal conductivity. Longitudinal: 108 W/m·K (62.4 Btu/ft·h·°F) at 18 °C (64 °F) (approx)

Chemical Properties

General corrosion behavior. Excellent resistance to atmospheric corrosion. Typical corrosion penetration outdoors is 2.5 μm/yr (64 μin./yr) in Palmerton, PA and 11.0 μm/yr (280 μin./yr) in New York City

Fabrication Characteristics

Suitable forming methods. Suited to forming by drawing, bending, roll forming, spinning, swaging, impact extrusion
Precautions in forming. Keep temperature above 21 °C (70 °F). Use pure soapy water or noncorrosive mineral oil as lubricant.
Standard finishes. Precautions in finishing. Air dried coatings are preferred over baked finishes because many enamels must be baked at temperatures high enough to impair mechanical properties of the zinc.
Joining. Soft solder with Pb-Sn, acidulated zinc chloride flux
Hot working temperature. 120 to 225 °C (250 to 435 °F)
Hot-shortness temperature. 300 to 420 °C (570 to 785 °F)

Table 2 Typical tensile properties of commercial rolled zinc (Zn-0.06Pb-0.06Cd) strip (a)

Orientation	Tensile strength MPa	ksi	Elongation, %
Hot Rolled			
Longitudinal	150	21.0	52
Transverse	170	25.0	30
Cold Rolled			
Longitudinal	150	22.0	40
Transverse	200	29.0	30
(a) Finish rolled to 0.6 mm (0.024 in.).			

Commercial Rolled Zinc Zn-0.3Pb-0.3Cd

Chemical Composition

Composition limits. Strip: 0.25 to 0.50 Pb, 0.25 to 0.45 Cd, 0.02 max Fe, 0.005 Cu, 0.001 Sn, 0.001 max Al
Consequence of exceeding impurity limits. Iron increases hardness and decreases ductility; copper changes the chemical characteristics; tin causes hot shortness; aluminum promotes intergranular corrosion.

Applications

Typical uses. Plates and strip for soldered battery cans, photoengraver's and lithographer's sheet
Precautions in use. Deforms under continuous load, particularly at elevated temperatures (52 °C [125 °F] and higher)

Mechanical Properties

Tensile strength. See Table 3.
Hardness. Hot rolled strip, 47 HB
Fatigue strength. Rotating beam at 10^8 cycles; hot rolled strip, 28 MPa (4.1 ksi)

Mass Characteristics

Density. 7.14 Mg/m^3 (0.258 lb/in.^{3}) at 21 °C (70 °F)

Thermal Properties

Melting point. 419 °C (786 °F)
Coefficient of thermal expansion. Linear, at 20 to 98 °C (70 to 210 °F): 39.9 μm/m·K (22.2 μin./in.·°F) longitudinal, 23.4 μm/m·K (13 μin./in.·°F) transverse
Specific heat. 395 J/kg·K (0.094 Btu/lb·°F) at 20 to 100 °C (68 to 212 °F)

Chemical Properties

General corrosion behavior. Excellent resistance to atmospheric corrosion. Typical corrosion penetration outdoors is 2.5 μm/yr (64 μin./yr) in Palmerton, PA and 11.0 μm/yr (280 μin./yr) in New York City.

Fabrication Characteristics

Formability. Suited to forming by drawing, bending, roll forming, swaging, impact extrusion
Precautions in forming. Keep temperature above 21 °C (70 °F). Use pure soapy water or noncorrosive mineral oil as lubricant.
Annealing temperature. 105 °C (220 °F)

Standard finishes. See Commercial Rolled Zinc, Zn-0.08Pb.
Joining. Soft solder with Pb-Sn, acidulated zinc chloride flux
Hot working temperature. 120 to 225 °C (250 to 435 °F)
Hot-shortness temperature. 275 to 420 °C (525 to 785 °F)

Table 3 Typical tensile properties of commercial rolled zinc (Zn-0.3Pb-0.3Cd) strip (a)

| Orientation | Tensile strength | | Elonga- |
	MPa	ksi	tion, %
Hot Rolled			
Longitudinal	160	23.0	50
Transverse	200	29.0	32
Cold Rolled			
Longitudinal	170	25.0	45
Transverse	210	31.0	28
(a) Finish rolled to 0.6 mm (0.024 in.).			

Copper-Hardened Rolled Zinc Zn-1Cu

Commercial Names

Previous trade name. Zilloy-40

Chemical Composition

Composition limits. Strip: 0.85 to 1.25 Cu, 0.10 max Pb, 0.012 max Fe, 0.005 max Cd, 0.001 max Sn, 0.001 max Al
Consequence of exceeding impurity limits. Iron or cadmium increases hardness and reduces ductility; tin causes hot shortness; aluminum promotes intergranular corrosion

Applications

Typical uses. Weatherstrip, name plates, ferrules, and drawn, formed or spun articles requiring stiffness
Precautions in use. Deforms under heavy, continuous load, particularly above 52 °C (125 °F)

Mechanical Properties

Tensile strength. See Table 4.
Hardness. Hot rolled strip, 52 HB; cold rolled strip, 60 HB

Fatigue strength. 42.0 MPa (6.1 ksi) at 10^8 million cycles
Creep-rupture characteristics. Creep rate at 83 MPa (12 ksi) and 25 °C (77 °F): longitudinal test, 0.15 days/%; transverse, 0.27 days/%

Mass Characteristics

Density. 7.17 Mg/m³ (0.259 lb/in.³) at 21 °C (70 °F)

Thermal Properties

Liquidus temperature. 422 °C (792 °F)
Solidus temperature. 419 °C (786 °F)
Boiling point. 907 °C (1665 °F)
Specific heat. 401 J/kg·K (0.0957 Btu/lb·°F) at 20 to 100 °C (68 to 212 °F)

Electrical Properties

Electrical conductivity. Volumetric, approximately 28% IACS at 20 °C (68 °F)
Electrical resistivity. 62.2 nΩ·m at 20 °C (68 °F), hot rolled

Chemical Properties

General corrosion behavior. Excellent resistance to atmospheric corrosion. Typical corrosion penetration is 2.5 μm/yr (64 μin./yr) in Palmerton, PA and 11.0 μm/yr (280 μin./yr) in New York City.

Fabrication Characteristics

Formability. Suited to forming by drawing, bending, roll forming, spinning and swaging
Precautions in forming. Keep temperatures above 21 °C (70 °F). Use pure soapy water or noncorrosive mineral oil as lubricant.
Precautions in finishing. Air-dried coatings are preferred over baked finishes because many enamels must be baked at temperatures high enough to impair mechanical properties of the zinc.
Joining. Soft solder with Pb-Sn, acidulated zinc chloride flux
Annealing temperature. 175 °C (347 °F)
Hot working temperature. 175 to 300 °C (445 to 570 °F)
Hot-shortness temperature. 300 to 419 °C (570 to 785 °F)

Table 4 Typical tensile properties of copper hardened rolled zinc (Zn-1Cu) strip

| Orientation | Tensile strength | | Elonga- |
	MPa	ksi	tion, %
Hot Rolled			
Longitudinal	170	24	50
Transverse	210	30	35
Cold Rolled			
Longitudinal	210	31	40
Transverse	280	40	25

Rolled Zinc Alloy Zn-1Cu-0.010Mg

Commercial Names

Previous trade name. Zilloy-15

Chemical Composition

Composition limits. Strip: 0.85 to 1.25 Cu, 0.006 to 0.016 Mg, 0.15 max Pb, 0.015 max Fe, 0.001 max Sn, 0.001 max Al, 0.04 max Cd
Consequence of exceeding impurity limits. Iron or cadmium increases hardness and reduces ductility; tin causes hot shortness; aluminum promotes intergranular corrosion.

Applications

Typical uses. Corrugated roofing and flat, drawn or mildly formed articles requiring maximum stiffness
Precautions in use. Deforms under heavy continuous load, particularly at elevated temperatures (52 °C [125 °F] and higher)

Mechanical Properties

Tensile strength. See Table 5.
Hardness. Hot rolled strip, 61 HB; cold rolled strip, 80 HB
Fatigue strength. Rotating beam at 10^8 cycles; 47 MPa (6.8 ksi)

Mass Characteristics

Density. 7.17 Mg/m³ (0.259 lb/in.³) at 21 °C (70 °F)

Thermal Properties

Liquidus temperature. 422 °C (792 °F)
Solidus temperature. 419 °C (786 °F)
Boiling point. 907 °C (1665 °F)
Coefficient of thermal expansion. Linear, at 20 to 100 °C (68 to 212 °F):

34.8 μm/m·K (19.3 μin./in.·°F) longitudinal, 21.1 μm/m·K (11.7 μin./in.·°F) transverse

Specific heat. 401 J/kg·K (0.0957 Btu/lb·°F) at 20 to 100 °C (68 to 212 °F)

Thermal conductivity. Longitudinal: 105 W/m·K (60.6 Btu/ft·h·°F) at 18 °C (64 °F)

Electrical Properties

Electrical conductivity. Volumetric: approximately 27% IACS at 20 °C (68 °F)

Electrical resistivity. 63.1 nΩ·m at 20 °C (68 °F)

Chemical Properties

General corrosion behavior. Excellent resistance to atmospheric corrosion. Typical corrosion penetration outdoors is 2.5 μm/yr (64 μin./yr) in Palmerton, PA and 11.0 μm/yr (280 μin./yr) in New York City.

Fabrication Characteristics

Formability. Suited to forming by drawing, bending and roll forming

Precautions in forming. Keep temperature about 21 °C (70 °F). Use pure soapy water or noncorrosive mineral oil as lubricant.

Standard finishes. Precautions in finishing. Air-dried coatings are preferred over baked finishes because many enamels must be baked at temperatures high enough to impair mechanical properties of zinc.

Joining. Soft solder with Pb-Sn, acidulated zinc chloride flux

Annealing temperature. 175 °C (345 °F)

Hot working temperature. 175 to 300 °C (475 to 570 °F)

Hot-shortness temperature. 300 to 420 °C (570 to 785 °F)

Table 5 Typical tensile properties of rolled zinc (Zn-1Cu-0.010Mg) strip (a)

Orientation	Tensile strength MPa	ksi	Elongation, %
Hot Rolled			
Longitudinal	200	29	20
Transverse	276	40	10
Cold Rolled			
Longitudinal	248	36	25
Transverse	317	46	10
(a) Finish rolled to 0.6 mm (0.024 in.).			

Zn-Cu-Ti Alloy Zn-0.8Cu-0.15Ti

Chemical Composition

Composition limits. Strip: 0.50 to 1.50 Cu, 0.12 to 0.50 Ti, 0.10 max Pb, 0.012 Fe, 0.05 Cd, 0.001 max Sn, 0.001 max Al

Consequences of exceeding impurity limits. Iron or cadmium increases hardness and reduces ductility; tin causes hot shortness; aluminum promotes intergranular corrosion.

Applications

Typical uses. Corrugated roofing, leaders, and gutters and formed articles requiring maximum creep resistance

Precautions in use. Creep resistance decreased with increasing temperature of use. Should be heat treated after cold working for maximum creep resistance.

Mechanical Properties

Tensile strength. See Table 6.

Creep rupture characteristics. Creep rate at 83 MPa (12 ksi) and 25 °C (77 °F): longitudinal test, 0.15 days/%; transverse, 0.27 days/%

Mass Characteristics

Density. 7.17 Mg/m³ (0.259 lb/in.³) at 21 °C (70 °F)

Thermal Properties

Liquidus temperature. 422 °C (792 °F)

Solidus temperature. 419 °C (786 °F)

Boiling point. 907 °C (1665 °F)

Coefficient of thermal expansion. Linear, at 20 to 100 °C (68 to 212 °F): 24.9 μm/m·K (13.8 μin./in.·°F) longitudinal, 19.4 μm/m·K (10.7 μin./in.·°F) transverse

Specific heat. 400 J/kg·K (0.096 Btu/lb·°F) at 20 to 100 °C (68 to 212 °F)

Thermal conductivity. Longitudinal: 105 W/m·K (60 Btu/ft·h·°F) at 18 °C (64 °F)

Electrical Properties

Electrical conductivity. Volumetric, approximately 27% IACS at 20 °C (68 °F)

Electrical resistivity. 62.4 nΩ·m at 20 °C (68 °F)

Chemical Properties

General corrosion behavior. Excellent resistance to atmospheric corrosion. Typical corrosion penetration is 2.5 μm/yr (64 μin./yr) in Palmerton, PA and 11.0 μm/yr (280 μin./yr) in New York City

Fabrication Characteristics

Formability. Suited to forming by drawing, bending and roll forming

Annealing temperature. High creep resistance can be restored after cold working by annealing for 45 min at 250 °C (480 °F) (metal at temperature for 45 min)

Joining. Soft solder with Pb-Sn, acidulated zinc chloride flux

Hot working temperature. 150 to 300 °C (300 to 570 °F)

Hot-shortness temperature. 300 to 420 °C (570 to 785 °F)

Table 6 Typical tensile properties of Zn-Cu-Ti alloy (Zn-0.8Cu-0.15Ti) strip (a)

Orientation	Tensile strength MPa	ksi	Elongation, %
Hot Rolled			
Longitudinal	221	32	38
Transverse	290	42	21
Cold Rolled			
Longitudinal	200	29	44
Transverse	260	37	60
(a) Finish rolled to 0.6 mm (0.024 in.).			

Superplastic Zinc Alloy

Commercial Names

Trade name. Super · Z 300, Formetal 22 Alloy, Korloy 2684 (other grades also available)

Chemical Composition

Nominal composition. 22 Al, 0.5 Cu, 0.01 Mg, rem Zn

Applications

Typical uses. Supplied as sheet for thermal forming. Especially useful for low-volume applications, where tooling costs must be kept low. Used for electronic enclosures, cabinets and panels, business machine parts, and medical and other laboratory instruments and tools

Precautions in use. Subject to creep if highly stressed and/or stressed at elevated temperature

Mechanical Properties

Tensile properties. See Table 7.
Hardness. See Table 7.
Impact strength. As-rolled, and as-annealed at 315 °C (600 °F) and air cooled:

Temperature		Test direction			
		Longitudinal		Transverse	
°C	°F	J	ft·lb	J	ft·lb
20	68	27	20	9.5	7
0	32	26.5	19.5	9	6.5
−5	23	26	19	8	6
−40	−40	15.5	11.5	4	3
−50	−58	8	5	2	1.5

Creep characteristics. See Table 7.

Mass Characteristics

Density. 5.20 Mg/m^3 (0.188 lb/in.^{3}) at 20 °C (68 °F)

Thermal Properties

Coefficient of thermal expansion. Linear at 20 to 100 °C (68 to 212 °F). Longitudinal: as rolled, 22.0 µm/m·K (12.2 µin./in.·°F); annealed, 26.6 µm/m·K (14.8 µin./in.·°F). Transverse: as rolled, 21.5 µm/m·K (11.9 µin./in.·°F); annealed, 26.8 µm/m·K (14.9 µin./in.·°F)

Electrical Properties

Electrical conductivity. Volumetric, as rolled, 32% IACS at 20 °C (68 °F); annealed, 28% IACS

Table 7 Typical room temperature mechanical properties of superplastic zinc

Property	As-rolled	Annealed(a)
Tensile strength, MPa (ksi)	310 (45)	400 (58)
0.2% yield strength, MPa (ksi)	255 (37)	350 (51)
Elongation, %	27	11
Hardness, HR15T	70	84
Creep strength, MPa (ksi)(b)	20 (3)	40-50 (6-7)

(a) Annealed at 315 °C (600 °F) and air cooled; properties are similar to those of as-thermal-formed sheet. (b) Stress to produce a creep rate of 0.01% in 1 000 h (1% in 11.4 yr).

The Use of Zinc in Corrosion Service

By Ernest W. Horvick
Zinc Institute, Inc.

MORE THAN 40% of all zinc consumed in the United States is used in controlling corrosion of iron and steel. The good resistance of zinc to atmospheric corrosion is the property most broadly exploited in zinc applications. In zinc and zinc alloy products, this characteristic serves throughout the useful life of the product. On galvanized steel, the corrosion resistance afforded by the zinc coating continues well into the service life of the part. Even when the underlying steel is exposed, the ability of zinc to protect the steel sacrificially continues. Zinc is by far the most effective of the common sacrificial coatings for steel. Zinc is also used in the form of anodes applied to steel structures (painted or unpainted) to protect them against electrolytic corrosion.

Corrosion Rates of Zinc and Iron. Zinc is much more resistant to corrosion than steel, except in very dry climates where the rate of attack is extremely low for both and in certain highly corrosive atmospheres where the rate of attack is quite high for both. The comparative corrosion rates vary widely depending on the conditions of exposure. The average corrosion rate of

iron is 25 times that of zinc. The comparative weight losses of zinc and iron in industrial, urban and rural areas is shown in Fig. 1.

Effect of Composition. The composition of a zinc product seldom has a significant effect on its rate of corrosion in atmospheric exposure. In any given

outdoor environment, all commercial forms of zinc probably exhibit corrosion rates within ±10% of each other. Data from an ASTM test program reveal that the atmospheric corrosion rates of three grades of zinc are similar (Fig. 2).

Anodic Coatings. Zinc can be anod-

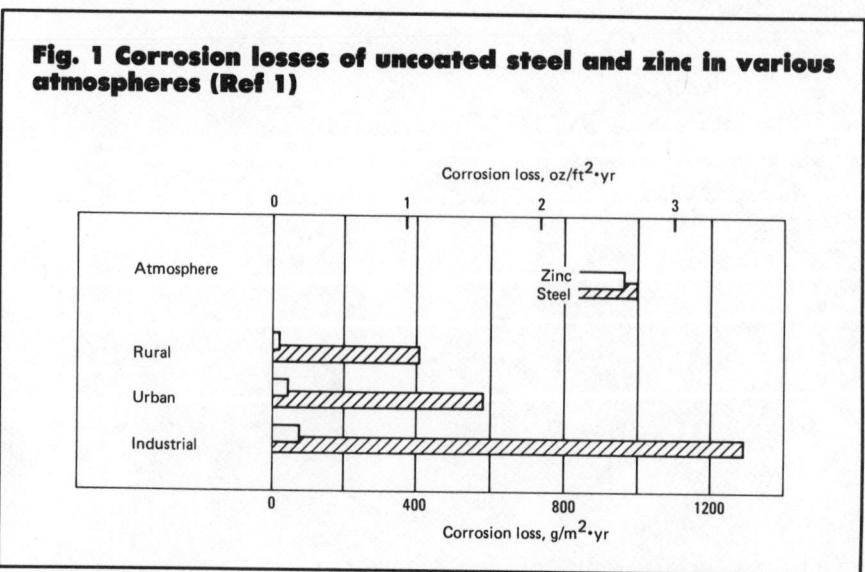

Fig. 1 Corrosion losses of uncoated steel and zinc in various atmospheres (Ref 1)

Fig. 2 Corrosion losses of three grades of zinc in various atmospheres (Ref 2 and 3)

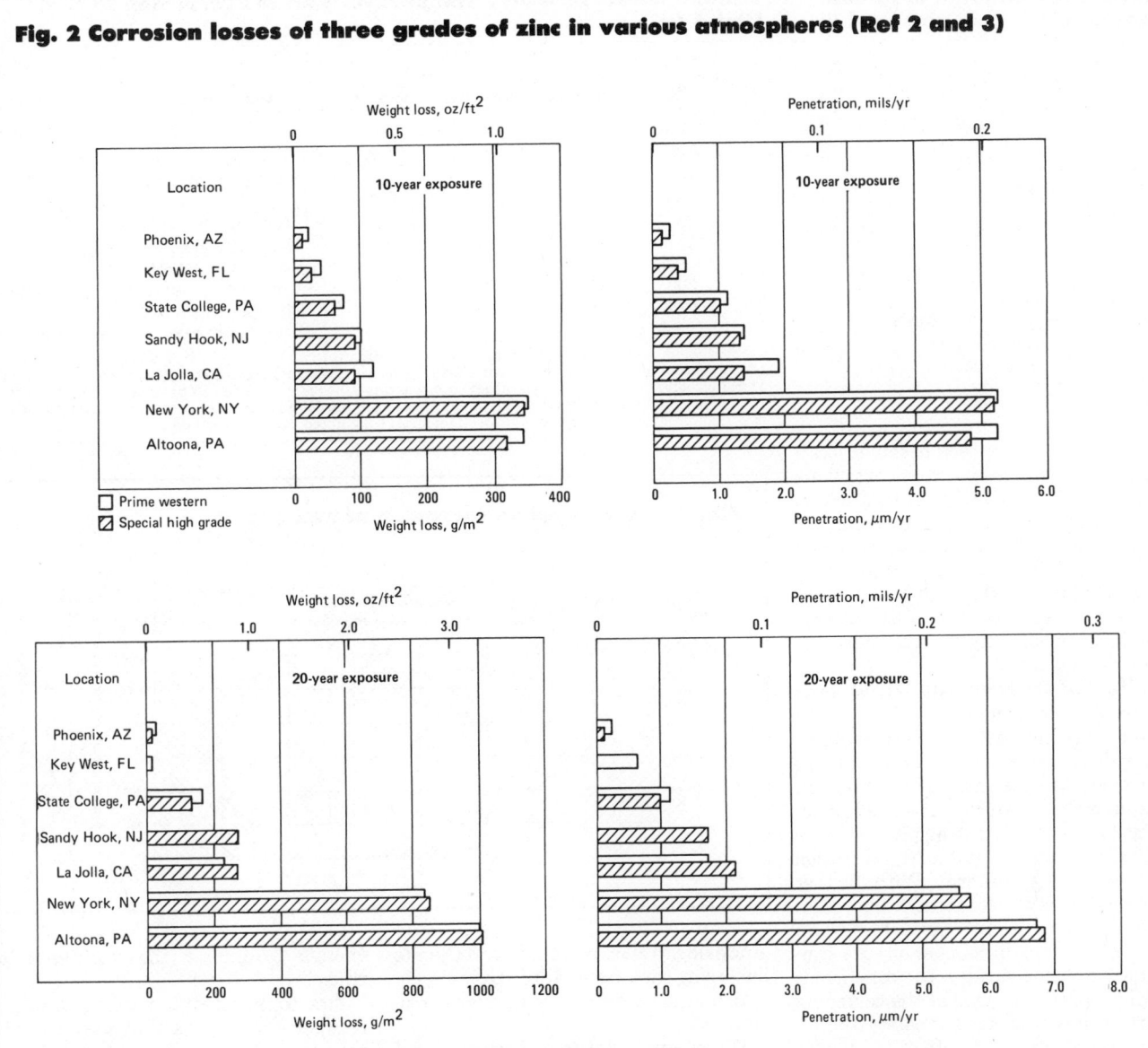

Panels 0.13-mm (0.05-in.) thick of three grades of zinc were exposed in selected test locations representative of highly industrial atmospheres (Altoona and New York), rural atmospheres (State College and Phoenix), and seacoast atmospheres (Key West, La Jolla and Sandy Hook). Corrosion data for Prime Western and Special High Grade types are shown above. Corrosion data for High Grade zinc (not shown) fall approximately midway between the values shown for these two types.

ized to enhance its abrasion and corrosion resistance. The anodic coating provides a good base for paint or can be used as is with its green-gray coloration. The process is approved for military and government agency applications under MIL-A-81801. Table 1 shows results of corrosion and abrasion testing.

Aqueous Corrosion

The corrosion rate of zinc in natural waters, as in atmospheric air, is less than that of iron under comparable conditions. Hence, zinc coatings are widely used in contact with water. As with other common metals, the rate of attack varies greatly with variations in exposure conditions. With zinc, the corrosion rate depends on temperature, composition and pH of the water, and the concentration of dissolved gases. The corrosion rate increases with increasing oxygen and carbon dioxide content. It is higher in soft water than in hard water, which often forms protective films. Weight losses ranging from 0.3 g/m^2 per day in hard water to 2.7 g/m^2 per day in distilled water have been observed with high-grade rolled zinc. The normal corrosion product formed on zinc in water is zinc carbonate.

Effect of Water Temperature.
The corrosion rate of zinc first increases and then decreases as the temperature of the water in which it is immersed is raised (Table 2). In the higher temper-

Table 1 Properties of green anodic coating on zinc (Ref 4)

Corrosion resistance
  Acetic acid salt
    spray (ASTM
    B287) . . . . 10-16 days
  Cass test (ASTM
    B368) . . . . 150 h
  Neutral salt spray
    (ASTM B117) . . . . 1500+ h (a)
Abrasion resistance
  Taber abrasion
    (500-g load) . . . . . . 1000 cycles
  BNF jet abrasion . . 35-50 g
Dielectric strength . . 300-400 V
Coating thickness . . . 0.032 mm (1.25 mils)
(a) May be as much as 8000 h.

Table 2 Effect of water temperature on the corrosion of zinc (Ref 5)(a)

Temperature		Corrosion rate			
°C	°F	g/m²·d(b)	µm/yr(c)	mils/yr(c)	Appearance of corrosion product
20	68 . . .	0.39	19.8	0.78	Definitely gelatinous; very adherent
50	122 . . .	1.37	71	2.8	Slightly less gelatinous; adherent
55	131 . . .	7.62	380	15	Mostly granular; nonadherent
65	149 . . .	57.7	3050	120	Decidedly granular; flaky and compact; nonadherent
75	167 . . .	46.0	2340	92	Decidedly granular flaky and compact; nonadherent
95	203 . . .	5.87	300	12	Compact, dense and flaky; adherent
100	212 . . .	2.35	120	4.7	Grayish white to black, very dense, resembling enamel; very adherent

(a) Specimens were rolled High Grade zinc, 80 by 120 by 0.76 mm (3.2 by 4.7 by 0.030 in.) in size. Surface preparation of specimens is not known. Corrosive medium was distilled water, approximately 15 L (4 gal). Duration of test was 15 days. Type of aeration was unwashed air (other details not given). Specimens were rotated in a horizontal position at 56 rpm on about a 50-mm (2-in.) radius. (b) Determined after removal of corrosion products (average of two specimens). (c) Calculated from weight loss.

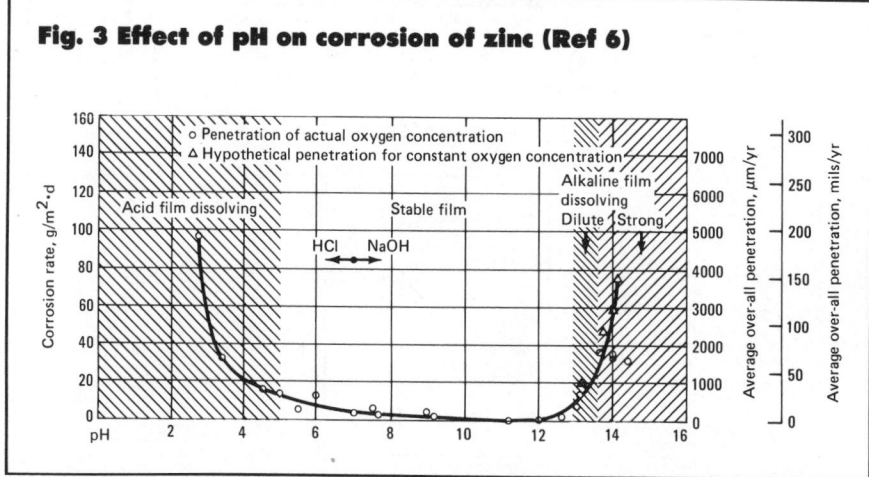

Fig. 3 Effect of pH on corrosion of zinc (Ref 6)

ature ranges, the attack becomes localized and pitting may occur, particularly in soft water. As shown in Table 2, increasing temperature causes changes in the reaction product film, resulting in pronounced changes in the corrosion rate of zinc. At the higher temperatures, the decreased solubility of oxygen in water usually causes a decrease in corrosion rate.

Effect of Oxygen and Other Gases in Water. Under conditions where the oxygen content cannot be replaced as quickly as it is consumed by the corrosion process, such as in quiescent water, a zinc product is attacked rapidly at local areas, creating pits. As more oxygen is made available, the corrosion becomes more uniform. With further increase in oxygen content of the water, the corrosion rate increases. For example, when thin films of moisture condense on a zinc surface, the concurrent rapid supply of oxygen at the corroding surface has a decided accelerating effect on the corrosion rate.

Attack in aerated water and attack in quiescent water both can be minimized by use of chromate conversion films. Data from experiments in which test pieces were immersed in water through which oxygen was bubbled show that corrosion occurs about eight times as fast in aerated water as in water boiled to remove gases and then cooled out of contact with air. Effects of temperature and dissolved oxygen concentration on corrosion of zinc in distilled water are shown in Table 3.

Aerated water invariably contains carbon dioxide and air and may also contain some sulfur dioxide and hydrogen sulfide. Carbon dioxide in normal concentrations reacts with the corrosion product and transforms it from zinc oxide to zinc hydroxide and then to basic zinc carbonate ($4ZnO \cdot CO_2 \cdot 4H_2O$), which forms a barrier to further corrosion.

Dissolved Acids, Bases and Salts. Zinc is not used in contact with acidic solutions or highly alkaline solutions because it corrodes rapidly in such mediums. As shown in Fig. 3, zinc is useful in aqueous solutions only in the pH range from 6.5 to 12.5. Its corrosion rate is high in acidic and highly alkaline solutions. The lowest corrosion rates are encountered in weakly alkaline solutions in the pH range mentioned above. The pH range of most natural waters used for domestic purposes is approximately 5.8 to 8.5.

Very dilute concentrations of acids accelerate corrosion rates beyond the limits of usefulness. Alkaline solutions of moderate strength are much less corrosive than acids of corresponding concentrations but are still corrosive enough to impair the usefulness of zinc.

Zinc-coated steel is used in handling refrigeration brines that may contain calcium chloride. Here, the corrosion rate is kept under control by adding sufficient alkali to bring the pH into the mildly alkaline range and by adding inhibitors such as sodium chromate. Certain salts, such as dichromates, borates and silicates, act as inhibitors of aqueous corrosion of zinc.

Corrosion Inhibitors. When zinc is used in contact with water in a closed system, inhibitors are frequently employed to minimize corrosion. Various inorganic inhibitors are available for use with zinc, such as sodium dichromate, sodium silicate, borax and hexametaphosphate. Mechanical exclusion films provided by adsorptive-type organic compounds such as lanolin are also useful for inhibiting corrosion of zinc. For most purposes, adjustment of the pH to the mildly alkaline range and addition of sodium dichromate is the preferred method. However, there is some danger of intensified pitting when an insufficient amount of inhibitor of this type is added.

Zinc-coated articles, if stored for significant periods of time under conditions where moisture may be trapped between closely adjacent surfaces, may become defaced by the appearance of a white reaction-product powder on the surface. This is known as "wet storage stain" or "white rust". Although it affects appearance, this powder generally is not harmful to the zinc coating. If unchecked, however, it can shorten the service life of the coating.

To avoid wet storage stain, it is essential to store zinc-coated material under conditions of free circulation of air and to avoid large periodic temperature changes that cause condensation of moisture on the metal. The reaction that causes wet storage stain does not occur in zinc-coated products in normal use once they have "aged" at the surface. However, it is important that new articles be carefully stored to avoid this problem.

Corrosion in Soils

As in natural waters, the corrosion rate of zinc in soils depends on factors such as dissolved salts, gases and pH of the ground water. In most soils, zinc forms an adherent protective coating, which slows the corrosion rate with time. Zinc anodes placed in the ground for cathodic protection are generally placed in a controlled environment (backfill) to prevent eventual polarization caused by this protective coating.

Nonaqueous Corrosion

Liquids. Many organic liquids that are nearly neutral in pH and substantially free of water do not attack zinc. Because of this, zinc and zinc-coated products are commonly used with gasoline, glycerine and inhibited trichlorethylene. The presence of free water may cause local corrosion because of lack of access to oxygen. When water is present, zinc may function as a catalyst in the decomposition of solutions like trichlorethylene, resulting in acid attack. Some organic compounds that contain acid impurities, such as low-grade glycerine, attack zinc. While neutral soaps do not attack zinc, some formation of zinc soap compounds may occur in dilute soap solutions.

Gases. Zinc may be used safely in contact with most common gases at normal temperatures, provided water is not present. Moisture content stimulates attack. Dry chlorine does not affect zinc, and hydrogen sulfide is also harmless because it reacts with zinc to form a surface layer of insoluble zinc sulfide. On the other hand, sulfur dioxide and chlorides are corrosive because they form water-soluble and hygroscopic salts.

Indoor Exposure. Zinc corrodes very little in ordinary indoor atmospheres of moderate relative humidity. Generally, a tarnish film develops slowly, starting at spots where dust particles are present on the surface. Moisture in the air up to 70% relative humidity has little effect on the tarnish rate. At and above this amount, however, zinc corrosion products absorb enough moisture to stimulate the attack to a perceptible rate, and thus the degree of corrosion is related to relative humidity.

Rapid corrosion can occur when surface temperature drops so that visible moisture condenses on the metal and dries slowly. Corrosion occurs because thin moisture films easily maintain high oxygen content due to the small volume of water and large water-air interface area. Considerably accelerated corrosion then can take place accompanied by formation of a corrosion-product film sufficiently thick to be objectionable. Chromate protective films are used to a considerable extent to prevent attack where accidental or limited contact with water is expected.

Atmospheres inside industrial buildings can be corrosive, particularly where condensation of heated moisture and gases such as SO_2 takes place near cool roofs. Typical data for five indoor locations are presented in Table 4. In these tests, the test racks were suspended near, but not in contact with, the roofs.

Outdoor Atmospheric Corrosion

The rate at which zinc corrodes in outdoor exposures is governed mainly by the frequency and duration of moisture contact, the rate of drying and the extent of industrial pollution in the air. The rates of corrosion in desert and rural locations are very low. The most harmful conditions are those in which frequent wetting of the metal occurs with dews and fogs in industrial environments where the condensed moisture is distinctly acidic. Formation of protective basic corrosion product films is interfered with when the metal is attacked by this acidic moisture. Rainfall is less harmful because it is seldom acidic. In fact, rainwater is helpful when it leaches out or washes off harmful products, such as acid dust particles in industrial areas and chlorides at seacoast locations.

The rate of drying is of great importance because the high oxygen content of thin films of moisture promotes rapid attack. Conditions conducive to slow drying apparently accelerate corrosion. In normal exposures, drying is rapid. However, in locations exposed to the atmosphere but sheltered from direct contact with rain, condensed moisture, such as dew, dries slowly.

Seacoast atmospheres are less corrosive to zinc than industrial atmospheres. Seacoast atmospheres contain many variables that may influence the

Table 4 Corrosion of rolled zinc in indoor industrial atmospheres(a)

Test atmosphere	g/m²·d(b)	Corrosion rate µm/yr(c)	mils/yr(c)
General atmosphere(d)	0.015	0.76	0.03
Heat, moisture, SO₂(c)	0.4-0.5	2.0-2.5	0.08-0.1
Industrial gases, cement dust(f)	0.05-0.1	0.25-0.51	0.01-0.02
Heat, gases, moisture, cement dust(g)	0.05-0.1	0.25-0.51	0.01-0.02
Industrial gases, moisture, coal dust(h)	0.1-0.15	0.51-0.76	0.02-0.03

(a) Specimens were 100 by 150 by 0.5 mm (4 by 6 by 0.02 in.). They were held in porcelain insulators on painted wooden racks, two specimens in the horizontal and two in the vertical position for each metal. Specimens were washed with ether and alcohol prior to test, which was of 10-year duration. (b) Corrosion criterion was loss in weight after chemical removal of corrosion products by 3-min immersion in 200 g/L CrO₃ at 80 °C (175 °F). (c) Calculated from weight loss. (d) Pyrometer shed, ZnO furnace building. (e) ZnO furnace building over furnaces. (f) Cement crusher building. (g) Cement kiln furnace building. (h) Coal breaker building. A steam locomotive frequently entered the building. Source: The New Jersey Zinc Co.

Table 5 Weight loss of iron coupled with other metals (Ref 7)(a)

Test location	Normal	Aluminum	Zinc	Weight loss of iron, %(b) Coupled with Copper	Lead	Nickel	Tin
Sandy Hook.	7.6	6.8	1.2	7.2	3.4	5.8	9.4
Key West.	5.8	4.0	0.1	10.9	2.1	6.5	4.3
Pittsburgh.	14.2	15.2	0.3	24.8	15.8	14.9	20.2
Altoona	9.5	5.4	4.3	13.2	9.6	6.4	12.8
Rochester.	11.0	11.6	(c)	16.2	15.2	12.5	16.2
New York City. . . .	12.0	7.8	(c)	14.6	11.7	8.5	13.4
State College.	3.7	2.2	1.2	5.0	2.3	3.9	4.4
Phoenix	1.1	0.3	0.1	1.2	0.8	1.1	1.5
La Jolla	74.7	39.3	5.6	58.4	39.0	(d)	52.8

(a) Test results after 7 year exposure. (b) To nearest 0.1%. (c) Corrosion was negligible; less than 0.06%. (d) Disk was missing.

rate of corrosion of a metal—for instance, the amount of rainfall, the prevailing relative humidity, the range of ambient temperature, the frequency and extent of cloudiness, the amount of industrial gas contamination in the air, and distance from the ocean. Sharp differences exist between rates of attack in exposures immediately adjacent to the shoreline and rates a few hundred yards inland.

Atmospheric Corrosion of Rolled Zinc. Valuable corrosion data on rolled zinc are available as a result of comprehensive studies and outdoor exposure tests initiated in 1931. Three grades of rolled zinc were used in these tests: Metal AA (Prime Western), Metal BB (High Grade) and Metal HH (Special High Grade). Four 230 by 305 mm (9 by 12-in.) weighed panels (two permanent, two removable) of each of the three grades of zinc were exposed at seven test sites representing a wide range of atmospheric conditions:

- *Altoona, PA*—Highly industrialized atmosphere

- *State College, PA*—Typical northeastern rural atmosphere; no industries within several miles of the test site
- *New York City*—Highly industrialized atmosphere, having some of the characteristics of a seashore location; considerable fog at some seasons of the year
- *Sandy Hook, NJ*—A hundred yards from the shoreline; originally chosen as a typical northern seacoast location, but in recent years the atmosphere has become considerably contaminated with industrial fumes
- *Key West, FL*—Fairly near the water; typical southern seacoast atmosphere; little industrial contamination
- *Phoenix, AZ*—Typical of semiarid and rural conditions in the southwestern United States
- *La Jolla, CA*—On a slight bluff, very near the edge of the water and subjected to ocean spray during storms and to very heavy nightly dewfall

Corrosion of zinc in outdoor exposure

at these various locations was generally quite smooth, essentially without pitting, and the weight change data for duplicate specimens were closely consistent. Therefore, the corrosion rates were reported in terms of average weight loss per unit area and average penetration. Results of the tests on Prime Western and Special High Grade rolled zinc slabs are given in Fig. 2. These data confirm that rolled zinc corrodes most rapidly in moist industrial atmospheres and least rapidly in dry pure atmospheres.

The outdoor corrosion rate of rolled zinc does not change with time, provided the nature of the atmosphere does not change. The main difference between the arid and rural areas lies in the moisture present in the latter as humidity, rainfall and, most important, dew formation. Industrial pollution is the greatest factor in determining the corrosion rate of rolled zinc in the atmosphere. Tests show that such pollution causes a 36-fold increase in corrosion rate over the rate at Phoenix, and a six-fold increase over that at State College. This substantiates other studies of the atmospheric corrosion of zinc and zinc coatings.

Corrosion of zinc in the atmosphere is controlled principally by four characteristics of this metal: (*a*) zinc is attacked by acids at rates that decrease as pH increases; (*b*) it is capable of forming insoluble basic salts when appropriate pH levels are reached; (*c*) it has a high hydrogen overvoltage; and (*d*) it is anodic to its metal impurities, to hydrogen, and to most surface contaminants.

Corrosion Fatigue

Corrosion fatigue is the weakening that occurs in a metal as a result of repeated stressing in the presence of corrosive agents. Although the fatigue strength of a metal has a fairly definite value in the absence of any form of corrosive influence, actual conditions are never so ideal. Frequently, the total amount of corrosion involved in corrosion fatigue is extremely small. It is important to prevent corrosion from the start because, once it has begun, cyclic stressing may lead to early failure even though further attack is prevented.

These considerations are significant in determining what preventive measures to adopt in practice and serve to explain why sacrificial or electrochemical protection is of major importance.

Fig. 4 Atmospheric corrosion results for corrugated galvanized sheet (Ref 8 and 9)

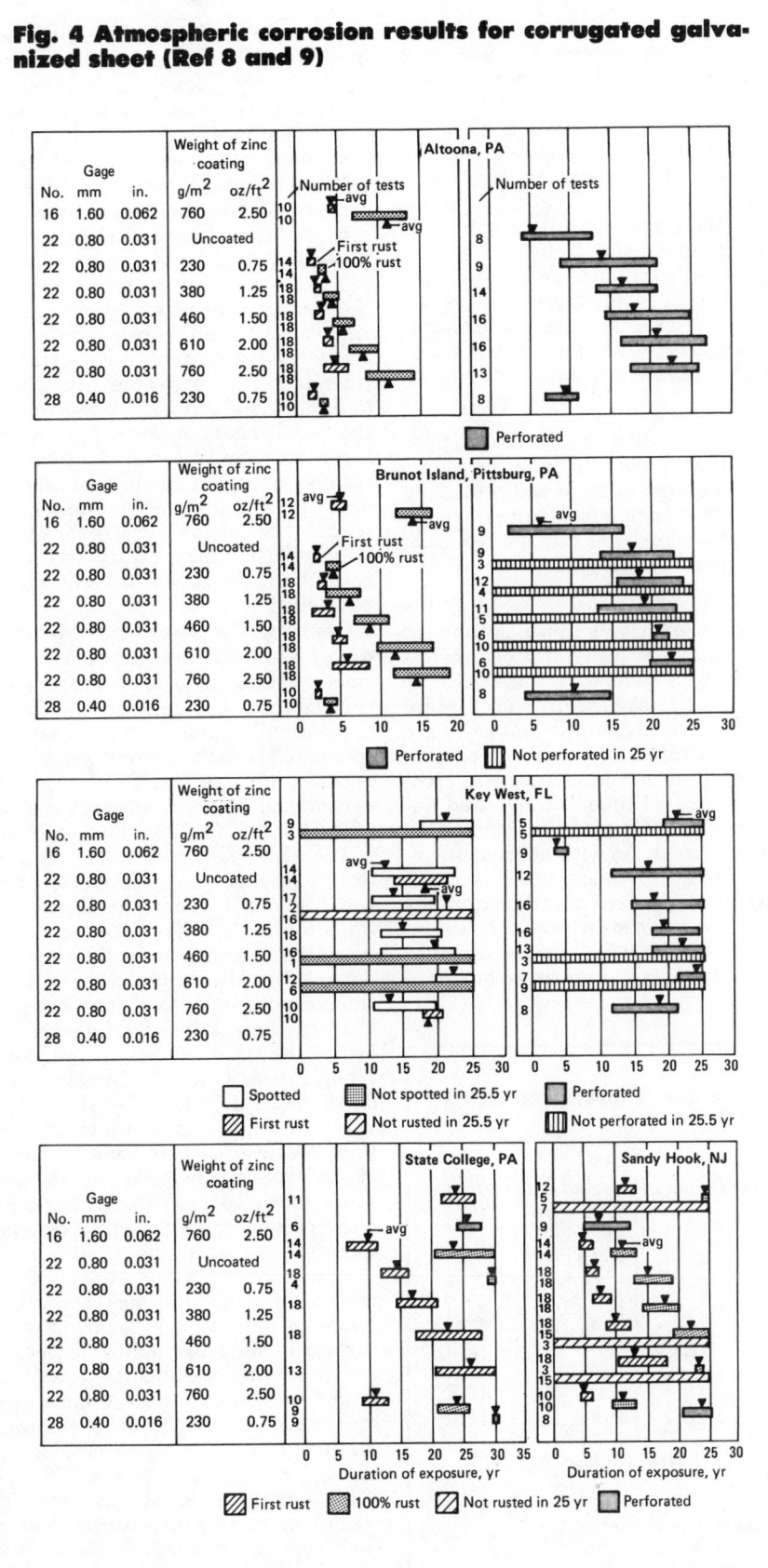

Because corrosion-fatigue cracks may be difficult to detect until they have reached dangerous proportions, a practical safeguard is to protect steel components throughout their lives by means of sacrificial zinc coatings, which give them complete cathodic protection. When zinc coatings are used, small- to medium-size imperfections in the coatings are of relatively little importance, and corrosion is prevented even in the microscopically small crevices that otherwise could lead to failure by corrosion-fatigue.

Any type of coating that covers the surface completely and retards the onset of corrosion may help, but if it is not anodic to steel it cannot provide cathodic protection at coating defects. In many general uses the fact that nonanodic coatings cannot prevent corrosion at small defects may be unimportant. When fatigue failure is not a consideration, a very small amount of corrosion has no significant effect.

Wherever corrosion fatigue can occur, however, the situation is very different. Coatings that cannot protect sacrificially may, as a result of discontinuities, cause severe pitting and, by concentrating the corrosion in small areas, stimulate growth of fatigue cracks. The important effects of corrosion acting together with fatigue were exemplified in England during the Second World War when steel wires for use in seawater failed after a comparatively short period of service. These failures were caused by vibration, which set up cyclic stresses in the presence of seawater from which the wires were not adequately protected. When galvanized wires were used, no corrosion-fatigue failures occurred.

In tests conducted at Melville, RI, the U.S. Navy determined that galvanizing combined with use of sacrificial zinc anodes prolonged the service lives of torpedo nets that, when unprotected, required repair for corrosion fatigue after only six months. In a typical test, one net so protected was found to be in good service condition after 620 days. Galvanizing also protected nets stored ashore in air.

Alloy additions and heat treatments considerably increase the tensile and fatigue strength of steel but have only minor effects on corrosion-fatigue behavior. Generally, the corrosion-fatigue behavior of low-alloy steel is little or no better than that of ordinary low-carbon steel unless the metal is provided with a protective coating such as zinc.

It was once assumed that in the presence of acidic rain or condensate, contact with zinc coatings might increase the hydrogen content of steel and thus offset the advantage of such coatings in combatting corrosion fatigue. It was found, however, that galvanized steel, with high hydrogen absorption, became brittle enough to snap when bent beyond its yield point, but had a long life when the stress was held within the elastic range. In acid mediums, contact with zinc greatly increased the life of steel at the higher acid concentration where the adverse effect of hydrogen would be most pronounced.

Electrochemical Corrosion

When two dissimilar metals are in electrical contact in the presence of an electrolyte, galvanic corrosion of one of the metals takes place while the other metal is protected. Metals can be arranged in a galvanic series according to their relative reactivities in a medium such as seawater, as shown in the article entitled "Corrosion Characteristics of Copper and Its Alloys", in this volume.

Zinc is high enough in the galvanic series to act as an anode when coupled with most of the common metals. It can therefore be used to protect other metals below it in the series, especially iron and steel. Generally, contact between small areas of zinc and large areas of a more noble or cathodic metal should be avoided. On the other hand, contact of small areas of cathodic metals with large areas of zinc is relatively harmless.

Zinc coatings protect iron and steel from corrosion (a) by affording protection as a continuous envelope that prevents contact of corrosive moisture with the basis metal and (b) by electrochemical action when the envelope is not continuous. If the coating is continuous, then the rate at which it is attacked is the only factor. If the coating is not continuous or becomes discontinuous due to weathering, pores, flaws or cracks, then the electrochemical property or anodic nature of zinc becomes effective in providing protection.

It is significant that on exposure to other than highly acidic atmospheric pollution, zinc forms a self-protecting film of fairly impermeable basic corrosion products. This film protects the metal from further attack.

Data from seven-year tests of the corrosion behavior of iron disks coupled with equal-size disks of other metals are given in Table 5. In these tests, at least partly successful efforts were made to ensure metallic contact in the couples. Ordinarily, such perfection is not obtained in assembled metal products fabricated commercially, and anticipated possible galvanic attack often does not occur. As with aqueous corrosion, contact between small areas of zinc and large areas of cathodic metals should be avoided, whereas it is relatively safe for small areas of other metals to contact large areas of zinc.

Zinc Coatings

For more than 100 years, zinc coatings have been used to protect iron and steel. Hot dip galvanizing is by far the most common process for zinc coating. Other methods include electrodeposition, sherardizing and metal spraying.

Alloy Layer. Coatings applied by conventional or hot dip galvanizing consist of three layers of iron-zinc alloy phases covered by an outer layer of relatively pure zinc. The phases in these layers progressively decrease in zinc content, and correspondingly increase in iron content, from surface to interface. The aggregate alloy layer also provides corrosion protection to the basis metal; when the pure zinc layer has finally corroded away, the reddish corrosion product of the exposed layer sometimes encountered is not to be mistaken for rust of the underlying steel. On the contrary, this is the optimum stage at which paint should be applied.

Coating Thickness and Uniformity. Although the corrosion rate of zinc in atmospheres is low, it must be remembered that galvanized coatings are normally quite thin. For example, commercial quality coating weight G90 is 275 g/m^2 (0.90 oz/ft^2) of sheet (as determined by the triple-spot test described in ASTM B90); thus, each side of the sheet is coated with 138 g/m^2 (0.45 oz/ft^2), or a layer a little over 19 μm (0.75 mil) thick. Therefore, it is important to specify a coating thickness that is sufficient for the intended application. Also, uniformity of coating thickness is important, especially in light coatings. When coatings are uneven, the thinner areas fail even though there is ample zinc elsewhere on the surface. With a heavy coating, variations in thickness are less important, because the sacrificial protection of the exposed steel is more effective. Thus increases in coating thickness often provide more than proportional increases in coating life.

The weights of zinc coatings are expressed in terms of grams per square metre or ounces per square foot of surface, except for galvanized sheet, for which coating weights are given in grams per square metre or ounces per square foot of sheet (that is, the total coating on both surfaces or sides of sheet). Thus, for a coating of 275 g/m^2 (0.90 oz/ft^2) on sheet, the average weight of coating per square foot of each side of the sheet is 138 g (0.45 oz).

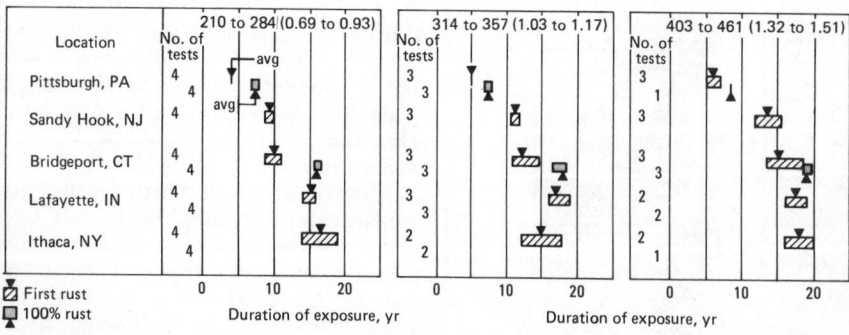

Fig. 5 Atmospheric corrosion results for zinc-coated wire strands (Ref 10)

Numbers at top of each chart indicate weight of zinc coating in g/m² (oz/ft²).

The protective value of different zinc coatings may be expressed in terms of the average thickness or weight per unit area of zinc removed in a given area. Suggested units are micrometres per square metre per day and mils (thousandths of an inch) per year.

To provide more complete information, coating-weight figures should be accompanied by equivalent thickness values. One hundred grams of zinc per square metre of surface is equivalent to a coating thickness of 14 μm; 1 oz of zinc per square foot of surface is equivalent to a coating thickness of 1.7 mils (0.0017 in.). From these equivalents, any other coating thickness can be calculated if the weight is given, or vice versa. The weights of coating for many galvanized products are covered by ASTM specifications. These are not maximum desired coating weights for every application (or maximum obtainable coating weight). ASTM specifications should be consulted for a description of the methods of properly measuring the thickness of the coating.

Criteria For Coating Life. There are different concepts of what constitutes termination of the useful life of a zinc coating. These include: (a) the first visible evidence of rust of the basis metal, (b) perforation of the basis metal, and (c) breakage of the galvanized item. Each concept has its pertinence. Useful life based on perforation of the basis metal or breakage of the item is controlled by thickness and corrosion resistance of the basis metal. Another method of evaluating life or degree of corrosion protection of a zinc coating is measurement of the rate at which the coating loses weight or thickness in a given environment.

The most practical way of expressing the corrosion resistance (response to different exposure conditions) of zinc coatings is in terms of the area on which rust forms due to exposure of the underlying steel. The protective value of the zinc coatings in the ASTM tests discussed below is expressed in terms of percentage of surface exhibiting rust. This method of evaluation is realistic for several reasons. Because of influences such as nonuniformity of coating thickness, nonuniformity of wetting and drying of different parts of a specimen, exposure of bare steel as a result of shearing, and presence of an alloy layer in the coating which may corrode at a rate different from that of the relatively pure zinc surface, corrosion initially takes place at the most vulnerable spots and then spreads. When depicted graphically, the curves showing progress of rust with time are S-shaped.

Galvanized Sheet. As a result of the large-scale and long-term test programs on galvanized steel sheet inaugurated in 1926 by the American Society for Testing and Materials, extensive information is available on the corrosion properties of this product. The specimens used were made from corrugated galvanized sheets, 0.66 by 0.76 m (26 by 30 in.), with five different weights of coating ranging from 230 to 760 g/m^2 (0.75 to 2.50 oz/ft^2), and uncoated sheets for comparison. These specimens were made from sheet rolled to three thicknesses by each manufacturer represented, all the sheet from each manufacturer being rolled from the same lot of steel.

The specimens were exposed at the following sites, which varied in corrosivity: Altoona, PA (industrial inland); Brunot Island, Pittsburgh, PA (severe industrial inland); State College, PA (rural); Sandy Hook, NJ (northern seacoast); and Key West, FL (southern seacoast). Recorded for each specimen were the numbers of years required for rusting to begin, for rust to cover 100% of the surface, and in some instances for perforation of each specimen to occur. The tests at Pittsburgh, Sandy Hook, and Key West were abandoned in 1951 and 1952 for various reasons, but the test results from these locations were essentially complete. Corrosion had proceeded so far that almost no additional information would have been obtained by continuing the tests. Many of the galvanized sheets exposed at Pittsburgh had not perforated when the test was discontinued.

A summary of inspection results from this test program are given in Fig. 4, showing years to first rust and years to 100% rust. Perforation data for the various test locations are also summarized.

Some of the uncoated 22-gage (0.80-mm, or 0.031-in.) sheets tested at Pittsburgh were perforated in 2 years. Time to first rust for the sheet coated with 230 g/m^2 of zinc (0.75 oz/ft^2) was 2 years, and average time to 100% rust was about 4 years; yet these sheets did not become perforated until after 14 years, and some still were not perforated after 25 years. This anomaly is explained by the fact that the zinc on the groundward exposed surface does not corrode as fast as that on the skyward surface. In atmospheres of this type, uncoated steel sheets exposed 30° to the horizon corrode about 50% more on the groundward side than on the skyward side. However, the basis metal could not rust until the zinc had been removed from the groundward surface. Therefore, time to perforation for a lightly galvanized sheet is much longer than that for an uncoated sheet of the same composition.

With reference to the test at State College, it is noteworthy that none of the 28-gage (0.40-mm, or 0.016-in.) sheets that originally had coatings of 230 g/m^2 (0.75 oz/ft^2) were perforated after 28 years. Six of the ten uncoated 22-gage (0.80-mm, or 0.031-in.) sheets were perforated. It is evident that in these tests basis-metal composition had little or no effect on corrosion rates of zinc coatings; but after the zinc had corroded away, the inherent resistance of the steel, as determined by its composition, was the main factor determining time to perforation.

Galvanized Wire and Wire Products

In 1936, ASTM inaugurated exposure tests on galvanized wire products such as farm fence, barbed wire, wire strand, chain link fence and unfabricated wires of various compositions. These tests were comparable to the tests on galvanized steel sheet begun in 1926 (see preceding section). The galvanized wire products were exposed at 11 locations in the U.S., including industrial, seacoast and rural sites. The rural sites chosen were widely distributed and thus were representative of rural atmospheric conditions prevailing throughout the country. Five different gages of wire, with zinc coatings ranging from 75 to 915 g/m^2 (0.25 to 3.00 oz/ft^2), were tested.

Unstranded wire and wire strand specimens were exposed at all test locations. Corrosion data for wire strand specimens at five test sites representative of industrial, rural and seacoast conditions are given in Fig. 5. Data on corrosion performance of zinc-coated unstranded wire at 11 test sites are given in Fig. 6. As with sheet, coating thickness is more important than basis-metal composition. Service life is proportional to coating thickness. The corrosion performance of any individual specimen of barbed wire or farm field fence can be compared directly with that of specific unstranded wire specimens.

Specimens of unstranded zinc-coated wire were exposed at Pittsburgh,

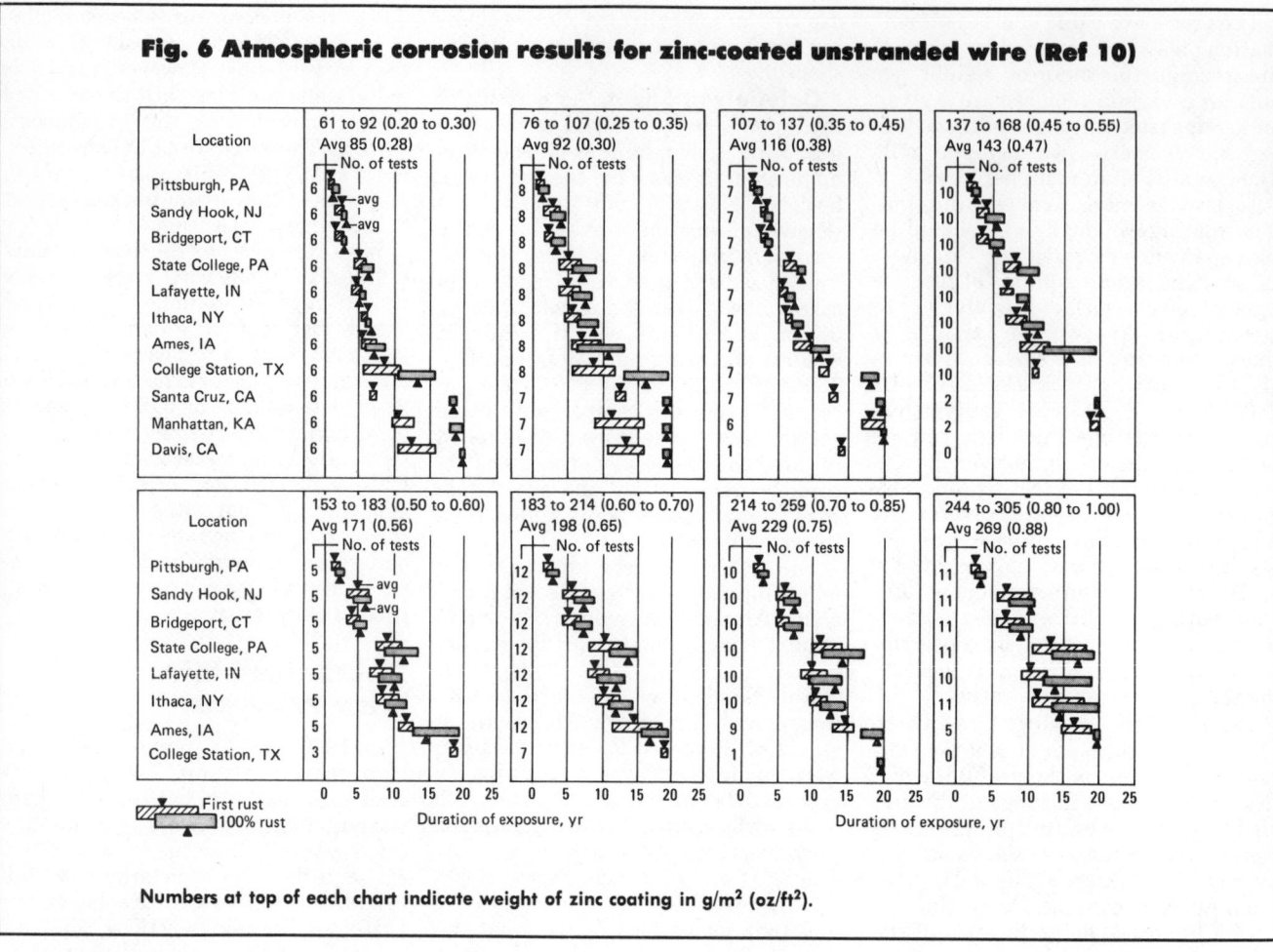

Fig. 6 Atmospheric corrosion results for zinc-coated unstranded wire (Ref 10)

Numbers at top of each chart indicate weight of zinc coating in g/m² (oz/ft²).

Sandy Hook and State College to determine weight loss of the zinc coating. Weight-loss values for these locations were as follows: Pittsburgh, 10.5 g/yr (0.369 oz/yr); Sandy Hook, 3.3 g/yr (0.117 oz/yr); and State College, 1.7 g/yr (0.060 oz/yr).

Cathodic Protection With Zinc Anodes

If a structure requiring protection is not, or cannot be, coated with zinc, corrosion still can be controlled by using zinc as a sacrificial galvanic anode in a cathodic-protection system.

Cathodic protection is effective in such diverse mediums as salt water, fresh waters and most soils. The structures most frequently protected by the cathodic method are underground pipelines and ship hulls. The source of the protective cathodic current can be either impressed (rectified ac or specially generated dc) or galvanic.

When a metal surface is in contact with an electrolyte, differences in electrical potential develop between local areas of the surface corresponding to their respective electrochemical reactivities; a more reactive (less noble) area is referred to as being "anodic" to a less reactive (more noble, or cathodic) area. The difference in potential between two areas causes current to flow from the cathodic area to the anodic area through the metal, and from the anodic area to the cathodic area through the electrolyte to complete the circuit. Where the current enters the electrolyte, metal ions go into solution, causing anodic corrosion.

Such corrosion can be prevented, and the metal protected, if the metal (steel, for example) is connected electrically by a wire or rod to zinc that also is in contact with the electrolyte. Under these conditions, current flows from the external zinc anode through the electrolyte to the steel surface, and this current is of sufficient magnitude to afford cathodic protection to the steel.

For best results, the zinc must be of proper composition. The composition most frequently used is in accordance with MIL-A-18001, which specifies maximum percentages as follows: 0.006 Pb, 0.0014 Fe, 0.06 Cd, 0.005 Cu, and 0.1 to 0.3 Al. Iron content may be increased to 0.003% when cadmium content is at least 0.025%.

REFERENCES

1. C. E. Richards, "The Corrosion Resistance of Galvanized Steel", European General Galvanizers Association, 1957
2. "On Atmospheric Corrosion Tests of Non-ferrous Metals and Alloys", Report of Subcommittee VI of Committee B-3, *ASTM Proceedings*, Vol 44, 1944, p 229
3. E. A. Anderson, "Corrosion of Rolled Zinc in the Outdoor Atmosphere", *Symposium on Atmospheric Exposure Tests on Nonferrous Metals*, STP 175, American Society for Testing and Materials, Philadelphia, p 9

4. R. E. Marce and L. M. Weisenberger, Paper No. G-T75-141, presented at the 8th SDCE International Die Casting Congress and Exposition, Detroit, MI, March 17-20, 1975

5. G. L. Cox, *Industrial Engineering Chemistry,* Vol 23, 1931, p 902-904

6. B. E. Roetheli, G. L. Cox and W. B. Littreal, *Metals and Alloys,* Vol 3, 1932, p 73

7. "On Galvanic and Electrolytic Corrosion (Plate III)", Report of Subcommittee VIII of Committee B-3, *ASTM Proceedings,* Vol 39, p 252-253

8. "On Inspection of Black and Galvanized Sheets", Report of Subcommittee XIV of Committee A-5, *ASTM Proceedings,* Vol 54, 1954, p 110-122

9. "Report of Committee A-5 on Corrosion of Iron and Steel", *ASTM Proceedings,* Vol 56, 1956, p 131

10. "On Atmospheric Exposure Tests of Wire and Wire Products", Report of Subcommittee XV of Committee A-5, *ASTM Proceedings,* Vol 55, 1955, p 126, 127

Precious Metals

Precious Metals and Their Uses

By E. D. Zysk
Engelhard Minerals & Chemicals
  Corp.

THE PRECIOUS NOBLE METALS are of inestimable value to modern civilization. Their functions in jewelry, coins and bullion, and as catalysts in devices to control auto exhaust emissions are widely understood. But in certain other applications, their functions are not as spectacular and, although vital to the application, are largely unknown except to the users. Their use in industry has grown until it can be said that they are vital to the support of our economy. There is hardly a facet of daily life that precious metals and their alloys do not influence. For instance, precious metals are used in dental restorations or dental fillings. Precious-metal solders are used in dentistry and in the jewelry and electronics industries. Much of our clothing today is produced with the aid of precious metals—in spinnerettes for producing synthetic fibers and in catalysts in various processes; for example, foodstuffs depend more and more on the use of fertilizer produced with the aid of a platinum-rhodium alloy catalyst woven in the form of gauze. Electrical contacts containing palladium are essential to telephone communications. Certain organometallic compounds containing platinum are significant drugs for cancer chemotherapy.

Resources and Consumption

Metal specialists at the U.S. Bureau of Mines continually survey the market in silver, gold and platinum-group metals to determine present availability and usage and to forecast future trends. *Mineral Commodity Profiles* and *Mineral Industry Surveys* covering these metals are issued periodically.

Much of the information in the following paragraphs was obtained from Ref 1 to 3. For the latest available information concerning these metals, it is recommended that the most recent issues of these publications be consulted.

Silver. In recent years, the U.S. has been a net importer of silver. Imports, including unrefined silver, supplied the equivalent of about 37% of U.S. industrial demand of 154 million troy ounces in 1977. The U.S. supply is obtained from primary and secondary sources. About one-third of primary silver is obtained from predominantly silver ores; the remaining primary silver is a by-product of the refining of copper, lead, zinc and other metals.

Five smelting and refining companies produce the major portion of domestic primary silver. The smelters and refineries treat ores, concentrates, residues and precipitates from company mines and plants in addition to materials purchased from other sources. Silver scrap is recycled by several primary smelters and a considerable number of small secondary refineries. In addition, secondary silver is recovered by several trading and fabricating companies and is recycled by end-product manufacturers.

Within recent years, government regulations relating to the environment and to control of emissions of hazardous compounds have limited the operation of some base-metal smelters that recover silver as a by-product. Silver in its ionic form, as in waste discharges from electroplating plants, is considered a potential source of pollution and a health hazard. Because of this, governmental agencies may be expected to increase their efforts to minimize discharges from processing plants.

The following table, which presents statistics for 1977, gives a general indication of the annual demand pattern for silver.

End use	Demand, millions of troy ounces
Silverware	23.5
Jewelry and artwork	8.1
Photography	53.7
Refrigeration	8.0
Coinage	0.1
Appliances	10.0
Batteries	5.8
Electrical equipment	18.8
Electronic components	12.5
Commemorative objects, including medallions and coins	4.3
Other	8.9
Total U.S. demand	153.7

Source: U.S. Bureau of Mines

Several factors can affect the supply/demand relationship for silver. One of these is the appreciable requirement for monetary purposes and another is the speculative or investor market in refined silver bars and sacks of domestic coins. In addition, there has been some interest in the collection of commemorative medallions and limited-use objects fabricated from silver. Whether the latter end use will continue to grow or regress in the future is a matter of considerable interest. In any event, depending on prices, a large potential secondary supply of silver is available in the form of coins, silverware, jewelry and commemorative objects.

Gold. Because of its esthetic beauty and enduring physical properties, gold is important not only to industry and the arts, but also as a commodity having long-term value.

In the past, gold was considered to be mainly a monetary metal. However, starting in the late 1950's, more gold was used by manufacturers and investors than was used for monetary purposes. Since 1968, gold has become, to a considerable extent, a free-market commodity with prices free to adjust to supply and demand. Despite this open market, almost half of all gold mined in the world, estimated at 1.3 billion troy ounces, is in various government vaults tied down by agreements between large industrial nations.

The United States mines and reclaims from scrap about two million ounces of gold per year. This amount is less than half the amount required by U.S. industry. The requirements for bullion and coins are almost equal to those of industry. Due to this general demand for gold, the net inflow of this metal from foreign sources is large. In 1977, for example, 8.1 million troy ounces were imported, mostly in the form of refined metal.

About 60% of domestic primary production comes from gold ores, and the remainder from by-products of the refining of copper and other base metals. U.S. refinery production includes gold from domestic mines, from imported ores and base bullion, and from domestic and foreign scrap. In recent years about 5 to 10% of U.S. refinery production has been derived from foreign ores, base bullion and scrap.

About 5 to 10% of total U.S. supply of gold comes from "old scrap", which is defined as metal discarded after use. New scrap, generated during manufacturing, is usually reclaimed by the fabricator and is not considered part of the market supply.

The U.S. has appreciable gold resources, some of which are marginally profitable to recover. The price is now high enough to encourage growth of production at a modest rate, but environmental restraints on placer mining and the high cost of developing lode deposits currently dictate that the United States will continue to import most of its gold.

The largest foreign producer of gold, the Republic of South Africa, produced about 30 million ounces of the estimated 1978 world production of 51 million ounces. Other important sources are the USSR, Canada, South America, Asia and Oceania. According to the U.S. Bureau of Mines, world resources are adequate to meet demand forecast for this metal to the year 2000.

The table that follows presents data for 1977, compiled by the U.S. Bureau of Mines, and illustrates the general pattern of consumption of gold in fabricated products.

End use	Demand, thousands of troy ounces
Jewelry and arts	2658
Dental supplies	728
Industrial	1205
Investment products	268
Total demand (1977)	4859

The use of gold in jewelry appears to be the highest requirement, approximately 55% of the gold consumed. Dental uses generally amount to 10 to 15% of annual demand. Industrial requirements are generally centered in the electronic industry. In the latter case, although the total number of end uses continues to be high, considerable emphasis has been directed to reducing the use of gold in present applications, because of its increasing value. About 3 to 6% of annual demand goes into small bars, medallions, coins and related products, which are bought mainly as a means of investment.

Platinum-Group Metals. Six closely related metals in the platinum group commonly occur together in nature. These transition elements are near neighbors in periods five and six of group VIII in the periodic table. Ruthenium, rhodium and palladium have a density of approximately 12 Mg/m³, and osmium, iridium and platinum have a density of around 22 Mg/m³.

The platinum-group metals are among the scarcest of metallic elements, and for this reason their cost is high. They occur as native alloys or in mineral compounds in placer deposits, sometimes together with gold, as well as in lode deposits in basic or ultrabasic rocks where they may be found together with nickel and copper. At the present time, most of the world's supply of platinum group metals is extracted from lode deposits from the Republic of South Africa, the USSR and Canada. Another major source, which is increasing in importance with broadening industrial use, is old scrap obtained from obsolete equipment, spent catalyst and discarded jewelry. Material of each type is often concentrated by a number of different methods, then refined by any of several chemical processes that conclude with a heating step to convert precipitates to metal in a porous, somewhat powdery form called "sponge". In commercial metal transactions, sponge is a common form, although ingots and shot are also traded.

In lode deposits, platinum-group metals are often associated with nickel and copper sulfide, which may be the principal product of mining (as in Canada and the USSR) or an important coproduct (as in the Republic of South Africa). In the ores, proportions of the six platinum-group metals vary from one lode deposit to another. Canadian deposits in the Sudbury District contain about equal amounts of platinum and palladium; USSR ores contain half as much platinum as palladium; South African deposits contain more than twice as much platinum as palladium. Generally, platinum and palladium together account for about 80 to 85% of the platinum-group metals present in any given ore, followed by ruthenium, rhodium, iridium and osmium, usually in that order.

The composition of placer deposits differs somewhat from lode deposits. Placer deposits are characterized by the nearly complete absence of palladium and the common presence of gold. It appears that palladium, and to a certain extent platinum, rhodium and ruthenium, are dissolved away during placer formation; thus in well-established, old placers such as the one in Witwatersrand, South Africa, only osmium and iridium are present (as the alloy osmiridium). The only economically important placers today may be found in Colombia, in South Africa and in the USSR, and together account for about 2% of total world production.

The United States depends almost totally on foreign sources of platinum-group metals. In 1977, for example, U.S. demand for platinum-group metals was 1 495 000 troy ounces, and the sources of supply were: Republic of South Africa, 46%; USSR, 23%; United Kingdom, 8%; other countries, 16%; domestic secondary production, 7%; and domestic mine production, 0.2%.

As a point of information, platinum-group metals are recovered from copper ores in the U.S. by three major refineries. In addition, about 30 to 40 smaller refineries handle or in some way process domestic scrap. However, most of them treat only platinum and palladium, and only three or four refine all six metals.

U.S. mine production and reserves are small; U.S. resources appear to be large but not well explored. The heavy U.S. dependence on foreign sources for these critical metals has strategic implications as well as a substantial impact on the U.S. balance of payments. The need for exploration and development of U.S. resources is apparent.

The U.S. Bureau of Mines forecasts that U.S. demand for platinum-group metals will grow at an annual rate of about 3% for the rest of this century to about 3.1 million ounces in the year 2000. In addition, demand in the rest of the world is forecast to grow more slowly than in the U.S. and is expected to reach about 9.7 million ounces in the year 2000. World reserves and resources appear to be more than adequate to meet this demand.

The demand patterns for 1977 for three of the most-used platinum-group metals are presented in the following table.

End use	Demand, thousands of troy ounces		
	Platinum	Palladium	Rhodium
Automotive	354	126	···
Chemicals	84	164	20
Petroleum refining	74	8	···
Ceramics and glass	59	···	14
Electrical	81	199	4
Dental supplies	21	80	···
Jewelry and ores	33	15	2
Other	62	44	5
Total demand	768	636	45

Source: U.S. Bureau of Mines

The United States and Japan currently use about 60% of the platinum-group metals produced. Western Europe and the USSR essentially divide the remaining 40%. Automotive (Pt/Pd catalyst) requirements began in 1974 and presently are substantial (32% of total consumption in 1977). In Japan about three-fourths of the platinum goes into jewelry, whereas in the U.S. and Western Europe about 5 to 15% of this metal is used in jewelry.

Trade Practices

The precious noble metals are bought and sold in troy ounces or, in markets where the metric system is used, in kilograms. One kilogram is equal to 32.15 troy ounces. The troy system of weights is based on the troy ounce of 480 grains, 31.1 grams or 20 pennyweight. One troy ounce is also equal to 1.097 avoirdupois ounces.

In respect to silver and gold, "fineness" refers to the weight portion of either metal in an alloy, expressed in parts per thousand. For example, 1000 fine silver (also fine silver) is pure silver, or 100% silver, and 1000 fine gold is 100% pure gold. Gold bullion that is commercially traded is at least 995 fine or higher. Sterling silver is 925 fine, or 925 parts (also 92.5%) silver and 75 parts (or 7.5%) copper. Furthermore, until 1964 the U.S. coin silver was an alloy of 90% silver (900 fine) and 10% copper. In respect to silver bullion that is traded, the silver content ranges from 999 fine to 999.9 fine. Gold or copper are silver impurities in any fineness of silver bullion.

Another way of indicating purity is by "karat" which is expressed in 24ths. In this system 24-karat (24K) gold is 1000 fine or pure gold, while 10K gold is 10/24ths or 41.7% gold. However, there are many different alloy combinations denoted by 10K gold, 14K gold,

etc. In each category (10K for example) they differ from each other in the number, type and proportions of the base-metal additions. Gold alloys used in jewelry are always called karat golds, while those used in dentistry and for electronic purposes are designated by percentage. Depending on the alloying constituents (silver, copper, zinc, nickel and sometimes platinum and palladium) yellow, red, green or white karat golds may be formed.

U.S. trade practice rules for the jewelry industry, set by the Federal Trade Commission, require that any article labeled gold contain at least 10K gold, with a tolerance of ½K. In gold clad (to base metal stock) the ratio of the weight of the gold alloy cladding to the total weight of the material is indicated, together with the karat of the cladding. For example, one-tenth 12K "gold filled" stock is a base metal surfaced with a layer or layers of 12K gold alloy that weighs 10% of the weight of the composite article. Such an article, if assayed in toto, would contain 5% gold.

The designation "gold filled" is limited by stamping regulations to articles in which the weight of the coating is at least one twentieth of the total. Lower ratios may be stamped "rolled gold plate". The quality mark cannot be applied to articles surfaced with an alloy less than 10K. In the cases of the gold filled and rolled gold plate materials, the karat gold is bonded to the base metal substrate by soldering, brazing, welding or mechanical means.

In respect to the platinum-group metals, platinum and palladium are traded on the New York Mercantile Exchange in respective lots of 50 and 100 troy ounces, while on the Chicago Mercantile Exchange units of 100 troy ounces are traded. On the N.Y. Mercantile Exchange, the metal may be in the form of bar or sheet.

The purity of platinum required may vary according to the end use or application. Although commercial-grade platinum must be at least 99.8% pure, platinum of at least 99.9% purity is used for alloying, laboratory ware and contacts. Higher-purity platinum sometimes with controlled impurities, is used for other specialized applications such as thermocouples and resistance thermometers. The present U.S. thermometric standard platinum, designated Pt 67, is 99.999% pure.

Federal regulations require that an article of trade may be marked platinum if it contains at least 98.5% platinum-group metals with no more than 5% of the other platinum-group metals (93.5% platinum minimum). Special stamping provisions cover some jewelry alloys. Furthermore, all platinum jewelry sold in the United Kingdom must be hallmarked.

Alloys used for dental purposes are rather complex in composition, and metallurgical considerations are dominant in their design. Various specifications have been established by the American Dental Association and by the federal government, but these do not cover all of the dental alloys.

Special Properties

The precious noble metals have unusual combinations of properties which are often superior to those of other materials, or which at times make them the only materials that can meet the specialized requirements of advanced technology or industry. The initial investment in these metals or their alloys may be high but is offset by long, reliable service and by ease and high recovery in refining. This relative completeness of recovery is of particular significance. The precious noble metals are virtually indestructible and because recovery is so high, their use in industry in many instances is economical as well as efficient.

Although the precious noble metals have a number of similar properties which distinguish them from other metals or alloys, some of which are corrosion resistance, good electrical conductivity, catalytic activity and excellent reflectivity, nevertheless, they have other distinctive individual characteristics.

Silver is a bright white metal. It is very soft and malleable. It does not oxidize at room temperature, but is attacked by sulfur. Nitric, hydrochloric and sulfuric acids attack silver, but it is resistant to many organic acids and to sodium and potassium hydroxide.

In all commercial applications, the special chemical properties, superior thermal and electrical conductivity, high reflectivity, malleability, ductility and/or corrosion resistance justify the high initial cost of the metal. In addition, uses have been established in photography, brazing, batteries, medicines, dentistry, mirror backings (silver-backed mirrors may become a more significant use as solar energy is developed), bearings, catalysts, coinage and nuclear control rods.

The use of silver in photography is based on the ability of exposed silver halide salts to undergo a secondary image-amplification process called development. In silver solders, the controlling factor is the rather low melting temperature of the alloys and the ability to wet various base metals at temperatures below the melting points of the metals to be joined. Such alloys do not dissolve or attack steel in normal usage, are ductile, have sufficient strength over a wide range of temperatures, and are capable of joining a wide variety of materials.

Silver that contains varying amounts of dispersed cadmium oxide (up to 20% CdO) is used in medium- and heavy-duty electrical contacts. In this composite material, silver imparts its good electrical and thermal conductivity as well as low surface contact resistance while the dispersed CdO improves resistance to sticking and welding plus good resistance to arc erosion (good arc quenching). The susceptibility of fine silver to sulfidation precludes the use of silver contacts in low-current, low-voltage and low-contact-force applications. In general, they should not be used below 10 V (except at high currents) or where a voltage drop of 0.2 V will be troublesome or in low-level audio circuits because of the electrical noise they would introduce.

Silver is used in engine bearings because it has good lubrication properties as well as moderate hardness, good thermal conductivity and low solubility in iron.

The good mechanical properties of certain silver-tin-mercury and silver-tin-copper-mercury alloys, and the small dimensional changes that occur during setting of these alloys, are the basis for the extended use of silver in dental amalgams.

Sterling silver retains its long-established position in uses where elegant appearance is of paramount importance. For jewelry and tableware, high reflectivity makes silver particularly attractive. Much work has been done in attempting to develop a nontarnishing sterling silver, but no such alloy appears to exist. Various thin protective coatings, such as rhodium, have been used on silver objects not likely to be scratched.

Silver-clad copper, brass, nickel and iron are produced for a variety of uses, ranging from electrical conductors and contacts to components for chemical equipment. Silver is also used in various chemical processes, including catalytic applications such as production of formaldehyde or oxidation of ethylene.

Silver coatings are applied to glass and ceramics by spreading on a special silver paste and then warming to red heat. These coatings are widely used in electronic devices and automotive applications. Chemical methods for applying conductive coatings to plastics and glass are also used as the base for electroplates. Metalloorganic solutions containing silver are applied and fired in the production of conductors, electrical grounds and shields, resistance heaters, electrode terminals and conductive bases for electroplating.

The rapid diffusion of oxygen through silver at elevated temperatures can be an advantage or disadvantage depending on the application. This phenomenon has been used to good advantage in the internal oxidation of base-metal alloying constituents (such as cadmium, rare earths, cerium or calcium) in silver alloys. The resulting silver composite containing fine, well-dispersed oxide particles has been used in electrical-contact applications.

Electrodeposited silver is used widely for electrical, electronic, industrial and decorative applications. Heavy electrodeposits can be used for surfacing chemical equipment and for bearings.

Gold is a bright, yellow, soft, and very malleable metal. Its special properties include corrosion resistance, good reflectance, resistance to sulfidation and oxidation, freedom from ionic migration, ease of alloying with other metals to develop special properties, and high electrical and thermal conductivity.

Because gold is easy to fashion, has a bright pleasing color, is nonallergenic and remains tarnish free indefinitely, it is much used in jewelry. For much the same reasons, it has long been used in dentistry in inlays, crowns, bridges and orthodontic appliances, as described in the article in this volume entitled "Gold in Dentistry".

Gold is used to a considerable extent in electronic devices, particularly in printed circuit boards, connectors, keyboard contactors and miniaturized circuitry. Because electronic devices employ low voltages and currents, it is important that the listed components remain completely free from tarnish films and remain chemically and metallurgically stable for the life of the equipment.

Gold is a good reflector of infrared radiation, and for this reason gold films are used in radiant heating and drying devices as well as in thermal-barrier windows for large buildings. A relatively recent and much publicized use of gold reflective coatings has been for protecting space-vehicle components and space suits from excessive solar radiation which could raise temperatures substantially.

Fired-on gold organometallic compounds are used to decorate porcelain and glassware. Chemically inert gold rupture discs are used in chemical process equipment. Because of good resistance to corrosion and wear, the gold alloy 70Au-30Pt has been used in the perforated "spinnerettes" through which cellulose acetate fibers are extruded. Gold has also been used in other industrial applications, such as sliding electrical contacts, fine-wire gold connectors (semiconductor industry), vacuum-deposited films or coatings (interconnecting links in thin-film integrated circuits), gold brazing alloys (for joining jet-engine components), and gold alloys in thermocouples for both cryogenic service (down to liquid helium temperature) and high-temperature use (up to 1300 °C).

Among the gold alloys, the Au-Ag-Cu-Pt-Pd alloys are used in dentistry because of their good mechanical properties, response to age-hardening treatments, nobility and moderate melting point. Social custom, available colors, relatively good resistance to tarnishing and corrosion, and adequate mechanical properties of the Au-Ag-Cu (yellow) golds and Au-Ni-Cu-Zn (white and suntan) golds account for their use for jewelry, eyeglass frames and rings; these alloys also are used for certain rubbing contacts in small electrical devices. Gold-silver alloys containing about 70% Au, generally with a few percent platinum, resist both oxidation and sulfidation, and have other properties useful for low-current electrical contacts.

Pure gold is readily electrodeposited and this metal, as well as rhodium and palladium, is used for surfacing certain high-frequency conductors for use in environments where silver corrodes. A substantial quantity of electrodeposited gold is being employed in surfacing plug-type electrical connectors. Platinum metals may be required at higher temperatures, to minimize diffusion and adhesion (sticking). Gold alloys are also electrodeposited on jewelry and other items where appearance is important. In some instances, several layers of gold and other metals are deposited successively, and the article is subsequently heated to produce an alloy by diffusion. Pure gold has high reflectivity in the red and infrared spectral ranges and therefore is sometimes used for surfacing infrared reflectors.

Although pure gold resists nitric, sulfuric and hydrochloric acids and many other corrosives, the use of pure gold in the chemical field is limited because of its susceptibility to attack by halogens, its softness and relatively low melting point and to some extent, its cost. However, gold sometimes is used as a lining for small calorimeter bombs and as a corrosion-resistant solder. The hard 70Au-30Pt alloy has been used for rayon spinnerettes, but generally has been replaced by Pt-10Rh.

Platinum-Group Metals. All six platinum-group metals are closely related and commonly occur together in nature. Their most distinctive trait in the metallic form is their exceptional resistance to corrosion. Of the six metals, platinum has the most outstanding properties and is the most used. Second in industrial importance is palladium, which is the lightest metal of the group.

Rhodium sometimes is fabricated in the unalloyed form but is more commonly used as an alloying element with platinum and to a lesser extent with palladium. In the unalloyed form, iridium is fabricated into large crucibles that are used in production of single crystals of YAG (yttrium-aluminum garnet) and GGG (gadolinium-gallium garnet, a substrate for bubble-memory devices). It also finds considerable use as an alloying element for platinum and rhodium. Ruthenium is mainly used as an alloying element for platinum and palladium. When alloyed with platinum and palladium, rhodium, iridium and ruthenium are increasingly effective hardening agents in the order shown. Osmium forms a toxic oxide at ambient temperature and is therefore a difficult metal to utilize. A naturally occurring alloy of osmium and iridium called osmiridium is very hard and has been used for fountain-pen tips and phonograph needles.

The important properties of the platinum-group metals are outlined so that additional uses may be determined.

- *Platinum* is a white, very ductile metal that remains bright in air at all temperatures up to its melting point. Platinum has the following engineering characteristics:

1 High melting point (1769 °C or 3216 °F)
2 It is readily strengthened by alloying with compatible noble metals.
3 It can be electroplated.
4 It is virtually nonoxidizable.
5 It resists molten glass and molten salts in oxidizing atmospheres.
6 Low vapor pressure
7 Low electrical resistivity and, conversely, high temperature coefficient of electrical resistivity, which makes it eminently suited for measuring elements in resistance thermometers
8 Stable electrical-contact resistance
9 Stable thermoelectric behavior (the Pt-10Rh vs Pt thermocouple is a defining instrument on the International Practical Temperature Scale of 1968)
10 High thermionic work function
11 Special magnetic properties when alloyed with cobalt
12 High thermal conductivity
13 High resistance to spark erosion (hence its use in spark plugs)
14 Excellent catalytic activity
15 Coefficient of thermal expansion matching that of common glass.

Platinum resists practically all chemical reagents and is soluble only in acids that generate free chlorine, such as aqua regia.

- *Palladium* is a white, very ductile metal with properties similar in many respects to those of platinum. Palladium has the following engineering characteristics:

1 It has a density of 12.02 Mg/m^3—approximately 56% that of platinum and 63% that of gold—and can be used in place of lower-cost gold alloys without sacrificing the good corrosion resistance of gold.
2 High melting point (1554 °C or 2829 °F)
3 Excellent ductility
4 It is easily cold worked.
5 Outstanding ability to form extensive ductile solid solutions with other metals.
6 It can be electroplated, electroformed and deposited electrolessly.

7 It is an effective whitener for gold.

8 Good catalytic activity.

Palladium resists tarnishing in ordinary atmospheres, but does tarnish slightly on outdoor exposure to sulfur-contaminated environments. When palladium is heated in air to 400 to 800 °C (750 to 1475 °F), a thin oxide film is formed; this film decomposes at higher temperatures, leaving the metal bright. Hydrochloric acid and sulfuric acid attack palladium slightly; nitric acid, ferric chloride and moist halogens attack it readily. Palladium absorbs hydrogen, which will diffuse at a relatively rapid rate when the metal is heated; this is the basis for laboratory apparatus for purifying hydrogen.

- *Rhodium* is a hard, white metal. It is fairly ductile when cold and quite ductile when hot. Rhodium is the whitest platinum-group metal and remains bright under all atmospheric conditions at ordinary temperatures. It resists most common acids, even at moderate temperatures. It resists hot aqua regia. High oxidation resistance and a high melting point (1963 °C or 3565 °F) permit rhodium to be used for fabricating items for use at high temperature. Rhodium can be electroplated on other metals to provide a hard, wear-resistant, permanently bright surface. Rhodium has high specular reflectivity and the highest electrical and thermal conductivities of all platinum-group metals.

- *Iridium* is a white metal of limited malleability at room temperature. It can be worked at elevated temperatures. Iridium oxidizes visibly when heated in air 600 to 1000 °C (1100 to 1850 °F), but remains bright at higher temperatures. Acids or aqua regia do not attack it; molten salts do. Iridium has exceptional corrosion resistance, and this property, coupled with a high-temperature strength comparable to that of tungsten up to 1650 °C (3000 °F) and high melting point (2447 °C or 4437 °F), permits its use in crucibles for melting nonmetallic substances at temperatures as high as 2100 °C (3700 °F). Iridium has a high modulus of elasticity (75 × 10^6 psi); is the only known metal that can be used for short periods of time at temperatures up to 2000 °C (3650 °F) in air without undergoing catastrophic failure; is catalytically active; and is the heaviest of all metals.

Table 1 Industrial applications of precious metals

Application	Special requirements	Metal or alloy
Electrical and Electronic Devices		
Spark plug electrodes	Resistance to corrosion and erosion	Thoriated Pt-4W
Jet engine glow plugs	Relight on flameout	Thoriated Rh-Pt
Leads for thermistors	Freedom from oxidation	Pt and Ag + binder
Transistor junction	Doping contact	Au and doping alloy
Transistor junction	Nondoping contact	Ir-Pt
Resistors and potentiometers　Resistance wire	High resistivity, low temperature coefficient and low contact resistance	8W-Pt; 5Mo-Pt; 10Ru-Pt; Au-Pd-Fe; dental-type alloys
Resistance film	High resistivity, low temperature coefficient and low contact resistance	Au-Pd-Pt
Electrode for ceramic condensers	Applicability, nonoxidizing, solderability	Ag or Pt, with bonding agent
Electrode for air condensers	Corrosion resistance	Ag and Au
Conductor in printed circuits	Corrosion resistance, solderability, wear resistance (Rh)	Ag, Au, Rh, Pd (Ag may lead to ionic shorting)
Connectors (such as terminals, lugs and tabs)	Low contact resistance, solderability	Ag, Au, Pd electro- or electroless plate
High-temperature wiring	Conductivity, oxidation resistance, low contact resistance	Pt-clad base metal, solid Ag, Ag-Mg-Ni
Fuse	High conductivity and oxidation resistance	Ag-Au
Solid lead in mercury contact devices	Negligible solubility, freedom from oxidation	Pt where wetting required; also 10 It-Pt. Ir where no wetting desired. Rh-plated steel for collector rings
Bonding in vacuum devices requiring vacuum-tight low-vapor-pressure seals	Desired melting point and low vapor pressure	28Cu-72Ag, 20Cu-80Au, 40Ni-60Pd, Au-Pd
Brazing alloy for tungsten	Ductility, high melting point, vapor pressure	Platinum
Instrument Applications		
Sensing element for resistance thermometer	Stable and known resistance, high temperature coefficient	Ultrapure Pt
Thermocouples	Stable temperature relation	10Rh-Pt vs Pt, 6Rh-Pt vs 30Rh-Pt, 13Rh-Pt vs Pt, 5Rh-Pt vs 20Rh-Pt, Au-Pd vs Rh-Pt, Au-Pd vs Ir-Pt
Thermocouples	For sensing ultra-high temperature in oxygen-free atmosphere	Ir-Rh vs Ir
Thermocouples	High emf	Au-Pd vs Rh-Pt, Au-Pd vs Au-Pd-Pt
Thermocouple connectors	Low-resistance joints with base-metal wires	Platinum plate
Galvanometer suspensions	Corrosion resistance, strength and conductivity	40Cu-60Pd (slow cooled); 14-Kt Au; Ag-Cu
Galvanometer pivots	Hardness and corrosion resistance	60 Os. Ru alloy
Contact parts in low level switches	Low electrical contact resistance, good wear resistance	Rh electroplate; 69Au-6Pt, 25Ag; Pt, Pd and hard dental alloys
Slip rings, brushes for selsyns	Low contact resistance, good wear resistance and minimum friction	18-Kt gold; dental alloys; 60Pd-40Cu; Ag; Au electroplate, Rh electroplate
Sensing element for gas analyzer	Catalytic action proportional to gas content	Pd-Pt; platinum metal
Glass and Ceramics Industries		
Tanks and crucibles for optical glass	Insolubility, high melting point, noncontaminating	Pure platinum

(continued)

Table 1 (continued)

Application	Special requirements	Metal or Alloy
Bushings and valves for fiberglass	Insolubility, high strength	10Rh-Pt
Crucible for continuous melting-glass frit	Noncontaminating	Platinum
Crucibles for melting optical salt crystals	Insolubility, high melting point, noncontaminating	Platinum
Metalized glass and ceramics, metal film bonded to ceramic by heat	Nonoxidizing, desired color	Liquid-bright Au and Pt pastes
Metallized glass and ceramics, metal film by vacuum sublimation	Desired properties	Au, Pd, Rh, Ag and alloys
Heater windings for glass, ceramic and ferrite research	Nonoxidizing, high melting point, low vapor pressure	Pt. 20Rh-Pt and 40Rh-Pt
Chemical Industry		
Septum in a hydrogen-purification system	Selective transmission	Pd; 60Pd-40Ag
Catalyst for removal of oxygen from H_2	Activity at low temperature	Pd on alumina
Septum in an oxygen-purification system	Selective transmission	Pure silver
Catalyst for production of nitrogen or nitrogen-hydrogen heat treating atmosphere from ammonia	Activity and long life	Platinum metal
Catalyst for production of formaldehyde from methanol	Activity	Silver
Catalyst for production of ethylene oxide from ethylene	Activity	Silver
Catalyst for destruction of odoriferous or hazardous contaminants	Activity	Platinum metal
Catalyst for ammonia plus air to yield HNO_3	Long life, high efficiency	Rh-Pt
Catalyst for ammonia, air and methane to yield HCN	Long life, high efficiency	Rh-Pt
Rayon spinnerettes	Corrosion resistance, strength ductility	Rh-Pt, Pt-Au
High-temperature HCl containers	Corrosion resistance	Platinum
Electrochemical Applications		
Insoluble anode for electrolytic protection	Non-film-forming, high corrosion resistance	Platinum, 20Pd-Pt and 50Pd-Pt, Pt clad T2, Pt-coated T1, Ta
Insoluble anode for production of persulfates, perchlorates; and for electroplating	Corrosion resistance in chlorides, sulfates; proper anodic reaction	Platinum and 5Ir-Pt
Positive plates in primary and secondary batteries	Corrosion resistance, conductivity and depolarization	$Ag-Ag_2O_2$
Fuel cell electrodes	Catalytic activity, corrosion resistance	Platinum metals
Container for tantalum capacitors	Corrosion resistance, high conductivity	Silver
Tomer-milliwatt meter anode	Insolubility	Platinum
Aerospace Applications		
Brazing alloy in stainless steel systems for handling rocket fuels and oxidizers	Corrosion resistance, compatibility	Au-Cu-Ni, Au-Ni-Cr

(continued)

- *Ruthenium* is a very hard, white metal that cannot be worked cold. It can be worked only after being heated to a fairly high temperature, and then only with extreme difficulty. Ruthenium resists common acids, including aqua regia, at temperatures up to 100 °C (212 °F). Ruthenium has high resistance to contamination by lead. Like iridium, it is principally used as a hardener for platinum and palladium. Ruthenium has a high melting point (2310 °C or 4190 °F); is exceptionally hard; has a high elastic modulus (60×10^6 psi); exhibits good resistance to attack by molten lithium, sodium, potassium, copper, silver and gold in the absence of oxygen; has low electrical-contact resistance at temperatures up to 600 °C (1100 °F) and resists any tendency of the contacts to weld together at these temperatures.
- *Osmium* is a white, hard metal that is not malleable at either room or elevated temperatures. It forms a toxic oxide at ambient temperatures. Osmium is used as an alloying element to provide other precious metals with extreme hardness and resistance to corrosion. Osmium has the highest melting point of all the platinum-group metals, 3045 °C (5513 °F).

Commercial Forms and Uses

Semifinished Products. Silver, gold, platinum, palladium and rhodium can be drawn to rod and wire as small as 25 μm (0.001 in.) in diameter. Iridium may be drawn to diameters as small as 75 μm (0.003 in.). Some of the platinum alloys containing iridium or rhodium may be drawn to diameters of 7.5 μm (0.0003 in.).

Sheet, strip, ribbon and foil in a broad range of alloys, sizes and thicknesses can be produced. Silver, gold, platinum and some of its alloys can be rolled to thicknesses as low as 2.5 μm (0.0001 in.), but cannot be guaranteed retain the greater strength of the base metals. Clad materials can be obtained as wire, sheet, strip and formed parts with a great variety of substrate materials.

Tube is manufactured in a wide range of sizes and in round, half-round and square sections. Seamless tube made of platinum, palladium, gold and most alloys of these metals is manufactured in sizes ranging from 0.4-mm 16-in.) OD by 0.1-mm (0.004-in.)

Table 1 (continued)

Application	Special requirements	Metal or Alloy
Special Uses		
Crucible for molten lead	Insolubility and high melting point	Ir under oxygen-free atmosphere
Crucible for molten bismuth	Insolubility and high melting point	Ru under oxygen-free atmosphere
Crucible for molten NaOH	High corrosion resistance	Silver
Container for high temperature sulfur and sulfur gases	High corrosion resistance	Gold
Container for high temperature SO_2	Corrosion resistance, ductility	Pure Pt, pure Au, Au-Pt alloy
Container for high temperature (1000 °C) H_2S	Corrosion resistance, ductility	Gold, platinum
Container for S and H_2S below 1000 °C	Corrosion resistance, ductility	Gold
Neutron absorber	High absorption cross-section	Iridium
Intense gamma-ray source	Radiation energy; moderate half-life	Iridium
Magnet	Highest known energy product and corrosion resistance, ductility	23Co-Pt
Laboratory ware	Corrosion and heat resistance	Platinum, 6Ir-Pt, 3.5Rh-Pt
Reflectors		
Visible and infrared reflecting surface	High efficiency	Ag where protected; Rh where exposed
Ultraviolet and infrared reflecting surface	High and uniform reflectivity	Rhodium
Red and infrared reflecting surface	High long-wave reflectivity	Gold
Safety Devices		
Over-pressure protector (frangible disk)	Reproducible tensile properties, corrosion resistance	0.6Ir-Pt, Ag, Au
Fuse wire for temperature-limiting fuse	Required and constant melting point oxidation resistance	Gold

wall up to 44-mm (1.750-in.) OD by 5-mm (0.200-in.) wall. Tube in larger sizes or made of less ductile materials (such as platinum alloyed with 25% or more rhodium or with over 25% iridium) is manufactured only as seamed tube 3-to-75-mm (1/8-to-3-in.) ID by 0.25-to-2.5-mm (0.010-to-0.100-in.) wall. Pure rhodium and pure iridium are usually furnished as seamed tube 3-to-40-mm (1/8-to-1½-in.) ID by 0.25-to-0.6-mm (0.010-to-0.025-in.) wall and single lengths about 150 mm (6 in.) long. Base-metal tube with outer cladding or inner lining of platinum, gold, silver or any of the commercial precious metal alloys is available.

Noble metal powders are produced for a wide range of electronic and industrial applications.

Electronic powders are chemically precipitated to produce particle sizes less than 10 μm and tend to be high in surface area. They are used as inks in hybrid circuits. Flake powders tend to produce shinier, smooth films; spherical particles more often result in dull-appearing surfaces. Platinum, palladium, 40Pt-20Pd-40Au, 10Pt-20Pd-70Au, 7.5Pt-22.5Pd-70Au and 75Au-25Pd powders have been used for electronic purposes. Trials are usually necessary to determine the most suitable powder from the standpoint of both cost and performance.

Powders intended for industrial uses are composed of mixtures of particles that range in size from about 2 to 3 μm to as large as 840 μm (20 mesh), depending on the size required. These powders are suitable for use in powder metallurgy parts, as protective coatings against hostile industrial environments, as raw materials in alloy manufacture and for various other uses.

Industrial Uses. Requirements and materials for about 65 industrial applications of precious metals are cited in Table 1. Additional information on selection and application of precious metal contacts is included in the article on electrical contact materials, which will be found in Volume 3 of the 9th Edition of this Handbook, or on pages 801–816 of Volume 1 of the 8th Edition.

Coatings. Several cladding or coating processes are used in producing composite articles with precious metal surfaces. Table 2 lists the most important of these for precious metals, with characteristics, common thickness ranges and typical applications of each.

Jewelry. Gold, the first jewelry metal, still is the most popular. Tradition, a distinctive color, and the karat mark aid in maintaining gold's popularity. Color and karat are factors to be considered in selection. Yellow is the most popular color, but red, green and white karat golds also are available. The 14K alloy is most popular in the United States, although significant quantities of all kinds of jewelry are made of 18K gold. At the same time, there is significant use of 10K gold, especially for rings set with synthetic colored stones. Gold–plated jewelry generally is produced for mass-market jewelry lines rather than for fine jewelry lines.

Hand crafting is usual where only a few exclusive creations are made, but where many duplicates will be required, die forming or casting is appropriate. For simple rings, mechanical forming methods are justified where more than a thousand units are required; below this quantity, casting is less costly. When complex shapes such as watchcases are produced from clad or filled stock, intricate and expensive dies are required to maintain uniformity of the cladding.

Platinum is frequently used to make settings for the finest jewelry. In addition to its high intrinsic worth, its workability and strength ensure reliable retention of jewels, and its white color enhances the brilliance of diamonds.

Manufacturers use 10% Ir platinum for either wrought or cast items; in some instances, 5% Ru platinum may be used. The 15 and 20% Ir alloys are preferred for some of the more delicate pieces, such as small chain. Where weight is objectionable, as in earrings, palladium is preferred to platinum because of its lower density.

Palladium has platinum-like characteristics which have led to increased use, particularly in quality jewelry.

Table 2 Precious metal coatings

Method	Characteristics	Thickness range	Examples of applications
Mechanical and Thermal Bonding (Cladding)			
Brazing, hot pressing, hot and cold rolling, puddling, casting	100% density, good adhesion, high wear resistance, uniform thickness	2.5 μm (0.1 mil) and up	Precious-metal-clad base metals for jewelry, electrical contacts, chemical apparatus, or other industrial uses; applicable for all malleable precious metals and alloys
Vacuum Coating			
Vacuum metallizing	Fairly uniform coating, transparent layers, good adhesion	0.025 to 12.5 μm (1 to 500 μin.)	For decorative purposes, reflectors (rhodium on glass), condensers for electronic devices (mostly metals on paper, plastic or lacquered surfaces); applicable for Ag and Au. Nucleation with Ag required prior to applying Zn on plastic condensers
Cathode sputtering	Very even coating, good adhesion, high density	1.2 to 125 μm (0.05 to 5 mils)	For improved corrosion resistance, silver in surgical gauzes, gold on thin Al-alloy foils, diaphragms, mirrors
Electrochemical and Chemical Coating			
Electroplating	Reasonably dense and usually well adhering deposits; mechanical and physical properties depend greatly on plating conditions	0.15 to 125 μm (6 to 5000 μin.)	Decorative uses, improved corrosion and wear resistance, electrical contacts; applicable to a wide range of elemental precious metals and some of their alloys
Fired on Films			
Formulated metalloorganic solutions, thermal decomposition	Thin, well-adhering film	50 to 200 μm (2 to 10 μin.)	Ceramic and electronic uses, printed circuits, decorations; applicable to bright Au, Ir, Pt, Pd and Ag, mostly on nonmetallic surfaces
Resins containing very fine suspended metal particles with a low-melting inorganic glass flux	Thick, adhering films	12 to 40 μm (0.5 to 1.5 mils)	Electronic applications
Chemical decomposition coating	Thin, well-adhering film	Usually very thin	Mirrors

Sterling is the standard silver jewelry alloy in spite of its tendency to tarnish.

All of the white metals—platinum, palladium, and white gold—are frequently finished with rhodium plate for whiteness and wear resistance.

REFERENCES

1. H. J. Drake, Silver, *Mineral Commodity Profiles,* MCP-24, Sept 1978, U.S. Department of the Interior, Bureau of Mines
2. W. C. Butterman, Gold, *Mineral Commodity Profiles,* MCP-25, Oct 1978, U.S. Department of the Interior, Bureau of Mines
3. J. M. Jolly, Platinum-Group Metals, *Mineral Commodity Profiles,* MCP-22, Sept 1978, U.S. Department of the Interior, Bureau of Mines

Corrosion Resistance of Precious Metals

By W. Z. Friend (retired)
Formerly, Supervisor
Chemical Industry Applications
The International Nickel Co., Inc.

PRECIOUS METALS are well known for their excellent resistance to corrosion in ordinary environments, but they also find increasing application in construction of chemical process equipment for handling highly corrosive materials. Because of their high cost, precious metals normally are used only where corrosion of less expensive materials is excessive or product purity is essential. The trend in chemical processing has been toward use of higher and higher temperatures and pressures, resulting in increases in severity of corrosion problems. Thus, despite their high initial cost, precious metals solve some corrosion problems more economically than do other materials.

Platinum

The corrosion resistance of platinum in several highly oxidizing environments is matched by few, if any, other materials. Because of its excellent resistance to oxidation in air at high temperatures, platinum is used in making furnace heating coils. For continuous operation at temperatures above about 1260 °C (2300 °F), the only suitable materials for winding coils are molybdenum and Pt-10Rh alloy. Mo-

lybdenum, however, is subject to rapid oxidation in air and can be used only in a protective nonoxidizing atmosphere; hence, furnaces with molybdenum coils are relatively complicated and costly to operate. By use of Pt-10Rh coils, continuous working temperatures as high as 1480 °C (2700 °F) can be maintained. With alloys higher in rhodium content, furnaces can be built for working temperatures up to 1760 °C (3200 °F).

In production of optical glass and other high-purity glass, platinum alloys are used for linings of melting furnaces, pouring gates, and other parts for which corrosion resistance is critical. These alloys are the only materials that can be exposed in air to temperatures of about 1400 °C (2550 °F), which are required for melting glass. Bushings and orifices of Pt-10Rh are commonly used for producing glass fibers at 1400 °C.

Platinum is resistant to nitric acid at all temperatures and concentrations, including the supertemperatures encountered with boiling acid under pressure. In these environments, it is greatly superior to the stainless steels commonly used with nitric acid at lower temperatures, and titanium is the only competitive material. How-

ever, under some conditions, titanium may form explosive compounds with fuming nitric acid. Platinum and a Pt-Rh alloy are commonly used in the form of fine gauze catalysts in synthesis of nitric acid by oxidization of ammonia at temperatures up to 955 °C (1750 °F). There are similar applications of Pt-Rh alloy gauze catalysts in synthesis of hydrocyanic acid from air, ammonia and methane, at temperatures as high as 1300 °C (2375 °F).

Perchlorates and persulfates are other examples of highly oxidizing chemicals that are corrosive to most of the common metals and alloys. The high resistance of platinum to these chemicals is indicated by the fact that platinum, or platinum-clad copper or silver, is commonly used for insoluble anodes in electrolytic production of persulfates, perchlorates and bromates. Platinum also is highly resistant to strong hypochlorite and chlorine dioxide bleaching solutions, which attack most common metals and alloys except titanium. Because these solutions are highly corrosive, they usually are contained in glass or ceramic.

Platinum is one of the most resistant of all metals to nonoxidizing hydrofluoric acid solutions, but platinum, as well as palladium, gold and osmium, are strongly attacked by mixtures of hydrofluoric acid with nitric acid or with hydrogen peroxide.

Platinum is resistant to sulfuric acid at most temperatures and pressures and for many years was used in construction of fairly large vessels for concentration of sulfuric acid. A long-established application of Pt-Rh alloys is spinnerets for extrusion of alkaline viscose rayon fibers into hardening baths containing sulfuric acid, zinc sulfate and hydrogen sulfide.

Several chemical processes are based on reactions of organic and inorganic materials with various halogen gases at high temperatures. Platinum has shown a high degree of resistance to hydrogen chloride gas, probably better than most other available metals and alloys. It is considered useful in this gas at temperatures up to 1095 °C (2000 °F), and possibly as high as 1205 °C (2200 °F). Typical applications of platinum-lined equipment include autoclaves for manufacture of ethyl chloride and tubes for vapor-phase reactions with hydrogen chloride gas at temperatures up to about 1010 °C (1850 °F).

Platinum has poor resistance to chlorine at elevated temperatures. It can be used freely in chlorine at 100 °C (212 °F), but above this, up to 250 °C (480 °F)—particularly when the gas may not be dry—consideration of specific conditions is required.

Platinum shows high resistance to hydrogen fluoride and to most fluoride compounds at high temperatures. Platinum-lined Inconel tubes are used for pyrolysis reactions in production of organic fluorine compounds in a reaction system operating continuously at about 705 °C (1300 °F). Platinum containers are used in production of large synthetic optical crystals from molten lithium fluoride and from molten sodium chloride.

Autoclaves lined with platinum or gold have been used to handle mixtures of phosphates and fluorides at temperatures up to 500 °C (930 °F) and pressures up to 55 MPa (8.0 ksi).

Sulfurous gases at high temperatures are destructive to most metals, including the common gold alloys (although gold itself is not attacked) and all silver alloys, but are resisted well by platinum. Sulfur dioxide appears to have no effect on platinum, and hydrogen sulfide only a slight effect. Unalloyed platinum exposed to SO_2 at 800

°C (1474 °F) for 1 h exhibited no change in weight or appearance, and 1 h at 1000 °C (1830 °F) caused a weight loss of only 130 mg/m². A 12-h exposure to H_2S at 1000 °C caused a weight gain of 2.4 g.m², but no loss in ductility. Gold would be recommended for service in sulfurous vapor below 1000 °C. In sulfur vapor, platinum alloys are attacked more than unalloyed platinum, and it has been noted that Pt-10Rh suffers intergranular sulfidation at 1095 °C (2000 °F).

Palladium

Palladium generally is less resistant to corrosion than platinum. For example, palladium, unlike platinum, has poor resistance to highly oxidizing chemical environments. However, palladium (as well as gold and rhodium) is reported to perform satisfactorily in equipment used for converting urea to melamine at 275 to 595 °C (525 to 1100 °F), where platinum is rapidly attacked. One application of palladium in oxidizing atmospheres is trays for firing of phosphors. Most of the palladium employed in the chemical industry is used for catalysts in hydrogenation or dehydrogenation processes.

Iridium

Iridium is highly resistant to corrosion by a wide variety of severe corrosives, including boiling aqua regia, hot hydrochloric acid and hot chlorine. Unfortunately, this metal is very difficult to forge or fabricate, so it is not ordinarily used by itself. Consequently, in most of its applications it is used as an alloying element in platinum, to which it imparts added corrosion resistance to all of the above environments.

Rhodium

Rhodium is somewhat similar to iridium in its resistance to a large number of severe corrosives, including boiling aqua regia, and in the fact that it is difficult to fabricate. Rhodium finds much of its application as an alloying addition to platinum, to which it imparts added corrosion resistance in such media as aqua regia, hot hydrochloric acid, boiling 10% ferric chloride, and wet chlorine. A 63Ni-37Rh alloy has better general corrosion resistance than 14-karat yellow gold.

Gold

Gold is resistant to several highly oxidizing environments, but not to some of the halogens. It has good resistance to boiling nitric acid solutions below 1.42 specific gravity, but is rapidly attacked by mixtures of nitric acid with hydrochloric, hydrobromic or hydriodic acid, and also by hot mixtures of nitric and sulfuric acids. However, gold is reported to have useful resistance to mixtures of nitric and hydrofluoric acids.

Gold is resistant to sulfuric acid at most temperatures and concentrations in the absence of oxidizing salts. Considerable gold-lined equipment once was used for evaporation and concentration of sulfuric acid. Although gold is superior in corrosion resistance to any other unalloyed precious metal, some gold-platinum alloys are equally corrosion resistant and have better mechanical properties, which make them preferable for constructional applications.

Gold is resistant to nonoxidizing phosphoric acid and phosphates at all temperatures and is used for lining autoclaves that handle phosphate mixtures at temperatures up to 500 °C (930 °F).

In the absence of air or other oxidizing agents, solutions of hydrochloric or hydrofluoric acid are not corrosive to gold, even at their boiling temperatures. However, at high temperatures under air pressure, HCl can cause considerable attack.

Unalloyed gold is not attacked by oxygen, sulfur, sulfur dioxide or selenium, but it does react with tellurium at elevated temperatures.

Gold is second only to platinum in resistance to dry hydrogen chloride gas at high temperatures, giving satisfactory service at temperatures up to about 900 °C (1650 °F). However, it exhibits poor resistance to chlorine at temperatures above about 80 °C (175 °F). Gold is resistant to dry fluorine up to about 310 °C (590 °F), and to hydrogen fluoride and other fluoride compounds at considerably higher temperatures. Reactions involving hydrochlorination or hydrofluorination at elevated temperatures are among the applications of gold-lined equipment in the chemical industries. Gold has good resistance to dry iodine vapor at 480 °C (900 °F) and has been used for closure gaskets in equipment for production of zirconium by the iodide method.

Electroplated gold 1.3 µm (0.05 mil) thick, and electroplated rhodium 0.38 µm (0.015 mil) thick, have been used as stop-offs in ferric chloride and chromic acid etching of printed circuits. Because of the porosity of such thin coatings, it is advantageous to employ an intermediate nickel coating 5.1 to 7.6 µm (0.2 to 0.3 mil) thick under the gold or rhodium plate. This improves subsequent soldering of the gold and provides increased resistance to wear and tarnishing. The solderability of thin gold plating is impaired somewhat by exposure as an etching stop-off.

Silver

Silver has been used for many years in production of chemical and food products to which it has a high degree of corrosion resistance or for which very high product purity is required. In most of these applications, fine silver has been used as solid sheet, in tubular form or as a lining over copper, nickel or carbon steel plate. The silver linings can be homogeneously bonded to the basis-metal plate by brazing or by solid-phase bonding. Silver-lined tubing frequently is made by expanding silver tubing inside the basis-metal tubing. High-purity silver also is used as an electroplated coating over one of the above basis metals.

Silver plating 3.8 µm (0.15 mil) thick on internal surfaces and 12.7 µm (0.5 mil) thick on external surfaces, plus rhodium plating 0.18 µm (0.007 mil) thick on all surfaces, has been used on copper alloy waveguide equipment for 15 years with excellent results. Table 1 gives values of microwave attenuation in precious-metal-plated waveguides before and after salt-spray testing.

Silver shows a higher degree of resistance to high-temperature caustic alkalies than do most other metals. It has been used for evaporating pans for concentration of sodium hydroxide during production of chemically pure grades of caustic soda. In the vacuum process of continuous evaporation of NaOH or KOH to the anhydrous condition at temperatures around 325 °C (615 °F), silver-lined nickel or Inconel evaporator tubes are used, particularly where the caustic alkalies contain chlorates that may increase the corrosion rates of nickel alloys at these temperatures. Molds, stirrers and ladles of silver also are used in handling and casting high-purity caustics.

Silver has excellent resistance to hot concentrated organic acids such as acetic, formic, citric, lactic, fumaric, phthalic and benzoic acids, fatty acids, and phenol. It is commonly used for evaporation and concentration of these organic acids where very high product purity is required. Silver is used in vacuum pans, evaporators, condensers and storage vats. Fruit and beverage syrups, essential oils and some pharmaceuticals frequently are concentrated in silver equipment to avoid effects on the flavor or stability of the products. Silver-lined steam-heated boiling pans, mixing vessels, autoclaves and vacuum pans are used. Hormones and vitamins generally are prepared in silver equipment. Silver-plated copper centrifuges are used for separating a crystalline vitamin from a solution containing hydrochloric acid. Large silver-lined autoclaves are used in synthesis of urea from carbon dioxide and ammonia at temperatures up to 200 °C (390 °F) and pressures up to 20 MPa (3.0 ksi).

Silver is employed extensively in applications where halogens and halogen acids are encountered. One of the most important of these is handling of wet chlorine gas in water-purification installations. Control and metering equipment for this purpose, as well as tubing, are constructed of fine silver or of an Ag-7.5Cu alloy, often in association with plastics. Silver also is used in contact with aqueous solutions of hydrochloric acid, particularly when the acid is associated with organic hydrocarbon liquids, such as in pressure vessels and process equipment. Resistance of silver to dilute hydrochloric acid usually is best where there is little likelihood that the protective chloride film will be removed.

Silver shows a high degree of resistance to boiling hydrofluoric acid solutions of all concentrations when they do not contain any sulfur compounds or sulfuric acid. Silver tubes are used in condensers for condensing 70% hydrofluoric acid from the hot HF vapors formed during hydrofluorination of uranium compounds. Both open and closed vessels with capacities up to 760 L (200 gal), with either bonded linings or solid silver inner pans, are used for evaporation of fluoride solutions and preparation of fluorine compounds.

Fluophosphoric acid is prepared by reaction of phosphoric anhydride with anhydrous hydrogen fluoride, or with a concentrated aqueous solution of hydrofluoric acid, in a silver reactor at elevated temperature. Fluophosphoric salts are made by fusion of fluophosphoric acid with the desired base in silver vessels.

Silver-lined evaporators are used for preparation of monobasic and dibasic sodium phosphates.

Silver is resistant to anhydrous hydrogen fluoride gas at considerably elevated temperatures. It resists hydrogen chloride gas up to about 225 °C (435 °F), but has minimal resistance to chlorine gas above room temperature.

Table 1 Effect of salt spray on microwave attenuation in precious-metal-plated waveguides

Plating material	KU band attenuation(a) At 15 GHz Before salt spraying	After salt spraying 50 h A	After salt spraying 50 h B	At 30 GHz Before salt spraying	After salt spraying 50 h A	After salt spraying 50 h B
Silver(b)	0.14	0.15	0.13	0.15	0.13	0.15
Silver(b) + rhodium(c)	0.13	0.14	0.15	0.14	0.14	0.14
Gold(d)	0.29	0.29	0.27	0.26	...	...
Gold(e)	0.28	0.29	0.29	0.27	...	...
Gold(f)	0.38	0.41	0.43	0.39	...	...
Silver(b) + gold(c)	0.16	0.16	0.16	0.22	...	...
Silver(b) + gold(d)	0.22	0.21	0.20	0.26	...	...
Black oxide on copper plus fungicidal varnish	0.29	0.28	0.38	0.25	0.25	0.29

(a) Attenuation, in dB/m, of plated copper alloy waveguides 15.8 by 7.90 mm (0.622 by 0.311 in.) in inside dimensions. Data in columns marked A and B are results of measurement in two directions. (b) 12.7 µm (0.5 mil) thick on external surfaces and 3.8 µm (0.15 mil) thick on internal surfaces. (c) 0.18 µm (0.007 mil) thick. (d) 0.38 µm (0.015 mil) thick. (e) 0.76 µm (0.030 mil) thick. (f) 1.27 µm (0.050 mil) thick.

Properties of Silver and Silver Alloys

Commercially Pure Silver

Compiled by C. D. Coxe (deceased),
A. S. McDonald
and
G. H. Sistare, Jr.
Handy & Harman

Applications

Typical uses. The largest single use for commercially pure silver is for photographic emulsions. The second largest use is in the electrical and electronic industries, for electrical contacts in the medium to high current and voltage categories for conductors and in primary batteries. Silver is deposited on glass to form mirrors, and particles of metallic silver about 1 to 5 μm in diameter are used in pastes for metallizing other nonconducting materials. In the chemical industry, silver is used as a catalyst for the dehydrogenation of methanol to make formaldehyde, and in the oxidation of ethylene to ethylene oxide. Silver may also be used to line reactors and vessels, particularly caustic evaporators or crystallizers. Silver also is used as a liner in heavy duty journal bearings.

Mechanical Properties

Tensile properties. Typical. Tensile strength, 130 MPa (18.2 ksi) for 5-mm (0.2-in.) diam wire annealed at 600 °C (1050 °F); yield strength (divider method), 55 MPa (7.9 ksi). See also Fig. 1.

Hardness. Research on the effect of oxygen on the hardness of annealed silver of various purities indicates that oxidation of impurities during oxidizing anneals generally causes a substantial increase in surface hardness and restrains grain growth, effects that are absent in spectroscopically pure silver. Very pure silver had a hardness of 25 HV after a hydrogen anneal at 650 °C (1200 °F), and 27 HV after annealing in air at 650 °C.

Poisson's ratio. 0.37 for annealed material; 0.39 for hard drawn material

Elastic modulus. Tension, 71 GPa $(10.3 \times 10^6$ psi)

Mass Characteristics

Density. 10.49 Mg/m³ (0.379 lb/in.³) or 5.527 troy oz/in.³) at 20 °C (68 °F); density is lowered by cold work and probably by oxygen

Thermal Properties

Melting point. For oxygen-free silver, 960.8 °C (1760.9 °F)

Coefficient of thermal expansion. Linear, 19.68 μm/m·K (10.93 μin./in.·°F) at 0 to 100 °C (32 to 212 °F), 20.61 μm/m·K (11.45 μm/in.·°F) at 0 to 500 °C (32 to 930 °F)

Specific heat. 234 J/kg·K (0.056 Btu/lb·°F) at 0 °C (32 °F), 238 J/kg·K (0.0568 Btu/lb·°F) at 100 °C (212 °F)

Thermal conductivity. 4186.8 W/m·K (2419 Btu/ft·h·°F) at 0 °C (32 °F)

Electrical Properties

Electrical conductivity. Effect of percentage reduction for extremely pure 5-mm (0.091-in.) diam wire at 20 °C (68 °F):

Reduction, %	% IACS
Annealed	102.8
10.2	102.2
20.0	101.0
37.0	99.7
48.6	99.5
60.0	99.4
68.5	98.4
74.0	98.1

The electrical conductivity of commercial drawn wire may be much lower than 98-99%.

Electrical resistivity. Annealed 2.3-mm (0.091-in.) diam wire, 17.7 $n\Omega \cdot m$ at 20 °C (68 °F). Temperature coefficient, 0.041 $n\Omega \cdot m$ per K (0.025 $n\Omega \cdot m$ per °F) from 0 to 100 °C (32 to 212 °F)

Chemical Properties

General corrosion behavior. Silver does not appear to oxidize at room temperature in air and thus differs from copper, but it is attacked and blackened by ozone. Silver oxide, however, does exist and has extremely high resistivity. Sulfur attacks silver rapidly, as it does copper, and the rate of tarnishing of silver in indoor atmospheres is determined by the supply of sulfur atoms, because the coating is nonprotective. This sulfide decreases the reflectivity of silver and also increases the electrical contact resistance, particularly at low currents, because it is nonohmic in character. The rate of sulfidation of silver indoors in a large city is of the order of 7 $mg/m^2 \cdot d$. Much work has been done searching for a tarnish-resistant high-silver alloy, but it appears that substantial additions of noble metals are required to achieve this goal, about 50% Pd or 70% Au being needed for complete resistance. Various protective plates have been used to protect silver from tarnishing. Of these, rhodium plate applied over a very thin nickel plate is the most successful and maintains a pleasing appearance but is little used.

Resistance to specific corroding agents. Silver is resistant to acetic acid and has been used for condensers handling this acid. It is also resistant to various other organic acids and foods that are free from sulfur. It shows good resistance to phenol and to hydrofluoric and phosphoric acids, provided that these also are substantially free from sulfur.

Silver is attacked by all the low-melting molten metals, such as mercury, sodium and potassium and their mixtures, lead, tin, indium and bismuth: consequently, the use of silver in heat exchangers and other devices that employ liquid-metal heat-transfer mediums should be avoided.

It is resistant to sodium and potassium hydroxides, is used in the laboratory for caustic fusions, and has also been considered for large equipment. However, silver creeps at the

Fig. 1 Tensile properties of commercial fine silver, 2.3-mm (0.091-in.) diam wire

Cold drawn after annealing

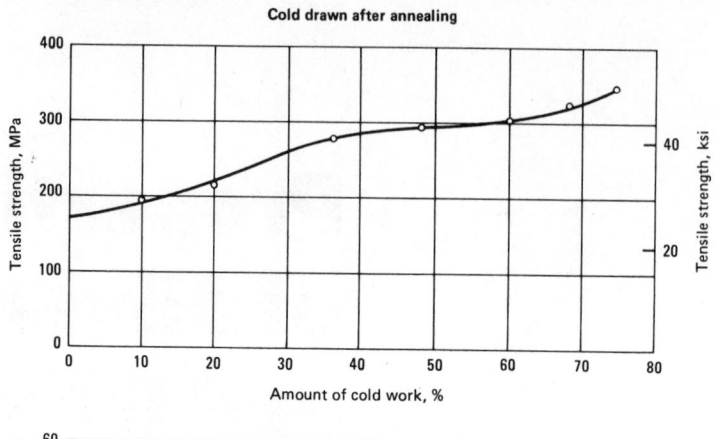

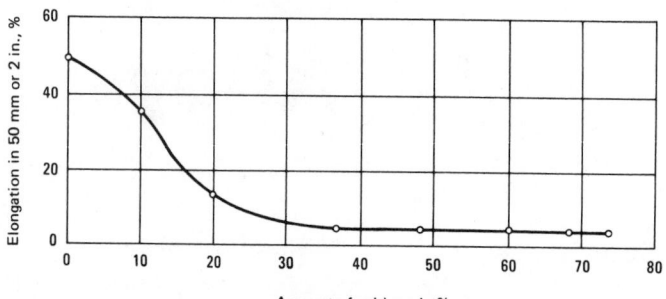

Cold drawn 49% before annealing

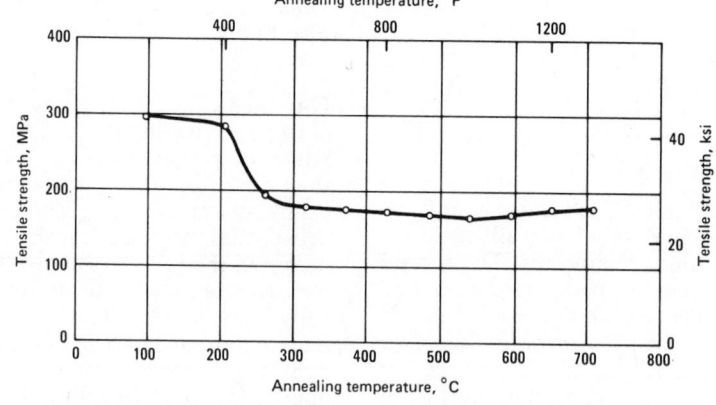

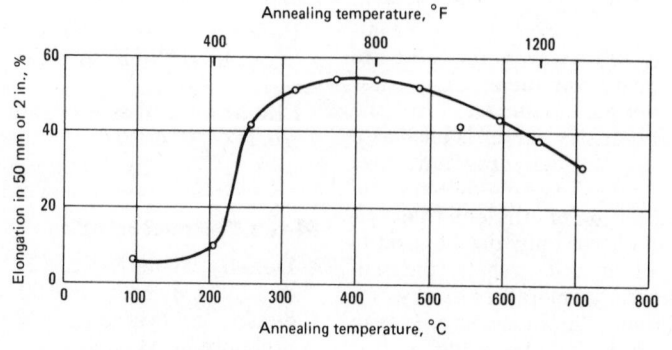

Fig. 2 Tensile properties and electrical conductivity of silver-copper alloys

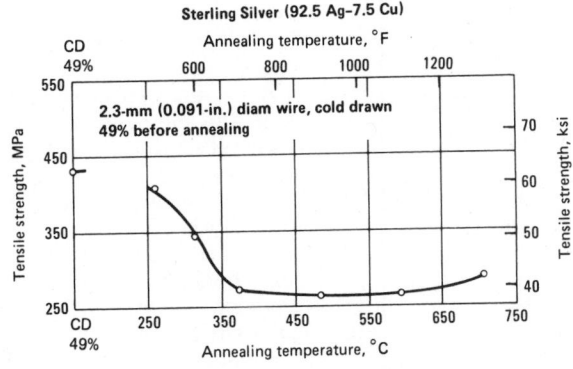

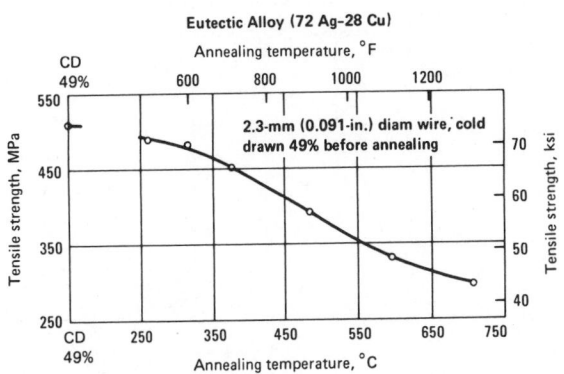

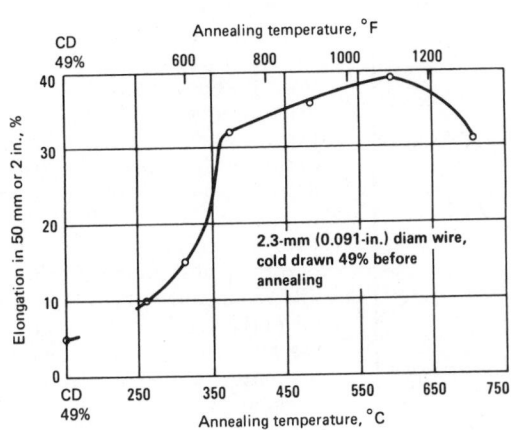

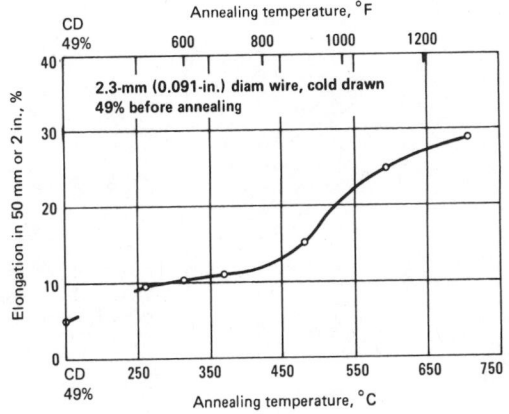

fusion temperatures of these hydroxides and its use for large equipment would require supporting vessels. It is attacked by moist bromine, iodine and chlorine and vigorously by HCl, HI and HBr. Alkaline cyanides, in the presence of air or other oxidizing agents, dissolve silver rapidly. Nitric acid that contains traces of nitrous acid attacks silver vigorously, as does hot concentrated sulfuric acid. Hot dilute sulfuric acid also attacks silver.

Fabrication Characteristics

Recrystallization temperature. 20 to 200 °C (68 to 392 °F)

Silver-Copper Alloys

Compiled by C. D. Coxe (deceased), A. S. McDonald and G. H. Sistare, Jr.
Handy & Harman

Commercial Names

Common names. Sterling silver (92.5% Ag min), coin silver (90Ag–10Cu)

Chemical Composition

Composition limits. Sterling silver must contain at least 92.5% Ag. The remainder is unrestricted but is nor-

mally copper because, in general, other metals have proved less desirable and are less effective hardeners. Coin silver is 90% Ag and 10% Cu. The eutectic alloy contains 28.1% Cu.

Applications

Typical uses. Silver-copper alloys have been used for thousands of years. Copper is effective in hardening silver, but lowers the melting point considerably and lowers the electrical and thermal conductivities appreciably. Sterling silver is used for flat and hollow tableware and various items of jewelry. Coin silver

Fig. 2 (continued)

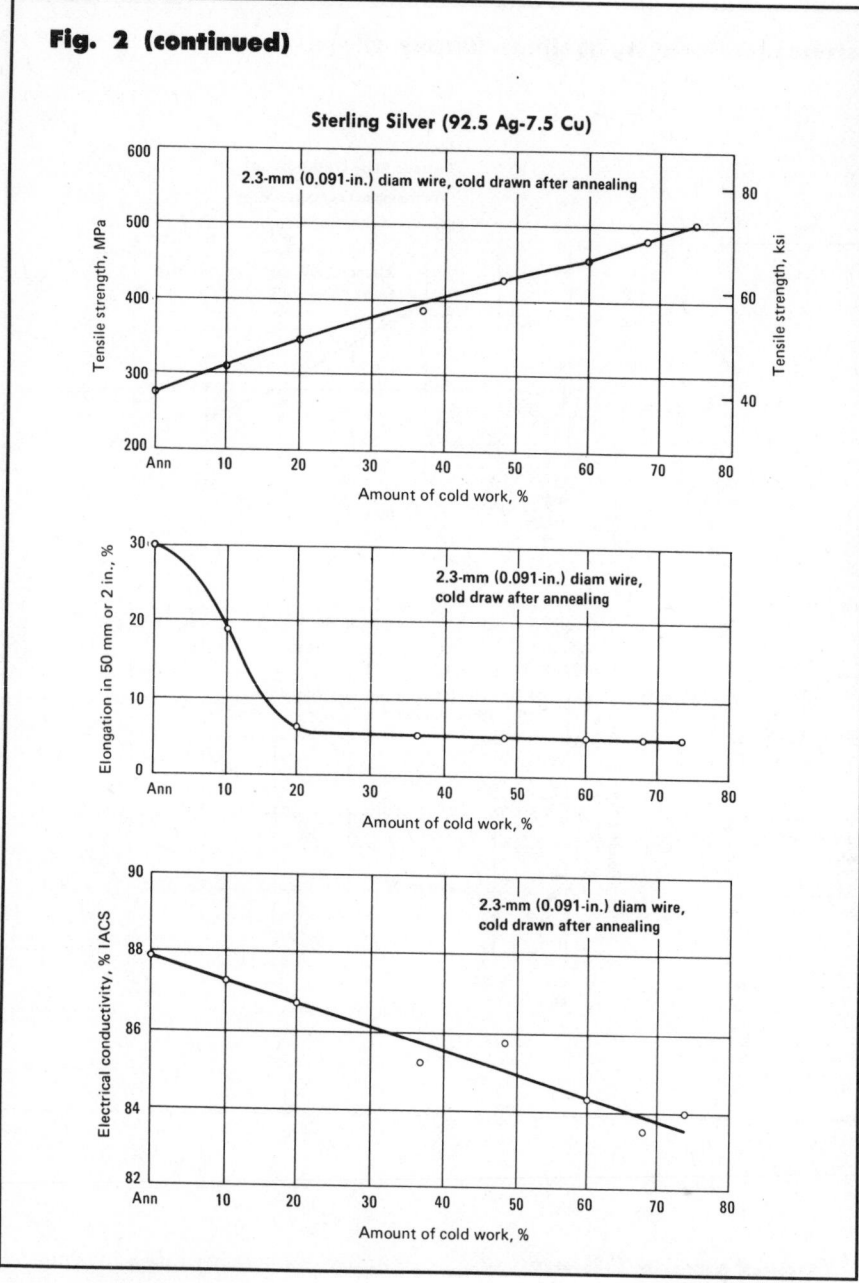

Sterling Silver (92.5 Ag-7.5 Cu)

2.3-mm (0.091-in.) diam wire, cold drawn after annealing

2.3-mm (0.091-in.) diam wire, cold draw after annealing

2.3-mm (0.091-in.) diam wire, cold drawn after annealing

with 10% Cu and 90% Ag was used for U.S. silver coins and is used for electrical contacts operating under service conditions where pure silver is considered too soft and is more likely to pit. The 28% Cu eutectic alloy finds some use as a brazing or soldering alloy. With heavy cold work, it is quite strong and is used for spring-type electrical contacts.

Mechanical Properties

Tensile properties. See Fig. 2 and 3.

Hardness. See Fig. 3.

Electrical Properties

Electrical conductivity. See Fig. 2.

Electrical resistivity. The addition of copper to silver raises the resistivity to a greater extent if the copper is held in solution by quenching and to a lesser extent if it is precipitated by aging or slow cooling. In Fig. 3, the curve labeled "Commercial wire, annealed" shows resistivity typical of commercial phosphorus-deoxidized wire, annealed between 480 and 540 °C (900 and 1000 °F) and cooled to room temperature in 1 h. The de-

crease in resistivity between 20 and 28% Cu is not significant; deviations of this amount can be expected from lot to lot of the same composition.

Chemical Properties

General corrosion behavior. At ordinary temperatures, the presence of copper in solid solution in silver will have little effect on the resistance of the metal to corrosion. The presence of small areas of the slightly less noble copper-rich phase might be expected to cause difficulty because of electrolytic effects, but apparently the difference between the potentials is small enough for the duplex alloys to behave satisfactorily in their usual applications. In seawater, however, and in similar electrolytes, some selective attack may be anticipated. At slightly elevated temperatures, copper oxidizes selectively. This behavior is of some consequence in electrical contacts, since it necessitates higher contact pressure. At approximately 600 °C (1100 °F), oxidizing atmospheres will cause rapid oxidation of the copper, and oxygen will diffuse to a considerable depth, forming a substance called "fire". One hour of exposure to air at this temperature will oxidize the 7.5% Cu alloy to a depth of 0.003 in. This was formerly very troublesome but with the production of well-deoxidized alloys and the use of nonoxidizing atmospheres, effects from the oxidation of copper have been minimized.

Resistance to specific corroding agents. Because sulfur tarnishes the silver-copper alloys in about the same way it tarnishes silver, sulfur must be excluded or the silver protected by appropriate coatings or wrapping if appearance is to be maintained without polishing.

Fabrication Characteristics

Processing. In melting silver-copper alloys, oxygen content should be brought to a low level before pouring at 1050 to 1100 °C (1920 to 2000 °F). Where the electrical conductivity is not important, final deoxidation with 0.025% P is convenient. Cadmium has been used as a partial deoxidizer; 0.5% or more is required. Lithium also is being used with success. Melting the material under a cover of broken graphite and pouring the alloy through a reducing flame during casting also gives good results. The alloy may be reduced approximately

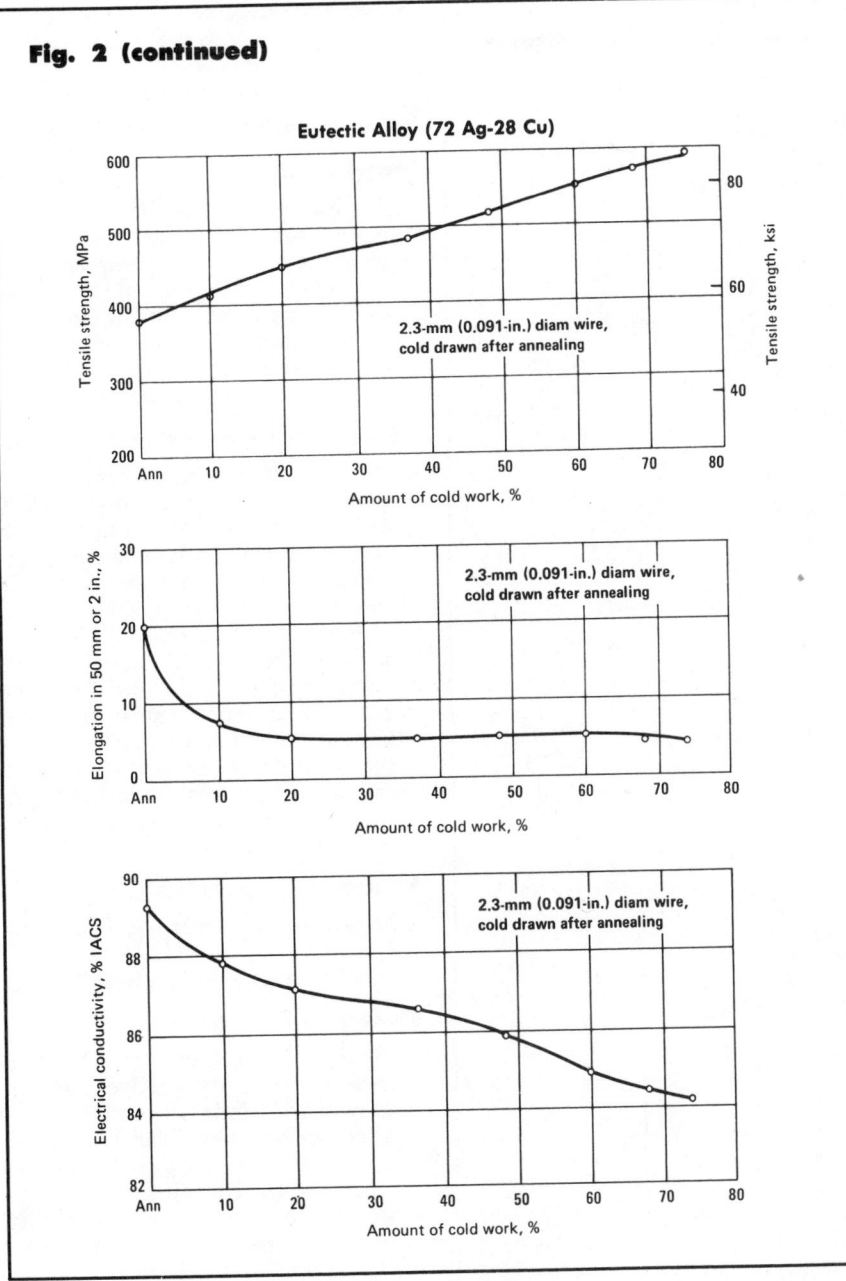

Fig. 2 (continued)

Eutectic Alloy (72 Ag-28 Cu)

2.3-mm (0.091-in.) diam wire, cold drawn after annealing

tation of the copper-rich phase and, if prolonged, increase the electrical conductivity considerably. Coin silver will remain duplex after any annealing treatment, and ages in much the same manner as the 7.5% Cu alloy. Both alloys respond to an aging treatment of 2 h at about 280 °C (535 °F) or 1 h at 300 °C (575 °F). The mechanical properties of coin silver are virtually the same as those of copper sterling after the usual annealing treatments at about 650 °C (1200 °F), because the composition of the silver-rich phase will be the same. Alloys containing 20 to 30% Cu have much more of the copper-rich phase and show less age hardening. In practice, relatively little deliberate use is made of the precipitation-hardening phenomenon in the silver-copper alloys. The solution temperature 700 to 730 °C (1300 to 1350 °F) is rather close to the solidus temperature 780 °C (1435 °F) and requires better temperature control than is available to many artisans who work with these alloys. In alloys heavily deoxidized with phosphorus, incipient melting in the grain boundaries may occur at 700 to 730 °C (1300 to 1350 °F) and when this happens the piece is likely to crack during quenching. Furthermore, these alloys are extremely soft at the solution temperature and are easily damaged. On the other hand, when soldering is done, the metal surrounding the joint may be heated to the solution temperature, and air cooling will cause some hardening. This counteracts the softening that would otherwise result from the soldering of work hardened metal.

60% in rolling, and short anneals are suitable at 540 to 650 °C (1000 to 1250 °F) in a steam atmosphere or a salt bath. Where the higher temperature is used, quenching is required for producing full softness. Alternation of oxidizing and reducing atmospheres is very damaging. Where light oxidation has occurred, pickling in a hot sulfuric acid solution (5 to 10%) is suitable. Heavy oxidation or "fire", can be eliminated only by removing considerable metal, either mechanically or chemically.

Heat treatment. Sterling silver can be age hardened without difficulty, and part of the merit of this composition in providing acceptable properties after miscellaneous treatments results from some hardening on cooling in air. The solubility of copper in silver at 650 °C (1200 °F) is about 4%, and at 730 °C (1350 °F) about 6%, so sterling silver processed at these temperatures is duplex with small amounts of the copper-rich phase scattered through the silver-rich matrix. Aging treatments cause precipi-

Silver-base Brazing Alloys

Compiled by C. D. Coxe (deceased), A. S. McDonald and G. H. Sistare, Jr. Handy & Harman

Commercial Names

Common name. Silver brazing alloys
Former name. Silver solders

Fig. 3 Effect of copper content on properties of silver-copper alloys

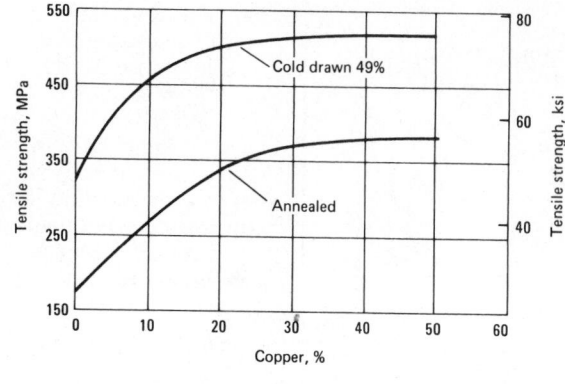

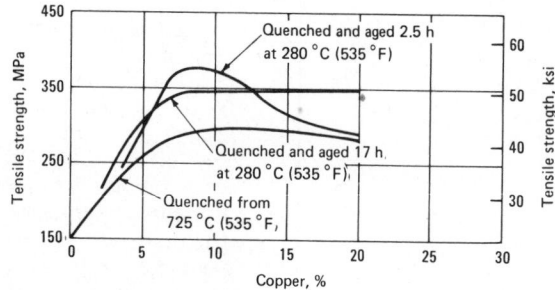

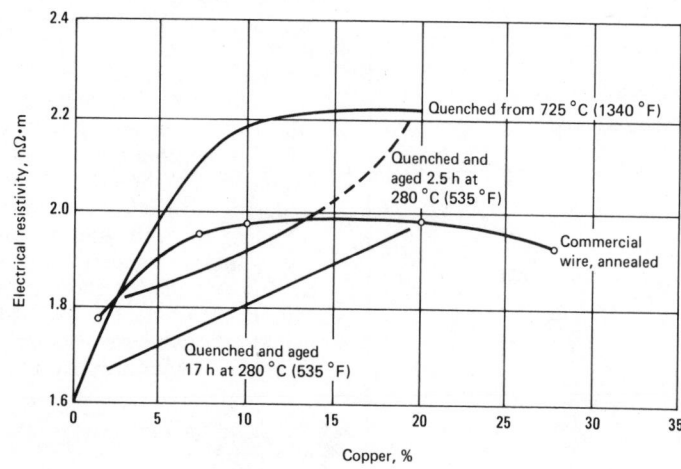

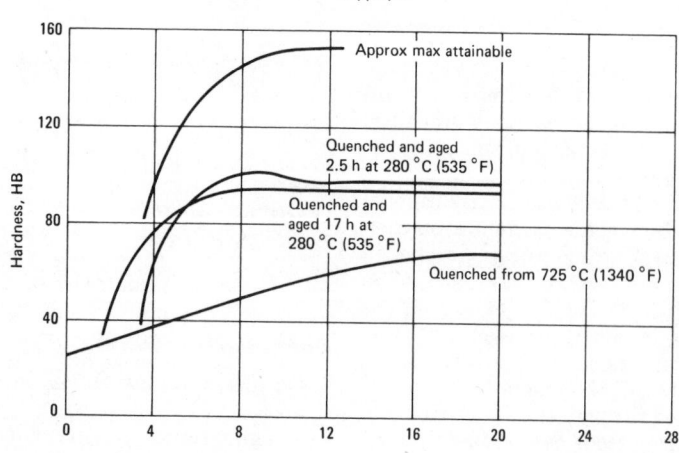

Specifications

AWS. A5.8

Chemical Composition

Composition limits. See Table 1.

Applications

Typical uses. Filler metal for brazing copper, nickel and cobalt alloys, tool steels, stainless steels and precious metals. Filler metal for brazing carbide tips onto cutting tools, or wear resisting tips made of tungsten or molybdenum onto copper alloy resistance welding electrodes

Mechanical Properties

Tensile properties. Typical. Tensile strength (approximate range), 280 to 410 MPa (40 to 60 ksi). The strength of silver brazing alloys declines rapidly at elevated temperatures. Short-time tests on the alloy 50Ag-15.5Cu-16.5Zn-18Cd indicate that the loss in strength at 220 °C (400 °F) will approximate 20 to 30% of the strength at room temperature and 50% at 280 °C (500 °F). The alloys that contain 5% or less of zinc and cadmium have better elevated-temperature strengths than those high in zinc and cadmium.

Thermal Properties

Liquidus temperature. See Table 1.
Solidus temperature. See Table 1.

Electrical Properties

Electrical conductivity. The electrical conductivity of silver brazing alloys varies from 10 to 80% IACS. Alloys with higher silver content and lower zinc content have the highest conductivity. The silver-copper eutectic (72Ag-28Cu) has conductivity of approximately 77% IACS. The high-silver, low-zinc Ag-Cu-Zn alloys also find some use in electrical contacts.

Chemical Properties

General corrosion behavior. The corrosion resistance of silver-base brazing alloys is better than that of most of the nonferrous-base metals alloy with which they are used.

Fabrication Characteristics

Formability. Silver-base brazing alloys are malleable and ductile and can be fabricated into sheet and wire with 50% or greater reductions between anneals.

Table 1 Silver-based brazing alloys

AWS classification	Composition(a), % Ag	Cu	Zn	Others	Solidus temperature °F	°C	Liquidus temperature °F	°C	Brazing temperature °F	°C
BAg-1	44.0-46.0	14.0-16.0	14.0-18.0	23.0-25.0 Cd	1125	607	1145	618	1145-1400	618-760
BAg-1a	49.0-51.0	14.5-16.5	14.5-18.5	17.0-19.0 Cd	1160	627	1175	635	1175-1400	635-760
BAg-2	34.0-36.0	25.0-27.0	19.0-23.0	17.0-19.0 Cd	1125	607	1295	702	1295-1550	702-843
BAg-2a	29.0-31.0	26.0-28.0	21.0-25.0	19.0-21.0 Cd	1125	607	1310	710	1310-1550	710-843
BAg-3	49.0-51.0	14.5-16.5	13.5-17.5	15.0-17.0 Cd, 2.5-3.5 Ni	1170	632	1270	688	1270-1500	688-816
BAg-4	39.0-41.0	29.0-31.0	26.0-30.0	1.5-2.5 Ni	1240	671	1435	779	1435-1650	779-899
BAg-5	44.0-46.0	29.0-31.0	23.0-27.0	· · ·	1250	677	1370	743	1370-1550	743-843
BAg-6	49.0-51.0	33.0-35.0	14.0-18.0	· · ·	1270	688	1425	774	1425-1600	774-871
BAg-7	55.0-57.0	21.0-23.0	15.0-19.0	4.5-5.5 Sn	1145	618	1205	652	1205-1400	652-760
BAg-8	71.0-73.0	Rem	· · ·	· · ·	1435	779	1435	779	1435-1650	779-899
BAg-8a	71.0-73.0	Rem	· · ·	0.25-0.50 Li	1410	766	1410	766	1410-1600	766-871
BAg-13	53.0-55.0	Rem	4.0-6.0	0.5-1.5 Ni	1325	718	1575	857	1575-1775	857-968
BAg-13a	55.0-57.0	Rem	· · ·	1.5-2.5 Ni	1420	771	1640	893	1600-1800	871-982
BAg-18	59.0-61.0	Rem	· · ·	9.5-10.5 Sn, .025 max P	1115	602	1325	718	1325-1550	718-843
BAg-19	92.0-93.0	Rem	· · ·	0.15-0.30 Li	1435	779	1635	891	1610-1800	877-982
BAg-20	29.0-31.0	37.0-39.0	30.0-34.0	· · ·	1250	677	1410	766	1410-1600	766-871
BAg-21	62.0-64.0	27.5-29.5	· · ·	5.0-7.0 Sn, 2.0-3.0 Ni	1275	691	1475	802	1475-1650	802-899

(a) Total maximum allowable impurities in each alloy is 0.15%.

Silver-Magnesium-Nickel Alloys Ag-0.25Mg-0.2Ni

Compiled by C. D. Coxe (deceased), and
L. Godfrey
Handy & Harman

Chemical Composition

Approximate composition. Type A: 0.24% Mg, 0.2% Ni. Type B: 0.20% Mg, 0.2% Ni

Applications

Typical uses. Used where high electrical and thermal conductivity are desired together with hardness that will not be affected by annealing during brazing or soldering, or by high service temperatures. Typical uses include: (a) electrical contacts that are to be affixed by brazing without loss of hardness, (b) high-thermal-conductivity spring clips for miniature vacuum tubes, (c) instrument and relay springs requiring high electrical conductivity or operation at high ambient temperature, and (d) electrical parts requiring drastic forming and subsequent hardening.

Precautions in use. These alloys have low creep rates (about 1/10 that of silver), and under creep conditions they fracture with negligible elongation. In elevated-temperature service, failure will usually occur by breaking rather than by relaxation or extension. The softer grade, type B, is less troublesome in this respect; it is less brittle. It is the preferred grade except where maximum hardness is required.

Mechanical Properties

Tensile properties. Typical. Tensile strength: type A, 470 to 510 MPa (68 to 74 ksi); type B, 400 to 470 MPa (58 to 68 ksi). Yield strength: type A, 360 to 400 MPa (52 to 58 ksi); type B, 310 to 380 MPa (45 to 55 ksi). Elastic limit: type A, 300 to 330 MPa (43 to 48 ksi); type B, 260 to 290 MPa (38 to 43 ksi). Elongation: type A, 2 to 12% in 50 mm or 2 in.; type B, 9 to 21% in 50 mm or 2 in.

Hardness. Type A, 160 to 180 HK; type B, 140 to 160 HK

Elastic modulus. Tension: type A, 83 GPa (12×10^6 psi); type B, 83 GPa (12×10^6 psi)

Creep-rupture characteristics. See Table 2.

Electrical Properties

Electrical conductivity. Volumetric: types A and B, 75% IACS at 20 °C (68 °F)

Chemical Properties

General corrosion behavior. Similar to fine silver

Fabrication Characteristics

Formability. Silver-magnesium-nickel alloys, when heated in air or oxygen, harden by internal oxidation. The magnesium, originally present in solid solution, is precipitated as submicroscopic MgO by oxygen diffusing from the surface inward at a faster rate than the magnesium will diffuse outward. The nickel, present as dispersed particles, is added to inhibit grain growth at the hardening temperature. In the annealed condition, these alloys are quite soft, slightly harder than fine silver, and are readily formable. Hardening by oxidation more than doubles the strength, and the hardened alloys cannot thereafter be restored to their original condition by annealing. Long-time heating at a high temperature in a reducing environment will cause some softening. These alloys may be hardened at

Table 2 Rupture stress of oxidation-hardened Ag-Mg-Ni alloys

Temperature °C	°F	Rupture stress, MPa (ksi) 2 h	100h
Type A			
263	500	200(30)	150(22)
399	750	130(19)	100(14)
538	1000	62(9)	40(6)
816	1500	7(1.0)	5.5(0.8)
Type B			
263	500	210(31)	180(26)
399	750	130(19)	80(12)
538	1000	70(10)	40(6)

temperatures between 650 and 800 °C (1200 and 1475 °F) in air (not combustion gases). The lower temperatures require longer times. Strip 0.4 mm (0.015 in.) thick requires 1 h at 745 °C (1375 °F) to harden through; 0.8-mm (0.030-in.) strip requires 4 h. Hardening time varies as the square of the thickness.

Dental Amalgam 50Hg-34.5Ag-13Sn-2Cu-0.5Zn

Compiled by R. M. Waterstrat
and
G. Dickson
National Bureau of Standards

Commercial Names

Common name. "Silver" filling

Specifications

ANSI. MD 156.1
Government. For alloy to be mixed with mercury: U.S.–350a
Foreign. ISO R1559

Chemical Composition

Composition limits. 65 min Ag, 29 max Sn, 6 max Cu, 3 max Hg, 2 max Zn
Consequence of exceeding impurity limits. Deviation from the above composition limits may produce excessive dimensional changes, reduced strength, and/or poor corrosion resistance.

Applications

Typical uses. Restoring lost tooth structure

Precautions in use. Small amounts of moisture (from the hand or mixing equipment) added during amalgamation cause excessive expansion of alloys containing zinc. Excess mercury in the alloy after packing causes the alloy to expand and flow or creep excessively. Excessive working of the alloy during amalgamation or packing reduces or eliminates setting expansion. Improper condensation (packing) increases the rate of corrosion and decreases physical properties.

Mechanical Properties

Tensile properties. Tensile strength, 10 to 17% of compressive strength
Compressive properties. Compressive strength, 280 to 350 MPa (40 to 50 ksi) after 5 days
Hardness. 90 HK
Elastic modulus. Tension, 60 GPa (8.7×10^6 psi)
Creep-rupture characteristics. All amalgams flow or creep when subjected to loading. Loads of 10% of the compressive strength may result in a creep rate of 2 μm/m·s at room temperature

Mass Characteristics

Density. 11 Mg/m^3 (0.397 lb/in.^{3})
Solidification shrinkage. Expands or contracts 0 ± 0.2% during hardening by a diffusion reaction at room temperature

Thermal Properties

Solidus temperature. Begins to sweat mercury and to break down at about 75 °C (167 °F)

Coefficient of thermal expansion. Linear, 22 to 28 μm/m·K (12 to 16 μin./in.·°F) near body temperature

Electrical Properties

Standard electrode potential. –0.5 V vs gold in normal sodium chloride solution

Optical Properties

Color. Silvery white

Chemical Properties

General corrosion behavior. Slight tarnishing or corrosion may occur in the oral environment. Electrolytic corrosion and pitting may result from contact with other metals.
Resistance to specific corroding agents. The amalgam is attacked readily by inorganic acids.

Fabrication Characteristics

Alloying. Amalgam is made by mixing at room temperature, 5 parts of the 65% Ag alloy (as particles) with 5 to 8 parts of mercury. Sufficient mercury is squeezed out during packing to reduce the mercury content to approximately 50%. Contamination with moisture must be avoided during amalgamation.
Compacting pressure. 10 to 30 MPa (1.5 to 4.5 ksi)
Forming. May be formed by packing and by carving with suitable tools or instruments while in the plastic form before hardening occurs. After hardening, the alloy is rather brittle and can be shaped by grinding.

Properties of Gold and Gold Alloys

Commercial Fine Gold

Compiled by J. A. Bard
Matthey Bishop, Inc.

Commercial Names

Common name. Called "proof gold" if more than 99.99% Au

Applications

Typical uses. The usual grade of refined gold contains from 99.95 to 99.98% Au and is suitable for most purposes, including dental and jewelry alloys. Metal that contains 99.5% Au is acceptable for international exchange and by the U.S. Mint without a refining penalty. Coin gold containing only copper as a hardener with 89.9 to 91.7% Au may also be acceptable.

Gold of high purity is employed for decorative and dental uses, for surfacing china and glass, as a thin film on glass for selective light filters stable over a wide range of temperature, for thermal limit fuses to protect electric furnaces, as a target in x-ray apparatus, as a freezing-point standard, as a high-melting solder to produce vacuum-tight pressure welds, for the lining of chemical equipment, and clad on phosphor bronze or nickel silver for contact springs in radio-frequency circuits.

Besides decorative uses and for infrared reflectors, electroplated gold has wide electrical applications in waveguides to provide a coating resistant to corrosion and tarnishing, on grid wires to suppress secondary emission, on variable-resistor terminals to give low-noise internal contact, for adhesion and flexibility of coating on vibrating and flexing components; on contacts for low and stable contact resistance, low cathode-glow discharge and capacitive-current weight loss and low-rms noise voltage. It is also used as a stop-off in electroplating. Gold is evaporated or sputtered onto selected areas of solid state electronic devices such as silicon transistors and integrated circuit chips to provide electrical terminals for these devices, and onto which small diameter fine gold wires may be thermo-compression bonded for electrical connection to the lead frames or other external circuits. Gold and silicon, or germanium, make low melting eutectics and this may be done *in situ* simply by heating pure gold in contact with silicon to produce a solder that bonds the semiconductor to its base, or other terminals. The low melting gold tin eutectic (prealloyed) is also used for similar purposes. Other solders may include gold together with antimony for "N" type semiconductors and indium for "P" type.

Precautions in use. Since January 1, 1975, a license is not required to buy or sell gold, but future transactions should take into account any changes in federal regulations. In melting, avoid contamination with base metals, particularly lead, bismuth and the like. Keep atmosphere oxidizing during melting. Avoid contact with hydrochloric acid containing free chlorine; aqua regia; concentrated sulfuric acid containing oxidizing agents; arsenic and phosphoric acids; and alkali cyanides, particularly in the presence of oxygen.

Mechanical Properties

Tensile properties. See Table 1.
Hardness. See Table 1.
Poisson's ratio. 0.42 (form not known)
Elastic modulus. See Table 1.
Fatigue strength. 31.7 MPa (4.6 ksi) at 10^7 cycles of reversed bending

Table 1 Mechanical properties of proof gold (99.99 + % Au)

Condition	Tensile strength		Yield strength (0.2% offset)		Elonga-tion(a), %	Hard-ness, HB	Modulus of elasticity	
	MPa	ksi	MPa	ksi			GPa	10⁶ psi
Cast.....................	125	18	...	...	30	33	74.5	10.8
Wrought, annealed.........	130	19	Nil	Nil	45	25	79.9	11.6
60% reduction	220	32	205	30	4	58	79.3	11.5
(a) In 50 mm or 2 in.								

Mass Characteristics

Density. 19.32 Mg/m³ (0.698 lb/in.³) at 20 °C (68 °F)

Thermal Properties

Melting point. 1064 °C (1948 °F)
Coefficient of thermal expansion. Linear, 14.2 μm/m·K (7.9 μin./in.·°F) at 20 °C (68 °F)
Specific heat. 131 J/kg·K (0.0312 Btu/lb·°F) at 18 °C (64 °F)
Thermal conductivity. 300 W/m·K (170 Btu/ft·h·°F) at 0 °C (32 °F)

Electrical Properties

Electrical conductivity. Volumetric, 73.4% IACS at 20 °C (68 °F)
Electrical resistivity. 21.9 nΩ·m at 0 °C (32 °F), 23.5 nΩ·m at 20 °C (68 °F)
Relative attenuation. 1.19 (copper = 1); in waveguide 10.16 × 22.86 mm (0.400 × 0.900 in.) ID (λ = 32 mm), 0.139 db/m; in waveguide 4.32 × 10.67 mm (0.170 × 0.420 in.) ID (λ = 12.5 mm), 0.6 db/m. *Skin depth:* λ = 10 mm, 0.45 μm (18 μin.); λ = 100 mm, 1.43 μm (57 μin.); λ = 1 m, 4.53 μm (180 μin.). *Noise voltage:* 0.6 μV rms at 0.5 to 200 Hz (gold ring, graphite brush; pressure, 1.08 kPa; speed, 0.35 m/s). *Contact erosion:* cathode glow discharge in air: weight loss, 0.886 (platinum = 1). Capacitive current: weight loss, 1.14 (platinum = 1)

Fabrication Characteristics

Formability. Suited to forming by all methods
Weldability. Torch braze with silver solder, no flux, any flame; oxyacetylene weld with gold, no flux, any flame; resistance weld by any method
Annealing temperature. 300 °C (575 °F), but usually no annealing is required
Hot working temperature. Can be worked at any temperature below the melting point
Casting temperature. 1100 to 1300 °C (2000 to 2370 °F)

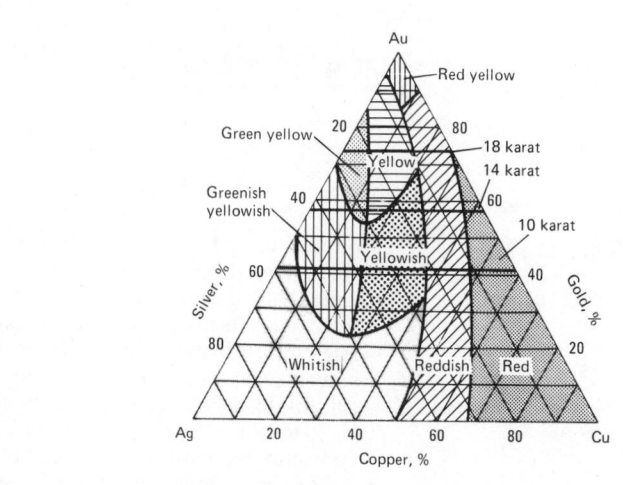

Fig. 1 Color chart for gold-silver-copper alloys for jewelry and dental applications

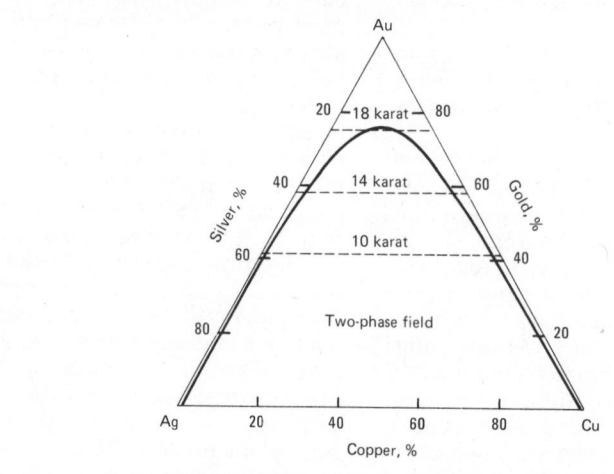

Fig. 2 Isothermal section of the gold-silver-copper ternary phase diagram at 371 °C (700 °F)

Fig. 3 Variation of hardness with silver content for gold-silver-copper alloys

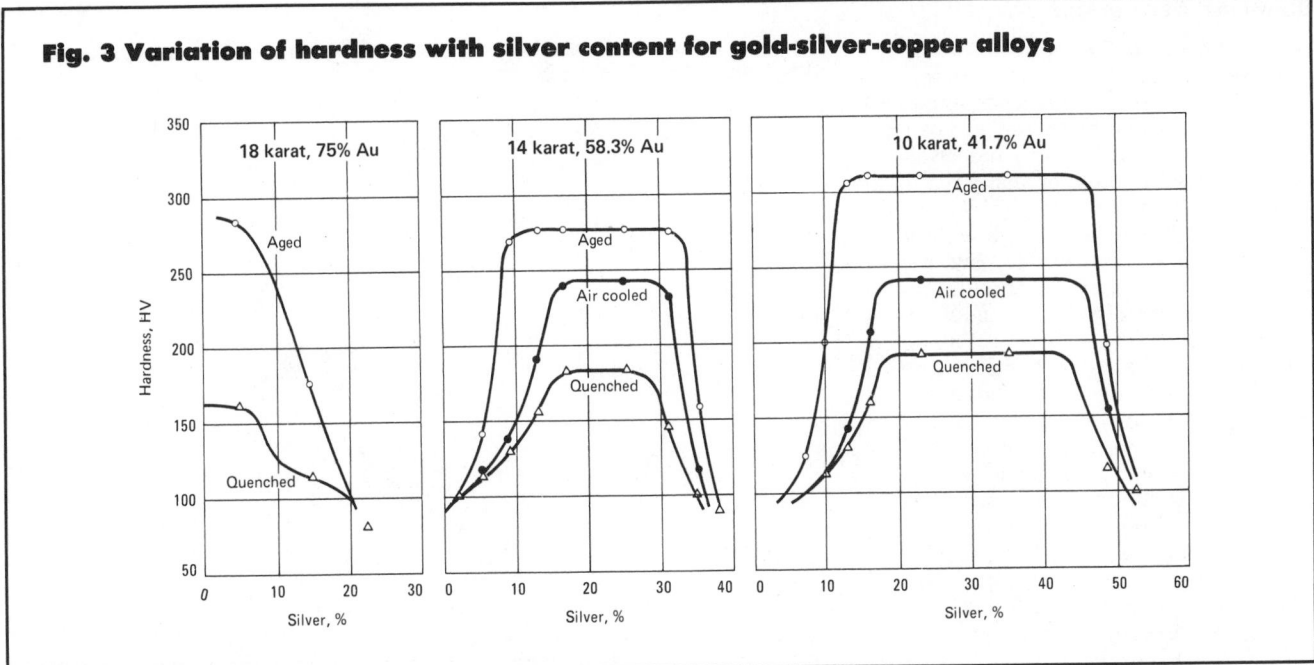

Gold-Silver-Copper Alloys

Compiled by George H. Sistare, Jr. and
Allen S. McDonald
Handy & Harman

Commercial Names

Common names. Green, yellow and red golds

Chemical Composition

Alloy types. Most of the commercially important colored alloys for jewelry and dental applications are based on the gold-silver-copper system (see Fig. 1)—frequently modified by the addition of zinc, and sometimes of nickel, for jewelry alloys, and of palladium and platinum for dental alloys.

In the ternary-phase diagram, the two-phase field of the silver-copper system extends well in toward the gold corner of the diagram (see Fig. 2). Alloys in the single-phase solid solution area on both the silver-rich and copper-rich sides are generally soft and not hardenable, except for the order-hardening gold-copper alloys containing approximately 75% Au. Two-phase alloys near the single-phase limit at 370 °C (700 °F) are quite soft when annealed and may be precipitation hardened by solution annealing, quenching, and aging at 260 to 315 °C (500 to 600 °F). Alloys lying farther into the two-phase region are harder in the annealed condition.

Figure 3 shows the effect of composition on hardness of gold-silver-copper alloys at three karat levels. The colors indicated in the triangular diagram may be modified by additions of other metallic elements. Zinc is frequently added to gold-silver-copper alloys: (a) as a deoxidizer, (b) to lighten the color (it makes reddish alloys more yellow), (c) to lessen the hardening that may occur on air cooling and (d) to lower the melting temperature for gold solder.

Where it is desirable to reduce the grain size of cast gold-silver-copper alloys, fractional percentages of iridium or rhodium, plus ruthenium, have been used, particularly in the dental field, to refine the structure. In the wrought jewelry alloys, additions are occasionally desired to reduce the rate of grain growth, and a very small amount of cobalt or, less desirably, iron can be used to accomplish this; nickel has some effect in this direction but its solubility is relatively high, particularly in the low-silver alloys. The addition of considerable percentages of nickel lightens the color and increases the solid-solution hardness. Iron may cause inclusions, and cobalt and nickel will form low-solubility phases with some deoxidizers.

The 18 karat gold-silver alloy has a good green color but is too soft for general use, except as a finishing plate, while the red 18 karat gold-copper alloy is troublesome to work because of ordering transformation in the solid state. The 18 karat gold alloys of the gold-silver-copper type are yellow in color. A wider range of colors is available in the 14 and 10 karat alloys.

Because the properties of 10 and 14 karat alloys are controlled largely by the ratio of silver to copper, regardless of the gold content, alloys can be converted from one gold content to another gold content having similar properties, by addition of pure gold to low-karat alloys or by addition of standard base alloy to high-karat golds.

Applications

Typical uses. These alloys are used mostly in jewelry, but sometimes are used for slip rings and brushes on electrical instruments. Parts may be cast to shape, made from rolled or drawn stock, or made from clad material comprising a layer of gold on one or both sides of a core made of nickel silver, pure nickel, brass or bronze. Certain of these alloys can be electrodeposited, and sometimes coatings are produced by electrodeposition followed by heating to cause diffusion with the underlying metal.

Chemical Properties

Resistance to specific corroding agents. The 18 karat alloys contain 75% Au by weight, corresponding to about 50 to 60 at. % Au, and are very resistant to tarnishing in indoor atmospheres, although probably not to outdoor urban exposures. They resist nitric, sulfuric and hydrochloric acids, and ferric chloride, but are attacked rapidly by mixtures of nitric and hydrochloric acids (aqua regia) and slowly by solutions of alkali cyanides plus air or other oxidizing agents. They are attacked when made anodic in hydrochloric acid or cyanide solutions. Fourteen-karat alloys contain 58.33% Au by weight, or about 30 to 40 at. % Au, and thus approach the region where the nobility decreases rapidly. They tarnish slowly in air but because of the color this is not obvious. If free from stress, they are quite resistant to nitric acid or ferric chloride, but if highly stressed, these alloys may be cracked by various corrosives or may crack even in the apparent absence of a corrosive. Mercury is also likely to crack 14 karat alloys when they are under stress. Decreasing the gold content still further to 12 and 10 karats (about 20 at. % Au) results in a rather sharp decrease in resistance to acids and an increase in the susceptibility to tarnish and sensitivity to stress-corrosion cracking. The resistance to corrosion of some of the alloys may be lowered, or occasionally raised, by aging.

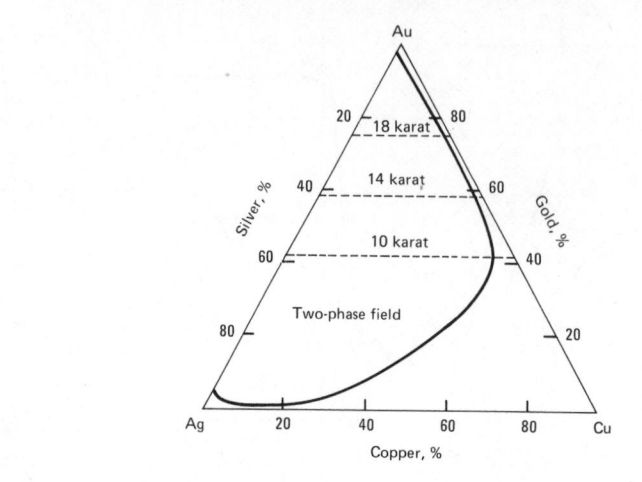

Fig. 4 Isothermal section of the gold-copper ternary phase diagram at 316 °C (600 °F)

Gold-Nickel-Copper Alloys

Compiled by George H. Sistare, Jr.
and
Allen S. McDonald
Handy & Harman

Commercial Names

Common name. White gold

Chemical Composition

Alloy types. Historically, white golds developed from a patented gold-nickel-zinc alloy containing about 80% of gold that was offered as a substitute for platinum jewelry. In extending the concept to conventional karat levels, it was immedi-

Table 2 Typical compositions and fire cracking tendencies of white golds

Alloy	Historical sequence	Au	Ni	Cu	Zn	Fire cracking tendency
18 kt	Early	75	17.30	2.23	5.47	Marked
14 kt	Early	58.33	15.17	18.04	8.46	Marked
14 kt	Intermediate	58.33	10.82	22.08	8.77	Moderate
14 kt	Modern	58.33	12.21	23.47	5.99	Slight
10 kt	Early	41.67	21.24	25.25	11.84	Marked
10 kt	Intermediate	41.67	15.12	30.96	12.25	Moderate
10 kt	Modern	41.67	17.08	32.85	8.40	Slight

ately discovered that some copper was essential for workability at 18 karat. Subsequently, it was found that sizeable amounts of copper were essential if 14 and 10 karat alloys were to be worked at all. Modern white golds are based on alloys in the gold-nickel-copper system to which 5 to 12% zinc is added.

The gold-nickel-copper system is similar to the gold-silver-copper system in that both are dominated by a two-phase immiscibility gap in the solid state ternary field. In the gold-nickel-copper system, the immiscibility gap in the gold-nickel binary system extends into the ternary field. In the gold-silver-copper system, the silver-copper binary eutectic generates

an immiscibility gap that extends into the ternary. Note however, that in one case (gold-nickel-copper), the two-phase field is opposite the copper corner, while in the other (gold-silver-copper), the two-phase field is opposite the gold corner. Compare the gold-nickel-copper isothermal section in Fig. 4 with the gold-silver-copper isothermal section in Fig. 2. It follows that in constant karat pseudo-binary sections (i.e., sections at fixed gold content) the two-phase field is asymmetrical in the gold-nickel-copper system and symmetrical in the gold-silver-copper system. Thus, white golds based on fixed nickel-copper ratios are dissimilar at different karat levels, whereas yellow golds based on fixed silver-copper ratios, by and large, are readily converted from one karat to another.

Gold-nickel-copper-based white golds work harden faster and are harder after annealing than gold-silver-copper-based yellow golds. All gold-nickel-copper-based white golds lie within the two-phase region at room temperature. The 18 and 14 karat alloys can be homogenized at elevated temperatures. The 10 karat alloys do not homogenize with any practical heat-treatment. The 18 karat alloys can be age hardened but the 14 or 10 karat alloys are not age hardenable modified with an addition of cobalt. It takes about 0.75% Co at 14 karat and 1.75% Co at 10 karat. In both instances, the cobalt is substituted for copper. These modifi-

cations are mostly used as casting alloys and usually can be aged as cast without first solution annealing.

Gold-nickel-copper-based white golds are very susceptible to fire cracking. They firecrack when they are given a full anneal after light cold working (reductions of less than 50%). Expedients such as stress relieving prior to annealing are ineffective in preventing firecracking. White gold compositions are the result of a compromise between color and fire cracking tendency. Increasing the copper content reduces the fire cracking tendency, but offsets the whitening effect of nickel. This can be compensated for with the addition of zinc, which tends to decolorize the copper and enhance the whitening effect of nickel. Unfortu-

nately, it also enhances the fire cracking tendency. This is illustrated in Table 2, which lists typical compositions of gold-nickel-copper-zinc white golds in order of decreasing karat. At 18 karat, little can be done to reduce fire cracking tendency; with 75% gold, any significant replacement of nickel with copper would result in an unacceptable color. At 14 karat, considerable amounts of copper can be added, and at 10 karat, even greater amounts of copper are tolerable. Note in the "Historical sequence" in Table 2 how the increase in copper and decrease in zinc contents have progressively led to a decrease in fire cracking tendency, while maintaining an acceptable color.

Gold-Platinum Alloy 70Au-30Pt

Compiled by D. J. Accinno
Engelhard Industries, Inc.

Applications

Typical uses. Because of its high corrosion resistance, this alloy is used for spinnerets in the numerous methods of rayon production that employ highly corrosive chemicals. The alloy is also used as a high-melting-point platinum solder.

Mechanical Properties

Tensile properties. Typical. Tensile strength, 639 MPa (92.7 ksi); reduction in area, 50%. See also Table 3.

Hardness. See Table 3.

Table 3 Typical mechanical properties of gold-platinum alloy (70Au-30Pt)

Condition	Proportional limit MPa	ksi	Yield strength MPa	ksi	Hardness(a), HB
Annealed(b)	200	28.8	245	35.4	130
Annealed(c).........	...	...	...	...	114
Hard(d)	450	64.9	570	82.5	169

(a) 500-Kg10ad. (b) 1090 °C (2000 °F), air cooled. (c) 1000 °C (1830 °F), quenched. (d) 66% reduction.

Elastic modulus. Tension, 113.8 GPa (16.51×10^6 psi)

Mass Characteristics

Density. Annealed, 19.92 Mg/m^3 (0.720 lb/in.^{3}) at 20 °C (68 °F)

Thermal Properties

Liquidus temperature. 1450 °C (2642 °F)
Solidus temperature. 1228 °C (2242 °F)

Electrical Properties

Electrical resistivity. 340 nΩ·m (quenched); 220 nΩ·m (aged) at 20 °C (68 °F). Temperature coefficient, 0.0059 nΩ·m per K at 0 to 1200 °C (32 to 2190 °F)

Fabrication Characteristics

Joining. Braze with gold solder; no flux; any flame
Annealing temperature. 1090 °C (2000 °F) in air

Gold in Dentistry

By John P. Nielsen
Professor of Metal Science
Polytechnic Institute of New York;
Director of Research
J. F. Jelenko & Co.

DENTAL RESTORATION TECHNIQUES rely heavily on precious metals in general, and gold in particular, for filling of cavities and construction of orthodontic appliances, crowns, bridges and partials. Various material requirements are essential for alloys used in dentistry. Alloys must be nontoxic and biocompatible and must have sufficient nobility for the oral environment, sufficient fabricability for production of the desired shapes, and those physical properties necessary for good resistance to service stresses and wear. In addition, there are numerous special requirements that depend on the type and color of the appliance desired, the fabricating technique, and the skill of the dental technician and dentist. Examples include low melting temperature to facilitate melting with a gas-air torch, good solderability for components to be joined by soldering, good ductility for adjustment to final fit and for hand-tool burnishing, good ceramic-to-metal bonding properties for appliances made by the ceramic-metal technique, and good dimensional stability. For the most part, gold color is preferred over platinum color.

* The alloy numbers in Tables 3, 4, 6 and 7 are only for purposes of identification in this article and do not have significance in any specification.

Direct Filling of Cavities (Ref 1)

Pure gold foil and mat gold still are used for direct filling of cavities by dentists skilled in use of these materials. Stronger fillings can be obtained by use of platinum foil sandwiched between layers of gold foil, but this combination of materials requires even more skill than gold foil alone. Recently, powdered gold produced by chemical precipitation or atomization also has been used for direct filling of cavities. The physical properties of golds for direct filling are given in Table 1.

Wrought Alloys (Ref 2)

Wire and foil made of gold and other precious metals are used in dentistry mainly for fabricating orthodontic appliances. American Dental Association (ADA) Specification No. 7 designates two grades of wire—one high in precious-metal content and one low (see Table 2). Composition ranges and limits (and colors) of eight precious-metal dental wire alloys* are given in Table 3. Table 4 presents required physical and mechanical properties of the alloys whose compositions are given in Table 3.

Cast Alloys (Ref 1, 2)

By far the largest use of precious metals in dentistry is in restorations involving metal casting. Such applications include cover inlays, crowns, multiple-unit bridges and partials. The lost wax method of molding is universally used; most melting is done by torch, but use of resistance and induction melting is increasing. Castings are small, ranging in weight from 1.5 to 40 g (1 to 25 pennyweight).

Each wax pattern is made individually and is used for making only one casting. No casting shrinkage allowance is built into the pattern, and thus a mold investment that expands for this compensation (about 1.25% in its

Table 1 Properties of golds for direct filling

Material	Density, Mg/m³	Tensile strength		Hardness, HK
		MPa	ksi	
Gold foil	15.8 to 15.9	269 to 292	1855 to 2015	69
Mat gold	14.3 to 14.7	159 to 166	1095 to 1145	52 to 62
Powdered gold	14.4 to 14.9	153 to 187	1055 to 1290	55 to 64

Table 2 Composition and minimum properties of wrought gold alloy wire for dentistry (a)

Alloy type	Gold plus platinum-group metals, % (min)	Tensile strength (oven-cooled wire) MPa	ksi	Yield strength (oven-cooled wire) MPa	ksi	Elongation in 50 mm or 2 in., % Quenched wire	Oven-cooled wire	Fusion temperature °C	°F
I-High precious metal	75	930	135	860	125	15	4	950	1740
II-Low precious metal	65	860	125	690	100	15	2	870	1600

(a) American Dental Association Specification No. 7. See *Guide to Dental Materials and Devices*, 5th Ed.; American Dental Association, Chicago, 1971.

Table 3 Compositions and colors of precious-metal alloys for high-strength dental wires

Alloy (a)	Au	Pt	Pd	Composition (b), % Ag	Cu	Ni	Zn	Color
1	25 to 30	40 to 50	25 to 30	...	...	...	...	Platinum
2	54 to 60	14 to 18	1 to 8	7 to 11	11 to 14	1 max	2 max	Platinum
3	45 to 50	8 to 12	20 to 25	5 to 8	7 to 12	...	1 max	Platinum
4	62 to 64	7 to 13	6 max	9 to 16	7 to 14	2 max	1 max	Light gold
5	64 to 70	2 to 7	5 max	9 to 15	12 to 18	2 max	1 max	Gold
6	56 to 63	5 max	5 max	14 to 25	11 to 18	3 max	1 max	Gold
7	10 to 28	25 max	20 to 37	6 to 30	14 to 21	2 max	2 max	Platinum
8	...	1 max	42 to 44	38 to 41	16 to 17	1 max	...	Platinum

(a) Numbers are for identification in this article only. (b) Fractional percentages of iridium, indium and rhodium are omitted here.

linear dimensions) is required. Numerous casting alloys are available, partly because of the wide range of requirements for the various types of restorations, and partly because increases and fluctuations in the prices of precious metals (especially since 1968) have made it necessary to consider cost to the consumer more carefully. Table 5 gives composition and hardness ranges for the four types of gold casting alloys designated by ADA Specification No. 5, while Table 6 gives composition ranges and limits for seven gold casting alloys that are representative of a majority of those currently on the market.

The first four alloys in Table 6 correspond to ADA types I, II, III and IV; they are listed in order of increasing hardness. The softest, alloy 9 (type I), is used for inlays subjected to minimal amounts of biting stress. Alloy 10 (type II) is used for inlays at the biting surface. Alloy 11 (type III) is a general-purpose alloy used for inlays as well as for crowns and bridges. Alloy 12 (type IV) is for larger appliances such as saddles, clasps, one-piece partial dentures and long-span bridges.

The copper and silver in these alloys are added for solid-solution, two-phase, and precipitation hardening, the copper being the more potent of the two. Platinum and palladium increase potential for precipitation hardening. Palladium preserves the nontarnishing characteristic as gold content is lowered and silver content is increased. Zinc is a deoxidizer. These alloys gener-

ally are inoculated (Ref 3) with metals such as iridium and ruthenium to decrease grain size of castings from 300 to 50 μm.

The metal-ceramic restoration technique has had a revolutionary effect on restorative dentistry. In this technique, which is described in the Coleman patent of 1961 (Ref 4), the Weinstein patent of 1962 (Ref 5) and the references cited in these patents (dating back to 1927), a metal crown or bridge modified to act as a substrate structure is prepared by the usual lost wax method, and the facing side is veneered with a tooth-simulating porcelain. This technique requires that the alloy have a coefficient of thermal expansion closely matching that of the porcelain, a high enough solidus temperature to permit the porcelain to bake at about 1040 °C (1900 °F), and no elements (such as copper) that bleed colored oxides into the molten porcelain during baking. Alloy 13 in Table 6 is the basic gold-colored alloy for the metal-ceramic technique. The combined palladium and platinum content serves simultaneously to raise the solidus temperature and lower the coefficient of thermal expansion. Alloy 14 also is a gold-base alloy for the metal-ceramic technique but is platinum in color; it is stronger than alloy 13 and hence better suited for long-span bridges. Most of the alloys for metal-ceramic restorations contain indium, tin and iron in small quantities, and these additions assist in hardening.

Alloy 15 is one of a new series of casting alloys—the so-called "borderline" or "low-gold" alloys for crowns and bridges and for harder inlays (Ref 6). The properties of these alloys are similar to those of ADA type III or type IV alloys (see Table 7). These low-gold alloys were introduced in response to the large increase in the price of gold in the early 1970's. When properly balanced in composition, they are tarnish resistant in the oral environment and are yellow in color. However, in low-gold alloys, the compositional balance required for dental tarnish resistance is delicate, and such alloys should be clinically evaluated for this property before being marketed as dental alloys.

All of the casting alloys discussed above, with the exception of type I and type II alloys, are markedly age hardenable. Appliances made of these age-hardenable alloys are easily adjusted for good fit while in the soft, quenched state. After fitting, appliances are given aging heat treatments that increase strength and hardness considerably and thus provide good resistance to service stresses and wear. (Because these alloys differ widely in response to heat treatment, details of heat treating schedules for specific alloys should be obtained from the manufacturer.) In some instances, these alloys have been cast in heated investment molds and aged by slow cooling in the mold. However, when this practice is followed, the soft, quenched state is bypassed, and adjustment for fit is much more difficult.

The mechanical properties given in Table 7 were determined by the methods adopted by the Technical Committee of the Dental Gold Institute (Ref 7).

Solders

Solders for dental appliances are alloys of gold, silver, copper, tin and zinc (see Table 8). Copper and silver are varied to control color and working characteristics. Some of the copper may be replaced by nickel to produce solders with colors approximately matching those of white or platinum-colored wire.

Dental alloys are almost always soldered in an open flame, and thus dental solders are formulated to ensure desirable properties under these conditions.

Applications of gold solders in dentistry are of two broad classes: orthodontia, and crown and bridge work. In orthodontia, the high-strength precious-metal wires previously described are joined. In crown and bridge work, parts made of 22-karat wrought gold and high-strength wires are assembled and soldered. In many instances, the solder not only serves to join the parts but also is used in bulk as part of the structure itself. Complicated appliances may require assembly in steps by successive soldering operations, making it necessary to use a series of solders with successively lower melting points.

Compositions. The gold contents of dental solders vary from 40 to 85%, but most of the solder used contains 58 to 65%. Lower-karat solders often are preferred for orthodontic soldering, because high temperatures may adversely affect the physical properties of the wires used in constructing appliances. Tarnishing and corrosion are not significant factors, because these appliances are not worn for extended periods of time. High-melting, usually high-karat, solders are required where stepwise assembly is practiced. High-karat solders also are useful where close matching of solder color with the color of high-karat gold is desired.

When gold, tin and zinc contents of a general-purpose or high-karat solder are held constant and copper and silver contents are varied, significant differences in working characteristics are obtained with only small changes in melting range. Changes in color also are produced; higher-copper alloys have rich goldlike colors, whereas higher-silver alloys are lighter and less like gold in appearance. Melting ranges of high-silver solders are narrower than those of high-copper solders. When flowed onto high-karat gold surfaces, such as ADA type I or type II dental castings, high-silver solders spread freely and cause minimum attack of the substrate. The high-copper solders, however, attach themselves to the substrate before they are entirely melted; on continued heating, they spread over the substrate, but also readily alloy with it and finally "burn through". Burnthrough results when the alloy formed by diffusion of the solder into the substrate melts at a lower temperature than the solder; thus, the surface of the solid substrate is melted away, and the substrate can be "burned through" in time. When high-karat gold alloys are soldered, the alloy formed has a melting temperature higher than that of high-silver solders and lower than that of high-copper solders. Spreading and burnthrough also result when the common dental-wire alloys are soldered.

The high-copper solders are useful when it is necessary to add material to build up a part that is deficient in size because of wear, accident or improper design.

When a general-purpose solder is used, composition is chosen so that neither ease of buildup nor ease of flow is overemphasized. Instead, a material is chosen that is a useful compromise in these properties.

Table 4 Physical and mechanical properties of high-strength precious-metal dental wires (a)

Alloy(b)	Tensile strength(c) Soft (d) MPa	ksi	Hard (e) MPa	ksi	Proportional limit Soft MPa	ksi	Hard MPa	ksi
1	860-1240	125-180	(g)	(g)	550-1035	80-150	(g)	(g)
2	760-895	110-130	1105-1280	160-186	495-705	72-102	895-1040	130-151
3	965-1035	140-150	1105-1170	160-170	760-825	110-120	895-965	130-140
4	620-795	90-115	825-1140	120-165	380-550	55-80	585-965	85-140
5	565-825	82-120	895-1140	130-165	365-505	53-73	710-960	103-139
6	580-690	84-100	660-1080	96-157	360-400	52-58	485-855	70-124
7	660-1020	96-148	1035-1325	150-192	415-795	60-115	760-1105	110-160
8	690-760	100-110	895-1170	130-170	435-600	63-87	740-875	107-127

Alloy(b)	Elongation, % (8-in. gage) Soft	Hardened	Hardness, HB(f) Soft	Hardened	Fusion temp (wire method) °C	°F	Specific gravity, Mg/m³
1	14-15	(g)	200-245	(g)	1500-1530	2730-2790	16.9-17.6
2	12-22	5-10	150-190	240-285	1005-1100	1840-2010	15.0-18.5
3	8-10	7-9	210-230	250-270	1065-1120	1950-2050	15.5-15.8
4	14-26	2-8	166-195	240-295	945-1015	1730-1860	14.5-15.6
5	14-20	1-3	135-200	230-290	900-930	1650-1710	14.1-15.2
6	20-28	1-2	138-170	220-280	875-900	1610-1650	13.7-14.0
7	9-20	1-8	150-225	180-270	940-1080	1725-1975	11.5-15.6
8	16-24	8-15	150-200	235-270	1045-1075	1910-1970	10.7-11.2

(a) See Table 3 for chemical compositions. (b) Numbers are for identification in this article only. (c) Tension tests on wires 1.0 mm (0.040 in.) in diam. Most elongation data on 8-in. gage lengths. (d) Quenched from 705 to 870 °C (1300 to 1600 °F) depending on type of alloy. (e) Cooled slowly and uniformly from 450 to 250 °C (840 to 480 °F) in 30 min. This is a severe hardening treatment used in testing to determine the behavior of wire under adverse conditions. Manufacturers recommend hardening treatments for specific uses. (f) Brinell hardness numbers obtained with the "Baby Brinell" testing machine. (g) Not appreciably affected by heat treatment.

REFERENCES

1. R. W. Phillips, *Science of Dental Materials*, 7th Ed., W. B. Saunders, Philadelphia, PA, 1973
2. E. M. Wise, *Gold*, D. Van Nostrand, Princeton, NJ, 1964
3. J. P. Nielsen and J. J. Tuccillo, Grain Size in Cast Gold Alloys, *Journal of Dental Research*, Vol 45, 1966, p 964-969
4. R. L. Coleman, P. S. Cecil and J. A. Kerpel, U.S. Patent No. 2 980 998, 1961
5. M. Weinstein et al, U. S. Patents No. 3 052 982 and 3 052 983, 1962
6. J. P. Nielsen and J. J. Tuccillo, U. S. Patents No. 3 424 577 (1969) and 3 767 391 (1973)
7. R. C. Brumfield, *Journal of the American Dental Association*, Vol 49, July 1954

Table 5 Composition and hardness ranges of gold casting alloys for dentistry (a)

Alloy	Gold plus platinum-group metals, % (min)	Hardness (b), HV	Au	Ag	Composition, % Cu	Pd	Pt	Zn
I (soft) 83		50 to 90	80.2 to 95.8	2.1 to 12.0	1.6 to 6.2	0.0 to 3.6	0.0 to 1.0	0.0 to 1.2
II (medium) 78		90 to 120	73.0 to 83.0	6.9 to 11.6	5.8 to 10.5	0.0 to 5.6	0.0 to 1.2	0.0 to 1.4
III (hard).............. 78		120 to 150	71.0 to 79.8	5.2 to 13.1	7.1 to 12.6	0.0 to 6.5	0.0 to 7.5	0.0 to 2.0
IV (extra hard) 75		greater than 150	62.1 to 71.9	8.0 to 17.1	8.6 to 15.1	0.0 to 10.1	0.2 to 8.2	0.0 to 2.7

(a) American Dental Association Specification No. 5. See *Guide to Dental Materials and Devices*, 5th Ed., American Dental Association, Chicago, 1971.

Table 6 Composition limits (by weight) of alloys used in dental castings

Alloy (a)	Type	Au	Ag	Cu	Pd	Pt	Zn
9	I(b)-soft................. 79 to 92.5		3 to 12	2 to 4.5	0.5 max	0.5 max	0.5 max
10	II(b)-medium .... 75 to 78		12 to 14.5	7 to 10	1 to 4	1 max	0.5
11	III(b)-hard .... 62 to 78		8 to 26	8 to 11	2 to 4	3 max	1
12	IV(b)-partial............ 60 to 71.5		4.5 to 20	11 to 16	5 max	8.5 max	1 to 2
13	Porcelain .... 83 to 88		0 to 1.3	...	4.5 to 6	4.0 to 16	...
14	Porcelain .... 49 to 52.5		12 to 16	...	25 to 30	...	...
15	Low gold................ 42 to 56		25 to 46.5	6.5 to 20	3.5 to 10	...	1

(a) Numbers are for identification in this article only. (b) American Dental Association Specification No. 5. See Table 5.

Table 7 Physical and mechanical properties of gold-alloy castings used in dentistry (a)

Alloy (b)	Treatment(c)	Hardness, HB	Tensile strength MPa	ksi	Proportional limit MPa	ksi	Elongation, %	Liquidus temperature °C	°F
9	Q 45-70		205-310	30-45	55-105	8-15	20-35	950-1050	1740-1920
10	Q 80-90		310-380	45-55	140-170	20-25	20-35	930-970	1705-1780
11	Q 95-115		330-395	48-57	160-205	23-30	20-25	950-1000	1740-1830
11	A 115-165		415-565	60-82	200-400	29-58	6-20	950-1000	1740-1830
12	Q 130-160		415-515	60-75	240-325	35-47	4-25	970-985	1600-1805
12	A 210-235		690-825	100-120	415-635	60-92	1-6	870-985	1600-1805
13	Q 155-175		460-515	67-75	415-485	60-70	4-6	1150-1190	2100-2175
13	A 175-190		...	...	...	...	...	1150-1190	2100-2175
14	Q 190-210		620-760	90-110	515-585	75-85	8-12	1240-1280	2265-2335
14	A 212-230		...	...	...	...	...	1240-1280	2265-2335
15	Q 160-190		485-620	70-90	345-485	50-70	20-35	900-940	1650-1725
15	A 180-200		550-895	80-130	415-825	60-120	1-3	900-940	1650-1725

Note: Modulus of elasticity of these alloys ranges from 76 to 125 GPa (11 to 18×10^6 psi). (a) See Table 6 for compositions of alloys. (b) Numbers are for identification in this article only. (c) Q, quenched; A, aged.

Table 8 Compositions and properties of three classes of precious-metal solders

Class	Composition (a) Au	Ag	Cu (b)	Hardness, HB (c)	Melting range (d), °C
Low-karat solder..............	45	30 to 35	15 to 20	140	816 to 691
General-purpose solder	60	12 to 22	12 to 22	110	835 to 724
High-karat solder	80	3 to 8	8 to 12	80	871 to 746

(a) Solders of all three classes contain 2 to 3% Sn and 2 to 4% Zn. (b) May be replaced in part by nickel (see text). (c) Values are for as-cast material and represent class averages. (d) From lowest solidus to highest liquidus.

Properties of Platinum and Platinum Alloys

Commercially Pure Platinum 99.95% Pt

Compiled by Edward D. Zysk
Engelhard Minerals & Chemicals
Corp.

Applications

Typical uses. Of the platinum group metals, platinum is the least rare, and it is the most widely used because of its general corrosion resistance, high melting point, appearance and ductility. Platinum of the highest purity is required for use in resistance thermometers and thermocouples. Various alloying elements such as rhodium, ruthenium and iridium and, for special purposes, other hardeners are employed to develop higher mechanical properties or to protect against special corrosion conditions. Platinum or its alloys are used for the cathodic protection of ship hulls, for electrical contacts, brushes, precision potentiometer wire, chemical production, laboratory ware, spinnerettes for synthetic fibers, anodes in both solid and clad form, and for jewelry. It is also used as a crucible liner for producing high purity optical glass or as a bushing in the extrusion of fiberglass. Platinum is an outstanding catalyst for oxidation as in the production of H_2SO_4 and HNO_3; for hydrogenation as in the production of vitamins and other chemicals; and in the petroleum reforming process as in the production of high octane gasolines. Certain organo-metallic compounds containing platinum have significant antitumor activity.

Mechanical Properties

Tensile properties. Typical. Annealed at 700 °C (1290 °F): tensile strength, 124 to 165 MPa (18 to 24 ksi); proportional limit, < 13.8 MPa

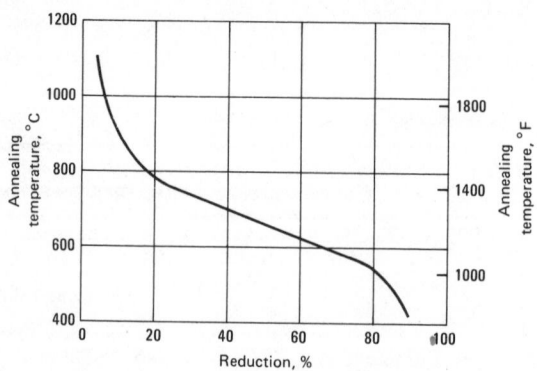

Fig. 1 Hardness of cold rolled and annealed platinum as a function of reduction during rolling (Ref 2)

Samples about 99.99% pure were cold rolled to various reductions, then annealed 15 min at temperatures indicated by the curve in the lower graph.

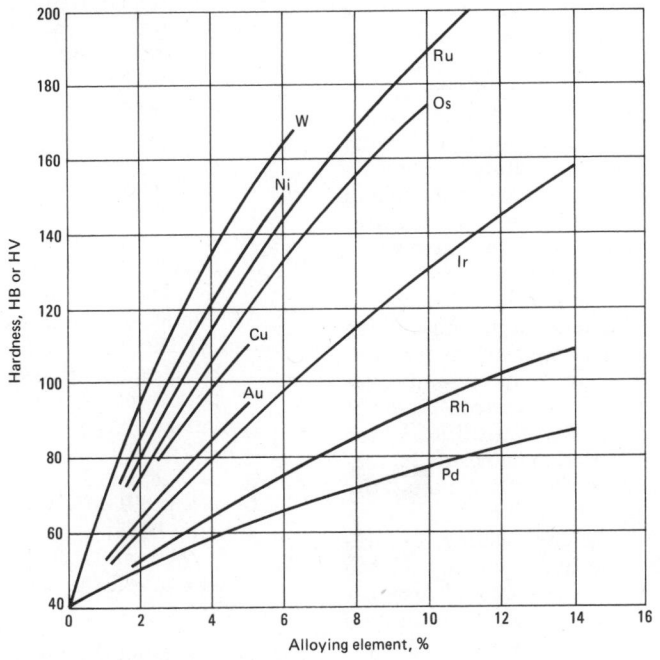

Fig. 2 Effect of various alloying additions on the hardness of annealed platinum

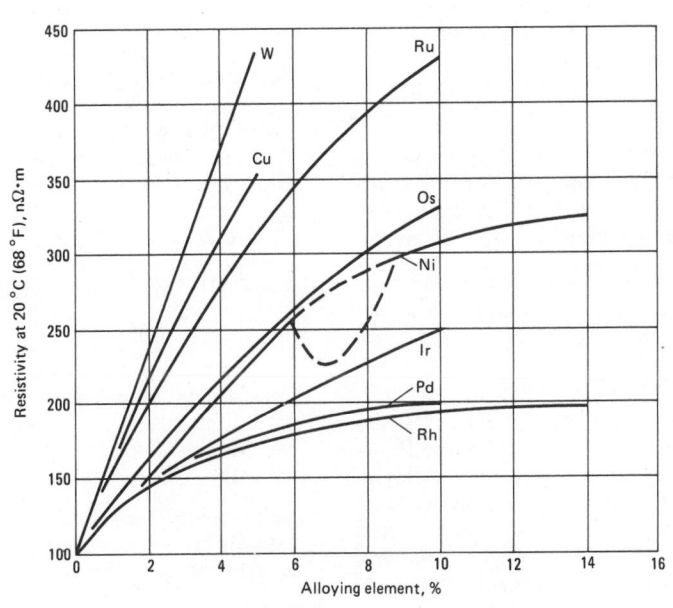

Fig. 3 Effect of various alloying additions on the electrical resistivity of platinum (Ref 9)

(< 2 ksi); elongation, 30 to 40% in 50 mm or 2 in. Hard drawn, 50% cold worked: tensile strength, 207 to 241 MPa (30 to 35 ksi); elongation, 1 to 3% in 50 mm or 2 in.

Effect of low temperature. Coarse grain material: 63.4 MPa (8.2 ksi) at room temperature; 283 MPa (41.0 ksi) at −195 °C (−317 °F); 565 MPa (82.0 ksi) at −253 °C (−425 °F). Fine grain material: 124 MPa (18.0 ksi) at 21 °C (70 °F); 448 MPa (65.0 ksi) at −195 °C (−317 °F) (Ref 1)

Effect of elevated temperature. Annealed thermocouple quality (about 99.99% pure): 143 MPa (20.7 ksi) at room temperature; 90 MPa (13.0 ksi) at 400 °C (750 °F); 55 MPa (8.0 ksi) at 800 °C (1470 °F); 34 MPa (5.0 ksi) at 1000 °C (1830 °F); 21 MPa (3.0 ksi) at 1200 °C (2190 °F). Tensile strength of 99.98% pure material of two grain sizes is given in Ref 1 for 20 K to 827 °C (−423 to 1521 °F)

Hardness. Annealed at 700 °C (1290 °F): 37 to 42 HV; hard drawn, 50% cold work: 90 to 95 HV. Electrodeposited: approx 600 HV. Effect of cold rolling, see Fig. 1. Effect of alloying, see Fig. 2.

Poisson's ratio. 0.39 (Ref 3)

Elastic modulus. At 20 °C (68 °F), annealed at 700 °C (1290 °F). Tension: static, 171 GPa (24.8 × 10⁶ psi); dynamic, 169 GPa (24.5 × 10⁶ psi). Hard drawn, 50% cold work, tension: static, 156 GPa (22.6 × 10⁶ psi)

Creep-rupture characteristics. For Pt and Pt-Pd alloys, see Ref 4.

Mass Characteristics

Density. 21.46 Mg/m³ (0.775 lb/in.³) at 25 °C (77 °F)

Thermal Properties

Melting point. 1769 °C (3217 °F) (Ref 5)

Coefficient of thermal expansion. Linear, 9.1 μm/m·K (5.1 μin./in.·°F) from 20 to 100 °C (68 to 212 °F) (Ref 6)

Specific heat. 132 J/kg·K (0.0314 Btu/lb·°F) at 0 °C (32 °F) (Ref 7)

Latent heat of fusion. 113 kJ/kg

Thermal conductivity. 71.1 W/m·K (41 Btu/ft·h·°F) at 0 °C (32 °F) (Ref 8)

Electrical Properties

Electrical resistivity. 98.5 nΩ·m at 0 °C (32 °F); 106 nΩ·m at 20 °C (68 °F). Temperature coefficient: 0.0039 per K from 0 to 100 °C (32 to

212 °F). Effect of alloying, see Fig. 3.

Optical Properties

Color. Silver white
Spectral reflectance. Bulk: 70.1% at 589 nm. Electrodeposited: 58.4% at 441 nm; 59.1% at 589 nm; 59.4% at 668 nm (Ref 10, 11, 12)

Chemical Properties

General corrosion behavior. See the article in this volume, "Corrosion Resistance of Precious Metals". See also Ref 13.
Resistance to specific corroding agents. Resistant to reducing or oxidizing acids at room temperature; attacked by aqua regia (a mixture of nitric and hydrochloric acids); attacked slowly by hydrochloric acid plus other oxidizing agents. Resistant to ferric chloride at room temperature; hydrobromic acid plus bromine attacks at room temperature. All of the free halogens attack at elevated temperatures; hydrochloric acid in the absence of oxidizing agents does not attack, and platinum is useful against this normally active gas up to 1090 °C (2000 °F). Sulfur dioxide does not attack even at 1090 °C (2000 °F) (Ref 14).

As an anode, platinum is outstanding and is used commercially in sulfuric and persulfuric acids, various sulfate-chloride plating electrolytes, and in chlorates with very little corrosion. If electrolyzed with alternating current, chlorides may attack, a characteristic exploited in etching platinum and platinum alloys.

Platinum is highly resistant to acid potassium sulfate, sodium carbonate, potassium nitrate at moderate temperatures and to sodium carbonate at 800 to 900 °C (1475 to 1650 °F) under nonoxidizing conditions. Although attacked vigorously by molten alkali cyanides and polysulfides, it is quite resistant to the normal sulfides plus alkali. Certain phosphates attack at high temperatures and care must be taken to avoid reducing conditions, particularly when compounds of arsenic, phosphorus, tin, lead or iron are present. Resistant to molten glasses, especially to those low in lead and arsenic. Platinum, even in the form of thin leaf, is resistant to corrosion and tarnishing on exposure to the atmosphere, including urban sulfur.

Fabrication Characteristics

Annealing. Annealing temperature depends on the purity of the material and amount of prior cold work. Fig. 1 shows the effect of reduction during rolling of 99.99% pure platinum; grain size after annealing platinum of this purity is determined almost entirely by prior reduction; virtually no grain growth occurs on the usual short anneals. It is probable, however, that platinum free from oxygen in solution will show grain growth after recrystallization and may have a still lower annealing temperature.

Air is the preferable atmosphere for annealing platinum; hot reducing atmospheres, particularly where silica, iron or easily reduced oxides are nearby, are almost certain to result in contamination. Annealing at too frequent intervals can result in substantial grain growth causing the "orange peel effect" during subsequent working or polishing. To prevent the formation of orange peel, the reduction in cross sectional area before annealing should not be less than 30%. Annealing for too long a length of time as well as at too high a temperature can result in thermal etching (grains of metal become clearly visible).
Precautions in working. The maintenance of oxidizing conditions throughout processing is essential to avoid contamination. To remove iron, pickling in hot hydrochloric acid after rolling and before annealing is essential for high-purity wire and sheet.

REFERENCES

1. R. P. Carreker, Jr., General Electric Research Lab Report 55-RL-1413, Schenectady, NY, 1955
2. E. Gruneisen, Annalen der Physik, Vol 25, 1908, p 825
3. W. Koster and J. Scherb, Zietschrift für Metallkunde, Vol 49, 1958, p 501
4. E. P. Sadowski, H. J. Albert, D. J. Accinno and J. S. Hill, "Stress Rupture Properties of Some Platinum and Palladium Alloys", AIME Metallurgical Society Conference, "Refractory Metals and Alloys", Vol II, M. Semchysen and J. J. Harwood (Eds.), Interscience Publishers, NY, 1961
5. The International Practical Temperature Scale of 1968 Amended Edition of 1975, Metrologia, Vol 12, 1976, p 7-17
6. P. Hidnert and W. Sander, NBS Circular 486, U.S. Dept. of Commerce, National Bureau of Standards, Washington, D.C., 1950
7. F. N. Jaeger and E. Rosenbohm, Physics, Vol 6, 1939, p 1123
8. R. W. Powell, R. P. Tye and M. J. Woodman, Platinum Metals Review, Vol 6, 1962, p 138
9. R. F. Vines and E. M. Wise, "Platinum Metals and Their Alloys", International Nickel Co., 1941
10. P. Drude, Annalen der Physik, Vol 39, 1890, p 481
11. W. Meier, Annalen der Physik, Vol 31, 1910, p 1017
12. G. Hass and L. Hadley, "Optical Properties of Metals", in American Institute of Physics Handbook, 2nd Ed., New York, 1965, p 6-107 to 6-118
13. Corrosion Handbook, John Wiley & Sons, 1948
14. E. M. Wise and J. T. Eash, Transactions of AIME, Vol 128, 1938, p 282

Platinum-Palladium Alloys

By J. Hafner and R. Volterra
Metals and Controls Div.
Texas Instruments, Inc.
Reviewed for this volume by
J. A. Bard
Matthey Bishop, Inc.

Applications

Typical uses. Platinum-palladium alloys are used in place of pure platinum for jewelry in Europe; in the United States, stamping laws do not provide for this type of alloy. Platinum-palladium alloys, with or without additions of other metals, are used for electrical contacts. Platinum with up to 20% Pd is used as an insoluble anode in seawater, and low-palladium platinum alloys have received consideration in the glass industry.

Mechanical Properties

Tensile properties. See Fig. 4.
Hardness. See Fig. 5.

Structure

Microstructure. Platinum and palladium form a continuous series of

Fig. 4 Tensile strength of annealed platinum-palladium alloys as a function of palladium content

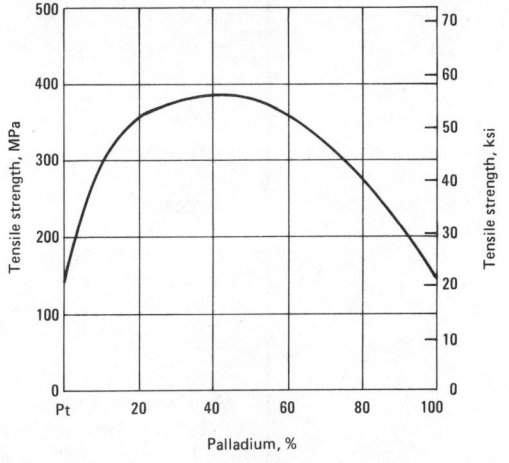

Fig. 5 Hardness of platinum-palladium alloys as a function of palladium content

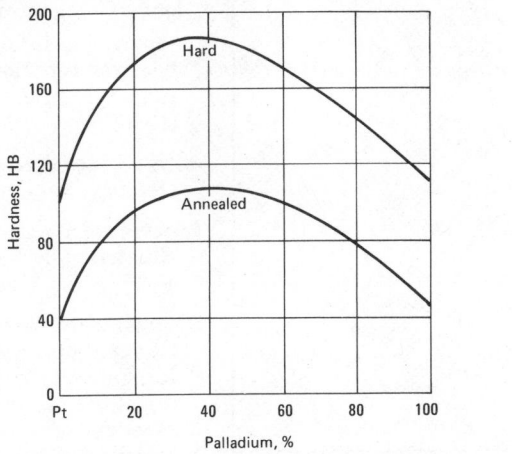

Fig. 6 Electrical resistivity of platinum-palladium alloys as a function of palladium content

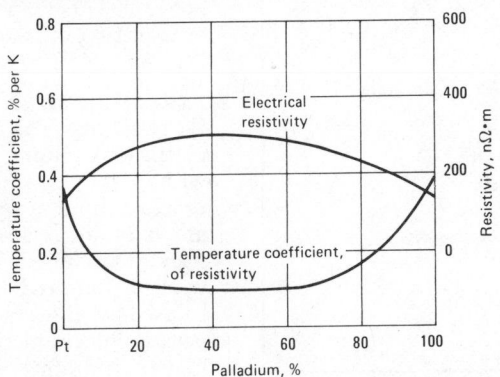

solid solutions. Solidus and liquidus curves are close together; the maximum interval between them is about 60 °C, or about 110 °F (50 Pt—50 Pd). No transformations in the solid state have been reported.

Electrical Properties
Electrical resistivity. See Fig. 6.
Temperature coefficient of resistivity. See Fig. 6.

Chemical Properties
General corrosion resistance. The alloys containing less than 25% Pd perform the same as pure platinum in most chemical mediums. The resistance to nitric acid decreases as the palladium content increases, but an alloy with only 2% Pt is as resistant to this reagent as a 14-karat gold alloy. The platinum-rich alloys do not discolor when heated in air, but the palladium-rich alloys darken between 400 and 750 °C (750 and 1380 °F), because of the formation of palladium oxide, which is stable in that interval of temperature but which decomposes at higher temperature. Prolonged heating at high temperature causes slight weight loss by volatilization; however, at 900 °C (1650 °F) in oxygen, the loss in weight is less than that of pure platinum. Presumably this is caused by the adsorption of oxygen in palladium. The solubility of hydrogen in palladium is reduced significantly by the addition of platinum. At over 34% Pt, only adsorption can be observed. The 90 Pt—10 Pd alloy has the lowest corrosion rate in seawater.

Fabrication Characteristics
Workability. All platinum-palladium alloys can be cold worked. Those with high palladium content should be annealed in inert or nitrogen atmospheres. Recommended annealing temperature is 899 °C (1740 °F).

Platinum-Iridium Alloys

By J. Hafner and R. Volterra
Metals and Controls Div.
Texas Instruments. Inc.
Reviewed for this edition by
J. A. Bard
Matthey Bishop, Inc.

Applications
Typical uses. Platinum-iridium alloys are used in the electrical, elec-

Fig. 7 Tensile strength of platinum-iridium alloys as a function of iridium content

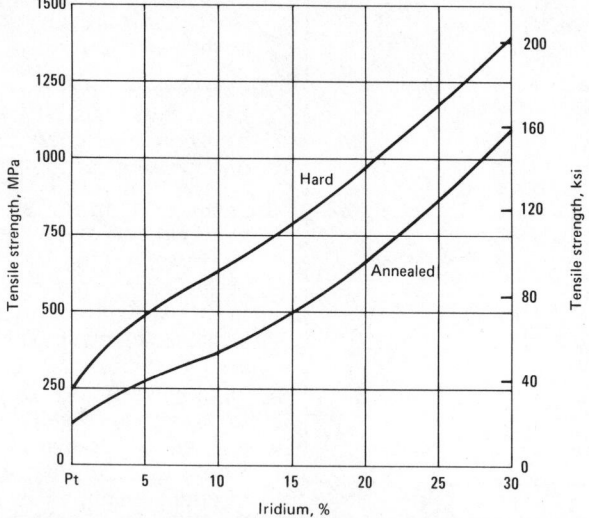

Fig. 8 Hardness of platinum-iridium alloys as a function of iridium content

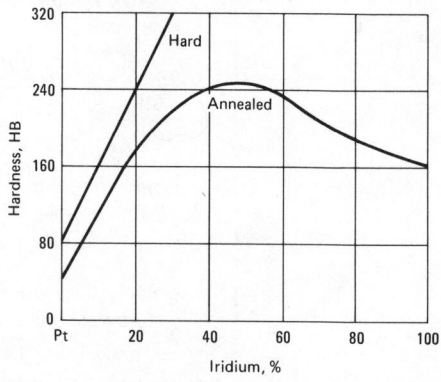

Fig. 9 Electrical resistivity of platinum-iridium alloys as a function of iridium content

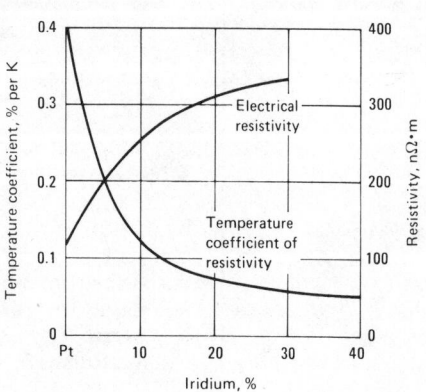

trochemical, chemical, medical and jewelry field. The Pt-10% Ir alloys are also used for standards of length and weight, because of their permanence. Some of the application of different alloys are:

Application	Iridium, %
Laboratory ware	0.4 to 0.6
Jewelry	5 to 15
Medical	10
Electrical contacts	10 to 25
Electrodes for electrochemical processes	10
Tubing for pens, hypodermic needles, spring elements	25 to 30

Mechanical Properties

Tensile properties. See Table 1 and Fig. 7.

Hardness. See Table 1 and Fig. 8.

Structure

Microstructure. The platinum-iridium alloys solidify in a complete series of solid solutions. Below about 995 °C (1825 °F), a very sluggish separation into two solid solutions has been reported. The duplex region ranges from 7 to 99% Ir at 700 °C (1290 °F)

Mass Characteristics

Density. See Table 1.

Electrical Properties

Electrical resistivity. See Table 1 and Fig. 9.

Chemical Properties

General corrosion resistance. The resistance to corrosion and tarnishing of the platinum-iridium alloys is excellent. The resistance to aqueous solutions of halogens and aqua regia increases with iridium content. Because iridium forms a volatile oxide at high temperatures, there is a noticeable loss by evaporation when these alloys are annealed in air above 900 °C (1650 °F). The loss by evaporation increases with temperature and iridium content.

Fabrication Characteristics

Workability. The workability of the platinum-iridium alloys decreases with increasing iridium content. The practical limit of workability for cast alloys is about 40% Ir. Alloys with about 25 to 30% Ir can be hot or cold worked. Cold reductions to 75% can be used for alloys with 20% Ir or less; permissible reductions for alloys of higher iridium content are lower.

Annealing temperatures between 1000 and 1200 °C (1830 and 2190 °F) are suitable for alloys to 10% Ir. For alloys of higher iridium content, temperatures near 1400 °C (2550 °F) are recommended. Platinum alloys with 10 to 90% Ir have been reported to be susceptible to precipitation hardening. Noticeable changes in physical properties can be obtained by heating at approximately 800 °C (1475 °F) after quenching from about 1700 °C (3090 °F)

Table 1 Typical properties of platinum-iridium alloys

% Iridium and temper	Tensile strength MPa	ksi	Hardness, HB	Density Mg/m³	lb/in.³	Electrical resistivity, nΩ·m	Temperature coefficient(a), per °C
5% annealed....	275	40	90	21.49	0.777	190	0.00188
5% hard........	485	70	140				
10% annealed...	380	55	130	21.53	0.778	250	0.00126
10% hard	620	90	185				
15% annealed...	515	75	160	21.57	0.780	285(b)	0.00102
15% hard	825	120	230				
20% annealed...	690	100	200	21.61	0.781	310	0.00081
20% hard	1000	145	265				
25% annealed...	860	125	240	21.66	0.783	330	0.00066
25% hard	1170	170	310				
30% annealed...	1105	160	280	21.70	0.784	350	0.00058
30% hard	1380	200	360				
35% annealed...	···	···	···	21.79	0.787	360	0.00058
35% hard	···	···	···				

(a) Of electrical resistivity at 0 to 160 °C (32 to 320 °F). (b) By interpolation.

Platinum-Rhodium Alloys (3.5 to 40% Rh)

By R. B. Green
Radio Corporation of America
Reviewed for this edition by
A. R. Wroblewski
Engelhard Minerals and Chemicals Corp.

Applications

Typical uses. Rhodium is the preferred addition to platinum for most applications at high temperatures under oxidizing conditions because, unlike most other hardeners, rhodium is not selectively volatilized. The 10% Rh alloy is used more than any of the other alloys in this series. In the production of nitric acid, it is used as a catalyst for the oxidation of ammonia by air. This alloy, with its composition controlled closely, is the positive element in the standard thermocouples (Pt—10 Rh versus platinum) that are used to define the International Practical Temperature Scale of 1968 in the range from 630.74 °C to the gold point (1064.43 °C). Some use is made of the Pt—13 Rh versus platinum thermocouple in instruments that are calibrated for it.

The following couples are accepted internationally and standard temperature/electromotive tables are available (for further information see NBS Monograph 125 and latest addition of ASTM Monograph 565).

Pt-10Rh vs Pt (Type S)
Pt-13Rh vs Pt (Type R)
Pt-30Rh vs Pt-6Rh (Type B)

Table 2 Typical properties of platinum-rhodium alloys

% Rhodium and temper	Tensile strength MPa	ksi	Elongation(a) %	Hardness, HB	Density Mg/m³	lb/in.³	Electrical resistivity(b), nΩ·m	Temperature coefficient(c), per °C
3.5% annealed .	170	25	35	60	20.90	0.755	166	0.0022
3.5% hard(d) ...	415	60	···	120				
5.0% annealed .	205	30	35	70	20.65	0.746	175	0.0020
5.0% hard (d) ..	485	70	···	130				
10% annealed ..	310	45	35	90	19.97	0.722	192	0.0017
10% hard(d)....	620	90	2	165				
20% annealed ..	485	70	33	120	18.74	0.677	208	0.0014
20% hard(d)....	895	130	2	210				
30% annealed ..	540	78	30	132	17.62	0.637	194	0.0013
30% hard(d)....	1060	154	0.5	238				
40% annealed ..	565	82	30	150	16.63	0.601	175	0.0014
40% hard(d)....	1255	182	0.5	290				

(a) In 50 mm or 2 in. (b) At 20 °C (68 °F). (c) Of electrical resistivity at 20 to 100 °C (68 to 212 °F). (d) Hard, as cold worked, 75% reduction.

Types R and S couples are generally used to 1400 °C (2552 °F) for extended service, while the Type B couple may be used to 1600 °C (2912 °F). The preferred atmosphere for the use of these thermocouples is air. For emf stability, high purity alumina insulators and protection tube should be used.

Pt-3.5 Rh alloy is used as a crucible and shows very little loss in weight at high temperatures. Platinum-rhodium alloys are used in the glass industry; they stand up well on contact with molten glass. The Pt-10Rh alloy is used for feeder dies and in handling glasses of high melting point. The Pt-10Rh alloy also is used for rayon spinnerettes. Pt-10Rh and

Pt-20Rh alloys are used as windings in high-temperature furnaces that operate under oxidizing atmospheres. The 40% Rh alloy has been used as a winding in furnaces operating between 1500 and 1800 °C (2800 and 3275 °F)

Mechanical Properties
Tensile properties. See Table 2.
Hardness. See Table 2.

Mass Characteristics
Density. See Table 2.

Electrical Properties
Electrical resistivity. See Table 2.
Temperature coefficient of electrical resistivity. See Table 2.

Fabrication Characteristics

Workability. These alloys can be worked hot or cold. Alloys with a higher rhodium content may require more hot work before they can be cold worked successfully. The hot working temperature range is between 900 and 1200 °C (1650 and 2190 °F), and the annealing temperature range is between 900 and 1000 °C (1650 and 1830 °F). If proper melting procedures are followed, reductions up to 90% between anneals are possible.

Platinum-Ruthenium Alloys

By F. E. Carter
Engelhard Industries, Inc.
Minerals and Chemicals Corp.
Reviewed for this edition by
Edward D. Zysk
Engelhard Minerals and Chemicals Corp.

Applications

Typical uses. The alloy that contains 5% Ru is used in jewelry and has properties essentially the same as the 10% Ir alloy—the so-called "hard platinum" of the jewelry trade. The same alloy is also used for laboratory electrode stems and for certain other chemical equipment, but it is not completely suitable for service at high temperature under strongly oxidizing conditions. Platinum-ruthenium alloys are frequently employed as electrical contacts—the 5% Ru alloy in the medium-duty field, the 10% alloy in aircraft magnetos, and the 14% alloy for heavy-duty contacts. Complex platinum-base alloys that contain 4 to 5% Ru are being used to some extent for spark plug electrodes in aircraft. The 10 and 11% Ru platinum alloys are about equally in demand, for electrical contacts and hypodermic needles.

Mechanical Properties

Tensile properties. Tensile strength: 5% Ru: annealed, 415 MPa (60 ksi); hard, 795 MPa (115 ksi). 10% Ru: annealed, 570 MPa (83 ksi); hard, 1035 MPa (150 ksi). Elongation in 50 mm or 2 in.: 5% Ru: annealed,

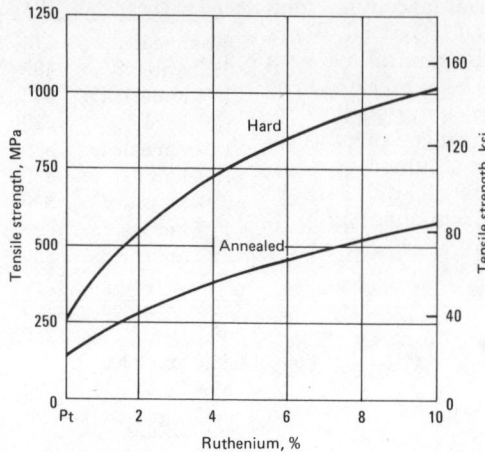

Fig. 10 Tensile strength of platinum-ruthenium alloys as a function of ruthenium content

Initially reduced by 75%, then annealed 15 min.

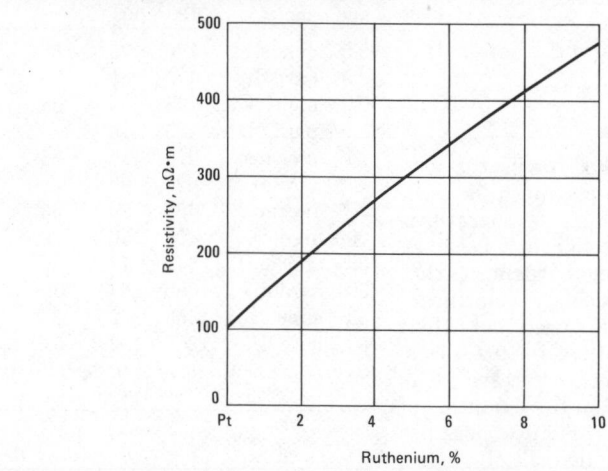

Fig. 11 Electrical resistance of platinum-ruthenium alloys as a function of ruthenium content

34%; hard, 2%; 10% Ru: annealed, 31%; hard, 2%. See also Fig. 10.
Hardness. 5% Ru: annealed, 130 BHN, RB 70; hard, BHN 210, RB94. 10% Ru: annealed, BHN 190, RB86; hard, BHN 280. 14% Ru, annealed, VHN 240

Mass Characteristics

Density. 5% Ru, 20.67 Mg/m^3 (0.747 lb/in.^{3}); 10% Ru, 19.94 Mg/m^3 (0.720 lb/in.^{3})

Electrical Properties

Electrical resistivity. 5% Ru, 315 nΩ·m; 10% Ru, 430 nΩ·m; 14% Ru, 460 nΩ·m. See also Fig. 11.

Temperature coefficient of resistivity. 5% Ru, 0.0009 per °C at 0 to 1000 °C (32 to 1830 °F); 10% Ru, 0.0008 per °C at 0 to 1000 °C (32 to 1830 °F); 14% Ru, 0.00036 per °C at 0 to 100 °C (32 to 212 °F)

Fabrication Characteristics

Workability. The 5% Ru alloy is hot worked between 900 and 1200 °C (1650 and 2200 °F); and the 10% Ru alloy, between 1000 and 1300 °C (1825 and 2375 °F). Annealing temperature for the 5% alloy is about 1000 °C (1825 °F); and for the 10%

alloy, about 1100 °C (2000 °F). Maximum reduction between anneals should not exceed 90% for the 5% Ru alloy, or 75% for the 10% alloy. The 14% Ru alloy approaches the practical limit of workability. The atmosphere for high-temperature anneals should be only slightly oxidizing, since excessively oxidizing atmospheres cause loss of ruthenium in the same manner as with iridium alloys.

79Pt-15Rh-6Ru

Compiled by J. D. Mitilineos
Sigmund Cohn Corp.

Commercial Names

Trade name. Alloy no. 851

Chemical Composition

Composition limits. 78.9 to 80.1 Pt, 14.9 to 15.1 Rh, 5.9 to 6.1 Ru
Consequence of exceeding impurity limits. Values of electrical resistivity, temperature coefficient of electrical resistivity, and tensile strength will not meet specifications.

Applications

Typical uses. This alloy has remarkably high tensile strength and hardness, combined with excellent corrosion resistance, weldability, stability, and shelf life. In solid solution form it is used for wire-wound potentiometers, galvanometers, suspension strips, explosive bridgewires, contacts, strain gages, catalytic glow plugs, and coratron wire for copiers.

Mechanical Properties

Tensile properties. Typical. Tensile strength, 2070 MPa (300 ksi); yield strength, 1515 MPa (220 ksi); elongation, 2% in 254 mm or 10 in.
Hardness. 371 HK
Elastic modulus. Tension, 205 GPa (30 × 10⁶ psi)

Mass Characteristics

Density. 18.6 Mg/m³ (0.67 lb/in.³)

Thermal Properties

Solidus temperature. 1880 °C (3415 °F)
Coefficient of thermal expansion. Linear, 15.6 μm/m·K (8.69 μin./in.·°F)
Thermal electromotive force vs Cu. 3.24 mV/°C

Electrical Properties

Electrical resistivity. 308 nΩ·m at 0 °C (32 °F); temperature coefficient, 0.0006 nΩ·m per K at 0 to 100 °C (32 to 212 °F)

Chemical Properties

General corrosion behavior. Excellent resistance to corrosion
Resistance to specific corroding agents. Very slowly attacked even by aqua regia. Can only be put into solution with a caustic fusion

Fabrication Characteristics

Weldability. Excellent

Platinum-Nickel Alloys

Compiled by J. D. Mitilineos
Sigmund Cohn Corp.

Chemical Composition

Nominal compositions. Platinum-nickel alloys range in composition from 0 to 20% nickel.

Applications

Typical uses. These alloys have long been used for their strength at high temperatures. Taut band strips for electrical meters are made from 10% Pt-Ni alloys. In addition to high tensile strength, this alloy is remarkably free of hysteresis. Trace amounts of Ni are added to Pt electrical contacts to increase hardness.
Precautions in use. Selective oxidation of nickel limits the use of Pt-Ni alloys at high temperature under oxidizing conditions.

Mechanical Properties

Tensile properties. Tensile strength:

Nickel, %	Tensile strength	
	MPa	ksi
Annealed		
5	640	93
10	815	120
15	910	130
20	910	130
Hard, 90% reduced		
5	1240	180
10	1550	225
15	1690	245
20	1725	250

See also Fig. 12.

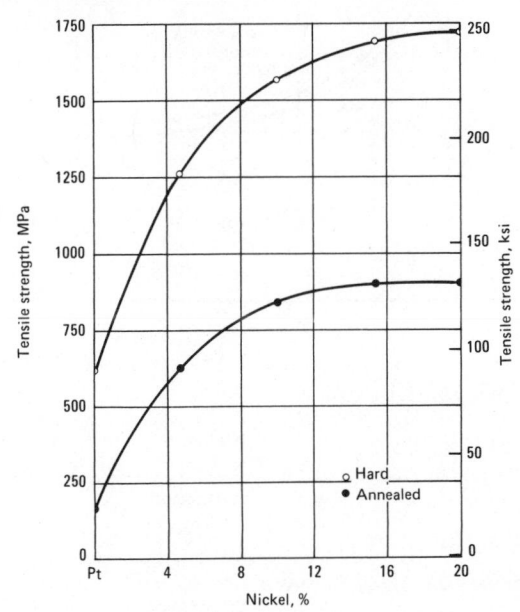

Fig. 12 Tensile strength of platinum-nickel alloys as a function of nickel content

Fig. 13 Hardness of platinum-nickel alloys as a function of nickel content

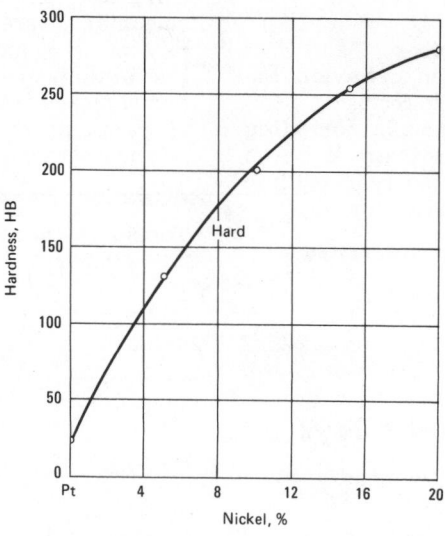

Fig. 14 Electrical resistivity of platinum-nickel alloys as a function of nickel content

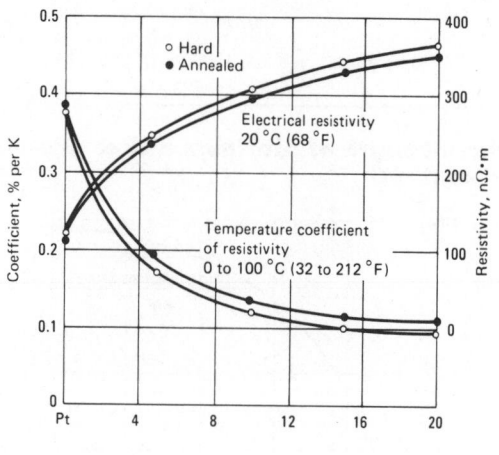

Hardness. 5% Ni, 130 HB; 10% Ni, 200 HB; 15% Ni, 255 HB; 20% Ni, 280 HB. See also Fig. 13.

Elastic modulus. Tension for 10% Pt-Ni alloy: 170 GPa (25×10^6 psi)

Electrical Properties

Electrical resistivity. At 20 °C (68 °F):

Nickel, %	Resistivity, nΩ·m
Annealed	
5	236
10	298
15	330
20	350
Hard	
5	244
10	304
15	440
20	360

See also Fig. 14.

Temperature coefficient of electrical resistivity. At 0 to 100 °C (32 to 212 °F):

Nickel, %	Coefficient, Ω/Ω per K
Annealed	
5	0.00179
10	0.00135
15	0.00114
20	0.00102
Hard	
5	0.00170
10	0.00125
15	0.00105
20	0.00094

See also Fig. 14.

Platinum-Tungsten Alloys

Compiled by J. D. Mitilineos
Sigmund Cohn Corp.

Chemical Composition

Nominal compositions. Four platinum-tungsten alloys are commonly used: 2%, 4%, 6% and 8% W.

Applications

Typical uses. The 4% W platinum alloy was originally developed for spark plug electrodes in aircraft engines. Its resistance to lead contamination is superior in the hard drawn condition. It was also used for grids in radar tubes because of reduced electron emission. These uses have diminished over the years.

The Pt-8% W alloy is used for potentiometer wire because it has excellent wear resistance and low noise characteristics. Its high strength at elevated temperatures makes it suitable for strain gages. (But see **Precautions in use** below). It is often used in vacuum or inert atmospheres to lessen the problem of volatilization. Pt-8% W has also been used for glow wires in copiers, and in reticles, bridgewire, diode switches, heater elements, and cladding for isotope fuel.

Fig. 15 Mechanical properties of platinum-tungsten alloys as a function of tungsten content

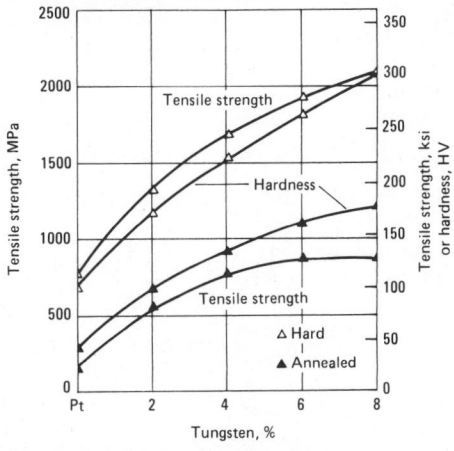

Fig. 16 Electrical resistivity of platinum-tungsten alloys as a function of tungsten content

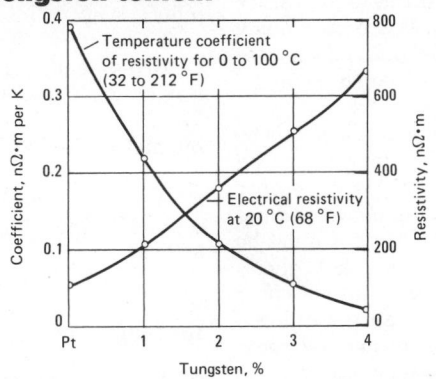

Fig. 17 Thermal electromotive force of platinum-tungsten alloys vs platinum as a function of tungsten content

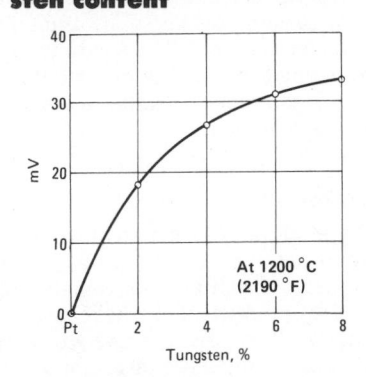

Electrical Properties

Electrical resistivity. At 20 °C (68 °F) annealed: 2% W, 215 nΩ·m; 4% W, 360 nΩ·m; 6% W, 530 nΩ·m; 8% W, 665 nΩ·m. See also Fig. 16.

Temperature coefficient of electrical resistivity. Annealed 0 to 100 °C (32 to 212 °F): 2% W, 0.0022 per K; 4% W, 0.0011 per K; 6% W, 0.0006 per K; 8% W, 0.00025 per K. See also Fig. 16.

Thermal electromotive force vs Pt. At 1200 °C (2190 °F), cold junction at 0 °C (32 °F): 2% W, 19 mV; 4% W, 26.5 mV; 6% W, 31.5 mV; 8% W, 34 mV. See also Fig. 17.

Precautions in use. The Pt-8% W alloy is not recommended for use at high temperature under oxidizing conditions because of the selective oxidation of tungsten.

Mechanical Properties

Tensile properties. Tensile strength, typical:

Tungsten, %	Tensile strength	
	MPa	ksi
Annealed at 1200 °C (2190 °F)		
2	570	82
4	770	110
6	860	125
8	995	130
Hard, 99.8% reduced		
2	1340	195
4	170	245
6	1930	280
8	2070	300

See also Fig. 15.

Hardness. Sheet:

Tungsten, %	Hardness, HV
Annealed at 1200 °C (2190 °F)	
2	100
4	133
6	158
8	180
Hard, 50% reduced	
2	170
4	220
6	260
8	300

See also Fig. 15.

Thermal Properties

Liquidus temperature. 8% W, 1910 °C (3470 °F)

Solidus temperature. 8% W, 1870 °C (3400 °F)

Platinum-Cobalt Permanent Magnet Alloy

Compiled by A. R. Robertson
Engelhard Minerals & Chemicals Corp.

W. J. Jellinghaus, in 1936, discovered that certain platinum-cobalt alloys had an unusually high coercive force (Ref 1). To date, coercivities up to 540 kA/m have been observed in these alloys at room temperature. The most valuable data on these alloys have been published in a series of papers by J. B. Newkirk, R. Smoluchowski, A. H. Geisler and D. L. Martin; the last of the series is cited below (Ref 2). The platinum-cobalt alloys used for permanent magnets are near the 50 at. % composition (23.3 wt % Co). In this region the alloys are disordered face-centered cubic at high temperature and ordered face-centered tetragonal at low temperature. Alloys near 30 at. % Co form a superlattice similar to Cu_3Au, face-centered cubic with cobalt atoms on the corner sites and platinum atoms on the face centers. The strain induced by dimensional change in a rigid alloy when it is ordered results in considerable hardening.

Applications

Typical uses. When an alloy of platinum and cobalt is properly processed, it exhibits extremely high energy products. For this reason, this material is used in areas where the length-to-thickness ratio of most other magnetic materials would be unfavorable. In many critical applications that require minimum space or weight and stable temperature performance without fear of attrition, platinum-cobalt magnets may be the best choice. Typical uses include focusing magnets, hearing aid magnets, magnetic phonograph cartridges, electric watch magnets, rotors in miniature motors and gyro bearings. Other applications include Pt-Co films for use in digital magneto-optic recording, medical implants for atrial stimulation and sensing, as well as in a system for delivery of magnetic emboli via a guided catheter to specific cerebral arteries.

Properties and Fabrication

The magnetic characteristics of the platinum-cobalt alloys can be varied by adjusting composition within the range from 40 to 60 at. % Co, and by varying the heat treatment. Curves of magnetic properties as functions of time and temperature of aging are given in papers by D. L. Martin (Ref 3, 4). According to Martin:

1 The disordered phase has a higher saturation $(B_i)_P$ and residual induction B_r than the ordered alloy.
2 The coercive force H_c and H_{ci}, and the related energy product $(BH)_m$, increase to a maximum and then decrease with additional aging time.
3 Peak values of H_c, H_{ci} and $(BH)_m$ were obtained in alloys aged at 600 °C (1110 °F). The coercive force reaches its maximum before ordering is complete.
4 The effect of cobalt on the magnetic properties was not established precisely. However, alloys with 49 to 50 at. % Co have the highest coercive force and the highest $(BH)_{max}$.
5 Magnetic properties depend not only on time and temperature of aging but also on the temperature at which the alloys are previously heated for disordering and the rate at which they are cooled from the disordering temperature.
6 The platinum-cobalt alloys can be prepared either by melting or by powder metallurgy; with proper techniques, they can be worked hot or cold, but sulfur content must be controlled to prevent hot shortness. The disordered phase is softer and more ductile than the ordered phase and can be retained at room temperature by quenching.

REFERENCES

1. W. J. Jellinghous, Zeitschrift fur Technische Physik, Vol 17, 1936, p 33
2. J. B. Newkirk and R. J. Smoluchowski, Applied Physics, Vol. 22, 1951, p 290
3. D. L. Martin, Effects of Temperature on Remanence Magnetics, Conference on Magnetism and Magnetic Materials, American Institute of Electrical Engineers, 1957, p 188
4. D. L. Martin, Processing and Properties of Cobalt Platinum Permanent Magnet Alloys, Transactions of the Metallurgical Society of AIME, Vol 212, Aug 1958, p 478-485

ADDITIONAL REFERENCES

H. J. Albert and L. R. Rubin, "Magnetic Properties of Platinum Group Metals and Their Alloys", in Platinum Group Metals and Their Compounds—Advances in Chemistry, Series #98, edited by Robert L. Gould, American Chemical Society, 1971

M. S. Walmer, Engelhard Industries Technical Bulletin, 1961-1962

J. C. Chaston, British Patent 849, 1960 p 505

P. Brissoneau, A. Blanchard and H. Bartholin, Institute of Electrical and Electronic Engineers Transactions on Magnetics, Vol 2, 1966, p 479

R. A. McCurrie and P. Gaunt, Philosophical Magazine, Vol 13, 1966, p 567

P. Gaunt, Philosophical Magazine, Vol 13, 1966, p 579

D. Treves, J. T. Jacobs and Savatzky, Journal of Applied Physics, Vol 46, No. 6, June 1975

J. Driller and Parsonnet, Institute of Electrical and Electronic Engineers Transactions on Magnetics, Vol 9, No. 3, Sept 1973, p 444-447

R. A. McCurrie, Nature, Vol 216, October 14, 1967, p 149

Properties of Palladium and Palladium Alloys

Commercially Pure Palladium 99.85% Pd

Compiled by E. M. Wise and R. F. Vines
The International Nickel Co., Inc.
Reviewed for this edition by
J.A. Bard
Matthey Bishop, Inc.

Applications

Typical uses. Palladium resembles platinum in appearance, ductility and strength. Its nobility and melting point are somewhat lower than those of platinum, but its lower cost per unit weight and lower density provide economic advantage. The major use of palladium is for contacts in light-duty electrical relays, where its freedom from tarnishing provides extreme reliability and noise-free transmission, required in voice circuits. Its effectiveness as a catalyst accounts for its use in the removal of oxygen from atmospheres used for heat treatment, the recombination of hydrogen and oxygen, the hydrogenation of terpines, and also the manufacture of organics such as vitamins. Hydrogen will diffuse selectively through a palladium septum, yielding pure gas. However, the gas must be initially free from sulfur. Palladium hardened with ruthenium

provides an "all precious metal" white jewelry alloy that sets off diamonds to advantage. Generally a solid solution former, palladium is employed as the major or auxiliary element in dental, electrical contact and special resistance alloys. Palladium is an important constituent in the high-temperature solders that have low vapor pressures, excellent wettability and minimum penetration into austenitic alloys. Palladium with a minimum purity of 99.8%

(UNS P03980) meeting ASTM B589 and MIL-E-46065 (MR) is available commercially. Contamination with low-melting-point metals causes embrittlement, with base metals causes hardening and decreased corrosion resistance, and with silicon causes loss of hot strength.

Mechanical Properties

Tensile properties. Typical. 1.3-mm (0.05-in.) wire. After annealing at

Fig. 1 Tensile properties of cold drawn deoxidized palladium as a function of annealing temperature

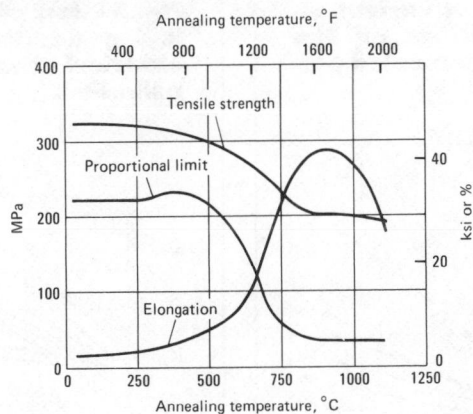

Annealed 5 min.

Fig. 2 Hardness of cold rolled palladium as a function of reduction during rolling

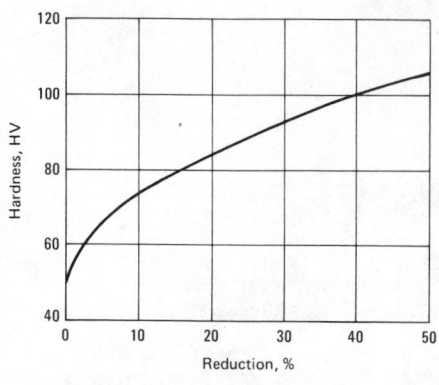

Fig. 3 Hardness of palladium as a function of annealing temperature

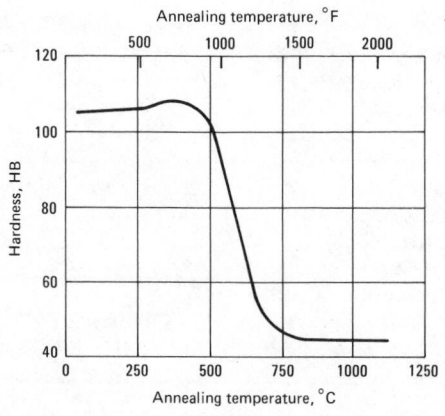

Fig. 4 Effect of various alloying additions on the hardness of annealed palladium

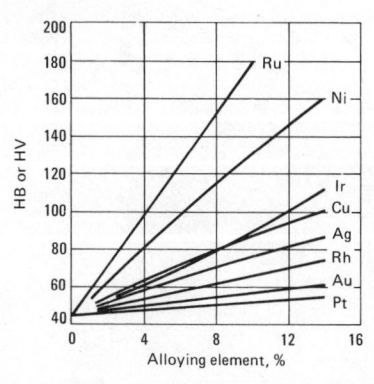

Fig. 5 Effect of various alloying additions on the electrical resistivity of palladium

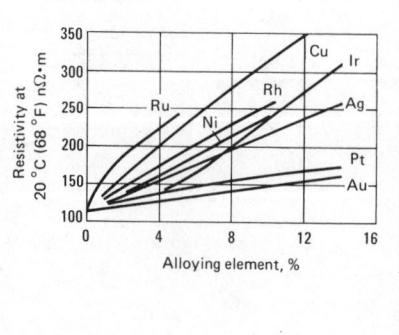

Fig. 6 Reflectance of palladium and Pd-5%Ru as a function of wave length

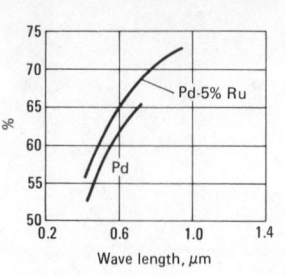

high temperature: tensile strength, 145 MPa (21 ksi); elongation, 24%. After annealing at 800 °C (1470 °F): tensile strength, 172 MPa (25 ksi); elongation, 30%. Tensile strength at various temperatures, material annealed at 1100 °C (2000 °F): at room temperature, 193 MPa (28.0 ksi); at 400 °C (750 °F), 125 MPa (18.1 ksi); at 800 °C (1470 °F), 57 MPa (8.3 ksi); at 1000 °C (1830 °F), 26 MPa (3.8 ksi)

The tensile properties of palladium are mildly sensitive to the kind and amount of residual deoxidizing agents present. Palladium containing such residuals may have a tensile strength of 172 to 207 MPa (25 to 30 ksi) as annealed and 324 MPa (47 ksi) after cold drawing 50%. The tensile properties of deoxidized palladium, initially reduced 50% by cold drawing and annealed for 5 min at various temperatures, are shown in Fig. 1. The optimum anneal for this material is about 800 °C (1470 °F)

Hardness. Rolled and annealed: 37 to 42 HV. Effect of cold rolling, see Fig. 2. Effect of annealing, see Fig. 3. Electrodeposited palladium is much harder than wrought material, ranging from 190 HV for metal from the chloride bath, to about 400 HV for deposits from the complex nitrite baths. For jewelry and other applications where a stronger material is desired, palladium usually is hardened with ruthenium. The relative effects of the various additions of the hardness of annealed material are given in Fig. 4.

Elastic modulus. Tension: 112 GPa (16.3 × 10⁶ psi)

Mass Characteristics

Density. 12.02 Mg/m³ (0.434 lb/in.³ at 20 °C)

Thermal Properties

Melting point. 1552 °C (2826 °F)
Coefficient of thermal expansion. Linear: 11.76 µm/m·K (6.53 µin./in.·°F) at 20 °C (68 °F)
Specific heat. 245 J/kg·K (0.0584 Btu/lb·°F) at 0 °C (32 °F)
Thermal conductivity. 76 W/m·K (44 Btu/ft·h·°F) at 18 °C (64 °F)

Electrical Properties

Electrical conductivity. 16% IACS at 20 °C (68 °F)
Electrical resistivity. 108 nΩ·m at 20 °C (68 °F). Effect of alloying, see Fig. 5.

Optical Properties

Reflectance. 62.8% in white light. Increases slightly in going from red to blue (see Fig. 6)

Chemical Properties

General corrosion behavior. Generally, palladium is less corrosion resistant than platinum, but more corosion resistant than silver. In ordinary atmospheres palladium is resistant to tarnish, but some discoloration may occur during exposure to moist industrial atmospheres that contain sulfur dioxide. Adding palladium to gold or silver alloys improves the tarnish resistance.
Resistance to specific corroding agents. At room temperature palladium is resistant to corrosion by hydrofluoric, perchloric, phosphoric and acetic acids. It is attacked slightly by sulfuric, hydrochloric and hydrobromic acids, especially in the presence of air; and it is attacked readily by nitric acid, ferric chloride, hypochlorites, and moist chlorine, bromine and iodine. Palladium is resistant to molten sodium or potassium nitrate, but not to sodium peroxide, sodium hydrate or sodium carbonate.

Fabrication Characteristics

Precautions in melting. Torch melting of small lots of palladium alloys is common in jewelry and dental applications. For larger melts, palladium is best inductively melted under an argon or lean hydrogen-nitrogen gas cover with care to prevent contamination with silicon, which causes hot shortness. The melt is deoxidized with 0.05% Al or calcium boride just before pouring.
Hot working temperature. 760 to 1100 °C (1400 to 2000 °F)

Recrystallization temperature. 600 °C (1100 °C)
Annealing. Nitrogen-hydrogen mixtures, nitrogen, argon or steam provide suitable annealing atmospheres. The pure metal may be annealed at about 800 °C (1470 °F); 1000 to 1100 °C (1830 to 2010 °F) is required for some of the harder alloys. Slow cooling of palladium from 815 °C (1500 °F) to 425 °C (800 °F) will cause a blue oxide coating to form. To avoid this, the metal should be quenched in water or cooled in a nitrogen atmosphere. Cooling in hydrogen will cause a phase change to occur, with accompanying distortion.
Joining. Palladium can be melted with an oxy-hydrogen torch or welded with plasma or gas tungsten-arc welding equipment. An oxidizing oxyacetylene flame is desirable for soldering palladium with platinum solders melting from 1100 to 1300 °C (2000 to 2375 °F). A gas-air torch and lower-melting white gold solders are employed for soldering palladium jewelry.

Palladium-Silver Alloys

By J. Hafner and R. Volterra
Metals and Controls Div.
Texas Instruments, Inc.
Reviewed for this edition by
J.A. Bard
Matthey Bishop, Inc.

Applications

Typical uses. Alloys with 1, 3, 10, 40, 50 and 60% Pd are employed for electrical contacts, the lower-palladium alloys showing less transfer than silver, and the higher-palladium alloys providing surety of contact, due to the absence of a tarnish film.

The 60 Pd—40 Ag alloy, which has electrical resistivity of 42 microhm-cm (252 ohms per cir mil-ft) at 20 °C (68 °F) and a temperature coefficient of resistivity of only 0.00003 per °C between 0 and 100 °C, is used for precision resistance wires.

Palladium-silver alloys are used for brazing stainless steel, Inconel and other heat-resistant alloys; the 90 Ag—10 Pd alloy has a flow point of 1065 °C (1950 °F) and is much less likely to dissolve or penetrate the

base metal than are nickel-base brazing alloys.

Mechanical Properties

Tensile properties. See Fig. 7.
Hardness. See Fig. 8.
Modulus of elasticity. Maximum at about 35% Pd content

Structure

Microstructure. Palladium and silver form a continuous series of solid solutions. Solidus and liquidus curves are close, with a maximum separation of about 60 °C (about 110 °F) for the 30% Pd alloy. No transformations in the solid state have been determined.

Mass Characteristics

Density. Can be calculated from the ratio of the components and the values of the same properties of the unalloyed metals

Thermal Properties

Coefficient of thermal expansion. Can be calculated from the ratio of the components and the values of the same properties of the unalloyed metals

Electrical Properties

Electrical resistivity. See Fig. 9.

Optical Properties

Color. The color of the alloys varies with palladium content; alloys with as little as 15 to 20% Pd have the color of palladium

Chemical Properties

General corrosion resistance. Alloys with more than about 50% Pd resist tarnishing. Nitric acid dissolves all the alloys, but they are quite resistant to hydrochloric acid, except in the presence of oxidizing agents. Cyanides attack all the alloys, particularly those rich in silver. Metallographic etching can be performed with the Jewett-Wise etch (10% KCN, 10% $NH_4S_2O_8$). The high-temperature corrosion in oxygen decreases with increasing palladium content; the 25% Pd alloy has only half as much weight loss as fine silver at 1650 °F. Addition of silver to palladium increases the solubility of hydrogen up to 30 to 40% silver, decreasing it to zero at about 75% Ag. The 60% Pd—40% Ag alloy has been suggested as a selective diffusion septum for the separation of hydro-

Fig. 7 Tensile strength of palladium-silver alloys as a function of silver content

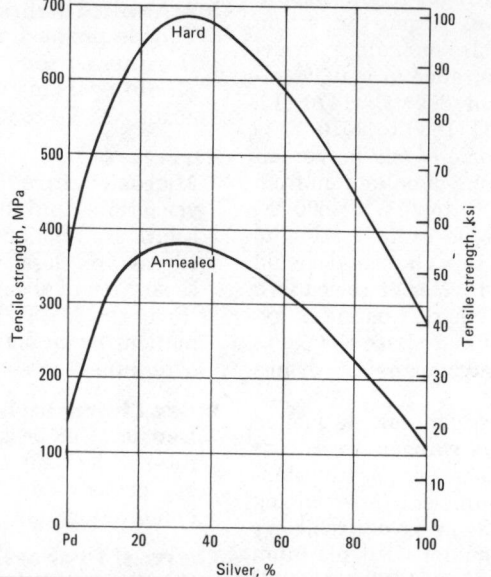

Fig. 8 Hardness of palladium-silver alloys as a function of silver content

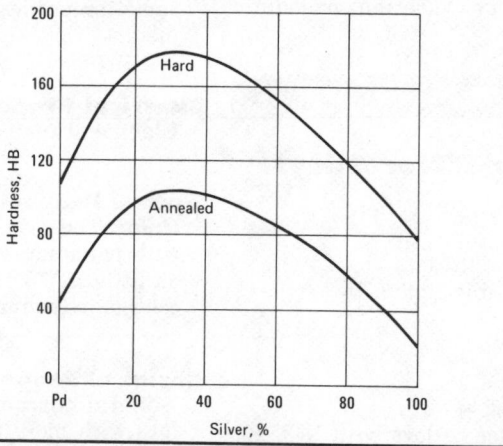

Fig. 9 Electrical resistivity of palladium-silver alloys as a function of silver content

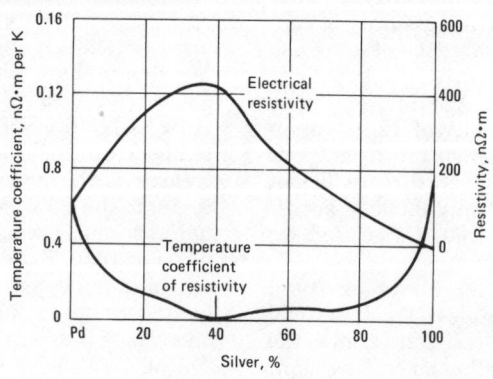

gen from other gases. The separation of high-purity hydrogen also has employed a 75 Pd-25 Ag alloy. (U. S. Patent 2,773,561)

Fabrication Characteristics

Workability. All the alloys can be cold worked. Annealing, preferably at 850 °C (1560 °F), must be in inert or nitrogen atmospheres, to prevent oxidation.

60Pd-40Cu

Compiled by H. T. Reeve
Bell Telephone Laboratories
Reviewed for this edition by
Edward D. Zysk
Engelhard Minerals and Chemicals
  Corp.

Applications

Typical uses. 60Pd-40Cu alloy was devised for electrical contacts in the milliampere range, in circuits containing sufficient capacity so that a considerable rush of current occurs on closure. It is also used where a hard material is required for slip rings running against brushes of the same material, and is generally given a final heat treatment to convert it to the ordered, highly conducting condition. This conversion is accelerated greatly by cold work or by the presence of hydrogen in solution.

Mechanical Properties

Tensile properties. Typical. Tensile strength: annealed, 515 MPa (75 ksi); hard drawn, 1330 MPa (193 ksi)

Mass Characteristics

Density. 10.6 Mg/m^3 (0.383 lb/in.^{3}) at 20 °C (68 °F)

Thermal Properties

Liquidus temperature. Approx 1224 °C (2235 °F)
Solidus temperature. Approx 1196 °C (2185 °F)

Electrical Properties

Electrical resistivity. Annealed and quenched, 350 nΩ·m; 35 nΩ·m ordered, which is best produced by heating cold worked material at 300 °C (570 °F)

Temperature coefficient of electrical resistivity. Annealed, 0.00032 per K; ordered, 0.00224 per K at 20 to 100 °C (68 to 212 °F)

Palladium-Silver-Copper Alloys

Compiled by P. J. Cascone
J. F. Jelenko & Co.

Applications

Typical uses. Alloys that contain about 45% Pd with sufficient copper to make them age hardenable to the desired level are used in dentistry, as are variants that contain a small percentage of platinum or gold. These alloys are used also for electrical contacts subjected to sliding wear or for applications that require good spring properties. About 1% Zn may be present, and this, as well as platinum, appears to accelerate age hardening. The alloys containing 10 to 25% Pd are used for high-strength brazed joints.

Mechanical Properties

Hardness. The hardening response of this system is quite good as shown in Fig. 10. Although the alloy 40Pd-30Ag-30Cu can be readily worked in the quenched condition, upon age hardening it attains a hardness in excess of 450 HV.

Thermal Properties

The liquidus and solidus features of the Pd-Ag-Au system follow the general pattern of the gold-silver-copper system. Palladium increases the liquidus and solidus temperatures of the silver-copper alloys much more rapidly than gold. The silver-copper eutectic persists into the ternary liquidus diagram, terminating at about 30Pd-45Ag-25Cu. The eutectic decomposition on the silver-copper side degenerates into a dome-shape two-phase region in the ternary diagram enabling most of the alloys to be age hardenable (Ref 1). Below 600 °C (1110 °F) the ordered phases PdCu$_5$ and Pd$_3$Cu$_5$ appear (Ref 2). These phases extend across the ternary diagram to the silver rich side of the immiscibility field.

Chemical Properties

Resistance to tarnishing. The silver-palladium alloys that contain 50 to 60 weight or atomic % Pd, have good resistance to tarnishing, but that of the corresponding copper-palladium alloys is not quite so good. If substantial age hardening is required, the palladium content should not exceed about 45%, but small amounts of gold or platinum can be added without impairing the hardening, and may actually increase it along with the resistance to tarnishing. As a result, a whole series of useful quaternary alloys exists between the Cu-Ag-Pd and Cu-Ag-Au systems.

Fabrication Characteristics

Melting and working. At high temperatures most of the ternary alloys are solid solutions and all are workable after quenching. The alloys must be melted in such a manner that oxygen, silicon and sulphur are low in the metal when it is cast. Final deoxidation with a few hundredths per cent of calcium or calcium boride often is useful. Annealing in nitrogen at about 800 °C (1475 °F) is suitable for most of the alloys, and the age hardenable ones should be cooled fairly rapidly for full softness. If much zinc is present, oil quenching may be required, to prevent partial hardening during cooling. Treatment for 1/4 to 5 h between 400 and 450 °C (750 and 850 °F) hardens many alloys effectively.

REFERENCES

1. W. M. Guertler, Ed. *Konstitution der Ternären Metallischer Systeme*, No. 11, Rotadruck Ernst Jaster, Berlin, 1960
2. R. P. Elliott, Ed., *Constitution of Binary Alloys, First Supplement*, McGraw-Hill, NY, 1965, p 378

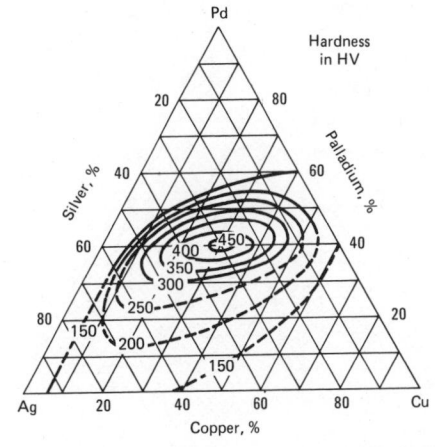

Fig. 10 Maximum hardness of aged palladium-silver-copper alloys

Palladium-Silver-Gold Alloys 40Ag-30Pd-30Au

Compiled by P. J. Cascone
J. F. Jelenko & Co.

Applications

Typical uses. The palladium-silver-gold alloys that can be very easily clad to other metals are used when high resistance to chemical corrosion is needed and when other material (e.g., tantalum) may present fabrication difficulties. The addition of indium or tin, within solid solubility limits, results in a series of useful dental alloys to which porcelain is fused. The palladium-gold-silver alloys that are made susceptible to precipitation hardening by additions of small amounts of other metals, such as copper, are the base of a group of dental alloys. However, these are heterogeneous two-phase alloys and may be less corrosion-resistant than the solid solution alloys, which have no such additions.

Fig. 11 Hardness of annealed palladium-silver-gold alloys

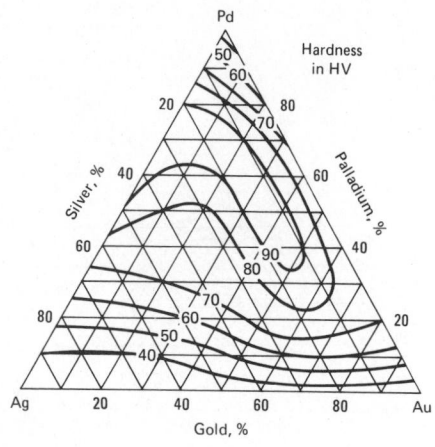

Mechanical Properties

Tensile properties. 30Pd-40Ag-30Au alloy: typical tensile strength, 640 MPa (92.5 ksi)

Hardness. 30Pd-40Ag-30Au alloy: 130 HB after cold rolling. See also Fig. 11.

Thermal Properties

This single phase system exhibits a narrow melting range throughout. There are no solid-state transformations. Various properties of this system are given in Ref 1.

Chemical Properties

General corrosion behavior. In binary alloys, the addition of 20% Au to palladium or about 60% Au to silver results in substantial resistance to nitric acid, and the ternary alloys from these points to the gold corner of the ternary diagram are very resistant to nitric acid. These alloys are particularly useful where the presence of halogen acids precludes the use of even the best of the corrosion-resistant base metal alloys. The alloy 40Pd-30Au-30Ag shows a loss of 0.06 g/m²·d in the atmosphere formed by boiling 20% hydrochloric acid and air. The alloys containing 50 to 60% Pd or about 70% Au are tarnish-resistant. At high temperatures, the alloys do not oxidize in air but they dissolve oxygen, and any subsequent treatment in a reducing atmosphere will produce surface imperfections.

Fabrication Characteristics

Workability. These alloys can be cold worked without difficulty but must be annealed in an inert atmosphere (e.g., nitrogen) with a low dew point to avoid palladium oxidation.

Annealing temperature. 850 °C (1560 °F)

REFERENCE

1. W. M. Guertler, Ed. *Konstitution der Ternären Metallischen Systeme,* No. 11, Rotardruck Ernst Jaster, Berlin, 1960

95.5Pd-4.5Ru

Compiled by D. J. Accinno
Engelhard Minerals and Chemicals Corp.

Applications

Typical uses. Jewelry, electrical contacts, reduction of nitric oxide by monolithic supported Pd-Ru alloys. This alloy is standard for palladium jewelry in the United States; both wrought and cast forms used. This alloy, including richer ruthenium alloys (containing up to 12% ruthenium) is used for electrical contacts. The latter percentage is normally accepted as the limit of commercial workability

Mechanical Properties

Tensile properties. Typical. Annealed: tensile strength, 380 MPa (55 ksi); elongation, 30% in 50 mm or 2 in. Cold drawn: tensile strength, 560 MPa (81 ksi); elongation, 3% in 50 mm or 2 in. See also Fig. 12 and Table 1.

Hardness. See Table 1.

Elastic modulus. 141 GPa (20.4 × 10^6 psi)

Mass Characteristics

Density. Annealed: 12.07 Mg/m³ (0.436 lb/in.³); as cast: 11.62 Mg/m³ (0.420 lb/in.³) at 20 °C (68 °F)

Electrical Properties

Electrical resistivity. 242 nΩ·m at 20 °C (68 °F)

Temperature coefficient of electrical resistivity. 0.0013 per °C at 0 to 100 °C (32 to 212 °F)

Thermal electromotive force vs Pt:

Temperature		
°C	°F	mV
200	390	+ 2.58
400	750	+ 5.73
600	1110	+ 9.00
800	1470	+12.12
1000	1830	+15.30
1200	2190	+18.36

Fabrication Characteristics

Melting and casting. Melting with city gas-oxygen torch and the use of MgO or high purity Al₂O₃ lined crucibles is common practice, although

Table 1 Typical mechanical properties of 95.5Pd-4.5Ru

Condition	Proportional limit		Yield strength		Hardness(a), HV
	MPa	ksi	MPa	ksi	
Annealed(b)	270	39	350	51	152
Annealed(c)	...	...	...	...	126
Hard (60% red)	...	...	...	...	286

(a) 5 kg. (b) 1000 °C (1830 °F), air cooled. (c) 1000 °C (1830 °F), quenched.

Fig. 12 Effect of annealing temperature on the tensile properties of 95.5Pd-4.5Ru

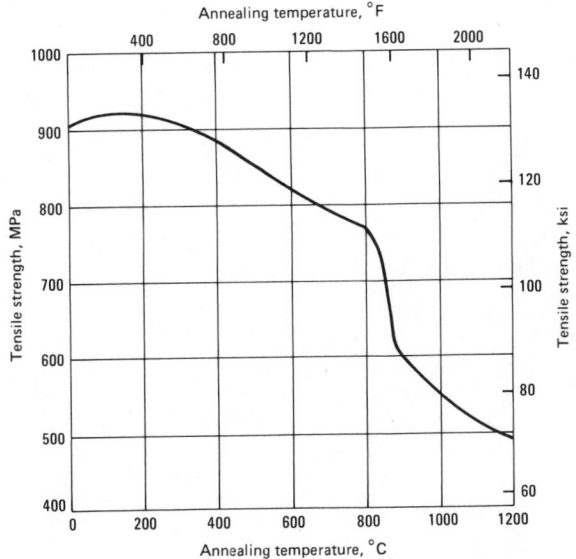

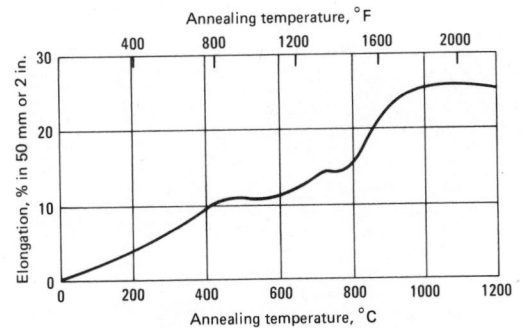

the use of a zirconia crucible yields superior results and is preferred. To obtain sound castings, deoxidation by 0.05 to 0.1% Al just prior to casting is advised. When using high frequency induction to melt the charge, magnesia or preferably zirconia lined crucibles with an argon covering atmosphere are the best approach. Final deoxidation of this melt with a small amount of calcium boride is generally recommended.

Annealing temperature. 900 °C (1650 °F) is suitable for annealing, based on elongation and tensile strength (see Fig. 12). Annealing in nitrogen plus 3 to 7% hydrogen is desirable for high-quality jewelry. Torch annealing should be done in an oxidizing flame.

Joining. Soldering and melting can be readily accomplished with an oxyacetylene torch (oxidizing flame). To avoid hot tearing of cast jewelry, care must be exercised to avoid possible contamination with oxygen or sulfur and particularly silicon and phosphorus.

SELECTED REFERENCES

Aldred, A. T., *Transactions of the AIME,* Vol 224, 1962, p 1082-1083

Darling, A. S. and J. M. Yorke, *Platinum Metals Review,* Vol 4, 1960, p 104-110

Wise, E. M., *Palladium Recovery, Properties and Uses,* Academic Press, NY, 1968, p 68

Atkinson, R. H. and G. P. Gladis, *Transactions of the AIME,* Vol 166, 1946, p. 426-435

Atkinson, R. H., *Transactions of the AIME,* Vol 206, 1956, p 1029-1035

Pure Metals

Preparation and Characterization of Pure Metals

By G. T. Murray
Associate Professor
Metallurgical and Welding
    Engineering
California Polytechnic State
    University

AS A RESULT of the constant quest for the true values of physical and chemical properties of metals, there has been continual improvement in the purity levels attainable and in the accuracy and completeness of techniques for measuring these purity levels. Therefore, the property values reported for "pure metals" in this section of the Handbook, which were determined at different times and in different laboratories, vary considerably in meaningfulness from one metal to another and from one property measurement to another.

The rapidly growing electronic microcircuit industry also has placed severe demands on metal suppliers for metals of the highest reproducible purity attainable. Trace impurity elements in concentrations below 1 ppm can prevent proper functioning of certain electronic devices.

The need for "ultrapure" metals for measurement of physical and chemical properties and for the electronic microcircuit industry poses two important problems: how to obtain such purity and how best to measure levels of trace impurity elements.

Preparation Methods

What is commonly referred to as "commercial-purity" metal normally is used as the starting material in ultrapurification operations. Depending on the metal in question, "commercial purity" usually means a purity between 99.0 and 99.95%. Commercial-purity metal can be prepared by a variety of processes, of which electrolytic processes such as electrowinning and electrorefining are among the most common. In both electrowinning and electrorefining, metal is deposited by electroplating from a bath. In electrowinning, the starting material usually is in the form of a concentrated ore or compound; in electrorefining, it is in metallic form. Many different types of baths are employed. For titanium and vanadium, fused salt baths are used, whereas chromium sometimes is produced by electrolysis of an aqueous solution of chromium-alum or chromic acid. For applications such as semiconductors, material produced by electrolytic processes is of insufficient purity and must be subsequently ultrapurified by one of the methods described below.

Fractional crystallization is a liquid-phase method that relies on differences in solubility in a liquid solvent among the various solid phases present in the impure metal. Basically, the metal to be purified is dissolved in a hot, often organic, solvent. The solvent selected is such that the metal is much more soluble at elevated temperatures than at lower temperatures but that impurities are fairly soluble even at the lower temperatures. On subsequent cooling of the solvent, the "pure" metal precipitates out of solution whereas most of the impurities remain. This process can be repeated many times, using a new batch of solvent each time. Gallium has been purified to the 99.9999% level by this method. This purity is required for manufacture of semiconducting gallium arsenide, which is employed for light-emitting diodes in items such as small hand-held calculators, digital watches and a variety of electronic displays.

Ultrapure silver, gold, palladium and platinum also are produced by fractional crystallization. In some instances, the metal being refined is precipitated and impurities are left in the solvent (as described above); in others, the impurities are precipitated (as compounds). Maximum purity in these metals, however, is obtained by zone refining (see below) following fractional crystallization.

Zone refining, also a liquid-phase technique, probably is the most widely used of all preparation methods. The classic zone refining experiments by Pfann (Ref 1) led to production of germanium of sufficient purity for development of the first transistor.

In zone refining, a molten zone is made to move slowly from one end of a bar of impure metal to the other. During this "zone pass", redistribution of impurities takes place because of differences between the solubility limits of impurity elements (limiting impurity concentrations) in the liquid phase of the metal and the corresponding limits in the solid phase. Under equilibrium conditions, the resulting distribution is measured by the coefficient K_o, which is defined as follows:

$$K_o = C_s/C_l \qquad \text{(Eq 1)}$$

where C_s is impurity concentration in the just-freezing solid phase and C_l is impurity concentration in the liquid phase. In practically all instances of freezing, equilibrium is not attained, and it is more appropriate to use an effective distribution coefficient, K_e, which is a function of freezing velocity, impurity diffusion and thickness of the diffusion layer, as well as the ratio C_s/C_l. When K_e is less than one, as in most instances, and with slow movement of the zone (for example, 10 mm/h or 0.39 in./h), the impurity concentration in the solid phase, C_s, at distance x from the starting end after a single pass of a liquid zone of length l, is as follows:

$$C_s/C_o = 1 - (1 - K_e) \exp [-K_e x/l] \qquad \text{(Eq 2)}$$

where C_o is initial concentration in the liquid phase. Additional passes of the zone in the same direction cause further concentration of impurities at one end of the bar. After many zone passes, this end is removed and discarded.

Metals and semiconductors were first zone refined by placing a bar of the material in a long boat-type crucible. Later to be introduced was the floating-zone technique (Ref 2), in which the metal is suspended in a vertical position and the molten zone is held in place by its own surface tension; heat sources commonly used for this technique include an electron beam and an induction coil. In the floating-zone technique, the diameter of the bar is limited to approximately 15 mm (0.59 in.), but this method has the distinct advantage that the material being refined is not contacted, and thus not contaminated, by a crucible. This is particularly advantageous for the high-melting-temperature reactive metals such as titanium, zirconium, niobium, tantalum, vanadium, tungsten and molybdenum, whereas metals such as gold, silver, copper, aluminum, zinc, lead, tin and bismuth, which melt below about 1200 °C (2190 °F), are zone refined in a boat.

Vacuum Melting. Zone refining of materials often is conducted in a dynamic vacuum in order to enhance the degree of purification. However, many metals—particularly those with high melting points—can be purified to a significant degree by the vacuum melting process alone. Although vacuum melting may not produce the degree of purity attainable by zone refining, it is less expensive and yields material of sufficient purity for a wide variety of applications.

In vacuum melting, purification occurs by degassing—that is, removal of oxygen, nitrogen and hydrogen, as well as CO or CO_2 formed by side reactions of oxygen with carbon—and by vacuum distillation of high-vapor-pressure impurity elements.

Degassing takes place because the solubility of gaseous elements in the liquid decreases when the partial pressure of the same elements in the surrounding gaseous medium is decreased. This was experimentally verified for partial pressures of about 10 to 100 kPa (75 to 750 mm of mercury) in the early experiments of Seiverts, which led to the well-known relationship:

$$S \propto \sqrt{P} \qquad \text{(Eq 3)}$$

where S is the solubility of a gas in the liquid phase and P is the partial pressure of the same gas in the surrounding medium.

This purification process is dependent on (a) the ability of the vacuum system to maintain a sufficiently low gas partial pressure near the molten surface, (b) diffusion of gas atoms through the liquid to the surface, (c) the presence or absence of any stirring action that might enhance transport of gas atoms in the liquid phase, and (d) the composition of the starting material.

Vacuum distillation during melting is a purification process based on preferential evaporation of solute. The degree of purification is dependent on the ratio of the vapor pressure of the solute to that of the solvent. For a high degree of purification, it is essential that solute vapor pressure be high relative to solute partial pressure in the gaseous medium in the immediate vicinity of the molten surface. As the solute content at the liquid/vapor interface becomes diminished, a concentration gradient is set up within the liquid. At this time, which may be very early in the melting operation, material transport in the liquid phase becomes the rate-controlling process. Thus, provided that vapor pressures are favorable and that the pumping speed of the vacuum system is sufficient to maintain a low partial pressure of the solute element, purification should proceed at a rate that depends on the diffusivity of the solute in the liquid.

Distillation. Straight distillation (in which heated material changes from solid to liquid to vapor), like vacuum-melting distillation, is an important vapor-phase purification process. If the distillation is conducted under conditions of near-equilibrium between the liquid and vapor phases, impurity elements will concentrate in the liquid phase. The vapor, or distillate, is then allowed to condense to form a solid of higher purity than that of the starting material. The most common distillation method is fractional distillation, in which the metal is repeatedly vaporized and condensed to liquid on a series of plates placed in a vertical column. A high reflux ratio (ratio of amount of liquid returning to the column from the condenser to the amount of vapor removed to the condenser) is desirable. Some metals, however, are purified in a single stage by simply condensing all the vapor produced by the still; this process has been used for alkali metals such as barium, calcium, lithium and sodium (Ref 3). Distilled magnesium is further purified by zone refinement.

A variation of straight distillation is sublimation, in which the metal passes directly from the solid phase to the vapor phase. Only metals that have high vapor pressure when in the solid state are suited to this process. Such a metal has a higher vapor pressure than most impurity metals, so that such impurities are left to concentrate in the remaining solid while the vapor is condensed, forming higher-purity metal.

Chemical Vapor Deposition. In purification by chemical vapor deposition, the starting material is reacted to form a compound, and the compound is subsequently decomposed in the vapor state. The metal vapor then is condensed to form a solid higher in purity than the starting material.

One of the more popular of the chemical vapor deposition processes is the iodide process, which has been used extensively to purify titanium, zirconium and chromium (Ref 4). For all three of these metals, the starting charge of the metal is reacted to form a volatile metal-iodide compound, which in turn is thermally decomposed to liberate iodine vapor. The pure metal is allowed to condense onto a suitably heated substrate (glass tubes and wires of the base metal have been used), while the iodine returns to the metal charge to form more iodide compound. Hence, the iodine acts as a carrier of the metal from the charge to the substrate.

In this process, some impurities are almost always carried over to the vapor phase along with the metal being purified. However, if a proper temperature is maintained, oxygen, nitrogen, hydrogen and carbon, as well as many metallic impurities, will not be carried over. Typical purities obtained are about 99.96% for titanium, 99.98% for zirconium (plus hafnium, which is present at about the 200-ppm level) and 99.995% for chromium. In all cases, the starting metal has a purity on the order of 99.9%. Chromium has been purified to its highest state to date by this method. Only iron is carried over with these metals to a significant extent. Thus, if a low-iron starting metal is used, the condensed vapor will approach a purity level of 99.999%.

Other metals that have been purified by chemical vapor deposition include hafnium, thorium, vanadium, niobium, tantalum, molybdenum and many less commercially important metals (Ref 4).

Characterization of Purity

The traditional system of describing metal purity is based on measurement of total impurity-element content and subtraction of this number from 100%. The result is reported in terms of "number of nines"— for example, "five nines", which indicates a purity of 99.999%. This system is quite adequate for many applications; however, unless the method of measurement and its sensitivity are reported, the system is meaningless and unacceptable for some fast-growing technical fields. In many analyses, certain elements (the gaseous elements, for example) are not measured, and often the method employed is not sensitive enough to detect impurity levels near the low end of the parts-per-million range. These impurity elements, referred to as "trace" elements, now can be detected by a variety of analytical methods.

Trace-Element Analysis. The sophisticated analytical methods now available include electron-beam microprobe analysis and the chromatographic methods. However, the most commonly used methods are:

1 Emission spectroscopy—for simultaneous determination of metallic elements in the parts-per-million range and greater
2 Mass spectroscopy—for determination of metallic elements in the parts-per-billion range; the analytical results are as accurate as the reference standard used
3 Neutron activation—for determination of metallic elements, and particularly oxygen, in the parts-per-million range
4 Atomic absorption—for sequential determination of metallic elements in the parts-per-billion range
5 Vacuum fusion—for determination of hydrogen and nitrogen in the parts-per-million range
6 Conductometry—for determination of carbon in the parts-per-million range

Emission spectroscopy is the most common analytical method and normally is used for detecting trace elements in concentrations of 10 to 1000 ppm. It is relatively inexpensive and yields results for all metallic elements in one analysis. Mass spectroscopy is more sensitive (and more expensive) than emission spectroscopy, and can easily detect impurity levels as low as 0.01 ppm. However, accuracy is dependent on good standards, and the technique is not accurate above the 50- to 100-ppm level for some elements. Generally, it is the best method for verification of purity at the five nines level and for obtaining information on all residual elements in the sample. Neutron activation analysis is more sensitive than mass spectroscopy, but is unable to detect many elements because of their inherent radioactive characteristics. It is an expensive method, but for certain elements that are difficult-to-identify, it is extremely sensitive and accurate. Atomic absorption is excellent for concentrations of 0.1 to 10 ppm, when only a few elements are present. However, the specific elements being sought must be known, which generally requires that atomic absorption analysis be preceded by emission spectroscopic analysis. All factors considered, mass spectroscopy is the preferred method for measuring trace elements in ultra-high-purity metals.

In summary, the only way of describing purity that is both accurate and meaningful is to state the entire list of possible impurities, the findings, and the limits of detection applicable to the specific analytical procedure used for each element.

Resistance-Ratio Test. The amount of resistance to passage of electrons through a sample of high-purity metal, particularly at low temperatures, is extremely sensitive to the amount of trace elements present in the sample. This fact gives rise to the "resistance-ratio" test, which is an extremely sensitive qualitative method of measuring purities of 99.999% and higher. This test is valuable not only because it is extremely sensitive but also because measurement of electrical resistivity is relatively simple.

Making a resistance measurement at a single (low) temperature would require accurate dimensional measurements. To avoid this requirement, resistance measurements are made both at the low temperature and at room temperature, and the ratio of the room-temperature value to the low-temperature value is reported. Unless otherwise stated, it is assumed that the low-temperature measurement was made at liquid-helium temperature (4.2 K).

The electrical resistivity of a metal can be conveniently divided into three parts:

$$\rho_T = \rho_{th} + \rho_d + \rho_i \qquad \text{(Eq 4)}$$

where ρ_T is total resistivity, ρ_{th} is resistivity due to thermal vibrations of the lattice, ρ_d is resistivity due to lattice imperfections (consisting primarily of vacancies, dislocations and grain boundaries) and ρ_i is resistivity due to impurity atoms. Variation of ρ_T with

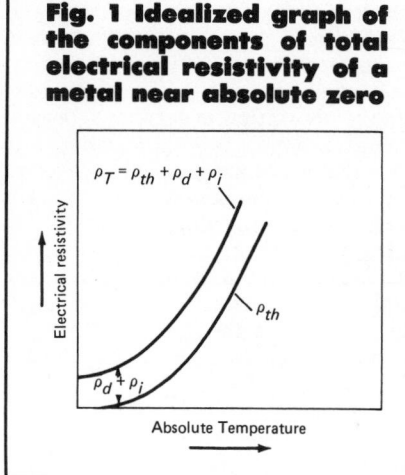

Fig. 1 Idealized graph of the components of total electrical resistivity of a metal near absolute zero

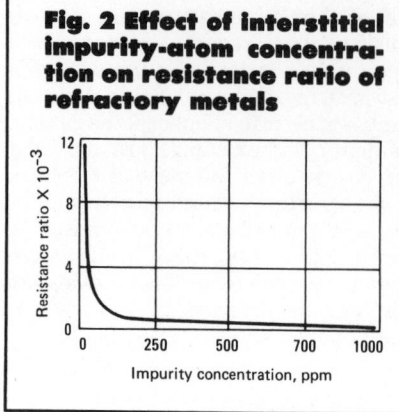

Fig. 2 Effect of interstitial impurity-atom concentration on resistance ratio of refractory metals

Table 1 Resistance ratios of samples of zone-refined metals(a)

Metal	Resistance ratio
Aluminum	40 000
Gold	2 000
Molybdenum	14 000
Nickel	3 000
Niobium	2 000
Rhenium	45 000
Tantalum	7 000
Tungsten	90 000
Vanadium	300
Zirconium	200

(a) See Table 2 for impurity contents of these samples.

temperature in terms of the components ρ_{th} and $(\rho_d + \rho_i)$ is depicted in Fig. 1. The sum $\rho_d + \rho_i$ is essentially temperature independent, whereas ρ_{th} is strongly temperature dependent ($\rho_{th} \propto T^5$), approaching zero at absolute zero temperature. Thus, resistivity near absolute zero affords a measure of $\rho_d + \rho_i$.

Point defects (vacancies) contribute to resistivity to about the same extent as impurity atoms. However, well-annealed metals contain far fewer point defects than impurity atoms. Dislocations of a typical density of 10^{11} per m^2 contribute an insignificant amount to the resistance ratio. In the highest-purity metals obtained to date (impurity concentrations of 10^{-5} to 10^{-6} atomic %), the contribution of ρ_d is still small compared with that of ρ_i. Total resistivity near 0 K, therefore, is a good measure of the ρ_i contribution.

Caution should be exercised in using the resistance ratio as a characterization of purity for several reasons. Most important is the fact that in resistance-ratio testing the impurity element in question is not determined (different impurity elements have vastly different effects on ρ_i). In addition, because only impurity atoms in solid solution are effective electron-scattering centers, nothing is learned about the impurity content in precipitate (compound) form. Finally, even for impurity atoms in solid solution, the resistance ratio is a sensitive measure of purity only when the impurity level is about 100 ppm or less. This is illustrated in Fig. 2, which is an estimate of variation in resistance ratio with concentration of interstitial atoms (O_2, N_2 and C) in refractory metals. This graph shows that the impurity level can be reduced from 500 to 250 ppm without appreciably affecting the ratio, whereas a reduction from 5 to 2.5 ppm has a marked effect.

Resistance ratios of zone-refined metals are listed in Table 1. Their corresponding chemical compositions, as measured by mass spectroscopy for metallic elements, vacuum fusion for gaseous elements and conductometric analysis for carbon, are listed in Table 2. Ratios higher than those shown in Table 1, and ratios for other metals, have been reported (Ref 5); however, they were not accompanied by chemical analyses. In fact, some of the ratios were so large that the impurity concentrations they indicated were too low to be detected by methods currently available.

REFERENCES

1. W. G. Pfann, *Transactions of the American Institute of Mining, Metallurgical and Petroleum Engineers*, Vol 194, 1952, p 861
2. H. C. Theuerer, *Transactions of the American Institute of Mining, Metallurgical and Petroleum Engineers*, Vol 206, 1956, p 1316
3. P. A. Schmidt, *Journal of the Electrochemical Society*, Vol 113, 1966, p 201
4. R. F. Rolsten, *Iodide Metals*, Wiley, New York, 1961
5. W. G. Pfann, *Zone Melting*, 2nd Ed., published jointly by Wiley, New York, and by the Cryogenics Laboratory, National Bureau of Standards, Boulder, CO

Table 2 Impurity-atom concentrations of samples of zone-refined metals(a)

Impurity element	Concentration(b), ppm of impurity of zone-refined:									
	Al	Au	Mo	Ni	Nb	Re	Ta	W	V	Zr
C........	5	<1	10	40	8	5	10	5	57	20
H	<1	<1	0.9	0.2	0.4	0.2	<0.1	0.1	3	3
O........	5	2	4.3	25	23.4	0.5	3.5	0.8	250	200
N	<1	1	0.5	10	4	1	2.3	0.1	3	2
Ag	<0.08	4	<0.7	<0.002	<0.3	<0.001	<0.004	<0.12	<0.002	<0.4
Al........	···	ND	<0.03	0.3	0.15	0.05	0.05	0.07	0.1	3
As........	<0.05	ND	<0.01	<0.04	<0.01	<0.002	<0.002	<0.005	<0.05	<0.01
Au	<0.02	···	<0.02	<0.15	<0.03	<0.15	<0.2	<0.3	0.6	<0.2
Bi	<0.03	ND	<0.02	<0.01	<0.01	<0.12	<0.04	<0.12	<0.02	<0.007
Ca........	0.15	ND	0.04	0.1	0.02	0.05	<0.008	0.02	0.1	0.04
Cd........	<0.05	ND	<1.0	<0.08	<0.5	<0.02	<0.007	<0.025	<0.03	0.5
Cl........	0.4	ND	0.4	0.1	0.03	0.1	0.01	0.2	0.1	2
Co........	<0.1	ND	<0.06	<0.1	<0.01	0.06	0.3	0.1	<0.15	<0.007
Cr........	0.2	ND	0.1	1.5	0.05	0.08	0.2	<0.001	<5	0.5
Cu........	0.2	1	<0.02	<0.04	<0.01	0.005	0.02	0.005	<0.3	0.01
Fe........	0.5	2	12	12	0.12	3	0.3	0.01	<20	30
Ga	<0.1	ND	<0.02	<0.4	<0.01	<0.004	<0.003	<0.01	20	<0.02
Ge........	<0.1	ND	<0.02	<0.7	<0.01	<0.02	<0.005	<0.04	<0.6	<0.03
Hf........	<0.06	ND	<0.03	<0.03	<0.02	<0.01	<0.4	<0.04	<0.03	40
In	<0.1	ND	<1	<1	<0.07	<0.2	<0.002	<0.03	<0.03	<0.08
Ir	<0.06	ND	<0.03	<0.02	<0.01	<0.2	<0.2	<0.15	<0.06	<0.03
K	0.4	ND	1	0.2	<0.04	0.01	0.02	<0.02	0.4	0.004
Li	<0.03	ND	<0.02	<0.02	<0.01	0.004	0.001	<0.001	<0.02	<0.001
Mg	<0.1	ND	<0.25	0.02	<0.05	0.02	0.006	0.15	<0.25	<0.05
Mn	<0.1	ND	0.06	0.03	0.03	0.02	0.01	0.03	<0.15	<0.03
Mo	<0.1	ND	···	0.5	<0.7	4	0.2	<0.1	0.08	<0.6
Na	<0.1	ND	<1	<0.04	<0.03	ND	0.015	<0.01	<0.05	<1
Nb	<0.04	ND	1	<0.02	···	1.2	25	<1	0.8	<0.5
Ni........	<0.1	ND	0.1	···	0.15	0.2	1.5	<0.02	12	1.5
P........	<0.06	ND	<0.03	<2	<30	0.02	<0.05	<0.05	0.2	0.1
Pb........	<0.1	0.5	<0.03	<0.02	<0.02	<0.25	<0.08	<0.25	<0.003	<0.015
Pd........	<0.1	ND	<1	<0.03	<0.5	<0.02	<0.08	<0.25	15	<0.8
Pt........	<0.06	ND	<0.06	<0.03	0.02	<0.3	<0.4	<0.5	<0.04	<0.2
Re........	<0.04	ND	<0.4	<0.02	<0.02	···	<0.2	<1	<0.5	<0.3
Rh	<0.04	ND	<0.1	<0.01	<0.06	<0.005	<0.002	<0.06	<0.5	<0.2
Ru	<0.1	ND	<0.3	<0.03	<0.4	<0.04	<0.006	<0.2	<0.5	<0.6
S........	0.4	ND	1	<0.12	<0.07	1	0.02	0.07	0.1	<1
Sb........	<0.03	ND	<0.2	<2	<0.04	<0.004	<0.004	<0.03	<0.02	<0.15
Si	<1	0.5	0.08	<0.2	0.6	0.5	<0.02	0.3	20	1.5
Sn........	<0.04	ND	<0.4	<0.4	<0.3	<0.02	<0.006	<0.02	<0.03	<0.25
Ta........	<0.02	ND	2	<0.5	50	3	···	5	<0.3	<0.2
Ti	<0.1	ND	<1	0.15	<0.02	<0.07	0.01	<0.01	6	1
V........	<0.1	ND	<0.02	<0.01	<0.8	<0.001	0.01	<0.01	···	0.05
W	0.07	ND	20	1.5	6.4	15	1.2	···	7	<0.7
Zn........	1	ND	<0.02	<0.4	<0.02	0.005	<0.004	<0.02	<0.4	<0.5
Zr........	<0.08	ND	<0.04	<0.15	<0.3	<0.003	<0.1	<0.6	<0.12	···
RE(c)	<0.5	<0.1	<1	<0.5	<0.5	<0.1	<0.2	<0.2	<1	<0.5

(a) See Table 1 for resistance ratios of these samples. (b) ND, not detectable (<0.01 ppm). (c) Rare earths.

Properties of Pure Metals*

Actinium

See table at end of this section

Aluminum (Al)

Compiled by H. Y. Hunsicker
Aluminum Company of America
L. F. Mondolfo
Consultant
and
P. A. Tomblin
The De Havilland Aircraft Co. of
Canada, Ltd.

Structure

Crystal structure. Face-centered cubic; $a = 0.404958$ nm at 25 °C
Slip plane. (111)
Slip direction. [110]
Twinning plane. (111)

Mass Characteristics

Atomic weight. 26.98154
Density. 2.6989 Mg/m^3 at 20 °C.
Effect of temperature, see Fig. 1.
Effect of deformation, 0.1 to 0.3%

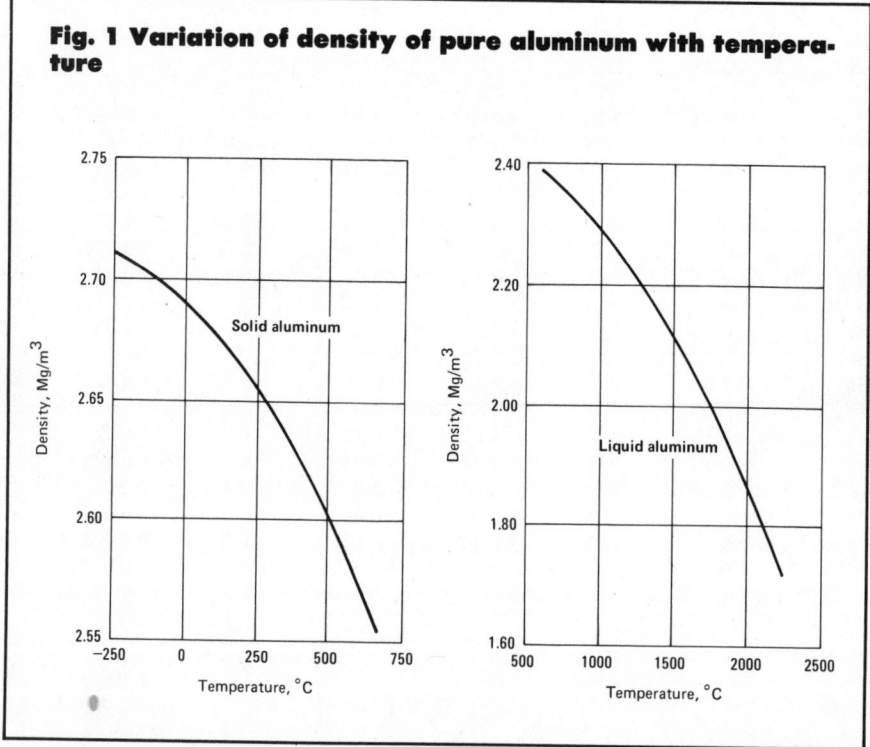

Fig. 1 Variation of density of pure aluminum with temperature

*The purity of the sample used in determining any of the properties listed in these compilations varies considerably from one metal to another and from one property to another. Furthermore, the technique used to measure a given property may vary. As a result, the accuracy and precision of the reported values vary. However, the values listed in these compilations are regarded by the compilers to be the best values available. Additional data may be found in the compilations for commercially pure grades of metals, but these additional values may not apply to samples of laboratory purity. All temperatures related to thermodynamic data have been corrected to the International Practice Temperature Scale of 1968, unless otherwise noted. For a discussion of the 1948 and 1968 temperature scales, see T. B. Douglas, *Journal of Research of the National Bureau of Standards*, Vol 73A, 1969, p 451–470.

Table 1 Tensile properties of aluminum

Amount of cold work	Tensile strength, MPa	Yield strength, MPa	Elongation, %
Annealed	40 to 50	15 to 20	50 to 70
40%	80 to 90	50 to 60	15 to 20
70%	90 to 100	65 to 75	10 to 15
90%	120 to 140	100 to 120	8 to 12

Fig. 2 Hardness and strength of aluminum as function of purity

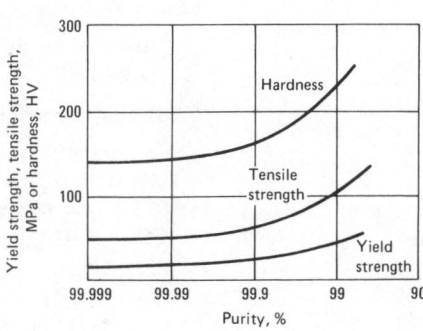

decrease at 90 to 99% deformation
Volume change on freezing. 6.5% contraction

Thermal Properties

Melting point. 660.4 °C
Boiling point. 2494 °C
Thermal expansion:

Temperature range, °C	Average coefficient $\mu m/m \cdot K$
−200 to 20	18.0
−150 to 20	19.9
−100 to 20	21.0
−50 to 20	21.8
20 to 100	23.6
20 to 200	24.5
20 to 300	25.5
20 to 400	26.4
20 to 500	27.4

Specific heat. At 25 °C, 900 J/kg·K; at 660.4 °C (liquid), 1.18 J/kg·K
Latent heat of fusion. 397 kJ/kg
Latent heat of vaporization. 10.78 MJ/kg
Heat of combustion. 31.05 MJ/kg Al
Thermal conductivity. At 25 °C, 247 W/m·K; at 660.4 °C (liquid), 90 W/m·K

Electrical Properties

Electrical conductivity. Volumetric, 64.94% IACS
Electrical resistivity. 26.55 nΩ·m at 20 °C Temperature coefficient, 114.5 pΩ·m per K at 20 °C
Electrolytic solution potential. −0.84 V vs 0.1N calomel electrode in NaCl-H_2O_2 solution
Temperature of superconductivity. 1.2 K

Magnetic Properties

Magnetic susceptibility. Mass: 7.88×10^{-9} mks

Optical Properties

Spectral reflectance. 90% for white light from a tungsten filament, 86 to 87% for λ = 220 to 250 nm, 96% for λ = 1.0 μm, 97% for λ = 1.1 to 10 μm
Spectral hemispherical emittance. At 9.3 μm: 3% at 25 °C, 15 to 20% at 660 °C (liquid)

Nuclear Properties

Neutron cross section. For neutron energy of 0.02 V, 0.2 b/atom; for 100 MV, 0.6 to 0.7 b/atom

Mechanical Properties

Tensile properties. See Table 1 and Fig. 2.
Hardness. See Fig. 2.
Modulus of elasticity. 62 GPa
Elastic modulus. Shear, 25 GPa at 25 °C
Velocity of sound. At 25 °C, 6200 m/s at 660 °C (liquid), 4650 m/s

Americium

See tables at end of this section

Antimony (Sb)

Compiled by S. C. Carapella, Jr.
ASARCO Incorporated

Antimony is used as an alloying element in lead alloys (for battery grids, printers' type, solder, bearings, cable sheathing and ammunition) and in tin alloys (such as pewter and costume jewelry alloys). Trace quantities are added to copper-base alloys, to prevent dezincification; to ductile iron, to assist in forming nodular graphite; and to gray iron, where the antimony acts as a powerful pearlite former. In the form of Sb_2O_3, antimony is used in enamels, glass, pigments, catalysts and flame retardants. Antimony is used as a component of III-V semiconductors such as InSb, AlSb and GaSb, and as an alloying ingredient in thermoelectric alloys.

Structure

Crystal structure. Hexagonal (rhombohedral equivalent); a = 0.4307 nm, c = 1.1273 nm

Mass Characteristics

Atomic weight. 121.75
Density. 6.697 Mg/m³ at 26 °C

Thermal Properties

Melting point. 630.7 °C
Boiling point. 1587 °C
Coefficient of thermal expansion. Linear, 8 to 11 μm/m·K at 20 °C
Specific heat. 207 J/kg·K at 25 °C
Latent heat of fusion. 163.17 kJ/kg
Latent heat of vaporization. 1602 kJ/kg

Thermal conductivity. 25.9 W/m·K

Electrical Properties

Electrical resistivity. 370 nΩ·m at 0 °C

Magnetic Properties

Magnetic susceptibility. Volume: -10.2×10^{-6} mks

Optical Properties

Spectral reflectance. 70% for $\lambda = 58.9$ μm

Chemical Properties

Resistance to specific corroding agents:

Corroding agent	Resistance
Air	Moderate general attack when air is moist and light is present
Alkalis and alkali salts	General attack
Ammonia	Resistant
Aqua regia	Severe general attack
Carbon dioxide	Resistant
Chlorine	Severe general attack
Hydrochloric acid	Moderate attack in presence of air
Hydrofluoric acid	Resistant
Nitric acid	Severe general attack
Sulfuric acid	Severe general attack by warm concentrated acid; resistant to cold or dilute acid

Mechanical Properties

Tensile strength. 11.40 MPa
Hardness. 30 to 58 HB
Elastic modulus. Tension, 77.759 GPa; shear, 19 GPa

Arsenic (As)

Compiled by S. C. Carapella, Jr.
ASARCO Incorporated

In quantities of 0.5 to 2%, arsenic improves the sphericity of lead shot. Small percentages are added to lead alloys for battery grids and cable sheathing to improve their hardness. In amounts up to 3%, arsenic improves the properties of lead-base bearing alloys. Minor additions improve corrosion resistance and increase recrystallization temperature of copper and stabilize pearlite in ductile and gray cast irons. Arsenic is a component of III-V semiconductors such as GaAs, GaAsP and InAs.

Arsenic is normally available in the α-metallic and β-amorphous form. Other allotropes of arsenic have been reported but supportive evidence is meager.

The toxicity of arsenic is related to its chemical state and can be extremely high. The degree of toxicity of arsenic in the elemental state is relatively low. Metallic arsenic on exposure to the atmosphere will develop an oxide coating and for this reason care must be taken not to ingest or handle it. Inhalation of dust and fumes are to be avoided. Controlled exhaust ventilation of work area is required to comply with OSHA standard of 10 μg of arsenic per cubic metre of air.

Structure

Crystal structure. Hexagonal (rhombohedral equivalent); at 26 °C, $a = 0.3760$ nm and $c = 1.0548$ nm

Mass Characteristics

Atomic weight. 74.9216
Density. 5.778 Mg/m³ at 26 °C

Thermal Properties

Melting point. 816 °C at 3.91 MPa
Boiling point. Sublimes at 615 °C
Coefficient of thermal expansion. Linear, 5.6 μm/m·K at 20 °C
Specific heat. 328 J/kg·K at 25 °C
Latent heat of fusion. 370.3 kJ/kg
Latent heat of sublimation. 426.77 kJ/kg

Electrical Properties

Electrical resistivity. α-metallic form, 260 nΩ·m at 0 °C
Electrochemical equivalent. Valence 3, 0.15254 mg/C; valence 5, 0.25876 mg/C

Magnetic Properties

Magnetic susceptibility. Volume: -3.9×10^{-6} mks at 18 °C

Mechanical Properties

Hardness. 3.5 Moh

Barium (Ba)

Compiled by G. C. Carter
(Currently with National Academy of Science)
National Bureau of Standards

Structure

Crystal structure. Body-centered cubic, $cI2$ ($Im3m$); $a = 0.5025$ nm at 26 °C. High-pressure phases: Ba II, hexagonal, $hP2$ ($P6_3/mmc$), formed at 5.9 GPa; BA III, structure not determined (Ref 1).

Mass Characteristics

Atomic weight. 137.3
Density. 3.5 Mg/m³ at 20 °C (Ref 2)

Thermal Properties

Melting point. 729 ± 2 °C (Ref 3)
Boiling point. 1637 °C (calculated from vapor pressure data) (Ref 2, 4)
Specific heat:

K	°C	J/kg·K
2	−271	0.1232
4	−269	0.7805
10	−263	14.06
20	−253	67.47
30	−243	109.5
40	−233	134.7
50	−223	151.1
100	−173	177.1
200	−73	192.1
298	+25	204.7

Electronic coefficient (γ). 19 ± 3 mJ/kg·K² (Ref 3)
Latent heat of fusion. 56.4 ± 7.6 kJ/kg (Ref 3)
Temperature of superconductivity. Ba II, ~1.3 K at 5.5 GPa; Ba III, 3.05 K from 8.5 to 8.8 GPa; Ba III, ~5.2 K for pressures >14.0 GPa (Ref 5)

Electrical Properties

Work function. 1.7 to 2.55 eV (0.27 to 0.41 aJ), depending on conditions and techniques of the experimental determination (Ref 6, 7, 8)

Magnetic Properties

Magnetic susceptibility. Molar: 0.254 mks at 20 °C (Ref 9)

Nuclear properties

Stable isotopes. [130]Ba, isotope mass 129.90628, 0.10% abundant; [132]Ba, isotope mass 131.90505, 0.095% abundant; [134]Ba, isotope mass 133.90449, 2.4% abundant; [135]Ba, isotope mass 134.90567, 6.5% abundant; [136]Ba, isotope mass 135.90456, 7.8% abundant; [137]Ba, isotope mass 136.90582, 11.2% abundant; [138]Ba, isotope mass 137.90524, 71.9% abundant (Ref 10)

Unstable isotopes. ^{122}Ba, 12 m; ^{123}Ba, 12 m; ^{124}Ba, ~24 m, 11 m; ^{125}Ba, 8 m, ~6.5 m; ^{126}Ba, 97 m; ^{127}Ba, 10 m, 18 m; ^{128}Ba, 2.42 d; ^{129}Ba, 2.5 h, 2.1 h; ^{130}Ba, 9 ms; ^{131}Ba, 14.3 m, 11.7 d; ^{133}Ba, 38.9 h, 10.4 y; ^{135}Ba, 28.7 h; ^{136}Ba, 0.308 s; ^{137}Ba, 2.55 m; ^{139}Ba, 83.3 m; ^{140}Ba, 12.79 d; ^{141}Ba, 18.3 m; ^{142}Ba, 10.7 m; ^{143}Ba, 13.6 s; ^{144}Ba, 11 s; ^{145}Ba, 6.2 s; ^{146}Ba, 2.2 s (Ref 10)

Mechanical Properties

Modulus. Bulk, 10.30 GPa; isothermal compressibility, 97.1 μm^2/N (Ref 12)

REFERENCES

1. P. Eckerline, H. Kandler and A. Stegherr, Landolt-Börnstein Tables, New Series III/6, *Structure Data of Elements and Intermetallic Phases*, edited by K.-H. Hellwege and A. M. Hellwege, Springer-Verlag, NY, 1971
2. R. J. Elliott, *Constitution of Binary Alloys, First Supplement*, McGraw-Hill, NY, 1965
3. R. Hultgren, P. D. Desai, D. T. Hawkins, M. Gleiser, K. K. Kelley and D. D. Wagman, *Selected Values of the Thermodynamic Properties of the Elements*, American Society for Metals, Metals Park, OH, 1973. [Additional thermodynamic data are available in this compilation.]
4. K. A. Gschneider, Jr., Physical Properties and Interrelationships of Metallic and Semimetallic Elements, *Solid State Physics*, Vol 16, edited by F. Seitz and D. T. Turnbull, Academic Press, NY, 1964, p 275
5. B. W. Roberts, "Superconductive Materials and Some of their Properties", NBS Technical Note 724, Superintendent of Documents, U.S. Government Printing Office, Washington, DC
6. V. S. Fomenko, *Handbook of Thermionic Properties - Electronic Work Functions and Richardson Constants of Elements and Compounds*, edited by G. V. Samsonov (translation from the Russian), Plenum Press Data Division, NY, 1966
7. G. A. Haas and R. E. Thomas, Thermionic Emission and Work Function, chapter 2 in *Measurements of Physical Properties*, edited by E. Passaglia, in Vol 6, part 1 of *Techniques of Metals Research*, edited by R. F. Bunshah, Interscience, 1972, p 91
8. H. B. Michaelson, *Handbook of Chemistry and Physics*, 55th Ed., edited by R. C. Weast, CRC Press, Cleveland, OH, 1974, E-81
9. Landolt-Börnstein Tables, II Band, 9. Teil, *Magnetische Eigenschaften 1*, K.-H. Hellwege and A. M. Hellwege, Eds., Springer-Verlag, NY, 1962
10. N. E. Holden and F. W. Walker, Chart of the Nuclides, Knolls Atomic Power Laboratory, United States Atomic Energy Commission, distributed by Educational Relations, General Electric Co., Schenectady, NY. [Revised to April 1972, dated October 1972. Additional nuclear data are available on this chart.]
11. W. B. Pearson, *Handbook of Lattice Spacings and Structures of Metals*, Pergamon Press, NY, Vol 1, 1958; Vol 2, 1967

Berkelium

See tables at end of this section

Beryllium (Be)

Compiled by John P. Denny
Kawecki Berylco Industries, Inc.

Beryllium is used as an alloying addition to copper and nickel to produce an age-hardening alloy used for springs, electrical contacts, spot welding electrodes and nonsparking tools. Beryllium is also added to aluminum and magnesium to achieve grain refinement and oxidation resistance. Unalloyed beryllium is used in weapons, spacecraft, nuclear reactor reflector segments, neutron sources, windows for x-ray tubes and radiation detection devices, rocket nozzles, aircraft brake discs, precision instruments and mirrors. Inhalation or ingestion of beryllium and its compounds should be avoided. Users should comply with occupational safety and health standards applicable to beryllium in Title 29, Part 1910, Code of Federal Regulations.

Structure

Crystal structure. α-phase, close-packed hexagonal; $a = 0.22858$ nm, $c = 0.35842$ nm at room temperature. β-phase, body-centered cubic; $a = 0.255$ nm at 1270 °C

Slip planes. Primary: (0002), (10$\bar{1}$0). Secondary: (10$\bar{1}$1)

Twinning planes. Primary: (10$\bar{1}$2). Secondary (10$\bar{1}\bar{1}$), (10$\bar{1}$2), (10$\bar{1}$3)

Mass Characteristics

Atomic weight. 9.0122
Density. 1.848 Mg/m^3 at 20 °C

Thermal Properties

Melting point. 1290 °C
Boiling point. 2770 °C
Phase-transformation temperature. 1250 °C
Coefficient of thermal expansion. Linear:

Temperature range, °C	Average coefficient, μm/m·K
25 – 100	11.6
25 – 300	14.5
25 – 600	16.5
25 – 1000	18.4

Specific heat. 1.886 kJ/kg·K at 20 °C

Latent heat of fusion. 1.30 MJ/kg

Thermal conductivity. 190 W/m·K

Electrical Properties

Electrical conductivity. 38 to 43% IACS

Electrical resistivity. 40 nΩ·m at 20 °C

Temperature coefficient of electrical resistivity. 0.025 $\Delta\rho/\rho_0$/K

Magnetic Properties

Magnetic susceptibility. Mass: -1×10^{-2} mks at 20 °C

Optical Properties

Color. Steel gray
Spectral hemispherical emittance. 61% for $\lambda = 650$ nm

Nuclear Properties

Thermal neutron cross section. 0.01 b

Chemical Properties

General corrosion behavior. Beryllium, a highly reactive metal, forms stable compounds with most other elements. It has excellent corrosion resistance at room tempera-

Table 2 Tensile properties of beryllium

Temperature, °C	Tensile strength, MPa	Yield strength(a), MPa	Elongation, %
Vacuum Hot-pressed Block			
Room temperature	228-352(b)	186-262	1-3.5
200........................	207-297	...	6-15
400........................	152-186	...	19-40
600........................	138-152	...	15-25
800........................	48	...	7-8
Hot Extruded Billet			
Room temperature	483-690	310	5-20
Cross-rolled Sheet			
Room temperature	483-621	345-414	10-40
Hot-isopressed Billet (High Purity)			
Room temperature	345-414	242-276	3-6

(a) 0.2% offset. (b) Instrument grades have minimum tensile strength of 345 MPa with microyield strength of (35 to 83 MPa).

ture (except for certain acids and alkalies) because of a thin protective oxide coating.

Resistance to specific corroding agents. Beryllium reacts appreciably with oxygen and nitrogen above 760 °C. Impure beryllium containing carbide or chloride reacts with moist air at room temperature. Beryllium does not react with hydrogen at any temperature; it reacts with fluorine at room temperature and with C1, B, I, HCl, HF and CO_2 at elevated temperatures. Beryllium has excellent resistance to pure water up to 300 °C if carbide and chloride are absent and if grain size is fine. Beryllium is attacked by dilute HF, HCl, H_2SO_4, and HNO_3 at room temperature.

Fabrication Characteristics

Machinability. Powder metallurgy material behaves well in most machining operations, provided carbide tools are used. Chip formation is similar to that of cast iron. All machining, especially careless grinding, produces a damaged surface layer that must be removed by etching a minimum of 0.05 mm per surface for critical, highly stressed applications.

Recrystallization temperature. 725 to 900 °C, depending on amount of cold work and annealing time

Hot-working temperature. 800 to 1100 °C

Mechanical Properties

Tensile properties. Especially elongation, depend strongly on preferred orientation and grain size. Beryllium mill products are produced chiefly by powder metallurgy using different consolidation procedures. Wrought products are produced either from cast or powder metallurgy materials. Extreme anisotropy in elongation occurs in wrought material; hence, uniaxial tensile results are valueless as an indication of behavior under conditions of complex stress. Tensile properties of different forms may vary extremely, with the greatest uniformity occurring for vacuum hot-pressed powder. See Table 2 for some representative values.

Hardness. 75 to 85 HRB

Poisson's ratio. 0.02 to 0.075

Elastic modulus. 275 to 300 GPa

Impact strength. 1.4 to 5.4 J

Plane-strain fracture toughness. 9 to 13 MPa$\sqrt{m}$

SELECTED REFERENCES

A. R. Kaufmann, P. Gordon and D. W. Lillie, The Metallurgy of Beryllium, *Transactions of the American Society for Metals,* Vol 42, 1950, p 785

W. W. Beaver and K. G. Wikle, Mechanical Properties of Beryllium Fabricated by Powder Metallurgy, *Transactions of the American Institute for Mechanical Engineers,* Vol 200, 1954, p 559

Reactor Handbook, 2nd Ed., Vol 1, Materials, Interscience Publishers, NY, 1960

G. E. Darwin and J. H. Buddery, *Beryllium,* Academic Press, NY, 1960

H. H. Hausner, *Beryllium: Its Metallurgy and Properties,* University of California Press, Berkeley, CA, 1965

Bismuth (Bi)

Compiled by S. C. Carapella, Jr.
ASARCO Incorporated

Bismuth is used extensively in the production of fusible alloys (low-melting-point alloys) as a carbide stabilizer in the manufacture of malleable iron, and as an additive to low-carbon steel and aluminum to improve machinability. Compounds of bismuth are used for catalysts, in pharmaceutical and for semiconductor applications.

Structure

Crystal structure. Hexagonal (rhombohedral equivalent); at 25 °C, $a = 0.4546$ nm and $c = 1.1860$ nm

Mass Characteristics

Atomic weight. 208.980

Density:

°C	Mg/m³
25	9.808
271	1.0067
300	1.003
400	0.991
600	0.966
802	0.940
962	0.920

Thermal Properties

Melting point. 271.4 °C

Boiling point. 1564 °C

Coefficient of thermal expansion. Linear, 13.2 μm/m·K at 20 °C

Specific heat:

°C	J/kg·K
25	122
217.4	146
327	141
427	137
527	134

Latent heat of fusion. 53.976 kJ/kg

Latent heat of vaporization. 854.780 kJ/kg

Thermal conductivity:

°C	W/m·K
0	8.2
300	11.3
400	12.3
500	13.3
600	14.5

Vapor pressure:

°C	kPa
893	0.1013
1053	1.013
1266	10.13
1564	101.3

Electrical Properties

Electrical resistivity:

°C	nΩ·m
0	1050
300	1289
700	1535

Mechanical Properties

Hardness. 7.0 HB; 2.5 Moh
Elastic modulus. Tension, 32 GPa
Liquid surface tension:

°C	mN/m
300	376
400	370
500	363

Magnetic Properties

Magnetic susceptibility. Volume: -1.68×10^{-5} mks

Nuclear Properties

Stable isotopes. ^{209}Bi
Thermal neutron cross section. For 2.2 km·s neutrons: absorption, 0.034 ± 0.002 b; scattering, 9 ± 1 b

Boron (B)

Compiled by James C. Schaefer
ESB, Inc.

Hot-wire boron is prepared by reduction of boron halides on a hot tungsten wire. The boron may form as wire or small chunks. It is black, the purest form of boron. Electrolytic boron, prepared by molten salt electrolysis, is a black powder of 40 to 325 mesh. Electrolytic boron has a purity of 99% and above. Magnesium-reduced boron, prepared by reduction of boric oxide (B_2O_3) with magnesium, results in a light brown powder in the purity range of 95 to 97%.

Boron formed on hot tungsten wire is used for reinforcement of metals and plastics and weight reduction. Electrolytic boron powder is used for preparation of borides for deoxidation of alloys. Isotopic B-10 and its compounds are used for neutron absorption. Boron in compound form is used for medicinal and cleaning purposes, as rocket fuels, diamond substitutes, and additives to aluminum alloys to improve electrical and thermal conductivity, and for grain refining of aluminum alloys.

Boron is nontoxic. No precautions are required for wire form boron. Electrolytic product contains fine dust, which slowly oxidizes. Large particles are not affected, but the fine dust should be kept in closed containers and under a protective gas such as argon. Magnesium-reduced boron is a powder that requires only the normal precautions of air filtration and facial masks for the workers. Borides require no handling or storage precautions, but boron hydrides are very sensitive to shock and can detonate easily, and boron halides are corrosive and toxic.

Structure

Crystal structure. Material prepared at about 800 °C and below: amorphous. Prepared between about 800 and 1100 °C: α phase, rhombohedral, $R\bar{3}m$; $a = 0.506$ nm, $\alpha = 58°4'$; unit cell contains a single B_{12} icosahedron. Prepared between about 1100 and 1300 °C: γ phase, tetragonal, $P4_2/nnm$; $a = 0.875$ nm, $c = 0.506$ nm; the 50 atoms per unit cell are distributed among four equivalent B_{12} icosahedra of required symmetry $2/m$ and two tetrahedral positions $42m$. Prepared above about 1300 °C: β phase, rhombohedral, $R\bar{3}m$; $a = 1.012$ nm, $\alpha = 65°28'$; unit cell contains $108 \pm$ atoms, values differ slightly according to investigator

Mass Characteristics

Atomic weight. 10.81
Density. Amorphous, 2.3 Mg/m^3 at <800 °C; α phase, 2.46 Mg/m^3 at 800 to 1100 °C; γ phase, 2.37 Mg/m^3 at 1100 to 1300 °C; β phase, 2.35 Mg/m^3 at >1300 °C

Thermal Properties

Melting point. Approximately 2300 °C
Boiling point. Approximately 2550 °C
Phase transformation temperature. Unknown
Coefficient of thermal expansion. Linear, 1.1 to 8.3 μm/m·K in temperature range from 20 to 750 °C
Specific heat:

K	°C	kJ/kg·K
82–195	−191– −78	0.0297
197–273	−76–0	0.754
273–373	0– +100	1.285
373	100	1.620
773	500	1.976
1173	900	2.135

Enthalpy:

K	°C	kJ/kg
400	127	120
600	327	416
800	527	786
1000	727	1200

Entropy. 604 J/kg·K
Latent heat of fusion. 22 000 kJ/kg
Latent heat of vaporization. 34 900 kJ/kg
Heat of combustion. 5.4 J/kg

Electrical Properties

Electrical resistivity:

°K	°C	Resistivity
123	−150	4×10^5
263	−10	4×10^4
273	0	3×10^4
300	+27	6.5×10^3
373	100	4×10^2
443	170	30
593	320	0.4×10^{-1}
793	520	1.2×10^{-2}
873	600	2×10^{-3}

Electrochemical equivalent. 37 μg/C
Standard electrode potential. At 25 °C, 0.87 V vs standard hydrogen electrode
Ionization potentials:

I	8.296 V
II	23.98 V
III	37.75 V
IV	258.1 V
V	338 V

Semiconductor properties. p-type dopant for silicon and germanium. Intrinsic current carrier concentration, 5×10^{20} per m^3 at 160 °C to 9×10^{25} per m^3 at 850 °C
Dielectric constant. Approximately 12

Optical Properties

Color. Crystalline is black; amorphous is brown
Refractive index. 2.5 using Hg line (579 nm)

Nuclear Properties

Stable isotopes. B-10, atomic weight 10.01294, 19.9% abundant; B-11, atomic weight 11.00931, 80.1% abundant
Neutron absorption. B-10, 3850 b; B-11, 0.05 b; natural boron, 755 b
Unstable isotopes:

Isotope	Atomic weight	Half life	Particles emitted
B-8.....	...	0.78 s	β^+
B-9.....	9.01333	3×10^{-19} s	P, (2α)
B-12.....	...	0.019 s	β^-
B-13.....	...	0.035 s	β^-

Chemical Properties

Effects of specific corroding agents. Reactivities and conditions for reaction of boron with several materials are
Fluorine, instantaneous at room temperature
Chlorine, above 500 °C
Bromine, above 600 °C
Iodine, about 900 °C
Hydrochloric acid, none
Hydrofluoric acid, none
Nitric acid (hot, concentrated), slow
Oxygen, slight at room temperature, rapid above 1000 °C
Hydrogen iodide, explosive
Hydrogen, above 840 °C
Nitrogen, bright red heat
Sodium hydroxide, no reaction at room temperature, slow at 500 °C
Boron nitride, none
Metals, CAUTION: above 900 °C, many metals react rapidly with boron, and the reactions are exothermic.

Mechanical Properties

Tensile properties. Tensile strength: 98.8% pure, amorphous, 1.6 to 2.4 MPa; fibers, 2.6 to 3.1 MPa
Compressive properties. Compressive strength: with B_2O_3 present, up to 0.5 MPa
Hardness. 99.9% crystalline: 3300 HK (with 100-g load), 9.3 moh
Elastic modulus. Tension: amorphous, 440 MPa

Cadmium (Cd)

Compiled by S. C. Carapella, Jr.
ASARCO, Incorporated

Cadmium is used for electroplating steel to improve its corrosion resistance, in powder form for mechanical plating of fasteners and other parts for corrosion protection, in low-melting-point alloys, brazing alloys, bearing alloys, nickel-cadmium batteries, and nuclear control rods, and as an alloying ingredient to copper to improve hardness. It also is used in the manufacture of pigments, plastic stabilizers, phosphors and semiconductor compounds. Cadmium with a minimum purity of 99.90% is available commercially; cadmium of this purity is covered by ASTM B440. Care must be taken to avoid creating toxic dust or fume; melting or handling conditions that do create dust or fume require capture at the source by an exhaust ventilation system. When capture of dust or fume is not feasible, approved NIOSH respiratory protective equipment must be worn. Maximum threshold limit values for cadmium dust is 0.2 mg/m^3 and for cadmium oxide fume 0.1 mg/m^3.

Structure

Crystal structure. Close-packed hexagonal, D_6^4h ($P6_3/mmc$); $a = 0.29793$ nm and $c = 0.56181$ nm at 26 °C
Slip plane and direction. (0001), [1120]
Twinning plane. (1012)
Distance of closest approach. 0.2973 nm

Mass Characteristics

Atomic weight. 112.40
Density:

°C	Mg/m³
26	8.642
330 (liquid)	8.020
400	7.930
600	7.720

Volume contraction on freezing. 4.74%

Thermal Properties

Melting point. 321.1 °C
Boiling point. 767 °C
Coefficient of thermal expansion. Linear, 31.3 μm/m·K at 20 °C
Specific heat. 230 J/kg·K at 20 °C; 264 J/kg·K at 321 to 700 °C
Latent heat of fusion. 55 kJ/kg
Latent heat of vaporization. 887 kJ/kg
Thermal conductivity. 98 W/m·K at 0 °C
Vapor pressure:

°C	kPa
382	0.1013
437	1.013
595	10.13
767	101.3

Electrical Properties

Electrical conductivity. Volumetric, 25% IACS at 20 °C
Electrical resistivity. 72.7 nΩ·m at 22 °C; 341 nΩ·m at 400 °C; 348 nΩ·m at 600 °C, 358 nΩ·m at 700 °C
Electrochemical equivalent. Valence +2, 582.4 Mg/C
Electrode reduction potential. 0.40 V for the reaction Cd = Cd^{++} + 2e$^-$, where potential for H$_2$ = 0.0 V

Magnetic Properties

Magnetic susceptibility. Volume: -2.2×10^{-6} mks

Optical Properties

Color. Silver-gray
Spectral reflectance:

λ, nm	%
410	78
474	74
518	72.5
554	73

Refractive index. 1.8 at λ = 578 nm
Absorptive index. 1.17 at λ = 578 nm

Nuclear Properties

Stable isotopes:

Atomic weight	Relative abundance, %
106	1.22
108	0.88
110	12.39
111	12.75
112	24.07
113	12.26
114	28.86
116	7.58

Thermal neutron cross section. At 2200 m/s, $2450 \pm 50 \times 10^{-28}$ b

Chemical Properties

General corrosion behavior. As a protective coating on steel and cast iron parts, cadmium offers corrosion protection in marine atmospheres, under alkaline conditions, and in damp indoor applications. Cadmium-plated steel fasteners resist galvanic attack when used with aluminum parts.

Mechanical Properties

Tensile strength. 71 MPa
Elongation. 50% in 1 in.
Hardness. 16 to 23 HB
Poisson's ratio. 0.33 at room temperature
Elastic modulus. Tension, 55 GPa; shear, 19.2 GPa
Liquid surface tension. 0.564 N/m at 330 °C; 0.611 at 450 °C

Calcium (Ca)

Compiled by J. F. Smith
Ames Laboratory
U. S. Department of Energy
Iowa State University

Metallic calcium is used as a reducing agent in the preparation of thorium, zirconium, uranium, chromium, vanadium and the rare earths. It is also used as a deoxider, decarburizer or desulfurizer for various ferrous and nonferrous alloys. Calcium is used as an alloying or modifying agent for aluminum, beryllium, copper, lead, tin and magnesium alloys. Other uses for calcium include getters for residual gases in high vacuums and vacuum-tube applications and reagents for purification and scavenging of inert gases. Calcium reacts readily with atmospheric components, particularly water vapor, and is not inert to nitrogen. To avoid contamination it must be handled in a dry inert gas atmosphere or in a vacuum.

Structure

Crystal structure. α-phase, face-centered cubic, $cF4$ *(Fm3m); a =* 0.5588 nm at 26.6 °C. β-phase, body-centered cubic, $cI2$ *(Im3m); a =* 0.4480 nm at 467 °C. Minor amounts of hydrogen stabilize a hexagonal form, and a low-symmetry form of undetermined structure results from contamination by nitrogen and/or carbon.
Minimum interatomic distance. 0.3952 nm at 25 °C

Mass Characteristics

Atomic weight. 40.08
Density. Solid, 1.55 Mg/m^3 at 25 °C; liquid, 1.37 Mg/m^3 at 839 °C
Density vs temperature. Solid, $\Delta d/d_0 \cdot K = -66.9 \times 10^{-6}$ at 0 to 400 °C; liquid, $\Delta d/d_0 \cdot K = -221 \times 10^{-6}$ at 839 to 1382 °C
Volume change on freezing. 4.7% contraction
Volume change on phase transformation. β- to α-phase, 0.04% contraction

Thermal Properties

Melting point. 839 °C at 1 atm; $\Delta T_m/\Delta P = 170$ μK per Pa
Boiling point. 1484 °C
Phase-transformation temperature. 448 °C; ΔT trans/$\Delta P = 33$ μK/Pa
Coefficient of thermal expansion. Linear, α-phase. Up to −267 °C: $\Delta l/l_0 \cdot K = 5 \times 10^{-11}T + 81 \times 10^{-12}T^3$, where T is in K; 1 μm/m·K at −253 °C; 3.3 μm/m·K at −243 °C; 14.16 μm/m·K at −198 °C; 22.15 μm/m·K at 10 °C; 22.3 μm/m·K (avg) for 0 to 400 °C. β-phase, 33.6 μm/m·K for 467 to 603 °C. Liquid, 73.7 μm/m·K for 839 to 1382 °C.
Specific heat:

°C	J/kg·K
25	631.5
127	654.9
327	737.5
448 (fcc)	807.9
448 (bcc)	732.0
627	917.4
839 (bcc)	1136
839 (liquid)	730.8
1027	730.8

Enthalpy. $H_{298} - H_0 = 142.4$ kJ/kg; entropy, $S_{298} = 1.03$ kJ/kg·K
Heat of fusion. 212.9 kJ/kg
Latent heat of transformation. α- to β-phase, 22.95 kJ/kg
Latent heat of sublimation. 4.447 MJ/kg aT 25 °C

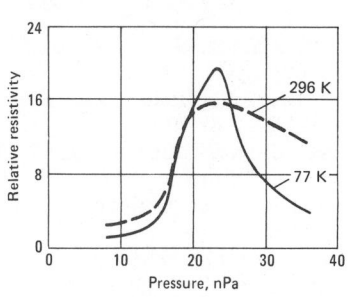

Fig. 3 Pressure dependence of the electrical resistivity of calcium

Relative resistivity is the ratio of the resistivity at the high pressure to the resistivity at the same temperature, but at one atmosphere pressure.

Heat and free energy of formation of oxide. CaO: $\Delta H^0_{298} = -11.32$ MJ/kg Ca; $\Delta G^0_{298} = -10.77$ MJ/kg Ca
Thermal conductivity. W/m·K: $\lambda = 190 - 0.22T$ from 150 to 360 °C, $\lambda = 0.31T - 1.5$, from 360 to 600 °C, where T is in K
Vapor pressure. α-phase, $\log P = 10.77 \times 9260/T$; β-phase, $\log P = 10.38 \times 8980/T$; liquid, $\log P = 9.67 - 8190/T$; where T is in K and P is in Pa
Diffusion coefficients. At 500 to 800 °C:

Element	$D_0 m^2/s$	H, KJ/mol
Ca	8.3 x 10^{-4}	161
C	2.7 x 10^{-7}	97.5
Fe	3.2 x 10^{-9}	125
Ni	1.0 x 10^{-9}	121
U	1.1 x 10^{-9}	146

Electrical Properties

Electrical conductivity. Volumetric, 49.6% IACS
Electrical resistivity. α-phase, 31.6 nΩ·m at 0 °C; liquid, 330 nΩ·m at 839 °C
Temperature coefficient of electrical resistivity. 4.02×10^{-3} per K at 0 °C
Pressure dependence of electrical resistivity. Unusual and currently believed to result from a manifestation of the electronic band structure

whose degree of overlap has been calculated to be highly sensitive to interatomic spacing. See Fig. 3.

Thermoelectric potential. vs Cu: β-phase, 9.6 μV/K at the melting point; liquid, 9.9 μV/K at the melting point

Electrochemical equivalent. 0.20762 mg/C

Electrolytic solution potential. vs H_2, −2.87 V at 20 °C

First ionization potential. 6.11 V

Hall coefficient. −0.228 nV·m/A·T (independent of temperature from −193 to 27 °C)

Work function. Thermionic, 0.359 aJ; photoelectric, 0.46 aJ

Magnetic Properties

Magnetic susceptibility. Volume: 2.71×10^{-5} mks

Magnetic permeability. 1.0000271

Optical Properties

Color. A fresh clean surface is a lustrous silvery white but darkens upon exposure to the atmosphere

Dielectric constant. Ellipsometry has been used to determine the real (ϵ_1) and imaginary (ϵ_2) parts of the complex dielectric constant as a function of vacuum wavelength (λ):

λ, nm	$-\epsilon_1$	ϵ_2
1771.2	65.2	...
1549.8	51.8	11.7
1377.6	41.8	8.58
1239.8	34.2	7.08
1033.2	24.2	4.71
885.6	17.5	3.40
774.9	12.8	2.48
688.8	9.6	1.95
619.9	7.5	1.62
563.6	5.8	1.39
516.6	4.5	1.22
476.9	3.6	1.09
442.8	2.8	1.01
413.3	2.2	1.01
364.7	1.4	1.02
326.3	...	0.95

Reflectivity. At normal incidence (r), refractive index (n), and adsorptive index (A) may be generated from these quantities through the following relations: $\epsilon_2 = 2nK$; $\epsilon_1 = n^2 + K^2$; $A = 4\pi K/\lambda$; $r = [(n-1)^2 + K^2] /[(n+1)^2 + K^2]$

Nuclear Properties

Stable isotopes:

Mass No.	Abundance, %
40	96.97
42	0.64
43	0.145
44	2.06
46	0.0033
48	0.185

Unstable isotopes. Isotopes with mass numbers 37, 38, 39, 41, 45, 47, 49 and 50 have been produced. No. 45 has the longest half-life (180 days); it is used in tracer experiments.

Chemical Properties

General corrosion behavior. Extremely poor corrosion resistance

Fabrication Characteristics

Recrystallization temperature. Below 300 °C; even at room temperature, X-ray diffraction patterns show no broadening or distortion after extensive deformation.

Mechanical Properties

Tensile properties. Annealed: tensile strength, 48.0 MPa; yield strength, 13.7 MPa; elongation, 51 to 53%; reduction in area, 58 to 62%. As rolled: tensile strength, 115 MPa; yield strength, 84.8 MPa; elongation, 7%; reduction in area, 35%

Hardness. Annealed: 16 to 18 HB

Poisson's ratio. 0.31

Elastic modulus. Tension, 19.6 GPa; shear, 7.38 GPa; bulk, 15.2 GPa

Compressibility. For 0 to 3900 MPa, $\Delta V/V_0$ Pa $= -6.578 \times 10^{-5} + 7.732 \times 10^{-11} P - 4.9 \times 10^{-13} P^2$, where P is in MPa

Liquid surface tension. For 839 to 1000 °C, $\gamma = 0.472 - 10^{-4}T$, where T is in K and γ is in N/m

Californium

See tables at end of this section

Cerium (Ce)

Compiled by K. A. Gschneidner, Jr. and B. J. Beaudry
Ames Laboratory
U.S. Department of Energy
Iowa State University

Cerium is used as an alloying additive to ferrous alloys to scavenge sulfur, oxygen, etc., and to nodulize cast iron; improves high temperature oxidation resistance of super-alloys. It is also used in glass polishing compounds, petroleum cracking catalysts, lighter flints, glass decolorizing agents, carbon arc lights, ceramic capacitors, $CeCo_5$ permanent magnets and pyrophoric ordnance devices. Cerium readily oxidizes at room temperature in air. It should be stored in vacuum or inert atmosphere; storage in oil is not recommended. Turnings can be ignited easily and burn white hot. Finely divided cerium should not be handled in air.

Structure

Crystal structure. α-phase, face-centered cubic, $Fm3m$ O^5_h; $a = 0.485$ nm at 77 K. β-phase, double close-packed hexagonal, $P6_3/mmc$ D^4_{6h}; $a = 0.36810$ nm, $c = 1.1857$ nm at 24 °C. α-phase, face-centered cubic, $Fm3m$ O^5_h; $a = 0.51610$ nm at 24 °C. δ-phase, body-centered cubic, $Im3m$ O_h; $a = 0.411$ nm at 768 °C

Minimum interatomic distance. γ-phase, 0.171 nm at 77 K; β-phase, $r_a = 0.18405$ nm, $r_c = 0.18237$ nm, radius $CN_{12} = 0.18321$ nm at 24 °C; γ-phase, 0.18247 nm at 24 °C

Mass Characteristics

Atomic weight. 140.12

Density. α-phase, 8.160 Mg/m^3 at 77 K; β-phase, 6.6893 Mg/m^3 at 24 °C; γ-phase, 6.7704 Mg/m^3 at 24 °C; δ-phase, 6.70 Mg/m^3 at 768 °C; liquid, 6.679 Mg/m^3 at 804 °C

Volume change on freezing. 1.1% expansion

Volume change on phase transformation. γ- to α-phase, 16.0% volume contraction on cooling at 110 K; γ- to β-phase, 1.2% volume expansion on cooling at 273 K; δ- to γ-phase, 0.3% volume expansion on cooling

Thermal Properties

Melting point. 798 ± 3 °C

Boiling point. 3433 °C

Phase-transformation temperature. γ- to δ-phase: 726 °C; γ- to β-phase: $M_s = 237$ to 278 K, $M_f = ?$. β- to γ-phase: $A_s = 373$ to 451 K, $A_f = 420$ to > 451 K. γ- to α-phase: $M_s =$

89 to 116 K, $M_f \cong$ 4.2 K. α- to γ + β-phases: A_s = 158 to 180 K, A_f = 190 to 210 K. β- to α-phase: M_s = 45 K, M_f = 15 K; α- to β-phase: A_s = 125 K, A_f = 200 K

Coefficient of thermal expansion. At 24 °C. Linear: γ-phase, 6.3 μm/m·K. Linear along crystal axes. γ-phase, 6.3 μm/m·K along a axis. Volumetric: γ-phase, 18.9×10^{-6} per K

Specific heat. 192.3 J/kg·K at 25 °C

Entropy. At 25 °C, est, 495.6 J/kg·K

Latent heat of fusion. 38.97 kJ/kg

Latent heat of transformation. γ- to δ-phase, 21.34 kJ/kg

Latent heat of vaporization. 3.016 MJ/kg at 25 °C

Heat of combustion. For cubic CeO_2 at 25 °C: ΔH_c^0 = 7.78 MJ/kg Ce; ΔG_f^0 = −7.35 MJ/kg Ce

Recrystallization temperature. About 325 °C

Thermal conductivity. γ-phase, 11.3 W/m·K at 25 °C

Vapor pressure. 0.001 Pa at 1290 °C; 0.101 Pa at 1554 °C; 10.1 Pa at 1926 °C; 1013 Pa at 2487 °C

Electrical Properties

Electrical resistivity. β-phase, 828 nΩ·m at 25 °C; 41 nΩ·m at 2 K. γ-phase, 744 nΩ·m at 25 °C. Liquid, 1270 nΩ·m at 800 °C

Ionization potential. Ce(I): 5.47 V; Ce(II): 10.85 V; Ce(III): 20.198 V; Ce(IV): 36.758 V

Hall coefficient. +0.181 n nV·m/A·T at 25 °C

Temperature of superconductivity. Bulk cerium not superconducting down to 0.25 K at atmospheric pressure; α-phase becomes superconducting at 1.5 K at 5 GPa

Magnetic Properties

Magnetic susceptibility. Volume, mks units. β-phase: 1.50×10^{-3} at 25 °C, obeys Curie-Weiss law from 50 to 320 K with an effective moment of 2.61 Bohr magnetons and θ = −41 K. γ-phase: 1.38×10^{-3} at 25 °C, obeys Curie-Weiss law above 0 °C with an effective moment of 2.52 Bohr magnetons and θ = −50 K

Magnetic transformation temperature. β-phase: Néel temperatures at 12.4 K (cubic sites) and 13.7 K (hexagonal sites)

Optical Properties

Color. Metallic silver

Spectral hemispherical emit- tance. 30.9% for λ = 645 nm at 850 to 1225 °C

Nuclear Properties

Thermal neutron cross section. 0.7 b

Chemical Properties

General corrosion behavior. Cerium oxidizes readily in air at room temperature. Oxidation rates increase with temperature. Interstitial impurities increase the oxidation rate, while some solid solution additives, like scandium, decrease the oxidation rate. Hydrogen will react with cerium at room temperature.

Resistance to specific corroding agents. Cerium reacts vigorously with dilute acids. Cold water slowly attacks cerium, hot water reacts faster. The presence of the fluoride ion retards acid attack by the formation of CeF_3 on the surface of the metal.

Mechanical Properties

Tensile properties. β-phase: yield strength, 87 MPa; reduction in area, 24% at 24 °C. γ-phase: tensile strength, 117 MPa; yield strength, 28 MPa; elongation, 22%; reduction in area, 34% at 24 °C

Hardness. 22 HV

Poisson's ratio. γ, 0.248

Strain-hardening exponent. 0.3

Elastic modulus. γ-phase at 27 °C: tension, 30.0 GPa; shear, 12.0 GPa; bulk, 19.8 GPa

Kinematic liquid viscosity. 0.479 mm²/s at 804 °C

Liquid surface tension. 0.72 N/m at 804 °C

Cesium (Cs)

Compiled by John H. Madaus
Callery Chemical Co.

Cesium ignites immediately on contact with air if poured or sprayed and reacts explosively with water. Cesium may form peroxide compounds if allowed to oxidize in the absence of water normally present in atmosphere. The resulting peroxides may be shock sensitive with easily reduced compounds in the same manner that potassium superoxide is shock sensitive with mineral oil. Cesium metal must be contained under vacuum, inert gas, or anhydrous liquid hydrocarbons protected from oxygen or air exposure. Safety and handling information is available from suppliers.

Structure

Crystal structure. Body-centered cubic; Im3m, cI2; a = 0.613 nm at −10 °C

Mass Characteristics

Atomic weight. 132.9054

Density. 1.903 Mg/m³ at 0 °C, 1.892 Mg/m³ at 18 °C (Ref 1), 1.827 Mg/m³ at 40 °C (liquid) (Ref 2)

Thermal Properties

Melting point. 28.64 ± 0.17 °C (Ref 3)

Boiling point. 670 °C (Ref 4,5)

Specific heat. 201.6 J/kg at 20 °C (solid), 239.5 J/kg at 670 °C (liquid) (Ref 6), 155.7 J/kg at 670 °C (vapor) (Ref 6)

Heat of fusion. 16.38 kJ/kg (Ref 3)

Heat of vaporization. 611.3 kJ/kg (Ref 6)

Thermal conductivity. 18.42 W/m·K at 28.64 °C (liquid) (Ref 7), 4.6×10^{-3} W/m·K at 670 °C (vapor) (Ref 6)

Vapor pressure. From −23 to 28.64 °C (solid), $\log P + \dfrac{-4120}{T} - 1.0 \log T + 8.32$; from 28.64 to 377 °C (liquid), $\log P = \dfrac{-4042}{T} - 1.4 \log T + 9.05$, where P is in Pa and T is in K (Ref 8)

Electrical Properties

Electrical conductivity. 4.5 MS/m at 28.64 °C (solid)

Electrical resistivity. 200 nΩ·m at 20 °C

Ionization potential. 3.893 V (Ref 6)

Magnetic Properties

Magnetic susceptibility. Volume: 167×10^{-6} mks (Ref 7)

Optical Properties

Color. 99.99% pure material is bronze colored

Mechanical Properties

Viscosity. 0.686 mPa·s at 28.64 °C (Ref 6)

Surface tension. 0.0394 N/m at 28.64 °C (Ref 6)

REFERENCES

1. L. Losana, Gazzetta Chimica Italiana, Vol 65, 1935, p 855
2. M. Eckardt and E. Graefe, Zietschrift für Anorganische und Allgemeine Chemie, Vol 23, 1900, p 385
3. K. Clusius and H. Stern, Zeitschrift für Angewandte Physik, Vol 6, 1954, p 194, and Chemical Abstracts, Vol 48, 1954, p 6869a
4. O. Ruff and O. Johannsen, Chemische Berichte, Vol 38, 1905, p 3608
5. "Cesium", *Gmelins Handbuch der Anorganischen Chemie,* Vol 25, 8th ed., Verlag Chemie, Berlin, 1955
6. W. D. Weatherford, Jr., J. C. Tyler and P. M. Ku, Properties of Inorganic Fluids and Coolants for Space Applications, WADC Tech Report 59-598, Southwest Research Institute, San Antonio, TX, 1959
7. C. A. Hampel, "Rubidium and Cesium", *Rare Metals Handbook,* 2nd ed., Reinhold Publishing Corp. New York, NY, 1961, p 434–440
8. J. W. Mellor, *Comprehensive Treatise on Inorganic and Theoretical Chemistry,* Vol 2, Supplement 3, John Wiley & Sons, New York, NY, 1963

Chromium (Cr)

Compiled by H. C. Aufderhaar
Union Carbide Corp.

Chromium, also known as chrome metal or electrolytic chromium, is an alloying agent used in steel and various nickel-base and cobalt-base superalloys, aluminum-base alloys, electrical resistance alloys, hard facing grains and powders, and for electroplating. No special precautions need be taken when using chromium. Chromium metal, produced by the electrolytic or pyrometallurgical processes, has chromium content in the range of 99.0 to 99.5%, with carbon at 0.050 max. In addition, an iodide process chromium is available with typical purity of 99.99% Cr.

Structure

Crystal structure. Body-centered cubic, cI2 (Im3m): a = 0.28844 to 0.28848 nm at 20 °C (Ref 1, 2, 3).

Table 3 Tensile properties of recrystallized, swaged, arc-cast electrolytic chromium(a) (Ref 3)

Temperature, °C	Tensile strength, MPa	Proportional limit, MPa	Elongation in 25 mm, %	Reduction in area, %	Modulus of elasticity, GPa
20	83	...	0	0	0.248
200	234	...	0	0	...
300	154(b)	11.7	3	4	0.290
350	197	105	6	8	0.168
400	225(c)	132	51	89	0.227
500	...	...	30	75	...
600	242	69	42	81	0.200
700	203	...	33	85	...
800	180	97	47	92	0.255

(a) Recrystallized at 1200 °C in hydrogen. Strain rate of testing, 0.017 m/m per min. (b) Yield strength, 131 MPa. (c) Yield strength, 140 MPa.

Above 1840 °C: face-centered cubic; a = 0.38 nm (Ref 1)

Mass Characteristics

Atomic weight. 51.996
Density. 7.19 Mg/m^3 at 20 °C (Ref 2, 4)

Thermal Properties

Melting point. 1875 °C (Ref 2, 3, 4)
Boiling point. 2680 °C (Ref 2)
Phase-transformation temperature. 1840 °C (Ref 1)
Coefficient of thermal expansion. Linear: 6.2 μm/m·K (Ref 2)
Specific heat. 459.8 J/kg·K at 20 °C (Ref 2, 3)
Entropy. From Ref 1:

°C	kJ/kg·K Solid	Gas
25	0.46	3.35
227	0.71	3.56
727	1.09	3.83
1227	1.36	4.00
1727	1.59	4.13
2227	...	4.24
2727	...	4.31

Latent heat of fusion. 258 to 283 kJ/kg (Ref 2, 3)
Latent heat of vaporization. 6168 kJ/kg (Ref 2, 3)
Thermal conductivity. 67 W/m·K at 20 °C, 76 W/m·K at 426 °C, 67 W/m·K at 760 °C (Ref 1)
Vapor pressure. From Ref 1:

°C	Pa
965	3.2×10^{-4}
1093	2.8×10^{-3}
1197	2.7×10^{-2}
1288	2.4×10^{-1}
1875	$9.9 \times 10^{+2}$

Electrical Properties

Electrical conductivity. 13% IACS at 20 °C (Ref 2, 3, 4)
Electrical resistivity. From Ref 1:

°C	nΩ·m
−260	5
20	130
152	180
200	200
407	310
600	400
652	470
1000	660

Temperature coefficient of electrical resistivity. At 0 °C, 0.03 nΩ·m per K (Ref 1)
Electrochemical equivalent. Valence 3, 0.17965 mg/C; valence 6, 0.08983 mg/C (Ref 3)
Electrolytic solution potential. For valence 3, 0.5 V vs hydrogen electrode (Ref 11)
Hydrogen overvoltage. 0.38 V (Ref 11)
Temperature of superconductivity. 0.08 K (Ref 3)
Work function. 0.7337 aJ (Ref 1)

Magnetic Properties

Magnetic susceptibility. Volume: 4.5×10^{-5} mks (Ref 2)

Optical Properties

Color. Steel gray
Reflectance. 67% at λ = 300 nm; 63% at λ = 1000 nm; 70% at λ = 500 nm; 88% at λ = 4000 nm (Ref 2)
Refractive index. 1.64 to 3.28 for λ from 257 to 608 nm (Ref 2)
Absorptive index. 3.69 to 4.30 for λ from 257 to 608 nm (Ref 7)

Nuclear Properties

Stable isotopes. ^{50}Cr, 4.31% abundant; ^{52}Cr, 83.76% abundant; ^{53}Cr, 9.55% abundant; ^{54}Cr, 2.38% abundant (Ref 1)

Unstable isotopes. From Ref 1:

Isotope no.	Half-life
48	23-24 h
49	41.7-41.9 min
51	27.5-27.9 d
55	3.52-3.6 min
56	5.9 min

Chemical Properties

Resistance to specific corroding agents. (A 10% solution at 12 °C was used unless otherwise noted.) Chromium is resistant to the following acids: acetic, aqua regia, benzoic (saturated), butyric, carbonic, citric, fatty, formic, hydrobromic, hydroiodic, lactic, nitric, oleic, oxalic, palmitic, phosphoric, picric, salicylic, stearic and tartaric. Chromium is not resistant to hydrochloric acid or other halogen acids. Chromium is resistant to the following agents: acetone, air, ethyl and methyl alcohol, higher alcohols, aluminum chloride, aluminum sulfate, ammonia, ammonium chloride, barium chloride, beer, benzyl chloride (saturated and 100%), calcium chloride, carbon dioxide, carbon disulfide, carbon tetrachloride (saturated and 100%), dry chlorine, chlorobenzene (saturated and 100%), chloroform, copper sulfate, ferric chloride, ferrous chloride, foodstuffs, formaldehyde, fruit products, glue, hydrogen sulfide (100%), magnesium chloride, milk, mineral oils, motor fuels, crude petroleum products, phenols, photographic solutions, printing ink, sodium carbonate, sodium chloride, sodium hydroxide, sodium sulfate, sugar, sulfur (100%), sulfur dioxide (100%), chlorinated, distilled or rainwater, zinc chloride and zinc sulfate (Ref 1, 3, 8, 9)

Mechanical Properties

Tensile properties. Iodide chromium at room temperature, as-swaged: tensile strength, 413 MPa; 0.2% yield strength, 362 MPa; elongation, 44%; reduction in area, 78% (Ref 2, 10). Iodide chromium at room temperature, swaged and recrystallized: tensile strength, 282 MPa; elongation, 0%; reduction in area, 0% (Ref 2, 10). Electrolytic chromium, see Table 3.

Hardness. As cast, forged: room temperature, 125 HB; 700 °C: 70 HB. Electrodeposited, annealed: 500 to 1250 HB, depending on the amount of hydrogen in the deposit. Electrodeposited and annealed: 70 to 90 HB. Extruded, annealed at 1100 °C: 110 HV. Extruded, annealed, rolled at 400 °C: 160 HV (Ref 1)

Elastic modulus. Tension, see Table 3.

Impact strength. Unnotched Charpy, as arc-cast electrolytic chromium: room temperature, 2 J; 400 °C: 160 J (Ref 3)

REFERENCES

1. J. C. Bailar, et al (Eds.), *Comprehensive Inorganic Chemistry*, Vol 3, Pergamon Press, 1973, p 624 & ff

2. H. F. Mark, et al (Eds.), *Encyclopedia of Chemical Technology*, Vol 6, 3rd Ed., Wiley and Sons, NY 1979, p 54 & ff

3. C. A. Hampel (Ed.), *The Encyclopedia of the Chemical Elements*, Reinhold Book Corporation, NY 1968, p 145 & ff

4. C. J. Smithells (Ed.), *Metals Reference Book*, Vol III, Plenum Press, NY, 1967, p 685

5. W. W. Coblentz and R. Stair, National Bureau of Standards Research Paper, Vol 39, 1928, p 352

6. P. Hidnert, *Journal of Research for the National Bureau of Standards*, Vol 27, 1941, p 113

7. Freederickaz, Annalen der Physik, Vol 34, 1911, p 780

8. J. E. Hosdowich, *Material and Methods*, Vol 24, 1946, p 896

9. McKay and Worthington, *Corrosion Resistance of Metals and Alloys*, Reinhold, NY, 1936

10. Sully, Brandeis and Mitchell, *Journal of the Institute of Metals* Vol 81, 1952-53, p 585

11. H. S. Taylor, *Treatise on Physical Chemistry*, Vol 1, 1931, p 354

Cobalt (Co)

Compiled by D. J. Maykuth
Metals and Ceramics Information Center
Battelle Memorial Institute

Cobalt is used as an alloying element in (a) permanent and soft magnetic materials, (b) superalloys—high-temperature creep-resistant alloys, (c) hard facing and wear-resistant alloys, (d) sintered carbide cutting tools, (e) steels—high speed, tool, and others, (f) cobalt-base tool materials, (g) electrical-resistant alloys, (h) high-temperature spring and bearing alloys, (i) magnetostrictive alloys. and (j) special-expansion and constant-modulus alloys

Structure

Crystal structure. α-phase, close-packed hexagonal, $hP2$ ($P6_3/mmc$); a = 0.25071 nm, c = 0.40686. β-phase, face-centered cubic, $cF4$ ($Fm\,3m$); a = 0.35441

Minimum interatomic distance. β-phase, 0.25061 nm

Mass Characteristics

Atomic weight. 58.9332

Density. At 20 °C: 8.832 Mg/m³ for α-phase; 8.80 Mg/m³ for β-phase

Volume change on phase transformation. β-1 to α-phase (cooling), −0.3% (approx)

Thermal Properties

Melting point. 1495 °C

Boiling point. 2900 °C (approx)

Phase-transformation temperature. β- to α-phase (cooling), 417 °C; a transformation near 1120 °C has not been confirmed

Coefficient of thermal expansion. Linear, 13.8 μm/m·K near room temperature; 14.2 μm/m·K at 200 °C; see also Fig. 4.

Specific heat. 414 J/kg·K at 20 °C

Latent heat of fusion. 292 kJ/kg

Latent heat of vaporization. 7.209 MJ/kg

Thermal conductivity. 69.04 W/m²·K at 20 °C

Electrical Properties

Electrical conductivity. 27.6% IACS at 20 °C

Electrical resistivity. 52.5 nΩ·m at 20 °C; temperature coefficient, 5.31 nΩ·m per K

Thermoelectric force. See Fig. 5.

Electrochemical equivalent. Valence +2, 0.03050 mg/C

Magnetic Properties

Magnetic permeability. Initial, 68; maximum, 245

Coercive force. 708.3 A/m for H_{max} = 0.1 T

Saturation magnetization. 1.87 T ($4\pi\,I_s$)

Residual induction. 0.49 T for H_{max} = 0.1 T

Hysteresis loss. 690 J/m³·cycle for B_{max} = 0.5 T

Curie temperature. 1121 °C

Fig. 4 Linear thermal expansion of cobalt (relative to 30 °C)

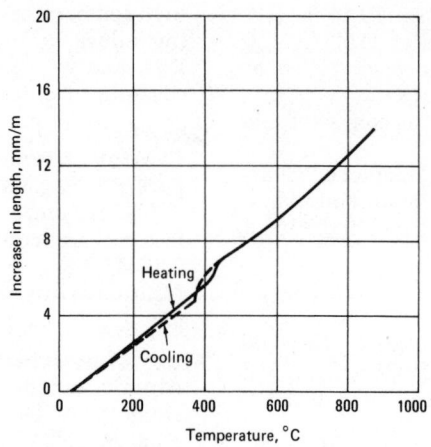

Fig. 5 Thermoelectric force of cobalt

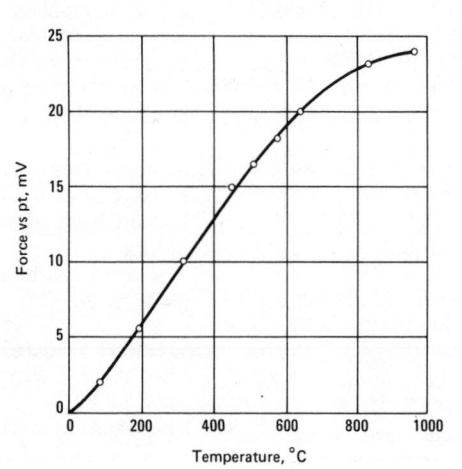

Table 4 Mechanical properties of cobalt

Form and purity	Tensile strength, MPa	0.2% yield strength, MPa	Compessive yield strength MPa
As cast (99.9)	234.4	...	291.0
Annealed (99.9)	255.1	...	386.8
Swaged (99.9)	689.5	...	...
Zone refined (99.8)	944.6	758.5	...

Optical Properties
Spectral reflectance:

λ, nm	%
200	37
1060	67.5
6750	92.7
12030	96.6

Mechanical Properties
Tensile properties. See Table 4.
Poisson's ratio. 0.32
Elastic modulus. Tension, 211 GPa; shear, 826 GPa; compression: 183 GPa
Velocity of sound. 457.2 m/s

Columbium

See *Niobium*

Copper (Cu)

Compiled by A. W. Blackwood
and
J. E. Casteras
ASARCO Incorporated

Structure

Crystal structure. Face-centered cubic, structure symbol, A1; $Fm3m$; $cF4$. Lattice parameter, 0.361509 ± 0.000004 nm at 25 °C (Ref 1, 2)
Twinning planes. (111) twin plane, [1$\bar{1}$2] twin direction; (11$\bar{1}$) twin plane, [112] twin direction (Ref 3)
Cleavage planes. None (Ref 4)
Minimum interatomic distance. 0.2551 nm (Ref 2)
Electronic structure. Band 1:

Fermi-surface description	Orbit description	Magnetic field direction
Sphere with necks touching [111]	Neck	[111]
	Belly	[111]
Brillouin zone faces	Dog's bone	[100], [110]
	Four-cornered rosette	[100]

Table 5 Relative volume vs pressure for pure copper at 25 °C (Ref 1)

Pressure, GPa	Relative volume, V/V₀
0.0	1.000
0.5	0.996
1.0	0.993
1.5	0.990
2.0	0.986
2.5	0.983
3.0	0.980
3.5	0.977
4.0	0.974
4.5	0.971
5.0	0.968
6.0	0.962
7.0	0.956
8.0	0.951
9.0	0.945
10.0	0.940
12.0	0.930
14.0	0.921
16.0	0.912
18.0	0.904
20.0	0.896
22.0	0.889
24.0	0.881
26.0	0.874
28.0	0.868
30.0	0.861
32.0	0.855
34.0	0.849
36.0	0.843
38.0	0.838
40.0	0.832
42.0	0.827
44.0	0.822
46.0	0.817
48.0	0.812
50.0	0.808
55.0	0.797
60.0	0.786
65.0	0.777
70.0	0.768
75.0	0.759
80.0	0.751
85.0	0.743
90.0	0.736
95.0	0.729
100.	0.722
120.	0.697
140.	0.677
160.	0.658
180.	0.642
200.	0.627
250.	0.596
300.	0.571
350.	0.550
400.	0.532
450.	0.516

Table 6 Mean coefficient of linear thermal expansion for pure copper (Ref 1)

Temperature, K	Mean coefficient, μm/m·K
2	0.0006
4	0.0025
6	0.0074
8	0.016
10	0.030
12	0.052
14	0.083
16	0.128
18	0.186
20	0.26
25	0.6
50	3.8
75	7.6
100	10.5
150	13.6
200	15.2
250	16.1
293	16.7
350	17.3
400	17.6
500	18.3
600	18.9
700	19.6
800	20.4
1000	22.4
1200	24.8

Mass Characteristics

Atomic weight. 63.54
Density. (Ref 1):

°C	Mg/m³
20	8.93
Melting point	7.940
1100	7.924
1200	7.846
1300	7.764

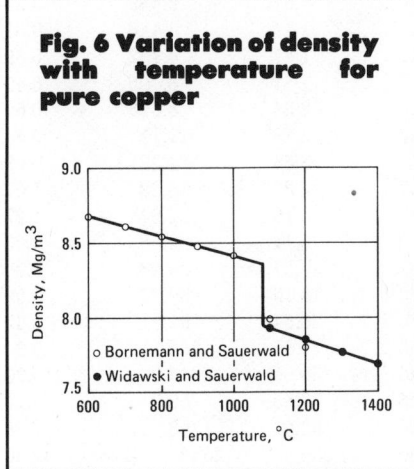

Fig. 6 Variation of density with temperature for pure copper

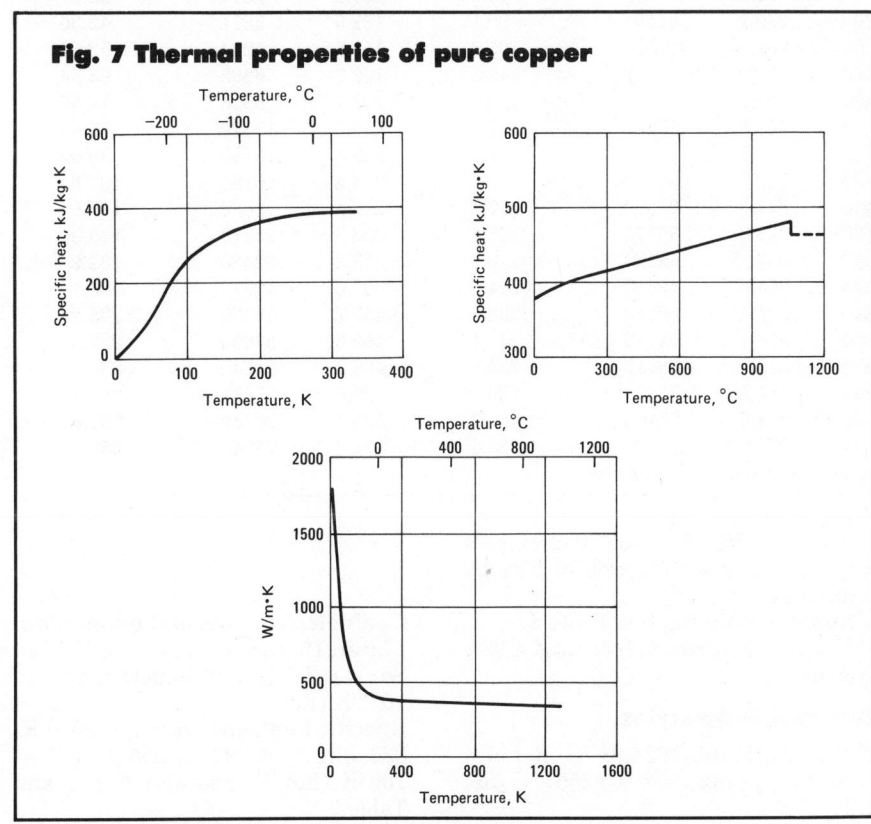

Fig. 7 Thermal properties of pure copper

Table 7 Thermodynamic properties of copper (Ref 1)

Temperature, K	C_p^0 J/kg·K	$H_T^0 - H_0^0$(a) J/kg	$(H_T^0 - H_0^0)/T$ J/kg·K	S_T^0 J/kg·K	$-(G_T^0 - H_0^0)$ J/kg	$(G_T^0 - H_0^0)/T$ J/kg·K
1	0.0117	0.00565	0.00565	0.0112	0.00552	0.00552
2	0.0278	0.0249	0.0124	0.0239	0.0228	0.0114
3	0.0530	0.0644	0.0214	0.0395	0.0543	0.0181
4	0.0916	0.135	0.0338	0.0596	0.103	0.0258
5	0.148	0.253	0.0507	0.0859	0.176	0.0351
6	0.228	0.439	0.0733	0.120	0.277	0.0463
7	0.335	0.717	0.103	0.162	0.417	0.0596
8	0.474	1.120	0.140	0.216	0.606	0.0757
9	0.651	1.684	0.187	0.282	0.853	0.0947
10	0.873	2.439	0.244	0.360	1.174	0.117
11	1.14	3.446	0.313	0.456	1.57	0.144
12	1.47	4.752	0.395	0.570	2.09	0.175
13	1.87	6.405	0.493	0.703	3.51	0.209
14	2.34	8.513	0.607	0.858	2.72	0.250
15	2.89	11.11	0.741	1.039	4.45	0.297
16	3.54	14.32	0.895	1.245	5.59	0.349
17	4.30	18.22	1.072	1.481	6.96	0.409
18	5.16	22.94	1.275	1.747	8.56	0.475
19	6.14	28.58	1.504	2.061	10.46	0.551
20	7.27	35.28	1.763	2.392	12.68	0.634
25	15.15	89.75	3.59	4.80	30.17	1.21
30	26.64	192.8	6.42	8.51	62.87	2.09
35	41.51	361.8	10.34	13.71	117.8	3.37
40	58.86	612.0	15.30	20.36	202.4	5.07
45	77.55	952.7	21.17	28.36	323.7	7.19
50	96.84	1388.	27.78	37.53	488.0	9.757
60	135.3	2549	42.50	58.60	965.9	16.10
70	170.9	4084	58.35	82.18	1668	23.83
80	202.2	5955	74.42	107.1	2614	32.67
90	229.1	8115	90.17	132.5	3811	42.35
100	251.9	10520	105.2	157.8	5264	52.64
110	271.0	13140	119.5	182.9	6968	63.34
120	287.2	15940	132.8	207.1	8918	74.32
130	300.9	18870	145.2	230.7	11110	85.43
140	312.7	21950	156.7	253.4	13530	96.62
150	322.7	25130	167.4	275.2	16180	107.8
160	331.3	28390	177.5	296.5	19030	118.9
180	345.3	35170	195.5	336.3	25370	140.9
200	356.1	42190	210.9	373.3	32460	162.2
220	364.6	49400	224.6	407.6	40270	183.1
240	371.4	56760	236.5	439.7	48750	203.2
260	376.7	64240	247.1	469.6	57850	222.5
273.15	379.7	69210	253.4	488.2	64140	234.8
280	381.1	71820	256.5	497.6	67530	241.1
298.15	384.6	78760	264.2	521.7	76780	257.5
300	384.9	79490	265.0	524.0	77740	259.2

(a) H_0^0 is enthalpy at 0 K and 1 atm.

See also Fig. 6. Density decreases 0.028% with a reduction of 50% by drawing

Specific volume. See Table 5.
Volume change on freezing. 4.92% contraction

Thermal Properties

Melting point. 1084.88 °C (Ref 5)
Boiling point. 2595 °C; 2567 °C (Ref 1)

Coefficient of thermal expansion. Linear, 16.5 μm/m·K at 20 °C (Ref 1). See also Table 6. Volumetric, 49.5 × 10^{-6}/K (Ref 2)
Specific heat. 494 J/kg·K at 2000 K; 386 J/kg·K at 293 K; 255 J/kg·K at 100 K (Ref 1). See also Fig. 7 and Table 7.

Table 8 Thermal conductivity of pure copper (Ref 1, 2)

Temperature, K	Conductivity, W/m·K
0	0
1	2870
2	5730
3	8550
4	11300
5	13800
6	15400
7	17700
8	18900
9	19500
10	19600
11	19300
12	18500
13	17600
14	16600
15	15600
16	14500
18	12400
20	10500
25	6800
30	4300
35	2900
40	2050
45	1530
50	1220
60	850
70	670
80	570
90	514
100	483
150	428
200	413
250	404
273	401
300	398
350	394
400	392
500	388
600	383
700	377
800	371
900	364
1000	357
1100	350
1200	342
1300	334(a)
1373	160
1773	172
1973	176
2273	177

(a) Extrapolated value.

Enthalpy, entropy. See Table 7.
Latent heat of fusion. 205 J/kg (Ref 1); 204.9 J/kg (Ref 5); 206.8 J/kg (Ref 2)
Latent heat of vaporization. 4729

Table 9 Radioactive tracer diffusion data for copper

Solute (tracer)	Crystalline form(a)	Purity %	Temperature range, °C	Form of analysis(b)	Activation energy (Q), kJ/mole	Frequency factor (D_0), mm²/s
^{110}Ag	S, P	...	580-980	RA	195	61
^{76}As	P	...	810-1075	RA	176.3	20
^{198}Au	S, P	...	400-1050	SS	178	3
^{115}Cd	S	99.98	725-950	SS	191	93.5
^{141}Ce	P	99.999	766-947	RA	115.5	21.7×10^{-7}
^{51}Cr	S, P	...	800-1070	RA	224	102
^{60}Co	S	99.998	701-1077	SS	226	193
^{67}Cu	S	99.999	698-1061	SS	211	78
^{152}Eu	P	99.999	750-970	SS, RA	112.4	11.7×10^{-6}
^{59}Fe	S, P	...	460-1070	RA	218	136
^{72}Ga		...	...	...	192.1	55
^{68}Ge	S	99.998	653-1015	SS	187.4	39.7
^{203}Hg	P	...	...	...	184	35
^{177}Lu	P	99.999	857-1010	RA	109.5	43×10^{-8}
^{54}Mn	S	99.99	754-950	SS	383	10^9
^{95}Nb	P	99.999	807-906	RA	251.4	204
^{63}Ni	P	...	620-1080	RA	225	110
^{102}Pd	S	99.999	807-1056	SS	227.6	171
^{147}Pm	P	99.999	720-955	RA	115	36.2×10^{-7}
^{195}Pt	P	...	843-997	SS	157	48×10^{-3}
^{35}S	S	99.999	800-1000	RA	206	23×10^2
^{124}Sb	S	99.999	600-1000	SS	176	34
^{113}Sn	P	...	680-910	...	188	11
^{160}Tb	P	99.999	770-980	RA	114.9	89.6×10^{-8}
^{204}Ti	S	99.999	785-996	SS	181	71
^{170}Tm	P	99.999	705-950	RA	101.1	72.8×10^{-8}
^{65}Zn	P	99.999	890-1000	SS	198.8	73

$$D_T = D_o \exp (Q/RT), \text{ where T is in K}$$

(a) P, polycrystalline; S, single crystal. (b) RA, residual activity; SS, serial section.

Fig. 8 Typical annealing curves for pure copper

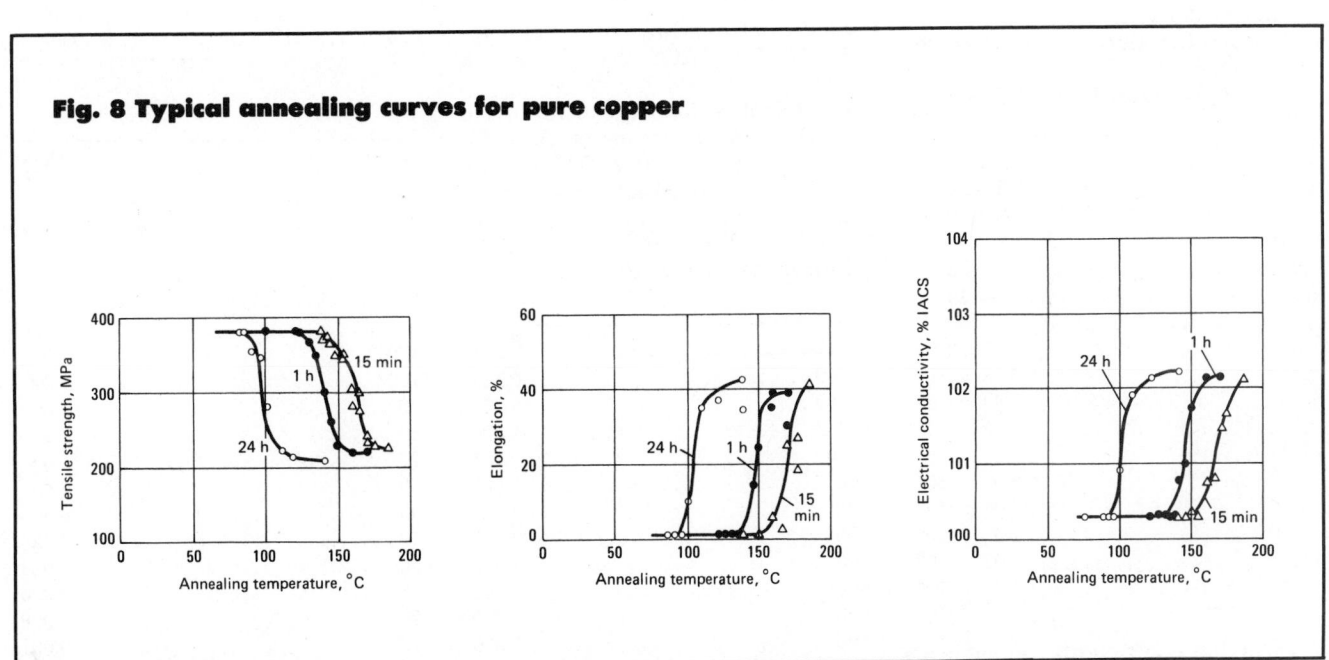

Fig. 9 Electrical properties of pure copper

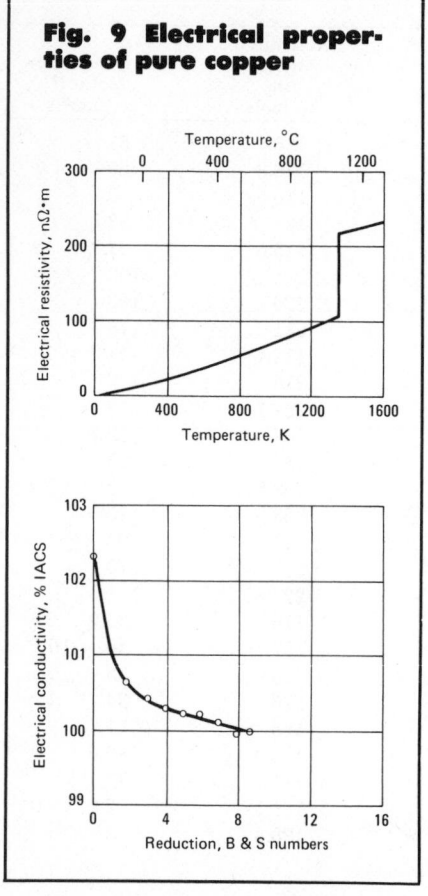

Fig. 10 Variation of tensile properties with amount of cold reduction for pure copper wire

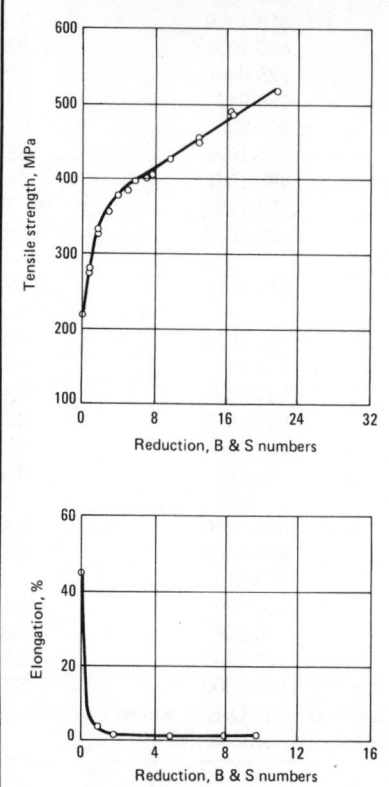

kJ/kg (Ref 1); 4726 kJ/kg (Ref 2); 4793 kJ/kg (Ref 6)

Thermal conductivity. 398 W/m·K at 27 °C (Ref 1). See also Fig. 7 and Table 8 (Ref 2, 1)

Recrystallization temperature. See Fig. 8.

Vapor pressure. From Ref 1:

°C	Pa
946	1.3×10^{-3}
1035	1.3×10^{-2}
1141	1.3×10^{-1}
1273	1.3
1432	13
1628	130
1879	1330
2067	5.33×10^3
2207	1.33×10^4
2465	5.33×10^4
2595	1.01×10^5
2760	2.02×10^5
3010	5.07×10^5
3500	1.01×10^6
3640	2.02×10^6
3740	4.05×10^6

Diffusion coefficient. See Table 9.

Electrical Properties

Electrical conductivity. Volumetric, 103.06% IACS. See also Fig. 8 and 9.

Electrical resistivity. 16.730 nΩ·m at 20 °C (Ref 1); temperature coefficient, 0.068 nΩ·m at 20 °C; pressure coefficient, −0.228 aΩ·m/Pa for pressure range 100 kPa to 9.8 GPa. See also Fig. 9. Effects of impurities dealt with in Ref 7, 8, 9, 10

°C	R_T/R_O
−200	0.151
−100	0.557
0	1.000
+100	1.431
200	1.862
300	2.299
400	2.747
500	3.210
600	3.695
800	4.750
1000	5.959

Table 10 Normal-incidence reflectance of freshly evaporated mirror-coating copper (Ref 2)

Wavelength, μm	Reflectance, %
0.220	40.1
0.240	39.0
0.260	35.5
0.280	33.0
0.300	33.6
0.315	35.5
0.320	36.3
0.340	38.5
0.360	41.5
0.380	41.5
0.400	47.5
0.450	55.2
0.500	60.0
0.550	66.9
0.600	93.3
0.650	96.6
0.700	97.5
0.750	97.9
0.800	98.1
0.850	98.3
0.900	98.4
0.950	98.4
1.0	98.5
1.5	98.5
2.0	98.6
3.0	98.6
4.0	98.7
5.0	98.7
6.0	98.7
7.0	98.7
8.0	98.8
9.0	98.8
10.0	98.9
15.0	99.0

K	nΩ·m
250	14.0
220	12.0
200	10.6
180	9.2
160	7.75
140	6.35
120	4.90
100	3.50
90	2.80
80	2.15
70	1.53
60	0.95
50	0.50
40	0.22
30	0.063
25	0.025
20	0.008
15	0.001

Resistivity ratio. From Ref 2:
Thermoelectric potential for Cu vs Pt. (Ref 1, 2):

°C	mV
−200	−0.19
−100	0.37
0	0
100	+0.76
200	1.83
300	3.15
400	4.68
500	6.41
600	8.34
700	10.47
800	12.81
900	15.37
1000	18.16

Electrochemical equivalent. 0.3294 mg/C for Cu^{+2}; 0.6588 mg/C for Cu^+

Electrolytic solution potential. All vs standard hydrogen electrode (Ref 1):

$Cu^{+2} + e^- \rightleftarrows Cu^+$, 0.158 V

$Cu^{+2} + 2e^- \rightleftarrows Cu$, 0.3402 V

$Cu^+ + e^- \rightleftarrows Cu$, 0.522 V

Ionization potential. Cu(I), 7.724 V; Cu(II), 20.29 V; Cu(III), 36.83 V (Ref 1)

Hydrogen overvoltage. In $1N$ H_2SO_4. $\eta = a + b (\log i)$; where η is overvoltage in V, i is current density in A/cm^2 constant a is 0.80 V and constant b is 0.115 V (Ref 6)

Hall effect. Hall voltage, -5.24×10^{-4} V at 0.30 to 0.8116 T; Hall coefficient, -5.5 mV·m/A·T (Ref 2)

Electron emission. Secondary electron emission: 1.3 max secondary electron yield; 600 eV primary electron energy for max yield; 200 eV for E(I) crossover; 1500 eV for E(II) crossover

Work function. From Ref 2:

Work function, eV	Conditions	Method of determination
4.5	1160-1200 K	Thermionic
4.6	1350 K	Thermionic
4.4	1100-1300 K	Thermionic
4.76	⋯	Photoelectric
4.86	⟨111⟩	Photoelectric
5.61	⟨110⟩	Photoelectric
4.60	⋯	Contact potential
4.51	⋯	Contact potential

Magnetic Properties

Magnetic susceptibility. Determined largely by the quantity of iron

Table 11 Optical properties of copper (Ref 2)

Wavelength, μm	Index of refraction	Extinction coefficient	Reflectance (calculated)
Bulk Copper			
0.3650	1.0719	2.0710	0.5004
0.4050	1.0769	2.2890	0.5491
0.4360	1.0707	2.4610	0.5860
0.5000	1.0308	2.7843	0.6528
0.5500	0.7911	2.7177	0.7013
0.5780	0.3250	2.8923	0.8716
0.6000	0.1491	3.2867	0.9508
0.6500	0.1074	3.9104	0.9740
0.7500	0.1034	4.8847	0.9835
1.0000	0.1471	6.9334	0.9881
Single Crystal Copper			
0.4400	1.1070	2.5565	0.5965
0.4600	1.0942	2.6320	0.6131
0.4800	1.0618	2.7124	0.6341
0.5000	1.0836	2.7684	0.6390
0.5200	1.0438	2.7784	0.6490
0.5400	0.9324	2.7348	0.6674
0.5600	0.6470	2.7200	0.7440
0.5800	0.2805	2.9764	0.8931
0.6000	0.1360	3.3464	0.9565
0.6200	0.1040	3.6525	0.9714
0.6400	0.0972	4.0692	0.9798
0.6600	0.0897	4.0692	0.9798
Evaporated Copper			
0.1025	1.05	0.70	0.098
0.1113	0.95	0.73	0.115
0.1215	0.95	0.78	0.137
0.1306	0.96	0.83	0.148
0.1392	1.00	0.91	0.165
0.1500	1.02	1.02	0.192
0.1603	0.98	1.04	0.219
0.1700	0.94	1.12	0.254
0.1800	0.90	1.21	0.296
0.1900	0.88	1.36	0.335
0.2000	0.94	1.51	0.378
0.500	0.88	2.42	0.625
0.600	0.186	2.980	0.928
0.700	0.150	4.049	0.966
0.800	0.170	4.840	0.973
0.900	0.190	5.569	0.977
1.000	0.197	6.272	0.981
1.35	0.45	7.81	0.971
1.69	0.58	9.96	0.977
2.28	0.82	13.0	0.981
3.00	1.22	17.1	0.984
3.4	1.53	20.3	0.985
3.97	1.94	23.1	0.986
4.87	2.86	28.9	0.987
5.0	2.92	27.45	0.985
5.8	3.71	34.6	0.988
7.00	5.25	40.7	0.988
7.3	5.79	43.2	0.988
8.35	7.28	49.2	0.988
9.6	9.76	57.2	0.988
10.25	11.0	60.6	0.988
10.8	12.6	64.3	0.988
12.25	15.5	71.9	0.989

present as an impurity. If the copper is free from oxygen, the iron is present in solid solution and has a small effect. The presence of oxygen results in the precipitation of Fe_3O_4 and a greater effect on magnetic properties. The measurements below were probably made on oxygen-bearing coppers.

Temperature	Volumetric mks
18 °C	-1.08×10^{-6}
1080 °C	-0.97×10^{-6}
1090 °C	-0.68×10^{-6}
−259 to −253 °C	-1.22×10^{-6}
296 K	-68.61×10^{-6}

Optical Properties

Color. Reddish metallic (Ref 1)
Spectral reflection coefficient. For incandescent light, 0.63 (Ref 1)
Reflectance. Polished or electroplated surfaces, see below (Ref 2); mirror coatings, see Table 10; calculated, see Table 11.

Wavelength, μm	Reflectance (a), %
0.25	25.9
0.30	25.3
0.35	27.5
0.40	30.0
0.50	43.7
0.60	71.8
0.70	83.1
0.80	88.6
1.0	90.1
2.0	95.5
4.0	97.3
6.0	98.0
8.0	98.3
10.0	98.4
12.0	98.4

(a) From polished surface at close-to-normal incidence.

Nominal spectral emittance. 0.15 for polished Cu at λ = 655 nm and 1080 K
Refractive index. See Table 11.
Absorptive index. Coefficient of absorption of solar radiation, 0.25 (Ref 1)

Nuclear Properties

Stable isotopes:

Isotope	Atomic weight	Natural abundance, %
^{63}Cu	62.9298	69.09
^{65}Cu	64.9278	30.91

Unstable isotopes:

Isotope	Lifetime	Modes of decay	Mean decay energy, MeV
^{58}Cu	3.20 s	β^+	8.569
^{59}Cu	82.0 ± 0.4 s	β^+,EC(a)	4.8
^{60}Cu	23.0 ± 0.3 min	β^+,EC(a)	6.12
^{61}Cu	3.41 h	β^+,EC(a)	2.242
^{62}Cu	9.8 min	β^+,EC(a)	3.939
^{64}Cu	12.9 h	β^-	0.573
		β^+,EC(a)	1.677
^{66}Cu	5.10 ± 0.02 min	β^-	2.633
^{67}Cu	61.88 ± 0.11 h	β^-	0.576
^{68}Cu	30 s	β^-	4.6

(a) EC, electron capture.

Chemical Properties

General corrosion behavior. Insoluble in hot and cold water (Ref 1)
Resistance to specific corroding agents. Soluble in HNO_3 and in hot H_2SO_4. Slightly soluble in HCl and NH_4OH (Ref 1)

Mechanical Properties

Tensile properties. Tensile strength: annealed, 209 MPa; cold drawn, 344 MPa (Ref 2); See also Fig. 8 and 10. Yield strength at 0.5% extension, under lead: annealed, 33.3 MPa; cold drawn, 333.4 MPa (Ref 2). Elongation: annealed, 60%; cold drawn, 14% (Ref 2); see also Fig. 8. Reduction in area; annealed, 92%; cold drawn, 88% (Ref 2).
Hardness. Cold drawn, 37 HRB
Poisson's ratio. 0.308 calculated from elastic modulus; annealed, 0.343 (Ref 5); cold drawn, 0.364 (Ref 2)
Strain-hardening exponent. Annealed, 0.54 (Ref 3)
Elastic modulus:

Tension, GPa	Ref
128	2
112 (cold drawn)	2
125 (annealed)	3
129.8	5

Shear, GPa	Ref
46.8	2
46.4 (annealed)	3
48.3	5

Bulk, GPa	Ref
140	2
137.8	5

Elastic modulus along crystal axes. Tension:⟨100⟩, 68 GPa; ⟨111⟩, 21 GPa. Shear: ⟨100⟩ 77 GPa (Ref 3)
Specific damping capacity. log decrement: 3.2×10^{-3} (Ref 2)
Dynamic liquid viscosity. From Ref 1:

°C	mPa·s
1085	3.36
1100	3.33
1150	3.22
1200	3.12

Liquid surface tension. 99.99999% purity, in vacuum: 1.300 N/m at the melting point. 99.999% purity, in N_2: 1.341 N/m at 1100 °C; 1.338 N/m at 1150 °C; 1.335 N/m at 1200 °C. 99.997% purity, at the melting point: 1.355 N/m in He or H_2; 1.358 N/m in Ar; 1.352 N/m in vacuum
Coefficient of friction. Static. Cu on Cu: 4.0 in H_2 or N_2; 1.6 in air or O_2; 1.4 (clean); 0.8 in paraffin oil (Ref 1); 0.7 in paraffin oil plus 1% lauric acid (Ref 2)
Velocity of sound. 4759 m/s for longitudinal bulk waves; 3813 for irrotational rod waves; 2325 for shear waves; 2171 for Rayleigh waves

REFERENCES

1. Weast (Ed.), *CRC Handbook of Chemistry and Physics,* 55th ed., CRC Press, Cleveland, 1974
2. *American Institute of Physics Handbook,* 3rd ed., McGraw Hill, NY, 1972
3. Tegart, W. J. McGregor, *Elements on Mechanical Metallurgy,* Mac-Millan, NY, 1966
4. Tetalman, A. S., and McEvily, A. J., *Fracture of Structural Materials,* Wiley & Sons, NY, 1967
5. Coates, P. B. and Andrews, J. W., A Precise Determination of the Freezing Point of Copper, *Journal of Physics, F: Metal Physics,* Vol 8, No. 2, 1978
6. *Table of Physical and Chemistry Constants,* 14th ed., compiled by Kaye, G. W. C. and Laby, T. H., Longman Group, Ltd., London, 1973
7. Smart, J. S., Smith, A. A. and Phillips, A. J., Preparation and

Some Properties of High Purity Copper, *Transactions of the AIME,* Vol 143, 1941

8. Smart, J. S. and Smith, A. A., Effect of Iron, Cobalt, and Nickel on Some Properties of High Purity Copper, *Transactions of the AIME,* Vol 147, 1942

9. Smart, J. S. and Smith, A. A., Effect of Certain Fifth-Period Elements on Some Properties of High Purity Copper, *Transactions of the AIME,* Vol 152, 1943

10. Smart, J. S. and Smith, A. A., Effect of Phosphorus, Arsenic, Sulfur, and Selenium on Some Properties of High Purity Copper, *Transactions of the AIME,* Vol 166, 1946

Curium

See tables at end of this section

Dysprosium (Dy)

Compiled by K. A. Gschneidner, Jr. and B. J. Beaudry
Ames Laboratory
U.S. Department of Energy
Iowa State University

Dysprosium is used as a control rod in nuclear reactors, and in phosphors, catalysts and garnet microwave devices. It is also used to measure neutron fluxes. Dysprosium will remain shiny in air at room temperature. However, turnings can be ignited and will burn white hot. Finely divided metal should not be handled in air.

Structure

Crystal structure. α' phase, orthorhombic, *Cmcm* D^{17}_{2h}; $a = 0.3596$ nm; $b = 0.6183$ nm; $c = 0.5678$ nm at 85 K. α phase, close-packed hexagonal, $P6_3/mmc$ D^4_{6h}; $a = 0.35915$ nm; $c = 0.56501$ nm at 24 °C. β phase, body-centered cubic, $Im3m$ O^9_h; $a = 0.398$ nm at 1395 °C
Slip planes. At 24 °C: primary {1010}, secondary {0002}
Twinning planes. At 24 °C: primary {1121}, secondary {1012}
Minimum interatomic distance. r_a = 0.17958 nm; r_c = 0.17522 nm; radius CN_{12} = 0.17740 nm at 24 °C

Mass Characteristics

Atomic weight. 162.50
Density. α phase, 8.551 Mg/m³ at 24 °C; β phase, 8.56 Mg/m³ at 1395 °C
Volume change on freezing. Approx 4.5% contraction

Thermal Properties

Melting point. 1412 °C
Boiling point. 2567 °C
Phase-transformation temperature. α' to α phase, 86 K; α to β phase, 1381 °C
Coefficient of thermal expansion. At 24 °C. Linear: 9.9 μm/m·K. Linear, along crystal axes: 7.1 μm/m·K along a axis, 15.6 μm/m·K along c axis. Volumetric: 29.8×10^{-6} per K
Specific heat. 173.0 J/kg·K at 25 °C
Entropy. 460.9 J/kg·K at 25 °C
Latent heat of fusion. 68.06 kJ/kg
Latent heat of phase transformation. α to β phase, 25.62 kJ/kg
Latent heat of vaporization. 1.787 MJ/kg at 25 °C
Heat of combustion. For cubic Dy_2O_3 at 25 °C: $\Delta H^0_c = 5.72$ MJ/kg Dy; $\Delta G^0_f = -5.45$ MJ/kg Dy
Recrystallization temperature. About 550 °C
Thermal conductivity. 10.7 W/m·K at 25 °C
Vapor pressure. 0.001 Pa at 804 °C; 0.101 Pa at 988 °C; 10.1 Pa at 1252 °C; 1013 Pa at 1685 °C

Electrical Properties

Electrical resistivity. 926 nΩ·m at 25 °C; 24 nΩ·m at 4 K. Along crystal axes, at 25 °C: 1110 nΩ·m along a axis, 766 nΩ·m along c axis. Liquid: 2100 nΩ·m at 1414 °C
Ionization potentials. Dy(I), 5.93 V; Dy(II), 11.67 V; Dy(III), 22.80 V; Dy(IV), 41.15 V
Hall coefficient. Along crystal axes, at 20 °C: -0.03 nV·m/A·T along b axis; -0.37 nV·m/A·T along c axis
Temperature of superconductivity. Bulk dysprosium is not superconducting down to 0.45 K at atmospheric pressure.

Magnetic Properties

Magnetic susceptibility. Volume (mks units), at 27 °C: $\chi_a = 0.0717$ and $\chi_c = 0.0511$; obeys Curie-Weiss law above 250 K with an effective moment of 10.64 Bohr magnetons and $\theta_a = 169$ K and $\theta_c = 121$ K
Saturation magnetization. 3.71 at 0 K along <1120>
Magnetic transformation temperatures. Curie temperature 85 K, Néel temperature 178.5 K

Optical Properties

Color. Metallic silver
Spectral hemispherical emittance. 29.7% for λ = 645 nm at 1413 to 1437 °C

Nuclear Properties

Thermal neutron cross section. 1100 b

Chemical Properties

General corrosion behavior. Remains shiny in air at room temperature. The rate of oxidation is slow even at 1000 °C due to the formation of a dark, tightly adhering oxide on the surface. The presence of water vapor increases the rate of oxidation.
Resistance to specific corroding agents. Dysprosium does not react with cold or hot water but will react vigorously with dilute acids. It is attacked slowly by concentrated sulfuric. The presence of the fluoride ion retards acid attack due to the formation of DyF_3 on the surface of the metal.

Mechanical Properties

Tensile properties. Tensile strength, 132 MPa; yield strength, 39 MPa; elongation, 23%; reduction in area, 22%
Hardness. 44 HV
Poisson's ratio. 0.238
Strain-hardening exponent. 0.35
Elastic modulus. At 27 °C: tension, 63.0 GPa; shear, 25.4 GPa; bulk, 40.1 GPa
Elastic constants along crystal axes. At 27 °C: $c_{11} = 74.3$ GPa; $c_{12} = 25.3$ GPa; $c_{13} = 20.8$ GPa; $c_{33} = 79.0$ GPa; $c_{44} = 24.7$ GPa

Einsteinium

See table at end of this section

Erbium (Er)

Compiled by K. A. Gschneidner, Jr.
and B. J. Beaudry
Ames Laboratory
U.S. Department of Energy
Iowa State University

Erbium is used in lasers and in phosphors, garnet microwave devices, ferrite bubble devices and catalysts. Erbium will remain shiny in air at room temperature. However, turnings can be ignited and will burn white hot. Finely divided metal should not be handled in air.

Structure

Crystal structure. Close-packed hexagonal: $P6_3/mmc$ D^4_{6h}; a = 0.35592 nm; c = 0.55850 nm at 24 °C

Slip planes. Primary $\{10\bar{1}0\}$, secondary $\{0002\}$ at 24 °C

Twinning planes. Primary $\{11\bar{2}1\}$, secondary $\{10\bar{1}2\}$ at 24 °C

Minimum interatomic distance. r_a = 0.17796 nm; r_c = 0.17335 nm; radius CN_{12} = 0.17566 nm at 24 °C

Mass Characteristics

Atomic weight. 167.26
Density. 9.066 Mg/m³ at 24 °C
Volume change on freezing. Approx 9.0% contraction

Thermal Properties

Melting point. 1529 °C
Boiling point. 2868 °C
Coefficient of thermal expansion. At 24 °C. Linear: 12.2 μm/m·K. Along crystal axes: 7.9 μm/m·K along a axis, 20.9 μm/m·K along c axis. Volumetric: 36.7×10^{-6} per K
Specific heat. 167.8 J/kg·K at 25 °C
Entropy. 437.5 J/kg·K at 25 °C
Latent heat of fusion. 119.0 kJ/kg
Latent heat of vaporization. 1.896 MJ/kg
Heat of combustion. For cubic Er_2O_3 at 25 °C: ΔH^0_c = 5.68 MJ/kg Er; ΔG^0_f = −5.40 MJ/kg Er
Recrystallization temperature. About 520 °C
Thermal conductivity. 14.5 W/m·K at 25 °C
Vapor pressure. 0.001 Pa at 908 °C; 0.101 Pa at 1113 °C; 10.1 Pa at 1405 °C; 1013 Pa at 1896 °C

Electrical Properties

Electrical resistivity. 860 nΩ·m at 25 °C; 47 nΩ·m at 4 K. Along crystal axes, at 25 °C: 945 nΩ·m along a axis, 603 nΩ·m along c axis. Liquid: 2260 nΩ·m at 1531 °C
Ionization potentials. Er(I), 6.10 V; Er(II), 11.93 V; Er(III), 22.74 V; Er(IV), 42.48 V
Hall coefficient. Along crystal axes, at 20 °C: +0.03 nV·m/A·T along b axis; $R_{H,c}$ = −0.36 nV·m/A·T along c axis
Temperature of superconductivity. Bulk erbium is not superconducting down to 0.03 K at atmospheric pressure.

Magnetic Properties

Magnetic susceptibility. Volume (mks units): χ_a = 0.0314 and χ_c = 0.0353 at 25 °C; obeys Curie-Weiss law above 195 K with an effective moment of 9.9 Bohr magnetons and θ_a = 32.5 K and θ_c = 61.7 K
Saturation magnetization. >3.33 T at 4.2 K along $<10\bar{1}0>$ and $<11\bar{2}0>$; 3.33 T at 4.2 K along $<0001>$
Magnetic transformation temperatures. Curie temperature, 19.6 K; a spin rearrangement at 53 K; Néel temperature, 85 K

Optical Properties

Color. Metallic silver
Spectral hemispherical emittance. 37.2% for λ = 645 nm from 1027 to 1587 °C

Nuclear Properties

Thermal neutron cross section. 170 b

Chemical Properties

General corrosion behavior. Erbium stays shiny in air at room temperature. The rate of oxidation is slow even at 1000 °C due to the formation of a dark, tightly adhering oxide on the surface of the metal.
Resistance to specific corroding agents. Erbium does not react with cold or hot water, but will react vigorously with dilute acids. The attack by concentrated sulfuric is slow. The presence of the fluoride ion retards acid attack due to the formation of ErF_3 on the surface of the metal.

Mechanical Properties

Tensile properties. Tensile strength, 139 MPa; yield strength, 37 MPa; elongation, 14%; reduction in area, 14%
Hardness. 42 HV
Poisson's ratio. 0.250
Strain-hardening exponent. 0.25
Elastic modulus. At 27 °C: tension, 65.9 GPa; shear, 28.3 GPa; bulk, 46.7 GPa
Elastic constants along crystal axes. At 27 °C: c_{11} = 83.67 GPa; c_{12} = 29.29 GPa; c_{13} = 22.22 GPa; c_{33} = 84.45 GPa; c_{44} = 27.53 GPa

Europium (Eu)

Compiled by K. A. Gschneidner, Jr. and
B. J. Beaudry
Ames Laboratory
U. S. Department of Energy
Iowa State University

Europium is used as control rods in nuclear reactors and as phosphors, especially as the red component in color television screens. Europium oxidizes rapidly in air at room temperature; therefore, it should be handled and stored under an inert atmosphere, storage in oil is not recommended. Finely divided europium can ignite spontaneously in air.

Structure

Crystal structure. Body-centered cubic: $Im3m$ O^9_h; a_o = 0.45827 nm at 24 °C
Minimum interatomic distance. 0.19844 nm at 24 °C; radius CN_{12} = 0.20418 nm

Mass Characteristics

Atomic weight. 151.96
Density. 5.244 Mg/m³ at 24 °C
Volume change on freezing. 4.8% contraction

Thermal Properties

Melting point. 822 °C
Boiling point. 1529 °C
Coefficient of thermal expansion. At 24 °C. Linear: 35.0 μm/m·K Volumetric: 105×10^{-6} per K
Specific heat. 182.0 J/kg·K
Entropy. At 25 °C: 512.1 J/kg·K
Latent heat of fusion. 60.63 kJ/kg
Latent heat of vaporization. 0.9522 MJ/kg at 25 °C

Latent heat of combustion. For monoclinic Eu_2O_3 at 25 °C: ΔH^0_c = 5.42 MJ/kg Eu; ΔG^0_f = −5.13 MJ/kg Eu

Recrystallization temperature. 300 °C

Thermal conductivity. Est, 13.9 W/m·K at 25 °C

Vapor pressure. 0.001 Pa at 399 °C; 0.101 Pa at 515 °C; 10.1 Pa at 685 °C; 1013 Pa at 963 °C

Electrical Properties

Electrical resistivity. 900 nΩ·m at 25 °C; 6 nΩ·m at 4 K; liquid, 2440 nΩ·m at 822 °C

Ionization potential. Eu(I): 5.67 V; Eu(II): 11.25 V; Eu(III): 24.66 V; Eu(IV): 42.28 V

Hall coefficient. + 2.44 nV·m/A·T

Temperature of superconductivity. Bulk europium is not superconducting down to 0.03 K at atmospheric pressure

Magnetic Properties

Magnetic susceptibility. Volume: 0.0134 mks at 25 °C; obeys Curie-Weiss law above 100 K with an effective moment of 8.3 Bohr magnetons

Saturation magnetization. >0.72 T at 4 K

Magnetic transformation temperature. Néel temperature, 89 K

Optical Properties

Color. Metallic silver when free from surface contamination

Nuclear Properties

Thermal neutron cross section. 4300 b

Chemical Properties

General corrosion behavior. Europium is the most air reactive of the rare earth metals, especially in moist air. In dry air, a dark coating is formed which retards oxidation. Hydrogen reacts with europium at about 250 °C

Resistance to specific corroding agents. Europium reacts vigorously with cold water and dilute acids

Mechanical Properties

Hardness. 17 HV

Poisson's ratio. 0.167

Elastic modulus. Tension, 18.2 GPa; shear 7.8 GPa; bulk, 8.3 GPa at 27 °C

Gadolinium (Gd)

Compiled by K. A. Gschneidner, Jr. and B. J. Beaudry
Ames Laboratory
U.S. Department of Energy
Iowa State University

Gadolinium is used as a burnable poison in shields and control rods in nuclear reactors, in host materials for rare earth phosphors, catalysts and garnet microwave devices. Gadolinium will tarnish slightly in air. Turnings can be ignited and burn white hot. Finely divided gadolinium should not be handled in air.

Structure

Crystal structure. α phase, close-packed hexagonal, $P6_3 \cdot mmc$ D^4_{6h}; a = 0.36336 nm, c = 0.57810 nm at 24 °C. β phase, body-centered cubic, $Im3m$ O^9_h; a = 0.406 nm at 1260 °C

Slip planes. At 24 °C: primary {1010}, secondary {0002}

Twinning planes. At 24 °C: primary {1121}, secondary {1012}

Minimum interatomic distance. At 24 °C: r_a = 0.18168 nm; r_c = 0.17858 nm; radius CN_{12} = 0.18013 nm

Mass Characteristics

Atomic weight. 157.25

Density. α phase, 7.901 Mg/m³ at 24 °C; β phase, 7.80 Mg/m³ at 1265 °C

Volume change on freezing. 2.0% contraction

Thermal Properties

Melting point. 1313 °C

Boiling point. 3273 °C

Phase-transformation temperature. α to β phase, 1235 °C

Coefficient of thermal expansion. At 100 °C. Linear: 9.4 μm/m·K at 100 °C. Linear, along crystal axes: 9.1 μm/m·K along a axis, 10.0 μm/m·K along c axis. Volumetric: 28.2×10^{-6} per K

Specific heat. 235.7 J/kg·K at 25 °C

Entropy. 432.1 J/kg·K at 25 °C

Latent heat of fusion. 63.94 kJ/kg

Latent heat of transformation. 24.88 kJ/kg

Latent heat of vaporization. 2.5286 MJ/kg at 25 °C

Heat of combustion. For monoclinic Gd_2O_3 at 25 °C: ΔH^0_c = 5.79 MJ/kg Gd; ΔG^0_f = −5.50 MJ/kg Gd

Recrystallization temperature. About 500 °C

Thermal conductivity. 10.5 W/m·K at 25 °C

Vapor pressure. 0.001 Pa at 1167 °C; 0.101 Pa at 1408 °C; 10.1 Pa at 1760 °C; 1013 Pa at 2306 °C

Electrical Properties

Electrical resistivity. Solid: 1310 nΩ·m at 25 °C; 24 nΩ·m at 4 K; liquid: 1950 nΩ·m at 1315 °C. Along crystal axes, at 25 °C: 1351 nΩ·m along a axis, 1217 nΩ·m along c axis

Ionization potentials. Gd(I), 6.14 V; Gd(II), 12.1 V; Gd(III), 20.48 V; Gd(IV), 43.86 V

Hall coefficient. −0.448 nV·m/A·T at 350 °C; along crystal axes, at 20 °C: −1.0 nV·m/A·T along a axis; −5.4 nVm/A·T along c axis

Temperature of superconductivity. Bulk gadolinium is not superconducting down to 0.37 K at atmospheric pressure.

Magnetic Properties

Magnetic susceptibility. Volume: 0.117 mks at 77 °C; obeys Curie-Weiss law above 77 °C with an effective moment of 7.98 Bohr magnetons and $\theta_a = \theta_c$ = 317 K

Saturation magnetization. 2.63 T at 0 K

Magnetic transformation temperatures. Curie temperature, 293.2 K

Optical Properties

Color. Metallic silver

Spectral hemispherical emittance. Solid: 33.7% for λ = 645 nm at 1025 to 1313 °C; 34.2% for λ = 645 nm at 1313 to 1600 °C

Nuclear Properties

Thermal neutron cross section. 40 000 b

Chemical Properties

General corrosion behavior. Gadolinium tarnishes slightly in air at room temperature. Even at 1000 °C the oxidation rate is slow because of the formation of a dark, tightly adhering oxide on the surface. The presence of water vapor increases the rate of oxidation. After heating to 550 °C in vacuum, hydrogen will react at 250 °C.

Resistance to specific corroding agents. Gadolinium does not react with cold or hot water, but will react vigorously with dilute acids. It is attacked slowly by concentrated sulfuric. The presence of the fluoride ion retards acid attack due to the formation of GdF_3.

Mechanical Properties

Tensile properties. Tensile strength, 122 MPa; yield strength, 17 MPa; elongation, 47%; reduction in area, 58%

Hardness. 37 HV for polycrystalline; 23 HV for {10$\bar{1}$0} prismatic face; 69 HV for {0001} basal plane

Poisson's ratio. 0.254

Strain-hardening exponent. 0.37

Elastic modulus. At 27 °C: tension, 55.8 GPa; shear, 22.2 GPa; bulk, 37.9 GPa

Elastic constants along crystal axes. At 27 °C: c_{11} = 67.83 GPa; c_{12} = 25.59 GPa; c_{13} = 20.73 GPa; c_{33} = 71.23 GPa; c_{44} = 20.77 GPa

Liquid surface tension. 0.81 N/m^2 at 1313 °C

Gallium (Ga)

Compiled by H. Clinton Snyder
Aluminum Company of America
and
R. Frankena
Ingal International Gallium, GmbH
Schuandorf, West Germany

Gallium is used predominantly in the electronics industry where it is combined with elements of group III, IV or V of the periodic table to form semiconducting materials; most often, it is combined with arsenic and/or phosphorus for uses in light emitting diodes, laser diodes, solar cells, transistors, etc. Combined as the oxide with other oxides in garnets for magnetic bubble domain devices; as the metal for heat transfer medium, eutectic alloys, liquid seals, high temperature lubricant; in superconducting compounds such as GaV_3 and as compounds in organic reactions. Commercially available metal ranges from 99.5% pure to 99.9999 + %. The most common impurities are Hg, Pb, Sn, Zn and Cu. If certain impurity limits of high-purity gallium are exceeded, the optoelectric properties of electronic materials are degraded or destroyed. Gallium is tested for purity using emission spectrography, mass spectrography and by residual resistivity measurement. Gallium ordinarily is not considered to be hazardous but in compounds or alloys, it may be toxic, depending upon nature of the other components or ions. Gallium in aluminum causes severe intergranular corrosion of the aluminum.

Structure

Crystal structure. Orthorhombic, *Cmca*: a = 0.45258 nm; b = 0.45186 nm; c = 0.76570 nm at 24 °C

Minimum interatomic distance. 0.2437 nm

Mass Characteristics

Atomic weight. 69.72

Density:

°C	Phase	Mg/m^3
20	Solid	5.907
29.65	Solid	5.9037
29.8	Liquid	6.0947
32.4	Liquid	6.093
200	Liquid	5.972
500	Liquid	5.779
600	Liquid	5.720
1010	Liquid	5.492
1100	Liquid	5.445

Volume change on freezing. 3.2% expansion

Thermal Properties

Melting point. 29.78 °C

Boiling point. 2403 °C; some sources list 2237 °C as the boiling point, but this is reported to be an error caused by gallium suboxide pressure

Coefficient of thermal expansion. Linear, along crystal axes, from 0 to 20 °C: 11.5 μm/m·K along a axis, 31.5 μm/m·K along b axis, 16.5 μm/m·K along c axis (Ref 1). Volumetric: solid from 0 to 29.7 °C, 58 000 mm^3/m^3·K; liquid at 100 °C, 120 000 mm^3/m^3·K; liquid at 900 °C, 97 000 mm^3/m^3·K

Specific heat. See Table 12.

Latent heat of fusion. 80.16 kJ/kg

Latent heat of vaporization. 3893 kJ/kg

Heat of combustion. Ga_2O_3: − 15 480 kJ/kg Ga

Table 12 Specific heat of gallium

K	°C	Phase	J/kg·K
4.3	− 268.9	Solid	0.122
16.1	− 257.1	Solid	19.25
60.1	− 213.1	Solid	175.7
273-297	0-24	Solid	372.3
285.7-473	12.5-200	Liquid	397.6

Thermal conductivity. Polycrystalline, at 29.8 °C: 33.49 W/m·K. Along crystal axes at 20 °C; 40.82 W/m·K along a axis, 88.47 W/m·K along b axis, 15.99 W/m·K along c axis. Liquid at 77 °C: 28.68 W/m·K along a axis, 34.04 W/m·K along b axis, 38.31 W/m·K along c axis

Vapor pressure:

°C	Pa
800	7.866×10^{-4}
1000	1.093×10^{-1}
1200	3.999
1400	5.999×10^1
1600	5.066×10^2
1800	2.799×10^3
2000	1.147×10^4
2200	3.733×10^4
2403	1.013×10^5

Electrical Properties

Electrical resistivity. Polycrystalline, at 20 °C: 150.5 nΩ·m. Along crystal axes, at 20 °C: 174 nΩ·m along a axis, 81 nΩ·m along b axis, 543 nΩ·m along c axis (Ref 2). Supercooled liquid: at 0 °C, 252 nΩ·m: at 20 °C, 256.1 nΩ·m. Liquid: at 40 °C, 260 nΩ·m; at 600 °C, 378 nΩ·m

Electrochemical equivalent. Valence + 3: 0.241 mg/C

Electrolytic solution potential. vs H_2: − 0.56 V at 25 °C

Hydrogen overvoltage. Near melting point: solid, − 0.31 V; liquid, − 0.44 V

Temperature of superconductivity. 8.4 K

Magnetic Properties

Magnetic susceptibility. Volume (mks units): solid at 80 K (− 193 °C), 3.07×10^{-4}; solid at 17 °C, 2.71×10^{-4}; liquid at 40 °C, 0.31×10^{-4}

Optical Properties

Color. Liquid metal is silvery white; solid metal is silvery with a bluish cast

Reflectance. Solid: 75.6% for λ = 436 nm; 71.3% for λ = 589 nm; liquid: 88.8% for λ = 435 nm; 88.4% for λ = 546 nm; 88.6% for λ = 691 nm

Nuclear Properties

Stable isotopes. ^{69}Ga, isotope mass 68.9257, 60.4% abundant; ^{71}Ga, isotope mass 70.9249, 36.6% abundant

Chemical Properties

General corrosion behavior. Liquid gallium oxidizes rapidly to form a protective layer of oxide. Gallium reacts with mineral acids depending on concentration and temperature. Gallium reacts with caustic—especially in the presence of iron metal. The rate of gallium corrosion is inversely related to the purity as is the tendency to super cool. At elevated temperatures, gallium is a corroding agent for many metals (Ref 3). At room temperature diffusion of gallium into many metals takes place, with formation of an often low-melting compound in grain-boundaries and grains of the corroded metal

Mechanical Properties

Hardness. 1.5 to 2.5, Mohs scale
Elastic modulus. Compressibility, at 20 °C: 0.021 $nm^3/m^3 \cdot Pa$ between 15 and 50 MPa
Fracture behavior. Polycrystalline masses shatter easily
Kinematic liquid viscosity. 287 m^2/s at 30 °C; 183 m^2/s at 500 °C
Liquid surface tension. In vacuum: 0.709 N/m at 30 °C; 0.712 N/m at 100 °C; 0.718 N/m at 200 °C; 0.743 N/m at 500 °C

REFERENCES

1. R. W. Powell, Electrical resistivity of gallium and some anisotropic properties of the metal, *Proceedings of the Royal Society* A, Vol 209, 1951, p 525
2. R. W. Powell, M. J. Woodman and R. P. Tye, Further measurements relating to the anisotropic thermal conductivity of gallium, *British Journal of Applied Physics*, Vol 14, 1963, p 432-435
3. L. R. Kelman, W. D. Wilkinson and F. L. Yaggee, Resistance of materials by the attack of liquid metals, USAEC-Report, Argonne National Laboratory ANL-4417, 1950

Germanium (Ge)

Compiled by C. D. Thurmond
Bell Laboratories

Structure

Crystal structure. Face-centered cubic (diamond); $a = 0.565754$ nm at 25 °C (Ref 1)

Mass Characteristics

Atomic weight. 72.59
Density. 5.3243 Mg/m^3 at 25 °C (Ref 2)

Thermal Properties

Melting point. 937.4 °C (Ref 3)
Boiling point. 2834 °C (Ref 3)
Coefficient of thermal expansion. Linear, 5.722 $\mu m/m \cdot K$ at 25 °C (Ref 4)
Specific heat. 321.7 $J/kg \cdot K$ at 25 °C (Ref 3)
Entropy. 428.3 $J/g \cdot K$ for solid at 25 °C (Ref 3)
Latent heat of fusion. 509.0 kJ/kg (Ref 3)
Latent heat of vaporization. 4558.7 kJ/kg (Ref 3)
Thermal conductivity. 58.6 $W/m \cdot K$ at 25 °C (Ref 5)
Vapor pressure. 0.140 mPa at 937.4 °C (Ref 3)

Electrical Properties

Electrical resistivity. Intrinsic, 0.45 $\Omega \cdot m$ at 25 °C (Calculated from the intrinsic carrier concentration and the carrier mobilities.)
Carrier density. Intrinsic, 2.1 μm^3 at 25 °C (Ref 6)
Forbidden energy gap. 0.7437 eV at 0 K; 0.6642 eV at 25 °C (Ref 6)
Drift mobility. Upper limit at 25 °C: electrons, 0.48 $m^2/V \cdot s$; holes, 0.20 $m^2/V \cdot s$ (Ref 7)

Mechanical Properties

Elastic constants. At 25 °C: $c_{11} = 128.53$ GPa, $c_{12} = 48.26$ GPa, $c_{44} = 67.98$ GPa (Ref 8)

REFERENCES

1. A. S. Cooper, *Acta Crystallographica*, Vol 15, 1962, p 578
2. M. E. Straumanis and E. Z. Aka, *J Appl Phys*, Vol 23, 1952, p 330
3. R. Hultgren, P. D. Desai, D. T. Hawkins, Molly Gleiser, K. K. Kelley and D. D. Wagman, *Selected Values of the Thermodynamic Properties of the Elements*, American Society for Metals, 1973, p 204
4. A. S. Bhalla and E. W. White, *Physica Status Solidi*, Vol 5, 1971, p K51
5. J. A. Carruthers, T. H. Geballe, H. M. Rosenberg and J. M. Ziman, *Royal Society of London Proceedings A*, Vol 238, 1957, p 502
6. See C. D. Thurmond, *J Electrochem So*, Vol 122, 1975, p 1133
7. S. M. Sze and J. C. Irvin, *Solid State Electronics*, Vol 11, 1968, p 599
8. H. J. McSkimin and P. Andreatch, Jr., *J Appl Phys*, Vol 34, 1963, p 651

Gold (Au)

Compiled by S. C. Carapella, Jr.
ASARCO, Incorporated

Structure

Crystal structure. Face-centered cubic: $a = 0.40786$ nm
Minimum interatomic distance. 28.78 nm

Mass Characteristics

Atomic weight. 196.9665
Density. 19.302 Mg/m^3 at 25 °C

Thermal Properties

Melting point. 1064.43 °C
Boiling point. 2857 °C
Coefficient of thermal expansion. Linear. At 20 °C, 14.2 $\mu m/m \cdot K$; from 0 to 950 °C, $L_t = L_0 [1 + (14.103t + 0.001628t^2 + 0.000001145t^3) \times 10^{-6}]$, where t is in °C
Specific heat:

°C	J/kg
25	128
227	133
627	142
1027	163
1063	170
1127	166
1227	159

Latent heat of fusion. 62.762 kJ/kg
Latent heat of vaporization. 1.6987 kJ/kg
Thermal conductivity. 317.9 $W/m \cdot K$ at 0 °C; 1749 $W/m \cdot K$ at 4.2 K
Vapor pressure:

°C	kPa
1770	0.1013
2036	1.013
2383	10.13
2857	101.3

Diffusion coefficients. At 20 °C:

Fig. 11 Temperature dependence of the electrical resistivity of gold

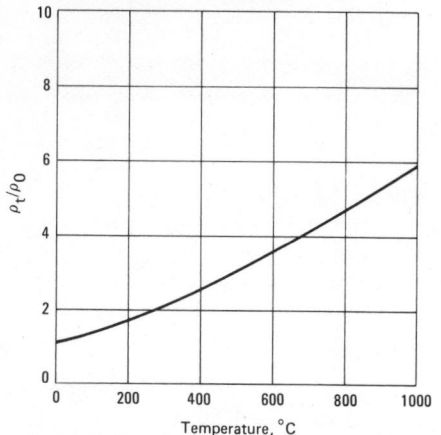

Fig. 13 Reflectance of gold as a function of wave length

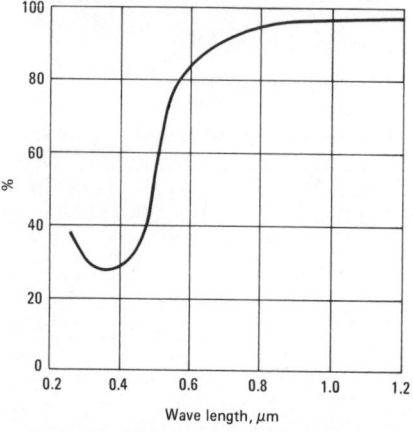

Fig. 14 Total hemispherical emittance of gold as a function of temperature

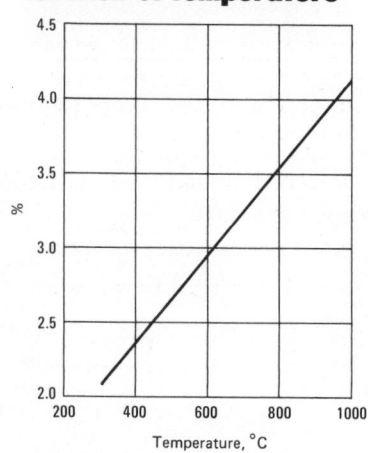

Fig. 12 Temperature dependence of the thermal electromotive force of gold vs platinum

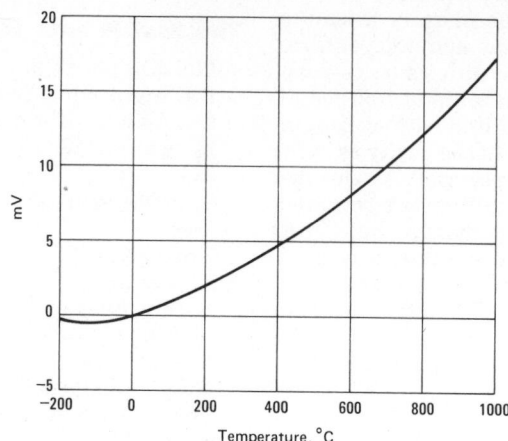

Positive values indicate Au is positive to Pt.

Element	Matrix	Diffusion, m^2/s
Fe	Au	3×10^{-26}
Ni	Au	1×10^{-30}
Cu	Au	6×10^{-34}
Pd	Au	2×10^{-35}
Au	Au	2×10^{-40}
Pt	Au	1×10^{-45}
Au	Pb	1×10^{-15}
Au	Cu	5×10^{-24}
Au	Ag	5×10^{-30}
Au	Pd	5×10^{-35}
Au	Pt	1×10^{-35}

Electrical Properties

Electrical conductivity. 73.4% IACS at 20 °C
Electrical resistivity. 20.1 nΩ·m at 0 °C; 23.5 nΩ·m at 20 °C. Temperature coefficient. From 0 to 100 °C, 0.004 per K. See also Fig. 11.
Thermal electromotive force. vs Pt, see Fig. 12.
Effect of alloying elements on resistivity. Annealed condition:

Element	Resistivity, nΩ·m	Increase in resistivity, %
1% Ag	28.2	28.2
1% Pd	29	31.8
1% Cd	30.7	39.5
1% Pt.	33	50.0
1% Cu	35.9	63.3
1% In.	45	104.0
1% Zn	49.3	124.0
1% Ni	51	132.0
1% Sn	76	245.0
1% Co	178	710.0
1% Fe	269	1220.0

Magnetic Properties

Magnetic susceptibility. Volume: -1.79×10^{-6} mks

Optical Properties

Reflectance. See Fig. 13.
Emittance. See Fig. 14.

Mechanical Properties

Tensile properties. Tensile strength, 103 MPa (annealed wire). Elongation, 30%
Elastic modulus. Tension, 78 GPa
Liquid surface tension. At 1200 °C, 1070 mN/m; at 1300 °C, 1020 mN/m

Holmium (Ho)

Compiled by K. A. Gschneidner, Jr. and B. J. Beaudry
Ames Laboratory
U.S. Department of Energy
Iowa State University

Holmium is used in phosphors and ferrite bubble devices and will remain shiny in air at room temperature. Turnings can be ignited and will burn white hot. Finely divided holmium should not be handled in air.

Structure

Crystal structure. Close-packed hexagonal: $P6_3/mmc$ D^4_{6h}; $a = 0.35778$ nm, $c = 0.56178$ nm at 24 °C

Slip planes. At 24 °C: primary {1010}, secondary {0002}

Twinning planes. At 24 °C: primary {1121}, secondary {1012}

Minimum interatomic distance. At 24 °C: $r_a = 0.17889$ nm; $r_c = 17433$ nm; radius $CN_{12} = 0.17661$ nm

Mass Characteristics

Atomic weight. 164.9304
Density. 8.795 Mg/m³ at 24 °C
Volume change on freezing. Approx 7.4% contraction

Thermal Properties

Melting point. 1474 °C
Boiling point. 2700 °C
Coefficient of thermal expansion. At 24 °C. Linear: 11.2 μm/m·K. Linear, along crystal axes: 7.0 μm/m·K along a axis, 19.5 μm/m·K along c axis. Volumetric: 33.6×10^{-6} per K

Specific heat. 164.6 J/kg·K at 25 °C
Entropy. At 25 °C: 454.9 J/kg·K
Latent heat of fusion. 102.28 kJ/kg
Latent heat of vaporization. 1.824 MJ/kg at 25 °C
Latent heat of combustion. For cubic Ho_2O_3 at 25 °C: $\Delta H^0_c = 5.70$ MJ/kg Ho; $\Delta G^0_f = -5.43$ MJ/kg Ho
Thermal conductivity. 16.2 W/m·K at 25 °C
Recrystallization temperature. About 520 °C
Vapor pressure. 0.001 Pa at 845 °C; 0.101 Pa at 1036 °C; 10.1 Pa at 1313 °C; 1013 Pa at 1771 °C

Electrical Properties

Electrical resistivity. 814 nΩ·m at 25 °C; 70 nΩ·m at 4 K. Along crystal axes, at 25 °C: 1015 nΩ·m along a axis, 605 nΩ·m along c axis. Liquid: 2210 nΩ·m at 1476 °C
Ionization potentials. Ho(I), 6.02 V; Ho(II), 11.80 V; Ho(III), 22.80 V; Ho(IV), 42.35 V
Hall coefficient. Along crystal axes, at 20 °C: +0.02 nV·m/A·T along b axis; −0.32 nV·m/A·T along c axis
Temperature of superconductivity. Bulk holmium is not superconducting down to 0.38 K at atmospheric pressure.

Magnetic Properties

Magnetic susceptibility. Volume (mks units): $\chi_a = 0.0500$ and $\chi_c = 0.0466$ at 25 °C; obeys Curie-Weiss law above 140 K with an effective moment of 11.2 Bohr magnetons and $\theta_a = 88$ K and $\theta_c = 73$ K
Saturation magnetization. 3.87 T at 4.2 K along <1120> and 3.80 T at 4.2 K along <1010>
Magnetic transformation temperatures. Curie temperature, 20 K; Néel temperature, 132 K

Optical Properties

Color. Metallic silver

Nuclear Properties

Thermal neutron cross section. 64 b

Chemical Properties

General corrosion behavior. Holmium stays shiny in air at room temperature. The rate of oxidation is slow even at 1000 °C due to the formation of a dark, tightly adhering oxide on the surface.
Resistance to specific corroding agents. Holmium does not react with cold or hot water but will react vigorously with dilute acids. It is attacked slowly by concentrated sulfuric. The presence of fluoride ion retards acid attack due to formation of HoF_3 on the surface of the metal.

Mechanical Properties

Tensile properties. About the same as dysprosium and erbium
Hardness. 46 HV
Poisson's ratio. 0.237
Elastic modulus. At 27 °C: tension, 65.2 GPa; shear, 26.4 GPa; bulk, 41.3 GPa
Elastic constants along crystal axes. At 27 °C: $c_{11} = 77.2$ GPa; $c_{12} = 26.1$ GPa; $c_{13} = 21.5$ GPa; $c_{33} = 81.1$ GPa; $c_{44} = 26.1$ GPa

Indium (In)

Compiled by S. C. Carapella, Jr.
ASARCO Incorporated

The major application of indium is in solders and fusible alloys (low-melting-point alloys). Other applications are in the manufacture of atomic reactor control rods, bearings, low-pressure sodium lamps and as a surface coating on aluminum wire conductors for making low-resistance contact and terminal joints. Some lesser applications are in dental alloys, semiconductors, radiation detector badges, surface lubricants and for protective finishes.

Structure

Crystal structure. Tetragonal; at 26 °C, $a = 0.32517$ nm and $c = 0.49459$ nm

Mass Characteristics

Atomic weight. 114.82
Density:

°C	Mg/m³
25	7.286
164	7.026
194	7.001
228	6.974
271	6.939
300	6.916

Volume change on freezing. 2.5% contraction

Thermal Properties

Melting point. 156.63 °C
Boiling point. 2073 °C
Coefficient of thermal expansion. Linear, 32.1 μm/m·K at 20 °C
Specific heat:

°C	J/kg·K
25	233
127	252
156.63 (solid)	264
156.63 (liquid)	257
227	256
327	255
427	254

Latent heat of fusion. 28.42 kJ/kg
Latent heat of vaporization. 2015.8 kJ/kg
Thermal conductivity. 86.6 W/m·K at 0 °C
Vapor pressure:

°C	kPa
1215	0.1013
1421	1.013
1693	10.13
2073	101.3

Electrical Properties

Electrical resistivity:

°C	nΩ·m
0	80
154	291
181	301
222	319
280	348

Electrochemical equivalent. Valence 3, 396.4 µg/C

Magnetic Properties

Magnetic susceptibility. Volumetric: 7.0×10^{-6} mks

Nuclear Properties

Stable isotopes. 113, 115
Thermal neutron cross section. For 2.2 km·s neutrons: absorption, 190 + 10 b; scattering, 2.2 ± 0.5 b

Mechanical Properties

Tensile strength. 2.62 MPa
Compressive strength. 2.14 MPa
Hardness. 0.9 HB
Elastic modulus. Tension, 10.8 GPa

Iridium (Ir)

Compiled by Leonard Bozza
Engelhard Minerals & Chemicals Corp.

Small crucibles made of iridium have been used for studying high-temperature reactions. A major use for iridium is crucibles for producing large, pure, defect-free man-made crystals for electronic and industrial applications. Single crystals so formed are used as substrates in magnetic bubble memory devices, solid-state lasers, insulating substrates for semiconductors, monoclinic filters and substitutes for natural gemstones in jewelry. Iridium also is used as an alloying element to harden platinum, electrodes in spark plugs for severe operating conditions such as jet engine igniters, thermocouple elements, and radioactive isotopes for industrial applications and cancer therapy.

Mass Characteristics

Atomic weight. 192.9
Density. 22.65 Mg/m³ at 20 °C

Thermal Properties

Melting point. 2447 °C (Ref 1)
Boiling point. 4500 °C (Ref 2)
Coefficient of thermal expansion. Linear, 6.8 µm/m·K at 20 °C (Ref 3)
Specific heat. 130 J/kg·K at 20 °C (Ref 4)

Thermal conductivity. 147 W/m·K at 0 to 100 °C (Ref 5)

Electrical Properties

Electrical resistivity. 47.1 nΩ·m at 0 °C, 53 nΩ·m at 20 °C (Ref 5). Temperature coefficient, 0.00427 nΩ·m per °C at 0 to 100 °C (Ref 8)
Thermal electromotive force. Pt67 (Reference junction at 0 °C): +3.626 mv at 400 °C; +6.271 mv at 600 °C; +12.741 mv at 1000 °C (Ref 6)

Magnetic Properties

Magnetic susceptibility. Mass: 0.19×10^{-8} mks at 18 °C (Ref 10)

Optical Properties

Reflectivity. 64% at λ = 0.45 µm; 70% at λ = 0.55 µm; 78% at λ = 0.75 µm (Ref 7)
Emissivity. 0.30 at 0.65 µm for solid unoxidized metal (Ref 8, 9)

Chemical Properties

General corrosion behavior. Iridium is the most corrosion resistant element. It is not affected by common acids, including hot sulfuric acid. It is slightly attacked by sodium hypochlorite solutions but not by aqua regia at ordinary temperatures. However, at elevated temperatures and pressures, aqua regia does attack iridium, and this method may be used for dissolving iridium and its refractory alloys for analysis. Iridium is virtually insoluble in lead even at high temperatures, and use is often made of this fact in preliminary steps in chemical analysis.

Fabrication Characteristics

Working data. Iridium can be arc melted (inert gas cover), electron beam melted or consolidated by powder metallurgy techniques. It is hot worked using these procedures similar to those used for tungsten. Final working is done at warm temperatures, which produces a fibrous structure. Iridium has limited malleability at room temperature.

Mechanical Properties

Tensile properties. (See also Table 13.) Properties of 0.5 mm wire. Tensile strength: annealed at 1000 °C, 1100-1240 MPa; hot drawn, 2070-2480 MPa. Elongation: annealed, 20-22%; hot drawn, 13-18%. (Ref 13)
Hardness. Annealed at 1000 °C, 200-240 HV; as cast, 210-240 HV; hot drawn, 600-700 HV (Ref 13)

Modulus of elasticity. Tension: static, 517 GPa; dynamic 527 GPa. Compression: 210 GPa (Ref 14)
Poisson's ratio. 0.26

Table 13 Tensile properties of iridium annealed at 1500 °C (Ref 16)

Temperature, °C	Tensile strength, MPa	0.2% yield strength, MPa	Reduction in area, %
24	623	234	6.8
500	530	234	12.7
750	450	142	51.0
1000	331	43.4	80.6

REFERENCES

1. International Practical Temperature Scale of 1968, Amended Edition of 1975, *Metrologia*, Vol 12, 1976, p 7-17
2. R. F. Hampson, Jr., and R. F. Walker, *J Research Nat Bur Standards*, Vol 65A, 1961, p 289
3. P. Hidnert and W. Souder, NBS Circular 486, U.S. Dept of Commerce, 1950
4. F. M. Jaeger and E. Rosenbohn, *Proc Acad Sci Amsterdam*, Vol 34, 1931, p 808
5. R. W. Powell *et al.*, *Platinum Metals Review*, Vol 6, 1962, p 138
6. G. F. Blackburn and F. R. Caldwell, *J Research Nat Bur Standards*, Vol 66C, 1962, p 1
7. M. Auswarter, *Z Tech Physik*, Vol 18, 1927, p 457
8. D. L. Goldwater and W. Danforth, *Phys Review*, Vol 103, 1956, p 871
9. R. C. Weast (Ed.), *Handbook of Chemistry and Physics*, 58th Ed., CRC Press, Cleveland, 1977, p E-230
10. K. Honda, *Annalen der Physik*, Vol 32, 1910, p 1027
11. *Corrosion Handbook*, John Wiley & Sons, NY, 1948
12. D. W. Rhys and E. G. Price, *Engelhard Industries Technical Bulletin*, Vol V, No. 2, Sept 1964
13. *Engelhard Industries Technical Bulletin*, Vol VI, No. 3, Dec 1965
14. R. I. Jaffee *et al.*, "High Temperature Properties and Alloying Behavior of the Platinum Group Metals", ONR Contract No. 2547(00), NRO 39-067

Fig. 15 Phase diagram for iron (Ref 1)

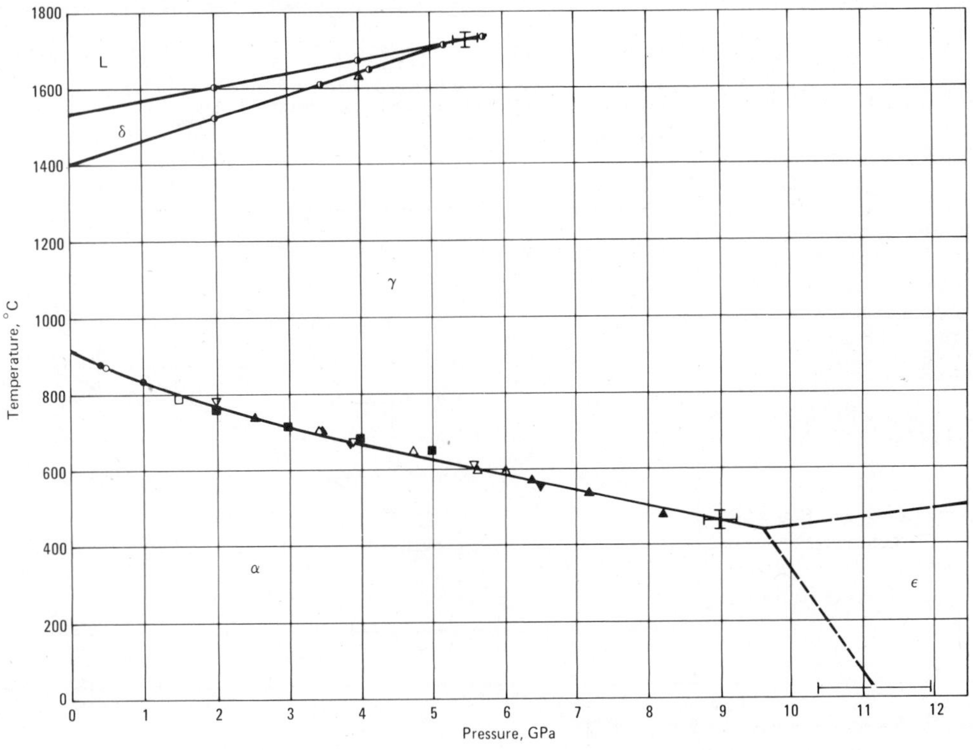

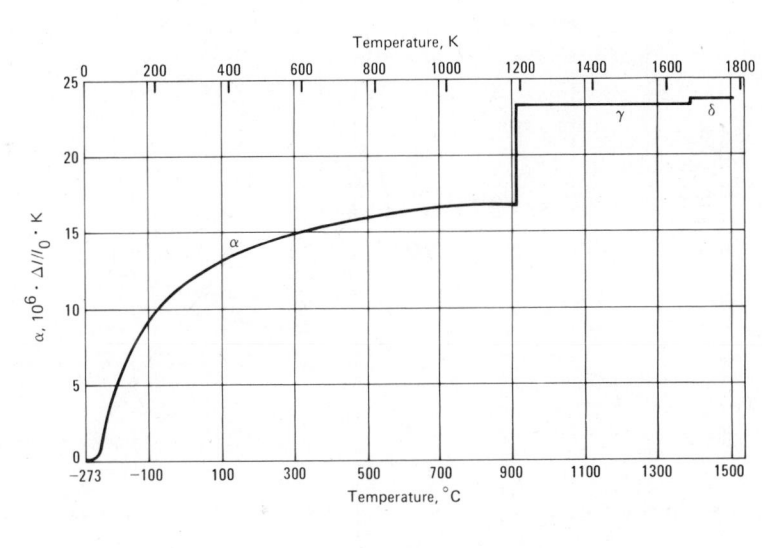

Iron (Fe)

Compiled by L. R. Smith
Ford Motor Company
(formerly with University of
Michigan)
and
W. C. Leslie
University of Michigan

Iron of sufficient purity so that its properties are essentially those of the element is commonly called high-purity iron. Iron is *not* an article of commerce; instead, it is employed almost exclusively in research. Iron of very high purity can be prepared by a variety of methods, but the last stage of the process is purification by floating zone refining, often combined with treatment in oxidizing and reducing atmospheres. In order to maintain this purity, it is essential to avoid contamination of the iron, which can occur by reactions with the atmosphere or with containers.

Structure

Crystal structure. The various

Fig. 16 Coefficient of linear thermal expansion for iron (Ref 12)

phases of iron are shown as a function of temperature and pressure in Fig. 15. The crystal symmetry and the space group for each phase are listed below.

α-iron and δ-iron: bcc, $Im3m$
γ-iron: fcc, $Fm3m$
ε-iron: cph, $P6_3/mmc$

Lattice parameter. As a function of temperature, see Table 14. Reference 3 gives the pressure dependence of the lattice parameters of the bcc and cph phases of iron at 23 ± 3 °C as:

$$a(\text{bcc}) = 0.2866 \, (1 + P/27.5)^{-0.056}$$
$$a(\text{cph}) = 0.2523 \, (1 + P/32.5)^{-0.033}$$

where c/a (cph) = 1.603 ± 0.001 and is independent of pressure; P is the pressure in GPa; and a(bcc), a(cph) and c are lattice parameters in nanometres.

Slip Plane and Direction:

Phase	Slip direction	Slip plane
α-iron	$\langle 111 \rangle$	{110}, {112}, {123}*
γ-iron	$\langle 101 \rangle$	{111}
ε-iron	$\langle 1120 \rangle$	{1010}

*It is generally considered that slip in bcc iron at room temperature and above can occur on any plane containing $\langle 111 \rangle$.

Twinning plane. 112, direction $\langle 111 \rangle$
Cleavage plane. 100
Minimum interatomic distance:

α-iron: 20 °C, 0.24825 nm; 907 °C, 0.25119 nm
γ-iron: 950 °C, 0.25815 nm; 1361 °C, 0.26029 nm
δ-iron: 1390 °C, 0.25388 nm; 1508 °C, 0.25458 nm

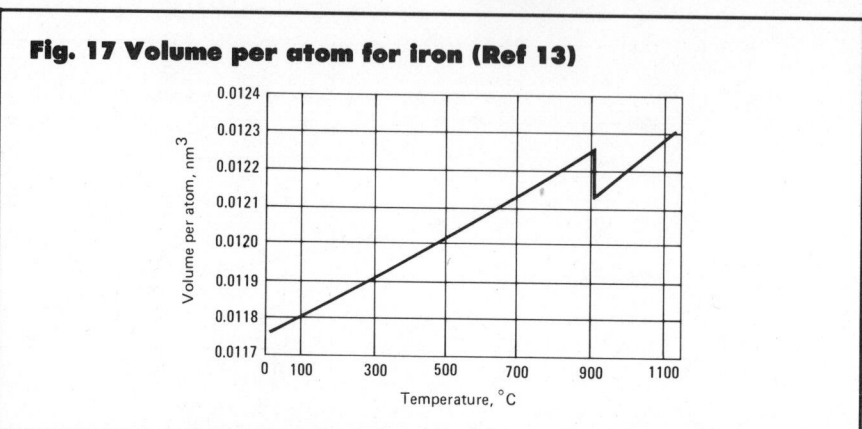

Fig. 17 Volume per atom for iron (Ref 13)

Fig. 18 Specific heat of iron (Ref 6 and 15)

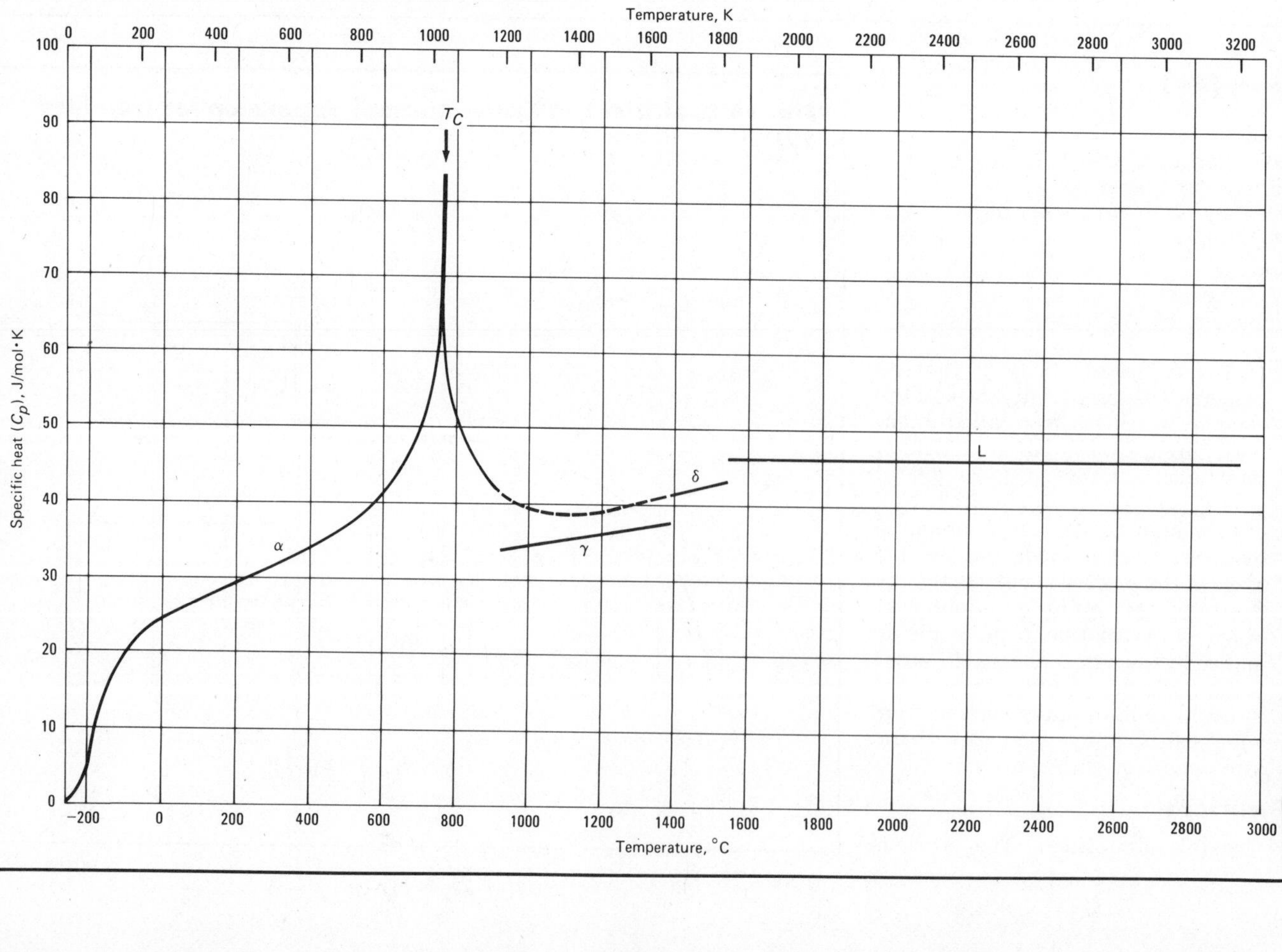

Fig. 19 Thermal conductivity of iron (Ref 17)

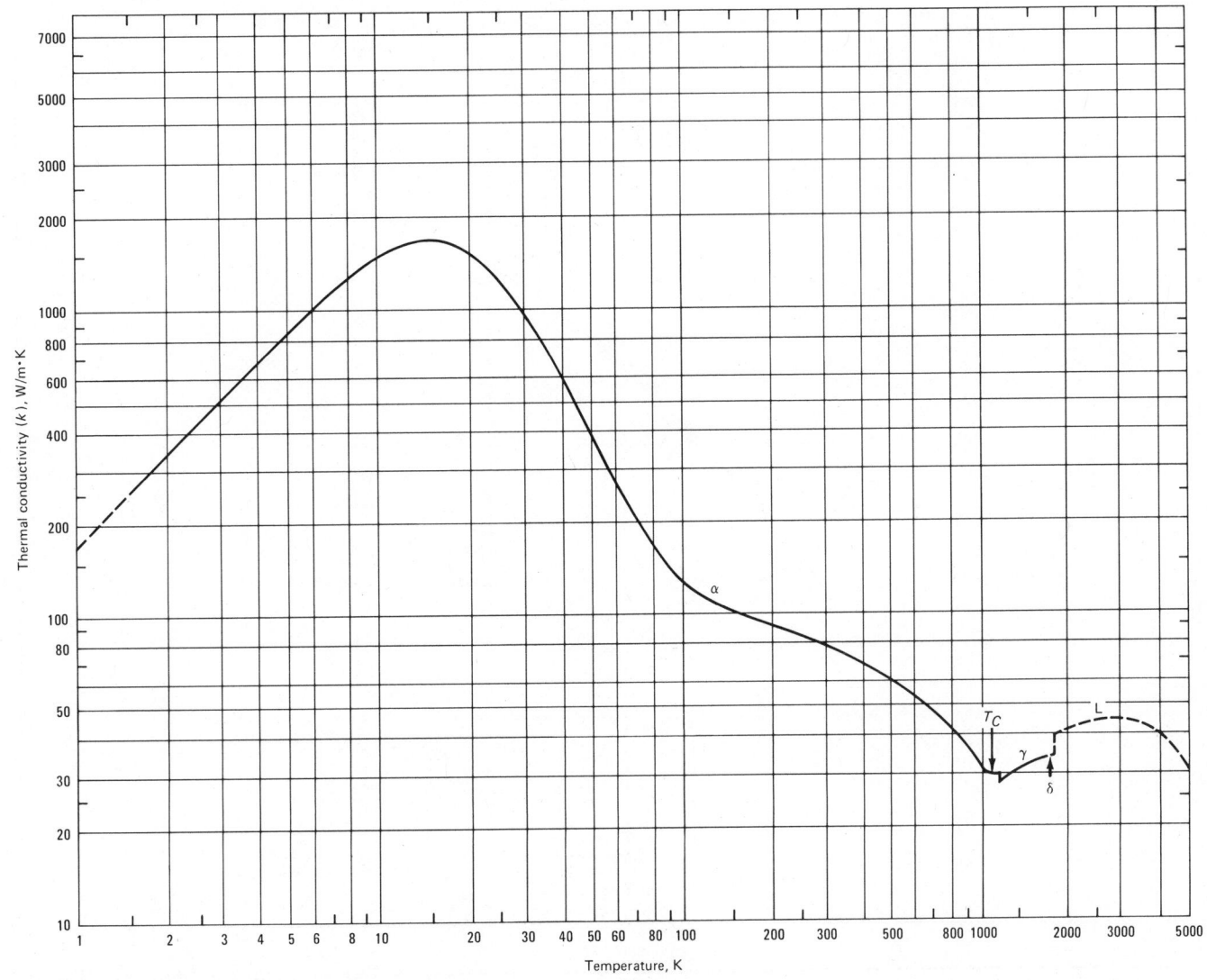

Microstructure. Zone-refined iron shows an essentially featureless, equiaxed grain structure.

Fracture behavior. Zone-refined iron can exhibit considerable ductility at −269 °C (4.2 K) (reduction in area of almost 100%), whereas less pure irons become brittle at temperatures below −153 °C (120 K) (Ref 4 and 5). Impure irons show either cube-face cleavage or conchoidal grain-boundary fracture, depending on grain size and impurities present (Ref 6).

Mass Characteristics

Atomic weight. 55.847 (based on $^{12}C = 12$, International Union of Pure and Applied Chemistry, 1961)

Density. α-Fe at 20 °C, $\rho = 7.870$ Mg/m³; γ-Fe at 912 °C, 7.694 Mg/m³; δ-Fe at 1394 °C, 7.406 Mg/m³ (Ref 7). Liquid Fe at melting point (1538 °C), $\rho = 7.035$ Mg/m³; at 1550 °C, 7.01 ± 0.03 Mg/m³; at 1564 °C, 7.00 Mg/m³ (Ref 6)

Density vs temperature. α-Fe (to 912 °C) $10^5 \cdot \Delta\rho/\rho_o K = 4.3$, γ-Fe (912 to 1394 °C), 6.7; δ-Fe (1394 to 1539 °C), 4.8 (Ref 8)

Density vs deformation. The density of high-purity α-iron will decrease very slightly with increasing cold work. The decrease in density can be estimated from the Stehle-Seeger relation (Ref 9). $N = (\Delta\rho/\rho_o)/2b^2$, where N is the dislocation density and b is the Burgers vector. The appropriate value of N at a strain of 0.20, taken from Ref 10, yields an estimate of $\Delta\rho/\rho_o = 2.5 \cdot 10^{-5}$.

Volume change on freezing. L to δ, −3.4%

Volume change on phase transformation. δ to γ (1394 °C), −0.52%; γ to α (912 °C), +1.0%; computed from changes in lattice parameter

Thermal Properties

Melting point. 1538 °C (Ref 11)
Boiling point. 2870 °C (Ref 11)
Phase-transformation temperatures. α to γ, 912 °C; γ to δ, 1394 °C
Coefficient of linear thermal ex-

Fig. 20 Vapor pressure of iron (Ref 19)

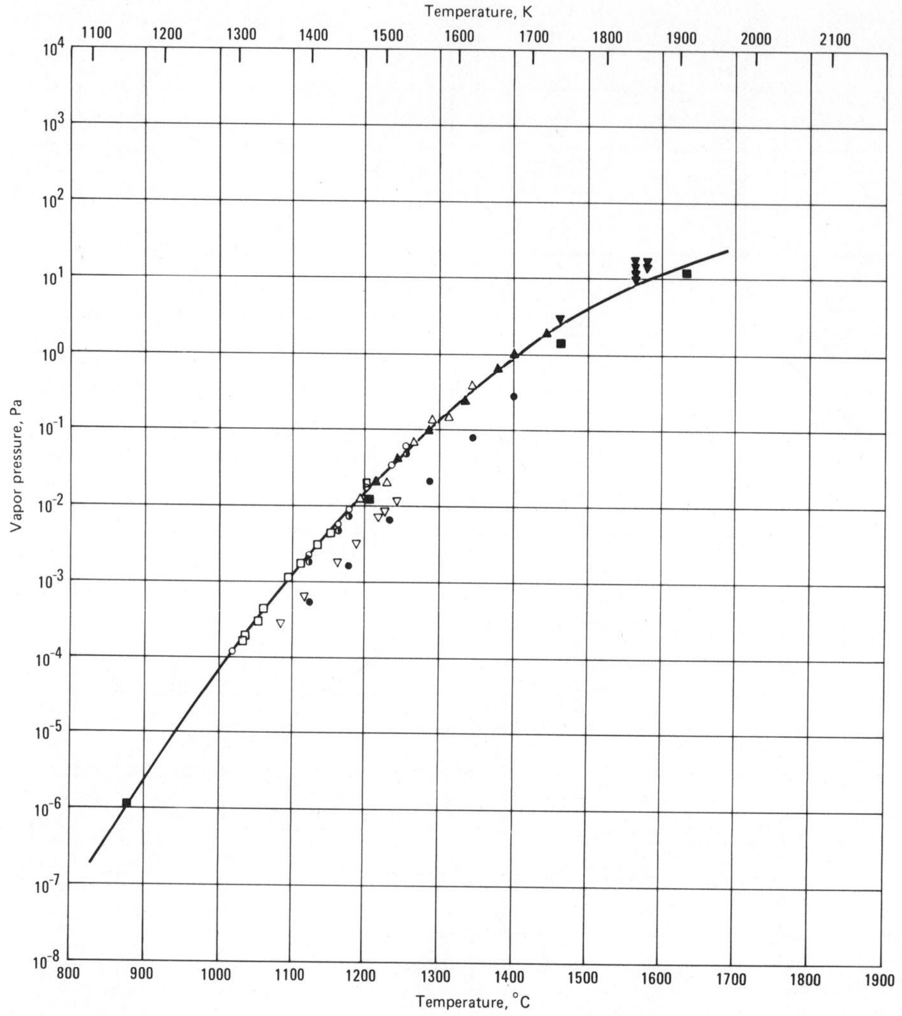

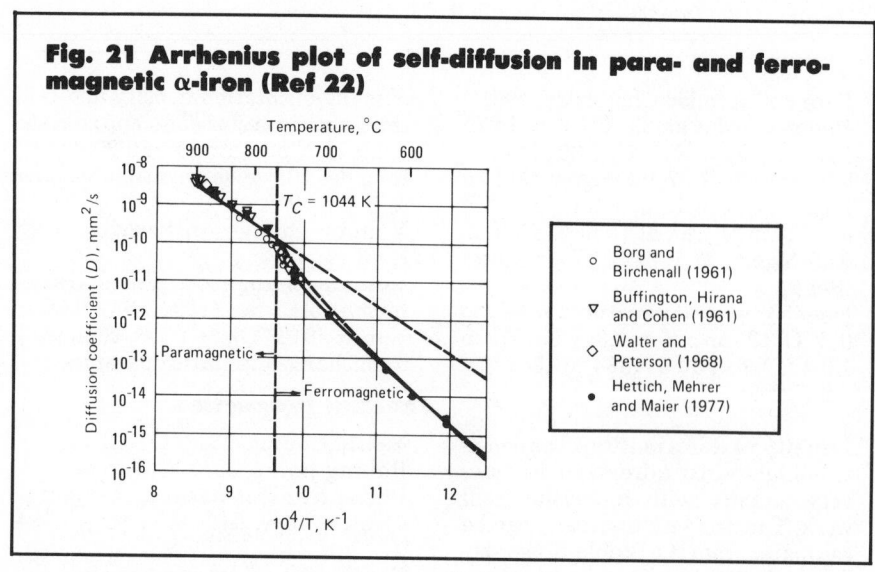

Fig. 21 Arrhenius plot of self-diffusion in para- and ferromagnetic α-iron (Ref 22)

pansion. Values recommended in Ref 12 are shown in Fig. 16 and Table 15. These values are considered accurate to within ±3% at temperatures below 627 °C, ±5% below 912 °C, and ±20% above 912 °C. The volume per atom as a function of temperature is shown in Fig. 17.

Coefficient of volumetric thermal expansion (liquid). The value of $10^6 \cdot \Delta v/v_o \cdot K = 140$ has been reported in Ref 14 (somewhat greater than three times the linear coefficient in the δ range).

Specific heat. The specific heats at constant pressure of α, γ, δ, and liquid iron, as a function of temperature, are shown in Fig. 18 and Table 16. Data for temperatures below 27 °C are primarily taken from the compilation reported in Ref 6. Data for

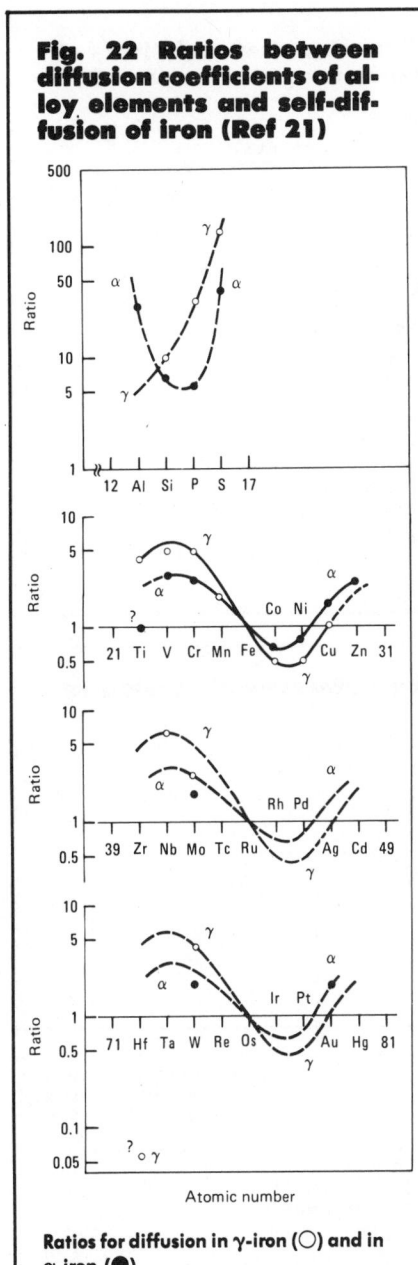

Fig. 22 Ratios between diffusion coefficients of alloy elements and self-diffusion of iron (Ref 21)

Ratios for diffusion in γ-iron (O) and in α-iron (●).

Fig. 23 Electrical resistivity of iron

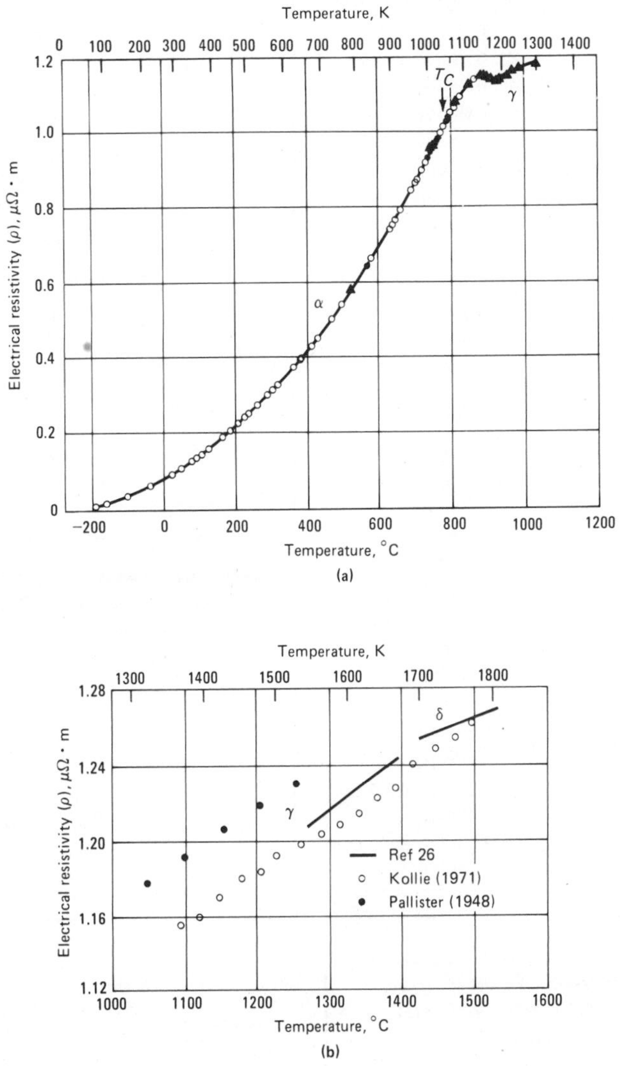

(a) −273 to +1027 °C (Ref 24 and 25). (b) 1027 to 1527 °C (Ref 26).

temperatures above 27 °C are taken from Ref 15.

Enthalpy, entropy and free-energy function. Values taken from Ref 15 are given in Table 16.

Latent heat of fusion. An adopted average value, taken from Ref 15, is 247 ± 7 kJ/kg for α-iron at the melting point (1538 °C).

Latent heat of phase transformation. Selected experimental averages, reported in Ref 15, are

Transformation	Temperature, K	Latent heat (ΔH°_t), kJ/kg
α to γ......	1185	16
γ to δ......	1667	15

Values at other temperatures can be derived from Table 16.

Latent heat of vaporization. Reference 16 gives a value of 7018.3 kJ/kg at the boiling point of iron (2870 °C).

Thermal conductivity. As a function of temperature is shown in Fig. 19. Data, given in Table 17, are

taken from Ref 17. The recommended values are for well-annealed high-purity iron, and those below −73 °C apply only to iron having a residual electrical resistivity of 0.143 nΩ·m. The values are accurate to within ±5% below −173 °C, ±3% from −173 °C to room temperature, ±2% from room temperature to +727 °C, and ±3 to 8% at higher temperatures.

Heat of combustion and free energy of formation. Reference 18 gives the following values for 25 °C:

Table 14 Lattice parameter of iron (Ref 62)

Temperature, °C	Lattice parameter, nm	
20	0.28665	α-Fe
53	0.28676	
154	0.28708	
248	0.28750	
315	0.28775	
378	0.28806	
451	0.28840	
523	0.28879	
563	0.28882	
588	0.28890	
642	0.28922	
660	0.28920	
706	0.28923	
730	0.28935	
754	0.28940	
764	0.28940	
772	0.28943	
799	0.28946	
862	0.28988	
898	0.29012	
907	0.29005	
950	0.36508	γ-Fe
1003	0.36535	
1076	0.36599	
1167	0.36660	
1249	0.36720	
1361	0.36810	
1390	0.29315	δ-Fe
1439	0.29346	
1480	0.29378	
1508	0.29396	

Fig. 25 Pressure dependence of the electrical resistivity of iron (Ref 30)

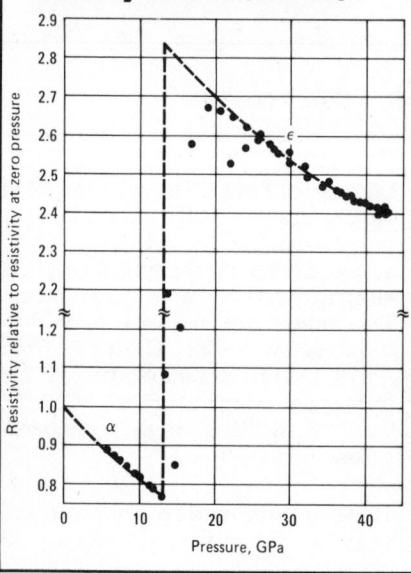

Fig. 24 Temperature coefficient of the electrical resistivity of iron in the neighborhood of the Curie temperature (Ref 25)

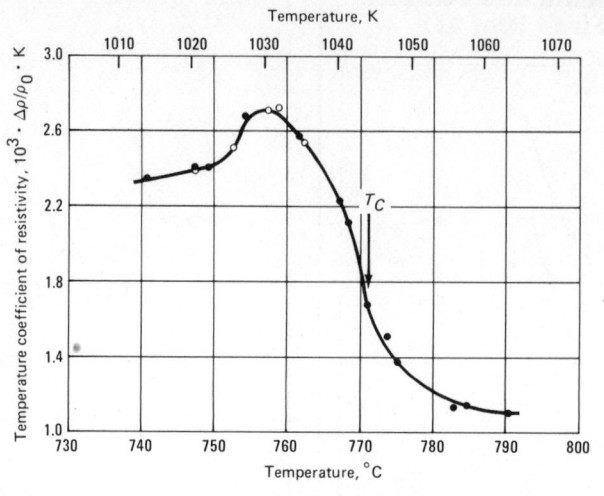

Fig. 26 Resistivity change of iron deformed in tension at −196 °C and at +25 °C (Ref 31)

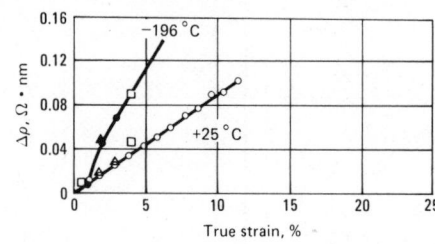

Oxide	Molecular weight	$-\Delta H°_f$, kJ/kg Fe	$-\Delta F°_f$, kJ/kg Fe
Fe$_{0.947}$O.....	68.89	4758 ± 14	4357 ± 18
Fe$_2$O$_3$.......	159.70	14700 ± 75	13245 ± 11 3
Fe$_3$O$_4$.......	231.55	19990 ± 150	18146 ± 16 5

	Diffusion constant (D_0), mm²/s	Activation energy (Q), MJ/kg
In γ-Fe	70	5.12
In paramagnetic α-Fe	16	4.30
In ferromagnetic α-Fe	50	4.30

The diffusion constant, D, can be calculated using the Arrhenius equation:

$$\ln D = \ln D_0 - \frac{Q}{RT}$$

In ferromagnetic iron, the Arrhenius plot of D cannot be approximated by a straight line; instead, it shows strong curvature (see Fig. 21). For practical purposes, the diffusion coefficients for alloying elements differ from the self-diffusion of iron by factors that are independent of temperature. These factors, shown in Fig. 22, can be used to determine the diffu-

Vapor pressure. Data reported in Ref 19 are shown in Fig. 20. For the temperature range 1178 to 1394 °C, Ref 20 reports that

$$\log p = m/T + b$$

where m is $-20\,908 \pm 109$; b is 12.161 ± 0.070; T is in K and p is in Pa.

Diffusion coefficient. Reference 21 gives the following values for self-diffusion of iron:

Fig. 27 Thermoelectric power of iron (Ref 32)

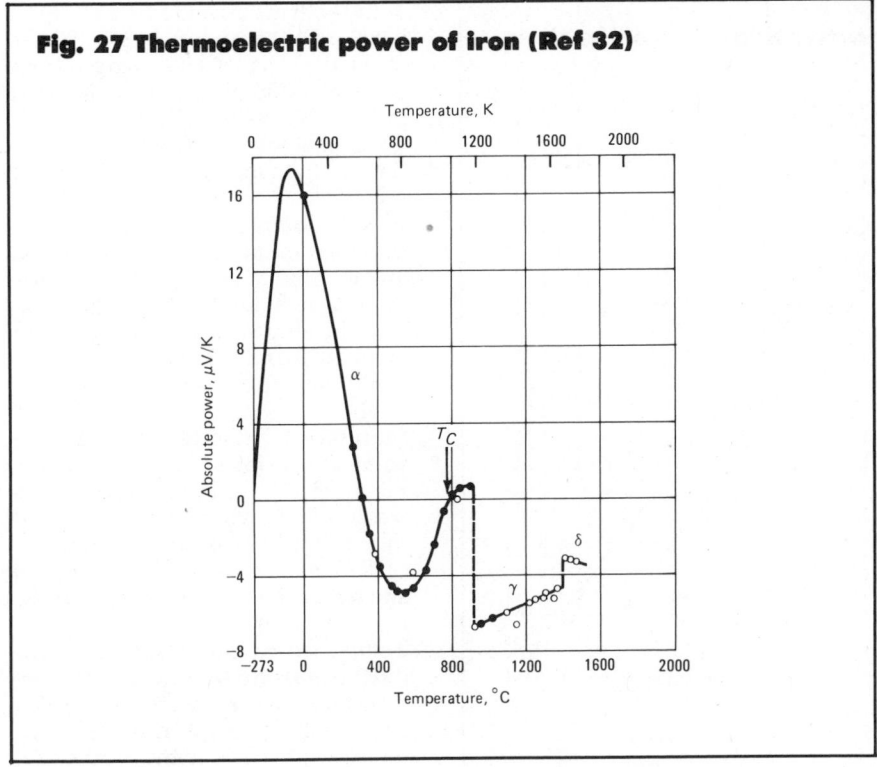

Table 15 Linear thermal expansion of iron (Ref 12)

Temperature		.Change in length, %(a)	Coefficient(α), $10^6 \cdot \Delta 1/1_o \cdot K$
°C	K		
−273	0	· · ·	0
−268	5	−0.204	0.01
−248	25	−0.203	0.20
−223	50	−0.203	1.3
−173	100	−0.184	5.6
−73	200	−0.102	10.1
+20	293	0.000	11.8
127	400	0.134	13.4
227	500	0.274	14.4
327	600	0.421	15.1
427	700	0.575	15.7
527	800	0.735	16.2
627	900	0.899	16.4
727	1000	1.065	16.6
827	1100	1.230	16.7
912(b)	1185(b)	1.370	16.8
912(b)	1185(b)	0.993(c)	23.3
927	1200	1.028(c)	23.3
1127	1400	1.494(c)	23.3
1327	1600	1.960(c)	23.3
1377	1650	2.077(c)	23.3
1394(d) 1667(d) to 1502 to 1775		· · ·	23.6

(a) Change from length at 20 °C. (b) α to γ phase transition. (c) Typical values. (d) γ to δ phase transition.

Fig. 28 Hydrogen overvoltage (activation) of iron in aqueous solution of NaCl (Ref 35)

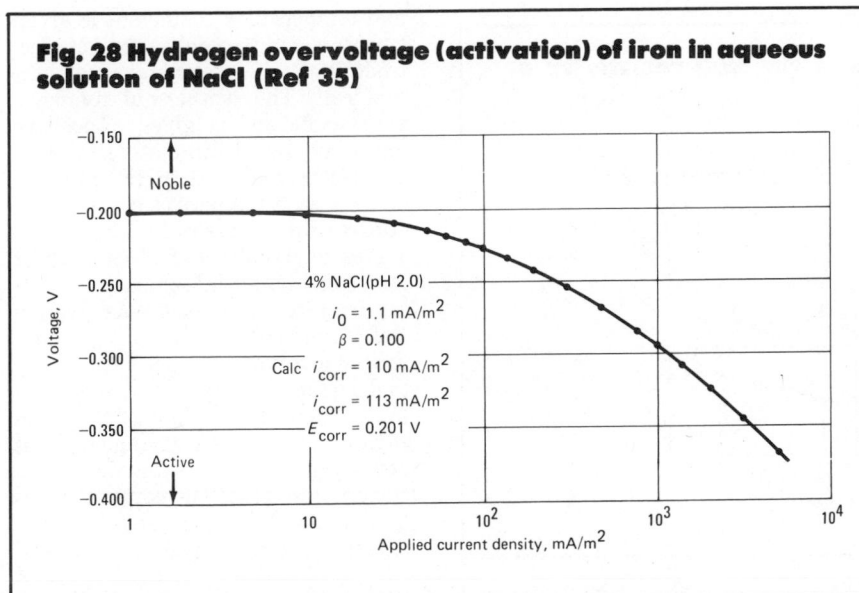

	Enthalpy (H), aJ	
	Formation	Migration
In γ-Fe..........	0.245 ±0.024	0.147
In paramagnetic α-Fe.....	0.257 ±0.024	0.202
In ferromagnetic α-Fe.....	0.256 ±0.024	0.202 ±0.040

The significance of these values is that monovacancy migration occurs only at temperatures well above ambient (> 450 K, or > 180 °C).

Electrical Properties

Volumetric electrical conductivity. 17.59% IACS at 25 °C

Electrical resistivity. The variation with temperature for solid iron is shown in Fig. 23(a) and (b). Data are from Ref 23 (−273 to +73 °C), Ref 25 (73 to 1027 °C), and Ref 26 (1027 to 1527 °C). The value of 1.39 $\mu\Omega$·m is given in Ref 6 for liquid iron at the melting point (1536 °C).

Temperature coefficient of electrical resistivity. From the data in Ref 27, the fundamental coefficient is

$$(\rho_{100 \, °C} - \rho_{0 \, °C})/\rho_{0 \, °C} = 0.00616$$

sion constants for other elements in γ-iron or α-iron. Grain-boundary diffusion in γ-iron and α-iron, of iron and some alloying elements, can be described by

$$D_0 = \frac{0.054 \; \mu m^3/s}{\delta}$$

where δ represents the thickness of the grain boundary in μm. For grain-boundary diffusion in both α-iron and γ-iron, Q is 2.78 MJ/kg.

Vacancy formation and migration. Reference 23 gives the following enthalpies for formation and migration of monovacancies in iron:

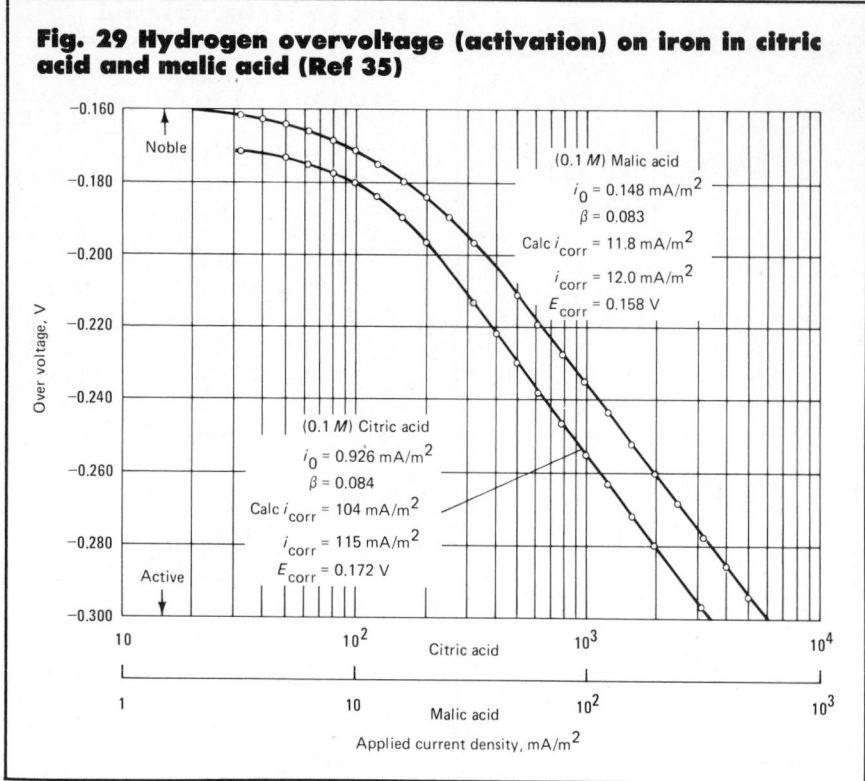

Fig. 29 Hydrogen overvoltage (activation) on iron in citric acid and malic acid (Ref 35)

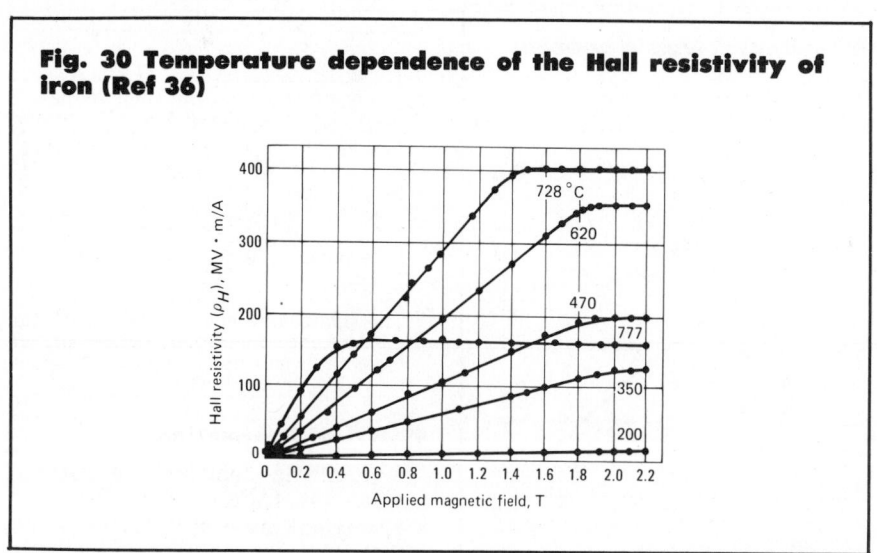

Fig. 30 Temperature dependence of the Hall resistivity of iron (Ref 36)

or 0.616% increase per K. The temperature coefficient in the neighborhood of the Curie temperature is shown in Fig. 24. The resistivity ratio $\rho_{300 K}/\rho_{4 K}$ of several high-purity irons is tabulated in Table 18. This ratio shows a fair correlation with purity, provided measurements are made in the proper longitudinal magnetic fields (Ref 29).

Pressure coefficient of electrical resistivity. The variation of resistivity with pressure at 25 °C is shown in Fig. 25. From this figure, the pressure coefficient is −0.18 per Pa. The resistance increased at the α to ε transition (at 13 GPa) by 366%.
Electrical properties vs deformation. The change in resistivity with true strain for iron deformed at −196

°C ($\rho = 5.8$ nΩ·m) and at +25 °C ($\rho = 0.098$ μΩ·m) is shown in Fig. 26. At 25 °C, the conductivity changed from 17.593% IACS (no reduction in area) to 17.577% IACS (9.5% RA).
Thermoelectric power. The variation with temperature is shown in Fig. 27.
Electrochemical equivalent. Based on an atomic weight of 55.85 and a value of 96 495 coulombs per gram equivalent weight for the Faraday, the electrochemical equivalents are 0.1929 and 0.2893 mg/C (C, coulomb) for Fe^{+3} and Fe^{+2}, respectively.
Standard electrode potential. The value of −0.4402 V for the reaction $Fe = Fe^{++} + 2e^-$ at 25 °C, where iron would be the negative terminal in a cell whose second electrode is SHE (standard hydrogen electrode) and where the Fe^{++} activity is unity, is given in Ref 33.
Temperature coefficient of standard electrode potential. The thermal temperature coefficient, $(dV°/dT)_{th}$, at 25 °C is given in Ref 33 as +0.923 mV/K, and the isothermal temperature coefficient, $(dV°/dT)_{iso}$, at 25 °C as +0.052 mV/K. The thermal temperature coefficient is given a positive value because the hot electrode is the (+) terminal in a thermal cell. The isothermal temperature coefficient is given a positive value as the electromotive force of the isothermal cell, SHE//iron, increases with temperature.
Ionization potential. The value 1.27aJ is given in Ref 34 as the first ionization potential for iron.
Hydrogen overvoltage. As given in Ref 35, the relationship between hydrogen overvoltage (activation), η_a, and current density, i, is $\eta_a = -\beta \log (i/i_0)$, where β (the slope of the Tafel region) and i_0 (the exchange current density) are constants that depend on the environment. The current density consists of contributions from the external applied current density and the local action current density. When the external applied current is zero, the overvoltage equals the corrosion potential, E_{corr}, and the local action current density equals the corrosion current, i_{corr}. The variation in hydrogen overvoltage of pure iron in 4% NaCl (2.0 pH) is shown in Fig. 28. In 4% NaCl, the overvoltage is essentially constant for pH 1 to 4. Hydrogen overvoltage of pure iron in 0.1 M citric acid and 0.1 M malic acid is

shown in Fig. 29. Overvoltage constants are given with the figures. Because there is a significant variation in overvoltage constants for different crystal orientations of pure iron, the data in the figures should be considered as average values for random orientation.

Hall effect. The Hall resistivity of polycrystalline ferromagnetic metals may be empirically expressed by

$$\rho_H = R_0 H + R_1 M$$

where R_0 and R_1 are the respective coefficients of ordinary and extraordinary Hall effect, H is the applied magnetic field, and M is the magnetization. The temperature dependence of ρ_H is shown in Fig. 30. Figures 31 and 32 show the temperature dependence of ordinary and extraordinary Hall coefficients. For single crystals, Ref 38 reports the extraordinary Hall coefficient to be dependent on the orientation of the crystal (see Fig. 33). A detailed discussion of galvanomagnetic effects in iron whiskers (at −273 to 27 °C) is given in Ref 39 and 40.

Electrical resistivity vs alloying. The electrical resistivity of some binary iron alloys at −269 °C (4.2 K) versus solute concentration in atom percent is shown in Fig. 34. Results from the iron-nickel system overlap those of the iron-palladium system; the iron-tungsten system data are identical with the results for the

Fig. 31 Temperature dependence of the ordinary Hall coefficient of iron (Ref 37)

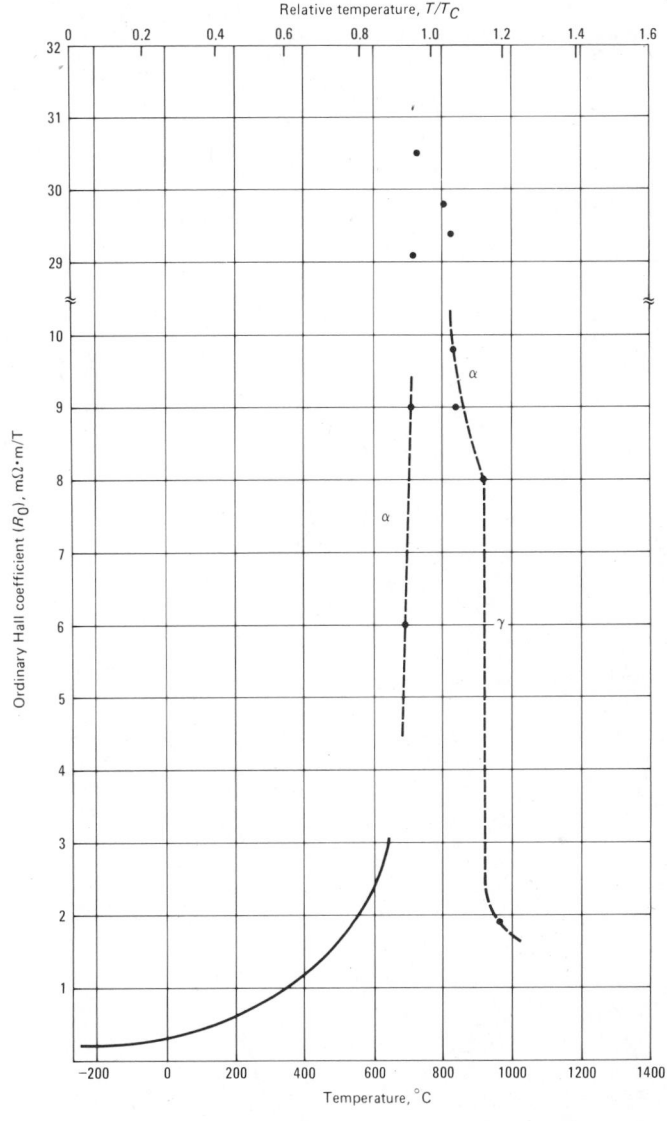

Fig. 32 Temperature dependence of the extraordinary Hall coefficient of iron (Ref 36)

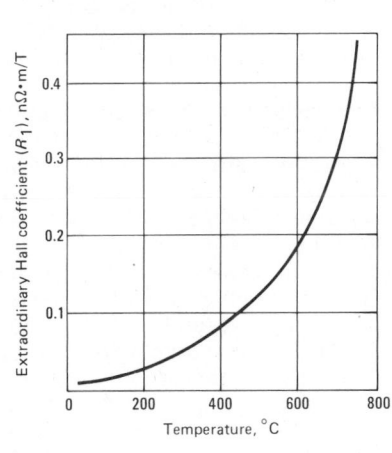

iron-manganese system. For clarity, low concentration points for some alloys have been omitted from the figure.

Electron emission. Reference 42 gives values of 260 kA/m²·K² for α-iron and 15 kA/m²·K² for γ-iron.

Work function of high-purity iron is sensitive to impurities and surface condition. Using iron that was electropolished and then cleaned through repeated cycles of ion bombardment (with argon) and annealing, the value of the work function for the (100) plane of α-iron was found to be 4.67 ± 0.03 eV (0.748 ± 0.005 aJ) (Ref 43). Using positive-ion-emission and electron-emission data, the change in work function at the α to γ transformation (ϕ_γ/ϕ_α) was found to be between +0.06 and +0.09 eV (+0.010 to +0.014 aJ) (Ref 44).

Magnetoresistance. The variation at −269 °C (4.2 K) of residual resistance in longitudinal and transverse magnetic fields is shown in Fig. 35. This effect can also be seen as a dependence of the residual electrical resistance upon measuring current density (see Fig. 36).

Table 16 Thermal properties of iron (Ref 6, 15 and 16)

Temperature °C	K	Specific heat (a), J/mol · K	Enthalpy (b), kJ/ mol	Entropy (c), J/ mol · K	Free-energy function (d), J/ mol · K
α and δ Phases					
− 273.15	0	0	− 4.498	− 27.280	Infinite
− 271.7	1.5	0.00749			
− 271.2	2.0	0.01017			
− 269.2	4.0	0.02134			
− 267.2	6.0	0.03452			
− 265.2	8.0	0.0502			
− 263.2	10.0	0.0682			
− 261.2	12.0	0.0916			
− 259.2	14.0	0.1188			
− 257.2	16.0	0.1556			
− 255.2	18.0	0.1987			
− 253	20	0.2573			
− 243	30	0.753			
− 233	40	1.57			
− 223	50	3.01			
− 213	60	4.81			
− 203	70	6.74			
− 193	80	8.62			
− 183	90	10.42			
− 173	100	12.05	− 4.067	− 21.150	46.800
− 163	110	13.56			
− 153	120	14.90			
− 143	130	16.11			
− 133	140	17.20			
− 123	150	18.12			
− 113	160	18.91			
− 93	180	20.33			
− 73	200	21.46	− 2.301	− 9.280	29.505
− 53	220	22.38			
− 33	240	23.18			
− 13	260	23.89			
+ 25.00	298.15	24.98	0	0	27.280
27	300	25.02	0.046	0.1544	27.281
127	400	27.36	2.665	7.673	28.290
227	500	29.71	5.518	14.031	30.273
327	600	32.05	8.606	19.654	32.590
427	700	34.60	11.934	24.778	35.009
527	800	37.90	15.548	29.560	37.405
577	850	40.15	17.498	31.923	38.167
627	900	43.03	19.573	34.291	39.823
677	950	47.15	21.819	36.717	41.030
727	1000	54.27	24.331	39.261	42.210
747	1020	59.77	25.466	40.385	42.698
757	1030	64.76	26.087	40.989	42.492
769	1042 (T_C)	83.51	26.978	41.847	43.236
777	1050	54.70	27.430	42.278	43.434
787	1060	51.55	27.964	42.795	43.694
807	1080	48.45	28.957	43.709	44.177
827	1100	46.37	29.904	44.578	44.673
912	1185 ($T_{\alpha-\gamma}$)	41.34	33.597	47.810	46.738
1394	1667 ($T_{\gamma-\delta}$)	41.03	52.601	61.244	56.970

(continued)

Magnetic Properties

Magnetic properties vs treatment and composition. Magnetic properties can be varied over a wide range by such factors as impurity content (particularly of carbon, sulfur, nitrogen and oxygen), impurity distribution (high-temperature solution anneal or low-temperature precipitation or aging), grain size, grain orientation, and strain or cold work.

Magnetic susceptibility. Temperature variations of the reciprocal of the mass paramagnetic susceptibility of iron are shown in Fig. 37.

Magnetic permeability. For polycrystal, H_2-treated iron, the value 0.0176 H/m for the initial permeability, and 0.314 to 0.352 H/m for the maximum permeability at room temperature were reported in Ref 6. For an iron single crystal, the maximum permeability in the [100] direction was reported as 1.80 to 1.82 H/m.

Coercive force. Reference 47 reports that for zone-refined iron the coercive force, H_c, follows the relation

$$H_c = 1.83 + 4.14/Q^{1/2}$$

where Q is the grain size in mm². After heat treatment for 10 h at 880 °C in H_2 with a furnace cool, the coercive force at room temperature was reported to be 10.74 A/m. A different treatment (60 h at 1300 °C in H_2, followed by 20 h at 870 °C in H_2 then furnace cooling) gave a coercive force of 1.35 A/m.

Saturation magnetization. Reference 48 gives the value of 2.158 T for room temperature. Reference 49 gives the magnetization per atom at 0 K (M_0) as 2.216 μ_B (Bohr magnetons), or 2.055×10^{-23} J/T.

Residual induction. The value of 1.183 T is given in Ref 48.

Hysteresis loss. The value per cycle of 15 to 19 J/m³ is given in Ref 6.

Magnetostriction of single-crystal and polycrystalline annealed electrolytic iron is shown in Fig. 38(a) and (b). If the directions of magnetization and strain measurement relative to the [100] direction for a cubic crystal are φ and ψ, respectively, then the variation of the magnetostriction of a self-saturated domain (which depends on the angular position of the vector **M** at saturation) can be expressed by five constants, as shown in the following formula:

Table 16 (continued)

Temperature °C	K	Specific heat (a), J/mol·K	Enthalpy (b), kJ/mol	Entropy (c), J/mol·K	Free-energy function (d), J/mol·K
1427	1700	41.36	53.963	62.052	57.589
1527	1800	42.36	58.149	64.443	59.418
1538	1811(T_M)	42.47	58.626	64.706	59.614
γ Phase					
912	1185($T_{\alpha-\gamma}$)	33.81	34.497	48.546	46.715
927	1200	33.93	35.010	48.976	47.081
1027	1300	34.77	38.442	51.717	49.426
1127	1400	35.60	41.957	54.319	51.630
1227	1500	36.44	45.559	56.800	53.707
1327	1600	37.27	49.247	59.176	55.677
1394	1667 ($T_{\gamma-\delta}$)	37.81	51.762	60.715	56.944
Liquid Phase					
1538	1811(T_M)	45.91	72.433	72.290	59.574
1627	1900	45.91	76.511	74.493	61.504
1727	2000	45.91	81.104	76.847	63.575
1827	2100	45.91	85.698	79.087	65.558
1927	2200	45.91	90.291	81.223	67.462
2027	2300	45.91	94.884	83.264	69.290
2127	2400	45.91	99.477	85.218	71.049
2227	2500	45.91	104.066	87.092	72.476
2327	2600	45.91	108.659	88.893	74.381
2427	2700	45.91	113.247	90.625	75.962
2527	2800	45.91	117.841	92.295	77.489
2627	2900	45.91	122.429	93.906	78.969
2727	3000	45.91	127.018	95.462	80.403
2827	3100	45.91	131.606	96.968	81.794
2927	3200	45.91	136.195	98.425	83.144

(a) C_p. (b) $H°_T - H°_{298}$. (c) $S_T - S°_{298}$. (d) $-(G°_T - H°_{298})/T$.

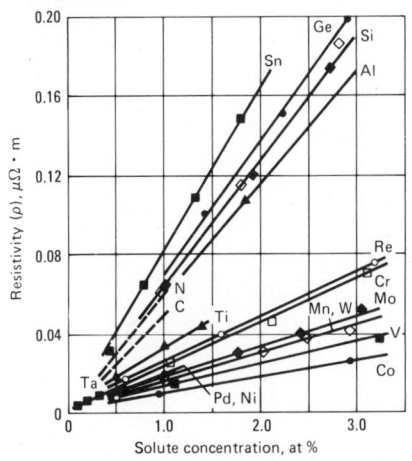

Fig. 34 Electrical resistivities of some binary iron alloys at −269 °C (4.2 K) (Ref 41)

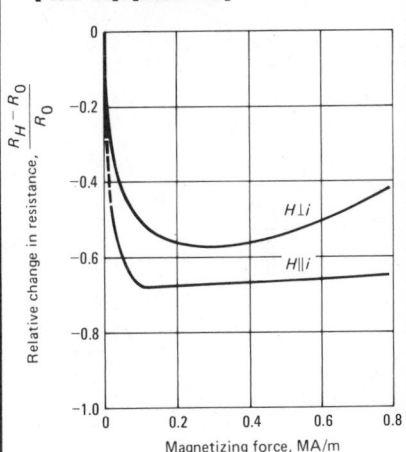

Fig. 35 Magnetoresistance of iron at −269 °C (4.2 K) (Ref 45)

Shown are curves for application of the magnetic field strength (H) parallel and perpendicular to the electric current (i). (R_H = resistance in field H; R_0 = resistance in no field. R_0 at 293 K is 258 times R_0 at 4.2 K.)

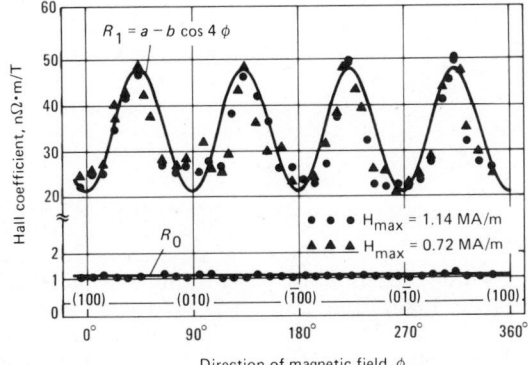

Fig. 33 Effect of magnetic field direction on Hall coefficients of iron at 27 °C (Ref 38)

φ is the angle between the magnetic field and the [100] axis when current is passed along the [001] axis.

$\Delta l/l_0 =$
$\{1/8 (h_1 - h_2 - h_3) + 5/48\ h_4\}$ cos 2ϕ
$+\{1/64 (-6h_3 + h_4 - 2h_5)$ cos 4ϕ
$+\{3/8\ h_1 + 1/8\ h_2 + 5/16\ h_4\}$ cos 2ϕ cos 2Ψ
$+\{1/2\ h_2 + 1/8\ h_5\}$ sin 2ϕ sin 2Ψ
$+\{3/64\ h_4 + 1/32\ h_5\}$ cos 4ϕ cos 2Ψ
$+\{-1/16\ h_5\}$ sin 4ϕ sin 2Ψ

Reference 51 reports the following

Table 17 Thermal conductivity of iron (Ref 17)

Solid			Solid			Liquid (a)		
Temperature °C	K	Conductivity, W/m·K	Temperature °C	K	Conductivity, W/m·K	Temperature °C	K	Conductivity, W/m·K
−273	0 0		−23	250	86.5	1537	1810	40.3(b)
−272	1 .... 171(b)		0	273	83.5	1600	1873.2	41.3(b)
−271	2 342		+25	298.2	80.4	1627	1900	41.5(b)
−270	3 511		27	300	80.2	1700	1973.2	42.3(b)
−269	4 677		50	323.2	77.4	1727	2000	42.6(b)
						1800	2073.2	43.2(b)
−268	5 839		77	350	74.4	1900	2173.2	43.9(b)
−267	6 993		100	373.2	72.0	1927	2200	44.1(b)
−266	7 1140		127	400	69.5	2000	2273.2	44.6(b)
−265	8 1270		200	473.2	63.4	2127	2400	45.0
−264	9 1390		227	500	61.3	2200	2473.2	45.2(b)
−263	10 1480		300	573.2	56.4	2327	2600	45.5(b)
−262	11 1560		327	600	54.7	2400	2673.2	45.6(b)
−261	12 1630		400	673.2	50.4	2527	2800	45.8(b)
−260	13 1670		427	700	48.8	2600	2873.2	45.9(b)
−259	14 1690		500	773.2	44.8	2727	3000	45.8(b)
						2800	3073	45.8(b)
−258	15 1700		527	800	43.3	2927	3200	45.6(b)
−257	16 1690		600	873.2	39.4	3000	3273	45.4(b)
−255	18 1630		627	900	38.0	3127	3400	45.1(b)
−253	20 1540		700	973.2	34.2	3327	3600	44.2(b)
−248	25 1270		727	1000	32.8	3527	3800	43.0(b)
−243	30 1000		786	1059	29.7	3727	4000	41.5(b)
−238	35 788		800	1073.2	29.8	4227	4500	36.8(b)
−233	40 623		827	1100	29.8	4727	5000	30.8(b)
−228	45 499		900	1173.2	30.0	5227	5500	23.3(b)
−223	50 405		910	1183(α)	30.0	5727	6000	14.7(b)
						6227	6500	5.1(b)
−213	60 285		912	1185(γ)	28.0			
−203	70 216		927	1200	28.3			
−193	80 175		1000	1273.2	29.6			
−183	90 150		1027	1300	30.0			
−173	100 134		1100	1373.2	30.9			
−150	123.2 115		1127	1400	31.2			
−123	150 104		1200	1473.2	31.9			
−100	173.2 .... 99.1		1227	1500	32.1			
−73	200 .... 94.0		1300	1573.2	32.7			
−50	223.2 .... 90.4		1327	1600(γ)	33.0			
			1400	1673.2(δ)	33.5(b)			
			1427	1700	33.8(b)			
			1500	1773.2	34.3(b)			
			1527	1800	34.5(b)			
			1537	1810	34.6(b)			

(a) Values for liquid iron are provisional. (b) Extrapolated or estimated values.

room temperature values for the constants in this formula:

$$h_1 = 36.1 \pm 2.1$$
$$h_2 = -34.5 \pm 1.4$$
$$h_3 = -1.2 \pm 0.5$$
$$h_4 = 3.3 \pm 0.7$$
$$h_5 = 0.8 \pm 0.3$$

The variation of these constants with temperature is treated in Ref 52.
Magnetic transformation (Curie) temperature. The value of 1044 ± 2 K is given in Ref 53.

Optical Properties

Color. Silvery white, resembling platinum more than ingot iron or steel.
Reflectance. The normal spectral reflectance of polished iron varies from 65% at a wavelength of 1.5 μm to 97% at 15 μm (Ref 19 and 54).
Absorptance. The normal spectral absorptance of polished iron varies from about 0.33 at a wavelength of 1.5 μm to 0.03 at 15 μm (Ref 54).
Emittance. The normal spectral emittance of polished iron at about 927 °C varies from 35% at a wavelength of 0.65 μm to 26% at 1.5 μm and 11% at 15 μm (Ref 54).

Nuclear Properties

Stable isotopes. Reported in Ref 55 are:

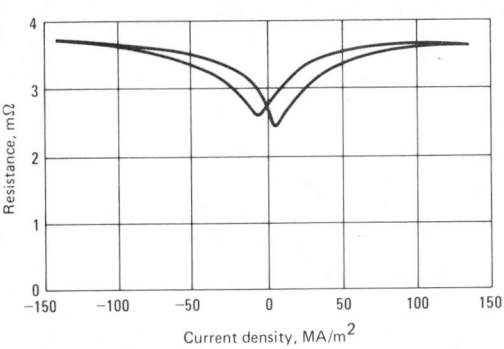

Fig. 36 Residual resistance of iron at −269 °C (4.2 K) as a function of measuring current density (Ref 45)

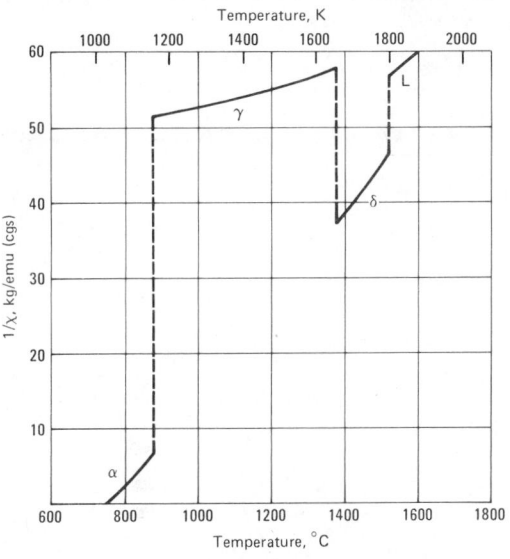

Fig. 37 Temperature dependence of the reciprocal of the mass paramagnetic susceptibility of iron (Ref 46)

To change susceptibility in cgs units to mks units, multiply by 4π.

Table 18 Resistivity ratios of iron (Ref 25) (a)

Sample	Impurity (ppm)	Ratio
Arajs "B" (1965)	200-500	27
Rosenberg (1965)	~ 100	50, 83
Kempt *et al* (1959)	~ 100	110
Arajs (1964)	~ 80	250
Arajs "A" (1965)	~ 25	300
Badiali *et al* (1963)	~ 25	380
Takaki and Kimura (1973)(b)	···	>5000

(a) ρ 300 K/ρ₄ₖ. (b) From Ref 28, rather than Ref 25.

Effects of neutron irradiation. On properties of materials are principally due to lattice defects produced. These effects involve structural, mechanical, electronic (associated with trapping of charge), and diffusion-controlled properties (see Ref 56). The effect of irradiation on yield strength and reduction in area of vacuum-melted iron (0.003% C, 0.0055% O_2, 0.0005 N) is shown in Fig. 39. For a review of mechanical properties of irradiated iron and iron alloys, see Ref 58. Corrosion effects are discussed in Ref 59.

Chemical Properties

General corrosion behavior. Of iron in aqueous solutions is shown schematically in Fig. 40. Irons of high purity show a remarkably high resistance to corrosion, sometimes remaining untarnished in laboratory atmospheres for months or years (Ref 6). It was shown in Ref 61 that zone-refined iron corrodes at the same rate in hydrochloric acid whether cold worked or annealed and is not affected by any heat treatment schedule. Results reported in Ref 35 indicate an orientation effect, as certain crystal faces are attacked more than others.

Effects of specific corroding agents. On zone-refined iron at 25 °C are as follows:

Acid (oxygen free)	pH	Corrosion rate, g/m²·day	Ref
0.1*M* citric	2.06	2.9	62
0.1*M* malic	2.24	0.3	62
4% NaCl	1 to 4	3.0	35
0.12*N* HCl.....	1.01	2.0	61

Stability (Pourbaix) diagrams. The regions of stability of various species of iron in water at 25 °C are shown as a function of potential (relative to a standard hydrogen electrode) and pH in Fig. 41.

Isotope	Atomic weight*	Percent of total
Fe⁵⁴	53.9396	5.82
Fe⁵⁶	55.9349	91.66
Fe⁵⁷	56.9354	2.19
Fe⁵⁸	57.9333	0.33

*Relative to C¹²

Unstable isotopes. Reported in Ref 55 are:

Isotope	Half-life	Decay mode	Particle energy, fJ
Fe⁵²	....8.2 h	β⁺, EC†	130
Fe⁵³	....8.5 min	β⁺, EC	450, 380, 260
Fe⁵⁵	....2.6 yr	EC	···
Fe⁵⁹	....45.1 ± 0.5 days	β⁻	252.0, 76.1, 43.7
Fe⁶⁰	....3(10⁵) yr	β⁻	≤ 22
Fe⁶¹	....6.0 min	β⁻	450

†EC, electron capture

Fig. 38 Longitudinal magnetostriction in (a) single crystals of iron and (b) polycrystalline iron (Ref 50)

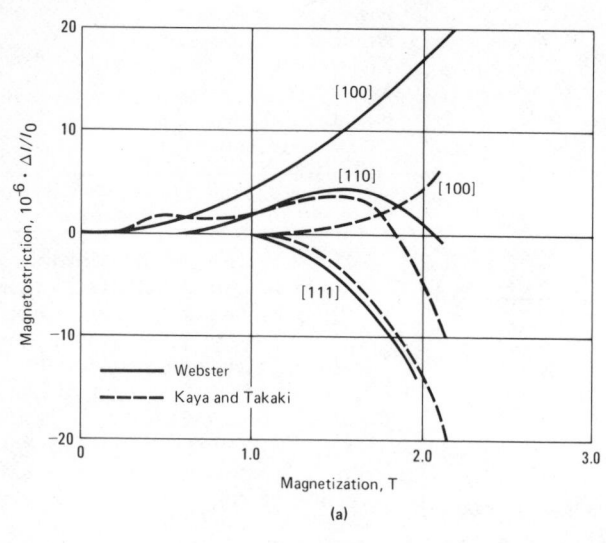

(a)

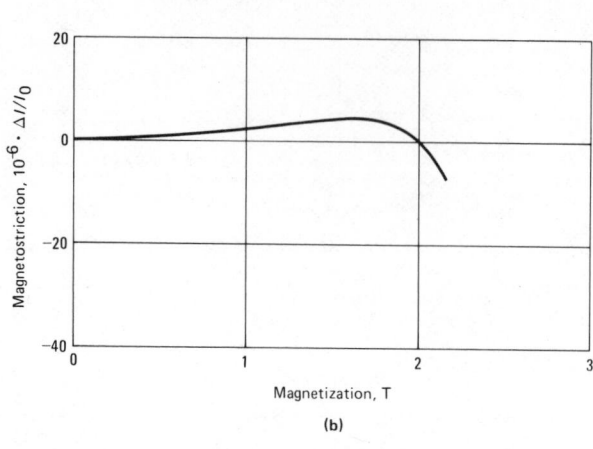

(b)

Fig. 39 Effect of temperature on yield strength and reduction in area of iron before and after irradiation (Ref 57)

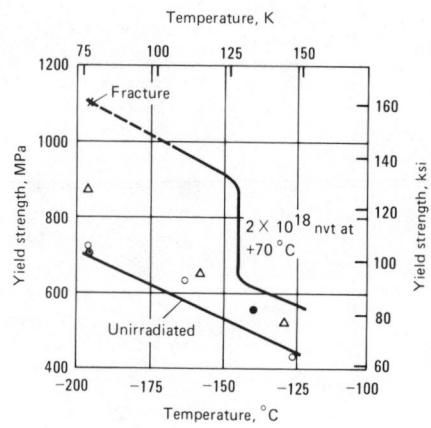

○ 6×10^{17} nvt at $-70\ ^\circ$C

△ 2×10^{18} nvt at $-70\ ^\circ$C

● 1×10^{18} nvt at $+150\ ^\circ$C

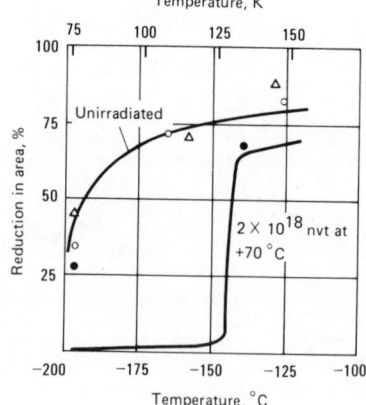

○ 6×10^{17} nvt at $-70\ ^\circ$C

△ 2×10^{18} nvt at $-70\ ^\circ$C

● 1×10^{18} nvt at $+150\ ^\circ$C

Note: nvt is the neutron dose and is equivalent to the number of neutrons per square centimetre.

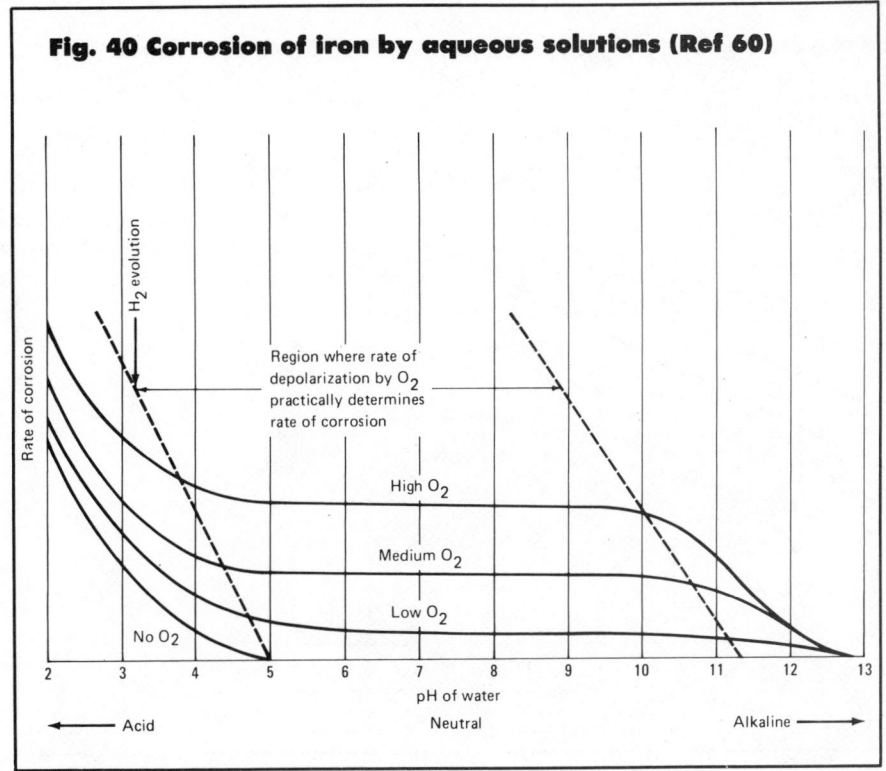

Fig. 40 Corrosion of iron by aqueous solutions (Ref 60)

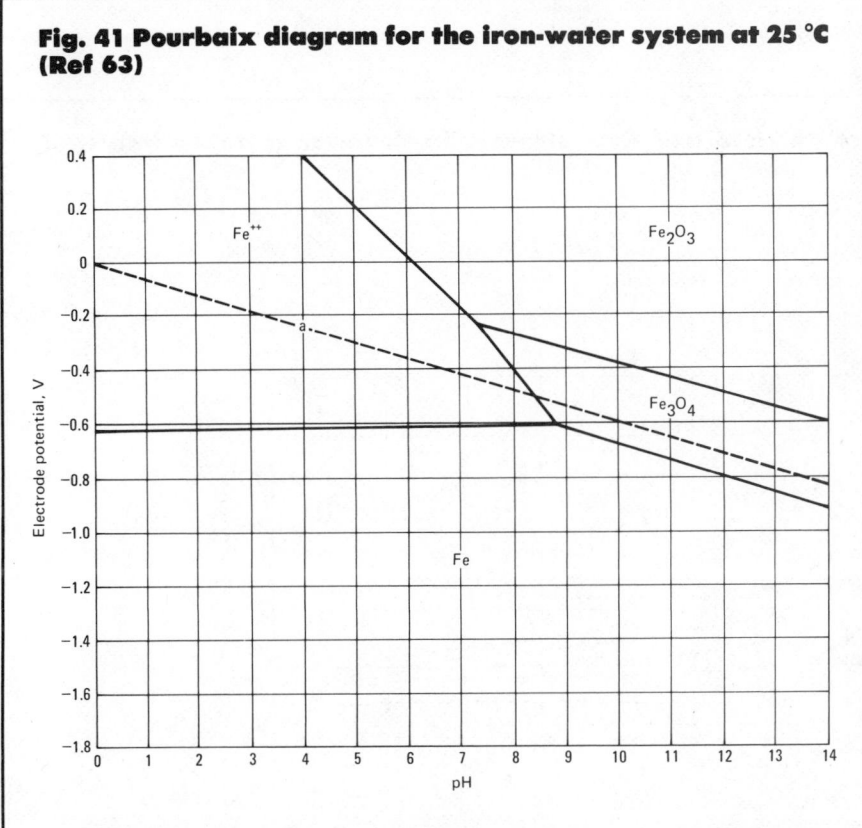

Fig. 41 Pourbaix diagram for the iron-water system at 25 °C (Ref 63)

Fe, Fe₃O₄ and Fe₂O₃ are solid substances; water is stable above line *a*, H₂ gas below.

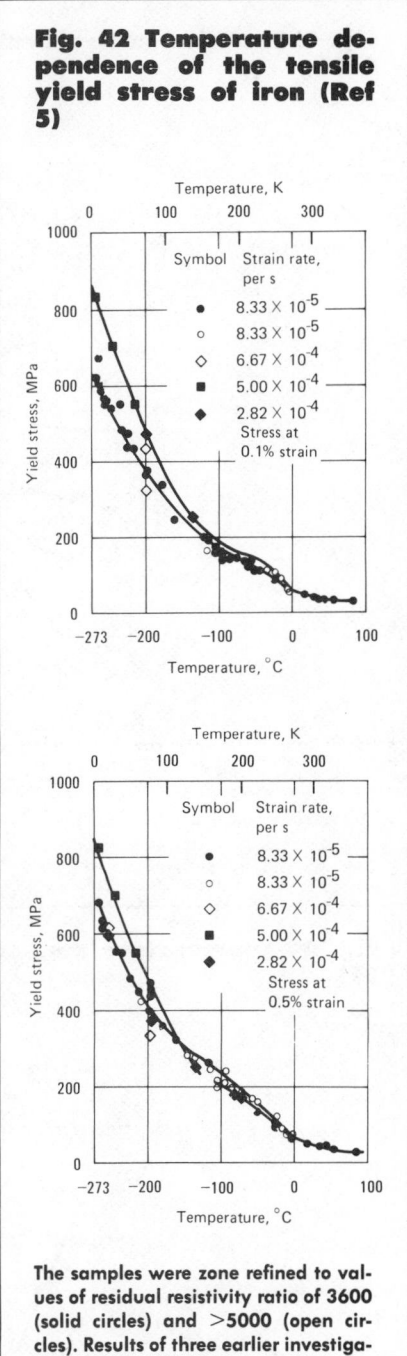

Fig. 42 Temperature dependence of the tensile yield stress of iron (Ref 5)

The samples were zone refined to values of residual resistivity ratio of 3600 (solid circles) and >5000 (open circles). Results of three earlier investigations are also shown.

Mechanical Properties

Tensile properties. The data for zone-refined iron have been summarized in Ref 5. Plots of the temperature dependence of yield stress are shown in Fig. 42. The very strong dependence shown is an inherent characteristic of α-iron. The yield

Fig. 43 The effect of strain rate on the strength of polycrystalline iron at room temperature (Ref 66)

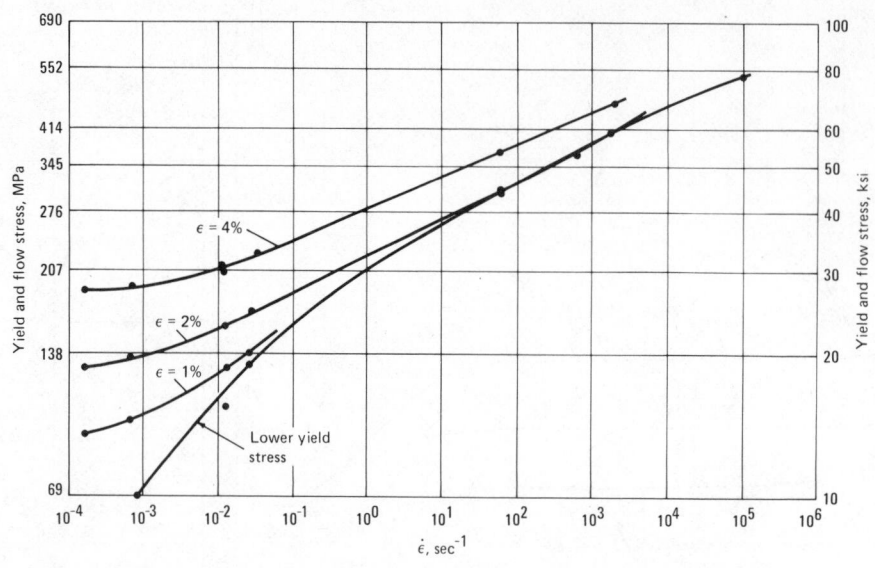

Fig. 44 Temperature dependence of yield and flow stresses in titanium-gettered iron (Ref 67)

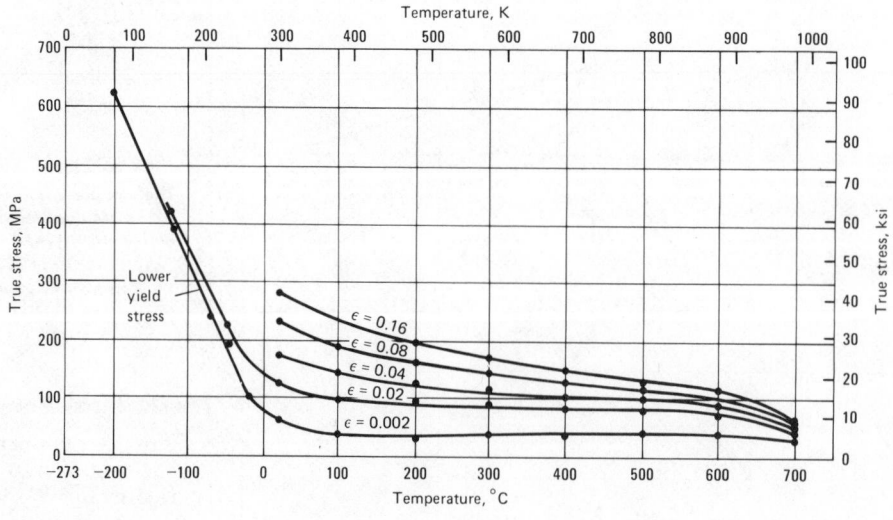

$\dot{\epsilon} \approx 2.5 \times 10^{-4}$ sec^{-1} grain size ASTM 5 to 6.

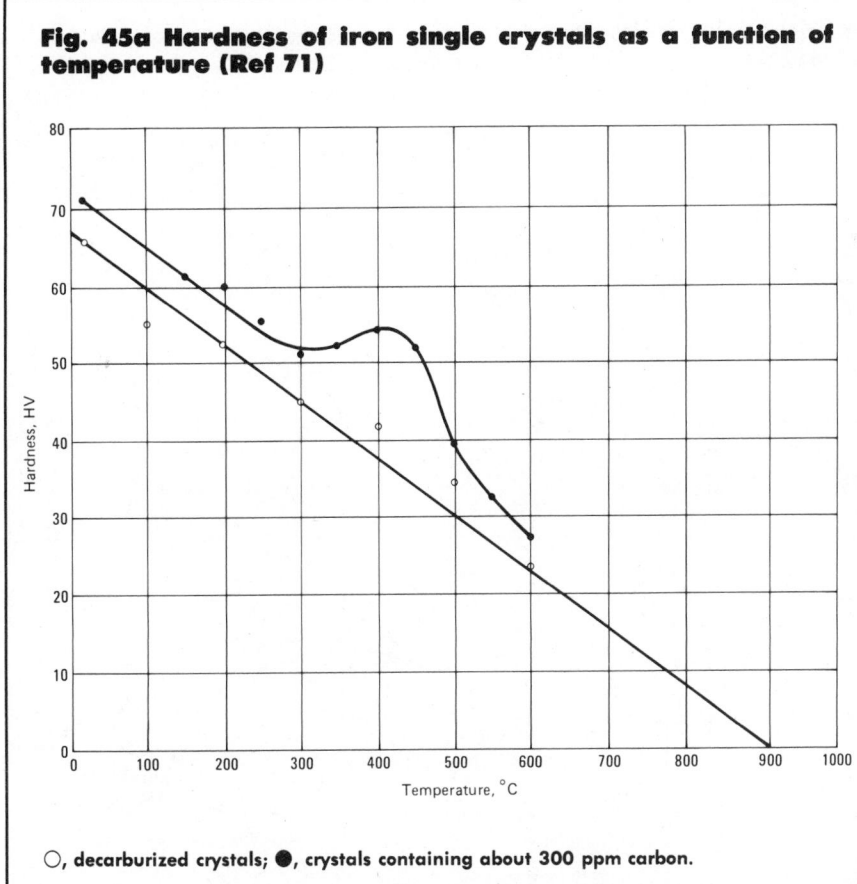

Fig. 45a Hardness of iron single crystals as a function of temperature (Ref 71)

○, decarburized crystals; ●, crystals containing about 300 ppm carbon.

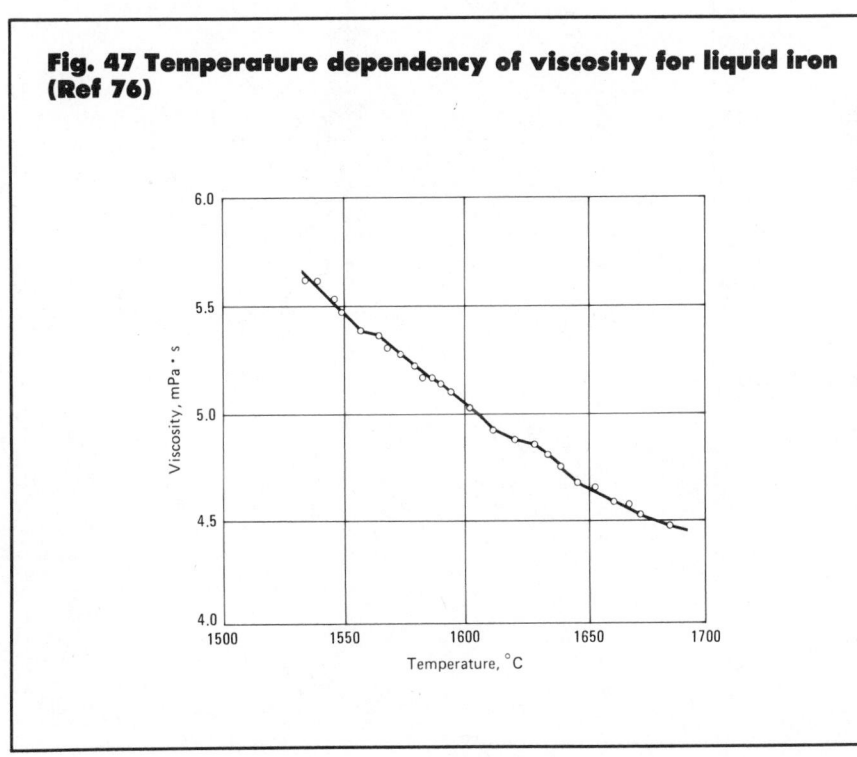

Fig. 47 Temperature dependency of viscosity for liquid iron (Ref 76)

stress, however, is not a smooth function of temperature; instead, there is a concave-downward region in the plot, centered about −30 °C (about 240 K), that has been observed by several workers (see Ref 4, 64 and 65). The scatter in these data for zone-refined iron is due principally to the single-crystal or "bamboo-structure" specimens that were employed. The dependence of the yield stress and flow stress of interstitial-free iron of grain size ASTM 5 to 6 on strain rate and temperature is shown in Fig. 43 and 44. The grain-size dependence of the yield stress is discussed in Ref 68 and 69.

Compressive properties. The yield strength in compression is the same as in tension. For a further discussion see Ref 70.

Hardness. Vickers hardness as a function of temperature for pure iron single crystals, and crystals containing 300 ppm carbon, is shown in Fig. 45(a). The anisotropy of the hardness of pure iron single crystals is shown in Fig. 45(b).

Poisson's ratio. The value of 0.291 at room temperature is given in Ref 69.

Strain-hardening exponent. About 0.3

Elastic moduli. The following room temperature values are reported in Ref 72:

- Young's modulus: 208.2 GPa
- Bulk modulus: 166.0 GPa
- Shear modulus: 80.65 GPa
- Compressibility: 6.024 TPa^{-1}

First, second and third order elastic stiffnesses are given in Ref 73.

Elastic moduli along crystal axes. The directional dependence of Young's modulus, the shear modulus, and Poisson's ratio are shown in Fig. 46.

Properties of single crystals. Plastic flow characteristics of single crystals are described in Ref 75. Three-stage hardening was observed at room temperature for all crystal orientations. At −130 and −196 °C (143 and 77 K) the critical resolved shear stress law was not obeyed. The critical resolved shear stress was greater in an anti-twinning direction than in a twinning direction.

Ductile-brittle transition temperature. For interstitial-free iron (Charpy V-notch, ½ size), Ref 67 gives −34 °C for ASTM grain size 4 to 5, and −29 °C for ASTM grain size 0 to 2.

Fig. 45b Hardness of decarburized crystals of iron as a function of crystal orientation (Ref 71)

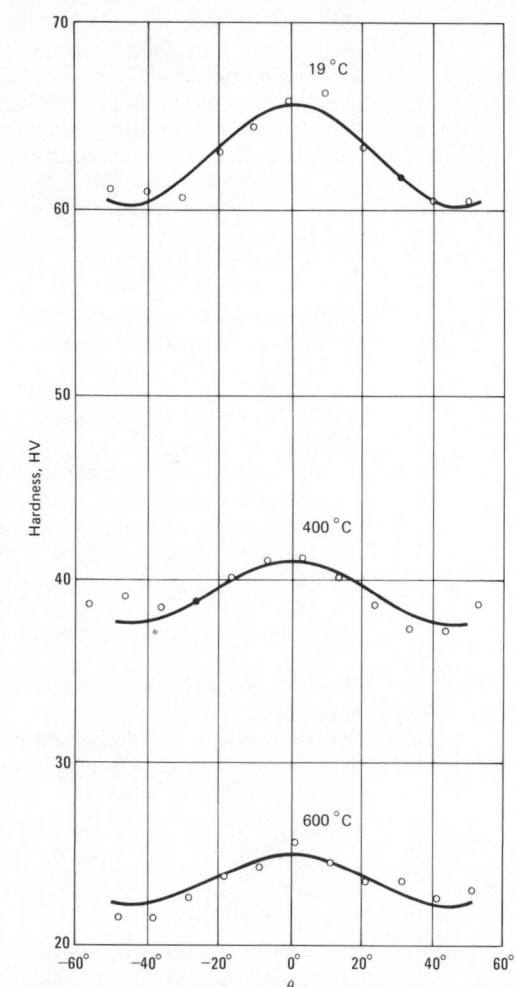

At θ = 0, the diagonal of the hardness impression is the projection of the <100> direction on the crystal surface.

Fig. 48 Surface tension of pure iron (Ref 77)

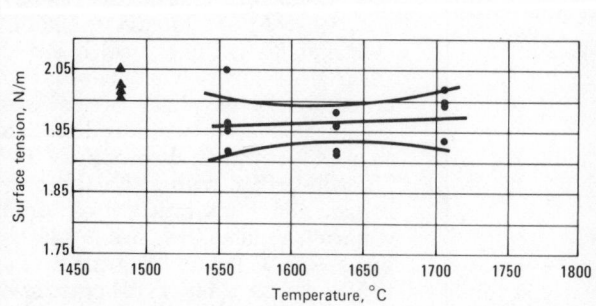

Fig. 46 Direction dependence of the elastic properties of iron in the (100) plane (Ref 74)

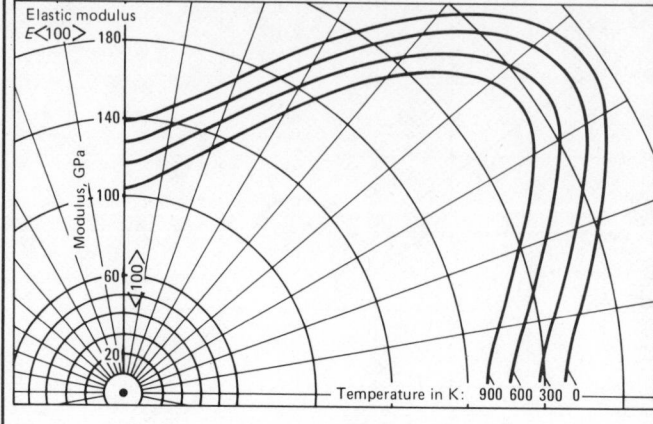

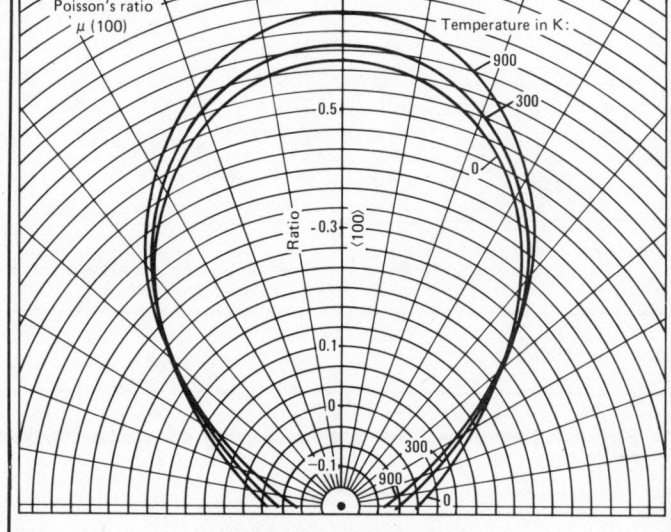

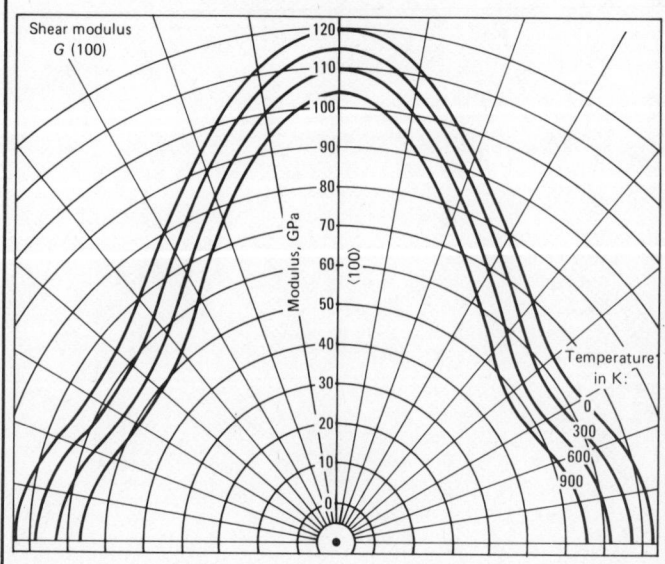

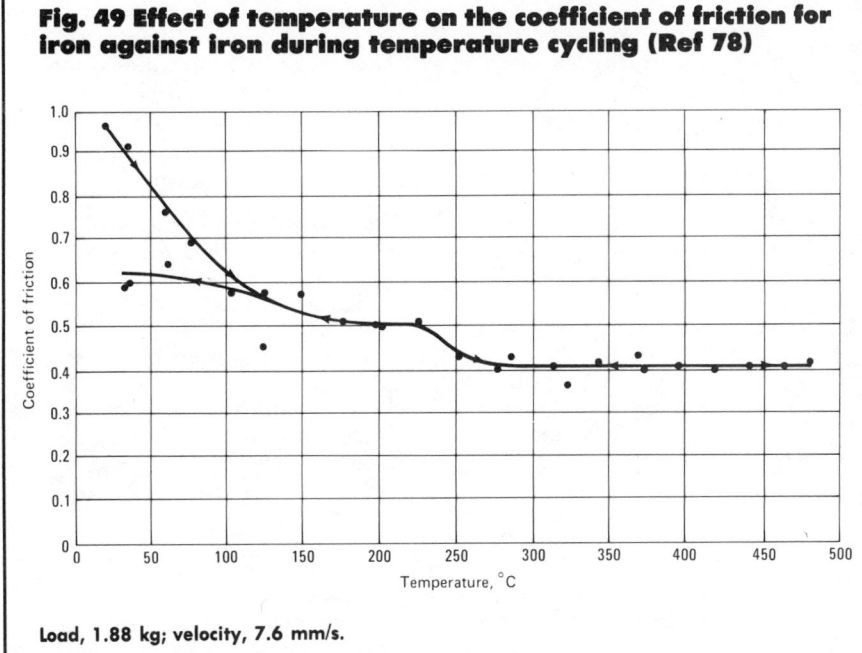

Fig. 49 Effect of temperature on the coefficient of friction for iron against iron during temperature cycling (Ref 78)

Load, 1.88 kg; velocity, 7.6 mm/s.

Viscosity of liquid. See Fig. 47.

Liquid surface tension. See Fig. 48.

Coefficient of friction. See Fig. 49.

Velocity of sound. Reference 79 gives the longitudinal and transverse velocities at room temperature as 5952 and 3222 m/s, respectively.

REFERENCES

1. J. F. Cannon, *J Phys Chem Ref Data,* 3(3), 1974, p 781
2. U. A. Kohlaas, P. Dünner, and N. Schmitz-Pranghe, *Z Angew Phys,* 23(4), 1967, p 245
3. H. Mao, W. A. Bassett, and T. Takahashi, *JAP,* 38(1), 1967, p 272
4. J. R. Low, Jr., The Deformation and Fracture of Iron, in *Iron and Its Dilute Solid Solutions,* C. W. Spenser and F. E. Werner (Eds.), John Wiley, NY, 1963
5. H. Matsui, S. Moriya, S. Takaki and H. Kimura, *Trans JIM,* 19, 1978, p 163
6. G. A. Moore and T. R. Shives, in *Metals Handbook,* Vol 1, American Society for Metals, Metals Park, OH, 1961 p 1206
7. L. Zwell, G. R. Speich, and W. C. Leslie, *Met Trans,* 4, 1973, p 1990
8. W. Hume-Rothery, Z. S. Basinski, and A. L. Sutton, *Proc Roy Soc (London),* A229, 1955, p 459
9. H. Stehle and A. Seeger, *Z Phys,* 146, 1956, p 217
10. A. S. Keh and S. Weissmann, *Electron Microscopy and the Strength of Crystals,* Interscience, NY, 1963, p 231
11. J. Chipman, *Met Trans,* 3, 1972, p 55
12. Y. S. Touloukian, R. K. Kirby, R. E. Taylor, and P. D. Desai, *Thermophysical Properties of Matter,* Vol 12, Thermal Expansion, Plenum, NY, 1970
13. H. Stuart and N. Ridley, *JISI,* 204, 1966, p 711
14. S. Watanabe and T. Saito, *Trans Japan Inst Met,* 14, 1973, p 120
15. R. L. Orr and J. Chipman, *Trans AIME,* 239, 1967, p 630
16. JANAF Thermochemical Tables, 2nd Ed., U.S. Government Printing Office, June 1971
17. C. Y. Ho, R. W. Powell, and P. E. Liley, *J Phys Chem Ref Data,* 3, 1974, p 1
18. J. F. Elliott and M. Gleiser, *Thermochemistry for Steelmaking,* Addison-Wesley, Reading, MA, 1960
19. Y. S. Touloukian, *Thermodynamic Properties of High Temperature Solid Materials,* Vol 1, MacMillan, NY, 1967, p 604
20. K. M. Myles and A. T. Aldred, *J Phys Chem,* 68(1), 1964, p 65
21. J. Fridberg, L. Törndahl, and M. Hillert, *Jernkont Ann,* 153, 1969, p 273
22. G. Hettich, H. Mehrer and K. Maier, *Scripta Met,* 11, 1977, p 795
23. H.-E. Shaefer, K. Maier, M. Weller, D. Herlach, A. Seeger and J. Diehl, *Scripta Met,* 11, 1977, p 803
24. S. Soffer, J. A. Dreesen, and E. M. Pugh, *Phys Rev,* 140 (2A), 1965, p A668
25. D. S. Miller and S. Arajs, *Mem Sc Rev Met,* 65, 1968, p 103
26. A. Cezairliyan and J. L. McClure, *J Res NBS,* 78A(1), 1974, p 1
27. J. G. Hust and P. J. Giarratano, NBS Spec Pub 260-50, 1975, 32 p
28. S. Takaki and H. Kimura, *Scripta Met,* 10, 1976, p 701
29. S. Arajs, B. F. Oliver, and J. T. Michalak, *J Appl Phys,* 38, 1967, p 1676
30. A. S. Balchan and H. G. Drickamer, *Rev Sci Inst,* 32(3), 1961, p 308
31. K. Tanaka and T. Watanabe, *Japanese JAP,* 11(10), 1972, p 1429
32. M. Shimizu and M. Sakoh, *J Phys Soc Jap,* 36(4), 1974, p 565
33. A. J. deBethune, T. S. Licht, and N. Swendeman, *J. Electrochem Soc,* 106(7), 1959, p 616
34. G. V. Samsonov, *Handbook of the Physiochemical Properties of the Elements,* Plenum, NY, 1968
35. M. Stern, *J Electrochem Soc,* 102(12), 1955, p 609
36. I. A. Tsoukalas, *Phys Stat Sol (a),* 22, 1974, p K59
37. T. Okamoto, H. Tange, A. Nishimura, and E. Tatsumoto, *J Phys Soc Jap,* 17, 1962, p 717
38. A. A. Hirsch and Y. Weissman, *Phys Let,* 44A(4), 1973, p. 239
39. P. N. Dheer, *Phys Rev,* 156(2), 1967, p 637
40. R. W. Klaffky and R. V. Coleman, *Phys Rev,* 10, 1974, p 2915
41. S. Arajs, F. C. Schwerer, and R. M. Fisher, *Phys Stat Sol,* 33, 1969, p 731
42. V. S. Fomenko, *Handbook of Thermoionic Properties,* Plenum, NY, 1966
43. K. Ueda and R. Shimizu, *Japanese JAP,* 11, 1972, p 916
44. R. V. Hill, E. K. Stefanakos, and R. F. Tinder, *JAP,* 42(11), 1971, p 4296
45. J. Frühauf and F. Günther, *Phys Stat Sol (a),* 23, 1974, p 399

46. Y. Nakagawa, *J Phys Soc Jap*, 11, 1956, p 855

47. A. Hoffman, *Arch Eisenhuettenw*, 40(12), 1969, p 999

48. H. E. Cleaves and J. M. Heigel, *J Res NBS*, 28(643), 1942, RP1472; J. G. Thompson and H. E. Cleaves, J Res *NBS*, 16(105), 1936, RP860

49. H. Danan, A. Herr, and A. J. P. Meyer, *JAP* 39(2), 1968, p 669

50. F. Brailsford, *Physical Principles of Magnetism*, D. Van Nostrand, NY, 1966, p 147

51. R. D. Greenough, C. Underhill, and P. Underhill, *Physica*, 81B, 1976, p 24

52. G. M. Williams and A. S. Pavlovic, *JAP*, 39(2), 1968, p 571

53. S. Arajs and R. V. Colvin, *J Appl Phys*, 35, 1964, p 2424

54. Y. S. Touloukian and D. P. Dewitt, *Thermophysical Properties of Matter*, Vol 7, Thermal Radiative Properties, Plenum, NY, 1970

55. R. Weast (Ed.), *Handbook of Chemistry and Physics*, 55th Ed., CRC Press, Cleveland, OH, 1974

56. C. O. Smith, *Naval Eng J*, 78(5), 1966, p 789

57. S. B. McRickard and J. G. Y. Chow, *Acta Met*, 14, 1966, p 1195

58. J. G. Y. Chow, S. B. McRickard, and D. H. Gurinsky, *ASTM Spec Tech Pub*, No. 341, 1963, p 46

59. V. I. Spitsyn, *Rec Chem Prog*, 31(1), 1970, p 27

60. F. L. LaQue and N. R. Copson, *Corrosion Resistance of Metals and Alloys*, Reinhold, NY 1963

61. Z. A. Foroulis and H. H. Uhlig, *J Electrochem Soc*, 111(5), 1964, p 522

62. M. Stern, *J Electrochem Soc*, 102(12), 1955, p 663

63. M. Pourbaix, *Atlas of Electrochemical Equilibria in Aqueous Solutions*, Pergamon Press, London, 1966

64. D. Tseng and K. Tangri, *Scripta Met*, 11, 1977, p 719

65. I. J. Diehl, M. Schreiner, S. Staiger and S. Zwiesele, *Scripta Met*, 10, 1976, p 949

66. W. C. Leslie, R. J. Sober, S. G. Babcock, and S. J. Green, *Trans ASM*, 62, 1969, p 690

67. W. C. Leslie, *Met Trans*, 3, 1972, p 5

68. W. B. Morrison and W. C. Leslie, *Met Trans*, 4, 1973, p 379

69. N. Nagata, S. Yoshida, and Y. Sekino, *Trans ISIJ*, 10, 1970, p 173

70. T. L. Altshuler and J. W. Christian, *Phil Trans Roy Soc London*, A261, 1967, p 1121

71. T. Takeda, *Japanese JAP*, 12(7), 1973, p 974

72. G. R. Speich, A. J. Schwoeble, and W. C. Leslie, *Met Trans*, 3, 1972, p 2031

73. H. M. Ledbetter and R. P. Reed, *J Phys Chem* Ref Data, 2(3), 1974, p 531

74. H. H. Wawra, *Arch Eisenheuttenw*, 45(5), 1974, p 317

75. W. A. Spitzig and A. S. Keh, *Acta Met*, 18, 1970, p 611

76. Y. Ogino, F. O. Borgmann, and M. G. Frohberg, *Trans ISIJ*, 14, 1974, p 84

77. R. Murarka, W-K. Lu, and A. E. Hamielec, *Met Trans*, 2, 1971, p 2949

78. M. B. Peterson, J. J. Florek, and R. E. Lee, *ASLE Trans*, 3, 1960, p 101

79. K. H. Schramm, *Z Metallkde*, 53(11), 1962, p 729

Lanthanum (La)

Compiled by K. A. Gschneidner, Jr.
and
B. J. Beaudry
Ames Laboratory
U.S. Department of Energy
Iowa State University

Lanthanum is used as an alloying additive to ferrous alloys to scavenge sulfur, oxygen, etc.; it improves high temperature oxidation resistance of super-alloys. Lanthanum is also used in optical lenses, petroleum cracking catalysts, carbon-arc lights, lighter flints and ceramic capacitors. Lanthanum readily oxidizes at room temperature in air. It should be stored in vacuum or inert atmosphere; storage under oil is not recommended. Turnings can be ignited easily and burn white hot. Finely divided lanthanum should not be handled in air.

Structure

Crystal structure. α-phase, double close-packed hexagonal, $P6_3/mmc$ D^4_{6h}; at 24 °C: $a = 0.37740$ nm; $c = 1.217$ nm. β-phase, face-centered cubic, $Fm3m$ O^5_h; $a = 0.5303$ nm at 325 °C. γ-phase, body-centered cubic, $Im3m$ O^9_h; $a = 0.426$ nm at 865 °C **Minimum interatomic distance.** At 24 °C: $r_a = 0.18870$ nm; $r_c = 0.18712$ nm; radius $CN_{12} = 0.18791$ nm

Mass Characteristics

Atomic weight. 138.9055

Density. α-phase, 6.146 Mg/m³ at 24 °C; β-phase, 6.190 Mg/m³ at 325 °C; γ-phase, 5.97 Mg/m³ at 865 °C; liquid, 5.949 Mg/m³ at 922 °C

Volume change on freezing. 0.6% contraction

Volume change on phase transformation. α- to β-phase, 0.5% volume contraction on heating; β- to γ-phase, 1.3% volume expansion on heating

Thermal Properties

Melting point. 918 °C

Boiling point. 3464 °C

Phase-transformation temperature. α- to β-phase. $A_s = 330$ °C, $A_f = 336$ °C. β- to α-phase: $M_s = 251$ °C, $M_f = 247$ °C. β- to γ-phase: 865 °C

Coefficient of thermal expansion. At 24 °C. Linear: 12.1 µm/m·K. Linear, along crystal axes: 4.5 µm/m·K along a axis; 27.2 µm/m·K along c axis. Volumetric: 36.2 × 10⁻⁶ per K

Specific heat. 195.2 J/kg·K at 25 °C

Entropy. At 298.15 K: 409.6 J/kg·K

Latent heat of fusion. 44.61 kJ/kg

Latent heat of transformation. α- to β-phase, 2.6 kJ/kg; β- to γ-phase, 22.47 kJ/kg

Latent heat of vaporization. 3.103 MJ/kg at 25 °C

Heat of combustion. For hexagonal La_2O_3 at 25 °C: $\Delta H^0_c = 6.44$ MJ/kg La; $\Delta G^0_f = -6.16$ MJ/kg La

Recrystallization temperature. About 300 °C

Thermal conductivity. α-phase, 13.4 W/m·K at 25 °C

Vapor pressure. 0.001 Pa at 1301 °C; 0.101 Pa at 1566 °C; 10.1 Pa at 1938 °C; 1013 Pa at 2506 °C

Electrical Properties

Electrical resistivity. α-phase; 615 nΩ·m at 25 °C, 3 nΩ·m at 7 K. Liquid: 1350 nΩ·m at 922 °C

Ionization potential. La(I): 5.577 V; La(II): 11.06 V; La(III): 19.1774 V

Hall coefficient. −0.035 nV·m/A·T at 25 °C

Temperature of superconductivity. α-phase, 5.10 K; β-phase, 6.00 K

Magnetic Properties

Magnetic susceptibility. Volume (mks units): at 24 °C: α-phase, 5.33 × 10^{-5}; β-phase, 5.87 × 10^{-5}

Optical Properties

Color. Metallic silver
Spectral hemispherical emittance. 28.2% for λ = 645 nm at 920 to 1220 °C

Nuclear Properties

Thermal neutron cross section. 8.9 b

Chemical Properties

General corrosion behavior. Lanthanum oxidizes readily in air at room temperature, and oxidation rates increase with temperature. Hydrogen will react with lanthanum at room temperature.
Resistance to specific corroding agents. Lanthanum reacts vigorously with dilute acids. Cold water slowly attacks lanthanum, hot water reacts faster. The presence of the fluoride ion retards acid attack by the formation of LaF_3 on the surface of the metal.

Mechanical Properties

Tensile properties. Similar to neodymium
Hardness. 28 HV
Poisson's ratio. 0.288
Elastic modulus. At 27 °C: tension, 38.4 GPa; shear, 14.9 GPa; bulk, 30.3 GPa
Kinematic liquid viscosity. 0.445 mm²/s at 922 °C
Liquid surface tension. 0.71 N/m at 922 °C

Lead (Pb)

Compiled by J. F. Smith
Lead Industries Assoc.
and
A. T. Balcerzak
St. Joe Lead Co.

Lead is used in lead-acid storage batteries, ammunition, cable sheathing, pipe, sheet, counterweights, bearings, ballast, gaskets, type metal, low-melting alloys, steel coatings and foil. Applications include

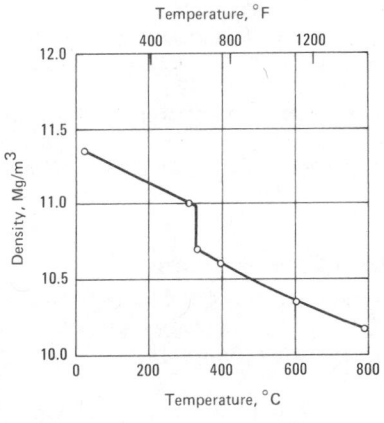

Fig. 50 Temperature dependence of the density of lead

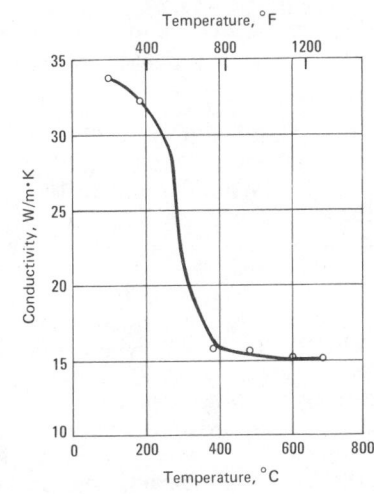

Fig. 51 Temperature dependence of the thermal conductivity of lead

sound and vibration control and x-ray shielding. Lead also is used as an alloying ingredient in steel and copper alloys to improve machinability, and in many chemicals.

CAUTION: lead presents a health hazard, and should not be used to conduct very soft water for drinking, or in contact with foods. Inhalation of lead dust and fumes should be avoided.

Structure

Crystal structure. Face-centered cubic, $a = 0.49489$ nm

Minimum interatomic distance. 0.3499

Mass Characteristics

Atomic weight. 207.19
Density. See Fig. 50.
X-ray absorption characteristics. See Fig. 2 in data compilations for lead and lead alloys.

Thermal Properties

Melting point. 327.4 °C
Boiling point. 1750 °C
Coefficient of thermal expansion. Linear, 26.5 μm/m·K at −190 to +19 °C, 29.3 μm/m·K at 17 to 100 °C.
Specific heat and enthalpy:

°C	C_p, J/kg·K	ΔH, kJ/kg
25	128.7	0
127	132.0	13.24
227	136.8	26.67
327.4(s)	142.1	40.76
327.4(l)	147.9	63.94
427	146.5	78.53
527	144.9	93.16
627	143.3	107.54
727	140.4	121.81
827	139.0	135.94

Latent heat of fusion. 22.98 to 23.38 kJ/kg
Latent heat of vaporization. 945.34 kJ/kg
Thermal conductivity. See Fig. 51.
Vapor pressure:

°C	kPa
957	0.1013
1140	1.013
1389	10.13
1750	101.3

Electrical Properties

Electrical resistivity:

°C	nΩ·m
20	206.43
100	270.21
200	363.78
300(a)	479.38
340(b)	978.67
400	1014.18

(a) Solid. (b) Liquid.

Electrochemical equivalent. Valence +2, 1.0736 mg/C; valence +4, 0.5368 mg/C

Standard electrode potential. 0.122 V vs standard hydrogen electrode.

Temperature of superconductivity. 4 K (-269 °C)

Magnetic Properties

Magnetic susceptibility. Volume: -1.5×10^{-6} mks

Optical Properties

Spectral reflectance. 62% at $\lambda = 589$ nm

Refractive index. Solid, 2.01 in yellow light; molten 0.415 for $\lambda + 602$ nm

Absorptive index. Solid, 3.48 in yellow light

Chemical Properties

Resistance to specific corroding agents. See Table 2 in data compilations for lead and lead alloys.

Mechanical Properties

Damping capacity. See Fig. 1 in data compilations for lead and lead alloys.

Dynamic liquid viscosity. 1.67 mPa·s

Liquid surface tension. 438 kN/m at 400 °C

Velocity of sound. 1.227 mm/s at 18 °C

Lithium (Li)

Compiled by J. E. Selle
Oak Ridge National Laboratory

Lithium is used as a scavenging agent for inert gases; as an alloying element with aluminum, magnesium, zinc and lead; in heat transfer applications; in tritium breeding; in the synthesis of organic compounds and in battery anode material. Compounds containing lithium are used as refrigerant dryers and catalysts, high temperature lubricants and as reagents in the ceramic and chemical industries. Lithium is very reactive and care must be taken to avoid reaction with air, water vapor or other reactive gases. Airtight containers should be used for containment. Niobium, tantalum and molybdenum containers are preferred for temperatures above 600 °C. Ferrous alloys

are not recommended for long term use above 550 °C but are satisfactory below this temperature.

Structure

Crystal structure. α-phase, close-packed hexagonal: $a = 0.3111$ nm; $c = 0.5093$ nm. β-phase, body-centered cubic: $a = 0.35089$ nm

Minimum interatomic distance. Distance of closest approach, β-phase: 0.3039 nm. Goldschmidt atomic radii, 12-fold coordination: 0.157 nm

Mass Characteristics

Atomic weight. 6.939

Density. 0.534 Mg/m³ at 20 °C

Density vs temperature. From 227 to 1027 °C: $\rho = 562 - 0.100 \times T$, where T is in K and ρ is in kg/m³

Expansion on melting. 1.5% of solid volume

Expansion on phase transformation. 0.12% (calculated)

Thermal Properties

Melting point. 180.7 °C

Boiling point. 1342 °C

Phase-transformation temperature. α- to β-phase (cooling), -193 °C

Coefficient of thermal expansion. Linear: 56 μm/m·K at 20 °C

Specific heat. 3.3054 kJ/kg·K at 20 °C; 3.5146 kJ/kg·K from 0 to 100 °C; 4.2258 kJ/kg·K from 180 to 500 °C

Enthalpy. $H_{st} - H_0 = 666.88$ kJ/kg at 25 °C; liquid: $H_1 - H_{273} = -7.517 \times 10^5 + 4.169 \times 10^3 T$ J/kg (T in K); solid: $H_s - H_{273} = -1.030 \times 10^6 + 3.7799 \times 10^3 T$ J/kg (where T is in K)

Entropy. $S_{st} = 4.2189$ kJ/kg·K

Latent heat of fusion. 432.32 kJ/kg

Latent heat of phase transformation. 6.452 kJ/kg

Latent heat of vaporization. 22.73 MJ/kg

Thermal conductivity. 44.0 W/m·K at 180.7 °C

Thermal conductivity vs temperature. $k = 21.874 + 0.056255\, T - 1.8325 \times 10^{-5} T^2$ W/m·K (where T is in K)

Vapor pressure. $\log p = 10.015 - \dfrac{8064.5}{T}$ where T is in K and p is in Pa

Self diffusion coefficient. $D = 1.41$ (± 0.12) $\times 10^{-7} \left(\exp \dfrac{2825 \pm 90}{RT} \right)$ m²/s for 195 to 450 °C

Table 19 Unstable isotopes of lithium

Isotope	Half-life, s	Decay	Energy, MeV
5(a)	10^{-21}	p	...
8	0.85	β⁻	13
9	0.17	β⁻	13.5 (75%); 11 (25%)

(a) Isotope mass 5.0125.

Electrical Properties

Electrical resistivity. Solid: 93.5 nΩ·m at 20 °C; liquid: 250 nΩ·m at 180.7 °C

Temperature dependence of electrical resistivity. $R = 22.56 - 0.6665\, T - 4.255 \times 10^{-4} T^2 + 1.398 \times 10^{-7} T^3$ nΩ·m, where T is in K

Thermoelectric potential. vs Pt (reference junction, 0 °C)

Hot junction temperature, °C	Thermal emf, nV
-200	-1.12
-100	-1.00
$+100$	$+1.82$

Ionization potential. Li (I), 5.39 V; Li (II), 75.619 V; Li (III), 122.419 V

Magnetic Properties

Magnetic susceptibility. Volume: 2.242×10^{-3} mks

Nuclear Properties

Stable isotopes. ⁶Li, isotope mass 6.01512, 7.42% abundance; ⁷Li, isotope mass 7.01600, 92.58% abundance

Unstable isotopes. See Table 19.

Thermal neutron cross section. ⁶Li, 45 ± 10 mb; ⁷Li, 37 ± 4 mb

Chemical Properties

General corrosion behavior. Lithium tarnishes quickly in oxygen, nitrogen and moist air. Solubilities of various metallic elements are very sensitive to the lithium purity.

Mechanical Properties

Hardness. 0.6 (Mohs scale)

Dynamic viscosity. $\mu = 0.645$ MPa·s at 180.7 °C; 0.140 MPa·s at 1335 °C; $\log \mu = -3.080 + \dfrac{57.63}{T} - 5.172 \times 10^{-4} T$ MPa·s, where T is in K

Liquid surface tension. $\sigma = 0.396$ N/m at 180.7 °C; $\sigma = 0.240$ N/m at

1335 °C; $\sigma = 0.447 - 1.07 \times 10^{-4} T - 1.35 \times 10^{-8} T^2$ N/m, where T is in K

Velocity of sound. For 185 to 827 °C: $v = 4784.5 - 0.591 T$ m/s, where T is in K

Lutetium (Lu)

Compiled by K. A. Gschneidner, Jr. and B. J. Beaudry
Ames Laboratory
U.S. Department of Energy
Iowa State University

Lutetium is used in ferrite bubble devices. It will remain shiny in air at room temperature. Turnings can be ignited and will burn white hot. Finely divided lutetium should not be handled in air.

Structure

Crystal structure. Close-packed hexagonal: $P6_3/mmc$ D^4_{6h}; $a = 0.35052$ nm, $c = 0.55494$ nm at 24 °C

Minimum interatomic distance. At 24 °C: $r_a = 0.17526$ nm; $r_c = 0.17172$ nm; radius $CN_{12} = 0.17349$ nm

Mass Characteristics

Atomic weight. 174.967
Density. 9.841 Mg/m³ at 24 °C
Volume change on freezing. Approx 3.6% contraction

Thermal Properties

Melting point. 1663 °C
Boiling point. 3402 °C
Coefficient of thermal expansion. At 24 °C. Linear: 9.9 μm/m·K. Linear, along crystal axes: 4.8 μm/m·K along a axis, 20.0 μm/m·K along c axis. Volumetric: 29.6×10^{-6} per K
Specific heat. 153.1 J/kg·K at 25 °C
Entropy. 291.3 J/kg·K at 25 °C
Latent heat of fusion. 106.6 kJ/kg
Latent heat of vaporization. 2.444 MJ/kg at 25 °C
Heat of combustion. For cubic Lu_2O_3 at 25 °C: $\Delta H^0_c = 5.37$ MJ/kg Lu; $\Delta G^0_f = -5.11$ MJ/kg Lu
Recrystallization temperature. About 600 °C
Thermal conductivity. 16.4 W/m·K at 25 °C
Vapor pressure. 0.001 Pa at 1241

°C; 0.101 Pa at 1483 °C; 10.1 Pa at 1832 °C; 1013 Pa at 2387 °C

Electrical Properties

Electrical resistivity. 582 nΩ·m at 25 °C; 45 nΩ·m at 4 K. Along crystal axes, at 25 °C: $\rho_b = 766$ nΩ·m; $\rho_c = 347$ nΩ·m
Ionization potentials. Lu(I), 5.426 V; Lu(II), 13.9 V; Lu(III), 20.9596 V; Lu(IV), 45.19 V
Hall coefficient. At 20 °C. -0.0535 nV·m/A·T. Along crystal axes: 0.045 nV·m/A·T along a axis, -0.26 nV·m/A·T along c axis
Temperature of superconductivity. Bulk lutetium is not superconducting down to 0.03 K at atmospheric pressure; it becomes superconducting at 0.3 K and 11 GPa

Magnetic Properties

Magnetic susceptibility. Volume (mks units). At 27 °C: 1.293×10^{-4}. Along crystal axes: 1.353×10^{-4} along a axis, 1.173×10^{-4} along c axis

Optical Properties

Color. Metallic silver

Nuclear Properties

Thermal neutron cross section. 108 b

Chemical Properties

General corrosion behavior. Lutetium stays shiny in air at room temperature. Even at 1000 °C, the rate of oxidation is slow due to the formation of a pink, tightly adhering oxide on the surface of the metal.
Resistance to specific corroding agents. Lutetium does not react with cold or hot water but reacts vigorously with dilute acids. Concentrated sulfuric slowly attacks lutetium. The presence of the fluoride ion retards acid attack due to the formation of LuF_3 on the surface of the metal.

Mechanical Properties

Tensile properties. About the same as erbium
Hardness. 44 HV
Poisson's ratio. 0.261
Elastic modulus. At 27 °C: tension, 68.4 GPa; shear, 27.1 GPa; bulk, 47.6 GPa
Elastic constants along crystal axes. At 27 °C: $c_{11} = 86.23$ GPa; $c_{12} = 32.0$ GPa; $c_{13} = 28.0$ GPa; $c_{33} = 80.86$ GPa; $c_{44} = 26.79$ GPa

Magnesium (Mg)

Compiled by S. C. Erickson
The Dow Chemical Co.

Primary magnesium has a minimum purity of 99.8% and must meet definite specifications limiting individual impurities. This purity is sufficient for most chemical and metallurgical uses. Most of the pure magnesium sold is produced electrolytically as primary magnesium. For applications requiring a minimum of specific impurities, special grades of electrolytic magnesium are available. Silicothermic magnesium is produced by thermal reduction of magnesium oxide. High-purity sublimed magnesium is produced by sublimation of primary electrolytic magnesium under vacuum. Typical analyses are shown in Table 20.

Unless otherwise indicated, the properties listed here for pure magnesium were determined on metal of 99.98 + % purity.

Alloyed with small amounts of aluminum, manganese, rare earths, thorium, zinc or zirconium, magnesium yields alloys with high ratios of strength to weight at both room and elevated temperatures. The alloys have unexcelled machinability, workability by all common methods, stability in many atmospheres, and high damping capacity. Magnesium is an active chemical element and reacts with many common chemical oxidizing agents. A number of metals such as thorium, titanium, uranium and zirconium are prepared by thermal reduction with magnesium. As a catalyst, magnesium is useful for promoting organic condensation, reduction, addition and dehalogenation reactions. It is useful for the synthesis of complex and special organic compounds by the Gringnard process. Its use in pyrotechnics is well established. Magnesium powder can be dispersed in hydrocarbons and mixed in solid propellants for high-energy fuels. Magnesium alloyed with other metals such as aluminum, copper, cast iron, lead, nickel and zinc improves their properties. It also deoxidizes copper and brass, desulfurizes iron and nickel, and debismuthizes lead. As a galvanic anode, it provides effective corrosion protection for water heaters, underground pipe lines, ship hulls, ballast tanks and other underground and under-

Table 20 Typical chemical compositions of pure magnesium available commercially

Designations	Al	Ca	Cu	Fe	Mn	Ni	Pb	Si	Sn	Other metallics Each	Total	Mg(a)	Ref
Primary electrolytic......	0.005	0.0014	0.0014	0.029	0.06	<0.0005	0.0007	0.0015	<0.0001	<0.05	<0.13(b)	99.87	1
Magnesium 2 .	...	...	<0.02	<0.05	<0.01	<0.001	<0.01	...	<0.01	<0.05	<0.10	99.90	1
Magnesium 3 .	<0.004	<0.003	<0.005	<0.03	<0.01	<0.001	<0.01	<0.005	<0.005	<0.01	<0.08	99.92	1
Magnesium 4 .	<0.002	<0.003	<0.004	<0.03	<0.004	<0.001	<0.005	<0.005	<0.005	<0.01	<0.07	99.93	1
Magnesium 5 .	...	...	<0.003	<0.003	<0.004	<0.001	<0.005	<0.005	<0.005	<0.01	<0.05	99.95	1
Silicothermic..	0.007	0.004	<0.001	0.001	0.002	<0.0005	0.001	0.006	0.001	<0.01	<0.04	99.96	2, 3
High-purity sublimed(c) ..	0.0004	0.001	0.0002	0.0007	<0.001	<0.0005	<0.0005	<0.001	<0.001	<0.01	<0.02	99.98	3

(a) Magnesium by difference. (b) 0.006 H, 0.0025 N, 0.0022 O; hydrogen, nitrogen and oxygen not reported for other grades. (c) Not available commercially.

Fig. 52 Temperature dependence of the density of magnesium

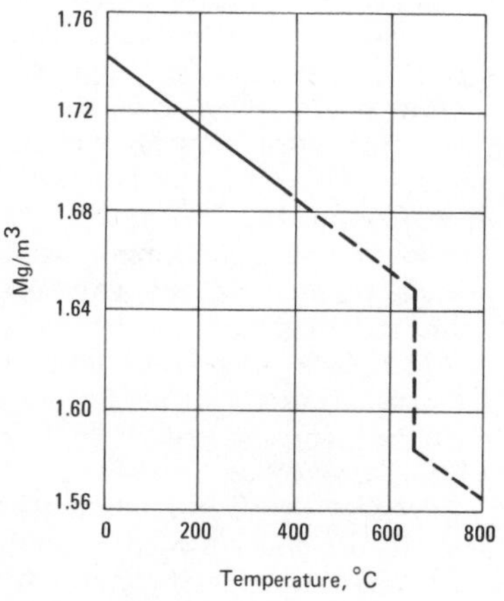

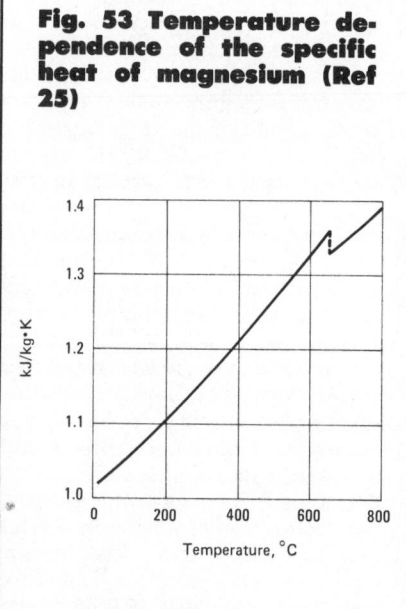

Fig. 53 Temperature dependence of the specific heat of magnesium (Ref 25)

Table 21 Typical mechanical properties of magnesium at 20 °C

Form and section	Tensile strength, MPa	0.2% tensile yield strength, MPa	0.2% compressive yield strength, MPa	Elongation(a), %	Hardness HRE	HB(b)
Sand cast, 13-mm (½-in.) diam...	90	21	21	2-6	16	30
Extrusion, 13-mm (½-in.) diam...	165-205	69-105	34-55	5-8	26	35
Hard rolled sheet	180-220	115-140	105-115	2-10	48-54	45-47
Annealed sheet	160-195	90-105	69-83	3-15	37-39	40-41

(a) In 50 mm or 2 in. (b) 500-kg load; 10-mm diam ball.

Fig. 54 Temperature dependence of the electrical resistivity of magnesium

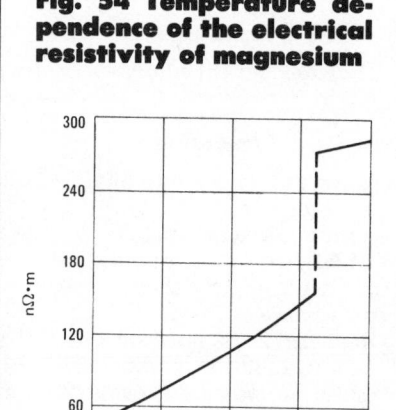

water structures. Small light-weight high-current-output primary batteries use magnesium alloy as the anode. Low capture cross section for thermal neutrons and low-level retention of induced radioactivity point to varied uses for magnesium in atomic energy.

Structure

Crystal structure. Close-packed hexagonal. At 25 °C: $a = 0.32087 \pm 0.00009$ nm; $c = 0.5209 \pm 0.00015$ nm; $c/a = 1.6236$ (Ref 4, 5)

Slip planes. Primary, (0001), $\langle 11\bar{2}0 \rangle$; secondary, $\{10\bar{1}0\}$, $\langle 11\bar{2}0 \rangle$; $\{10\bar{1}1\}$, $\langle 11\bar{2}0 \rangle$ at elevated temperatures (Ref 6 to 9)

Twinning planes. Primary, $\{10\bar{1}2\}$; secondary, $\{30\bar{3}4\}$; $\{10\bar{1}3\}$ at elevated temperatures (Ref 10, 11)

Cleavage plane. No definite cleavage plane (Ref 6, 12)

Minimum interatomic distance. 0.3196 nm

Fracture type. See Ref 6 and 12.

Mass Characteristics

Atomic weight. 24.312

Density. 1.738 Mg/m³ at 20 °C; solid: approx 1.65 Mg/m³ at 650 °C; liquid: approx 1.58 Mg/m³ (Ref 13, 14). See also Fig. 52 (Ref 13, 14, 15)

Volume change on freezing. 4.2% shrinkage

Volume change during cooling. From 650 (solid) to 20 °C: 5% shrinkage

Thermal Properties

Melting point. 650 °C (Ref 16, 17)

Boiling point. 1107 ± 10 °C (Ref 3, 17, 18)

Thermal expansion. Polycrystalline at 20 °C: 25.2 μm/m·K (Ref 19 to 23). Values for all magnesium alloys

Fig. 55 Effect of alloying additions on the electrical resistivity of magnesium

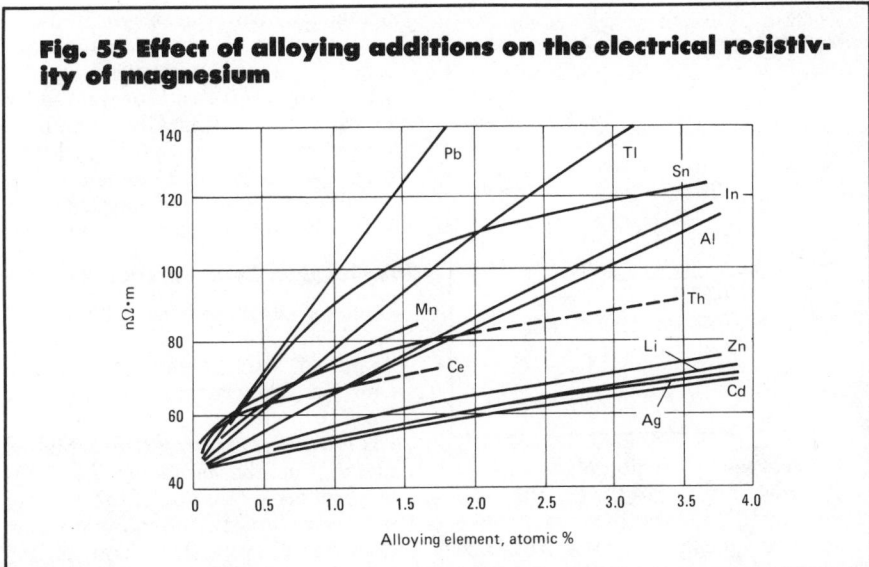

Fig. 56 Temperature dependence of the modulus of elasticity of magnesium

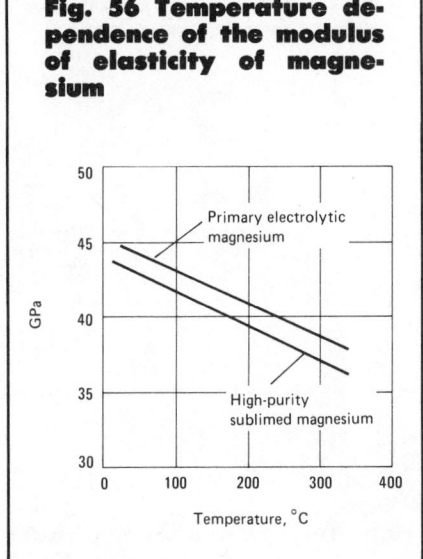

Fig. 57 Damping capacity of magnesium as a function of strain

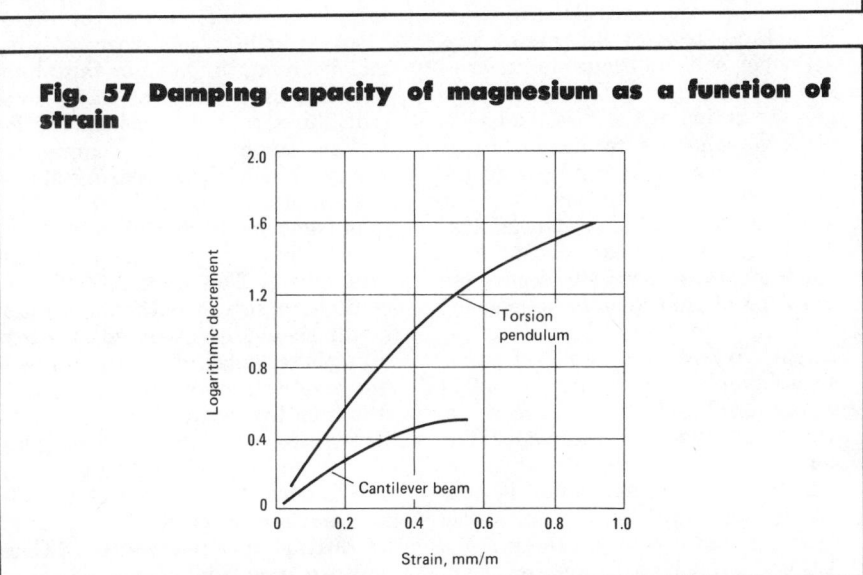

Fig. 58 Minimum creep rate of magnesium as a function of stress and temperature

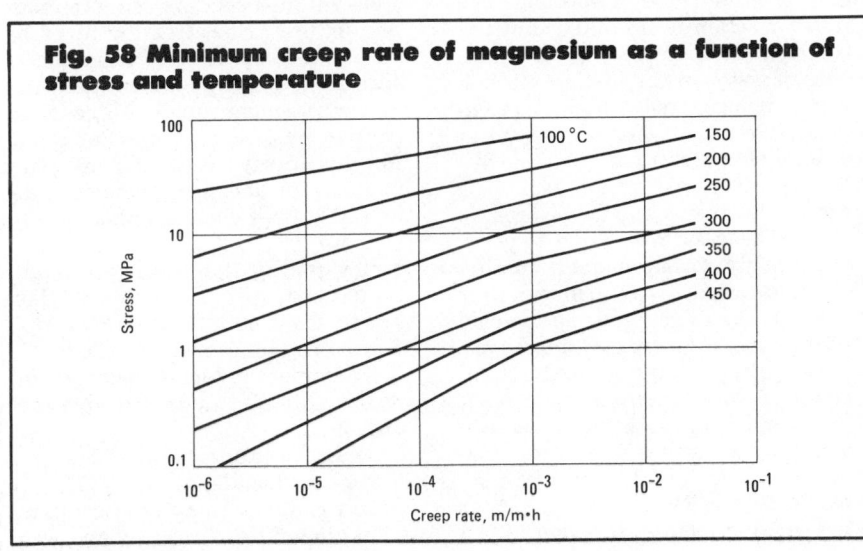

are approximately the same. $L_t = L_0 [1 + (24.8t + 0.00961t^2) \times 10^{-6}]$, where t is in °C. Along crystal axes, from 15 to 35 °C: 27.1 μm/m·K along a axis; 24.3 μm/m·K along c axis (Ref 24)

Specific heat. 1.025 kJ/kg·K at 20 °C (Ref 25). See also Fig. 53.

Latent heat of fusion. 360-377 kJ/kg (Ref 25)

Latent heat of sublimation. 6113-6238 kJ/kg at 25 °C (Ref 17)

Latent heat of vaporization. 5150-5400 kJ/kg (Ref 17)

Thermal conductivity. 418 W/m·K at 20 °C (Ref 26). Temperature dependence: $K/T = 0.017 + 2.26 \times 10^{-5}/\rho$, where ρ is in nΩ·m, T is in K and K is in W/m·K (Ref 2)

Heat of combustion. 24 900 – 25 200 kJ/kg Mg

Electrical Properties

Electrical conductivity. 38.6% IACS

Electrical resistivity. Polycrystalline at 20 °C: 44.5 nΩ·m (Ref 20, 27). Liquid at 650 °C: 247 nΩ·m (Ref 28 to 30). Along crystal axes, at 20 °C: 44.8 nΩ·m along a axis; 37.4 nΩ·m along c axis (Ref 31). Temperature coefficient: polycrystalline at 20 °C: 0.165 nΩ·m per K (Ref 27). Along crystal axes, at 20 °C: 0.165 nΩ·m per K along a axis; 0.143 nΩ·m per K along c axis (Ref 31). Effect of temperature, see Fig. 54 (Ref 28 to 30, 32 to 34). Effect of alloying, see Fig. 55 (Ref 14, 27)

Contact potential. + 0.44 mV vs Pt

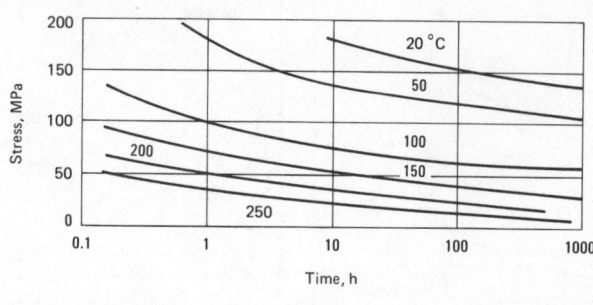

Fig. 59 Stress-rupture life of magnesium as a function of stress and temperature

at 0 to 100 °C (Ref 35); − 0.222 mV vs Cu at 27 °C (Ref 27)
Electrochemical equivalent. 126 mg/C
Electrolytic solution potential. 1.63 mV vs saturated calomel electrode at 25 °C in aerated 3% NaCl solution (Ref 36)

Magnetic Properties

Magnetic susceptibility. Mass: 0.00627 − 0.00632 mks (Ref 37)
Magnetic permeability. 1.000012

Optical Properties

Reflectivity. 72% at λ = 0.500 μm; 74% at λ = 1.00 μm; 80% at λ = 3.0 μm; 93% at λ = 9.0 μm (Ref 38)
Refractive index. 0.37 at λ = 0.589 μm (Ref 39)
Absorption constant. 4.42 at λ = 0.589 μm (Ref 39)
Color. Bright silvery white
Emissivity. 0.07 at 22 °C (Ref 40)

Nuclear Properties

Thermal neutron absorption cross section. 0.063 ± 0.004 b (Ref 41)

Chemical Properties

General resistance to corrosion. Dependent on surface film formation; the rate of formation, solution, or chemical change of the film varies with the medium to which it is exposed and also with the alloying elements or impurities present in the metal. Magnesium has good resistance to both indoor and outdoor atmospheres and, in the absence of galvanic couples, even shows resistance to more aggressive environments such as seawater. Indoor tarnishing is controlled largely by the relative humidity. In mild marine and industrial inland atmospheres,

the degree of corrosion resistance far exceeds that of mild steel. In stagnant distilled water at room temperature, magnesium forms a protective film that stops action (Ref 42).
Resistance to specific agents. The action of salt solutions on magnesium is dependent on both the anion and the cation of the dissolved salt. Neutral solutions of heavy metal salts will generally cause severe attack. Magnesium suffers little, if any, attack in alkalies, chromates, fluorides, nitrates or phosphates; more vigorous corrosion occurs in solutions of chlorides, bromides, iodides and sulfates. Mineral acids, except hydrofluoric and chromic acids, dissolve magnesium rapidly. Aqueous solutions of organic acids attack magnesium, whereas fatty acids, hot or cold, dry or containing water, do not. Magnesium is not affected by aliphatic and aromatic hydrocarbons, ketones, ethers, glycols and alcohols, with the exception of anhydrous methyl alcohol. The latter reaction is inhibited, but not completely suppressed, by the presence of water in the methyl alcohol. Pure halogenated organic compounds do not attack magnesium at ordinary temperatures, but at elevated temperatures, or if water is present, corrosion can be severe. No marked reaction was found to occur between magnesium and methyl chloride, carbon tetrachloride, or chloroform, even after prolonged heating under increased pressures. Lower alkyl halides, up to amyl derivatives, have been shown to react with magnesium only under pressure and at temperatures in excess of 270 °C, but higher alkyl halides are reported to react with magnesium at their boiling points. In general, the

presence of water greatly stimulates the reaction between magnesium and halogenated compounds at elevated temperatures. Fluorinated hydrocarbons are generally without action on magnesium when dry (Ref 42) (See also the separate article on the corrosion resistance of magnesium in this volume.)

Fabrication Characteristics

Casting temperature. 705 to 760 °C
Type of flux. Open-pot melting, Dow No. 250; crucible melting, Dow No. 310
Precautions in melting. Molten metal must be protected from the atmosphere by the use of inert gas or protective fluxes. Molten magnesium does not react with carbon, silicon carbide, or combinations of these. There is little, if any, reaction with molybdenum, tungsten or tantalum. Low-carbon (welded or cast) steel crucibles are used as containers for molten magnesium of commercial purity—avoid nickel-bearing steels. Use a protective agent (Dow No. 181) to prevent magnesium from burning when it is being poured in an open atmosphere. The usual safety precautions observed with any molten metal should be observed. Preheat all tools or metal introduced in molten magnesium. Keep pot settings free from iron scale.
Precautions in fabrication. A supply of an approved extinguishing agent should be readily accessible to any machining, grinding or similar operations on magnesium. Good housekeeping and sharp machine tools are the best deterrents to magnesium fires. Heat treating furnaces should have a protective atmosphere, such as SO_2 or BF_3, when operating at high temperatures. Magnesium powder must be kept dry (Ref 43)
Machinability index. For pure magnesium and all magnesium alloys, 500 (free cutting brass = 100) (Ref 44)
Hot working temperature range. 93 to above 482 °C for 99.98% Mg; 177 to above 482 °C for 99.80% Mg
Annealing temperature. 150 to 200 °C. Maximum reduction between anneals, 50 to 60% under suitable conditions
Forming temperature. 150 to 200 °C for best results
Joining. Rivet composition, aluminum alloy 5056. Oxyacetylene weld

with pure magnesium welding rod, magnesium welding flux and neutral flame. Resistance welding is satisfactory. Helium-arc or argon-arc welding is preferred. Use pure magnesium welding rod and no flux

Recrystallization temperature. 93 °C for 1-h anneal after 30% cold reduction (99.98% Mg); 177 °C for 1-h anneal after 30% cold reduction; 93 °C for 1-h anneal after 60% cold reduction (99.80% Ma) (Ref 19)

Mechanical Properties

Tensile properties. See Table 21; see also Ref 45

Compressive properties. See Table 21.

Hardness. See Table 21.

Elastic modulus. Tension at 20 °C. 99.98% Mg: dynamic, 44 GPa; static; 40 GPa. 99.80% Mg: dynamic, 45 GPa; static, 43 GPa. See also Fig. 56 (Ref 46 to 49)

Damping capacity. See Fig. 57 (Ref 50, 51)

Creep-rupture characteristics. See Fig. 58 and 59 (Ref 52, 53)

Dynamic liquid viscosity. At 650 °C, 1.23 mPa·s; at 700 °C, 1.13 mPa·s (approx values)

Liquid surface tension. At 681 °C, 0.563 N/m; at 894 °C, 0.502 N/m (Ref 54)

Coefficient of friction. 0.36 at 20 °C, magnesium vs magnesium

REFERENCES

1. F. J. Krenske, J. W. Hays and D. L. Spell, *Journal of Metals,* AIME, January 1958, p 28
2. "High-Purity Magnesium", Dominion Magnesium, Ltd., Bulletin TIB 551, Toronto, Canada
3. W. Leitgehel, *Zeitschrift fur Anorganische und Allgemeine Chemie,* Vol 202, 1931, p 305
4. R. S. Busk, *Transactions of AIME,* Vol 188, 1950, p 1460
5. F. W. Batchelder and R. F. Raeuckle, *Physical Review,* Vol 105, 1957, p 59
6. F. E. Hauser, P. R. Landon and J. E. Dorn, *Transactions of the American Society for Metals,* Vol 48, 1956, p 986 and *Transactions of AIME,* Vol 206, 1956, p 589
7. A. R. Chaduri, H. C. Chang and N. J. Grant, *Transactions of AIME,* Vol 203, 1955, p 682
8. Technical Report 55-241, Wright Air Development Center, Dow Chemical Company, August 1955
9. R. E. Reed-Hill and W. D. Robertson, *Journal of Metals,* AIME, Vol 209, April 1957, p 496
10. S. L. Couling and C. S. Roberts, *Acta Crystallographica,* Vol 9, 1956, p 972
11. R. E. Reed-Hill and W. D. Robertson, *Acta Metallurgica,* Vol 5, 1957, p 717
12. R. E. Reed-Hill and W. D. Robertson, *Acta Metallurgica,* Vol 5, 1957, p 728
13. R. S. Busk, *Transactions of AIME,* Vol 194, 1952, p 207
14. Adolf Beck, *The Technology of Magnesium and Its Alloys,* F. A. Hughes and Company, Ltd., London, 1943
15. H. Grothe and C. Mangelsdorff, *Zeitschrift fur Metallkunde,* Vol 29, 1937, p 352
16. F. D. Rossini, D. D. Wagman, E. H. Evans, S. Levine and I. Jaffe, National Bureau of Standards Circular 500, 1952
17. D. R. Stull and G. C. Sinke, *Thermodynamic Properties of the Elements,* Advances in Chemistry Series, No. 18, American Chemical Society, Washington, D.C., 1956, p 124
18. H. Hartman and R. Schneider, *Zeitschrift fur Anorganische und Allgemeine Chemie,* Vol 180, 1929, p 275
19. R. A. Townsend, *Metals Handbook,* American Society for Metals, 1948, p 1013
20. P. Hidnert and W. T. Sweeney, Journal of Research of the National Bureau of Standards, Vol 1, 1928, p 771
21. K. Scheel, *Zeitschrift fur Physik,* Vol 5, 1921, p 167
22. J. B. Austin, *Physics,* Vol 3, 1932, p 240
23. H. Esser and H. Eusterbrock, *Archiv fur das Eisenhuttenwesen,* Vol 14, 1941, p 341
24. P. W. Bridgman, *Proceedings of the American Academy of Arts and Sciences,* Vol 67, 1932, p 27
25. R. A. McDonald and D. R. Stull, *American Chemical Society,* Vol 77, 1955, p 5293
26. W. Bungardt and R. Kallenbach, *Metallwirtschaft—Metallwissenschaft und Technik,* Vol 4, 1950, p 317
27. E. J. Salkovitz, A. J. Schindler and F. W. Kammer, *Physical Review,* Vol 105, 1957, p 887
28. F. H. Harn, *Physical Review,* Vol 84, No. 2, 1951, p 855
29. E. Scala and W. D. Robertson, *Transactions of AIME,* Vol 197, 1953, p 1141
30. A. Roll and H. Motz, *Zeitschrift fur Metallkunde,* Vol 48, No. 5, May 1957, p 272
31. J. L. Nichols, *Journal of Applied Physics,* Vol 26, No. 4, 1955, p 470
32. R. W. Powell, *Philosophical Magazine,* Series 7, Vol 27, No. 185, 1939, p 677
33. G. Grube and E. Schiedt, *Zeitschrift fur Anorganische und Allgemeine Chemie,* Vol 194, 1930, p 190
34. G. Grube, L. Mohr and R. Bornhak, *Zeitschrift fur Elektrochemie,* Vol 40, 1934, p 160
35. *Temperature, Its Measurement and Control in Science and Industry,* American Institute of Physics, Reinhold Publishing Company, NY, 1941, p 1308
36. R. E. McNulty and J. D. Hanawalt, *Transactions of the Electrochemical Society,* Vol 81, 1942, p 429
37. M. Gaber, Michigan State University, private communication, 1958
38. W. W. Coblenz, *Journal of the Franklin Institute,* Vol 170, p 169, and Bulletin of the National Bureau of Standards, Vol 2, 1906, p 457 and Vol 7, 1911, p 197
39. P. Drude, *Annalen Der Physik,* Vol 39, 1890, p 481
40. *Handbook, American Institute of Physics,* McGraw-Hill Book Co., NY, 1957
41. Chart of the Nuclides, Knolls Atomic Power Laboratory, General Electric Company, 1956
42. L. Whitby, "Magnesium and Its Alloys", *Corrosion Resistance of Metals and Alloys,* McKay and Worthington ACS Monograph, Reinhold Publishing Corporation, NY
43. Standard for Magnesium, No. 48, National Fire Protection Association, Boston, MA, 1957
44. Report of Independent Research Committee on Cutting Fluids, ASTE, *Automotive Industry,* Vol 88, No. 8, 1943, p 48
45. M. W. Toaz and E. J. Ripling, *Transactions of AIME,* Vol 206, 1956, p 936
46. J. R. Frederick, PhD dissertation, University of Michigan, 1947
47. J. R. Frederick and C. H. Church, University of Michigan, private communication, 1957

48. D. W. Levinson and W. Graft, Armour Research Foundation, private communication, 1957
49. R. W. Fenn, Jr., Proceedings of the American Society for Testing and Materials, Vol 58, 1958, p 826
50. R. E. Maringer, Battelle Memorial Institute, private communication, 1956
51. W. A. Babington and G. F. Weissman, Bell Telephone Laboratories, private communication, 1957
52. C. S. Roberts, *Transactions of AIME*, Vol 197, 1953, p 1121
53. J. L. Bernard, R. Caillat and R. Darras, *Progress in Nuclear Energy, Metallurgy and Fuels*, Vol 2, Pergamon Press, NY, 1957
54. V. G. Givov, Aluminum-Magnesium Institute (Russia) Vol 14, 1937, p 99

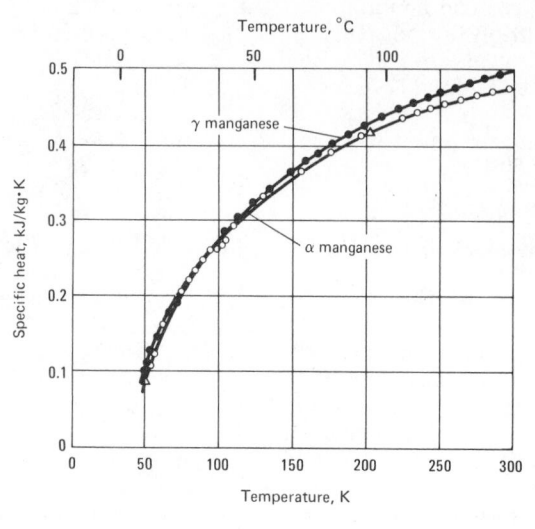

Fig. 60 Effect of temperature on the specific heat of manganese

Manganese (Mn)

Compiled by Manganese Centre Paris, France

Structure

Crystal structure:

Phase	Structure	a, nm	c, nm
α	Cubic	0.8931	...
β	Cubic	0.6326	...
γ	Tetragonal	0.3784	0.940

(a) Stabilized at room temperature.

Mass Characteristics

Atomic weight. 54.9380
Density. Solid at 20 °C: α phase, 7.47 Mg/m³; β phase, 726 Mg/m³; γ phase, 7.21 Mg/m³ (Ref 1). Liquid at melting point, 5.73 Mg/m³ (Ref 2)
Volume change on freezing. 1.7%

Thermal Properties

Melting point. 1244 ± 4 °C
Boiling point. 2095 °C (Ref 3)
Phase-transformation temperature. α to β, 710 °C; β to γ, 1079 °C; γ to δ, 1143 °C (Ref 4)
Coefficient of thermal expansion. Linear: α phase, 22.8 μm/m·K at 0 to 100 °C and 24.05 μm/m·K at 100 to 200 °C; β phase, 24.9 μm/m·K at 0 to

100 °C; γ phase, 14.75 μm/m·K at 0 to 100 °C (Ref 5)
Specific heat. At 25 °C: α phase, 477 J/kg; β phase, 645 J/kg; γ phase, 502 J/kg. See also Fig. 60.
Latent heat of fusion. 267 kJ/kg
Latent heat of transformation. Cooling: α to β, 40.8 kJ/kg; β to γ, 41.5 kJ/kg; γ to δ, 32.8 kJ/kg
Latent heat of vaporization. 4.09 MJ/kg

Electrical Properties

Electrical conductivity. Volumetric, α phase: 0.9% IACS
Electrical resistivity. (largely dependent on gas absorbed) α phase, 1440 nΩ·m at 22 °C (Ref 6); β phase, 910 nΩ·m at 20 °C (Ref 7); γ phase, 400 nΩ·m (Ref 8)
Standard electrode potential. 1.134 v

Magnetic Properties

Magnetic susceptibility. Volume: α phase, 1.24×10^{-4} mks at 25 °C; γ phase, 1.07×10^{-4} mks at 1079 °C (Ref 9)
Influence of treatments. Manganese is ferromagnetic after certain treatments—for example, after absorption and expulsion of nitrogen, which leaves the lattice expanded.

Mechanical Properties

Tensile properties. α and β phases are brittle. γ phase (extrapolated):

tensile strength, 496 MPa; yield strength, 241 MPa; elongation 40%
Hardness. γ phase, 35 HRC (extrapolated)
Elastic modulus. Tension: γ phase, 191 GPa at 17 °C (extrapolated). Bulk: α at 17 °C, 1.08×10^{-11} m²/N (Ref 10)

REFERENCES

1. *Metals Reference Book*, 5th Ed., Butterworths, London, 1976
2. N. A. Vatolin and O. A. Yesin, *Physics of Metallurgy and Metallography*, Vol 16, No. 6, 1963, p 129
3. E. Rapoport and G. C. Kennedy, *Journal of Physics and Chemistry of Solids*, Vol 27, 1966, p 93
4. G. T. Meaden, *Metallurgical Reviews*, No 125, p 98; Stal, U.S.S.R., No 12, 1977, p 1098
5. Sully, *Manganese*, Butterworths, 1955
6. G. T. Meaden and P. Pelloux Gervais, *Cryogenics*, Vol 5, 1965, p 227
7. F. Bunke, *Annalen der Physik*, Vol 21, 1934, p 139
8. H. D. Efling, *Annalen der Physik*, Vol 37, 1940, p 162

9. C. T. Kriesman and McGuire, *Physical Review*, Vol 98, 1955, p 936

10. Rosen, *Physical Review*, Vol 165, 1968, p 357

Mercury (Hg)

Compiled by M. Nowak
Vice President
Troy Chemical Corporation

Table 22 Standard enthalpy, entropy and Gibbs free energy of mercury

Temperature, K	$H_l - H_{st}$ kJ/kg	Liquid $S_T - S_{st}$ kJ/ kg·K	$G_T - H_{st}$ T kJ/kg·K	$H_T - H_{st}$ kJ/kg	Gas(a) $S_T - S_{st}$ kJ/ kg·K	$G_T - H_{st}$ T kJ/kg·K
298.15	0	0.0000	−0.3786	0	0.0000	−0.8723
350	7.180	0.0222	−0.3803	5.385	0.0166	−0.8736
400	14.05	0.0406	−0.3840	10.561	0.0305	−0.8764
450	20.87	0.0566	−0.3888	15.738	0.0427	−0.8800
500	27.66	0.0709	−0.3942	20.935	0.0536	−0.8841
550	34.42	0.0838	−0.3999	26.111	0.0635	−0.8883
600	41.20	0.0956	−0.4056	31.308	0.0725	−0.8927
630	45.27	0.1022	−0.4090	34.419	0.0776	−0.8952
650	47.99	0.1064	−0.4113	36.485	0.0808	−0.8970
700	54.77	0.1165	−0.4169	41.661	0.0885	−0.9013
750	61.57	0.1259	−0.4224	46.859	0.0957	−0.9055

(a) Ideal.

Mercury is the only common metal that is liquid at room temperature. It is rarely found in the free and uncombined state in nature, most often occurring as the ore *cinnabar* (HgS). Mercury is widely used for thermometers, barometers, diffusion pumps and other laboratory instruments. It is used commercially in mercury-vapor lamps, in lamps and lamp tubes for advertising signs, in switches for instruments and control devices, in dental preparations, and in batteries. Mercury chemicals are widely used for making pesticides, antifouling paints, high-grade paint pigments, explosives and medicines. Mercury cells are used for production of caustic chlorine.

Prime virgin mercury as commonly obtained by refining directly from mercury ores has a purity of at least 99.9%, and in many instances 99.99%. Metal of lesser purity is generally obtained by reclaiming discarded mercury.

Precautions in use. *Health hazard.* Mercury vapor is readily absorbed through the respiratory tract, the gastrointestinal tract or unbroken skin. Mercury acts as a heavy-metal poison, but its effect becomes known only after prolonged exposure. Acute poisoning from mercury vapor is extremely rare. Mercury absorbed from vapor is eliminated from the human body fairly quickly through the urinary and fecal tracts. Mercury levels resulting from exposure to mercury vapor or to inorganic mercury compounds (including aryl mercury compounds such as phenylmercuric acetate) are rapidly reduced and do not accumulate. On the other hand, mercury levels resulting from exposure to alkyl mercury compounds such as methyl mercury or ethyl mercury compounds cannot be eliminated quickly, and tend to accumulate.

Mercury is a very volatile element, and dangerous levels of mercury vapor are readily attained at room temperature in enclosed spaces that are not adequately ventilated. The present toxicity limit for mercury vapor in air is 0.1 mg/m³. At 20 °C, air saturated with mercury vapor contains a concentration more than 100 times this limit.

Because of the toxic nature of mercury and many mercury compounds, certain precautions are mandatory during handling and disposal. Containers should be securely covered. All operations involving mercury metal should be carried out in a well-ventilated area or in a closed system to prevent accumulation of mercury vapor in the workspace; this is of utmost importance if the operation involves heating mercury above room temperature. Workspaces should be continually monitored with special electronic instruments to detect any rise of mercury-vapor concentration above the established safe working limit. Workers should be provided with masks or special breathing devices, and the level of mercury in the body of every worker should be periodically monitored by specially trained medical personnel. Any spills of liquid, or escape of vapor from a closed heated system, must be countered by immediate decontamination of the affected workspace. Disposal is ordinarily accomplished by sending impure mercury or concentrated mercury compounds to reclamation centers where purified metal is produced from the discards. Mercury compounds such as methyl mercury are dangerous pollutants and are required to be removed from effluents before they are discharged into natural waters. Sludges and other solid wastes that are contaminated with small concentrations of mercury are sometimes buried at approved sites.

Structure

Crystal system. Rhombohedral below −39 °C. Structure symbol, *A*1o; space group, $R\bar{3}m$; *hR*1

Mass Characteristics

Atomic weight. 200.59
Density. Solid: 14.193 Mg/m³ at −39 °C. Liquid: 14.43 at melting point; 13.595 at 0 °C; 13.546 at 20 °C; 13.352 at 100 °C; 13.115 at 200 °C; 12.881 at 300 °C
Compressibility. Volumetric, from 1 to 493 atm: 4×10^{-6} per atm at 20 °C

Thermal Properties

Boiling point. 356.58 °C at 1 atm
Freezing temperature. −38.87 °C
Vapor pressure:

Temperature, °C	Vapor pressure
−20	2.41 mPa
0	24.6
20	160.0
50	1.689 Pa
100	36.40
150	374.2
200	2.305 kPa
220	4.284
240	7.580
260	12.839
280	20.914
300	32.904

320	50.173
340	74.381
356.58	101.3
360	107.49
380	151.77
400	209.86

Critical point. 1677 °C and 74.2 MPa (732 atm). Critical density: 3.56 Mg/m^3

Coefficient of thermal expansion. Volumetric, liquid: 182×10^{-6} at 20 °C

Specific heat:

Temper- ature, K	C_p, J/kg·K	
	Liquid	Gas(a)
234.28	142.1(b)	...
234.28	142.0	...
298.15	139.6	103.7
350	137.7	103.7
400	136.8	103.7
450	136.0	103.7
500	135.6	103.7
550	135.36	103.7
600	135.38	103.7
630	135.5	103.7
650	135.6	103.7
700	136.1	103.7
750	136.7	103.7

(a) Ideal. (b) Solid.

Enthalpy, Entropy. See Table 22.
Latent heat of fusion. 11.8 kJ/kg
Heat of vaporization. 61.42 kJ/kg at 25 °C.
Latent heat of vaporization. 272 kJ/kg
Thermal conductivity. 8.21 W/m·K at 0 °C; 9.67 at 60 °C; 10.9 at 120 °C; 11.7 at 160 °C; 12.7 at 220 °C

Electrical Properties

Electrical resistivity:

Temperature, °C	Resistivity, nΩ·m
20	958
50	984
100	1032
200	1142
300	1275
350	1355

Temperature coefficient, 0.9 nΩ·m/K at 20 °C
Thermoelectric potential. Hg vs Pt: -0.60 mV for 100 °C hot junction and 0 °C cold junction

Standard electrode potential. vs hydrogen electrode at 20 °C: 0.851 V for $Hg^{++} + 2e^- \rightleftarrows Hg$; 0.7961 V for $2Hg^{++} + 2e^- \rightleftarrows 2Hg^+$; 0.905 V for $2Hg^{++} + 2e^- \rightleftarrows (Hg_2)^{++}$
Ionization potential. Hg(I), 10.43 V; Hg(II), 18.75 V; Hg(III), 34.30 V; Hg(IV), 72 V; Hg(V), 82 V
Hydrogen overvoltage. 1.06 V
Contact potential. Hg vs Sb, -0.26 V; Hg vs Zn, $+0.17$ V

Magnetic Properties

Magnetic susceptibility. Volume: -1.9×10^{-6} mks at 18 °C

Optical Properties

Color. Filtered mercury has a bright, clean, silvery appearance if the metal contains less than 1 ppm of impurities.
Spectral reflectance. 71.2% for $\lambda =$ 550 um
Refractive index. 1.6 to 1.9 at 20 °C

Nuclear Properties

Thermal neutron cross section. Capture, 420 b; scattering 5 to 15 b

Mechanical Properties

Dynamic viscosity. 1.55 mPa·s at 20 °C
Surface tension. 0.465 N/m at 20 °C; 0.454 N/m at 112 °C; 0.436 N/m at 200 °C; 0.405 N/m at 300 °C; 0.394 N/m at 354 °C. Angle of contact on glass: 128° at 18 °C

Mischmetal (MM)

Compiled by K. A. Gschneidner, Jr. and
B. J. Beaudry
Ames Laboratory
U. S. Department of Energy
Iowa State University

Mischmetal is used as an alloying additive in ferrous alloys to scavenge sulfur, oxygen and other substances. Mischmetal is also added to magnesium-base alloys to improve high temperature strength and to ductile irons to nodularize graphite. Other uses of mischmetal include lighter flints, pyrophoric ordnance devices and mischmetal cobalt ($MMCo_5$) per-manent magnets. Mischmetal oxidizes at room temperature in air. Turnings can be ignited easily and burn white hot. Finely divided mischmetal should not be handled in air. Because mischmetal is an indefinite mixture of rare earth metals, the properties of a particular mischmetal depend on its composition, which in turn depends on the mineral source for the mixture. Listed below are the properties of two of the most common mixtures. All values are estimated.

Bastnasite-derived Mischmetal

Specifications

UNS number. E21000

Chemical Composition

Composition limits. Total mixed rare earths: 99.0 min; mixture consists of 50.0 Ce; 38.0 La; 12.0 Nd; 4.0 Pr; 1.0 other rare earths

Structure

Crystal structure. α-phase, probably double close-packed hexagonal, $P6_3/mmc$ D^4_{6h}; $a = 0.371$ nm; $c = 1.195$ nm, at 24 °C. β-phase, probably body-centered cubic
Minimum interatomic distance. At 24 °C: $r_a = 0.186$ nm; $r_c = 0.184$ nm; radius $CN_{12} = 1.85$ nm

Mass Characteristics

Atomic weight. 140.3
Density. 6.5 Mg/m^3 at 24 °C
Volume change on freezing. 0.2% expansion

Thermal Properties

Melting point. 875 °C
Boiling point. Expected to evaporate incongruently with the initial loss of a major constituent (Nd) by boiling at approx 3100 °C
Phase-transformation temperature. α- to β-phase, 795 °C
Coefficient of thermal expansion. At 24 °C. Linear: 8.7 μm/m·K. Volumetric: 26×10^{-6} per K
Specific heat. 193 J/kg·K at 25 °C
Entropy. 467 J/kg·K at 25 °C
Latent heat of fusion. 22 kJ/kg
Latent heat of transformation. 3.1 kJ/mol
Latent heat of combustion. For hexagonal R_2O_3 at 25 °C: $\Delta H^0_c = 6.4$ MJ/kg MM; $\Delta G^0_f = -6.0$ MJ/kg MM

Recrystallization temperature. Approx 350 °C

Thermal conductivity. 13 W/m·K at 25 °C

Electrical Properties

Electrical resistivity. Solid: 800 nΩ·m at 25 °C. Liquid: 1300 nΩ·m at 880 °C

Magnetic Properties

Magnetic susceptibility. Volume 1.6 × 10⁻³ mks at 25 °C

Optical Properties

Color. Metallic silver

Spectral hemispherical emittance. Liquid, 30%

Chemical Properties

General corrosion behavior. Mischmetal oxidizes in air at room temperature. Oxidation rates increase with increasing temperature.

Resistance to specific corroding agents. Mischmetal reacts vigorously with dilute acids. The presence of the fluoride ion retards acid attack by the formation of rare earth fluoride, RF_3, on the surface of the metal

Mechanical Properties

Tensile properties. At 24 °C: tensile strength, 138 MPa; yield strength, 48 MPa; elongation, 25%; reduction in area, 50%

Hardness. 35 HV

Poisson's ratio. 0.27

Elastic modulus. At 27 °C: tension, 35 GPa; shear, 14 GPa; bulk, 25 GPa

Kinematic liquid viscosity. 0.46 mm²/s at 875 °C

Liquid surface tension. 0.70 N/m at 875 °C

Monazite-derived Mischmetal

Specifications

UNS number. E31000

Chemical Composition

Composition limits. Total mixed rare earths, 99.0 min; mixture consists of 45.0 Ce; 20.0 La; 19.0 Nd; 6.0 Pr; 4.0 Sm; 2.0 Gd; 2.0 Y; 2.0 other rare earths

Structure

Crystal structure. α-phase, double close-packed hexagonal, $P6_3/mmc$ D^4_{6h}; a = 0.369 nm; c = 1.188 nm at 24 °C. β-phase, probably body-centered cubic

Minimum interatomic distance. At 24 °C: r_a = 0.184 nm; r_c = 0.183 nm; radius CN_{12} = 0.184 nm

Mass Characteristics

Atomic weight. 140.5

Density. 6.7 Mg/m³ at 24 °C

Volume change on freezing. 0.1% expansion

Thermal Properties

Melting point. 920 °C

Boiling point. Expected to evaporate incongruently with the initial loss of a major constituent (Nd) by boiling at ~3100 °C

Phase-transformation temperature. α- to β-phase, 830 °C

Coefficient of thermal expansion. At 24 °C. Linear: 8.6 μm/m·K. Volumetric: 26 × 10⁻⁶ per K

Specific heat. 195 J/kg·K at 25 °C

Entropy. At 298.15 K, 477 J/kg·K

Latent heat of fusion. 46 kJ/kg

Latent heat of transformation. 22 kJ/kg

Latent heat of combustion. For hexagonal R_2O_3 at 25 °C: ΔH^0_c = 6.4 MJ/kg MM; ΔG^0_f = −6.0 MJ/kg MM

Recrystallization temperature. Approx 350 °C

Thermal conductivity. 13 W/m·K at 25 °C

Electrical Properties

Electrical resistivity. Solid: 800 nΩ·m at 25 °C; liquid, 1300 nΩ·m at 925 °C

Magnetic Properties

Magnetic susceptibility. Volume: 5.2 × 10⁻³ mks at 25 °C

Optical Properties

Color. Metallic silver

Spectral hemispherical emittance. Liquid, 30%

Chemical Properties

General corrosion behavior. Mischmetal oxidizes in air at room temperature. Oxidation rates increase with increasing temperature.

Resistance to specific corroding agents. Mischmetal reacts vigorously with dilute acids. The presence of the fluoride ion retards acid attack by the formation of rare earth fluo-

ride, RF_3, on the surface of the metal

Mechanical Properties

Tensile properties. At 24 °C: tensile strength, 138 MPa; yield strength, 48 MPa; elongation, 25%; reduction in area, 50%

Hardness. 35 HV

Poisson's ratio. 0.27

Elastic modulus. At 27 °C: tension, 37 GPa; shear, 15 GPa; bulk, 27 GPa

Kinematic liquid viscosity. 0.47 mm²/s at 920 °C

Liquid surface tension. 0.71 N/m at 920 °C

Molybdenum (Mo)

Compiled by J. Z. Briggs
Reviewed for this edition by
Robert Q. Barr
Climax Molybdenum Co.

Molybdenum is used as alloying additions and for electrical and electronic parts; missile and aircraft parts, high-temperature furnace parts; die-casting cores; hot working tools; boring bars; thermocouples; nuclear-energy applications; corrosion-resistant equipment; equipment for glass melting furnaces; metallizing. Molybdenum is not suitable for continued service at temperatures above 500 °C in an oxidizing atmosphere unless protected by adequate coating

Structure

Crystal structure. Body-centered cubic, a = 0.31468 nm at 25 °C

Slip planes. {112} at 20 °C; {110} at 1000 °C

Slip direction. [111]

Interatomic distance. 0.27252 nm min

Metallography. Electrolytic polishing is preferred. Etching: (1) 10 g NaOH + 30 g K_3Fe $(CN)_6$ + 100 litres water; (2) 1 g NaOH + 35 g K_3Fe $(CN)_6$ + 600 litres water; (3) Murakami solution

Mass Characteristics

Atomic weight. 95.94

Density. At 20 °C: 10.22 Mg/m³; see also Fig. 61.

Compressibility. At 293 °C: 36 μm²/N

Fig. 61 Density of various products of molybdenum

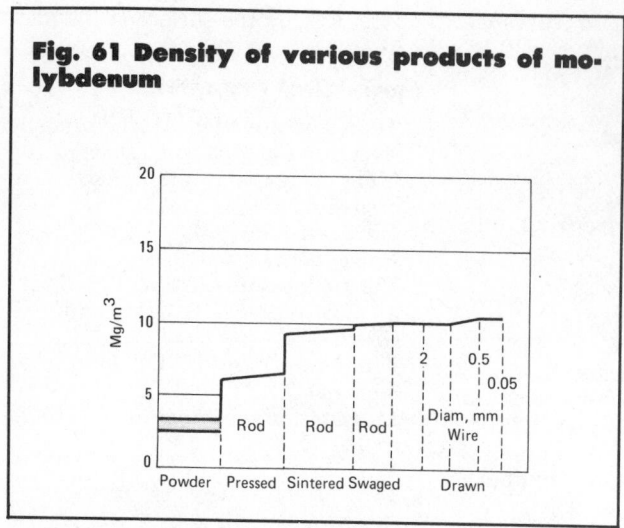

Fig. 64 Temperature dependence of the thermal conductivity of molybdenum

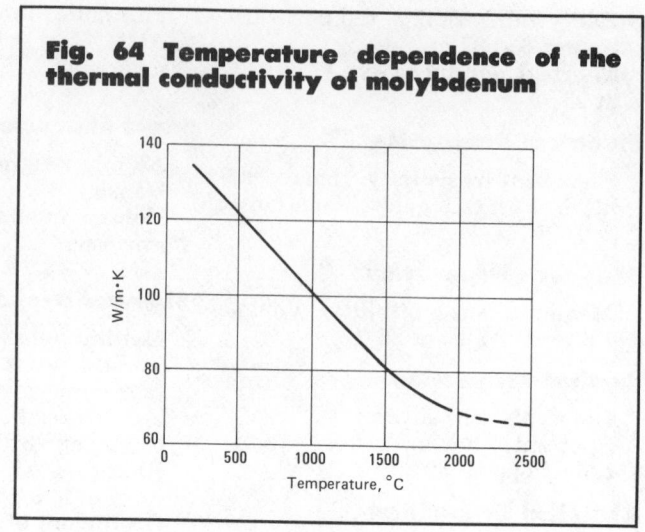

Fig. 62 Linear thermal expansion of molybdenum

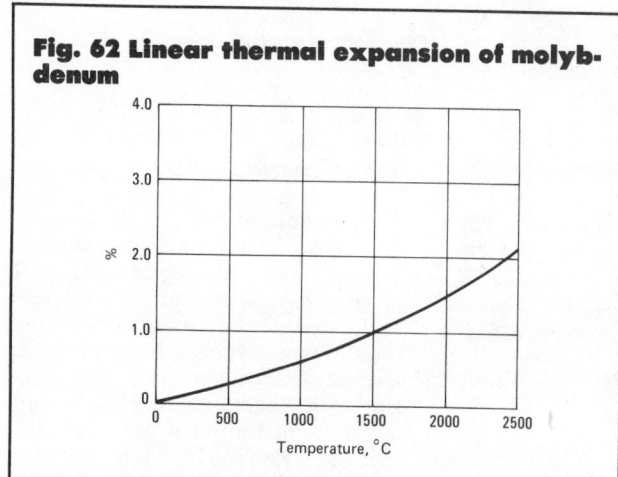

Fig. 65 Temperature dependence of the electrical resistivity of molybdenum

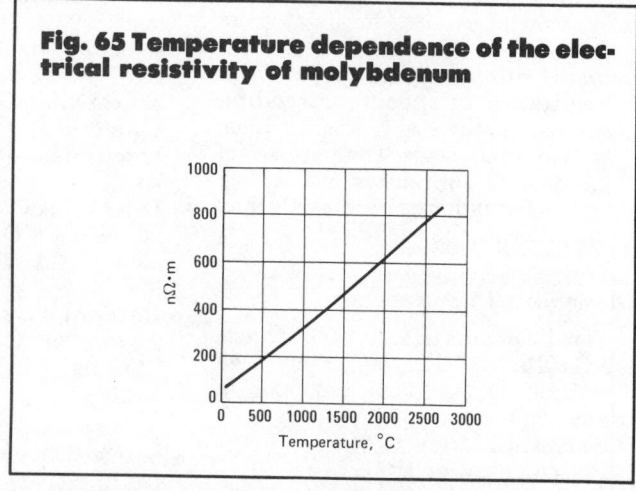

Fig. 63 Temperature dependence of the specific heat of molybdenum

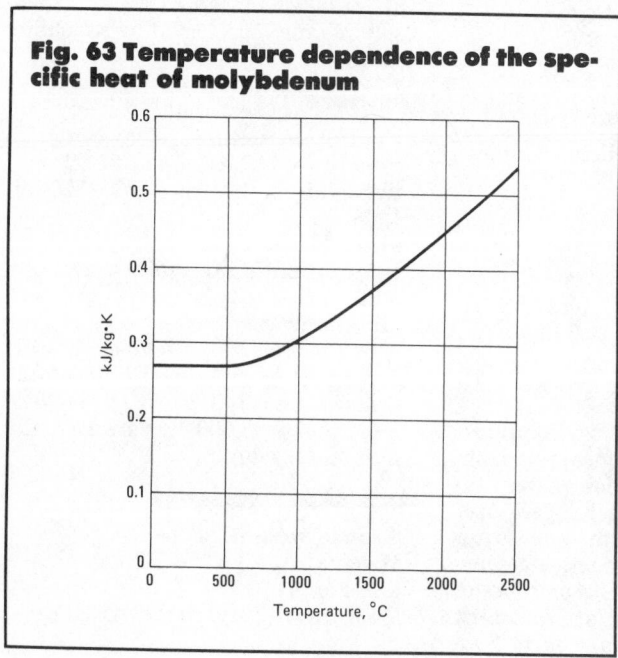

Fig. 66 Temperature dependence of the total normal emittance of molybdenum

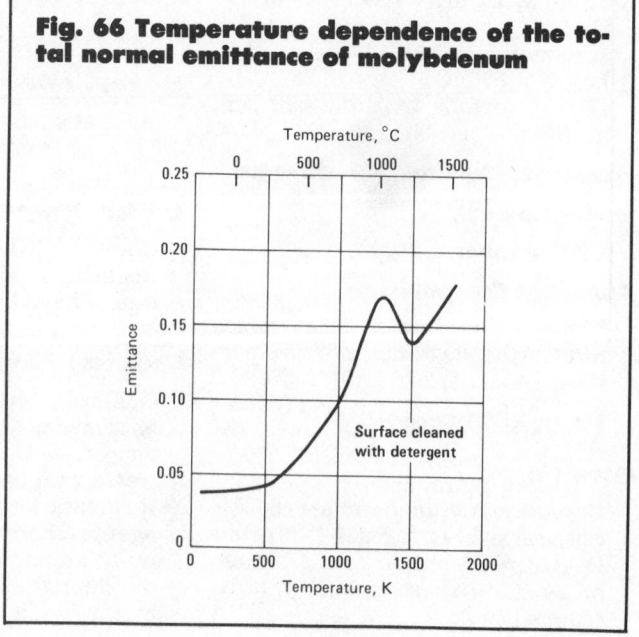

Fig. 67 Temperature dependence of the tensile strength of molybdenum

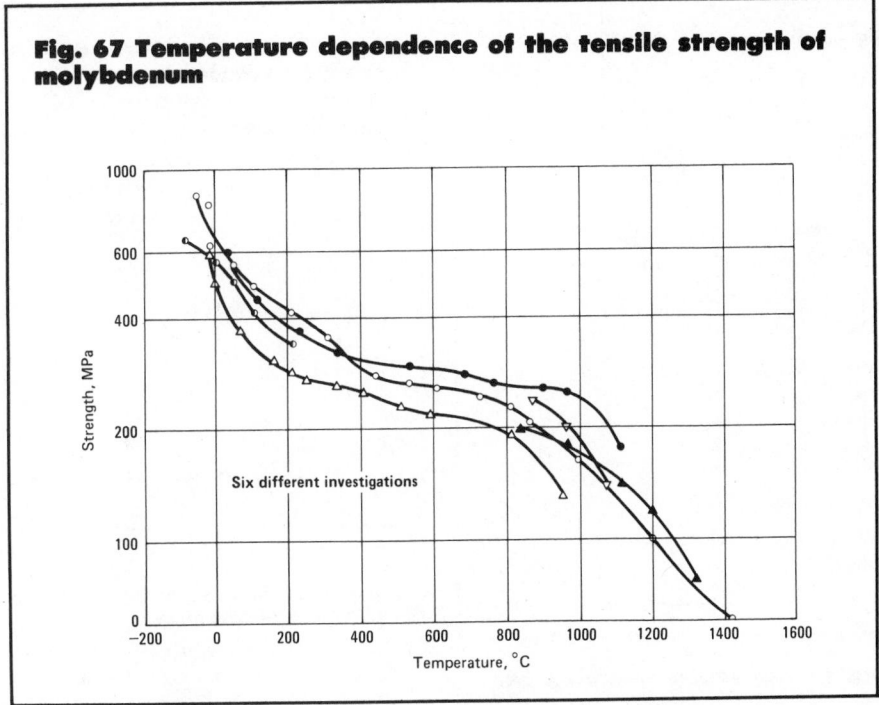

Six different investigations

Fig. 68 Temperature dependence of the yield and fracture strengths of molybdenum

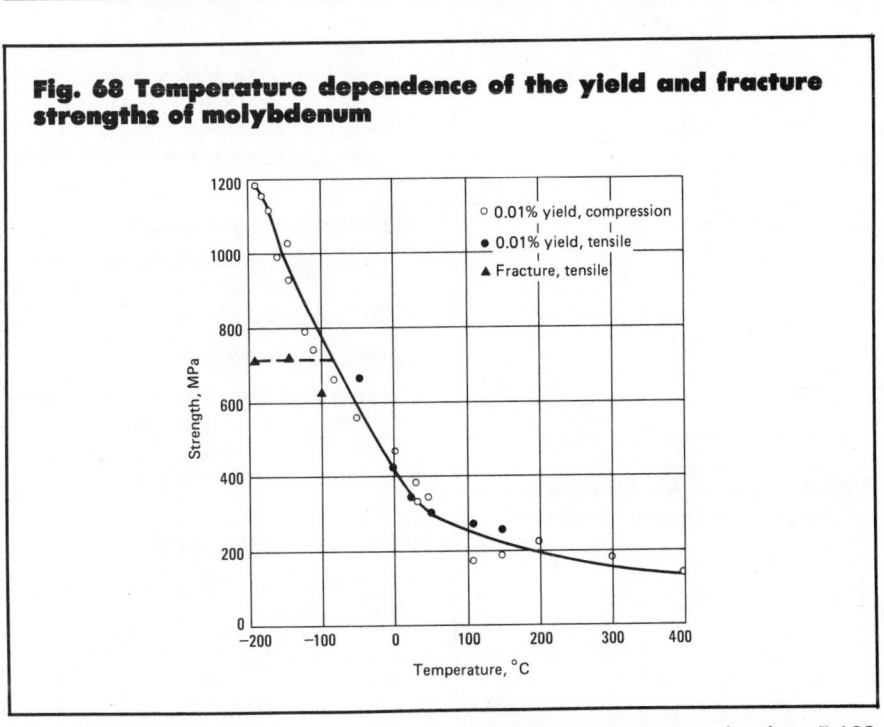

- ○ 0.01% yield, compression
- ● 0.01% yield, tensile
- ▲ Fracture, tensile

Fig. 69 Effect of product form on the tensile properties of molybdenum

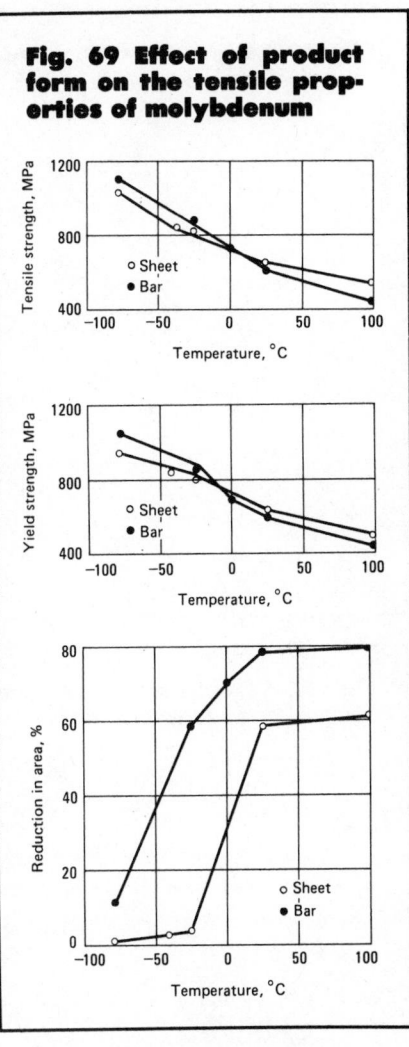

- ○ Sheet
- ● Bar

Fig. 71 Temperature dependence of the hardness of molybdenum

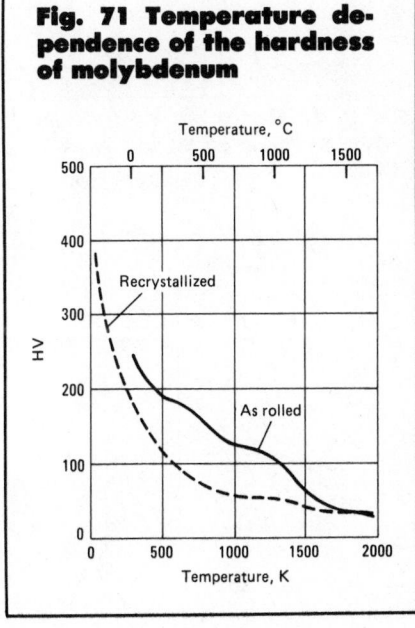

Recrystallized

As rolled

Thermal Properties

Melting point. 2610 °C
Boiling point. 5560 °C
Thermal expansion. See Fig. 62.
Specific heat. At 20 °C: 276 J/kg. See also Fig. 63.
Latent heat of fusion. 270 kJ/kg (est)

Latent heat of vaporization. 5.123 MJ/kg
Thermal conductivity. At 20 °C: 142 W/m·K. See also Fig. 64.
Heat of combustion. 7.58 MJ/kg Mo
Recrystallization temperature. 900 °C min; commercial products normally require higher temperatures

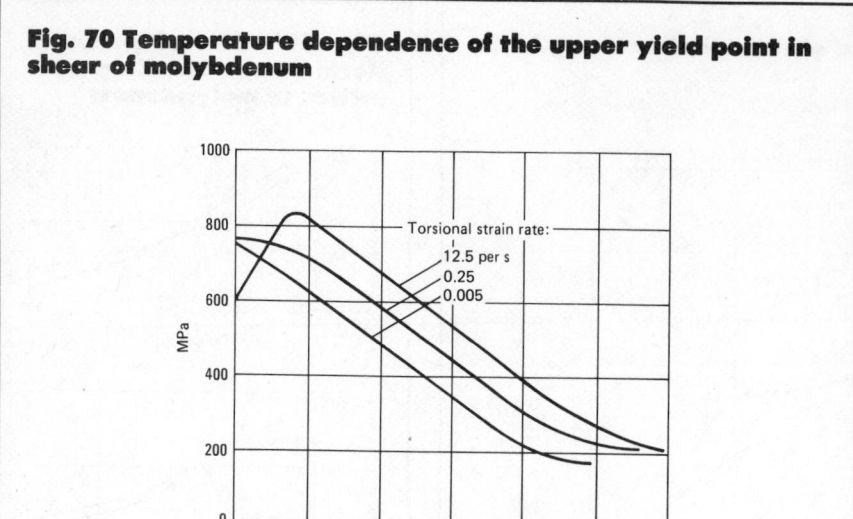

Fig. 70 Temperature dependence of the upper yield point in shear of molybdenum

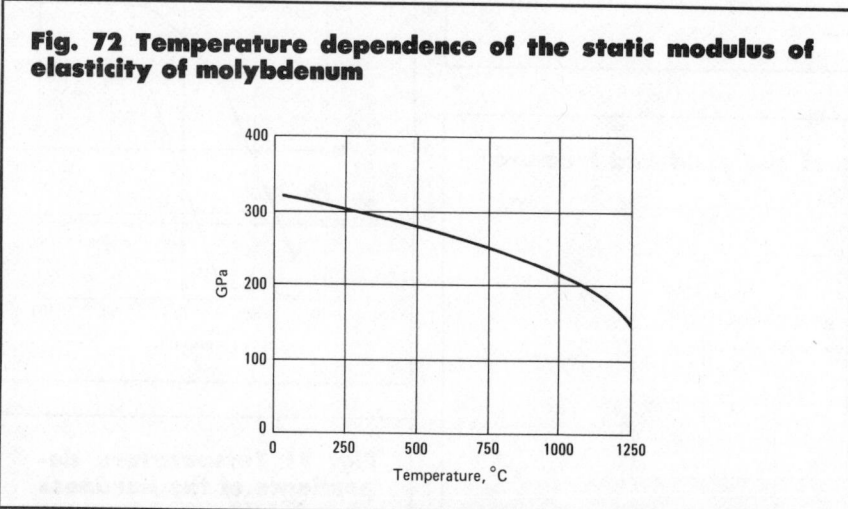

Fig. 72 Temperature dependence of the static modulus of elasticity of molybdenum

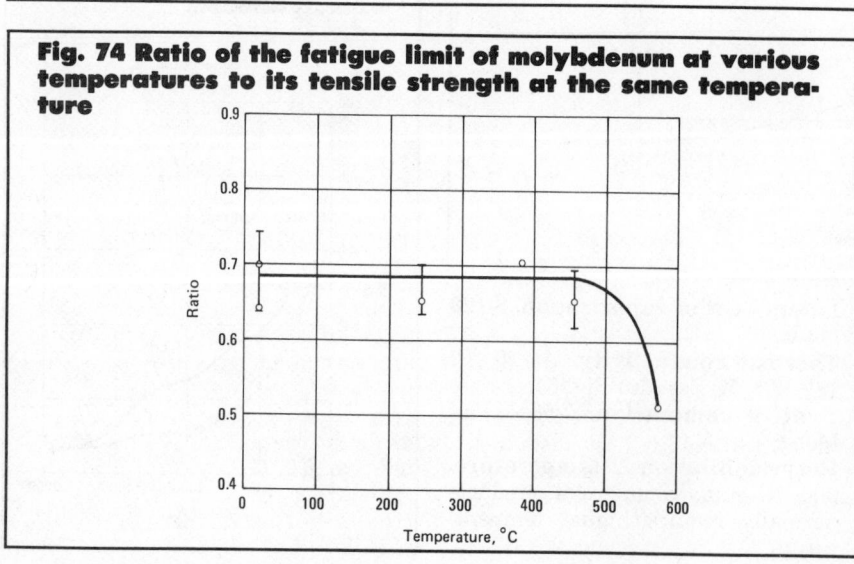

Fig. 74 Ratio of the fatigue limit of molybdenum at various temperatures to its tensile strength at the same temperature

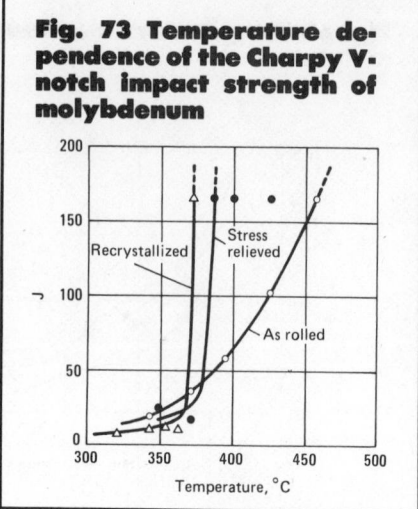

Fig. 73 Temperature dependence of the Charpy V-notch impact strength of molybdenum

Electrical Properties

Electrical conductivity. At 0 °C: 34% IACS

Electrical resistivity. At 0 °C; 52 nΩ·m. See also Fig. 65.

Thermal electromotive force. vs Pt, 0 to 100 °C: 1.45 mV

Electrochemical equivalent. Valence 6, 0.1658 mg/C

Hydrogen overpotential. At 100 A/m^2: 0.44 V

Magnetic Properties

Magnetic susceptibility. Mass: 1.17×10^{-8} mks at 25 °C; 1.39×10^{-8} mks at 1825 °C

Optical Properties

Reflectivity. 46% at 500 nm, 93% at 10 000 nm

Color. Silvery white

Total normal emittance. See Fig. 66.

Mechanical Properties

Tensile properties. See Fig. 67, 68 and 69.

Shear properties. See Fig. 70.

Hardness. See Fig. 71.

Elastic modulus. See Fig. 72.

Impact strength. See Fig. 73.

Fatigue strength. See Fig. 74.

Creep-rupture characteristics. See Fig. 75 and 76.

Directional properties. If not cross rolled, the tensile strength of molybdenum sheet may be as much as 20% greater in the direction of rolling than when the inclination of the direction of tension to that of rolling is between 45 and 90°

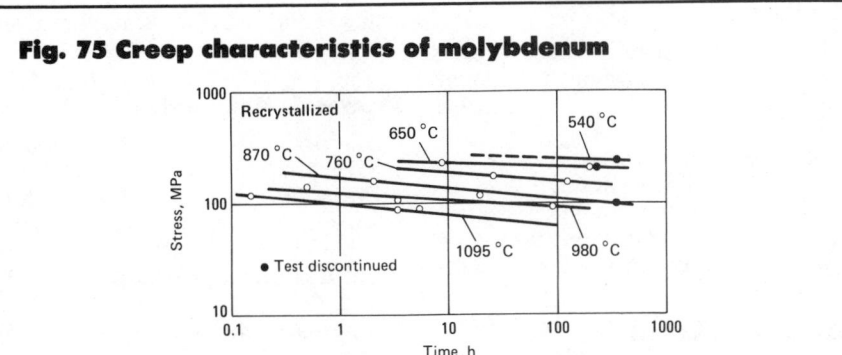

Fig. 75 Creep characteristics of molybdenum

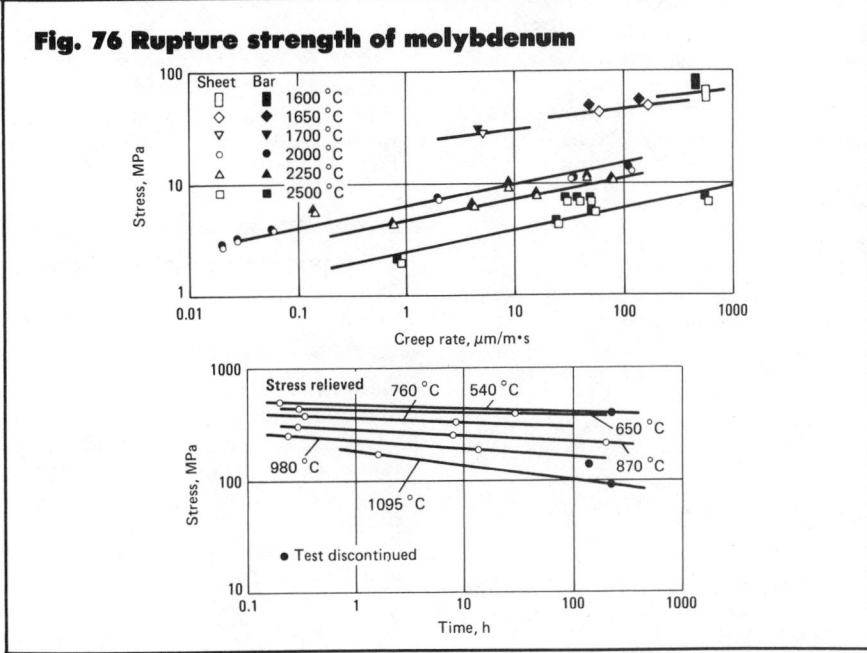

Fig. 76 Rupture strength of molybdenum

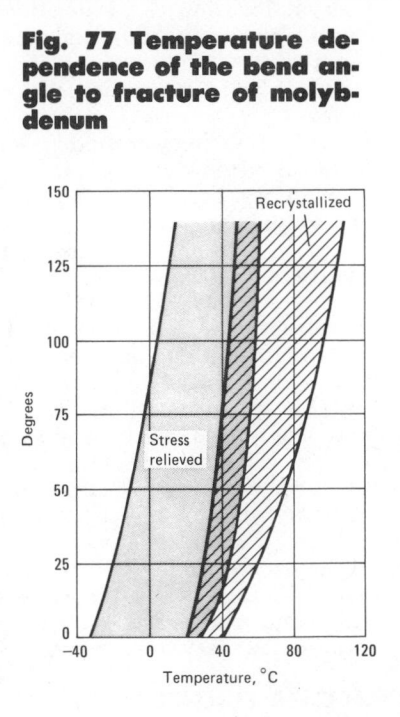

Fig. 77 Temperature dependence of the bend angle to fracture of molybdenum

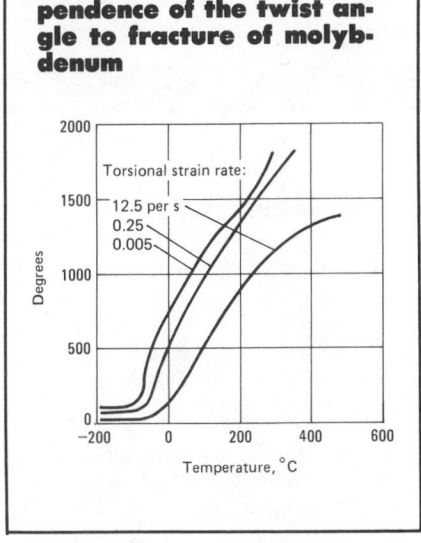

Fig. 78 Temperature dependence of the twist angle to fracture of molybdenum

Properties of single crystals. Tensile strength, 350 MPa

Velocity of sound. Longitudinal wave, 6370 m/s; shear wave, 3410 m/s; "thin rod", 5500 m/s

Chemical Properties

General corrosion behavior. Molybdenum has particularly good resistance to corrosion by mineral acids, provided oxidizing agents are not present. It is also resistant to many liquid metals and to most molten glasses. In inert atmospheres, it is unaffected up to 1760 °C by refractory oxides. Molybdenum is relatively inert in hydrogen, ammonia and nitrogen up to about 1100 °C, but a superficial nitride case may be formed in ammonia or nitrogen.

Resistance to specific corroding agents:

Agent	Corrosion, mils/year
20% hydrochloric acid, boiling, unaerated	0.90
49% hydrofluoric acid, room temperature, unaerated	0.14
85% phosphoric acid, 100 °C, unaerated	0.29
40% sulfuric acid, boiling, unaerated	1.5
Liquid lithium, sodium and sodium-potassium, up to 900 °C	<1
Liquid bismuth, 980 °C	nil

Fabrication Characteristics

Consolidation. In most instances, molybdenum is consolidated from powder by compacting under pressure followed by sintering in the range from 1650 to 1900 °C. Some molybdenum is consolidated by a vacuum-arc-casting method in which a pre-formed electrode is melted by arc formation in a water cooled mold.

Hot working temperature. Generally forged between 1180 and 1290 °C down to 930 °C

Annealing temperature. Normal stress-relieving temperature is 870 to 980 °C

Recrystallization temperature. Depends on prior working and condition; 1180 °C for full recrystallization in 1 h of a 5/8 in. bar reduced 97% by rolling

Suitable forming methods. Conventional methods

Formability. See Fig. 77 and 78.

Precautions in forming. Must be heated to the proper temperature relative to its thickness and forming speed

Heat treatment. Not hardenable by heat treatment but only by work hardening

Suitable joining methods. Can be brazed or joined mechanically, as well as welded by arc, resistance, percussion, flash and electron-beam methods. Arc-cast molybdenum is preferred to a powder metallurgy product for welding. Absolute cleanliness of surface is essential. Fusion welding must be carried out in closely controlled inert atmosphere

SELECTED REFERENCES

Molybdenum Metal, Climax Molybdenum Co., 1960

Molybdenum Mill Products, Climax Molybdenum Company of Michigan

Neodymium (Nd)

Compiled by K. A. Gschneidner, Jr. and
N. J. Beaudry
Ames Laboratory
U. S. Department of Energy
Iowa State University

Neodymium is used as an alloying additive to ferrous alloys to scavenge sulfur, oxygen and other elements and to strengthen magnesium alloys. It is also used as a laser material, glass coloring agent and in petroleum cracking catalysts, carbon-arc lights, lighter flints, and ceramic capacitors. Neodymium oxidizes at room temperature in air. It should be stored in a vacuum or inert atmosphere; storage in oil is not recommended. Turnings can be ignited easily and will burn white hot. Finely divided neodymium should not be handled in air.

Structure

Crystal structure. α-phase, double close-packed hexagonal, $P6_3/mmc$ D^4_{6h}; $a = 0.36582$ nm, $c = 1.17966$ nm at 24 °C. β-phase, body-centered cubic, $Im3m$ O^9_h; $a = 0.413$ nm at 880 °C

Minimum interatomic distance. At 24 °C: $r_a = 0.18291$ nm; $r_c = 0.18137$ nm; radius $CN_{12} = 0.18214$ nm

Mass Characteristics

Atomic weight. 144.24

Density. α-phase, 7.008 Mg/m³ at 24 °C; β-phase, 6.80 Mg/m³ at 8800 °C; liquid, 6.67 Mg/m³ at 1035 °C

Volume change on freezing. 0.9% contraction

Volume change on phase transformation. α- to β-phase, 0.1% volume expansion on heating

Thermal Properties

Melting point. 1021 °C

Boiling point. 3074 °C

Phase-transformation temperature. α- β-phase, 863 °C

Coefficient of thermal expansion. At 24 °C. Linear: 9.6 μm/m·K. Linear, along crystal axes: 7.6 μm/m·K along a axis, 13.5 μm/m·K along c axis. Volumetric: 28.7×10^{-6} per K

Specific heat. 190.0 J/kg·K at 25 °C

Entropy. At 25 °C: 492.9 J/kg·K

Latent heat of fusion. 49.51 kJ/kg

Latent heat of transformation. 21.00 kJ/kg

Latent heat of vaporization. 2.271 MJ/kg at 25 °C

Thermal conductivity. 16.5 W/m·K at 25 °C

Heat of combustion. Hexagonal Nd_2O_3 at 25 °C: $\Delta H^0 c = 12.5$ MJ/kg Nd; $\Delta G^0_f = -5.96$ MJ/kg Nd

Recrystallization temperature. 400 °C

Vapor pressure. 0.001 Pa at 955 °C; 0.101 Pa at 1175 °C; 10.1 Pa at 1500 °C; 1013 Pa at 2029 °C

Electrical Properties

Electrical resistivity. 643 nΩ·m at 25 °C; 68 nΩ·m at 4 K. Liquid: 1550 nΩ·m at 1022 °C

Ionization potential. Nd(I), 5.49 V; Nd(II), 10.72 V; Nd(III), 21.85 V; Nd(IV), 40.32 V

Hall coefficient. $+0.0971$ nV·m/A·T at 25 °C

Temperature of superconductivity. Bulk neodymium is not superconducting down to 0.25 K at atmospheric pressure

Magnetic Properties

Magnetic susceptibility. Volume: 3.62×10^{-3} mks at 25 °C; obeys Curie-Weiss law above 35 K with an effective moment of 3.45 Borh magneton and $\theta_a = +5$ K and $\theta_c = 0$ K

Saturation magnetization. >0.59 T at 2 K along $\langle 11\bar{2}0 \rangle$

Magnetic transformation temperature. Néel temperatures at 7.5 K (cubic sites) and 19.0 K (hexagonal sites)

Optical Properties

Color. Metallic silver

Spectral hemispherical emittance. 28.0% for $\lambda = 645$ nm from 1021 to 1300 °C

Nuclear Properties

Thermal neutron cross section. 46 b

Chemical Properties

General corrosion behavior. Neodymium oxidizes in air at room temperature but at a slower rate than lanthanum or cerium. Oxidization rates increase with increasing temperature; interstitial impurities increase the rate of oxidation. Hydrogen will react with neodymium at room temperature.

Resistance to specific corroding agents. Neodymium reacts vigorously with dilute acids and slowly with concentrated sulfuric. The presence of the fluoride ion retards acid attack due to the formation of NdF_3 on the surface of the metal

Mechanical Properties

Tensile properties. Tensile strength, 169 MPa; yield strength, 71 MPa; elongation, 28%; reduction in area, 72%

Hardness. 18 HV

Poisson's ratio. 0.279

Strain-hardening exponent. 0.28

Elastic modulus. At 27 °C: tension, 42.2 GPa; shear, 16.5 GPa; bulk, 31.8 GPa

Elastic constants along crystal axes. At 27 °C: $c_{11} = 54.77$ GPa; $c_{12} = 24.60$ GPa; $c_{13} = 16.56$ GPa; $c_{33} = 60.80$ GPa; $C_{44} = 15.01$ GPa

Liquid surface tension. 0.69 N/m at 1021 °C

Neptunium

See tables at end of this section

Nickel (Ni)

Compiled by Donald L. Pasquine
The International Nickel Company,
Inc.

Structure

Crystal structure. Face-centered cubic; $a = 0.35167$ nm at 20 °C

Mass Characteristics

Atomic weight. 58.71
Density. 8.902 Mg/m^3 at 25 °C

Thermal Properties

Melting point. 1453 °C
Boiling point. Approx 2730 °C
Coefficient of thermal expansion. Linear, 13.3 μm/m·K at 0 to 100 °C
Specific heat. 471 J/kg·K at 100 °C
Recrystallization temperature. 370 °C
Thermal conductivity. 82.9 W/m·K at 100 °C

Electrical Properties

Electrical conductivity. Volumetric, 25.2% IACS at 20 °C
Electrical resistivity. 68.44 nΩ·m at 20 °C; temperature coefficient, 69.2 nΩ·M per K at 0 to 100 °C

Magnetic Properties

Magnetic susceptibility. Ferromagnetic
Magnetic permeability. μmax = 1240 at $B = 1900$ G
Coercive force. 167 A/m (from $H = 4$kA/m)
Saturation magnetization. 0.616 T at 20 °C
Residual induction. 0.300 T
Hysteresis loss. 685 J/m^3 at $B = 0.6$ T
Curie temperature. 358 °C

Optical Properties

Color. Grayish-white
Spectral reflectance. 41.3% for λ = 0.30 μm

Nuclear Properties

Effect of neutron irradiation. Results in small increase in tensile strength but large increase in yield strength

Chemical Properties

General corrosion behavior. Nickel is not an active element chemically, does not readily evolve hydrogen from acid solutions and usually requires the presence of an oxidizing agent for significant corrosion to occur. Generally, reducing conditions retard corrosion whereas oxidizing conditions accelerate corrosion of nickel in chemical solutions. However, nickel may also form a protective corrosion-resistant, or passive, oxide film on exposure to some oxidizing conditions.

Mechanical Properties

Tensile properties. Typical. Tensile strength, 317 MPa; 0.2% offset yield strength, 59 MPa; elongation, 30% in 50 mm or 2 in.
Hardness. 64 HV (annealed)
Poisson's ratio. 0.31 at 25 °C
Elastic modulus. Tension, 207 GPa; shear, 76 GPa; compression, 207 GPa
Velocity of sound. 4.7 km/s at 40 °C

SELECTED REFERENCES

E. M. Wise and R. H. Schaefer, The Properties of Pure Nickel, *Metals and Alloys,* Vol 16, 1942
W. A. Wesley, Preparation of Pure Nickel by Electrolysis of a Chloride Solution, *Journal of the Electrochemical Society,* Vol 103 (No. 5), 1956
G. W. P. Rengstorff, "High-Purity Metals", Report 222, Defense Metals Information Center, 1966
"Nickel and Its Alloys", National Bureau of Standards Monograph 106, 1968
J. Crangle and G. M. Goodman, The Magnetization of Pure Iron and Nickel, *Proceedings of the Royal Society* (Series A), London, Vol 321, 1971
R. Hultgen, *et al, Selected Values of the Thermodynamic Properties of the Elements,* American Society for Metals, 1973

Niobium (Columbium) (Nb)

Compiled by E. S. Bartlett
Battelle Memorial Institute
Reviewed for this edition by
L. H. Belz
Kawecki Berylco Industries, Inc.

Niobium is used as an alloying element in nickel- and cobalt-based superalloys as well as some grades of stainless and low alloy steels. It is also used as an alloy base for various combinations with zirconium, hafnium, tungsten, tantalum and molybdenum to increase high temperature mechanical properties. Niobium, niobium-titanium alloys and niobium-tin alloys are used as superconductors. Niobium oxidizes and becomes contaminated with absorbed oxygen rapidly above about 400 °C in oxygen-containing atmospheres, including atmospheres normally considered neutral or reducing; absorbs hydrogen at temperatures between about 250 and 950 °C from hydrogen-containing atmospheres. Contamination by interstitial elements results in loss of ductility at ambient temperature. Consequences of high impurity levels include impaired fabricability, increased ductile-to-brittle transition temperature, considerable low-temperature strengthening with attendant loss in ductility, intensified strain-aging effects at slightly elevated temperature, and slight strengthening at higher temperature.

Structure

Crystal structure. Body-centered cubic, $a = 0.3294$ nm, atomic diameter, 0.294 nm
Slip plane. 110
Metallography. (1) Grind through 000 emery; (2) rough polish with coarse diamond in kerosine; (3) standard finish polish with alumina; (4) etchant (all acids in parts by volume of laboratory reagent grades): 30 lactic—10 nitric—5 hydrofluoric acid solution (more HF for alloys); (5) chemical polish (for freedom of distortion, if required): 30 lactic—30 nitric—1 to 2 hydrofluoric; (6) electrolytic etch (for particularly uniform grain-boundary definition): 90H$_2$SO$_4$-10HF at 2 V

Mass Characteristics

Atomic weight. 92.9064
Density. At 20 °C: 8.57 Mg/m^3

Thermal Properties

Melting point. 2468 °C
Boiling point. 4927 °C
Coefficient of thermal expansion. Linear:

Fig. 79 Temperature dependence of the specific heat of niobium

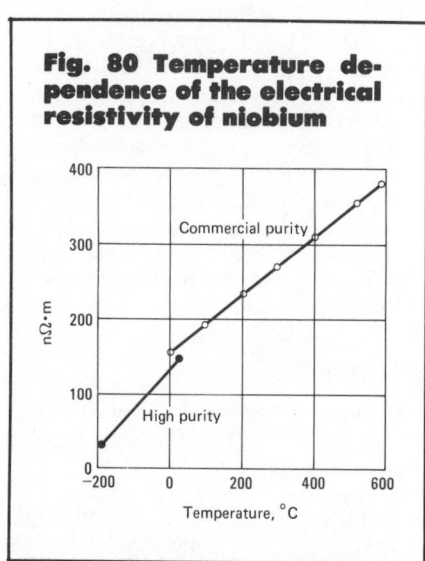

Fig. 80 Temperature dependence of the electrical resistivity of niobium

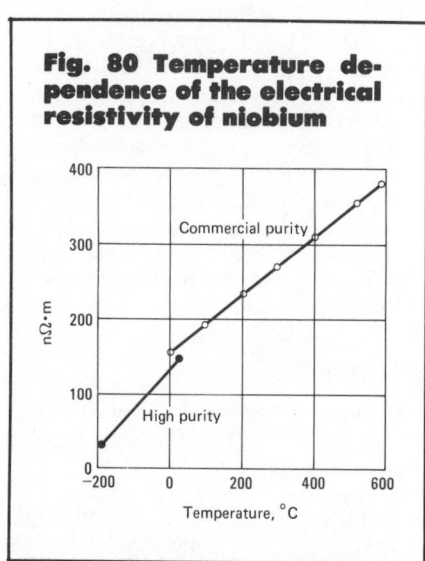

Fig. 81 Temperature dependence of the thermal electromotive force of niobium vs platinum

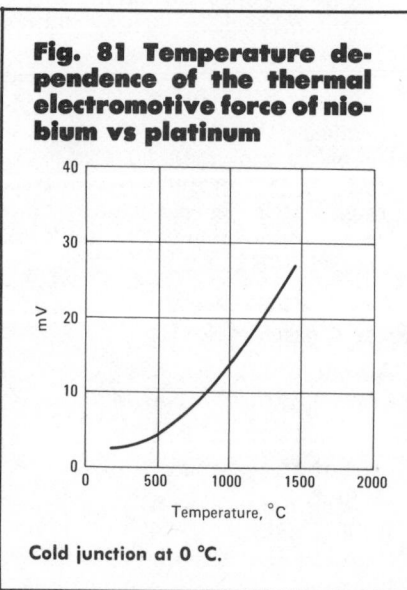

Cold junction at 0 °C.

Fig. 82 Temperature dependence of the emittance of niobium

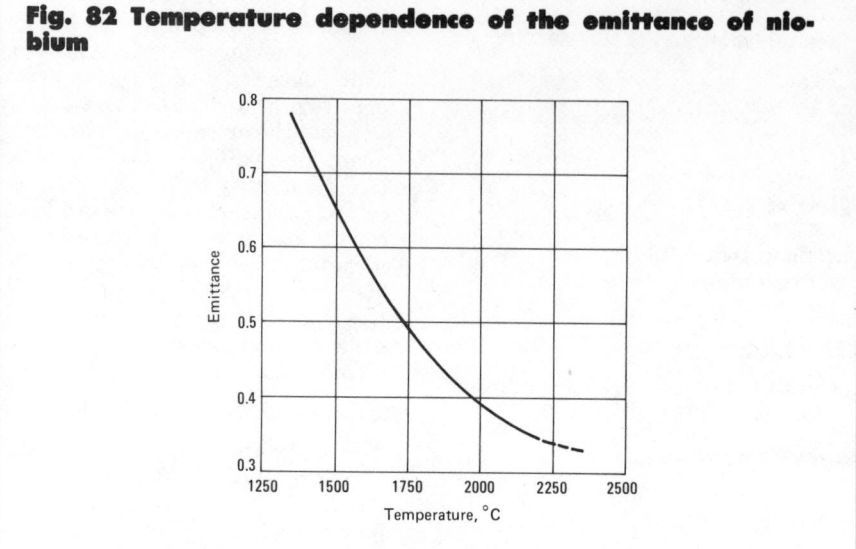

Fig. 83 Temperature dependence of the tensile strength of niobium

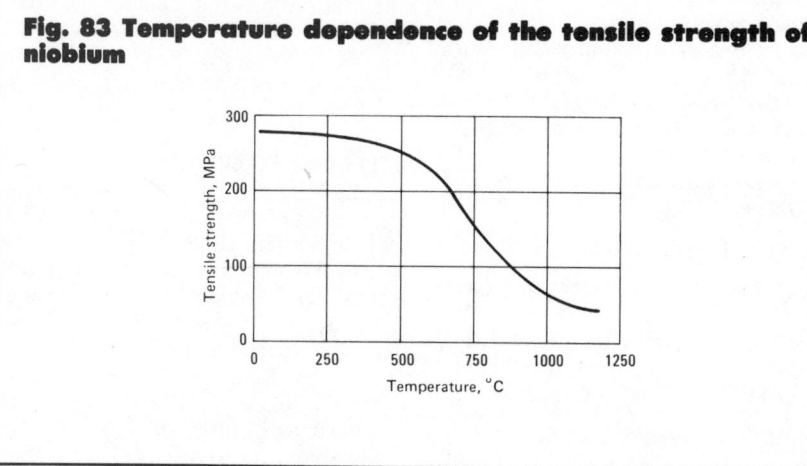

°C	Coefficient, μm/m·K	
	Average(a)	Instantaneous
300........ 7.31		7.38
400........ 7.39		7.54
500........ 7.47		7.61
600........ 7.56		7.87
700........ 7.64		8.03
800........ 7.72		8.20
900........ 7.80		8.37
1000........ 7.88		8.52

(a) Mean value between 18 °C and indicated temperature.

$\Delta l/l_0 = 6.892 \times 10^{-6} T + 8.17 \times 10^{-10}T^2$, where T is in K

Specific heat. See Fig. 79.
Latent heat of fusion. 290 KJ/kg
Latent heat of vaporization. 7490 KJ/kg
Thermal conductivity:

°C	W/m·K
0 .... 52.3
100 .... 54.4
200 .... 56.5
300 .... 58.6
400 .... 60.7
500 .... 63.2
600 .... 65.3

Electrical Properties

Electrical conductivity. At 18 °C: 13.2% IACS
Electrical resistivity. See Fig. 80.
Temperature coefficient. 0 to 600 °C: 0.395 nΩ·m per K
Thermal electromotive force. vs Pt, see Fig. 81.

Table 23 Creep and creep-rupture behavior of niobium at various temperatures

Temperature, °C	Stress for rupture in 1 h, MPa	10-h stress for:				100-h stress for:				1000-h stress for:		
		0.05% creep	0.1% creep	0.2% creep	Rupture	0.05% creep	0.1% creep	0.2% creep	Rupture	0.05% creep	0.1% creep	0.2% creep
Wrought Products												
400............	...	160	276	360	...	83	140	200	...	45	66	107
500............	...	186	214	230	...	121	140	160	...	80	93	110
700............	...	...	...	...	...	...	20	25	...	...	17	12
870............	62	...	...	...	55	...	...	...	48	...	...	...
980............	48	...	...	...	45	...	...	...	42	...	...	...
1200	35.8	...	...	...	32	...	...	...	28	...	...	...

Electrochemical equivalent. 0.1926 mg/C

Hall coefficient. 0 to 900 K, 0.09 nV·m/A·T

Magnetic Properties

Magnetic susceptibility. Volume: at 25 °C: 28×10^{-6} mks

Optical Properties

Refractive index. 1.80

Spectral emittance. $\lambda = 650$ nm; see also Fig. 82.

Nuclear Properties

Thermal neutron cross section. At 2200 m/s: 1.1 b

Chemical Properties

General corrosion behavior. Niobium is moderately to highly resistant to corrosion in most aqueous mediums that are usually considered highly corrosive, such as dilute mineral acids, organic acids and organic liquids. Notable exceptions are dilute strong alkalis, hot concentrated mineral acids, and hydrofluoric acid, which attack the metal rapidly. Gaseous atmospheres at high temperature attack niobium rapidly, primarily by oxidation, although oxygen contents may be very low. Niobium and its alloys are remarkably resistant to corrosion by certain liquid metals, notably lithium and NaK, and to high temperatures (900 to 1010 °C); this coupled with low-capture cross section for thermal neutrons renders niobium materials most attractive for reactor application.

Mechanical Properties

Tensile properties. Highly dependent on purity, particularly the content of interstitial elements. Values listed are for material of good commercial purity (only 100 to 200 ppm interstitial contaminants).

Wrought: tensile strength, 585 MPa; elongation, 5%.

Annealed: tensile strength, 275 MPa; yield strength, 207 MPa; elongation, 30%; reduction in area, 80%. See also Fig. 83.

Hardness. Annealed: 80 HV. Wrought: 160 HV

Poisson's ratio. At 25 °C: 0.38

Strain-hardening exponent. 0.24, similar to low-carbon steel

Elastic modulus. At 25 °C: tension, 103 GPa; shear, 37.5 GPa. At 870 °C: tension, 90 GPa.

Ductile-to-brittle transition temperature. < 147 K; increases sharply with lower purity

Creep-rupture characteristics. See Table 23.

Fabrication Characteristics

Alloying practice. High-vacuum powder metallurgy techniques may be utilized effectively. Consumable-electrode vacuum arc melting and electron-beam furnace melting may be used for alloying purposes. High-vacuum techniques purify niobium at temperatures above 1980 °C, through volatilization of NbO_2.

Precautions in melting. Exclude atmospheric contaminants as completely as possible. Cold hearth techniques are required to prevent crucible reaction.

Recrystallization temperature. Material cold reduced 70 to 80% completely recrystallizes in 1 h at 1090 °C

Hot working temperature. 800 to 1100 °C may be necessary to break down the ingot structure of columbium. This requires conditioning of the breakdown product to remove the contaminated surface layer. Subsequent working is done cold.

Maximum reduction between anneals. Virtually unlimited.

Precautions in forming. Because of the high probability of seizure and galling, selection of lubricant and die material is important in extreme-pressure methods. Carbon tetrachloride (for machining) or sulfonated tallow or proprietary waxes (for spinning and drawing) are preferred lubricants. Polished aluminum bronze has been recommended as a die material for extreme-pressure processes.

Suitable joining methods. Welding processes capable of excluding interstitial contaminants from the hot zone are satisfactory.

Osmium (Os)

Compiled by H. J. Albert
Engelhard Industries, Inc.
Reviewed for this volume by
J. A. Bard
Matthey Bishop, Inc.

Osmium and its alloys are useful for their hardness and resistance to wear and corrosion. The resistance to rubbing wear is more than would be expected on the basis of hardness; alloys of equal hardness are inferior to osmium. Osmium is used in fountain nibs, phonograph needles, electrical contacts, and in instrument pivots. Osmium should not be heated in the presence of oxygen, since it has a toxic oxide OsO_4 that boils off at 130 °C

Structure

Crystal structure. Close-packed hexagonal, $a = 0.27341$ nm and $c = 0.43197$ nm at 26 °C (Ref 1). The space group is $D^4_{6h} - C6/mmc$ (Ref 2)

Mass Characteristics

Atomic weight. 190.2

Density. At 26 °C, calculated from

lattice constants: 22.583 Mg/m³ (Ref 1); 22.57 Mg/m³ obtained directly on an arc-melted button of osmium

Thermal Properties

Melting point. Approx 2700 °C
Boiling point. Approx 5500 °C (Ref 3)
Coefficient of thermal expansion. 3.2 μm/m·K at 50 °C parallel to c axis; 2.2 μm/m·K at 50 °C parallel to a axis (Ref 4); mean value, 2.6 μm/m·K at 50 °C
Specific heat. At 0 °C, 0.12973 kJ/kg·K; at 100 °C, 0.131 kJ/kg·K; at 1600 °C, 0.161 kJ/kg·K. From 25 to 2727 °C, $c_p = 0.125 + 0.0190\ T$, where T is in K and c_p is in kJ/kg·K (Ref 5)

Electrical Properties

Electrical resistivity. Approx 95 nΩ·m at 20 °C (Ref 6); temperature coefficient, 0 to 100 °C: 0.0042 nΩ·m per K (Ref 7)

Magnetic Properties

Magnetic susceptibility. Approx 0.93×10^{-6} mks (Ref 8)

Chemical Properties

Resistance to specific corroding agents. Easily oxidized and forms a tetroxide boiling at 130 °C. It is rapidly attacked by HNO_3 and aqua regia at room temperature. It is not attacked by H_2SO_4 at room temperature or 100 °C, nor by 36% HCl or 40% HF at 20 °C (Ref 9)

Fabrication Characteristics

Working data. Completely unworkable. May be shaped by melting, powder metallurgy and grinding

Mechanical Properties

Hardness. Approx 800 HV, arc-melted button
Elastic modulus. 560 GPa (est) (Ref 10)

REFERENCES

1. Swanson, Fuyat and Ugrinic, *Standard X-ray Diffraction Powder Patterns*, National Bureau of Standards Circular 539, Vol 4, p 8-9
2. Barth and Lunde, *Zeitschrift für Physikalische Chemie*, Vol 121, 1926, p 78-102
3. D. Richardson, *Spectroscopy in Science and Industry*, Wiley and Sons, NY, 1938, p 64
4. Owen and Roberts, *Zeitschrift für Kristallographie, Kristallgeometrie, Kristallphysik, Kristallchemie*, Vol 69A, 1937, p 497-498
5. Jaeger and Rosenbohm, *Proceedings of the Academy of Sciences of Amsterdam*, Vol 34, 1931, p 85
6. Blau, *Elektrotechnische Zeitschrift*, Vol 25, 1905, p 198
7. Lombardi, *Elektrotechnische Zeitschrift*, Vol 25, 1902, p 42
8. Honda and Sone, *Science Reports of the Tohoku Imperial University*, Vol 2, 1913, p 26
9. Wise, *Corrosion Handbook*, Wiley and Sons, NY, 1948, p 311-312
10. W. Koster, *Zeitschrift für Electrochemie*, Vol 49, 1943, p 233

Palladium (Pd)

Compiled by E. M. Wise and R. F. Vines
The International Nickel Co., Inc.
Reviewed for this edition by
J. A. Bard
Matthey Bishop, Inc.

Structure

Crystal structure. Face-centered cubic: $a = 0.38902$ nm at 20 C (Ref 1)

Mass Characteristics

Atomic weight. 106.4
Density. 12.02 Mg/m³ at 20 °C (Ref 2)

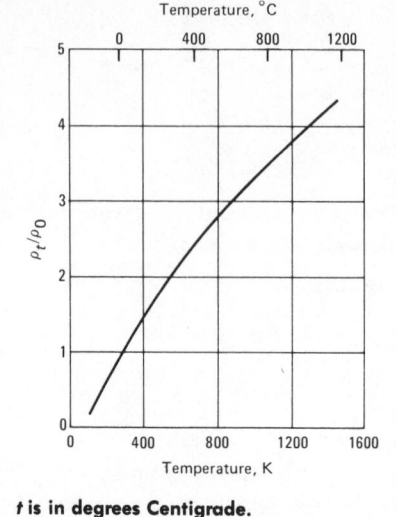

Fig. 84 Ratio of the electrical resistivity of palladium at various temperatures to the resistivity at 0 °C

t is in degrees Centigrade.

Thermal Properties

Melting point. 1552 °C (Ref 3, 4, 5)
Boiling point. Approx 3980 °C
Coefficient of thermal expansion. 11.76 μm/m·K at 20 °C; $L_t = L_0 (1 + 1.167 \times 10^{-5}t + 2.187 \times 10^{-9}t^2)$ where t is in °C (Ref 6)
Specific heat. 245 J/kg·K at 0 °C; 296 and 311 J/kg·K at 1000 °C (Ref 7, 8)

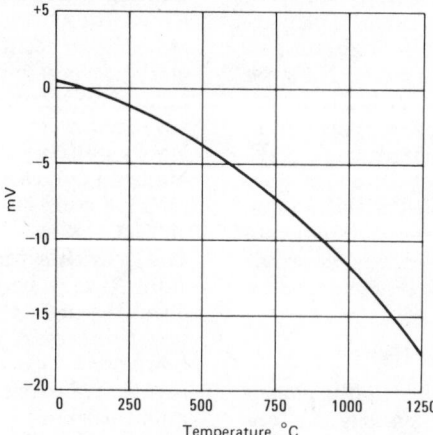

Fig. 85 Temperature dependence of the thermal electromotive force of palladium vs platinum (Ref 11)

Cold junction at 0 °C.

Thermal conductivity. 70 W/m·K at 18 °C (Ref 9)

Vapor pressure. At 1000 °C, 1.53 × 10^{-3} Pa; at 1500 °C, 8.23 Pa; at 1554 °C, 15.7 Pa

Electrical Properties

Electrical conductivity. 16% IACS at 20 °C

Electrical resistivity. 108 nΩ·m at 20 °C; 100 nΩ·m at 0 °C. Temperature coefficient: 0.00377 per K (Ref 10). See also Fig. 84.

Thermal electromotive force. See Fig. 85.

Magnetic Properties

Magnetic susceptibility. Mass, at 18 °C: approx 7.3 × 10^{-8} mks

Optical Properties

Reflectance. 62.8% in white light; increases slightly in going from blue to red

Emittance. Solid, 0.33 and liquid, 0.37 at λ = 0.65 μm (Ref 13)

Chemical Properties

General corrosion behavior. At room temperature palladium is resistant to corrosion by hydrofluoric, perchloric, phosphoric and acetic acids. It is attacked slightly by sulfuric, hydrochloric and hydrobromic acids, especially in the presence of air; and it is attacked readily by nitric acid, ferric chloride, hypochlorites, and moist chlorine, bromine and iodine. In ordinary atmospheres palladium is resistant to tarnish, but some discoloration may occur during exposure to moist industrial atmospheres that contain sulfur dioxide. Adding palladium to gold or silver alloys improves the tarnish resistance.

REFERENCES

1. C. S. Barrett, *Structure of Metals*, McGraw-Hill, NY 1952
2. E. A. Owen and E. L. Yates, *Philosophical Magazine*, Vol 15, 1933, p 472
3. C. O. Fairchild, W. H. Hoover and M. F. Peters, *Journal of Research of the National Bureau of Standards*, Vol 2, 1929, p 931
4. F. H. Schofield, *Proceedings of the Royal Society* (London), Section A, Vol 155, 1936, p 301
5. L. D. Morris and S. R. Scholes, *Journal of the American Ceramic Society*, Vol 18, 1935, p 359
6. L. Holborn and A. L. Day, *Annalen der Physik*, Vol 4, 1901, p 104
7. F. M. Jaeger and W. E. Veenstra, *Proceedings of the Academy of Sciences of Amsterdam*, Vol 37, 1934, p 280
8. H. Holtzmann, "Festschrift 50 jahriger", *Siebert G.M.B.H.*, Hanau, Germany, 1931, p 147
9. W. Jaeger and H. Diesselhorst, *Wissenschaftliche Abhandlungen der Physikalisch-Technischen Reichsanstalt*, Vol 3, No. 269, 1900, p 415
10. R. F. Vines and E. M. Wise, *"Platinum Metals and Their Alloys"*, International Nickel Company, 1941
11. L. Holborn and A. L. Day, *Sitzungsberichte der Akademie der Wissenschaften* in Berlin, 1899, p 694; and Annalen der Physik, Vol 2, 1900, p 505
12. M. Auwater, *Zeitschrift fur Technische Physik*, Vol 18, 1927, p 457
13. W. F. Roeser and H. T. Wensel, *Temperature—Its Measurement and Control in Science and Industry*, Reinhold Publishing Corporation, NY, 1941, p 1293

Platinum (Pt)

Compiled by Edward D. Zysk
Engelhard Minerals & Chemicals Corp.

Structure

Crystal structure. Face-centered cubic, a = 0.39231 nm at 25 °C (Ref 1)

Mass Characteristics

Atomic weight. 195.09

Density. 21.45 Mg/m³ at 20 °C, calculated from lattice parameter (Ref 2). 21.46 Mg/m³ at 25 °C (measured)

Thermal Properties

Melting point. 1769 °C (Ref 3)
Boiling point. 3800 °C (Ref 4)
Coefficient of thermal expansion. Linear: 9.1 μm/m·K from 20 to 100 °C (Ref 5)
Specific heat. 132 J/kg·K at 0 °C (Ref 6)
Latent heat of fusion. 113 kJ/kg
Thermal conductivity. 71.1 W/m·K at 0 °C (Ref 7)

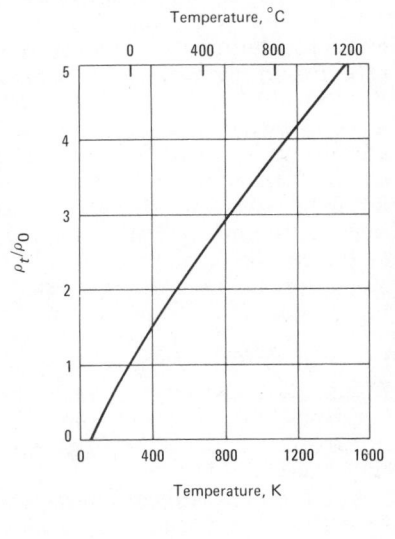

Fig. 86 Ratio of the electrical resistivity of platinum at various temperatures to the resistivity at 0 °C (Ref 20)

t is in degrees Centigrade.

Vapor pressure. For 1377 to 1767 °C: $\rho = 11.767 - 27\,575/t$, where t is in °C and ρ is in Pa (Ref 4, 8). See also Ref 5.

Oxide evaporation rate. Platinum has exceptional resistance to oxidation. Upon heating in air at all temperatures to the melting point, it remains untarnished. At temperatures above about 750 °C, slight weight loss occurs due to volatilization of the metal itself and the formation of a volatile oxide of platinum. This oxide is essentially PtO_2, although the presence of both PtO_2 and PtO has been noted (Ref 9); mass spectrometric techniques indicate that the main oxide molecules are PtO_2 and PtO_3 (Ref 10). As higher temperatures are reached, loss to volatilization of the oxide species becomes greater than that of the metal. In determining weight loss of platinum at elevated temperatures, it is important that some consideration be given not only to the kinetic aspects of oxide formation, but also to the equilibrium between the metal and oxide species. The loss due to oxide formation is influenced by such factors as oxygen pressure, rate of gas flow over the metal surface, degree of saturation of the surrounding area with the oxide species and the geometry of the system. Should the evaporating species be removed, the evaporation rate is increased (Ref 11). Equilibrium vapor pressures of the six platinum group metal oxides are dealt with in Ref 12; rate of evaporation of platinum in vacuum in low pressure oxygen or air are covered in Ref 13, 14, 15; air only in Ref 16, 17, 18; air, nitrogen, argon, hydrogen and oxygen in Ref 19. Care should be taken in applying research data relating to oxide vaporization, as indicated in the cited references, to actual commercial applications because operating conditions may not be similar to those in the experiments.

Electrical Properties

Electrical resistivity. 98.5 nΩ·m at 0 °C; 106 nΩ·m at 20 °C. Temperature coefficient: 0.003927 K from 0 to 100 °C; see also Fig. 86.

A laboratory standard platinum resistance thermometer is the standard instrument used on the International Practical Temperature Scale of 1968 from 13.81 K to 630.74 °C. The thermometer resistor must be strain free, annealed, pure platinum (at least 99.998% Pt) to achieve a mandated temperature coefficient of at least 0.03925 nΩ·m per K. Below 0 °C, the resistance temperature relation of the thermometer is determined by reference function and specified deviation equations. From 0 to 630.74 °C, two polynomial equations provide the resistance temperature relation (see Ref 3). For data on the effect of trace impurities on the temperature coefficient of electrical resistivity, see Ref 21.

Thermal electromotive force. Very pure platinum, U. S. Thermometric Standard Pt 67 (supplied as a standard reference material by the U. S. National Bureau of Standards), is used as the reference electrode in comparing the thermoelectric behavior of individual metals and alloys. Measurements of the thermal emf of samples of platinum (with the "joined" hot end at 1200 °C and the two free ends at 0 °C) against Pt 67 are useful for estimating the purity of different lots of platinum. Small amounts of impurities (except gold) in solid solution in platinum make the slightly impure platinum thermoelectrically positive to the plati-

num without the additions. Iron is troublesome in this respect. For data on the effect of trace impurities on the thermal emf of platinum see Ref 21.

Work function. 0.852 to 0.876 aJ (5.32 to 5.47 eV) (Ref 22)

Magnetic Properties

Magnetic susceptibility. Mass: 0.012204 mks (Ref 23)

Optical Properties

Color. Silver white
Reflectance. Bulk: 70% at 589 nm. Electrodeposited: 58.4% at 441 nm; 59.1% at 589 nm; 59.4% at 668 nm (Ref 24, 25, 26)
Spectral emittance. At 650 nm: solid, 0.30; liquid, 0.38 (Ref 27)
Total hemispherical emittance. From Ref 27:

°C	Emittance
25	0.037
100	0.047
500	0.096
1000	0.152
1500	0.191

Chemical Properties

General corrosion behavior. See the article in this volume entitled "Corrosion Resistance of Precious Metals". See also Ref 28.
Resistance to specific corroding agents. Platinum is resistant to ferric chloride at room temperature. Hydrobromic acid plus bromine attacks it at room temperature. All of the free halogens attack platinum at elevated temperatures but hydrochloric acid in the absence of oxidizing agents does not attack it and platinum is useful against this normally active gas up to 2000 °F. Sulfur dioxide does not attack platinum even at 2000 °F (Ref 29)

As an anode, platinum is outstanding and is used commercially in sulfuric and persulfuric acids, various sulfate-chloride plating electrolytes, and in chlorates with very little corrosion. If electrolyzed with alternating current, chlorides may attack it, and this is a characteristic exploited in etching platinum and platinum alloys.

Platinum is quite resistant to acid potassium sulfate, sodium carbonate, potassium nitrate at moderate temperatures and to sodium carbonate at 1475 to 1650 °F under nonoxidizing

conditions. Although it is attacked vigorously by molten alkali cyanides and polysulfides, it is quite resistant to the normal sulfides plus alkali. Certain phosphates attack it at high temperatures and care must be taken to avoid reducing conditions, particularly when compounds of arsenic, phosphorus, tin, lead or iron are present. Platinum is resistant to molten glasses, especially to those low in lead and arsenic.

Platinum, even in the form of thin leaf, is resistant to corrosion and tarnishing on exposure to the atmosphere, including urban sulfur-bearing atmospheres

REFERENCES

1. H. E. Swanson and E. Tatge, *National Bureau of Standards Circular 539*, 1953
2. H. E. Swanson and E. Tatge, *National Bureau of Standards Circular 539*, U. S. Dept. of Commerce, National Bureau of Standards, Washington, D. C., Vol 1, 1953, p 21, 31, 531
3. The International Practical Temperature Scale of 1968 Amended Edition of 1975, *Metrologia*, Vol 12, 1976, p 7-17
4. R. F. Hampson, Jr. and R. F. Walker, *Journal of Research of the National Bureau of Standards*, Vol 65A, 1961, p 289
5. P. Hidnert and W. Sander, *NBS Circular 486*, U. S. Dept. of Commerce, National Bureau of Standards, Washington, D. C., 1950
6. F. N. Jaeger and E. Rosenbohm, *Physics*, Vol 6, 1939, p 1123
7. R. W. Powell, R. P. Tye and M. J. Woodman, *Platinum Metals Review*, Vol 6, 1962, p 138
8. L. H. Dreger and J. L. Margrave, *Journal of Physical Chemistry*, Vol 64, 1960, p 1323
9. J. H. Norman, H. G. Staley and W. E. Bell, *Journal of Physical Chemistry*, Vol 71, 1967, p 3886
10. A. Olivei, *Journal of Less Common Metals*, Vol 29, 1972, p 18
11. G. C. Fryburg and H. M. Petrus, *Journal of the Electrochemical Society*, Vol 108, 1961; p 496
12. C. B. Alcock and G. W. Hooper, *Proceedings of the Royal Society (A)*, Vol 254, 1960, p 557
13. G. K. Burgess and R. G. Waltenberg, National Bureau of Standards Scientific Paper 254 (1916)
14. T. Kubaschewski, Zeitschrift fur Electrochemie, Vol 49, 1943, p 446

15. E. K. Rideal and O. H. Wansborough-Jones, *Proceedings of the Royal Society (A)*, 1929, Vol 123, p 202 (1933)

16. W. Betteridge and D. W. Rhys, "The High Temperature Oxidation of the Platinum Metals and Their Alloys", in *First International Congress on Metallic Corrosion 1961*, L. Kenworth (ed.), Butterworths, London, 1962, p 185.

17. E. Raub and W. Plate. Zeitschrift fur Metallk unde, Vol 48, 1957, p 529

18. R. W. Douglass, C. A. Krier and R. I. Jaffee. "Summary Report on High Temperature Properties and Alloying Behavior of the Refractory Platinum-Group Metals". NR 039-067, Battelle Memorial Institute, Columbus, OH, August 1961

19. J. S. Hill and H. J. Albert, Engelhard Industries Technical Bulletin, Vol 4, No. 2, 1963, p 59-63

20. W. P. Roeser and H. T. Wensel, in *Temperature, Its Measurement and Control in Science and Industry*, Reinhold Publishing Corp., NY, 1941, p 1312

21. J. Cochrane, Englehard Industries Technical Bulletin, 1970, Vol XI, No. 2, p 58-73

22. A. Ertel, Physical Review, Vol 78, 1950, p 353

23. F. E. Hoare and J. C. Walling, Proceedings of the Physical Society, Section B, Vol 164, 1951, p 337

24. P. Drude, Annalen der Physik, Vol 39, 1890, p 481

25. W. Meier, Annalen der Physik, Vol 31, 1910, p 1017

26. G. Hass and L. Hadley, "Optical Properties of Metals" in *American Institute of Physics Handbook,* 2nd ed., NY, 1965, p 6-107 to 6-118.

27. W. F. Roeser and W. T. Wensel in *Temperature, Its Measurement and Control in Science and Industry*, Reinhold, NY, 1941, p 1313-1314

28. *Corrosion Handbook,* John Wiley & Sons, 1948

29. E. M. Wise and J. T. Eash, *Transactions of AIME,* Vol 128, 1938, p 282

Plutonium (Pu)

Compiled by M. B. Brodsky
Argonne National Laboratory

The term plutonium usually implies [239]Pu of at least 95% purity (generally 99.7 to 99.99 wt%). Small amounts of δ-phase stabilizers, such as 0.1 wt% Al, may cause retention of the δ-phase at room temperature. The term plutonium, however, also implies [239]Pu sufficiently free of δ-phase stabilizers so that only the α-phase is present at room temperature.

Typical uses of plutonium include nuclear weapons, nuclear fuel, neutron sources, heat sources for thermoelectric generators, production of higher isotopes and transplutonic elements. Plutonium is a highly radioactive alpha emitter, extremely poisonous, and is properly handled in gloveboxes. It is about twice as poisonous as radium to the human system. The maximum permissible body burden is 0.6 μg. When handling quantities in excess of 300 g, the possibility of nuclear criticality must be considered.

Structure

Crystal structure. See Table 24 (Ref 1, 4, 5)

Slip planes. Alpha phase ($\bar{1}02$), (112), (111), ($10\bar{1}$), ($\bar{4}11$) and (118) (Ref 3)

Minimum interatomic distance. (Ref 1, 6):

Phase	Minimum distance, nm
α	0.25
β	0.297
γ	0.3026
δ	0.3279
δ'	0.3249
ε	0.3149

Metallography. Standard grinding procedures using kerosene or carbon tetrachloride as a lubricant. Polishing is done on microcloth charged with gamma alumina or 1 μm diamond. Suitable etches contain tetraphosphoric acid, water and 2-ethoxyethanol in the following proportions: 7:36:57 for low temperature phases; 12:33:55 for delta phase; 2:3:5 for long etching. Tetraphosphoric acid may be replaced by orthophosphoric acid and less water.

Fracture behavior. Although the metal exhibits some toughness, it fails by brittle fracture

Mass Characteristics

Atomic weight. 239.052
Density. See Table 24.
Compressibility. Alpha plutonium: approximately 0.2 per Pa at atmospheric pressure to 0.05 per Pa at 10 GPa; the volume at 10 GPa is about 90% of the volume at atmospheric pressure (Ref 1, 5). Beta plutonium, at 200 °C: 0.23 ± 0.10/Pa in the range 0 to 200 MPa

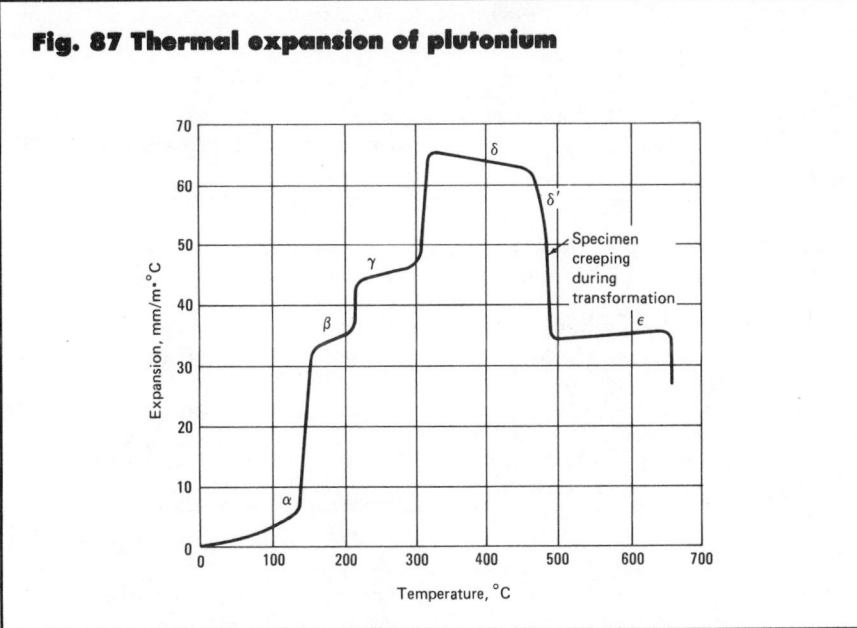

Fig. 87 Thermal expansion of plutonium

Table 24 Crystal structure and density of various phases of plutonium

Phase	Lattice symmetry	a_0, nm	Lattice constants b_0, nm	c_0, nm	β (deg)	Density, Mg/m³	Atoms per unit cell
α....	Monoclinic	0.6183(21 °C)	0.4822	1.0963	101.79	19.86	16
β....	Monoclinic	0.9284(190 °C)	1.0463	0.7859	92.13	17.70	34
γ....	Orthorhombic	0.3159(235 °C)	0.5768	1.0162	...	17.14	8
δ....	Face-centered cubic	0.46371(320 °C)	...	...	...	15.92	4
δ'....	Body-centered tetragonal	0.334(465 °C)	...	0.444	...	16.00	2
ε....	Body-centered cubic	0.3636(490 °C)	...	...	...	16.51	2

Table 26 Coefficient of linear thermal expansion for plutonium

Phase	From dilatometric data Coefficient, μm/m·K	Temperature, °C	From x-ray data Coefficient, μm/m·K		Temperature, °C
α	67	80-120	α_1 perp. to c axis	66	21-104
			α_2 parallel to b axis	73	
			α_3 parallel to c axis	29	
			Average	56	
β	41	160-200	α_1	94	93-190
			α_2 parallel to b axis	14	
			α_3 perp. to (101)	18	
			Average	42	
γ	35	220-280	α_1 parallel to a axis	-19.7 ± 1.0	210-310
			α_2 parallel to b axis	39.5 ± 0.6	
			α_3 parallel to c axis	84.3 ± 1.6	
			Average	34.7 ± 0.7	
δ	-8.6	340-440	α	-8.6 ± 0.3	320-420
δ'	-596(a)	470	α_1 parallel to a axis	444.8 ± 12.1	450-479
			α_2 parallel to c axis	-1063.5 ± 18.2	
			Average	-57.9 ± 10.1	
ε	15	490-550	α	36.5 ± 1.1	490-550
Liquid	50 ± 25(b) at 665		...		...

(a) Value exaggerated by creep during residual transformation. (b) Volumetric.

Table 25 Transformation properties of plutonium (Ref 1, 5, 7)

Transformation	Temperature, °C	Volume change, %	Heat of transformation, kJ/kg
α to β....	120	9	17.6
β to γ....	210	2.5	2.6
γ to δ....	315	6.9	2.5
δ to δ'....	452	-0.4	0.4
δ' to ε....	480	-2	7.4
ε to liquid...	640	-1 to -2	13.1

Electrochemical equivalent. Valence 3, 0.8256 mg/C; valence 4, 0.6142 mg/C

Hall coefficient. $+35$ pV·m/A·T at room temperature (Ref 3)

Superconductivity. None found at 1.3 K in metal which is 99.99% pure

Magnetic Properties

No magnetic ordering has been found in any of the phases of plutonium, whether pure or alloyed. However, compounds of plutonium are often magnetic, especially when the Pu-Pu distance increased beyond 0.34 nm

Magnetic susceptibility. See Table 29 (Ref 1, 5)

Optical Properties

Color. White. When slightly oxidized, yellow tarnish; when heavily oxidized, green black

Nuclear Properties

Unstable isotopes. See Table 30.
Thermal neutron cross section. See Table 30.

Chemical Properties

General corrosion behavior. Plutonium is a highly reactive metal, similar in reactivity to the rare earths

Resistance to specific corroding agents. Relatively inert to dry air but corrodes rapidly if traces of moisture are present (Ref 1). Reacts slowly with water at room temperature

Fabrication Characteristics

Machinability. Similar to 3003 Al
Recrystallization temperature. Approximately 120 °C
Casting temperature. In vacuum: 800 to 900 °C
Alloying practice. Low melting elements are commonly added as pure metals to the molten bath. Alloys of

Volume change on transformation. See Table 25.

Thermal Properties

Melting point. 640 °C
Boiling point. 3235 °C (Ref 5)
Phase-transformation temperature. See Table 25.
Coefficient of thermal expansion. See Table 26 and Fig. 87 (Ref 6)
Specific heat. α-phase, 33.9 kJ/kg·K at 25 °C; β-phase, 41.4 kJ/kg·K at 160 °C; γ-phase, 46.0 kJ/kg·K at 280 °C; δ-phase, 45.6 kJ/kg·K at 350 °C; δ'-phase, 55.3 kJ/kg·K at 455 °C; ε-phase, 43.5 kJ/kg·K at 500 °C (Ref 1)
Latent heat of fusion. See Table 25.
Latent heat of phase transformation. See Table 25.
Latent heat of vaporization. 336.9 kJ/mol·
Latent heat of combustion. 1058.7 kJ/mol Pu
Recovery temperature. 109 °C
Thermal conductivity. 6.5 W/m·K at room temperature
Vapor pressure. $\log P = -17587/T + 10.02$, where P is in Pa and T is in K (Ref 5)
Chemical diffusion. See Table 27 (Ref 3). Self-diffusion coefficient, ε-phase: 1.2×10^{-7} cm²/s at 500 °C

Electrical Properties

Electrical resistivity. See Table 28 (Ref 1, 5)
Thermodynamic potential. See Table 28 (Ref 2, 6)

Table 27 Chemical diffusion of plutonium

Composition, at. % Pu	Diffusion, coefficient, cm²/s	Activation energy, kJ/mol
Mg-Pu System		
0.045	1×10^4	150
0.562	2.45×10^{-2}	118.6
1.124	1.05×10^{-2}	118.5
1.686	3.6×10^{-4}	93.70
Pu-Zn System (δ-Pu)		
38.5	7.70×10^{-6}	98.39
46.2	3.56×10^{-7}	63.01
61.6	8.86×10^{-9}	49.40
69.3	9.60×10^{-9}	52.96
U-Pu System		
1.75	0.14×10^{-7}	56.1
3.50	0.15×10^{-7}	57.4
5.25	0.18×10^{-7}	59.0
7.00	0.28×10^{-7}	63.6
8.75	0.44×10^{-7}	68.2
10.50	0.88×10^{-7}	74.9
12.25	1.18×10^{-7}	78.7
14.00	2.00×10^{-7}	83.7
15.75	2.57×10^{-7}	86.2

Table 28 Electrical properties of plutonium

Phase	Temperature, °C	Electrical resistivity uΩ·cm	Temperature coefficient of resistivity, 10^4/°C	Thermoelectric potential vs Pt(a) mV
α	−223	128.0	+184.05	...
	107	141.4	−2.08	1.44
β	147	108.5	−0.62	2.23
γ	232	107.8	−0.50	3.81
δ	352	100.4	+0.72	5.92
δ'	462	102.1	+4.43	7.63
ε	501	110.6	0.00	8.31

(a) Reference junction at 20 °C.

Table 29 Magnetic susceptibility of plutonium

Allotrope	Temperature, °C	Mass susceptibility, emu/g	Temperature range, °C	Mean temperature coefficient
α	20	0.0280	20-118	-1.8×10^{-5}
β	132	0.0290	132-198	-16.4×10^{-5}
γ	224	0.0280	224-302	-11.5×10^{-5}
δ	358	0.0268	358-446	-12.3×10^{-5}
δ'	464	0.0266	464-477	$+36.3 \times 10^{-5}$
ε	488	0.0270	488-570	-12.5×10^{-5}

Table 30 Nuclear properties of plutonium

Isotope	Half-life, yr	Emitted particles	Cross sections, Capture	10^{-24} cm² Fission
238	86.4	α(5.49, 4.45 MeV), γ	403	16.8
239	2.4×10^4	α(5.15 MeV), γ	315	746
240	6.6×10^3	α(5.16 MeV), γ	250	0.03
241	13.2	β⁻(0.021 MeV), γ	390	1010
242	3.8×10^5	α(4.90 MeV), γ	19	<0.2

refractory metals are added as master alloys

Deoxidizers. Cerium has been used in deoxidizing plutonium. Plutonium melts have been made under potassium chloride-sodium chloride covers

Melting practice. High-vacuum furnaces are commonly used to melt plutonium alloys. Magnesia and coated graphite crucibles are used to 1200 °C, and thoria crucibles are used to 1500 °C. Tantalum can be used to 1000 °C to contain molten plutonium. Magnesia, graphite and copper are suitable mold materials.

Hot working temperature. Can be worked readily in the delta (fcc) temperature range, 312 to 458 °C. Beta plutonium is ductile and can be worked (Ref 7)

Heat treatment. Plutonium is given a cold treatment at −23 °C to complete the beta to alpha transformation (Ref 1)

Mechanical Properties

Mechanical properties depend heavily on microstructure, and are especially sensitive to the presence of microcracks caused by the large volume change associated with the β-to-α phase transformation

Tensile properties. Typical for cast alpha at 25 °C (Ref 1, 2, 3): tensile strength, 415 MPa; yield strength, 275 MPa; elongation, 0.2 to 0.5%; proportional limit, 160 MPa

Compressive properties. Typical for cast alpha at 25 °C (Ref 1, 2, 3): compressive strength, 830 MPa; compressive yield strength, 415 MPa

Hardness. 250 to 283 HV, 10-kg load (Ref 3)

Poisson's ratio. 0.15 to 0.21

Elastic modulus. Tension, 107 GPa; shear, 45 GPa

Fatigue strength. Typical, rotating beam: 90 MPa at 10^8 cycles

Liquid surface tension. 0.5 N/m

Viscosity. Dynamic, molten Pu: 7.4 mPa·s at 650 °C; 6.2 mPa·s at 750 °C

REFERENCES

1. W. N. Miner et al, Plutonium, in *Rare Earth Metals Handbook,* 2nd ed., Reinhold, NY, 1961
2. E. Grison and W. P. H. Lord, Eds., *Second International Conference on Plutonium Metallurgy,* Cleaver-Hume Press, London, 1960
3. J. H. Kittel et al, Plutonium and Plutonium Alloys as Nuclear Fuel Materials, in *Nuclear Design and Engineering.* C. F. Bonilla and T. A. Jaegger, Eds., North Holland Publishing, Amsterdam, 1971
4. A. S. Coffinberry and W. N. Miner, Eds., *The Metal Plutonium,* American Society for Metals, Cleveland, 1961
5. A. S. Coffinberry and M. B. Waldron, Eds., The Physical Metallurgy of Plutonium, *Progress in Nuclear Energy,* Vol I, Series V, Pergamon Press, London, 1956
6. E. L. Francis, *Plutonium Data Manual,* I.G.R. 161 (RG/R), Industrial Group Headquarters, Risley, Warrington, Lancashire, 1959
7. W. D. Wilkinson, Ed., *Extractive and Physical Metallurgy of Plutonium and Its Alloys,* Interscience Publishers, NY, 1960

Potassium (K)

Compiled by J. R. Keiser
Oak Ridge National Laboratory

Few uses have been found for potassium metal, though an alloy of sodium and potassium is used as a heat transfer medium. Because potassium is an essential element in plant growth, potassium or K_2O is a main component of plant fertilizer. Potassium is also used as the super oxide, KO_2, to produce oxygen in gas masks. Potassium is highly reactive and must be handled with great care. Use of dry and oxygen-free inert gas atmosphere is essential if reactions are to be avoided.

Structure

Crystal structure. Body-centered cubic, type $A2$; $a = 0.5344$ nm at 20 °C

Mass Characteristics

Atomic weight. 39.09
Density. (Ref 1)

°C	Mg/m³
−273	0.909
−173	0.894
−73	0.873
+20	0.855
100	0.820
200	0.797
300	0.774
400	0.751
600	0.702
800	0.653
1000	0.602

Volume change on freezing. 2.41% contraction (Ref 2)

Thermal Properties

Melting point. 63.2 °C (Ref 1)
Boiling point. 756.5 °C (Ref 1)
Coefficient of thermal expansion. Linear: 83 µm/m·K for 0 to 95 °C; volumetric, (liquid): $V/V_0 = 1 + 2.58 \times 10^{-4}T + 13.08 \times 10^{-8}T^2 + 1.98 \times 10^{-12}T^3$ for 63.2 to 1250 °C, where T is in °C (Ref 1)
Specific heat. 770.45 J/kg·K at 20 °C
Specific heat vs temperature. Solid, from −173 to 63.2 °C: $C_p = 538.07 + 0.8004T$ J/kg·K, where T is in K. Liquid, from 63.2 to 1150 °C: $C_p = 839.14 - 0.3675T + 4.594 \times 10^{-4}T^2$ J/kg·K, where T is in °C (Ref 1)
Enthalpy. Where T is in °C and H_{0c} is the enthalpy of the solid at 0 °C: solid, $H_c - H_{0c} \times 710.62T + 1.0388 T^2$ J/kg; liquid, $H_l - H_{0c} = 56\,178 + 841.01T - 0.1585T^2 + 1.0502 \times 10^{-4}T^3$ J/kg (Ref 1)
Entropy. Where T is in K and S_{0c} is the entropy of the solid at 0 °C: solid, from 0 to 63.2 °C: $S_c - S_{0c} = 329.47 \log T + 2.0776T - 13\,703$ J/kg·K; liquid, from 63.2 to 800 °C: $S_l - S_{0c} = 2189.9 \log T + 0.4864T + 1.5723 \times 10^{-4}T^2 - 50\,481$ J/kg·K (Ref 1)
Latent heat of fusion. 59.45 kJ/kg (Ref 1)
Latent heat of vaporization. 1985 kJ/kg (Ref 1)
Thermal conductivity. 108.3 W/m·K at 293 K (Ref 1)
Thermal conductivity vs temperature. Solid: $k = 1.26 \times 10^2 - 6.03 \times 10^{-2}T$ W/m·K; liquid: $k = 43.8 - 2.22 \times 10^{-2}T + \dfrac{3950}{T + 273.2}$ W/m·K (T in °C) (Ref 1)
Vapor pressure. $\text{Log} \dfrac{p}{p_0} = \dfrac{-4625.3}{T} + 6.59817 - 0.700643 \log T$ (T in K; $p_0 = 101.325$ kPa $= 1$ atm); $p = 3.95$ kPa at 773 K; $p = 6.24$ kPa at 1273 K (Ref 1)

Table 31 Diffusion characteristics of potassium (Ref 3)

Solute	Temperature range, °C	Activation energy, kJ/kg	Frequency factor, mm²/s
¹⁹⁸Au	5.6 to 52.5	345.89	0.129
⁴²K	−52.0 to 61.0	1002.3	16
²²Na	0 to 62.0	797.80	5.8
⁸⁶Rb	0.1 to 59.9	940.21	9.0

Table 32 Unstable isotopes of potassium (Ref 6)

Isotope	Half-life	Decay mode	Energy, pJ
37	1.2 s	β⁻	0.82
38	7.71 min	β⁺, γ	0.43, 0.35
38ₘ	0.95 s	β⁻	0.80
40(a)	1.28 x 10⁹ yr	β⁻ EC, γ	0.21 0.23
42	12.4 h	β⁻, γ	0.56, 0.32
43	22.4 h	β⁻, γ	0. 3, 0.07
44	22.0 min	β⁻, γ	0.42, 0.85, 0.64
45	16 min	β⁻, γ	0.34, 0.18, 0.64

(a) Isotope mass 39.974, 0.0012% abundance.

Diffusion characteristics. See Table 31.

Electrical Properties

Electrical resistivity. Solid: 61 nΩ·m at 0 °C (Ref 4). Liquid: 142.7 nΩ·m at 100 °C; 238.2 nΩ·m at 250 °C; 1096.7 nΩ·m at 1000 °C; or $\rho = 79.898 + 0.6371T - 1.3959 \times 10^{-4}T^2 + 5.3020 \times 10^{-7} T^3$ nΩ·m (T in °C) for 93 to 1093 °C (Ref 1)
Thermoelectric potential. vs Pt: 1.83×10^{-7} V/K at 25 °C; 1.988×10^{-6} V/K at 250 °C; 1.0168×10^{-5} V/K at 800 °C
Electrochemical equivalent. For K^+: 0.4052 mg/C (Ref 1)
Electrolytic solution potential. vs H_2: −2.922 V (Ref 2)
Ionization potential:

Degree of ionization	Potential, V
I	4.339
II	31.81
III	46
IV	60.9
V	82.6
VI	99.7
VII	118
VIII	155

Hall coefficient. −4.9 aV·m/A·T (Ref 1)
Work function. 0.359 aJ (Ref 2)

Magnetic Properties

Magnetic susceptibility. Volume (mks units): at 30 °C, 4.94×10^{-6}; at 100 °C, 4.72×10^{-6}; at 250 °C, 4.61×10^{-6} (Ref 1)

Optical Properties

Color. Silver white
Refractive index. 0.392 for $\lambda = 313$ nm; 0.924 for $\lambda = 134$ nm; 0.964 for $\lambda = 128$ nm (Ref 5)

Nuclear Properties

Stable isotopes. ³⁹K, isotope mass 38.96371, 93.10% abundant; ⁴¹K, 6.88% abundant (Ref 6)
Unstable isotopes. See Table 32.

Chemical Properties

General corrosion behavior. Potassium is a highly reactive metal and consequently is found only in a combined state. It reacts vigorously with water to form the hydroxide and for this reason must be kept in a moisture-free environment. Potassium reacts with many other materials including hydrogen, oxygen, sulfur, nitrogen, bromine and graphite. It also forms alloys with many metals.

Mechanical Properties

Elastic constants. At 295 °C: C_{11}, 3.715 GPa; C_{12}, 3.153 GPa; C_{44}, 1.88 GPa (Ref 7)
Kinematic liquid viscosity. 0.00628 mm²/s at 69.6 °C; 0.00328 mm²/s at 250 °C; 0.00254 mm²/s at 400 °C (Ref 4)
Liquid surface tension. σ (N/m) = 0.1157 − 6.4 × 10^{-5} T (T in °C) (Ref 1)

REFERENCES

1. O. J. Foust (Ed.), *Sodium—NaK Engineering Handbook,* Vol. 1, Gordon & Breach, NY, 1972 p 10-89

2. T. P. Whaley, Sodium, Potassium, Rubidium, Cesium, and Francium, *Comprehensive Inorganic Chemistry,* Pergamon, NY, 1973, p 369-381

3. R. C. Weast (Ed.), *Handbook of Chemistry and Physics,* 55th ed., CRC Press, Cleveland, 1974, p F-65

4. E. A. Schoeld, Potassium, *The Encyclopedia of the Chemical Elements,* Reinhold, NY, 1968, p 552-561

5. J. C. Sutherland and E. T. Arakawa, *Journal of the Optical Society of America,* Vol 58 (No. 8), 1968, p 1080-1083

6. R. C. Weast (Ed.), *Handbook of Chemistry and Physics,* 55th ed., CRC Press, Cleveland, 1974, p B-253-54.

7. S. K. Sangal and P. K. Sharma, *Czechoslovak Journal of Physics,* Vol B19, 1969, p 1098

Praseodymium (Pr)

Compiled by K. A. Gschneidner, Jr. and B. J. Beaudry
Ames Laboratory
U. S. Department of Energy
Iowa State University

Praseodymium is used as an alloying additive to ferrous alloys to scavenge sulfur, oxygen, etc. It is also used as a glass and ceramic coloring agent, and in petroleum cracking catalysts, carbon-arc lights and $PrCo_5$ permanent magnets. Praseodymium oxidizes at room temperature in air. It should be stored in vacuum or inert atmosphere; storage in oil is not recommended. Turnings can be ignited easily and will burn white hot. Finely divided praseodymium should not be handled in air.

Structure

Crystal structure. α-phase, double close-packed hexagonal, $P6_3/mmc$ D^4_{6h}; a = 0.36721 nm, c = 1.18326 nm at 24 °C. β-phase, body-centered cubic, Im^3m O^9_h; a = 0.413 nm at 815 °C
Minimum interatomic distance. At 24 °C: r_a = 0.18360 nm; r_c = 0.18197 nm; radius CN_{12} = 0.18279 nm

Mass Characteristics

Atomic weight. 140.9077
Density. α-phase, 6.773 Mg/m³ at 24 °C; β-phase, 6.64 Mg/m³ at 815 °C; liquid, 6.609 Mg/m³ at 935 °C
Volume change on freezing. 0.02% contraction
Volume change on phase transformation. α- to β-phase, 0.5% volume expansion on heating

Thermal Properties

Melting point. 931 °C
Boiling point. 3520 °C
Phase-transformation temperature. α- to β-phase, 795 °C
Coefficient of thermal expansion. α-phase, at 24 °C. Linear: 6.7 μm/m·K. Linear, along crystal axes: 4.5 μm/m·K along a axis, 11.2 μm/m·K along c axis. Volumetric: 20.2 × 10^{-6} per K
Specific heat. 194.8 J/kg·K at 25 °C
Entropy. At 25 °C, 524.7 J/kg·K
Latent heat of fusion. 48.88 kJ/kg
Latent heat of transformation. 22.48 kJ/kg
Latent heat of vaporization. 2.524 MJ/kg at 25 °C
Heat of combustion. For cubic Pr_6O_{11} at 25 °C: ΔH^0_c = 6.73 MJ/kg Pr; ΔG^0_f = −6.32 MJ/kg
Recrystallization temperature. About 400 °C
Thermal conductivity. 12.5 W/m·K at 25 °C
Vapor pressure. 0.001 Pa at 1083 °C; 0.101 Pa at 1333 °C; 10.1 Pa at 1701 °C; 1013 Pa at 2305 °C

Electrical Properties

Electrical resistivity. 700 nΩ·m at 25 °C; 22 nΩ·m at 4K. Liquid: 1130 nΩ·m at 932 °C
Ionization potential. Pr(I): 5.42 V; Pr(II) 10.55 V; Pr(III): 21.624 V; Pr(IV): 38.98 V; Pr (V): 57.45 V
Hall coefficient. + 0.0709 nV·m/A·T at 25 °C
Temperature of superconductivity. Bulk praseodymium is not superconducting down to 0.25 K at atmospheric pressure

Magnetic Properties

Magnetic susceptibility. Volume: 3.34 × 10^{-3} mks at 25 °C; obeys Curie-Weiss law above 100 K with an effective moment of 3.56 Bohr magnetons and $\theta \cong 0$ K
Saturation magnetization. >0.53 T at 4 K along $\langle 11\bar{2}0 \rangle$
Magnetic transformation temperature. Single crystal, strain free praseodymium does not order magnetically, most polycrystalline samples order at various temperatures below 25 K

Optical Properties

Color. Metallic silver
Spectral hemispherical emittance. 29.4% for λ = 645 nm from 931 to 1225 °C

Nuclear Properties

Thermal neutron cross section. 11 b

Chemical Properties

Corrosion behavior. Praseodymium oxidizes in air at room temperature but at a lower rate than lanthanum or cerium. Oxidation rates increase with temperature. Interstitial impurities in the metal increase the rate of corrosion in air. Hydrogen will react with praseodymium at room temperature.
Resistance to specific corroding agents. Praseodymium reacts vigorously with dilute acids. It reacts slowly with concentrated sulfuric. The presence of the fluoride ion retards acid attack due to the formation of PrF_3 on the surface of the metal

Mechanical Properties

Tensile properties. Approximately the same as neodymium
Hardness. 37 HV
Poisson's ratio. 0.289
Elastic modulus. At 27 °C: tension,

37.9 GPa; shear, 14.7 GPa; bulk, 28.9 GPa

Elastic constants along crystal axes. At 27 °C: c_{11} = 49.35 GPa; c_{12} = 22.95 GPa; c_{13} = 14.3 GPa; c_{33} = 57.40 GPa; c_{44} = 13.59 GPa

Kinematic liquid viscosity. 0.431 mm^2/s at 935 °C

Promethium (Pm)

Compiled by K. A. Gschneidner, Jr. and B. J. Beaudry
Ames Laboratory
U. S. Department of Energy
Iowa State University

Promethium is used as a lightly shielded radioisotope power source. It is also a highly radioactive β emitter (^{147}Pm)

Structure

Crystal structure. α-phase, double close-packed hexagonal, $P6_3/mmc$ D^4_{6h}; a = 0.365 nm, c = 1.165 at 24 °C. β-phase, probably body-centered cubic

Minimum interatomic distance. At 24 °C: r_a = 0.1825 nm; r_c = 0.1797 nm; radius CN_{12} = 0.1811

Mass Characteristics

Atomic weight. 147
Density. 7.264 Mg/m^3 at 24 °C

Thermal Properties

Melting point. 1042 °C
Boiling point. Est, 300 °C
Phase-transformation temperature. 890 °C
Coefficient of thermal expansion. At 24 °C. linear: est, 11 μm/m·K. Linear. Along crystal axes, est: 9μm/m·K along a axis, 16 μm/m·K along c axis. Volumetric: 33 × 10^{-6} per K
Specific heat. Est, 186 J/kg·K at 25 °C
Entropy. At 25 °C: est, 487 J/kg·K
Latent heat of fusion. Est, 52 kJ/kg
Latent heat of transformation. Est, 20 kJ/kg
Latent heat of vaporization. Est, 2.37 MJ/kg at 25 °C
Thermal conductivity. Est, 15 W/m·K at 27 °C

Heat of combustion. For monoclinic Pm$_2$O$_3$ at 25 °C: est, ΔH^0_c = 6.19 MJ/kg Pr; ΔG^0_f = −5.88 MJ/kg Pm
Recrystallization temperature. Est, 400 °C

Electrical Properties

Electrical resistivity. Est, 750 nΩ·m at 25 °C
Ionization potential. Pm(I), 5.55 V; Pm(II), 10.90 V; Pm(III), 22.04 V; Pm(IV), 40.69 V

Magnetic Properties

Magnetic susceptibility. Probably strongly paramagnetic with a susceptibility somewhat greater than that of cerium at 24 °C
Magnetic transformation temperature. Probably exhibits two Néel temperatures which fall between those observed in neodymium and samarium

Optical Properties

Color. Metallic silver

Chemical Properties

General corrosion behavior. About the same as neodymium
Resistance to specific corroding agents. About the same as neodymium

Mechanical Properties

Tensile properties. About the same as neodymium
Hardness. 63 HK
Poisson's ratio. Est, 0.28
Elastic modulus. At 27 °C: est, tension, 46 GPa; shear, 18 GPa; bulk, 33 GPa

Protactinium

See tables at end of this section

Rhenium (Re)

Compiled by Warren R. Knipple
Cleveland Refractory Metals
Rhenium is used for electrical contacts, thermocouples, filaments for electronic devices, and to increase ductility in alloys of molybdenum and tungsten for use in electronics, thermocouples and welding rods. Rhenium begins to generate a white

nonpoisonous vaporous oxide, Re$_2$O$_7$, at about 600 °C when heated in air.

Structure

Crystal structure. Close-packed hexagonal, a = 0.2760 nm, c = 0.4458 nm
Minimum interatomic distance. 0.2746 nm

Mass Characteristics

Atomic weight. 186.207
Density. 21.04 Mg/m^3 at 20 °C

Thermal Properties

Melting point. 3180 °C
Boiling point. 5900 °C
Specific heat:

°C	kJ/kg·K
25	25.7
500	25
1000	30
1500	33
2000	37

Latent heat of fusion. 178 kJ/kg
Latent heat of vaporization. 3417 kJ/kg
Recrystallization temperature. 1200 to 1500 °C for 1 h, depending on purity of metal and amount of cold work
Thermal conductivity. 71.2 W/m·K at 20 °C
Vapor pressure:

°C	mPa
1525	1 x 10^{-6}
2000	0.004
2200	0.11
2400	1.9
2600	.16
2800	170
3000	1100

Electrical Properties

Electrical conductivity. 9.3% IACS
Electrical resistivity. 193 nΩ·m at 20 °C; see also Table 33.
Temperature coefficient of electrical resistivity. 0 to 100 °C, 0.00395 per K; see also Table 33.
Thermoelectric force. See Table 33.
Temperature of superconductivity. 1.699 K
Thermionic work function. 4.80 eV at Richardson constant of 0.52 MA/m^2·K^2

Table 33 Electrical properties of rhenium

Temperature, °C	Electrical resistivity, nΩ·m	Temperature coefficient of resistivity, per K	Thermoelectric potential vs platinum, mV
20	193	. . .	0
100	254	0.00395	0
300	400	0.00383	0.61
500	526	0.00358	2.31
700	630	0.00333	4.9
900	725	0.00313	8.8
1100	805	0.00294	13.0
1300	870	0.00274	19.4
1500	930	0.00258	26.1
1700	985	0.00244	35.1
1900	1030	0.00231	. . .
2100	1065	0.00217	. . .
2300	1090	0.00204	. . .

Magnetic Properties

Magnetic susceptibility. Volume: 863×10^{-6} mks

Optical Properties

Spectral hemispherical emittance. 42% for $\lambda = 655$ nm from 0 to 2000 °C

Nuclear Properties

Stable isotopes. Re 185, atomic weight 184.953007, 37.398% abundant; Re 187, atomic weight 186.955791, 62.602% abundant

Chemical Properties

General corrosion behavior. Rhenium is unattacked by hydrochloric acid, resistant to sulfuric acid, and dissolves readily in nitric acid. In general, rhenium can be solvated by alkalies and fused salts. It is highly resistant to attack by molten tin, zinc, silver, copper and aluminum.

Mechanical Properties

Tensile properties. True stress, B, at unit strain is 2.53 GPa; strain hardening exponent, n, is 0.353; see also Table 34 and Fig. 88.
Elastic modulus. At 20 °C, 460 GPa
Hardness. Arc-melted button, 135 HK; annealed rod, 270 HK; rod swaged 40% in cross-sectional area, 825 HK
Poisson's ratio. 0.49
Strain-hardening exponent. 0.353
Creep-rupture characteristics. See Fig. 89.

Fig. 88 Effect of temperature on the tensile properties of rhenium

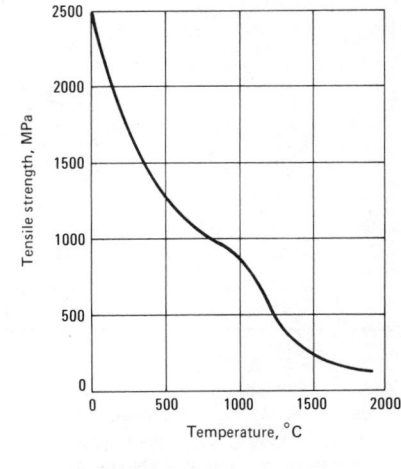
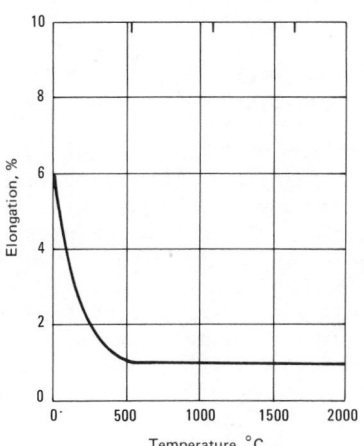

Sheet 1.3 mm thick, reduced 15%.

Fig. 89 Creep characteristics of rhenium and rhenium alloys

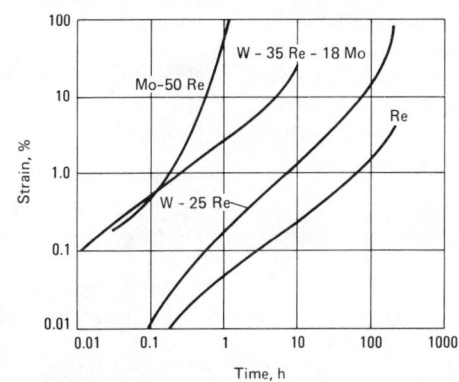

Tests were conducted at 41.4 MPa and 1600 °C in hydrogen after a 2 h annealing cycle at that temperature.

Table 34 Tensile properties of rhenium at room temperature

Product	Tensile strength, GPa	Yield strength(a), GPa	Elongation, %	Reduction in area, %
Sheet, 0.25-mm Thick				
Annealed	1.16	0.93	28	30
Reduced 30.7%	2.22	2.14	2	<1
Wire, 1.3-mm Diam				
Annealed	1.17	. . .	10	16
Reduced 15%	2.32	. . .	. . .	2
Rod, 3.2-mm Diam				
Annealed	1.13	0.32	24	22

(a) 0.2% offset.

Fabrication Characteristics

Consolidation. Rhenium can be consolidated by powder metallurgy techniques, inert-atmosphere arc-melting, and thermal decomposition of volatile halides. The powder metallurgy product is usually made by pressing bars at 200 MPa, followed by vacuum presintering at 1200 °C and hydrogen sintering at 2700 °C

Working. Rhenium is usually fabricated from sintered bar by cold working and annealing, since it is likely to be hot short at normal hot working temperatures. Reductions of 10 to 20% can be taken with intermediate anneals for 1 to 2 h at 1700 °C required. Primary working is by rolling, swaging, or forging. Wire-drawing has been done. Strip and wire as thin as 2 mils are possible.

Suitability for forming. Excellent ductility at room temperature allows forming of complex shapes.

SELECTED REFERENCE

The Study and Use of Rhenium Alloys, edited by E. M. Savittskii and M. A. Tylkina, Moscow, 1975. Translated from Russian and published by Amerind Publishing Co., Pvt. Ltd., 1978. Available through the U.S. Bureau of Interior and the National Science Foundation, Washington, D.C.

Rhodium (Rh)

Compiled by Leonard Bozza
Engelhard Minerals & Chemical Corp.

Pure rhodium is used principally where maintained high and uniform reflectivity is essential. This includes mirrors, principally those made by electrodeposition of rhodium on metal, plus some made by subliming rhodium on glass. Various types of light filters can be made by applying very thin coatings by subliming. A substantial amount of rhodium is electrodeposited as a nontarnishing finishing plate on jewelry articles, including white gold, silver, and other metals. Some use is also made of rhodium for the plating of sliding electrical contact surfaces; for this purpose, rhodium is sometimes applied over a heavier plate of palladium. Some pure wrought rhodium is used, but rhodium is most important as an alloying element with platinum. These materials find much use at elevated temperatures for crucibles, furnace windings, glass-working equipment, thermocouples, and particularly for catalysts.

Structure

Crystal structure. Face-centered cubic; lattice parameter 3.796 kX at 20 °C (Ref 1)

Mass Characteristics

Atomic weight. 102.905
Density. 12.41 Mg/m^3 at 20 °C

Thermal Properties

Melting point. 1963 °C (Ref 2)
Boiling point. Approx 3700 °C (Ref 3)
Coefficient of thermal expansion. Linear: 8.3 μm/m·K from 20 to 100 °C (Ref 4)
Specific heat. 247 J/kg·K (Ref 5)
Thermal conductivity. 150 W/m·K at 0 to 100 °C (Ref 6)

Electrical Properties

Electrical resistivity. 45.1 nΩ·m at 20 °C. (Ref 6) Temperature coefficient, 0.043 per K from 20 to 100 °C (Ref 11)

Magnetic Properties

Magnetic susceptibility. Volume: 14.3×10^{-6} mks at 18 °C (Ref 7, 8)

Optical Properties

Reflectance. See Fig. 90.
Emittance. At λ = 0.65 μm: solid, 0.24; liquid, 0.30 (Ref 2)

Chemical Properties

General corrosion behavior. Rhodium remains bright and untarnished during atmospheric exposure, and its general resistance to corrosion is exceptionally high. It is even resistant to boiling aqua regia. However, rhodium is attacked somewhat by hydrobromic acid, particularly when hot; by moist iodine; and by sodium hypochlorite. Hot sulfuric acid also attacks it, and sufficient corrosion occurs on electrolysis with alternating current in sulfuric acid to make this a useful metallographic etch.

Mechanical Properties

Tensile properties. Tensile strength: annealed, 951 MPa; hard, 2068 MPa
Hardness. 101 HB, 122 HV (Ref 13)
Elastic modulus. Hard wire, 293 GPa; 135 HB, 380 GPa (Ref 14)

REFERENCES

1. N. E. Swanson *et al.*, NBS Circular 539, U. S. Dept of Commerce, 1954
2. International Practical Temperature Scale of 1968, Amended Edition of 1975, *Metrologia*, Vol 12, 1976, p 7-17
3. R. F. Hampson, Jr. and R. F. Walker, *Journal of Research of the National Bureau of Standards*, Vol 65A, 1961, p 289
4. W. H. Wanger, *Journal of Research of the National Bureau of Standards*, Vol 3, 1929, p 1029
5. F. M. Jaeger and E. Rosenbohm, *Proceedings of the Academy of Sciences of Amsterdam*, Vol 24, 1931, p 85
6. R. W. Powell *et al.*, *Platinum Metals Review*, Vol 6, 1962, p 138

Fig. 90 Reflectance of rhodium as a function of wave length (Ref 9, 10)

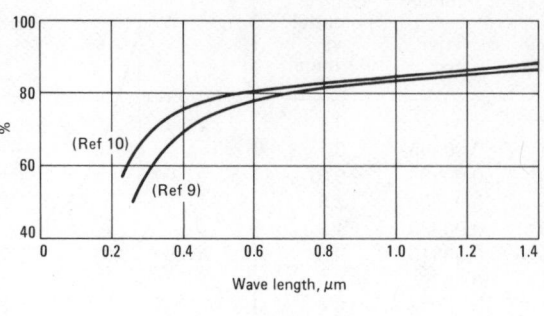

7. K. Honda, *Annalen der Physik,* Vol 32, 1910, p 1027
8. K. Honda, *Science Reports of Tohoku Imperial University,* Vol 1, 1912, p 1
9. W. W. Coblentz and R. Stair, *Journal of Research of the National Bureau of Standards,* Vol 22, 1939, p 93
10. M. Auwarter, *Journal of Applied Physics,* Vol 10, 1939, p 705
11. E. G. Price and B. Taylor, *Nature,* Vol 195, 1962, p 272
12. W. F. Roeser and H. T. Wensel, *Temperature, Its Measurement and Control in Science and Industry,* Reinhold Publishing Corporation, NY, 1941, p 1293
13. W. Köster, *Zeitschrift für Electrochemie,* Vol 49, 1943, p 233
14. J. S. Acken, *Journal of Research of the National Bureau of Standards,* Vol 12, 1934, p 249

Rubidium (Rb)

Compiled by J. R. Keiser
Oak Ridge National Laboratory

Current uses of rubidium are limited but include such applications as vacuum tubes and photoelectric cells. Potential uses include use as a heat transfer medium and as a fuel for ion propulsion engines. Chemical behavior of rubidium is similar to that of potassium; it reacts vigorously with air or water and should be handled and stored in a dry-inert environment

Structure

Crystal structure. Body-centered cubic, A2; $a = 0.562 \pm 0.003$ nm at -173 °C; $a = 0.570$ nm at 0 °C (Ref 1)

Mass Characteristics

Atomic weight. 85.467
Density. From Ref 1 and 2:

°C	Mg/m³
20	1.532
39	1.475
50	1.47
150	1.46
220	1.45

Volume change on melting. 2.5% increase (Ref 1)

Thermal Properties

Melting point. 38.89 °C (Ref 3)
Boiling point. 688 °C
Coefficient of thermal expansion. Linear: 90 μm/m·K at 20 °C; 340 μm/m·K at 40 to 100 °C (Ref 4). Volumetric, 3.38×10^{-4} m/m·K (Ref 1)
Specific heat. 334.89 J/kg·K at 0 °C; 379.7 J/kg·K at 50 °C (Ref 5)
Enthalpy. Solid: $H_T - H_{298} = 160.16\ T + 0.32080\ T^2 - 76259$. Liquid: $H_T - H_{298} = 384.47\ T - 88846$. T is in K and $H_T - H_{298}$ is in J/kg (Ref 6)
Entropy. Solid: $S = 368.84 \log T + 0.64162\ T - 288.96$. Liquid: $S = 885.45 \log T - 1297.9$. T is in K and S is in J/kg (Ref 6)
Latent heat of fusion. 25.535 kJ/kg (Ref 4)
Latent heat of vaporization. 887.46 kJ/kg (Ref 4)
Thermal conductivity. From Ref 7:

°C	Phase	W/m·K
0	Solid	58.3
25	Solid	58.2
38.89	Solid	58.1
38.89	Liquid	33.3

Vapor pressure. From Ref 5:

°C	KPa
294	0.1333
387	1.333
519	13.33
569	26.66
628	53.33

Electrical Properties

Electrical resistivity. Solid: 116 nΩ·m at 0 °C. Liquid: 231.5 nΩ·m at 50 °C; 274.7 nΩ·m at 100 °C (Ref 8). Temperature coefficient: 0.06 nΩ·m per K (Ref 4)
Electrochemical equivalent:

Valence	Equivalent, mg/C
1	0.8858
2	0.4429
3	0.2214
4	0.1107

Electrolytic solution potential. vs H_2: -2.924 V at 25 °C (Ref 8)
Ionization potential. Rb (I), 4.176 V; Rb (II), 27.5 V; Rb(III), 40 V (Ref 9)
Work function. 0.335 aJ (Ref 2)
Photoelectric threshold. 7.3×10^{-4} C (Ref 1)

Magnetic Properties

Magnetic susceptibility. Volume: 2.6×10^{-3} mks at 30 °C (Ref 10)

Optical Properties

Color. Silvery white

Nuclear Properties

Natural isotopes. [85]Rb: isotope mass 84.9117, 72.15% abundant. [87]Rb: 27.85% abundant, decays with half-life of 5×10^{11} year (Ref 11)
Unstable isotopes. In addition to [87]Rb, there are 25 other unstable isotopes of rubidium

Chemical Properties

General corrosion behavior. Rubidium is very active: it ignites in air at room temperature and will react explosively in oxygen. Rubidium can easily dissociate water to form the hydroxide and will react spontaneously with gaseous chlorine and fluorine and explosively with liquid bromine

Mechanical Properties

Hardness. 0.3 Mohs scale (Ref 5)
Modulus of elasticity. 2.35 GPa (Ref 1)
Kinematic liquid viscosity. From Ref 1:

°C	mm²/s
38	0.4573
39	0.4561
40	0.4528
50	0.4267
99.7	0.3359
140.5	0.2904
179	0.2586
220.1	0.2332

Liquid surface tension. From Ref 2:

°C	N/m
39	0.0847
100	0.080
550	0.051
632	0.0468

REFERENCES

1. M. A. Filyand and E. I. Semenova, *Handbook of the Rare Elements,* Vol 1 (M. E. Alferieff, Trans, and Ed.) Boston Technical Publishers, Cambridge, MA, 1968, p 219-229
2. T. P. Whaley, Sodium, Potassium, Rubidium, Cesium, and Francium, *Comprehensive Inorganic Chemistry,* Trotman-Dickenson *et al,* Eds., Pergamon, NY, 1973, p 369-381
3. *CRC Handbook of Chemistry and Physics,* 55th ed., CRC Press, Cleveland, 1974, p B-28
4. C. A. Hampel, Rubidium and Cesium, *Rare Metals Handbook,* Reinhold, NY, 1961, p 434-440
5. C. E. Mosheim, Rubidium, *The Encyclopedia of the Chemical Elements,* C. A. Hampel, Ed., Reinhold, NY, 1968, p 604-610
6. *CRC Handbook of Chemistry and Physics,* 55th ed., R. C. Weast, Ed., CRC Press, Cleveland, 1974, p D-57
7. *CRC Handbook of Chemistry and Physics,* 55th ed., R. C. Weast, Ed., CRC Press, Cleveland, 1974, p E-15
8. F. M. Perel'man, *Rubidium and Cesium,* R.G.P. Towndrow, Trans., and R. W. Clarke, Ed. Pergamon Press, NY, 1965, p 13
9. *CRC Handbook of Chemistry and Physics,* 55th ed., R. C. Weast, Ed., CRC Press, Cleveland, 1974, p E-68
10. *CRC Handbook of Chemistry and Physics,* 55th ed., R. C. Weast, Ed., CRC Press, Cleveland, 1974, p E-124
11. *CRC Handbook of Chemistry and Physics,* 55th ed., R. C. Weast, Ed., CRC Press, Cleveland, 1974, p 265-266

Ruthenium (Ru)

Compiled by R. H. Atkinson, retired, The International Nickel Co., Inc.
Reviewed for this edition by Edward W. Zysk
Engelhard Minerals & Chemicals Corp.

Ruthenium is used as a hardener for platinum and palladium for jewelry and other applications, including 10 to 11% Ru platinum for aircraft magneto contacts and similar contacts. Ruthenium can be used as an electric contact at temperatures to around 500 °C, because its oxide is conductive. Hard, complex high-ruthenium alloys are also used for pen tipping and the like. Some of these alloys also contain osmium. A significant application for ruthenium is in the thick-film paste systems used for printed circuit resistance elements.

Structure

Crystal structure. Close-packed hexagonal; $a = 2.6987$ kX; $c = 4.2728$ kX

Mass Characteristics

Atomic weight. 101.07
Density. At 20 °C, 12.45 Mg/m³; 12.37 Mg/m³ at 25 °C, computed from lattice constants

Thermal Properties

Melting point. 2310 ± 20 °C
Boiling point. Approx 4080 ± 100 °C
Coefficient of thermal expansion. Linear, 5.05 μm/m·K at 20 °C
Specific heat. 240 J/kg·K at 0 °C

Electrical Properties

Electrical resistivity. 76 nΩ·m at 0 °C
Electrochemical equivalent. Valence 2, 0.527 mg/C
Temperature of superconductivity. 2.04 K

Magnetic Properties

Magnetic susceptibility. Mass: 1.12×10^{-8} mks

Optical Properties

Reflectivity. 63% avg in the visible range

Chemical Properties

General corrosion behavior. Approaches that of iridium; unaffected by common acids, including aqua regia, at temperatures up to 100 °C, or by sulfuric acid up to 500 °C; moderately attacked by aqueous solutions of alkaline hypochlorites. Ruthenium exhibits good resistance to attack by certain molten metals. For example, at 200 °C above the melting points of the respective metals (under argon cover), it is not attacked by lithium, sodium, potassium, gold, silver, copper, lead, bismuth, tin, tellurium, indium, cadmium, calcium and gallium. However, gold, silver and copper wet the surface of ruthenium.

Fabrication Characteristics

Metallography. Etching: use a-c electrolytic
Consolidation. Powder from refining process may be consolidated by powder metallurgy techniques, or by orgon-arc melting.
Hot working temperature. 1500 to 2400 °C
Forming. Very difficult to work, but with care it can be forged at temperatures above 1500 °C. Ruthenium can be consolidated by powder metallurgy methods plus sintering above 1450 °C
Compacting pressure. 275 MPa
Formability. Monocrystalline zone-refined rod can be bent over a small radius but, if worked, recrystallizes on annealing and becomes relatively brittle

Mechanical Properties

Tensile properties. Tensile strength for compact powder bar, hot rolled 50%; 540 MPa at room temperature, 246 MPa at 3650 °C

Samarium (Sm)

Compiled by K. A. Gschneidner, Jr. and B. J. Beaudry
Ames Laboratory
U. S. Department of Energy
Iowa State University

Alloyed with cobalt, samarium is used as a permanent magnet, $SmCo_5$. Samarium is also used in nuclear reactors as a burnable poison, as a phosphor and in catalysts and ceramic capacitors. This metal oxidizes slowly in air at room temperature. Storage in an inert atmosphere or vacuum is recommended. Turnings can be ignited easily. Finely divided samarium should not be handled in air.

Structure

Crystal structure. α-phase, rhombohedral, $R\bar{3}m\ D^5_{3d}$; $a = 0.89834$ nm, $\alpha = 23.311°$ (hexagonal parameters, $a = 0.36290$ nm, $c = 2.6207$ nm at 24 °C). β-phase, close-packed hexagonal, $P6_3/mmc\ D^4_{6h}$; $a = 0.3663$ nm; $c = 0.5845$ nm at 450 °C. γ phase, body-centered cubic, $Im3m\ O^9_h$; $a = 0.407$ nm at 920 °C

Minimum interatomic distance. At 24 °C: r_a = 0.18145 nm; r_c = 0.17937 nm; radius CN_{12} = 0.18041 nm

Mass Characteristics

Atomic weight. 150.4
Density. α-phase, 7.520 Mg/m³ at 24 °C; β-phase, 7.352 Mg/m³ at 450 °C; γ-phase, 7.400 Mg/m³ at 926 °C
Volume change on freezing. 3.6% contraction

Thermal Properties

Melting point. 1074 °C
Boiling point. 1794 °C
Phase-transformation temperature. α- to β-phase, 734 °C; β to α-phase, 727 °C; β- to γ-phase, 922 °C
Coefficient of thermal expansion. At 24 °C. Linear: 12.7 μm/m·K. Linear, along crystal axes: 9.6 μm/m·K along a axis 19.0 μm/m·K along c axis. Volumetric, 38.1 × 10^{-6} per K
Specific heat. 196.4 J/kg·K at 25 °C
Entropy. At 25 °C: 428.9 J/kg·K
Latent heat of fusion. 57.31 kJ/kg
Latent heat of transformation. 20.70 kJ/kg
Latent heat of vaporization. 1.374 MJ/kg at 25 °C
Heat of combustion. For monoclinic Sm_2O_3 at 25 °C: ΔH^0_c = 6.05 MJ/kg Sm; ΔG^0_f = −5.78 MJ/kg Sm
Recrystallization temperature. About 440 °C
Thermal conductivity. 13.3 W/m·K at 25 °C
Vapor pressure. 0.001 Pa at 508 °C; 0.101 Pa at 642 °C; 10.1 Pa at 835 °C; 1013 Pa at 1150 °C

Electrical Properties

Electrical resistivity. 940 nΩ·m at 25 °C, 67 nΩ·m at 4 K
Ionization potential. Sm(I), 5.63 V; Sm(II), 11.07 V; Sm(III), 23.20 V; Sm(IV), 41.01 V
Hall coefficient. −0.021 nV·m/A·T at 25 °C
Temperature of superconductivity. Bulk samarium is not superconducting down to 0.37 K at atmospheric pressure

Magnetic Properties

Magnetic susceptibility. Volume: 8.03 × 10^{-4} mks at 17 °C, does not obey Curie-Weiss law
Saturation magnetization. >0.042 T at 4 K along ⟨0001⟩ and >0.034 T at 4 K along ⟨11$\bar{2}$0⟩
Magnetic transformation temper- ature. Ordering temperatures at 14 K (cubic sites) and 106 K (hexagonal sites)

Optical Properties

Color. Metallic silver
Spectral hemispherical emittance. Solid Sm, 43.7% for λ = 645 nm from 852 to 1077 °C

Nuclear Properties

Thermal neutron cross section. 5600 b

Chemical Properties

General corrosion behavior. Samarium oxidizes very slowly at room temperature in air. The rate of oxidation increases with temperature. Hydrogen will react at about 250 °C with samarium metal
Resistance to specific corroding agents. Samarium reacts vigorously with dilute acids but only slowly with concentrated sulfuric. The presence of fluoride ion retards acid attack due to the formation of SmF_3 on the surface of the metal

Mechanical Properties

Tensile properties. Tensile strength, 157 MPa; yield strength, 69 MPa; elongation, 22%; reduction in area, 34%
Hardness. 39 HV
Poisson's ratio. 0.282
Strain-hardening exponent. 0.23
Elastic modulus. At 27 °C: tension, 50.0 GPa; shear, 19.5 GPa; bulk, 37.7 GPa

Scandium (Sc)

Compiled by K. A. Gschneidner, Jr. and
B. J. Beaudry
Ames Laboratory
U. S. Department of Energy
Iowa State University

Scandium is used as a neutron window or filter in reactors. It is also used in high intensity lamps because of the multilined spectrum of incandescent scandium vapor. Turnings of scandium can be ignited and will burn white hot. Finely divided scandium should not be handled in air. Ingots of pure scandium can be stored in air.

Structure

Crystal structure. α-phase, close-packed hexagonal $P6_3/mmc$ D^4_{6h}; a = 0.33088 nm; c = 0.52680 nm at 24 °C. β-phase structure probably body-centered cubic; lattice parameters not determined
Minimum interatomic distance. At 24 °C: r_a = 0.16544 nm; r_c = 0.16269 nm; radius CN_{12} = 0.16407 nm

Mass Characteristics

Atomic weight. 44.9559
Density. α-phase, 2.989Mg/m³ at 24 °C

Thermal Properties

Melting point. 1541 °C
Boiling point. 2836 °C
Phase-transformation temperature. α- to β-phase, 1337 °C
Coefficient of thermal expansion. At 24 °C. Linear: 10.2 μm/m·K. Linear, along crystal axis: 7.6 μm/m·K along a axis: × 15.3 μm/m·K along c axis. Volumetric: 30.5 × 10^{-6} per K
Specific heat. 567.4 J/kg·K at 25 °C
Entropy. 773.6 J/kg·K at 25 °C
Latent heat of fusion. 313.6 kJ/kg
Latent heat of transformation. 89.2 kJ/kg
Latent heat of vaporization. 8.404 MJ/kg at 25 °C
Heat of combustion. For cubic Sc_2O_3 at 25 °C: ΔH_c^0 = 21.2 MJ/kg Sc; ΔG_f^0 = 20.2 MJ/kg Sc
Recrystallization temperature. About 550 °C
Thermal conductivity. 15.8 W/m·K at 25 °C
Vapor pressure. 0.0010 Pa at 1036 °C; 0.101 Pa at 1243 °C; 10.1 Pa at 1533 °C; 1013 Pa at 1999 °C

Electrical Properties

Electrical resistivity. 514 nΩ·m at 25 °C; 31 nΩ·m at 4 K. Along crystal axes, at 25 °C: 642 nΩ·m along a axis; 287 nΩ·m along c axis
Ionization potential. Sc(I): 6.54 V; Sc(II): 12.80 V; Sc(III): 24.76 V; Sc(IV): 73.47 V; Sc(V): 91.66 V; Sc(VI): 111.1 V
Hall coefficient. −0.013 nV·m/A·T at 25 °C
Temperature of superconductivity. Bulk scandium is not superconducting down to 0.032 K at atmo-

spheric pressure, but is superconducting at 0.032 K and 21 GPa.

Magnetic Properties

Magnetic susceptibility. Volume (mks units): at 24 °C: 2.471×10^{-4}. Along crystal axes: 2.490×10^{-4}; along a axis; 2.419×10^{-4} along c axis

Optical Properties

Color. Metallic silver

Nuclear Properties

Thermal neutron cross section. 24 b

Chemical Properties

General corrosion behavior. Scandium remains shiny in air at room temperature; discoloration starts at about 300 °C. Oxidation proceeds slowly to completion at 1000 °C
Resistance to specific corroding agents. Scandium reacts readily with most acids. When fluoride ions are present, forming ScF_3, they retard the attack by nitric, hydrochloric and other acids.

Mechanical Properties

Tensile properties. Tensile strength, 256 MPa; yield strength, 174 MPa; elongation, 5%; reduction in area, 8%
Hardness. 50 HK, 36 HV, 40 HB, 85 HRH
Poisson's ratio. 0.279
Elastic modulus. At 27 °C: Tension, 75.2 GPa; shear, 29.4 GPa; bulk, 56.7 GPa
Elastic constants along crystal axes. At 27 °C: $c_{11} = 99.3$ GPa; $c_{12} = 45.7$ GPa; $c_{13} = 29.4$ GPa; $c_{33} = 106.9$ GPa; $c_{44} = 27.7$ GPa

Selenium (Se)

Compiled by S. C. Carapella, Jr.
ASARCO Incorporated

Selenium is used in rectifiers, in photovoltaic cells, in xerographic drums, as a colorizing and decolorizing agent in glass, as a color pigment used in paints, ceramics and plastics, as an additive to improve machinability of low carbon steels, stainless steels, copper alloys and invar, as an additive to lead-antimony battery grid metal to improve properties, and as a vulcanizing agent to improve temperature and abrasion resistance of rubber.

Structure

Crystal structure. γ phase, hexagonal; at 20 °C, $a = 0.43640$ nm and $c = 0.49594$ nm. α and β phases, monoclinic

Mass Characteristics

Atomic weight. 78.96
Density:

Form	°C	Mg/m³
γ phase	25	4.809
α phase	25	4.389
β phase	25	4.470
Vitreous	20	4.280
Liquid	217	3.975
	267	4.060
	305	4.020
	406	3.910

Thermal Properties

Melting point. γ phase, 217 °C; vitreous softens at 40 °C
Boiling point. 684.9 °C
Phase-transformation temperature. α to β unknown; β to γ, 209 °C (?)
Coefficient of thermal expansion. Linear at 20 °C: γ phase, 49 μm/m·K; vitreous, 37 μm/m·K
Specific heat. γ phase, 317 J/kg·K at 25 °C; vitreous, 462 J/kg·K at 22 °C
Latent heat of fusion. 84.93 kJ/kg
Latent heat of vaporization. 1213.3 kJ/kg
Thermal conductivity. At 25 °C: γ phase, 2.48 W/m·K; vitreous, 0.51 W/m·K
Vapor pressure:

°C	kPa
344	0.1013
431	1.013
540	10.13
684.9	101.3

Electrical Properties

Electrical resistivity. At 25 °C: γ phase, 100 MΩ·m; vitreous, 100 GΩ·m
Electrochemical equivalent. Valance +6, 136.4 μg/C

Magnetic Properties

Magnetic susceptibility. Volume: -3.9×10^{-6} mks

Optical Properties

Refractive index. Vitreous at wave length of 1.152 μm, 2.4969. γ phase single crystals at 23 °C:

Wave length, μm	Index Ordinary	Index Extraordinary
1.06	2.790	3.608
1.15	2.737	3.573
3.39	2.65	3.46
10.6	2.64	3.41

Nuclear Properties

Stable isotopes. 74, 76, 77, 78, 80, 82
Thermal neutron cross section. For 2.2 km/s neutrons: absorption, 11.8 ± 0.4 b; scattering, 11 ± 2 b

Mechanical Properties

Hardness. 2.0 moh
Elastic modulus. Tension, 53.82 GPa; shear, 6.46 GPa
Surface tension:

°C	mN/m
220	105.5
250	100.5
280	98.0
310	98.2

Silver (Ag)

Compiled by S. C. Carapella, Jr.
ASARCO Incorporated and
D. A. Corrigan
Handy & Harman

Structure

Crystal structure. Face-center cubic at 25 °C, $a = 0.408621$ nm

Mass Characteristics

Atomic weight. 107.868
Density. At 20 °C: 10.49 Mg/m³.
Liquid:

°C	Mg/m³
960.5	9.30
1000	9.26
1092	9.20
1195	9.10
1300	9.00

Volume change on freezing. 5% contraction

Thermal Properties

Melting point. 961.9 °C. Freezing

Fig. 91 Temperature dependence of the electrical resistivity ratio of silver (Ref 14-16)

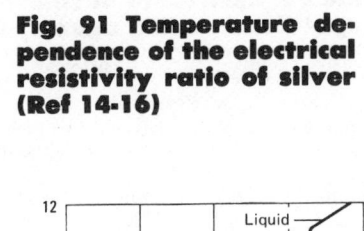

ρ_t/ρ_0 at 20.4 K is about 0.10; and at 1.3 K, 0.0068 (Ref 16)

Fig. 92 Reflectance of silver as a function of wave length (Ref 19, 20)

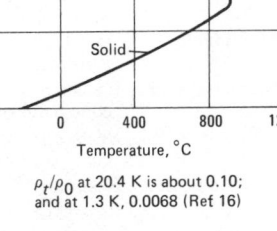

point in approximate equilibrium with the oxygen in the atmosphere (partial pressure of oxygen 20 kPa) is about 950 °C (Ref 1, 2). Freezing point is not lowered by carbon.

Boiling point. 2163 °C (Ref 8)

Coefficient of thermal expansion. Linear: at 20 °C, 19.0 μm/m·K; −190 to 0 °C, 17.0 μm/m·K (Ref 3, 4, 5); 0 to 900 °C, $L_t = L_0 (1 + 19.494 \times 10^{-6} t + 1.0379 \times 10^{-9} t^2 + 2.375 \times 10^{-12} t^3)$, where t is in °C (Ref 6); 0 to 100

°C, 19.68 μm/m·K (Ref 7); 0 to 500 °C, 20.61 μm/m·K

Specific heat. Solid: at 25 °C, 235 J/kg·K; 127 °C, 239 J/kg·K; 527 °C, 262 J/kg·K; 961 °C, 297 J/kg·K (Ref 8). Liquid: 961 to 2227 °C, 310 J/kg·K (Ref 8)

Latent heat of fusion. 104.2 kJ/kg (Ref 8)

Latent heat of vaporization. 2.63 MJ/kg (Ref 8)

Recrystallization temperature. 20 to 200 °C (68 to 392 °F), depending on purity and amount of cold work (Ref 9)

Thermal conductivity. 428 W/m·K at 20 °C; 356 W/m·K at 450 °C (Ref 10)

Vapor pressure. Liquid (Ref 8):

°C	Pa
1304	1.013×10^2
1510	1.013×10^3
1783	1.013×10^4
2163	1.013×10^5

Liquid: $\log p = \dfrac{-13350}{T} + 10.486$

(Ref 11);

solid: $\log p = \dfrac{-14020}{T} + 11.012$

(Ref 12); where T is in K and p is in Pa. High rate of volatilization at high temperatures and with oxidizing gases rather than under reducing gases (Ref 13)

Electrical Properties

Electrical conductivity. 108.4% IACS for extremely pure silver (105% referred to very pure copper)

Electrical resistivity. 14.7 nΩ·m at 0 °C. Temperature coefficient: from 0 to 100 °C, 0.0041 per K. Temperature dependence: see Fig. 91.

Cold working of silver considerably increases resistivity: 5% for 90% reduction (Ref 17). Annealing commercially pure silver successively in air and hydrogen disrupts grain boundaries and increases resistance.

Tension reduces resistivity slightly as does hydrostatic pressure: 12 000 kg/cm² causes 4% reduction (Ref 18).

Thermal electromotive force vs Pt. +0.74 mV; cold junction at 0 °C, hot junction at 100 °C (Ref 19).

Magnetic Properties

Magnetic susceptibility. Volume: -2.27×10^{-6} mks

Optical Properties

Color. As a result of the high and fairly uniform reflectance in the visible range, silver is considered white, but if human eyes were sensitive to a slightly shorter wave length region, silver would be colored.

Reflectance. For clean silver, high in the visible and infrared, but low in near-ultraviolet; see Fig. 92.

Emittance. Solid silver at 0.65 μm: extremely low and not known accurately; values of 0.044 at 940 °C and 0.072 at 980 °C have been observed for liquid silver (Ref 21). Other experiments showed no discontinuity at the melting point of silver, the emissivity being about 0.055 at about 700 °C (Ref 22).

Mechanical Properties

Tensile properties. Considerable spread in values for tensile strength and hardness of high purity silver. Average tensile strength, 125 MPa for 5-mm wire annealed at 600 °C (Ref 23)

Hardness. High purity silver: hydrogen anneal 650 °C, 25 HV; air anneal at 650 °C, 27 HV; electrodeposited silver (higher electrical resistivity than wrought silver): 100 HV

Elastic modulus. Strained 5%, then heated 0.5 h at 350 °C, 71.0 GPa (Ref 24)

Poisson's ratio. Annealed: 0.37; hard drawn, 0.39 (Ref 25)

Liquid surface tension. 0.923 N/m at 995 °C

REFERENCES

1. W. F. Roeser and A. I. Dahl, *Journal of Research of the National Bureau of Standards,* Vol 10, 1933, p 661
2. N. P. Allen, *Journal of the Institute of Metals,* Vol 49, 1932, p 49
3. R. Buffington and W. M. Latimer, *Journal of the American Chemical Society,* Vol 48, 1926, p 2305
4. H. Ebert, *Zeitschrift für Physik,* Vol 47, 1928, p 712
5. F. C. Nix and D. MacNair, *Physical Review,* Vol 61, 1942, p 74
6. H. Esser and H. Eusterbrock, *Archiv furdas Eisenhüttenwesen,* Vol 14, 1941, p 341
7. B. A. Rodgers, I. C. Schoonover and L. Jordan, Silver: Its Proper-

ties and Industrial Uses, National Bureau of Standards Circular, C412, 1936. L. Addicks (Ed.), *Silver in Industry*, Reinhold, NY, 1940

8. Hultgren *et al*, *Selected Values of the Thermodynamic Properties of the Elements*, ASM, Ohio, p 17-21

9. A. Butts and C. D. Coxe, *Silver-Economics Metallurgy and Use*, Van Nostrand, 1967, p 146-151

10. W. Hume-Rothery and P. W. Reynolds, *Proceedings of the Royal Society*, Vol A167, 1938, p 25

11. P. L. Woolf, G. R. Zellars, E. Foerster and J. P. Morris, *U.S. Bureau of Mines Report*, Investigation No. 563, 1960, 10 pages

12. H. M. Schadel, Jr. and C. E. Birchenall, *Transactions of the Metallurgical Society*, Vol 188, 1950, p 1134-1138

13. I. N. Plaksin and A. Y. Brechsted, *Zhurnal Prikladnoi Khimil*, Vol 11 (No. 12), 1938, p 1055, 1158, 1262, 1556

14. A. Butts and C. D. Coxe, *Silver—Economics, Metallurgy and Use*, Van Nostrand, 1967, p 112

15. E. F. Northrup, *Journal of the Franklin Institute*, Vol 178, 1914, p 85

16. F. Pawlek and D. Rogalla, *Cryogenics*, Vol 6, 1966, p 14

17. G. Tammann and K. L. Dreyer, *Annalen der Physik*, Vol 5, 1933, p 16, 111

18. P. W. Bridgman, *Proceedings of the American Academy of Arts and Science*, Vol 52, 1917, p 573

19. W. W. Coblentz and R. Stair, *Journal of Research of the National Bureau of Standards*, Vol 2, 1929, p 343

20. M. Auwarter, *Zeitschrift für Technische Physik*, Vol 18, 1937, p 457

21. G. K. Burgess and R. G. Waltenberg, *Bulletin of the National Bureau of Standards*, Vol 11, 1915, p 605

22. C. C. Bidwell, *Physical Review*, Vol 3 (No. 2), 1914, p 439

23. F. Saeffel and G. Sachs, *Zeitschrift für Metallkunde*, Vol 17, 1925, p 353

24. J. McKeown and O. Hudson, *Journal of the Institute for Metals*, Vol 60, 1937, p 109

Silicon (Si)

Compiled by H. C. Aufderhaar
Union Carbide Corp.

Silicon is used mainly in the primary and secondary aluminum industry, in silicones and in steels. The purity of the silicon for these purposes varies from 96.7 to 98.5 Si, 0.10 to 0.75 Al, and 0.03 to 0.40 Ca. The remainder is chiefly iron

Structure

Crystal structure. Face-centered cubic, A4; a = 0.54307 nm (Ref 1)
Minimum interatomic distance. 0.235 nm (Ref 1)

Mass Characteristics

Atomic weight. 28.08
Density. 2.330 Mg/m^3 at 25 °C

Thermal Properties

Melting point. 1410 °C (Ref 2)
Boiling point. 3280 °C (Ref 2)
Coefficient of thermal expansion. Linear: at 20 °C, 2.8 to 7.3 μm/m·K (Ref 3); from 0 to 1400 °C, 2.9 μm/m·K (Ref 4); see also Ref 5, 6.
Specific heat. At 20 °C: 677.96 J/kg·K (Ref 7). Temperature dependence: from 0 to 900 °C, C_p = 855 + 0.0919 T − 1.50 × 10^7 T^{-2}, where T is in K and C_p is in J/kg·K (Ref 7)
Latent heat of fusion. 1807.9 kJ/kg (Ref 2)
Latent heat of vaporization. 10 606 kJ/kg (Ref 8)
Heat of combustion. 31 350 kJ/kg Si (Ref 9)
Thermal conductivity. 83.680 W/m·K at 20 °C

Electrical Properties

Electrical resistivity. 1 nΩ·m at 0 °C (Ref 5). It is not certain whether the temperature coefficient of electrical resistivity of silicon is positive or negative. Pressure dependence: −1.224 nΩ·m per GPa (Ref 10)
Electrochemical equivalent. 0.07269 mg/C (Ref 11)
Electrolytic solution potential. vs H$_2$: −0.453 V (Ref 12)
Hydrogen overvoltage. vs Pt: 0.192 ± 0.002 V (Ref 13)
Hall effect. 4100 V·m/A·T at 20 °C (Ref 3)

Magnetic Properties

Magnetic susceptibility. Volume: −1.63 × 10^{-6} mks (Ref 3)

Table 35 Resistance of silicon to specific corroding agents

Corrosive agent	Resistance
Air	Resistant
Ammonia	Resistant; reacts with vapors at bright red heat
Bromine	Resistant; burns at 930 °F
Carbon dioxide	Resistant
Chlorine	Resistant; burns at 640 °F
Copper sulfate	Resistant (10% solution)
Ferric chloride	Resistant (10% solution)
Hydrochloric acid	Resistant (dilute or conc, cold or boiling)
Hydrofluoric acid	Resistant (dilute or conc, cold or boiling)
Hydrogen sulfide	Resistant
Iodine	Resistant
Nitric acid	Resistant (dilute or conc, cold or boiling)
Oxygen	Resistant
Potassium hydroxide	Attacked
Sodium hydroxide	Attacked
Sulfur	Resistant; reacts at elevated temperatures
Sulfur dioxide	Resistant
Sulfuric acid	Resistant (dilute or conc, cold or boiling)
Water, distilled	Resistant
Water, rain	Resistant

Optical Properties

Color. Dark steel gray
Reflectance. 39 to 29% from 200 to 130 nm (Ref 14). Wavelength dependence, see Ref 10.
Refractive index. 3.87 at λ = 578 nm; 4.24 at λ = 589 nm; 407 at λ = 1000 nm (Ref 3)
Absorptive index. 0.12 at λ = 578 nm; 0.114 at λ = 589 nm; 0.095 at λ = 1000 nm (Ref 3)

Chemical Properties

Resistance to specific corroding agents. See Table 35.

Mechanical Properties

Compressive strength. Chill cast specimens, 25 by 25 by 90.2 mm: 92.87 MPa (Ref 15)
Elastic modulus. Tension. Chill cast specimens, 25 by 25 by 90.2 mm: 112.7 GPa (Ref 15). Chill cast specimens, 6.5 by 6.5 by 76 mm: 106.8 GPa (Ref 4)
Bending properties. Chill cast

specimens, 6.5 by 6.5 by 76 mm on 75-mm span: deflection under 53-N load, 0.0457 mm; modulus of rupture, 62.37 MPa; breaking strength, 156 N (Ref 4)

REFERENCES

1. W. B. Pearson, *A Handbook of Lattice Spacings and Structures of Metals and Alloys,* Pergamon Press, NY, 1958
2. Kubaschewski, Evans and Alcock, *Metallurgical Thermochemistry,* 4th Ed., Pergamon Press, 1967
3. *International Critical Tables,* McGraw-Hill, NY 1926
4. Kinzel and Cunningham, *Metals Technology,* Vol 6, 1939, p TP1138
5. A. E. van Arkel, *Reine Metalle,* J. Springer, Berline, 1939
6. Hobling, *Zeitschrift für Angewandte Chemie,* Vol 40, 1927, p 655
7. K. K. Kelly, *Bureau of Mines Bulletin 371,* 1934, p 43
8. L. L. Quill, *The Chemistry and Metallurgy of Miscellaneous Materials,* McGraw-Hill, NY, 1950
9. J. F. Elliot and M. Gleiser, *Thermochemistry for Steelmaking,* Vol I, Addison-Wesley, 1960
10. J. W. Mellor, *A Comprehensive Treatise of Inorganic Chemistry,* Vol VI, 1925, p 152
11. *G. A. Rousch, Transactions of the Electrochemical Society,* Vol 70, 1938, p 293
12. G. W. Akimow and A. S. Oleschko, *Korrosion und Metallschutz,* Vol 10, 1934, p 134
13. Thiel and Hammerschmidt, *Zeitschrift für Anorganische und Allgemeine Chemie,* Vol 132, 1923, p 15
14. Johnson, *Proceedings of the Physics Society (London),* Vol 53, 1941, p 53
15. Templin, *Metals and Alloys,* Vol 3, 1932, p 136

Sodium (Na)

Compiled by J. R. Keiser and J. H. DeVan
Oak Ridge National Laboratory

Sodium is used as a liquid metal heat transfer medium, a working fluid for evaporative heat pipes, and an electrical conductor in homopolar generators. It is also used in vapor lamps for highway lighting, as an alloying addition for lead, zinc, and aluminum and as a reactant for deoxidation of metals and for reduction of metal fluorides. Sodium is highly reactive with water; hydrogen released by the reaction is potentially explosive. Molten sodium will burn in ambient air. Iron-based alloys are usually selected as containers in transport of liquid sodium. Argon, helium and nitrogen are used as cover gases to minimize sodium oxidation. Sodium fires are best extinguished by closing off air accesses or by blanketing with either nitrogen or inert solids, such as carbon granules. Commercial extinguishing media include sodium chloride, sodium carbonate and calcium phosphate. Carbon tetrachloride and solid carbon dioxide extinguishers should not be used on sodium fires.

Structure

Crystal structure. β-phase, body-centered cubic, type $A2$; $a = 0.42906$ nm at 20 °C. On cooling below 36 K, sodium partially transforms to α-phase, close-packed hexagonal, type $A3$; $a = 0.3767$ nm at 5 K (Ref 1)

Mass Characteristics

Atomic weight. 22.9898
Density. 0.9674 Mg/m³ at 25 °C; 0.9270 Mg/m³ at 100 °C; 0.7113 Mg/m³ at 1000 °C (Ref 2, 3)
Density vs temperature. ρ (kg/m³) $= 972.5 - 0.2011T - 1.5 \times 10^{-4} T^2$ for 0 to 96.6 °C (T in °C); ρ (kg/m³) $= 950.1 - 0.22976T - 1.460 \times 10^{-5} T^2 + 5.638 \times 10^{-9} T^3$ for 98 to 1370 °C (T in °C) (Ref 4)
Volume change on melting. $+2.71\%$ (Ref 4)

Thermal Properties

Melting point. 97.82 °C (Ref 4)
Boiling point. 881.4 °C (Ref 4)
Phase-transformation temperature. Incomplete transformation to α-phase occurs on cooling (below 36 K) or on deforming (below 51 K) (Ref 5)
Coefficient of thermal expansion.

Linear: 68.93 μm/m·K; $\frac{l}{l_0} = 1 + 6.893 \times 10^{-5}T + 0.63 \times 10^{-7}T^2$ for 0 to 96.6 °C (T in °C) (Ref 4); volumetric, $2.418 \times 10^{-4}/°C; \frac{V}{V_0}$ $= 1 + 2.4183 \times 10^{-4}T + 7.385 \times 10^{-8}T^2 + 15.64 \times 10^{-12}T^3$ for 97.83 to 1350 °C (T in °C) (Ref 4)

Specific heat. $C_p = 1.2220$ kJ/kg·K at 25 °C (solid); $C_p = 1.3210$ kJ/kg·K at 250 °C (liquid); $C_p = 2.5100$ kJ/kg·K at 1000 °C (vapor) (Ref 6, 7)
Specific heat vs temperature. Solid: $C_p = 1198.72 + 0.64894 T + 0.010527T^2$ J/kg·K for 0 to 97.8 °C (T in °C); liquid: $C_p = 1436.1 - 0.58026T + 4.6208 \times 10^{-4} T^2$ J/kg·K for 97.8 to 900 °C (T in °C) (Ref 4)
Enthalpy. Where H_{0c} is enthalpy of the solid state at 0 °C: solid, $H_c - H_{0c} = 1199.26T + 0.3247T^2 + 3.510 \times 10^{-3}T^3$ J/kg for 0 to 97.8 °C (T in °C); liquid $H_l - H_{0c} = 98960 + 1436.7T - 0.29025T^2 + 1.5410 \times 10^{-4}T^3 + 2.400 \times 10^7 \times e^{-(13600/(T + 273)}$ J/kg; for 97.8 to 900 °C (T in °C) (Ref 4)
Entropy. Where S_{0c} is entropy at 0 °C: solid, $S_c - S_{0c} = 4162.42 \log(T) - 5.1036T + 0.0052658T^2 - 9140.2$ J/kg·K for 0 to 97.8 °C (T in K); liquid, $S_l - S_{0c} = 3752.6 \log T - 0.8330T + 2.3112 \times 10^{-4} T^2 - 8673.9$ J/kg·K for 97.8 to 900 °C (T in K) (Ref 4)
Latent heat of fusion. 113 kJ/kg (Ref 4)
Latent heat of vaporization. 3.874 MJ/kg (Ref 4)
Thermal conductivity. 131.4 W/m·K at 25 °C; 79.6 W/m·K at 250 °C (Ref 4)
Thermal conductivity vs temperature. k (W/m·K) $= 135.6 - 0.167T$ for 0 to 95 °C (T in °C); k (W/m·K) $= 91.8 - 0.049T$ for 104 to 832 °C (T in °C) (Ref 4)
Vapor pressure. (Ref 4)

°C	Pa
100	1.43 x 10⁻⁵
200	1.81 x 10⁻²
300	1.85
500	5.19 x 10²
700	1.40 x 10⁴
900	1.20 x 10⁵

also, $\log P_s$ (Pa) $= 6.438 \times 10^5 - \dfrac{5.641 \times 10^8}{T} - (5 \times 10^4) \log T$ for 100 to 877 °C (T in K)
Diffusion characteristics coefficient. See Table 36.

Electrical Properties

Electrical resistivity. Solid: 46.9

nΩ·m at 20 °C; liquid: 96.4 nΩ·m at 97.8 °C; 221.4 nΩ·m at 400 °C; 463.5 nΩ·m at 800 °C; 737.6 nΩ·m at 1100 °C (Ref 4)

Temperature dependence of electrical resistivity. r_s (nΩ·m) = 42.9 + 0.1993T + 9.848 × 10^{-5} T^2 for −223 to 97.8 °C (T in °C); r_s(nΩ·m) = 61.44 + 0.3504T + 5.695 × 10^{-5} T^2 + 1.667 × 10^{-7} T^3 for 130 to 1090 °C (T in °C) (Ref 4)

Thermoelectric potential. vs Pt: (Ref 4)

°C	mV
25	2.9 × 10^{-2}
50	5.3 × 10^{-2}
100	8.4 × 10^{-2}
200	1.36 × 10^{-1}
300	2.26 × 10^{-1}
400	3.90 × 10^{-1}
500	6.15 × 10^{-1}
600	9.96 × 10^{-1}
700	1.47
800	2.02
900	2.63

Electrochemical equivalent. For a valence of 1, 0.238 mg/C
Electrolytic solution potential. vs H_2: −2.711 V at 25 °C (Ref 8)
Ionization potential. (Ref 8):

Ionization state	Potential, V
I	5.138
II	47.29
III	71.715
IV	98.88
V	138.37
VI	172.09
VII	208.444
VIII	264.155

Hall coefficient. −2.5 aVm/A·T (Ref 4)
Work function. 0.365 aJ (Ref 8)

Magnetic Properties

Magnetic susceptibility. (Ref 4):

°C	Volume susceptibility, mks units
30	7.29 × 10^{-6}
95	7.24 × 10^{-6}
150	7.04 × 10^{-6}
250	6.95 × 10^{-6}

Optical Properties

Color. Silver
Refractive index. Liquid, 0.0045 for

Table 36 Diffusion characteristics of sodium (Ref 8)

Solute	Temperature range, °C	Activation energy, kJ/mol	Frequency factor, m²/s
^{198}Au	1 to 77	9.25	3.34 x 10^{-8}
^{42}K	0 to 91	35.3	0.08 x 10^{-8}
^{22}Na	0 to 98	42.2	0.145 x 10^{-8}
^{86}Rb	0 to 85	35.5	0.15 x 10^{-8}

Table 37 Unstable isotopes of sodium

Isotope	Half-life	Decay mode	Particle energy, pJ
20	~0.39 s	β^+, α	...
21	23 s	β^+	0.402
22	2.58 yr	β^+, EC	0.087
24^m	~0.02 s	IT, β^-	...
24	15 h	β^-	0.223
25	60 s	β^-	0.61
26	1.0 s	β^-	1.07

λ = 589.3 μm; solid, 4.22 for λ = 589.3 μm (Ref 8)

Nuclear Properties

Stable isotopes. ^{23}Na, isotope mass 22.9898, 100% abundance (Ref 4)
Unstable isotopes. See Table 37. (Ref 4)

Chemical Properties

Resistance to specific corroding agents. At 25 °C, sodium passivates in dry O_2 but oxidizes in moist air to form Na_2O, NaOH, and finally Na_2CO_3. Sodium is highly pyrophoric in air at or above 125 °C. Sodium reacts with CO_2 above 200 °C to form Na_2O, C, and possibly $Na_2C_2O_4$. Below 320 °C, water (gas or liquid) reacts with liquid sodium to produce NaOH (solid) and H_2. (H_2 is potentially explosive if O_2 is present.) Above 320 °C, products of the H_2O reaction include NaOH (liquid) NaH, Na_2O, and H_2. A useful technique for removing sodium residues is by reaction with water vapor in nitrogen or noble gas at 70 °C followed by water rinsing. Sodium undergoes metallic dissolution in anhydrous liquid ammonia, the solution ultimately converting to sodium amide. Sodium reacts with alcohols to form sodium alcoholates and hydrogen. N-butyl alcohol can be used to slowly dissolve sodium at 25 °C. Ethyl alcohol is much more reactive and can be ignited if sodium comes in contact with air. Solid sodium is relatively inert toward dry hydrocarbons that do not have an active hydrogen or acetylene hydrogen component. In contact with molten sodium, alkyne hydrogen atoms are liberated, and aryl hydrocarbons can be polymerized or decomposed.

Mechanical Properties

Kinematic liquid viscosity. (Ref 4)

°C	mm²/s
100	0.7338
200	0.5001
300	0.3921
400	0.3323
500	0.2955
600	0.2568
700	0.2313
800	0.2134

Liquid surface tension. 0.192 N/m at 97.8 °C; 0.161 N/m at 400 °C; 0.146 N/m at 550 °C; 0.113 N/m at 881.4 °C; also, σ (N/m) = 0.2067 − 1 × 10^{-4}T for 97.8 to 881.4 °C (T in °C) (Ref 4)
Velocity of sound. V(m/s) = 2577.25 − 0.524 T (T in °C) (Ref 2)

REFERENCES

1. A. Taylor and Brenda J. Kagle, *Crystallographic Data on Metals and Alloy Structures*, Dover, NY, 1963

2. M. Sittig, *Sodium, Its Manufacture, Properties and Uses*, Chapter 9, Physical and Thermodynamic Properties of Sodium, G. W. Thomson and E. Garelis (Eds.), American Chemical Society Monograph Series 133, Reinhold Publishing Corp., NY, 1956

3. J. P. Stone, et al., High Temperature Properties of Sodium, NRL-6241, Naval Research Laboratory, 1965

4. O. J. Foust, Ed., *Sodium-NaK Engineering Handbook*, Vol 1, Physical Properties, H. J. Bomelburg and C. R. F. Smith, Gordon and Breach, 1972, p 1-88

5. C. S. Barrett, X-ray Study of the Alkali Metals at Low Temperature, *Acta Crystallographica*, Vol 9, 1956, p 671

6. D. C. Ginnings, et al., Heat Capacity of Sodium Between 0° and 900 °C, The Triple Points and Heat

of Fusion, *Journal of Research for the National Bureau of Standards*, Vol 45, 1950, p 23

7. G. H. Golden and J. G. Tokar, Thermophysical Properties of Sodium, ANL-7323, Argonne National Laboratory, 1967
8. R. C. Weast (Ed.), *Handbook of Chemistry and Physics*, 55th ed., CRC Press, Cleveland, 1974

Strontium (Sr)

Compiled by G. C. Carter
(Currently with National Academy of Science)
National Bureau of Standards

Structure

Crystal structure. α phase, face-centered cubic, cF4 (*Fm3m*); *a* = 0.60849 nm at 25 °C. β phase (commonly referred to as γ-phase in earlier literature), body-centered cubic, cI2 (*Im3m*); *a* = 0.485 nm at 614 °C (Ref 1, 2). High-pressure phase: body-centered cubic, cI2 (*Im3m*) (Ref 2).

Mass Characteristics

Atomic weight. 87.62
Density. α phase, 2.6 Mg/m³ at 20 °C (Ref 3); β phase, 2.55 Mg/m³ at 614 °C (calculated from x-ray data) (Ref 1)

Thermal Properties

Melting point. 768 °C (Ref 4)
Boiling point. 1099 °C (calculated from vapor pressure data) (Ref 3, 5)
Phase transformation temperature. α to β, 557 °C (Ref 4)
Specific heat:

K	°C	J/kg·K
1	−272	0.0477
2	−271	0.139
4	−269	0.612
10	−263	8.46
20	−253	54.4

Electronic coefficient (γ), 41.5 ± 0.3 mJ/kg·K (Ref 4)
Latent heat of fusion. 104.7 kJ/kg (calculated from binary phase diagram data) (Ref 4)

Electrical Properties

Work function. 2.1 to 2.74 eV (0.34 to 0.44 aJ), depending on conditions and techniques of the experimental determination (Ref 6, 7, 8)

Magnetic Properties

Magnetic susceptibility. Molar: 1.16 mks at 22 °C (Ref 9)

Nuclear Properties

Stable isotopes. ⁸⁴Sr, isotope mass 83.913431, 0.56% abundant; ⁸⁶Sr, isotope mass 85.909276, 9.9% abundant; ⁸⁷Sr, isotope mass 86.908894, 7.0% abundant; ⁸⁸Sr, isotope mass 87.905628, 82.6% abundant (Ref 10)
Unstable isotopes. ⁷⁸Sr, 31 m; ⁷⁹Sr, 8.1 m; ⁸⁰Sr, 1.7 h; ⁸¹Sr, 2.5 m; ⁸²Sr, 25.0 d; ⁸³Sr, 32.4 h; ⁸⁵Sr, 67.7 m, 65.2 d; ⁸⁹Sr, 50.5 d; ⁹⁰Sr, 29 y; ⁹¹Sr, 9.48 hr; ⁹²Sr, 2.71 h; ⁹³Sr, 7.5 m; ⁹⁴Sr, 1.29 m; ⁹⁵Sr, 26 s; ⁹⁶Sr, 4.0 s; ⁹⁷Sr, ≤ 0.2 s; ⁹⁸Sr, ~ 0.85 s, ⁸⁹Sr to ⁹⁸Sr mode of decay by negative electron (Ref 10)

Mechanical Properties

Modulus. Bulk, 11.61 GPa; isothermal compressibility, 86.1 μm²/N

REFERENCES

1. P. Eckerlin, H. Kandler and A. Stegherr, Landolt-Börnstein Tables, New Series III/6, *Structure Data of Elements and Intermetallic Phases*, edited by K.-H. Hellwege and A. M. Hellwege, Springer-Verlag, NY, 1971
2. W. B. Pearson, *Handbook of Lattice Spacings and Structures of Metals*, Pergamon Press, NY, Vol 1, 1958; Vol 2, 1967
3. R. J. Elliott, *Constitution of Binary Alloys, First Supplement*, McGraw-Hill, NY, 1965
4. R. Hultgren, P. D. Desai, D. T. Hawkins, M. Gleiser, K. K. Kelley and D. D. Wagman, *Selected Values of the Thermodynamic Properties of the Elements*, American Society for Metals, Metals Park, OH, 1973. [Additional thermodynamic data are available in this compilation.]
5. K. A. Gschneidner, Jr., Physical Properties and Interrelationships of Metallic and Semimetallic Elements, *Solid State Physics*, Vol 16, edited by F. Seitz, and D. T. Turnbull, Academic Press, NY, 1964, p 275
6. V. S. Fomenko, *Handbook of Thermionic Properties—Electronic Work Functions and Richardson Constants of Elements and Compounds*, edited by G. V. Samsonov (translation from the Russian), Plenum Press Data Division, NY, 1966
7. G. A. Haas and R. E. Thomas, Thermionic Emission and Work Function, chapter 2 in *Measurements of Physical Properties*, edited by E. Passaglia, in Vol 6, part 1 of *Techniques of Metals Research*, edited by R. F. Bunshah, Interscience, 1972, p 91
8. H. B. Michaelson, *Handbook of Chemistry and Physics*, 55th ed., edited by R. C. Weast, CRC Press, Cleveland, OH, 1974, E-81
9. Landolt-Börnstein Tables, II Band, 9. Teil, *Magnetische Eigenschaften 1*, K.-H. Hellwege and A. M. Hellwege, Eds., Springer-Verlag, NY, 1962
10. N. E. Holden and F. W. Walker, Chart of the Nuclides, Knolls Atomic Power Laboratory, United States Atomic Energy Commission, distributed by Educational Relations, General Electric Co., Schenectady, NY. [Revised to April 1972, dated October 1972. Additional nuclear data are available on this chart.]

Tantalum (Ta)

Compiled by Mortimer Schussler
Fansteel, Inc.

Tantalum provides a combination of properties not found in many refractory metals—excellent fabricability, low ductile-to-brittle transition temperature and high melting point. The largest use of tantalum at this time is in electrolytic capacitors. Sizeable quantities of tantalum also are used in chemical process equipment (such as heat exchangers, condensers, thermowells and lined vessels), notably for handling nitric, hydrochloric, bromic and sulfuric acids, and combinations of these acids with many other chemicals. Spinnerettes for extruding man-made fibers constitute another important application of tantalum. Because of its high melting point, tantalum is used for heating elements, heat shields and other components of vacuum furnaces. Tantalum has been used in specialized aerospace and nuclear applications. Tantalum also is used in prosthetic devices in contact with body fluids and as an alloy component in superalloys, and tantalum carbide is an important constituent of cemented carbide cutting tools made from

Fig. 93 Temperature dependence of the entropy of tantalum

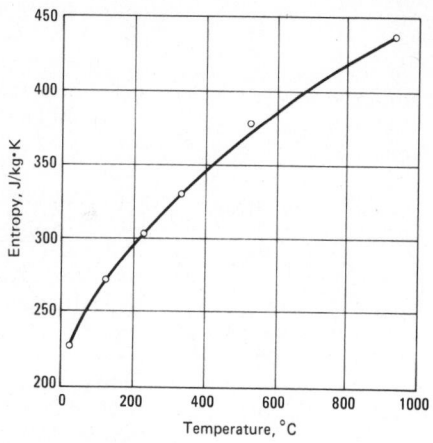

Fig. 94 Electrical resistivity of tantalum at low temperatures

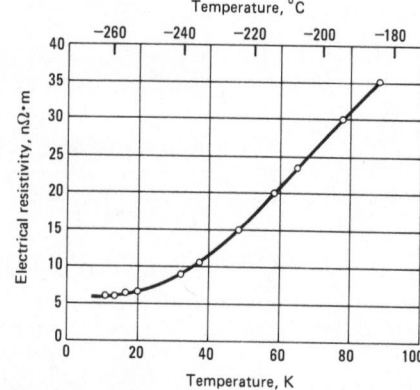

Unannealed rod, 99.98% pure tantalum.

Fig. 96 Temperature dependence of the total emittance of commercially pure tantalum

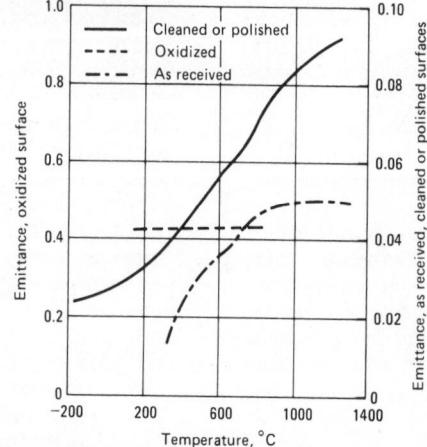

mixtures of titanium, tungsten and tantalum carbides. Yield and ultimate strengths are increased, and ductility is reduced, by increases in the amount of interstitial elements (oxygen, nitrogen, carbon and hydrogen). Embrittlement of the tantalum can occur if contamination by these elements is sufficiently severe. High-purity tantalum (99.90% min) is available commercially with the following maximum impurity limits, in ppm: 500 Nb, 300 W, 100 to 200 O, 100 Fe, 100 Mo, 50 to 75 C, 50 to 75 N, 50 Ni, 50 Si, 50 Ti, 10 H.

Structure

Crystal structure. Body-centered cubic, $I\frac{4}{m}\bar{3}\frac{2}{m}$; $a = 0.33026$ nm at 20 °C
Slip planes. {110}
Cleavage planes. {110}
Minimum interatomic distance. 0.2854 nm

Mass Characteristics

Atomic weight. 180.948
Density. At 20 °C, 16.6 Mg/m³

Thermal Properties

Melting point. 2996 °C
Boiling point. 5427 °C
Coefficient of thermal expansion. Linear: 6.5 μm/m·K near 20 °C. Temperature dependence: $\alpha = 6.5 + 0.34 \times 10^{-3} T + 0.12 \times 10^{-6} T^2$, where T is in °C and α is in μm/m·K
Specific heat. At 0 °C, 139.1 J/kg·K. Temperature dependence: $C_p = 139.04 + 1.757 \times 10^{-2} T + 1.375 \times 10^{-6} T^2$, where T is in K and C_p is in J/kg·K
Entropy. At 25 °C, 229 J/kg. See also Fig. 93.
Latent heat of fusion. 145 to 174 kJ/kg
Latent heat of vaporization. 4160 to 4270 kJ/kg
Heat of combustion. 5634 to 5772 kJ/kg Ta
Thermal conductivity:

°C	W/m·K
−73	56.1
20	54.4
127	59.9
527	66.6
927	72.9
1327	77.0
1727	80.8
2127	83.7
2527	85.8

1500°C   0.51 +0.27   0.78 ×10⁻⁶
12%

Vapor pressure:

°C	mPa
2351	0.6298
2365	0.7488
2487	4.019
2566	9.820
2615	17.20
2652	24.40
2675	37.03

Electrical Properties

Electrical conductivity. 13% IACS

Electrical resistivity. At 20 °C: 135.0 nΩ·m. See also Fig. 94 and 95. Temperature coefficient: From 0 to 100 °C, 0.0038 per K. See also Fig. 95. Pressure coefficient:

Gage pressure, kPa	Resistivity ratio(a)
0	1.000
98	0.984
200	0.968
390	0.941
590	0.918
780	0.898
980	0.882

(a) Ratio of resistivity at pressure to resistivity at zero pressure.

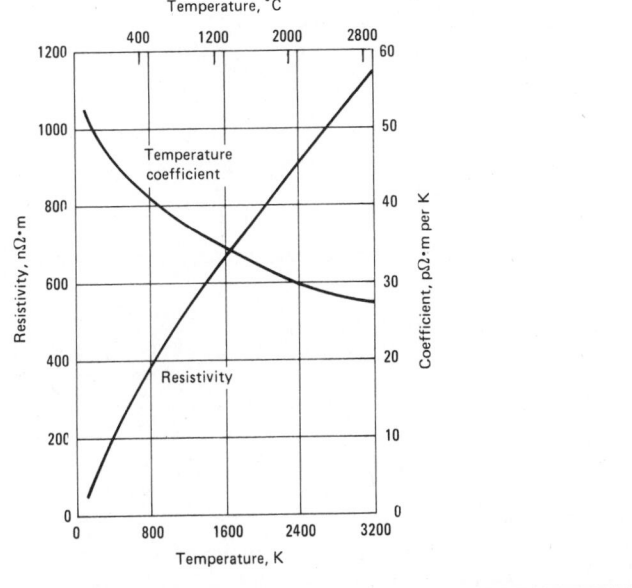

Fig. 95 Electrical resistivity of tantalum and its temperature coefficient

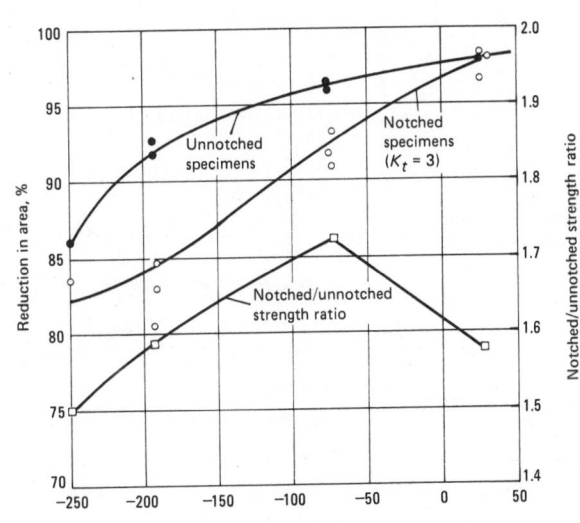

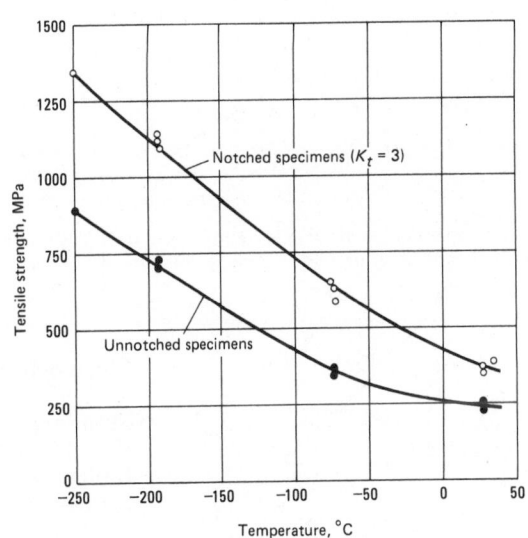

Fig. 97 Low-temperature tensile properties of electron-beam-melted tantalum bar

Sample impurities: <0.003% C, <0.003% O, 0.0008% N, <0.08% others. Bar was annealed for 3 h at 1200 °C: hardness, 83 HV; grain size, ASTM 5. Crosshead speed: unnotched specimens, 0.5 mm/min; notched specimens, 0.13 mm/min

Fig. 98 Elevated-temperature tensile properties of 1-mm-thick electron-beam-melted tantalum sheet

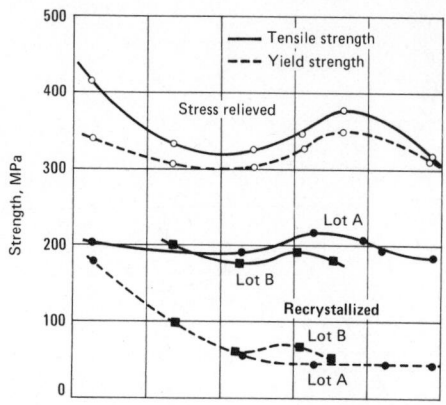

Sample impurities (both lots): 0.0030% C, 0.0016% O, 0.0010% N, <0.040% others. Stress-relieved sheet was cold rolled 95% and stress-relieved for ¼ h at 730 °C; recrystallized sheet was cold rolled 75% and recrystallized by heating for 1 h at 1200 °C. Crosshead speed, 1.3 mm/min

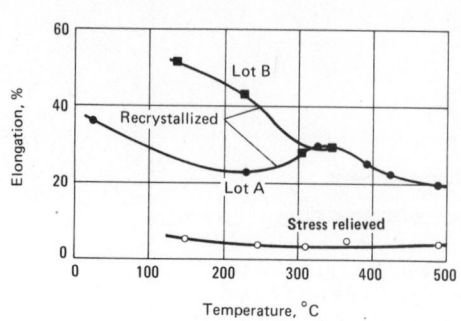

Fig. 99 High-temperature tensile strength of tantalum

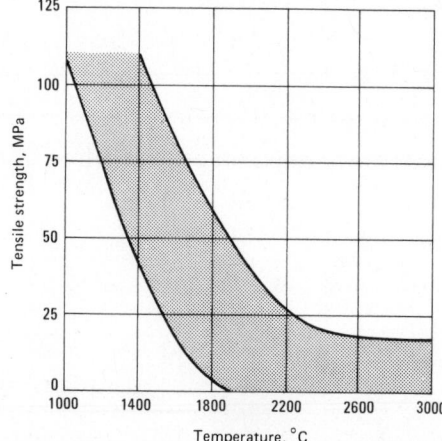

Note that upper portion of curve is characterized by high strain rates and high interstitial content whereas lower portion of curve is characterized by low strain rates and low interstitial content.

Thermal electromotive force. vs Pt:

Temperature at hot junction(a), °C	Thermal emf, mV
−200	+0.21
−100	−0.10
100	+0.33
200	0.93
600	5.95
1000	15.20
1200	21.41

(a) Cold junction at 0 °C.

Electrochemical equivalent. Valence 5: 0.3749 mg/C
Standard electrode potential. vs H_2: 1.12 V
Ionization potential. 7.3 ± 0.3 V
Hall coefficient. + 0.095 nV·m/A·T (virtually independent of temperature from 100 to 900 K)
Temperature of superconductivity. 4.38 K
Electron emission. 600 kA/m²·T^2, where T is in K
Work function. 0.657 aJ (4.10 eV)

Positive-ion emission. 1.60 aJ (10.0 eV)
Dielectric constant. For Ta_2O_5 layer: approx 400 kV per mm (10 kV per mil)
Thickness of anodic oxide film. At 0 °C: 1.6 nm/V; at 100 °C: 2.5 nm/V; at 200 °C: 3.0 nm/V

Magnetic Properties

Magnetic susceptibility. At 25 °C: 10.4×10^{-6} mks

Optical Properties

Spectral emittance. 0.49 for λ = 650.0 nm
Total hemispherical emittance. At 1400 °C: 0.20; at 1500 °C: 0.21; at 2000 °C: 0.25. See also Fig. 96.
Total radiation. At 1300 °C: 73.0 kW/m²; at 1530 °C: 128.0 kW/m²; at 1730 °C: 212.0 kW/m²

Nuclear Properties

Natural isotopes. 181
Thermal neutron cross section. 21.3 b

Chemical Properties

General corrosion behavior. Tantalum oxidizes in air above 300 °C. It has excellent resistance to corrosion by a large number of acids, by most aqueous solutions of salts, by organic chemicals and by various combinations and mixtures of these agents. Also, tantalum exhibits good resistance to many corrosive as well as

Fig. 100 Effects of temperature and oxygen content on Charpy V-notch impact energy of wrought electron-beam-melted tantalum

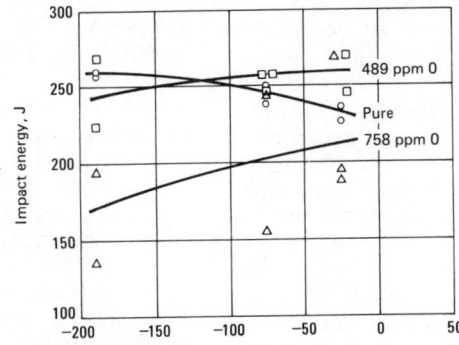

Sample impurities: <44 ppm C + N

Fig. 101 Rotating-beam fatigue strength of wrought electron-beam-melted tantalum

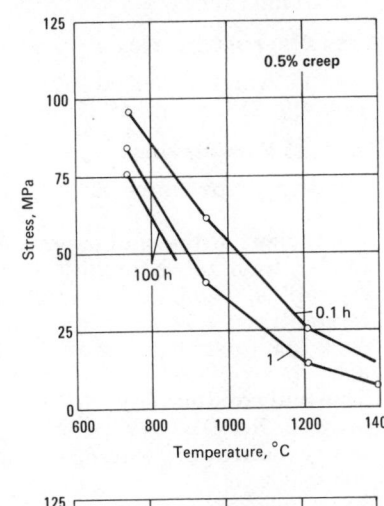

Sample impurities: <44 ppm C + N

Fig. 102 Creep characteristics of 1-mm-thick electron-beam-melted tantalum sheet

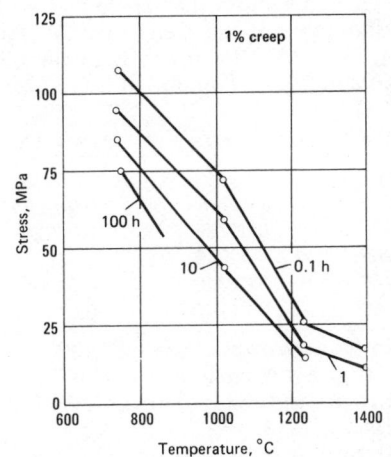

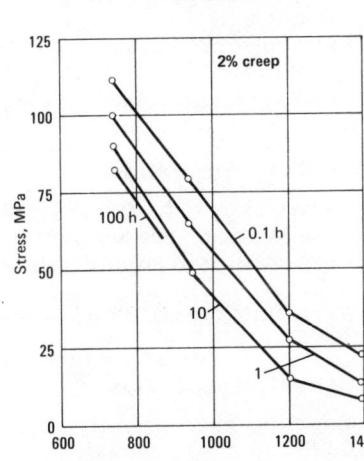

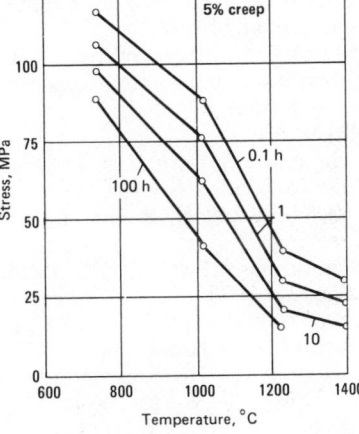

Sample impurities: 0.0030% C, 0.0016% O, 0.0010% N, <0.040% others. Sheet was cold rolled 75% and recrystallized by heating for 1 h at 1200 °C.

common gases and to many liquid metals.

Resistance to specific corroding agents. Tantalum is attacked by hydrofluoric acid, by fuming sulfuric acid and by strong alkalis. The presence of salts that hydrolyze to form hydrofluoric acid or strong alkalis also can lead to attack of tantalum. Tantalum may become embrittled by hydrogen if it is the cathode in a galvanic couple exposed to an acid environment or to a hydrogen-containing atmosphere at elevated temperature. Halogen gases can attack tantalum: fluorine causes attack at both room and elevated temperatures, chlorine at and above 250 °C, bromine at and above 300 °C, and iodine at somewhat higher temperatures. Bromine plus methanol also attacks tantalum.

Fabrication Characteristics

Precautions in melting. Exclude oxygen, hydrogen, nitrogen and carbon. Melt in vacuum or inert atmosphere.

Hot working temperature. None; it is worked cold

Annealing temperature. Above 1050 °C in high vacuum for complete recrystallization, with resulting grain size as follows:

Final annealing temperature(a), °C	Average ASTM grain size(b)
1200	5 to 6
1300	4
1400	3 to 4
1425	3 to 4
1600	2
1700	1
1800	0 to 1

(a) Material cold rolled 75% after intermediate annealing, then annealed 1 h at indicated temperature. (b) Determined by comparison with ASTM grain-size chart at 100×.

Maximum reduction between anneals. Greater than 95%

Suitable forming methods. Tantalum can be formed by spinning, deep drawing, bulging, bending, blanking, punching and stretch forming using conventional methods, equipment and tooling normally found in shops fabricating heat-resistant alloys.

Compacting pressure. 10 to 85 MPa depending on the physical properties of the powder

Sintering temperature. 2300 to 2600 °C in high vacuum will essentially remove all detrimental impurities contained in the powder

Machinability. Fully recrystallized unalloyed tantalum has machinability similar to that of soft copper. Use chlorinated hydrocarbons, light oil or water-soluble oil as a cutting fluid, and high speed tool steel or cemented carbide tools. Tantalum can be successfully turned, bored, drilled, tapped, reamed, shaped, milled, sawed and ground to desired tolerances and surface finishes.

Joining. Gas tungsten-arc, gas metal-arc, resistance and electron beam welding can be used for joining tantalum. High-purity inert gas (argon or helium) or vacuum must be used in fusion welding. Resistance spot and seam welding can be done in air or under water with proper precautions. Silver brazing alloys, copper, and several specially developed refractory metal brazing alloys can be used to braze tantalum to itself or to dissimilar metals such as stainless steels. Brazing is done in vacuum or under an inert atmosphere (high-purity argon or helium). Tantalum also can be bonded to dissimilar metals by explosive cladding, and in some instances by roll bonding.

Cleaning. To avoid contamination of tantalum by interstitial elements and metallic impurities, it is mandatory that the material be chemically cleaned before any heating operation (such as annealing or welding). Such cleaning involves thorough degreasing (detergent or solvent); chemical etching in 20 vol % HF, 20 vol % H_2SO_4 and 60 vol % HNO_3; hot- and cold-water rinsing (deionized water recommended); and spot-free drying. The etching solution may be strengthened (by adding HF) or weakened (by adding water) to achieve the amount of stock removal necessary to ensure cleanness.

Mechanical Properties

Tensile properties. See Fig. 97 to 99.
Hardness. Electron-beam melted, 110 HV; P/M compact, 120 HV
Poisson's ratio. 0.35 at 20 °C
Elastic modulus. Tension: 186 GPa at 20 °C, 159 GPa at 750 °C. Shear: 69 GPa at 20 °C. Compressibility:

MPa	$\Delta V/V_0$ 99.9% Ta	99.95 + % Ta
490	0.00244	0.00243
980	0.00488	0.00485
1470	0.00728	0.00726
1960	0.00969	0.00967
2450	0.01208	0.01208
2940	0.01447	0.01448

Impact strength. See Fig. 100.
Fatigue strength. See Fig. 101.
Creep-rupture characteristics. See Fig. 102.
Damping characteristics. In polycrystalline sheet vibrated at a frequency of 0.65 Hz, maximum damping occurs at 1100 °C.
Friction characteristics. Galls against itself and against type 18-8 stainless steel

REFERENCES

F. T. Sisco and E. Epremian, *Columbium and Tantalum*, Wiley, 1963
F. F. Schmidt and H. R. Ogden, "The Engineering Properties of Tantalum and Tantalum Alloys", DMIC Report 189, Sept 13, 1963
J. G. Sessler and V. Weiss, "Aerospace Structural Metals Handbook—Volume 11A, Nonferrous Heat Resistant Alloys", AFML-TR-68-115 (Vol. 11A), Jan 1968
D. R. Mash, D. W. Bauer and M. Schussler, Fabricating the Refractory Metals, *Metal Progress*, Feb-Mar-Apr 1971

Technetium (Tc)

Compiled by C. C. Koch
Oak Ridge National Laboratory

Technetium is used as a radioactive tracer in medicine with potential uses arising from its favorable corrosion-inhibiting properties and its high superconducting transition temperature. There is contamination hazard due to its radioactivity. Classed as moderately toxic. All sample preparation, etc., which could disperse solid ^{99}Tc must be carried out in glove-box facilities. The data that follow are for ^{99}Tc only

Structure

Crystal structure. Close-packed hexagonal; A3; $a = 0.2735$ nm, $c = 0.4388$ nm, $c/a = 1.604$ at 25 °C (Ref 1)
Minimum interatomic distance. 0.2716 nm (Ref 1)

Mass Characteristics

Atomic weight. 99.0000
Density. 11.5 Mg/m^3 at 25 °C

Thermal Properties

Melting temperature. 2200 °C (Ref 2)
Coefficient of thermal expansion. Linear, from 150 K to 25 °C. Polycrystalline: 7.05 μm/m·K. Along crystal axes: 7.04 μm/m·K along a axis: 7.06 μm/m·K along c axis (Ref 3)
Thermal conductivity. 50.2 W/m·K at 25 °C (Ref 4)
Recrystallization temperature. 700 to 800 °C
Enthalpy of sublimation. 6.68 kJ/kg (Ref 5)
Vapor pressure. See Fig. 103 (Ref 5)

Electrical Properties

Electrical resistivity. 185.0 nΩ·m at 25 °C (Ref 6). See also Fig. 104.
Temperature of superconductivity. 7.8 K (Ref 7)
Work function. 0.782 aJ (4.88 eV) (Ref 8)

Magnetic Properties

Magnetic susceptibility. Volume: 1.63×10^{-4} mks units (Ref 9)

Nuclear Properties

Isotopes. See Table 38.
Thermal neutron cross section. 22 b at 2200 m/s

Fig. 103 Vapor pressure of ^{99}Tc

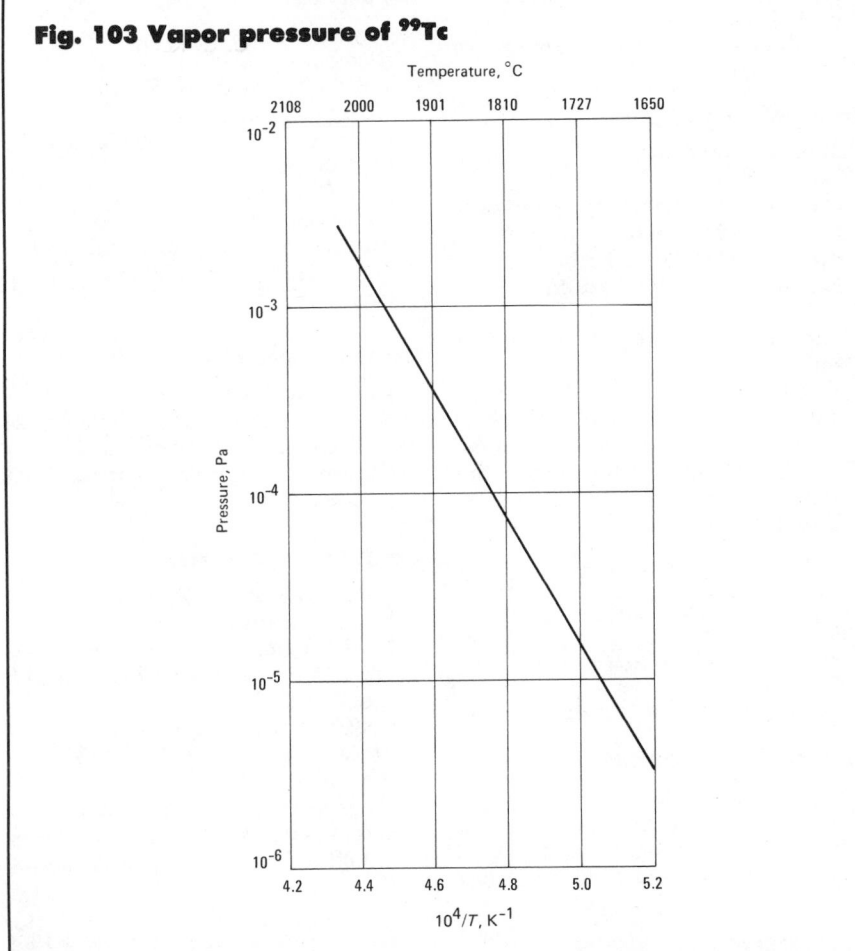

Fig. 104 Temperature dependence of the electrical resistivity of ^{99}Tc

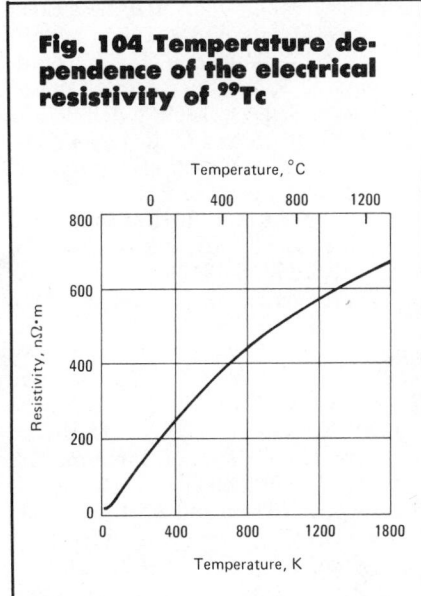

oxidizing acids such as nitric, sulfuric and aqua regia but not in hydrochloric acid. The pertechnetate ion has been shown to be an efficient anticorrosion agent in solution

Mechanical Properties

Tensile properties. Tensile strength. 1510 MPa as-rolled (46% reduction); 798.6 MPa after annealing 10 min at 950 °C; 0.2% offset yield strength, 1290 MPa as-rolled (46% reduction); 319 MPa after annealing 10 min at 950 °C (fully recrystallized); elongation in 1 in.: 4% as-rolled (46% reduction); 30% after annealing 10 min at 950 °C (Ref 10)

Hardness. 46% cold worked, 394 HV, 442 HB; annealed at 950 °C: 151 HV, 112 HB (Ref 10)

Poisson's ratio. 0.31 (Ref 11)

Elastic modulus. Tension, 322 GPa; shear, 123 GPa; bulk, 281 GPa (Ref 11)

Velocity of sound. At 25 °C: longitudinal velocity, 6220 m/s; shear velocity, 3270 m/s

REFERENCES

1. W. B. Pearson, *A Handbook of Lattice Spacings and Structures of Metals and Alloys,* Pergamon Press, NY, 1958
2. E. Anderson, R. A. Buckley, A. Hellawell, and W. Hume-Rothery, *Nature,* Vol 188, 1960, p 48-9
3. J. A. C. Marples and C. C. Koch, *Physics Letters* (Sec A, Vol 41, 1972, p 307-8

Table 38 Isotopes of technetium

Isotope	Half-life	Decay	Energy J × 10^{13}	MeV
92	4.4 min	β$^+$, ε	3.96	2.47
93	2.7 h	ε, β$^-$	1.31	0.82
94	293 min	ε, β$^+$	1.31	0.816
95	20 h	ε	...	...
96	4.3 days	ε	...	...
97	2.6 × 10^6 yr	ε	...	...
98	1.5 × 10^6 yr	β$^-$	0.48	0.30
99	2.14 × 10^5 yr	β$^-$	0.468	0.292
100	17 s	β$^-$	5.41	3.38
101	14.0 min	β$^-$	2.11	1.32
102	4.5 min	β$^-$	3.5	2.2
103	50 s	β$^-$	3.5	2.2
104	18 min	β$^-$	7.4	4.6
105	7.7 min	β$^-$	5.4	3.4
106	37 s	β$^-$	~10	~6.5

Chemical Properties

General corrosion behavior. Technetium metal tarnishes slowly in air and burns in oxygen. At 400 °C it reacts with fluorine to form the hexafluoride and with chlorine to form a mixture of the hexachloride and tetrachloride. Technetium dissolves in

4. D. E. Baker, *Journal of Less-Common Metals,* Vol 8, 1965, p 435

5. O. H. Krikarion, J. H. Carpenter and R. S. Newbury, *High Temperature Science,* 1969, p 313-330

6. C. C. Koch and G. R. Love, *Journal of Less-Common Metals,* Vol 12, 1967, p 29-35

7. S. T. Sekula, R. H. Kernohan and G. R. Love, *Physical Review,* Vol 155, 1967, p 364-369

8. S. Trasatti, *Surface Science,* Vol 32, 1972, p 735-738

9. C. C. Koch, W. E. Gardner and M. J. Mortimer, *Low Temperature Physics-LT 13,* Vol 2, edited by K. D. Timmerhaus, W. J. O'Sullivan and E. F. Hammel, Plenum, NY, 1974, p 595-600

10. R. G. Nelson and D. P. O'Keefe, Concluding Progress Report, *A Study of Tungsten-Technetium Alloys,* Battelle Memorial Institute Pacific Northwest Laboratory, BNWL-865, 1968

11. G. R. Love, C. C. Koch, H. L. Whaley and Z. R. McNutt, *Journal of Less-Common Metals,* Vol 20, 1970, p 73.75

Tellurium (Te)

Compiled by S. C. Carapella, Jr.
ASARCO Incorporated

Tellurium is used as an additive to steel and copper to improve machinability, an additive to cast iron to control depth of chill, in the production of malleable cast iron as a carbide stabilizer, and in lead base alloys to improve their properties. It is an important constituent of thermoelectric alloys. It is also used in fuses for explosives, as a vulcanizing agent in rubber, a catalyst in chemical reactions, as a glass forming agent in glasses, and as a colorizing agent in glass and ceramics.

Structure

Crystal structure. Hexagonal; at 25 °C, $a = 0.44565$ nm and $c = 0.59268$ nm

Mass Characteristics

Atomic weight. 127.60
Density. 6.237 Mg/m^3 at 25 °C. Temperature dependence: solid (0 to 450 °C), $6.250 - 0.000261T$; liquid

(450 to 1000 °C), $6.170 - 0.000777T$
Volume change on freezing. 5% contraction

Thermal Properties

Melting point. 449.5 °C
Boiling point. 988 °C
Coefficient of thermal expansion. Linear, 18.2 μm/m·K at 20 °C
Specific heat. 201 J/kg·K at 25 °C
Latent heat of fusion. 86.113 kJ/kg
Latent heat of vaporization. 446.43 kJ/kg
Thermal conductivity. Polycrystalline, between 5.98 and 6.02 W/m·K at about 20 to 28 °C. Single crystals: 3.3 W/m·K ∥ to c axis, 2.1 W/m·K ⊥ to c axis
Vapor pressure:

°C	kPa
505	0.1013
617	1.013
768	10.13
988	101.3

Electrical Properties

Electrical resistivity. Polycrystalline, between 1 and 50 mΩ·m at 25 °C. Single crystals, at 20 °C: 5 mΩ·m ∥ to c axis, 1.5 mΩ·m ⊥ to c axis

Mechanical Properties

Tensile strength. 10.8 to 11.25 MPa
Hardness. 25 HB, 2.3 moh
Elastic modulus. Tension (single crystals): 42.57 GPa ∥ to c axis, 20.45 GPa ⊥ to c axis. Shear (polycrystalline), 15.16 GPa
Liquid surface tension. 186 mN/m at 450 °C

Terbium (Tb)

Compiled by K. A. Gschneidner, Jr. and B. J. Beaudry
Ames Laboratory
U.S. Department of Energy
Iowa State University

Terbium is used as a phosphor and in catalysts. It will remain shiny in air at room temperature. Turnings can be ignited and will burn white hot. Finely divided terbium should not be handled in air.

Structure

Crystal structure. α' phase, orthorhombic, $Cmcm\ D^{17}_{2h}$; $a = 0.360$ nm, $b = 0.625$ nm, $c = 0.571$ nm at −73 °C. α phase, close-packed hexagonal, $P6_3/mmc\ D^4_{6h}$; $a = 0.36055$ nm, $c = 0.56966$ nm at 24 °C. β phase, body-centered cubic, $Im3m\ O^9_h$; $a = 0.402$ nm at 1290 °C
Minimum interatomic distance. $r_a = 0.18028$; $r_c = 0.17639$; radius $CN_{12} = 0.17833$ nm

Mass Characteristics

Atomic weight. 158.925
Density. α phase, 8.230 Mg/m^3 at 24 °C; β phase, 8.12 Mg/m^3 at 1290 °C
Volume change on freezing. 3.1% contraction

Thermal Properties

Melting point. 1356 °C
Boiling point. 3230 °C
Phase-transformation temperature. α' to α phase, −73 °C; α to β phase, 1289 °C
Coefficient of thermal expansion. At 24 °C. Linear: 10.3 μm/m·K. Linear, along crystal axes: 9.3 μm/m·K along a axis, 12.4 μm/m·K along c axis. Volumetric: 31.0×10^{-6} per K
Specific heat. 181.9 J/kg·K at 25 °C
Entropy. 463.7 J/kg·K at 25 °C
Latent heat of fusion. 67.92 kJ/kg
Latent heat of transformation. 31.59 kJ/kg
Latent heat of vaporization. 2.446 MJ/kg at 25 °C
Heat of combustion. For cubic Tb_2O_3 at 25 °C: $\Delta H^0_c = 5.90$ MJ/kg Tb; $\Delta G^0_f = -5.60$ MJ/kg Tb
Recrystallization temperature. 500 °C
Thermal conductivity. 11.1 W/m·k at 25 °C
Vapor pressure. 0.001 Pa at 1124 °C; 0.101 Pa at 1354 °C; 10.1 Pa at 1698 °C; 1013 Pa at 2510 °C

Electrical Properties

Electrical resistivity. 1150 nΩ·m at 25 °C; 35 nΩ·m at 4 K. Along crystal axes, at 25 °C: 1235 nΩ·m along a axis, 1015 nΩ·m along c axis. Liquid: 1930 nΩ·m at 1358 °C
Ionization potentials. Tb(I), 5.85 V; Tb(II), 11.52 V; Tb(III), 21.71 V; Tb(IV), 3950 V
Hall coefficient: Along crystal axes at 20 °C: −0.10 nV·m/A·T along b axis; −0.37 nV·m/A·T along c axis

Temperature of superconduction. Bulk terbium is not superconducting down to 0.37 K at atmospheric pressure.

Magnetic Properties

Magnetic susceptibility. Volume (mks units): $\chi_a = 0.129$ and χ_c 0.0738 at 27 °C, obeys Curie-Weiss law above 240 K with an effective moment of 9.77 Bohr magnetons and $\theta_a = 239$ K and $\theta_c = 195$ K
Saturation magnetization. 3.39 T at 0 K along $\langle 11\bar{2}0 \rangle$
Magnetic transformation temperatures. Curie temperature, 221 K; Néel temperature, 195 K

Optical Properties

Color. Metallic silver

Nuclear Properties

Thermal neutron cross section. 44 b

Chemical Properties

General corrosion behavior. Terbium stays shiny in air at room temperature. The rate of oxidation is slow even at 1000 °C due to the formation of a dark, tightly adhering oxide on the surface. Water vapor increases the rate of oxidation. After heating to 550 °C in vacuum, hydrogen will react at 250 °C.
Resistance to specific corroding agents. Terbium does not react with cold or hot water but will react vigorously with dilute acids. It is slowly attacked by concentrated sulfuric. The presence of the fluoride ion retards acid attack due to the formation of TbF_3.

Mechanical Properties

Tensile properties. About the same as gadolinium
Hardness. 38 HV for polycrystalline; 30 HV for {10$\bar{1}$0} prismatic face; 80 HV for {0001} basal plane
Poisson's ratio. 0.255
Elastic modulus. At 27 °C: tension, 57.1 GPa; shear, 22.7 GPa; bulk, 38.9 GPa
Elastic constants along crystal axes. At 27 °C: $c_{11} = 69.24$ GPa; $c_{12} = 24.98$ GPa; $c_{13} = 21.79$ GPa; $c_{33} = 74.39$ GPa; $c_{44} = 21.75$ GPa

Thallium (Tl)

Compiled by S. C. Carapella, Jr.
ASARCO Incorporated

Thallium is used in alloying to lower the freezing point of certain metals, such as mercury in arctic thermometers and low temperature mercury switches. It is also a component of fusible alloys, glass, and may be used as an additive in the counter-electrode metal for selenium rectifiers. Compounds of thallium are used for catalysts and semiconductor applications. Because of the high toxicity of thallium, skin contact and inhalation of dust and fumes are to be avoided. Impervious gloves and aprons should be worn and dust and fumes in work areas are to be controlled by exhaust ventillation.

Structure

Crystal structure. Close packed hexagonal below 230 °C; $a = 0.34560$ nm, $c = 0.55248$ nm. Body centered cubic above 230 °C; $a = 0.3874$ nm

Mass Characteristics

Atomic weight. 204.37
Density. 11.872 Mg/m^3 at 20 °C
Volume change on freezing. 3.23% contraction

Thermal Properties

Melting point. 303 °C
Boiling point. 1473 °C
Coefficient of thermal expansion. Linear, 28 $\mu m/m \cdot K$ at 20 °C
Specific heat. 150 J/kg·K for liquid, 130 J/kg·K for solid
Latent heat of fusion. 20.27 kJ/kg
Latent heat of vaporization. 802.833 kJ·kg
Thermal conductivity. 47 W/m·K at 0 °C
Vapor pressure:

°C	kPa
818	0.1013
972	1.013
1179	10.13
1473	101.3

Electrical Properties

Electrical resistivity. 150 nΩ·m at 0 °C (solid), 740 nΩ·m at 303 °C (liquid)
Electrochemical equivalent. Valence +3, 706.01 μg/C
Standard electrode potential. 0.336 V vs standard hydrogen electrode

Magnetic Properties

Magnetic susceptibility. Volume: -3.1×10^{-6} mks

Optical Properties

Color. Dull gray

Nuclear Properties

Stable isotopes. 203, 205
Thermal neutron cross section. For 2.2 km/s neutrons: absorption, 3.3 ± 0.5 b; scattering, 14 ± 2 b

Mechanical Properties

Tensile strength. 8.9 MPa
Elongation. 40% in 5 in.
Hardness. 2 HB
Liquid surface tension. 467 mN/m at 303 °C, 450 mN/m at 450 °C

Thorium (Th)

Compiled by J. F. Smith
Ames Laboratory
U. S. Department of Energy
Iowa State University

Thorium, as a solid or fluid in elemental, intermetallic, or oxide form, is used as a fuel for nuclear reactors since it is a fertile material for the generation of fissionable uranium-233. The oxide form of thorium is used for gas mantles. Thorium oxide additions control grain size in tungsten filaments and strengthen nickel alloys (TD nickel). Thorium metal is used as an alloying addition in magnesium technology and is used as a deoxidant for molybdenum, iron and other metals. Thorium has a variety of applications in electronic technology. Thorium is radioactive. Pure, fresh thorium is a weak α-emitter, but old thorium, with accumulated decay products, also emits β-particles and penetrating γ-rays. Thorium is also chemically quite reactive. In finely divided form, thorium can be pyrophoric, and in dust form, it may be explosive. Chemical toxicity of thorium and its compounds is generally low.

Structure

Crystal structure. α-phase: face-centered cubic, A1, cF4 (Fm3m); a =

Fig. 105 Thermal conductivity of thorium

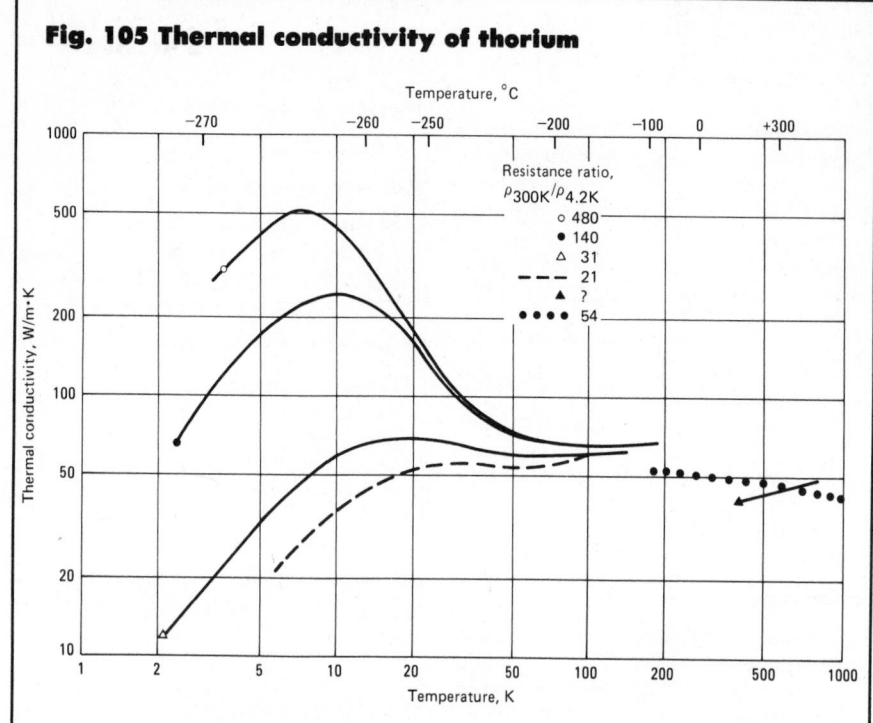

Fig. 106 Thermoelectric power of thorium

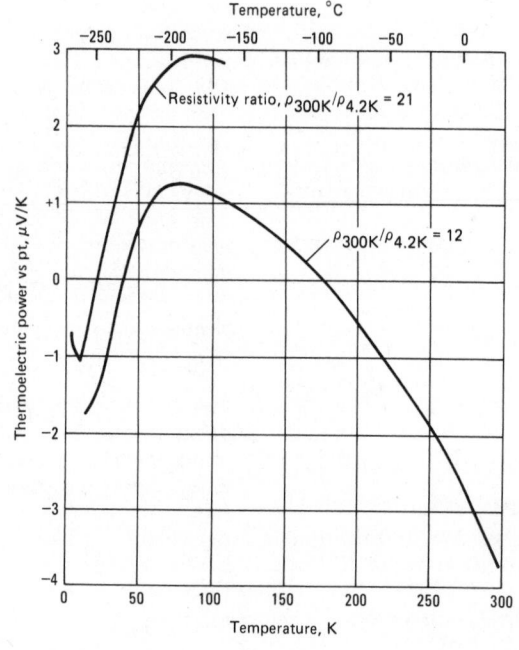

0.5086 nm at 25 °C. β-phase: body-centered cubic, A2, cI2 (Im3m); a = 0.411 nm at 1450 °C

Slip planes. Deformation textures imply that {111} slip planes are active throughout the temperature range of study, −196 to 900 °C

Minimum interatomic distance. 0.3596 nm at 25 °C

Microstructure. Common inclusions in thorium metal are gray ThO₂, a so-called "white phase" of debated identity which often surrounds cast grains of calcium-reduced thorium, gold-colored nitrides, and an occasional massive particle of tungsten in arc-melted material. A fine, Widmanstätten-like microstructure with cream-colored needles and angular inclusions, which turn a deep blue after exposure to air for 1 day, can be produced by melting in graphite crucibles.

Mass Characteristics

Atomic weight. 232.038

Density. Solid: 11.8 Mg/m³ at −273 °C; 11.72 Mg/m³ at +25 °C; 10.89 Mg/m³ at 1755 °C. Liquid: 10.35 Mg/m³ at 1755 °C

Density vs temperature. $\Delta d/d_o \cdot K = -34.2 \times 10^{-6}$ at 25 °C

Volume change on freezing. −5%

Thermal Properties

Melting point. 1755 °C

Boiling point. ~4800 °C

Phase-transformation temperature. β- to α-phase (cooling), 1345 °C

Coefficient of thermal expansion. Linear: 10.9 μm/m·K at −193 °C; 11.4 μm/m·K at +25 °C; 12.6 μm/m·K at 600 °C; 13.3 μm/m·K at 750 °C; 14.0 μm/m·K at 850 °C; 14.9 μm/m·K at 950 °C

Specific heat. 117.78 J/kg·K at 25 °C

Specific heat vs temperature:

°C	J/kg·K
−253	19.95
−223	72.992
−173	98.850
−123	107.56
−73	112.1
−23	115.26
27	117.89
127	123.5
327	134.3
427	139.9
527	145.3
727	156.4
827	161.7
927	167.3

Fig. 107 Temperature dependence of the tensile strength of thorium

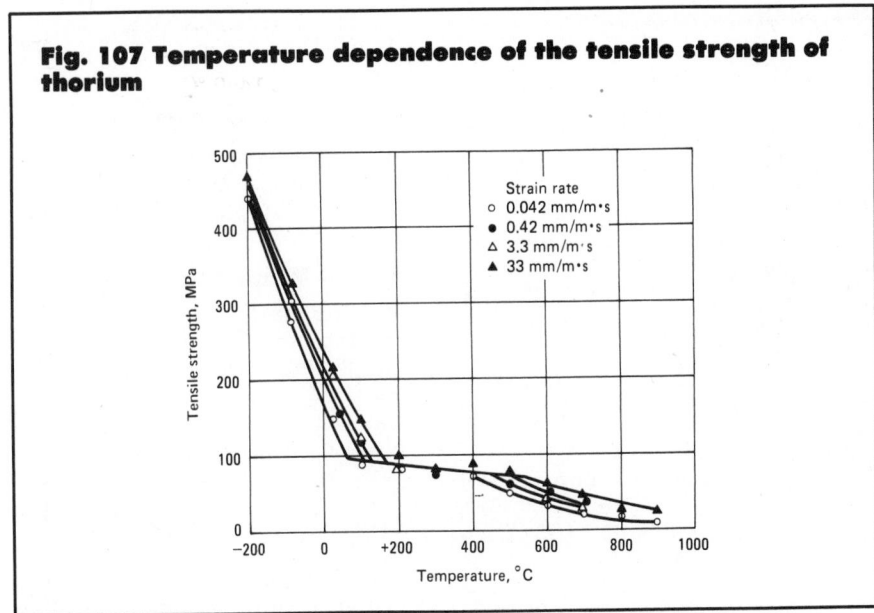

Fig. 109 Temperature dependence of the hardness of thorium

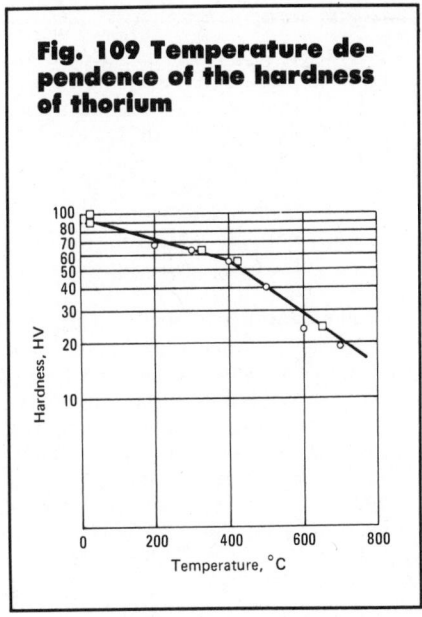

Fig. 108 Temperature dependence of the yield strength of thorium

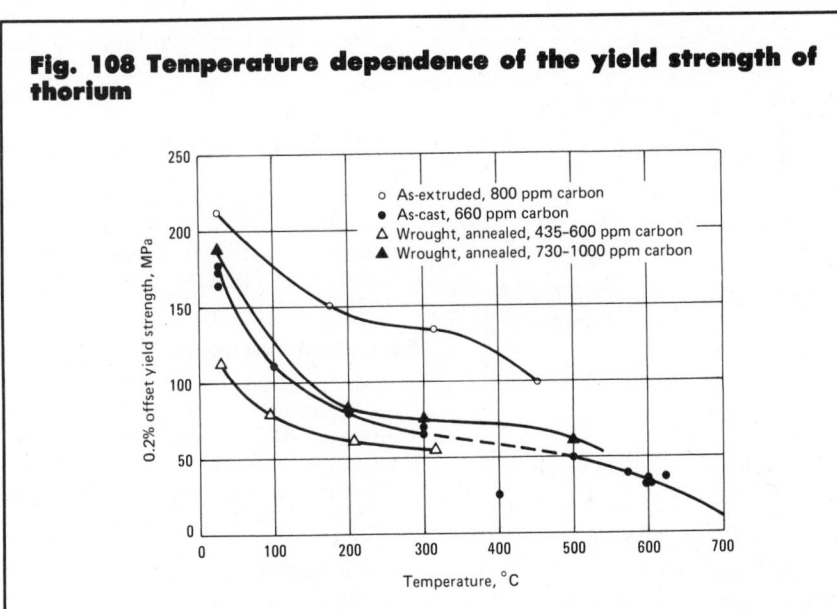

Enthalpy. $H_{298} - H_0 = 28.06$ kJ/kg at 25 °C; entropy, $S_{298} = 230.1$ J/kg·K at 25 °C

Latent heat of fusion. 59.47 kJ/kg

Latent heat of phase transformation. 15.6 kJ/kg

Latent heat of sublimation. 2.539 MJ/kg at 25 °C

Enthalpy of oxide formation. (ThO_2) -4.6460 MJ/kg at 25 °C

Free energy of oxide formation. (ThO_2) -4.4274 MJ/kg at 25 °C

Thermal conductivity. 77 W/m·K at 25 °C

Thermal conductivity vs temperature. See Fig. 105.

Vapor pressure. Solid: $\log P$ (Pa) = $-(28780 \pm 620)/T + 10.997 \pm 0.333$ at 1484 to 1683 °C; liquid: $\log P$ (Pa) = $-(29770 \pm 220)/T + 11.030 \pm 0.098$ at $1747 - 2187$ °C, where T is in K

Diffusion coefficient. (D in m²/s, activation energy in kJ/mol, 1 in K) Self-diffusion: $D = 1.7 \times 10^{-4}$ exp $(-327/RT)$ for α-phase; $D = 0.5 \times 10^{-4}$ exp $(-230/RT)$ for β-phase. Hydrogen in α-phase: $D = 2.92 \times 10^{-7}$ exp $(-40.8/RT)$ at infinite dilution. Carbon in β-phase: $D = 2.2 \times 10^{-6}$ exp $(-113/RT)$. Nitrogen in β-phase: $D = 3.2 \times 10^{-7}$ exp $(-71/RT)$. Oxygen in β phase: $D = 1.3 \times 10^{-7}$ exp $(-94.1/RT)$

Electrical Properties

Electrical conductivity. Volumetric, 11% IACS

Electrical resistivity. 157 nΩ·m at 25 °C

Temperature coefficient of electrical resistivity. 0.560 nΩ·m per K

Pressure coefficient of electrical resistivity. 5.7 aΩ·m per GPa

Thermoelectric potential. See Fig. 106.

First ionization potential. 6.08 V

Hall coefficient. -0.088 to -0.13 nV·m/A·T at 25 °C

Temperature of superconductivity. 1.390 K at zero field

Increase of electrical resistivity with carbon addition. 55 pΩ·m/ppm carbon by weight

Work function. 0.559 aJ

Magnetic Properties

Magnetic susceptibility. Volume: 60.7×10^{-6} mks at 25 °C

Magnetic permeability. 1.0000607 at 25 °C

Optical Properties

Color. A fresh surface exhibits a bright silvery luster, which darkens after prolonged exposure to air

Emissivity. Total: 30% at 1127 °C; 30.5% at 1227 °C; 31% at 1327 °C; 32% at 1427 °C; 34% at 1527 °C. Change in emissivity at the α–β transition <0.5%

Nuclear Properties

Unstable isotopes. Isotopes from Th^{223} through Th^{235} are known and all are radioactive with half-lives ranging from 0.9 s to 1.4×10^{10} years. This latter half-life is associated with Th^{232}, which is the isotope constituting essentially 100% of the natural abundance; Th^{232} emits α-particles with energies of 6.38×10^{-13} J

Effect of neutron irradiation. Dimensional changes in irradiated thorium are essentially isotropic, relatively small, and are associated with the increase in volume due to the accumulation of fission products in the material. Tensile strength has been found to increase 75% after a neutron exposure of 10^{19} nvt and an additional 30% at double that exposure

Chemical Properties

Resistance to specific corroding agents. In air between 100 and 900 °C, corrosion is principally oxidation and follows a linear reaction rate; at 800 °C, a weight gain of 2.88 kg/m^2 per day in air is typical. Above 850 °C, reaction with oxygen follows the parabolic law. Reaction with nitrogen also follows the parabolic law in the range 671 to 1490 °C, and in purified nitrogen, typical weight gain at 800 °C is 0.96 kg/m^2 per day. Thorium corrodes in water to form thorium oxide and hydrogen and loses weight by spallation; typical weight loss of unalloyed thorium in high-purity water at 178 °C is 0.109 kg/m^2 per day

Fabrication Characteristics

Machinability. Thorium is readily machined. However, high-purity thorium tends to be gummy; its softness causes continuous turnings. Lower-purity material contains abrasive oxides, which cause rapid tool wear. It is advisable to use a water-soluble coolant when machining thorium.

Recrystallization temperature. Depends upon purity and amounts of prior cold work, but is approx 650 °C

Sintering temperature. 1100 to 1200 °C

Annealing temperature. Initial recovery: 525 °C. Recrystallization: 650 °C

Hot-working temperature. 750 to 900 °C

Mechanical Properties

Tensile properties. As-cast thorium (0.02 to 0.08 wt% carbon): tensile strength, 219 MPa; yield strength (0.2% offset), 144 MPa; elongation, 34%; reduction in area, 35%. See also Fig. 107 and 108.

Compressive properties. Values closely comparable to tensile values

Hardness. 56 to 114 HV with 20-kg load; see also Fig. 109.

Poisson's ratio. 0.27

Strain-hardening exponent. 0.18 at 25 °C

Elastic modulus. Tension, 72.4 GPa; shear, 27.6 GPa; bulk, 57.7 GPa

Elastic properties along [100] crystal axis. Tension modulus, 60.1 GPa, shear modulus, 47.8 GPa, Poisson's ratio, 0.394

Impact strength. 13 J (approx) for calcium-reduced thorium at 25 °C

Fatigue strength. Endurance limit 97 MPa (approx) for calcium-reduced thorium, reversed bending test

Specific damping capacity. 3×10^{-3} in range 0 to 300 °C

SELECTED REFERENCE

J. F. Smith *et al*, *Thorium: Preparation and Properties,* Iowa State University Press, Ames, IA 1975

Thulium (Tm)

Compiled by K. A. Gschneidner, Jr. and B. J. Beaudry
Ames Laboratory
U.S. Department of Energy
Iowa State University

Thulium is used in phosphors, ferrite bubble devices and catalysts. Irradiated thulium (^{169}Tm) is used as a portable radiographic source. Thulium will remain shiny in air at room temperature. Turnings can be ignited and will burn white hot. Finely divided thulium should not be handled in air; due to its high vapor pressure at its melting point, thulium should not be arc melted.

Structure

Crystal structure. Close-packed hexagonal: $P6_3mmc$ D^4_{6h}; $a = 0.35375$ nm, $c = 0.55540$ nm at 25 °C

Minimum interatomic distance. At 24 °C: $r_a = 0.17688$ nm; $r_c = 0.17236$ nm; radius $CN_{12} = 0.17462$ nm

Mass Characteristics

Atomic weight. 168.9342
Density. 9.321 Mg/m^3 at 24 °C
Volume change on freezing. Approx 6.9% contraction

Thermal Properties

Melting point. 1545 °C
Boiling point. 1950 °C
Coefficient of thermal expansion. At 24 °C. Linear: 13.3 $\mu m/m \cdot K$. Linear, along crystal axes: 8.8 $\mu m/m \cdot K$ along a axis, 22.2 $\mu m/m \cdot K$ along c axis. Volumetric: 39.8×10^{-6} per K
Specific heat. 160.0 $J/kg \cdot K$ at 25 °C
Entropy. At 25 °C: 438.1 $J/kg \cdot K$
Latent heat of fusion. 99.68 kJ/kg
Latent heat of vaporization. 1.374 MJ/kg at 25 °C
Latent heat of combustion. For cubic Tm_2O_3 at 25 °C: $\Delta H^0_c = 5.59$ MJ/kg Tm; $\Delta G^0_c = -5.33$ MJ/kg Tm
Recrystallization temperature. About 600 °C
Thermal conductivity. 16.9 $W/m \cdot K$ at 25 °C
Vapor pressure. 0.001 Pa at 599 °C; 0.101 Pa at 748 °C; 10.1 Pa at 964 °C; 1013 Pa at 1300 °C

Electrical Properties

Electrical resistivity. 676 $n\Omega \cdot m$ at 25 °C; 56 $n\Omega \cdot m$ at 4 K. At 25 °C: 880 $n\Omega \cdot m$ along a axis, 472 $n\Omega \cdot m$ along c axis
Ionization potentials. Tm(I), 6.18 V; Tm(II), 12.05 V; Tm(III), 23.68 V; Tm(IV), 42.55 V
Hall coefficient. -0.18 $nV \cdot m/A \cdot T$ at 20 °C
Temperature of superconductivity. Bulk thulium is not superconducting down to 0.35 K at atmospheric pressure.

Magnetic Properties

Magnetic susceptibility. Volume (mks units): $\chi_a = 0.0160$ and $\chi_c = 0.0195$ at 25 °C; obeys Curie-Weiss law above 55 K with an effective moment of 7.61 Bohr magnetons and $\theta_a = 17$ K and $\theta_c = 41$ K
Saturation magnetization. 2.79 T at 4.2 K along <0001>
Magnetic transformation temperatures. Curie temperature, 25 K; a spin rearrangement at 42 K, Néel temperature, 58 K

Optical Properties

Color. Metallic silver

Nuclear Properties

Thermal neutron cross section. 125 b

Chemical Properties

General corrosion behavior. Thulium stays shiny in air at room temperature. Even at 1000 °C the rate of oxidation is slow due to the formation of a dark, tightly adhering oxide on the surface of the metal.

Resistance to specific corroding agents. Thulium does not react with cold or hot water, but reacts vigorously with dilute acids. The attack by concentrated sulfuric is slow. The presence of the fluoride ion retards acid attack due to the formation of TmF_3 on the surface of the metal.

Mechanical Properties

Tensile properties. About the same as erbium

Hardness. 48 HV

Poisson's ratio. 0.217

Elastic modulus. At 27 °C: tension, 74.0 GPa; shear, 30.4 GPa; bulk, 44.4 GPa

Tin (Sn)

Compiled by Joseph B. Long
Tin Research Institute, Inc.

Under certain specific conditions at low temperature, tin can transform from the normal tetragonal metal (β-tin or white tin) to a cubic form (α-tin or gray tin) that has entirely different properties. Because this transformation is accompanied by an increase in volume, the resultant expansion causes disintegration of the metal to coarse powder or to local "warts". The equilibrium temperature of transformation is 13.2 °C. In practice, it is extremely difficult to initiate the change, and even after transformation has started, the rate is slow. Moreover, common impurities such as bismuth, antimony and lead inhibit the change. Fear of failure at low temperatures of fabricated products made of tin or high-tin alloys or of the disintegration of tin in storage is largely unfounded.

The existence of "gamma tin," a brittle modification that is sometimes mentioned in technical literature, has been disproved (Ref 1). Abrupt changes in some properties at elevated temperatures, such as high temperature ductility, are ascribed more accurately to impurities. Many of the properties listed here were determined on tin of 99.95% purity.

Structure

Crystal structure. α-phase: face-centered cubic, A4, cF8 (Fd3m); a = 0.64912 nm (Ref 2,3). β-phase: body-centered tetragonal, A5, tI4 ($I4_t/amd$); a = 0.58314 nm, c = 0.31815 nm (Ref 4 to 9)

Slip elements. β-phase (Ref 10, 11):

	20 °C		150 °C	
Slip plane	Slip direction		Slip plane	Slip direction
(110)	[001] [111]		(110)	[$\bar{1}$11]
(100)	[001] [010]			
(101)	[101]			
(121)				

Twinning plane. (301) (Ref 3)
Interatomic distances. (Ref 12):

Table 39 Effect of current density on hydrogen overvoltage of tin

Current density, A/m²	Overvoltage (V) in Ref:						
	1	1	52	53	53	54	55
1	...	...	...	...	...	...	0.50
5	...	...	...	...	...	...	0.65
10	0.85	0.66	0.73	0.59	0.57	...	0.73
20	...	...	...	...	...	0.71	0.89
50	...	...	...	...	...	0.87	1.09
100	0.97	0.85	0.89	0.72	0.71	0.98	1.29
200	...	...	...	...	...	1.04	...
500	...	...	...	...	...	1.13	...
1000	0.98	0.89	0.99	0.86	0.83	1.19	...
2000	...	...	...	...	...	1.37	...
5000	...	...	1.00	...	...	...	...
10 000	...	0.98	...	...	...	0.88	...

Table 40 Optical properties of tin at λ = 546.1 nm (Ref 60 to 63)

Surface	Reflectance	Refractive index	Absorptive index
Film, 42-200 nm thick(a)	0.70	2.4	1.9
Film, 2.5 nm thick(a)	...	3.0	0.17
Bulk solid	0.80	1.0	4.2
Liquid	0.80	1.7	3.1

(a) Vacuum evaporated film.

	Interatomic distance, nm	Number of neighbors
α tin	0.279	4
	0.456	8
β tin	0.302	4
	0.318	2
	0.376	4
Liquid tin (250 °C)	0.338	10

Microstructure. Because tin is an extremely soft metal, it is often difficult to obtain an unworked surface free from scratches, and the low crystallization temperature of tin may result in false structures if distortion occurs during polishing. To overcome these difficulties, it is therefore necessary to take special precautions during both mounting and polishing.

Fracture behavior. Ductile

Mass Characteristics

Atomic weight. 118.69

Density. α-phase, 5.765 Mg/m³ at 1 °C (Ref 13). β-phase: 7.2984 Mg/m³ at 15 °C (Ref 14), 7.168 Mg/m³ at 20 °C (Ref 15, 16). Liquid:

°C	Mg/m³	Ref
298	6.94	17
409	6.840	15, 16
474	6.789	15, 16
523	6.761	15, 16
538	6.77	17
574	6.729	15, 16
602	6.711	15, 16
648	6.671	15, 16
816	6.62	17
1093	6.45	17
1371	6.29	17
1573	6.16	17

Table 41 Unstable (radioactive) isotopes of tin (Ref 64)

Mass No.	Half life	Energy of radiation, MeV Particles	γ rays
?	......4.5, 3 h		
108?	......4 h		
111?	......35 min	β⁻1.45, 1.51	
113	......30-33, 25 min 112, ~100, 70 days	β⁻1.2	~0.09, 0.85, ?
117	......14.5, 14, 13 days		0.175, 0.17, 0.159, 0.162, 0.152, 0.157
119	......≥100, 279, ~250, 245 days		0.069, 0.064
121	......27.0, 26.4, 27.5 h long, >400 days	β⁻0.383, 0.35, ~0.4 β⁻0.41, 0.42	
123	......130, 136 days 39.5, 41, 40 min	β⁻1.42, 1.3, ~1.5 β⁻1.26, 1.12, 1.32, ~1.7	0.394, ? 0.153, ~0.17, ~0.4
124	......0.4-0.9 × 10¹⁶ yr?	β⁻1.0-1.5, ?	
125	......10.0, 9.9, 11, 9.4 days	β⁻2.38, 2.34, 2.1 β⁻2.6, 2.33, 2.06 β₁⁻2.37 β₂⁻0.40	?, ~1.9
	9.5, 9.8, 10 min	β₁⁻2.04, 2.05, β₂⁻2.2, 2.06 1.17 β₃⁻0.51, ?, 0.5, ~0.5	γ₁0.326, 0.36 γ₂1.86, >1 γ₃1.37
126	......70, 80 min	β⁻0.7, 2.7	~1.2

Volume change on freezing. 2.8% contraction

Volume change on phase transformation. β-phase to α-phase, 27% expansion

Thermal Properties

Melting point. 231.9 °C (Ref 18)

Effect of pressure on melting point. From Ref 19:

Pressure, MPa	Temperature, °C
51	232.26
76	233.09
101	233.89
151	235.47
203	237.18

Boiling point. 2770 °C (Ref 20)

Phase-transformation temperature. β-phase to α-phase, 13.2 °C (Ref 21)

Coefficient of thermal expansion. For α-phase (Ref 22,23):

Temperature, °C	Linear, μm/m·K	Volumetric, mm³/m·K
−200	13.5	40 600
−150	16.6	49 900
−100	18.1	54 300
−50	19.2	57 500
0	19.9	59 800
50	23.1	69 200
100	23.8	71 400
150	26.7	80 200

Volumetric for liquid phase (Ref 16,17,24):
106 000 mm³/m·K for 232 to 400 °C, 105 000 mm³/m·K for 400 to 700 °C, 100 000 mm³/m·K for 232 to 1600 °C

Coefficients of linear expansion along crystal axes. From Ref 22,25:

Temperature, °C	Coefficient, μm/m·K Parallel to c-axis	Perpendicular to c-axis
−200	21.3	9.4
−150	24.1	12.8
−100	25.7	14.3
−50	27.0	15.2
0	28.4	15.8
50	32.9	16.6
100	35.7	17.9
150	38.4	19.2
200	40.4	19.9

Specific heat. From Ref 20,26: α-phase at 10 °C, 205 J/kg·K; β-phase at 25 °C, 222 J/kg·K, β-phase from 25 to 231 °C, $155 + 0.22 \times T$ J/kg·K

(where T is temperature in K); liquid phase from 232 to 1000 °C, 257 J/kg·K

Latent heat of fusion. 59.5 kJ/kg (Ref 27)

Latent heat of phase transformation. 17.6 kJ/kg (Ref 27)

Latent heat of vaporization. 2.4 MJ/kg (Ref 20, 28, 29, 30)

Thermal conductivity. From Ref 31 to 34:

Phase	°C	W/m·K
β	−170	80.8
β	0	62.8
β	100	60.7
β	200	56.5
Liquid	232 to 332	32.6

Vapor pressure. From Ref 30:

°C	Pa
727	9.9 × 10⁴
927	0.16
1027	1.1
1127	5.9
1227	23
1327	89
1527	746
1727	4.0⁸ × 10³
1827	8.41 × 10³
2027	2.9 × 10⁴
2127	5.13 × 10⁴
2227	8.51 × 10⁴

Diffusion coefficient. Selfdiffusion along crystal axes in β-phase (Ref 35):

Temperature, °C	Coefficient, μm²/s c-axis	a-axis
180.5	0.0111	0.00527
197.0	0.0148	0.00601
210.7	0.0213	0.00930
223.1	0.0265	0.00929

Electrical Properties

Electrical conductivity. 15.6% IACS

Electrical resistivity:

°C	μΩ·m
0	0.110
100	0.155
200	0.200
231	0.220

For liquid phase (Ref 43):

°C	μΩ·m
232	0.450
300	0.468
400	0.490
500	0.515
600	0.540
700	0.563
800	0.587
900	0.612

Electrical resistivity along crystal axes. β-phase (Ref 44, 45): 0.120 μΩ·m parallel to c-axis, 0.092 μΩ·m perpendicular to c-axis

Pressure coefficient of electrical resistance. -9.51×10^{-6} at 30 °C between 0 to 2.9 MPa (Ref 46)

Thermoelectric power. Between solid and liquid tin (liquid is at the higher potential) (Ref 47), 1.6 μV/K
Electrochemical equivalent. Valence +2, 615.03 g/C; valence +4, 307.51 g/C (Ref 48)

Standard electrode potential. Versus standard hydrogen electrode:
$$\text{Sn/Sn}^{2+} \quad -0.14 \text{ V}$$
$$\text{Sn}^{2+}/\text{Sn}^{4+} \quad -0.15 \text{ V}$$
(calculated) (Ref 49,50)

Ionization potential. 7.297 V, spectrographic (Ref 51)
Hydrogen overvoltage. See Table 39.

Hall effect. -0.02 nV·m/A·T at 0.4 T and room temperature (Ref 56 to 58)
Temperature of superconductivity. 3.73 K (Ref 58)
Photoelectric work function. 464 eV (Ref 59)

Magnetic Properties
Magnetic susceptibility. Mass: α-phase, -39×10^{-11} mks; β-phase at 18 °C, $+34 \times 10^{-11}$ mks; liquid phase at 250 °C, -45×10^{11} mks

Optical Properties
Color. White, with bluish tinge
Reflectance: Refractive and absorption indices. See Table 40.
Emittance. 0.04 at 50 °C (Ref 14)

Nuclear Properties
Stable isotopes. Results from three measurements (Ref 64):
Unstable (radioactive) isotopes. See Table 41.

Mass number	Abundance, %		
112	1.01,	0.90,	0.94
114	0.68,	0.61,	0.65
115	0.35,	0.33,	?
116	14.28,	14.07,	14.36
117	7.67,	7.54,	7.51
118	23.84,	23.98,	24.21
119	8.68,	8.62,	8.45
120	32.75,	33.03,	33.11
122	4.74,	4.78,	4.61
124	6.01,	6.11,	5.83

Mechanical Properties
Poisson's ratio. 0.33
Elastic modulus. Room temperature (Ref 65): as-cast (coarse grained), 41.6 GPa; self-annealed (fine grained), 44.3 GPa. Effect of temperature (Ref 66):

	Percentage of value at 16 °C	
°C	Poly-crystalline	Single crystals
25	98	99
50	92	96
75	86	93
100	79	90
125	72	87

Elastic modulus along crystal axes. Room temperature: 84.7 GPa along (001) plane (maximum value), 26.3 GPa along (110) plane (minimum value). Effect of temperature (Ref 67 to 69): 53.9 MPa/K at -180 to 0 °C, 75.5 MPa/K at 0 to 100 °C, 121.6 MPa/K at 100 to 200 °C
Dynamic liquid viscosity. From Ref 70 to 75:

°C	mPa·s
232	2.71
250	1.88
300	1.66
400	1.38
500	1.18
600	1.05
700	0.95
800	0.87

Liquid surface tension. From 400 to 800 °C: $700 - 0.17 \times T + (25 + 0.015 \times T)$ mN/m (where T is temperature in K) (Ref 76)
Velocity of sound. In solid: 2.60 km/s at 18 °C (Ref 3); in liquid, 2.27 km/s at 232 °C and 12 MHz (Ref 77)

REFERENCES

1. A. Hickling and F. W. Salt, Transactions of the Faraday Society, Vol 37, 1941, p 333
2. L. D. Brownlee, Nature, Vol 166, 1950, p 482
3. C. J. Smithells, *Metals Reference Book*, Butterworths, 1949
4. A. J. C. Wilson, ed., "Structure Reports for 1949", International Union of Crystallography, Vol 12
5. R. Clark, G. B. Craig and B. Chalmers, Acta Crystallographica, Vol 3, 1950, p 479
6. A. Ievins, M. Straumanis and K. Karlsons, Zeitschrift für Physikalische Chemie (B), Vol 40, 1938, p 347
7. L. W. McKeehan and H. J. Hoge, Zeitschrift für Kristallographie, Vol 92, 1935, p 476
8. W. Stenzel and J. Weertz, Zeitschrift für Kristallographie, Vol 84, 1932, p 20
9. E. R. Jette and F. Foote, Journal of Chemical Physics, Vol 3, 1935, p 605
10. E. Schmid and W. Boas, *Plasticity of Crystals*, E. A. Hughes, 1950
11. K. Brausch, Zeitschrift für Physik, Vol 93, 1935, p 479
12. C. Gamertsfelder, Journal of Chemical Physics, Vol 9, 1941, p 450
13. H. Endo, Bulletin of the Chemical Society of Japan, Vol 2 1927, p 131
14. C. L. Mantell, *Metals Handbook*, 1939 edition, p 1714
15. Hess, Berichte der Deutschen Physikalischen Gesellschaft, Vol 11, No. 3, 1905, p 403
16. K. Bornemann and P. Siebe, Zeitschrift für Metallkunde, Vol 14, 1922, p 329
17. A. L. Day, R. B. Sosman and J. C. Hostetter, American Journal of Science, No IV, Vol 37, 1914, p 1
18. P. G. J. Gueterbock and G. N. Nicklin, Journal of the Society of Chemical Industry (London), Vol 44, 1925, p 370T
19. J. Johnston and L. H. Adams, American Journal of Science, Vol 31, 1911, p 501
20. K. K. Kelley, U. S. Bureau of Mines Bulletin 383, 1935
21. C. E. Homer and H. C. Watkins, Metal Industry (London), Vol 60, 1942, p 364
22. H. D. Erfling, Annalen der Physik (V) (Leipzig), Vol 34, 1939, p 136

23. F. L. Uffelman, Philosophical Magazine, Vol 10, 1930, p 633

24. T. R. Hogness, Journal of the American Chemical Society, Vol 43, 1921, p 1621

25. B. G. Childs and S. Weintroub, Proceedings of the Physical Society (B), Vol 63, 1950, p 267

26. E. Cohen and K. D. Dekker, Zeitschrift für Physikalische Chemie, Vol 127, 1927, p 214

27. O. Kubaschewski, Zeitschrift für Electrochemie, Vol 54, 1950, p 275

28. A. W. Searoy and R. D. Freeman, Journal of the American Chemical Society, Vol 76, 1954, p 5229

29. L. Brewer and R. F. Porter, Journal of Chemical Physics, Vol 21, 1953, p 2012

30. E. C. Baughan, Quarterly Reviews, Chemical Society, Vol 7, 1953, p 103

31. M. Jakob, Zeitschrift für Mettallkunde, Vol 16, 1924, p 353

32. W. B. Brown, Physical Review, Vol 22, 1923, p 171

33. S. Konno, Science Reports of the Tohoku Imperial University, Vol 8, 1919, p 169

34. C. H. Lees, Philosophical Transactions of the Royal Society, Vol 208, 1908, p 381

35. P. J. Fensham, Australian Journal of Scientific Research (A), Vol 4, 1951, p 229

36. E. Scala and W. D. Robertson, Journal of Metals, Vol 5, 1953, p 1141

37. F. Forester and G. Tschenke, Zeitschrift für Metallkunde, Vol 32, 1940, p 191

38. "High Purity Tin", Capper Pass, Ltd., 1940

39. W. B. Pietenpol and H. A. Miley, Physical Review, Vol 34, 1929, p 1588

40. Y. Matuyama, Science Reports of the Tohoku Imperial University, Vol 16, 1927, p 447

41. E. F. Northrup and R. G. Sherwood, Journal of the Franklin Institute, Vol 182, 1916, p 477

42. E. F. Northrup and V. A. Suydam, Journal of the Franklin Institute, Vol 175, 1913, p 153

43. "Properties of Tin", Tin Research Institute, 1965

44. B. Chalmers and R. H. Humphrey, Philosophical Magazine, Vol 25, 1938, p 1108

45. P. W. Bridgeman, Proceedings of the American Academy of Arts and Sciences, Vol 68, 1933, p 95

46. P. W. Bridgeman, Proceedings of the American Academy of Arts and Sciences, Vol 72, 1938, p 157

47. F. Cirkler, Zeitschrift für Naturforschung (A), Vol 8, 1953, p 646

48. G. A. Roush, Transactions of the Electrochemical Society, Vol 73, 1938, p 285

49. W. M. Latimer, *Oxidation Potentials,* Prentice Hall, 1952

50. M. M. Haring and J. C. White, Transactions of the Electrochemical Society, Vol 73, 1938, p 211

51. S. Tolansky, Proceedings of the Royal Society (A), Vol 144, 1934, p 574

52. J. O'M. Bockris and S. Ignatowicz, Transactions of the Faraday Society, Vol 44, 1948, p 519

53. A. G. Pecherskaya and V. V. Stender, Zhurnal Prikladnoi Khimii, Vol 19, 1946, p 1303

54. H. Hunt, J. F. Chittum and H. W. Ritchey, Transactions of the Electrochemical Society, Vol 73, 1938, p 299

55. G. Schmid and E. K. Stoll, Zeitschrift für Elektrochemie, Vol 47, 1941, p 360

56. P. Räthjen, Physikalische Zeitschrift, Vol 25, 1924, p 84

57. G. Busch, J. Wieland and H. Zoller, Helvetica Physica Acta, Vol 24, 1951, p 49

58. D. Shoenberg, *Superconductivity,* Cambridge University Press, 1952

59. R. Hischberg and E. Lange, Naturwissenschaften, Vol 39, 1952, p 131

60. P. L. Clegg, Proceedings of the Physical Society (B), Vol 65, 1952, p 774

61. D. G. Avery, Philosophical Magazine, Vol 41, 1950, p 1018

62. C. V. Kent, Physical Review, Vol 14, 1919, p 459

63. P. Erochin, Annalen der Physik (IV), Vol 39, 1912, p 213

64. "Nuclear Data", U.S. National Bureau of Standards Circular 499 and Supplements 1, 2, 3, 1951–2

65. J. W. Cuthbertson, Journal of the Institute of Metals, Vol 64, 1939, p 209

66. L. Rotherham, A. D. N. Smith and G. B. Greenough, Journal of the Institute of Metals, Vol 79, 1951, p 439

67. W. Koster, Zeitschrift für Metallkunde, Vol 39, 1948, p 1

68. Y. L. Yousef, Philosophical Magazine, Vol 37, 1946, p 490

69. S. Aoyama and T. Fukuroi, Science Reports of the Tohoku Imperial University, Vol 28, 1940, p 423

70. T. P. Yao and V. Kondic, Journal of the Institute of Metals, Vol 81, 1952, p 17

71. A. J. Lewis, Proceedings of the Physical Society, Vol 48, 1936, p 102

72. K. Gering and F. Sauerwald, Zeitschrift für Anorganische und Allgemeine Chemie, Vol 223, 1935, p 204

73. V. H. Stott, Proceedings of the Physical Society, Vol 45, 1933, p 530

74. F. Sauerwald and K. Topler, Zeitschrift für Anorganische und Allgemeine Chemie, Vol 157, 1926, p 117

75. M. Plüss, Zeitschrift für Anorganische und Allgemeine Chemie, Vol 93, 1915, p 1

76. D. V. Atterton and T. P. Hoar, Journal of the Institute of Metals, Vol 81, 1952–53, p 541

77. O. J. Kleppa, Journal of Chemical Physics, Vol 18, 1950, p 1331

Titanium (Ti)

Compiled by W. Stuart Lyman
Manager, Technical and Market Services
Copper Development Association, Inc.
(formerly with Battelle Memorial Institute)
Reviewed for this edition by Douglas H. Wilson
RMI Company

Designations

Common name. Iodide titanium, electrolytic titanium

Typical uses. Experimentation and research; commercial applications requiring freedom from interstitial alloying elements (oxygen, nitrogen, carbon and hydrogen)

Structure

Crystal structure. α-phase: close-packed hexagonal; $a = 0.295030$ nm, $c = 0.468312$, $c/a = 1.5873$. β-phase: body-centered cubic, $a = 0.332$ nm at 900 °C

Mass Characteristics

Atomic weight. 47.9

Density. α-phase: 4.507 Mg/m³ at 20 °C. β-phase: 4.35 Mg/m³ at 885 °C (from indirect measurements)

Chemical Composition

Impurity element	Typical concentration, %	
	Electro-lytic Ti	Iodide crystal bar Ti
Fe	0.009	0.002
Si...........	0.002	0.005
Ca	...	0.003
Cu	0.007	<0.001
Mg..........	<0.001	0.003
Mn	<0.001	0.003
Sn	<0.020	0.001
Zr	<0.001	0.050
C	0.008	0.001
O..........	0.037	0.03-0.06
N..........	0.004	0.002
Cl	0.073	...
Ti (by difference) ..	99.837	99.90-99.87

Thermal Properties

Melting temperature. 1668 ± 10 °C
Boiling point. 3260 °C (estimated)
Vapor pressure. From 1587 to 1698 K.

$$\log P = 7.7960 - \frac{24\,644}{T} - 0.000227T$$

Phase-transformation temperature. α to β, 882.5 °C
Coefficient of thermal expansion. At 20 °C, 8.41 × 10⁻⁶/c at 1000, 10.1 × 10⁻⁸/c (estimated)
Thermal expansion in crystallographic directions (calculated from lattice parameters):

Temperature, °C	Direction	Expansion, μm/m·K
20 to 400 ...	Perpendicular to c axis	10.2
20 to 700 ...	Perpendicular to c axis	11.0
20 to 700 ...	Along c axis	12.8

Specific heat. Below 13 K:

$$C_p = 0.0706 + 5.43 \times 10^{-4}T^3$$

Above room temperature:

$$C_p = 669.0 - 0.037188\,T - 1.080 \times 10^7 T^{-2}$$

where C_p is in J/kg·K and T is in K

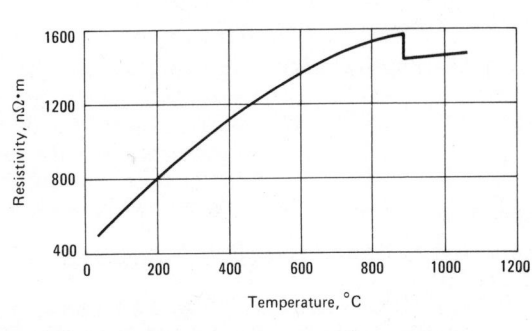

Fig. 110 Electrical resistivity of 99.9% pure titanium

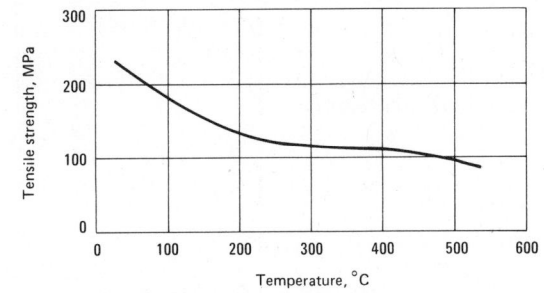

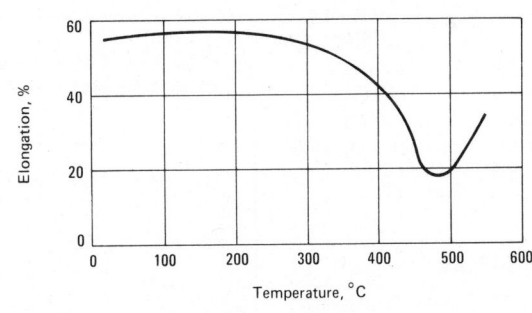

Fig. 111 Typical tensile properties of 99.9% pure titanium

Temperature, K	C_p, J/kg·K
50	99.3
75	210.0
100	300.2
125	363.2
150	409.4
175	440.8
200	465.1
225	484.1
250	499.4
275	512.6
298.15	522.3
300	537.8
350	567.8
400	586.6
450	598.9
500	607.2
550	612.8
600	616.7
650	619.3
700	620.9
750	621.9
800	622.4
850	622.4
900	622.3
950	621.7
1000	621.0
1050	620.2
1100	619.2
1150	618.1

Latent heat of fusion. 440 kJ/kg (estimated)

Latent heat of transformation. 91.8 kJ/kg (estimated)

Latent heat of vaporization. 9.83 MJ/kg (estimated)
Thermal conductivity. 11.4 W/m·K at −240 °C

Electrical Properties

Electrical resistivity. 420 nΩ·m at 20 °C. See also Fig. 110.
Superconductivity. Critical temperature: 0.37 to 0.56 K

Magnetic Properties

Magnetic susceptibility. Volume, at room temperature: 180 ($\pm$1.7) × 10^{-6} mks

Optical Properties

Total hemispherical emittance. 0.30 at 710 °C

Nuclear Properties

Stable isotopes:

Mass number	Natural abundance, %	Cross section, barns
46	7.95	0.6
47	7.75	1.6
48	73.43	8.0
49	5.51	1.8
50	5.34	0.2

Chemical Properties

General corrosion behavior. Greater resistance than that of commercial grades of unalloyed Ti.

Mechanical Properties

Tensile properties. Typical, at room temperature: tensile strength, 235 MPa; 0.2% yield strength, 140 MPa; elongation in 50 mm, 54%. See also Fig. 111.
Minimum bend radius. Less than 1t.

Hardness. Ingot melted from: electrolytic titanium, 70 to 74 HB; iodide titanium, 65 to 72 HB
Velocity of sound. 4970 m/s

Tungsten (W)

Compiled by Stephen W. H. Yih
Consultant

Structure

Crystal structure. α-phase: body-centered cubic, $cI2$ ($Im3m$); a = 0.316522 $\pm$ 0.00009 nm at 25 °C. β-phase: occurs only in the presence of oxygen and is probably W_3O; stable below 630 °C and is of type $A15$ or $cP8$ ($Pm3n$); a = 0.5046 nm at 25 °C
Minimum interatomic distance. α-phase, 0.274116 nm; β-phase, 0.252 nm

Mass Characteristics

Atomic weight. 183.85
Density. 19.254 Mg/m³

Thermal Properties

Melting point. 3410° $\pm$ 20 °C (Ref 1)

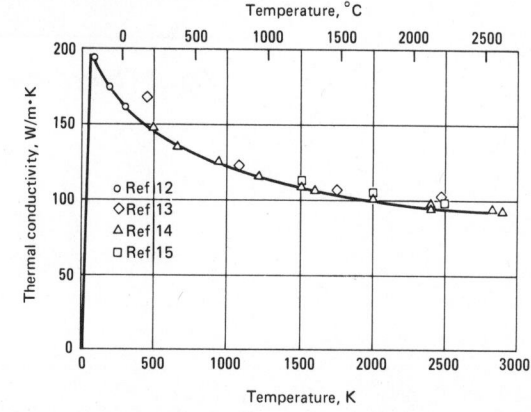

Fig. 112 Temperature dependence of the thermal conductivity of tungsten (Ref 11)

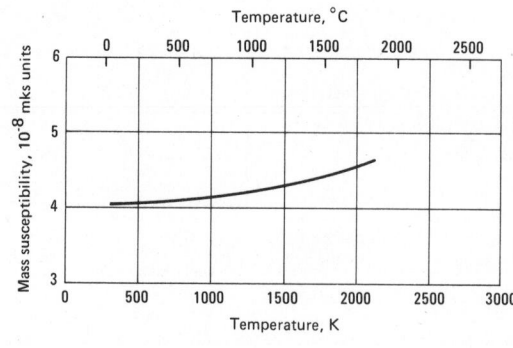

Fig. 113 Temperature dependence of magnetic susceptibility of tungsten (Ref 17)

Table 42 Values of coefficients for calculation of thermal expansion of tungsten from 25 to 2500 °C (Ref 6)

Material	Coefficient A_0	A_1	A_2
Powder metallurgy rod	-8.69×10^{-3}	3.83×10^{-4}	7.92×10^{-8}
Powder metallurgy sheet	-4.58×10^{-3}	3.65×10^{-4}	9.81×10^{-8}
Arc-cast sheet	-6.76×10^{-3}	3.91×10^{-4}	8.98×10^{-8}

Fig. 114 Temperature dependence of the tensile strength of tungsten (Ref 11)

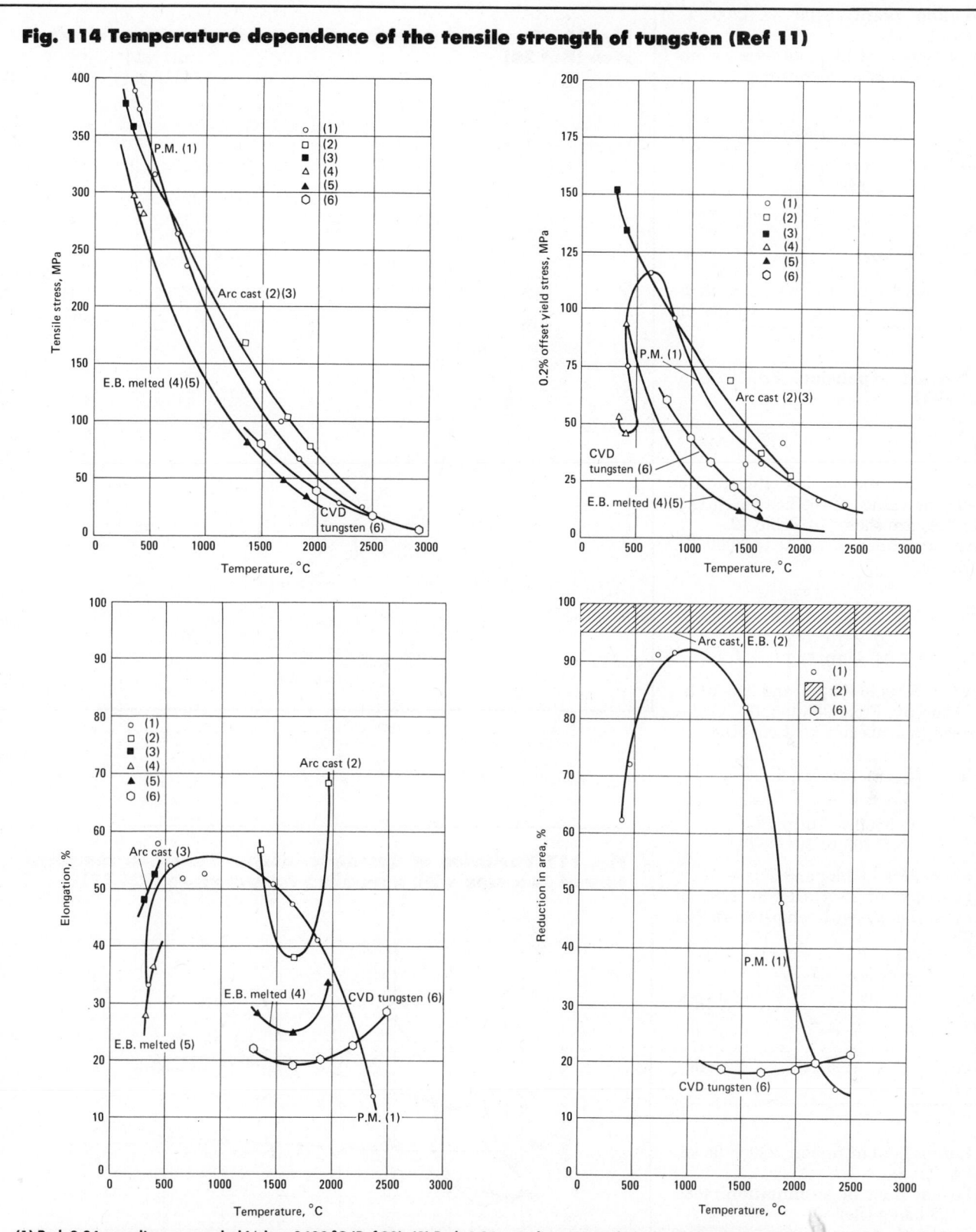

(1) Rod, 2.36-mm diam, annealed ½ h at 2400 °C (Ref 20). (2) Rod, 4.06-mm diam, annealed 1 h at 1982 °C (Ref 21). (3) Rod, 4.06-mm diam, annealed 1 h at 1648 °C (Ref 21). (4) Rod, 4.06-mm diam, annealed 1 h at 1371 °C (Ref 22). (5) Rod, 4.06-mm diam, annealed 1 h at 1982 °C (Ref 22). (6) Rod, 4.06-mm diam, annealed 1 h at 2845 °C (Ref 23).

Boiling point. 5700° ± 200 °C (Ref 2)

Coefficient of thermal expansion. Linear at low temperatures:

Temperature K	°C	Coefficient, μm/m X-ray spacing data (Ref 3)	Extensometer data
180	−93	3.7	3.9 (Ref 4)
140	−133	3.2	3.4 (Ref 4)
100	−173	2.3	2.7 (Ref 4)
80	−193	1.8	2.2 (Ref 5)
60	−213	1.1	1.5 (Ref 5)
40	−233	0.4	0.6 (Ref 5)

Thermal expansion. From 25 to 2500 °C:

$$\frac{L - L_{25°C}}{L_{25°C}} \times 100 = A_0 + A_1 T + A_2 T_2$$

where T is the Celsius temperature and the values of coefficients A_0, A_1 and A_2 are shown in Table 42.

Specific heat. From 0 to 3000 °C (Ref 7):

$$C_p = 135.76 \left(1 - \frac{4805}{T^2} + (9.1159 \times 10^{-3}) T + (2.3134 \times 10^{-9}) T^3\right)$$

where C_p is in J/kg·K and T is in K

Enthalpy. From 935 to 2975 °C (derived from specific heat equation:

$$H_T - H_{298} = 135.76 \left(T + \frac{26.14}{T}\right) - 4.266 \times 10^3 + (4.5569 \times 10^{-3}) T^2 + (5.78205 \times 10^{-10}) T^4$$

where H is in J/kg and T is in K

Entropy. At 25 °C (Ref 8): 178.3 J/kg·K. Change with temperature (Ref 9):

°C	J/kg·K
25	0
127	37
727	163
1627	262

Latent heat of fusion. 220 ± 36 kJ/kg (Ref 2)

Latent heat of sublimation. 4680 ± 25 kJ/kg (Ref 10)

Thermal conductivity. See Fig. 112.

Vapor pressure. From 2327 to 2827 °C (Ref 10):

$$\log_{10} P = \frac{-45385}{T} + 2.865$$

where P is in Pa and T is in K

Fig. 115 Temperature dependence of the hardness of tungsten (Ref 24)

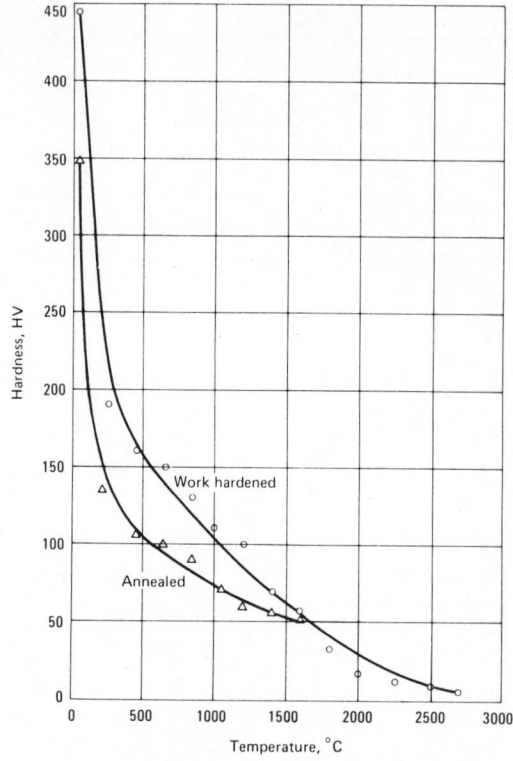

Fig. 117 Variation of ductile-to-brittle transition temperature of tungsten with annealing temperature (Ref 11)

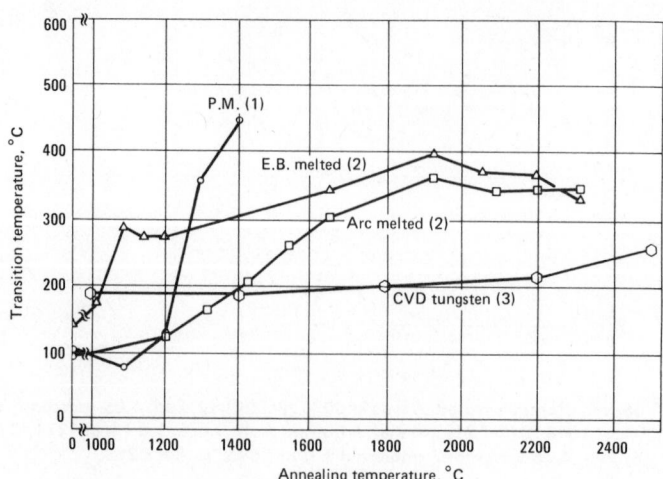

Ductile-to-brittle transition temperature determined by 4t bend for tungsten sheet: (1) Ref 27, (2) Ref 22, (3) Ref 28.

Fig. 116 Temperature dependence of Poisson's ratio and the elastic modulii of tungsten (Ref 11)

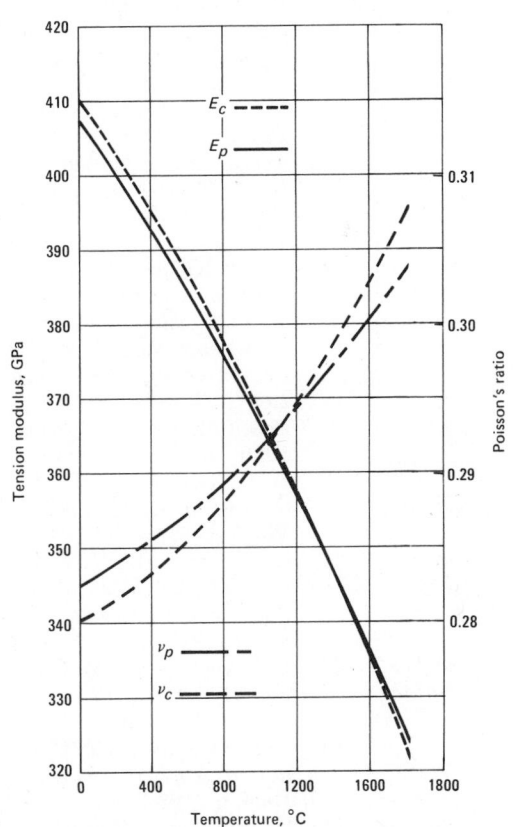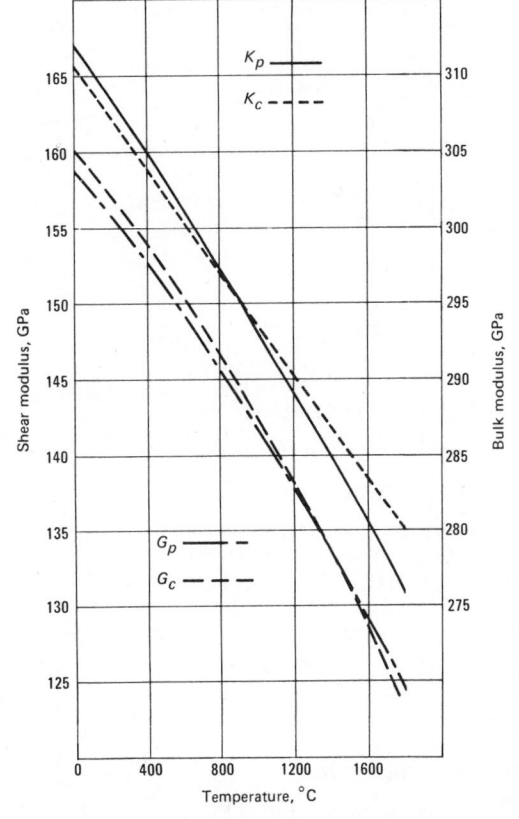

Poisson's ratio and elastic moduli as calculated from single-crystal elastic constants (ν_c E_c G_c and K_c) (Ref 25) and from measurements on poly-crystalline tungsten. (ν_p, E_p, G_p and K_p) (Ref 26)

Fig. 118 Temperature dependence of the 1-h rupture strength of tungsten (Ref 11)

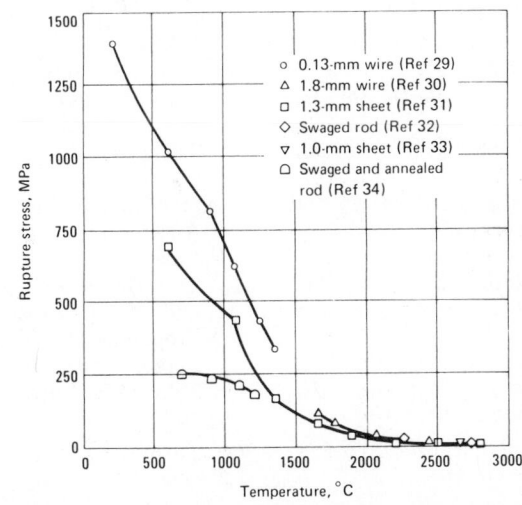

- ○ 0.13-mm wire (Ref 29)
- △ 1.8-mm wire (Ref 30)
- □ 1.3-mm sheet (Ref 31)
- ◇ Swaged rod (Ref 32)
- ▽ 1.0-mm sheet (Ref 33)
- ◠ Swaged and annealed rod (Ref 34)

Electrical Properties

Electrical resistivity. 53 nΩ·m at 27 °C. From 27 to 967 °C (Ref 15):

$$\rho = (4.33471 \times 10^{-14})\,T^2 + (2.19691 \times 10^{-10})\,T - (1.64011 \times 10^{-8})$$

where ρ is in Ω·m and T is in K
Hall constant. 10.55 µV·m/A·T (Ref 2)
Temperature of superconductivity. 0.016 K (Ref 16)

Magnetic Properties

Magnetic susceptibility. See Fig. 113.

Optical Properties

Total emissivity. From Ref 18:

°C	Total emissivity
127	0.042
527	0.088
1327	0.207
1727	0.260
2127	0.296
2527	0.323
2927	0.341

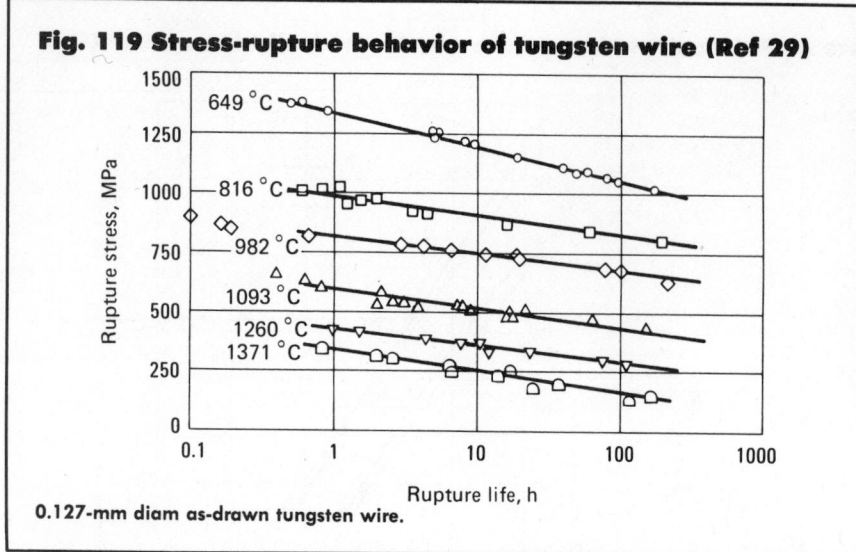

Fig. 119 Stress-rupture behavior of tungsten wire (Ref 29)

0.127-mm diam as-drawn tungsten wire.

Nuclear Properties

Stable isotopes. [180]W, isotope mass 179.9470, 0.14% abundant; [182]W, isotope mass 181.9483, 26.41% abundant; [183]W, isotope mass 182.9503, 14.40% abundant; [184]W, isotope mass 183.9510, 30.64% abundant; [186]W, isotope mass 185.9543, 28.41% abundant (Ref 19)

Mechanical Properties

Tensile properties. See Fig. 114.
Hardness. See Fig. 115.
Poisson's ratio. See Fig. 116.
Elastic modulus. See Fig. 116.
Ductile-to-brittle transition temperature. See Fig. 117.
Creep-rupture characteristics. See Fig. 118 and 119.

REFERENCES

1. A. Cezairliyan, *High Temperature Science,* Vol 4 (No. 3), 1972, p 248-252
2. G. D. Rieck, *Tungsten and Its Compounds,* Pergamon Press, Oxford, 1967
3. J. S. Shah and M. E. Straumanis, *Journal of Applied Physics,* Vol 42 (No. 9), 1971, p 3288
4. F. C. Nix and D. McNair, *Physical Review,* Vol 61, 1942, p 74
5. R. J. Corruccini and J. J. Gniewek, *National Bureau of Standards Monograph 29,* 1961
6. J. B. Conway and A. C. Losekamp, *Transactions of TMS-AIME,* Vol 236, 1966, p 702-709
7. M. Hoch, *High Temperature, High Pressure,* Vol 1, 1969, p 531-542
8. K. Clusius and P. Franzosini, *Zeitschrift für Naturforschung,* Vol 14, 1959, p 99
9. U. Schmidt, O. Volmer and R. Kohlhass, *Zeitschrift für Naturforschung,* Vol 25, 1970, p 1258-1264
10. E. R. Plante and A. B. Sessoms, *Journal of Research of the National Bureau of Standards,* Vol 77A (No. 2), 1973, p 237-242
11. S. W. H. Yih and C. T. Wang, *Tungsten: Sources, Metallurgy, Properties and Applications,* Chapter 6, Plenum Press, New York, 1979
12. N. G. Backlund, *Journal of Physics and Chemistry of Solids,* Vol 28, 1967, p 2219-2223
13. B. E. Neimark and L. K. Voronin, *High Temperature* (USSR; English translation), Vol 6, 1968, p 999-1010
14. R. E. Taylor, F. E. Davis and R. W. Powell, *High Temperature, High Pressure,* Vol 1, 1969, p 663-673
15. V. A. Vertogradskii and V. Ya. Chekhovskoi, *Teplofizika Vysokikh Temperatur,* Vol 8 (No. 4), 1970, p 784-788
16. B. B. Triplett *et al, Journal of Low Temperature Physics,* Vol 12 (No. 5/6), 1973, p 499-518
17. C. Kittel, *Introduction to Solid State Physics,* 3rd Ed., Wiley and Sons, New York, 1971
18. D. E. Gray, *American Institute of Physics Handbook,* 3rd Ed., McGraw-Hill, New York, 1972, p 6-79
19. R. C. Weast (Ed.), *Handbook of Chemistry and Physics,* CRC Press, Cleveland, OH, 1977
20. H. G. Sell, W. R. Morcom and G. W. King, "Development of Dispersion Strengthened Tungsten Base Alloys", AFML-TR-65-407, Part II, Westinghouse Lamp Division, Bloomfield, NJ, 1966
21. W. D. Klopp and P. L. Raffo, "Effects of Purity and Structure on Recrystallization, Grain Growth, Ductility, Tensile and Creep Properties of Arc-Melted Tungsten", NASA-TND-2503, Lewis Research Center, Cleveland, OH, 1964
22. W. D. Klopp and W. R. Witzke, "Mechanical Properties and Recrystallization Behavior of Electron-Beam-Melted Tungsten Compared with Arc-Melted Tungsten", NASA-TND-3232, Lewis Research Center, Cleveland, OH, 1966
23. J. L. Taylor and D. H. Boone, *Journal of Less-Common Metals,* Vol 6, 1964, p 157-164
24. G. S. Pisarenki, V. A. Borisenko and Yu. A. Kashtalyan, *Soviet Powder Metallurgy and Metal Ceramics,* Vol 5, 1962, p 371-374
25. R. Lowrie and A. M. Gonas, *Journal of Applied Physics,* Vol 38, 1967, p 4505-4509
26. R. Lowrie and A. M. Gonas, *Journal of Applied Physics,* Vol 36, 1965, p 2189-2192
27. H. R. Ogden, "Refractory Metals Sheet-Rolling Program", DMIC Report 176, Battelle Memorial Institute, Columbus, OH, 1962
28. A. C. Schaffhauser, "Low Temperature Ductility and Strength of Thermochemically Deposited Tungsten and Effects of Heat Treatment", AFML-TR-179, Oak Ridge National Laboratories, Oak Ridge, TN, 1966
29. D. L. McDanels and R. A. Signorelli, "Stress-Rupture Properties of Tungsten Wire from 1200° to 2500 °F", NASA-TND-3467, NASA, 1966
30. J. K. Y. Hum and A. Donlevy, "Some Stress Rupture Properties of Columbium, Molybdenum, Tantalum and Tungsten Metals and Alloys Between 2400°F and 5000 °F", Report 354D, Society of Automotive Engineers, New York, 1961
31. C. A. Drury, R. C. Kay, A. Bennett and M. J. Albom, "Mechanical Properties of Wrought Tungsten", Report ASD-TDR-63-585, Vol 2, Marquardt Corp., Van Nuys, CA, 1963
32. W. V. Green, *Transactions of*

AIME, Vol 215, 1959, p 1057-1060

33. E. C. Sutherland and W. D. Klopp, "Observations of Properties of Sintered Wrought Tungsten Sheet at Very High Temperatures", NASA-TND-1310, 1963

34. J. W. Pugh, *Proceedings of ASTM*, Vol 57, 1957, p 906-916

Uranium (U)

Compiled by Ronald D. Nelson
Battelle Pacific Northwest
Laboratories

Natural uranium nominally contains 0.006% ^{234}U, 0.72% ^{235}U, and the remainder ^{238}U. The term "enriched uranium" designates uranium containing higher than normal ^{235}U; "depleted uranium" designates lower than normal ^{235}U. Other designations include alpha uranium, beta uranium and gamma uranium for the three polymorphic forms.

The most common use is as enriched uranium in nuclear reactors. It is used either in unalloyed metallic form or as uranium oxide. In the latter case, reactor fuel elements usually are made of a mixture of uranium oxide and plutonium dioxide. Enriched uranium is also used as a nuclear explosive. ^{233}U, which is produced by neutron irradiation of ^{232}Th, is also fissionable and so can be used in nuclear reactors.

Massive uranium metal offers no substantial problem in handling and storage. It oxidizes slowly in air and forms an adherent oxide. Finely divided uranium metal, on the other hand, is pyrophoric and care must be exercised to prevent fires. Uranium is a controlled nuclear material and emits weak alpha radiation. It is also a heavy-metal poison and, like other heavy-metal poisons, must be processed under controlled conditions to avoid ingestion of fumes or dusts by workers.

Structure

Crystal structure. Alpha: orthorhombic (*Cmcm*; *oC*4); $a = 0.2854$ nm, $b = 0.5869$ nm, $c = 0.4955$ nm at 298 K. Beta: complex tetragonal (*P*4$_2$/*mnm*; *tP*30); $a = 1.0748$ nm, $c = 0.5652$ nm at 950 K. Gamma: body-centered cubic (*Im3m*; *cI*2); $a = 0.3535$ nm at 1100 K

Slip planes. At 300 to 875 K, primary slip in α-U is (010) [100], which cross slips onto (001)

Twinning planes. The most frequently observed type of twinning occurs on (130) and less frequently on (172) and (176) in alpha uranium

Mass Characteristics

Atomic weight. 238.029

Density. Alpha phase: 19.05 Mg/m^3 at 298 K from x-ray data; 18.7 to 19.05 Mg/m^3 for wrought metal. Beta phase: 18.13 Mg/m^3 at 973 K. Gamma phase: 17.91 Mg/m^3 at 1173 K. Liquid: 17.25 Mg/m^3 at 1410 K. Temperature coefficients: alpha phase, 0.001 Mg/m^3·K; beta phase, 0.0009 Mg/m^3·K; gamma phase, 0.0012 Mg/m^3·K; liquid, 0.0016 Mg/m^3·K

Volume change on melting. 2.2% expansion

Volume change on phase transformation. On cooling: beta to alpha, 1.0% contraction; gamma to beta, 0.6% contraction

Thermal Properties

Melting point. 1406 K

Boiling point. 4091 K

Phase-transformation temperature. Alpha to beta, 934 K; beta to gamma, 1042 K

Coefficient of thermal expansion. The thermal expansion of wrought alpha uranium is highly anisotropic and depends on fabrication history and the resultant preferred orientation. Linear: quenched alpha phase, 12 μm/m·K at 298 K and 28 μm/m·K at 900 K; beta phase, 28 μm/m·K at 1000 K; gamma phase, 20 μm/m·K between 1175 and 1400 K. Coefficients of linear thermal expansion along crystal axes. Alpha phase between 50 and 923 K:
$^\alpha[100] = 2.422 \times 10^{-5} - 9.83 \times 10^{-9} T + 4.602 \times 10^{-11} T^2$; $^\alpha[010] = 3.07 \times 10^{-6} + 3.47 \times 10^{-9} T - 3.845 \times 10^{-11} T^2$; $^\alpha[001] = 8.72 \times 10^{-6} + 3.704 \times 10^{-8} T + 9.08 \times 10^{-12} T^2$, where T is temperature in K.
Beta Phase: $^\alpha[100]$, 25 μm/m·K; $^\alpha[001]$, 5 μm/m·K

Specific heat:

Phase	Temperature, K	Specific heat, J/kg
α	300	117
α	600	145
α	800	172
α	900	190
β	940	179
β	1040	179
γ	1050	160
γ	1300	160

The equation for specific heat of α-U vs temperature is

$$C_p = 103.6 + 0.0180\,T + 8.49 \times 10^{-5}\,T^2$$

where C_p is in J/kg·K and T is in K

Latent heat of fusion. 38.72 kJ/kg at 1406 K

Latent heat of phase transformation. Alpha to beta, 12.3 kJ/kg at 943 K; beta to gamma, 20.1 kJ/kg at 1042 K

Latent heat of vaporization. 2.069 kJ/kg at 1406 K

Thermal conductivity:

Phase	Thermal conductivity, W/m·K	Temperature, K
α	9.8	10
α	15.8	20
α	2.17	100
α	27.6	300
α	31.7	600
α	41.3	900
β	43.9	1000
γ	46.3	1100

Recrystallization temperature. Generally between 650 and 750 K, but is highly dependent on purity and fabrication history

Vapor pressure. 1 μPa at 1500 K and 17.5 mPa at 2000 K

Enthalpy. At 298 K: 26.74 kJ/kg; entropy, 211 J/kg·K at 298 K

Electrical Properties

Electrical resistivity. Alpha phase, 300 nΩ·m at 300 K; beta phase, 560 nΩ·m at 1000 K; gamma phase, 540 nΩ·m at 1100 K; liquid, 66 mΩ·m at 1200 K. Temperature coefficient: alpha phase, 0.021 per K at 27 °C and 0.039 per K at 627 °C. Electrical resistivity along crystal axes for alpha phase at 273 K: [100], 390 nΩ·m; [010], 240 nΩ·m; [001], 262 nΩ·m

Hall coefficient. 380 nVm/A·T at 300 K

Temperature of superconductivity. <0.5 K

Work function. 0.58 aJ

Magnetic Properties

Magnetic susceptibility. Volume: 390×10^{-6} mks at 300 K

Optical Properties

Spectral reflectance. 73.5% for $\lambda = 660$ nm
Spectral hemispherical emittance. 26.5% for $\lambda = 660$ nm

Nuclear Properties

Unstable isotopes:

Isotope	Abundance, %	Half-life, years
^{234}U	0.0055	2.47×10^5
^{235}U	0.720	7.1×10^6
^{238}U	99.274	4.51×10^9

Isotope	Particle emitted	Energy, MeV
^{234}U	Alpha	4.77, 4.72
^{235}U	Alpha	4.58, 4.47, 4.40, 4.2
^{238}U	Alpha	4.18

Thermal neutron cross section. For 0.025 eV neutrons. Natural uranium: capture, 6.7 b. ^{238}U: capture, 1.6 b. ^{235}U: capture, 100 b; fission, 580 b

Mechanical Properties

Wide variations exist in all mechanical properties of alpha uranium and depend markedly on a large number of parameters, most notably preferred orientation, grain size, fabrication history, heat treatment, and type and distribution of impurities. For instance, fracture stress decreases from approximately 600 MPa for a grain size of 1 μm to 130 MPa for a grain size of 10 μm

Tensile properties. At 293 K (approximate). As cast: tensile strength, 400 MPa; 2% offset yield strength, 200 MPa; elongation, 4%; reduction in area, 10%. Beta annealed (grain size, 500 μm): tensile strength, 615 MPa. Wrought alpha uranium: tensile strength, 1150 MPa; yield strength, 740 MPa; elongation, 7%; reduction in area, 14%

Hardness. Coarse-grained α-U, 185 HV at 300 K; fine-grained α-U, 250 HV at 300 K. β-U, 30 HV at 950 K. γ-U, 1 HV at 1100 K

Vanadium (V)

Compiled by I. Drangel
and
G. L. Martin
Materials Research Corp.

Commercial applications for pure vanadium are limited. The primary present use is as an alloying agent for steel. Other alloy applications of interest are in electronics, superconductivity, and nuclear power. Typical composition of pure vanadium is: 99.7 V (by difference); impurities: 0.03 Fe, 0.04 Al, 0.10 Si, 0.01 Ti, 0.03 Mo, 0.02 O, 0.02 Ni, 0.02 C, 0.001 H. Increased impurity levels particularly of silicon, oxygen, nickel, carbon and hydrogen have adverse effects on hardness and ductility.

Structure

Crystal structure. Body-centered cubic; lattice parameters: calcium reduced, 0.30278 nm; iodide, 0.30258 nm

Mass Characteristics

Atomic weight. 50.941
Density. 6.1 Mg/m^3

Thermal Properties

Melting point. 1900 $\pm$ 25 °C
Coefficient of thermal expansion. Linear:

Temperature range, °C	Average coefficient, μm/m·K
23 to 100	8.3
23 to 500	9.6
23 to 900	10.4
23 to 1100	10.9

Specific heat. 498 J/kg·K at 0 to 100 °C
Thermal conductivity. 31.0 W/m·K at 100 °C

Electrical Properties

Electrical resistivity. 248 to 260 nΩ·m at 20 °C

Nuclear Properties

Thermal neutron cross section. 4.7 $\pm$ 0.02 b

Chemical Properties

Resistance to specific corroding agents. At room temperature, vanadium and its alloys have excellent resistance to corrosion in salt water and dilute hydrochloric acid; good corrosion resistance in sodium hydroxide solutions; and poor corrosion resistance in nitric acid solutions. Resistance to attack by liquid alkali metals is good.

Mechanical Properties

Tensile properties. Typical at 1025 °C: tensile strength, 53 MPa; elongation, 37% in 1 in.
Hardness. 72 HB, electron beam ingot. See also Table 43.
Poisson's ratio. 0.36
Elastic modulus. Tension, 124 to 137 GPa; shear, 46.4 GPa
Creep-rupture characteristics. Limiting creep stress, 4.63 MPa for 1% deformation in 24 h at 1000 °C. Stress/density ratio at 1000 °C, 110

Fabrication Characteristics

Recrystallization temperature. 800 to 1010 °C
Standard finishes. The machining of vanadium metal is similar to that of stainless steel and presents no

Table 43 Typical mechanical properties for vanadium metal at room temperature

Condition	Tensile strength, MPa	Yield strength, MPa	Elongation in 50 mm, %	Reduction of area, %	Hardness HRA	Hardness HRB	Cold bend, deg
Bar, 25.4-mm Diam(a)							
Hot rolled	472	439	27.0	54.4	···	85	···
Wire, 3.9-mm Diam(b)							
Vacuum annealed	538	463	25.0	87.5	48	···	180
Cold drawn 80%	910.8	765	6.8	76.5	54	···	180
Sheet, 1.9-mm Thick(c)							
Vacuum annealed	536	454	20.0	53.0	···	83	180
Cold rolled 84%	828	776.3	2.0	40.6	···	100	180

(a) Specimen size: 12.8-mm diam by 51 mm. (b) Specimen size: 3.9-mm diam by 51 mm. (c) Specimen size: 1.9 mm by 12.7 mm by 51 mm.

special problem except where the metal surface has been severely contaminated with oxygen and nitrogen.

Suitable joining methods. Satisfactory electric welding of vanadium requires adequate protection of the weld pool and heat-affected zone with a neutral gas, such as argon or helium, to prevent or minimize contamination with oxygen, hydrogen and nitrogen. Flame welding is not practical because of the reactivity of any combustion gas mixture with molten vanadium. Vanadium can be joined by welding to ferritic and austenitic stainless steels, titanium and titanium alloys, as well as to low-carbon steel.

Ytterbium (Yb)

Compiled by K.A. Gschneidner, Jr. and B. J. Beaudry
Ames Laboratory
U.S. Department of Energy
Iowa State University

Ytterbium is used for phosphors, ceramic capacitors, ferrite devises and catalysts. Ytterbium ([170]Yb), which has been formed by neutron irradiation of thulium ([169]Tm), is used as a portable radiograph source; ytterbium foils are used to measure pressure and as stress tranducers. Ytterbium will tarnish slightly at room temperature in air. Massive ytterbium can be handled in air, but should be stored in an inert atmosphere or vacuum. Finely divided ytterbium should not be handled in air.

Structure

Crystal structure. α phase, close-packed hexagonal, $P6_3/mmc$ D^4_h; a 0.38799 nm, $c = 0.63859$ nm at 24 °C. β phase, face-centered cubic, $Fm3m$ O^5_h; $a = 0.54848$ nm at 24 °C. γ phase, body-centered cubic, $Im3m$ O^9_h; $a = 0.444$ nm at 798 °C
Minimum interatomic distance. 0.19392 nm at 24 °C

Mass Characteristics

Atomic weight. 173.04
Density. α phase, 6.902 Mg/m^3 at 23 °C; β-phase, 6.966 Mg/m^3 at 24 °C; γ phase, 6.56 Mg/m^3 at 798 °C; liquid, 6.292 Mg/m^3 at 824 °C

Volume change on freezing. 5.1% contraction
Volume change on phase transformation. β to γ phase, 0.1% volume contraction on heating

Thermal Properties

Melting point. 819 °C
Boiling point. 1196 °C
Phase-transformation temperature. α to β phase: $A_s = 280$ K; β to α, $M_s \cong 260$ K; β to γ, 795 °C
Coefficient of thermal expansion. At 24 °C. Linear: 26.3 μm/m·K. Linear, along crystal axes: 26.3 μm/m·K along a axis. Volumetric: 79.0 × 10^{-6} per K
Specific heat. 154.5 J/kg·K at 25 °C
Entropy. 345.8 J/kg·K at 25 °C
Latent heat of fusion. 44.36 kJ/kg
Latent heat of transformation. 10.11 J/kg
Latent heat of vaporization. 87.90 kJ/kg at 25 °C
Heat of combustion. For cubic Yb$_2$O$_3$ at 25 °C: $\Delta H^0_c = 5.23$ MJ/kg Yb; $\Delta G^0_f = -5.00$ MJ/kg Yb
Recrystallization temperature. About 300 °C
Thermal conductivity. 38.5 W/m·K at 25 °C
Vapor pressure. 0.001 Pa at 301 °C; 0.101 Pa at 400 °C; 10.1 Pa at 541 °C; 1013 Pa at 776 °C

Electrical Properties

Electrical resistivity. 250 nΩ·m at 25 °C; 10 nΩ·m at 4 K; liquid, 1080 nΩ·m at 821 °C
Ionization potentials. Yb(I), 6.254 V; Yb(II), 12.17 V; Yb(III), 25.03 V; Yb(IV), 43.66 V
Hall coefficient. +0.377 nV·m/A·T at 20 °C
Temperature of superconductivity. Bulk ytterbium is not superconducting down to 0.015 K at atmospheric pressure.

Magnetic Properties

Magnetic susceptibility. Volume: 8.8 × 10^{-6} mks at 2 °C

Optical Properties

Color. Metallic silver

Nuclear Properties

Thermal neutron cross section. 36 b

Chemical Properties

General corrosion behavior. Ytterbium tarnishes slightly in moist

air. It oxidizes slowly at elevated temperatures and reacts readily with hydrogen at 250 °C.
Resistance to specific corroding agents. Ytterbium does not react with cold water, but will tarnish in hot water; reacts vigorously with dilute acids

Mechanical Properties

Tensile properties. Tensile strength, 59 MPa; yield strength, 6.9 MPa; elongation, 42%; reduction in area, 90%
Hardness. 17 HV
Poisson's ratio. 0.207
Strain-hardening exponent. 0.62
Elastic modulus. At 27 °C: tension, 23.9 GPa; shear, 9.9 GPa; bulk, 13.5 GPa
Kinematic liquid viscosity. 0.424 mm^2/s at 824 °C

Yttrium (Y)

Compiled by David T. Peterson
Ames Laboratory
U. S. Department of Energy
Iowa State University

Yttrium is used in magnesium alloys and oxidation resistant alloys, and it is also used in garnets and ferrites for electronic components. Yttrium is a host material for rare earth phosphors, including the red color (Eu) in color television screens. Some simulated diamonds (yttrium aluminum garnets) contain yttrium. Yttrium tarnishes slowly in air at room temperature. Turnings can be ignited quite easily and burn with great evolution of heat. Finely divided yttrium should be handled with great care and should be kept away from air and oxidizing agents.

Structure

Crystal structure. α-phase, close-packed hexagonal, $P63/mmc$; $a = 0.3648$ nm, $c = 0.5732$ nm at 25 °C. β-phase, body-centered cubic, $Im3m$; $a = 0.410$ nm above 1478 °C
Slip planes. $[10\bar{1}0]<[1\bar{2}10]>$ from -196 °C to 224 °C; $[0002]<[1\bar{2}10]>$ from -196 °C to 224 °C
Twinning planes. $[11\bar{2}1]<1\bar{1}\bar{2}6>$ at 25 °C
Minimum interatomic distance. α-phase, 0.3557 nm at 25 °C; β-phase, 0.355 nm at 1478 °C

Fracture behavior. Primarily ductile

Mass Characteristics

Atomic weight. 88.9059
Density. Solid, 4.469 Mg/m³ at 25 °C

Thermal Properties

Melting point. 1522 °C
Boiling point. 3338 °C
Phase-transformation temperature. 1478 °C
Coefficient of thermal expansion. Linear: 10.6 μm/m·K. Linear, along crystal axes: 6.0 μm/m·K along a axis; 19.7 μm/m·K along c axis. Volumetric; 31 700 mm³/m³·K
Specific heat. 298.4 J/kg·K at 25 °C
Entropy. At 298.15 K: 499.7 J/kg·K
Latent heat of fusion. 128.19 kJ/kg
Latent heat of transformation. cph → bcc, 56.15 kJ/kg
Latent heat of vaporization. 4.777 MJ/kg at 25 °C
Heat of combustion. −10.72 MJ/kg Y at 25 °C and constant pressure
Recrystallization temperature. 550 °C
Thermal conductivity. 17.2 W/m·K at 25 °C
Vapor pressure. 0.001 Pa at 1220 °C; 0.101 Pa at 1458 °C; 10.1 Pa at 1809 °C; 1013 Pa at 2356 °C

Electrical Properties

Electrical resistivity. 596 nΩ·m at 25 °C; 32 nΩ·m at 4 K. Along crystal axes at 25 °C: 725 nΩ·m, along a axis; 355 nΩ·m along c axis
Ionization potentials. Y(I), 6.38 V; Y(II), 12.24 V; Y(III), 20.52 V; Y(IV), 61.8 V
Hall coefficient. $R_{H,b} = -0.027$ nV·m/A·T and $R_{H,c} = -0.16$ nV·m/A·T at 25 °C

Magnetic Properties

Magnetic susceptibility. Volumetric at 27 °C (mks units): 1.186×10^{-4}. Along crystal axes: 1.233×10^{-4}; along a axis; 1.109×10^{-4} along c axis

Nuclear Properties

Thermal neutron cross section. 1.3 b

Optical Properties

Color. Steel gray
Spectral hemispherical emittance. Solid: 37.4% at 950 to 1522 °C, liquid: 37.0% at 1522 to 1675 °C

Chemical Properties

General corrosion behavior. Yttrium metal oxidizes very slowly in air at temperatures up to 450 °C
Resistance to specific chemical agents. Yttrium metal reacts vigorously with hydrochloric and nitric acids. It does not react with hydrofluoric acid or with HCl or NNO₃ containing fluoride ion

Mechanical Properties

Tensile properties. At 25 °C, annealed rod: tensile strength, 186 MPa; yield strength, 27 MPa; elongation, 17% in 1 in.; reduction in area, 24%
Hardness. 40 HV
Poisson's ratio. 0.24
Strain-hardening exponent. $n = 0.22$
Elastic modulus. Tension, Young's, 63.6 GPa; shear, 25.5 GPa; bulk, 40.7 GPa
Elastic modulus along crystal axes. $c_{11} = 77.9$ GPa; $c_{12} = 28.5$ GPa; $c_{13} = 21.0$ GPa; $c_{33} = 76.9$ GPa; $c_{44} = 24.3$ GPa

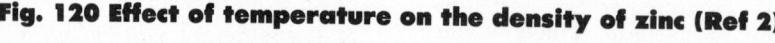

Zinc (Zn)

Compiled by Ernest W. Horvick
The Zinc Institute, Inc.

Structure

Crystal structure. Close-packed hexagonal; $a = 0.26648$ nm, $c = 0.49470$ nm; tests made with spectroscopically pure zinc (Ref 1)
Slip planes. Primary (00.1) at 25 °C
Twinning planes. (10.2)
Cleavage planes. (00.1)
Minimum interatomic distance. 0.26594 nm (Ref 2)
Fracture type. Basal cleavage

Mass Characteristics

Atomic weight. 65.38
Density. 7.133 Mg/m³ at 25 °C; also see Fig. 120.
Volume change on freezing. 7.28% between 469 °C and 0 °C (Ref 2)

Thermal Properties

Melting point. 420 °C (Ref 3,4)
Boiling point. 906 °C (Ref 5, 6, 7)
Coefficient of thermal expansion. Linear, single crystals at 0 to 100 °C: 15 μm/m·K along a axis, 61.5 μm/m·K along c axis. Polycrystalline solid at 20 to 250 °C: 39.7 μm/m·K; temperature effect, $L_t = L_0 (1 + 35.4 \times 10^{-8} t + 1 \times 10^{-8} t^2)$ (Ref 8). Liquid at 500 to 600 °C, 60 μm/m·K (Ref 2)
Specific heat. 382 J/kg at 20 °C (Ref 5,9); also see Fig. 121.
Latent heat of fusion. 100.9 kJ/kg (Ref 5)
Latent heat of vaporization. 1.782 MJ/kg (Ref 5)
Thermal conductivity. 113 W/m·K at 25 °C (Ref 3); also see Fig. 122.
Heat of combustion. −341 MJ/kg Zn (Ref 5)

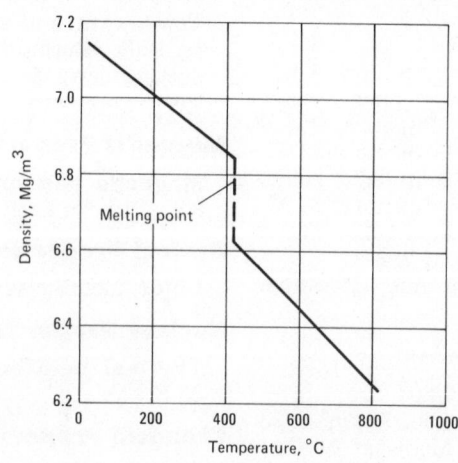

Fig. 120 Effect of temperature on the density of zinc (Ref 2)

Fig. 121 Effect of temperature on the specific heat of zinc (Ref 5, 9)

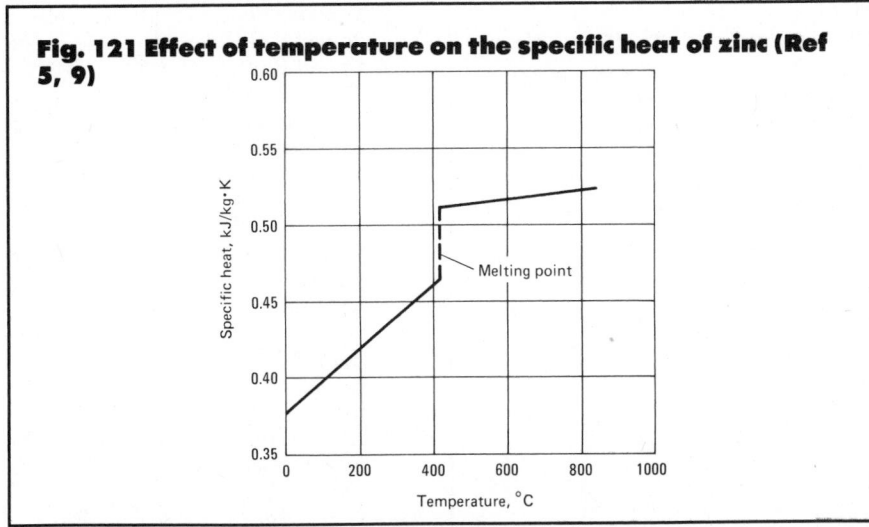

Fig. 122 Effect of temperature on the thermal conductivity of zinc (Ref 3, 10, 11)

Fig. 123 Effect of temperature on the electrical resistivity of zinc

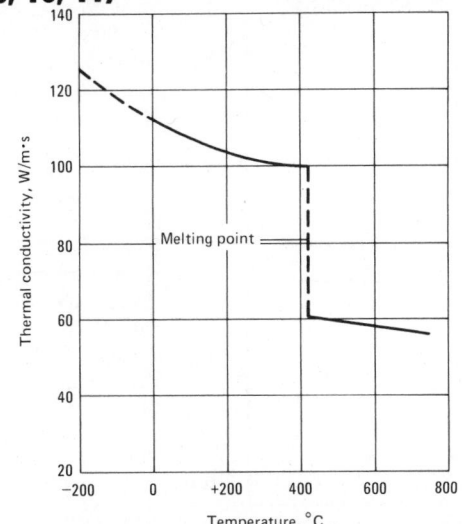

Composition:
99.993% Zn
0.005% Fe
0.0004% Pb
0.0018% Cd
Traces of As and S

Electrical Properties

Electrical conductivity. 28.27% IACS (Ref 12)

Electrical resistivity. Single crystals at 20 °C: 58.9 nΩ·m along a axis, 61.6 nΩ·m along c axis (Ref 13). Polycrystalline solid: 59.16 nΩ·m at 20 °C (Ref 3); temperature coefficient at 0 to 100 °C (Ref 3), 0.0419 nΩ·m per K; also see Fig. 123; pressure coefficient at room temperature (calculated), -25 nΩ·m per TPa at 100 kPa to 300 MPa; see also Fig. 124.

Electrochemical equivalent. 338.8 μg/C (Ref 15)

Electrolytic solution potential. -0.7618 V vs standard hydrogen electrode (Ref 3)

Hydrogen overvoltage. 0.75 V at 108 A/m^2 for metal rubbed with fine emery (Ref 16)

Temperature of superconductivity. 0.84 ± 0.05 K (Ref 17)

Magnetic Properties

Magnetic susceptibility. Volume: -123×10^{-6} mks (Ref 3)

Optical Properties

Color. Blue white

Spectral reflectance. 74.7% at $\lambda = 0.5000$ μm; 69.9% at $\lambda = 0.8000$ μm; 53.3% at $\lambda = 1.0100$ μm; 70.0% at $\lambda = 1.1300$ μm (Ref 18); also see Fig. 125.

Refractive index. 1.19 in white light ($\lambda = 0.5500$ μm) $\rho = 70°$; $2\psi = 74° 39'$ (Ref 18)

Absorptive index. 3.71 in white light ($\lambda = 0.5500$ μm) $\rho = 70°$; $2\psi = 74° 39'$ (Ref 18)

Mechanical Properties

Elastic properties. Compressibility, see Fig. 126.

Coefficient of friction. 0.21, rolled zinc vs rolled zinc

Surface tension. Liquid, 0.755 N/m at 450 °C (Ref 3)

Velocity of sound. 3.67 km/s at room temperature (shape and size of specimen wire unknown) (Ref 20)

REFERENCES

1. E. R. Jette and F. Foote, Journal of Chemical Physics, Vol 3, 1935, p 605
2. Erich Pelzel and Franz Sauerwald, Zeitschrift für Metallkunde, Vol 33, 1941, p 229
3. "Zinc and Its Alloys", National Bureau of Standards Circular No. 395, 6 Nov, 1931

Fig. 124 Effect of pressure on the electrical resistivity of zinc at 21 °C (Ref 14)

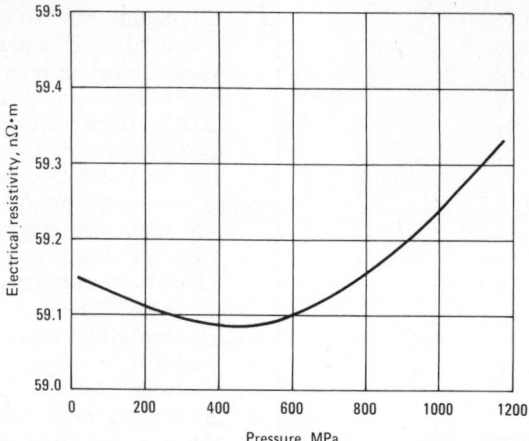

Fig. 125 Effect of wave length on the spectral reflectance of zinc (Ref 18, 19)

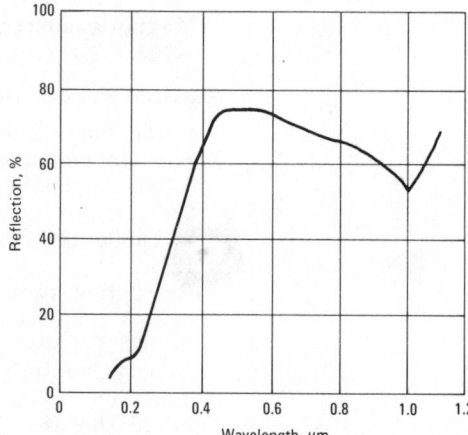

Fig. 126 Effect of temperature on the compressibility of zinc

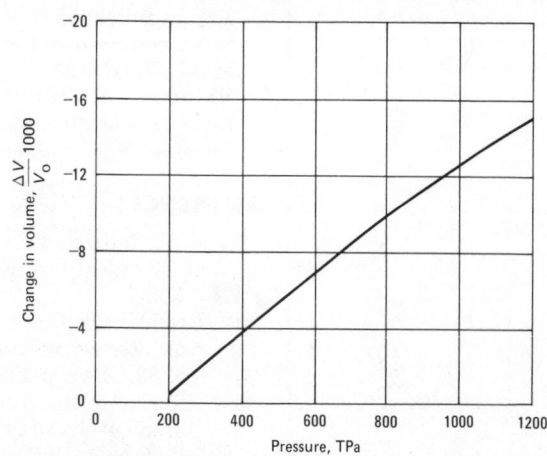

4. William Roeser and H. T. Wensel, Journal of Research of the National Bureau of Standards, Vol 14, 1935, p 247
5. C. G. Maier, U.S. Bureau of Mines Bulletin, 1930 p 324
6. J. Fischer, Zeitschrift für Anorganische und Allgemeine Chemie, Vol 219, 1934, p 367
7. W. Leitgebel, Zeitschrift für Anorganische und Allgemeine Chemie, Vol 202, 1931, p 305
8. A. Schulze, Physikalische Zeitschrift, Vol 22, 1921, p 403
9. K. K. Kelley, U.S. Bureau of Mines Bulletin, 1934, p 371
10. L. C. Bailey, Proceedings of the Royal Society (London) A, Vol 134, 1931, p 51
11. C. C. Bidwell, Physical Review, Series II, Vol 58, 1940, p 561
12. "Rolled Zinc", The New Jersey Zinc Co. Bulletin, 1929
13. W. J. Poppe, Physical Review, Vol 46, 1934, p 815
14. International Critical Tables, Vol 6, p 136
15. H. J. Creighton and W. A. Koehler, *Electrochemistry,* John Wiley & Sons, New York, 1944
16. C. L. Mantell, *Industrial Electrochemistry,* McGraw-Hill, New York, 1931, p 52
17. D. Shoenberg, Proceedings of the Cambridge Philosophical Society, Vol 36, Issue 1, 1940, p 84
18. J. Bor, A. Hobson and C. Wood, Proceedings of the Physical Society, Vol 51, 1939, p 932
19. G. B. Sabine, Physical Review, Series II, Vol 55, 1939, p 1064
20. G. Gerosa, Atti della Reale Accademia Nazionale dei Lencei Tendiconti, Issue IV, Vol 4, 1888, p 127

Zirconium (Zr)

Compiled by R. T. Webster
Teledyne Wah Chang Albany

Zirconium is nontoxic and, consequently, does not require serious limitations on its use because of health hazards. (Ref 1).

Zirconium is pyrophoric because of its heat-producing reaction with oxidizing elements such as oxygen. Large pieces of sheet, plate, bar, tube and ingot can be heated to high temperatures without excessive oxidation or burning, but small pieces

Fig. 127 Temperature dependence of mean coefficient of linear thermal expansion of zirconium (Ref 3)

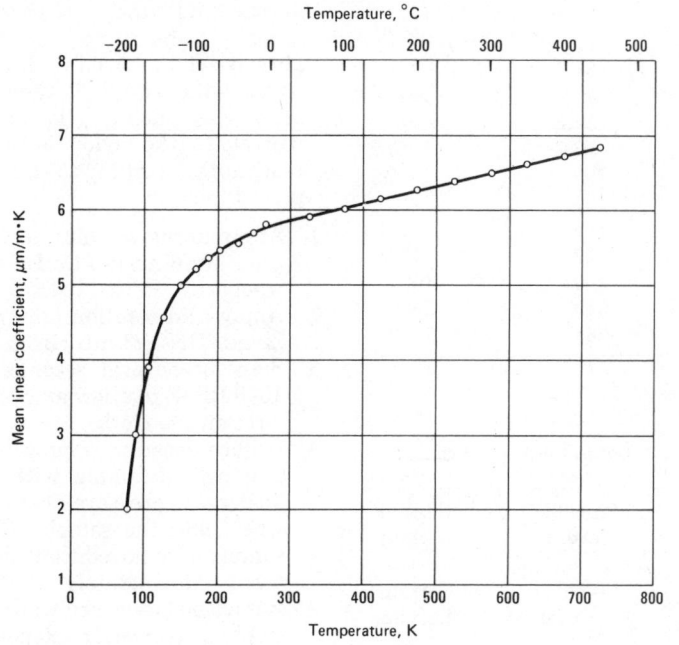

with a high surface area to mass ratio, such as machine chips and turnings, are easily ignited and burn at extremely high temperatures. It is recommended that large accumulations of chips and other finely divided material be avoided. Also, in storing the chips and turnings care should be taken to place the material in non-flammable containers and isolated areas. One storage method that works quite well is to keep the material covered with water in the containers and in turn, use oil on the water to keep it from evaporating. If a fire accidentally starts in zirconium, do not attempt to put it out with water or ordinary fire extinguishers, but use dry sand, powdered graphite, or commercially available Met-L-X powder. Large quantities of water can be used to control and extinguish fires in other flammables in the vicinity of a zirconium fire.

When zirconium is exposed to highly corrosive attack by concentrated acids, such as red fuming nitric acid, it is possible that in time exposed surfaces of the zirconium will be converted to a finely divided powder, which can ignite, possibly with explosive force. If the proper balance between water vapor and nitrogen dioxide above the liquid is maintained, this hazard can be eliminated.

Fig. 128 Temperature dependence of specific heat of zirconium (Ref 3)

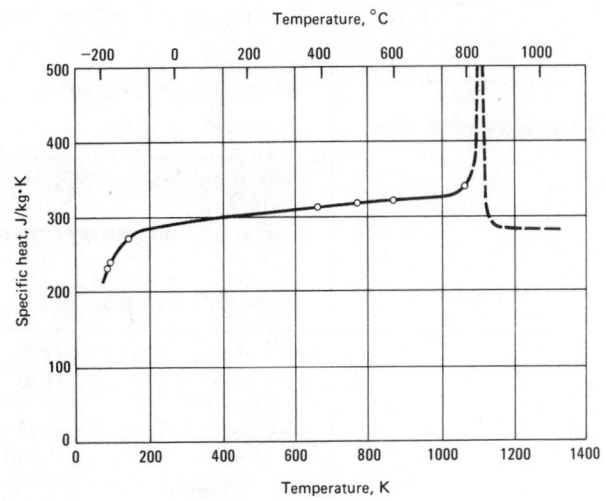

Structure

Crystal structure. (Ref 2) α-phase, close-packed hexagonal; at 20 °C: $a = 0.323115$ nm; $c = 0.51477$ nm; $c/a = 1.5931$. β-phase, body-centered cubic; at 862 °C: $a = 0.36090$ nm
Slip planes. $\{10\bar{1}0\}$ at 20 °C
Twinning planes. $\{10\bar{1}2\}$ $\{11\bar{2}1\}$ $\{11\bar{2}2\}$ $\{11\bar{2}3\}$ at 20 °C
Cleavage planes. α-phase, $\{1000\}$. β-phase, $\{100\}$
Minimum interatomic distance. α-phase at 20 °C, $d_1 = 0.316$ nm, $d_2 = 0.312$ nm; β-phase, $d_1 = 0.322$ nm
Microstructure. Polishing and etching zirconium to observe the microstructure is not difficult when the proper techniques are used. Due to the tendency of zirconium to smear during polishing, an attack-polish technique is used. A solution of alumina is used in conjunction with a dilute acid solution to attack the

Table 44 Optical and electronic properties of zirconium (Ref 4)

Temperature		Brightness temperature (λ-652 nm)		Total radiation, Kw/m² A/m²	Electron emission,
K	°C	K	°C		
1000	727	967	694	16.8	...
1100	827	1059	786	22.7	...
1200	927	1151	878	30.3	...
1300	1027	1242	969	40.6	...
1400	1127	1332	1059	54.0	...
1500	1227	1423	1150	72.0	0.2
1600	1327	1513	1240	100	1.8
1700	1427	1602	1329	134	13
1800	1527	1691	1418	175	84
1900	1627	1779	1506	222	405
2000	1727	1866	1593	280	1600
2100	1827	1952	1679	345	5200
2130	1857	1980	1707	365	7200

Table 45 Unstable isotopes of zirconium (Ref 1)

Mass No.	Half life	Mode of decay and radiation	Energy of radiation, MeV	Correctness of mass No.	Existence of element
86	17 h	K(a)		Probable	Certain
87	1.6 h	β⁻	2.10	Certain	Certain
		γ	0.6, 0.3		
88	85 days	K(a) γ	0.41	Probable	Certain
89(c)	4.4 min	IT(b)	0.59	Certain	Certain
		β⁻	0.9, 2.4		
		γ	1.5		
89(c)	78 h	K(a)	0.91	Certain	Certain
		β⁻	0.92	Radiation emitted by short-lived daughter	
		γ			
93(d)	~5 × 10⁶	β⁻	0.06	Probable	Certain
95(d)	yr	β⁻	0.39, 1.0; e^-(e)	Certain	Certain
	65 days	γ	0.73, 0.92		
97(d)	17 h	β⁻	1.91	Certain	Certain
		γ	0.75	Radiation emitted by short-lived daughter	

(a) K = K-electron capture. (b) IT = isomeric transition. (c) Nuclide 89 exists in two isometric states. (d) The nuclides 93, 95 and 97 have been identified as products of fusion of ^{235}U induced by slow neutrons. (e) e^- = internal conversion electron.

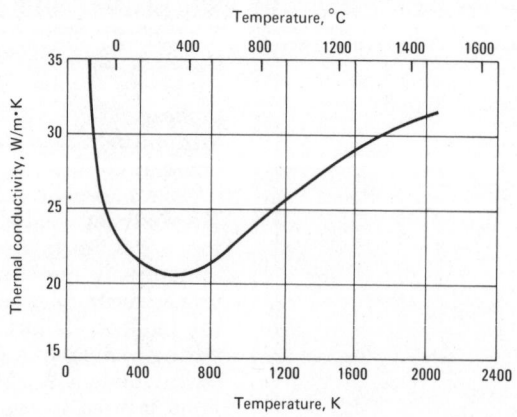

Fig. 129 Temperature dependence of the thermal conductivity of zirconium (Ref 4)

sample surface chemically at about the same rate that it is being removed by abrasion. The right combination of wheel speed, hand pressure, acid solution, and abrasive will result in a true undisturbed microstructure. *Sanding*—the sample is sanded on abrasive cloth down to a 3/0 grit. *Polishing*—the sample is then polished on a wheel using nylon cloth over met cloth (the nylon is fairly acid resistant), according to the following procedure:

1 Apply abrasive solution (5 g of 3 μm alumina/150 mL H_2O) to wheel
2 Apply acid solution (250 mL H_2O/ 22 mL HNO_3/3 mL HF) to wheel
3 Spin for several seconds (approx 1000 RPM) to allow an even film to form on the cloth
4 Reduce speed to approx 550 RPM and polish sample with light to moderate pressure between the wheel and the sample. When the sample is removed from the wheel, wash immediately with H_2O (squirt bottle works well) or over-etching will occur. Repeat above polishing as necessary to obtain a surface with no disturbed metal

To maintain the proper acid concentration on the wheel, the polishing cloth should be thoroughly rinsed with water after 1 to 2 min use. New alumina and acid should be applied to continue polishing.

Etching—swab the sample surface with acid solution (22 mL H_2O/22 mL HNO_3/3 mL HF) for 5 to 10 s, rinse in water to prevent overetching

Mass Characteristics

Atomic weight. 91.22
Density. α-phase, 6.505 Mg/m³ (low in hafnium) to 6.574 Mg/m³ (high in hafnium); β-phase at 979 °C, 6.046 Mg/m³ (high purity crystal bar zirconium)

Thermal Properties

Melting point. 1852 ± 2 °C
Boiling point. 3700 °C
Phase-transformation temperature. 862 ± 5 °C
Coefficient of thermal expansion. Linear: 5.85 μm/m·K at 20 °C for heterogeneously oriented polycrystals; temperature dependence, see Fig. 127; along crystal axes, 5.65 μm/m·K perpendicular to c-axis and 6.96 μm/m·K parallel to c-axis. Volumetric: 17.68 μm/m·K for 0.005 at. %

Fig. 130 Dependence of the electrical resistivity of zirconium on interstitial impurities (Ref 5)

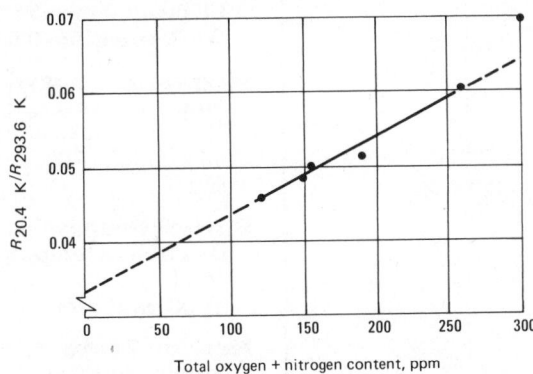

Fig. 131 Thermoelectric force of zirconium-platinum thermocouple (Ref 3)

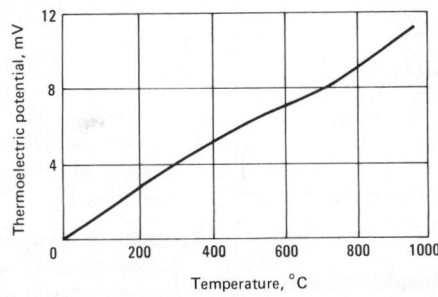

Fig. 132 Effect of irradiation and subsequent annealing on the ductility of sponge zirconium (Ref 7)

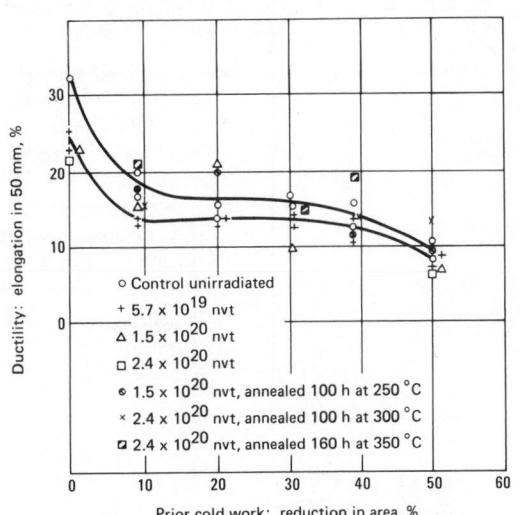

Hf; 17.47 μm/m·K for 1.2 atom% Hf
Specific heat. See Fig. 128.
Enthalpy. $H_T - H_{298}$:

°C	kJ/kg·K
100	23.1
200	52.2
300	80.3
400	112.6
500	146.4
600	182.0

Entropy. 426.1 J/kg·K at 25 °C
Latent heat of fusion. 25 kJ/kg
Latent heat of phase transformation. 42.2 kJ/kg
Latent heat of vaporization. 6520 kJ/kg
Heat of combustion. Heat of formation of ZrO_2, 5940 kJ/kg Zr
Thermal conductivity. (Ref 3) At 25 °C, 21.1 W/m·K; 100 °C, 20.4 W/m·K; 200 °C, 19.6 W/m·K. Temperature dependence:
$K = 30.8 (\sigma - 0.000327) T + 3.81$
where K is thermal conductivity, W/m·K; σ is electrical conductivity in reciprocal nΩ·m and T is temperature in K (see also Fig. 129).
Vapor pressure:

°C	Pa
1574	1.013×10^{-5}
1690	1.013×10^{-4}
1822	1.013×10^{-3}
1976	1.013×10^{-2}
2156	1.013×10^{-1}
2367	1.013
2620	1.013×10
2926	1.013×10^2
3304	1.013×10^3
3783	1.013×10^4
4409	1.013×10^5

Diffusion coefficient. At 800 °C: 2 $\times 10^{-3}$ mm²/s for hydrogen; 2 $\times 10^{-7}$ mm²/s for oxygen; 1 $\times 10^{-7}$ mm²/s for nitrogen

Electrical Properties
Electrical conductivity. Volumetric, 4.1% IACS
Electrical resistivity. 450 nΩ·m; temperature coefficient, $44 \pm 1 \times 10^{-4}$ per K at 0 to 200 °C; pressure coefficient, average reduction of 0.2% in resistance for each 980 MPa; dependence on plastic deformation, 2 to 5% increase with cold reductions of 64 to 96%; dependence on impurities, see Fig. 130.

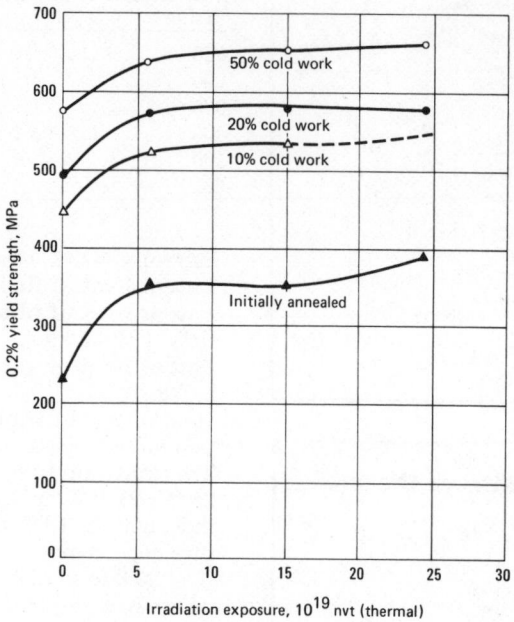

Fig. 133 Effect of irradiation on the yield strength of sponge zirconium (Ref 7)

Exposure temperature, 50 to 60 °C

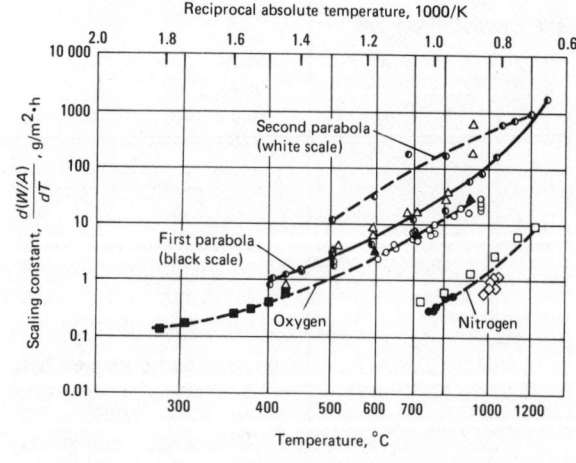

Fig. 134 Scaling rate of zirconium at elevated temperature (Ref 7)

77 K (-196 °C), 0.126 nV·m/A·T at 300 K ($+27$ °C) (Ref 6)

Temperature of superconductivity. 0.63 K (-272.52 °C)

Electron emission. See Table 44.

Work function. 0.656 aJ

Magnetic Properties

Magnetic susceptibility. Volume: 16×10^{-6} mks units at 25 °C, 19.2 mks units at 700 °C, 24 mks units at 860 °C

Optical Properties

Brightness temperature. See Table 44.

Total radiation. See Table 44.

Nuclear Properties

Stable isotopes. From Ref 3:

Mass No.	Abundance, %	Thermal neutron absorption cross section, barn
90	51.5	0.1
91	11.2	1.5
92	17.1	0.2
94	17.4	0.07
96	2.8	0.05

Unstable isotopes. See Table 45.

Effect of irradiation on properties. See Fig. 132 and 133.

Chemical Properties

Resistance to specific agents. Although zirconium is able to maintain a bright surface permanently at room temperature, to form a stable oxide of high melting point, and possibly to form a continuous oxide surface, oxidation resistance in air or oxygen at moderately high temperatures is poor, as indicated in Fig. 134.

Mechanical Properties

Tensile properties. See Fig. 132 and 133.

Poisson's ratio. 0.35 at room temperature

Elastic properties. Tension, 99.284 GPa. Pressure dependency of specific volume (compressibility) (Ref 8):

Pressure, MPa	Relative specific volume
0.10	1.000
2940	0.967
3920	0.956
5880	0.937
6860	0.929
7850	0.922
8830	0.916
9810	0.910

Thermoelectric potential. See Fig. 131.

Electrochemical equivalent. 0.2363 mg/coulomb

Ionization potential. 34.33 eV

Hydrogen overvoltage. 0.83 V at 10 A/m²

Hall coefficient. 0.118 nV·m/A·T at

Specific damping capacity. 17×10^{-4} at 25 °C, decreasing to 6.2×10^{-4} at 260 °C with a sharp increase to 1×10^{-2} at 610 °C

Coefficient of friction. For zirconium sliding on zirconium, 0.42 at 20 °C (Ref 9)

Velocity of sound. 4.62 km/s

REFERENCES

1. H. Loevenstein and H. L. Gilbert, "Zirconium: A Review and Summary of Published Data", Technical Information Service Extension, Oak Ridge, TN, Oct 1958
2. A. Taylor and Brenda J. Kagle, "Crystallographic Data on Metal And Alloy Structures", Dover Publication, Inc., New York, 1963
3. G. L. Miller, "Zirconium", Academic Press, New York, 1957
4. C. Y. Ho, R. W. Powell, and P. E. Liley, "Thermal Conductivity of Selected Materials", NSRDS-NBS16, Feb 1968
5. D. L. Douglass, "The Metallurgy of Zirconium", Atomic Energy Review Supplement, 1971
6. Ted G. Berlincourt, "Hall Effect, Resistivity, and Magnetoresistivity of Th, U, Zr, Ti, and Nb", Atomic International, Division of North American Aviation, Incorporated, Canoga Park, CA., *Physical Review*, Vol 114, No. 4, May 15, 1959
7. B. Lustman and K. Kerze, Jr., "The Metallurgy of Zirconium", McGraw-Hill Book Co., New York, 1955
8. John M. Wash, Melvin H. Rice, Robert G. McQueen, and Frederick L. Yarger, "Shock-Wave Compressions of Twenty-Seven Metals, Equation of State of Metals", *Physical Review*, Vol 108, No. 2, Oct 15, 1957
9. Donald H. Buckly and Robert L. Johnson, "Relation of Lattice Parameters to Friction Characteristics of Beryllium, Hafnium, Zirconium, and Other Hexagonal Metals in Vacuum", NASA TND-2670, March 1965

Properties of the Actinide Metals

Polymorphic modifications, transformation temperatures and structural data for the actinide metals

Metal	Lattice symmetry	Temperature range of stability, °C	Lattice constants				Density(a), Mg/m³	Atoms per unit cell
			a, nm	b, nm	c, nm	β, deg		
Actinium	Face-centered cubic	···	0.5 331	···	···	···	10.07	4
Thorium	α, face-centered cubic	Below 1400	0.5 0843(25 °C)	···	···	···	11.724	4
	β, body-centered cubic	1400-1750	0.4 11(1450 °C)	···	···	···	11.10	2
Protactinium...	α, body-centered tetragonal	Below 1165	0.3 929	···	0.3 241	···	15.37	2
	β, body-centered cubic	1165-1575	0.3 81(1186 °C)	···	···	···	13.87	2
Uranium	α, orthorhombic	Below 666	0.2 8537(298 °C)	0.5 8695	0.4 9548	···	19.07	4
	β, tetragonal	666-774	1.0 763 ± 5(720 °C)	···	0.5 652 ± 5	···	18.11	30
	γ, body-centered cubic	774-1132	0.3 524 ± 2(805 °C)	···	···	···	18.06	2
Neptunium ...	α, orthorhombic	Below 279	0.6 663 ± 3	0.4 723 ± 1	0.4 887 ± 2	···	20.45	8
	β, tetragonal	279-574	0.4 897 ± 2(313 °C)	···	0.3 388 ± 2	···	19.36	4
	γ, body-centered cubic	574-637	0.3 518(600 °C)	···	···	···	18.04	2
Plutonium	α, monoclinic	Below 118	0.6 183(21 °C)	0.4 822	1.0 963	101.79	19.86	16
	β, monoclinic	−118-−200	0.9 284(190 °C)	1.0 463	0.7 859	92.13	17.70	34
	γ, orthorhombic	−200-312	0.3 159(235 °C)	0.5 768	1.0 162	···	17.14	8
	δ, face-centered cubic	312-458	0.4 6371(320 °C)	···	···	···	15.92	4
	δ′, body-centered tetragonal	458-480	0.3 34(465 °C)	···	0.4 44	···	16.00	2
	ε, body-centered cubic	480-640	0.3 636(490 °C)	···	···	···	16.51	2
Americium....	α, close-packed double hexagonal	Below 1074	0.3 4(20 °C)	···	1.1 2	···	13.6	4
	β, face-centered cubic	1074-1176	0.4 894 ± 5	···	···	···	13.65	4
Curium.......	α, close-packed double hexagonal	Below 1176	0.3 49	···	1.1 33	···	13.5	4
	β, face-centered cubic	1176-1330		···	···	···		4
Berkelium	α, close-packed double hexagonal	Below 910	0.3 416 ± 5	···	1.1 068	···	14.79	4
	β, face-centered cubic	910-983	0.4 999 ± 5	···	···	···	13.24	4
Californium...	α, close-packed double hexagonal		0.4 002	···	1.2 804	···	9.31	4
	β, face-centered cubic	Below 940	0.5 74	···	···	···	8.72	4
Einsteinium...	α, close-packed double hexagonal		7	···	7	···	7	4
	β, face-centered cubic	Below 820	0.5 75	···	···	···	8.84	4

(a) Determined by x-ray methods.

Properties related to $N(E_F)$ for α-phase actinide metals

Metal	$\rho_{300} - \rho_4$(a), n$\Omega \cdot$m	χ_{300}(b), 10^4 emu/mol	γ(c), mJ/(mol$\cdot$K^2)	n(d)	U_0/W(e)	Magnetic ordering or superconducting(f)
Thorium.....	140	0.8	4.3	3	0.3	S
Protactinium.	180	2.7	...	2.8	(0.4)	S
Uranium	310	3.9	9.1	1-4(g)	0.6	S
Neptunium ..	1230	5.5	12.4	2	0.7	...
Plutonium ...	1380	5.1	17.0	2	1.8	...
Americium ..	670	7.8	6.0	5	50	S
Curium	860	119.0	...	2	>50	M
Berkelium...	...	...	...	...	>50	M
Californium..	...	...	...	...	>50	M

(a) Electrical resistivity (difference in values at 300 and 4 K) $\propto N(E_F)$. (b) Magnetic susceptibility at 300 K (27 °C). (c) Coefficient of the electronic term in the specific heat, $\gamma \propto N(E_F)$. (d) Low temperature value in $\rho - \rho_0 = aT^n$. (e) Ratio of polar-state formation energy to band-width (for $U_0/W \leq 1$, bands are formed). (f) M, orders magnetically; S, superconductor. (g) A function of crystallographic direction.

Typical mechanical properties of pure actinide metals(a)

Metal	Tensile strength, MPa	Yield strength, MPa	Elongation, % in 50 mm	Hardness, HV
Thorium(b)..........	120	48	36	150
Uranium(c)	585	240	10	220
Plutonium(d)	525	300	8	250
For comparison:				
Nickel(e)............	315	60	30	70

(a) Measured at room temperature on high- or commercial-purity, polycrystalline material after prior plastic deformation and annealing. (b) Iodide (high-purity) grade; forged, hot and cold rolled, then annealed at 850 °C. (c) Magnesium-reduced (commercial purity) grade; hot rolled as α phase, annealed in β phase and water quenched. (d) Electrolytic (high-purity) grade; prior treatment consisted of extrusion at 108 °C without subsequent annealing. (e) Electrolytic grade; cold rolled, then annealed at 550 °C.

Periodic Table of the Elements

Metals ← | → Nonmetals

Key to chart:
Atomic Number → **50** → Oxidation States (+2 +4)
Symbol → **Sn**
Atomic Weight → 118.69
→ Electron Configuration (−18−18−4)

Transition Elements

Group	Ia	IIa	IIIb	IVb	Vb	VIb	VIIb	VIII	VIII	VIII	Ib	IIb	IIIa	IVa	Va	VIa	VIIa	0	Orbit
1	**1** H, +1 −1, 1.0079, 1																	**2** He, 0, 4.00260, 2	K
2	**3** Li, +1, 6.939, 2−1	**4** Be, +2, 9.0122, 2−2											**5** B, +3, 10.81, 2−3	**6** C, +2 +4 −4, 12.011, 2−4	**7** N, +1 +2 +3 +4 +5 −3, 14.0067, 2−5	**8** O, −2, 15.9994, 2−6	**9** F, −1, 18.998403, 2−7	**10** Ne, 0, 20.179, 2−8	K−L
3	**11** Na, +1, 22.9898, 2−8−1	**12** Mg, +2, 24.312, 2−8−2											**13** Al, +3, 26.98154, 2−8−3	**14** Si, +2 +4 −4, 28.08, 2−8−4	**15** P, +3 +5 −3, 30.97376, 2−8−5	**16** S, +4 +6 −2, 32.06, 2−8−6	**17** Cl, +1 +5 +7 −1, 35.453, 2−8−7	**18** Ar, 0, 39.948, 2−8−8	K−L−M
4	**19** K, +1, 39.09, −8−8−1	**20** Ca, +2, 40.08, −8−8−2	**21** Sc, +3, 44.9559, −8−9−2	**22** Ti, +2 +3 +4, 47.9, −8−10−2	**23** V, +2 +3 +4 +5, 50.941, −8−11−2	**24** Cr, +2 +3 +6, 51.996, −8−13−1	**25** Mn, +2 +3 +4 +6 +7, 54.9380, −8−13−2	**26** Fe, +2 +3 +4 +7, 55.847, −8−14−2	**27** Co, +2 +3, 58.9332, −8−15−2	**28** Ni, +2 +3, 58.71, −8−16−2	**29** Cu, +1 +2, 63.54, −8−18−1	**30** Zn, +2, 65.38, −8−18−2	**31** Ga, +3, 69.72, −8−18−3	**32** Ge, +2 +4, 72.59, −8−18−4	**33** As, +3 +5 −3, 74.9216, −8−18−5	**34** Se, +4 +6 −2, 78.96, −8−18−6	**35** Br, +5 −1, 79.904, −8−18−7	**36** Kr, 0, 83.80, −8−18−8	−L−M−N
5	**37** Rb, +1, 85.467, −18−8−1	**38** Sr, +2, 87.62, −18−8−2	**39** Y, +3, 88.9059, −18−9−2	**40** Zr, +4, 91.22, −18−10−2	**41** Nb, +3 +5, 92.9064, −18−12−1	**42** Mo, +6, 95.94, −18−13−1	**43** Tc, +4 +6 +7, 98.9062, −18−13−2	**44** Ru, +3, 101.07, −18−15−1	**45** Rh, +3, 102.905, −18−16−1	**46** Pd, +2 +4, 106.4, −18−18−0	**47** Ag, +1, 107.868, −18−18−1	**48** Cd, +2, 112.40, −18−18−2	**49** In, +3, 114.82, −18−18−3	**50** Sn, +2 +4, 118.69, −18−18−4	**51** Sb, +3 +5 −3, 121.75, −18−18−5	**52** Te, +4 +6 −2, 127.60, −18−18−6	**53** I, +1 +5 +7 −1, 126.9045, −18−18−7	**54** Xe, 0, 131.30, −18−18−8	−M−N−O
6	**55** Cs, +1, 132.9054, −18−8−1	**56** Ba, +2, 137.3, −18−8−2	**57*** La, +3, 138.9055, −18−9−2	**72** Hf, +4, 178.49, −32−10−2	**73** Ta, +5, 180.948, −32−11−2	**74** W, +6, 183.85, −32−12−2	**75** Re, +4 +6 +7, 186.207, −32−13−2	**76** Os, +3 +4 +6 +8, 190.2, −32−14−2	**77** Ir, +3 +4, 192.9, −32−15−2	**78** Pt, +2 +4, 195.09, −32−16−2	**79** Au, +1 +3, 196.9665, −32−18−1	**80** Hg, +1 +2, 200.59, −32−18−2	**81** Tl, +1 +3, 204.37, −32−18−3	**82** Pb, +2 +4, 207.19, −32−18−4	**83** Bi, +3 +5, 208.980, −32−18−5	**84** Po, +2 +4, (209), −32−18−6	**85** At, (210), −32−18−7	**86** Rn, 0, (222), −32−18−8	−N−O−P
7	**87** Fr, +1, (223), −18−8−1	**88** Ra, +2, 226.0254, −18−8−2	**89**** Ac, +3, (227), −18−9−2	**104** Rf, +4, (261), −32−10−2	**105** Ha, (262), −32−11−2	**106** (263)													−O−P−Q

Orbit labels: K, K−L, K−L−M, −L−M−N, −M−N−O, −N−O−P, −O−P−Q, −N−O−P, −O−P−Q

***Lanthanides**

58 Ce, +3 +4, 140.12, −20−8−2	**59** Pr, +3 +4, 140.9077, −21−8−2	**60** Nd, +3, 144.24, −22−8−2	**61** Pm, +3, 147, −23−8−2	**62** Sm, +2 +3, 150.4, −25−8−2	**63** Eu, +2 +3, 151.96, −25−8−2	**64** Gd, +3, 157.25, −25−9−2	**65** Tb, +3, 158.925, −27−8−2	**66** Dy, +3, 162.50, −28−8−2	**67** Ho, +3, 164.9304, −29−8−2	**68** Er, +3, 167.26, −30−8−2	**69** Tm, +3, 168.9342, −31−8−2	**70** Yb, +2 +3, 173.04, −32−8−2	**71** Lu, +3, 174.967, −32−9−2

****Actinides**

90 Th, +4, 232.038, −18−10−2	**91** Pa, +4 +5, 231.0359, −20−9−2	**92** U, +3 +4 +5 +6, 238.029, −21−9−2	**93** Np, +3 +4 +5 +6, 237.0482, −22−9−2	**94** Pu, +3 +4 +5 +6, 239.052, −24−8−2	**95** Am, +3 +4 +5 +6, (243), −25−8−2	**96** Cm, +3, (247), −25−9−2	**97** Bk, +3 +4, (247), −27−8−2	**98** Cf, +3, (251), −28−8−2	**99** Es, (254), −29−8−2	**100** Fm, (257), −30−8−2	**101** Md, +2 +3, (258), −31−8−2	**102** No, +2 +3, (259), −32−8−2	**103** Lr, +3, (260), −32−9−2

Numbers in parentheses are mass numbers of most stable isotope of that element

Système Internationale d'Unités (SI)

SI Base Units

Quantity	Unit	Symbol	Quantity	Unit	Symbol	Quantity	Unit	Symbol
Length	metre	m	Amount of substance	mole	mol	Plane angle(a)	radian	rad
Mass	kilogram	kg				Solid angle(a)	steradian	sr
Time	second	s	Luminous intensity	candela	cd	(a) Supplementary unit		
Electric current	ampere	A						
Thermodynamic temperature	kelvin	K						

SI Derived Units(a)

Quantity	Unit	Symbol	Formula	Quantity	Unit	Symbol	Formula
Frequency (of a periodic phenomenon)	hertz	Hz	s^{-1}	Capacitance	farad	F	C/V
				Electric resistance	ohm	Ω	V/A
Force	newton	N	$kg \cdot m/s^2$	Conductance	siemens	S	A/V
Pressure, stress	pascal	Pa	N/m^2	Magnetic flux	weber	Wb	$V \cdot s$
Energy, work, quantity of heat	joule	J	$N \cdot m$	Magnetic flux density	tesla	T	Wb/m^2
Power, radiant flux	watt	W	J/s	Inductance	henry	H	Wb/A
Quantity of electricity, electric charge	coulomb	C	$A \cdot s$	Luminous flux	lumen	lm	$cd \cdot sr$
				Illuminance	lux	lx	lm/m^2
Electric potential, potential difference, electromotive force	volt	V	W/A	Activity (of radionuclides)	becquerel	Bq	s^{-1}
				Absorbed dose	gray	Gy	J/kg

(a) Derived units in this list include only those units for which special names and symbols have been approved by the General Conference on Weights and Measures (CGPM)

SI Prefixes(a)

Prefix	Multiplication factor	Symbol	Prefix	Multiplication factor	Symbol
exa	$1\ 000\ 000\ 000\ 000\ 000\ 000 = 10^{18}$	E	deci(b)	$0.1 = 10^{-1}$	d
peta	$1\ 000\ 000\ 000\ 000\ 000 = 10^{15}$	P	centi(c)	$0.01 = 10^{-2}$	c
			milli	$0.001 = 10^{-3}$	m
tera	$1\ 000\ 000\ 000\ 000 = 10^{12}$	T			
giga	$1\ 000\ 000\ 000 = 10^{9}$	G	micro	$0.000\ 001 = 10^{-6}$	μ
mega	$1\ 000\ 000 = 10^{6}$	M	nano	$0.000\ 000\ 001 = 10^{-9}$	n
			pico	$0.000\ 000\ 000\ 001 = 10^{-12}$	p
kilo	$1\ 000 = 10^{3}$	k			
hecto(b)	$100 = 10^{2}$	h	femto	$0.000\ 000\ 000\ 000\ 001 = 10^{-15}$	f
deka(b)	$10 = 10^{1}$	da	atto	$0.000\ 000\ 000\ 000\ 000\ 001 = 10^{-18}$	a

(a) Used to form multiples and decimal fractions of the base and derived SI units. (b) Normally avoided. (c) Use not recommended.

Abbreviations and Symbols

A area

a crystal lattice edge length along *a* axis

AC air cooled

ADA American Dental Association

AMS Aerospace Material Specification (of SAE)

ANSI American National Standards Institute, Inc.

ASME American Society of Mechanical Engineers

ASTM American Society for Testing and Materials

avg average

AWG American wire gage

AWS American Welding Society

b barn (neutron capture cross section)

b crystal lattice edge length along *b* axis

bcc body-centered cubic

Btu British thermal unit

B&S Brown and Sharpe (gage)

c crystal lattice edge length along *c* axis, elastic constant (for single crystals)

CDA Copper Development Association

cgs centimetre-gram-second system of units

cph close-packed hexagonal

CN coordination number

C_p specific heat at constant pressure

D die casting

D diameter, diffusion constant

d day

d diameter

DA duplex annealed

di didymium (praseodymium plus neodymium)

diam diameter

E elastic modulus, electrical potential

EC electron capture

emf electromotive force

EPA Environmental Protection Agency

Eq equation

est estimate

F free energy

FC furnace cooled

fcc face-centered cubic

ft foot

G modulus of rigidity, torsion modulus, Gibb's free energy

gal gallon

GMAW gas metal-arc welding

GTAW gas tungsten-arc welding

H enthalpy, magnetic field strength, coercive force

h hour

HB Brinell hardness

HRB Rockwell "B" hardness

HRC Rockwell "C" hardness

HRE Rockwell "E" hardness

HRF Rockwell "F" hardness

HR15T Rockwell "15T" superficial hardness

HV Vickers hardness

IACS International Annealed Copper Standard (of electrical conductivity)

ILZRO International Lead Zinc Research Organization

I radiation intensity

i electrical current

in. inch

IT isomeric transition

K_c critical plane-stress stress-intensity factor

k_{Ic} plane-strain fracture toughness

ksi kips (1000 lb) per square inch

L litre, longitudinal, liquid

l length

lb pound

log common logarithm (base 10)

LT long transverse

M molar concentration, magnetization

max maximum

MIG metal inert-gas (welding)

min minimum, minute

mks metre-kilogram-second system of units

MM mischmetal

N normal concentration

n strain hardening exponent

NACE National Association of Corrosion Engineers

OD outside diameter

OSHA Occupational Safety and Health Administration

oz ounce

P permanent mold casting

P pressure

pH negative logarithm of hydrogen-ion activity

P/M powder metallurgy

ppb parts per billion

ppm parts per million

psi pounds per square inch

Q activation energy, grain size

R ratio of minimum stress to maximum stress, electrical resistivity, universal gas constant, Hall effect coefficient

r radius, minimum interatomic distance, plastic strain ratio

RE rare earth (elements)

rem remainder

rev revolution

RWMA Resistance Welder Manufacturer's Association

S sand casting

S entropy

s second, solid

SAE Society of Automotive Engineers

SHE Standard Hydrogen Electrode

SMAW shielded metal-arc welding

STA solution treated and aged

T temperature

t tonne (metric ton, or 1000 kg)

t time, thickness

T_c Curie temperature

TIG tungsten inert-gas (welding)

UNS Unified Numbering System (ASTM-SAE)

vs versus

WC water quenched

yr year

° degree (angular measure)

°C degree Celsius (centigrade)

°F degree Fahrenheit

Ω ohm

> greater than

< less than

% percent

+ plus, in addition to, including, positive ion charge

− minus, negative ion charge

± maximum deviation

× multiplied by, diameters (magnification)

· multiplied by

α alpha phase, alpha particle

$\beta+$ positron

$\beta-$ electron

γ, σ surface tension

Δ difference

δ grain-boundary thickness

ϵ strain, dielectric constant

$\dot{\epsilon}$ strain rate

λ wave length

μ viscosity

ρ density, electrical resistivity

χ magnetic susceptibility

Index

Numbered Alloys

10 alloy. See *Copper alloys, specific types, C17500.*
25 alloy. See *Copper alloys, specific types, C17200.*
50 alloy. See *Copper alloys, specific types, C17600.*
165 alloy. See *Copper alloys, specific types, C17000.*

A

Actinium, pure 714, 832-833
Adhesive bonding
  aluminum 201-202
  copper metals 456
  magnesium 547-549, 550
Admiralty metals
  antimonial. See *Copper alloys, specific types, C44400.*
  arsenical. See *Copper alloys, specific types, C44300.*
    inhibited. See *Copper alloys, specific types, C44300, C44400 and C44500.*
  phosphorized. See *Copper alloys, specific types, C44500.*
Age hardening. See also *Precipitation hardening.*
  aluminum alloys 30-32, 33, 36-38, 40-42, 43
  copper alloys 256
Alclad products
  core and cladding combinations 211
  corrosion resistance 210-211
Alcology. See *Copper alloys, specific types, C68800.*
Alloy M25. See *Copper alloys, specific types, C17300.*
Aluminum. See also *Aluminum alloys.*
  pure 714-715
  solubility in magnesium 525
  U.S. shipments 4, 17
  world production 4
  zone refined, impurity concentration 713
Aluminum alloys
  adhesive bonding 201-202
  annealing 28-29, 30
  applications 16-23
    building and construction 16-17
    consumer 21-22
    container and packaging 17-18
    electrical 20-21
    machinery and equipment 22
    transportation 18-20
  brazing
    dip brazing 199-200
    filler metals 199-200
    sheet for 29, 201
    summary of procedures 200
    torch brazing 199-200
    vacuum furnace brazing 199-200

casting alloys
  alloy systems 140-141
  casting processes 143-148
  characteristics 143, 144, 145
  designation system ... 141-143, 144, 145
  mechanical properties 148-151
  modification 149, 150
  quality of castings 148
castings 9-10
corrosion resistance
  alclad products 210-211
  anodized products ... 225-226, 229, 232
  atmospheric corrosion 219-228
  cathodic protection 210-211
  chemical products, packaging 228-229, 231-233
  composition, effect of 206-209
  corrosion fatigue 219, 220
  deposition corrosion 211-212
  erosion-corrosion 219
  exfoliation corrosion 218-220
  food, packaging 228-229, 231
  galvanic corrosion ... 207, 209-210
  high purity waters 222-223
  intergranular corrosion 212
  microstructure, effect of 206-209
  natural waters 223
  nonmetallic building materials 226-228
  oxide film protection 204-205
  pharmaceuticals, packaging ... 228-229, 231-233
  pitting 204-206
  ratings 209-211, 213
  seawater 223-225, 228-232
  soil 22
  solution potentials 206-207
  stress-corrosion cracking 210-211, 212-218
extrusions 5-6
fabrication characteristics 14-16
fasteners 202-203
forgeability 6
forgings 5-9
forming 14-16
foundry products 140-151
heat treatment 28-43
  aging 29-33, 36-38, 40-42, 43
  precipitation hardening ... 29-33, 36-38, 40-42, 43
  castings 32, 33
  cold work, effect after quenching 38, 39
  corrosion resistance, effect of quenching 32-35
  dimensional change during 39-42, 43
  dimensional stability 42-43
  quality control 32
  quenching 32-35, 40-41, 43
  refrigeration, effect on aging ... 35-38
  solution heat treatment 31, 32, 35, 38-40

wrought alloys 30-32
impacts 8-10, 13
joining 16, 191-202
machining 5, 6, 13, 15, 187-190
  chip characteristics 189-190
  machinability ratings 188-189
  surface finish 190
  tool wear 187-189
mill products 4-5, 44-62
powder metallurgy parts 10-13
product forms 4-14
properties. See also data compilations for specific alloys. 3-4
soldering 200-201
stampings 4, 13-14
sheet, stamping 13, 14, 180-186
  biaxial stretching 180
  characteristics 183-184
  deep drawing 180
  flanging 180, 186
  forming-limit diagrams 182-183
  material properties 180-181
  pure bending 180
  shape analysis 184, 185
  stretch bending 180, 181, 186
  stretch/draw 184-186
  tests 180-182
temper designations 24-27
weldbonding 202
welding
  filler metals 193-194
  joint types 193-194
  joint preparation 193-195
  processes 196-199
  weldability 192-193
  finishing 195-196
  weld strength 195, 197-199
wrought alloys
  bar 51-52
  designation system 44-51
  extrudability 54
  extrusions, interconnecting 55, 57-58
  flat rolled products 51
  mechanical properties ... 55, 58, 59-62
  physical properties 53-54, 58
  elevated-temperature properties 56, 57-58, 62
  low-temperature properties 62
  rod 51-52
  shapes 52-57
  tubular products 52
  wire 51-52
Aluminum alloys, powder metallurgy, specific types. See also *Aluminum alloys, wrought, specific types; Aluminum casting alloys, specific types*

201 AB
  as-sintered properties 13
  fatigue curves 15
  tensile strength, effects of density and thermal condition on 14

601 AB
  as-sintered properties 11
  fatigue curves 14
  tensile strength, effects of density
    and thermal condition on 13
Aluminum alloys, wrought, specific types.
  See also *Aluminum alloys, powder*
  metallurgy, specific types; Aluminum
  casting alloys, specific types
1050 63-64
  composition 45
  mechanical properties 59
  physical properties 53
  product forms 45
1060 64-65
  annealing temperature 29
  applications 46
  composition 45
  corrosion resistance 46, 209
  fabrication characteristics 46
  mechanical properties 59
  physical properties 53
  product forms 45
  solution potential 207
  stress-corrosion cracking
    resistance 209
  weldability 193
1100 ....65-66, 67
  annealing curves 28, 30
  annealing temperature 29
  anodic polarization curve 205
  applications 46
  atmospheric corrosion
    resistance 221, 222
  composition 45
  corrosion in chemical solutions 233
  corrosion pit density vs anodic
    coating thickness 229
  corrosion resistance 46, 209, 223,
          224, 230, 231, 232
  fabrication characteristics 46
  forming-limit diagram 183
  impacts, mechanical properties 10
  impacts, minimum wall thickness 9
  machinability rating 188
  mechanical properties 59
  physical properties 53
  product forms 45
  properties, expected for welds 197
  properties, gas metal-arc
    welded plate 198
  solution potential 207
  stress-corrosion cracking
    resistance 209
  weldability 193
1135
  atmospheric corrosion resistance 224
1145 66-67
  composition 45
  physical properties 53
  product forms 45
1188
  atmospheric corrosion resistance 224
1199 67-68
  atmospheric corrosion resistance 224
  composition 45
  physical properties 53
  pitting potential, effect of
    chloride-ion activity 206
  product forms 45
  seawater corrosion resistance 230
1350 ....68-69, 70
  annealing temperature 29
  applications 46
  composition 45
  corrosion resistance 46, 209
  fabrication characteristics 46
  machinability rating 188
  mechanical properties 59
  physical properties 53

  product forms 45
  stress-corrosion cracking
    resistance 209
  weldability 193
  wire conductors 20
2011 ....69-70, 71
  applications 46
  composition 45
  corrosion resistance 46, 209
  fabrication characteristics 46
  machinability rating 188
  mechanical properties 59
  physical properties 53
  product forms 45
  stress-corrosion cracking
    resistance 209, 213
  weldability 193
2014 70-73
  aging characteristics at low aging
    temperatures 36-38
  annealing temperature 29
  applications 46
  composition 45
  corrosion resistance ...46, 209, 224, 225
  fabrication characteristics 46
  fatigue curves 15
  forgeability, relative 6
  forging, relative cost vs
    forging weight 5
  impacts, mechanical properties 10
  impacts, minimum wall thickness 9
  machinability rating 188
  mechanical properties 59
  physical properties 53
  product forms 45
  quenching stresses 40, 43
  solution potential 207
  stress-corrosion cracking
    resistance 209, 213
  weldability 193
2017
  annealing temperature 29
  corrosion resistance 209, 223
  galvanic corrosion with
    magnesium 607
  machinability rating 188
  stress-corrosion cracking
    resistance 209
2018
  corrosion resistance 209
  stress corrosion cracking
    resistance 209
2024 ....72-75, 76-78
  aging characteristics at low aging
    temperatures 31, 36-38
  annealing temperature 29
  applications 46
  composition 45
  corrosion rate affected by
    quenching rate 34
  corrosion resistance 46, 209, 224,
          225, 226, 232
  fabrication characteristics 46
  fatigue characteristics 35
  galvanic corrosion with
    magnesium 607
  machinability rating 188
  mechanical properties 59
  natural aging curve 31
  physical properties 53
  product forms 45
  properties, elevated
    temperatures 56, 62
  quenching rate, effect on yield
    strength after aging 34
  solution potential 207
  stress-corrosion cracking
    resistance 209, 213
  tensile properties affected by
    cold work before aging 39

  weldability 193
alclad 2024 ....72-75, 76-77
  galvanic corrosion with
    magnesium 607
2025
  corrosion resistance 209
  forging, relative cost vs
    forging weight 5
  stress-corrosion cracking
    resistance 209
2036 75-76
  annealing temperature 29
  applications 46
  composition 45
  corrosion resistance 46, 209
  fabrication characteristics 46
  machinability rating 188
  mechanical properties 59
  mechanical properties, sheet 181
  physical properties 53
  product forms 45
  solution potential 207
  stress-corrosion cracking
    resistance 209
2048 ....77-78, 80-82
  composition 45
  mechanical properties 59
  physical properties 53
  product forms 45
  stress-corrosion cracking
    resistance 213
2117
  annealing temperature 29
  corrosion resistance 209
  stress-corrosion cracking
    resistance 209
2124 ....78-79, 82-84
  annealing temperature 29
  composition 45
  mechanical properties 59
  physical properties 53
  product forms 45
  stress-corrosion cracking
    resistance 213
2218 ....79, 85
  applications 46
  composition 45
  corrosion resistance 46, 209
  fabrication characteristics 46
  mechanical properties 59
  physical properties 53
  product forms 45
  stress-corrosion cracking
    resistance 209
  weldability 193
2219 ....86-88, 89, 91
  annealing temperature 29
  applications 46
  composition 45
  corrosion resistance 46, 209
  fabrication characteristics 46
  machinability rating 188
  mechanical properties 59
  physical properties 53
  product forms 45
  solution potential 207
  stress-corrosion cracking
    resistance 209, 213
  weldability 193
2319
  physical properties 53
  product forms 45
2618 ....88-90, 92, 93
  applications 47
  composition 45
  corrosion resistance 47, 209
  fabrication characteristics 47
  forgeability, relative 6
  machinability rating 188
  mechanical properties 59

Aluminum Alloys, Wrought (cont.)
physical properties 53
product forms 45
stress-corrosion cracking
  resistance 209
weldability 193
3003 .... 90-92, 94, 95
annealing temperature 29
applications 47
atmospheric corrosion
  resistance 221, 222
composition 45
corrosion resistance 47, 209, 223,
    224, 225, 230, 231, 232
fabrication characteristics 47
fatigue characteristics,
  weldments 195, 199
galvanic corrosion with
  magnesium 607
impacts, mechanical properties of 10
machinability rating 188
mechanical properties 59
mechanical properties, sheet 181
physical properties 53
product forms 45
properties, expected for welds 197
properties, gas metal-arc
  welded plate 197
solution potential 207
stress-corrosion cracking
  resistance 209
weldability 193
alclad 3003 .... 90-92, 94
3004 .... 92-94, 96, 97
annealing temperature 29
applications 47
atmospheric corrosion
  resistance 221, 222
composition 45
corrosion in distilled water 205
corrosion resistance 47, 209, 224,
    225, 232
fabrication characteristics 47
machinability rating 188
mechanical properties 60
physical properties 53
product forms 45
properties, expected for welds 197
solution potential 207
stress-corrosion cracking
  resistance 209
weldability 193
alclad 3004 92-94
3105 .... 94-95, 97
annealing temperature 29
applications 47
composition 45
corrosion resistance 47, 209
fabrication characteristics 47
mechanical properties 60
physical properties 53
product forms 45
stress-corrosion cracking
  resistance 209
weldability 193
4032 .... 95-96, 98
applications 47
composition 45
corrosion resistance 209
fabrication characteristics 47
forgeability, relative 6
mechanical properties 60
physical properties 53
product forms 45
stress-corrosion cracking
  resistance 209
weldability 193
4043 97-98
atmospheric corrosion resistance 204
composition 45

mechanical properties 60
physical properties 53
product forms 45
5005 98-99
annealing temperature 29
applications 47
composition 45
corrosion resistance 47, 209, 224
fabrication characteristics 47
machinability rating 188
mechanical properties 60
physical properties 53
product forms 45
properties, expected for welds 197
stress-corrosion cracking
  resistance 209
weldability 193
5050 99-100
annealing temperature 29
applications 47
composition 45
corrosion resistance 47, 209, 224, 225
fabrication characteristics 47
machinability rating 188
mechanical properties 60
physical properties 53
product forms 45
properties, expected for welds 197
properties, gas metal-arc
  welded plate 197
solution potential 207
stress-corrosion cracking
  resistance 209
weldability 193
5052 101-102
annealing curves 28, 30
annealing temperature 29
applications 47
composition 45
corrosion resistance 47, 209, 224,
    225, 230, 231
fabrication characteristics 47
galvanic corrosion with
  magnesium 607
machinability rating 188
mechanical properties 60
mechanical properties, sheet 181
physical properties 53
product forms 47
properties, expected for welds 197
properties, gas metal-arc
  welded plate 197
solution potential 207
stress-corrosion cracking
  resistance 209
weldability 193
5056 102-103
annealing temperature 29
applications 47
composition 45
corrosion resistance ... 47, 209, 230, 231
fabrication characteristics 47
galvanic corrosion with
  magnesium 607
machinability rating 188
mechanical properties 60
mechanical properties, sheet 181
physical properties 53
product forms 45
solution potential 207
stress-corrosion cracking
  resistance 209
weldability 193
alclad 5056 102-103
5083 .... 103, 104
annealing temperature 29
applications 47
composition 45
corrosion resistance 47, 209,
    224, 230, 232

cruciform weldment, mercury
  cracking of 212
fabrication characteristics 47
forgeability, relative 6
machinability rating 188
mechanical properties 60
physical properties 53
product forms 45
properties, expected for welds 197
properties, gas metal-arc
  welded plate 197
solution potentials 207
stress-corrosion cracking
  resistance 209, 216-217
weldability 193
5086 104-105
annealing temperature 29
applications 48
composition 45
corrosion resistance ... 48, 209, 224, 232
fabrication characteristics 48
machinability ratings 188
mechanical properties 60
mechanical properties, sheet 181
physical properties 53
product forms 45
properties, expected for welds 197
properties, gas metal-arc
  welded plate 197
solution potential 207
stress-corrosion cracking
  resistance 209
weldability 193
alclad 5086 104-105
5154 .... 105, 106
annealing temperature 29
applications 48
composition 45
corrosion resistance 48, 209,
    224, 230, 232
fabrication characteristics 48
fatigue characteristics,
  weldments 195, 199
machinability rating 188
mechanical properties 60
physical properties 53
product forms 45
properties, gas metal-arc
  welded plate 197
solution potential 207
stress-corrosion cracking
  resistance 209
weldability 193
5182 106-107
annealing temperature 29
composition 45
machinability rating 188
mechanical properties 60
mechanical properties, sheet 181
physical properties 53
product forms 45
solution potential 207
5252 107
applications 48
composition 45
corrosion resistance 48, 209
fabrication characteristics 48
mechanical properties 60
mechanical properties, sheet 181
physical properties 53
product forms 45
stress-corrosion cracking
  resistance 209
weldability 193
5254 108-109
annealing temperature 29
applications 48
composition 45
corrosion resistance 48, 209
fabrication characteristics 48

mechanical properties 60-61
physical properties 53
product forms 45
stress-corrosion cracking
  resistance 209
weldability 193
5356 109
  composition 45
  microstructure, effect on susceptibility
    to stress-corrosion
    cracking 213-215
  physical properties 53
  product forms 45
5357
  atmospheric corrosion resistance 224
5454 109-110
  annealing temperature 29
  applications 48
  composition 45
  corrosion resistance ... 48, 209, 224, 232
  fabrication characteristics 48
  machinability rating 188
  mechanical properties 61
  physical properties 53
  product forms 45
  properties, expected for welds 197
  solution potential 207
  stress-corrosion cracking
    resistance 209
  weldability 193
5456 110-111
  annealing temperature 29
  applications 48
  composition 45
  corrosion resistance ... 48, 209, 224, 232
  fabrication characteristics 48
  machinability rating 188
  mechanical properties 61
  physical properties 48
  product forms 45
  properties, expected for welds 197
  solution potential 207
  stress-corrosion cracking
    resistance 209
  weldability 193
5457 111-112
  annealing temperature 29
  applications 48
  composition 45
  corrosion resistance 48, 209, 232
  fabrication characteristics 48
  machinability rating 188-189
  mechanical properties 61
  physical properties 53
  product forms 45
  stress-corrosion cracking
    resistance 209
  weldability 193
5652 112-113
  annealing temperature 29
  applications 48
  composition 45
  corrosion resistance 48, 209
  fabrication characteristics 48
  mechanical properties 61
  physical properties 53
  product forms 45
  stress-corrosion cracking
    resistance 209
  weldability 193
5657 113
  applications 48
  composition 45
  corrosion resistance 48, 209
  fabrication characteristics 48
  machinability rating 189
  mechanical properties 61
  physical properties 53
  product forms 45

stress-corrosion cracking
  resistance 209
  weldability 193
6005 113-114
  annealing temperature 29
  applications 48
  composition 45
  corrosion resistance 48
  mechanical properties 61
  physical properties 53
  product forms 45
6009 114-115
  annealing temperature 29
  composition 45
  machinability rating 189
  mechanical properties 61
  mechanical properties, sheet 181
  physical properties 54
  product forms 45
  solution potential 207
6010 115
  annealing temperature 29
  composition 45
  machinability rating 189
  mechanical properties 61
  mechanical properties, sheet 181
  physical properties 54
  product forms 45
  solution potential 207
6051
  corrosion resistance 223, 230, 231
6053
  annealing temperature 29
  corrosion resistance 209, 224,
                                 231, 232
  stress-corrosion cracking
    resistance 209
6061 115-117
  aging characteristics at low aging
    temperatures 31, 36-38
  annealing temperature 29
  applications 48
  composition 45
  corrosion resistance 48, 209, 224,
                           225, 230, 231, 232
  fabrication characteristics 48
  fatigue characteristics,
    weldments 195, 199
  fatigue curves 14
  galvanic corrosion with
    magnesium 607
  forgeability, relative 6
  forging, relative cost vs forging
    weight 5
  forming, change from alloy 5052 to
    eliminate cracking during 16
  forming-limit diagram 183
  impacts, mechanical properties of 10
  impacts, minimum wall thickness 9
  machinability rating 189
  mechanical properties 61
  mechanical properties, sheet 181
  natural aging curve 31
  physical properties 54
  product forms 45
  properties, expected for welds 197
  properties, gas metal-arc welded
    plate 197
  quenching rate, effect on yield
    strength after aging 34
  solution potential 207
  stress-corrosion cracking
    resistance 209, 213
  weldability 193
alclad 6061 115-117
6063 117-118
  annealing temperature 29
  applications 49
  composition 45

corrosion resistance 49, 209, 226,
                              231, 232
fabrication characteristics 49
galvanic corrosion with
  magnesium 607
machinability rating 189
mechanical properties 61
physical properties 54
product forms 45
properties, expected for welds 197
solution potential 207
stress-corrosion cracking
  resistance 209
weldability 193
6066 118-119
  annealing temperature 29
  applications 49
  composition 45
  corrosion resistance 49, 209
  fabrication characteristics 49
  mechanical properties 61
  physical properties 54
  product forms 45
  weldability 193
  stress-corrosion cracking
    resistance 209
6070 119
  applications 49
  composition 45
  corrosion resistance 49, 209, 232
  fabrication characteristics 49
  mechanical properties 61
  physical properties 54
  product forms 45
  stress-corrosion cracking
    resistance 209
  weldability 193
6101 119-120
  application 49
  composition 45
  corrosion resistance 49, 209
  fabrication characteristics 49
  mechanical properties 61
  physical properties 54
  product forms 45
  stress-corrosion cracking
    resistance 209
  weldability 193
6151 120-121
  applications 49
  composition 45
  corrosion resistance 209
  fabrication characteristics 49
  forgeability, relative 6
  forging, relative cost vs forging
    weight 5
  mechanical properties 61
  mechanical properties, sheet 181
  physical properties 54
  product forms 45
  quenching stress 40, 41, 43
  solution potential 207
  stress-corrosion cracking
    resistance 209
6201
  applications 49
  composition 45
  corrosion resistance 49, 209
  fabrication characteristics 49
  mechanical properties 61
  physical properties 54
  product forms 45
  stress-corrosion cracking
    resistance 209
  weldability 193
6205 121-122
  composition 45
  mechanical properties 61
  physical properties 54
  product forms 45

Aluminum Alloys, Wrought (cont.)
6262 122
  applications 49
  composition 45
  corrosion resistance 49, 209
  fabrication characteristics 49
  machinability rating 189
  mechanical properties 61
  physical properties 54
  product forms 45
  stress-corrosion cracking
    resistance 209
  weldability 193
6351 122-123
  composition 45
  impacts, mechanical properties 10
  mechanical properties 62
  physical properties 54
  product forms 45
  properties, expected for welds 197
  seawater corrosion resistance 232
  solution potential 207
6463 123
  applications 49
  composition 45
  corrosion resistance 49, 209
  fabrication characteristics 49
  machinability rating 189
  mechanical properties 62
  physical properties 54
  product forms 45
  stress-corrosion cracking
    resistance 209
  weldability 193
7001
  annealing temperature 29
  corrosion resistance 209
  stress-corrosion cracking
    resistance 209
7005 123-125
  annealing temperature 29
  composition 45
  mechanical properties 62
  physical properties 54
  product forms 45
  solution potential 207
  stress-corrosion cracking
    resistance 213
  weldability 193
X7016, solution potential 207
7021, mechanical properties, sheet 181
X7021, solution potential 207
7029, mechanical properties, sheet 181
X7029, solution potential 207
7039 ....
  seawater corrosion resistance 232
  stress-corrosion cracking
    resistance 213, 232
7049 125-126
  annealing temperature 29
  composition 45
  machinability rating 189
  mechanical properties 62
  physical properties 54
  product forms 45
  solution potential 207
  stress-corrosion cracking
    resistance 213
7050 126-128
  aging characteristics at room
    temperature 36-38
  annealing temperature 29
  composition 45
  machinability rating 189
  mechanical properties 62
  physical properties 54
  product forms 45
  quenching rate, effect on yield
    strength after aging 34
  solution potential 207

stress-corrosion cracking
    resistance 213
7072 128-129
  composition 45
  mechanical properties 62
  physical properties 54
  product forms 45
  seawater corrosion resistance 230
  solution potential 207
7075 129-132
  aging characteristics at low aging
    temperatures 36-38
  annealing temperature 29
  applications 49
  composition 45
  corrosion rate affected by quenching
    rate 34
  corrosion resistance 49, 209, 224,
              225, 226, 230, 232
  exfoliation corrosion resistance 219
  fabrication characteristics 49
  forgeability, relative 6
  forged part, mechanical properties ... 13
  forging, relative cost vs forging
    weight 5
  galvanic corrosion with
    magnesium 607
  impacts, mechanical
    properties 10, 13
  impacts, minimum wall thickness 9
  machinability rating 189
  mechanical properties 62
  natural aging curve 31
  physical properties 54
  precipitation-hardening
    curves 31, 36-38, 43
  product forms 45
  properties, elevated
    temperatures 56, 58, 62
  quenching rate, effect on yield
    strength after aging 34
  solution potential 207
  stress-corrosion cracking
    resistance 209, 213-216
  weldability 193
alclad 7075 129-132
  galvanic corrosion with
    magnesium 607
  mechanical properties
7079
  annealing temperature 29
  atmospheric corrosion
    resistance 224, 232
  forged parts, variation in
    mechanical properties 8, 9
  forging, relative cost vs forging
    weight 5
  stress-corrosion cracking
    resistance 213
7146, mechanical properties, sheet 181
X7146, solution potential 207
7149, stress corrosion cracking
    resistance 213
7175 131-134
  applications 49
  composition 45
  corrosion resistance 49
  fabrication characteristics 49
  mechanical properties 10, 62
  physical properties 54
  product forms 45
  stress-corrosion cracking
    resistance 214
7178 134-135
  annealing temperature 29
  applications 49
  composition 45
  corrosion resistance 49, 209
  fabrication characteristics 49
  machinability rating 189

  product forms 45
  solution potential 207
  stress-corrosion cracking
    resistance 209, 214
  weldability 193
alclad 7178 134-135
7475 135-139
  annealing temperature 29
  composition 45
  machinability rating 489
  mechanical properties 62
  physical properties 54
  product forms 45
  solution potential 207
  stress-corrosion cracking
    resistance 214
Aluminum bronzes. See also specific
  types, under *Bronzes*.
  alpha aluminum bronzes, heat
    treating 259
  corrosion in various
    media 390-391, 468-469
**Aluminum casting alloys, former
  ASTM designations**
  C4A. See *Aluminum casting alloys, specific
    types, 295.0*
  CN42A. See *Aluminum casting alloys,
    specific types, 242.0*
  CS43A. See *Aluminum casting alloys,
    specific types, 208.0*
  G4A. See *Aluminum casting alloys,
    specific types, 514.0*
  G8A. See *Aluminum casting alloys,
    specific types, 518.0*
  G10A. See *Aluminum casting alloys,
    specific types, 500.0*
  GH70B. See *Aluminum casting alloys,
    specific types, 535.0*
  S5A. See *Aluminum casting alloys, specific
    types, B443.0*
  S5B. See *Aluminum casting alloys, specific
    types, 443.0*
  S5C. See *Aluminum casting alloys, specific
    types, C443.0*
  S12A. See *Aluminum casting alloys,
    specific types, A413.0*
  S12B. See *Aluminum casting alloys,
    specific types, 413.0*
  SC51A. See *Aluminum casting alloys,
    specific types, 355.0*
  SC51B. See *Aluminum casting alloys,
    specific types, C355.0*
  SC64A. See *Aluminum casting alloys,
    specific types, 308.0*
  SC64D. See *Aluminum casting alloys,
    specific types, 319.0*
  SC84A. See *Aluminum casting alloys,
    specific types, A380.0*
  SC84B. See *Aluminum casting alloys,
    specific types, 380.0*
  SC92A. See *Aluminum casting alloys,
    specific types, 354.0*
  SC102A. See *Aluminum casting alloys,
    specific types, 383.0*
  SC114A. See *Aluminum casting alloys,
    specific types, 384.0*
  SG70A. See *Aluminum casting alloys,
    specific types, 356.0*
  SG70B. See *Aluminum casting alloys,
    specific types, A356.0*
  SG91A. See *Aluminum casting alloys,
    specific types, 359.0*
  SG100A. See *Aluminum casting alloys,
    specific types, A360.0*
  SG100B. See *Aluminum casting alloys,
    specific types, 360.0*
  SN122A. See *Aluminum casting alloys,
    specific types, 336.0*
  ZC81A.B. See *Aluminum casting alloys,
    specific types, 713.0*

ZG61A. See *Aluminum casting alloys, specific types, 712.0*
ZG71B. See *Aluminum casting alloys, specific types, 771.0*
Aluminum casting alloys, specific types. See also *Aluminum alloys, powder metallurgy, specific types; Aluminum alloys, wrought, specific types.*
13. See *Aluminum casting alloys, specific types, 413.0.*
A13. See *Aluminum casting alloys, specific types, A413.0.*
40E. See *Aluminum casting alloys, specific types, 712.0.*
43. See *Aluminum casting alloys, specific types, 443.0, A443.0, B443.0.*
A43. See *Aluminum casting alloys, specific types, C443.0.*
100.1
   corrosion resistance 211
   stress-corrosion cracking
      resistance 211
108. See *Aluminum casting alloys, specific types, 208.0.*
A108. See *Aluminum casting alloys, specific types, 308.0.*
A132. See *Aluminum casting alloys, specific types, 336.0.*
142. See *Aluminum casting alloys, specific types, 242.0.*
150.1
   corrosion resistance 211
   stress-corrosion cracking resistance . .211
170.1
   corrosion resistance 211
   stress-corrosion cracking resistance . .211
195. See *Aluminum casting alloys, specific types, 295.0.*
B195. See *Aluminum casting alloys, specific types, 296.0.*
201.0 152-154
   composition 142
   tensile properties, test bars 149
206.0 154-155
   composition 142
   tensile properties, test bars 149
A206.0 154-155
   composition 142
   tensile properties, test bars 149
208.0 155
   characteristics 144
   composition 142
   corrosion resistance 210, 227
   machinability rating 188
   solution potential 207
   stress-corrosion cracking
      resistance 210
   tensile properties, test bars 149
   weldability 193
213.0
   characteristics 144
   weldability 193
214. See *Aluminum casting alloys, specific types, 514.0.*
218. See *Aluminum casting alloys, specific types, 518.0.*
A218. See *Aluminum casting alloys, specific types, A535.0.*
B218. See *Aluminum casting alloys, specific types, B535.0.*
220. See *Aluminum casting alloys, specific types, 520.0.*
222.0
   characteristics 144
   weldability 193
224.0
   corrosion resistance 210
   stress-corrosion cracking
      resistance 210

238.0
   solution potential 207
240.0
   corrosion resistance 210
   stress-corrosion cracking
      resistance 210
242.0 155-157
   characteristics 144
   composition 142
   corrosion resistance 210, 211
   machinability rating 188
   stress-corrosion cracking
      resistance 210, 211
   tensile properties, test bars 149
   weldability 193
A242.0
   corrosion resistance 210
   stress-corrosion cracking
      resistance 210
249.0
   corrosion resistance 210
   stress-corrosion cracking
      resistance 210
295.0 157-158
   characteristics 144
   composition 142
   corrosion resistance 210, 227
   machinability rating 188
   solution potential 207
   stress-corrosion cracking
      resistance 210
   tensile properties, test bars 149
   weldability 193
B295.0 See *Aluminum casting alloys, specific types, 296.0.*
296.0 158-159
   characteristics 144
   composition 142
   solution potential 207
   tensile properties, test bars 149
   weldability 193
308.0 159
   characteristics 144
   composition 142
   corrosion resistance 211
   machinability rating 188
   solution potential 207
   stress-corrosion cracking
      resistance 211
   tensile properties, test bars 149
   weldability 193
319.0 159-160
   characteristics 144
   composition 142
   corrosion resistance 210, 211, 227
   machinability rating 188
   solution potential 207
   stress-corrosion cracking
      resistance 210, 211
   tensile properties, test bars 149
   weldability 193
319 Allcast. See *Aluminum casting alloys, specific types, 319.0.*
328.0
   characteristics 144
   weldability 193
332.0
   characteristics 144
   corrosion resistance 211
   stress-corrosion cracking
      resistance 211
   weldability 193
A332.0 See *Aluminum casting alloys, specific types, 336.0.*
333.0
   characteristics 144
   weldability 193
336.0 160-161
   characteristics 144
   composition 142

corrosion resistance 211
stress-corrosion cracking
   resistance 211
   tensile properties, test bars 149
   weldability 193
354.0 . . . . .161, 162, 163
   characteristics 144
   composition 142
   corrosion resistance 211
   machinability rating 188
   stress-corrosion cracking
      resistance 211
   tensile properties, test bars 149
   weldability 193
355.0 161-165
   applications 145
   composition 142
   corrosion resistance 210, 211, 227
   machinability rating 188
   solution potential 207
   stress-corrosion cracking
      resistance 210-211
   tensile properties, test bars 149
   weldability 193
C355.0 161-165
   applications 145, 146
   characteristics 144
   corrosion resistance 210, 211
   machinability rating 188
   mechanical properties 151
   stress-corrosion cracking
      resistance 210, 211
   tensile properties, test bars 149
   weldability 193
356.0 164-167
   applications 145
   characteristics 144
   composition 142
   corrosion resistance 210, 211, 227
   machinability rating 188
   microstructure, effect of sodium
      modification 150
   precipitation hardening 32, 33
   solution potential 207
   stress-corrosion cracking
      resistance 210, 211
   tensile properties, test bars 149
   weldability 193
A356.0 164-167
   applications 145, 146
   characteristics 144
   composition 142
   corrosion resistance 211
   machinability rating 188
   microstructure, effect of sodium
      modification 150
   stress-corrosion cracking
      resistance 211
   weldability 193
F356.0
   corrosion resistance 211
   stress-corrosion cracking
      resistance 211
357.0 167
   applications 146
   characteristics 144
   composition 142
   machinability rating 188
   tensile properties, test bars 149
   weldability 193
A357.0 . . . . .167, 168
   applications 145
   characteristics 144
   composition 142
   corrosion resistance 211
   machinability rating 188
   stress-corrosion cracking
      resistance 211
   tensile properties, test bars 149
   weldability 193

Aluminum Casting Alloys (cont.)
358.0
corrosion resistance ... 211
stress-corrosion resistance ... 211
359.0 ... 167-168, 169
characteristics ... 144
composition ... 142
corrosion resistance ... 211
machinability rating ... 188
stress-corrosion cracking
resistance ... 211
tensile properties, test bars ... 149
weldability ... 193
360.0 ... 168-169
applications ... 145
characteristics ... 145
composition ... 142
corrosion resistance ... 211
machinability rating ... 188
stress-corrosion cracking
resistance ... 211
tensile properties, test bars ... 149
A360.0 ... 168-169
characteristics ... 145
composition ... 142
corrosion resistance ... 211
machinability ratings ... 188
stress-corrosion cracking
resistance ... 211
tensile properties, test bars ... 149
364.0
corrosion resistance ... 211
stress-corrosion cracking
resistance ... 211
380.0 ... 169, 170
applications ... 145
characteristics ... 145
composition ... 142
corrosion resistance ... 211
machinability rating ... 188
stress-corrosion cracking
resistance ... 211
tensile properties, test bars ... 149
A380.0 ... 170
characteristics ... 145
composition ... 142
corrosion resistance ... 211
machinability rating ... 188
stress-corrosion cracking resistance . 211
tensile properties, test bars ... 149
383.0 ... 170
characteristics ... 145
composition ... 142
corrosion resistance ... 211
stress-corrosion cracking
resistance ... 211
tensile properties, test bars ... 149
384.0 ... 170-171
characteristics ... 145
composition ... 142
corrosion resistance ... 211
stress-corrosion cracking
resistance ... 211
tensile properties, test bars ... 149
A384.0 ... 170-171
tensile properties, test bars ... 149
390.0 ... 171-172
composition ... 142
corrosion resistance ... 211
machinability rating ... 188
stress-corrosion cracking
resistance ... 211
tensile properties, test bars ... 149
A390.0 ... 171-172
composition ... 142
machinability rating ... 188
tensile properties, test bars ... 149
392.0
corrosion resistance ... 211
stress-corrosion cracking
resistance ... 211
413.0 ... 172-173
applications ... 145
characteristics ... 145
composition ... 142
corrosion resistance ... 211
machinability rating ... 188
stress-corrosion cracking
resistance ... 211
tensile properties, test bars ... 149
weldability ... 193
A413.0 ... 172-173
characteristics ... 145
composition ... 142
corrosion resistance ... 211
stress-corrosion cracking
resistance ... 211
tensile properties, test bars ... 149
443.0 ... 173-174
composition ... 142
corrosion resistance ... 210, 227
machinability rating ... 188
solution potential ... 207
stress-corrosion cracking
resistance ... 210
tensile properties, test bars ... 150
weldability ... 193
A443.0 ... 173-174
composition ... 142
weldability ... 193
B443.0 ... 174-174
applications ... 145, 146
characteristics ... 144
composition ... 142
corrosion resistance ... 211
stress-corrosion cracking
resistance ... 211
tensile properties, test bars ... 150
weldability ... 193
C443.0 ... 173-174
characteristics ... 145
composition ... 142
corrosion resistance ... 211
stress-corrosion cracking
resistance ... 211
tensile properties, test bars ... 150
A444.0
corrosion resistance ... 211
stress-corrosion cracking
resistance ... 211
512.0
characteristics ... 144
corrosion resistance ... 210
stress-corrosion cracking
resistance ... 210
weldability ... 193
513.0
applications ... 145
characteristics ... 144
corrosion resistance ... 210, 211
stress-corrosion cracking
resistance ... 210, 211
weldability ... 193
514.0 ... 174
characteristics ... 144
composition ... 142
corrosion resistance ... 210
machinability rating ... 188
solution potential ... 207
stress corrosion cracking
resistance ... 210
tensile properties, test bars ... 150
weldability ... 193
518.0 ... 174-175
applications ... 145
characteristics ... 145
composition ... 142
corrosion resistance ... 211
machinability rating ... 188
stress-corrosion cracking
resistance ... 211
tensile properties, test bars ... 150
520.0 ... 175
applications ... 146
characteristics ... 144
composition ... 142
corrosion resistance ... 210, 227, 232
machinability rating ... 188
solution potential ... 207
stress-corrosion cracking
resistance ... 210
tensile properties, test bars ... 150
weldability ... 193
535.0 ... 175-176
characteristics ... 144
composition ... 142
corrosion resistance ... 210
stress-corrosion resistance ... 210
tensile properties, test bars ... 150
weldability ... 193
A535.0 ... 175-176
composition ... 142
B535.0 ... 175-176
composition ... 142
corrosion resistance ... 210
machinability rating ... 188
stress-corrosion cracking
resistance ... 210
D612. See *Aluminum casting alloys, specific types, 712.0*
613 Tenzaloy. See *Aluminum casting alloys, specific types, 713.0.*
705.0
characteristics ... 144
corrosion resistance ... 210, 227
stress-corrosion cracking
resistance ... 210
weldability ... 193
707.0
characteristics ... 144
corrosion resistance ... 210, 211, 227
stress-corrosion cracking
resistance ... 210, 211
weldability ... 193
710.0
characteristics ... 144
corrosion resistance ... 210, 227
stress-corrosion cracking
resistance ... 210
weldability ... 193
711.0
characteristics ... 144
corrosion resistance ... 211, 227
stress-corrosion cracking
resistance ... 211, 227
weldability ... 193
712.0 ... 176-177
composition ... 142
corrosion resistance ... 210, 227
machinability rating ... 188
stress-corrosion cracking
resistance ... 210
tensile properties, test bars ... 150
weldability ... 193
D712.0 See *Aluminum casting alloys, specific types, 712.0.*
713.0 ... 177-178
applications ... 146
characteristics ... 144
composition ... 142
corrosion resistance ... 210, 211, 227
machinability rating ... 188
stress-corrosion cracking
resistance ... 210, 211
tensile properties, test bars ... 150
weldability ... 193
750. See *Aluminum casting alloys, specific types, 850.0.*
771.0 ... 178

characteristics .144
composition 142
corrosion resistance 210
stress-corrosion cracking
resistance .210
tensile properties, test bars 150
weldability 193
850.0 178-179
characteristics .144
composition 142
corrosion resistance 210, 211
machinability rating 188
stress-corrosion cracking
resistance 210, 211
tensile properties, test bars 150
851.0
characteristics .144
corrosion resistance 210
stress-corrosion cracking
resistance .210
852.0
characteristics .144
corrosion resistance 210
stress-corrosion cracking
resistance .210
Almag 25. See *Aluminum casting alloys,
specific types, 535.0.*
Precedent 71A. See *Aluminum casting
alloys, specific types, 771.0.*
AMAX-LP copper. See *Copper alloys,
specific types, C10800.*
Americium, pure 715, 832-833
Ampco A1. See *Copper alloys, specific types,
C95200.*
Ampco B2. See *Copper alloys, specific types,
C95300.*
Ampco C3. See *Copper alloys, specific types,
C95400.*
Ampco D4. See *Copper alloys, specific types,
C95500.*
Ampco 21. See *Copper alloys, specific types,
C62500.*
Ampcoloy 495. See *Copper alloys, specific
types, C95700.*
AMSIL copper. See *Copper alloys, specific
types, C10400, C10500 and C10700.*
AMS specifications. See also listings in data
compilations for individual alloys.
copper tubular products 264
Amzirc Brand copper. See *Copper alloys,
specific types, C15000.*
Annealing
aluminum alloys 28, 29
copper metals 253-255
copper wire 272
Anneal-resistant electrolytic copper. See
Copper alloys, specific types, C11100.
Anodized aluminum, corrosion
resistance 225-226, 229-232
Anti-acid metal. See *Copper alloys, specific
types, C93800.*
Antimony, pure 715, 716
Applications. See also specific metals and
alloys.
aluminum alloys 46-49
copper metals .466
copper tubular products 261, 262
gold in dentistry 684-687
lead 495-498
magnesium 525
tin powder 616
zinc 629-637
Arsenic, pure 716
ASME specifications. See also listings in data
compilations for individual alloys.
copper tube and pipe 264
ASTM specifications. See also listings in data
compilations for individual alloys.
aluminum casting alloys, former
designations. See cross references to
specific former designations under
*Aluminum casting alloys, former ASTM
designations.*
copper casting alloys 384-386
copper tube and pipe 264
copper wire .266-273
magnesium alloys
designations 525-526, 527, 528
temper designations, copper
metals 248-251

B

Barium, pure 716-717
Bearing applications, copper alloy
castings 392-393
Bearing materials, tin alloy 614-615
Bending stress, magnesium structures . . .552
Berkelium, pure 717, 832-833
Berylco 10. See *Copper alloys, specific types,
C17500.*
Berylco 165. See *Copper alloys, specific
types, C17000.*
Beryllium copper. See *Copper alloys, specific
types, C17000, C17200, C17300 and
C17600.*
Beryllium coppers
10C. See *Copper alloys, specific types,
C82000.*
20C. See *Copper alloys, specific types,
C82500.*
30C. See *Copper alloys, specific types,
C82200.*
50C. See *Copper alloys, specific types,
C81800.*
70C. See *Copper alloys, specific types,
C81400.*
165C. See *Copper alloys, specific types,
C82400.*
245C. See *Copper alloys, specific types,
C82600.*
275C. See *Copper alloys, specific types,
C82800.*
Be-modified chrome copper. See *Copper
alloys, specific types, C81400.*
casting alloy 10C. See *Copper alloys,
specific types, C82000.*
casting alloy 30C. See *Copper alloys,
specific types, C82200.*
casting alloy 35C. See *Copper alloys,
specific types, C82200.*
casting alloy 53B. See *Copper alloys,
specific types, C82200.*
casting alloy 165C. See *Copper alloys,
specific types, C82400.*
casting alloy 245C. See *Copper alloys,
specific types, C82600.*
casting alloy 275C. See *Copper alloys,
specific types, C82800.*
grain-refined casting alloy 21C. See
*Copper alloys, specific types, beryllium
copper 21C.*
heat treating 256-257, 258
standard casting alloy. See *Copper alloys,
specific types, C82500.*
Beryllium cupro-nickel. See *Copper alloys,
specific types, C96600 and beryllium
copper nickel 72C.*
Beryllium, pure 717-718
Bismuth, pure 718-719
solution potential 207
Boron, pure .719-720
Brasses
56-2-10-12. See *Copper alloys, specific
types, C97300.*
63-1-1-35. See *Copper alloys, specific
types, C85700 and 85800.*
67-1-3-29. See *Copper alloys, specific
types, C85400.*
70-30. See *Copper alloys, specific types,
C26000.*
72-1-3-24. See *Copper alloys, specific
types, C85200.*
76-2½-6½-15. See *Copper alloys, specific
types, C84800.*
81-3-7-9. See *Copper alloys, specific types,
C84400.*
82-4-14. See *Copper alloys, specific types,
C87500 and C87800.*
85-5-5-5. See *Copper alloys, specific types,
C83600.*
Admiralty brass. See *Copper alloys,
specific types, C44300, C44400 and
C44500.*
cartridge brass, 70%. See *Copper alloys,
specific types, C26000.*
cast, corrosion ratings 390-391
clock brass. See *Copper alloys, specific
types, C34200 and C35300.*
engraver's brass. See *Copper alloys,
specific types, C34200 and C35300.*
extra-high leaded brass. See *Copper alloys,
specific types, C35600.*
extra quality brass. See *Copper alloys,
specific types, C26000.*
forging brass. See *Copper alloys, specific
types, C37700.*
free-cutting brass. See *Copper alloys,
specific types, C36000.*
free-cutting tube brass. See *Copper alloys,
specific types, C33200.*
free-cutting yellow brass. See *Copper
alloys, specific types, C36000.*
free-turning brass. See *Copper alloys,
specific types, C36000.*
galvanic corrosion with magnesium . . .607
heavy-leaded brass. See *Copper alloys,
specific types, C34200 and C35300.*
high brass. See *Copper alloys, specific
types, C33000.*
high copper yellow brass. See *Copper
alloys, specific types, C85200.*
high-leaded brass. See *Copper alloys,
specific types, C34200, C35300 and
C36000.*
high-leaded brass (tube). See *Copper
alloys, specific types, C33200.*
high strength yellow brass. See *Copper
alloys, specific types, C86100, C86200,
C86300 and C86500.*
leaded high strength yellow brass. See
Copper alloys, specific types, C86400.
leaded nickel brass. See *Copper alloys,
specific types, C97300.*
leaded red brass. See *Copper alloys,
specific types, C83600.*
leaded semi-red brass. See *Copper alloys,
specific types, C84400 and C84800.*
leaded yellow brass. See *Copper alloys,
specific types, C85200, C85400, C85700
and C85800.*
low brass, 80%. See *Copper alloys, specific
types, C24000.*
low-leaded brass. See *Copper alloys,
specific types, C33500.*
low-leaded brass (tube). See *Copper
alloys, specific types, C33000.*
medium-leaded brass, 62%. See *Copper
alloys, specific types, C35000.*
medium-leaded brass, 64.5%. See *Copper
alloys, specific types, C34000.*
naval brass, antimonial. See *Copper alloys,
specific types, C46600.*
naval brass, arsenical. See *Copper alloys,
specific types, C46500.*
naval brass, high leaded. See *Copper
alloys, specific types, C48500.*
naval brass, inhibited. See *Copper alloys,
specific types, C46500, C46600 and
C46700.*

Brasses (cont.)
naval brass, leaded. See *Copper alloys, specific types, C48200 and C48500.*
naval brass, medium leaded. See *Copper alloys, specific types, C48200.*
naval brass, phosphorized. See *Copper alloys, specific types, C46700.*
naval brass. See *Copper alloys, specific types, C46400, C46500, C46600 and C46700.*
naval brass, uninhibited. See *Copper alloys, specific types, C46400.*
No. 1 yellow brass. See *Copper alloys, specific types, C85400.*
plumbing goods brass. See *Copper alloys, specific types, C84800.*
red brass, 85%. See *Copper alloys, specific types, C23000.*
silicon brass. See *Copper alloys, specific types, C87500 and C87800.*
silicon red brass. See *Copper alloys, specific types, C69400.*
spinning brass. See *Copper alloys, specific types, C26000.*
spring brass. See *Copper alloys, specific types, C26000.*
temper designations 248-249
tin brass. See *Copper alloys, specific types, C41900.*
white manganese brass. See *Copper alloys, specific types, C99700.*
yellow brass. See *Copper alloys, specific types, C26800, C27000 and C33000.*
yellow brass, 65%. See *Copper alloys, specific types, C27000.*
yellow brass, 66%. See *Copper alloys, specific types, C26800.*
Brazing
aluminum ...., 199-201
copper alloys 449-453
Britannia metal. See *Tin alloys, specific types, pewter.*
Bronzes
64-4-4-8-20. See *Copper alloys, specific types, C97600.*
66-5-2-2-25. See *Copper alloys, specific types, C97800.*
70-5-25. See *Copper alloys, specific types, C94300.*
75-3-8-2-12 manganese aluminum bronze. See *Copper alloys, specific types, C95700.*
78-7-15. See *Copper alloys, specific types, C93800.*
79-6-15. See *Copper alloys, specific types, C93900.*
80-10-10. See *Copper alloys, specific types, C93700.*
81-4-4-11 aluminum bronze. See *Copper alloys, specific types, C95500.*
83-4-6-7. See *Copper alloys, specific types, C83800.*
83-7-7-3. See *Copper alloys, specific types, C93200.*
84-10-2½-0-3½. See *Copper alloys, specific types, C92900.*
85-4-11 aluminum bronze. See *Copper alloys, specific types, C95400.*
85-5-9-1. See *Copper alloys, specific types, C93500.*
86½-12-0-0-1½. See *Copper alloys, specific types, C91700.*
87-8-1-4. See *Copper alloys, specific types, C92300.*
87-11-1-0-1. See *Copper alloys, specific types, C92500.*
88-3-9 aluminum bronze. See *Copper alloys, specific types, C95200.*
88-6-1½-4½. See *Copper alloys, specific types, C92200.*

88-8-0-4. See *Copper alloys, specific types, C90300.*
88-10-0-2. See *Copper alloys, specific types, C90500.*
88-10-2-0. See *Copper alloys, specific types, C92700.*
89-1-10 aluminum bronze. See *Copper alloys, specific types, C95300.*
89-6-5. See *Copper alloys, specific types, C87200.*
92-4-4. See *Copper alloys, specific types, C87200.*
95-1-4. See *Copper alloys, specific types, C87200.*
444 bronze. See *Copper alloys, specific types, C54400.*
alpha nickel aluminum bronze. See *Copper alloys, specific types, C95800.*
aluminum bronze, 5%. See *Copper alloys, specific types, C60600 and C60800.*
aluminum bronze, 7%. See *Copper alloys, specific types, C61300 and C61400.*
aluminum bronze, 8%. See *Copper alloys, specific types, C61000.*
aluminum bronze, 9%. See *Copper alloys, specific types, C62300.*
aluminum bronze, 11%. See *Copper alloys, specific types, C62400.*
aluminum bronze 9A. See *Copper alloys, specific types, C95200.*
aluminum bronze 9B. See *Copper alloys, specific types, C95300.*
aluminum bronze 9C. See *Copper alloys, specific types, C95400.*
aluminum bronze 9D. See *Copper alloys, specific types, C95500.*
aluminum bronze A. See *Copper alloys, specific types, C60600.*
aluminum bronze D. See *Copper alloys, specific types, C61400.*
aluminum bronze E. See *Copper alloys, specific types, C63000.*
architectural bronze. See *Copper alloys, specific types, C38500.*
bearing bronze. See *Copper alloys, specific types, C54400.*
bearing bronze 660. See *Copper alloys, specific types, C93200.*
bushing and bearing bronze. See *Copper alloys, specific types, C93700.*
cast, corrosion ratings 390-391
commercial bronze, 90%. See *Copper alloys, specific types, C22000.*
"G"-bronze. See *Copper alloys, specific types, C90300.*
high-conductivity bronze. See *Copper alloys, specific types, C40500.*
high leaded tin bronze. See *Copper alloys, specific types, C93200, C93500, C93700, C93800, C93900 and C94300.*
high-silicon bronze. See *Copper alloys, specific types, C65500.*
high-silicon bronze A. See *Copper alloys, specific types, C65500.*
hydraulic bronze. See *Copper alloys, specific types, C83800.*
jewelry bronze, 87½%. See *Copper alloys, specific types, C22600.*
leaded commercial bronze. See *Copper alloys, specific types, C31400.*
leaded commercial bronze, nickel-bearing. See *Copper alloys, specific types, C31600.*
leaded Navy "G" bronze. See *Copper alloys, specific types, C92300.*
leaded nickel bronze. See *Copper alloys, specific types, C97600 and C97800.*
leaded nickel-tin bronze. See *Copper alloys, specific types, C92900.*

leaded tin bronze. See *Copper alloys, specific types, C92300, C92500, C92600 and C92700.*
low-silicon bronze. See *Copper alloys, specific types, C65100.*
low-silicon bronze B. See *Copper alloys, specific types, C65100.*
manganese aluminum bronze. See *Copper alloys, specific types, C95700.*
manganese bronze (60 000 psi). See *Copper alloys, specific types, C86400.*
manganese bronze (65 000 psi). See *Copper alloys, specific types, C86500.*
manganese bronze (90 000 psi). See *Copper alloys, specific types, C86100 and C86200.*
manganese bronze (100 00 psi). See *Copper alloys, specific types, C86300.*
medium bronze. See *Copper alloys, specific types, C94500.*
Navy "M" bronze. See *Copper alloys, specific types, C92200.*
nickel aluminum bronze. See *Copper alloys, specific types, C63000 and C63200.*
nickel gear bronze. See *Copper alloys, specific types, C91700.*
penny bronze. See *Copper alloys, specific types, C40500.*
phosphor bronze, 1.25% E. See *Copper alloys, specific types, C50500.*
phosphor bronze, 5%. A. See *Copper alloys, specific types, C51000.*
phosphor bronze, 8% C. See *Copper alloys, specific types, C52100.*
phosphor bronze, 10% D. See *Copper alloys, specific types, C52400.*
phosphor bronze B-2. See *Copper alloys, specific types, C54400.*
phosphor bronze, free-cutting. See *Copper alloys, specific types, C54400.*
phosphor gear bronze. See *Copper alloys, specific types, C90700.*
propeller bronze. See *Copper alloys, specific types, C95800.*
silicon bronze. See *Copper alloys, specific types, C87200.*
soft bronze. See *Copper alloys, specific types, C94300.*
steam bronze. See *Copper alloys, specific types, C92200.*
stem manganese bronze. See *Copper alloys, specific types, C86400.*
tin bronze. See *Copper alloys, specific types, C90300 and C90500.*
tin bronze, 65. See *Copper alloys, specific types, C90700.*
Bunch stranded copper conductors 266, 272

C

Cable. See also *Wire.*
copper 265-274
Cadmium, pure 720-721
solution potential 207
Cadmium copper. See *Copper alloys, specific types, C16200.*
Cadmium copper, deoxidized. See *Copper alloys, specific types, C14300.*
Cadmium in copper 241-243
Calcium, pure 721-722
Californium, pure 832-833
Castings. See *specific metals and alloys.*
Cathodic protection, zinc anodes 654-655
Centrifugal casting. See also *Castings, Foundry products.*
aluminum alloys 146-147
copper alloys 384
Cerium, pure 722-723
Cesium, pure 723-724

Chemical analysis, trace elements in pure metals 711
Chemical vapor deposition for purifying metals 711
Chip characteristics, in machining aluminum 189-190
Chrome copper. See *Copper alloys, specific types, C81500.*
Chrome Copper 999. See *Copper alloys, specific types, C18200.*
Chromium, pure 724-725
   solution potential 207
Chromium copper. See *Copper alloys, specific types, C18200, C18400, C18500 and C81500.*
Chromium coppers, heat treating 257, 259
Coatings
   copper metals, corrosion protection ...465
   copper wire 272
   tin 613-614
   zinc on iron and steel 651-653
Cobalt, pure 725-726
Cobron. See *Copper alloys, specific types, C66400.*
Cold drawing, for sizing of tubular products 264
Cold working
   aluminum alloys, relation to heat treatment 32, 38, 40-42
   copper metals 241
Common desilverized lead. See *Leads and lead alloys, specific types, corroding lead.*
Composite-mold casting. See also *Castings, Foundry products.*
   aluminum alloys 147
Composition
   copper casting alloys 383-384
Composition metal. See *Copper alloys, specific types, C83600.*
Concentric-lay stranded copper conductors 266, 268
Conductivity, copper casting alloys 393
Continuous casting. See also *Castings, Foundry products.*
   aluminum alloys 147
Copper 239-247
   annealing 241
   anneal-resistant 241, 242, 243
   corrosion resistance 239-240
   decorative colors and finishes 240
   galvanic corrosion with magnesium ...607
   hydrogen embrittlement 239-240
   mechanical working 241
   solution potential 207
   stress-corrosion cracking 239-240
Copper alloys 239-274
   age-hardenable 242
   alloy families 241-242
   decorative colors and finishes 240
   deoxidizers 243
   insoluble elements, effect on machinability 242-243
   solid solution 242
Copper alloys, castings
   applications 387, 389-390
   ASTM specifications 384, 386
   centrifugal casting 384
   composition 383-384
   conductivity 393
   corrosion resistance 390-391
   cost considerations 393-394
   dimensional tolerances 388-389
   machinability 388
   mechanical properties 385-387
   permanent mold casting 384
   plaster mold casting 384-385
   sand casting 384
   solidification control 385

Copper alloys, corrosion environments
   acetic acid 473, 475-476
   alkalis 477
   ammonium hydroxide 477
   anhydrous ammonia 477
   atmospheric 467, 470
   beer 480
   benzol 480
   biofouling 472
   carbon dioxide 480-481
   carbon monoxide 480-481
   creosote 479
   dry oxygen 482-483
   fatty acids 476
   fresh water 470-471
     copper aluminum alloys 471
     copper nickels 471
     copper-silicon alloys 471
     copper-zinc alloys 470
   gasoline 479, 483
   halogen gases 475, 481
   hydrochloric acid 473-675
   hydrocyanic acid 476-477
   hydrofluoric acid 475
   hydrogen 481-483
   hydrogen sulfide 481
   linseed oil 480
   oleic acid 476
   organic compounds 478, 482
   oxidizing salts 478-479
   phosphoric acids 473-474
   salts 477-479
   salt water 471-472
   selection for specific environment 466-467, 468-469
   soil 467, 470
   steam 471
   stearic acid 476-477
   sugar 480, 483
   sulfur compounds 480
   sulfur dioxide 481, 483
   sulfuric acid 473, 474, 475
   tartaric acid 477
Copper alloys, corrosion resistance
   aluminum brasses 464
   aluminum bronzes 465
   brasses 461, 463, 464
   copper nickels 464-465
   copper-silicon 464, 465
   inhibited brasses 464
   nickel silvers 464, 465
   phosphor bronzes 464, 465
   tin brasses 463-464
Copper alloys, corrosion service ...466-483
   tubes, condenser 472
   tubes, heat exchange 472
Copper alloys, heat treating
   aluminum bronzes 259
   annealing temperatures 256
   beryllium coppers 256-257, 258
     aging 257, 258
     solution treating 256-257
   chromium coppers 257-258, 259
   copper-nickel-phosphorus alloys 257
   zirconium copper 258-259
Copper alloys, specific types
   beryllium copper 21C 438
   beryllium copper nickel 72C 438-439
   C10200
     stress relaxation 484-487
     tubes, applications 262
     tubes, mechanical properties 263
   C11000
     acetic acid-acetic enhydride, corrosion in 475
     acetic acid, corrosion in 476
     alcohols, corrosion in 482
     aldehydes, corrosion in 481
     amine-system, corrosion in 478
     ammonia, corrosion in 477

     atmospheric corrosion 470
     beet-sugar, corrosion in 483
     $CaCl_2$ refrigeration brine, corrosion in 478
     composition 467
     contaminated naphtha, corrosion in 483
     ester solutions, corrosion in 479-480
     ethers, corrosion in 481
     ethylene glycol, corrosion in 482
     fasteners 443
     hydrocyanic acid, corrosion in 477
     hydrogen, corrosion in 482, 483
     hydrogen cyanide, corrosion in 476
     isopropyl ether-acetic acid, corrosion in 476
     ketones, corrosion in 481
     sodium chloride brine, corrosion in 478
     stearic acid, corrosion in 476-477
     stress relaxation 484-487
     sulfuric acid, corrosion in 474
     tubes, piercing temperature 263
   11600, stress relaxation 486, 488
   C12000
     atmospheric corrosion 470
     ketones, corrosion in 481
     steam condensate, corrosion in 471
     stress relaxation 485, 488
     tubes, used for 471
   C12200
     composition 467
     tubes, applications 262
     tubes, mechanical properties 263
     tubes, piercing temperature 263
   C12300, tubes 472
   C13400, stress relaxation 486, 488
   C13700, stress relaxation 485, 488
   C14200
     steam condensate, corrosion in 471
     sulfuric acid, corrosion in 474
   C15710 298
   C15720 299
   C15735 ....299, 300
   C16200 300
     stress relaxation 486, 488
   C17000 301-302
     aging, properties corresponding to 258
   C17200 ....303-304, 305
     aging, properties corresponding to 258
     stress relaxation 487
   C17300 ....303-304, 305
   C17500 306-307
     aging, properties corresponding to 258
     stress relaxation 488
   C17600 308
   C18200 309
   C18400 309
   C18500 309
   C18700 310
   C19000 488
   C19200 311
     tubes, applications 262
     tubes, mechanical properties 263
   C19400 ....311-312, 313
   C19500 313
   C21000 ....256, 316
   C22000 ....317, 318
     composition 467
     stress-relieving temperatures 256
     tubes, piercing temperature
   C22600 ....318, 319
   C23000 ....320-321, 322
     alcohols, corrosion in 482
     atmospheric corrosion 470
     composition 467
     contaminated naphtha, corrosion

Copper Alloys (cont.)
in 483
  galvanic corrosion 459
  steam condensate, corrosion in 471
  stress-relieving temperatures 256
  sulfur compounds, corrosion in 480
  tartaric acid, corrosion in 477
  tubes, applications 262
  tubes, mechanical properties 263
  tubes, piercing temperature 263
C24000 ....256, 322, 323
C26000 ....323-326, 327, 481
  ammonia, corrosion in 477
  atmospheric corrosion 470
  composition 467
  ethylene glycol, corrosion in 482
  fasteners 444
  ketones, corrosion in 481
  oleic acid, corrosion in 476
  stress-relieving temperatures 256
  stearic acid, corrosion in 476-477
  sulfuric acid, corrosion in 474
  tartaric acid, corrosion in 477
  tubes, applications 262
  tubes, extrusion 261-262
  tubes, mechanical properties 263
  tubes, piercing temperature 263
C26800 327-328
C27000 327-328
  dezincification 461
  fasteners 443
  stress-relieving temperatures 256
C28000 ....328, 329-330
  composition 467
  contaminated naphtha, corrosion
    in 483
  fasteners 444
  stress-relieving temperatures 256
  sulfur compounds, corrosion in 480
  tubes, piercing temperature 263
C31400 330
C31600 331-331
C33000 331
  tubes, applications 262
  tubes, mechanical properties 263
C33200 332
C33500 333
C34000 333-334
C34200 ....333, 334
C34900 ....335, 336
C35000 336
C35300. ....334, 335
C35600 336-337
C36000 337-338
  composition 467
  fasteners 444
  stress-relieving temperatures 256
  tubes, applications 262
  tubes, extrusion 261-262
C36500 338-339
C36600 338-339
C36700 338-339
C36800 338-339
C37000 339
C37700 340-341
C38500, composition 342, 467
C40500 342
C40800 343
C41100 ....343, 344
C41500 344-345
C41900 345
C42200 345
C42500 346
C43000 ....346, 347
C43400 347
C43500 347-348
  tubes, applications 262
  tubes, mechanical properties 263
C44200
  atmospheric corrosion 470

amine-system, corrosion in 478
contaminated naphtha,
  corrosion in 483
ethylene glycol, corrosion in 482
C44300 348-349
  beet-sugar, corrosion in 483
  composition 467
  gasoline, corrosion in 483
  impingement attack 460
  steam condensate, corrosion in 471
  stress-relieving temperatures 256
  tubes 472
  tubes, applications 262
  tubes, extrusion 261-262
  tubes, mechanical properties 263
C44400 348-349
  alcohols, corrosion in 482
  beet-sugar, corrosion in 483
  composition 467
  steam condensate, corrosion in 471
  tubes 472, 480
  tubes, applications 262
  tubes, mechanical properties 263
C44500 348-349
  beet-sugar, corrosion in 483
  composition 467
  intercrystalline corrosion 461
  steam condensate, corrosion in 471
  tubes 472, 480
  tubes, applications 262
  tubes, mechanical properties 263
C46200
  fasteners 443
C46400 349-351
  composition 467
  contaminated naphtha,
    corrosion in 483
  fasteners 443, 444
  tubes, applications 262
  tubes, mechanical properties 263
  tubes, piercing temperature 263
C46500 349-351
  composition 467
  fasteners 444
  tubes, applications 262
  tubes, mechanical properties 263
C46600 349-351
  composition 467
  fasteners 444
  tubes, applications 262
  tubes, mechanical properties 263
C46700 349-351
  composition 467
  fasteners 444
  tubes, applications 262
  tubes, mechanical properties 263
C48200 351-352
C48500 352-353
  fasteners 444
C50500 353-354
C51000 354-355
  aldehydes, corrosion in 481
  composition 467
  ester solutions, corrosion in 480
  ethylene glycol, corrosion in 482
  fasteners 443
  oleic acid, corrosion in 476
  paper-mill vapor, corrosion 483
  stress relaxation 489
  stress-relieving temperatures 256
  sulfuric acid, corrosion in 474
C51100 355
C52100 355-356
  atmospheric corrosion 470
  composition 467
  paper-mill vapor, corrosion in 483
C52400 356
C54400 356-357
C60600 357-358
C60800 358

tubes, applications 262
tubes, mechanical properties 263
C60800
  ester solutions, corrosion in 479
  ethylene glycol, corrosion in 482
C61000 358-359
  atmospheric corrosion 470
C61300 ....359-360, 361
  composition 467
  oleic acid, corrosion in 476
C61400, fasteners 359, 360-361, 443
C61500 362
C61800
  ethylene glycol, corrosion in 482
  paper-mill vapor, corrosion in 483
C62300 362-363
  ester solutions, corrosion in 479
C62400 364
C62500 364-365
C63000 365-366
  ethylene glycol, corrosion in 482
  fasteners 443
C63200 ....366, 367
C63600 367
  ester solutions, corrosion in 479
C63700, composition 467
C63800 367-369
C64200, fasteners 443
C65100 369
  composition 467
  fasteners 443, 444
  tubes, applications 262
  tubes, mechanical properties 263
C65500 369-370
  acetic acid, corrosion in 476
  acetic acid-acetic enhydride,
    corrosion in 475
  alcohols, corrosion in 482
  aldehydes, corrosion in 481
  amine-system, corrosion in 478
  atmospheric corrosion 470
  composition 467
  ester solution, corrosion in 479
  ethers, corrosion in 481
  ethylene glycol, corrosion in 482
  fasteners 443, 444
  hydrocyanic acid, corrosion in 477
  hydrogen cyanide, corrosion in 476
  ketones, corrosion in 481
  oleic acid, corrosion in 476
  stearic acids, corrosion in 476-477
  sulfuric acid, corrosion in 473
  tubes, applications 262
  tubes, mechanical properties 263
C65800
  hydrochloric acid, corrosion in 475
  paper-mill vapor, corrosion in 483
C66100
  fasteners 443
C66400 ....370, 371
C67500
  fasteners 443
C68700
  composition 467
  tubes, used for 472
  tubes, applications 262
  tubes, mechanical properties 263
C68800 370-371
C69000 372
C69400 372
C70400 373
C70600 373-374
  composition 467
  ethylene glycol, corrosion in 482
  tubes, applications 262
  tubes, mechanical properties 263
  tubes, used for 472
C71000 374-375
  beet-sugar, corrosion in 483
  fasteners 443

steam condensate, corrosion in 471
tartaric acid, corrosion in 477
C71300
 tartaric acid, corrosion in 477
C71500 375-376
 composition 467
 amine-system, corrosion in 478
 beet-sugar, corrosion in 483
 ester solutions, corrosion in 479
 ethylene glycol, corrosion in 482
 fasteners 443
 gasoline, corrosion in 483
 hydrofluoric, corrosion in 475
 stress relieving temperatures 256
 tubes, applications 262
 tubes, mechanical properties 263
 tubes, used for 472
C71900 376-377
C72200 377-378
C72500 378
 stress relaxation 489
C73200
 paper-mill vapor, corrosion in 483
C74500 378-379
 fasteners 444
C75200 379-380
 composition 467
 paper-mill vapor, corrosion in 483
 stress-relieving temperatures 256
C75400 380
C75700 380-381
C77000 381
 paper-mill vapor, corrosion in 483
C78200 382
C81100 395
C81400 395-396
 composition 393
 mechanical properties 393
C81500 396
 composition 393
 mechanical properties 393
C81800 396-397
 composition 393
 mechanical properties 393
C82000 397-398
 composition 393
 mechanical properties 393
C82200 398-399
 composition 393
 mechanical properties 393
C82400 399-400
C82500 400-402
 composition 393
 mechanical properties 393
C82600 402-403
C82800 403-405
 composition 393
 mechanical properties 393
C83600 404-406
 applications 392
 composition 384
 foundry properties, sand casting 385
 machinability 389
 mechanical properties 386
 working stress, ASME Code
  castings 390
C83800 406-407
 applications 392
 composition 384
 machinability 389
 mechanical properties 386
C84400 407
 applications 392
 composition 384
 foundry properties, sand casting 385
 machinability 389
 mechanical properties 386
C84800 407-408
 applications 392
 composition 384

foundry properties, sand casting 385
machinability 389
mechanical properties 386
C85200 408
 applications 392
 composition 384
 machinability 389
 mechanical properties 386
C85400 408-409
 applications 392
 composition 384
 foundry properties, sand casting 385
 machinability 389
 mechanical properties 386
C85700 409
 composition 384
 mechanical properties 386
C85800 409
 composition 384
 foundry properties, sand casting 385
 mechanical properties 386
C86100 409-410
C86200 409-410
 compositions 384
 mechanical properties 386
C86300 410-411
 composition 384
 foundry properties, sand casting 385
 machinability 389
 mechanical properties 386
C86400 411-412
 composition 384
 machinability 389
 mechanical properties 386
C86500 412-415
 composition 384
 foundry properties, sand casting 385
 machinability 389
 mechanical properties 386
C86700
 composition 384
 mechanical properties 386
C87200 416
 composition 384
 foundry properties, sand casting 385
 mechanical properties 386
C87400
 composition 384
 mechanical properties 386
C87500 416-417
 composition 384
 foundry properties, sand casting 385
 mechanical properties 386
C87600
 composition 384
 mechanical properties 386
C87800 416-417
 composition 384
 mechanical properties 386
C87900
 composition 384
 mechanical properties 386
C90300 417-418
 composition 384
 foundry properties, sand casting 385
 machinability 389
 mechanical properties 386
C90500 418
 composition 384
 machinability 389
 mechanical properties 386
C90700 418
C91100
 applications 394
 composition 384
C91300
 applications 394
 composition 384
C91700 418-419
C92200 419-421

composition 384
foundry properties, sand casting 385
machinability 389
mechanical properties 386
working stress, ASME Code
 castings 390
C92300 422
 composition 384
 machinability 389
 mechanical properties 386
C92500 422
C92600 422-423
C92700 423
C92900 423
C93200 424
 applications 394
 composition 384
 machinability 389
 mechanical properties 386
C93500 424
 applications 394
 composition 384
 machinability 389
 mechanical properties 386
C93700 424-427
 applications 394
 composition 384
 foundry properties, sand casting 385
 machinability 389
 mechanical properties 386
C93800 427-428
 applications 394
 composition 384
 machinability 389
 mechanical properties 386
C93900 428
C94100
 composition 384
C94300 428
 applications 394
 composition 384
 foundry properties, sand casting 385
 machinability 389
 mechanical properties 386
C94400, composition 384
C94500 429
 composition 384
C94700
 composition 384
 mechanical properties 386
C94800
 composition 384
 mechanical properties 386
C94900
 composition 384
 mechanical properties 386
C95200 429-431
 composition 384
 machinability 389
 mechanical properties 386
C95300 430-431
 composition 384
 foundry properties, sand casting 385
 machinability 389
 mechanical properties 386
C95400 431-433
 composition 384
 machinability 389
 mechanical properties 386
C95500 433-434
 composition 384
 machinability 389
 mechanical properties 386
C95600, machinability 389
C95700 434
C95800 434-435
 foundry properties, sand casting 385
C96400 435
C96600 435-436
C97300 436

Copper Alloys (cont.)
composition 384
machinability 389
mechanical properties 386
C97600 436-437
composition 384
foundry properties, sand casting 385
mechanical properties 386
C97800 437
composition 384
foundry properties, sand casting 385
mechanical properties 386
C99400 437
C99700 437-438

Copper and copper alloys
stress relaxation 484-490
stress relaxation, mechanical
components 487-489

Copper cable 265-274

Copper metals 243-246
applications 239, 240-241, 246-247, 466
corrosion potentials 458-459, 460
corrosion fatigue 459, 463
crevice corrosion 459, 460
dealloying 459, 461
deposit attack 458, 459
fabricated mill products 245-246
fretting 459, 460-461
galvanic corrosion 458-459
general corrosion 458, 459
impingement attack 459, 460-461
intercrystalline corrosion 459, 461-462
mining and refining 243-245
pitting 459-461
stress-corrosion cracking 462-463
sulfur corrosion 463, 464
water-line attack 459, 460

Copper metals, annealing
continuous strand 255
grain size 253, 254, 255
grain-size stabilized alloys 253
hydrogen embrittlement 255
mechanical properties, correlation
with 253-254
pretreatment, effect of 255
recrystallization 253, 255
temperatures 253
testing 254
time, effect of 255

Copper metals, brazing
atmospheres 450-452
beryllium coppers 450
cadmium-bearing copper 450
chromium copper 450
copper-aluminum alloys 450
copper-nickel-zinc alloys 450
copper-silicon alloys 450
copper-tin alloys 450
copper-zinc alloys 450
deoxidized coppers 449
dissimilar metals 450, 451
filler metals 450, 451
fluxes 450-452
joint clearance 452
oxygen-free coppers 449
processes 452-453
postbraze treatment 453
surface preparation 452
tough pitch coppers 449
zirconium coppers 450

Copper metals, cast, corrosion
ratings 390-391

Copper metals, heat treating
age hardening 256
homogenizing 252-253
martensitic transformation 259
precipitation hardening 256
spinodal decomposition 259-260
stress relieving 255-256

Copper metals, joining. See also *Copper
metals, brazing; Copper metals,
soldering; Copper metals, welding*
adhesive bonding 456
diffusion bonding 456
electroplating 457
mechanical joining 440-443, 444
process selection 440, 441, 442
roll bonding 457

Copper metals, soldering
advantages 443, 445
coated copper alloys 448
fluxes 446-447
mechanical properties,
joints 448-449, 450
methods 447-448
solders 444-446
surface preparation 447
testing 448

Copper metals, welding
aluminum bronzes 455
beryllium coppers 453, 454
brasses 454-455
cadmium-coppers 453, 454
chromium-copper 453, 454
copper-nickels 456
deoxidized coppers 453
electron beam 457
filler metals 453, 454
friction 457
gas metal-arc 453, 454
gas tungsten-arc 453, 454
laser 457
nickel silvers 456
oxygen-free coppers 453
resistance 453, 455
silicon bronzes 455-456
tin brasses 455
tin bronzes 455
tough pitch coppers 453
ultrasonic 457
zirconium coppers 453, 454

Copper nickels
10%. See *Copper alloys, specific types,
C70600.*
20%. See *Copper alloys, specific types,
C71000.*
30%. See *Copper alloys, specific types,
C71500.*
70-30. See *Copper alloys, specific types,
C96400.*
chromium-bearing. See *Copper alloys,
specific types, C71900 and C72200.*
tin-bearing. See *Copper alloys, specific
types, C72500.*
corrosion in various media 468-469

Copper-nickel-phosphorus alloys,
heat treating 242, 257
Copper-nickel-silicon alloys 242
Copper, pure 726-733
Copper resources 246, 247
Coppers, corrosion in various
media 468-469
Coppers, electrical 240-241
Coppers, specific types
C10100 275-278
C10200 275-278
C10300 279
C10400 280-281
C10500 280-281
C10700 280-281
C10800 281-282
C11000 282-290
C11100 291
C11300 .... 291-292, 293
C11400 .... 291-292, 293
C11500 .... 291-292, 293
C11600 .... 291-292, 293
C12500 .... 292-293, 294
C12700 .... 292-293, 294

C12800 .... 292-293, 294
C12900 .... 292-293, 294
C13000 .... 292-293, 294
C14300 .... 294, 295
C14310 .... 294, 295
C14500 295
C14700 295-296
C15000 .... 296-297, 298
stress relaxation 485-488
Copper, temper designations 248-251
Copper tube shells, production
extrusion 261-262
rotary piercing 262, 263
Copper tubes, production
cold drawing 264
reducing 264
Copper, tubular products 261-264
applications 261, 262
joints 261
mechanical properties 261, 263
Copper wire 265-274
characteristics 267-273
coating 272
rectangular wire 266, 272
round wire 266, 267
square wire 266
stranded wire 266-273
tin coated 266, 273
Copper wire, drawing
annealing 272
flat wire 272
processes 271-272
rod preparation 271
Copper wire, materials
electrical bronzes 265-266
high-conductivity 265-266
high-copper alloys 265-266
Copper wire rod, fabrication
continuous casting 269-270
Hazelett process 271
Outokumpu process 271
Properzi system 270
rolling 266-269
Southwire system 270
wirebar 266

Coronze. See *Copper alloys, specific types,
C63800.*
Corrosion, copper metals 458-465
Corrosion environments, zinc, behavior
in various 648-650
Corrosion fatigue
aluminum alloys 219, 220
copper metals 459, 463
Corrosion in water, aluminum
alloys 220-222, 205, 208
Corrosion protection
copper metals 459, 465
magnesium alloys 602, 603
Corrosion resistance
aluminum 204-236
aluminum alloys, effect of
quenching 32-35
copper 239-240
copper alloys 463-465
copper casting alloys 390-391
gold 669-670
iridium 669
magnesium alloys 596-609
palladium 669
platinum 668-669
rhodium 669
silver 670
zinc 646-655
Corrosion testing
magnesium alloys 603, 604
Cupronickels. See *Copper nickels.*
Curium, pure 733, 832-833

D

Dairy metal. See *Copper alloys, specific types, C97600.*
Dealloying, copper metals 459, 461
Dealuminification. See *Dealloying.*
Deep drawing, magnesium 543, 545
Deep drawing zinc. See *Zinc alloys, specific types, commercial rolled zincs.*
Degassing for purifying metals 710
Dental alloys, tin, use in 615
Dental amalgam 678
Dentistry, use of gold 684-687
Deposition corrosion, aluminum
  alloys 211-212
Design, die casting, zinc 633-634
Dezincification. See *Dealloying.*
Die casting. See also *Castings, Foundry products.*
  aluminum alloys 140, 143, 145, 147
  copper casting alloys 384
Die castings, zinc 630-634
Diffusion bonding, copper metals 456
Dimensional control, copper alloy
  castings 388-389
Directional solidification. See also
  Solidification. Aluminum
  castings 150-151
Distillation for purifying metals 710
Dysprosium, pure 733

E

Einsteinium, pure 733
Electrochemical corrosion, zinc ... 650, 652
Electrode potentials. See also *Solution potentials.* Second-phase constituents in
  aluminum alloys 207
Electrolytic solution potential. See *Solution potential*
Electrolytic tin. See *Tin alloys, specific types, commercially pure tins.*
Electrolytic tough pitch copper. See *Copper alloys, specific types, C11000.*
Electrolytic tough pitch copper, anneal resistant. See *Copper alloys, specific types, C11100.*
Erbium, pure 734
Erosion-corrosion, aluminum alloys 219
Europium, pure 734-735
Everdur. See *Copper alloys, specific types, C87200.*
Exfoliation corrosion, aluminum
  alloys 218-220
Extrusions
  aluminum alloy 4, 5-6
  aluminum alloys 51-58
  magnesium 527-528, 534-535, 537
EZDA 12. See *Zinc alloys, specific types, zinc foundry alloy ZA-12.*

F

Fabrication
  copper wire rod 266-271
  wrought zinc 636-637
Fasteners, copper, mechanical
  joining 440-443
Fastening, aluminum 202-203
Fatigue, zinc 650-652
Fatigue strength, magnesium 531, 532
Federal specifications. See also *listings in data compilations for individual alloys.*
Fire refined copper. See *Copper alloys, specific types, C12500, C12700, C12800, C12900 and C13000.*
Fire refined tough pitch copper. See *Copper alloys, specific types, C12500.*
Fire refined tough pitch copper with silver. See *Copper alloys, specific types, C12700, C12800, C12900 and C13000.*

Floating zone technique for purifying
  metals 710
Fluxes
  brazing 450-452
  soldering 446-447
Foreign specifications. See listings in data compilations for individual alloys.
Forgings
  aluminum alloy 5-14
  magnesium 528-529, 537-538, 539,
                      540, 541
Formability, magnesium 540-541, 542
Formetal 22 Alloy. See *Zinc alloys, specific types, superplastic zinc alloy.*
Forming, aluminum sheet 180
Forming-limit diagrams, aluminum
  alloy sheet 182-183
Foundry products, aluminum. See also *Castings* and specific casting processes such as *Sand Casting, Die casting.*
Fractional crystallization for purifying
  metals 709-710
Fractional distillation for purifying
  metals 710
Free-machining copper. See *Copper alloys, specific types, C14500, C14700 and C18700.*
Fusion welding, copper metals 441, 442

G

G5. See *Copper alloys, specific types, C95400.*
Gadolinium, pure 735-736
Gallium, pure 736-737
Galvanic corrosion
  aluminum alloys 209-211, 213-214
  aluminum, coupled with dissimilar
    metals 203
  copper metals 458-459
  magnesium alloys 604-609
Galvanized coatings, zinc 651-653
Gases, copper alloys, corrosion
  rate 480-483
Fresh water corrosion, copper
  alloys 470-471
Germanium, pure 737
Gilding metal, 95%. See *Copper alloys, specific types, C21000.*
Gold
  corrosion resistance 669
  jewelry 666-667
  production 660
  special properties 662-663
Gold, commercial fine 679-680
Gold, green. See *Gold-silver-copper alloys.*
Gold in dentistry 684-687
  cast alloys 684-686, 687
    applications 685
    composition 684-685, 687
    hardness ranges 685, 687
    mechanical properties 686, 687
    physical properties 687
  cavity filling, materials for 684
    gold foil, physical properties 684
    mat gold, physical properties 684
    powdered gold, physical
      properties 684
  copper addition 685
  lost wax process 684-685
  metal-ceramic technique 685, 687
  silver addition 685
  solders
    applications 686
    composition 686
  wire alloys
    color 684, 685
    composition 684, 685
    mechanical properties 684, 686
    physical properties 684, 686
  zinc addition 685

Gold-nickel-copper alloys 682-683
Gold-platinum alloy 683
Gold, proof. See *Gold, commercial fine.*
Gold, pure 737-738
Gold, red. See *Gold-silver-copper alloys.*
Gold-silver-copper alloys 680-682
Gold, white. See *Gold-nickel-copper alloys.*
Gold, yellow. See *Gold-silver-copper alloys.*
Gold, zone refined, impurity
  concentration 713
Gomak-3. See *Zinc alloys, specific types, AG40A.*
Gomak-5. See *Zinc alloys, specific types, AC41A.*
Gun metal. See *Copper alloys, specific types, C90500.*

H

Herculoy. See *Copper alloys, specific types, C87200.*
High-strength modified copper. See *Copper alloys, specific types, C19400.*
Holmium, pure 738-739
Hard lead (94-6). See *Leads and lead alloys, specific types, 6% antimonial lead.*
Hard lead (96-4). See *Leads and lead alloys, specific types, 4% antimonial lead.*
Heat treating
  aluminum alloys 28-43
  copper metals 252-260
High-zinc brasses, corrosion in various
  media 468-469
Homgenizing, copper metals 252-253
Hot working, copper metals 241
HSM copper. See *Copper alloys, specific types, C19400.*
Hydrogen embrittlement
  copper 239-240
  copper annealing 255

I

Impacts, aluminum alloy 8-10
Indium, pure 739-740
Inserts, load bearing, in magnesium
  parts 538-540, 541, 542
Insulation, copper wire and cable 274
Intergranular corrosion, aluminum
  alloys 212
Investment casting. See also *Castings, Foundry products.* Aluminum
  alloys 147
Iodide process for purifying metals 711
Iridium, corrosion resistance 669
Iridium, pure 740
Iron, pure 741-760

J

Joining
  aluminum 191-203
  copper metals 440-457

K

Kayem 12. See *Zinc alloys, specific types, zinc foundry alloy ZA-12.*
Korloy 2570. See *Zinc alloys, specific types, zinc foundry alloy ZA-12.*
Korloy 2684. See *Zinc alloys, specific types, superplastic zinc alloy.*

L

Lanthanum, pure 760-761
Lead
  galvanic corrosion with magnesium ...607
  solution potential 207
Lead and lead alloys
  applications 495-499
  corrosion resistance 495, 511-522
    acids 515-522
    atmospheric 512-513
    chemicals 515-522

Lead and lead alloys (cont.)
differential aeration 514
galvanic corrosion 513-514, 515-516
soil 514-515
underground ducts 513-514
water 511-512
grades of lead 494
lead-base alloys. See also *Leads and lead alloys, specific types.* 494
lead-base solders 497, 505-506
pig leads 494
plumtum series 498-499
products 495-499
properties of lead. See also *Lead, pure and Leads and lead alloys, specific types.* 494-495
refining of lead 493-494
sound-control materials 498, 499
sources of lead 493
Leaded copper. See *Copper alloys, specific types, C18700.*
Lead in copper 242-243
Lead, pure 761-762
Leads and lead alloys, specific types
acid-copper lead 494
arsenical lead 494, 502-503
calcium lead, Pb-0.065Ca-0.7Sn 504
calcium lead, Pb-0.065Ca-1.3Sn ... 504-505
calcium lead, Pb-0.07Ca 503
calcium lead, Pb-0.09Ca-0.3Sn 503
calcium lead, Pb-0.09Ca-0.5Sn ... 503-504
calcium lead, Pb-0.09Ca-1.0Sn 504
chemical lead 494, 501-502
common lead 494
copper-bearing lead. See *Leads and lead alloys, specific types, acid-copper lead.*
corroding lead 494, 500, 501
lead-base babbitt (alloy 7) 508-509
lead-base babbitt (alloy 8) 509
lead-base babbitt (alloy 13) 509
lead-base babbitt (alloy 15) 510
silver-lead solder 505
1% antimonial lead 506
4% antimonial lead 506-507
6% antimonial lead 507
8% antimonial lead 508
9% antimonial lead 508
5-95 solder 505
20-80 solder 505
50-50 solder 506
Lithium, pure 762-763
Low-beryllium copper. See *Copper alloys, specific types, C17500.*
Low-carbon steel, galvanic corrosion with magnesium 607
Low temperature properties, magnesium 531-532, 533, 534
Low-zinc brasses, corrosion in various media 468-469
Lubaloy. See *Copper alloys, specific types, C41100.*
Lubronze. See *Copper alloys, specific types, C42200.*
Lusterloy. See *Copper alloys, specific types, C61500.*
Lutetium, pure 763

M

Machinability
aluminum alloys 187-190
copper casting alloys 388-389
magnesium alloys 549, 551
Machining
aluminum, chemical milling vs mechanical milling 14, 16
aluminum part, relative cost 13, 15
Magnesium 525-552
applications 525

production 525, 526
solution potential 207
standard designations 525-526, 527, 528
Magnesium alloys 525-552
adhesive bonding 547-549, 550
bearing strength 528, 530
bending 552
casting alloys 526-527
castings 533-534, 536
compressive strength 528, 529-530, 531
deep drawing 543, 545
elevated temperature properties 532-533, 534, 535
extrusions 527-528, 534-535, 537, 546
fatigue strength 531, 532
forgings 528-529, 537-538, 539, 540, 541
formability 540-541, 542, 546
hardness 528, 530-531
impact extrusions 535, 537
inserts 538-540, 541, 542
joining 546-549
low-temperature properties 531-532, 533, 534
machinability 549, 551
mechanical properties 529-533
plate buckling 550-552
riveting 549
shear strength 528-530
sheet and plate 529, 541-543, 544
spot welds 547, 548
stress relieving 547
stretch forming 543, 545-546
weight reduction 551-552
welding 546-547, 548
Magnesium alloys, corrosion resistance 596-609
acids 600
alkalis 600
at elevated temperature 601-602
atmospheric 597-598, 599, 600
cold working, effects of 597
composition, effects of 596-597
corrosion protection 602, 603, 607, 609
corrosion testing 598, 603-604, 607
fresh water 598
galvanic corrosion 604-609
gases 601
heat treatment, effects of 597, 598
organic compounds 600-601
salt solutions 599-600, 601
soils 601
temperature, effects of 602
Magnesium alloys, specific types
AM60A 569
composition 526, 528
mechanical properties 526, 528
AM100A 570, 571
atmospheric corrosion 599
composition 526, 528
mechanical properties 526, 528
AS41A 571
composition 526, 528
mechanical properties 526, 528
AZ10A 553
composition 528
mechanical properties 528
AZ21X1 554
applications 528
composition 528
mechanical properties 528
AZ31B 554-555
atmospheric corrosion 599, 600, 604, 606, 607
composition 528, 529
corrosion rate in 3% NaCl 598

electrode potential in 3% NaCl 598
galvanic corrosion 605-607
marine corrosion 600, 603-604, 606, 607
mechanical properties 528, 529
AZ31C 554-555
composition 528
mechanical properties 528
AZ61A 555-556
atmospheric corrosion 599
composition 528
corrosion in salt solutions 601
corrosion potential in 3% NaCl 598
electrode potential in 3% NaCl 598
galvanic corrosion 607
mechanical properties 528
AZ63A 571-572
composition 528
corrosion potential in 3% NaCl 598
electrode potential in 3% NaCl 598
marine corrosion 603-604
mechanical properties 528
AZ80A 556, 557
atmospheric corrosion 599
composition 528, 529
forgeability 528, 529
mechanical properties 528, 529
AZ81A 573, 574
composition 526, 528
AZ91A 574, 575
composition 526, 528
mechanical properties 526, 528
AZ91B 574, 575
AZ91C 574, 575, 576
composition 526, 528
marine corrosion 603, 604
mechanical properties 526, 528
AZ92A 577-578
atmospheric corrosion 599
composition 526, 528
corrosion potential in 3% NaCl 598
electrode potential in 3% NaCl 598
mechanical properties 526, 528
EZ33A 578, 579, 580, 581
composition 527, 528
mechanical properties 527, 528
HK31A, cast 557, 558, 559, 560, 561
HK31A, wrought 557, 558, 559, 560, 561
composition 528, 529
corrosion potential in 3% NaCl 598
electrode potential in 3% NaCl 598
mechanical properties 528, 529
HM21A 561-562, 563, 564, 565
composition 528, 529
corrosion potential in 3% NaCl 598
electrode potential in 3% NaCl 598
forgeability 528, 529
mechanical properties 528, 529
HM31A 565, 566
composition 528
mechanical properties 528
HZ32A 584, 585-586
composition 527, 528
mechanical properties 527, 528
K1A 587
composition 526, 528
mechanical properties 526, 528
M1A 567, 568
atmospheric corrosion 599
composition 528
corrosion potential in 3% NaCl 598
electrode potential in 3% NaCl 598
mechanical properties 528
PE 555
composition 528, 529
mechanical properties 528, 529
QE22A 587-588, 589
composition 527, 528
mechanical properties 527, 528

QH21A 589, 590, 591
  composition 527, 528
  mechanical properties 527, 528
ZE41A 591-592
  composition 526, 527, 528
  mechanical properties 526, 527, 528
ZE63A 592-593
  composition 526, 528
  mechanical properties 526, 528
ZH62A 593
  composition 526, 528
  mechanical properties 526, 528
ZK21A 568
  composition 528
  mechanical properties 528
ZK40A 568
  composition 528
  mechanical properties 528
ZK51A 594
  composition 526, 528
  mechanical properties 526, 528
ZK60A 568-569
  composition 528, 529
  corrosion potential in 3% NaCl 598
  electrode potential in 3% NaCl 598
  forgeability 528, 529
  mechanical properties 528, 529
ZK61A 595
  composition 526, 528
  mechanical properties 526, 528
ZM21A
  composition 528
  mechanical properties 528
Magnesium, pure 763-768
Manganese, pure 768
Martensitic transformation, copper
  alloys 259
Mazak-3. See *Zinc alloys, specific types,*
  AG40A.
Mazak-5. See *Zinc alloys, specific types,*
  AC41A.
Mechanical properties. See also listings of
  specific properties in data compilations
  for individual metals and alloys.
Mechanical properties
  aluminum alloys 55, 58, 59-62
  aluminum casting alloys 148-151
  copper alloy castings 385-389, 387
  magnesium 529-533
Mercury, pure 769-770
Metal, purity characteristics
  resistance-ratio test 711-712, 713
  trace-element analysis 711
Metals, purification methods
  chemical vapor deposition 711
  degassing 710
  distillation 710
  floating zone technique 710
  fractional crystallization 709-710
  iodide process 711
  sublimation 710
  vacuum melting 710
  zone refining 710
Mild steel
  solution potential 207
Mill products, aluminum 44-62
MIL specifications. See also listings in data
  compilations for individual alloys.
Mischmetal 770-771
Modern pewter. See *Tin alloys, specific
  types, pewter.*
Molybdenum, pure 771-776
Molybdenum, zone refined, impurity
  concentration 713
Monel, galvanic corrosion with
  magnesium 607
Muntz metal. See *Copper alloys, specific
  types, C28000.*
  antimonial leaded. See *Copper alloys,
  specific types, C36700.*

arsenical leaded. See *Copper alloys,
  specific types, C36600.*
free-cutting. See *Copper alloys, specific
  types, C37000.*
inhibited leaded. See *Copper alloys,
  specific types, C36600, C36700 and
  C36800.*
leaded. See *Copper alloys, specific types,
  C36500, C36600, C36700 and C36800.*
phosphorized leaded. See *Copper alloys,
  specific types, C36800.*
uninhibited leaded. See *Copper alloys,
  specific types, C36500.*

N

N-4 alloy. See *Copper alloys, specific types,
  C15000.*
Navy Tombasil. See *Copper alloys, specific
  types, C87200.*
NDZ. See *Copper alloys, specific types,
  C99400.*
Neodymium, pure 776
Neptunium, pure 777, 832-833
Nickel in copper 242
Nickel, pure 777, 833
Nickel, solution potential 207
Nickel, zone refined, impurity
  concentration 713
Nickel silvers
  55-18. See *Copper alloys, specific types,
  C77000.*
  65-10. See *Copper alloys, specific types,
  C74500.*
  65-12. See *Copper alloys, specific types,
  C75700.*
  65-18. See *Copper alloys, specific types,
  C75200.*
  65-15. See *Copper alloys, specific types,
  C75400.*
Nickel silvers, corrosion in various
  media 468-469
Nickel silvers, specific types
  12%. See *Copper alloys, specific types,
  C97300.*
  20%. See *Copper alloys, specific types,
  C97600.*
  25%. See *Copper alloys, specific types,
  C97800.*
Niobium (Columbium), pure 777-779
Niobium, zone refined, impurity
  concentration 713
Noble metals. See *Precious metals.*
Nondezincification alloy. See *Copper alloys,
  specific types, C99400.*
Novoston. See *Copper alloys, specific types,
  C95700.*
Number 3 Die Casting Alloy. See *Zinc alloys,
  specific types, AG40A.*
Number 5 Die Casting Alloy. See *Zinc alloys,
  specific types, AC41A.*

O

Organic compounds, copper alloys,
  corrosion rate 479-480
Orthodontic appliances, gold 684-687
Osmium, pure 779-780
Ounce metal. See *Copper alloys, specific
  types, C83600.*
Oxygen-free copper. See *Copper alloys,
  specific types, C10100 and C10200.*
Oxygen-free electronic copper. See *Copper
  alloys, specific types, C10100.*
Oxygen-free extra-low-phosphorus copper.
  See *Copper alloys, specific types,
  C10300.*
Oxygen-free low-phosphorus copper. See
  Copper alloys, specific types, C10800.
Oxygen-free silver copper. See *Copper alloys,
  specific types, C10400, C10500 and
  C10700.*

P

Palladium
  corrosion resistance 669
Palladium, commercially pure 699-701
Palladium-copper alloy 702-703
Palladium, pure 780-781
Palladium-ruthenium alloy 704-705
Palladium-silver alloys 701-702
Palladium-silver-copper alloys 703
Palladium-silver-gold alloys 703-704
Permanent magnet alloys. See
  Platinum-cobalt permanent magnet alloy.
Permanent mold casting. See also *Castings,
  Foundry products.*
  aluminum alloys 144, 145, 147
  copper alloys 384
Pewter 614
Pewter. See *Tin alloys, specific types, pewter.*
Phospher bronzes, corrosion in various
  media 468-469
**Phosphorus-deoxidized, tellurium-bearing
  copper.** See *Copper alloys, specific types,
  C14500.*
Physical properties, aluminum
  alloys 53-54, 58
Piercing, rotary 262-263
Pipe. See *Tubular products.*
Pitting
  aluminum alloys 204-206
  copper metals 459-461
Plaster casting. See also *Castings, Foundry
  products.*
  aluminum alloys 146
Plaster mold casting, copper
  alloys 384-385
Plate buckling, magnesium 550-552
Platinum, corrosion resistance 668-669
Platinum, commercially pure 688-690
**Platinum-cobalt permanent
  magnet alloy** 697-698
Platinum-group metals
  iridium 664
  jewelry 666-667
  osmium 665
  palladium 663-664
  platinum 663
  production 660-661
  rhodium 664
  ruthenium 664-665
  special properties 660-661
Platinum-iridium alloys 691-693
Platinum-nickel alloys 695-696
Platinum-palladium alloys 690-691
Platinum, pure 781-783
Platinum-rhodium alloys 693-694
Platinum-rhodium-ruthenium alloy 695
Platinum-ruthenium alloys 694-695
Platinum-tungsten alloys 696-697
Plumbum series 498-499
Plutonium, pure 783-785, 832-833
Potassium, pure 786-787
Powder metallurgy parts, aluminum
  alloy 10-13
Praseodymium, pure 787-788
Precious metals 659-667
  coatings 666, 667
  commercial forms 665-667
  gold 660
  industrial uses 664-666
  jewelry 666-667
  platinum-group 660-661
  silver 659-660
  special properties 662-665
  trade practices 661-662
Precious metals, corrosion resistance
  gold 669-670
  iridium 669
  palladium 669
  platinum 668-669

Precious metals (cont.)
rhodium 669
silver 670
Precious metals, industrial applications
ceramics 664-665
chemical 664-666
coatings 666, 667
containers 666
crucible 665-666
electrical/electronic 664
electrochemical 665
glass 664-665
instruments 664
powder 666
reflectors 666
safety devices 666
Precipitation hardening. See also *Aging.*
aluminum alloys 29-30, 34,
38-39, 40-42, 43
copper alloys 256
Promethium, pure 788
Properties, elevated temperature,
aluminum alloys 56, 57-58, 62
Properties, low-temperature,
aluminum alloys 62
Protactinium, pure 788, 832-833
Pure metals 709-713
characterization 711-712, 713
preparation 709-711
Pure metals. See index entries under
individual elements.

Q

Quenching, aluminum
alloys 32-35, 40-41, 43

R

**Resistance-ratio test for pure
metals** 711-712
Resistance welding, copper
metals 441, 442
Rhenium, pure 788-790
Rhenium, zone refined, impurity
concentration 713
Rhodium, corrosion resistance 669
Rhodium, pure 790-791
Riveting, magnesium 549
Rolled zinc, atmospheric
corrosion 647, 650
**Rope-lay stranded copper
conductors** 266, 269, 270-271
Rubidium, pure 791-792
Ruthenium, pure 792

S

SAE specifications. See also listings in data
compilations for individual alloys.
Salts, copper alloys, corrosion
rate 477-479
Salt water corrosion, copper
alloys 471-472
Samarium, pure 792-793
Sand casting. See also *Castings, Foundry
products.*
aluminum alloys 144, 145-146, 147
copper alloys 384
Scandium, pure 793-794
Selenium in copper 242-243
Selenium, pure 794
Sheet, zinc, galvanized 651, 653
Sheet and plate,
magnesium 529, 541-543, 544
Shell mold casting, aluminum alloys. See
also *Castings, Foundry products.* ...146
Silicon bronzes, corrosion in various
media 468-469
Silicon in copper 242
Silicon, pure 796-797

Silver
corrosion resistance 670
production 659-660
solution potential 207
special properties 662
Silver-base brazing alloys 675-677
Silver-bearing tough pitch copper. See
Copper alloys, specific types, C11300,
C11400, C11500 and C11600.
Silver, coin. See *Silver-copper alloys.*
Silver, commercially pure 671-673
Silver-copper alloys 673-675
Silver in copper 242, 243
Silver-magnesium-nickel alloys 677-678
Silver, pure 794-796
Silver solders. See *Silver-base brazing alloys.*
Silver, sterling. See *Silver-copper alloys.*
Sodium, pure 797-799
Soft undesilverized lead. See *Leads and lead
alloys, specific types, corroding lead.*
Soil corrosion, aluminum alloys 225
Soils, copper corrosion
resistance 467, 470
Soldering, aluminum 200-201
Soldering, copper metals 443-449, 450
Solders. See also *Solders, specific types.*
applications 201, 445
compositions for soldering
aluminum 201
compositions for soldering copper 445
dental appliances, gold 686, 687
lead-base 497, 505-506
melting range 201, 445
tin alloy 614
Solders. See specific types, such as *soft solder*
and *tin-silver solder,* under *Tin alloys,
specific types.*
Solidification, aluminum castings 150-151
Solidification, copper casting alloys 385
Solid-state welding, copper
metals 441, 442
Solution heat treatment, aluminum
alloys 31, 32, 35, 38-40
Solution potentials, aluminum
alloys 206-207
Sound-control materials 498, 499
Special brasses, corrosion in various
media 468-469
Specifications. See specific type of
specification, such as *ASTM, AMS,
SAE, Federal, MIL, Foreign.* See also
listings of specifications in data
compilations for individual alloys.
Spinodal decomposition, copper
alloys 259-260
Stainless steel
galvanic corrosion with magnesium ...607
solution potential 207
Stamping, aluminum alloy sheet 180-186
Stampings, aluminum alloy 13-14
Steam corrosion, copper alloys 471
Steel
cadmium-plated, galvanic corrosion
with magnesium 607
zinc-plated, galvanic corrosion with
magnesium 607
Steel wire, zinc galvanized 652-654
Straits tin. See *Tin alloys, specific types,
commercially pure tins.*
Stranded copper conductors 266, 268-273
Strescon. See *Copper alloys, specific types,*
C19500.
Stress-corrosion cracking
aluminum alloys 210-211, 212-218
copper 239-240
copper metals 459, 462-463
Stress relaxation, copper and copper
alloys 484-490
Stress relieving
copper alloys 255-256

copper metals 462-463
Stretch forming, magnesium ...543, 545-546
Strip, copper, temper
designations 248-249
Strontium, pure 799
Sublimation for purifying metals 710
Sulfur-bearing copper. See *Copper alloys,
specific types,* C14700.
Sulfur copper. See *Copper alloys, specific
types,* C14700.
Sulfur corrosion, copper metals ... 463, 464
Superalloys. See *Nickel alloys, specific types*
Superstone 40. See *Copper alloys, specific
types,* C95700.
Super·Z 300. See *Zinc alloys, specific types,
superplastic zinc alloy.*

T

Tantalum, pure 799-804
Tantalum, zone refined, impurity
concentration 713
Tellurium in copper 242-243
Technetium, pure 804-806
Tellurium, pure 806
Temper designations, aluminum and
aluminum alloys 24-27
Terbium, pure 806-807
Terne metal 497-498
Testing, aluminum stamping
alloys 180-182
Thallium, pure 807
Thorium, pure 807-810, 823-833
Thulium, pure 810-811
Tin 613-616
solution potential 207
unalloyed 614
Tin alloys
battery-grid 615
bearing materials 614-615
cast iron, addition to 615
copper-tin bronzes 615
dental alloys 615
pewter 614
solders 614
titanium, addition to 615
type metals 615
use in organ pipes 614
zirconium, addition to 615-616
Tin alloys, specific types
antimonial tin solder 619
bearing alloy 617, 623
casting alloy 617, 623
commercially pure tins 617, 618-619
hard tin 619
pewter 617, 624-625
soft solder, 60Sn-40Pb 620-621
soft solder, 63Sn-37Pb 620
soft solder, 70Sn-30Pb 620
tin babbitt alloy 1 ....617, 621-622
tin babbitt alloy 2 .... 617, 622
tin babbitt alloy 3 .... 617, 622, 623
tin die-casting alloy 623
tin foil 623-624
tin-silver solder 619
white metal 624
Tin chemicals, applications 616
Tin, coatings
electroplating 614
hot dip 614
tinplate 613-614
Tin in copper 242
Tin powders, applications 616
Tin, pure 811-814
Titanium, galvanic corrosion with
magnesium 607
Titanium alloys, tin addition 615
Titanium, pure 814-816
TLW 3-833
Tombasil. See *Copper alloys, specific types,*
C87500 and C87800.

Tool life, in machining aluminum ... 187-189
Tough pitch copper with silver. See *Copper alloys, specific types, C11300, C11400, C11500 and C11600.*
Trace-element analysis for pure metals 711
Trace elements, chemical analysis in pure metals 711
Tube, copper, temper designations .. 249-251
Tube reducing 264
Tubes. See *Tubular products.*
Tubes, copper alloys 472
Tubular products, copper 261-264
Tungsten, pure 816-821
Tungsten, zone refined, impurity concentration 713
Type metals ....,.. 496

U

Unbreakable Metal. See *Zinc alloys, specific types, zinc-base slush-casting alloys.*
Unilay stranded copper conductors 266
Uranium, pure 821-822, 832-833

V

Vacuum melting for purifying metals
Valve metal. See *Copper alloys, specific types, C84400.*
Vanadium, pure 822-823
Vanadium, zone refined, impurity concentration 713
Vapor deposition. See *Chemical vapor deposition.*

W

Wear applications, copper alloy castings 392-393
Wearite 4-13. See *Copper alloys, specific types, C62500.*
Weight reduction, magnesium 551-552
Weldbonding, aluminum 202
Welding
  aluminum 191-199
  copper metals 453-456
  magnesium 546-547
Welding processes, aluminum 196-199
White Tombasil. See *Copper alloys, specific types, C99700.*
Wire, copper 265-274
Wirebar, copper 266
Wiredrawing, copper 271-273
Wire, copper, temper designations 248
Wire rod, copper 266-271
Wrought alloys
  aluminum 44-62
  zinc 635-637

Y

Ytterbium, pure 823
Yttrium, pure 823-824

Z

ZA-12. See *Zinc alloys, specific types, zinc foundry alloy ZA-12.*
Zamak-3 (die casting). See *Zinc alloys, specific types, AG40A.*
Zamak-5 (die casting and sand casting). See *Zinc alloys, specific types, AC41A.*
Zilloy-15. See *Zinc alloys, specific types, rolled zinc alloy.*
Zilloy-40. See *Zinc alloys, specific types, copper-hardened rolled zinc.*
Zinc
  galvanic corrosion with magnesium ... 607
  solution potential 207
Zinc alloys, applications 630-637
Zinc alloys, gravity casting, applications 635
Zinc alloys, specific types
  12%-Al
    applications 635-636
  27%-Al
    applications 635-636
  AC41A alloy 639
    composition 630-631
    creep data 634-635
    designations 630-632
    mechanical properties 633
    physical properties 633
  AG40A alloy 638
    composition 631
    creep data 634-635
    designations 630-631
    mechanical properties 633
  alloy 7
    composition 630-631
    designations 630-631
    mechanical properties 633
    physical properties 633
  commercial rolled zincs 641-643
  copper-hardened rolled zinc 643
  ILZRO 16 640-641
    composition 632
    designations 631-632
    mechanical properties 633
    physical properties 633
  rolled zinc alloy 643-644
  superplastic zinc alloy 644-645
  zinc-base slush-casting alloys 639-640
  zinc foundry alloy ZA-12 640
  Zn-27Al 641
  Zn-Cu-Ti alloy 644
Zinc alloys, wrought
  characteristics 636
  classification 636
Zinc anodes, cathodic protection 654
Zinc, applications 629-637
Zinc, corrosion environments
  acids 648
  aqueous 647-648
  atmospheric 649-650
  gases 649
  indoor exposure 649-650
  inhibitors, use of 648-649
  nonaqueous liquids 649

oxygen in water, effect of 648-649
salts 648
seacoast 649-650
soils 649
Zinc, corrosion resistance
  anodic coatings, effect of 646-647
  composition, effect of 646-647
  electrochemical corrosion 650, 652
  fatigue 650-652
  rate, compared to iron 646
Zinc, corrosion service 646-655
Zinc, die castings
  aging 631
  alloys 630-633
  applications 632-633
  assembly 631
  finishing 631-632
  heat treatment 631-632
Zinc, die castings engineering
  die design 633-634
  casting design 634
Zinc dust
  applications 629-630
Zinc foundry alloy ZA-12. See *Zinc alloys, specific types, zinc foundry alloy ZA-12.*
Zinc foundry alloy ZA-27. See *Zinc alloys, specific types, Zn-27Al.*
Zinc, galvanized
  alloys layer 652
  coating thickness 652-653
  coating uniformity 652-653
  life of coating 653
  sheet 651, 653
  steel wire 652-654
Zinc in copper 242-243
Zinc oxide, applications 630
Zinc, pure 824-826
Zinc, rolled, atmospheric corrosion, tests 647, 650
Zinc, slab
  composition 629
  grades 629
  production 629-630
  superplastic 629
Zinc, solubility in magnesium 525
Zinc sulfide, applications 629
Zinc, wrought
  classification 636-637
  extrusions 636
  fabrication 636-637
  finishing 637
  machining 637
  mechanical properties 636-637
  rolled products 636
  soldering 637
  welding 637
  wire drawing 636
Zirconium copper. See *Copper alloys, specific types, C15000.*
Zirconium copper, heat treating ... 258-259
Zirconium, pure 826-831
Zirconium, zone refined, impurity concentration 713
Zirconium alloys, tin addition 615-616
Zone refining for purifying metals 710